ENCYCLOPEDIA OF CAVES

SECOND EDITION

Editors

WILLIAM B. WHITE
The Pennsylvania State University

DAVID C. CULVER
American University

AMSTERDAM • BOSTON • HEIDELBERG • LONDON NEW YORK • OXFORD
PARIS • SAN DIEGO • SAN FRANCISCO • SYDNEY • TOKYO
Academic Press is an imprint of Elsevier

Academic Press is an imprint of Elsevier
225 Wyman Street, Waltham, MA 02451, USA
The Boulevard, Langford Lane, Kidlington, Oxford, OX51GB, UK
Radarweg 29, PO Box 211, 1000 AE Amsterdam, The Netherlands
525 B Street, Suite 1900, San Diego, CA 92101-4495, USA

First edition 2005
Second edition 2012

Copyright © 2012 Elsevier Inc. All rights reserved

No part of this publication may be reproduced, stored in a retrieval system or transmitted in any form or by any means electronic, mechanical, photocopying, recording or otherwise without the prior written permission of the publisher.

Permissions may be sought directly from Elsevier's Science & Technology Rights Department in Oxford, UK: phone (+44) (0) 1865 843830; fax (+44) (0) 1865 853333; email: permissions@elsevier.com. Alternatively you can submit your request online by visiting the Elsevier web site at http://elsevier.com/locate/permissions, and selecting *Obtaining permission to use Elsevier material*.

Notice
No responsibility is assumed by the publisher for any injury and/or damage to persons or property as a matter of products liability, negligence or otherwise, or from any use or operation of any methods, products, instructions or ideas contained in the material herein.

Library of Congress Cataloging-in-Publication Data
Encyclopedia of caves/editors David C. Culver, William B. White. – 2nd ed.
  p. cm.
  Includes bibliographical references and index.
  ISBN 978-0-12-383832-2 (alk. paper)
  1. Speleology–Encyclopedias. 2. Caves–Encyclopedias. I. Culver, David C., 1944- II. White, William B. (William Blaine), 1934-GB601.E534 2012
  551.44′703–dc23

2011039751

British Library Cataloguing in Publication Data
A catalogue record for this book is available from the British Library

ISBN: 978-0-12-383832-2

For information on all Academic Press publications
visit our website at www.elsevierdirect.com

Printed and bound in China
12 13 14   10 9 8 7 6 5 4 3 2

**Working together to grow
libraries in developing countries**

www.elsevier.com | www.bookaid.org | www.sabre.org

ELSEVIER   BOOK AID International   Sabre Foundation

Contents

Contents by Subject Area   ix
List of Contributors   xi
Guide to the Encyclopedia   xiii
Preface   xv

A

Adaptation to Low Food   1
Kathrin Hüppop

Adaptive Shifts   9
Francis G. Howarth and Hannelore Hoch

Anchihaline (Anchialine) Caves and Fauna   17
Boris Sket

Ancient Cavers in Eastern North America   25
Patty Jo Watson

Asellus aquaticus: A Model System for Historical Biogeography   30
Rudi Verovnik

Astyanax mexicanus: A Model Organism for Evolution and Adaptation   36
William R. Jeffery

B

Bats   45
Thomas H. Kunz, Susan W. Murray and Nathan W. Fuller

Beetles   54
Oana Teodora Moldovan

Behavioral Adaptations   62
Jakob Parzefall

Breakdown   68
Elizabeth L. White

Burnsville Cove, Virginia   74
Gregg S. Clemmer

C

Camps   85
Gregg S. Clemmer

Castleguard Cave, Canada   89
Derek Ford

Cave Dwellers in the Middle East   94
Paul Goldberg and Ofer Bar-Yosef

Cave Ecosystems   99
Kevin S. Simon

Cave, Definition of   103
William B. White and David C. Culver

Cavefish of China   107
Li Ma and Ya-hui Zhao

Chemoautotrophy   125
Annette Summers Engel

Clastic Sediments in Caves   134
Gregory S. Springer

Closed Depressions in Karst Areas   140
Ugo Sauro

Coastal Caves   155
John E. Mylroie

Contamination of Cave Waters by Heavy Metals   161
Dorothy J. Vesper

Contamination of Cave Waters by Nonaqueous Phase Liquids   166
Caroline M. Loop

Cosmogenic Isotope Dating of Cave Sediments   172
Darryl E. Granger and Derek Fabel

Crustacea   177
Horton H. Hobbs III

D

Dinaric Karst: Geography and Geology   195
Nadja Zupan Hajna

Diversity Patterns in Australia   203
William F. Humphreys

Diversity Patterns in Europe   219
Louis Deharveng, Janine Gibert and David C. Culver

Diversity Patterns in the Dinaric Karst   228
Boris Sket

Diversity Patterns in the Tropics   238
Louis Deharveng and Anne Bedos

Diversity Patterns in the United States   251
Horton H. Hobbs III

Documentation and Databases   264
Peter Matthews

E

Ecological Classification of Subterranean Organisms   275
Eleonora Trajano

Entranceless Caves, Discovery of   277
Nevin W. Davis

Entrances   280
William B. White

Epikarst   284
Michel Bakalowicz

Epikarst Communities   288
David C. Culver, Anton Brancelj and Tanja Pipan

Evolution of Lineages   295
Eleonora Trajano and Marina Cobolli

Exploration of Caves—General   304
William B. White

Exploration of Caves—Underwater Exploration Techniques   310
Jill Heinerth

Exploration of Caves—Vertical Caving Techniques   314
Mark Minton and Yvonne Droms

F

Folklore, Myth, and Legend, Caves in   321
Paul Jay Steward

Food Sources   323
Thomas L. Poulson

Friars Hole System, West Virginia   334
Stephen R.H. Worthington and Douglas M. Medville

G

Gammarus minus: A Model System for the Study of Adaptation to the Cave Environment   341
Daniel W. Fong

Geophysics of Locating Karst and Caves   348
Barbara Anne am Ende

Glacier Caves   353
Jason D. Gulley and Andrew G. Fountain

Guano Communities   357
Pedro Gnaspini

Gypsum Caves   364
Alexander Klimchouk

Gypsum Flowers and Related Speleothems   374
William B. White

H

Helictites and Related Speleothems   379
Donald G. Davis

Hydrogeology of Karst Aquifers   383
William B. White

Hydrothermal Caves   391
Yuri Dublyansky

I

Ice in Caves   399
Aurel Perşoiu and Bogdan P. Onac

Invasion, Active versus Passive   404
Dan L. Danielopol and Raymond Rouch

J

Jewel Cave, South Dakota   411
Mike Wiles

K

Karren, Cave   419
Joyce Lundberg

Karren, Surface   425
Joyce Lundberg

Karst   430
William K. Jones and William B. White

Kazumura Cave, Hawaii   438
Kevin Allred

Krubera (Voronja) Cave   443
Alexander Klimchouk

L

Lampenflora   451
Janez Mulec

Lechuguilla Cave, New Mexico, U.S.A.   456
Patricia Kambesis

Life History Evolution   465
David C. Culver

M

Mammoth Cave System, Kentucky   469
Roger W. Brucker

Mapping Subterranean Biodiversity   474
Mary C. Christman and Maja Zagmajster

Marine Regressions   482
Claude Boutin and Nicole Coineau

Maya Caves 486
Andrea Stone and James E. Brady

Microbes 490
Annette Summers Engel

Minerals 499
Bogdan P. Onac

Modeling of Karst Aquifers 508
Georg Kaufmann, Douchko Romanov and Wolfgang Dreybrodt

Mollusks 512
David C. Culver

Morphological Adaptations 517
Kenneth Christiansen

Multilevel Caves and Landscape Evolution 528
Darlene M. Anthony

Mulu Caves, Malaysia 531
Tony Waltham and Joel Despain

Myriapods 538
David C. Culver and William A. Shear

N

Natural Selection 543
Peter Trontelj

Neutral Mutations 549
Horst Wilkens

Niphargus: A Model System for Evolution and Ecology 555
Cene Fišer

Nitrate Contamination in Karst Groundwater 564
Brian G. Katz

Nullarbor Caves, Australia 568
Julia M. James, Annalisa K. Contos and Craig M. Barnes

P

Paleoclimate Records from Speleothems 577
Victor J. Polyak and Rhawn F. Denniston

Paleomagnetic Records in Cave Sediments 585
Ira D. Sasowsky

Paleontology of Caves 590
Blaine W. Schubert and Jim I. Mead

Passage Growth and Development 598
Arthur N. Palmer

Passages 603
George Veni

Population Structure 608
Valerio Sbordoni, Giuliana Allegrucci and Donatella Cesaroni

Postojna–Planina Cave System, Slovenia 618
Stanka Šebela

Protecting Caves and Cave Life 624
William R. Elliott

Q

Quartzite Caves of South America 635
Augusto S. Auler

R

Recreational Caving 641
John M. Wilson

Rescues 648
John C. Hempel

Responses to Low Oxygen 651
Frédéric Hervant and Florian Malard

Root Communities in Lava Tubes 658
Fred D. Stone, Francis G. Howarth, Hannelore Hoch and Manfred Asche

S

Salamanders 665
Špela Gorički, Matthew L. Niemiller and Danté B. Fenolio

Saltpetre Mining 676
David A. Hubbard Jr.

Scallops 679
Phillip J. Murphy

Shallow Subterranean Habitats 683
Tanja Pipan and David C. Culver

Show Caves 690
Arrigo A. Cigna

Siebenhengste Cave System, Switzerland 698
Pierre-Yves Jeannin and Philipp Häuselmann

Sinking Streams and Losing Streams 707
Joseph A. Ray

Sistema Huautla, Mexico 712
C. William Steele and James H. Smith, Jr.

Soil Piping and Sinkhole Failures 718
Barry Beck

Solution Caves in Regions of High Relief 723
Philipp Häuselmann

Solution Caves in Regions of Moderate Relief 733
Arthur N. Palmer

Species Interactions 743
David C. Culver

Speleogenesis, Hypogenic 748
Alexander Klimchouk

Speleogenesis, Telogenetic 765
Franci Gabrovšek

Speleothem Deposition 769
Wolfgang Dreybrodt

Speleothems: General Overview 777
William B. White

Spiders and Related Groups 786
James R. Reddell

Springs 797
William B. White

Stalactites and Stalagmites 805
Silvia Frisia and Jon D. Woodhead

Sulfuric Acid Caves 810
Arthur N. Palmer and Carol A. Hill

T

Tiankeng 821
Xuewen Zhu, Weihai Chen and Yuanhai Zhang

U

Ukraine Giant Gypsum Caves 827
Alexander Klimchouk

Underwater Caves of the Yucatán Peninsula 833
James G. Coke, IV

Uranium Series Dating of Speleothems 838
Christoph Spötl and Ronny Boch

V

Vertebrate Visitors—Birds and Mammals 845
Nikola Tvrtkovic

Vicariance and Dispersalist Biogeography 849
John R. Holsinger

Vjetrenica Cave, Bosnia and Herzegovina 858
Ivo Lučić

Volcanic Rock Caves 865
Stephan Kempe

W

Wakulla Spring Underwater Cave System, Florida 875
Barbara Anne am Ende

Water Chemistry in Caves 881
Janet S. Herman

Water Tracing in Karst Aquifers 887
William K. Jones

Wetlands in Cave and Karst Regions 897
Tanja Pipan and David C. Culver

White-Nose Syndrome: A Fungal Disease of North American Hibernating Bats 904
Marianne S. Moore and Thomas H. Kunz

Worms 910
Elzbieta Dumnicka

Index 917

Contents By Subject Area

TYPES OF CAVES

Anchihaline caves and fauna
Cave, definition of
Coastal caves
Entranceless caves, discovery of
Glacier caves
Gypsum caves
Hydrothermal caves
Multilevel caves and landscape evolution
Quartzite caves of South America
Show caves
Solution caves in regions of high relief
Solution caves in regions of moderate relief
Sulfuric acid caves
Volcanic rock caves

CAVE FEATURES

Breakdown
Clastic sediments in caves
Entrances
Ice in caves
Karren, cave
Paleomagnetic records in cave sediments
Passage growth and development
Passages
Scallops

SURFACE KARST FEATURES

Closed depressions in karst areas
Dinaric karst — geography and geology
Karren, surface
Karst, general overview
Sinking streams and losing streams
Soil piping and sinkhole features
Springs
Tiankeng
Wetlands in cave and karst regions

HYDROLOGY AND HYDROGEOLOGY

Contamination of cave waters by heavy metals
Contamination of cave waters by nonaqueous phase liquids
Epikarst
Hydrogeology of karst aquifers
Modeling karst aquifers
Nitrate contamination in karst groundwater
Passages growth and development
Speleogenesis, hypogenetic
Speleogenesis, telogenetic
Sinking streams and losing streams
Water chemistry in caves
Water tracing in karst aquifers

SPELEOTHEMS AND OTHER CAVE FORMATIONS

Gypsum flowers and related speleothems
Helictites and related speleothems
Karren, cave
Minerals
Paleoclimate records from speleothems
Speleothem deposition
Speleothems — general overview
Stalactites and stalagmites

CAVE AGES AND PALEOCLIMATE

Cosmogenic isotope dating of cave sediments
Multilevel caves and landscape evolution
Marine regression
Paleomagnetic record in cave sediments
Uranium series dating of speleothems

EXCEPTIONAL CAVES

Burnsville Cove, Virginia
Castleguard Cave, Canada
Friars Hole Cave system, West Virginia
Jewel Cave, South Dakota
Kazumura Cave, Hawaii
Krubera (Voronja) Cave,
Lechuguilla Cave, New Mexico, USA
Mammoth Cave system, Kentucky
Mulu Caves, Malaysia
Nullarbor Caves, Australia
Postojna-Planina Cave system, Slovenia
Quartzite caves of South America
Siebenhengste Cave system, Switzerland
Sistema Huautla, Mexico
Ukraine giant gypsum caves
Underwater caves of the Yucatan Peninsula
Vjetrenica Cave, Bosnia and Hercegovina
Wakulla Spring underwater cave system, Florida

GROUNDWATER CONTAMINATION AND LAND-USE HAZARDS IN CAVE REGIONS

Contamination of cave waters by heavy metals
Contamination of cave waters by nonaqueous phase liquids
Nitrate contamination in karst groundwater
Soil piping and sinkhole failures

HISTORICAL USE OF CAVES

Ancient cavers in eastern North America
Cave dwellers in the Middle East
Folklore, myth, and legend, caves in
Maya caves
Saltpetre mining

CONTEMPORARY USE OF CAVES

Lampenflora
Mapping subterranean biodiversity
Protecting caves and cave life
Recreational caving
Show caves

EXPLORATION OF CAVES

Camps
Documentation and databases
Entranceless caves, discovery of
Exploration of caves — general
Exploration of caves — underwater exploration techniques
Exploration of caves — vertical exploration techniques
Recreational caving
Rescues

BIOLOGY OF PARTICULAR ORGANISMS IN CAVES

Asellus aquaticus — a model system for historical biogeography
Astyanax mexicanus — a model system for evolution and adaptation
Bats
Beetles
Cavefish in China
Crustacea
Gammarus minus — a model system for the study of adaptation to the cave environment
Microbes
Molluscs
Myriapods
Niphargus — a model system for evolution and ecology
Paleontology of caves
Salamanders
Spiders and related groups
Vertebrate visitors — birds and mammals
White Nose Syndrome — a fungal disease of North American hibernating bats
Worms

COMMUNITIES AND HABITATS

Cave ecosystems
Ecological classification of cave organisms
Epikarst communities
Guano communities
Lampenflora
Root communities in lava tubes
Shallow subterranean habitats

ECOLOGY

Chemoautotrophy
Food sources
Niphargus — a model system for evolution and ecology
Population structure
Responses to low oxygen
Species interactions

BIOGEOGRAPHY AND DISPERSAL

Asellus aquaticus — a model system for historical biogeography
Invasion, active versus passive
Marine regression
Vicariance and dispersalist biogeography

DIVERSITY

Diversity patterns in Australia
Diversity patterns in Dinaric karst
Diversity patterns in Europe
Diversity patterns in the tropics
Diversity patterns in the United States
Mapping subterranean biodiversity

EVOLUTION AND ADAPTATION

Adaptation to low food
Adaptive shifts
Astyanax mexicanus — a model system for evolution and adaptation
Behavioral adaptations
Evolution of lineages
Gammarus minus — a model system for the study of adaptation to the cave environment
Life history evolution
Morphological adaptation
Natural selection in caves
Neutral mutation theory

List of Contributors

Giuliana Allegrucci Tor Vergata University, Italy
Kevin Allred Hawaii Speleological Survey
Barbara Anne am Ende Deep Caves Consulting
Darlene M. Anthony Roane State Community College
Manfred Asche Museum für Naturkunde, Germany
Augusto S. Auler Instituto do Carste, Brazil
Michel Bakalowicz Université Montpellier 2, France
Craig M. Barnes University of Sydney, Australia
Ofer Bar-Yosef Harvard University
Barry Beck P.E. LaMoreaux & Associates, Inc.
Anne Bedos Museum National d'Histoire Naturelle, France
Claude Boutin Université Paul Sabatier, France
James E. Brady California State University, Los Angeles
Anton Brancelj National Institute of Biology, Slovenia
Roger W. Brucker Cave Research Foundation
Donatella Cesaroni Tor Vergata University, Roma, Italy
Weihai Chen Chinese Academy of Geological Sciences, China
Kenneth Christiansen Grinnell College
Mary C. Christman University of Florida
Gregg S. Clemmer Butler Cave Conservation Society, Inc.
Marina Cobolli University of Rome "La Sapienza", Italy
Nicole Coineau Observatoire Océanologique de Banyuls, France
James G. Coke, IV Texas
Annalisa K. Contos University of Sydney, Australia
David C. Culver American University
Dan L. Danielopol Austrian Academy of Sciences, Austria
Nevin W. Davis Butler Cave Conservation Society, Inc.
Donald G. Davis National Speleological Society
Louis Deharveng Muséum National d'Histoire Naturelle, France
Rhawn F. Denniston Cornell College
Joel Despain National Park Service
Wolfgang Dreybrodt University of Bremen, Germany
Yvonne Droms U.S. Deep Caving Team
Yuri Dublyansky Innsbruck University, Austria
Elzbieta Dumnicka Polish Academy of Sciences, Poland
William R. Elliott Missouri Department of Conservation
Derek Fabel University of Glasgow, Scotland

Danté B. Fenolio Atlanta Botanical Garden
Cene Fišer University of Ljubljana, Slovenia
Daniel W. Fong American University
Derek Ford McMaster University, Canada
Andrew G. Fountain Portland State University
Silvia Frisia The University of Newcastle, Australia
Nathan W. Fuller Boston University
Franci Gabrovšek Karst Research Institute at ZRC SAZU, Slovenia
Janine Gibert Université Claude Bernard Lyon I, France (deceased)
Pedro Gnaspini University of São Paulo, Brazil
Paul Goldberg Boston University and Harvard University
Špela Gorički University of Maryland
Darryl E. Granger Purdue University
Jason D. Gulley University of Texas
Philipp Häuselmann Swiss Institute for Speleology and Karst Studies (SISKA), Switzerland
Jill Heinerth Heinerth Productions, Inc.
John C. Hempel EEI Geophysical
Janet S. Herman University of Virginia
Frédéric Hervant Université Claude Bernard Lyon 1, France
Carol A. Hill University of New Mexico
Horton H. Hobbs III Wittenberg University
Hannelore Hoch Universität zu Berlin, Germany
John R. Holsinger Old Dominion University, Norfolk, Virginia
Francis G. Howarth Bernice P. Bishop Museum
David A. Hubbard Jr. Virginia Speleological Survey
William F. Humphreys Western Australian Museum
Kathrin Hüppop Institute for Vogelforschung, Germany
Julia M. James University of Sydney, Australia
Paul Jay Steward Cave Research Foundation
Pierre-Yves Jeannin Swiss Institute for Speleology and Karst Studies (SISKA), Switzerland
William R. Jeffery University of Maryland
Patty Jo Watson Washington University
William K. Jones Karst Waters Institute
Patricia Kambesis Cave Research Foundation
Brian G. Katz U.S. Geological Survey

Georg Kaufmann Free University of Berlin, Germany
Stephan Kempe University of Technology Darmstadt, Germany
Alexander Klimchouk Ukrainian Institute of Speleology and Karstology, Ukraine
Thomas H. Kunz Boston University
Caroline M. Loop Groundwater Management Associates
Ivo Lučić Speleological Association Vjetrenica, Bosnia and Herzegovina
Joyce Lundberg Carleton University, Ottawa, Canada
Li Ma Department of Biology, University of Maryland
Florian Malard Université Claude Bernard Lyon 1
Jim I. Mead East Tennessee State University
Douglas M. Medville West Virginia Speleological Survey
Mark Minton U.S. Deep Caving Team
Marianne S. Moore Boston University
Janez Mulec Karst Research Institute, at ZRC SAZU, Slovenia
Phillip J. Murphy University of Leeds, UK
Susan W. Murray Boston University
John E. Mylroie Mississippi State University
Matthew L. Niemiller University of Tennessee
Bogdan P. Onac University of South Florida, Tampa, U.S.A., and "Emil Racovita" Institute of Speleology, Romania
Arthur N. Palmer State University of New York Oneonta
Jakob Parzefall University of Hamburg, Germany
Aurel Perşoiu University of Suceava, Romania
Tanja Pipan Karst Research Institute at ZRC SAZU, Slovenia
Victor J. Polyak University of New Mexico
Thomas L. Poulson Jupiter, Florida
Joseph A. Ray Crawford Hydrology Laboratory, Western Kentucky University
James R. Reddell The University of Texas at Austin
Douchko Romanov Free University of Berlin, Germany
Raymond Rouch Centre National de la Recherche Scientifique, France (Retired)
Ira D. Sasowsky University of Akron

Ugo Sauro University of Padova, Italy
Valerio Sbordoni Tor Vergata University, Italy
Blaine W. Schubert East Tennessee State University
Stanka Šebela Karst Research Institute at ZRC SAZU
William A. Shear Hampden-Sydney College
Kevin S. Simon The University of Auckland, New Zealand
Boris Sket Univerza v Ljubljani, Slovenia
James H. Smith, Jr. Environmental Protection Agency
Gregory S. Springer Ohio University
C. William Steele Boy Scouts of America
Andrea Stone University of Wisconsin-Milwaukee
Fred D. Stone University of Hawai'i at Hilo
Annette Summers Engel University of Tennessee
Oana Teodora Moldovan Emil Racovitza Institute of Speleology, Romania
Eleonora Trajano Universidade de São Paulo, Brazil
Peter Trontelj University of Ljubljana, Slovenia
George Veni National Cave and Karst Research Institute
Rudi Verovnik University of Ljubljana, Slovenia
Dorothy J. Vesper West Virginia University
Tony Waltham British Cave Research Association, U.K.
Elizabeth L. White Hydrologic Investigations
William B. White The Pennsylvania State University
Mike Wiles Jewel Cave National Monument
Horst Wilkens University of Hamburg, Germany
John M. Wilson Marks Products, Inc.
Jon D. Woodhead The University of Melbourne, Australia
Stephen R.H. Worthington Worthington Groundwater, Canada
Maja Zagmajster University of Ljubljana, Slovenia
Yuanhai Zhang Chinese Academy of Geological Sciences, China
Ya-hui Zhao Institute of Zoology, Chinese Academy of Sciences, China
Xuewen Zhu Chinese Academy of Geological Sciences, Guilin, China
Nadja Zupan Hajna Karst Research Instituteat ZRC SAZU

Guide to the Encyclopedia

The *Encyclopedia of Caves* is a complete source of information on the subject of caves and life in caves, contained within a single volume. Each article in the *Encyclopedia* provides an overview of the selected topic to inform a broad spectrum of readers, from biologists and geologists conducting research in related areas, to students and the interested general public.

In order that you, the reader, will derive the maximum benefit from the *Encyclopedia of Caves*, we have provided this Guide. It explains how the book is organized and how the information within its pages can be located.

SUBJECT AREAS

The *Encyclopedia of Caves* presents 128 separate articles on the entire range of speleological study. Articles in the *Encyclopedia* fall within 17 general subject areas, as follows:

- Types of Caves
- Cave Features
- Surface Karst Features
- Hydrology and Hydrogeology
- Speleothems and Other Cave Deposits
- Cave Ages and Paleoclimate
- Exceptional Caves
- Biology of Particular Organisms in Caves
- Communities and habitats
- Ecology
- Cave Invasion
- Biogeography and Dispersal
- Evolution and Adaptation in Caves
- Evolution and Adaptation in Caves
- Diversity
- Contemporary Use of Caves
- Historical Use of Caves
- Ground Water Contamination and Land Use Hazards in Cave Regions

ORGANIZATION

The *Encyclopedia of Caves* is organized to provide the maximum ease of use for its readers. All of the articles are arranged in a single alphabetical sequence by title. An alphabetical Table of Contents for the articles can be found beginning on page v of this introductory section.

So they can be more easily identified, article titles begin with the key word or phrase indicating the topic, with any descriptive terms following this. For example, "Invasion, Active versus Passive" is the title assigned to this article, rather than "Active versus Passive Invasion," because the specific term *Invasion* is the key word.

You can use this alphabetical Table of Contents by itself to locate a topic, or you can first identify the topic in the Contents by Subject Area on page x and then go to the alphabetical Table to find the page location.

ARTICLE FORMAT

Each article in the *Encyclopedia* begins with introductory text that defines the topic being discussed and indicates its significance. For example, the article "Behavioral Adaptations" begins as follows:

Animals living in darkness have to compete for food, mates, and space for undisturbed reproduction just as their epigean conspecifics do in the epigean habitats, but there is one striking difference: In light, animals can use visual signals. Thus, important aspects of their behavior driven by visual signals cannot apply in darkness. The question arises, then, of how cave dwellers compensate for this disadvantage in complete darkness. This article uses several examples to compare various behavior patterns among cave dwelling populations with epigean ancestors.

Major headings highlight important subtopics that are discussed in the article. For example, the article "Beetles" includes these topics: Adaptations, Colonization and Geographical Distribution, Systematics of Cave Beetles, Ecology, Importance and Protection.

CROSS-REFERENCES

Cross-references appear within the *Encyclopedia* as indications of related topics at the end of a particular article. As an example, a cross-reference at the end of an article can be found in the entry "Camps." This article concludes with the statement:

See Also the Following Articles

Recreational Caving • Exploration of Light Sources

This reference indicates that these related articles all provide some additional information about Camps.

BIBLIOGRAPHY

The Bibliography section appears as the last element of the article. This section lists recent secondary sources that will aid the reader in locating more detailed or technical information on the topic at hand. Review articles and research papers that are important to a more detailed understanding of the topic are also listed here. The Bibliography entries in this Encyclopedia are for the benefit of the reader to provide references for further reading or additional research on the given topic. Thus they typically consist of a limited number of entries. They are not intended to represent a complete listing of all the materials consulted by the author or authors in preparing the article. The Bibliography is in effect an extension of the article itself, and it represents the author's choice as to the best sources available for additional information.

INDEX

The Subject Index for the *Encyclopedia of Caves* contains more than 4600 entries. Within the entry for a given topic, references to general coverage of the topic appear first, such as a complete article on the subject. References to more specific aspects of the topic then appear below this in an indented list.

Preface

Throughout history, caves have always been of at least some interest to almost everyone. During the past few centuries, caves have been of passionate interest to at least a few people. The number of those with a passionate interest has been continuously growing. The core of the cave enthusiasts are, of course, the cave explorers. However, scientists of various sorts, mainly geologists and biologists, have also found caves useful and fascinating subjects for scientific study.

There have always been cave explorers. Some, such as E.A. Martel in France in the late 1800s, achieved amazing feats of exploration of deep alpine caves. In the United States, the number of individuals seriously interested in the exploration of caves has grown continuously since the 1940s. Cave exploration takes many forms. Some cavers are interested in caving simply as a recreational experience, not intrinsically different from hiking, rock climbing, or mountain biking. But many pursue genuine exploration. Their objective is the discovery of new cave passages never before seen by humans. As the more obvious entrances and the more accessible caves have been explored, cave exploration in the true sense of the word exploration, has become more elaborate and more difficult. To meet the challenge of larger, more obscure and more difficult caves, cavers have responded with the invention of new techniques, new equipment, and the training required to use them. To meet the challenge of long and difficult caves, cavers have been willing to accept the discipline of project and expedition caving and to accept the arduous tasks of surveying caves as they are explored. The result has been the accumulation of a tremendous wealth of information about caves that has been invaluable to those studying caves from a scientific point of view.

In the early years of the twentieth century, a few geologists became interested in the processes that allowed caves to form. Biologists were interested in the unique habitats and the specialized organisms that evolved there. In both sciences and in both Europe and the United States, the interest was in the caves themselves. The study of caves was focused inward and some proposed the study of caves to be a separate science called speleology. In Europe, largely due to the influence of the Romanian biologist Emil Racoviţă, other subterranean habitats were included in the study of cave life. In the latter decades of the twentieth century, there was a gradual change in perspective, and the study of caves came to be seen as important for its illumination of other realms of science.

In the past few decades, the geological study of caves has undergone a tremendous expansion in point of view. The caves themselves are no longer seen as simply geological oddities that need to be explained. Caves are repositories and are part of something larger. As repositories, the clastic sediments in caves and the speleothems in caves have been found to be records of past climatic and hydrologic conditions. Cave passages themselves are recognized as fragments of conduit systems that are or were an intrinsic part of the groundwater system. Active caves give direct insight into the hydrology and dry caves are records that tell something of how drainage systems have evolved. Techniques for the dating of cave deposits have locked down events much more accurately in the caves than on the land surface above. Caves then become an important marker for interpreting the evolution of the surface topography. Even the original, rather prosaic, problem of explaining the origin and development of caves has required delving into the chemistry of groundwater interactions with carbonate rocks and on the fluid mechanics of groundwater flow.

Cave biology has likewise evolved from an exercise in taxonomy—discovering, describing, and classifying organisms from caves—to the use of caves as natural laboratories for ecology and evolutionary studies. The central question that has occupied the attention of biologists at least since the time of Lamarck is how did animals come to lose their eyes and pigment. The question gets answered each generation using the scientific tools available and its most contemporary form is a question of the fate of eye genes themselves. Cave animals have also served as models for the study of adaptation because of their ability to survive in the harsh environments of caves. There are also interesting biological questions about the evolutionary history of cave animals that are being unraveled using a variety of contemporary techniques. Finally, there is increasing concern about the conservation of cave animals. Nearly all have very restricted ranges and many are found in only a single cave. The past two decades have seen a phenomenal growth in the understanding of how to manage cave and karst areas to protect the species that depend on them.

One should not suppose that caves are of interest only to geologists and biologists. Caves are repositories of archaeological and paleontological resources. Ancient art has been preserved in caves. Caves appear in folk tales, legends, mythology, and in the religions of many peoples. Caves appear frequently in literature, either as an interesting setting for the story or as a metaphor. The latter has a history extending at least back to Plato.

In planning the content of the first edition of the *Encyclopedia of Caves*, the editors were faced with this great variety of "clients" with their highly diverse interests in caves. Several decisions were made. One was that we would address the interests of as many "clients" as possible given the limitations of space. Thus, the Encyclopedia, in addition to the expected articles on biology and geology, also contains articles on exploration techniques, archaeology, and folklore. A second decision was to allow authors a reasonable page space so they could discuss their assigned subject in some depth. As a result of this decision, the Encyclopedia contains a smaller number of articles and thus a smaller number of subjects than might be expected. The object was to provide a good cross-section of contemporary knowledge of caves rather than attempt an entry for every possible subject.

The level of presentation was intended to be at the college level. In this way, the articles would have sufficient technical depth to be useful to specialists but would still be accessible to the general reader. Some of the subjects are intrinsically more technical than others, but we have attempted to keep to a minimum the specialist jargon and in particular the obnoxious acronyms that turn many technical subjects into a secret code known only to insiders.

The editors maintained that same criteria and guidelines in constructing the second edition. Authors of articles in the first edition were invited to revise and update their articles and most of them did. Many of the entries from the first edition have been completely revised and expanded; others received only minor updates. Many new articles have been added. The first edition contained 107 articles; the second edition contains 128. In addition, a few old entries were dropped and new articles with new authors were added.

The selection of authors was made by the editors. We attempted to select contributors who we knew were expert in the subject being requested of them. For many subjects there was certainly a choice of potential experts and our selection was to some degree arbitrary. We sincerely hope that no one is offended that some other person was selected rather than them.

We take this opportunity to thank the authors for their hard work. The Encyclopedia is a collective effort of many people in many disciplines. We are particularly appreciative of everyone's efforts to communicate with cave enthusiasts outside of their particular discipline.

William B. White
David C. Culver
May, 2011

A

ADAPTATION TO LOW FOOD

Kathrin Hüppop

Helgoland

INTRODUCTION

Subterranean environments are characterized not only by continuous darkness but also by a reduced variability in the number of specific abiotic conditions such as moisture, temperature, and water chemistry, as well as by isolation and restriction in space. Additionally, hypogean systems are relatively energy-limited compared to photosynthetically based epigean systems. As a response, many cave animals share numerous adaptations to the food scarcity of their environment. They not only show morphological and behavioral adaptations but also have evolved several special physiological characters, most notably *energy economy*, a reduction in energy consumption, which has a high selective advantage in cave animals and has been observed in numerous species in a variety of phyla (Poulson, 1963; Culver, 1982; Hüppop, 2000). Factors concerning adaptation to food scarcity in caves are illustrated in the causal network shown in Figure 1.

On one hand, the high environmental stability in caves, including darkness and sometimes predator scarcity, allows the evolution of characters; on the other hand, it requires character changes. In fact, most characteristics of adaptation to food scarcity can only be realized in ecologically stable and, above all, predator-poor caves (Fig. 1). Food scarcity acting as a selective force in caves requires adaptations. Possible adaptations of cave animals to survival in caves low in food are an improved food-finding ability, an improved starvation resistance, a reduced energy demand, and life history characters changed toward more *K*-selected features (Hüppop, 2000). Further, a higher food utilization efficiency, dietary shift, and feeding generalism may be realized. Many of these characteristics have evolved coincidentally, depending on the kind of food scarcity.

As a consequence, most real cave animals show no or only minor signs of malnutrition despite the low food availability in their environment.

TYPES OF FOOD SCARCITY

Not only the intensity but also the quality of the food scarcity and the duration of this selective force determine the degree of adaptation. Food scarcity in caves can have three facets: general food scarcity, periodic food supply, and patchy food scarcity. General food scarcity holds for nearly all caves and occurs especially in caves with no or low but continuous food input. Additionally, many caves are not stable throughout the year. Periodic food supply characterizes caves that are flooded periodically (normally several times during the rainy season) or caves with periodic food input by animals visiting the cave regularly. Seasonally flooded caves are subject to severe changes regarding food input, water quality, oxygen content, temperature, and competitors or predators. During the wet season, food supply can be very high and even abundant for some weeks or months. After exhaustion of these food reserves, animals in such caves suffer food scarcity like animals in generally food-poor caves. Some cave animals have to cope with patchy food scarcity. This means that food is not necessarily limited but is locally concentrated and difficult to find and exploit. Under such conditions, cave organisms can be observed aggregated at patchy food resources.

WHAT TO FEED IN CAVES

Food Input

The basic food resource in most caves is organic matter from external origin. Wind, percolating surface water, flooding, and streams provide input of many kinds of organic matter, such as detritus, microorganisms, feces, and accidental or dead animals. Some caves are visited

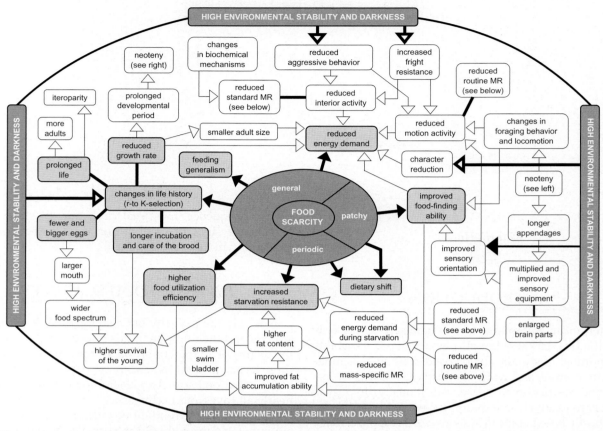

FIGURE 1 Network of characters observed in many cave animals, their causes, consequences, and interrelations. Dark gray: selection factors; light gray: main characters influenced by food scarcity; close arrowheads require something, open arrowheads allow something. *Redrawn after Hüppop, 2000.*

actively by epigean animals for shelter or reproduction. Such caves are much richer in food than are more isolated ones, because the visitors provide an additional food input in the form of their feces or their carcasses. Bat guano can present an immense source of food for guanobionts. Bacteria and above all microfungi decompose detritus and guano, thus building the basis for a food pyramid in caves. Lava tubes can be rich in food due to exudates from roots growing through the ceilings into the caves (Poulson and Lavoie, 2000).

Chemoautotrophy

As the only primary producers in caves, a few species of chemoautotrophic bacteria may support the survival of cave animals, especially in caves that have no natural entrance and where the absence of water infiltration from the surface excludes the input of photosynthetic food (Sarbu, 2000). However, these chemoautotrophic systems are quantitatively important in only a few exceptional caves, the best known example being the Movile Cave in Romania.

Influence of Cave Type

The amount of food supply in caves depends on the cave type, on surface connections, and on the geographic location. Generally, the food supply in tropical and subtropical caves is greater than in temperate ones because the biomass in the tropical epigeum is greater and its production is mostly uninterrupted (Poulson and Lavoie, 2000). As a consequence, selection pressure can be expected to be weaker, the evolutionary rate slower, and the appearance of troglobites not as fast in such caves compared to caves with low energy input such as temperate ones. In fact, troglobites are far more abundant in temperate zones than in the tropics, and species richness in caves is often correlated with the amount of available energy.

FOOD-FINDING

A variety of morphological and physiological adaptations and changes in foraging behavior are the basis for more efficient foraging and increased food-finding ability of cave animals. Such alterations are only advantageous under food scarcity and in darkness. In the case of high food–prey density or in light conditions, cave species are inferior to competing epigean relatives.

Appendages and Sensory Equipment

The most obvious morphological alterations in cave animals are longer legs, antennae, fins and barbels, or enlarged or flattened heads. If these body parts bear sensory organs, their enlarged surface can be correlated to an increased number of chemosensitive or mechanosensitive organs. As a consequence, an increased sensitivity to chemical and mechanical stimulants and changes in foraging behavior are possible. Cave animals can detect the food faster and at a greater distance from their bodies than can epigean ones and, as a side effect, spend less energy for food searching. All of these characters have been observed in a broad variety of taxa, from amphipods to crayfish, isopods, spiders, beetles, fish, salamanders, and more. Cavefish have been studied most intensively in this respect, mainly those of the famous North American cavefish family the Amblyopsidae, the Mexican characid fish *Astyanax fasciatus*, and cave salamanders.

In the Amblyopsidae, a positive trend in several of the specified troglomorphic features progresses from an epigean species over four gradually more cave-adapted species (Poulson, 1963). Adaptive alterations to the cave conditions are correlated with enlarged associated brain parts, whereas smaller optic lobes reflect the reduction of eyes as a consequence of darkness and uselessness. For cave salamanders, the most likely function of the elongated limbs is to raise the body and particularly the head above the cave floor to increase efficiency of the lateral-line system. They also permit the salamanders to search a larger area per unit of energy expended and thus increase feeding efficiency. In interstitial species, the evolution of appendage length is different than in cave species. Due to the small size of the interstitial gaps, they tend to have shorter appendages and a more worm-like appearance (Coineau, 2000).

Behavior

Changes in foraging behavior can also increase the food-finding ability. In the darkness of caves, a food-searching behavior concentrated on only the two-dimensional bottom or other surface areas can be much more economic in time and cost than a food search in a three-dimensional space, as exhibited by most surface animals in light and what they also try to do in darkness. Several cave animals have abandoned the shoaling or grouping behavior and adopted a continuous moving mode as a consequence of darkness and food scarcity in the cave habitat. They compensate for the optically orientated and spatially limited food-searching mode of epigean relatives by covering a greater area using chemo- and mechanosensors. The amblyopsid cavefish have developed a different swimming behavior, referred to as glide-and-rest swimming. This behavior, also enabled by the larger fins, not only conserves energy but also results in a reduction of interference noise for neuromast receptors, thus improving prey detection.

Other Factors

Most cave animals cope with food scarcity by taking a wide range of food or exhibiting a different food preference compared to surface relatives. Sometimes they show a dietary shift if one food source becomes scarce. A higher food utilization efficiency in cave animals as adaptation to the food scarcity is still not proven. Finally, the improvement of one feature sometimes may have more than one positive effect on the cave animals. The elaboration of the antennae in amphipods not only enhances food-finding ability, and thus survivorship, but also improves the mate-finding ability in populations with often low densities. Elongated bodies presumably facilitate movement through an interstitial medium.

Back to the Network

An improved, that is, a more efficient, food-finding ability is adaptive predominantly in patchily food-limited cave habitats and may be realized through changes in foraging behavior and improved ability for sensory orientation. The latter includes improvements not only in taste and smell senses but also in spatial orientation, as is required in the darkness of caves. A multiplied and improved sensory equipment can be causally connected with longer appendages and this can be a result of neoteny. An improved food-finding ability itself can reduce the general energy demand of cave animals or improve the ability of fat accumulation.

GENERAL ENERGY DEMAND

A reduced general energy demand is highly adaptive in the food scarcity of caves. It reflects a resistance not only to starvation during periodic food limitation or to general food scarcity but also to food patchiness, low oxygen content, or other abiotic factors in the cave environment. The energy demand of an animal is usually quantified by its metabolic rate. Meaningful information on the metabolic rate is given by the measurement of oxygen consumption of the entire organism or parts of it. In addition, indirect parameters such as respiratory frequency, resistance to anoxia, ability to survive starvation periods, body composition, growth rate, gill area, or turnover rate of adenosine triphosphate (ATP) have been used to compare metabolic rates, and in most investigations the metabolic rates of the hypogean species were found to be more or less lower than those of their epigean relatives.

Aquatic Cave Animals

Not only in caves but also in interstitial habitats, aquatic animals above all practice striking energy economy. Several hypogean amphipod, isopod, decapod, and fish species have been shown to live with metabolic rates much lower than those of surface relatives. The most detailed analysis of cave adaptation in fish (Poulson, 1963) demonstrates a decreasing trend in the metabolic rate from the epigean species in the Amblyopsidae over the troglophilic to gradually more cave-adapted ones. High fat reserves together with low metabolic rates explain the long survival time of the most troglobitic amblyopsid species when starved. However, high fat contents may lead to misinterpretations of metabolic rates. Because fat tissue is known to have a relatively low maintenance metabolism compared to other tissues or organs, lean body mass or bodies with comparable fat contents should be preferred as a metabolic reference to avoid misinterpretations of the metabolic rate. In a variety of the Mexican cavefish *A. fasciatus* a very high fat content (Table 1) suggested a reduced metabolic rate compared to the epigean relative (Hüppop, 2000; see Table 1). The recalculation of the metabolic rate, taking fat content into account, resulted in nearly identical values in both varieties of the fish species. Although obviously adapted to a periodically low energy environment (see "Periodic Starvation" below), as can be seen from the high fat content, the hypogean *A. fasciatus* were not yet able to reduce their energy turnover rate in adaptation to a general food scarcity.

Terrestrial Cave Animals

Only a few investigations on metabolic rates of terrestrial cave animals exist. Although food scarcity generally is even greater in the terrestrial than in the aquatic cave environment, only a few cave arthropod species were found to show a tendency toward energy economy.

Activity

Every activity increases the energy consumption of animals. The standard metabolic rate (*i.e.*, the lowest oxygen consumption rate that can be measured during a test) excludes motion activity and is a measure of the physiological adaptation of cave animals to food scarcity. However, the routine metabolic rate (*i.e.*, the mean metabolic rate over 24 hours, which includes spontaneous activity) is a more appropriate index of actual energy expenditures in nature; it actually may have the highest rank among the parameters determining adaptation to food scarcity in cave animals. The routine metabolic rate may be reduced in cave animals due to minimized motion activity, to changed motion patterns (temporal as well as morphological), to reduced or no-longer-practiced aggressive and territory behavior, or to reduced fright reactions. Actually, in most cave animals activity is reduced. Although an increase in food-finding ability in cave animals often seems to go along with an increase in food-searching activity, changed motion patterns result in a reduction of energy expenditure, sometimes to a fantastic extension. For example, in the most cave-adapted species of the amblyopsid fish in North America, over 90% of the total energy savings by adaptations are based on the reduced activity (Poulson, 1985).

Excitement and Aggression

Metabolic rates are definitely elevated by an animal's reaction to disturbance (excitement) and by aggressive behavior. The standard metabolic rate is elevated by interior activity or by excitement without expression in motion activity. The routine metabolic rate increases due to external activity, including motion activity resulting from excitement or aggression. In the Mexican characid fish *Astyanax fasciatus* the reduction of

TABLE 1 Fat Contents and Condition Factors of the Epigean and One Hypogean Variety (from La Cueva de El Pachon) of the Mexican Characid Fish *Astyanax fasciatus* in Relation to the Number of Experimental Starvation Days

		Days of Starvation		
		0	109	174
Fat content (% wet body mass)	epigean	9	2	
	hypogean	37	28	27
Fat content (% dry body mass)	epigean	27	8	
	hypogean	71	63	62
Condition factor (100 g cm^{-3})	epigean	2.0	1.6	
	hypogean	2.9	2.5	1.9

Source: *Adapted from Hüppop (2000).*

aggressive behavior as a consequence of the loss of vision in the darkness of caves was proven. An increased resistance to disturbance has been shown to be important for energy economy in the amblyopsid fish (Poulson, 1963). The generally low standard and routine metabolic rates of cave amblyopsids and their resistance to disturbance are interpreted not only as adaptations to the reduced food supply, by a factor of about 100 compared to the surface, but also as a by-product of relatively stable cave conditions and a general lack of predators in the amblyopsid cave environment.

The Conflicts of Body Size

The energy reserves of larger animals last longer and are more resistant to food shortage than are small animals because the metabolic rate of animals is not directly proportional to body mass but is related to mass by the following equation

$$\text{metabolic rate} = am^b$$

where a = intercept, m = wet body mass, and b = mass exponent/slope smaller than 1 (Withers, 1992). Consequently, subterranean animals have to resolve the conflict between two advantages: (1) to be larger with a lower energy demand per unit mass but a higher one per individual; or (2) to be small, thus requiring more energy per unit mass but less per animal, and/or being able to live in crevices.

A special case of subterranean habitat is the interstitial. In addition to food scarcity it is constrained by the grain size of the substrate. Interstitial animals are limited in size and shape due to the small size of the interstitial gaps and often have shortened appendages, excluding the posterior appendages, which tend to be elongated (Coineau, 2000). However, although they have a comparatively higher mass-specific routine metabolic rate than their surface relatives, their routine metabolic rate per individual is smaller. Thus, many interstitial forms may have reduced their body size not only to fit better into the small crevices but also to cope better individually with food scarcity in their special habitat. That the motion activity, necessary for foraging within the small crevices, and thus the routine metabolic rate can in turn be increased in interstitial animals (Danielopol et al., 1994) has a counterproductive effect.

Ectothermy and Neoteny

Troglobites are exclusively ectotherms. The generally very low metabolic rates of ectotherms (only 10 to 20% or even less that of similar sized endotherms) are the basis for their success in zones characterized by limited resource supplies, such as shortages in food, oxygen, or water. Ectotherms can utilize energy for reproduction that endotherms are forced to use for thermoregulation. Finally, ectotherms are able to exploit a world of small body sizes unavailable to endotherms. Body sizes less than 2 grams are not feasible for endotherms because the curve relating metabolism to body mass becomes asymptotic to the metabolism axis at body masses lower than 2 grams (Withers, 1992).

Within the ectothermic vertebrates, only fish and amphibians evolved cave species. Because they are the largest animals in cave communities, they usually represent the highest trophic level in the cave food web and can survive in large populations only in relatively food-rich caves. All troglobitic salamanders are aquatic throughout their life or at an early stage of their development. They often show the retention of larval characters, known as neoteny, which enables them to survive in a relatively less food-scarce aquatic cave habitat compared to the terrestrial cave habitat (Culver, 1982). Finally, suppression of the energetically expensive metamorphosis in hypogean salamanders can be interpreted as an adaptation to general food scarcity.

Hypoxic Conditions

Besides food scarcity, numerous cave or interstitial species have to cope with temporary, permanent, or patchy hypoxic conditions. Also, this character of some cave environments forces reduced metabolic rates and has been proven in crustaceans and fish. High amounts of fermentable fuels result in a more sustained supply for anaerobic metabolism, and glycogen utilization rates and lactate production rates are significantly lower in hypogean crustaceans. In contrast to surface species, several hypogean species have no sharp break in the oxygen uptake lines under depleting oxygen concentrations. This absence of a discontinuity in the oxygen uptake line is called *oxyregulation* and is considered to be adaptive in environments characterized by variable oxygen conditions (Danielopol et al., 1994).

Character Reduction

Many features become reduced during the evolution of cave animals. This regressive evolution is generally described as the reduction of "functionless" characters in cave animals. It not only concerns structural but also behavioral and physiological traits. Character reduction, that is, saving energy from not building or not maintaining useless characters when living in strongly food-limited cave environments, should be advantageous for cave animals: they can transfer the saved energy to the development or support of other characters or to growth, reproduction, or survival during starvation periods. There exist a few hints among beetles and spiders of this possible strategy of cave animals to adapt to a food-restricted cave environment. Nevertheless, more often the reduction of characters in cave animals seems to be the result of accumulated neutral mutations.

Back to the Network

A reduced energy demand is adaptive mainly in those caves that are generally low in food. A lowered interior activity, meaning a reduced standard metabolic rate, can be the result of an increased fright resistance, a lowered aggressive behavior, and changes in biochemical mechanisms. Reduced motion activity (*i.e.*, a lowered routine metabolic rate) can be achieved by means of reduced body movement to escape (a reduced number of) predators or for aggression, changes in foraging behavior and locomotion, and an improved sensory orientation, resulting in fewer movements for food searching. In the end, reduced metabolic rates result in a higher availability of energy for growth and/or a greater resistance to starvation. Additionally, character reduction, reduced growth rates, and smaller adult body size have the ability to reduce energy demand in cave animals. The reduced energy demand in cave animals can have two effects. Under the aspect of metabolic span (see "Longevity" below), a reduction per time is correlated with a prolonged lifetime combined with iteroparity. On the other hand, the reduction per individual life enables higher survival rates of individuals or even the increase of population size.

LIFE HISTORY CHARACTERS

The extremes of the spectrum of life history adaptations are characterized as *r*- and *K*-selection. Whereas *r*-selection (*r* being the slope of the population growth curve) means a trend toward high population growth rate under temporarily good conditions in relatively unpredictable and changing habitats, *K*-selection (*K* being the carrying capacity of the habitat) can be realized only in more predictable and stable habitats, and the appropriate fitness measure is the maximum lifetime reproduction. *K*-selected species are characterized by low or no population growth; they have reached a maximal *K*. This situation is connected with fewer but larger and more nutrient-rich eggs, increased time required for hatching, prolonged larvae stage, generally decreased growth rate, delayed and perhaps infrequent reproduction, increased longevity, and parental care. Many cave animals show a couple of these characters, demonstrating a trend toward more *K*-selection in cave species. The life history of cave animals has been the subject of some reviews; the main investigations were done on a variety of invertebrates, particularly crustaceans and arthropods, and on fish and salamanders (Hüppop, 2000).

Egg Size

Bigger eggs with more energy-rich yolk release larvae that are bigger at the time of yolk depletion when they have to start feeding on external food. These larvae have a bigger head with larger mouth, so they can start external feeding on a wider spectrum of food particles and may have a better chance of survival. For example, in the Mexican characid fish *Astyanax fasciatus* the eggs of several hypogean varieties and the larvae of the most adapted variety have significantly more yolk than those of the commensurate epigean conspecific relative (Fig. 2 and Fig. 3; Hüppop, 2000). The cave larvae live longer on their yolk reserve (4.5 versus 3 days), are longer when yolk is depleted (5 versus 4 mm), and start storage of adipose tissue already when fed *ad libitum* some days earlier than epigean larvae do (6 versus 9 days).

Bigger larvae may also have a higher resistance to starvation and a higher mobility for food searching and for effective escape reactions. In extreme, the trend toward fewer but bigger eggs in cave animals can result in a single larva per reproductive season that possibly never feeds, as in a cave beetle species.

Growth Rate

A reduced growth rate in cave animals is adaptive to food scarcity because it means a reduced energy demand per time. More animals can live on a defined amount of food, or a defined group of animals can survive longer on it. A reduction in energy demand per unit time through a lower metabolic rate, together with a reduction of absolute and relative costs of reproduction, can make possible an increase of population density and hence an increase in the number of females actually breeding per year.

Longevity

There is evidence that the total metabolic turnover in a lifetime not only of endotherms but also in ectotherms is the product of the energy turnover rate and the duration of life, called the *metabolic span*. Generally, lower metabolic rates and slower growth rates (that is, "slower living") are tied to increased longevity. Because the reproductive success of an animal might be defined by the ability to live long enough to survive the gap between good years, an increased lifetime in cave animals is advantageous in a generally food-scarce environment or in caves where relatively food-rich reproductive seasons occur irregularly. The increased longevity of cave animals connected with a delay in maturity and a trend from semelparity to iteroparity means that the population is less likely to disappear in years when food supply is too low to allow females to produce offspring.

A Case Study

Extremely prolonged lifetimes of more than 150 years are known among North American cave crayfish. However, the amblyopsid cavefish, intensively investigated by Poulson (1963), are the best-known example of how cave animals adapt their life history to food scarcity. Within this group of fish species it is obvious how cave animals with increasingly slower energy turnover rates have increasingly prolonged life cycles connected with many increasingly K-selected features, such as bigger and fewer eggs with prolonged developmental time, prolonged branchial incubation time (=parental care), bigger larvae at first external feeding, reduced growth rate, delayed maturity, and multiplied chances to reproduce with increasing cave adaptation (Table 2). Population growth rate and population density decrease with increasing phylogenetic age of the cave species, and the population structure shifts toward adults (Poulson, 1963).

Back to the Network

Life history changes toward more K-selected characters are correlated with a prolonged lifetime (and consequently iteroparity) and/or with a shift toward more adults in the population. Additionally, more K-selection may include bigger and fewer eggs and longer incubation and brood care, giving the offspring a higher chance of survival. A reduced growth rate and a smaller adult size may save energy. A prolonged developmental period may result in neoteny which in turn can have an influence on food-finding ability through appendage lengthening and changes in foraging behavior.

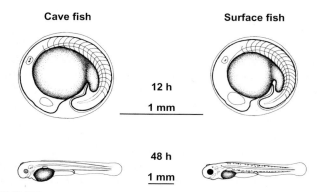

FIGURE 2 Egg size at an age of 12 hours and larvae size at an age of 48 hours of a cave variety (from La Cueva de El Pachon) and the epigean relative variety of the Mexican characid fish *Astyanax fasciatus*. After Hüppop (2000).

PERIODIC STARVATION

Besides a general food scarcity, many cave animals are faced with temporal periodicity of food; hence, they need an improved ability to survive long periods of starvation. Seasonality in caves, as already mentioned, is based on periodic flooding or on animals visiting the cave periodically, such as bats. Normally, this results in annual cycles.

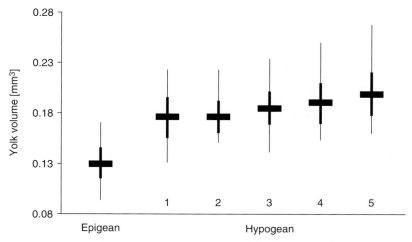

FIGURE 3 Yolk volume (mean, standard deviation, and range) at an age of 12 hours of five hypogean varieties of the Mexican characid fish *Astyanax fasciatus* (from La Cueva Chica, El Sotano de Yerbaniz, La Cueva del Rio Subterraneo, El Sotano de Las Piedras, and La Cueva de El Pachon) and the epigean relative. After Hüppop (2000).

TABLE 2 Some Life History Characters of the Epigean Amblyopsid Fish *Chologaster cornuta* Compared to the Troglobitic *Amblyopsis rosae*

	Ch. cornuta	A. rosae
Egg size (mm)	0.9 to 1.2	1.9 to 2.2
Developmental time	few weeks	5 to 6 months
Larval size (mm) when hatching	3	5
when leaving gill cavity	8	12
Adult size (mm)	23 to 55	36 to 48
Female body mass (g)	0.93	1.25
Adult maturity (year)	1	3
Maximal life span (years)	1	5 to 6
Number of mature ova per female	98	23
Average number of reproduction per lifetime	1	0.6
Maximum number of reproduction per lifetime	1	3

Source: *Adapted from Poulson (1963, 1985) and Culver (1982).*

Fat Accumulation

The main way to improve the survival capacity in periodically food-scarce cave environments is the accumulation of large amounts of adipose tissue during food-rich seasons. High lipid contents have been observed not only in cave animals but also in many surface species subjected to seasonal changes in food supply. The energy content per gram of lipids is roughly twice that of proteins or carbohydrates; therefore, fat accumulation is the best way to store energy. This may be achieved through excessive feeding, increased feeding efficiency, or improved metabolic pathways favoring lipid deposition. Cave animals build up fat reserves during the food-rich season and store them in their abdominal cavity, in subdermal layers, intra- or extracellularly in the hepatopancreas (decapods) or in the muscles, within the orbital sockets of the reduced eye, or in the more or less reduced swim bladder (as some fish do). Hypogean amphipods, decapods, remipedes, collembola, beetles, and several fish species all over the world have been observed to be able to survive starvation periods better than their epigean relatives. Several of them accumulate fat deposits, sometimes up to huge amounts. They are able to survive starvation periods from several weeks to one year (as proven with fish), probably even more. By producing eggs with more yolk, cave animals can enable their young to better resist short starvation periods.

A hypogean variety of the Pyrenean salamander *Calotriton asper* exhibited a hypermetabolism and higher glycogen in liver and muscles (+25%) and triglyceride stores in muscles (+50%) in the fed state than the epigean ones (Issartel et al., 2010). During experimental fasting the energetic reserves always remained higher in the hypogean individuals.

In the North American amblyopsid cavefish species the increasing ability to resist starvation is correlated with the different stages of various morphological adaptations (Poulson, 1963).

Individuals of a hypogean variety of the Mexican characid fish *Astyanax fasciatus* fed *ad libitum* in the laboratory were able to accumulate fat up to 71% of dry body mass compared to only 27% in the conspecific epigean fish variety (Hüppop, 2000). Not until after a starvation period of almost half a year did the condition factor of individuals of the hypogean variety of *A. fasciatus* fall below the condition factor of the epigean fish in an unstarved condition (Table 1).

Energy Demand

A lowered energy demand during starvation can mainly be attained by the reduction of the routine metabolic rate in minimizing the motion activity. But also the reduction of the standard metabolic rate in lowering internal activity is possible; even changes in biochemical mechanisms, for example, changes in electron transport system activity have been shown (Mezek et al., 2010).

In some hypogean crustaceans, the locomotory, ventilatory, and respiratory rates were drastically lowered during long-term starvation, whereas surface species showed lower decreases in these rates and responded by a marked and transitory hyperactivity (Hervant et al., 1997). In a state of temporary torpor the hypogean species were able to survive starvation periods largely longer than 200 days.

A hypogean variety of the Pyrenean salamander *Calotriton asper* exhibited a 20% decrease in oxygen consumption under experimental fasting, whereas epigean individuals experienced no significant change (Issartel et al., 2010).

Back to the Network

An augmented fat content is extremely useful in cave animals confronted with periodic food scarcity in that it increases their resistance to starvation. Cave animals may be able to improve their capacity of fat accumulation by an improved food-finding ability, perhaps supported by a higher food utilization efficiency. Lowered metabolic rates reduce energy demand during starvation and thus increase starvation resistance. An increased starvation resistance possibly also has an influence on the survival rate of the young.

See Also the Following Articles

Natural Selection
Food Sources

Bibliography

Coineau, N. (2000). Adaptations to interstitial groundwater life. In H. Wilkens, D. C. Culver, & W. F. Humphreys (Eds.), *Subterranean ecosystems* (pp. 189–210). Amsterdam: Elsevier.

Culver, D. C. (1982). *Cave life: Evolution and ecology*. Cambridge, MA: Harvard University Press.

Danielopol, D. L., Creuzé des Châtelliers, M., Mösslacher, F., Pospisil, P., & Popa, R. (1994). Adaptation of crustacea to interstitial habitats: A practical agenda for ecological studies. In J. Gibert, D. L. Danielopol & J. A. Stanford (Eds.), *Groundwater ecology*. New York: Academic Press.

Hervant, F., Mathieu, J., Barré, H., Simon, K., & Pinon, C. (1997). Comparative study on the behavioral, ventilatory, and respiratory responses of hypogean and epigean crustaceans to long-term starvation and subsequent feeding. *Comparative Biochemistry and Physiology*, A118(4), 1277–1283.

Hüppop, K. (2000). How do cave animals cope with the food scarcity in caves? In H. Wilkens, D. C. Culver, & W. F. Humphreys (Eds.), *Subterranean ecosystems* (pp. 189–210). Amsterdam: Elsevier.

Issartel, J., Voituron, Y., Guillaume, O., Clobert, J., & Hervant, F. (2010). Selection of physiological and metabolic adaptations to food deprivation in the Pyrenean newt *Calotriton asper* during cave colonisation. *Comparative Biochemistry and Physiology*, A155(1), 77–83.

Mezek, T., Simi, T., Arts, M. T., & Brancelj, A. (2010). Effect of fasting on hypogean (*Niphargus stygius*) and epigean (*Gammarus fossarum*) amphipods: A laboratory study. *Aquatic Ecology*, 44(2), 397–408.

Poulson, T. L. (1963). Cave adaptation in amblyopsid fishes. *American Midland Naturalist*, 70(2), 257–290.

Poulson, T. L. (1985). Evolutionary reduction by neutral mutations: Plausibility arguments and data from amblyopsid fishes and linyphiid spiders. *National Speleological Society Bulletin*, 47(2), 109–117.

Poulson, T. L., & Lavoie, K. H. (2000). The trophic basis of subsurface ecosystems. In H. Wilkens, D. C. Culver & W. F. Humphreys (Eds.), *Subterranean ecosystems* (pp. 231–249). Amsterdam: Elsevier.

Sarbu, S. M. (2000). Movile Cave: A chemoautotrophically based groundwater ecosystem. In H. Wilkens, D. C. Culver, & W. F. Humphreys (Eds.), *Subterranean ecosystems* (pp. 319–343). Amsterdam: Elsevier.

Withers, P. C. (1992). *Comparative animal physiology*. Fort Worth: Saunders College Publishing.

ADAPTIVE SHIFTS

Francis G. Howarth and Hannelore Hoch

Bishop Museum, Honolulu
Museum für Naturkunde, Berlin

THEORY OF ADAPTIVE SHIFT

For over a century, the evolution of obligate cave species (troglobites) was assumed to be restricted to continental regions that had been influenced by glacial events and that the ecological effects resulting from the severe climatic fluctuations provided the isolation and necessary elements to facilitate their evolution. Because most troglobites appear to have no surviving ancestors, it was assumed that they evolved allopatrically (*i.e.*, as a consequence of extinction of their closely related epigean species); however, it is possible that many temperate-zone troglobites evolved parapatrically through adaptive shifts and that their current isolation from surface relatives occurred after cave adaptation.

In the adaptive shift model, populations of epigean species invade subterranean habitats to exploit novel resources, and the drastic change in habitat is considered to be the driving force for genetic divergence and speciation. The presence of exploitable food resources provides the evolutionary "incentive" for adaptation. Evolutionary theory predicts that adaptation and new species are more likely to arise from large, expanding populations to exploit marginal resources at the edge of their habitat and to make adaptive shifts to exploit new environments than are small populations.

Adaptive shifts proceed in a three-step process. First, a new habitat or resource becomes available for exploitation. This opening of an ecological niche can occur either by an organism expanding its range or by geological or successional processes creating an opportunity for a resident population. Second, there is a shift in behavior to exploit the new habitat or food resource. Third, if the behavioral shift successfully establishes a new population, natural selection fosters adaptation. In a classic comparative study of cave and surface springtails, Christiansen (1965) demonstrated that a behavioral shift to exploit a new resource or survive a new environmental stress occurs first, and, if it is successful, morphological, physiological, and additional behavioral changes follow. Even if the populations remain in contact, the selection pressures imposed by the different environments and augmented by environmental stress can force the two populations to diverge.

Speciation by adaptive shift is parapatric in that adaptive differentiation, accompanied by a reduction in gene flow, proceeds across a steep environmental gradient within a contiguous area. If the colonization is successful, the founding population reproduces and expands into underground habitats. Maintenance of the barrier to gene flow could be facilitated by selection against hybridization (*e.g.*, decreased hybrid viability), assortative mating, or spread of the incipient cave population away from the narrow hybrid zone, thereby reducing the effect of introgression from epigean individuals. It has been hypothesized for cave species that selection for novel mating behaviors may be the principal origin of isolation (Hoch and Howarth, 1993). Many adaptive shifts undoubtedly fail; that is, the founding population dies out. Conversely, interbreeding and selection following a shift can produce a single more adaptable population capable of exploiting both environments.

FACTORS UNDERLYING ADAPTIVE SHIFTS INTO CAVE HABITATS

The trigger for an adaptive shift into caves is the availability of suitable habitat and exploitable resources. Only a small percentage of epigean groups have representatives inhabiting caves, and several factors appear to be involved in determining whether or not an organism can take advantage of the opportunity to exploit caves. These factors often act in concert but for clarity can be considered either *intrinsic* or *extrinsic*. Intrinsic factors are characteristics inherent in the organism that allow it to make the shift—for example, the ability to live in damp, dark habitats. Extrinsic factors are those imposed on the organism by the environment.

Intrinsic Factors

Preadaptation

The role of preadaptation has been recognized for more than a century as one of the principal factors explaining which taxa have successfully adapted to live in caves. Most preadaptations simply result from the correspondence of an organism's preferred environment with that found in caves. That is, organisms that have characteristics that allow them to live in damp, dark, wet-rock microhabitats on the surface have a better chance of surviving in caves than do organisms that do not possess these traits. Similarly, survival is enhanced if their normal food resource naturally occurs in caves. Nymphs of epigean cixiid planthoppers are admirably preadapted to caves as they feed on plant roots and therefore already have a suite of behavioral, morphological, and physiological adaptations to survive and feed underground. In the cave species, these preadapted nymphal characters have been retained into adulthood so that the adults are also able to live underground.

Genetic Repertoire

In order to adapt to caves, a population must have the ability to change. For example, if an organism is genetically hard-wired to require light, temperature change, or other environmental cues to complete its life cycle and reproduce, it is unlikely to colonize caves. In cixiids, salamanders, and a few other groups, the adaptive shift to living in caves was greatly facilitated by neoteny or the retention of nymphal characters into adulthood. Development of neoteny in salamanders is believed to involve relatively small changes in their genes that regulate development. This demonstrates that a small genetic change can have a large effect on phenotype.

Founder Events and Subsequent Population Flush and Crash Cycles

Because only a few individuals from the parent population make the initial shift, only a subset of the genetic diversity found in the parent population can be carried into the founding population. Certain alleles may be lost, while certain rare ones may become more abundant. Inbreeding, response to environmental stress, and expression of previously rare alleles may result in coevolved blocks of loci becoming destabilized, thus allowing new combinations of genes. If the new colony expands rapidly into the new habitat (*e.g.*, due to abundant food or reduced competition and predation), the relaxed selection pressure allows the survival of new mutations and unusual recombinants that ordinarily would be lost due to low fitness. Thus, the genetic variability may be quickly reestablished in the founding population; however, by this time there may be significant genetic divergence from the parent population. A subsequent population crash (as when the new colony exceeds the carrying capacity) places extreme new selection pressures on the population. These cycles of expansions and crashes in founding populations can facilitate adaptation to a new environment (Howard and Berlocher, 1998).

In parapatric divergence, occasional backcrosses with the parent population can increase genetic diversity and provide additional phenotypic variability on which natural selection can act. Mixing of the two gene pools by interbreeding becomes less likely as the new population expands its range beyond that of the parent population and adapts to the stresses of the new environment.

Response to Stress

Environmental stresses in a novel habitat may facilitate adaptive shifts (Howarth, 1993). Organisms living under environmental stresses often experience higher mutation rates and display greater phenotypic and genetic variation. In nature, selection and the high energetic costs of stress usually reduce this variation; however, where sufficient exploitable food energy compensates for the extra costs of survival in a novel habitat, some of the enhanced variation may survive and in time result in a new population adapted to cope with the stresses. The higher energetic costs required to cope with stress augment both natural and relaxed selection pressures to favor the loss of unused characters, such as eyes and bodily pigment of troglobites. The lower metabolic and fecundity rates possessed by troglobites may in large part be an adaptation to cope with stresses found in subterranean environments.

Mating Behavior and Hybridization

An important constraint among organisms colonizing caves is the ability to locate mates and reproduce underground. The normal cues used by a species may not be present or the mate recognition signal system may be confusing in the unusual environment. For example, sex pheromones would not disperse in the same way in caves as they do on the surface. Also, an animal may have difficulty following the pheromone plume in a three-dimensional dark maze. The rarity of troglobites that use airborne sounds to locate and choose a mate attests to the difficulty of using audible sound underground and suggests that muteness in crickets is a preadaptation for colonizing caves. In addition, environmental stress can reduce sexual selection and disrupt mating behaviors. Founder events can exacerbate these effects and the resulting release of sexual selection pressure may allow hybridization with relatives. The maintenance of limited gene flow between the diverging populations is believed to be important in providing the additional genetic variability to accelerate adaptation; however, the evolution of more appropriate mating behaviors may be the principal isolating mechanism separating a newly established cave population from its surface relatives.

Extrinsic Factors

Presence of Cavernicolous Habitats

Obviously, there must be an opportunity for a surface population to colonize caves; that is, the surface population must live in contact with cavernous landforms. Geological processes (such as lava flows) can create new cave habitats, ecological succession or the immigration of a new organism can introduce suitable new food resources into caves, or a pre-adapted surface population can migrate into the cave area. The Hawaiian Islands formed sequentially in line, and the successful colonization of each island after it emerged from the sea had to proceed in an orderly fashion. Phytophagous insects had to wait for their hosts to establish and so on. Terrestrial obligate cave species are not likely to have crossed the wide water gaps between each island. Even though suitable lava tubes and cave habitats were available from the beginning, the evolution of cave species had to wait for their surface ancestors to become established and for appropriate food resources to accumulate in caves. These circumstances provided tremendous advantages to the first organisms that could exploit a newly available resource and preempt the resource from subsequent colonists.

Presence of Exploitable Food Resources

An exploitable food resource is the paramount prerequisite that allows individuals of a population to shift into a new habitat or lifestyle. In fact, in most examples of adaptive shift, both the cave and surface species feed on the same food resource. Sparse soil and organic litter with areas of barren rock on the surface are characteristic of cavernous landforms, including those created by limestone, lava, and talus. Where an interconnected system of cave and cave-like subterranean voids occur, organic material does not accumulate on the surface but is washed underground by water, falls underground by gravity, or is carried underground by living organisms (Howarth, 1983). Many surface organisms will follow their food underground and attempt to exploit it. Even though their food may be abundant in caves, most organisms adapted to living in surface habitats may become lost in deep cave passages because their normal cues are reduced or absent or because they cannot cope with the harsh environment. Thus, much of this sinking material is outside the reach of most surface organisms; however, it becomes a rich reward for any organism that can adapt to exploit it.

Environmental Stresses

The environment in deep caves and cave-like voids is highly stressful for most surface animals, and relatively few surface species can survive for long underground. Even though food may be adequate, it is much more difficult to find and exploit in the complex three-dimensional dark maze, especially because many environmental cues (*e.g.*, light/dark cycles, temperature changes, and air currents) are absent. Furthermore, the atmosphere is characteristically above the equilibrium humidity of body fluids, making respiration and water balance difficult to sustain (Howarth, 1980). The substrate is moist, and the voids occasionally flood. Rock-lined pools are often pitfall traps for small animals. Carbon dioxide and oxygen concentrations can reach stressful levels. In many cave regions, radiation from decaying heavy isotopes may exceed hazardous levels. Because of the maze-like space, escape to a less stressful environment is virtually impossible, and cave animals must either adapt to cope with these and other stresses *in situ* or die out.

Ancestral Habitats

Troglobites originated from a variety of damp surface habitats. Specifically, the ancestral habitats include soil, leaf litter, mosses, and other damp microhabitats in forests; damp, wet-rock habitats (such as marine littoral, riparian, and cracks in barren lava flows); and guano-inhabiting species living in caves and animal burrows. These source habitats corroborate the concept of preadaptation.

CASE STUDIES

Hawai'i

Twelve arthropod lineages have independently adapted to caves on two or more islands in Hawai'i, indicating that cave adaptation is a general phenomenon fostered by ecological and evolutionary factors. On the young (less than one million years) Hawai'i Island, some taxa have undergone (or are undergoing) adaptive shifts into both caves and new surface habitats in a phenomenon called *adaptive radiation*. Many of these troglobites still have extant close surface relatives living in nearby habitats. The better known among these are listed in Table 1 and discussed below. The number of examples will increase as more taxa are studied. Cave species on the older islands tend to be relicts today; however, they also are believed to have originally evolved by adaptive shifts.

Cixiid Planthoppers (Hemiptera: Fulgoromorpha: Cixiidae)

Cixiid planthoppers exemplify how adaptive shifts might occur (Fig. 1). The nymphs of nearly all species in the family live in or close to the soil and feed on the xylem sap of roots. Cixiid nymphs of surface species molt to adults, emerge above ground, and live, feed, and mate on vegetation. Plants living on or near cavernous landforms often send their roots through the interconnected system of voids deep underground following nutrients and percolating groundwater. Thus, roots are often an important source of food energy in caves, especially in the tropics and lava tubes. Nymphs of surface species may migrate along roots deep into caves from their normal shallow habitat. These nymphs can exploit deep roots, as the deep habitat is similar to their normal one except for the perpetually saturated atmosphere, and for this stress they already have mechanisms to excrete excess water as xylem sap is very dilute. When they molt, the adults may have little chance to find their way to the surface to reproduce and would eventually die from the stressful environment; however, if host plant roots were abundant and large numbers of nymphs continually made their way underground, an adult might occasionally find a mate and reproduce fortuitously. Most of these early chance matings, if successful at all, would produce normal surface adults that would also become lost. Eventually, some offspring might acquire traits from random mutation or genetic recombination that are passed on and that allow the offspring to survive and reproduce underground more easily. If this incipient colony successfully establishes an underground population, it could more fully exploit the deeper root resources. In fact, a whole new habitat and resource would be opened to the expanding population as it could move throughout

TABLE 1 Parapatric Cave and Surface Species-Pairs Occurring on Hawai'i Island

Cave Species	Common Name	Surface Relative	Ancestral Habitat
Littorophiloscia species	Slater, Wood louse	*Littorophiloscia hawaiiensis*	Marine littoral
Oliarus makaiki	Cixiid planthopper	*Oliarus koanoa*	Mesic forest
Oliarus polyphemus*	Cixiid planthopper	*Oliarus* species	Rain forest
Oliarus lorettae	Cixiid planthopper	*Oliarus* species	Dry shrub land
Schrankia howarthi*	Moth	*Schrankia howarthi**	Rain forest
		Schrankia altivolans*	Dry to wet forests
Nesidiolestes ana*	Thread-legged bug	*Nesidiolestes selium*	Cryptic habitats in rain forest
Caconemobius varius*	Rock cricket	*Caconemobius fori**	Barren lava flows
		C. sandwichensis	Marine littoral
Anisolabis howarthi	Earwig	*Anisolabis maritima*	Marine littoral
		Anisolabis hawaiiensis	Barren lava flows?
Lycosa howarthi	Wolf spider	*Lycosa* cf *hawaiiensis**	Barren lava flows

*Represented by several distinct populations or species. Polymorphic cave populations may represent separate invasions or could result from divergence of a single lineage after cave adaptation.

the interconnected voids in the cave region and leave its surface relatives behind. Adaptation to the new environment might be rapid in the invasive expanding phase. Subsequently, the newly established subterranean population will acquire the troglomorphic traits characteristic for obligate cavernicoles: reduction of eyes, wings, and bodily pigment. The surface environment on cavernous landforms provides little suitable habitat for epigean planthoppers, so that the subterranean population is often larger than surface populations over caves. This is especially true on volcanoes where eruptions can create vast new habitats for cave populations. In Hawai'i, the principal pioneering plant on young lava flows is the endemic tree *Metrosideros polymorpha* (Myrtaceae), the roots of which are the hosts of *Oliarus polyphemus s.l.* (*sensu lato*) nymphs. Even widely scattered small trees on 100-year-old flows send abundant roots deep into caves. *O. polyphemus* can migrate underground to colonize new caves from neighboring older lava flows, and in fact they have been found in caves less than 25 years old.

In Hawai'i, the genus *Oliarus* has undergone extensive adaptive radiation in surface habitats after the successful colonization of a single ancestral species, and about 80 endemic species and subspecies have been described from all major islands. Eighteen surface species are known from the island of Hawai'i, where at least three independent adaptive shifts into subterranean habitats have occurred (Table 1). The clearest case for adaptive shift among cave cixiids involves *O. makaiki* and *O. koanoa*. *O. makaiki* has reduced wings, eyes, and bodily pigment and is known from a single 1500- to 3000-year-old lava tube from Hualālai Volcano. Its male genitalia (which usually provide an excellent means for distinguishing species in the group) are virtually identical to the fully winged and eyed epigean *O. koanoa*. *O. koanoa* is widespread in mesic forests on Hawai'i. Its nymphs are frequent in shallow caves and even occur near the entrance in the same cave with *O. makaiki*. Adults of *O. koanoa* may be accidental in caves, as accumulations of body parts from thousands of dead individuals have been found in dead-end passages in some caves (Hoch and Howarth, 1999).

The highly troglomorphic *O. polyphemus s.l.* (Fig. 2) also occurs in the same cave with *O. makaiki*, as well as in virtually every suitable cave containing its host roots, *Metrosideros polymorpha*; however, each cave system appears to harbor a unique population of this troglobite. Each population differs only slightly in morphology, but the mating calls are highly distinctive and probably sufficient to reduce hybridization if the populations were to come in contact (Hoch and Howarth, 1993). The ancestor of *O. polyphemus s.l.* has not been determined, although it certainly belongs in the endemic Hawaiian group of species. It is also unknown whether *O. polyphemus s.l.* represents a single colonization event in caves with subsequent divergence underground or whether some populations represent separate invasions by the same or closely related surface species.

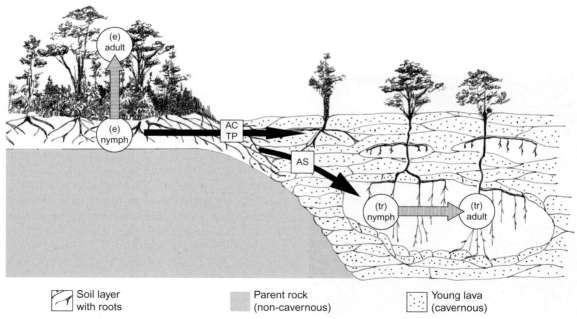

FIGURE 1 Diagrammatic view of an adaptive shift from surface to underground environments by a cixiid planthopper. AC: accidental; AS: adaptive shift; (e): epigean; TP: troglophile; (tr): troglobite.

Adult planthoppers use substrate-borne sounds to recognize and locate their mates. Among surface species, females feed on the host plant while the males fly from host to host and initiate a species-specific mating song. If a receptive female hears an appropriate song, she will answer with her own song. The male orients to the call, and the two alternately sing and listen until the male finds the female, who remains *in situ*. Studies have shown that Hawaiian cave *Oliarus* essentially use the same mate recognition system: substrate-borne vibrations, with living roots being an excellent transmitting medium. In laboratory studies on the cave cixiid *Oliarus polyphemus s.l.*, the flightless females feed on the host root and occasionally call. Males are also flightless and wander from root to root, listening. When a male hears a female, he answers, and if the song is appropriate the two sing and listen alternately until the male locates the stationary female. Although the behavior has not been confirmed in natural settings in caves, it is believed that this switch in behavior occurred as an adaptation to the subterranean environment. Singing is energetically costly, and males expend extra energy finding roots in the dark, three-dimensional maze. Also, predators can use these songs to locate prey, and males would be especially vulnerable if they had to sing on every root they found; therefore, males would be more likely to find a female if they did not have to initiate calling. Females conserve energy by staying in place and feeding, and they produce fine wax filaments that deter predators (Hoch and Howarth 1993).

Terrestrial Isopods (Crustacea: Isopoda: Philosciidae)

Species in the genus *Littorophiloscia* are marine littoral and inhabit tropical shorelines worldwide. One epigean species, *L. hawaiiensis*, is endemic to Hawai'i and is so far known only from littoral habitats on the islands of Hawai'i and Laysan. Surprisingly, an undescribed troglobitic species also lives on Hawai'i Island. The cave form is remarkably similar in morphology to *L. hawaiiensis*, differing only in the troglomorphic characters (reduced eyes and pigment) displayed by the cave population. The males even share the same distinctive sexual characters. Results from a recent molecular phylogenetic study showed that *L. hawaiiensis* and the cave *Littorophiloscia* are distinct species that diverged from each other most likely by an adaptive shift (Rivera *et al.*, 2002). The initial adaptive shift must have been from salty marine littoral habitats to freshwater terrestrial habitats, possibly along shorelines of coastal springs in young lava flows. Suitable freshwater shorelines are not well developed on Hawai'i Island, but lava tubes would offer a vast new habitat. The complex three-dimensional system of anastomosing voids characteristic of young basaltic lava flows provides an immeasurably large habitat for cave animals. In contrast, *L. hawaiiensis* is restricted to a narrow band along the coast where it lives under rocks resting on soil. Thus, an adapting cave population would quickly move away from contact with its ancestral population and greatly exceed it in size.

A possible alternative explanation is the classical model of cave adaptation in which the coastal species became extinct, stranding an incipient cave population; however, this model still requires that the adaptive shift to freshwater (and possibly to caves) had to occur before extinction, and that circumstance is precluded in the classical model. This scenario also requires a subsequent colonization by the halophilic littoral species that secondarily reestablished parapatry.

Crickets (Orthoptera: Gryllidae)

On the island of Hawai'i, the marine littoral rock cricket *Caconemobius sandwichensis* has colonized both unvegetated lava flows and caves, both of which are barren-rock habitats like its ancestral home. Several distinct populations of rock crickets inhabit each of these inland environments, but it is unknown whether these represent multiple adaptive shifts from the marine littoral habitat or result from divergence after colonization of the new habitat, or both. Even though the marine littoral and lava flow cricket (*C. fori*) species are distinct morphologically, a hybrid zone between the two was found around a pool of brackish water in a deep crack where a young lava flow entered the sea, suggesting that the adaptive shift is ongoing. *C. fori* is a nocturnal scavenger on very young (*i.e.*, one month to about a century old) lava flows on Kīlauea, disappearing when plants colonize the flow. It hides in deep

FIGURE 2 Adult female of *Oliarus polyphemus*. Photo by W.P. Mull. Used with permission.

cracks and caves during the day where it overlaps with several distinct cave populations or species. Additional recognizable populations of lava crickets, resembling *C. fori*, occur allopatrically on Mauna Loa and Hualālai volcanoes, also on Hawai'i Island.

Up to three morphologically, behaviorally, and physiologically distinct populations of cave crickets can be found in larger caves, and individual caves may harbor a unique population of one or more of these forms. The total diversity of cave-adapted rock crickets on Hawai'i is astounding, especially given the young age of the island. These multiple simultaneous adaptive shifts into such different habitats from the same ancestor further corroborate the contention that cave adaptation via adaptive shifts is a general phenomenon.

A possible scenario for the origin of troglobitic rock crickets and wolf spiders on the island of Hawai'i might proceed like this. Marine littoral crickets could have been among the earliest colonists of newly emerging islands, because all that they required was already present: a rocky shoreline and ocean-derived flotsam. Barren lava was the original terrestrial habitat, and sea birds would have sought these new islands for nesting, presenting an inviting habitat for the littoral crickets to shift to the land where they could feed on wind-borne debris. Subsequent accumulation of organic material underground would have allowed the colonization of caves by either the seacoast or lava cricket preempting some resources, possibly before many plants and potential competitors arrived on the island.

Wolf Spiders (Arachnida: Lycosidae)

The big-eyed wolf spiders (family Lycosidae) live on barren landscapes in many parts of the world and are good long-distance dispersers. They would also have been among the earliest colonists on emerging islands in Hawai'i. There are several barren-ground populations of an unidentified species of *Lycosa* (possibly *L. hawaiiensis*) on Hawai'i Island that differ in behavior, morphology, and color pattern, but surprisingly they can hybridize. They occupy a range of habitats from hot, dry barren lava on coastal plains to the freezing stone deserts above 4000 meters on Mauna Kea and Mauna Loa. On young lava flows, they prey on *Caconemobius* lava crickets. One member of this group followed its prey underground, becoming the anomalous small-eyed, big-eyed wolf spider (*Lycosa howarthi*).

The lycosid wolf spiders are characterized by the possession of four huge eyes (for spiders) and four smaller eyes, and they are among the better-sighted spiders. The troglobitic species has six to eight vestigial eyes, yet differs from the lava flow species only by the characters associated with cave adaptation. As noted, both the epigean spider and cricket were early colonists on Hawai'i, and possibly the spider was able to preempt the cave habitat because it was among the first predators to have the opportunity to exploit the young caves on the island.

Moths (Lepidoptera: Noctuidae)

Recent molecular analysis of mitochondrial (mtDNA) and nuclear (nDNA) of cave and surface *Schrankia* has documented another example of ongoing adaptive shifts into caves (Medeiros et al., 2009). Based largely on wing color pattern and external morphology, five endemic species had been described from surface habitats in Hawai'i. Three species were recorded from both Hawai'i and Maui islands. In addition, two distinct forms were known from caves on Hawai'i and Maui islands: a pale, small-winged, and vestigial-eyed cave form and a drably patterned, larger winged, sighted form found in the twilight and entrances of caves. The two were believed to be separate species, but intermediate forms occurred in larger caves, suggesting hybridization. The DNA analysis revealed a different and more interesting story. All five surface species were shown to belong to a single highly polymorphic species *Schrankia altivolans*, and all cave populations belong to another polymorphic species *S. howarthi*. The two species are closely related but can also be separated by the form of the male genitalia as well as wing pattern. The DNA sequences confirm that hybridization occurs between the cave and twilight zone morphs. However, they also show that *S. howarthi* is segregating into distinct geographic populations. For example, the Maui *S. howarthi* form a distinct group within the Hawai'i Island group, indicating that *S. howarthi* arose on Hawai'i and subsequently dispersed to Maui where it independently became cave-adapted. The west Hawai'i and montane Mauna Loa populations each also form distinct groups, whereas populations from southeast Hawai'i are more diverse and include members related to the other groups. The latter happenstance might be explained by its youth as the caves sampled are all less than 700 years old, and nearly the entire surface of Kīlauea is probably less than 10,000 years old. Surprisingly, the montane *S. howarthi* group is closer to *S. altivolans* than it is to the remaining *S. howarthi*. This might result from lineage sorting or past hybridization, but it also suggests independent adaptation to caves.

Schrankia larvae feed on succulent new growth of roots and occasionally on rotting organic matter, and thus are preadapted to live in caves. Like *Oliarus*, cave *Schrankia* have adapted to exploit plant roots in the vast system of voids in young lava flows. However, *Schrankia* use close-range pheromones to locate mates, and unlike *Oliarus*, cave-adapted *Schrankia* are not fully isolated reproductively from their surface relatives. Volant males would be able to find mates by following the female scent plume at least in larger cave passages. Pheromone communication probably would be less effective within the smaller voids. By necessity, all specimens used in the DNA study came from accessible cave passages. Moths living in the more isolated, inaccessible smaller voids might include a greater proportion of cave-adapted individuals than the populations sampled.

Other Islands

Adaptive shifts are best known from islands in large part because of their youth and isolation. Thus, many diverging species pairs are still extant, and their evolutionary history has not been obscured by geological or ecological events. Examples of troglobite evolution by adaptive shift are known from the Galapagos, Canary, and Greater Antilles islands. In the Galapagos, at least ten species of troglobitic terrestrial invertebrates still live parapatrically with their putative epigean ancestors (Peck and Finston, 1993). At least seven extant parapatric pairs of troglobitic and epigean species are known from the Canary Islands (Oromí et al., 1991). Some of these, such as the *Dysdera* spiders (Arnedo et al., 2007), *Trechus* ground beetles (Contreras-Díaz et al., 2007), and cockroaches, represent multiple invasions into caves, in parallel with Hawai'i.

Continents

Clear examples of adaptive shifts in continental caves are relatively rare, possibly because the great age of these systems has obscured the evolutionary history of the taxa involved or because the phenomenon has not generally been considered; nevertheless, there are a few cases. Perhaps the best documented is the detailed study (Culver et al., 1995) done on the spring- and cave-inhabiting amphipod *Gammarus minus*, which occurs in freshwater basins in temperate eastern North America. That study demonstrated that several cave populations were derived from independent invasions from springs and that some hybridization had occurred; however, the constraints imposed by the dynamics of water flow and other factors may have isolated some populations from each other. In spite of considerable genetic and morphological diversity, these populations are all considered one species. In parallel with some Hawaiian forms, the definition of what constitutes a species is sometimes not clear. Adaptive shifts have been proposed to explain the origins of cave species in North Queensland, Australia, but the few phylogenetic analyses completed to date have not supported the contention.

CONCLUSIONS

In cavernous regions, organic energy is continually being transported into an extensive system of subterranean voids, as evidenced by the characteristic presence of areas with a barren exposed rocky surface. Epigean species can exploit food resources on and near the surface, but increased environmental stress levels and absent or inappropriate cues prohibit their access to this resource in the deeper voids. Animals following their normal food deeper may eventually become lost and be unable to return to the surface. Surface species continue to be frequent accidentals in caves and provide a large proportion of the food to the cave-adapted predators and scavengers. Over time, a few of these accidentals may eventually survive and establish a new population. A major initial factor facing a new cave population may be the ability to locate a mate and reproduce underground. Once a population is able to live and reproduce underground, a large new habitat would be opened up to it, and the population could expand rapidly. Evolution of troglobites by adaptive shift may be a common phenomenon; however, many examples of adaptive shift remain unrecognized, especially because the phenomenon is not generally considered. Other scenarios (such as allopatric evolution, which appears to be supported for many temperate troglobites) are also conceivable. The adaptive shift model and the classical theory of troglobite evolution may not be mutually exclusive.

THE FUTURE

Adaptive shifts offer extraordinary opportunities for research. Comparative studies of closely related pairs of species adapted to such different environments as caves and surface habitats should provide better understanding not only of the processes of cave adaptation but also of selection and species formation in general. Molecular systematics will clarify the evolutionary history of taxonomic groups as well as elucidate the genetic basis of adaptive shifts. Phylogenetic analyses can elucidate the evolutionary transformation of specific morphological, behavioral, physiological, and ecological traits that occurred during the process of adaptation to novel habitats. The remarkable divergence from their epigean relatives displayed by troglobites makes such studies especially interesting. Comparative physiological experiments can potentially dissect out for study the specific mechanisms involved in coping with a variety of environmental stresses. For example, terrestrial cave species live in an air-filled aquatic environment, while the close relatives of some live in desert-like environments. These species pairs are ideal models for studies on the physiological mechanisms involved in regulating water balance.

See Also the Following Article

Root Communities in Lava Tubes

Bibliography

Arnedo, M. A., Oromí, P., Múrria, C., Macías-Hernández, N. E., & Ribera, C. (2007). The dark side of an island radiation: Systematics and evolution of troglobitic spiders of the genus *Dysdera* (Araneae, Dysderidae) in the Canary Islands. *Invertebrate Systematics, 21*, 623–660.

Christiansen, K. (1965). Behavior and form in the evolution of cave Collembola. *Evolution, 19*, 529–537.

Contreras-Díaz, H. G., Moya, O., Oromí, P., & Juan, C. (2007). Evolution and diversification of the forest and hypogean ground-beetle genus *Trechus* in the Canary Islands. *Molecular Phylogenetics and Evolution, 42*, 687–699.

Culver, D. C., Kane, T. C., & Fong, D. W. (1995). *Adaptation and natural selection in caves: The evolution of Gammarus minus.* Cambridge, MA: Harvard University Press.

Hoch, H., & Howarth, F. G. (1993). Evolutionary dynamics of behavioral divergence among populations of the Hawaiian cave-dwelling planthopper *Oliarus polyphemus* (Homoptera: Fulgoroidea: Cixiidae). *Pacific Science, 47*, 303–318.

Hoch, H., & Howarth, F. G. (1999). Multiple cave invasions by species of the planthopper genus *Oliarus* in Hawaii (Homoptera: Fulgoroidea: Cixiidae). *Zoological Journal of the Linnean Society, 127*, 453–475.

Howard, D. J., & Berlocher, S. H. (Eds.), (1998). *Endless forms: Species and speciation* New York: Oxford University Press.

Howarth, F. G. (1980). The zoogeography of specialized cave animals: A bioclimatic model. *Evolution, 34*, 394–406.

Howarth, F. G. (1983). Ecology of cave arthropods. *Annual Review of Entomology, 28*, 365–389.

Howarth, F. G. (1993). High-stress subterranean habitats and evolutionary change in cave-inhabiting arthropods. *American Naturalist, 142*, S65–S77.

Medeiros, M. J., Davis, D., Howarth, F. G., & Gillespie, R. (2009). Evolution of cave living in Hawaiian *Schrankia* (Lepidoptera: Noctuidae) with description of a remarkable new cave species. *Zoological Journal of the Linnean Society, 156*, 114–139.

Oromí, P., Martin, J. L., Medina, A. L., & Izquierdo, I. (1991). The evolution of the hypogean fauna in the Canary Islands. In E. C. Dudley (Ed.), *The unity of evolutionary biology* (Vol. 1). Portland, OR: Dioscorides Press. pp. 380–395.

Rivera, M. A. J., Howarth, F. G., Taiti, S., & Roderick, G. K. (2002). Evolution in Hawaiian cave-adapted isopods (Oniscidea: Philosciidae): Vicariant speciation or adaptive shift? *Molecular Phylogenetics and Evolution, 25*, 1–9.

Peck, S. B., & Finston, T. (1993). Galapagos islands troglobites: The questions of tropical troglobites, parapatric distributions with eyed-sister-species, and their origin by parapatric speciation. *Mémoires de Biospéologie, 20*, 19–37.

ANCHIHALINE (ANCHIALINE) CAVES AND FAUNA

Boris Sket

Univerza v Ljubljani, Slovenia

Anchihaline (or anchialine) habitats are water bodies in hollows along the sea coasts where the influence of the sea may be felt and which are inhabited by some subterranean species. This usually indicates an underground connection of the cave or pool with the sea. Such a habitat may contain seawater, but it primarily has layers of different brackish salinities. In exceptional cases, it has freshwater but is inhabited by some animals of the recent marine provenance (Sket, 1996). We must, however, exclude genuine sea caves, "spaces ... containing a high number of speleophilous forms of the rocky littoral with exclusion of any speleobiotic species" (Riedl, 1966).

In principle, anchihaline habitats can occur everywhere along the coast, but hardly any have been noted and investigated outside the tropical or warm moderate climatic zones. Geologically, two very different types of coasts are often of interest. Some anchihaline caves are in karstified limestones, while others are in lava fields. Because the sea level along most coasts during the Ice Ages was approximately 100 meters lower than it is now, the coasts may be hollowed out to such a depth. They often contain dripstone formations (such as stalactites and stalagmites), indicating their own previous continental and terrestrial nature. The limestones may be of different ages; some being old and primarily compact, with others being quite recently formed and still primarily porous coral limestones or little transformed coral reefs. Similarly, anchihaline habitats are also formed in lava flows, either in systems of cracks or in the tunnel-shaped lava tubes.

GEOGRAPHY AND SOME HISTORICAL DATA

Although well known since ancient times, the so-called blue caves in the Mediterranean (*e.g.*, Capri, Italy; Biševo, Croatia) were never investigated speleobiologically, as their morphology does not promise extensive anchihaline habitats in them. A number of localities with anchihaline caves have been discovered and investigated along tropical seas and in the Mediterranean. Since the middle of the nineteenth century, brackish anchihaline habitats of some fish and shrimp species in Cuba have been known. Another historical site of this kind is Jameos del Agua, a segment of a lava tube on the island of Lanzarote (Islas Canarias). The twentieth century began with the discovery of interesting fauna in some south Italian caves. Important anchihaline caves were later sampled in Libya and in islands of the Caribbean.

Riedl (1966), who systematically studied sea caves of the Mediterranean, conceived the concept of the *marginal cave* (*Randhöhle* in German), which was similar to the present-day concept of the anchihaline cave. Riedl defined the marginal cave by its position at the sea coast, the presence of brackish water, and the presence of troglobiotic species in it, but he also supposed that the origin of such caves was due to wave action and their gradual loss of connection to the sea. Thus, they should be a vector of the stygobiotic fauna originating in the sea and making it limnic.

An important study by Holthuis (1973) followed Riedl's studies and dealt with a number of shrimp species, including the now well-known red shrimps, as well as with a new concept of *anchihaline pools* that was free of suppositions about the past and the future of the habitat and therefore easier to apply. However, Holthuis obtained his animals from open pools in Hawaii and on the Red Sea coast, which contained some troglomorphic animals along with a very scarce selection of normal marine animals. In the period that followed, Stock and his numerous colleagues investigated a number of anchihaline caves in the Caribbean, mainly in regard to their faunistic and taxonomic features but also biogeographically. This region appeared to be particularly rich in such caves and fauna (*e.g.*, Stock, 1994). Stock suggested and substantiated the spelling change of *anchialine* to *anchihaline*.

Sket studied the caves along the Adriatic coast and gave the first detailed quantitative data about ecological conditions in them (Sket, 1986). Later, he also specified the concept of the anchihaline cave habitat. Iliffe (2001) is noted for his sampling of a great number of caves in the Caribbean region as well as in the tropical islands of the Indo-Pacific. His studies revealed a high number of new animal species representing an array of groups. In the meantime, a number of other field researchers and taxonomists were either sampling caves more locally or describing and studying collected animals.

Yager (1981) discovered the new crustacean group Remipedia, probably the most astonishing discovery in anchihaline waters. About the time of this discovery, research intensified into the blue holes and inland anchihaline caves in the Bahamas. Por, Fosshagen, and Boxshall studied Copepoda and Tantulocarida; Kornicker and Danielopol studied Ostracoda; Stock and Holsinger studied Amphipoda; Holthuis and others studied Caridea; and Stock and Botosaneanu studied Isopoda. Jaume noted euhaline anchihaline caves in the Mediterranean (*e.g.*, Jaume and Boxshall, 1996), and Por described hyperhaline pools in the Arabian Peninsula. A number of authors discussed the models and hypotheses regarding the colonization of subterranean habitats by animals in which the anchihaline caves played a non-negligible role.

CAVE MORPHOLOGY AND HYDROLOGY

(See Por, 1985; Sket, 1986, 1996; Iliffe, 2001.) The morphology of the anchihaline caves and other cavities containing anchihaline habitats is of low importance for their fauna; however, it may be extremely important for its accessibility (particularly for the time before SCUBA techniques were introduced) and to gain an understanding of its composition and distribution, as well as relations between species. Horizontal caves that are either filled with water along the entire corridor or that exhibit series of pools with different connections between themselves and with the sea are comparatively scarce. Such caves can be either karst caves or lava tubes. Horizontal coastal caves may, however, be very extensive inundated cave systems, such as those being investigated along the Mexican coasts.

More common are shallow vertical caves. They may be of small dimensions like the natural wells on the Adriatic islands in Croatia, which are sometimes artificially elaborated as watering places for sheep. Similar are *grietas* in some Caribbean islands or cracks in lava fields of the Galapagos; however, accessible vertical parts are always connected with horizontal corridors or fissures that may connect groups of caves. *Cenotes* have large dimensions and are found mainly in Mexico's Yucatán Peninsula; these, again, are often extended by systems of horizontal corridors. An important characteristic of such habitats is that a significant part of the aqueous body is illuminated by daylight.

Even more open are anchihaline pools in lava or in coral reefs. Such pools must be connected to the sea and may be interconnected by systems of more or less dark corridors or fissures, although the open space may prevail. Such pools are particularly well developed in Hawaii.

Due to a U-shaped tube effect (phreatic loop), in caves close to coasts under the influence of tides, the tides may also be felt underground, although with a certain delay and decline. This means that in the mild cases, such as in the Adriatic coastal caves, the entire stratified column of water moves twice a day up and down, carrying the floating organisms with itself and remarkably changing the life conditions for substratum-bound animals, particularly in the reach of the halocline. The tides in the open Adriatic are only approximately 0.5 meters, but more extreme conditions may occur elsewhere. In the Bahamian blue holes, the effects of very important tidal currents (dangerous for a diver) have not yet been studied for their influence on living conditions. In the horizontal St. Paul's Underground River corridor (Palawan, Philippines), the seawater moves with the tides for some kilometers in and out along the bottom. It is difficult, however, to imagine that any anchihaline fauna exists in such a cave.

The aqueous body is seldom purely marine (euhaline) but such are present in some deeply set submarine caves that have to be very extensive to contain any troglobionts. However, at least the water in lower portions of the cave systems in the underground of the

Bahamas (Porter, 1993), as well as some caves in the Mediterranean islands (Jaume, 1996) is euhaline. When influenced by freshwater, the situation may be very different for a cave on a small island than one that is on the mainland coast. In any case, the relation between the fresh and marine part of the column is governed by the Ghyben–Herzberg principle. In a porous aquifer, if the freshwater overlies seawater, the interface will be below the sea level approximately 40 times the height of the water above sea level. On an island, the freshwater body will take on the form of a lens (Ghyben–Herzberg lens), the depth of which depends on the precipitation at the site and the speed of the water outflow toward the sea. The shape of the freshwater lens and the height of the underlying layers of seawater will change slowly during the year and from year to year. Less predictable and more complex is the situation on the coast of the continent (or of a larger island), where canalized (thus concentrated) water inflow from inland may be the result of net-like systems of corridors. This may complicate the salinity distribution and will sometimes change the salinity relations very rapidly.

The transition from euhaline to limnic salinities in the column is never perfectly gradual. The depth of very gradually stratified or even homogeneous waters is regularly divided by a halocline that can be a few decimeters to a couple of meters thick. The halocline is a layer where the salinity abruptly changes (sometimes from polyhaline to nearly limnic). In genuine caves—such as in caves in general—the water temperature may be constant and equal throughout the column, approximately equal to the local yearly average. If the water body is exposed to the surface temperature influences, this may also cause temperature stratification, changing its direction throughout the year. During the winter in moderate climates, the surface cooling in freshwater lakes causes the loss of stratification and penetration of cooler (*i.e.*, heavier) and well-aerated surface water toward the bottom. In anchihaline waters, the density changes caused by varying temperatures are much lower than the density differences caused by a small salinity difference. With cooling from $20°C$ to $4°C$, the density of pure water changes from 998 to $1000 \, g \, L^{-1}$, while with the increase of salinity from 10 to 20 parts per thousand, the density changes from 1009 to $1019 \, g \, L^{-1}$. This makes the water column in mixohaline anchihaline caves very stable and prevents mixing of layers, with consequences as explained below.

Although the source of organic input can appear rather obvious, some sources have also been traced by sophisticated isotope methods (Pohlman *et al.*, 2001). In caves of Mexico, the soil particles may originate from a tropical rain forest, while in the Mediterranean these particles may be from the poor remains of soil in deep karren. The use of the holes as watering places, however, may cause the introduction of very high quantities of debris which is deposited on the bottom. Another source of organic matter is production in illuminated parts of caves. Rich planctonic algal communities may be present there, while rocky walls can also be overgrown by microscopic algae. The virtual contribution of autotrophic bacteria has only been supposed. Rich sources of oxidizable substances can be used in such a way. Some nitrate production (which may be by nitrite oxidation) has been noted within the halocline in a Mexican cave. On the other hand, remarkable hydrogen sulfide (H_2S) concentrations in many anchihaline caves are a strong basis for chemoautotrophy; clouds of bacteria within the halocline and mats covering the bottom have been observed in some caves, and the sulfide-oxidizing mixotrophic bacteria of the genera *Beggiatoa* and *Thiothrix* have been identified.

Wherever the input of organic substances is remarkable, they are deposited on the bottom where they putrefy, often causing total consumption of oxygen and an accumulation of hydrogen sulfide. If communication with the open sea is feeble, a thick lower layer of the water may be infested by this gas. A 250-$mmol \, L^{-1}$ concentration was established in a Mexican cave, while such remarkable concentrations have only been confirmed by smell in Adriatic caves. Such an accumulation can be identified up to the water surface in extreme cases.

The oxygen that is consumed in cave waters must be renewed from air on the surface, by an influx of precipitation water and water jets from inland, at the surface, or by mixing or diffusion from the open sea in deeper layers. Therefore, the surface layer is usually comparatively well aerated while the oxygen concentration diminishes toward the halocline; below the halocline it may rise again or decrease further. It is the rule that dissolved oxygen does not reach its saturation in anchihaline waters. Its depletion may be remarkable, probably total, although no analysis until now has been accurate enough to establish this. Such are the conditions close to the debris-covered bottom under a deep layer of hydrogen sulfide–rich water. In some other caves, the oxygen saturation is the lowest or close to zero within the halocline, the most stable water layer. Characteristically, the pH has also always been lowest in the halocline when measured.

NONCRUSTACEAN GROUPS IN THE ANCHIHALINE FAUNA

Sponges are represented very richly in true marine caves, sometimes with some deep-sea elements, such as the calcified *Calcarea pharetronida*. This is particularly evident in some sack-like Mediterranean caves that capture and hold cold water throughout the year and also harbor small glass sponges (Hexactinellida) and the exceptional carnivorous Cladorhizidae (*Asbestopluma hypogea* Vacelet and Boury-Esnault) (Vacelet et al., 1994). Some species are claimed to be true troglobionts of anchihaline caves, such as the haplosclerid *Pellina penicilliformis* Van Soest et Sass in marine caves in the Bahamas and the axinellid *Higginsia ciccaresei* Pansini et Pesce in southern Italy; the latter is a typical anchihaline animal living in brackish water.

Very little is known about gastropods in anchihaline caves. The presence of the tiny and mostly interstitial marine *Caecum* spp. is most probably in connection with some sand deposits within the cave. Only some of the apparently hydrobioid snails in Adriatic caves have yet been identified.

Among the few filter feeders are some tube worms occurring quite regularly at high salinities, such as *Filogranula annulata* (O.G. Costa) in Adriatic caves. Some stygobiotic errant polychaetes have been described, such as, for example, *Gesiella jameensis* (Hartmann—Schroeder) from the Canaries. Probably more sediment-dependent are the tiny members of the family Nerillidae. It is remarkable that the Dinaric cave tube worm (*Marifugia cavatica* Absolon and Hrabe) has never been found in a cave within the influence of the sea. Some oligochaetes occurring in such caves do not seem to be closely related to them.

Some species of sessile chaetognaths (family Spadellidae) occur in euhaline anchihaline waters, and at least the eyeless and pigmentless *Paraspadella anops* Bowman et Bieri from the Bahamas is supposed to be stygobiotic; some other species from caves are oculated.

Fishes are among the longest known anchihaline animals; approximately 10 anchihaline species are known today. The family Ophidiidae is exapted for such habitats by their shy behavior, nocturnal habits, and/or deep-sea prevalence. The stygobiotic *Lucifuga subterraneus* Poey and *Stygicola dentatus* Poey have been recognized in Cuban *grietas* and similar localities since the middle of the nineteenth century. New, related species have recently been found in the region. The stygobiotic and euryhaline *Ogilbia galapagosensis* (Poll et Leleup) from underground waters of the Galapagos Islands has variably reduced eyes and a closely eyed relative in the species *O. deroyi* (Van Mol), which dwells in littoral crevasses. In Yucatán caves, the ophidiid *Ogilbia pearsei* (Hubbs) is a neighbor of *Ophisternon infernale* (Hubbs), representing another family of often cryptically living fishes, Synbranchidae. The synbranchid *Ophisternon candidum* (Mees), at up to 370 mm in length, is probably the longest cave animal.

It inhabits brackish waters in northwestern Australia, accompanied by the gobiid *Mylieringa veritas* Whitley. It is remarkable that related (congeneric) species occur in anchihaline caves in mutually very remote areas.

ANCHIHALINE CRUSTACEA

Generally, Crustacea are by far the richest and most diversely represented group in subterranean waters, and the same holds for anchihaline cave waters. The predominance of crustaceans may also be felt within a particular cave where the specimen number or the biomass relation between crustaceans and noncrustaceans may be 10:1 or higher. Among Crustacea, by far the richest in species are Amphipoda and Copepoda, although Decapoda (due to their size) are probably the most obvious. For Remipedia, the anchihaline caves are the only known habitat, and it seems to be the central one also for Thermosbaenacea.

Remipedia are crustaceans with by far the most plesiomorphic trunk and appendages but with a highly specialized (and diversified) anterior end of its body, a very nice case of mosaic evolution. They have up to now only been found in anchihaline caves in euhaline salinities, or close to them, usually below the halocline in poorly aerated water (typically less than 1 mg L^{-1} oxygen). Remipedia are hermaphrodites. Up to 45-mm-long, centipede-shaped swimmers have been found in the tropical belt connecting Mexico, the Bahamas, and the Canaries and along the northwestern Australian coast; they seem to be absent in the Mediterranean. In some Bahamian caves, a number of species may occur together; *Speleonectes lucayensis* Yager was the first found and described.

Copepods are the richest group of *Entomostraca*, which include approximately 25% of crustacean species either in the whole or in subterranean habitats, but only 14% of those in anchihaline waters. In subterranean freshwaters, Harpacticoida are several times richer than Cyclopoida, while Calanoida are negligible (approximately 440:150:10 as known in 1986); in anchihaline waters, the harpacticoids are not so important, while calanoids are comparatively well represented. Each of these groups has some ecologically endemic families, or nearly so, such as Superornatiremidae in Harpacticoida, Epacteriscidae in Calanoida, and Speleoithonidae in Cyclopoida. According to existing data, a family may be also geographically extremely limited, such as the calanoid Boholinidae in the Philippines. Of particular interest is the small and relatively plesiomorphic group Misophrioida, an essentially (but not exclusively) deep-sea group discovered recently in Atlantic and Indo-Pacific tropical anchihaline caves.

Ostracods are easily recognizable as tiny crustaceans with an extremely shortened body hidden between two valves. The order Halocyprida is well represented with species of very diversely shaped and sculptured valves. Among them, the genus *Danielopolina* is interesting for its presence in both, the anchihaline waters and in the deep sea.

Leptostracans are small malacostracans with a bivalved carapace on the anterior part of the body and stalked eyes. *Speonebalia cannoni* Bowman, Yager, and Iliffe from a cenote-like cave in the Caicos Islands is eyeless.

Decapods are the best-known crustaceans represented in anchihaline caves, primarily by different kinds of shrimps. Members of the small family Procarididae have only been found in such habitats in the tropical Atlantic and in Hawaii; Agostocarididae seem to be restricted to the Caribbean region. Atyidae are a huge family of freshwater shrimps, represented at the sea nearly exclusively by some genera in anchihaline waters. The most remarkable is the genus *Typhlatya*, which is found with some closely related genera exhibiting a range of polyhaline to limnic species distributed circumtropically; they also exhibit different degrees of troglomorphism. The atyid *Antecaridina lauensis* (Edmondson) alone is very widely distributed—Madagascar, Red Sea, Hawaii—which is testament to the dispersion possibilities of some shrimps outside the anchihaline habitat. This has been also demonstrated for Palaemonidae, represented by the genus *Macrobrachium*, for example, which repeats some distributional characteristics of the family Atyidae. The palaemonid *Typhlocaris lethaea* Parisi from Libya is probably the largest cave invertebrate at all. A series of anchihaline genera belong to the essentially marine family Alpheidae. The alpheid *Barbouria cubensis* (Von Martens) in the Caribbean region has been known since the beginning of the twentieth century; the Indo-Pacific *Parhippolyte uveae* Borradaile, again one of the famous red shrimps, has been known from the end of the nineteenth century. Some Grapsidae crabs, such as *Orcovita* spp., have been found recently, although they are mostly not very troglomorphic. The only anomuran, the historical *Munidopsis polymorpha* Koelbel (family Galatheidae), from the inundated lava tube in the Canary Islands, belongs to an essentially deep-sea genus.

Thermosbaenaceans are small (up to 5 mm long) and delicate crustaceans with a short carapax under which the female broods her eggs; all are stygobiotic. Although the first species was found in hot (approximately 45°C) freshwater springs, and some species inhabit fresh continental waters, most are nevertheless anchihaline. The group is distributed in the Tethyan fashion, which will be described later. As an example, *Halosbaena acanthura* Stock is widely spread in the Caribbean, inhabiting interstitial and cave waters of brackish to hyperhaline salinities. Related species are known from the Canaries and northwestern Australia in almost freshwater. *Monodella halophila* S. Karaman, found in caves along the northeastern Adriatic coast, often most densely populates the least aerated water layers of a polyhaline salinity.

Mysids are small and delicate shrimp-like crustaceans with a ventral brood chamber. It is essentially a marine group that inhabits anchihaline caves circumtropically, with some exceptional species in southern Italy. Such is also the distribution of the genus *Spelaeomysis*, which resembles the atyid shrimp *Typhlatya* in its geographical and ecological distribution and the span of morphological adaptations. The widely distributed and nontroglomorphic species *S. cardisomae* Bowman seems to penetrate the entire variably brackish part of the groundwater body in the Caribbean island of San Andres, occurring even in the putrefied and probably deoxygenated waters in the land crab burrows.

Mictaceans slightly resemble the thermosbaenaceans. This is a small group of some deep-sea and two marine cavernicolous species in the Bermudas (*Mictocaris halope* Bowman et Iliffe) and the Bahamas.

Isopods are one of the major groups of Malacostraca, characterized mainly by their foliaceous abdominal appendages (pleopods); they are very diverse in shape and size. Since early in the twentieth century, some *Typhlocirolana* spp. from the Mediterranean basin have been known; some are anchihaline while some are from freshwaters. Similarly old is the history of *Annina lacustris* Budde—Lund from east Africa, which is a very euryoecious, nontroglomorphic animal that also occurs in mixohaline anchihaline caves. All of these animals belong to the family Cirolanidae, which is essentially marine, and the great majority of freshwater species is stygobiotic. A number of genera and species have been found in anchihaline waters, particularly in the Caribbean (such as *Bahalana mayana* Bowman in Mexico and the Yucatán), but also in the Mediterranean and in Indo-Pacific areas and elsewhere, such as *Haptolana pholeta* Bruce and Humphreys in western Australia. Similar is the frame habitat selection of the genus *Cyathura* (Anthuridae), although these animals are strongly sediment bound and burrow, while cirolanids are free walkers and even occasional swimmers. The stygobiont anthurids are only present in the tropics.

The presence of tiny species of the families Microcerberidae and Microparasellidae in anchihaline waters such as artificial wells is mostly due to sand deposits, as they are essentially interstitial animals. Genuine cave animals are evidently some members of the related (*i.e.*, asellote) families of Janiridae (*Trogloianiropsis lloberai* Jaume) in the Balearic Islands; Stenetriidae (*Neostenetroides schotteae* Ortiz, Lalana, et Perez) in Cuba; and Gnathostenetroididae (*Stenobermuda mergens*

Botosaneanu et Iliffe) in the Bahamas, which were found in Caribbean to Bermudian anchihalines, as well as in Mediterranean caves. The family Atlantasellidae seems to be endemic to anchihaline caves of the tropical Atlantic (Bermuda and the Caribbean).

Amphipoda seem to be by far the richest anchihaline group (more than twice the species number of any other crustacean group or three times all noncrustacean species), although the effect of numerous researchers cannot be excluded. The family level taxonomy of this group (counted to 6000 or even 8600 world species) is not yet settled, but it has to be used for the sake of surveying. The most characteristic group is comprised of Hadziidae and Melitidae with particularly high diversities of genera in the Caribbean region (e.g., *Metaniphargus, Bahadzia*) and Mediterranean (e.g., *Hadzia*) and with a number of scattered taxa elsewhere (e.g., *Liagoceradocus* spp.); Western Australia seems to represent the third such diversity center (with the genus *Nedsia*, for example). The group as a whole, as well as some genera, are represented by species of different salinity preferences, from euhaline to limnic. On a smaller scale and limited to the Mediterranean and Atlantic, is the genus *Pseudoniphargus*; the majority of its species are freshwater but never very far from the sea. Taxonomically remote is the family Ingolfiellidae; although including some marine species, it also includes ancient freshwater species in the center of Africa, for example. This family also has some anchihaline species.

Some taxonomically and geographically isolated species seem to be more or less direct descendants of marine relatives, such as *Antronicippe serrata* Stock and Iliffe of Pardaliscidae, from anchihaline caves in the Galapagos Islands.

At least most species or groups mentioned above are primarily marine, conquering continental waters with more or less success. Different are some amphipods of the Dinaric (i.e., Adriatic) anchihaline caves. Some species of *Niphargus* are present there; *N. hebereri* Schellenberg is moderately widely distributed, although not obligate, while the narrowly endemic *N. pectencoronatae* Sket seems to be an obligate anchihaline species (Fig. 1). Niphargids are a large group of essentially freshwater animals and only secondarily did they adapt to the brackish water; *N. hebereri* is very euryhaline and also in other respects a very tolerant species.

BIOGEOGRAPHY

Different patterns of geographical distribution can be seen in anchihaline cave faunas. One of the most remarkable patterns is the so-called Tethyan distribution (Humphreys, 2001), if historical grounds are emphasized, or circumtropical distribution, if we suppose ecological connections. The Tethys Sea encircled the globe in the Mesozoic period, when the supercontinent Pangaea split into a northern Laurasia and southern Gondwanaland. It was only in the Tertiary when southern and northern continents came again so close to each other that they disconnected this circumtropical seaway; note that even the Mediterranean was completely dry for some period in the Miocene. A number of animal groups, even genera and some species, have distribution ranges that reflect the possible circumglobal distribution within the Tethys. We may presume that particularly in the tropical belt a number of Tethyan inhabitants have survived and many of them succeeded in colonizing marginal waters, avoiding in this way competition with modern sea fauna. These include the Remipedia, Thermosbaenacea, hadziid amphipods, the *Typhlatya*-related group of shrimp genera, and even the mysid genus *Spelaeomysis*, the thermosbaenacean genus *Halosbaena*, and some others. However, it is difficult to believe that some circumtropically spread species, such as the shrimp *Antecaridina lauensis*, could conserve its morphological unity without a recent or nearly recent gene flow among populations; its genetic unity, however, has not been proven. An open question is also what occurred during the Messinian salinity crisis. Did the Tethyan elements temporarily change for freshwaters? Maybe this is connected with the existence of some freshwater populations of some elements such as *Hadzia fragilis* S. Karaman in the Adriatic region (Fig. 2).

A certain number of those Tethyan elements characteristically skip the Mediterranean in their west—east distribution, such as Remipedia and the genus Halosbaena; therefore, it is also possible that some Tethyan elements disappeared from the dried-out Messinian Mediterranean, which was later colonized again by only some of them. The third possibility is their disappearance during the colder climates in place after the Tertiary.

Other distribution patterns are more locally restricted. An instructive one is the paralittoral distribution

FIGURE 1 The amphipod *Niphyargus pectencoronatae*. Photo by Boris Sket.

FIGURE 2 The amphipod *Hadzia fragilis*. Photo by Boris Sket.

area along the northeastern (Dinaric) coast of the Adriatic. Typical anchihaline taxa (such as *Hadzia fragilis* and *Monodella halophila*) are present along the coast, including the islands, in brackish waters. They are, however, absent in such waters within the gulf of Kvarner (Quarnero) which was not yet marine prior to the Pleistocene. These species are replaced in the Kvarner anchihaline caves by generally continental species, such as *Niphargus arbiter* G. Karaman, while *Hadzia* is represented there by freshwater populations. This is clear evidence of a historical—not a purely ecological—basis for the species distribution in anchihaline habitats.

BIOLOGY AND ECOLOGICAL DISTRIBUTION OF INHABITANTS

Unfortunately, for most anchihaline species the ecological inclinations are virtually unknown. Some groups are evidently represented only by euhaline or at least polyhaline species which are all unquestionable immigrants from the deep sea or close relatives of deep-sea animals: the glass sponges and carnivorous sponges in the Mediterranean sea caves, the anomuran *Munidopsis polymorpha* in the Canary Islands, and the mictacean *Mictocaris halope* in the Bermudas. There are also unique representatives of some marine groups such as chaetognaths or leptostracans (*Paraspadella anops* and *Speonebalia cannoni*); however, the completely anchihaline group Remipedia also seems to be polyhaline.

Most anchihaline animals are to some degree euryhaline, with different optima more or less expressed. The distribution of shrimp species in a series of pools in Hawaii seems to depend mostly on different salinities. They are able to adapt their position in the water body to other local conditions, particularly the presence of competitors or predators. The physically delicate thermosbaenacean *Monodella halophila*, along the Adriatic coast, is limited to comparatively high salinities, although its very close relative *M. argentarii* Stella inhabits freshwaters in the Italian Monte Argentario. The Adriatic caves are inhabited by some (most of them by two) amphipod species which may be predatory on *Monodella*. This might also explain the (supposed) absence of *Monodella* in some Caribbean islands where caves were probably not investigated meters deep and where hadziid amphipods were regularly found. Some essentially limnic species, such as the copepod *Diacyclops antrincola* Kiefer or the shrimp *Troglocaris* sp. may be present at higher salinities than the supposed marine derivatives *Hadzia fragilis* and *Salentinella angelieri* Ruffo et Delamare—Deboutteville, while the primarily limnic *Niphargus hebereri* Schellenberg reaches 30 parts per thousand salinity, which is close to marine values.

It was observed in some Adriatic anchihaline waters that at least some highly troglomorphic animals do not avoid sunlit parts of the water body if they are not forced to do so by competing surface animals. The population of *Niphargus hebereri* was the most dense at the water surface in a shaft-like cave. Such a behavior (as well as similar situations in continental cave waters) teaches us that light and dark relations are not of direct importance for all cave species. This also excuses us when we treat the open anchihaline pools jointly with anchihaline caves. The so-called red shrimps are very characteristic inhabitants of those pools, and they show quite different relations toward sunshine. The atyid *Halocaridina rubra* Holthuis also appears in open anchihaline pools during sunshine, while *Antecaridina lauensis* is more photophobic but is able to change its color from red to white translucent and back in response to light and disturbance. The hippolytid *Parhippolyte uveae* Borradaile hides in cracks after sunset and during cloudy days, while it reappears in ponds in masses when the sun strikes the water; it is also able to rapidly change its coloration. One has to add, however, that these observations were made in single localities, and we do not know if such was the behavior of the species in any conditions.

Again, very characteristic is the sometimes apparently positive (in fact, probably neutral) relation of anchihaline animals toward low oxygen concentrations in water. This discussion has unfortunately some very serious limitations. First, due to some analysis problems at very low oxygen concentrations, we do not know if cave animals have been found in totally anoxic or just deeply suboxic water; however, the presence of H_2S in a thick layer speaks in favor of a total lack of oxygen. Second, particularly high densities of some animal colonies within the anoxic (or suboxic) water layers are most probably a result of competition (or predation) avoidance; however, this speculative conclusion has never been subjected to an experimental verification. Anyway, at extremely low oxygen concentrations and in the presence of H_2S (not quantitatively evaluated) euhaline Remipedia were found commonly or regularly; the mesohaline *Monodella halophila*, often; and the originally limnic *Niphargus hebereri*, quite so. Resistance to extremely low oxygen concentrations has been experimentally studied and verified in some freshwater cave crustaceans; it may be explained by their characteristically low metabolism.

Thus, based on the evidence available, anchihaline animals are in general very euryecious (generalist) toward a number of ecological parameters. One of the most durable among stygobionts is *Niphargus hebereri*, which has been found in limnic conditions (kilometers far inland) and in mixohaline waters of up to 30 parts per thousand salinity; in very food-poor habitats and in deposits of decaying organic debris; in well-oxygenated waters as well as in apparently anoxic layers; in waters smelling strongly of H_2S; and in a sink polluted with mineral oil.

TROPHIC RELATIONS

Supposedly, the richest source of food is input from the Earth's surface, while some food may be produced in the cave itself, as mentioned above. It is safe to say that the currents reaching into a euhaline marine cave are bringing food particles as well as pelagic larvae. Therefore, such caves are usually populated with a depauperate marine fauna, and the stygobionts are to some degree excluded by competition. As the most modest users of organic matter and as ecological generalists (as shown above), stygobionts are able to populate the rest of the coastal underground. Some interesting studies of the organic matter transport within genuinely marine and anchihaline caves have been made, but generalizations are still not possible. The sessile suspension (filter) feeders are nearly absent; some sponges and tube worms are on the border between the marine and anchihaline realm. Very few observations have been done about the behavior of those crustaceans that possess putative filtering structures, such as atyid shrimps and thermosbaenaceans. They seem to be predominantly on the bottom and collect organic detritus particles or bacteria from the sediment. Probably more pelagic is the rich assortment of anchihaline calanoid copepods. On the other hand, some inhabitants are proven predators, such as the procaridid shrimp *Procaris ascensionis* Chace, that feed on other shrimps. The maxillipeds of Remipedia are explicitly predatory, while on the fang tip of their maxilla I, a poison gland discharges. Most probably, the majority of anchihaline amphipods are generalistic, feeding on detritus and carcasses as well as on live animal prey.

THEORETICAL IMPORTANCE OF ANCHIHALINE HABITATS

The position of anchihaline habitats on the doorstep between the epigean and hypogean realms, as well as between marine and freshwater environments, makes them very intriguing for theoretical exploitation. Several theories about the colonization of the underground include anchihaline habitats, either implicitly or explicitly. However, the first question that has to be answered regards the origin of the anchihaline fauna itself. The marine provenance of the majority is out of the question if we look at the taxonomic composition. Only one local fauna is known for which a higher percentage of members are freshwater by origin (anchihaline fauna along the Adriatic coast). The fact that the band inhabited by it was out of the sea during the last glaciation is not peculiar to that area. Probably, it was the particularly rich continental cave fauna that succeeded to rule anchihaline waters and thus prevented intrusion of other biota.

A much-discussed question has been the relation of anchihaline fauna to the deep sea. Their phylogenetic proximities have been discussed previously. At one time it was thought that anchihaline species were directly derived from deep-sea species. Later, data about deoxygenation of deeper sea layers during the geological past cast doubts on this possibility, and some researchers came to regard the deep-sea species and anchihaline species as two parallel lineages originating from littoral ancestors. However, cladistic analyses of some ostracod and some copepod groups have revealed that at least some anchihaline species in some regions may originate from deep-sea faunas (Danielopol, 1990).

The anchihaline habitat may also be regarded as a door into continental waters. The regression model and the two-step model suggest that a marine benthic animal colonizes the marine cave, adapts to the cave environment and to the less saline water, and becomes isolated from the sea after the sea regresses, which results in a limnic stygobiont. These and other models differ in some details which become theoretically negligible (making all models complementary) if we consider the ecologically and genetically sound supposition of an active immigration. Active immigration into a new habitat would be furthered by a diversified gene pool and wide ecological potential of the species and would be forced to occur only by its own normally excessive reproduction and population expansion. Sea regression (stranding) in this context is not an act of repression; it implies only interruption of the gene flow (with the marine or surface part of the population) and therefore faster specialization. Of course, this is equally important for understanding the origin of the surface freshwater fauna, which may have occurred the same way on the surface, and the origin of continental cave fauna, which occurred by immigration from surface freshwater underground.

ANCHIHALINE FAUNA AND HUMANS

It has always been a challenge to humans to enter the dark underworld; sometimes such a venture reveals the beauty of speleothems or offers the promise of a cool atmosphere in the heat of summer. So, many karst and lava caves have been exploited for tourism, and those that are close to coastal tourist resorts are particularly subject to such a possibility. Some of them contain anchihaline pools.

Šipun cave in Cavtat (near Dubrovnik, Croatia) was adapted for such visits decades ago, with the anchihaline pool at its end being shown as a special gem. Its comparatively very rich fauna of tiny crustaceans is not particularly interesting for most nonprofessionals but has nonetheless been greatly endangered by pollution produced by the visitors. Anchihaline caves play an important role in tourism and the economy of Bermuda; nevertheless, a number of them have been damaged or destroyed—along with their diverse fauna—by pollution or by quarrying activities. On the other hand, the lava cave Jameos del Agua in Lanzarote (Islas Canarias, Spain) has also been exploited for tourism and its *jameito* or *cangrejo ciego* (*Munidopsis polymorpha*) is an official symbol of the island that is used in various promotional campaigns. The gorgeous hypogean estuary St. Paul's Underground River (Palawan, Philippines) is the central attraction of St. Paul's Subterranean National Park; unfortunately, there are no data on anchihaline fauna in it. Very imposing is the 7-cm-long anchihaline shrimp *Typhlocaris lethaea* from the cave Giok-Kebir, close to Benghazi (Libya); this might be the world's largest troglobiotic invertebrate and could be an interesting subject for interested tourists. Although smaller, the beautifully colored anchihaline red shrimps can be remarkable enough to be given local vernacular names such as *opaeula* (*Halocaridina rubra*) in Hawaii. They may be a subject of local taboos, such as *pulang pasayan* (*Parhippolyte uveae*) in the Philippines, or even of a kind of worship, such as *ura buta* (the same species) on the Fiji Island of Vatulele. No doubt, red shrimps might also become an attraction for ecotourists.

Apart from species protection to serve tourist needs or religious habits, modern humans have become aware of the biological importance of such habitats. A system of anchihaline pools on the Hawaiian island of Maui has been protected as a natural reserve particularly for the rich anchihaline fauna. A flooded crack in the Sinai Peninsula has (or used to have) a similar status.

See Also the Following Article

Coastal Caves

Bibliography

Danielopol, D. L. (1990). The origin of the anchialine fauna: The "deep sea" versus the "shallow water" hypothesis tested against the empirical evidence of the Thaumatocyprididae (Ostracoda). *Bijdragen tot de Dierkunde, 60*(3–4), 137–143.

Holthuis, L. B. (1973). Caridean shrimps found in land-locked salt-water pools at four Indo-west Pacific localities (Sinai Peninsula, Funafuti Atoll, Maui and Hawaii Islands), with the description of the new genus and four new species. *Zoologische Verhandelingen, 128*, 1–48.

Humphreys, W. F. (2001). Relict faunas and their derivation. In H. Wilkens, D. C. Culver & W. F. Humphreys (Eds.), *Subterranean ecosystems* (pp. 417–428). Amsterdam: Elsevier.

Iliffe, T. M. (2001). Anchialine cave ecology. In H. Wilkens, D. C. Culver & W. F. Humphreys (Eds.), *Subterranean ecosystems* (pp. 59–73). Amsterdam: Elsevier.

Jaume, D., & Boxshall, G. A. (1996). The persistence of an ancient marine fauna in Mediterranean waters: New evidence from misophrioid copepods living in anchialine caves. *Journal of Natural History, 30*, 1583–1595.

Pohlman, J. W., Cifuentes, L. A., & Iliffe, T. M. (2001). Food web dynamics and biogeochemistry of anchialine caves: A stable isotope approach. In H. Wilkens, D. C. Culver & W. F. Humphreys (Eds.), *Subterranean ecosystems* (pp. 345–356). Amsterdam: Elsevier.

Por, F. D. (1985). Anchialine pools: Comparative hydrobiology. In G. M. Friedman & W. E. Krumbein (Eds.), *Hypersaline ecosystems: The gavish sabkha. Ecological studies: Analysis and synthesis* (Vol. 53, pp. 136–144). New York: Springer.

Porter, M. (1993). The formation of biota of inland and oceanic blue holes. *Pholeos, 13*(2), 1–8.

Riedl, R. (1966). *Biologie der meereshöhlen*. Hamburg/Berlin: Parey.

Sket, B. (1986). Ecology of the mixohaline hypogean fauna along the Yugoslav coast. *Stygologia, 2*(4), 317–338.

Sket, B. (1996). The ecology of the anchihaline caves. *Trends in Ecology and Evolution, 11*(5), 221–225.

Stock, J. H. (1994). Biogeographic synthesis of the insular groundwater faunas of the (sub)tropical Atlantic. *Hydrobiologia, 287*, 105–117.

Vacelet, J., Boury-Esnault, N., & Harmelin, J. G. (1994). Hexactinellid cave, a unique deep-sea habitat in the scuba zone. *Deep-Sea Research, 41*(7), 965–973.

Yager, J. (1981). Remipedia, a new class of Crustacea from a marine cave in the Bahamas. *Journal of Crustacean Biology, 1*(3), 328–333.

ANCIENT CAVERS IN EASTERN NORTH AMERICA

Patty Jo Watson

Washington University

INTRODUCTION

The longest cave in the world (more than 625 km, and still being explored and mapped) is the Mammoth

Cave System in west central Kentucky in the United States. Beginning about 4000 years ago, the people living in the Mammoth Cave region began exploring some portions of this immense subterranean labyrinth. Indigenous peoples continued to use parts of the Mammoth Cave System until about 2000 years ago. The materials discarded or lost underground many millennia ago preserve detailed information, not only about prehistoric activities in the cave, but also about subsistence practices and other lifeways aboveground.

CAVE ARCHAEOLOGY IN EASTERN NORTH AMERICA

The Mammoth Cave System was not the only cave entered by aboriginal inhabitants of the Americas. To the contrary, any cave opening large enough for a person to squeeze into was probably explored, and well known. The earliest trip documented far into the dark zone of any cave in the Western Hemisphere is that of a 45-year-old man whose *ca.* 8000-year-old skeletal remains were preserved in Hourglass Cave, high in the Southern Rocky Mountains (Mosch and Watson, 1997). In Eastern North America, the oldest detailed archaeological record of dark zone cave exploration dates to approximately 5000 years ago when eight or nine people made one or two trips far back inside an 11-km-long cave in what is now northern Tennessee. They left a scattering of charcoal from their cane torches, and—in the damp clay floor of one passageway—multiple impressions of their feet (Watson *et al.*, 2005).

For many millennia prior to the time of European contact, indigenous peoples of the Americas entered caves, explored them, and used them for a variety of purposes: for example, as storage places, as sources of valuable minerals or of magical powers, as shrines, as cemeteries, or as places to contact the spirit world. The archaeology of the Mammoth Cave System illustrates several of those activities for the two millennia between 4000 and 2000 years ago.

ARCHAEOLOGY OF THE MAMMOTH CAVE AREA

Like many other regions of the eastern United States, west central Kentucky was first inhabited by Paleo-Indian hunter-gatherers, who were present by approximately 14,000 years ago. Archaeological evidence for this time period (*ca.* 14,000 to 11,500 years before present) is sparse, but data from here and elsewhere in the midcontinent indicate to archaeologists that Paleo-Indian communities were small, and were dispersed fairly widely across the landscape.

In archaeological terminology, the Paleo-Indian period in Eastern North America is followed by the Archaic period (*ca.* 11,500 to 3000 years ago), which is succeeded by the Woodland period (3000 to 1100 years ago), and—in some places—the Mississippian period (1100 to 500 years ago). The explorers who left their footprints in that Tennessee cave 5000 years ago are spoken of by archaeologists as Archaic people. The earliest explorers of the Mammoth Cave System were also Archaic folk, as were the first people to enter Fisher Ridge Cave just east of Mammoth Cave National Park, Lee Cave and Bluff Cave (both within Mammoth Cave National Park but not known to be connected to the Mammoth Cave System), and Adair Glyph Cave some distance east of Mammoth Cave (Crothers *et al.*, 2002).

During most of the Archaic period, all societies in the Eastern Woodlands of North America hunted, fished, and gathered the abundant resources of the forests where they lived. By 3000 years ago, however, at the end of the Archaic and beginning of the Early Woodland period, some local groups had become farmers growing several species of small-seeded plants: sunflower, a relative of sunflower called *sumpweed* or *marshelder, goosefoot, maygrass* (also known as *canary grass*), bottle gourd, and a gourd-like form of squash (Browman *et al.*, 2009). These Early Woodland folk were among the first agriculturists in Eastern North America, and some of them were also highly skilled cavers who confidently made their way through many miles of passages in what is now known to be the world's longest cave, the Mammoth Cave System. Since the 1950s, archaeologists and speleologists have studied remains left by these early cavers to discover where they went and what they did underground, and also to investigate the evidence they left of their early farming economy.

PREHISTORIC ARCHAEOLOGY IN THE WORLD'S LONGEST CAVE

Dry cave passages are excellent repositories of archaeological and paleontological remains because they lack factors of weathering and decay that quite rapidly alter or destroy all perishable materials aboveground. Some of the earliest Euroamerican explorers of the Mammoth Cave System noted well-preserved traces of prior human presence there. When the National Park Service acquired the land above Mammoth Cave, Salts Cave, and a series of other caves (known since 1972 to form a single cave system; see Brucker and Watson, 1987), archaeologist Douglas Schwartz at the University of Kentucky was asked to prepare a series of reports on Mammoth Cave area archaeology. Subsequently, other investigators affiliated with the Cave Research Foundation and the National Park Service took up the work, which is still continuing (Crothers, 2001; Crothers *et al.*, 2002; Watson, 1969, 1997). Results are summarized below.

Chronology

Several dozen radiocarbon dates have been obtained on a wide variety of material from various parts of the prehistoric archaeological record in the Mammoth Cave System (Crothers *et al.*, 2002, Table 1 and Figure 1 therein). As already noted, the dates span some 2000 years, from a little before 4000 years ago to a little after 2000 years ago. Most of them, however, cluster around 2500 years before present, plus or minus a few hundred years. Two archaeological periods are represented: Late Archaic and Early Woodland. Both Archaic and Early Woodland cavers explored several miles of passages in the upper and lower levels of Mammoth Cave and Salts Cave, but the Early Woodland record is much more abundant and much more extensive than is the Archaic one.

Archaeological Evidence

The most prominent and pervasive category of prehistoric material in dark zone cave passages is torch and campfire debris. Everywhere the ancient cavers traveled, they left a scatter of cane and dried weed stalks, as well as charred fragments of these torch materials. Charcoal smudges left by torches on cave walls, ceilings, and breakdown boulders are also commonly visible. River cane seems to have been the preferred torch material, but dried stems of goldenrod and false foxglove (Fig. 1) were also used. In addition to the torch remains, fragmentary bindings and cordage made from plant fibers or the inner bark of certain trees are strewn along the passage floors.

Less abundant items include wooden and gourd containers, mussel shell scrapers or spoons, basketry and vegetal fiber artifacts (including remains of bags, cordage, and moccasin-like footwear), digging sticks, scaling or climbing poles, prints of human feet in dust or damp clay floored passages (Fig. 2), dried human excrement (paleofeces), and the physical remains ("mummies," actually simply desiccated bodies) of two of the ancient cavers themselves.

FIGURE 1 Torch of dried weed stalks (*Gerardia* sp., false foxglove) with inner-bark tie, *in situ* on the floor of Indian Avenue, Salts Cave, Mammoth Cave National Park, Kentucky. *Cave Research Foundation photo by Pete Lindsley. Used with permission.*

FIGURE 2 (A) Outline of a prehistoric caver's foot in the dust covering a breakdown boulder in Indian Avenue, Salts Cave, Mammoth Cave National Park, Kentucky. (B) Imprint of a prehistoric caver's foot in mud floor of the Upper Crouchway, Unknown Cave, Mammoth Cave National Park, Kentucky. *Cave Research Foundation photos by William T. Austin. Used with permission.*

In most places within the dry cave passages entered by the Early Woodland people there is evidence that they removed cave minerals from the walls, breakdown boulders, ceilings, and cave sediments. These minerals consist of various sulfate compounds, most prominently gypsum (hydrous calcium sulfate) and mirabilite (hydrous sodium sulfate), with epsomite (hydrous magnesium sulfate) probably also included, although it is much less abundant in the cave than are gypsum and mirabilite.

Archaeological excavation in the entrance areas of both Mammoth Cave and Salts Cave yielded a considerable quantity of charred plant material, and of animal bone (deer, turkey, raccoon, opossum, squirrel, rabbit, turtle, fish, rodents), as well as tools made of bone, chipped stone, and ground stone. The Salts Cave excavations also recovered a considerable quantity of human bone (Robbins, 1997). Physical remains of prehistoric people are known from a few cave sites where they died by accident (*e.g.*, an adult man in Mammoth Cave and a young boy in Salts Cave), or were purposely buried (Crothers *et al.*, 2002). Use of sinkholes and vertical shafts as mortuary facilities, into which the dead were lowered or dropped, was much more common than cave interments across the Midwest, Midsouth, and Southeast.

Interpretations

The somewhat scant evidence dating to the earliest human presence in Mammoth Cave, Salts Cave, Lee Cave, and Fisher Ridge Cave, as well as the Tennessee footprint cave, indicates that Archaic people who ventured into these underground locales seem to have been exploring rather than working or spending long periods of time there. Adair Glyph cave is quite different, however (DiBlasi, 1996). Here the ancient people used sticks or their fingers on a mud-floored passage to make a dense series of curvilinear and rectilinear markings whose meanings are unknown to us, but whose creation must have taken considerable time and thought, and probably more than one trip into the cave. At any rate, it is obvious that those responsible for the markings were not simply exploring this room, which is about one kilometer from the entrance. At Fisher Ridge Cave the archaeological evidence primarily indicates exploration, but there is a lattice or checkerboard design (scratched onto a large breakdown boulder) that was present when the first Euroamerican cavers entered that passage during the early 1980s. Finally, several geometric and representational figures (some scratched, some drawn with charcoal) documented in both Salts Cave and Mammoth Cave are probably prehistoric, either Late Archaic or Early Woodland, or both (Crothers, 2001; DiBlasi, 1996).

In Wyandotte Cave, Indiana, and 3rd Unnamed Cave, Tennessee, Late to Terminal Archaic people quarried and worked chert derived from underground sources. At 3rd Unnamed Cave they also engraved rectilinear, curvilinear, and representational figures onto the limestone ceilings, walls, and breakdown boulders (Franklin, 2008), some of which are reminiscent of the mud images in Adair Glyph Cave. Thus, it appears that the first people to enter dark zones of several big caves in Eastern North America were indeed exploring these complexes, but they were also obtaining chert from some of them, and in several they were carrying out ritual activities as well.

During the subsequent, Woodland period (3000 to 1200 years ago), caves were important to many communities across the Midsouth and Southeast as ceremonial and mortuary foci. Dozens of examples are known in Kentucky, Tennessee, the Virginias, Alabama, Georgia, Florida, and Texas, but the prehistoric archaeological record in the Mammoth Cave System ceases 2000 years ago. For the centuries between about 2800 and 2300 years BP, however, Early Woodland cavers spent considerable amounts of time in both Salts Cave and Mammoth Cave mining sulfate minerals. They were obviously very familiar with several miles of complicated passage systems in each of these caves, and were quite capable of navigating their way through multiple levels of walking passages, crouchways, and crawlways. They battered the gypsum crust off interior cave surfaces; they dug gypsum crystals (satin spar and selenite) from crevices in cave walls, and/or from sediment on cave floors and ledges; they scooped up mirabilite from those areas where it forms so abundantly that it piles up in and on the breakdown and on passage floors; they brushed mirabilite crystals off cave walls into gourd, wood, or basketry containers; and they removed mirabilite-bearing cave sediments, presumably so that the salty, medicinal sulfate deposits could be leached from them outside the cave.

The technology used for all this work was simple but effective, consisting of cane and weed stalk torches; twigs and small branches for campfire fuel; digging sticks, climbing poles (Fig. 3) and ladders; baskets (Fig. 4), fiber bags, and gourd and wood containers. The minerals removed from the cave were probably sought for their medicinal (mirabilite, and perhaps epsomite as well) or magical attributes. Battering the gypsum crust off cave walls produces gypsum powder, which can be mixed with water or grease to make white paint, known ethnographically to be important in ritual contexts. Gypsum crystals may have been thought to have special powers that would enable curers and wisdom-keepers to diagnose and treat illness, to foretell the future, or to exercise other

Woodland and Mississippian periods up to the time of European contact (Simek, 2008), are clear evidence that many, if not all, activities underground took place within settings that were very special to the ancient cavers. We cannot know exactly what they thought about the world underground, but there is information from ethnographic and ethnohistorical records indicating general concepts that help us speculate about prehistoric belief systems in Eastern North America (*e.g.*, Diaz-Granados, 2008; Hudson, 1976; Simek and Cressler, 2008). Many North American Indians believed in a multilayered universe, wherein there was another world (or several, the details differ from group to group) below the surface world, and another (or several) above the surface world. Each layer was inhabited by an array of creatures specific to it. The ancestors of humankind, having originated in the underground world, had climbed onto the surface world and established themselves there in the primordial past. Sky beings (*e.g.*, falcon or owl spirits) and subterranean beings (the most powerful and fearsome being a kind of griffin, or dragon-like creature known variously as the Uktena, Michibichi, the Water Panther, and the Long Tailed Beast) could be contacted and engaged in human enterprises by specially gifted individuals (*e.g.*, shamans) who knew how to carry out the dangerous procedures required: in particular, how to journey to those realms, negotiate with their inhabitants, obtain magical substances and objects, and return safely to the surface world. It seems likely that the enigmatic cave pictographs, petroglyphs, and mud glyphs were created by specialists in the esoteric knowledge necessary to ensure safe journeys to and from portions of the world underground.

FIGURE 3 Prehistoric scaling pole, Indian Avenue, Salts Cave, Mammoth Cave National Park, Kentucky. Radiocarbon age of 2760 ± 40 years (uncalibrated; Beta-87915). *Cave Research Foundation photo by François-Marie Callot and Yann Callot. Used with permission.*

FIGURE 4 Prehistoric split-cane basket in Ganter Avenue, Mammoth Cave, Mammoth Cave National Park, Kentucky. Radiocarbon age of 2630 ± 55 years (uncalibrated; Beta-47292). *Cave Research Foundation photo by Roger W. Brucker. Used with permission.*

magical abilities (all these activities are attested ethnographically among indigenous North American populations). Early Woodland mining of cave sulfates is most extensive and best known in Salts Cave and Mammoth Cave, but evidence of gypsum mining has also been documented recently in Hubbards Cave, Tennessee, and probably took place in Big Bone Cave, Tennessee, as well (Crothers *et al.*, 2002).

The pictographs and mud glyphs created in some caves as early as the Late Archaic period, and present in many other subterranean locales dating from

CONCLUSION

The first human inhabitants of America were also the first North American cavers. They found and explored hundreds of caves, especially in the midcontinental karst region that extends from the Midwest through the Midsouth to the southeastern United States, including what is now known to be the longest cave in the world. Archaeologists and speleologists have investigated only a small sample of indigenously utilized caves, but the resulting information makes it clear that subterranean landscapes were well known to aboriginal communities who lived in these karstic areas. As was the case with other features of their surroundings, prehistoric people were familiar with the physical and spiritual aspects of the world underground, knew how to get there, and once there how to travel to their objectives. They also knew how to negotiate with the supernatural beings that inhabited these realms of darkness, and what to do to ensure safe return to the surface world. Because of the size and complexity of the caves they frequented, it can quite justifiably be claimed that they were the best cavers in the world, a distinction held by Archaic and Woodland peoples of Eastern North America until less than a century ago.

See Also the Following Articles

Cave Dwellers in the Middle East
Maya Caves

Bibliography

Browman, D. L., Fritz, G. J., Watson, P. J., & Meltzer, D. J. (2009). Origins of food producing economies in the Americas. In C. Scarre (Ed.), *The human past* (pp. 306–349). London: Thames and Hudson.

Brucker, R. W., & Watson, R. A. (1987). *The longest cave* (2nd ed.). Carbondale: Southern Illinois University Press.

Crothers, G. M. (2001). Early Woodland mineral mining and perishable remains in Mammoth Cave, Kentucky. In P. Drooker (Ed.), *Fleeting identities: Perishable material culture in archaeological research* (pp. 314–334). Carbondale: Southern Illinois University Press.

Crothers, G. M., Faulkner, C. H., Simek, J. F., Watson, P. J., & Willey, P. (2002). Woodland cave archaeology. In D. Anderson & R. Mainfort (Eds.), *The woodland southeast* (pp. 502–524). Tuscaloosa: The University of Alabama Press.

Diaz-Granados, C. (2008). Picture cave. In D. H. Dye (Ed.), *Cave archaeology in the Eastern Woodlands* (pp. 203–215). Knoxville: University of Tennessee Press.

DiBlasi, P. J. (1996). Prehistoric expressions from the central Kentucky karst. In K. C. Carstens & P. J. Watson (Eds.), *Of caves and shell mounds* (pp. 40–47). Tuscaloosa: University of Alabama Press.

Franklin, J. D. (2008). Big cave archaeology in the East Fork Obey River Gorge. In D. H. Dye (Ed.), *Cave archaeology of the Eastern Woodlands* (pp. 141–155). Knoxville: University of Tennessee Press.

Hudson, C. (1976). *The southeastern indians*. Knoxville: University of Tennessee Press.

Mosch, C. J., & Watson, P. J. (1997). An ancient Rocky Mountain caver. *Journal of Cave and Karst Studies, 59*, 10–14.

Robbins, L. M. (1997). Prehistoric people of the Mammoth Cave area. In P. J. Watson (Ed.), *Archeology of the Mammoth Cave area* (pp. 137–162). Reprint of the 1974 Academic Press publication, with a new preface by the editor. St. Louis, MO: Cave Books.

Simek, J. F. (2008). Afterword: Onward into the darkness. In D. H. Dye (Ed.), *Cave archaeology of the Eastern Woodlands* (pp. 261–270). Knoxville: University of Tennessee Press.

Simek, J. F., & Cressler, A. (2008). On the backs of serpents. In D. H. Dye (Ed.), *Cave archaeology of the Eastern Woodlands* (pp. 169–191). Knoxville: University of Tennessee Press.

Watson, P. J. (Ed.), (1969). *The prehistory of Salts Cave, Kentucky* Springfield: Illinois State Museum.

Reprint of the 1974 Academic Press publication, with a new preface by the editor. Watson, P. J. (Ed.), (1997). *Archeology of the Mammoth Cave area*. St. Louis, MO: Cave Books.

Watson, P. J., Kennedy, M. C., Willey, P., Robbins, L. M., & Wilson, R. C. (2005). Prehistoric footprints in Jaguar Cave, Tennessee. *Journal of Field Archaeology, 30*, 25–43.

ASELLUS AQUATICUS: A MODEL SYSTEM FOR HISTORICAL BIOGEOGRAPHY

Rudi Verovnik
University of Ljubljana

INTRODUCTION

The isopod *Asellus aquaticus* is a common and widespread freshwater crustacean in most parts of Europe. It prefers slow-flowing streams and lakes and is often found in ponds and ditches shaded over by trees—the decaying leaves being its favorite food, but it is extremely eurytopic and eurythermic, surviving even in brackish waters. Young specimens are often found in interstitial hyporheic habitats, possibly due to avoidance of predation. Being so eurytopic the species can be considered as preadapted to the subterranean life. Thus, many hypogean populations are found scattered throughout Europe (Fig. 1). In the northern part of its range these habitats are commonly of anthropogenic origin including mines, water supply pipes, and waterworks, indicating very recent hypogean invasion. Indeed the specimens from such habitats as well as from caves in northern Europe are only slightly troglomorphic with their slight body and sometimes eye depigmentation. Highly troglomorphic populations are confined only to two relatively small areas in southeastern Europe.

The first such area is the vicinity of Mangalia in the Dobrogea region in eastern Romania with the famous Movile Cave. This region is known for its highly sulfurous mesothermal waters (21°C) and well-established subterranean food chain based on chemoautotrophic bacteria. From here the troglomorphic subspecies *A. aquaticus infernus* was described. It is commonly found in the hand-excavated wells and sulfurous springs in the karstic aquifer around Mangalia, but much sparser in the Movile Cave itself. It is characterized by several troglomorphic traits including *K*-strategy traits.

The second region is in the northwestern part of the Dinaric karst, a well-known "hot spot" of subterranean biodiversity. Here several caves are inhabited by partially or highly troglomorphic populations, which were described as separate subspecies. In addition, several karst poljes with isolated surface water drainages are also inhabited by the morphologically distinct subspecies, *A. aquaticus carniolicus*. The common trait linking these surface populations with subterranean subspecies is the presence of a large respiratory *area* on posterior pleopods—the unsclerotized part of the expodite (Fig. 2). The respiratory *area* is much smaller in all other populations throughout Europe, including the *A. a. infernus*. Due to this common feature, it has been presumed that the surface populations in karst poljes are the ancestors of the subterranean populations. As in most troglobionts the epigean ancestors are either extinct or unknown, and this gives us a unique opportunity to study cave invasions through historical biogeography.

DEVELOPMENT OF TROGLOMORPHIES

Even those subterranean populations of *A. aquaticus* that are not considered extremely troglomorphic can be easily distinguished from the surface populations by their lack of body pigmentation and reduction of eyes and some cuticular structures (Fig. 3). The progressive troglomorphies present in *A. a. cavernicolus* include enlargement of the body and elongation of appendages, in particular of the antennae II which can be linked with enhancement of the extraoptical sensory structures. *K*-strategy consequences in this subspecies seem to be limited to different paedomorphoses including more isometric articulation and more symmetric shape of some appendages. In subspecies *A. a. infernus*, the smaller number of eggs produced by females compared to the surface-dwelling subspecies is the only observed adaptation in reproductive biology, indicating that *A. aquaticus* is a relatively young immigrant to the subterranean waters.

Even from this brief description of the troglomorphic traits, it is evident that they differ in different subterranean populations. It is expected that subterranean populations in Mangalia and those from the Dinaric karst do not share a common ancestor, being separated by more than 1000 km distance. It is, however, much more surprising to see that differences occur also among subterranean populations in a small area of the northwestern part of the Dinaric karst and even within the same cave system. In the Planina Cave, populations from the Rak River and the Pivka River channels differ in troglomorphic traits. The Rak channel population is considered to be more troglomorphic, characterized by multiplication and shortening of setae. This implies that even single caves were invaded on several occasions, strongly supporting the hypothesis of active cave invasions of *A. aquaticus*.

POSTOJNA-PLANINA CAVE SYSTEM

The main theater of the historical biogeographic studies is the Postojna-Planina Cave System (PPCS), an approximately 25-km-long system comprising several caves with accessible river passages of hypogean flow of the Pivka and Rak rivers. Upstream from the PPCS the plain of the Postojna basin is inhabited by the nominotypic subspecies *A. a. aquaticus*, while the short but spectacular Rakov Škocjan Polje is inhabited by the

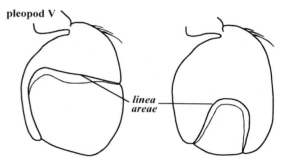

FIGURE 2 The shape of the nonsclerotized respiratory *area* of the fifth pleopod of *Asellus aquaticus*. The large respiratory *area* form is distributed along the Dinaric karst.

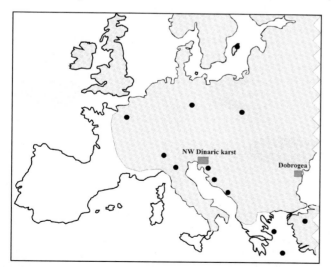

FIGURE 1 Approximate distribution of *Asellus aquaticus* in Europe. Black dots denote nonhomogeneous troglomorphic populations; the squares, areas with specialized troglomorphic populations.

FIGURE 3 Troglomorphic and surface populations of *Asellus aquaticus* are easy to distinguish. *Photo courtesy of Boris Sket. Used with permission.*

large respiratory *area* subspecies *A. a. carniolicus*. In the hypogean Pivka River the first cave is the famous touristic Postojna Cave where only slightly depigmented specimens of the *A. a. aquaticus* are found at the entrance. The Pivka River then flows through caves Črna jama and Pivka jama where a mixture of fully pigmented, partially depigmented, and fully depigmented and eyeless specimens is present. Pivka Cave is also the type locality of *A. a. cavernicolus*. Morphometric studies revealed that the type specimens collected 80 years ago are more troglomorphic than the recent inhabitants of the Pivka Cave. This strange phenomenon could perhaps be explained by the eutrophication of the Pivka River due to organic pollution, which was particularly severe in the second half of the twentieth century. This enabled several epigean species, including *A. a. aquaticus*, to penetrate deeper into the cave system. Downstream from the Pivka Cave, there are several kilometers of inaccessible river passages after which the river reappears in the Pivka channel of Planina Cave. Highly troglomorphic *A. a. cavernicolus* is abundant here; however, single pigmented specimens can also be found after floods. In the Rak channel only *A. a. cavernicolus* is present. Downstream of the PPCS the epigean river Unec in the large Planina Polje is stretching, inhabited by the large respiratory *area* subspecies *A. a. carniolicus*.

In order to understand the connections between these (sub)populations and subspecies and to put them into the phylogeographic context, different molecular studies were performed. With an RAPD (Random Amplified Polymorphic DNA) study several interesting patterns emerged: first, the grouping of the populations is congruent with subspecific status; second, the mixed population from the Pivka Cave is genetically homogeneous despite morphological heterogeneity, and is more closely related to the upstream surface populations than to downstream *A. a. cavernicolus*; third, the genetic diversity was significantly lower in the Pivka Cave and the Pivka channel of the Planina Cave, possibly due to bottlenecks caused by the recent organic pollution of the river. The gene flow estimates confirm the discontinuity between the Pivka Cave and the Pivka channel (Fig. 4); however the differences between gene-flow estimates among populations in the PPCS were not significant. Molecular analyses also confirmed clear separation between *A. a. cavernicolus* populations from the Pivka and the Rak channels which is in line with observed morphological differences. The results of RAPD analyses have currently been corroborated also by a preliminary microsatellite study where the gene-flow estimate across 10 microsatellites showed the lowest gene flow between these two populations. All above-mentioned analyses confirm separate status of all populations of *A. aquaticus* in PPCS, but due to the anonymous nature of RAPD DNA fragments, no historical biogeographic conclusions can be drawn.

To add the historical component to the molecular differentiation, a wider scale mt-DNA sequence study was employed. One of the most important issues in phylogeographic studies is the calibration of the molecular clock: the speed with which the changes in DNA sequences evolved. To do the calibration, either well-developed series of fossil records or good timing geological events that have caused the split of the populations and prevented gene flow between them up to present are required. None of these was available in *A. aquaticus*, as the only geological information that could be potentially useful is the estimate of the beginning of the karstification and the speleogenesis in the region of the study. This proved to be broadly estimated between 1 and 5 myr, thus providing only a general timeframe for the evolution of troglomorphic populations. Luckily, a molecular clock of a related subterranean aquatic isopod *Stenasellus* spp. calibrated by the separation of the Tyrrhenian Islands could be used.

At the level of the PPCS populations, the sequence of the first subunit of cytochrome oxidase mitochondrial gene (COI) revealed no hydrogeographic or taxonomic grouping, thus negating the RAPD-based conclusions. The groups that appeared did confirm a higher level of

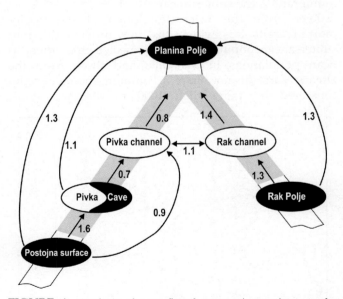

FIGURE 4 Analysis of gene flow between *A. aquaticus* populations along two confluent sinking rivers in the PPCS. The values on the arrows are estimated number of effective migrants per generation. Black symbols indicate populations with pigmented individuals belonging to *A. a. aquaticus* (Planina surface) and *A. a. carniolicus* (Planina and Rak Polje). White symbols indicate troglomorphic populations belonging to *A. a. cavernicolus*; the black-and-white symbol stands for the morphologically mixed population. The darker area represents the subterranean river sections.

isolation of the Rak channel population where two population-specific COI haplotypes were dominant. On the other hand, a closer relationship between the Pivka Cave and the Pivka channel was also revealed opposing the RAPD-based conclusions. However, these results should be treated with caution as the presence of widespread and common haplotypes in the PPCS indicates incomplete lineage sorting due to recent splits and possibly also due to sporadic introgression. Since the mt-DNA is transmitted only maternally, even a single introgression event involving a female from a different population could be retained in the observed haplotype diversity of the receiver population. To understand the phylogeographic patterns, a broader geographic scale, which in this case provides also a wider timeframe, should be addressed.

BROADER VIEW

In order to explain the results at a broader geographic scale, further details on the presence of cave populations in the northwestern part of the Dinaric karst should be discussed. Apart from the PPCS, the subspecies *A. a. cavernicolus* is also distributed upstream from the polje Rakov Škocjan and also downstream from the Planina Polje. Additionally, a disjunct population of this troglomorphic subspecies was considered to be present in the Labodnica/Trebiciano Cave in hypogena Reka River which is a part of the Adriatic drainage; all other studied rivers in the northwestern Dinaric karst drain toward the Danube. Other less troglomorphic populations of the subspecies *A. a. cyclobranchialis* are present in the Krka drainage farther east from the PPCS. The large respiratory *area A. a. carniolicus* is present also at the polje of the Cerknica Lake east of the PPCS. Two more subspecies are described from isolated karst poljes farther eastward, both sharing the large respiratory *area* character with *A. a. carniolicus*.

At this broader scale, including the northeastern Dinaric karst, mt-DNA analysis reveals structuring that could be linked to recent and also to the presumed historical hydrography of the region. It also provides additional confirmation of separate cave invasions by *A. aquaticus* in three geographically and currently also hydrologically separate areas: the Krka drainage, the Ljubljanica drainage (PPCS included), and the Reka drainage. In the Ljubljanica drainage there is a further split between the PPCS and the cave populations downstream from the Planina Polje, providing evidence of an additional separate cave invasion.

With the use of the stenasellid molecular clock the splits identified by the phylogenetic and phylogeographic analyses can be traced down to the geographic isolation of the north western Dinaric karst populations between 2.7 and 4.3 mya, thus within the supposed timeframe of the speleogenesis in the region. Soon after that, the first fragmentation followed within the Dinaric karst area separating the Reka River population from the rest (Fig. 5). At that time the historical drainage of the Ljubljanica River was dominant, receiving water from the regions much farther eastward than at present. This situation changed approximately 1 mya, when the current extent of the Ljubljanica drainage was established. At that time the eastern karst poljes were still connected, draining more directly toward north. The present fragmented hydrology of the eastern drainage was established around 0.8 mya also with a shift of part of the drainage to the Krka River (Fig. 5). These splits also provide the approximate ages of the cave invasions, as most of the changes of drainages can be linked to the ongoing karstification. Thus the first invasion of the subterranean environment was possibly completed in the Reka River drainage, followed by the Ljubljanica and Krka River drainages. This corresponds well with the decreasing troglomorphy of the populations, the one from the Reka River in the Labodnica/Trebiciano Cave being the most troglomorphic with slender and elongated appendages and highly paedomorphic shape of the pleopods.

ASELLUS KOSSWIGI

The troglomorphic population from the Labodnica/Trebiciano Cave in the Reka River splits basally within the northwestern Dinaric karst group in mt-DNA phylogeny; therefore its previous designation as belonging to the subspecies *A. a. cavernicolus* was not supported. This is confirmed also by a large number of population-specific RAPD fragments found in the Reka River population, indicating a long period of isolation. In addition, the troglomorphic population from the Reka River is characterized by a distinct nuclear 28S rDNA sequence, which is otherwise highly conserved throughout the European range of the species (Fig. 6). The phylogeny based on 28S rDNA puts the Reka River population firmly into the northeastern Dinaric karst clade, making the historical scenario described in the previous chapter more plausible.

The solution of the ambiguous taxonomic position of the Reka River troglomorphic population came with the discovery of the syntopic occurrence of the troglomorphic and nontroglomorphic specimens of *A. aquaticus* in the resurgence of the Reka River near Monfalcone in Italy. Despite syntopic occurrence, no intermediate forms were found, indicating that there is an effective reproductive barrier in place. The subterranean population was therefore described as separate

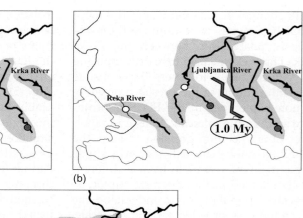
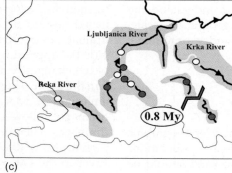

FIGURE 5 Graphical representations of a hypothetical chain of events that best explains the current distribution of haplotypes of *Asellus aquaticus* in the northwestern Dinaric karst. A zigzag line indicates fragmentation events that separated drainages. The pale line denotes subterranean river passages; white-filled circles subterranean populations; and brown-filled circles isolated karst polje surface populations. The maximum age of the events was estimated by applying the COI clock rate of stenasellids.

species *Asellus kosswigi*, honoring the first discoverer of this cave population. The species status was confirmed also by molecular studies, as no traces of gene exchange between the cave and the surface populations was observed, both for mitochondrial and nuclear genes. One should note that there are several troglobiotic species described within genus *Asellus*, but all are from the Far East, mostly Japan. *Asellus kosswigi* is currently the second species of the genus recognized in Europe and illustrates the common side effect of phylogeographic studies: description of new species.

CONTINENTAL SCALE

One must understand that the processes within the range of *Asellus aquaticus* in the northwestern part of the Dinaric karst were not isolated, and therefore an even wider continental scale phylogeographic study was performed. To get an insight into the historical biogeography at the continental scale, more than 70 populations were sampled throughout the species range in Europe, also covering the rest of the Dinaric karst. Based on the COI molecular clock calibration for stenasellids, the expansion of the species to central Europe was estimated

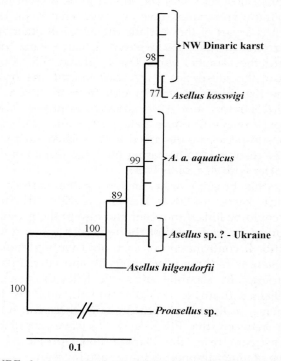

FIGURE 6 A simplified 28S rDNA phylogeny of the European populations of the *Asellus aquaticus*. The posterior probability of each node (in percent) is shown on branches.

at approximately 9 mya when the southern Russia region was connected with the Pannonian Basin by the partially brackish Parathetys Sea, which provided dispersal routes for the ancestral *A. aquaticus*. The Pannonian basin was the origin of the first expansion in Europe and its southwestern edge still hosts the highest genetic diversity, indicating the presence of a very old refugium in this region (Fig. 7). Between 4 and 8 mya the species dispersed throughout most of its current southern range, forming secondary refugia in the Balkan Peninsula, Apennine Peninsula, and in the newly forming Dinaric karst. All these refugia are confirmed by the presence of geographically defined groups of the COI phylogeny and host large haplotype diversity (Table 1, Fig. 7).

In the COI phylogeny the northwestern Dinaric karst clade is positioned as a sister group of the central European clade and an isolated clade from the Baške Oštarije karst polje in the central Dinaric karst. This provides a reasonable assumption that the Dinaric karst region was colonized from the Pannonian basin. Due to karstification, the populations in the northwestern parts of the Dinaric karst, and the Baške Oštarije became isolated from the rest of the species range, which is also evident from the 28S rDNA phylogeny (Fig. 6). As the 28S rDNA nuclear gene is highly repetitive, homogenization through concerted evolution over most of the species range could be achieved with very limited gene flow. Due to long-term isolation, the northwestern Dinaric karst escaped that process, retaining several specific 28S rDNA sequences. The highly troglomorphic population from Mangalia in Romania was found to be identical in its 28S rDNA sequence to the other populations in Europe, and had COI haplotypes similar to those from the nearby surface populations.

Through distribution of COI haplotype diversity, several additional questions were addressed. The first was whether the species survived Pleistocene glaciations in the presumed permafrost areas north of the Alps. Several populations with high haplotype diversity, similar to the observed diversity in the populations in southern refugia, were found as far north as the Netherlands and northern Germany, indicating that such microrefugia did exist (Table 1). The second comparison was between habitat types where large rivers were found to be most diverse, providing the core habitat of the species. It was more surprising to find that haplotype diversity in the subterranean populations was twice as high as in the karstic springs, the presumed origin of some of the cave invasions (Table 1). Several hypotheses were formulated to explain that, including relaxed selection in caves due to reduced competition and predation. The second is that caves

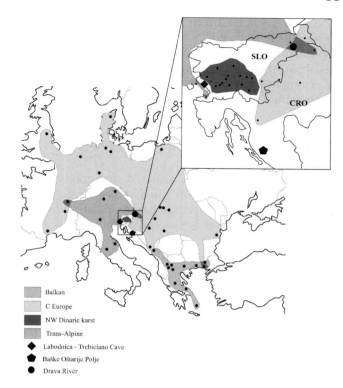

FIGURE 7 Geographic distribution of clades of the COI mitochondrial DNA haplotypes in European *Asellus aquaticus*. The northwestern Dinaric karst region is magnified for clarity. The possibly oldest refugium and origin of dispersion situated in the western Pannonian region (Drava River) is denoted by a black circle. Areas of overlapping distribution of clades are indicated by darker color.

TABLE 1 ANOVA (Analysis of Variance) Design for Two Hypotheses Explaining the Geographic Pattern of Nucleotide Diversity (π) of *Asellus aquaticus* in Europe

Hypothesis	ANOVA groups	N	π	SD
Classic refugia	Dinaric karst	17	0.00314	1.03E-05
	Apennine	6	0.01259	0.000506
	Pannonian	14	0.01012	0.000147
	S. Balkan	12	0.00965	0.000332
	N. Europe	12	0.01129	0.000299
Habitat	river	21	0.01736	0.00039
	stream	17	0.00185	6.65E-06
	lake	5	0.00189	3.79E-06
	subterranean	7	0.00902	0.00011
	spring	11	0.00487	8.86E-05

served as shelters during Pleistocene climatic oscillations, which reduced or even eliminated the surface populations during extremely cold or dry conditions. The most obvious explanation is that caves in general are a much larger habitat compared to springs, therefore hosting larger and more stable populations.

PROSPECTS FOR THE FUTURE

Many questions were answered and many stories unfolded, but there is still much more to be discovered using the model species *Asellus aquaticus*. One of the unresolved issues is the taxonomic status of the isolated populations from the northwestern Dinaric karst. Despite all evidence, the process of speciation within this group seems to be at such an early phase that the isolation and number of separate species could not be unequivocally resolved by the studied molecular markers. Microsatellites could provide these answers as they allow much more accurate detection of gene flow between populations. This will provide opportunities to test *A. aquaticus* as a model species for parallel speciation, possibly of a common speciation process in many subterranean taxa. It is also not clear whether the formation of the troglomorphic populations was a result of a complete isolation from the surface source population or it could have been achieved in sympatry through ecological speciation. These are certainly very promising prospects for the future and *A. aquaticus* still has many important questions to answer.

See Also the Following Article

Invasion, Active versus Passive

Bibliography

Avise, J. C. (2000). *Phylogeography: The history and formation of species.* Cambridge, MA: Harvard University Press.

Coen, E., Strachan, T., & Dover, G. (1982). Dynamics of concerted evolution of ribosomal DNA and histone gene families in the *melanogaster* species subgroup of *Drosophila*. *Journal of Molecular Biology, 158*, 17–35.

Culver, D. C. (1982). *Cave life.* Cambridge, MA: Harvard University Press.

Culver, D. C. & Sket, B. (2000). Hotspots of subterranean biodiversity in caves and wells. *Journal of Cave and Karst Studies, 62*, 11–17.

Ketmaier, V., Argano, R., & Caccone, A. (2003). Phylogeography and molecular rates of subterranean aquatic Stenasellid isopods with a peri-Thyrrenian distribution. *Molecular Ecology, 12*, 547–555.

Lynch, M. & Milligan, B. G. (1994). Analysis of population genetic structure with RAPD markers. *Molecular Ecology, 3*, 91–100.

Prevorčnik, S., Blejec, A., & Sket, B. (2004). Racial differentiation in *Asellus aquaticus* (L.) (Crustacea: Isopoda: Asellidae). *Archiv für Hydrobiologie, 160*, 193–214.

Sket, B. (1994). Distribution of *Asellus aquaticus* (Crustacea: Isopoda: Asellidae) and its hypogean populations at different geographic scales, with a note on *Proasellus istrianus*. *Hydrobiologia, 287*, 39–47.

Turk-Prevorčnik, S. & Blejec, A. (1998). *Asellus aquaticus infernus*, new subspecies (Isopoda: Asellota: Asellidae), from Romanian hypogean waters. *Journal of Crustacean Biology, 18*, 763–773.

Verovnik, R., Sket, B., Prevorčnik, S., & Trontelj, P. (2003). Random amplified polymorphic DNA diversity among surface and subterranean populations of *Asellus aquaticus* (Crustacea: Isopoda). *Genetica, 119*, 155–165.

Verovnik, R., Sket, B., & Trontelj, P. (2004). Phylogeography of subterranean and surface populations of water lice *Asellus aquaticus* (Crustacea: Isopoda). *Molecular Ecology, 13*, 1519–1532.

Verovnik, R., Sket, B., & Trontelj, P. (2005). The colonization of Europe by the freshwater crustacean *Asellus aquaticus* (Crustacea: Isopoda) proceeded from ancient refugia and was directed by habitat connectivity. *Molecular Ecology, 14*, 4355–4369.

ASTYANAX MEXICANUS: A MODEL ORGANISM FOR EVOLUTION AND ADAPTATION

William R. Jeffery
University of Maryland

EVOLUTION, ADAPTATION, AND MODEL ORGANISMS

Biological evolution is the change in inherited traits over successive generations in populations of organisms. Evolutionary modification of traits occurs when variation is introduced into a population by gene mutation or genetic recombination or is removed by natural selection or genetic drift. Adaptation is a key evolutionary process in which the fitness of traits and species are adjusted by natural selection to become better suited for survival in specific ecological habitats. The environment acts to promote evolutionary change through changes in development, but little is known about the underlying mechanisms. One of the reasons is that there are only a few suitable model organisms that can shed light on this topic. A model organism is a species that is studied extensively with anticipation that the results can be applied to biological phenomena in general. Cave animals can serve as excellent models to study the relationships between the environment, evolution, adaptation, and development because of the relative simplicity of their habitat. Troglomorphic (cave-related) traits, such as elongated appendages, lowered metabolism, specialized sensory systems, and loss of eyes and pigmentation have evolved as a response to the effects of perpetual darkness. Here, we describe the characid fish *Astyanax mexicanus*, one of the few model organisms that is available for studying the molecular, developmental, and genetic basis of evolution and adaptation.

ASTYANAX MEXICANUS AS A MODEL ORGANISM

Astyanax mexicanus is a single species with a surface-dwelling form (surface fish) and many different cave-dwelling forms (cavefish) (Fig. 1A). The divergence between surface fish and cavefish is relatively recent, having occurred less than three million years ago. Cavefish have evolved constructive features, such as new methods of feeding and physical orientation, which adapt them to life in caves. They have lost or substantially reduced their eyes, pigmentation, and other regressive traits whose evolutionary significance is not completely understood (Table 1).

The biology of *Astyanax mexicanus* makes it an excellent model organism. It is well suited for laboratory research because it can be raised on a simple diet, produces a large number of embryos, and has a relatively short generation time, about 4–6 months. Surface fish and cavefish embryos are large, clear, and suitable for experimental manipulations, such as tissue transplantations. Gene inhibition and overexpression methods have been developed in *Astyanax mexicanus*, and phenotypic changes can be studied by powerful genetic analysis (Jeffery, 2009a). Surface fish and cavefish are interfertile. Moreover, different cavefish populations can be crossed to perform genetic complementation tests. Therefore, cavefish can be compared to surface fish in essentially the same way that mutants are compared to wild-type phenotypes in model organisms such as *Drosophila melanogaster*. Finally, some cavefish populations have been derived independently from surface fish ancestors, providing a rich foundation for studying convergence or parallel evolution.

NATURAL HISTORY OF ASTYANAX MEXICANUS

The ancestors of *Astyanax mexicanus* surface fish first entered Central America after the Isthmus of Panama was formed about 3 million years ago. Since then their range has gradually expanded north into Mexico and the southwestern United States. During the Pleistocene epoch there was probably periodic northern expansion and contraction of this range due to temperature fluctuations. Surface fish entered limestone caves through entrapment by stream capture during this period but permanent colonization probably occurred only in a

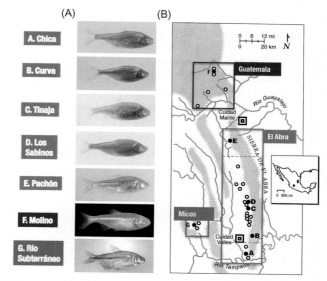

FIGURE 1 *Astyanax mexicanus* cavefish populations in northeastern Mexico. (A) Photographs of cavefish populations named on the left and located according to letters on the map. (B) Map showing the distribution of 29 cavefish populations in Tamaulipas and San Luis Potosí, Mexico. Each sphere represents one cavefish population. Filled spheres labeled with a letter represent populations shown and named on the left. Boxes indicate the location of independently evolved Guatemala (blue box), El Abra (red box), and Micos (green box) cavefish populations. Dashed lines representing zones of cavefish populations that may have evolved some troglomorphic features separately subdivide the El Abra box. Inset: Outline map of Mexico showing northeastern region in B as a rectangle and a filled sphere indicating the location of the Granadas cavefish population in Guerrero.

TABLE 1 Constructive and Regressive Troglomorphic Traits in *Astyanax mexicanus*

Constructive Traits	Regressive Traits
Jaws	Eyes
Taste buds	Melanophores
Feeding posture behavior	Melanin
Forebrain	Optic tectum
Hypothalamus	Rib-bearing vertebrae
Orbit bones	Aggressive behavior
Superficial neuromasts	Alarm behavior
Fat deposition	Growth control
Maxillary teeth	Scales
Nares	Fin rays
Olfactory interneurons	
Vibration attraction behavior	

restricted portion of their range. The hotspot of cavefish distribution is the Sierra Madre Orientale (Sierra de El Abra, Sierra de Guatemala, and Micos regions) in the northeastern Mexican states of Tamaulipas and

San Luis Potosí (Fig. 1B; Mitchell et al., 1977). There is a single outlying cavefish population located across the continental divide in the southern Mexican state of Guerrero. No troglomorphic cavefish have been discovered in other areas of Mexico or Central America. This suggests that the caves of northeastern Mexico may have unknown features that are ideal for cavefish evolution.

Originally only three Mexican characid cavefish populations were known, and these were classified as members of a distinct genus consisting of three species: *Anoptichthys jordani*, *A. antrobius*, and *A. hubbsi*. Based on molecular studies and the ability to produce fertile offspring, all cavefish populations and nearby surface fish are now considered to be a single species: *Astyanax mexicanus*. Thirty *Astyanax mexicanus* cavefish populations with varying degrees of troglomorphic features have been identified (Fig. 1). At least four separate cavefish lineages were established in the El Abra, Guatemala, Micos, and Guerrero areas and evolved troglomorphic traits independently (Jeffery, 2009a). Further troglomorphic evolution occurred as the founders of these lineages dispersed underground and became isolated in separate cave systems (*e.g.*, El Abra, Fig. 1B).

The 30 *Astyanax mexicanus* cavefish populations are named after their caves of origin (Mitchell et al., 1977). For example, Pachón, Molino, Rio Subterráneo, and Granadas cavefish are found in La Cueva de El Pachón (Tamaulipas), El Sótano de Molino (Tamaulipas), La Cueva de la Rio Subterráneo (San Luis Potosí), and La Cueva de las Granadas (Guerrero), respectively (Fig. 1). The Rio Subterráneo and Granadas cavefish are considered as examples of phylogenetically young populations, based on minimal reductions in eye size and amount of pigmentation, whereas the Pachón and Molino cavefish are considered phylogenetically old, with eyes and pigmentation regressed to the maximal extent. Troglomorphic traits range between these extremes in other cavefish populations.

In some cases, most famously Chica cavefish, there has been subsequent introgression between cavefish and surface fish to form hybrid populations. La Cueva Chica is located near a large surface stream, Rio Tampáon, from which surface fish enter the cave via underground conduits and hybridize with cavefish. Chica cavefish have been collected and sold in pet stores as blind *Astyanax* cavefish, but are not recommended for most research purposes because of their recent hybrid background.

TROGLOMORPHIC TRAITS

Troglomorphic traits are the sum of constructive and regressive changes (*e.g.*, gains and losses) that have evolved in cavefish (Table 1, Fig. 3). The most conspicuous troglomorphic traits are the loss or reduction of eyes and pigmentation. Eye and pigment regression can be considered developmental defects. Although the eye primordia and body pigment cell precursors initially develop in cavefish embryos, they fail to complete differentiation and eventually degenerate (Fig. 2). Some parts of the cavefish embryonic eye are more degenerate than others, suggesting a degree of independence in their troglomorphic evolution. The lens is reduced to a tiny vestige and the cornea and ciliary body are absent. Although a layered retina is initially formed, retinal growth eventually ceases and the layering becomes disorganized. Photoreceptor cells initially differentiate but subsequently disappear and are not replaced in adults. Neurites grow from the retinal ganglion cells to the

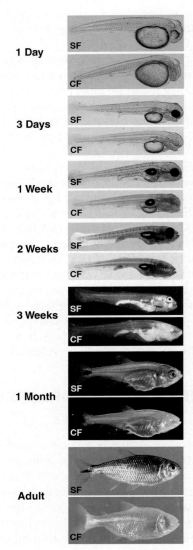

FIGURE 2 *Astyanax mexicanus* surface fish (SF) and Pachón cavefish (CF) shown from top to bottom at selected developmental stages and as adults.

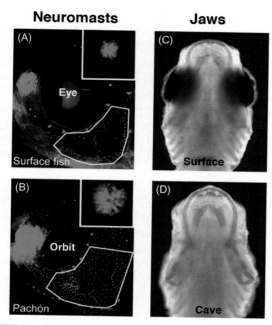

FIGURE 3 Constructive traits. (A,B) The numbers and diameters (insets) of superficial neuromasts (green dots) are increased in a region below the orbit (white outlined areas) in Pachón cavefish (B) relative to surface fish (A). Views are of the head from a lateral side. (C,D) Jaw span and protrusion is increased in Pachón cavefish (D) relative to surface fish larvae (C). Views are of the head from the ventral side with jaw and pharyngeal cartilage stained blue. *Reprinted with permission from Yamamoto et al. (2009).*

Constructive traits include those that are beneficial for feeding, such as larger jaws (Fig. 3C,D), more taste buds, and additional maxillary teeth, or to meet the challenges of navigation in the dark, hallmarked by an enhanced sensory neuromast system (Fig. 3A,B). Changes in jaw shape promote feeding or prey capture in different ways: small jaws are useful for biting, which occurs in surface fish, whereas large protruding jaws are used for sucking, which is prevalent in cavefish. Taste buds are employed to distinguish food from nonfood items as cavefish graze on the sediment of cave pools. There is also a dramatic change in posture that is related to substrate feeding. Cavefish glide smoothly across the substrate at a 45° angle using their shovel-like jaws to efficiently scoop and sample detritus. In contrast, surface fish feed primarily in the water column guided by vision. When surface fish are forced to search for food on the bottom of a darkened laboratory tank, they approach the substrate at a 90° angle, awkwardly swivel on their body axis, and pick up food inefficiently. Competition between surface fish and cavefish for limited food in lighted or darkened tanks has shown that surface fish can outcompete cavefish in the light, but cavefish can obtain food more successfully in the dark.

In northeastern Mexico, there is a rainy season, when food supply in caves may be abundant, and a dry season, when food may be minimal. Probably to survive such fluctuations, cavefish have accumulated fat deposits under the skin throughout the body and especially in the empty eye sockets. Thus, the ability to detect and consume different types of food and to store energy may be required for adaptation to the cave environment.

Navigation without vision in a dark cave is accomplished by increasing the size and number of motion-detecting neuromasts. As in other teleosts, *Astyanax mexicanus* has canal neuromasts located in the lateral lines and superficial neuromasts distributed freely in the head and body (Fig. 3A,B). An increase in superficial neuromasts is the cause of a behavior that attracts cavefish to water vibrations (Yoshizawa et al., 2010). Vibration attraction behavior is strong in cavefish but weak and rarely seen in surface fish, probably because of the risk in attracting large predators. Attraction to vibrations confers an advantage for feeding in darkness and could have been one of the first evolutionary changes that resulted in adaptation to the cave habitat. An enhanced neuromast system also allows cavefish to "memorize" their complex physical environment. *Astyanax mexicanus* perform exploratory swimming behaviors when challenged with new environments that later allow them to avoid obstacles, and this spatial orientation behavior is enhanced in cavefish.

The ventral forebrain and hypothalamus are enlarged in cavefish compared to surface fish (Rétaux et al.,

optic tectum through the optic nerve but they also probably disappear in adults. In concert with visual decay, cavefish have smaller optic tecta, the paired regions of the brain that transmit visual information from the eyes. There are also differential effects on three types of pigment cells: black melanophores, silver iridophores, and orange xanthophores. The number of melanophores is reduced to various extents in different cavefish populations. Furthermore, black melanin pigment is absent in most Pachón and Molino cavefish, resulting in albinism. Changes in iridophores and xanthophores, if any, are less severe than those affecting melanophores.

Other regressive traits in cavefish are a reduced metabolic rate, which is presumably beneficial for energy conservation, reductions in the size of scales and number of fin rays, and loss of one or two rib-bearing vertebrae. Fewer vertebrae lead to a shorter body axis relative to surface fish. The advantage of a shorter body is not clear. Finally, some of the behaviors typical of surface fish are absent in cavefish. These include schooling behavior, which is based on vision, the scattering (or alarm) reaction, which is triggered by a body secretion warning conspecific fish schools of predator approach, and aggressive behavior. The absence of large predators in caves could facilitate these regressive behavioral changes.

2008). One of the reasons for this increase may be to match gustatory processing centers in the central nervous system with the development of additional taste buds. Ventral forebrain changes also involve an increase in the number of GABAeric olfactory interneurons, which are crucial for olfactory learning and discrimination. This constructive change in olfaction is coupled with a greater size, and presumably receptor capacity, of cavefish nasal areas.

The adaptive significance of some constructive characters is more difficult to understand. For example, the surface fish craniofacial skeleton includes six flat bones surrounding each orbit. Cavefish have a different craniofacial organization: their orbital bones expand over the empty eye sockets and are subdivided into numerous smaller bones. The changes in craniofacial structure are due to the absence of a fully differentiated eye, which negatively influences orbital bone growth in surface fish. Surface fish-like craniofacial features can be restored in cavefish by swapping its embryonic lens for that of a surface fish (see below). As a by-product of eye degeneration, craniofacial reorganization may have no adaptive significance.

DEVELOPMENTAL BASIS OF TROGLOMORPHIC TRAITS

The primary focus of developmental studies in the *Astyanax mexicanus* model system has been to determine the molecular mechanisms of eye and pigment degeneration (Jeffery, 2009b). These studies have mostly been done with Pachón cavefish.

As in other teleosts, the embryonic eyes of *Astyanax mexicanus* consist of two major parts: the lens, which is derived from the surface ectoderm, and the optic cup, an out-pocketing of the brain. The optic cup is the precursor of the retina and its melanin-pigmented epithelium (the retinal pigment epithelium or RPE). Eyes with a smaller lens and optic cup relative to surface fish initially develop in cavefish embryos but they subsequently cease growth, degenerate, and disappear into the orbits (Fig. 2). The *pax6* gene, which encodes a transcription factor, is a key regulator of eye development in vertebrates. The small cavefish optic cup shows reduced *pax6* expression relative to surface fish, which is caused by upregulation of the *sonic hedgehog* (*shh*) gene along the anterior embryonic midline. *Shh* does not affect *pax6* expression directly. Instead *Shh* proteins diffuse from the midline, resulting in upregulation of the *pax2* and *vax1* genes, which in turn negatively regulate *pax6* in the optic cup.

The expression of *pax6* is also downregulated in the cavefish lens. The cavefish lens is formed just as in surface fish, but lens fiber cell differentiation does not occur. Instead of differentiating, the cavefish lens undergoes programmed cell death. The αA-*crystallin* gene, which encodes a cell survival protein, is downregulated, and the *heat shock protein 90α* (*hsp90α*) gene, which encodes a cell death-promoting protein, is upregulated in the cavefish lens. Together, these changes in gene expression trigger lens cell death in cavefish. The lens normally induces the anterior parts of the eye, including the cornea and ciliary body. Therefore, in the absence of a functional lens these optic components are missing in cavefish.

Differentiation of the retina and RPE are also initiated in cavefish. Neural and glial cells differentiate normally during retinal development but this is not the case for rod photoreceptor cells. Cavefish rod cells are formed with abnormally shortened outer segments, but they subsequently disappear and are not replaced. Parts of the cavefish retina also undergo cell death and its layers become extremely disorganized. Melanin pigmentation appears in the RPE in some cavefish populations, although their RPE does not seem to differentiate normally.

Eye growth continues throughout life in teleosts due to the division of stem cells in the lens and the outer edge of the retina/RPE (the ciliary marginal zone or CMZ). The rate of cell proliferation in the cavefish lens, however, is insufficient to replace the dying cells, and consequently the lens becomes progressively smaller during larval development. In contrast, there is a high rate of cell division in the cavefish CMZ. However, the absence of net growth in the retina is due to death of the newly born cells before they can contribute to its structure. Why the cavefish retina continues to produce new cells, presumably at high energetic cost, only to subsequently destroy them, is a paradox. The final size of the vestigial (internalized) eye is variable both within and between different cavefish populations.

The critical role of *Shh* in eye degeneration has been demonstrated by *shh* gene overexpression in surface fish. Similar to cavefish, *Shh* upregulation causes increases in *pax2* and *vax1* expression and a decrease in *pax6* activity, reducing the size of the surface fish optic cup (Yamamoto et al., 2004). In addition, *Shh* overexpression induces cell death in the surface fish lens. Thus, a surface fish mimic of cavefish eye degeneration can be made by *Shh* overexpression.

The process of eye regression in cavefish is summarized as follows. First, *Shh* is over expressed along the embryonic midline. Second, the increase in *Shh* signaling down regulates *pax6*, which reduces the size of the optic cup and induces lens cell death. Third, the nonfunctional lens is either not able to induce other eye tissues or to protect them from cell death and thus allows them to mount a sufficient growth rate to match the increasing body size.

There are three ways in which the cavefish eye can be rescued from degeneration. First, eyes and vision can be restored in hybrids produced by crossing members of independently evolved cavefish lineages. Rescue by genetic complementation shows that some of the genes responsible for eye degeneration are different in these lineages. Second, inhibition of lens cell death with the Hsp90 antagonist radicicol can partially rescue eye development, indicating that the cell death program is reversible. Third, the cavefish eye can be rescued by lens transplantation (Yamamoto and Jeffery, 2000). A surface fish lens transplanted into the cavefish optic cup evades cell death and survives to influence the development of other eye parts, including the cornea, ciliary body, and retinal photoreceptor cells. Moreover, in the presence of a surface fish lens, cell death is prevented in the cavefish retina, allowing it to resume growth and form a large external eye, indicating that cavefish have retained the instructions for optic development.

Reduction or elimination of melanophores is another major regressive trait. Melanophores and other body pigment cells are formed from the neural crest, which originates in surface ectoderm adjacent to the neural plate. During neural tube formation, neural crest cells leave the surface and migrate long distances through the embryo to eventually differentiate into many different adult tissues, including spinal ganglia, the peripheral nervous system, facial bones, and body pigment cells. It seems unlikely that such an important cell type could be modified without lethality. Accordingly, no defects in neural crest cell formation or migration have been discovered in cavefish embryos. Instead, the block in pigmentation is beyond this point at the stage of melanophore differentiation.

During melanophore differentiation, melanin is synthesized within the melanosome, a membrane-bound organelle, via a series of well-known biochemical reactions (Fig. 4A). The amino acid L-tyrosine is transported into the melanosome, where it is converted to L-DOPA by tyrosinase. L-DOPA is then converted into melanin by a cascade of enzymatic reactions, catalyzed either by tyrosinase or other enzymes. In albino cavefish, melanin synthesis can be rescued by supplying exogenous L-DOPA (Fig. 4B), but not L-tyrosine. Therefore, cavefish melanophores lack melanin not because they are deficient in tyrosinase or downstream enzymes but instead because they do not contain L-tyrosine. A defect at the first step in the melanin synthesis pathway has evolved independently in several different cavefish populations.

Developmental studies of constructive traits have focused on jaws and taste buds, which require shh gene expression (Yamamoto et al., 2009). During mouth development, shh is first expressed throughout the oral epithelium and later attenuated to the developing taste buds. Experimental downregulation of shh in cavefish or surface fish embryos results in smaller jaws with fewer taste buds. Reciprocally, shh upregulation results in larger jaws with more taste buds. Thus, the pleiotropic shh gene functions in both feeding apparatus enlargement and eye degeneration, revealing an important antagonistic linkage between these constructive and regressive traits.

INHERITANCE AND GENETIC BASIS OF TROGLOMORPHIC TRAITS

Eye degeneration is controlled by mutations in 6–12 different genes, each with a small role in the process (Jeffery, 2009a). Quantitative trait locus (QTL) analysis has been used to map eye regression traits and to identify candidate genes. One of the eye candidate genes is αA-crystallin (see above) whose downregulation may affect the survival of the cavefish lens. Other candidate genes for eye degeneration are ROM1, which is required for normal development of rod photoreceptor cells, and Shroom2, which regulates melanosome localization in the RPE. The over expressed shh gene is not

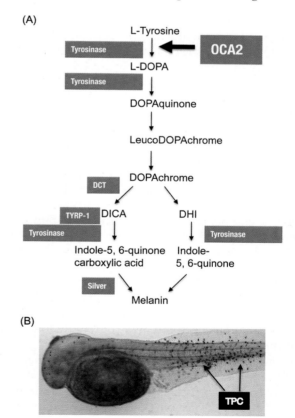

FIGURE 4 (A) Melanin biosynthetic pathway showing substrates and positions of enzyme and Oca2 function. (B) Albino Pachón cavefish embryo provided with exogenous L-DOPA substrate rescues melanin-containing pigment cell precursors (TPC).

represented by a QTL, suggesting that upstream regulatory genes, rather than *shh* itself, have mutated in cavefish (Protas et al., 2008). The identification of additional candidate genes remains challenging in the absence of a sequenced *Astyanax mexicanus* genome.

Pigmentation changes are inherited independently from eye regression, as indicated by segregation of these two traits in the F2 progeny of a cavefish × surface fish cross (Jeffery, 2009a). Multiple genes with additive effects also decrease melanophore number. Similar to eyes, the genes responsible for melanophore reduction can be different in various cavefish populations. The number of genes controlling changes in pigmentation is potentially high: 18 different QTL affect melanophore pigmentation (Protas et al., 2007). Included in the many QTL and predicted pigmentation genes are two traits with relatively simple inheritance: albinism and the brown eye phenotype.

Cavefish albinism is inherited as a recessive trait controlled by *oculocutaneous albinism 2* (*oca2*), a single gene that also has a major role in mouse and human albinisms (Protas et al., 2006). Loss of *oca2* function has occurred in both Pachón and Molino cavefish, but through different mutations: large deletions encompassing two distinct regions in the protein coding sequence. Thus, for an unknown reason, the *oca2* gene is repeatedly targeted for mutation, resulting in independent origins of albinism. The Oca2 protein may control accessibility of L-tyrosine to tyrosinase for conversion to L-DOPA, perhaps by serving as a transport channel in the melanosome membrane.

Mutations in the *brown* gene decrease the number of melanophores and change pigmentation from black to brown in the (degenerate) eyes of many different cavefish populations. Genetic analysis predicted a single gene responsible for the *brown* phenotype, and a single QTL was identified (Gross et al., 2009). The *brown* gene encodes the melanocortin 1 receptor (Mc1r), a component of the Melanosome Stimulating Hormone-α (MSH-α) signaling system, which is involved in changing pigmentation in many different vertebrates. Furthermore, different mutations in the *Mc1r* gene, either in the coding or regulatory regions, are responsible for the brown eye phenotype in various cavefish populations, and *Mc1r* is another example of a gene that has been mutated repeatedly to generate the same troglomorphic phenotype.

Several mutated genes are responsible for taste-bud amplification but as in the case of eyes, none of these is likely to be *shh* (Protas et al., 2008). This indicates that the effects of *Shh* on taste-bud development also occur downstream of the mutated genes. In contrast to the large number of genes involved in eye and pigment degeneration, only three gene loci, one major gene and two modifiers, are required to change feeding behavior. This is surprising because of the complexity thought to be associated with behavioral traits.

EVOLUTION OF TROGLOMORPHIC TRAITS

Constructive troglomorphic traits are likely to have evolved by natural selection. The best example is the attraction to vibrations, which is rare in surface fish but robust in cavefish. Vibration attraction behavior has a genetic basis and is beneficial for feeding, allowing it to be influenced by natural selection. The evolutionary forces controlling regressive traits are less clear. Two major hypotheses have been proposed: (1) accumulation of neutral mutations and genetic drift; and (2) negative selection based either on energy conservation or indirect pleiotropy involving a tradeoff with constructive traits.

In the case of eye loss, selection seems to be supported over neutral mutation, although all cavefish researchers do not agree with this conclusion. All known eye QTL decrease eye size (Protas et al., 2007). In contrast, the accumulation of random neutral mutations might be expected to increase as well as decrease the size of the eyes. However, decreases in eye function could also be explained by the fact that most natural mutations are recessive. An additional consideration is that genes such as *shh* and *hsp90α* increase rather than decrease expression activity during cavefish eye development. Particularly in the case of *shh*, eye loss could be the result of indirect selection involving a tradeoff for the development of larger jaws and more taste buds at the expense of eyes. Because many genes are mutated to give rise to eye loss, it is also possible that one or more evolutionary forces are acting in concert on this complex troglomorphic trait.

Reduced pigment cell number is probably caused by neutral mutation and genetic drift (Protas et al., 2008). Genetic analysis shows that individual QTL can increase or decrease melanophore abundance. When albinism is considered, however, the picture becomes less clear. On one hand, neutral mutations and subsequent genetic drift could have occurred if *oca2* is not pleiotropic and thus does not affect other developmental pathways. On the other hand, the position of Oca2 function at the beginning of the melanin biosynthesis pathway suggests another possibility (Fig. 4A). This is a critical branch point in which L-tyrosine can be converted to either (1) L-DOPA and melanin or (2) to dopamine, which could have adaptive significance in changing feeding behaviors. Therefore, inhibition of melanin synthesis at a step that would provide more L-tyrosine substrate for dopamine synthesis could be another example of the role of indirect natural selection via pleiotropic tradeoff as a driving force for evolution and adaptation.

See Also the Following Articles

Adaptation to Low Food
Adaptive Shifts
Gammarus minus: A Model System for the Study of Adaptation
Niphargus: A Model System for Evolution and Ecology

Bibliography

Gross, J. B., Borowsky, R., & Tabin, C. J. (2009). A novel role for Mc1r in the parallel evolution of depigmentation in independent populations of the cavefish *Astyanax mexicanus*. *PLoS Genetics, 5* (1), e10000326.

Jeffery, W. R. (2009a). Regressive evolution in *Astyanax* cavefish. *Annual Reviews of Genetics, 43*, 25–47.

Jeffery, W. R. (2009b). Evolution and development in the cavefish *Astyanax*. *Current Topics in Developmental Biology, 86*, 191–221.

Mitchell, R. W., Russell, W. H., & Elliot, W. R. (1977). Mexican eyeless characin fishes, genus *Astyanax*: Environment, distribution, and evolution. *Special Publication of the Museum of Texas Tech University, 12*, 1–89.

Protas, M., Hersey, C., Kochanek, D., Zhou, H., Wilkens, H., Jeffery, W. R., et al. (2006). Genetic analysis of cavefish reveals molecular convergence in the evolution of albinism. *Nature Genetics, 38*(1), 107–111.

Protas, M., Conrad, M., Gross, J. B., Tabin, C. J., & Borowsky, R. (2007). Regressive evolution in the Mexican cave tetra, *Astyanax mexicanus*. *Current Biology, 17*(5), 452–454.

Protas, M., Tabansky, I., Conrad, M., Gross, J. B., Vidal, O., Tabin, C. J., et al. (2008). Multi-trait evolution in a cave fish, *Astyanax mexicanus*. *Evolution & Development, 10*(2), 196–209.

Rétaux, S., Pottin, K., & Alunni, A. (2008). *Shh* and forebrain evolution in the blind cavefish *Astyanax mexicanus*. *Biology of the Cell, 110*(3), 139–147.

Yamamoto, Y. & Jeffery, W. R. (2000). Central role for the lens in cave fish eye degeneration. *Science, 289*(5479), 631–633.

Yamamoto, Y., Stock, D. W., & Jeffery, W. R. (2004). Hedgehog signalling controls eye degeneration in blind cavefish. *Nature, 431* (7010), 844–847.

Yamamoto, Y., Byerly, M. S., Jackman, W. R., & Jeffery, W. R. (2009). Pleiotropic functions of embryonic *sonic hedgehog* expression link jaw and taste bud amplification with eye loss during cavefish evolution. *Developmental Biology, 330*(1), 200–211.

Yoshizawa, M., Gorički, Š., Soares, D., & Jeffery, W. R. (2010). Evolution of a behavioral shift mediated by superficial neuromasts helps cavefish find food in darkness. *Current Biology, 20* (18), 1631–1636.

B

BATS

Thomas H. Kunz, Susan W. Murray, and Nathan W. Fuller

Boston University

INTRODUCTION

Bats (order Chiroptera) are an ecologically diverse and geographically widespread mammalian taxon. With 1232 extant species currently described (Simmons, personal communication), members of two suborders (Yinpterochiroptera and Yangochiroptera) constitute approximately one-fifth of all living mammals. They range in size from the tiny 2-gram Kitti's hog-nosed bat (*Craseonycteris thonglogyai*) to the large Malayan flying fox (*Pteropus vampyrus*) weighing about 1200 grams, with a wingspan of approximately 1.5 meters. Bats are characterized by several life-history traits that make them unique among mammals. For example, compared to small terrestrial species, bats of similar size have very few young and long periods of pregnancy and lactation, and some are known to live up to 42 years in the wild (Kunz and Parsons, 2009). Differences in life-history traits between bats and other small mammals are often attributed to the evolution of flight and echolocation (Crichton and Krutzsch, 2000; Kunz and Fenton, 2003).

Bats are the only mammals that have evolved powered flight. Thus, along with echolocation, these two traits have made it possible for bats to exploit many different types of roosting habitats (*e.g.*, foliage, tree cavities, caves, rock crevices) that are generally inaccessible to most terrestrial mammals. Echolocation and flight have also made it possible for bats to exploit an unprecedented variety of food sources, including fruit, nectar, pollen, leaves, seeds, invertebrates (*e.g.*, insects, spiders, crustaceans, and scorpions), small vertebrates (*e.g.*, fish, frogs, lizards, birds, and other small mammals), and blood (Kunz and Fenton, 2003).

Although echolocation has evolved independently in birds and mammals, and even within different groups of mammals (*e.g.*, bats, shrews, cetaceans), the most sophisticated and diversified form of echolocation can be found in laryngeal-emitting species of bats. Echolocation is used for prey detection and capture, for navigation, and in some instances for communication. In additon to echolocation, several species may rely on a combination of vision, olfaction, and prey-generated sounds to locate food (Kunz and Fenton, 2003).

The roosting habits of bats are often highly specialized, with different species occupying tree cavities; spaces beneath exfoliating bark; unmodified foliage; leaves modified into so-called tents; abandoned ant, termite, and bird nests; large and small caves; rock crevices; and a wide variety of manmade structures, including mines, old and new buildings, ancient stone ruins, and bridges (Kunz, 1982; Kunz and Fenton, 2003). Caves alone provide a variety of structural substrates for roosting, including crevices, cavities, textured walls and ceilings, massive rooms, rock outcrops, and rock rubble on floors. In addition, the microclimates of caves occupied by bats can vary enormously, depending on latitude, altitude, depth, and volume, as well as the number, size, and position of openings to the outside. These variables can influence the amount of airflow, the presence of flowing and standing water, and daily and seasonal variations in atmospheric pressure, temperature, and humidity. Thus, the environmental conditions within caves may be hot, cold, dry, humid, still, or windy, although different species appear to select parts of caves and cave-like structures that minimize energy expenditure and water loss, and that facilitate an array of social interactions.

This article highlights the biology of bats that roost in caves and cave-like structures. Specifically, we briefly discuss why bats live in caves, where they are found, their roosting requirements, threats from moving wind turbines and emerging diseases, as well as conservation and management issues important to protecting cave-roosting species.

CAVE BATS AND THEIR DISTRIBUTION

Cave bats are defined as trogloxenes, species that complete part of their life cycle outside of caves. Their ability to fly and echolocate has allowed many species to exploit caves and similar subterranean habitats for roosts and to forage for food often at considerable distances from these structures. All bats that belong to the suborder Yangochiroptera and orally echolocating species belonging to the suborder Yinpterochiroptera are able to navigate in dark spaces and to feed in habitats ranging from open fields, high altitudes, and dense forested areas to over lakes and streams. Rousette fruit bats (family Pteropodidae; suborder Yinpterochiroptera) may also roost in caves, but instead of using vocal cords to produce echolocation calls, they rely on tongue clicks to produce audible sounds to facilitate navigation in the dark. The handful of other pteropodids found roosting in caves are restricted to areas where enough ambient light makes it possible for them to navigate in and out, as they largely rely on vision and not echolocation while in flight (Kunz and Fenton, 2003).

Bats are virtually ubiquitous; they are known on all continents, except Antarctica, and from many archipelagos and oceanic islands (Kunz, 1982; Fleming and Racey, 2010). Most bats occur in tropical regions, where they are often the most diverse and abundant mammals present. The diversity of bats generally increases as one travels from the poles toward the equator, a pattern that is largely attributable to an increase in habitat complexity as latitude decreases (Kunz and Fenton, 2003).

The geographic distribution of cave-roosting bats not only depends on the presence of caves (Palmer and Palmer, 2009), but is also a consequence of specific roosting requirements (Kunz, 1982). For example, although the ghost bat (*Macroderma gigas*), the orange leaf-nosed bat (*Rhinonycteris aurantius*), and the large bent-wing bat (*Miniopterus schreibersii*) are all found in Australian caves, their roosting requirements and hence geographic distributions are markedly different. The orange leaf-nosed bat selects caves that are extremely hot and humid (28 to 30°C and greater than 94% relative humidity) and are known from only 10 caves in Australia. In contrast, the large bent-wing bat can be found roosting at a broader range of temperature and humidity and has one of the widest reported distributions of cave-roosting bats, encompassing southern

FIGURE 1 Townsend's big-eared bat (*Corynorhinus townsendii*). Photo by J.S. Altenbach. Used with permission.

Europe, Africa, Southeast Asia, Japan, and Australia (Zubaid et al., 2006). Similarly, different species of bats in the West Indies select caves whose temperatures may range from 18 to 40°C (Fleming and Racey, 2009). The presence and abundance of caves with characteristics allowing the creation of gradients in temperature will strongly influence the presence and abundance of different bat species.

Bats are found almost everywhere subterranean spaces exist. The distributions of cave-dwelling bats are determined largely by species-specific roosting requirements that vary depending on their ecology and evolutionary history. Local and global distributions and densities of bats that rely on caves for at least part of their life cycle are in turn determined largely by the distribution, quantity, and characteristics of available caves (Palmer and Palmer, 2009). For example, the Townsend's long-eared bat (*Corynorhinus townsendii*), endemic to North America (Fig. 1), almost exclusively roosts in caves and cave-like structures (Kunz, 1982).

FUNCTIONS OF CAVE ROOSTS

Bats roost in caves for a variety of reasons, including courtship and mating, raising young, and hibernation. They seek shelter during the day and disperse from these sites to forage for food at night. During the day, bats typically sleep, rest, groom, or interact with their roost-mates. For example, a typical day for lactating female lesser long-nosed bats (*Leptonycteris curasoae*) involves sleeping or resting quietly for up to 12 hours, interspersed with periodic grooming and nursing. Although female *L. curasoae* usually roost together in caves during the day, they seldom interact with males, except during a brief mating season in November and December (Muñoz-Romo and Kunz, 2009). The common vampire bat (*Desmodus rotundus*) forms long-term social bonds, and individuals groom one another as they interact socially while occupying both cave and tree roosts (Crichton and Krutzsch, 2000; Kunz and Fenton, 2003). In addition, many insectivorous species retreat to caves between feeding bouts, where they may cull the wings and heads of insects that were captured while foraging. Frugivorous species sometimes transport large fruits to caves where they cull soft pulp and where they can reduce the risks of predation. Some species that roost in foliage or tree cavities during warm months typically hibernate in caves and cave-like structures during the winter (Kunz, 1982; Kunz and Fenton, 2003; Zubaid et al., 2006).

Courtship and Mating

Several types of mating systems have been described for bats (Crichton and Krutzsch, 2000). Mating systems of bats and other mammals are often classified into three general categories: promiscuity, polygyny, and monogamy. Mating systems of bats, however, cannot be easily categorized into one of these groups, as they often exhibit a continuous spectrum of mating behaviors.

Promiscuity is a type of mating system in which both males and females have multiple partners. However, such a system is almost always highly structured, with some males siring more offspring than others. Promiscuity is common among temperate cave-roosting species, possibly because of the limited time available for mating in autumn before individuals enter hibernation. Males of most temperate species generally do not roost with females during warm months but instead roost alone or in small groups. Assemblages of males and females from temperate regions often gather at caves and mines in the autumn (referred to as swarming behavior), a time that also coincides with courtship and mating behavior. During the swarming period, bats are active in caves and mines at night, where males can often be observed displaying and chasing females. Male and female greater horseshoe bats (*Rhinolophus ferrumequinum*) have a mating system in which males establish territorial sites inside caves and mines in early autumn. Females gather at these sites and selectively visit and mate with different males in their territories (Crichton and Krutzsch, 2000).

Polygyny, a mating system that is most common in bats, is characterized by one male mating with several females (Crichton and Krutzsch, 2000). An example of this type of mating system can be observed in the cave-roosting greater spear-nosed bat (*Phyllostomus hastatus*). In this species, females roost in small stable groups, often remaining together for 10 years or more. Females often form discrete roosting groups in solution cavities or "potholes" on cave ceilings, where a dominant male typically defends a group of females from intrusions by other males. By defending the females, or the roost cavity, a so-called harem male is often able to mate with several females. Sometimes these harem males are accompanied by a subordinate male who positions himself in the harem to assume a dominant role if the harem male should become injured or die. The risks and costs associated with mate-guarding behavior can be substantial. For example, a harem male greater spear-nosed bat may incur some injuries while defending the females or roost cavity and may only sire 60 to 90% of the young born to those females. A similar pattern of mate guarding and courtship has been observed in the Jamaican fruit bat (*Artibeus jamaicensis*), which commonly roosts in caves on many of the islands in the West Indies and throughout Central and South America (Crichton and Krutzsch, 2000).

Monogamy occurs when males and females form long-term pair bonds. This type of mating system has been described for only a few species of bats. Two examples are the African false vampire bat (*Cardioderma cor*) and the American false vampire bat (*Vampyrum spectrum*), both of whom are carnivorous, sit-and-wait predators. An extended period of parental involvement in which males provision both females and young may have contributed to the evolution of monogamy in these and other species. To our knowledge, lek mating behavior, in which males assemble and display simultaneously to attract females, has not been observed in cave-roosting bats (Crichton and Krutzsch, 2000).

Rearing Young

During the pregnancy and lactation periods, females form maternity colonies, which are often located in separate places from roosts used by males. In most species of bats, the responsibility of raising young lies solely with females. Pregnancy and lactation are both energetically expensive events; thus females and their young can benefit from the heat generated when they form dense clusters in partially enclosed spaces often found in tree cavities, buildings, caves, and cave-like structures. Roosting together in large clusters may reduce the energy expenditure of some individuals by up to 50%. When lactating females disperse from roosts in the evening to feed, they often leave their pups in a warm, incubator-like environment. Females incur high energy costs when they forage and return to the roosts one to three times each night to find and suckle their dependent young. Thus, assembling in warm places can help reduce the energy needed by small bats to remain euthermic (Kunz, 1982; Crichton and Krutzsch, 2000; Kunz and Fenton, 2003). The formation of multispecies assemblages of bats may also contribute to reduce energy expenditure. By roosting in the same cave, species with different activity patterns can form larger assemblages, and thus contribute to a higher ambient temperatures, and reduced energy expenditure (Fleming and Racey, 2009).

Each spring, Brazilian free-tailed bats (*Tadarida brasiliensis*) migrate from Mexico to the southwestern United States to form large maternity colonies in caves and sometimes other structures such as buildings and bridges. This species is thought to form some of the largest aggregations of mammals known to humankind, where a single colony may exceed several million individuals in single cave (Fig. 2). Each time a female Brazilian free-tailed bat returns from a feeding bout to suckle her young, she faces the daunting task of finding her own pup among the millions of babies that are left on the ceilings and walls of the cave. A mother bat begins this adventure using spatial memory by returning to the area in the cave where she left her pups before emerging to feed. She then uses vocal and olfactory cues to ultimately identify her own pup among the thousands or more present in the immediate surroundings (Crichton and Krutzsch, 2000). Hungry pups will sometimes attempt to nurse from almost any female, although lactating females usually guard against milk stealing from unrelated individuals. The investment that a mother bat makes in her pups is substantial, often requiring quantities of food intake equal to her entire body mass each night during peak lactation (Kunz and Parsons, 2009).

Young Brazilian free-tailed bats grow rapidly from a diet of energy-rich milk. Mothers nourish their young with milk for several weeks, because young bats cannot fly and feed on their own until their wings have almost reached adult dimensions. Within 6 weeks of birth, young Brazilian free-tailed bats are able to fly and forage on their own. In contrast to most other mammals that typically wean their young at about 40% of adult size, most insectivorous bat species suckle their young until they are about 90% of adult size (Crichton and Krutzsch, 2000; Kunz and Fenton, 2003; Kunz and Parsons, 2009).

HIBERNATION

Bats have evolved behavioral and physiological mechanisms to avoid long periods of adverse weather and low food or water availability. Some species migrate to more suitable areas where food is readily available, but others use daily torpor, a controlled lowering of body temperature to conserve energy when food resources are limited or unavailable. Only temperate species in the families Vespertilionidae and Rhinolophidae are known to hibernate in caves and mines (Fig. 3) (Kunz, 1982; Kunz and Fenton, 2003).

Hibernating bats rely on stored fat as their primary energy source during hibernation and are sustained on these reserves for upwards of 6 to 8 months. Hibernation is an energy-saving strategy that is strongly influenced by the ambient conditions in a cave. When a bat is hibernating, low ambient temperatures lead to a decrease body temperature and metabolic rate. When the ambient temperature is too cold or too warm, bats typically arouse and move to other parts of a hibernaculum. It is important for hibernating bats to occupy caves and mines that provide a variety of temperatures, because individuals often change roosting positions as the season progresses (Kunz and Fenton, 2003; Zubaid *et al.*, 2006).

During hibernation, bats lower their body temperature to within a few degrees of the ambient temperature, but individuals arouse periodically by employing nonshivering thermogenesis to increase their body temperature. Bouts of hibernation can last anywhere from a few days to several weeks. In areas with moderate winters, for example, bats such as the greater horseshoe bat (*Rhinolophus ferrumequinum*) may arouse

FIGURE 2 Emerging Brazilian free-tailed bats (*Tadarida brasiliensis*) from a maternity cave in south-central Texas. *Photo by M.D. Tuttle, Bat Conservation International. Used with permission.*

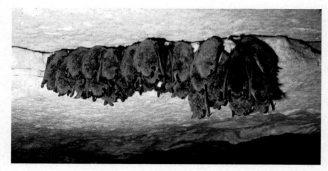

FIGURE 3 Hibernating cave bats (*Myotis velifer*) in a gypsum cave in south-central Kansas. *Photo by T.H. Kunz.*

periodically and feed on insects on warm winter nights. Arousals from deep hibernation are energetically costly, with a single arousal expending the energy equivalent of a bat spending 30 days in deep torpor. Thus, if hibernating bats arouse too often, either because the microclimate is suboptimum or from human disturbance, they may not have enough fat reserves to survive the winter (Kunz and Fenton, 2003; Zubaid et al., 2006).

COSTS AND BENEFITS OF LIVING IN CAVES

The decision about where to roost is critical to the survival and reproductive success of bats. The type of roost that a bat selects is influenced by its morphology, ecology, and physiological requirements and often reflects a compromise between the costs and benefits associated with a particular type of roost (Kunz, 1982). For cave-roosting species, the benefits of living in a cave usually outweigh costs that they may otherwise incur. In the following section, we discuss the major costs and benefits considered critical for the selection of roosts by cave-dwelling bats. It is important to note that roosting requirements and relevant costs and benefits are not uniform for all species and may even vary intraspecifically, depending on geographic location, reproductive condition, and/or season (Kunz and Fenton, 2003; Zubaid et al., 2006).

Benefits

Caves offer a wide range of benefits to bats including structurally and climatically stable environments, and protection from predators and adverse weather. Microclimate, specifically temperature and relative humidity, is arguably the most important factor in roost selection by cave-dwelling bats (Kunz, 1982; Zubaid et al., 2006). Different bat species roost in a variety of microclimates within caves and mines, and this variation is often correlated with body size, diet, phylogeny, and their ability to enter torpor. Metabolically conservative species on islands are also likely to derive benefits from the stable thermal microclimate of caves, and hurricane-prone regions appear to favor the prevalence of cave-dwelling bats (Fleming and Racey, 2009).

Compared to nonvolant mammals, bats experience high rates of evaporative water and heat loss, due in large part to their relatively high surface area-to-volume ratio, enhanced by the large surface of their naked wing membranes. At low relative humidity (<20%), bats may lose up to 30% of their body mass per day from evaporative water loss alone. This rapid dehydration can be lethal. Thus, many bat species select caves that have high relative humidity to help conserve water during the day (Kunz and Fenton, 2003).

Bats use at least four different strategies for conserving energy while in their roosts (Kunz and Fenton, 2003; Zubaid et al., 2006). Some species select roosts with ambient temperatures within their thermal neutral zone. For example, the California leaf-nosed bat (*Macrotus californicus*) often exploits geothermally heated mines to conserve heat during the winter. Other species form large colonies in areas in caves that have little airflow (e.g., high-domed ceilings or dead-end chambers), which contribute to an increase in roost temperatures as the metabolic heat generated by the bats becomes trapped. The lesser long-nosed bat (*Leptonycteris curasoae*), the large bent-wing bat (*Miniopterus schreibersii*) the sooty mustached bat (*Pteronotus quadridens*), and the Brazilian free-tailed bat (*Tadarida brasiliensis*) are examples of cave-dwelling species that form colonies large enough to substantially increase the temperature of their roost environment. Mormoopids, natalids, and many phyllostomids in the West Indies form multispecies assemblages that increase the temperature of the caves in which they roost (Fleming and Racey, 2009). During the winter months, hibernating species, such as big brown bats (*Eptesicus fuscus*), little brown myotis (*Myotis lucifugus*), Indiana myotis (*M. sodalis*), and the gray bat (*M. grisescens*), select colder roost environments that allow them to reduce their body temperature and thus become torpid. Torpor not only reduces the amount of energy a bat expends, but also helps reduce water loss. Finally, some species form dense clusters in winter that buffer individuals from changes in ambient temperature, a behavior that also reduces their energy expenditure (Kunz, 1982; Zubaid et al., 2006; Fleming and Racey, 2009).

In addition to the energy savings that cave-roosting bats may experience, many species also benefit from social interactions. For example, the environmental stability of caves often facilitates social interactions such as finding, attracting, and guarding mates; information transfer; and interactions that evolve through kin selection and/or reciprocal altruism (Crichton and Krutzsch, 2000). Females that roost together sometimes share information about food resources, such as the location of flowering and fruiting trees. Information transfer, presumably, is facilitated in the greater spear-nosed bats (*Phyllostomus hastatus*) through vocal contact, which may help females coordinate efforts to defend food patches from conspecifics. Also, bats often emerge synchronously from caves and cave-like structures, which may decrease an individual's risk of predation (Fleming and Racey, 2009) as a consequence of the so-called selfish herd effect (Kunz, 1982).

The common vampire bat (*Desmodus rotundus*), another very social species, has evolved a system of sharing blood with both relatives and unrelated female roost-mates. Vampire bats must obtain a blood meal at least once every three days or they will invariably die of starvation and dehydration. Females often share blood with roost-mates that are at risk of starving, but this sharing occurs only among individuals with whom they are closely associated. This behavior is referred to as reciprocity (or reciprocal altruism) and occurs when the cost to the individual performing the altruistic act is less than the benefit to the recipient when such an act is later reciprocated (Crichton and Krutzsch, 2000).

Costs

There are several potential costs associated with living in caves, most of which are related to living in large groups. Large numbers of bats that live in close physical contact with others may be more prone to transmit certain diseases or increase the risk of parasitic infestations. High mite infestations on a bat, for example, may cause an increase in the amount of time an individual spends grooming, and thus increase is daily energy expenditure (Kunz, 1982). Large groups of bats are more likely to transmit infectious pathogens (Kunz and Fenton, 2003; Kunz and Parsons, 2009), including fungal spores associated with white-nose syndrome (Frick et al., 2010). Large concentrations of bats in poorly ventilated chambers may also affect the balance of certain gasses in a cave, requiring additional physiological mechanisms to cope (Kunz, 1882; Fleming and Racey, 2009).

Birds of prey, such as owls, hawks, and falcons, are known to swoop down into columns of bats that emerge nightly from caves and mines. Some predatory birds, such as the bat hawk (*Machaerhamphus alcinus*) in Africa and the bat falcon (*Falco rufigularis*) in Central and South America even specialize on bats. Wintering merlins (*Falco columbarius*) regularly congregate around large bat colonies in the West Indies, influencing nightly departure patterns of bats during the season when merlins are important predators (Rodríguez-Durán and Lewis, 1985). Most predatory birds are territorial, so their numbers at any one cave are usually rather small.

Sometimes only a single mating pair of birds is associated with one colony of bats, thus the impact on local populations is probably minimal. Other animals that prey on cave-roosting bats include snakes, raccoons, skunks, opossums, and other bats; even frogs have been observed preying on bats. Notwithstanding, few studies have evaluated the impact that predators have on bat colonies in caves.

The distribution of caves in most terrestrial landscapes is highly variable (Palmer and Palmer 2009), and some may be located at considerable distances from the food resources on which bats depend. Bats that roost alone or in small groups in tree cavities and in foliage can often take advantage of food resources located near their roosts, but cave bats, especially those that form large aggregations, more often must commute considerable distances to foraging sites. Because flight is energetically expensive, bats must make compromises between colony size and the amount of energy spent commuting to feeding sites and the energy that is conserved by selecting roosts that have microclimates conducive to energy and water conservation (Kunz, 1982).

Local food resources may not be sufficient to support the energy and nutrient budgets of all individuals that form large cave colonies; thus some individuals must disperse considerable distances in order to secure their daily energy and nutrient requirements. For example, many maternity colonies of the Brazilian free-tailed bat (*Tadarida brasiliensis*) number in the millions, requiring some individuals to fly upwards of 50 km each night to obtain their food. This is the case for other species that form large colonies in caves, such as the large bent-wing bat (*Miniopterus schreibersii*), the lesser long-nosed bat (*Leptonycteris curasoae*) (Kunz and Fenton, 2003), and most members of the family Mormoopidae in the West Indies (Fleming and Racey, 2009).

CONSERVATION AND MANAGEMENT

In recent years, reductions in the numbers of cave bat populations have increasingly concerned conservation biologists (Kunz and Racey, 1998; Kunz and Fenton, 2003). One of the major problems that places bat populations at risk is that they have relatively low reproductive rates and thus are unable to recover quickly from population declines (Kunz and Fenton, 2003). Cave bats face a variety of human threats that may vary in different regions of the world. Some threats reflect differences due to socio economic conditions, habitat types, and cultural attitudes toward bats. Notwithstanding, several successful approaches in recent years, such as habitat restoration and cave protection, have been employed to protect bats, their roosts, and their food resources. Increasingly, most if not all geopolitical units (cities, states, countries) are faced with issues related to increased surface and subsurface mining operations, spelunking, ecotourism, vandalism, sealing of caves and mines for safety reasons, and deliberate killing of bats. Other local threats include guano mining and over-collection of bats for scientific research, each of which can have adverse affects on cave populations.

Bats are often portrayed by the media as being evil, vile, or vectors of disease, but these negative attitudes can best be overcome through increased educational efforts. Lack of basic information about the natural history of most cave-roosting bats is a concern in many regions of the world. Shared knowledge about the ecology and behavior of different species is essential for developing sound conservation strategies, and for informing natural resource managers and decision makers who will improve the well-being of bats and the food and roost resources on which they depend (Kunz and Fenton, 2003; Kunz et al., 2011).

ECOSYSTEM SERVICES PROVIDED BY CAVE-ROOSTING BATS

Many cave-dwelling bats provide essential ecosystem services by helping to maintain forest diversity by dispersing seeds and pollinating flowers (Kunz et al., 2011). Changes in bat diversity or abundance due to forest fragmentation or roost destruction can lead to the dysfunction of forest ecosystems. In addition to plant-visiting bats that disperse seeds and pollinate flowers, many insectivorous bats that roost in caves consume vast quantities of insects. Some insectivorous species feed on insects that cause significant damage to agricultural crops (Boyles et al., 2011). The Brazilian free-tailed bat (*Tadarida brasiliensis*), for example, is known to feed on insects that cause millions of dollars in damage to corn and cotton crops in the United States each year. In addition, nearly everywhere that large quantities of guano have accumulated in caves, local communities have discovered its value as a fertilizer. In some parts of the world, guano is mined locally and sold commercially for fertilizer (Kunz and Fenton, 2003; Kunz et al., 2011).

The organic input from bat guano (feces and urine) is essential for sustaining the health of cave ecosystems (Palmer and Palmer, 2009). Many cave-dependent organisms (*e.g.*, fungi, arthropods, fish, salamanders) depend on bats to produce guano and thus provide critical food resources in an environment where other sources of organic nutrients are relatively scarce. Typically, organic nutrients are provided to caves via water flow. Subterranean streams wash coarse terrestrial detritus into caves, which is increased significantly following heavy rains, and groundwater drips provide a source of dissolved organic compounds. These processes can be rather infrequent or occur at a relatively slow rate, whereas the seasonal nutrient deosition with bat colonization can be very rapid.

The influx of organic nutrients from bat guano provides a major source of carbon, nitrogen, and water, especially into otherwise dry caves. In addition to carbon and nitrogen, bat guano also contains important minerals, including sodium, potassium, phosphorous, iron, calcium, and magnesium, which are often limiting nutrients in many ecosystems. The amount of energy and inorganic nutrient that is provided by bat guano depends on the energetic content of the food that bats consume and how well the bats are able to assimilate ingested food. For example, there are a number of caves in south-central Texas that house large (400,000 to 1,000,000 individuals) maternity colonies of Brazilian free-tailed bats. Within these caves, each lactating female bat may produce 0.12 g dry mass of feces per night, which contains 3.64 kJ of free energy. Thus, a colony of Brazilian free-tailed bats comprised of 1,000,000 individuals will produce enough feces to provide a cave ecosystem with 3,640,000 kJ of energy in a given year (J.D. Reichard, unpublished data).

Bat guano provides the nutrient base on which complex food webs develop in caves (Palmer and Palmer, 2009). In particular, arthropod communities are the most successful colonizers of bat guano deposits in caves and form surprisingly rich communities comprised of several trophic levels. Some arthropod groups feed directly on bat guano or on the microorganisms that colonize the guano deposits. These groups commonly include various species of mites, coleopterans (beetles), diplopods (millipedes), bat ectoparasite larvae and adults (fleas), collembolans (springtails), isopods (woodlice), cockroaches, crickets, and silverfish, among many others. These detritivores are in turn fed on by various species of predators including chilopods (centipedes), hemipterans (true bugs), arachnids, pseudoscorpions, carnivorous beetles and mites, and vertebrates, including fish and salamanders, some of which have become entirely dependent on the cave fauna and flora derived from the organic input from bats (Palmer and Palmer, 2009).

Most populations of bats in temperate regions do not use caves for the entire year, using them for only a few months while they are raising young or during hibernation. For example, hibernating bats have little impact on cave ecosystems because they only deposit small amounts of guano; however, bats that form maternity colonies incaves have considerable impact on the cave ecosystems owing to high fecal deposition associated with increased energetic demands of lactation and parental care. The effects of seasonal bat colonies on cave ecosystems can be observed in organisms that directly depend on guano for nutrients and energy. For example, populations of dermestid beetles, which feed on fallen bats and guano, closely track the presence of bats. When bats are present, the beetles mostly feed on bat carcasses. After the bats migrate from the cave, dermestid beetles shift their diet to the accumulated guano; lack of new guano causes the population of adult beetles to decline, leaving only eggs that will hatch when the bats return in the spring. Cave-roosting bats in tropical regions discard culled wings of insects, seeds, and bits of fruit and leaves dropped from feeding roosts, and thus provide a year-round supply of energy and nutrients for a variety of cave organisms (Kunz and Racey, 1998; Kunz et al., 2011).

Interestingly, organisms that are not typically associated with bat guano also depend on this nutrient-rich organic source. For example, the larval stages of the cave-adapted salamanders (*Eurycea spelaea*), which inhabit caves in the Ozark Plateau that house large maternity roosts of gray bats (*Myotis grisescens*), consume guano. Guano contains nearly the same amount or more nutrients as the salamander's usual food (cave amphipods). Thus, salamander populations increase significantly in caves where bats are present for a portion of the year (Palmer and Palmer, 2009).

Not only do bats provide nutrients to caves through defecation, urination, and deposits of discarded food items, they also influence the cave microclimate. Metabolic heat is generated by bats and trapped in some caves, although additional heat is generated from the metabolic activities of microbes in guano deposits. This heat can substantially increase the temperature of caves beyond what can be expected from the influence of outside ambient temperatures. Relative humidity, which is typically high in most cave systems, can also be affected by bats (Kunz and Fenton, 2003; Fleming and Racey, 2009). Water from feces and urine and exhaled air also provide important sources of moisture that is especially vital in dry caves.

With the appearance of white-nose syndrome (WNS) in North America, cave ecosystems may face threats in the near future as a result of widespread mortality among cave-dwelling species (see below). If the mortality of other hibernating species follows the effects of WNS observed in the little brown myotis (*Myotis lucifugus*), then there could be major implications for cave organisms that depend on the nutrients that bats import into caves. Initially, the death of bats in a cave could result in a nutrient boom from the accumulation of carcasses that decompose on the cave floor. However, such a period of nutrient richness will be followed by a significant decrease in nutrient input as fresh guano deposits vanish with declining bat populations. While the cessation or decline of guano deposition may not necessarily lead to extinction of cave specialists, the loss of nutrient input will most likely result in significant population declines, loss of genetic diversity, and at worse the collapse of some cave ecosystems.

Human Disturbance

Humans enter caves for various reasons, including scientific research exploration, shelter, tourism, mining, and even sometimes for collecting bats to eat. Whatever the intentions might be, these activities can have adverse consequences for bats (Kunz and Racey, 1998; Kunz and Fenton, 2003). Disturbing bats during maternity periods, whether they are handled or not, can cause pregnant females to abort their young or cause young to fall to the floor, leading to injury or death. Hibernating bats are particularly vulnerable to disturbances from human activities inside caves. Additionally, when humans disturb hibernating bats, they often respond by arousing, which is energetically expensive. Nontactile stimuli, such as light and noise, can also increase the activity of bats that hibernate in caves and thus lead to unintended arousals and the depletion of critical energy reserves (Kunz and Fenton, 2003).

Habitat Destruction and Alteration

Surface and subterranean mining activities can have adverse impacts on cave-roosting bats, because such activities can modify the physical structure and microclimate of caves. Because many bats have very specific roosting requirements, such changes may cause bats to completely abandon these sites. Just as important is the fact that mining operations frequently use toxic chemicals. For example, in some regions, modern gold-mining methods use cyanide to extract gold from ore, and as a consequence such practices have unexpectedly killed enormous numbers of cave bats, by contaminating water sources from which they derive their food and drinking water. While increases in the number of abandoned mines may have increased the abundance and distribution of some cave-roosting species, reclamation of mines and the closing of others have also led to an increase in bat mortality when these structures are closed without verifying their presence (Kunz and Fenton, 2003).

The habitat surrounding caves can be just as important for bats, as is the roosting environment within the cave itself. Many hibernating species, such as the endangered Indiana myotis (*M. sodalis*) typically roost beneath exfoliating bark during the warm months but hibernate in caves during winter. To survive a prolonged period of hibernation, bats must be able replenish their fat reserves following autumn migration and before entering hibernation, thus productive foraging habitats located near hibernacula are essential for their success. The vegetation around a cave not only supports source populations of insect prey but may also buffer the interior of the cave from severe changes in wind flow and temperature.

Threats to Cave Bats from White-Nose Syndrome

Several pathogens and putative pathogens have been linked to cave bats (Kunz and Parsons, 2009), some of which are zoonotic (*e.g.*, can be transmitted to humans). Among the zoonotic viruses, rabies and Ebola are perhaps the best known to the general public, but there is little to no evidence that these viruses have caused mass mortality in bats. Recently, the emerging fungus (*Geomyces destructans*, Gd) associated with white-nose syndrome (WNS) has led to the death of over a million hibernating bats in the northeastern United States, and since its first discovery in New York State in 2006 (Blehert et al., 2009; Frick et al., 2010), has contributed to one of the most precipitous declines of wildlife in recorded history. Within four years of being reported, this putative pathogen has spread from New York to 16 other states, south to North Carolina, west to Missouri and Oklahoma, and northward to the provinces of New Brunswick, Ontario, and Quebec in Canada. Based on an analysis of reported mortality in the little brown myotis (*M. lucifugus*), the species most affected by this condition, regional extinction has been predicted in this species within 20 years (Frick *et al.*, 2010). A strain of *Gd*, the same fungal pathogen that has been linked to WNS in North America, has also been reported from hibernating bats in European caves but, to date, there is no evidence of WNS-related symptoms or mass mortality in these species (Wibbelt et al., 2010).

Threats to Cave Bats from Wind Turbines

The proliferation of wind energy facilities throughout the world also places bats at risk of being killed from being struck directly by turbines blades or indirectly from barotrauma (*e.g.*, damage to the lungs on being exposed to negative pressure associated with moving turbine blades). To date, most bat fatalities attributed to wind turbines in North America have been tree-roosting species, but up to 24% of the bat fatalities at some wind energy facilities have been cave-roosting species (Kunz et al., 2007; Arnett et al., 2008), including the big brown bat (*Eptesicus fuscus*) the little brown myotis (*M. lucifugus*), the northern long-eared myotis (*M. septentrionalis*) the tricolored bat (*Perimyotis subflavus*), and the Brazilian free-tailed bat (*Tadarida brasiliensis*). Several species of cave-roosting bats in Europe are also being killed by wind turbines (Rydell *et al.*, 2010), a pattern that is likely to prevail throughout the world where wind-energy facilities are being installed and operated. Most importantly, wind-energy facilities should not be installed near caves that serve as maternity roosts, swarming sites, or hibernacula.

Fatalities to cave-roosting and tree-roosting bats from wind turbines can be markedly reduced operationally by increasing the rotor cut in speed of turbines to 6.5 m s^{-1} (Arnett *et al.*, 2010). The rationale for this mitigation effort at sites where wind turbines have been inappropriately located is that aerial insects, and hence aerial feeding insectivorous bats, are most active at low wind speeds, typically less than 6 m s^{-1}. Additionally, if wind turbines blades can be operationally feathered at certain time of year (*e.g.*, especially during fall migration), fatalities can be reduced significantly (Cryan and Barclay, 2009).

The Paradox of Vampire Bats

An endless source of myths and legends, vampire bats offer a valuable lesson about the need to learn more about these and other bat species before it is too late to protect them from extinction. Three species of vampire bats, each of which use caves as roosts, range from northern Mexico through South America, but only the common vampire bat (*Desmodus rotundus*) is sufficiently abundant to be considered a nuisance to humans and their livestock. All three species depend on a diet of blood, but they feed on a variety of different animals. Most of our knowledge of vampire bats comes from the common vampire bat, which specializes on mammalian blood as its primary source of food. Populations of vampire bats increased sharply throughout Latin America following the introduction of livestock by European settlers over 500 years ago. Because the common vampire bats feed on cattle and occasionally on humans, it is considered a pest in most parts of Central and South America. The economic loss due to cattle dying from bat-transmitted rabies alone is a major concern in many regions of Latin America, and thus massive eradication efforts are often implemented, leading to the death of other species as well.

Lack of education and misguided attempts to control vampire bat populations have led to the mass destruction of nontargeted species. Nonselective killing techniques, such as fire and gas (fumigating caves), have been used either because local landowners are often unaware of the differences between vampire bats and other species or because they are uninformed about the ecological value of bats in general. Poisons, such as strychnine, or anticoagulants are often applied to the wounds on livestock because vampire bats return to wounds that they made the previous night. Selective approaches that concentrate on controlling vampire bats should be used whenever possible. Recent discoveries by researchers indicate that chemicals present in the saliva of common vampire bats have important medical benefits (*e.g.*, reducing the risks of stroke and heart attacks in humans). Thus, a bat species that is considered a nuisance or public health threat by some segments of society may offer enormous benefits to others.

Many local, national, and international organizations have become engaged in efforts to support research on bats and have helped to educate the public about the benefits of these flying mammals to humankind. Many caving organizations have joined this effort to protect bats in general, and cave-roosting bats in particular. Television programs, newspaper articles, the Internet, and other media must be used to promote the ecological value of bats and the importance of caves for sustaining bat populations on a worldwide scale. Organizations such as Bat Conservation International (www.batcon.org), Bat Conservation Trust (www.bats.org.uk), the Lubee Bat Conservancy (www.lubee.com), the Organization for Bat Conservation (www.batroost.com), and the National Speleological Society (www.caves.org) are among a growing number of nongovernment organizations that are contributing to these efforts. Notwithstanding, additional efforts are needed to help promote and protect the 1232 species of living bats known worldwide. Given the threats to cave-roosting bats from anthropogenic disturbances, emerging diseases, misinformation, and human ignorance, management practices that conserve cave-roosting species are needed to ensure the survival of the rare and unique cave ecosystems that these species help to create and maintain.

Acknowledgments

We wish to acknowledge support from the American Society of Mammalogists (S.W.M.), Bat Conservation International (S.W.M. and N.W.F.), National Geographic Society (T.H.K.), National Science Foundation (T.H.K.), National Speleological Society (T.H.K. and N.W. F.), Lubee Bat Conservancy (T.H.K.), and U.S. Fish and Wildlife Service (T.H.K.) for funding our research on bats. We also thank Tigga Kingston, Wendy R. Hood, and Armando Rodríguez-Durán for making helpful suggestions on earlier drafts of this manuscript.

See Also the Following Article

White-Nose Syndrome

Bibliography

Arnett, E. B., Brown, W. K., Erickson, W. P., Fiedler, J. K., Hamilton, B. L., Henry, T. H., et al. (2008). Patterns of bat fatalities at wind energy facilities in North America. *Journal of Wildlife Management*, 72(1), 61–78.
Arnett, E. B., Huso, M. M. P., Schirmacker, M. R., & Hayes, J. P. (2010). Altering turbine speed reduces bat mortality at wind-energy facilities. *Frontiers of Ecology and the Environment*, 9, 209–214. doi:10.1890/100103.
Blehert, D. S., Hicks, A. C., Behr, M., Meyer, C. U., Berlowski-Zier, B. M., Buckles, E. L., et al. (2009). Bat white-nose syndrome: An emerging fungal pathogen? *Science*, 323(5911), 227.
Boyles, J. G., Cryan, P. M., McCracken, G. F., & Kunz, T. H. (2011). Economic importance of bats in agriculture. *Science*, 332(6025), 41–42.
Crichton, E. G., & Krutzsch, P. H. (Eds.), (2000). *Biology of bat reproduction*. San Diego: Academic Press.
Cryan, P. M., & Barclay, R. M. R. (2009). Causes of bat fatalities at wind turbines: Hypotheses and predictions. *Journal of Mammalogy*, 90(6), 1330–1340.
Fleming, T. H., & Racey, P. A. (Eds.), (2009). *Island bats: Evolution, ecology, and conservation*. Chicago: University of Chicago Press.
Frick, W. F., Pollock, J. F., Hicks, A., Langwig, K., Reynolds, D. S., Turner, G., et al. (2010). A once common bat faces rapid extinction in the northeastern United States from a fungal pathogen. *Science*, 329(5992), 679–682.
Kunz, T. H. (Ed.) (1982). *Ecology of bats*. New York: Plenum Press.
Kunz, T. H., & Fenton, M. B. (Eds.), (2003). *Bat ecology*. Chicago: University of Chicago Press.
Kunz, T. H., & Parsons, S. (Eds.), (2009). *Ecological and behavioral methods for the study of bats*. Baltimore: Johns Hopkins University Press.
Kunz, T. H., & Racey, P. A. (Eds.), (1998). *Bat biology and conservation*. Washington, DC: Smithsonian Institution Press.
Kunz, T. H., Arnett, E. B., Erickson, W. P., Johnson, G. D., Larkin, R. P., Strickland, M. D., et al. (2007). Ecological impacts of wind energy development on bats: Questions, hypotheses, and research needs. *Frontiers of Ecology and the Environment*, 5, 315–324.
Kunz, T. H., Braun de Torrez, E., Bauer, D. M., Lobova, T. A., & Fleming, T. H. (2011). Ecosystem services provided by bats. *Annals of the New York Academy of Sciences*, 1223, 1–38.
Muñoz-Romo, M., & Kunz, T. H. (2009). Dorsal patch and chemical signaling in males of the long-nosed bat, *Leptonycteris curasoae* (Chiroptera: Phyllostomidae). *Journal of Mammalogy*, 90(5), 1139–1147.
Palmer, A., & Palmer, M. (Eds.), (2009). *Caves and karst of the USA*. Huntsville, AL: National Speleological Society.
Rodríguez-Durán, A., & Lewis, A. R. (1985). Seasonal predation by Merlins on sooty mustached bats in western Puerto Rico. *Biotropica*, 17, 71–74.
Rydell, J., Bach, L., Dubourge-Savage, M.-J., Green, M., Rodrigues, L., & Hedenstrom, A. (2010). Mortality of bats at wind turbines links to nocturnal insect migration. *European Journal of Wildlife Research*, 56, 823–827.
Wibbelt, G., Kurth, A., Hellmann, D., Weishaar, M., Barlow, A., Veith, M., et al. (2010). White-nose syndrome fungus (*Geomyces destructans*) in bats, Europe. *Emerging Infectious Diseases*, 16, 1238–1242.
Zubaid, A., McCracken, G. F., & Kunz, T. H. (Eds.), (2006). *Functional and evolutionary ecology of bats*. New York: Oxford University Press.

BEETLES

Oana Teodora Moldovan

Emil Racovitza Institute of Speleology, Cluj Department, Romania

INTRODUCTION

As part of the Insect class, representatives of the order of Coleoptera usually have a sclerotized body with sclerotized forewings that are leathery or horny and modified to act as rigid covers (*elytra*) over the membranous, reduced, or even absent hindwings. The mouthparts are adapted for cutting, nibbling, and chewing, and the antennae have usually 8 to 11 articles. The male genitalia are retractable, and the females do not possess an ovipositor. Beetles are the most numerous of all insects, with more than 400,000 described species—almost 40% of all known animals (Hammond, 1992). Big or small, they are everywhere, occurring in all environments, including caves, lava tubes, cracks and fissures in a massif in mesovoid shallow substratum (MSS) in limestone, or in other different rocks such as schist, gneiss, granodiorites, basalts, quartzites, grits, and so on.

Of the 166 families of the order, only very few have species in the underground world: the aquatic Dryopidae, Dytiscidae (predaceous diving beetles), Elmidae, Hydrophilidae (water scavenger beetles), and Noteridae, as well as the terrestrial Carabidae (ground beetles), Curculionidae (weevils), Histeridae, Leiodidae, Pselaphidae, Staphylinidae (rove beetles), Endomychidae, Ptiliidae, Scydmaenidae, and Tenebrionidae (darkling beetles) (Fig. 1). Even underground they are the best represented of all animals, with more than 2000 species.

It is not surprising that one of the first discovered cave animals was a beetle, in Postojnska Jama (Slovenia). In 1831, Čeč observed specimens that looked very much like ants on the beautiful stalagmites. Count Franz von Hohenwart sent the material to the Austrian naturalist Schmidt, who described *Leptodirus hochenwartii*, an amazing terrestrial species displaying a high degree of adaptation to cave life (Fig. 2).

ADAPTATIONS

Several adaptations of cave beetles to darkness, low and heterogeneous in time and space food input (at least in temperate regions, where the food is brought inside caves through the cracks network, only in some seasons, and mostly by water), and a relatively constant climate are characteristic of cave life, but the degree of morphological, anatomical, behavioral, and physiological adaptation is not similar in all species. One of the

FIGURE 2 *Leptodirus hochenwartii* (Schmidt, 1832). *Photo courtesy of Valika Kuštor, Slovenia. Used with permission.*

first morphological changes to occur among beetles during colonization of caves is a loss of pigmentation. The cuticle becomes thinner, and the color of the cave beetles given by the benzoquinones is a red brown. The most evident morphological change is the reduction or complete lack of eyes. The ocelles and rhabdomeres can disappear completely, and reduction can also affect the optical center, much more pronounced than for stygobitic fishes and decapod crustaceans.

Elongation of the body and antennae, which become slender and longer, is linked to their preferred habitat (the network of cracks) and serves as compensation for the lack of eyes. Longer antennae mean also longer mechanical, gustatory, olfactory, and hygrosensitive sensilla, as well as enhancement of the receptory surface. Cave beetles have a larger reception surface compared with their epigean relatives.

Life in caves makes the use of the wings impossible. Wings are completely lost in some species, and the elytra are fused together. Moreover, in highly specialized species, under the elytrae is located a compartment containing air for humidity regulation that causes a false physiogastry, similar to the bulging abdomen of ants and termites filled with lipid reserves.

The internal anatomical modifications are especially due to the scarcity of food and its uneven distribution throughout the year; therefore, adapted species have developed a fat body containing huge vesicles filled with fat, proteins, and glycogen that allow survival during several months of fast. There is also a modification of the exocrine glands, observed on some Leiodidae. The

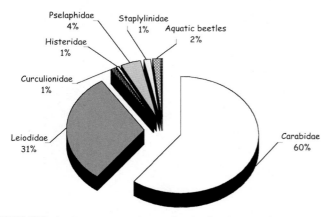

FIGURE 1 Proportion of cave beetle families in the group of Coleoptera. *Source: Adapted from Juberthie and Decu (1998).*

soil species have a big sternal exocrine gland (secreting pheromones) that disappears completely in cave relatives, being replaced by smaller unicellular glands. The secreted substances switch from a mixture of very volatile and less volatile ones for soil species to less volatile ones, perceived only at small distances (the special cuticular hydrocarbons), in subterranean species. In caves, the presence of food can be an attractant at long distances, while the small-distance pheromones act in place, releasing the mating behavior and even the laying of eggs. This behavior saves the energy needed for the production of offspring. Even the larvae save energy by not feeding during the development, as they are protected by small "houses" of clay.

Several breeding experiments were done in the Pyrenean Cave laboratory of Moulis (France) (Deleurance-Glaçon, 1963) on beetles with different degrees of adaptation. The more adapted have low fecundity with a reduction in the number of ovarioles. The females lay few eggs or only one that is bigger, with more vitellinic reserves; also, the time required for egg hatching is longer. The larval stages and larval life are reduced, and the time spent as a pupa increases (Fig. 3). The life cycle varies; French species spend 4 to 5 years as adults, and the American *Ptomaphagus* species spend only 2 years (Peck, 1986).

The activity of epigean species depends very much on the day/night and seasonal alternations. For cave beetles in the absence of light and other 24-hour environmental cues, periods of activity and rest do not have a daily rhythm.

Tropical caves with important food input are inhabited by less adapted beetles, especially those caves with large deposits of guano. The process of adaptation of guanophilic and guanobiotic species is slowed down. In these regions, at low altitude the species are troglomorphic, and true troglobitic beetles typically inhabit high mountain caves (Sbordoni et al., 1977).

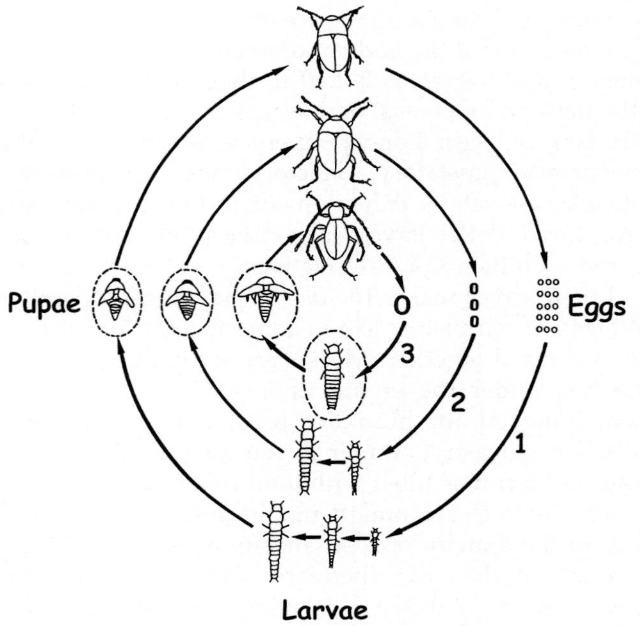

SYSTEMATICS OF CAVE BEETLES

The family Carabidae includes the Trechinae, representing more than 2000 subterranean species. These have colonized all the subterranean habitats, from soil to caves. The cavernicolous species have been classified into two morphological types (anophthalm and aphenopsian), corresponding to how advanced they are in the adaptive evolutionary processes (Fig. 4). The anophthalm type characterizes the endogean and some cave species: depigmented, reduced or no eyes, anterior body and appendages slightly elongated. The other type, aphenopsian, is a very evolved species that is eyeless with a very long body, antennae, and legs, and very thin cuticle. Typical for this last group are *Aphaenops* in Europe and *Neaphaenops* and *Mexaphaenops* in North America. But one of the most amazing adaptations concerns the sensory equipment, with development of olfaction through increased numbers of antennae receptors and lengthening of the mechanical trichobotries on the ellytrae. Some Trechinae are polyphagous predators, such as the French *Aphaenops*, which has a diet consisting of adults and larvae of other beetles, springtails, flies, sometimes

FIGURE 3 Development cycles of underground beetles with different levels of adaptations: 1, endogean; 2, less adapted; 3, very adapted.

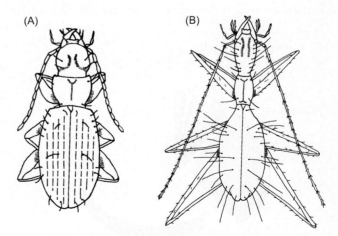

FIGURE 4 Different Trechinae adaptations to caves: (A) anophthalm and (B) aphenopsian. *Source: Adapted from Ginet and Decou (1977).*

millipedes, crickets, diplurans, and so on, and which hunts in caves and MSS (Juberthie and Decu, 1998). Others are very specialized on a single prey, such as *Rhadine subterranea* from Great Onyx Cave (United States), which feeds exclusively on the eggs and young of the cricket *Hadenoecus subterranea*. The anatomy and behavior of this beetle are surprisingly adapted for locating and stealing the eggs deposited one at a time beneath the surface of silt. Chemical substances left by the cricket release the search behavior of the beetle, and when the prey is found a hole is dug and the contents of the eggs are devoured (Mohr and Poulson, 1966). Some species can be rare in caves, only sometimes being found on the sandy banks of subterranean rivers or during flooding seasons. The Trechinae are very mobile and active in the search for food, probably covering large areas in fissured massifs. In this group, the steps of evolution and the way in which adaptation has occurred can be observed very well thanks to the presence of many species outside, in moist or dark habitats (such as under rocks or moss) and deep in the soil. The troglomorphic features of the cave Carabidae are very diversified; some of them are explained by adaptation to different underground compartments, but others represent the original contribution of some phyletical line or the years spent as cave inhabitants (Fig. 5).

Another rich family is the Leiodidae, especially the group of Leptodirini, with almost 200 genera and more than 800 species. Leptodirini occur only in European caves, being replaced in North America by smaller but very interesting other subfamilies, such as the Ptomaphagini. The Leiodidae have colonized all the subterranean habitats, but they consume organic or decomposed matter and are not predators. Cave clay or moonmilk that contains bacteria, fungi, and algae is important to their diet, and these beetles can sometimes be found in huge numbers on these substrates. Species of this group have different morphologies according to their degree of adaptation to cave life; four morphological types are accepted: (1) *bathyscioid*, for humicolous, endogenous, and some less specialized cave species with more or less globular forms of body and short appendages; (2) *pholeuonid*, characterizing specialized forms with longer, slender bodies and appendages and false physiogastry; (3) *leptodiroid*, for highly specialized species, such as *Leptodirus*, with extremely elongated legs and antennae and a very small anterior part of the body; (4) *scaphoid* (from *scaphe*, Greek for "boat"), which are also highly specialized and very similar to the previous group but with a different form of the body like a boat (Fig. 6). Not enough data are available regarding predators of these beetles. It is generally accepted that they do have as many as their epigean relatives, but some cases of predation from harvestmen and pseudoscorpions have been reported and it is probably more pronounced on the eggs and larvae than on the adults. Other Coleoptera families have few troglobitic representatives but do have interesting adaptations to subterranean life.

It was considered that the aquatic beetles were not as successful in colonizing hypogean habitats as the terrestrial ones. The first stygobiotic beetle was mentioned in France at the beginning of the twentieth century. Until recently, 31 species were known to inhabit cave streams, springs, or wells (Fig. 7A), usually in warmer climates. Today, their number has increased. In Australia alone are now more than 100 known species of diving beetles (Dytiscidae) inhabiting the calcretes of the arid parts of Australia (Watts and Humphreys, 2006). Besides the typical adaptations of cave beetles, aquatic species have other adaptations, such as switching between swimmers and crawlers, different methods of obtaining air through cuticular respiration or tracheated elytral respiration, pupation at the bottom of subterranean waters, and smaller size (from 1.1 to 4.5 mm).

The Curculionidae is the best represented family of beetles on the surface, but there are few cave inhabitants, with no typical cave adaptation. These species display different degrees of specialization to a deep soil environment, such as the eyeless *Troglorhynchus monteleonii* from a cave in Central Italy (Osella, 1982). It is not a question of adaptation for cave life but rather one of a different degree of specialization to a deep soil environment (Juberthie and Decou, 1998). The cave in this case is more a trap.

Troglophilic and guanobiotic specialization (many myrmecophilous and termitophilous species) of the Histeridae can explain the low number of cave species; the first cave specimen was discovered in Turkey (Fig. 7B). More than 150 taxa of Pselaphidae are subterranean inhabitants (Fig. 7C). The origin of the temperate troglobitic pselaphids can be traced to the Tertiary, under similar conditions as of the humid and relatively

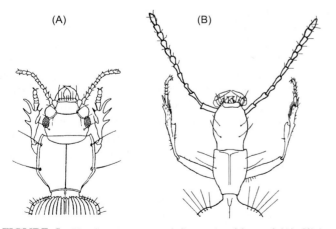

FIGURE 5 Head, antennae, and first pair of legs of (A) *Clivina subterranea* and (B) *Italodytes stammeri* showing the differences, respectively, between new colonizers and old colonizers of caves.

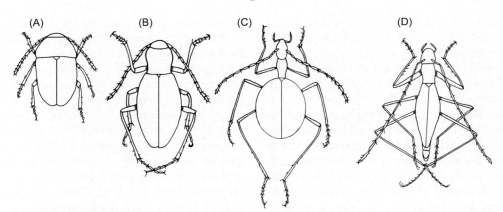

FIGURE 6 Different types of Leptodirini as adaptations to caves: (A) bathyscioid; (B) pholeuonid; (C) leptodiroid; (D) scaphoid. *Source: Adapted from Ginet and Decou (1977).*

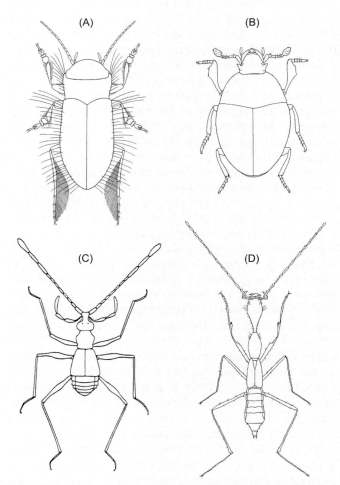

FIGURE 7 Cave beetles: (A) aquatic *Morimotoa gigantea* (Uéno, 1957) (from Japan); (B) histerid *Spelaeabraeus agazzli* (Moro, 1957) (from Italy); (C) pselafid *Decumarellus sarbui* (Poggi, 1994) (from Romania); (D) staphylinid *Domene vulcanica* (Oromi and Hernandez, 1986) (from Canary Islands).

cold forests of the African mountains where forms with small eyes and no wings are largely spread in the humus.

Most Staphylinidae are from the Mediterranean region (Morocco, Algeria, Spain, Italy) or nearby (Romania and the Canary and Madeira Islands), but generally they are troglophilic (Fig. 7D).

Other families of cave beetles are associated with guano deposits and therefore do not share the same morphological adaptations as the cave species.

GEOGRAPHICAL DISTRIBUTION AND COLONIZATION

The subterranean beetles spread on all continents and on some islands, not only in karstic areas (Fig. 8). The richest regions are the temperate ones. The glaciated areas usually do not have cave beetles, but some troglobitic species have recolonized the underground habitats. There are parts of the world where the beetle fauna is very diversified, such as the Mediterranean karst. The reasons for the separation of different populations with wide ranges of distribution can be paleogeographical, paleoclimatical, or ecological. European and North American species are the best known, given the number of specialists and long biospeleological tradition, but also the favorable climatic conditions. The extreme north and south areas were repeatedly covered with ice, and the limits of the glaciers are very well reflected today in the limits of the troglobitic beetle distribution.

The beetles that have colonized the underground world were preadapted or exapted; they were terrestrial and nocturnal, preferring moist habitats, frequently the

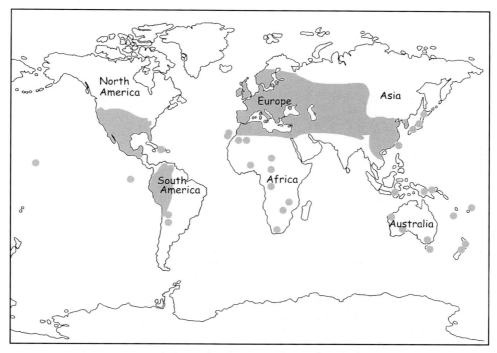

FIGURE 8 World distribution of cave beetles (in gray). *Source: Adapted from Juberthie and Decu (1998).*

fissures of soil, bark, or entrances to caves. The moment of cave occupation is still debated, but it is often related with the Quaternary glacial−interglacial periods (the climate relict hypothesis) or with the natural processes of empty environment colonization (the habitat shift hypothesis). One of these hypotheses can be verified by using morphological features, characteristics of the male genitalia, or molecular phylogenies.

Peck (1981) proposed the following scenario for colonization of caves in the Grand Canyon (between 1160- and 1580-m altitude) by *Ptomaphagus hirtus*. The first species arose during the interglaciary period after the Illinois glaciation (235,000 to 185,000 BP) and dispersed underground during the next glaciation. Then, again, new cave species appeared in the next interglaciary period (150,000 to 90,000 BP). Colonization with troglophilic or troglobitic species was followed by isolation during the interglaciary periods which separated the populations, and new cave species appeared. This hypothesis has been tested and validated with alloenzyme paleodating and paleotemperature measurements.

More recently, molecular analyses on west Mediterranean cave beetles brought interesting explanations about the origin and the speciation of cave inhabitants (Ribera *et al.*, 2010). The origin of Leptodirini (Leiodidae) in this area is represented by perfectly adapted ancestors to the subterranean domain. By combining obtained phylogenies with a calibrated molecular clock, ages of the major events in the phyletical line history, the origin of species, and species diversification can be obtained (Fig. 9). For the Mediterranean region one important geological event was used to calibrate the molecular clock together with estimation of the mutation rate for the studied genes. Detachment of the Catalonian and Sardinian continental plates was the determinant of the subterranean species vicariance. Phylogeny and molecular clock on subterranean Trechini from the above-mentioned geographical area have shown that the origin of the subterranean clade is placed at 10 Ma BP and separation events occurred in the same geographical area on species of the same morphological type exclusively within the Pliocene and Pleistocene (Faille *et al.*, 2010). These two examples sustain the habitat shift hypothesis.

Nevertheless, the opposing hypothesis (the climatic relict) is sustained by phylogenies of diving beetles from aquifers beneath the arid parts of Australia. Lineages populated groundwaters during the late Miocene or Early Pliocene and were caused by drastic climatic changes. Humid habitats were replaced with dry habitats, as can be seen today in most parts of this continent (Leys *et al.*, 2003).

ECOLOGY

Cave beetles generally live in relatively stable climates with constant temperature, an atmosphere saturated with

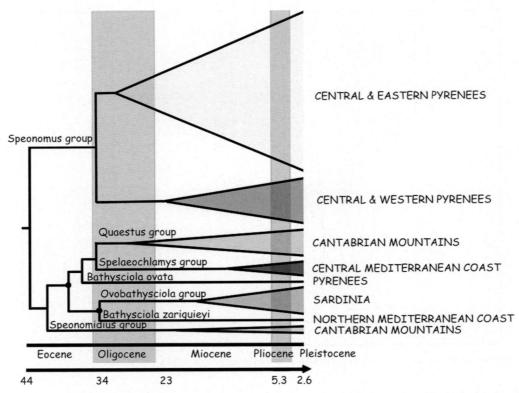

FIGURE 9 Phylogenetic tree of Leptodirini from the Western Mediterranean obtained from combined mitochondrial and nuclear sequences. *Source: Modified from Juan and Emerson (2010).*

water vapor, and no air currents. There is a link between the presence of beetles in caves and food input in spring and autumn (for temperate regions). It has been observed that populations migrate between the network of cracks and the caves, depending on the presence of food and on climatic parameters. The fissures and cracks offer a more stable habitat than the big passages and rooms that have large volumes of air and are more or less in direct contact with the natural entrances. Very sensitive to any change in the conditions of their environment, cave beetles quickly initiate a behavior, even if it is only to run.

Interesting examples are provided by studies from Transylvania (Romania), France, and the United States. Variations of the number of individuals at different times during a year were determined to be influenced by air temperature or the level of the subterranean stream. The example in Figure 10 relates the increase of water level to the withdrawal of the local population of *Pholeuon* (Racoviţă, 1971).

Determining the size of cave beetle populations has been a concern of coleopterologists. Individuals captured at a given time have been marked and then recaptured after a period of time. Estimates can then be made by comparing the number of recaptured marked individuals with the total number of all marked ones. For example, the *Neaphaenops tellkampfi* in Mammoth Cave has been estimated to be 750,000 individuals (Barr and Kuehne, 1971). In France, the population of catopid *Speonomus* in an MSS station was estimated at 1,000,000 individuals for two species, and the population of trechine *Aphaenops* was estimated to be 100,000 individuals (Juberthie and Decu, 1998).

The presence of two or more beetle species in the same cave usually means that one of them is dominant, especially if they compete for the same food resources. The species can choose different places in the underground system and the predators' different niches. In the Transylvanian caves, two genera of Leptodirinae coexist: *Pholeuon* and *Drimeotus*. With very few exceptions, *Drimeotus* is limited primarily near the entrances and in small numbers, or in the network of cracks, while the *Pholeuon* are numerous in the deep cave passages. In Mammoth Cave, beside the numerous *Neaphaenops tellkampfi*, there are other two dominant trechine with different distributions: *Pseudoanophthalmus menetriesi*, which is found in the upper level, and *P. striatus*, which is found along the subterranean river. Mammoth Cave has one of the largest and most complex subterranean biological communities of all known caves, with no less than six trechine species, one pselaphid, and one leiodid. In White Cave, the animal's entire life is dependent on crickets and to a lesser extent

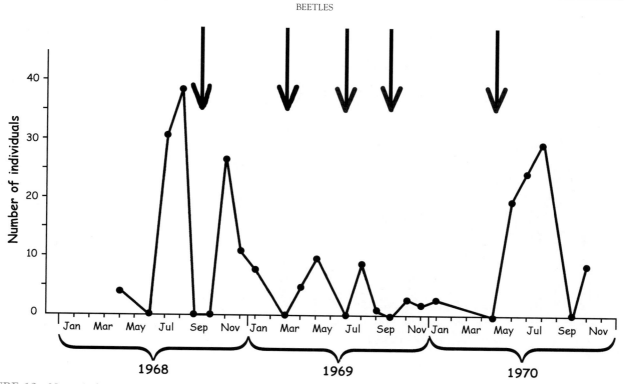

FIGURE 10 Numerical variation of *Pholeuon moczaryi* in Vadu Crişului Cave (Romania), in the period 1968 to 1970 as a consequence of flooding periods (indicated by arrows). *Source: Adapted from Racoviţă (1971).*

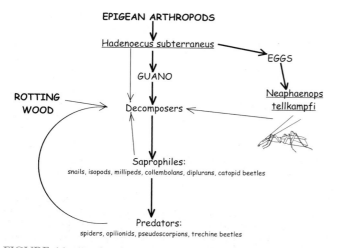

FIGURE 11 Food web in White Cave (Kentucky). *Source: Adapted from Barr and Kuehne (1971).*

on a few pieces of rotten wood (Barr and Kuehne, 1971); a generalized food web for this community is shown in Figure 11. On the partially decomposed cricket guano live saprophytic snails, isopods, millipedes, springtails, diplurans, and leiodid beetles. These are eaten by spiders, harvestmen, pseudoscorpions, and trechine beetles (it has not been determined whether or not the predators are highly specialized).

IMPORTANCE AND PROTECTION

The diversity of species makes this group of cave invertebrates one of the best for testing many of the hypotheses concerning adaptation strategies during colonization of empty places on Earth. Cave beetles are also very precious to the natural biodiversity of the world. One species can populate one cave or one massif, so the degree of endemism is very high. On the other hand, in many regions the biospeleological explorations are only just beginning, and the task of the coleopterologist is to find and describe newly discovered species. The importance of their studies lies primarily in the development of knowledge regarding conservation management measures to be taken for more vulnerable or rare species. Experiments carried out in a Romanian show cave indicate that the presence of tourists has eliminated the troglobiotic beetles from visited parts of the cave and has also influenced the yearly dynamics of the leptodirine population (Fig. 12) in the nonvisited part.

Rare and having a strange morphology, cavernicolous species have attracted the attention of collectors, especially in Europe; however, laws to prevent over-collection of specimens from caves seem to have had little effect on this trade.

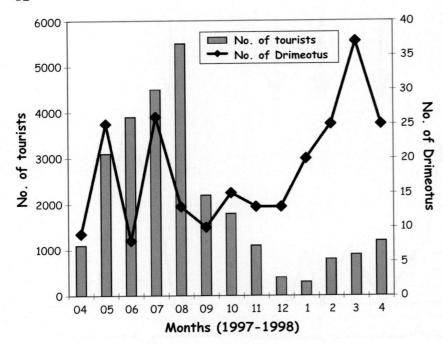

FIGURE 12 Variation in the number of *Drimeotus* in Urşilor Cave (Romania) as a function of tourist periods. *Source: Modified after Moldovan et al. (2003).*

Bibliography

Barr, T. C., & Kuehne, R. A. (1971). Ecological studies in the mammoth cave system of Kentucky. II. The ecosystem. *Annales de Spéologie, 26*(1), 47–96.

Deleurance-Glaçon, S. (1963). Recherches sur les coléoptères troglobies de la sous-famille des Bathysciinae. *Annales des Sciences Naturelles-Zoologie, 1*(5), 1–172 (in French).

Faille, A., Ribera, I., Deharveng, L., Bourdeau, C., Garnery, L., Quéinnec, E., & Deuve, T. (2010). A molecular phylogeny shows the single origin of the Pyrenean subterranean Trechini ground beetles (Coleoptera: Carabidae). *Molecular Phylogeny and Evolution, 54*(1), 97–106.

Ginet, R., & Decou, V. (1977). *Initiation à la biologie et à lécologie souterraines*. Paris: Delarge.

Hammond, P. M. (1992). Species inventory. In B. Groombridge (Ed.), *Global biodiversity. Status of the Earth's living resources*, (pp. 19–36). London: Chapman & Hall.

Juan, C., & Emerson, B. C. (2010). Evolution underground: Shedding light on the diversification of subterranean insects. *Journal of Biology, 9*(3), 17.

Juberthie, C., & Decu, V. (Eds.), (1998). *Encyclopaedia biospeologica* (Vol. II). Paris: Société de Biospeologie.

Leys, R., Watts, C. H. S., Cooper, S. J. B., & Humphreys, W. F. (2003). Evolution of subterranean diving beetles (Coleoptera: Dytiscidae: Hydroporini, Bidessini) in the arid zone of Australia. *Evolution, 57*(12), 2819–2834.

Mohr, C. E., & Poulson, T. L. (1966). *The life of the cave*. New York: McGraw-Hill.

Moldovan, O. T., Racovitza, G., & Rajka, G. (2003). The impact of tourism in Romanian show caves: The example of the beetle populations in the Ursilor cave of Chiscau (Transylvania, Romania). *Subterranean Biology, 1*(56), 73–78.

Osella, G. (1982). I curculionidi cavernicoli italiani (riassunto). *Lavori Societa Italiana Biogeographia, 7*, 337–338 (in Italian).

Peck, S. B. (1981). Evolution of the cave Cholevinae in North America (Coleoptera: Leiodidae). In *Proceedings 8th international congress of speleology* (Vol. 2, pp. 503–505). Kentucky: Bowling Green.

Peck, S. B. (1986). Evolution of adult morphology and life-history characters in cavernicolous Ptomaphagus beetles. *Evolution, 40*(5), 1021–1030.

Racoviţă, G. (1971). La variation numérique de la population de *Pholeuon (Parapholeuon) moczaryi* Cs. De la grotte de Vadu-Crişului. *Travaux de l'Institut de Spéologie "E. Racovitza," 10*, 273–278.

Ribera, I., Fresneda, J., Bucur, R., Izquierdo, A., Vogler, A. P., Salgado, J. M., & Cieslak, A. (2010). Ancient origin of a Western Mediterranean radiation of subterranean beetles. *BMC Evolutionary Biology, 10*(1), 29.

Sbordoni, V., Argano, R., Vomero, V., & Zullini, V. (1977). Ricerche sulla fauna cavernicola del Chiapas (Messico) e delle regioni limitrofe: Grotte esplorate nel 1973 e nel 1975. Criteri per una classificazione biospeleologica delle grotte. In *Subterranean fauna of Mexico* (Part III). Quaderni Accademia Nazionale Lincei, 71 (pp. 5–74).

Watts, C. H. S., & Humphreys, W. F. (2006). Twenty six new Dytiscidae (Coleoptera) of the genera Limbodessus Guignot and Nirripirti Watts and Humphreys, from underground waters in Australia. *Transactions of the Royal Society of South Australia, 130*(1), 123–185.

BEHAVIORAL ADAPTATIONS

Jakob Parzefall

Biozentrum Grindel, Zoologisches Institut, und Zoologisches Museum der Universität Hamburg

Animals living in darkness have to compete for food, mates, and space for undisturbed reproduction just as their epigean conspecifics do in the epigean habitats, but there is one striking difference: in light, animals can use visual signals. Thus, important aspects of behavior driven by visual signals cannot apply in darkness. The question arises, then, of how cave dwellers compensate for this disadvantage in complete darkness. This article uses several examples to compare various behavior patterns among cave-dwelling populations with epigean ancestors.

COMPARISON OF BEHAVIOR PATTERNS IN CAVE-DWELLING ANIMALS AND THEIR EPIGEAN RELATIVES

Potential cave dwellers must have the sense organs and behavior necessary to find food and to reproduce in caves. Such animals may be said to be preadapted to cave life, and in fact some of these animals can survive in the darkness without behavioral adaptations and can reduce behavioral characters not necessary in the caves. In contrast, some cave dwellers have improved sense organs and have acquired behavior adapted to their extreme habitats.

Food and Feeding Behavior

Suitable food sources and quantity vary from cave to cave. In general, cave animals depend upon food brought in from outside and are omnivorous. With some exceptions, most caves do not have an abundance of food compared to above ground habitats. Food sources can be widely distributed or concentrated in patches, and their occurrence is mostly unpredictable; therefore, food-finding abilities have to be improved, and food must be stored to ensure survival during long starvation periods.

When the blind fish population of *Astyanax fasciatus* from Pachon Cave, Mexico, was studied in competition experiments conducted in darkness, it was found that they retrieved 80% of small pieces of meat distributed on the bottom of an aquarium, whereas the epigean fish got only 20% (Hüppop, 1987). Among North American amblyopsid fish, which comprise six species in four genera, the ability to detect invertebrates at low prey densities in the dark is much better for the cave-living species *Amblyopsis spelea* than the troglophile *Chologaster agassizi*. When one *Daphnia* was introduced into a 100-L tank, the *A. spelea* found the prey hours before the *C. agassizi* did. In addition, the maximal prey detection distance is greater in cave species. *Daphnia* was detected by *Typhlichtys subterraneous* within 30 to 40 mm and by *C. agassizi* within 10 mm (Poulson, 1963). Density-dependent cannibalism in subterranean amblyopsids seems to account for most of the mortality of the young (Poulson, 1963).

Cannibalism was also observed in Brazilian cave fishes apparently subject to severe food shortage (Parzefall and Trajano, 2010).

The reaction to prey by salamanders has been studied in the facultative cave-living Pyrenean salamander *Euproctus asper*, which has fully developed eyes, and the blind Dinaric salamander *Proteus anguinus*. Both species react to living and dead chironomids. Even in light, where *E. asper* can use its visual sense, *P. anguinus* required less time to initiate the first snapping response to dead prey. When the time between the start of an experiment and the first snap at prey was divided into pre-approach and approach phases, it turned out that the difference found could be attributed to the pre-approach phase (Fig. 1A). Living prey were detected more quickly than dead prey in both species, but *E. asper* needed more time in the darkness than did *P. anguinus* (Fig. 1B,C). These data show that *P. anguinus* is well adapted to search prey on the basis of chemical and mechanical information. In contrast, *E. asper* demonstrated a more directed, visually dominated approach behavior with live prey in light and can switch to a more active, widely foraging mode with live prey in darkness and dead prey in light. This young cave colonizer seems less adapted to the dark but is capable of foraging successfully in both epigean and hypogean habitats.

Compensating for the unpredictability of food quality and quantity also results in physiological adaptations. Cave animals are able to survive for long periods without food—nearly one year for invertebrates and up to several years for caves fishes and salamanders (Vandel, 1964). In *A. fasciatus*, the cavefish are able to build up enormous fat reserves. A 1-year-old cavefish fed *ad libitum* had a mean fat content of 37% fresh body mass compared to 9% in epigean fish under the same conditions (Hüppop, 1987).

Reproductive Behavior

Having found enough food to reach sexual maturity, the next problem to be solved by cave dwellers is finding a sexual partner in the darkness. Subsequently, they need behavior patterns that provide effective fertilization in the absence of any visual orientation. In species with high population densities, it is easy for the male to find a female. The male of the galatheid crab *Munidopsis polymorpha* of the marine cave Jameos del Aqua on Lanzarote in the Canary Islands receives a chemical signal sent by the molting female ready for reproduction (Parzefall and Wilkens, 1975).

In terrestrial invertebrates, a comparable situation has been observed: The females of the cave crickets *Hadenoecus subterraneus* and *H. cumberlandicus* that are ready to mate release an olfactory attractant. Normally,

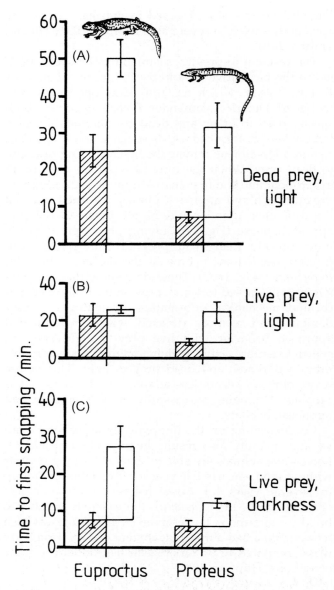

FIGURE 1 Snapping response in *Euproctus asper* and *Proteus anguinus*. The mean time interval between the start of the experiments and the first snap at prey was divided into a pre-approach phase (hatched bars) and an approach phase (open bars) in three different experimental treatments. Standard errors of the mean are shown on top of the bars. *Source: From Uiblein et al. (1992). Used with permission.*

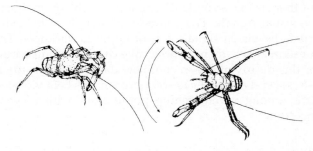

FIGURE 2 The male *Munidopsis polymorpha* (right) displays with cheliped shaking in front of a female. *Source: From Parzefall and Wilkens (1975). Used with permission.*

several males reach the attractive female at the same time. They can transmit information about their high reproductive fitness by sending tactile signals through the air by their elongated antennae, as is done by both of the cricket species (Hubbell and Norton, 1978). In *M. polymorpha* the male emits rhythmic water waves with chelipeds (Fig. 2). They must repeat these signals several times. The female eventually decides which male is good, accepts the sperm transfer (or not), and escapes.

A comparable situation has been found in the livebearing poeciliid fish *Poecilia mexicana*, which lives in a high-population-density cave habitat. The males check conspecific females by nipping at the genital region. The females ready for reproduction produce a species-specific chemical signal and attractant for about 3 days during a cycle of about 28 days. A female accepting a male stops swimming and allows the copulation. Normally, bigger males are preferred on the basis of visual signals. Only the cavefish female is able to perform this behavior in darkness (Fig. 3); she does so by switching from the visual system to a lateral line system (the fish have one lateral line system only) (Parzefall, 2001). The females of the cave population exhibited also a preference for well-nourished males even under completely dark conditions (Plath et al., 2005).

In the above-mentioned crab and cricket species, data about sexual behavior in their epigean relatives are lacking, so we cannot determine whether the reproductive behavior has changed in adaptation to the dark habitat. For *Poecilia mexicana*, comparative data on epigean conspecifics reveal that in the epigean habitat visually orientated sexual displays are lacking, in contrast to other species of the genus such as *P. velifera*, *P. latipinna*, and *P. reticulata*. So, P. mexicana seems to be preadapted to cave life and has improved its reproductive fitness in the dark by means of a special female choice behavior based on a lateral line system (the fish have one lateral line system only).

Species with lower population densities in cave habitats, such as the characid fish *Astyanax fasciatus* or the salamander *Proteus anguinus*, attract conspecifics from chemical signals transmitted in the water. Comparative studies with the epigean proteid *Necturus maculosus* have demonstrated that this information is species specific (Parzefall et al., 1980). The animals also constantly deposit a substance while in contact with the substrate and at communal resting places. This substance is individual specific but does not provide any detailed information about sex or reproductive state; it merely brings members of the species

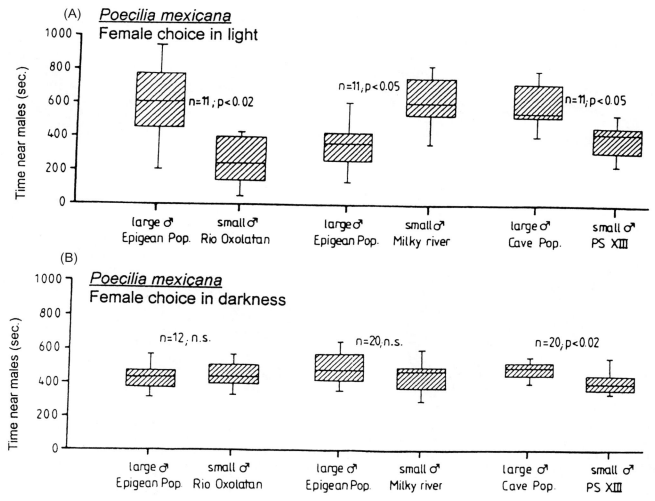

FIGURE 3 Female choice behavior in different populations of *Poecilia mexicana* in (A) light and (B) darkness. The female had the choice to swim to the compartment of a big or a small mature male, and the time spent in a male compartment was measured. A clear partition prevented direct contact. The middle line in the box plot represents the median; the upper end of the box, the 75% value; the lower end of the box, the 25% value. The whiskers represent the 90% and 10% values, respectively. *Source: From Parzefall (2001). Used with permission.*

together. For recognition of sex and reproductive state, *Proteus anguinus* requires direct contact. When sexually motivated, a male establishes a territory that a female may enter only after direct body contact. The male sends a chemical signal by fanning his tail against the female and from time to time begins to walk away. The female follows and nips at the genital region of the male. After a short walk, they stop and the male deposits a spermatophore, which the female retrieves and places in her cloacal region. Unfortunately, comparative data in the epigean salamander *Necturus maculosus* are lacking.

Aggressive Behavior

Aggressive behavior consists of different patterns of threatening postures and attacks followed by fights. This behavior has various functional aspects and is absolutely necessary in darkness; therefore, it has to be adapted to cave conditions.

Food competition results in food territories for groups, pairs, or solitary animals. Within a group, a limited food supply can lead to aggressive encounters. In general, defending food resources is only adaptive when the costs are not higher than the potential incoming energy of the food. The majority of data available regarding aggression among cave and epigean populations are for the characid fish *Astyanax fasciatus*. The epigean form is widely distributed in Mexico. When undisturbed, the epigean fish defends small territories of 10 to 20 cm, depending on body size, by fin spreading, snake swimming, and ramming (Fig. 4). In the laboratory, epigean fish of both sexes display the entire

FIGURE 4 Aggressive patterns in the epigean *Astyanax fasciatus*. (A) Aggressive fine erection; the head-down position of the fish on the right expresses a higher aggressive motivation. (B) Snake swimming is shown by the fish on the right and aggressive fin erection by the fish on the left. (C) The fish on the left is ramming against the one on the right. (D) Both fish show circling and tail beating. *Source: From Parzefall and Hausberg (2001). Used with permission.*

aggressive pattern. The subdominant fish demonstrates submission by a head-up position and trying to hide or escape. The fights can be very strong; in smaller aquaria that offer no place to hide, the death of the subdominant fish can result.

Aggressive behavior depends on optical releasers. Using dummies of different types, it has been found that natural shape and locomotion are important visual signals. Tests with infrared video have shown that the epigean fish is not able to perform aggressive patterns in complete darkness (Hausberg, 1995). They do not establish territories at all. From such data we can conclude that epigean fish, when colonizing caves, were no longer able to perform aggressive encounters. In the blind populations of the Pachon, Piedras, and Yerbaniz caves, Hausberg (1995) noted a high percentage of fish with injuries on fins and scales. In an experiment with the Pachon population, the number of injured fish increased in the absence of food. The aggressive behavior observed within this fish includes defending small territories of a few centimeters by biting, circling, and tail beating (Fig. 5). Also in these experiments, a striking difference in swimming behavior was observed; the fish that were regularly fed glided slowly through the water of the entire aquarium without initiating aggressive encounters against conspecifics. When food was lacking, the locomotor activity decreased. The fish mostly hovered at the bottom and rhythmically flicked their fins; fish entering the small area of a few centimeters were attacked. The territory size was correlated with the aggressiveness of the fish, and the aggressive patterns differed from those shown in epigean fish. The cavefish has developed an aggressive behavior with signals that are only effective in close body contact.

Among cave-living invertebrates, the galatheid crab *Munidopsis polymorpha* of the marine cave Jameos del Aqua on Lanzarote in the Canary Islands feeds mainly on diatoms on lava rocks. The animals keep a minimal distance from one another according to the length of their second antenna (Fig. 6). Any closer than this distance, and *Munidopsis* attacks with its extended chelipeds and by snapping its pincers. This behavior does not

FIGURE 5 Ramming and circling in the Pachon cave population of *Astyanax fasciatus*. Source: From Parzefall and Hausberg (2001). Used with permission.

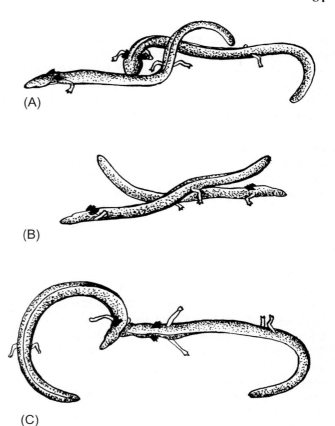

FIGURE 7 Aggressive behavior of a *Proteus anguinus* male against an intruder in his territory: (A) body contact (chemical identification); (B) tail beating; (C) biting. Source: From Parzefall (1976). Used with permission.

FIGURE 6 Aggressive interactions among *Munidopsis polymorpha*. Source: From Parzefall and Wilkens (1975). Used with permission.

depend on optical releasers but on water movements (Parzefall and Wilkens, 1975). The aggressive patterns of *Munidopsis* are very similar to the one described for the deep-water, bottom-living, epigean galatheid *Munida sarsi*. The author believes that, despite their naturally dim environment, vision is still the primary sense involved in the aggressive behavior of this species, so it appears that the aggressive behavior of galatheids is effective in light and in darkness, with no striking differences.

In the blind cave salamander *Proteus anguinus*, studies in the laboratory have revealed that males show aggressive behavior and territoriality for only a very short reproductive period. Normally, the animals rest under stones in groups of both sex without any aggressive reaction. When a male became sexually active, it begins to control its conspecifics by contacts with his snout and allows only females in the reproductive state to remain in the hiding place. Intruders will be attacked by tail beating, ramming, and biting in close body contact (Fig. 7). After being attacked in a particular territory, *Proteus* avoids that territory for several days on the basis of chemical cues on the substrate. In the Poeciliid fish *Poecilia mexicana*, males use aggressive behavior to establish a size-dependent rank order within a mixed school. The females have a reproductive cycle of about 28 days and are attractive to males within the first 3 days of the cycle. The dominant male controls the females by nipping in response to an attractive female. In the field, the pair separates from the shoal and become more or less sedentary. The male nips and tries to copulate while also defending the female. During aggressive encounters with more or less equal-sized males, small males use a female-like body coloration to try to sneak copulations (Parzefall, 1979). The population of *P. mexicana* that colonized a limestone cave in Tabasco (Mexico) does not school, and the males do not fight. In laboratory studies with epigean fish and cavefish having functional eyes, a quantitative, genetically based reduction of aggressive patterns and schooling has been demonstrated (Parzefall, 1979). The reaction is highly variable within the population. Some of the cavefishes tested seemed unable to understand the attacks and answered by nipping and copulation attempts. It seems that aggression in these cavefishes is a disadvantage, because fighting males risk losing the opportunity for contact with an attractive female in darkness.

DISCUSSION

Studies of behavior in cave dwellers have revealed complex systems of responses to visual, chemical, and tactile stimuli. Many animals can survive in complete darkness with no visual signals. The use of weak electric signals among cave dwellers has not been detected. In some cases, an existing behavior (such as the aggression exhibited by *Astyanax fasciatus*) has changed to a more effective behavioral system. These changes are always based on existing above ground behavior, and no completely new behavioral character has been found in cave animals.

Acknowledgments

The author would like to thank the DFG (German research board) for financial support over many years, the Mexican government for providing research and collecting permits over many years, and M. Hänel (Zoological Institute Hamburg) for drawing the figures.

See Also the Following Articles

Adaptation to Low Food
Morphological Adaptations
Natural Selection in Caves

Bibliography

Hausberg, C. (1995). Das aggressionsverhalten von *Astyanax fasciatus* (Characidae, Teleostei): Zur ontogenie, genetik, und evolution der epigäischen und hypogäischen form. Ph.D. dissertation, University of Hamburg, Germany.

Hubbell, T. H., & Norton, R. M. (1978). The systematics and biology of the cave-crickets of the North American tribe Hadenicini (Orthoptera, Saltatoria, Ensifera, Rhaphidophoridae, Dolichopodinae). *Miscellaneous Publications of the Museum of Zoology of the University of Michigan, 156*, 1–124.

Hüppop, K. (1987). Food finding ability in cave fish (*Astyanax fasciatus*). *International Journal of Speleology, 16*, 59–66.

Parzefall, J. Z. (1976). Die Rolle der chemischen Information im Verhalten des Grotten-olms *Proteus anguinus* LAUR. (Proteidae, Urodela). *Tierpsychol, 42*, 29–49.

Parzefall, J. (1979). Zur genetik und biologischen bedeutung des aggressionsverhaltens von *Poecilia sphenops* (Pisces, Poeciliidae): Untersuchungen an Bastarden ober- und unterirdisch lebender Populationen. *Zeitschrift für Tierpsychologie, 50*, 399–422.

Parzefall, J. (1989). Sexual and aggressive behaviour in species hybrids of *Poecilia mexicana* and *Poecilia velifera* (Pisces, Poeciliidae). *Ethology, 82*, 101–115.

Parzefall, J. (2001). A review of morphological and behavioural changes in the cave molly *Poecilia mexicana*, from Tabasco, Mexico. *Environmental Biology of Fishes, 62*, 263–275.

Parzefall, J., & Hausberg, C. (2001). Ontogeny of the aggressive behaviour in epigean and hypogean populations of *Astyanax fasciatus* (Characidae, Teleostei) and their hybrids. *Mémoires de Biospéologie, 28*, 153–157.

Parzefall, J., & Trajano, E. (2010). Behavioural patterns in subterranean fishes. In E. Trajano, M. E. Bichuette & B. G. Kapoor (Eds.), *Biology of subterranean fishes* (pp. 81–114). Enfield, NH: Science Publishers.

Parzefall, J., & Wilkens, H. (1975). Zur ethologie augenreduzierter tiere: Untersuchungen an *Munidopsis polymorpha* koelbel (Anomura, Galatheidae). *Annales de Spéléologie, 30*, 325–335.

Parzefall, J., Durand, J. P., & Richard, B. (1980). Chemical communication in *Necturus maculosus* and his cave-living relative *Proteus anguinus* (Proteidae, Urodela). *Zeitschrift für Tierpsychologie, 53*, 133–138.

Plath, M., Heubel, K. U., de Leon, F. G., & Schlupp, I. (2005). Cave molly females (*Poecilia mexicana*, Poeciliidae, Teleostei) like well-fed males. *Behavioral Ecology and Sociobiology, 58*, 144–151.

Poulson, T. L. (1963). Cave adaptation in amblyopsid fishes. *American Midland Naturalist, 70*, 257–290.

Uiblein, F., Durand, J. P., Juberthie, C., & Parzefall, J. (1992). Predation in caves: effects of prey immobility and darkness on the foraging behaviour of two salamanders, *Euproctus asper* and *Proteus anguinus*. *Behavioural Processes, 28*, 33–40.

Vandel, A. (1964). *Biospéologie la biologie des animaux cavernicoles*. Paris: Gauthier-Villars.

BREAKDOWN

Elizabeth L. White

Hydrologic Investigations

Caves as seen by cave explorers are remarkably stable. Cavers very rarely encounter rockfalls or evidence of recent rockfalls. Yet caves are frequently littered with the piles of rock fragments known as *breakdown*. The scale of breakdown ranges from small isolated loose rocks to ceiling collapses that may completely block the passage. To understand the presence of breakdown it is necessary to consider both breakdown mechanics and also the geological processes that could set the stage for breakdown.

BREAKDOWN MORPHOLOGY

Piles of breakdown are unsorted and highly permeable. Layering is undistinguishable or nonexistent. Breakdown is an essential part of the cave landscape. One can distinguish small-scale breakdown features that are the various types of breakdown blocks themselves and large-scale features that are cavern features consisting of (or generated by) breakdown processes.

Small-Scale Features

Breakdown can be classified by the relationship of individual blocks to the bedding of the parent bedrock:

> *Block breakdown* consists of masses of rock with more than one bed remaining as a coherent unit.
> *Slab breakdown* consists of fragments of single beds.
> *Chip breakdown* consists of small rock chips and shards derived from the fragmentation of individual beds.

This scheme has the advantage that breakdown observed in the field can be properly classified without speculation as to its origin; however, it has the disadvantage of also being a function of the limestone lithology. Thus, limestone fragments of a given intermediate size might be blocks if derived from a thin-bedded limestone or slabs if derived from a massively bedded limestone. In general, however, this classification has been found useful for the areas studied.

The classification scheme given above is very broad and general. For any specific investigation, it is possible to derive much more detailed classification schemes based on the characteristics of the fragment surface, whether the fragments are dissolutional or fracture, the fragment shape, the predominant mode of origin, the morphological class of trench block, fault wedge, pendant cluster or other form, and the position of the breakdown features in relationship to the cave passage (Jameson, 1991).

Block breakdown can be massive; blocks measure up to tens of meters on a side and are usually bounded by bedding planes along the bedding and by joint planes across the bedding. Slab breakdown has a plate shape, with slab thickness being controlled by the thickness of the beds; the width of individual slabs varies from tens of centimeters to many meters. Chip breakdown ranges in size from centimeters to tens of centimeters; the shape of chip breakdown is variable and dependent on the process that created the breakdown. Crystal wedging, frost pry, and closely spaced joints produce very angular chunks, whereas pressure-induced spalling and mineral replacement produce flatter, more irregular shards.

In Mammoth Cave, slab breakdown is the most common and is distributed through all levels of the cave. Block breakdown occurs where major roof collapse has taken place and where dividing walls have fallen between coalescing vertical shafts. An extensive breakdown has occurred in the upper gallery of the Great Salts Cave Section. This passage is floored with block and slab breakdown to a depth of 12 or more meters for a distance of more than a kilometer. The largest breakdown block so far observed in this passage is a single block 19 m long, 4.5 m wide, and 1 m thick.

Large-Scale Features

Although breakdown blocks form a variety of features in caves ranging from a few scattered blocks to major collapsed passages, it is useful to describe two types of features: terminal breakdown and breakout domes. *Terminal breakdown* occurs at the end of collapsed major cave passages. Eroding valleys on the surface have the effect of causing collapse (breakdown) in the caves below. Massive breakdown that completely occludes a cave passage is referred to as terminal breakdown. In a number of caves, artificial entrances have been created where the cave passage would have intersected the surface topographic valley. In the Central Kentucky Karst, a terminal breakdown is the most common terminator of passages. Often the terminal breakdown contains sandstone as well as limestone fragments where the collapse has extended upward to the overlying caprock. Major trunk passages beneath the sandstone-capped plateau were once continuous feeder conduits carrying groundwater from the Sinkhole Plain to the south and east of the plateau to Green River in the north. These formerly continuous passages have been truncated by ceiling collapse. Some are actual intersections of the passages with the surface; others have collapsed at depth. The present-day configuration of the cave system is due in large part to these random features of collapse. Similar terminal breakdown occurs in many other caves with and without caprock.

Breakout domes, among the most remarkable of cavern features, are the huge rooms that form as a result of major ceiling collapse. Some of these, such as Chief City in Mammoth Cave, have floor dimensions of more than 100 m and ceiling heights of 30 m. The Rumble Room in the Rumbling Falls Cave System under Fall Creek Falls State Park in Tennessee has a ceiling about 20 stories tall. This breakout dome is the largest in the eastern United States and the second-largest breakout dome known in the United States. Other breakout domes include the large rooms in Camps Gulf Cave (Tennessee) (Fig. 1), Rothrock's Cathedral in Wyandotte Cave (Indiana), the entrance room in Hellhole Cave (West Virginia), the entrance room in Marvel Cave (Missouri), Devil's Sinkhole (Texas), and Salle de la Verna in Pierre Saint-Martin (France). The details of the

FIGURE 1 Breakout dome in Camps Gulf Cave, Tennessee. Note figures for scale. *Photo courtesy of Peter and Ann Bosted. Used with permission.*

enlargement mechanism are less clear in Devil's Sinkhole than in the others mentioned, although it also has the beehive-shaped room and the gigantic debris cone typical of all breakout domes.

Careful examination of many breakdown areas reveals a continuum of sizes, from very large breakdown rooms to small, roughly circular or elliptical breakdown areas in cave ceilings. The features at the small end of the scale are sometimes only 3 m in diameter and involve only one or two beds. The morphological term *breakout dome* describes all such features, regardless of their size.

Debris piles vary in size from dome to dome, but in those domes that are accessible the volume of debris is much smaller than the enclosing volume of the dome. Because the bulk density of the debris cone is considerably less than that of the original bedrock, it is apparent that large quantities of material must have been removed. Large breakout domes must therefore have formed at a time when water was actively circulating near their base. The dome could then enlarge by dissolution action of fallen blocks with concurrent stoping of the sides. The dome itself is usually circular or elliptical in contour. The top is often capped by a single massive bed.

BREAKDOWN MECHANICS

Most mechanisms advanced for cave breakdown have been drawn from the rock mechanics of mining. However, caves are not mines. Mines are fresh excavations in the bedrock and it is the stress pattern around the opening that determines such hazards as rock bursts, spalling, and roof collapse (Brady and Brown, 1993). Caves are thousands of years old and were excavated by slow dissolution of the bedrock rather than by blasting. The horizontal strains in the cave would be expected to have annealed out and the only residual strain is that due to gravitational loading of the cave ceiling by the overlying beds.

It is not known for certain that all horizontal strains caused by tectonic deformation of the strata have been annealed. Massive ceiling failures in limestone mines are thought to be due to compressive stresses parallel to the bedding (Esterhuizen *et al.*, 2007). The circular or elliptical pattern and dome-shaped residual cavity seen in mine collapses have some similarity to breakout domes in caves (Fig. 1) and it is possible that similar mechanisms are responsible.

The simplest model is that of brittle fracture of fixed ceiling beams under gravitational loading (White and White, 1969). The concept dates back to the work of Davies (1951), who based his model on the mining literature. The model assumes a rectangular passage formed in near-horizontal limestone. A small amount of elastic sag of the unsupported roof beds causes these beds to separate slightly. Each bed offers some support to the beds above it so that the stress pattern closes to an arch or stress dome in the rock above the cave passage. The gravitational load due to the rock above the top of the stress dome is distributed onto the rock that makes up the wall of the cave passage and does not contribute to passage instability. Cave passages at great depths are not less stable than more shallow passages. It is only if surface erosion or other geological processes breach the stress dome that the passage stability is reduced.

Figure 2 shows the parameters of the fixed-beam model and the dome shape of the stress distribution. The beds act as fixed beams across the cave passage. For the ceiling to be stable, the bending strength of the beams must be greater than the gravitational force acting on the weight of the unsupported span. Thicker beds are stronger than thin beds. There will be a critical thickness (t_{Crit}) for any given passage width at which the strength of the bed is just sufficient to support its weight. When the mechanics are worked out, only the length of the beam (L) and the critical bed thickness (t_{Crit}) remain. The width of the beam cancels out. The beam length (L) is set equal to the passage width, while the extent of the ceiling bed along the axis of the cave passage does not enter the calculation. Figure 3 shows the roof stability according to the fixed-beam model. For a fixed beam, the critical thickness (m) is:

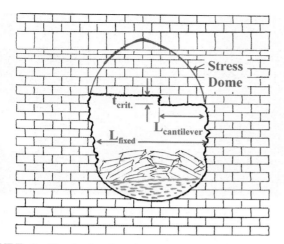

FIGURE 2 Sketch showing the parameters for the fixed beam model of cave breakdown. Note that the stress arch is a sketch, not the result of a calculation.

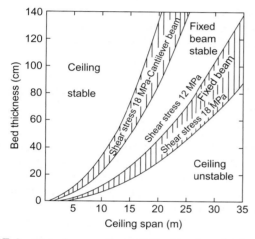

FIGURE 3 Plot of cave ceiling stability based on the fixed beam model and a range of bed shear strengths as indicated.

$$t_{Crit} = \frac{\rho L^2}{2 S \cos \theta} \quad (1)$$

where ρ is the density of the bedrock (in kg/m^3), θ is the bedding dip (in degrees), and S is the flexural strength (in MPa). If the ceiling beds are not supported at both sides of the cave passage, they are treated as cantilever beams for which the critical thickness is:

$$t_{Crit} = \frac{3 \rho L^2}{2 S \cos \theta} \quad (2)$$

The fixed-beam model implies a completely elastic response of the ceiling beds. It does not allow for plastic deformation and long-term creep that could lead to bed failure in the absence of any geologic triggers. It also has no time dependence; a stable ceiling would remain stable until some geologic process destabilized it.

If the fixed beam, brittle fracture model is complete, explanations must be found for any recent breakdown. Numerous occurrences of recent breakdown have been reported. Some have occurred near entrances where freeze–thaw cycles may be responsible. Some have occurred deep in the cave but in areas of active vadose water. However, a fraction of recent breakdown occurrences are in dry passages with no obvious triggering mechanism. There have been four documented roof failures in Mammoth Cave in the past century. Three were massive rockfalls, but the fourth involved a plastically deformed ceiling slab that had been mapped in detail. The slab displayed extensive plastic deformation in the 1960s. Sometime in the early 1970s it fell.

A more comprehensive model allows for inelastic creep (Tharp and Holdrege, 1994; Tharp, 1995). Materials break through a mechanism of crack propagation. Tharp's model is based on the propagation of microcracks, which allow deformation and creep. The crack propagation velocity is given by:

$$v = c \left(\frac{K_I}{K_{Ic}}\right)^n \quad (3)$$

where K_I is the stress intensity (MPa $m^{0.5}$), K_{Ic} is the fracture toughness of the bedrock, and c is a constant related to the activation energy for crack movement. The parameter c is given by:

$$c = V_0 \, e^{-\frac{H}{RT}} \quad (4)$$

where H is the enthalpy of activation = 67–147 kJ/mole; R is the gas constant = 8.3145 J/K; T is temperature in kelvins and V_0 and n are fitting constants.

Propagation of microcracks allows inelastic deformation and also a time-to-failure. Using the Tharp model, the time-to-failure is the time scale of crack propagation through beds of nominal thickness. The timeframe can range from thousands of years to as much as 1 million years. Thus, the time-to-failure can be in the same range as the age of the cave passage, implying that breakdown can occur at any time, even in the absence of geologic triggering processes.

The Tharp model introduces fracture toughness as another parameter in addition to the flexural strength for determining whether particular beds will collapse. The Paleozoic limestones of the eastern United States (where many of the breakdown investigations have been made) are dense, fine-grained rocks. Coarse fracturing occurs along joints and bedding plane partings. But, within the rock mass there is little to inhibit crack propagation, and these rocks break mainly by brittle fracture. It is for this reason that the fixed-beam model has worked so well.

Porous and vuggy rocks, such as the Tertiary limestone beds of the Caribbean, may have a lower flexural strength but they have higher fracture toughness, because pores and vugs inhibit crack propagation. The caves on Mona Island in Puerto Rico have large, relatively flat chambers with little breakdown because of the toughness of the porous, young limestone beds. Because of the inability of the limestone beds to propagate cracks, nearly-flat roof spans of 30 m or more are found throughout these caves.

GEOLOGIC INFLUENCES ON BREAKDOWN PROCESSES

Geologic Processes That Initiate Breakdown

Within the context of either the fixed-beam or Tharp model, any geologic process that lengthen the beams or convert fixed beams into cantilever beams can move the ceiling beds from stable to unstable configurations. Beam thickness, flexural strength, and fracture toughness are properties of the bedrock and do not change during evolution of the cave passage. A list of triggering processes, not necessarily complete, is as follows:

Passage enlargement below the water table. Phreatic passages continue to enlarge as water flows through them. If the hydrologic conditions are such that the passage is not drained, it may continue to enlarge until it becomes mechanically unstable.
Removal of buoyant support. By Archimedes' principle, the ceiling beds of a water-filled cave passage are buoyed upward by a force proportional to the ratio of the density of water to the density of the bedrock. For limestones, with typical densities in the range of 2.65 g/cm^3, 35% to 42% of the buoyant support of the ceiling is lost when the cave is drained.
Effects of base-level back flooding. During the time when the emergent cave passage is in the floodwater zone, rises and falls in base level alternately fill and drain the cave passage. Additional dissolution at this time, particularly dissolution along ceiling joints, can turn fixed beams into cantilever beams and destabilize the ceiling.
Action of vadose water. Formation of vertical shafts, solution chimneys, and solutionally enlarged fractures by the action of undersaturated vadose water often has the effect of cutting ceiling beds, thus changing fixed beams into cantilever beams (Fig. 4).
Ice wedging. Caves that draw in cold winter air can have freezing conditions some distance inside. When water moving through joints and bedding plane partings freezes, the expansion creates enough force to fracture the bedrock.

FIGURE 4 Collapsed ceiling slabs due to a ceiling channel cutting the bedding perpendicular to the cave passage. Pohl Avenue, Mammoth Cave, Kentucky. *Photo by the author.*

Crystal wedging. Replacement of calcite in the bedrock by other minerals can exert a wedging effect. Because gypsum has a greater volume than the calcite it replaces, enough force is generated to fracture the bedrock.

Crystal Wedging or Limestone Replacement to Initiate Breakdown

Many breakdown areas in caves with extensive sulfate minerals (primarily gypsum) suggest that crystal wedging and replacement of limestone by gypsum and possibly other sulfate minerals are important factors in this type of cavern collapse. Features that are characteristic of mineral-activated breakdown are (1) walls and ceilings fractured in irregular patterns, often with visible veins of gypsum following the fractures; (2) breakdown consisting of thin, irregular splinters and shards of bedrock (Fig. 5); and (3) curved plates of bedrock ranging in size from a few centimeters to more than a meter hanging from the ceiling at steep angles cemented only by a thin layer of gypsum. Microscope examination of thin sections of the bent beds shows that the sagging and bending are due to the direct replacement of limestone by gypsum. Another characteristic feature is the collapses that take the form of symmetrical mounds with coarse, irregular blocks at the base grading upward into a rock floor at the top.

Crystal wedging produces a subset of chip breakdown. Chip breakdown consists of rock fragments that are smaller than individual bedding plane slabs and can result from many processes, including purely mechanical ones. Crystal wedging breakdown appears to be of two types. Type I consists of angular rock fragments broken on sharp planes that cut the bedding planes. Type I breakdown, with fractures filled with gypsum, results from mechanical wedging due to crystallization of the gypsum. Similar rock fragments are found near cave

FIGURE 5 Shards of breakdown created by replacement of limestone by gypsum. Turner Avenue, Mammoth Cave, Kentucky. *Photo by the author.*

entrances, where they result from frost action. Type II breakdown is more complex. The fragments and plates are angular and sharp and are fractured across the usual zones of weakness—bedding planes and joints (Fig. 5). Many of the fragments, only a few centimeters on a side and less than a millimeter thick, crush like broken glass when walked upon. These irregular plates are the signature of the crystal replacement process. The limestone bedrock is shattered and intermixed with gypsum so that the passage walls become piles of rubble. Unlike most breakdown, which results from purely mechanical processes, Type II breakdown involves a complex chemistry of mineral replacement, the details of which are still under investigation (White and White, 2003).

ROLE OF BREAKDOWN IN SPELEOLOGICAL PROCESSES

Both geologic triggering and slow creep of beds under load assure that breakdown can occur at any time during the evolutionary history of a cave passage; however, breakdown processes are most active during the enlargement phase of cave development and during the decay phase of cave development. The role of breakdown in the enlargement phase includes the following:

Breakdown during the enlargement phase exposes more limestone surfaces and thus increases the rate of dissolution.
Upward stoping by breakdown processes can create large chambers, if actively circulating water removes the breakdown blocks at floor level.
Upward stoping along fracture zones with removal of fallen blocks results in the formation of stoping shafts; Sótano de las Golondrinas in Mexico is an outstanding example, as are the tiankengs of China.

Breakdown in master conduits, particularly during the floodwater stage, can provide a support structure for groundwater dams. Silt and clay that deposit behind the blockage seal the dam, raise hydraulic heads upstream, and thereby generate a hydraulic gradient for the formation of new tap-off passages.

Breakdown continues to play a role during the stagnation and decay phases of cave development as follows:

Breakdown processes can stope upward to interconnect previously isolated cave levels into an integrated system of passages.
Truncation of cave passages by the formation of terminal breakdown is a dominant process in the breakup of continuous conduits into the fragments characteristic of the decay stage of cave development. The lowering land surface intersects the stress dome, thus destabilizing the underlying cave passage.
The final phase in the decay of caves is the passage collapse that takes place when the eroding land surface intersects the stress dome in the rocks of the cave ceiling.

The final residue of a cave is a rubble zone consisting largely of breakdown.

See Also the Following Article

Clastic Sediments in Caves

Bibliography

Brady, B. H. G., & Brown, E. T. (1993). *Rock mechanics for underground mining*. London: Chapman and Hall.

Davies, W. E. (1951). Mechanics of cavern breakdown. *National Speleological Society Bulletin*, 13, 36–43.

Esterhuizen, G. S., Dolinar, D. R., Ellenberger, J. L., Prosser, L. J., & Iannacchione, A. T. (2007). Roof stability issues in underground limestone mines in the United States. In S. S. Peng, C. Mark, G. Finfinger, S. Tadolini, A. W. Khair, K. Heasley & Y. Luo (Eds.), *Proceedings of the 26th international conference of ground control in mining* (pp. 336–343). Morgantown, WV: West Virginia University.

Jameson, R. A. (1991). Concept and classification of cave breakdown: An analysis of patterns of collapse in Friars Hole Cave System, West Virginia. In E. H. Kastning & K. M. Kastning (Eds.), *Proceedings of the Appalachian karst symposium* (pp. 35–44). Huntsville, AL: National Speleological Society.

Tharp, T. M. (1995). Design against collapse of karst caverns. In B. F. Beck (Ed.), *Karst geohazards* (pp. 397–406). Rotterdam: A. A. Balkema.

Tharp, T. M., & Holdrege, T. J. (1994). Fracture mechanics analysis of limestone cantilevers subject to very long term tensile stress in natural caves. In P. P. Nelson & S. E. Laubbach (Eds.), *First North American rock mechanics symposium proceedings* (pp. 817–824). Rotterdam: A. A. Balkema.

White, E. L., & White, W. B. (1969). Processes of cavern breakdown. *National Speleological Society Bulletin, 31*, 83–96.

White, W. B., & White, E. L. (2003). Gypsum wedging and cavern breakdown: Studies in the Mammoth Cave System, Kentucky. *Journal of Cave and Karst Studies, 65*, 43–52.

BURNSVILLE COVE, VIRGINIA

Gregg S. Clemmer

Butler Cave Conservation Society, Inc.

INTRODUCTION

For more than half a century, cavers have pushed, explored, and mapped the caves of Virginia's Burnsville Cove. Located near the Bath/Highland County border approximately 80 km west of Staunton, Virginia, this sparsely populated area is known for its rural, scenic character. The systematic study and mapping of Breathing Cave by Nittany Grotto of the Pennsylvania State University, beginning in 1954, marked the first organized effort to chart what was then the largest and best known cave in Burnsville Cove. Initial work produced an overland survey, passage cross sections and longitudinal profiles, and an analysis of the cave's formation relative to stratigraphic folds and faults.

Bevin Hewitt's dramatic aqualung dive into the Mill Run spring in 1956 and his discovery of Aqua Cave beyond fueled interest in finding more caves in the Cove. Ike Nicholson's discovery of Butler Cave two years later (Clemmer, 2001) and the rapid discovery of more than 16 km of large cave passages there, including the Sinking Creek trunk, confirmed that Burnsville Cove possessed vast underground secrets.

The 1982 Burnsville Cove Symposium summarized the geology and cave description of the Burnsville Cove reported to that date. Nicholson and Wefer (1982) described five caves—Boundless, Breathing, Butler, Better-Forgotten, and Aqua—as being part of an underground integrated karst drainage. These caves, although not connected by human transit, comprised the Butler Cave–Sinking Creek Cave System, described at that time as containing approximately 35 km of mapped passages.

In the past three decades, additional discoveries and mappings have greatly expanded the extent and understanding of this cave system. With more than 102 km of mapped cave passages in Burnsville Cove, this isolated corner of the Old Dominion ranks as one of the primary karst regions in the United States (Fig. 1).

PHYSICAL AND GEOLOGIC SETTING

Burnsville Cove sits astride the Highland/Bath County line, 20 km southwest of the village of McDowell, Virginia. Bordered by Jack Mountain on the west and Tower Hill Mountain to the east, the Cove is a broad,

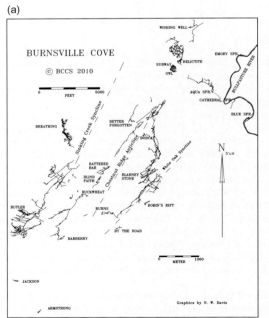
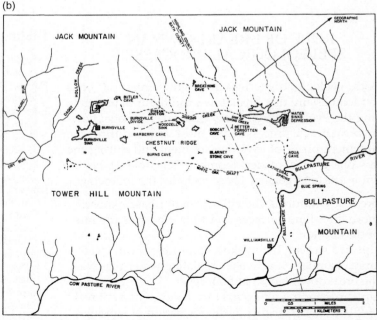

FIGURE 1   (A) Map of Burnsville Cove showing the drainage system, location of caves, and large surface depressions. (B) Outline map showing principal caves in Burnsville Cove in relation to each other. *Maps courtesy of the Butler Cave Conservation Society.*

dual-synclinal valley, split longitudinally by a folded anticlinal ridge known as Chestnut Ridge. The region measures approximately 8 km long and 5 km wide, the geologic structure plunging to the northeast.

The Helderberg limestones of Silurian-Devonian age make up the exposed karst of the Cove. The Helderberg rocks are not a uniform calcareous sequence, a fact that significantly impacts the speleogenesis of the area (White and Hess, 1982). The stratigraphic labeling of the carbonate rocks has varied over time and with different investigators. The sequence, bottom to top, consists of the Tonoloway, Keyser, New Creek, Corriganville, and Licking Creek limestones. Below the carbonate sequence is the Wills Creek Shale and above it the Oriskany Sandstone. Thicknesses of individual beds vary significantly over short distances. Of these members, the Tonoloway and Keyser each contain three thin layers of insoluble sandstone that significantly impact cave development in Burnsville Cove. With water rarely able to breach these layers, cave passages are often floored and/or roofed by these sandstones for long distances, lending a characteristic flat ceiling to many of the galleries.

The geologic structure of Burnsville Cove is not straightforward. The two synclinal valleys are separated by the folded anticline of Chestnut Ridge, which extends northeast from the little village of Burnsville. Sinkholes dot the pastures and woods, particularly the upper (southwest) portion of Burnsville Cove, where Burnsville Sink harbors the entrances to Butler Cave. At the Water Sinks 5.5 km to the northeast, a dramatic stream insurgence flows into Water Sink Cave (the Subway). Water drainage in almost all other circumstances in Burnsville Cove is subsurface (Davis and Hess, 1982), discharging at several springs on the Bullpasture River 8 km northeast of Burnsville.

Aqua Cave is the primary resurgence for Sinking Creek Valley, a western synclinal valley containing Boundless, Butler, Breathing, Better Forgotten, Barberry, Basswood, Buckwheat, Blind Faith, Battered Bar, Water Sink, Owl, Helictite, and other smaller caves, as well as the Burnsville Turnpike/Black Canyon portion of the Chestnut Ridge Cave System. Emory Spring, a road-covered karst spring 1.5 km upriver from the Aqua Cave resurgence, contains no known cave but may be the resurgence for waters in Helictite Cave as well as the recently opened Wishing Well Cave. Cathedral Spring, located on the Bullpasture River 800 m downstream of the Aqua Cave resurgence, is the primary outflow for the caves of the eastern synclinal valley: Robins Rift, By-the-Road, and the Burns-Blarney Stone portion of the Chestnut Ridge Cave System. Of key note here is the drainage divide in the Chestnut Ridge Cave System made possible by the North/South Trunk passage cutting transversely through the Chestnut Ridge anticline. Waters in the Burnsville Turnpike (possibly the largest contiguous underground chamber known in Virginia, averaging 20 m wide by 15 m high and extending more than 1200 m, with the most massive segment measuring 45 m wide and 425 m long) continue into the narrow confines of Black Canyon to the 622 Sump only to reappear in Aqua Cave and flow into the Bullpasture River. Waters in the Cyanide Canyon section and all of the Blarney Stone and Burns sections of the system emerge at Cathedral Spring.

HISTORY OF EXPLORATION

Well-known Breathing Cave was mined for saltpeter during the Civil War. It gained prominence after World War II as a popular sport cave with members of the newly formed National Speleological Society (NSS). Interest spread among NSS members and other cavers to investigate Mill Run Spring on the Bullpasture River. Bevin Hewitt's dive into Aqua Cave in 1956 amply demonstrated that more cave waited to be found, and Ike Nicholson's discovery of Butler in 1958 attracted a considerable number of experienced cavers to Burnsville Cove. The August 1958 Sand Canyon camp expedition (Nicholson and Wefer, 1982) "recon mapped" 4500 m of new cave. Dozens of leads abounded. Nittany Grotto joined the survey in November 1958 and quickly mapped more than 8 km of intricate passages off the upstream and downstream trunk. Upstream discoveries in 1959 took cavers beyond Penn State Lake to the distant, joint-controlled passages of what later became known as Mbagintao Land. Far downstream, two parallel sumps—Last Hope and Rats' Doom—blocked Butler's advance to the northeast, but four years later, an obscure lead in Kutz Pit Junction revealed the muddy, remote galleries of Marlboro Country.

Overland surveys from Butler to Breathing plus the added data from both cave maps pointed to a possible connection. By 1967, a host of cavers were pushing from both caves to connect (Nicholson and Wefer, 1982). Despite new discoveries in Breathing and repeated efforts at digging in both caves, no connection has been found and the caves remain more than 150 m apart.

On Chestnut Ridge, a small pit discovered in 1959 was pushed in hopes of extending Butler downstream, but it quickly degenerated into a vertical crawl of daunting challenges. Ten years later, cavers returned to the Better Forgotten Cave and hammered through the Vertical Crawl to find more than 300 m of a large trunk passage ending in a terminal sump. Upstream portions of the cave remained choked in breakdown.

As the 1960s ended, a number of cavers, hoping to protect the pristine nature of Butler Cave, formed the Butler Cave Conservation Society (BCCS). In 1975, the BCCS purchased the 65-acre tract of land containing the Nicholson entrance to Butler Cave and today continues as the nation's oldest, private cave conservation preserve.

Exploration in Burnsville Cove slowed in the 1970s. The Robins Rift dig was a dynamic project in the relatively caveless eastern synclinal valley and eventually led to approximately 600 m of discovery, but surface digs in the Cove yielded little significant cave. Remapping in Butler took priority and despite a kilometer of virgin cave discovered in Mbagintao Land, some began to feel that Burnsville Cove had yielded most of its secrets.

In 1979, members of the Shenandoah Valley Grotto visited an obscure blowing cave on Chestnut Ridge first located by David Nicholson in 1957. In a series of gritty, exhausting trips commencing in early 1982, they dropped deep into the heart of the ridge, following good air and a small, contorted stream passage in this cave they called Bobcat. Their discovery of the North/South Trunk in 1983 led to new, exciting discoveries of extensive, large caves. Finding day trips too short, highly fatiguing, and vastly inefficient for mapping in such remote passages, they began camping in the cave. From 1984 to 1990, 27 separate underground camps explored and surveyed more than 14.5 km of virgin cave, establishing Bobcat at the time as Virginia's deepest cave.

In 1989, Ron Simmons conducted a series of cave dives into the constricted fissures of Cathedral Spring. After widening the cherty, underwater conduit, Simmons mapped a larger, descending passage to a depth of 46 m. With 290 m surveyed and the cave continuing its plunge, Simmons discontinued the exploration in the face of serious hypothermia, extended decompression times, and special gas requirements.

In March of 1991, ridge cavers dug into a small sink on the eastern flank of Chestnut Ridge and found Blarney Stone Cave. A twisting, wet passage similar to that encountered in Bobcat led to an extensive, multilevel, decorated gallery of cave passages exceeding 6 km in length. Blarney Stone and Bobcat were connected in a dual team effort in August 1994, thus forming the 22.5-km Chestnut Ridge Cave System. Protracted, in-cave excavation of an air-blowing lead in Upper Ghost Hall led to the 2003 discovery of more than 4 km of passages extending under the eastern valley and into the flank of Tower Hill Mountain. Aid climbs up air-blowing Pigeon Tooth Dome and at the far reaches of infeeding Falling Water hint at much more discovery. Off the main trunk passage, descending Duane's Drop led to a huge, treacherously floored-in-breakdown trunk passage dubbed the Boulder Dash that soon cleared into a sandy, potential camp room named Phantom Ranch. A classic dry sump eventually intersects the rarely seen Cathedral River which sumps both up and downstream. An air-blowing lead on a high wall overlooking the Boulder Dash—Opportunity Knocks—was pushed in a series of ever brutal trips until the team of Mike Ficco, Ed Kehs, and Jon Lillestolen dug and mapped a physical connection with Burns Chestnut Ridge Cave in December 2005, establishing the Chestnut Ridge System as Virginia's third longest cave system.

Digging for new caves continued, and in 1993 Barberry Cave (Schwartz, 1999) was discovered in an obscure pasture sinkhole 1 km east of Burnsville. A second entrance dug the following summer provided a more comfortable, safer entry to a cave that had now been mapped to a length of nearly 2 km. In November 1994, Ben Schwartz and Mike Ficco pushed an extremely low water crawl more than 30 m to find a large trunk passage headed to the northeast. Survey here added 3 km of cave to Barberry, but the Barberric Crawl soon had everyone considering a second series of cave camps. The first campers into Barberry, however, became temporarily trapped for several days when high water completely sealed the Crawl. Despite a media frenzy, the group engineered an intrepid self-rescue only to have the entrance later placed off-limits by the landowner.

Undeterred, cavers resumed digging once again, this time on the land of caver Nevin W. Davis. After drilling a 25-cm hole 21 m down into the ceiling of the large Barberry trunk passage beneath the Davis farm, cavers began excavating and shoring to install a large metal tank endwise. Further drilling and blasting, accompanied with a full complement of spectacular setbacks, eventually opened the Big Bucks Pit entrance to Barberry Cave in 1996. Meanwhile, an ongoing dig on the slopes above the Water Sinks broke into virgin cave in March 1996. Named Helictite Cave for its pretty display of such speleothems in the entrance area, the cave led to a rabbit warren of joint-controlled passages that currently exceeds 11 km.

Strong air also lured cavers into Burns Chestnut Ridge Cave. Extremely arduous trips to "bottom" this tight cave had exhausted and frustrated a succession of caving teams for more than three decades, but a concerted effort led by Gregg Clemmer, Nevin W. Davis, and Tom Shifflett finally revealed a sizable stream passage 200 m below the entrance. Subsequent trips pushed the depth to 240 m, surpassing the Chestnut Ridge Cave System as Virginia's deepest. As previously stated, connection of Burns to the Chestnut Ridge Cave System in

December 2005 established the system as Virginia's third longest (32+ km) and second deepest (245 m).

Given this success, digging for caves in Burnsville Cove accelerated. Buckwheat Cave was opened after an hour's effort in March 1998. Pushing low crawls and following air brought mappers to within 4 m of connecting with the far downstream end of Barberry. Cave diggers found Blind Faith the following March after digging down 4 m into a blind sinkhole. The cave has been mapped to more than 1 km of passage and drains toward Woodzell Sink.

A dig in April 2000 yielded an improbable entrance to the surprisingly extensive Battered Bar Cave. Atop a narrow karst saddle on the west flank of Chestnut Ridge, the cave currently exceeds 2100 m in mapped length. Two months after this discovery, a utility lineman climbed up a roadside bank 1.5 km east of Burnsville to check a telephone repeater and discovered a blowing hole. This By-the-Road Cave is now gated and managed by the BCCS in a unique agreement with the Virginia Department of Transportation, which owns the entrance.

Recent digging has opened Basswood (named for the Virginia state champion basswood tree shadowing the entrance) in 2002, Fuhl's Paradise Pit in 2005, Buckets of Smoke and Water Sink (the Subway) in 2007, and Wishing Well in 2010.

DESCRIPTIONS OF CAVES

Breathing Cave

Breathing Cave is one of the best known and most visited noncommercial caves in Virginia. Until the discovery of Butler Cave, it was the largest known cave in the state (Douglas, 1964). Partially mapped several times after 1945, the cave was more completely surveyed by Nitttany Grotto of the Pennsylvania State University in the late 1950s who charted it to a length of 7.3 km (Holsinger, 1975). A recent survey by the Gangsta Cavers now charts the cave to 9.9 km.

Breathing was mined for saltpeter during the Civil War. A challenging, heavily joint-controlled, parallel maze cave, it divides 30 m inside the large sinkhole entrance into the Historic Section or saltpeter mine on the left and the Main Section on the right. Here, to the right, several kilometers of parallel, interconnected passages trend southeast, ending in a series of very low, narrow, wet crawls one surveyor termed *pseudopsyphons*. At the extreme terminus, the cave approaches the Good News Passage in Butler Cave (Nicholson and Wefer, 1982). The cave is a registered National Natural Landmark.

Aqua Cave

Aqua Cave was discovered by Bevin Hewitt in July 1956 by diving 8 m horizontally into Mill Run Spring. A low airway was subsequently blasted from the left ceiling of the spring, giving cavers without dive gear a sporting access to nearly 2 km of bracing, fairly large river cave. The 1984 discovery of an extensive upper level, the Big Brother section, pushed the cave to a length of almost 3 km. Ron Simmons' dive of the stream's in-cave resurgence charted flooded passages headed into Chestnut Ridge, but did not discover any air-filled continuations.

Butler Cave

With almost 27 km of mapped passages, Butler Cave ranks as the second longest cave in Burnsville Cove. Formed mostly in the Tonoloway limestone, extensive portions of the cave are sandwiched between two of the sandstones, giving some passages a distinctive flat ceiling over great expanses (White and Hess, 1982). The cave underlies the western synclinal valley of the Cove, a fact that in large part determines its passage layout (Fig. 2). Entering the spacious, central Trunk Channel at Sand Canyon, visitors follow the axis of the syncline as they walk up- or down stream. Infeeder branch passages intersect primarily from the west. A number of wet-weather streams course through these side passages to the Main Trunk. All waters in Butler Cave have been dye-traced to their resurgence at Aqua Cave (Davis and Hess, 1982). Although not as spectacularly decorated as other caves in the Cove, Butler awes with its sheer volume. It remains a cave of great challenges, be they a novice's first glimpse of the Moon Room or a veteran's long trek to the remoteness of such exotic destinations as Djibouti or the Doom Room. Like Breathing Cave, Butler Cave is a registered National Natural Landmark. The cave is owned and managed by the Butler Cave Conservation Society. A second entrance to the system, accessed by an angled 29° culvert pipe, was dug in 1998.

Better Forgotten Cave

Better Forgotten is an aptly named, tight, muddy, multidrop cave on the west flank of Chestnut Ridge near the Bath/Highland County line. Its 12-m pit entrance leads to a series of narrow vertical drops as the cave develops down-dip. This eventually intersects a 580-m-long section of stream trunk passage that ends downstream in a sump (Holsinger, 1975). The cave is 1200 m long and reaches a depth of 130 m. The stream has been dye-traced to its resurgence at Aqua Cave on the Bullpasture River. Better Forgotten Cave is owned by the BCCS.

FIGURE 2 The Glop Slot is a very narrow squeeze at the bottom of the entrance pit at the Nicholson entrance to Butler Cave. Until the opening of the SOFA entrance, all who entered the cave had to pass through it. *Photo from BCCS Oscar Estes collection. Used with permission.*

Boundless Cave

Boundless Cave opens in a small sink 800 m southwest of the entrance of Butler Cave. Trending northeast, the passage is characteristically very low and filled with sand and cobbles. A small stream has been traced to its resurgence in Aqua Cave more than 6 km to the northeast (Holsinger, 1975).

Robins Rift

Opened by digging a large, cold air sink located at the western base of Tower Hill Mountain, Robins Rift quickly developed a notorious reputation for entrance instability. At least four separate cave digs over the past 35 years have attempted to keep this cave open to visitation. With more than 600 m of passage surveyed, it stood as the largest cave in the eastern synclinal valley of the Burnsville Cove until the discovery of Blarney Stone Cave in 1991. Access is currently blocked by yet another entrance collapse. Water in Robins Rift has been traced to the Cathedral Spring resurgence (Davis and Hess, 1982).

Bobcat Cave (Chestnut Ridge Cave System)

Bobcat Cave was previously known as Chestnut Ridge Blowing Cave (Douglas, 1964). The two entrances to Bobcat are situated about 10 m apart near the top of Chestnut Ridge just south of the Bath/Highland County line. The 550-m-long entrance series is a muddy, contorted, plunging slot following a small stream that had been dye-traced to Cathedral Spring. This stream is finally intersected by Tombstone Alley, a dry, paleo-overflow segment that leads to the North/South Trunk, a passage ranging from 6 to 21 m wide and up to 12 m high (Rosenfeld and Shifflett, 1995).

As one of three large trunk passages in the cave, the North/South Trunk winds through large breakdown and exceptional displays of aragonite trees (Figs. 3 and 4). To the north, pits interrupt the trunk passage, which appears to terminate at voluminous SVG Hall. A 30-m lead climb here, however, leads to the blowing Porpoise Passage, which crosses over the Chestnut Ridge anticline. Beyond, the 6-m Mud Piton Climb leads to the 24-m Damart drop and a second pitch of 11 m, Polypro Pit. This drops into the second trunk passage consisting of the Sixth of July Room to the south and the highly decorated Jewel Cave/Big Bend area to the north. Maret's Lead out of Sixth of July rambles through big rooms floored in slippery mud and challenging down-climbs for 600 m, ending in a sump 190 m beneath the entrance. This has been dye-traced to the Aqua Cave resurgence. A small, blowing infeeder here leads to Black Canyon, 800 m of washed, scrambling stream cave in very dark limestone. A sporting up-climb through a cascade intersects the beginning of the Burnsville Turnpike, the third trunk and by far the largest. Extending more than 1200 m with widths approaching 44 m and heights up to 30 m, the Turnpike is one of the biggest and most remote underground passages in Virginia. The Turnpike ends in up-trending breakdown, with a stream entering from a lower level of breakdown.

The North/South Trunk south of its intersection with Tombstone Alley is a pleasant walk of 300 m toward the Camp Room. Off the southeast side of the Camp Room, a steeply plunging down-dip lead drops into the Shamrock Dome area of the cave. A series of muddy slots and down-climbs leads to Satisfaction Junction and the 722 Sump. Southwest from the Camp Room, but on the same relative level, the North/South Trunk continues as a series of rambling, up- and down-climbs. A tight slot with air can be followed through collapsed breakdown to the South Lead

FIGURE 3 An anthodite speleothem called the Elk Horn in Bobcat Cave; the white vertical piece at the bottom is about 25 cm long. *Photo by Ron Simmons.*

FIGURE 4 Crystal speleothems found in a section of Bobcat Cave called the North/South Trunk. *Photo by Ron Simmons.*

Terminus, a comfortable room ending in more breakdown. A small hole, dug under a ledge, drops into the Blarney Stone continuation of the system. This cave is considered one of the most demanding in the Old Dominion.

Blarney Stone Cave (Chestnut Ridge Cave System)

Blarney Stone Cave was discovered by digging open an obscure sink on the eastern slope of Chestnut Ridge in March 1991. Its muddy fissure entrance series—four short rope drops and a series of challenging downclimbs—is a shorter version of the Bobcat entrance series. After 300-m, a small overflow tube leads to larger cave. On an upper level, a 5-m aid climb across a 14-m-deep shaft accesses an impressive paleotrunk named Ghost Hall. Sporting highly decorated stalagmites, stalactites, totem poles, and columns, Ghost Hall leads south to expansive Upper Ghost Hall. Black Diamond crawl, a small stream crawl with black gravel, also exits the south end of Ghost Hall. Black Diamond intersects a pit that descends to the Pearly Gates, named for a beautiful and prolific display of cave pearls. Moon River extends upstream and downstream below this point for nearly 600 m. Numerous side passages abound here, the most notable being Stairway to Heaven, an extensive series of challenging up-climbs that rise more than 150 m. North of Ghost Hall, the cave winds through the totem poles of Leprechaun Forest. Extraordinary crystalline white chandeliers decorate a delicate section of the wall. A large walking passage rambles to the north, eventually finding the obscure lead beyond the Earthworks to the cave's connection with Bobcat Cave. Very promising leads remain in the southeastern section of the cave that connects with Burns, particularly climbs at the end of Falling Water and Pigeon Tooth Dome.

Burns Chestnut Ridge Cave (Chestnut Ridge Cave System)

Before its 2005 connection with the Chestnut Ridge Cave System, Burns was the deepest cave in Burnsville

Cove at 240 m. The entrance series of low, sinuous, muddy crawls; tight, body-sized cracks; and plunging slot canyons to the bottom of the cave is one of the most arduous 500 m of cave in the United States. An impressive stream canyon below the 198-m level soon sumps upstream but flows north to a series of cascades at the 213-m level. Here, a high lead some 12 m above the stream leads to nearly 2-km of walking passage, eventually giving access to the rarely visited Cathedral River, and ultimately its connection with the Blarney Stone section of the Chestnut Ridge Cave System. This passage was mapped downstream for 335 m to a point where the water came to within 10 cm of the ceiling. The downstream cave approaches Robins Rift to within 300 m.

Barberry Cave

Barberry Cave has three excavated entrances, but only one open for access. The entrance to Big Bucks Pit is an excavated 21-m shaft dropping into an impressive 23-m-high trunk passage. At the bottom of the pit, this decorated, spacious stream trunk extends 670 m to the north and 550 m to the south and is aligned with the south end of the Burnsville Turnpike 1.6 km to the north (Fig. 5). The stream ends in a sump, but the trunk passage continues another 150 m to massive breakdown and in very close proximity to Buckwheat Cave. The WOWay, a sizable side lead entering the main trunk passage from the west, extends via watery passages to within 120 m of Butler Cave. An air-blowing lead at the end of this very tortuous passage holds promise of a connection with Butler.

FIGURE 5 Morphine Waterfalls. The main stream in Barberry Cave, Bath County, Virginia, is flowing down the bedding before cascading over a 2-m waterfall. The main passage in this area is 12 m wide, 20 m high, and has massive flowstone decorating the walls. *Photo by Philip C. Lucas. Used with permission.*

Buckwheat Cave

Buckwheat Cave, dug open in 1998, plunges as a walking stream passage into a series of low water crawls, blocked by massive breakdown. Coming to within 4 m of portions of the north extensions of Barberry Cave, Buckwheat drains a small part of the western flank of Chestnut Ridge. To date, 670 m of cave have been mapped in Buckwheat to a depth of 42 m. A connection to Barberry Cave is proven via air and barometric studies but remains elusive.

Blind Faith Cave

Blind Faith was discovered the year after Buckwheat Cave by digging a 4-m shaft in a small sink in the next wooded valley 600 m north of Buckwheat. A series of crawls and challenging down-climbs eventually drop into a going stream passages. This degenerates downstream in an extremely low, down-trending passage. Upstream, the cave winds for several hundred meters along the western flank of Chestnut Ridge but stops well short of connecting to nearby Buckwheat. More than 1000 m of cave has been charted in Blind Faith to a depth of 48 m.

Battered Bar Cave

Battered Bar Cave is located about 450 m north of the Blind Faith entrance on the edge of a deep sink corresponding to the terminal end of the Burnsville Turnpike in the Chestnut Ridge Cave System. A narrow, 18-m pit leads to an even tighter slot that slopes down 8 m to the top of a slippery 12-m-deep shaft. A steep, muddy up-climb leads to a third drop of 4 m. Beyond, through massive breakdown blocks, the cave opens up dramatically. In the first big room, a fissure leads down to the Ramp, a steeply inclined 30-m-long chute-way floored with breakdown lingering at the angle of repose. Two walking passages extend to the south from the bottom of the Ramp, the left-handed passage dividing after some 125 m into left and right branches. The left branch approaches the downstream end of Blind Faith Cave, hinting at a possible connection and ending near an unusual folded limestone feature called the Stone Rainbow. The right branch continues south 200 m, plunging dramatically to a sump in a passage covered with pure white sand. The right trunk at the bottom of the Ramp snakes through breakdown, then extends for 300 m into a maze of small passages ending in breakdown with some leads blowing air. Underneath the Ramp, a stream can be followed for 150 m into massive breakdown, including a pool that may be diveable. Any connection with the Burnsville Turnpike is approximately 250 m beyond. Battered Bar has been surveyed for more than 2080 m with a measured depth of 119 m.

Helictite Cave

Helictite Cave is formed near the top of the Helderberg limestone sequence, dissolved mainly out of the New Creek and Licking Creek limestones. A vast maze cave of tubes and canyons with one major paleostream passage, Helictite possesses dramatic examples of dogtooth spar, cave pearls, helictites, and slickenslides (Figs. 6–8). With more than 11 km of cave mapped since its dug-open discovery in 1996, Helictite is not a typical Burnsville Cove Cave and drains to Emory Spring instead of Aqua.

By-the-Road Cave

Found in June 2000, the entrance is a recent sink collapse located in the eastern synclinal valley 800 m southwest of Robins Rift. Strategically placed to offer access to the sumped, upstream portion of the Burns section of the Chestnut Ridge Cave System, By-the-Road is currently mapped to an ongoing in-cave dig. Considerable air blows from the cave in hot weather. The cave is gated and managed by the Butler Cave Conservation Society at the request of the Virginia Department of Transportation.

Basswood Cave

Dug open in July 2002 after several months of excavation, Basswood Cave is accessed by a 7-m ladder and 4-m climb that lead to a series of low, narrow crawls with additional, short pitches that end in a seasonally wet, very low crawl. In 2008–2009, cavers spent considerable time widening an air-blowing crawl midway through the cave to discover an additional 140 m of cave passages, only to have the cave, still blowing strong air, shrink to less than passable dimensions.

FIGURE 6 A cluster of helictites, the signature of Helictite Cave, Highland County, Virginia. When this cave was first entered some of the first speleothems seen were helictites. As exploration and surveys continued it was realized that this cave was far richer in this type of speleothem than other Virginia caves. This photo is part of a large cluster of helictites located just 30 meters inside the entrance. *Photo by Arthur N. Palmer. Used with permission.*

FIGURE 7 Pool spar, Helictite Cave, Highland County, Virginia: one of the crystal forms of calcite sometimes called *dogtoothed spar*. In Helictite Cave, the dogtoothed spar in this photo is located in a pool that is about 6 m long and 2 m wide. *Photo by Philip C. Lucas. Used with permission.*

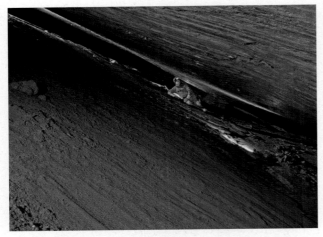

FIGURE 8 The Slickenslides Room. Helictite Cave has areas where faults are intersected by passage development. Here the slickenslides of a fault have dropped into the cave passage below. This makes for a dramatic passage where the scrapings or slickenslides are easily recognized on the ceiling and floor of this passage. *Photo by Philip C. Lucas. Used with permission.*

Although physically near Barberry and Buckwheat, Basswood Cave remains unconnected.

Fuhl's Paradise Pit

Discovered in the spring of 2005, Fuhl's was found by excavating a stone-filled sink that always melted snow in winter. Its 7-m entrance pit leads to 160 m of mostly dry, walking passage with no obvious continuations, hinting that Fuhl's is an isolated segment of ancient fossil trunk passage.

Water Sinks Cave (The Subway)

The discovery of this voluminous cave in 2007, now mapped to 4.5 km, revealed virgin cave beneath long-known Owl Cave. Although the two have no known connection, both caves are prominent Cove insurgences that drain to Aqua.

Buckets of Smoke Cave

This 5-m pit sits astride the near crest of Chestnut Ridge and is one of the highest entrances in Burnsville Cove. Discovered in 2007, Buckets of Smoke's marginal drainage over its 240-m mapped length has not yet been traced to either the Cove's eastern or western synclines.

Wishing Well Cave

Wishing Well Cave was opened in the Summer of 2010 by a caver team led by Phil Lucas, digging over a 3-year period at the bottom of a 16-m shaft at the northern end of Burnsville Cove. Initial exploration and mapping follow a strong current of air channeled through fractured layers of sandstone atop deeper layers of limestone. This latest discovery in Burnsville Cove possesses considerable potential not only as a new cave but for its possible connection to nearby Helictite as well as its potential connection into the Emory Spring drainage.

SPECIAL ATTRIBUTES

A preliminary report on the cave fauna of Burnsville Cove was published in the *Burnsville Cove Symposium* (Holsinger, 1982) and identified 11 invertebrates and 8 vertebrates. Two species, the amphipod *Stygobromus conradi* and a beetle (*Pseudanophtalmus*), are endemic to Burnsville Cove. A more recent report (Hershler *et al.*, 1990) noted the discovery of a new species of aquatic snail, *Fontigens morrisoni*, citing Butler Cave as one of only two locations. An updated report describes the discovery of a new species of springtail (*Arrhopalites*) from Butler Cave, since found in another Virginia cave outside the Cove. The appearance of unknown animal tracks deep in the North/South Trunk of Bobcat Cave and the finding of a "raccoon-like" skeleton beneath the Camp Room proved a startling discovery. Photographs and castings were subsequently identified as belonging to *Martes pennanti,* which is a fisher unreported in Virginia for almost two centuries.

A preliminary report on cave mineralogy of the Burnsville Cove was also published in the *Burnsville Cove Symposium* (White, 1982) and characterized the secondary mineral deposits as sparse though widely dispersed. Recent discoveries in selected areas of the Chestnut Ridge System, Barberry, and Helictite caves reveal an astounding array of helictites, aragonite trees, anthodites, cave pearls, and moon milk. Sediment studies in Butler Cave have found evidence of magnetic reversal as well as iron fixing filamentous bacteria in a brownish-yellow layer of goethite.

After more than a half-century of systematic investigation, it is now understood that the caves of the Burnsville Cove are hydrologically connected. Despite the physical barrier of the Chestnut Ridge anticline, the improbable presence of the North/South Trunk in the Chestnut Ridge System provides a key underground connection between the Butler Cave—Sinking Creek drainage of the western syncline and the caves of the less understood eastern syncline of the Burns—eastern Bobcat—Blarney Stone—Cathedral Spring drainage. Despite the connection of Bobcat and Blarney Stone caves in 1994, and their subsequent connection to Burns in 2005, connections between other caves of Burnsville remain elusive. With dedicated cavers mapping an average of 1.6 km of new cave a year over the past half century, Burnsville Cove continues to be recognized as one of the world's preeminent cave/karst frontiers.

Bibliography

Clemmer, G. S. (2001). That cave just had to be there. Ike Nicholson and the discovery of Butler Cave. In J. A. Campbell (Ed.), *Virginia Cavalcade, 50*(1), 52–63. Richmond, VA: The Library of Virginia.

Davis, N. W., & Hess, J. W. (1982). Hydrogeology of the drainage system, Burnsville Cove, Virginia. *National Speleological Society Bulletin, 44,* 78–83.

Douglas, H. H. (1964). *Caves of Virginia*. Falls Church, VA: Virginia Cave Survey.

Hershler, R., Holsinger, J. R., & Hubricht, L. (1990). A revision of the North American freshwater snail genus *Fontigens* (Prosobranchia: Hydrobiidae). *Smithsonian Contributions to Zoology, 509,* 49.

Holsinger, J. R. (1975). *Descriptions of Virginia caves*. Bulletin No. 85. Charlottesville, VA: Virginia Division of Mineral Resources.

Holsinger, J. R. (1982). A preliminary report on the cave fauna of Burnsville Cove, Virginia. *National Speleological Society Bulletin, 44,* 98–101.

Nicholson, I. K., & Wefer, F. W. (1982). Exploration and mapping of the Sinking Creek System. *National Speleological Society Bulletin, 44,* 48–63.

Rosenfeld, J. R., & Shifflett, T. E. (1995). The caves of Burnsville Cove, Virginia. In C. Zokaites (Ed.), *Underground in the Appalachians: A guidebook for the 1995 NSS convention* (pp. 9–19). Huntsville, AL: National Speleological Society.

Schwartz, B. (1999). Exploring Barberry Cave. *NSS News, 57,* 268–275.

White, W. B. (1982). Mineralogy of the Butler Cave–Sinking Creek System. *National Speleological Society Bulletin, 44,* 90–97.

White, W. B., & Hess, J. W. (1982). Geomorphology of Burnsville Cove and the geology of the Butler Cave–Sinking Creek System. *National Speleological Society Bulletin, 44,* 67–77.

C

CAMPS

Gregg S. Clemmer
Butler Cave Conservation Society, Inc.

The thorough exploration and survey of an extensive cave system demand that all participants "push the limits" of the cave to its "bitter end." Such idealistic, oft-used phrases employed by cavers reflect a deeply held philosophy—an ethic some would say—that goal-oriented, expedition-style caving requires careful preparation, long-term dedication, and extensive stamina. "Push trips" to the bottom or to the far reaches of vast, complex cave systems challenge all three requirements. Planning entails cooperation, competency, specific goals, a mountain of gear, and the occasional kilometer of rope. Participation impacts everything from bank accounts and vacation time to jobs and marriages.

It is a given that everyone enters the cave in superb physical and mental condition. But what happens when the mountain of gear is consumed, when the kilometer of rope is rappelled, when the strongest caver is exhausted, and the cave still goes down and down, on and on? In the years since the founding of the National Speleological Society, Inc. in 1941, U.S. cavers have continued to push the limits. In the first half-century of the Society's existence, for a variety of reasons—personal comfort, novelty and intrigue, or simply because the cave went on and on—cavers in a few instances resorted to underground camps to pursue their respective goals. Cavers in Europe confronted the same challenges. Exploration in Switzerland's Hölloch expanded to an underground camp in 1949. After four cavers were trapped by high water for 9 days in 1952, all exploration camps in Hölloch became wintertime endeavors (Courbon et al., 1989). The widely acclaimed 1952 descent into Pierre Saint-Martin employed a 5-day underground camp but ended tragically with the death of Marcel Loubens (Tazieff, 1953). Gouffre Berger became the first caver to break 1000 m in depth, a feat realized in 1956 which owed much to the staged underground camps of 1954–1955 at 494 m and 860 m (Cadoux, 1957). A 1955 expedition into the Cigalere also employed staged cave camps, but, instead of pushing the cave ever deeper, explorers confronted a daunting series of waterfalls as they ascended into the mountain (Casteret, 1962).

The evolving European model for underground expedition camping utilized advance supply teams to rig pits, lay phone lines, transport mountains of gear, and establish camps. Single-rope techniques being unknown, pits were negotiated by cable and rope ladders backed up by belay. The early exploration of Utah's Neff Canyon Cave in October 1953 followed this European model and utilized a support crew to aid four cavers on a 33-hour trip to the bottom of the cave. Engaged in a rivalry with a local climbing club, the group carried in a large amount of gear—sleeping bags, cable ladders, 150 m of rope, field phones, and coils of wire—to support their effort. But, after enduring a "fitful sleep in cold, cramped quarters" in their unsuccessful attempt to find the cave's deepest point, the explorers emerged "completely exhausted." One member of the support crew spent a week in the hospital suffering "utter fatigue." They chalked up their failure to "bulky packs and unmanageable gear in the narrow, jagged passageways" (Green and Halliday, 1958).

A few months later, Floyd Collins' Crystal Cave Expedition (C-3 Expedition) electrified the caving community with a sensational attempt in Flint Ridge to push the far reaches of Kentucky's most extensive cave. With movie cameras rolling and backed by a full complement of sponsors, fawning reporters, and radio broadcasters, the cavers entered the cave with ambitious goals for a week underground. Metal "Gurnee cans" protected gear through tight, rocky crawlways. Field phones connected remote sections of the cave with the surface support crew. Experts in cave biology, geology, hydrology, medicine, and meteorology accompanied the hard-charging explorers, all eager to measure the cave, the cavers, and the phenomena therein.

The expedition ended with numerous official reports detailing everything from sleeplessness and mild shocks from ring voltages in the phone system to the morale-boosting effects of candy and tobacco being delivered from the surface. The festive, self-congratulatory tone at the expedition's end ignored the large amounts of trash buried or burned in the cave. Several kilometers of abandoned telephone wire would litter the passages for decades (Brucker and Lawrence, 1955). Thankfully, the "success" of the C-3 Expedition was never repeated, but in a subsequent report 2 years later one participant recommended "simplicity in all phases of trip organization" as a future goal, warning that too often "success is judged by size instead of actual accomplishment" (Smith, 1956).

In contrast, the Butler Cave Camp of August 1958 commenced as a closely held secret. Not eager to get "scooped" and exploring mostly in blue jeans and wool shirts, the seven-man crew eagerly pushed deep into this Virginia discovery without field phones, surface support crew, or scientific agenda. At the end of an exciting, tiring week, they exited with 5 km "recon-mapped" and a collection of superb color slides, but they buried their trash and spent carbide in the cave.

Youthful exuberance and naiveté characterized the August 1962 cave camp in Indiana's Sullivan cave. With lofty goals to map the cave, sample the soil for microbes, conduct psychological surveys on participants, and clean up extensive vandalism, the teenagers elected to spend 2 weeks camping in the cave despite its relatively close proximity to the entrance. Field phones connected them to the surface. Equipped with sleeping bags on cots and supplied with a double-burner Coleman stove, canned goods, rye bread, and even fresh vegetables (celery, carrots, and lettuce), the explorers endured a miserable, cold existence in wet, muddy clothes despite five complete changes of underground wardrobe. Although they mapped 2 km of passage, the young "Sullivaneers" discovered only half of it to be virgin cave. One participant characterized part of their mapping as "a comedy of errors."

Vastly more significant discoveries rewarded a two-person, week-long camp in Ellison's Cave, Georgia, in 1969. Despite the dramatic failure of their only stove at base camp more than 250 m below the surface, this man-and-woman team stomached cold food and dank conditions to survey almost 4 km of cave without field phones and surface crew, all with minimal impact to the cave (Smith, 1977).

A bizarre example of cave camping occurred in 1972 in Midnight Cave, Texas, when one man spent 6 months underground. Dismissed as nothing more than a publicity stunt by some, the venture did garner enormous attention, including a feature article in *National Geographic*. The subject entered the cave ostensibly to investigate the long-term psychological and physiological effects of solitary confinement and sensory deprivation. Amenities included a canopied sleeping area on a wooden platform, extensive incandescent lighting, field phone, books, and record player. At the end of his time underground, the relieved cave dweller declared his trial a success as the "longest beyond time experiment in history" (Siffre, 1975).

Ongoing exploration of Wind Cave in South Dakota employed an underground camp in 1972. Situated near the Master Room, the relatively comfortable camp was supported by a surface crew, stocked by supply teams, and connected to the surface by field phones. To thwart hypothermia, participants toyed with the novel notion of running heat lamps in the camp on 480 V piped down the telephone cable. Despite the discovery of major extensions to the cave, no one favored a second underground camp the following year. "The logistics of running a base camp, although successful, were very difficult and time-consuming," wrote one organizer. Henceforth, the survey reverted to "long, single-day trips from the surface" (Scheltens, 1988).

All of this experience gained was lost on organizers of Project SIMMER when 118 cavers descended on Simmons-Mingo Cave in West Virginia in October 1973. Ambitiously planned much like a military operation with a chain of command, mess tent, and administration tent, the expedition ultimately consumed 10 hours of preparation to every hour actually spent underground. Planners managed to lay more than 15 km of wire for field phones, then touted their work as the "world's largest in-cave communication network." The Gurnee can of the C-3 Expedition morphed into a "Carts can," a stovepipe and plywood contraption used to haul gear into the cave. Plagued by poor sleeping bags, wet clothes, ringing phones, and a miserable camp spot, the crews mapped less than a kilometer of cave. Project SIMMER had profited nothing from the C-3 experience of 20 years earlier and never issued a final expedition report. Participants even abandoned the phone wire in the cave. Yet, beyond these disappointments, the overall underground manager opined at the end of the experience that for deep or remote cave exploration, "the small camping party [would be] more efficient than the larger, more formally organized group."

American deep-caving expeditions to Mexico also began camping underground. European participants with extensive expedition experience contributed a wealth of knowledge toward maintaining a comfortable, efficient subterranean camp. Prolonged underground stays were begun in the mid-1960s and by the late 1970s had pushed the reaches of vast, deep, technically difficult caves (such as Sotano de San Agustin, La Grieta, Sumidero Yochib, Sotano del Rio Iglesia, and Sistema Purificacion) far

FIGURE 1 A meal being prepared at camp 3 in Sotano de San Agustin.

beyond the reaches of conventional day-trip caving (Stone, 1978). (See Figs. 1 and 2.)

Outside the warmer climes of Mexico, occasional cave camping in the temperate latitudes of the United States generated little appeal for second attempts. Despite some glowing declarations of expedition success, the cave campers of Neff Canyon, C-3, Butler Cave, Sullivan Cave, Ellison Cave, Wind Cave, Project SIMMER, and even a successful, comfortable camp deep in Fern Cave in Alabama brokered little enthusiasm to repeat their adventures. Pushing caves to their limits in the chillier continental 48 states went back to being brutal day-trip endeavors. Remote underground camps were best left for the warmer caves south of the border. The 1983 discovery of a large cave system under Chestnut Ridge, near Burnsville, Virginia, provided the impetus for yet another group to try an extended underground camp in the United States. More than 20 km of challenging, decorated, virgin galleries rewarded those who endured the cold, sloppy, tortuous entrance series of Bobcat Cave (see Figs. 3 and 4). Yet exhaustion and the real threat of hypothermia limited all efforts to safely extend exploration via increasingly longer day trips. With no other choices, cavers with decades of grueling experience grudgingly confronted the possibility that camping underground was the only feasible way to continue the survey. Given the Spartan experiences a generation earlier at nearby Butler Cave, few relished the idea. No cave in the United States had ever been continuously pushed and mapped in such a manner. Nevertheless, over the next 10 years, more than 15 km of passageways were explored and surveyed via 27 underground camps in Bobcat Cave, culminating in the 1994 connection with 7-km-long Blarney Stone Cave.

To be fair, few caves offer the isolation and daunting physical challenges that justify camping underground, but Bobcat Cave did, and once the decision to camp had been made the question became one of how to thwart

FIGURE 2 Cooking area of the White Lead Room in La Nita, which is part of Sistema Huautla in Oaxaca in southern Mexico.

FIGURE 3 Bobcat Cave; a view of the cooking area after several days of use.

hypothermia, obtain adequate nourishment, maintain endurance, and still safely and efficiently get the cave competently explored and mapped. Custom nylon coveralls were the first big difference from previous camps. Until the early 1980s, experienced American cavers—with rare exceptions—went underground clothed primarily in cotton and wool. "Farmer John" coveralls ruled the day. Wet suits were tight and uncomfortable but battled hypothermia and provided protection far better than

FIGURE 4 Bobcat Cave; one caver's area in the main camp chamber. He is using a hammock instead of a sleeping pad on the ground.

blue jeans and corduroy jackets or the wool sweaters and flannel shirts of earlier times. Comfort counted, and the eternal cold of soggy cotton and smelly wool when still kilometers from the entrance begged for garments promising warmth, agility, and the ability to stay dry despite the wearer's body heat.

Nylon did just that. Participants in the first Bobcat camp purchased yards of the fabric, adapted a borrowed coverall pattern, and sewed their own. Cave packs evolved the same way. The bulky, battered metal towing cans of the C-3 Expedition and Project SIMMER never even came under consideration. Instead, long, cylindrical, flexible nylon duffel bags, also self-sewn, performed admirably. With a tether on one end for upright attachment to seat harnesses when ascending or descending drops, a handle in the middle for grasping in crawls and crevices, and back straps for carrying over long distances, the "camp duff" proved invaluable for getting gear and food into camp. Double or triple thicknesses of trash bags protected food, clothes, and sleeping bags from devastating leaks. Sucking out the air from such packed bags before tying them off provided additional space. Two decades later, nylon packs and coveralls enjoy almost universal use in caving and even exude their own fashion statements, having spawned a cottage industry in custom cave gear, vertical rigs, and personalized repair using a variety of incredibly durable fabrics. Changing into dry, warm camp clothes on reaching camp boosted morale, especially if one's body heat aided the process; thus, wickers and polypropylene replaced cotton and wool undergarments.

Sleeping underground, though, had always been a prolonged struggle against chill and dampness. Cotton or down sleeping bags were dismal failures, but lightweight fiberfill or synthetic bags worked nicely when laid on a foam pad (or, for example, a Therm-a-Rest pad) atop a reflective ground cloth. A stocking cap kept head and ears warm all night. Some campers even wore gloves. A dry change of socks, bound up in small plastic bags, assured dry feet even when moving about camp in wet, muddy cave boots. An extra polypropylene top and bottom, properly bagged, provided the luxury of a pillow. Hammocks, although favored by some on Mexican expeditions, were quickly abandoned after a fitful night tossing in the damp, 48°F chill of Bobcat Cave. The camp site itself needed to be relatively level, spacious enough for sleeping quarters and a community kitchen and eating area, and fairly close to reliable water. A drop of iodine per gallon of water accomplished water purification. The latrine was located in respectable proximity to camp, dug into a clay bank.

Eating revolved around breakfast and dinner, supplemented during the day by personal preferences (energy bars, gorp, cheese, candy, premade sandwiches, even a baked potato). Freeze-dried food covered most menus, being far tastier than the wretched examples of the past and significantly lighter than canned goods. (*Note*: In desert caves or where water is scarce, canned goods could be a significant supply for both food and water.) Tea, coffee, sugar, salt, oatmeal, dried fruit, pepper and other spices, and even luxury condiments were easily stuffed in zip-lock bags and buried in the depths of packed sleeping bags.

Aside from ropes, climbing gear, bolt kits, and survey gear, community camp gear on the initial trip included a small white gas stove with repair kit; several full, secured fuel bottles; a cooking pot for hot water; three or four collapsible plastic gallon jugs; first-aid kit; and trowels and toilet paper, all divided among the participants. Outside of replenishment items, these were secured in the cave from camp to camp. Luxury items ranged from washcloths and personal journals to cards and a harmonica. Carbide provided 90% of the lighting, with candles around camp adding an intimate touch and saving acetylene.

Cavers know that adaptability and incentive remain a vital part of pushing the limits. Future camps may very well embrace caving LED lamps, for which rumored 50-hour burn times on one set of four D cells would surely lighten camp duffs of pounds of bulky calcium carbide on the way in and spent carbide on the way out. Every Bobcat expedition entered the cave as a small, self-contained team. Never did more than nine cavers (three teams of three) participate; six proved the average. Surface crews lounging in administration tents fielding phones attached to kilometers of wire strung through near-virgin cave were never an option and would probably violate the conservation ethic of today's cavers. Instead, with a safety contact just a few kilometers from the entrance, the expedition entered the cave with competent associates on the surface aware and available. In the years since, cave camping has remained a seldom-used tool of American cavers.

The recent multisump isolation of camping on a tarp suspended above water deep in Mexico's Sistema Huautla is surely the extreme (Stone et al., 2002), but the continued success of camps in making new discoveries such as in Kentucky's Fisher Ridge, New Mexico's Lechuguilla, and Virginia's Omega System is a tribute to cavers' adaptability to the challenging extremes of the caves they continue to push.

See Also the Following Article

Exploration of Caves—General

Bibliography

Brucker, R. W., & Lawrence, J., Jr. (1955). *The caves beyond: The story of Floyd Collins' Crystal Cave exploration.* New York: Funk & Wagnall's.

Cadoux, J. (1957). *One thousand meters down.* London: George Allen & Unwin.

Casteret, N. (1962). *More years under the Earth.* London: Neville Spearman.

Courbon, P., Chabert, P., Bosted, P., & Lindsley, K. (1989). *Atlas of great caves of the world.* St. Louis, MO: Cave Books.

Green, D. J., & Halliday, W. R. (1958). America's deepest cave. *National Speleological Society Bulletin, 20,* 31–37.

Scheltens, J. (1988). 50 miles beneath the wind. *NSS News, 46,* 5–14; 28–37.

Siffre, M. (1975). Six months alone in a cave. *National Geographic, 147*(3), 426–435.

Smith, M. O. (1977). *The exploration and survey of Ellison's Cave, Georgia.* Birmingham, AL: Smith Print and Copy Center.

Smith, P. (1956). Seven principles of effective expedition organization. *National Speleological Society Bulletin, 18,* 46–49.

Stone, B. (1978). Underground camps for deep caves. *AMCS Activities Newsletter, 8,* 37–45.

Stone, B., Am Ende, B., & Monte, P. (2002). *Beyond the deep: The deadly descent into the world's most treacherous cave.* New York: Warner Books.

Tazieff, H. (1953). *Caves of adventure.* New York: Harper & Brothers.

CASTLEGUARD CAVE, CANADA

Derek Ford

McMaster University

Castleguard Cave is the longest cave system currently known in Canada (21 km) and the foremost example anywhere of a cavern extending underneath a modern glacier (Fig. 1). It displays many striking features of interactions between glaciers and karst aquifers, a complex modern climate, rich mineralization, and a troglobitic fauna that has possibly survived one or more ice ages beneath deep ice cover in the heart of the Rocky Mountains (Ford (1983); Ford et al., (2000); Muir and Ford (1985)).

GEOGRAPHICAL SETTING

The cave is located in the northwest corner of Banff National Park, Alberta, very close to the Continental Divide. The region is one of rugged alpine mountains with many horn peaks, cirques, and U-shaped valleys typical of intensive glacial erosion, plus a few small but high plateaus. The range of elevation is from 1500 m asl in the floors of trunk valleys to summits at 3500 m. Mean annual temperatures are 0 to $-14°C$ across this height range. Natural boreal forests extend up to ~2100 m, passing into grass and low shrub tundra and then alpine desert generally above 2400 m. The Columbia Icefield is a plateau ice cap 320 km^2 in area and 200 to 300 m thick, the largest remaining ice mass in the Rocky Mountains. Valley glaciers radiate up to 10 km out from it today. Ice thickness and extent were much greater during the major glaciations, when the glaciers extended 100 km or more from the icecap, with only the mountain peaks protruding as nunataks.

The karst rocks are resistant carbonates of Cambrian age. The Cathedral Formation (>560 m thick) is massively bedded, very resistant crystalline limestone that contains the cave. Above it, the Stephen Formation (80 m) is a limestone shale that can block much descending groundwater but leaks readily through some major fractures (i.e., it is an aquitard). It is overlain by further thick-bedded limestones and dolostones, the Eldon and Pika Formations. The summit strata are mechanically weaker shales, sandstones, and dolostones. Beneath the Icefield, around the cave and north of it, these rocks dip regularly south-southeast at 4 to 6°. South of the cave and parallel to it there is a sharp downfold in the Cathedral rocks that caused some slippage of bedding planes (thrusting) to the north. A valley is excavated along the downfold, with a glacier from the Icefield at its head and the Castleguard River starting at the glacier snout. Castleguard karst groundwater drainage reaches the river via 60 or more springs.

On the surface, the Cathedral limestones host a suite of small but typical alpine karst landforms such as karren, solution and suffosion dolines, and vertical shafts. They are particularly well seen in the Meadows, a broad, shallow valley north of the cave mouth (Fig. 1). Many of these features were overridden and lightly eroded by glaciers during a minor re-advance—the "Little Ice Age," which occurred during the past 500 years. The glaciers are now receding. Meltwater streams sink underground around their edges or in the Meadows. At places, streams can be heard cascading down shafts still concealed beneath the flowing ice.

MORPHOLOGY AND GENESIS

Castleguard Cave is a textbook example of a meteoric water dissolutional cave in limestone. Cavers can enter it only at its downstream end at 2010 m asl in the north wall of Castleguard River valley, more than 300 m above

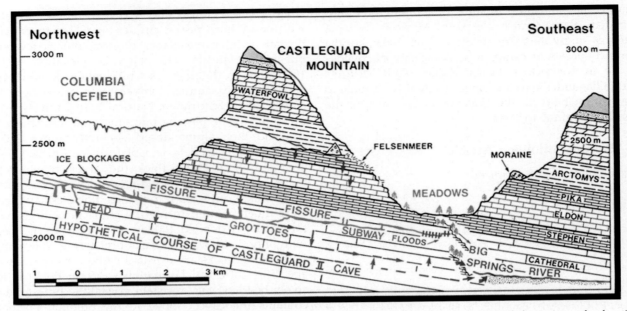

FIGURE 1 Schematic section through Castleguard Mountain, Alberta, Canada, showing the geological formations, the location of Castleguard Cave, Castleguard II, and the Meadows.

the valley floor. From there, the cave ascends 386 m to terminations underneath the Icefield, where explorers are farther from their only entrance (and exit) than in any other known cave. There are three distinct morphologic sections. The *Headward Complex* is comprised of inlet passages beneath the modern Icefield that were created by repeated glacial blockage and rerouting of sinking waters in the past. The passages are plugged by glacier ice or debris today. Younger vadose shafts pass down through them and become blocked by constrictions or debris below. The *Downstream* or *Entrance Complex* includes low tunnels in two major bedding planes created by flooding and obstructions by glacier ice in Castleguard Valley in the past. Finally, the *Central Cave* is a sequence of remarkably long, straight conduits created where one master bedding plane is intersected by a pair of vertical joints that are linked by a sedimentary dike (Grottoes Dike) crossing them (Fig. 2A); in the bedding plane there is some evidence of crushing and shearing, indicating that differential slip opened it up a little, permitting groundwater to penetrate at its juncture with the joints and dike.

The cave possibly originated as a single phreatic loop beneath the Stephen impermeable cover rocks that descended more than 370 m below a paleowatertable and then re-ascended to ancient springs just below the Meadows. More certainly, as Castleguard Valley was entrenched below the Stephen Formation the cave became enlarged to nearly its modern dimensions, as shown in Figure 2B. It was then comprised of two shallow, principal loops with vadose canyon entrenchments up to 20 m deep at their upstream ends, grading downstream into phreatic tubes 4 to 5 m in diameter and of beautiful circularity (Fig. 3). The downstream loop discharged into the Helictite Passage in the Entrance Complex by a vertical lift (phreatic shaft) of 24 m. Following further entrenchment of the Valley, the main cave headwaters were diverted into a lower cave (Castleguard II) and residual waters drained through constricted undercapture passages in the bottoms of the loops (Fig. 2C). The undercaptures channel local invasion waters passing through the leaky Stephen rocks today, but the Central Cave and Headward Complex are essentially abandoned hydrologic relics. The Downstream Complex, however, can be flooded with terrifying rapidity when waters pour out of another, quite independent lifting shaft within it (Boon's Blunder), which fills the first kilometers of low passages entirely and discharges the waters through the explorers' entrance (Fig. 4).

Modern Hydrology

Modern waters drain underground from the glacier soles, alpine karst, and Meadows to a set of springs extending 3 km downstream from the Big Springs, which are a trio of dramatic overflows 15 to 40 m above the valley floor (Fig. 1). The waters flow through the putative series of inaccessible caves (Castleguard II), as illustrated in Figure 5. The Artesian Spring, the lowest in elevation, is perennial. As the annual melt season progresses, upstream springs such as Gravel and Watchman springs begin to flow. The Big Springs,

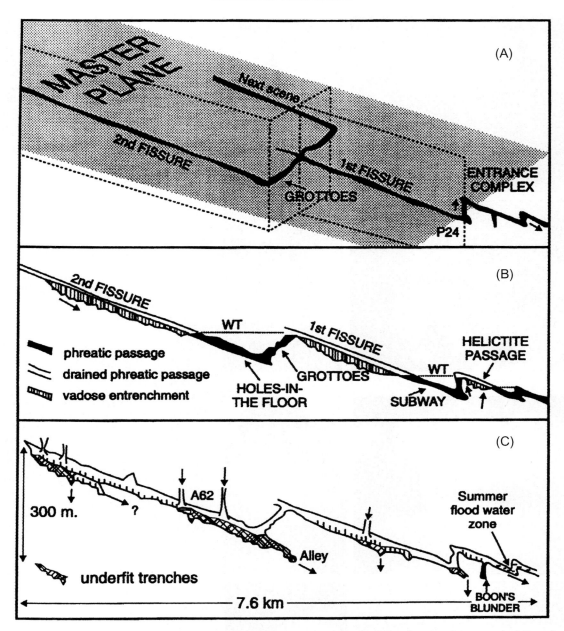

FIGURE 2 (A) The initial phreatic passages in the Central Cave and Downstream Complex showing the master bedding plane and intersecting vertical fractures that guided them. They discharged into the Entrance Complex bedding planes via P24, a vertical shaft up a small fault. (B) The cave at the close of the principal enlargement phase. Drawdown vadose canyons supplied water to a succession of shallow phreatic loops. (C) The modern cave; small invasion vadose streams have cut shafts and underfit trenches in the drained galleries and are lost into impenetrably small, undercapture passages continuing on down into Castleguard II.

100 m higher than the Artesian Spring, have a maximum discharge $>7~m^3~s^{-1}$ and handle average summer floods. Their capacity is exceeded when there is very strong melting at the head of the Meadows and/or on the Icefield. Groundwater then backs up in the aquifer until it floods Boon's Blunder and the Downstream Complex of the cave, 270 m above the Big Springs. Cave discharges can be $>5~m^3~s^{-1}$. Tracer dye placed in glacier edge sinkholes can reach the Big Springs 4+ km distant and 750 m lower in as little as 3.5 hours. This is again a textbook example—here, of a dynamic alpine karst aquifer. The large number and great height range of its springs are attributed to repeated disruptions such as debris plugs during the glaciations.

The Boon's Blunder flooded shaft was dived for the first time in 1987, when it was found to be a phreatic

FIGURE 3 The architecture of phreatic looping in the Central Cave: (A) In the Central Grotto, a cross-section on the master bedding plane (Fig. 2A) and aligned on a Neptunian dike (on left). Water flowed gently upslope here to spill down into (B) the First Fissure, a vadose entrenchment below the master plane (seen in roof), and then into (C) the Subway, a textbook example of the ideal phreatic passage. At the downstream end of the Subway, the water then flowed upward again under hydrostatic pressure, creating (D) the 24-m Shaft, a phreatic lifting element.

FIGURE 4 (A) The Ice Crawls; the residue of summer floodwaters is frozen solid in the exploring season, April. (B) The Ice Crawls end at this floodwater phreatic lift, a 7-m shaft where the violent flow creates a shingle beach at the foot. (C) Scene 100 meters from the entrance in June, with the 7-m shaft in far rear. The winter airflow in has reversed and moist flow out of the cave has coated its frozen walls with hoarfrost. The first flood will remove this. (D) The cave entrance at the end of winter, overlooked by the Watchman peak in the background.

lifting shaft like the 24-m shaft into the Central Cave. It reaches the same master bedding plane (Fig. 2A), where a bedding passage trends northwest up the stratal dip and subparallel to the First Fissure. In a series of dives in 2009 and 2010 the British cave diver, Martin Groves, swam this upstream for 845 m before reaching the air-filled subway-type passage that remains to be explored. From his estimates of the passage dimensions the volume of water in perched storage here is ~ 6000 m^3, which compares very well with

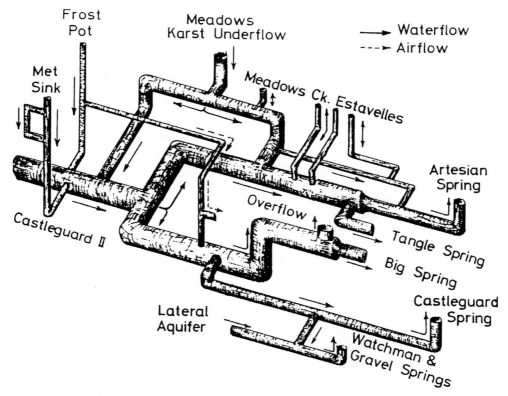

FIGURE 5 The proportional flow model of the inaccessible Castleguard II cave system, its feeder invasion vadose shafts, and discharge springs that Smart (1983) derived from quantitative hydrological studies. Cross-sectional areas of individual passages are proportional to their share of the total groundwater flow measured in the system, and their complicated pattern of interconnections is deduced from dye tracing and flood overflow behavior.

an estimate of 5400 m^3 that Smart (1983) derived from change of D/H ratio in the waters over the course of the first summer flood event, that is, that the water is being expelled by simple piston flow.

Cave Sediments, Speleothems, and Dating

Subglacial boulders, gravel, and sand were swept or bulldozed into the head of the cave. Many have been partially cemented by calcite and later eroded, indicating a long history of filling and removal. Throughout the Central Cave are remnants of three partial fillings with varved silts and clays, separated by phases of erosion or calcite deposition. They are deposits of glacier flour, settled out of suspension on occasions when the cave became backflooded with subglacial waters because the Castleguard Valley was filled with flowing ice.

Despite its location beneath glaciers or alpine desert, the cave beyond the modern entrance flood zone is well decorated with speleothems, chiefly very pure, white calcite. There has been much speculation about mechanisms for its deposition in the absence of any sources of soil CO_2 overhead (see Atkinson, 1983). There is one double layer of cave pearls that are all edge-rounded cubes 5 to 7 mm in diameter, a unique deposit. In the warmer central sector (temperature of +2 to +3°C, relative humidity of ≥95%), there are small evaporative aureoles of aragonite, huntite, hydromagnesite, gypsum, mirabilite, and epsomite; evaporation in these extreme conditions is due to strong drafts blowing through the cave when it is not in flood.

Although most speleothems appear to postdate the latest phase of varve deposition, there are many remains of older stalactites and stalagmites that suffered erosion during one or more varve floods. Uranium series dating and paleomagnetic studies of some of the oldest examples show that the cave became relict (*i.e.*, Castleguard II was already well developed) more than 780,000 years ago. The varved clays are younger. This antiquity is typical of multilevel alpine caves.

THE CLIMATE OF THE CAVE

Castleguard Cave is a chilly place; however, because it passes through a big mountain (Fig. 1) the geothermal heat flux is able to warm the central, most insulated parts around the Grottoes to approximately 3°C. In winter, this is much warmer than outside temperatures. The cave then functions as a chimney, with the warmer central cave air flowing upstream into the sole of the Icefield and dense, cold air pouring in through the explorers' entrance to replace it. The cold, dry air freezes residual pools of summer floodwater in the Entrance Complex, giving cavers a 300-m belly crawl over dusty ice. The first liquid water is encountered only 1 km inside the cave. There is a subsidiary daily effect because the draft is strongest and coldest immediately before dawn, not a good time to be coming out through the ice crawls. In summer, the dynamic situation is reversed as cold, moist air from the interior flows down and out through the explorers' entrance into the warm exterior. Thick hoarfrost is deposited onto the chilled entrance walls until the first floods of summer arrive to remove it.

An unexpected feature of the 2009 and 2010 dives was that the temperature of the perched water body in Boon's Blunder was gauged at 5°C, whereas air and seepage water temperatures in the Central Cave a few hundred meters away at the same elevation do not rise above 3.0 to 3.5°C; the heart of the mountain is chilled below its geothermal balance by the winter airflow here.

THE FAUNA OF CASTLEGUARD

Harsh as conditions are, the cave has a small population of animals that prefer it or are entirely dependent on its protection. The first, troglophiles, are packrats nesting above the floodwater limits in the Entrance zone. They have their own private entrances from the Meadows overhead that are too small for human cavers. Two species of blind and unpigmented (troglobitic or fully adapted) crustaceans live in pools in the Central Cave, where the water apparently never freezes. An isopod, *Salmasellus steganothrix*, is known elsewhere in the Canadian Rockies. An amphipod, *Stygobromus canadensis*, is known only in Castleguard. It is possible that the cave served as a subglacial refuge for these species during the last glaciation or longer. A packrat nest and other organic material were found in the Headward Complex in 1983 and carbon-dated to ~9000 years BP, indicating that in early postglacial times the Icefield had receded from at least part of it, thus providing some food for the troglobites downstream.

CONCLUSIONS

Castleguard Cave has proved to be a fine laboratory for the study of dissolutional cavern genesis and of groundwater flow under glacier cover. It appears that most passages in the Cave that are large enough for humans have now been found and mapped, except perhaps in the Headward area and the new discoveries beyond Boon's Blunder. The underlying, very dynamic complex of Castleguard II that drains much of the Columbia Icefield today continues to defy all attempts at physical exploration. Under the Meadows and around the summit massif of the mountain more than 100 mapped sinkholes plus unmeasured subglacial streams feed water into Castleguard III, another almost entirely unknown system that may contain both extensive fossil and active galleries. There is great potential here for future generations of speleologists.

Bibliography

Atkinson, T. C. (1983). Growth mechanisms of speleothems in Castleguard Cave, Columbia Icefields, Alberta, Canada. *Arctic and Alpine Research*, 15(4), 523–536.

Ford, D. C. (Ed.), (1983). Castleguard Cave and Karst, Columbia Icefields Area, Rocky Mountains of Canada: A symposium. *Arctic and Alpine Research*, 15(4), 425–554.

Ford, D. C., Lauritzen, S.-E., & Worthington, S. (2000). Speleogenesis of Castleguard Cave, Rocky Mountains, Alberta, Canada. In A. Klimchouk, D. C. Ford, A. N. Palmer, & W. Dreybrodt (Eds.), *Speleogenesis: Evolution of karst aquifers* (pp. 332–337). Huntsville, AL: National Speleological Society.

Muir, R. D., & Ford, D. C. (1985). *Castleguard*. Gatineau, Quebec: Parks Canada Centennial Publication, Department of Supply and Services.

Smart, C. C. (1983). The hydrology of the Castleguard Karst, Columbia Icefields, Alberta, Canada. *Arctic and Alpine Research*, 15(4), 471–486.

CAVE DWELLERS IN THE MIDDLE EAST

Paul Goldberg and Ofer Bar-Yosef

Boston University and Harvard University

This article offers information about prehistoric caves in the Middle East. Caves are a source of fascination for the public as well as many scientists. In this short entry we will attempt to briefly describe and summarize some of the discoveries in Middle Eastern caves and what those discoveries reveal about human behavior.

INTRODUCTION

Caves figure strongly in the archaeological record, through most archaeological time periods and across most continents. Caves not only served as loci of habitation where daily activities such as food processing and sleeping took place but also functioned as gathering places for spiritual and religious activities.

Prehistoric caves—particularly those in the Middle East—are special sedimentary environments (Fig. 1). In

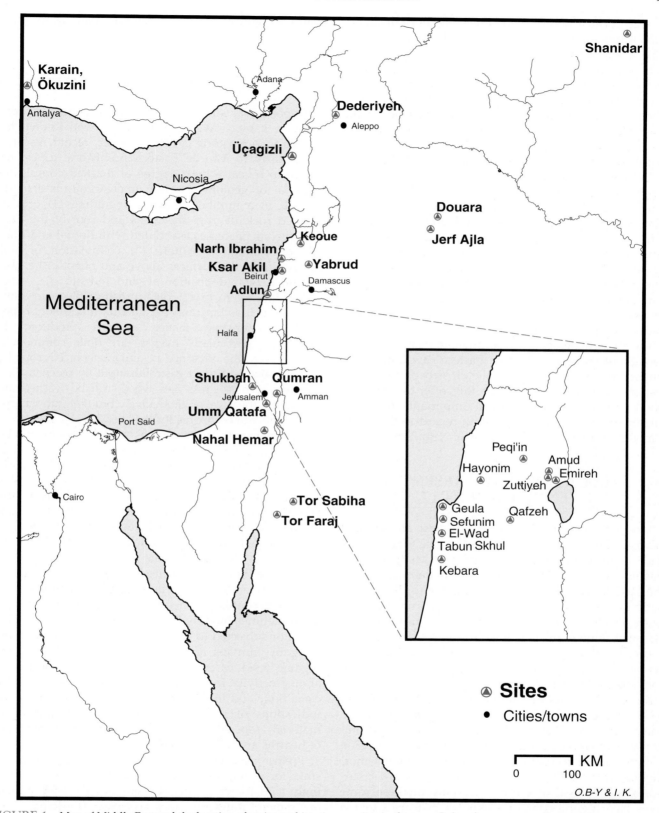

FIGURE 1 Map of Middle East and the location of major prehistoric cave sites in the area. Only a few caves are discussed here.

essence, because of the overall low energies of the depositional processes, caves serve as excellent sedimentary traps: whatever is brought into the cave tends to stay there. As a result, caves can preserve faithful records of past environments, as well as past human activities. Environmental information is conveyed by the presence of macro- and microfaunal remains—particularly the latter. Plant remains, an additional environmental indicator, are scarce, although microbotanical remains such as pollen and phytoliths can be found. Phytoliths tend to be better preserved and abundant and provide insights not only about past local and regional conditions but also about human activities, such as the gathering of plants for fuel, bedding, and food.

In addition, anthropogenic deposits also reveal the nature of past human activities at a cave by incorporating animal bones and stone artifacts. These objects inform us about hunting and butchering practices, the manufacture and use of curated stone tools, and the use and function of space (*e.g.*, working areas, fireplaces, dumping, and sleeping zones).

The above statements are pertinent to many cave sites throughout the world. Middle Eastern caves, however, benefit from their location at the crossroads to human past and present migrations and thus offer the opportunity to monitor and document important phases in human evolution, particularly with regard to the origin of modern humans and the demise of Neanderthals.

LOCATION AND PRESENT ENVIRONMENT

Most of the caves in the Middle East are situated either in the Mediterranean climatic zone (Fig. 1), particularly in a belt close to the present Mediterranean coastline, or in the drier, steppic and desertic region. The Mediterranean climate, which includes distinct climatic gradients from west to east and from north to south, is characterized by warm, dry summers and cool, wet winters in which 500 to more than 1000 mm of precipitation can fall. Similarly, the transition from steppe to desert is marked by reductions in rainfall from 500 to less than 100 mm.

Most of the caves outlined in this section developed in limestone and formed during the late Tertiary/Quaternary period under phreatic conditions and commonly as enlargements along joints. Vadose expansion is more recent, resulting in a vaulted or domed morphology with vertical chimneys that extend to the surface above several of the caves. Many of these same caves have depressions or sinkholes that are situated beneath the chimneys. In any case, modern-day karstic activity is generally negligible, and the remains of formerly more extensive and well-developed dripstone features signify karstic conditions that were much different from what occurs today.

CAVE DEPOSITS AND PROCESSES

Depositional and post-depositional processes operating in karstic caves are relatively well known but less so in prehistoric caves; however, recent research, particularly in Middle Eastern prehistoric caves, has begun to reveal a high degree of marked complexity. Material of geological origin may accumulate in the interior in a number of ways, including (1) gravity-derived rockfall from the walls and roof of the cave; (2) aqueous processes associated with fluvial and phreatic deposition, or runoff; (3) colluviation of soils derived from the surfaces above and outside the cave; and (4) Aeolian deposition of sand and silt.

Middle Eastern caves contain distinct biogenic contributions, such as bird and bat guano. In addition, coprolites and organic matter remains produced by carnivores—particularly hyenas—are quite common. In many caves, the presence of millimeter-sized rounded stones indicates distinct gastrolith input by pigeons. Plant remains, such as grass and wood, can be washed and blown into the cave, or alternatively brought into the cave by past human occupants for fuel, bedding, or shelter.

Anthropogenic contributions to cave sedimentation in this region are noteworthy and tend to be unnoticed in many prehistoric cave settings throughout the world. These accumulations consist of primarily bone and shell remains, as well as the buildup of ashes, organic matter, and charcoal that are associated with fireplaces and burning activities. Additional traces of fine-grained soil and sediment that were tracked into the cave by humans and other animals are subtle but important.

Once within the cave itself, deposited material is commonly modified by a number of processes that are often penecontemporaneous with deposition. Deposits are sometimes modified by wind, runoff, and dripping water or subjected to burrowing by animals or to trampling by humans and other occupants. Mineral and organic residues are sometimes moved by human or animal activity from their original location within the cave. Bone and lithic discard evidence of activities and indications of removal, dumping, and trampling of material associated with the cleaning or maintenance of hearths are not uncommon.

Prehistoric caves in this region are damp and act as sinks for water, organic matter, and guano and accordingly tend to act as chemical engines whereby numerous mineral transformations take place (Karkanas et al., 2000; Schiegl et al., 1994, and Schiegl et al., 1996). These secondary alterations commonly include the precipitation (and dissolution) of carbonates that form

FIGURE 2B Tabun Cave, Mount Carmel. This photograph is from the upper part of the cave, above the uppermost ledge in Fig. 2A. In contrast to the lower deposits, these are largely anthropogenic in nature and consist of lighter and darker bands of ash and red clay, respectively. These anthropogenic layers are overlain by massive reddish clay that contains decimeter-sized blocks of roof fall. The latter attests to an enlargement and eventual opening of the chimney that leads up to the surface of Mount Carmel. It is through this chimney that the clayey *terra rosa* soils were washed into the cave.

FIGURE 2A Tabun Cave, Mount Carmel. Shown in this photograph (taken during the 1969 season) are bedded sandy and silty sediments that were blown into the cave from the adjacent coastal plain. These contain Lower and Middle Paleolithic industries, including the Upper Acheulian (UA), Acheulo-Yabrudian (AY), and Mousterian (M). Note the strong dip of the lowermost sediments that plunge into a subsurface swallow hole. This subsidence took place during the earlier phases of deposition, as the bulk of the sediments from the middle part of the photograph upward are roughly horizontal. The uppermost part of the photograph (see Fig. 2B) is composed of interbedded clay and ashy deposits, punctuated with numerous blocks of roof fall. The dating of the deposits here is a subject of debate, but estimates using thermoluminescence (TL) dating on burned flints suggest that the deposits down to the top of the large hole on the left are close to 350,000 years old (Mercier *et al.*, 1995).

speleothems, layered travertines, or the so-called cave breccia, which represent calcite-cemented clastic sediments typically consisting of inwashed soil material. Another characteristic type of mineral alteration involves the formation of several types of phosphate minerals, along with the formation of opal and the breakdown of clays.

Paleoenvironmental and paleoclimatic changes expressed in the faunal records, as well as by depositional and post-depositional accumulations and removals, are registered in a rather coarse chronology. Speleothems in Nahal Soreq Cave and a cave near Jerusalem provided detailed records of paleoclimatic fluctuations of the past 170,000 years. These karstic caves were rarely penetrated by humans, and these oscillations follow the pattern known from other localities of the Northern Hemisphere.

THE PREHISTORIC AND HISTORICAL SEQUENCE OF CAVE OCCUPATIONS IN THE MIDDLE EAST

Prehistoric caves in the region produced the basic archaeological sequence of at least the past 400,000 years (Fig. 1); several examples are illustrated in the following text, with comments related to the themes discussed above. The oldest known cave deposits were exposed in Umm Qatafa, Tabun (Fig. 2A,B), and the Yabrud IV rockshelter, where the lowermost layers—more than 400,000 years old—contained a core-and-flake industry, a phenomenon that intersects the Acheulian sequence. The Late (or Upper) Acheulian is known from several caves. This industry is characterized by hand axes or bifaces, but these tools are not quantitatively the most dominant tool type. The Late (or Upper) Acheulian is followed by the Acheulo–Yabrudian, where the combination of bifaces and scrapers, often shaped on thick flakes, appear in Tabun, Hayonim, Qesem, Zuttiyeh, and the Yabrud I rockshelter, as well as in open-air sites in the El-Kowm basin (northeast Syria). This entity occupies the Levant between the Acheulian sequence and the

FIGURE 3 Qafzeh Cave, Lower Galilee. This cave is famous for its series of Middle Paleolithic human burials in layers XVII to XXI. The morphological analyses identified the remains as archaic modern humans and are therefore considered as coming out of Africa. These layers, dated by thermoluminescence (TL) from about 95,000 to 114,000 calendar years BP (Valladas et al., 1988), consist of fine-grained angular rock fall that has been reworked by slopewash and colluvial processes. The darker band in layer XXI results from manganese enrichment associated with a subsurface spring that was operational during that time.

FIGURE 4 Kebara Cave, Mount Carmel. Situated about 12 km south of Tabun, Kebara is considerably younger, dating to about 60,000 to 70,000 calendar years BP. This view of the southeast corner of the cave shows mostly Upper Paleolithic (UP) deposits in the walls on the right and left and Middle Paleolithic (MP) sediments on the floor of the excavation. The Upper Paleolithic deposits are comprised of finely bedded and laminated silt and sand-sized material (S) that has washed back into the cave via runoff. It commonly includes sand-sized aggregates that have been reworked from older sediments near the entrance. A large piece of limestone roof fall (LS) in the wall to the left shows that some gravity deposition also takes place. Interestingly, the other portions of this block have been subjected to diagenesis in which the original dolomite block has been transformed into a number of phosphate minerals, including francolite, crandallite, montgomeryite, and leucophosphite (Weiner et al., 1993). Punctuating these geogenic deposits are isolated hearths (H) representing some anthropogenic material within the mostly geogenic sediments. The Middle Paleolithic deposits, on the other hand, are mostly anthropogenic and consist of hearths and hearth products (charcoal, organic matter, and ash). Some of these hearths are revealed by a white area (see the base of the photograph).

Mousterian or Middle Paleolithic. Its geographic distribution indicates an origin in the northern Levant.

Cave sites with Middle Paleolithic (250,000–270,000 to 48,000–46,000 calendar years BP) remains have produced a wealth of evidence, as well as a large number of human burials and isolated human bones. The earlier deposits contain skeletal remains identified as archaic modern humans (also known as the Skhul-Qfazeh group; Fig. 3). Occurring stratigraphically above these human remains are those of southwestern Asia Neanderthals that were found in Kebara (Fig. 4), Amud, Dederiyeh, and Shanidar. Based on the evidence, both human populations demonstrated good hunting skills, the use of fire, and the ability to procure raw materials for making stone tools from a radius of 5 to 20 km around the sites. Mobility between lowland and highland areas has been easier to trace in Lebanon, south Jordan, and the Zagros because of the greater topographic relief in these areas.

Among Upper Paleolithic and Epi-Paleolithic (48,000–46,000 to 11,500 calendar years BP) cave occupations, the best known are (1) Ksar 'Akil, with an unusual 18-m sequence of Upper Paleolithic layers; (2) the few Levantine Aurignacian sites (e.g., Yabrud II, El-Wad, Kebara, and Üçagizili); and (3) those attributed to the Natufian culture that immediately preceded the earliest farming villages. The Natufians camped in caves, built rooms inside the main chambers, and buried their dead. Their use of caves was probably more intense than that of their predecessors. During the prehistoric periods, caves were often used for camping by entire groups and rarely used as stations for performing specific tasks. One of the best examples is the use of caves at higher altitudes for short-term camps by hunters.

Since the Neolithic period (11,500 to 7500 calendar years BP) the use of caves changed dramatically. Several caves served as camps for special artisans, but in more than one case the caves became sacred and were utilized, as in Nahal Hemar and Nahal Qana caves, for storage of paraphernalia or other sacred objects. Larger caves could have served as locations for ceremonies, as indicated by a few caves in Turkey.

During the Chalcolithic and Bronze ages (7500 to 3200 calendar years BP), caves were employed for various purposes. Certain karstic caves, such as Peqi'in, served as burial grounds, while others functioned as storage facilities, animal pens, and even refugia. During the Bronze Age, caves continued to be used in similar ways, and a unique example is the warrior burial in a cave near the Jordan Valley. Finally, a common use of most caves during the last millennium BC and the first two millennia AD was by shepherds who often spent the late fall and winter months in these protected shelters.

See Also the Following Articles

Ancient Cavers in Eastern North America
Maya Caves

Bibliography

Ayalon, A., Bar-Matthews, M., & Kaufman, A. (2002). Climatic conditions during marine oxygen isotope stage in the eastern Mediterranean region from the isotopic composition of speleothems of Soreq Cave, Israel. *Geology, 30*(4), 303–306.

Bar-Matthews, M., Ayalon, A., & Kaufman, A. (1997). Late Quaternary paleoclimate in the eastern Mediterranean region from stable isotope analysis of speleothems at Soreq Cave, Israel. *Quaternary Research, 47*, 155–168.

Bar-Matthews, M., Ayalon, A., Kaufman, A., & Wasserburg, G. J. (1999). The eastern Mediterranean paleoclimate as a reflection of regional events: Soreq Cave, Israel. *Earth and Planetary Science Letters, 166*, 85–95.

Frumkin, A., Ford, D. C., & Schwarcz, H. P. (1999). Continental oxygen isotopic record of the last 170,000 years in Jerusalem. *Quaternary Research, 51*, 317–327.

Karkanas, P., Bar-Yosef, O., Goldberg, P., & Weiner, S. (2000). Diagenesis in prehistoric caves: The use of minerals that form in situ to assess the completeness of the archaeological record. *Journal of Archaeological Science, 27*, 915–929.

Karkanas, P., Goldberg, P., 2010. Phosphatic Features. In: G. Stoops, V. Marcelino, F. Mees (Eds.), *Interpretation of Micromorphological Features of Soils and Regoliths*. Elesevier, Amsterdam, pp. 522–542.

Mercier, N., Valladas., H., Valladas, G., & Reyss, J.-L. (1995). TL dates of burnt flints from Jelinek's excavations at Tabun and their implications. *Journal of Archaeological Science, 22*, 495–509.

Schiegl, S., Lev-Yadun, S., Bar-Yosef, O., Goresy, A. E., & Weiner, S. (1994). Siliceous aggregates from prehistoric wood ash: A major component of sediments in Kebara and Hayonim caves (Israel). *Israel Journal of Earth Science, 43*, 267–278.

Schiegl, S., Goldberg, P., Bar-Yosef, O., & Weiner, S. (1996). Ash deposits in Hayonim and Kebara Caves, Israel: Macroscopic, microscopic and mineralogical observations, and their archaeological implications. *Journal of Archaeological Science, 23*, 763–781.

Valladas, H., Reyss, J. L., Joron, J. L., Valladas, G., Bar-Yosef, O., & Vandermeersch, B. (1988). Thermoluminescence dating of Mousterian "Proto-Cro-Magnon" remains from Israel and the origin of modern man. *Nature, 331*(6157), 614–616.

Weiner, S., Goldberg, P., & Bar-Yosef, O. (1993). Bone preservation in Kebara Cave, Israel, using on-site Fourier transform infrared spectrometry. *Journal of Archaeological Science, 20*, 613–627.

CAVE ECOSYSTEMS

Kevin S. Simon
The University of Auckland, New Zealand

DEFINITION AND BOUNDARIES

Caves are unique among ecosystem types in their physical structure and biological function. Embedded in rock and lacking light, the geological setting defines the bounds of cave ecosystems and dictates how energy and matter move through them. As humans, we typically think of the bounds of cave ecosystems as the walls of the passages that we enter and explore, ranging from dimly lit cave entrances to deep, dark passages. As it turns out, the cave environment that humans experience is really just one habitat within a broader ecosystem; we simply see only the portion of the ecosystem we fit into. Beyond those passages are voids, some quite small and some filled with water, which are interconnected and through which move organisms, energy, and matter. If the cave ecosystem extends beyond the cave walls, where then do we draw the line that bounds the ecosystem? The first real attempt at defining the bounds of a cave ecosystem was made by French biologist, Raymond Rouch. In his research at the Baget Basin (summarized in Rouch, 1986) he realized that when he collected animals in caves for study, he was really only sampling a small portion of a bigger system. By sampling at sinking streams and at springs that drained different portions of the aquifer, he found a much greater abundance and diversity of animals than he observed from sampling only in caves. He realized that the true boundaries of the ecosystem extended well beyond caves, to where water first entered the subsurface to where it returns to the surface at springs. Rouch argued that the true ecosystem was not really a "cave" ecosystem at all; rather it was a "karst" ecosystem, that is, the entire drainage basin and the highly interconnected and diverse habitats within it (Rouch, 1977). The edges of the ecosystem are not cave walls; rather they are the hydrologic boundaries that determine where water flows on and under the landscape. As it turns out, the same physical and chemical force, flowing water, that creates caves also defines the boundaries of the ecosystem, the various habitats within the ecosystem, and how energy and matter move through it.

PHYSICAL ENVIRONMENT AND HABITAT ZONES

Rainwater that falls on the surface of a karst landscape follows gravity, percolating through soils and into a network of small fractures just below the soil surface or sinking rapidly through relatively large openings as sinking streams (Fig. 1). That water flows through cave streams to the water table and ultimately it returns to the surface at springs. The entire land area and aquifer volume that drain water to the point of resurgence are the drainage basin that comprises the karst ecosystem. Within that basin are definable habitats that differ in physical characteristics and biological communities. This zone of contact between soils and bedrock, the epikarst, is the interface zone between the

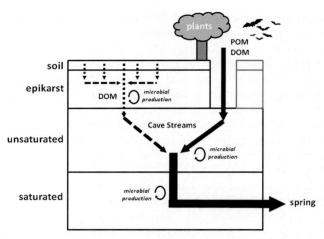

FIGURE 1 Diagram of energy flux among karst ecosystem zones. DOM and POM are dissolved and particulate organic matter, respectively.

surface and caves. It is a unique environment comprised of a heterogeneous network of small fissures that are intermittently wetted, with some spaces permanently wetted while others are sometimes or always dry. Water that percolates through the epikarst sometimes drips into larger cave passages and into small pools and cave streams. These streams coalesce with other streams that are fed by larger conduits which are directly connected to the surface through relatively large openings, visible as sinking streams at cave entrances. While epikarst and conduits are both interfaces, or ecotones, with the surface, they differ dramatically in physical structure, biological communities, and capacity to deliver energy and matter to the subsurface. The small, heterogeneous fractures in the epikarst result in relatively slow water flow rates and permit only very small particles or dissolved material to pass from surface to subsurface. Further, the epikarst represents a large area of interface within the basin. In contrast, the much larger conduit openings, such as cave entrances, can support much higher and more variable flow and pass larger particles and organisms from the surface to the subsurface. While the conduits are larger, they are generally restricted to a few specific focal points within the basin.

Water derived from epikarst and conduit flow paths pass through streams which are characterized by the same types of physical structures (riffles and pools) seen in surface streams. Depending on the nature of connection to the surface, cave streams can differ dramatically in temperature, flow, and chemistry, with conduit-fed streams generally much more variable in all of these aspects. These streams and passages below the epikarst, but above the water table, comprise the unsaturated (or vadose) zone of the ecosystem. The unsaturated zone contains both aquatic and terrestrial habitats, and it is the component of ecosystems that most humans are familiar with as caves. Eventually, cave passages and fractures in the rock intersect the water table, they become completely filled, and the system is comprised exclusively of aquatic habitat. This habitat is generally only accessible via diving or through wells driven below the water table. By nature of the amount of time the water has spent underground, this zone of the aquifer is more stable in terms of temperature and chemistry than the epikarst or unsaturated zone. Ultimately, water from the saturated zone resurges at springs, which are the downstream interface with the adjacent surface ecosystem.

The terrestrial component of cave ecosystems follows the same fracture system as water, and indeed represents conduits that are currently or historically channels for water flow. It is, however, restricted to the unsaturated portion of the aquifer and has only been studied in areas directly accessible by humans. Howarth (1983) categorized the terrestrial environment of caves into five zones: (1) entrance, (2) twilight, (3) transition, (4) deep, and (5) stagnant air. These zones are in fact a gradient from a surface in which light declines and temperature and humidity become less variable. This framework applies only to the larger conduit openings, but there is likely a terrestrial habitat within the epikarst, but it is virtually impossible to access and characterize. The terrestrial and aquatic habitats within karst ecosystems should not be viewed as separate and they are likely extremely connected, as is the case for surface ecosystems. For example, organic matter carried into caves by streams is deposited into the terrestrial habitat by dynamic overflooding and drying cycles. Animals and organic matter from terrestrial zones can in turn be deposited into streams. However, the nature of this connection has not been well studied.

ENERGY FLUX AND LIMITATION

Darkness is probably the most defining feature of cave ecosystems, and it has perhaps had the greatest influence on organisms in caves. The effect of no light is manifest through energy flux in food webs. In nearly all ecosystems, light-driven photosynthetic production by plants is the key source of energy that fuels all other living organisms in the ecosystem. How can food webs persist in caves where there is no light? There are essentially two options. The first is to fix energy derived from something other than light into organic molecules. When the chemical conditions are right, microbes can use an array of metabolic strategies that extract energy from chemical sources to create microbial biomass

(Engel, 2007). This chemical-driven microbial production, chemoautotrophy, can support diverse and fascinating food webs in caves (Sarbu et al., 1996). It turns out though, that most caves that humans can access lack the unique chemistry needed to support substantial chemoautotrophic production. There are, however, some large aquifers such as the Edward Aquifer in Texas, United States, that appear to have very large, deep systems driven by chemoautotrophy.

In the absence of light and chemoautotrophy, the only other option for sustaining a food web is dead organic matter from plants, or detritus, which is imported from the surface. This detritus ranges from dissolved compounds to large particles such as leaves and wood (Simon et al., 2007). That detritus is delivered to the subsurface through two key vectors: flowing water and organisms. As water flows from the surface, it carries with it detritus that is either dissolved or entrained in the water. In the case of water percolating through soils and into the epikarst, only dissolved organic matter and very fine particles that pass through soils and fractures in the rock are transmitted. That material can be used by microbes which are in turn consumed by small animals, notably copepods, which can achieve very diverse communities in the epikarst. In the case of larger conduits connected to the surface, water can carry not only dissolved detritus, but also much larger particles ranging from leaves to logs. That material is carried through caves, and deposited in streams or on stream banks, especially during floods. Thus, the nature of the physical structure of the aquifer dictates what type and how much energy is made available to various habitats in caves. Animals and plants can also deliver organic matter to caves. Well-known examples include bats and crickets, which forage on the surface and return to the subsurface where their feces, eggs, and bodies are consumed by microbes and animals in caves. Indeed, entire specialized communities (guanobionts) rely on bat and bird feces. This is well demonstrated for terrestrial habitats, but less clear in aquatic habitats in caves. In other cases, those animals that exit caves and forage on the surface are themselves energy within caves. For example, cave cricket eggs are preyed on by specialized cave beetles. In a few cases, plant roots can penetrate down into cave passages where material from those roots can be consumed by cave animals.

If one French scientist, Rouch, was the pioneer of defining the ecosystem, it was another French ecologist, Janine Gibert, who pioneered the study of how energy and matter move through karst ecosystems. Gibert did this by characterizing the amount and type of organic matter moving from overlying soils, out of the epikarst and through the vadose zone in cave streams, and ultimately out of the saturated zone via a spring at the Dorvan-Cleyzieu basin (Gibert, 1986). Among Gibert's most important findings were that organic matter dissolved in water was a much larger flux than was the flux of organic particles such as pieces of leaves and wood and the bodies of surface animals. Those fluxes were extremely variable over time and tightly linked to patterns of high and low flow as driven by rainfall. Finally, she noted that microbes were likely to be key players in mediating the transfer of energy from detritus to animals in karst.

More recent research at Organ Cave in the United States has confirmed that bacterial productivity fuels cave stream food webs (Simon et al., 2003). In Organ Cave, some streams are fed only by epikarst water while others receive water from sinking streams and shafts. Leaves placed in the streams did not move far before being retained and eventually decomposed. Leaf decomposition was rapid only in those streams that contained a particular stygophilic amphipod which shreds and consumes leaves readily. In the other streams with only stygobitic animals, leaves were not readily consumed and were only slowly degraded by microbes in streams where particulate organic matter was scarce. A series of experiments subsequently showed that carbon from bacterial production was used by all animals as a food source in both types of streams. Interestingly, it appears that dissolved organic matter, especially that derived from soils, rather than large particles such as leaves is the main energy source for food webs even when leaves and wood are available. While some cave animals can use the larger particles directly as food, many seem to rely almost exclusively on microbes that use dissolved organic compounds. The relationship between microbes and animals is not unidirectional, as the animals also strongly influence the productivity of the microbes that they consume (Cooney and Simon, 2009). These issues of patterns of energy transport and use are important as energy scarcity has been considered the key driving force in the evolution of cave organisms. For example, there is a great deal of indirect evidence such as reduced metabolic rate, larger but fewer eggs, and increased longevity of subterranean animals that is consistent with the hypothesis of energy limitation of karst ecosystems.

NUTRIENTS

Availability of detritus, or carbon, may not be the only factor that limits energy availability in karst ecosystems. Karst ecosystems based on chemoautotrophy are clearly limited by the availability of inorganic elements (especially forms of iron, nitrogen, and sulfur;

see Engel, 2007) other than carbon. The physical factors that dictate the dynamics of these elements have strong influence on the magnitude and location of the energy base of food webs in those systems, with interface zones such as the water–air interfaces in Movile Cave (Sarbu et al., 1996) acting as hotspots for productivity in karst ecosystems. The importance of inorganic nutrients in limiting productivity in other karst ecosystems is possible as microbes that use detritus can be limited by nutrients such as nitrogen and phosphorus. This issue is almost completely unstudied in karst ecosystems and the few data available are contradictory.

CONNECTIVITY AND DISTURBANCE IN KARST ECOSYSTEMS

Caves present an interesting dichotomy in terms of stability and disturbance that is determined in large part by the nature of hydrologic connectivity with the surface. On one hand, caves seem like rather stable, benign places. They are always dark and generally have stable temperature and humidity. On the other hand, caves can be spectacularly harsh places to live. Food is scarce and, during floods, some cave streams become raging torrents. The reality is that, depending on place and time, caves can vary dramatically in stability and harshness. This is dictated in large part by the nature of connectivity to the surface. Streams and pools fed by epikarst water are much more stable in flow, temperature, and chemistry than are streams connected to the surface by large openings. The geological framework of the aquifer and the climate the system is located in also strongly influence the nature of disturbance and stability in caves. High flow during storms, in particular, is a key disturbance in karst systems as it refills voids, inundates terrestrial habitat, increases delivery of energy and matter from the surface and scours microbial films. Ironically, while it was recognized that floods are likely to be quite important for cave animals (Hawes, 1939), few studies have systematically examined how floods influence the animal and microbial communities in caves.

Disturbance by humans is now a component of most karst ecosystems. This disturbance can take the form of pollution by chemicals and sediments that have had detrimental effects on karst ecosystems. Of particular importance is the role humans play in changing energy availability in caves via pollution. In particular, humans almost invariably increase the input of organic matter and nutrients to caves. This comes from a diverse array of sources ranging from septic system waste to bits of cotton clothing left by cavers. Often the organic matter humans introduce is readily used by microbes and it can indirectly change the physical environment by reducing oxygen availability and smothering habitat. One common outcome of this perturbation is a shift in abundance of cave animals such that a few tolerant taxa proliferate and become abundant while other more sensitive taxa are reduced in abundance or extirpated. A second common outcome is the increased abundance of surface animals in caves at the expense of cave-adapted taxa. It appears that the increased energy availability lessens the competitive advantage that cave animals have in low productivity environments. In some cases humans can reduce energy input to caves by severing connectivity to the surface. Examples include human-induced loss of bats and alteration of surface vegetation, soils, and hydrology in ways that cut off organic matter input to the subsurface. Finally, human activity can change the magnitude and pattern of water flow, the lifeblood of most karst ecosystems, by altering surface vegetation, soils, and streams or through extraction of water from within aquifers.

See Also the Following Articles

Food Sources
Chemoautotrophy

Bibliography

Cooney, T., & Simon, K. S. (2009). Influence of dissolved organic matter and invertebrates on the function of microbial films in groundwater. *Microbial Ecology, 58*(3), 599–610.

Engel, A. S. (2007). Observations on the biodiversity of sulfidic karst habitats. *Journal of Cave and Karst Studies, 69*(1), 187–206.

Gibert, J. (1986). Ecologie d'un systeme karstique jurassien. Hydrogéologie, dérive animale, transits de matières, dynamique de la population de Niphargus (Crustacé Amphipode). *Memoires de Biospeologie, 13*(40), 1–379 (in French).

Hawes, R. S. (1939). The flood factor in the ecology of caves. *Journal of Animal Ecology, 8*, 1–5.

Howarth, F. G. (1983). Ecology of cave arthropods. *Annual Review of Entomology, 28*, 365–389.

Rouch, R. (1977). Considerations sur l'ecosystem karstique. *Compte Rendus Academie des Sciences de Paris Serie D, 284*, 1101–1103 (in French).

Rouch, R. (1986). Sur l'ecologie des eaux souterraines dans la karst. *Stygologia, 2*(4), 352–398 (in French).

Sarbu, S. M., Kane, T. C., & Kinkel, B. F. (1996). A chemoautotrophically based cave ecosystem. *Science, 272*(5270), 1953–1955.

Simon, K. S., Benfield, E. F., & Macko, S. A. (2003). Food web structure and the role of epilithic films in cave streams. *Ecology, 84*(9), 2395–2406.

Simon, K. S., Pipan, T., & Culver, D. C. (2007). A conceptual model of the flow and distribution of organic carbon in caves. *Journal of Cave and Karst Studies, 69*, 279–284.

CAVE, DEFINITION OF

William B. White and David C. Culver[†]*

The Pennsylvania State University, [†]American University

DEFINITIONS AND POINTS OF VIEW

Humankind has been involved with caves for millennia. Caves were places of shelter for early humans in many parts of the world. They have served as places of worship for many societies in many times and places. Caves have been used as storehouses, as munitions factories, and as resting places for the dead. Caves play a prominent role in the myths and legends of many cultures throughout recorded history. Caves are secret places. Small children make "caves" by draping blankets over furniture. In contemporary society, caves frequently appear in movies and in cartoons. Show caves continue to draw thousands of visitors each year. Caves are also of great interest to scientists and explorers. Because caves are voids in rock, they are considered geological features, and indeed many textbooks (*e.g.*, Ford and Williams, 1989) firmly defend this point of view. However, caves are more than their geology because of their interaction with people and with organisms. One textbook (White, 1988) recognizes their human appeal by defining caves as "a natural opening in the Earth, large enough to admit a human being, and which some human beings choose to call a cave."

CAVES AS PLACES FOR EXPLORATION

Caves are enticing, awaking an interest in many to see "where it goes." Although the first cave explorers are lost in the mists of history, cave exploration as a specialized human activity dates from the middle of the nineteenth century (Shaw, 1992). Earlier authors include discussions of caves in general travelogues and descriptions of regions. The most famous of these is Valvasor's 1656 book, *The Glories of the Duchy of Carniola* (an area that is part of present-day Slovenia). In North America, Horace C. Hovey, Luella Owen, and a few others wrote popular accounts of their expeditions. The first modern speleologist is said to be Adolf Schmidl, who explored many Slovenian caves in the mid-nineteenth century. Somewhat later came Edouard A. Martel in France who is considered the father of cave exploration. Unlike his American counterparts, Martel organized serious caving expeditions into the large caves and deep pits of the Pyrenees, the Alps, and many other places in Europe, an activity that also required inventing the technology of cave exploration as he went along.

Cave exploration by organized caving societies was well under way in Europe in the early twentieth century. Organized exploration came later in the United States when several caving groups formed the National Speleological Society in 1941. During the past 50 years, cave exploration has blossomed into a recreational activity for thousands of individuals throughout the world and is organized into hundreds of caving clubs, national societies, and specialized scientific organizations. Most cavers take their explorations seriously and spend substantial time in preparing maps and writing detailed reports. Exploration and surveying of caves are among the few remaining activities where useful contributions to knowledge can be made by nonprofessionals.

Much of the allure of caves resides in their remoteness and wilderness character even beneath urban sprawl. This is, in part, because the underground landscape with its total darkness and unusual shapes of rock and mineral deposits is so alien compared with the familiar surface landscape. Caves are remote in the sense of the time and effort required to explore them. The farthest reaches of a large cave system may be only a few kilometers from the entrance as the crow flies and may be no more than 10 or 15 km as the caver crawls. And, of course, the outside world is only a few tens or hundreds of meters away, vertically, through solid rock. However, reaching the farthest corners of a large cave system, doing a bit of exploring and surveying, and returning to the entrance may require 24 to 36 hours. Or, it may require several days and an underground camp. In the same time one could have traveled comfortably across the continent on a jet plane, attended a conference on the opposite coast, returned home, and been clean and well rested to boot. The remoteness of the cave arises not from distance but from the time needed to traverse it, from the obstacles that must be overcome, and from the sense of the strange and the bizarre in the cave landscape, a landscape duplicated nowhere on the Earth's surface. Caving, from this point of view, is truly an esthetic or wilderness experience. It requires solitude, a leisurely pace, and a sense of absorption into the environment. The emergence of the caver into the misty air under the bright stars of a summer's night is indeed the return from another world.

CAVES AS GEOLOGICAL REPOSITORIES AND CAVES AS PARTS OF GROUNDWATER FLOW PATHS

The cave environment may be described as dark, wet, neutral to mildly alkaline, and oxidizing. The variations of these parameters are much less than the variations of similar parameters on the Earth's surface. Chemical reactions under these very precisely controlled conditions permit the growth of unusual minerals and the growth of crystals of exceptional size. Mineral deposits take on the form of stalactites, stalagmites, flowstone, and other forms known collectively as *speleothems*. Because these deposits are nourished by water seeping down from the surface, changes in the climate and vegetation on the surface leave their signatures in the growth bands of the speleothems. The deposits of caves have become an important source of paleoclimatic information.

Because caves are void spaces, they tend to fill up with various materials collapsed or washed in from the surface. Debris piles near cave entrances preserve archaeological and paleontological deposits. Streamborne deposits of clays, silts, sands, and gravels record past flood conditions. Because the filling of caves takes place slowly over very long periods of time and because still older deposits are preserved in the high abandoned levels of cave systems, cave deposits are a history book for the ice ages.

Caves, of course, do not form in isolation. Every cave is related to a drainage system that now or at some time in the past carried water on its way from an inlet point to an outlet at a big spring. The underground pathways in limestone through which water moves from sinkholes and sinking streams to the outlets at springs are known as *conduits*. Conduit systems are very long but most of them are invisible. Conduits may be water-filled or may have no humanly accessible entrances. Another definition of caves is that they are those conduit fragments that are accessible to human exploration. Careful inspection of the size, shape, and patterns of caves as well as details of solutional sculpturing on the cave walls provides much information on the present or past behavior of the groundwater flow system.

Much of the value of caves to the geological sciences is that they preserve records over long periods of time. Streams erode and deepen surface valleys and in so doing destroy the stream channels, floodplains, and valley shapes that were there before. In contrast, caves deepen by forming new passages at lower levels, leaving the old levels high, dry, and well preserved.

CAVES AS HABITAT

We can also look for a definition of caves by considering what animals consider caves to be dwelling places. As habitats, caves have several distinct environmental properties. In temperate zones in the summer, they tend to be cooler; conversely, they tend to be warmer in the winter. This characteristic of temperature buffering is shared not only by what we would call true caves by any definition, but also by rock overhangs and shelters as well. Some of the fauna of true cave entrances such as phoebes and swallows are found in rock overhangs as well. If we define caves as an area of temperature buffering, then not only will rock overhangs be included but so will many manmade structures such as cellars and the underside of bridges.

A more appropriate environmental parameter to consider is darkness. Species isolated in the darkness of caves evolve a characteristic morphology, including loss of eyes and pigment. The presence of eyeless, depigmented species in a habitat would be one way to define a cave from an organism's point of view. What would be the characteristics of such a definition? First, the cavity would have to be large enough to have a zone of darkness. This would eliminate many short cavities. That is, the length of the habitat must be great enough relative to the diameter of the opening so that sunlight does not penetrate to the far reaches of the habitat. In practical terms, this length ranges from a few meters to hundreds of meters. This definition would exclude many open-air pits, which are formed the same way caves are—by dissolution of carbonate rock through the action of carbonic and sulfuric acid.

The second characteristic of such a habitat would be that it had been around long enough for animals either to evolve *in situ* or for animals to migrate into the habitat. It seems likely that most caves and karst areas are old enough to have an eyeless, depigmented fauna. Some of the youngest caves known are on carbonate islands such as San Salvador Island in the Bahamas. These caves, less than 125,000 years old, have an eyeless, depigmented cave fauna. In glaciated regions, the caves may be quite old but have only been ice free for perhaps 10,000 years. In this case, there are few if any terrestrial species that are either depigmented or eyeless. There are typically eyeless, depigmented aquatic species. These may have survived underneath the ice in unfrozen freshwater or have colonized from unglaciated areas. In any case, 10,000 years is almost certainly not long enough for species to evolve eyelessness *in situ*, a process that probably takes between 100,000 and 1,000,000 years (Culver *et al.*, 1995). Caves in evaporites (gypsum) and lava form much more quickly (and disappear more quickly). There are few reports of eyeless, depigmented species in gypsum caves, but lava tubes often have a rich fauna. The reasons for this are complex, but one of the factors is that the lava fields themselves are considerably older than a lava tube. Species found in lava tubes only a few thousand years old almost certainly migrated there from other lava tubes and cavities in the lava.

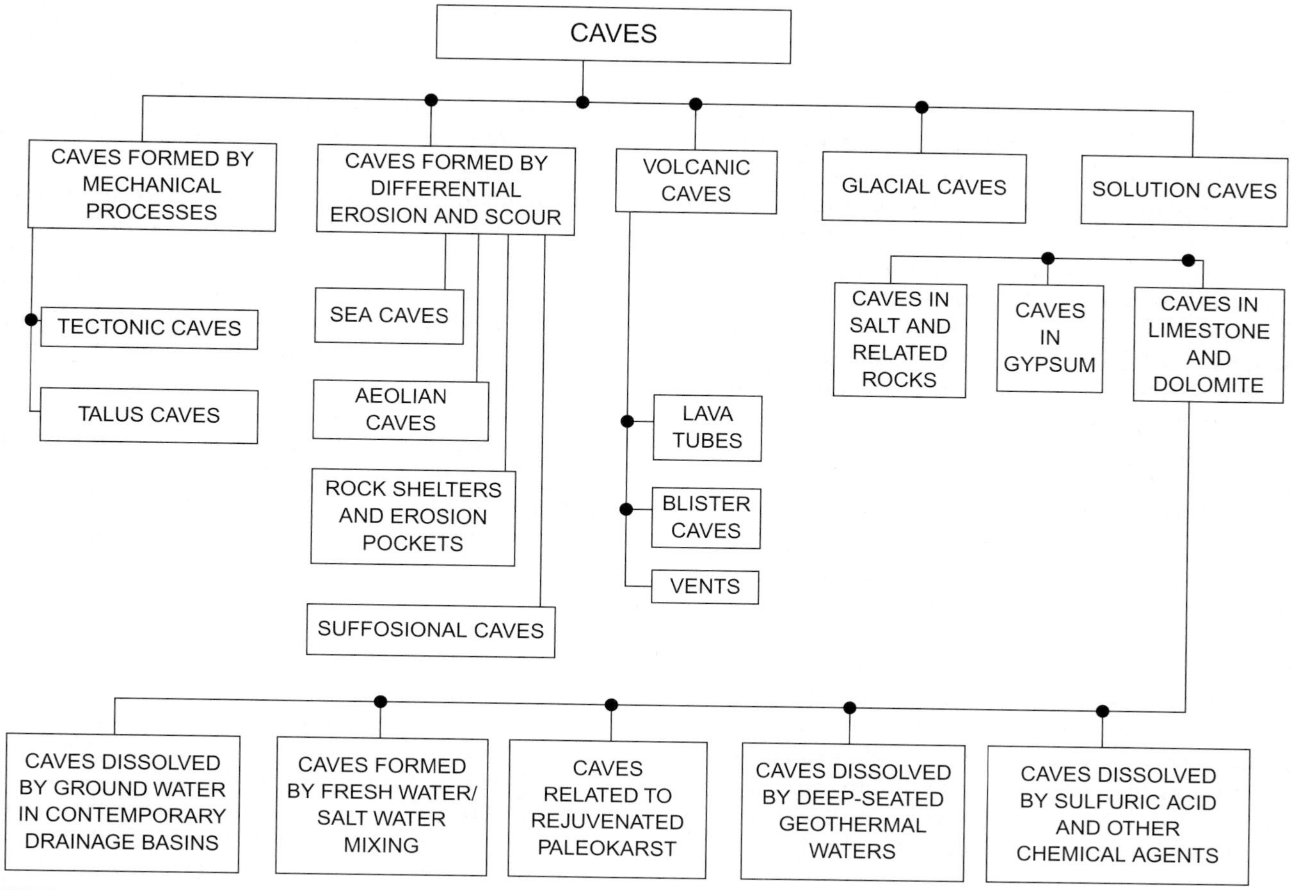

FIGURE 1 Types of caves.

A third characteristic of such a habitat is that, if darkness combined with the presence of eyeless species is how it is delineated, then it will include a wide variety of aquatic habitats in darkness that have no connection with our intuitive idea of a cave. These include the underflow of rivers, marine and freshwater beaches, and any subsurface water. Many of these habitats, collectively termed *interstitial* or alternatively *permeable and small cavities* (as opposed to permeable large cavities of caves), have a rich subterranean fauna of eyeless, depigmented species. At least in principle, we can recognize the differences between species in interstitial habitats and species in caves. While they share a lack of pigment and eyes, interstitial species tend to have shorter appendages, smaller body size, and a more worm-like appearance than do cave species (Coineau, 2000). In practice, it is difficult to distinguish species from many taxonomic groups. For example, snails occur in both habitats but there is little to distinguish the two groups morphologically. In areas with carbonate rocks, we can also distinguish the two by the size of the cavity. The simple, nonscientific definition of caves as natural cavities in a rock that can be entered by people does not help in this case. Cave animals can obviously thrive in cavities that are too small to be entered by people. Cavity diameters exist in a continuum, with a breakpoint between laminar and turbulent flow cavities of approximately 1 cm. The transition between laminar and turbulent flow is also an important biological transition, and one that in a general sense separates interstitial and cave habitats.

Thus, we can define a cave from a biological point of view as a cavity, at least part of which is in constant darkness, with turbulent water flow and with eyeless, depigmented species present.

TYPES OF CAVES

When we define caves in terms of human access rather than in terms of geologic setting, caves can be found in many different geologic settings and have been formed by many kinds of processes (Fig. 1). Caves can be formed by purely mechanical processes. Bulk masses of rock can be fractured and shifted by tectonic processes such as faults. Likewise, rock masses can break along fractures and then pull apart by the rock masses sliding under the influence of gravity. In either case, void spaces formed between the rock masses are called *tectonic caves*. Tectonic caves tend to form in hard, massive, brittle rocks such as sandstones. They tend to be small, typically a few tens of meters, although some reach lengths of hundreds of meters.

Boulder piles, if the boulders are sufficiently large to allow humans to explore the pore spaces between the boulders, can form caves. These openings are called *talus caves*. Because talus caves are simply the interstices between a pile of boulders that may or may not be completely roofed over, talus caves really do not have a definable length. Some talus caves in the Adirondack Mountains have been claimed to be several kilometers in length, although this requires a fairly generous definition of a cave. Rocks may be differentially eroded to produce cave-sized openings. The categories of erosional caves are defined by the erosion process.

Sea caves are formed by wave action on sea cliffs. Fractures in the rock produce zones of weakness that focus the attack of the waves. Sea caves are formed in many kinds of resistant rocks in many parts of the world. Fingal's Cave on the Isle of Staffa in the Scottish Hebrides (made famous by Mendelssohn in his *Hebrides Overture*) is in columnar basalt. Sea caves form in granite on the coast of Maine. Many sea caves are found on Santa Cruz and Anacapa Islands off the coast of Southern California (Bunnell, 1988, 1993). Sea caves extend for distances from a few meters to a few hundred meters into the sea cliff. Frequently, the caves consist of an inner chamber that is much larger than the connection passage to the open sea. Access to a sea cave may be possible at low tide, while the cave may be flooded at high tide. Some sea caves have blow holes in the back that spurt water when tides or storm surges rush into the cave.

Aeolian caves are formed by wind action. These are common in the arid regions of the American Southwest. Wind-blown sand scours sandstone cliffs and sculpts out chambers a few meters to a few tens of meters in diameter. A typical aeolian cave is a bowl-shaped chamber carved in solid rock. Ceiling heights vary from 1 to 2 meters. Often the entrance is a smoothly carved hole in the cliff much smaller than the chamber inside.

Rock shelters are borderline caves. In places where a resistant bed of rock, typically sandstone, overlies weaker beds of rock, typically shales, the weaker rock can be eroded away, leaving the resistant rock beds to form a roof. Rock shelters are usually small, a few meters to a few tens of meters in depth, although they can be tens to hundreds of meters wide. As caves, rock shelters usually do not extend to total darkness; however, such shelters were habitats for early humans. They are often rich repositories of archaeological material and are sometimes referred to as caves by archaeologists.

Fine-grained, poorly consolidated sediment can be swept away by stormwaters. In such locations as Badlands National Monument in western South Dakota, sediments flushed by stormwaters produce small caves in the loosely consolidated silts and clays. These are known as *suffosional caves*.

Of great interest to cave explorers are a variety of caves in volcanic rocks. Lava conduits (also known as tubes, tunnels, or pyroducts) form on the sides of volcanoes within streams of lava by a process called *inflation* or by the crusting over of channels. Internal erosion by the flowing lava often enlarges these conduits to result in open passages with diameters of meters to tens of meters and lengths up to many kilometers. The processes that form these primary caves are entirely different from the processes that form secondary solution caves in limestone and other rock but they are caves in the same sense that other natural openings of human size are caves. Lava caves are also habitat for organisms. Volcanic areas that have discharged fluid basaltic lavas usually have systems of lava conduits including volcanic islands such as Hawaii, Galapagos, the Canaries and Azores, Jeju Island (South Korea), Iceland, and intracontinental, hot-spot derived lava plateaus such as southern Oregon, Northern California, Kenya, Syria, Jordan, and Saudi Arabia, as well as southern Australia and Undara/Queensland. Lava Beds National Monument in Northern California provides easily accessed examples (Waters *et al.*, 1990).

When glaciers melt, the meltwater drains down into the glacier through fissures. This water moves along the base of the glacier and, because it is slightly warmer than the freezing point of water, it gradually carves out long tunnels that open at the front of the glacier, forming the rivers that drain the glaciers. Glacier caves are ice tunnels with floors of rock and walls and ceilings of ice. When the surface of the glacier is below freezing, the tunnels drain and become open to exploration. When the glacier is melting, the tunnels are often filled with water.

Most important are the solution caves. These form by chemical dissolution of the bedrock by circulating groundwater. They come in a great variety depending on the type of rock and the source and chemistry of the water that does the dissolving. Most solution caves are formed in limestone, a rock consisting largely of calcium carbonate, or dolomite, a rock consisting largely of calcium magnesium carbonate. These would be called limestone caves or dolomite caves. In arid regions, the more soluble rock, gypsum and calcium sulfate, is exposed at the land surface, and one finds gypsum caves, for example, in West Texas, western Oklahoma, and New Mexico, where gypsum rock occurs at the land surface. Caves easily form in salt and a few other highly soluble materials, but such caves are rare because salt does not survive at the land surface except in a few extremely arid regions.

Figure 1 shows the variety of sources for the water responsible for the development of caves. Most of the caves are dissolved by the movement of groundwater in contemporary drainage basins. In coastal regions, the mixing of fresh groundwater with saltwater produces an aggressive solution that can dissolve out caves. Some caves (for example, the large caves of the Black Hills of South Dakota) are formed from hot water rising up from deep within the rock. Carlsbad Caverns in New Mexico and other caves of the Guadalupe Mountains have been formed by sulfuric acid derived from the oxidation of hydrogen sulfide migrating upward from the oil fields to the east.

In summary, caves form in a great variety of rocks by a great variety of geological and chemical processes. Each has its importance to geology. However, the common theme that binds this diverse collection of cavities together is their interest to human explorers and their use as habitat by cave-adapted organisms.

Bibliography

Bunnell, D. E. (1988). *Sea caves of Santa Cruz Island*. Santa Barbara, CA: McNally and Loftin.

Bunnell, D. E. (1993). *Sea caves of Anacapa Island*. Santa Barbara, CA: McNally and Loftin.

Coineau, N. (2000). Adaptations to interstitial groundwater life. In H. Wilkens, D. C. Culver, & W. F. Humphreys (Eds.), *Subterranean ecosystems* (pp. 189–210). Amsterdam: Elsevier.

Culver, D. C., Kane, T. C., & Fong, D. W. (1995). *Adaptation and natural selection in caves*. Cambridge, MA: Harvard University Press.

Ford, D. C., & Williams, P. W. (1989). *Karst geomorphology and hydrology*. London: Unwin Hyman.

Shaw, T. R. (1992). *History of cave science*. Sydney, Australia: Sydney Speleological Society.

Waters, A. C., Donnelly-Nolan, J. M., & Rogers, B. W. (1990). Selected caves and lava-tube systems in and near Lava Beds National Monument, California. *U.S. Geological Survey Bulletin*, 1673.

White, W. B. (1988). *Geomorphology and hydrology of karst terrains*. New York: Oxford University Press.

CAVEFISH OF CHINA

Li Ma and Ya-hui Zhao

*Department of Biology, University of Maryland, U.S.A.
Key Laboratory of the Zoological Systematics and Evolution, Institute of Zoology, Chinese Academy of Sciences, Beijing, China*

THE DISTRIBUTION AND DIVERSITY OF CHINESE CAVEFISH

Cavefishes, or hypogean fishes, are a distinctive group of fishes. Their life history binds them to be located in caves and other subterranean waters. In 1854, Schiner divided all creatures that live in subterranean and underground waters into three types: troglobites, troglophiles, and trogloxenes. Troglobites have typical adaptive characteristics allowing them to complete their entire life cycles in caves; troglophiles lack special adaptations to caves but nonetheless rely heavily on subterranean waters during their life cycles; and trogloxenes enter the cave accidentally and their life cycles and characteristics do not necessarily require caves.

Globally, most species of cavefish are distributed in the tropics and subtropics, with 75% of all known populations predominantly located in Southeast Asia or Central and South America, where the landscape is composed of limestone and other soluble rock types (also known as a karst landscape). The southern and western parts of China have more than 620,000 km^2 of karst (Huang et al., 2008), which provides suitable conditions for cavefish evolution. The earliest record of cavefish in Chinese history was in 1436 when a local doctor, Mao Lan, recorded cavefish in South Yunnan. This was the well-known golden-line fish, which lives mostly in Dian Lake near caves in Yunnan Province. This golden-line fish, now recognized as *Sinocyclocheilus grahami*, is a troglophilic, partially cave-dwelling fish. The earliest paper report on a troglobitic fish is also in China. In 1540, Yingjing Xie, a local governor of Guangxi, went to A'lu Cave and recorded "there was a kind of transparent fish coming out if the subterranean river rose very much" in his travel notes on A'lu Cave. This transparent (also blind) fish still lives in the same cave and was described as *Sinocyclocheilus hyalinus* (Fig. 1) in 1994 (Chen et al., 1994).

At the end of 2010, China had 95 described species of hypogean (cave and phreatic) fishes which belong to three families: Cyprinidae, Cobitidae, and Balitoridae (Romero et al., 2009; Table 1). There are 55 species of Chinese hypogean cyprinids that include 4 genera: *Onychostoma*, *Sinocrossocheilus*, *Sinocyclocheilus*, and *Typhlobarbus*. *Onychostoma macrolepis* is a cavefish that was found to "hibernate" during winter. This is the only hypogean species in China found north of the Yangtze River. The family Cobitidae is a family characterized by a wormlike or fusiform body, whose members are mostly bottom dwellers in freshwater. Two troglomorphic species of the genus *Protocobitis* have been described and both are from China. The family Balitoridae is another freshwater family of the order Cypriniformes, which is characterized by having three or more pair of barbels. Some species are scaleless. Many tend to hide underneath rocks. The Chinese hypogean fishes of this family include 7 genera: *Heminoemacheilus*, *Oreonectes*, *Paracobitis*, *Paranemacheilus*, *Schistura*, *Triplophysa*, and *Yunnanilus*.

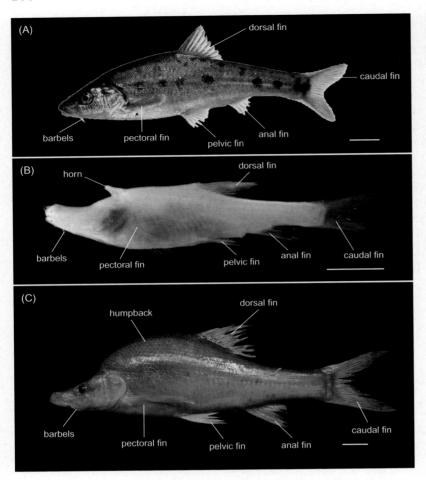

FIGURE 1 A comparison between the general morphologies of (A) adult *S. grahami*, (B) *S. hyalinus*, and (C) *S. brevibarbatus*. Scale bars = 1 cm.

The genus *Sinocyclocheilus* (Cyprinidormes, family: Cyprinidae), established by Fang Bingwen in 1936, is endemic to China and can be found in the karst cave waters and surface rivers or lakes in Yunnan-Guizhou Plateau and the surrounding region (including east of Yunnan Province, south central of Guizhou Province, and northwest of Guangxi Zhuang Autonomous Region) (Fig. 2). Almost all *Sinocyclocheilus* species live in caves for at least part of their life cycles. Yunnan has the most *Sinocyclocheilus* species, which is mainly attributed to an abundance of troglophiles, while Guangxi and Guizhou contain about half of the troglobitic cavefish in China. The genus *Sinocyclocheilus* is mainly distributed near 25°N. In river systems, most species of *Sinocyclocheilus* are found in the Qianjiang River north tributary of Xunjiang River, upstream of the largest Xijiang River tributary of the Zhujiang River system, and more upstream of the Hongshuihe Valley (including the associated underground river). *Sinocyclocheilus* may be the biggest cyprinid genus in China with more than 60 nominal species currently described. At least 50 species are valid among them, and 25 (50%) are troglobites (Zhao and Zhang, 2009).

BIOLOGY OF CHINESE CAVEFISH

Chinese cavefish vary greatly in their size. For example, the maximum length of adult cavefish is quite variable. Some species are large, like *Sinocyclocheilus altishoulder* and *S. hugeibarbus*, where the largest recorded individual is 200 mm (standard length); whereas in *Oreonectes anophthalmus*, the average adult length is around 30 mm.

Due to the absence of photosynthesis in caves, almost all Chinese cavefish are predators. Based on stomach contents, their diet seems to consist of fish scales, mollusks, and insects. The diet also can include algae (mostly diatoms), insect larvae, and other less defined organic materials.

Reproduction also varies considerably in the genus *Sinocyclocheilus*. Some species show continuous reproductive activity in all four seasons, a phenomenon that is seldom seen at other cyprinid fish. Normally most fish in the subtropical zone of the Northern Hemisphere multiply during the first half of each year. Perhaps because these areas have four distinct seasons, and the winter temperature is low, fish in these areas

TABLE 1   All the Valid Species of Chinese Hypogean Fishes

Species	Namer	Year	Distribution
Family: Cyprinidae			
Genus Onychostoma	Günther	1896	
1. *Onychostoma macrolepis*	Bleeker	1871	Beijing, Hebei, Shandong, Shanxi, Henan
Genus Sinocrossocheilus	Wu	1977	
2. *Sinocrossocheilus bamaensis*	Fang	1981	Guangxi, Guizhou
3. *Sinocrossocheilus liuchengensis*	Liang	1987	Guangxi
4. *Sinocrossocheilus megalophthalmus*	Chen, Yang, and Cui	2006	Guangxi
Genus: Sinocyclocheilus	Fang	1936	
5. *Sinocyclocheilus altishoulderus*	Li and Lan	1992	Guangxi
6. *Sinocyclocheilus anatirostris*	Lin and Luo	1986	Guangxi
7. *Sinocyclocheilus angularis*	Zheng and Wang	1990	Guizhou
8. *Sinocyclocheilus angustiporus*	Zheng and Xie	1985	Guizhou, Yunnan
9. *Sinocyclocheilus anophthalmus*	Chen and Chu	1988	Yunnan
10. *Sinocyclocheilus aquihornes*	Li and Yang	2007	Yunnan
11. *Sinocyclocheilus bicornutus*	Wang and Liao	1997	Guizhou
12. *Sinocyclocheilus brevibarbatus*	Zhao, Lan, and Zhang	2009	Guangxi
13. *Sinocyclocheilus brevis*	Lan and Chen	1992	Guangxi
14. *Sinocyclocheilus broadihornes*	Li and Mao	2007	Yunnan
15. *Sinocyclocheilus cyphotergous*	Dai	1988	Guizhou
16. *Sinocyclocheilus donglanensis*	Zhao, Watanabe, and Zhang	2006	Guangxi
17. *Sinocyclocheilus furcodorsalis*	Chen, Yang, and Lan	1997	Guangxi
18. *Sinocyclocheilus grahami*	Regan	1904	Yunnan
19. *Sinocyclocheilus guilinensis*	Ji	1982	Guangxi
20. *Sinocyclocheilus guishanensis*	Li	2003	Yunnan
21. *Sinocyclocheilus huaningensis*	Li	1998	Yunnan
22. *Sinocyclocheilus hugeibarbus*	Li and Ran	2003	Guizhou
23. *Sinocyclocheilus hyalinus*	Chen and Yang	1994	Yunnan
24. *Sinocyclocheilus jii*	Zhang and Dai	1992	Guangxi
25. *Sinocyclocheilus jiuxuensis*	Li and Lan	2003	Guangxi
26. *Sinocyclocheilus lateristritus*	Li	1992	Yunnan
27. *Sinocyclocheilus lingyunensis*	Li, Xiao, and Luo	2000	Guangxi
28. *Sinocyclocheilus longibarbatus*	Wang	1989	Guangxi
29. *Sinocyclocheilus longifinus*	Li and Chen	1994	Yunnan
30. *Sinocyclocheilus luopingensis*	Li and Tao	2003	Yunnan
31. *Sinocyclocheilus macrocephalus*	Li	1985	Yunnan
32. *Sinocyclocheilus macrolepis*	Wang	1989	Guizhou, Guangxi
33. *Sinocyclocheilus macrophthalmus*	Zhang and Zhao	2001	Guangxi

(Continued)

TABLE 1 (Continued)

Species	Namer	Year	Distribution
34. *Sinocyclocheilus macroscalus*	Li	1992	Yunnan
35. *Sinocyclocheilus maculatus*	Li	2000	Yunnan
36. *Sinocyclocheilus maitianheensis*	Li	1992	Yunnan
37. *Sinocyclocheilus malacopterus*	Chu and Cui	1985	Yunnan
38. *Sinocyclocheilus microphthalmus*	Li	1989	Guangxi
39. *Sinocyclocheilus multipunctatus*	Pellegrin	1931	Guangxi, Guizhou
40. *Sinocyclocheilus oxycephalus*	Li	1985	Yunnan
41. *Sinocyclocheilus purpureus*	Li	1985	Yunnan
42. *Sinocyclocheilus qiubeiensis*	Li	2002	Yunnan
43. *Sinocyclocheilus qujingensis*	Li, Mao, and Lu	2002	Yunnan
44. *Sinocyclocheilus rhinocerous*	Li and Tao	1994	Yunnan
45. *Sinocyclocheilus robustus*	Chen and Zhao	1988	Guizhou
46. *Sinocyclocheilus tianlinensis*	Zhou, Zhang, and He	2003	Guangxi
47. *Sinocyclocheilus tileihornes*	Mao, Lu, and Li	2003	Yunnan
48. *Sinocyclocheilus tingi*	Fang	1936	Yunnan
49. *Sinocyclocheilus wumengshanensis*	Li, Mao, and Lu,	2003	Yunnan
50. *Sinocyclocheilus xunlensis*	Lan, Zhao, and Zhang	2004	Guangxi
51. *Sinocyclocheilus yangzongensis*	Tsü and Chen	1977	Yunnan
52. *Sinocyclocheilus yaolanensis*	Zhou, Li, and Hou	2009	Guizhou
53. *Sinocyclocheilus yimenensis*	Li and Xiao	2005	Yunnan
54. *Sinocyclocheilus yishanensis*	Li and Lan	1992	Guangxi
55. *Typhlobarbus nudiventris*	Chu and Chen	1982	Yunnan
Family: Cobitidae—loaches			
Genus Protocobitis	Yang	1994	
56. *Protocobitis polylepis*	Zhu, Lu, Yang, and Zhang	2008	Guangxi
57. *Protocobitis typhlops*	Yang, Chen, and Lan	1994	Guangxi
Family: Balitoridae			
Genus Heminoemacheilus	Zhu and Cao	1987	
58. *Heminoemacheilus hyalinus*	Lan, Yang, and Chen	1996	Guangxi
59. *Heminoemacheilus zhengbaoshani*	Zhu and Cao	1987	Guangxi
Genus Oreonectes	Günther	1868	
60. *Oreonectes anophthalmus*	Zheng	1981	Guangxi
61. *Oreonectes furcocaudalis*	Zhu and Cao	1987	Guangxi
62. *Oreonectes macrolepis*	Huang, Chen, and Yang	2009	Guangxi
63. *Oreonectes microphthalmus*	Du, Chen, and Yang	2008	Guangxi
64. *Oreonectes retrodorsalis*	Lan, Yang, and Chen	1995	Guangxi
65. *Oreonectes translucens*	Zhang, Zhao, and Zhang	2006	Guangxi
Genus Paracobitis	Bleeker	1863	

(Continued)

TABLE 1 (Continued)

Species	Namer	Year	Distribution
66. *Paracobitis maolanensis*	Li, Ran, and Chen	2006	Guizhou
67. *Paracobitis posterodarsalus*	Ran, Li, and Chen	2006	Guangxi
Genus Paranemacheilus	Zhu	1983	
68. *Paranemacheilus genilepis*	Zhu	1983	Guangxi
Genus Schistura	McClelland	1838	
69. *Schistura dabryi microphthalmus*	Liao and Wang	1997	Guizhou
70. *Schistura lingyunensis*	Liao and Luo	1997	Guangxi
Genus Triplophysa	Rendahl	1933	
71. *Triplophysa aluensis*	Li and Zhu	2000	Yunnan
72. *Triplophysa gejiuensis*	Chu and Chen	1979	Yunnan
73. *Triplophysa longibarbatus*	Chen, Yang, Sket, and Aljancic	1998	Guizhou
74. *Triplophysa nandanensis*	Lan, Yang, and Chen	1995	Guangxi
75. *Triplophysa nasobarbatula*	Wang and Li	2001	Guizhou
76. *Triplophysa qiubeiensis*	Li and Yang	2008	Yunnan
77. *Triplophysa rosa*	Chen and Yang	2005	Chongqing
78. *Triplophysa shilinensis*	Chu and Yang	1992	Yunnan
79. *Triplophysa tianeensis*	Chen, Cui, and Yang	2004	Guangxi
80. *Triplophysa xiangshuingensis*	Li	2004	Yunnan
81. *Triplophysa xiangxiensis*	Yang, Yuan, and Liao	1986	Hunan
82. *Triplophysa yunnanensis*	Yang	1990	Yunnan
83. *Triplophysa zhenfengensis*	Wang and Li	2001	Guizhou
Genus Yunnanilus	Nichols	1925	
84. *Yunnanilus bajiangensis*	Li	2004	Yunnan
85. *Yunnanilus beipanjiangensis*	Li, Mao, and Sun	1994	Yunnan
86. *Yunnanilus discoloris*	Zhou and He	1989	Yunnan
87. *Yunnanilus longidorsalis*	Li, Tao, and Lu	2000	Yunnan
88. *Yunnanilus macrogaster*	Kottelat and Chu	1988	Yunnan
89. *Yunnanilus macrolepis*	Li, Tao, and Mao	2000	Yunnan
90. *Yunnanilus nanpanjiangensis*	Li, Mao, and Lu	1994	Yunnan
91. *Yunnanilus niger*	Kottelat and Chu	1988	Yunnan
92. *Yunnanilus obtusirostris*	Yang	1995	Yunnan
93. *Yunnanilus paludosus*	Kottelat and Chu	1988	Yunnan
94. *Yunnanilus parvus*	Kottelat and Chu	1988	Yunnan
95. *Yunnanilus pulcherrimus*	Yang, Chen, and Lan	2004	Guangxi

finish their breeding activity as early as possible, so that the young fry have time to grow before the next cold season. Thus, it is likely that breeding of most *Sinocyclocheilus* species occurs between the spring and summer. Since the streams in all caves inhabited by *Sinocyclocheilus* connect with surface rivers, their water levels are affected by regional precipitation and can change radically within a short time period. These

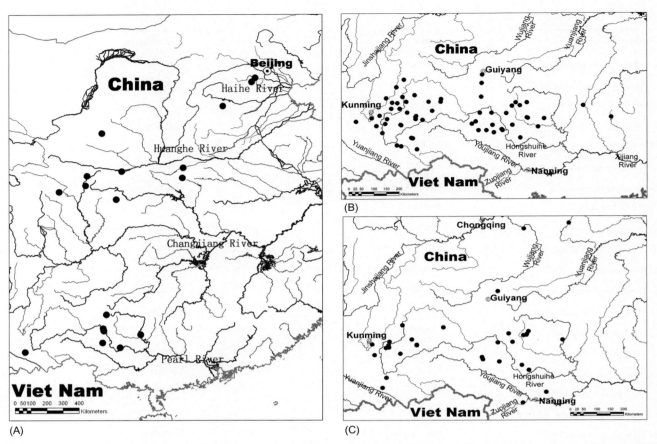

FIGURE 2 Distribution map of Chinese hypogean fishes. (A) Distribution of hypogean Cyprinidae (excluding the genus *Sinocyclocheilus*). (B) Distribution of the species in the genus *Sinocyclocheilus*. (C) Distribution of hypogean Cobitidae and Balitoridae.

radical changes in water levels may trigger reproductive activity. The sex ratio of *S. tingi* varies from 1.31 to 4.2, with females being more abundant than males. The egg diameter of *S. macrolepis* is 1.5~2.0 mm, whereas the diameter of mature *S. bicormutus* and *S. angularis* eggs is 1.5~2.0 mm, and the mature egg of *S. tingi* averages 2.1 mm. According to dissection and observation, it was found that the absolute brood size of *S. macrolepis* is about 2000 and the average brood size of *S. tingi* is 2595~3087. In both species, embryos develop synchronously, suggesting that ovulation occurs simultaneously. The fecundity of 2-year-old *S. yangzongensis* is 3585, and increases to 22,000 eggs after 5 years.

THE CHARACTERISTICS OF CHINESE CAVEFISH

Chinese cavefish have five main characteristics that make them unique among the cavefish of the world.

1. China hosts the greatest and most diverse number of troglobitic fish species. So far, at least 46 species of troglobites have been found in China with the number still increasing with the discovery of new species.

2. Although China has the richest variety of troglobite fish species, they mainly belong to two genera from Cypriniformes, *Sinocyclocheilus* and *Triplophysa*, which together account for 71.7% of the Chinese troglobitic cavefish.

3. The range of preferred habitat is relatively narrow. Chinese troglobitic fish, excluding *Triplophysa xiangxiensis* which lives in Hunan Province and *T. rosa* that lives in Chongqing, are all distributed in Yunnan Province, Guizhou Province, and the Guangxi Zhuang Autonomous Region. In particular, the Guangxi Zhuang Autonomous Region has the largest number of troglobitic cavefish (22 species, accounting for 48% of all Chinese troglobite fish). Chinese troglobitic fish species are most concentrated in the karst areas of the Yunnan-Guizhou Plateau. Most of these species are located only in one independent water body, such as one small river, one brook, even one cave. For example, *Oreonectes anophthalmus* is unique to Taiji Cave of Wuming, Guangxi.

4. Intense speciation is seen among Chinese cavefish, despite the narrow area of habitat in which they live. The genus *Sinocyclocheilus* is a good example. Its range reaches from Yunnan Yimen in the west to Guangxi Fuchuan in the east, and north from Guizhou Huaxi to Yunnan Yanshan in the south. The distance between east and west is 900 km, but the distance between south and north is only 300 km. In this relatively small area, 50 species of *Sinocyclocheilus* have been recorded, making *Sinocyclocheilus* the largest genus of Chinese Cyprinidae, as well as the genus with the greatest amount of adaptive morphology. The large diversity of *Sinocyclocheilus* species is probably the result of strong selective pressures of isolated cave environments, combined with the unique environmental effects that might be present in the Qinghai-Tibet Plateau uplift. It is interesting to note that this intense speciation is seldom seen in the other areas of the world where cavefish are distributed.
5. Each species of Chinese cavefish tends to have a small population size. Limited food resources are probably responsible for this. A good example is *S. hyalinus*, of which only very few specimens were found in their only habitat, A'lu cave in Guangxi Zhuang Autonomous Region.

MORPHOLOGY AND ADAPTATION

Cave habitats are unique environments characterized by permanent darkness, the absence of green plants, and seasonal scarcity. Chinese cavefish have evolved a series of constructive and regressive morphological changes to survive in these harsh conditions. Constructive features often include a protruding jaw, an increase in the number of taste buds, over-developed barbels, and various specialized appendages, while regressive changes mainly include eye degeneration, reduction or loss of pigmentation, and the disappearance of scales. In addition to these common adaptations, Chinese cavefish have other unique features, such as the development of a humped back, a horn, and a head drape. Below we discuss the morphology and adaptation of Chinese cavefish from the perspective of their constructive and regressive structures.

Changes in Body Shape

Cavefish often have a very different outward appearance compared to their surface counterparts. There are three different body shapes of cavefish, the fusiform type, the humpback type, and the head-horn type. In *Sinocyclocheilus*, the head shape of the surface species is very similar to the normal shape of other fishes (*e.g.*, *S. grahami*), in which the length of the head is longer than the height of the body; the snout is slightly pointed; the mouth is subinferior; and the upper jaw is elongated (Fig. 1). In some cave-dwelling fish, the head is duckbilled, with the anterior half-depressed and the posterior half-raised, and the jaws are wider and more protruding than their related surface species. For instance, the jaws of *S. hyalinus* and *S. anophthalmus* are wider and more protruding than *S. grahami* (Fig. 3). The pleat on the head ridge that is found in *S. hyalinus*

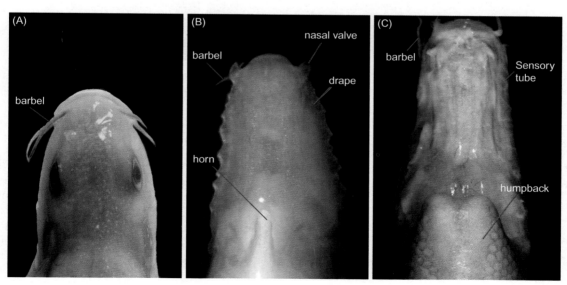

FIGURE 3 Differences in head shape between (A) *S. grahami*, (B) *S. hyalinus*, and (C) *S. anophthalmus*. The jaws of *S. hyalinus* and *S. anophthalmus* adults are wider and more protruding than the jaw of *S. grahami*.

FIGURE 4 Examples of the different species of *Sinocyclocheilus*. (A) *S. tianlinensis*; (B) *S. microphthalmus*; (C) *S. furcodorsalis*; (D) *S. broadihornes*; (E) *S. tileihornes*; (F) *S. hugeibarbus*; (G) enlarged image of *S. hugeibarbus* head; (H) *S. rhinocerous*; (I) enlarged image of *S. rhinocerous* head.

is also a very unique morphological feature (Fig. 3), and may have the function of increasing sensitivity to fluctuations in water flow.

Humpback and Horn

One of the prominent constructive structures of *Sinocyclocheilus* is the humpback and horn, which can be very well developed in some of the troglobitic species. The humpback is free of bone and mainly consists of adipose tissue (Wang et al., 1995), while the horn consists of the frontal and parietal bones. The horn structure is found in both sexes of the same species, and the shape of the horn also varies between different *Sinocyclocheilus* species: some are small (*S. angularis*), some are forked (*S. bicornutus*), some show a thin protrusion (*S. rhinocerous*), and several show a tile shape (*S. tileihornes*) (Fig. 4).

The histology of the *S. hyalmus* head horn has been extensively studied (Fig. 5). Overall, the structure of horn can be divided into three segments, from anterior

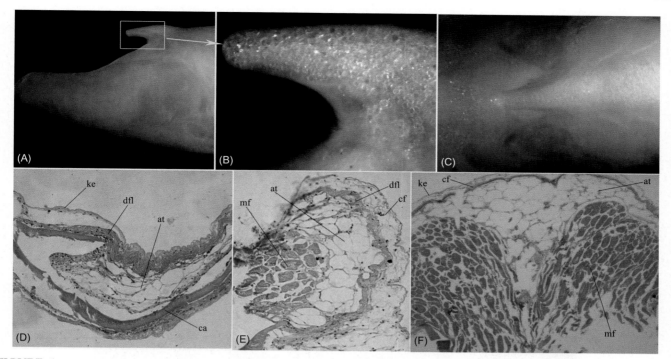

FIGURE 5 Horn structure of *S. hyalmus*. (A) Lateral view of head; (B) magnified image of horn; (C) dorsal view of horn. The horn is divided into three segments; (D), (E), and (F) show the histological structure of horn. ke, keratinized epithelium; dfl, dense fibrous layer; at, adipose tissue; mf, muscle fibers; cf, collagen fibrils; ca, cartilage.

to posterior: the first is the apical section, the second is the middle section, and the third is the basal section. The apical section is characterized by a thick layer of keratinized epithelium containing thick collagen fibrils (Fig. 5D). The middle section is composed of keratinized epithelium, lying over a compact fibrous layer, adipose tissue, and muscle fibers. A thick dermal reticular layer forms the compact fibrous layer under the thick keratinized epithelium. There are adipose tissue and muscle fibers under the fibrous layer (Fig. 5E), and the basal section has a much thinner layer of collagen fibrils and a compact fibrous layer under the keratinized epithelium, which is characterized by the presence of very thick adipose tissue and muscle fibers under the fibrous layer (Fig. 5F). Neither nerve ending nor glands are present in the horn. These characteristics suggest that the function of the horn may be to store fat for nourishment of the adjacent brain.

The horn may also be used for protecting the brain from bumping into rock walls. Li and Tao (1994) studied horn structure in *S. rhinocerous*, and found that its bony part is composed of three pieces, each shaped like a right-angled triangle (Fig. 6A). As shown in the schematic diagram (Fig. 6B), "A" stands for the tip of the horn and "BCD" stands for the base of horn. Thus when "A" hits a rocky surface, the pressure will transmit downward to line "BE," "CE," and "DE" and then disperse and decrease greatly, thereby protecting

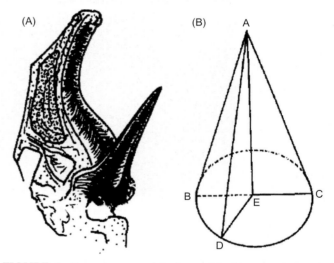

FIGURE 6 Bone structure of the horn of *S. rhinocerous* (A) with a schematic diagram (B).

the brain during swimming in rocky caves (Li et al., 1997). Although the bony horn of *S. rhinocerous* can't completely protect the body from injuries produced by bumping into rock walls, it could disperse pressure equally throughout the head surface to reduce the magnitude of the force. There are also some other hypotheses on functions of horn-like character, but real answers still need further studies.

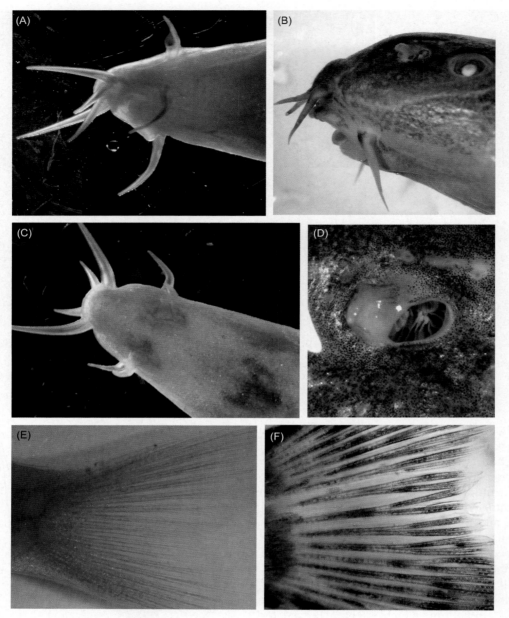

FIGURE 7 The barbels, nostril, and caudal fin of *Triplophysa*.

The humpback is found in different clades of *Sinocyclocheilus*, demonstrating that it is the result of convergent evolution, and likely to be adaptive. Although the precise function of the horn-like structure is still unknown, it is found only among some hypogean species of the genus *Sinocyclocheilus*.

Sensory Apparatus

Sensory structures such as barbels, taste buds, nostrils, and the lateral line system have been augmented in cavefish. Most of the Chinese cave-dwelling species have a more sensitive sense of smell, with the nostrils more anterior than their surface relatives, and in some species the nasal valve is hyper-developed and takes on a beard shape (Fig. 7). At the same time, cave species often have an increased number of taste buds: for example, *S. hyalinus* has more taste buds than *S. grahami*.

The barbel has a tactile function, and also has taste buds on it, implying both mechanosensory and gustatory functions. Fish can use barbels to detect the surrounding environment and amplify the search for food. There are abundant neurons and sinuses in the barbels of *Sinocyclocheilus anatirostris*, showing that the barbels have a prominent sensory function. All *Sinocyclocheilus* have two pairs of well-developed barbels; however, the extent of barbel development differs according to the level of exposure to light. Among those Chinese cave-dwelling species that live in both cave and non-cave environments with abundant light, most have barbels that are moderately developed and extend from the anterior edge of the eyes to the preopercular bone. Those partially cave-dwelling species that live in shaded bodies of water have more developed barbels that can extend to the trailing edge of the opercle, and may even extend to the starting point of the pectoral fin, such as is the case in *S. longibarbatus*. There are several cave-restricted species of Chinese cavefish in which barbel augmentation is often less than that found in partially cave-dwelling species. The least developed barbels do not reach the leading edge of the eye, as in *S. cyphotergous*, while the most developed barbels among the cave-restricted species just touch the preopercular bone, as in *S. microphthalmus*. This shows that the development of the barbels is inversely correlated with eye development: the barbels are long when the eye is smaller, a possible tradeoff for loss of eyesight. This is a very interesting phenomenon if one takes into account that cave-restricted species also have a very developed sensory tube and a special projection on the head. Therefore, it is reasonable to conclude that enhanced barbels is a primitive feature compensating for the loss of eyesight, whereas the sensory tube and projection are more advanced compensatory forms.

The lateral line system of *Sinocyclocheilus* is very developed and specialized. In addition to possessing the tubal system of Cyprinidae, both sides of the lateral line canal also have many short branches named *sensory tubes*. These are especially abundant in the head (Fig. 3C). In species such as *S. cyphotergous*, which shows the most extensive sensory tube network, sensory tubes can even be found under the surface of the skin on both sides of lateral line at the anterior part of the trunk. These special sensory tubes are found in all species of *Sinocyclocheilus*, which suggests that this feature was present in their common ancestor. It is estimated that the initial evolution of the sensory tube system occurred in a common ancestor living in an environment that may have had intimate connections with a karst cave environment. During the course of evolution and adaptation, this common ancestor acquired the trait first, and then passed it to its descendants.

Specialized Appendages

Troglobitic species often have more developed appendages, such as modified pectoral and pelvic fins, than troglophilic species (Fig. 1). The overdeveloped appendages could diminish the energy consumption of the animal and improve the efficiency of movement, as seen in *Sinocyclocheilus*. In *Sinocyclocheilus*, the degeneration of the opsin system directly weakens the motor skills of the fish and results in a decreased ability to respond quickly to environmental changes. At the same time, the enhanced pectoral fin increases the fish's ability to balance itself. The morphological changes in the pectoral girdle and pectoral fin are detailed as follows. Overall, the pectoral girdle of *Sinocyclocheilus* cave species tends to be more elongated and narrower than surface species. Corresponding to a shorter surface for muscle adherence, the muscle attachment surface composed of the cleithrum and corcoideum has a more flattened concave shape. In some partially cave-dwelling species, the cleithrum and corcoideum are greatly reduced and in some cases they disappear completely. Additionally, the coracoid of surface species is wide, and higher than the cleithrum, with shallow ridges to accommodate muscle attachments. In *Sinocyclocheilus*, the coracoid develops into a triangle with the tip positioned anterior to the plane of the cleithrum.

In partially cave-dwelling species, the pectoral fin is short, with a rear projection that does not reach the starting point of the ventral fin, and an average number of 15 branches in the fin ray. In contrast, the pectoral fin of the cave-restricted species is long, with a rear protraction that extends past the starting point of the ventral fin, and an average of 13 branches in the fin ray.

Adipose Storage

Chinese cavefish restricted to caves store large amounts of adipose tissue in various parts of the body, such as the forehead, the horn, the base of the dorsal fin, the base of the caudal fin, and the sides of the body (Fig. 8). There is also a considerable amount of adipose tissue stored in the eye socket (Fig. 9F). Fat deposition in the eye sockets should not be dismissed simply as padding, because the storage of fat is very important to cave animals and allows them to survive during seasons in which food does not enter caves. The lack of a primary food source that can be produced in darkness may cause a food limitation and thus require improvements in food-finding and energy storage capabilities. For this reason cave animals store fat wherever they have space, such as in the empty eye sockets and in the horn. Storing fat locally in the horn may function to provide adjacent tissues, such as the brain, nourishment when there is low food input into the cave.

Eye Degeneration

The adaptive changes in the optic system of Chinese cavefish are mainly seen as changes in the size of

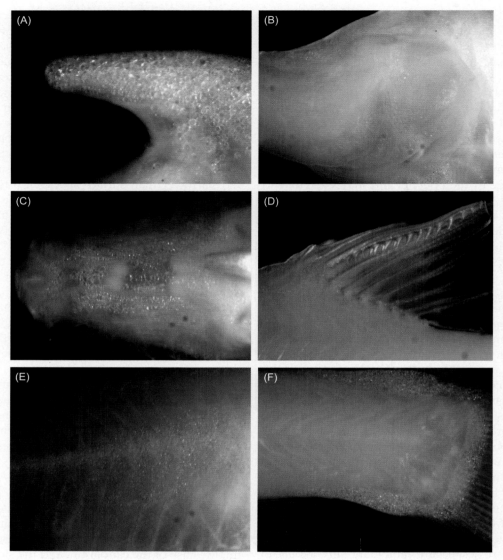

FIGURE 8 The distribution of adipose tissue in *S. hyalinus*. *S. hyalinus* store fat in the horn (A), forehead (B,C), base of the dorsal fin (D), lateral face of the body (E), and at the base of the caudal fin (F).

the eye, the degeneration of the eye's structure, and in some species its complete disappearance. There are three different eye structures seen in the genus *Sinocyclocheilus*: normal, dot-eye, and blind (Fig. 4). These variations are strongly correlated with differing amounts of sunlight each species is exposed to in the caves they inhabit. Those troglophilic species that are sometimes active at the surface of the water in brighter areas have eyes that are clear but underdeveloped when compared to surface-dwelling species. The circumorbital structures of these fish are also modified, such that the lacrimal and supraorbital bones retain the shapes found in surface-restricted species, while the infraorbital and postorbital bones regress to a tubular shape. In some species, the supraorbital bone and jawbones have a tendency to fuse, such as in *S. tingi*. In species that live in half-enclosed caves with weak light, such as *S. macrophthalmus*, eyes often tend to be larger, which may be beneficial for detecting weak light. Structural changes are even more pronounced in troglobitic Chinese cavefish species that spend their entire life cycle in deep caves or underground streams and are only active near the mouth of these caves at night, returning to the depths before daybreak. The eyes of these fish are so small that they are barely visible under the surface of the skin that grows over the eye socket, and in many cases they even disappear completely. The infraobital and postorbital bones either form tubes or disappear as well, while the supraorbital bone and jawbone fuse together. The sides

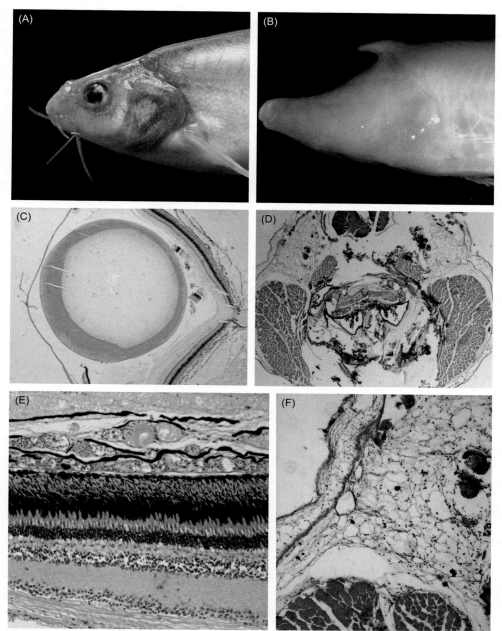

FIGURE 9 Comparison of the eye structures of *S. grahami* (A,C,E) and *S. hyalinus* (B,D,F). (A) Lateral view of *S. grahami* with normal eyes; (B) lateral view of *S. hyalinus* without eyes; (C) section of a *S. grahami* eye; (D) transversal section through the eye region of *S. hyalinus* showing the region containing fat tissue and a flap of skin; (E) retinal structure of *S. grahami*; (F) enlarged image of *S. hyalinus* eye region.

of the jawbone are widened and the rima oculi narrowed. The skin of the jaw has elongated and fused with the inferior skin to seal up the rima oculi completely. In addition, adipose tissue is deposited around the eye as it degenerates and can even replace it completely, as in *S. anatirostris* and *S. hyalmus* (Fig. 9).

Here we use *S. grahami* and *S. hyalinus* to demonstrate the details of the various changes in the eyes of Chinese cavefish (Fig. 9). On examination of the external features of these fish, it can be seen that the *S. grahami* has fully developed eyes located toward the front section of the head (Fig. 9A), while the eyes of *S. hyalinus* seem to disappear (Fig. 9B). The eye structure of *S. grahami* is complete and includes the cornea, iris, pupil, and retina, with its fully developed neural structure (Fig. 9C,E). In *S. hyalinus*, the eye disappears completely, to the point that only adipose tissue is found in the eye orbit when the head is serially sectioned (Fig. 9D,F). In contrast, *S. anophthalmus* does have small eyes buried deep within the orbits of the skull, which are also reduced in size, as seen in skeletal comparisons between species (Fig. 10D). Sometimes, the level of degeneration is different even within same species, as is known to occur in *S. anophthalmus* and *Triplophysa tianeensis*. Most individuals of these species have lost eyes completely; however, a few individuals have vestigial eyes buried under the skin (Fig. 11B) on one side or both sides. Histological sections of *S. rhinocerous* showed only a single rod cell and no cone cells, thus demonstrating that their eyes have lost light sensitivity.

Loss of Pigmentation

Body pigmentation in teleosts is due to three types of dermal chromatophores: black melanophores, which contain melanin; silver iridophores, which contain purines; and yellow xanthophores, which contain pteridines. There is a dramatic decline in the total number of melanophores in Chinese cavefish, as well as a strong reduction in the ability of these cells to synthesize melanin. Those cavefish species that live exclusively in subterranean streams where there is no light are often completely albino, whereas those species that live around cave entrances and are only partial cave dwellers often have light black-brown or dark brown pigment on their bodies. This is seen in all three of our sample species (Fig. 1): the surface species *S. grahami*, which has a deep yellow body color; the cave-dwelling species *S. hyalinus*, which is an albino with no scales; and *S. brevibarbatus*, which is normally semitransparent with milky white fins. Of greater interest is the fact that individuals of the same species, captured in a cave environment, are often lightly colored or colorless, while those captured outside of the cave often have a much darker body color (*S. microphthalmus*). Some Chinese cavefish can gain pigmentation when they enter a bright environment. It seems likely that the disappearance of pigmentation in Chinese cavefish is mainly caused by the inhibition of gene expression in the pigment production pathway, rather than by the total loss of pigmentation genes.

Disappearance of Scales

The development of scales in cave-adapted *Sinocyclocheilus* takes on an obvious regressive trend,

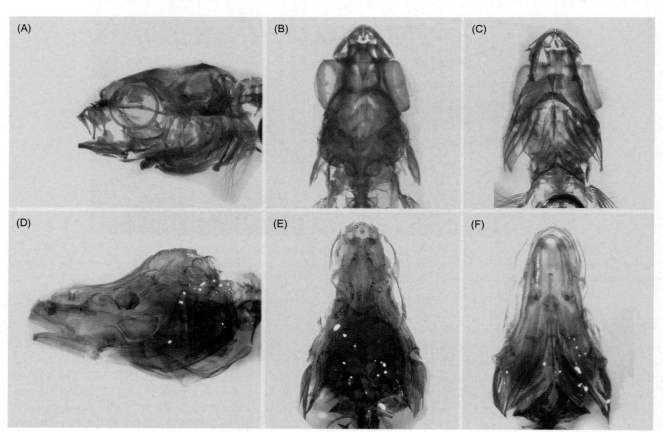

FIGURE 10 Comparison between the head skeleton of *S. grahami* (A,B,C) and *S. anophthalmus* (D,E,F). (A,D) lateral views; (B,E) dorsal views; (C,F) ventral views.

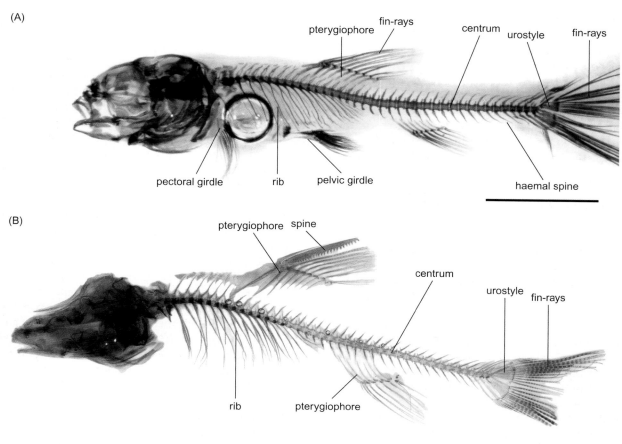

FIGURE 11 Whole-mount staining of bone and cartilage of *S. grahami* (A) and *S. anophthalmus* (B). The cartilage was stained with Alcian blue and the bone was stained with Alizarin red. Scale bars = 50 mm.

as seen in their reduced distribution over the body and the relative size of the lateral line scales. Those species that live in abundant light, such as *S. grahami*, have large well-developed scales that are overlapping over the entire body. The most extreme example of scale regression is seen in cave-restricted species such as *S. tianlinensis*, which have no scales on the body at all. We can only find lateral line pores on the sides.

The rate of scale regression the lateral line due to adaptation to the cave environment is much slower than the speed at which general body scales are lost. As previously mentioned, even those *Sinocyclocheilus* species that have completely lost their body scales still retain the scales of the lateral line, although they are underdeveloped. For those species that have partial regression of body scales, the lateral line scales retain a larger shape than those general body scales above and below them, while in species that have well-developed body scales, those of the lateral line are just as well developed or even larger. This indicates that the lateral line scales of Chinese cavefish are more resistant to regressive evolution during cave adaptation.

PHYLOGENESIS, SPECIATION MECHANISMS, AND BIOGEOGRAPHY

Evolution is at the center of all Chinese cavefish research. Due to the isolated nature of caves, it is possible to combine geological data with the existing records of cavefish distribution to reveal the synergic relationship between cavefish evolution and cave development. The origin of the genera *Sinocyclocheilus* and *Triplophysa* is attributed to the ancient geology and climate of China. Current theory suggests that the primitive ancestor of *Sinocyclocheilus* may have originally lived on the Yunnan-Guizhou Plateau during the late Tertiary period. During the Quaternary period, the Qinghai-Xizang Plateau underwent a sudden upward shift that caused the geological environment to change greatly, while the Yunnan-Guizhou Plateau also underwent an upheaval but settled at a different elevation. At the same time, the temperature of the Earth began to decline, such that the size of the polar ice caps greatly increased. As a result of these drastic environmental changes, the *Sinocyclocheilus* common ancestor was forced to live in caves and eventually took on such adaptations as tetraploidy of chromosomes, an increase in the number of body scales, and a smaller body size in order to survive.

Sinocyclocheilus is a monophyletic group with four clades, given the names *jii*, *angularis*, *cyphotergous*, and *tingi*, named according to their most representative species. The geographical distribution of these different clades displays very little overlap. The phylogenetic tree of *Sinocyclocheilus* based on combined morphological and molecular data indicates that different cavefish species invaded individual cave waters multiple times and acquired their troglomorphic traits independently (Xiao et al., 2005; Fig. 12). The hypothesized troglophilic ancestor of modern *Sinocyclocheilus* species was presumably distributed throughout the karst region of Yunnan and Guizhou at the beginning of the plateau uplift. After the violent sudden upheaval of the plateau approximately 3.4 Ma BP, also known as Act A of the Qinghai-Xizang Movement, the eastern population became isolated from the other groups and evolved into today's *jii* clade. Phylogenetic trees based on maximum parsimony, Bayesian analysis, or morphology, all show that jii is the ancestral clade. Soon afterward, some *Sinocyclocheilus* in the western region became true troglobites, dwelling exclusively in caves and/or subterranean rivers. These became the ecologically isolated angularis clade, which became separated from other troglophilic relatives partially in response to the various climate changes and structural changes of the karst landscape. The angularis clade emerged from this ancestral group. Most species of angularis clade have a horn on the head and a humpback. In addition to the appearance of the horn and humpback, the species in this clade are all troglobites, possessing the usual regressive characteristics typical of caves; their eyes are either reduced in size or absent; and their pigmentation is decreased. Almost at the same time, further upheaval of the land during Act B (2.6 Ma BP) of the Qinghai-Xizang Movement served to isolate the remaining two clades: tingi and cyphotergous, which are sister groups. These clades share a common characteristic in not forming the parietal bone projection on the head and back. The tingi clade, which with one exception exclusively exhibits troglophilic levels of adaptation to hypogean life, became limited to the watershed of the Nanpanjiang River. This clade mainly shows a troglophilic life cycle and will forage out of the cave, although not far away from cave entrances. The troglophilic and troglobitic members of the cyphotergous clade are currently distributed among the drainage sites of the Hongshuihe River. The cyphotergous clade is characterized by a humpback and a reduced lateral line scale number, normally 55 scales. Whether a given *Sinocyclocheilus* species emerged as troglobite or a troglophile can be closely tied to the geological characteristics of the Yunnan-Guizhou Plateau as it exists today. Troglobitic species are more heavily concentrated in the central area of the distribution range, where violent upheavals have produced steep slopes and a complicated karst environment. In contrast, troglophilic species tend to be found closer to the eastern and western edges of the overall distribution area, which exhibit less elevation change. Vicariance is the primary mechanism for speciation in the genus *Sinocyclocheilus*, with diversification mainly resulting from genetic drift in isolated populations (Zhao and Zhang, 2009).

Geographical isolation is the main cause of species diversity in *Sinocyclocheilus*. The ancestor of *Sinocyclocheilus* was likely to be a troglophilic cavefish. During its subsequent evolutionary history, some troglophilic cave *Sinocyclocheilus* became isolated in caves, which gradually produced special characteristics adaptive to the dark environment, and thus troglobitic cavefish species appeared. The isolation of cave or subterranean waters prohibited genetic communication between the fish groups living in different caves. In the cave-rich southern areas, which are affected by high temperatures, abundant precipitation, and surface erosion, cave collapse or underground river diversion was frequent, further enhancing the geographical isolation of species that once had a continuous range. Thus, different populations isolated by underground water evolved in different directions, finally resulting in the formation of independent species.

A very interesting phenomenon not often encountered in cave-adapted animals is that different *Sinocyclocheilus* species can be spatially distributed at the same underground river or same cave at the same time. For example, *S. anatirostris*, *S. microphthalmus*, and *S. lingyunensis* all live in Guangxi Lingyun caves. This phenomenon may be attributed to abundant rainfall in these areas, which enters caves and isolates existing fish populations spatially, thus eventually leading to the formation of separate species in the same cave.

Although geographical isolation is the most important mechanism for *Sinocyclocheilus* speciation, small population sizes also have a role in this process. In small isolated populations there are powerful effects of *genetic drift* as well as natural selection. In small populations derived from a single founding population, genetic differentiation and speciation can occur more rapidly due to the absence of migration and gene pool exchange. In summary, speciation in this genus is attributed to several different factors, including isolation and small population sizes, which result in

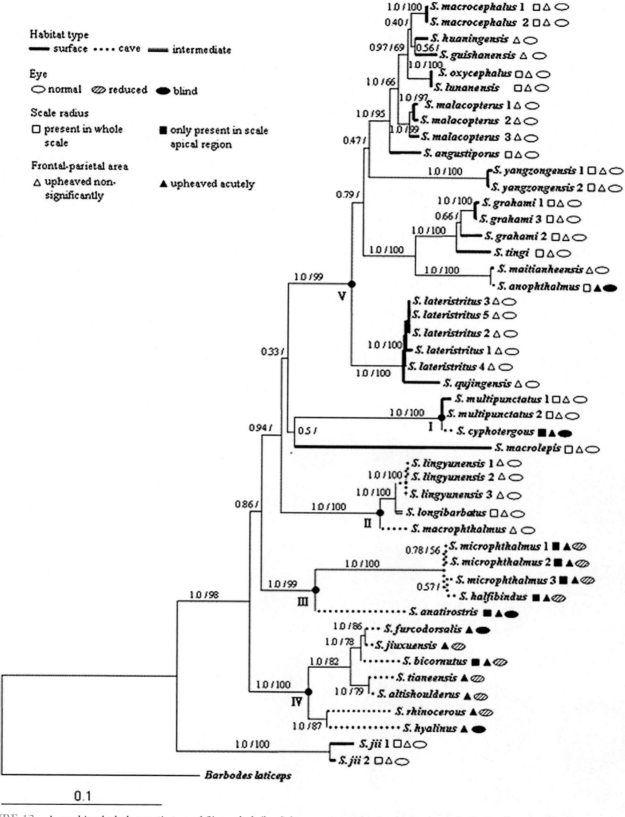

FIGURE 12 A combined phylogenetic tree of *Sinocyclocheilus* fishes constructed using both morphological and molecular data.

inbreeding, rapid genetic differentiation, and fixation. For example, *S. anophthalmus* and *S. maitianheensis* have morphological variation, but their genetic difference is small (only 0.36%), suggesting that enormous and sudden selective pressures led to rapid evolution of adaptive features. Small population speciation mechanisms probably accelerate and amplify this process, and have resulted in a particularly rich *Sinocyclocheilus* species diversity.

The distribution of *Sinocyclocheilus* in karst environments is a consequence of drainage system distribution, climate conditions, and karst development. As described above, geographical isolation was probably the most important factor in speciation; however, the distribution of the tingi clade in the upstream parts of the Nanpanjiang River may be the result of diffusion in this area after the third diversity. And the ψ-type structure in Guangxi karst landform, a Cenozoic downfaulted basin and valley, has had a significant effect on this pattern. The distribution of *Sinocyclocheilus* was separated coincidentally by both flanks and extension line of Guangxi ψ-type structure at different areas. The jii clade occupies the east part of the Guangxi ψ-type structure's east flank, including the Guijiang river and Hejiang river; and no distribution records of *Sinocyclocheilus* exist between the middle and east wing of Guangxi ψ-type structure, which is due to the absence of caves in this area. In contrast, the cyphotergous clade distributes to the east and midline west areas of the Guangxi ψ-type structure. And the distribution of *Sinocyclocheilus* has an obvious relationship to the vertical altitude of karst from Yunnan to Guangxi. The tingi clade is mainly distributed in eastern Yunnan Province, the central eastern plateau, and the upstream parts of the Nanpanjiang River at altitudes between 1500 m and 2000 m. Downstream, in the middle parts of the Nanpanjiang River, the altitude of most areas is between 1000 m and 1500 m, and the angularis clade is found in this region. The cyphotergous clade is located in the far downstream parts of the river, where elevation falls to between 500 m and 1000 m. Finally, the altitude where the jii clade is found is only about 200 m. Thus, the geographical distribution of *Sinocyclocheilus* clades has a vertical basis.

RESEARCH AND CONSERVATION

Hypogean fishes are susceptible to the threats from habitat degradation, hydrological manipulations, environmental pollution, overexploitation of resources, and introduction of alien species. Because most Chinese cavefish are endemic to small areas and have small populations, any of these threats

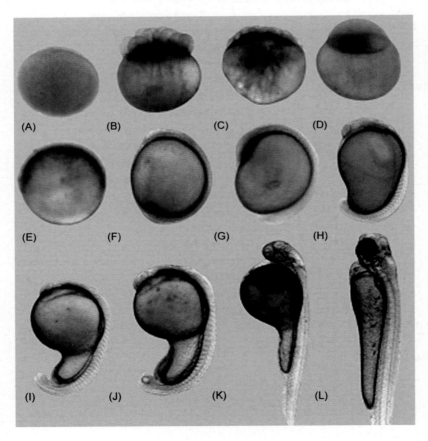

FIGURE 13 The embryonic development of *S. grahami*. (A) The zygote, a few minutes after fertilization; (B) 16-cell stage; (C) 64-cell stage; (D) sphere stage; (E) 50% epiboly stage; (F) bud stage; (G) 5-somite stage; (H) 22-somite stage; (I) 24-somite stage; (J) 28-somite stage; (K) 42-somite stage; (L) long-pec stage.

could have serious consequences. Currently, some populations have been considered "threatened" or "vulnerable." In fact, most Chinese cavefish are currently placed in the "vulnerable" status. In 2004, 14 species cavefish were placed on the "Chinese red list" (Wang Song, Xie Yan (2004). China species red list. Beijing: Higher Education), including 9 species of Cyprinidae (8 species of *Sinocyclocheilus*) and 5 species of Cobitidae.

Chinese cavefish are also threatened by the rapid economic growth, which has depleted their habitats for living and reproduction. To protect these valuable and unique species of hypogean fish, more research should focus on the present habitats that still exist, while practical and effective protective measures must be established for the karst caves and deep pools where they live. Protection should also be put into place so that predation by other animals and humans is prevented. On April 3, 2008, the first "autonomous district level cave rare fish natural area" was set up in Guangxi Lingyun. This was the first natural reserve area to focus on preserving the unique and rare cavefish of China, and it covers an area 684 hectares wide, including areas along an underground river and six distinct caves. Even more promising is the success of researchers at the Yunnan Endemic Species Breeding Center, who in 2007 were able to stimulate artificial reproduction in *Sinocyclocheilus grahami* (Fig. 13). The study of Chinese cavefish has drawn more people's attention to the urgent need for conservation measures to protect this fascinating group of highly endangered animals.

See Also the Following Article

Astyanax mexicanus—A Model Organism for Evolution and Adaptation

Bibliography

Chen, Y., Yang, J., & Zhu, Z. (1994). A new fish of the genus *Sinocyclocheilus* from Yunnan with comments on its characteristic adaptation (Cypriniformes: Cyprinidae). *Acta Zootaxonomica Sinica*, 19, 246–253.

Huang, Q., Cai, Y., & Xing, X. (2008). Rocky desertification, antidesertification, and sustainable development in the karst mountain region of southwest China. *Ambio*, 37, 390–392.

Li, W., & Tao, J. (1994). A new species of Cyprinidae from Yunnan —*Sinocyclocheilus rhinocerous* sp. nov. *Journal of Zhanjiang Ocean University*, 14, 1–3.

Li, W., Wu, D., Chen, A., & Tao, J. (1997). Histological study on the horn-like projection of the head of *Sinocyclocheilus rhinocerous*. *Journal of Yunnan University*, 19, 426–428.

Romero, A., Zhao, Y., & Chen, X. (2009). The hypogean fishes of China. *Environmental Biology of Fishes*, 86, 211–278.

Wang, D., Huang, Y., Liao, J., & Zheng, J. (1995). Taxonomic revision of the genus *Gibbibarbus* Dai. *Acta Academiae Medicinae Zunyi*, 18, 166–168.

Wang, S., & Xie, Y. (2004). *China species red list*. Beijing: Higher Education.

Xiao, H., Chen, S., Liu, Z., Zhang, R., Li, W., Zan, R., & Zhang, Y. (2005). Molecular phylogeny of *Sinocyclocheilus* (Cypriniformes: Cyprinidae) inferred from mitochondrial DNA sequences. *Molecular Phylogenetics and Evolution*, 36, 67–77.

Zhao, Y., & Zhang, C. (2009). *Endemic fishes of* Sinocyclocheilus *(Cypriniformes, Cyprinidae) in China—Species diversity, cave adaptation, systematics and zoogeography*. Beijing: Science Press.

CHEMOAUTOTROPHY

Annette Summers Engel

University of Tennessee

INTRODUCTION

Photosynthesis is not possible in the dark zone of a cave. Consequently, the assumption has been that nearly all life on Earth, and especially life in caves, depends on organic carbon and energy derived from photosynthesis. But, reactive rock surfaces and mineral-rich groundwater provide energy sources for specialized microorganisms that gain cellular energy from chemical transformations of inorganic compounds— such as hydrogen, reduced iron, or hydrogen sulfide, present in groundwater or sediments—and convert inorganic carbon sources into organic carbon. Essentially, microorganisms utilize the available chemical energy that might otherwise be lost to a system. In this manner, chemosynthesis provides a rich alternative energy source for organisms, with the results being that ecosystem biodiversity and population densities are higher compared to some nonchemosynthetically based ecosystems that rely on inconsistent and limited inputs of organic carbon.

This article focuses on chemosynthesis and chemosynthetically based ecosystems in caves and karst. In general, our knowledge of the evolution and metabolism of major chemosynthetic microbial groups has increased in recent years, partly due to advances in molecular genetics methods, but also because of expanded efforts to investigate habitats where chemosynthetic microbes exist. Thanks to research from cave and karst settings, as well as from research done on significant chemosynthetic populations at deep-sea hydrothermal vents (Deming and Baross, 1993) and within the deep terrestrial subsurface (Stevens and McKinley, 1995), the "world is green" view is beginning to change. The article concludes with a discussion of the importance of subsurface chemosynthetically based ecosystems and suggests relevant aspects of future research in cave and karst systems.

TERMINOLOGY EXPLAINED

The physiological mechanisms for capturing chemical energy during chemosynthesis are diverse, and there are several descriptive qualifiers that define an organism based on its carbon and energy sources. Microbial metabolism terminology is reviewed in this book's article "Microbes." For the purposes of this article, microbes that gain energy through chemosynthesis and fix inorganic carbon are *chemolithoautotrophs* (literally "self-feeding rock-eaters"), and energy is gained by transferring electrons from one chemical (electron donor) to another (electron acceptor). Typically, chemolithoautotrophs use compounds present in rocks or groundwater. Chemical electron donors include, but are not limited to, molecular hydrogen, reduced sulfur compounds, metals, and so on. Organisms that gain cellular energy from chemical transformations but use organic carbon compounds for their carbon source are *chemoorganotrophs*, and *heterotrophs* use organic carbon for cellular energy and carbon sources. Several studies have shown that chemolithoautotrophs can grow if organic carbon is present as *mixotrophs*, in which both chemolithoautotrophy and heterotrophy are expressed simultaneously. Microorganisms also have oxygen requirements, and can respire aerobically, anaerobically, or ferment, all of which relates to electron acceptor utilization. Oxygen is the terminal electron acceptor for aerobic metabolic processes (Table 1). In reducing environments, microbes that do not require oxygen (*anaerobes*) use a variety of alternative electron acceptors for respiration in a sequence of energetic, reduction reactions that occur along thermodynamic (and redox) gradients, from nitrate to carbon dioxide (Table 1). There are several other energetically favorable, chemolithoautotrophic pathways that occur in the absence of oxygen, including

TABLE 1 Examples of Chemolithoautotrophic Energy Reactions and Carbon Fixation Pathways

Energy Reaction	Metabolic Process	Major Genera	Electron Donor	Electron Acceptor	Carbon Pathway[1]
AEROBES					
$H_2 + \frac{1}{2}O_2 \rightarrow H_2O$	Hydrogen oxidation	*Alcaligenes, Aquifex*	H_2	O_2	Calvin, rTCA
$H_2S + 2O_2 \rightarrow SO_4^{2-} + 2H^+$ or $2S^0 + 3O_2 + 2H_2O \rightarrow 2H_2SO_4$	Sulfur oxidation	*Sulfolobus, Thiobacillus*	H_2S, S^0	O_2, NO_3^-	rTCA, Calvin, A-CoA
$2FeS_2 + 7O_2 + 2H_2O \rightarrow 2FeSO_4 + 2H_2SO_4$	Sulfur oxidation	*T. ferrooxidans*	S^{2-}, FeS_2	O_2	Calvin
$4Fe^{2+} + O_2 + 10H_2O \rightarrow 4Fe(OH)_3 + 8H^+$	Iron oxidation	*T. ferrooxidans*	Fe^{2+}	O_2, NO_3^-	Calvin
$2Mn^{2+} + O_2 + 2H_2O \rightarrow 2MnO_2 + 4H^+$	Manganese oxidation	*Shewenella*	Mn^{2+}	O_2	Calvin
$NH_4^+ + 2O_2 \rightarrow NO_3^- + 2H^+ + H_2O$	Ammonium oxidation	*Nitrobacter, Nitrosomonas*	NH_4^+, NO_2^-	O_2	Calvin
$CH_4 + 2O_2 \rightarrow CO_2 + 2H_2O$	Methane oxidation	*Methanomonas*	CH_4	O_2	RMP
ANAEROBES					
$4H_2 + CO_2 \rightarrow CH_4 + 2H_2O$	Methanogenesis	*Methanoseata, Methanococcus*	H_2	CO_2	A-CoA
$2CO_2 + 4H_2 \rightarrow CH_3COOH + 2H_2O$	Acetogenesis	*Acetobacterium*	H_2	CO_2	A-CoA
$4H_2 + SO_4^{2-} + 2H^+ \rightarrow H_2S + 4H_2O$	Sulfate reduction	*Archeoglobus*	H_2	$SO_4^{2-}, S_2O_3^{2-}$	A-CoA
$4H_2 + SO_4^{2-} + 2H^+ \rightarrow H_2S + 4H_2O$	Sulfate reduction	*Desulfovibrio, Desulfobacter*	H_2	SO_4^{2-}	A-CoA, rTCA
$FeOOH + 2H^+ + \frac{1}{2}H_2 \rightarrow Fe^{2+} + 2H_2O$	Iron reduction	*Geobacter, Shewenella*	H_2	Fe^{3+} (as FeOOH)	Calvin
$NH_4^+ + NO_2^- \rightarrow N_2 + 2H_2O$	Ammonium oxidation	*Kuenenia*	NH_4^+	NO_2^-, SO_4^{2-}	A-CoA
$2 NO_3^- + 5H_2 \rightarrow N_2 + 4H_2O + 2OH^-$	Denitrification	*T. denitrificans, Pyrolobus*	H_2, H_2S, S^0	NO_3^-	Calvin, A-CoA

[1] *A-CoA = Acetyl-CoA pathway; rTCA = reductive tricarboxylic acid cycle; Calvin = Calvin–Benson cycle; RMP = Ribulose monophosphate cycle*

sulfide oxidation via nitrate reduction and anaerobic ammonium oxidation (anammox) (Table 1).

CARBON FIXATION PATHWAYS

Chemolithoautotrophy has been crucial in the transformation of chemical energy from one reservoir to another during Earth's history. With photoautotrophy, these two processes are the only mechanisms by which the inorganic carbon is incorporated into the biosphere (referred to as *primary productivity*). Most extant chemolithoautotrophy is accomplished by one of at least four pathways: the Calvin–Benson–Bassham (CBB) cycle (also known as the reductive pentose phosphate cycle), the acetyl coenzyme A (acetyl-CoA) (also known as the Wood–Ljungdahl) cycle, the reductive tricarboxylic acid cycle (rTCA) (or Arnon), and the 3-hydroxypropionate cycle. Based on evolutionary relationships of the 16S (small-subunit) ribosomal RNA (rRNA) gene and the subsequent "Tree of Life" (Woese, 1987), autotrophic carbon fixation pathways are distinctly related to the phylogenetic position of the respective organisms. Only members from the domains of Bacteria and Archaea are capable of chemolithoautotrophy (*i.e.*, plants are photoautotrophs).

Some researchers have speculated that the last common ancestor between Bacteria and Archaea used either the acetyl-CoA pathway or the rTCA cycle, with the rTCA cycle most closely resembling what is considered to be the first autotrophic pathway. Moreover, chemosynthesis probably preceded photosynthesis on Earth because none of the ancestral lineages are phototrophic (Schidlowski, 2001). Although both the acetyl-CoA and rTCA pathways are used by diverse Bacteria and Archaea, such as some sulfate-reducers and acetogenic bacteria, as well as methanogens (Archaea), little is known about the ecology and evolution of the rTCA cycle and the microbes capable of this important pathway. Instead, some of the most deeply branching bacteria on the "Tree of Life" are thermophilic bacteria Aquificales and Thermotogales, including the genera *Aquifex* and *Hydrogenobacteria*, and fix carbon via the CBB cycle, whereas the archaeal thermophiles, such as *Thermoproteus* and *Pyrodictium*, fix carbon via the rTCA cycle. The CBB cycle is found in many obligate and facultative autotrophic microbes and is the dominant pathway in the Eukarya (*e.g.*, plants). Many researchers have postulated that the key enzyme in the CBB pathway, ribulose-bisphosphate carboxylase (RuBisCo), is the most abundant protein on Earth. The 3-hydroxypropionate cycle was considered to be relegated to only one phylum of bacteria, the *Chloroflexi*, but now it has been detected from Archaea and seems to be more important than previously considered.

MAJOR CHEMOLITHOAUTOTROPHIC MICROBIAL GROUPS

Electron donors are essential for chemolithoautotrophic growth. For microbes other than chemolithoautotrophs (*e.g.*, heterotrophs, chemoorganotrophs), organic carbon molecules are used as electron donors to gain cellular energy. One of the most energetic chemolithoautotrophic electron donors is molecular hydrogen (H_2). Microbes that use H_2 also use CO_2 as an electron acceptor, either to make methane (methanogenesis) or acetate (acetogenesis). Sulfate-reducing microbes also compete for H_2, but some can use small organic molecules as electron donors. Hydrogen- and ammonia-oxidizing bacteria utilize H_2 aerobically. Additional electron donors include reduced sulfur compounds (*e.g.*, hydrogen sulfide, elemental sulfur), ammonia (NH_4^+), and reduced iron (Fe^{2+}) and manganese (Mn^{2+}). The following sections describe chemolithoautotrophic microbial groups based on required electron donors (Table 1).

Hydrogen

H_2, hydrogen gas, accumulates from the anaerobic breakdown of organic molecules by fermentative (heterotrophic) bacteria, and serves as an important energy source for aerobic and anaerobic archaeal and bacterial chemolithoautotrophs. In most cases, H_2 is rapidly consumed under anaerobic conditions by methanogens and sulfate-reducers. However, if anaerobic growth is slower than production of H_2 by heterotrophs or chemoorganotrophs, then H_2 will diffuse into the aerobic environment where it can be used by aerobic H_2-utilizing microorganisms. Chemolithoautotrophic aerobic H_2-oxidizing microbes oxidize H_2 as the electron donor by a membrane-bound hydrogenase (enzyme), oxygen as the electron acceptor, and fix CO_2 via the CBB or rTCA cycle (Table 1). A wide variety of gram-negative and gram-positive bacteria are H_2-oxidizers, including *Alcaligenes, Hydrogenobacter, Pseudomonas*, and *Aquifex*.

The anaerobic microbial groups that require H_2 as an electron donor follow predictable redox chemistry regarding the utilization of available terminal electron acceptors. Reduction of nitrate, iron, or manganese during anaerobic respiration is typically done by heterotrophs or chemoorganotrophs, but iron or manganese reduction can also be accomplished by some chemolithoautotrophs. Ferric iron is used as an exclusive electron acceptor by dissimilatory iron-reducing microbes that use H_2 as their electron donor, although some require organic carbon compounds, such as acetate or formate. One of the most studied iron-reducers, *Shewanella putrefaciens*, is a chemoorganotroph, and can also use MnO_2 as a sole electron acceptor. Biological iron reduction in aquatic sediments is an important anaerobic process.

Dissimilatory sulfate-reducing microbes are obligate anaerobes that reduce sulfate to hydrogen sulfide (H_2S) (Table 1). Many sulfate-reducers are heterotrophic or chemoorganotrophic, divided into two broad physiological subgroups. Group I dissimilatory sulfate-reducers, including the genera *Desulfovibrio*, *Desulfomonas*, and *Desulfobulbus*, use lactate, pyruvate, ethanol, or other fatty acids as carbon sources. Group II genera oxidize acetate, and include *Desulfococcus*, *Desulfosarcina*, *Desulfobacter*, and *Desulfonema*. These bacterial groups are found within the *Deltaproteobacteria* class of the Proteobacteria phylum, although there are sulfate-reducing Archaea that belong to the genus *Archeoglobus*. Certain species are also capable of growing chemolithoautotrophically with H_2 as the electron donor, sulfate as the electron acceptor, and CO_2 as the sole carbon source. Chemolithoautotrophs use the acetyl-CoA pathway, while the rTCA cycle is used by chemoorganotrophs. If sulfate concentrations are high, then sulfate-reducing bacteria completely oxidize fermentation by-products to CO_2. In low-sulfate anaerobic environments, however, sulfate-reducing bacteria compete with methanogens for H_2 and organic compounds. The reduction of elemental sulfur is also possible, and chemolithoautotrophic sulfur-reducing Archaea include *Thermoproteus*, *Acidianus*, and *Desulfurolobus*, which are common in acidic environments. Although not exclusively a chemolithoautotrophic pathway, the disproportionation of elemental sulfur is also possible, whereby elemental sulfur is split into two different compounds, typically sulfite and thiosulfate, to produce both H_2S and sulfate. *Desulfocapsa* is a common sulfur-disproportionating bacterial genus.

Acetogenesis results in less overall cellular energy than methanogenesis, but both metabolic groups are found in similar habitats. Acteogenic bacteria, such as *Clostridium* and *Acetobacterium*, are obligate anaerobes that form acetate from the oxidation of H_2 using the acetyl-CoA pathway for CO_2 fixation. These organisms are known as *homoacetogens*. H_2 is the common electron donor, but donors can also be from sugars, organic acids, and amino acids. Many acetogens also reduce nitrate and thiosulfate, and are tolerant of low pH, thereby making them more versatile than other anaerobes.

Methanogens are exclusively Archaea, and are one of the most common anaerobic microbes in highly reducing conditions in close association with decomposing organic material. Methanogens oxidize H_2 as the electron donor while reducing CO_2 to methane using the acetyl-CoA pathway (Table 1). These methanogens are usually found in close association (syntrophy or interspecies hydrogen transfer) with fermenting microbial groups because of the need for the continuous H_2 supply provided through fermentation. Chemolithoautotrophic methanogenesis is less efficient than those utilizing formate, acetate, methanol and other alcohols, carbon monoxide, or even elemental iron, as alternative electron donors. The physiologies of these different types of methanogens (chemolithoautotrophic versus chemoorganotrophic) are distinct, and there are seven major groups of methanogens based on this physiology, including the genera *Methanobacterium*, *Methanosaeta*, *Methanococcus*, and *Methanolobus*.

Reduced Inorganic Sulfur Compounds

High concentrations of reduced sulfur compounds are toxic to most organisms. H_2S gas, in particular, reacts with operational biomolecules to form nonfunctional complexes that inhibit respiration. However, H_2S, elemental sulfur, thiosulfate, polythionates, metal sulfide, and sulfite can serve as electron donors for chemolithoautotrophic sulfur-oxidizing bacteria and Archaea (note: for the purposes of this discussion, any microbe capable of oxidizing any reduced sulfur compound is referred to as a sulfur-oxidizer) (Table 1). One difficulty for sulfur-oxidizing microbes is competing with chemical oxidation of reduced sulfur compounds. As a result, most sulfur-oxidizing bacteria occupy aerobic/anaerobic interfaces where sulfide and oxygen meet, and are therefore chemotactic. Interface or gradient growth is important in some ecosystems where sulfur-oxidizing bacteria form symbiotic relationships with animals, such as at deep-sea hydrothermal vent sites.

The earliest microbiological research regarding chemo-lithoautotrophic metabolism was done with sulfur-oxidizing bacteria in the late 1880s. Some sulfur-oxidizers thrive in low pH environments (acidophiles), while others require neutral pH conditions. Sulfur-oxidizing bacteria can form large microbial mats or thick biofilms in a range of habitats, including acid mine drainage and mine tailings, sulfidic thermal springs, marine sediments, and sewage sludge. There are several different types of sulfur-oxidizers, including those belonging to the genus *Thiobacillus* and the "morphologically conspicuous sulfur bacteria" such as *Beggiatoa*, *Thiothrix*, *Thiomicrospira*, and *Thiovulum*, among others. Virtually all of these colorless sulfur bacteria (as compared to the "green" and "purple" sulfur bacterial groups that are anoxygenic phototrophs) are gram-negative and deposit intracellular elemental sulfur when grown on H_2S. Although most sulfur-oxidizing bacteria are chemolithoautotrophic, using either the CBB or rTCA cycles for carbon fixation, some can also be chemoorganotrophic, heterotrophic, or mixotrophic with respect to their carbon source. Typically, molecular oxygen is the terminal electron acceptor, but some sulfur-oxidizers use nitrate as an electron acceptor (e.g., *Thiobacillus denitrificans*).

Ammonia and Nitrite

Nitrification includes the oxidation of both ammonia and nitrite, and results in the formation of nitrite and nitrate, respectively. The formation of nitrate requires a two-step process in which ammonia is first oxidized to nitrite by ammonia-oxidizers, and then nitrite is oxidized to nitrate by nitrite-oxidizers (Table 1). The two groups of nitrifying bacteria, belonging to the family *Nitrobacteraceae*, typically work synchronously in a habitat as chemolithoautotrophs that use the ammonia or nitrite as electron donors and CO_2 as the sole source of carbon. In some rare instances, small organic compounds can also be used. Genera of ammonia-oxidizers have the prefix *Nitroso-*, including *Nitrosomonas* and *Nitrosospira*, whereas genera of nitrite-oxidizers have the prefix *Nitro-*, including *Nitrobacter* and *Nitrospira*. Nitrification occurs at the aerobic/anaerobic interface, but nitrifying bacteria have a high affinity for oxygen. Microbiological ammonia and nitrite oxidation can occur over a broad range of conditions, including in acid soils, and at low and high temperatures.

The oxidation of ammonia anaerobically, referred to as anammox, involves coupling ammonia oxidation to nitrite reduction under anaerobic conditions (Table 1). The end product is nitrogen gas (N_2), but an intermediate by-product of this reaction is hydrazine (rocket fuel) that is contained in unique anammoxasomes, or specialized intracellular organelles. If ammonia is limited, these microbes can use organic acids to convert nitrate and nitrite into N_2. Anammox microbes fix carbon via the acetyl-coA pathway and have been identified from marine and freshwater habitats. So far, anammox microbes belong to the Planctomycetes phylum with four provisionally described bacterial genera from freshwater (*Brocadia*, *Kuenenia*, *Anammoxoglobus*, and *Jettenia*), and *Scalindua* from marine habitats.

Iron and Manganese

As with sulfur oxidation, microbes can also gain energy from ferrous iron (Fe^{2+}) or manganous manganese (Mn^{2+}) oxidation. In the past, Fe^{2+}- and Mn^{2+}-oxidation was attributed to many different microbes based on the accumulation of iron or manganese minerals associated with cellular material. However, recent investigations have found that most of these microbes do not gain energy from mineral buildup. Iron rapidly oxidizes to ferric (Fe^{3+}) iron at neutral pH, so successful iron-oxidizers live in low pH environments, such as acid mine drainage, acid springs, mine tailings, or acid soils containing sulfide minerals such as pyrite. Chemolithoautotrophic iron-oxidizing bacteria include the acidophilic aerobe *Thiobacillus ferrooxidans*. Despite research attempting to isolate other chemolithoautotrophs, the most common iron-oxidizers, *Leptospirillum* and *Gallionella*, and some manganese-oxidizers, such as *Arthrobacter* and *Hyphomicrobium* (except *H. manganoxidans*), are chemoorganotrophs. Ecologically, iron- and manganese-oxidizers are important for detoxifying the environment by lowering the concentration of dissolved toxic metals.

One-Carbon Compounds

There has been some controversy regarding whether or not microbial processes based on reduced one-carbon (C-1) compounds, as the sole carbon and energy source, represent true chemolithoautotrophy; C-1 compounds used as electron donors are biogenic, thereby not fulfilling the *litho* component to chemolithoautotrophy. However, cycling C-1 compounds by potential chemolithoautotrophs is gaining appeal because understanding CO_2 and CH_4 dynamics is important for carbon budgets of aquatic ecosystems in surface and subsurface environments.

Methanotrophs and methylotrophs utilize methane or methanol, respectively, for cellular energy and growth under aerobic conditions (Table 1). Aerobic methanotrophs sequentially oxidize methane to a series of intermediates, including methanol, formaldehyde, and formate, before generating CO_2 enzymatically with methane monooxygenase. Some of the intermediates can be incorporated into the cell, which is one of the reasons why C-1 metabolism is not considered to be strictly autotrophic. There are two physiological groups of methanotrophic bacteria, both obligate aerobes. Type I methanotrophs include the genera *Methylomonas*, *Methylosphaera*, and *Methylomicrobium* within the *Gammaproteobacteria*, and fix carbon by assimilating formaldehyde in the ribulose monophosphate pathway. Type II methanotrophs fix formaldehyde in the serine pathway, and genera include *Methylocystis* and *Methylosinus* within the *Alphaproteobacteria*. Methanotrophy had been previously associated with neutrophilic pH conditions, but bacteria from the phylum Verrucomicrobia have been described recently that oxidize methane below pH 5 and optimally around pH 2.

Methane can also be oxidized anaerobically by a syntrophic consortium of bacteria and anaerobic methanotrophic (ANME) Archaea, referred to as anaerobic oxidation of methane or AOM. Sulfate is the electron acceptor for sulfate-reducing bacteria. ANMEs are closely related to methanogenic Archaea, and AOM is the enzymatic reverse of methanogenesis. This process has been identified from marine and freshwater lake sediments and water columns. Anaerobic oxidation of methane with nitrite as the electron acceptor has recently been identified by the new microbe "Candidatus Methylomirabilis oxifera" in the absence of a syntrophic association.

Growth on carbon monoxide by carboxidobacteria (carbon monoxide-oxidizing bacteria) occurs with a range of electron acceptors, including oxygen, elemental sulfur, sulfate, nitrate, or ferric iron. Members of the genera *Pseudomonas*, *Bacillus*, *Alcaligenes*, and *Clostridium* are known for carbon monoxide oxidation.

CHEMOLITHOAUTOTROPHY IN CAVES AND KARST SETTINGS

Cave ecosystems have been considered to be both energy- and nutrient-limited because organic matter, assumed to be derived primarily from photosynthetically produced material on the surface, needs to be brought into the subsurface. Knowledge of chemolithoautotrophic microbial processes, however, has changed our understanding of the types of ecosystem-level processes that can occurs when surface input of organic matter is limited. Studies of chemolithoautotrophic processes in caves have been done since the 1960s, with intensive efforts since the mid-1990s.

At a glance, Table 2 shows that the sulfur cycle in karst has been intensively studied. Research efforts have been sustained for decades for some systems, including the Movile Cave (Romania), Frassasi Caves (Italy), and Lower Kane Cave (Wyoming, U.S.A.), but several new systems have been discovered and extensively investigated (Table 2). To be fair, much of the recent research has been done in systems most likely to harbor chemolithoautotrophic microbes (Table 2), such as those having sulfidic or methanogenic waters, metal-laden sediments, or obvious microbial mats and biofilms formed under extreme geochemical conditions.

In general, most studies have provided a census of microorganisms and their metabolic processes, with very few ecosystem-level studies that attempt to integrate the activities of different microorganisms with respect to each other, or studies of cave or karst systems where chemolithoautotrophy may not be a dominant metabolism. From recent studies, it is clear that more diverse chemolithoautotrophic microbial groups are present than previously considered. The following section summarizes our current knowledge of chemolithoautotrophic microbial groups in cave and karst settings, divided into the sulfur, nitrogen, methane, and metal (Fe, Mn) cycles.

Sulfur Cycle

Caves with active microbial sulfur cycling typically have H_2S-rich springs that discharge into passages, termed *sulfidic caves*. Because of H_2S oxidation, either abiotically or biotically, sulfuric acid is generated and promotes local limestone bedrock dissolution. This cave formation process is known as *sulfuric acid speleogenesis*, described in detail from Lower Kane Cave by Egemeier (1981) (Fig. 1). These types of caves, the microbial communities, and geochemical and geological processes have been reviewed recently by Engel (2007).

Prior to and just after the 1986 discovery of the Movile Cave ecosystem, there were few studies describing sulfur-based microbial populations in caves. Sulfur-oxidizing bacteria occur in both aqueous and subaerial habitats, with filamentous groups forming thick microbial mats, rope-like structures, biofilms, and webs within cave spring orifices, pools, or streams. In Movile Cave, there is a thin microbial mat that floats at the water surface (Fig. 2) and an anaerobic biofilm mixed with clay that covers the submerged cave floor. For most studies, the main goals have been to understand the microbial diversity and spatial distribution of microbes, but rates of primary productivity have been quantified from a few systems (Table 2). Recent advances in integrative isotopic, molecular, and microscopy techniques are revealing how higher level organisms in the cave ecosystems are dependent on, and even in potential symbiotic associations with, chemolithoautotrophic sulfur-oxidizing microbes (e.g., Dattagupta et al., 2009).

16S rRNA gene sequence surveys have certainly increased our understanding of the microbial diversity from these systems. Phylogenetic analyses from several of the cave systems reveal that sulfur-oxidizing bacteria closely related to the genera *Thiovulum*, *Thiothrix*, *Thiobacillus*, *Beggiatoa*, and *Thiomicrospira* occur. However, unclassified (due to the inability to culture these groups to date) microbial groups belonging to the *Epsilon-proteobacteria* class of the Proteobacteria phylum dominate many of the microbial communities in the sulfidic systems. Sequences from many other different bacterial lineages, including candidate divisions, have also been retrieved from sulfidic karst systems, suggesting a richer microbial diversity than previously considered. Sulfate-reducing and sulfur-disproportionating microbes have also been identified from 16S rRNA gene surveys and culture-based methods of water and sediments (Table 2). Sulfate-reducers are important for recycling sulfur compounds by generating H_2S that sulfur-oxidizers can use. In cave streams, this interactions among microbes across redox gradients in microbial mats, which are cycling elements from oxidized to reduced forms (e.g., sulfate to

TABLE 2 Chronological Summary of Studies Describing Chemolithoautotrophic Microorganisms Found in Various Caves Worldwide

Cave or Karst System	Date(s) of Study[1]	Chemolithoautotrophic Metabolic Processes[2]	Approach[3]
Caves in France and Iowa (U.S.A.)	1963, 1987	Iron oxidation	M, C
Florida Aquifer karst, U.S.A.	1965, 1992–1999; 2006	Sulfur oxidation; methanogenesis and methanotrophy	M, C, DNA, BA, SI
Bungonia Caves, Australia	1973–1994	Iron reduction, sulfate reduction, sulfur oxidation	M, C
Mammoth Cave, Kentucky, U.S.A.	1977; 2009	Ammonium and nitrite oxidation	C
Edwards Aquifer, Texas, U.S.A.	1981, 1986; 2005–2011	Sulfur oxidation; methanogenesis and methanotrophy; sulfate reduction, ammonium oxidation, nitrite oxidation, anaerobic ammonium oxidation	SI, DNA, FG
Movile Cave, Romania	1986–2011	Sulfur oxidation, methanotrophy, methanogenesis, ammonium oxidation, nitrite oxidation	M, C, SI, DNA, FG, BA, RI, nanoSIMS
Cesspool Cave, Virginia, U.S.A.	1986, 2001; 2009	Sulfur oxidation; sulfate reduction	M, C, DNA, RI
Grotta Azzurra, Italy	1986–2007	Sulfur oxidation, sulfate reduction	C, M, SI
Parker Cave, Kentucky, U.S.A.	1988, 1998	Sulfur oxidation	M, DNA
Karst springs, Spitsbergen	1994	Sulfate reduction, sulfur oxidation	M, SI
Cueva de Villa Luz,[4] Mexico	1994–2007	Sulfur oxidation, sulfate reduction	M, C, DNA
Kugitangtou Caves, Turkmenistan	1994, 1997, 2001	Sulfur oxidation, sulfate reduction, iron oxidation	M, C, DNA
Zoloushka Cave, Ukraine	1994, 2001	Sulfate reduction, denitrification, sulfur oxidation, iron oxidation	C, SI
Frasassi Caves,[5] Italy	1996–2011	Sulfur oxidation, sulfate reduction	M, C, SI, DNA, RI, nanoSIMS
Anchihaline caves, Mexico	1997	Ammonium oxidation, sulfur oxidation, methanotrophy	SI
Maltravieso Cave, Spain	1997	Ammonium oxidation	C
Bundera Sinkhole, Australia	1999	Sulfur oxidation	SI
Lower Kane Cave, Wyoming, U.S.A.	1999–2011	Sulfur oxidation, sulfate reduction, iron reduction, methanogenesis, iron oxidation	M, C, SI, DNA, FG, BA, RI, FISH
Glenwood Cavern, Colorado, U.S.A. (and nearby mine adit)	2005–2007	Sulfur oxidation, hydrogen oxidation, ammonium oxidation	M, DNA, FG
Ayalon Cave, Israel	2006–2008	Sulfur oxidation, methanotrophy, methanogenesis	M, SI
Acquasanta Terme caves, Italy	2009–2011	Sulfur oxidation, sulfate reduction	DNA, FISH
Sistema Zacatón, Mexico	2007–2011	Sulfur oxidation, sulfate reduction, methanogenesis, anaerobic ammonium oxidation	M, DNA, FG
Coldwater Cave, Iowa and Minnesota (U.S.A.)	2010	Ammonium oxidation, nitrite oxidation, anaerobic ammonium oxidation	C, DNA, FG

[1] *Representative years for which published results can be found in the literature, thus far.*
[2] *Listed in order of dominance, if known.*
[3] *Approach used to identify and to verify chemolithoautotrophic processes. M = microscopy; C = culture techniques; SI = stable isotopes; DNA = phylogenetic relationships based on 16S rRNA analyses; FG = functional gene surveys; BA = biomarker assays; RI = radioisotope experiments; FISH = fluorescence in situ hybridization; nanoSIMS = nano-secondary ion mass spectroscopy. For a description of the approaches, refer to this book's article "Microbes."*
[4] *Also known as Cueva del Azuffre or Cueva de las Sardinas.*
[5] *Includes the Grotte di Frasassi and Grotta Sulfurea systems.*

FIGURE 1 Main trunk passage in Lower Kane Cave, Wyoming (U.S.A.). White, filamentous microbial mats dominated by sulfur-oxidizing bacteria colonize shallow sulfidic water, beginning at the lower right corner (water flows from the lower right to upper left). This microbial mat extends for approximately 20 m, with an average thickness of 5 cm. Piles of gypsum surround the stream (especially on the left), formed from the replacement of limestone during sulfuric acid speleogenesis.

FIGURE 2 Partially flooded passage in Movile Cave, Romania. A white filamentous microbial mat floats on the surface of the water. A grid was constructed to encourage additional mat growth.

FIGURE 3 Discontinuous cave-wall biofilms and crusts (dark patches) from Lower Kane Cave, Wyoming. Crusts form on gypsum (light areas), and condensation droplets and mucus-like drops are suspended from the crusts. Elemental sulfur is typically associated with the crusts.

sulfide, then sulfide to sulfur to sulfate, or CO_2 to bicarbonate to organic carbon and back to CO_2), as well as involving abiotic and biotic transformation processes, has been proposed as an example of nutrient spiraling of sulfur (and carbon).

Biofilms on subaerial cave-wall surfaces, referred to as microbial draperies, cave-wall biofilms, or snottites, have been described from the several active sulfidic caves. Although some of the longest draperies reportedly reach up to 10 cm in length or more, cave-wall biofilms typically occur as discontinuous patches of insoluble crusts with subcentimeter long, mucus-like droplets suspended from the crusts (Fig. 3). Condensation droplets associated with the biofilms have pH values between 0 and 3.

Acidity is maintained, in part, because the replacement of carbonate with gypsum in the sulfuric acid speleogenesis process that physically separates the cave-wall solutions from the underlying carbonate that would buffer pH to more neutral conditions, and also because of continued microbial sulfur oxidation. Laboratory isolates and phylogenetic studies demonstrate that microbes are closely related to extremophiles, including those belonging to the genera *Acidithiobacillus*, *Sulfobacillus*, and *Acidimicrobium*, as well as the *Thermoplasmales*, actinomycetes, and the bacterial candidate lineage TM6. Filamentous fungi and protists have also been described from cave-wall biofilms.

Nitrogen Cycle

Nitrogen enters the cave in various forms, including as atmospheric gas, nitrates from fertilizers, organic matter in soil, and from bat and rat guano. Ammonium is produced by nitrogen-fixing bacteria, which would be present in soils overlying limestone. The ammonium could be brought into the caves via bedrock fissures or washed in with stream sediments. Many anaerobic microbes (*e.g.*, sulfate-reducers) can also fix nitrogen. *Nitrosomonas* oxidizes ammonia to nitrite, and *Nitrobacter* oxidizes nitrite to nitrate. This two-step process can influence nitrate mineral precipitation, such as niter and nitrocalcite (saltpeter). Whether microbes are involved in the precipitation of these minerals or not is still debated. Nitrate minerals are typically found in dry cave sediments, such as in the southeastern United States. Isolates of *Nitrobacter* have been obtained from Mammoth Cave (Kentucky) sediments rich in saltpeter and suggest a geomicrobiological link, but more work is needed to confirm this.

Nitrification is an important chemolithoautotrophic pathway in nitrogen-limited freshwater streams and lakes because it enhances overall nitrogen availability. Limited work has been done from caves. One study based on stable nitrogen isotope ratio analysis from anchihaline caves in Mexico indicates that nitrification may be the primary energy source to that ecosystem, in addition to sulfur- and methane-oxidation. Recently, from research done in a mine adit at Glenwood Springs, Colorado, ammonium oxidation by Crenarchaea may be occurring (Table 2). Characterization of nitrogen cycling from sediments impacted by atrazine (sourced from agricultural land runoff) in Coldwater Cave, Iowa and Minnesota, indicates that atrazine may negatively affect nitrogen cycling, and cause an increase in nitrification at the expense of nitrogen fixation (Iker et al., 2010).

Methane Cycle

Methane cycling is beginning to interest karst researchers because of its potential significance to ecosystem development and primary productivity in some cave systems. Methanotrophy and methanogenesis, while expected to occur in groundwater have only been studied from a few cave and karst systems, likely because of a limited understanding of what the dissolved methane concentrations are. Studies, particularly in Movile Cave, have been based on enrichment and isolation techniques for methanotrophs and methanogens, $^{13}CH_4$-labelling incorporation, and DNA analysis following separation of ^{12}C from ^{13}C (a process referred to as stable isotope probing) (Table 2). Very low carbon isotope values for dissolved inorganic and organic carbon in Movile Cave, groundwater from the Floridan Aquifer (Opsahl and Chanton, 2006), and groundwater from the Edwards Aquifer in Central Texas (Longley, 1981), are suggestive of active methanotrophy. Moreover, higher trophic levels (e.g., animals) have isotopic values that indicate their food sources were organic carbon produced from these microbial processes.

Iron and Manganese Bacteria

According to Caumartin (1963), iron plays a significant role in the microbiology of a cave, and most cave sediment contains iron bacteria. Indeed, microbes associated with iron and manganese cycling have been found in cave sediments and speleothems based on microscopy and culturing, with chemolithoautotrophic *Gallionella*, and heterotrophic *Leptothrix* and *Crenothrix* being the most common genera described. Fossil or encrusted sheaths and stalks of these bacteria have been observed in stalactites, sediments, and corrosion crusts from several caves, including caves from Iowa, the Black Hills (South Dakota), Lechuguilla Cave (New Mexico), Kugitangtou caves (Turkmenistan), Grand Caymen Island caves, the Nullarbor Plain caves in Australia, as well as others. However, it is difficult to distinguish between chemical and biological iron/manganese precipitates, and it is not evident from mineralization, fossil or active, whether an organism was chemolithoautotrophic. Therefore, the ecology of metal cycling remains speculative and the roles of these microbial groups as potential primary producers remain unclear.

CHEMOLITHOAUTOTROPHICALLY-BASED CAVE ECOSYSTEMS

Although chemolithoautotrophic metabolism was described from sulfidic springs more than 100 years ago, it has been estimated that chemolithoautotrophs do not produce efficient energy, and therefore, could not be significant primary producers for an ecosystem when compared to photosynthetically produced energy. However, in the subsurface, photosynthesis is impossible. For research done in deep-sea hydrothermal vents, the deep terrestrial subsurface, and caves and karst aquifers, which can also be restricted from an influx of organic matter from the surface, the significance of chemolithoautotrophy to ecosystem-level processes is being revised.

Part of the revision is because of the extremely rich biodiversity in these ecosystems. From marine environments, chemolithoautotrophically based ecosystems include deep-sea hydrothermal vents (Deming and Baross, 1993), hydrocarbon cold seeps, and estuarine sediments. *Gammaproteobacteria* and *Epsilonproteobacteria* are in chemosymbiotic associations with animals, such as the vestimentiferan tube worms (*Riftia pachypila*) and clams (*Calyptogena magnifica*). Deep-sea sediments exposed to hydrocarbon seeps have mussels with methanotrophic symbionts, and some mussels (*Bathymodiolus thermophilus*) also have sulfur- and methane-oxidizing bacterial symbionts. Clams, including *Solemya* and *Thyasira*, living in brackish water and estuarine sediments have sulfur-oxidizing symbionts. The relationships between chemolithoautotrophic microorganisms and animals are beneficial to both organisms, as the animals provide protective habitat and the bacteria supply nutrients and energy.

From caves and karst aquifers, our knowledge of chemosymbiotic associations in freshwater systems is limited. But, the recent discovery of a potential chemosymbiotic association between the sulfur-oxidizing bacteria *Thiothrix* spp. (*Gammaproteobacteria*) and *Niphargus* spp. (amphipod) in the Frasassi Caves (Italy) (Dattagupta et al., 2009) provides an incentive to keep investigating other possible associations. The highly diverse Movile Cave ecosystem, with 33 new cave-adapted taxa identified from 30 terrestrial invertebrate species (24 are endemic) and 18 species of aquatic animals (9 endemic), is another cave system to investigate possible chemosymbiotic associations (Sarbu et al., 1996). The recent discovery of the Ayalon Cave ecosystem in Israel, with possibly a new sulfur-based, chemosynthetic ecosystem and eight invertebrates new to science, is testimony that substantial and unknown subterranean diversity in karst still exists. Other opportunities to expand our knowledge of chemolithoautotrophically based karst systems include the Edwards Aquifer, where the role of chemolithoautotrophy is just beginning to be investigated, as well as the Upper Floridan aquifer.

CHEMOLITHOAUTOTROPHY IN THE FUTURE

Caves are important and relatively accessible habitats to study chemolithoautotrophic metabolism. In the absence of organic carbon compounds, reactive rock surfaces and mineral-rich groundwater in the subsurface provide an assortment of potential energy sources for chemolithoautotrophic microorganisms. In comparison to the research done with macroscopic organisms in caves, the distribution of chemolithoautotrophic microbial groups and associated biogeochemical processes occurring in caves have infrequently been addressed, although the methodology exists. Of those investigations describing cave microbial communities, many have been simply observational and the physiological capabilities of microorganisms were inferred based on the utilization of a specific substrate in cultures or from 16S rRNA gene sequence phylogenetic affiliation. To date, there have been few stable isotope or radioisotope studies done to describe chemolithoautotrophy in cave ecosystems, and combined microscopy and molecular studies have just recently been initiated that should continue to expand our understanding of chemolithoautotrophic processes in the subsurface. From these investigations, potential chemosymbiosis between microbes and animals provide exciting future research opportunities.

See Also the Following Articles

Microbes
Cave Ecosystems

Bibliography

Caumartin, V. (1963). Review of the microbiology of underground environments. *National Speleological Society Bulletin*, 25(1), 1–14.

Dattagupta, S., Schaperdoth, I., Montanari, A., Mariani, S., Kita, N., Valley, J. W., et al. (2009). A novel symbiosis between chemoautotrophic bacteria and a freshwater cave amphipod. *The ISME Journal*, 3(8), 935–943.

Deming, J., & Baross, J. (1993). Deep-sea smokers: Windows to a subsurface biosphere? *Geochimica et Cosmochimica Acta*, 57(14), 3219–3230.

Egemeier, S. J. (1981). Cave development from thermal waters. *National Speleological Society Bulletin*, 43, 31–51.

Engel, A. S. (2007). Observations on the biodiversity of sulfidic karst habitats. *Journal of Cave and Karst Studies*, 69(1), 187–206.

Iker, B. C., Kambesis, P., Oehrle, S. A., Groves, C., & Barton, H. A. (2010). Microbial atrazine breakdown in a karst groundwater system and its effect on ecosystem energetic. *Journal of Environmental Quality*, 39(2), 509–518.

Longley, G. (1981). The Edwards Aquifer—Earth's most diverse groundwater ecosystem. *International Journal of Speleology*, 11 (1–2), 123–128.

Opsahl, S. P., & Chanton, J. P. (2006). Isotopic evidence for methane-based chemosynthesis in the Upper Floridan aquifer food web. *Oecologia*, 150(1), 89–96.

Sarbu, S. M., Kane, T. C., & Kinkle, B. K. (1996). A chemoautotrophically based cave ecosystem. *Science*, 272(5270), 1953–1955.

Schidlowski, M. (2001). Carbon isotopes as biogeochemical recorders of life over 3.8 Ga of Earth history: Evolution of a concept. *Precambrian Research*, 106(1–2), 117–134.

Stevens, T. O., & McKinley, J. P. (1995). Lithotrophic microbial ecosystems in deep basalt aquifers. *Science*, 270(5235), 450–455.

Woese, C. R. (1987). Bacterial evolution. *Microbiological Reviews*, 51(2), 221–271.

CLASTIC SEDIMENTS IN CAVES

Gregory S. Springer

Department of Geological Sciences, Ohio University

BASIC PHYSICAL PROPERTIES

Clastic sediments are broken fragments of preexisting rocks that have been transported and redeposited. Clastic sediments can be derived from any rock type, including rocks enclosing caves and originating outside of caves (Fig. 1), as well as secondary deposits such as travertine. Clastic sediments are volumetrically the most common deposits in caves. The majority of clastic particles in cave sediments are detrital grains eroded from land surfaces and carried into caves by streams, mass movements, wind, wave action, and ice (Table 1). Geomorphologists use clastic cave sediments to understand the processes that shape caves and landscapes, including system responses to climate change, while hydrologists may focus on the driving and resisting forces that affect sediment movement through karst systems. In contrast, archaeologists and paleontologists seek to know how objects came to rest in a cave, why they were buried, and whether the objects have been disturbed or altered since emplacement. The diversity of sediments and their uses means that research goals vary markedly from one study to another, but all cave studies share a commonality in that subterranean research techniques are modified versions of techniques developed for surface deposits and landforms. Analyses begin with sediment description.

Scientists studying cave sediments focus on their stratigraphy and are typically interested in mineralogy, median grain sizes, sorting, and sedimentary structures. The mineral content of sediment grains is largely a function of their source. Grains eroded from cave surfaces or the overlying landscape will contain minerals found in the surrounding rocks (*e.g.*, calcite or chert). These locally derived sediments are defined as *autochthonous* and contrast with allochthonous grains derived from neighboring nonkarst landscapes. *Allochthonous sediments* commonly include insoluble silicate minerals, such as quartz, feldspar, or mica, which are easily distinguished from soluble carbonate and evaporite minerals. Perhaps surprisingly, clay-sized sediments in caves draining silicate rocks are typically dominated by quartz and contain very little to virtually no carbonate grains. Allochthonous sediments can be used to identify cave stream sources if local streams drain rocks with distinctive mineral compositions.

The median grain size is a diameter for which 50% of all grains are larger and 50% of all grains are smaller (Fig. 2). Sorting refers to the relative proportions of different grain sizes. As examples, a well-sorted sand deposit may consist almost entirely of 1-mm grains, but a poorly sorted deposit may contain similar proportions of clay, silt, sand, and pebbles. Cumulative grain size distributions are useful for understanding sediment textures, which are functions of size and sorting, and how sediments were deposited. For instance, the middle flat regions in the two quietwater curves in Figure 2 mean the sample has significant amounts of sand and significant amounts of clay, but very few grains of intermediate sizes. This is the result of attrition, whereby a sandy limestone ceiling was slowly dissolved and quartz sand grains fell onto clays being deposited by slow-moving water.

Sediment grains are often organized into distinctive patterns, called *sedimentary structures*, during deposition. For instance, sand ripples contain thin inclined

FIGURE 1 Clastic sediments slumping into a short cave passage connecting a sinkhole and stream passage. The cave stream occasionally pushes rounded stream cobbles uphill and out of the tube, but cobbles are falling back into the cave along with angular blocks shed by cliffs and hillsides in the sinkhole. Together, the sediments are a gravity deposit.

TABLE 1 Types of Clastic Sediments Commonly Found in Caves

	Surface Equivalent
Water-Borne (Stream) Facies Class: Detrital Sediments Transported by Flowing Water	
Bedload: Comprised of grains mobilized by tractive forces at the base of stream flows	Channel lag deposit
Slackwater: Comprised of grains carried in suspension and deposited floodwaters	Overbank deposit
Quietwater: Comprised of grains carried in suspension and deposited within standing bodies of water	Lacustrine (lake) deposit
En mass: Comprised of grains transported as semicoherent units by floodwaters (sliding beds) or debris flows	Debris flow deposit
Gravity Facies Class: Sediments or Blocks Transported by Gravity, but Independent of Fluid Flow	
Collapse: Comprised of blocks detached from cave ceilings, walls, and entrance areas	Talus and scree
Infiltrates: Comprised of detrital grains transported by gravity through discrete, subvertical pathways	
Residual Facies Class: Sediments Derived from Weathering of Primary or Secondary Deposits	
Residuum: Bedrock weathering products preserved *in situ* with primary grains or sedimentary structures observable	Saprolite

Sources: Springer *et al.* (1997); Bosch and White (2007).

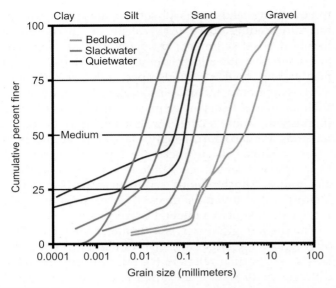

FIGURE 2 Seven example grain-size distributions plotted by particle diameter (mm) and cumulative percent finer. Two curves are derived from ancient stream bedload deposits and 50% of their grains are sand size or smaller (orange lines). In contrast, slackwater sediments deposited by floodwaters have median grain sizes in the silt range (green). The floodwaters were moving more quickly than those responsible for the quietwater sediments (blue), but the quietwater sediments have comparatively large median grain sizes. This unexpected finding reflects sand grains falling from a sandy limestone ceiling onto clays being deposited in a permanently flooded cave passage; median grain size is not always a reliable indicator of past water velocities.

FIGURE 3 Sediments deposited since the 1940s in a West Virginia cave. White lines separate three major stratigraphic packages. The basal gravel bed was deposited prior to a large logjam forming just downstream of this photograph. The logjam created a large pool during floods, wherein slackwater sediments consisting of silt and organic detritus were deposited to form the middle stratigraphic unit. Eventually, the cave floor aggraded to the top of the logjam and this allowed gravel to move once again. The resulting channel deposit (behind student) was deposited along with silt banks.

layers (laminae) arranged as cross-bedding (Fig. 3) formed as grains rolled off ripple crests. Cross-bedding is evidence of deposition by moving water or air, but many other sedimentary structures exist and can be used to infer other sediment transport and deposition processes. A bed lacking sedimentary structures is generically referred to as *massive* and this too may have meaning, perhaps recording sediments having been transported chaotically within gravity flows (*e.g.*, mudflows and other landslides).

STRATIGRAPHY

Stratigraphy describes how individual beds of sediment are arranged relative to one another, including their stacking orders and ages. Each bed represents a different chapter in cave passage evolution, so deciphering how and why beds are associated with one another is valuable for understanding cave and landscape histories. For instance, the stratigraphy seen in Figure 3 includes streambed gravels (*bedload*) overlain by finer sediments that accumulated in a large pool behind a logjam. By slowing floodwater velocities, the logjam prevented gravel from being moved and created the conditions necessary for finer sediments to settle out. The channel bottom *aggraded* (filled with sediment) over several decades until it was nearly level with the top of the logjam, allowing a new gravel layer to be deposited a meter above the old one. The stream is now incising through the accumulated sediments because cave explorers removed the logjam.

Caves can possess multiple generations of deposits scattered among passages whose ages may differ by thousands or millions of years. Long geomorphic records can be constructed by combining the stratigraphic records of many individual deposits. This requires placing deposits in chronological sequence, which is typically accomplished through paleomagnetic, radiometric, cosmogenic isotope, and relative dating techniques. Cave sediments generally adhere to the three laws of traditional stratigraphic studies, including the law of superposition stating that the oldest sediments are at the bottom of a deposit. The law of original horizontality states that all sedimentary beds are deposited flat-lying, but clay and fine silt commonly adhere to inclined cave surfaces to create similarly inclined lamina (thin layers) in a process called *parallel accretion* (Fig. 3). So, tilted lamina in caves should not be automatically assumed disturbed. The law of cross-cutting relationships is violated in unique ways and objects cutting through a deposit may be older than the sedimentary deposit. This occurs when sediments are deposited around projections (*e.g.*, travertine) on cave surfaces. Care is needed with regard to the three laws, but violations are not so frequent as to void standard stratigraphic techniques.

A widely encountered stratigraphic sequence is the abandonment suite, which is comprised of stream sediments that become finer toward the top of a section. Active stream tiers commonly carry ample bedload, which accumulates to varying depths on passage floors. A passage ceases to receive bedload when a lower level passage forms. Floodwater velocities in the older, higher tier decrease because it mostly experiences backflooding or overflow from the lower passage. In this lower velocity setting, suspended sediments are deposited atop the older sediments, hence from the bottom to top, a typical abandonment succession is gravel, sand, silt, clayey silt. Abandonment suites can be mistaken for channel infill created during aggradation of a streambed and the silts and clayey silts can be deposited long after a passage has been abandoned, so fining-upward sequences should be interpreted carefully.

Long-term deposition (*aggradation*) may be recorded as banks of sediments on channel margins and thick deposits beneath streambeds. This can occur when cave passages are obstructed (Fig. 3), climate changes, or excess sediment is supplied by tributary streams, glaciers, or landslides. Aggradation can be inferred where the evidence is consistent with paragenesis or ascending erosion. Paragenesis occurs when clastic sediments shield cave floors from erosion and ceiling erosion creates vertical accommodation space. Over time, streambed and ceiling elevations increase (ascend) and a thick aggradational deposit is created. The process is unique to caves and leaves distinctive evidence, including ceiling pendants and meander notches that propagated upstream during passage growth. Lateral erosion can detach pendants from ceilings and leave them otherwise intact and upright in an aggradational sequence, which is convincing evidence of long-term erosion and sediment transport, rather than simple infilling as part of passage abandonment.

The *stratigraphic facies* concept is widely used for reconstructing depositional relationships and geomorphic histories (Table 1). Facies are distinctly unique bodies of sediment identifiable on the basis of appearance, composition, texture, or sediment sizes. A *depofacies* possesses unique sedimentary structures and stratigraphic successions whose origins are derived from a common depositional process or environment. Examining Figure 3, the uppermost gravel layer is channel fill adjacent to a silt-dominated point bar. The two stratigraphic units have very different textures and appearances, but fluvial (stream) processes deposited them more or less simultaneously, so they represent a stream depofacies traceable throughout much of the cave. In contrast, organic-rich sediments below the gravel layer were deposited in a large, intermittent pool and lack cross bedding (Fig. 3). Hence, a lacustrine facies is recognized (Table 1), but the facies cannot be traced throughout the cave because it is restricted to the area upstream of the logjam. We can estimate the extent backflooding created by the logjam by mapping the extent of pool slackwater sediments.

SEDIMENT PRODUCTION

Clastic sediments are produced by the physical disaggregation of preexisting rocks during weathering and mechanical erosion. Chemical weathering weakens rocks by altering mineral compositions and by removing the minerals cementing them together. Physical weathering involves fragmentation due to tensional or compressive stresses, including expansion of clay minerals within rocks, frost shatter, and root wedging. The latter two are frequently sources of angular rock fragments, including breakdown, in and near cave entrances (Fig. 1). Breakdown blocks are clastic sediment and they easily exceed the total volume of all other clastic deposits in a cave, occurring where passage widths exceed the tensile strength of ceiling or overhanging rock beds.

Cave streams mechanically erode conduit walls through a combination of *plucking* and *corrasion*. Plucking adds sediment to karst streams when intense hydraulic forces pull or wedge blocks of rock from conduit surfaces. Corrasion occurs when sediment grains carried by floodwaters strike bedrock surfaces with sufficient force to break off fragments. This process is generally referred to as *abrasion* when the result is a smooth or polished cave surface and *percussion* when it results in obvious flaking and rock breakage. Percussion generally implies impact of large grains, such as cobbles, whereas abrasion is most commonly associated with "sand blasting" by suspended sediment. As reported by Malcolm Newson (1971), solid carbonate grains created by abrasion can collectively weigh more than the dissolved solids in those same cave-fed floodwaters. Whether corrasion plays a major role in cave development elsewhere is a matter of ongoing research, but it is undoubtedly a nontrivial source of sediment to high-energy cave streams where large grains impact cave walls.

Carbonates may contain chert or other low-solubility minerals, which enter cave passages as the surrounding rock is eroded. Where abundant, chert can be the major or only significant autochthonous bedload supplied to cave streams. In other cases, comparatively high concentration of autochthonous grains may accumulate in clays and silts when sedimentation rates are otherwise very slow (Fig. 2), perhaps recording protracted pipe-full conditions in a quietwater setting (many hundreds of years) (Table 1).

SEDIMENT TRANSPORT AND DEPOSITION

Stream Sediments

Sediment moves through cave streams in suspension, whereby turbulence keeps grains aloft in the water column with little or no contact with the streambed, and as bedload. The latter moves by rolling, toppling, sliding, and saltation (similar to bouncing). Grains moving in suspension and as bedload are collectively referred to as the stream's *sediment load*, a quantity often expressed in kilograms per second. Load increases during floods and little or no load is carried between floods because stream waters are moving too slowly to entrain (put in motion) grains.

Intuitively, high water velocities are needed to entrain and transport large sediment grains, but this is also true of clay and silt-sized particles because they are cohesive. The relationship between size and the velocities needed to entrain particles is shown in Hjulstrom's Diagram (Fig. 4A). Particles will continue to be transported at velocities less than what was needed to entrain them and coarse sediments are deposited first because of their greater weights. Clay and silt are deposited last because they are easily suspended. The velocity-dependent deposition of sediment grains causes them to be sorted into beds containing similar-sized particles. This is often expressed as a fining upward sequence consisting of a basal gravel bed beneath a sand bed capped with a silt bed.

Hjulstrom's Diagram is useful for presenting idealized relationships between grain sizes and sediment transport. But the mechanics of sediment transport are complex and a number of factors must be considered, including those describing conditions at the scale of individual grains (Fig. 5). These individual particles shield one another from the full brunt of floodwaters, normal forces pinning particles in place, and friction and gravity. Those resisting forces compete against buoyancy, drag, lift, and shear stress. All submerged objects are buoyant and the effective weight of a submerged quartz grain is only 62% of its dry weight, a significant difference, although not enough for it to float. Lift and drag are created by vertical and horizontal pressure differences. Fluid pressure is inversely proportional to velocity, so higher pressures are reached on the upstream and undersides of grains (Fig. 5C). Much like an airplane wing, the pressure difference pulls grains toward lower pressure regions. In contrast, shear stress is generated by differences in

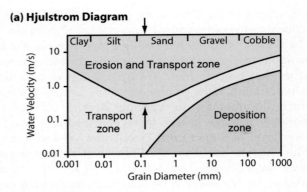

FIGURE 4 Hjulstrom's Diagram is based on laboratory experiments and is divided up into three zones. Sediments will be entrained when water velocities reach the uppermost dividing line and, as indicated by arrows, fine sand is the easiest to move because finer sediments are cohesive. Sediments will continue to move even after velocities fall into the transport zone because their inertia has been overcome, but grains will eventually be deposited when velocities cross the lower dividing line. However, sediments finer than sand are easily suspended and may remain in motion for long time periods (left side of diagram). The Shield's Diagram is more complex, but is preferred by sedimentologists. The Reynolds shear stress and Shield's parameter are calculated based on a variety of variables discussed in the text and take into account fluid and grain properties.

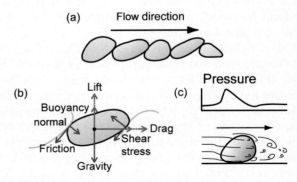

FIGURE 5 (A) Shows imbricated sediment grains on a channel bed. The forces acting on a grain are shown in (B) with purple indicating a resisting force and green representing driving forces. The latter act to entrain particles. Lift and drag are the result of pressure differences with higher pressures forming at the upstream and undersides of particles (C). Shear stress is derived from the momentum of moving water and transferred through the water's viscosity.

water velocity above grain surfaces. The water on grain surfaces has an effective velocity of zero, but the water's viscosity helps transmit the momentum of floodwaters to grain surfaces and accelerates them in a downstream direction. Greater differences in velocity translate to larger shear stresses.

The Shield's Diagram incorporates many more factors than Hjulstrom's and is widely used in bedload studies. The Shield's parameter accounts for the competing influences of particle weight and shear stress. Motion is expressed as a function of the Reynolds shear stress and Shield's parameter (Fig. 4B). The former is calculated using particle diameter, kinematic viscosity, and shear velocity. The latter incorporates additional variables, including shear stress exerted on the streambed, and altogether the calculations help express the effects of turbulence around a grain. The importance of turbulence at grain and macroscopic scales should not be underestimated because it creates large lift and drag forces by driving fast-moving water close to grain surfaces, thus creating large velocity gradients. So, together the Shield's parameter and Reynolds shear stress take into account the driving and resisting forces discussed previously.

Gravity Sediments

A gravity deposit results from sediment moving as a semicoherent or highly viscous mass of solid particles, as opposed to a large volume of water moving a much smaller volume of sediment. Gravity deposits are easily distinguished from stream sediments (alluvium) by a lack of sorting. Breakdown is, technically, a gravity deposit, but many other gravity deposits are found in caves, including landslides of many sizes. Clastic sediments and highly degraded rock can collapse to form fast-moving landslides or be slowly churned as they creep downslope. Debris flows, similar to rapidly moving liquid concrete, are known to enter caves in mountainous regions, but hillslope-derived gravity deposits (colluvium) are most common in caves. Colluvium may enter a cave via impassable openings and spread out to form alluvial fans made of many different sizes of material (Fig. 1).

GEOMORPHIC PERSPECTIVE

As major contributors to landscape denudation, streams carry solutes and sediment produced during weathering and erosion. Sediment quantities and calibers (sizes) are sensitive to changes in climate, tectonics, and base-level elevations, which make old clastic sediments useful for understanding past changes in earth systems, especially when considered together with passage morphologies and elevations. This is well documented by studies of the Mammoth Cave System in Kentucky, U.S.A., and its local base level, the Green River. Drs. Darryl Granger, Art Palmer, and Vic Schmidt age-dated clastic sediments in Mammoth Cave to reconstruct landscape responses to the Pleistocene Ice Ages (Schmidt, 1982; Granger et al., 2001). Early in the cave's history, the Green River was apparently incising slowly because cave streams had time to create large, low gradient trunk passages. Smaller but steeper canyon passages formed soon after glaciers rearranged river paths farther to the north and caused the Green River to incise rapidly, increasing local hydraulic gradients. Subsequently, cycles of sediment infilling and incision occurred in Mammoth Cave as the Green River alternately aggraded and incised in response to Ice Age climate fluctuations. We would know considerably less about the effect of Ice Ages on the region's largest rivers without the unique ability of caves, such as Mammoth, to protect clastic sediments for millions of years.

Clastic cave sediments can also be used to reconstruct flood histories of surface rivers where caves are found in riverbanks. Paleoflood studies are most practical when a surface river is the only significant source of water and sediment to a cave because this simplifies the identification and interpretation of slackwater sediments (Springer and Kite, 1997). Once identified, a minimum floodwater depth can be inferred from slackwater sediment elevations and this stage can in turn be used to estimate flood discharge. Such studies of clastic sediments offer the potential to significantly extend flood records and improve flood recurrence interval calculations. Clastic cave sediments have many other potential applications and the bibliography includes references to a broad range of studies.

See Also the Following Articles

Breakdown
Cosmogenic Isotope Dating of Cave Sediments

Bibliography

Bosch, R. F., & White., W. B. (2007). Lithofacies and transport of clastic sediments in karstic aquifers. In I. D. Sasowsky & J. Mylroie (Eds.), *Studies of cave sediments: Physical and chemical records of paleoclimate* (pp. 1–22). Dortrecht: Kluwer. doi:10.1007/978-1-4020-5766-3_1.

Granger, D. E., Fabel, D., & Palmer, A. N. (2001). Pliocene-Pleistocene incision of the Green River, Kentucky, determined from radioactive decay of cosmogenic ^{26}Al and ^{10}Be in Mammoth Cave sediments. *Geological Society of America Bulletin*, 113(7), 825–836. doi:10.1130/0016-7606(2001)113<0825:PPIOTG>2.0.CO;2.

Newson, M. D. (1971). A Model of Subterranean Limestone Erosion in the British Isles Based on Hydrology. *Transactions of the Institute of British Geographers, 54*, 55—70.

Schmidt, V. A. (1982). Magnetostratigraphy of Sediments in Mammoth Cave, Kentucky. *Science, 217*(4562), 827—829. doi:10.1126/science.217.4562.827.

Springer, G. S., & Kite, J. S. (1997). River-derived slackwater sediments in caves along Cheat River, West Virginia. *Geomorphology, 18*(2), 91—100. doi:10.1016/S0169-555X(96)00022-0.

Springer, G. S., Kite, J. S., & Schmidt, V. A. (1997). Cave sedimentation, genesis, and erosional history in the Cheat River Canyon, West Virginia. *Geological Society of America Bulletin, 109*(5), 524—532. doi:10.1130/0016-7606(1997)109<0524:CSGAEH>2.3.CO;2.

CLOSED DEPRESSIONS IN KARST AREAS

Ugo Sauro

Department of Geography, University of Padova

TYPES OF CLOSED DEPRESSIONS

Closed depressions are the dominant and most distinctive forms of the karst landscape. The following types of closed depressions can be observed in karst areas (listed according to increasing size):

1. *Rain pits* or *rain craters* (1 cm to about 3 cm in size)
2. *Solutions pans* or *kamenitza* (centimeter- to meter-scale features)
3. *Dolines*; also called *sinkholes* (diameters ranging from some meters to over 1 km)
4. *Compound depressions* (hundreds of meters to a few kilometers)
5. *Poljes* (reaching up to some tens of kilometers)

Dolines, compound depressions, and poljes link the surface and the underground fissure and cave systems.

SOLUTION DOLINES AS THE DIAGNOSTIC FORMS OF KARST SURFACES

The solution doline (also called a *sinkhole*) is considered to be the diagnostic form of the karst landscape and is also defined as the index form (White, 1988; Ford and Williams, 2007). Dolines are the most common closed depressions in the Carso di Trieste, which is known as the classical karst because it is regarded as the most typical karst region. Dolines commonly have circular to subcircular plan geometry, and a bowl- or funnel-shaped concave profile. Their depths range from a few decimeters to a few hundred meters, and their inner slopes vary from subhorizontal to nearly vertical.

The Slav term *doline* means "depression" and in a broad sense includes channels and hollows of different types, such as fluvial valleys, dry valleys, blind valleys, uvalas, poljes, and karst dolines. The name *dolina* was first applied in 1848 by Morlot to a circular, closed, karst depression. Subsequently, Cvijic, in 1895, extended use of the term to all circular, closed depressions in karst areas. Gams (2000) further suggested that the word *doline* could be replaced by the term *kraska*, to highlight its importance as a morphological marker of the karst landscapes.

Scientific investigation of dolines began in the middle of the nineteenth century, during construction of the Southern Railway in the Austro-Hungarian Empire through the Karst of Trieste. At that time, the most common theory was that dolines resulted from the collapse of cave roofs. If that were true, the natural hazard for a railway crossing the karst was high. The railway engineers, on the other hand, came to a different conclusion: that most dolines originated through dissolution.

Dolines link the surface and the underground drainage systems. The absence of still water, such as lakes or ponds, at the bottoms of most dolines reveals that water is lost to the underground karst system, most commonly to caves. Through the sole observation of surface forms, a doline seems very simple: it is a bowl- or funnel-like depression with one or a few swallowing cavities at the bottom which are covered by soil. Cross sections outcropping in road cuts and quarries, however, reveal that dolines are actually rather complex. To understand how dolines and, in general, all large, closed karst depressions develop, it is necessary to examine (1) their morphology and size; (2) their locations and relationships with the topographic and geomorphologic settings; (3) their structures; (4) their hydrological behavior and related solution processes; (5) other processes that play a role in their evolution; (6) examples of evolution; and (7) peculiar morphologies that occur under specific environmental conditions. In the following sections, all of these aspects will be discussed.

Morphology and Size of Dolines

Most dolines show a circular or slightly elliptical plan geometry (commonly, the main axis/minor axis ratio is <1.5 and the depth/width ratio is about 0.1). The three-dimensional form of dolines may be compared to a bowl (nearly hemispherical depression), a funnel (nearly conical depression), or a well (nearly cylindrical depression). In some doline populations, more complex forms, such as truncated conical depressions with flat bottoms and star shapes, are common. The bowl shape is the most typical doline shape and is four to ten times more common than funnel-shaped dolines (Cvijic, 1895). The well-shaped doline is uncommon. Soil-covered forms show an almost circular plan geometry. Depressions that developed on bare rock are more irregular and commonly show sharp contours that follow the main fracture systems.

The dolines described by Cvijic as "holotypes" were subsequently carefully analyzed by Šušteršič (1994), who recognized that "... the dolines ... are not as regular as many simple morphometric methods presume... some of these irregularities are probably due to the roughness of the neighboring terrain... the others are probably due to the greater dynamism of the whole doline which does not permit the 'bowl' to achieve a regular shape." Šušteršič (1994) also identified the presence of three distinct concentric areas within each single doline: (1) a flat central area, which is generally covered by soil; (2) a ring of steep slopes; and (3) an outer belt with a gentler slope.

A population of dolines observed in a large-scale topographic map rarely shows the entire spectrum of possible sizes attained by these karst depressions, because smaller forms are not represented. The morphometric parameters of the dolines measured in the Carso di Trieste indicate that there is not a continuous, asymptotic size distribution; therefore, it is possible to recognize different subpopulations. The smaller forms are relatively shallow, and it is difficult to identify them through field observation alone. Within a doline population, there are a few forms that can be distinguished from the typical doline of the population because of some particular characteristics, such as elongation, asymmetry, steepness of the slopes, depth, and so on. Morphometrical methods and analyses of various doline populations are illustrated in a number of papers (Bondesan et al., 1992).

Locations and Relationships with the Topographical and Geomorphological Settings

Dolines commonly occur in populations with variable numbers of individuals and with different densities, ranging from a few to over 200 per square kilometer. Within a population, the density depends on the surface slope. On subhorizontal or gentle slopes, the density of dolines is higher than on steep slopes. Dolines are commonly lacking on very steep slopes. Two major types of karst plateaus can be distinguished on the basis of the spatial distribution of dolines: one that is characterized by isolated dolines (the classical karst type) and another that is characterized by honeycomb systems of dolines. In many karst plateaus, intermediate situations can be observed. With respect to the distribution of dolines, it is possible to recognize the influence of major fracture systems. The center of the doline is commonly set at the intersection between two or more fractures. Rows of dolines commonly follow fault lines, especially when the dislocations are linked to extensional tectonics that bring different rock units into contact. In karst plateaus where a net of dry valleys can be distinguished, dolines are aligned along the valley bottoms.

The Structure of Dolines

In the bare karst, such as at elevations above the timberline, it has been observed that the structure of closed depressions is controlled by the macro- and micro-structural properties of the soluble rocks. For example, in Monte Canin (Julian Alps, between Italy and Slovenia) the properties of the rock yield a high density of steep-sided, rocky, nival depressions, or *kotlici*. In the typical bowl- or funnel-shaped dolines, the rock is normally covered by soil and other surface deposits. The structure of the dolines in the soil-covered karst is visible only through artificial cuts. Some basic characteristics of dolines can be inferred from the analysis of many cross-sections. In particular, it can be clearly observed that the doline structure begins to develop before the onset of the surface depression. Many small depressions in the hard rock, which are completely filled by soil and surface deposits, go undetected if one looks only at the surface. These small forms have been termed *embryonic cryptodolines* or *subsoil dolines* (Nicod, 1975). Some of these structures host paleosols, beside the present-day soils. The presence of paleosols in the filling sediments indicates that the development of cryptoforms took place over a long time and that cryptodolines have acted as sediment traps since the beginning of their formation. In the early phases, the trapping action was effective for soils, which show less erosion than in the surrounding area in correspondence to the cryptodepression. When the cryptodolines evolved as surface forms, the trapping action also became efficient for other materials, such as rock debris from the slopes, soil sediments, aeolian dust, volcanic ashes, and so on. In the Mediterranean karst, these fillings are commonly red (*terra rossa*, or "red soil"), and the color is the result of weathering processes. In the dolines of the classical and dinaric karst, the filling commonly consists of a slightly reddish brown silt, which is primarily composed of weathered, loess-like sediment of aeolian origin. The filling may be up to 10 m thick. The profile of some dolines exhibits multilayered filling.

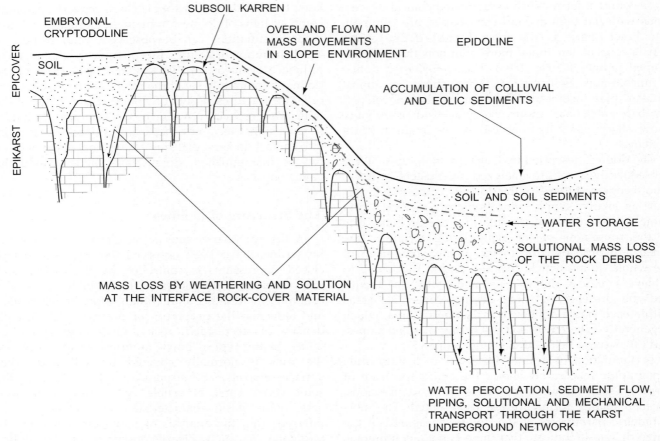

FIGURE 1 Schematic showing the structures of a doline and a cryptodoline. The epikarst in the host rock, the fillings, and the soil are outlined. The secondary porosity of the soluble rock is greater at the bottom of the dolines. The main morphodynamic and hydrologic processes are also provided.

The sketch of the structure of an ideal middle latitude doline (Fig. 1) shows that the solid rock/filling materials interface is not a simple boundary, but rather a complex, involute surface that consists of many rounded bulges in the soluble rock separated by crevices and fissures that become narrower inward. These covered solution forms, called *rounded karren*, develop through the slow flow of the water solution at the rock/filling interface. The structures of dolines in different karst areas show large variability depending on the role that is played by different factors, such as lithology, structural conditions of the rock, topography, climate conditions, and features inherited from previous morphodynamical events.

Hydrological Behavior and Related Solution Processes in Dolines

A simple doline is a hydrological form that can be compared to a first-order valley segment. Its concavity is an expression of the convergence of water toward the bottom which is a transitional point between surface and underground hydrologic networks. The physical structure of the doline is the framework of its hydrological functioning and, because this functioning is also related to the genesis and evolution of the form, the doline may be considered the surface expression of a peculiar three dimensional hydrostructure. The geomorphological setting and the structure of the dolines allow the distinction of three major hydrostructural types (Fig. 2): (1) *point-recharge doline*, (2) *drawdown doline*, and (3) *inception doline*.

A point-recharge doline begins to form when a fluvial net loses water to the cavities of a soluble rock. It is necessary for protocaves to begin to form inside the rocky mass, connecting surface recharge points to an underground network. Once developed, a protocave focuses both drainage and solute removal at the points of transition from the surface to the underground network, thus causing the development of a surface depression. The point-recharge doline is commonly located along a

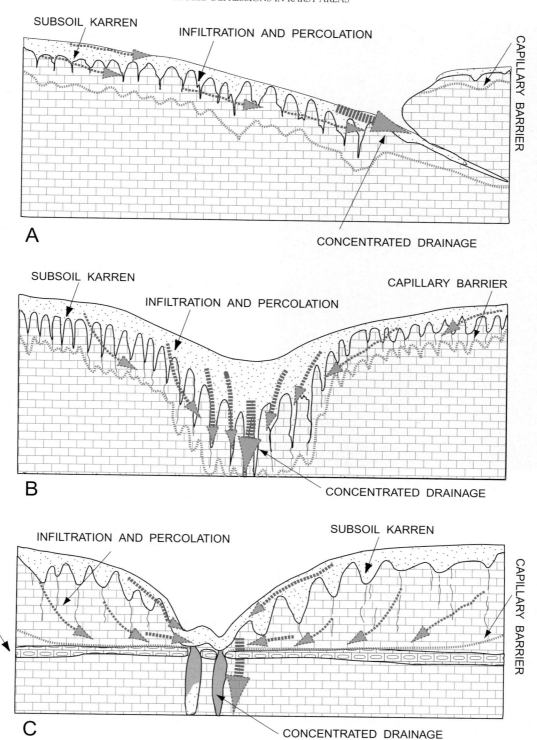

FIGURE 2 The three main types of solution dolines. (A) In the point-recharge doline, the strongly asymmetric hollow is linked by one main sink, fed primarily by surface and soil water. (B) In the drawdown doline, the depression results from focusing of the dissolution inside the water infiltration zone of the rock through centripetal convergence of the mainly subsurface water held inside the epikarst (*i.e.*, the upper zone of the soluble rock presents a greater secondary porosity). (C) In the inception doline, the depression also originates from a centripetal convergence of water, but this occurs inside a preexisting hydrogeological structure and is triggered by a change of hydraulic conductivity of the rocky mass, influenced by lithological and structural factors.

FIGURE 3 A line of funnel shaped dolines in the Monti Lessini along the bottom of a dry valley (Venetian Pre-Alps, Italy). These inception dolines develop just above a lithology change.

increases centripetally toward each leakage point, and more water molecules are brought into contact with the rock surface near the major drain. There, the rock is dissolved more rapidly, secondary porosity increases faster than in the peripheral zone, and the hydraulic conductivity increases. Consequently, a depression tends to form at each leakage zone, the center of which deepens faster than in the surrounding areas. A drawdown doline is, therefore, the product of the positive feedback resulting from the interrelations between the hydrological and the solute processes acting inside the outer layer of a fractured soluble rock.

Gams (2000) calls the dissolution process at the central part of a doline *accelerated corrosion*. The hydrostructure corresponding to this central part is comparable to the funnel-shaped depression created by a pumping well in the water table of a phreatic aquifer. The outer rock layer, which is characterized by high secondary porosity, is shallower in the peripheral areas and thicker in the central parts of the dolines. It hosts a hanging aquifer, which is nearly saturated during wet climate periods and nearly empty during dry periods. The existence of the epikarst with the characteristics of a water reservoir was recognized through the study of the hydrological regime of karst springs.

The inception dolines develop from the interception by the epikarst of a hydrogeological structure that developed inside the rock and previously triggered by a change in hydraulic conductivity of the rocky mass. A change of lithology or the presence of an impermeable layer, such as clay or chert, may cause the formation of a hanging aquifer. If connections such as fractured zones exist and allow water to overcome an obstacle, a hydrostructure similar to the drawdown dolines develops, yielding a focused dissolution both above and below the bottleneck (Fig. 3). Above, the fractures of the rock are enlarged to fissures; below, small shafts develop.

This type of doline differs from the drawdown doline because of the more marked lithological and structural control and because of an early evolution of the hydrostructure, which is not necessarily dependent on the epikarst.

The models illustrated in Figure 2 do not consider the role of surface deposits or filling materials. These components are important, because the water is absorbed by soil, surface materials, and filling sediments before it reaches the fissures in soluble rock. The fillings may host small aquifers nested above the epikarst and act as pads that slowly release water downward, thus influencing with their physical character both the flow velocity at the cover–rock interface and the water regime of the epikarst and the main karst springs.

rather dry hydrographic net and shows transitional features between the depression of a blind valley and a common drawdown doline (Fig. 2A). It is often elongated and skewed in the direction of the flow, with the upstream side being gentler than the downstream side.

The drawdown doline is the most common and typical karst form. To understand the drawdown dolines it is necessary to consider the behavior of the water in the rock zone just beneath a nearly horizontal or gently sloping surface. Near the surface, the rock fractures tend to open by tensional relaxation; therefore, the soil water seeps through the fractures and enlarges them through dissolution. With time, secondary porosity develops inside the shallow rock layer near the surface which then becomes saturated in wet periods. If, in some points of this layer, the water is able to find and enlarge routes to an underground network, flow paths converging to the leakage zones are thus established. As a consequence, the water velocity

Other Processes in Doline Evolution

Even if the major process of doline formation is dissolution through a differential mass wasting of the rock, the following processes also play roles in the shaping of the depression: soil forming processes and other weathering processes, slope processes, overland flow processes, the capacity of the closed depression to trap different types of sediments, and the processes of evacuation of filling materials.

The soil acts as a filter and as an insulating layer with respect to the outer environment. Soil releases solutions rich in humic acids and air with a high content of carbon dioxide. It also yields fracture-filling material.

Frost shattering of the rock is a very important weathering process. In the fillings of most midlatitude dolines, a variable amount of angular rock debris is produced through this mechanism. During the cold phases of the Pleistocene, some dolines were completely filled with rock debris. Slope processes influenced by the gravitational force, such as creep and solifluction, are responsible for soil thinning, for the formation of a debris cover along the slopes, and for its thickening in the bottom areas. The overland flow processes are responsible for erosion, the washing of soils and loose material, and for the deposition of colluvial sediments in the bottom area of the depression. Once formed, a closed depression acts as a trap for different types of materials that are carried by wind, rain, and so on.

In the midlatitude karst, the doline filling contains variable amounts of loess-like deposits that were transported by the wind during cold Pleistocene phases, volcanic ashes, sands, and silt deposited by rain, and so on. The fine-grained material is commonly deposited on the entire karst surface and then accumulates by overland flow into the central part of the depression. Many authors consider the filling material to be a residue of limestone dissolution. Even if in some karst areas this may be true, most of the midlatitude doline fillings consist of allochthonous materials, with the exception of angular rock debris due to *in situ* frost shattering. The filling material may be evacuated through the karst conduits through different processes, such as mass wasting of the rock debris by solution, subsurface flow, liquefaction of the loose material, or piping.

Examples of Evolution

Some examples will help to understand the evolution of dolines. Point-recharge dolines are present in the gypsum karst of the Santa Ninfa Plateau (Sicily, Italy). There, chains of dolines follow the pattern of a fluvial network that developed on the impermeable cover and was overprinted on the gypsum. Inside some chains, the bottom of the upstream doline is at a lower altitude than that of the progressively downstream doline. The last upstream doline marks the end of a blind valley. It is clear that each closed depression became less active after the development upstream of a new swallow hole. So, the speed of bottom deepening is strictly linked to the activity of the swallow hole.

In the Waitomo district of New Zealand, populations of point-recharge dolines occur in the framework of interstratal karstification. It is possible to reconstruct the transitions from a fluvial network in an impervious rock to a honeycomb system of dolines that developed in the underlying limestone units and the subsequent recovery to fluvial network in the impervious rock unit previously overlain by the limestones, following their chemical erosion. Gunn (1986) proposed a model based on five sample areas, each representing a different stage of the transition. The model outlines the important roles played by both allogeneic recharge and the location of previous subterranean drainage structures.

A small, nearly horizontal area of the classical karst, which was surveyed at very high resolution (scale 1:1000; contour interval 1 m; spot elevation, at the bottoms of some depressions, checked at the resolution of 0.1 m), is characterized by drawdown dolines, including some very small and shallow forms (Fig. 4A). The main morphometric parameters reveal three main subpopulations represented by the following types of dolines: (1) small and shallow dolines less than 20 m in diameter and 0.4 to 2 m in depth; (2) medium-sized dolines 12 to 50 m in diameter and 2 to 7 m in depth; and (3) large dolines, 50 to 120 m in diameter and 8 to 15 m in depth. From the ratios between the numbers of individuals of each group it is possible to infer that most of the embryonic surface dolines will probably abort prior to becoming medium-sized forms. Their hydrological functioning is at the limit between triggering or not the positive feedback that would allow them to evolve into typical dolines. The coexistence of the three subpopulations also suggests different ages for the forms, which are probably related to the starting times of their evolution during favorable morphoclimatic phases. The smaller and shallower dolines could be the result of the karst morphogenesis that began at the end of the last cold period, which was triggered by the preexistence of cryptodolines.

Seven rock terraces were cut in the Montello (Venetian Pre-Alps of Italy) neotectonic anticlinal morphostructure consisting of Upper Miocene conglomerate during the tectonic uplift. The terraces subsequently became the sites of doline morphogenesis (Fig. 4B). Depending on the ages of the terraces, it is possible to analyze populations of drawdown dolines that developed in very similar geomorphological

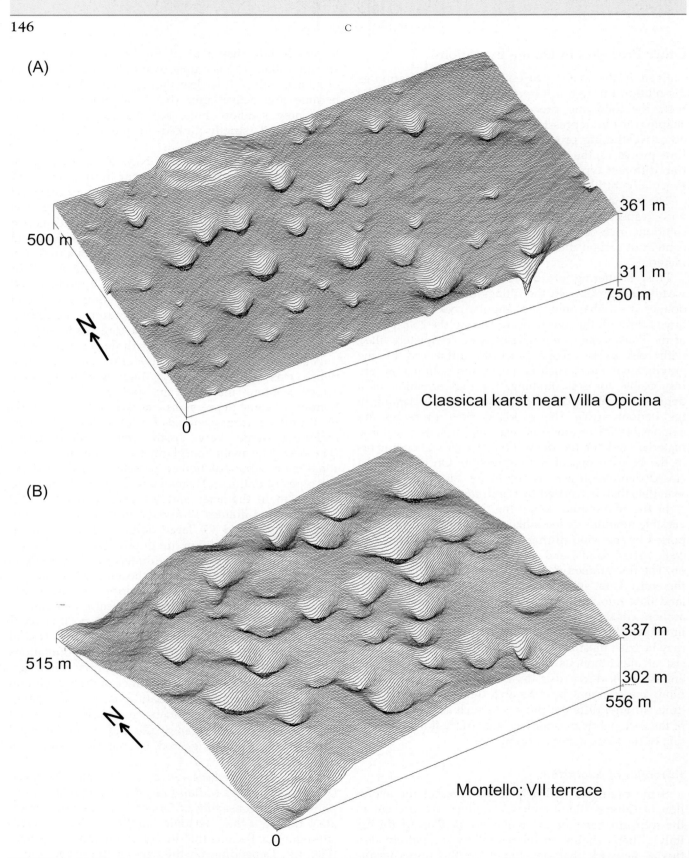

FIGURE 4 Digital elevation model of two doline areas. (A) Classic Karst near Opicina (Triest). (B) Seventh terrace in Montello Hill (Venetian Pre-Alps). Morphometric parameters and distribution of dolines are more homogeneous in the Montello area than in the Opicina example. In the latter, the dolines show a larger variability of morphometric values. *Modified from Ferrarese and Sauro, 2001.*

environments and now represent different evolutionary stages. A "standard" doline for each paleosurface was reconstructed from morphometrical analysis of the dolines of each terrace (with the exception of the lowest one, where no recognizable dolines exist) (Fig. 5). The series show that the standard forms are not the expression of a regular and linear growth, probably because of the different influence of some local factors on each terrace. Nevertheless, there is a general trend toward increasing sizes and depths of the dolines, from the lower and younger terraces to the upper and older ones. What is clear is that the deepening process was faster than the widening process. What was increasing in a perfectly linear way was the sum of the volumes of the dolines for an area unit. The depressions tended to evolve from bowl-like to dish-like forms. The flat bottoms of the older and larger depressions result from thick fillings that formed inside the hollows. In the lower terraces, the dolines are isolated spots on a nearly flat surface, but on the upper terrace their boundaries are shared and a honeycomb karst morphology is almost reached (Fig. 4B).

In the southern Monte Baldo (Venetian Pre-Alps of Italy) are dolines that differ in size and character. Many host very thick fillings; some open dolines filled up to the rim look like amphitheaters with flat bottoms, which are open on one side. The fillings commonly consist of angular rock debris supported by a loess-like silt matrix. Some relicts of large egg-carton-like structures show that a population of big dolines, which probably developed in the late Tertiary, had been largely dismantled by the periglacial processes during the cold phases of the Pleistocene. During the Interglacials, the karst morphogenesis prevailed against other competitive processes, and so, many dolines survived up to the present day, even though their morphology has been strongly modified (Sauro, 1995).

The Velebit Mountains of Croatia have different types of dolines. In the central plateau, the dolines are very large and deep and are funnel shaped, almost without covers and fillings and with open shafts on the slopes and at the bottom. In the southern plateau, the depressions are smaller and partially filled by rock debris and loess-like deposits. The main difference between the two groups is represented by the lithology of the host rocks: the larger rocky dolines developed in a massive limestone breccia characterized by low sensitivity to frost shattering, while the smaller and partially filled dolines developed in a limestone more susceptible to gelivation. The larger depressions remained, therefore, filling free, and during the winter they trapped large amounts of wind-transported snow. The large amount of meltwaters accelerated the evolution of the forms, which reach huge sizes.

In the Eyre Peninsula of South Australia, clusters and lines of dolines developed on a stack of fossil coastal fore-dunes of Middle and Late Pleistocene age, now consisting of calcarenites. Some dolines occur high in the local topography and are also aligned in groups. Their distribution is influenced by the diversion of groundwaters in fractures that cut into the pre-Pleistocene basement (granites) and the predominant action of solution processes in the limestone above such zones. This is probably related to an earlier mass wasting of the sand during its diagenesis in the open fissures of the underlying rock with a local increase of porosity of the newly formed rock above. The influence of a buried surface on the development of hydrostructures and related forms is referred to as *underprinting* by Twidale and Bourne (2000).

The above examples demonstrate the important roles played by various factors and processes such as the morphostructural setting of the soluble rock, its lithology and density of discontinuities, its susceptibility to frost shattering, the qualities of the rock units above and below, the characters of the erosional surface on which the karst forms began to develop, the morphoclimatic system and its relative changes, and the occurrence of such events as falling volcanic ashes, among others.

Populations of Dolines Linked to Specific Environmental Conditions

In humid tropical climates, some populations of closed depressions show peculiar characteristics that distinguish them from doline populations of the midlatitude karst. In particular, the closed depressions are often larger than typical dolines and in the topographical maps show a star-shaped figure. The cartographic representation of a tropical karst landscape with closed depressions resembles a classical karst landscape with dolines, but with an inverted relief. In fact, most of the rounded, concentric figures drawn by the map contour lines are conical or tower-like hills encircling the concentric star-shaped contours of a closed depression (Fig. 6). The egg-carton shape of the basin is characterized by a great difference between minimum and maximum depths (differences in elevation between the bottom of the depression and the lowest and highest points of the perimeter). This type of tropical doline, called a *cockpit* from their Jamaican name, has very steep slopes, usually from 30 to 40°, and is coated by a thin soil cover. Often, the rock outcropping on the steeper slopes shows evidence of biokarstic weathering, such as honeycomb alveolar cavities, and locally of carbonate deposition. At the bottom, the fillings are shallow, and sometimes it is possible to observe a flat, sponge-like rocky bottom, blackened by biokarstic weathering, and a sharp change of gradient at the contact with

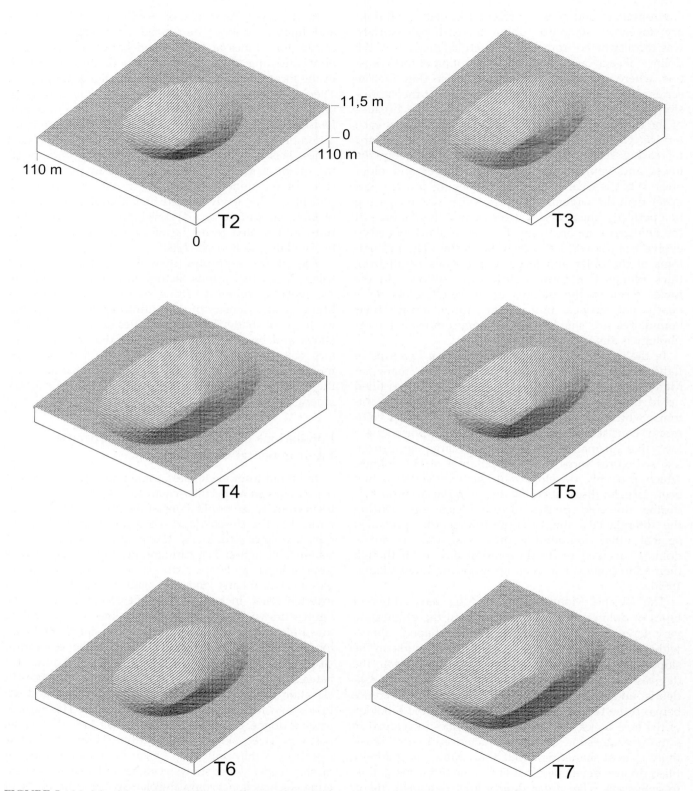

FIGURE 5 Models of the standard doline for every Montello terrace, ordered from the youngest (T2) to the oldest (T7). The model is drawn using the average value of each morphometric parameter. The horizontal scale is the same for all drawings, while the vertical scale is slightly decreasing from T2 to T7.

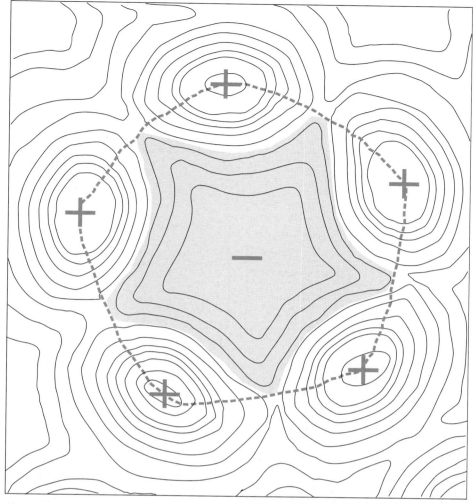

FIGURE 6 Sketch of a tropical cockpit. The star-shaped depression is encircled by conical hills separated by dry valleys and gaps. The planar shape of the watershed line is polygonal.

the surrounding slopes. During the wet season, the shallow holes at the bottom of some depressions behave like springs, and temporary ponds may form.

The morphological differences between the normal solution dolines and the tropical cockpits are mainly due to the presence of the valley-like depressions entrenched on the slopes of the cockpits, which determine the star-like form of these sinks (Williams, 1985). These incisions in the slopes, even if similar to some morphological types of the fluvial landscapes, are solution forms. Their presence in the tropical dolines, and not in the midlatitude dolines, is due to such different factors as: (1) the larger dimensions of the slopes of the tropical dolines; (2) the poorer development and role of the epikarst; (3) the minor influence of limestone debris and soil cover; and (4) the more active overland flow and biokarstic weathering that act on the slopes.

Similar to tropical dolines, some populations of midlatitude dolines, such as those of the coastal belt of Dalmatia (Velebit, Croatia) and of Cantabria (Spain), may be interpreted as inherited forms, which first developed in subtropical humid conditions during the late Tertiary. The survival of inherited forms was possible where the dismantling processes competing with the solution processes, such as frost shattering, have played a secondary morphogenetic role due to both peculiar environmental conditions and a low susceptibility of the rock to gelifraction. Also similar to the tropical dolines, some very large dolines of the Central Velebit plateau can be explained as old, inherited forms that were not considerably altered by the periglacial processes of the cold Pleistocene phases because they developed in a limestone breccia resistant to frost shattering.

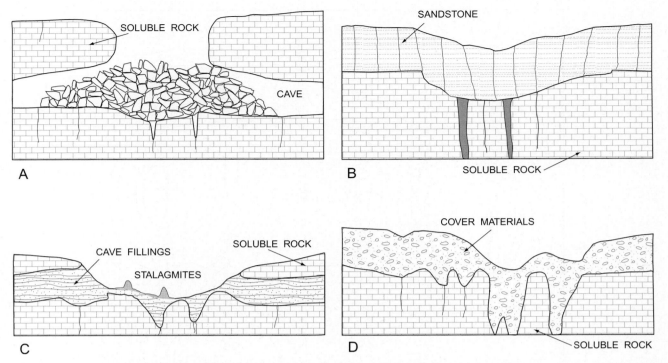

FIGURE 7 The four main types of dolines that do not directly originate by solution of rock at or near the surface: (A) collapse doline, resulting from the collapse of the roof of a cave; (B) subsidence doline, formed by the settling of an insoluble rock following solution of the underlying soluble rock; (C) intersection doline, originating from the emptying of fillings of an old fossil cave due to intersection with the topographical surface; (D) cover doline, which has developed in incoherent rocks burying a soluble rock or partially filling a karst depression.

Summary of Solution Dolines

The solution doline may be considered the most typical karst geo-ecosystem that links the surface and the underground networks. A doline is both a three-dimensional form, and a multicomponent geosystem consisting of several elements of variable thickness, such as soil, surface deposits, fillings, and epikarst. If the development of a doline is triggered by a positive feedback in the synergical interactions of hydrological and solute processes, other cooperative and competitive processes may also play an important role in its evolution. In particular, the soil plays a cooperative function by enriching the water with carbon dioxide. The fillings, on the contrary, commonly act as a competitive factor by slowing water seepage into the epikarst and, if carbonates are present in their composition, by increasing the water pH, thus decreasing potential dissolution of the underlying solid rock. Without the presence of soil cover it is not possible for a doline to attain a typical bowl- or funnel-like shape, as documented by the irregular depressions of the bare high mountain karst. Soil and other surface deposits, therefore, play a fundamental role in the evolution of the forms. Dissolution is the dominant morphogenetic process in the cockpits of the tropical karst. If the soil cover is very thin on the steep slopes, its role can be replaced by a rock layer that is strongly corroded by biological processes.

OTHER TYPES OF DOLINES

In karst areas it is possible to find other types of dolines that do not directly originate in the solution of rock at or near the surface. The main types are (1) *collapse dolines*; (2) *subsidence dolines*; (3) *intersection dolines*; and (4) *cover dolines* (Fig. 7).

Collapse Dolines

Collapse dolines result from the collapse of a cave. They show different shapes, with vertical or overhanging walls. Collapse dolines that allow entry into a subterranean system are also called *karst windows*, because they open into the underground environments. The best known karst windows are the dolines of Skocjanske Jame in Slovenia (Fig. 8), where the Reka River enters into a large cave feeding the aquifer of the classical karst and the spring of Timavo 40 km away.

FIGURE 8 The karst windows on the first part of the subterranean course of the Reka river in the Skocjanske Jame system (Slovenia).

FIGURE 9 A huge collapse doline called Red Lake near Imotski (Croatia). The depth from the rim to the lake is over 250 m, and the depth of the lake itself is greater than 250 m.

The range of sizes of the collapse dolines varies from a few meters to hundreds of meters. The Velika dolina (large doline) of Skocjanske is about 500 m in diameter and 164 m in depth. The Blue Lake and Red Lake dolines of Imotski (Croatia) are among the most spectacular collapse dolines in the world (Fig. 9). The main axis of the Blue Lake doline is about 1.5 km; the diameter of the Red Lake doline is about 400 m at the lake surface, and its total depth, including the submerged part, is 520 m.

The Chinese name of the largest of the collapse dolines is *tiankeng*. A tiankeng is a collapse doline at least 100 m long, wide and deep, and with perimeter walls that are close to vertical (Zhu and Waltham, 2006). The greatest of these depressions exceed 100 million m^3 in volume and have developed in correspondence of large cave chambers where underground rivers are flowing.

These forms are not the result of a unique collapse episode, but of a number of collapses often related to oscillations of the water table (Beck, 1984). Most of the collapse dolines do not provide access to a subterranean system because of collapse debris at the cavity bottom.

A particular type of collapse doline is the *cenotes*, open depressions in the coastal karst belts giving access to the water table and resulting from the collapse of the roof of submerged cavities. The best-known cenotes are in Yucatán and in the Caribbean Islands. The evolution of cenotes and of the related cave systems is often due to both efficient dissolution by the brackish waters of the coastal belts and oscillations of the water table. Some cenote populations could be interpreted to be intersection dolines.

Subsidence Dolines

Subsidence dolines are closed depressions caused by the subsiding of an area. They may be due to upward propagation of an initial collapse of the roof of a deep cavity with development of a breccia-pipe structure; when this structure reaches the surface a closed depression with an unstable bottom is formed. Subsidence closed basins are also found on insoluble rocks overlying soluble rocks. If the underlying rocks are eroded by solution processes, the rocks above subside. These hollows, which may host lakes, are frequent on rocks overlying very soluble minerals, such as gypsum and salt.

Intersection Dolines

Intersection dolines are depressions formed when an old cave system, partially or totally filled with sediments, is cut by the topographic surface as a consequence of the lowering of such a surface by chemical erosion. The opening of such fossil caves reactivates the old hydrostructures, leading to evacuation of the fillings and development of closed depressions. These forms are relatively common in the classical karst, where chains of depressions or slender depressions also some hundred meters long exist. On the bottom of such a doline, it is easy to find relicts of cave fillings and stalagmites. The recent building of a highway has cut many of these forms and they have been labeled *roofless caves* by Mihevc (2001); here, the term *intersection dolines* is preferred because they are no longer underground forms.

Cover Dolines and Other Types

Cover dolines are the closed depressions formed in incoherent materials such as alluvial deposits, glacial drift, superficial deposits, soil sediments, and so on. These forms, also commonly called *alluvial dolines*, have different characteristics according to the type, continuity, and thickness of the deposits and the presence of a well-developed karst relief under the deposits or of a still incipient karst hydrology, among other factors. The alluvial plains of a large river and the flat bottoms of large solution dolines, retaining filling deposits, may be considered as end members of the various development environments for this type of doline. The influence of buried karst hydrostructures on the formation of cover dolines is a clear example of underprinting. Various processes may play a role in the evolution of these forms, such as water infiltration, washing and piping, suffosion (in the sense of erosion from below), suction of the sediments caused by oscillations in the water table, liquefaction of the sediments and the linked mud flows, and changes in volume caused by freeze–thaw cycles.

The two main models for the development of cover dolines are (1) upward migration of a cavity and (2) subsidence, as determined by the gradual rearrangement (a sort of inner creeping) of the cover material induced by its mass wasting through the karst net. If the cover material is sufficiently cohesive, the mass wasting by the water infiltrating underground produces a cavity with an arched roof; the arch gradually migrates upward, where it reaches the surface and causes the sudden opening of an ephemeral cover doline with nearly vertical walls. The subsidence cover dolines form in less cohesive sediments.

In addition to the forms of dolines described above are other types originating by different processes, such as dolines along fault lines caused by seismotectonic movements and anthropogenic dolines. Also, craters similar to dolines were created by shelling during World War I in the chalky limestone of the Venetian Pre-Alps and are an interesting anthropogenic type. The crater-like depressions behave as drawdown dolines because of the rock fracturing caused by the explosions..

UVALAS AND COMPOUND AND POLYGENETIC SINKS

In the old literature, a large closed depression that does not show a doline morphology is also referred to as an *uvala*. This name is one of the first classical ones applied to a type of karst depression that has been abandoned because it was erroneously utilized by some authors in combination with the theory of the karst cycle. A recent review about the uvala concept, based on the analysis of a population of large karst basins of the dinaric karst, revalues this term (Calic, 2009). Indeed, there are forms that match the original concept such as closed basins mostly of elongated and irregular plan shape ranging in size from approximately one up to several kilometers without a flat bottom and never flooded, generated mainly by the karst process, and so without clear evidences of fluvial and glacial shaping. In the dinaric karst, development of such depressions of uvala type is mostly guided by tectonically broken zones of regional extension, in which corrosion is intensified due to rock fracturing. So, from the genetic point of view, uvalas could be considered forms of accelerated corrosion triggered by local tectonic settings and dynamics.

Beside the uvalas, both compound and polygenetic closed depressions can be observed in the karst landscapes. A compound hollow is a form that originated from the fusion of simpler forms. If more dolines coalesce together, an irregular depression develops, sometimes with a lobate perimeter.

A polygenetic sink is a closed depression that clearly evolved through both the karst process and another morphogenetic process. The most frequent types are the (1) *tecto-karstic hollows*; (2) the *fluvio-karstic hollows*; and (3) the *glacio-karstic hollows*.

A tecto-karstic hollow is a closed basin that developed inside a tectonic depression that also evolved through karst processes. A fluvio-karstic hollow may be considered the closed ends of blind valleys that evolved through both fluvial and karstic processes. Glacio-karstic basins are common forms in the alpine high mountain karst. These forms were formed by both karst processes and glacial abrasion and are of two main types: (1) the high plateau type, often elongated and similar to a bathtub; and (2) the glacial cirque type, which developed in the bottom of a glacial cirque. Larger glacio-karstic depressions can be over 1 km long. Their bottoms are often occupied by rocky hummocks, drumlins and till deposits..

DEPRESSIONS IN KARST AREAS

Poljes are the largest closed depressions observed in the karst terrains. In 1895, Cvijic defined a polje as a "large karst depression, with a wide, flat and nearly horizontal floor, completely enclosed between steep slopes." This simple definition is insufficient to explain the genesis of this type of form and the role of the karst processes in its characterization. To understand this form it is necessary, as for dolines, to consider both the structure of the form and its dynamics; in other words, to consider the entire geosystem expressed by the form. Poljes present a large variability, and the development of most of them cannot be explained through the karst process only, because they are polygenetic forms, resulting from the combination of a number of processes.

Typical of this form are the poljes of the dinaric karst, which show large, flat, and nearly horizontal floors, from one to some tens of kilometers long. The sides, usually sloping at about 30°, connect with the bottoms at a sharp angle. Closed depressions of this type are also present in other karst regions of the world and are referred to by various names, such as *campo* or *piano* in Italy, *plans* in France, and *hojos* in Cuba.

The main environmental peculiarity of this type of form is the common absence of a permanent lake inside the basin; an ephemeral lake without surface outlets may form and disappear during the seasonal cycle in relation to the precipitation regime. The lake, if present, represents the transition point between the surface and the underground hydrology and serves as a window to the underground aquifer. In other words, the polje floor is affected by the oscillations of the local water table. Although the lake may be fed by both surface runoff and underground circulation, the draining of it occurs exclusively through the subterranean karst network (Fig. 10). The dynamics of the lower part of the basin is strictly linked with the seasonal cycle of water input, throughput, and output. The temporary lake is responsible for the solution of both the floor and the base of the slopes, leading to the planation of the bottom and to its lateral enlargement by marginal corrosion. The forming, standing, and dissipation of the water body is also the cause of a redistribution and leveling of the filling materials and of their volumetric reduction by solution and erosion via the subterranean circulation system (Fig. 11).

In the structure of the poljes it is possible to recognize a large variability in the thickness of the filling deposits, ranging from a few decimeters to more than 100 m. Swallowing cavities and springs may open at both the bases of the slopes and inside the floor area. Of these, some are real caves, and others are cover dolines developed in the fillings above cavities in the underlying rock. From the hydrological point of view, some behave permanently as springs or as swallowing cavities, and others may invert their functioning from springs to swallow holes and *vice versa*. This last type is referred to as an *estavelle* in France.

FIGURE 10 A typical blind valley: the Biogradski ponor (Bosnia-Herzegovina).

FIGURE 11 A typical karst polje in the dinaric range: the Dabarsko Polje (Bosnia-Herzegovina).

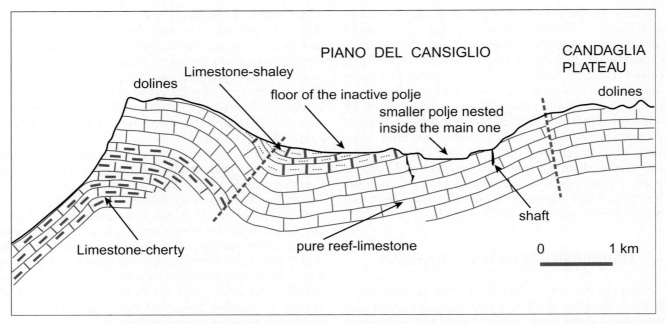

FIGURE 12 The structural polje of Piano del Cansiglio in the Venetian Pre-Alps (Italy). The polje is now inactive and hosts smaller depressions nested inside the main floor. This form has been described as a lithological contact polje.

The classification of poljes may be based on both the geomorphological character and the hydrodynamics. Gams (1994) used these two criteria to distinguish five main types of poljes: (1) *border poljes*, (2) *piedmont poljes*, (3) *peripheral poljes*, (4) *overflow poljes*, and (5) *base-level poljes*. A border polje is located at the transition between a nonkarstic and a karstic area. The recharge of the basin is, therefore, mainly due to surface streams and the escape of water through stream-sinks. A piedmont polje is located downslope of a mountain area from where a large amount of debris has been received, filling the depression and hosting a local aquifer. Here, the recharge is also in the main made by surface streams. A peripheral polje is a depression fed by a large internal area of impermeable rocks with a centrifugal stream network. The sinks are located around the periphery of the inlier. An overflow polje is underlain by a belt of relatively impermeable rocks that act as a hydrological barrier to water that emerges at springs on one side of the polje floor and escapes via stream-sinks on the other side of the basin. A base-level polje is a polje with a floor cut entirely across karst rocks and is affected by the vertical oscillations of the water table; consequently, it is inundated during high-level periods.

Based on the books of Ford and Williams (1989, 2007), it is possible to reduce the categories to three basic types: (1) border poljes, (2) structural poljes, and (3) base-level poljes. From the geomorphological point of view, most poljes correspond to tectonic depressions such as graben, faultangle depressions, pull-apart basins, and so on. These dislocations often lead to contact between rocks with different permeabilities, thus leading to conditions of lithological contact. Many structural poljes have also been described as lithological contact poljes (Fig. 12), a situation corresponding, at least partially, to that of the border poljes.

Some structural poljes have trapped hundreds of meters of sediments hosting local aquifers. An example is the Piano del Fucino in the Central Apennines (Italy), the bottom of which, during most of the Holocene, was occupied by a lake subject to numerous vertical variations in the level. In early Roman and modern times tunnels were excavated to drain the lake and to reclaim the floor for agriculture. Today, the intensive agriculture is supported by overexploitation of the alluvial aquifer.

Some poljes have developed along typical fluvial valleys. A specific example is represented by Popovo Polje in Bosnia-Herzegovina, which is 40 km long and only 1 to 5 km wide. The floor slopes gently throughout its length as demonstrated by the course of the river Trebinjcica. Before the construction of a draining tunnel, the bottom was occupied from October to May by a seasonal lake.

The best examples of true karst poljes can be found in tropical karst regions, such as in southern China. Here, the floors of certain populations of large cockpits are in the vertical oscillation zone of the water table. The base-level floors of these cockpits expand by marginal corrosion, leading to the gradual dismantling of the ridges and to fusion of the floors. The resulting compound depressions represent true karst poljes, probably a unique type of polje created by karst processes only.

See Also the Following Article

Soil Piping and Sinkhole Failures

Bibliography

Beck, B. F. (1984). *Sinkholes: Their geology, engineering and environmental impact*. Rotterdam: A. A. Balkema.
Bondesan, A., Meneghel, M., & Sauro, U. (1992). Morphometric analysis of dolines. *International Journal of Speleology, 21*, 1–55.
Calic, J. (2009). *Uvala—Contribution to the study of karst depressions, with selected examples from Dinarides and Carpatho-Balkanides*. PhD thesis, Postojna.
Cvijic, J. (1895). *Karst*. Beograd: Geografska Monografija.
Ferrarese, F., & Sauro., U. (2001). Le doline: Aspetti evolutivi di forme carsiche emblematiche [The doline: Evolution aspects of the emblematic karst form]. *Le Grotte d'Italia, 5*(2), 25–38 (in Italian).
Ford, D., & Williams, P. W. (1987). *Karst geomorphology and hydrology*. London: Unwin Hyman.
Ford, D., & Williams, P. W. (2007). *Karst hydrogeology and geomorphology*. Chichester, U.K.: Wiley.
Gams, I. (1994). Types of poljes in Slovenia, their inundation and land use. *Acta Carsologica, 23*, 285–300.
Gams, I. (2000). Doline morphogenetical processes from global and local viewpoints. *Acta Carsologica, 29*, 123–138.
Gunn, J. (1986). Solute processes and karst landforms. In S. T. Trudgill (Ed.), *Solute processes* (pp. 363–437). New York: John Wiley & Sons.
Mihevc, A. (2001). *Speleogeneza divaskega Kraza [The speleogenesis of the Divaca Karst]*. Ljubljana, Slovenia: Zalozba ZRC, ZRC SAZU.
Nicod, J. (1975). Corrosion de tipe crypto-karstique dans le karst méditerranéen. *Bulletin Association Geographique François, 428*, 284–297 (in French).
Sauro, U. (1995). Highlights on doline evolution. In I. Barany-Kevei (Ed.), *Environmental effects on karst terrains* (Vol. 34, pp. 107–121). Szeged, Hungary: Universitatis of Szegediensis.
Šušteršič, F. (1994). Classic dolines of classical sites. *Acta Carsologica, 23*, 123–156.
Twidale, C. R., & Bourne, J. A. (2000). Dolines of the Pleistocene dune calcarenite terrain of western Eyre Peninsula, South Australia: A reflection of underprinting? *Geomorphology, 33*, 89–105.
White, W. B. (1988). *Geomorphology and hydrology of carbonate terrains*. New York: Oxford University Press.
Williams, P. W. (1985). Subcutaneous hydrology and the development of doline and cockpit karst. *Zeitschrift für Geomorphologie, 29*, 463–482.
Zhu, X., & Waltham, T. (2006). Tiankeng: Definition and description. *Speleogenesis and Evolution of Karst Aquifers, 4*(1), 1–8.

COASTAL CAVES

John E. Mylroie
Mississippi State University

INTRODUCTION

Caves that form in coastal environments are controlled by different factors from caves that form in traditional inland settings. The first and most obvious factor is the physical and chemical power of waves and salt water acting on coastal rocks. Second, and less obvious, is the fact that sea level can change, and with that change, the position of the coastline moves. Therefore the position of cave development by coastal processes will also move. Sea level can change in a variety of ways, but there are two ways that are of particular importance to cave formation on coasts. First, sea level can change on a global scale, called *eustatic* sea level change. The most common reason for this change is the amount of ice on the continents. From 10,000 to 2,600,000 years ago—the Pleistocene Epoch—the Earth underwent a series of ice advances called *glaciations* (the "Ice Ages") and a series of ice retreats called *interglacials*. As ice sheets grew, evaporated seawater falling as snow on land was trapped as ice, and sea level dropped worldwide. When the ice melted as an interglacial occurred, sea level rose as the meltwater flowed back into the ocean basins. Eustatic sea level change of this type is called *glacioeustatic* sea level change. Evidence indicates that the Earth went through at least 15 of these glacial cycles, and hence sea level changes, in the Pleistocene. Second, sea level position can also shift as isolated events at specific locations because the land is either subsiding or being uplifted. Such sea-level change is called *local* sea level change, as only that local area is affected, and is commonly caused by tectonic movements of the Earth, or compaction and subsidence of sediments, as in river deltas.

Caves found in coastal areas fall into two major categories, karst caves and pseudokarst caves. The latter can form in almost any rock material by a variety of mechanisms. The most common pseudokarst cave in a coastal environment is the sea cave, produced by wave action. Sea caves are usually single chambers or a small collection of chambers and fissures open to the sea. The precise scientific name is *littoral cave*, meaning a cave formed within the range of tides. Coastal weathering and erosive conditions commonly create coastal cliffs, which in turn allow production of other pseudokarst cave types, such as tafoni (subaerial weathering pockets), talus caves (boulder and block rubble deposits), and fissure caves (cliff failure fractures). While tafoni are only isolated pockets, talus and fissure caves can sometimes be extensive. However, talus and fissure caves are a direct result of coastal erosion, and as a result are in an environment of continued modification and destruction, which results in loss of these cave types through time. Sea caves, being in solid bedrock, persist longer.

SEA CAVES

The most common of the coastal caves are sea caves, and they are found the world over. Sea caves are caves that form by wave erosion in coastal areas that contain exposed bedrock. They can develop in almost any type of bedrock, with wave energy utilizing fractures and other preexisting weaknesses in the rock to quarry out voids by mechanical action. The chemical action of salt water can also exploit rock weakness. The compression of air caused by water flowing forcefully into cracks and fractures in the rock can break rock, including rock above sea level. In addition to exploiting weaknesses in the rock, waves also interact with themselves and the sea floor to create constructive interference patterns that can focus wave energy at specific sites along a uniform coast. As a result, seas caves can form in rock where no obvious weakness exists.

Sea caves can vary from arches and small voids only a few meters across to very large chambers up to 100 m deep and wide (Fig. 1). The sea caves seen on coastlines today have formed rapidly, as sea level has only been at its present elevation for perhaps 3000 to 5000 years, following melting of the large continental glaciers at the end of the last glaciation. In areas such as Alaska and Norway, where the Earth's crust was depressed by large masses of ice during glaciation, the shoreline is now rising as the crust rebounds to a stable position following melting of the ice. In so doing, sea caves formed many thousands of years ago have now been carried high above modern sea level.

The Earth is currently in an interglacial, or between glaciations, so glacial ice is at a minimum and sea level is high. The last interglacial occurred 131,000 to 119,000 years ago, and during that time the ice melted back a bit more than present conditions, and sea level was about 6 m higher than it is today. On some rocky coasts, sea caves produced at that time are still visible, 6 m above the ocean, if more recent erosion has not obliterated them.

Sea caves have had a long history of interaction with people, especially wherever sailors have used the ocean on rocky coastlines. Sea caves were particularly favored by smugglers to hide stolen goods, and to also hide the small, fast sailing ships that carried such cargo out of reach of the taxman. Pirates allegedly buried treasure in sea caves, but most sea caves are in an active erosional environment, and anything buried would not survive long. Pirates who chose old sea caves above modern sea level would have had better success. Some sea caves formed during a past sea level higher than today contain significant archaeological and paleontological remains.

Sea caves are ubiquitous on the rocky coasts of the world. Fingal's Cave in Scotland, the Blue Grotto of Capri in the Mediterranean, Sea Lion Cave on the coast of Oregon, and Arcadia Cave on the coast of Maine are well-known sea caves visited by tourists on a regular basis. Many organisms use sea caves as a refuge, particularly seals, sea lions, and other marine mammals, as well as birds, which roost in the ceiling ledges above the reach of waves. From the viewpoint of cave exploration, sea caves are not of major category, primarily because they are short in length. In areas where other types of caves are rare, such as in Southern California, sea caves offer the best cave exploration option. Occasionally, sea caves can have spacious chambers and over 500 m of passages. Exploration of sea caves can be very dangerous for those not experienced in handling strong waves, tides, and currents.

FLANK MARGIN CAVES

The coastal environment creates a very unusual type of cave when limestones are present. The interaction of freshwater and seawater produces a unique geochemical situation that allows caves to form by dissolution that are very different from both sea caves, made by mechanical wave action, and most limestone caves located in the interior of continents, which are usually underground stream conduits formed by freshwater dissolution.

FIGURE 1 Left, looking out of a large sea cave on Eleuthera Island, Bahamas. Right, Sea arches and waves, Barbados.

Freshwater is slightly less dense than seawater, because of the extra salt dissolved in seawater. Average freshwater has a density of 1.0 g/cm^3 and average seawater has a density of 1.025 g/cm^3. The difference in density is only one part in 40, but it is sufficient that when freshwater flows toward the ocean inside an aquifer, it floats on top of the seawater that has invaded the aquifer from the ocean. The boundary between the freshwater and salt water is called the *halocline* ("halo" meaning salt, "cline" meaning boundary) if it is a sharp boundary. If the boundary is broad, containing water of brackish salinity, it is called a *mixing zone*. The freshwater flows toward the ocean because rainfall infiltrates the land behind the coast, piling up in the aquifer until there is sufficient slope to drive the water toward the ocean. Where the freshwater is piled up inland, because of buoyancy, the halocline sinks downward into the seawater, much like a piece of wood floating in water. Because the difference in density is one part in 40 for each centimeter (or each meter) that the freshwater piles up above sea level, it sinks 40 cm (or 40 m) into the seawater. As the freshwater flows down this small slope toward the ocean, its elevation above sea level decreases, and in buoyant response, the halocline rises up toward sea level, a 40-cm rise for each 1 cm of elevation loss of the water table. At the coast, the freshwater discharges to the sea as a thin sheet. This configuration of freshwater over seawater is called the *freshwater lens*, because seen in cross section in an island (where the water discharges to coasts on either side), the freshwater body is seen to have the shape of a lens, similar to a lens in a magnifying glass. Figure 2 shows this relationship in a diagrammatic fashion, with vertical exaggeration. Understanding the freshwater lens is critical to successful exploitation of fresh groundwater in islands and coastal areas.

When the freshwater lens is formed within a limestone aquifer in a coastal region, a very unique type of cave, called a *flank margin cave*, can develop. Limestones are made up of $CaCO_3$, either as the mineral calcite, or its close polymorph (*i.e.*, alternate crystal structure), *aragonite*. Seawater is usually saturated with $CaCO_3$, and cannot dissolve limestone very well. Freshwater that has had a long residence time in a limestone aquifer is also commonly saturated with $CaCO_3$ and also cannot dissolve more limestone. However, because the freshwater and seawater became saturated with $CaCO_3$ under different initial conditions, when they mix they are capable of doing more dissolution, a process called *mixing corrosion* or *mixing dissolution*. Where the freshwater lens meets the seawater at the halocline, it is possible to dissolve out large voids, or caves that otherwise would not be able to form.

The top of the freshwater lens is also a place where waters can mix. In this case, the freshwater at and below the water table (in this case, the top of the lens) is called *phreatic* water, and the water descending from the ground surface above is called *vadose* water. It is common for both the phreatic and vadose waters to be saturated with respect to $CaCO_3$, but as with the case of mixing seawater and freshwater, the phreatic and vadose waters are saturated at different initial conditions, so that when they mix, the water can dissolve more $CaCO_3$. Therefore both the top and bottom of the freshwater lens are favorable environments for the dissolution of $CaCO_3$.

The top of the lens (the water table) and the bottom of the lens (the halocline) represent density interfaces. Organic particulate material transported by the vadose water flow from the land surface commonly floats on the top of the water table. Some of this organic material may then become waterlogged, and work its way to the bottom of the lens where it floats on denser seawater at the halocline. The decay of the organic matter at these interfaces creates CO_2, which dissolves in the water to make carbonic acid that promotes $CaCO_3$ dissolution. In both situations, if the amount of organic material becomes too much, its decay will use up the local oxygen supply to create anoxic conditions. If the anoxic conditions persist, anaerobic bacteria will create H_2S, which can later encounter water with oxygen in it to create H_2SO_4 or sulfuric acid, a very powerful acid which can dissolve even more $CaCO_3$.

The mixing, and the possible anoxic conditions, that promote dissolution can occur at the top and bottom of the lens throughout the lens area. However, these environments are superimposed on each other at the edge, or margin of the freshwater lens where the top of the lens slopes down to sea level, and the bottom on the lens rises up to sea level. Because the lens is thin at this point, its flow velocity is high because the entire lens discharge is being forced through a thinning wedge. The combination of increased flow and

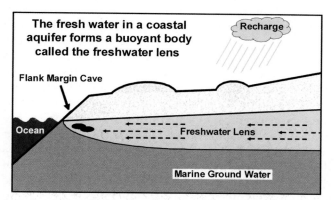

FIGURE 2 Cartoon diagram of a freshwater lens in a limestone island, showing flank margin cave development in the lens margin, under the flank of the enclosing landmass. The lens is drawn with vertical exaggeration.

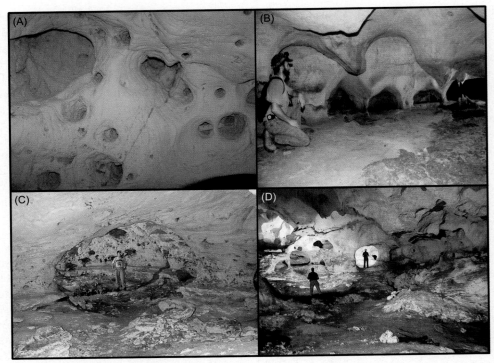

FIGURE 3 Interior of flank margin caves, showing the complex dissolutional morphology. (A) Small pores, vugs, and pockets (ruler in lower right is 10 cm long), Cumulus Cave, Crooked Island, Bahamas. (B) Small pockets bordering a chamber in Cueva del Agua Sardinera, Isla de Mona, Puerto Rico. (C) Chamber with wall pockets and pores, Ten Bay Cave, Eleuthera, Bahamas. (D) Large chamber with pockets, arches, and tubes, 1702 Cave, Crooked Island, Bahamas.

superposition of the favorable geochemical environments for $CaCO_3$ dissolution results in large voids forming very rapidly at the margin of the lens. As noted earlier, this discharge occurs at the flank of the land, so the caves thus developed are called *flank margin caves*. In addition, in tropical islands and coasts, the limestone is commonly very young. Unlike the ancient limestones of the mid-continent regions, these young limestones have not been buried, squeezed, or greatly altered. The rocks still have a high degree of primary porosity, and water moves through them easily. Such rock is said to be *eogenetic*. True conduit flow is difficult to develop, as the rock has a high degree of permeability, meaning that the water has many flow routes to choose from. So how do caves form in this setting?

Flank margin caves are not true conduits, like caves formed by sinking streams in continental interiors. They are instead mixing chambers. Conduit caves form by turbulent water flow, but flank margin caves develop in the laminar (nonturbulent) flow of a porous, eogenetic limestone aquifer. The freshwater enters the developing flank margin cave as diffuse flow and exits, after mixing with seawater, as diffuse flow. Flank margin caves form without human-sized entrances. As a result of their nonconduit origin, flank margin caves do not have long tunnels or a dendritic pattern found in most conduit caves. The flank margin caves are a series of oval rooms from pore size up to large chambers, which tend to be extensive in the horizontal direction, but limited in the vertical direction, a result of developing in the thinning margin of the freshwater lens (Fig. 3). The chambers can connect in a somewhat random manner, creating caves that are unpredictable in their pattern. Maze-like areas are common, indicating regions where chamber development did not go to completion. As the caves were growing, the mixing zone advanced into them, such that the back wall of the cave (the wall farthest from the ocean) is the youngest. Complex cross-connections between chambers can develop, and the caves can be quite complex despite their simple mode of development. The typical flank margin cave consists of one or more large chambers located just inside the edge of the island. The cave may trend parallel to the coast for some distance, but rarely penetrates very far inland, as its development site was restricted to the margin of the lens (Fig. 4). The caves form entrances when the erosion of the hillside that contains them breaches into the underlying cave. Initially, this entrance may be a small opening, but through time it can enlarge as more of the outer wall of the cave erodes away. Throughout up to 14,000 m³ in volume developed in that time frame, vindicating how rapidly this mixing dissolution process can occur (Fig. 3). On stable limestone coasts around the world are many flank margin caves that developed during the last interglacial sea level highstand.

Isla de Mona, halfway between Puerto Rico and the Dominican Republic, is a small island that has been uplifted by tectonic forces. It has huge flank margin caves that formed almost 2 million years ago (Fig. 4). These caves are very large because they developed in a freshwater lens before the Pleistocene glaciations began and so had a longer time to dissolve before sea-level fluctuation caused the freshwater lens to change position. When the initial glaciation began early in the Pleistocene, these caves were drained and speleothems such as stalactites and stalagmites developed. When a subsequent interglacial sea level highstand occurred, sea level rose and the caves were reinvaded by the freshwater lens, partially dissolving the speleothems produced during the dry phase. The caves were then uplifted by tectonics well beyond any further glacioeustatic sea level changes, and have been preserved for exploration today. In this case, local sea level change affected only this one island. Some of the cave chambers are over 400,000 m³ in volume, much wider than they are high, but with many complex connections with adjacent chambers. The chambers have ancient speleothems much modified by attack from an invading freshwater lens, and more modern speleothems that have grown since the last uplift event, and which are in pristine condition. The longest flank margin cave in the world, Sistema del Faro, is located on Isla de Mona, and has over 19 km of survey on its map (Fig. 4). Flank margin caves commonly have numerous entrances, and in tropical settings are warm and friendly. The pleasant conditions and many entrances make them easy to explore. When movies and TV dramas show pirates in caves, it seems they always have enough light, and they move easily through large open passages. Flank margin caves actually do look a bit like this show-biz characterization (Fig. 5).

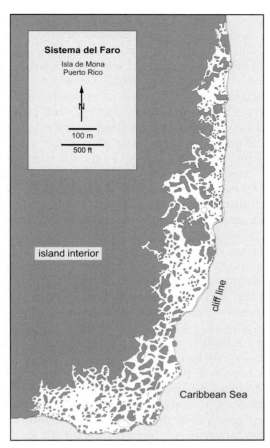

FIGURE 4 Map of Sistema del Faro on Isla de Mona, Puerto Rico, a very large flank margin cave. Note how the cave wraps around the edge of the island, following the margin of a past freshwater lens. *Cartography by Marc Ohms.*

the Bahamas and other carbonate island areas, there are flank margin caves in all states of erosional destruction, from almost intact to almost entirely removed. On islands such as Bermuda, where weathering and erosion of the limestone hills is more rapid than in the Bahamas, most flank margin caves have been entirely removed.

Because they develop in the freshwater lens, flank margin caves are sensitive to sea level change. If sea level falls, the caves are drained as the freshwater lens follows sea level downward, and cave development ceases. If sea level goes up, the lens is pushed higher, the caves become flooded with pure seawater from below, and dissolution and cave enlargement stop.

The Bahama Islands are tectonically stable, meaning they are not rising or falling because of tectonic forces. The dry flank margin caves that explorers enter there today resulted from the freshwater lens being 6 m higher than today 131,000 to 119,000 years ago, during the last interglacial. That sea level highstand lasted only about 12,000 years, but caves with individual chambers

BLUE HOLES

Blue holes are names for large, deep pits that form on islands and lagoons in tropical waters. They are named blue holes because their great depth gives them a very dark blue color (Fig. 6). They commonly connect to cave systems at depth. The name *blue hole* was first published in 1725, and later appeared on British Admiralty charts from the Bahama Islands in the 1840s. Blue holes became popularized in the early 1970s when cave divers began to make the first serious investigations of their depths. Since that time, blue holes have been the subject of a number of major scientific investigations, including the discovery in 1979 of a new class of crustaceans.

Blue holes are defined as "subsurface voids that are developed in carbonate banks and islands; are open to the Earth's surface; contain tidally-influenced waters of fresh, marine, or mixed chemistry; extend below sea level for a majority of their depth; and may provide access to submerged cave passages" (Mylroie et al., 1995, p. 225). Blue holes can be additionally characterized as being found in two settings: (1) *ocean holes*, which open directly into the present marine environment and contain marine water, usually with tidal flow; and (2) *inland blue holes*, which are isolated by present topography from marine conditions, open directly onto the land surface or into an isolated pond or lake, and contain tidally influenced water of a variety of chemistries from fresh to marine (Fig. 6).

While blue holes are best known from the Bahama Islands, they are found in a wide variety of tropical coasts and islands. Their origin is tied to the coast, island, and lagoon environments where they are found, which means that their development has been influenced by glacioeustatic sea level changes of the Pleistocene. Blue holes commonly contain many stalactites and stalagmites (known as *speleothems*) that are now under water, indicating that the blue holes were drained by glacioeustatic sea level lowstands, allowing the speleothems to form from drip water, then were flooded by return of sea level as the ice sheets melted on the continents at the end of the last glaciation. Some of these speleothems are more than 350,000 years old, indicating that the blue holes containing them are very old, and have undergone repetitive sea level lowstands and highstands.

There are three main hypotheses of how blue holes form: 1) drowning of surface karst features such as pits and sinkholes; 2) collapse of deep-seated phreatic dissolution voids; and 3) bank margin fracturing. Blue holes come in a variety of morphologies, and may represent features of polygenetic (*i.e.*, many origins) development, in which case a combination of the above hypotheses may be correct.

Exploration of blue holes generally involves cave diving to great depths. Such exploration is at the leading edge of technology and stamina, requiring the use of mixed gases, long decompression stops, total darkness, the danger of silt-out (stirring up silt so that the way out cannot be seen), and tight passages. Cave diving in blue holes is extremely dangerous, and many cave divers and scientists have lost their lives trying to penetrate into the unknown. Unlike many other types of exploration, however, there is no substitute for direct human exploration. The blue holes and their associated caves, and the contents of those caves, cannot be viewed, measured, or sampled without someone going there.

FIGURE 5 Hamilton's Cave, Long Island, Bahamas. Spacious, well-lit chambers like this one are common in flank margin caves.

FIGURE 6 Blue holes, so named for their deep blue color. Left, an inland blue hole, Andros Island, Bahamas. Right, an ocean hole, Dean's Blue Hole, the world's deepest at 200+ m, Long Island, Bahamas.

CONCLUSION

Coastal caves are important to science as they contain information about present and past sea level conditions. Their utilization as a habitat makes them important for many organisms over the breadth of the Animal Kingdom. The speleothems contained within blue holes and flank margin caves contain within their layers evidence about changes in the Earth's climate over hundreds of thousands of years. Flank margin caves reveal past conditions of the freshwater lens, an essential component of island habitation. While flank margin caves are generally easy to explore, sea caves can be dangerous for the unwary, and blue holes exceptionally dangerous even for the well trained.

See Also the Following Articles

Anchihaline Caves and Fauna
Underwater Caves of the Yucatán Peninsula

Bibliography

Bunnell, D. (1988). *Sea caves of Santa Cruz Island*. Santa Barbara, CA: McNally & Loftin.

Mylroie, J. E., Carew, J. L., & Moore, A. I. (1995). Blue holes: Definition and genesis. *Carbonates and Evaporites, 10*(2), 225–233.

Mylroie, J. E., & Mylroie, J. R. (2007). Development of the Carbonate Island karst model. *Journal of Cave and Karst Studies, 69*(1), 59–75.

Palmer, R. (1997). *Deep into blue holes*. Nassau, Bahamas: Media Publishing.

Vacher, H. L. (1988). Dupuit-Ghyben-Herzberg analysis of strip-island lenses. *Geological Society of America Bulletin, 100*(4), 580–591.

Waterstrat, W. J., Mylroie, J. E., Owen, A. M., & Mylroie, J. R. (2010). Coastal caves in Bahamian eolian calcarenites: Differentiating between sea caves and flank margin caves using quantitative morphology. *Journal of Cave and Karst Studies, 72*(2), 61–74.

CONTAMINATION OF CAVE WATERS BY HEAVY METALS

Dorothy J. Vesper

West Virginia University

Heavy metals are ubiquitous throughout nature, including within caves and karst environments. Evaluating the accumulation and transport of metals in cave waters requires understanding the governing physical and chemical processes. While the presence of heavy metals in speleothems and cave deposits has been investigated in some detail, the general metal cycling through the karst system is less well known but can be inferred from analogous investigations in surface systems. The term *heavy metals* is poorly defined and has been used inconsistently through time and in the scientific literature. In the context of this discussion, the metals and metalloids discussed are those defined as potentially toxic by the U.S. Environmental Protection Agency and the World Health Organization (Table 1). It should be noted that many of these metals, while toxic in large quantities, are essential nutrients in small quantities.

NATURAL AND ANTHROPOGENIC SOURCES

Metals are omnipresent in atmospheric, marine, and terrestrial settings. In karst environments, they are most likely to be found in three primary compartments: soils, the matrix–fracture–conduit system, and springs (Fig. 1). Within conduits and caves, heavy-metal rich minerals may be found incorporated into speleothems, coatings, fillings, rinds, and other cave deposits. Additionally, both caves and springs may have metals present in water or associated with suspended and bed sediments.

Metals may be part of the natural background or anthropogenic (Table 2). Spectacular deposits of metal-rich speleothems can occur when caves exist in proximity to natural geologic sources. For example, Cupp–Coutunn Cave in Turkmenistan has speleothems rich in manganese, iron, lead, and zinc due to the presence of overlying bituminous coal and subsequent hydrothermal alternation. Mbobo Mkula Cave in South Africa also boasts unusual metal-rich speleothems thanks to the presence of overlying ore minerals and a sulfide-rich black shale. The host rock itself may contribute to the metal load. Low levels of trace metals in spring waters in Nevada and California have been attributed to the paleomarine chemistry at the time of carbonate deposition. Given that the groundwater feeding the springs was thousands of years old, the water chemistry was attributed to dissolution of the carbonates. Metals may also be found as detrital material within the host rock. The manganese source for Jewel Cave in South Dakota has been attributed to such detritus.

Anthropogenic sources of metals are widespread (Table 2). Sources may be diffuse, such as emissions from fuel combustion, or more localized, such as point-source discharges from manufacturing facilities. Acid mine drainage (AMD) is a common metal source in many karst regions (the Appalachians, Kentucky, and Tennessee in the United States and in southern China). Studies of metals in AMD in Tennessee have shown that Fe and Mn concentrations decrease in water when introduced to karst systems. The most likely explanation is that metals precipitate in the presence of alkalinity. In cave stream/spring water, the drop in metal concentrations is associated with the production of flocs of Fe and Mn hydroxides. There is some suggestion that the metals may precipitate as a metal armor on the host rock, thereby limiting the interaction between water and rock. Manufacturing sources also contribute to metal contamination. Horse Cave in Kentucky and its associated aquifer and springs have been impacted by discharge from a metal-plating factory that began operations in 1970. Although some dilution occurred, elevated concentrations of Cr, Ni, and Cu were found in both cave and spring waters within the Hidden River basin.

ENVIRONMENTAL METAL CHEMISTRY

General Chemistry

Once metals are introduced into karst settings, their storage and transport depend on physical processes, the specific metal chemistry, and the chemistry of the surrounding environment. Metal speciation influences solubility and the likely mode of transport through the aquifer. Metal speciation also controls the bioavailability and toxicity of the metal. Metal chemistry can be complex and depends on many competing variables. Although a brief description is provided herein, the reader is referred to the bibliography, which lists a few of the excellent texts on the topic.

In general, metals are present in three forms: mineral, otherwise associated with solids, or dissolved (Fig. 2). Mineral-bound metals exist as source materials in soils in bedrock, as secondary cave deposits, and as detrital material throughout karst systems. Heavy metals may also be associated with the surfaces of solids and are often associated with suspended and bed sediments. Metal interactions at solid surfaces range in intensity from exchangeable (loosely bound) to incorporation into the near-surface mineralogy (tightly bound). Metals may also be incorporated into surface coatings (both organic and inorganic) or attached to insoluble organic matter. It is difficult to distinguish between metals that are electrostatically bound to the surface, specifically adsorbed, or coprecipitated with the surface mineral or coating; therefore, they are often loosely referred to in combination as *sorbed* metals. Metals are present in water as free metal ions, soluble metal complexes, or associated with colloids. Metals may also form complexes with soluble or colloidal organic compounds and thereby increase their solubility dramatically. Copper, in particular, has a strong affinity for organic compounds. Metals may move between these compartments via chemical reactions such as dissolution, sorption, or desorption or via physical processes such as deposition and entrainment (Fig. 2).

Two key variables that control metal solubility and chemical form are pH and reduction–oxidation (redox)

TABLE 1 Potentially Toxic Metals and Their Abbreviations

Metal	Abbreviation	Metal	Abbreviation
Arsenic	As	Nickel	Ni
Beryllium	Be	Lead	Pb
Cadmium	Cd	Antimony	Sb
Chromium	Cr	Scandium	Sc
Cobalt	Co	Selenium	Se
Copper	Cu	Titanium	Ti
Iron	Fe	Thallium	Tl
Mercury	Hg	Vanadium	V
Molybdenum	Mo	Zinc	Zn
Manganese	Mn		

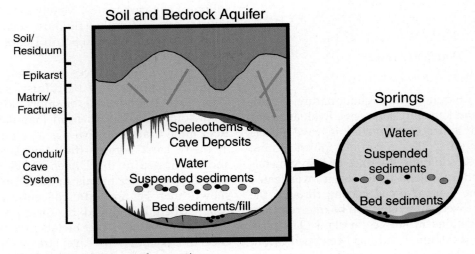

FIGURE 1 Locations for heavy metal storage in karst settings.

TABLE 2 Sources of Heavy Metals

Type of Source	Examples of Specific Sources
Natural	Original marine deposition; detrital materials in bedrock; ore bodies; hydrothermal deposits; black shales; coals
Anthropogenic	Mining and mineral processing; agriculture (fertilizer, pesticides, preservatives, irrigation); emissions and solid wastes from fossil fuel combustion; sewage and solid wastes; manufacturing (metallurgical, electronic, ceramic, chemical, pharmaceutical); sport and military shooting; breakdown of metal alloys and paints

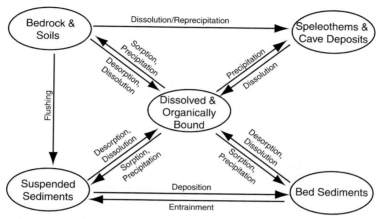

FIGURE 2 Storage compartments for heavy metals in karst settings. Arrows indicate some of the chemical reactions and physical processes that transfer metals between compartments.

state. Many metals are more soluble in acidic waters than in neutral waters and thus tend to precipitate in well-buffered karst waters. Redox reactions occur when electrons are transferred between metal species. For example, ferric iron (Fe^{3+}) is reduced to ferrous iron (Fe^{2+}) by the addition of one negatively charged electron. In actuality, free electrons do not exist, and the reaction occurs by combination with another "half reaction" in which an electron is lost. Redox reactions are extremely important because they control the solubility of many metals. While some metals are mobile in reducing conditions (As, Mn, Fe, Mo), others are mobile in oxidizing conditions and immobile in reducing environments (Zn, Cu, Hg, due to sulfide mineralization in reduced settings). For many metals, both the pH and the redox conditions must be known to predict what species should be present.

Redox states change spatially and through time. Spatial changes may occur on a microscale, and temporal changes may occur on a scale as short as hourly (e.g., with storms and changing hydraulic conditions). The degree to which metal redox reactions in caves are mediated and catalyzed by microorganisms is a topic of increasing interest. Microbes can be either electron acceptors or electron donors and can therefore either oxidize or reduce metals as part of their metabolic process. Hence, both precipitation and dissolution of metals can be induced by microbial action. Much of the microbial action may occur along redox gradients, such as those found at the edges of caves.

Other factors may influence the redox state, such as introduction of oxygen during a storm and degradation of organic compounds. The breakdown of either natural organic matter or organic contaminants in the overlying soil, karst aquifer, or spring sediments can create locally reducing and acidic environments capable of solubilizing and mobilizing metals.

Iron and Manganese

Iron and manganese can be found in many locations in caves. While oxide and hydroxide forms are the most common, heavy-metal carbonates, phosphates, nitrates, and sulfates have also been identified in caves. In general, while reduced iron (Fe^{2+}) is soluble in water, when it oxidizes to Fe^{3+} it forms oxide and hydroxide precipitates such as goethite (FeOOH) and ferrihydrite ($Fe(OH)_3$). *Limonite* is a generic term that refers to both mineral and amorphous forms of iron oxides and hydroxides. Likewise, reduced manganese (Mn^{2+}) is soluble, but its oxidized forms are not. Fully oxidized Mn often forms the common cave mineral birnessite, which can be poorly crystallized (δ-MnO_2). In reality, however, the chemistry is far more complicated than this. The reactions also depend on metal concentration, solution pH, CO_2 partial pressure, and presence of organic compounds. Both reduced and oxidized metals can be mobilized by organic complexation, and reduced metals can precipitate into metal–carbonate minerals.

Trace and Contaminant Metals

Iron and manganese oxides and solids also play a key role in trace-metal chemistry because of their ability to scavenge trace metals from solution and because—given their relative concentrations—they often control the overall redox chemistry of the solution or sediment. Iron oxides are commonly associated with As, Cu, Ni, Mn, and Zn; manganese oxides commonly contain Co, Fe, Ni, Pb, and Zn. Therefore, the transfer of Fe and Mn between mobile and immobile forms ultimately controls the trace-element chemistry, even if the trace elements are themselves not redox sensitive. For example, while precipitation of Fe and Mn oxides may remove trace and contaminant metals from solution, dissolution of the same oxides may rerelease trace metals back into solution.

Studies of soils and marine and lake sediments have shown that Fe and Mn oxides are typically the controlling factor in determining trace-metal concentrations. Trace metals may also be associated with organic compounds (both soluble and insoluble) and inorganic coatings on particulates.

METAL STORAGE AND TRANSPORT

Schematic Scenarios

Heavy metals are introduced into karst aquifers from surface runoff and spills, reactions in the soil zone, and dissolution of overlying geologic units and the host rock (Fig. 1). The metals can be either dissolved or associated with colloids and particulates (Fig. 2). Dissolved metals may be transported through the system or may change chemical form and precipitate as speleothems, onto walls or onto sediments. The fate of the dissolved metals may depend on the type of recharge. If dissolved metals are introduced slowly from fractures in the cave ceiling they may—given the right chemical conditions—form speleothems. If the metals arrive as part of a cave stream, they may be more likely to be transported through the system or to form coatings on stream sediments. Metals introduced in particulate form can either travel through the system as suspended sediments and be discharged at a spring or can be deposited within the aquifer.

Metals deposited within the fractures and conduits of the aquifer may be stored for extended periods. It is also possible, however, that they will either be later dissolved (minerals and coatings) or re-entrained (sediments) and be flushed from the system. One of the final possible storage locations on the flowpath is spring sediments. Depending on spring morphology and hydrology, it is possible for some metals to accumulate in the springbed sediments.

Speleothems and Cave Deposits

Coatings of Fe and Mn are not uncommon in caves. Although far less common, extensive decorations incorporating Fe, Mn, and trace metals do occur. Examples of manganese deposits exist as powders and coatings in Matts Black Cave in West Virginia and Rohrer's Cave in Pennsylvania, as stream-cobble coatings in Butler Cave in Virginia, and as floor fills in Jewel Cave in South Dakota. Examples of iron-rich speleothems also exist in Rohrer's Cave, which has stalactites, stalagmites, and columns of limonite. Spectacular Fe and Mn speleothems exist in Mbobo Mkula Cave in South Africa. The metals in this cave are mobilized when acidic water, produced in the overlying shales, infiltrates through chert layers in the limestone. Trace metals are often found in association with the Fe and Mn deposits. Samples from Rohrer's Cave have been shown to contain up to 20% heavy metal oxides. Deposits of trace-metal speleothems also exist, although they are more rare. For example, malachite, a copper-carbonate, has been observed in cave crusts and speleothems. Recent studies indicate that metal incorporation into speoleothems may be an indicator of water flow. Speleothems in Grotta di Ernes in Italy have distinct layers of metals in stalagmite laminations; their source has been interpreted as associated with organic-colloids during infiltration.

Suspended and Bed Sediments

Given the near-neutral pH of karst water and the input of oxygenated surface waters, neither Fe nor Mn is likely to be present in appreciable concentrations as dissolved metals in water. This is one factor that distinguishes the transport of heavy metals from the more commonly studied alkaline earth metals (e.g., Ca, Mg). The alkaline earth metals, present from the dissolution of the carbonate host rock, are almost solely present in the dissolved form. If dissolved heavy metals are present, they are likely to be oxidized and precipitated as cave deposits or directly onto particulates in the sediments. Once deposited in a cave or spring, sediment Fe and Mn may be influenced by small-scale transitions in redox state. This is analogous to metal behavior in lake and marine sediments where redox gradients, and the associated Fe and Mn chemistry, change over short vertical distances.

The importance of particulate metal transport has long been established in surface-water systems where it has been shown that nearly all heavy metals are transported in association with solids. Groundwater studies in the karst aquifer at the Oak Ridge National Laboratory have also shown that most of the metals are associated with colloidal or particulate matter. Metal and radionuclide transport in granular and fractured aquifers is often attributed to colloidal transport. Karst aquifers, however, are able to transport much larger sizes of particles; therefore, metal transport is not limited by the size of particulate to which the metal is adhered or incorporated. Data from springs in Kentucky and Tennessee, in which concentrations of digested and filtered samples were compared, demonstrated that Fe, Mn, and the trace elements were present in particulates larger than 0.45 μm.

Storm-Enhanced Transport of Sediment-Associated Metals

The ability of the aquifer to transmit sediment is largely a function of flow velocity; therefore, the flushing of particles from the overlying soil, entrainment of sediments already within the aquifer, and the deposition of suspended sediments within conduits and in springs are all key physical processes controlling metal transport and are closely linked to the groundwater velocity.

Groundwater velocity is not constant through time in karst aquifers. Systems that rapidly transmit recharge water exhibit increases in velocity during storm events. Research has shown that the sediment transport that occurs during storms controls the transport of heavy metals as well. Storm-event samples collected from springs in Kentucky and Tennessee demonstrated clearly that the heavy-metal concentrations increased dramatically during storm conditions (Fig. 3). Data from different-sized storms and different types of springs in the same area suggest that this relationship is consistent and that total metal transport is episodic and enhanced by storms.

While some "dissolved" (less than 0.45 μm) heavy metals were present throughout the storm events, their concentrations were relatively constant in comparison to the particulate metals. This suggests that "dissolved" metal transport through the aquifer occurs continuously at low concentrations while particulate metal transport occurs primarily during storms.

Data from Matts Black Cave indicates that the distribution of manganese coatings may also be influenced by the same physical flow processes. The coatings are much thicker on the ceilings (up to 10 mm) than on the walls (up to 2 mm) or along the stream (poorly coated), suggesting that metal deposition and stream erosion are in competition; therefore, the thickness and the distribution of metal coatings within caves may change temporally with storm events.

SUMMARY

Heavy metals are present throughout caves and karst systems due to both natural and anthropogenic sources. The solubility and transport of iron and manganese, two of the most common heavy metals, are controlled by pH and redox conditions. Although typically soluble in water in reducing conditions, the oxidized metals tend to precipitate as cave deposits

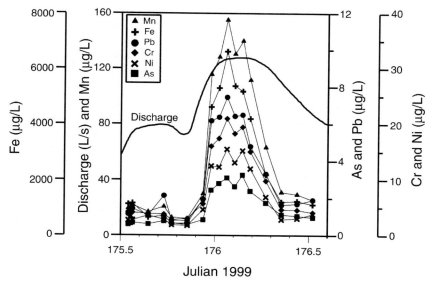

FIGURE 3   Concentrations of heavy metals in spring water (Beaver Spring). Symbols indicate digested sample concentrations and represent the total metal transport (both dissolved and particulate components). *1999 data from Fort Campbell Army Base, Kentucky/Tennessee.*

or on sediments. While speleothems of Fe and Mn are unusual, coatings of the same are not. Many metals are never present in the dissolved state and are introduced, stored, and transported through the system as colloids or larger particles. Particulate metal transport is enhanced during storms when high groundwater velocities permit the metals to be entrained and suspended. The transport and storage of trace and contaminant metals are closely linked to the iron and manganese chemistry. Although some trace metals are not sensitive to redox conditions, their behavior is governed by their association with iron and manganese so they are impacted by redox conditions.

Bibliography

Allard, B. (1994). Groundwater. In B. Salbu & E. Steinnes (Eds.), *Trace elements in natural waters* (pp. 151–176). Boca Raton, FL: CRC Press.

Borsato, A., Frisia, S., Fairchild, I. J., Somogyi, A., & Susini, J. (2007). Trace element distribution in annual stalagmite laminae mapped by micrometer-resolution X-ray fluorescence: Implications for incorporation of environmentally significant species. *Geochimica et Cosmochimica Acta, 71*, 1494–1512.

Hill, C. A. (1982). Origin of black deposits in caves. *National Speleological Society Bulletin, 44*(1), 15–19.

Hill, C. A., & Forti, P. (1997). *Cave minerals of the world* (2nd ed.). Huntsville, AL: National Speleological Society.

Horowitz, A. J. (1991). *A primer on sediment–trace element chemistry* (2nd ed.). Chelsea, MI: Lewis Publishers.

McCarthy, J. F., & Shevenell, L. (1998). Processes controlling colloid composition in a fractured and karstic aquifer in eastern Tennessee, USA. *Journal of Hydrology, 206*(3–4), 191–218.

Northrup, D. E., & Lavoie, K. H. (2001). Geomicrobiology of caves: A review. *Geomicrobiology Journal, 18*(3), 199–222.

Salomons, W., & Förstner, U. (1984). *Metals in the hydrocycle*. Berlin: Springer-Verlag.

Sasowsky, I. D., & White, W. B. (1993). Geochemistry of the Obey River Basin, north-central Tennessee: A case of acid mine drainage in a karst drainage system. *Journal of Hydrology, 146*, 29–48.

Siegel, F. R. (2002). *Environmental geochemistry of potentially toxic metals*. Berlin: Springer-Verlag.

Vesper, D. J., & White, W. B. (2003). Metal transport to karst springs during storm flow: An example from Fort Campbell, Kentucky/Tennessee, U.S.A. *Journal of Hydrology, 276*, 20–36.

White, W. B., Vito, C., & Scheetz, B. E. (2009). The mineralogy and trace element chemistry of black manganese oxide deposits from caves. *Journal of Cave and Karst Studies, 71*, 136–143.

CONTAMINATION OF CAVE WATERS BY NONAQUEOUS PHASE LIQUIDS

Caroline M. Loop

Groundwater Management Associates

Due to their unique properties, nonaqueous phase liquids (NAPLs) are widely used as solvents, insulators, and fuels. Some of the same valuable properties, however, make NAPLs toxic and difficult to remove from soils and groundwater once they have been spilled. In the heterogeneous karst subsurface, the rate of movement of NAPL contamination can vary by orders of magnitude. Tailoring a conceptual model to a particular NAPL release and subsurface characterization is necessary to develop multiple working hypotheses and better guide detection and monitoring techniques.

NAPL CHARACTERISTICS AND SOURCES

Four characteristic chemical properties that influence the behavior of NAPLs in the environment are solubility, density, vapor pressure, and viscosity. Nonaqueous phase liquids must, by definition, have limited solubility in water, allowing them to remain in a separate phase (Fig. 1). Thus, the dissolved, or aqueous phase, concentration of a nonaqueous phase liquid is moderate to low. Unfortunately, even the limited aqueous phase concentration is often much higher than the maximum concentration level established by the U.S. Environmental Protection Agency (Table 1). Solubilities are typically less than 5000 mg/L for chlorinated solvents, less than 2000 mg/L for most gasoline components, and less than 1.0 mg/L for polychlorinated biphenols (PCBs).

Nonaqueous phase liquids are divided between those that are less dense than water (LNAPLs) and those that are more dense than water (DNAPLs). Hydrocarbons, including benzene and ethylbenzene, are common LNAPLs, as are vinyl chloride and styrene. LNAPLs are often used in the synthesis of plastics, and as gasoline. With a density less than water, they will be present as a separate phase above the water table (Fig. 2). The water table will be depressed in proportion to the thickness and density of the LNAPL immediately above it. DNAPLs include some very toxic chemicals such as perchloroethylene (PCE), PCBs, and insecticides and herbicides such as lindane and atrazine. PCE and its breakdown products were commonly used as solvents, especially in dry cleaning. (Their use, however, is decreasing due to their highly toxic nature.) PCBs, for example, the Aroclor formulations, were commonly used as liquid insulators in capacitors and transformers. DNAPLs will sink below the water table and continue to migrate deeper in an aquifer until they meet sufficient resistance (Fig. 3).

NAPLs can be characterized based on their vapor pressure, which is a measure of the tendency of the compound to evaporate from a pure liquid of the compound. NAPLs are either volatile, with a vapor pressure greater than 10^{-4} atm, or semivolatile, with a vapor pressure between 10^{-4} and 10^{-11} atm. Benzene

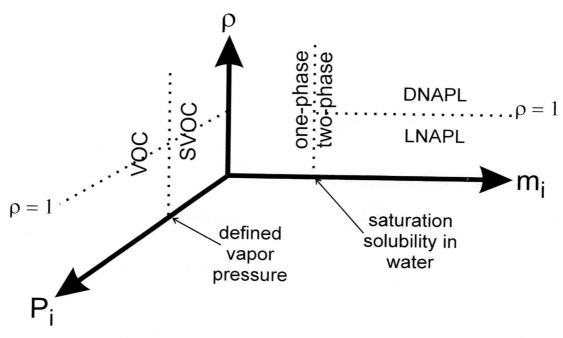

FIGURE 1 Diagram of important properties for characterizing organic contaminants. ρ = density (g/cm^3), m_i = solubility of contaminant i in water, and P_i = vapor pressure of contaminant i.

TABLE 1 Density, Vapor Pressure, and Solubility Values for Common NAPLs

Compound	Density (g/cm³)	Vapor Pressure (kPa)	Solubility (mg/L)	U.S. EPA MCL (mg/L)
Toluene	0.86	3.75	500	1
Benzene	0.88	12.4	1800	0.005
o-Xylene	0.88	0.88	175	10
Ethylbenzene	0.90	1.24	150	0.7
Vinyl chloride	0.91	334	2800	0.002
Styrene	0.91	0.62	310	0.1
Aroclor 1242	1.4	5.29×10^{-5}	0.24	0.0005
Trichloroethylene (TCE)	1.5	9.63	1100	0.005
Carbon tetrachloride	1.6	14.9	760	0.005
Tetrachloroethylene (PCE)	1.6	2.48	200	0.005
Lindane	1.9	8.19×10^{-6}	7.5	0.0002

Note: Equilibrium aqueous phase solubility concentrations are orders of magnitude larger than the U.S. Environmental Protection Agency's maximum concentration level (MCL).

has a high vapor pressure, as can be seen when gasoline fumes rise in the summer, whereas PCBs are almost wholly semivolatile. Because many NAPLs have a high vapor pressure, they had sometimes historically been thought to fully evaporate when poured on the ground. This belief turned out to be incorrect and, in fact, such practice significantly contaminated underlying soils and aquifers. As discussed later, vapor pressure may be helpful in locating NAPL contaminants in karst.

Viscosity influences the subsurface mobility of nonaqueous phase liquids in that the less viscous the liquid, the farther it can migrate into pores and fractures. Chlorinated solvents such as trichloroethylene (TCE) and PCE have very low viscosities. Creosote

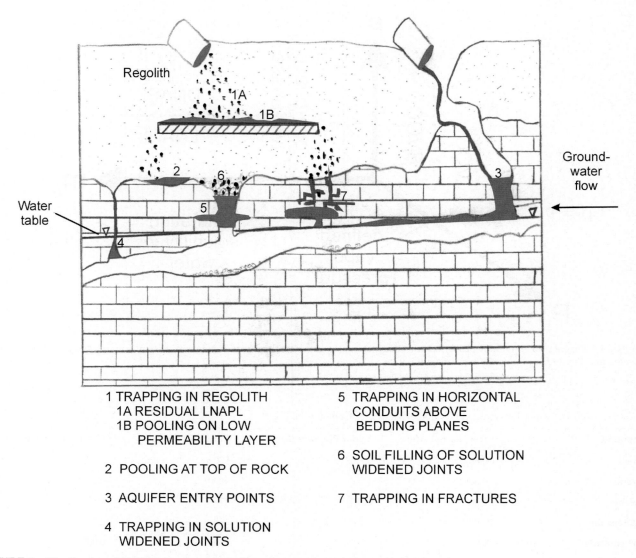

FIGURE 2 Distribution of LNAPLs in a hypothetical karst setting. *Source: After Wolfe et al. (1997).*

wood-treatment compounds behave as DNAPLs and are characterized by a high viscosity. Fluids can increase in viscosity with time, called *weathering* in the petroleum industry, as they lose more volatile components. It is easier for flowing water, such as might be found in a cave stream, to entrain less viscous fluids, and move them farther from the source.

Although petroleum compounds are formed and present in the subsurface naturally, they are rarely found in near-surface aquifers and soils unless they have been spilled. Other NAPLs, including PCBs, chlorinated solvents, and agricultural chemicals such as lindane and atrazine, have been developed in laboratories and tailored to industrial uses. Freon 11, also known as trichlorofluoromethane, is one such chemical that was used extensively until the toxic effects of spills and leaks was found to exceed its practical use. The distribution of NAPLs in karst soils and aquifers is nearly always the result of human misuse, whether by transportation accidents, leaking tanks, or improper disposal.

TRANSPORT INTO KARST AQUIFERS

Both the volume and rate of release can play a key role in the movement and distribution of NAPL contaminants. For example, a gasoline spill from a 10-gallon bucket will move through and be sorbed to soil and regolith differently than a slow leak from a 1000-gallon tank that can build up a pressure head of product, which drives it farther into the surface. On the extreme

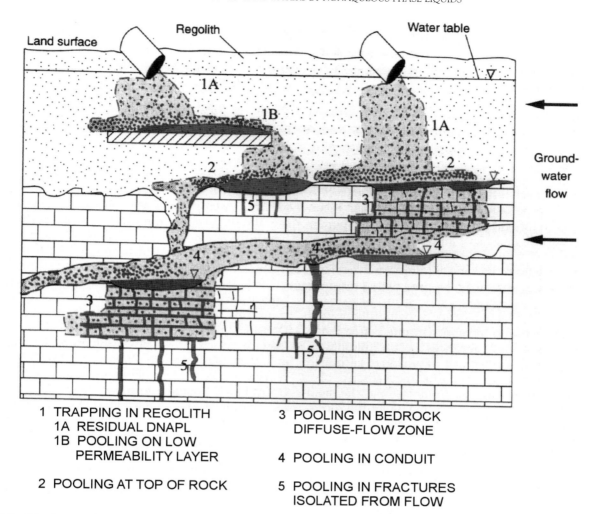

FIGURE 3 Distribution of DNAPLs in a hypothetical karst setting. *Source: From Wolfe et al. (1997).*

end of this scale is the gasoline tanker truck that turns over and releases thousands of gallons quickly onto the ground surface, which then flows overland and into a sinkhole or sinking stream where it can more quickly be transported into a bedrock aquifer.

Epikarst

Climate can also impact the rate and volume of NAPL entering the karst subsurface. Karst soils are usually thin, and they overlie the regolith and irregular bedrock of the epikarst. In a wet climate three things may occur to move the NAPLs more quickly through an unsaturated soil and possibly epikarst zone. First, when water coats soil, organic, or regolith particles, if NAPL is spilled, it travels as a nonwetting fluid and will be held less securely between particle grains because it is, in a sense, lubricated by the aqueous phase.

Second, and not inconsequentially, soil and regolith pores may periodically be flushed by precipitation. In this case, pressure head builds up from the accumulation of recent precipitation, and even light nonaqueous phase contaminants can be forced to migrate to deeper pores. If a less permeable layer is present in the soil, the influx of rainwater may cause an NAPL that was temporarily pooled above the layer to migrate horizontally to an area where the less permeable layer is no longer present. At that point, the NAPL may resume its vertical flow toward the water table. While the NAPL is held in soil or regolith pores, it may dissolve into the aqueous phase, which may then potentially be flushed farther through the system during precipitation events. Thus, NAPLs held in the soil or regolith can continue to dissolve and be long-term sources of aqueous phase contamination to an aquifer with which the NAPL is not in direct contact.

A third influence of climate on the transport of NAPLs into karst aquifers is the impact on sinkhole formation and collapse. Sinkholes form localized catchments for storm runoff in karst regions. Some sinkholes are plugged with soil so that NAPLs held within them may migrate downward only slightly more quickly than in nonsinkhole soils, with the only difference being a comparatively larger volume of precipitation percolating through. On the other end of the spectrum, some sinkholes have open drains so that any runoff captured will be immediately carried into the subsurface. In some cases, however, a soil plug may itself be subject to piping failure, especially during large storms. When this occurs, both the soil and the NAPL are transported into the subsurface and leave an open hole for further contaminants to more quickly enter the system.

The top of bedrock in karst is often an irregularly sculptured surface that may be undulating or may have deep crevices along joints and fractures separated by intermediate pinnacles. Water, moving downward from overlying soil into the epikarst, must often move laterally for substantial distances before finding an open fracture or shaft that will permit vertical movement into the unsaturated zone of the bedrock. Like groundwater, infiltrating contaminants may be held for extended periods of time, pooling along the top of relatively impermeable bedrock. While still in the vadose zone, pools resting along the top of the bedrock can be described as *microphase pools*, which have collected in pores between regolith and other particles. The pools should be distinguished from continuous macrophase pools that may exist floating on water (LNAPLs) or below water (DNAPLs), but not distributed within pore space.

In the event that the top of the limestone or dolomite bedrock is below the water table, LNAPL pools will most likely resist resting on the bedrock; rather they will tend to remain in the soil or regolith directly above the water table. Wells intersecting a few feet of LNAPL above the water table are not uncommon in contaminated areas. DNAPLs, in contrast, are able to migrate below the water table, and may pool on less permeable layers, such as the bedrock surface. In this case, water will flow in the direction of hydraulic gradient; however, DNAPLs often flow in the direction of bedrock dip. In one specific case, an aqueous phase plume developed in one direction, whereas the DNAPL itself migrated down-dip in a different direction.

Fractures, Open Drains, and Sinking Streams

Karst bedrock is often dissected by vertical or near-vertical fractures and joints, many of which have been solutionally widened. These provide fast paths for NAPLs into the subsurface aquifer once the critical height for entry into the fracture has been exceeded. The critical height of a DNAPL pool is a function of the interfacial tension between the NAPL and water, the wetting angle between the DNAPL and the solid surface in the presence of pore water, the density difference between the NAPL and water, and the fracture aperture. Both LNAPLs and DNAPLs will migrate in similar ways through fractures above the water table. One exception is that the higher density of the DNAPLs will cause them to move more quickly under gravity. Viscosity will also play an important role in the ability to move through a given aperture. Storm flow can, again, add to the pressure above the non-aqueous fluid, causing it to migrate farther down the fracture.

Once at the water table, LNAPLs and DNAPLs will behave very differently, with fractures acting as temporary traps for LNAPLs and preferential flowpaths for DNAPLs. In a scenario where NAPL is moving down a fracture, which farther down intersects the water table, and then joins with a phreatic conduit, LNAPL will remain in the fracture above the water table, whereas DNAPL will continue to travel down the fracture into the conduit, as long as the critical DNAPL pool height is exceeded (Figs. 2 and 3). If drought conditions lower the water table so that water flowed like a surface stream in the conduit, the same LNAPL could potentially travel down the fracture and float on the water in the free surface stream down the length of the conduit. When the water level rises, the LNAPL will again get caught in a fracture or a sump and again be trapped, possibly even below the water table.

In this same scenario, there will be much less difference in the transport of the DNAPL with a variation in water level. Because it is denser than water, the DNAPL, with sufficient pool height, can move through the fracture and into the conduit. An air-filled fracture would likely offer less resistance than a water-filled one, yet in both cases, the DNAPL will not be held in the fracture indefinitely. In the event that the critical pool height is not exceeded, the NAPL will diffuse into the rock matrix over time. Once the NAPL is depleted, the contaminated aqueous phase will act to reverse the concentration gradient and dissolve into freshwater undersaturated with the NAPL, moving through the fractures.

Open drains, such as those sometimes found in sinkholes, as well as sinking streams, offer a very quick path to the underlying karst aquifer. NAPLs can be poured into the open drains. They can also either float in the case of LNAPLs, or be pushed along in the case of DNAPLs in surface streams. Sinking streams and open drains allow contaminants direct access to quick subsurface flow and can allow NAPLs to move at a rate of kilometers per hour. This is very important to take into consideration in an area where a spring acts as the local water-supply source.

STORAGE IN KARST AQUIFERS

Once NAPLs have entered a karst aquifer, they can be a source of contamination for thousands of years, given their low solubility values and the difficulty often encountered when attempting to retrieve them. Natural degradation is faster for hydrocarbons than chlorinated compounds such as chlorinated solvents and PCBs, but for a large spill may take many decades. Little is known about the ability of microbes present in karst systems to degrade any of the NAPL compounds. Hence, it is important to consider the locations in which NAPL might be stored in the karst aquifer.

Pools in Conduits

Macrophase NAPL pools—free surface pools not confined to pore spaces—can be present in conduits. DNAPL pools could be present at the bottom of a flowing subsurface stream, whereas LNAPL pools would float on top of the water. Pools such as these could also be present in dry cave passages, but in all cases require a large volume of contaminant, quickly injected to prevent either (1) seepage into sediments or (2) horizontal movement to a more permeable material. In turbulent flow, which characterizes most flow in subsurface streams, NAPL can be entrained or carried as an emulsion until the water velocity again decreases. Macrophase pools have a low surface-area-to-volume ratio, so dissolution is limited. When exposed to air, NAPL compounds may volatilize within the cave passages.

Fractures

Fractures may intersect conduits in all directions. As discussed earlier, fractures extending from the roof of a conduit may act to indefinitely trap LNAPLs, but fractures in walls may also act to isolate NAPLs. DNAPLs are especially likely to migrate into subaqueous fractures in conduit floors (Fig. 3). Once they reach a depth where the critical pool height is no longer exceeded, they diffuse into the fracture matrix, from which they may act as a long-term source for aqueous phase contamination.

Matrix and Vugs

NAPLs can be stored in the bedrock matrix or in larger vugs. Diffusion of NAPLs into and out of the matrix occurs on a relatively slow scale. The quickest period of activity in the matrix may be the result of the flushing of pores during storm activity, as when the potentiometric surface changes slope from down to the conduit during baseflow, to a mound above the conduit during intense storm flow. Storage in bedrock matrix and vugs may represent the portion of a contaminant spill most resistant to short-term remediation.

Sediment

NAPLs may sorb to or be present in pores between sediments. In the first case, the NAPL may travel on the sediment, ready to desorb once it reaches a less concentrated aqueous environment. In the second case, once the sediment pile is disrupted, the NAPL is again free to move with the water, or down-dip in the case of DNAPLs. NAPL can be held in the sediment pile by capillary forces or a microphase pool. With storm movement, a fresh flush of contaminants may be remobilized when NAPL is associated with sediment piles in karst conduits.

DETECTION OF NAPL IN KARST

Springs and Caves

In an area where springs and caves are accessible, monitoring should begin by sampling these locations during storm and baseflow conditions. Springs are good monitoring sites as they are often down-gradient from spills or leaks, and if previous dye traces exist, they can help to define a groundwater basin and evaluate the risk to nearby populations. Spring sediments should be observed. Conductivity and temperature are inexpensive tools for estimating the timing of a spring's response. In caves, one must be careful if NAPLs are a suspected contaminant. In one instance in the 1960s, a carbide lamp ignited a gasoline spill in a cave and killed several people, some by flames, but others by asphyxiation. Less dramatic is the instance in which a volatile NAPL produces toxic air within a cave.

Wells

As mentioned previously, wells can intersect meters of LNAPL product above the water table. Fracture trace analysis can help to site wells in some instances, and wells can also be used for dye trace studies. In karst, wells must be carefully considered, especially with the slow diffusion into and out of the bedrock matrix as compared to the rate at which other transport processes may be operating in the aquifer. DNAPLs are likely more difficult to find using wells, because they may have migrated into deeper fractures in the aquifer, and they are harder to direct by altering the hydraulic gradient. With either type of NAPL, but especially with DNAPLs, one must be very careful to case wells properly in a contaminated area. An open borehole is an excellent way to transmit dissolved contamination or DNAPL to an underlying, formerly uncontaminated aquifer.

Soils

When an NAPL is spilled, it often passes through soil, which can be collected for quantitative laboratory analysis. Soil sampling may be useful for verifying the type of NAPL and whether any former spills took place at the site. Soil sampling in a sinkhole is not recommended, due to possible piping failure. Soil gas sampling may be very useful, especially for LNAPLs. Contaminants can volatilize, with the resulting gas moving up through the soil. Depending on the season, the contaminant, and the subsurface configuration, soil gas sampling may be a helpful technique for identifying and locating an NAPL product.

SUMMARY

Due to the heterogeneous nature of the karst subsurface, the rate of either water or NAPL transport through karst aquifers is highly variable. Both the quantity and timing of NAPL releases are important for understanding how the pollutant might be trapped in the aquifer. Individual NAPL characteristics such as density, solubility, vapor pressure, and viscosity are also key to recovering a contaminant from soils or groundwater. Over time, NAPL held in the epikarst or matrix can dissolve into the aqueous phase. Aqueous concentrations can be toxic and persist for many years, especially in the case of chlorinated compounds, which are naturally degraded more slowly than hydrocarbons. However, NAPL from large spills may move through a conduit on the order of kilometers per hour. All information about a specific karst system, including dye traces, spring response, depth to and shape of the bedrock surface, is important for evaluating the potential of NAPLs to be held in and transported through the subsurface. The study of NAPL contamination in karst aquifers is a relatively new aspect of karst science, and in the future will certainly be enhanced by additional case studies and research.

Bibliography

Black, D. F. (1966). Howard's cave disaster. *National Speleological Society News*, 24, 242–244.

Crawford, N. C., & Ulmer, C. S. (1994). Hydrogeologic investigations of contaminant movement in karst aquifers in the vicinity of a train derailment near Lewisburg, Tennessee. *Environmental Geology*, 23(1), 41–52.

Ewers, R. O., Duda, A. J., Estes, E. K., Idstein, P. J., & Johnson, K. M. (1991). *The transmission of light hydrocarbon contaminants in limestone (karst) aquifers. Proceedings of the third conference on hydrogeology, ecology, monitoring, and management of ground water in karst terranes.* Dublin, OH: Association of Ground Water Scientists & Engineers, National Ground Water Association.

Jancin, M., & Ebaugh, W. F. (2002). Shallow lateral DNAPL migration within slightly dipping limestone, southwestern Kentucky. *Engineering Geology*, 65, 141–149.

Krothe, N., Fei, Y., McCann, M. R., & Cepko, R. P. (1999). Polychlorinated biphenyl (PCB) contamination of a karst aquifer in an urban environment, central Indiana, USA. In J. Chilton (Ed.), *Groundwater in the urban environment: Selected city profiles.* Rotterdam: A. A. Balkema.

Loop, C. M., & White, W. B. (2001). A conceptual model for DNAPL transport in karst ground water basins. *Ground Water*, 39(1), 119–127.

Mercer, J. W., & Cohen, R. M. (1990). A review of immiscible fluids in the subsurface: Properties, models, characterization, and remediation. *Journal of Contaminant Hydrology*, 6, 107–163.

Pankow, J. F., & Cherry, J. A. (1996). *Dense chlorinated solvents and other DNAPLs in groundwater.* Waterloo, Ontario, Canada: Waterloo Press.

Schwarzenbach, R. P., Gschwend, P. M., & Imboden, D. M. (1993). *Environmental organic chemistry.* New York: John Wiley & Sons.

Wolfe, W. J., Haugh, C. J., Webbers, A., & Diehl, T. H. (1997). Preliminary conceptual models of the occurrence, fate, and transport of chlorinated solvents in karst regions of Tennessee, U.S. Geological Survey Water-Resources Investigations Report 97-4097. Reston, VA: U.S. Geological Survey.

COSMOGENIC ISOTOPE DATING OF CAVE SEDIMENTS

Darryl E. Granger and *Derek Fabel*[†]

*Purdue University, [†]University of Glasgow

INTRODUCTION

Natural curiosity prompts both cave explorers and first-time visitors to wonder "How old is this cave?" and "Why is it here?" Scientists have more specific reasons to study and date cave sediments. For example, geomorphologists use caves to learn about landscape evolution, or the sequence of events that shaped the rivers, hills, and valleys around us. Paleontologists study fossils in cave sediments to learn about animal and plant evolution and about the ecological communities that lived long ago. Paleoanthropologists study our ancestors' bones that are found in caves—whether from cave dwellers, or often those that were eaten in the caves. These fossils and their dates help teach us about our own human origins. Archaeologists search for clues about human use of caves. Some scientists also study caves for their own sake, to learn about how water flows through rock, and how the spectacular and labyrinthine underground environment is formed. All of these various fields require information about the age of the cave and its contents.

Caves are important across so many fields of science because the conditions underground are so protected and stable that minerals, rocks, and fossils can be preserved in exquisite condition for millions of years. Sediments and fossils on the ground surface are gradually but constantly weathered and eroded over time. The landscape on the surface changes slowly but surely as hillslopes are worn down, rivers incise or fill their beds with sediment, and forests grow and recede. By contrast, caves and their sediments can be nearly pristine, with delicate minerals, fossils, and sediments still intact. Although the hill or mountain that a cave is formed in may change over time, the cave itself is contained in solid rock, so it can maintain its original shape until the entire mountain itself is eventually eroded away.

Because knowing the age of a cave or its sediments is critical for learning about the past, several techniques for dating cave sediments and minerals have been developed. Each dating scheme has its own advantages and limitations. Some of the dating techniques such as paleomagnetism, uranium-series disequilibrium, and radiocarbon dating have become well established and widely used. This article concerns another, relatively new, dating technique that employs radioactive nuclides to date when sediment was brought into a cave. The method is called *cosmogenic nuclide burial dating* (Granger and Muzikar, 2001).

RELATIVE VERSUS ABSOLUTE DATING TECHNIQUES

When attempting to date a particular cave or its contents, there are several possible techniques to consider, depending on the age and the particular fossils or minerals in question. Some of these dating techniques are relative, indicating whether one thing is older or younger than another but not the exact age of either. For example, characteristic fossils in a cave can be used to place it relative to other deposits. Other dating techniques are absolute, meaning they give a numeric age that does not depend on correlations with any other site. Absolute ages are defined using some sort of "clock" that operates at a known and constant rate. By far the most widely used and reliable clock is radioactive decay.

To understand radioactive decay, it is helpful to first review the basic structure of the atomic nucleus. A nucleus is made of protons and neutrons. The number of protons in a nucleus determines to a large degree the way that an atom behaves; in fact, the elements of the periodic table are defined by the number of protons they have. Sometimes two different atoms may have the same number of protons, but different numbers of neutrons. In this case, the atoms are of the same element, but they have different masses. These two atoms are called *isotopes*. For a given element, some isotopes are stable, remaining unchanged over time. Other isotopes are radioactive, in which case the nucleus spontaneously breaks apart, losing mass and energy in radioactive decay. Radioactive decay occurs at a constant rate for any given isotope. If a certain amount of a radioactive isotope is contained in a rock, then half of that amount will decay in a characteristic time called the *half-life*. Half of the remaining half will decay after another half-life and so forth *ad infinitum*. Here is the clock for dating radioactive materials. If the original amount of a radioactive isotope is known, and the remaining amount can be measured, then the difference indicates the amount of time that has passed. The trick for cave scientists is to find a material with a known initial amount of radioactive isotopes. Cosmogenic nuclides provide just such a case.

BURIAL DATING WITH COSMOGENIC NUCLIDES

One way to date cave sediments is by determining the radioactive loss of cosmogenic nuclides. (The term *nuclide* refers to atoms regardless of their element, as opposed to *isotope*, which always refers to atoms of the same element.) Cosmogenic nuclides are produced by cosmic rays—energetic particles coming from outer space that constantly bombard Earth. Most of the cosmic rays are absorbed in the atmosphere, producing particles such as ^{14}C that is used for radiocarbon dating. Some secondary cosmic rays reach the ground surface and cause nuclear reactions inside rocks and minerals found within a few meters of the surface. During these nuclear reactions, the nuclei inside the mineral grains are broken apart, forming lighter nuclides. By chance, some of the products of these reactions are radioactive. For example, we can consider reactions in the mineral quartz. Quartz has a chemical formula SiO_2. Silicon nuclei each have 14 protons, and most of them have 14 neutrons. The common silicon nucleus thus has a mass of 28, written ^{28}Si. An incoming cosmic ray particle will occasionally break apart a silicon nucleus. If a proton and a neutron are lost, then the ^{28}Si is converted into ^{26}Al. Fortunately for dating, ^{26}Al is radioactive, with a half-life of 700,000 years. Another reaction that occurs in quartz is the conversion of ^{16}O to ^{10}Be through the loss of 4 protons and

2 neutrons. Beryllium-10 is also radioactive, with a half-life of about 1.4 million years. These two different radioactive nuclides are produced in the same quartz grain, and are the key to dating sediment burial in caves. A diagram of the reaction producing ^{26}Al is shown in Figure 1.

Over thousands of years, as rocks are exposed to cosmic rays they build up an inventory of ^{26}Al and ^{10}Be. Many repeated measurements of quartz grains on the ground surface have shown that ^{26}Al is produced about 6.8 times faster than ^{10}Be. Since both the production rates and the half-lives are known, the concentrations of ^{26}Al and ^{10}Be in rocks exposed to cosmic rays can be calculated. For most rocks at the ground surface, the ^{26}Al:^{10}Be ratio is 6.8:1. If quartz is brought into a cave, though, the grains are shielded from cosmic rays so ^{26}Al and ^{10}Be are no longer produced.

After the quartz-bearing sediment is brought into the cave, radioactive decay gradually lowers the concentrations of both ^{26}Al and ^{10}Be. Aluminum-26 decays faster than ^{10}Be, so the original ^{26}Al:^{10}Be ratio decreases over time. After 700,000 years half of the ^{26}Al is gone, but only 30% of the ^{10}Be has decayed. The original ratio of 6.8:1 has thus been lowered to about 4.8:1. The ^{26}Al:^{10}Be ratio provides the radioactive clock that we can use to date cave sediments, with an original ratio of 6.8:1 that decreases exponentially over time. Figure 2 shows the decay of the two nuclides, and the ^{26}Al:^{10}Be ratio as a function of time. As the concentrations of ^{26}Al and ^{10}Be get smaller and smaller over time, they become more difficult to measure. The practical limit to measurement usually occurs after about 5 million years of burial for this reason.

Aluminum-26 and ^{10}Be in sediment can only be measured using a very sensitive technique called *accelerator mass spectrometry* (AMS). This is because the concentrations of the cosmogenic nuclides are extremely small. For example, only 5 atoms of ^{10}Be may be produced in a gram of quartz in an entire year, and a sample

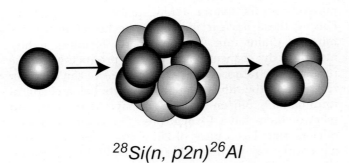

FIGURE 1 An example of a spallation-type nuclear reaction, in which an incoming cosmic ray neutron (left) impacts a ^{28}Si nucleus (center), to knock off two neutrons and a proton (right), making radioactive ^{26}Al. Neutrons are indicated by the darker-colored spheres, protons by the lighter-colored spheres.

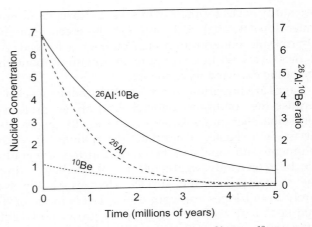

FIGURE 2 A graph of the concentrations of ^{26}Al and ^{10}Be (arbitrary units), and the ^{26}Al:^{10}Be ratio in quartz grains over time. The grains are washed into a cave with an original ^{26}Al:^{10}Be ratio of 6.8:1. Because ^{26}Al decays faster than ^{10}Be, the ^{26}Al:^{10}Be ratio decreases over time. The ^{26}Al:^{10}Be ratio can thus be used to date when the sediment was deposited in the cave.

may contain less than a million atoms of ^{10}Be. AMS is capable of measuring an isotope ratio (e.g., ^{10}Be/^{9}Be or ^{26}Al/^{27}Al) as low as 10^{-16}. That is, if there are 10^{16} atoms of the common isotope ^{27}Al, then AMS can detect a single atom of cosmogenic ^{26}Al.

Requirements for Burial Dating

As with any dating technique, it is important to consider the circumstances for which dates will be reliable, and those for which dates will be unreliable or impossible to obtain. Burial dating has several rather strict requirements. First, the sediment must have washed into the cave from outside, that is, it must be allochthonous. Otherwise, there would be none of the cosmogenic nuclides to begin with. Second, the sediment must contain the mineral quartz, because that is the mineral for which we know the production rates of ^{26}Al and ^{10}Be. Quartz is not always common in cave-forming bedrock, so even if there is allochthonous sediment in the cave it may not be datable. Third, the sample must be buried underground by at least 20 meters for the technique to be reliable over millions of years. It is possible to date more shallowly buried samples, but this introduces higher uncertainties. Fourth, there are limitations on the burial times. The sediment cannot have been buried more than about 5 million years, or the ^{26}Al and ^{10}Be are no longer detectable. Uncertainties in measuring ^{26}Al and ^{10}Be are usually 3–5%, making it difficult to achieve burial dates more precise than about 100,000 years. So the sediment must have been buried for at least 100,000 years. Finally, the sediment must come into the cave without a prior history of burial. If the sediment were, for instance, buried at the bottom of a doline for a million years and then washed into the cave, then the burial age would account for the total time spent buried, not just that in the cave. Although these uncertainties limit the application of burial dating somewhat, there are many situations for which the technique is ideal. We describe two examples below.

Example 1: The Development of Mammoth Cave, Kentucky

Mammoth Cave, the longest known cave in the world, has developed alongside the Green River in Kentucky. It is an example of a water-table cave, or one that has developed nearly horizontal passages that are closely controlled by the level of groundwater flow (Palmer, 1981). At Mammoth Cave, the water level is in turn controlled by the elevation of the Green River. Rainfall on the nearby Pennyroyal plateau quickly infiltrates the karst bedrock until it reaches cave passages that are filled or nearly filled with water. These underground streams then flow toward the Green River, passing through a sandstone-capped limestone plateau to discharge as springs on the Green River. In addition to the large recharge area that captures abundant rainfall, the sandstone-capped plateau is a major reason why Mammoth Cave is so long. The sandstone is a rock made of cemented quartz sand that is very resistant to erosion. Over time, as the Green River has cut through the sandstone and into the underlying limestones, cave passages have formed at successively lower levels. In many landscapes, the old cave passages above would be destroyed by erosion as new cave passages were formed below. However, at Mammoth Cave the sandstone is so resistant to erosion that the older passages have not eroded away. The sandstone forms ridges beneath which are preserved stacks of cave passages, the oldest passages near the top and the youngest passages at modern river level.

The preservation of old passages at Mammoth Cave provides a wonderful opportunity for studying how the cave has developed, and how the Green River has incised and aggraded over time. Geologist and hydrologist Art Palmer has spent many years carefully working out the sequence of events that are encrypted within Mammoth Cave's passages. It is very difficult, though, to decipher the history of the cave without dates that can be used to tie passages together across the cave system, and to match episodes of cave development with other geologic events. Burial dating with cosmogenic nuclides has provided a set of dates that helps show how the development of the Mammoth Cave System has been strongly influenced by climate change and the growth of ice sheets across North America. Although ice sheets did not reach Mammoth Cave itself, they did impact the Green River, which alternately incised and aggraded, forming sets of passages beneath the ridges at Mammoth Cave.

Burial dating works at Mammoth Cave because quartz pebbles from conglomerates within the sandstone upland are carried into the cave by sinking streams. These streams carry the pebbles through the cave system and into the Green River. When the river incises deeper, new passages are formed at lower elevations. The old passages are no longer occupied by streams, so whatever quartz pebbles were being carried through the cave are left in place, to sit within the now-abandoned cave passages. These packages of quartz-bearing sediment can be found throughout nearly the entire cave system. It is important to realize that the sediments indicate not when the passage formed, but when the passage was abandoned. It is only through careful analysis of the cave that the abandonment of one passage can be linked to the growth of another.

Granger *et al.* (2001) dated sediments from throughout the Mammoth Cave System. These samples reach ages up to 3.5 million years old in the uppermost levels of the cave, and tell an interesting story of how the cave developed over time. First, the dates show that the Mammoth Cave System is quite old. If the sediments that fill the cave are up to 3.5 million years old, then the cave itself must be significantly older than that! The upper levels of the cave system substantially predate the ice ages, which began roughly 2.5 million years ago. The cave, then, reveals how the landscape of central Kentucky responded to this major climate change. The initial response to climate change seems to be that the entire cave system filled up with sediment. Visitors to Mammoth Cave will notice telltale signs of sediment everywhere, even in nooks and crannies on the passage ceilings. These sediments show that most of the cave was filled up at about 2.4 million years ago, which in turn indicates that the Green River valley must be filled up as well. The landscape response to climate change was river aggradation, perhaps due to increased hillslope erosion that would have supplied more sediment than the river could carry. The next chapter in the story of Mammoth Cave is river incision and cave development at lower levels. Mammoth Cave is developed in levels, indicating that river incision was episodic. Each new level represents a pulse of river incision. Granger *et al.* (2001) found that these incision pulses correlate with large glaciations that covered most of eastern Canada and the northeastern United States, advancing as far as the northern edge of Kentucky. These large ice sheets completely reorganized river systems that were either buried beneath the ice or blocked by great ice dams. In fact, the modern courses of rivers such as the Ohio River, the Missouri River, and the northern Mississippi River were shaped along the edges of ice sheets. Each time that the Ohio River or Mississippi River incised due to glaciation or sea-level lowering, the Green River followed closely behind. Mammoth Cave thus records in its passages a key to the history of eastern North America over the past 3.5 million years.

Example 2: Caves and Human Evolution

Caves have always held an important place in human evolution, because so many fossils have been found in them. This is not necessarily because early humans lived in caves, but because the bones are so well-preserved, and often because predators or scavengers carried the bones there. Cave deposits, though, have been difficult to date beyond a few hundred thousand years. Cosmogenic nuclide burial dating has provided a new tool for dating these old fossils.

One example is found at the caves of Atapuerca, Spain. These are multilevel caves formed as the nearby Arlanzon river incised. A railroad cut through the limestone breached into the caves, and revealed that at several places the cave had filled with sediment from above, where shafts connected the passages to the surface. These sedimentary infills contain an abundance of fossils, including human ancestors.

The sedimentary infill at Atapuerca known as Sima del Elefante fills a vertical canyon 18 meters deep. The lower part of the infill contains some of the oldest fauna found at Atapuerca, including weasels, beavers, cave bears, and European jaguar. In 2007, a partial jaw and teeth from a human ancestor were discovered, as were several stone tools. Two samples of sediment were taken from very near the fossil for cosmogenic dating. The two samples agree closely, and give an average burial age of 1.18 ± 0.12 million years. This makes the bone the oldest dated fossil of a human ancestor in Europe (Carbonell *et al.*, 2008).

A second example comes from Zhoukoudian, near Beijing, China. This is a vertical cave that was completely filled with sediment. The site is famous for the discovery of "Peking Man" in the late 1920s. Peking Man was among the first discovered human ancestors, and played an important role in the early development of the science of paleoanthropology. The age of Peking Man has been debated for many years, with most people believing that the fossil is about 500,000 years old. Cosmogenic nuclide burial dating provides a way to determine the age absolutely.

Shen *et al.* (2009) collected six samples of quartz sediment from the cave infill, and also four stone tools made of quartz. Out of these ten samples, six were used to date Peking Man. These samples have an average age of 0.77 ± 0.08 million years, substantially older than previously thought. They were able to correlate sedimentation of the cave with global climate cycles, and together with the cosmogenic burial ages suggested that Peking Man occupied northern China during a cold phase, when Beijing was cooler and drier than today. This has important implications for the types of environments that early humans were capable of colonizing.

SUMMARY

These two examples of burial dating with cosmogenic nuclides show only the beginning of what the dating technique can do. The sediments in the Mammoth Cave System were an integral part of how the cave was formed. The sediments reveal the evolution of the cave system, and how cave development is tightly coupled to river incision and aggradation. In this case, Mammoth Cave was ideal because it was a water-table cave that carried quartz from local bedrock. In contrast, Atapuerca and Zhoukoudian are sedimentary infills where sediment (and animals) fell into a preexisting cavity. Such cave infills are the norm in archaeology and paleoanthropology because they collect bones and artifacts over long periods of time. In this case, the cosmogenic nuclides dated the sedimentary infill rather than the cave itself. There are many more situations where geomorphologists, paleoanthropologists, and other scientists can benefit from dating cave sediments over the past 5 million years.

See Also the Following Articles

Clastic Sediments in Caves
Multilevel Caves and Landscape Evolution
Mammoth Cave

Bibliography

Carbonell, E., Burmúdez de Castro, J. M., Parés, J. M., Pérez-González, A., Cuenca-Bescós, G., Ollé, A., et al. (2008). The first hominin species of Europe. *Nature*, 452(7186), 465–470.

Granger, D. E., Fabel, D., & Palmer, A. N. (2001). Pliocene-Pleistocene incision of the Green River, Kentucky, determined from radioactive decay of cosmogenic ^{26}Al and ^{10}Be in Mammoth Cave sediments. *Geological Society of America Bulletin*, 113(7), 825–836.

Granger, D. E., & Muzikar, P. F. (2001). Dating sediment burial with cosmogenic nuclides: Theory, techniques, and limitations. *Earth and Planetary Science Letters*, 188(1–2), 269–281.

Palmer, A. N. (1981). *A geological guide to Mammoth Cave National Park*. Teaneck, NJ: Zephyrus Press.

Shen, G. J., Gao, X., Gao, B., & Granger, D. E. (2009). Age of Zhoukoudian *Homo erectus* determined with ^{26}Al/^{10}Be burial dating. *Nature, 458*(7235), 198–200.

CRUSTACEA

Horton H. Hobbs III
Wittenberg University

INTRODUCTION

Crustaceans are one of the oldest and most diverse arthropods as well as one of the most successful groups of invertebrates on Earth, with approximately 40,000 extant species described and some 150,000 species recognized. They have been extremely successful in aquatic habitats, yet some species have become adapted on land as well. Their fossil record indicates that they are an ancient group, having occupied marine environments since the lower Cambrian period yet very early in their evolutionary history they invaded freshwater habitats. Although about 90% of the currently recognized taxa are widespread in marine systems, the remaining 10% are found in diverse inland waters and assume important roles in various ecosystem processes of many surface and subterranean lotic and lentic habitats (Hobbs 2000). This article focuses on the hypogean members, specifically on those crustaceans that are highly adapted to dwelling in groundwater ecosystems and generally referred to as *stygobionts* (obligate hypogean aquatic forms).

In addition to being taxonomically diverse, crustaceans are anatomically disparate, having evolved an assortment of body forms accomplished by developing highly specialized body segments and appendages as well as by fusing various segments. As a group, crustaceans are bilateral, having internal and external segmentation, and an open hemocoel. They have a rigid, chitinous exoskeleton composed of a thin proteinaceous epicuticle and a thick multilayer procuticle that in many groups is hardened by small inclusions of calcium carbonate. Their bodies are generally divided into the cephalon (head), thorax, and abdomen (with the former two sometimes combined as the cephalothorax). The many jointed appendages are biramous (or secondarily uniramous), may occur in all regions of the body, and these arthropods possess paired antennules (uniramous in all crustaceans except malacostracans), antennae, mandibles, and maxillae.

Crustaceans have invaded the hypogean realm, occupying interstitial and other groundwater habitats, including anchialine waters (inland groundwater with subsurface marine connections harboring unique fauna) in karst (see article entitled "Anchialine Caves" in this volume) as well as in other landforms (*e.g.*, lava). Some species of amphipod and isopod crustaceans have abandoned the groundwater and have been successful in the terrestrial hypogean environment. Most species dwelling in subterranean environments exhibit a suite of characteristic traits that are adaptive for life in such extreme ecosystems. Examples include reduction or loss of eyes and pigments, elongation of appendages, increased chemical and tactile sensitivity, degeneration of circadian rhythms, lowered fecundity and metabolic rates, and increased longevity and ovum volume. These embody behavioral, ecological, morphological, and physiological modifications that include both the reduction or loss of characters (regressive evolution) as well as the augmentation of others (constructive evolution). These various adaptations combine to generate the convergence characteristic of most obligate, cave-adapted organisms that is referred to as *troglomorphy* (see Culver and Wilkens 2000).

Crustacean taxonomy continually undergoes reevaluation and revision and the classification structure used herein (generally based on Martin and Davis, 2001) reveals five classes having subterranean representatives: Branchiopoda, Remipedia, Maxillopoda, Ostracoda, and Malacostraca (Table 1); a brief discussion of the classes and their hypogean representatives (predominantly stygobionts) is presented next.

SYNOPSIS OF STYGOBIOTIC CRUSTACEAN TAXA

Class Branchiopoda

Branchiopods are relatively small heterogeneous crustaceans that share few characteristics, including small to vestigial head appendages with similar mouthparts, flattened leaflike thoracic legs called *phyllopods* that usually decrease in size posteriorly, and a pair of spines or claws on the ultimate body segment. Classification of branchiopods has undergone numerous revisions and, of the currently recognized orders, only Diplostraca has subterranean members. There are about 100 species of the suborder Cladocera (450 total species) that occupy the subterranean environment (Table 1).

They are known from subterranean waters (especially the interstitial/hyporheos) on all continents, but especially well in Bosnia and Herzegovina, France, Romania, Slovenia, and Spain. These small transparent crustaceans have a carapace that is laterally compressed and attached dorsally to the body around which it is wrapped, excluding the head. Troglomorphic adaptations are very minor except in a few species of the genera *Alona* and *Spinalona*, evident by a lack of eyes

TABLE 1   Abbreviated Classification of Crustacea, Inclusive Only of Those Groups Dwelling in Subterranean Habitats[a]

Subphylum Crustacea (4570+)
  Class Branchiopoda (~100)
    Subclass Phyllopoda (~100)
      Order Diplostraca (~100)
        Suborder Cladocera (79)
          Family Daphniidae (20)
          Family Moinidae (5)
          Family Bosminidae (2)
          Family Macrothridicae (8)
          Family Eurycercidae [=Chydoridae] (44)
        Infraorder Ctenopoda (2)
          Family Sididae (2)
    Infraorder Haplopoda (1)
      Family Leptodoridae (1)
  Class Remipedia (24)
    Order Nectiopoda (24)
      Family Godzilliidae (5)
      Family Micropacteridae (1)
      Family Speleonectidae (18)
  Class Maxillopoda (1108+)
    Subclass Tantulocarida (1)
    Subclass Mystacocarida (<20)
    Subclass Copepoda (1087+)
      Order Platycopioida (11)
      Order Calanoida (39)
      Order Misophrioida (17)
      Order Cyclopoida (231)
      Order Gelyelloida (2)
      Order Harpacticoida (787)
  Class Ostracoda (~1000)
    Subclass Myodocopa (235)
      Order Myodocopida (8)
      Order Halocyprida (29 strictly anchihaline and approximately 200 marine interstitial species)
    Subclass Podocopa (~365)
      Order Platycopida (1)
      Order Podocopida (120+, of which approximately half are anchihaline)
    Subclass Mystacocarida (<20 intertidal and subtidal interstices)
  Class Malacostraca (2338)
    Sublcass Phyllocarida (1)
      Order Leptostraca (1)
    Subclass Eumalacostraca (2337)

Superorder Syncarida (249)
   Order Bathynellacea (228)
   Order Anaspidacea (21)
Superorder Peracarida (1866)
   Order Spelaeogriphacea (4)
   Order Thermosbaenacea (33)
   Order Mysida (45)
   Order Mictacea (1)
   Order Bochusacea (3)
   Order Amphipoda (784)
     Suborder Gammaridea (784)
   Order Isopoda (973)
     Suborder Phreatoicidea (12)
     Suborder Anthuridea (24)
     Suborder Microcerberidea (64)
     Suborder Flabellifera (144)
        Superfamily Cymothoidea (Cirolanidae: 93)
        Superfamily Sphaeromatoidea (Sphaeromatidae: 51)
     Suborder Asellota (448)
     Suborder Caloabozoidea (1)
     Suborder Oniscidea (280)
   Order Tanaidacea (7)
   Order Cumacea (16)
Superorder Eucarida (222)
   Order Decapoda (222)
     Suborder Pleocyemata (222)
       Infraorder Stenopodidea (1)
       Infraorder Caridea (111)
       Infraorder Astacidea (41)
       Infraorder Anomura (2)
       Infraorder Brachyura (67)

[a]Number = approximate number of described subterranean species.

and a carapace that is translucent and sparsely pigmented. Additionally, some have conserved a suite of primitive characters (*e.g.*, setation of valve rims), which suggests that the protected constancy of the hyporheic has allowed for the survival of some old taxa.

Class Remipedia

The discovery of remipedes in the subterranean waters of Lucayan Cavern on Grand Bahama Island in 1979 presented a major surprise. On one hand these blind crustaceans possess characteristics that are very primitive (*e.g.*, long homonomous body, paddle-like antennae, double ventral nerve cord), yet they have attributes that are traditionally considered advanced (*e.g.*, maxillipeds and biramous limbs that are not platelike). These small (<1–4 cm) translucent crustaceans lack a carapace but a cephalic shield covers the head, appendages of which include a pair of rodlike processes anterior to antennules and prehensile mouthparts; the head and first trunk segment comprise the cephalothorax; and they have an elongate trunk of up to 32 unfused segments, each bearing a pair of similar, laterally directed biramous limbs (Fig. 1).

They have been observed by divers primarily within submerged caves below the density interface between the overlying fresh or slightly brackish water and the underlying dense saltwater, although one species, *Speleonectes epilimnius* Yager and Carpenter, is found above the density gradient. Most swim ventral side up as the result of synchronized beating of the trunk appendages and apparently stay below this interface in the oxygen-poor (as low as 0.08 mg/L) saltwater layer. These predators likely feed in the overlying, well-oxygenated freshwater lens and locate their food by chemosensory means. Remipedes are small (up to 45 mm total length) and are commonly associated with other stygobiotic crustaceans such as caridean shrimps, cirolanid isopods, haziid amphipods, mysids, ostracods, and thermosbaenaceans. Twenty-four species reside in three families containing eight genera (Tables 1 and 2) and are known only from anchialine caves in the Cape Range Peninsula in Western Australia, Bahamas, Canary Islands, Cuba, Turks and Caicos, and the Yucatán Peninsula (Mexico).

Class Maxillopoda

Maxillopods are mostly small crustaceans although barnacles (Thecostraca: Cirripedia) are conspicuous deviations. They have a reduced abdomen and lack a full complement of legs. Treated below are those subclasses having subterranean representatives: Tantulocarida, Mystacocarida, and Copepoda.

Subclass Tantulocarida

These small (<0.5 mm long), ectoparasitic crustaceans are restricted to crustacean hosts. The only stygobiotic tantulocarid (*Stygotantulus stocki* Boxshall and Huys 1989) is known from an anchialine lava pool on Lanzarote in the Canary Islands where it is parasitic on two families of harpacticoid copepods.

Subclass Mystacocarida

This small group (most species <0.5 mm in total length; two genera with <20 described species and subspecies; Table 1) of marine interstitial crustaceans is characterized by an elongate body that is divided into a head and a 10-segmented trunk. The carapace and compound eyes are lacking; the first trunk segment has one pair of maxillipeds not fused to the head; and it has a telson bearing large, pincer-like furca. Because of the retention of primitive head segmentation, the lack of fusion of the cephalon and maxillipedal trunk segment, simplicity of mouth appendages, and absence of trunk compartmentalization, this is likely one of the most primitive of all crustaceans. (Others argue that these features may be related entirely to pedomorphosis and adaptation for interstitial habitats.)

FIGURE 1 Dorsal view of the remipede, *Micropacter yagerae* Koenemann, Iliffe, and van der Ham, from Old Blue Hill Cave, Caicos Island, Bahamas. *Photo courtesy of Stefan Koenemann. Used with permission.*

TABLE 2 Occurrence of Remipedia

Family	Genus	Species	Distribution
Godzilliidae	*Godzilliognomus*	2	Bahamas
	Godzillius	1	Turks and Caicos
	Pleomothra	2	Bahamas
Micropacteridae	*Micropacter*	1	Turks and Caicos
Speleonectidae	*Cryptocorynetes*	3	Bahamas, Dominican Republic
	Kaloketos	1	Turks and Caicos
	Lasionectes	2	Australia, Turks and Caicos
	Speleonectes	12	Bahamas, Canary Islands, Cuba, Dominican Republic, Mexico (Yucatán)

Little is known concerning the life history of this group. Apparently eggs are laid free and likely they have up to six naupliar stages. Their small size and worm-like body are adaptations for interstitial life, using head appendages to aid in crawling among sand grains where they glean detritus and microorganisms from the surfaces of sediment particles (see Zinn 1986). Although probably more widespread, they demonstrate a patchy distribution within the littoral and sublittoral sands in southern and western Africa, Australia, Brazil, Chile, southern Europe, the Gulf of Mexico, the Mediterranean Sea, and the east coast of the United States. Species of the genus *Ctenocheilocaris* are restricted to the Neotropical region, whereas *Derocheilocaris* spp. inhabit the Nearctic, Palearctic, and Ethiopian regions.

Subclass Copepoda

The subclass Copepoda is a very large and diverse group of crustaceans (approximately 220 families, 2300 genera, and 14,000 species) and, because they can attain incredibly high densities, are considered to be the most abundant metazoans on Earth (more individuals but fewer species than the insects). Lacking compound eyes and a carapace, the basic body plan consists of a head with well-developed mouthparts and antennae; a six-segmented thorax bearing swimming appendages, with the first segment fused to the head with maxillipeds; and a five-segmented abdomen, lacking appendages but including a telson. Development occurs within a few days to three weeks from fertilized eggs that hatch into a larval stage called a *nauplius*. Six naupliar stages are followed by six copepodid stages, the last of which is the adult (no additional molts). Six of the 10 orders have subterranean representatives, most of which demonstrate varying degrees of troglomorphy, and are treated next briefly (Table 1).

The Platycopioida order consists of one family (Platycopiidae), four genera, and 11 species that have retained numerous primitive characters. They are considered to be the first order to diverge from the main lineage of copepods. The eight species of the genus *Platycopia* are known from the benthos in sea and coastal waters (Africa, Bahamas, northern Europe, Japan, United States). *Sarsicopia polaris* Martinez-Arbizu is found in muddy sediments of the Arctic Ocean, and two anchialine genera and species are endemic to a single cave in Bermuda: *Antrisocopia prehensilis* Fosshagen and *Nanocopia minuta* Fosshagen, one of the smallest known copepods.

The Calanoida order contains approximately 2400 species in about 250 genera, yet only 39 species assigned to 23 genera and seven families are described from subterranean waters (Tables 1 and 3). Calanoids have biramous antennae and antennules that are greatly elongated, and the point of major body articulation occurs between the thorax and abdomen, which is marked by a distinct narrowing of the body. This primarily filter-feeding, planktonic group is geographically widespread, having been found in hypogean settings in Australia, Bahamas, Balearic Islands, Barbuda, Belize, Bermuda, Canary Islands, Caroline Islands, southeast China, Cuba, Dinaric Alps, Fiji, southern France, Galapagos Islands, Herzegovina, Istria, Italy, Madagascar, Mediterranean, Mexico, Philippines, and Russia.

The Misophrioida order is represented by 16 genera and 34 widely distributed copepods that are planktonic, hyperbenthic, and deep-sea, open-ocean, and hypogean water-dwellers. The subterranean species demonstrate particularly disjunct distributions, which are summarized in Table 4.

Members of the Cyclopoida order are free-living, planktonic, and associated with various substrates in benthic or littoral habitats. They detect their prey with the aid of mechanoreceptors on their first antennae and grasp their food with precision with their first maxillae or, occasionally, with their second maxillae and maxillipeds. Most are voracious predators although some are parasitic. Antennules and uniramous antennae are moderately long (never as long as antennules of harpacticoids; see below); a fifth pair of legs is

TABLE 3 Occurrence of Described Stygobiotic Calanoid Copepods in Anchihaline, Cave, or Well Habitats

Family	Genus	Species/Subspecies	Occurrence
Boholiniidae	2	2	Anchihaline
Diaptomidae	8	11	Freshwater
Epacteriscidae	3	14	Anchihaline; freshwater
Fosshageniidae	1	1	Anchihaline
Pseudocyclopiidae	3	4	Anchihaline; marine
Ridgewayiidae	5	5	Anchihaline
Stephidae	1	2	Anchihaline
Total	23	39	

TABLE 4 Subterranean Misophrioida Copepoda Distributions and Habitats

Genus	Species	Distribution	Habitat
Boxshallia	*bulboantennulata*	Canary Islands	Anchihaline
Dimisophria	*cavernicola*	Canary Islands	Anchihaline
Expansophria	4	Canary Islands; Galapagos Islands; Italy; Sardinia; Palau	Anchihaline
Huysia	*bahamensis*	Bahamas	Anchihaline
Misophria	*kororiensis*	Atlantic	Anchihaline
Palpophria	*aestheta*	Canary Islands	Anchihaline
Protospeleophria	*lucayae*	Bahamas	Anchihaline
Speleophria	4	Angaur Islands, Palau; Balearic Islands; Bermuda	Anchihaline, cave
Speleophriopsis	2	Balearic Islands; Canary Islands	Anchihaline
Stygomisophria	*kororiensis*	Koror Island, Palau	Anchihaline

highly reduced, the sixth pair vestigial. Cyclopoid copepods have not been well studied in subterranean environments. Nearly 750 species are recognized, but only about 231 species and subspecies are reported from freshwater caves, interstitial habitats, and wells (Table 1) (*e.g., Acanthocyclops, Diacyclops, Eucyclops, Halicyclops, Idiocyclops, Kieferiella, Metacyclops, Neocyclops, Speleoithona,* and *Speocyclops*) from Africa, Asia, Australia, Bahamas, Cuba, Europe, Madagascar, and North and South America. A diverse and abundant copepod fauna is actively being studied inhabiting epikarstic drip pools (Pipan 2005). About 60 additional species and subspecies are known from marine and brackish water interstices.

The Gelyelloida order is represented by a single genus and two species inhabiting European freshwater karst systems. *Gelyella droguei* Rouche and Lescher-Moutoué and *G. monardi* Moeschler and Rouch are restricted to karst waters in Montpellier, France, and in the Swiss Jura, respectively. This copepod order is characterized by a distinct combination of gnathostomous mouthparts and unusual derived features.

The Harpacticoida order is characterized by a body that is generally worm-like and cylindrical, with anterior segments not much larger than posterior ones; antennules and biramous antennae are quite short and the point of major body articulation occurs between the fifth and sixth thoracic segments; locomotory pereiopods are reduced, consistent with loss of swimming ability and use of these appendages as levers against sand grains. This group of copepods occurs virtually in all aquatic environments; most are benthic and are well suited to move through interstices feeding on detritus and on microorganisms (*e.g.*, bacteria, diatoms, and protozoa). The harpacticoid copepods comprise 42 families with approximately 375 genera containing about 3000 species. At least 12 families (about 445 species) are known from freshwater caves, interstitial groundwaters, and wells (*e.g., Chappuisius, Elaphoidella, Neochinophora, Parastenocaris, Spelaeocamptus, Stygonitocrella*). Some 342 species and subspecies in 63 genera and 13 families are found in the marine interstitial habitat (*e.g., Leptocaris, Nitocra, Novocrinia, Psammotopa, Stygolaophonte*).

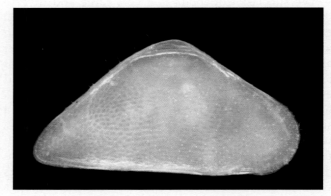

FIGURE 2 A candonid ostracod (subclass Podocopa). *Photo courtesy of Stuart Halse. Used with permission.*

Class Ostracoda

Ostracods are small (usually <1 mm, rarely 2 cm) crustaceans, with short, oval bodies encased within bivalved shells hinged dorsally (Fig. 2). This is a very diverse group (5700 species) and although most are found in the marine environment (to depths of 7000 m) they are abundant worldwide in all aquatic systems. Most species are benthic, many are planktonic, some are commensal on echinoderms or various malacostracan Crustacea (*e.g.*, Sphaeromicolinae on Isopoda, Entocytherinae on cambarid Decapoda), and a few are terrestrial in moist habitats.

Although some species are known only from caves, most hypogean ostracods are found in anchialine habitats, interstitial waters, springs, and wells. Examples of ostracods occupying various subterranean waters are *Cavernocypris lindbergi* Hartman (Afghanistan cave), *Pseudocandona jeanneli* Klie (cave in Indiana, United States), *Mixtacandona juberthieae* Danielopol (cave in southern France), and *Pseudolimnocythere hartmanni* Danielopol (well in Greece). These stygobionts display various troglomorphic adaptations including reduction or loss of eye structure, pigments, number of setae, and some have a very elongate or trapezoidal-shaped carapace. Most of the approximately 1000 subterranean species are assigned to the order Podocopida (Fig. 2) although the order Myodocopida is dominated by ostracods living in marine interstitial habitats (Bermuda, Galapagos Islands, Jamaica) and a coastal sea cave on Niue in the central Pacific (*Dantya ferox* Kornicker and Iliffe) (Table 1). The order Halocyprida is represented by 29 troglomorphic species dwelling in anchialine habitats, blue holes, and caves: *Danielopolina* (11 species in the Bahamas, Canary Islands, Cuba, Galapagos Islands, Jamaica, Yucatán Peninsula of Mexico, and the Cape Range Peninsula of Western Australia), *Deeveya* (7 species in the Bahamas and the Turks and Caicos islands), *Euconchoecia* (single species in Palau), and *Spelaeoecia* (11 anchialine species in the Bahamas, Bermuda, Cuba, Jamaica, and the Yucatán Peninsula); approximately 200 are marine interstitial species.

Of interest, the wide, irregular distribution, primitive nature, and troglomorphic adaptations of these taxa point to an extended history in a suitable cave environment. However, during the most recent period of Pleistocene glaciation, the sea level was lowered at least 100 m, resulting in coastal, anchialine caves becoming dry and then reinundated (substantiated today by the presence in these submerged caves of stalactites and stalagmites and other speleothems that were formed only in air by dripping or flowing water). This suggests that present-day amchialine ostracod fauna are recent invaders (within the past 15,000–18,000 years) and that they likely used an alternate, deeper habitat as refuge for considerable periods of time.

Class Malacostraca

This diverse group (about 29,000 species) far exceeds the species richness of any other crustacean and is divided into three subclasses, of which two have species dwelling in hypogean environments: Phyllocarida and Eumalacostraca. The body fundamentally comprises a five-segmented cephalon, eight-segmented thorax, six-segmented abdomen (seven in leptostracans), and telson; the carapace may be absent, reduced, or may cover part or all of the thorax and even several abdominal segments; many have none to three pairs of maxillipeds; antennules and antennae are usually biramous; the abdomen generally bears five pairs of biramous pleopods and one pair of biramous uropods; and they are mostly gonochoristic.

Subclass Phyllocarida

The subclass Phyllocarida is represented by the single order Leptostraca, which is the most primitive member of Malacostracans and is characterized by a head with movable, articulated rostrum; biramous antennules and uniramous antennae; absent maxillipeds; phyllopodous thoracopods; a laterally compressed bivalved carapace (lacking hinge) covering the thorax; and an elongate abdomen consisting of seven free pleomeres plus telson. The order is represented by fewer than 20 species assigned to six genera, most of which are small (5–15 mm long), but one species is nearly 4 cm in length. Most are epibenthic, are suspension feeders, and occur in low-oxygen marine environments. A single stygobiont, *Speonebalia cannoni* Bowman, Yager, and Iliffe (1985) (Nebaliidae), is known from two caves on Providenciales, Caicos Islands (Fig. 3).

Subclass Eumalacostraca

Members of this subclass have head, thorax, and abdomen; up to three thoracomeres are fused with the head, appendages of which are usually modified as maxillipeds; most groups have a well-developed carapace; and they possess a telson and paired uropods. There are three of four superorders with subterranean representatives.

SUPERORDER SYNCARIDA

This freshwater group, derived from marine stock, demonstrates the most primitive living body

FIGURE 3 The stygobiotic leptostracan, *Speonebalia cannoni* Bowman, Yager, and Iliffe from Providenciales, Caicos Islands. *Photo courtesy of Tom Iliffe. Used with permission.*

FIGURE 4 The parabathynellid syncarid, *Hexabathynella* spp. Photo courtesy of Jane McRae. Used with permission.

architecture of any eumalacostracan and many of the rather uniform trunk segments lack appendages. They do not have a carapace; telson with or without furcal lobes; some pereiopods are biramous; and the pleopods are variable. They either crawl or swim and, although little is known about most species, some are likely omnivorous. Unlike most other crustaceans, which transport eggs and thus carry early embryos, syncarids lay their eggs or release them into the water subsequent to copulation. Approximately 260 species have been described that are placed into two orders (Table 1): Bathynellacea and Anaspidacea; 95% of these are stygobionts.

The primitive Bathynellacea order lacks maxillipeds, is by far the more diverse group of syncarids, and is worldwide in distribution. Approximately 228 species are assigned to 60 genera that are placed within three families: Bathynellidae (*e.g.*, *Bathynella chappuisi* Delachaux, Switzerland; *Bathynella primaustraliensis* Schminke, Australia), Leptobathnelidae (*e.g.*, *Acanthobathynella knoepffleri* Coineau, Ivory Coast, Africa), and Parabathynellidae (*e.g.*, *Hexabathynella* spp., Africa, Australia, New Zealand, North and South America (Fig. 4); *Notobathynella williamsi* Schminke, Australia; *Parabathynella stygia* Chappuis, Croatia, Slovakia). They typically inhabit freshwater interstitial media in epigean, cave, and well hypogean habitats in Africa, Asia, Australia, Europe, Japan, Madagascar, Malaysia, New Zealand, and North and South America.

Order Anaspidacea (five families) has one pair of maxillipeds, is diverse in Australia, and species richness is particularly high on the island of Tasmania (numerous species awaiting formal descriptions). The Anaspididae family (*e.g.*, *Anaspides tasmaniae* (Thomson) (Fig. 5)) inhabits various freshwater environments in Tasmania, including caves. However, most other syncarids are interstitial dwellers or live strictly in subsurface groundwater, including caves and springs (*e.g.*, the exclusively stygobiotic family Psammaspididae—*Psammaspides* spp.—particularly diverse in New South Wales, Australia, and *Eucrenonaspides* spp. in Tasmania). Psammaspidae is represented only by *Eucrenonaspides oinotheke*, a spring-dwelling syncarid endemic to Tasmania. Family Koonungidae is

FIGURE 5 The anaspidacean syncarid, *Anaspides tasmaniae* Thomson, from Tasmania. Photo courtesy of Dr. Stefan Eberhard (www.subterraneanecology.com.au). Used with permission.

found in sediment interstices, open water, and sinkholes (*e.g.*, *Koonunga crenarum* Zeidler in southeast Australia and Tasmania). Family Stygocarididae (historically considered a separate order) is represented by fewer than 10 species assigned to four genera: *Oncostygocaris*, *Parastygocaris*, *Stygocarella*, and *Stygocaris*. These tiny species inhabit interstitial waters in epigean and hypogean environments in New Zealand, Tasmania, and South America.

SUPERORDER PERACARIDA

Crustaceans placed into this superorder demonstrate a trend toward reduction of the carapace. They possess one (rarely two to three) maxillipeds; gills are thoracic or abdominal; unique thoracic coxal endites (oostegites) form a ventral brood pouch or marsupium in females; they lack true larval stages; the young hatch as mancas, a prejuvenile stage lacking the last pair of thoracopods; pleopods lack appendix interna; and the telson is without caudal rami.

This highly successful group of malacostracans (approximately 11,000 species) is divided into 10 orders; all but the Lophogastrida have known subterranean species (Table 1). Most are marine but many occupy freshwater and terrestrial habitats. Peracarids are diverse in their habits and size, ranging from a few millimeters to 44 cm in length and some are symbionts as well as stygobionts.

The Spelaeogriphacea order is limited to fresh groundwater habitats in South Africa, South America, and Western Australia (Gondwana plates). This group was initially represented by a single species, *Spelaegriphus lepidops* Gordon, which was described from pools and a stream in Bats Cave, Table Mountain (now known from a second cave), South Africa, where it was observed swimming swiftly using rapid undulations of the body. Evidence suggested that it feeds largely on detritus.

In 1987 and again in 1998 two genera were described, one from a lake in a freshwater cave in Brazil (*Potiicoara brasiliensis* Pires—now known from two other caves) and the other from the Millstream aquifer in arid northwestern Australia (*Mangkurtu mityula* Poore and Humphreys; a second species, *M. kutjarra* Poore and Humphreys, has been added to this genus (Fig. 6) (Poore and Humphreys 1998, Humphreys and Harvey 2001).

These small, blind, unpigmented stygobionts possess a short, saddle-like carapace fused with the first thoracomere and, anteriorly, produced into a broadly triangular rostrum; they have one pair of maxillipeds; pereiopods 1 through 7 are simple, biramous, with shortened exopods; the exopods on pereiopods 1, 2, and 3 are modified for producing currents, and on pereiopods 4 through 7 they are modified as gills; the abdomen is elongated, often exceeding half of the total body length; and pleopods 1 through 4 are biramous and natatory, and pleopod 5 is reduced.

Order Thermosbaenacea is a group of small (2–5 mm), aquatic crustaceans that have a short carapace fused with the first thoracic somite (remaining seven thoracic segments free) extending posteriorly over two to three additional segments. In females, the carapace provides a brood pouch; there is a single pair of maxillipeds; pereiopods are biramous, simple, and lack epipods; there are two pairs of uniramous pleopods; the uropods are biramous; and the telson is free or forms a pleotelson with the last pleonite.

At least 33 stygobiotic species have been assigned to seven genera and four families (one additional species of *Limnosbaena* not formally described) (Table 5; see Wagner, 1994, for a revision of the order). They are known from anchialine habitats, caves, cenotes, various interstices, cold springs, thermal springs (45°C), and wells in fresh, to oligohaline, to hypersaline waters. These tethyan relicts demonstrate a very large geographic range in those areas once covered by the shallow Tethys Sea or along its former coastlines.

Order Mysida is represented by more than 1000 species that are widespread over all continents where they inhabit coastal and open ocean waters as well as continental fresh waters and various groundwater habitats. Some species are intertidal and burrow into the sand during periods of low tides; most of these shrimp-like crustaceans swim with the aid of thoracic exopods and are omnivorous suspension feeders that eat algae, zooplankton, and suspended detritus. They range in length from approximately 2 mm to 8 cm and display a well-developed carapace covering most of the thorax. Compound eyes are stalked, sometimes reduced; they have one to two pairs of maxillipeds not associated with cephalic appendages; the abdomen is elongated; pereiopods are biramous although the last pair is sometimes reduced; pleopods are reduced or modified; and the statocyst is usually located in each uropodal endopod.

Of the two suborders, only the Mysida has stygobiotic representatives and is divided into four families, three of which have stygobiotic and stygophilic species (Table 6). At least 45 stygobiotic species are recognized, most of which are endemics. The current distribution of the majority of these suggests that most colonized ground waters as a consequence of uplifting and stranding of their marine ancestors, which resulted from regressions of the Tethys and Mediterranean seas. Clearly other taxa have invaded ground waters more recently, most being stygophiles. Adaptations for subterranean existence seem to be limited to reduction or loss of body pigments (red pigments often retained) as well as reduction or loss of eyes but not eyestalks that contain endocrine organs.

Mictaceans are stygobiotic crustaceans, lacking body pigmentation and without visual elements in reduced eyestalks. The head narrows anteriorly into a triangular

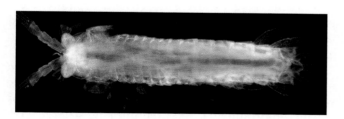

FIGURE 6 The spelaeogriphacean, *Mangkurtu katjarra* Poore and Humphreys, from groundwater in limestone in the upper Fortescue River in Western Australia. *Photo courtesy of Jane McRae. Used with permission.*

TABLE 5 Occurrence of Subterranean Thermosbaenaceans

Family	Genus	Species	Occurrence
Halosbaenidae	3	5	Caves, marine interstitial
Monodellidae	2	24	Anchihaline, artesian wells, brackish wells, caves, coastal springs, interstitial, phreatic, thermal springs, wells
Thermosbaenidae	1	1	Thermal springs
Tulumellidae	1	3	Anchihaline, caves, cenotes
Total	7	33	

TABLE 6 Occurrence of Described Subterranean Mysids (Order Mysida)

Family	Genus	Species	Occurrence
Lepidomysidae	2	10	Anchihaline, caves, crab burrows, interstitial, phreatic, prawn culture field
Mysidae	16	28	Anchihaline, caves, coral reef caves, land crab burrows, marine caves
Stygiomysidae	1	7	Anchihaline, caves, phreatic
Total	19	45	

rostrum and is fused posteriorly with the first thoracomere. The carapace is not developed but small lateral carapace folds produce a head shield laterally covering bases of mouthparts; gills are lacking; pereiopods are simple; pleopods are reduced and uniramous; and uropods are biramous, with two to five segmented rami.

A recent revision of the order Mictacea has resulted in its being monotypic (Table 1). The single species, *Mictocaris halope* Bowman and Iliffe, assigned to the family Mictocarididae, is known from four marine caves on Bermuda (Fig. 7). This small species (up to 3.5 mm in total length) swims, using exopods of thoracopods 2 through 6, and rarely rests or walks on the substrate. Mouthparts suggest that it is not a predator but probably procures food by scraping and/or filtering fine particulates from the water column and benthos.

Order Bochusacea is the most recently established peracarid order, which was erected to accommodate three species placed in two genera belonging to the family Hirsutiidae (initially placed in the order Mictacea; two deep-sea benthic species: *Hirsutia bathylis* Sanders, Hessler, and Garner off Surinam in northeast South America, and *Hirsutia sandersetalia* Just and Poore from Bass Strait, Australia). With additional material it became clear that a new order, Bochusacea, should be erected to house the hirsutiids (Gutu and Iliffe 1998). Additionally, a third species, *Thetispelecaris remex* Gutu and Iliffe, from three marine/anchialine caves in the Bahamas, was placed in the order. Although similar to the mictaceans, the bochusaceans possess different pereiopod forms, among other features. They too are swimmers and filter feeders; pereiopod 1 is specialized for feeding and not for locomotion.

The Amphipoda order is characterized by the absence of a carapace; the body is divided into head, thorax, and abdomen, each bearing appendages; the head bears two pairs of antennae; it has a single pair of maxillipeds, seven pairs of uniramous pereiopods, with the first, second, and often others modified as chelae or subchelae; periopodal coxae are expanded as lateral side plates; gills are thoracic; the abdomen consists of two regions of three segments each, an anterior pleon with pleopods, and posterior urosome with appendages modified as uropods; and the telson is free or fused with the last urosomite. This group of crustaceans is generally slender and laterally compressed although some are dorsoventrally flattened (e.g., *Heterophlias*). They range in length from 1 mm to giant deep ocean benthic forms reaching 25 cm.

FIGURE 7 The stygobiotic mictacean, *Mictocaris halope* Bowman and Iliffe, from a cave in Bermuda. *Photo courtesy of Tom Iliffe. Used with permission.*

Amphipod crustaceans are commonly found in numerous aquatic ecosystems around the globe where they have invaded freshwater, brackish, and marine environments, often comprising a large portion of the biomass in a habitat. A few species also dwell in various terrestrial ecosystems (Talitridae; e.g., supralittoral sandy beaches, moist forest litter) and they exhibit a great diversity of feeding strategies including carnivory, herbivory, parasitism, scavenging, and suspension feeding. Approximately 7000 species are assigned to three suborders: Gammaridea, Caprellidea, and Hyperiidea. Stygobiont amphipods are known only from the very large (>5700 species) Gammaridea suborder where some 784 troglomorphic species are assigned to approximately 159 genera and 29 families (Table 7). The following families demonstrate the richest subterranean biodiversity: Bogidiellidae, Crangonyctidae, Hadziidae, Melitidae, and Niphargidae. The amphipod suborder is

TABLE 7 Occurrence of Subterranean Amphipods (Suborder Gammaridae)

Family	Genus	Species	Occurrence
Allocrangonyctidae	1	2	Caves: United States (Missouri, Oklahoma)
Aoridae	1	1	Marine interstitial: Ile de Batz (Atlantic Ocean)
Bogidiellidae (including Artesiidae)	33	105	Caves, hyporheos, interstices of sandy beaches and sublittoral sands, cold mountain springs, wells: nearly worldwide (not from continental Africa south of equator, or continental Australia)
Calliopiidae	3	3	Groundwater, springs: New Zealand (North and South islands)
Crangonyctidae	6	158	Caves, springs, wells: Holarctic region (mostly North America)
Gammaridae	17	38	Caves, springs, wells: Canary Islands, Eurasia, Mediterranean region, North America, Solomon Islands
Hadziidae	26	78	Caves: Australia, southern Europe, Bahamas, Caribbean, Fiji, Hawaiian Islands, Mexico, United States
Hyalellidae	2	2	Caves: Australia, Brazil, Venezuela
Hyalidae	1	3	Subterranean freshwater: Comores and Zanzibar islands (off coast of Eastern Africa)
Ingofiellidae	3	37	Caves, interstitial (marine), springs, wells: Africa, Caribbean, Mediterranean regions, South America
Liljeborgiidae	1	1	Marine cave: Bermuda
Lysiannasidae	1	1	Anchihaline cave: Galapagos Islands
Melitidae	23	54	Anchihaline, caves, freshwater and marine interstitial, and wells: Australia, Balearic, Canary, Caribbean, Galapagos, Hawaiian, Pacific, and Philippine islands
Metacrangonyctidae	2	15	Hyporheos, springs, wells: Canary Islands, Mediterranean region
Metaingolfiellidae	1	1	Well: southern Italy
Neoniphargidae	2	2	Shallow groundwaters: Victoria and Tasmania (?), Australia
Niphargidae	8	221	Caves, hyporheos, interstitial, wells: Asia, Europe
Paracrangonyctidae	2	3	Interstices, wells: Chile, Kerguélen Island (in Indian Ocean), New Zealand
Paramelitidae	9	10	Caves, groundwater habitats: South Africa, Western Australia, southern Victoria, and Tasmania
Pardaliscidae	2	3	Anchihaline caves: Bahamas, Galapagos, Lanzarote, Turks and Caicos
Perthiidae	1	1	Caves: southwestern Australia
Phreatogammaridae	2	2	Caves, wells: south Australia, New Zealand (South Island), Spain
Pontogeneiidae	1	2	Cave: Japan
Plustidae	2	3	Freshwater caves and wells: Japan
Pseudocrangonyctidae	2	11	Caves, springs, wells: northeastern Asia
Salentinellidae	2	14	Freshwater and brackish water caves, hyporheos, wells: Mediterranean region of northern Africa and southern Europe
Sebidae	1	2	Artesian well: Texas (Hays County)
Sternophysingidae	1	8	Caves, springs: southern Africa
Talitridae	3	3	Caves: Corsica, Isla de la Palma (Canary Islands), Kauai (Hawaiian Islands), Sardinia
Total	159	784	

widespread globally with the greatest taxonomic diversity of troglomorphic species occurring in eastern and southern North America, the Caribbean/West Indian region, central and southern Europe, and the Mediterranean region. These stygobionts evolved from surface ancestors, moving into hypogean ground waters from both freshwater and marine environments involving both active and passive dispersal.

Order Isopoda is quite a diverse group with more than 11,000 described aquatic and terrestrial species.

They inhabit nearly all environments where most are free living yet some are partly (Flabellifera) and others are exclusively (Epicaridea) parasitic. Their feeding habits extend from herbivores to omnivorous scavengers, detritivores, predators, and parasites. They range in length from about 0.5 mm to 4.4 cm (the largest is the benthic *Bathynomus*). This diverse group of crustaceans is dorsoventrally flattened; they lack a carapace; the first thoracomere is fused with the head; they have a single pair of maxillipeds and seven pairs of uniramous pereiopods which are modified as ambulatory, natatory, or prehensile limbs; pleopods are biramous and modified as natatory or for gas exchange (gills in aquatic taxa, pseudotracheae in terrestrial Oniscidea); and the telson is usually fused with pleonites 1 through 6.

Representatives of subterranean isopods are known from seven suborders: Phreatoicidea, Anthuridea, Microcerberidea, Flabellifera, Asellota, Calabozoidea, and Oniscidea (Tables 1, 8) and are global in distribution.

Suborder Phreatoicidea is the most ancient group of isopods and is derived from marine ancestors. These isopods are currently restricted mostly to freshwaters of South Africa, Australia, India, and New Zealand. Most are epigean; however, stygobiotic species (at least 12 species in nine genera), known from caves and wells, are extant on all of these continental Gondwana fragments except South Africa.

Anthuridea is principally a marine group consisting of four families; however, at least 24 troglomorphic species assigned to two families and three genera dwell in the interstitial sediments of anchialine habitats, bays, beaches, caves, and wells in the Canary Islands, the Caribbean and Indian Ocean islands, Indonesia, Mexico, New Zealand, and South America.

The Microcerberidea consists of two families. The Atlantasellidae is represented by a single species, *Atlantasellus cavernicolous* Sket, which is restricted to anchialine caves in Bermuda. Clearly its recent origin is from marine stock. The Microcerberidae are slender, stygobionts of which approximately 64 species in six genera inhabit interstices of caves, marine beaches, and wells in the coastal regions of southern and western Africa, Southeast Asia, the Caribbean, Indian Ocean islands, Japan, the Mediterranean, and western North America.

Suborder Flabellifera is represented by 18 families, only 2 of which have subterranean members: Cirolanidae and Sphaeromatidae. The predominantly marine family Cirolanidae has species dwelling in subterranean environments ranging from fresh water to salinities near that of seawater. Approximately 93 species assigned to 29 genera are currently recognized and range from less than 3 mm in length (*Arubolana parvioculata* Notenboom) to around 33 mm (*Speocirolana bolivari* (Rioja)). Most are blind, lack body pigments, are somewhat convex dorsally, and are opportunistic benthic scavengers and predators. Their geographical distribution includes East Africa, the western Atlantic, and North America. (Cave and phreatic habitats are occupied by *Speocirolana* in Mexico and Texas, and by *Antrolana* and *Cirolanides* in Virginia and West Virginia, and Texas, respectively, and by *Kagalana* and *Haptolana* in Western Australia.)

The Sphaeromatidae is mostly a marine group with four genera of stygobionts represented by 43 species inhabiting subterranean waters in southern Europe and 8 species in hot springs in the southwestern United States and north-central Mexico. The very diverse genus *Monolistra* (at least 35 species) occupies karst waters along the Dinaride and Italo-Dinaride systems.

The Asellota are divided into a number of superfamilies (not treated herein) and nine families have subterranean representatives. The family Asellidae is a large benthic group (about 18 genera and 266 species) that is found primarily in caves, springs, and wells in northwestern Africa, Europe, Japan, and Central and North America. The Stenasellidae represent an ancient group that demonstrates primitive characters and that are blind and often orange or pink in body color. At least 71 species are assigned to 10 genera and these crustaceans inhabit caves, the hyporheos, cool and thermal springs, and wells in much of Africa, Southeast Asia, eastern, southern, and western Europe, Indonesia, Malaysia, Mexico, and the Edwards Plateau in North America. The small Stenetriidae isopods are known only from mixohaline waters in a cave on Curaçao and from a blue hole on Andros Island, Bahamas. The Janiridae is represented by 13 species assigned to five genera of marine origin. They are known from caves, springs, and wells in east Asia, western Europe, Italy, Japan, and North America (California). Microparasellidae are small, quite diverse (79 species in four genera), and widespread. They are known from eastern Asia, Australia, the Caribbean, Indian Ocean islands, Japan, and the Mediterranean. The Paramunnidae is represented by a single species inhabiting subterranean waters. Although the genus *Munnogonium* currently contains seven species, only *M. somersensis* Kensley inhabits a marine cave in Bermuda and shows no obvious troglomorphic adaptations. The gnathostenetroidid isopods are small crustaceans of marine origins and the seven species distributed evenly over three genera demonstrate a disjunct distribution in anchialine caves in the Turks and Caicos islands and Bermuda (the eyed *Stenobermuda*), an anchialine cave in the Bahamas (*Neostenetroides*), and the intertidal zone in Japan and a thermal spring on Ischia, a small coastal island

TABLE 8 Taxonomic Treatment of Subterranean Isopods

Suborder	Family	Genus	Species/Subspecies
Phreatoicidea	Amphisopidae	3	5
	Amphisopodidae	1?	1?
	Nichollsidae	1	2
	Phreatoicidae	3	4
Total	**4**	**8**	**12**
Anthuridea	Anthuridae	1	20
	Paranthuridae	2	4
Total	**2**	**3**	**24**
Microcerberidea	Atlantasellidae	1	1
	Microcerberidae	6	63
Total	**2**	**7**	**64**
Flabellifera	Cirolanidae	29	93
	Sphaeromatidae	4	51
Total	**2**	**33**	**144**
Asellota	Asellidae	18	266
	Stenasellidae	10	71
	Stenetriidae	1	2
	Janiridae	5	13
	Microparasellidae	4	79
	Paramunnidae	1	1
	Gnathostenetroididae	3	7
	Protojaniridae	4	9
Total	**8**	**46**	**448**
Calabozoidea	Calabozoidae	1	1
Total	**1**	**1**	**1**
Oniscidea	Armadillidae	—	—
	Armadillidiidae	—	—
	Berytoniscidae	—	—
	Cylisticidae	—	—
	Eubelidae	—	—
	Ligiidae	—	—
	Mesoniscidae	—	—
	Oniscidae	—	—
	Philosciidae	—	—
	Plathyarthridae	—	—
	Porcellionidae	—	—
	Scleropactidae	—	—
	Scyphacidae	—	—

(*Continued*)

TABLE 8 (Continued)

Suborder	Family	Genus	Species/Subspecies
	Spelaeoniscidae	–	–
	Styloniscidae	–	–
	Trachelipidae	–	–
	Trichoniscidae	54	–
Total	17	99	280
GRAND TOTAL	35	197	973

TABLE 9 Occurrence of Tanaidaceans in Subterranean Habitats

Family	Genus	Species	Occurrence
Apseudidae	*Apseudes*	*bermudeus*	Anchihaline: Bermuda
		bowmani	Marine: Koror Island (Palau)
		orghidani	Anchihaline: Bermuda
	Pugiodactylus	*agartthus*	Marine cave: Niue Island (South Pacific)
Metapseudidae	*Calozodion*	*propinquus*	Anchihaline, open ocean: Bermuda
Nototanaidae	*Nesotanais*	*maclaughlinae*	Anchihaline: Eli Malk Island (Palau)
Parapseudidae	*Swireapseudes*	*birdi*	Anchihaline: Bahamas

southwest of Naples, Italy (*Caecostenetroides*). Nine species belonging to four genera are assigned to the Protojaniridae family and are quite separate from other members of the Asellota suborder. They are found in subterranean freshwaters in South Africa, Argentina (South America), southern India, islands of the Indian Ocean, and Namibia.

The monotypic suborder Calabozoidea (Calabozoidae (*Calabozoa pellucida* Van Lieshout)) is known only from wells in the environs of the small town of Calabozo in northern Venezuela. This transparent, blind isopod appears to be restricted to phreatic waters.

The suborder Oniscidea is generally a terrestrial group (*e.g., Sinoniscus cavernicolous* Schultz, China) characterized by extreme reduction of the first pair of antennae and by biramous pleopods modified into pseudotracheae. Most members of the suborder are troglobionts yet some are stygobionts; some 276 aquatic and terrestrial species assigned to 99 genera and 17 families are recognized. Examples of aquatic species are *Abebaioscia troglodytes* Vandel from the Nullarbor Plain, Australia; *Haloniscus searlei* Chilton, a scyphacid from Australia in saline lakes up to 159 parts per thousand; a scyphacid isopod, *Haloniscus* sp. from Australia; the styloniscid *Thailandoniscus annae* Dalens, from Thailand; the trichoniscids *Mexiconiscus laevis* (Rioja), which is terrestrial as a juvenile and aquatic as an adult, and *Titanethes albus* (Koch), an amphibious species from Italy and Slovenia.

The order Tanaidacea is known worldwide from benthic marine habitats and only a few species live in brackish or nearly fresh water. They often dwell in burrows or tubes and are known from all ocean depths. Approximately 850 species are recognized and, being small (0.5–2 cm long), many are suspension feeders, some detritivores, and others are predators. The carapace is fused with the first two thoracic segments; the first and second thoracopods are maxillipeds, with the second one being chelate; and thoracopods 3 through 7 are simple, ambulatory pereiopods.

Very little is known about subterranean species, mostly due to very minimal sampling. Currently seven species, assigned to the families Apseudidae, Netapseudidae, Nototanaidae, and Parapseudidae, are known from anchialine/marine caves in the Bahamas, Bermuda, Eli Malk and Koror islands (Palau) (Gutu and Iliffe 1989), and Niue Island (South Pacific) (Table 9).

The order Cumacea is represented by rather strange-looking crustaceans having a large, bulbous anterior end and elongated posterior, a carapace that is fused to and covers the first thoracic segments, and three pairs of maxillipeds. Pereiopods 1 through 5 are simple and ambulatory, and pleopods are present in males, but usually absent in females.

TABLE 10 Subterranean Cumaceans

Family	Genus	Species	Occurrence
Bodotriidae	*Cyclaspis*	2	Submarine caves: Bahamas, Jamaica
Diastylidae	*Oxyurostylis*	*antipai*	Submarine cave: Jamaica
Nannastacidae	*Campylaspis*	2	Blue hole: Andros Island, Bahamas; submarine cave: Bermuda
	Cumella	9	Anchihaline: Bermuda; blue hole: Andros Island, Bahamas; submarine cave: Jamaica
	Schizotrema	2	Anchihaline, caves: Bermuda
Total	5	16	

They are distributed worldwide and about 850 species are recognized. These small peracarids (0.5–2 cm in length) are mostly marine (a few brackish and freshwater species) and are generally benthic. Subterranean species are not well studied and only 16 stygobionts placed in five genera and three families are known from blue holes and submarine caves on Andros Island (Bahamas), Bermuda, and Jamaica (see Petrescu et al. 1994) (Table 10).

SUPERORDER EUCARIDA

Eucarida is the final superorder treated herein and is characterized by crustaceans having a carapace covering and fused dorsally with the head and all thoracomeres (cephalothorax); they usually have stalked compound eyes; gills are thoracic; and the telson lacks caudal rami.

Order Decapoda is the only eucarid found in subterranean environments. This order is amazingly diverse with nearly 15,000 species known from marine, brackish, freshwater, and terrestrial habitats. Although widespread in distribution, these crustaceans are significant subterranean players in temperate and tropical regions of the world dominated by karst or volcanic terrains and are well represented in the western Atlantic, Caribbean region, Central and North America, and numerous Pacific Ocean islands. They are characterized by a well-developed carapace enclosing the branchial chamber, are unique among decapods in possessing three pairs of maxillipeds, and have five pairs of functional uniramous or weakly biramous pereiopods (thus the derivation of the order name). The order is split into two suborders, the Dendrobranchiata (about 450 species, mostly penaeid and sergestid shrimps) and the Pleocyemata (all remaining decapods as well as all stygobionts). The only dendrobranchiate decapod that occupies subterranean waters is a stygoxenic penaeid shrimp, *Penaeus indicus* H. Milne Edwards, found in Mangapwani Cave in East Africa. The suborder Pleocyemata contains all of the remaining decapods and is geographically widespread globally. It is commonly divided into 7 infraorders, only 5 of which have stygobiotic representatives: Stenopodidea (Fig. 8), Caridea, Astacidea, Anomura,

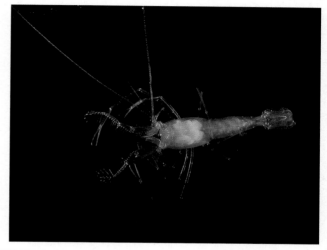

FIGURE 8 The stygobiotic stenopodid, *Macromaxillocaris bahamaensis* Alvarez, Iliffe, and Villalobos, from an anchihaline cave in the Bahamas. *Photo courtesy of Tom Iliffe. Used with permission.*

and Brachyura. The two remaining pleocyemate infraorders are represented in subterranean environments by the stygophilic *Jasus edwardsii* (Hutton) from an intertidal marine cave in southeastern New Zealand (Palinura), by the stygoxenic *Stenopus spinosus* in a lava tube in the Canary Islands, by *Odontozona addaia* from a marine cave on the island of Minorca, Spain (Stenopodidea), and by a probable stygophile *Naushonia manningi* Alvarez, Villalobos, and Iliffe from an anchialine cave on Acklins Island, Bahamas (Thalassinidea).

Approximately 2800 species of natant decapods are assigned to the infraorder Caridea. These caridean shrimps have chelate first or second pereiopods (no chelation in the genus *Procaris*); first pleopods are somewhat reduced; and the pleuron of the second abdominal tergite is enlarged, overlapping that of the first and third. Global in distribution, only some 111 species in 40 genera and six families are known from anchialine habitats, blue holes, caves, cenotes, springs, and wells in some temperate but mostly tropical latitudes (Table 11).

TABLE 11 Occurrence of Stygobiotic Caridean Shrimps

Family	Genus/Subgenus	Species	Occurrence
Procarididae	1	3	Anchihaline
Atyidae	15	55	Anchihaline; marine caves; cenote; coral limestone pools; saline caves, pools, and wells; wells (see Sket and Zaksek 2009)
Agostocarididae	1	2	Anchihaline, blue hole, cenote
Palaemonidae	11	30	Anchihaline, caves, coral rock pools, fissures, submarine caves, subterranean water, wells
Alpheidae	6	12	Anchihaline, caves, coral rock pools, lava pools, lava tubes (marine), submarine caves (see Anker 2008)
Hippolytidae	6	9	Anchihaline, blue holes, caves, cenotes, pools in coral or lava, sea caves
Total	40	111	

TABLE 12 Stygobiotic Crayfishes (Astacoidea: Cambaridae) in the United States, Cuba, and Mexico

Genus	Subgenus	Species/Subspecies	Occurrence
Cambarus	*Aviticambarus*	3	United States: Alabama, Tennessee
	Erebicambarus	hubrichti	United States: Missouri
	Jugicambarus	6	United States: Arkansas, Florida, Georgia, Missouri, Oklahoma
	Puncticambarus	nerterius	United States: West Virginia
Orconectes	*Orconectes*	9	United States: Alabama, Indiana, Kentucky, Missouri, Tennessee
Procambarus	*Austrocambarus*	5	Cuba; Mexico: Oaxaca, Veracruz
	Leconticambarus	1	United States: Florida
	Lonnberguis	2	United States: Florida
	Ortmannicus	11	Mexico: San Luis Potosí; United States: Florida
	Remoticambarus	pecki	United States: Alabama
Troglocambarus		maclanei	United States: Florida

The infraorder Astacidea, which includes about 900 species of crayfishes and chelate lobsters (>500 of those are crayfishes), is global in distribution (all continents except Africa), and the first three pairs of pereiopods are chelate. Crayfishes are the only astacideans that are stygobiotic and, of these, only the superfamily Astacoidea is represented in the subterranean environment. Some 41 species from a single family (Cambaridae) have been described as obligate cave-dwelling species and subspecies; about 50 additional pigmented, stygoxenic, or stygophilic crayfishes commonly invade subterranean waters around the globe (Astacidae, Cambaridae, Parastacidae). The following four genera are assigned stygobiotic species and subspecies from Cuba, Mexico, and North America (primarily in the Appalachian, Florida Lime Sink, Interior Low Plateau, and Ozark karst regions): *Cambarus* (11 species), *Orconectes* (9 species and subspecies), *Procambarus* (20 species and subspecies), and the monotypic *Troglocambarus* (Table 12). Most of these are troglomorphic and are opportunistic omnivores (see Hobbs et al. 1977).

Although diverse, the infraorder Anomura has only two stygobiotic species assigned to two families (Table 1). They possess a variably shaped carapace; the first pair of pereiopods is chelate, the third pair never, and the fifth pair reduced (never used for walking) and function as gill cleaners. The pleopods are reduced or absent. The superfamily Galatheoidea is represented by two families having subterranean species: Aeglidae and Galatheidae. *Aegla cavernicola* Türkay is a poorly studied aeglid, known only from a cave in São Paulo, Brazil. The aggressive galatheid *Munidopsis polymorpha* Koelbel has received much attention (life history and agonistic, feeding, and reproductive behaviors) and is restricted to a single cave (lava tube) and wells (marine groundwater) on Lazarote, Canary Islands.

TABLE 13 Occurrence of Stygobiotic Brachyurans

Family	Genus	Species	Occurrence
Hymenosomatidae	3	3	Caves: Indonesia
Trichodactylidae	2	2	Caves: Chiapas and Tabasco, Mexico
Goneplacidae	1	1	Caves: New Britain, Indonesia
Xanthidae	1	1	Anchihaline cave: Ecuador
Potamidae	5	7	Caves: Sarawak, Borneo; Laos; Palau Tioman, Malaysia; Thailand; Ryukyu Islands
Pseudothelphusidae	5	20	Caves: Belize; Chiapas, Mexico; Colombia and Venezuela, South America; Guatemala; Cuba
Gecarcinucidae	5	6	Caves: Sarawak, Borneo; Thailand; Ryukyu Islands
Parathelphusidae	2	10	Caves: Irian Jaya, Indonesia (stygophile?); Philippines
Sundathelphusidae	4	7	Caves: New Britain, Indonesia; Papua New Guinea; Philippine Islands
Hydrothelphusidae	1	1	Cave: Madagascar
Grapsidae	2	7	Anchihaline, caves: Guam; Jamaica; Java, Nusa Lain Island, Indonesia; Niue Island, Polynesia; Papua New Guinea: Solomon Islands
Sesarmidae	1	1	Cave: Jamaica
Varunidae	1	1	Anchihaline cave: Ryukyu Islands
	33	67	

The cephalothorax of the infraorder Brachyura (true crabs) is dorsoventally flattened and commonly expanded laterally; the abdomen is reduced, symmetrical, and flexed beneath the thorax; uropods are usually absent; first pereiopods are chelate; and the males lack third, fourth, and fifth pairs of pleopods.

Approximately 7000 crabs are recognized, most of which are marine; however, freshwater and tropical terrestrial species occur in epigean and hypogean habitats. At least 67 crabs assigned to 33 genera and 13 families are categorized as stygobionts (Table 13). Additionally, the following families are represented as facultative cavernicoles in primarily tropical subterranean settings (mostly caves and anchialine habitats; numbers are approximate): Trichodactylidae (1 genus, 2 species), Xanthidae (2, 2), Potamidae (7, 9), Potamonautidae (2, 2), Pseudothelphusidae (10, 21), Gecarcinucidae (2, 2), Parathelphusidae (1, 5), Sundathelphusidae (5, 5), Hydrothelphusidae (1, 1), Gecarcinidae (3, 5), and Grapsidae (5, 12).

See Also the Following Articles

Asellus aquaticus: *A Model System for Historical Biogeography*

Gammarus minus: *A Model System for the Study of Adaptation*

Bibliography

Anker, A. (2008). A worldwide review of stygobiotic and stygophilic shrimp of the family Alpheidae (Crustacea, Decapoda, Caridea). *Subterranean Biology, 6*, 1–16.

Botosaneanu, L. (Ed.), (1986). Stygofauna mundi, *a faunistic, distributional, and ecological synthesis of the world fauna inhabiting subterranean waters (including the marine interstitial)*. Leiden: E. J. Brill.

Boxshall, G. A., & Huys, R. (1989). New tantulocarid, *Stygotantulus stocki*, parasitic on harpacticoid copepods, with an analysis of the phylogenetic relationships within the Maxillopoda. *Journal of Crustacean Biology, 9(1)*, 126–140.

Culver, D. C., & Wilkens, H. (2000). Critical review of the relevant theories of the evolution of subterranean animals. In H. Wilkens, D. C. Culver, & W. F. Humphreys (Eds.), *Ecosystems of the world 30: Subterranean ecosystems* (pp. 381–398). Amsterdam: Elsevier Press.

Gutu, M., & Iliffe, T. M. (1989). Description of two new species of Tanaidacea (Crustacea) from the marine water caves of the Palau Islands (Pacific Ocean). *Travaux du Muséum d'Histoire Naturelle Grigore Antipa, 30*, 169–180.

Gutu, M., & Iliffe, T. M. (1998). Description of a new hirsutiid (n.g., n.sp.) and reassignment of this family from order Mictacea to the new order, Bochusacea (Crustacea, Peracarida). *Travaux du Muséum d'Histoire Naturelle Grigore Antipa, 40*, 93–120.

Hobbs, H. H., Jr., Hobbs, H. H., III, & Daniel, M. A. (1977). *A review of the troglobitic decapod crustaceans of the Americas*. Smithsonian Contributions to Zoology, 244, 1–183.

Hobbs, H. H., III (2000). Crustacea. In H. Wilkens, D. C. Culver, & W. F. Humphreys (Eds.), *Ecosystems of the world 30: Subterranean ecosystems* (pp. 95–107). Amsterdam: Elsevier Press.

Humphreys, W. F., & Harvey, M. S. (Eds.), (2001). *Subterranean biology in Australia 2000* Records of the Western Australian Museum (Supplement No. 64).

Martin, J. W., & Davis, G. E. (2001). *An updated classification of the recent Crustacea* (Science Series 39). Los Angeles, CA: Natural History Museum of Los Angeles County, 1–24.

Petrescu, I., Iliffe, T. M., & Sarbu, S. M. (1994). Contributions to the knowledge of cumaceans (Crustacea) from Jamaica. II. Five new species of the genus *Cumella*. *Travaux du Muséum d'Histoire Naturelle Grigore Antipa, 34*, 347–367.

Pipan, T. (2005). *Epikarst—A promising habitat: Copepod fauna, its diversity and ecology: A case study from Slovenia (Europe)*. Carsologica 5, 1–101.

Poore, G. C. B., & Humphreys, W. F. (1998). First record of Spelaeogriphacea from Australasia: A new genus and species from an aquifer in the arid Pilbara of Western Australia. *Crustaceana, 71*(7), 721–742.

Sket, B., & Zakšek, V. (2009). European cave shrimp species (Decapoda: Caridea: Atyidae), redefined after a phylogenetic study: Redefinition of some taxa, a new genus and four new *Troglocaris* species. *Zoological Journal of the Linnean Society, 155*(4), 786–818.

Wagner, H. P. (1994). A monographic review of the Thermosbaenacea (Crustacea: Peracarida): A study on their morphology, taxonomy, phylogeny and biogeography. *Zoologische Verhandelingen (Leiden), 291*, 3–338.

Zinn, D. J. (1986). Mystacocarida. In L. Botosaneanu (Ed.), Stygofauna mundi, *a faunistic, distributional, and ecological synthesis of the world fauna inhabiting subterranean waters (including the marine interstitial)* (pp. 385–388) Leiden: E. J. Brill.

D

DINARIC KARST: GEOGRAPHY AND GEOLOGY

Nadja Zupan Hajna

ZRC SAZU *Karst Research Institute*

TRAITS OF THE DINARIC KARST

There are many important karst regions around the world, but the Dinaric karst remains the "classical" karst for many reasons. A large limestone region with similar style of landscape, it is the type-site for many features and phenomena. The term *karst* (*kras*) is derived from the Kras plateau (the northwest part of the Dinaric karst). From the region originates international terms such as *polje*, *uvala*, *doline*, *kamenitza*, and *ponor*. The Dinaric karst is also the landscape where *karstology* and *speleology* as sciences were born. Already three World Heritage properties from the region, Plitvice Lakes (Croatia), Škocjanske Jame (Slovenia), and Durmitor National Park (Montenegro), are inscribed on the UNESCO list.

The Dinaric karst is geographically and geologically the carbonate part of the Dinaric Mountains (Dinarides), which form the western part of the Balkan Peninsula and the entire Adriatic Sea littoral belt, between 42–46°N and 14–21°E. The name of the mountain belt derives from Dinara Mountain, extending in a northwest–southeast direction for a length of 84 km between Croatia and Bosnia, with Troglav (1913 m) as the highest peak. The Dinarides (Dinaric Alps) from a tectonic view include all of the External (Outer) Dinarides and Internal (Central) Dinarides. According to some authors the Dinarides also include the Southern Limestone Alps in Slovenia, but for various reasons they are considered part of the Alps.

The boundary line of the Dinaric karst differs depending on whom you consult (*e.g.*, Roglič, 1965; Gams, 1974) and more or less follows the boundary of carbonate rocks toward the Pannonian basin (Fig. 1). Present in the region are some noncarbonate rocks, such as the Eocene flysch rocks in the External Dinarides and clastic sediments and igneous rocks in the Internal Dinarides. The western border of the Dinaric karst is represented by the Adriatic Sea. The northern border is the border between the Alps and the Dinarides and follows the Soča, Idrijca, and Sava rivers. The eastern border is not very clear, because there carbonate rocks alternate with noncarbonates of the Internal Dinarides, the Dinaridic ophiolite zone, and the Panonian Basin (Fig. 2). The eastern border is more or less a line from Samobor and Karlovac in Croatia, Banja Luka in Bosnia, and Peč in Kosovo. According to Roglič (1965), carbonate areas between Sarajevo, Užice, and Durmitor are also included in the Dinaric karst because, from a geological point of view, these are Adria-derived thrust sheets (*e.g.*, Durmitor, Sandzak plateau). The southern border is also not very distinctive; it more or less follows the last carbonates of Prokletije mountain, northwest of the river Drim/Drin in Albania.

From the northwest to the southeast, the Dinaric karst is about 650 km long and up to 150 km wide. Covering more than 60,000 km^2 this is the paramount karst region in Europe. The highest peaks are more than 2000 m high and some of the karst features (*e.g.*, caves with speleothems) are now below the present sea level. According to different geological, hydrological, climate, and geomorphic characteristics, the whole Dinaric karst can be divided into three belts parallel to the Adriatic Sea: low coastal Adriatic karst, high mountain karst, and low continental interior karst.

The Dinaric karst is situated in a humid temperate climate zone, with prevailing west winds. Along the Adriatic Sea and along river valleys toward Dinaric Mountains there is a Mediterranean climate with dry and hot summers and wet and fresh winters. The average precipitation there is about 800 mm per year. On the south sides of the Dinaric Mountains a strong downslope northeast wind burja (bora) is typical. The

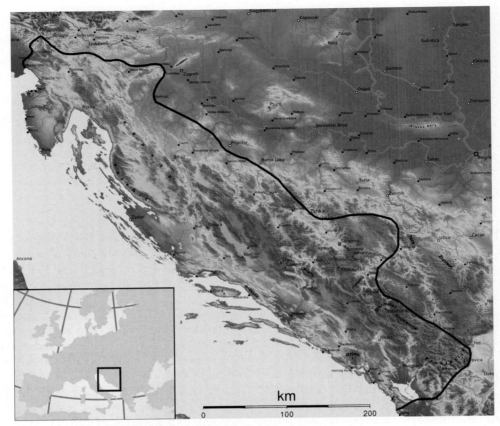

FIGURE 1 Approximate borders of the Dinaric karst. *After Roglič (1965) and Gams (1974).*

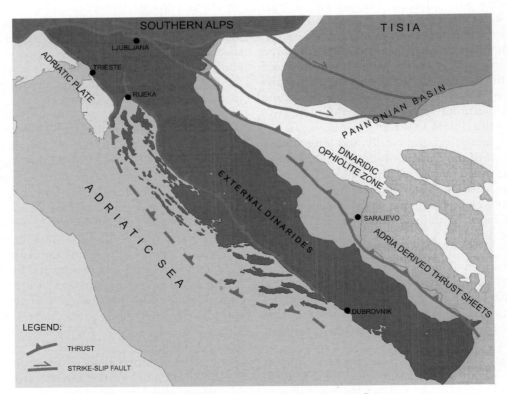

FIGURE 2 Simplified tectonic map of the Dinaric Mountains. *After Schmid et al. (2004) and Šumanovac et al. (2009).*

inner parts of the Dinaric karst have a mountainous climate, especially on higher elevations of karst plateaus; a moderate continental climate is present in the intermountain basins and in transition areas toward the Pannonian basin. The Dinaric Mountains act as an orographic barrier where the highest amount of precipitation falls (an average of about 3000 mm per year). The highest annual precipitation is above Boka Kotorska bay (about 5000 mm per year).

The vegetation corresponds to the climate and the soil cover, which on the karst is not very abundant. The Dinaric karst is also well known as limestone desert; the bare rocky landscape was and is the result of climate conditions, and intense land use over the centuries. There is almost no vegetation on the islands (Fig. 3) nor along the coast where salt water sprayed by bora impedes all growth. There are large grasslands with plant species that vary from sub-Mediterranean to alpine floral elements. Forest is present on higher parts of the plateaus and mountains (*e.g.*, Trnovski gozd, Snežnik, Velebit, Kapela). From the coast toward the high peaks of the mountains, the typical vegetation is as follows. From the coast toward the mountains are degraded forests of pubescent oak and hornbeam; they are gradually followed by forests of pubescent oak and flowering ash. Vertically, between 700 and 1300 m, are coastal beech forests with the indigenous forests of the black pine. The continental slopes of mountains at elevations between 600 and 900 m are covered with alpine beech forests. They extend quite low on the continental side, sometimes descending to the edges of karst plains. Elevations between 1200 and 1400 m are inhabited by beech and fir forests (*Abieti-Fagetum dinnaricum*) which are also the most common forests of the Dinaric karst. Higher up (above 1600 meters), there is normally a subalpine beech forest. In different areas are well-preserved spruce forests, which inhabit altitudes over 1400 m. The highest parts of the mountains are covered with the indigenous virgin forest of dwarf mountain pine.

The hydrological characteristics of the Dinaric karst are the result of lithology, tectonic structures, climate, and geomorphic evolution. Water in the Dinaric karst is drained toward the Adriatic and Black Sea. There are only a few surface streams in spite of high amounts of precipitation. In the areas with prevailing dolomites, fluviokartic drainage is developed. On karst poljes are sinking rivers, usually flowing underground from one polje to another. Underground connections between different karst poljes of the Dinaric karst are known, such as the underground water connections between Gatačko, Nevesinjsko, Fatničko, Dabarsko, and Popovo poljes (Bosnia and Herzegovina) and Ombla spring at Dubrovnik (at sea level, Croatia). Another example is the river Ljubljanica in Slovenia, draining 1100 km^2 (a mean discharge of 56 m^3 s^{-1}), which crosses four karst poljes. A few large rivers cross the carbonate belts along the Dinaric karst: Kolpa, Zrmanja, Krka, Cetina, Neretva, Morača, Una, Vrbas, Bosna, and Drina are the largest. Most of the rivers form deep canyons (*e.g.*, Cetina, Pliva, Tara). On some of the rivers tufa deposits are abundant (*e.g.*, Krka, Zrmanja, Una, and Pliva). Some of the more famous bodies of water are the Plitvice lakes with dams and waterfalls on the Korana river (Croatia); they were added to UNESCO's World Heritage List in 1979. Huge karst springs are typical of the region, with up to a few hundred m^3 s^{-1} maximum discharge (*e.g.*, Buna (Fig. 4), Bunica, Trebišnjica, and Ljuta). Some of the springs along the Adriatic Sea are located under present sea level (*e.g.*, Donjobrelska Vrulja below Biokovo, and Zečica below Velebit); locally they are called *vrulje*.

FIGURE 3 Bare rocky landscape on Pag Island in Croatia. In the foreground is thick-bedded, well-fissured Cretaceous limestone, and in the distance are thin-bedded Paleogene limestones on which stone walls were made on the boundaries of plots. *Photo by the author.*

FIGURE 4 Efficacious karst spring of Buna in Bosnia. *Photo by the author.*

STRUCTURAL GEOLOGY OF THE DINARIDES

The geological evolution of the Dinarides is closely associated with the history of the Tethys Ocean, which closed during the Mesozoic and Cenozoic as a result of the convergence of the African and Eurasian plates. Intermediate microplates played an important role in ocean shortening. The present geological structures resulted from post-collision processes in the Alpine orogenic system, which started before 35 Ma (Vrabec and Fodor, 2006). The main period of thrusting and folding of the area is post-Eocene—the result of post-collision processes between the Africa and Europe plates. The latest tectonic phase in the region started by counterclockwise rotation of the Adria microplate approximately 6 Ma BP. The rotation caused reactivation of already existing Dinaric faults, which were the consequence of the aforementioned thrusting, as dextral strike-slip faults.

Different opinions about the geologic evolution of the Dinaric system in the northeastern Adriatic region reflect its complexity. The main tectonostratigraphic units of the Dinarides (Tari, 2002) are related to Early and Middle Triassic rifting and Late Jurassic-to-present-day compression.

Early and Middle Triassic rifting was marked by strong magmatism and horst and graben-related deposition overlying the Variscian basement. Along the "eastern" Apulian margin toward the Sava-Vardar, ocean rifting was followed by subduction-generated extension from the Early Triassic until the Late Jurassic. Subduction-related attenuation of the continental crust along the eastern margin of Apulia caused the formation of a back-arc basin in the oceanic crust. The remnants of these active continental margin lithologies are found within the eastern thrust belt as the ophiolite mélange of the Internal Dinarides Ophiolite Belt.

Late Jurassic-to-present-day compression generated (1) the eastern thrust belt, foredeep and foreland; (2) the northern Dinarides accretionary wedge; (3) the western thrust belt, foredeep and foreland; (4) the eastern Adria imbricated structures; and (5) wrenching and tectonic inversion. The Dinaridic carbonate platform toward the west presented the foreland of the generally west-directed thrusting. During the Early Cretaceous, compressional stresses began to be transmitted westward through the Dinarides, causing the migration of the foredeep basin and regional uplift of the eastern thrust belt. Subduction of the oceanic plate along the northern margin of the Dinarides resulted in the accumulation of this accretionary wedge from the Maastrichtian to the Eocene. From the end of the Cretaceous until the Early Eocene the entire carbonate platform was uplifted. During the Eocene, the Dinaridic carbonate platform was finally buried under the flysch deposits in the broad foredeep basin of the western thrust belt. At the beginning of the Oligocene, collision and progressive subduction of the Adria below the Dinarides created the imbricate structures of the Adria provenance in front of the western thrust belt. The structural style of the Dinaridic thrust belt is a result of the polyphase tectonic compression and the competence of the sedimentary units involved. The competent carbonate rocks are the strongest influencing factor on the structural style of the thrust belt. The compression started with ramping along the deep decollement from the root zone with a southwestern tectonic transport. In this way, by progressive overstepping of the thrust faults, various structural forms were created along the eastern and western thrust belt: fault bend folds, tear fault-related folds, and folded thrust structures reworked by footwall deformations. The northeast–southwest striking system of the dextral strike-slip faults during the Oligocene to Miocene was followed by northwest–southeast striking and wrenching in the Early and Middle Miocene, affecting the South Pannonian Basin, the western thrust belt, and the Adriatic foreland. This activity is reflected in the large flower structures of the Dinaridic thrust belt.

The main structures of the Dinaric karst relief have a northwest–southeast direction, in the so-called Dinaric direction. This structural predisposition is visually reflected in the direction of high plateaus and karst poljes from south Slovenia to Montenegro. From a distant view, the Dinaric karst has expressive structural relief dependent on overthrust structures and on dextral strike-slip faults (both in Dinaric direction).

LITHOLOGY OF THE DINARIC KARST

The Dinaric karst denotes an area confined mostly to Mesozoic and Cenozoic carbonates of External Dinarides and Mesozoic carbonates of Internal Dinarides. Carbonate sedimentary rocks along the Dinarides were formed on carbonate platform from the Upper Paleozoic to the Paleocene. The youngest limestones are present in the belt along the Adriatic coast, from Eocene, Paleocene, to Cretaceous in age; more or less Cretaceous and Jurassic limestone and dolomites are present in the middle (External Dinarides) and northeastern part (Internal Dinarides); Triassic limestones and dolomites are mainly parts of the middle (External Dinarides) and northeastern outer parts (Internal Dinarides) of the Dinaric karst. Geological data (*e.g.*, Vlahović *et al.*, 2002) indicate that the External Dinarides were formed by the destruction of a single, yet in morphological terms highly variable, shallow water carbonate platform: the Adriatic Carbonate Platform. There is confusion in the scientific literature due to the different names used for the same shallow-water carbonate platform and ideas of two existing carbonate platforms at the same time. The platform was very dynamic in some periods because of its paleogeographic position during the Mesozoic, especially during the Late Cretaceous. The final disintegration of the platform area culminated in the formation of flysch trough(s) in the Late Cretaceous and Paleogene and the subsequent uplift of the Dinarides.

Along the Dinaric karst, karst features are favored especially in massive and thick bedded Cretaceous limestones and Jelar Breccia of the Oligocene or younger age. Upper Cretaceous massive limestones are exceptionally well karstified because of their properties (very pure). In Jelar Breccia solutional karst features such as karren, dolines, conical hills, and caves are also well developed. Along the south and southwest parts of Velebit, where breccia is located, ridges of conical hills (*kuk* in the local language) are numerous. Between these hills, big and deep karst depressions (duliba in the local language) are developed. Among the most well known are Hajdučki and Rožanski kukovi near Zavižan (Northern Velebit National Park); Dabarski kukovi near Baške Oštarije; Bojinac (Fig. 5) above Paklenica Canyon (National Park Paklenica); and Tulove grede on the south edge of the Velebit. The thick-bedded Jelar Breccia, comprised predominantly of angular, poorly sorted clasts of Cretaceous limestones and dolomites, Triassic carbonates, and Paleogene limestones, covers large areas along the northeastern Adriatic coast on the southwest rim of Velebit Mountain (western Croatia). The largest outcrop is >100 km long, 2–10 km wide, and thicker than 500 m in places. The Jelar Breccia is usually considered a result of the disintegration of frontal parts of the major thrusts. The large leveled surfaces and caves are developed also in Eocene–Oligocene Promina Beds (Promina Formation) which consist of noncarbonate and carbonate parts. Carbonate conglomerates may be seen, for instance, on mountain Promina, at Sjevernodalmatinska zaravan (northern Dalmatian Plain; the area between Velebit, Novigradsko more and river Krka).

During the Pleistocene, Dinaric Mountains with elevation above 1100 m were glaciated. Stagnant glaciers were covering the top of the plateaus with small tongues flowing over plateau edges. Valley glaciers reshaped the original fluvial valleys into typical U-shaped valleys and deposited vast amounts of sediments. There is a number of characteristic glacial features such as hanging valleys, remains of moraines, fluvioglacial sediments, erratic blocks, and striated cobbles and boulders. The remains of Pleistocene glaciations are visible on mountains and plateaus from Slovenia to Montenegro; for example, Snežnik, Risnjak, Velebit, Orjen, Lovčen, and Durmitor.

KARST FEATURES OF THE DINARIC KARST

The most characteristic relief forms of the Dinaric karst are high karst plateaus, numerous poljes in the central part stretched in northwest–southeast direction, enormous leveled surfaces, various dolines, large and deep caves, sinking rivers, and abundant springs. Karst forms are surface and subsurface karst features formed mainly by solutional processes of carbonate rocks of the region which are extremely well folded and fractured due to tectonic events. Linear faulting is rare in carbonate rocks; usually carbonates are fractured to different degrees. According to the degree of deformities of the bedrock, we distinguish in carbonates fissured (fissures), broken (blocks of rock), and crushed zone (clay, silt, breccia). The presence of fractured zones is very important for karstification and formation of karst features usually favor defined ones; for example, fissure zones are important for formation and location of karren fields, giant grikelands, and strings of dolines.

On various limestones, dolomites, and so on, several kinds of karren are developed, especially on massive and thick-bedded limestone. Due to its specific mechanical and petrographic properties, the Jelar Breccia is characterized by an abundance of dissolution features on exposed surfaces (*e.g.*, solutional runnels, heelprints, corrosional steps, kamenitzas).

The most typical karst features of the Dinaric karst are depressions of various sizes: dolines, collapse dolines, uvalas, and poljes. Dolines are closed

FIGURE 5 Karst features in Jelar Breccia on Bojinac, South Velebit in Croatia. *Photo by the author.*

karst depressions of different sizes and genesis (Ford and Williams, 2007). Solution dolines form where vertical karst drainage is present and supported with suitable lithology, geological structures, slope inclination, and long enough time. In some areas their distribution is very dense, up to 200 dolines per km^2. Where dolines totally pock some parts of the surface and occupy almost all the space, we are dealing with typical polygonal karst (egg-box-like topography). Landscapes displaying polygonal karst can be found in many places in the Dinaric karst; for example, on Biokovo above Makarska, on plateaus above Boka Kotorska bay, and on high plateaus in Herzegovina. Where the principal process of the formation of a doline is the collapse above underlying caves, it is called a *collapse doline*. Collapse dolines are common features of the Dinaric karst and are especially numerous above known large caves. More than 15 large collapse dolines are located above the underground stream of river Reka which sinks to Škocjanske Jame in southwest Slovenia. The biggest collapse dolines are near Imotski (Croatia). With a depth of 528 m, Crveno Jezero is the deepest of them; its bottom water is about 280 m deep. On the Dinaric karst many dolines are inherited forms of primary caves and shafts (Mihevc, 2007), which, due to surface denudation, become part of the karst relief. Many large and deep closed depressions are located along the high Dinaric plateaus (*e.g.*, Snežnik, Velebit). They are named *konta*, *draga*, or *duliba* and their origin as solutional or collapse dolines is not always clear. During the Pleistocene they were significantly remodeled by glaciation and frost action and in some of them moraines are found.

Uvalas are polygenetic-formed karst closed depressions, which are connected to the geological structure (Čalić, 2009), usually in broken (well-fractured) zones of regional extension. They are larger than dolines (km-scaled), they are mainly disconnected with the karst water table, and they can look like a small polje without a surface stream. Uvalas are frequent relief features of the Dinaric karst but they are not developed at low elevations and on leveled surfaces.

Karst poljes are large karst depressions with flat bottoms, which are the result of solutional processes and karst water drainage. Due to different geological and hydrological factors Gams (1978) distinguished five types of karst poljes: *border polje, piedmont polje in alluvial valley, peripheral polje, overflow polje,* and *baselevel polje*. Most of large karst poljes of the Dinaric karst are baselevel poljes in their origin even if they have some noncarbonate rocks at their bottom. Water on the surface causes base-leveled corrosion, lateral corrosion (marginal corrosion), and consecutive leveling of the rocky bottom at the water table. Oscillation of the water table causes hydrological phenomena such as springs, ponors, estavelas, floods, and intermittent lakes.

The karst poljes of the Dinaric karst are structurally controlled and almost all of them are stretched along overthrust edges and fault zones of regional importance in so-called Dinaric direction (northwest—southeast). Between parallel dextral strike-slip faults or between normal faults, separate blocks are sunk to different elevations (*e.g.*, pull-apart basin formation; graben formation). Some of them are just on the water table level, or they were in the past. These blocks of carbonates at the water table are, or were, leveled and sometimes also laterally enlarged by corrosion and karst poljes were formed. Some parts of the poljes sank, or they are still sinking, and these immersed parts are usually filled up with sediments. Many of the karst poljes along the Dinaric karst contain fluvial, lacustrine, or glacial Neogene and Quaternary sediments up to few hundred meters deep (*e.g.*, Nevesinjsko, Gacko, Glamočko, and Livanjsko poljes). The proofs for existing lakes in sinking basins during the Miocene and Plio-Pleistocene are also fossil fauna and coal-bearing deposits in some of the karst poljes (*e.g.*, Kočevsko, Sinjsko, Duvanjsko, Livansko, and Gacko poljes) from Slovenia to Montenegro. Older sediments are in karst poljes in the northeast and central area of the Dinaric karst, while in the bottoms of the poljes on the southwest side of the highest karst plateaus (*e.g.*, the southwest part of Slovenia, Istria, and Velebit, and along the Adriatic Sea) sediments are in thinner layers or they do not exist (*e.g.*, Planinsko polje in Slovenia). It is also seen that in some of the Dinaric fault zones in the area, karst poljes are not developed yet (blocks are not developed or sink to the water table; *e.g.*, along the Raša fault in southwest Slovenia). The age of the sediments in the karst poljes (whether Miocene, Pliocene, Pleistocene, or Holocene) decline in age from east to west and the presence of the developed karst poljes along fault zones is in connection with the youngest dextral strike-slip tectonic influence by recent Adria microplate movements. Below the sediments, more or less leveled rocky bottoms of karst poljes exist. At different elevations of karst polje bottoms or at their edges (*e.g.*, polje of Lika, Nevesinjsko polje, and Fatničko polje) or even between poljes (*e.g.*, between Fatničko and Dabarsko poljes) are leveled surfaces, which are remnants of a time when polje bottoms were corroded at the water table as one single unit and now they are as separate blocks left behind (they are no longer sinking). On some of these older leveled blocks (surfaces) dolines are already formed (*e.g.*, the north central part of Nevesinjsko polje and the south edge and southeast part of Fatničko polje).

Leveled surfaces are often connected with baselevel corrosion on karst poljes, but they can be also produced after a long period of denudation on the input margin (Ford and Williams, 2007). The complex of processes producing corrosional plains in karst has been called *lateral solution planation* or *corrosional planation* and involves a combination of vertical dissolution, lateral undercutting of hillsides, and spring head sapping. Larger leveled plains are found along the Dinaric karst in south Slovenia, in Istria, between Karlovac (Croatia) and Bihać (Bosnia), Dalmatia (Fig. 6), and in Bosnia and Herzegovina. Corrosional plains cut different geological structures (*e.g.*, folds, lithology) in the larger area. In local languages, leveled surfaces are named *ravnik*, *podolje*, and *zaravan*. In some of them rivers formed canyons, for example, at Krka in the Sjevernodalmatinska zaravan (north Dalmatian karstic plain) or at river Cetina in the Zadvarska zaravan; some of old ones are already dissected by dolines and/or conical hills (*e.g.*, Kras plateau in Slovenia).

Contact karst develops on the contact between permeable (karst) and nonpermeable (nonkarst) rocks, for example, on the contact between carbonates and Paleozoic and Mesozoic clastic rocks or Eocene flysch rocks on the outer borders of the Dinaric karst or within its borders. Contact karst is a landform associated with allogeneic input. Different karst features are limited with the border belt of carbonate areas and they are influenced by sinking rivers. There are many big sinking rivers along Dinaric karst (*e.g.*, Reka sinks in Škocjanske Jama, river Pivka sinks in Postojnska Jama, and Zalomka River sinks to the Biogradski ponori). Remnants of old river streams are dry valleys, which lack an active watercourse over all or parts of their length for at least part of the year, because water is lost in karst or is in a much lower position. For instance, there are two dry valleys crossing the Kras plateau in southwest Slovenia: the largest one in Slovenia is the 400-m-deep Čepovanski dol between Trnovski gozd and Banjščice plateaus, and another big dry valley is situated in upper part of river Bregava between Dabarsko polje and Stolac in Herzegovina. A contact karst feature is also the blind valley, which is a valley cut by a stream flowing from insoluble rocks onto soluble rocks and sinking in a ponor. A series of blind valleys is developed in Matarsko podolje in southwest Slovenia, where along geological contact between flysch and limestones 17 brooks sink into the karst edge. Blind valley bottoms are widened by baselevel and lateral corrosion (Mihevc, 2007).

More than 20,000 caves are known in the Dinaric karst; they are located from the high mountains to below present sea level in the form of vertical shafts or horizontal passages. Some of the caves are active and some of them are relict parts of karst drainage systems. Caves are developed in various carbonate rocks of the region and their formation was influenced by the local presence of different geological structures and position of water gradient (base level). The deepest explored caves are known from the North Velebit in Croatia; for example, shafts Lukina Jama-Trojama (−1392 m), Slovačka Jama (−1320 m), Velebita cave system (−1026 m) with a 513-m underground vertical shaft, and Meduza (−679 m). Several significant caves were discovered in the Jelar Breccia and Liassic limestone of the Crnopac area (South Velebit). Their main morphological characteristic is a network of multiphase cave passages, some of them with very large cross-sectional dimensions. The most important caves of Crnopac massif are Munižaba (5993 m long, −437 m) and Kita Gaćešina (10,603 m long, −456 m). There are many horizontal caves linked to the ponors such as Postojnska Jama (a 20,570-m-long system of 5 caves) and Škocjanske Jama (6200 m long) in Slovenia, and Đulin ponor-Medvedica (16,396 m long) in Croatia. Some of the horizontal caves are managed for cave tourism; notably Postojnska Jama, Cerovačke pećine (3800 m long) in Croatia, Vjetrenica (6700 m long) on Popovo polje in Bosnia and Herzegovina, and Lipska pećina (3500 m) in Montenegro.

Twenty years ago the prevailing thinking was that the surface of the Dinaric karst was formed in the Middle Pliocene (Habič, 1991). Before then karst was covered and contained by impermeable rocks, and waters flowed superficially over impounded carbonate rocks. Later tectonic movements contributed to surface transformation and to karst dissection and surface streams disappeared into karstic underground. The evidence of the surface water flows were clastic (fluvial) sediments found on the karst surface. With

FIGURE 6 Sjevernodalmatinska zaravan (north Dalmatian karstic plain in Croatia) cuts different geological elements; in the distance is Promina Mountain. *Photo by the author.*

FIGURE 7 Unroofed cave near Grotta Gigante at the Italian part of Kras plateau. *Photo by the author.*

different methods the denudation rate in humid temperate zone was calculated to about 20–50 m per million years (Gams, 1974). With this rate of surface denudation many caves were already breached and their ceilings were removed—this is how roofless caves were formed (Mihevc, 2007; Fig. 7)—and from some caves only their insoluble clastic sediments remain on the karst surface. Unroofed caves are calculated to be older than 5 Ma BP. The paleomagnetic and magnetostratigraphic research of karst sediments in Slovenia (Zupan Hajna et al., 2008) has been carried out since 1997. More than 2500 samples were taken and analyzed in 36 different profiles at 21 locations in caves and on the surface. The results were in some sites calibrated by the thorium-uranium (Th-U) method and by paleontological and geomorphological analyses. Calibrated data contributed to reconstruction of speleogenesis, deposition in caves, and indirectly to evolution of karst surfaces and succession of tectonic movements. The evolution of caves in Slovenia took part within one post-Eocene karstification period. The period contains distinct phases of massive deposition in now-existing caves with still preserved sediments dated to about 5.4–4.1 Ma BP (Miocene–Pliocene), 3.6–1.8 Ma BP (Pliocene–Quaternary), and Quaternary. From the age of the cave sediments it follows that the caves have to be even older.

LAND USE

The Dinaric karst has a long history of human impact such as deforestation and transformation into stony semidesert, and also a long history of reforestation (Gams and Gabrovec, 1999). Originally the karst was completely forested. When the human economy changed to stockbreeding and farming, the process of anthropogenic deforestation began. In the history there have been two main reasons for deforestation (Kranjc, 2008): (1) economic (the requirements of new land, pastures, timber use, and trade) and (2) social (local increases in population, mass migration, wars, and raids). Nowadays dense natural forests, extensive forest plantations, dry karst shrublands, and also completely barren karst areas can all be found on the Dinaric karst.

The main regions of the Dinaric karst, the Kras plateau, the islands, and the low plateaus of the Dalmatian coast, present the types of land use typical of Mediterranean countries which produce changes in the surface (Nicod, 2003): the deforestation and adaptation of the karst surface for agriculture; the stone clearing effects; the use of extracted stones for dry-stone walls and hillslope terraces; and the land reclamation and management in the dolines, uvalas, and poljes. Dry stone wall terraces were built around hillsides in order to limit soil erosion and to provide small fields for agriculture. Stones removed from the soil have mostly been accumulated in dry walls. In some areas the only agriculture surfaces were cultivated dolines. With the declining role of agriculture, increased social mobility, and an aging and insufficient agricultural workforce, terraces and small gardens in the dolines have lost their former role and now they have been almost entirely replaced by meadows and bushes.

The environmental problems, on which many factors interfere, were of interest to various and multidisciplinary researchers. During the Austro-Hungarian Empire and Yugoslavia kingdom, surface streams were regulated, ponors were widened, and screens were installed in front of them to hold back deposits, all to shorten the flood time in the karst poljes. After World War II the construction of dams on rivers and streams for water reservoirs started. The managements in the large Dinaric poljes pose very intricate and difficult problems (e.g., water tightness of artificial lakes and draining of natural habitats) depending on the structural and hydrogeological conditions. They have been studied in many works (e.g., Bonacci, 1987; Milanović, 1981).

See Also the Following Article

Diversity Patterns in the Dinaric Karst

Bibliography

Bonacci, O. (1987). *Karst hydrology: With special reference to the Dinaric karst.* Berlin: Springer-Verlag.

Čalić, J. (2009). *Uvala: Contribution to the study of karst depressions, with selected examples from Dinarides and Carpatho-Balkanides* (Doctor thesis). Nova Gorica, Slovenia: University of Nova Gorica.

Gams, I. (1974). *Kras.* Ljubljana, Slovenia: Slovenska matica.

Gams, I. (1978). The polje: The problem of definition. *Zeitschrift für Geomorphologie,* 22(2), 170–181.

Gams, I., & Gabrovec, M. (1999). Land use and human impact in the Dinaric karst. *International Journal of Speleology, 28B*(1—4), 55—77.

Habič, P. (1991). Geomorphological classification of NW Dinaric karst. *Acta Carsologica, 20*, 133—164.

Kranjc, A. (2008). History of deforestation and reforestation in the Dinaric karst. *Geographical Research, 47*(1), 15—23.

Mihevc, A. (2007). The age of karst relief in West Slovenia. *Acta Carsologica, 36*(1), 35—44.

Milanović, P. T. (1981). *Karst hydrology*. Littleton, CO: Water Resources Publications.

Nicod, J. (2003). Understanding environmental problems in Dinaric karst. *Dela, 20*, 27—41.

Roglić, J. (1965). The delimitations and morphological types of the Dinaric karst. *Naše jame, 7*(1—2), 79—87.

Schmid, S. M., Fügenschuh, B., Kissling, E., & Schuster, R. (2004). Tectonic map and overall architecture of the Alpine orogen. *Eclogae geologicae Helvetiae, 97*, 93—117.

Šumanovac, F., Orešković, J., Grad, M., & ALP 2002 Working Group (2009). Crustal structure at the contact of the Dinarides and Pannonian basin based on 2-D seismic and gravity interpretation of the Alp07 profile in the ALP 2002 experiment. *Geophysical Journal International, 179*, 615—633.

Tari, V. (2002). Evolution of the northern and western Dinarides: A tectonostratigraphic approach. *EGU Stephan Mueller Special Publication Series, 1*, 223—236.

Vlahović, I., Tišljar, J., Velić, I., & Matičec, D. (2002). The karst Dinarides are composed of relics of a single Mesozoic platform: Facts and consequences. *Geologia Croatica, 55*(2), 171—183.

Vrabec, M., & Fodor, L. (2006). Late Cenozoic tectonics of Slovenia: Structural styles at the northeastern corner of the Adriatic microplate. In N. Pinter, G. Grenerczy, J. Weber, S. Stein & D. Medak (Eds.), *The Adria microplate: GPS geodesy, tectonics and hazards (NATO Science Series, IV, Earth and Environmental Sciences, 61)* (pp. 151—168). Dordrecht, Germany: Springer.

Zupan Hajna, N., Mihevc, A., Pruner, P., & Bosák, P. (2008). *Paleomagnetism and magnetostratigraphy of karst sediments in Slovenia.* (Carsologica, 8, Založba ZRC) (p. 266). Ljubljana, Slovenija. Slovenia: Založba ZRC, Postojna.

DIVERSITY PATTERNS IN AUSTRALIA

William F. Humphreys

Western Australian Museum, University of Adelaide and University of Western Australia

INTRODUCTION

The diversity of subterranean fauna in Australia, and tropical areas worldwide, has not long been recognized. Until recent decades, Australia was thought to be deficient in overtly cave-adapted (troglomorphic) animals. This circumstance was considered to have resulted from a number of causes: (1) the relative sparseness of carbonate rocks in Australia, as found in other Gondwanan fragments, compared with the world average (Fig. 1); (2) the general aridity of the

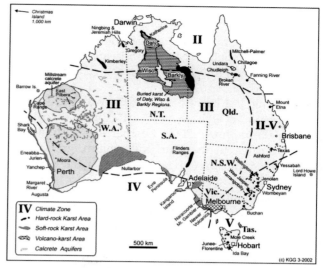

FIGURE 1 Karst areas of Australia and the bioclimatic zones: II, tropical; III, subtropical dry; IV, transitional zone with winter rain; V, warm temperate; II—IV warm temperate/tropical transition zone. The two pink areas within the common outline represent the Pilbara (north) and Yilgarn (south) cratons and their associated orogens that together comprise the western shield. *After Hamilton-Smith and Eberhard, 2000. Graphic by K. G. Grimes, modified.*

continent—it is the most arid inhabited continent, two-thirds of which receives less than 500 mm of rain annually—generally resulting in both dry caves and low input of food energy into the underground voids; (3) the global lack of cave-adapted animals in tropical areas; and (4) the lack of widespread and repeated glaciations, which was perceived to be the main factor driving the evolution of troglobites in the Northern Hemisphere, then the focus of research on subterranean animals. Concomitantly, in Australia there was perceived to be a high proportion of animals found only in caves but not specialized for cave life, that is, lacking overt troglomorphisms. Although not articulated, these arguments would have applied also to stygofauna, the inhabitants of underground waters in both karstic and alluvial aquifers.

Understanding the biogeography of an area is reliant on having a broad spatial and taxonomic sample of the biota, a comprehensive taxonomy, a well-developed systematic and paleoclimate framework, and a fully developed geographical understanding (especially of paleodrainage and plate tectonics). There are serious deficiencies in information on most of these fields of endeavor in Australia. The taxonomic and systematic framework is very patchy and many groups of interest to hypogean questions remain largely unstudied (*e.g.*, Thysanura, Collembola, Diplura, Oligochaeta), or are just beginning to be studied, so it is still too early for them to contribute in detail to biogeographical understanding (*e.g.*, many higher taxa in Oligochaeta,

Copepoda, Ostracoda, Amphipoda, Diplura, Gastropoda). Hence, the focus here will be on some higher taxa for which there is more adequate information, and on some systems, such as the groundwater inhabitants of the smaller voids (mesovoids), for which there is a useful body of data.

During the last two decades of the twentieth century, more focused, as well as more widespread, exploration of caves (Humphreys 2000; Eberhard and Humphreys 2003) and later, in the third millennium, groundwater and nonkarst substrates, has shown that the Australian tropics and arid zones contain especially rich subterranean faunas (Humphreys and Harvey 2001; Austin et al., 2008). However, no area of Australia has been well studied for its hypogean life, the distribution of effort has been very uneven across the country, and many areas remain effectively unexplored for subterranean fauna. Detailed examination of subterranean biology in Australia is sparse and studies have been largely restricted to faunal surveys. Prominent karst areas, such as the Barkley and Wiso regions, have barely been examined because of their remoteness from population centers. Other remote areas, such as the Nullarbor, in which there has been a long history of cave research, have proved to have sparse hypogean assemblages, especially among the stygofauna, but interesting occurrences of the misophrioid copepod *Speleophria*, and heavily sclerotized troglobitic species such as the cockroach *Trogloblatella* and the mygalomorph spider *Troglodiplura* which has South American affinities, and very diverse chemoautotrophic microbial communities form mantles in the saline groundwater (Holmes et al., 2001). Even within those relatively well-surveyed areas, the taxonomic effort is seriously underdeveloped. For example, in one compilation, 63% of the stygofauna from New South Wales was undescribed. Where species have been described, there are many oddities, not yet well placed within their lineage and thus contributing poorly to understanding the biogeography of the Australian hypogean biota. The main knowledge base and most active research are from the western shield and South Australia.

GEOGRAPHIC FACTORS

In contrast to the widespread glaciation that directly influenced many of the classical karst areas in the Northern Hemisphere, Australia has not been subjected to extensive glaciation since the Permian. The biogeography of the hypogean fauna of Australia has been influenced by the continent's past connections with Pangaea and Gondwana, as well as having formed the eastern seaboard of the Tethys Ocean during the Mesozoic. Australia is a fragment of Gondwana together with Africa, India, Madagascar, South America, and Antarctica. Gondwana itself fragmented and Eastern Gondwana (India, Antarctica, and Australasia) became isolated from South America and Africa by 133 million years ago. By the Upper Cretaceous (ca. 80 million years ago), Australia was joined only to Antarctica and it formed the eastern seaboard of Tethys. These lands shared a Gondwanan flora and fauna, and when the final separation between them occurred (45 million years ago), both lands were well watered and supported cool temperate and subtropical forests.

The separation of Australia from Antarctica, and its subsequent rapid northward drift toward Southeast Asia, has been the most significant factor that has shaped the Australian subterranean fauna in the Tertiary. It resulted in the formation of the Southern Ocean seaway that led to the development of the circum-Antarctic Ocean winds and currents that markedly altered the climate of the Southern Hemisphere, causing Australia to become much drier. The formation of the Antarctic ice cap 15 million years ago saw the beginning of a series of marked climatic fluctuations that have greatly stressed the Australian (and other Gondwanan) flora and fauna. Warm and wet interglacial periods alternated with very dry, cool, and windy glacial stages, but only a small area of the Eastern Highlands and Tasmania were subject to extensive ice cover. These cyclic fluctuations, superimposed on a generally increasing and spreading aridity, provided conditions under which subterranean refugia played an important role. Most of the detailed molecular phylogenetic studies of Australian subterranean fauna indicate an origin of subterranean species associated with this developing aridity.

Shield Regions and the Cretaceous Marine Transgressions

Australia has several major shield regions—parts of the Earth's crust little deformed for a prolonged period—that have been emergent since the Paleozoic. The largest is the *western shield*, which includes the Pilbara and Yilgarn cratons and associated orogens (Fig. 1). These stable, truly continental areas of Australia have a nonmarine, presumably freshwater history extending through several geological eras. The Cretaceous marine inundation, at *circa* 120 Ma BP, would have eliminated nonmarine life in the submerged areas (Fig. 2) and only 56% of the current land area of the continent remained above sea level. This has important implications for lineages with poor dispersal ability, as is typical of subterranean fauna. The distribution of ancient lineages, both epigean and subterranean, may be expected to reflect this marine incursion in two

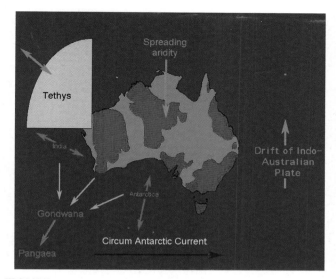

FIGURE 2 Deep history events that have influenced the biogeography of Australian subterranean faunas (see text). The shaded continental areas have not been covered by the oceans since the Paleozoic.

ways. First, ancient terrestrial and freshwater lineages may have survived on these continually emergent landmasses. Second, marine ancestors may have become stranded along the shores as the Cretaceous seas retreated and today may be represented as relictual marine lineages now far inland.

Caves and other subterranean habitats can remain as relatively stable environments over long periods of time because they are well insulated from the climatic perturbations that profoundly affect surface environments and surface animals. There, a number of ancient geographical and phylogenetic relictual groups have survived (Spelaeogriphacea, Remipedia, Thermosbaenacea, *etc.*) (Fig. 3). Owing to their limited potential for dispersal, their present distributions may contain a great deal of information about past geography and climates. The ghost of Cretaceous and earlier marine transgressions is probably reflected in the distribution of Phreatoicidea, an ancient group of isopods, in both their epigean and subterranean forms, the latter being restricted to the tropics, and it has been well documented in the crangonyctoid amphipods (see Box 1).

In this respect aquatic subterranean faunas hold a special significance because, unlike terrestrial troglobionts (troglobites), the aquatic troglobiont (stygobites) fauna contain many relict species that are only distantly related to surface forms. These lineages provide the most compelling evidence that the distribution of some relict fauna occurred through rafting on tectonic plates moved by seafloor spreading. Recently a number of notable discoveries of such relict fauna have been made in Australia whose geographical distribution and lifestyles suggest origins variously in Pangaea, Gondwana, Eastern Gondwana, and Tethys.

Cave Atmosphere

The latitudinal position and general aridity of Australia make cave atmosphere a significant biogeographic determinant in Australia. Cave environments have traditionally been separated into different zones—the entrance, twilight, transition, and deep zones—with characteristics related to the remoteness from the surface environment, such as more stable temperature and humidity and reduced light and food energy input. On the basis of research in the Undara lava tube, Howarth and Stone (1990) developed the concept of a fifth zone, the *stagnant-air zone*, which is characterized by elevated carbon dioxide and depressed oxygen levels. Only in such areas were highly troglomorphic species found in cave passages. However, in other tropical areas, such as arid Cape Range, highly troglomorphic species occur in caves that have unremarkable concentrations of oxygen and carbon dioxide, some even occurring in sunlight near cave entrances, but only where the air is saturated, or nearly saturated, with water vapor.

Howarth (1987) also addressed the importance of water content in the cave atmosphere, largely from his Australian studies. Both tropical and temperate cave systems lose water when the outside air temperature (strictly, the outside water vapor pressure) drops below that in the cave. In the tropics, where average seasonal temperature differences are less than in temperate regions, caves tend to be warmer than the surface air at night and cooler during the day. Even if both air masses are saturated with water, the cave will tend to dry out as water vapor leaves the cave along the vapor pressure gradient—the so-called *tropical winter effect*.

Owing to widespread aridity, this concept has particular relevance to Australia and also in tropical areas where the general form of many caves (giant grikes, small and shallow caves) and/or low subterranean water supply make them vulnerable to drying. Within this context the extent of the deeper cave zones (transition and deep) will fluctuate as the boundary of threshold humidity levels migrates with the changing atmospheric conditions farther into or out of the cave. Such changes occur in ecological time, associated with daily and seasonal fluctuations in air density and humidity, and through evolutionary time, in response to climatic cycles and long-term climatic trends. Such changes should have little effect on groundwater or on troglobites in deep caves, which are extensive enough

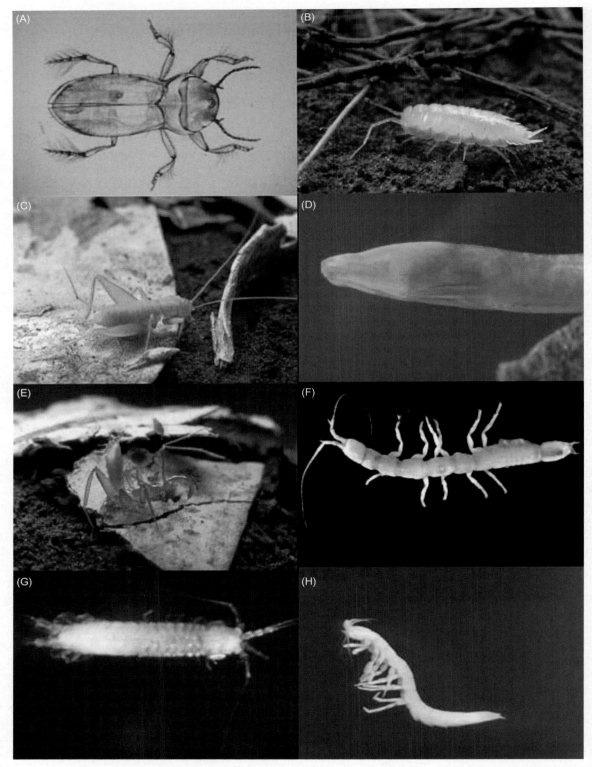

FIGURE 3 Subterranean animals: (A) *Limbodessus eberhardi* (Dytiscidae), one of 100 species of blind diving beetles from calcrete aquifers in the Australian arid zone; (B) unnamed blind philosciid isopod; (C) *Ngamarlanguia luisae* (Gryllidae: Nemobiinae) from Cape Range, the only troglobitic cricket in Australia; (D) head of *Ophisternon candidum* (Synbranchiformes), one of two Australian cavefish; (E) *Draculoides vinei* (Schizomida), one of seven species of microwhip scorpions known from Cape Range; (F) the phreatoicidean isopod *Phreatoicoides gracilis*; (G) *Mangkurtu mityula* (Spelaeogriphacea), a subterranean family that is known from only four species, two in Australia, and Africa and Brazil; (H) *Pygolabis humphreysi*, from Ethel Gorge calcrete, belongs to a family of flabelliferan isopods, the Tainisopidae, known only from groundwater in the Kimberley and Pilbara regions of Western Australia. *Photographs (B—E) Douglas Elford, Western Australian Museum; (A and G) William Humphreys, former from a painting by Elyse O'Grady; and (F and H) GDF Wilson, Australian Museum.*

BOX 1

AMPHIPODS

Australia is a major center of amphipod diversity and much of this diversity is represented by stygobitic species. In a checklist of the Australian inland aquatic amphipods, Bradbury and Williams (1999) reported 74 species belonging to nine different families, of which 29 species in five families were reported from subterranean waters. Ten years later the number of stygobitic amphipod species has significantly increased, particularly because of systematic surveys in the Pilbara and Yilgarn region of Western Australia and various regions in South Australia. The data provided here are based on taxonomic descriptions and on (partly unpublished) molecular assessments. At present, there is evidence for more than 144 stygobitic amphipod species: more than 37 species (8 genera) in the Paramelitidae; more than 24 species (9 genera of which 3 are undescribed) in the Neoniphargidae; more than 51 species (several undescribed genera) in the Chiltoniidae; more than 30 species (4 genera) in the Melitidae; and 4 species (2 genera) in the Bogidiellidae. An interesting pattern of amphipod distribution and diversity is emerging. The majority of the species have distributions that are restricted to individual aquifers or tributaries, as is expected for taxa that are confined to groundwater and that have low dispersal abilities. Much of the diversity is occurring in groundwaters of the arid region. Whereas some families appear to be restricted to the temperate southeast and southwest of the continent (Neoniphargidae), others are much more widespread and encompass parts of the arid zone (Chiltoniidae, Melitidae, Paramelitidae) and tropical areas (Melitidae, Paramelitidae). Some families are more restricted, such as the Bogidiellidae, which are known only from coastal (anchihaline) habitats in the Cape Range and Barrow Island of Western Australian and Eyre Peninsula of South Australia. Notably, while taxa in southern areas comprise both stygobitic and epigean species, northern taxa, in the arid tropics and subtropics, comprise only stygobitic taxa.

Speciation and adaptation to subterranean habitats probably involved marine transgressions in the context of the Melitidae and Bogidiellidae because species in these families are all found near present or historical shorelines. The evolution of the stygobitic species in the Paramelitidae, Neoniphargidae, and Chiltoniidae, which occur predominantly in the areas that were not inundated by the high sea level stand in the Cretaceous, probably was driven by aridification, forcing epigean freshwater species to seek refuge in subterranean habitats.

---

to contain the entire change. But, in the shallow caves common in the Australian tropics, such changes are likely to cause large areas of cave systems to dry out. Such processes may lead to the extinction of certain cave fauna, or impede movement through the epikarst and thus could promote speciation between different karst areas. The high diversity of Schizomida in the Pilbara and Cape Range, and of the paradoxosomatid millipede *Stygiochiropus* in arid Cape Range, are candidates for such analysis.

Humid caves within the arid zone have permitted the survival of a diverse troglobitic fauna in arid Cape Range, the affinities of which lie with the inhabitants on the floor of the rainforest, both temperate and tropical, habitats now thousands of kilometers distant. While the fauna is now geographically relict, the driving force resulting in the initial invasion of the caves is unknown—species may have established in caves coincident with the onset of aridity to escape the surface drying, or they may have established in caves seeking resources unrelated to the onset of aridity. This question cannot be resolved for Cape Range because the aridity has been sufficiently intense to extirpate entirely close relatives at the surface. Other tropical areas, such as North Queensland, offer greater prospect of resolving such issues because contemporary lineages occur with surface and cavernicolous species exhibiting various degrees of troglomorphy. However, the recent finding of diverse troglobiont communities also throughout the arid western shield supports the hypothesis that aridity has driven the fauna to seek more humid habitat underground and where it has evolved troglomorphic traits, because the lineages to which they belong are not commonly subterranean in more humid regions with suitable subterranean voids.

Another area where a resolution of the causes of colonization of the hypogean environment may be resolved is in the groundwater calcrete deposits (see Box 2) in the arid zone. There, many different lineages of diving beetles (Dytiscidae) have invaded the groundwater and become eyeless and flightless (Fig. 3). Each calcrete body has a unique dytiscid assemblage, there is no overlap in species between different calcretes, and speciation appeared to have taken place *in situ* because multiple species (1 to 4 of different sizes in many cases being sister species) are common among the 100 stygal

> **BOX 2**
>
> **GROUNDWATER CALCRETES**
>
> The long period of emergence and the ensuing erosion down to the Archaean basement has resulted in classical karst terrain being absent from the western shield. However, thin carbonate deposits are widespread throughout the arid zone and are well developed as groundwater (valley) calcretes (hereafter termed *calcrete*) which occur widely in Australia but in isolated, though sometimes extensive, pockets usually associated with paleodrainage lines (Fig. 5). Calcretes are carbonate deposits forming from groundwater near the water table in arid lands as a result of concentration processes by near-surface evaporation. They form immediately upstream of salt lakes (playas), chains of which form a prominent part of the landscape in the more arid parts of Australia. The playas, which reflect the groundwater base level, are the surface manifestation of paleodrainage channels incised into Precambrian basement rocks by rivers that largely stopped flowing when the climate changed from humid to arid in the Paleocene. Hence, the paleovalleys predate the fragmentation of Gondwana.
>
> Calcretes are especially important in the Australian context as they form in arid climates (annual rainfall <200 mm) with high potential evaporation (>3000 mm per year). Although quite thin (10–20 m thick) the groundwater calcretes often develop typical karst features and within them easy movement of groundwater reflecting open conduits. Groundwater salinity may vary markedly owing to the episodic recharge characteristic of the arid zone (Humphreys 1999, 2001) and the calcrete aquifers have been described as subterranean estuaries (Humphreys *et al.*, 2009).
>
> Because they are deposited at intervals from the groundwater flow, the calcrete masses are separated by habitat that is unsuitable for stygofauna, namely, Tertiary valley-fills, largely clays, and salt lakes. Consequently, they form isolated karst areas along the numerous major paleovalleys, some of which may date from the Permian. The sediments filling the paleochannels are mostly Eocene or later but the age of the calcretes is poorly defined. The extensive alluvial fan calcretes and some of the river valley calcretes formed in the Oligocene (37–30 Ma BP) may have followed the onset of the continental aridity. Many of the calcrete areas, especially those north of 31°S, are being actively deposited and the others have probably been remobilized and redeposited, attributes that make the dating of calcrete deposits using standard radiometric methods problematic. However, molecular phylogenies of the diverse diving beetle, oniscidean isopod, parabathynellid, and amphipod faunas, the numerous species of which are each restricted to a single major calcrete area or paleodrainage tributary (*e.g.*, Leys *et al.*, 2003; Finston *et al.*, 2007; Guzik *et al.*, 2009), indicate that the calcretes have been present for at least 5–8 million years.

species in the arid zone. Molecular studies suggest that numerous lineages invaded the calcrete aquifers during the constrained time period (8–4 Ma BP), which suggests that it occurred in response to a widespread factor, such as might be expected from spreading aridity.

Within the third millennium there has been a recognition that nonkarst and nonalluvial substrates, such as pisolites and fractured rocks, support extremely diverse subterranean faunas in both aquatic and terrestrial habitats, particularly in the western shield region of Western Australia. This recognition has resulted from the requirement in Western Australia to include subterranean fauna as a component of environmental assessment for major mining projects in this mineraliferous region. At a local scale this troglobiont diversity is comparable to that found in the richest karst areas, such as Bayliss Cave in Queensland. Regionally, consultancy companies report between 150 and 300 troglobiont species from their projects within the 220,000 km^2 Pilbara region, about the same size as the massive Nullarbor karst of southern Australia where 27 troglobiont species are known. The proportion of species in common to the consultancy companies is currently unknown but is likely to be low, as many troglobiont species occupy small geographic areas.

Energy Supply

Energy enters subterranean systems largely mediated by water, but also by animals and plant roots. Because these elements themselves are not uniformly distributed across Australia they have the potential to influence Australian cave biogeography. The carriage of organic matter in surface water is strongly affected by seasonal rainfall and plant growth. The episodic rainfall, characteristic of the arid zone, means that some areas potentially have unpredictable energy supplies. Evaporation greatly exceeds rainfall in arid zones and groundwater recharge—which carries dissolved organic matter, the predominant energy supply in subterranean waters—occurs only following episodic large rainfalls, sometimes years apart.

Plants provide the raw material that is transported by water into the subterranean realm, but they also directly transport energy into hypogean habitats by means of sap transport within the roots and by root growth to depths of up to 76 m. Roots, especially tree roots, are an important and reliable source of energy for troglobitic cixiid and meenoplid fulgoroid Homoptera. These occur in the lava tubes of tropical North Queensland and similar fauna are found in karst across the tropics, into the Kimberley and down the arid west coast, to the south of Cape Range. They are also found on roots in calcrete, pisolite, and fractured rock substrates on the western shield. Tree roots are also utilized by cockroaches throughout the country (*e.g.*, in the Nullarbor, *Trogloblatella nullarborensis*).

Tree root mats also represent a reliable food supply for elements of the rich communities of aquatic invertebrates, including some exhibiting troglomorphisms, occurring in some shallow stream caves of Western Australia. They provide habitat, and probably food, for stygofauna in the Nullarbor, in calcrete aquifers of the western shield, and in anchihaline caves in Cape Range and Christmas Island where they are associated with a diverse fauna largely comprising crustaceans.

Roots, like guano, often provide copious quantities of energy to cave communities, which may be quite diverse. Roots in the Undara lava tube, Queensland, and the Tamala Limestone of Western Australia, support diverse cave communities. However, whereas the former contain numerous highly troglomorphic species, the latter has few stygomorphic species, many being indistinguishable for surface species. Superficial subterranean systems are often supported by roots but to what extent is unknown, as, for example, the cave and groundwater faunas of groundwater calcretes in the arid zone.

Animals may transport energy into cave systems and deposit it there as excreta, exuvia, carcasses, and eggs. In Australia such troloxenic agents exhibit marked latitudinal differences. In the south, raphidophorid crickets are the most conspicuous trogloxenic agents, whereas bats, while not diverse, are locally abundant where they form breeding colonies. In the tropics bats are widespread, diverse, and important producers of guano, as, to a lesser degree, are swiftlets in more humid areas.

Guano is usually intermittently distributed in both space and time because it is dependent on the seasonal biology of the birds and bats. In consequence, the cave communities associated with guano are highly specialized and differ markedly from the cave fauna not dependent on guano. Markedly troglomorphic species are not commonly found in the energy-rich, but temporally unstable, guano communities. However, immense populations of mites (>100 cm^{-2}) are seasonally present and numerous, mostly rare, species (50 +) may occur, occupying distinct parts of the cave system.

STYGOFAUNA AND CRUSTACEA

Stygofauna are discussed in the context of Crustacea, which comprise the overwhelming majority of stygofauna, but the Dytiscidae example above introduced the insect component.

The magnitude of the biodiversity present in subterranean waters globally has only recently been given prominence. Australia, especially the northwestern and southeastern parts, has unexpectedly come to the attention of stygobiologists and systematists on account of its diverse regional groundwater fauna (stygofauna). Recently, these have been determined to include a number of higher order taxa variously new to science (*i.e.*, the flabelliferan isopod family Tainisopidae from Kimberley and Pilbara), new to the Southern Hemisphere (Thermosbaenacea, Remipedia, Epacteriscidae), or new to Australia (Spelaeogriphacea, Pseudocyclopiidae). Many of these taxa occur near coastal and anchihaline waters and are widely interpreted as comprising a relictual tethyan fauna, although molecular phylogeographic support has yet to be demonstrated. Several of these lineages have congeneric species, which are known elsewhere only from subterranean waters on either side of the North Atlantic—the northern Caribbean region and the Balearic and Canary archipelagos (see Box 3).

Syncarida

The Syncarida are crustaceans now entirely of inland waters. The Anaspidacea are confined to Australia, New Zealand, and southern South America. In southeastern Australia they are often large and mostly surface living, although several stygomorphic species occur in cave streams and groundwater, and an undescribed family has been reported that is restricted to caves. In contrast, both families of Bathynellacea have a global distribution, often even at the generic level, and are widespread in Australia. *Bathynella* (Bathynellidae) is found from Victoria to the Kimberley and elsewhere the genus occurs globally. Genera within the Parabathynellidae known from Australia exhibit different regional affinities. *Chilibathynella* and *Atopobathynella* are known from

> **BOX 3**
>
> ## ANCHIHALINE HABITATS—TETHYAN RELICTS?
>
> *Anchihaline* (or *anchialine*) habitats comprise near-coastal mixohaline waters, usually with little or no exposure to open air and always with more or less extensive subterranean connections to the sea. They typically show salinity stratification and may usefully be considered groundwater estuaries. They typically occur in volcanic or limestone bedrock and show noticeable marine as well as terrestrial influences. The water column is permanently stratified with a sharp thermohalocline separating a surface layer of fresh or brackish water from a warmer marine, oligoxic water mass occupying the deeper reaches. They have a significant amount of autochthonous primary production, via a sulfide-based chemoautotrophic bacterial flora, as well as receiving advected organic matter from adjacent marine or terrestrial epigean ecosystems. Anchihaline habitats are mostly found in arid coastal areas and are circumglobally distributed in tropical/subtropical latitudes.
>
> Anchihaline habitats support specialized subterranean fauna (Fig. 6), predominantly crustaceans representing biogeographic and/or phylogenetic relicts. These specialized anchihaline endemics are largely restricted to the oligoxic reaches of the water column below the thermohalocline. The structure of these assemblages is predictable, and, remarkably, however remote an anchihaline habitat, this predictability frequently extends to the generic composition.
>
> In continental Australia, anchihaline systems occur adjacent to the North West Shelf (Cape Range and Barrow Island) and on Christmas Island (Indian Ocean), an isolated seamount 360 km south of Java but separated from it by the Java Trench.
>
> Cape Range supports a fauna comprising atyids, thermosbaenaceans, hadziid amphipods, cirolanid isopods, remipeds, thaumatocypridid ostracods (*Danielopolina*), and an array of copepods such as epacteriscid and pseudocyclopiid calanoids, and speleophriid misophrioids.
>
> Some are the only known representatives of higher taxa in the Southern Hemisphere (Class Remipedia; Orders Thermosbaenacea, Misophrioida), and several genera are known elsewhere from anchihaline systems on either side of the North Atlantic (*Lasionectes, Halosbaena, Speleophria, Danielopolina* (Fig. 6)) and the atyid *Stygiocaris* (Fig. 6) is closely related to the amphi-Atlantic genus *Typhlatya*. These obligate stygal lineages are thought to have poor capability or opportunities for dispersal, and this attribute, combined with their distributions which closely match areas covered by the sea in the late Mesozoic, suggest that their present distributions could have resulted by vicarianace as a result of the movement of tectonic plates (Fig. 2).
>
> Anchihaline systems on oceanic islands support a different group of fauna, but the structure of these assemblages is similarly predictable, even between oceans. Christmas Island is a seamount and supports an anchihaline fauna characterized by the stygobitic shrimp *Procaris* (Decapoda), which belongs to the primitive, highly aberrant family, Procarididae which appears globally to be restricted to anchihaline caves. This family has been reported elsewhere only from other isolated seamounts, namely, Bermuda and Ascension Island in the Atlantic Ocean, and Hawaii in the Pacific. In each case, as with Christmas Island, the procaridids are associated with alpheid, hippolytid, and atyid shrimp. These co-occurrences of two primitive and presumably ancient caridean families support the contention that crevicular habitats have served as faunal refuges for long periods of time. Recently, an endemic species of *Danielopolina* has been described from Christmas Island, the first juxtaposition of a member of the anchihaline faunas characteristic of epicontinental and seamount island anchihaline faunas. There is no coherent theory as to their distribution to remote seamounts such as Christmas Island.

Chile and southeastern Australia, while the latter is also found in India and throughout northwestern Australia, including Barrow Island and Cape Range, and the paleodrainage channels of the arid center. *Notobathynella* is found across Australia and New Zealand, while *Hexabathynella*, from the eastern Australian seaboard, has a more global distribution, being found in New Zealand, southern Europe, Madagascar, and South America. Endemic genera include *Kimberleybathynella* from the Kimberley, *Brevisomabathynella* and *Billibathynellla* from the western shield, and *Octobathynella* from New South Wales. As 30 described species are endemic to single calcrete masses the overall diversity is expected to be very high. Bathynellacea are small stygobites, mostly inhabitants of interstitial freshwater environments, although an undescribed genus of large, free-swimming parabathynellid occurs in water close to marine salinity (>30 000 mg L^{-1} TDS) in the Carey paleodrainage systems of the arid zone, where it is associated with a number of maritime copepod lineages such as Ameiridae (Harpacticoida) and *Halicyclops* (Cyclopoida).

Copepoda

Remarkably little work had been conducted on non-marine copepods in Australia until this millennium. Recent work on groundwater copepods, largely from groundwater calcretes of the western shield, and the near coastal, especially anchihaline systems of the northwest, has revealed higher taxa not previously described from Australia, in some cases even from the Southern Hemisphere.

Numerous species of copepods are being described from Australian groundwaters, largely from the western shield, including ten new genera of Cyclopoida and Harpacticoida, and several genera are reported for the first time from Australia [*Nitocrella* Ameiridae (Eurasia), *Parapseudoleptomesochra* Ameiridae (global), *Nitocrellopsis* Ameiridae (Mediterranean) *Haifameira* Ameiridae (depth of Mediterranean Sea), *Pseudectinosoma* Ectinosomatidae (Europe) and the family Parastenocarididae (Pangaea, freshwater)]. The broader distribution of these lineages within Australia awaits investigation but some of them have been found in tropical Queensland.

The occurrence of about 17 near-marine lineages (*e.g.*, *Halicyclops*) in the center of the western shield alongside lineages considered to be ancient freshwater lineages (*Parastenocaris*: Parastenocarididae) is notable. It may reflect both the salinity stratified, often hypersaline groundwater in the more southerly paleodrainage systems, as well as the ancient origins of the fauna. Interestingly, *Halicyclops* is almost completely absent from marine interstitial in Australia, although it is the world's most speciose element in this habitat. It was probably replaced here by the genus *Neocyclops*, which is very diverse, testifying to an ancient invasion of inland waters by *Halicyclops*. *Mesocyclops* has a mostly tropical distribution; *Metacylcops* (*trispinosus* group) and *Goniocyclops* have an Eastern Gondwanan distribution; and the limits to the distributions of newly described genera of Ameiridae, Canthocamptidae, and Cyclopinae await confirmation.

Ostracoda

Ostracods recorded from Australian inland waters are mainly from the families Limnocytheridae, Ilyocypridae, and Cyprididae. In the Murchison, ostracods from the families Candonidae, Cyprididae, and Limnocytheridae have been recorded in open groundwater but stygophilic species occur only in the Limnocytheridae and Candonidae, the latter including the globally widespread genus *Candonopsis* (subfamily Candoninae), which occurs widely and in a wide variety and age of substrates. Species are known from Pleistocene syngenetic dune karst (Tamala Limestone), several species from Tertiary (probably Miocene) groundwater calcretes on the western shield, and from the Kimberley (Devonian Reef Limestone). In Europe there are only a few, mostly hypogean species that are considered to be Tertiary relicts with surface relatives today occurring in tropical and subtropical surface waters; they are especially diverse in Africa. The subfamily Candoninae (family) Candonidae are common elements of stygofauna globally but recent finds from the Pilbara describe about 25% of the world's genera but these are more closely related to the South American and African Candoninae than to those of Europe.

The thaumatocypridid genus *Danielopolina*, previously unreported in the Southern Hemisphere, occurs as a tethyan element in the anchihaline system at Cape Range and on Christmas Island, Indian Ocean, an isolated seamount island where it occurs with the first living representative of *Microceratina* (Cytheruridae). Fossils in marine cave facies in the Czech Republic suggest that the former lineage already inhabited marine caves in the Jurassic.

In the five years to 2010 there was major progress in the describing and understanding of the subterranean ostracods from Australia. One of the hotspots of their diversity is the Pilbara region from where 86 species have been described belonging to 12 genera in the subfamily Candoninae. Only one of the species and one genus is not endemic to this region, while all the other species have a very limited distribution, comprising short-range endemics. The tribe Humphreyscandonini was described to accommodate eight Candoninae genera endemic to the Pilbara. This tribe shows some close similarities with the fossil, Tertiary fauna of the Northern Hemisphere. Other genera from this region have a clear Gondwanan connection, and even the genus *Candonopsis*, of the tribe Candonopsini, is here represented by a distinct lineage, and a separate subgenus. This subgenus is the only taxon connecting the Pilbara's candonins with those from the Kimberley and Murchison regions. Candoninae fauna from the Murchison is less rich, but also comprises only endemic species. Even though it has been quite thoroughly explored, the Murchison Candoninae comprises only five described species. East Australia also has a unique Candoninae fauna, but only one monotypic genus have been described from Queensland, while many more await description. The east Australian Candoninae belongs to the tribe Candonopsini and they are more closely connected with the South and Central American Candoninae than to the rest of Australia. Few taxa other than

Candoninae have been described from Australian subterranean waters, and most belong to the genus *Gomphodella* (family Limnocytheridae) and are endemic to the Pilbara and Murchison regions, and represent a very old lineage of freshwater ostracods. Most of the species are known from fossils, and some close relatives today live in the subterranean waters of the Balkan Peninsula, and ancient African Lakes. One species of the family Cyprididae has been described from the Kimberley region, but, as all the other Cyprididae species found in the wells and occasionally bores around Australia, probably represents only a stygophile, rather than a true stygobiont.

The western shield is considered to have been a single landmass continuously emergent from the sea since at least the Paleozoic. If this is the case then there is no clear hypothesis to account for the extraordinary disparity (Table 1) in the gross composition of the subterranean fauna between contiguous parts of the western shield, loosely termed the Pilbara (to the north) and the Yilgarn (to the south) and the associated orogens.

Isopoda

Phreatoicidean isopods (Fig. 3) have a Gondwanan distribution and occur widely across southern Australia (and in tropical Arnhem Land and the Kimberley) in surface habitats that have permanent water—surface expressions of groundwater—usually as cryptic epigean species. Their distribution is strongly associated with the areas of the continent not submerged by Cretaceous seas (see also Fig. 4). About 76 species in 30 genera are described from Australia, of which 10 species in 8 genera are hypogean (cavernicolous or spring emergents). They are under active revision and numerous taxa are being described. Five hypogean species occur on the Precambrian western shield. The family Hypsimetopidae is represented by the genera *Pilbarophreatoicus* in the Pilbara and *Hyperoedesipus* in the Yilgarn regions (separate cratons of the western shield). These genera are closely related to the hypogean genus *Nichollsia* found in the Ganges Valley of India and a new genus from caves of Andrah Pradesh (East-Central India). These occurrences suggest that Hypsimetopidae were hypogean prior to the separation of Greater Northern India from the western shore of Australia (*ca.* 130 My BP). *Crenisopus*, a stygobitic genus occurring in a sandstone aquifer in the Kimberley and another new species on Koolan Island off Western Australia, is a link between African and Australasian lineages of phreatoicideans. The genus is basal to most families in the Phreatoicidea, suggesting divergence after they entered freshwater but prior to the fragmentation of East Gondwana during the Mesozoic era.

The family Cirolanidae with three species and two genera occurs on Cape Range, Barrow Island, and in the Pilbara of Western Australia. The family Tainisopidae, endemic to northwestern Australia, occurs in the exposed and greatly fragmented Devonian Reef system throughout the western Kimberley as well as in remote outcrops of this fossil reef in northeastern Kimberley. A second clade of this family (Fig. 3) with five described species inhabits groundwater calcretes in the Pilbara, from which the Kimberley was separated by the Cretaceous marine incursions. The location and distribution of this family is indicative of ancient marine origins. Recent cladistic analysis suggests that this family is related to the cosmopolitan marine Limnoriidae but at a basal level, suggesting

TABLE 1 The Distribution of the better known higher taxa of subterranean invertebrates between the adjacent Pilbara and Yilgarn regions on the Australian Western Shield that has been continuously emergent since the Paleozoic

Taxon	Pilbara	Yilgarn	% in Common
Candonine ostracod genera	13	1	8
Candonine ostracod species	58	5	0
Copepoda genera	43	30	4
Copepoda species	25	15	21
Spelaeogriphacea species	2	0	0
Tainisopidae species	5	0	0
Schizomida species	26	0	0
Dytiscidae species	0	89	0
Nocticolidae species	9	0	0

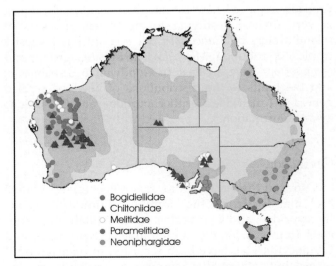

FIGURE 4 The distribution of subterranean amphipod families in relation to long emergent land areas (deeper shading) in Australia. Map by Remko Leijs.

the family is older than the more derived families like the Cirolanidae, which have an extensive Cretaceous fossil record. Among Asellota, the Janiridae occur widely across southern Australia and Tasmania and the genus *Heterias*, which also occurs in New Zealand, is likely to be highly speciose. Protojaniridae occur in the Northern Territory and this family has a Gondwanan distribution similar to the Phreatoicidea.

Terrestrial isopods (Oniscidea) are a prominent component of cave fauna throughout Australia, as elsewhere in the world, yet there are few described highly troglomorphic Oniscidea (Fig. 3) and these are placed in the genera *Abebaioscia*, *Andricophiloscia*, and *Haloniscus* (Philosciidae), and *Troglarmadillo* (Armadillidae) that occur widely on the western shield. Numerous other troglobitic species are recognized belonging to the genera *Styloniscus* (Stylonisciidae), an undescribed genus (Stenonisciidae), *Laevophiloscia* and *Haloniscus* (Philosciidae), *Hanoniscus* (Oniscidae), and *Buddelundia* and *Troglarmadillo* (Armadillidae). Their distribution seems to reflect the general aridity that developed following the separation of Australia from Antarctica, rather than to suggest more ancient relictual distributions. So, in the more humid southern regions Styloniscidae are a prominent component of cave fauna, as they are in the wet forest of the surface, but they are troglophilic. Armadillidae, Ligiidae, and Scyphacidae are also common in Tasmanian caves but none is troglomorphic. On the mainland, Olibrinidae, Philosciidae, and Armadillidae are prominent among subterranean fauna. In the drier areas of Australia, where armadillids are such a prominent part of the surface fauna, they appear in caves more frequently, and many have overt troglomorphies. These troglobites are known from the Nullarbor, North Queensland (Chillagoe), Cape Range, and Kimberley and the western shield. The troglobitic Philosciidae, Platyarthridae, and Oniscidae from Western Australia are undescribed.

A single epigean species of *Haloniscus searlei*, an aquatic oniscidean isopod, is known from salt lakes (playas) across southern Australia. Numerous stygobitic species of *Haloniscus* occur in groundwater calcrete deposits of the Yilgarn region of the western shield sometimes in saline waters, and from Ngalia Basin northwest of Alice Springs in Central Australia.

Spelaeogriphacea

In Australia, this order of stygal crustaceans is known from two species, each in a separate lacustrine calcrete deposit in the Fortescue Valley, a major ancient paleovalley of the northern Pilbara region. The four extant species of spelaeogriphaceans occur with very circumscribed distributions in subterranean freshwater habitats of Africa (Table Mountain, South Africa), South America (western Mato Grosso, Brazil), and Australia, all fragments of Gondwana. The supposition of a Gondwanan origin is possibly refuted by a marine fossil, *Acadiocaris novascotia*, from shallow marine sediment of Carboniferous age in Canada, which has been attributed to this order. All living spelaeogriphaceans occur in or above geological contexts that are earliest Cretaceous or older. The colonization of Gondwanan freshwater is likely to have occurred after the retreat of the Gondwanan ice sheet (after 320 Ma BP) and prior to the dissolution of Gondwana (142–127 Ma BP).

Decapoda

Atyid shrimps (Decapoda) are widespread in surface waters throughout the tropics. They may have colonized Australia from Asia via the Indonesian archipelago, but their presence as stygobitic species in caves and groundwaters of the Canning Basin suggest a more

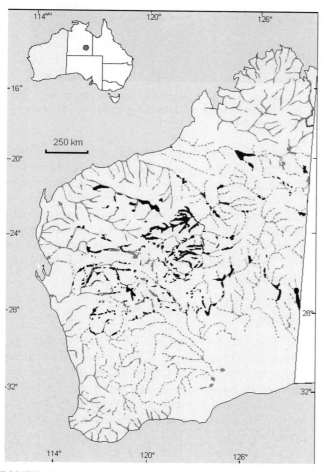

FIGURE 5 The distribution of groundwater calcretes in Western Australia, which occur throughout the arid land north of 29°S. Most occur immediately upstream of salt lakes (playas) within paleodrainage channels (dotted lines).

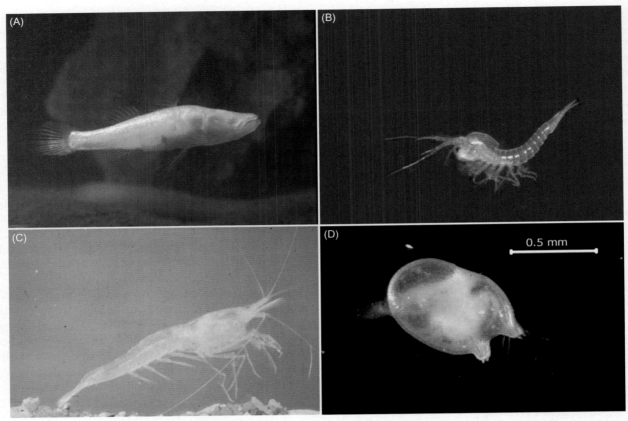

FIGURE 6 Stygal animals from Australian anchihaline waters. (A) *Milyeringa veritas* (Eleotridae); (B) *Halosbaena tulki* (order Thermosbaenacea); (C) *Stygiocaris stylifera* (Decapoda: Atyidae); (D) *Danielopolina kornickeri* (Ostracoda: Thaumatocyprididae). *Photographs (A) and (B) by Douglas Elford; (C) and (D) by William Humphreys; both of the Western Australian Museum.*

ancient origin. The genera *Parisia, Pycnisia, Pycneus,* and *Caridina,* the first three being endemics, form a paraphyletic clade and occur as species with small ranges in northern and central Australia. A second clade includes *Antecaridina* and the endemic genus *Stygiocaris* (Fig. 6) from anchihaline systems of an isolated seamount (Christmas Island, Indian Ocean) and the epicontinental system of the northwest, respectively. *Stygiocaris* is most closely related to *Typhlatya* found in subterranean waters on either side of the North Atlantic.

CHELICERATA

The chelicerates globally comprise a biodiverse component of cave communities and the class is represented in the Australian cave fauna by the orders Acarina, Palpigradi, Amblypygi, Opiliones, Pseudoscorpiones, Schizomida, Scorpiones, and Araneae.

Acarina

The mite family Pediculochelidae (Acariformes) was first recorded in Australia from a dry cave in Cape Range where a specimen was attributed to *Paralycus lavoipierrei* that is described from California. *Tiramideopsis* (Mideopsidae) occurs in the Millstream aquifer, a genus previously known from similar habitats of India and suggesting ancient links (*cf.* Phreatoicidea). Generally, the poorly known mesostigmatid mite fauna of Australian caves does not appear to constitute a distinctive cave fauna or exhibit any of the morphological characteristics of deep-cave arthropods.

Palpigradi

The first native species, *Eukoenenia guzikae*, was described in 2008, collected from a borehole in calcrete substrate in the southern western shield. Since then other species have been recovered from the Pilbara, Barrow Island, and Kimberley of Western Australia, all from boreholes in karst and fractured rock substrates, or caves.

Amblypygi

Troglophilic species of *Charon* are found on Christmas Island and in the Northern Territory.

Opiliones

Cavernicolous species of Triaenonychidae in Tasmania and New South Wales often show depigmentation, attenuation of pedipalps and legs, a reduction in size (but not loss of) eyes, reduced sclerotization, and other troglomorphic features. The cave fauna of Tasmania, unlike continental Australia, have distribution patterns more like those of other periglacial areas of the world in which profound environmental changes were associated with Cainozoic glacial cycles. The distribution of the opilionid genus *Hickmanoxyomma*, which is exclusively cavernicolous, appears to have resulted from the ablation of surface forms in the south and east of Tasmania, where the effect of glaciation was most intense, and the occurrence of some sympatric species suggests that there may have been multiple phases of cave invasion. In contrast, in the coastal lowlands to the north and northeast, where periglacial conditions were less extreme, surface-dwelling species of *Hickmanoxyomma* are present. A cavernicolous assamiid with reduced eyes, but not strongly troglomorphic, and the strongly troglomorphic *Glennhuntia glennhunti* ("Phalangodidae") from arid Cape Range, are both probably rainforest relicts although the wider affinities of both families are unknown. Several species of unknown affinity have recently been collected from boreholes in the arid Pilbara region.

Pseudoscorpiones

The worldwide family Chthoniidae is most commonly represented among troglomorphic species in Australia. The genera *Tyrannochthonius*, *Pseudotyrannochthonius*, and *Austrochthonius* are widespread with cave populations in eastern and western parts of the continent. Syarinidae, which occur in the rainforests of Africa, Asia, and the Americas, occur widely in Australia and as a troglophile in Cape Range and the Pilbara. The Hyidae, known from India, Madagascar, and southeast Australia, are represented in Australia in the Kimberley and by markedly troglomorphic species of *Indohya* from arid ecosystems.

Schizomida

Schizomids are essentially a tropical forest element that occurs across the top of the continent, as far south as the humid caves in the arid Cape Range and subterranean systems throughout the Pilbara. Troglobitic species are known from various genera, of which the most troglomorphic belong to the genera *Draculoides* (Fig. 3) and *Paradraculoides*.

Scorpiones

A troglobitic scorpion (*Liocheles*: Liochelidae) occurs on rainforest-covered Christmas Island, Indian Ocean. *Aops oncodactylus*, a genus endemic to arid Western Australia, is the first troglobitic urodacid and the first troglobitic scorpion recorded from continental Australia. Only one other troglobitic scorpion is known outside the Americas, from Sarawak (Malaysia).

Araneae

Troglodiplura, which has South American affinities, is the only troglobitic mygalomorph spider in Australia, and occurs in caves in the arid Nullarbor region. Like the sympatric cockroach *Trogloblatella*, it is heavily sclerotized, suggesting a more drying atmosphere than is generally associated with troglomorphic animals found elsewhere, such as in the arid zone caves of Cape Range, which have affinities with rainforest floor communities. The primitive araneomorph (true) spider *Hickmania troglodytes* from Tasmania is a troglophilic member of Austrochilidae, a family that also occurs in Chile and Argentina.

Large lycosoid spiders occur widely in the arid areas, one of which, *Bengalla bertmaini* (Tengellidae), from Cape Range, is highly troglomorphic, lacking eyes and pigment.

Symphytognathidae occur as troglobitic elements in the tropical caves of arid Cape Range and monsoonal Northern Territory (Katherine) as *Anapistula*, found as epigean elements in the wet tropics of Australia, Malaysia, and Indonesia.

Cavernicolous Filistatidae occur throughout Australia (*Wandella*) but the monotypic *Yardiella* from Cape Range has relatives in northeast India and the family has a generally Gondwanan distribution.

A blind troglobitic *Tetrablemma* (Tetrablemmidae) occurs in the Pilbara, presumably a relict from past humid environments. Likewise *Desognanops* (Trochanteriidae), from the Pilbara, is a blind rainforest relict.

Among the Pholcidae, *Wugigarra* occurs along the eastern seaboard and to the southeast of the continent while the western three-quarters of the continent contains old elements of the pholcid fauna. If the distribution of the genus were restricted by current ecological conditions, then the genus would be expected to be found in the west and other refugia, but this apparently is not the case. This distribution may be due to the marine subdivision of the continent by the Cretaceous marine transgression. The genus *Trichocyclus* occurs as a cavernicolous element throughout much of the rest of the continent from the Nullarbor to the tropics. Both genera are considered to be Gondwanan relicts.

MYRIAPODA

Diplopoda

Troglobitic millipedes have rarely been reported in Australia. The Cape Range karst hosts a number of paradoxosomatid millipedes of the genus *Stygiochiropus*, three of which are declared rare fauna being known from single small caves, while *S. communis* comprises a species swarm occurring through the range in three regions each with a major genetic cluster. Caves between regions differ by a mean of 25.6% fixed allelic differences (allozymes), while caves within regions show 0.92–3.06% fixed differences. The first recorded troglobitic spirobolid millipede is *Speleostropus nesiotes*, a genus endemic to Barrow Island.

Chilopoda

The scolopendrid centipede *Cryptops* (*Trigonocryptops*) *roeplainsensis* occurs along the coastal boundary of the Nullarbor karst. Large scutigerid centipedes of the genus *Allothereura* are seen in the more superficial parts of both humid and dry caves in the arid areas but are not troglobitic.

INSECTS

Diplura

Numerous undescribed species belonging to Campodeidae, Japygidae, Projapygidae, and Parajapygidae have been collected from limestone caves and boreholes in karst and fractured rock aquifers.

Zygentoma

Among the Nicoletiidae, *Trinemura* is represented in caves in the west, while *Metrinura* is found in the caves of the northeast. Numerous species of Nicoletiidae and Ateluridae have recently been collected from boreholes in the western shield, especially the Pilbara region.

Collembola

The composition of collembolan fauna changes between the south and north of Australia. Caves in the south of the continent contain up to five genera of troglobitic Collembola, while those in tropical areas have only two genera. The genera *Adelphoderia* and *Arrhopalites* are not recorded as troglobites in tropical caves, but because the former is known from both temperate and tropical rainforest litter it seems likely to occur in tropical caves. This apparent trend in diversity may well reflect the greater sampling effort in southeast Australia. *Oncopodura* occurs in southeast Australia and in the Northern Hemisphere.

Planthoppers: Relicts or Invaders?

There is continuing debate as to whether cave fauna result from active colonization or occur as relicts as a result of the extirpation of surface populations by adverse conditions (*e.g.*, glaciation, aridity). The cave fauna on arid Cape Range are clearly relictual in that they are now remote from the humid forest from which the fauna were sourced. However, the aridity is sufficiently intense to have obliterated all close surface relatives and so the process by which it became relictual cannot be resolved. By contrast, in a grossly similar fauna in Far North Queensland, the troglobitic cixiid and meenoplid planthoppers have some members with surface relatives and many intermediate forms. These lineages show many reductive, but no progressive trends, and this has been interpreted as support for the active colonization of the subterranean realm, rather than as a process of relictualization (Hoch and Howarth, 1989).

In North Queensland seven evolutionary lines of planthoppers (Fulgoroidea) of the families Cixiidae (genera *Solonaima*, *Undarana*, *Oliarus*) and Meenoplidae (*Phaconeura*, a continent-wide genus) are found. *Solonaima* (Cixiidae) exhibits four independent invasions of the caves and shows a full range of adaptations to cave life, from epigean to troglobitic, together with intermediate stages. This lineage provides an excellent model for the stepwise evolution of cave forms and the reconstruction of the historic process of cave adaptation—the loss of eyes and pigmentation, reduction of wings and tegmina, and increased phenotypic variation, such as wing venation, even within same species, suggesting a relaxation of selection pressure. To Hoch and Howarth (1989) this suggested that there had been fragmentation of the rainforest owing to the drying climate during the Miocene. This model, argued on other evidence, has also been suggested for the arid Cape Range region on the west coast of the continent.

Blattodea

Cockroaches represent a widespread and common element of many Australian caves, particularly those where the predominant energy source is guano from bats or swiftlets where. *Paratemnopteryx* and related genera (*Gislenia*, *Shawella*) are prominent. *Paratemnopteryx stonei* exhibits significant morphological variation in seven tropical caves spread over a 150-km distance in

North Queensland, such variation being consistent with molecular variation (Slaney and Weinstein, 1996). The genus *Neotemnopteryx* is widespread on the east coast and is represented by 14 species, of which five species are cavernicolous, but troglobitic species occur in the Nullarbor and the southwest coast: *N. wynnei* and *N. douglasi*, respectively. In the Nullarbor, where caves are relatively dry, the large, eyeless but highly sclerotized *Trogloblatella nullarborensis* is found. In contrast, the Nocticolidae occur widely in the Old World tropics and a number of cave species occur throughout the Australian tropics, down to arid Cape Range where *Nocticola flabella* is found, the world's most troglomorphic cockroach, which is distinguished by its pale, fragile, translucent appearance. In contrast, a more robust monotypic troglobite, *Metanocticola*, is found on Christmas Island. Numerous species of troglobitic *Nocticola* (and Blattidae) are being collected from baited traps inserted down holes bored in pisolites and fractured rocks of the arid Pilbara region. The genus *Nocticola* also occurs in the Philippines, Vietnam, Ethiopia, South Africa, and Madagascar.

Orthoptera

Many cave crickets (Rhapdidophoridae), which occur in cave and bush habitats across southern Australia, are trogloxenes, like some bats. During the day these moisture-loving insects tend to congregate in relatively cool, moist, and still air to avoid desiccation. In the evening, part of the cricket population moves outside the cave entrance to feed but they return underground before dawn and so transport organic matter into the cave. Rhaphidophoridae have a disjunct global distribution in the temperate zones of both hemispheres. The Macropathinae are considered to be the basal group and these have a circum-Antarctic distribution, suggesting a Gondwanan origin. Generic diversity is much greater in Australia and New Zealand than elsewhere. Four genera are restricted to Australian temperate zones and a further three genera to Tasmania itself. The remaining three subfamilies inhabit the Boreal zone, suggesting vicariance owing to the Mesozoic dissolution of Pangaea.

In contrast, the only truly troglobitic cricket in Australia is the pigmy cricket *Ngamalanguia* (Nemobiinae: Gryliiidae) (Fig. 3), a monotypic genus endemic to Cape Range that lacks eyes, ocelli, tegmina, wings, and auditory tympana; is pale; and has exceptionally long antennae.

Coleoptera Dystiscidae

Globally, beetles are by far the most intensively studied cave animals. Chief among them are the trechine carabid beetles, of which more than 2000 species have been described. Of these, more than 1000 species are troglomorphic, inhabiting caves from periglacial areas of Australia and New Zealand (25 species), eastern Palearctic (*ca.* 250 species), western Palearctic (*ca.* 600 species), and Nearctic and Neotropical (*ca.* 200 species).

Unlike mainland Australia, Tasmanian caves support a distinctive cave fauna of carabid beetles from the tribes Trechini (a strongly hydrophilous group forming a dominant element of cave fauna of the periglacial areas of Europe, North America, New Zealand, and Japan) and Zolini (confined to Australasia) each containing two genera with troglobitic species. In the periglacial areas of Tasmania, vicariant patterns similar to those for opilionids may be deduced for the trechine and zoline carabid beetles, which form such a prominent part of the Tasmanian cave fauna. Harpalinae, a globally widespread and predominantly phytophagous group, typical of dry country, are considered unsuitable for cave colonization, and yet many genera are represented in caves in Australia. Two genera of the Calleidini occur in guano caves in Australia, which suggests, because these beetles are typically arboreal, the possibility of a reversal from the arboreal habit typical of this tribe, to an edaphic or subterranean life.

Although the Cholevidae is well represented in the more humid parts of Australia, the tribe Leptodirinae (Bathysciinae), which comprises the predominant component of the rich cholevid beetle fauna of the Northern Hemisphere, is entirely missing from Australia and the rest of the Southern Hemisphere. In the Snowy Mountains area of the mainland, where periglacial conditions also persisted, is found the only troglomorphic psydrinid beetle known globally. Numerous other families of beetles occur in caves throughout Australia, in both the humid and arid areas, but most seem to be accidentals. The Australian troglobitic fauna, especially those that associate with periglacial areas, differ from those in the Northern Hemisphere, owing to the composition of the surface fauna, rather than due to different evolutionary trends. Recent collections from baited traps lowered into bores drilled in karst, pisolites, calcretes, and fractured rock have yielded a sparse but diverse troglomorphic beetle fauna awaiting study including members of the families Carabidae, Curculionidae, Staphylinidae (Pselaphinae), and Tenebrionidae. In addition, Trogidae are commonly associated with guano deposits derived from cave roosting bats and swiftlets.

Australia hosts more species of stygobitic diving beetles (Dytiscidae) than the rest of the world combined. More than 100 species of Hydroporinae (and the only two stygobitic species of Copelatinae) are found mainly in shallow calcrete aquifers in the Yilgarn, western shield, and in Ngalia Basin, Northern Territory. Several species occur in alluvia of New South Wales and South Australia.

VERTEBRATES

Caves in the wet-dry (monsoonal) tropics commonly provide refuge to vertebrates during the dry season and clearly this temporary habitation has an impact on the trophic relations of these caves. Among them are tree frogs (e.g., *Litoria caerula*), which are also abundant in uncapped boreholes, and fish, such as the common eel-tail catfish, *Neosilurus hyrtlii*, and the spangled perch or grunter, *Leiopotherapon unicolor*. In the dry season, the fish may survive in caves and underground water systems and from there they would contribute to the repopulation of the seasonally inundated floodplains.

Australia has two described highly troglomorphic fishes which are sympatric where they occur at Cape Range. The blind gudgeon, *Milyeringa veritas* (Eleotridae) (Fig. 6), is of unknown affinity but inhabits water ranging from seawater to freshwater in a largely anchihaline system in Cape Range (a second, undescribed species is known from Barrow Island). Swamp eels (Synbranchidae) are represented in Australia by two species of *Ophisternon*, of which *O. candidum* is a highly troglomorphic species (Fig. 3). Stygobitic specimens of unknown affinity were recorded from Barrow Island and the Pilbara during 2010. The genus occurs widely in the coastal wetlands of the Indo-Malayan region, with one other troglomorphic species inhabiting caves in Quintana Roo, Mexico.

Snakes are commonly seen in caves, especially in the tropical regions where they predate bats (e.g., the banded cat-snake *Boiga fusca ornata*). The blind snake, *Ramphotyphlops longissimus*, from the Barrow Island karst has apparent troglomorphies and may represent the first troglobitic reptile.

Birds are rarely represented in Australian caves other than as superficial components inhabiting cave openings. The exceptions are swiftlets (*Collocalia* species) that build their nests in the dark zone, on smooth concave walls high above the cave floor in some tropical caves in Far North Queensland and Christmas Island (Indian Ocean). The nests of some species are intensively harvested for the gourmet delicacy "birds' nest soup" in Southeast Asia and India. The Christmas Island glossy swiftlet (*Collocalia esculenta natalis*) is endemic to Christmas Island where, in the absence of cave bats, they are the prime source of guano in caves. A number of other species of *Collocalia* occur in the Indian Ocean, Southeast Asia, and Queensland, mostly nesting in caves. The nests detach from the cave walls in dry air, a factor that may account for their absence from the drier tropical areas, such as the Kimberley. The various subspecies inhabit few of the caves available, being known from only five caves on Christmas Island, whereas the white-rumped swiftlet (*Collocalia spodiopygus chillagoensis*) occurs in less than 10% of approximately 400 caves at Chillagoe in Queensland.

Bats comprise nearly a third of the Australian mammalian fauna. Seven families of bats, comprising about 30% of the Australian bat fauna, are found in caves. The 17 species of cave-dwelling bats in Australia are largely restricted to the tropics and comprise insectivorous and frugivorous bats and vertebrate predators (ghost bats, *Macroderma gigas*). Six species are restricted to the Cape York peninsula and 11 species occur across the northern part of the continent, two of them extending along the west coast to the arid Pilbara region. Only four species are restricted largely to the center of the continent, two being restricted to the western plains of Queensland and New South Wales.

CONCLUSIONS

In a global context, the most striking features of the subterranean fauna of Australia are (1) the apparent age of the lineages present in subterranean environments, and (2) the high proportion of geographic relicts present in the subterranean systems that are widely separated from their near relatives. Although much remains to be done to establish consistent patterns, numerous independent examples suggest similar processes but at a range of spatial and temporal scales.

In the southeast, there is evidence that Pleistocene glaciation influenced the cave fauna. But, over most of mainland Australia, the overwhelming influence seems to have been relict distributions resulting from increasing aridity during the Tertiary, particularly in the Miocene. Numerous terrestrial and aquatic lineages have affinities with Gondwana, or with Western Gondwana, often at the generic level. In terrestrial lineages, these are commonly associated with rainforests. Numerous crustaceans, often lineages entirely comprising stygal species, and even a fish lineage, have distributions throughout the area of the former Tethys Ocean. Many lineages from northwestern Australian anchihaline waters comprise species congeneric with those inhabiting caves on either side of the North Atlantic.

Acknowledgments

I thank G.D.F. Wilson (non-oniscidean Isopoda), S. Taiti (Oniscidea), R. Leijs (Amphipoda), T. Karanovic

(Copepoda), I. Karanovic (Ostracoda), M.S. Harvey (Chelicerata), S. Halse, and S. Eberhard for their comments and/or unpublished information, and K. Grimes for the basis of Figure 1.

See Also the Following Articles

Diversity Patterns in the Dinaric Karst
Diversity Patterns in Europe
Diversity Patterns in the Tropics
Diversity Patterns in the United States

Bibliography

Austin, A. D., Cooper, S. J. B., & Humphreys, W. F. (Eds.), (2008). Subterranean connections: Biology and evolution in troglobiont and groundwater ecosystems. *Invertebrate Systematics*, 22(2), 85–310.

Bradbury, J. H., & Williams, W. D. (1999). Key to and checklist of the inland aquatic amphipods of Australia. *Technical Reports of the Australian Museum*, 14, 1–121.

Eberhard, S. E., & Humphreys, W. F. (2003). The crawling, creeping and swimming life of caves. In B. Finlayson & E. Hamilton-Smith (Eds.), *Beneath the surface: A natural history of Australian caves* (pp. 127–147). Sydney: University of New South Wales Press.

Finston, T. L., Johnson, M. S., Humphreys, W. F., Eberhard, S., & Halse, S. (2007). Cryptic speciation in two widespread subterranean amphipod genera reflects historical drainage patterns in an ancient landscape. *Molecular Ecology*, 16(2), 355–365.

Guzik, M. T., Cooper, S. J. B., Humphreys, W. F., & Austin, A. D. (2009). Fine-scale comparative phylogeography of a sympatric sister species triplet of subterranean diving beetles from a single calcrete aquifer in Western Australia. *Molecular Ecology*, 18(17), 3683–3698.

Hamilton-Smith, E., & Eberhard, S. (2000). Conservation of cave communities in Australia. In H. Wilkens, D. C. Culver & W. F. Humphreys (Eds.), *Subterranean ecosystems* (pp. 647–664). Amsterdam: Elsevier.

Hoch, H., & Howarth, F. G. (1989). The evolution of cave-adapted cixiid planthoppers in volcanic and limestone caves in North Queensland, Australia (Homoptera: Fulgoroidea). *Mémoires de Biospéologie*, 16, 17–24.

Holmes, A. J., Tujula, N. A., Holley, M., Contos, A., James, J. M., Rogers, P., & Gillings, M. R. (2001). Phylogenetic structure of unusual aquatic microbial formations in Nullarbor caves, Australia. *Environmental Microbiology*, 3(4), 256–264.

Howarth, F. G. (1987). The evolution of non-relictual tropical troglobites. *International Journal of Speleology*, 16, 1–16.

Howarth, F. G., & Stone, F. D. (1990). Elevated carbon dioxide levels in Bayliss Cave, Australia: Implications for the evolution of obligate cave species. *Pacific Science*, 44(3), 207–218.

Humphreys, W. F. (1999). Relict stygofaunas living in sea salt, karst and calcrete habitats in arid northwestern Australia contain many ancient lineages. In W. Ponder & D. Lunney (Eds.), *The other 99%: The conservation and biodiversity of invertebrates* (pp. 219–227). Mosman, New South Wales: Transactions of the Royal Zoological Society of New South Wales.

Humphreys, W. F. (2000). The hypogean fauna of the Cape Range peninsula and Barrow Island, northwestern Australia. In H. Wilkens, D. C. Culver & W. F. Humphreys (Eds.), *Subterranean ecosystems* (pp. 581–601). Amsterdam: Elsevier.

Humphreys, W. F. (2001). Groundwater calcrete aquifers in the Australian arid zone: The context to an unfolding plethora of stygal biodiversity. *Records of the Western Australian Museum*, (Suppl. 64), 63–83.

Humphreys, W. F., & Harvey, M. S. (Eds.), (2001). Subterranean biology in Australia 2000. *Records of the Western Australian Museum*, (Suppl. 64).

Humphreys, W. F., Watts, C. H. S., Cooper, S. J. B., & Leijs, R. (2009). Groundwater estuaries of salt lakes: Buried pools of endemic biodiversity on the western plateau, Australia. *Hydrobiologia*, 626, 79–95. Erratum: *Hydrobiologia*, 632, 377.

Leys, R., Watts, C. H. S., Cooper, S. J. B., & Humphreys, W. F. (2003). Evolution of subterranean diving beetles (Coleoptera: Dytiscidae: Hydroporini, Bidessini) in the arid zone of Australia. *Evolution*, 57(12), 2819–2834.

Slaney, D. P., & Weinstein, P. (1996). Geographical variation in the tropical cave cockroach *Paratemnopteryx stonei* Roth (Blattellidae) in North Queensland, Australia. *International Journal of Speleology*, 25(1–2), 1–14.

DIVERSITY PATTERNS IN EUROPE

Louis Deharveng,[*] Janine Gibert,[†] and David C. Culver[‡]

[*]Muséum National d'Histoire Naturelle, Paris, France,
[†]Université Lyon I, France, [‡]American University, U.S.A.

INTRODUCTION

To an extent that is unusual in most branches of zoology systematics, Europe is both a hotspot of subterranean biodiversity and a hotspot of research into subterranean biology, both historically and at present.

The scientific study of cave life can be traced back to Johann von Valvasor's comments in 1689 on the European cave salamander *Proteus anguinus*. This species, the only stygobiotic salamander in Europe, reaches a length of more than 25 cm, making perhaps the largest stygobiont known anywhere. It occurs throughout the Dinaric mountains in northeast Italy, Slovenia, Croatia, and Bosnia and Herzegovina. During the late eighteenth century and much of the nineteenth century, living Proteus were collected and delivered to many scientists throughout Europe. It was this animal more than any other cave animal that played a formative role in the emerging theories of evolution of Lamarck and Darwin. The first invertebrate was described in 1832, also from Slovenia, as *Leptodirus hochenwartii*, a bizarre appearing beetle with an enlarged abdomen and long spindly appendages.

Besides the caves of the Dinaric region, the cave fauna of the French and Spanish Pyrenees began to attract attention, and the Pyrenean fauna began to be described by the mid-nineteenth century. In 1907, the Romanian zoologist E.G. Racovitza published the enormously influential "Essai sur les problèmes biospéologiques," which set the agenda for biospeleological research in the coming decades. Together with the French entomologist René

Jeannel, as well as Pierre-Alfred Chappuis and Louis Fage, he established in 1907 an association named Biospeologica. This association had three objectives: (1) to explore caves and look for subterranean species; (2) to obtain identifications and descriptions from specialists for all material sampled; and (3) to publish results in a series entitled *Biospeologica*. Ultimately, Biospeologica was responsible for the inventory of the fauna of more than 1500 caves, mostly in Europe. More than 50 monographic treatments of the taxonomy and distribution of European cave fauna were published between 1907 and 1962 (Racoviţă, 2005).

Even with more than a century and a half of description and cataloging of the European subterranean fauna (Juberthie and Decu, 1994–2000), both species descriptions and inventories are far from complete. The most important large-scale inventory project of recent years, the European project PASCALIS (Protocol for the Assessment and Conservation of Aquatic Life in the Subsurface) ran from 2002 to 2004. It developed common protocols (Gibert, 2001) for the comparison of subterranean aquatic species diversity at six sites in five countries (France, Spain, Belgium, Italy, and Slovenia), and brought to light a large amount of new taxa (Deharveng et al., 2009). Individual country assessments of subterranean biodiversity are active as well, and at the beginning of the PASCALIS project the most advanced of these was in Italy, where there were more than 6000 records for 899 subterranean species (Stoch, 2001).

DIVERSITY COMPARISONS TO OTHER CONTINENTS

As of 2000, approximately 5000 obligate subterranean aquatic (stygobionts) and terrestrial (troglobionts) species from Europe had been described. By contrast, 1200 had been described from Asia, 500 from Africa, and 1000 from North America (Gibert and Culver, 2009). The dominance of Europe in known subterranean species is the result of several factors. First, Europe has been better studied than the other continents (Deharveng et al., 2000). This is particularly evident in the status of biodiversity assessment in non-cave subterranean habitats. About half of the stygobiotic species known from Europe are not primarily cave-dwellers, but rather live in other subterranean habitats such as the underflow of streams and epikarst (Pipan, 2005). In North America, these habitats are little studied and account for less than a fifth of the known stygobionts. A peculiar yet widespread terrestrial habitat, the MSS (*milieu souterrain superficiel*), which is constituted of small size voids among scree covered with soil, hosts a number of troglobionts in Europe; this habitat is unsampled in North America. In the cases of Asia and Africa, large geographic areas have not been studied with respect to subterranean species, both cave and non-cave (Culver et al., 2001).

Second, subterranean biodiversity in Europe is actually higher than on other continents. There is some empirical evidence to support this. On a worldwide basis, Culver and Sket (2000) list 20 caves and wells that are known to have a total of 20 or more stygobionts and troglobionts. Of these, 13 are in Europe, 3 are in North America, and the rest are scattered elsewhere. This pattern is retrieved on a broader regional basis. Among 7 large karst areas of Europe and North America that are among the best documented for their terrestrial subterranean fauna, the top rankings were two European sites (Fig. 1). In several regions, especially the western Balkans (northeast Italy, Slovenia, Croatia, Bosnia and Herzegovina, and Serbia) and the Pyrenees (France and Spain), even the casual observer can note the large number of stygobionts and troglobionts present, relative to caves on other continents. A possible explanation for the increased diversity of stygobionts in the western Balkans is the complex biological and geological history of the Dinaric mountains (Sket, 1999). The amount of available subterranean habitat in these mountains is large and has had a long history of cave development and evolution. In addition, the complex history of the Mediterranean Sea, including the fact that it almost dried up about 6 million years ago (the Messinian crisis), may have resulted in a greater invasion rate of the subterranean realm from marine waters. Moreover, multiple invasions may have occurred from surface continental waters. In the case of terrestrial species, it is possible that invasion of European caves was enhanced during interglacials of the Pleistocene relative to North America. This is because mountain ranges in Europe are largely east–west oriented, whereas mountain ranges in North America are largely north–south oriented. Thus, in

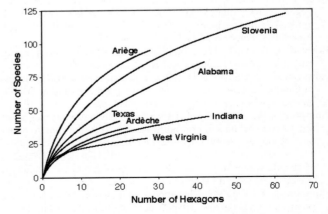

FIGURE 1 Species accumulation curves. Incidence functions of number of troglobionts plotted against the number of 100-km² hexagons in 3 regions of Europe and 4 regions of North America. *From Culver et al., (2006).*

North America, species could escape hot summers either by migrating north or by invading caves. In Europe, the orientation of the mountains reduced the migration potential, perhaps resulting in increased invasion rates of caves and other subterranean habitats.

TAXONOMIC RICHNESS OF THE EUROPEAN SUBTERRANEAN FAUNA

The subterranean diversity pattern is better known in Europe than in other continents, yet it is not always well documented. This is obvious from the extensive study of stygobiotic fauna conducted during the recent project PASCALIS on six sites across Western Europe that resulted in the discovery of at least 109 species new to science, that is, more than 40% of the collected species. Patterns are likely to be similar for terrestrial fauna as reflected in the large number of new taxa described every year from Europe, though no synthesis is available.

The distribution of the biota is controlled by a combination of factors, including geological history, physicochemical variables, aquifer structure, connectivity between karst units, and biological interactions resulting in complex biodiversity patterns, with an obvious hierarchical dimension (Malard et al., 2009). Local (α) and regional (γ) diversity is better known than the increase in diversity over spatial scales (β), although β-diversity is likely to be particularly significant across spatial scales (aquifer, basin, and region). Put another way, the diversity in a single cave is always much less than the overall diversity of caves even within a small area as a result of reduced opportunities for migration and invasion—as far as endemism is significant inside the area of interest.

The taxonomic diversity of the European fauna, with more than 5000 stygobionts and troglobionts, is too extensive and complex to be reviewed here, but several highlights are especially noteworthy. The first of these are the beetles in the terrestrial habitats. Two families of beetles dominate in temperate zone caves: the Carabidae and the Leiodidae. Both of these families reach their zenith of subterranean species richness in southern Europe, while southern China is emerging since the 1990s as a second hotspot of biodiversity for Carabidae. Among Carabidae, troglobiotic species are distributed in 40 genera of the subfamily Trechinae (Fig. 2), including seven genera with 18 or more cave species: *Anophthalmus, Aphaenops, Duvalius, Geotrechus, Hydraphaenops, Neotrechus,* and *Orotrechus*. The genus *Duvalius* is especially speciose, with more than 250 species. In contrast, there is only one carabid genus in North America with more than 20 cave species—*Pseudanophthalmus*—although it has more than 200 species.

Among the Leiodidae, a total of 28 genera of the tribe Leptodirini are known from the Pyrenees. Among them, *Bathysciola*, a European genus, has more than 20 species in this mountain range, often relatively widely distributed and living mostly in soil habitats. In contrast, the 27 other genera are almost exclusively constituted of subterranean species, all oligospecific, often troglomorphic and narrowly distributed (Fig. 3F). In the Dinaric mountains, more than 50 Leiodidae genera are only known from caves, many exhibiting the highest levels of troglomorphy known among beetles worldwide, with *Antroherpon* containing 27 species. Once again, the contrast with temperate North America is instructive. There are only three genera, a single troglomorphic species (*Glacicavicola bathyscioides*), and *Ptomaphagus* has 19 cave obligate species—all oculated.

Among aquatic species, two groups are especially noteworthy. One is the amphipod family Niphargidae. There are nine genera and *Niphargus* alone has more than 200 species. Taxonomically, the diversity of the Niphargidae is rivaled by that of the Crangonyctidae in North America (Holsinger, 1993). Especially speciose is the genus *Stygobromus*, with nearly 200 species. However, the crangonyctids seem to be less diverse ecologically than the niphargids, both in terms of morphological variation and habitat variation. The final group worthy of special note are the mollusks in the Dinaric Mountains. In the Slovenian part of the Dinaric Mountains, for example, there are 37 aquatic obligate cave snails, one aquatic cave clam (Fig. 4C), and 11 species of terrestrial obligate cave snails. The entire Dinaric Mountain region has several times that many species.

TROGLOMORPHY AND RELICTNESS

Aside from its high taxonomic richness, the European subterranean fauna includes a variety of taxa outstanding by their degree of troglomorphy, their phyletic isolation, or their unique adaptive features (Figs. 2, 3, 4).

Subterranean evolution has produced more troglomorphy in Europe than in any other continent. The highest troglomorphic representatives of the most diversified groups of terrestrial cave fauna—beetles, springtails, woodlice, aquatic Crustacea, spiders, pseudoscorpions—are almost all European, Eastern Asia ranking second. As an example, all Collembolan families present in temperate caves, except Paronellidae and Arrhopalitidae, have their most modified troglomorphic species in Europe (*i.e.*, with thinner claws or longer antennae, in addition to depigmentation and anophthalmy): *Ongulogastrura longisensilla* (Pyrenees) in Hypogastruridae, *Ongulonychiurus colpus* (Cantabric range in Spain) in Onychiuridae, *Tritomurus*

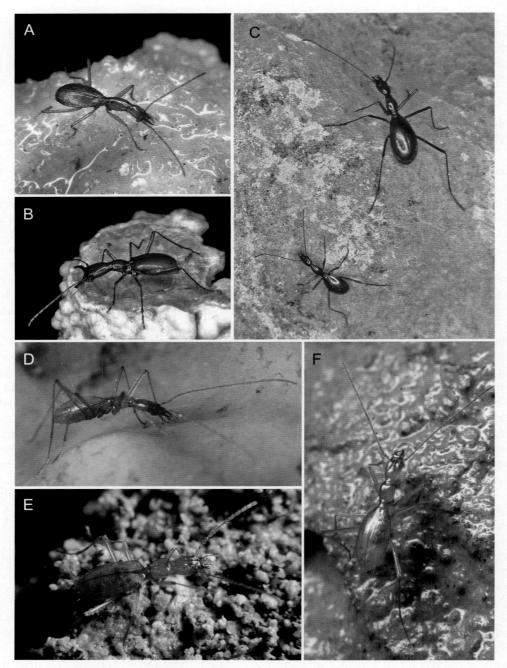

FIGURE 2 Troglomorphic Trechinae (Carabidae) of Western Europe. (A) *Lessinodytes pivai* from Italian Alps; (B) *Italaphaenops dimaioi* from Italian Alps, the largest cave Trechinae in the world; (C) *Sardaphaenops supramontanus grafittii* and *Sardaphaenops adelphus* from Sardinia; (D) *Aphaenops pluto* from French Pyrenees; (E) *Trichaphaenops obesus* from French Alps; (F) *Aphaenops cerberus* from French Pyrenees. *Photos (A) to (C) courtesy of Enrico Lana and (D) to (F) courtesy of Christian Vanderbergh. Used with permission.*

falcifer (Pyrenees) and *T. veles* (Croatia) in Tomoceridae, *Isotomiella unguiculata* (Jura) in Isotomidae, *Pseudosinella cabidochei* (Pyrenees) and *Verhoeffiella longicornis* (Dinaric karst; Fig. 3E) in Entomobryidae, *Oncopodura lebretoni* (Pyrenees) in Oncopoduridae, *Galeriella liciniana* (Dinaric karst) in Sminthuridae, and *Megalothorax tuberculatus* (Pyrenees) and *Neelus klisurensis* (Dinaric karst) in Neelidae. No explanation has been provided so far for the prevalence of such high levels of troglomorphy in southern Europe compared to other regions on Earth.

Caves host more relicts, that is, species without close phyletic relatives in other habitats of their

FIGURE 3 Some remarkable subterranean species of Europe. (A) *Titanobochica magna*, a recently described, highly troglomorphic pseudoscorpion from Portugal; (B) *Belisarius xambeui*, a relictual troglophilic scorpion of Catalonia, only representative in Europe of the family Troglotayosicidae; (C) *Stenasellus virei hussoni*, a frequent stygobiont of central Pyrenees (Isopoda); (D) *Thaumatoniscellus speluncae* from Croatian caves, one of the two species of *Thaumatoniscellus* in the monotypic subfamily Thaumatoniscellinae (Isopoda); (E) *Verhoeffiella longicornis*, a highly troglomorphic Entomobryidae Collembola from Croatia; (F) *Trocharanis mestrei* from Pyrenees, a troglobiotic beetle of the monospecific genus *Trocharanis*. Photo (A) courtesy of Sofia Reboleira; photos (B), (C), (F) courtesy of Christian Vanderbergh; and photos (D), (E) courtesy of Jana Bedek. Used with permission.

distribution area, than any other habitat, and Europe has more relicts than any other continent. Almost all are narrow endemics, most are troglobiotic and highly troglomorphic (Fig. 3), but some among the most remarkable (such as the millipede *Galliobates gracilis* of the monotypic family Galliobatidae, or the scorpion *Belisarius xambeui* from Catalonia; Fig. 3B) are only troglophilic, being found both in caves and soil. Though most relicts are concentrated in the Dinaric range and the Pyrenees, a few can be found as far north as the northern Swiss Alps (*Niphatrogleuma wildbergeri*, a millipede) and as far south as southern Spain (*Dalyat mirabilis*, the only European beetle of the South African–American subfamily Promecognathinae).

At least several suprageneric taxa have their sole subterranean representatives in Europe. The Dinaric karst is particularly rich in this respect, with serpulids, sponges, bivalves (Fig. 4), and the Cnidarian *Velkovrhia enigmatica*. Extensive karstic outcrops such as the Dinarids exposed to large bioclimatic fluctuations during long geological periods offer repeated opportunities for subterranean colonization and may account for this richness in unusual cave taxa, often derived from marine ancestors.

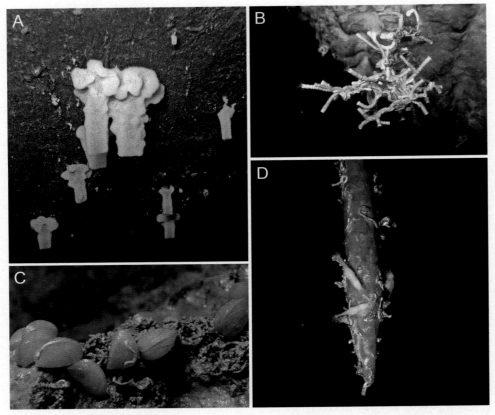

FIGURE 4 Some remarkable stygobionts of the Dinaric karst in Croatia. (A) *Eunapius subterraneus*, the only subterranean freshwater sponge in the world; (B) *Marifugia cavatica*, the only freshwater serpulid in the world (Polychaeta); (C) *Congeria kusceri*, the only stygobiotic bivalve in the world, and *Marifugia cavatica*; (D) *Troglocaris anophthalmus* (Decapoda Atyidae) and *Marifugia cavatica* on a stalactite. *Photos (A), (B), (D) courtesy of Branko Jalzic and photo (C) courtesy of Helena Bilandzija. Used with permission.*

GEOGRAPHIC PATTERNS WITHIN EUROPE

While the inventory of known subterranean species is far from complete, existing data provide some important information about diversity patterns.

The first general pattern is that there is a gradient in species richness with diversity dropping off to the north. A comparison of Italy and the United Kingdom, two European countries thoroughly inventoried, shows this in a striking manner. In the United Kingdom, a total of 12 troglobionts and 15 stygobionts, with a further 19 interstitial forms have been recorded. In Italy, a total of 265 stygobionts, 321 troglobionts, and a few more from the MSS were known in 2001. A difference of the same magnitude between southwestern countries of Europe and Belgium was highlighted during the PASCALIS project, and was additionally expected to grow with increasing sampling effort. Of course, part of this difference is a result of the covering of the United Kingdom and Belgium by Pleistocene ice sheets or permafrost, but undoubtedly other factors are at work to produce this diversity gradient. As one moves north, mean annual temperature declines and, all other things being equal, so does net primary productivity. While little or no primary production occurs in subterranean habitats, they are dependent on surface primary production for food. Thus, food availability decline with increasing latitude is reflected in the diversity pattern (Gibert and Deharveng, 2002).

It was recently shown that the north–south increase in diversity follows the average annual temperature gradient up to midlatitude only. To the south, diversity decreases again. Actually, there is a midlatitude ridge between 42° and 46° north in Europe where biodiversity peaks (Fig. 5). It roughly corresponds to parts of the mountain ranges of the northern Mediterranean region. In addition to the Coleoptera, Amphipoda, and Gastropoda mentioned earlier, many other groups reach their highest diversity along this ridge, including isopods, spiders, and springtails. Historical circumstances, especially the extensive shallow embayments in the Tertiary and the Mediterranean salinity crisis of

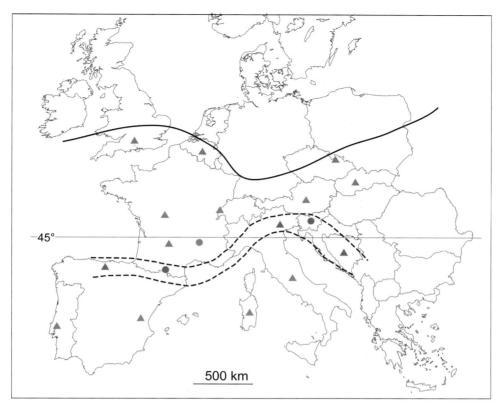

FIGURE 5 The midlatitude biodiversity ridge of European troglobionts (redrawn and updated from Culver et al., 2006). Red circles are the diversity hotspots in Slovenia and Ariege. Red triangles are other probable diversity hotspots. Orange circle is the Ardèche karst; orange triangles are other moderately rich karsts. Blue triangles are karsts with very few or no troglobionts. The boundary of the Pleistocene ice sheet is shown as a scored solid line. A pair of dashed lines indicates the hypothesized position of the high diversity ridge.

the Miocene, as well as a large variety of karstic subterranean habitats, may account for the exceptional diversification observed today. The presence of extensive karsts moderately impacted by Pleistocene glaciations along the ridge area and the long-term high productivity in surface habitats at these latitudes may have allowed the persistence of this striking biodiversity pattern.

The second striking pattern is that of hotspots nested inside hotspots; along the midlatitude ridge, subterranean diversity and endemism are not homogeneous, but peak in particular areas. These areas, long known as exceptionally rich, are all located in mountainous and humid parts of the ridge. They exhibit an unusual concentration of cave relictual endemics and highly troglomorphic species across various taxonomic groups, both aquatic and terrestrial. The richest area is the Dinaric range, and the Pyrenees are second (Fig. 6), while several other hotspots, less rich but also less documented, exist in various areas (southern Italy, southern Massif Central in France, southern Catalonia and Cantabrian Cordillera in Spain, Southern Alps in France).

The third and the fourth patterns highlighted below are formally not specific to Europe, but are best documented in this continent.

The third pattern that is evident is that the ratio of troglobionts to stygobionts declines with increasing latitude from the biodiversity ridge northward (Fig. 7). The reduction in troglobionts may be largely the result of Pleistocene effects. No subterranean terrestrial habitats existed, as far as we know, within the ice sheets. On the other hand, some subterranean aquatic species may have survived the Pleistocene in running water habitats underneath the ice cap. Additionally, it may be that recolonization has been more rapid among aquatic species (for example, through fluvial corridors and due to the hydrological continuum and the connectivity between aquifers). Stygobionts can be found in almost all parts of Europe with suitable groundwater habitats, even though these habitats are often not cave habitats, but rather other subterranean habitats such as the underflow of streams. To the south of the biodiversity ridge, available evidence suggests that the ratio of troglobionts to stygobionts does not increase any more or may decrease significantly, but data remain insufficient.

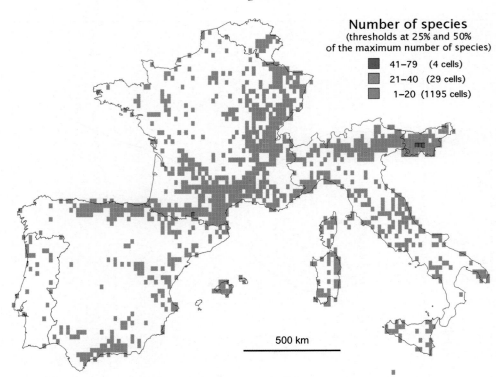

FIGURE 6 Hotspots of richness in stygobionts. Each cell across southwestern Europe is 0.2° × 0.2°. *From Deharveng et al., 2009.*

The fourth pattern is that of local changes in subterranean fauna resulting of habitat differences. For instance, Copepoda, the most diverse group of stygobionts, were shown to contribute almost equally to the overall species richness in karst and porous aquifers and were usually slightly more diverse in nonsaturated compared to saturated habitats. Karst aquifers studied during the PASCALIS project were often richer in stygobionts than porous aquifers and unsaturated zones of the karst often richer than saturated zones. Overall, however, correlation of habitat types and stygobiotic species richness did not exhibit consistent patterns across sites (Table 1). Nevertheless, even when species richness was similar, taxonomic composition often differed among habitats of the PASCALIS sites, so that, except in Wallonia and Cantabria, increase in species richness for a given number of samples was clearly more significant when several habitats and both aquifer types were included in the analyses. This might reflect the fact that increased regional richness is correlated on average to stronger habitat specialization.

At a finer scale, and more than any other factor, habitat granularity generates a clear-cut pattern of stygobiont distribution. The large voids in karstic channels clearly enable a large size range of aquatic organisms to occur while very tiny spaces in porous aquifers represent a barrier to larger animals. For example, in the Vidourle basin in southern France, the amphipod *Niphargus virei* and the decapod atyid *Gallocaris inermis*, are limited to the karstic aquifer, while most micro-Crustacea are found in both porous and karstic aquifers of this site.

The diversity patterns in subterranean terrestrial communities are less well known regarding habitat types, but seem to be basically different, because, for unknown reasons, interstitial animals of the deep soil (euedaphobionts, equivalent to the stygobionts of porous aquifers) only exceptionally colonize caves. Biodiversity of large subterranean voids is therefore mostly based on relatively large size invertebrates, while edaphobiont species are, in fact, more diverse. This claim is based on compilation of heterogeneous and scattered datasets from the literature. Actually, the fauna living in "porous" terrestrial habitats of caves has been very scarcely studied, and comparative data between cave and soil fauna at the same site are lacking.

ASSESSING EUROPEAN BIODIVERSITY PATTERNS FOR CONSERVATION

The current incomplete state of knowledge of groundwater biodiversity in Europe remains a major constraint on successful implementation of its conservation. The PASCALIS project elaborated a toolbox for the assessment of this diversity that includes validated methods

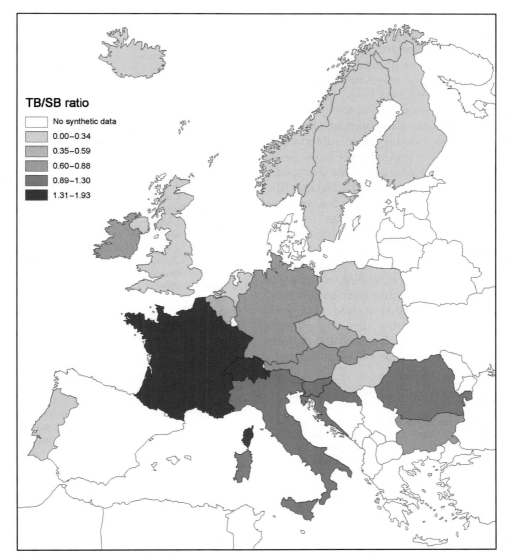

FIGURE 7 Ratio of troglobionts (TB) to stygobionts (SB) for western and central European countries where synthetized data are available *(from Gibert and Culver, 2005)*. The pattern includes non-cave stygobionts.

TABLE 1 Number of Species Collected in Different Habitats of the Six Sites of the PASCALIS Project

	Aquifers		Habitats Inside Aquifers			
	karst	porous	u	s	h	p
Jura (France)	39	*41*	33	32	32	*34*
Roussillon (France)	26	*43*	18	17	*35*	31
Lessinia (Italy)	*58*	48	*39*	30	38	21
Krim (Slovenia)	*64*	59	34	42	41	*48*
Cantabria (Spain)	52	32	*46*	28	26	18
Wallonia (Belgium)	*19*	16	15	*17*	10	15

Karst habitats: upper unsaturated zone (u) and lower saturated zone (s); porous analogues: hyporheic zone (h) and phreatic zone (p). In bold italics are highest values for habitat and aquifer at each site *(from Dole-Olivier et al., 2009)*.

for (1) determining the reliability of patterns of regional biodiversity revealed by mapping of existing data; (2) obtaining by means of a standardized field sampling method an unbiased estimate of groundwater biodiversity in regions for which data were insufficient; and (3) predicting overall species richness based on biodiversity indicators in regions with incomplete data sets. These considerations led to the development of a common database that allowed zooming in on subterranean landscapes in order to consider groundwater biodiversity at different spatial as well as ecological scales (European, country, region, local, aquifer types, and habitats). The project devised a specific action plan for the conservation of groundwater biodiversity at a European level by selecting priority regions for conservation using complementarity approaches (Michel et al., 2009), by identifying the spatial scale of relevance for conserving biodiversity within these regions, and by formulating a series of appropriate measures for maintaining their biodiversity. Finally, shifting to an aquifer-based instead of a grid-cell–based approach for evaluating diversity patterns and habitat selection is the next challenge to significantly increase the relevance of conservation policy in Europe.

Acknowledgments

This work was supported in part by the European project PASCALIS No. EVK2-CT-2001-002121. We thank Michel Perreau and Arnaud Faille for information on Leiodidae and Carabidae beetle biodiversity, and Sofia Reboleira and Mark Judson for information on Pseudoscorpiones and relictual taxa.

See Also the Following Articles

Diversity Patterns in Australia
Diversity Patterns in the Dinaric Karst
Diversity Patterns in the Tropics
Diversity Patterns in the United States

Bibliography

Culver, D. C., Deharveng, L., Bedos, A., Lewis, J. J., Madden, M., Reddell, J. R., et al. (2006). The mid-latitude biodiversity ridge in terrestrial cave fauna. *Ecography, 29*(1), 120–128.

Culver, D. C., Deharveng, L., Gibert, J., & Sasowsky, I. D. (Eds.), (2001). *Mapping subterranean biodiversity—Cartographie de la biodiversité souterraine.* Charles Town, WV: Karst Waters Institute (Special Publication 6).

Culver, D. C., & Sket, B. (2000). Hotspots of subterranean biodiversity in caves and wells. *Journal of Cave and Karst Studies, 62*(1), 11–17.

Deharveng, L., Dalens, H., Drugmand, D., Simon-Benito, J. C., da Gama, M. M., Sousa, P., et al. (2000). Endemism mapping and biodiversity conservation in Western Europe: An arthropod perspective. *Belgian Journal of Entomology, 2*(1), 59–75.

Deharveng, L., Stoch, F., Gibert, J., Bedos, A., Galassi, D. M. P., Zagmajster, M., et al. (2009). Groundwater biodiversity in Europe. *Freshwater Biology, 54*(4), 709–726.

Dole-Olivier, M.-J., Castellarini, F., Coineau, N., Galassi, D. M. P., Martin, P., Mori, N., et al. (2009). Towards an optimal sampling strategy to assess groundwater biodiversity: Comparison across six European regions. *Freshwater Biology, 54*(4), 777–796.

Gibert, J. (2001). Protocols for the assessment and conservation of aquatic life in the subsurface (PASCALIS): A European project. In D. C. Culver, L. Deharveng, J. Gibert & I. D. Sasowsky (Eds.), *Mapping subterranean biodiversity—Cartographie de la biodiversité souterraine* (pp. 19–21). Charles Town, WV: Karst Waters Institute (Special Publication 6).

Gibert, J., & Culver, D. C. (2009). Assessing and conserving groundwater biodiversity: An introduction. *Freshwater Biology, 54*(4), 639–648.

Gibert, J., & Culver, D. C. (2005). Diversity patterns in Europe. In D. C. Culver & W. B. White (Eds.), *Encyclopedia of caves* (pp. 196–201). Amersterdam, The Netherlands: Elsevier Press.

Gibert, J., & Deharveng, L. (2002). Subterranean ecosystems: A truncated functional biodiversity. *Bioscience, 52*(6), 473–481.

Holsinger, J. R. (1993). Biodiversity of subterranean amphipod crustaceans: Global patterns and zoogeographical implications. *Journal of Natural History, 27*, 821–835.

Juberthie, C., & Decu, V. (Eds.), (1994–2001). *Encyclopaedia biospeologica* (Vols. 1–3). Moulis, France: Société Internationale de Biospéologie (in French).

Juberthie, C., & Juberthie-Jupeau, L. (1975). La réserve biologique du laboratoire souterrain du C.N.R.S. à Sauve (Gard). *Annales de Spéléologie, 30*, 539–551 (in French).

Malard, F., Boutin, C., Camacho, A. I., Ferreira, D., Michel, G., Sket, B., et al. (2009). Diversity patterns of stygobiotic crustaceans across multiple spatial scales in Europe. *Freshwater Biology, 54*(4), 756–776.

Michel, G., Malard, F., Deharveng, L., Di Lorenzo, T., Sket, B., & De Broyer, C. (2009). Reserve selection for conserving groundwater biodiversity. *Freshwater Biology, 54*(4), 861–876.

Pipan, T. (2005). *Epikarst—A promising habitat.* Ljubljana, Slovenia: ZRC Publishing.

Racoviţă, G. (2005). L'entreprise Biospeologica. Sa creation, son activité et ses realizations. *Endins, 28*, 25–34 (in French).

Sket, B. (1999). High biodiversity in hypogean waters and its endangerment—The situation in Slovenia, Dinaric Karst, and Europe. *Crustaceana, 15*, 125–139.

Stoch, F. (2001). Mapping subterranean biodiversity: Structure of the database and mapping software SKMAP and report of status for Italy. In D. C. Culver, L. Deharveng, J. Gibert & I. D. Sasowsky (Eds.), *Mapping subterranean biodiversity—Cartographie de la biodiversité souterraine* (pp. 29–35). Charles Town, WV: Karst Waters Institute (Special Publication 6).

DIVERSITY PATTERNS IN THE DINARIC KARST

Boris Sket

Univerza v Ljubljani, Slovenia

WHAT IS THE DINARIC KARST?

The Dinarides or Dinaric Alps are a part of the Alpine system along the east coast of the Adriatic, the western part of the Balkan Peninsula in southern Europe (Fig. 1). In the extreme northeast of Italy and the west of Slovenia they are detached from the Southern Calcareous Alps by a narrow belt of the so-called isolated karst. The Dinaric

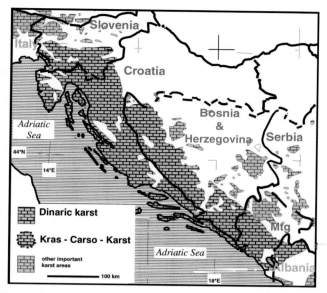

FIGURE 1 Position of the Dinaric karst. Mtg = Montenegro. *After Sket (1997).*

karst with the surrounding belt of the isolated karst patches covers approximately the southeastern halves of Slovenia, Croatia, and Bosnia and Herzegovina, and the whole of Montenegro, extending into southwestern Serbia and Kosovo. In northwestern Albania it extends up to Hellenides.

The Dinaric land appeared atop the Adriatic microplate out of the sea approximately 30 million years ago as an island that separated the sea of Paratethys from the then still existing narrow part of the Tethys Ocean. Its flora and fauna originated and developed in comparative isolation from the rest of Europe. After orogenic faulting and erosion of the flysch layer, karstification began—most probably at the end of the Pliocene. It caused the origin of cave habitats, hydrographic changes, and splintering of many biotic distribution areas.

Tropical or subtropical climates during the Tertiary were followed by intermittent glaciations in the Pleistocene. However, the closest extensive glacier in the region was in the Julian Alps in northwestern Slovenia, whereas in the Dinarides proper, only some of the highest peaks were glaciated. The dry land extended during the last glacials far south into today's Adriatic Sea, enabling the continental biota to move away from their former proximity of the coldest points. It has been supposed that after glaciations, 10,000 years ago, the whole Dinaric karst, except for its highest elevations and a narrow coastal strip, before man's impact, had been covered by extensive forests.

Humans began to colonize the territory (Sket, 1997) at least by the end of the Riss glacial or 65,000 years BP. With the onset of agriculture and animal husbandry (Mesolithic, 10,000–4300 BC), more forest clearing occurred. Illyric tribes depended primarily on domestic animals, migratory grazing (transhumance), and storage of hay. Wintering of large herds caused higher pressure on forests' existence. The Copper to Iron Ages (3000–200 BC) meant additional forest devastation for mining and the primitive metallurgy. In the Halstatt period, southern Slovenia was already quite densely inhabited. Grave constructions testify to the existence of bare rocky grounds in Lika (Croatia) in 500 BC and at the coast in Bronze and Roman times. Up until 60 years ago, before the reforestation started to be successful, very large parts of the Dinarides were barren rock or nearly so. The measures that are responsible for comparatively green karst sites nowadays were the introduction of the European black pine (*Pinus nigra*) for reforestation, a ban on goats, and a remarkable reduction in the human population in karst areas; the population moved to the towns and to developing industrial centers as well as to the fertile fields of the Pannonian Plain.

Today's Dinaric karst extends to more than 56,000 km^2; it is an approximately 100-km-wide and 600-km-long belt between the Adriatic Sea and the northeast hills and plains. It includes most Adriatic islands and drowned caves below sea level, and reaches some peaks having 2000-m elevations. Mountain ridges generally run parallel to the Adriatic coast (northwest–southeast). One of the specific characteristics of this karst are *poljes*, which are large, closed, periodically flooded depressions, typically without a surface in- and outflow.

Thick carbonate layers, mainly Triassic limestones (but also dolomites and limestones of other ages), are highly porous; in the approximately 7000-km^2 area of the Dinaric part of Slovenia only, close to 7000 caves have been registered and investigated.

In the outer Dinarides (toward the Adriatic) the climates are Mediterranean or sub-Mediterranean. The mean yearly temperatures are between 12° and 17°C. Precipitation is mainly about 1000–2000 mm yr^{-1}, but is unevenly distributed and more than 50% in the cold half-year; this causes regular droughts in summer and washing away of soil in winter. The climate of the inner Dinarides is more continental and cooler and there is less precipitation. The nonfluctuating temperatures of underground habitats approximately equal mean yearly temperatures of the region, which are 8–16°C higher than winter temperatures even in subtropical regions. The floods in rainy periods cause some temperature fluctuations and enrich the underground with fresh food resources.

The comparatively isolated position of the Dinaric territory along with high precipitation, its tropical past, and very dynamic geomorphologic changes allowed development of a very diverse flora and fauna. This is attested to by the fossil molluscan faunas of Pliocene-Pleistocene lakes as well as by the diversity of survivors. The territory's position on the edge of formerly glaciated regions allowed it to be a refuge for thermophile elements during the Pleistocene. Particularly useful refuges are the thermally conditioned hypogean habitats, where animals are not exposed to unfavorable winter temperatures.

COMPOSITION AND DIVERSITY OF SUBTERRANEAN FAUNA

History

Disregarding the remote and isolated first mention of a cavefish in China in 1540, a series of zoological discoveries in caves started in the extreme northwest part of the Dinarides. The amphibian *Proteus anguinus* was first mentioned by Valvasor in 1689 and scientifically described by Laurenti in 1768; however, it was first seen in a cave in 1797. The next was the most unusual cave beetle, found by the local cave worker Luka Čeč in the famous cave Postojnska Jama, described by Schmidt in 1832 as *Leptodirus hochenwartii*. These and a great number of other discoveries in the nineteenth century were in the Kras (= Carso = Karst) of Slovenia, which gave its name to the geomorphologic phenomenon of the *karst*. Discoveries of cave animals in other parts of the world (first in the Caucasus, Appalachians, Pyrenees, and New Zealand) occurred in the middle of the century, and in other parts of the Dinarides even toward its end. Thanks to such a history, the "classical Karst" won the appellation of the "cradle of speleobiology (biospeleology)." Hamann's (1896) book *Europäische Höhlenfauna*, which even at that time already cited 400 references, was still devoted mainly to the cave species from Slovenia.

The most recent census of the Dinaric obligate subterranean fauna (Sket et al., 2004) revealed more than 900 species, 600 terrestrial and 330 aquatic troglobionts. If we compare this to 930 and 420 troglobionts in the whole of North America, the importance of such numbers is evident. Among regions of approximately the same size class in the world, the Dinaric karst is by far the richest in the obligate subterranean fauna. This does not seem to be a consequence of the state of investigations. To the troglobionts, one should add some regularly occurring eutroglophiles as well as a high number of normal surface animals (trogloxenes or accidentals) that occasionally, but with a higher or lower regularity, happen to occur in caves. Some representatives of the Dinaric fauna are shown in Figure 2.

Terrestrial Cave Fauna

By far the most diverse group of troglobionts are beetles (Coleoptera), represented by more than 200 species. This should not cause wonder, because Coleoptera are by far the richest animal group in the world; nevertheless, its representation underground is very biased; it is represented mainly by members of only two families. In addition, the other insect groups—which are still extremely diverse on the surface—are hardly represented underground at all. The most important in Dinaric caves is the family Cholevidae with its subfamily Leptodirinae (known also as Catopidae: Bathysciinae). Approximately 175 species are distributed in as many as 50 genera. Although this is originally an edaphic (*i.e.*, soil-inhabiting) group, only a few cave-inhabiting species are not specialized to the cave habitat. And only few Dinaric genera are not endemic to this region: the troglophilous *Phaneropella lesinae* Jeannel exhibits a transadriatic distribution and its congenerics occur also far in the Asia Minor. The edaphic and troglophilous *Bathyscia montana* Schioedte may serve as an archetype of edaphic leptodirines: it is less than 2 mm long, egg shaped, with short legs and antennae. Some troglobionts are similarly built, but most of the others attained different degrees of troglomorphism. They are regularly bigger and at least slightly elongated.

The higher troglomorphic species may be *pholeuonoid*, which means that they have a spindle-shaped abdomen and a neck-shaped prothorax. Such species include *Parapropus* spp. in the northwestern part of the Dinarides; similarly shaped cave beetles may occur also in other karst regions (such as *Pholeuon* in Romania). Only in the Dinaric karst have the beetles attained the highest, that is, the *leptodiroid*, degree of troglomorphism. In the name-giving *Leptodirus hochenwartii* Schmidt from Slovenia and Croatia, the legs and antennae are extremely elongated, the "neck" (prothorax) is nearly cylindrical, and high-vaulted elytrae make the abdomen egg shaped or nearly globular (Fig. 2F). Of similar shape are more than 25 member species of the genus *Antroherpon* in the southeastern part of the Dinaric karst. The high number of species probably indicates the effectiveness of such a body transformation. Nowhere else in the world has this group of beetles (Cholevidae) achieved such diversity as in the Dinaric caves.

The other beetle group in Dinaric caves is the family Carabidae with 12 genera and 80 troglobiotic species of its subfamily Trechinae. The carabid beetles in surface soil are predators; they are twice as big and longer legged than their scavenger cholevid neighbors. The northwestern *Anophthalmus* spp. or the southeastern *Neotrechus* spp. are, however, larger, more slender, and with longer appendages than the epigean or

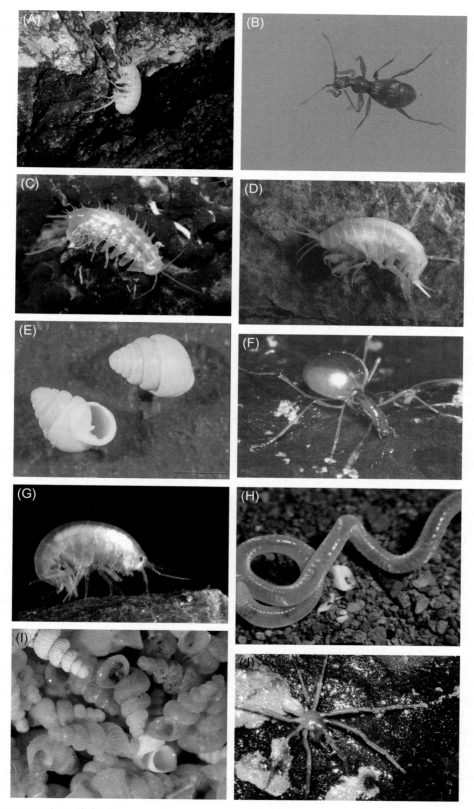

FIGURE 2 Some representatives of the Dinaric subterranean fauna. (A) *Typhlogammarus mrazeki* (Amphipoda), (B) *Machaerites spelaeus* (Coleoptera: Pselaphidae), (C) *Monolistra monstruosa* (Isopoda: Sphaeromatidae), (D) *Niphargus orcinus* (Amphipoda: Niphargidae), (E) *Zospeum kusceri* (Gastropoda: Carychiidae), (F) *Leptodirus hochenwartii* (Coleoptera: Cholevidae), (G) *Synurella ambulans* (Amphipoda: Crangonyctidae), (H) *Delaya bureschi* (Oligochaeta: Haplotaxidae), (I) *Stalita taenaria* (Araneae: Dysderidae), (J) *Lanzaia vjetrenicae* (Gastropoda: Hydrobioidea).

troglophilous *Trechus* spp. Some other troglobionts such as the *Aphaenopsis* spp. are even more troglomorphic, but this transformation never reached the degree it has in cholevids. The Dinaric Trechinae are also less exceptional; their diversity is approached by other cave faunas, such as those in North America. To the same family, but its subfamily Pterostichinae, belong also two comparatively large, approximately 10-mm-long species of *Laemostenes*; although also living outside caves, they are worth mentioning because they are present in nearly all caves in the whole Dinaric karst and far outside it.

The family Pselaphidae (Fig. 2B) is represented by only 20 troglobiotic species. These are tiny beetles with very slender thoraxes and wide abdomens with shortened elytrae. Their surface relatives are also soil-dwellers. The related extensive family Staphylinidae, with a nearly tap-shaped body with very short elytrae, resulted in the Dinarides in no troglobiotic species, but a number of species occur as troglophiles or occasional guests in the entrances of the caves. Examples include some tiny species of *Atheta* and some larger *Glyptomerus* spp. Some tiny weevils (Curculionidae) are also known from Dinaric caves. Some species of the genus *Troglorrhynchus* are often regarded as troglobionts. However, they depend on tree roots that are not as common and extensively developed in karst caves as they are in lava tubes, instead they are shallow underground. *Troglorrhynchus* spp. are in fact, as soil inhabitants, most probably merely guests in caves.

Most troglobiotic beetles in the Dinarides are eyeless and have reduced hind wings but none is really pigmentless; although translucent, most are dark brown, whereas some are yellowish brown in color. Although carabid beetles are more or less evenly distributed along the Dinaric area, slightly more diversified in the northwest, the specific cholevids become more diverse toward the southeast; the southern Herzegovina, with neighboring parts of Croatian Dalmacija and Montenegro, by far being the richest.

The next richest groups of troglobionts are the spiders and false scorpions. The genus *Troglohyphantes* is very characteristic among spiders (Araneae). Its small representatives of mostly less than 4 mm in body length span their small sheet webs either in the forest litter, in micromammalian burrows (they are *microcavernicole*), in entrance parts of the caves, or deeper in them. Only the latter (deep in caves) are mostly pigmentless and eyeless; there are approximately 25 troglobiotic species. In contrast, the always more than 5-mm-long spiders of the family Dysderidae hunt without webs. Among 25 species of 7 mainly fully troglobiotic genera belongs also the first scientifically described cave spider *Stalita taenaria* Schioedte. These are shiny brown to orange in the cephalothorax and legs and dirty white on the abdomen. Not less than 7 spider families are represented by troglobiotic species in the Dinarides. The strangest might be the family Anapidae; its species *Pseudanapis relicta* Kratochvil from Montenegro measures only 1.4 mm in length and its relatives are geographically very far away.

Similarly numerous are false scorpions (Pseudoscorpiones), which, however, are less diverse, representing only three families, which have similar body shapes. The most specialized troglobiotic species are those with extremely slender and elongated pincers on the second pair of appendages and an up to 8-mm body length. They are very different from their relatives that inhabit forest soil. *Neobisium spelaeum* Schioedte was the first species described from caves.

The other groups represented by troglobiotic species are (in descending order) millipedes (Diplopoda), woodlice (Isopoda: Oniscidea), snails (Gastropoda), centipedes (Chilopoda), harvestmen (Opiliones), and planarians (Turbellaria: Tricladida). Remarkable millipedes are species of the genus *Apfelbeckia* that are most probably subtroglophile or even trogloxene. At approximately 10 cm they are the largest European millipedes. During the hot and dry Mediterranean summers, they are found only at the entrances of caves, giving those rooms a sweetish scent because of their defense glands. Among snails, widely distributed are tiny species of *Zospeum* (Fig. 2E), a genus present also in the Alps and Pyrenees; their shells are about 2 mm high, ovoid, may be seen on wet cave walls or may be washed out of the earth from the cave bottom. In the southeastern parts of the Dinarides, bigger snail-shells may be seen, such as those of *Aegopis spelaeus* A.J. Wagner, which can be up to 20 mm in diameter. Terrestrial is also the small prosobranch *Pholeoteras euthrix* Sturany with a hairy shell. The pigmentless species of terrestrial planarians from Slovenia has not yet been described.

Like everywhere, bats are an important group of subtroglophile cave inhabitants. However, very large colonies hardly exist anymore—nowadays, a colony of 1000 specimens is a very high number for the area. One such year-round cave species is the long-fingered bat (*Myotis capaccinii* Bonaparte). Particularly bound to caves is the Schreibers' bat (*Miniopterus schreibersii* Kuhl). The most often seen species, but more solitary, are different horseshoe bats (*Rhinolophus* spp.). Bats are important vectors of food resources between the surface and the underground. Individuals of other mammal species may also be present deep in caves, including the rodent the edible dormouse (*Glis glis* Linne), which was first scientifically described just from the beech forests of the Dinarides, and some martens (*Martes foina* Erxleben). No birds enter the totally dark parts of Dinaric caves. However, the twilight areas near the entrances used to be common nesting places for large colonies of rock doves (*Columba livia* Gmelin), giving a number of caves the name Golobina, Golubinka, and so on (*golob* or *golub* is the Slovene or Croatian name for *Columba* spp.). Unfortunately, their populations have been heavily decimated or dispersed by hunters or by other human actions. On the other hand, groups of feral pigeons joined their wild relatives in some caves.

Aquatic Subterranean Fauna

As is generally true throughout the world, by far the prevailing group in Dinaric hypogean waters is the crustaceans (Sket, 1999). Peculiar, however, to the Dinarides is their richness in snails (Gastropoda); therefore, we discuss them first here.

With approximately 130 species, the Dinaric aquatic gastropods represent close to half of the world's known aquatic troglobiotic snail fauna. They are also a very common appearance in Dinaric cave waters and their empty shells may accumulate in more or less pure piles (thanatocenoses), even some meters long. Nearly all of them belong to the group Hydrobioidea whose formal subdivision is still actively changing. These are snails with a discoid, conical to nearly rod-shaped shell, with a simple or highly elaborated, ear-like mouth and a smooth to strongly ribbed whirl. The mostly 2- to 3-mm-high shells are approximately twice as big as those of their North American relatives, which is a difficult-to-explain curiosity. The shells of *Hadziella* species (from the northwest Dinarides) are flat; *Hauffenia* are mostly widely conical, *Iglica* are very slender (*iglica* in Slavonic languages means "small needle"); the *Lanzaia* (Fig. 2J) and *Plagigeyeria* spp. from the southeast exhibit very diversely ornamented, ribbed shells with elaborate mouths. A common member of pulmonate snails (Pulmonata: Ancylidae) is the generalist, genetically/cytologically diverse and therefore widely distributed species *Ancylus fluviatilis* O.F. Mueller with a cap-shaped shell without a spiral. It may also be present in caves and such a cave population might be clinally (gradually) troglomorphized: skin pigmentation gradually disappears and eyes become more and more reduced along the sinking stream in the Postojna-Planina Cave System. *Acroloxus tetensi* (Kuščer), with a similarly cap-shaped shell, is a troglobiotic species of Acroloxidae.

Among crustaceans, the most numerous are copepods and amphipods, with approximately 60 species each. Of amphipods (Amphipoda), not less than 45 species are of the genus *Niphargus* (Fig. 2D). This is the most diverse genus of the group, spread over almost all of Europe (and in Iran and the Arabian Peninsula in Asia). It inhabits all freshwater habitats, including forest ditches, brooks, caves, and interstitial waters. Specimens of different species may be 2 to 30 mm long and extremely diverse in their body shapes. However, all of these species are absolutely without eyes. The same variability span is achieved in the *Niphargus* species of the Dinarides. One of the smallest species, *N. transitivus* Sket, inhabits interstitial and karst waters in the extreme northwest. The slender and up to 20-mm-long *N. stygius* Schioedte, the type species of its genus, inhabits mainly percolation waters in Western Slovenia; the very stout and large *N. orcinus* Joseph and its relatives are spread throughout larger cave water bodies along the Dinaric karst. Particularly interesting is the long-legged *N. balcanicus* Absolon with a densely spiny back, from Herzcegovina; similar is *N. dolichopus* Fišer, Trontelj, and Sket with even longer pereopods, from Bosnia. The related *Niphargobates orophobata* Sket from the fissure system in Slovenia has the only close relative in the Greek island of Kriti (Crete). Another relative of *Niphargus*, *Carinurella paradoxa* Sket, which inhabits interstitial waters within the northwest Dinaric karst, is able to roll into a ball. The up to 25-mm-long and stout *Typhlogammarus mrazeki* Schaeferna (Fig. 2A) is an omnipotent raptor that can catch a shrimp swimming or climb up a vertical wall with a thin layer of water. The extremely fragile *Hadzia fragilis* S. Karaman has particularly numerous relatives in the Mediterranean and in the Caribbean. It is a mainly anchihaline species but a few populations are present in continental freshwaters.

Copepods (Copepoda) are represented by some generalist and a number of troglobiotic species, many coming from the genera *Diacyclops* and *Acanthocyclops*. Cyclopoids are particularly richly represented in larger bodies of water, where they may even be accompanied by the explicitly planctonic diaptomids (Calanoida: Diaptomidae); four Dinaric cave species is an enormous number for freshwater Calanoida. *Troglodiaptomus sketi* Petkovski exhibits a holodinaric distribution; the others are more endemic. Harpacticoids are particularly well represented in the percolation waters in fissure systems. Their genera *Elaphoidella* with 20 and *Parastenocaris* with 15 species are the leading groups, and numbers are still growing.

Isopoda are not numerous, but are very diversely represented. The asellids (family Asellidae) are represented by the troglomorphic races of *Asellus aquaticus* Linne and by some troglomorphic *Proasellus* species, such as *P. slovenicus* Sket in the northwest or the diversified species *P. anophthalmus* S. Karaman in the southeast. Very characteristic is the rich group of aquatic pill-bugs *Monolistra* (family Sphaeromatidae, Fig. 2C), some of them, for example, *M. spinosissima* Racovitza, with long spines on their backs. The largest isopods are cirolanids (Cirolanidae) *Sphaeromides*, distributed close to the

Adriatic coast but without any known ecological or historical connection to the sea. The water fleas (Cladocera), although with only three tiny (less than 0.5-mm-long) troglobiotic *Alona* species, are relatively richly represented. The generally interstitial Microparasellidae, such as *Microcharon* spp. or *Microparasellus* sp., are not numerous, but they also occur in caves or in gravel beds outside them.

Thermosbaenacea are represented by an anchihaline coastal species, *Monodella halophila* S. Karaman, and a mainly interstitial freshwater species *Limnosbaena finki* Meštrov and Lattinger. All troglobiotic decapod shrimps belong to the family Atyidae (Sket & Zakšek, 2009); they are very common in Dinaric caves and up to three species may occur in the same cave (in Vjetrenica). The most widely distributed species aggregate *Troglocaris (Troglocaris) anophthalmus* Kollar is holodinaric and split into a number of independent species. Members of subgenera *Troglocaridella* and *Spelaeocaris* are southeastern merodinaric; five species have been recognized in the latter.

Also worth mentioning are 13 epizoic to parasitic species of Turbellaria Temnocephalida, belonging to the family Scutariellidae and living mainly on atyid cave shrimp. Besides these troglobiotic species, only one epigean species is from Europe, again from the same, Dinaric, region.

Three species of true filter feeders should be mentioned, because these types of animals are very rare in caves. Besides some troglozene or eutroglophile species, the only troglobiotic freshwater sponge (Porifera: Spongillidae), *Eunapius subterraneus* Sket et Velikonja, inhabits Dinaric caves near Ogulin (Croatia); it is of a softer consistency than the surface species and forms only a few gemmulae. The only troglobiotic clam (Bivalvia: Dreissenidae), *Congeria kusceri* Bole, is similar to and related to the well-known zebra clam; it fastens itself to rocks by means of byssal threads. The only troglobiotic tube worm (Polychaeta: Serpulidae), *Marifugia cavatica* Absolon et Hrabě, fastens itself to rocks with its less than 1-mm-wide calcareous tubes. It is particularly interesting that generations of tubes may accumulate to build meter-thick tufa-like layers. The only troglobiotic cnidarian (Hydrozoa: Bougainvilliidae), *Velkovrhia enigmatica* Matjašič et Sket, is ecologically similar. A colonial species, it attaches to the substratum by means of stolons, with sessile medusoids still distinctly developed. All these putative species exhibit a holodinaric distribution.

The only troglobiotic vertebrate in Europe is the usually approximately 20-cm-long cave salamander *Proteus anguinus* Laurenti. Its distribution area is holodinaric with some localities in the small Italian part of the Dinaric karst and with the extreme southeastern reliable localities in southeast Herzegovina. At present, two subspecies are formally described, although this does not match its whole diversity. All proteus populations exhibit outer gills, only three toes on the first legs, and two on the hind legs. The holodinarically distributed troglomorphic race *P. a. anguinus* has a more or less colorless skin and an elongated head with reduced eyes, hidden below the skin. The nontroglomorphic *P. a. parkelj* Sket & Arntzen is limited to a tiny area in the extreme southeastern part of Slovenia. It is very darkly (sometimes black) pigmented, has normally developed eyes, and different body proportions. However, both races are obligate cave-dwellers and both come out of springs on some nights.

In the central and southern parts of the Dinaric karst, a number of endemic cyprinid fishes (Pisces: Cyprinidae) with small distribution areas are present. Most of them are known as regular periodical colonizers of caves (subtroglophiles), leaving the open waters of poljes before the retreat of the water and appearing again soon after a flood. They were observed to do both directions of migrations actively, not because they were directly forced to do so by hydrological events. In ancient times some species were economically important; special constructions were placed in caves to catch them in masses since they were a highly prized export article of the otherwise poor karst regions. A number of species used to be attributed to the genus *Phoxinellus*, earlier called *Paraphoxinus*. They exhibit different degrees of scale reduction; *P. alepidotus* Heckel possesses only a few scales along the lateral line. These fishes have been recently mainly displaced to *Telestes* and *Delminichthys*; taxonomically the most particular species is *Aulopyge huegelii* Heckel.

BIOGEOGRAPHICAL PATTERNS IN THE DINARIC UNDERGROUND

Widely Spread Taxa

Species or species groups may inhabit very different, wide areas within the Dinaric karst or—in a few cases—beyond it. The latter may be particularly instructive, possibly explaining to us the noncontiguous distribution areas of even highly specialized cave species. Such is the case with the isopod *Asellus aquaticus*, amphipod *Synurella ambulans* F. Mueller (Fig. 2G), and gastropod *Ancylus fluviatilis*. They are highly generalist species of approximately European distribution that penetrate underground particularly along sinking rivers. In the Dinaric karst, some cave populations of these species became troglomorphic and troglobiotic. Similar but less well known is the situation with some generalist cyclopoid copepods.

Transdinaric Distribution

Some groups of species inhabit most of the Dinaric karst, but occur also in areas to the east or to the west. Such is the shrimp genus *Troglocaris*; relatives of the handful of Dinaric species are present in Georgia (Sakartvelo) in the east. Dinaric species of *Sphaeromides* have close relatives in Bulgaria and eastern Serbia. For both groups, molecular analysis excluded the French species from being congeneric. The rich genus *Monolistra* covers the whole Dinarides and reaches far into the southern calcareous Alps.

Among terrestrial animals, transdinaric are the gastropods *Zospeum*, a genus reaching from the southeastern Dinarides, through the southern Alps, far into the Pyrenees.

Holodinaric Distribution

Also the scattered holodinaric distribution of some taxa may be a consequence of multiple (*i.e.*, polytopic and/or polychronous) immigration with a possible subsequent extinction of surface populations. In some cases, the immigration was followed by speciation; in fact, in all cases studied by DNA analysis, a split into distinct (biological) species could have been shown. In some other cases, the holodinaric area might be inhabited by a homogeneous species. The "holodinaric" area, starting with the northwestern border of the Dinaric karst, finishes with the southeasternmost Herzegovina. In all holodinaric species (or genera), the taxon is never contiguously spread over most of the area; it is instead split into groups of populations in hydrographically isolated karst areas. Such a distribution exhibits the most unique cavernicolous representatives of their higher groups: the amphibian *Proteus anguinus*, the clam *Congeria kusceri*, the tube worm *Marifugia cavatica*, and the cnidarian *Velkovrhia enigmatica*. Other cases are particular representatives of bigger groups, for example, one gastropod species, *Zospeum amoenum* Frauenfeld, the spider *Parastalita stygia* Joseph, the amphipod *Niphargus steueri* Schellenberg and some others.

Merodinaric Distribution Patterns

The northwestern merodinaric elements are, for example, the small beetle genus *Leptodirus*, the flat-shelled hydrobioid snails *Hadziella*, and the monolistrine (Isopoda) subgenus *Microlistra*. Their approximate counterparts in the southeastern merodinaric area are the beetle genus *Antroherpon* with numerous leptodiroid species, and hydrobioid snail genera *Lanzaia* and *Plagigeyeria*. No explanation for this bipolarity of the continental Dinaric cave fauna has been presented until now. One has to emphasize that both continental merodinaric areas, when combined, do not correspond to the holodinaric area. A paralittoral merodinaric distribution along the Adriatic coast is exhibited by the anchihaline amphipod *Hadzia fragilis* S. Karaman, the thermosbaenacean *Monodella halophila* S. Karaman, and in a big part of the belt also by the amphipod *Niphargus hebereri* Schellenberg. These predominantly anchihaline species are not present in the brackish subterranean waters of the Kvarner (Quarnero) Gulf, which was inundated by the sea only late in the Pleistocene; this shows us clearly that also the paralittoral distribution pattern has a historic—not purely ecological—background.

Smaller Distribution Areas

Most species and subspecies, but also some genera, exhibit smaller distribution areas within one of the merodinaric areas. The most interesting fact is that most of these areas do not follow the recent hydrological divides. One species (or subspecies) may either cross the borders of a divide or may be limited just to a part of the actual drainage area. This has been more properly studied for species of the isopod genus *Monolistra* and of some hydrobioid snails. It has been supposed that such distribution areas had been achieved in geologically past drainage areas (sometimes still prekarstic, on the surface) and maintained by competition between related species today. Most other small distribution areas are not specifically explainable; they may be either relics of a formerly wider area or results of locally limited immigrations underground. A number of animal species is known from only one locality, either a cave or a spring; the most striking is the amphipod *Niphargobates orophobata* Sket from just one jet of percolating water in Planinska Jama (Slovenia).

Endemism

Only a negligible number of obligate subterranean species cross the borders of the Dinaric region. Such is the case of some amphipods, snails, and beetles crossing the slightly indistinct border between the Dinaric and South Alpine regions, and such is the case of some copepod species, which may even exhibit wider distribution areas. On the other hand, also comparatively high is the endemism between regions within the Dinarides. At the moment, the only distribution data for administrative regions (states) that are known definitely do not reflect the natural biogeographical units. Nevertheless, the number of endemic species (for countries) is mostly 40–60%, sometimes even much higher. Consideration of nominal subspecies would increase the endemicity remarkably as has been studied in some large genera, while even a number of endemic genera exist, scattered to different taxonomic groupings; they are particularly numerous in the cholevid beetles. Only 5 of the 44 present cholevid genera are not endemic for the Dinaric karst and at least 23 genera seem to be endemic for one of the states (Slovenia, Croatia, or Bosnia and Herzegovina). Because political borders are crossing regions with homogeneous faunistic assemblages, the endemism degree would certainly be much higher in natural biogeographic provinces.

HOTSPOTS WITHIN THE HOTSPOT

In this highly biotically diverse area, some cave systems have particularly rich cave faunas. When searching for caves with more than 20 obligate cave inhabitants, 20 such caves or cave systems were traced and as many as 6 of them were in the Dinaric karst (Culver and Sket, 2000). The Postojna-Planina Cave System (PPCS) is, with 99 such species, the richest among them. The system consists of 17 km and 6 km of passages connected by 3 km of flooded corridors not yet mastered by cave divers. The dry parts of the Postojnska Jama is one of the oldest (since 1818) and most famous tourist caves in the world, with a small railway in it; some parts of the system are used extensively by tourists, but there are still many "wild" parts that are rich in fauna. This is the type locality for a number of "first cave" animals, including the first described troglobiont, the beetle *Leptodirus hochenwartii*, the first cave spider *Stalita taenaria*, and many others. Altogether nearly 60 animal species or subspecies were first found and described from this system, and the European cave salamander, *Proteus anguinus*, was first seen in its natural habitat in Črna Jama in 1797. It is also a site of long-term ecological studies. Dry parts of the system are inhabited by 37 species, 9 of which are beetles. There are 62 aquatic troglobiotic species; particularly numerous are crustaceans, snails, and oligochaetes. The main artery of the system is the sinking river Pivka, which sinks at the Postojnska Jama and resurges from the Planinska Jama; there is approximately 10 km of underground bed in-between. A mixed fauna of troglobiotic, eutroglophile, and trogloxene species inhabits it. The last ones are mainly insects; aquatic larvae of some of them are present all along the course. Populations of some eutroglophile species exhibit a clinaly (gradually) increased troglomorphy along the stream. These caves are in Slovenia, which is the richest in the world for aquatic subterranean fauna. The next four richest caves are also in Slovenia: Križna Jama, Logarček, Šica-Krka System, and Grad.

In Herzegovina, which seems to be the richest area in the world for terrestrial cave fauna, evidence is growing that a number of additional caves will be catapulted to the status of the richest ones. Until now, only Vjetrenica in Popovo Polje seems to reach the PPCS in number of troglobiotic species. This complex cave system has 6 km of passages, which include a number of small streams, pools, and trickles of water. The surface animals (trogloxenes) are very few here. There is no sinking stream, but water jets after rains import organic debris through crevices and shafts by which the rich fauna are being fed. Among troglobionts there are 10 species of amphipods, some of which are very large, and three species of decapod shrimps. There are approximately 10 species of beetles. Particularly noteworthy are the amphibious catopid beetle *Hadesia vasiceki* J. Mueller and the amphipod *Typhlogammarus mrazeki* Schaeferna, which occupy the cave hygropetric (described below).

SPECIAL ASSEMBLAGES

In an area with such a high number of faunal elements and, therefore, comparatively high pressure for interspecific competition, one can expect further ecological specialization of species. Therefore, besides a number of more or less "trivial" communities and synusiae, composed of species that sometimes merely by chance are locally combined, a small number of characteristic groupings with particular ecological demands or abilities exist.

Ice caves are scattered all over the area. They are usually sack-shaped and surrounded by forest in moderately higher elevations and bear ice deposits during the whole year. Particularly wet soil with close to zero temperature is the ecological characteristic of the proximity of such a subterranean glacier in summer. This environment is usually particularly rich in troglobiotic and troglophilic (e.g., *Nebria* spp.) Coleoptera. However, the highly troglomorphic leptodirine beetles of the genus *Astagobius* seem to have specialized to such an environment.

The *cave hygropetric* is a rocky wall, usually sinter, overflown by a thin layer of slipping water. A number of leptodirine beetles seem to be specialized for climbing in such a semiaquatic environment and collecting organic particles brought by water. All of them—although not closely related—have a pholeuonoid body shape, with very strong claws, and with particularly hairy mouth apparatus. The best-known representative of such beetles is *Hadesia vasiceki*. These obligate inhabitants may join some nonspecialized semiterrestrial or semiaquatic guests using such environments as hunting grounds or simply as distribution paths.

A number of aquatic and probably some terrestrial species are nearly limited to the percolation-water-filled *crevice systems*, which include the *epikarst*; however, such species are primarily found in percolation-water-fed rimstone pools or puddles on cave walls or bottoms. Morphologically adapted to such narrow places seem to be the harpacticoid copepods of the genera *Parastenocaris* and *Elaphoidella* in particular, while the tiny amphipod *Niphargobates* exhibits strong and curved legs and claws appropriate for climbing in less narrow crevices.

Thermal waters are a trophically particularly inhospitable environment. They force higher metabolism intensities and, because they are purified as they make their way through greater earth depths, they offer particularly little food and little oxygen. In hypothermal waters of 15–28°C on the northwestern edges of the Dinaric karst, some stenasellid isopods, such as the thermophile race of *Protelsonia hungarica* Mehely and tiny gastropods *Hadziella thermalis* Bole, are present. In Slovenia, close to the Pleistocene Alpine glacier, stenasellids were found only in such an environment. It seems, in fact, that biota are limited to trophically less inhospitable springs, because the filtration of huge quantities of water from depths gave no results.

Sinking streams are streams that, after some course on the surface, flow underground. The subterranean bed of a sinking stream is, in fact, an ecotone environment. Ecological conditions in them are more fluctuating, are richer in food resources, and more diverse than in "autogenous" waters. Some surface species may penetrate along such streams for shorter or longer stretches underground. Among them are adaptation-prone eutroglophile species such as *Asellus aquaticus*. Along such a subterranean bed, a rich assemblage of selected surface and subterranean animals is formed; the fauna changes its quantitative and qualitative composition with gradual ecological changes leaving the sinking point and the surface influences. Although some troglobionts, for example, *Troglocaris* shrimp and even *Proteus* and some *Niphargus* spp., find such a rich environment particularly inviting, others strictly avoid it. On the other hand, this is the only cave environment where a number of insect larvae (of Ephemeroptera, Plecoptera, and Trichoptera) may be present.

POLLUTION AND PROTECTION

A number of threats face the Dinaric cave fauna. In the nineteenth and early twentieth centuries, some animals had been caught in large numbers for trading purposes. The amphibian *Proteus anguinus* was reportedly a popular pet in parlor aquaria, while the rare cave beetles won high prices among amateur entomologist-collectors. Some dealers organized genuine chains of collectors all along the Dinaric karst. Because they were using baited traps they were potentially able to affect some populations.

In the decades of economic development of the traditionally poor and economically passive Dinaric karst after the Second World War, a number of hydrotechnical projects have been accomplished. Surface and underground dams and artificial tunnels changed the hydrology of some parts of the Dinaric karst remarkably. We may expect that some previously isolated population-races might come in contact and fuse again or at least that some populations were "polluted" by repeated input of foreign genes. New competition assemblages may have formed, threatening the existence of some species. The hydrological regime of the underground in wide surroundings of those constructions definitely changed, also changing the living conditions (and sometimes threatening the existence) of cave populations and species. For example, the live deposits of the *Marifugia*-tufa with its builder and many interstitial guests is now dead, due to the lack of regular long-lasting floods in Popovo Polje and in the sink-cave Crnulja. This unique phenomenon has been destroyed. Such is also the destiny of the massive colony of the clam *Congeria kusceri* in Žira Jama.

Of course, the most serious threat to the cave biota is the omnipresent and diverse pollution. The net-shaped underground hydrological connections make the direction of pollutant outflows unpredictable and their effect very wide-reaching. In the case of the sinking river Pivka, a moderate amount of organic pollution may enable surface animals to increase their success in competitive situations to the detriment of troglobionts. In this way, pollution may indirectly extinguish cave faunas. However, spills of generally poisonous materials are not a rare event in the karst and only the fact that such spills are also detrimental to the precious resource of potable water allows us to efficiently protect the subterranean fauna.

However, as early as 1920, the Section for Nature and Natural Monuments Conservation of the Slovene Museum Society wrote and published a memorandum that outlined a very complete and detailed nature conservation plan. Beside the Alpine sites and biota, one of its main subjects was specifically karst caves as a whole as well as the particularly interesting cave fauna. A paragraph about a total ban on the commercial exploitation of cave fauna was included, saying that it should only be exploited for scientific and educational purposes. This was—very generally—realized in 1922. This early attempt at nature protection, initiated in a high degree by cave fauna, was followed by a number of species protection acts and other environmental legislation in all Dinaric countries. So, the legal background for the many-sided protection of this hotspot exists—but so do many practical hindrances for its practical implementation.

See Also the Following Articles

Dinaric Karst: Geography and Geology
Diversity Patterns in Australia
Diversity Patterns in Europe
Diversity Patterns in the Tropics
Diversity Patterns in the United States

Bibliography

Culver, D. C., & Sket, B. (2000). Hotspots of subterranean biodiversity in caves and wells. *Journal of Cave & Karst Studies, 62*, 11–17.
Hamann, O. (1896). *Europäische Höhlenfauna*. Jena: Hermann Costenoble.
Sket, B. (1997). Biotic diversity of the Dinaric karst, particularly in Slovenia: History of its richness, destruction, and protection. In I. D. Sasowsky, D. W. Fong & E. L. White (Eds.), *Conservation & Protection of the Biota of Karst* (pp. 84–98). Charles Town, WV: Karst Waters Institute (Special Publication 3).
Sket, B. (1999). High biodiversity in hypogean waters and its endangerment—The situation in Slovenia, Dinaric karst, and Europe. *Crustaceana, 72*(8), 767–779.
Sket, B., Paragamian, K., & Trontelj, P. (2004). A census of the obligate subterranean fauna in the Balkan Peninsula. In H. I. Griffiths & B. Krystufek (Eds.), *Balkan Biodiversity. Pattern and Process in Europe's Biodiversity Hotspot* (pp. 309–322). Dordrecht, The Netherlands: Kluwer Academic Publishers.
Sket, B., & Zakšek, V. (2009). European cave shrimp species (Decapoda: Caridea: Atyidae), redefined after a phylogenetic study; redefinition of some taxa, a new genus and four new *Troglocaris* species. *Zoological Journal Linnean Society, 155*(4), 786–818.
Valvasor, J. W. (1689). *Die Ehre des Herzogthums Crain*. Nuremberg, Germany: W. M. Endtner.

DIVERSITY PATTERNS IN THE TROPICS

Louis Deharveng and Anne Bedos

Museum National d'Histoire Naturelle de Paris

The intertropical belt has an ecological definition that roughly matches its geodesic limits (*i.e.*, between 23°27′N, the tropic of Cancer, and 23°27′S, the tropic of Capricorn; Fig. 1). Tropical climate, ranging from lowlands to the nival belt, can be characterized by a low thermal amplitude across seasons and usually high humidity, though deserts are also included in the belt (parts of Saudi Arabia, southern Africa, Sahara, Australia, and India). These climatic characteristics of humid tropics are also those of cave microclimate, aside from darkness. Caves are actually largely azonal habitats, with a very similar environment across all latitudes and all macroclimates. Beyond local variations, temperature is the most basic physical factor distinguishing abiotic environments in tropical versus nontropical caves.

HISTORICAL CONTEXT

As early as 1914, Jeannel and Racovitza claimed that troglobionts, though rare, were well-diversified in the tropics. This statement was overlooked by other authors for almost 60 years. Until the 1970s, authors who worked on the cave fauna repeatedly stressed that true subterranean animals were rare or absent in the tropics (Annandale et al. 1913; Leleup, 1956). Why, they argued, would animals develop adaptations to an environment that is similar to that outside the caves in its weak thermal amplitude and permanent high humidity? In the same vein, it was speculated that higher energy input in tropical caves would have lessened the "selection pressures for energy-economizing troglobite adaptations" (Mitchell, 1969), accounting for the scarcity of some troglomorphic traits among cave invertebrates of the tropics.

The idea that tropical troglobionts are scarce and mostly limited to aquatic species was based on surveys of a few tropical caves, mostly from Malaysia (Batu Caves), Burma (Farm Caves), and Congo. The rediscovery that a rich troglobiotic community including highly troglomorphic species might exist in the tropics was provided by Howarth (1973) based on a study of the fauna of Hawaiian lava tubes. Since then, the presence of troglobionts has been confirmed in other parts of the tropics, in volcanic as well as limestone caves (Chapman, 1980; Humphreys, 2000; Gnaspini and Trajano, 1994; Deharveng and Bedos, 2000; Deharveng et al. 2009).

CURRENT STATE OF KNOWLEDGE

Cave-restricted species (troglobionts and guanobionts) continue to be discovered in virtually every newly surveyed tropical karst, along with a profusion of troglophilic and guanophilic forms. This rapid progress should not hide the fact that most tropical karsts have not yet been biologically investigated. Enormous geographic (Figs. 1,2) and taxonomic gaps still hamper our global understanding of tropical cave biodiversity. Unevenness in taxonomic coverage is illustrated, for instance, by the Mulu cave fauna in Sarawak, which probably includes the largest number of named troglobionts in the tropics, but which has barely been surveyed for aquatic and terrestrial microarthropods. Another striking example is the giant rhaphidophorid crickets of Southeast Asia which, in spite of being the

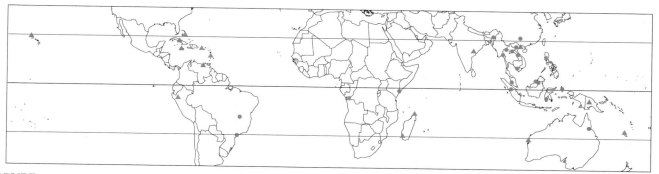

FIGURE 1 Main documented spots of tropical subterranean biodiversity, anchihaline and interstitial habitats excluded. Cave systems cited in Tables 1 and 2 shown as circles, others as triangles; in orange, more than 10 troglobiotic species; in blue, less than 10 troglobiotic species. *After Juberthie and Decu (1994–2001), updated.*

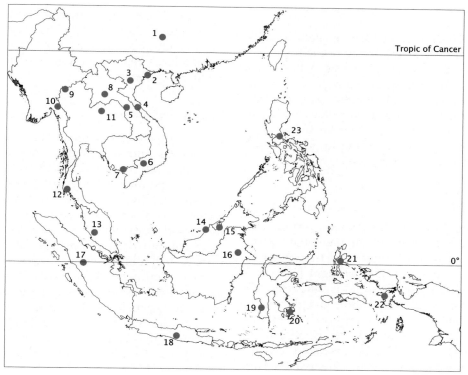

FIGURE 2 Main records of obligate subterranean species in Southeast Asia (*after Deharveng and Bedos, 2000, updated*). Red spots: best studied caves or karsts, with more than four obligate subterranean species. 1: Masan-dong (Mulun karst); 2: Halong Bay Caves; 3: Pu Luong karst; 4: Ke Bang karst; 5: Tham Thon (Khammouane karst); 6: Tan Phu lava tubes; 7: Hon Chong–Kompong Trach karst; 8: Vang Vieng; 9: Tham Chiang Dao; 10: Farm Caves; 11: Tham Phulu; 12: Phang Nga Caves; 13: Batu Caves; 14: Niah Cave; 15: Air Jernih system (Mulu karst); 16: Baai system (Sangkulirang karsts); 17: Sangki system (Gunung Seribu); 18: Gunung Sewu karst; 19: Gua Salukkan Kallang (Maros karst); 20: Muna karst; 21: Batu Lubang (Halmahera); 22: Fakfak karst; 23: Montalban.

most important invertebrates in food webs of tropical caves (Deharveng and Bedos 2000), remain undescribed in most cases. More generally, and in contrast with temperate cave fauna, the majority of tropical species listed in the literature are undescribed and their ecological status is often uncertain, due to the scarcity of faunistic surveys in surface habitats. Moreover, although the sampling of cave faunas is enough to derive rough biogeographic patterns, it is usually insufficient for the detection of evolutionary radiations among tropical cave groups. Each time fine-grain studies have been carried out, however, such radiations associated with narrow endemism among species have been recognized. If confirmed to be general, such a pattern would lead to a considerable increase in estimates of species richness in tropical caves. In all these aspects, we are obviously at the dawn of the biological exploration of tropical cave biodiversity.

SUBTERRANEAN HABITATS

A subterranean fauna is encountered in any kind of consolidated or unconsolidated rocks, as long as obscure, climatically stable, interconnected, and humid voids are available, together with minimal food input from the surface. The combination of void size available for fauna—small versus large—and their contents—air, freshwater, or salty water—allows the recognition of the basic types of subterranean habitats. In addition, terrestrial cave habitats can be further split into low-energy and high-energy, based on the nature and amount of food resources.

Traditionally, terrestrial interstitial habitats, that is, the deep soil—the richest compartment of the hypogean domain—is not dealt with by subterranean biologists. All other habitats (terrestrial low-energy and high-energy cave, aquatic interstitial, aquatic cave, marine/anchihaline interstitial, marine/anchihaline cave) are relevant to subterranean studies. They host their own fauna, which has diversified according to patterns that differ in temperate and tropical regions.

Aquatic Interstitial

Aquatic interstitial communities may be very diverse and are richer in species than the noninterstitial aquatic fauna of caves. Of particular interest is the epikarstic component of this fauna, which is brought into caves by dripping water from stalactites and is thus relatively easy to study in terms of composition and dynamics. The interstitial fauna is mostly composed of minute microcrustaceans, with copepods being dominant. In the tropics, however, most of our knowledge on this fauna comes from nonkarstic, unconsolidated lowland terrains, and the caves most intensively surveyed for subterranean invertebrates have been at most barely surveyed for their interstitial fauna.

Anchihaline and Marine Caves

Littoral karsts of the tropics and subtropics often host a characteristic and rich fauna of anchihaline (or anchialine) and marine cave species, which is much more diversified than in temperate regions (Iliffe 2000). When present, anchihaline/marine species may locally account for the largest part of subterranean biodiversity, especially for oceanic islands. At the extreme, the Bahamas probably do not have a single terrestrial troglobiont, whereas more than 45 anchihaline stygobionts are listed from its submarine passages, including the majority of the 24 known species of Remipedia. Globally, however, the anchihaline fauna does not represent the most important component of tropical subterranean biodiversity, because of its narrow geographic range. Many favorable coastal karsts of Southeast Asia (Phang Nga Bay in Thailand, Halong Bay in Vietnam), have no or very few anchihaline/marine species described, whereas their freshwater cave fauna is rich. So far, the anchihaline fauna has been mostly sampled in neotropical islands and, more recently, in Cape Range (Australia), but it remains poorly investigated in Southeast Asia, the Pacific, and Africa. Several taxa linked to anchihaline/marine caves of the tropics are remarkable by their phyletic isolation (Remipedia, Mictacea), or affinities with deep-sea species (*e.g.*, *Thetispelecaris remex*, from the Bahamas, in the order Bochusacea). Other anchihaline stygobionts of the tropics include fish and a number of Crustacea, of which several have representatives both in anchihaline and freshwater habitats.

Cave Freshwater

Cave freshwater habitats typically host large-sized stygobionts and stygophiles. Microcrustaceans, which are typical of interstitial habitats, are also represented, either as permanent populations or brought in from the epikarst by percolating water. The largest freshwater stygobionts have been well studied in several tropical regions, such as Cuba, Thailand, and South Sulawesi, but data are lacking for most karsts of the tropics, including Central America and several large areas of Southeast Asia. In the tropics, this fauna is often rather elusive, but may be found locally as high-diversity assemblages. As a rule, phreatic water in cave passages hosts the richest communities of stygobionts, in contrast to cave puddles and streamlets (even endogenous) that are often devoid of large stygobionts. The richest stygobiotic fauna of continental Southeast Asia is found in the food-poor phreatic waters of Tham Phulu (Thailand), which host no less than six highly troglomorphic species that can co-occur in a single puddle: *Dugesia deharvengi*, *Heterochaetella glandularis*, *Theosbaena cambodjana*, *Aequigidiella aquilifera*, *Stenasellus rigali*, and *Siamoporus deharvengi*).

Aquatic habitats rarely show the striking contrast between high- and low-energy habitats observed in tropical terrestrial habitats, because guano that falls in running water is readily dispersed. When present, however, guano-rich puddles may host a few stygobionts, but there is no evidence that any are restricted to this habitat.

A characteristic feature of aquatic subterranean habitats in temperate regions is the overwhelming dominance of Crustacea and low representation of insects: the same is seen in the freshwater of tropical caves. However, the dominant taxonomic groups are different. In particular, stygobiotic crabs are totally absent in temperate caves, whereas they are frequent and diversified in the tropics (Guinot 1994).

Tropical freshwater stygobionts are often related to anchihaline or marine taxa. The recently described, highly troglomorphic freshwater hymenosomatid crabs of Kalimantan and Sulawesi (Fig. 3) belong to an essentially marine family, and the same pattern is seen in many tropical caves Caridea. In other cases, such as *Stenasellus* isopods, the marine origin is more remote and all members of the family are freshwater stygobionts, ranging from Southeast Asia to Africa and Europe. At least a few species of oniscid isopods (*Thailandoniscus annae* from Thailand and several undescribed *Sinoniscus* from southern China) and several freshwater beetles are secondary stygobionts, having colonized freshwater from terrestrial ancestors.

Oligotrophic Terrestrial Habitats and Troglobionts

Food-poor habitats are extensive in tropical caves. These include all deep cave passages that are free of bat and bird colonies, and out of reach of flooding by exogenous streams. As in temperate regions, these habitats are typically home to troglobiotic invertebrates, with a low representation of troglophilic species and an absence of guanobionts. Food resources consist of scattered feces of bats, birds, and crickets; roots; organic matter brought down by percolating water; and old debris from exceptional flooding episodes.

Cave-restricted troglobionts, though usually rare, are better known than other categories of tropical cave fauna because their unusual morphology, odd biology, frequent relictual status, and rarity attract the interest of biologists.

The tropical faunas of troglobionts exhibit striking local variations in their taxonomic composition, but they also share common characteristics across the whole intertropical belt. Local variations concern all groups, in particular the dominant ones: millipedes, woodlice, springtails, and crickets. Haplodesmidae millipedes (Fig. 4) are, for instance, highly diversified and often abundant in Southeast Asia caves, but absent from tropical America and Africa caves, where no millipede radiation has been recorded so far. Crickets, which often constitute the basis of invertebrate food webs in the tropics, exhibit a clear-cut distribution pattern: Rhaphidophoridae (Gryllacridoidea) in Southeast Asia and Phalangopsidae (Grylloidea) in Africa and tropical America. Patterns in troglobiotic springtails are also sharply contrasted, but at a finer level; in Southeast Asia the dominant troglobiotic groups are distributed as a complex mosaic: *Sinella* (Entomobryidae) in southern China and northeastern Thailand, *Lepidonella* (Paronellidae) in Vietnam, *Troglopedetes* (Paronellidae) in western Thailand, *Cyphoderopsis* (Paronellidae) from southern Thailand to Java, and *Pseudosinella* (Entomobryidae) in Sulawesi and Papua. Needless to say, the cause of these complex biogeographical patterns is unknown.

On the other hand, several groups have troglobiotic representatives across large parts of the tropics, whereas they are absent or very rare in temperate caves. Examples include scorpions and various poorly studied spider families among the predators, nocticolid cockroaches (Fig. 5) and paronellid Collembola among the decomposers, or fulguromorph Hemiptera among the root-feeders. The absence or rarity of groups that are highly diversified in northern temperate regions is a further characteristic of

FIGURE 3 *Sulaplax ensifer* Naruse, Ng, and Guinot, 2008 from a cave of Muna island (Sulawesi). *Photo courtesy of Tohru Naruse. Used with permission.*

FIGURE 4 Radiation of the millipede genus *Eutrichodesmus*, illustrated by four recently described species from Vietnamese caves (Golovatch et al., 2009): (A) *Eutrichodesmus regularis*; (B) *E. armatocaudatus*; (C) *E. asteroides*; (D) *E. aster*. Photos by Louis Deharveng.

FIGURE 5 Nocticolidae cockroach from Saripa Cave in Maros (Sulawesi). *Photo courtesy of Jean-Yves Rasplus. Used with permission.*

the tropical troglobiotic fauna, which is shared with southern temperate faunas. Examples of such groups are Onychiuridae and Hypogastruridae among springtails, and several groups of beetles.

At a finer scale, the level of geographic diversification of troglobiotic taxa across fragmented karst landscapes is of prime importance for biodiversity evaluation and conservation. No study has directly addressed this issue for the tropics, but indirect information is available from various taxonomic and faunistic studies concerning the Hawaiian cave fauna as well as radiations of Trechinae beetles, Telemidae spiders, Haplodesmidae and Glomeridae millipedes in southern China, and those of *Troglopedetes* springtails in western Thailand. All these data indicate that levels of endemism are probably higher in tropical than in temperate caves. Besides high levels of endemicity, two characteristics of troglobiotic taxa interest biologists: their often highly modified morphology (troglomorphy) and their frequent phyletic isolation (relictness). Both features are discussed separately below.

Guano and Guanobionts

A considerable proportion of cave terrestrial communities in the tropics are dependent on guano, produced by bats and locally by swiftlets (Fig. 6) or other birds. Troglobionts exploit scattered bat or swiftlet droppings, or even small guano patches, but they avoid guano accumulations, which are inhabited by guanophiles or guanobionts, often referred to as *troglophiles* on a morphological basis. Guanobionts complete their whole life cycle in or around piles of guano, and should be rather considered as a category of troglobionts as suggested by Chapman (1986) and Gnaspini (this book), even if they usually lack troglomorphic characters.

The arthropods foraging on and around guano piles are diverse in terms of their size and abundance. Giant arthropods (tailless whip-scorpions, Sparassidae spiders (Fig. 7), scutigerid centipedes, crickets, cockroaches) are the most conspicuous and, though not strictly linked to guano, they are frequent on the walls of most high-energy caves in the tropics. The guano microarthropods, comprised of a multitude of medium to small species (mites, springtails, millipedes, beetles, flies, and moth larvae), are present as huge populations foraging on and in guano piles, where guanobionts and more opportunistic species are mixed. The guano fauna includes both widespread species (often pantropical, *e.g.*, the springtail *Xenylla yucatana*, or even cosmopolitan, *e.g.*, the cockroach *Periplaneta americana*) and species with more restricted ranges. The latter category might be more important than previously thought, as shown by the recent discovery of a huge radiation of typically guanobiotic Cambalopsidae millipedes in Southeast Asia, each karst having one or two endemic species (Fig. 8).

Though several of the dominant groups in guano-related habitats are known across the whole tropics (Gnaspini, this book), regional differences exist for others. In tropical Asia for instance, the dominant medium-size arthropods on guano piles are Cambalopsidae millipedes, and giant Sparassidae spiders (Fig. 7) are frequent on cave walls of food-rich caves, whereas both taxa are absent from the guano caves of Africa and tropical America.

Troglophilic and Stygophilic Species

The most abundant and most conspicuous animals of tropical caves are species that are not strictly linked to subterranean environments but spend part of their life in caves, either seasonally or daily (subtroglophiles), or that have both cave and non-cave populations (eutroglophiles). These troglophiles are present in all the habitats listed above. Many taxa cited in the *Encyclopaedia Biospeologica* (Juberthie and Decu, 1994–2001) for tropical countries are troglophiles, accounting for the largest part of the fauna encountered in caves, far ahead of troglobionts.

First among troglophiles are bats (often classified as trogloxenes), with many species restricted to the tropics. Tropical caves commonly host very large bat colonies; two million individuals of *Tadarida plicata* live, for instance, in Khao Chong Pran Cave in Thailand. However, most bat species live as scattered individuals or small groups in caves. Taxonomic and functional diversity of cave bats are much higher in the tropics than in temperate regions. In addition to the classical insectivorous bats, several large fruit-bats

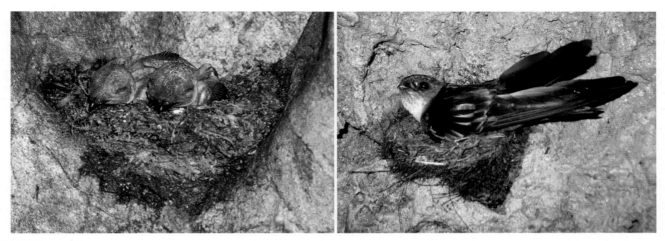

FIGURE 6 Swiftlets. Left: young, Ngalau Indah near Lintau Buo (Sumatra) (*Photograph by Louis Deharveng*). Right: female in its nest, Santo (Vanuatu). *Photo courtesy of Bernard Lips. Used with permission.*

FIGURE 7 *Heteropoda* sp. eating a giant cave cockroach *Miroblatta baai* Grandcolas (Baai Cave, Kalimantan). *Photo by the authors.*

often roost in tropical caves; blood-sucking and fish-eating species are even known in caves of tropical America. Morphological diversity is also striking. Among others, tropical bats include one of the smallest mammals in the world (*Craseonycteris thonglongyai*, from Thailand) and the strange naked bat *Cheiromeles torquatus*, from Sunda islands.

Bats are not the only troglophilic flying vertebrates. Several birds are linked to caves in the tropics. The most regular and conspicuous ones are different species of swiftlets in Southeast Asia, and the oilbird (*Steatornis caripensis*) in tropical America. Like bats, swiftlets (Fig. 6) may live in large colonies, which are often managed and exploited for their nests by local villagers in a remarkably sustainable way. Non-flying troglophiles are much rarer among vertebrates, the most conspicuous and frequent in Southeast Asia being the bat-eating snake Orthriophis taeniurus ridleyi.

Bats and birds can provide the basis of cave food webs, as they import, through their feces, most of the food resources exploited by terrestrial and, to a lesser extent, aquatic cave invertebrates. Like other cavernicoles, troglophilic species exploit this widespread resource, as well as debris brought into caves by flooding. Several of the largest species of crickets, tailless whip-scorpions, and giant Sparassidae spiders are known (or assumed) to leave caves every night for feeding, but this is certainly not the case for populations living in deep recesses of the caves. Actually, in spite of the widespread occurrence and conspicuousness of these troglophiles, very little is known about their biology. Regarding seasonal subtroglophiles, information is practically nonexistent. Small-sized species encountered in caves rather qualify as eutroglophiles; they are often abundant on cave guano or organic debris, as well as in surface soils. They include numerous springtails and mites, various beetles and bugs, but also larger species such as millipedes, spiders, woodlice, cockroaches, and ants.

Stygophilic species are also numerous in tropical caves, including fish and various invertebrates. Large-sized and big-eyed *Macrobrachium* (Palaemonidae) are the most conspicuous aquatic animals in many caves of Southeast Asia; they are common even far inside caves, usually excluding stygobionts from the habitats they occupy.

BROAD-SCALE PATTERNS OF TROGLOBIONT SPECIES RICHNESS

Geographical Patterns of Species Richness

The uneven sampling of tropical caves and biased taxonomic coverage of their fauna result in a heterogeneous data set, from which some general patterns of species richness can however already be drawn

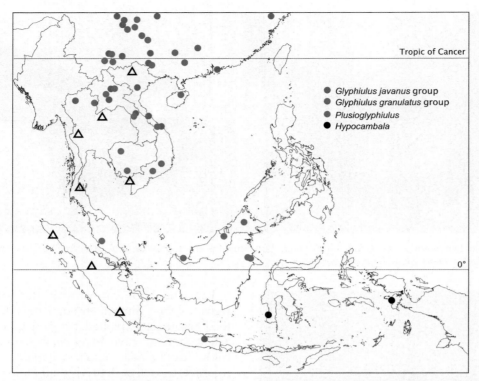

FIGURE 8 The distribution of Cambalopsidae millipedes in Southeast Asia (synthetized from the 2007–2009 works of Golovatch and collaborators). Each dot represents a locality (usually a guano cave, rarely a surface habitat locality) with one or two endemic species. Distribution data for the tramp species *Glyphiulus granulatus* are not displayed on the map. Empty triangles correspond to unidentified species.

1. Richness in troglobionts or guanobionts is rarely very low in the tropics, and never null as in temperate regions affected by glaciations. However, it varies considerably from place to place, and allows the recognition of both hot- and cold-spots of species richness (Table 1).
2. The terrestrial fauna appears to be more diverse in the Oriental and Australian regions than in the Neotropics or Africa. This is obvious from the list of cave species given by country in the *Encyclopaedia Biospeologica* (Juberthie and Decu, 1994–2001), and from updated data for the richest tropical caves, summarized in Table 1. Paucity in troglobionts is observed in many countries of tropical America, with the possible exceptions of highlands (Guatemala, Venezuela) and some subtropical regions of Mexico (Reddell 1981). The pattern is different for stygobionts, with comparatively higher richness in the caves of tropical America, Africa, and Australia than in Southeast Asia. Uneven sampling effort might, however, explain much of the difference, as illustrated by recent findings such as remarkable stygobiotic Hymenosomatidae crabs of Kalimantan and Sulawesi (Fig. 3).
3. Locally, large karst blocks tend to be richer in troglobionts than smaller ones, having a statistically wider diversity of habitats. In Sarawak, the Air Jernih system in the Mulu karst has significantly more cave-restricted species than the Niah Cave in the small isolated hill of Gunung Subis (Table 1). However, this does not necessarily apply to guanobionts, as illustrated by their high diversity in caves developed in small limestone outcrops such as the Batu Caves or Niah Cave in Malaysia.
4. "Most diversity in caves is expressed regionally rather than locally" (Culver and Sket, 2000). This low alpha- versus high beta-diversity reflects the high degree of micro-endemism among subterranean groups. The available evidence suggests that the impact of fragmentation on speciation might be stronger in tropical than in temperate karsts. A good illustration is the radiation of the guanobiotic Cambalopsidae millipedes that has been recently found in Southeast Asia caves (Fig. 8) with more than 50 species. The absence of MSS ("Milieu Souterrain Superficiel") in the tropics, except in some volcanic terrains, certainly contributes to accelerating diversification by limiting gene flow between populations of different karst units.
5. The richest hotspot of the tropical subterranean fauna today is the Maros karst in South Sulawesi, with 28 troglobionts and stygobionts in the

TABLE 1 Biodiversity and Faunistic Composition of Some Well-Studied Tropical Caves

		SKT System (SUL)	Air Jernih System (MAL)	Sangki System (SUM)	Ma San Dong (CHI)	Tham Chiang Dao (THA)	Batu Lubang (HAL)	Areias System (BRA)	Tham Non (LAO)	Tham Thon (LAO)	Hang Mo So (VIET)	Hang Nuoc (VIET)	Kulumuzi Cave (TAN)	Siju Caves (ASS)	Batu Caves (MAL)	Olhos d'Agua Cave (BRA)	Farm Caves (MYA)	Niah Great Cave (SAR)
Mollusca																		
Hirudinea				1									1					
Arachnida	Acari	1		1	1	1	1		1	1								
	Araneae	3	5	2	1	2	4		1	2	3		1					
	Opilionida	1	1	1	1	1	1		2	1	1	1	1	1		1		
	Palpigradida	1				1												
	Pseudoscorpionida	1		1		1		1			1		1			1		
	Schizomida	1		1		1				1			1	1			1	
	Scorpionida		1															
Myriapoda	Chilopoda												1					
	Diplopoda Glomeridesmida		1															
	Diplopoda Glomerida		1		1				1	1								
	Diplopoda Julida	1							1					1	1		1	
	Diplopoda Spirostreptida			1	1						1							
	Diplopoda Polydesmida	2	2	2	2	1	2	3	2	2	1				1	1		
	Diplopoda Stemmiulida						1			1								
	Diplopoda Callipodida											1						
Crustacea	Isopoda Oniscidea	3	2	3	2	2	3	2	2	1	1	3	1	2				
Collembola	Entomobryomorpha	2	1	2	2	2	2	3	1	2	2	3		1	1	1	2	
	Poduromorpha	1								2								
	Symphypleona	1		1		1		1	1			2						
Diplura	Campodeidae	1			1				1		1				1			
Zygentoma	Nicoletiidae									1								
Insecta	Blattodea	1	1	1		1	1			1	1		1					

(Continued)

TABLE 1 (Continued)

	SKT System (SUL)	Air Jernih System (MAL)	Sangki System (SUM)	Ma San Dong (CHI)	Tham Chiang Dao (THA)	Batu Lubang (HAL)	Areias System (BRA)	Tham Non (LAO)	Tham Thon (LAO)	Hang Mo So (VIET)	Hang Nuoc (VIET)	Kulumuzi Cave (TAN)	Siju Caves (ASS)	Batu Caves (MAL)	Olhos d'Agua Cave (BRA)	Farm Caves (MYA)	Niah Great Cave (SAR)
Coleoptera Carabidae	1			1													
Coleoptera Pselaphidae			1				1	2		1							
Coleoptera Leiodidae				1													
Dermaptera																	1
Hemiptera												1					
Hymenoptera Formicidae									1								
Orthoptera				1				1	1				1				
Troglobionts	**21**	**15**	**17**	**15**	**15**	**15**	**13**	**16**	**15**	**14**	**12**	**10**	**7**	**4**	**4**	**4**	**1**
Turbellaria	1	1	1														
Mollusca				1			1										
Crustacea																	
Amphipoda (Bogidiellidae)	1	1	1			1	1										1
Copepoda														1			
Decapoda Anomura (Aegiidae)							1										
Decapoda Brachyura	1	1							1								1
Decapoda Natantia	3	1										1					
Isopoda Anthuridae		1															
Isopoda Oniscidea				1											2		
Isopoda Stenasellidae		1	1														
Ostracoda														1			
Syncarida					1									1			
Pisces	1						1								1		
Stygobionts	**7**	**6**	**3**	**2**	**1**	**1**	**3**	**0**	**1**	**0**	**1**	**0**	**1**	**3**	**3**	**0**	**2**
Troglobionts + stygobionts	**28**	**21**	**20**	**17**	**16**	**16**	**16**	**16**	**16**	**14**	**13**	**10**	**8**	**7**	**7**	**4**	**3**

ASS: Assam (India); BRA: Brasil; CHI: Guangxi (China); HAL: Halmahera (Indonesia); LAO: Laos; MAL: Malaya (Malaysia); MYA: Myanmar; SAR: Sarawak (Malaysia); SUL: Sulawesi (Indonesia); SUM: Sumatra (Indonesia); TAN: Tanzania; THA: Thailand.

Salukkan Kallang-Tanette (SKT) system alone (Deharveng and Bedos 2000). Why such a high diversity in a lowland cave, whereas the best-known surrounding regions (*e.g.*, Java) are only moderately rich? The unique conjunction of four independent factors may explain this exceptional richness: seasonal local climate, proximity to the sea (several species present in Salukkan are considered of marine origin), situation at the foot of relatively high mountains (a potential source of cave colonizers), and intensive sampling. Surveys of other tropical karsts will tell us in the future whether the latter was the predominant factor.

Troglomorphy along Environmental Gradients

Gradients that structure cave fauna operate at different scales. At the local scale, inside a single underground system or karst unit, physicochemical factors (light, salinity, substrate granulometry, and food resources) are associated with striking changes in assemblage composition and troglomorphy level, operating on very short ecotones (often a few meters or less). There is strong evidence in particular that troglomorphy is linked to scarcity of food resources. Even more obviously than in temperate caves, high degree of troglomorphy and high diversity of troglomorphic species are only observed in oligotrophic habitats while high-energy habitats are only exploited by non- or weakly troglomorphic species.

Elevation is a major structuring gradient for the surface fauna. Because deep caves of the same system have a very similar microclimate whatever their altitude, the expectation is that the cave fauna will be less dependent on altitudinal gradient. In fact, the decline of bats and guanobionts with elevation is paralleled by important changes in the balance between ecological categories of cave fauna, and an increase in troglomorphic traits among troglobionts. Thus, outside subtropical China, troglomorphic beetles are exceptionally encountered in lowland of tropical regions but regularly found in highland caves, such as *Mayaphaenops sbordonii* at 3000 m in Guatemala, *Speleodesmoides raveloi* at about 3200 m in Venezuela, and several species in highlands of New Guinea.

Latitudinal gradients of seasonality and aridity are the main drivers of changes in surface ecosystems. Subterranean environments, because of their buffered microclimate, seem to be relatively isolated from these changes. Cross-checking of taxonomic and faunistic information allows us to recognize that troglomorphy increases with increasing seasonality, especially along a latitudinal gradient. This pattern is observed in various zoological groups and concerns both the number of troglomorphic species and the degree of modification of their morphological traits. A good example is that of subterranean beetles in Southeast Asia, which are few and weakly modified morphologically in the lowland humid tropics (two species in Indonesia, none in Malaysia), more numerous and more troglomorphic in eastern and northern Thailand (four or five species), and finally reaching the highest levels of troglomorphy and diversity in southern China, just north of the tropical belt. A similar pattern of increasing troglomorphy is seen among the most characteristic and frequent terrestrial cave groups of the region, that is, Rhaphidophoridae crickets, Cambalopsidae millipedes, and springtails. Interestingly, fish follow the same gradient but stygobiotic invertebrates do not: reduced-eyed and long-legged species of crabs and shrimps are no more frequent in the subtropics than near the equator (Guinot 1988).

Tropical versus Temperate Subterranean Biodiversity

The tropics are known to have the highest biodiversity on Earth for surface ecosystems, but this might not be true for subterranean habitats. Culver and Sket (2000) compared the specific richness of 20 caves and wells worldwide having 20 or more obligate subterranean species and came to the conclusion that "the scarcity of high diversity caves in the tropics is still a puzzle" (p. 16). Their study included two tropical caves, the SKT system in South Sulawesi, and the Bayliss lava tube in northern Australia. These caves ranked 12 and 15 out of 20 for total species richness, far behind the temperate cave systems of the Dinaric karst. The data we provide here (Table 2) concern a larger array of tropical caves from various regions, including the best-studied and richest ones, and it seems to confirm Culver and Sket's observation. However, aquatic microinvertebrates (Copepoda, Oligochaeta), which contribute the most biodiversity in temperate caves, have not been studied in SKT nor in most tropical caves. A more relevant comparison, based on terrestrial fauna alone, places these two caves much closer to the richest temperate caves (ranking 4 and 5 out of 21 in Table 2). They would rank even higher if the obligate subterranean guanobiotic species, which are considerably more diverse in the tropics than in temperate areas, were taken into account.

RELICTUAL VERSUS NONRELICTUAL TAXA

Distributional relicts represent a precious component of biodiversity for the paleobiogeographic and phyletic

TABLE 2 Tropical versus Temperate Cave Biodiversity

		Bioclimate	Aquatic	Terrestrial	Total	*Ranked on Terrestrial*	*Ranked on Total*
Europe: Pyrenees							
	Baget system (FR)*	Temperate	24	9	33	*18*	*8*
	Goueil di Herr (FR)*	Temperate	14	12	26	*15*	*10*
Europe: Italy							
	Busso del Rana*	Temperate	15	5	20	*20*	*14*
	Grotta dell'Arena *	Temperate	6	14	20	*13*	*14*
Europe: Dinaric karst							
	Grad (SLO)*	Temperate	17	3	20	*21*	*14*
	Jama Logarcec (SLO)*	Temperate	28	15	43	*10*	*5*
	Krisna Jama (SLO)*	Temperate	29	16	45	*8*	*4*
	Sica-Krka system (SLO)*	Temperate	27	7	34	*19*	*7*
	Sistem Postojna-Planina (SLO)*	Temperate	48	36	84	*1*	*1*
	Vjetrenica Jama (Bosnia)*	Temperate	39	21	60	*5*	*2*
Europe: Romania							
	Pestera de la Movile*	Temperate	18	29	47	*2*	*3*
USA							
	Mammoth Cave*	Temperate	15	26	41	*3*	*6*
	Shelta Cave*	Temperate	12	12	24	*15*	*11*
China							
	Feihu Dong	Sub-temperate	4	16	20	*8*	*14*
	Ma San Dong	Sub-tropical	2	15	17	*10*	*19*
Australia							
	Bayliss Cave*	Tropical	0	24	24	*4*	*11*
Southeast Asia							
	Sangki system (SUM)	Tropical	3	17	20	*7*	*14*
	Air Jernih system (SAR)	Tropical	6	15	21	*10*	*13*
	SKT system (SUL)	Tropical	7	21	28	*5*	*9*
Africa							
	Kulumuzi Cave (TAN)	Tropical	0	10	10	*17*	*21*
America							
	Areias system (BRA)	Tropical	3	13	16	*14*	*20*

Data are indicative, as ecological status of troglobiont may vary slightly between authors. Abbreviations are as in Table 1. FR: France; SLO: Slovenia.
*data from *Culver and Sket (2000)*.

information they carry. They are more numerous in caves than in any surface habitat, as the buffered microclimate of subterranean ecosystems is assumed to have protected cave biota from environmental fluctuations that often led surface fauna to extinction. Most cave relicts are troglobiotic species and most are found in temperate regions. From the observation that cave species of Hawaii are derived from the surface fauna present on the islands, Howarth (1973) hypothesized that the absence of relicts might be characteristic of tropical caves. Hawaii, however, as a young archipelago, did not experience drastic ecological changes that would have eliminated the surface fauna, thus turning cave species into relicts. Actually, evidence is accumulating that climatic relicts are also present in tropical caves, often in large numbers. The Cape Range of western Australia, thoroughly studied by Humphreys, is well known in this respect, but many other karsts host relictual species, particularly but not exclusively in seasonal regions of the tropics such as Laos or the Maros karst in South Sulawesi.

There are various degrees and kinds of relicts. In the tropics, marine relicts (*i.e.*, species secondarily adapted to freshwater) often represent the largest part of freshwater cave fauna. For instance, the crab *Cancrocaeca xenomorpha* of Maros (South Sulawesi) or the isopod *Cyathura (Stygocyathura) chapmani* of the Mulu caves are likely to have arisen recently from marine ancestors, but they are now totally disconnected spatially and ecologically from marine habitats. Among the terrestrial fauna, disjunct distributions also point to a relictual status. Trechinae, for instance, have extensively radiated in caves of southern China to produce more than 50 species, but their potential surface relatives are now hundreds of kilometers away, in the Himalayas. More often, relict species belong to oligo- and monospecific genera. Thus, a troglobiotic millipede from a cave of southern Vietnam, *Eostemmiulus caecus*, was shown to belong to Stemmiulidae, a large family previously unknown from Southeast Asia, but well diversified in the rest of the tropics. Another example is the recent finding, in the large Kebang-Khammouane karst across Laos and Vietnam, of two troglomorphic species that belong to two monospecific genera of Pseudochactidae, a family of scorpions described from central Asia in 1998 based on a single species (Fig. 9). It is actually no surprise that tropical karsts, affected by bioclimatic changes and sea-level fluctuations, have generated various relictual taxa. The overall picture is, however, of a lower diversity of relicts than in temperate caves, in line with the lower intensity of bioclimatic fluctuations in the tropics, as stressed by Chapman (1986).

FIGURE 9 *Vietbocap canhi* Lourenço and Pham (Pseudochactidae), a blind scorpion discovered in 2010 from a cave in the Phong Nha-Ke Bang National Park (Vietnam). *Photo courtesy of Elise-Anne Leguin. Used with permission.*

CONSERVATION ISSUES SPECIFIC TO TROPICAL CAVES

Karst and cave fauna are seriously threatened by human activities in the tropics, from pollution to uncontrolled tourism and limestone exploitation. Like other ecosystems, they have to face a growing demand on natural resources generated by the rapid development of southern countries, in a context of less stringent environmental regulations than in western countries. These threats are aggravated by two characters of tropical cave fauna: the importance of bats, and the high proportion of narrow endemics.

Bats, Guano, and Biodiversity Conservation

It has long been more or less implicitly admitted that guano hosted less biodiversity, or less interesting biodiversity than low-energy habitats, and in particular a low number of narrow endemics. This view is, however, challenged by the recent discovery of a high Cambalopsidae radiation in Southeast Asia caves (Fig. 8).

Mining, guano exploitation, hunting, and tourism all have a detrimental effect on bat and swiftlet populations. Indirectly, guano deposits, and hence their associated fauna, are affected. As common bats such as *Tadarida plicata* in tropical Asia are the largest providers of cave guano in subterranean ecosystems, threats on their colonies are paradoxically more worrying for overall biodiversity than threats on rarer non-colonial species.

When disturbance leads to bats abandoning a roosting site, nonphoretic guanobionts will not survive the disappearance of guano, as they cannot escape by active dispersal, and contrary to troglobionts, they cannot find refuge in deeper cracks of the rocks because there is no guano there. Given the vulnerability of bats and the huge contribution of guano species to overall cave diversity in the tropics, bat disturbance is probably today the most worrying threat to tropical subterranean biodiversity.

Endemism and Vulnerability

Caves host a larger proportion of endemics than any surface habitat, and cave endemics are on average more narrowly distributed than surface endemics. However, this remains poorly documented for tropical karsts. Rapid weathering of limestone under tropical climate often results in dramatic landscape of more or less isolated karst blocks. During this process, cave communities are fragmented and may genetically diverge. Recent studies have shown that even very small limestone blocks, when separated from each

FIGURE 10 A critical issue in karst biodiversity conservation: the Hon Chong hills in southern Vietnam (from Google map, photo dated 2006). Limestone hills are outlined in white; non-limestone hills are not outlined. Nui Khoe La and Nui Bai Voi will be quarried flat, except in the extreme north of Nui Bai Voi which is being developed as a tourist spot. Both hills host strictly endemic arthropods, especially in their cave systems. Of the other hills to the south, only Nui Hang Tien will remain unexploited.

other by a few kilometers of unfavorable habitat, may host narrowly endemic species in their underground arthropod fauna and very local relicts differing between karst blocks. More relictual and phyletically isolated taxa of arthropods have been recorded in the tiny hills of the Hon Chong karst in southern Vietnam, for instance, than in any other large karst of the country. Quarrying such small hills to the ground, as is being done in several regions of Malaysia and Vietnam (Fig. 10), and probably elsewhere, is the fastest way to lead to the extinction of a large number of the rarest subterranean microendemic species of invertebrates. The selection of quarrying and mining sites ought first to take into account the size and isolation of the ecosystem that will be impacted, as no remediation will be possible afterwards. This should be the cardinal rule for a sustainable exploitation of limestone in the tropics in terms of its impact on biodiversity.

See Also the Following Articles

Diversity Patterns in Australia
Diversity Patterns in the Dinaric Karst
Diversity Patterns in Europe
Diversity Patterns in the United States

Bibliography

Annandale, N., Brown, J. C., & Gravely, F. H. (1913). The limestone caves of Burma and the Malay Peninsula. *Asiatic Society of Bengal*, 9, 391–423.

Chapman, P. (1980). The biology of caves in the Gunung Mulu National Park, Sarawak. *Transactions of the British Cave Research Association*, 7, 141–149.

Chapman, P. (1986). Non-relictual cavernicolous invertebrates in tropical Asian and Australasian caves. In Ibynsa C./Badajoz 145–147 Barcelona (eds): Proc. 9th Int. Cong. Speleol., Barcelona (Spain), 2: 161–163.

Culver, D. C., & Sket, B. (2000). Hotspots of subterranean biodiversity in caves and wells. *Journal of Cave & Karst Studies*, 62, 11–17.

Deharveng, L., & Bedos, A. (2000). The cave fauna of Southeast Asia. Origin, evolution and ecology. In H. Wilkens, D. C. Culver & W. F. Humphreys (Eds.), *Ecosystems of the world: 30 subterranean ecosystems* (pp. 603–632). Amsterdam: Elsevier.

Deharveng, L., Bedos, A., Le, C. K., Le, C. M., & Truong, Q. T. (2009). Endemic arthropods of the Hon Chong hills (Kien Giang), an unrivaled biodiversity heritage in Southeast Asia. In C. K. Le, Q. T. Truong & N. S. Ly (Eds.), *Beleaguered hills: Managing the biodiversity of the remaining karst hills of Kien Giang, Vietnam* (pp. 31–57). Ho Chi Minh City, Vietnam: Nha Xuat Ban Nong Nghiep.

Gnaspini, P., & Trajano, E. (1994). Brazilian cave invertebrates, with a checklist of troglomorphic taxa. *Revista Brasileira Entomologia*, 38, 549–584.

Golovatch, S., Geoffroy, J. J., Mauriès, J. P., & van den Spiegel, D. (2009). Review of the millipede family Haplodesmidae Cook, 1895, with descriptions of some new or poorly-known species (Diplopoda, Polydesmida). In S. I. Golovatch & R. Mesibov (Eds.), *Advances in the Systematics of Diplopoda I. Zookeys* 7, pp. 1–53.

Guinot D (1994). Decapoda Brachyura. In C. Juberthie & V. Decu (Eds.), Encyclopaedia Biospeologica, vol. I, pp. 1991–2006. Société de Biospéologie, Moulis, France.

Howarth, F. G. (1973). The cavernicolous fauna of Hawaiian lava tubes. I. Introduction. *Pacific Insects*, 15, 139–151.

Humphreys, W. F. (2000). Relict faunas and their derivation. In H. Wilkens, D. C. Culver & W. F. Humphreys (Eds.), *Ecosystems of the world: 30 subterranean ecosystems* (pp. 417–432). Amsterdam: Elsevier.

Iliffe, T. M. (2000). Anchialine cave ecology. In H. Wilkens, D. C. Culver & W. F. Humphreys (Eds.), *Ecosystems of the world: 30 subterranean ecosystems* (pp. 59–76). Amsterdam: Elsevier.

Jeannel, R., & Racovitza, E. G. (1914). Biospeologica XXXIII. Enumération des grottes visitées 1911–1913 (5 ème série). *Arch. Zoologie Expérimentale et Générale*, 53, 325–558.

Juberthie, C., & Decu, V. (Eds.), (1994–2001). *Encyclopaedia biospeologica* (Vols. I–III).). Moulis, France: Société de Biospéologie.

Leleup, N. (1956). La faune cavernicole du Congo belge et considérations sur les coléoptères reliques d'Afrique intertropicale. *Annales du Musée du Congo Belge-Sciences Zoologiques*, 46, 1–170.

Mitchell, R. W. (1969). A comparison of temperate and tropical cave communities. *Southwestern Naturalist*, 14, 73–88.

Naruse, T., Ng, P. K. L., & Guinot, D. (2008). Two new genera and two new species of troglobiotic false spider crabs (Crustacea: Decapoda: Brachyura: Hymenosomatidae) from Indonesia, with notes on Cancrocaeca Ng, 1991. *Zootaxa*, 1739, 21–40.

Reddell, J. R. (1981). Review of the cavernicole fauna of Mexico, Guatemala, and Belize. *Texas Memorial Museum Bulletin*, 27, 1–327.

DIVERSITY PATTERNS IN THE UNITED STATES

Horton H. Hobbs III
Wittenberg University

INTRODUCTION

The scientific study of cave fauna (cavernicoles) in the United States had its inception in 1842 with the description by DeKay of the blind amblyopsid cavefish, *Amblyopsis spelaea*, from Mammoth Cave, Kentucky. By 1888 the number of obligate cave-dwellers, that is, troglobionts (terrestrial) and stygobionts (aquatic species), had increased to more than 50, as reported by Alpheus Packard in his compendium of North American cave fauna (Packard, 1888); numerous additional descriptions of cave-restricted organisms occurred through the 1940s and 1950s. By 1960 Brother Nicholas listed 334 species (Nicholas, 1960), in 1998 Stewart Peck estimated 1353 cave species (including interstitial and undescribed species and subspecies; Peck, 1998), and in 2000 Culver *et al.* reported that a total of 973 species and subspecies were described from caves in the 48 contiguous states (Culver *et al.*, 2000). Currently, 1138 species and subspecies are described (inclusive of Alaska and Hawaii) and are assigned to 239 genera and 112 families (Table 1). Nearly all have evolved from the independent invasion of surface organisms into the subterranean realm where physical isolation and thus cessation of gene flow with surface congeners has led to speciation. In the stressed subterranean environment where there is perpetual darkness and extremely limited food, selective pressures have led to the evolution of parallel and convergent regressive forms (profound morphological alterations) that are characteristic of troglobionts and stygobionts worldwide. This combination of characteristics, which includes reduction or loss of eyes and pigments, gracilization and elongation of appendages, increased chemical and tactile sensitivity, degeneration of circadian rhythms, lowered fecundity and metabolic rates, and increased longevity, is postulated to be adaptive to life in such extreme ecosystems and is termed *troglomorphy*.

This article summarizes the ecological, taxonomic, and geographic patterns of biodiversity of cave-inhabiting fauna in the 50 United States, focusing on troglobionts and stygobionts (separated because the two environments have very different environmental characteristics and faunas), but not ignoring cave visitors (trogloxenes/stygoxenes) and "cave lovers" (troglophiles/stygophiles). Specific numbers of organisms known from various cave areas (see below) are

TABLE 1 Summary of the Obligate Subterranean Phyla/Classes/Orders, Families, Genera, and Species/Subspecies Described from U.S. Caves

Phylum/Class/Order	Family	Genus	Species/Subspecies
Turbellaria	4	6	28
Oligochaeta	4	6	13
Mollusca	5	16	29
ARACHNIDA			
Acari	7	13	22
Aranaea	8	22	98
Opiliones	6	13	45
Pseudoscorpiones	9	29	152
Schizomida	1	1	1
Scorpiones	1	1	1
CRUSTACEA			
Copepoda	2	3	7
Ostracoda	2	5	14
Bathynellacea	1	1	1
Thermosbaenacea	1	1	1
Amphipoda	11	18	137
Isopoda	6	15	101
DECAPODA			
Shrimps	2	2	5
Crayfishes	1	4	35
Crabs	1	1	1
Chilopoda	3	4	4
Diplopoda	12	20	64
HEXAPODA			
Thysanura	1	2	2
Diplura	2	3	8
Collembola	6	12	77
INSECTA			
Orthoptera	1	2	3
Dermaptera	1	1	1
Hemiptera	2	2	2
Homoptera	1	1	7
Coleoptera	7	24	261
Diptera	1	1	1
Osteichtheys	2	5	6
Amphibia	1	5	11
Totals	**112**	**239**	**1138**

derived mostly from a cave biota database that is available on the World Wide Web at *www.karstwaters. org*. No attempt is made herein to summarize the protozoans or other microbial communities of caves although clearly these are incredibly important in cave ecosystems as sources of food (primary and secondary production), in the deposition of minerals, and in the process of speleogenesis. Additionally, attention is directed toward the vulnerability as well as to the challenges of conservation of this fauna and associated habitats.

CAVE ECOLOGY

The obvious feature of caves (Hobbs, 1992) is perpetual darkness, which results in the absence of green plants (producers of carbon-based molecules and thus "food" for organisms). Other characteristics of these food-poor cavities are low variances in temperature and humidity (usually near saturation), which make for a predictable, less variable environment. Evolutionarily, as an organism makes the transition to troglobiont or stygobiont, isolation is another key feature of that environment. These are unique combinations of characteristics that set caves apart from other ecosystems and that make this extreme environment inhospitable to most organisms.

The zonal variation in biological, chemical, and physical properties influences the distribution and abundance of fauna occurring in caves. The *entrance* area of caves is the ecotone between the epigean and hypogean worlds and has received only minimal study. Clearly, here there is more diversity and greater environmental variability than in any other cave zone. On both horizontal and vertical scales, entrances provide a transition of characteristics (*e.g.*, temperature, humidity, light) that may provide conditions for entry or survival for preadapted species or for relictual species that otherwise are rare or have become extinct on the surface due to climatic changes. They are important windows into the subterranean realm through which pass migrating trogloxenes (*e.g.*, bats, crickets) and can be the point of entry of important organic material. Farther into the cave but still within the limits of light penetration (the *twilight zone)*, the influence of surface conditions is apparent and variation of meteorological conditions is significantly less than at the surface and entrance areas. The *dark zone* initially demonstrates considerable influence of surface conditions but as distance increases from the entrance, it grades into a much less variable environment that is far from "constant" but significantly reduced in fluctuations of such parameters as atmospheric and water temperatures. In the dark, deep interior of caves that is characterized by more environmental constancy, virtually no food is produced and organisms are thus dependent on input of carbon from the surface that energetically supports most cave ecosystems (plant debris, bat and cricket guano). Exceptions to these generalizations about food occur in caves where chemoautotrophic production by sulfur-oxidizing microbial organisms (*e.g.*, sulfur bacteria: *Achromatium, Beggiatoa, Thiothrix*) occurs, resulting in sufficient energy to support and sustain complex cave ecosystems. These types of caves are dominated geochemically by reduced sulfur compounds and are rare occurrences (*e.g.*, Cesspool Cave in Allegheny County, Virginia, and Lower Kane Cave, Big Horn County, Wyoming).

Cave organisms can be separated on the basis of habitat and/or resource base: *terrestrial riparian communities* found on stream banks with a resource base of allochthonous particulate organic matter deposited by stream fluctuations; *terrestrial transitory organic matter (dung) communities* usually living within a few hundred meters of the surface with a resource base of organic matter (often fecal material) that is derived from the activities of animals (*e.g.*, bats, crickets, raccoons) moving in and out of caves; *terrestrial epikarst communities* living primarily in the network of small, air-filled cavities above the cave but below the surface; *aquatic stream communities* living primarily in cave streams ultimately dependent on dissolved and particulate organic matter derived from the surface; *aquatic phreatic communities* found in the permanent groundwater at or below the cave itself, including the hyporheos; and *aquatic epikarst communities* living primarily in the network of small flooded or partially flooded cavities above the cave but below the surface and most easily sampled within the cave in drip pools.

DISTRIBUTION OF KARST, CAVES, AND CAVERNICOLES

Within the continental United States, available and appropriate habitat for subterranean fauna is not continuous; thus cave-inhabiting organisms are found in distinct areas (mostly karst), some being widely distributed and isolated (Table 2, Figs. 1 and 2). Due to these separations and a unique geology, history, and climate, it should not be surprising that there are distinct differences in regional fauna, variances expressed in diversity, population densities, as well as taxonomic groups. Isolation, due in part to folding of strata, has resulted in high species richness (particularly of troglobiotic beetles) in the Appalachians, whereas in the Interior Lowlands this is not the case, likely due to cave connectivity. Most cave-adapted species are found south of the southern limits of the

TABLE 2 Comparison of Species Richness and Characteristics of 10 U.S. Karst Regions[a]

Major Karst Regions of the United States	Number of Troglobionts	Number of Stygobionts	Total Species	Number of Caves	Area of Karst (km^2)
Appalachians	185 (2)	91 (1)	276 (2)	7,441 (2)	37,268 (5)
Black Hills	2 (9)	0 (10)	2 (10)	160 (9)	7,272 (8)
Driftless Area	11 (8)	2 (8)	13 (9)	615 (7)	25,222 (7)
Edwards/Balcones	105 (3)	58 (3)	163 (3)	2,011 (4)	65,586 (2)
Florida Lime Sinks	0 (10)	24 (5)	24 (6)	627 (6)	27,338 (6)
Guadalupes	13 (7)	1 (9)	14 (8)	1,379 (5)	43,522 (4)
Hawaii (lava)	35 (4)	6 (6)	41 (5)	?	?
Interior Lowland Plateau	256 (1)	63 (2)	319 (1)	11,928 (1)	60,612 (3)
Mother Lode	20 (6)	3 (7)	23 (7)	179 (8)	390 (9)
Ozarks	31 (5)	51 (4)	82 (4)	6,964 (3)	110,125 (1)
Total	658	299	957	31,304	377,335

[a]Ranks from highest to lowest are presented in parentheses (modified from Culver and Hobbs, 2002). An additional 180 species are from other cave and karst regions throughout the contiguous United States and one described stygobiotic amphipod occurs in Alaska, bringing the total number of described obligate cavernicoles in the United States to 1138.

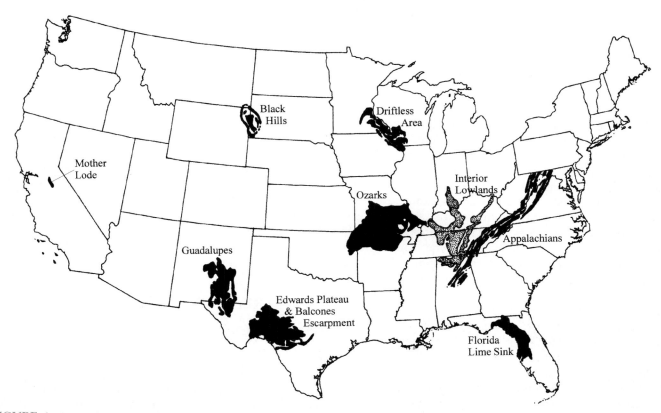

FIGURE 1 Map of major karst regions treated in this article (excluding the Hawaiian Islands, which are lava). The Interior Lowland Plateau is stippled in order to differentiate it from the Appalachians (extension into Canada truncated herein at Maryland–Pennsylvania border) and the Ozarks. *Modified from Culver et al., 2003.*

Pleistocene glacial ice sheet, yet some were able to survive subglacial conditions in the northern states (*e.g.*, New York, Wisconsin) as well as in Canada (*e.g.*, Alberta, British Columbia) and are represented primarily by groundwater amphipods and isopods. In the Driftless Area (parts of Illinois, Iowa, Minnesota, and Wisconsin) and other smaller karst regions north of the glacial boundary, low biodiversity in caves is

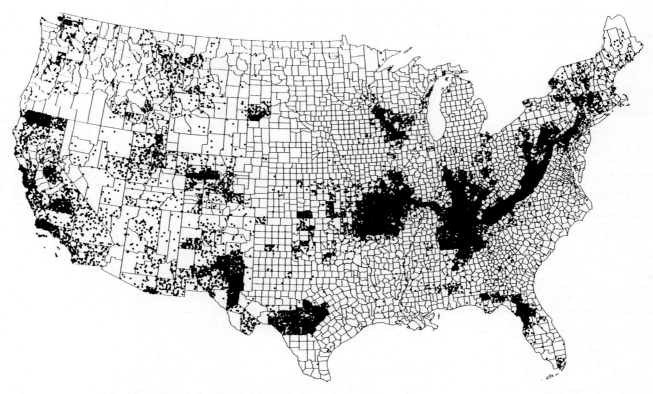

FIGURE 2 Dot map of the distribution of caves in the United States. Each dot represents one cave. *Modified from Culver et al., 1999a.*

attributed to the lack of time for colonization and isolation due to the effects of glaciers. Also, the Guadalupe Mountains of New Mexico have a depauperate cave fauna likely resulting from the aridity of the region. Although high densities of lava tubes occur in the western United States and limestone caves have developed in numerous smaller karst areas, this article focuses primarily on caves formed in soluble rock, particularly limestone, and follows the nine geologically defined cave regions discussed by Culver and Hobbs (2002) (Fig. 1). These are the Appalachians, Black Hills, Driftless Area, Edwards Plateau and Balcones Escarpment, Florida Lime Sinks, Guadalupe Mountains, Interior Lowland Plateaus, Mother Lode, and the Ozarks. Also, a tenth, the Hawaiian region, of volcanic origin (thus many islands are "young" ecologically), is treated.

Caves of the coterminous United States are shown in Figure 2 resulting from a plot of nearly 45,000 caves. Clearly the distribution of cavernicoles is influenced by the distribution and abundance of cavities available for habitation. Indeed, although it is useful to examine the number of cavernicoles in states (Fig. 3) or within various karst areas (Fig. 4), Christman and Culver (2001) and Culver *et al.* (2003) demonstrated that the availability of habitat expressed by the number of caves in a region (Table 2) is the best predictor of the biodiversity of cave-dwelling organisms. Also, the distribution of cavernicoles within caves is patchy and nonrandom and is usually associated with concentrations of food sources. Whereas troglobiont diversity within caves is likely affected not only by resource availability but also by variety (organic plant debris, bat or cricket guano, mammalian scat), stygobionts tend to be less diverse in part due to the preponderance of feeding generalists and the lack of specialization on food type.

TREATMENT OF THE FAUNA: INVERTEBRATES

Class Turbellaria

This diverse group of stygobiotic flatworms (28 species) is assigned to four families and six genera. The largest genus, *Sphalloplana*, is represented by 16 species found in subterranean settings primarily in the Appalachians and Interior Lowlands major karst regions.

Class Oligochaeta

This group of aquatic worms is represented by 13 species belonging to six genera and four families

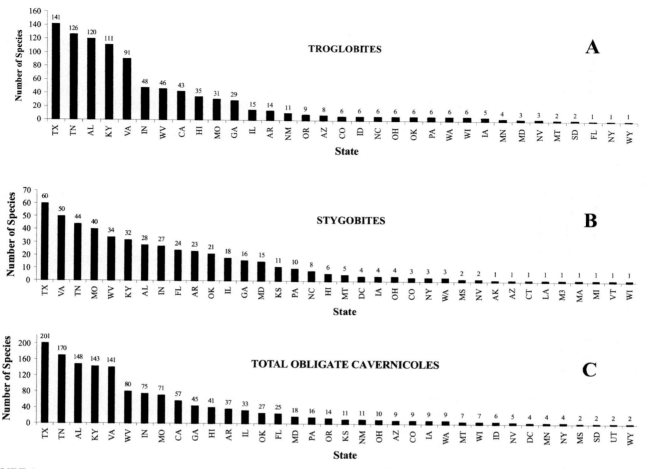

FIGURE 3 Frequency histograms by state of number of described obligate subterranean species. (A) Troglobiotic biodiversity by state, (B) stygobiont biodiversity by state, and (C) combined cavernicolous biodiversity by state.

mostly from the Appalachians and Interior Lowlands. They are poorly known from cave soils and streams as well as from interstitial waters, the latter habitat only rarely sampled. Of interest, seven species of the genus *Cambarnicola* (Branchiobdellidae) are restricted as ectosymbionts to cambarid crayfishes occupying subterranean waters. Tubificid worms are common to subterranean streams containing much organic material (or that are receiving sewage effluent!), but no stygobiotic forms are known.

Class Mollusca

Both aquatic and terrestrial snails inhabit caves, the diversity of stygobionts being nearly six times greater than that of troglobionts. Twenty-four stygobionts are placed in 12 genera in two families and three families contain four genera and five species of generally small, translucent troglobiotic gastropods. Numerous species of terrestrial epigean snails are deposited in caves on debris transported in by floods but apparently are not able to maintain populations for many generations. Yet a few species are troglobiotic (mostly from the Appalachians and Interior Lowlands karst regions, *e.g., Helicodiscus barri* Hubricht, *Carychium stygium* Call), feeding on guano and decaying allochthonous material (mainly plant) and associated microbes. Hydrobiid snails have been particularly successful in subterranean waters with 23 stygobionts described primarily from the Appalachians, Edwards Plateau and Balcones Escarpment, Interior Lowlands (*e.g., Antroselates spiralis* Hubricht), and Ozarks. Some of these snails are site endemics (*e.g., Antrobia culveri* Hubricht, known from a stream in a single cave in Taney County, Missouri) or are restricted to a single drainage (*e.g., Fontigens turritella* Hubricht, from two caves in the Greenbrier River drainage in Greenbrier County, West Virginia). *Physella spelunca* Turner and Clench is unique in that it is a site-endemic species restricted to a thermal cave stream (Lower Kane Cave; see above) draining into Big Horn River, Wyoming.

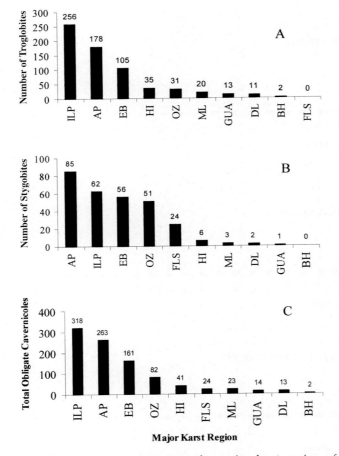

FIGURE 4 Frequency histograms by major karst region of described obligate cavernicoles. (A) Troglobionts, (B) stygobionts, and (C) total. AP, Appalachians; BH, Black Hills; DL, Driftless Area; EB, Edwards Plateau and Balcones Escarpment; FLS, Florida Lime Sink; GUA, Guadalupes; HI, Hawaii; ILP, Interior Lowland Plateau; ML, Mother Lode; OZ, Ozarks.

Class Arachnida

The cave-adapted arachnids are represented by a very diverse group of mostly terrestrial arthropods that are widely distributed throughout karst terrains of the United States. Mites (Order Acari) are small arachnids that are dominated by terrestrial species (11 genera and 19 species) with only three stygobionts placed in two genera. Mites are undoubtedly underrepresented in these totals due primarily to inadequate sampling. Seven families are represented, with the rhagidiids being the most diverse. Although most of the known mites are predaceous and commonly observed moving rapidly among organic debris in moist areas of caves, some species are parasites on bats and harvestmen. Stygobionts are known only from the Interior Lowlands (Indiana), and the greatest diversity of troglobionts resides in the Appalachians and Interior Lowlands; at least one species is endemic to lava tubes on Hawaii. No troglobiotic ticks are known but in some localities ticks are common, particularly in bat caves.

Spiders (Araneae) have been quite successful in the terrestrial cave environment with 98 species in 22 genera and 8 families. These troglobionts are known from all the major karst areas except the Black Hills and the Florida Lime Sinks. The genus *Cicurina* is particularly diverse in Texas (49 species) and *Phanetta subterranea* (Emerton) and *Porhomma cavernicola* (Keyserling) are widely dispersed throughout the Appalachians and Interior Lowlands, and less widely within the Ozarks. Five species assigned to five genera and three families are known from lava tubes in Hawaii. In addition to the troglobionts, a significant number of facultative species (*e.g., Gaucelmus augustinus* Keyserling, troglophile; *Meta ovalis* (Latreille), trogloxene; *Nesticus carteri* Emerton, troglophile) are common inhabitants of caves in the major karst regions.

The Order Opiliones (harvestmen or "daddy longlegs") are widely distributed throughout the karst regions of the United States but are not overly diverse except in the Mother Lode (13 species) and Edwards Plateau and Balcones Escarpment (13 species). A total of 44 species has been assigned to 13 genera and 6 families that are scattered over the Appalachians, Interior Lowlands, and other lesser karst areas, as well as those mentioned above. The genus *Banksula* is diverse in the Mother Lode and environs (9 species) and *Texella* is represented by 11 species in the Edwards Plateau and Balcones Escarpment region as well as an additional three species in California and the Guadalupes. Large populations of trogloxenic harvestmen are often encountered in ceiling pockets or beneath ledges, out of the desiccating effects of air movements where they form dense, undulating mats (*e.g., Leiobunum townsendii* Weeds in caves in the Edwards Plateau, Texas, and *L. bicolor* Wood in southern Ohio caves). These trogloxenic species are typically scavengers often associated with organic matter washed into caves, while the troglobiotic forms are more commonly predators feeding on microarthropods, such as collembola.

Pseudoscorpions (Order Pseudoscorpiones) are minute predators that have pincer-like pedipalps (as do scorpions), but lack the tail and sting characteristic of scorpions. Some are trogloxenic/troglophilic and are often associated with bat guano communities, whereas others are highly specialized troglobionts. This highly diverse group of small arachnids (152 species within 29 genera and nine families) is particularly abundant in Alabama caves (41 species) and also demonstrates high diversity in Texas (25 species), and California and Virginia (13 species each). They are found in all major karst regions except the Florida Lime Sinks, the Black Hills, and the Driftless Area. Two families are particularly widespread, Chernetidae and Chthoniidae, with the latter being incredibly diverse (95 species). The chernetids are dominated by the extensive genus *Hesperochernes*, which is found in the Appalachians, the Edwards Plateau and Balcones Escarpment, Interior Lowlands, Mother Lode, and the Ozarks. The chthoniids are heavily influenced by the genera *Apochthonius* (15 species), *Kleptochthonius* (31 species), and *Tyrannochthonius* (37 species) and are found in all the major karst regions except the Black Hills, Florida Lime Sinks, and the Driftless Area. Two species of the genus *Tyrannochthonius* are found in lava tubes on the Hawaiian Islands (Hawaii and Oahu).

The Order Schizomida is represented by a single terrestrial species, *Hubbardia shoshonensis* Briggs and Hom, from a cave in Inyo County, California. Although similar to true whip scorpions, they have a much shorter telson as well as other anatomical features that separate the two orders. They reside in leaf litter and under stones and are predators on small invertebrates.

Scorpions (Order Scorpiones) are believed to be one of the most ancient terrestrial arthropods and the most primitive arachnid. All are terrestrial predators characterized by two large pedipaps terminating in chelae as well as by the stinging apparatus derived from the telson and bearing a sharp barb called the *aculeus*. Although numerous pigmented species are found in U.S. caves, only one species is troglobiotic: *Uroctonus grahami* Gertsch and Soleglad, known only from Shasta County, California. At least 11 species are troglobiotic to the south in Mexican caves.

Subphylum Crustacea

Class Maxillopoda

The Subclass Copepoda (see "Crustacea" article, this volume) is not well sampled in the cave environment in the United States and this is reflected in the low species richness reported. Currently two families containing three genera and seven species and subspecies of these crustaceans are known from the Edwards Plateau and Balcones Escarpment in Texas; the Interior Lowlands in Illinois, Indiana, Kentucky, and Tennessee; and from Orange County, North Carolina. One obligate parasitic copepod, *Cauloxenus stygius* Cope, is associated with the Northern Cavefish, *Amblyopsis spelaea*, in southern Indiana caves. Recent work in Slovenian caves and a few caves in West Virginia has shown that copepods are important constituents in the epikarst community.

Like the copepods, the Class Ostracoda has not been studied thoroughly in U.S. caves. Even so, 14 species belonging to 5 genera and 2 families are stygobionts in the Appalachians, Edwards Plateau and Balcones Escarpment, Florida Lime Sinks, Interior Lowlands, and the Ozarks. Twelve species of the family Entocytheridae are commensal on cambarid crayfishes inhabiting caves and two species of Cyprididae are free living in cave streams and lakes.

Class Malacostraca

The Order Bathynellacea (see "Crustacea" article, this volume) is characterized by lacking maxillipeds and a carapace. These small crustaceans are not well studied, particularly in the United States, where only one species, *Iberobathynella bowmani* Delamare Deboutteville, Coineau, and Serban, is known from two localities in Dickens and San Saba counties, Texas. These reclusive organisms are interstitial or reside in groundwater, those habitats not having been adequately sampled.

The Order Thermosbaenacea (see "Crustacea" article, this volume) is represented by a single species, *Monodella texana* Maguire, which is known from a cave and wells in Bexar, Hays, and Uvalde counties, Texas. This is the sole North American representative of the Thermosbaenacea and is endemic to the Edwards Aquifer.

The Order Amphipoda is quite diverse with 134 stygobiotic species and subspecies in 18 genera and 10 families. An additional species is a talitrid troglobiont, *Spelaeorchestia koloana* Bousfield and Howarth, which is a rare endemic that feeds on plant roots and organic debris in a few caves on the island of Kauai, Hawaiian Islands. Most amphipods are scavenger/predators and are laterally compressed, giving them a shrimplike appearance. Amphipods occur in all major karst regions except the Black Hills, Driftless Area, and Guadalupes. Crangonyctidae is the most diverse and widely distributed family, with the genera *Bactrurus* and *Crangonyx* housing 7 and 12 stygobionts, respectively, yet the genus *Stygobromus* (98 species) is approximately five times as diverse as those two genera combined. In addition to most of these being found in unglaciated regions, *S. quatsinensis* Holsinger and Shaw is endemic to caves and springs north of the southern limits of Pleistocene glaciation in British Columbia, Canada, and southeastern Alaska. Also, eyed and pigmented stygophilic amphipods commonly inhabit cave and karst spring waters of the contiguous states and are major players in the trophic dynamics of some systems (*e.g.*, *Gammarus* spp. and *Hyalella azteca* Saussure).

The Order Isopoda (see "Crustacea" article, this volume) is represented in caves by 78 stygobionts in 11 genera and 4 families. The family Asellidae is immense with all but 8 species of the order assigned to this family of dorsoventrally flattened crustaceans. Geographically, these stygobionts are widespread, absent only from the Black Hills, Driftless Area, Guadalupes, and the Mother Lode major karst regions. The genus *Caecidotea* is represented by at least 60 species and subspecies and widely dispersed over these regions, whereas other species are much more restricted (*e.g.*, *Remasellus parvus* (Steeves) in Florida and *Salmasellus howarthi* Lewis in Washington). The genus *Lirceus* (also an asellid) has two stygobiotic species from southwestern Virginia (*L. culveri* Estes and Holsinger and *L. usdagalun* Holsinger and Bowman from Scott and Lee counties, respectively) and numerous stygophilic species (*e.g.*, *L. fontinalis* Rafinesque) that occupy springs and caves. The predominantly marine family Cirolanidae is represented in the United States by three dome-shaped and elliptical stygobitic species, *Cirolanides texensis* Benedict and *Speocirolana hardeni* Bowman, both from Texas, and the widely disjunct *Antrolana lira* Bowman from Virginia and West Virginia (Appalachians).

The troglobiotic isopods are represented by 19 species and subspecies in four genera. The 15 species in the contiguous United States belong to three genera (*Amergoniscus, Brackenridgia, Miktoniscus*) within the Trichoniscidae and these "pillbugs" are found in the Appalachians and Interior Lowlands as well as in Oklahoma, Oregon, and Texas. Four species of the genus *Hawaiioscia* are endemic to the Hawaiian Islands of Kauai, Maui, Molokai, and Oahu.

North American stygobiotic members of the Order Decapoda (see "Crustacea" article, this volume) are limited to 41 crayfishes, 5 shrimps, and 1 crab. Crayfishes in the United States belong to the family Cambaridae and the following 35 are stygobiotic: 11 species of *Cambarus* in the Appalachians, Interior Lowlands, and the Ozarks; 9 species and subspecies of *Orconectes* in the Interior Lowlands and Ozarks; 14 species and subspecies of *Procambarus* in the Florida Lime Sinks and Interior Lowlands; and a single species of *Troglocambarus* in the Florida Lime Sinks. Shrimps belong to two families: Atyidae, 2 species of *Palaemonias* in the Interior Lowlands, and Palaemonidae, three species of *Palaemonetes* in the Florida Lime Sinks and Edwards Plateau and Balcones Escarpment. The single crab (Grapsidae; *Hemigrapsus oregonensis* Dana, formerly *Hemigrapsus estellinensis* Creed) is endemic, and probably extinct, to a deep artesian hypersaline spring in Hall County, Texas. Approximately 50 stygophilic/stygoxenic crayfishes utilize caves and, of special note and concern, is the impact of introduced epigean species that utilize cave streams (*e.g.*, *Procambarus* (*Scapulicambarus*) *clarkii* (Girard) in a cave in San Diego County, California, and likely introduced into the area from the southern Gulf states).

Class Chilopoda

Troglobiotic centipedes are rare in caves, fast moving, usually small, slender, and voracious predators. They are found in organic debris, on clay, and burrowing in silt. Four genera and species from three families are known from the Guadalupes and the Appalachians as well as from north-central California and the Edwards Plateau and Balcones Escarpment.

Troglobiotic millipedes, Class Diplopoda, are slow-moving grazers, feeding on organic detritus and associated microbes. Typically found in food-poor caves, some 64 species and subspecies in 20 genera are known primarily from the Appalachians, Driftless Area, Guadalupes, Interior Lowlands, and Edwards Plateau and Balcones Escarpment. *Tetracion, Scoterpes*, and *Pseudotremia* are the three most important genera in North American caves and the largest genus, *Pseudotremia* (Cleidogonidae), has at least 33 troglobiotic species in Appalachian and Interior Lowland caves. In the west, the genus *Amplaria* is represented by at least six species in California, Idaho, Oregon, and Washington. Numerous troglophiles are noted in all of these karst regions as well.

Class Hexapoda

The Class Hexapoda is restricted to the primitive orders of insects lacking wings (apterygote) and, in caves, is represented by three orders: Thysanura, Diplura, and Collembola.

Troglobiotic *thysanurans* (bristletails) are unusual in caves. A single species, *Texoreddellia texensis* (Ulrich), a member of the Nicoletiidae family, is known from caves in 18 counties in Texas and *Speleonycta ozarkensis* Espinasa, Furst, Allen, and Slay, also a nicoletiid, occurs in several caves on the Ozark Plateau in Oklahoma and Arkansas. Trogloxenic species (*e.g.*, *Pedetontus* sp.) are not uncommon in entrance areas in many parts of the United States.

Troglobiotic *diplurans* are small and lack a median caudal filament but possess two cerci ("tails") that are long, segmented filaments in the Campodeidae and short, thick forceps-like appendages in the Japygidae (see Reddell 1983). Eight species are assigned to three genera and two families, with the campodeid *Litocampa* being the most diverse genus with six species. This poorly studied group is known from the Appalachians, Interior Lowlands, the Edwards Plateau and Balcones

Escarpment, and a single species from Lincoln County, Nevada (*Condeicampa langei* Ferguson). These are observed on wet flowstone, cricket guano, and on silty substrata. Additionally, at least seven species belonging to the genus *Haplocampa* are known from lava tubes in Northern California, Idaho, Oregon, and Washington.

Troglobiotic springtails (Order Collembola) are small, jumping insects that can occur in very high densities, particularly where organic matter has accumulated. Often they are observed moving about on the surface of small drip pools as well as on larger stream pools. Not only do they provide food for many small predators but they also graze on bacteria and fungi. They are a diverse group with 75 troglobiotic species assigned to 12 genera and are known from all the major karst regions, except the Florida Lime Sinks, as well as other minor karst areas. Entomobryidae is the largest family, with the greatest diversity in the genus *Pseudosinella* (25 species); *Sinella* has 13 species. Fifteen species make up the widely distributed genus *Arrhopalites* (Sminthuridae). Numerous troglophilic species (*e.g.*, *Sinella cavernarum* (Packard)) contribute to the terrestrial communities of many caves. Six troglobiotic species assigned to three genera and two families are found in lava tubes on the islands of Hawaii, Maui, and Oahu.

Class Insecta

The Class Insecta is characterized by those insects possessing wings (pterygote) or secondarily with reduced wings and is represented in caves by troglobiotic/stygobiotic members of five orders. Troglobiotic members of the Order Orthoptera are known only from Hawaiian lava tubes. Three species of flightless, mute crickets are assigned to two genera and belong to the family Gryllidae. *Caconemobius varius* Gurney and Rentz is known from five caves on Hawaii Island and displays small eyes, reduced body pigment, and a translucent exoskeleton. It is omnivorous or perhaps a scavenger and appears to be most abundant in low, wet caves. *C. howarthi* Gurney and Rentz has been collected from four caves on the island of Maui. It has been observed only in the dark zone of caves and appears to be more cave-adapted than *C. varius*. *Thaumatogryllus cavicola* Gurney and Rentz is known from approximately 10 small, generally shallow caves on the island of Hawaii. In some of these lava tubes it has been observed on dangling roots of grasses and trees that have penetrated the ceilings in characteristically wet areas. In the contiguous states, pigmented rhaphidophorid crickets often are very numerous in caves, particularly within the first 100 to 200 m of the entrance, and these trogloxenes are represented by species of the genera *Ceuthophilus*, *Hadenoecus*, and *Euhadenoecus*. These crickets not only migrate to the surface to forage for food during spring, summer, and fall (and thus provide food in the form of guano to caves), but they also use caves as refugia during cold periods. Some crickets (*e.g.*, *Ceuthophilus silvestris* Bruner in small caves in glaciated Wyandot County, Ohio) interact with caves only seasonally, entering them solely during the winter months.

A single endemic, troglobiont (*Anisolabis howarthi* Brindle) from Hawaii is the only known member of the Order Dermaptera (earwigs) to inhabit caves. Characterized by the heavily sclerotized posterior forceps (cerci), they use these for predation and are considered to be scavenging omnivores as well.

Only two troglobiotic "bugs" (Order Hemiptera—Heteroptera) are known, both endemic to lava tubes on the island of Hawaii. The thread-legged bug, *Nesidiolestes ana* Gagne and Howarth (Reduviidae), is a predator of arthropods, and the lava tube water-treader, *Speleovelia aaa* Gagne and Howarth, sucks on rotting fluids of deceased arthropods.

The Order Homoptera is represented by seven troglobitic planthoppers endemic to the islands of Hawaii, Maui, and Molokai. All belong to the Cixiidae family and are placed in the genus *Oliarus*. Often they are observed on roots that have penetrated ceiling cracks of lava tubes.

The Order Coleoptera has been incredibly successful in caves with 261 species in 21 genera and five families. Only three species are aquatic, two of which are dryopids (Oregon and Texas) and one a dytiscidae from Texas (*Haideoporus texanus* Young and Longley, the only stygobiotic predaceous diving beetle known in the United States). Beetles are major components of cave communities particularly in the Appalachians and Interior Lowlands. Troglobionts are dominated by the carabids, particularly the genus *Pseudanophthalmus* which is inordinately abundant in Kentucky and Texas caves, 62 and 67 species, respectively. Two carabids are endemic to lava tubes on the Hawaiian island of Maui. Troglobiotic ground beetles are represented by the genus *Rhadine* (17 species), which is restricted to Texas, particularly dense in the Edwards Plateau and Balcones Escarpment karst region. In certain Texas caves, *Rhadine s. subterraneanea* (Van Dyke) and *R. noctivaga* Barr feed exclusively on cave cricket eggs (*Ceuthophilus* spp.) that have been buried in fine-grained, calcareous deposits. This feeding behavior has also been demonstrated in Mammoth and other caves in the Pennyroyal Plateau of western Kentucky (Interior Lowlands) for the carabid *Neaphaenops tellkampfii* (Erichson) preying on eggs and nymphs of the cricket *Hadenoecus subterraneus*, and on the Cumberland Plateau of eastern Kentucky the cricket is *Hadenoecus cumberlandicus* and its predator is *Darlingtonea kentuckensis* Valentine. Leiodids (*Ptomaphagous* spp.) and pselaphids (predominantly *Batrisodes* spp.) make up the remaining troglobionts. Numerous troglophiles (*e.g.*, Carabidae, Leiodidae, Leptodiridae,

Ptinidae, Staphylinidae, Tenebrionidae) also are common members of cave communities. In addition to predators, the cavernicolous beetles are scavengers and opportunistic saprophiles.

A single, widely distributed troglobiotic fly (Order Diptera) is known from caves in the Appalachians, the Interior Lowlands, and the Ozarks: *Spelobia tenebrarum* (Aldrich), Sphaeroceridae. A variety of flies utilize caves periodically and include various culicids (*e.g., Culex pipiens* Linnaeus), heleomyzids (*e.g., Amoebaleria defessa* (Osten-Sacken)), mycetophilids (*e.g., Macrocera nobilis*), phorids (*e.g., Megaselia cavernicola* (Brues)), and sphaerocerids (*e.g., Leptocera tenebrarum* (Aldrich)).

Additional insects are observed in caves but are either troglo/stygoxenes or troglo/stygophiles and include lepidopterans (*e.g., Scoliopteryx libatrix* (Linnaeus)), psocopterans (*e.g., Psyllipsocus ramburi* Sélys-Longchamps), and siphonapterans (*e.g., Myodopsylla insignis* (Rothschild)). A particularly disturbing example of surface insects entering the subterranean world is the invasion of central Texas caves during the summer months by the trogloxenic, exotic red fire ant, *Solenopsis invicta* (Hymenoptea).

TREATMENT OF THE FAUNA: VERTEBRATES

A number of vertebrates spend varying times of their life in subterranean environments but most are not considered cavernicoles (*e.g.*, some snakes and mammals are adapted to a fossorial mode of life in the soil). Other vertebrates visit caves for shelter during times of unfavorable surface environmental conditions (*e.g.*, amphibians, snakes), to avoid predation (*e.g.*, pack rats, birds), or for breeding purposes (*e.g.*, birds, bats). Generally these cave-dwellers are classified as accidentals or as trogloxenes/stygoxenes; fishes and salamanders are the only vertebrates in which troglophilic/stygophilic or troglobiotic/stygobiotic species have evolved.

Class Osteichthyes

Approximately 85 stygobiotic fishes are known worldwide, however only six cave-obligate species are recognized from North American subterranean waters and belong to one of two families, Amblyopsidae and Ictaluridae. The amblyopsid cavefishes are stygobiotic except for one stygophilic species, *Forbesichthys agassizii* (Putnam), which inhabits caves and springs of Illinois, Kentucky, and Tennessee yet migrates to the surface at night to exploit food resources. Four additional species are confined to subterranean waters: *Amblyopsis rosae* (Eigenmann) from the Ozarks, *Amblyopsis spelaea* from the Interior Lowlands, *Speoplatyrhinus poulsoni* Cooper and Kuehne from a single cave in the Interior Lowlands, and *Typhlichthys subterraneus* Girard from the Interior Lowlands and the Ozarks. The ictalurid subterranean representatives, *Satan eurystomus* Hubbs and Bailey and *Trogloglanis pattersoni* Eigenmann, are known only from deep phreatic waters of the Edwards Aquifer in Bexar County, southwest Texas, and have never been observed in accessible caves. Members of both of these families are predatory feeders primarily of crustaceans except the toothless *T. pattersoni*, which is likely a grazer taking up detritus.

Other fishes occupy subterranean waters (some regularly demonstrate migratory patterns) and some of the families represented are the following: Anguillidae (*e.g., Anguilla rostrata* (LeSueur)), Centrarchidae (*Lepomis* spp.), Cottidae (*Cottus* spp.), Cyprinidae (*Notropis* spp.), Ictaluridae (*Ictalurus catus* (Linnaeus)), Mugilidae (*Mugil cephalus* Linnaeus), and except for *Cottus* sp., show no troglomorphic adaptations for living in this environment.

Class Amphibia

Oddly, no stygophilic or stygobiotic frogs are known from caves (possibly because they cannot complete their life history in the subterranean environment), but they certainly utilize caves (*e.g., Rana palustris* LeConte), primarily to retreat from surface climatic extremes. Globally, only 13 stygobiotic salamanders are known from hypogean waters and are assigned to two families, Proteidae and Plethodontidae, the former restricted to limestone caves and karst springs in Bosnia and Herzegovina, Croatia, Italy, and Slovenia. The plethodontids are represented by 11 species and subspecies in North America and are known from the Appalachians, Edwards Plateau and Balcones Escarpment, Interior Lowlands Plateau, and the Ozarks as well as from caves in the Dougherty Plain of southwestern Georgia and adjacent panhandle of Florida (Tables 2 and 3). Numerous additional species of plethodontid salamanders are stygophilic (*e.g., Pseudotriton ruber* (Latreille)) and troglophilic (*e.g., Eurycea lucifuga* (Rafenesque)).

Class Reptilia

Although there are no known obligate reptiles from caves in the United States, these habitats are used often by snakes for short durations or even significant amounts of time. Most tend to remain near entrances and use the cave as a refuge from extreme surface conditions. Some snakes prey on bats during their flights out of the cave (*e.g.*, the colubrid *Elaphe*

TABLE 3 Obligate Subterranean Salamanders (Plethodontidae) of the United States

Genus	Species/Subspecies	State	Global Conservation Status (Nature Serve)
Eurycea	latitans	Texas	G3
	rathbuni	Texas	G1
	robusta	Texas	G1
	spelaea	Arkansas, Kansas, Missouri, Oklahoma	G4
	tridentifera	Texas	G3
	troglodytes	Texas	G1
Gyrinophilus	gulolineatus	Tennessee	G1
	palleucus palleucus	Alabama, Georgia, Tennessee	G2
	palleucus necturoides	Tennessee	G2
	subterraneus	West Virginia	G1
Haideotriton	wallacei	Florida, Georgia	G2

obsoleata (Say)) and certain lizards are observed in the entrance areas of caves as well (*e.g.*, the geckonid *Coleonyx variegatus brevis* (Steineger) in Texas). In Florida the American alligator (*Alligator mississipiensis* (Daudin)) and the cooter turtle (*Pseudemys floridana* (LeConte)) are seen in the entrances to many submerged caves and will venture short distances into these cavities that serve often as the source of large karst spring runs.

Class Aves

The most noted of all birds that utilize caves is probably the troglophilic oilbird of Cueva del Guacharo in Venezuela, South America (*Steatornis caripensis* Humboldt), for centuries prized by locals as a source of cooking and lighting oil. Although the eyes of this bird are large and well developed for use in dim light, they utilize echolocation (as do bats) when in the complete darkness of caves. In the United States, no such highly specialized species occur in caves yet swallows (*e.g.*, *Hirundo* spp.), the canyon wren (*Catherpes maxicanus* (Swanson), and the eastern phoebe (*Sayornis phoebe* (Latham)) are common inhabitants of cave entrances. Various species inhabit the twilight zone of caves, including the eastern (common) screech-owl (*Otus asio* (Linnaeus)) and the turkey vulture (*Cathartes aura* (Linnaeus)) that nest in this semidark zone of caves.

Class Mammalia

The occurrence of mammals in caves is quite common although none is obligate and most are temporary residents/visitors, including numerous rodents such as pack rats (*Neotoma* spp.), mice (*Peromyscus* spp.), beaver (*Castor canadensis* Kuhl), the wood-chuck (*Marmota monax* (Linnaeus)), as well as carnivores such as bears (*Ursus* spp.) and raccoons (*Procyon lotor* Linnaeus). Clearly no bats are obligate cave-dwellers yet 37 of the 50+ species and subspecies found in the United States occupy caves at least occasionally. They are assigned to four families (Table 4) and most are insectivorous (*e.g.*, the endangered Indiana Bat, *Myotis sodalis* Miller and Allen), but some feed on fruits, pollen, and nectar (*e.g.*, Mexican long-tongued bat, *Choeronycteris mexicana* Tschudi). During the winter most hibernate in caves (or mines) often in large, dense clusters of up to several thousand individuals and the summers are spent in trees or buildings. A few species (*e.g.*, gray bat, *Myotis grisescens* Howell) may live in caves throughout the year, although different ones are utilized in winter and summer. Some species, like the Brazilian free-tailed bat (*Tadarida brasiliensis* (Saussure)), occupy caves in very large numbers and contribute an immense amount of organic material (guano, *i.e.*, bat droppings) to the cave ecosystem. During the summer approximately 20,000,000 individuals of this particular species occupy a single cave near San Antonio, Texas (Harvey et al. 1999), and, at that time, the population represents the largest concentration of mammals in the world. Guano produced by bats living in caves not only supports very large communities of guanophiles (*e.g.*, bat fleas, dermestid beetle larvae, gnats, and pseudoscorpions), but also this nitrogen-rich material has been mined for use as fertilizer as well as for producing saltpeter (potassium nitrate), which is an ingredient of gunpowder.

TABLE 4 Cave-Dwelling Bats of the United States

Family	Genus	Species/Subspecies	Common Name	Global Status
Molossidae	*Tadarida*	*brasiliensis*	Brazil free-tailed bat	G5
Mormoopidae	*Mormoops*	*megalophylla*	Ghost-faced bat	G4
Phyllostomidae	*Choeronycteris*	*mexicana*	Mexican long-tongued bat	G4
	Leptonycteris	*nivalis*	Mexican long-nosed bat	G3
	Leptonycteris	*yerbabunae*	Lesser long-nosed bat	G4
Vespertilionidae	*Antrozous*	*pallidus*	Pallid bat	G5
	Corynorhinus	*rafinesquii rafinesquii*	Rafinesque's big-eared bat	G3, G4
	Corynorhinus	*r. macrotis*	Big-eared bat	G3, G4
	Corynorhinus	*townsentii ingens*	Ozark big-eared bat	G4
	Corynorhinus	*townsendii townsendii*	Townsend's Western big eared bat	G4
	Corynorhinus	*t. pallescens*	Pale lump-nosed bat	G4
	Corynorhinus	*t. virginianus*	Virginia big-eared bat	G4
	Eptesicus	*fuscus*	Big brown bat	G5
	Idionycteris	*phyllotis*	Allen's big-eared bat	G3, G4
	Lasiurus	*borealis*	Eastern red bat	G5
	Lasiurus	*cinereus*	Hoary bat	G5
	Lasiurus	*c. semotus*	Hawaiian hoary bat	G5
	Myotis	*auriculus*	Southwestern Myotis	G5
	Myotis	*austroriparius*	Southeastern Myotis	G3, G4
	Myotis	*californicus*	Californian Myotis	G5
	Myotis	*ciliolabrum*	Western small-footed Myotis	G5
	Myotis	*evotis*	Long-eared Myotis	G5
	Myotis	*grisescens*	Gray Myotis	G3
	Myotis	*keenii*	Keen's bat	G2
	Myotis	*leigii*	Eastern small-footed Myotis	G3
	Myotis	*lucifugus*	Little brown Myotis	G5
	Myotis	*septentrionalis*	Northern Myotis	G4
	Myotis	*sodalis*	Indiana Myotis	G2
	Myotis	*thysanodes*	Fringed Myotis	G4, G5
	Myotis	*thysanodes vespertinus*	Pacific fringe-tailed bat	G4, G5
	Myotis	*t. pahasapensis*	Fringe-tailed Myotis	G4, G5
	Myotis	*velifer*	Cave Myotis	G5
	Myotis	*v. brevis*	Southwestern cave Myotis	G5
	Myotis	*volans*	Long-legged Myotis	G5
	Myotis	*yamanensis*	Yuma Myotis	G5
	Parastrellus	*hesperus*	Western pipistrelle bat	G5
	Perimyotis	*subflavus*	Eastern pipistrelle bat	G5

During the winter of 2006–07 disturbing observations were made of hibernating bat species in eastern New York. Affected bats demonstrated unusual behavior and displayed a white fungus on their muzzles and wings and were subsequently diagnosed with "White Nose Syndrome" (WNS) associated with an infection of the fungus *Geomyces destructans*. The number of mines and caves in which bats infected with the fungus has risen sharply and the estimates of losses exceed one million individuals. Infected populations with individual losses up to more than 90% have been confirmed in Connecticut, Delaware, Massachusetts, New Hampshire, New Jersey, Indiana, New York, Ohio, Pennsylvania, Tennessee, Vermont, Virginia, and West Virginia and is likely in Missouri and Oklahoma. Currently the nature of the relationship of the fungus to this widespread mortality is unclear.

DISCUSSION

The biodiversity of cavernicoles in the United States is summarized in Table 1 and their distribution patterns are demonstrated in Tables 2 and 3 and Figs. 3 and 4. Clearly the more mobile troglobionts have been very successful, with more than twice their numbers invading subterranean habitats when compared to stygobionts. Also, insects (361 species and subspecies including apterygous and pterygous forms; 32% of all obligate species), arachnids (319; 28%), and crustaceans (302; 26%) dominate the species richness summaries, with terrestrial arachnids making up nearly one-third of all subterranean species. Much variation is shown among major karst regions, with those to the north (Black Hills and Driftless Area) having few obligate species. The number of stygobionts ranges from zero (Black Hills) to 91 (Appalachians), and the number of troglobionts varies from zero (Florida Lime Sinks) to 256 (Interior Lowlands), and even at the generic level, overlap among regions is low. Karst regions with the greatest total biodiversity are the Interior Lowlands, Appalachians, and the Edwards Plateau and Balcones Escarpment (Fig. 4C), although troglobionts show somewhat different patterns than do stygobionts (Figs. 4A,B). States with the highest total species richness are Texas, Tennessee, Alabama, Kentucky, and Virginia. Stygobionts show a different pattern: Texas, Virginia, Tennessee, Missouri, and West Virginia (Fig. 3). Most assuredly, "hotspots" do exist where concentrations of fauna occur and northeastern Alabama is one of these centers of biodiversity for troglobionts and stygobionts Culver et al. 1999b. Of particular note, Jackson County has 1526 caves and 66 obligate cavernicoles, which translates into more than three times the number of caves and nearly twice as many species as any other Alabama county. This shows the strong relationship between the number of caves observed and the number of species reported ($r^2 = 0.81$).

Troglobionts and stygobionts make up slightly more than 50% of the imperiled U.S. fauna that is tracked in the central databases of the Natural Heritage Program. There are far too many potential (and realized) disturbances to discuss herein that threaten these out-of-sight organisms and the reader is referred to Elliott (2000) for an excellent review of them. Suffice it to say that the ultimate long-term survival of subterranean karst communities depends on appropriate management and protection of the cave, the groundwater, and the entire catchment area.

It becomes clear that these subterranean species are geographically concentrated in a small percentage of the landscape, with more than 50% of cave-inhabiting species occurring in less than 1% of the land. Hence, it is much easier to preserve a large percentage of at-risk species by focusing habitat conservation efforts in those areas of high concentrations of obligate cave fauna, or *hotspots*. Protecting and conserving karst habitats and their biodiversity is a challenging but most important task for modern and future speleologists. The subterranean biodiversity of the United States is globally significant but extremely vulnerable and little is known of the long-term impacts of WNS on bats or on entire cave ecosystems.

See Also the Following Articles

Diversity Patterns in Australia
Diversity Patterns in the Dinaric Karst
Diversity Patterns in Europe
Diversity Patterns in the Tropics

Bibliography

Christman, M. C., & Culver, D. C. (2001). The relationship between cave biodiversity and available habitat. *Journal of Biogeography, 28* (3), 367–380.

Culver, D. C., Christman, M. C., Elliott, W. R., Hobbs, H. H., III, & Reddell, J. R. (2003). The North American obligate cave fauna: Regional patterns. *Biodiversity and Conservation, 12*(3), 441–448.

Culver, D. C., & Hobbs, H. H., III (2002). Patterns of species richness in the Florida stygobitic fauna. In J. B. Martin, C. M. Wicks & I. D. Sasowsky (Eds.), *Hydrogeology and biology of post-Paleozoic carbonate aquifers* (pp. 60–63). Charles Town, WV: Karst Waters Institute (Special Publication 7).

Culver, D. C., Hobbs, H. H., III, Christman, M. C., & Master, L. L. (1999a). Distribution map of caves and cave animals in the United States. *Journal of Cave and Karst Studies, 61,* 139–140.

Culver, D. C., Hobbs, H. H., III, & Mylroie, J. E. (1999b). Alabama: A subterranean biodiversity hotspot. *Journal of the Alabama Academy of Sciences, 70,* 97–104.

Culver, D. C., Master, L. L., Christman, M. C., & Hobbs, H. H., III (2000). Obligate cave fauna of the 48 contiguous United States. *Conservation Biology, 14*(2), 386–401.

Elliott, W. R. (2000). Conservation of the North American cave and karst biota. In H. Wilkins, D. C. Culver & W. F. Humphreys (Eds.), *Ecosystems of the world 30: Subterranean ecosystems* (pp. 665–689). Amsterdam: Elsevier Press.

Harvey, M. J., Altenbach, J. S., & Best, T. L. (1999). *Bats of the United States (Arkansas Game & Fish Commission)*. Asheville, NC: U.S. Fish and Wildlife Service.

Hobbs, H. H., III (1992). Caves and springs. In C. T. Hackney, S. M. Adams & W. H. Martin (Eds.), *Biodiversity of the southeastern United States: Aquatic communities* (pp. 59–131). New York: John Wiley & Sons.

Nicholas, B. G. (1960). Checklist of macroscopic troglobitic organisms of the United States. *American Midland Naturalist, 64*(1), 123–160.

Packard, A. S. (1888). The cave fauna of North America, with remarks on the anatomy of the brain and the origin of the blind species. *Memoirs of the National Academy of Science, 4*, 1–156.

Peck, S. B. (1998). A summary of diversity and distribution of the obligate cave-inhabiting faunas of the United States and Canada. *Journal of Cave and Karst Studies, 60*(1), 18–26.

Reddell, J. R. (1983). A checklist and bibliography of the Japygoidea (Insecta: Diplura) of North America, Central America, and the West Indies. *The Pearce-Sellards Series (Texas Memorial Museum), 37*, 1–41.

DOCUMENTATION AND DATABASES

Peter Matthews

Informatics Commission, International Union of Speleology

INTRODUCTION

Where would we be without cave and karst documentation? Perhaps a cliché but it ranges from describing the contents of one cave to plotting the karst areas of Europe against existing protective reserves, and is the foundation upon which much of cave exploration, research, protection, and management is based.

This article offers a brief outline of what is involved in cave documentation, its issues and techniques. The reader is directed to the bibliography and to the Informatics Commission [22] of the International Union of Speleology's (UIS) website for further detail and examples via the bracketed numbers within this article. The plan for the UIS website is to continually add more and updated examples and practical help with cave and karst documentation. Readers who know of further examples that should be on the website, please contact the author. Any reference in this article to companies or products does not constitute endorsement, merely examples.

This article first considers cave documentation in general, and then expands on the database aspects. By *documentation*, we mean the recording of descriptive information in the broadest sense about specific caves and karst areas. As scientific documentation can be quite project-specific, we focus mainly on the more general type generated by cavers and speleologists. This is usually done on a volunteer basis and with a very low budget. Although many cavers see their activity primarily as a sport, they realize it is also useful to record what they find, especially as the information is otherwise not readily obtainable.

There are two broad categories of people passionate about caves and karst: amateur cavers and professional scientists. These quite different groups can be of significant benefit to each other. Caving groups over time gather a large amount of information about a large number of caves. While often not scientific experts, cavers are usually able to mine their data to flag up caves that may be of interest to scientists for deeper study but which would otherwise remain unknown. Scientists, of course, can help cavers understand the caves better and perhaps even guide further exploration.

Three relatively recent technical innovations have made a significant difference to the effectiveness of cave documentation: computers, the Internet, and global positioning systems (GPS). Computers and databases have revolutionized data storage space, retrieval, and presentation; the Internet and the web have revolutionized communication and helped reveal the existence of cave information from anywhere in the world, and permitted it to be remotely accessed and updated; and the GPS has simplified the positioning of caves, even where there are no landmarks at all, such as on the 1100-km Nullarbor Plain in Australia.

DOCUMENTATION

History

In the early days datasets traditionally consisted of an index card box for summaries and cartons or multiple filing cabinets for the detailed reports. Impressive datasets were built up by caving groups in many countries. Based on these, printed cave lists started to be published, such as *Pennsylvania Caves* in 1930 [1] and the Australian Speleological Federation's (ASF) national listing in *Speleo Handbook* in 1968 [2].

By the late 1960s, computerized cave databases had started to appear and then were helped along by the appearance of Digital Equipment's minicomputer and the desktop personal computer, both of which brought free computing within the reach of more cavers. Cave surveying and mapping programs started to appear, helped also by Hewlett Packard's small pen plotters.

In the 1980s came *geographic information systems* (GIS), and in the 1990s the web started to take off, enabling easy access to centralized databases. Today we are working with GPS, laser cave profile measurers, hand held in-cave map plotters, and even walk-through cave mapping using GPS principles with underground radio [10]. And since 2007 the *Karst Information Portal* (KIP) initiative [3] has provided a major advance in web access to cave information from around the world.

In 1986 UIS instituted its *Informatics Commission* (UISIC), whose role was to "encourage and facilitate the systematic collection and responsible use of cave, karst and related data on an international basis." To this end the Commission has operated a website since 1994 containing various aids to cave documentation, and has various projects in the pipeline. The Commission's *Cave Survey Working Group,* ably led by Philipp Häuselmann, has established standards for survey and mapping.

An example of a development from paper to computer database to publishing is ASF's *Australian Karst Index*, a project by this article's author. In 1972 I was trying to work out for ASF how we could automate storing, retrieving, and publishing cave information when I came across, in the United States, the *National Speleological Society's* (NSS) then defunct 80-column punch-card system, which had coded values for the field associated with each card column. Eureka! It sounds obvious now, but coded values and a fixed vocabulary for each field enabled me to design a tick-the-box-style 3-page form and a database with numerically coded values for most fields. With the help of free computing from my employer and a lot of hard work by caving clubs all around Australia we created a comprehensive national cave database which was then used to generate the national cave list by translating the codes into English and generating the masters for publication of the *Australian Karst Index 1985* [4]. In 1999 I converted this database to a fully relational one which ran on PCs, and since then it has been further developed by ASF for the web where it has been accessible online since 2001 and updateable since 2005, using ASF's open-source software [5].

Cave Documentation Issues

In managing cave documentation, the following issues commonly arise:

- *The difficulty of getting information at all*. To overcome human nature's inertia it is important to minimize impediments to reporting. For example, the Karst Institute in Slovenia was successful in receiving an impressive amount of information over the years by providing a post-paid, postcard-sized reporting form which cavers could simply fill in and drop in a postbox on their way home from a trip.
- *Cave identification*. To assign information to the correct cave on an ongoing basis there needs to be a method of identifying each cave and cave entrance. To avoid having to devise a unique name for every little cave, a cave numbering scheme is usually used and is usually hierarchical by state, area, and serial number, for example, 3B-5. This number is often marked at the cave entrance to positively identify the cave, and may also be the starting point for the cave survey.
- *Report types, filing, and retrieval*. In terms of filing and indexing paper documents there are basically three types: cave reports, trip reports, and the rest (books, scientific papers, newsletter and newspaper articles, theses, websites, *etc.*). The question is how to know that the information is there and how to access it in these documents.
- *Secrecy, trust, and publication*. Cavers generally don't like cave information to become available to the public. Do they trust the custodian? What is safe to publish? Get it wrong and few people will contribute data.
- *Safety of records*. Are they safe against loss by destruction and unauthorized dispersion?
- *Access policy*. To help ensure safety, some kind of published *Records Access Policy* is needed, and perhaps a *Data Use Agreement* for digital data.
- *Credibility*. If wrong or unreliable information gets into a database its credibility suffers and it therefore becomes much less useful. Data verification is important.
- *Standards*. What standards will be used for recording, indexing, storing?
- *Data transfer*. Invariably there comes a time when digital data needs to be transferred into or out of a system. What format will be used? How difficult will it be for the recipient to understand what they are receiving?

Recording

There are basically two scenarios: field recording of raw data from one cave or one trip, and then the back-office work consolidating, summarizing, and indexing accumulated data.

Recording can be helped by the availability of forms. For example, ASF has six forms which helped with its KID project: three field forms (cave report, trip report, sketch sheet) and three back-office summary forms (cave summary, area summary, and map summary); and the Victorian Speleological Association (VSA) has developed a form for comprehensively indexing any cave-related document. UISIC has a project to make forms available, including a cave summary form with customizable fields; these will eventually appear on the website. Recording directly into a database may be possible in some situations, such as into a hand held computer in the field for later uploading into a database, but paper forms are usually still convenient for initial recording and for consolidating information from various sources prior to data entry. It also helps others verify the data after data entry.

It is best to encourage recording at least the cave data while still on the trip. The longer the delay after arriving home, the less likely the data will be supplied.

Where data are not going into a database, one useful further recording step is at least the indexing of the documents, so that you know which documents to look in for details of a particular cave or speleological aspect. The simplest index would be a list of the trip reports or other documents with their date, title, and which areas and caves were mentioned.

Expedition recording is a whole subject in itself. Some of the issues are:

- Coordination, such as GPS data, camera timestamps, recording methods, and interfacing to members' instruments such as various GPS models. It is recommended that before an expedition a GPS datum be agreed on, all locations be recorded against that datum, all cameras be synchronized against local GPS time, and then the associated GPS date/time screen be photographed to allow accurate later auto-synchronizing of photos against GPS track points
- Consolidating data from the various expedition members' fieldbooks and instruments
- Issuing updated and consolidated data back to members to aid the next stage of exploration and mapping
- Publication of the expedition summary results
- The rational consolidation, storage, and distribution of the mass of detailed expedition results, such as cave descriptions, cave numbers, locations, photographs, specimens, participants, trip journal, *etc.*
- There are external issues as well, such as what expeditions by others are planned or have already been to the area, and the ethics to be observed when exploring in a foreign country
- Things that may help [6]:
  - Cambridge University Caving Club's (CUCC) *Troggle* expedition web database
  - The planned *International Caving Expeditions* (ICE) database, a combined European Speleological Federation (FSE) and UIS project

- VSA is developing a method to make all data from an expedition linked and accessible using an offline web browser, and with "KML" files for use with Google Earth
- UIS *Code of Ethics for Cave Exploration and Science in Foreign Countries* (available in seven languages)

Information Storage

Datasets can take many physical forms, ranging from paper files to spreadsheets to databases to GIS systems.

The traditional method of paper files will be with us for a while yet as that is how most of the data will be provided. But they do present a storage and retrieval problem. A lot of the information requires extraction or indexing before filing, so this can easily get behind in a volunteer environment. (By *extraction* is meant extracting the report's actual data into a database, and by *indexing* is meant cataloguing the report as to what type it is, what caves and/or areas are mentioned, and what kind and level of data are included.) The use of a separate sheet per cave after a trip (cave reports) can help, because then that sheet can simply be dropped into that cave's folder; thus all the data for a particular cave can be more readily located without any extra work. Obviously also extracting that data would provide even better access. Trip reports, on the other hand, often contain data on several caves and therefore some level of indexing is mandatory if the reference to each cave is not to be buried.

Any form of data extraction is subject to human error, so it can be an advantage to be able to easily refer to the original document if necessary. Scanning of the original documents is therefore a useful step, as well as it creating a backup copy. If the scans are checked carefully afterwards, the paper copy can then be discarded, so saving considerable storage space.

Today data extraction or indexing of original paper documents is likely to be in digital form, such as listed below. Because of the increasing level of skill and time involved, not all data may get to the database or GIS stage:

- Scans of original documents which are identified only by their filename
- Indexes of original documents listing their existence and preferably also their types of content
- Spreadsheets for simple data tables
- Databases allowing complex searching and reporting

- Survey data, being the measurements made in the cave or surface area, details of the people, instruments, and methods used, and usually also the resultant reduced data used for the subsequent map
- Cave maps drawn using solely computer methods and symbol libraries, some using the *Scalable Vector Graphics* (SVG) standard [27]; example vector software tools: CorelDraw, Adobe Illustrator, Inkscape (free SVG editor)
- GIS systems allowing the searching, manipulation, and presentation of combined spatial and nonspatial data [26]

Summary data and scans can also be stored in hand held computers (usually now with GPS) for taking back into the field. This can help record keepers when identifying and recording *new* caves.

To help the international use of data across a variety of spoken languages, UISIC has been promoting the storing of data as inherently language-independent numeric codes rather than as text in a specific language. These codes are part of extendable fixed vocabularies for each field, for example, each *rock type* would have a specific numeric code. When presented on the screen or in a report, the numeric code would be automatically translated into the chosen national language. UISIC has so far defined almost 700 fields and their codes on its website, some of them in German as well as in English [7].

The use of standards and guidelines can be a real benefit. Although not obligatory, their use can help with how to document caves, and also how to interpret documentation done by others. Here are some examples [8]:

- *UISIC existing*: Cave-mapping symbols, surface-mapping symbols, survey and mapping grades, database table designs, speleological subject classifications, multilingual word list [21] (340 speleo concepts in 12 languages)
- *UISIC in progress or planned*: Cave data field definitions and field value codes, extended subject classifications, data transfer and archiving format, minimum documentation standards
- *ASF existing*: Cave survey and mapping, cave naming, terminology, cave and karst numbering, map numbering

Retrieval

The only reason for storing data is to be able to retrieve it and present it later, often in specific relationships to other data. So retrieval ability is just as important as storage. For example, "show me the name, location and phone number of all owners of caves that contain protected biological species within 200m of this proposed highway route."

Paper-based retrieval of this kind of information can obviously be very tedious if not impossible in practice, hence the trend towards extracting as much information as possible into digital databases. But there is more to it than just the digitizing of the data if the retrieved data are to be usefully understood outside the immediate environment of its own database. For example:

- Differing understandings of apparently the same data
- Different formats and software for storing the data which inhibits transferring or consolidating data between systems
- Data spread over local differing databases, making the analysis of data over a wider area much more tedious
- Language problems when interpreting data internationally, for instance, when the data are in a foreign language

UISIC is working to alleviate these problems by promoting standard definitions for cave and karst data fields, numerically coded field values to help the language problem, and a common transfer format to assist data consolidation, exchange, and archiving [19].

The KIP, an international consortium including UIS, is working to permit international web-based access to cave and karst data from around the world, both by linking to existing databases and by creating new ones where the need arises [3]. Using portal techniques it is now possible to query and report on data across several independent databases in different locations in the world, though the more dissimilar they are the more set-up work is involved, of course.

A useful new way of displaying retrieved information is to show it overlaid on Google Earth by embedding the data in a KML file (Keyhole Markup Language). In addition to positional and extent information, this can also display other data values and photographs [25].

Surveying and Mapping

Surveying and mapping, both surface and underground, are major aspects of cave documentation. Maps can be used to provide clues for further exploration; structural detail for geomorphologists; locational detail for biological, cultural, or other cave contents (cave inventories); to assist with landowner issues and engineering works; and so on. They can also make good wall decoration of course.

Maps could be on traditional paper sheets, multi-sheet map books, computer screens, solid models, or GIS systems [24] (we're still waiting for full-size holograms). Computer screens could show them as plans, rotatable perspective 3D diagrams, walk-throughs, or deep, tiled zooms (such as on Google Earth) that show passage fine detail as well as an overview of hundreds of kilometers of passage. Laser stations can allow cave wall detail to be recorded and shown [9]. GIS systems allow the combining of spatial data with nonspatial data, and the visual presentation of the results of spatial queries on that data, for example, thematic maps [26]. Maps drawn using a computer and symbol libraries for wall and passage detail are also becoming common. The surveying and mapping standards mentioned above help with interpretation.

Traditionally, and also to a large extent today, surveying was done using a compass, clinometer, tape or equivalent, and fieldbook because these were cheap, robust, and readily available. But increasing use is being made of modern equipment such as laser rangefinders and even GPS-style radio location [10]. In-cave processing and plotting are now possible using handheld computers, for example, *Auriga* software [11]. The advantage of in-cave plotting to scale, whether by hand or computer drawn, is that measurement blunders are more immediately detected and so can be corrected on the spot. Something not often appreciated by surveyors is that much more time is spent back in the office producing the map than is spent doing the survey, so taking the extra time to get the survey right usually pays off—who wants to go back into a distant cave just to sort out a survey problem?

A common approach is to survey accurately on the surface by using precise survey instruments, existing detailed maps, or GPS to accurately locate the entrance(s) of a cave, and then adjust the typically less accurate underground survey to fit the accurate entrance positions. Where there is only one entrance, radio location equipment can sometimes be used to relatively accurately locate on the surface where the end of the cave is. These are part of the common technique of *closing the loop* to distribute the random unavoidable measurement errors around the survey.

Today several cave survey computer programs are in common use [20]. However, one issue is the different format each uses for storing the survey data, though to some extent they can convert another program's data. Another issue is the different data structure used in Europe and the United States. No commonly used data transfer format yet exists, though several attempts have been made [12]. This issue inhibits the consolidation of surveys done over a long period, and the long-term archiving of survey data independently of any particular software program.

A diagram on the UISIC data exchange website shows the various data items involved in a cave survey project and the relationships between them [13].

Data Safety

A significant issue with cave documentation is the safety of the data:

- The physical safety of the data—is there only one copy?
- Preventing the release of sensitive data, which could be to the detriment of the cave, its values, its contents, or relations with the landowner, manager, or data providers

Physically, there needs to be an up-to-date copy held at a different site in order to guard against total loss by fire, flood, burglary, or the like. Both locations must be secure, also against the dispersal of sensitive data, as there will be a reluctance to provide any data at all to a central repository if it is not perceived to be safe. While digital data are relatively easy to back up and store offsite, strict routines must still be established to ensure smooth and effective management of regular backups, and the restoration procedures in the event of loss.

A problem with digital data is that the whole data store is more easily stolen as it is so compact and easy to carry away, leaving no trace that it has been illicitly copied. Encryption can be a help. Holding the location data separately, and splitting it into several separately encrypted files, say, one per cave area, can also reduce the risk of complete dispersal.

Paper files are more difficult to back up, of course. Photocopying is expensive, can be unclear, and doubles the required storage space. Microfiche was a step forward in that the space taken was much more manageable, but it was still not cheap. But now color scanning is very cheap and practical, image resolutions can be chosen to show up obscure detail more clearly than the original, and the images can be used in day-to-day access as well as being a second copy for offsite backups.

Map backups are sometimes overlooked. Modern maps are often not a problem as they may already be digital, or at least drawn on an A4 or A3 sheet that can readily be scanned. However, sometimes there will be only one copy of a large-sheet map, perhaps painstakingly drawn many years ago using only manual techniques. Some possibilities include a commercial print or photocopy, a careful color photograph of the map with a high-resolution quality-lens digital camera, perhaps in several partial images later stitched together, or better, a commercial scan on a large scanner.

Sometimes government agencies will request data. Despite any iron clad assurances that the data will not be released to anyone else, experience has shown that this cannot always be relied upon. With staff changeovers and department reorganizations, the original agreements can soon be forgotten. They may be in writing, but after data escape it is too late: you gave your data to the Hydrology Department but now the caves have shown up on a public topo map by the Survey Department! It is certainly a dilemma—the information is there to be used, not hidden. One technique is not to provide bulk exact location data, only coarse locations, and to provide exact locations for specific caves only when actually required.

Publication

The cave and karst information collected is only useful if people know of its existence, so publication or at least reference to it in some form is desirable. Three useful examples that list published speleological material are shown in [14]. UIS also has a Documentation Centre [23], hosted by the Swiss Speleological Society, and with a network of co-operating libraries around the world.

Caves of ... books have always been popular with cavers, researchers, and land managers, but have also been contentious. Published locations have been the main issue. Some people just will not supply any information if it is to be published. A compromise that has worked is:

- Restrict sales to only the speleo and related community
- Publish only coarse locations, for example, to the nearest 10 km
- Provide information about whom to contact if more detailed information is needed

The web has made the publishing of cave data much more feasible and low cost. Web access can be either open to the public or graded by means of usernames and passwords. A web *database* can be particularly useful by allowing:

- Distributed updating of a central repository with automatic backups
- Wide and graded access
- Complex queries of the data
- A range of presentation alternatives
- With suitable coded data storage and programming, auto-translation of output into various national languages
- Online linking with other systems, for example, via the KIP

An example of a national web database is the KID in Australia. It follows the proposals of the UISIC field standards, and its software is open-source, free for others to use and develop further [15].

DATABASES

Introduction

Cave and karst databases have been referred to above but do require more elaboration. Basically, a database is the computer-based systematic storage of pieces of information in such a way that its contents can be queried, and the pieces of information that match the query can then be presented and consolidated in useful ways. For example, the results could be presented as a sorted table with just the required columns, or even as a generated thematically-colored map if the database was part of a GIS system.

Databases have become much more common in recent years and have generally been produced as a flow-on from earlier paper records, being a computer-searchable extraction and consolidation from them. But they are now also combining with the spatial capabilities of GIS systems so that we are seeing a convergence of factual data systems with mapping systems, with all their data capable of being queried, analyzed, and presented in much more useful ways. The UISIC website contains a list of cave datasets around the world, but is seriously out of date, so countries are encouraged to update their entries [16].

A spreadsheet is a common example of a database but a very rudimentary one: its columns represent the fields containing the different types of data such as *cave length* or *depth*; the rows represent different sets of those data for different items, for example, a row per cave; and the whole spreadsheet is, of course, a table representing a collection of one type, for example, caves. While a spreadsheet is convenient and easily available, as soon as the data relationships start to get more complex, such as further entities related to the caves such as maps, people, or photographs, or where fields need to be multivalued such as *pitch lengths*, then the spreadsheet becomes inadequate and a full *database management system* (DBMS) is needed.

Some examples of database software:

- Commercial DBMS: Access, Oracle, Filemaker, Paradox
- Free DBMS: MySQL, OpenOffice Base
- Commercial GIS: ArcGIS, MapInfo
- Free GIS: GRASS, QGIS

Components

When using databases it often proves handy to have at least a basic idea of their main components:

- *Entities:* An *entity* is a thing or event which we are collecting data about, such as a cave or a trip.
- *Fields:* Fields or *elements* are notionally labeled places in the computer where a particular type of data is stored, for example, the *cave name*, the *cave length*, *discovery date*, *owner type*, and so on. Where a table is being displayed, the fields are usually represented by the columns.

- *Records:* A *record* is like a row in a table, and represents a collection of field values for just one item, such as for one cave. The first field of each row is usually a special type of field called a *key field*, and serves to uniquely identify and distinguish that row or instance of the entity from all the other rows.
- *Tables:* As expected, a *table* is a collection of rows and columns, that is, a set of fields for a number of instances of the particular entity, for example, for a group of caves.
- *Linkages:* Sometimes the fields for one entity will be spread across several separate tables, perhaps for storage efficiency or perhaps to accommodate multivalued fields. The corresponding rows in each table are correlated by all having the same value for their identifying key field. These linkages are stored in the database and serve to assist in linking tables during various queries and views of the data.
- Where the database contains more than one entity, such as caves, areas, trips, maps, and people, these will also invariably be related to each other via linkages. For example, one of the fields in a cave's data might be specifying a particular map by containing the map's ID. This type of field in the cave's table is known as a *foreign key*, and serves to link the cave and map entities together.
- *Forms:* A *form* is a layout of fields displayed on the computer screen, perhaps combining fields from several tables and several entities to present a particular topic in the data. Forms are typically used to enter data into the database, as well as to display its data.
- *Queries and reports:* A *query* is a search of the database looking for records whose field values satisfy particular criteria. The result would typically be a table of matching records showing only the wanted fields. A *report* is a specification of usually a more formal layout for the presentation of the results of a query, often for a printed output. Once a query or report has been designed, and some are quite complex, it is usually possible to store them in the database for later reuse.

Database Issues

The following are some of the issues with using cave databases. Some apply to paper datasets as well, but they apply particularly to computerized databases.

- *Data accuracy and credibility:* When data are in a computer there is a tendency to believe that they are correct, yet there is plenty of scope for data to be wrong. Great care must be taken to verify that the data have been correctly copied or extracted from paper records, and that they consist only of facts, not opinions, guesses, or interpretations. If they do contain opinions, then the status of this information needs to be clearly evident to any user. Also: *when in doubt, leave it out*. Once a database loses credibility and cannot be relied on, its value will be much decreased.
- *Safety of the data:* The compactness of digital data makes unauthorized copying easier, even copying of the whole data store. There is a case for encryption and splitting up of the data.
- *Access policy:* As with paper files, there needs to be an access policy, but it also needs to take into account the ease of unauthorized copying. There may also be need for a *Data Use Agreement* to cover the use of the digital data.
- *Ownership and attribution:* This might become an issue in some circumstances, such as when data are being supplied to a third party for a license fee. If several clubs have contributed data to a centralized database, is the money to be split, and if so, how? One method would be to implement data attribution, that is, keep a tally of which organization supplied what data. UISIC has produced fields and example table structures to allow this [17].
- *Trust:* There is an almost universal desire among cavers not to let detailed cave information become public, so for a database to be successful in getting data contributions, cavers must trust that the custodian will safeguard it in accordance with agreed policies.
- *Specifications:* What fields will be included in the database? The fields and definitions being used need to be known by the users if they are to make effective use of it. There is a good case for using already defined standard fields to help understanding and data transfer [7].
- *Central repository:* Many databases have been initiated, designed, and operated by enthusiasts in local clubs. This means that for a while at least, they do have someone keen to look after them. However, the disadvantages are that (1) it depends on one person's continued availability, and (2) anyone wanting to analyze data over a wider region needs to separately negotiate and work with multiple databases, probably all of different designs. So, should the data instead be held in a central repository such as for a state, or better, for a whole country? This too can have disadvantages such as loss of local control and local enthusiasm, the "all eggs in one basket" syndrome, and harder staffing problems. One solution used in Austria is for the central national database to hold only summary data, periodically uploaded from local databases, and with the detailed data held only locally. This

also reduces the amount of central work needed. A similar approach has been used in Australia, though it is not yet working as well as in Austria.
- *Language:* Not everybody speaks English, of course, so UISIC has been encouraging the storage of data as language-independent numeric codes wherever possible.
- *Unique identifiers:* Each record in a database should have a globally unique identifier as its key field to prevent clashes and to identify the data if they are transferred to another database. In the case of cave records, this should be distinct from the normal public cave number as these are not necessarily unique on a global scale or permanent.
- *Data transfer:* The chances are that at various times data will need to be transferred into and/or out of a database. Unless transfer systems are set up, this can be a tedious, labor-intensive task. The obvious commonly used mechanism is the export and import of comma-separated-value files. However, when the databases are different, this needs investigation and manual manipulation, usually at the receiving end. This issue is discussed in more detail below.

Data transfer and archiving

There is often a need to transfer data between databases such as copying data from a club database into a state wide database, or supplying data to a government agency. Also needed is the archiving of data independently of any particular database software.

For a valid data transfer we need three things: an agreed meaning of the data, a unique identifier for each record, and an agreed transfer format:

- *Meaning:* The data must mean the same thing at each end of the transfer. For example, each end might have a field called *length*, but at one end the value might mean meters while at the other it might mean feet.
- *Identifier:* To avoid a possible clash at the target end, each end needs to use globally unique record identifiers. UISIC is recommending that these consist simply of a unique organization code identifying the original creating organization, followed by a serial number. They do not also need to indicate the type of record. The organization code consists of the ISO-3166 two-letter country code followed by a unique 3-letter organization code within that country [18].
- *Format:* So that the recipient can interpret the data received, it needs to be in an agreed format, for example, a special transfer format or markup language, or specified comma-separated-values.

UISIC has a project to define standards that permit these three requirements [19]. Part of this is a cave and karst data fields library. Although it currently contains almost 700 field definitions, many more fields are needed, so to be successful it will need the cooperation of cavers and karst scientists from the various disciplines to contribute any extra fields needed for their work.

The transfer format is planned to be a markup language called CaveXML [12] utilizing international XML (Extensible Markup Language) standards. It is expected to be compatible with the existing international standard Geographic Markup Language (GML).

Database Advantages and Disadvantages

Advantages

- A compact, convenient and systematic way of storing a large amount of cave and karst data
- Easily updated over the years without increasing the bulk
- Allows easy and tailored retrieval and presentation of the data
- Allows easy backup of the data
- Facilitates consolidation of the data to allow queries that could span a state, a country, or the whole world
- Allows a range of output formats for retrieved data
- Allows easy interfacing to spatial data in GIS systems
- A web database can allow easy remote access and remote updating, such as local clubs updating a state or national database

Disadvantages

- Requires a greater level of technical expertise to set up and operate
- Can increase the risk of data dispersal
- Is a secondary source, a step removed from the original documents, so is susceptible to transcription errors
- Requires periodical upgrading of the system to counteract the inevitable obsolescence of software and hardware

WHERE TO FROM HERE?

We still have a long way to go:

- To enable easy access to information from around the world we need further development and support of the KIP.
- To help with better recording of cave and karst data we need better communication and sharing of cave documentation know-how.
- To make cave data more usable by more organizations we need greater use of standards.

- To enable easy transfer of data between different organizations and research institutions we need our library of standard field definitions, our multilingual controlled vocabularies, and our standard transfer format.
- To build up our library of standard field definitions we need the cooperation of cavers and karst scientists for the various disciplines.
- To help UISIC implement its many projects to aid cave documentation we need more volunteers.
- We need accurate positions and altitudes for all our caves to assist geomorphological, hydrological, and other scientific studies.
- We need much greater uptake of GIS systems by caving clubs to make their hard-won information even more useful.
- We need GIS systems showing the karst areas in each country superimposed on existing protective reserves to see where further karst protection is needed.
- We need better access to information about expedition data.
- We need a greater level of awareness, understanding, and active cooperation between scientists and cavers to get maximum benefit from the scientists' knowledge and the vast amount of data that cavers collect.

These are just some of the challenges ahead.

Acknowledgments

My thanks first to Keith Wheeland who wrote the excellent first edition of this article, focusing mainly on the database aspects—most of Keith's points are included in this edition; he has also kindly perused this expanded version. Thanks also to all the cave documentation people in many countries who have shown me such hospitality and taken the time to explain their systems since my first *Cave Documentation Safari* in 1972. Thanks also to all the members of ASF clubs whose hard work and cooperation enabled cave documentation ideas to be turned into reality. And last, thanks to the editors of the *Encyclopedia of Caves* for the opportunity to include this cave and karst documentation article.

Bibliography

References to further detail are given by numbers cited in the text. Because of the large number of website citations, this nonstandard method of citation is used.

1. Stone, R. W. (1968). Pennsylvania caves. *Commonwealth of Pennsylvania Department of Internal Affairs, Topographic and Geologic Survey, Bulletin G3*, 1–63.
2. Matthews, P. (Ed.), (1968). *Speleo handbook*. Sydney: Australian Speleological Federation.
3. *Karst Information Portal (KIP)*. www.karstportal.org.
4. Matthews, P. (Ed.), (1985). *Australian karst index 1985*. Sydney: Australian Speleological Federation.
5. Lake, M., & Matthews, P. (2005). An open-source web-based national cave database. *Proceedings of the 14th International Congress of Speleology, 21–28 August 2005, Kalamos, Greece* (Vol. 2, pp. 473–475). Athens: Hellenic Speleological Society.
6. Expedition aids:
   - *The FSE/UIS International Caving Expeditions (ICE) database project*. www.eurospeleo.eu/index.php?option = com_content&view = article&id = 24&Itemid = 33.
   - Curtis, A. (2009). Troggle: A novel system for cave exploration information management. *Proceedings of the 15th International Speleological Congress, Kerrville, Texas, USA, July 19–26, 2009* (Vol. 1, pp. 431–436). International Union of Speleology. www.code.google.com/p/troggle.
   - *Victorian Speleological Association's (AU) web browser method* (in progress). Contact: matthews@melbpc.org.au.
   - UIS code of ethics for cave exploration and science in foreign countries. www.uis-speleo.org/ethic-en.html.
7. *UISIC's draft field definitions and numeric codes*. www.uisic.uis-speleo.org/exchange/atenlist.html.
8. Cave documentation standards and guidelines:
   - *General*: www.uis-speleo.org/guides.html.
   - *Survey*: www.uisic.uis-speleo.org/wgsurmap.html.
9. Canevese, E. P., Tadeschi, R., & Forti, P. (2009). Laser scanning use in cave contexts: The cases of Castellana (Italy) and Naica (Mexico). *Proceedings of the 15th International Speleological Congress, Kerrville, Texas, USA, July 19–26, 2009* (Vol. 3, pp. 2061–2067). International Union of Speleology.
10. Wenger, R., & Jeannin, P.-Y. (2009). Towards a positioning system for the subterraneous world (U-GPS). *Proceedings of the 15th International Speleological Congress, Kerrville, Texas, USA, July 19–26, 2009* (Vol. 1, pp. 612–617). International Union of Speleology.
11. LeBlanc, L. (2009). The Sierra Negra in a PDA: Expedition-wide electronic cave surveying. *Proceedings of the 15th International Speleological Congress, Kerrville, Texas, USA, July 19–26, 2009* (Vol. 3, pp. 2095–2099). International Union of Speleology. www.speleo.qc.ca/auriga.
12. Cave survey data transfer format attempt examples:
    - Dotson, D., & Wookey (1993). *HTO Survey Data Interchange Format*. www.chaos.org.uk/survex/cp/CP02/CPoint02.htm#Art6.
    - Lake, M. (2001). *CaveScript*. www.speleonics.com.au cavescript.
    - UISIC (2001). *CaveXML*. www.uisic.uis-speleo.org wgsurmap.html#exchange.
    - van Ieperen, T., & Petrie, G. (1997). *RosettaStal* (a format convertor). www.resurgentsoftware.com/rosettastal.htm.
13. Survey data model. www.cavexml.uis-speleo.org/task-sdm.html#diagram.
14. Some indexes to published material:
    - *UIS/SSS Speleological Abstracts/Bulletin Bibliographique Spéléologique (SA/BBS)*. www.ssslib.ch/bbs.
    - *Karst Information Portal*. www.karstportal.org.
    - Northup, D. E., Diana E. Northup, Emily Davis Mobley, Kenneth L. Ingham III, & William W. Mixon. (Eds.), (1998). *A guide to speleological literature of the English language: 1794–1996*. St. Louis: Cave Books.
15. Example national web database: *Australian Karst Index Database (KID)*. http://kid.caves.org.au/kid.
16. Country datasets: www.uisic.uis-speleo.org/contacts.html.
17. Example fields and table structure for data attribution: www.uisic.uis-speleo.org/exchange/histables.html.
18. *UISIC proposal for unique record identifiers*: www.uisic.uis-speleo.org/exchange/exchprop.html#rident & www.uisic.uis-speleo.org/exchange/exchprop.html#dident.
19. *UISIC data exchange project*: www.uisic.uis-speleo.org/exchange/exchprop.html.
20. Lists of cave survey programs:
    - www.resurgentsoftware.com/winkarst.html#Software%20Links/.
    - www.survex.com/related.html.

21. Multilingual word list and cave/karst glossaries:
    - Multilingual word list: www.uisic.uis-speleo.org/lexuni.html.
    - Glossary list: www.uisic.uis-speleo.org/lexintro.html#refs.
22. *UIS Informatics Commission*: www.uisic.uis-speleo.org.
23. *UIS Documentation Centre*: www.ssslib.ch/bbs/public/anglais/centreDoc.htm.
24. Example of mapping a long cave: Horrocks, R. D., & Austin, D. C. (2009). Making a digital map of Wind Cave, Wind Cave National Park, South Dakota. *Proceedings of the 15th International Speleological Congress, Kerrville, Texas, USA, July 19–26, 2009* (Vol. 3, pp. 2091–2094). International Union of Speleology.
25. KML for Google Earth: Wernecke, J. (2009). *The KML Handbook: Geographic visualization for the web*. Boston, MA: Addison-Wesley, Pearson Education Inc. Also: www.earth.google.com/support/bin/answer.py?hl = en&Also:answer = 148119.
26. Some GIS references:
    - General GIS introduction: www.rockyweb.cr.usgs.gov/outreach/articles/nss_gis_article.pdf.
    - Hose, L. (Ed.) (2002). *Journal of Cave and Karst Studies* (Special issue: Intro to cave and karst GIS), *64*(1). Huntsville, AL: National Speleological Society.
    - Examples of cave/karst GIS using ArcGIS: www.esri.com/industries/cavekarst.
    - MapInfo Professional: www.pbinsight.com/welcome/mapinfo.
    - *GRASS GIS*: www.grass.osgeo.org/ (free, open-source).
    - *Quantum GIS (QGIS)*: www.qgis.org/ (free, open-source).
27. Sample cave maps using SVG: www.carto.net/neumann/caving/svg.

E

ECOLOGICAL CLASSIFICATION OF SUBTERRANEAN ORGANISMS

Eleonora Trajano

Universidade de São Paulo, Brasil

Classifying, that is, grouping entities according to shared characteristics, is part of the human nature, constituting a first step to understand the surrounding world at different intellectual and emotional levels. Therefore, it is no surprise that the early researchers on cave faunas were concerned about classifying those organisms, not only into general systematic systems (reflecting their evolutionary relationships), but also into ecological systems, in view of the particularities of the selective regime in subterranean habitats, reflected in the morphology, distribution, and biology of organisms found there.

The hypogean (subterranean) domain contrasts greatly with the epigean one because of its permanent absence of light and, thus, of both photoperiodicity and photoautotrophy, tending toward food scarcity and environmental stability. Two different processes representing distinct steps in the evolution of subterranean lineages are *colonization*, which leads to the establishment of hypogean populations that spend at least part of their life cycles in this habitat, and *genetic isolation*, which may give rise to exclusively subterranean taxa. These two processes may be concurrent, but more often isolation postdates colonization and derives from independent events. A minimum requirement for successful colonization of the hypogean habitat is that colonizers not be dependent on light as a primary source of energy or for orientation cues. If the energy sources available in the hypogean habitat are sufficient for the physiological requirements of the species and the individuals are able to find food and mates for reproduction nonvisually, then they can complete their life cycle without leaving the subterranean habitat.

Subterranean populations are subject to selective regimes more or less distinct from those in the epigean environment, including the cessation of typically epigean selective pressures such as those related to light. If genetically isolated, such populations may differentiate due to the interruption of genetic flow from epigean individuals, causing cessation of indirect influence from surface conditions.

As soon as humans became aware of cave organisms, they noticed the existence of very different, "bizarre" animals living there together with "normal" (generally similar to epigean) animals. This led to the first classification of cave animals, published by J.C. Schiodte in 1849, with four categories based on habitat, namely, lighting and substrate: animals, respectively, in shadow, twilight, obscure zones, and on flowstone in the obscure zone ("animaux des ombres, crépusculaires, des regions obscures, et des regions obscures à concretions stalagmitiques"). A similar classification based on habitat was proposed in 1882 by G. Joseph, including: inhabitants of cave entrances, illuminated regions, and with variable temperature; inhabitants of middle, twilight regions; and, inhabitants of deep, aphotic regions with constant temperature (Racovitza, 2006).

In 1854, J.R. Schiner proposed a classification based on distribution and ecology: (1) *occasional visitors* (*hôtes occasionels*)—animals found in caves, but also at surface, everywhere "where they encounter conditions proper to their way of living"; (2) *troglophiles* (*troglophiles*)—animals living in regions where daylight still penetrates, which exceptionally may be found at surface or which have only light-loving representative forms; (3) *troglobites* (*troglobies*)—exclusively cavernicoles, never found in epigean habitats besides exceptional events such as floods (Racovitza, 2006).

This classification was modified in 1907 by E. Racovitza, who substituted "occasional visitor" with "trogloxene," redefining these categories: (1) *trogloxenes* (*trogloxènes*) are either lost or occasional hosts, the latter attracted by humidity or by food, but they do not live here always and not reproduce here; (2)

troglophiles (*troglophiles*) inhabit continuously the subterranean environment, but preferentially in superficial regions; they reproduce here often, but can be encountered also outside; (3) *troglobites* (*troglobies*) have as habitat exclusively the subterranean environment and stay preferentially in its deepest parts; they are very modified and offer the most profound modifications to life in dark. Troglophiles provide in all periods the main source of troglobites and they are the first colonizers when a new region of the subterranean environment is offered (Racovitza, 2006).

The *Schiner–Racovitza* classification is widely used by speleobiologists up to the present, being the basis for new proposals in the past century, which differed mainly in terminology and conceptualization of the third category. In fact, trogloxenes have been the main point of disagreement, including from animals with a well-defined ecological relationship with subterranean habitats, such as bats, to accidentals surviving there for variable periods. Some of the twentieth-century classifications were quite complicated, such as that by T. Pavan, in 1944 (Sket, 2008), which considered seven classes (eu- and subtrogloxenes, aphyletic and phyletic trogloxenes, sub- and eutroglophiles, and troglobites). Sket (2008) clearly stated the minimum criteria for an acceptable, biologically meaningful classification as being ecologically and evolutionarily adequate; applicable in association with the current state of biological knowledge; close to the traditionally used vocabulary; and having the greatest concordance with the historical priorities.

Some of the early classifications included the notion of "true" versus "false" cavernicoles, the former with features presenting a certain adaptation to the life in darkness and the latter without such characteristics, not differing from closely related epigean forms.

A first step to apply any classification is to define subterranean animals (encompassing populations in non-cave habitats). Trajano (2005) considered subterranean organisms those regularly found in the subterranean biotope and for which this is part of or the whole natural habitat. In contrast, *accidentals* may be introduced into caves by mishap (by being washed into caves or falling through upper openings, for instance) or when entering in search of a mild climate; although accidentals can survive temporarily, their inability to properly orient and to find food leads to their eventual demise.

From the evolutionary point of view, subterranean organisms (cavernicoles *sensu latu*) may be defined as evolutionary units responding to subterranean selective regimens. Subterranean habitats would provide resources—for example, food, shelter, substrate, climate—which affect survival and reproductive rates. Such units have a historical connectivity and therefore may be classified as systematically meaningful biological systems.

In contrast, accidentals are evolutionary *culs-de-sac*. From an ecological point of view, accidentals are potential resources for subterranean organisms (food, substrate, *etc.*). Resources *per se* have no historical connectivity; therefore, when an organism becomes a resource, it makes no sense to classify it further, identifying species or higher categories (a biologically meaningful classification of food would be according to its accessibility, such as availability, protection against predation, *etc.*), nutrition value, and so on. Moreover, accidentals are grouped by a negative trait (they are not subterranean organisms, as herein defined); therefore they do not constitute a biologically meaningful group. In conclusion, it is clear that the "accidental" concept has a different nature and therefore should not be included in the same Schiner–Racovitza system.

The North American speleobiologist T.C. Barr provided objective and clear definitions, followed by many other specialists such as Holsinger and Culver (1988): (1) *trogloxenes* are regularly found in the subterranean habitat, but must leave it during some period(s) in order to complete their life cycles (usually because hypogean food sources are insufficient for the species requirements); (2) *troglophiles* are able to complete their life cycles both in the hypogean and in the epigean environment, forming populations in both habitats, with individuals commuting between them and maintaining genetic flow between these populations; and (3) *troglobites* are species restricted to the subterranean domain and are usually characterized by apomorphic character states related to the hypogean life (*troglomorphisms*) such as reduction until loss of eyes and dark pigmentation. A fourth category, that of accidentals, was also mentioned. Sket (2008) suggests, for semantic reasons, that *troglobiont* is preferable to *troglobite*.

Some authors use distinct sets of words for subterranean aquatic species—*stygoxene*, *stygophiles*, and *stygobites* (*stygobionts*). This terminology has been proposed by Gibert *et al.* (1994) as descriptive of groundwater fauna, because it includes organisms living in nonkarst subterranean habitats, such as porous aquifers.

Another version, used by some Europeans, is based on Pavan–Ruffo's classification (Sket, 2008), which differs basically in terminology: *eutroglophile* instead of *troglophile* in Barr's classification, *subtroglophile* instead of *trogloxene*, and *trogloxene* instead of *accidental*. These terms are misleading because they tend to suggest an evolutionary progression or incomplete adaptation. Subtroglophiles are not quasi-eutroglophiles. At a given moment, each population is either a troglophile or a trogloxene (in Barr's sense). Because food is frequently the limiting factor, individuals may leave hypogean habitats periodically to eat (thus becoming trogloxenes), under special circumstances, and in highly eutrophic caves (*e.g.*, bat caves), individuals of such species may form troglophilic populations. However, this corresponds to a very particular ecological situation, not representing an evolutionary progression. In fact, as stated by Sket (2008), the boundary between epigean and hypogean habitats and between competing surface and cavernicolous animals may change (move) if trophic resources are enhanced. Other ecological traits, such as differences in population densities, are also irrelevant, because they represent responses to specific environmental conditions. It is noteworthy that, probably due to higher competition pressures, for a given species the densities of epigean populations are frequently lower than those of the troglophilic ones.

One of the most interesting and useful recent concepts in speleobiology is the distinction between source and sink populations: a sink population, if cut off from all other migrants, eventually becomes extinct, whereas a source population has excess production and continues to grow if isolated (Fong, 2004). Sink populations are habitat-level phenomena, corresponding to stranded groups of individuals in habitats less than suitable (in terms of space, food, and other resources necessary for self-sustained, source populations). Therefore, their presence in such habitats is unpredictable.

When applied to the Schiner–Racovitza classification, this conceptualization leads to clearer and biologically meaningful definitions of the three categories classically recognized: (1) *troglobites* (stygobites) correspond to exclusively subterranean source populations; sink populations may be found in surface habitats; (2) *troglophiles* (stygophiles) include source populations both in hypogean and epigean habitats, with individuals regularly commuting between these habitats, promoting the introgression of genes selected under epigean regimes into subterranean populations (and *vice versa*); (3) *trogloxenes* (stygoxenes) are instances of source populations in epigean habitats, but using subterranean resources (in the so-called obligatory trogloxenes, all individuals are dependent on both surface and subterranean resources). It is noteworthy that sink populations do not fit the Schiner–Racovitza scheme, unless one considers stranded troglobitic or stygobitic individuals as part of the subterranean source population from which they originated.

Bibliography

Fong, D. W. (2004). Intermittent pools at headwaters of subterranean drainage basins as sampling sites for epikarst fauna. In W. K. Jones, W. V. Coord, D. C. Culver & J. S. Herman, *Epikarst* (pp. 114–188). Proceedings of the symposium held October 1 through 4, 2003, Shepherdstown, Special Publication 9. Charles Town, WV: Karst Waters Institute.

Gibert, J., Stanford, J. A., Dole-Olivier, M. J., & Ward, J. V. (1994). Basic attributes of groundwater ecosystems and prospects for research. In J. Gibert, D. A. Danielopol & J. A. Stanford (Eds.), *Groundwater ecology*. San Diego, CA: Academic Press.

Holsinger, J. R., & Culver, D. C. (1988). The invertebrate cave fauna of Virginia and a part of eastern Tennessee: Zoogeography and ecology. *Brimleyana, 14*, 1–162.

Racovitza, E. G. (2006). Essay on biospeleological problems: French, English, Romanian. [Facsimile of the publication *Essai sur les problèmes biospéologiques* (1907), translated by D. C. Culver & O. Moldovan.] Cluj-Napoca, Romania: Institut de Speologie "Emil Racovitza."

Sket, B. (2008). Can we agree on an ecological classification of subterranean animals? *Journal of Natural History, 42*(21–22), 1549–1563.

Trajano, E. (2005). Evolution of lineages. In D. C. Culver & W. B. White (Eds.), *The encyclopedia of caves* (pp. 230–234). San Diego, CA: Academic Press.

ENTRANCELESS CAVES, DISCOVERY OF

Nevin W. Davis

Butler Cave Conservation Society, Inc.

INTRODUCTION: HUMANKIND'S NEED FOR EXPLORATION AND DISCOVERY

Humankind has always had the need to explore and use caves. Early Americans entered caves in the Virginias, Kentucky, and elsewhere. Polynesian Hawaiians have explored lava tubes on the Big Island of Hawaii from early times to the present. The Hawaiians used the tubes for ceremonial purposes, collecting water, and burials, but from the evidence left by charcoal from cane torches in remote places, they also must have enjoyed exploration for its own sake. Today there is a large group of cavers whose exploration instinct is satisfied with the traversing of known passages and the mapping or remapping of passages that are already "in the book" or "previously enjoyed" (slang terms for cave passages that are well known by the caving public), but there is also a select group of cavers who are satisfied only with exploration and mapping of "virgin" passages. So the question arises "What is a virgin cave passage?" Purists would say that a virgin cave passage is a passage that has never been seen by humans. A looser definition would be a passage that has never been recorded or described even though it may have been viewed years before or by aboriginal man. In this article we use the first definition.

A virgin cave passage can be found in known caves. Many times exploration will yield a passage that appears to have never before been entered. A passage that has no footprints or scrape marks is a good candidate for being virgin but there is no assurance in today's world of well-explored caves unless some obstacle is passed that would preclude others from having been there. These include but are not limited to the following:

1. Water-filled sumps that have no written or oral record of being dived
2. Very remote passages in a project cave that no casual explorer would have gotten to
3. Passages that are blocked by breakdown or dirt fill that, when removed, reveals extensions
4. Deep pits

Virgin cave passages can also be found in another interesting way. Suppose the cave is totally unknown and has no entrance. (An entrance is an opening large enough for a human to enter.) Absent an entrance a cave is very likely to be virgin. In Virginia the number of entrances per cave varies widely. One cave has 16 entrances, Clark's Cave has 8 entrances, most caves have 1 entrance, but some of the best known caves originally had no passable entrance. They were dug open. One of the best known of these is Butler Cave—Sinking Creek System with 26.8 km of passage that was dug open in May 1958. Coincidentally, in the same year Rane Curl published a paper entitled "A Statistical Theory of Cave Entrance Evolution" (Curl, 1958). In it he came up with a statistical way to predict the number of entranceless caves in a region with a distribution of caves with 0 through 4 entrances. For example, in West Virginia at that time, out of a population of 257 caves with lengths greater than 30 m, 228 caves had 1 entrance. His theory gave the number of entranceless caves in this area as 2405. This is an impressive number, but the theory also leads to a distribution function of lengths of caves within this set with 0 through 4 entrances. Unfortunately, 50% or 1200 of the entranceless caves would have passage lengths of less than 55 m (1200 would also have lengths greater than 55 m), while 200 of the single-entrance caves would have a length greater than 55 m. This means that any successful effort to dig open a cave would be met with a large number of short caves!

Fortunately, the search for entranceless caves in the real world is quite different from the theoretical approach. In the real world, some intelligence can be applied to the effort to better the odds of finding something significant. Here are some recent numbers from the Virginia Speleological Survey database. In Virginia there are 2601 caves with a length greater than 9 m. Of this population, 2066 caves have one entrance and 251 caves were dug open. The average length of the single-entrance caves is 239 m, while the average length of the formerly entranceless caves is 809 m. Clearly something is happening here! Cavers are not digging for short caves or are only reporting their successes.

ENVIRONMENTAL SETTINGS OF ENTRANCELESS CAVES

The discussion that follows is based on experience in the Appalachians in limestone terrains. Insight that applies to the Appalachian Mountains in the eastern United States should also apply to many other limestone terrains with significant rainfall. The environmental variables that produce significant long and deep known caves obviously likewise apply to entranceless caves. These include the following:

1. Large upland catchment areas directing the runoff into sinkholes and losing streams. Catchment areas can vary from under 260 hectares to more than 26,000 hectares.
2. Entrenched rivers with large karst springs.
3. Thick, contiguous beds of limestone.
4. Insoluble caprock that overlies some portion of the limestone, protecting it from surface solution. This condition occurs only in small areas in the folded Valley and Ridge Province of the Appalachians but is found over large areas in the Cumberland Plateau and in Central Kentucky near Mammoth Cave.
5. Moderately high relief. Relief of 30–60 m, as is found in the valleys of central Pennsylvania, leads to much shorter caves than are found in the Burnsville Cove in west central Virginia, where the relief from the highest cave entrance to the spring exceeds 244 m.

All of these environmental variables need not be in their most desirable range to produce long and deep caves. Hosterman Pit in central Pennsylvania sits under a hill in the center of Penns Valley. Nearby Pine Creek is only entrenched 60 m below the hilltop, the karst springs run at less than 30 liters/sec, and there is no caprock or large catchment area; however, the cave has more than 1.6 km of passage and is 59 m deep.

Another consideration in searching for caves is entrance lifetime. Caves are long-term features under the landscape with lifetimes measured in millions of years, whereas entrances to them are fleeting features with lifetimes measured in millennia. One factor affecting entrance lifetime in a temperate climate is airflow. Bedrock in any climate will assume the mean temperature of the region. This, of course, means that the air temperature in the depths of the caves surrounded by that bedrock also has this temperature. In the winter, outside air temperatures are often below freezing in the Appalachians while the cave temperatures are near 10°C. Almost all air movement in caves is driven by the chimney effect, that is, warm air rises. This means that warmed cave air exits the upper entrances of caves to draw the cold outside air into lower entrances. Eventually, the temperature of the rock at lower entrances drops below freezing causing any water in the rock to freeze, expand, and fracture the rock. This can also occur in caves with only one entrance because significant quantities of air can penetrate the soil, covering potential high entrances including domes and high ceiling fissures. The large surface area over a cave means that air velocities through the soil can be insignificant, while the velocities in the smaller cave passages can be meters per second. The upshot of all of this is that lower entrances to caves have very short lifetimes.

Another consideration in the Appalachians where most of the ridges are covered with deciduous trees is that leaves and twigs will soon cover and block small vertical entrances. Pits less than a meter in diameter can be totally blocked in one season. Leaves blocking a small entrance are soon followed by roots and more leaves and it is not long before all traces of an entrance are gone. Humankind does not help this condition much since it seems that a natural instinct is to cover holes with poles (for safety?) and the cycle of covering is repeated in an unnatural way, unless of course someone steps on the hidden, rotten poles in a few years and falls into the pit.

Humankind also has had an effect on valley and ridge margin sinks, all potential cave entrances. Any depression extends an invitation to be filled with trash. Many cave entrances have been lost this way. In a sinkhole dig in Burnsville Cove, Virginia, a domestic pig skull was found under 2 m of soil with some 50- to 70-year-old domestic trash near the surface. Because this sinkhole receives no drainage from nearby fields, the 2 m of compacted soil covering the skull took no more than 200 years to accumulate from the hill above. This gives an idea of rates of infilling of small sinks.

STALKING THE ELUSIVE ENTRANCELESS CAVE

The same moving air that can destroy a lower entrance to a cave can help find an upper entrance. In the winter when the ground is covered with snow, potential cave entrances can be found by looking for melted spots. Remember that air does not necessarily need a large opening to move through. It can move through a soil covering as well as loose rocks. The most sensitivity to this effect can be had with early season snows before the ground freezes. This will also lead to more false positives because rotting stumps and other such sources of heat can also melt the snow. The best condition to search for snow melt is with a new snowfall in midwinter with an overcast sky, since sunlight can also give false positives by shining through shallow snow cover onto rocks and melting the snow. This is a tried-and-true method that has led to countless new caves.

Stick to looking in smaller sinks. The smaller the better because larger sinkholes have more material blocking a potential entrance and will require a more protracted dig. Give preference to sinks with a bedrock outcrop. These are more likely to lead to a cave if for no other reason than that there is at least one solid wall to follow when digging!

A straight 1.5-m digging bar can be a helpful tool to probe suspected sites. In a small sinkhole, if the bar sinks easily into loose, organic material this is a potential dig site. Many times a 1.5-m bar will punch through into space at the top of a dome pit or fissure. If the bar indicates that the fill in a sink is consolidated material without organic content, it is probably a sign to look elsewhere.

The topography in the Valley and Ridge Province of the Appalachians features major parallel ridges of clastic rocks and oftentimes karstic valleys. This means that the clastic ridges serve as the catchment areas to direct recharge into the valley margin sinkholes. These sinks, although they are potential entrances to the caves that underlie the valleys, are poor choices for dig sites unless they are known to have been entrances in the past. The reason for this is that the bed load in the sinking streams tends to choke any entrances. An effort to open entrances in these locations may require equipment such as a trackhoe to get anywhere. At the base of Jack Mountain near Burnsville Cove in Virginia, Cycle Sink provides an example of this. This sink was a collapse in the bed of a sinking stream that reached 6 m depth without digging. Digging efforts there resulted in very little progress until a storm completely removed the collapse, filling it and leaving no sign that it ever existed. The stream now flows past this area before it sinks.

In a similar situation where the surface is composed of limestone outcrops, cavers did not dig at the stream sink. Instead, at a Giles County site in Virginia, the dig was 9 m from the stream sink and above it. This was a place where some air exited. The dig is now the entrance to a significant Virginia cave.

Traditionally, the way to find caves is to talk to landowners, hunters, ginseng hunters, and other outdoor people. The problem often is asking the right questions, particularly when trying to find entranceless caves. Ask about snow melt spots or places that may have entrances which are now covered over. If all else fails in the search for the elusive entranceless cave, go where there are already known long and/or deep caves. Search nearby or in an adjacent drainage and success will surely follow.

EXAMPLES

The Burnsville Cove in west central Virginia consists of a synclinal valley surrounded by ridges composed of clastic rocks. The valley is bisected by Chestnut Ridge, an anticlinal structure. Butler Cave–Sinking Creek System and its drainage lie on the western side of this ridge. The entrance to Bobcat Cave, a significant cave at 15 km in length, lies high on the western side of the ridge but the cave has passage on both sides of the ridge. Of special interest to cavers is the drainage on the eastern side of the ridge, especially to the southwest where the recharge area and the majority of cave passage must be. Bobcat Cave ends in breakdown and tight fissures to the southwest. Ridge-walking by the author over the years has led to the discovery of innumerable snow melt spots on Chestnut Ridge. In March 1991 a crew dug at one of these, a 1.2-m-diameter sink high on the eastern side. In 20 minutes a narrow fissure had been opened. Below was a 9-m drop followed by a second drop into a room. Two more drops lead to Strychnine Canyon, a tight, crawly passage over 274 m long that eventually ends in a large phreatic passage at Ghost Hall. This is the beginning of 7.5 km of passage that eventually connects to Bobcat Cave. The ultimate goal of finding a series of passages upstream to the recharge area has not been met, but 20 minutes of work to open access to more than 7.5 km of cave passage isn't bad!

Buckwheat Cave is a small, pleasant, 670-m-long cave in a ravine on the western side at the base of Chestnut Ridge. The entrance is 4.6 m above the base of the ravine on the north side. An ephemeral stream collects in the ravine and only flows past the entrance in the wettest of times. This cave is part of the population of entranceless caves that were opened by digging. With this knowledge cavers looked at the next ravine to the north. Topographically, it is very similar and even has a small sink 1.5 m above the bottom of the ravine on the north side. Stratigraphically, the sink is located in the same limestone unit as Buckwheat Cave. The sink is 2.4 m in diameter and only 0.6 m deep. It had no snow melt and the fill was rather tight and inorganic. Optimistic cavers spent one day digging there in March 1999 and after sinking a 4-m shaft, opened aptly named Blind Faith Cave. The cave is now 1019 m long and 48 m deep and comes very close to connecting to another excavated cave, Battered Bar Cave.

The first example described here is an illustration of the benefits of following the air. The use of snow melt in finding new caves is a powerful tool that can lead to minimum effort in opening them. The second example shows that intimate knowledge of a particular area can lead to discovery even if the signs are not otherwise favorable.

See Also the Following Articles

Entrances
Geophysics of Locating Karst and Caves

Bibliography

Curl, R. L. (1958). A statistical theory of cave entrance evolution. *National Speleological Society Bulletin, 20,* 9–21.

ENTRANCES
William B. White
The Pennsylvania State University

If a cave is to be accessible to human exploration it must have an entrance. Entrances, however, are rarer than might be expected. For the most part, the processes that form caves do not demand that they have entrances. If a cave entrance was created as part of the original cave-forming process, the entrance can easily be destroyed by later collapse and slumping. Most entrances are formed late in the history of caves by random processes of erosion and collapse. Some caves may have no entrances at all (White, 1988).

LOCATIONS OF ENTRANCES

The search for caves by cave explorers is a search for entrances. Some entrances, of course, are obvious—the storybook picture of a gaping hole in the hillside. These caves have generally been known for centuries. Other entrances are small, obscure, and often go unrecognized until the systematic combing of an area is undertaken. Still other entrances have been opened by human activities. They are found in road cuts, rock quarries, and by deliberate excavation. The search for cave entrances has become more intense as the more obvious openings have been discovered and the caves explored, surveyed, and described. Much exploration is now focused on finding concealed entrances. These are places where cave passages intersect the Earth's surface but where collapses, slumping, and other occlusions have erased all traces on the land surface. At such locations, a relatively modest amount of excavation may expose the cave passage.

TYPES OF CAVE ENTRANCES

Cave entrances come in three broad categories: (1) those that are an intrinsic part of the cave-forming process so they are the same age as the cave; (2) those that formed by later processes of truncation and collapse and thus are younger than the cave; and (3) those that formed by human activities such as quarrying and road building—these entrances are the youngest of all.

1. *Entrances formed as an intrinsic part of the cave-forming process:*
   Entrances at sinking streams
   Entrances at springs
   Pit and fissure entrances
2. *Entrances formed by later processes of truncation and collapse:*
   Cave passages truncated by valley deepening
   Entrances formed by sinkhole collapse
3. *Entrances created by human activity:*
   Entrances in quarries and road cuts
   Excavated entrances

Whether or not a sinking stream results in a cave entrance depends to some extent on the magnitude of the sinking stream. Large streams tend to produce large entrances. However, large streams can also close up cave entrances. Many stream entrances are blocked with stream sediment as well as logs and debris that have been washed into the entrance by floods. The stream flows through the permeable debris pile but explorers cannot follow. Stream sinks are often regarded as possible sites for digging but there may be tens of meters of flood debris between the surface stream and the underlying cave passage. Of the very large number of sinking streams, only a few are open cave entrances.

Some springs issue from open cave mouths, providing an entrance that is easy to enter (Fig. 1). Other springs are flooded but may be penetrated by divers. Often the passage flooding occurs only near the spring and a short distance upstream is the open cave passage. Other springs are blocked with rubble, permeable to the water but not to would-be explorers. Some dry entrances to caves are, in fact, abandoned spring mouths.

Pits and shafts are sometimes an integral part of the underlying cave system and sometimes are younger features that formed from infiltrating surface water more recently than the primary portion of the cave. Although pits and shafts are dissolved by seepage water clinging to the walls of the shaft and do not require streams of water for their development, some shafts are the sink points for small surface streams. Some shafts are open to the surface and some are not.

FIGURE 1 Overholt Blowing Cave, Pocahontas County, West Virginia. A cave entrance at a spring mouth. *Photo by the author.*

Open shafts can provide a natural entrance to the underlying cave system. However, because the development of the shaft is by independent water sources, the shaft may be connected to the underlying cave only by a small and sometimes impenetrable drain. Other drains may be blocked by loose rock and other debris that have fallen down the shaft.

Caves tend to form as near-horizontal passages directed toward major surface valleys. As the valleys deepen and widen, the caves are gradually shortened as valley walls retreat. With time, tributary valleys deepen and may cut directly through underlying cave passages. Rubble associated with the collapse of the tributary valley floor will cut the cave into two segments and may completely block both ends. Continued downcutting of the tributary valley can remove some or all of the rubble so that one or both of the cave passage segments become exposed on the valley wall.

Once cave passages are formed, they remain in place beneath the landscape. The landscape does not remain in place. Continuous erosion and valley deepening lowers the land surface so that the amount of rock above the cave passage gradually decreases. The roof of the cave loses the ability to support its own weight. Descending water from the land surface can enlarge joints and further weaken the roof until the cave ceiling collapses. The collapse also truncates the underlying passage. One or both ends may remain open on the sides of the sinkhole (Fig. 2).

Cave entrances are found at many size scales from the very large (Fig. 3) to moderate human-size to the very small (Fig. 4). There is no relationship between the size of the entrance and the size of the cave. For this reason it is necessary to check out every opening with a size sufficient to admit a human being (Fig. 5).

FIGURE 2 Tytoona Cave, central Pennsylvania. The entrance is formed by the collapse of the roof of a large passage carrying an underground stream. *Photo by the author.*

FIGURE 4 A small entrance: Roadside Pit, Pocahontas County, West Virginia. A small crevice in a road ditch opens into the top of a 20-m pit, which in turn opens into a large cave. *Photo by the author.*

FIGURE 3 A large entrance: Fortress of Ponor, Apuseni Mountains, Romania. Note human figures in lower left of the photograph. *Photo by the author.*

FIGURE 5 Solution conduit in a quarry wall. Every opening must be checked. *Photo by the author.*

Many cave passages are blocked, and thus do not provide entrances, because of the soil slumping and collapse that is a normal part of the weathering and retreat of valley walls. Such blockages are often quite shallow. Human activities that scrape away the weathered zone are, therefore, quite likely to produce entrances and expose the underlying cave passages. It is for this reason that so many cave entrances occur in quarries, road cuts, and other excavations. Figure 4 shows an entrance in the ditch of a country road that was opened by building the road. This innocuous opening is the top of a 20-meter pit that opens into a cave of substantial size.

STATISTICS OF CAVE ENTRANCES

Most caves have only one entrance but some have two entrances and a few have more than two entrances. Because entrance formation is primarily a stochastic process, one might expect a statistical relationship between the number of caves with a given number of entrances and the number of entrances and, indeed, this is the case (Fig. 6). For three populations of caves in Pennsylvania, West Virginia, and Alabama, the number of caves falls off systematically with the number of entrances. This leads to very interesting speculation. Because the number of caves with one entrance is

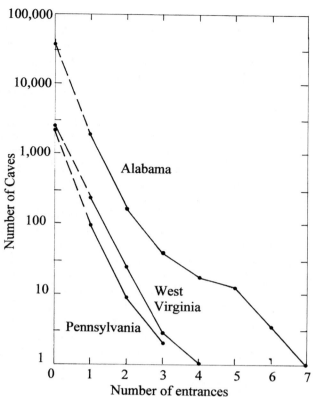

FIGURE 6 Number of caves as a function of the number of entrances. Data for Alabama are based on 1973 data of the Alabama Cave Survey. Data for Pennsylvania and West Virginia are from Curl (1958). The curves are extrapolated to predict the number of caves with zero entrances.

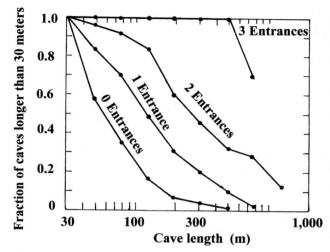

FIGURE 7 Distribution of caves in West Virginia as a function of number of entrances and cave length. The number of all caves decreases as their length increases (that is, there are many more small caves than there are large caves, regardless of number of entrances). Overall, caves with multiple entrances tend to be longer. Caves with zero entrances are predicted to be, on the average, smaller but also with fewer and fewer caves of greater length. After Curl (1958).

much larger than the number of caves with more than one entrance, the curves can be extrapolated to predict the number of caves that have zero entrances. There is predicted to exist a very large number of caves with no human access from the surface. A few of the previously entranceless caves have been exposed by human excavations. Consideration of the large distances between stream sinks and their resurgent springs suggest that many caves must exist that do not have entrances.

Because the most important mechanism for the formation of entrances is random truncation and collapse of cave passages, larger caves are more likely to have entrances than smaller caves. This hypothesis is supported by the distribution of caves by length (Fig. 7). Caves with larger numbers of entrances tend to be the larger caves. The hypothetical population of entranceless caves is likely to consist mostly of small caves.

ENTRANCES AS HABITAT

Although entrances might appear to be a sharp boundary between the outside surface environment and the inside cave environment, they are, in fact, transitional. Ecologists call such transitions *ecotones*. Outside is the surface environment—a sun-drenched hillside, a forest, a deep sheltered valley, or a desert. Inside is the cave environment—complete darkness, high humidity, and constant temperature. The boundary between inside and outside is often taken by cave surveyors as the *drip line*, that is, the line marking the last spot where rainfall reaches the ground. In the outer transitional zone, cool, moist cave air creates a microclimate that is cooler in the summer and warmer in the winter than the surrounding terrain. Ferns, mosses, and flowering plants, some of them rare, grow in these sheltered locations. The inner transitional zone is a *twilight zone* where enough daylight penetrates to permit the growth of some plants. The width of the transitional zone depends on the local setting. It may vary from a meter or less for a crawlway entrance on a cliff face to hundreds of meters for a large cave entrance at the head of a narrow valley.

The cave entrance zone is, in effect, a continuous sequence of microclimates, each segment of which can provide habitat for organisms. Organisms can migrate from the surface environment through the outer and inner transitional zones into the deep cave environment.

ENTRANCES AS PALEONTOLOGICAL AND ARCHAEOLOGICAL SITES

Cave entrance areas provide shelter for animals and provided shelter for early humans, who made their camps, either migratory or semipermanent, in appropriately shaped entrances. The best sites were entrances on south-facing slopes that were large enough for easy movement. Early humans typically left their debris—remnants of meals, scraps of clothing and tools, campfires, and sometimes burials. Predatory animals often dragged their prey into cave entrances. The bones remained behind.

Cave entrances are not stable. Because of freeze/thaw cycles and other hillside erosion, there is a continuous, if sparse, rain of soil and rock fragments into the sheltered area of the cave entrance. This pile of weathered material is known as *entrance talus*. Because of movement of material downslope, the entrance not only migrates deeper into the hill but also migrates upward as rocks break away above to contribute to the talus pile below. Entrance talus piles can vary from very small to depths approaching 100 m. The continuous accumulation of entrance talus buries animal remains and human debris, which are entombed in rough stratigraphic sequence as the talus pile enlarges. The sediments in cave entrance areas are, therefore, often excellent sites for paleontological and archaeological investigations.

See Also the Following Article

Entranceless Caves, Discovery of

Bibliography

Curl, R. L. (1958). Statistical theory of cave entrance evolution. *National Speleological Society Bulletin, 20*, 9–22.

White, W. B. (1988). *Geomorphology and hydrology of karst terrains.* New York: Oxford University Press.

EPIKARST

Michel Bakalowicz

Hydrosciences, Université Montpellier 2, France

DEFINITION

The word *epikarst* (in French: *épikarst*) gradually came into use during the 1990s following the definition by Mangin (1973) of the *epikarstic aquifer* (in French: *aquifère épikarstique*). According to Mangin, the epikarstic aquifer is the perched saturated zone within the superficial part of the karst that stores a part of the infiltrated water. The term *epikarst* is a generalization of the concept of the epikarstic aquifer. It is the shallow, superficial part of karst areas, in which stress release, climate, tree roots, and karst processes fracture and enlarge rock joints and cracks, creating a more permeable and porous zone over the massive carbonate rock in which only few open vertical joints and fine cracks occur (Fig. 1). The epikarst overlies the infiltration zone itself, which is intersected by occasional enlarged vertical fractures and karst conduits, so that the base of the epikarst acts as an aquitard and may contain a local perched water table, the so-called *epikarstic aquifer*. *Epikarst* generally corresponds to the *karren* (or *lapiaz*) zone of geomorphologists and speleologists (White, 1988). According to Williams (1983), dolines should be considered as part of it.

Instead of epikarst, Williams (1983) preferred *subcutaneous karst*, which, according to him, was an expression translated from the French *karst sous-cutané* and proposed by Birot to describe the shallow part of karst morphology in tropical regions. However, Birot did not use that expression, but instead used either *superficial*, *subsuperficial*, or *subepidermic*. Prior to this, Ciry employed the term *karst cutané* (cutaneous karst), referring specifically to the shallow caves and karst features of Burgundy, supposedly developed during glacial periods, when the permafrost limited karst processes during summer melting. It is now known that no karst processes occurred during glacial periods and that the *karst cutané* in Burgundy is what remains of ancient karst phases after the reduction of thickness through surface erosion.

The expression *subcutaneous karst* is no longer used, and *epikarstic zone*, *epikarstic aquifer*, and *epikarst* are now widely accepted.

EPIKARST, A NECESSARY CONCEPT

In Ecology

The concept of epikarst was first proposed by groundwater ecologists (Rouch, 1968), who considered that perched saturated zones must exist within the temporary percolation zone, that is, the infiltration zone. They observed aquatic microfauna, mainly Copepoda, in water dripping from stalactites, in shallow caves. According to Rouch, the observed diversity and abundance of the fauna imply the existence of a local and permanent perched phreatic zone, a few meters below ground surface. In nonkarstic rocks, ecologists had already described such a zone, rich in groundwater fauna, which was named by Mestrov the *hypotelminorheic zone*. The concept was applied to most rock types and was considered to refer to the "skin" of the subsurface.

Later, Juberthie *et al.* (1980) extended the concept to terrestrial underground fauna, by defining the

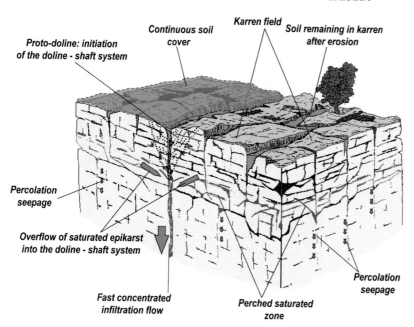

FIGURE 1 The epikarst is represented on its right part without a significant soil cover, showing a karren at its surface. The local saturated zone is shown. Arrows indicate (1) the percolation seepage, slow and dispersed in the fine cracks; (2) the rapid and concentrated infiltration flow through enlarged fractures and vertical conduits; and (3) the overflow of the local saturated zone into the doline—shaft system. Source: From Mangin (1975), modified.

underground shallow medium (in French *milieu souterrain superficiel* or MSS). The shallow zone of the karst is expressed as the "reservoir" of life, the place where exchanges occur between surface and subsurface, where surface animals discover and colonize the underground environment. A general paradigm (Rouch, 1986) explains the existence of fauna adapted to living in the karst and the underground environment by a permanent and active colonization from the ground surface of the epikarst, that essential medium for life, which is distinct from the cavernous environment.

In Hydrogeology

Drogue, in his unpublished doctorate thesis, mentions for the first time the hydrogeological role played by the epikarst and named it the *superficial karst*. He observed a perched, discontinuous, and temporary water table in the karstified limestones of Languedoc (southern France), drained by a fracture network to the saturated zone.

To explain the observed homogenization of isotopic characteristics in three karst aquifers of the French Pyrenees, Bakalowicz *et al.* (1974) invoked the existence of shallow storage in karst aquifers and suggested its generalization to all karst aquifers. The role of the epikarst is also shown in the process of *evapotranspiration*, that is, the consumption of groundwater by plants. Numerous karst scientists, mainly in France, were to show though hydrodynamic and hydrogeochemical studies that the epikarst is necessary to explain the functioning of the karst aquifer.

Moreover, in one of the earliest attempts to model a karst aquifer by a grid approach, Kiraly (1975) showed that the successful simulation of karst spring hydrographs is dependent on the introduction of a shallow layer of elevated porosity and permeability representing the epikarst.

In Karstology

The epikarst represents the vertical extension of the soil, and as such can act as a reservoir for the accumulation of organic matter. The decomposition of organic matter within this layer produces carbon dioxide, CO_2, which is the main agent of carbonate rock solution, and of karst processes, when dissolved in groundwater. The epikarst is the key site for carbonate rock solution. Karst depressions (dolines) are initiated by drainage of epikarst storage via vertical conduits. Karren areas develop in poorly drained areas of the epikarst.

EPIKARST, A NOT YET FULLY AGREED CONCEPT

Some authors (see Klimchouk, 2004) consider that dolines and more usually the whole shallow part of karst belong to epikarst. Bakalowicz (2004) and others (see the final discussion in Jones *et al.*, 2004) restricts the epikarst to the parts where shallow water storage may occur, considering that epikarst should be defined not only by the surface morphology, that is, karren fields and their possible soil or sediment cover, but overall by its hydrological functioning and the typical processes occurring in it. The main processes are related to water stored in the uppermost part of karst, because of a vadose seepage at its base: infiltration is delayed, soil CO_2 is directly used for carbonate solution, organics are dissolved and weathered, and dissolved salts are concentrated because of water consumption by evapotranspiration.

HOW DOES THE EPIKARST WORK?

The first approaches to studying the epikarst were indirect, in that its behavior was deduced from that of the whole karst aquifer. It soon became evident that a more direct approach to investigating the shallow zone of the karst aquifer was required. Several field test sites were designed for analyzing the epikarst. The use of tracer tests and artificial rainfall complemented the classical hydrodynamic, chemical, and isotopic investigations (Bakalowicz, 1995). According to the first interpretations by Mangin and by Bakalowicz, these investigations showed the existence of various infiltration modes through the shallow part of karst. The rainwater recharges the karst aquifer and emerges at its springs by the following pathways:

1. A part of the water infiltrates directly and quickly through wide fractures and vertical conduits, from dispersed infiltration at the karst surface, or from point infiltration through sinkholes.
2. The other part is stored in the epikarst where it contributes to different processes: a part is consumed by plants (evapotranspiration), another part percolates slowly through the fine cracks and rock porosity (slow infiltration), and, during the heavy rains, the last part is flushed away into the vertical conduits of the infiltration zone (the doline–shaft system; Fig. 2), from epikarst overflow (delayed infiltration).

Point and direct fast infiltration waters do not flow through the epikarst. Dolines and shafts allow direct vadose flow to rapidly reach the conduit system, acting as holes breaking the epikarst, and simultaneously working as overflow drains of epikarst when it is filled up with water. Figure 2 describes how infiltration water is distributed in karst aquifers according to the different types of underground flows.

PLACE OF EPIKARST IN KARST EVOLUTION AND MORPHOLOGY

The epikarst is completely involved in CO_2 production, storage and dispersion in karst aquifers. The epikarst works as a CO_2 reservoir, gently recharging the infiltration zone by means of the slow infiltration. As a

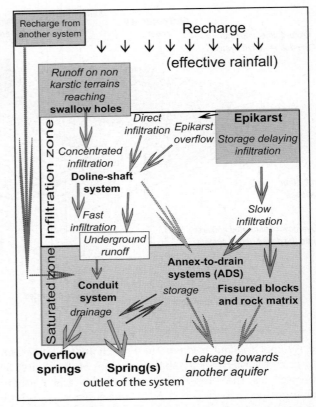

FIGURE 2 Functional scheme of a karstic system showing the distribution of infiltration water according to the different types of underground flows. The functions are written in italic; the "organs" are written in bold.

matter of fact, this slow infiltration induces mixing of air and water during the slow percolation in fine cracks and pores. This process drives the CO_2-rich air present in soil down into the system and disperses it through the whole infiltration zone. In that way, the fast infiltration water, from direct dispersed and point recharge, may dissolve CO_2 and consequently carbonate rock in the depth of the karst, at the top of the phreatic zone. Consequently, storage in the epikarst is an essential mechanism for karst development, at the surface as well as at depth.

The functioning of epikarst determines the spatial distribution of carbonate rock solution, not only at the ground surface, but also between the surface and the karst at depth. Vertical conduits are enlarged near the surface as closed depressions, then constituting the doline–shaft system. They allow for the introduction at depth of fast-infiltrated aggressive waters, which dissolve the rock at various depths, in such a way that fractures may be enlarged into conduits, which is a prerequisite for the development of caves.

At the top of the system, water stored in the epikarst gently dissolves the rock in the few meters below the soil, deepening fractures and cracks. The enlarged fractures can stock detritic sediments, residue of the dissolution of carbonate rocks or from displacement by rain, wind, and surface runoff—this is the karren field, which may or may not be covered with soil and plants. When climate and plant cover conditions are stable for a long time, karren, and then epikarst, may develop into spectacular landforms, such as in the Stone Forest in Yunnan, China, or the tsingy in Madagascar.

Recent developments in geophysics have unveiled the existence of the epikarst at depth. Two methods are particularly effective: ground penetrating radar (GPR) which shows the vertical structure of epikarst (Al-Fares et al., 2002), and magnetic resonance sounding (MRS), which demonstrates the presence of water stored in the epikarst. GPR has shown the existence of an extensive fractured zone within the first 10 m below the ground surface of a Mediterranean karst. The zone approximately follows the ground surface, independently of strata dipping. Occasional major vertical fractures are the origin of vertical conduits, which are in turn the origin of the conduit network which characterizes the karst aquifer.

THE EPIKARST, THE SKIN OF THE KARST

Finally, epikarst is an essential interface between the biosphere and the karst itself. Karst develops and evolves as a result of the epikarst, which distributes the infiltrated water and the rock solvent (CO_2 + water) in such a way that a characteristic landscape is created at the surface (closed depressions, karren) as well as at depth (the conduit network and caves). Therefore epikarst can be compared to the skin of a living being.

In the same way, the epikarst is a very fragile, sensitive medium. It may easily be eroded. During glacial periods of the Quaternary, the freeze/thaw alternation and the ice- and snow-cover movements totally destroyed the epikarst of cold regions of Europe and North America. Most of the slope debris that has accumulated at the foot of limestone cliffs originated from such a process. In Mediterranean areas, heavy rains and flash floods act in the same way, but the erosion is limited to steep slopes, and the accumulation occurs in dejection cones. Karst of the high mountains of Turkey, Lebanon, and Syria are often coated with a thick cover of carbonate debris which were not yet washed out. This debris cover works as a continuous buffer, filtering seasonal variations in infiltration flow, and is what may regulate the spring discharge.

Humans are also an efficient erosive agent. We deforest and cultivate the rich but generally thin soils. Some well-known examples in Ireland and England show preserved soil and epikarst under dolmen in areas otherwise devoid of such features. As observed in Mediterranean karst areas and in southern China, even when soils were once thick, they have progressively disappeared, revealing karren with rounded shapes, typical of formations more normally found under soil or a sediment cover. These examples show how much the epikarst is fragile and may be easily eroded. Moreover, during the construction of highways or ski stations the epikarst is systematically destroyed, either because from a geotechnical point of view it provides too fragile of a foundation for buildings, or because it provides a useful source of rubble.

Once the epikarst is eroded, a uniform rock surface, the pavement, may be observed, interspersed with only a few cracks and occasional large vertical conduits. To recreate the necessary conditions for karst evolution, soil and a plant cover must develop, but the process requires several thousand years in a temperate, humid climate.

See Also the Following Article

Epikarst Communities

Bibliography

Al-Fares, W., Bakalowicz, M., Guérin, R., & Dukhan, M. (2002). Analysis of the karst aquifer structure by means of a Ground Penetrating Radar (GPR). Example of the Lamalou area (Hérault, France). *Journal of Applied Geophysics, 51*, 97–106.

Bakalowicz, M. (1995). La zone d'infiltration des aquifères karstiques. Méthodes d'étude. Structure et fonctionnement. *Hydrogéologie, 4*, 3–21 (in French).

Bakalowicz, M. (2004). The epikarst, the skin of karst. In W. K. Jones, D. C. Culver, & J. S. Herman (Eds.), *Epikarst* (9, pp. 16–22). Leesburg, VA: Karst Water Institute. Special Publication.

Bakalowicz, M., Blavoux, B., & Mangin, A. (1974). Apports du traçage isotopique naturel à la connaissance du fonctionnement d'un système karstique. Teneurs en oxygène 18 de trois systèmes des Pyrénées, France. *Journal of Hydrology, 23*, 141–158 (in French).

Jones, W.K., Culver, D. & Herman, J.S., ed. (2004). Epikarst. Special Publication 9, Karst Water Institute, Leesburg, VA.

Juberthie, C., Delay, B., & Bouillon, M. (1980). Extension du milieu souterrain en zone non calcaire: Description d'un nouveau milieu et de son peuplement par les Coléoptères troglobies. *Mémoires de Biospéologie, 7*, 19–52 (in French).

Kiraly, L. (1975). Rapport sur l'état actuel des connaissances dans le domaine des caractères physiques des roches karstiques. In A. Burger & L. Dubertret (Eds.), *Hydrogeology of karstic terrains* (No. 3, pp. 53–67). Paris, France: International Union of Geological Sciences. Series B.

Klimchouk, A. (2004). Towards defining, delimiting and classifying epikarst: Its origin, processes and variants of geomorphic evolution. In W. K. Jones, D. C. Culver & J. S. Herman (Eds.), *Epikarst* (9, pp. 23–35). Leesburg, VA: Karst Water Institute. Special Publication.

Mangin, A. (1973). Sur la dynamique des transferts en aquifère karstique. *Proceedings of the 6th International Congress of Speleology, Olomouc, III, Canada*, 157–162 (in French).

Rouch, R. (1968). Contribution à la connaissance des Harpacticides hypogés (Crustacés, Copépodes). *Annales de Spéléologie, 23*(1), 5–167 (in French).

Rouch, R. (1986). Sur l'écologie des eaux souterraines dans le karst. *Stygologia*, 2(4), 352–398 (in French).

White, W. B. (1988). *Geomorphology and hydrology of karst terrains*. New York: Oxford University Press.

Williams, P. W. (1983). The role of subcutaneous zone in karst hydrology. *Journal of Hydrology*, 61(1), 45–67.

EPIKARST COMMUNITIES

David C. Culver,[*] Anton Brancelj,[†] and Tanja Pipan[**]

[*]*American University, U.S.A;* [†]*National Institute of Biology, Slovenia;* [**]*Karst Research Institute at ZRC SAZU, Slovenia*

INTRODUCTION

The *epikarst*, or subcutaneous zone, is the uppermost part of karstified rock, a perched aquifer, and an ecotone between surface and subterranean environments. It has been variously defined but in general it is the boundary between soil and rock in karst, honeycombed with small fractures and solution pockets. Its vertical extent varies from nearly zero to a few meters. It is a zone where environmental parameters (*e.g.*, temperature) can change significantly over short distances and where there are seasonal and even daily oscillations. The cavities and cracks in the rock in the epikarst zone may or may not be well integrated in the horizontal direction. Sometimes water percolating through the epikarst moves laterally at the base of it for substantial distances before finding a pathway deeper into the vadose zone. The epikarst consists of a series of small cavities and crevices, some of which are water-filled, some of which are filled with organic material, humus, and insoluble material, and some of which are air-filled. Whereas hydrogeologists often stress the water storage capacity of the epikarst zone, biologists usually stress the vertical movement of water in the zone, and the term *percolating water* is often used in this connection. Epikarst is one of the major aquatic shallow subterranean habitats.

SAMPLING TECHNIQUES

The habitat has rarely, if ever, been sampled directly. Instead, biologists have had to rely on indirect samples (Pipan, 2005). Direct methods for sampling fauna in the epikarst are still being developed. In theory, drilling vertically into the epikarst zone and collecting water from voids there is possible. In principle, a metal or plastic tube, closed at the bottom and with a series of holes some centimeters above the sealed end that is left in position for some time should act as a pitfall trap for fauna there—but that's in theory! In practice, there are two possibilities for sampling the fauna. The most direct is to sample percolating water from the trickles, especially in permanent trickles (Fig. 1). By means of a funnel, water from a trickle is directed into a plastic container with one or several holes on the sides. The hole is covered with a mesh (mesh size of 60–100 μm; Fig. 1). Containers need to be emptied periodically, perhaps once a month, to minimize losses due to predation and other factors.

The second method is to collect and filter water from small pools on calcareous slopes or on the bottom of galleries, which are filled with water from the trickles. The volume of the pools can vary from a few milliliters to a liter or more. Water from the pools can be collected by means of a pump (Fig. 2), which is very efficient in small rimstone pools or in the deep and narrow cracks on stalagmites that are filled with water. During the sampling, vigorous agitation of the water is recommended to collect particles from the bottom of the pools where most animals are attached. While sampling drip pools is much easier than sampling drips, the sampling from pools is a biased sample of drip water, and typically has fewer stygobionts, species specialized for subterranean life (Pipan *et al.*, 2010). They are sink populations in the ecological terminology of source–sink populations. Nonetheless, they are important collection sites for the epikarst fauna, especially when the first survey of fauna is made in a particular cave, and can yield a very diverse fauna (*e.g.*, Brancelj, 2002).

ENVIRONMENTAL CONDITIONS

Epikarst water is highly heterogeneous in fluxes. Measurement of discharge of 35 drips over one year in six Slovenian caves (Pipan, 2005) revealed that fluxes are extremely dependent on precipitation. They range from less than 0.5 mL min^{-1} to nearly 1400 mL min^{-1}. Measurements in the Slovenian cave Velika Pasjica showed that discharge can increase from 0.2 mL min^{-1} to more than 870 mL min^{-1} in less than 6 hours and drop back to minimal flow again in 24 to 48 hours. The relationship of discharge in a cave to precipitation depends on saturation of soil, cracks, and voids with water. If they are saturated with water, reaction is quick (within an hour). After a long dry period, drips react with long delays or even do not react at all, as rainwater is retained in soil, cracks, and voids and percolates downward very slowly. In general, drip water shows the same seasonal pattern as surface precipitation, but with a delayed response.

Temperature of water in the epikarst is relatively constant with some oscillations related to seasons and weather events. In principle, the water temperature in the epikarst equals the mean air temperature of the location, however it rarely, if ever, reaches this equilibrium. In her six study caves in Slovenia, Pipan found that temperature in dripping water varied from 0.4°C to 13.7°C in a region where the mean annual temperature was approximately 9°C.

Chemically, epikarst water is highly enriched in calcium ions with a concentration 10 to 50 times that of rainwater, indicating that it is actively dissolving calcium carbonate. Measurements of calcium concentrations in percolation water from six caves in Slovenia indicate high values. On average, the concentration of Ca was about 38 mg Ca L^{-1} (ranging from 10 to 100 mg Ca L^{-1}). Conductivity was also high—the average value was 355 μS cm^{-1} (ranging from 170 to 730 μS cm^{-1}), the result of the length of time the water has been in contact with bedrock, which may be weeks or even months. Measured values of oxygen concentration are usually between 60 and 100% of saturation, but this may be misleading since these measurements were based on epikarst water exposed to the air, which may replenish the oxygen.

Epikarst water contains varying amounts of coarse and fine particulate organic matter (POC) and dissolved organic carbon (DOC) as a result of feeding and metabolic activities in the soil and the epikarst itself. Epikarst is a source of organic input as well as the location of the exchange of surface and subsurface fauna. Input of DOC and POC as a source of nutrients for bacteria and invertebrates may arrive in the karst basin via two different pathways: openings (sinks and pits) or infiltration through soils. Water percolating through soils into the epikarst carries both DOC and POC, but POC is mostly filtered by soils. Once transported into the karst underground, POC and DOC are used or processed to different forms. An important element for DOC and POC transformation are microbial films on rocks (epilithion).

The DOC concentration in the epikarst drips in the Postojna-Planina Cave System (Slovenia) and Organ Cave (West Virginia, U.S.A.) was around 1 mg C L^{-1} (Simon et al., 2007). Organic carbon concentrations in sinking streams, cave streams, and the resurgence of the Postojna-Planina Cave System were higher, but in Organ Cave organic carbon concentrations of epikarst, cave stream, and resurgence waters were similar. In other caves, organic carbon concentrations of epikarst water can be much higher than 1 mg C L^{-1}. This highlights the need for careful measurement of organic carbon in different systems. Compared to cave stream water, the organic carbon in epikarst is low in aromaticity and humification, indicating that it was more readily assimilated into the biotic community (Simon et al., 2010).

FIGURE 1 Device for collecting fauna from percolating water. By means of a funnel, water is collected in a container with mesh-covered sides. The lower rim of the holes in the side of the container is positioned some centimeters above the bottom, allowing a small pool of water to form in the bottom of the container.

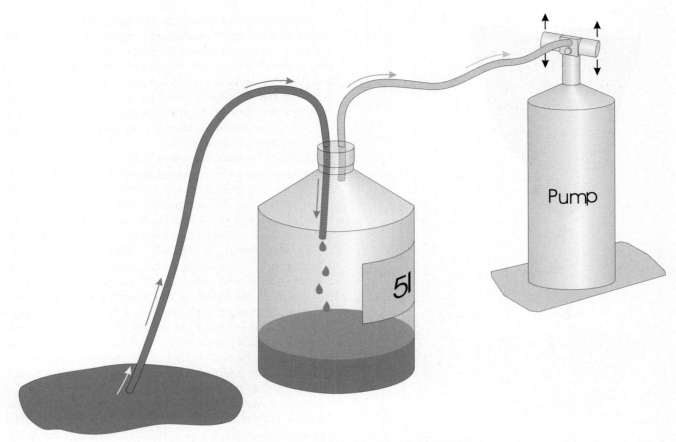

FIGURE 2 Device for collecting fauna from small amounts of percolating water in pools and cracks.

TAXONOMIC COMPOSITION

A wide variety of invertebrate species are known primarily or exclusively from epikarst. In the compendium of subsurface aquatic species *Stygofauna mundi* (Botosaneanu, 1986), over 10 species of oligochaetes, 280 species of crustaceans, 2 species of beetles, and 30 species of flatworms are reported from percolating waters of caves. Eight genera of crustaceans have 10 or more species that are found in the epikarst (Table 1). In North America, the amphipod *Stygobromus* and the isopod *Caecidotea* are common epikarst species, whereas in Europe the isopod *Proasellus*, the syncarid *Iberobathynella*, the cyclopoid copepods *Speocyclops* and *Diacyclops*, and the harpacticoid copepods *Bryocamptus*, *Elaphoidella*, and *Parastenocaris* predominate.

Intensive sampling of epikarst since 2000 in different countries all over the world (Brazil, China, Romania, Slovenia, Spain, Thailand, U.S.A.) has revealed rich fauna there. A rather long list of species from epikarst exists, with Copepoda (Cyclopoida and Harpacticoida), Ostracoda, and Syncarida (Bathynellaceace) as the most common representatives, followed by Amphipoda and Oligochaeta. Other groups, including Gastropoda, are less common in samples, although their empty shells are rather frequent in the pools filled by dripping water.

Only eight genera have been found in the epikarst (Table 2). They include the harpacticoid copepod from genera *Morariopsis* (the Dinaric Mountains), and

TABLE 1 Genera of Crustaceans with Ten or More Epikarst Species

Order	Genus	Total No. of Subterranean Species	No. of Epikarst Species
Cyclopoida	*Speocyclops*	41	28
	Diacyclops	34	10
Harpacticoida	*Elaphoidella*	91	34
	Parastenocaris	167	15
Isopoda	*Caecidotea*	57	13
	Proasellus	115	29
Amphipoda	*Stygobromus*	100	45
Syncarida	*Iberobathynella*	18	10

(Data from various sources).

TABLE 2   Aquatic Genera Likely Found Only in Epikarst

Class	Order	Genus	Number of Species
Crustacea	Harpacticoida	*Morariopsis*	3
	Harpacticoida	*Paramorariopsis*	3
	Syncarida	*Batubathynella*	1
	Amphipoda	*Niphargobates*	2
Insecta	Coleoptera	*Trogloguignotus*	1
	Coleoptera	*Troglelmis*	1

(Data from various sources)

Paramorariopsis (the Alps and Dinaric Mountains), the amphipod *Niphargobates* (the Balkans), the syncarid crustacean *Batubathynella* (Malaysia), the beetle *Trogloguignotus* (Venezuela), and the beetle *Troglelmis* (the Republic of Congo). Several of these species are shown in Figure 3. Since the publication of *Stygofauna mundi* in 1986, many more aquatic epikarst species have been described but new exclusively epikarst genera were recorded only within Copepoda (Harpacticoida), which is the most common and abundant stygobiotic group in epikarst.

Many epikarst species are found in genera where most of the other species are from interstitial habitats such as shallow alluvial aquifers and the underflow of streams (*i.e.*, hyporheic zone). Examples include the amphipod genus *Bogidiella*, the isopod genus *Microcharon*, and the Syncarida in general. These taxonomic distribution patterns suggest an affinity between interstitial and epikarst habitats. At the same time there are very few species which inhabit both habitats, interstitial and epikarst. In North America, the exclusively subterranean genus *Stygobromus* is found in a wide variety of subterranean habitats, including cave streams, interstitial habitats, deep groundwater, and seepage springs. In eastern North America, 28 of 56 species are found exclusively in the epikarst, and three others were found in epikarst as well as in other subterranean habitats.

Because part of the epikarst is air-filled, one would expect terrestrial species as well. To our knowledge, no one has sampled these species directly, but Gibert (1986) and Pipan *et al.* (2008) did collect considerable numbers of terrestrial species in nets under ceiling drips. They found a total of 12 and 18 taxa, respectively, including one representative of Collembola, one Diplura, and one Coleoptera, known only from caves. Pipan *et al.* (2008) found a snout beetle (weevil), *Troglorhynchus anophtalmus*, a very specialized coleopteran, feeding on the tree roots that penetrate from soil into epikarst and even caves. It is interesting that the most common terrestrial species Gibert found was the collembolan *Arrhopalites secondarius*, a globular-shaped species with morphology typical of deep soil species. Indeed, it would be surprising if the terrestrial epikarst fauna did not share affinities with the deep soil fauna. Little else is known about terrestrial epikarst species except for tantalizing hints scattered in the literature. For example, there is a very unusual carabid beetle (Coleoptera) in its own genus (*Horologion*), found only once, in 1938 in a small, shallow cave in West Virginia. In spite of repeated visits to the cave and nearby caves, it has never been collected again. Given that it was found in a very shallow cave and that it has not been seen since, it seems likely that this species and others collected in similar circumstances are in fact epikarst species. The terrestrial snail *Zospeum*, whose shells are common in many caves of the Dinaric karst, is likely an epikarst specialist. Although reported from other parts of the caves, especially from rotten wood near the entrances, its main habitat is small crevices in the ceiling and walls of the caves. Other parts of the caves are probably inhabited by "sink" populations.

ESTIMATING TOTAL SPECIES RICHNESS OF EPIKARST COPEPODS

Because of the possibility of continuous sampling, the epikarst fauna can be more thoroughly and accurately studied than fauna in other subterranean habitats. The epikarst copepod fauna is a significant part of the aquatic cave fauna, contributing about 20% of the species at the regional level. An example is from caves in Dinaric Karst in Slovenia, where 35 drips in six caves were continuously sampled for one year. Species accumulation curves and estimates of total copepod species richness were calculated. At the scale of individual drips, three to four months' sampling appeared to be sufficient to collect most of the species. At the scale of the individual cave, five drips sampled for one year appeared to be sufficient to collect most of the species, and at the regional scale, five caves were sufficient to collect 90% of all epikarst species of Copepoda (Pipan and Culver, 2007a). In the case of epikarst Copepoda, the relatively rapid saturation of accumulation curves and the relative completeness of samples is the result of the fine scale of heterogeneity and possibly intensive movement of individuals compared to other subterranean habitats. The epikarst is a finely structured habitat at a small scale (meters or tenths of meters) in which a relatively high number of species coexist because of habitats' patchiness and different environmental conditions.

Some copepod species in epikarst have very local distributions. In a series of caves in southwest Slovenia and West Virginia a significant difference in fauna

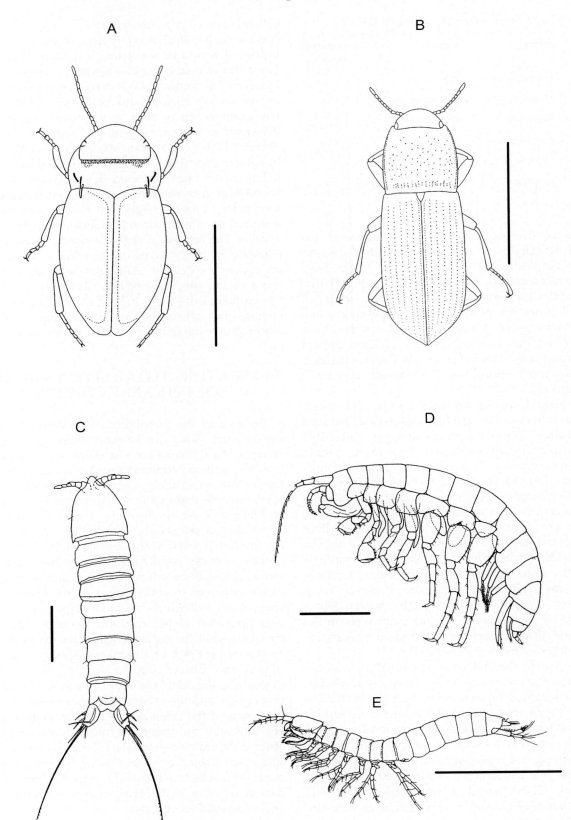

FIGURE 3 Examples of species from five genera endemic to epikarst (see Table 2). (A) *Trogloguignotus concii*, (B) *Troglelmis leleupi*, (C) *Morariopsis dumonti*, (D) *Niphargobates orophobata*, and (E) *Allobathynella malaya* (scale bar: 1 mm for (A), (B), (D), and (E); 0.1 mm for (C)).

composition of Copepoda over a distance of 100 m appeared (Pipan and Culver, 2007b). In general, the pattern of distribution of different taxa within a cave follows the same pattern as for distribution between the caves. This suggests a pattern of extremely restricted lateral flow under normal hydrological conditions. Under high flow conditions lateral flow may increase what is often the case during a toxic spill. Frequent occurrence of copepods in dripping water, knowledge of their local distribution within a cave, as well as their small size make them a useful tool in tracing the potential route of pollutants and potential water tracers within a small karst system.

MORPHOLOGICAL FEATURES

Compared to obligate cave-dwelling species, epikarst species tend to be smaller in size. Groups with larger sized species that are absent in the epikarst include fish, salamanders, crayfish, and shrimp. On the other hand, groups with very small species, usually less than 2 mm, predominate, both in terms of abundance and number of species. Primarily, these are cyclopoid and harpacticoid copepods. Even within a genus, epikarst species tend to be smaller. The North American amphipod genus *Stygobromus* is exclusively subterranean, and in the eastern United States occurs in epikarst, seepage springs, cave streams, and deep groundwater. Body sizes were smallest for epikarst species, averaging 5.3 mm for 28 species. In cave streams, sizes averaged 9.0 mm for 9 species (Culver et al., 2010). On the other hand, there were no differences in antennal lengths (corrected for body size) in the different habitats.

Epikarst copepods have some specific adaptations that make them easily recognizable as epikarst specialists. In Cyclopoida specific adaptations are, apart from reduction of segments on legs P1 to P4, the presence of very strong and robust spines on exopodites of the first pair of legs. In addition, they have very strong spines at the base of each endopodite on the first pair of legs oriented backward. Those spines prevent them from being washed away by strong current, as they act as anchors. In Harpacticoida the most characteristic are modifications in body shape, which is elongated and cylindrical. Furcal rami are divergent (up to 45°) and equipped with robust and stiff terminal setae. There are also short and strong or (more commonly) robust setae on the baseoendopodite of leg P5. Spines on exopodites, especially on the first pair of legs, are very robust and also curved. Short and robust spines and setae enable specimens to obtain better grip against water current not to be washed out from the epikarst (Brancelj, 2009).

BIOGEOGRAPHY OF EPIKARST SPECIES

Local species richness in the epikarst can be quite high compared to other subterranean environments. Gibert (1986) lists 12 species as occurring in the percolating waters of Cormoran Cave in France, among them four obligate subterranean species, including two harpacticoid and one cyclopoid crustacean species. In the 100-m-long Slovenian cave Velika Pasjica, Brancelj (2002) found an amazingly rich epikarst copepod fauna: 11 harpacticoids and 1 cyclopoid in drip pools. Pipan (2005), sampling both drips and pools in Županova Jama in Slovenia, found even more species—14 harpacticoids and 2 cyclopoids. In both cases, all species but one were stygobionts. In addition, intensive sampling by Pipan in the epikarst zone in six caves in Slovenia revealed a total of 37 taxa of Copepoda. Twelve of them are ubiquitous, the rest are stygobiotic, and 14 of them are endemic to Slovenia. The high species richness of Copepoda in Slovenian epikarst is in accord with the high richness of Copepoda also in other subterranean habitats. At the moment there are 51 cyclopoid and 56 harpacticoid species known from Slovenia of which 18 cyclopoid and 27 harpacticoid species are stygobionts and 2 cyclopoid and 13 harpacticoid species are restricted to epikarst only (*i.e.*, 15% of all recorded species in Slovenia).

Most epikarst species have been found in caves throughout the world very recently, although the very first ones, such as *Speocyclops infernus* and *Morariopsis scotenophila*, were described in 1930 and were collected from pools filled with drip water. The best studied karst region is the Karst region of Slovenia where seven caves have been intensively sampled by Pipan and Brancelj. The caves are situated in a rectangle of about 2400 km^2 and the distance between caves varied from about 1 km to approximately 60 km. Several patterns characteristic of epikarst species emerged. First, endemism is high. Among 37 taxa collected in the seven caves, 14 taxa are endemic to Slovenia. Second, within the 7 caves studied, distribution of species was patchy. Only one species, *Speocyclops infernus*, a cyclopoid copepod, was collected in all 7 caves. In contrast, one-third of taxa were present only in 2 caves (11 taxa) or in 1 cave (13 taxa). This indicates that the majority of stygofauna in epikarst zone are (1) very localized on a microscale, and (2) very diverse on a macroscale. This supports the hypothesis that communication between voids in the horizontal direction is limited and thus distribution/migration of individuals, including genetic material, is limited.

In spite this high level of heterogeneity and local endemism, some patterns did emerge. First, and most remarkably, the frequency of occupation of habitats at one scale (drips within a cave) was a good predictor of frequency of occurrence in different caves (Culver et al., 2009). This in turn suggests that the distribution of epikarst copepods is a very dynamic process, and that dispersal ability of different copepod species is an important determinant of distribution pattern. It also suggests that there are some scale invariant (fractal) characteristics of the distribution of epikarst copepods. Second, the contiguous distribution of many species of epikarst copepods within a cave is on the order of 100 m, suggesting that they can be used a tracers of the lateral movement of epikarst water (Pipan and Culver, 2007b).

ECOLOGY OF EPIKARST SPECIES

Pipan et al. (2006) analyzed the relationship between the epikarst copepod fauna collected from drips in five Slovenian caves and a variety of physical (drip rate, temperature, conductivity, ceiling thickness, and surface precipitation in the preceding month) and chemical (chloride, nitrate, sulfate, sodium, potassium, calcium, and magnesium) parameters. Canonical Correspondence Analysis (CCA) allows the simultaneous representation of environmental variables and species' preferences (Fig. 4). The axes of the resulting two-dimensional graph show the linear combination of environmental variables that explain the greatest possible amount of the variance of all the environmental variables taken together. Species lying close to one of the lines for the environmental variables is strongly associated with that variable. For example, *Moraria varica* is found in waters with higher concentrations of NO_3^-, and *Bryocamptus balcanicus* is associated with drips with higher flow rates (Fig. 4). Each species niche can be similarly defined by its position with respect to the canonical axes. Species with nearby positions have similar niches. Overall differences among the caves with respect to the environmental variables were found, and these differences help explain faunal differences among the caves.

The variables that best explained the variation in the 29 copepod species were thickness of the cave ceiling, and concentrations of Na^+, NO_3^-, and K^+. Nitrate is a macronutrient and the reason for a correlation with sodium and potassium is not clear, but it may be

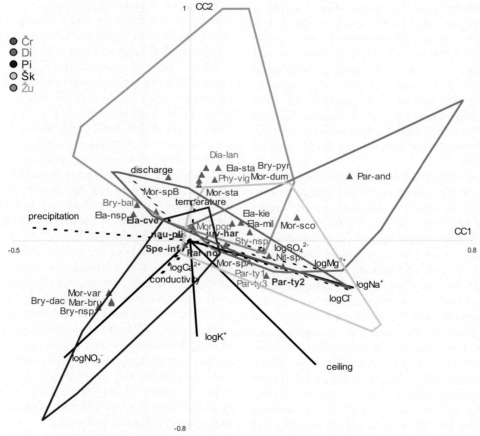

FIGURE 4 Ordination diagram based on species composition and abundance data of copepods in drips in relation to 12 environmental variables in five caves (Črna Jama (Čr), Dimnice (Di), Pivka Jama (Pi), Škocjanske Jame (Šk), and Županova Jama (Žu)). CC1 and CC2 are the first two canonical axes and together account for 51.9% of the total variance. See Pipan (2005) for further details.

connected with surface pollution. The negative correlation with ceiling thickness, resulting in the richest fauna from the thinnest ceilings, suggests that the ceiling, the zone of percolation, below the epikarst acts as a filter for fauna falling out of the epikarst.

The fauna of Copepoda in trickles and pools in "fossil" galleries (and probably all stygofauna collected there) is a result of accidental downward drift. The abundance of copepods collected under drips is positively correlated with flow rates, which is to be expected if the appearance of animals is the result of accidental downward drift. The extreme heterogeneity in the numbers of animals collected in drips is consistent with the hypothesis that occasionally all of the copepods get washed out from a small solution pocket. The presence of all life stages in the drip collections is also consistent with this hypothesis.

Based on the reduced frequency of nauplii and immature individuals, and the relative rarity of stygobionts in drip pools relative to drips, Pipan et al. (2010) conclude that drip pools are sink habitats for epikarst copepods. That is, if the continuing flow of copepods from drips were cut off, the copepod populations would mostly disappear. The flux of copepods can be surprisingly large. In Organ Cave, West Virginia, an average of one copepod per day was found in dripping water, and the yearly flux of copepods into pools often exceeds the number of copepods in the pool. Of course, very large pools may harbor source populations that are self-sustaining. This seems to be the case for an artificially enlarged drip pool in Škocjanske jame. Some species may also thrive in pools. Considering the high number of specimens of *Speocyclops infernus* collected in drip pools in Slovenian caves, it appears that the species either reproduces in pools or is easily washed out from epikarst. It commonly appears as the only species in some pools, articularly in pools with clay bottoms, where it frequently coexists with *Niphargus*.

See Also the Following Articles

Epikarst
Shallow Subterranean Habitats

Bibliography

Botosaneanu, L. (Ed.), (1986). *Stygofauna mundi* Leiden, The Netherlands: E. J. Brill.

Brancelj, A. (2002). Microdistribution and diversity of Copepoda (Crustacea) in a small karstic cave in central Slovenia. *Hydrobiologia*, 477(1–3), 59–72.

Brancelj, A. (2009). Fauna of unsaturated karstic zone in Central Slovenia: Two new species of Harpacticoida (Crustacea: Copepoda), *Elaphoidella millennia* n.sp., & *E. tarmani* n.sp., their ecology and morphological adaptations. *Hydrobiologia*, 621(1), 85–104.

Culver, D. C., Holsinger, J. R., Christman, M. C., & Pipan, T. (2010). Morphological differences among eyeless amphipods in the genus *Stygobromus* dwelling in different subterranean habitats. *Journal of Crustacean Biology*, 301(1), 68–74.

Culver, D. C., Pipan, T., & Schneider, K. (2009). Vicariance, dispersal, and scale in the subterranean aquatic fauna of karst regions. *Freshwater Biology*, 54(4), 918–929.

Gibert, J. (1986). Ecologie d'un système karstique Jurassien. Hydrogéologie, derive animale, transits de matiéres, dynamique de la population de *Niphargus* (Crustacé, Amphipode). *Mémoires Biospéologie*, 13(40), 1–380 (in French).

Pipan, T. (2005). *Epikarst—A promising habitat*. Ljubljana: ZRC Publishing.

Pipan, T., & Culver, D. C. (2007a). Regional species richness in an obligate subterranean dwelling fauna—Epikarst copepods. *Journal of Biogeography*, 34(5), 854–861.

Pipan, T., & Culver, D. C. (2007b). Copepod distribution as an indicator of epikarst system connectivity. *Hydrogeology Journal*, 15(4), 817–822.

Pipan, T., Blejec, A., & Brancelj, A. (2006). Multivariate analysis of copepod assemblages in epikarstic waters of some Slovenian caves. *Hydrobiologia*, 55(1), 213–223.

Pipan, T., Holt, N., & Culver, D. C. (2010). How to protect a diverse, poorly known, inaccessible fauna: Identification and protection of source and sink habitats in the epikarst. *Aquatic Conservation: Marine and Freshwater Ecosystems*, 20(7), 748–755.

Pipan, T., Navodnik, V., Janžekovič, F., & Novak, T. (2008). Studies of the fauna of percolation water of Huda Luknja, a cave in isolated karst in northeast Slovenia. *Acta Carsologica*, 37(1), 33–43.

Simon, K. S., Pipan, T., & Culver, D. C. (2007). A conceptual model of the flow and distribution of organic carbon in caves. *Journal of Cave and Karst Studies*, 69, 279–284.

Simon, K. S., Pipan, T., Ohno, T., & Culver, D. C. (2010). Spatial and temporal patterns in abundance and character of dissolved organic matter in two karst aquifers. *Fundamental and Applied Limnology*, 177, 81–92.

EVOLUTION OF LINEAGES

Eleonora Trajano[*] *and Marina Cobolli*[†]

[*]*University of São Paulo, Brazil,* [†]*University of Rome "La Sapienza", Italy*

Subterranean ecosystems are distinguished by the presence of species that may considerably differ from their epigean (surface) relatives in terms of morphology, physiology, behavior, and ecology. To investigate which organisms colonize the subterranean realm to establish more or less self-sustained populations (and why others do not succeed in this), how some of these lineages differentiate, and what mechanisms are causing this evolutionary change, are the main challenges and sources of debate for speleobiologists throughout the world.

COLONIZATION OF SUBTERRANEAN HABITATS

Subterranean organisms illustrate well the importance that preadaptations (there are authors who prefer "exaptations") may have for the adoption of a new way of life. (*Preadaptations*, as defined here, are character states conferring performance advantage in a given selective regime, but which have been selected in another, independent previous regime.) Classical preadaptations to the subterranean life include nocturnal activity, selecting for development of mechanosensory and/or chemosensory traits, and opportunistic, generalist feeding, allowing survival in a food-poor environment that has scattered, variable, and frequently unpredictable food sources. Therefore, subterranean communities represent a subsample of the epigean ones that is strongly biased toward nocturnal, ecologically generalist taxa, including those currently living in the area as well as those that had lived there, colonized the hypogean habitat, and then become extinct in the surface (see discussion of relicts below).

There are several possible colonization routes to hypogean habitats. In the case of aquatic organisms, marine faunas are a source of colonizers for both marine and freshwater hypogean habitats. The latter may be reached through invasion of aquifers along interstitial and other pathways (*e.g.*, anchihaline habitats, submarine karst springs) by marine organisms progressively adapting to freshwater conditions. Subterranean freshwater species may also evolve directly from marine ancestors by stranding of founder populations in gradually freshening groundwaters during marine regressions. Among hypogean populations derived directly from epigean freshwater ancestors, stream-dwellers dependent on lotic conditions (fast-flowing, well-oxygenated waters) may access the subterranean environment mainly through stream sinkholes and resurgences. Those adapted or at least tolerant of lentic conditions (slow-moving, low-oxygen waters), and especially small-sized organisms, may also penetrate deep into the water table, colonizing phreatic habitats both laterally and vertically (for instance, through hyporheic zones), and reaching caves through this route. In the first case, the boundary between epigean and hypogean habitats is clearer and, due to the fragmented nature of streams, disruption of populations is easier to achieve. Phreatic habitats, on the other hand, may be continuous throughout wide areas and encompass different kinds of subterranean habitats where troglomorphic taxa may be found. Routes of colonization of the subterranean realm by terrestrial organisms are basically limited by their size. A vertical colonization of caves through deep soil and spaces (cracks, crevices) in the overlying bedrock is probable for small organisms; large animals require larger contacts between the subterranean habitat and the surface, and large cave entrances may be a frequently used route. The importance of forest-floor litter fauna as a source of colonizers for hypogean habitats has long been noted for many karst areas in the world.

The colonization of subterranean habitats by preadapted taxa forming troglophilic populations could occur at any time, but it is predicted that it happens mainly during the periods most favorable to these preadapted species in the epigean environment, when reproductive success and survival rates are at their highest. Therefore, the generality of the notion of colonization constrained by unfavorable climatic conditions and immediately followed by speciation resulting in troglobites, which is central to the paleoclimatic model (see below) according to several modern authors, is highly questionable. Dispersion is a natural tendency of all organisms, which tend to occupy all suitable and ecologically accessible habitats. The gradient between epigean and hypogean environments is generally restricted to the proximity of their contacts. Thus, except perhaps for organisms living deep in a progressively drying soil or sediment over karst bedrock, epigean animals cannot be "forced" to enter subterranean habitats during stressful periods because they would not know in which direction to move in order to find access to the sheltered hypogean environment. Animals do not present purpose-oriented behaviors—they just survive where they are already living, to which they are adapted. On the other hand, organisms may be washed or dragged into the subterranean habitat at any time (by floods, for instance, in a case of *passive colonization*) and survive or not depending on the presence of preadaptations or simply by chance.

In conclusion, animals colonize subterranean habitats through cave entrances in the human sense (openings large enough for humans to cross) to much smaller contacts between the epigean and the hypogean realms, such as the MSS (mesovoid shallow substratum), epikarst, deep soil, and so on. Colonization may be horizontal, as in the case of animals following water courses until they reach sinkholes or resurgences, to vertical, for example, animals passing through soil and/or fractured rocky layers and reaching larger subterranean spaces as caves. The latter may be accelerated by climate change, for instance, due to progressive drying of the surface: individuals able to survive in deeper layers would follow changing gradients. However, this colonization by constraint is restricted to a few situations, where gradient changes follow a continuous, undisrupted homogeneous route. In the many cases, habitat heterogeneities (such as variations in depth along streams resulting in isolated pools during dry seasons or epochs) would prevent this mode of colonization.

MODELS OF GENETIC DIFFERENTIATION AND ORIGIN OF TROGLOBITES

As for epigean taxa, genetic differentiation is probably more frequently achieved by geographic isolation leading to allopatric speciation of troglobitic species. In view of the vertical spatial continuity, which allows permeability between the epigean and the hypogean compartments of the biosphere and which is higher for smaller organisms, the most plausible mechanism of geographic isolation for most subterranean populations is through local extinction of related epigean populations. A classical model for temperate regions, which also applies to the tropical ones, is based on local extinction due to paleoclimatic fluctuations, rendering the epigean environment unsuitable for many taxa. Because the buffered condition of the subterranean environment results in the maintenance of environmental variables within the tolerance range of several species, troglophilic populations in their hypogean habitats could survive periods of climatic stress above and could differentiate in the absence of genetic flow from their (locally extinct) epigean conspecifics.

The majority of hypogean terrestrial populations belongs to hydrophilic taxa that are preadapted to life in the generally damp subterranean environment. Thus, the colonization of hypogean habitats and establishment of troglophilic populations would take place mainly during humid periods (glacial or interglacial, depending on the region), and the geographic isolation in drier periods, when humid vegetation types are replaced by drier ones and the epigean drainage may be disrupted. During these drier periods, which are climatically stressful for hydrophilic and aquatic taxa, a decrease in food input is expected, causing a decline in subterranean population sizes. This would result in bottleneck effects and favor genetic drift leading to rapid differentiation. Under such conditions, one can predict strong selection for improved efficiency in finding food and mates (as indeed is observed in many troglobites), which, in turn, could allow a new increase in population sizes.

On the other hand, geological and hydrological barriers, such as the presence of insoluble, impermeable rocks interposed between the surface and habitable subterranean habitats, obliteration of subterranean conduits by collapse, siltation or chemical deposition, karst drainage divides, and stream capture, could also provide isolation events for specific areas and taxa without extinction of epigean populations.

In addition to the derivation of troglobitic species from well-established troglophilic populations, a possible instance of concomitant colonization and isolation is a consequence of stream capture. This is a noncyclical, geomorphologic model proposed for the intensively studied Mexican tetra characin fishes, genus *Astyanax*. Stream capture occurs when a surface stream running on an impervious stratum beneath which there is soluble rock (usually limestone) meets at an intersection with a vertical joint in the latter, which is gradually enlarged by dissolution and erosion to the point that all of the stream water sinks into the capturing pit. At this point, the stream reach downstream of the sinkhole and its fauna would be completely subterranean, isolated from the reach upstream by waterfalls. Nevertheless, it has been observed that surface fish have constant access to the caves from both above and below; thus the applicability of this model to the origin of the cave *Astyanax* species is questionable.

Models of parapatric and sympatric speciation have also been proposed and may account for the origin of some troglobitic species. These models postulate the occurrence of genetic differentiation without geographic isolation, dependent on intrinsic biological properties of the taxa. A particular explanation proposed initially for tropical terrestrial troglobites is the adaptive-shift model (Howarth, 1993), which is based on the notion that the hypogean habitat is so stressful for epigean organisms that those managing to colonize it are prone to rapid genetic reorganization and differentiation. The adaptive-shift model is not in accordance with the observation that most troglobites belong to preadapted taxa, for which the subterranean life would not represent such a new, highly stressful ecological challenge. As a matter of fact, troglophilic populations, especially in tropical areas, are far more frequent than predicted by this model, according to which every subterranean terrestrial population would tend to become troglobitic with time.

It is important to note that troglophiles, which by definition are as well adapted to the subterranean as to the epigean way of life, are not merely an "intermediate" or "preparatory" step in the evolution of troglobites. Only when extrinsic (climatic, geological, hydrological) or intrinsic (genetic) barriers interrupt or greatly decrease the gene flow between hypogean and epigean individuals does a troglophilic population have the potential to originate troglobitic species. Otherwise, troglophilic populations may remain as such forever, and the high proportion of troglophiles, at least in tropical caves, indicates that this is frequently the case.

SINGLE VERSUS MULTIPLE ORIGIN OF TROGLOBITES

By definition, troglobites cannot survive long or complete their life cycles in the epigean environment, either because conditions in the latter became unsuitable for the species (first step in the climatic model) or, further on, because they acquired specializations that preclude the epigean life, even when the original environmental conditions are restored. Therefore, their dispersion is basically subterranean, and most troglobitic species have small geographic distributions, restricted to limited areas or even to a single cave or cave system. Nevertheless, troglobites have been assigned to single species that present wide distributions and that may encompass superficially discontinuous karst areas.

In any case, but especially for the latter, the question arises of whether these species originated after a single (Fig. 1) or multiple (Fig. 2) independent colonization event by the same epigean ancestor, followed or not followed by subterranean dispersion. Because selective forces and constraints are generally similar throughout the subterranean environment, different populations of the same ancestral species isolated independently in the hypogean habitat tend to evolve in the same direction, acquiring similar apomorphic traits (e.g., reduced eyes and pigmentation) and frequently resulting in very similar, morphologically indistinguishable taxa—a classical case of parallel evolution. In many cases, such cryptic species resulting from independent speciation events can only be recognized by genetic methods.

Troglobitic species distributed in superficially discontinuous karst areas are strong candidates to represent examples of multiple colonization and independent speciation. However, keep in mind the possibility of a continuity through deep karst or that a previously continuous karst area (for instance, limestone on top of an anticlinal structure), harboring a single troglobitic species, was fragmented by erosion and karst denudation. This illustrates well the need for integration of

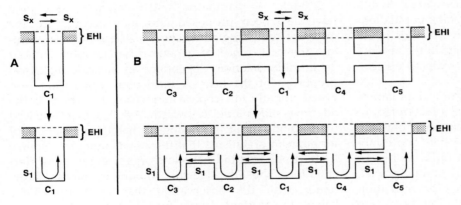

FIGURE 1 Generalized scheme for a single colonization event of cave system C_1 by an epigean ancestor (S_x), resulting, after genetic isolation, in a troglobitic species (S_1) restricted to that cave system (as shown in (A)) or also occurring in other cave systems (S_{2-5}) due to subterranean dispersal of S_1 (as shown in (B)). EHI, epigean/hypogean interface. Arrows indicate gene flow between populations. *Extracted from Holsinger, 2000.*

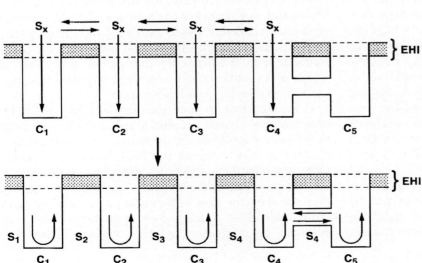

FIGURE 2 Generalized scheme for multiple colonization events by a single, widespread epigean ancestor (S_x), establishing independent populations in cave systems C_{1-4}, each one originating a troglobitic species (S_{1-4}) after genetic isolation; subsequently, S_4 expanded its range to cave system C_5 by subterranean dispersal. EHI, epigean/hypogean interface. Arrows indicate gene flow between populations. *Extracted from Holsinger, 2000.*

morphological, ecological, genetic, and geological data for the elucidation of this and many other relevant questions in biospeleology.

TROGLOMORPHIC TRAITS

Variation and Time of Isolation: "Recent" versus "Ancient" Troglobites

Since Racovitza's time and even before, degree of troglomorphism, usually reduction of eyes and pigmentation and elongation of appendices, has been used as a main measure of phylogenetic age, leading to the distinction between "ancient" and "recent" troglobites. Within the paradigm of gradual evolution, troglobitic species which are totally anophthalmic and depigmented, with long appendages, especially without known epigean relatives, have been considered as ancient troglobites, living in subterranean habitats much longer than those with partially regressed eyes and pigmentation and/or showing intrapopulation variation in such characters. For the latter, the isolation time would not be enough for fixation of transformed character states throughout the whole population, whereas species isolated for a long time and/or subject to fast differentiation conditions might differ considerably from epigean-related taxa, to the point that such relatives are not recognized anymore, either because the subterranean taxon has diverged too much or because the surface ones had became extinct. Subterranean taxa belonging to lineages extinct in the epigean environment are referred as phylogenetic relicts (*e.g.*, the crustacean order Spelaeogriphacea, with only three extant species, all troglobites showing a Gondwanaland distribution). Those belonging to epigean groups that have disappeared from a particular area but which still exist somewhere else are geographic or distributional relicts.

However, a straight correlation between degree of troglomorphism and phylogenetic age may be questioned, as pointed out by Trajano (2007) and exemplified for cixiid planthoppers by Wessel *et al.* (2007), who did not find a correlation between troglomorphy and age of caves. First, it is based on the assumption of fairly constant evolutionary (differentiation) rates among characters in more or less related taxa. The occurrence of different mosaic of character states in closely related species, as observed in several Brazilian cavefishes (*e.g.*, *Ancistrus cryptophthalmus*), suggests different mechanisms acting at different rates in each population. Such mosaics may encompass many characters, morphological, physiological, and/or behavioral, as seen in North American amblyopsid (Niemiller and Poulson, 2010) and Thai balitorid fishes (Parzefall and Trajano, 2010), hampering any attempt to rank such species according to their degree of specialization to the subterranean life (= troglomorphy). Moreover, intravariable populations may be a result of secondary introgression between fully differentiated troglobites and closely related epigean, eyed, and pigmented populations, as proposed for Mexican cavefishes.

Genetic studies, associated with morphological and physiological data, brought into light the complexity of these systems. There are two independent developmental units within the eyes, which may regress independently in closely related populations, as observed for Mexican cavefishes (Wilkens, 2010), and depigmentation may be achieved either morphologically, due to progressive loss of melanophores (mutations in a polygenic system), and/or physiologically, by loss of ability to synthesize melanin (through at least two different, apparently monogenic systems; Felice *et al.*, 2008)—these two modes of pigmentation loss may coexist in a population, but degenerative processes affecting monogenic systems are likely to proceed faster.

In addition to the specific mechanisms of character transformation, two factors may influence rates of divergence: population sizes and life-cycle strategies. Small populations tend to differentiate faster due to phenomena such as genetic drift. Population sizes are extremely influenced by ecological factors, and may vary among closely related species due to differences in habitat. Progressive specialization to the subterranean life may also affect population sizes, through an increasing efficiency in the use of resources allowing population increments, which would slow down differentiation rates. Moreover, K-selected life strategies imply lower differentiation rates due to delayed ages for first maturation and low reproductive rates. In conclusion, there is a complex balance between different genetic, ecological and biological factors producing the actual divergence rates, which may differ among related taxa and even among different characters (Trajano, 2007).

Constructive versus Regressive Traits: Definitions and Mechanisms

Two kinds of apomorphic traits of troglobites related to the hypogean life (troglomorphisms) can be distinguished: constructive and regressive traits. Constructive character states correspond to developed traits in relation to those observed in epigean relatives (*e.g.*, elongation of antennae in arthropods), for which natural selection, usually related to feeding and reproductive efficiency, may be evoked. Classical examples are the developed mechanosensory and chemosensory organs of many troglobites. Regressive characters are transformation series in which derived states correspond to a decrease until loss (character reversal) of structures, physiological, and behavioral components selected in the epigean environment occurs. Regressive traits, particularly reduced eyes and pigmentation, are the most conspicuous and distinguishing features of troglobites. These were understandably the first aspects of subterranean biology to attract the attention of researchers and are the central focus of most studies on hypogean organisms.

One of the most debated questions about subterranean evolution refers to the mechanisms underlying character regression, and dozens of hypotheses, more or less distinct, have been proposed since the nineteenth century, which included neo-Lamarckian (role of disuse) and orthogenetic (age of phyletic lines) theories. The main models elaborated within the neo-Darwinian paradigm fall into two groups: those directly or indirectly involving natural selection (adaptationists) and those involving adaptively neutral mutations. All of them, however, are based on the observation that regressing structures and behaviors are generally those that become useless in the subterranean habitat, that is, those directly or indirectly related to light.

Because nutrient scarcity is one of the most striking features of subterranean ecosystems and seems to be the logical explanation for development of many structures in troglobites, it is not surprising that economy of energy was one of the first Darwinian explanations evoked for character regression as well. According to this adaptationist model, mutant individuals for underdeveloped eyes and pigmentation, normally excluded from epigean populations by stabilizing selection, would have an advantage over normally eyed and pigmented individuals in the subterranean habitat because they could save extra energy (otherwise spent in useless structures) for survival and reproduction. Although appealing, this model finds no good support from empirical data and lacks generality—not every troglobite showing regressive traits lives in a food-poor environment, and some caves may be quite rich in nutrients, especially those harboring large populations of trogloxenes such as bats. Moreover, regression of eyes and pigmentation is also observed in some epigean organisms such as those living in deep soil, turbid large rivers, and also endoparasites, for which food is not a limiting factor. On the other hand, at least in theory, economy of energy could be in some cases associated with other genetic models, accelerating differentiation.

Currently, there are two main theories aiming to explain character regression in troglobites, based respectively on neutralism, that is, the accumulation of neutral mutations, and on pleiotropic effects of selection for constructive characters, as developmental tradeoffs. In the 1940s, Kosswig made a case for the accumulation of neutral mutations as the basis of regression of eyes and pigmentation in troglobites, as characters that became useless in a new environment. From that time on, many genetic studies, especially on the Mexican cavefish, genus *Astyanax*, were designed to put forward evidence for the neutral model, such as those by Wilkens (2010). The neutralism theory postulates the accumulation of mutations that become neutral after a change in way of life as a major cause of regression of any trait that loses its function under the new selective regime. There is genetic evidence that such mutations, excluded by stabilizing selection under the previous selective regime, would no longer be eliminated and, because mutations are more frequently deleterious than constructive, would accumulate and cause disorganization of structures and loss of physiological and behavioral components.

More recently, pleiotropic effects of development of antennae causing regression of eyes due to competition for neural centers have been hypothesized for North American freshwater amphipods *Gammarus minus* (Culver et al., 1995). Jeffery and colleagues, after experimental studies of eye development in epigean and troglobitic Mexican characins, *Astyanax* spp., have shown that the Hh midline signaling is a positive regulator of many developmental events, such as taste buds, whose number is increased in the cavefish. However, the enhanced signaling has a negative effect on eye development. Thus, selection for traits positively controlled by the Hh midline pathway would result in eye regression by default (Jeffery, 2005).

As a general model for eye regression, the pleiotropy hypothesis requires a particular association between a regressive and a constructive character in every troglomorphic species, which is highly unlikely in view of its wide occurrence among so different, distantly related taxa, from planarians, gastropods, and arthropods, to salamanders and fishes. It is improbable, for instance, that different mechanisms of sensorial compensation, affecting basically the mechanosensorial system in amblyopsid fishes, but the chemosensorial one in siluriforms such as *Rhamdia* catfishes, have the same collateral output, namely, the regression of eyes. It is relevant that different brain centers and embryonic placodes are involved in these sensorial systems in fish. The model proposed for *Astyanax* is also not consistent with the late regression of eyes observed for siluriforms, such as the armored catfish *Ancistrus cryptophthalmus*, among other studied Brazilian catfishes. As predominantly chemoriented fishes, it is expected that sensory compensation involves an increment in the taste bud number as in *Astyanax*, and that eyes should also regress at an early ontogenetic stage. However, this is contrary to the observed. Wilkens (2010) has shown that the neutral model does apply to the Mexican cavefish and that constructive and regressive developmental modules are inherited independently in this fish; thus eye regression would not be a spin-off effect of the improvement of constructive characters through pleiotropy.

Whatever the mechanism(s) leading to character regression, it does not operate in the same way and rate for all characters in each troglobitic species or population, as discussed above.

It is noteworthy that autapomorphies of troglobites do not necessarily encompass reproductive isolation. Reproductive characters are conservative and it is predicted that several other traits may change before isolation reproductive mechanisms are fixed throughout the population. Such is the case with the Mexican tetra characins, genus *Astyanax*, which encompass at least 19 cave troglomorphic populations, some of which have been shown to introgress with epigean, eyed, and pigmented tetras to produce fertile hybrids.

Within the paradigm of irreversibility of evolution (*Dollo's Law*), regressed characters have been viewed as the last possible state in the character transformation series, and troglobites as evolutionary dead-ends, unable to (re)colonize epigean habitats. Nevertheless, some authors question this notion. Based on genetic evidence, Culver *et al.* (1995) hypothesized that eyed amphipods, *Gammarus minus*, living in karst windows are derived from troglomorphic cave populations. According to these authors, the reacquisition of large eyes indicates that eye reduction in the original cave populations resulted from selection for degenerative mutations occurring at regulatory *loci* governing eye expression while the structural loci remain relatively intact. More recently (Dillman *et al.*, 2011) applied parsimony to a phylogenetic hypothesis for amblyopsid fishes, with epigean and troglophilic species as apical taxa inside the cluster of troglomorphic species. This implies accepting the origin of these species from troglobitic populations, therefore supporting the hypothesis of reevolution of eyes and melanic pigmentation, associated with the recolonization of epigean habitats.

The same reasoning is valid for the black proteus, *P. anguinus parkelj*, which is restricted to a few caves in a small area within the wide range of highly troglomorphic *P. anguinus*; in a recent molecular phylogeny, the black proteus was positioned as an apical taxa within several distinct populations of *P. anguinus anguinus*. Since vertebrates are born with eyes and melanic pigmentation, which regress along ontogeny, a truncation of the degenerative process due to blocking of genes responsible by triggering such process could result in the so-called character reversal. Clearly, this would not apply to invertebrates.

AN OVERVIEW OF MOLECULAR DATA

Since Darwin, cave organisms have stimulated the interest of biologists as they offer unique opportunities to comparatively unravel evolutionary processes in an ecologically simplified and temporally stable environment. Pioneer studies based on morphology have helped in gaining a better understanding of evolution in caves revealing, at the same time, that convergence is way more frequent in subterranean organisms than in their surface counterparts. Hence, it became evident that molecules would have provided an alternative, independent source of information, less prone to convergence and exhibiting variation appropriate to the different evolutionary questions posed.

In the early 1960s, molecular methods for examining genetic variation at the protein and nucleic acid levels started being adopted by evolutionary biologists. Since then an increasing amount of genetic data has been amassed for terrestrial and aquatic cave taxa and used to test a variety of hypotheses. The next paragraphs present an overview of the different methods and markers that have been placed at the disposal of cave biologists and briefly describe, through a few selected examples, how these markers have been used to describe evolutionary processes in caves.

Historically, assaying protein polymorphisms via electrophoresis has been the first and most widespread molecular method to describe allele frequency changes in species. Protein data have broadened our understanding of genetic variability of populations, gene flow, recognition of species boundaries, and phylogenetic relationships. While relatively cost-effective, the method has implicit limitations as it requires the use of fresh tissues (ethanol preserved samples and/or minimal invasive sampling are not applicable), and the number of loci and alleles that can be resolved tends to be relatively small.

Cave crickets belonging to the eastern Mediterranean genus *Dolichopoda* (Rhaphidophoridae) have been extensively studied on micro-, meso-, and macro-geographical scales (Sbordoni *et al.*, 2000 and references therein). Allozymes proved informative to discriminate geographically distant populations, to recognize processes of cryptic speciation, to assess levels and patterns of gene flow within species, and, more generally, to describe the microevolutionary phenomena of adaptation and speciation related to colonization of caves.

In the case of subterranean aquatic organisms, levels and patterns of allozymic divergence in the exclusively stygobiont isopods of the genus

Stenasellus accurately reflect the geological evolution of the peri-Tyrrhenian area, emphasizing the role played by the tectonic evolution of the area in shaping the diversification of the group and allowing the calibration of a specific allozymic molecular clock (Ketmaier *et al.*, 2000).

In the early 1970s DNA kinetics became an increasingly popular venue of research and evolutionary biologists started looking at methods based on DNA/DNA hybridization as a promising source of information alternative to protein polymorphisms. The rationale behind this approach is based on the double-stranded nature of DNA; the firmer the in-vitro bonds between previously single-stranded DNA fragments of two different species the higher their genetic affinity because of the relatively low number of mismatches between the heterospecific DNA fragments. This technique provides information on the degree of similarity among species and, hence, on phylogenetic relationships. Critiques have been raised that, while accurate at deeper phylogenetic levels, DNA/DNA hybridization tends to be more prone to experimental errors when closely related species are compared.

Caccone and Powell (1987) presented one of the very few study cases centered on cave organisms based on this technical approach. DNA/DNA hybridization data were used to reconstruct phylogenetic relationships within and between the North American cave crickets genera *Euhadenoecus* and *Hadenoecus*. They found a large degree of genetic divergence among populations and species, used the yielded phylogeny to calibrate a molecular clock, and concluded that these insects had, at the time when the article was published, one the fastest rates ever reported for invertebrates.

Popularity of protein electrophoresis and DNA/DNA hybridization peaked in the late 1970s/early 1980s; however, the number of studies based on these approaches progressively waned in the mid to late 1980s and virtually disappeared from top-ranked journals by the early 2000s. But the amount of molecular data gathered on cave organisms has certainly not decreased in the past two decades; rather the opposite. Central to this quantum leap in the use of molecular markers has been development of applications of the polymerase chain reaction (PCR) for investigating variation in DNA on a large scale. The possibility to easily and cost-effectively amplify a large number of genes also from old and/or badly preserved samples resulted in the production of increasingly large data sets on DNA sequence variation between and within species. The versatility of PCR and the fact that sequences produced worldwide have to be deposited in open access databases (*e.g.*, GenBank) have allowed scientists to pinpoint with extreme accuracy genes with levels of variation appropriate to the case under study. Genes of the mitochondrial genome (mtDNA) have played a prominent role in the past two decades, as they are relatively easy to access, do not recombine (and hence coalesce faster), and are (with a few exceptions) maternally inherited. Sequencing of mtDNA has proved instrumental in the fields of biogeography, phylogeography, and speciation among closely related species.

Allegrucci *et al.* (2009) sequenced multiple mtDNA genes to reconstruct the phylogeny and phylogeography of the genus *Dolichopoda* at the scale of the eastern Mediterranean area. Relationships among species resulted to reflect essentially a geographical pattern, with taxa coming from nearby areas being also genetically more closely related. The biogeographic scenario hypothesized to explain patterns of divergence foresaw dispersal as the main force promoting divergence followed by dispersal when the paleoclimatic conditions were favorable for surface dispersal.

A combination of vicariance and dispersal was also invoked by Ketmaier *et al.* (2003) to explain the current distribution and mtDNA phylogeography of the isopods belonging to the genus *Stenasellus*. The authors also took advantage of geological data to calibrate rates of substitutions for the cytochrome oxidase subunit I mtDNA gene and to compare them to those available for other invertebrates.

Leys *et al.* (2003) presented mtDNA evidences supporting a scenario of multiple independent origin and allegedly sympatric subterranean speciation in a group of diving beetles from calcrete aquifers in Western Australia. These subterranean systems host at least 12–13 sister species pairs or triplets of diving beetles, which can be readily discriminated from one another on both genetic and morphological bases. Species occurring sympatrically have nonoverlapping body sizes; this suggests a potential mechanism of sympatric speciation by niche partitioning of species within each calcrete.

MtDNA data need to be complemented with the analysis of more conserved nuclear (nuc) genes when the aim is to disentangle ancient cladogenetic events or to understand ecomorphological evolutionary processes.

Allegrucci *et al.* (2010) used a selection of mt- and nucDNA genes to unravel tempo and mode of evolution in a group of cave crickets belonging to the family Macropathinae (Rhaphidophoridae). This group shows a clear Gondwanan distribution and hence represents an ideal system to test whether molecular-based phylogenetic hypotheses mirror the geological events that led to the breakup of Gondwana. Most of the splits in the obtained phylogeny fit to a vicariance scenario based on the ancient tectonic events related to the

Gondwana fragmentation. However, some relationships remained controversial and could be explained either in light of sporadic events of long distance dispersal or as an artifact due to incomplete taxon sampling.

North American Sclerobuninae harvestmen (*Opiliones*) include three genera; each of these genera comprises series of taxa ranging from the surface ones with normal pigmentation and eye development to those inhabiting caves and showing different levels of troglomorphism, including reduction or lack of tegumentary pigments, elongated legs and palps, and reduced or absent eyes. Derkarabetian *et al.* (2010) coupled the sequencing of multiple mt- and nucDNA genes with the morphometric analysis of a suite of characters to gain insights into morphological homoplasy and rates of morphological changes in the framework of temporal evolution of the cave systems inhabited by these harvestmen. Phylogenetically unrelated taxa evolved troglomorphism multiple times convergently; in one circumstance a troglomorphic facies was found in specimens from a high-elevation stony debris habitat, suggesting that troglomorphism can potentially evolve also in non-cave habitats. A strong positive and linear relationship between degree of troglomorphism and divergence time was also found, making it possible to predict taxon age from morphology.

At the other tip of the scale (*i.e.*, at the intraspecific level), nuclear microsatellite loci are the markers of choice to obtain a finer description of the population genetic structure of a given species.

The increasing demand for low-cost sequencing has led to the development of technologies that parallelize the sequencing technology, resulting in the production of thousands of sequences at once. Parallel (or next-generation) sequencing was launched about ten years ago and machines started being marketed at reasonable prices less than five years ago. The decreasing costs of this new technology and the humongous amount of data produced in just a fraction of a single run are promoting a gradual but decisive transition from the genetic to the genomic level of investigation. Scanning a large portion of the genome is becoming increasingly feasible also for nonmodel organisms (hence including cave species) and will allow scientists to shed light on processes that were considered out of reach only a few years ago. Among others, certainly the most intriguing are those aimed at understanding the genetic basis of adaptive and regressive evolution in cave species with different degrees of troglomorphism. As to our knowledge a study focusing on subterranean organisms and using next-generation sequencing has yet to be published but this certainly represents a fascinating and very promising area of research that will attract the attention and efforts of cave biologists in the near future.

See Also the Following Articles

Asellus aquaticus: *A Model System for Historical Biogeography*
Niphargus: *A Model System for Evolution and Ecology*

Bibliography

Allegrucci, G., Rampini, M., Gratton, P., Todisco, V., & Sbordoni, V. (2009). Testing phylogenetic hypotheses for reconstructing the evolutionary history of *Dolichopoda* cave crickets in the eastern Mediterranean. *Journal of Biogeography*, 36(9), 1785–1797.

Allegrucci, G., Trewick, S. A., Fortunato, A., Carchini, G., & Sbordoni, V. (2010). Cave crickets and cave weta (Orthoptera, Rhaphidophoridae) from the southern end of the world: A molecular phylogeny test of biogeographical hypotheses. *Journal of Orthoptera Research*, 19(1), 121–130.

Caccone, A., & Powell, J. R. (1987). Molecular evolutionary divergence among North American cave crickets. II. DNA-DNA hybridization. *Evolution*, 41, 1215–1238.

Culver, D. C., Kane, T. C., & Fong, D. W. (1995). *Adaptation and natural selection in caves*. Cambridge, MA: Harvard University Press.

Derkarabetian, S., Steinmann, D. B., & Hedin, M. (2010). Repeated and time-correlated morphological convergence in cave-dwelling harvestmen (Opiliones, Lanatores) from Montane Western North America. *PLos One*, 5, e10388. doi:10.1371/journal.pone.0010388.

Dillman, C. B., Bergstrom, D. E., Noltie, D. B., Holtsford, T. P., & Mayden, R. L. (2011). Regressive progression, progressive regression or neither? Phylogeny and evolution of the Percopsiformes (Teleostei, Paracanthopterygii). *Zoologica Scripta*, 40(1), 45–60.

Felice, V., Visconti, M. A., & Trajano, E. (2008). Mechanisms of pigmentation loss in subterranean fishes. *Neotropical Ichthyology*, 6(4), 657–662.

Holsinger, J. R. (2000). Ecological derivation, colonization, and speciation. In H. Wilkens, D. C. Culver, & W. F. Humphreys (Eds.), *Subterreanean Ecosystems*. San Diego, Elsevier.

Howarth, F. G. (1993). High-stress subterranean habitats and evolutionary change in cave-inhabiting arthropods. *The American Naturalist*, 142(suppl), 65–77.

Jeffery, W. R. (2005). Adaptive evolution of eye degeneration in the Mexican blind cavefish. *Journal of Heredity*, 96(3), 185–196.

Ketmaier, V., Argano, R., & Caccone, A. (2003). Phylogeography and molecular rates of subterranean aquatic Stenasellid isopods with a peri-Tyrrhenian distribution. *Molecular Ecology*, 12(2), 547–555.

Ketmaier, V., Messana, G., Cobolli, M., De Matthaeis, E., & Argano, R. (2000). Biochemical biogeography and evolutionary relationships among the six known populations of *Stenasellus racovitzai* (Crustacea, Isopoda) from Tuscany, Corsica and Sardinia. *Archiv für Hydrobiologie*, 147, 297–309.

Leys, R., Watts, C., Cooper, S. J. B., & Humphreys, W. F. (2003). Evolution of subterranean diving beetles (Coleoptera: Dytiscidae: Hydroporini: Bidessini) in the arid zone of Australia. *Evolution*, 57, 2819–2834.

Niemiller, M. L., & Poulson, T. L. (2010). Subterranean fishes of North America: Amblyopsidae. In E. Trajano, M. E. Bichuette, & B. G. Kapoor (Eds.), *Biology of subterranean fishes* (pp. 169–280). Enfield, NH: Science Publishers.

Parzefall, J., & Trajano, E. (2010). Behavioral patterns in subterranean fishes. In E. Trajano, M. E. Bichuette, & B. G. Kapoor (Eds.),

Biology of subterranean fishes (pp. 81–114). Enfield, NH: Science Publishers.

Sbordoni, V., Allegrucci, G., & Cesaroni, D. (2000). Population genetic structure, speciation and evolutionary rates in cave-dwelling organisms. In H. Wilkens, D. C. Culver, & W. F. Humphreys (Eds.), *Ecosystems of the world 30: Subterranean ecosystems* (pp. 450–483). San Diego: Elsevier.

Trajano, E. (2007). The challenge of estimating the age of subterranean lineages: Examples from Brazil. *Acta Carsologica, 36*, 191–198.

Wessel, A., Erbe., P., & Hoch., H. (2007). Pattern and process: Evolution of troglomorphy in the cave-planthoppers of Australia and Hawaii—Preliminary observations (Insecta: Hemiptera: Fulgoromorpha: Cixiidae). *Acta Carsologica, 36*, 199–206.

Wilkens, H. (2010). Genes, modules and the evolution of cave fish. *Heredity, 105*(5), 413–422.

EXPLORATION OF CAVES—GENERAL

William B. White

The Pennsylvania State University

The skills, training, and equipment necessary to explore caves vary tremendously with the size and complexity of the cave. In common, however, are the environmental realities that caves are dark, usually wet and muddy, and have temperatures that are close to the regional average. The terrain is rugged with low crawls, narrow fissures, pits, puddles, pools, flowing streams, and piles of loose rocks. Only rarely does a cave passage have a level floor. The equipment needed by the explorer includes reliable light sources, clothing that provides protection from the environment, and any additional equipment needed for vertical and water-containing parts of the cave system.

SKILL LEVELS IN CAVE EXPLORATION

Small, Near-Horizontal Caves

In regions of low to moderate relief, caves tend to form as near-horizontal passages although there may be many small-scale irregularities in the form of short climbs, drops, and piles of broken rock. Exploration of these caves requires a certain amount of equipment and a certain skill level, but generally nothing that requires extensive training. Even in small, near-horizontal caves, there are certain minimum requirements for exploration.

Above all, caves are dark. Beyond the twilight zone at the entrance, darkness is absolute. Failure of light sources is the single most common risk in caving. Maneuvering one's way through even a small cave by touch alone is exceedingly difficult and for a cave of even moderate complexity, probably impossible. *Requirement (1)*: The explorer must understand the necessity of reliable light sources, backup sources, and the expected lifetime of light sources.

Cave temperatures are usually constant and near the mean annual temperature of the region. Caves in temperate climates are in the range of 8–12°C. Caves at high latitudes and/or high altitudes are colder, often in the range of 0–4°C. Caves tend to be wet and humid. The combination of low temperature and high humidity accelerates heat loss with risk of hypothermia. Tropical caves may pose the opposite problem. High temperatures, 20–25°C, combined with high humidity restricts heat loss with the resultant risk of overheating and possibly heat exhaustion. Some caves contain ponded water and/or flowing streams, which require crawling, wading, or swimming in water that is also at or close to cave temperature. *Requirement (2)*: The explorer must dress appropriately both for the expected environmental conditions and also for the expected activity. A fast-moving exploration team has different heat management requirements than a team engaged in survey or photography.

The cave terrain is rugged. Passages are irregular often scaling from dimensions of tens of meters to crawlways at the limit of human penetration. Ceilings are irregular and protruding ledges are common. Passage floors are often muddy. Loose rocks (breakdown) occur as isolated fragments and as piles tens of meters high. Passages may be discontinuous with vertical offsets that require climbing up or down. Some passages are traversed on ledges or by hopping along canyons that have been incised into the main passage floor. *Requirement (3)*: The explorer must be prepared with foot, head, and body protection against the abuse required for traversing cave passages.

Vertical Caves

Vertical caves are those that have substantial elevation differences between the lowest and highest points in the cave. To be considered a vertical cave, the elevation difference should occur over a relatively short horizontal distance. There exist caves, some volcanic caves, for example, with elevation differences between the top and bottom of more than 1000 m but with only gently sloping passages. The vertical component of caves may consist of pits, sequences of pits sometimes interconnected by small passages, and vertical crevices of various sizes, depths, and shapes. Pits are often pathways for descending water ranging from trickles down walls to large waterfalls.

Vertical caves may be entered from the top, from the bottom, or from various points between. The requirements for exploring vertical caves are more rigorous than those for small horizontal caves and usually require special training on the rigging and use of ropes and on techniques for descending and ascending ropes.

Large Cave Systems

What constitutes a "large" cave as distinguished from a "small" cave is to some extent in the eye of the beholder. As used in this article, a large cave is one that cannot be explored in its entirety in a single day. Exploration of large caves requires developing techniques for *push caving*, for project caving and for expedition caving possibly including cave camps. Push caving is done by a team that simply drives itself to continue exploration for 20- to 30-hour stretches. Requirements are mainly superb physical conditioning and the mental stamina to keep going to exhaustion and beyond. Project caving has an organizational structure. Teams return to the same cave over and over. Some type of organizational structure must be assembled. Records must be kept. The advances in exploration and survey are recorded so that each exploratory push adds to the previously explored cave in a systematic way. In addition to the standard caving skills, the requirement for an expedition caver is a willingness to accept a measure of discipline and control. The term *expedition caving* is applied to caves that are either or both difficult of access and difficult to reach the limits of exploration. Expedition caving may require an expedition to reach the cave and may also require extensive rigging and advance camps within the cave to advance the exploration.

Underwater Caves

It has, of course, been known for a long time that caves exist below regional base levels or below sea level. Until the past 20–30 years, the exploration of these caves was very limited because they were full of water and the necessary equipment did not exist. Although a few heroic explorations were made going back into the nineteenth century using pressurized suits and air compressors, it was the invention of self-contained underwater breathing apparatus (SCUBA) that allowed explorations to be extended to underwater caves.

Requirements for underwater exploration are the most formal and the most extensive. Formal training on the equipment is needed along with formal certification of cave-diving qualifications. A distinction can also be made between clear water diving like that found in many of the large spring caves in Florida and the Yucatán of Mexico and what may be called northern sump diving. The water-filled passages of many temperate and northern climate caves are lined with silt, such that movement of a diver's body quickly stirs up the silt and reduces visibility to near zero.

BASIC EQUIPMENT

Light Sources

The most basic items of equipment are the light sources. Without light, no exploration is possible and if all lights of an exploration party are lost, the party can expect to wait in the dark for rescue. Cavers are advised to each carry three sources of light, which are called the primary, secondary, and tertiary sources. New technology, particularly the development of white light-emitting diodes (LEDs), has greatly enlarged the caver's choices of light sources.

For many years the primary light source was the miner's carbide lamp. Most American cavers used a helmet-mounted lamp that consisted of a lower chamber containing lumps of calcium carbide, an upper chamber containing water, and an adjustable valve that allowed water to drip into the carbide producing acetylene gas. The acetylene produced in the lower chamber was ejected through a nozzle where it was burned to produce a bright yellow flame. A version more popular in Europe used a belt-mounted carbide and water chamber connected by a hose to the helmet-mounted nozzle and reflector. The chemical reaction is

$$CaC_2 + 2H_2O \leftrightarrow C_2H_2 + Ca(OH)_2$$

Carbide lamps had many advantages. They were nearly indestructible. A small repair kit with spare nozzles, filters, gaskets, and other parts allowed replacement of failed parts in the cave. A waterproof bottle holding 500 g of carbide would keep the lamp going for tens of hours. The large flame made a great deal of light. The lamp could also be used to warm a meal. The sooty flame was often used to mark survey stations. A carbide lamp combined with a plastic garbage bag tent offered some protection against hypothermia. The downside of carbide lamps was that the spent carbide is a white, pasty solid consisting mostly of calcium hydroxide but also containing all of the impurities that were present in the carbide. Spent carbide must be carried out of the cave because it is strongly alkaline and toxic to cave organisms. A further complication is that security authorities regard carbide as an explosive material and there are extensive restrictions on carrying or shipping the material.

Carbide lamps have been to a large extent replaced by battery operated light sources. These are available in a great variety of forms. The miner's Wheat lamp features a helmet-mounted fixture that holds the bulb and reflector with power provided by a belt-mounted rechargeable battery. Other belt-mounted lamps are used. Other primary light sources are helmet-mounted lamps with a small battery pack attached to the back of the helmet. These use disposable alkaline batteries, typically the AA size. They are much lighter than other lamps but require the caver to carry a sufficient supply of spare batteries.

Filament-type lamps are rapidly being replaced by arrays of white light-emitting diodes (LEDs). LEDs are more rugged than filament bulbs and are much more efficient, producing light for a longer time on a set of batteries. Some cavers carry dual light sources on their helmets—an LED array as their primary source and a filament lamp that can be focused to provide a sharp bright beam for probing high ceilings and possible side passages. The disadvantage to helmet-mounted lamps is that they provide flat lighting on an irregular, generally monochrome landscape. A hand-held lamp can provide better contrast on irregular cave floors.

The secondary source is often a flashlight of some kind. Small, rugged, waterproof flashlights such as the aluminum-bodied Maglites are widely used. Secondary sources are needed to provide light when it is necessary to refuel or repair the primary source. However, it should be remembered that the secondary will become the primary source if the original primary fails completely. For this reason the secondary source is often a second lamp similar to the primary source.

The tertiary source is the backup to the backup. In the early days, candles and a waterproof container of matches were often used. Escaping from a cave with a candle is not easily done. The present-day choices are the small LED devices, which are available in a variety of forms, some of which are even designed for key rings. The single LED will provide marginally enough light to find one's way through the cave and it will work for 30 hours or more on a single battery.

Hard Hats

The most important item in the caver's outfit is the hard hat. Caves have low and irregular ceilings; there are projecting ledges and other objects on which to bang one's head. Loose rocks get knocked down pits. Hard hats are essential. Two styles are in common use: miners' hats with a snout, and rock-climbers' helmets, which are rounded (Fig. 1). In either style, there is an outer shell of tough polymer that provides the actual protection and an inner harness that keeps the shell from coming into contact with the head. Hard hats should have chin straps with safety release features. Hard hats without straps can be easily knocked off during climbing. However, cavers have been known to become trapped by their hard hats when the caver slid down a narrow fissure only to discover that the hard hat would not fit; thus the necessity for a release feature on the chin strap. Hard hats should have mounting brackets for lamps, whether carbide or

FIGURE 1 A comparison of caving helmets. Left: a rock-climbing helmet fitted with a lamp bracket. Middle: a modern caving helmet with attached LED headlamp and battery pack on the back. Right: a miner's helmet with carbide lamp.

electric. Helmet-mounted light sources leave the hands free for climbing or other activities.

Clothing

Beginning cavers in small caves may wear any clothing that is rugged and that they don't mind getting dirty. More advantageous is a coverall type of outfit that protects the entire body, keeps mud off the inner clothing, and can be peeled off after exiting the cave. Specialty coveralls of tough nylon are available in a variety of bright colors. The desired clothing is dictated by the cave, its temperature, and presence of water and mud.

Underclothing should be selected to preserve heat, particularly if the cave is wet. Wool has this property. Cotton does not. Special clothing of polypropylene has become available for serious cave exploration.

Footwear should provide ankle support and have a good gripping sole. In the United States, ankle-high boots with cleated soles are often favored. European cavers have a preference for high-topped rubber boots ("Wellingtons"). These have the advantage of allowing the caver to wade calf-deep water with dry feet. Wool or polypropylene socks are recommended.

Cave Packs

Because of climbing, crawling, and the frequent necessity for squeezing through tight places, it is unwise for the caver to carry very much in his or her pockets. Most cavers carry packs in which to transport food, water, spare batteries or spare carbide for their lights, a first aid kit, repair kits for lights, surveying equipment, and anything else that might be needed underground. Most needed personal gear can be carried in a small side pack, but if the trip involves vertical caving, larger packs for ropes, ladders, and other climbing equipment will be needed. Packs must be made of tough material that will not tear as the pack is pushed, dragged, or rolled through the cave. It must remain tightly closed so that the contents are not lost. Straps work, whereas zippers tend to become clogged with mud. Tough waterproof packs are used to transport camping gear when advanced camps are needed

VERTICAL CAVING

Vertical caves range in depth from a drop of just a few meters into a cave entrance to deep alpine and high plateau systems where sequences of pits have reached depths of 1500 m and more. The techniques for descending and ascending vertical caves have evolved and improved greatly during the past several decades. Specialized equipment is well-established. The use of it requires training and practice, which is best accomplished under the guidance of experienced vertical cavers.

Going Down

The early days of pit exploration involved winches and lowering lines along with rope and cable ladders of various types. In early twenty-first century caving, the only surviving technique from the early days is the cable ladder. The ladder consists of two strands of woven cable as the legs of the ladder. The rungs are short (typically 15 cm) lengths of aluminum tubing, crimped to the cable a convenient distance apart. Thirty meters of cable ladder can be rolled to a diameter of 30 cm or less and stuffed in a pack. Sections of ladder can be strung together for long drops but today are rarely used for this purpose. Ladders are mainly used for short drops to avoid the nuisance of rigging climbing gear and for tight crevices where single-rope techniques are awkward. Ladder climbs should be belayed because of the possibility of slipping off the ladder.

Nearly all descents of vertical caves are made with what are known as *single-rope techniques*. The ropes themselves have undergone substantial innovation with older woven (laid) ropes being replaced by sheaf ropes. Sheaf ropes have a core of twisted polymer to provide strength surrounded by a braided sheaf. The sheaf allows the rope to slide smoothly through descending devices without the twisting experienced on the older laid ropes. Caving ropes are strong but are sensitive to abrasion. They must be rigged where they do not rub against the rock. This often requires that the rope be attached to a set of secondary anchors bolted to the wall along the drop. These are known as *rebelays* and require the caver to learn how to pass them, both descending and ascending. In part, the extensive use of rebelays has been made possible by the invention of the battery-powered drill which allows setting bolts in awkward places without excessive effort.

Descending a pit is a question of connecting oneself to the rope and sliding down it in a controlled manner, a procedure known as a *rappel*. Many rappelling devices have been devised. One of the most popular is the rappel rack, a U-shaped steel rod. Across the U are placed a series of brake bars, usually of aluminum. These have holes at one end for threading over the U while the other ends are notched so that the brake bars can be hinged away from the rack, thus allowing bars to be added or removed according to the weight of rope hanging below. The rope is threaded through the brake bars in such a way that the bars are locked against the U-rod of the rack and friction between the rope and the brake bars controls the speed of descent. The caver wears a harness of nylon webbing that is attached to the rack by a carabiner.

Going Up

Pits can be ascended by cable ladder, which works equally well going up or going down. However, ascending a pit or sequence of pits using single-rope techniques requires some means for climbing the rope. Prusik knots are a form of slip knot that will clamp around the main rope when tension is put on them and release when the tension is released. A set of three Prusik knots, two connected to the climber's feet and the third attached to the chest harness, provides a slow and tedious method of rope climbing but is useful as an emergency measure. Most cavers use ascenders of some type. There are several designs but all are based on a moving toothed cam that presses against the main rope when the climber's weight is on it. It releases and can be moved upward when the climber's weight is released. A number of ways have been devised for arranging the ascenders, each with their adherents. The overall result is that the caver essentially walks up the rope.

Some caves are explored from the bottom, which means that explorers need to ascend drops or pits without benefit of a rope. Standard rock-climbing techniques are used but are more limited because of poor lighting, wet and muddy surfaces, and the presence of weak or rotten rock. Most vertical ascents require drilling holes in the rock, setting bolts, and attaching anchors that support the climber while new and higher bolts are set.

LARGE SYSTEMS: EXPEDITION CAVING AND PROJECT CAVING

It is useful to make a distinction between *caving* and *cave exploring*. *Caving* is simply traversing the cave, whether for recreation, scientific observations, photography, or other objectives. *Cave exploring* is the systematic probing of all passages to obtain a complete description of the cave. For present-generation explorers, this implies that the passages will be surveyed and a map produced. Systematic exploration and survey is much more time-consuming than a casual trip through the cave and, for caves of even modest size, may require multiple trips or even multiple years of effort; thus, the terms *expedition caving* or *project caving*.

The approach to expedition caving depends on whether or not the frontier of exploration can be reached from some entrance in a reasonable period of time. If so, the expeditions can be based at the surface. Exploration parties enter the cave, proceed quickly to the point at which exploration is to begin, explore and survey, and then return to the surface to eat and sleep. The alternative is a cave in which reaching the frontier of exploration requires so much time that there is no time left for exploration, much less returning to the surface. Such caves require setting up underground camps.

With either type of large cave, complete exploration and survey requires a sustained effort and thus an organizational structure. The structure may take the form of a formal organization or a loosely knit "project." In either case, some leadership is necessary to coordinate expeditions, data collection, and data storage. Survey results need to be deposited in central files and between-expedition activity is necessary to plot draft maps and prepare objectives for the next expedition.

Lengths and organization of expeditions also depend on the accessibility of the cave. Many caves in the United States and Europe are within easy access of roads. The style is to conduct frequent weekend or weeklong expeditions. Expeditions to caves in inaccessible regions or that require extensive travel such as the deep caves of Mexico or the giant caves of Malaysia often extend for periods of months. For these expeditions, arrangement for supplies and their transport to the expedition site becomes an important part of expedition management.

Perhaps the most elaborate effort at project caving is the Cave Research Foundation. Although it has greatly expanded its original objectives, the Cave Research Foundation was organized to explore and survey the complex Mammoth Cave System in Kentucky. It has been at the task for more than 50 years and the end is not yet in sight. In the course of this 50-year effort, the length of the system was expanded from something like 40 km to almost 600 km. Few places in the Mammoth Cave System are exceptionally difficult; it's just that there is so much of it.

WATER AND UNDERWATER CAVING

Big-River Caves

Most caves are wet and many contain pools and flowing streams. Most of these caves can be traversed only by getting wet. However, a few caves exist that contain large, high-gradient rivers not different in kind from white water rivers on the surface. Exploration of such caves requires techniques beyond those of basic caving.

Cave exploration that requires total immersion in water requires protective clothing beyond the usual warm underclothes and coveralls. Wet suits are tight-fitting full-body outfits made of polymer foam with a protective smooth outer polymer layer. As the name implies, the wet suit allows water to soak through but provides an insulating layer to preserve body heat. Wet suits are widely used in any cave exploration that requires extensive exposure to water. They are often worn under coveralls for protection against snags and abrasion. Less common are dry suits, which are full-body outfits of water-impermeable polymer. They keep water out, but they also keep perspiration in. Dry suits work best for waterfalls and whitewater streams where the caver may be immersed for only short periods. Once out of the stream, the suit can be opened to allow perspiration and waste heat to escape.

Exploration of big-river caves is one of the most hazardous types of cave exploration. The distinction between a stream cave and a big-river cave can be made on the basis of whether or not a caver can wade the stream and not be swept away. Exploration of big-river caves requires much the same techniques as exploring vertical caves. Anchors must be placed and ropes played out to prevent the explorers from being washed away. Waterfalls must be rigged with ropes. Deep pools may require swimming. Great care must be taken to avoid being pinned by the force of flowing water.

Underwater Caves

Many caves end in sumps. Sumps are places where the ceiling of the passage dips below the surface of standing water. Some sumps are short and shallow. A dive of a few meters or a few tens of meters brings the explorer back up into an air-filled passage.

Some sumps are longer and deeper. Some caves are completely underwater. Cave systems in Florida, the Yucatán Peninsula, and the Bahamas have been drowned by the sea-level rise that took place at the end of the ice ages. Underwater caves can be quite long. Six caves are known in Florida with lengths in excess of 5 km. The underwater cave Ox Bel Ha in the Mexican state of Quintana Roo has a surveyed length of 70.65 km and a dozen other underwater caves are known in the Yucatán with lengths exceeding 5 km.

Both sump diving and underwater cave exploration require complex SCUBA equipment, extensive training, and, above all, experience. Formal training and certification is offered by the Cave Diving Section of the National Speleological Society. Underwater caving is far less forgiving of mistakes and equipment failures than any other type of caving. Rescues are rare. Body recoveries are all too common.

The horizontal limits of underwater exploration are dictated by air supply. The distance a diver can travel before one-third of the air supply is consumed was for a long time limited by how far the diver could swim. A new development is the use of underwater propulsion systems (scooters), which allows divers to go much farther with the same air supply. Most of the air in SCUBA tanks is wasted; it is breathed once and then exhaled as a stream of bubbles. Another development, however, is the invention of the rebreather, a device that recycles the air by extracting carbon dioxide. This ability greatly extends the diver's range.

The depth limits of underwater exploration are limited by the need for decompression. Pressure increases with depth, causing an increased solubility of gases, particularly nitrogen, in the diver's blood. Ascent from an underwater cave must be done slowly, lest nitrogen form bubbles in the blood, bringing on a painful condition known as the *bends*. Special gas mixtures have been devised to partially alleviate the problem.

SAFETY

Whether recreational caving in relatively small, near-horizontal caves or undertaking expedition caving in deep vertical caves, caving parties must take responsibility for their own safety. Rescue of disabled persons from caves is an extremely complex and expensive operation. Most rescue teams are volunteers. Their services should be reserved for serious accidents and not wasted on people who simply forgot to take extra batteries for their flashlights.

For beginning cavers in relatively small, near-horizontal caves, the rules are simple: (1) Never cave alone. A party of three works well. If one caver has accident, another can go for help while the third remains with the injured caver. (2) Carry the necessary light sources and know how many hours of service they will provide. Be out of the cave before the lights go out. (3) Always notify someone about where you are going and when you expect to return. If the entire party does not return, there is someone to call for help. (4) Be careful. Any injury that incapacitates a caver will require a massive rescue effort.

Most countries with a large number of caves or a large number of cavers have some type of formal cave rescue organization. In the United States, it is the National Cave Rescue Commission (NCRC). The NCRC is on call throughout the United States and maintains close communication with police and other fire and rescue organizations. It also offers formal training courses for those who might wish to participate in cave rescue activities.

CONSERVATION

Caves are fragile. Time moves slowly underground. Any damage done to a cave—broken speleothems, graffiti, litter—will take a very long time for natural processes to repair. Organisms that live in caves are also easily disrupted. There is an ethic among cavers to do their explorations with the minimum possible impact on the cave and its inhabitants. Anyone who visits caves, no matter how casually, should make the maximum effort to preserve the underground environment.

CLOSING COMMENTS

This article is a summary, not a technique manual. Those who are contemplating becoming cave explorers are urged to join experienced explorers and learn from their experience. Most countries have formal cave exploring organizations. In the United States the organization is the National Speleological Society with headquarters in Huntsville, Alabama. The bibliography lists some books on techniques and also books that describe some of the great explorations that have occupied cavers for many years.

See Also the Following Articles

Exploration of Caves—Vertical Exploration Techniques
Exploration of Caves—Underwater Exploration Techniques
Entranceless Caves, Discovery of
Recreational Caving

Bibliography

Brucker, R. W., & Watson, R. A. (1976). *The longest cave*. New York: Alfred A. Knopf.
Damon, P. H., Sr. (1991). *Caving in America*. Huntsville, AL: National Speleological Society.
Dasher, G. R. (1994). *On station. A complete handbook for surveying and mapping caves*. Huntsville, AL: National Speleological Society.
Exley, S. (1994). *Caverns measureless to man*. St. Louis, MO: Cave Books.
Farr, M. (1991). *The darkness beckons*. London, England: Diadem Books.
Hanwell, J., Price, D., & Witcombe, R. (2010). *Wookey Hole: 75 years of cave diving and exploration*. Wells, Somerset, U.K: Cave Diving Group.
Judson, D. (1984). *Caving practice and equipment*. Newton Abbot, U.K: David & Charles.
Marbach, G., & Tourte, B. (2002). *Alpine caving techniques*. Allschwil, Switzerland: Speleo Projects.
NSS Caver Training Committee (1992). *Caving basics*. Huntsville, AL: National Speleological Society.
Padgett, A., & Smith, B. (1987). *On rope*. Huntsville, AL: National Speleological Society.
Steele, C. W. (1985). *Yochib: The river cave*. St. Louis, MO: Cave Books.
Steele, C. W. (2009). *Huautla: Thirty years in one of the world's deepest caves*. Dayton, OH: Cave Books.
Stone, W., & Am Ende, B. (2002). *Beyond the deep*. New York: Warner Books.

EXPLORATION OF CAVES— UNDERWATER EXPLORATION TECHNIQUES

Jill Heinerth

Heinerth Productions, Inc.

When some cavers reach a water-filled sump, a sigh of disappointment rises in their throats—the end of accessible exploration. Yet to cave divers, this is the portal to a new frontier of discovery. In the past 70 years, the underwater exploration of caves has evolved from a high-risk adventure for self-described gear inventors to an evolved application of new technology driven by human endurance.

The evolution has been rapid and extensive and world record penetration dives of two decades past are now being conducted as routine excursions in a cave-diving class. Exploration has moved from wide open passages into tight, secluded sites, highly technical expeditions to remote locations, and sump dives to previously unimaginable depths. Beyond stretching the limits of human endeavor, cave-diving exploration has joined the ranks of serious "technical diving."

Three factors have contributed to increased range and productivity within cave-diving exploration: wider availability of training and mentoring, improved quality and availability of cave-diving equipment, and advanced technology.

Currently, cave diving educational programs are available from training agencies on every continent. The distinction between for-profit and nonprofit training organizations is quickly disappearing. From Finland to China to Australia, organizations and social networks have united people with cave-diving interests and the desire to seek general and specialized educational opportunities. For the most part, that explosion of availability and interest has been relatively recent. After forward progress was halted by a water-filled sump in the Mendip Hills of the United Kingdom in 1935, cavers organized mentoring programs and equipment for the exploration in the flooded caves of Somerset. The goal of the British Cave Diving Group (CDG) was to teach good cavers how to dive as opposed to training divers how to explore underwater caves. In North America, training emerged much later. American cave diving pioneer Sheck Exley published *Basic Cave Diving: A Blueprint for Survival* in 1977, taking a serious look at accident analysis and creating the foundation for formalized training. Since that time, numerous international organizations have used his basic rules as the centerpiece for teaching safe cave diving. In the past decade, entry-level recreational diver certifications may have leveled off, but participation in technical diving programs is still growing. In a recent survey conducted by the SDI/TDI Group (Scuba Diving International™ and Technical Diving International™) almost four out of ten divers indicated they were "diving tech" or they were interested in taking a technical diving program of some sort within 12 months (Lewis, 2011). Cave diving represents a large segment of the technical diving community, which includes deep-wreck divers and anyone diving beyond the range of direct and immediate ascent in an emergency situation.

The earliest training programs focused on essential cave-diving skills. These days, specialized training by accredited and insured instructors is available in activities including side-mount and no-mount exploration, deep mixed-gas cave diving, diver propulsion vehicles, photography, surveying, cartography, and the use of closed-circuit rebreathers in caves. Classes for many of these pursuits are available in Florida, Mexico, the Bahamas, France, Italy, Spain, Scandinavia, Australia, South Africa, Brazil, and Russia, with more destinations, instructors, and infrastructure growing around the world.

Equipment development for cave divers was aggressive in the 1970s, driving many of the safety enhancements that we take for granted in diving today. Valve manifolds, octopus second stages, buoyancy devices, and reels were developed and refined in this decade. Accurate underwater cave maps matured correspondingly, headed in America by Sheck Exley, Bob Friedman, and Frank Martz. Lighting instruments were also improved and new "Goodman" light-head handles enabled the use of two hands for reel work and survey. Additional tanks, called *stage bottles*, permitted divers to extend their penetrations significantly and the mapped underwater cave passages in north Florida were extended more than 30 miles. However, with increased activity came higher mortality rates. In 1974, fatalities peaked with 28 individuals perishing in water-filled caves (Bozanic and Halpern, 1974).

In the early 1980s, cave diving became the subject of the mainstream media. A television program, *Descent into Darkness*, introduced the sport to wider audiences. During this period, the first *NSS-CDS Cave Diving Manual* was published. Toward the end of the 1980s, cave diving became truly technical in nature. Closed-circuit rebreathers (CCRs) (Fig. 5), exotic gases, and diver propulsion vehicles (DPVs) increased the range of individual divers and groups, including the United States Deep Caving Team and the Woodville Karst Plain Project. In 1987, Dr. Bill Stone made the first 24-hour cave excursion using a CCR (Fig. 4), ushering in the dawn of a new age of extremely technical dives.

Currently, CCRs, DPVs, and small, high-powered HID and LED lights are facilitating remarkable exploration. Advanced dive computers, capable of multigas and multialgorithm profiling, have made decompression diving safer and more efficient. Decompression algorithms are traded on iPhones and BlackBerrys instead of being held secretly in the hands of elite teams. Time-worn, popular caves such as Ginnie Springs (Devil's Ear and Eye), Peacock Springs State Park, and other sites are enjoying a new era of exploration, as young divers penetrate beyond the ends of lines, often in passages that were previously thought to be too small for continued exploration. In some cases, completely new complex base levels and connections to other openings are being reported and surveyed.

One of the tools supporting new cave exploration is the side-mount diving technique (Fig. 1). Divers have come to recognize that this versatile configuration is one of the most comfortable, stable, and safe ways to enter smaller passages that were previously overlooked. Although this style of cave diving is not new, off-the-shelf equipment solutions from trusted manufacturers, new textbooks, and training opportunities have expanded the practice of side-mount diving. At the time of writing, several international training agencies have specialized programs in side-mount diving, including the International Association of Nitrox and Technical Divers (IANTD), National Association for Cave Diving (NACD), and National Speleological Society/Cave Diving Section (NSS-CDS). The following agencies provide training and certification in side-mount open water diving: International Association of Nitrox and Technical Divers (IANTD) and the Professional Association of Diving Instructors (PADI).

The first side-mount cave exploration took place beneath the hills of northern England in the early 1960s. British cave diver Mike Boon explored the sumps of Hardrawkin Pot in Yorkshire, England, by slinging the cylinder on his side with a bandolier-style harness.

FIGURE 1 Brian Kakuk explores the beauty of "The Badlands" of Dan's Cave in Abaco, Bahamas. His efforts are bringing the country ever closer to preserving this unique region, South Abaco Blue Holes Conservation Area, as a part of the National Park system. *Photo by the author.*

The impetus for side-mount diving in North America came from a much darker episode. In the late 1970s, a diver fatally pinned himself in an impossibly narrow crevice in Royal Spring in north Florida. Divers Sheck Exley and Wes Skiles were unable to reach him in their traditional back-mounted cylinders. Exley, holding a tank at his side, slid into the crevice beside the dead diver, breaking him loose from the cave. Skiles, seeing this configuration in action, quickly realized the potential for exploration of smaller passages, and thus, side-mount cave exploration began to expand through the United States. Other pioneering explorers including Woody Jasper, Forrest Wilson, Court Smith, Mark Long, Tom Morris, and Lamar Hires (Fig. 2) continued to evolve and refine the system, lending to its popularity today.

Perhaps the greatest paradigm shift in underwater cave exploration is the recent adoption of rebreather technology. In the late 1990s, Dr. Bill Stone pioneered the development and marketing of the first fully electronic CCRs that were targeted to noncommercial divers and explorers. At Wakulla Springs, the U.S. Deep Caving Team, using Stone's Cis-Lunar MK5P rebreathers, along with his advanced sonar-based mapper and FatMan scooters, accomplished the first accurate three-dimensional model of an underwater cave environment (Fig. 3). Some of the most impressive deep cave penetrations of the decade were logged by members of the Woodville Karst Plain Project (WKPP) using a purpose-built semi-closed circuit rebreather (SCR). In 2007, a seven-mile traverse by Jarrod Jablonski and Casey McKinley (Kernagis *et al.*, 2008) linked two separate openings and had the cave-diving community wondering whether they had reached the boundaries of human endurance. Yet, in the past decade, supported by the wide commercial availability of CCRs, a new wave of exploration is under way. AP Valves (Inspiration and Evolution), Innerspace Research (Megalodon), KISS, and others have established a viable worldwide commercial rebreather market. Several organizations support rebreather training programs, giving cave divers access to safe, standardized instruction. Once a rare curiosity, rebreathers are now common technology at most technical diving sites.

Rebreather diving is not a new idea. Rebreathers were available long before traditional open circuit scuba equipment. Open-circuit divers use either back- or side-mounted high-pressure cylinders affixed to buoyancy control devices (BCD) called *wings*. A regulator attached to the cylinder lowers the pressure of the supply gas so that it can be delivered to the diver on demand as she inhales. As the diver exhales, the resulting bubbles are vented directly into the water column. Since humans only use a very small portion of the oxygen molecules in an inhaled breath for metabolism, they exhale vital leftovers and create

FIGURE 2 Lamar Hires demonstrates the perfect trim afforded by the side-mount diving system. *Photo by the author.*

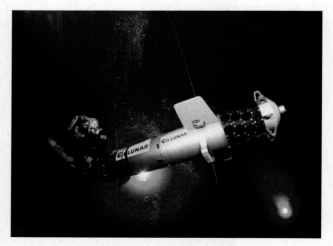

FIGURE 3 The author driving the Wakulla Mapper and using the Cis-Lunar MK5P rebreather during the United States Deep Caving Team explorations of the late 1990s. *Photo by Wes Skiles/United States Deep Diving Team. Used with permission.*

FIGURE 4 A side-mount diver may easily remove a single tank to negotiate a restriction and minimize impact on the cave itself. *Photo by the author.*

FIGURE 5 The author diving in a Sentinel Expedition rebreather, which offers not only a gaseous carbon dioxide monitor, but also a large heads-up display with lights and vibrating buzzer that alerts the diver to life-threatening scenarios. *Photo by the author.*

waste. A rebreather, on the other hand, takes advantage of the exhaled bubbles by recapturing them, scrubbing them clean of carbon dioxide, and returning them to the diver with the addition of supplemental oxygen, maintaining an appropriate level for metabolism. The first rebreathers were pure oxygen rebreathers which could only be used at very shallow depths, but allowed much longer, bubble-free operation for divers.

As interest grew in rebreathers and applications demanded technology for deeper depths, SCRs were developed. SCRs create some intermittent bubbles. They inject a steady stream of oxygen-enriched air (nitrox) into a set of breathing bags. As the diver's exhaled breath is recaptured in the system, a few bubbles are vented off and fresh nitrox is injected to make up the difference. The equipment is carefully calibrated so that the injection stream makes up the correct volume of oxygen lost through bubbles and diver metabolism. SCRs act as gas extenders and provide longer no-decompression times over standard-air diving, but have the disadvantage of venting gas in the form of bubbles. Various advances over the years enhanced this technology to reduce wastage through bubbling and increase monitoring capability.

Fully closed circuit rebreathers offer the greatest advantages over standard scuba. A CCR only vents gas on ascent. It has the highest efficiency and nets divers the longest no-decompression times. This technology can be up to 200 times more efficient than standard open-circuit scuba, dependent on depth, the ratio of efficiency improving at deeper ranges. CCRs are now developing into true life support devices that better protect the diver from human error and undetected dangers such as carbon dioxide buildup. The Sentinel rebreather from VR Technology brings to the market the first application of an advanced onboard carbon dioxide monitoring system. The three-pronged approach monitors the volume of oxygen that passes through the solenoid valve estimating the production of carbon dioxide, tracks the thermal profile of the reaction zone within the bed of scrubber material, and monitors gaseous carbon dioxide within the breathing loop using infrared technology, netting a reasonable expectation of scrubber life as well as detecting errors in preparation of the scrubber or failures of the system (Gurr, 2010).

Carbon dioxide is thought to be the least understood and greatest risk to rebreather divers. Difficult to track and impossible to detect at autopsy, carbon dioxide poisoning may turn out to be the silent killer that has taken the lives of many dozens of rebreather divers in the past decade. Professor Simon Mitchell from the Department of Anaesthesiology at the University of Auckland and Auckland City Hospital, leads the industry in researching the issues of carbon dioxide buildup and retention in divers. His paper, *Hyperbaric Conditions*, gives a comprehensive account of relevant physiology (Doolette and Mitchell, 2011). Mitchell describes that the act of diving itself alters respiratory control. He describes that sensitivity to CO_2 may be reduced by an increased partial pressure of oxygen, increased partial pressure of nitrogen, and increased work of breathing—all aspects of rebreather diving. In particular, he believes that an increased work of breathing results in varying degrees of CO_2 tolerance which may subsequently lead to the diver failing to recognize and correct a potentially life-threatening scenario (Mitchell, 2010).

Cave-diving fatalities have declined in the past 30 years, yet rebreather diving deaths are on the rise. The industry is at its pioneering stage where instructors and divers are working out common codes of practice, manufacturers are unregulated in many regions, and equipment testing is sometimes undertaken with human divers using gear that has no independent oversight. In ever more litigious times, it will be incumbent on the industry to self-police through the adoption of policies such as CE or equivalent third-party testing and documentation. Barring this, product liability lawsuits and outright banning of the use of such equipment at dive sites seem likely. In the dawning days of this exciting technology, it is clear we still have a long way to go to improve procedures and equipment safety. (CE EN14143: Essentially any person or company in Europe that is importing, manufacturing, selling, or using rebreathers requires a valid CE marking on the equipment. If it does not have a valid CE marking, then they are breaking the laws or regulations in their country, and the enforcement body in that country can ban the import, obtain court injunctions to stop manufacture or sale, and seize the goods. If there are accidents from the use of such equipment, it may become a criminal liability for the directors of the company involved. EN14143 is a testing and documentation protocol for all PPO2 monitors for use with a rebreather that is imported, manufactured, sold, or used in work in Europe.)

In addition to the diving apparatus itself, the tools for cave survey are also undergoing a renaissance. Traditional underwater cave survey is conducted using knotted survey line or fiberglass tape and compass. These traditional tools can only net extremely rudimentary surveys that are dependent on myriad imaginative additions by true artists of cartography. Underwater cave maps take years to assemble in even a basic form and are often fraught with errors and generalities. Accurate digital compasses, such as those provided on Suunto diving computers, add significantly to accuracy, but the future lies in far more accurate tools.

After proving the value of the Wakulla 3D Mapper in the late 1990s, Dr. Bill Stone went back to the drawing table to make the next huge leap in exploration technology. In 2007, DepthX made a historic dive in La Pilita cenote in Tamaulipas, Mexico, accurately mapping the sink in three dimensions, autonomously. It appears that the real future of cave mapping may not include humans. The untethered device dropped to a maximum depth of 100 m while compiling 340,000 sonar wall hits during the course of an approximately 1-hour-duration mission. The device's sonar arrays "scanned" the wall in 360° and even noted the presence of a 20-m-diameter tunnel leading off from the western wall of the chamber at a depth of approximately 40 m. As Stone continues to miniaturize this apparatus, it appears the future of underwater cave mapping will not only be more accurate but also safer.

Numerous challenges and rewards await the next generation of cave-diving explorers. Physical and technological advances are increasing the depths and penetrations that are humanly possible. The coming era may feature exciting new technologies from hard suits to miniature robots to thermal imaging and new decompression strategies. With education, mentoring, and experience, there are few limitations that stand before the imagination of those willing to push the limits of underwater cave exploration.

Bibliography

Bozanic, J., & Halpern, R. (1974). *Cave diving fatalities. A summary*. Workshop presentation for the NSS-CDS Annual Conference, Marianna, Florida.

Doolette, D. J., & Mitchell, S. (2011). Hyperbaric conditions. *Comparative Physiology*, 11–39.

Exley, S. (1986). *Basic cave diving: A blueprint for survival* (5th ed.). Huntsville, AL: National Speleological Society.

Exley, S. (1994). *Caverns measureless to man*. St. Louis, MO: Cave Books.

Gurr, K. (2010). *Technical diving from the bottom up*. VR Technology.

Heinerth, J. (2010). *The essentials of cave diving*. High Springs, FL: Heinerth Productions, Inc.

Heinerth, J., & Oigarden, W. (2008). *Cave diving: Articles and opinions*. High Springs, FL: Heinerth Productions, Inc.

Kakuk, B., & Heinerth, J. (2010). *Side mount profiles*. High Springs, FL: Heinerth Productions, Inc.

Kernagis, D., McKinlay, C., & Kincaid, T. (2008). Dive logistics of the Turner to Wakulla Cave Traverse. In P. Brueggeman & N. W. Pollock (Eds.), *Diving for science*, Proceedings of the American Academy of Underwater Sciences 27th Symposium. Dauphin Island, AL.

Lewis, S. (2011). *The six skills and other discussions: Creative solutions for technical divers*. Techdiver Publishing.

Mitchell, S. (2010). CO_2: *Big, bad and hard to measure*. Presentation at Eurotek, October 2010, Birmingham, England.

Stone, W. (2007). DEPTHX: Zacaton—Mission 1 begins. <www.stoneaerospace.com/news-/news-zacaton-mission4.php/>.

EXPLORATION OF CAVES—VERTICAL CAVING TECHNIQUES

Mark Minton and Yvonne Droms

U.S. Deep Caving Team

Disclaimer: Vertical caving is a potentially hazardous activity. The following discussion is not intended to teach proper vertical technique or to cover all details of vertical caving. Get proper training before entering any vertical cave.

INTRODUCTION

Although many caves can be explored by simply crawling, walking, and climbing over rocks and ledges, a great many require the use of ropes and related equipment for ascending and descending vertical drops. Even caves that are predominantly horizontal may have occasional pits, while others are almost entirely vertical with dozens of drops and relatively little horizontal extent. Individual pits can range from a few meters to hundreds of meters in depth. The world's deepest cave (Krubera) is over 2000 meters deep and contains scores of drops.

The simplest vertical cave passages can be negotiated without equipment by chimneying and other free-climbing techniques, although this can be dangerous since there is no safety net if one slips or if handholds break. For slightly more difficult drops one may get by with a simple handline or a ladder. However, serious vertical caves require specialized equipment, training, and practice to explore safely. Ropes, and the associated hardware to use them, are the only practical method for getting up and down the vast majority of vertical caves. The current state of the art is called single rope technique, or SRT. We shall confine our discussion to the most commonly used variants of SRT, although there are many different systems with ardent adherents and specialized applications.

ROPE

Modern ropes for caving are made of nylon or polyester and use kernmantle construction, which consists of a core of long, multistranded, twisted fibers surrounded by a tightly woven sheath. Unlike dynamic ropes for rock climbing, which are designed to stretch in order to absorb the shock of a fall, static caving ropes are designed to have as little stretch as possible. This contributes to the efficiency of climbing the rope, especially on long drops where stretch can be significant. For many years the standard diameter for caving rope was 11 mm, and this is still widely used in the United States. However, nowadays many cavers are opting for smaller diameters: 10-mm rope is common and advanced exploration may use 9-mm or even 8-mm rope. These smaller-diameter ropes are lighter and large amounts can be packed into a relatively small space for transport, but one must be increasingly careful to avoid abrasion when rigging them.

RIGGING

In the early days of vertical caving it was common practice to simply tie the rope to whatever was handy and throw it down the pit. This was easy to do but often resulted in dangerous abrasion of the rope as well as the possibility that loose rocks could be dislodged from the wall. It also frequently made it difficult to get over the lip at the top of the drop. Especially when using small-diameter rope, these conditions are unacceptable. Rig points should be chosen such that the rope hangs as freely as possible down the pit. Since this requires the rope to hang at least a short distance beyond the edge of the drop, an approach line is usually added for safety in getting onto and off of the main rope.

Natural rig points (bedrock bridges, large stalagmites, firmly wedged boulders, *etc.*) are preferable from a conservation point of view, but more often than not there is no suitable natural anchor available. In that case artificial anchors consisting of rock-climbing chocks, pitons, or, most often, expansion bolts are used. With the advent of lightweight battery-powered hammer drills, placing bolts has become very easy. One should use care to place bolts properly since there are usually limited spots available, and no one likes to see a proliferation of bolts at the top of a pit. When bolts are in high-traffic areas they should be made of stainless steel and ideally labeled in a manner that lets future visitors know the embedment depth, date of placement, and identity of the rigger. Another alternative is the use of glue-in bolts, which are more permanent and have less tendency to work loose with repeated use. All rig points should be backed up. Do not depend on a single bolt, piton, or chock. An additional advantage of using multiple rig points is that one can almost always construct a free-hanging rig by using Y-hangs (Fig. 1).

As important as a good rig point is a secure knot to tie off the rope. There are several knots that vertical

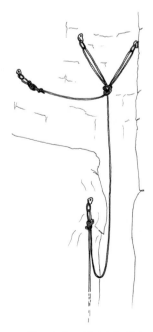

FIGURE 1 Y-hang with rebelay. *Sketch by Bob Alderson.*

cavers should know. These include the bowline, bowline on a bight, figure eight, double figure eight or rabbit-ears knot, figure nine (used with thinner rope), alpine butterfly, and double fisherman's knot. The double fisherman's knot is used to join two ropes together and is relatively easy to pass. One should also be familiar with the water knot for nylon webbing. There are a great many other knots that may come into play in various situations, but those mentioned here should serve in the majority of situations.

Even when the rope is placed properly at the top of a drop, there may be ledges, speleothems, waterfalls, and so on, that must be avoided farther down. In this case the rope may be secured at intermediate spots to eliminate rub points or change its orientation. Such intermediate rig points are called *rebelays* and *redirects* or *deviations*. A rebelay is a hard tie-off, where the rope is secured to an anchor similar to one used at the top of a drop (Fig. 1). This essentially divides the drop into two (or more) independent pitches, which, depending on circumstances, may be climbed simultaneously by two or more cavers. A redirect or deviation is not an actual anchor, but simply a rerouting of the rope through a carabiner tethered on a piece of cord or nylon webbing such that it pulls the main rope off of its original path and clear of obstacles. Both of these techniques require additional skill and training to negotiate safely.

Although rarely necessary with careful rigging, occasional rub points are unavoidable or impractical to eliminate. In such cases rope pads may be used. Any relatively abrasion-proof, flexible material may be used to protect the rope from rubbing against the rock. This includes pieces of carpet, heavy cloth, wide nylon webbing, slit rubber hose, or, temporarily, even cave packs or items of clothing. Rope pads are easiest to pass if they are tied off independently and placed such that the rope will simply run across them. When this is not possible an alternative is to tether the pad directly to the rope. This is a more secure arrangement in that the rope cannot slip off of the pad, but it is more difficult to pass.

SIT HARNESS

For all vertical work in caves, one should wear a properly fitting sit harness designed specifically for this purpose (Fig. 2A). The harness holds the body in a comfortable and secure upright position on rope even when there are no walls or floor within reach. A typical sit harness is made of wide nylon webbing and consists of a waist belt and two leg loops. It is closed by a lockable metal carabiner or quick link, often called a *maillon*, which provides a central attachment point to which various pieces of hardware may be connected. Frequently there are auxiliary gear loops for stowing vertical gear when it is not in use and for carrying other pieces of equipment as well. A harness designed for caving typically has a lower central attachment

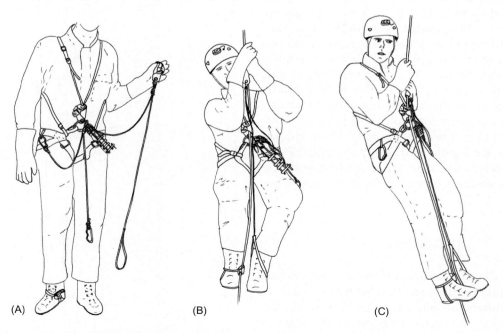

FIGURE 2 The frog system for ascending rope. (A) Frontal view showing harness. (B) Frontal view, climbing. (C) Side view, climbing. *Sketches by Bob Alderson.*

point than one for rock climbing, as this optimizes ascending efficiency.

While on rope it is more efficient to suspend a cave pack from the central link of the sit harness with a tether than to wear it on one's back. This prevents the weight of the pack from pulling the caver away from the rope. It also supports the weight of the pack when one is resting. The tether should be just long enough not to interfere with climbing, but not so long as to allow the pack to get caught or entangled beyond the caver's reach.

DESCENDING

While cable ladders may occasionally be used to negotiate vertical drops in caves, by far the most common method for going down pits is *rappelling* (also called *abseiling*) on a single rope. Many devices have been designed that apply friction to the rope in a variable and controllable fashion. Of these, the only ones in common use for caving today are the rack, bobbin, and figure eight. An older technique now rarely used is body rappelling, where the rope is wrapped around the caver's body to provide the friction necessary to control descent. Body rappelling is less comfortable and more dangerous than using a mechanical descender because the caver is not physically attached to the rope. For safety in approaching a drop, see the discussion of cow's tails below.

The figure eight is the simplest mechanical descender and consists of a stiff metal frame roughly in the shape of the numeral 8. The rope is threaded through the frame to provide friction. Disadvantages to the eight are that it gives only moderate control and must be removed from the harness in order to be attached to the rope, which could allow it to be dropped. It also applies a twist to the rope which many cavers do not like. The figure eight is only suitable for relatively short, nontechnical drops.

Rappel racks consist of a steel frame, basically in the shape of the letter U, which is attached to the central link of the sit harness. A quick link as opposed to a locking carabiner should always be used to attach a rack to the sit harness because a carabiner is far more prone to accidental unlocking. Perpendicular to the long axis of the U are mounted a number of steel or aluminum brake bars, which may either be fixed or pivot to open and close onto the other side of the rack. The rope is threaded under and over these bars to provide friction. An advantage of the rack is that friction can be varied based on both the number of bars in use and their spacing. Bars may be added or removed in the course of a rappel if necessary. A rack can be locked off by wrapping the rope around it so that the descent stops and the caver may use both hands for other tasks. Some cavers do not like aluminum brake bars because they wear more quickly and leave a gray residue on the rope.

A bobbin consists of two metal plates which pivot open to reveal two metal spools or bobbins around which the rope is threaded to provide friction. Unlike a rack, a bobbin should always be attached to the sit harness link with a locking carabiner rather than a quick link, because the latter, due to its smaller diameter, can sometimes inadvertently pry open the safety latch. A separate carabiner on the main link is used to help control friction. Auto-lock variants of the bobbin also have a safety lever which must be held in a specific position in order to proceed. If the lever is released, and also in some cases if it is squeezed, descent will be halted, for example, in the event that the rappeller becomes unconscious. It is more difficult to change the amount of friction in the course of a rappel with a bobbin than it is with a rack, although if one knows a rope will be slow the bobbin can be rigged differently at the beginning of the rappel to provide less friction. A bobbin can also be locked off by wrapping the rope around it. Bobbins work better on 10-mm or smaller diameter ropes.

ASCENDING

Going up a rope requires considerably more effort than going down, since one is fighting gravity. In addition, cavers more often ascend ropes at the end of a caving trip when energy reserves are already lower. Keeping this in mind is important so that fatigue does not lead to a dangerous situation. All mechanical ascending devices employ some sort of one-way cam that will slide up the rope freely and then lock and not slide back down when weighted. Proper arrangement and fitting of ascending devices and good, well-practiced technique are essential to efficient rope-climbing. A great number of different systems have been developed for ascending ropes, but we will cover only the most universal, the *frog system*. Frog is the world standard, used almost exclusively everywhere except in parts of the United States. One reason for its popularity is ease of crossing rebelays. Frog is less efficient on long free drops where the rope walker system may be preferred, but for general vertical caving it is ideal.

A basic frog system consists of a chest ascender, an upper ascender, and cow's tails (Fig. 2A). The chest ascender, which actually rides below the sternum, is designed to lie flat against the caver's belly and is attached directly to the sit harness at its central link. It is held up by a chest harness, whose only purpose is to hold the chest ascender in a vertical position; it does not support the climber. The chest harness may thus be made of very light webbing or cord since it does not bear any weight, although it should be adjustable for comfort when not in use.

The upper ascender is fitted with a long piece of cord or nylon webbing with one or two foot loops at the end to stand in. For safety it is attached by a shorter piece of webbing to the central link of the sit harness. Some cavers use the longer cow's tail (see below) for this, but that makes for a less versatile system overall. The lengths of these attachments are critical to efficient climbing and will vary with the height and technique of the climber. In use, the upper ascender is placed on the rope above the chest ascender and typically has a handle for easy gripping, although lighter and more compact versions without a handle are preferred by some cavers.

To climb a rope, the caver sits so that his/her weight is held by the chest ascender. The feet are tucked under the buttocks and the upper ascender is simultaneously slid up the rope as high as possible. The caver then stands up in the foot loop(s). While doing this the chest ascender will automatically slide upward. One then repeats this sit-stand process. Although simple in principle, it takes coordination and practice to climb efficiently (Fig. 2B,C).

Cow's tails are an essential safety element in any modern vertical system. They are used to attach a caver to a safety line or directly to an anchor point while still allowing some freedom of movement in order to perform the various maneuvers needed to gain access to rig points and to cross rebelays. Cow's tails are best constructed from knotted lengths of dynamic rope; those made from sewn webbing have been shown to produce significantly higher shock loads in the event of a fall. Typical cow's tails are in the form of a V with one side longer than the other for versatility. A nonlocking carabiner is attached to each arm of the V using a knot that will hold it stiffly in place. A knot at the bottom of the V is attached to the central sit-harness link. In addition to their use in vertical work, cow's tails make handy tethers for other items that might be dropped or lost.

For longer climbs, especially if one is carrying a heavy load, a small additional ascender may be attached directly to one foot. The Petzl Pantin is made especially for this purpose. Having an ascender directly on the foot helps hold the climber's body in a more upright position. It also provides an additional foothold for getting past rebelays and pitch heads, especially if the passage is constricted.

One additional type of "ascender" that is rarely used but can come in handy in certain situations is the Prusik knot. This is a simple knot tied with a loop of smaller-diameter rope or cord that will slide up the main rope when unweighted but will tighten and hold fast when weighted. A set of knots is very light and compact and could be used to climb a short drop in a remote location where it might not be worthwhile to carry normal climbing gear. In an emergency, Prusik knots can also be fashioned by cutting off the bottom of the main rope.

CHANGING OVER

For various reasons one may need to change direction of travel while on rope. It is possible to change from descending to ascending or vice versa, although the former is easier than the latter. When descending it is a relatively simple matter to attach one's ascenders and transfer weight to them. The descending device can then be removed and the caver can climb up the rope. When ascending it is possible to change direction by simply moving one's ascenders in reverse. By briefly holding open the cams and alternately sliding the ascenders downward, one can walk down the rope, but this is a slow and inefficient process not suitable for long distances. It is more complicated to successfully transfer one's weight from ascenders to a descending device and then be able to remove all of the ascenders. These techniques should be practiced on the surface where provisions have been made to lower the climber in the event of a problem.

A similar situation occurs when one must pass a knot in the rope. This may occur either because the original rope was not long enough and an additional piece was tied on, or because the rope was damaged and the bad section was isolated by a knot. In either event, a vertical caver must be prepared to pass an unexpected obstacle on rope while going either up or down.

TRAVERSES AND TYROLEANS

Occasionally one needs to cross a void horizontally rather than vertically. This can occur when a passage continues on the opposite side of a pit and it is more convenient to go straight across rather than down and back up, or when it is desirable to avoid deep water. If there is a small ledge to provide footing, a rope may be rigged a short distance above the ledge and tied off periodically so that there is never much slack between anchor points. One may then cross the ledge safely by clipping one or both cow's tails onto the rope and walking across, using the rope for support. If there is some elevation gain, an ascender may also be used to provide an easier grip. At each anchor, one cow's tail at a time is moved past the knot so that the caver remains securely attached at all times. Traverses may also be done when there is no ledge to walk on, although with considerably more difficulty. In such cases a separate rope may be rigged for the feet to provide support and balance.

If the space one needs to cross is not near a wall, then a rope may be rigged directly across the void. This is called a *Tyrolean* or *Tyrolean traverse* (Fig. 3). In this case the rope should be as tight as possible to minimize sag. Such rigging places extra force on the anchor points and on the rope and should be done carefully with this in mind. Many times a Tyrolean will be slightly inclined, in which case it is easy to slide down in one direction but is significantly more difficult going up the other way. Even an essentially horizontal Tyrolean will inevitably have some sag, so at some point one must travel uphill. To cross a Tyrolean, the caver clips in a cow's tail and upper ascender (without using the foot loops) as well as a carabiner at the waist. The caver hangs underneath the rope and may throw one or both legs over the rope for support. Using the ascender, one pulls along the rope to the other side. On a long or steep Tyrolean, a chest and/or foot ascender may be used as well. If the Tyrolean is highly inclined, a second rope may be rigged more loosely beneath the first. On descent, a rappel device may be used on the lower rope while a cow's tail is used for support on the upper one.

An alternative to a Tyrolean or even some wall traverses is to rig a rope loosely across the void in a large U shape. The caver rappels down one side of the rope until the other side is within reach. Ascenders are then attached to the rope going up, and the rack is removed as soon as it is unweighted. One then climbs in the usual fashion. Even though this requires more rope and more distance traveled, it is sometimes easier and more efficient than doing a traverse or a Tyrolean.

VERTICAL FROM THE BOTTOM UP

The majority of vertical caves are rigged from the top down. In other words, one enters the cave and descends various drops to get to the bottom. This is certainly the easiest approach, because pitches may be rigged with the aid of gravity. However, occasionally one encounters a vertical drop from the bottom with no ready access to the top. In this case more involved procedures are required to get up the drop. A few very deep caves have been explored almost exclusively from the bottom up. The most famous is Austria's Lamprechtsofen, which was explored to a depth (height) of over 1000 meters before an upper entrance was connected.

Many techniques have been used to climb vertical drops from the bottom, some quite ingenious. The simplest drops may be scaled by free climbing or chimneying. Sometimes one caver can boost another up into a high lead. When the climb is relatively short, a ladder can be brought in to gain access. A related technique involves a scaling pole, where a segmented metal pole is carried into the cave and assembled at the climb. A rope or cable ladder is suspended from the top of the pole and climbed to gain access to the upper passage. If there are appropriately located projections or formations, a rope may be used to lasso an anchor or a grappling hook can be used. In a few cases helium balloons have been used to drape a line over an anchor so that a rope could be pulled up. These techniques are rather risky because one cannot examine the anchor to see if it is strong enough or if the rope is securely in place.

The most reliable method of reaching a high lead from the bottom is bolt climbing. This used to be a laborious process using hand-driven expansion bolts, but with modern, lightweight hammer drills it has become almost routine, so much so that now any dome can be considered a viable lead. To bolt up a drop, one sets an expansion bolt in the wall from which is suspended a short section of webbing ladder called an *etrier*. An adjustable daisy chain is used to pull the climber's waist up to the bolt and keep him/her upright while climbing the etrier. A quickdraw is also attached to the bolt, through which runs a dynamic belay rope, much the same as in rock climbing. A separate person controls the belay, which will catch the climber in the event that a bolt fails. By repeating this process with another etrier, daisy chain, and quickdraw the climber advances up the wall, alternately moving the etriers and daisy chains upward. Domes over 100 meters high have been climbed by this method, and much passage has been discovered that would never have been seen otherwise.

To retreat from a climb that does not go or is no longer needed, one may either downclimb the drop in reverse, retrieving protective gear along the way, or rig a pull-down, which usually involves leaving some sort of anchor behind. The former is more efficient in

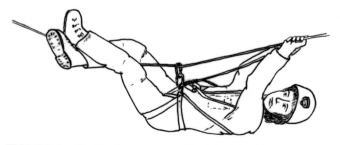

FIGURE 3   The Tyrolean traverse. *Sketch by Bob Alderson.*

terms of hardware, but is more time-consuming. In a pull-down, the main rope is doubled through the top anchor and the caver rappels on one side with the other end tied off or blocked, or on both of the ropes simultaneously. Once on the bottom of the drop, the caver pulls one end of the doubled rope through the anchor until the rope falls to the ground. Some sort of hardware at the top of the drop is generally sacrificed in this case.

Bibliography

Judson, D. (1995). *Caving practice and equipment*. British Cave Research Association and Birmingham, Alabama: Menasha Ridge Press.

Marbach, G., & Tourte, B. (2002). *Alpine caving techniques*. Allschwil, Switzerland: Speleo Projects.

Smith, B., & Padgett, A. (1996). *On rope*. Huntsville, Alabama: National Speleological Society.

Warild, A. (2007). *Vertical*. Newtown, Australia: Alan Warild.

F

FOLKLORE, MYTH, AND LEGEND, CAVES IN

Paul Jay Steward

Cave Research Foundation

Since the beginning of time man has fantasized about caves and the mysteries they hold. They are a doorway to the Underworld and have played a vital role in the narrative of cultures throughout the globe. Many of the stories associated with caves conjure up images of a dark and mysterious place inhabited by gods and demons and associated with everything from resurrection, fertility, worship, sacrifices, and the beginning of life, to a resting place for the dead. Others envision the Earth as hollow, teeming with lost worlds and civilizations.

The most popular of these stories are the ones handed down from ancient mythology. The Greeks believed the Underworld to be the kingdom of the dead ruled by Hades, a greedy god whose only concern was increasing the number of souls in his dark world. In Northern Ireland there is a cave called Saint Patrick's Purgatory. It is used as a place of worship and is believed to be the entrance to Hell.

Although it was a one-way trip for most, getting to the Underworld was not an easy chore. Between the world of the living and the dead, five rivers must be crossed: the Acheron (river of woe), the Cocytus (river of lamentation), the Phlegethon (river of fire), the Styx (river of hate, which surrounds the Underworld with nine loops), and the Lethe (river of forgetfulness).

Hermes, the god of travelers, was born in a cave on Mount Cyllene. He was quite familiar with the trail and would lead the souls of the dead through the darkness. Erebus was the god of darkness in the Underworld. The last river to cross was the Acheron. Charon would lead you across Acheron in a ferry, provided you were buried with a coin placed under your tongue for payment. Those who could not pay were trapped between the two worlds forever.

Across the river loomed the gates to the Underworld. Cerberus, a three-headed dog with a dragon tail, guarded the gates and would allow the souls to enter but never leave. Once through the gates, the souls appeared before a tribunal consisting of Hades, Aeacus, Minos, and Rhadamanthys. Souls who had committed crimes against the gods were sent to Tartarus, while the souls of the just were sent to Elysium Fields.

Many of the caves located on the island of Crete are associated with mythology. Zeus, the king of the gods, and brother to Hades, was born in a cave located on Mount Aegeum. Today it is called the Cave of Psychro. Amnisos Cave is also located on Crete and is believed to be the sanctuary of Artemis, the goddess of fertility. Aeolos, the god of wind and air, kept his winds locked in a cave and only let them out as instructed by the gods of Olympus. Somnus, the god of sleep, resided in a cave. The twins, Romulus and Remus, were found in a basket along the banks of a river and brought to the Cave of Lupercal where they were suckled and raised by a she-wolf. Many of the gods from the oceans lived in caves and grottoes. Even the goddess of music loved to play her instruments in a cave.

The Underworld was not just reserved for gods and souls of the dead. Cunning little elves and dwarfs spent much of their time living and hiding in caves. Harpies were filthy, winged monsters, with the faces of women and the bodies of vultures, that lived in caves. Hydra, the many-headed serpent, and Cacus, a fire-breathing troll, lived in caves. Polyphemus, a Cyclops, lived in a cave near Mount Aetna and survived on meals of human flesh. The serpent Python dwelled in a cave. Typhon, the scourge of mankind, lived in Kilikian Cave, and the famed Medusa lived in the secret Cave of the Gorgons.

Sibyl's Cave is located in the ancient town of Cumae, near Naples, and was the home of Sibyl, a famous female prophet. Many of her prophecies were written on leaves, which she placed at the mouth of her cave for people to find. These "Sibylline Leaves" were bound into books and consulted by the Romans in times of crisis. It is also through this cave that Sibyl was said to have led Aeneas to the Underworld.

Makua Cave, also known as Kaneana Cave, is located on the western shore of the Hawaiian Island of Oahu. The cave is named after the Hawaiian god Kane, the god of creation. It is believed that the cave is the womb from which mankind emerged and spread throughout the Waianae Coast. In ancient times Kahunas were forbidden to enter the cave for they believed it was the home of Nanaue, the shark man of Kaneana. It was said this half-man, half-shark beast would drag his victims to the back of the cave and devour them.

Located on the high slopes of Mount Amarnath, deep in the Himalayan Mountains, lies Amarnath Cave. This is one of the holiest sites in the Hindu religion. The cave contains several ice stalagmites that grow from a natural spring. The largest of these is believed to be a representation of the Hindu god Shiva. The smaller ones represent the Hindu gods Ganesha, Parvati, and Bhairava.

Every year during the months of July and August, tens of thousands of pilgrims bear the harsh climate and rugged terrain to trek to the cave to worship these phallus-shaped formations. According to the legend, in the cave Shiva disclosed the secrets of creation and immortality to his companion, Parvati, and those who make the difficult journey to the cave are promised salvation. Also in the cave with Parvati was a pair of mating doves that overheard the conversation and learned the secrets. The doves made the cave their eternal home and have been reborn every year since. Many pilgrims report seeing a lone pair of doves during their trek.

To the Maya, caves and cenotes were sacred places used for rituals and sacrifice. They called their lower world Xibalba, or Place of Phantoms. Many of their temples were built over cave openings, and much of their water came from the caves. The Maya believed they were born from the Earth goddess and originated out of seven caves.

Caves are mentioned throughout the bible. They are used for shelters, as tombs to bury the dead, as places to hide from one's enemies, and a place for quiet worship.

Lot and his two daughters left the city of Zoar and found shelter in a mountain cave. Abraham purchased Machpelah Cave, also called Cave of the Patriarchs, from Ephron and used it as a family tomb. This cave is one of the holiest sites known in the Jewish religion. The Patriarchs Abraham, Isaac, and Jacob, along with the Matriarchs Sarah, Rebecca, and Leah, are buried in the cave. It is also believed the cave is a passageway to the Garden of Eden.

Seeing their armies defeated by Joshua, the five kings of the Amorites fled to Makke'dah Cave. After being captured and killed, their bodies were returned to the cave and the cave sealed with large stones. David found refuge in Adullam Cave after fleeing from Saul. David wrote many of the Psalms in this cave. Also in this immense cavern, David trained his outlaw army of 400 men. In Engedi Cave, David found Saul asleep and cut off a part of his robe. Obadiah hid Elijah and 100 prophets in caves on Mount Carmel.

Caves were frequently used as stables that also contained rooms for people to sleep. Many believe that Jesus was born in one of these caves. After dying on the cross, Jesus was brought to a cave and buried.

Several historical books written by Egyptians several hundred years before the birth of Christ contain stories about the lives of Adam and Eve. It is believed that after being banned from the Garden of Eden, God sent Adam and Eve to live in a strange broad land, covered in sand, strewn with stones, and void of all vegetation. Here, God commanded them to live in the Cave of Treasures. The cave became a family shrine and many generations would live, die, and be buried in there.

Another cave associated with Adam and Eve is Lilith Cave. According to Jewish folklore, Adam's first wife was Lilith, not Eve, as commonly accepted. Unfortunately, it was not a match made in heaven. Adam and Lilith fought and argued constantly. Any attempts by Adam to mate with Lilith were met with rejection. Lilith refused to lie under Adam in the standard missionary position. She felt they were equal and therefore wanted to make love in equal positions. Adam wanted no part of this arrangement. In a rage, Lilith left the Garden and took up residence in a cave along the shores of the Red Sea in an area fraught with demons. In the cave, she made love to many demons and populated the world with thousands of demon children. She became known as the Mother of Demons, and is said to still reside in her cave.

Ti-Tsang Wang is the Chinese god of mercy. He wanders through the caverns of Hell searching for souls to save and help escape from the Underworld.

Amaterasu is the Japanese Sun goddess. One day her mischievous brother, Susanoo, destroyed her beautiful garden. Amaterasu was so mad she hid herself in a cave and covered the entrance with a large rock. This made the world dark and cold. For many weeks she hid in the cave. One day she heard laughter, music, and singing coming from outside her cave. Amaterasu was curious as to how all the other gods could sound so happy with the world so dark and cold. They told Amaterasu another god more beautiful than she had come and brought light to the world. As she peered out from the cave she saw her reflection in a mirror. Quickly, she was pulled out from the cave and told that the light was her own reflection. Amaterasu had never seen herself before and was amazed at her own beauty. Afterward she returned to her throne in the heavens to warm and light the Earth.

No talk of caves would be complete without a few ghost stories. The quiet, isolation, and absolute darkness found underground create the perfect setting for imaginations to run wild. Many of the caves across the world have horrible tales of ghosts tormenting unsuspecting visitors. One of the oldest and most documented cases in the annals of American haunted places is the Bell Witch Cave.

The cave is along the south side of the Red River, in the rural farming town of Adams, Tennessee. Above the cave is a known burial ground for the Choctaw Indians. While many believe it is Indian ghosts that haunt the cave, others believe it is a troubled spirit known as the Bell Witch.

Numerous stories abound of unusual happenings in the cave. Researchers have discovered high energy levels and evidence of paranormal activity in and around the cave. Visitors have reported strange sounds, eerie feelings, and even sightings of apparitions. In the cave, camera equipment is known to jam, stop working, and rewind for no apparent reason. Photographs taken in the cave sometimes contain white or orange glowing orbs floating about. Orbs are a common occurrence in locations associated with paranormal activity. Others have reported seeing a dark-haired woman drifting throughout the cave. The current owners of Bell Witch Cave allow tours from May 1 to October 31. They, too, tell of strange sounds and happenings in the cave. It is not unusual for a tour to be cut short by a disturbed guest wanting to return to the surface and vowing never to return.

FOOD SOURCES

Thomas L. Poulson

Jupiter, Florida

USE OF THE SCIENTIFIC METHOD, CAVEATS, AND TRADEOFFS

This article uses the scientific method to hypothesize that organisms in caves are food-limited and the adaptations of different species to test predictions about the degree of their limitation by food supply. There are a few cave types where organisms are not very food-limited and these are exceptions that prove the rule. The species in these food-rich caves show few or no adaptations to food scarcity. At one extreme are caves with abundant food from organic pollution, sulfur-based chemoautotrophy, natural deposits of fossil fuel, or extensive bat guano. At an intermediate scale are areas with nonsoluble ridges above limestone valleys where streams sink at the contact and can carry large amounts of decomposing plant material into caves. At the other extreme are caves in glaciated areas where food supply is variable, but there are few or no cave-adapted species. If they had been present, most became extinct when glaciers covered the caves.

Trade-offs for supply and kinds of food in caves are examined. These include quantity versus quality, quantity versus risk of injury, and quality versus unpredictability. In addition, predictability of food influences whether a species can be a specialist or generalist (another trade-off), and in terrestrial habitats the usability of a food type depends on the moisture of the substrate.

LIMITED FOOD SUPPLY IN MOST CAVES

Observations and First Principles

The absence of light in caves leads to the hypothesis of food limitation; no light means no photosynthesis. During growing seasons the above ground world is overwhelmingly green; the amount of all animal life combined is only about 20% of live and dead plant material plus dead plant parts. This is because animals that eat plants or dead plant parts are only about 15% efficient, and animals that eat plant eaters are on average only 20% efficient. This inefficient transfer of energy along a food chain is due to costs of doing business. Costs can include hunting, pursuit, capture, killing, eating, digestion, assimilation, making specific proteins and fats and carbohydrates, and reproduction. Organ system costs are circulation, respiration, excretion, neural processing, and metabolism. Then there are losses of feces and urine. Because the sum of costs goes up with steps in the food chain, it should be no surprise that the mass of organisms must be progressively reduced from plants to herbivores to first-order carnivores. In addition, no second- or third-order predators live exclusively in caves, especially not endothermic birds and mammals that have very high costs due to heat production and high metabolic rates.

Actually the food limitation in caves is even greater than the above plant-based food chain argument suggests, because food chains in caves mainly start with losses from the above ground food chains. These losses include death, feces, and urine with further losses to the fungi bacteria that decompose waste products. Even though fungi and bacteria have relatively low costs of doing business, there is still further energy lost with each species along a food chain of decomposers.

The food limitation in caves should be still greater because the amount, and especially sizes, of organic matter that gets into caves is progressively reduced with depth below the surface. This is because most caves do not have entrances of greater than 1 cm dimension, and those that have bigger entrances have very few of them relative to cave length. Why is this true? Even without the usual soil filter over the limestone, the soil moving downward with water tends to plug most of the openings (look at a road cut in limestone). The very small vertical joint cracks in the limestone layers and horizontal bedding plane cracks do not change much in size with depth, except in the rare greater than 1-cm pathways that may enlarge and become a cave (Palmer, 2007).

The restriction on entry of organic matter is even greater because the pathways of greater than 1 cm are convoluted. It is more like a series of pinball machines than a vertical series of sieves of decreasing pore size. Thus we should expect mostly millimeter-sized particulate organic matter and mostly dissolved organic matter to be brought into caves by percolating water.

The small amounts of tiny particulate and dissolved organic matter that percolate into most caves are likely to be low in food quality and very slowly used by microorganisms that are the main base of cave food chains. The logic is that the easily decomposed organic matter is eaten, lost by leaching, and used by microorganisms as it moves from the surface through the soil and through decreasing sizes and numbers of opening in the limestone. Soil detritivores, such as earthworms, and decomposers, such as fungi, "eat" the particulate organic matter. Bacteria use the easily metabolized sugars and other simple organic molecules leached out of particulate organic matter by water. Thus most dissolved organic matter to reach caves without entrances are large and complex molecules, such as fulvic and humic acids, that are metabolized very slowly. The colonies of *Actinomycete* bacteria, which use complex organic molecules and give caves their characteristic musty smell, grow less than 1 mm over decades. As is often true for slow-growing microorganisms, *Actinomycetes* are known to produce antibiotics such as streptomycin, which protect them from being eaten.

Even the rare particulate organic matter that is washed or that percolates into caves, with the usual centimeter to decimeter scale entrances, is so thoroughly leached that it is of low food quality and only slowly used. In Mammoth Cave streams, leached and black particulate organic matter does not disappear over decades between rare 25- to 100-year floods that wash in new organic matter (Poulson, 1992). However, fresh leaf and wood litter entering caves via sinking streams decomposes just as fast as in a surface desert stream. Even in caves with very small entrances, fresh sawdust or newly broken pieces of lumber are decomposed so rapidly by fungi and bacteria that they use up all of the dissolved oxygen in the water and so can kill even specialized cave organisms with low metabolic rates.

The relations of coarse particulate, fine particulate, dissolved organic matter, and biofilms to input by sinking streams and ceiling drips has been extensively studied by K. Simon and colleagues (chapter 2 in Culver and Pipan, 2009).

Species Adaptations Predicted from Hypotheses of Limited Food

If the hypotheses about extreme food limitation in caves are correct then one prediction is that cave species with the highest costs of doing business should be both relatively rare and should have many adaptations to cope with a shortage of food. Table 1 lays out predictions of decreasing adaptations to scarce food with decreasing costs of doing business, decreasing size, and decreasing complexity of organization of organ systems and nervous systems from fish and salamanders to protozoa and bacteria. For each cost, bacteria are given a score of 1 and estimates of the increasing costs are given. The least increases are for cellular functions, which should be the same, plus anabolism, where the number of proteins, fats, and carbohydrates increase from bacteria to fish and salamanders. The greatest increases are for locomotion and foraging where the costs increase dramatically with increasing size and complexity. Summing estimated costs gives a range from 4 to 68 for predictions of increased specializations to cope with limited food supply.

Table 2 shows that the predicted specializations from Table 1 are matched by observed increased specializations to food limitation from bacteria to fish and salamanders in caves. Among species that have evolved from troglophiles to troglobites, from snails and flatworms to fish and salamanders, all appear white and eyeless, but they are increasingly different from their troglophilic relatives and surface ancestors in their external morphology. Their body builds are less robust with lower weight per length and lower metabolic cost per length, their legs and sensory appendages are longer, and their macroscopically visible sense organs on body and appendages are larger, fancier, more numerous, and more dense. Their internal anatomy, where we can assess it macroscopically, also indicates

TABLE 1 Energetic Costs[1] of "Doing Business"

Organisms	Locomotion Foraging	"Capture" Ingestion Digestion	Cell and Circulation Costs	Respiration Excretion	Sum of Costs
Fish and salamanders	30	20	3	15	68
Arthropod predators	20	20	3	10	53
Arthropod detritivores	15	15	3	8	41
Snails and flatworms	6	5	2	4	17
Protozoa	3	2	1	2	8
Bacteria	1	1	1	1	4

[1]Costs are estimated relative to bacteria, which are given a score of 1 in each category. Organisms are listed in order of decreasing size and decreasing complexity of organization. The sum of costs is a predictor of the degree of food limitation. The adaptations listed in Table 2 support these predictions.
Based on these costs, it is hypothesized that fish and salamanders should be most limited by scarce food in caves and bacteria should be least limited.

TABLE 2 Troglomorphic specializations of kinds of organisms to food scarcity. These in caves support predictions from Table 1, based on complexity of organization and costs of "doing business." Specialization are adaptations in terms of efficiency, metabolic economy, and increased chance of successful reproduction. The number of pluses indicates the level of adaptation

Specialization	Fish and Salamanders	Arthropod Predators	Arthropod Detritivores	Snails and Flatworms	Protozoa and Bacteria
Foraging					
Search pattern	++++	+++	++	+	0
Reduced cost of transport	+++	++	++	0	0
Sensory appendage length	+++	++	++	+	0
Sensory organ density	+++	++	++	+?	0
Sensory organ kinds	+++	++	++	+?	0 NA
Brain center 1[1]	++++	++	++	+?	0 NA
Brain centers 2/3[2]	++++	+?	+?	0?	0 NA
Assimilation and growth	+++	++	+	0?	0
Metabolic					
Routine metabolic rate	+++	++	++	+	0
Starvation resistance	++++	+++	++	++	0?
Reproduction					
Egg size	++++	++++	+++	++?	0 NA
Parental care	+++	0	0	0	0
Multiple breeding attempts	++++	++	+	+?	0
Sum of adaptation	45	27	22	11	0
Sum of costs (Table 1)	68	53	41	17	0

[1]The first level of projection of sensory information, e.g., vagal (mouth taste), facial (skin taste), and somato sensory (lateral line touch).
[2]Processing and translation of sensory information from brain center 1.

greater ability to find food from arthropods. Refer to Mohr and Poulson (1966) or to Poulson and White (1969) for drawings of sense organs, such as smell and movement and touch and taste, and primary and secondary processing parts of the brain that process sensory information in troglophiles versus troglobites of Amblyopsid cavefish and of crayfish. Troglobite foraging behavior is also more efficient at finding and utilizing scarce food that is patchily distributed in space. Where we have data, troglobites that are hungry move long distances between turns and so increase their chances of encountering an area with food. Once they find a food item their rate of movement slows, their rate of turning increases, and they remain in the area even after they eat a food item because there are likely to be more food items in the same area.

Once troglobites encounter a food item, their efficiency at ingesting it and using it also increases from flatworms to fish. Bacteria simply absorb molecules across a cell membrane, and flatworms may catch live prey using their mucus and ingest it by extruding a pharynx. Among arthropods, some troglobitic spiders use a series of webs like a trap line to increase chances of capturing some prey (Poulson, 1992). It is suggested that they will have especially potent poisons to subdue the largest possible prey. Among fishes, Amblyopsids use many motion sensors on a very large head to detect prey, and have dense touch receptors to contact prey, huge mouths to suck in especially large prey items, and extremely efficient digestion with very fast growth in the rare cases where large prey items are captured and eaten.

Aquatic and Terrestrial Differences in Food Supply

Because water dissolves simple compounds faster as it flows, there are major differences in the variability and predictability of food, in time and space, in terrestrial and aquatic habitats in caves. This is true for food input by abiotic agents such as flowing water, falling into a pit, or blowing into a large entrance. It is also true for input by biotic agents such as commuting bats or cave crickets. The main difference is in whether the food is diluted and dispersed widely in aquatic habitats or remains concentrated in a local area in terrestrial habitats. In terms of relative food limitation this means that aquatic cave species will be more limited and more specialized for energy economy than terrestrial species.

A corollary prediction is that aquatic detritivore species should be slower growing and live longer than terrestrial species. Indeed the data show that the differences in longevity are incredibly large with the maximum life span of terrestrial detritivore troglobites (millipedes) on the order of a decade and the maximum life span of aquatic species (crayfish) on the order of a century. The selection for long life is that renewal of the detritus at the base of the aquatic food chain can be rare in time, on the order of decades (*e.g.*, statistically 10-, 20-, or even 50-year floods). With the same food base of plant detritus, terrestrial species also get rare pulses of food, but it is left by floods on stream banks and ceilings, so is not progressively leached. Thus it will remain approximately the same relatively high food quality for many years. This means that the most food-limited terrestrial species, such as millipedes and bristletails, can grow and complete their life cycles in less than a decade.

Exceptions That Prove the Rule of Food Limitation in Caves

There are past and present situations that are both much less food-limited and much more food-limited than most of the limestone caves in today's temperate zones of the Earth. These are considered at decreasing scales in time and space. The corollary, from earlier logic, is that this should affect the degree of species specialization to food limitation.

At a long time scale of glacial cycles there were probably times when less food washed into caves than at present, especially for aquatic species that may have been isolated in caves for tens of millions of years. During both cold−wet periods with glacial advances and during dry−warm periods with glacial retreats, there would have been less food input into caves. Because the more specialized troglobites are presumed to date from preglacial times, present adaptations to food scarcity are a complex result of fluctuating food supplies, with times of greatest food supply "crunches" being most important.

At an intermediate scale across space and over time are contrasts between caves in flat-bedded limestones in temperate and tropical regions, caves in flat-bedded and steeply folded limestones in temperate regions, and flat-bedded limestone caves and lava tube caves. It is argued that wet tropical caves should be less food-limited than temperate zone caves based on less seasonality, greater rainfall, year-round occupancy by bats in a higher percentage of caves, and faster rates of cave development (with more entrances) in the wet tropics. This is a strong hypothesis, but a Brazilian scientist (Eleonora Trajano) has argued that the tropics are not uniform. For example, glacial cycles have had great effects in the tropics, and there are both wetter and more arid zones present within the tropics. It is likely that data on degree of adaptation to food limitation within a taxonomic group are not yet adequate to support or reject those predictions based on less food limitation in the wet tropics than in temperate zone caves.

The caves in steeply folded and thin strata of the Appalachian Valley and Ridge Province of the United States should be less food-limited than caves in the flat-bedded and thick limestones of the Interior Low Plateau Province. This prediction is supported mainly by aquatic species with a greater number of species, greater density of individuals, and higher proportion of troglophile to troglobite species in caves of the

Appalachian Valley and Ridge Province (Culver and Pipan, 2009). The logic for this hypothesis is that the geology results in more large cave entrances and more regular and greater input of particulate organic matter to valley and ridge caves where the ridge tops are insoluble rocks, such as sandstone and shale, and the valley sides and bottoms are soluble limestone with caves. Water flowing off the insoluble ridge tops dissolves entrances where it encounters the limestone, so large amounts of unleached plant organic matter are washed into the caves almost every year. The trade-offs for cave species are that the food supply is reliable (+) and plentiful (+), but that the risk of being dislodged from rock refuges in streams and killed by abrasion or being washed into pools with predators is high (−). Under these circumstances short-lived troglophiles, which can take fast advantage of good times and recover quickly after bad times, coexist with more energy-efficient troglobites that have longer life cycles (especially isopod and amphipod species).

The contrast of limestone caves versus lava tube caves is even more striking and even more strongly supports the overarching hypothesis of the selective influence of low food supplies in caves (Poulson and Lavoie, 2000). Over geological time the changes in food input and limitation are reversed in limestone caves and in lava tubes. As erosion continues over thousands to hundreds of thousands of years, limestone caves become more open to food input and so less food-limited. But unlike lava tubes, we cannot readily date their time of initial formation. The long times prevent us from assessing how adaptations to food scarcity change. In contrast we can exactly date the time of lava tube formation and follow what happens in lava tubes of different ages. Lava tubes are most open to food input soon after sudden formation when they are colonized both by troglophiles and troglobites and population densities are at the highest. Over only tens to hundreds of years both wind-borne and water-borne sediments fill the larger voids above the lava tubes and abiotic agents introduce less food. Population densities de-cline and only those species adapted to food scarcity persist.

On a short time scale of one to tens of years, organic matter in the form of green "plants" (including blue-green bacteria) and lint in terrestrial parts of commercialized caves can greatly increase. None of the human-caused input of organic matter such as lint, which can be widespread, or algae and moss, which is localized around lights, seems to be eaten by specialized cave species or above ground native species but alien millipedes occasionally invade and graze algae and mosses around lights even when dense congregations of camel crickets are nearby and do not come to the moss and algae. Why are neither outside species, such as mice, or any cave species not attracted to the plant growth or to lint-covered surfaces? One hypothesis is that these kinds of organic matter are of poor food quality. Certainly blue-green bacteria and mosses are heavily defended because few or no herbivores eat them above ground, and much of lint is synthetic fibers that are complex and either not decomposed or slowly decomposed. A weaker hypothesis is that extensive modern human visitation of caves is only about 100 years old, so no species have had time to evolve to utilize either lint or green photosynthesizers associated with extensive human visitation. Still another, not mutually exclusive hypothesis, is that long-lived troglobites avoid areas because their predators are also attracted. Long-lived terrestrial troglobites must spread their risk of reproductive failure over many years and so anything that might lower the chances of successful reproduction, such as increased predation risk, is avoided.

Pollution of aquatic habitats by decomposable organic matter can occur across a large spatial scale, everything downstream of the input point, and its effects of decreasing the diversity of cave species occur on a short time scale, often within weeks or months. This is the same pattern seen outside of caves where a huge abundance of only a few species of short-lived and low-oxygen-tolerant species replaces a much lower abundance and much higher diversity of species that are not tolerant of low oxygen. Whether above- or below- ground, the dominant food base with organic pollution is colonial sewage bacteria and the only animal may be Tubificid worms. Oxygen is low because the biological oxygen demand is high when bacteria and fungi rapidly decompose large quantities of organic matter. This may not explain why energy-efficient troglobites do not still occur, because they have very low metabolic rates and so should be able to live where oxygen is low though probably not at the extremely low oxygen concentrations where sewage bacteria survive.

Other relevant observations are of the path of community recovery when input of excess decomposable organic matter, such as human or animal wastes, is stopped. First the sewage bacteria and worms are replaced by somewhat longer lived and more energy-efficient troglophiles such as crayfish and isopods. Then, after sewage bacteria and worms disappear, the area is recolonized by long-lived and energy-efficient troglobites, the troglophiles become rare, and the original diversity of species is restored. The temporary dominance of troglophiles suggests that either the demographic pressure of faster reproducing species or direct interference competition by the troglophiles (low-metabolic-rate troglobites should win with indirect resource competition based on energy efficiency) may explain the dominance of troglophiles when oxygen is not limiting. Still, another hypothesis is that pollution by decomposable organic matter is often associated with pollution by toxins and that, compared to troglophiles, troglobites are especially compromised by bioaccumulation of toxins over their much longer life spans and biomagnifcation of toxins if they are farther along food chains. Whatever the explanation, the changes with organic pollution underscore the usual food limitation in caves and the needs for adaptation to scarce food by troglobites.

FOOD SOURCES AND FOOD TYPES, ESPECIALLY IN TERRESTRIAL HABITATS

This section deals with the differences between input by abiotic agents and biotic agents and the more striking results of different kinds of biotic agents of input in terrestrial than in aquatic habitats. Considered first are aquatic versus terrestrial contrasts for the same food sources and types, then terrestrial extremes of food types, and the section ends with a detailed look at different fecal types of biotic input in terrestrial habitats.

Aquatic versus Terrestrial Contrasts for Extremes of Food Quality

As Table 3 shows, hypothesized differences in variability and predictability of food on land versus in water depend on whether there is abiotic input by water or biotic input by trogloxenes. There are opposite and common trade-offs for land and water for quantity versus quality of abiotic and biotic inputs. On the minus side for plant detritus, a combination of surface decomposers and leaching by water usually removes all but the compounds most recalcitrant to digestion, so bacterial and fungal decomposers are also much reduced. On the plus side with 10- to 100-year floods, especially in a few places near cave entrances, renewal of unleached detritus may "feed" the cave communities for another 10 to 100 years. Even leached wood, twigs, and leaves decompose slowly in caves because they are mostly lignins and cellulose that only some bacteria and fungi can digest, and very slowly at that. Once the detrital pieces get small enough to be eaten by worms, amphipods, and isopods, the rate of decomposition is enhanced. Cycles of bacteria-fungi-protozoa biofilm buildup, ingestion of pieces and digestion of the biofilm by animals, egestion of now smaller particles, and recolonization of pieces by biofilm organisms all decrease the size of the pieces and so the surface area to volume increases and decomposition slowly accelerates.

For biotic input by trogloxenes the quantity of food is less (−) but the quality of food is much greater (+). In addition, biotic input is much more predictable (+) both in time, every year and every month, and in space, usually near entrances and in the same spots. However, these advantages are much less when inputs are over water, as seen in the second column of Table 3.

In water, leached feces of trogloxenes are not much higher quality than leached plant debris, and both

TABLE 3 Trade-Offs (+ and −) of Food Variability and Predictability in Caves: How Quantity and Quality Vary in Space and Time

Trade-Off Category	Aquatic	Terrestrial
Abiotic Input[1]		
Occurrence	+ Gets everywhere	− Localized
Quantity	− Low densities	+ Locally high
Quality (leaching)	− Continued leaching	+ No further leaching
Renewal predictability	= / − Not every year	− Rare
Successional communities	− None	+ Some
Biotic Input[2]		
Regional occurrence	± Not everywhere	+ Localized
Occurrence in a cave	± Not everywhere	+ Localized
Quantity	− Dilution	+ Stays in place
Quality	− Only leached feces	+ Feces and others
Renewal predictability in a cave	± Moderate	+ Yearly

[1] Abiotic input is by flowing water, wind, and gravity.
[2] Biotic input is by trogloxenes such as bats, wood rats, and cave crickets. Inputs are feces, dead bodies, and eggs.

TABLE 4 Extreme Food Types in Terrestrial Caves: How Food Traits Influence the Ecology and Life History of Species That Eat Each Type and How This Influences Species Diversity

	Twigs and Leaves	**Feces**[1]	**Dead Bodies**	**Cricket Eggs**
Food Traits				
Caloric density	Low to very low	Moderate	High	Very high
Digestibility[2]	Low to very low	Moderate	High	Very high
Occurrence				
Among caves	High	Moderate	Low	Very low
Within a cave	High	Variable	Low	High
Amount per area[3]	Very high	Variable	Low	High
Renewal rate	Very low	High	Low	High
Monopolized by single species?	No	Variable	If small	Yes
Species Hypotheses				
Specialization?	No	Variable	No	Yes
Life span	Long	Variable	Short	Short
Species diversity	High	Variable	Low	Very low

[1] As shown in Table 5, there are many kinds of feces, so many of the categories are variable.
[2] Digestibility is largely determined by the N:C ratio.
[3] Amount per area is at locations where the food type is present, i.e., nonzero values.

become diluted and dispersed over wide areas. The only advantage of this reduction in quality is that a little food occurs everywhere and is easy to find. Consumers of this low-quality food must be very efficient digesters and are thus slow-growing and are probably long-lived. They also must be opportunistic omnivores. Species that have predatory ancestors, especially fish and salamanders, will get some benefit from eating detritus with a biofilm if their preferred food of copepods, isopods, and amphipods is especially scarce. Species that have herbivorous or detritivorous ancestors, such as crayfish and isopods, will greatly benefit from eating occasional live prey. For all of these reasons there is little opportunity for specialization of relatively few species in aquatic cave habitats.

Terrestrial Extremes of Food Types: Leaves, Feces, Dead Bodies, and Cricket Eggs

Many more species are on land because they have opportunities for specialization and almost no minus trade-offs (Table 3, last column). With less leaching, food quality is higher. Food quality varies from low with partly leached litter, to feces, to dead bodies, to very high with eggs. With no dilution, input stays where it is deposited, and this makes the food input more predictable in time and space. The consequences of these on-land advantages are explored in the next section, but suffice it to summarize here that a high diversity of terrestrial cave species occurs for four reasons (Table 4): (1) There are many food types with a wide range of quality and quantity; (2) within many food types there is a heterogeneity of digestibility for different components and so there can be successional replacement of species during decomposition; (3) predictability in time and/or space at, within, and among cave scales allows some species to specialize (with narrower niches more species can coexist); and (4) there is different local usability of the same food types because of interactions among microhabitat and moisture.

Table 4 contrasts aspects of food type and aspects of the species that are expected to use each type. Quality of food includes both caloric density and digestibility as indexed by carbon to nitrogen ratio. Predictability of food type includes both frequency of occurrence within and among caves and amounts where it occurs. These food type trait differences lead to hypotheses about the likelihood that species will specialize to use each type. At one extreme are variably leached leaves and twigs that are low in quality but high in amount; for this type consumers with little specialization and long life spans are expected. At the other extreme are cricket eggs that are high in quality but low in amount; for this type consumers with great specialization and short life spans. Corollaries of these primary predictions are secondary predictions about the diversity of species likely to use each food type. The highest diversity of consumer species is predicted for litter because it is most heterogeneous, can undergo very slow successional decomposition by different species, and has a low renewal rate. Its lowest quality components can last for many years, often decades, and cannot be monopolized by one or a few species. The lowest diversity of species is predicted for cricket eggs because they are homogeneous, have no successional use, have a yearly renewal rate, and can be monopolized by one species. This is a more specific case of the generalizations discussed above for cave species in general. Thus if food is of low quality and low renewal rate, in our example litter we expect species to be generalized with the plus and minus tradeoffs of a jack-of-all-trades but a master of none. We expect the opposite trade-offs for a specialist that is favored by food of high quality and regular renewal, such as our example of cricket eggs. These predictions are best supported by extensive data on the most specialized cave species known, namely Carabid beetles that specialize on eating cave cricket eggs (Griffith and Poulson, 1993).

A cricket egg is a real prize that can be quickly eaten by one beetle of about the same size as the egg, the ultimate in monopolization of and energy return from a single food item. A beetle spends only 10% of its time digging up eggs but 90% of its energy digging, so it is not surprising that a beetle defends its ongoing investment of digging given the high caloric reward of an egg. In fact the deeper the hole the more aggressively a beetle defends the hole. Once a beetle digs up an egg it runs with it until it is in a place safe—under a rock, up the wall, or on the ceiling—from other beetles that will try to steal the egg because this is much less expensive than digging up its own egg.

Given the high payoff it is not surprising that beetle predators of cricket eggs have sophisticated searching behaviors. First, they slow their rate of walking and increase their rate of turning when on the silt or sand substrates where cave crickets lay the overwhelming majority of their eggs. Second, beetles dig in spots where crickets have raked silt or sand into tiny mounds over the egg-laying site and may avoid the holes that a cricket makes to test the suitability of a site by inserting its ovipositor. The beetles may recognize an egg mound not only by texture but also because the raking of the cricket ovipositor removes beetle scents left as they forage for eggs. Third, the beetle digs only to a depth equal to that of the cricket's ovipositor, because if there is no egg at that depth there will be none deeper.

An even stronger support for these predictions of specialization to eat cave cricket eggs is the convergent evolution of this habit and in body size among at least five different species of Carabid beetles that eat eggs of at least three different species of cave crickets. In each case the cricket egg is about 5 mg and in most cases closely matched to the weight of the beetle. In some cases the ancestral species was smaller and in some cases larger than about 7 mg. Also, each beetle species chooses a substrate habitat of silt or sand closely matched in moisture content to that best for cricket egg development. In each case the beetle's life history is closely matched to the seasonal cycle of cricket egg laying. As predicted in each case, the egg can be monopolized by one individual beetle.

As with cricket eggs, another food item that is high in quality and can potentially be monopolized is dead bodies. Corpses that are too large to be carried away by scavengers are usually at the bottom of deep pits where surface species can blunder in and fall to their death. This is rare within and among caves, and we have no data to suggest that any cave species even has this food item as part of its regular diet. On the other hand, under maternity colonies of bats that number in the thousands (*e.g.*, gray bats) to millions (free-tailed bats), baby bats regularly fall from the ceiling and are quickly eaten by hordes of small scavengers, especially Dermestid beetles and camel crickets. Tiny corpses that can be carried away are rarely seen in caves because they are carried away by scavengers (see Mohr and Poulson, 1966). In fact, if we incapacitate or kill a small beetle or cricket and watch, we usually see that either beetles or crickets of the same species as the food item almost immediately find it and carry it away. Thus in this case the dying or dead species does not support any more species to the cave community; it merely enlarges the feeding niche of existing cave species that we predicted should be opportunistic feeders.

At the opposite extreme are food types that cannot be monopolized and have a high diversity of consumers, especially leaf and twig litter of very low quality and large piece size. Recall that in Table 4 it was predicted that the low quality and low predictability of renewal with very long times for decomposition, often decades, would prevent monopolization or specialization and might support a relatively high diversity of cave species. Also, relatively undecomposed litter is carried into caves only by infrequent floods (*e.g.*, 20- to 50-year events) and avoids continued leaching only if deposited at the extreme high watermark on ceilings or high on stream banks. Only there is it available as a moderately high quality food source every year for terrestrial cave species. In such situations many generalist species make up the cave community. Among eaters of fungi and bacteria, either grazed on or digested off of as a biofilm, are primary consumers such as springtails, bristletails, and millipedes. Predator generalists that eat the primary consumers include daddy longlegs, spiders, pseudoscorpions, Carabid beetles, and various mites. Springtails live for several years and millipedes may live for a decade. Primary and secondary consumer predators such as Carabid beetles live for several years and Linyphiid spiders for more than a decade.

So we see that the predictions for the extremes of terrestrial food types (Tables 3 and 4) are upheld, but the real test of the generality of these predictions is to look at an intermediate food type that has a wide range of food qualities and quantities, that is, feces.

A PLETHORA OF FECAL TYPES

In terrestrial habitats, of all the biotic inputs of food to caves by trogloxenes, their feces, and the usability of their feces, are most different and offer the greatest number of feeding niches for cave animals. It can be said that you are what you eat and your feces reflect even more what you eat. Consider, as a human omnivore, your own feces when you eat hot spicy food compared to salad compared to meat and potatoes (the cave community would be incredibly richer if humans lived in caves). In addition, the usability of any one fecal type can vary with its predictability in time and space and with the moisture of the substrate where it is deposited. Is it, as in Goldilocks, too wet, too dry, or just right? Thus if, like Doctor Doolittle, we could question cave species about food they would tell us that cave species have an incredibly rich vocabulary for feces. Their vocabulary for feces is every bit as rich as the vocabulary of Inuit Eskimos for frozen water, which includes words for different forms and amounts of snow and ice. As Eskimo survival is related to snow and ice, cave animals would tell us how different forms and amounts of feces are critical for niche separation of cave species that feed on the feces of raccoon, pack rat, bat, cave cricket, and cave sand beetle (Table 5).

TABLE 5 In Terrestrial Habitats in Caves, There Can Be a Diversity of Fecal Types and a Diversity of Species Eating Feces

	Raccoon	Pack Rat	Bat (insectivorous)	Cave Cricket	Cave Beetle
Fecal Traits					
Quality					
Caloric density	High	Moderate	Low	Very low	High
% easily digestible	50	30	20	60	70
Low digestibility	Hair, seeds	Hair, wood	Chitin	Little	None
Diet heterogeneity	High	Moderate	Low	Low	Very low
Quantity Where it Occurs					
Turd size mg.	13,700	60	20	5	0.1
Turds per day	10	20	50	5	5
Spatial dispersion	Random	Latrine	Roost to flight paths	Roost to walk-by	Under rocks
Predictability					
Among caves	20%	30%	10%	90%	70%
Area within cave	10%	5%	30%	40%	15%
Seasonality	Moderate	Little	Great	Little	Little
Drying risk	Winter	Usually	Little	Rarely	None
Leaching risk	None	None	None	Moderate	None
Predictions/Tests					
Species specialization					
To fecal type	No	A little	Yes	Yes	Yes
To microhabitat	No	Rocks, dirt	No	Rocks, dirt	Sand, dirt
Species succession	Yes, a lot	Yes, with depth	Yes, with depth	A little	None
Species diversity	++	+++	++++	+++	++

Note: For each fecal type, there should be increasing specialization of feeding with increasing predictability and quality of the feces but decreasing diversity of species. These data, predictions, and tests are from the Interior Low Plateau of east central United States and from Texas and New Mexico.

In Table 5, quality, quantity, and predictability of feces are considered, and how they lead to hypotheses and predictions about whether cave species specialize on one fecal type and whether distinct communities of species are associated with each fecal type. For quality, the most important component is ease of digestibility and the heterogeneity of the fecal producer's diet. For quantity important variables are where feces occur, turd size, numbers of turds per day, and spatial dispersion (*e.g.*, random or clumped). Predictability includes spatial and temporal occurrence within and between caves as well as abiotic constraints on whether the feces are dried or leached, both making them less useable. Quality, quantity, and predictability of feces together lead to opposite predictions for whether a species will specialize on a single fecal type and whether a diverse community of species will be associated with a single fecal type. The best predictors of species specialization are high quality, homogeneity of quality, and high predictability in time and space, for example, fresh bat feces at the top of a pile of guano under maternity roosts. The best predictors of species diversity are heterogeneity of quality, which leads to successional replacement of species, and large quantities, which leads to different species using different densities of feces, for example, from the top of a bat guano pile to areas beyond the periphery of a guano pile under a maternity roost. Because maternity colonies of bats are rare, the overall specialization of species to using bat feces is relatively low. What of other fecal types? We will now give a summary for each type. For details about communities associated with each fecal type, especially bats with completely different diets (fruit and blood in the tropics in addition to insects in the temperate zone), see Poulson and Lavoie (2000).

Raccoon feces should not support specialist feeders but its heterogeneity in quality and low predictability should allow use by a wide variety of species at the community scale. Raccoons rarely wander into caves and are only rarely found far from entrances. On this basis we should predict only generalist opportunistic feeders on feces. In fact, in early stages of decomposition in summer there can be domination of the most easily digestible components by dense squirming masses of maggots of Sphaerocerid flies that usually have a wide feeding niche and are only locally abundant. With fresh raccoon feces, the large size of individual turds and a tangling of turds in piles allows maggots to do best due to their collective facilitative behavior with an abundance of external digestive enzymes and increased rate of development allowed by generation of body heat. As successional decomposition proceeds, with fewer and fewer digestible components available, larvae of scavenger and fungal-eating species become the most common. At the end of decomposition, when the only remaining components are nearly undigestible (*e.g.*, raccoon hair or seeds) there is an even larger diversity but a very low density of very small generalist feeders. These can use tiny and scattered amounts of remaining digestible components and include springtails, mites, and scavenger beetle adults (Leiodids).

In winter the community using raccoon feces has few, if any, animal species but all major size classes of fungi. Low temperatures and dry conditions just inside cave entrances preclude use by most animals that in summer eat fungal hyphae or disrupted hyphal networks in the process of feeding. The result is that fungi cannot accumulate enough energy to produce increasingly large fruiting bodies from Phycomycetes to Ascomycetes to the mushrooms of Basidiomycetes.

In contrast to the random deposition of raccoon feces in place and time, pack rats have fecal latrines that can be in the same spot for decades, centuries, and even millennia. Consequently we might predict that many cave species could specialize on this fecal resource; however, they do not. Let us consider why not. The problem is that pack rats occur in few caves, regionally and locally, and where they do occur their latrines are usually in very dry entrances where, at best, only fungi can do well as users of feces. Because pack rats urinate elsewhere, for example, as one way of scent marking trails in and out of the cave, there is no mitigation of dry conditions by urination as can occur with bats that almost always urinate over their fecal piles in caves.

In cave entrances that have moist microclimates, pack rat fecal latrines can have a rich community of fecal users. The piling of pellets helps to create a stable microclimate which buffers variation in the cave microclimate and collectively creates a resource with low surface area to volume. The largest latrines have a spatial zonation, from pile top where fresh feces are deposited daily to pile edges where both fresh and old feces are scattered as a mosaic. A large species of fungus gnat grazes fungi on the freshest feces and a smaller species of fungus gnat eats a combination of fungi and bits of feces deeper in the pile and toward the periphery. Still deeper in the pile the feces are only powder consisting of mostly hard-to-decompose lignin and cellulose. Use of partially intact feces at the periphery depends on whether the substrate is rock, mud, or sand. On rock there is a smelly slurry of soluble nutrients with domination by saprophytic mites that can tolerate the low oxygen. On mud there can be large numbers of two species of springtail and a few predatory pseudoscorpions. On sand most nutrients are leached and there are more species with no dominance. Thus the overall diversity of species using piles of rat pellets depends on depth in the center of the pile and the degree of leaching into the substrate at the periphery of the pile.

As with rat feces, sand beetle (e.g. Neaphaenops) feces are localized and their use depends on the substrate; the difference is that sand beetle feces are much more predictable because there are so many areas where cave crickets lay eggs eaten by sand beetles. Thus we might predict some specialization in use of their feces. As explained previously, a 5-mg cricket egg is a huge meal for a 7-mg sand beetle, and the successful beetle usually retreats under a rock to prevent another beetle from stealing the egg. Under the rock the beetle ingests the egg and urinates and defecates as it digests the egg over a 1- to 2-week period. The community of consumers and their predators that use the feces depends on whether the rock is on a sand or silt–mud substrate. On silt the main consumers are scavenger beetles and springtails, and both mites and pseudoscorpions are predators. On sand the main consumers are bristletails, springtails, and mites and spiders are the main predators.

The feces of insectivorous bats can be like rat feces in occurring in piles, unlike rat feces in having no substrate constraints, and like raccoon feces in having rare places in which piles occur. In the rare caves with heat traps and adequate moisture where free-tailed or gray bat maternity colonies exist, the piles are so large that they create their own microclimate and the substrate is not an important constraint on use by cave animals. The microclimate requirements are so exacting that maternity colonies occur in fewer caves than rare hibernating roosts or transient colonies. The maternity colonies have stayed in the same caves for decades and probably centuries, and so it has been possible for some cave animals to specialize on this locally super-abundant food resource. When mother bats are nursing in a maternity colony the input of feces and urine is prodigious, and the surface of the guano pile can be a seething mass of varying combinations of mites, springtails, and camel crickets. These feed both directly on feces and urea from urine and also indirectly on the bacteria and fungi that metabolize feces and urine. As detailed in Poulson and Lavoie (2000), the animals using feces in huge piles of bat guano vary with depth in the pile and with decreasing density of feces from the center of the pile to the periphery. This zone blends into a fly-by scattering of fecal pellets analogous to the walk-by scattering of fecal splotches of cave crickets away from their entrance roosts.

Cave cricket and bat feces share a huge spatial variation in local density from areas under roosts to walk- and fly-by scatterings of turds on the way to and from near-entrance roosts. In both cases there is a gradient to higher diversity and more adaptation to food scarcity from the periphery of roosts to the much less dense feces and lower renewal rates in fly- and walk-by areas. The difference from bats is that cave crickets occur in almost every cave within their geographic range so a more predictable community is associated with the cricket's scattering of feces.

Unlike the huge piles of bat feces under maternity roosts, the veneers of feces under cave cricket roosts are another example of the Goldilocks principle for substrate moisture; it should be not too wet or too dry to be used by the highest diversity of cave animals. If the substrate is steeply sloped with lots of drip water from speleothems, then leaching removes what little nutrients are in the feces. If the substrate is too dry, for example, due to air movement in cold and dry winter microclimates, then only fungi can use the feces. The most diverse communities with highest densities of each species are found on flat rocky areas with the locally deepest veneers of fresh feces, such as under protected ceiling microhabitats of mini domes. In these areas both a common snail generalist and a rare snail specialist are found as well as other specialists including a millipede and two mites as consumers and a Pselaphid beetle and web worm as predators. On both mud and rock substrates Oligochaete worms may preempt the feces, and on sand the feces can be used only if rates of fecal input are high enough to create crusts that buffer the local moisture microclimate (Poulson, 1992).

Cave Crickets as Key Industry Species

Cave crickets are key species by virtue of their high frequency of occurrence, high densities where they occur, and high impact per individual as the largest terrestrial cave invertebrate. Within their geographic range they are found in virtually every cave with high densities in roosts just inside cave entrances. They are the basis for three separate cave communities. At entrances they go outside to feed about every 10 days and return to their roosts where they digest crop contents that may be 100–200% of their body weight. As they digest they defecate and one community depends on this fecal input. Adult crickets walk from entrance roosts to reproductive areas deep in the caves and defecate along the way. The walk-by feces they deposit, at very low densities with infrequent renewal, are the most important and reliable energy source for the most energy-efficient terrestrial species in the cave. The eggs they lay in sand and silt areas are eaten by Carabid sand beetles, and beetle feces are the basis for the third cave community that depends on cave crickets. Cave crickets are not only key species in several of the karst areas of the United States but also in parts of Europe and perhaps in Asia and South Africa.

See Also the Following Articles

Adaptation to Low Food
Guano Communities
Chemoautotrophy

Bibliography

Culver, D. C., & Pipan, T. (2009). *The biology of caves and other subterranean habitats*. New York: Oxford University Press.

Griffith, D. M., & Poulson, T. L. (1993). Mechanisms and consequences of intra-specific competition in a Carabid cave beetle. *Ecology, 74*(5), 1373–1383.

Mohr, C. E., & Poulson, T. L. (1966). *The life of the cave. Living World of Nature Series*. New York: McGraw-Hill.

Palmer, A. N. (2007). *Cave geology*. Dayton, OH: Cave Books.

Poulson, T. L. (1992). The Mammoth Cave ecosystem. In A. Camacho (Ed.), *The natural history of biospeleology* (pp. 569–611). Madrid, Spain: Museo Nacional de Ciencias Naturales.

Poulson, T. L., & Lavoie, K. L. (2000). The trophic basis of subsurface ecosystems. In H. Wilkens, D. Culver, & W. Humphreys (Eds.), *Ecosystems of the world 30: Subterranean ecosystems* (pp. 231–250). Amsterdam: Elsevier.

Poulson, T. L., & White, W. B. (1969). The cave environment. *Science, 165*(3897), 971–981.

FRIARS HOLE SYSTEM, WEST VIRGINIA

Stephen R.H. Worthington* and Douglas M. Medville[†]

*Worthington Groundwater
[†]West Virginia Speleological Survey

The Friars Hole System, in Greenbrier and Pocahontas counties, West Virginia, is the longest cave, in terms of surveyed passage (73.3 km, or 45.5 miles), in both the Valley and Ridge and Appalachian Plateaus provinces in the eastern United States. The cave's geological setting, in moderately faulted and gently dipping Mississippian limestones, is in the midpoint of a range of Appalachian cave settings, from those found in flat bedded unfaulted carbonates to the west, in the Cumberland Plateau, for example, to those in the more highly faulted and folded limestones in the Valley and Ridge Province to the east; Burnsville Cove in Virginia, for example.

As such, the Friars Hole System is an excellent example of a large cave, the development of which has been influenced by minor structures such as low angle thrust faults and the availability of competing flow paths to two regional base level springs 20 km apart. As a result, the cave has a complex hydrological history and a plan view that reflects both abandoned flow paths and currently active internal drainages within the cave system.

SETTING OF THE CAVE

Location

The Friars Hole System is located close to the boundary between the Valley and Ridge and Plateau provinces of the Appalachian Mountains. Friars Hole itself is a 150-m-deep dry valley between the summits of Droop Mountain and Jacox Knob, which are 850 m asl. Friars Hole extends for 13 km from the sinks of Bruffey Creek and Hills Creek in a southerly and southwesterly direction to Spring Creek (Fig. 1A). Sandstones, shales, and minor limestones of the Mississippian Mauch Chunk Series underlie these summits and the valley sides, but there are a series of inliers of Union Limestone along the valley bottom (Fig. 2). Streams flow onto the limestone sink soon after reaching it, and the entrances to Friars Hole System are found there. A number of these sinking streams feed active stream passages within the cave, which underlies a 6-km length of Friars Hole. Much of the cave lies under the nonkarstic flanks of the valley, with passages lying up to 185 m below the surface.

Hydrology and Hydrogeology

Hills Creek is the largest sinking stream along Friars Hole and its hydrology is complicated. At low flow it sinks completely in its gravel bed and is seen in succession in Cutlip Cave, Clyde Cochrane Cave, and Friars Hole System, where it forms Rocky River, the largest stream in the cave (Fig. 2). At high flow that part of Hills Creek that does not sink in its bed continues on the surface for 500 m beyond the main sink, and enters Hills-Bruffey Cave, which also takes the high flow discharge from Bruffey Creek. Hills-Bruffey Cave drains to the east to Upper Hughes Cave, Lower Hughes Cave, Martha's Cave, General Averell Pit, and finally Locust Creek Cave, which is the cave at Locust Spring on the east side of Droop Mountain. In very high flow conditions the water backs up at the entrances of Hills-Bruffey Cave and overflows on the surface to its final sinkpoint at Cutlip Cave, from which it flows south through Clyde Cochrane Cave and then through the Friars Hole System via Rocky River and rises at JJ Spring on Spring Creek, 19 km to the southwest of the Cutlip Cave entrance (Fig. 1B). The next sinking stream to the south of Hills Creek is Rush Run and, like Hills Creek and Bruffey Creek, has part of its discharge flowing to Locust Spring and part to JJ Spring.

The streams that sink directly into the Friars Hole System all drain to the southwest to JJ Spring. This spring drains some or all of the flow from an area of 265 km^2, and the streams draining via the known cave

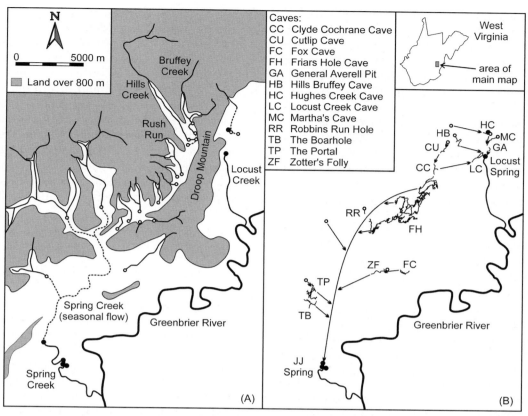

FIGURE 1 Topography and hydrology of the Friars Hole area. Friars Hole is the dry valley between the sinks to Hills Creek and Spring Creek (A). Mapped caves in the drainage area of Friars Hole. Arrows represent traced flow paths of cave streams (B).

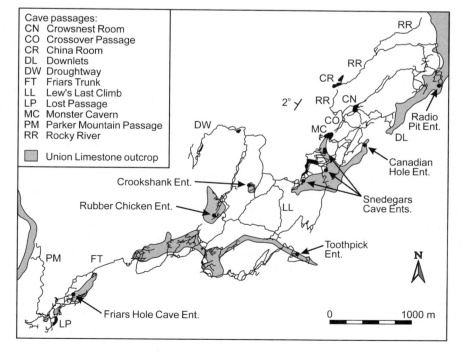

FIGURE 2 Cave passages of Friars Hole System, showing the relationship to surface outcrops of limestone.

in the Friars Hole System account for only a third of this discharge (Fig. 1). JJ Spring is located on Spring Creek, which flows across the Union and Pickaway limestones for several kilometers. At low flow both Spring Creek and its tributaries sink in their beds on reaching the Union Limestone, and the bed of Spring Creek is dry for 13 kilometers. At high flow the sinks are overwhelmed and there is flow along the length of Spring Creek (dashed line in Fig. 1A). There must be an extensive cave system underlying Spring Creek and its tributaries. One fragment of this is seen in Robbins Run Hole, most of which lies 25 m below the bed of Robbins Run. A second, flooded, fragment has been explored at Circulating Cenote, one of the springs for the Friars Hole System. Several other caves have been explored between the Friars Hole System and JJ Spring, including Fox Cave, Zotter's Folly, the Portal, and the Boarhole, but these are all located on the flanks of Spring Creek rather than directly under the creek (Fig. 1B).

Although JJ Spring is the perennial spring for Friars Hole, there are also three nearby overflow springs, the Cannon Hole, the Circulating Cenote, and Dale's Spring. The Cannon Hole is in fact an estavelle since at low flow it absorbs the surface flow of Spring Creek. The complexity seen at the springs of the Friars Hole System is common in karst aquifers, where distributaries are common where conduit systems discharge to the surface. Furthermore, the pattern of one perennial spring plus several intermittent overflow springs is also common. It is likely that the drainage to JJ Spring and the associated springs follows a single large conduit for most of the distance between Friars Hole and the springs, as shown in Figure 1B.

Geology

Almost all of the Friars Hole System is formed in the 48-m-thick Union Limestone and the underlying 30-m-thick Pickaway Limestone, which both belong to the Mississippian Greenbrier Group. The Union Limestone is primarily a pure limestone, with 91% of the section at Friars Hole being composed of oosparites and biosparites. However, there are four calcilutites, and these range in thickness from 50 cm to 3 m and have insoluble fractions that range from 47% to 63%. The Pickaway Limestone is much less pure than the Union Limestone, and consists of a regular alternation of calcilutites and calcarenites of up to 3 m in thickness. The underlying Taggard Formation is 7 m thick, and is composed predominantly of shale. It is only seen in the cave in the passages underlying the Crowsnest Room (Fig. 2), and JJ Spring is perched on the Taggard Shale.

The strata dip is to the northwest at 2° within the Friars Hole System and the dip is similar along the flow path to the springs. However, to the east of the cave underneath Droop Mountain there is a monocline which probably has thrust faulting associated with it. Thrust faulting is also seen at many locations within the cave (Fig. 3A). Many thrust faults in the cave have a strike of around N25°E and dip to the west, though there is much variability. The major joint set in the cave strikes at N60°E to N70°E. Although a majority of passages in the Friars Hole System are developed along joint sets and solutionally enlarged bedding plane partings that are parallel or subparallel to the strike, several substantial passage segments follow joints and bedding partings down-dip to the northwest (Fig. 2).

HISTORY OF EXPLORATION

Although the Snedegars entrance to the Friars Hole System was used for saltpeter mining during the U.S. Civil War, contemporary exploration of the cave began in the early 1950s with the descent of the 30-m Crookshank Pit, followed by its connection to Snedegars Cave in June 1964. The entrance to Friars Hole Cave, 4 km to the southwest of Snedegar Cave, was dug open in August 1964 and led to over 6 km of passage in a separate cave.

In March 1976, following excavation of a doline containing a sinking stream 1800 m to the northeast of the Friars Hole entrance, 7 km of passage were surveyed in Rubber Chicken Cave. Within two months, connections between it and the nearby Friars Hole and Snedegar-Crookshank caves were established, resulting in a single cave (the Friars Hole System) with a combined length of 24 km of surveyed passage.

In the summer of 1976, a major extension was found in Canadian Hole, a short, multipitch cave 600 m to the north of the Snedegar Saltpetre entrance. Over the next year, over 8 km of passage were surveyed, culminating in November 1977 with a connection to the Friars Hole complex to the south. A final connection to another nearby cave (Toothpick Cave) was made in November 1978, resulting in over 42 km of surveyed passages in the entire complex.

Continued work in the Friars Hole System since then has increased the total surveyed length of the cave to 72.7 km by the summer of 2002. A significant extension to the cave took place in 1995–2000 where an upper passage (Fig. 2, PM) entering the Friars Hole trunk was followed for 2.2 km to the west and northwest, ending beneath the Robbins Run valley 600 m to the west of the Friars Hole valley and establishing Robbins Run as a paleo infeeder to the Friars Hole System. Additional surveys since 2002 have increased the surveyed length of the system to 73.3 km.

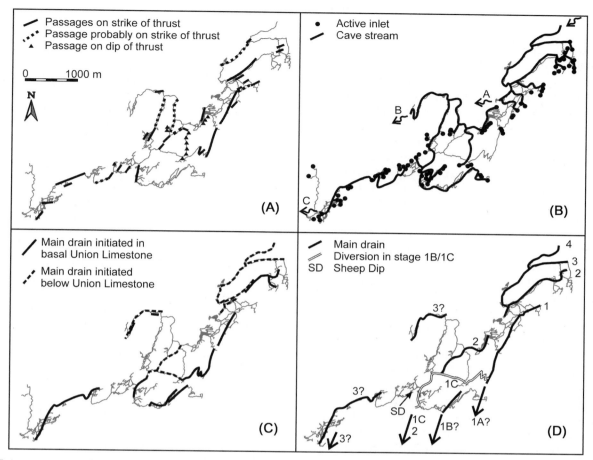

FIGURE 3 Cave passages of Friars Hole System, showing passages following thrust faults (A), active inlets and cave streams (B), stratigraphic control of main drains (C), and a possible sequence of development of the cave (D).

HYDROLOGY OF THE CAVE

Seven of the entrances of the Friars Hole System (Friars Hole Cave, Rubber Chicken Cave, Toothpick Cave, Crookshank Cave, Canadian Hole, and the Staircase and North Entrances to Snedegars Cave) are active stream sinks (Figs. 2 and 3B). The remaining three entrances, Radio Pit and the Saltpetre and Maze entrances to Snedegars Cave are abandoned stream sinks. In addition to the active stream sink entrances to the cave there are more than 100 other inlet passages where streams currently enter the cave. These range from small flows to large sinking streams.

In the cave, many of the inlet passages can be followed upstream to sediment blockages quite close to the surface, and thus these locations are clustered underneath the surface inliers of the limestone (Figs. 2 and 3B). These passages are typically tall vadose canyons that range from less than 50 cm up to several meters in width. They usually descend joints through the Union Limestone, but also often flow down-dip for short distances when they reach one of the impure limestone beds. Most inlet passages descend through the Union Limestone within a few tens of meters, though a few are longer. The longest passage in the Union Limestone is Lew's Last Climb, which was initiated along a low-angle fault, descending over a distance of 600 m before reaching the Pickaway Limestone; subsequently a tall, narrow canyon passage has downcut below the fault (Figs. 2 and 3A).

There is some integration of stream passages in the Union Limestone, but most takes place in the Pickaway Limestone, where the stream passages have much lower gradients. Within the known cave the major streams integrate into three major drainages (Fig. 3B). The eastern drainage has the largest cave stream, Rocky River, and drains some or all of the flow from a catchment area of 45 km^2. Rocky River is a large strike-oriented cave stream that can be followed in the cave for more than 2 km to a boulder blockage (Fig. 3B, point A). There are several long tributaries to Rocky River, such as the one sinking at the Canadian Hole entrance. This descends through the Union Limestone with three major drops (11 m, 15 m, and 9 m), and then flows down-dip in the upper Pickaway Limestone until it joins the strike-oriented Rocky River.

The central drainage in the cave includes the streams sinking at the Toothpick, Snedegar Staircase, Crookshank, and Rubber Chicken entrances to the cave. The longest stream is the Toothpick stream, which can be followed almost continuously for 5.5 km, ending at a sump, which is the lowest point in the cave (Fig. 3B, point B). The path of the Toothpick stream is sinuous. After dropping rapidly through the Union Limestone by two 10-m drops, most of the passage is in the basal Union Limestone or uppermost Pickaway Limestone, and much probably formed at the intersection of major bedding planes and thrust faults (Figs. 3A and 3C).

The western drainage in the cave is aligned along the strike and closely follows the base of the Friars Hole valley. Consequently, only short tributaries join the main stream. The cave stream is last seen in an aqueous passage called Water World (Fig. 3B, point C). Along most of its path the stream follows Friars Trunk, which is one of the largest passages in the cave, averaging 15 m in width and 8 m in height.

The three separate cave streams seen in the cave join together before reaching JJ spring (Fig. 1B). The gradient from the lowest point in the cave to the spring is only 0.002, and it seems likely that most of this strike-aligned pathway is along a perennially flooded conduit below the level of Spring Creek.

PALEOHYDROLOGY

Surface Paleohydrology

The Friars Hole System has a long and complex history, and the evolution of the cave is intimately related to the evolution of the surface overlying it. The progressive capture of the surface creek along Friars Hole has left the valley in a largely relict situation, allowing the evolution of the surface drainage to be inferred.

Hills Creek and its tributaries originally flowed along the length of the Friars Hole valley and formed a tributary to Spring Creek. As the creek eroded downward it eventually reached the Union Limestone and was captured underground. With continued surface erosion the sink point for Hills Creek has retreated upstream a distance of more than 4 km, and is now 40 m lower than the original sink point. This has formed a long inlier of Union Limestone along the eastern part of Friars Hole, which terminates at the original sink point, which fed the Downlets in the Friars Hole System (Fig. 2). The elevation of the top of the Union Limestone at the Downlets is 780 m asl, which is the highest elevation of this contact along Friars Hole, and this is the reason that the original sink point was at this location.

Flow from the original sink point of Hills Creek was to the southwest through the Friars Hole System, but Locust Spring is a much closer spring than JJ Spring, and offers a much steeper hydraulic gradient. As the sink point of Hills Creek retreated to the north its location became closer to Locust Spring (Fig. 1B), and east-flowing cave streams such as the one in Locust Creek Cave must have captured at least some of the flow. At present much of the flood discharge from Hills Creek drains into Hills-Bruffey Cave and then east under Droop Mountain to Locust Spring, but the low-flow sink point of Hills Creek 500 m to the west of the flood sink is gradually capturing the flow back toward the Friars Hole System and to JJ Spring on Spring Creek.

The underground capture of Hills Creek at the Downlets bisected the drainage along Friars Hole. Upstream, the 54 km^2 Hills Creek watershed fed this large sinking stream. Downstream, the catchment area of 20 km^2 along the remainder of Friars Hole fed a much smaller creek, which eventually eroded down to the limestone contact, and was captured underground. The process of fragmentation of the surface drainage by capture underground has continued to the present day, resulting in Friars Hole now being a dry valley. However, small creeks continue to flow off the clastic rocks on the flanks of Friars Hole, but now sink soon after reaching the Union Limestone, which outcrops along the axis of Friars Hole (Figs. 1 and 2).

Cave Paleohydrology

The Friars Hole System has a long and complex history, which is not yet fully understood. Despite the survey of almost 75 km of passages, the exploration of many has been stopped by sediment fill, boulder blockages, or water-filled passages. Furthermore, the currently explored cave only spans one-third of the distance between the major sink at Hills Creek and the springs for the cave (Fig. 1). Clearly, there are many tens of kilometers of unexplored cave, including both currently active stream passages and relict passages.

Despite the only partial knowledge of the underground drainage in the area, a number of factors are helpful in deducing the paleohydrology of the cave. We assume that the fossil drainage was guided by the same principles that determine present flow patterns. Determining factors include stratigraphy, structure, and the water table elevation. Modern cave streams descend rapidly through the pure Union Limestone, and have much lower gradients as they follow impure horizons subhorizontally in the basal Union or upper Pickaway. Low angle thrust faults are common in the cave (Fig. 3A), with at least 5 km of major passages being initiated along the intersection of a thrust fault with a prominent bedding plane.

The water table elevation is important in determining flow direction, with generally down-dip flow in the vadose zone and strike-oriented flow in the phreatic zone. Detailed analysis of cave streams has shown that some were initiated solely on bedding planes and followed a down-dip course, and others were initiated along joints or along the intersection of a bedding plane and a joint or fault. The currently known cave is all above the present water table, but the low gradient between the lowest point in the cave and the springs indicates that most of the intervening conduit, which closely follows the stratal strike, is below the water table.

The largest passages in the cave tend to follow the strike of the strata. These main drains are inferred to be the principal passages at each stage in the history of the Friars Hole System, and are presumed to have connected the major sink into the cave (Hills Creek) with the spring or springs. A number of these main drains can be identified in the cave, and are usually found in the lower Union Limestone or upper Pickaway Limestone (Fig. 3C). These main drains range in elevation over 80 m, with the higher ones being found in an up-dip location in the southeast part of the cave, and lower ones parallel and down-dip to the northwest. This pattern is most clearly seen in the easternmost part of the cave, where four major, parallel main drains can be identified, and are numbered 1–4 in Figure 3D. The first three of these are abandoned, and the fourth is Rocky River, the present underground course of Hills Creek.

Solutional scallops on passage walls are commonly used to infer flow direction and discharge, but are rarely preserved in the major passages of the Friars Hole System, where exfoliation has been common in the impure limestones of many passage walls. However, scallops are well preserved in the critical Sheepdip area (Fig. 3D, SD), where there are a series of round, phreatic tubes in the Union Limestone. These are unique in the cave because all other passages in the cave in the Union Limestone are vadose canyons. Scallop measurements in the Sheepdip area showed that the paleoflow was to the east, rather than to the southwest, which is the current flow direction. The paleo water table at this time can be identified from the transition from vadose to phreatic passages, and was at 702 m, and this eastward flow descended to at least 22 m below the water table.

Scuba diving in flooded cave passages in the area surrounding Friars Hole has shown that most sumps descend just a few meters below the water table. The interpretation of paleoflow shown in Figure 3D assumes that most conduits were initiated below the water table and within a few meters of it. Undulating profiles along some of the main drains confirm that some sections must have been below the water table. Conversely, tall canyons have developed along some main drains, indicating these carried vadose streams at one time.

The initial sink point of Hills Creek was at the Downlets, and the underground course of that creek was a large strike-oriented conduit (Fig. 3D, main drain 1). The subsequent main drains are further down-dip, have an average vertical spacing of 17 m, and are associated with later sinkpoints of Hills Creek. The parallel courses of the main drains in the eastern part of the cave make interpretation there straightforward. However, the interpretation is only tentative for the remainder of the cave. For instance, flow to the south at 1A and 1B is less well supported than at 1C (Fig. 3D), and any of these three paths imply a major unexplored passage to the south, underneath the clastic caprock, and leading to a spring on Greenbrier River, where the Union Limestone would have been exposed earlier than on Spring Creek. The relationships shown for main drains 2 and 3 in Figure 3D are even less certain, but are consistent with passage sizes, orientations, and elevations. What is certain is that there are unexplored continuations to many passages, though many are undoubtedly blocked by sediment or are water-filled.

Age of the Cave

An indication of the age of the cave can be inferred from speleothem samples taken from it. The ages of 40 speleothems from the cave have been determined by uranium series dating, and 10 of these gave ages beyond the 350,000-year limit of the dating method. A sample from the Crossover Passage, a stage 2 passage (Figs. 2 and 3D), was paleomagnetically reversed and thus is more than 700,000 years old. A sample from the Lost Passage, a stage 3 passage at the cave's southwest end (Figs. 2 and 3D), was more than 350,000 years old and was magnetically normal, but had a $^{234}U/^{238}U$ ratio in secular equilibrium, a process that takes more than 1 million years. This sample probably dates from the Olduvai subchron between 1.72 and 1.88 million years ago. If this is true, and assuming a steady rate of base-level lowering, then the oldest passages in the cave are 4.1 million years old.

Bibliography

Dougherty, P. H., Jameson, R. A., Worthington, S. R. H., Huppert, G. N., Wheeler, B. J., & Hess, J. W. (1998). Karst regions of the eastern United States, with special emphasis on the Friars Hole Cave System, West Virginia. In D. Yuan & Z. Liu (Eds.), *Global karst correlation* (pp. 137–155). Beijing, China: Science Press.

Medville, D. M. (1981). Geography of the Friars Hole Cave System, U.S.A. In B. F. Beck (Ed.), *Proceedings of the 8th International Congress of Speleology, Bowling Green, Kentucky* (pp. 412–413). Huntsville, AL: National Speleological Society.

Medville, D. M., & Medville, H. E. (1991). Structural controls on drainage beneath Droop Mountain, West Virginia. In E. H. Kastning & K. M. Kastning (Eds.), *Appalachian karst* (pp. 11–18). Huntsville, AL: National Speleological Society.

Sasowsky, I. D., White, W. B., & Medville, D. M. (1989). The remarkably constant longitudinal profile of Toothpick Stream, Friars Hole Cave System, West Virginia, U.S.A. In *Proceedings of the 10th International Congress of Speleology* (pp. 284–286). Budapest: Hungarian Speleological Society.

Storrick, G. D. (1992). *The caves and karst hydrology of Southern Pocahontas County and Upper Spring Creek Valley*. Morgantown, WV: West Virginia Speleological Survey.

G

GAMMARUS MINUS: A MODEL SYSTEM FOR THE STUDY OF ADAPTATION TO THE CAVE ENVIRONMENT

Daniel W. Fong

American University

ADAPTATION TO THE CAVE ENVIRONMENT

A diverse assemblage of animal groups, ranging from vertebrates such as fish and salamanders, to crustaceans such as crayfish and scuds, to insects such as beetles and springtails, to even clams and sponges, contain species that are adapted to subterranean habitats such as caves and are never found on surface habitats. Most of these species are strongly convergent in a suite of morphological, physiological, and behavioral characteristics, termed *troglomorphy*, that are clearly related to the dominant environmental factor in caves: the absence of light (Culver and Pipan, 2009). Examples of morphological features include reduced to complete loss of eyes and body pigment, elaborated extra-optic sensory structures such as longer antennae and bigger or higher density of taste buds, elongated walking appendages, and more slender body forms compared to related surface species. The elaborated and elongated structures are generally assumed to be adaptations to total darkness, resulting from directional selection for enhanced chemosensory and tactile performance as well as for increased efficiency of movement, although this assumption is seldom tested. The reduction to loss of eyes and body pigment is part of a general phenomenon of evolutionary loss of features common to most, if not all, phyletic lineages; the loss of hearing in moths on islands without bats is but one of many examples (Fong et al., 1995). Therefore the evolution of elaborated and of even reduced features is not unique to cave organisms. The study of the evolutionary mechanisms leading to troglomorphy in cave species is interesting, however, because both elaborated and reduced features are exhibited by the same organism, have evolved independently and repeatedly across many higher taxonomic groups, and are clearly related to the consequences of a single overriding environmental constraint: total darkness. It is possible that different species have evolved troglomorphy through different mechanisms. But part of the allure of the study of troglomorphy is that it seems probable that one universal mechanism is responsible for the fact that such a diverse syndrome of superficially similar characteristics is exhibited by a wide array of distantly related taxa.

GENERAL HYPOTHESES ON THE MECHANISM OF ADAPTATION TO THE CAVE ENVIRONMENT

The search for the mechanism behind the evolution of troglomorphy has focused mainly on the reduction to loss of eyes and body pigment, historically referred to as *regressive evolution*. The lack of emphasis on the elaborated features results from the assumption that these are obviously adaptations to darkness, or simply more examples of evolution by means of natural selection. The mechanism of regressive evolution is more difficult to elucidate, however, not because it is difficult to explain but because many of the proposed explanations are difficult to frame into testable hypotheses (see Fong et al., 1995). Hypotheses on the mechanism of regressive evolution in cave species fall into three types: nonadaptive, directly adaptive, and indirectly adaptive.

The nonadaptive hypothesis invokes as its mechanism neutral mutations and genetic drift, and attributes no role to natural selection. Simply put, features such as eyes and body pigment serve no function in darkness; consequently, mutations affecting structural or regulatory genes underlying these traits are invisible to natural selection, or neutral, and would accumulate over time. The consequence of accumulating such mutations will eventually lead to the breakdown of the feature. An adaptation is by definition a feature that is an evolved response to natural selection, and strong convergence among distantly related taxa such as the dorsal fin in sharks and dolphins is usually a textbook example of evidence of selection, or similar responses to similar selective pressures. Under the neutral mutation scenario, interestingly, strong convergence for regressive evolution among diverse taxa in caves is not considered as similar responses to the common selective constraint of total darkness. The critical assumption of this nonadaptive hypothesis is independence of the genetic bases governing the development and expression of the sets of elaborated and reduced features: the elaborated features are adaptations, but the reduced features are not.

The directly adaptive hypothesis suggests that selection favors individuals with smaller eyes or less pigment *per se* in darkness. The selective advantage of having smaller eyes or less pigment is not always clear, but is usually couched in terms of the energy or material saved by spending less of it on a useless structure. The nature of this energy or material, however, is nebulous, and in any case, even if such energy or material savings can be obtained, the selective advantage can only be realized if the unspent energy or material can be spent on features that increase fitness in darkness, such as extra-optic sensory structures; therefore the directly adaptive hypothesis is mechanistically indistinguishable from the indirectly adaptive hypothesis (Fong, 1989).

The indirectly adaptive hypothesis assumes an evolutionary tradeoff, or a negative genetic relationship, among components between the sets of elaborated and reduced features through antagonistic pleiotropy. Under this scheme, the evolutionary reduction in features is a correlated response to selection for elaborated features, that regressive evolution is a direct consequence of selection for enhanced features in darkness. Under this hypothesis, components of the sets of reduced and elaborated features are genetically dependent, evolving as one integrated unit.

CRITERIA FOR THE STUDY OF ADAPTATION TO THE CAVE ENVIRONMENT

A model system for the study of adaptation to the cave environment must satisfy several criteria (Culver and Pipan, 2009). The first is that phylogenetic information must exist to indicate the direction of evolution. In other words, there is independent evidence that the troglomorphic features are derived rather than ancestral character states. Simply relying on a correlation between the different morphologies of different species in different habitats can be problematical. For example, large eye in surface species has long been assumed the ancestral state and small eye in cave species the derived state among fish lineages of the family Amblyopsidae in North America. A recently constructed phylogeny based on nucleotide sequence variation, however, indicates that large eye in the genus *Chologaster* is likely a derived condition within the *Typhlicththys* clade, with absence of eyes as the ancestral condition (Dillman et al., 2010). This criterion is extremely important because it rules out a majority of species as model systems because most troglomorphic species in caves are highly endemic with no close surface relatives for comparison (Barr and Holsinger, 1985), and because a relatively small number of phylogenies including closely related cave and surface taxa currently exist. Second, there should be data providing genetic and selective perspectives on population structure. Data on genetic population structure informs on the extent of genetic isolation between surface and cave counterparts as well as within and among cave populations. The structure of the selective environment reflects the extent of heterogeneity between the surface and cave habitats, and equally important, heterogeneity within the cave habitat. Third, there should be evidence that both reduced and elaborated characters show heritable variation in the cave and the related surface taxa. If selection has driven the presumed adaptations to fixation in cave populations, it is difficult to demonstrate active selection on these characters. This criterion excludes most highly troglomorphic species because all individuals lack any manifestation of eyes and of body pigment: there is simply no variation. In any case, selection must act first on the ancestral character states on initial colonization of the cave environment, and it is the features found in the ancestral surface taxa that show the presumptive ancestral character states. Ideally, these three criteria are satisfied by a single species with troglomorphic and nontroglomorphic populations occupying different habitats and with data supporting troglomorphy as the derived condition. Finally, there should be evidence of multiple, independent evolution of troglomorphy in different cave populations. The concordance or discordance of patterns observed among independently derived populations of the same species is extremely informative to our ability to generalize conclusions across a large number of species belonging to a diverse assemblage of taxonomic groups. The amphipod *Gammarus minus* is an excellent model organism to study the evolution of troglomorphy in light of these criteria.

GENERAL ECOLOGY OF GAMMARUS MINUS

The freshwater amphipod crustacean *Gammarus minus* was first described by the naturalist Thomas Say in 1818 from specimens obtained from a spring run in southern Pennsylvania. The species has an extensive geographic range, occupying a narrow band along the Appalachian Mountains from southern Pennsylvania to northwestern Alabama and the northern Mississippi, plus much of the east central United States including most of Tennessee and Kentucky and parts of southern Ohio, southern Indiana, and western Illinois, as well as across the Mississippi River west to central Missouri, northern Arkansas, and northeastern Oklahoma (Fig. 1). Within its range *G. minus* is a common inhabitant of carbonate springs (Glazier et al., 1992) where groundwater resurge onto the surface, especially in karst areas. It is also a common inhabitant of cave streams throughout its range, but troglomorphic populations occur only in some caves in two widely separated, limited geographic areas (Fig. 1). *G. minus* also occurs in the streams of some karst windows, and some karst window populations in these two areas may exhibit morphologies that are intermediate between spring and troglomorphic cave populations. Figure 2 shows an idealized subterranean drainage basin and the hydrological connection among the spring, cave stream, and karst window habitats.

Populations of *G. minus* in all habitats probably feed similarly on organic detritus but derive their nutrients mainly from the associated microbial fauna and flora. Troglomorphic cave populations may also derive much of their energy from grazing on the biofilm coating the substratum of cave streams. Compared to spring populations, they are also more likely to prey on other organisms such as small worms and smaller crustaceans such as isopods or other amphipods, and show a stronger tendency toward cannibalism, especially on smaller or injured individuals. Density of spring populations may exceed 200 m^{-2}, but is usually an order of magnitude lower for troglomorphic populations in cave streams. The densities of nontroglomorphic populations in cave streams and in karst windows are extremely variable but may reach similarly high values as spring populations. Similar to most amphipods in the family Gammaridae, *G. minus* is sexually dimorphic in size: the body length of

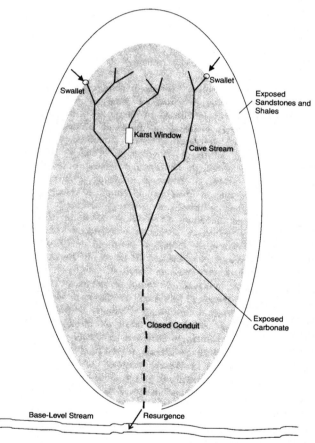

FIGURE 2 Aerial view of an idealized subterranean drainage basin. Headwater cave streams originate at swallets (indicated by arrows) where surface water sinks at the contact of nonsoluble rock and carbonate rock (stippled area) or from percolating water dripping into cave passages. Cave streams are occasionally exposed at karst windows where parts of the cave ceiling have collapsed to the surface. Water flows through a closed conduit before exiting permanently at a resurgence, or spring, and continues as a spring run before emptying into a surface stream. Populations of *Gammarus minus* may occupy all three habitats within a basin: resurgence, cave stream, and karst window. *From Culver et al., 1995.*

FIGURE 1 Distribution of *Gammarus minus*. Solid areas indicated by arrows are regions with troglomorphic populations: Greenbrier Valley (GV) in West Virginia and Wards Cove (WC) in Virginia. *Modified from Culver et al., 1995.*

mature females is about 67% of mature males. During reproduction, males and females form precopulatory pairs in amplexus. Although amplexing pairs are observed throughout the year, their numbers tend to peak between December and March in spring populations, but do not show a consistent temporal pattern in troglomorphic cave populations, with some showing a peak in the winter months and some not. Fertilization is external. Females deposit eggs into a ventral marsupium formed by overlapping cuticular plates, where they are fertilized when a male deposits a sperm packet. There is no larval stage; fertilized eggs develop directly within the marsupium into miniature amphipods, about 1 mm in length. The newly hatched amphipods are released from the marsupium into the environment when the female molts. The life cycle is about one year in spring populations and probably in nontroglomorphic cave populations as well as karst window populations, but may be longer in troglomorphic cave populations. Specimens collected from any of the habitats can live for two to three additional years in the laboratory.

VARIATION IN GAMMARUS MINUS

The spring habitats of *Gammarus minus* are characterized by hard, alkaline water of pH 6 or higher, conductivity greater than $100\,\mu s\,cm^{-1}$, and temperatures at or near the average annual ambient temperature for the latitude (10–12°C in the states of West Virginia and Virginia). The occurrence of *G. minus* in cold, carbonate springs is ubiquitous. Springs with the appropriate physical and chemical properties but without *G. minus* are extremely rare, at least in West Virginia and Virginia. Within its spring habitat, *G. minus* is usually the dominant macroinvertebrate species in terms of density and probably biomass. Sexually mature individuals of spring populations have large compound eyes with about 40 ommatidia, a first pair of antennae at about 45–50% of body length, and brownish body pigmentation (Fig. 3). Body size is extremely variable among spring populations, with body length of mature males ranging from 6 to 11 mm. The variation in body size among springs may be inversely correlated with the intensity of size-selective predation by small fish predators such as sculpins (*Cottus* sp.).

Gammarus minus is also commonly observed in cave streams throughout its range. The cave streams within a subterranean drainage basin resurge onto the surface at a spring, and the morphology of most cave populations is identical to or only slightly different from that of hydrologically connected spring populations. However, populations of *G. minus* occurring in some caves in only two small geographic areas, one in the Greenbrier Valley of southeastern West Virginia and one in the Wards Cove area of southwestern Virginia (Fig. 1), are morphologically differentiated from spring populations. These morphologically distinct cave populations of *G. minus* exhibit the classic troglomorphic syndrome common to species highly adapted to the subterranean environment (Jones *et al.*, 1993). They differ from spring populations and nontroglomorphic cave populations in having consistently large body size with mature males attaining 10–11 mm in length, pale to bluish instead of brownish body coloration, longer first antennae that are at least 65% of body length, and greatly reduced compound eyes with only a few to no discernable ommatidia (Fig. 3; Holsinger and Culver, 1970).

Gammarus minus also occurs in the streams of some karst windows. A karst window forms when a portion of the ceiling of a cave passage collapses through to the surface; thus the stream in a karst window was once a subterranean cave stream. Some karst window

FIGURE 3 (A) A live, adult specimen of *Gammarus minus* from a spring population, 7-mm body length. (B) Specimen from a troglomorphic cave population, 10-mm body length. Notice the difference in eye size, antennae length, and body coloration. *Photo by the author.*

populations of *G. minus* are located in subterranean drainage basins with a troglomorphic population in a cave stream and a nontroglomorphic spring population at the resurgence. Among five such karst window populations in five different subterranean drainage basins, one is morphologically identical to the troglomorphic cave population, one is identical to the spring population, and three are intermediate between the cave and spring populations (Culver et al., 1995), suggesting the interesting possibility that some of the karst window populations are evolving from the troglomorphic phenotype back toward the nontroglomorphic phenotype.

POPULATION STRUCTURE OF GAMMARUS MINUS

Although water from springs empties into base level surface streams, often after flowing for only short distances along spring runs, the physical and chemical properties of the spring water differ greatly in magnitude and show much less daily and annual variation than those of the surface stream water. In addition, the density of fish predators, such as trout, darters, and sculpins, is much higher in the streams than in the springs. Thus migration of *G. minus* individuals among even geographically proximate springs situated along small surface streams may be restricted, but may be possible during winter when the temperature of the surface streams drops to values matching those of springs. Gene flow among geographically distant spring populations separated by large streams, such as rivers of the fourth order or higher, appears highly improbable. Clearly, spring populations of *G. minus* exhibit classic isolation by distance, with distance measured as hydrological distance rather than Euclidean distance. Among *G. minus* populations situated along small streams and a large river in Greenbrier Valley, West Virginia, significant proportions of the sequence variation in the mitochondrial gene COI (35%) and in the nuclear gene ITS-1 (17.2%) are explained by hydrological distance (Carlini et al., 2009).

Both of the geographic areas that contain troglomorphic populations of *G. minus* have extensive cave development and large cave systems exceeding 20 km in passage length. Culver *et al.* (1995) hypothesized that such large cave systems are necessary for the genetic isolation of cave populations from spring populations to allow for the evolution of the troglomorphic traits. Genetic isolation of these *G. minus* populations probably arises from the fragmented nature of the habitats. The cave streams in these subterranean drainage basins are separated from their spring resurgences by phreatic sediments and porous rock when the cave stream passes below the water table (Ford and Williams, 2007). The resultant closed conduit channels are barriers to gene flow for organisms such as *G. minus*. The physical barrier for *G. minus* between the cave stream and the resurgence can be rather abrupt, even in the absence of such a closed conduit. For example, the cave stream in McClung-Zenith Cave in Monroe County, West Virginia, flows underneath a large breakdown pile of a collapsed portion of the former cave ceiling at the cave entrance for about 25 m and emerges on the other side as a spring. This spring harbors a typical surface population of *G. minus* at a density of over 100 m^{-2}, yet *G. minus* has not been collected in the cave stream just a short distance away. Consequently, troglomorphic cave populations are physically and genetically isolated from hydrologically connected spring populations. Because different subterranean drainage basins are hydrologically connected only through their respective resurgences and their base level streams, troglomorphic cave populations are also effectively genetically isolated from each other. This is supported by estimates of divergence among populations as indicated by F_{ST} values obtained from sequence variation in the COI gene (Carlini et al., 2009). For example, water in Persinger Cave resurges at Davis Spring. Divergence between the Persinger Cave and Davis Spring populations is 0.78, similar in magnitude to divergence between the Persinger Cave population and cave populations located in other subterranean drainage basins each associated with a different spring: the Hole Cave at 0.83, Organ Cave at 0.69 and 0.86, and Fallen Rock Cave at 0.88.

ORIGIN OF TROGLOMORPHIC CAVE POPULATIONS OF GAMMARUS MINUS

There is strong evidence indicating cave populations of *G. minus* originated via colonization at some time in the past by individuals dwelling at the spring where the cave water resurges, and that cave populations in different drainage basins are each independently derived from a different spring population. On average, cave populations show significantly lower within-population sequence variation in the COI and ITS-1 genes than spring populations, and that despite the substantial morphological difference between cave and surface populations of *G. minus*, cave populations are on average more closely related to hydrologically proximate surface populations than to cave populations in other drainages (Kane et al., 1992; Carlini et al., 2009). Cave populations in general show significantly smaller

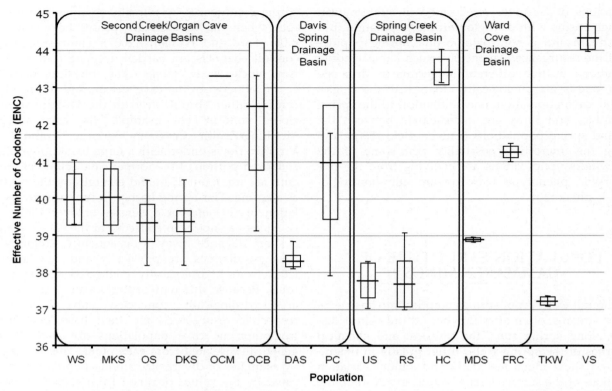

FIGURE 4 The effective number of codons (ENC; Wright, 1990) in COI sequences from 15 populations of *Gammarus minus*. High ENC indicates small effective population size and low ENC indicates large effective population size. The mean (horizontal line), standard deviation (box), and range (vertical bar) of ENC are indicated for each population (except for OCM where all individuals were fixed for one sequence). Populations are grouped by drainage basin and habitat type (spring populations are designated by the letter S, cave populations by the letter C, and the single karst window population by the letters KW). The TKW and VS populations were the only sites sampled within their respective drainage basins, and so are not grouped with any other population. *Reproduced from Carlini et al., 2009.*

effective population sizes as measured by levels of codon bias compared to surface populations in the same drainage (Fig. 4; Carlini et al., 2009), indicating they may have suffered genetic bottlenecking during initial colonization of the cave habitat, or more probably from periodic episodes of genetic bottlenecking events since initial colonization such as lengthy dry spells, or both. This is because, within a drainage basin, effects of environmental perturbation of the physical and chemical parameters such as discharge or temperature would be more severe at upstream locations and be dampened farther downstream with increasing volume of flow, and the most downstream portion of a subterranean basin is the resurgence at the spring. The evidence clearly shows that the troglomorphic features are the derived character states in *G. minus*. Interestingly, this also suggests the possibility that the nontroglomorphic features of karst window populations of *G. minus* are secondarily derived character states from the ancestral troglomorphic state in each basin. An examination of the structure of the vestigial compound eyes shows that individuals in four different cave populations have consistently lost different parts of the compound eye (Fig. 5; Culver et al., 1995). Therefore, troglomorphy has evolved independently in different populations of *G. minus* occupying different subterranean drainages.

There is some evidence indicating genetic structuring of *G. minus* individuals within a subterranean basin. Within Organ Cave, sequence variation in the COI gene was compared between specimens from a headwater stream at the top of the drainage (OCM) and from the large river near the lowest accessible point of the drainage (OCB). Although the OCB individuals showed only a small amount of variation, all OCM individuals were fixed for a single sequence (Carlini et al., 2009). The lack of genetic variation at OCM may be explained by genetic drift. Small headwater streams at the top of a subterranean drainage are more likely than larger, lower level streams to periodically dry up during prolong droughts, leading to extinctions of localized populations of *G. minus* followed by recolonization by a small subset of the main population from lower level streams when the headwater streams resume their flow.

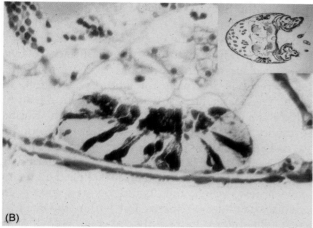

FIGURE 5 The compound eye structures of two cave populations of *Gammarus minus*. (A) Vestigial compound eye from an Organ Cave specimen. (B) Vestigial compound eye from a Benedict Cave specimen. The inset in each panel shows the cross-section of the entire head through the eyes. Vestigial eyes from other cave populations are not shown, but different cave populations have lost different parts of the compound eye. *Photo by the author.*

GAMMARUS MINUS AS A MODEL SYSTEM

In summary, *Gammarus minus* is an excellent model system to study adaptation to the cave environment. It is a single species with troglomorphic cave populations and nontroglomorphic surface populations. Cave populations, in addition to surface populations, show heritable, individual variation in the reduced as well as the elaborated structures (Fong, 1989). Hydrological and genetic evidence (see Culver et al., 1995) clearly indicate that troglomorphic populations invaded cave streams from surface spring habitats, thus showing the troglomorphic traits are the derived condition. The selective environments of both surface and cave populations are well characterized, and data exist on the genetic population structure, both among and within populations and habitats. Evidence from allozyme variation and morphological variation (see Culver *et al.*, 1995) as well as from DNA sequence variation (Carlini *et al.*, 2009) strongly indicates that multiple cave populations have evolved troglomorphy independently.

See Also the Following Articles

Asellus aquaticus: *A Model System for Historical Biogeography*
Niphargus: *A Model System for Evolution and Ecology*
Astyanax mexicanus: *A Model System for Evolution and Adaptation*
Natural Selection
Morphological Adaptations

Bibliography

Barr, T. C., Jr., & Holsinger, J. R. (1985). Speciation in cave faunas. *Annual Review of Ecology and Systematics, 16,* 313–337.

Carlini, D. B., Manning, J., Sullivan, P. C., & Fong, D. W. (2009). Molecular genetic variation and population structure in cave and surface populations of the freshwater amphipod *Gammarus minus*. *Molecular Ecology, 18*(9), 1932–1945.

Culver, D. C., Kane, T. C., & Fong, D. W. (1995). *Adaptation and natural selection in caves: The evolution of* Gammarus minus. Cambridge, MA: Harvard University Press.

Culver, D. C., & Pipan, T. (2009). *The biology of caves and other subterranean habitats*. Oxford, UK: Oxford University Press.

Dillman, C. B., Bergstrom, D. E., Noltie, D. B., Holtsford, T. P., & Mayden, R. L. (2010). Regressive progression, progressive regression or neither? Phylogeny and evolution of the Percopsiformes (Teleostei, Paracanthopterygii). *Zoologica Scripta, 40*(1), 45–60.

Fong, D. W. (1989). Morphological evolution of the amphipod *Gammarus minus* in caves: Quantitative genetic analysis. *American Midland Naturalist, 121,* 361–378.

Fong, D. W., Culver, D. C., & Kane, T. C. (1995). Vestigialization and loss of nonfunctional characters. *Annual Review of Ecology and Systematics, 26*(1), 249–268.

Ford, D., & Williams, P. (2007). *Karst hydrogeology and geomorphology*. New York: John Wiley and Sons.

Glazier, D. S., Horne, M. T., & Lehman, M. E. (1992). Abundance, body composition and reproductive output of *Gammarus minus* (Amphipoda, Gammaridae), in ten cold springs differing in pH and ionic content. *Freshwater Biology, 28*(2), 149–163.

Holsinger, J. R., & Culver, D. C. (1970). Morphological variation in *Gammarus minus* Say (Amphipoda, Gammaridae), with emphasis on subterranean forms. *Postilla*(146), 1–24.

Jones, R. T., Culver, D. C., & Kane, T. C. (1992). Are parallel morphologies of cave organisms the result of similar selection pressures? *Evolution, 46,* 353–365.

Kane, T. C., Culver, D. C., & Jones, R. T. (1992). Genetic structure of morphologically differentiated populations of the amphipod *Gammarus minus*. *Evolution, 46,* 272–278.

Wright, F. (1990). The "effective number of codons" used in a gene. *Gene, 87*(1), 23–29.

GEOPHYSICS OF LOCATING KARST AND CAVES

Barbara Anne am Ende

Deep Caves Consulting South Riding, Virginia

It can be quite challenging to find caves that have entrances, as the openings may be small, obscure, or plugged. Detecting caves with no openings to the surface at all is that much more difficult. In addition to explorers, developers and landowners have a profound need to understand the subsurface in karst terrains for public safety concerns to avoid or mitigate collapse issues (Fig. 1).

On occasion, karst cavities in bedrock are accidentally encountered during well drilling. However, most techniques are based on methods that rely on some measurable physical property that can be sensed remotely. It is the variations in the properties of a void, whether it is air-, water-, or sediment-filled, such as reflections of energy or resistivity to current flow, that suggest the presence of a void.

Principally, the techniques used are cave radio location, electrical resistivity techniques, microgravity, and ground-penetrating radar. Less useful, but possible under limited conditions, are seismic studies, thermal variations, and interferometry. Some people may consider dowsing to be a viable technique, but it has no scientific validity.

For most of these techniques, there are tradeoffs. Each setting may be interpreted better using one method versus another. Then there are choices to be made, for instance, electrode configurations, frequencies, spacing of detectors, and so on. Generally, low frequencies penetrate deeper, but do not give the resolution of high-frequency energy systems. Careful study of these tradeoffs is necessary before one goes to the expense and effort of doing a geophysical study.

MICROGRAVITY

Gravity is the attraction of one body of mass to another. With respect to the Earth, gravitational attraction decreases with increasing distance from the center of the Earth. All else aside, the pull of gravity is less on a mountaintop than in the neighboring valley. The *geoid* is the surface that would represent sea level, should it be represented across the globe. The force of gravity is always perpendicular to the geoid surface. *Bouguer gravity anomalies* are the variations in the gravity field after corrections are made for latitude, elevation, and terrain. These are the result in variation in density of the rock. If a cave is present beneath the Earth's surface, the gravity field would be less than the surrounding area, as karst features filled with air, sediment, or water are less dense than bedrock (Fig. 2). Considering the radius of the Earth is more than 6000 km, the effect of a cave a few meters tall would be very small, and therefore such measurements fall within the realm of *microgravity* studies. Indeed, the "average" gravitational measurement is about 980 gal, whereas the anomaly measured over passages in Maxwelton Cave, West Virginia, ranged between 20 and 30 μgal for passages ~3 m in diameter and 10–30 m deep.

A challenge with microgravity studies is that a large cavity at a deep depth may be impossible to distinguish from a small void at a shallow depth. Nevertheless, microgravity studies have successfully predicted cave locations and studies have been conducted in karst terrain throughout the world.

ELECTRICAL RESISTIVITY IMAGING (ERI)

The basic technique for ERI is to insert a line of metal electrodes into the ground. Constant current is injected into two, and the voltage difference is measured sequentially between the remaining pairs of electrodes. This application of current is repeated through the rest of the pairs of electrodes and an apparent resistivity is calculated for each measurement. These data are then inverted to create a resistivity model of the subsurface generally using an iterative smoothness-constrained least-squares technique (Fig. 3). Software, such as the

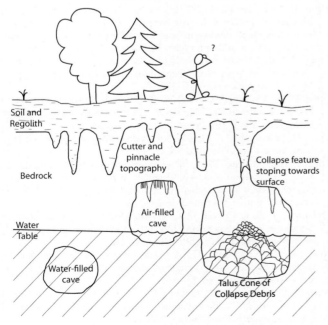

FIGURE 1 Various subsurface targets to be detected and distinguished by geophysical methods.

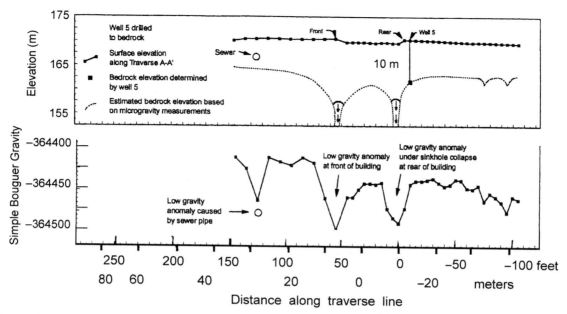

FIGURE 2 A microgravity traverse beneath a building showing the anomalies due to an underlying cavity. Traverse conducted by the Center for Cave and Karst Studies, Western Kentucky University, Bowling Green, KY.

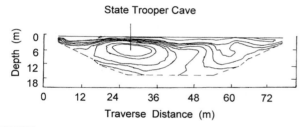

FIGURE 3 Resistivity profile over State Trooper Cave, Bowling Green, Kentucky. Traverse constructed by the Center for Cave and Karst Studies, Western Kentucky University, Bowling Green, KY.

freeware program RES2DINV, can be used for this processing. Caveats about possible errors in the interpretation are discussed in McLean and Luke (2006). In their extensive RSI imaging at Ft. Stanton Cave, New Mexico, transects were conducted with measurements to ~50 m. Several passages were predicted and later discovered.

High resistivity values are considered bedrock. A water-filled void would have low resistivity, as the natural water would conduct the current relatively easily and salt water would conduct the current even better. An air-filled cavity would have a very high resistivity. However, it should be noted that alternate explanations can be interpreted from a line of measurements. It may be impossible to distinguish between a single large void, or several smaller, closely spaced cavities.

Prikryl et al. (2009) performed experiments using ERI in karst terrain and found that after testing 5 electrode array configurations, dipole sizes, and numbers of electrodes, the pole–dipole configuration increased the depth that was imaged versus dipole–dipole. The Wenner–Schlumberger configuration penetrates deeper (>30 m) relative to dipole–dipole, but with deeper penetration there is loss of signal strength and resolution. Further, they found interpretations of the data are less ambiguous with 3D versus 2D surveys, although 3D surveys take longer and are therefore more costly.

ERI has been used at a great number of sites throughout the world to determine where cave passages might lie. The technique is subject to interpretation, but generally considered as reliable as any other method available for predicting the unknown.

AUDIO-MAGNETOTELLURIC SOUNDINGS (AMT)

AMT is a technique related to EMI; instead of introducing a current, natural-source, multifrequency electromagnetic signals from lightning or atmospheric disturbances produce the current. Karst features must be at least 250 m across to be detected using AMT (Weary and Pierce, 2009). Despite this limitation, there are several characteristics in AMT's favor. The system works to a depth of about 1 km. The equipment is portable; 2 ground electrodes and 2 magnetic coils cover an area of 50 m × 50 m, which contrasts favorably with traditional ERI studies which require electrode deployment in line lengths 4–5 times traditional ERI surveys. Two people can do 6–8 soundings per day.

GROUND-PENETRATING RADAR (GPR)

Radio detection and ranging (RADAR) is the technique for sending radio frequency energy out, measuring the time for the return, and interpreting reflections as objects of interest. Originally, it was developed for locating objects, such as aircraft or ships, above ground. GPR is the application of the technique to the subsurface.

Electromagnetic waves, generally in the VHF frequencies (30–300 MHz), are used for GPR. A transmitter sends a pulse of energy into the ground. The waves are reflected at contacts between materials of different physical properties. Typically this would occur at bedding planes, but in the case of karst, it could be the wall of a cavity, whether it is air-, water-, or sediment-filled, as each situation would result in a reflection of the waves. An antenna receives the reflections and records the elapsed time from transmission.

If a velocity is assumed for the bedrock/soil/sediment, the two-way travel time of the waves can be converted to depth (Fig. 4). Errors can result from velocity changes associated with variations in moisture content and rock type. Calibrating the system with known features or similar rock types can improve accuracy.

The devices used for conducting surveys are varied. Some are pulled by autos, others can be dragged behind a person in a harness. Commonly, GPR units are configured much like a lawn mower with four wheels and an upright handle, but heavier. Other units use flexible antenna several meters long which can be dragged behind the operator. Generally a larger antenna allows deeper penetration. Most of these units can only be used on moderately level and unforested terrain. The presence of clay in the subsurface can degrade the ability to use GPR successfully.

SEISMIC MODELING TECHNIQUES

Petroleum companies use seismic studies of bedrock to image the subsurface. Conceptually, the technique is very similar to GPR, but elastic waves are produced, for example, through explosions or *Vibroseis trucks*. Geophones are implanted in the ground along arrays to measure the time of returns of the waves as they are reflected off discontinuity surfaces such as bedding planes or potentially the walls of karst features. Typically, low frequencies are used and the subsurface is imaged at depths from 2 to more than 10 km. Because any feature less than the wavelength of the seismic signal is unresolvable, this technique holds limited use for finding caves. Paleokarst horizons may be identified as chaotic zones due to their irregular nature crossing bedding planes. Geotechnical engineers use modified techniques with different wavelengths and have the potential for detecting smaller caves at depths less than a few hundred meters.

INTERFEROMETRY

Synthetic aperture radar (SAR) is a technique of imaging where a microwave beam (0.3–30 cm) is used from an overhead platform. Relative motion between the antenna and ground-based target is processed such that it produces finer spatial resolution than basic radar.

Interferometric SAR (InSAR) records Earth surface deformation by recording the difference between two or more InSAR images (Fig. 5). By using the phase of the returned waveform, this technique can measure centimeter-scale (or possibly millimeter-scale) changes in land surface height. Currently, only the European Space Agency has InSAR-capable satellites, though airborne sensors also exist.

THERMAL VARIATION

Thermal infrared imaging has been used to recognize cave entrances by the thermal contrast between the more uniform cave temperature and the ground

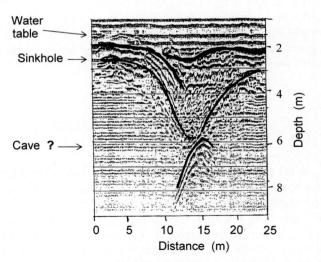

FIGURE 4 Ground-penetrating radar images of cutter and pinnacle topography at a site in central Florida.

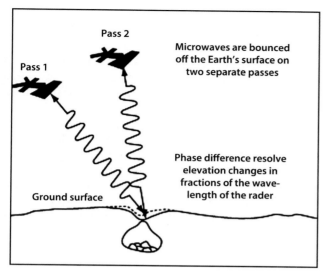

FIGURE 5 Interferometry relies on two passes of an overhead SAR sensor and the difference in phase to calculate centimeter-scale changes in land surface elevation.

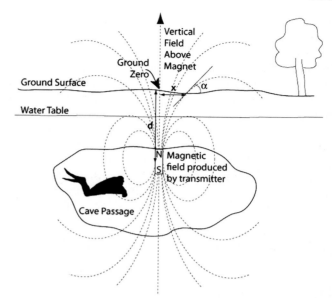

FIGURE 6 When deployed, the transmitter acts like a bar magnet and creates a magnetic field (dashed lines). Ground zero lies on the surface directly above the transmitter. The depth to the transmitter, d, may be calculated using distance from ground zero, x, and the angle of the field, α, at a distance, x, from GZ.

surface with much wider diurnal and seasonal variations (am Ende, 2009). For this technique to be used for caves without entrances, a cave would have to extend upward to a very shallow depth, probably less than a meter, for the cave air temperature to create an anomaly on the ground surface.

This technique was used by Gunn *et al.* (2004) in the United Kingdom where very old coal mines with their entrances covered by fill were environmental hazards due to potential collapse, noxious gas leaks, and acid mine drainage. Airborne thermal imaging, both daytime and predawn, resulted in the identification of known shafts and revealed sites that might represent covered mines.

CAVE RADIOLOCATION

If a cave can be entered at a remote location and traversed underground to a distant site, the location can be determined even if there is no entrance in the vicinity. For that reason, cave radio will be considered in this discussion of geophysics, although an entrance does exist somewhere. The primary purpose for cave radio is to find the exact spot on the surface of the Earth that overlies the location underground where a transmitter has been placed (Gibson, 2010). This can be used to correct errors with in-cave surveying by using GPS or a better survey on the surface. Also cave radio locations have been used to locate where a well should be drilled to reach water or to drill or dig a new entrance to a cave.

A transmitter is taken into the cave where a battery produces an alternating current in a coil of wire at a frequency between about 100 Hz and 10 kHz. Traditionally, the coil is ranged in a large loop, about 0.7 m in diameter. This setup produces an alternating magnetic field with lines of force much like those around a bar magnet (Fig. 6). The coil must be precisely leveled so that the vertical field is, in fact, directly above the transmitter. In air-filled caves, this is not generally a problem. Underwater, however, the cave floor sediment can be soft and it is difficult to deploy the transmitter horizontally. To solve this challenge, Adrian Richards and Ken Smith developed a "pinger" that is a 60 cm × 5 cm PVC tube in which a specially designed, narrow coil is placed. Underwater, the sealed tube, with some air inside, floats and naturally places itself in a vertical orientation, thus producing a magnetic field in the proper orientation.

While the current is flowing and the magnetic field is present, a person must search over the ground surface with a receiving antenna to locate "ground zero" exactly above the site where the transmitter is located. An amplifier converts the amplitude of the magnetic field into sound, and the person on the surface wears headphones while carrying the receiver. The signal can be heard within about 50 m of ground zero and with experience (and open ground), the site can be located within about five minutes. Ground zero is the spot where the receiver reads null in all directions. Depth to transmitter can be calculated mathematically when a 45° field location is found at a measured distance from ground zero.

Examples of using cave radio include Frank Reid's location of a point inside Honey Creek Cave, Texas. A shaft was drilled, successfully intersecting the cave passage at a depth of 44 m in a direct hit. Brian Pease pushed the limits of cave radio technology by doing location at Culverson Creek, West Virginia. The cave passage was approximately 200 m deep and it was deeper than Pease could accurately calculate. Later, Pease provided the technology to conduct 44 underwater cave radio locations during the Wakulla 2 expedition, most of the passages lying at a depth of 90 m.

TRANSIENT ELECTROMAGNETICS (TEM)

Using the *transient electromagnetics* technique, typically, a square-shaped transmitter loop (which may be many meters across) is placed on the ground and a steady current applied to the loop produces a primary magnetic field. When pulsed off, the decaying secondary magnetic field is measured by sensor coils on the ground surface. The voltage curve induced in the receiver coil is called the *transient*. Resistivity in the bedrock controls the diffusion and dissipation of current. Generally, a larger loop size is used for deeper targets and the useful depths for this technique are one to three loop diameters. Longer decay times come from deeper structures within the Earth. Traverses are generated by moving the system of transmitter and sensors along a series of positions.

Interpreting the data collected through TEM is complicated. The recorded transient must be normalized for many parameters such as current intensity, loop size, received gain, and comparison with modeled, uniform resistivity. The transient data may be interpreted through curve matching or computer inversion. TEM is best suited for mapping conductive targets, such as saltwater-saturated sediments.

DOWSING

Dowsing (also known as *divining* or *witching*) is a technique principally used for finding underground water, particularly underground rivers. The technique is performed by a person walking across the ground holding a forked stick or two thin metal rods bent into an L shape. When the person walks over underground water, the forked stick bends downward or the metal rods cross. No mechanism for this process is known and its veracity is highly disputed. The scientific community has no evidence the technique is repeatable and testable (Enright, 1995) and therefore dismisses dowsing as without merit.

Water is not the only matter that is considered possible to find through dowsing. Historically, minerals were dowsed for and recently explosives in war zones are considered fair game. Caves may be full of water and thus prime suspects for detecting. On the other hand, with no clear mechanism and a handful of anecdotal successes, even air-filled caves are considered by some as subject to detection through dowsing.

Believers in dowsing have anecdotal evidence from numerous successes. Detractors point out that the successes are no more significant than from random chance or from extraneous information. Despite the use of dowsing for centuries, we are no closer to understanding the technique and it remains within the realm of believers.

CONCLUSION

Cave radiolocation has been used successfully to establish surface locations above caves many times. For unknown cave and karst features, a variety of techniques have been used. Knowing when to use any particular method may require a series of tradeoffs depending on local circumstances, including what equipment is available. Electrical resistivity studies are, perhaps, the most commonly used technique to locate karst features. In some studies, multiple techniques are used, which can be useful as one technique may validate or refine the results from another.

Bibliography

am Ende, B. (2009). Discrimination between caves, overhangs, and large vugs using long-wave infrared imaging. In W. B. White (Ed.), *Proceedings of the 15th international congress of speleology* (pp. 571–579). Kerrville, TX.

Enright, J. T. (1995). Water dowsing: The *Scheunen* experiments. *Naturwissenschaften, 82*, 360–369.

Gibson, D. (2010). *Cave radiolocation*. Raleigh, NC: Lulu Enterprises.

Gunn, D., Ager, G., Marsh, S., Raines, M., Water, C., McManus, K., et al. (2004). The development of thermal imaging techniques to detect mineshafts. *British Geological Survey Internal Report.* IR/01/24.

McLean, J., & Luke, B. (2006). Electrical resistivity surveys of karst features near Fort Stanton, Lincoln County, New Mexico. In L. Land, V. W. Leuth, W. Raatz, P. Boston & D. L. Love (Eds.), *Caves and karst of southeastern New Mexico* (pp. 227–232). Socorro, NM: New Mexico Geological Society (Guidebook, 57th Field Conference).

Prikryl, J., McGinnis, R., & Green, R. (2009). Characterization of karst solutional features using high-resolution electrical resistivity surveys. In W. B. White (Ed.), *Proceedings of the 15th international congress of speleology* (pp. 597–602). Kerrville, TX.

Weary, D. J., & Pierce, H. A. (2009). Geophysical prospecting for a major spring conduit in the Ozark karst system (Missouri, USA) using audio-magnetotelluric (AMT) soundings. In W. B. White (Ed.), *Proceedings of the 15th international congress of speleology* (pp. 604–610). Kerrville, TX.

GLACIER CAVES

Jason D. Gulley* and Andrew G. Fountain†

*University of Texas Institute of Geophysics,
†Portland State University

Glaciers are some of the most rapidly evolving and dynamic karst aquifers on the planet. Glacier caves have frequently been classified as pseudokarst by many geomorphologists because glacier caves form by melting instead of dissolution. Considering that the vast majority of glacier caves are by-products of self-organizing hydrological systems, they are, at least hydrologically speaking, karst aquifers (Gulley et al., 2009). Over the past 40 years, processes of hydrological self-organization and cave formation in glaciers have become the subject of considerable scientific interest because of the discovery that hydrological processes, including cave formation, can modulate the speed of glaciers and ice sheets (including Greenland and Antarctica). Understanding these complicated links between glacier hydrology and glacier motion are important for understanding and predicting the rate of global sea level rise. In addition to ice melt, the mechanical transfer of ice as icebergs can also be influenced by hydrological processes and make important contributions to sea level rise.

INTRODUCTION TO GLACIERS

By definition, a glacier is a perennial accumulation of snow or ice that moves. Glaciers are sedimentary formations of seasonal snow that accumulates on the upper elevations of a glacier and buries the previous season's accumulation. As the annual layers become buried in subsequent years, they become compacted, increase in density, and eventually metamorphose into glacier ice (Benn and Evans, 2010). In most alpine environments within the United States, Europe, and South America, ice is relatively close to its melting temperature. Under these conditions, ice is mechanically soft, easily deformable, and readily flows downhill. Therefore, snow that accumulates at high elevations is transferred downhill as ice, to lower, warmer, elevations where it is removed by melting. Rates of ice motion and deformation are controlled by glacier thickness and slope, and can be greatly enhanced by meltwater that is either produced at or delivered to glacier beds. In subpolar and temperate regions of the globe, ice at the melting temperature allows a film of water to exist at the bottom of a glacier. Additionally, surface meltwater can penetrate to the glacier bottom where it may pressurize the glacier bed. In both instances, water decreases friction between the ice and its substrate, allowing the glacier to move faster by sliding than without water. The volume of water at glacier beds is an important factor in determining the degree of lubrication and the amount of sliding. In polar regions, where temperatures are extremely cold and little or no melting occurs, glacier motion is very slow and caused entirely by ice deformation.

Those unfamiliar with glaciers frequently assume that glaciers are very cold, freezing any water on contact. In fact, many glaciers in the world are considered "warm," or "temperate," meaning the ice is at the melting temperature. For glaciers in the western United States and Europe, for example, it is common to see streams flowing across the glacier surface before plunging into the ice interior via an opening in the ice such as a crevasse (fracture on the ice surface) or a moulin (melt-enlarged vertical shaft in a glacier). Under these conditions water coexists with ice and the two phases are close to thermodynamic equilibrium and even small additions of heat, such as the flux of geothermal heat from the Earth, can melt the ice at the bottom of glaciers. Temperate glaciers exist in regions where the average annual temperatures are not much below 0°C. These glaciers exist in the temperate and tropical zones of the Earth, including Europe, North America (exclusive of the Canadian Arctic and the Brooks Range, Alaska), and the glaciers of Africa, South America, New Zealand, New Guinea, and most in central Asia including the Hindu Kush-Himalaya region.

In areas where average annual temperatures are well below freezing, the temperature of glacier ice is also below freezing for much of the ice thickness. These glaciers are restricted to the polar regions (e.g., Antarctica) or high altitudes (e.g., Alaskan peaks). If the entire thickness of the glacier is below freezing, the glacier is frozen to its bed. These glaciers are called *polar* or *cold-based glaciers* and they flow more slowly than their temperate counterparts because all motion is accomplished by internal ice deformation. Although the surface may melt for a short period in summer, water is found only in or under the ice in restricted circumstances (see "Subglacial Caves" section, below). In subpolar conditions, such as Arctic Sweden, polythermal glaciers may form where the upper portion of the ice is below freezing, but ice near the bed is warm. Another condition exists for the two ice sheets (Greenland and Antarctica) whereby the entire thickness of the ice is below freezing except at the very bottom where water may be present. This is partly due to the insulation against the polar cold provided by the great thickness of the ice, geothermal heat coming from the Earth's interior, and heat generated by ice deformation. Also, the melting temperature of the ice is lower than 0°C due to the great pressure of the overlying mass.

PROCESSES OF GLACIER CAVE FORMATION

Caves are common to all glaciers that experience melting. Caves that are entirely enclosed in the main body of the glacier are classified as englacial caves, whereas caves at the glacier bed are subglacial caves. Regardless of where they are located, caves primarily form in glaciers by melting. While many of the most well-known glacier caves have formed in glaciers on the flanks of volcanoes as a result of geothermal heat and steam vents, most glacier caves are formed by a less obvious source of heat—friction. All flowing water generates heat through friction, but the rate of heat production is quite small relative to other heat sources, such as sunlight or air temperature, and self-heating is typically ignored in most hydrological applications. However, for water flowing in a glacier, there are no other sources of heat and because of the close thermodynamic equilibrium between ice and water, any frictional heat produced is consumed by melting ice. Consequently, passageways (caves) in ice enlarge in summer when surface melt is the greatest and the water discharge in the conduits is the largest. Opposing this enlargement is ice creep, acting under the weight of ice above the passage, which squeezes the passage closed. As long as the water can melt the walls, the passage remains open. Larger passageways drain water from smaller passageways and a network, much like a dendritic cave system, forms in and under temperate glaciers. In winter, the surface of the glacier is typically covered with a cold blanket of snow and the network is starved of water. The rate of ice creep dominates melt enlargement and the passageways close. Near the glacier surface, passageways close more slowly than those in deep ice because the ice pressure is reduced close to the surface.

The processes of cave formation in glaciers share many similarities with that in limestone. Caves form along hydrologically efficient flow paths, networks of fractures, or bedding planes that connect a water source to a discharge point (Gulley et al., 2009). In glaciers, a highly permeable feature may be a crevasse. Ultimately, the morphology of caves in glaciers is determined by the type of glacier (polar or temperate), the crevasses that are present, and the magnitude of meltwater discharge. Generally, increasing the number of crevasses on a glacier reduces the meltwater discharge in any one crevasse, thus reducing its capacity for possible enlargement. In the following section we discuss the broad categories of englacial caves that form in different glaciers and then discuss subglacial caves.

ENGLACIAL CAVES

The formation of englacial caves typically requires a network of fractures to melt-enlarge into caves. Fractures may include surface crevasses, thrust faults, or relict crevasses/faults that become separated from their zone of formation due to ice flow (Gulley et al., 2009). Surface crevasses drain meltwater and rain from the glacier surface into the glacier interior. Where a surface stream drains into a crevasse, a moulin forms. The crevasse may close due to the ice advecting into a region of compression, but as long as the stream flow is maintained, the moulin will survive. In thin regions of a glacier the moulin may reach the glacier bottom. French speleologists and glaciologists have made the deepest known descent of any moulin, reaching a depth of 205 m below the ice surface of the Greenland Ice Sheet (Badino, 2007).

Thrust faults can also be exploited by water. Such faulting occurs in glacier termini when subjected to extreme compression. A glacier thins toward the terminus, causing the ice to slow, and the thicker, consequently faster, ice upstream continues to push, causing compression. When the compression is sufficiently large, the ice fractures such that the upper ice moves over the deeper ice. Low wide caves can form along these faults as pressurized subglacial water discharges along the thrust plane. Caves have been mapped along thrust faults in the polythermal Aldegonda Glacier, Svalbard, Norway and the Matanuska Glacier, Alaska, U.S.A. (Fig. 1).

On glaciers where crevasses may be absent from large areas of melting ice, caves may form from the incision of surface streams. The streams initially form slot canyons. Provided the rate of stream incision

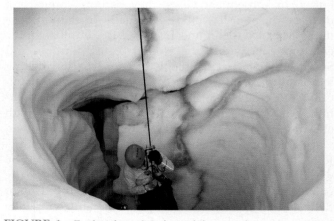

FIGURE 1 Englacial conduit formed from surface water exploiting a thrust fault, Matanuska Glacier, Alaska, U.S.A. *Photo by Jason Gulley.*

FIGURE 2 A shallow cut-and-closure cave in the debris-covered Khumbu Glacier, Khumbu Himal, Nepal. This portion of the passage is beneath approximately 20 m of ice. *Photo by Jason Gulley.*

FIGURE 3 A prominent canyon suture formed from ice pressure squeezing the canyon walls shut is evident in the passage ceiling. This portion of the passage is beneath about 15 m of ice. Longyear Glacier, Svalbard, Norway. *Photo by Jason Gulley.*

exceeds that of overall surface melt, closure of the upper parts of canyon walls by ice creep forms caves (Fig. 2). These types of caves are *cut-and-closure caves* and the environmental requirements for their formation generally restrict them to polar or polythermal glaciers or debris-covered glaciers. Polar glaciers have slow rates of ice motion and have few, if any, crevasses. Surface streams have large catchment areas and have high discharges per unit area where they are funneled through topographic depressions or along the margins of glaciers. Because stream discharge, and hence melt, is concentrated, streams can incise into the ice much more rapidly than the surface is lowered by melting. On debris-covered glaciers, surface debris insulates the underlying ice and melt is concentrated in stream channels. Creep closure of the canyon walls isolates the stream from the glacier surface and forms a cave. The effects of creep closure are more obvious at increasing ice depths and create a pronounced suture in the ceiling running parallel to the passage (Fig. 3). The depths of cut-and-closure caves are ultimately limited by the rate of creep closure relative to the rate of enlargement. If a cave incises to an ice depth that is capable of closing a conduit by creep over winter, the next season's meltwater will back up in the cave and discharge through any available holes in the cave roof. If no such holes are available, the cave will fill completely with water and freeze shut and a new cave will begin forming nearby. Cut-and-closure caves include the longest mapped englacial cave—in the Scott Turner Glacier, in Svalbard, Norway, >2 km long with a total elevation change >100 m. This cave is shallow, never more than a few tens of meters below the ice surface, but has incised to the glacier bed to create a subglacial passageway near the glacier terminus.

On stagnant ice, that which is no longer moving, glacier karst can readily develop because the deformation imposed by ice movement is absent and increases the lifetime of englacial and subglacial passageways. Sinking streams, closed depressions, and lakes that drain rapidly are common features of stagnant ice. On debris-covered glaciers, such as those in the

FIGURE 4 Phreatic-vadose transition in a cave formed by lake drainage in the Ngozumpa Glacier, Khumbu Himal, Nepal. This portion of the cave is beneath approximately 20 m of ice. *Photo by Jason Gulley.*

FIGURE 5 An N-channel incised into frozen till beneath the thin, cold-based Rieper Glacier, Svalbard, Norway. This portion of the cave is beneath 10–15 m of ice. *Photo by Jason Gulley.*

Himalayas, lakes may form in topographic depressions on the glacier surface and drain through caves. These caves are often nearly horizontal and form along permeable debris-filled structures in the ice that connect lake basins. Often these caves exhibit a *keyhole* cross-section (Fig. 4), indicative of initial water-filled conditions that melt a circular cross-section. Water empties from the lake and the cave no longer flows at roof-full condition. Water in the partly full cave no longer melts the roof and only melts the side and bottom, incising the cave downward until the lake empties.

Water-filled surface crevasses may penetrate deep into a glacier by hydrofracturing due to the density difference between water and ice. A water-filled crevasse exerts an outward pressure on the walls that is greater than the pressure exerted by the ice. Hydrofracturing forms deep, fan-shaped cracks that may later be altered into a more circular morphology by concentrating water flow in along one path and creep closure of marginal areas of the fracture exposed to less water flow. Exploration of hydrofractures in ice that is less than 150 m thick has led to the discovery of subglacial cave systems in several glaciers in Svalbard, Norway.

SUBGLACIAL CAVES

Subglacial caves form where englacial passageways reach the glacier bed or where surface stream or lakes sink along the ice-bed interface at ice margins. The common form of subglacial cave is one that is incised upward into the ice. These form like englacial caves through the frictional heating of the water that, in this case, melts the ice roof. Less common are those that are cut into the bedrock with an ice roof (Fig. 5).

Because much of the extent of subglacial caves is difficult to access physically, much of what is known has been inferred from indirect measurements such as dye tracing, geophysical measurements, and geochemistry. Dye tracing has shown that the spatial pattern of subglacial caves is dendritic. Meltwater sinks into the glacier surface at many points, primarily through crevasses or moulins, yet discharges from the glacier at a few discrete exits. Locally, however, caves may be anastomotic. Subglacial caves near the terminus, along the ice margin, and in rare instances where subglacial caves are accessible via moulins have been observed to form in the same locations, with very similar morphologies, each year.

Subglacial caves can form in polar glaciers that are frozen to their beds from cut-and-closure caves that have incised to the glacier bed. Under thin ice, subglacial caves formed by cut-and-closure caves that reach the bed are frequently incised in underlying till or soluble/erodible rock. Subglacial caves can also form from hot gases that escape from volcanoes covered by glaciers. Caves formed by escaping volcanic gases are concentrated around fumeroles where gas escapes and passageways are organized to efficiently discharge these escaping gases. Such caves are common to glaciated volcanic peaks, such as the Cascades (United States) and Antarctica, and to volcanic terrain in Iceland.

Ice margin caves are commonly seen by even the most casual tourist. The location where the glacial stream exits the glacier is an ice marginal cave. Ice marginal caves also form where streams from valley walls enter glaciers or where streams from the surface of the glacier sink along the contact of the ice and valley wall. In both cases, they provide some access to the glacier interior that is frequently limited by flooding or ice thickness.

EXPLORATION OF GLACIER CAVES

Glacier caves are typically entered soon after surface melting has ceased sufficiently to permit safe exploration. However, once melt-enlargement ceases, the caves begin to close rapidly such that for ice depths greater than 150 m the caves are frequently closed by ice creep before they become sufficiently dry. Caves at depths between 80 and 150 m require exploration as soon as surface melting has stopped.

There are many hidden hazards in glacier caves. The most obvious glacier cave entrances are often where glacier streams discharge from glacier fronts; however, these entrances are often wide and covered by thin ice, making them prone to frequent collapse. Rainfall or an increase in surface temperatures can cause glacier caves to flood rapidly and catastrophically. In cold glaciers, refreezing of the surface of these floodwaters can create dangerous false floors when the water level recedes again. Special caution should be used when exploring volcanic glacier caves because they can trap deadly gases and have been implicated in a number of deaths in ice caves. Glacier caves combine all of the hazards of glacier travel with all of the hazards of alpine caving. When approached with experience and a healthy amount of common sense, glacier cave exploration can be fairly safe.

Bibliography

Badino, G. (2007). *Caves of sky: A journey into the heart of glaciers.* : Graffice Tintoretto (TV).

Benn, D. I., & Evans, D. J. (2010). *Glaciers and glaciation.* New York: Oxford University Press.

Gulley, J., Benn, D., Screaton, E., & Martin, J. (2009). Mechanisms of englacial conduit formation and their implications for subglacial recharge. *Quaternary Science Reviews, 28,* 1984–1999.

GUANO COMMUNITIES

Pedro Gnaspini

University of São Paulo, Brazil

GUANO AND ITS IMPORTANCE TO CAVE COMMUNITIES AND TO SCIENCE

Guano is the accumulation of feces of particular animals in time and space. A commonly known example is seabird guano, which is the accumulation of droppings of large colonies of seabirds. This has been exploited for many years as a source of nitrogen compounds. In caves, guano also occurs, and it comes from three main different animal sources: birds, bats, and crickets. It should be stressed that the animals that add guano into the caves are always recognized as trogloxenic animals. Bat guano has also been exploited in the past, especially for the production of gunpowder.

The cave ecosystem lacks primary producers and completely depends on energy that comes from the outside; hence, it is a food-poor environment. Actually, the food supply in caves tends to be low or sporadic in quantity and poor or variable in quality. It is also clear that any accumulation of food inside caves (as is a guano pile) is of great attraction to detritivorous scavenger animals. Other common examples of energy sources in caves include dead animals and debris wasted in from the surface (by water, gravity, *etc.*).

There are some caves where the accumulation of guano is so large that the caves are considered to be food-rich places, the so-called eutrophic caves. These caves are more common in tropical countries but they occur in all continents. For example, many caves in Mexico and Brazil are also eutrophic, although in these countries there are many more caves that are food-poor (poecilotrophic). In some of these food-poor caves bat guano deposits are scattered and do not constitute the main or only base of the trophic chain. Nevertheless, it is a very important food source. Eutrophic caves tend to occur where there is an enlarged chance of bat aggregation due to the biology of the bats or to physical characteristics of the cave areas. For instance, in Brazil, limestone caves tend not to be eutrophic because they are generally larger and occur in large number in particular areas, offering more shelters for bats. In turn, sandstone caves are smaller and occur in smaller numbers leading to the concentration of bats in the same shelters, and, therefore, to the concentration of bat guano. Nevertheless, Poulson and Lavoie (2000) also considered that bat and cricket guano provide the most important biotic input of energy to hypogean environments in temperate zones.

Scattered feces (even of other vertebrates, such as cats, pumas, dingoes, foxes, otters, rats, agoutis, and didelphids) may also be used as food by cave animals. In some cases it may be preferred because such animals would avoid the high predation risk that they would be subjected to in guano deposits, where large communities may be established. An interesting example of association to feces of other vertebrates was provided by Calder and Bleakney (1965), who noticed that in Nova Scotia bats are not important, and the cave fauna depends mainly on porcupine dung.

In addition to the availability of food, bats may have a large influence on the cave climate. In many caves with a higher concentration of bats, especially considering large chambers connected to other points of the cave by narrow/low conduits, temperature tends to be very high, reaching (and sometimes surpassing) 40°C. Concentration of ammonia is also very high, making these places very difficult for human exploration. These are considered *bat caves*.

Because guano communities are somewhat isolated and, in general, present a simple structure, they are an important field for ecological studies. Moreover, the physical and nutritional properties of the guano vary according to the kinds of animals producing the piles; namely, insects, birds, or bats (in turn with several kinds of feeding preferences) providing an additional field for comparison of community structures.

Guano has been studied in different places and with different foci, including mineral and chemical analysis of bat guano, mathematical simulation of community dynamics, and extensive faunistic surveys. Guano communities have been the subject for ecological studies in several caves around the world, but more commonly in tropical than in temperate caves, contrary to what would be expected from the higher development of temperate speleobiology. Theoretical definitions about guano inhabitants are based mainly on papers by Decu and colleagues (focusing on Eastern European caves) and Poulson and colleagues (relative to North American caves). Both of these groups started their research on guano fauna in the 1960s while Gnaspini and colleagues (focusing on South American caves) started their research in the 1980s. Their ideas are summarized in papers by Poulson (1972), Decu (1986), and Gnaspini (1992), and a general review can be found in Gnaspini and Trajano (2000).

TYPES OF GUANO

Guano in caves comes from three main different animal sources: birds, bats, and crickets. Guano of rhaphidophorid crickets represents a large amount of food input for cave animals in temperate caves, especially those in the United States. Some trogloxenic crickets (*Hadenoecus subterraneus*) leave the caves for food periodically, and when they come back they deposit a large amount of feces based mainly on vegetal matter. Cricket guano may also be found in Asian caves, but, in this case, it should be considered a by-product of bat guano because these crickets feed mainly inside caves on the guano of other animals (Deharveng and Bedos, 2000).

Several birds, such as doves, owls, parakeets, swallows, swiftlets, and vultures, may nest near cave entrances, and their droppings may be used as a food supply in the twilight zone. In some tropical caves, however, bird droppings such as those from oilbirds (guacharos, *Steatornis caripensis*; northern South America and Caribbean) and swiftlets (*Aerodramus* sp.; tropical oriental caves) may be an important food source. They colonize the dark zones, and their droppings may form large piles. Swiftlets feed on insects and are common in Southeast Asia where their nests are exploited for a traditional dish in human cuisine. Oilbirds are common in northern South American caves, occurring in northern Bolivia, Colombia, Guyana, Peru, Venezuela, and on the island of Trinidad. They are known to navigate based on sound, like bats, but their clicks can be heard by humans, unlike those of bats. Therefore, they can be found deep inside caves. They feed mainly on palm fruits, which they carry inside the caves. Consequently their guano is comprised of both rejected seeds (with some pulp adhering) and droppings.

The most important source of animal droppings in caves throughout the world, however, is bat guano. In the northern temperate regions and other parts of the old world, most cave-dwelling bats have an insectivorous diet. This type of guano is generally finely granulated and consists of small pieces of insect cuticle, sometimes with some material still adherent.

In some parts of the old world, for instance, in African caves, frugivorous bats also deposit guano. This kind of guano is generally granulated and consists of seeds and decaying plant matter (Fig. 1). It is similar to oilbird guano regarding physical and nutritional characteristics. It is in the Neotropics, however, that the feeding habits of bats (especially within the Phyllostomidae) are very diversified. Frugivorous and insectivorous bats also occur. The most common in inhabited areas may be the vampire, or hematophagous, bats (especially *Desmodus rotundus*). Owing to the high availability of prey in the form of domestic animals, and roosts not being a limiting factor in cave areas, the populations of these bats are very high

FIGURE 1 Close-up of a frugivorous bat guano pile, showing a *Maxchernes* pseudoscorpion and the large number of seeds with adhered digested material. *From Gnaspini and Trajano, 2000.*

FIGURE 2 Overview of a somewhat old and dry hematophagous bat guano pile.

FIGURE 3 Close-up of a carnivorous bat guano pile, showing a large concentration of cholevine larvae (at least two species of the genus *Dissochaetus*). *From Gnaspini and Trajano, 2000.*

(Trajano, 1985). Their guano is a very dark and smelly paste (Fig. 2). Although they form small colonies (up to ten individuals), large carnivorous bats may also deposit large amounts of guano. These animals feed on birds and small mammals and occasionally on large insects, and their guano is formed of large, somewhat compact droppings full of digested animal material mixed with feathers and/or hairs (Fig. 3). Although nectarivorous bats (Glossophaginae) are relatively frequent in caves, guano piles from these bats have not been identified thus far, probably because the colonies are in general small and, on account of its liquid nature, the guano would mix with the substrate.

Sometimes, in the same cave one can find more than one kind of bat guano pile with some maybe even mixed with each other. The texture and nutritional characteristics differ among the various types of bat guano and these different characteristics will influence the associated communities.

Other important characteristics influencing the establishment of associated fauna are the substrate and the place where guano is deposited, because they would influence the availability of the organic matter (see Poulson and Lavoie, 2000). For instance, if guano is deposited on sandy substrates and/or under dripping water, nutrients may be washed away more rapidly.

THE FAUNA ASSOCIATED WITH GUANO DEPOSITS—DEFINITIONS

Considering that guano is an important source of food in caves, there are many animals that directly feed on this substrate and/or on microorganisms (*e.g.*, bacteria and fungi) that grow on guano. These animals are called *guanophages*. To complete the guano communities, there are also several predators that feed on the guanophages. Deharveng and Bedos (2000) also recognize two types of associated communities: the giant arthropod community (GAC) and the meso- and microinvertebrate community (MIC). The latter is always on or close to guano deposits and formed by large numbers of species of guanophages and their predators, and the former consists mainly on one species of guanophages and mostly on predators, which may live far from the guano piles and come closer to feed.

Due to the specificity of these communities to the guano, some authors (see, *e.g.*, Decu, 1986) have argued that these animals should not be considered cavernicoles, because they occur in caves just after this specific source of food, and not because of the cave environment. Therefore, these would not depend on caves from an ecological point of view. Because this would mean that (1) guano inhabitants colonize caves only because of guano, and would not be there if that food supply was missing, and (2) they do not depend on the cave ecosystem nor are they subjected to cave environmental constraints, Gnaspini (1992) discussed that, in order to reach and colonize the guano piles frequently deposited far from known cave entrances, organisms should be able to orient themselves and also survive inside the cave. In other words, they must have an ecological relationship with the caves and should be considered cavernicoles. In addition, if true, the same argument to exclude guano dwellers should have been used to exclude other communities associated with specific sources of food in caves. This would be the case, for instance, in the animals that feed specifically on roots in caves, which are an important part of the troglobitic fauna of Hawaii. Therefore, guano piles should simply be treated as a substrate, and, as any substrate inside a cave, it would be colonized by cavernicoles.

Because guano is an important source of food, and, sometimes, the only source of specific types of food (such as digested blood or vegetal matter, especially fruits and seeds), there are some cave-dwelling animals that are restricted to guano piles, and some that occasionally visit this substrate to feed directly on guano or to prey on guano dwellers. Therefore, following the same rationale of "affinity to the environment" used to classify cave animals into troglobites, troglophiles, and trogloxenes, cavernicolous guano dwellers can be classified as guanobites, guanophiles, and guanoxenes. It should be reinforced that this classification should be used only for animals inhabiting guano deposits inside caves.

Guanobites are organisms that, when in caves, exclusively inhabit guano deposits, and whose entire biological cycle takes place in this substrate. It should be noted that guano deposits are discontinuous in caves, and hence guanobites may be found on any cave substrate as they move through the cave environment to colonize guano piles. Guanobites, however, do not reproduce or feed in these substrates. Guanophiles may inhabit and reproduce both in guano piles and in other substrates of the cave environment. Guanoxenes may be found feeding and/or reproducing on guano deposits but depend on other substrate(s) in the caves to complete their biological cycle. Although theoretically possible, this third type seems to be unlikely in practice.

It should be noted that the relationships with the cave and with the guano are not necessarily interdependent. The degree of fidelity to, or dependence on, the cave environment is neither influenced by nor influences the relationships of the cavernicoles to the guano (i.e., their degree of fidelity to the guano environment). For example, an animal restricted, in caves, to guano piles (thus, a guanobite) may occur in the epigean environment as well, in this case not being a troglobite, as expected if the two classifications were correlated. Therefore, the guano inhabitants should be classified independently according to their cave and guano relationships. When analyzing the cave relationships of an organism, one must take into account its possible occurrence in the epigean environment, as well as the possible modifications acquired during colonization and isolation in the hypogean environment, in order to classify it as a trogloxene, a troglophile, or a troglobite. Independently, one may also consider whether or not an organism inhabits substrates other than guano inside the cave.

The entire biological cycle of a guanobite takes place in the guano piles inside caves, and, therefore, entirely in the cave environment. Thus, a guanobite may be either a troglobite or a troglophile, because both can complete their entire cycle inside caves. It can never be classified as a trogloxene, because this group must leave the caves (and therefore any substrate inside the caves) to complete its life cycle. This applies also to guanophiles, because at least the populations observed in the guano piles complete their life cycles inside the cave. Finally, guanoxenes can be trogloxenes, troglophiles, or troglobites. It is noteworthy that a classification combining troglo- and guano-related classifications is very useful because it gives a precise notion of the ecological-evolutionary relationships of the organisms studied.

Troglomorphisms generally are adaptive responses to the general low predictability and availability of food in the subterranean environment. Therefore, it would not be expected to find troglomorphisms in species associated with guano piles (especially not among guanobites), because these animals would be living under a large food availability. However, some cases seem to occur.

COMMUNITY STRUCTURE AND SUCCESSION IN BAT-GUANO PILES

Like all communities associated with the cave environment, guano communities are simple and composed of only a few trophic levels. Two levels can be recognized: the guanophages and their predators. In turn, the relationships between the animal components are very complex, because most predators would feed on any of the items in the guanophage level, and maybe even within the predator level. Therefore, the trophic relationship within this community should be viewed as a food web instead of a food chain.

In temperate regions, the seasonal activity of cave bats leads to seasonal changes in the structure of guano piles and, consequently, in the associated fauna (Poulson, 1972; Decu, 1986; Poulson and Lavoie, 2000). In general, these bats are active during the summer and hibernate (in the same or in other caves) in winter. When bats are present and active, fresh guano accumulates continuously. As bats reduce their activity or leave the sites, there is a progressive desiccation of the deposits, promoting chemical and microclimatic changes. The microclimatic properties of the guano depend on the depth and the distance from the center of the deposit. Because several taxa show microclimatic preferences, a faunistic succession is observed in the guano deposits. Guano piles are initially alkaline owing to the concentration of ammonia in the feces. They subsequently become acid as a result of the decomposition of alkaline compounds and fermentation of the piles.

On the other hand, Decu (1986), based on the literature and his own observations in Cuba, stated that bats are permanently active in tropical caves. As the guano input is continuous, there would be no seasonal differences in the structure and composition of the guano deposits and the associated fauna. In both cases guano deposition is predictable, but, whereas in temperate bat caves deposition is seasonal, in tropical bat caves such deposition is continuous. This led Decu and Tufescu (1976) to characterize caves as belonging to a temperate type when there is seasonal succession, and to a tropical type when there is not, irrespective of their geographic location. However, Poulson and Lavoie (2000) consider that the predictability of bat guano availability in temperate caves is much smaller than in tropical caves.

In the Brazilian karstic areas so far studied, continuous guano deposition usually does not occur because roost availability is high and the bats form itinerant colonies, frequently moving between and inside caves. Therefore, guano piles show temporal changes in these caves, passing through the same phases of deposition, and subsequent desiccation after the bat colony leaves the sites. In contrast to temperate caves, succession in Brazilian caves is not necessarily (if at all) seasonal. In fact, the temporal and spatial deposition of guano may be highly unpredictable. Neotropical bats show regular activity throughout the year, but can leave a specific roost at any time, causing the interruption of guano deposition and consequent succession. Therefore, from whichever cave region, piles of bat guano are "qualitatively and microclimatically nonuniform ... [and] ... look like a mosaic of microhabitats that shelter a mosaic of zoological communities in various evolutional stages" (Decu, 1986).

THE ASSOCIATED FAUNA—EXAMPLES

As expected, a large variety of animals inhabit guano piles. A huge list of papers citing communities associated with bat guano can be found in the literature, but it is not the goal of this chapter to present an extensive review of the available data. For a review and discussion of guano dwellers in tropical caves, please refer to Gnaspini and Trajano (2000) and Deharveng and Bedos (2000). For temperate caves, see, for example, Calder and Bleakney (1965), Negrea and Negrea (1971), Decu and Tufescu (1976), and the review by Poulson and Lavoie (2000). When a broad overview on the guano communities is made, we could recognize that some taxa occur almost everywhere in the world, constituting the basis of most guano communities. This is especially the case, of course, when the family level is analyzed: generic and specific levels would give more differences and would allow a better biogeographic comparison. These are the "basic" taxa (*sensu*, *e.g.*, Negrea and Negrea, 1971). In addition, there are some less common but still frequent groups, which could be called "accessory" taxa (also *sensu* Negrea and Negrea, 1971). Of course, faunal regionalism would also interfere with the taxa observed in the different places. For instance, although crickets are commonly associated with guano, rhaphidophorid crickets occur in North American, European, and Australian caves, whereas phalangopsid crickets occur in African and South American caves.

Good examples of basic taxa are laelapid, macrochelid, oribatid, and uropodid mites; chernetid pseudoscorpions; entomobriid, hypogastrurid, and isotomid springtails; tineid moths; carabid, cholevine, histerid, and staphylinid beetles; and drosophilid, fanniid, milichiid, and phorid flies. Examples of accessory taxa are planarians; earthworms; other pseudoscorpion families (Cheliferidae, Ideoroncidae, Neobisiidae); harvestmen; spiders; schizomids; isopods; millipedes; chilopods; psocopterans; cockroaches; crickets; several other beetle families (*e.g.*, Aderidae, Curculionidae, Dermestidae, Elateridae, Hydrophilidae, Nitidulidae, Ptiliidae, Tenebrionidae, Trogidae); and several other dipteran families (*e.g.*, Ceratopogonidae, Dolichopodidae, Empididae, Muscidae, Sciaridae, Sphaeroceridae, Stratiomyiidae).

As expected, mites (Acari) are by far the most diversified group of guano inhabitants. Guano-inhabiting mites include guanophages and predators. Argasid ticks may occasionally be found on guano, but they should not be considered to be true guano-inhabitants because they drop from their hosts (bats) when well fed, and stay there for a while before starting to look for new hosts. Springtails (Collembola) are, together with mites, the most speciose invertebrates in guano piles, and may form extremely large populations of guanophages. The Brazilian species *Acherontides eleonorae* (Fig. 4) is a good

FIGURE 4 A large concentration of *Acherontides eleonorae* collembolans, with some scattered mites. Close-up of the region pointed out in Fig. 2 where the white spots indicate the presence of these animals.

example of a guanobite and is noteworthy in view of the extremely large populations (hundreds of individuals cm^{-2}) found on old, dry piles of hematophagous bat guano from São Paulo caves (therefore concentrating on the borders of the guano piles through succession), and because they survive and breed when kept in closed vials with this substrate.

After Collembola and Acari, larvae of tineid Lepidoptera are among the most ubiquitous guanophages in many caves where they make their cases. Pyralid moths were also frequently recorded. Beetles (Coleoptera) of the families Dermestidae, Histeridae, Hydrophilidae, Scarabaeidae, Staphylinidae (including pselaphines), and Tenebrionidae are occasionally found associated with guano piles, sometimes in large populations. However, cholevine leiodids may be considered to be the commonest guano-inhabiting beetles (Fig. 3). Aderid beetles were frequently found on swiftlet guano in Southeast Asia, where they form guanobitic populations (Deharveng and Bedos, 2000).

The most important guanophage flies (Diptera) belong to the families Drosophilidae, Fanniidae, Milichiidae, Muscidae, and Phoridae, many of which are probably guanobites. In Brazil, they show a preference for piles of hematophagous guano; faniids and muscids are occasionally found associated with guano from carnivorous bats in Southern Brazilian caves (Gnaspini and Trajano, 2000). A striking example of a troglobite-guanobite is the African mormotomyiid *Mormotomyia hirsuta*, which is a highly modified species whose larvae live on guano.

In tropical America, bugs (Heteroptera) of the families Cydnidae and Lygaeidae are associated exclusively with guano from frugivorous bats or from oilbirds (due to the similar physical and nutritional characteristics), where they may establish very large breeding populations (see Gnaspini and Trajano, 2000). Lygaeids were also recorded in frugivorous bat guano from Asia and New Guinea. In caves these insects are restricted to guano and thus should be considered guanobites. Representatives of these families are generally plant eaters (and some are detritivores). This may explain why the cave species occur only in guano from frugivorous animals.

Cockroaches (Blattaria) are typical tropical cave-dwellers. These highly opportunistic insects are, as expected, found in several substrates, including different types of guano, and may establish huge populations. For instance, Braack (1989) recorded several species in a single cave, with the population of one species reaching 75,000 individuals and of two other species reaching, each, 40,000–50,000 individuals. Crickets (Ensifera) are omnivorous cavernicoles observed throughout the world, and may concentrate near and on guano piles. Isopods are also commonly recorded on guano, sometimes forming large populations. In Brazil, for instance, some groups show preference to particular kinds of guano: philosciids in guano of frugivorous bats from São Paulo caves, scleropactids in guano of insectivorous bats from one Pará cave, and armadillids in guano of carnivorous bats from Mato Grosso do Sul (Gnaspini and Trajano, 2000). Diplopods may also be found associated with guano, and some species may be considered troglobite-guanobites, because they are troglomorphic.

In addition, other higher taxa were occasionally collected in or near guano piles. Earthworms (Oligochaeta) may use guano as an extension of their endogean habitat; they need a moist and nutrient-rich substrate, such as that found in guano from hematophagous and frugivorous species (where large populations of enchytraeid worms were found in Brazil; Gnaspini and Trajano, 2000). Psocopterans feed on fungi and may be found in guano. Subulinid gastropods were recorded in some tropical American caves (see Gnaspini and Trajano, 2000), and may form guanobitic populations in Asia (Chapman, 1984). Harvestmen (Opiliones) may also be found near the piles and are observed occasionally feeding on guano.

Among the predators, pseudoscorpions may be found in any kind of guano, preying on the abundant springtails and mites. A few cases of large populations were recorded, for example, the Brazilian chernetid *Maxchernes iporangae* (Fig. 1), which seems to be restricted to a single cave where it exclusively inhabits frugivorous bat guano piles (see Gnaspini and Trajano, 2000). Spiders (Araneae) such as Pholcidae and *Loxosceles* spp. (Sicariidae) sometimes make their webs over or near guano piles, opportunistically preying on guanophages. Barychelid trapdoor spiders were observed associated to swiftlet guano in Asia (Chapman, 1984). Whip spiders (Amblypygi) were frequently recorded in guano from swiftlets in Southeast Asia (Deharveng and Bedos, 2000). Guanobitic geophilomorph chilopods were recorded in Asia (Chapman, 1984). Reduviid bugs (Heteroptera) are not generally directly associated to guano but occasionally feed on guanophages.

It is important to stress that aquatic communities may also use guano as food. For instance, veliid bugs were recorded on floating guano particles from Namibia; aquatic guanophages that feed on hematophagous bat guano seem to be the main food item of a troglobitic catfish from Northeastern Brazil; guano may be nutritionally important for troglobitic salamanders (Fenolio et al., 2006); and even whole aquatic communities may depend on bat guano.

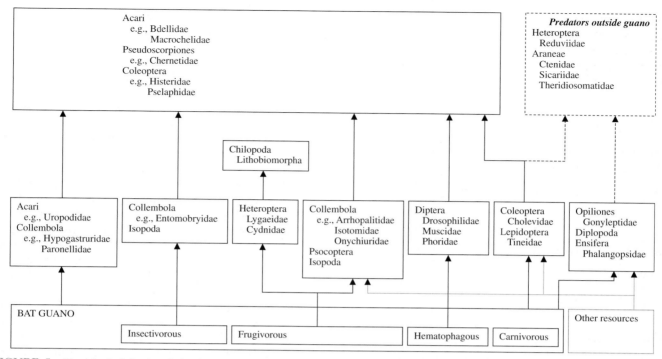

FIGURE 5 Hypothetical food web for bat-guano communities from São Paulo caves, Southeastern Brazil. Arrows indicate direction of energy flow.

THE FOOD WEB—EXAMPLES

As mentioned before, animals inhabiting guano piles may be guanophages and their predators and may also inhabit other substrates. Figure 5 shows an example of a Brazilian food web, where a large variety of types of guano may be found. Note that some taxa may be found in any kind of guano (*e.g.*, tineid larvae), whereas some are restricted to one kind (*e.g.*, flies in guano from hematophagous bats, and heteropterans and lithobiomorphans in guano from frugivorous species).

CONSERVATION

Guano mining causes an impact on the bat population, which can move away from the mining site, resulting in the interruption of guano deposition. This, in turn, would lead to the abandonment and/or death of guano inhabitants, especially when guanobites are considered. This also includes aquatic environments. It is also clear that any important change in the epigean environment may lead to the death and/or abandonment of bats (and/or birds and/or crickets), and, consequently, the guano inhabitants. Again, guanobites are important targets, and guanobites-troglobites may even be extinguished. For instance, logging or forest cutting may locally extinguish frugivorous bats and birds, interrupting the deposition of guano inside the local caves. One important example is the popular use of poisonous pastes against hematophagous bats in Neotropical countries because of their menace to cattle. The paste is applied on the fur of the bats, which, due to their social behavior, lick each other and die in large groups. In the following days and months there is a large increase in energy input due to the large amount of cadavers deposited in the caves. However, afterward, energy input drastically decreases. Pollution from metals may also represent a threat to bats.

CONCLUSIONS

In summary, guano is an important source of food in caves all around the world. There are some basic taxa that are almost always present in guano communities in caves, and accessory taxa, which are somewhat less common but still frequent groups. In addition, the composition of guano communities shows some differences related to the kind of guano (*e.g.*, bird vs. bat guano, guano of frugivorous vs.

insectivorous vs. hematophagous bats, *etc.*). Although there are large number of papers about guano communities in caves, there is still much research to be done and many important findings have been occurring in the past decades.

Acknowledgments

Sonia Hoenen and David Culver provided helpful criticisms to the manuscript. The author has a research fellowship from CNPq (Conselho Nacional de Desenvolvimento Científico e Tecnológico). Figures 1 and 3 were reprinted/modified from Gnaspini and Trajano (2000).

See Also the Following Articles

Food Sources
Bats
Diversity Patterns in the Tropics

Bibliography

Braack, L. E. O. (1989). Arthropod inhabitants of a tropical cave "island" environment provisioned by bats. *Biological Conservation, 48*, 77–84.

Calder, D. R., & Bleakney, J. S. (1965). Microarthropod ecology of a porcupine-inhabited cave in Nova Scotia. *Ecology, 46*, 895–899.

Chapman, P. (1984). The invertebrate fauna of the caves of Gunung Mulu National Park. *Sarawak Museum Journal, 51*, 1–22.

Decu, V. (1986). Some considerations on the bat guano synusia. *Travaux de l'Institut de Spéologie "Émile Racoviță," 25*, 41–51.

Decu, V., & Tufescu, M. V. (1976). Sur l'organisation d'une biocénose extrême: la biocénose du guano de la grotte "Pestera lui Adam" de Baile Herculane (Carpates Méridionales, Roumanie). *Travaux de l'Institut de Spéologie "Émile Racoviță," 15*, 113–132 (in French).

Deharveng, L., & Bedos., A. (2000). The cave fauna of Southeast Asia. Origin, evolution and ecology. In H. Wilkins, D. C. Culver & W. F. Humphreys (Eds.), *Subterranean ecosystems* (pp. 603–632). Amsterdam: Elsevier Press.

Fenolio, D. B., Graening, G. O., Collier, B. A., & Stout, J. F. (2006). Coprophagy in a cave-adapted salamander; the importance of bat guano examined through nutritional and stable isotope analysis. *Proceedings of the Royal Society B, 273*(1585), 439–443.

Gnaspini, P. (1992). Bat guano ecosystems. A new classification and some considerations, with special references to Neotropical data. *Mémoires de Biospéologie, 19*, 135–138.

Gnaspini, P., & Trajano, E. (2000). Guano communities in tropical caves. In H. Wilkins, D. C. Culver & W. F. Humphreys (Eds.), *Subterranean ecosystems* (pp. 251–268). Amsterdam: Elsevier Press.

Negrea, A., & Negrea, S. (1971). Sur la synusie du guano des grottes du Banat (Roumanie). *Travaux de l'Institut de Spéologie "Émile Racoviță," 10*, 81–122 (in French).

Poulson, T. L. (1972). Bat guano ecosystems. *Bulletin of the National Speleological Society, 34*(2), 55–59.

Poulson, T. L., & Lavoie, K. H. (2000). The trophic basis of subsurface ecosystems. In H. Wilkins, D. C. Culver & W. F. Humphreys (Eds.), *Subterranean ecosystems* (pp. 231–249). Amsterdam: Elsevier Press.

Trajano, E. (1985). Ecologia de populações de morcegos cavernícolas em uma região cárstica do sudeste do Brasil. *Revista Brasileira de Zoologia, 2*(5), 255–320 (in Portuguese).

GYPSUM CAVES

Alexander Klimchouk

Ukrainian Institute of Speleology and Karstology, Simferopol, Ukraine

OCCURRENCE OF GYPSUM AND TYPES OF GYPSUM KARSTS

Gypsum ($CaSO_4 \cdot 2H_2O$) is one of the most common evaporite rocks, that is, inorganic rocks formed by chemical precipitation in a concentrated solution. As gypsum is much more soluble than limestone, it is much less common in outcrops. This is one of the reasons why gypsum karst has received relatively little appreciation in the past in the mainstream of karstology, and has commonly been considered of limited significance. However, since the 1970s it has been increasingly realized that karst processes in gypsum operate extensively when it occurs below various types of cover beds within the upper few hundred meters of sedimentary sequences, that is, in intrastratal karst settings that are very common over cratonic regions (Klimchouk *et al.*, 1996). Various engineering problems, such as ground instability and construction damage due to collapse and subsidence, leakages from canals and reservoirs, and complications during mining operations readily arise in gypsum karst areas due to both the presence of preformed caves and fast development of new conduits.

The variety of depositional settings in which gypsum can form ranges from deep-water, to subaerial, to lacustrine, so the mode of occurrence of gypsum and the associations of lithofacies is quite diverse. Evaporites are most susceptible to recrystallization, dissolution, and replacement among common rocks. Gypsum deposited in the shallow subsurface is converted to anhydrite with burial, and back to gypsum as the sulfate reenters the shallow subsurface realm and is washed in cooler and fresher fluids. Gypsum and anhydrite occur widely as rare to frequent interbeds in sequences containing halite, carbonates, and siliciclastic sediments, but also as massive beds. Individual beds of gypsum with thicknesses of 10–40 m are fairly common, although sequences of evaporite/carbonate/siliciclastic rocks may exceed 1000 m and more in thickness. Further details of gypsum geology can be found in published comprehensive accounts (i.e. Warren, 2006).

The karst that develops in gypsum has many features in common with carbonate karst, but also has significant peculiarities imposed by the specific features of evaporite geology and dissolution processes

(Klimchouk, 2002; Klimchouk et al., 1996). In characterizing gypsum karst and caves, it is particularly important to refer to the overall settings of karst development, which are conveniently generalized by the evolutionary classification of karst types. Types of karsts are viewed as successive stages in geological/hydrogeological evolution, from deposition through burial to reemergence, between which the major boundary conditions, the overall circulation pattern, and extrinsic factors and intrinsic mechanisms of karst development appear to change considerably (see the Figure 3 in this book's article, "Speleogenesis, Hypogenic"). The different types of karsts are marked by characteristic styles of cave development, which are particularly distinct in gypsum.

Syngenetic karst, which commences in freshly deposited gypsum, is limited in extent and does not present any significant caves.

Intrastratal karst is considered to develop within rocks already buried by younger strata, where karstification is later than deposition of the cover rocks. It is by far the predominant karst type in gypsum. As a consequence of standard denudation and uplift on the continents, the buried rocks are shifted with time into progressively shallower positions. Karstification may be initiated at any of the stages of intrastratal development *en route* back to the surface. The entire sequence of intrastratal karst types (stages) includes (in the order as they potentially evolve) *deep-seated karst*, *subjacent karst*, *entrenched karst*, and *denuded karst*. The respective settings differ by the degree of separation of soluble units from the surface and by structural and hydrogeological conditions for karstification. In terms of groundwater dynamics, deep-seated karst is always confined (or semiconfined). In subjacent karst the major hydrogeological confinement is locally breached by erosion of the cover so that hydrodynamics changes through to vadose and water table. Further erosional incision leads to the establishment of entrenched karst, where major valleys incise below the bottom of the karst aquifer and largely or entirely drain the gypsum beds that remain capped with protective insoluble cover for the most part of an area. Denuded karst is former intrastratal karst that develops where the soluble rock has been almost completely exposed from beneath the cover, but with substantial inheritance of solution porosity from the preceding stages of intrastratal development.

Where poorly permeable clays or shales surround gypsum in a stratified sequence, intrastratal karst may not develop at all until the gypsum bed is exposed to the surface. *Open karst* is karst that commences only after exposure of the soluble rock, without any significant intrastratal karstification. While both open and denuded karst represents exposed settings, styles of cave development differ dramatically in these types of karsts. Paradoxically, conduit development in gypsum in open karst is somewhat hindered by rapid dissolution of gypsum. Overall, the true open karst type in gypsum is commonly characterized by limited karstification, while karst types of the intrastratal group frequently demonstrate a very high degree of karstification.

IMPLICATION OF EQUILIBRIUM CHEMISTRY AND DISSOLUTION KINETICS TO SPELEOGENESIS IN GYPSUM

Chemical equilibrium and reaction kinetics for gypsum dissolution differ from those of calcite, imposing some important consequences for speleogenesis. The solubility of gypsum and calcite depends on conditions different for each mineral. The solubility of gypsum is roughly one order of magnitude greater than the solubility of calcite under typical CO_2 pressures found in karst. However, this difference can be much greater under certain conditions (see Fig. 6B in this book's article, "Speleogenesis, Hypogenic"). The gypsum solubility can be considerably enhanced or renewed at the presence of other salts in groundwaters, by anaerobic reduction of sulfates in the presence of organic matter, by dedolomitization of intercalated dolomite layers, and by some other factors. This is particularly important for hypogene speleogenesis in intrastratal settings.

From the perspective of speleogenesis, the most important difference of dissolution processes between limestone and gypsum lies in kinetics. The dissolution of calcite, the principal constituent of carbonate rocks, is controlled mainly by the reaction at the rock surface. Experimental studies of dissolution rates for limestones suggest that relatively high rates drop rapidly when a solution reaches about 60–90% saturation because kinetic reaction changes its order in this region (White, 1977; Palmer, 1991). The nonlinear dissolution rate is low and this kinetic switch is of the ultimate importance for speleogenesis in carbonates as considerable dissolution can continue, although at slow rates, along entire flow paths. A positive feedback loop between discharge and the rate of fissure growth makes the mechanism self-accelerating. The rapid enlargement stage begins when the water is able to pass through the entire conduit while preserving considerable undersaturation. This represents the breakthrough event resulting in a boost of the growth rate, provided that hydrogeological settings permit increasing discharge.

In contrast to limestone, dissolution of gypsum is controlled mainly by diffusion across a boundary layer. It was long assumed that the dissolution reactions of gypsum follow a linear rate law and no switch to slow kinetics occurs with an increase in concentration. Hence slow but uniform enlargement throughout long initial flow paths cannot occur to open them up. However, recent experiments have shown that gypsum dissolution does slow at roughly 95% saturation (Jeschke et al., 2001), which allows some slow dissolution at very close to saturation. In any case, enlargement of initial tight flow paths in gypsum to permit a fast dissolution regime (the breakthrough conditions) seems to be much more difficult than in limestones. Caves may form only where gradients are high enough, or initial widths of openings are large enough, and/or where flow distances are short enough to enable breakthrough conditions within a feasible time. These conditions do not occur frequently in exposed settings. The result is that speleogenesis in open gypsum karst settings is limited, localized, and does not extend to considerable depth (Klimchouk, 2002).

Because of the high solubility and fast dissolution of gypsum (which also depends on flow velocity), early conduit development in this rock is extremely sensitive to variations of boundary conditions such as fissure aperture widths and hydraulic gradient (Birk, 2002). The range of variations, between which the speleogenetic initiation of conduits is attained either very fast or virtually never, is very narrow. This explains why, on one extreme, gypsum formations may remain virtually untouched by cave development in many cases, and on the other extreme conduits may develop fast along favorable flow routes.

High dissolution rates in gypsum and their dependence on flow velocity cause the flow/dissolution feedback in gypsum speleogenesis to be stronger than in limestones. Before the breakthrough event, disparity of dissolution rates between alternative flow paths in gypsum grows faster than in limestone. After breakthrough, enlargement rates in the successful conduits increase more dramatically in gypsum than in limestone. These differences determine that speleogenetic competition between alternative flow paths is much stronger in gypsum than in limestone and that a "runaway" development of the favorable paths is more pronounced (Klimchouk, 2002).

SPELEOGENESIS IN DIFFERENT TYPES OF KARSTS

The above-mentioned peculiarities of gypsum dissolution manifest themselves best in *open karst* settings. Precipitation that falls onto the exposed gypsum rocks contributes greatly to surface denudation but it has a limited potential to create caves because the surface runoff and dispersed infiltration get almost saturated very fast. No epikarst is known to form in open gypsum karst. Instead, a kind of sealing crust develops on exposed gypsum in arid and semi-arid climates due to recrystallization in the uppermost layers caused by repeated wetting and drying. In the saturated (phreatic) zone of unconfined aquifers, speleogenetic initiation of initial tight flow paths in gypsum is very slow and it may not occur within reasonable geological time. Speleogenesis in open gypsum karst settings depends on the availability of initial pathways wide enough for breakthrough (undersaturated thoughflow) to occur quickly. It finds and exploits a few favorable paths (if there are any), while other potential paths languish their growth. After breakthrough, passages quickly develop to accommodate the highest possible discharge, and water table or vadose conditions are quickly established. Linear or crudely branching caves form (Fig. 1), particularly where flashy concentrated recharge occurs, rapidly adjusting to the

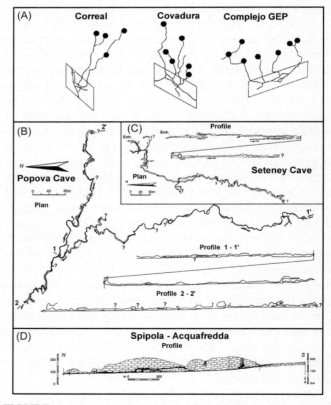

FIGURE 1 Typical sketches, plans, and profiles of epigene gypsum caves formed in unconfined settings: (A) in the open karst of Sorbas, Spain; feeding dolines are indicated by dots (*adapted from Calaforra, 1998*); (B) and (C) in the open karst of the Ekeptze-Gadyk massif, North Caucasus, Russia (*adapted from Ostapenko, 2001*); (D) the Spipola-Acquafredda system in the open karst of the Emilia Romagna, Italy (*adapted from Grimandi, 1987*).

contemporary geomorphic setting and available recharge. Most of the dissolution occurs where water first enters the gypsum, or is focused along streambeds during high-flow events. No deep phreatic conduits develop in unconfined gypsum aquifers. In the vadose zone, cave development concentrates along wide vertical percolation paths, forming vertical pitches, and along free stream courses often following minor insoluble interbeds. Caves of this type may have several levels of subhorizontal passages connected by vertical pits.

Open gypsum karsts are found mainly in arid and semi-arid environments, and where fairly thick gypsum units were surrounded by impervious sequences prior to exposure. The best examples of caves formed in open karst type are documented in the Central Apennines and Sicily in Italy (Neogene gypsum), in Sorbas (Neogene gypsum) and Vallada (Triassic gypsum) regions in Spain, in New Mexico and in Oklahoma, in the United States (Permian gypsum), and in the Pinega region of Russia (Permian gypsum). They include the longest gypsum caves of this type: Kulogorskaja-Troja (Pinega; 14.3 km long), Jester Cave (Oklahoma; 11.8 km long), Spipola-Aquafredda (Central Apennines; 11 km long and 118 m deep; Fig. 1D), the Gueva de Aqua (Sorbas; 8.35 km long), and the deepest gypsum cave in the world, Tunel dels Sumidors (Vallada; 210 m deep). Other significant gypsum caves of this type are explored in the Diebel Nador and Oranais areas in eastern and western Algeria; in the Ar Rabitat/Bir area in northwest Libya; in central and northern Somalia; in some mountain areas in the North Caucasus in Russia; and in Central Asia in Tajikistan.

In *intrastratal settings*, conditions and styles of speleogenesis change in the course of reburial, geomorphic development, and hydrogeological evolution. Caves can initiate and develop in confined conditions (deep-seated karst; hypogene speleogenesis). Subjacent karst represents transitional conditions from confined to unconfined ones, where hypogene and epigene speleogenesis may operate depending on the locally prevailing flow regime, but hypogene speleogenesis still dominates. The further development of preformed hypogene cavities and the contemporaneous epigene speleogenesis can continue through unconfined conditions (entrenched and denuded karst).

In *deep-seated gypsum karst*, there are several common favorable situations and respective mechanisms for cave initiation and development. The most common is transverse speleogenesis in a stratified sequence, where the gypsum bed is sandwiched between good aquifers such as clastic or dolomite beds in a multistory leaky confined system, with diffuse input from below and upwelling flow driven by head gradient across the gypsum bed (Fig. 2A). Where laterally connected fracture

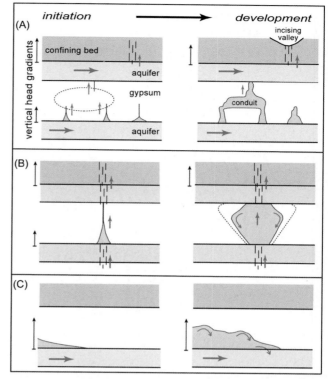

FIGURE 2 Initiation and development of hypogene conduits and cavities in gypsum (*adapted from Klimchouk, 2000*): (A) transverse speleogenesis in a fractured gypsum bed sandwiched between aquifers; (B) transverse speleogenesis along a single fracture zone cross-cutting an otherwise low-fissured bed; (C) lateral contact speleogenesis driven by natural convection in the case of a massive nonfractured bed underlain by a prolific aquifer.

systems exist in the gypsum, reticular maze caves can form by aggressive waters rising from the underlying aquifer through many discrete points (feeders; Fig. 3). Where fractures are arranged in several horizontal levels due to textural/structural inhomogeneities or the presence of minor interbeds, multistory networks develop (see Fig. 7A in the article "Speleogenesis, Hypogenic"). Reticular, rectilinear, or polygonal maze patterns guided by fractures are common in gypsum (Fig. 3), while anastomotic and spongework maze patterns are not.

The transverse hypogene speleogenesis across a soluble bed in a multiple aquifer system has an important mechanism that suppresses strong flow/dissolution feedback and allows many conduits to develop. When some favorable flow paths across the gypsum achieve the breakthrough, their growth accelerates, but this quickly diminishes the head gradient across the gypsum in the vicinity of the successful conduits. As the hydraulic resistance of the conduit becomes smaller than that of the adjacent aquifers, discharge through the conduit is now controlled by the hydraulic

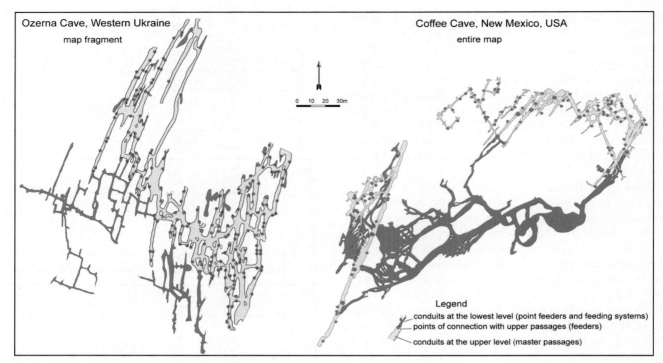

FIGURE 3 Distribution of point feeders through networks of master passages in transverse hypogene maze caves. Red dots indicate feeders; subvertical conduits at the lower story are shown in blue; connecting passages at the upper level are shown in yellow. Left: Ozerna Cave in the Neogene gypsum, Western Ukraine (*adapted from Klimchouk, 1990*). Right: Coffee Cave in the Permian gypsum, New Mexico, U.S.A. (*adapted from Stafford et al., 2008*). Lower level passages locally form feeding subsystems that connect to the master level through point feeders.

conductivity of the least permeable lithological unit in the system and by the boundary conditions (external hydraulic control), but no longer by the diameter of the conduit. Unless and until the boundary conditions change, the flow and enlargement rates in the conduit remain roughly constant at some level. The positive feedback loop is no longer the determinant of conduit development. Because the vertical head gradient between the aquifers is still maintained, although diminished, at some lateral distance apart from the successful transverse conduit, alternative conduits do not languish but continue to grow and eventually reach the breakthrough, either to the upper aquifer or laterally to the conduit that had been "broken through" earlier.

On the general background of upwelling hydraulic communication, flow pattern commonly acquires local lateral components at various levels because of vertical heterogeneities in hydraulic conductivity. These are caused by an imperfect connection of vertically superimposed fracture systems (see Fig. 7 in the article, "Speleogenesis, Hypogenic"), the presence of minor insoluble intercalations, and overall discordance in plan positions of ultimate recharge and discharge points. The lateral flow components cause merging individual transverse passage clusters into laterally extensive cave systems, giving a false impression of generally lateral cave-forming flow.

Outstanding examples of transverse hypogenic maze caves are the giant gypsum caves in the Neogene gypsum of Western Ukraine. Five of them occupy the top ranks in the list of the longest gypsum caves in the world (Table 1), totaling over 514 km of passages. The longest gypsum cave Optymistychna is currently mapped for 232 km and is the second longest cave in the world after the Mammoth–Flint-Ridge system in Kentucky, U.S.A. Other prominent examples include network caves encountered by mines in the Paleogene gypsum of the Paris basin, France; the maze caves in the Neogene gypsum of Estremera, Madrid basin, Spain; in the Permian gypsum of the Pecos Valley Region in New Mexico, U.S.A.; and in the Pinega and Fore-Ural regions of Russia.

Where cross-bed fractures in gypsum are scarce or where water rises into the gypsum not diffusely but through localized fracture zones in the underlying formation, isolated passages or clusters of a few passages form (Fig. 2B). Good examples are the first hypogenic caves in the Messinian (Neogene) gypsum in Italy, recently encountered by the Moncalvo underground quarry (Vigna *et al.*, 2010). Interception of these caves caused a massive inrush of the pressurized waters into

TABLE 1 The World's Longest (>2000 m) and Deepest (>100 m) Gypsum Caves

Name	Longest Gypsum Caves of the World		
	Length, m	Country, Region	Rock Age
Optymistychna	232,000	Ukraine, Western	Neogene
Ozerna	131,400	Ukraine, Western	Neogene
Zoloushka	92,000	Ukraine, Western	Neogene
Mlynki	36,000	Ukraine, Western	Neogene
Kristal'na	22,610	Ukraine, Western	Neogene
Kulogorskaja-Troja	16,250	Russia, Pinega	Permian
Jester	11,800	U.S.A., OK	Permian
Spipola-Aquafredda	10,500	Italy, Emilia-Romagna	Neogene
Olimpijskaja-Lomonosovskaya	9110	Russia, Pinega	Permian
Slavka	9100	Ukraine, Western	Neogene
Verteba	8550	Ukraine, Western	Neogene
Agua, cueva de	8350	Spain, Sorbas	Neogene
Cater Magara	7300	Syria	Neogene
Park's Ranch	6595	U.S.A., NM	Permian
Konstitutzionnaja	6130	Russia, Pinega	Permian
Kungurskaya Ledjanaja	5700	Russia, Fore-Ural	Permian
Mushkarova Jama	5050	Ukraine, Western	Neogene
Selman	4794	U.S.A., OK	Permian
Severny Sifon	4617	Russia, Pinega	Permian
Ordynskaya	4600	Russia, Fore-Ural	Permian
Kumichevskaja	4520	Russia, Pinega	Permian
Zolotoj Kljuchik	4380	Russia, Pinega	Permian
Covadura	4245	Spain, Sorbas	Neogene
Estremera, cueva de	4000	Spain, Madrid	Neogene
Crystal Caverns	3807	U.S.A., NM	Permian
Double Barrel	3724	U.S.A., NM	Permian
Scrooge	3700	U.S.A., NM	Permian
Umm al Masabih	3593	Libya	Jurassic
Gostry Govdy	3570	Ukraine, Western	Neogene
Simfonia	3240	Russia, Pinega	Permian
Pekhorovskaja	3200	Russia, Pinega	Permian
Carcass	3164	U.S.A., NM	Permian
Martin	3150	U.S.A., NM	Permian
Abisso LU.S.A.	3000	Italy, Emilia-Romagna	Neogene
Leningradskaja	2950	Russia, Pinega	Permian
Wimmelburger Schlotte	2840	Germany	Permian
Horseshoe Valley	2760	U.S.A., OK	Permian

(*Continued*)

TABLE 1 (Continued)

Name	Length, m	Country, Region	Rock Age
Longest Gypsum Caves of the World			
Pshashe-Setenay	2690	Russia, N. Caucasus	Jurassic
Pinezhskaja Terehchenko	2600	Russia, Pinega	Permian
Vodnaja	2600	Russia, Pinega	Permian
Jansill/Driftwood	2562	U.S.A., NM	Permian
River Styx	2562	U.S.A., TX	Permian
Fanning Ranch	2557	U.S.A., NM	Permian
Yubuleynaya	2555	Russia, Pinega	Permian
Atlantida	2525	Ukraine, Western	Neogene
Eras'kina 1-2	2500	Russia, Pinega	Permian
Ingh. Ca' Siere	2500	Italy, Emilia-Romagna	Neogene
Triple Engle Pit	2485	U.S.A., NM	Permian
10-years LSS	2450	Russia, Pinega	Permian
Bukovinka	2408	Ukraine, Western	Neogene
Nescatunga	2398	U.S.A., OK	Permian
Segeberger-Kalkhoehle	2360	Germany, S. Harz	Permian
Coffee Cave	2321	U.S.A., NM	Permian
Hyaenlabyrint	2310	Somalia	Paleogene
Pekhorovskij Proval	2262	Russia, Pinega	Permian
Geograficheskogo obshchestva	2150	Russia, Pinega	Permian
Ugryn'	2120	Ukraine, Western	Neogene
Re Tiberio	2110	Italy, Sicily	Neogene
Hay's	2037	U.S.A., NM	Permian
Kulogorskaja-5	2035	Russia, Pinega	Permian
Popova	2032	Russia, N. Caucasus	Jurassic
Michele Gortani	2015	Italy, Emilia-Romagna	Neogene

Name	Depth, m	Country, Region	Rock Age
Deepest Gypsum Caves of the World			
Tunel dels Sumidors	210	Spain, Vallada	Triassic
Abisso LU.S.A.	204	Italy, Emilia-Romagna	Neogene
Pozzo A	>200	Italy, Emilia-Romagna	Triassic?
Triple Engle Pit	134	U.S.A., NM	Permian
Corall, sima del	130	Spain, Sorbas	Neogene
Covadura	126	Spain, Sorbas	Neogene
Campamento, sima del	122	Spain, Sorbas	Neogene
Aguila, sima del	112	Spain, Vallada	Triassic
Rio Stella-Rio Basino	100	Italy, Sicily	Neogene
AB 6	100	Russia, N. Caucasus	Jurassic

the mine. After several months of pumping, the water level was lowered, giving access to 5–7-m- wide passages with distinct hypogene morphology.

In highly soluble aquifers such as gypsum, natural convection circulation and gravitational separation of water due to density differences play important roles in speleogenesis, especially in hypogenic speleogenesis. As forced-flow regimes in confined settings are commonly sluggish and water with lesser density enters the gypsum from below, free convection patterns powered by density gradients are widely operative in hypogenic speleogenesis in gypsum. Various morphological effects of buoyancy dissolution are well represented in hypogenic gypsum caves, including keyhole sections, rising-wall channels, ceiling half-tubes, cupolas, and domepits. In mixed convection systems, buoyancy dissolution effects are particularly pronounced during the mature stage of speleogenesis, where considerable conduit space has been created and vertical hydraulic gradients across the gypsum bed are diminished.

Speleogenesis at the base of the gypsum due to buoyancy dissolution may operate even without guiding cross-bed and forced transverse flow across the unit. Where the gypsum is underlain by a prominent aquifer, the aggressive water attacks the gypsum from the bottom, and when it gets loaded with dissolved sulfates, it is replaced by less dense waters, returns to the basal aquifer, and outflows with the regional flow (Fig. 2C). In this way vast isolated irregular chambers and passages, or ramiform cavities, can form at the bottom of thick gypsum sequences. Outstanding examples of this kind of speleogenesis include numerous giant Schlotten-type caves in the Zechstein (the Upper Permian) gypsum encountered by mines in the South Harz region of Germany at depths up to 400 m (Kempe, 1996).

Reemergence of soluble units to the shallow subsurface and local breaching of the hydrogeological confinement in the course of uplift and erosion signifies the stage of *subjacent karst*, marked by drastic changes of boundary conditions for speleogenesis. At the beginning of this stage discharge through preformed hypogenic caves increases in areas of the breaching, with respective increase of dissolution rates and volume of cavities. With the subsequent establishment of vadose zone and water-table conditions dissolution is localized along the water table and, where surface streams are swallowed, along underground streambeds. Dissolution along the water table is favored by distinct differences in density and aggressiveness between the top 5–10-cm layer and the bulk of water in partially submerged cave passages. This leads to expressed horizontal notching and widening of passages (Fig. 4A,B,C). In passages located below the

FIGURE 4 Modification of hypogene gypsum caves in unconfined (water table) conditions. (A) Basic unmodified fissure-like passage. (B) Passage with horizontal wall notches and inclined facets developed by dissolution along the water table. (C) Diagram illustrating natural convection flow pattern in a passage under water-table conditions (exclamation mark and red arrows indicate the most aggressive water at the top layer of water) and evolution of the cross section. (D) Map of the 92-km-long Zoloushka Cave in Western Ukraine; the dashed lines in the profile inset indicate the water-table positions before and after the quarrying (*map courtesy of the Chernivtsy Speleological Club; from Andrejchouk, 2008*). (E) A typical passage at the upper level in Zoloushka Cave showing horizontal notching due to dissolution at the water table. *Photos A and B courtesy of A. Kananovich; photo E by the author.*

water table, natural convection effects lead to the development of characteristic tip-down triangular cross-sections, with flat ceilings (*laugdecke* in German), combined with inclined wall facets.

A good example of the hypogene maze cave in the subjacent karst stage is the longest explored underwater gypsum cave Ordinskaya in the Fore-Ural, with over 4.5 km of passages mapped by cave divers (Fig. 5). Divers also mapped more than 100 active feeders at the bottom of passages through which upward recharge still occurs to the cave from the underlying limestone aquifer. Another outstanding example is the maze of Zoloushka Cave in Western Ukraine, in which 92 km of passages have been mapped since 1976 (Fig. 4D). The position of the water table in the area during the Holocene is marked by extensive horizontal notching and widening in passages at the upper story (Fig. 4E). With the open-pit quarry operation and groundwater withdrawal since the late 1950s the water table has dropped more than 20 m (Fig. 4D, inset profile), which made the huge maze cave accessible for direct exploration.

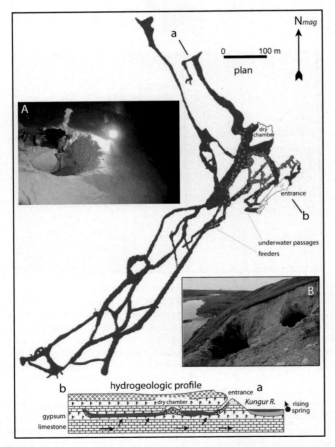

FIGURE 5 The map of Ordinskaya Cave in the Fore-Ural, the longest underwater gypsum cave. The hypogenic cave is presently under conditions of subjacent karst. Red circles indicate active feeders at the bottom of passages mapped in the central part of the cave (one of which is shown on the inset photo (A)). Inset photo (B) shows the entrance of the cave and the entrenching Kungur River. *Modified and compiled from the map and the photos by P. Sivinskykh (2009).*

Deepening of erosional valleys below the gypsum bottom signifies the stage of *entrenched karst*. Most preformed caves become relict, although those within the reach of high levels of nearby rivers can be subject to backflooding, which greatly contributes to further solution widening of passages. A characteristic feature of many entrenched gypsum karsts are vertical solution pipes (also known as *chimneys* or *comins*) that perforate gypsum sequences and often superimpose on the relict hypogene caves, being formed by aggressive percolation waters leaking through the vadose zone from perched aquifers above.

Surface denudation may lead to almost complete removal of the protective cover in some areas of instrastratal karst regions, which establishes denuded karst type. Caves inherited from the preceding stages progressively disintegrate through collapsing. This is why areas of denuded gypsum karst demonstrate much denser doline distribution than open karsts.

Contemporary cave development under unconfined conditions in subjacent, entrenched, and denuded karst types generate single-conduit or crudely branching caves, very much in the same way as in open karst. However, the presence of the insoluble cover provides more aggressive recharge from the surface via dolines and ponors, and the presence of inherited cave porosity facilitates more rapid epigene cave development. In the regions of intrastratal gypsum karst, contemporary epigene caves coexist with (and often superimpose onto) relict hypogene caves, demonstrating strikingly different patterns and mesomorphology (Fig. 6). Being evolutionary stages of intrastratal karst development, subjacent, entrenched, and denuded karst types often intersperse in an area or lie adjacent to each other and to deep-seated karst within the same aquifer in block-fault and monoclinal tectonic settings. Such variations are due to differentiated rates of neotectonic uplift. Remarkable examples include the Pinega and Pre-Ural regions in Russia, Western Ukraine, the South Harz in Germany, and the Gypsum Plain in New Mexico, U.S.A.

SUMMARY

Gypsum caves include some of the world's longest caves, although the deepest caves in this rock barely reach 200 m in depth. There are several characteristic patterns and styles of caves that form in gypsum, distinctly related to particular settings of karst development. These settings change regularly with the hydrogeological and geomorphological evolution.

In confined intrastratal settings (deep-seated karst) the following types of caves form in gypsum by

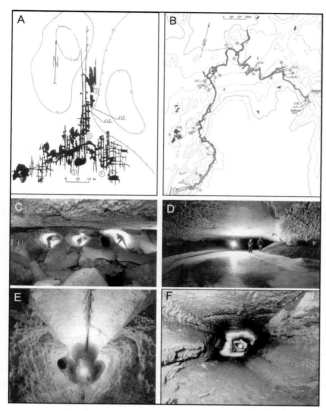

FIGURE 6 Patterns and morphology of hypogene (left column, (A), (C), and (E)) and epigene (right column, (B), (D), and (F)) gypsum caves in the Pinega region in north Russia. (A) Simfonia Cave (hypogene transverse maze cave); (B) Kumichevka-Vizborovskaya Cave (epigene stream cave); (C), (E) Simfonia Cave; (D) Lomonosovskaya Cave; (F) Golubinsky Proval Cave. *Cave maps from Malkov et al., 2001; photos by Yu. Nikolayev.*

different mechanisms of hypogene speleogenesis, depending on lithostratigraphic, structural, and hydrogeological conditions:

1. Reticular 2D and 3D (multistory) maze caves
2. Isolated passages along prominent disruptions
3. Isolated irregular chambers and passages at the base of gypsum

The morphology of these caves can be considerably modified on the subsequent stages of unconfined intrastratal karst development, particularly in the water-table conditions in subjacent and entrenched karsts.

The common patterns of gypsum caves that form contemporaneously in unconfined intrastratal settings (subjacent, entrenched, and denuded karst types) include:

4. Linear and crudely branching through-flow caves
5. Vertical solution pipes fed by perched aquifers in cover beds

In open karst, linear and crudely branching through-flow caves are the predominant type. They form close to the water table and in the vadose zone. No deep phreatic conduit development occurs in open gypsum karst.

See Also the Following Article

Ukraine Giant Gypsum Caves, Speleogenesis, Hypogenic

Bibliography

Andrejchouk, V. N. (2007). *Zoloushka cave*. Simferopol-Sosnowiec: Silezian University – Ukrainian Institute of Speleology and Karstology (in Russian).

Birk, S. (2002). Characterisation of karst systems by simulating aquifer genesis and spring responses: Model development and application to gypsum karst. *Tübingen Geowissenschaftliche Arbeiten, Reine C, 60*.

Calaforra, J. M. (1998). *Karstologia de yesos*. Almeria, Spain: Universidad de Almeria.

Grimandi, P. (1987). Grotta della Spipola. *Ipoantropo (Italy), 5*, 51–64.

Jeschke, A., Vosbeck, K., & Dreybrodt, W. (2001). Surface controlled dissolution rates of gypsum in aqueous solutions exhibit nonlinear dissolution kimetics. *Geochimica et Cosmochimica Acta, 65*(1), 27–34.

Kempe, S. (1996). Gypsum karst of Germany. In A. Klimchouk, D. Lowe, A. Cooper & U Sauro (Eds.), *Gypsum Karst of the World* (pp. 207–224) International Journal of Speleology Theme issue 25 (3–4).

Klimchouk, A. B. (2002). Evolution of karst in evaporites. In F. Gabrovšek (Ed.), *Evolution of karst: From prekarst to cessation* (pp. 61–96). Postojna-Ljubljana, Slovenia: Založba ZRC.

Klimchouk, A. B., Ford, D. C., Palmer, A. N., & Dreybrodt, W. (Eds.), (2000). *Speleogenesis: Evolution of karst aquifers* Huntsville, AL: National Speleological Society.

Klimchouk, A. B., Lowe, D., Cooper, A., & Sauro, U. (Eds.), (1996). *International Journal of Speleology 25*(3–4), 307.

Malkov, V. N., Gurkalo, E. I., Monakhova, L. B., Shavrina, E. V., Gurkalo, V. A., & Frants, N. F. (2001). *Karst and caves of the Pinezhje*. Moscow: Ecost Association (in Russian).

Ostapenko, A.A. (2001). *Underground karst forms in the sulphate deposits of the Western Caucasus*. PhD thesis, Krasnodar: Kubansky University (in Russian).

Palmer, A. N. (1991). Origin and morphology of limestone caves. *Geological Society of America Bulletin, 103*(1), 1–21.

Sivinskykh, P. (2009). Features of geological conditions of the Ordynskaya Cave, Fore-Urals, Russia. In A. B. Klimchouk & D. C. Ford (Eds.), *Hypogene speleogenesis and karst hydrogeology of artesian basins*. Simferopol: Ukrainian Institute of Speleology and Karstology.

Stafford, K. W., Land, L., & Klimchouk, A. (2008). Hypogenic speleogenesis within seven rivers evaporites: Coffee cave, eddy county, New Mexico. *Journal of Cave and Karst Studies, 70*(1), 47–61.

Warren, J. K. (2006). *Evaporites: Sediments, resources and hydrocarbons*. Berlin: Springer.

White, W. B. (1977). Role of solution kinetics in the development of karst aquifers. In J. S. Tolson & F. L. Doyle (Eds.), *Karst hydrogeology* (pp. 503–517). Huntsville, AL: UAH Press.

GYPSUM FLOWERS AND RELATED SPELEOTHEMS

William B. White

The Pennsylvania State University

Caves provide excellent environments for the growth of mineral crystals. Cave conditions of temperature and moisture remain constant for long periods of time. Some 256 different minerals have been identified from caves, although most are quite rare (Hill and Forti, 1997). Most of the speleothems found in caves are composed of calcium carbonate. However, minerals composed of sulfates of various kinds are found in dry caves. The crystal growth habits of the sulfate minerals are quite different from the habits of carbonate minerals, and the resulting speleothems have a quite different appearance. Sulfate mineral crystals tend to grow as large, curved fibrous bundles having the appearance of flower petals as well as other shapes. The most common of the sulfate minerals is gypsum, $CaSO_4 \cdot 2H_2O$, and the commonly appearing speleothems are known as *gypsum flowers*.

EVAPORITE MINERALS IN CAVES

Calcium sulfate occurs in three stages of hydration. Anhydrite is $CaSO_4$, the anhydrous salt. It is stable at temperatures above 58°C and at high salinities. Anhydrite is common in sedimentary deposits where the high ionic strength of seawater lowers the activity of water and stabilizes anhydrite structure. Anhydrite is not stable in the cave environment because of the lower temperature and the presence of freshwater. $CaSO_4 \cdot \frac{1}{2}H_2O$ is ordinary plaster but occurs only rarely in nature as the mineral bassanite. Gypsum, $CaSO_4 \cdot 2H_2O$, is the fully hydrated form. Gypsum is monoclinic with unit cell parameters: $a = 5.679$ Å, $b = 15.202$ Å, $c = 6.522$ Å, and $\beta = 118.4°$. Single crystals of gypsum are water-clear. These transparent masses are called *selenite*. Selenite can be easily cleaved into plates and slabs perpendicular to the b-crystallographic axis, the {010} cleavage plane.

Most crystals grow at different rates in different crystallographic directions. The fast growth direction in gypsum is along the crystallographic c-direction. Because of the differences in growth rate, gypsum often takes the form of long acicular crystals and fibers.

In addition to gypsum, two other sulfate minerals have similar crystal structures and also form crusts and "flowers": mirabilite, $Na_2SO_4 \cdot 10H_2O$, and epsomite, $MgSO_4 \cdot 7H_2O$. These minerals are highly water soluble, so they appear only in very dry caves. Mirabilite is unstable in the outside atmosphere. If mirabilite crystals are removed from the cave, the mineral loses its water of crystallization to become thenardite, Na_2SO_4. The large water-clear crystals of mirabilite collapse into a white powder. The behavior of epsomite is unusual. Unlike mirabilite, which loses all ten water molecules at once, epsomite loses water stepwise where the first step is $MgSO_4 \cdot 7H_2O$ (epsomite) $\leftrightarrow$ $MgSO_4 \cdot 6H_2O$ (hexahydrite) $+ H_2O$. The conditions of temperature and water vapor pressure for this reaction are very close to cave conditions, so small shifts in these parameters will cause the mineral to hydrate or dehydrate, thus changing mineralogy depending on small seasonal changes in the cave environment.

Other sulfate minerals that occur in caves include the double salts bloedite, $MgSO_4 \cdot Na_2SO_4 \cdot 4H_2O$, and eugsterite, $CaSO_4 \cdot Na_2SO_4 \cdot 2H_2O$. Celestine, $SrSO_4$, occurs as a minor mineral associated with gypsum deposits where seepage waters are rich in Sr^{2+}. Typically celestine appears as pale-blue crystals embedded in a white gypsum crust. In caves where the sulfate ion is derived from oxidation of sulfide minerals or from the oxidation of H_2S, there may occur aluminum-containing sulfate minerals such as alunite, $KAl_3(SO_4)_2(OH)_6$. Some caves are associated with ore deposits. The oxidation and hydration of the metal sulfides of the ore deposits by reaction with cave waters produces a complex suite of iron, copper, and other metal sulfate minerals (Alpers *et al.*, 2000).

SPELEOTHEMS

Gypsum and the other sulfate mineral speleothems exhibit a great variety of shapes depending on growth rate and the source of water.

A. Bulk crystals
   1. Soil gypsum
   2. Dentate gypsum
   3. Gypsum needles
B. Gypsum crusts
   1. Granular crusts
   2. Fibrous crusts
C. Fibrous gypsum
   1. Gypsum flowers
   2. Angel hair
   3. Cave cotton

Gypsum can grow in clastic sediments to form a variety of crystals which can reach dimensions of many centimeters. These are sometimes elongated and other times irregular masses. The surfaces are rough and rarely display crystal faces, although the interior may be clear selenite. The presence of some moisture films on gypsum in air-filled cave passages can result in large-faceted crystals. Bulk gypsum crystal clusters can reach dimensions of meters and have been termed *gypsum chandeliers*. Gypsum needles are straight, clear crystals that grow by extrusion from the soil in some caves (Fig. 1). At the base of each needle, where it is inserted into the soil, is a V-shaped notch. Polarized light microscopy shows that the needles are twins, with the {100} twin plane emerging from the tip of the V and forming the central plane of the needle. The V-notch, termed a *reentrant twin*, provides an optimum site for the deposition of new crystal from moisture in the soil. Thus, the gypsum needles grow from the base and are extruded from the soil as the crystals grow. Growth at the reentrant V maintains a linear form so that gypsum needles do not curve in the manner of many other gypsum speleothems. Gypsum needles have been observed with lengths of more than one meter.

Gypsum crusts are formed on cave walls from minute amounts of moisture seeping through pores in the limestone rock. Some crusts have a sugary appearance and are composed of sand-sized grains of gypsum. Others have a satin appearance in cross section and are composed of fibrous crystals arranged perpendicular to the crust. Crusts have been observed to grow over historical names that had been scratched on the cave walls. They grow irregularly because of the variation in the permeability of the bedrock. Locations where growth is faster produce blister-like bulges in the crusts, and these may erupt to form gypsum flowers. As crusts continue to grow, they often break free of the wall and peel away, exposing bare bedrock and later new crusts.

Most spectacular of the gypsum speleothems are the gypsum flowers (Fig. 2). Gypsum flowers range in size from a few centimeters to tens of centimeters, often with the "petals" elaborately curved. Under the optical microscope, the petals of the gypsum flowers are seen to be bundles of fibers, each fiber a separate gypsum crystal. Gypsum flowers grow from the same seepage waters that form gypsum crusts. The solutions move faster in more porous zones and thus the crystals grow faster near the center of the zone. The faster growing crystals tend to push past the slower growing fibers at the side, thus causing the overall mass of fibers to curve outward, producing the curved flower petals. The fibers that make up gypsum flowers are relatively coarse, perhaps fractions of a millimeter to a millimeter in diameter. Much smaller and longer fibers occur as the form known as *angel hair*. The fibers that make up angel hair appear to be single crystals with lengths of a meter or more with a cross section of a millimeter or less. Still smaller fibers occur in a speleothem called *cave cotton*. As the name implies, cave cotton is a mass of tangled fibers generally with fiber lengths of a few tens of centimeters but diameters of tens or hundreds of micrometers. The thin crystals flex easily so that the mass appears much like a ball of cotton.

Mirabilite, epsomite, and other sulfate minerals have similar crystal habits although different crystal structures on the atomic scale. As a result, "flowers" of mirabilite and epsomite are also found in extremely dry cave passages. However, the high solubility of these minerals means that they dissolve easily in any available seepage. Stalactites of epsomite (Fig. 3) and mirabilite (Fig. 4) occur when there is sufficient seepage to form liquid droplets. Otherwise, the minerals occur as curved bundles of fibers or fibrous crystals. It is difficult to distinguish the water-soluble minerals from gypsum by visual inspection only, although they can be distinguished by taste.

SOURCES AND DEPOSITIONAL MECHANISMS OF EVAPORITE MINERALS

There are multiple sources for the gypsum and other sulfate minerals that occur in limestone caves.

FIGURE 1 Gypsum needles, Cumberland Caverns, Tennessee. *Photo by the author.*

FIGURE 2 Gypsum flowers, Cumberland Caverns, Tennessee. *Photo by the author.*

FIGURE 3 Epsomite stalactites on mixed sulfate mineral crust, Cottonwood Cave, New Mexico. *Photo by the author.*

1. Gypsum beds that occur within the limestone
2. Gypsum derived from the oxidation of pyrite that occurs within the limestone
3. Gypsum derived from oxidation of pyrite and other sulfide minerals in overlying rock formations
4. Gypsum derived from oxidation of H_2S
5. Gypsum derived from the hydration of anhydrite pods within the limestone

The chemistry of gypsum deposition is exceedingly simple. Evaporation of a solution containing Ca^{2+} and SO_4^{2-} causes them to combine to form gypsum.

$$Ca^{2+} + SO_4^{2-} + 2H_2O \leftrightarrow CaSO_4 \cdot 2H_2O$$

The solubility of gypsum is approximately 2400 mg/L at 10°C making it relatively easy for gypsum to dissolve in one part of the groundwater flow system and be transported to a site of deposition elsewhere. Beds of primary gypsum occur in many limestone sequences; thus they can provide a source of gypsum that requires only transport to its final site of deposition.

Pyrite, FeS_2, occurs widely distributed in many sedimentary rocks. It is unstable in wet, oxidizing environments. Although the details are complicated, the overall reaction is:

$$FeS_2 + 7/2\ O_2 + H_2O \leftrightarrow Fe^{3+} + 2SO_4^{2-} + 2H^+$$

Ferric iron is highly insoluble in mildly alkaline karstic waters and precipitates as $Fe(OH)_3$ so that iron

FIGURE 4 Water-clear mirabilite stalactites hanging on massive gypsum crystals, Turner Avenue, Mammoth Cave, Kentucky. *Photo by the author.*

rarely migrates far from the site of the original pyrite. The reaction can be more complicated because Fe^{3+} can also break down more pyrite with the release of additional acidity. Microorganisms often catalyze these reactions. The sulfuric acid produced by pyrite oxidation can either react with limestone directly to produce gypsum which is then transported in solution, or the reaction can be delayed until the reactants reach the walls of the cave passage where gypsum is deposited *in situ*. In either case the combined reaction is:

$$H_2O + 2H^+ + SO_4^{2-} + CaCO_3 \leftrightarrow CaSO_4 \cdot 2H_2O + CO_2$$

Some limestones contain nodules of anhydrite incorporated within the limestone. When these are exposed by the formation of cave passages, the anhydrite reacts with water to form gypsum.

Some caves, particularly those in the Carlsbad area of New Mexico, have been formed by the dissolving action of sulfuric acid rather than by carbonic acid, which is the usual mechanism in shallow groundwater caves. The source of sulfuric acid is the oxidation of hydrogen sulfide, H_2S, which seeps upward from nearby petroleum fields until it encounters oxygen-rich groundwater. The sulfuric acid dissolves the limestone and also can deposit quantities of gypsum which can then be redistributed to form gypsum speleothems (Hill, 1987).

It is difficult to determine the details of gypsum deposition because the solutions from which the crystals grow seep so slowly from pores in the rock that no collectable liquid is observed. Evidence for relatively rapid growth is provided by the observation of gypsum crusts overlying historic signatures that had been scratched in the cave wall. Early Americans who inhabited the Mammoth Cave region mined Mammoth Cave and nearby Salts Cave for gypsum and other salts. In some places where the ceiling had been scraped down, the bedrock now shows growths of mirabilite that must have formed during the several thousand years since the paleomining operation.

Bibliography

Alpers, C. N., Jambor, J. L., & Nordstrom, D. K. (2000). Sulfate minerals: Crystallography, geochemistry and environmental significance. *Reviews of Mineralogy and Geochemistry, 40.* 608 pp.

Hill, C. A. (1987). Geology of Carlsbad Cavern and other caves in the Guadalupe Mountains, New Mexico and Texas. *New Mexico Bureau of Mines and Mineral Resources, 117.* 150 pp.

Hill, C., & Forti, P. (1997). *Cave minerals of the world* (2nd ed.). Huntsville, AL: National Speleological Society.

H

HELICTITES AND RELATED SPELEOTHEMS

Donald G. Davis

National Speleological Society

SUBAERIAL HELICTITES

Helictites are elongated speleothems that, unlike stalactites, may grow in any direction. Upward-growing helictites have sometimes been called *heligmites*, but there is little logical basis for this distinction, because helictites do not occur as separate "up" and "down" forms. Helictites may be straight, smoothly curving, or even spiral (*helical*, the root meaning of helictite), but in most cases they twist and turn erratically. Accordingly, the alternative names *erratics*, *eccentrics*, or *eccentric stalactites* have been used by some authors. Helictites are usually composed of calcite or aragonite, more rarely other minerals. They occur in a great range of sizes, from hair-thin and a fraction of a centimeter long to several centimeters wide and more than a meter in length. All share the common characteristic of a narrow central canal of capillary size.

Because helictites are so conspicuous and unusual in appearance, many (often fanciful) theories have been proposed for their origin. By growing artificial helictites of sodium thiosulfate, Huff (1940) demonstrated that hydrostatic pressure feeding capillary flow was the true mechanism. In natural carbonate helictites, the tip is extended by deposition of calcium carbonate around the central pore as the outflowing moisture evaporates or loses carbon dioxide. Moore (1954) subsequently explained helictite curvature by a combination of effects of impurities, crystallographic-axis rotation, and stacking of wedge-shaped crystals. These factors take precedence over gravity because the rate of flow is too slow to form a hanging drop at the tip. Increased flow can cause helictites to convert to soda-straw stalactites, and decreased flow, *vice versa*.

Varieties including *filiform* (hair-like), *vermiform* (worm-like), and *antler'* (forking) helictites have been defined on the basis of shape and size. Aragonite helictites may be beaded (Fig. 1), consisting of a string of conical beads of radiating fibrous crystals. The larger ends of the cones may face either the attached end of the helictite or the free end, but the orientation is usually consistent within each chain. This bizarre-looking beaded structure is relatively rare and has never been explained.

ANTHODITES

Anthodites (Fig. 2) were defined by Henderson (1949) from Skyline Caverns, Virginia, as "clusters of slender, branching, tubular formations originating near a common point and radiating outward and downward." The term *anthodite*, which simply means "flower rock," has come to be used in conflicting ways, with many writers applying it to branching aragonite sprays that lack central canals. For the latter speleothem, the term *frostwork*, which was used in the Black Hills (South Dakota, U.S.A.) before 1900, has much earlier priority. The first edition of *Cave Minerals of the World* called frostwork "acicular anthodites" and speleothems of the Skyline Caverns type "quill anthodites." The second edition (Hill and Forti, 1997) treated anthodites and frostwork as separate speleothem types (following a discussion exchange in the National Speleological Society Bulletin).

Anthodites, in Henderson's original sense, are not very clearly distinct from helictites. Occurrence in clusters in which they radiate or curve away from a common base seems to be the only distinct morphological trait characterizing "anthodites." Typical helictites are usually more independent and less smoothly curvilinear. These macroscopic differences reflect the growth habit of the original primary mineral. Anthoditic

FIGURE 2 Anthodites in Skyline Caverns, Warren County, Virginia. *Photo courtesy of William B. White. Used with permission.*

FIGURE 1 (A) Spectacular display of beaded helictites. (B) Detail of beaded helictites. *Photos courtesy of Kevin Downey. Used with permission.*

morphology is generally seen in examples that originated as aragonite; helictitic morphology in those originating as calcite. Alternation between mineralogies, causing overgrowth of one mineral by the other, as well as conversion of aragonite to calcite, results in intermediate or transitional forms that are difficult to classify and interpret.

FROSTWORK

Frostwork (Fig. 3) consists of sprays of acicular aragonite needles, often intricately branching. Like helictites, these needles are not gravitationally oriented and may grow in any direction. However, because it lacks internal feeder channels, frostwork is genetically distinct from helictites and from Henderson's anthodites, although all of these may be found in proximity to each other.

Frostwork apparently grows from seepage of a thin film of carbonate-bearing moisture over the crystal surfaces, although growth from aerosols has also been suggested. It is usually found in sections of caves below dolomitic bedrock and where evaporation prevails. These conditions encourage enrichment of magnesium in wall moisture, facilitating aragonite growth. Relatively high cave temperature is also favorable but not essential. When evaporation has concentrated the solution sufficiently that magnesium minerals can also precipitate, the aragonite needles may be tipped with blobs of hydromagnesite moonmilk. Individual frostwork "trees" may grow up to half a meter long. Composite masses can form even larger conical stalagmitic growths of intertwined aragonite, which are often hollow where drips have passed through the center. Such hollow aragonite stalagmites have been called *logomites* in the Black Hills caves. Where not soiled, frostwork is almost always snowy white, rarely incorporating colored contaminants. It is among the most aesthetically impressive cave decorations.

FIGURE 3 Aragonite frostwork tree in Lechuguilla Cave, New Mexico. *Photo courtesy of Ballman/Downey/Widmer. Used with permission.*

FIGURE 4 Active subaqueous helictites in a pool in Lechuguilla Cave, New Mexico. *Photo courtesy of Kevin Downey. Used with permission.*

CAVE SHIELDS

Shields, also called *palettes*, are disk-shaped speleothems consisting of paired circular or oval plates separated by a medial crack, which is an extension of a fracture in the rock substrate. Like helictites, they may be oriented in any direction. The largest are more than 3 m wide. Shields and helictites, though different in appearance, are closely related genetically. From a geometric viewpoint, shields can be regarded as flattened, biaxial "helictites" in which growth occurs along the margin of a plane rather than the end of a line. Shields and helictites are similar in being fed by seepage under hydrostatic pressure; an active shield in France was observed jetting five arcs of water from its edge. To sustain flow from a shield's extended margin requires a higher rate of discharge than in a helictite. Shields often grow helictites from their upper plates, and draperies from overflow from the lips of the lower plates.

The medial crack in a shield is usually of capillary size along the edge, but may be wider, up to several millimeters, internally. This may be caused by crystal-growth wedging along the active edge pushing the plates slightly apart. Some have suggested that flexing by Earth tides prevents the cracks from closing up; however, self-generated wedging may be all that is really required. Shields that have split open, showing the internal surfaces, display concentric growth rings, and in some cases, also closely spaced horizontal ribs, 2 mm or so wide, that presumably reflect former fluctuating water levels inside the shield.

Shields are sometimes confused with flowstone canopies undermined by sediment removal; canopies usually curve and have no medial crack. Shields are found in fewer caves than helictites, but are often abundant where they do exist. They are probably favored by low primary permeability in the substrate rock, forcing seepage to be localized along fractures; highly permeable walls are more likely to grow helictites. Welts are incipient shields—linear excrescences along cracks where flow has not become confined to particular sections.

SUBAQUEOUS HELICTITES

Until 1987, all helictites were assumed to grow surrounded by air. In that year, spectacular displays of spaghetti-like helictites were discovered in Lechuguilla Cave, New Mexico, United States, growing from the undersides of shelfstone and from subaqueous calcite crust in pools (Fig. 4). They are usually 2–5 mm wide, but sometimes more than 1 cm, and up to 30 or more centimeters long. By the year 2000, more than 30 sites with active or "dead" subaqueous helictites had been recorded in Lechuguilla Cave, which remains almost the only cave in which they are known (a few small ones have been found in Virgin Cave in the same area).

Like subaerial helictites, subaqueous helictites have tiny central canals, but the mechanism of capillary seepage under hydrostatic pressure, with CO_2 degassing or evaporation at the tip, does not apply in the pool environment. Instead, subaqueous helictites have invariably been found where blocks or crusts of gypsum are in contact with flowstone upflow from the pool basins. It is now generally accepted that the subaqueous helictites were produced by the common-ion effect, in which stringers of gypsum-enriched water flowed into pools already saturated with calcium bicarbonate, triggering deposition of the less soluble calcium-mineral species (calcite) where the calcium-rich solutions mixed. Water analyses at the type locality were consistent with this mechanism (Davis et al., 1990). The initial result is development of shelfstone and crust barriers between the incoming and ambient waters, with further inflow being restricted to small pores around which rings of calcite grow. Extension of these rings into the pool creates the helictites. Subaqueous helictites grow generally downward, as is appropriate to the greater density of the entering gypsum-rich solution, though some turn upward toward the ends.

Lechuguilla Cave is almost the only known subaqueous-helictite site because it is exceptionally well supplied with gypsum left over from its sulfuric-acid speleogenesis, some of it in contact with flowstone-growing seepage. This is a rare and unstable situation.

SUBAQUEOUS "HELICTITE BUSHES"

In certain passages of Wind Cave, South Dakota, U.S.A., there are clusters of intertangled helictitic calcite speleothems (Fig. 5), typically about 0.5 m or less long, but a maximum of 2 m. They almost always extend upward from discontinuities in floor crust, though the largest clump grows downward from a passage-end pinch. All yet observed appear to be inactive. These were informally called *helictite bushes* by the cavers who observed them, and were assumed for many years to be subaerial helictites of unusual form.

After the discovery of Lechuguilla Cave's subaqueous helictites, Davis (1989) examined the Wind Cave bushes and concluded, from several kinds of evidence, that they represent another type of subaqueous speleothem, developed below the water table at a time between the end of speleogenesis and the drainage of the passages. They are quite distinct from all other helictites, either subaerial or subaqueous. They are frequently strap-shaped and irregularly flattened in cross section (in contrast to the cylindrical outline typical of

FIGURE 5 Helictite bush of subaqueous origin in Wind Cave, South Dakota. Calcite rafts demonstrate former submergence. *Photo courtesy of Kevin Downey. Used with permission.*

helictites), and often branch dendritically. Their internal canals are not capillary tubes, but relatively large, irregular conduits that may be more than 1 cm wide and that more or less mirror the convolutions of the speleothem's cross section. It is unlikely that such structures grew upward in a subaerial environment.

The distribution of helictite bushes shows no correlation with sources of downward-seeping water. They are largely confined to a prominent dip-oriented passage trend that bisects the main part of the maze cave and are always near the lowest level in any particular section. The typical upward growth suggests that the incoming fluid was less dense than the surrounding ambient water (in contrast to the Lechuguillan case). Davis (1989, 1991) proposed that helictite bushes grew by interaction of near-surface groundwater, possibly via the common-ion effect, with thermal fluid rising from depth along a fracture set. In this scenario, helictite bushes would be more closely related to geysermites and submarine "white smokers" than to other helictite types. They may be low-energy analogs of the subaqueous tufa towers and "sand tufa" of Mono Lake, California (Davis, 2004).

Chemical and isotopic analyses of helictite-bush fragments by LaRock and Cunningham (1995) failed to confirm a deep-seated origin of the helictite-growing fluid, but did indicate growth in a cooling thermal environment at

33–42°C, approximately 200,000 years ago. Their specimens showed a multiple-layered structure: an inner tube of crystals with c-axes parallel to the speleothem's axis, an outer coating of "palisade" crystals at right angles, and a transitional layer between. The transition and palisade layers were repeated a number of times in some examples, perhaps recording cycles of emergence/submergence. Fossil microbial filaments, most abundant in the inner layer, suggest a microbiological role in bush growth.

LaRock and Cunningham (1995) found no analytic evidence for common-ion-effect involvement, and suggested that the calcite was deposited via CO_2 degassing at the overlying water surface. They proposed that the flow inside the bush branches was driven not by fluid rising from depth, but by water-table fluctuations, causing a pressure gradient toward the passage to develop in water trapped behind the passage-lining crust when the water level in the open passage dropped faster than that behind the crust. Their interpretation does not explain why the bushes are largely confined to a narrow passage trend, whereas the crust lining is much more extensive in the cave, or why the phenomenon has not been seen in other caves with subaqueous crust linings. These unique speleothems need further study.

Bibliography

Davis, D. G. (1989). Helictite bushes: A subaqueous speleothem? *National Speleological Society Bulletin, 51*(2), 120–124.

Davis, D. G. (1991). Wind Cave helictite bushes as a subaqueous speleothem: Further observations. *Geo^2, 19*(1), 13–15.

Davis, D. G. (2004). Helictite bushes, tufa towers, and inverted rimstone dams. *Geo^2, 31*(2), 13–19.

Davis, D. G., Palmer, A. N., & Palmer, M. V. (1990). Extraordinary subaqueous speleothems in Lechuguilla Cave. *National Speleological Society Bulletin, 52*(2), 70–86.

Henderson, E. P. (1949). Some unusual formations in Skyline Caverns, Virginia. *National Speleological Society Bulletin, 11*, 31–34.

Hill, C. A., & Forti, P. (1997). *Cave minerals of the world* (2nd ed.). Huntsville, AL: National Speleological Society.

Huff, L. C. (1940). Artificial helictites and gypsum flowers. *Journal of Geology, 48*(6), 648–659.

LaRock, E. J., & Cunningham, K. I. (1995). Helictite bush formation and aquifer cooling in Wind Cave, Wind Cave National Park, South Dakota. *National Speleological Society Bulletin, 57*(1), 43–51.

Moore, G. W. (1954). *The origin of helictites*, Occasional Paper No. 1. Huntsville, AL: National Speleological Society.

HYDROGEOLOGY OF KARST AQUIFERS

William B. White

The Pennsylvania State University

An *aquifer*, according to standard definitions, is a rock unit capable of storing and transmitting water. Both properties are necessary. The open spaces within the rock mass provide the storage space. These open spaces—pores, fractures and, in the case of karst aquifers, pipe-like conduits—must be interconnected so that water can flow through. In order to appreciate the special complications of aquifers that contain cave-size openings, it is necessary to first describe the more common porous media aquifers. We then proceed to the special properties of karst aquifers. These are presented as a *conceptual model*, that is, a picture or cartoon of the parts and pieces of a karst aquifer, how they interrelate, and how they function. The purpose of a conceptual model is to set a foundation on which more precise and quantitative descriptions of aquifer behavior can be built.

POROUS MEDIA AQUIFERS

In porous media aquifers, pores between mineral grains are the open spaces available for the storage of water. Porosity is defined as the ratio of the volume of pores to the volume of bulk rock and is usually expressed as a percentage. Pore space by itself is insufficient. There must also exist pathways for the water to move through the pores so that overall the rock will transmit water. The ability of a rock to transmit water is referred to as its *permeability*. It is possible to have a rock, for example, a volcanic pumice, that has a high porosity but an almost zero permeability. Many types of rocks can be effective aquifers, but not all of them. Shales typically have both low porosity and low permeability. Rocks that do not transmit water are called *aquicludes*. They frequently serve as barriers or confining layers in the movement of groundwater.

In addition to the pore spaces between mineral grains, many rock units have been fractured so that the bulk rock is laced with cracks, known as *joints*, and more sparsely with clusters of joints, known as *fractures* or *joint swarms*. In addition, there may be partings where the rock has separated along bedding planes. The term *fracture* is used loosely for all of these mechanical openings. Fractures also act as pathways for groundwater movement so that one can speak of *matrix permeability* referring to the interconnected pores and *fracture permeability* referring to joints, larger fractures, and bedding plane partings. In some rocks, for example, fractured granites and fractured massive basalts, fracture permeability completely dominates matrix permeability.

Aquifers are usually named with the name of the rock formation in which they occur. Rock formations extend indefinitely along the beds but the beds have a definite thickness. The aquifer thickness enters into various calculations but the lateral extent does not.

Some aquifers are local; others extend over hundreds of kilometers. Descriptions of the principal aquifers in the United States have been compiled into the *Ground Water Atlas* by the U.S. Geological Survey (United States Geological Survey, 1992–1998).

Groundwater in porous media aquifers moves very slowly, typically a few meters per year. The velocity is proportional to the hydraulic gradient according to Darcy's law

$$v = -K \frac{dh}{d\ell}$$

The proportionality constant, K, has units of velocity and is called the *hydraulic conductivity*. The hydraulic conductivity is related to properties of the rock, properties of the fluid, and the strength of gravity that provides the ultimate driving force moving the water. If the cross-sectional area, A, of the bit of aquifer being examined is specified, Darcy's law also describes the volume of water moving across the specified area per unit time. Combining these ideas

$$Q = vA = -A \frac{N d^2 \rho g}{\eta} \frac{dh}{d\ell}$$

The parameter d is the mean diameter of the mineral grains and N is a geometric factor to take account of the irregular shape of the grains. Nd^2 is the permeability, a property of the rock. Permeability and hydraulic conductivity are often confused. Because both measure the ability of the rock to transmit water, they are sometimes used—incorrectly—interchangeably. Hydraulic conductivity has units of velocity; permeability has units of area. Darcy's law is valid for other fluids (such as petroleum) as well as water. The properties of the fluids are described by their density, ρ, and viscosity, η, both functions of temperature. The gravitational force is described by g, the acceleration due to gravity.

Darcy's law is one of the most powerful statements in all of hydrology. It underlies most of the theory of groundwater flow. When combined with statements of the conservation of water within a unit volume of the aquifer, the result is a differential equation that describes the groundwater flow field. Many pages of standard textbooks on groundwater hydrology deal with the boundary conditions and solutions of these equations. The results take the form of flow fields, techniques for testing water wells, prediction of contaminant plumes, and many others. One of the most fundamental distinctions between karst aquifers and most porous media aquifers is that in important parts of the karst aquifer, Darcy's law does not work.

Another aspect of karst aquifers that differs remarkably from normal aquifers is in the relationship between groundwater and surface water. In most porous media aquifers, groundwater and surface water are disconnected. In a rainstorm, some water soaks into the ground as infiltration, I. Some water runs off over the land surface to enter creeks and rivers, which quickly drain the water away as runoff, R. Some water is lost back to the atmosphere by evaporation and by the transpiration through the leaves of plants. Generally, these factors are taken together as evapotranspiration, E. Over long time periods, these factors must balance

$$P - E = I + R$$

Surface water flows in streams and rivers at velocities of fractions of a meter per second whereas groundwater moves at velocities of meters per year. The two can be treated as almost separate entities. The same is not true in karst aquifers.

KARST AQUIFERS

The Triple Porosity Model

The internal structure of a karst aquifer contains three types of permeabilities and is thus sometimes described by what is called the *triple porosity* or *triple permeability* model.

1. Matrix permeability is due to the pore spaces within the bedrock. Matrix permeability in karstic aquifers is not intrinsically different from the matrix permeability in porous media aquifers.
2. Fracture permeability is due to joints, joint swarms, and bedding plane partings. Again, these are not intrinsically different from fractures in sandstone, granite or other massive rocks except for their tendency to be enlarged by dissolution of the bedrock.
3. Conduit permeability consists of solutionally generated caves or conduits in the aquifer bedrock. Flow of water in conduits is similar to flow in pipes. The flow regime may be turbulent. The flow velocities are in the range of centimeters per second compared with meters per year in porous media aquifers. Water is confined to the conduits with little dispersion. As a result, contaminants are not dispersed to form a plume. Instead, they are carried down the conduit to a spring with little dilution except for incoming tributary passages.

The properties of each component of the permeability are summarized in Table 1. Many Paleozoic limestones and dolomites have very little matrix porosity, while young limestones that have not been subject to the heat and pressure of deep burial are quite porous.

TABLE 1 Characteristics of the Three Components of the Triple-Permeability Model

Permeability	Aperture	Travel Time	Flow Mechanism	Guiding Equation	Distribution
Matrix	μm–mm	Long	Darcian flow field: laminar	$h_f = \frac{\eta v L}{\rho g (N d^2)}$	Continuous medium
Fracture	10 μm–10 mm	Intermediate	Cube law: mostly laminar, may be nonlinear components	$\frac{Q}{h} = \frac{C}{f} B^3$	Localized but statistically distributed
Conduit	10 mm–10 m	Short	Darcy–Weisbach: open channel and pipe flow turbulent	$h_f = \frac{f L v^2}{4 g r}$	Localized

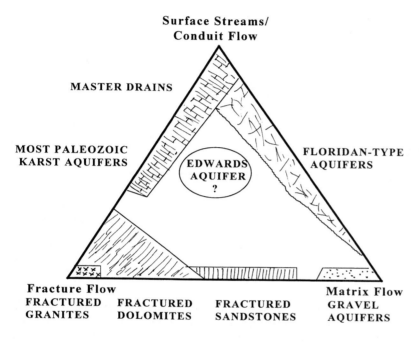

FIGURE 1 The three types of aquifer permeabilities expressed as a triangular diagram. Normal aquifers would lie along the bottom of the triangle. Karst aquifers are those in which conduit permeability is an important component.

Limestones and dolomites tend to be hard, brittle rocks and are often highly fractured.

All aquifers can be categorized by the triple porosity model, as illustrated in Figure 1. Various nonkarstic aquifers such as fractured sandstones, fractured granites, and gravel aquifers lie along the base of the triangle where there is little or no contribution from dissolution processes in the development of the aquifer. Dense, brittle carbonate rocks form aquifers that lie along the side of the triangle between conduit flow and fracture flow. Aquifers in young carbonates such as the Floridan aquifer contain flow systems that are a mix of matrix and conduit flow. There are aquifers, such as the important Edwards Aquifer in central Texas, that contain contributions from all three types of permeabilities.

Conduit Permeability

The feature of karst aquifers that makes them distinct from other aquifers is the presence of conduits, which act as networks of pipes carrying water rapidly through the aquifer. The degree of conduit development is extremely variable. The connection of the conduits to the other types of permeabilities is also highly variable. As a result, there is a range of aquifer types, as illustrated in Figure 2, with very different roles played by the conduit system.

At one end of the sequence (A) is the downstream reach of a sinking stream with its catchment mostly as surface streams on noncarbonate rock. In this case the conduits take the form of a drainage tunnel. Water moves through the tunnel at velocities similar to those in surface streams, in one end and out the other. There are only weak connections between the water flowing in the conduit and the fracture and pore water in the surrounding rock. When a larger fraction of the drainage basin is karstic, sinkhole drains and smaller conduits converge to form master trunk systems (B). Master trunk and branchwork conduits form where the flow of water is under moderate to high hydraulic gradients. These conduit systems behave hydraulically much like surface streams except that they happen to be underground.

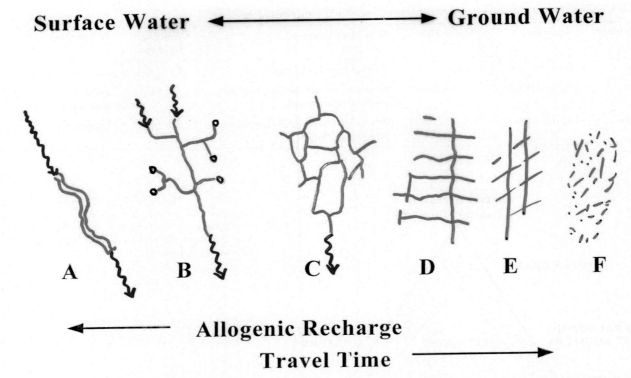

FIGURE 2 Sketches showing various types of conduit permeabilities: (A) Single conduit with mainly surface water input; (B) branchwork conduits with sinkhole and sinking stream inputs; (C) anastomotic maze of interconnected tubes developed along bedding plane partings; (D) network maze developed along vertical fractures; (E) permeability consisting of solutionally widened fractures; and (F) matrix permeability only. Flow is most localized at (A) and most diffuse at (F). Travel times increase left to right. The sequence moves from a surface stream with a roof at (A) to a porous-medium groundwater aquifer at (F).

When gradients are low or where the free flow of the conduit system is impeded, water circulates along multiple pathways, gradually dissolving out either networks of joints or multiple loop openings along bedding plane partings. Dissolution along fractures leads to network mazes (C); circulation along bedding plane partings leads to anastomotic mazes (D). If master trunk and branchwork conduit systems can be likened to surface streams, network and anastomotic mazes can be likened to swamps. Like swamps, much water can move through these multiple pathways but at very low velocities. The sequence continues with aquifers of less and less karstic character to fracture aquifers (E) and porous media aquifers (F). Real aquifers can be much more complicated depending on their geology and how they have evolved over time.

The flow through conduits is generally turbulent whereas the flow in fractures and pores is generally laminar. Flow velocity increases with conduit size. Under usual hydraulic gradients, the onset of turbulence occurs when conduit apertures reach about 1 centimeter. We may take one centimeter—either the width of a solutionally widened fracture or the diameter of a tubular conduit—as the threshold for conduit flow behavior. Continuous flow paths through the rock with an effective width of one centimeter or greater would be considered conduit permeability. Smaller openings would be fracture or matrix permeability.

Caves are not exactly equal to conduits. Caves are explored by humans, which gives them a minimum dimension of roughly 0.5 m whereas with a minimum aperture of 1 centimeter there are many conduits that are too small to be called caves (or, at least, are too small to be explored by humans). Conduits are organized to provide continuous flow paths because for them to function as permeability in the aquifer, water must be able to move continuously from input to output. Caves are often occluded and are discontinuous; caves are fragments of conduits. Active conduits are located near or below the water table. The dry caves that lie above the active zone of groundwater circulation are fragments of abandoned conduits.

Water flows through conduits either as open channel flow (a stream with a free air surface), or as pipe flow (a completely water-filled tunnel). A single conduit may have sections with open stream passage and sections that are water-filled. Exploration of cave sumps by scuba divers has shown that conduits

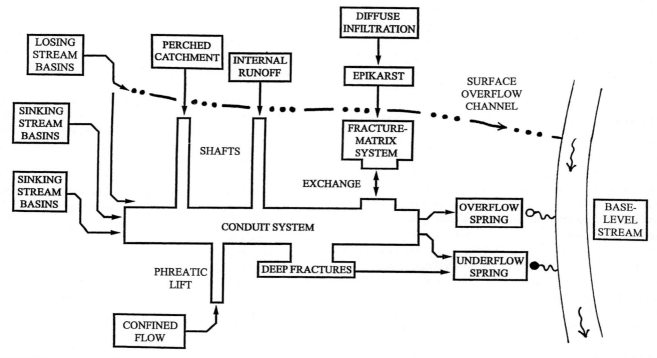

FIGURE 3 A conceptual model for a carbonate aquifer showing various inputs and flow paths. Any particular groundwater basin may or may not contain all of these features.

frequently undulate in the vertical plane with alternating free surface streams and sumped passages.

Karst Groundwater Basins and Groundwater Flow Systems

Because of the rapid transmission of water through conduits and because of the rapid drainage of surface water into the aquifer through sinkholes and surface streams, it is necessary to combine both surface water and groundwater characteristics. The result is the concept of a groundwater basin which includes both surface and subsurface components. Combining water sources with the internal characteristics of the aquifer and with the discharge of water back to surface routes produces the *standard model* of karstic aquifers illustrated in Figure 3.

The standard model contains the following sources of recharge:

1. *Dispersed (or diffuse) infiltration*. Rainfall onto the karst surface infiltrates through the soils, the epikarst, underlying fractured bedrock, to eventually reach the water table and the active conduit system. This component is not intrinsically different from infiltration in any aquifer.
2. *Allogeneic recharge from surface streams*. Mostly, these drain from nearby drainage basins on noncarbonate rocks. Allogeneic recharge enters the aquifer at well-defined swallets. Sometimes these are cave entrances, pits, or debris piles in blind valleys, and sometimes they are just sections of streambed where water is gradually lost. Flood flow in allogeneic basins injects large pulses of water into the aquifer on short time scales. There is little or no filtration where surface streams become groundwater. Dissolved contaminants, suspended contaminants, and even massive objects such as cans, old tires, household garbage, and dead animals can be swept into the aquifer during periods of high flow.
3. *Internal runoff*. The land surface above karst aquifers is often pocked with closed depressions of various sizes. Some of these are soil mantled, some have exposed bedrock, and a few expose an underlying underground stream—often called *karst windows*. Some closed depressions have collapses in the bottom where soil piping has occurred. During moderate storms, the rainfall infiltrates through the soils into the underlying aquifer through fractures and bedding plane partings. However, there are often no surface runoff channels in sinkhole topography. Overland flow during storms drains to the bottoms of the closed depressions to make its way into sinkhole drains and thus directly into the underlying aquifer. Storm runoff can pond in the

closed depressions and the seepage pressure can induce soil piping failures. Soil piping failures inject volumes of clastic sediment into the aquifer along with any contaminants that may have accumulated in the sinkhole.

Piracies and Spillover Routes

Unlike surface water basins, groundwater basins rarely have rigorously fixed boundaries. Escape routes along conduits that cross the basin boundaries into adjacent drainage basins are common. These are known as *piracy routes* and are most common in locations where the groundwater basin seems to be defined by the overlying surface water basin. Overflow routes on the land surface converge to a master stream that is often the location of the spring from which the groundwater discharges. However, certain surface streams that sink underground within the surface water basin may cross surface divides, sometimes under ridges or mountains, and appear in springs far outside the original basin. Such piracies can be identified by tracer studies, but these must be conducted with great care because dyes injected into swallow holes may reappear in completely unexpected places.

Karst aquifers evolve over time. New conduits are formed at depth and old conduits are drained and abandoned. The old conduits may have converged on a spring in a particular surface water basin. The new conduits may have shifted the flow path so that the present system drains to a new spring which may be in a different surface water basin. Under low-flow conditions, the present-day basin may be well behaved with all recharge draining to the active spring. During flood flow, however, rising water levels within the conduit system may activate some of the abandoned conduits. The result is that floodwaters are diverted into the old drainage system. Such spillover routes are also a common feature of karst aquifers.

The Karst Water Table

In a porous medium aquifer, rainfall infiltrates through the soil into the underlying bedrock. It moves vertically through the pore spaces in the rock, displacing the air that occupied the pores. This portion of the aquifer is known as the *vadose* or *unsaturated zone*. The infiltrating water continues to move vertically until it reaches a portion of the aquifer where all pore spaces are water-filled. This portion of the aquifer is known as the *phreatic* or *saturated zone*. The interface between the vadose and phreatic zones is the *water table*. In a porous medium aquifer, the water table is a surface that stands higher under topographic highs and lower under topographic lows. The water table slopes from topographic highs to surface streams which are the locations of groundwater discharge. The slope of the water table is supported by the hydraulic resistance for groundwater moving through the pore spaces of the rock.

In a karst aquifer, the open pipes of the conduit system have a very low hydraulic resistance. The presence of conduits creates a trough in the water table. The water table in the fracture and matrix permeability in the nearby bedrock slope toward the conduits rather than to the surface streams. The conduit system acts as a drain for the entire aquifer. Sinkholes and sinking streams allow very rapid input of storm water into the aquifer. The groundwater trough fills quickly, often flooding the conduit systems.

In terrain with low to medium relief, the karst water table is well defined. The location of the water table can be determined by measuring the elevation of standing water in wells. If there is a sufficient density of wells, the standing water elevations can be contoured to construct a map of the water table surface. In high relief terrain, the position and even the existence of the water table is more difficult to define. Recharge water from rainfall and snow melt in mountainous regions often descends through the vadose zone as waterfalls in shafts and steep gradient free-surface streams which may cross over each other. The presence of even a large free surface stream does not necessarily mark the location of the water table. Furthermore, sumped passages may occur that are perched above regional base levels, and these also do not necessarily indicate the location of the water table either.

CHARACTERISTICS OF KARST AQUIFERS

Discharge Characteristics: Hydrographs

Karst aquifers usually discharge through a relatively small number of large springs. The discharge, turbidity, and chemistry of water from karst springs provide information on the generally inaccessible processes upstream in the groundwater basin.

Surface drainage basins respond very rapidly to storms. A sudden intense storm will send a flood down the valley of the surface stream. Depending on the degree of development of the conduit system, groundwater basins may respond much the same way. If there is a well-developed conduit system, storm inputs through sinking streams and internal runoff into sinkholes will move rapidly through the system. The discharge at the spring will increase rapidly and

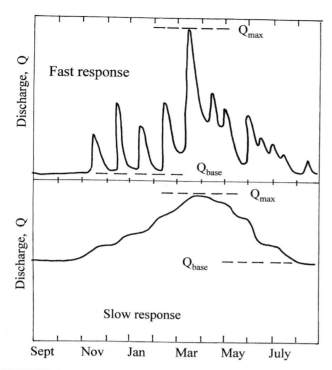

FIGURE 4 Sketches of (A) a fast response hydrograph showing individual storm peaks and (B) a slow response hydrograph showing only seasonal wet and dry periods. The horizontal scale is the water year, which begins in September in the eastern United States.

the water may become muddy. Other springs, draining groundwater basins with more poorly developed conduit systems, will have a more subdued response. These responses can be described by measuring the time variation of discharge at the spring and constructing spring hydrographs (Fig. 4). Fast response springs are those with open conduit systems so that the rise and fall of the storm hydrograph is short compared with the average spacing between storms. The storm flow from the spring rises, peaks, and falls back close to base flow before the next storm arrives. The discharge from slow response springs may increase during wet seasons and decrease during dry seasons, but the system responds too slowly for the spring to be impacted by individual storms. There are, of course, intermediate-response springs with some storm response but with the hydrographs of individual storms smeared together.

Spring hydrographs provide several parameters for characterizing the aquifers that feed the spring. The ratio of peak discharge to base flow discharge, Q_{max}/Q_{base}, is a measure of the "flashiness" of the aquifer. The recession of the hydrograph often has an exponential form. It can be fitted to a function of the form

$$Q = Q_0 \exp(-t/\tau)$$

where Q is spring discharge in volume per unit time, t is time, and τ is the response time of the aquifer. Plots of hydrograph recession curves for many highly karstic groundwater basins show two (or more) segments, a fast response time on the order of a few days, and a slow response time on the order of 20–30 days. One interpretation is that the fast response represents the draining of the conduits while the slow response is the draining of fractures and small openings that feed into the conduits.

Carrying Capacity of Conduit Systems

When surface streams flowing on noncarbonate rocks reach the contact with the carbonates, the amount of surface-stream flow lost to the carbonate aquifer depends on the carrying capacity of the conduit system. If the conduit system is poorly developed so that it can carry only a portion of the base flow, the stream will remain on the surface across the carbonate rock. The flow in the surface stream will be smaller than expected for the area of the basin because part of the flow will be lost into the carbonate aquifer. At the next stage, the conduit system can carry base flow but not storm flow. At this stage, the surface stream will go dry during the dry season. With further enlargement of the conduit system, small storms can also be accommodated. At this stage, the surface stream will be dry most of the time and will carry water only during flood flow. Because of disuse, the streambed will become overgrown with vegetation, banks will cave in, and sinkholes may develop along the stream channel. At the final stage, when the conduit system can carry the most extreme flood flows, the surface channel will eventually disappear. Upstream from the swallet, surface flow on noncarbonate rocks will continue to deepen the channel, resulting in a blind valley.

Base Flow/Area Relationships

The boundaries of the groundwater basins that subdivide karst aquifers are determined by a variety of evidence—geological constraints, tracer tests from sinkholes and sinking streams to springs, and explored cave passages—but the basin boundaries cannot be measured directly. In most regions, the base flow emerging from the spring that drains a groundwater basin is proportional to the basin area.

$$Q = kA$$

The proportionality constant, k, varies depending on the hydrogeology of the basin, particularly the development of the conduit system. Basins with well-developed conduit systems will drain effectively between storms so that base flows will be small. Basins with poorly developed conduit systems will store water more effectively so that base flows are maintained. A serious disagreement between basin area calculated from base flow and basin area determined from the interpreted boundaries is an indication of the presence of piracy routes or of serious error in interpreting the boundaries.

Clastic Sediments

The solutionally widened fractures and conduits in karst aquifers are sufficiently wide to permit the transport of clay, silt, sand, and gravel and even materials up to the size of cobbles and boulders. More importantly, the velocity of moving water is sufficient to drag solid particles through the aquifer.

Clastic sediments are transported mainly during flood flow episodes. Very small particles, mainly clay, are transported in suspension, making the water turbid or muddy. When water velocities decrease following a storm event, the sediments in suspension settle out, leaving a layer of mud. Coarser particles, sand and gravel, are transported as bedload. That is, the particles are rolled along the bottom of the conduit by the tractive force of the water flowing over them. Bedload transport requires a minimum velocity, which depends on particle size, to place the particles in motion. Banks of silt, sand, and gravel are frequently found in caves. Streambeds in caves are often armored with a layer of gravel and cobbles that move only during flood flows.

Transport of clastic sediment is an essential part of the functioning of the karst aquifer. In completely developed aquifers where all surface streams sink underground, the insoluble material carried by the sinking surface streams must be carried through the conduit system and washed out through the springs. If this did not happen, the conduit system would eventually become completely clogged with sediment and drainage would be forced to return to the surface.

WATER SUPPLY AND WATER QUALITY

Water Wells in Karst

Groundwater for domestic or industrial use is extracted by drilling wells. Pumping water from the wells lowers the water level in the well (a quantity known as *drawdown*), thus creating a local hydraulic gradient that forces water from the surrounding rock to flow toward the well. For any given aquifer, there will be a maximum quantity of water that can be pumped from the well determined by the balance between the pumping rate and the drawdown. Water supply wells are located to provide the maximum quantity of water with minimum drawdown.

The optimum placement of wells in karst aquifer is difficult to determine. A well drilled into a flooded conduit would provide large quantities of water with negligible drawdown, but the water might be of questionable quality. In contrast, wells drilled into unfractured carbonate rock are often dry holes or at best provide only small quantities of water. For those aquifers with fracture permeability consisting of vertical fractures, wells drilled on fractures or on fracture intersections provide the best yields. For those aquifers with fracture permeability consisting of bedding plane partings, these partings provide high permeability zones that will be tapped regardless of the detailed location of the well.

Water Quality

In the highly populated and developed countryside of today, one learns quickly not to drink from creeks and rivers. Municipal water companies draw water from surface sources, but these supplies are filtered, treated, and chlorinated before being released into public water supplies. Water from wells and springs, on the other hand, is often thought to be "pure." This statement is not true in general and it is exceptionally untrue when applied to karst aquifers. Karst springs are the outlets draining open conduit systems. Conduits receive portions of their water from sinkholes and sinking streams. There is little or no filtration. Any surface contaminants that enter the conduit system are flushed through to the springs with little dilution. Karst springs should be regarded with the same suspicion as surface streams.

Groundwater in carbonate aquifers tends to contain high concentrations of magnesium and calcium. These ions, Mg^{2+} and Ca^{2+}, make the water "hard." Hard water is not a health risk, but it does require more soap for washing and laundry. It tends to precipitate calcium and magnesium carbonates in hot-water heaters and other plumbing fixtures and may require water softeners for many uses. Wells drilled into fractured carbonate aquifers are also at risk. Open fractures can allow contaminants to migrate down from the land surface. Groundwater in the fractures generally flows down-gradient to the conduit system. Pumping on wells in fractures may reverse the hydraulic gradients and cause water to flow from the conduits to the wells, bringing contaminants to the wells.

Contaminant Transport

Because of high velocity flow restricted to the conduit system, contaminants in karst aquifers generally do not spread out into a plume as they do in normal aquifers. Different categories of contaminants have different modes of transport within karst aquifers. Types of contaminants are listed below:

1. *Water-soluble compounds*. Inorganic salts and some organic compounds are soluble in water. These move with the water in the aquifer and appear at the springs possibly in a somewhat diluted form. In the case of a spill, water-soluble compounds will reach the spring in about the same time as the travel time of the water, often a matter of hours or at most a few days.
2. *Light nonaqueous phase liquids (LNAPLs)*. Gasoline, fuel oil, home heating oil, and related hydrocarbons are less dense than water and are only slightly soluble. These compounds will float on the water table and will float on free-surface underground streams. However, they tend to pond behind sumps and tend to be trapped in pockets in the ceilings of water-filled conduits. Rising water levels can force LNAPLs upward along fractures where fumes can enter homes and other buildings.
3. *Heavy nonaqueous phase liquids (DNAPLs)*. Chlorinated hydrocarbons such as trichloroethylene (TCE) and perchloroethylene (PCE)—both used as solvents, degreasers, and dry cleaning agents—as well as polychlorinated biphenyls (PCB) and many other compounds are denser than water and are only slightly soluble. These materials are sometimes trapped in the epikarst but when they enter the aquifer, they tend to sink to the lowest water-filled passages or become trapped in the clastic sediments that occur in the conduits. As a result, spills of DNAPLs often never reappear at springs and remain trapped in the karst aquifer for long periods of time.
4. *Metals*. Metallic elements such as chromium, cadmium, lead, and mercury are highly toxic, while others such as copper and zinc are less so, and these can be carried into karst aquifers either as ionic species in solution or as solid particles of various sorts. Because of the alkaline chemistry of karst waters, some metals are precipitated, some are adsorbed on clay particles, and some are incorporated into the manganese and iron oxides/hydroxides that form coatings on cave stream sediments. Metal transport in karst aquifers, therefore, involves a very complex chemistry that is not easily generalized.
5. *Pathogens*. Viruses, bacteria, protozoa, and larger organisms are easily transported into karst aquifers because of the large solution openings and the absence of filtering. Most common of these are the fecal coliform family of organisms and fecal streptococci. These organisms are indications of contamination by sewage or animal waste. *Giardia lamblia* is of most concern among protozoa. It is released in a cyst form in animal feces and is present in many surface waters. Sinking streams carry the stable cysts into the subsurface. Die-off is slow underground so that karst waters remain contaminated far from surface inputs.

See Also the Following Articles

Modeling of Karst Aquifers
Sinking Streams and Losing Streams
Springs
Water Tracing in Karst Aquifers

Bibliography

Ford, D., & Williams, P. W. (2007). *Karst hydrogeology and geomorphology and hydrology*. Chichester, U.K.: John Wiley.
Milanović, P. T. (1981). *Karst hydrogeology*. Littleton, CO: Water Resources Publications.
Milanović, P. T. (2004). *Water resources engineering in karst*. Boca Raton, FL: CRC Press.
Palmer, A. N. (1991). Origin and morphology of limestone caves. *Geological Society of America Bulletin, 103*(1), 1–21.
Palmer, A. N. (2007). *Cave geology*. Dayton, OH: Cave Books.
United States Geological Survey (1992–1998). Ground water atlas of the United States. *Hydrologic investigations atlas*, 730-B-M, 12 folio volumes.
White, W. B. (1988). *Geomorphology and hydrology of karst terrains*. New York: Oxford University Press.
White, W. B. (2002). Karst hydrology: Recent developments and open questions. *Engineering Geology, 65*(2), 85–105.
White, W. B. (2007). Groundwater flow in karst aquifers. In J. W. Delleur (Ed.), *Handbook of groundwater engineering* (2nd ed., pp. 21-1–21-47). Boca Raton, FL: CRC Press.

HYDROTHERMAL CAVES

Yuri Dublyansky

Innsbruck University, Austria

DEFINITION

The term *hydrothermal karst* defines a process of dissolution of cavities in the rocks under the action of hot waters. This definition, though quite simple, is sometimes difficult to apply, because it requires another definition of which water should be called hot or thermal. In hydrological studies any water that is appreciably warmer (5°C or more) than the surrounding environment is called thermal. This type of definition is fairly satisfactory when it is applied to a still-active process in areas with a moderate climate. It is difficult to apply, however, to those settings where neither resurgence temperatures nor annual climatic averages are applicable (*i.e.*, to deep-seated waters tapped by boreholes or to fossil karst process). Conventionally, the temperature of 20°C is considered to be the lower limit of the hydrothermal environment. Although some meteoric karst systems in hot arid climates may exceed this limit without any thermal input, most hydrothermal cave systems relate to hypogenic sources of energy (*i.e.*, internal heat of the Earth). Formed due to the action of rising waters, hydrothermal systems typically lack any genetic relationships to recharge from overlying or directly adjacent surfaces. Caves must be uplifted with the rock and intersected by surface erosion by common karst, or by drilling, mining, or quarrying to be discovered and studied. Hydrothermal karst is a special case of hypogene karst.

The concept that some caves could have been formed by ascending thermal waters rather than by cold descending, gravity-driven ones was first put forth as early as in the mid-nineteenth century. It has been suggested that Pb-Zn ores in some of the European deposits in carbonate rocks were emplaced in dissolution cavities and that these cavities owe their existence to the same solutions from which, at later stages, the ores were deposited. In his "A Treatise on Metamorphism", Charles Van-Hise (1904) provided explanations for how and why hydrothermal solutions move advectively through the rocks, and what causes their aggressiveness. He conjectured that most hydrothermal solutions originate as common meteoric waters that become heated during their circulation deep in the Earth's crust. These early works laid a foundation for the concept of hydrothermal karst.

SETTINGS OF HYDROTHERMAL KARST

It is convenient to subdivide hydrothermal karst settings into the following three categories: endokarst, deep-seated hydrothermal karst, and shallow hydrothermal karst.

Endokarst

It is a well-known fact that the rate of karst development decreases with increasing depth. However, deep drilling for oil and gas reveals that solutional porosity of carbonate rocks at depths of 4–5 km may be as great as 18–28%, and the porosity of aluminosilicate rocks may be as great as 30–35%. The pressures of fluids at these depths are typically greater than the hydrostatic ones and may approach lithostatic values. At those levels, where the pressure exceeds the strength of the rock, pores and cavities may exist only if they are filled with high-pressure fluid, which prevents them from failure. The process of formation of such cavities is termed *endokarst* (Andreychouk et al., 2009). The size of endokarstic cavities does not normally exceed several centimeters. Though this type of karst apparently does not produce traversable caves, it may play a significant role in creation of deep-seated reservoirs for hydrocarbons.

Deep-Seated Hydrothermal Karst

This covers a range of depths (approximately 0.3 to 4.0 km) where the temperature gradients are relatively small, pressures are close to hydrostatic, and the influence of the temperature changes at the Earth's surface is practically absent. Processes of cave excavation and cave infilling occur in response to the change in physicochemical parameters of fluids moving toward the Earth's surface, such as the decrease in temperature and pressure. It is thought that dissolution related to the elevated content of carbonic acid is the leading factor in initiation and enlargement of caves in the deep setting. Recent studies, however, revealed the important role of dissolution processes involving sulfuric acid. At certain specific combinations of bedrock solubility and frequency of fractures, dissolution is accompanied by collapse. Characteristic collapse breccias are known in many fossil hydrothermal systems related to Pb-Zn ore deposits in dolomitic rocks (*e.g.*, in Silezia, Poland).

Shallow Hydrothermal Karst

The shallow setting describes processes developing near the free surface of the thermal water—both below and above it. In this zone the pressures are low (down to atmospheric) and temperatures may range from boiling to just slightly exceeding the ambient ones. The temperature gradients may be significant, which leads to the appearance of some specific and powerful processes, like thermal convection and condensation corrosion. Also, this is a zone where upwelling thermal waters meet colder oxidized meteoric waters and atmosphere. This may induce specific reactions and processes such as H_2S oxidation, mixing corrosion, and cooling corrosion. The caves formed in such a setting commonly exhibit extremely diverse morphologies. Characteristic dimensions of individual caves are appreciably greater than those of the deep-seated hydrothermal caves. Also, shallow hydrothermal karst speleothem types are much more varied than those of the deep-seated karst.

All enterable active hydrothermal caves (those containing hot waters) are examples of shallow hydrothermal karst. Examples of such caves are known in many places including Turkmenistan (Bakharden Cave, with a lake with temperature of 35–37°C), Italy (Grotta Giusti, 32–34°C), Mexico (Sistema Zacatón, 29–32°C), and Hungary (Molnár János Cave, 18–24°C).

CHEMISTRY OF FLUIDS AND PROCESSES OF CAVE EXCAVATION

Hydrothermal caves associated with the CO_2 and caves formed by waters containing H_2S are considered two major classes of hydrothermal (hypogene) caves.

Dissolution by Rising Thermal Water (CO_2)

Thermal waters rising from significant depth are commonly saturated with CO_2, which may originate from metamorphism of carbonate rocks and igneous activity. The solubility of CO_2 in water depends on both temperature and pressure. Water saturated with respect to CO_2 at deep levels (e.g., 2–4 km) becomes supersaturated as it rises toward the surface. Hence, CO_2 must exsolve in the gaseous phase and leave the system. Rising carbonic thermal waters also cool down. Due to inverse relationships between carbonate solubility and temperature, they may acquire and maintain aggressiveness, even at decreasing CO_2 levels. The solubility of $CaCO_3$ increases evenly along the ascending fluid path, but near the land surface (or water table) it drops drastically. Such nonlinear behavior leads to the appearance of two geochemical zones: a zone of carbonate dissolution at depth and a zone of carbonate precipitation closer to the surface.

Oxidation of Sulfides (H_2S)

Sulfuric waters become aggressive when their dissolved H_2S oxidizes on contact with oxygen-rich waters or air to form sulfuric acid. Conversion of H_2S to H_2SO_4 produces a sharp increase in dissolution. The effect is attenuated when CO_2 generated by the H_2SO_4-$CaCO_3$ reaction is degassed. In two settings, hydrogen sulfide oxidation is an important speleogenetic process. The first is subaqueous dissolution of carbonates near the water table. Large rooms of the Carlsbad Cavern in New Mexico, U.S.A., are believed to be formed this way. The second setting is dissolution of and subsequent replacement of calcite by gypsum and its consequent removal above the water table. The mechanism, termed *replacement corrosion*, was suggested for caves of the Big Horn Basin in Wyoming, U.S.A.

Dissolution Due to Mixing of Waters (CO_2 and H_2S)

Solutional aggressiveness can be renewed or enhanced by mixing of waters with contrasting chemistry, particularly those differing in CO_2 and H_2S content or salinity. Mixing of waters having different temperatures produces a similar effect, due to contrasts in CO_2 contents. This effect, known as *mixing corrosion*, is thought to be responsible for the development of network maze cave systems in Budapest, Hungary, for example, Pál-völgy, Szemlohegy, and Ferenc-hegy.

Hydrothermal Karst in Noncarbonate Rocks

The mechanisms described above pertain to the most common variety of hydrothermal karst developing in carbonate rocks. Besides, hydrothermal caves have been reported from silicate rocks (quarzite, scarn, jasperoid, quartz veins), sulfate rocks (gypsum, anhydrite), rock salt, and even from massive sulfide ores.

Hydrothermal Karst Related to Oxidation of Sulfide Ores

A substantial amount of heat can be released when infiltrating oxygen-rich waters react with sulfide ores. This may lead to both increased temperatures and enhanced carbonate aggressiveness of waters passing through ore bodies. Those thermally and chemically modified waters may then attack carbonate rocks to produce a specific type of hydrothermal karst. This type of hydrothermal karst is commonly triggered by mining activities that may facilitate access of waters to the ore bodies.

Dissolution in Subaqueous and Subaerial Settings

Many large hydrothermal caves and cave systems have formed in subaqueous conditions, that is, below the water table. In addition, significant development of caves can also occur in a subaerial setting, above the open surfaces of underground thermal lakes by a mechanism of *condensation corrosion*. Moisture evaporating from the lake surface moves upward and condensed on cooler bedrock. The resulting water film becomes aggressive by dissolving gaseous CO_2; the condensate attacks the bedrock and creates cavities with characteristic cupola-like morphology (Fig. 1). This speleogenetic mechanism can be very efficient, but requires a number of prerequisites, such as elevated temperature of water (promoting evaporation) and sustained gradient of temperatures between the

FIGURE 1 Solutional cupolas in an ascending channel presumably formed in a subaerial setting by condensation corrosion. Kraus cave, Austria. *Photo by L. Plan. Used with permission.*

vapor phase and cave wall (promoting condensation). Importantly, by this mechanism, cavities can develop above the underground lake water in which it is saturated with respect to bedrock and is not capable of dissolving it.

MORPHOLOGY OF HYDROTHERMAL CAVES

Hydrothermal karst produces a large variety of cave morphologies with cave sizes ranging from solution enlarged pores to extensive cave systems with total mapped passage length exceeding 100 km. The most common morphologic types of hydrothermal caves are discussed below.

Solution Porosity

Solution enlarged pores are commonly observed in borehole cores from depths up to 4–5 km in geothermal areas and oil fields. They may form extensive layers of rock in which solution voids account for as much as 5–15% of the entire rock volume. In these places such horizons become parts of oil and gas reservoirs. Layers of solution-enhanced transmissivity appear where the movement of the fluids is slow, of the order of a few millimeters per year or less, and where hydraulic structures which could concentrate the flow (faults and fissures) are absent. Such zones of enhanced porosity may become the inception horizons for future caves.

Isometric Rooms

Roughly spherical pockets or rooms with diameters ranging from 0.5 to 8.0 m were reported from Khod Koniom Cave in the Crimea. The cavities, lined with crystals of hydrothermal calcite (temperature of formation 40–85°C) and filled with red clay are truncated by a later vertical vadose cave at a depth of 80–200 m. Similar caves occur in Kighizstan, where medieval excavations in the Birksu mercury mine (Turkestan Range) uncovered a series of near-spherical rooms 3–8 m in diameter. The rooms coalesce in two- and three-dimensional clusters following the bedding planes and minor faults. Originally, the rooms were entirely filled by massive hydrothermal calcite containing veinlets of cinnabar, but the ore was removed by medieval miners.

Individual Chambers

Individual chambers are distinguished from isometric rooms discussed above by their larger dimensions, ranging from tens to hundreds of meters. Caves of this type are composed of one or several large individual chambers. The latter commonly have a length of 100–200 m, a width of 30–60 m, and a height of 80 m. Examples of caves belonging to this type are Bakharden Cave in Turkmenistan (Kopet-Dag Range), Novoafonskaya Cave in Abkhazia (Caucasus), Karani Cave in the Crimea, Ukraine, and Champignons Cave in Provence, France. Another spectacular example of individual chambers are cenotes Caracol and La Pilita (Mexico), which have nearly spherical shapes and diameters of 80 to 100 m and are filled with moderately thermal water (29.6 and 31.6°C).

Single-Conduit Caves

As implied by the name, the single-conduit caves are composed of a single long, typically tube-shaped passage, which can be aligned horizontally, parallel to the water table, or at a steep angle to it. Horizontal single conduit caves are exemplified by the hydrothermal Hellespont, Spence, and Kane Caves (Wyoming, U.S.A.), which represent nearly horizontal, tube-shaped conduits 60–600 m long. They have developed in a vadose setting, where ascending H_2S-bearing fluid comes in contact with the air, resulting in replacement corrosion. Thermal springs discharge through the three above-mentioned caves, and several inactive caves having similar morphology are known in the region. Steeply dipping single-conduit caves develop in a phreatic setting. Examples of such caves are Grotte de Chat (France), Pozzo del Merro (Italy; the deepest explored underwater cave of the world), and cenote Zacatón (Mexico).

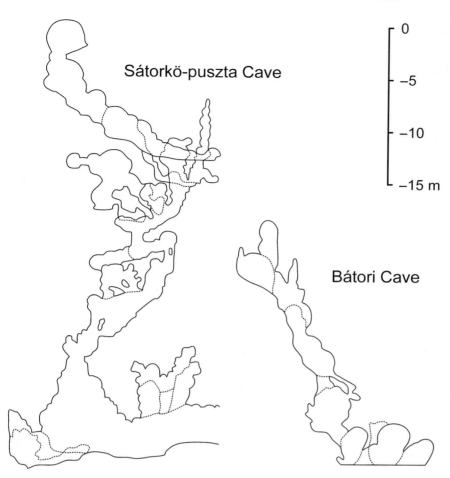

FIGURE 2 Examples of the presumably monogenic hydrothermal bush-like caves with cupolas: Sátorkö-puszta Cave in the Pilis Mountains and Bátori Cave in Buda Hills, Hungary. *Maps by M. Juhász, P. Borke, and L. Kárpat.*

Bush-Like Caves or Caves with Cupolas

Such caves typically consist of a basal chamber from which a branching pattern of rising passages develops. The branches are composed of coalesced spherical cupolas whose typical size is 0.5–1.5 m, and cupolas terminate most of the branches. Such caves are known in Hungary (*e.g.*, Sátorkö-puszta and Bátori caves; Fig. 2) and in the Azérous Mountains in northern Algeria. These types of caves are thought to be due to the delivery of hot water to a single input point at the base of carbonate rock having low fissure density. Bush-like caves are an example of monogenetic hydrothermal karst, although the exact mode of formation for such caves is not yet understood. One hypothesis ascribes their formation to the convective movement of water in the phreatic zone, whereas another model (preferred) invokes the convective movement of moist air above the hot-water table coupled with condensation corrosion.

Phreatic Maze Caves

Maze caves are the most common types of hydrothermal cave systems. Among them network caves, anastomotic caves, spongework caves, and ramiform caves are distinguished.

Network caves are angular grids of intersecting passages formed by widening of nearly all major fractures within favorable areas of soluble rocks. Two-dimensional rectilinear maze systems are created where rising water is trapped in densely jointed carbonate rock below a relatively impervious bed. Examples of such a pattern are Cserszegtomaji-kut and Acheron-kut caves in Hungary, developed in Triassic dolomite under the cover of Miocene sandstone.

Multistory rectilinear network caves are more common. Examples of these are caves of the Buda Hills, Hungary (Pál-völgy, Szemlo-hegy, Ferenc-hegy) and caves of the Black Hills in South Dakota (Wind and Jewel caves). Many of the carbonate-hosted lead-zinc ore deposits (the so-called Mississippi Valley type) exhibit a network pattern of solutionally widened fractures later filled—partly or entirely—with sulfides (Jefferson City Mine in Tennessee and Devil's Hole Mine in the United Kingdom).

Anastomotic caves consist of curvilinear tubes that intersect in a braided pattern. They usually form a two-dimensional array along a single favorable parting of low-angle fracture. Less common three-dimensional variants follow more than one geologic structure. An example of a supposedly hydrothermal three-dimensional anastomotic cave is the Pobednaya Cave in Kirhgizstan, with its 1.5 km of very narrow, tube-shaped crawlways.

Spongework caves consist of interconnected cavities of varied size in a seemingly random three-dimensional pattern. Such caves appear to form by the coalescing of intergranular pores and minor interstices. *Ramiform caves* consist of irregular rooms and galleries wandering three-dimensionally with branches extending outward from the main areas of development. Passage interconnections are common, producing a continuous gradation with spongework and network caves. József-hegy Cave in the Buda Hills, Hungary, is a good example of the ramiform pattern.

Small-Scale Morphology

Hydrothermal caves commonly exhibit characteristic morphology of cave elements (*e.g.*, cave passages, rooms, as well as yet smaller morphological features, such as rock relief of cave walls). Among those features Klimchouk (2007) has defined the *morphologic suite of rising flow*. The suite comprises three major components: (1) feeders (inlets); (2) transitional wall and ceiling features; and (3) outlet features. Feeders correspond to sites of input of rising water into hypogenic caves. Typical feeders are vertical or subvertical conduits, either individual or forming small networks. Transitional wall and ceiling features form with a considerable role of buoyant effects (upward-focused dissolution associated with rising limbs of free convection cells). They include such small morphological forms as rising wall channels, ceiling channels (half-tubes), and ceiling cupolas. Outlet features represent channels of varying morphology, connecting the cave to the next upper story or discharging water out of the cave-development zone. Typically they are represented by cupolas and domepits (vertical tubes) that rise from the ceilings of cave passages and rooms. The suite comprises morphological features created by rising water, which may or may not be thermal. Nevertheless, identification of these morphological elements may be an important first step in determining the hydrothermal origin of a cave.

CAVE DEPOSITS

Mineralogy

Calcite is the most common mineral of hydrothermal caves developed in carbonate rocks. In addition, such minerals as quartz, barite, fluorite, and sulfides are commonly reported from the deep-seated hydrothermal caves. (Ore-related hydrothermal karst, where the list of minerals can be quite large, is not considered here.) Shallow hydrothermal karst caves rarely contain any "exotic" minerals.

Character of Cave Deposits

Large euhedral (that is, having perfect crystallographic shape) calcite crystals, aggregates, thick crusts, and sediments reflecting stable hydrodynamic conditions are common in deep-seated hydrothermal caves. Individual calcite crystals can be as large as 10–30 cm and even 100 cm; gypsum crystals can be even larger, reaching in some cases the length of several meters (the most striking example is the 11-m-long gypsum crystals in Naica cave, Mexico; these largest-ever-found natural crystals were formed at 52°C).

The morphology of calcite crystals in deep-seated hydrothermal caves is normally simple, dominated by a scalenohedron (dogtooth spar) sometimes combined with an obtuse rhombohedron. In contrast, the deposits of shallow caves commonly reflect a more dynamic environment. Crusts are typically less thick (except in the immediate vicinity of the water table, where massive mammillary crusts called *cave clouds* may form) and euhedral crystals are rare. The size of crystals in aggregates ranges from several millimeters to a few centimeters. The dominant crystal morphology is a combination of a scalenohedron and a prism with the crystal tip blunted by obtuse rhombohedron (nailhead spar). Minerals might be contaminated by clay, which indicates that the paleowaters were dynamic enough to carry the particulate matter. In addition to subaqueous deposits, two more types of speleothems occur in shallow hydrothermal karst: waterline deposits (rafts, folia, cave cones), and subaerial deposits (*e.g.*, cave popcorn). These two types are also common in cold karst.

HYDROTHERMAL CAVE LIFE

A peculiar cave life is known to exist in some active hydrothermal caves, the most striking example being Movile cave (Romania). The cave hosts slightly thermal (20.9°C) water rich in dissolved hydrogen sulfide. The hydrogen sulfide is used as an energy source by bacteria to fix inorganic carbon, thus producing a food base for a highly evolved chemoautotrophic (*i.e.*, based on chemosynthesis rather than photosynthesis) ecosystem. As many as 48 species of cave-adapted terrestrial and aquatic invertebrates, 33 of which are endemic to this ecosystem, have been identified in Movile. The cave did not have a natural entrance: it was intercepted by an artificial shaft at 18 m below the surface. It is thought that life in the cave has been separated from the outside environment for the past 5.5 million years.

IDENTIFICATION OF HYDROTHERMAL CAVES

Unequivocal evidence of the hydrothermal origin of a cave can be provided by (1) the presence of thermal waters in a cave (obviously, this indicator is only relevant in presently active hydrothermal-karst caves); (2) the presence of hydrothermal minerals deposited on the cave walls (hydrothermal character of mineralization is inferred from mineralogical and/or geochemical evidence; it is also necessary to demonstrate that the two speleogenetic stages, cave development and mineralization, are related to the same process); and (3) the presence of isotopic alteration of the cave walls (character and degree of alteration must correspond to the water/rock interaction at elevated temperature, which is inferred on the basis of isotopic calculations).

Evidence suggesting the hydrothermal origin of a cave includes characteristic morphological features, indicating very slow water movement, free convection of water, and active enlargement of caves in a subaerial setting (air convection and condensation corrosion mechanism). Additional suggestive evidence includes a lack of morphological features and deposits indicating running water environment; a lack of association with the surface features of "conventional" karst; and spatial association with hydrothermal activity (including extinct one).

REGIONAL EXTENT

Hydrothermal origins have been established for many large cave systems of the world: for example, caves in the United States (South Dakota, Wyoming, Montana), Austria, England, France, Italy, Poland, Czech Republic, Slovak Republic, Hungary, Romania, Ukraine (Crimea), Caucasus, Kyrgyzstan, Israel, Iran, Algeria, Namibia, Mexico, Mauritania, and South Africa.

See Also the Following Article

Speleogenesis, Hypogenetic

Bibliography

Andreychouk, V., Dublyansky, Y., Ezhov, Y., & Lysenin, G. (2009). *Karst in the Earth's Crust: its distribution and principal types.* Sosnowiec (Poland) — Symferopol (Ukraine): University of Silesia — Ukrainian Institute of Speleology and Karstology.

Audra, P., Hoblea, F., Bigot, J.-Y., & Nobecourt, J.-C. (2007) The role of condensation-corrosion in thermal speleogenesis: Study of a hypogenic sulfidic cave in Aix-Les-Bains, France. *Acta Carsologica*, 36/2, 185—194.

Collignon, B. (1983). Spéléogenèse hydrothermale dans les Bibans (Atlas Tellien—Nord de l'Algérie). *Karstologia*, 2, 45—54 (in French).

Dublyansky, Y. (1997). Hydrothermal cave minerals. In C. Hill & P. Forti (Eds.), *Cave minerals of the world* (2nd ed., pp. 252—255). Huntsville, AL: National Speleological Society.

Dublyansky, Y. (2000). Hydrothermal speleogenesis: Its settings and peculiar features. In A. Klimchouk, A. Palmer, D. Ford & W. Dreybrodt (Eds.), *Speleogenesis: Evolution of karst aquifers* (pp. 292—297). Huntsville, AL: National Speleological Society.

Dzulynsky, S., & Sass-Gustkiewicz, M. (1985). Hydrothermal karst phenomena as a factor in the formation of the Mississippi Valley-type deposits. *Handbook of strata-bound and stratiform ore deposits, Vol. 13* (Pt. 4), 391—439.

Egemeier, S. J. (1981). Cavern development by thermal waters. *National Speleological Society Bulletin*, 43, 31—51.

Ford, D. C., & Williams, P. W. (1989). *Karst geomorphology and hydrology*. London: Unwin Hyman.

Ford, T. D. (1995). Some thoughts on hydrothermal caves. *Cave and Karst Science*, 22, 107—118.

Klimchouk, A. (2007). *Hypogene speleogenesis: Hydrogeological and morphogenetic perspective* (Special Paper 1). Carlsbad, NM: National Cave and Karst Research Institute.

Palmer, A. N. (1991). Origin and morphology of limestone caves. *Geological Society of America Bulletin*, 103, 1—21.

Sarbu, S. M., Kane, T. C., & Kinkle, B. K. (1996). A chemoautotrophically based cave ecosystem. *Science*, 272(5270), 1953—1955.

Takácz-Bolner, K., & Kraus, S. (1989). The results of research into caves of thermal water origin. *Karszt és Barlang (Hungary)* (Special Issue), 31—38.

Van-Hise, C.R. (1904). *A Treatise on Metamorphism*. Monographs of the United States Geological Survey, v. 47, Washington: Govt. Print. Off.

I

ICE IN CAVES

Aurel Perşoiu and Bogdan P. Onac
University of Suceava, Romania
University of South Florida

Ice in caves is a common occurrence in the Northern Hemisphere during the winter when the temperature drops to 0°C and below. These cold temperatures extend inside the cave for a few hundred meters, causing the drip water (more rarely flowing water) to freeze and form seasonal ice stalactites, stalagmites, and crusts. The life of these speleothems is short. The speleothems melt during summer when both geothermal heat and warmer outside airflow raise the cave temperature above 0°C. This freezing/thawing cycle rejuvenates every winter. Under favorable conditions and particular cave settings, however, ice survives the melting season, and accumulates into large masses of perennial ice (Fig. 1). In contrast to glacier caves (*i.e.*, caves within glaciers or at the contact between glaciers and the rock beneath them), rock caves hosting perennial ice are known as *ice caves*. However, in some places, caves with seasonal ice or even lacking ice but having a colder climate than the external environment, were inappropriately called ice caves.

Whereas seasonal ice forms a variety of speleothems, perennial ice occurs mostly in the form of stratified ice blocks (and to a minor extent, stalagmites and domes), in many ways similar, but not identical, to mountain glaciers. Thus, we consider that the term *glacier* can be extended to perennial ice accumulations in caves, provided that a series of conditions are met: large, stratified ice blocks (a few meters in thickness), well-marked seasonal mass balance fluctuations and active flow. The last point is especially important, as flow of cave ice has been rarely described. The cave walls can restrict flow (*e.g.*, Focul Viu Ice Cave, Romania; Velika ledena jama v Paradani, Slovenia), or in some cases, the flow of ice is merely an artifact of the ice lying on an inclined surface (*e.g.*, Eisriesenwelt, Austria). In some cases, however (*i.e.*, Scărişoara Ice Cave, Romania), active flow is present, with the ice moving under its own weight, generating glacier-specific features: push-moraines, ice tongues, folds (when obstacles are encountered), and so on. If choosing to name an underground ice accumulation *glacier*, one should make sure that the use of this term does not bring in the confusion between ice caves and glacier caves (Yonge, 2004).

Ice caves have been described from various climatic zones, extending from subpolar (Canada, Norway, and Siberia) to the Mediterranean, with a few occurrences in the hot arid climates of the southwestern United States and central Turkey. A peculiar combination of climatic, hydrologic, and (cave) morphologic factors is required for the genesis and preservation of ice in caves, in conditions that are otherwise not favorable for its formation (*i.e.*, too warm or too dry climates). A series of mechanisms have been proposed for accumulation of ice in caves, but often the mechanisms of *cave cooling* (a prerequisite for ice genesis) and those leading to *cave ice formation* were confused. Cooling of the cave can be achieved by (1) trapping of cold air (Fig. 2A); (2) unidirectional ventilation (Fig. 2B); (3) evaporative cooling; or (4) as a consequence of the cave's geographic setting in the subpolar climatic zone. Most of the cave ice forms by freezing of water, with sublimation of moisture and snow diagenesis playing less important roles.

Caves at high latitude (*e.g.*, Canada, Norway, and Siberia) or high altitude (*e.g.*, Altai Mountains) are often under permafrost conditions. As a consequence, the air temperature is below 0°C, which provides favorable conditions for ice preservation in the cave. Permafrost conditions, however, restrict water inflow and therefore only a limited amount of ice (in summer) can form by congelation. Perennial ice in caves, such as those in Canada's Yukon Territory, is formed by both congelation and desublimation processes. In summer, moist air entering the cave desublimates to form thick hoar ice deposits on the cave walls, which, depending on the cave's morphology and topoclimatic conditions can either melt, with the resulting water subsequently freezing to form congelation ice (*e.g.*, in the Tsi-tché-han Cave, Canada) or accumulate on the floor to form massive ice (*e.g.*, Serendipity Cave, Canada). Hoar deposits on cave walls can also be found in caves outside the permafrost zone (*e.g.*, Scărişoara Ice Cave and Dobsinská l'adová jaskyňa,

FIGURE 1 The upper part of the ice block in the Great Hall of Scărişoara Ice Cave and various ice speleothems (stalactites, domes). *Photo by Aurel Perşoiu.*

Slovakia), but the deposition mechanism is slightly different. The cold air entering these caves promotes sublimation of ice and desublimation of the resulting moisture on the undercooled walls and/or ice (Fig. 3A). This hoar ice melts in early summer and the resulting water drains away before the initiation of the freezing in autumn.

Trapping and diagenesis of snow have been proposed as mechanisms for ice formation in caves, but thus far, no detailed descriptions and studies have been put forward. Ice in Q5 Cave in Vancouver Island (Yonge, 2004) and in Avenul din Albele (Retezat Mountains, Romania) might have formed in this way. A number of caves throughout Europe have snow accumulations near their entrances, but the conversion to ice via snow metamorphosis and freezing of meltwater inside the snow pack was not reported. An explanation could be the reduced thickness of the snow layer (up to 4–5 m), which prevents snow diagenesis through weight compaction and also permits through-flow of water before freezing.

Freezing of liquid water in an undercooled cave environment is the most common mechanism of cave ice formation. As mentioned above, cooling can be achieved through different mechanisms (points (1)–(3)), but generally, it is a combination of them.

Trapping of cold air is the most common mechanism of cave cooling, leading to cave ice formation. It is most

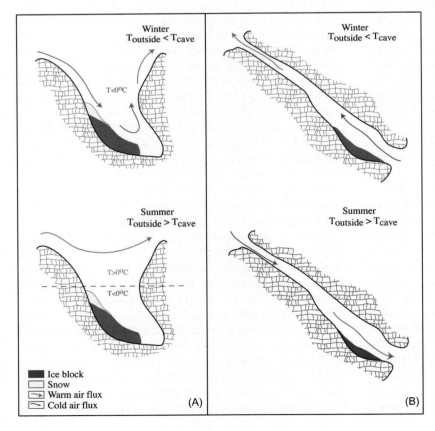

FIGURE 2 Schematic diagram showing the most common cave cooling mechanisms. (A) Trapping of cold air; (B) unidirectional ventilation.

FIGURE 3   Ice speleothems. (A) Hexagonal hoar ice crystals (*photo by Aurel Perşoiu*); (B) complex ice stalagmite morphology (*photo courtesy of Iosif Viehmann*); (C) cryogenic calcite on ice surface (*photo by Bogdan P. Onac*); (D) overview of the 18-m-high ice wall in Little Reservation of Scărişoara Ice Cave. *Photos courtesy of Claudiu Szabo and Gheorghe Fraţilă. Used with permission.*

common in caves with a single entrance (or multiple entrances situated in the upper part of the cave) and descending passages (Fig. 2A), in which cold air sinks in winter, displacing the warm air, which is pushed out through the same entrance (if the diameter is large enough; e.g., Scărişoara and Borţig ice caves, Romania) or through other entrances (e.g., Focul Viu Ice Cave). The direct cooling under the influence of sinking cold air is accompanied by *evaporative cooling*, induced by evaporation of moisture from the walls (Wigley and Brown, 1976). In summer, the air inside the cave is colder and thus heavier than the external air and is trapped inside the cave, hence the name of the phenomenon. However, "trapping" of cold air as the origin of ice caves is a rather misleading term, as the presence of the cold air inside the cave in summer is a consequence of the presence of ice itself, rather than a cause for it. Cooling begins in early autumn (when air temperature outside the cave becomes lower than inside and cold airflow propagates into the cave) and lasts until late spring (the actual length depends on the general climatic conditions outside the cave, being longer in colder climates and shorter in warmer ones). In summer, however, the inflow of cold air ceases, and the cave air temperature begins to rise, due to the heat conduction through the walls and air column in the entrance shaft(s) and the warmer percolating water.

Thus, the heat delivered to the cave leads to ice melting and hence cooling of air, so that cave atmosphere is maintained at 0°C and the cold air remains trapped inside the cave.

Cooling by ventilation (chimney effect) is the third most important mechanism of cave cooling below freezing point. It occurs in caves with (at least) two entrances situated at different elevations, as a consequence of temperature contrast between the cave and the exterior. These differences result in pressure variance at the lower entrance, which in turn triggers a dynamic ventilation of the cave. If the temperature inside the cave at the lower entrance is higher than outside, cold air will be advected to the cave and warm air will be pushed out through the upper entrance (in winter); while if the cave air temperature is below the outside one (in summer), warm air will move off the cave through the lower entrance and cold air will be advected through the upper one (Fig. 2B). These mechanisms of air circulation through the cave lead to cooling of the lower entrance and ice genesis by water freezing in winter, while in summer melting of ice will maintain negative air temperatures and thus prevent inflow of warm air.

The first cold air wave inside the cave usually promotes evaporation of water from the walls and condensation of moisture from air. Next, water freezes;

this is a multiple-stage process, forming different types of ice. Occasionally, liquid water (from the summer melting season), pooling on top of older ice (*e.g.*, in Scărişoara and Borţig ice caves) will freeze from top to bottom to form *lake ice*. In a second stage, if temperature is close to (but still below) 0°C and water has time to spread out, dripping water freezes on top of this lake ice to form a thinner layer of *floor ice*. When temperatures drop below −2°C, dripping water will freeze to form ice stalagmites and mounds. If the negative air temperatures persist for longer periods, freezing conditions will occur in the ceiling of the caves, with subsequent formation of ice stalactites and curtains. The freezing of carbonate-rich water causes the expulsion of CO_2 from the solution and subsequent precipitation of *cryogenic carbonates* (Fig. 3B). These carbonates are easy to diagnose as they show a marked shift toward extremely high $\delta^{13}C$ values (up to +18‰ VPDB). A mixture of cryogenic carbonates and other organic and inorganic impurities accumulates at the bottom of the lake ice layer, forming, together with the ice, an annual couplet, similar to varves in lakes. Most of the glaciers (see above) in caves in Europe are formed by the accumulation of such couplets, although the annual lamination is usually partly destroyed by prolonged and strong summer melting. Accumulation over longer periods of time leads to relatively thick *ice blocks* (up to 24 m in Scărişoara Ice Cave), whose bases are close or slightly below the pressure melting point of ice. As a consequence, ice becomes plastic and steadily flows, with rates varying between 0.5 and 30 mm per year. Differentiated basal melting rates also contribute to variations in the speed and direction of flow, as do constrictions imposed by the surrounding cave walls. Therefore, the ice can bend and fold or spill over steps in the bedrock to form ice tongues. Moving of ice is accompanied by rock polishing (*e.g.*, in Scărişoara Ice Cave a 3-m-high band of rock with different textures can be seen above the present-day ice level, indicating the former height of the ice block), moraine formation (mixture of rocks, cryogenic calcite, and tree trunk remains), frost shattering of the rock walls in the vicinity of the ice block, and patterned ground arrangements. A peculiar phenomenon associated with ice in caves is the formation of *cryogenic cave micropearls*, which form through precipitation of $CaCO_3$ in summer, from carbonate-rich waters, and also in winter, during the freezing of water and loss of CO_2, as described above.

Apart from the large ice blocks, the most interesting cave ice formations are the *ice stalagmites*. They form in the so-called *periglacial cave environment*, at a certain distance from the larger ice blocks, where the cold air halo centered on the ice block partly vanishes, leaving the air temperature near the cave ceiling above 0°C allowing the dripping water to enter the cave and subsequently freeze on the overcooled ground. Similar ice stalagmites can also form in the coldest parts of caves, before the freezing of the ceiling, as well as near the entrances of unglaciated cave. These stalagmites display a wide range of morphologies, from simple candle-style to more complex ones as a result of air temperature and variations in relative humidity (Fig. 3C).

A particular type of ice stalagmite, known as *bamboo* or *thermoindicator*, consists of successions of perfectly transparent ice knobs and narrow sections of opaque ice (Fig. 4A). The narrow portions are formed at lower temperatures when dripping water freezes instantly, favoring the growth in height. The knobs are created when higher temperatures at the apex of stalagmites allow dripping water to flow down the stalagmite before freezing, thus increasing their diameter. This growth mechanism was confirmed by daily measurements and observations performed on stalagmites in the Great Reservation of Scărişoara Ice Cave between February 9 and March 23, 1965 (Fig. 4B).

SCIENTIFIC SIGNIFICANCE OF ICE IN CAVES

Short descriptions and papers related to ice caves are scattered throughout various journals and books published since the beginning of the nineteenth century. However, it was only over the past 50 years that ice caves stimulated complex scientific investigations. Most of the earlier studies targeted the ice in caves or the caves *per se* (mostly climatic, morphologic, and glaciologic studies). The more recent investigations are aimed at an integrated assessment of the processes in ice caves. In most cases, the ultimate goal is obtaining *information on past climatic changes* from proxies in ice such as pollen, water stable isotopes, stratigraphy, chemical properties of ice and the impurities within, and so on (Racoviţă and Onac, 2000; Yonge, 2004).

The organic matter preserved within the ice layers of many caves is generally well-suited for radiocarbon age dating, so that a very precise, absolute chronology can be tied to the ice stratigraphy. With a few notable exceptions (*e.g.*, Eisriesenwelt, ~4000 years; Scărişoara, ~10,000 years (Perşoiu, 2011), the radiocarbon ages of all the other ice blocks (cave glaciers) from Europe and North America indicate they are younger than 1000 years. Supported by stratigraphic data, the existing ages thus far point toward a complex interaction between cave ice and climate over this time period.

Most of the past climate studies were examining the oxygen and hydrogen stable isotope variation in ice. The isotopic composition of ice mirrors that of meteoric water before freezing inside the cave. There is a

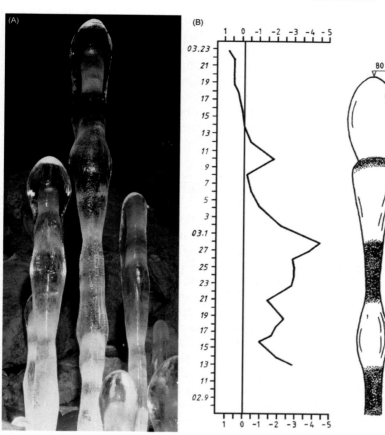

FIGURE 4 Bamboo ice stalagmites and their temperature record. *Photo by Aurel Perşoiu; graph adapted from Racoviţă and Onac (2000).*

strong correlation between outside temperature and the oxygen isotopic composition in the ice precipitation. By combining the ages of the ice at different depths with high-resolution stable isotope analyses along the core, a detailed record of climate changes in the cave's region (mean-annual temperature, source of moisture, *etc.*) since ice began its accumulation can be derived.

At Scărişoara Ice Cave, studies of pollen, micro- and macrocharcoal, and macrofossil recovered from cave ice provided valuable paleoecological data. The collected data offered an accurate picture of past vegetation dynamics and composition at both local and regional scale, along with clear signals of human impact over the past 1000 years.

Traces of industrial pollution were documented from ice caves in Romania and Montenegro. This field of study has great potential, although the great mobility of the chemical species in ice during melting/refreezing events may modify the original signal.

Studies in the fields of crystallography and mineralogy have revealed a wealth of data related to the presence of ice in caves along with the description of rare minerals such as monohydrocalcite or ikaite associated with these environments. Furthermore, isotopic and mineralogical investigations in ice caves enabled the identification of cryogenic cave carbonates (Fig. 3B).

A peculiar phenomenon associated with ice in caves is the formation of cryogenic cave micropearls, which form through the precipitation of $CaCO_3$ from carbonate-rich waters by either ordinary degassing of CO_2 when cave temperature is above 0°C or during freezing of water and loss of CO_2, as described above.

Cave tourism has rapidly developed in the past several years. Tourism in ice caves has increased the pressure on the fragile cave environment and the ice. An extreme example in this regard is the complete melting of ice in Kungur Ice Cave (Russia) in the mid-1980s. Since then, it has beautifully recovered due to adequate new management solutions. This led to the growth of cave management—related research in ice caves, mainly the study of microclimate and the interaction between visitors and the cave environment. Body heat from visitors and the output from lights contribute heat to the cave environment, which become greatly intensified when doors are added to a cave, causing a change in the ventilation regime. All these affect the melting of ice by sublimation or/and condensation processes. Such climate studies allow long-term predictions on how the presence of visitors and other physical modifications might impact the cave microclimate. Cave managers can use this information to better maintain the appropriate environmental conditions in their ice caves.

On Earth, caves are hosted by almost every major type of rock that has a soluble component. Caves are natural traps for sediments, unique life environments, and valuable repositories for various forms of paleoclimatic and paleoenvironmental information since their deposits are protected from destructive processes acting on the surface. Extraterrestrial caves on other planets (*e.g.*, lava tubes on the moon, Mars, *etc.*) may host ice deposits, which in turn represent radically different life conditions for some microorganisms (Boston, 2004). Understanding all aspects related to Earth's ice accumulation and preservation in caves (even when developing under deserts) and their particular microbiological setting could ultimately have a paramount importance in deciphering the history of celestial bodies. Investigating minerals, microorganisms, or other materials trapped within ice in caves could give valuable clues in the formation of these celestial bodies. Furthermore, a wide diversity of cold-adapted bacteria and fungi can live and grow while frozen in ice at temperatures approaching −30°C. Studying such microorganisms capable of metabolizing while under higher pressure and very low temperatures may help our understanding of ice caves as extreme life environments and shed light on possible similar life forms that could exist on Mars, Jupiter's moon Europa, or other extraterrestrial planets (Boston, 2004).

Bibliography

Boston, P. J. (2004). Extraterrestrial caves. In J. Gunn (Ed.), *Encyclopedia of cave and karst science* (pp. 355–358). London: Fitzroy-Dearborn Publishers.

Perşoiu, A. (2011) Palaeoclimate significance of perennial ice accumulations in caves: an example from Scărişoara Ice Cave, Romania. PhD thesis, University of South Florida, Tampa, USA.

Racoviţă, G., & Onac, B. P. (2000). *Scărişoara Glacier Cave*. Monographic Study. Cluj-Napoca, Romania: Editura Carpatica.

Wigley, T. M. L., & Brown, M. C. (1976). The physics of caves. In D. T. Ford & C. H. C. Cullingford (Eds.), *The science of speleology* (pp. 329–358). London: Academic Press.

Yonge, C. (2004). Ice in caves. In J. Gunn (Ed.), *Encyclopedia of cave and karst science* (pp. 435–437). New York: Fitzroy-Dearborn Publishers.

INVASION, ACTIVE VERSUS PASSIVE

Dan L. Danielopol[*] *and Raymond Rouch*[†]

[*]*Austrian Academy of Sciences*, [†]*Centre National de la Recherche Scientifique France (Retired)*

Why and how subterranean habitats were invaded by surface-dwelling animals is one of the themes that has fascinated generations of speleologists. The subterranean realm is colonized actively and/or passively by aquatic and terrestrial fauna. Ecological evidence and arguments for evolutionary scenarios that support various models of colonization are provided in this article. Modern evolutionary ecological research offers new perspectives to better understand the invasibility of subterranean systems by both terrestrial and aquatic fauna.

INTRODUCTION

The topic of subterranean life is closely related to the question of why and how organisms penetrate and further settle, sometimes in large numbers, in an apparently inimical environment, that is, where space and energy can be very constraining for the development of flourishing organismic populations. Caves accessible to humans were one of the first habitats to be intensively explored over the past 200 years. This led to the discovery of a diverse aquatic and terrestrial troglobitic fauna, sometimes without direct relationship to the living surface dwelling animals. This discovery stimulated naturalists to propose various explanations for the subsurface habitat colonization by surface-dwelling animals.

Students working on subterranean animals used the term *invasion* to stress that the subsurface realm is extensively and, in many cases, massively colonized by surface-dwelling animals which later adapt to subterranean life conditions. The invasion modalities are grouped in two major categories: (1) active colonization, which is mainly related to the environmental cues as perceived by the surface-dwelling organisms, and (2) passive colonization where animals arrive into and settle within the subsurface environment either by the force of or by the chance of the environmental dynamics.

Scenarios explaining the active and/or passive invasion of subterranean animals were proposed during the late nineteenth century as well as during a large part of the twentieth century by naturalists interested in the evolutionary processes that generated the troglobitic and/or stygobitic fauna. At the beginning of the twentieth century Neo-Lamarckians, such as E.G. Racovitza and R. Jeannel in Europe and A.M. Banta in North America, favored the view of an active invasion of the subterranean realm by animals that could not survive at the Earth's surface either because of climatic changes or because of strong predator pressures. These animals retreated into subterranean habitats which offered them a kind of stable refugium against epigean environmental constraints. A similar view is shared by A. Vandel (1965) who is an exponent of the organicist view of subterranean evolution, which considers that old species, like senescent individuals unable to live in an epigean dynamic environment, retire into subterranean habitats. An origin of troglobites due to passive invasion was proposed at the end of the nineteenth century by E.R. Lankester who considered that the troglobitic evolution could start with the organisms that accidentally colonized cave habitats. Most of these kinds of hypotheses were not followed by rigorous scientific tests.

During the second part of the twentieth century the authors assisted in an increased interest in the ecology of subterranean animals. Biologists proposed various scenarios explaining the origin of the diverse hypogean fauna. As compared to the previous generation of scenarios, these are more carefully documented and offer better possibilities to test their plausibility. In this article, a brief review of this information is subsequently presented within its evolutionary and ecological context.

EVOLUTIONARY SCENARIOS

Models Based on Active Dispersal

The Climatic-Relict Model

This assumes that surface-dwelling animals actively colonize subterranean habitats during periods when the environment changes drastically. Such events happened during the Pleistocene period in the Northern Hemisphere. Dry and/or cold climate at the soil surface, especially in karstified areas, is considered an important climatic stressor that constrained various animal groups to invade underground systems. The fraction of organisms that could not adapt to subterranean environments became extinct while the new troglobionts or stygobionts built a relictual fauna. Various terrestrial and aquatic hypogean animals from Europe, North America, Africa, and Australia are considered to fit this model.

The Adaptive-Shift Model

Preadapted surface-dwelling organisms are able to colonize unconstrained subsurface habitats where they progressively adapt to the new environment. The invasion of terrestrial fauna and their subsequent evolution in caves formed in volcanic regions such as the Hawaiian Islands is an example proposed by F.G. Howarth (1983). For marine anchihaline caves a similar scenario called the *zonation model* was proposed by T.M. Iliffe (1986).

The Active Colonization Model

This scenario, well documented by R. Rouch and D. L. Danielopol (1987), considers the possibility of colonization of subterranean environments by preadapted and/or generalist organisms. This is an active process which does not occur during periods of environmental stress. The authors consider this model as having a wide generality for the subterranean invasion. The process of invasibility following the scenario of Rouch and Danielopol is decoupled from the speciation process, while in the adaptive-shift model both processes play an additive role.

Colonization through Marine Shallow-Water Ecotones

The penetration route within the subterranean aquatic caves follows either directly through the entrance of the caves or through submarine karstic springs or porous clastic sediments along coastal areas.

Transit of Marine Fauna through Epigean Limnic Systems

This is an alternative for the colonization of caves and/or interstitial habitats by marine benthic animals. It occurs through the adaptation of euryhaline animals in a first step to surface inland waters and further by the invasion of subterranean waters. The latter scenario applies to crustaceans that colonized the Dinaric Karst.

Colonization from the Deep Sea

Bathyal and/or abyssal fauna displaying biological convergent traits with troglobitic fauna could colonize shallow marine caves, especially on volcanic islands. It was proposed that deep-sea animals—sponges, crustaceans, or fishes—could, during that time, penetrate through the crevices of volcanic or karstic rocks up to the anchihaline shallow caves.

Escape from Epigean Predators and/or Strong Competitive Pressure

The invasion of subterranean habitats by animals, especially small invertebrates that tried to escape to stronger predation and/or competition, is a scenario envisaged by generations of biospeleologists. It is hypothesized that many crustaceans invaded subterranean aquatic habitats in order to escape the predation and/or competition pressure of other animal groups such as insects. For instance, E.G. Racovitza (1907), at the beginning of the twentieth century, suggested that the troglobitic isopod *Spelaeoniscus debrugei* took refuge in caves because of the incomplete capacity to protect itself by rolling up into a ball when predators attack. Because the isopod's antennae are left out during the rolling up of the body (Fig. 1), it was assumed that these crustaceans were easily preyed on by epigean invertebrates.

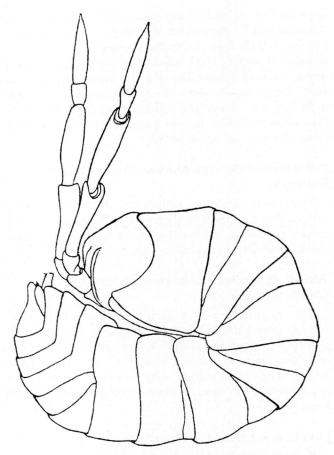

FIGURE 1 *Spelaeoniscus debrugei*, a troglobiont isopod from North Africa. From E.G. Racovitza (1907). Used with permission.

Models Based on Passive Dispersal

Passage of Epigean Animals during Massive Surface Water Infiltration into Subterranean Systems

There is a whole series of possibilities for passive migration of fauna into the subsurface. In alluvial plains along river channels during floods, epigean animals are transported by surface water which later infiltrates the subsurface, taking animals with it. One example is the periodic invasion of crustacean Cladocera from the backwaters of the Danube wetlands (in Austria) into the aquifer existing below the floodplain. Surface streams that flow into karstic systems, for example, through sinkholes or infiltrates through macro- and microchannels, allow the invasion of subterranean karst by a diverse fauna. The dispersal of this fauna can continue underground into large karstic systems through the complex drainage systems. It is hypothesized that the highly diverse stygobitic crayfish fauna of Florida originated through this process.

The Regression Model

Dutch biologist J. Stock developed a general model for the invasion of marine benthic fauna into inland subterranean waters based on the eustatic regressive sea level movements. Stock noticed that various crustacean groups with stygobitic representatives, such as the amphipod Ingolfiellidea, are presently known from sites located on ancient marine paleocoasts. He assumed also that once animals colonized subsurface coastal marine habitats such animals become thalassostygobionts that minimally dispersed into other geographic areas. During marine regressive phases, the coastal thalassostygobionts, instead of following the regression of the marine water, remained in place and progressively adapted to inland nonmarine subterranean water (brackish and later freshwater) which replaced the marine water. Such animals become limnostygobionts. In this way the invasion of subterranean waters happens through a so-called passive regression process and by vicariant events, that is, the inland originally marine form is stranded and respectively isolated from the marine population. Stock's model has since been extended to more detailed scenarios, such as the two-step evolutionary model of J. Notenboom, C. Boutin, and N. Coineau (Fig. 2) and the three-step model of J.R. Holsinger. They combine the active migration process of preadapted animals with a second passive process during marine regressions, and they include a third phase of ecological isolation in the new inland subterranean habitats. It is assumed that the three-step evolutionary model is able to explain also the invasion of both marine and nonmarine benthic animals into the subsurface environment. Despite the opinion that the passive regression models have a wide generality, they were never carefully tested for their basic ecological and biogeographical assumptions that the animals cannot follow the marine water during the regressive periods and do not further disperse inland far away from the original site (see the next section for additional arguments).

Rafting and Resettlement of the Fauna in New Subterranean Habitats

Interstitial marine animals can be transported passively on various objects with the whole sediment over various distances. Harpacticoids and ostracods could spread in this way.

Erosion of Sediment and the Drift of Subsurface Fauna with the Surface Water Flow

Surficial interstitial marine and even freshwater fauna are especially exposed during storms and floods to the resuspension of the sediment and the passive transport through the surface water column to other

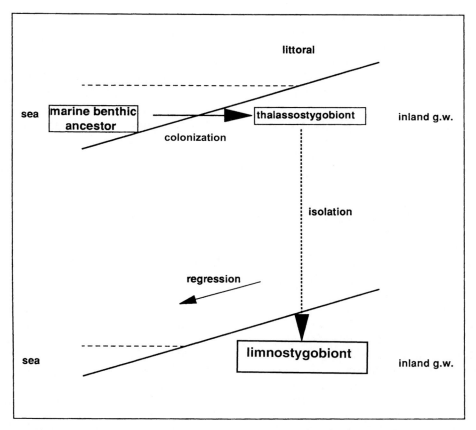

FIGURE 2 Schematic presentation of the two-step evolutionary model, which explains the invasion of inland subterranean habitats by marine fauna; g.w. = groundwater. *Adapted from J. Notenboom (1991). Used with permission.*

places. After the resettlement they are able to recolonize shallow subsurface habitats.

Unique versus Multiple Invasions

Various biospeleologists assumed that important invasions of a subterranean habitat, such as a cave, followed by the adaptation of the originally epigean animals to the hypogean life and the expansion of their geographic range, occurred during unique historical events such as the strong climate deterioration on the Northern Hemisphere during the Pleistocene era. Today this view is challenged by the evidence of multiple invasions of the subterranean environment by epigean animals preadapted over a wide geographic range. For instance, R. Rouch and D.L. Danielopol (1987) consider this a basic aspect of their active colonization model. Multiple colonization occurs either simultaneously or independent in time. The amphipod *Gammarus minus* independently invaded various caves of West Virginia at different times, taking advantage of the suitable subterranean habitats.

Ward and Palmer coined the term *interstitial highway* to describe the long-term dispersal routes below the soil surface. It applies especially to interstitial habitats along river channels and alluvial plains. Hypogean dwelling animals, such as the amphipods *Bogidiella albertimagni* and *Niphargopsis casparyi*, known from Western to Eastern Europe, expanded their geographic range along large river systems such as the Rhône, the Rhine, and the Danube.

THE ECOLOGICAL EVIDENCE

Evolutionary and/or historical scenarios can be checked for their plausibility by examining recent field ecological situations and/or by laboratory experimental observations. Active migration of animals is commonly observed for thigmotactic animals living on the bottom of running waters from where they penetrate the hyporheic interstitial systems without being constrained by some environmental stress. There are hundreds of species, insect larvae, crustaceans, water mites, and various worms observed to invade shallow subsurface habitats and adopt the hyporheal as their common life place. The colonization and further adaptation to groundwater habitats by crustaceans, for example, representatives of the harpacticoid genus *Elaphoidella*, with wide ecological valence in tropical, subtropical, and temperate zones, is in accordance with the active colonization scenario of R. Rouch and D.L. Danielopol (1987).

Laboratory experiments with the amphipod *Gammarus roeseli* demonstrated that this species actively migrates through the porous space of gravel sediments when exposed to oxygen gradients; respectively, they avoid hypoxic zones and look for normoxic water. New volcanic islands that allow the development of sandy beaches are rapidly colonized by marine epigean animals that settle into the interstitial habitats, for instance, the interstitial habitats of a sandy beach of Long Island near New Guinea. This is an island that emerged through a volcanic eruption and was colonized by two species of crustacean harpacticoids within the first 20 months of its existence.

Observation in the field and/or laboratory of the penetration of epigean animals in subsurface habitats because of predator pressure pointed to a more complex behavior than previously known. This behavior was used in the colonization under the constraint scenario mentioned above. For instance, observation of *Gammarus roeseli* exposed to fish predation shows that a fraction of the surface-dwelling animals migrate into the sediment, while others reduce their movement at the sediment surface. However, because animals are very thigmotactic, they penetrate into the sediment anyway, provided there is the necessary vital porous space.

Passive invasion of karstic systems by diverse epigean animals, both insect larvae and crustaceans, was thoroughly studied during long-term filtration of exurgencies by the ecologists working in France. Studies done on the Baget, Moulis, and Dorvan karstic drainage basins documented that a surprisingly high number of animals penetrate during the rainy period into the subsurface. Much of this fauna is further released outside at the exurgencies following the water that flows through the subsurface voids. In the experimental area Rhitrodat at the Biological Station in Lunz am See, Austria, it was documented that the hyporheic zone of a gravel stream can be recolonized within hours by various meio- and macrofauna such as rotifers, cyclopoids, and chironomids. A significant relationship between the volume of the entrapment of sediment and the number of animals caught pointed to the passive invasion of this environment by various animals.

The view that aquatic animals take refuge in subsurface habitats, mainly in the riverbed sediments, under the pressure of climatic drought was checked in the field and laboratory. Observation of two Northern California streams, intermittent and perennial, did not show an increase of the hyporheic fauna in either type of stream during the dry period. The reaction is more complex in a stream flowing through the Sonoran Desert in Arizona, where during the dry period the shallow interstitial habitats accumulate epigean benthic invertebrates which later disappear with the increasing sediment dryness. Also, observation in the laboratory using oligochaetes in sediment columns pointed out that the progressive dryness of the sediment determines a low fraction of the worms to follow the water into the depth of the column. Most of the animals remain entrapped in the unsaturated zone of the sediment. Marine interstitial fauna, especially in tidal littoral zones, are able to migrate tenths of centimeters during the seasons.

BRIDGING THE GAP—EVOLUTIONARY ECOLOGY AT WORK

Evolutionary biologists, such as R. Lewontin, point out that organisms are active subjects which are able to choose (or select) their habitat; in this way organisms define their niche. Of course organisms are also objects of the selective pressures imposed by the environment that they experience. Hence, the fact that organisms represent the subject and the object of selection forms an important biological principle which gives support to the active invasion scenarios. The success of invasibility as the penetration and the settlement of an organism in the explored environment depends on the level of resources available for its survival and reproduction. Organisms with low energetic requirements and unspecialized for their resource acquisition (the so-called generalists) have an advantage over more specialized organisms or those needing high energetic resources. At a premium are also those surface organisms already predisposed to colonize subsurface habitats. There are general ecological principles that apply to the colonization of the subterranean environment too.

Both population geneticists and ecologists dealing with metapopulation studies of plants and insects recognized that the success of invading new environments depends on the selective capacity of organisms for progressive adaptation during their dispersion paths. The process, called *adaptive infiltration*, supports some of the models that explain the active subterranean invasion, for example, the zonation model of T.M. Iliffe (1986). However, in environments that deteriorate rapidly, animals can become extinct before significant evolutionary changes have time to occur. This latter argument weakens the credibility of scenarios based on the principle of refugium under constraints, for example, the climatic-relict hypothesis. The generality of invasion by surface-dwelling organisms of subterranean habitats is a well-documented phenomenon that cannot be explained by local processes such as the predation and/or competition pressure of epigean animals (remember the scenario of the refugium of animals underground under the pressure of voracious surface predators).

The passive invasion scenario of J. Stock and its modern variants of bi- and triphase evolutionary models have a weakness; they do not explain why marine animals that colonize marine interstitial habitats should remain in place or migrate minimally during the regression phase of the sea. As mentioned above, marine interstitial animals display the capacity to migrate vertically or horizontally depending on the environmental conditions. Marine regressions proceed at a low rate, theoretically allowing animals to track the receding sea and to keep their marine ecological requirements. Obviously we need more ecophysiological information about the way marine littoral fauna adapt to inland subterranean habitats.

CONCLUSIONS

1. The examination of alternative hypotheses which could illuminate the problem of subterranean invasion, as advocated by L. Botosaneanu and J.R. Holsinger (1991), proved to be a successful scientific strategy during the past 10–20 years. It stimulated active research where the quality of information for various models could be compared.
2. The scenarios regarding which animals repeatedly colonized the subsurface realm during geologic time could be tested with field and laboratory data.
3. The evolutionary models following which epigean animals migrated into the subterranean realm under external environmental pressures such as climatic deterioration, predation, and competition pressure seems to have limited generality, applying (if real) more to local situations.
4. The passive invasion of marine animals as described by the scenario that considers migration in inland groundwater occurring during marine regressive phases (widely favored by biospeleologists) needs additional ecological research to keep it as a robust explanatory model.
5. The scenario describing which deep-sea animals were able to actively colonize anchihaline caves was seldom corroborated by recent research and various students who supported this hypothesis were converted to an alternative explanation—the migration of shallow marine organisms into submarine caves.

Note: Additional information is available in: Bodergat and Marmonier (1997); Botosaneanu and Holsinger (1991); Culver and Holsinger (1994); Culver et al. (1994); Danielopol et al. (1999); Gibert et al. (1994); Racoviță (1907); Uiblein et al. (1996); Wilkens et al. (2000).

Acknowledgments

Our research was financially supported over the years mainly by the F.W.F. Austria and the Austrian Academy of Sciences (grants attributed to D.L. Danielopol) and by the C.N.R.S., France (for R. Rouch). We thank D.C. Culver (American University, Washington, D.C.) for critical review of the manuscript.

See Also the Following Articles

Evolution of Lineages
Vicariance and Dispersalist Biogeography
Adaptive Shifts

Bibliography

Bodergat, A.-M., & Marmonier, P. (Eds.) (1997). Contraintes et instabilité de l'envirronement: Stratégies adaptatives des organismes récents et fossiles. In: *Geobios-Memoire Special* (Vol. 21). Lyon (in French): Editions Université Claude-Bernard.

Botosaneanu, L., & Holsinger, J. R. (1991). Some aspects concerning colonization of the subterranean realm—Especially of subterranean waters: A response to Rouch & Danielopol, 1987. *Stygologia*, 6, 11–39.

Culver, D. C., & Holsinger, J. R. (Eds.) (1994). Biogeography of subterranean crustaceans: The effects of different scales. *Hydrobiologia*, 287, 95–104.

Culver, D. C., Kane, T. C., & Fong, D. W. (1995). Adaptation and natural selection in caves. The evolution of *Gammarus minus*. Cambridge, MA: Harvard University Press.

Danielopol, D. L., Martens, K., & von Vaupel Klein, J. C. (Eds.) (1999). Crustacean biodiversity in subterranean, ancient lake and deep-sea habitats. *Crustaceana*, 72(8), 721–722.

Gibert, J., Danielopol, D. L., & Stanford, J. (Eds.). (1994). *Groundwater ecology*. San Diego, CA: Academic Press.

Giere, O. (1993). *Meiobenthology. The microscopic fauna in aquatic sediments*. Berlin: Springer-Verlag.

Howarth, F. G. (1983). Ecology of cave arthropods. *Annual Review of Entomology*, 28, 365–389.

Iliffe, T. M. (1986). The zonation model for the evolution of aquatic faunas in anchihaline caves. *Stygologia*, 2, 2–9.

Notenboom, J. (1991). Marine regressions and the evolution of groundwater dwelling amphipods (Crustacea). *Journal of Biogeography*, 18, 437–454.

Racoviță, E. G. (1907). Spelaeoniscus debrugei n.g., n.sp., isopode terrestre cavernicole d'Algerie. *Archivs Zoologie Expérimental et Génerale*, 7, 69–77 (in French).

Rouch, R., & Danielopol, D. L. (1987). L'origine de la faune aquatique souterraine. Entre le paradigme du refuge et le modèle de la colonisation active. *Stygologia*, 3, 345–372 (in French).

Uiblein, F., Ott, J., & Stachowitsch, M. (Eds.), (1996). Deep sea and extreme shallow-water habitats: Affinities and adaptations (Biosystematics and ecology series 11). Vienna: Austrian Academy of Sciences Press.

Vandel, A. (1965). *Biospeleology. The biology of cavernicolous animals*. Oxford, England: Pergamon Press.

Wilkens, H., Culver, D. C., & Humphreys, W. F. (Eds.). (2000). *Subterranean ecosystems* Amsterdam: Elsevier.

J

JEWEL CAVE, SOUTH DAKOTA

Mike Wiles

Jewel Cave National Monument

HISTORY

Early History

The first written record of Jewel Cave is a 1900 mining claim, the Jewel Tunnel Load, filed by Frank and Albert Michaud and Charles Bush. The "jewels" of Jewel Cave are the calcite crystals (Fig. 1) that line most of the cave's walls, ceilings, and floors. The cave's entrance was originally a blowing hole that was too small for human entry. It had to be enlarged to facilitate mining and to provide easy entrance for tourists. Little mining was actually done, however, and the cave failed as a major tourist attraction in the early years because it was too far off the beaten track. In 1908 Jewel Cave became one of America's first national monuments—the first one established for protection of a cave—and in 1933 it became part of the National Park Service.

Early Exploration

Until the late 1950s, little was known of the cave and only about 3 km of passages had been discovered. The situation changed dramatically when Dwight Deal, a young college student who was writing a master's thesis on the geology of Jewel Cave, invited Herb and Jan Conn—climbing friends in their late thirties—to accompany him in surveying and mapping the cave as an aid to his research. In the process they discovered many more miles of cave passages. They quickly realized that Jewel Cave was not the small cave it had been assumed to be for so many years.

Later Exploration

After Deal (1962) completed his work and moved on in pursuit of his career, the Conns continued exploring the cave for over 20 years. Through the careful and systematic documentation of a survey, they discovered over 100 km of cave passages before turning the effort over to younger explorers (Conn and Conn, 1981). They demonstrated that Jewel Cave was not a small cave; rather, it was one of the world's longest cave systems.

BAROMETRIC AIRFLOW

The Conns also recognized that the cave's pronounced airflow responded very closely to changes in the outside air pressure. In 1965, Herb Conn measured the airflow resulting from known barometric pressure changes, and developed a mathematical model to predict the total volume necessary to account for the behavior of the airflow at the entrance (Conn, 1966). Although no one can know for sure just how big the cave may ultimately be, 100 million cubic meters is an astonishingly large number and is strong evidence that the vast majority of the cave system is yet to be found. The wind has been known to blow barometrically at speeds over 56 km per hour through crawlways that are thousands of meters from the entrance. There are strong breezes even at the farthest known extents of the cave—an irresistible lure for any cave explorer.

PHYSICAL CHARACTERISTICS

Three-Dimensional Cave System

So far, more than 249.6 km of cave passages have been discovered beneath about 8 km^2 of surface area. Exploration has shown the cave to be a complex three-dimensional maze with passages occurring in at least

FIGURE 1 Calcite crystals. The largest crystal is about 5 cm long.

five different levels (Fig. 2). However, the passages are not distributed homogeneously throughout the subsurface. The cave is typically a maze of large passages where it is beneath a hill, but it bottles down to just a few crawlways where it crosses beneath the canyons—and it again becomes a maze of large passages beneath the next hill. The entire cave appears to have formed phreatically along joints and bedding planes, but development was most likely controlled by the same forces that produced the modern-day landscape. In order for the cave to avoid the canyons the way it does, it must have formed concurrently with the downcutting streams that created nearby Hell Canyon and Lithograph Canyon.

Loft Level

The uppermost loft level typically has smooth, domed ceilings and powdery weathered limestone covering flat floors. Passage development appears to have been controlled more by horizontal bedding planes than by joints, and is less likely to follow the parallel trends commonly exhibited by most of the other cave levels. The calcite-crystal coating so abundant in most of the cave is generally absent, but there are remnants suggesting it once coated the surfaces of the loft level passages and was later redissolved. These remnants often contain some of the largest calcite crystals found in the cave.

Chert Level

Chert level passages are found below the loft level, primarily in a 3- to 5-meter zone of alternating beds of chert and limestone. These passages often have flat ceilings and angular pieces of broken chert littering the floor.

Subchert Level

These passages are found immediately below the chert zone and usually have rounded surfaces and domed ceilings. The bedrock usually consists of punky, often brecciated limestone. A smooth, thin crystal coating typically thins to nothing as it nears the chert level.

Main Level

Main level passages lie beneath the chert level. They are often large, strongly joint-oriented, and usually breakdown modified. The ubiquitous deposits of manganese minerals are often found beneath the breakdown collapse. These manganese deposits are composed of a variety of manganese oxides and hydroxides; most are mineralogically unidentified and are simply referred to as "manganese."

Lower Level

The lower level occurs in bedded dolomitic limestone, and often exhibits some development along bedding planes. It is common to find the crystal coating separated from the underlying bedrock, presumably because it had been deposited phreatically onto a surface that had previously weathered to a punky consistency, and thus was not as firmly attached to the walls and ceilings as elsewhere in the cave. Manganese deposits tend to be thickest in the lower level. One deposit near the scenic tour route is more than a meter thick.

Basement Level

Only a few hundred feet of passages have been discovered in the basement level. It is more joint controlled than the lower level and usually has boxwork, but no crystal coating or manganese.

Breakdown Modification

Breakdown collapse is common throughout the cave system, and many of the larger rooms have resulted from the merging of two or more levels by collapse of the intervening rock. What would have been flat floors are now covered in piles of breakdown. This makes moving through the cave a constant up-and-down

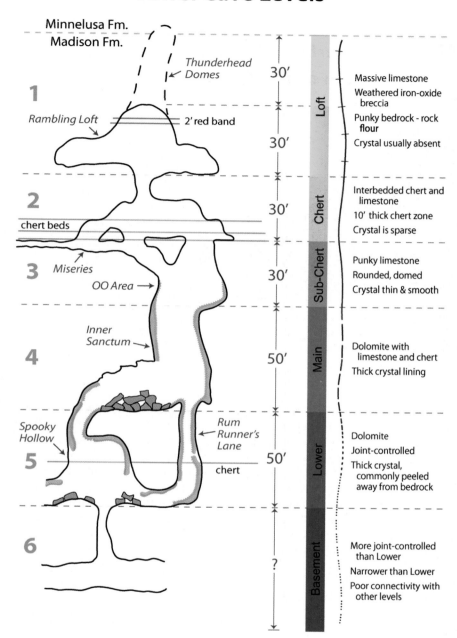

FIGURE 2 Cave levels.

experience. Even without the breakdown, no one level of cave passages is very continuous. Often the caver must cross *between* large passages via small holes and crawlways, and must constantly change levels to go from one end of the cave to the other.

Passage Sizes

Jewel Cave passages average 3 m wide and 5 m high, but range from 20 cm to 33 m in height, and 0.3–55 m in width. Most of the crawls are relatively short, but one area, named the Miseries, consists of nearly 550 m of hands-and-knees to belly-crawls. The Volksmarch is about 800 m of uninterrupted walking.

GEOLOGY

Jewel Cave was formed in the Mississippian Pahasapa Limestone. It is approximately 130 m thick in the Jewel Cave area, and all known cave passages are found in the upper 75 m of the unit. This relationship holds true even

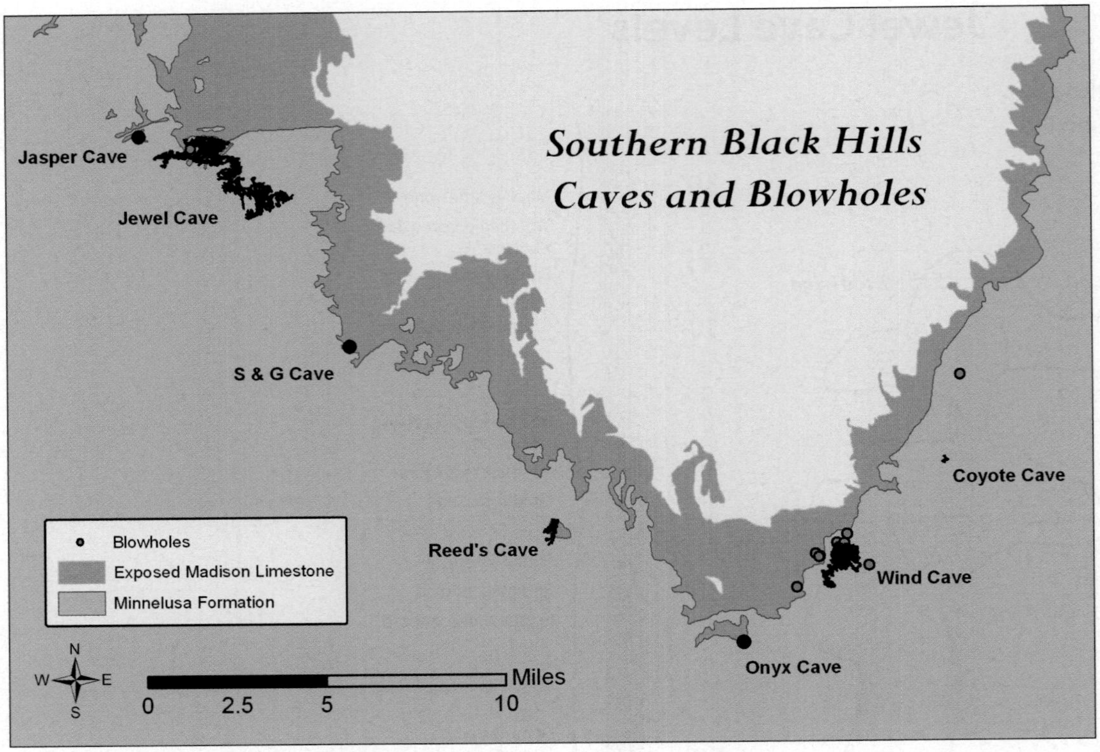

FIGURE 3 Relationship between caves and the Minnelusa cap.

at Wind Cave, where the limestone is only 107 m thick. Nearly the entire cave is in limestone that is overlain by the Pennsylvanian Minnelusa Formation.

Recent Geologic Mapping

Detailed geological mapping of the Jewel Cave Quadrangle (Wiles, in Fagnan, 2009) has identified six subunits within the Minnelusa: (1) 12 m of basal sandstone with thickness-compensatory red siltstone; (2) 15 m of thin-bedded limestone; (3) 37 m of sandstones; (4) 37 m of dolomite with interbedded sandstones; (5) 37 m of medium-to-coarse-grained sandstones; and (6) over 30 m of brecciated sandstone layers. Unit thicknesses were constant throughout the quadrangle, allowing the mapping of subtle faults and folds. The folds express themselves in extensive dip-slope topography and exhibit a strong correlation between synclines and surface valleys. Cross-cutting relationships indicate that the passages of Jewel Cave postdate all major folding and faulting events.

Relationship between Caves and the Minnelusa Cap

One of the most striking observations is the fact that Jewel Cave exists almost exclusively where the limestone is capped with the Minnelusa Formation. This relationship holds throughout the southern Black Hills to the point that, without exception, no cave over 60 m in length is known to exist within uncapped limestone (Fig. 3). Furthermore, there is no mappable paleotopographical relief within the quadrangle. Reported examples of paleokarst topography have turned out to be misinterpretations of previously unrecognized structures, often because the Pahasapa Limestone is so easily confused with the limestone subunits of the Minnelusa.

Cave Fill

Red fill material found within Jewel Cave has traditionally been called *paleofill*, based on the belief that a paleokarst topography had developed on top of the Pahasapa prior to Pennsylvanian time, and that subsequent deposition by the Minnelusa sea filled the associated "paleocave" passages; these were later intersected by the development of cave passages as they exist today. However, the absence of paleokarst topography casts doubt on the validity of this scenario. While the fill material is certainly from the overlying Minnelusa, the physical evidence suggests a more recent timeframe, when dissolution within the Pahasapa reached up into the Minnelusa, and Pennsylvanian material collapsed into the actively forming passage. Some of Jewel Cave's passages even contain blocks of sandstone that are identical to the basal Minnelusa sandstone exposed in nearby road cuts. If this scenario is correct, it would be more accurate to think of the "paleofill" as a "neofill," because it formed contemporaneously with the development of the cave—long after deposition and lithification of the Minnelusa.

HYDROLOGY

No part of the cave is known to intercept the water table, as is the case with the famous Lakes region of nearby Wind Cave. Water arrives only in the form of vadose dripping in areas that are near surface drainages where erosion has cut through the overlying Minnelusa Formation to within 30 m of the upper surface of the Pahasapa Limestone (Wiles, 1992); the lower subunits of the Minnelusa are quite permeable. Most of the cave is capped with stratigraphically higher subunits that restrict water flow. Areas of the cave below these subunits show virtually no evidence of dripping water, present or past. Only about 0.25% of the cave shows signs of dripping, and only half of that is presently active. Dripstone and occasional small pools can be found in the hydrologically active areas, but Jewel Cave shows no evidence of streams or rivers. Apparently, a higher water table had at one time supported surface flow, and then both the groundwater level and the surface flow had decreased significantly soon after the cave had formed. There was little opportunity for mechanical enlargement due to vadose flow.

ORIGIN OF JEWEL CAVE

Because recent observations have demonstrated a strong correlation between the cave passages and modern geological features, a new model is needed to explain how the cave could have formed as a result of the most recent processes that shaped the present-day structure and topography. More work is needed to address important features such as calcite spar, boxwork, and manganese deposits, but this hypothesis does provide a broad framework to account for the documented correspondence with modern geologic and topographic features, as well as provide a plausible source of recharge. Following is a description of proposed core events. Its purpose is not so much to say, "Here is the answer," but to spur the imagination to find better answers for the new, unexpectedly puzzling observations.

Working Hypothesis

In the southern Black Hills, caves longer than 60 m are only found in limestone that is presently capped with Minnelusa. This means that the Minnelusa must have played an essential role in the development of large caves, and that this development must have occurred in a geologically recent time. The base of the Minnelusa consists of a medium- to coarse-grained sandstone that readily allows vadose infiltration of rainwater into the cave today. In the past, this sandstone could have functioned as a local aquifer blanketing the Pahasapa Limestone, which itself has a low primary permeability. This aquifer would provide nonpoint-source recharge to the maze of fractures within the limestone, but there would be little circulation until down-cutting streams of nearby canyons intercepted the top of the aquifer, establishing a local gradient and causing groundwater to move toward the canyons and discharge into gaining streams. Some of the water would circulate phreatically as deep as 75 m below the top of the Pahasapa before its dissolutional capacity was depleted and it discharged into the canyons via fractures, rather than conduits. With groundwater approaching the canyons from both sides, there would be little cross-flow beneath the canyons, and consequently little cave development would occur in those areas. Water moving vertically across the Pahasapa/Minnelusa interface would enlarge fractures within the limestone, but the overlying Minnelusa would collapse and fill these conduits nearly as fast as they formed, resulting in deposits of "neofill." As the volume of the cave increased and the climate became drier, discharge would exceed recharge: the Minnelusa aquifer would dry up and the water in the cave would drain away. This must have happened relatively quickly, because there is virtually no evidence of vadose activity in Jewel Cave.

BIOLOGY

Cave Life

Jewel Cave is largely devoid of life except near the historic entrance. This is because of the relative lack of moisture, or a mechanism to carry organic material deep into the cave where it could serve as a food source for subterranean creatures. Animals that feed outdoors do use the entrance area for shelter, however. These are mostly bats and packrats and the parasites and other organisms that feed on these mammals and their excretions.

Bats

Though they almost certainly did not inhabit the cave prior to the enlargement of the long windy constriction that was the natural entrance of the cave, several species of bats now make the cave their home. Many *Myotis* use the cave as a day roost during the spring, summer, and fall, and can be found as far as 300 m from the entrance. They are joined in the winter by several hundred male *Corynorhinus townsendii* (Townsend's big-eared bats). Jewel Cave is presently the world's largest known hibernaculum for this species. Interestingly, the *C. townsendii* prefer to hibernate closer to the entrance, in the path of airflow, which varies widely in temperature and humidity when the cave is inhaling. When the temperature is coldest, they will hang in clusters of a hundred or more. When the air is warmer, many individuals will break out of the cluster and form smaller clusters nearby. The *C. townsendii* are frequently active throughout the winter. On the other hand, the *Myotis* move very little, and seem to prefer the warmer and more constant temperatures of rooms deeper in the cave and away from the airflow. They commonly roost singly or in pairs, and are occasionally found in clusters of up to a dozen individuals. From 2000 to 2009 the total wintering population averaged over 1300—a modest increase from the previous decade's average of around 1200.

Microbiology

Microbes are known to inhabit caves, even when there is little or no organic food source. A single preliminary study at Jewel Cave has identified such microbial life, but more research is needed to build a base of knowledge from this meager starting point.

SPELEOTHEMS

Some of the speleothems found in Jewel Cave include dripstone (stalactites, stalagmites, flowstone, draperies); helictites; gypsum formations (flowers, needles, and beards); aragonite; calcite crystals, coatings, and rafts; boxwork; helictite branches; pool fingers; popcorn; popcorn stalagmites (many of which are hollow); and hydromagnesite balloons.

Popcorn Stalagmites

The popcorn stalagmites are particularly curious. They range in size from 5 cm high and 5 cm in diameter, to 5 m high and 1.2 m in diameter, and many have holes down their center axis. Sometimes the hole is off-center, forming a slot along the side of the stalagmite. In some cases the hole is even deeper

FIGURE 4 Hydromagnesite balloon. The balloon is about 2.5 cm high.

than the stalagmite is high. Though almost certainly formed by the action of dripping water, none of these formations is found in areas where water is dripping today, nor do they occur where there is any sign of dripstone from past vadose activity. Clearly, they formed under significantly different circumstances than are observed today, and the nature of their origin remains a mystery.

Hydromagnesite Balloons

Hydromagnesite balloons (Fig. 4) were first discovered in Jewel Cave, and have since been found in half a dozen other caves throughout the world. These delicate bubbles of hydromagnesite typically have a pearly white luster and come in a variety of irregular shapes. They are no more than 4 cm across, and the shell is made up of layers of hexagonal plates with a total thickness of only $25\,\mu m$. The balloons appear to have been inflated. Both biological and chemical mechanisms have been proposed, but no firm conclusions have yet been reached. Several hundred balloons occur in two widely separated locations in the cave—both in passages that experience significant airflow.

OTHER BLACK HILLS CAVES

Jewel Cave is one of over 200 caves known in the Black Hills. Most are dry caves with joint-oriented passages, and many have a calcite crystal coating—though most often thinner and less extensive than that found at Jewel. Thick manganese deposits are only found in one other local cave. The majority of these caves are less than 150 m long. Five of them are 1.5–10 km long, and only a few have a barometric wind. One notable exception is Wind Cave, with a prominent barometric wind and over 219.76 km of surveyed passages. It has the same "feel" as most Black Hills caves, but has extensive boxwork development, and little in the way of a calcite crystal coating.

CONCLUSION

Jewel Cave is a unique cave system of immense proportions. Thus far, only a small part of it has been found, and there is much to be learned about its extent and speleogenesis. It has a great potential for future discovery, particularly in the areas of exploration, geology, and biological science.

Bibliography

Conn, H. (1966). Barometric wind in Wind and Jewel Caves, South Dakota. *National Speleological Society Bulletin, 28*(2), 55–69.

Conn, H., & Conn, J. (1981). *The jewel cave adventure*. St. Louis, MO: Cave Books.

Deal, D. E. (1962). *Geology of Jewel Cave National Monument, Custer County, South Dakota, with special reference to cavern formation in the Black Hills*. University of Wyoming.

Fagnan, B. A. (2009). Geologic map of the Jewel Cave Quadrangle, South Dakota. *7.5 Minute Series Geological Quadrangle Map 9*. South Dakota Geological Survey.

Wiles, M. E. (1992). *Infiltration at Wind and Jewel Caves, Black Hills, South Dakota*. South Dakota School of Mines & Technology.

K

KARREN, CAVE

Joyce Lundberg

Carleton University, Ottawa

The term *karren* indicates small-scale dissolutional features of rock surfaces of the order of millimeter to meter scale with an upper size limit of perhaps 5 m. Cave karren (sometimes called *speleogens*) are usually minor modifications to the main passage form and smaller than the passage, although some may dominate the passage form or even comprise the whole passage. While some are inherent to the initial speleogenetic process, most are subsequently superimposed on passage walls. General discussions on cave karren can be found in the following texts: Ford and Williams (2007); Klimchouk et al., (2000); Palmer (2007); Slabe (1995); White (1988).

INTRODUCTION

Cave karren develop where soluble rocks interact with solvent fluid (typically water, but it can be water vapor or aerosol) and typically where that fluid is moving. Flow may be in phreatic (pressure flow) or vadose (free flow) or alternating (floodwater) conditions. Aggressivity may be enhanced by organic activity or local acidification. Karren usually develop on bare rock but occasionally they can form under sediments or guano. Classification based on genesis is difficult because many similar forms are produced by dissimilar processes, for example, circular wall/ceiling cavities may be produced by phreatic eddies, or in mixing-water caves, or in air-filled caves from trickling vadose water, or where water vapor condenses on surfaces (perhaps in conditions as diverse as deep inside thermal caves or the entrance zones of tropical caves). Another example is the notch: some notches are simply dissolved out bedding planes while others mark water level; some form above sediment level (where sediment armors the rock against dissolution) while others form below sediment level (where sediment pore water is aggressive).

CONTROLS ON KARREN FORM

In most cases the shape of the karren feature relates to the properties of the flow rather than properties of the rock, for example, free-fall drips produce *pits*; film flow produces *longitudinal rills/flutes* or sometimes just etched surfaces; channel flow with turbulent eddies and rollers produces *dissolutional potholes*, *scallops* (probably the most ubiquitous cave karren feature), *transverse flutes*, *anastomoses*, *pendants*, *half-tubes*, and floor/wall/ceiling *pockets*. If flow is slow enough that turbulence does not dominate flow, then density and convection currents may govern the form, producing *facets*, *notches*, *bevels*, and convection *cupolas*. Condensation of vapor on surfaces produces corrosion *cusps/hollows* and, probably, *bellholes*.

In some cases the dissolutional features are modified strongly by rock properties. Karren are best developed (*i.e.*, have the smoothest, most regular shape) in homogeneous, fine-grained, pure rock. Rock heterogeneity results in differential dissolution where features may be preferentially etched out (Fig. 1) or the less soluble material may stand proud of the cave wall (*e.g.*, impure beds, chert nodules, veins, or fossils). *Boxwork* is the result of preferential dissolution of the more-soluble rock between intersecting sets of less-soluble veins. There is no standard term, other than *etch karren*, for the forms produced from this type of differential dissolution because there is no standard morphology; each shape is unique to the rock properties that govern it.

The production of cave karren may be complicated by local, in-cave biological activity and in some cases a unique suite of forms is produced. An example is *photokarren* (Fig. 2): suites of small tubes, rods, and cones etched into the rock surface in the dimly lit entrance zones of some caves (especially tropical caves); the diagnostic feature is that they are all oriented toward the

FIGURE 1 The influence of rock properties on cave karren: a complex stylolite is etched out of the wall, Clearwater Cave, Mulu, Sarawak. *Photo courtesy of Keith Christenson. Used with permission.*

FIGURE 2 The influence of local biological activity: photokarren near to the entrance of Clearwater Cave, Mulu, Sarawak, are angled toward the incident (low) light.

light. The control mechanism is not simple, but it appears that photosensitive biota colonize in patches, stimulating very local rock dissolution such that the surface immediately under the biota retreats at the angle of incidence of the light. Organic activity associated with guano produces its own suite of features, few of which have yet been documented.

Finally, karren form may be affected by clastic material carried in the water. Abrasional features (*scour marks*) are at the edge of the karren category, since they are not purely dissolutional. Similarly many cave potholes are largely abrasional in origin.

DESCRIPTIONS OF FEATURES

Drip pits occur where water drips freely onto bare rock (*e.g.*, into cave entrances, or from small vadose inlets), producing a suite of simple circular cavities, of semicircular to parabolic cross section, typically 1–5 cm in diameter and depth (Fig. 3A). Simple pits are common in wet caves, but in dry caves they usually get filled with sediment and modified into round-bottomed *dissolution basins*. These can be of any size, but are typically <1 m; they form in dry caves where a thin, patchy sedimentary or organic cover inhibits drainage, and aggressivity is often enhanced by organic activity (Fig. 3B).

Longitudinal rills/flutes are suites of vertical grooves that develop where a water film flows down a high angle rock face (Fig. 4A), for example, on vertical shafts from descending vadose water. They vary in size from ~1 cm (rills) up to ~30 cm (flutes) in width and perhaps many meters in length, often extending the whole length of the shaft. Although the bigger flutes may be somewhat interrupted or influenced by bedding planes and joints, the basic form is not governed by rock properties. Sometimes the flute wall may itself be made up of smaller-scale rills. Decantation flutes form where stored water drains over a rock face, for example, out of bedding plane partings or sediment or organic material (Fig. 4B).

Dissolutional potholes are sharp-edged, rounded-interior basins in stream beds (Fig. 5). The scale is typically 20–200 cm in diameter and depth. They usually have a close-to-perfectly-round circumference and are often deeper than they are wide. They develop in channels of steep gradient where flow is strongly turbulent and velocity is high. They may be entirely dissolutional or modified by abrasion when clastic particles are introduced. Some become abandoned when a drainage channel develops in the pothole wall; then a new pothole develops below, leading to a suite of potholes lining channel walls and floor.

Scallops are spoon-shaped hollows (Fig. 6) that develop under deep flowing water with subcritical turbulent flow (at velocities generally between 1 cm sec^{-1} and 3 m sec^{-1}). They develop in suites, overlapping and intersecting each other (Fig. 6B), often covering the whole of the wetted surface. They are markedly asymmetrical, the steep side facing downstream (Fig. 6C).

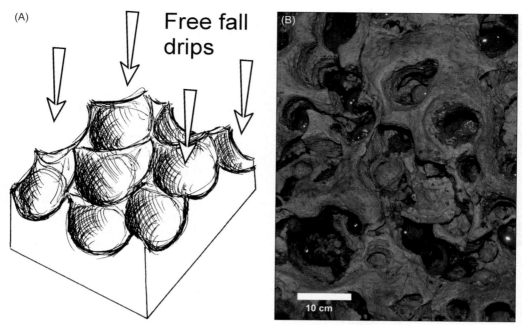

FIGURE 3 (A) Drip pits that develop in caves under free-falling drops are similar to subaerial rainpits (although usually larger because of the larger drop size). (B) Small dissolution basins developed in association with a thin sediment cover in the floor of Stonehorse Cave, Mulu, Sarawak.

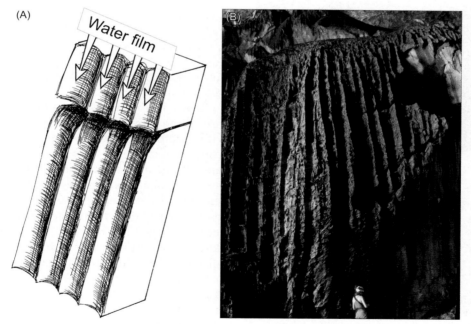

FIGURE 4 (A) Flutes develop on steep walls from film flow. (B) Extraordinarily large decantation wall flutes carved by aggressive waters draining from the high-level shelf of guano, Deer Cave, Mulu, Sarawak. *Photo by Donald McFarlane. Used with permission.*

Their wavelength decreases with increasing velocity (a log-normal length distribution), ranging from ~1.5 cm in flood flows of ~3 m sec^{-1} to ~100 cm in gentle flows of ~1 cm sec^{-1}. Thus, dominant water velocity and direction can be estimated from scallop wavelength. The scallop shape is cut where flow has separated from the rock surface, then become reattached (Fig. 6C), allowing aggressive water direct access to the rock face rather than the more usual indirect diffusion through the saturated boundary layer. The initial eddy may be caused by an initial inhomogeneity; one eddy then triggers others, so that the process becomes self-propagating.

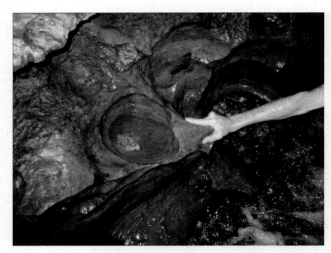

FIGURE 5 Dissolutional potholes in fine-grained limestones, Lagarta Cave, Ecuador.

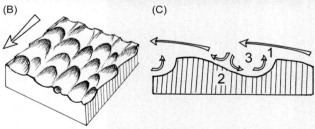

FIGURE 6 (A) Scallops in Cave of the Winds, Mulu, Sarawak (*photograph by Keith Christensen, used with permission*). (B) Block diagram shows typical packing pattern of scallops. (C) Longitudinal profile of scallop or flute showing separation of flow at 1, reattachment of flow at 2, and eddy at 3.

The frequency of detachment increases as velocity increases, reducing wavelength. For any one section of passage, the scallops of characteristic wavelength are formed by the dominant discharge. In vadose passages the upper level scallops, reached only by flood flows, are smaller.

Transverse flutes are essentially scallops of infinite width (normal scallops are typically 1–20 cm wide but in very stable flow conditions can reach ~2 m); they have the same cross section as scallops, but are laterally more extensive. They develop the same way as scallops, but require a laterally continuous separation of flow, that is, a transverse, roller eddy. Thus, they are rare.

Scour marks develop where clastic particles are carried at high velocities; the scallop form becomes elongated and polished. These are standard fluvial forms, but relatively rare in caves since they need both very high velocities (>3 m sec^{-1}) and silica sand.

Anastomoses, *pendants*, and *half-tubes* are commonly associated with confined flow. *Paragenesis*, or antigravitational erosion, occurs where sediments filling the passage confine flow against the ceiling, thus exposing the ceiling to pressurized flow and enhanced dissolution. Paragenetic features therefore develop upward.

The term *anastamose* refers to a reticulate pattern (such as in the veins of a leaf); in caves it applies to a pattern of interweaving channels (Fig. 7) cut into a fissure of low dip that is penetrated by water under pressure. The fissure may exist between two rock surfaces (bedding plane or joint), or between sediment fill and cave roof (*i.e.*, paragenesis). Anastomosing channels (typically a few centimeters up to a meter in width) are usually exposed to view only when the lower confining bed or fill is removed. If the relict channels are not truly anastomosing, then they may simply be called *drainage grooves*.

Floodwater can sometimes be forced into a complex three-dimensional, reticulate route through the rock, with interconnecting pockets and tubes at various scales; this creates a form of *spongework*. The scale can range from a few tens of centimeters up to tens of meters. Spongework can also be produced from locally acidified waters and in mixing water caves. In some cases spongework is produced where phreatic water is very slow moving: the very slow dissolution kinetics close to saturation allow differential dissolution to enhance subtle difference in rock properties.

Pendants are positive forms hanging from passage ceilings, typically ~10–100 cm in dimension. If they are remnant from removal of intervening rock through eddy dissolution they will have sinusoidal cross sections (like egg boxes). If they are remnant from anastomoses or paragenesis the cross section resembles an inverted flat-topped mesa (Fig. 7A). Paragenetic pendants can be complex and very attractive (Fig. 8). Positive forms can also be left on walls and floors, but, since they do not hang down, are not termed pendants; for example, *echinoliths* are remnant forms that protrude upward from the floor. These are often of complex shape, multifaceted with concave surfaces.

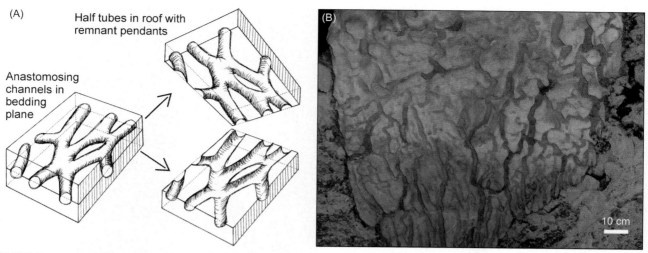

FIGURE 7 (A) block diagram of bedding plane anastomosing channels: when the lower block is removed the upper bedding planes' half-tubes with intervening remnant pendants are exposed. (B) Paragenetic anastomoses formed at junction of rock and former sedimentary fill in Cave of the Winds, Mulu, Sarawak. *Photo courtesy of Keith Christensen. Used with permission.*

FIGURE 8 Paragenetic pendants exposed by removal of sedimentary fill, Lagang Cave, Mulu, Sarawak. *Photo courtesy of Keith Christensen. Used with permission.*

A *half-tube* (Fig. 9) is a channel of semicircular cross section usually cut upward into a ceiling but sometimes laterally into walls, and usually meandering more sinuously than the channel it occupies. It is almost always paragenetic, the location, especially of the wall channels, explicable only by perching of the stream on a former fill or confinement against the roof. The paragenetic ceiling tube will become more entrenched as sediment continues to fill the passage, the meanders migrating downstream (Fig. 9A).

In some simple phreatic passages nonmeandering, ogive-shaped (gothic arch) ceiling channels follow the guiding joint: these have been explained as the result of mixing corrosion where water trickling through the joint intersects the main phreatic passage. A third suggested mode of origin is that CO_2 is trapped against the ceiling during flooding, enhancing aggressiveness.

If ceiling tube entrenchment cannot keep pace with sediment filling, then the blockage causes locally very high hydraulic potentials, and water is forced to adopt new routes above the main passage, exploiting any available fracture. The resultant *bypass tubes* and *corkscrews* often follow very tortuous routes.

Solution pockets are negative forms usually in ceilings but also developed in walls and floors; they are blind pockets of rounded cross section, often circular to elliptical in plan, sometimes multicuspate, and developed on a guiding fracture (Fig. 10A). They range in size from ~10 cm to many meters. If they are markedly taller than wide, they are called *chimneys* and sometimes *avens*. They may form through mixing corrosion where a joint introduces foreign water into a phreatic cave passage, or in vadose caves where water seeping from a joint renews its aggressivity by absorbing CO_2 from the cave atmosphere. Narrow, tapered solution pockets form in epiphreatic conditions where aggressive floodwater is forced under pressure into fractures. Where several of these form side by side the residual rock may be sharply bladed.

Bell holes are also blind pockets developed in ceilings; however, they are never associated with guiding joints or lithological control (Fig. 10B,C). The form has been called "negative stalagmite"-shaped; they are parabolic at top, flared at base, usually much deeper than wide, with perfectly vertical long axes. Typically they are ~1 m deep and 20–30 cm wide, and are usually arranged in suites. They are most often reported from the humid tropics. Theories of formation have ranged from phreatic eddying to bat-urine corrosion, but, since

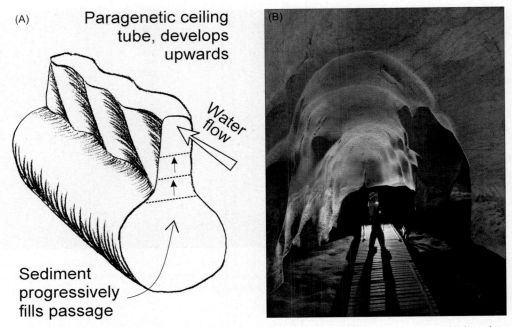

FIGURE 9 (A) Paragenetic ceiling half tube develops by upward entrenchment, meandering more sinuously than the passage it occupies. (B) Paragenetic ceiling half tube, Niah Caves, Miri, Sarawak *(photograph by Keith Christensen, used with permission)*. The main passage, still largely sediment-filled, is about 15 m wide.

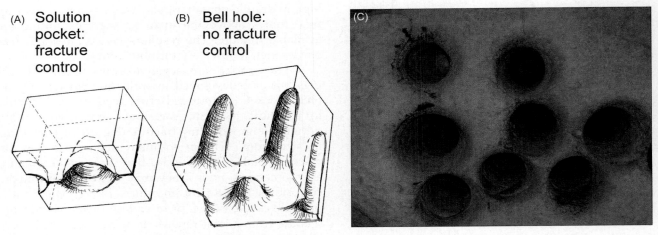

FIGURE 10 (A) A solution pocket has a guiding fracture and often lithological control. (B) A bell hole has no guiding fracture and no apparent lithological control. (C) A suite of bell holes in the roof of Fruit Bat Cave, Mulu, Sarawak. The photograph was taken after exodus of the bats roosting inside the bell holes.

many are currently active without phreatic conditions, recent research suggests that they may form by condensation corrosion associated with bat roosting.

Another blind pocket feature is the *cupola*. These are complex multicuspate alveolar pockets, often in hierarchical suites (Fig. 11A). They can range from small to passage-sized scale (at which size they would not be classified as karren). Cupolas are characteristic features of thermal caves. Thermal waters rising upward through rock are often very aggressive (from dissolved CO_2 or H_2S). Below water level, thermal convectional cells may be established in quasi-static water and carve out rounded, spherical cupolas. Alternatively, the aggressive medium may be steam from the warm waters condensing on cooler passage roofs, again circulating in a thermal convection cell and thus producing cupola forms. Cupolas may form part of a complex three-dimensional network similar to spongework, or they may be part of a branching dendritic network. Most cupolas have no guiding fractures.

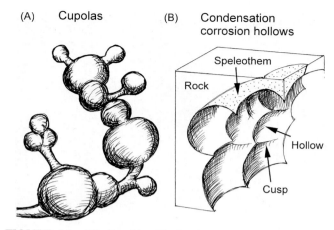

FIGURE 11 (A) Cupolas. (B) Condensation corrosion hollows and cusps, cutting equally through rock and speleothem (stippled).

Condensation from vapor also produces *condensation corrosion hollows* and *cusps*. These forms can be quite complex, but are typically large, shallow, scallop-like hollows, with somewhat sharp cusps where the hollows intersect. These are usually cut smoothly through all material regardless of geological structure and texture (Fig. 11B), for example, a single hollow may be cut through bedrock and speleothem without any obvious impact of the change in properties. Reprecipitation of calcite sometimes occurs downwind of the dissolution, on the lee side of the hollows. Sometimes *condensation rills/flutes* are produced. Condensation corrosion has long been known from H_2S-rich vapor in hydrothermal caves. While condensation rates are very much greater in hydrothermal caves, recent research suggests that condensation may be important in many nonthermal caves, for example, during summer afternoons in tropical cave entrances. Condensation requires significant changes in relative humidity, and corrosion requires dissolved gases such as CO_2 and H_2S. Condensation may be caused just by climatic effects, or by establishment of bat colonies. Corrosion is most obvious in entrance areas of caves, in narrow passages where an increase in airflow velocity causes a decrease in pressure and temperature, and in thermal caves directly above steam vents.

The final examples of cave karren occur where water (either meteoric or thermal) is almost static. Where open, vadose standing water becomes aggressive by absorbing CO_2 from the atmosphere, a sharp horizontal *waterline notch* develops, exactly at the water level and exactly horizontally. The notch often cuts straight across geological structure. The scale is typically tens of centimeters, but, where water level is stable over long periods, it may penetrate the rock face quite deeply, for example, up to a meter in normal meteoric water caves. Shallow notches often occur in suites corresponding to changing water levels. Horizontal *corrosion notches* can also form where density gradients are set up by the production of heavy ion pairs and solute ions; these sink, setting up cellular flows that carry fresh solute or H^+ to the water surface and to the rock wall. In this case notches are formed that taper off steeply below the water line. Maximum dissolution is focused on the point of first contact but the cellular flow carries some H^+ ions downward, resulting in declining dissolution with depth (with a linear or exponential decay). The form may then be an inclined face at about 45° (a *facet*) that meets the flat ceiling of the notch to produce an inverted-triangular cross section (the *laughöhle*). The best-studied examples are developed in gypsum, where 1–3-mm-thick density currents have been measured against the rock surface. Continued retreat of the inclined face under stable water conditions leaves a horizontal ceiling, termed a *laugdecke* or *corrosion bevel*. These are relatively rare because conditions are not often stable enough for long enough.

See Also the Following Article

Scallops

Bibliography

Ford, D. C., & Williams, P. (2007). *Karst hydrogeology and geomorphology* (especially pp. 248–265). London: Unwin Hyman.

Klimchouk, A. B., Ford, D. C., Palmer, A. N., & Dreybrodt, A. (2000). *Speleogenesis evolution of karst aquifers* (especially pp. 100–111, 407–426). Huntsville, AL: National Speleological Society.

Palmer, A. N. (2007). *Cave geology* (especially pp. 147–155). Dayton, OH: Cave Books.

Slabe, T. (1995). *Cave rocky relief and its speleological significance*. Ljubljana, Slovenia: Založba ZRC/ZRC Publishing.

White, W. B. (1988). *Geomorphology and hydrology of karst terrains* (especially pp. 91–102). New York: Oxford University Press.

KARREN, SURFACE

Joyce Lundberg

Carleton University, Ottawa

INTRODUCTION

Small-scale dissolutional features of rock surfaces or *karren* are usually of the order of millimeter to meter scale, but some extreme examples extend to almost 30 m. Dissolution of the surface of soluble rock produces various rills, channels, pits, and basins. Individually, these are called karren, each type with its own name; collectively, a mass of dissolution features is called a

FIGURE 1 The Pinnacles of Mulu, Sarawak; a karrenfield made up of ~30-m-tall pinnacles separated by deep narrow grikes.

karrenfield. General discussions on surface karren can be found in the following texts: Bögli (1981); Ford and Lundberg (1987); Ford and Williams (2007); Ginés (2004); Ginés et al., (2009); Lundberg (2012); Trudgill (1985); White (1988). These are important hydrologically because much of the recharge to karst areas occurs here. The most widespread of karst features, karren can form in a great variety of situations and present myriad forms that have spawned a plethora of jargon in several languages! Fresh rainwater is the most common agent of erosion, but snow melt is important in alpine areas, and coastal areas (lacustrine or marine) have their own suite of processes and karren forms. Most karren are the result of simple dissolution but many are mediated by biological activity, sometimes apparent only at the microscopic scale.

Most commonly the feature of interest is the negative form—the shape that has been removed—such as the various types of channels and basins; but some of the most dramatic karren landscapes are made of the positive forms—the emergent remnant features. Classic among these are the pinnacle karsts of the world, such as the Pinnacles of Mulu (Fig. 1), the Tsingy of Madagascar, and the Stone Forest of China, mostly impenetrable terrain of deep clefts and extremely sharp vertical blades. Although karren are so widespread, several important karst areas show little visible karren on the surface (e.g., the Sinkhole Plain of Kentucky).

BASIC CONTROLS

Karren develop where a solvent comes into contact with a soluble rock. The most common solvent is fresh meteoric water. The most common rocks are carbonates (limestone, dolostone, marble), but karren are very nicely displayed on evaporite rocks (gypsum, halite).

FIGURE 2 Karren of San Salvador Island, Bahamas. The smooth, rounded, white surfaces are only recently emergent from under the soil cover. The longer-exposed surfaces, detectable by their covering of gray epilithic cyanobacteria, are considerably sharpened.

Obviously, dissolution requires that the water be aggressive but even the very gentle acidification caused by absorption of atmospheric CO_2 in rainwater is sufficient for shallow karren to form on bare rock. Dissolution is much faster where water has picked up additional soil CO_2 or biogenic acids. Snow melt is more aggressive than simple rainwater because of the higher solubility of CO_2 in cold water. Seawater is generally not aggressive to carbonates but, in areas of restricted circulation with intense biological activity, can become aggressive.

One of the more important controls on karren morphology is the hydrodynamic control (i.e., how the fluid impinges on the rock surface, the amount of discharge, and the frequency of wetting). Bare rock can have direct raindrop impact, or, with greater discharges, sheet flow or channel flow; additionally, bare rock can quickly dry out as soon as rainfall has stopped. However, where rock has a cover (of snow, of vegetation, or of sediment) the fluid has to impinge on the surface by percolation, it is generally spread out more evenly, and it generally remains damp for much longer after cessation of rainfall. Covered-rock forms are therefore distinctly smoother than bare-rock forms (Fig. 2).

FIGURE 3 Dissolutional pavements in dolostones of Dodo Mountain, Mackenzie Mountains, northwest Canada. Dissolution, although slow in this permafrost region, is not overwhelmed by frost action so the joints have been opened by about 7 cm since retreat of Wisconsinon ice. Scale can be estimated from the 4–6-m-tall cliffs.

Biological activity is important indirectly in that it governs the amount of biogenic CO_2 and/or organic acids available. However, sometimes biological action is more direct, in the form of rock boring. Borings can be quite large (e.g., up to several centimeters for gastropods and echinoderms) to microscopic (e.g., many boring fungi, algae, and sponges may penetrate the rock surface by only a few millimeters).

The environmental setting is important since climate dictates precipitation (amount, intensity, and physical form—snow, dew), and temperature (which largely controls biological activity). Karren of the wet tropics are generally the largest, sharpest, and most dramatic, while karren of the arctic are generally subdued. However, if frost action is not dominant, karren can be well displayed even in the arctic (e.g., the Mackenzie Mountains, Canada, is a periglacial area, but shows nice *dissolutional pavements*; Fig. 3).

Finally, lithological factors also control the development of karren. Rock purity obviously governs solubility, but beyond this, lithological factors may dominate the morphology such that karren forms are not obvious. For example, they are not usually well

FIGURE 4 The impact of rock texture on small pinnacles in forest, Sarawak, Borneo. (A) Massively bedded and homogeneous limestone in Mulu. (B) Massively bedded but highly stylolitic limestone in Niah. Both regions have similar climate and vegetation.

displayed on highly fossiliferous rocks or very thinly bedded rocks. Figure 4 shows the effect of rock texture on the morphology of emergent subsoil pinnacles from Sarawak, Borneo.

DESCRIPTION OF SOME KARREN FEATURES

In this account non-English terms are included where they are regularly used, even by English speakers.

Negative forms of circular plan are called *pits* if they are deep and narrow, *basins* if they are round-bottomed and shallow, and *pans* if they are flat-bottomed and shallow (Fig. 5). There is no limitation on size but most are $< \sim 1$ m in diameter and depth. Most have some evidence of biological colonization of the surfaces and most retain water for some time after

FIGURE 5 (A) Flat-bottomed solution pan (or *kamenitza*) with vertical to slightly overhanging walls, in San Salvador Island, Bahamas. The floor retains some dried-out cyanobacteria and/or algae that rehydrate on wetting. The interpan remnant rock surface shows the small pits and spikes that are ubiquitous on bare karst rock.

FIGURE 6 Rillenkarren on salt; Cardona, Spain. These are deeper and wider than most rillenkarren on limestone.

rainfall. They are best displayed on bare rock, the subsoil surfaces generally simply showing rounded depressions rather than distinct basins or pans.

Negative features of linear plan that are governed strictly by hydrodynamic controls (*i.e.*, water flow) include rills, flutes, and channels. Most channels form when discharge becomes focused into a stream. However, rills (*rillenkarren*, sometimes called *solution flutes*) are the only type of channel that forms at the crest of a block, from the top down (Fig. 6), and from direct impact of raindrops. They give way downslope to a smooth nonchanneled surface. They are parabolic in cross section, just under 2 cm in width on limestone, always packed together in parallel suites, and separated by knife edges. Where rills develop to either side of a ridged crest, they meet at the top in a herringbone pattern.

Drainage channels from focused flow under gravity control, *solution runnels*, are sharp-edged and narrow on bare rock (*rinnenkarren*) and smoother and wider under soil cover (*rundkarren*). They develop toward the foot of slopes and extend upslope over time. These are the rock equivalent of a normal fluvial gully, and so, depending on slope, may have a dendritic plan like a river system, and will get deeper downslope as they collect more discharge. On bare rock they often meander. They are typically several decimeters wide and deep. The smooth runnels that form by focused drainage under soil are exposed to view by soil erosion (Fig. 7).

Another type of channel is the *decantation runnel or decantation flute*. If water is stored and its aggressivity is enhanced (*e.g.*, by biological activity) then the overflow will form a runnel that is deepest at the point of emergence and peters out downslope. The storage may be in bedding planes, or clumps of soil or vegetation. If the drainage emerges as a sheet then a suite of decantation flutes will form.

Although water is required for karren formation, many features show the influence of rock properties more than flow properties. A very common type of karren is the dissolutionally opened joint. The first stage of opening is the *splitkarren* (like a small tear). The more fully opened joint is called a *grike* (Fig. 8). Suites of grikes develop along intersecting sets of joints leaving the remnant emergent, flat-topped *clint*, or more pointed, round-topped *pinnacle*, whose shape and size

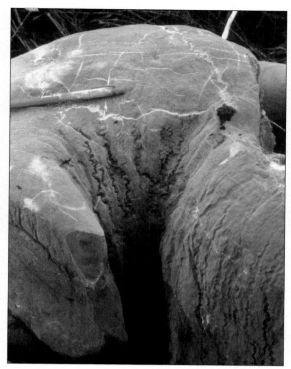

FIGURE 7 The rounded shape of this rundkarren from Vancouver Island, Canada, is indicative of its subsoil origin. Since soil erosion, the bare surface has become etched by tiny meandering channels (*rillenstein*).

FIGURE 9 Emergent rounded subsoil pinnacles in Wee Jasper, Australia, becoming modified by rillenkarren.

FIGURE 8 Clints and grikes of northwest Burren, County Clare, Ireland, making up a dissolutional pavement. The flat-topped clints are remnant from the glacially scoured bedding surface. The grikes are partly full of soil and vegetation.

vary with joint angles and frequency. The joints exposed since ice retreat in glacially scoured limestone landscapes are often now some 10–30 cm wide and up to a few meters deep. However, joints that have been opening without such a time limitation may be many meters wide and deep (*e.g.*, the grikes that separate the huge pinnacles in Fig. 1). The larger ones are called *bogaz* or *corridors* and make up *labyrinth karst* systems.

Dissolution under a cover of soil, vegetation, or even sometimes snow, can be many times faster than on bare rock, and hydrodynamic controls are quite different. Capillary forces prolong contact between solvent and solute. Where hydraulic gradient is low, this allows exploitation of subtle differences in rock properties resulting in *cavernous karren*, with curved intersecting holes rather like a sponge. *Subcutaneous* karren (*bodenkarren*) are always smoothly rounded. Steeply sloping rundkarren develop where hydraulic gradient is high. Under more acidic conditions features become more exaggerated. A *cutter* is a deep grike that has formed under an acidic soil cover, often with a complex curved cross section, some cavernous karren, and remnant rounded pinnacles. Where rounded subsoil pinnacles are exposed by soil erosion they become sharpened by development of rillenkarren and runnels (Fig. 9). Small sharp pinnacles are called *spitzkarren*, examples of which are well displayed in coastal karren (see below).

The junction between standing water or soil and rock is often marked by the development of a *corrosion notch*, whose form is not related to lithological controls. This may simply be a shallow horizontal indentation but

FIGURE 10 Cliff-foot swamp notch developed at Niah Caves, Sarawak, Borneo.

FIGURE 11 Coastal karren of Curacao, Netherlands Antilles. The basins and spitzkarren of the supra-littoral can be seen in the foreground, and the intertidal notch at the breaking wave in the distance.

swamp notches are often quite narrow vertically while extending into the rock by several meters (Fig. 10).

KARREN ASSEMBLAGES

Sometimes suites of karren are developed over large areas, creating landscape-scale karren assemblages or karrenfields. Examples are *alpine karrenfields* of bare rock above the treeline, showing features associated with snow melt (such as *heelprint karren—trittkarren*).

Pinnacle karst (Fig. 1) is another example of a karrenfield, made up of pinnacles separated by deep, narrow grikes. The sharpest and tallest pinnacle karst develops only on the purest and most mechanically strong rock. Most are inherited from subsoil pinnacles after regional uplift or base-level lowering.

Dissolutional pavements (Figs. 3, 8) are developed on the bedding planes of flat or gently dipping strata where joints have been opened up by dissolution. Typically the bedding plane was exposed by glacial scour (but any lateral planation process will do), so pavements are characteristic of the formerly glaciated regions. A pavement often degrades into rubble or vegetation patches. However, if the tops of the clints remain largely intact while the grikes enlarge, the end product is a *ruiniform karrenfield* of tall, narrow emergent clints.

A unique suite of karren forms develops in marine coastal regions where direct bioerosion is common. *Coastal karren* can be divided into three basic landforms: the *intertidal platform*; the *intertidal notch* (plus associated overhanging *visor*); and the *ramp* of basins and pinnacles. The first two are well displayed in tropical regions. The third is typical of tops of the visors, the more exposed parts of tropical coasts, and most of the temperate coasts (Fig. 11).

Bibliography

Bögli, A. (1981). Solution of limestone and karren formation. In M. M. Sweeting (Ed.), *Karst geomorphology* (pp. 64–89), Benchmark Papers in Geology 59. Stroudsburg, PA: Hutchinson Ross Publishing Company.

Ford, D. C., & Lundberg, J. (1987). A review of dissolutional rills in limestone and other soluble rocks. *Catena Supplement, 8*, 119–140.

Ford, D. C., & Williams, P. (2007). *Karst hydrogeology and geomorphology*. London: Unwin Hyman.

Ginés, A. (2004). Karren. In J. Gunn (Ed.), *Encyclopedia of caves and karst science* (pp. 64–89). New York, London: Fitzroy Dearborn.

Ginés, A., Knez, M., Slabe, T., & Dreybrodt, W. (Eds.), (2009). *Karst rock features—Karren sculpturing (Carsologica, 9)*. Ljubljana, Slovenia: Založba ZRC/ZRC Publishing.

Lundberg, J. (2012). Micro-sculpturing of solutional rocky landforms. In Frumkin A. (Ed.), *Karst geomorphology, 6*, Treatise on geomorphology, Elsevier.

Trudgill, S. T. (1985). *Limestone geomorphology*. London, New York: Longman.

White, W. B. (1988). *Geomorphology and hydrology of karst terrains*. New York: Oxford University Press.

KARST

William K. Jones[*] and William B. White[†]

[*]Karst Waters Institute,
[†]The Pennsylvania State University

INTRODUCTION

The wonderfully diverse landscapes that make up the surface of planet Earth are the result of a complex interplay between tectonic forces, erosive forces, and the mineralogical composition of the underlying bedrock. Tectonic forces elevate the land, thus providing the gravitational gradients needed to drive the erosive forces. The erosional carving of the landscape is accomplished by a combination of chemical reactions and mechanical transport. Chemical weathering breaks down the mineral grains and their cements, taking portions into solution and leaving the insoluble portions as regolith or soil. Mechanical transport involves gravity-driven downslope movement such as soil creep and landslides and transport by flowing water in storm runoff and flowing streams and sometimes by wind scour and by creeping glaciers. In rocks such as limestone, dolomite, and gypsum, chemical weathering dominates over mechanical transport, resulting in a unique set of landscapes known as *karst*. Because of the dominant removal of rock masses in solution, karst regions are those with internal drainage.

The distribution of karst landscapes over the Earth's surface to a large extent follows the distribution of carbonate (limestone and dolomite) and gypsum rocks (Fig. 1). As a percentage of total land area, a compilation by Paul Williams gives values for carbonate rock karst (Table 1).

The data in Table 1 are only karst areas on carbonate rocks. If areas of gypsum karst and a few rare occurrences of karst on salt and other rocks are included, the total comes into the 15–20% range widely cited as the fraction of the Earth's land area that is karst.

The objective of this article is to provide a broad overview of karst. There is a large literature describing individual karst regions. Regional descriptions include Europe and Russia (Herek and Stringfield, 1972); China (Sweeting, 1995); and North America (Palmer and Palmer, 2009).

ORIGIN OF THE WORD *KARST*

Location

The word *karst* is a Germanized form of the name of a carbonate plateau that is situated above the Adriatic Sea immediately to the east of Trieste, Italy, and covers about 750 km^2 (Fig. 2). The regional name for this area is Kras from the pre-Indo-European root *Ka(r)* meaning stone or rock (Gams, 2003). Variations on the name include *Carusadus*, *Karusad*, *Grast*, *Krs*, and *Carso*. Some researchers argue that the "classical" karst should be considered just the Kras plateau, with about one third of the area in Italy and two thirds in Slovenia. In as much as the Kras is the type locality for karst there is some justification for this rather narrow view. The idea of a "type locality" is that the location gives its name

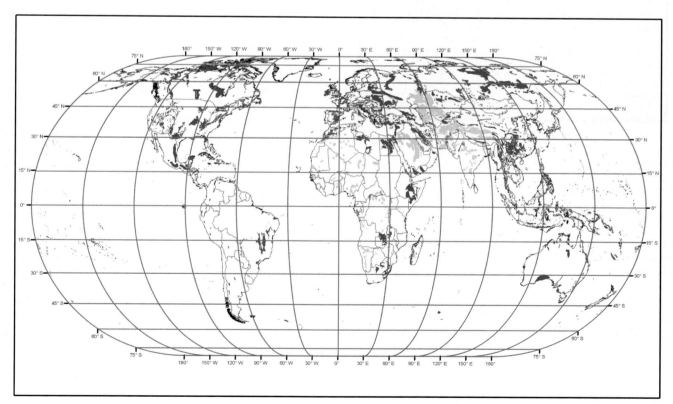

FIGURE 1 World map showing the distribution of karst regions. *Map courtesy of Professor Paul W. Williams. See also Williams and Ford, 2006.*

TABLE 1 World Distribution of Karst

	Percentage of Land Area	Karst area in million km²
Russian Federation	16.1	3.3
South America	2.1	0.37
Africa	9.2	2.8
North America (including Central America and the Caribbean)	18.3	4.1
Southeast Asia (including China, Malaysia, and Indonesia)	10.8	1.7
Middle East and Central Asia	23.0	2.6
Europe (excluding Iceland and Russia)	21.8	1.3
Australasia (including Pacific islands)	6.2	0.59
World total	12.5	16.7

FIGURE 3 Map showing the expanded view of the classical karst that includes the caves and karst poljes in the vicinity of Postojna. Map from R.A. Baroody after Gams, 2003.

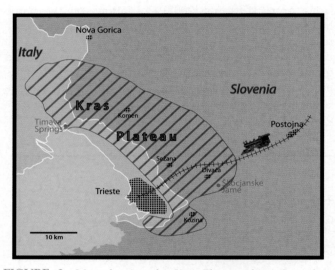

FIGURE 2 Map showing the Kras Plateau, the railway line, Škocjanske Cave, and the resurgence of water from the cave at Timava (Timavo) Springs. This is the region that gives karst its name and is sometimes referred to as the "karst of Trieste." Map from R.A. Baroody after Petrič, 2005.

to the geologic formation or feature being described, presumably for the first time in the scientific literature. The type locality is not necessarily the best example of a particular feature; it is just the location where that type of feature was first described. Note that "Kras" refers to a specific region and "karst" refers to a landform. To add to the possible confusion, in the Slovene language *Kras* is the region and *kras* is the landform.

A more realistic definition of the classical karst is an expanded area that extends eastward from Trieste at least to Postojna and south to Rijeka (Croatia). This is sometimes called the *karst triangle* and this area picks up some important features such as karst poljes that are not included in the landforms of the Kras plateau. Geologists from Vienna who gave us the first modern descriptions of karst landforms in the late 1800s probably considered the karst to include the karst of the Dinaric Mountains of Slovenia and much of the extensive karst landforms extending south through the former Yugoslavia (Fig. 3). Kranjc (1994) suggests that all of the Dinaric karst in present-day Slovenia be considered the classical karst. Many of the technical names for karst features also come from the Dinaric karst region. Terms in international usage include *doline*, *polje*, *ponor*, and *hum*. Of course many spectacular karst landforms such as the tower karst of south Asia were essentially unknown to Western geologists at the time, so regardless of how we define the "classical karst" it does not include all karst features and landscapes.

History

Slavic tribes followed the Roman occupation of the Kras and were established in the area by 800 A.D. The Romans used Trieste as a port and called it Tergeste. The present territory of Slovenia and Trieste, Italy, were part of the Austrian-Hungarian (Habsburg) empire from 1382 to 1919. Trieste became the major imperial port of the Habsburgs when it was declared a "free port" in 1719. Trieste and part of Slovenia as far east as Ljubljana were under Italian control from 1919 until 1945 when Yugoslavia emerged as an independent nation. The population of Trieste was almost evenly split between Italians and Slavs, but in 1954 the UN handed Trieste back to Italy and it is today a "border" city between Italy and Slovenia. Slovenia, with most of the classical karst, became an independent country in 1991 and is presently in the European Union. Many of the names and spellings have changed from the Slavic roots to German to Italian to Slovene, so it is sometimes confusing to read the earlier literature and keep track of the changing place names. For example, Postojna with its famous cave was known as Adelsberg in German and Postumia in Italian.

When the Archduke Ferdinand Maximilian crossed the Kras on a "highway" in 1850 he called the Kras a "cursed desert" and was apparently very happy to reach the Obelisk in Opicina above Trieste that marks the end of the Kras plateau above the city (Morris, 2001). The Kras landscape appeared sparsely populated, deforested, rocky, and waterless. The Kras "threshold" was a region dreaded by many travelers and traversed as quickly as possible. The Kras is now easily crossed on modern highways and has been extensively reforested, but it still strikes the traveler as an alien landscape in many respects.

Karst phenomena were well known to the ancient civilizations of the Mediterranean. Descriptions of caves and karst features are found throughout Greek and Roman mythology (Clendenon, 2009). Many of the Greek settlements along the Adriatic coast appear to be preferentially located on areas characterized by limestone rocks and karst springs (Crouch, 1993). Ideas and descriptions of the Dinaric karst had been written long before the science of geomorphology came into being. Posidonios of Apameia (Roman, 135–50 B.C.) studied the Škocjan Caves and the Timava (Timavo) Springs (Kranjc, 1994). In 1599 Father F. Imperato tried (unsuccessfully) to conduct a tracer test using "floaters" to establish the connection between Škocjan Caves (Fig. 4) and the springs at Timava some 35 km northwest of the cave (Shaw, 1992). An early "scientific" description of the Kras was a monograph by Janez Valvasor (1686) that describes in detail the polje of Cerknica and its periodic lake (Fig. 5) and Škocjanske Jame (cave). Hacquet published several papers around 1780 that describe karst landforms and hydrology in Slovenia and viewed karst as a special landform and a geological and hydrological specialty (Kranjc, 1994). The term *karst* had certainly appeared in papers before the mid-1850s, but it did not become established in the international literature until the late 1800s.

Karst

Caves and karst had been studied in many locations around the world for several thousand years, but the "modern" science of geomorphology dates to about 1858 (Roglic, 1972). The term *karst* became established in the international scientific literature as a result of an outstanding type locality in the Kras and the timing of events. Vienna became a center of geomorphological studies, and the Kras plateau with its many caves and dolines was part of the empire. Travel between the capital of the Habsburg Empire in Vienna and its port on the Adriatic was slow and tedious until the Austrian Southern Railway line was opened in 1857. After that, geomorphologists could easily visit Postojna and the Kras Plateau. Some of the early descriptions of dolines were from road cuts right along the railway tracks.

Professor Albrecht Penck from the Vienna School of Physical Geography published the first textbook on geomorphology in 1894, but it was his student Jovan Cvijić who really established the term *karst* in the geological literature when he published *Das Karstphänomen* in 1893. Cvijić was from Serbia and he described karst features throughout the Dinaric Mountains of Yugoslavia with many examples from the area of Postojna and Trieste. Many of the most influential geomorphologists of the late 1800s and early 1900s visited the *classical karst* and the name became firmly established, at least in Western geology circles.

Caves represent the subsurface aspect of a karst landscape and some researchers focused primarily on the subterranean features observed in caves rather than on the surface landscape. Understandably, speleologists sometimes formed different theories about the process that created the unique landforms on the surface. The French cave explorer E.A. Martel used a decidedly "speleological" approach to his studies of karst areas and he preferred the term *Phenomenes du calcaire* (Martel, 1894), which he probably felt put less emphasis on surface features. Examples of other terms for karst include *limestone geomorphology*, *limestone landscapes*, and *carbonate aquifers*. Nevertheless, the term *karst* (noun or adjective) and its many forms such as *karstic* (adjective) is firmly established in the international literature and is found in the title of many books and articles.

KARST LANDFORMS

Karst regions are characterized by a unique set of landforms as described below. Karst regions also differ from other geomorphic regions by the presence of subsurface landforms—caves—which are also a distinguishing characteristic. In general, surface karst features are more pronounced in regions where the caves have direct hydrologic connections to the land immediately overlying the caves and the surface and subsurface features develop more or less simultaneously. Some refer to the surface expression as *exokarst* and the subsurface expression as *endokarst*.

FIGURE 4 Entrance to Škocjanske Jame, Kras region, Slovenia. *Photo by W.K. Jones.*

Pavements and the Epikarst

Carbonate rocks react with rainfall and infiltrating water either at the exposed rock surface or at the soil–bedrock contact. Chemical attack on the rock tends to be concentrated on structural weaknesses such as joints and bedding plane partings. If the limestone is very pure or if the regolith has been removed, the exposed limestone surface becomes a *pavement karst*. Dissolution along joints produces deep crevices known as *grikes* (in the U.K.) or *cutters* (in the U.S.). The exposed bedrock is often sculptured into complex small solutional forms known as *karren*. A karst surface covered with regolith may appear featureless from above but the regolith/bedrock contact, which is where the active dissolution takes place, is likely to be highly irregular with deep regolith-filled crevices separated by high pinnacles. The regolith/bedrock interface is usually sharp with carbonate-free regolith on one side and unweathered limestone on the other side. The crevices and other solution features often disappear at depths of a few to a few tens of meters as the infiltrating waters consume their acidity and the dissolution process slows. The zone that includes the regolith and the solutionally sculptured bedrock is known as the *epikarst*.

Closed Depressions (Dolines, Sinkholes)

Karst drains internally through fractures in the bedrock. This internal runoff tends to concentrate along more transmissive zones in the fractures or at fracture

FIGURE 5 Cerkniško jezero, the intermittent lake in the Cerkniško Polje east of Postojna, was described in detail by Valvasor in 1689. Karst poljes are considered a typical karst feature of the Dinaric karst but are not represented in the Kras plateau of Trieste. *Photo by W.K. Jones.*

intersections and as a result dissolves a depression in the bedrock. Closed depressions, variously known as sinkholes or dolines, are one of the most characteristic features of karst landscapes. Bedrock dissolution is only one process contributing to the formation of closed depressions. Cave roof collapse can migrate upward to the surface, producing a collapse sinkhole (or doline). In karst regions with thick soils, piping of soil into the subsurface can produce a cover-collapse sinkhole with little modification of the bedrock.

Sinkholes are found in a very wide range of sizes from meter-size depressions to multikilometer diameters and in depths from fractions of a meter to hundreds of meters. Regions where closed depressions are the dominant landform are not typically dissected by surface streams and are known as *sinkhole plains* or *doline karst*. These landscapes tend to have a similar appearance in many parts of the world (Fig. 6).

Poljes

The largest known closed depressions are the *poljes* found in the Dinaric karst and identified in many parts of the world. The typical polje (a Slavic word meaning "field") has a flat floor which may represent a deep alluvial fill or a planated bedrock surface. The polje floors flood seasonally and are frequently the best agricultural land. Streams typically enter the poljes from springs, flow across the floor as surface streams, and sink on the opposite side in swallets known as *ponors*.

Cones and Towers

In nonkarstic landscapes, flowing streams are integrated from small creeks to large creeks to small rivers to large rivers and on to the sea. The result is a landscape of mountains, hills, and valleys all with open outlets for drainage. Internal drainage in karst sculptures the landscape into closed depressions with

FIGURE 6 (A) Doline karst in Slovenia. (B) Doline karst in West Virginia, USA. *Photos by W.K. Jones.*

FIGURE 7 (A) Cone and tower karst in Puerto Rico. (B) Cone and tower karst (fengcong) along the Lijiang, Guangxi Province, China. *Photos by W.B. White.*

individual drainage outlets on the bottom and the residual rock mass in-between. If the residual rock mass is dominant, the landscape is viewed as a surface pocked with closed depressions. If the closed depressions are dominant, the residual rock masses stand out in relief as what is known as *cone and tower karst* (Fig. 7).

Cone and tower karst takes many forms. There may be closed-spaced, steep-sided hills separated by chaotic gorges. There may be isolated towers separated by planated surfaces. There may be large closed depressions with roughly pyramidal hills at the intersections; this form is sometimes known as *cockpit karst*. All of these possibilities can be incorporated into the concept of polygonal karst (Williams, 1972) where the cones, towers, and closed depressions are analyzed in terms of the pattern of their drainage divides.

Caves

Although the chemistry of the dissolution process is similar, the degree of landform development on the surface and in the subsurface can be different. There are karst regions with some development of surface landforms but very few enterable caves. There are large and extensive cave systems with little expression of karst landforms on the surface above. The internal drainage in karst regions follows localized conduits from recharge points to discharge at karst springs. Explored caves are fragments of conduits or fragments of abandoned conduits that are sufficiently large for human exploration and which happen to have a connection with the land surface. The population of known caves is only a small sampling of the remaining unknown and inaccessible part of the conduit system.

LANDSCAPES OF MIXED ORIGINS

The original description of karst from the Dinaric mountains was of a region entirely underlain by carbonate rocks. Such regions were termed *holokarst*, meaning the region was entirely karst with little or no surface drainage. The Dinaric mountains are exceptional in that they contain thousands of meters of carbonate rock. In many karst regions the stratigraphic thickness of carbonate rocks is much less. Other processes will be operating, and the resulting landscape is of mixed origin.

Fluviokarst

Fluviokarst is the name applied to many landscapes where exposed karstic rocks make up part but not all of the drainage basins. In fluviokarst there are stream tributaries and headwater drainage basins on insoluble rocks. These surface streams typically sink at karstic rock contacts, leaving only dry channels or dry valleys until, downstream, the underground drainage reappears from large karst springs. The size of fluviokarst basins varies from less than 1 km^2 to hundreds of km^2. Details depend on the thickness and lithology of the karstic rocks, on the relief of the drainage basin and the placement of karstic and nonkarstic rocks within the basin, and on the structural characteristics of the basin.

Alpine Karst

The term *alpine karst* refers to landscapes developed in high mountain regions such as the Alps and the Pyrenees in Europe or the Rocky Mountains in the United States and Canada. Alpine karst is the result of both the setting and the processes. The alpine setting implies high relief, providing high hydraulic gradients for the movement of water, and usually structural complexity in the rocks with extreme folding and faulting. The dissolutional processes are also influenced by the extremes in climate, cold winters with thick snow pack and hot summers often with intense storms.

In many alpine regions, soils are thin or nonexistent with the only soil accumulations in the bottoms of closed depressions. The exposed bedrock is sculptured into a complex of karren forms. Deep shafts and deep widened crevices develop along open fracture systems. Caves usually have an extensive vertical component with many pits and waterfalls.

Glaciokarst

The time scale for the evolution of most karst areas is several million years. Those karst areas at high altitudes or moderate to high latitudes have, therefore, been exposed to the comings and goings of the Pleistocene glaciations. In some regions such as western Ireland and northern England, the glaciations have scraped off the soil, exposing dramatic pavement karst. In other regions such as the American Midwest, the carbonate rock surface has been covered with thick layers of glacial till. Much of alpine karst also bears the imprint of mountain glaciers.

COASTAL KARST, EOGENETIC KARST, AND MIXING-ZONE KARST

Several factors set coastal karst and island karst apart from karst developed in continental interiors. For karst development directly on the coast, continuous wetting by waves, tides, and storm surges makes the karstic rocks a suitable habitat for a variety of organisms. Algal growths provide organic acids to supplement the effects of carbonic acid. Boring organisms provide a mechanism for sculpturing the rock surface.

Near the coast is a zone where salt water mixes with freshwater from the interior. The mixing-zone waters become undersaturated even if both salt water and freshwater are themselves at saturation. Mixing-zone water dissolves out very characteristic caves known as *halocline caves* or *flank margin caves*. Mixing-zone karst is mainly a subsurface phenomenon except for cave entrances exposed by receding coastlines.

A third factor, not limited to coastal areas but most frequently found there, is the presence of highly porous limestone. The carbonate rocks that support most karst areas in the continental interiors have been strongly compacted by deep burial and tectonic deformation. As a result, the primary permeability is extremely low and groundwater movement is through fractures and bedding plane partings. Caves develop along these pathways and surface karst landforms are strongly controlled by them. Collectively, landforms developed in dense compacted rocks is known as *telogenetic karst*. In contrast, young limestones, typically found in coastal regions, are still undergoing diagenesis, have never suffered compaction or tectonic deformation, and as a result have a high primary permeability. Karstification and diagenetic processes are both at work, resulting in what is called *eogenetic karst*.

HYPOGENETIC KARST

Traditional karst, both surface and subsurface, forms by a top-down process with CO_2-containing meteoric water infiltrating through the epikarst and downward through the vadose and phreatic zones of the aquifer to eventually reappear at karstic springs. In recent decades, it has been realized that there are also bottom-up processes by which karst features are formed by up-welling deep waters. These features, mostly caves, are known as *hypogenetic karst*.

Hydrothermal Karst

Some deep-seated solutions are also primarily carbonate waters although often at high CO_2 pressures. These solutions are also at temperatures well above surface ambients and so the resulting dissolution features have been called *hydrothermal karst*.

Sulfuric Acid Karst

A completely different chemistry is responsible for some large cave systems such as Carlsbad Caverns and Lechuguilla Cave in New Mexico. The up-welling solutions contain H_2S which oxidizes to sulfuric acid when the solutions reach the oxygen-bearing shallow groundwater. Action of the sulfuric acid on the limestone in the reaction zone can produce very large cave chambers although there may be little expression of surface karst. Sulfuric acid caves often have characteristic three-dimensional patterns with large interconnected chambers. Residual masses of gypsum are also an indicator.

See Also the Following Articles

Closed Depressions in Karst Areas
Karren, Surface
Soil Piping and Sinkhole Failures
Tiankeng

Bibliography

Clendenon, C. (2009). *Hydromythology*. Lansing: Fineline Science Press.
Crouch, D. P. (1993). *Water management in ancient Greek cities*. New York: Oxford University Press.
Cvijić, J. (1893). Das Karstphänomen. Versuch einer morphologischen Monographie. *Geographische Abhandlungen, 5*(3), 218–329.
Gams, I. (2003). *Krasv Sloveniji*. Ljubljana: Zalozba ZRC, ZRC SAZU.
Herek, M., & Stringfield, V. T. (1972). *Karst: Important karst regions of the Northern Hemisphere*. Amsterdam: Elsevier.
Kranjc, A. (1994). About the name and the history of the region Kras. *Acta Carsologica, 23*, 83–89.
Martel, E. A. (1894). *Les abimes*. Paris: Delagrave.
Morris, J. (2001). *Trieste and the meaning of nowhere*. New York: Simon and Schuster.
Palmer, A. N., & Palmer, M. V. (2009). *Caves and karst of the USA*. Huntsville, AL: National Speleological Society.
Petrič, M. (2005). Hydrogeological characteristics of Kras. In A. Mihevc (Ed.), *Kras: Water and life in a rocky landscape* (p. 21). Ljubljana: Zalozba ZRC.
Roglič, J. (1972). Historical review of morphologic concepts. In M. Herak & V. T. Stringfield (Eds.), *Karst: Important karst regions of the Northern Hemisphere* (pp. 1–18). Amsterdam: Elsevier.
Shaw, T. R. (1992). *History of cave science*. Broadway, Australia: Sydney Speleological Society.
Sweeting, M. M. (1995). *Karst in China*. Berlin: Springer-Verlag.
Williams, P. W. (1972). Morphometric analysis of polygonal karst in New Guinea. *Geological Society of America Bulletin, 83*, 761–796.
Williams, P. W., & Ford, D. C. (2006). Global distribution of carbonate rocks. *Zeitschrift für Geomorphologie, Supplement, 147*, 1–2.

KAZUMURA CAVE, HAWAII

Kevin Allred
Hawaii Speleological Survey

INTRODUCTION

Among the many known lava tube caves presently explored throughout the world, Kazumura Cave, on the big island of Hawaii, stands foremost in length and vertical extent. It is also the longest linear cave in the world, extending farther over a straight-line distance than any other. This cave contains a wide range of speleothems and morphology with populations of cave-adapted invertebrates. Because much of the cave is located under rural subdivisions, it is a popular place to visit for locals and out-of-state cavers.

PHYSICAL SETTING

Location and Surface Relationships

Kazumura Cave was formed, along with many nearby caves, approximately 300–500 years BP in a large tholeiitic lava flow complex known as the Ai-laauShield Flow on the eastern side of the Island of Hawaii (Holcomb, 1987). It is thought that the Kazumura Cave flow was one of the last in the Ai-laau flow complex; however, black glassy-skinned lava has intruded into the cave in several places, showing that later flows occurred after the cave had cooled significantly. The flow originated from the summit of Kilauea Volcano at about 1189 m. The upper end of the cave was collapsed and filled from highway construction at the 1130-m elevation, but continuous passage can be traversed downslope for 41.8 km to the lower end at an elevation of 28 m (Fig. 1). It is, by far, the longest linear cave in the world, covering a straight-line distance of 32.2 km. The cave has an average slope of less than 2°. It has a temperature range from 15°C near the Kilauea caldera to 22°C at the lower end.

The surface slopes of the Ai-laau flows are presently undergoing dramatic transformation from a barren lava wasteland to lush tropical old growth jungle atop ever-thickening soils. In many cave passages, delicate tree roots penetrate the thin soils and then cracks of the roof supply an energy source for cave-adapted crickets, fantails (Fig. 2), and spiders (Howarth, 1973). Lava tube slime, a little-known-about organic growth, sometimes coats ceilings and walls with gold, white, or red.

Human Impacts

Because much of the cave is beneath developing rural subdivisions, it is being negatively impacted. Sometimes with a roof less than 3 m thick, it is prone to collapse from bulldozing, and is sometimes used for domestic sewage, graywater, and trash disposal. Recently, its increased notoriety has led to some industrial tourism resulting in many thousands of feet of passage modification. In these areas, it no longer retains its virgin natural appeal.

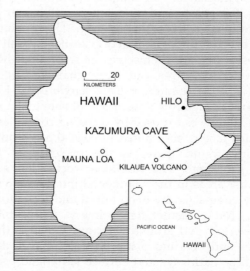

FIGURE 1 The location of Kazumura Cave on the Island of Hawaii.

HISTORY OF EXPLORATION

Local Visitation

Not much is known of the earliest exploration of Kazumura Cave. There is some evidence of use by prehistoric Hawaiians. Many parts of the cave also contain modern paraphernalia such as rotten rope, string, trash, and bits of worn-out shoes. In one section, someone used an innovative approach by placing pages of a *Reader's Digest* magazine along the floor to mark his or her way back to the nearest entrance.

FIGURE 2 A fantail clings to a tree root which has penetrated the roof of the cave. *Photo by Mike Shambaugh. Used with permission.*

Survey and Systematic Exploration

An entrance of Kazumura Cave was a designated civil defense shelter in the early 1960s, and its name originated from this. In August 1979, acting on reports of a major lava tube, a U.K. speleological expedition explored and surveyed 11.7 km of the cave with a vertical range of 261 m, making Kazumura Cave one of the three longest lava tube caves in the world. It was also visited and partially surveyed by Japanese cavers. Finally, 14 years later, members of the Hawaii Speleological Survey of the National Speleological Society began a more determined survey using profile views. The original British portion was extended to 12.5 km and various plugs and collapses were dug through to eventually extend the cave to 65.4 km with a vertical extent of 1101.8 m. The cave presently has 101 known entrances (Fig. 3).

DESCRIPTION OF KAZUMURA CAVE

Speleogenesis

Kazumura Cave is a well-developed "master tube," formed from the crusting over of a highly liquid pahoehoe lava flow issuing steadily for a long period from Kilauea Volcano. Because of the insulative nature of the lava tube, the Kazumura flow extended 39 km to the Pacific Ocean and significantly extended the coastline there. But the cave itself becomes low and pinches off in pooled lava 3.5 km from the beach. Although some portions of the cave developed by the crusting over of the main open lava stream, evidence indicates that

FIGURE 3 One of many entrances to Kazumura Cave. *Photo by Mike Shambaugh. Used with permission.*

FIGURE 4 Over time, the flowing lava cut down the floor of Kazumura Cave, creating deep canyon-like passages. Here, near an entrance (upper center) an intermediate roof formed, reinsulating the lava. *Photo by Mike Shambaugh. Used with permission.*

FIGURE 5 During the thermal erosion of Kazumura Cave, increased turbulence at the outside of bends has formed cutbanks (right). Slipbanks result on the inside of bends (left) where there is less thermal erosion. *Photo by Mike Shambaugh. Used with permission.*

generally, the cave passage first formed as a braided, interweaving delta-like pattern near the growing snout of the flow. The cave roof was thickened through accretion against the cool ceilings. Overflow or *breakout* further built up the roof. Kazumura Cave passages vary widely in shape from wide and low typical of the embryonic braided passages to deep and canyon-like, which indicates significant downcutting (Fig. 4).

Thermal Erosion

As the flow extended, one or two braids pirated the flow and downcutting occurred through a process known as *thermal erosion* (Hulme, 1973). This is a partial melting of the underlying substrate with an accompanying mechanical plucking away of unmelted crystals. Deepened canyon-like passages, lavafalls, and

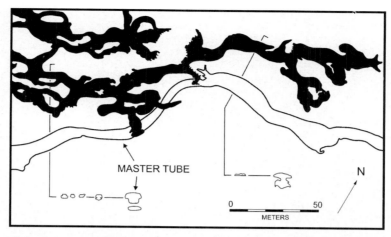

FIGURE 6 Early-formed braided tubes are in black. The developed master tube is in white. One or two braids pirated the Kazumura lava flow and were deepened by thermal erosion (note cross sections).

exposed beds of old pahoehoe and aa flows exposed behind ruptured wall linings are all evidence of this thermal erosion. More turbulent flow eroding aggressively into the substrate on the outside of bends created cutbanks (Fig. 5). The early formed braided passages that are not lava filled are commonly found at the ceiling level off the main passage (Fig. 6). As thermal erosion progressed, portions of the now unsupported ceiling collapsed into the free-flowing lava and this breakdown was carried away or wedged into upper areas during temporary flooding.

Extraneous Tubes

The most outstanding proof of thermal erosion is found where the downcutting lava stream encountered older lava tubes or large bubbles. Here, melting of the substrate was not possible due to the air-cooled voids. The stream was not able to cut into and invade these spaces unless the thin separating rind eventually collapsed. These rinds are as little as 10 cm thick (Allred and Allred, 1997). Several extraneous tubes have been identified in the cave. They are recognized as large, egg-shaped bulges protruding from walls. Most tubes ruptured from the contraction of hot walls during the cooling of the cave, but some of them remain unbroken; thus the extent of the voids within is unknown. The longest extraneous tube found in Kazumura Cave is nearly 1 kilometer long under an extraordinarily large hump in the floor which diverted the stream to the side. The thin separation finally collapsed during the cooling of Kazumura Cave, allowing access into a master tube which crosses nearly perpendicular to the Kazumura passage. This passage contains lava formations unlike any found in Kazumura Cave proper. Heligmites and miniature volcanoes erupted out of the floors.

FIGURE 7 Near entrances or in spacious passages, later-formed intermediate roofs insulated the lava stream.

Multilevel and Lavafall Development

Intermediate roofs are formed in some spacious passages and below entrances that were open during the eruption (Fig. 7). Cooling formed crusts which reinsulated the lava stream. Because of these

FIGURE 8 With increased turbulence in steeper slopes, the lava flow in Kazumura Cave cut back into the substrate, creating lavafalls and plunge pools. This lavafall is 9 m high. Its molten 14-m-wide plunge pool is estimated to have been 27 m deep. *Photo by Mike Shambaugh and Kevin Allred.*

secondary roofs, entrances in the cave system have not completely bisected the underground continuity of the cave. Up to four stacked levels are found in some places. Breakout overflows through skylights probably also occurred, but they probably did not leave much connected passage unless they happened during the early braiding stage. Lavafalls are well developed in the cave closer to the caldera where steeper slopes contributed to greater turbulence. It is thought that some headward backcutting occurred on many of the lavafalls until they reached a level, nonturbulent upstream floor (Allred and Allred, 1997). Lavafalls can be up to 12.9 m high and their solidified plunge pools up to 15.5 m in diameter (Fig. 8). By calculating the amount of crystallization contraction in the sunken plunge pools, some of these are estimated to have eroded 20–30 m in depth.

LAVA SPELEOTHEMS

Primary Speleothems

In many parts of the cave, fluctuations in the flowing stream against a ceiling have formed sharktooth stalactites composed of many layers of linings. Other primary speleothems are splashed stalactites, stretched lava, Pele's Hair, squeeze-ups, and lava blades which sometimes resemble rillenkarren on a bare carbonate karst. Boiling within partially solidified lava linings has resulted in extruded forms such as tubular lava stalactites above globular stalagmites, helictites, and blisters (Allred and Allred, 1998).

Secondary Speleothems

Secondary deposits of white crusts and crystals appear to be gypsum. They occur in dry places in the cave away from rain seepage through the porous ceilings.

See Also the Following Article

Volcanic Rock Caves

Bibliography

Allred, K., & Allred, C. (1997). Development and morphology of Kazumura Cave, Hawaii. *Journal of Cave Karst Studies, 59*(2), 67–80.

Allred, K., & Allred, C. (1998). Tubular lava stalactites and other related segregations. *Journal of Cave and Karst Studies, 60*(3), 131–140.

Holcomb, R. T. (1987). Eruptive history and long-term behavior of Kilauea Volcano. *U.S. Geological Survey Professional Paper, 1350,* 261–350.

Howarth, R. T. (1973). The cavernicolous fauna of Hawaiian lava tubes. 1. Introduction. *Pacific Insects, 15*(1), 139–151.

Hulme, G. (1973). Turbulent lava flows and the formation of lunar sinuous rilles. *Modern Geology, 4,* 107–117.

KRUBERA (VORONJA) CAVE

Alexander Klimchouk
Ukrainian Institute of Speleology and Karstology, Simferopol, Ukraine

INTRODUCTION

At the dawn of the new millennium, in January 2001, Krubera (Voronja) Cave in the Arabika Massif, Western Caucasus, became the deepest known cave in the world, with a depth of −1710 m (Klimchouk and Kasjan, 2001). For the first time, the deepest known cave was found to be outside Western Europe. In the article on Krubera Cave in the previous edition of this encyclopedia (written in 2003), when the explored depth of the cave was still at −1710 m, the present author wrote: "The future possibility of locating a 2000 m+ system in the area is exceptionally good." Discovering the first cave on the planet deeper than 2000 m had been a long-standing dream of cavers around the world, and this was set in 2001 as an official goal of the Call of the Abyss project of the Ukrainian Speleological Association (Ukr.S.A.), one of the most ambitious and successful exploration projects in the history of speleology. In October 2004 this goal was reached, when Krubera Cave was pushed to depth of 2080 m. In subsequent years, the Ukr.S.A. expeditions have explored the cave to depth of 2191 m in the main branch, and also explored the second branch in this cave, called Nekujbyshevskaja, to a depth of 1697 m.

THE ARABIKA MASSIF: LOCATION AND PHYSIOGRAPHY

The Arabika Massif is one of the largest limestone massifs of the Western Caucasus (Fig. 1A,B). It is located in Abkhazia, a republic that officially belongs to Georgia but since 1992 holds claim to being an independent state.

To the northwest, north, northeast, and east, Arabika is bordered by the deeply incised canyons of Sandripsh, Kuturusha, Gega, and Bzyb rivers (Fig. 1C). The Bzyb River separates Arabika from the adjacent Bzybsky Massif, another outstanding karst area with many deep caves, including the Snezhnaja-Mezhonogo-Iljuzia system (−1,753 m) and Pantjukhina Cave (−1508 m). To the southwest, Arabika is bordered by the Black Sea, with limestones dipping continuously below the sea level.

The Arabika Massif has a prominent high central sector with elevations above the tree line at ~1800−1900 m. This is an arena of classical Alpine-type karst, a glaciokarstic landscape with numerous glacial trough valleys and cirques, with ridges and peaks between them. The bottoms of the trough valleys and karst fields lie at elevations of 2000−2350 m, and ridges and peaks rise to 2500−2700 m. The highest peak is the Peak of Speleologists (2705 m) but the dominant summit is a typical pyramidal horn of the Arabika Mount (2695 m). Some middle- to low-altitude ridges covered with forest lie between the central sector and the Black Sea. A plateau-like middle-altitude outlier of the massif in its south sector is Mamzdyshkha, with part of the plateau slightly emerging above the tree line.

KRUBERA CAVE AND OTHER DEEP CAVES IN THE ARABIKA MASSIF

Among several hundred caves known in the Arabika Massif, 15 have been explored below 400 m and five below 1000 m (shown in Fig. 1C). Several are located within the Ortobalagan Valley, a perfectly shaped, relatively shallow, perched glacial trough of the sub-Caucasian stretch, which holds the advanced position in the central sector toward the seashore (Fig. 2). Since 1980, Ukrainian cavers have been undertaking systematic efforts in exploring deep caves in the Ortobalagan Valley including Krubera (Voronja) Cave (number 1 on Fig. 1C; −2191 m), and the Arabikskaja System (number 4 on Fig. 1C), which consists of Kujbyshevskaja Cave (−1110 m) and Genrikhova Bezdna Cave (−965 m to the junction with Kujbyshevskaja). Another deep cave in the valley, located in its very upper part, is Berchilskaja Cave (−500 m; number 11 on Fig. 1C) explored by Moldavian and Ukrainian cavers. The Ortobalagan Valley extends along the crest of the Berchilsky anticline, which dips gently northwest.

An open-mouthed 60-m shaft, the Krubera entrance, was first documented in the early 1960s by Georgian researchers, who named it after Alexander Kruber, a founder of karst science in Russia. The early exploration was stalled by an impassable squeeze in a meandering passage at −95 m which led off from the foot of the entrance shaft. During the 1980s, the main focus of the Ukrainian expeditions to the Ortobalagan Valley were Kujbyshevskaja Cave (−1110 m) and Genrikhova Bezdna Cave (−956 m) connected into a single system in 1987 (the Arabikskaya System). At the same time, these expeditions had pushed Krubera Cave from −95 to −340 m by breaking through a series of critically narrow meanders between successive vertical shafts. During this period the cave received its secondary name Voronja (Crow's Cave), owing to the number of crows nesting in the entrance shaft.

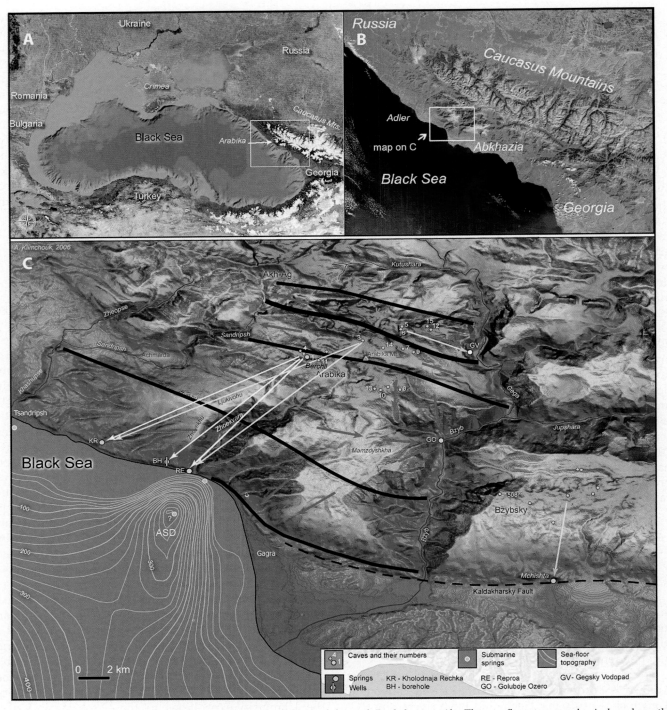

FIGURE 1 (A) and (B): Location and shaded maps of the Arabika and Bzubsky massifs. The sea-floor topography is based on the SRTM30plus data (NASA). (C) Deep caves are indicated by dots and numbers (explained in the text) and major springs are specified by two-character indexes (explained in the legend). Solid black lines show the crests of major anticline folds. Red arrows reflect previous ideas on groundwater basins and flow directions controlled by major fold structures. White and yellow arrows indicate the actual hydrologic connections established by dye-tracing experiments in 1984–1985.

From 1992–1999 the explorations in Arabika were suspended due to the Georgian-Abkhazian ethnic conflict and subsequent turmoil. In 1999, the Ukr.S.A. expedition recommenced work in Krubera Cave and made a major breakthrough by discovering and exploring two branches that stretched from the old series in different directions: the Main Branch to −740 m and the Nekujbyshevskaja Branch to −500 m. The Main Branch

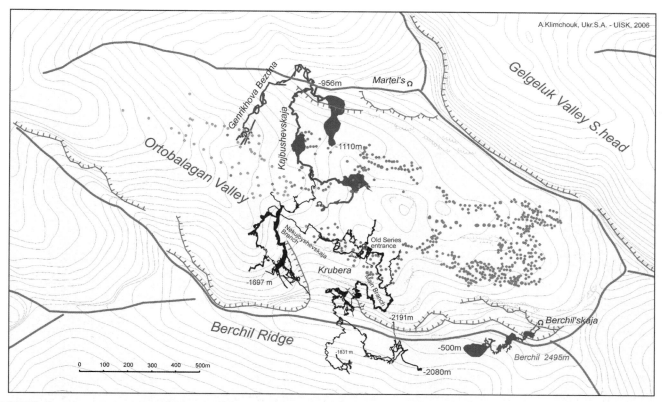

FIGURE 2   Major caves in the Ortobalagan Valley. Red dots indicate dolines.

was quickly pushed farther in 2000: in August to −1200 m and September to −1410 m. In January 2001, the Ukr.S.A. expedition explored the cave to −1710 m, establishing it as the new deepest cave in the world. In 2004, the cave was explored to depth of 1840 m in August, and to −2080 m in October. Further efforts in the Main Branch led to reaching the terminal sump at −2145 m in 2006. The current deepest point at −2191 m was reached by the Ukr.S.A. expedition in 2007 through diving to −46 m in the terminal sump. In the Nekujbyshevskaja Branch, systematic digging efforts in boulder chokes since 2004 resulted in a series of breakthroughs and the eventual exploration to the depth of 1697 m in 2010. Most of the Ukr.S.A. explorations in Krubera Cave since 1999 have been led by Yury Kasyan.

All the large caves of the Ortobalagan Valley belong to a single hydrologic system, developed in and near the crest zone of the Berchil'sky anticline. The direct connection of Krubera Cave with the Arabikskaja System, although not established yet, is a sound speleological possibility. The Main Branch and the Nekujbyshevskaja Branch in Krubera are largely independent, predominantly vertical, parts of the system which deviate from the Krubera old series at −220 to −240 m (Figs. 2 and 3). Their upper sections (to altitudes of about 900−800 m) cut through the thickly bedded and massive, often sandy Upper Jurassic limestones. The caves are predominantly combinations of vadose shafts and steep meandering passages, although in places they cut apparently old fossil passages at different levels (*e.g.*, at altitudes of 2070−2040 m in Kujbyshevskaja Cave, 1200−1240 m in the Main Branch of Krubera, and 980−1150 m in the Nekujbyshevskaja Branch of Krubera Cave, *etc.*). The antiquity of these passages is supported by the ages of speleothems falling beyond the ^{230}Th dating limit (>500 ka). Both branches of Krubera Cave are extremely vertical up to the depths of about 1600 to 1700 m (altitudes of about 750−650 m). The Main Branch below this level becomes more inclined, largely following the strata dip, but then it goes steeply down again to the depths of about 2050 to 2150 m (altitudes of about 200−100 m).

Cave development is strongly controlled by the block-fault structure. The cave entrances are aligned along the anticlinal crest (Figs. 2 and 3) but the cave passages and shafts are controlled by diagonal and orthogonal fractures and faults, and comprise complex winding patterns in the plan view, remaining largely within and near the anticlinal crest zone. Two dominant diagonal, SE−NW-stretching lines are recognizable in the plan pattern of Krubera Cave. The other diagonal direction, of SW−NE orientation, is also used by many smaller fragments of Krubera Cave and by the most part of Genrikhova Bezdna Cave. The south − north orthogonal direction is expressed in a large fossil passage in the

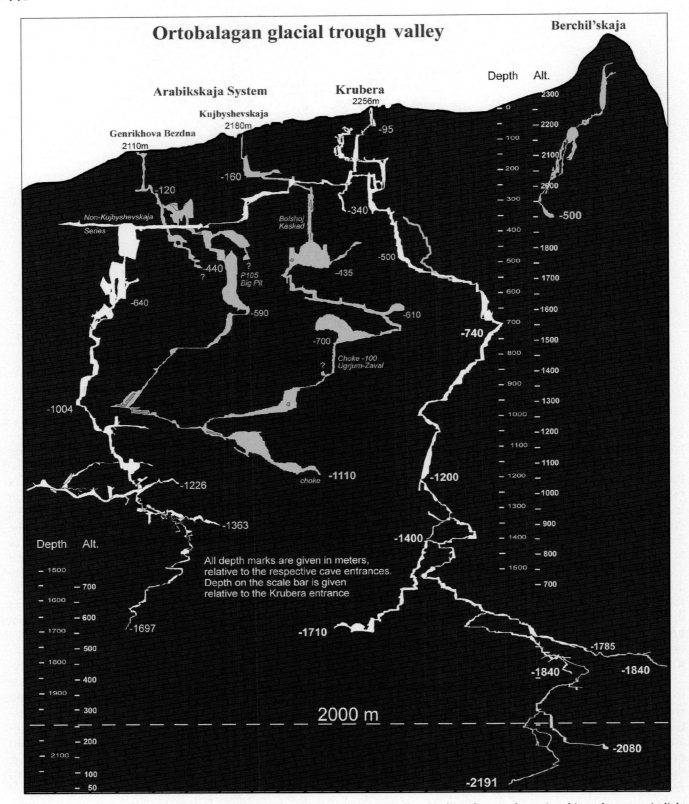

FIGURE 3 Combined cave profile projected on the axis of the Ortobalagan Valley. Krubera Cave is shown in white, other caves in light brown.

Nekujbyshevskaja Branch at altitudes of 980–1150 m, in the terminal sump passage in the Main Branch, and also in some parts of the nearby Kujbushevskaja Cave. The deep portion of Krubera display a more pervasive conduit pattern with a mixture of phreatic morphology, characteristic of the zone of high-gradient floods, and vadose downcutting elements that are observed even below the water table (in the terminal sump).

The cave hydrology is very variable, depending on season. During winter, flow through the cave is at a minimum as recharge from the surface is virtually absent. Late May and July are marked by the maximum flux due to massive snow melting. During this period of the highest flow, conduits in many parts of the system at altitudes below 1100–1000 m get filled with water above local obstructions up to local levels of 10–20 m (the "bottleneck" effect). In the summer and fall seasons, flow varies with the precipitation regime. During low-flow periods, a small permanent stream (up to 1 L s^{-1}) first appears in the Krubera Main Branch at a depth of about 340 m. The flow disappears and reappears at various levels, but never increases significantly, even in the deepest sections. There are only small permanent water flows in the Nekujbyshevskaya Branch, but at the depth of 1660 m a large influent stream with discharge of about 10 L s^{-1} comes from a tight side passage. This water most likely arrives from the nearby Kujbyshevskaya Cave, where the only stream of compatible discharge is known, disappearing in a boulder choke at the bottom at −1110 m.

The cave is very cold, with water temperatures of 1.0°C at −100 m depth, rising slowly up to 7.2°C at 2000 m depth. Air temperature is only a few tenths of a degree higher. The phreatic zone effectively intercepts the geothermal heat flux and drains it through springs. This leaves the vadose zone above under strong influence of cold percolation that mainly comes from snow melt at altitudes of about 2000–2300 m. This results in an expressed geothermal depression characteristic for alpine-type karst massifs.

Other deep caves in Arabika include the Iljukhina System (−1273 m; number 3 in Fig. 1C) located in the center of the massif, Dzou Cave (−1090 m; number 5 in Fig. 1C) and Moskovskaja Cave (−1125 m; number 6 in Fig. 1C) in the northeastern part, and Sarma Cave (−1760 m; number 13 in Fig. 1C) in the southeastern part of the high sector of Arabika. Sarma Cave, now the second deepest cave in the world, is located along the same anticline as Krubera but on its southern slope, at the similar advanced position relative to the seashore.

THE ARABIKA MASSIF: GEOLOGY

The Arabika Massif is composed of Lower Cretaceous and Upper Jurassic limestones that dip continuously southwest to the Black Sea and extend below the modern sea level (Fig. 4). In the central part of Arabika the Cretaceous cover is retained only in a few ridges and

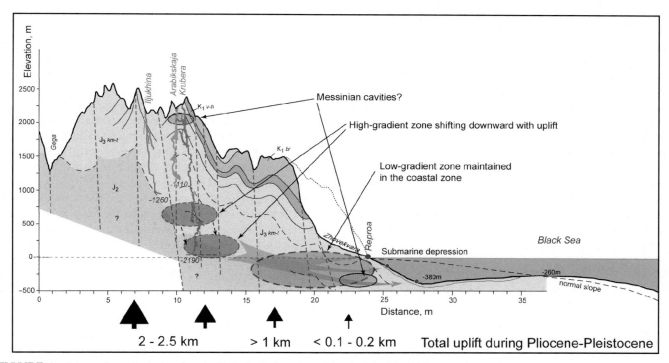

FIGURE 4 Schematic geological and speleohydrological section of the Arabika Massif. See text for explanations.

peaks, as well as in small patches within the trough valleys (Valanginian and Hauterivan limestones, marls, and sandstones). In the south and southeast, on the coastwise ridges, the Cretaceous succession is more continuous and includes Barremian and Aptian-Senomanian limestones and marly limestones with abundant concretions of black chert. The core part of the massif is composed of the Upper Jurassic succession resting on the Bajocian Porphyritic Series, which includes sandstones, clays, and conglomerates at the top, as well as tuff, tuff sandstones, conglomerates and breccia, porphyry, and lava. The Porphyritic Series forms the nonkarstic basement of Arabika, which is exposed only on the northern and eastern outskirts, locally in the bottoms of the Kutushara and Gega River valleys.

The Upper Jurassic succession begins from the bottom with thin-bedded Kimmeridgian-Oxfordian cherty limestones, marls, sandstones, and clays, which are identified in the lowermost part of Krubera Cave. Above lies the thick Titonian succession of thick-bedded limestones with marly and sandy varieties. Sandy limestones are particularly abundant through the upper 1000-m sections of deep caves of the Ortobalagan Valley.

The tectonic structure of Arabika is dominated by the large sub-Caucasian (NW–SE oriented) anticline, with the gently dipping southwestern megaflank, complicated by several low-order folds, and steeply dipping northeastern flank. The axis of the anticline roughly coincides with the ridge bordering the Gelgeluk Valley to the north. Located on the southwestern flank of the major anticline is another large one (Berchil), in which the crest is breached by the Ortobalagan Valley. There are several smaller subparallel anticlines and synclines farther southwest, between the Berchil anticline and the coast (Fig. 1C).

The plicative dislocation structure of the massif is severely complicated by faults, with the fault-block structure strongly controlling both cave development and groundwater flow. Major faults of the sub-Caucasian orientation delineate several large elongated blocks that experienced uplift with different rates during Pliocene and Pleistocene. This had a pronounced effect on the development of deep groundwater circulation and the karst system. Both longitudinal and transverse faults and related fracture zones play a role in guiding groundwater flow; the fracture zones guide flow across the strike of major plicative dislocations, from the central sector toward the Black Sea.

THE ARABIKA MASSIF: HYDROGEOLOGY

Major onshore karst springs with individual average discharges of $1-2.5\,m^3\,s^{-1}$ are located at altitudes ranging from 1 m (Reproa Spring) to 540 m (Gegsky Vodopad). Two of them are located in the shore area: these are Reproa (average discharge $2.5\,m^3\,s^{-1}$, alt. 1 m asl; RE in Fig. 1C) and Kholodnaja Rechka ($1.2\,m^3\,s^{-1}$, 50 m asl; KR in Fig. 1C). Other two major springs are located in the river canyons bordering Arabika to the east: Goluboe Ozero in the Bzyb canyon ($2.5\,m^3\,s^{-1}$, 90 m asl; GO in Fig. 1C) and Gegsky Vodopad in the Gega canyon ($1\,m^3\,s^{-1}$, 540 m asl; GV in Fig. 1C). There are also several smaller springs in the town of Gagra.

Some boreholes located along the shore of the Black Sea yield karstic groundwater from depths of 40–280 m below sea level. Georgian hydrogeologists reported that other much deeper boreholes tapped karstic waters low in total dissolved solids at depths of 500 and 1750 m in the Khashupse Valley near Tsandripsh and 2250 m near Gagra. This suggests the existence of a deep karst system and vigorous karst groundwater circulation at depth. Submarine springs are known in the Arabika area, emerging from the floor of the Black Sea in front of the massif. Shallow springs at depths of 5–7 m below sea level can be reached by free dive near Tsandripsh. Other submarine springs are known near the eastern part of Gagra at depths of 25–30 m. Offshore hydrochemical profiling revealed submarine discharge at depths up to 400 m. Recently, an outstanding feature of the sea-floor topography near Arabika has been recognized using the DEM based on the SRTM30Plus dataset (NASA). This is a huge submarine depression on the shelf in front of the Zhoekvara River mouth, which has dimensions of about 5×9 km and a maximum depth of about 380 m bsl. The Arabika Submarine Depression (ASD; shown on Figs. 1C and 4) is a closed feature with internal vertical relief of about 120 m (measured from its lowest rim) separated from the abyssal slope by the bar at a depth of about 260 m. It has steep northern and northeastern slopes (on the side of the massif) and gentle south and southwestern slopes. Its formation is apparently karstic. Presently ASD seems to be a focus of submarine discharge of the karst systems of Arabika. The existence of ASD, along with other lines of evidences, points to the possibility of much lower sea level positions in the past than is suggested by Pleistocene glacioeustatic oscillations.

The hydrogeologic model for the Arabika Massif that dominated before the 1980s did not allow a possibility of hydrologic connection between the central high sector of the massif and the coastal springs. The model implied the existence of elongated groundwater basins corresponding to synclines and separated by anticlines of the sub-Caucasian trend, with several superimposed aquifers in each basin vertically separated by minor "nonkarstifiable" beds within the carbonate succession (Kiknadze, 1972; 1979). According to this notion, groundwater recharged within the central part of the massif would flow northeast, beneath the nonkarstic cover, and southwest to the Goluboje Ozero spring, as shown by red arrows on Figure 1C. The recharge areas for the coastal and submarine springs were assumed to be only the proximal low-altitude ridges.

This model was disproved by speleological explorations and dye-tracing studies during the 1980s under the coordination of the Institute of Geological Sciences of the Ukrainian Academy of Science. A series of large-scale dye-tracing experiments was conducted in Arabika in 1984 and 1985. Tracers injected in the Kujbyshevskaja and Iljukhina caves were detected in the Kholodnaja Rechka and Reproa springs, proving groundwater flow to the south-southwest across major tectonic structures over a distance of 13–16 km as the crow flies (shown by white and yellow arrows on Fig. 1C). The tracer from Kujbyshevskaja was also detected in a borehole located between these two springs (BH on Fig. 1C), which yields groundwater from a depth of 200 m below sea level. This has been interpreted as an indication of the connection of the cave with the submarine discharge. The large Central Karst Hydrologic System, which encompasses most of the southeastern flank of the Arabika anticline, had been identified in this way (Klimchouk, 1990). These experiments have revealed the deepest karst hydrologic system in the world with its overall vertical range of about 2500 m (measuring to the borehole water-bearing horizon) or even 2700 m (measuring to the deepest reported submarine discharge points). Another tracer was injected in the Moskovskaja Cave (−970 m) and detected at the Gegsky Vodopad spring, indicating the presence of a karst hydrologic system comprising the northeastern flank of the Arabika anticline (the Northern System). No connections have been proven of any cave with yet another major spring, Goluboje Ozero in the Bzyb River canyon, although it apparently drains a large area of the eastern sector of the massif (the hypothetical Eastern Karst Hydrological System). It is not clear where Sarma Cave (−1760 m) drains to, Goluboje Ozero to the southeast or Reproa to the southwest, at the shore.

The results of the dye-tracing tests have radically changed notions of the hydrogeology of Arabika and revealed its outstanding speleological perspectives, strongly stimulating further efforts for exploration of deep caves. They demonstrated that groundwater flow is not subordinate to the fold structure but is largely controlled by faults that cut across the strike of major folds, and that the large part of the central sector of Arabika is hydraulically connected to the springs along the seashore and with submarine discharge points.

EVOLUTION OF THE DEEP CAVE SYSTEM IN ARABIKA

Krubera Cave has an extremely steep profile and reveals a huge thickness of the vadose zone. The lower boundary of the vadose zone (the top of the phreatic zone) is at an elevation of about 110 m at low flow, which suggests a low overall hydraulic gradient of 0.007–0.008. Groundwater low in dissolved solids is tapped by boreholes in the shore area at depths of 40–280, 500, 1750, and 2250 m below sea level, which suggests the existence of a deep flow system with vigorous flow. Submarine discharge along the Arabika coast is reported at depths up to ∼400 m bsl. A huge closed submarine depression is revealed at the sea floor next to Arabika, with the deepest point of ∼400 m bsl. It is difficult to interpret these facts in terms of the development of karst systems controlled by contemporary sea level, or even within the range of its Pleistocene fluctuations (up to −150 m).

A hypothesis is suggested that early karst systems in Arabika could have originated in response to the Messinian salinity crisis (5.96–5.33 Ma BP), when the Black Sea (Eastern Paratethys) could have almost dried up. The dramatic sea level drop of ∼1500 m is well established for the Messinian time in the adjacent Mediterranean. French karstologists have recently demonstrated (*e.g.*, Mocochain *et al.*, 2006) that the Messinian crisis played a great role in karst development in the Mediterranean region, where deep conduits had formed in response to the Messinian lowering of the base level and imposed a strong influence on subsequent karst evolution.

The hypothesis that the dramatic sea level drop could have taken place in the Black Sea basin during the Messinian time had been put forward long ago (Hsu and Giovanoli, 1979) mainly based on deep-sea drilling data (DSDP Sites 379, 380, 381). It has been strongly corroborated by recent studies of regional geology, including data from bio- and magnetostratigraphy of the key sedimentary sequences, seismic profiling (establishing the Messinian erosional surface in the Eastern Paratethys), studies of deep-water delta complexes, and so on.

Before the Late Miocene, the present coastal Western Caucasus was a low- to middle-altitude mountain terrain. Temporary desiccation of the Black Sea in the Messinian time established the base level at many hundreds of meters below the present level and caused conduit development deep in the Arabika Massif. These early systems were flooded after the Pliocene transgression. Uplifts of the Arabika area during Pliocene and especially Pleistocene were highly differentiated by elongate zones (blocks) of the sub-Caucasian stretch (parallel to the coast). The total uplift amounted to 2—2.5 km in the central sector of Arabika, whereas it was minimal (0.1—0.2 km) in the zone proximal to the coast. Hydraulic continuity was always maintained across the zones, between the main recharge area in the rising central sector of Arabika and the coastal zone and submarine springs. The presence of high conduit porosity of the Messinian origin in the coastal/submarine sector allowed the zone of high hydraulic gradient during uplift to be pushed far inland, beneath the rising central sector (Fig. 4). This created favorable conditions for the enhanced conduit development at depth in the central sector, quick adjustment of the water table to new uplift pulses, and eventual development of a huge vadose zone and extremely steep and deep cave systems such as Krubera Cave. This was further favored by recurring sea level drops up to −150 m during the Pleistocene, which caused the steepening of hydraulic gradients beneath the central sector of Arabika and enhancement of conduit porosity in the upper part of the present phreatic zone.

This evolution scenario is indirectly supported by ^{230}Th dating of speleothems from the deep parts of Krubera Cave. Stalagmites from depths of 1640 m and 1820 m (elevations 640 m and 436 m asl) have yielded ages older than 200 ka (max. 276 ka), which suggests that the deep vadose zone already existed in the Middle Pleistocene at the latest.

The above discussion demonstrates that it is not by chance that the deepest cave in the world with the exceptionally high vertical range of almost 2200 m has been discovered in the Arabika Massif. There were unique geological and paleogeographic preconditions for that.

Acknowledgment

This article is an outcome of enormous efforts of several generations of cavers who have explored the deep caves in Arabika over the past 30 years.

Bibliography

Hsu, K. J., & Giovanoli, F. (1979). Messinian event in the Black Sea. *Palaeogeography, Palaeoclimatology, Palaeoecology, 29*(1—2), 75—94.

Kiknadze, T. Z. (1972). *Karst of the Arabika Massif*. Tbilisi, Russia: Metzniereba.

Kiknadze, T. Z. (1979). *Geology, hydrogeology and activity of limestone karst*. Tbilisi, Russia: Metzniereba.

Klimchouk, A. B. (1990). *Karst circulation systems of the Arabika Massif*. Peschery (Caves). Perm: Perm University.

Klimchouk, A., & Kasjan, J. (2001). Krubera (Voronja): In a search for the route to 2000 meters depth: The deepest cave in the world in the Arabika Massif, Western Caucasus. *NSS News, 59*(9), 252—257.

Mocochain, L., Clauzon, G., Bigot, J. Y., & Brunet, P. (2006). Geodynamic evolution of the perimediterranean karst during the Messinian and the Pliocene: Evidence from the Ardèche and the Rhône Valley systems canyons, southern France. *Sedimentary Geology, 188—189*, 219—233.

L

LAMPENFLORA

Janez Mulec

Karst Research Institute, Scientific Research Centre of the Slovenian Academy of Sciences and Arts

DEFINITIONS

Lampenflora—originally a German term adopted in the English vocabulary, sometimes called *lamp flora*, or in French *la maladie verte*—designates the phenomenon—proliferation of phototrophic organisms near artificial light sources. Lampenflora grows at sites where, under natural circumstances, it would not appear. This alien flora can be found in the proximity of light sources in caves, mines, and other artificial and light-deprived environments, such as cellars and stores (Fig. 1). Besides artificial light, high humidity, typically around 90%, which sometimes approaches saturation, is also a characteristic for these environments. Humidity and, occasionally, seeping water, together with light, support the growth of phototrophic organisms. Various types of aerophytic cyanobacteria and algae, as well as some mosses and ferns, dominate. The term *aerophytic* designates living on air in terrestrial habitats, on rocks, stones, sediments, bark of trees, needing only the atmospheric moisture (humidity) to sustain life cycle or part of it. Several other terms are also used to define the ecology of lampenflora phototrophs, such as *subaerophytic*, indicating organisms living in aerophytic localities with an intermittent water supply. Sometimes in this context the term *subaerial* is used, describing the location of the habitat which is near the Earth's surface, that is, underground. To describe interactions between an organism and a solid substratum, the term *epilithic* is used for lampenflora as well. It is generally used for organisms attached to stones or rocks or their growth on stony substrata, aerophytic, or submerged under (the film of) water. The term *endolithic* denotes growth of biota under the surface of stones and rocks, including small microcavities and fissures. Organisms are usually strongly adhered to the substratum with their thalus. In the early phase of colonization and succession, cyanobacteria and eukaryotic algae usually play the most important role, while mosses and ferns appear later in succession. Vascular plants are sometimes found around lamps, but almost always only as germinating shoots. Using the plant's reserves some growth can occur, even in the absence of light (mixotrophy). Only lampenflora in show caves and tourist mines as a result of artificial lighting is discussed here.

Many caves are recognized because of natural beauty, amazing cave formations, and places of inspiration. Humans developed intimate bonds with caves early in history, as they offered temporary or permanent shelter for our ancestors and, later, caves became attractive for visiting and recreation. Especially interesting are those caves with prehistoric paintings, such as Lascaux Cave in France. A few dozen caves of natural and cultural importance are listed on the United Nations Educational, Scientific, and Cultural Organization (UNESCO) World Heritage List (World Heritage Convention). Cave tourism is considered to be one of the oldest forms of tourism. Nowadays in many of these places, artificial illumination has been installed to attract visitors. Illuminated areas such as rocky surfaces, sediments, and artificial materials around lamps quickly become colonized by phototrophic organisms. The sites of tourist interest, such as prehistoric paintings or extraordinary cave formations, are frequently well lit to present them to the public, and consequently they become suitable substrata for lampenflora. A clear distinction between naturally present aerophytic microscopic phototrophs and higher plants in cave entrances and lampenflora is that lampenflora organisms colonize natural and artificial substrata deep in the underground and are completely dependent on artificial lighting. Lampenflora in caves

FIGURE 1 Lampenflora in Postojnska Jama, Slovenia, 2008. Photograph taken by Svatava Kubešová.

can also be found attached to solid surfaces in illuminated submerged cave formations, such as cave ponds and riverbeds. Although lampenflora is a result of light eutrophication underground and a cause of biodeterioration or biofouling of various types of substrata in the underground, it has been preserved in some caves as a tourist attraction.

COMPOSITION AND ECOLOGY OF LAMPENFLORA

Every exposed solid surface is subject to colonization by organisms, and in this aspect the underground is no exception. With the introduction of light on a solid surface in light-deprived underground habitats, phototrophic organisms become competitive for this ecological niche. The problem with the growth of lampenflora appears with installation of permanent electric lighting in show caves and tourist mines. The use of carbide lamps or portable electric lights, such as LEDs (light-emitting diodes), during cave exploration does not sustain the growth of phototrophs. Generally, various abiotic parameters play important roles for development and community structure of phototrophs: light, temperature, characteristics of the substratum, carbon dioxide, oxygen, availability of nutrients, and, if present, the characteristics of surrounding water (streaming, pH, alkalinity, acidity, hardness, salinity, pollutants). Among biotic factors, the following are important: competition, parasitism, predation, and organic exudates.

Diversity of lampenflora depends mainly on lighting regime, quantity, and quality—light spectrum of available energy. Compared to sunlight, the artificial light source shows no oscillations in intensity. An incandescent lamp or a fluorescent lamp does not notably influence the community composition. When a cave is lit for a longer period, flora becomes more diverse and succession is faster. Diversity is further dependent on temperature, and quantity and routes of introduced cells, spores, or other viable propagules underground. At low photosynthetic photon flux densities (PPFDs), a small increase in temperature can result in a large increase of biomass yield. For example, the green alga *Chlorella* sp., frequently identified in lampenflora communities, increased biomass 30-fold at 2.5 μmol photons m^{-2} s^{-1} when the temperature increased from 9 to 11°C. At cave temperatures, the light saturation point is fast reached at low PPFDs (<10 μmol photons m^{-2} s^{-1}), and biosynthesis of accessory photosynthetic pigments below the light saturation point is considerably elevated. Lampenflora can be found at PPFDs lower than 0.2 μmol photons m^{-2} s^{-1} up to several several hundred μmol photons m^{-2} s^{-1}.

In a contrast to cave fauna that can move around, lampenflora uses different modes of dispersion in the underground. Three main modes of transport of viable propagules in the underground can be distinguished: air currents, water flow, and introduction by migratory animals and humans. Caves with mass tourism are subjected to high input of lint and other detritus from visitors which serve as a nutrient source. Air currents in caves appear due to meteorological changes of external climate (oscillations of temperature and air pressure, and seasonal dynamics) and consequently these changes are reflected in underground spaces. An important factor in spreading lampenflora inside show caves is a local air current, which is formed because of heating of cave air with strong, mainly halogen lamps. In such places lampenflora growth was observed in a distance more than 10 m from a lamp. An interesting observation came from Aggtelek Cave, Hungary, where green patches appeared after snow melt in spring, but when water evaporated, the green coating in the immediate vicinity of lamps disappeared. Extremely strong lighting dried the surface and restricted lampenflora growth. But in the zone more distant from the lamps, flora continued to grow.

Seeping water passing through crevices and cracks brings viable propagules from the epikarst zone above the caves. Vegetation in caves thrives better in places with seeping water. Abundance of algae is probably also dependent on air humidity. As mentioned above, in the immediate vicinity of incandescent or halogen lamps humidity drastically drops. An important source of allochthonous organisms entering underground is the formation of bioaerosols from fine mist or spray at a ponor due to splashing of a river. Microscopic allochthonous organisms, such as algae, are then further transmitted in a cave with air currents.

Solid surfaces subjected to colonization, which are covered with sediments, enable faster growth of lampenflora; in addition, rough surfaces with minor irregularities represent a better substratum for its attachment. Substratum characteristics at the micro level notably influence lampenflora growth and its distribution around lamps. This may be why there is often an absence of correlation between PPFDs and the concentrations of main photosynthetic pigment—chlorophyll a at individual lamps.

In comparison to naturally growing prototrophs in cave entrances, diversity of lampenflora is poor. Generally, in comparison with cave entrances, areas around lamps have more stable conditions with no extreme environmental oscillations, for example, freezing–thawing. Typical representatives are soil aerophytic cyanobacteria and algae. Organisms successful in colonizing a new ecological niche are cosmopolitan—ubiquitous in nature—having simple nutrition requirements and wide ecologic tolerance; they reproduce fast and are easily adaptable to new conditions. Taking only microscopic lampenflora into consideration, that is, cyanobacteria and algae, it is not possible simply to define their general distribution pattern. In the species succession, the first colonists are usually fast growing and are slowly ousted by the more persistent organisms. Although at first sight the community of lampenflora seems static, it is not so as the species succession process is rather dynamic. In the development of a phototrophic community, a random combination of species first occurs; vegetation zones are formed later, usually after many years. However, in this phase an important factor to initiate growth is the presence of appropriate inoculum.

On occasion, cyanobacteria are pioneering organisms in many harsh environments (Whitton and Potts, 2000), but the first colonizers are frequently fast-growing green algae which are gradually overgrown by cyanobacteria. On electrically lit sites with high humidity, or with condensed water on solid surfaces, and especially at sites with seeping water, diatoms slowly become more abundant in the community. Diatom adhesion can occur in the light or dark, but for successful adhesion energy is required. One of important factors influencing diatom adherence is the presence of Ca^{2+} ions. Greenish patches of microscopic cyanobacteria and algae are progressively becoming covered by islands of mosses and finally ferns. Cyanobacteria are usually the most abundant ($\sim 50\%$), followed by Chrysophyta or Chlorophyta. Among cyanobacteria, the following genera are frequently encountered: *Aphanocapsa*, *Aphanothece*, *Chroococcus*, *Gleocapsa*, *Leptolyngbya*, *Myxosarcina*, and *Synechocystis*. Among green algae are *Apatococcus lobatus*, *Chlorella* spp., *Chlorococcum* spp., *Coccomyxa* spp., *Klebshormidium flaccidum*, and *Stichococcus* spp. Among diatoms in lampenflora communities, researchers report on frequent presence of the genera *Achnanthes*, *Aulacoseira*, *Cocconeis*, *Diadesmis*, *Fragilaria*, *Hantzschia*, *Navicula*, and *Nitzschia*.

Most mosses identified in caves are not specific for this habitat and can exist also in other places with very low light irradiances. Low PPFDs permit bryophytes to live in places inhospitable to other plants. Bryophytes in caves sometimes express etiolation—adaptation to low quantity of available light by using their physiological capacity to expose more surface area to capture the few available photons. For example, in some species the number of chloroplasts and size of grana can increase in response to low light. Some bryophytes are able to grow over a relatively wide range of PPFDs (Glime, 2007). In caves, moss protonema frequently grow. Tufa-forming moss, *Eucladium verticillatum*, is a frequent dweller in poorly and powerfully lit places in caves; for example, in Slovenian show caves *E. verticillatum* had the highest ecological tolerance width based on the span of PPFDs, from 1.4 to 530.0 μmol photons m^{-2} s^{-1}. Sometimes in lampenflora communities even at low PPFDs, bryophytes can complete their life cycle by longer exposure to light irradiance. Such an example is *Cratoneuron filicinum* which lives in zones where lights are on 24 h d^{-1} with PPFD between 2.1 and 2.4 μmol photons m^{-2} s^{-1} (Mulec and Kubešová, 2010). Besides species succession of algae and cyanobacteria, succession in the bryophyte community should be taken into consideration as well. Ferns are also found in the lampenflora community, but their diversity is low, and frequently they are found only as prothalli.

ALTERATION OF UNDERGROUND HABITAT

The cave ecosystem is in dynamic equilibrium, which is disturbed when high amounts of nutrients and/or energy–light have been introduced underground. Newly formed lampenflora biomass is available for cave-adapted animals and other occasional dwellers of caves. Such new conditions indirectly influence cave fauna as they are adapted to (nearly) oligotrophic conditions. Along with easier available nutrients, newcomers become competitive for a new ecological niche and a population of troglobiotic animals can be affected. Terrestrial troglobiotic species are frequently observed grazing on lampenflora.

Vegetation and other heterotrophic microbial communities do not change just the aesthetic appearance of a cave, but also mechanically and biochemically cause the deterioration of cave paintings and speleothems. Organisms interact with rock, mobilize ions, and release compounds. Electron transfer, proton uptake, removal of cations from the substrate due to the formation of chelating complexes, and phototrophic processes with transfer of carbon dioxide, oxygen production, and alkaline reaction, all contribute to biodeterioration of the stony substrata (Warscheid and Braams, 2000). At the micro level, changes of pH are based on the lighting regime, increase of pH during photosynthesis, and lowering of pH during respiration or fermentation (Albertano et al., 2000). The mucilage surrounding cells acts as a slimy adhesive for airborne particles. In addition, these polysaccharides help to retain humidity, increase adhesion to the substratum, and stabilize dust particles.

A very important problem occurs when live or dead lampenflora becomes encrusted with calcium carbonate, irrespective of whether this carbonate is a result of abiotic or biotic precipitation. Limestone and the other calcareous substrata, which often occur in subterreanean environments, can promote the growth of calcifying species. Such an amorphous mix of dead phototrophs and carbonate irreversibly destroys speleothems or other objects of cultural value.

CONTROL AND RESTRICTION OF GROWTH

Cave management faces two urgent problems regarding lampenflora: (1) how to remove lampenflora without damaging the substratum, once it is established; and (2) how to prevent its growth and biodeterioration in the future (Fig. 2). Many soil algae, which are common in lampenflora, are known to have also heterotrophic metabolism, that is, organisms which use organic compounds for most or all of their carbon requirements. In addition to these, some cyanobacteria can even use fermentative biochemical pathways to obtain energy (Whitton and Potts, 2000). These physiological adaptations allow these organisms to survive periods with no available light, which is why the decreasing PPFDs or shortening of time of exposition to light irradiance is not a satisfactory and long-lasting solution. Even short but continuous pulses of light can drive the light phase of photosynthesis.

Problems connected with lighting and phototrophic biofilms are usually not properly solved. The main obstacle is that the cause, that is, light, remains on the site and new viable propagules are constantly arriving. The proper way to prevent lampenflora growth is still unknown. The simplest solution would be complete removal of existing phototrophic communities, cessation of lighting, and abolition of tourist visits, which in many cases is not acceptable for the cave management.

FIGURE 2 Lampenflora encrusted in flowstone. Contact between flowstone and algae coating. Image was taken with low vacuum scanning electron microscope 5500 LV JEOL (back scattered electron image). *Photograph taken by Alenka Mauko.*

Many different procedures have already been applied to control lampenflora, including physical and chemical approaches (Mulec and Kosi, 2009). To eliminate lampenflora from cave formations, mechanical removal with water and brushes is usually applied. This is not a sustainable approach because many fragile formations are destroyed. Introduction of ozone-producing lamps or UV, when tourists are not visiting, also may negatively affect cave biota.

In the past, in order to kill lampenflora several very toxic compounds were applied, such as DCMU (diuron, N-(3,4-dichlorophenyl)-N'-dimethylurea), bromine compounds, formalin, cupric ammoniac solution, which are nowadays totally unacceptable to introduce into the cave environment. To kill lampenflora nowadays, such places are frequently sprayed with 5% bleach solution (NaClO) that effectively kills lampenflora. From the NaClO solution, gaseous chlorine can be released which gives its characteristic smell in a cave. In the reaction of hypochlorite and nitrogenous compounds, toxic chloramines and even carcinogenic trihalomethanes are released. Low chlorine concentration in the cave environment can kill other cave biota. Less toxic, odor-free, and more environmentally friendly is hydrogen peroxide (H_2O_2) which is already applied in some caves; however, the pH of unbuffered 15% H_2O_2 solution is 4 and thus corrosive for limestone and other formations. Oxidized plant material later becomes a source of nutrients for bacteria and fungi, and for this reason dead material should be removed from a cave. Unfortunately, all the above-mentioned compounds are not specifically targeted against vegetation and have a biocidal effect for the cave fauna as well. In the future, more studies should go in this direction. There are reports on the potential use of selective herbicides such as Atrazine (6-chloro-N-ethyl-N'-(1-methylethyl)-1,3,5-triazine-2,4-diamine) and Simazine (6-chloro-N,N'-diethyl-1,3,5-triazine-2,4-diamine), but their widespread use underground is not suitable because the green coloring on the karst formations persists (Grobbelaar, 2000). The use of bleach or other nonspecific oxidizing compounds to control lampenflora in places with art is likely to damage features of significance, for example, inscriptions, prehistoric paintings, frescoes, stuccoes, and mosaics, and is thus strongly dissuaded for managing such places. The crucial step in limiting the adverse impacts of lampenflora is to start with treatment as soon as the lampenflora is visible.

Future perspectives regarding lampenflora management should undoubtedly include development of a procedure which would kill lampenflora with no or minimum impact for caves. Still, before applying any procedure to remove lampenflora it seems reasonable to check for their species composition, but such data can be of inventory value. Even when a new taxon is identified or a peculiar community structure revealed, one has always to keep in mind that this alien flora normally does not occur in underground environments, and its occurrence is directly linked to human intervention in the underground.

FIGURE 3 Barbarossahöhle, anhydrite show cave in Germany. Note color of water, lightened with cool white LED and additionally with red LED light. *Photograph taken by Alexander Chrapko.*

An important step in controlling lampenflora growth is an appropriate installation of lamps, housings, and modes of illumination: short-time lighting of individual sectors, avoidance of damp surfaces, places covered with sediments and easy accessibility for tourists, places with air currents, dripping, and seeping water. Lamps installed in hypogea should emit light with a spectrum out of the absorption maxima of chlorophyll a. If we strictly follow this direction, green lighting does not give a natural view of the underground (Roldán et al., 2006). Chlorophyll a has two peaks in the absorption of red and blue parts of a light spectrum. Another possible alternative is yellow lighting because algae do not absorb strongly in the yellow part of the light spectrum. The common observation in hypogea where colored filters are used to restrict lampenflora growth is that there is less growth compared to white lights. A very promising recent approach is the installation of LED lighting (Fig. 3) because of low-energy consumption, long-lasting LEDs, and potential to tune the desirable emission spectrum (Toomey et al., 2009). With recent advances in technology, the use of fiber optics in cave lighting seems another alternative. With optimization and more studies in cave lighting, such installations have a potential for enhanced control to reduce lampenflora growth.

See Also the Following Article

Show Caves

Bibliography

Albertano, P., Bruno, L., D'Ottavi, D., Moscone, D., & Palleschi, G. (2000). Effect of photosynthesis on pH variation in cyanobacterial biofilms from Roman catacombs. *Journal of Applied Phycology, 12*, 279–384.

Glime, J. M. (2007). Light: Cave mosses. In *Bryophyte ecology, physiological ecology* (Vol. 1, Chapter 9-5, pp. 43–52) (Ebook sponsored by Michigan Technological University and the International Association of Bryologists). Accessed on 3 October 2011 at http://www.bryoecol.mtu.edu/.

Grobbelaar, J. U. (2000). Lithophytic algae: A major threat to the karst formation of show caves. *Journal of Applied Phycology, 12*, 309–315.

Mulec, J., & Kosi, G. (2009). Lampenflora algae and methods of growth control. *Journal of Cave and Karst Studies, 71*, 109–115.

Mulec, J., & Kubešová, S. (2010). Diversity of bryophytes in show caves in Slovenia and relation to light intensities. *Acta Carsologica, 39*, 587–596.

Roldán, M., Oliva, F., Gónzales del Valle, M. A., Saiz-Jimenez, C., & Hernández-Mariné, M. (2006). Does green light influence the fluorescence properties and structure of phototrophic biofilms? *Applied and Environmental Microbiology, 72*, 3026–3031.

Toomey, R. S. III, Olson, R. A., Kovar, S., Adams, M., & Ward, R. H. (2009). Relighting Mammoth Cave's new entrance: Improving visitor experience, reducing exotic plant growth, and easing maintenance. In W. B. White (Ed.), *ICS 2009 15th International Congress of Speleology Proceedings* (Vol. 2, pp. 1223–1228). Kerrville, TX: Greyhound Press.

Warscheid, T., & Braams, J. (2000). Biodeterioration of stone: A review. *International Biodeterioration and Biodegradation, 46*, 343–368.

Whitton, B. A., & Potts, M. (Eds.), (2000). *The ecology of cyanobacteria: Their diversity in time and space* Dordrecht: Kluwer Academic Publishers.

LECHUGUILLA CAVE, NEW MEXICO, U.S.A.

Patricia Kambesis

Cave Research Foundation

Prior to 1986, Lechuguilla Cave was known as a minor historic site in the back country of Carlsbad Caverns National Park. At that time the cave contained 120 meters of dry, dead-end passages that were accessed via a 27 meter entrance shaft known as Misery Hole. In the 1950s cave explorers noted a significant amount of wind that issued from a rubble pile at the bottom of the known cave, indicating the potential for more cave passages, and over the years several attempts were made to dig out the rubble pile. In 1984, on a park-approved permit, cavers once again worked on finding a way through the rubble choke. In May of 1986, after two years of digging efforts, a team of explorers broke through into a major cave passage in Lechuguilla Cave. This began one of the most remarkable chapters in American cave exploration and speleology. The cave quickly grew in length and depth and became known as the deepest limestone cavern in the United States and ultimately fourth in length for U.S. caves. On the world scale it is the fifth longest.

Lechuguilla Cave has been called the "jewel of the underground," and rightfully so, with its incredible variety of spectacular speleothems that decorate many of the cave's massive chambers, circuitous tunnels, and confounding mazes. However, it is not just the cave's aesthetic beauty and explorer's appeal that define Lechuguilla as a world-class cave. Due in large part to the stewardship of the National Park Service and the foresight of the American caving community, science has been a part of the exploration efforts in the cave since the very beginning (Turin and Plumer, 2000). Consequently, important studies in geology, cave microbiology, microclimatology, mineralogy, geomicrobiology, and geochemistry have been conducted in the cave. Scientific discoveries have gone hand-in-hand with cave exploration and survey.

Geologically, Lechuguilla Cave is unusual in that its speleogenesis involves sulfuric acid and the degassing of hydrogen sulfide from a hydrocarbon-rich basin. The concept of basinal degassing is significant not just for cave formation in the area but also to the understanding of migration of hydrocarbons, the formation of petroleum reservoirs, the deposition of uranium ore, and the origin of Mississippi Valley–type lead-zinc deposits (DuChene and Hill, 2000). Lechuguilla Cave is one of the world's best-documented examples of a cave formed by sulfuric acid in a hypogene setting

Mineral residues and some of the more unusual speleothems in Lechuguilla Cave show indications of microbial influence on cave development. The significance of these geomicrobiological interactions is providing insight into the basic mechanisms of dissolution and precipitation by microorganisms and on the origin of specific types of speleothems (Northup et al., 2000). The microbe environment in Lechuguilla Cave is also being studied by NASA scientists who believe that it may be an analog to similar environments on other planets.

The exploration and concurrent survey of Lechuguilla Cave have influenced the modes and methods of cave mapping in the United States. The abundance and delicacy of formations, unusual sediments, and microbiology within the cave have caused both explorers and resource managers to reevaluate human impact on all caves in general and has raised the profile of cave conservation and restoration efforts.

PHYSICAL SETTING

Lechuguilla Cave is located within Carlsbad Caverns National Park, which is part of the Guadalupe Mountains

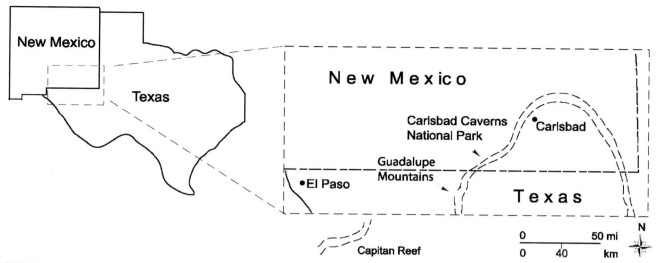

FIGURE 1 Location map of Carlsbad Caverns National Park, Eddy County, New Mexico.

of southeastern New Mexico (Fig. 1). These mountains are situated on the northern margin of the Chihuahan Desert. The area climate is semi-arid to arid with average winter temperatures of 7°C and summer temperatures of 27°C. The annual rainfall ranges from 20–50 cm with more than half of the precipitation coming during the summer months. Vegetation of the area includes cacti, succulents, and desert shrubs. The cave's name comes from one of the local flora, *Agave lechuguilla*, which grows in large clumps on limestone ledges and slopes, and is a common plant in Lechuguilla Canyon and at the entrance of the cave.

GEOLOGIC SETTING

Lechuguilla Cave is developed within the Capitan Reef Complex which forms a belt of Permian-aged carbonates 8 km wide and 650 km long and defines the northwest perimeter of the Delaware Basin, a major hydrocarbon basin in the southwest United States. The Delaware Basin lies between the structurally active Basin and Range Province and the non-structurally active Great Plains (Hill, 2000). The Guadalupe Mountains are a fault block that has been exhumed by erosion and evaporite solution-subsidence within the Delaware Basin.

Though the Guadalupe Mountains are considered a karst area, features typical of classic karst areas are relatively rare. There are some sinkholes, paleokarst, and a variety of springs both small and large, but there is no direct correlation between the surface topography and geomorphology and the extensive caves that lie within the mountains. This is a reflection of the cave's development in a hypogene setting. Cave entrances

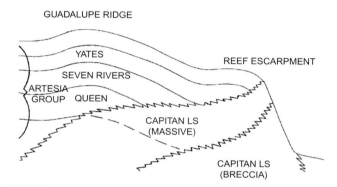

FIGURE 2 Rock units exposed in Lechuguilla Cave. *After Jagnow, 1977.*

are not past or present points of hydrologic recharge or discharge. Rather, they are fortuitous collapses that have exposed the caves to the surface. In all of its great length, Lechuguilla Cave has only one entrance.

STRATIGRAPHY AND STRUCTURE

Paleontologic and stratigraphic data from rock exposures within Lechuguilla Cave indicate that most of the cave is developed either in the back reef facies of the Permian reef or within the reef and forereef slope facies. The cave spans four stratigraphic units including (in descending order) the Yates Formation, Seven Rivers Formation, Queen Formation, and Capitan Formation (Fig. 2). Backreef facies include the Yates, Seven Rivers, and Queen formations. The reef and forereef units encompass the Capitan Formation. It has been speculated that the cave could extend into the upper reaches of the Goat Seep Formation, a unit that consists of massive reef and forereef talus of dolomite composition, deposited along the shelf margin. However, to date no Goat Seep exposures have been documented in the cave.

Rock units exposed in the cave include the Yates Formation which is composed of dolomites, calcareous quartz siltstones, and fine-grained sandstones. This unit is exposed at the entrance of Lechuguilla Cave, in the upper levels of the Chandelier Graveyard, in Tower Place, and in many of the other upper level passages. The Seven Rivers Formation consists of dolomites and, occasionally, quartz siltstones. This formation is evident throughout most of the upper parts of the cave from the entrance through the Rift. The Queen Formation is made up of dolomites interbedded with siltstones and fine sandstones. Within the cave this formation is exposed in Windy City, Sugarlands, and in the north and south ends of the Rift.

The Capitan Formation is divided into massive and breccia members. The massive member lacks bedding and consists of fossiliferous and dolomitic limestones. The breccia member is made up of reef limestone fragments mixed with lenses of siliciclastic material. The massive Capitan outcrops at depths of 175–275 m within the cave. The breccia member outcrops at depths below 250 m throughout the cave.

The major folds within the vicinity of Lechuguilla Cave include the Guadalupe Ridge anticline and the Walnut Canyon syncline. The cave lies within the structural limb shared by these two folds. Joint sets trend either parallel or perpendicular to these major folds. Cave passage orientations are controlled by these two major joint sets. Lechuguilla Cave is located within a structural trap for natural gas located just off the Guadalupe Ridge anticline. The cave is also located beneath the Yates Formation which functions as a stratigraphic trap for H_2S (Hill, 2000).

The development of Lechuguilla Cave was influenced by the arrangement of stratigraphic units, differential dolomitization of these units, joint patterns, and folds (Hill, 2000). All four controls first appeared in the Late Permian with some modification by later tectonic and diagenetic events (Hill, 1996). However, the overall layout and geometry of the cave is a function of regional groundwater flow and sulfuric acid production (Palmer and Palmer, 2000). Because the genesis and development of the cave is not coupled to surface hydrogeology, Lechuguilla Cave is classified as a *hypogene cave system*.

REGIONAL/LOCAL HYDROGEOLOGY

The permeable Capitan Limestone is the aquifer for the region and the upland surface of the Guadalupe Mountains is the recharge area. Water moves through the shelf rocks of the Artesia Group and drains down-dip into the Capitan Limestone aquifer. Water moves slowly through the reef and discharges into the Pecos River at Carlsbad Spring in the city of Carlsbad, New Mexico. The caves of the Guadalupe Mountains, including Lechuguilla Cave, show a combination of deep phreatic and water-table characteristics. Water-table conditions are responsible for horizontal level development of cave passages. In these areas, cave passages cut across bedding planes. The vertical development of the system is a function of deep phreatic conditions. Lechuguilla Cave was formed in a diffuse-flow aquifer regime. The cave may have functioned as a mixing chamber for hypogene-derived H_2S and meteoric derived freshwater (Hill, 2000).

The Aretesia Group (Yates, Seven Rivers, Queen, and Goat Seep formations) was probably the original reservoir where water containing hydrogen sulfide was stored prior to and during sulfuric acid speleogenesis. The aquifer also functioned as the avenue for eastward migration of hydrogen-sulfide water with oxygenated meteoric water (DuChene, 2000). The aquifer possibly also served as the site where the mixing of H_2S water with oxygenated water occurred and which ultimately formed the passages of Lechuguilla Cave. Hydrogen sulfide speleogenesis took place at or near the water table (Palmer and Palmer, 2000).

Over time, as the water table dropped, cave passage development occurred where H_2S was ascending to the water table in significant quantities (Jagnow, 1989). As the water table lowered during uplift and deformation, H_2S was episodically released, forming major rooms and passages (Palmer and Palmer, 2000). When H_2S occurred during times when the water table was stable, horizontal levels of cave passages developed. The massive rooms and long passage trends within Lechuguilla Cave reflect these events (Fig. 3).

The relationship of the ages of the mineral alunite with respect to elevations within the cave suggests that there was an 1100-meter decline in the water table from 12 million years ago to the present (Polyak et al., 1998). This suggests that the Capitan aquifer water table was relatively flat during the Late Miocene and Pliocene (Polyak & Provencio, 2000). It is speculated that the paleospring that would have been associated with Lechuguilla Cave has not been preserved because it has been truncated or obliterated by valley downcutting and erosion of the ancient Pecos River Valley (Hill, 2000).

Flowing water is relatively rare in caves of the Guadalupe Mountains. There is only one area in Lechuguilla Cave that has flowing water (Kambesis, 1989). This is the Lost Pecos River in the Far East Branch of the cave. A small stream issues from a crack

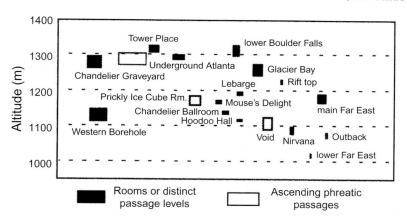

FIGURE 3 Elevation of rooms and passages in Lechuguilla Cave. The width of each block is proportional to the width of the room or passage. *After Palmer and Palmer, 2000.*

in the wall and feeds a series of pools in the area. This water probably originates from a small perched aquifer that the cave passage has intersected.

Drip and pool water is fairly common throughout the cave and is sourced from vadose meteoric water that has seeped down from the surface along joints, bedding planes and interconnected pores in the rock. The geochemistry of Lechuguilla Cave's pool waters is quite diverse. The water chemistry is a function of precipitation chemistry, bedrock, and the gypsum deposits that occur in close proximity (Turin and Plummer, 2000). The water is further modified by evaporation, speleothem deposition, and microbial activity. Recent research indicates that the water levels in cave pools respond to regional climate change rather than just random drip migration (Despain and Stock, 2008).

BIOLOGY

The megascopic biology of Lechuguilla Cave is somewhat limited. Cave crickets and arachnids are present in the entrance passages and to some extent past Boulder Falls, the first significant pit in the cave. Bat bones have been found throughout the system but no bats currently inhabit the cave, except at the entrance.

However, the microscopic biology is quite a different story. It is speculated that since the microbial systems within Lechuguilla Cave have no multicellular grazers or predators, and no meteoric influences, and have been in existence for long spans of time, the evolution of these microbial communities are independent of surface organisms (Boston, 1999). Kiym Cunningham was the first to recognize microbial evidence in corrosion residues and in some speleothems. Microbial research in the cave indicates that the microbe communities are chemolithotrophs. These types of bacteria metabolize sulfur, iron, and manganese minerals and may assist in enlarging the cave and determining the shapes of some unusual speleothems. Other studies indicate that some microbes may possess medicinal qualities that are beneficial to humans.

Lechuguilla Cave contains remnants of what could be past and perhaps present geomicrobiological interactions. These features include corrosion residues, moonmilk, pool fingers, webulites, and u-loops. Microorganisms of today have been found in association with carbonate and silicate speleothems, sulfur compounds, and iron and manganese oxides (Cunningham *et al.*, 1995).

Research on microbial ecology has revealed that the microorganisms in Lechuguilla Cave, though surviving in nutrient-poor conditions, are metabolically versatile and unexpectedly diverse. Preliminary research results indicate that microbial communities overcome the limitations of a nutrient-poor environment by cooperative and mutualistic associations (Barton and Jurado, 2007).

MINERALOGY

Lechuguilla Cave is well known for its unusual mineralogy and for the spectacular nature of its speleothems. Almost of the caves of the Guadalupe Mountains display many of the same features as Lechuguilla. However, the Lech displays almost features as large-scale, text book examples. Lechuguilla Cave contains some of the most spectacular examples and varieties of gypsum speleothems including massive selenite chandeliers, delicate clear gypsum crystals, gypsum needles, gypsum cotton, and crusts. Calcite speleothems, though not in as great abundance, are also spectacular in their shape and configuration and occur in places throughout the cave where meteoric waters seep into the underground. Large stalagmites have formed in Tower Place. Stunning arrays of soda straws, stalactites, and draperies festoon such places as Deep Secrets, the Oasis Pool, and the Pearlsian Gulf.

Large amounts of massive or laminated gypsum are common within the cave. These deposits accumulated where groundwater flow was not strong enough to carry away the sulfate produced by the sulfuric acid reaction with the limestone when the cave was enlarging (Davis, 2000). Some of the gypsum beds have been dissolved away by meteoric water that percolates into the cave or from freshwater reflooding. Many of these gypsum remnants occur in dry sections of the cave and in places resemble small glaciers (Glacier Bay, Prickly Ice Cube Room, Land of Fire and Ice, Blanca Navidad).

Elemental sulfur deposits occur as yellow masses or as stringers within larger masses of gypsum. The sulfur, which can be massive, granular, or platy, is exposed by dissolution of the surrounding gypsum (Davis, 2000). Locations where sulfur deposits have been observed include the Void, Ghosttown, Ghostbusters Balcony, and near Blanca Navidad.

Caves formed by sulfuric acid-bearing waters contain direct by-products of their origin including gypsum, hydrated halloysite, and alunite. Lechuguilla contains many of these minerals. The alunite is important because it contains potassium which gives it the potential to be dated by the K-Ar or $^{40}Ar/^{39}Ar$ dating methods (Polyak and Provencio, 2000). Polyak et al. (1998) used $^{40}Ar/^{39}Ar$ dated alunite to determine that Lechuguilla Cave is at least 5 million years old.

Lechuguilla Cave is rich in corrosion residues that contain bacterial and fungal communities, indicating that microbes could be active participants in limestone dissolution in conjunction with abiological processes (Cunningham et al., 1995). Gypsum and sulfur in the Guadalupe Mountains are significantly enriched in the light isotope of sulfur (Hill, 2000). The enrichment of light sulfur implies that hydrocarbons and sulfur bacteria were involved in speleogenesis.

Endelleite and montmorillonite fill solution pockets in many places in the cave. The minerals celestite and barite have also been found in many of the maze areas of the cave (Chandelier Maze).

Some speleothems have a biologic origin. Rusticles, which are observed in only two locations with in the cave, are composed on iron oxide minerals which contain fossil microbial casts (Davis, 2000). Pool fingers, which grew underwater, are calcite in composition but seem to have a microbial origin.

One of the "signature" speleothems of Lechuguilla Cave are underwater helictites. These are vermiform in shape and contain a central canal. These formations developed as a result of the common-ion effect (Davis, 2000). Other unusual mineralogies include folia, hydromagnesite balloons and fronds, raft accumulations and cones, splash rings, and silticles.

Though the wide variety of mineralogies and speleothems seem randomly diverse, they are related to each other by the unusual origin of the cave and/or from convective airflow loops that are driven by temperature gradients that result from the cave's great vertical relief (Davis, 2000).

PALEONTOLOGY

Vertebrate paleontologic remains are very rare. Pleistocene fossil remains have been found in fills near the entrance area. It is speculated that the fill is from an ancient sinkhole that may have trapped some local mammals. There are bat remains scattered throughout the system, some of them of paleontologic significance. There is only one other locality within Lechuguilla Cave that contains vertebrate fossil remains. A fully articulated ringtail cat skeleton was found in the Land of Awes, an area that is far removed from the current entrance and the surface. This specimen indicates the possibility of an ancient entrance that is no longer accessible.

Lechuguilla Cave has formed within the Capitan Reef complex. The primary frame-building organisms of the reef include calcareous sponges, calcareous algae, and bryozoans. These fossils are very common within the bedrock of Lechuguilla Cave. Many of the fossils found within the cave have been broken into sandstone size fragments that accumulated in voids in the Capitan Reef and in the backreef lagoon (DuChene, 2000). Fossils are also exposed in the bedrock walls that are not covered with mineral crusts and coatings. The most abundant fossil remains are those of invertebrate reef organisms including bivalves, brachiopods, bryozoans, crinoids, porifera, gastropods, urchin spines, corals, and cephlapods. These occur throughout all areas of the cave.

DESCRIPTION OF THE CAVE

The overall pattern of the chambers and passages of Lechuguilla Cave is characteristic of caves formed by sulfuric acid speleogenesis. These types of caves are characterized by ramiform passage morphology, large passage size, and horizontal passages connected by deep shafts and fissures. Irregular passage shapes are common, as are dead-end rooms and galleries and blind alcoves. The ramified shape and layout of Lechuguilla Cave (Fig. 4) are formed by the oxidation of rising H_2S, where the main passages and large domed rooms correspond to favored paths of H_2S movement (Palmer, 1996). The large passage sizes are a

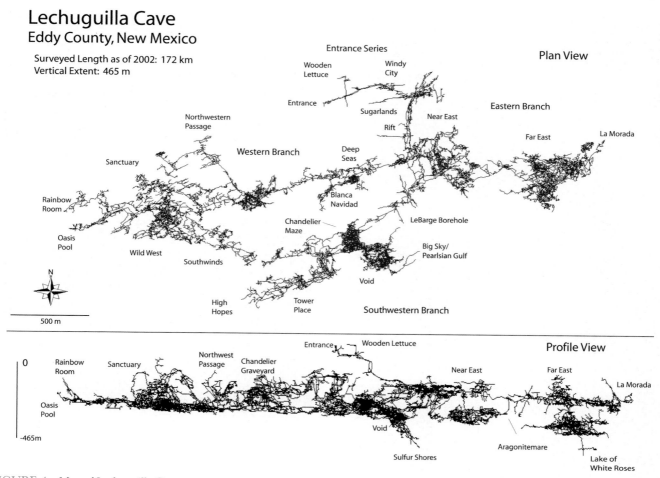

FIGURE 4 Map of Lechuguilla Cave.

function either of large H_2S inputs or H_2S inputs over longer periods of time.

Lechuguilla Cave contains a number of fissures (or rifts) that descend from the floors of large rooms or passages and pinch downward with bottoms that are clogged with calcite crust, carbonate sediment, or that terminate in water (Fig. 5). These rifts extend below the level of major cave rooms and may be inflow routes along which oxygenated and H_2S-rich water first began to mix (Palmer and Palmer, 2000). Such areas include the routes from the Chandelier Graveyard to Sulfur Shores and from the Outback to Lake of the White Roses.

Different levels within Lechuguilla Cave are connected via ascending passages such as the Great White Way, which connects the south Rift to Deep Secrets, and the Gulf of California fissure, which connects the mazes of Snow White's Passage to the LeBarge Borehole.

Ascending tubes with large cross-sections and great vertical extent may also be the result of the mixing of two major water sources (groundwater and meteoric water). Such an ascending tube rises 200 meters from the Rift to the cave's entrance (Palmer and Palmer, 2000). The Wooden Lettuce Passage (whose highest point is located approximately 6 meters above the current entrance) may have been a paleo-entrance.

Horizontal passages may represent former water table levels (Fig. 6). The Western Borehole is a 1.6 km-long linear trend of horizontal passages with areas of elliptical passage cross-section, which may have served as an outlet for water from zones of maximum cave origin (Palmer and Palmer, 2000). The Western Borehole ends abruptly without breakdown collapse or major passage extensions.

The walls of many of the large passages in Lechuguilla Cave are honeycombed with interconnected, nontubular solution cavities of various sizes and geometry forming three-dimensional mazes of spongework-type morphology (Fig. 7). These passages, called Boneyard, are ubiquitous throughout Lechuguilla Cave and in most other caves in the Guadalupe Mountains. They develop where many interconnected openings are solutionally enlarged at comparable rates. Mazes

FIGURE 5 Fissure descending into the LeBarge Borehole, Southwestern Branch. *Photo courtesy of Dave Bunnel. Used with permission.*

FIGURE 6 Large passage of the Great Western Borehole, Western Branch. *Photo by Dave Bunnel. Used with permission.*

FIGURE 7 Boneyard maze. *Photo by Dave Bunnel. Used with permission.*

originate in mixing zones where the aggressiveness of the water is high.

HISTORY OF EXPLORATION

Prior to May 1986, the known extent of Lechuguilla Cave consisted of a 27 meter entrance pit that dropped into a large three-level chamber containing several insignificant side passages. Surveyed passage length was 120 meters. Covering the floor was a thick veneer of guano which coated a large rubble pile. Of note was the fact that a tremendous volume of air issued from the rubble pile, sometimes strong enough to blow sediment up the entrance pit. Later, other visitors to the cave suggested it be named Misery Hole because of the great amount of dust that blew out of the entrance pit.

Lechuguilla Cave had been known at least since the early part of the twentieth century. The name and date of the cave's original discoverer is unknown but it is possible that guano miners may have made the first entrance into the cave. In 1914, John Ogle, Cad Ogle, and C. Whitfield filed a placer mining claim to mine bat guano from the cave. At that time guano mining was a major industry in the Guadalupe Mountains. Only a small amount of guano was actually removed from the cave and it appears that the project was abandoned after only preliminary mining efforts (Frank, 1988). After the initial mining activities, the cave saw a minimal amount of activity. Park rangers visited the cave in the early 1940s, and in the early 1950s Colorado Grotto members mapped the cave and published a descriptive report.

In the 1970s Cave Research Foundation (CRF) and park personnel made several trips to the cave. CRF remapped the cave and the map was published in D.J. Jagnow's *Cavern Development in the Guadalupe Mountains*, in January of 1979. CRF digging crews began working the lowest levels of the blowing rubble pile, having to shore it up as they went. Progress on the unstable rock pile was minimal and CRF abandoned the dig project.

In 1984 digging activities led by cavers from Colorado took a new approach on the breakdown pile using the wind and bedrock walls as their guide. Over the course of four digging expeditions, significant progress was made on the rubble pile. On May 25, 1986 the dig broke out on the side of a talus slope into a low, wide room. Continuing through a squeeze and following the wind, the diggers/explorers broke into a large-going passage.

This marked the beginning of one of the greatest North American cave discoveries of the twentieth century. Between 1986 and 1998, over 160 km of cave passages were discovered, explored, and documented.

After the breakthrough, exploration moved at breakneck speed, with 5–8 km of passages documented with each expedition. The first 48 months of exploration and survey work delineated the main framework of the cave and established its vertical extent. Explorers pushed and mapped the east–west trending entrance passages to a 43-meter pit named Boulder Falls. The cave continued at the base of the pit, through a series of large, gypsum-floored corridors and rooms (Colorado Room, Windy City, Glacier Bay, Sugarlands). The east–west trend terminated at the Rift, a large fissure that has a vertical extent of over 30 meters in places. At the south end of the Rift the cave splits into three separate east–west trending branches.

Explorers pushed a small hole in the rift which leads to a tall, slanting gypsum fissure named the Great White Way. Deep Secrets was the first significant room at the base of the fissure and it contained not only an abundance of gypsum coatings and assorted calcite speleothems but also some of the longest soda straws ever observed in the Guadalupe Mountains. A series of mazey rooms and climb-ups eventually broke into the Great Western Borehole, a 1.6 km-long, gypsum-coated trunk passage which branched to two terminations. The southern terminus ended at the calcite-encrusted Oasis Pool located at the base of a sheer wall. The northern terminus was pushed to the Rainbow Room, an ascending bedding plane passage. In other places in the Western Borehole, lofty avens led to upper-level segments including the Chandelier Graveyard and Pelucidar and Barsoom.

A major junction in the Rift (the EF Junction) led to more gypsum-coated mazes and yet another descending, gypsum-coated fissure. This proved to be the gateway to the Southwest Branch, one of the most spectacularly decorated sections of the cave. Exploration and survey documented lakes and pools in areas containing flowstone, dripstone, and gypsum formations (Lake LeBarge, Lake Chandelar, Yellow Brick Road, Lake Castrovalva). Long, tubular passages and convoluted boneyard mazes coated with all manner of gypsum mineralogy were discovered and mapped (LeBarge Borehole, Land of Awes, Chandelier Ballroom, Prickley Ice Cube Room, Chandelier Maze, Hoodoo Hall).

Passages trending to the southwest of the Chandelier Maze opened into more boneyard mazes (Land of the Lost), large austere chambers floored with breakdown (the Voids), high-vaulted rooms floored with pools, flowstone, and cave pearls (Big Sky and Pearlsian Gulf). Thick deposits of rock flour coat the floors of many of the rooms and passages. Another steeply descending fissure leads to Sulfur Shores and what may be the top of the water table. High-level fissures trending north from the Prickly Ice Cube Room lead to large shafts, upper-level rooms containing some of the largest flowstone formations in the cave (Underground Atlanta and Tower Place), and more perched lakes and boneyard mazes (High Hopes).

At the far end of the Rift is a pit leading to a series of ramps and fissures that are coated with dark corrosion residues that drop into the Eastern Branch of Lechuguilla Cave. This branch of the cave proved to be the most technically challenging, requiring the rigging of multiple pitches and traverses. More perched lakes, and large-scale mazes and rooms characterized this branch of the system (Lake of the Blue Giants, Nirvana, Megamaze, Moby Dick Room). Explorers were temporarily stopped by a tall, aragonite-coated fissure named the Aragonitemare. A series of climbing attempts finally reached the top of the fissure (which was measured at 60 meters) and led to a continuation of the Eastern Branch called the Far East Maze. Cavers documented east-trending maze passages, some with corrosion residues or massive aragonite trees covering the floors and other areas with perched pools coated with calcite rafts (Grand Guadalupe Junction, Silver Bullet Passage, Land of Enchantment, Boundary Waters). At the Grand Guadalupe Junction, upward-trending passages were explored in a series of complex upper mazes (La Morada Maze, Firefall Hall). Pits in the floor of Grand Guadalupe Junction dropped into the Outback, a sparsely decorated section of lower-level cave defined by more slanting fissures and long, low-pancake crawlways. In the middle of the Outback, the bottom of the passage literally dropped out to the deepest point of the cave known as Lake of the White Roses at 486 meters, which may also be the water table. By 1990, more than 80 km of passages had been discovered, explored, and documented.

Between 1991 and 1998 another more than 80 km of discoveries had been documented. The slowdown in pace of exploration was a function of the remoteness of many of the leads, the necessity of technical climbing in some areas (Davis, 1999), and because the NPS had, at various times, closed the cave to exploration trips for management purposes.

Since 1999, exploration and survey work has continued to fill in more maze passages that surround all major branches of the cave. New discoveries identified significant northern and southern extensions in the Western Branch of the cave. From the central Western Borehole, a technical climb led to the discovery of another one of the cave's major chambers, Hudson Bay. Northwest-trending passages led to the northernmost extent of Lechuguilla Cave. At the known limits of the Western Borehole, explorers extended the perimeters of the cave via breakouts to the Wild West, South Winds, the Sanctuary, and Emerald City. This section of the cave contains passages that head back toward the High Hopes block, suggesting a possible connection between the Southwest and Western branches of the cave. To date, that connection has not been made.

In 2005, a major extension named the Coral Sea was discovered off the eastern edge of the Outback in the Far East Branch of the cave. This section of cave has seen close to three kilometers of survey and contains a high density and variety of calcite and aragonite speleothems, numerous occurrences of active flowstone and dripstone, and over 100 lakes and pools. These features are in striking contrast to the Outback which, overall, is dry and speleothem-poor. The Coral Sea is the easternmost extension of Lechuguilla Cave.

The new explorations in the Southwest Branch of the cave have concentrated on pursuing climbing leads and upper-level passages, specifically in the Chandelier Maze. Though no significant passage extensions have been discovered as of yet, the various climbs have led to short sections of upper levels, some that connect back to known areas of the complex Chandelier Maze.

In addition to new exploration, significant efforts have been made in resolving survey issues and raising the overall quality of the survey work in Lechuguilla Cave. Loop closures have been improved, survey errors identified and resolved, and in-cave sketches upgraded to today's standards. The Cave Resource Office (CRO) at Carlsbad Caverns National Park oversees the cartographic effort in Lechuguilla Cave and a team of volunteer cartographers work on the series of quadrangle maps for the cave system.

As of 2010, Lechuguilla Cave has a surveyed length of 209 km. There is still good potential for more extensions in all sections and levels of the cave. Access to Lechuguilla Cave is limited to approved scientific researchers, survey and exploration teams, and National Park Service management-related trips.

See Also the Following Article

Sulfuric Acid Caves

Bibliography

Barton, H., & Jurado, V. (2007). What's up down there? Microbial diversity in caves. *Microbe*, 2(3), 132–138.

Boston, P. (1999). A bit of peace and quiet: The microbes of Lechuguilla. *NSS News*, 57(8), 237–238.

Cunningham, K., Northup, D., Pollastro, R. M., Wright, R., & LaRock, E. (1995). Bacteria, fungi and biokarst in Lechuguilla Cave, Carlsbad Caverns National Park, New Mexico. *Environmental Geology*, 25(1), 2–8.

Davis, D. G. (1999). Lechuguilla Cave: The second 50 miles. *NSS News*, 57(8), 228–238.

Davis, D. G. (2000). Extraordinary features of Lechuguilla Cave, Guadalupe Mountains, New Mexico. *Journal of Cave and Karst Studies*, 62(2), 147–157.

Despain, J., & Stock, G. (2008). Dated cave pool shelfstones as indicators of climate change. *Lechuguilla Cave Newsletter*(3), 38–39 (Winter 2007–2008).

DuChene, H. R. (2000). Bedrock features of Lechuguilla Cave, Guadalupe Mountains, New Mexico. *Journal of Cave and Karst Studies*, 62(2), 109–119.

DuChene, H. R., & Hill, C. A. (2000). Introduction to the caves of the Guadalupe Mountains Symposium. *Journal of Cave and Karst Studies*, 62(2), 53.

Frank, E. F. (1988). Cave history—Lechuguilla Cave. *NSS News*, 46(11), 378–379.

Hill, C. A. (1996). *Geology of the Delaware Basin, Guadalupe, Apache, and Glass Mountains, New Mexico and West Texas, Publication 96–39*. Midland, TX: Society of Economic Paleontologists and Mineralogists—Permian Basin Section.

Hill, C. A. (2000). Sulfuric acid, hypogene karst in the Guadalupe Mountains of New Mexico and West Texas, U.S.A. In A. B. Klimchouk, D. C. Ford, A. N. Palmer & W. Dreybrodt (Eds.), *Speleogenesis, evolution of karst aquifers* (pp. 309–316). Huntsville, AL: National Speleological Society.

Jagnow, D. H. (1977). *Geologic factors influencing speleogenesis in the Capitan Reef Complex, New Mexico and Texas*. Unpublished M.S. Thesis, University of New Mexico, Albuquerque.

Jagnow, D. H. (1979). *Caverns development in the Guadalupe Mountains*. Columbus, OH: Cave Research Foundation.

Jagnow, D. H. (1989). The geology of Lechuguilla Cave, New Mexico. In P. M. Harris & G. A. Grover (Eds.), *Subsurface and outcrop examination of the Capitan Shelf Margin, Northern Delaware Basin* (pp. 459–466). San Antonio, TX: Society of Economic Paleontologists and Mineralogists (Core Workshop 13).

Kambesis, P. (1989). Extending the limits, a major breakthrough in Lechuguilla Cave. *NSS News*, 47, 300–305.

Northup, D., Dahm, C., Melim, L., Spilde, M., Crossey, L., Lavoi, K., et al. (2000). Evidence for geomicrobiological interaction in Guadalupe Caves. *Journal of Cave and Karst Studies, 62*(2), 80–89.

Palmer, A. (1996). *Cave patterns in the Guadalupe Mountains* (pp. 51–52). National Speleological Society Convention, Salida, Program with abstracts.

Palmer, A., & Palmer, M. (2000). Hydrochemical interpretation of cave patterns in the Guadalupe Mountains, New Mexico. *Journal of Cave and Karst Studies, 62*(2), 91–108.

Polyak, V. J., McIntosh, W. C., Guven, N., & Provencio, P. (1998). Age and origin of Carlsbad Cavern and related caves from $^{40}Ar/^{39}Ar$ of alunite. *Science, 279*(5358), 1919–1922.

Polyak, V. J., & Provencio, P. (2000). Summary of the timing of sulfuric-acid speleogenesis for Guadalupe Caves based on the ages of alunite. *Journal of Cave and Karst Studies, 62*(2), 72–74.

Turin, J., & Plummer, M. (2000). Lechuguilla Cave pool chemistry, 1986–1999. *Journal of Cave and Karst Studies, 62*(2), 135–143.

LIFE HISTORY EVOLUTION

David C. Culver

American University

INTRODUCTION

The concept that life history characteristics, for example, age at first reproduction, number of offspring, and longevity, were not just a passive reflection of what the environment would allow, but rather that they were molded by natural selection gained great currency in the 1960s and 1970s. Since then a variety of models of life history evolution have been put forward to explain the multifaceted differences in life history characteristics of species. The subterranean environment in general and the cave environment in particular have characteristics that act as strong selective agents in the evolution of life histories. Foremost among these are extreme food scarcity and the reduction in seasonal cues for reproduction. Nearly all cave animals are characterized by increased longevity, greatly diminished reproductive rates, and increased investment in the offspring that are produced, often in the form of increased egg size (Culver, 1982).

There is perhaps no aspect of the biology of cave animals that is both more interesting and more difficult to study. Many undergraduate and graduate students have set out to study life history and population growth of cave animals, but given longevities that often measure in the decades and reproductive rates that are exceptionally low, very little change is often observed in populations over the course of a normal research project. This is further complicated in many cases by the inability to sample the subterranean habitat in times of high water, often the critical time in demography of subterranean populations. What we know about the evolution of life history of cave animals is based on a few long-term studies with bits and pieces of information from many shorter term studies (Trajano, 2001).

LIFE HISTORY EVOLUTION OF AMBLYOPSID FISH

The fish in the family Amblyopsidae are a small lineage of fish comprising six species, four of which are stygobionts in North American caves and one of which is a stygophile in North American caves (Niemiller and Poulson, 2010). The extensive work of Thomas Poulson (1963) on life history characteristics of these species is both the most extensive comparative study of cave and non-cave species in the same lineage and the most influential study of any cave organisms. Poulson was the first cave biologist to look at the life history of cave animals as an adaptation to the cave environment.

One species, *Chologaster cornuta*, lives in freshwater swamps in the Coastal Plain of the southeastern United States. The other species associated with caves, the stygophilic *Forbesichthys agassizi*, is found in springs and caves in the central United States, especially Kentucky. The other four species are all obligate cave dwellers. The genus *Amblyopsis* has two species, *A. rosae* in the Ozarks of Arkansas, Kansas, Missouri, and Oklahoma, and *A. spelaea* in the Interior Low Plateaus of Indiana and Kentucky. *Typhlichthys subterraneus* is found in a wide area from Missouri to Tennessee. *Speoplatyrhinus poulsoni* is one of the rarest vertebrates in North America, known from only a single cave in Alabama. Little is known about its life history and it will not be considered further.

When the stygobiotic species are compared to the other two species, they differ in a set of characteristics which represent a reduction in reproductive effort on the part of the cave species (Table 1). Compared to *Chologaster* and *Forbesichthys*, all the obligate cave-dwelling species show at least a 20% reduction in the number of eggs, at least a 100% increase in the age of first reproduction, at least a 50% reduction in the proportion of the population breeding at any given time, and at least a 40% reduction in growth rate. Poulson (1963) argues that all of these characteristics are an adaptation to a food-poor environment, and there is little doubt that he is correct in this assertion. It is important to note that all of the characteristics listed in Table 1 could also be the result of food stress on a population. The response of almost any population to food stress is an increase in the age at first reproduction, a reduction in the number of eggs, a reduction in the number of breeding females, and a reduction in growth rate. In a sense, these are imposed by the environment. However, in the case of cavefish, there is strong evidence that this is not just an environmental response,

TABLE 1 Life History Characteristics of Amblyopsid Fish That Are Consistent with a Response to Starvation, as Well as Adaptation to a Food-Poor Environment

Species	Age at First Reproduction (months)	Number of Eggs	Maximum Proportion of Females with Eggs	Growth Rate (mm per year)
Chologaster cornuta	12	93	1.00	2.4–3.8
Forbesichthys agassizi	12	150	1.00	1.7–2.2
Typhylichthys subterraneus	24	50	0.50	1.0
Amblyopsis spelaea	40	70	0.10	1.0
Amblyopsis rosae	37	23	0.20	0.9

TABLE 2 Life History Characteristics of Amblyopsid Fish That Are an Adaptation to a Food-Poor Environment

Species	Longevity (years)	Egg Diameter (mm)	Average No. of Broods	Maximum No. of Broods	Reproductive Effort per Brood (mm^3 gm^{-1} of female)	Maximum Lifetime Reproductive Effort (mm^3 gm^{-1} of female)
Chologaster cornuta	1.3	0.9–1.2	1.0	1	64	64
Forbesichthys agassizi	2.3	1.5–2.0	2.0	2	148	297
Typhylichthys subterraneus	4.2	2.0–2.3	1.5	3	452	903
Amblyopsis spelaea	7	2.0–2.3	0.5	5	52	260
Amblyopsis rosae	4.8	1.9–2.2	0.6	3	83	249

but rather it is a genetically programmed response to the cave environment which is the result of selection for a set of life history characteristics.

The strongest evidence for the adaptive nature of cavefish life histories comes from an additional set of life history characteristics that cannot be the simple result of the environmental pressure of starvation (Table 2). These include at least a doubling of life span, at least a 40% increase in egg size, and at least a 50% increase in the maximum number of broods as a result of increased longevity. One stygobiotic species, T. subterraneus, shows an increase in reproductive effort per brood as measured by total egg volume per gram of female. All stygobiotic species show an increase in maximum lifetime reproductive effort compared to the swamp-dwelling C. cornuta and the same or greater effort than the stygophilic F. agassizi. The maintenance and even a possible increase in lifetime reproductive effort is somewhat surprising, and may not simply be the result of increased longevity. This apparent anomaly will be addressed in a later section.

Differences in life history characteristics among the stygobiotic fish have two general explanations. One is that the species have had different amounts of evolutionary time in caves. Thus, the more "extreme" life histories of A. rosae and A. spelaea compared to T. subterraneus may be because the Amblyopsis species have been isolated in caves longer. There is no direct evidence for this and a second explanation seems more likely: the different species occupy different subterranean habitats which result in different selective pressures. For example, A. rosae occurs in areas near the surface while T. subterraneus occurs deeper in the groundwater.

GENERALITY OF LIFE HISTORY CHARACTERISTICS

While there is no other comparative study within a lineage of both cave- and surface-dwelling species that is as complete as that of the amblyopsid fish, available evidence from other groups indicates that the patterns observed in amblyopsid fish are widespread. Only a few cases will be mentioned here.

Longevity in cave animals, at least aquatic species, seems to be greatly increased relative to surface species. The French cave-dwelling amphipod Niphargus virei has a maximum age of 35 years and routinely reaches the age of 10 years, based on annual growth rates (Turquin and Barthelemy, 1985). There are several reports of cavefish living for more than ten years in a laboratory setting, including Pimelodella kronei (Brazil), Phreatichthys andruzzi (Somalia), and Caecobarbus geertsii (Congo).

Cave springtails (Collembola) often have double the life span of soil- and litter-dwelling species, and harpacticoid crustaceans living in the interstices of gravel aquifers also have double the life span of surface-dwelling species. In these and many other cases of reported longevity, it is impossible to determine if increased longevity was a direct result of selection in the subterranean environment or whether only those lineages with great longevity can successfully invade caves. The reason these causes cannot be separated is that it is often impossible to find a closely related surface-dwelling ancestor to compare to the subterranean species and thus lineage and selection effects are confounded.

The epitome of longevity in cave animals is from the study of the crayfish *Orconectes australis australis* in Shelta Cave, Alabama, by John Cooper (Cooper, 1975). Based on growth rates of marked individuals over a period of several years, the minimum age at sexual maturity was 15 years, and the best estimate, based on average growth rates rather than maximum growth rates, was 35 years. Furthermore, this was the estimated age of sexual maturity. The estimated age at first reproduction was over 100 years. Since Cooper's study was completed, the population of crayfish in Shelta Cave crashed to less than 10% of its former size. This was probably the result of an ill-advised gate to the cave that kept out bats, which probably were an important food supply to the crayfish in the form of guano or the organisms that fed on guano. Nevertheless, it is an impressive example of the ability of cave animals to survive for extended periods of time.

Increased egg size and reduced numbers of eggs is widely reported among subterranean animals, including harpacticoid copepods living in interstitial habitats and cave amphipods. One especially clear-cut example without the confounding effects of lineage differences is the amphipod *Gammarus minus*. Populations of this species (actually a species complex) have invaded subterranean drainages in extensive karst areas in Virginia and West Virginia. The average egg number in a cave population was 50% less than in a spring population less than 5 km away, for females of the same size. In contrast, the volume of an individual egg is about 40% larger for the cave population. Many other subterranean animals have very small numbers of eggs, but usually comparison with the appropriate surface species is unavailable.

One remarkable life history adaptation to cave life is shown by some leiodid beetles in the genus *Isereus* (Ginet and Decou, 1977). Two species, *I. colasi* and *I. serullazi*, have a nonfeeding larval stage that lasts nearly 100 days followed by a pupal stage lasting more than 115 days. The evolutionary advantage of the suppression of larval feeding is not entirely clear, but it is probably related to escaping predation, a factor whose importance is suggested by the length of the pupal period as well. If prey available to the prey are scarce, the loss from lack of feeding might be quite small.

One feature of the life history of subterranean organisms shows a great deal of variability, the seasonality or lack of seasonality of reproduction. For predators such as the carabid beetle, *Neaphaenops tellkampfi*, which feeds on the cricket eggs, reproduction is timed to the availability of cricket eggs. Because the crickets forage in surface habitats at night, their reproduction is seasonal. For most subterranean species, seasonality is much less clear-cut. It is a common (and frustrating) observation that a cave population that shows a clear seasonality one year fails to show the same pattern the next year. Work by the French biologist Marie-José Turquin and colleagues (Mathieu and Turquin, 1992; Turquin and Barthelemy, 1985) provides the basis for this variation. She points out that species living in cave streams are subject to floods year round. For the populations of the amphipod *Niphargus* that she studied in southern France, flood frequency reached a maximum in the summer. The impact of these floods was large: it caused significant mortality and it brought in significant amounts of food. Both mortality and reproduction were keyed to these events, which of course do not occur at the same time or intensity each year. Other populations that live in cave pools or live in the interstices of gravels in streams are not faced with the same hydrological regime. In these environments, floods do not occur. Instead their habitat can shrink due to drying. In these populations, ovigerous females and egg laying follow several months after the onset of the wet season. Of course, terrestrial populations would be subject to different temporal patterns, but Turquin's point that reproduction is keyed to events of high mortality and/or food input is likely to be a general one.

THE PARADOX OF HIGH REPRODUCTIVE EFFORT IN CAVE ANIMALS

Except for Turquin, few investigators have noted what seems to be a paradoxical aspect of the life history of stygobionts and troglobionts. Even though the number of eggs produced in any one brood is small and the possibility of enough resources to allow for reproduction is small, the potential for reproduction of cave animals is often quite large (Culver and Pipan, 2009). This is even the case for the amblyopsid cavefish. The reproductive effort per brood is highest in the stygobiont *T. subterraneus*, and the lifetime maximum reproductive effort of the other two stygobionts, *A. rosae* and *A. spelaea*, is much greater than that of the swamp-dwelling *C. cornuta* and nearly that of the spring-dwelling *C. agassizi*. How can we account for this paradox of long life span and a potentially high reproductive rate even in the amblyopsid cavefish, which are so exquisitely adapted to subterranean life? Turquin suggests that this paradox is in fact essential to life in caves. Although there are situations where there is a low but constant source of food such as production by chemoautotrophic bacteria, the more usual situation in caves is that food comes into the subterranean realm in pulses or spurts, which may be more or less predictable. Turquin suggests that the combination of a life span and the ability to expend significant reproductive effort make life in food-poor subterranean environments possible. If this idea is indeed correct, then it goes toward explaining some otherwise perplexing features of the life history of subterranean organisms. The high variability of reproduction between years makes sense because the animals are responding to opportunities for reproduction when there is food input that is not necessarily predictable from year to year. The increased longevity of cave animals allows them to "wait" for opportunities. It also fits in nicely with the idea that for most of the time, the growth rate (r of the standard growth equations) is slightly negative for cave animals. This was certainly the case for *O. australis australis* in Shelta Cave and was likely true for many of the fish populations studied by Poulson. Occasionally, growth rates must be positive or these populations would become extinct. The long life span and relatively high potential investment in reproduction allows stygobionts and troglobionts to take advantage of these opportunities for growth.

See Also the Following Articles

Natural Selection
Adaptation to Low Food

Bibliography

Cooper, J. E. (1975). *Ecological and behavioral studies in Shelta Cave, Alabama, with emphasis on decapod crustaceans*. Lexington, KY: University of Kentucky.

Culver, D. C. (1982). *Cave life. Evolution and ecology*. Cambridge, MA: Harvard University Press.

Culver, D. C., & Pipan, T. (2009). *The biology of caves and other subterranean habitats*. Oxford, U.K.: Oxford University Press.

Ginet, R., & Decou, V. (1977). *Initiation à la biologie et a l'écologie souterraines*. Paris: J.-P. Delarge.

Mathieu, J., & Turquin, M. J. (1992). Biological processes at the population level. II. Aquatic populations: Niphargus (stygobiont amphipod) case. In A. I. Camacho (Ed.), *The natural history of biospeleology* (pp. 263–293). Madrid: Monografias, Museo Nacional de Ciencias Naturales.

Niemiller, M. L., & Poulson, T. L. (2010). Subterranean fishes of North America: Amblyopsidae. In E. Trajano, M. E. Bichuette & B. G. Kapoor (Eds.), *The biology of subterranean fishes* (pp. 169–281). Enfield, NH: Science Publishers.

Poulson, T. L. (1963). Cave adaptation in amblyopsid fishes. *American Midland Naturalist*, 70(2), 257–290.

Trajano, E. (2001). Ecology of subterranean fishes: An overview. *Environmental Biology of Fishes*, 62, 133–160.

Turquin, M. J., & Barthelemy, D. (1985). The dynamics of a population of the troglobitic amphipod *Niphargus virei* Chevreux. *Stygologia*, 1, 109–117.

M

MAMMOTH CAVE SYSTEM, KENTUCKY

Roger W. Brucker
Cave Research Foundation

INTRODUCTION

Mammoth Cave System is located about 160 km (100 miles) south of Louisville, Kentucky, and about 56 km (35 miles) northeast of Bowling Green, Kentucky. Most of the cave lies within Mammoth Cave National Park, a World Heritage Site and a part of the United Nations program of International Biosphere Reserves. It is the longest cave in the world by a factor of 3, with about 631 km (392 miles) of surveyed passage (Fig. 1). Between its historical discovery in the late 1700s and 1957, it was considered a single cave 59 km (37 miles) long, although surrounded by several long caves.

RECENT AND ONGOING EXPLORATION

In 1957, the discovery of a connection between Floyd Collins' Crystal Cave and Unknown Cave in nearby Flint Ridge revealed the likelihood that additional discoveries of natural connections between these long caves would increase the length of Mammoth Cave. In 1972, a team of explorers led by John Wilcox discovered a natural passage connection between the integrated Flint Ridge Cave System (connecting Colossal Cave, Salts Cave, Unknown Cave, and Floyd Collins' Crystal Cave) and Mammoth Cave, a system then totaling 232 km (144 miles) in length (Brucker and Watson, 1987). Between 1972 and 1983, connections were found between Proctor Cave, Mammoth Cave, and Roppel Cave (Bordon and Brucker, 2000), extending the underground labyrinth outside the boundaries of Mammoth Cave National Park.

Exploration continues, and Mammoth Cave as of 2010 is separated from the 185 km (115 miles) Fisher Ridge Cave System by only 90 m (300 feet). Explorers familiar with the caves of the Central Kentucky Karst region have predicted that the Mammoth Cave System may eventually reach 1600 km (1000 miles) in length, although it might require cave diving to realize this potential. The park is visited by 750,000 annually. About half of the park visitors go on guided cave trips under the leadership of the National Park Service.

PREHISTORIC EXPLORERS AND MINERS

The first explorers of Mammoth Cave were indigenous inhabitants of Eastern North America. Radiocarbon dating of organic materials they discarded or lost underground indicates an activity span of about 2400 years, starting in 2250 BC. These explorers entered the caves originally for seasonal shelter and to explore the labyrinth of passages; later they mined mirabilite, gypsum minerals including selenite (euhedral and subhedral gypsum crystals), and epsomite. These ancient cavers explored about 19 km of passages using cane and dry weed-stock torches for light. Extensive archaeological investigations have led to a reconstruction of the eastern aboriginal diet, based primarily on plant materials recovered from human paleofeces preserved in the cave (Watson, 1997). A few desiccated aboriginal corpses (mummies) have been found, one of a prehistoric miner crushed when a heavy rock shifted.

Since 1999 several large passages have been found that were entered by the prehistoric Indians, but which were not found by later explorers. These discoveries include several trunk passages averaging 10 m wide and which extend for over 250 m each. They contain a rich assemblage of prehistoric soot markings, wall battering, torch fragments, and other organic materials with no evidence of historic or modern visitation. While the earliest prehistoric explorations may have been motivated by curiosity, it has been hypothesized

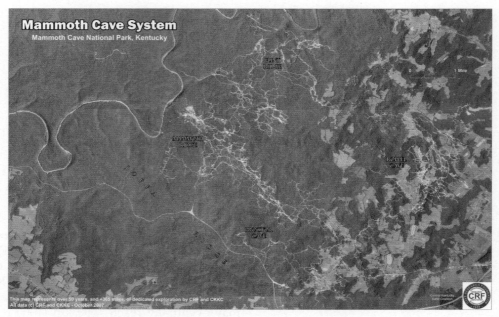

FIGURE 1 The Mammoth Cave System as compiled [C] 2007. The map shows 587 km (365 miles) of passages mapped by the Cave Research Foundation and Central Kentucky Karst Coalition. *Map courtesy of the Cave Research Foundation. Used with permission.*

that the later extensive mining of gypsum and other minerals was for ceremonial uses.

HISTORY OF MAMMOTH CAVE

The oral history of Mammoth Cave begins with long hunters, solitary woodsmen who roamed the forested Midwest in search of game in the late 1700s. The cave was mined for saltpeter, one ingredient of gunpowder, by a succession of owners, culminating in a commercial saltpeter mining operation with 70 slaves during the War of 1812. Wagonloads of saltpeter were shipped to Delaware and made into black powder by E. I. duPont. With the spread of population westward into Kentucky between 1795 and 1840, a road network and stagecoaches brought tourists to visit Mammoth Cave. Mammoth Cave was part of the "grand tour" by international visitors, along with Niagara Falls, NY (Olson, 2010). In 1838, Franklin Gorin, an owner of the cave, brought his slave, Stephen Bishop, to guide visitors through the cave. He proved to be one of the most personable and celebrated explorers and guides in a tradition of African-American guides that extended unbroken until 1941. His cave map shows about 40 km (25 miles) of the cave.

The building of the Louisville and Nashville Railroad in 1858 swelled tourism and created opportunities for entrepreneurs to exhibit nearby smaller caves. Commercial rivalry intensified as the fiercely competitive "cave wars" characterized the struggle for tourist dollars. The cave wars, fueled by the highly publicized

FIGURE 2 The Historic Entrance to Mammoth Cave. The Historic Entrance to Mammoth Cave is one of 26 entrances to the cave system. Only three are natural: Historic Entrance, Salts Cave Entrance, and Echo River Entrance (underwater.). *Photo courtesy of the National Park Service.*

entrapment and death of cave explorer Floyd Collins in Sand Cave in 1925 (Murray and Brucker, 1982), gradually ebbed as marginal caves went bust and the movement to create Mammoth Cave National Park gained momentum. The park was formally established in 1941 and dedicated in 1946. As of 2011, Mammoth Cave has 26 entrances (Fig. 2), a fact that enables explorers to reach unexplored passages relatively quickly compared with travel times in large one- or two-entrance caves.

GEOGRAPHIC AND GEOLOGIC SETTING

The Mammoth Cave System is developed in Mississippian limestone rocks on the west side of the Cincinnati Arch, primarily the Girkin limestone, Ste. Genevieve limestone, and St. Louis limestone. The Mammoth Cave Plateau, or Chester Cuesta, is separated from the Pennyroyal Plateau by the Chester Escarpment, which rises 60 m (200 feet) above the Sinkhole Plain to the Mammoth Cave Plateau. The Sinkhole Plain and Mammoth Cave Plateau contain the Mammoth Cave System, which is drained by the deeply incised and meandering Green River. The rocks dip gently northwest (Palmer, 1981).

The Central Kentucky Karst is an area of about 390 km^2 (150 square miles), consisting of 30 or more overlapping karst drainage basins. Sinkholes and sinking creeks funnel runoff and groundwater to caves. Rolling farmland with hundreds of closed depression sinkholes characterizes most of the area's topography. The Mammoth Cave Plateau topography is developed in a dendritic pattern of valleys and ridges oriented generally perpendicular to Green River. The upland represents a fluvial surface on the flat-lying Big Clifty Sandstone, with remnants of overlying Upper Mississippian bedrock on the highest ridges. The valley drainage breached the resistant sandstone and was diverted into sinkholes, creating karst valleys below the base of the sandstone (White *et al.*, 1970).

The ages of Mammoth Cave passage levels and their clastic fills generally correlate with the adjustment of Green River elevations over a period of 3.5 million years during the late Pliocene and Pleistocene (Granger *et al.*, 2001). Cosmogenic Al and Be in passage sediments indicate at least four periods of downcutting and at least three periods of aggradation, and the intervals are in agreement with surface strath terrace level evidence. Such events as ice sheet advances and retreats, drainage adjustments, and climate changes influenced the paleo-Teays River, Ohio River, and Green River water-table positions through time. Approximate ages of fills located in passages on tourist routes are Hippodrome 3.5 myr, Main Cave 2.35 myr, Violet City 2.27 myr, Forks of the Cave 1.78 myr, Cleaveland Avenue. 1.34 myr, and Buchanans Way 0.8 myr. The age of original passage development probably dates to as much as 10 myr prior to the deposit of the fills.

Unlike rivers on the surface where drainage divides can be delineated easily in the field, these karst basins and valleys are dynamic. The subsurface consists of a three-dimensional network of integrated conduits that preserve the underground former flow paths or paleo-drainage routes, cave passages that have migrated downward through time. Basin low water boundaries in the area have been determined by dye tracing, but heavy rains overwhelm the constricted conduit system and back up cave water stages to twice or more the base level river stage. Under these conditions, water may be pirated through earlier drains to adjacent basins. The interconnected conduits not only create spillover routes, but also serve as migration corridors for cave-dwelling animals and leads for cave explorers (White and White, 1989).

MAMMOTH CAVE PASSAGE PATTERNS

Passages in Mammoth Cave are generally arranged in a tree-like dendritic or angulate pattern on several levels. Truncations of passages and the development of piracy conduits complicate interpretation of the pattern. The farthest upstream ends are either narrow canyon-like passages or vertical shafts (Fig. 3), cylindrical voids that resemble the interior of grain storage silos up to 12 m (40 feet) in diameter and 60 m (200 feet) deep (Brucker *et al.*, 1972). Such shafts generally have small drains relative to their volume. Vertical shafts are often found at the edges of upland

FIGURE 3 Keller Well. A typical vertical shaft is cylindrical like the inside of a silo. Shafts short-circuit groundwater from sinkholes to small drains leading to the lowest levels. Many shafts contain waterfalls during heavy rains and thin films of aggressive descending water during drier times. A meter-long stick shows scale in Keller Well. *Photograph taken by the author.*

reentrants where water flows off the impermeable caprock and sinks into the exposed limestone. Shaft catchment areas may be hundreds of square meters of sandstone upland. High volumes of fast-moving water descend the shaft walls to produce a rapid rate of enlargement and material removal. Slower moving films of water descend the walls continuously. Vertically flowing groundwater in the shafts represents a karst "chain saw" in comparison to the dissolving capacity of slow-moving water flowing horizontally. Slope retreat gradually isolates the larger shafts from their catchment areas. Moving downstream, shaft drains and sinkhole drains join to form downcutting canyons that change to tubular passages with elliptical cross-sections as they reach a temporary base level (Fig. 4).

In their downstream extremities, the lowest active level trunk passages sump at the Green River level. An artificial pool of the Green River at 106 m (421 feet) mean sea level is behind Dam No. 6 located at Brownsville, KY. Built in 1893, the pool backflooded to raise the normal pool stage in the cave about 1 m. Cave divers have reported submerged elliptical cross-section tubes from 2 to 10 m deep at the resurgences of Turnhole Spring, Echo River Spring, and Pike Spring on Green River. Additionally, water discharges through root-like distributary systems below the pool stage into the river.

Upper levels of the flat-lying cave were once continuous drains from the karst valleys and Sinkhole Plain to the paleo-Green River. The highest elevation passages are cut into segments by surface valleys that break down passage ceilings and by vertical shafts that short-circuit surface runoff and groundwater to the lowest active levels of the cave.

FIGURE 4 Cleaveland Avenue. Cleaveland Avenue is an elliptical tube in cross-section, suggesting shallow phreatic origin. It diverted water from the higher original trunk passages. Abundant gypsum crystals adorn walls and ceilings. *Photo courtesy of the National Park Service.*

Some passages have been truncated by sinkholes or valleys that formed after the active stage has passed. Inside the cave, such segments are marked by terminal breakdowns or fills. A cave map may show a continuation of the original passage, confirmed by elevation data and stratigraphic correspondence. In the Mammoth Cave System, several of these segmented trunk passages can be traced for miles. One example of a sequence of segments is Kentucky Avenues, extending from the Frozen Niagara Entrance to a breakdown at Grand Central Station, a missing segment, and then continuing as Sandstone Avenue, another breakdown, followed by a long segment starting at Violet City Entrance and extending as Broadway to a distributary net near the Historic Entrance. One of the distributary passages is Houchins Narrows that leads to the Historic Entrance. Yet another segment is Dixon Cave that terminates at a former big spring at the paleo-Green River. One or more of the missing segments of this sequence have been destroyed or may await discovery in the future.

A primary reason why so many hundreds of miles of passages are present in Mammoth Cave is that the capping bed of clastic rock acts as a "roof" to the cave system. The resistant Big Clifty Sandstone and a thin bed of impermeable shale at its base constitute the caprock that has protected much of the cave system from erosion. A mapping program has revealed how protected passage fragments were part of a set of evolving drainage patterns that show how subterranean watercourses have developed through time. As can be seen in the lowest water-filled passages as well as at upper levels, drainage diversion or piracy is a ubiquitous occurrence. Diversion routes often develop as trenches cut in the bedrock floor of a tube or canyon. Their gradient deepens and they may depart from the original passage to become tubes or canyons leading to lower reaches of the system.

MAMMOTH CAVE MINERALS

Mammoth Cave is not noted for extensive displays of speleothems, such as stalactites, stalagmites, columns, and flowstone. These depositional formations are confined to parts of the cave system where the clastic capping bedrock is thin or has been removed and carbonate-saturated groundwater can descend. The Frozen Niagara section shown to visitors contains a significant display of these speleothems.

Various evaporative minerals can be seen in parts of the cave where the clastic caprock above it remains intact. These include gypsum crystals, needles, cotton, flowers, massive crusts, and loose deposits resembling drifted snow (Fig. 5). Less common are epsomite and

FIGURE 5 Gypsum flowers are found where the sandstone and shale caprock are intact over the cave passages. *Photo courtesy of the National Park Service.*

mirabilite crystals. The origin of the sulfate may be pyrite in the limestone or overlying beds, or dissolved gypsum from overlying beds. Investigators believe that aborigines gathered and used mirabilite as a laxative and perhaps as food seasoning, gypsum as a paint ingredient, and selenite crystals as ceremonial objects.

MAMMOTH CAVE ORIGIN

As a significant and complex cave, Mammoth Cave's origin in process terms has been studied since the 1920s. Early studies of the cave were mainly descriptive and based on only about 10% of the cave system known today. Inadequate cave maps also hampered early investigations. A.C. Swinnerton observed that horizontal passages in Mammoth Cave were formed in the "zone of discharge," a 30-m (100-foot) band located between the highest stage of the base level Green River and its lowest cave resurgence. At Mammoth Cave, this "shallow phreatic zone" extends from 9 m (30 feet) below pool stage to 21 m (70 feet) above pool stage. The regional dip is toward the Green River, but some passages veer across the strike. Hence, it appears that the dip is not the only structural control (Palmer, 1981).

BIOLOGY OF MAMMOTH CAVE

The caves of the Mammoth Cave area have a very diverse and well-studied biology. Currently, 41 species of cave-adapted organisms are known from the Mammoth Cave area. This number means that the park has one of the highest subterranean diversities worldwide. Scientific study of the cave biology of the Mammoth Cave System began in the middle of the 1800s and actively continues today. These studies have included taxonomic studies of specific organisms, ecological studies of terrestrial and aquatic systems, and evolutionary studies of the adaptation of cave animals. Poulson (1992) maintains that Mammoth Cave is the best-studied and best-understood cave ecosystem in the world. Notable species include the endangered, endemic Kentucky Cave Shrimp (*Palaemonias ganteri* Hay), two species of troglobitic fish, and five co-occurring cave trechine beetle species. The caves of the park are also home to two endangered species of bats. The caves of Mammoth Cave National Park are the type locality of over 25 cave-adapted species.

Two aspects of this high species diversity are explained here. First, the kinds of obligate and facultative organisms are especially high for the aquatic environment; that is, we find fish, crayfish, shrimp, isopods, amphipods, flatworms, copepods, worms, and snails. Because of the gradients in the amount, kind, and quality of organic input, and also water quantity, there are replacement of species in several groups, for example, four fish, two crayfish, two isopods, three amphipods, many copepods, and two flatworms. Second, in geographic scale terms, Mammoth Cave lies at the intersection of several cave and karst underground and surface species dispersal corridors along escarpments and sinkhole plains. Different species come from all directions. Striking examples are the four fish and seven Carabid beetles living in different parts of the cave.

Prominent among observable animals are cave guests like bats and pack rats, cave lovers like salamanders and cave crickets, and permanent cave dwellers like blindfish and beetles.

Since there is no light in the cave and hence no growth of green plants, cave organisms must rely on food brought in from above ground. Food chains in aquatic habitats start with dead plant material washed into the cave from cracks, sinkholes, and sinking streams. Sinkhole streams and sinks above the oldest vertical shafts, such as Mammoth Dome Sink and dry ravines, take coarse particulate organic matter during rare heavy rain events, but only fine particulate organic matter and mostly dissolved organic matter enter via the epikarst of the Sinkhole Plain and Mammoth Cave Plateau karst valleys. Such material is serially reduced in amount, size, and quality as it moves through the epikarst. Decline in food quality also occurs through deposition, leaching, and the use of decomposers from active vertical shafts to master shaft drains to base level rivers. This leads to a parallel decrease in density and hence change in aquatic species, for example, two different species of isopods, three of amphipods, two of flatworms. Downstream of master shaft drains, one can observe a change in species of cavefish, shrimp, and snail as passages become wider and deeper with lower velocity and more bottom silt.

Decomposers such as bacteria and molds do the initial breakdown of organic matters and animals eat the decomposers. The most important food chains in terrestrial habitats start with the feces and eggs of cave crickets. Cave crickets roost inside cave entrances and at night venture out to forage for bits of dead and decomposing matter. One community of organisms depends on the veneers of guano under the dense roosts of cave crickets. Another community depends on the feces of cave beetles that eat eggs that crickets lay in the sand away from the cave entrance. Cave beetles search vigorously for cricket eggs using touch and smell to locate the sand that crickets rake over their eggs.

From ceiling stains, it is evident that Mammoth Cave used to contain millions of hibernating bats, but today there are very few. Reasons may include the alteration of microclimate due to entrance modification and visitation by saltpeter miners and hundreds of thousands of modern tourists. Passages beneath the caprock are especially dry, and this influences the biology, mineralogy, and archaeology. Hydrated sulfate minerals, for example, epsomite, mirabilite, and gypsum, apparently "suck" the moisture from living and dead organisms. For the biologist these are the "Great Kentucky Desert," but for the archaeologist they are sites without decomposition and thus preserved the remnants of aboriginal people who used the caves.

As with any aquatic life, the organisms of Mammoth Cave are vulnerable to pollution. Several disastrous petroleum and chemical spills have confirmed the sensitivity of species in large streams, especially long-lived shrimp, crayfish, and fish at the end of food chains. An emerging area for research is bioaccumulation and biomagnification of toxins, such as mercury, that migrate up the food chain, affect the reproduction of long-lived animals, and are remobilized by floods. Another area of concern is incompletely treated human waste. This excess organic matter favors fast-reproducing species whose numbers overwhelm the usual cave species and so reduce species diversity. Awareness of pollution threats is focusing keen attention on violations of the Clean Air Act and Clean Water Act, and the effects of development on the Tier III protected waters of Mammoth Cave.

A MAGNET FOR INTERNATIONAL CAVE RESEARCH

The modern era of cave investigation at Mammoth Cave started in 1954 (Lawrence and Brucker, 2010) and with the organization of the Cave Research Foundation (CRF) in 1957. That group undertook a systematic exploration and mapping program as the basis for a scientific program in archaeology, biology, and hydrogeology. More than a dozen books, dozens of theses, and hundreds of scientific papers and cave maps have resulted from a cooperative partnership between the National Park Service and CRF scientists. University research and educational activities associated with Mammoth Cave include three organizations at Western Kentucky University, Bowling Green, KY: the Center for Cave and Karst Studies, the Hoffman Institute, and the Mammoth Cave International Center for Science and Learning.

Bibliography

Borden, J. D., & Brucker, R. W. (2000). *Beyond Mammoth Cave*. Carbondale, IL: Southern Illinois University Press.

Brucker, R. W., Hess, J. W., & White, W. B. (1972). Role of vertical shafts in the movement of ground water in carbonate aquifers. *Ground Water*, 10(6), 5–13.

Brucker, R. W., & Watson, R. A. (1987). *The longest cave*. Carbondale, IL: Southern Illinois University Press.

Granger, D. E., Fabel, D., & Palmer, A. N. (2001). Pliocene–Pleistocene incision of the Green River, Kentucky, determined from radioactive decay of cosmogenic ^{26}Al and ^{10}Be in Mammoth Cave sediments. *Geological Society of America Bulletin*, 113, 825–836.

Lawrence, J., Jr., & Brucker, R. (2010). *The caves beyond: The story of the Floyd Collins' Crystal Cave exploration* (reprint of the 1955 edition). Dayton, OH: Cave Books.

Murray, R. K., & Brucker, R. W. (1982). *Trapped! The story of Floyd Collins*. Lexington, KY: University Press of Kentucky.

Olson, C. O. (2010). *Mammoth Cave by lantern light: Visiting America's most famous cave in the 1800s*. Dayton, OH: Cave Books.

Palmer, A. N. (1981). *A geological guide to Mammoth Cave National Park*. Teaneck, NJ: Zephyrus Press.

Poulson, T. L. (1992). The Mammoth Cave ecosystem. In A. Camacho (Ed.), *The natural history of biospeleology* (pp. 569–611). Madrid: University of Madrid.

Watson, P. J. (Ed.), (1997). *Archeology of the Mammoth Cave area*. St. Louis, MO: Cave Books.

White, W. B., Watson, R. A., Pohl, E. R., & Brucker, R. W. (1970). The Central Kentucky Karst. *Geographical Review*, 160(1), 88–115.

White, W. B., & White, E. L. (Eds.), (1989). *Karst hydrology, concepts from the Mammoth Cave area*. New York: Van Nostrand Reinhold.

MAPPING SUBTERRANEAN BIODIVERSITY

Mary C. Christman[*] *and Maja Zagmajster*[†]

[*]*Department of Statistics, University of Florida,* [†]*Department of Biology, University of Ljubljana*

INTRODUCTION

The ability to describe and explain the spatial distribution of subterranean fauna is becoming increasingly more important as humans continue to expand their reach and encroach on traditionally inaccessible areas.

Knowing where and under what conditions various fauna are found is important from different ecological and management perspectives. To that end, it is important to understand the processes that have led to and sustain such distributions, and in the light of recent environmental changes, to help in understanding the impact of future environmental scenarios. Additionally, it is important for developing appropriate conservation plans (Adams et al., 2000). Hence, an increasingly urgent need is the ability to accurately map the geographic distribution of various species and ecosystems (and often as critical, the distribution of the required habitat). Mapping includes not only the graphical presentation of species occurrences and derived parameters such as densities and species richness over a region but, more importantly, it includes the statistical spatial modeling that provides both the map and the means of assessing its accuracy. Here we present some statistical approaches that provide such information which can be used in interpretation, hypothesis testing, and decision making. Statistical modeling provides the means for studying the spatial process that generates the observed geographic distribution of species and hence, it is useful both for research into the generating process as well as for identifying areas in which high species richness might be expected, although not yet reported.

DATA TYPES

To statistically model the spatial distribution of species or species diversity, one needs to have geo-referenced positions of the localities where the species occur; that is, their exact position on the Earth's surface as determined with coordinates. Accuracy of coordinates can differ among datasets, referring to the specific site (cave) at which species were observed or to the centroid of a large subregion of the study area in which the species is associated. Data that consist of the locations of occurrence of species or individuals are called a *point pattern*. Point patterns are usually analyzed for inference concerning whether the locations (points) tend to cluster in some meaningful way. Mapping of these data usually is in the form of locations of occurrence (presences) on a map.

More commonly, individual locations are not reported (often for purposes of confidentiality), yet such information can be used for exploration of subterranean biodiversity patterns. The data are spatially aggregated over subregions within the study area and the resulting counts or totals are reported at the level of the subregions. These areas can be artificially created, for example, grid cells constructed by overlaying a grid on the study region (Zagmajster et al., 2008a, 2010), or based on geopolitical boundaries such as counties in the United States (Christman and Culver, 2001a). These types of data, referred to as *lattice data* (Cressie, 1993), are often analyzed explicitly to determine if spatial autocorrelation exists, that is, to study the form of the correlation among neighboring areas. Hence, the level of spatial scale should be considered as a characteristic of the dataset used in analyses since it influences the methods used to statistically model the spatial distribution of the species or diversity measures.

The aforementioned presentations most often refer to presence of the species, and such statistics as species richness derived from the data. Adding information on sampling effort increases the value of biodiversity maps (Zagmajster et al., 2008a, 2010). In this chapter we focus on the mapping of quantitative data, especially species richness, either point-referenced or aggregated to subregions. Yet the general ideas are the same, whether one is mapping biodiversity or abundances of individual species or species assemblages.

DECOMPOSITION OF THE SPATIAL PROCESS

The process that generates the observed spatial distribution of species richness can be decomposed into several components (Cressie, 1993). First, there is the large-scale trend that can be explained due to environmental or geophysical features. These explanatory variables might include such attributes as cave length or depth, cave type, surface climate characteristics, altitude, latitude, distance from certain features such as the maximal extent of an ancient ice sheet or embayment, and others. For example, large-scale patterns of precipitation might help explain the spatial arrangement of caves in karst areas and, hence, of cave species.

Next is mesoscale or small-scale variation that is not explainable by the large-scale trend. This is local variability that is not typically captured over the geographic scale of the large-scale trend. Such variation might arise because of unknown factors that operate at smaller scales, that is, elevation or slope aspect. It might also be due to even finer spatial features such as connectivity of caves. The variations at this scale are often observed to have covariability, that is, adjacent areas often have similar numbers of species, possibly due to connectivity. As a result, this variation is often modeled as having spatial autocorrelation among nearby areas.

The final component is random error or the unexplained variability that is unaccounted for after the fitting of the other components. This variability could be due to sampling intensity or to factors not present in the model, such as size of the cave entrance opening. It is usually assumed that these variations are independent of one another, that is, the value at one site neither influences nor is influenced by a value at a nearby site.

Using this decomposition of the spatial variability, we can write that $Y(s)$, the number of species at site s, is a linear combination of the components:

$$Y(s) = \mu(s) + \delta(s) + \varepsilon(s)$$

where $\mu(s)$ is the large-scale trend at site s, $\delta(s)$ is the spatial correlation component describing the mesoscale variability, and $\varepsilon(s)$ is the random noise or unexplained variability. In order to do statistical analyses, it is assumed that the three components are independent of each other. When this assumption holds, the modeler can estimate the precision of each estimated model component.

An example of the decomposition of actual data is given in Figure 1. In this instance, Christman and Culver (2001a,b) modeled the number of species found in counties in the southeastern United States as a function of the number of caves in that county (large-scale trend), the autocorrelation between adjoining counties (mesoscale variability), and random error (unexplained variability). Counties without caves were constrained by the model to have zero species; hence, the flat regions on the graphs.

INVESTIGATING SPATIAL PATTERNS

When mapped, even visual inspection can offer new insight into the patterns observed, but having georeferenced data allows us to explore for patterns that may not be obvious in a map. Any statistical investigation

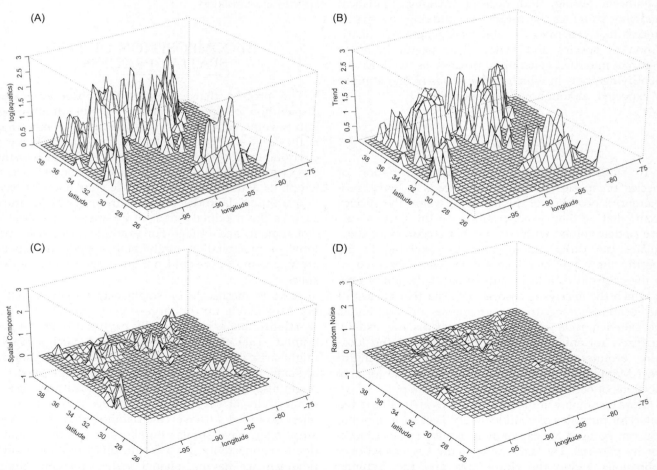

FIGURE 1 Perspective plot showing (A) the geographic distribution of the observed counts of stygobites (log-transformed) in counties in the southeastern part of the United States; (B) the fitted regression estimate of counts for the large-scale component of the model in which log-transformed number of stygobites is related to the number of caves (also log-transformed); (C) the spatial autocorrelation mesoscale component in which the log-transformed number of stygobites is predicted conditionally using the log-transformed counts of nearby surrounding counties; and (D) the residual or unaccounted for variation in the data.

considers the spatial relationships of data, where observations close in space are more similar than those farther apart. Causes range from emigration and immigration, common history or environmental features, or other possible reasons. Such proximity-dependent similarity in values of the variables of interest is called *spatial autocorrelation*. This autocorrelation is an important component of spatial processes and often drives the mesoscale variation described in the previous section. It makes spatial statistical analyses different from nonspatial approaches. When autocorrelation exists, localities are not considered independent and this must be taken into account when testing hypotheses concerning the causes of the observed distributions or whether changes have occurred in the distribution between two time periods. Exploratory analyses of the presence and strength of spatial autocorrelation use autocorrelation indices such as Moran's I (Moran, 1950), Geary's c (Geary, 1954), black-white join count statistics (Cliff and Ord, 1981), and the Mantel test (Mantel, 1967). Spatial autocorrelation can exist only at certain distances or range of distances, and not at others. These statistics can be applied at the different distances to identify the extent of spatial autocorrelation. Another approach that provides both a map and measures of the spatial autocorrelation involve *variogram estimation* and *kriging*.

INTERPOLATORS AND KRIGING

Besides point occurrence maps and/or presence within bigger units, a better visual presentation that can provide additional insight into the pattern observed is based on interpolation techniques. Numerical interpolators (Fig. 2) are used to smooth data so that contour plots or smooth three-dimensional perspective plots of the variable of interest can be constructed. Several kinds of interpolators are available, each providing a slightly different method for smoothing data so that a continuous surface is displayed. Techniques include simple methods such as moving averages, locally weighted smoothing, smoothing splines, and others.

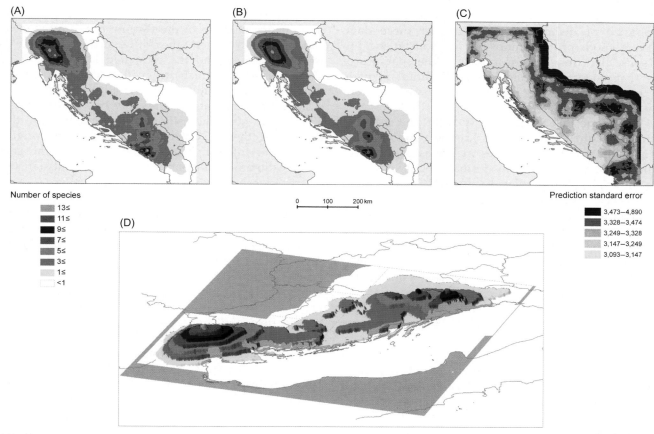

FIGURE 2 Different graphical representations of the same data, number of obligate subterranean beetles (from families Cholevidae and Carabidae) per 20 × 20-km quadrats in northwestern Balkans. (A) Filled contours based on inverse distance weighting; (B) filled contours based on ordinary kriging; (C) prediction standard error map of ordinary kriging; (D) a three-dimensional presentation of the ordinary kriging predictions. Scale refers to (A), (B), and (C). The first legend refers to (A), (B), and (D), and the second legend to (C). *Mapping procedure as in Zagmajster et al., 2008b.*

Moving-average smoothing involves estimating the mean for a location using the set of observed points within a given radius of the location to be predicted. Locally weighted smoothing uses a weighted mean instead of the arithmetic mean where the weights are usually some function of the distances of the observed points to the location to be predicted. A smoothing spline replaces the weighted mean with the value predicted by a local regression over the points within the region.

One needs to be aware that the type of the surface provided by the smoother depends on parameters that are specified in the smoothing algorithm. For example, at each point in the vertices of a grid overlying the study region, a moving average smoother uses the average of the $p\%$ data points closest to the point being estimated (p chosen before interpolating). When p is small, only data located very close to the prediction site are used; hence, the surface appears spiky and very similar to the original data. If p is large, then more of the surrounding data are used and the result is a much smoother surface. The choice of parameter values influences the final map display and thus, like any graphic, can be manipulated to emphasize or deemphasize features of these data. These methods provide maps but are nonstatistical in the sense that they are not models of the spatial process that generates these data and cannot provide any means of assessing the bias, precision, or accuracy of the map.

A technique that is statistically valid and uses the spatial structure of the data is kriging (Cressie, 1993; Haining, 1990) which provides measures of the error of the map generated from a small set of observed data. In kriging, the predicted value of $Y(s_0)$, the number of species at a prediction site, s_0, is the sum of the estimated large-scale trend plus the estimated autocorrelation component. In kriging with no large scale trend component, the optimal (in the sense of smallest variance) predictor for a new observation is a weighted mean (linear combination) of the variable values (Y) at the n sampled sites ($s_1, s_2, \ldots, s_n$):

$$p(Y(s_0)) = \sum_{i=1}^{n} \lambda_i Y(s_i)$$

where the weights, λ_i, are functions of the autocorrelation between the observed value of Y at s_i and the value of Y at the new site, s_0. Kriging provides a means of studying the fine-scale spatial structure of subterranean biodiversity because autocorrelation among sites can be ascribed to processes occurring at the spatial level of individual sites. In addition, it can be used to interpolate values at intervening locations between observed sites and hence, to map biodiversity, for example, over the whole study region. In fact, that is its most common use, because the kriging procedure will give a predicted value at an observed site equal to the observed value at that site. In kriging, one can also map the standard errors associated with the predicted values in order to gain some sense of the precision of the predictions.

Kriging requires some stringent assumptions be met in order to be a valid method for modeling the spatial distribution. These include stationarity and isotropy. A spatial process is said to be second-order stationary if the following is true: (1) the trend is constant over the region, that is, the mean $\mu(s) = \mu$ is constant for every site s in the region; and (2) the covariability between two sites may depend on the distance and direction between them but not on the location of the sites within the study region. The second point indicates that the variance of the Y variable is constant everywhere as well. A process is said to be isotropic if the covariability between two sites does not depend on the direction between them.

When a large-scale trend does exist (i.e., $\mu(s) \neq \mu$), one approach is first to model the large-scale trend and obtain estimates of the varying trend in the mean, call this $\hat{\mu}(s)$. Then the differences between the observed values and the estimated large-scale trend, $Y(s) - \hat{\mu}(s) = \varepsilon(s)$, which are called the *residuals*, are modeled using kriging. The results of kriging the residuals are estimates of the other two components of Y, $\delta(s)$, and $\varepsilon(s)$. The estimates $\hat{\mu}(s)$ and $\hat{E}(s)$ are then added together to obtain estimates for the original variable of interest. The standard error of the combined estimator can be approximated so that statistical inferences can be explored.

A cautionary note concerning the use of kriging for mapping subterranean biodiversity is required. The geologic conditions necessary for caves and other subterranean features needed by obligate subterranean species are not normally uniform over the study region. Kriging assumes that the study region is a convex polygon within which the sites at which data are taken represent a sample from the continuous spatial field and, hence, assumes that the conditions are continuous over the entire study area. As a result, it will predict nonzero estimates of species richness at sites within the study area, even if those sites do not have any appropriate habitat for cave fauna. In some instances, this is easily resolved by masking, which is the process of removing those locations from possible interpolation or kriging. This is effective only if the inappropriate area is small relative to the study region. When habitable areas are patchier, the correct modification to the modeling is not so clear. Some work has been done using a method called *zero-inflated spatial models*, but more research is required to determine the best method of resolving the spatial relationships under such situations. One technique that has been used is to aggregate the point data to another scale, for example, to sum up the number of caves and total number of species found in a given area.

MODELING ISSUES

Aggregation and Scale

One of the first issues that must be resolved when performing spatial modeling is the scale at which the analysis is to be done. One of the impediments can be the positional accuracy of the localities; they can be either poorly described in the literature (so determination of exact position is difficult) or public access can be restricted due to conservation. In that case, data can still be used, but at a different scale. They can be aggregated to the level of county, or into the units of a grid that overlays the study area. Scaling has both a positive and a negative impact on ecological studies. First, it may be that different processes are acting at different scales and the researcher should attempt to partition the effects of the different scales. For example, similarities between adjacent caves could be due to cave connectivity (a microscale process), altitude, or aspect of the hillside on which the cave is located (a mesoscale process), or the latitude at which the cave is found (a macroscale process). On the other hand, if data are aggregated in such a way that the aggregation scale does not mimic the scale at which a process is occurring, then it is not possible to study the relationships at the appropriate scales. The exploratory tests for spatial structure mentioned earlier (Moran's *I*, Mantel test, *etc.*) are useful for identifying possible spatial scales that are relevant to the problem at hand.

Figure 3 is an example of an analysis done at different spatial scales in order to identify the appropriate level of aggregation of the point-level data. The data on obligate subterranean beetles from the Dinaric karst region were mapped at different spatial scales of quadrats. Though generally the pattern of highest species richness remained the same, there were some differences. The quadrats of highest species richness were not always overlapping among scales. Additionally, the spatial autocorrelation among quadrats with species numbers changed as the scale changed. It increased up to the grid cell sizes of 20×20 km, and began to decrease as the aggregation units got larger. When the units are too big and cover too big areas, the similarities among the quadrats start to diminish. From this analysis it was suggested that the suitable unit for data aggregation, based on the data available, was at the scale of 20×20-km grid cells (Zagmajster *et al.*, 2008a).

The size and shape of the aggregation units can influence the analyses (Openshaw and Taylor, 1979). For example, when square grid cells are used, the neighboring quadrats are usually considered to be those with a common boundary—this precludes the grid cells on the diagonal. A possible solution to this problem of different neighbors is the use of hexagons, as developed by White, Kimerling, and Overton (1992). Additionally, hexagons have the advantage of retaining equal shape and equal area over the globe (White *et al.*, 1992), which makes them especially suitable for analyses when very large geographic areas are considered. They have been widely used in mapping biodiversity in the United States (Adams *et al.*, 2000).

Another problem of data aggregation is the choice of the most suitable size of the unit for the problem being studied. Relationships between variables can change as the scale of data aggregation is changed, a phenomenon called the *ecological fallacy* or the *modifiable areal unit problem* (MAUP). A good review of the issues can be found in Gotway and Young (2002).

When data are aggregated into blocks, the statistical models used to analyze them and estimate the spatial relationships differ as well. Now the model is based on analyzing relationships among discrete units rather than points. The decomposition remains the same but the model is more flexible in that isotropy is no longer required for analysis. In fact, it is common that the spatial autocorrelation is different along different compass points in many biological applications. Models for analyzing aggregated data are provided in Cressie (1993) and Haining (1990), among others. In order to make the model tractable, it is sometimes assumed that the spatial correlation is independent of direction between sites and location within the study region. Other simplifying assumptions include constraining the autocorrelation to exist only if the two sites are within K miles (or kilometers) of each other or setting the autocorrelation to be an inverse function of the distance between two sites, for example, setting the autocorrelation to some function of $d(s,t)^{-p}$ where $d(s,t)$ is a distance metric and p is the decay factor, indicating how fast the autocorrelation drops off as the two sites (s and t) become farther apart. Note that in this model, one does not attempt to estimate species richness at sites in between aggregated locations because, by definition, there are no such sites between aggregated blocks.

Figure 1 was generated using a lattice model with the mean number of species in any county being a

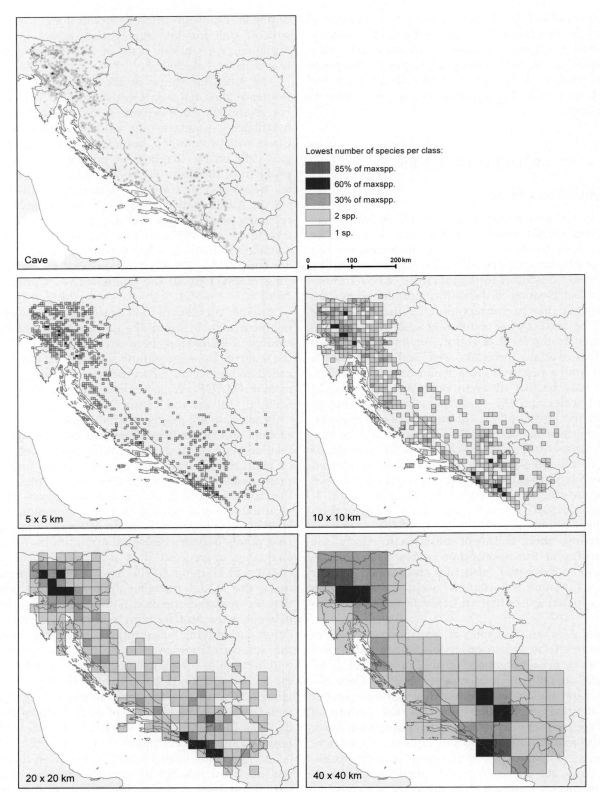

FIGURE 3 Number of obligate subterranean beetles (from families Cholevidae and Carabidae) in northwestern Balkans, mapped at different scales (marked in lower left corner of maps). The classes are delimited relatively according to maximal number of species per unit used (maxspp.). Only localities where positions determined with centroid of maximal 3-km radius were considered. *See also Zagmajster et al., 2008a.*

function of the number of caves in that county and of the number of species in surrounding counties whose centroids were within 56 km of the county being estimated (Christman and Culver, 2001a). The spatial correlation parameterization used was

$$\eta(s_i, s_j) = \begin{cases} 0, & \text{if } ||s_i - s_j|| > 56 \text{ km} \\ \dfrac{||s_i - s_j||^{-1}}{\max_{r,t}\{||s_r - s_t||^{-1}\}}, & \text{if } ||s_i - s_j|| \leq 56 \text{ km} \end{cases}$$

where $||s_i - s_j||$ is the Euclidean distance between the two locations s_i and s_j and $\max_{r,t} ||s_r - s_t||$ is the greatest distance between any two locations in the dataset. The ratio of the maximum distance to the local distance constrains the parameter η to fall within the interval $(-1, +1)$ and, hence, is interpretable as a correlation coefficient.

Excess Zeroes

Although the above models work well when the assumptions are met and there are few zeroes, a common problem with obligate subterranean data is the excess number of zeroes. This occurs either because there is no habitat available for cave species or because there are no records of species within individual caves or subregions. In the first case they can be regarded as "true" zeroes, while in the second the situation is less clear. It could be that the cave either has not been explored yet or has been explored but no organisms were found. In subterranean fauna, sampling can be difficult so that species detectability and catchability are often low and false negatives (species present, yet not recorded) are common. One way of dealing with the zeroes then would be to include information on sampling effort, even if (or especially if) "unsuccessful." Unfortunately, this is rarely or not recorded in the literature.

The problem of excess zeroes in mapping can be partially diminished when data are aggregated into bigger units, and not analyzed at the original scale of individual caves. In the case of lack of habitat, the lattice model can be adapted to require that any prediction be predicted on the existence of karst (or caves or whatever other variable might be appropriate). Some models have been used that allow for the fact that a zero may be either due to lack of available habitat or simply no data have been observed (*cf.* Ridout *et al.*, 2001). These models have been developed under the assumption that there is no spatial autocorrelation among sites and, hence, are applicable only if the assumption holds (and therefore has limited usability in studies of subterranean biodiversity). Some work has been done to incorporate excess zeroes in a kriging model (*cf.* Warren, 1997), but research is still ongoing.

See Also the Following Article

Documentation and Databases

Bibliography

Adams, J. S., Stein, B. A., & Kutner, L. S. (2000). Biodiversity: Our precious heritage. In B. A. Stein, L. S. Kutner & J. S. Adams (Eds.), *Precious heritage: The status of biodiversity in the United States* (pp. 3–18). New York: Oxford University Press.

Christman, M. C., & Culver, D. C. (2001a). The relationship between cave biodiversity and available habitat. *Journal of Biogeography*, 28(3), 367–380.

Christman, M. C., & Culver, D. C. (2001b). Spatial models for predicting cave biodiversity: An example from the southeastern United States. In D. C. Culver, L. DeHarveng, J. Gibert & I. D. Sasowsky (Eds.), *Mapping subterranean biodiversity: Proceedings of an international workshop held March 18 through 20, 2001, Laboratorie Souterrain du CNRS, Moulis, Ariege, France* (pp. 36–38). Charles Town, WV: Karst Waters Institute.

Cliff, A. D., & Ord, J. K. (1981). *Spatial processes: Models and applications.* London: Pion Limited.

Cressie, N. A. C. (1993). *Statistics for spatial data* (Rev. ed.). New York: Wiley & Sons.

Geary, R. C. (1954). The contiguity ratio and statistical mapping. *The Incorporated Statistician*, 5(3), 115–145.

Gotway, C., & Young, L. (2002). Combining incompatible spatial data. *Journal of the American Statistical Association*, 97(458), 632–648.

Haining, R. (1990). *Spatial data analysis in the social and environmental sciences.* New York: Wiley & Sons.

Mantel, N. (1967). The detection of disease clustering and a generalized regression approach. *Cancer Research*, 27(1), 209–220.

Moran, P. A. P. (1950). Notes on continuous stochastic phenomena. *Biometrika*, 37(1–2), 17–23.

Openshaw, S., & Taylor, P. J. (1979). A million correlation coefficients: Three experiments on the modifiable areal unit problem. In N. Wrigley (Ed.), *Statistical methods in the spatial sciences* (pp. 127–144). London: Pion Limited.

Ridout, M., Hinde, J., & Demetrio, C. G. B. (2001). A score test for testing a zero-inflated Poisson regression model against zero-inflated negative binomial alternatives. *Biometrics*, 57(1), 219–223.

Warren, W. G. (1997). Changes in the within-survey spatio-temporal structure of the northern cod (*Gadus morhua*) population, 1985–1992. *Canadian Journal of Fisheries and Aquatic Sciences*, 54(Suppl.), 139–148.

Zagmajster, M., Culver, D. C., & Sket, B. (2008a). Species richness patterns of obligate subterranean beetles (Insecta: Coleoptera) in a global biodiversity hotspot—Effect of scale and sampling intensity. *Diversity and Distributions*, 14, 95–105.

Zagmajster, M., Sket, B., & Culver, D. C. (2008b). The representation of subterranean beetle biodiversity with the use of interpolation methods. In D. Perko, M. Zorn, N. Razpotnik, M. Čeh, D. Hladnik, M. Krevs, T. Podobnikar, B. Repe & R. Šumrada (Eds.), *Geographical information systems in Slovenia 2007–2008* (pp. 237–245). Ljubljana, Slovenia: ZRC Publishing.

Zagmajster, M., Culver, D. C., Christman, M. C., & Sket, B. (2010). Evaluating the sampling bias in pattern of subterranean species richness: Combining approaches. *Biodiversity and Conservation*, 19(11), 3035–3048.

MARINE REGRESSIONS

Claude Boutin* and Nicole Coineau†

*Université Paul Sabatier, Toulouse, France,
†Observatoire Océanologique de Banyuls, Université Paris 6, France

BACKGROUND

The ocean milieu not only covers more than two-thirds of the Earth's surface, but very likely has played a major role in the appearance, the evolution, and diversification of the life on the whole planet, including the different terrestrial and aquatic habitats existing on the continents. Even caves and other subterranean aquatic biotopes were often colonized by animal species directly coming from marine habitats during the periods of marine regressions. Because of this, the present geographic distribution of some aquatic cave inhabitants is closely correlated with the past position of shorelines before a recession of the sea. In some cases cave or groundwater animal species are the sole evidence of past ocean presence, and could be used to date the marine regression period.

FROM THE SEA TO FRESHWATER AND LAND

Most biologists and paleontologists have long agreed that life appeared first in marine environments, and that all animal phyla (all the different lineages of the animal kingdom) which later colonized continents first existed and evolved in the sea. A good number of phyla are still exclusively marine; some of them are well known, such as the Echinodermata (sea urchins, starfishes) or the Cephalocordata (the amphioxus), while other phyla such as the Ctenophora or the Chaetognatha are known only by zoologists. In contrast, a great number of phyla have both marine branches and other branches living in continental freshwaters or in terrestrial habitats. This is the case of Arthropoda and Vertebrata, which include many marine species and many other species presently living in river and lake freshwaters or in terrestrial habitats of continents. While colonizing the freshwater habitats long ago, many groups continued to live in the sea; they are therefore represented both by marine and freshwater species belonging to the same order, often to the same family, and sometimes to the same genus. Such situations may be observed in many crustacean groups such as Decapoda, Isopoda, and Amphipoda. In these groups the colonization of continental freshwaters by marine ancestral populations may have occurred repeatedly at different periods, originating groups of continental species that were different from their distant marine ancestors.

ADAPTABILITY TO DIVERSE HABITATS

Since the 1970s, biologists interested in the development of life on Earth and in the diversity of living organisms have made significant advances in the field of historical biogeography and evolutionary ecology. Now we better understand why species occur where they are presently distributed. These researchers have also discovered the surprising ability of many species to colonize a great variety of habitats, including the most extreme and drastic environments. Such habitats occur in polar regions covered by ice caps; in deep parts of oceanic plains; around the hydrothermal vents that form sea bottom oases; in the deep part of continental caves; and in all subterranean waters present in continental caves, marine caves, and anchihaline caves (a kind of coastal cave related to the sea but filled with water also subjected to continental freshwater influences), as well as in continental and fresh groundwaters.

There is a general agreement for considering that the ancestral populations of many troglobionts (species living in caves and other terrestrial subterranean habitats) lived first in soils, humus, litter, or moss strata of forest ecosystems. They lived mainly during the Cenozoic and Pleistocene periods, in temperate regions, and probably at any period from the Mesozoic to the present in tropical regions. Thus it is shown that terrestrial subterranean species or populations are derived from surface ancestral species.

For the limnostygobionts, the subterranean freshwater species presently living in caves or in continental groundwaters, the question is less simple because their origin is double. Some species clearly belong to groups that presently live in surface freshwaters and appear to be derived from surface freshwater ancestral populations. But many other species of limnostygobionts are representatives of marine groups, totally absent in surface freshwaters, and showing evident phylogenetic affinities with marine species or groups belonging to the marine planktonic, benthic, or interstitial communities. With reference to their origin, the first group of limnostygobionts was called *limnicoid* and the second group *thalassoid* stygobionts.

MARINE REGRESSIONS AND THE BIOGEOGRAPHY OF THALASSOID STYGOBIONTS

The direct marine origin of the thalassoid limnostygobionts has been postulated and acknowledged for a long time by many stygobiologists who came first from France and then from many different countries in the world.

Among the thalassoid limnostygobionts occurring in continental subterranean habitats around the present marine shores, it is possible to mention some Protista such as the Foraminifera known from groundwaters of Central Asia, the Sahara, and Morocco and a high number of Metazoa (or Animalia phyla). Subterranean Porifera and Cnidaria are thalassostygobionts still living in littoral caves, but Platyhelminthes, Nematoda, Annelida (Oligochaeta and Polychaeta), and Gastrotricha, Mollusca, and Arthropoda (mainly representatives of many Crustacea orders but also Hydracarina) often occur within the freshwater thalassoid stygofauna.

The ecological and evolutionary processes that changed the habitats of some ancestral marine populations that originated the thalassoid limnostygobionts have interested stygobiologists for the past 30 years (Coineau and Boutin, 1992; Holsinger, 2000). J. Stock (1980) was the first to propose a *regression model evolution* which provided an understanding of the occurrence and distribution of stygobiontic amphipods living in groundwaters of several Mediterranean countries and Atlantic islands. The marine regression was considered the event responsible for the origin of subterranean thalassoid crustaceans. These species were simply viewed as descendants of marine populations "stranded" by the receding sea. Then several authors independently described a *two-step model of evolution and colonization* emphasizing the succession of two very different ecological changes. The first one, occurring in marine coastal habitats, is called the "vertical transition." It is the active colonization of subterranean milieus, either in marine caves or more often in interstitial and crevicular environments, correlated with an adaptation to the life in darkness and, in the latter habitats, in small size interstices. The second step, sometimes called a *horizontal transition*, is an adaptation to life in brackish water and progressively in freshwater. This may have occurred passively, first in the intertidal zone, and then continued on the spot in groundwater during and after the sea recession (Boutin and Coineau, 1990; Holsinger, 1994; and Fig. 1).

This way of viewing the origin of thalassoid limnostygobionts occurring in caves or in groundwaters, and the feeble ability for active dispersal observed in interstitial and subterranean biota, allows us to understand the correlation that may be observed between the extent of past marine gulfs and the present distribution area of continental thalassoid stygobionts (Fig. 2). Notenboom (1991) emphasized such correlations between the location of past marine gulfs and the distribution of present stygobiotic amphipods from Spain. Boutin (1994a) explained similar correlations occurring with the metacrangonyctid amphipods from Morocco. Many other distribution patterns of limnostygobiontic peracarid crustaceans including a number of thalassoid amphipods, isopods, and Thermosbaenacea could probably be well understood in the same way when considering the two-step model of evolution and colonization.

WHEN THALASSOID LIMNOSTYGOBIONTS ARE THE SOLE EVIDENCE OF A MARINE TRANSGRESSION–REGRESSION CYCLE

Usually subterranean biologists use the data provided by the studies of regional historical geology and paleogeography for the reconstruction of biogeographic scenarios that explain the observed species distribution pattern. However, an exactly opposite approach, using biological and paleogeographical data, is sometimes possible. Furthermore, it may happen that some thalassoid limnostygobionts, belonging to a well-known group of species, occur in a region characterized by an incompletely known paleogeographic history. This can occur when all the Mesozoic and Cenozoic rocks have been removed by recent erosion.

Such a situation occurs in southwestern Morocco where two locations of the amphipod *Longipodacrangonyx* are known in the groundwaters of the Tiznit region, about 40 km south of the known southern limits of the Eocene marine gulf of the Souss Valley. Farther south, near the city of Guelmim, another amphipod belonging to the most primitive lineage of the genus *Metacrangonyx* occurs in many sites. Other amphipod species of *Longipodacrangonyx* and species of the primitive lineage of *Metacrangonyx* are well known in many other regions of Morocco where they clearly exhibit a distribution respectively related to the past marine embayments of the Lutetian (Mid-Eocene) sea and with those of the Cenomano-Turonian (Mid-Cretaceous). Therefore they are considered the descendants of coastal ancestral populations which colonized the continental sediments during the regressions of the Late Lutetian (about 40 Ma BP) for *Longipodacrangonyx* and during the Turonian (about 90 Ma BP) for one group of *Metacrangonyx* species. The geological history of the regions of Tiznit and Guelmim is far from well known because of a recent uplift of the region resulting in the erosion of all marine sediments of Mesozoic and Cenozoic age. There are no data or marks of possible marine deposits in the

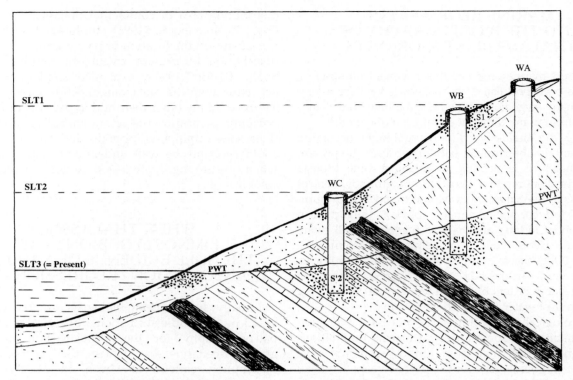

FIGURE 1 Origin of two present thalassoid limnostygobiontic species S'1 and S'2, in relation to two successive marine transgression-regression cycles, from time T1 to the present, T3. The sea level first reached shoreline SLT1 at time T1 and then the sea receded. Later a new transgression reached the shoreline SLT2 at time T2 and receded until the present shoreline SLT3. S1 was a marine species that colonized the coastal sediments at time T1 (vertical transition), S2 was a thalassostygobiont living during T2 period, and S3 is a present thalassostygobiontic species. The present limnostygobionts, S'1 and S'2, are derived from S1 and S2 during the sinking of the water table up to the present water table (PWT). WA, WB, and WC are three present wells. Two vicariant thalassoid stygobiontic species, S'1 and S'2, occur in WB and WC since their origin is related to the two different regressions that occurred after T1 and T2 (because of an impermeable black layer reducing the possible genetic fluxes between the groundwater populations of WB and WC).

region that are more recent than 400 Ma BP (Choubert and Faure-Muret, 1962; Andreu-Bousut, 1991). The occurrence and location of *Longipodacrangonyx* is an indication of an Eocene gulf more southerly extended than supposed from observation of preserved sediments, and the occurrence of *Metacrangonyx* east of the present shoreline is evidence of the presence of the Cretaceous sea during a period that ended with the Turonian marine regression. The presence of this ancient sea near Gelmim and Tiznit is not really surprising for geologists, since the Tarfaya Cretaceous Basin is well known about 200 km in the south, as well as the High Atlas Cretaceous Basin immediately north of the Souss Valley, near Agadir, where marine sediments are preserved. However, the continuity of the Turonian Sea from the Tarfaya Basin to the High Atlas was only a possibility before the discovery in this region of a species of *Metacrangonyx*, which clearly belongs to a group of species that settled in continental groundwaters during the regression of the Turonian shorelines.

Generally, after a marine regression, the presence of the sea is usually marked on the continent by the presence of sedimentary rocks containing classic fossils used for dating the transgression. But if a subsequent erosion occurs, all sediments with their fossils may be removed. In contrast, the thalassoid crustaceans—often called *living fossils*—settled during the marine regression and later still dwelled in subterranean groundwaters. They may survive such erosive destruction, as they are able to change habitat when groundwaters, first embedded in superficial sedimentary rocks, shift into deeper and older rocks during a regional uplift.

DATING MARINE REGRESSIONS OR LAND EMERSIONS FROM THE EXTANT STYGOFAUNA

It is noteworthy that the thalassoid limnostygobionts may provide not only the evidence of an ancient sea in the region where they occur, but that they can also, if they belong to a well-known monophyletic group of species, provide an indication about the date of the marine regression that allowed the colonization of

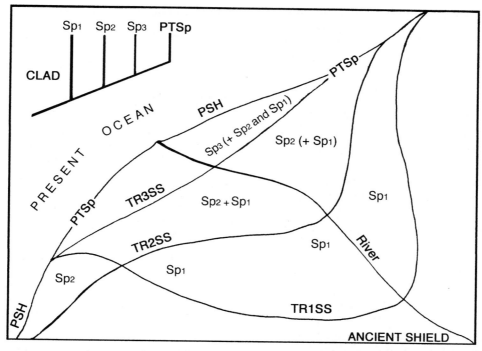

FIGURE 2 Schematic map showing the geographic distribution of three species of thalassoid limnostygobionts Sp1 to Sp3, in relation to three successive marine regressions. TR1SS to TR3SS are the seashore lines during the maximum of transgression at the periods T1, T2, and T3. Each transgression was followed by a marine regression as far as the present shore (PSH). PTSp is a present thalassostygobiontic species still living in a coastal interstitial or cave environment. CLAD is a simplified cladogram showing the phylogenetic relationships within the group of present stygobiontic species.

subterranean waters by their ancestral populations. The above example in southwestern Morocco allows the dating of two different marine regressions: one during the Eocene, south of the present Souss Valley, and the other during the Late Turonian in the Guelmim Province.

The date of the most ancient exposed lands of Israel and Palestine certainly is not Cenozoic, as generally proposed. A species of *Metacrangonyx* belonging to a lineage settled in continental subterranean waters during the marine regression of the Senonian (Late Cretaceous) occurs near the Dead Sea, as well as in the eastern Egyptian Sinai near the Eilat Gulf. Some geologists considered the possible emergence of isolated lands in this region during the Late Cretaceous from indirect evidence, but these emerged lands or islands are difficult to localize with precision, and in many parts of Israel marine sediments indicate the presence of the sea only from the beginning of the Cenozoic. The characteristics of the two species of *Metacrangonyx* known in the region indicate that these species are part of a derived group of species that accomplished the transition from a marine to a continental environment during the Senonian. The suggested conclusion is that the Late Cretaceous is the age of the most ancient parts of the Levant (Boutin, 1997).

The emergence and the growing of an island, when marine sediments are uplifted above the sea surface, is an ecological event very similar to a marine regression occurring on the margin of a continent. During the closure of the eastern Mediterranean, all the lands of the Levant (eastern Mediterranean countries) formed a continuous emerged land. This was marked on its margins by recent cycles of marine transgressions and regressions. It appeared first as islands, and the greatest age of the land in these countries is that of the most ancient emerged island.

Similarly, the age of the most ancient part of the Canary Archipelago has been dated from the Senonian, based on the study of the thalassoid amphipods occurring in wells of Fuerteventura Island (Boutin, 1994b). A part of Fuerteventura is not volcanic, unlike the remainder of the Archipelago; instead it is made of Cretaceous sedimentary marine rocks forming a first emerged island before the Cenozoic, while all studied volcanic rocks are Cenozoic. This island emergence is quite similar to a marine regression discovering lands above sea level and at the same time developing available groundwater habitats for marine crustaceans that are candidates for originating the thalassoid limnostygobiontic fauna. It is noteworthy that all the wells housing the Fuerteventura metacrangonyctid amphipod are dug in the metamorphic limestones of the "basal complex" and never in the Cenozoic volcanic rocks. *Metacrangonyx* is the sole evidence of the Senonian age of the emersion of the most ancient part of the island of Fuerteventura.

EVALUATING THE RATE OF EVOLUTIONARY CHANGES FROM THE MARINE REGRESSION CALENDAR

Many observations made by paleontologists as well as by neontologists suggest that the morphological change in animal lineages is extremely slow in habitats ecologically stable for millions of years, and rather important and rapid in unstable habitats subjected to repeated disturbances (Boutin and Coineau, 1991). As for the molecular evolution, it is generally described following the molecular clock model with the molecular changes supposed to occur regularly and proportionally to the duration. But the average rate of molecular changes fixed in generations forming a lineage is different according to the considered gene, or molecule. Moreover, it is difficult to be sure that, during a long period, the fixation of occurring mutations is always constant. This rate could be dependent on repeated "bottleneck" occurring in animal populations living in unstable habitats. Therefore it has been suggested (Boutin and Coineau, 2000) that it would be especially interesting and informative to study, comparatively, both at a molecular level and from a morphological point of view, some thalassoid limnostygobionts belonging to different lineages of a group that settled in continental groundwaters during different marine regressions. As a matter of fact, after the vertical transition the different lineages of interstitial limnostygobionts lived for a more or less long period in the changing and unstable coastal biotopes before the horizontal transition to the more stable continental groundwaters, which resulted from the repeated marine regressions. Probably such comparisons will permit the testing of molecular clocks and the respective value of molecular and morphological characters for the reconstruction of phylogenies. These studies would be possible when the dates of the successive marine regressions, which originated the different lineages within a thalassoid group of limnostygobionts, are well known (Danielopol, 1980). The first studies with this aim have just been performed and new results are expected by stygobiologists.

Finally, the biogeography and the history of the thalassoid stygobiontic fauna are closely related to the shifts of the shorelines, and the origin and some evolutionary aspects of these stygobionts result from marine regressions.

See Also the Following Articles

Vicariance and Dispersalist Biogeography
Invasion, Active versus Passive

Bibliography

Andreu-Boussut, B. (1991). Les ostracodes du Crétacé-moyen (Barrémien à Turonien) le long d'une transversale Agadir-Nador (Maroc). *Strata, 14*(Série 2), 1–405 (in French).

Boutin, C. (1994a). Phylogeny and biogeography of metacrangonyctid amphipods in North Africa. *Hydrobiologia, 287*, 49–64.

Boutin, C. (1994b). Stygobiology and historical geology: The age of Fuerteventura (Canary Island), as inferred from its present stygofauna. *Bulletin de la Société Géologique de France, 165*, 273–285.

Boutin, C. (1997). Stygobiologie et géologie historique: L'émersion des terres de Méditerranée orientale datée à partir des Amphipodes Metacrangonyctidae (Micro-Crustacés souterrains). *Geobios-Memoires Special, 21*, 67–74 (in French).

Boutin, C., & Coineau, N. (1990). Regression model, Modèle biphase d'évolution et origine des micro-organismes stygobies interstitiels continentaux. *Revue de Micropaléontologie, 33*, 303–322 (in French).

Boutin, C., & Coineau, N. (1991). Instability of environmental conditions and evolutionary rates. Example of thalassoid subterranean microcrustaceans in Mediterranean countries. *Bulletin de l'Institut de Géologie du Bassin d'Aquitaine, 50*, 63–69.

Boutin, C., & Coineau, N. (2000). Evolutionary rates and phylogenetic age in some stygobiontic species. In H. Wilkens, D. C. Culver & W. F. Humphreys (Eds.), *Ecosystems of the world, 30: Subterranean ecosystems* (pp. 433–451). Amsterdam: Elsevier.

Choubert, G., & Faure-Muret, A. (1962). Évolution du domaine atlasique marocain depuis les temps paléozoïques. In *Livre à la mémoire du Professeur Paul Fallot* (Tome 1) (pp. 447–527). Paris: Société géologique de France (in French).

Coineau, N., & Boutin, C. (1992). Biological processes in space and time: Colonization, evolution and speciation in interstitial stygobionts. In A. I. Camacho (Ed.), *The natural history of biospeleology* (Monografias 7) (pp. 423–451). Madrid: C.S.I.C.

Danielopol, D. L. (1980). An essay to assess the age of the freshwater interstitial ostracods of Europe. *Bijdragen tot de Dierkunde, 50*, 243–291.

Holsinger, J. R. (1994). Pattern and process in the biogeography of subterranean amphipods. *Hydrobiologia, 287*, 131–145.

Holsinger, J. R. (2000). Ecological derivation, colonization, and speciation. In H. Wilkens, D. C. Culver & W. F. Humphreys (Eds.), *Ecosystems of the world, 30: Subterranean ecosystems* (pp. 399–415). Amsterdam: Elsevier.

Notenboom, J. (1991). Marine regressions and the evolution of groundwater dwelling amphipods (Crustacea). *Journal of Biogeography, 18*, 437–454.

Stock, J. H. (1980). Regression model evolution as exemplified by the genus *Pseudoniphargus* (Amphipoda). *Bijdragen tot de Dierkunde, 50*, 105–144.

MAYA CAVES

Andrea Stone[*] *and James E. Brady*[†]

[*]*University of Wisconsin-Milwaukee,* [†]*California State University, Los Angeles*

Ancient Maya civilization extended across southern Mexico, Guatemala, Belize, and western Honduras

GENERAL CHARACTERISTICS OF ANCIENT MAYA CAVE UTILIZATION

The ancient Maya explored caves with the aid of wooden torches. Bundles of split pine torches have been recovered in cave entrances and burned remains recovered in deeper passages. Charcoal torch strikes on cave walls are also common. One of the few cave-specific tools developed by the Maya is a ceramic tube that served as a torch holder. Unlike Paleolithic cave explorers who penetrated equal depths by firelight, the ancient Maya never used oil lamps. Light appears to have been a limiting factor in their speleological exploits. Their record for penetration of a cave is 3 km, set at Actun Chek, Belize. The longest caves in the Maya area are the Chiquibul and Cave's Branch cave systems of Belize, where individual caves reach 15 km in length. Most caves explored by the ancient Maya are more modest; however, small size did not preclude their designation as sacred sites. Even the smallest caves, cenotes, and rockshelters were ritually utilized. Although the Maya generally explored the deepest parts of caves, the heaviest concentration of artifacts is usually found in the light to twilight zones near entrances. Forays into the dark zone may have been restricted to ritual specialists.

FIGURE 1 Map of the Maya area with some important archaeological cave sites.

(Fig. 1). Although this zone has a highland component, it is the limestone-covered Maya Lowlands that saw the rise of the great Classic cities, which are famous for their hieroglyphic inscriptions, fine architecture, and sculpture. The Lowland Maya offer an unusual case study of one of the world's great civilizations emerging in a tropical karst landscape. Moreover, the prevalence of caves in the environment shaped the development of the Classic Maya civilization. Although caves served some practical ends, for instance, sources of water in the Yucatán, they principally provided religious sanctuaries and theaters of ritual activity. They also figured prominently in the Maya's mental world of myth and symbol. In their highly stratified society, the Maya's ritual use of caves was an institution that cut across class boundaries. This widespread cultural practice also had great antiquity, accompanying the rise of sedentary villages, beginning in the Early Preclassic Period (ca. 1200 B.C.). Cave rituals became more elaborate during the Late Classic fluorescence (600–900 A.D.) and continued in northern Yucatán and the Maya Highlands through the Postclassic Period (900–1550 A.D.). The fact that cave ceremonialism continues today in the Maya area, despite a legacy of colonial persecution, shows the centrality of caves in indigenous religion at the time of the conquest.

ANTIQUITY OF MAYA CAVE USE

Evidence for cave utilization in the Maya area prior to the advent of sedentary villages, circa 1200 B.C., is extremely limited. The discovery at Loltun Cave, Yucatán, of stone tools associated with extinct megafauna offers some evidence for use during the Late Pleistocene. Charcoal from a hearth in Carwash Cenote, Quintana Roo, yielded a radiocarbon date of 8250 ± 80 years BP. The picture is equally dim during the Archaic period but begins to clarify during the Preclassic. At Copan, Honduras, ceramic evidence from Gordon's Cave #3 indicates its use as an ossuary as early as 1000 B.C. Major architectural modifications at the Cueva de las Pinturas, Guatemala, have been dated to the end of the Middle Preclassic (300 B.C.), and stone masonry at Naj Tunich, Guatemala, dates to the Late Preclassic. An elaborate figure carved at an entrance to Loltun Cave suggests that the site was appropriated for elite use by the Late Preclassic.

CAVE BURIAL

Human skeletal material in caves reflects both the disposal of sacrificial victims and the burial of loved ones. Rockshelters were often used as cemeteries in the southern Maya Lowlands from at least the Late Preclassic (300 B.C. to 300 A.D.) to the end of the Classic Period (900 A.D.). Important individuals were buried in cave alcoves that were closed off with crude stone walls. This practice appears to be confined to the Preclassic except at Naj Tunich, which is unique in having elaborate masonry tombs dating to the Classic period.

A number of cave ossuaries have been reported, generally in the Highlands, although the data are sketchy. On the eastern edge of the Maya area, Gordon's Cave #3 near Copan contains hundreds of cremated remains from roughly 1000 B.C. that appear to have been brought to the cave in cloth bags. On the western edge of the Maya area in Chiapas, small, walled-up caves yielded sealed ceramic vessels that contained cremated human remains dating to the Postclassic (900–1550 A.D.). Ethnohistorical sources from Chiapas mention the worship of bundles in caves containing the remains of ancestors, and ethnographic sources suggest that important males were interred in lineage caves up until the close of the nineteenth century. If, as these sources suggest, the ossuaries are associated with ancestor worship, it is interesting to note that many Maya still believe that ancestors live within the Earth.

CAVE MODIFICATIONS

Almost all Maya caves show extensive breakage of speleothems. Often broken formations have been moved or taken from the cave, and speleothems are frequently reported in cultural deposits at surface sites. Modern indigenous terms for speleothems identify them as congealed water or dripping water turned to stone. Because water is so essential to agriculture, speleothems are directly related to fertility, and today they are placed on household altars. Similar ideas were probably held by the Classic Maya.

The Maya deployed various types of architectural constructions to modify space within caves. Among the most common are stone walls defining enclosures. Walls with low doorways intentionally impede access by forcing visitors to enter on hands and knees. Retaining walls and dirt fill were used to create level spaces. Plaster floors and stone pavements also demarcated special areas. Altars, often nothing more than a crude stack of stones, are frequently reported (Fig. 2). Stairways leading down into caves appear most frequently in Yucatán and Quintana Roo. Dams have been reported in several caves with seasonal flooding. Finally, small architectural structures, probably reserved for the most private rituals, were occasionally built inside caves.

FIGURE 2   Altar-like construction from Actun Kabal, Belize. *Photo by Andrea Stone.*

CAVE ART

With about 60 known decorated caves, the Maya Lowlands have one of the most important cave art traditions in the world. Pigment-based art consists of paintings, drawings, and both positive and negative hand- and footprints, often utilizing charcoal and clays available in the cave. However, prepared mineral pigments, such as red (from hematite), blue, and yellow occur in rare instances. The most typical carvings, mainly found in Yucatán, Quintana Roo, and Belize, are deeply engraved petroglyphs showing frontal faces, meanders, and geometric elements. Another category of sculpted art is the modified speleothem, typically shaft-shaped speleothems with crudely carved faces (Fig. 3). These may represent spirit beings embodied in stone. Artistic modification of speleothems was widely practiced among the Classic Maya. Rare examples of painted petroglyphs and sculptures modeled from clay are also known.

Although the sculpted art tends to be crude, some of the painted art is refined and resembles elite art. This is epitomized by the paintings at Naj Tunich, Guatemala, dating to the seventh and eighth centuries. Both hieroglyphic inscriptions and depictions of human figures, which evince elite cave utilization, are present. The hieroglyphs reveal that political relations among regional sites were mediated by ritual cave use and that some caves, such as Naj Tunich, held higher status than others as regional pilgrimage destinations. The painted cave art has been the subject of technical analysis including AMS radiocarbon dating and multispectral imaging.

ETHNOHISTORY AND ETHNOGRAPHY

Colonial documents and modern ethnographic studies of the Maya aid in the reconstruction of Classic

Maya cave use. An important colonial source is the *Popol Vuh*, a K'iche' Maya text recounting the adventures of supernatural twins in an underworld realm called Xibalba, "place of fright," a thinly veiled allusion to a cave. The *Popol Vuh* reveals thematic associations of caves with danger, regeneration, and the ballgame.

In indigenous cosmology the first appearance of humans on the face of the Earth is from a cave, the womb of the Earth. Thus, caves are the primary symbol of human and world creation. Not surprisingly, caves play an important role in conceptions of sacred geography, definitions of territorial boundaries, and pilgrimage routes in today's Maya communities and probably did so in the past.

As penetrations into the sacred Earth, caves are the residences of ancestors and supernatural beings. The most important indigenous deity in the Maya Highlands today is known by a name that translates as "Hill-Valley." He/she controls the fruits of the Earth and is petitioned in caves. The cave's association with water is equally important among contemporary groups, many of whom believe that storms emanate from caves. In dry areas, such as Yucatán, the presence of a cave or cenote influenced the location of settlements. Cave water was ritually pure and in Yucatán was called *suhuy ha* or "virgin water." Archaeological evidence of ritual activity near pools of water in caves and the placement of ceramic vessels and stone troughs under drips, often in remote areas, are testimony to the antiquity of this belief.

CAVES AND COMMUNITY

The identity of Maya communities with caves is so close that the cave often gives its name to the community. In colonial rituals at the foundation of a new community, the presence of a cave was considered essential and lent authority to settlers' claims to occupy and govern a territory. Because the cave symbolized political legitimacy in Mesoamerica, it is not

FIGURE 4 Artificial cave dug in volcanic ash under the central plaza of the Pre-Columbian settlement of Utatlán, Guatemala. This cave is an important religious site among contemporary highland Maya. *Photo by James Brady.*

FIGURE 3 Modified speleothem from a cave in Guatemala. *Drawing by Anne Chojnacki.*

surprising that the pre-Hispanic Maya elite incorporated caves into their ritual activity and their elaborate architecture. Caves have been found to underlie building complexes, including pyramids, at such sites as Chichen Itza, Mayapan, Tulum, Polol, and Dos Pilas.

Caves were so important that, in the nonkarstic area of the Maya Highlands where caverns do not do occur naturally, artificial caves were excavated, often beneath pyramids or other architecture. Such tunnels have been found at the sites of Utatlán, Mixco Viejo, Zaculeu, Iximche, and La Lagunita in Guatemala (Fig. 4). The fact that artificial caves have also been reported at the Central Mexican sites of Teotihuacán, Xochicalco, and Acatzingo Viejo suggests that caves were a fundamental concern throughout Mesoamerica.

See Also the Following Articles

Ancient Cavers in Eastern North America
Cave Dwellers in the Middle East

Bibliography

Brady, J. E. (1997). Settlement configuration and cosmology: The role of caves at Dos Pilas. *American Anthropology, 99*(3), 602–618.

Brady, J. E. (1999). *Studies in Mesoamerican cave use. Sources for the study of Mesoamerican ritual cave use* (2nd ed. (Publication 1)). Los Angeles: California State University.

García-Zambrano, A. J. (1994). Early colonial evidence of Pre-Columbian rituals of foundation. In M. G. Robertson & V. Field (Eds.), *Seventh Palenque Round Table, 1989* (pp. 217–227). San Francisco, CA: Pre-Columbian Art Research Institute.

McNatt, L. (1996). Cave archaeology of Belize. *Journal of Cave Karst Studies, 58*(2), 81–99.

Stone, A. J. (1995). *Images from the underworld: Naj Tunich and the tradition of Maya cave painting.* Austin: University of Texas Press.

Stone, A. J. (1997). Pre-Columbian cave utilization in the Maya area. In C. Bonsall & C. Tolan-Smith (Eds.), *The human use of caves (BAR International Series 667)* (pp. 201–206). Oxford, England: Archaeopress.

Thompson, J. E. S. (1975). Introduction to the reprint edition. In H. C. Mercer (Ed.), *The hill-caves of Yucatán.* Norman: University of Oklahoma Press.

Vogt, E. Z. (1981). Some aspects of the sacred geography of Highland Chiapas. In E. P. Benson (Ed.), *Mesoamerican sites and world views* (pp. 119–142). Washington, DC: Dumbarton Oaks Research Library and Collection.

MICROBES

Annette Summers Engel

University of Tennessee

INTRODUCTION

The Earth's continental subsurface contains abundant and active microbial biomass that exceeds the biomass found in other parts of the biosphere (*e.g.*, Whitman *et al.*, 1998). Reactive interfaces among air, water, and rock offer suitable habitats for microbial activity, including within cave and karst landscapes. As such, cave and karst environments serve as reservoirs for diverse microbes over extended periods of time. This article describes the history of microbiological research in karst, current methodological approaches being used to understand microbial diversity and ecology, as well as the roles of microbes in important geologic and geochemical subsurface karst processes.

MICROBIAL DIVERSITY AND METABOLISM

Microorganisms in cave and karst settings have varying metabolism. Microbial metabolism refers to how a microbe obtains carbon, along with other nutrients, and gains energy (Fig. 1). Carbon for cellular growth originates by either converting inorganic carbon (CO_2, HCO_3^-) to organic carbon as an *autotroph*, or assimilating organic carbon initially produced by autotrophs. *Heterotrophs* use already existing organic carbon for cellular energy and their carbon sources. The physiological mechanisms for capturing chemical energy are diverse, and the distinction between a chemosynthetic

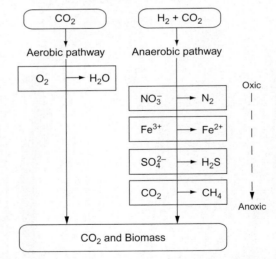

FIGURE 1 Schematic representation of energy-yielding metabolic reactions. Aerobic microorganisms, such as iron-oxidizing bacteria, use dissolved oxygen (O_2) to generate energy. In contrast, anaerobes use alternative terminal electron acceptors. For methanogenesis, which occurs under the most reducing conditions, CO_2 is the electron acceptor. CO_2 is required for all chemolithoautotrophic metabolic reactions, and H_2 is required by most anaerobes. Aerobic and anaerobic metabolism can occur simultaneously, although there is usually spatial and temporal separation of the reactions within a microbial habitat. For instance, aerobic microorganisms colonize uppermost portions at the air-water interface, while anaerobes occupy inner regions of a mat or water-sediment interface.

and a photosynthetic organism is based on whether the initial source of energy is from inorganic chemicals (*litho*) or light (*photo*). Microbes that gain energy through chemosynthesis and fix inorganic carbon are *chemolithoautotrophs* (literally "self-feeding rock-eaters"). During chemosynthesis, microorganisms gain energy by transferring electrons from one chemical (electron donor) to another (electron acceptor) that can come from groundwater or colonized rock and mineral surfaces. Chemical electron donors can include, but are not limited to, molecular hydrogen or reduced sulfur compounds; organic molecules (*organo*), such as acetate or formate, can also be used. Chemolithoautotrophs can also grow if organic carbon is present as *mixotrophs*, in which both chemolithoautotrophy and heterotrophy are expressed simultaneously. However, an organism is no longer classified as a chemolithoautotroph if organic compounds are used to gain energy or for a carbon source. Organisms that gain cellular energy from chemical transformations but use organic carbon compounds for their carbon source are *chemoorganotrophs*.

Microorganisms are well suited to live in geochemically or mineralogically extreme environments. Extremophiles can be adapted to live in conditions that are nutrient poor (*oligotrophic*), low pH (*acidophilic*), high pH (*alkaliphilic*), high temperature (*thermophilic*) or low temperature (*psychrophilic*), or sulfide-rich (*sulfidophilic*). Classification of microbial metabolism is also based on oxygen requirements and whether a microbe respires aerobically, anaerobically, or ferments, all of which relates to electron acceptor utilization (Fig. 1). Microorganisms that require oxygen are *aerobes*, and oxygen serves as the terminal electron acceptor for metabolic processes yielding energy, such as from the oxidation of reduced sulfur compounds (*e.g.*, hydrogen sulfide, thiosulfate), iron oxidation, or ammonia oxidation. In more reducing environments, where organic carbon is rapidly consumed, microbes that do not require oxygen (*anaerobes*) use alternative electron acceptors for respiration in a sequence of energetic, reduction reactions that occur along thermodynamic (and redox) gradients. After oxygen, electron acceptors are reduced in the following order: $NO_3^- \rightarrow Mn^{4+} \rightarrow Fe^{3+} \rightarrow SO_4^{2-} \rightarrow CO_2$.

Most anaerobes are obligate, but *microaerophilic* organisms require low oxygen concentrations, while *facultative* organisms grow in the presence or absence of oxygen, using different, and the most energetic, electron acceptors available. Some important anaerobic, redox reactions include nitrate or dissimilatory nitrogen reduction ($NO_3^- \rightarrow N_2$), ferric iron reduction ($Fe^{3+} \rightarrow Fe^{2+}$), dissimilatory sulfate reduction ($SO_4^{2-} \rightarrow H_2S$), and methanogenesis ($CO_2 \rightarrow CH_4$). Microorganisms involved in these processes can grow as chemolithoautotrophs or chemoorganotrophs. There are several other energetically favorable pathways that occur in the absence of oxygen, including sulfide oxidation via nitrate reduction and anaerobic ammonium oxidation (anammox).

HISTORY OF MICROBIOLOGY STUDIES

The earliest microbiological studies from caves used microscopy and culturing for specific metabolic groups. Microbes were found to be prevalent in sediments and water. By the early 1900s and through the 1940s, many cave deposits, such as cave nitrates (saltpeter or saltpetre) (*e.g.*, Faust, 1949), carbonate moonmilk, and other speleothems, were considered to have a microbiological origin. Chemolithoautotrophic, iron- and sulfur-oxidizing bacteria were identified and sulfur-oxidizing bacteria were implicated in the sulfuric acid-promoted dissolution of limestone (*e.g.*, Principi, 1931). Incidentally, it has only been within the past 5–10 years that modern methodology has been able to prove that some of these deposits and speleogenetic processes are the result of microbial activity.

By the mid-1960s, the ecological and geological roles of microbes in caves started to come into focus (*e.g.*, Caumartin, 1963; Symk and Drzal, 1964). However, based on what was known of microbial diversity by the 1960s through the 1980s, which was predominantly from culture-based approaches, microbial communities in the subsurface were considered to be a subset of surface, such as soil, communities that were flushed underground by meteoric drip waters, surface streams, or air currents, or were carried by animals into caves. All life in caves was thought to be dependent on photosynthetically produced organic matter, and therefore, life in caves was considered to be nutrient-limited. Consequently, the primary ecological function of microbes in caves was as heterotrophs, that degraded organic matter and were food sources for higher organisms. Because of oligotrophic conditions, microbial biomass was considered to be low and insufficient to impact most geological or geochemical processes, except in specialized habitats such as in sulfidic aquifers or metalliferous sediments. There was some recognition that laboratory enrichment and strain isolation by culturing was perhaps a problem because culturing selected for fast-growing rather than slow-growing groups, and most cave habitats were found to be oligotrophic.

Microbial primary production based on chemosynthesis, rather than photosynthesis, was introduced in the late 1980s from the discoveries of the deep-sea hydrothermal vents. This research coincided with the development and application of molecular genetics techniques in the 1980s and 1990s, and provided timely release from the earlier, one-sided ecological views of microbes in ecosystems. For cave and karst research, work in the early to mid-1990s from the Movile Cave, Romania (Sarbu et al., 1996), brought the significance of microbes in chemosynthetically based ecosystems into focus. Since then, it is clear that microorganisms can have tremendous biomass and metabolic diversity in the subsurface, and that many of the metabolic processes are not dependent on sunlight.

Cave and karst microbiological studies have kept pace with technology, and are now based on applying geochemical models to genetic diversity data resulting from automated DNA sequencing, high-throughput computing analyses of environmental rRNA and functional gene sequence phylogenies, or from oligonucleotide rRNA probing, microautoradiography, and stable-isotope probing. All of these methods evaluate biochemical diversity and the evolution of genes, enzymes, and metabolites (e.g., Hutchens et al., 2004; Chen et al., 2009; Dattagupta et al., 2009). Essentially, research efforts over the past 40 years have gone from answering the basic question, "Who's there?," to a more difficult question to answer, "What are they doing?"

METHODOLOGICAL APPROACHES

Techniques are rapidly changing and being improved. Virtually any method has some type of error associated with it, which should be considered when interpreting results. The approaches described herein are a sampling of possible strategies that could be used to understand "Who's there?" and "What are they doing?" Without question, a combination of different methods, while not always feasible, still provides the best results.

Culturing and Metabolic Assays

Culturing involves growing microbes in the laboratory by providing cells with all the essential nutrients needed for growth, including electron donors and acceptors. Even when the metabolic requirements are known (or guessed), however, enrichment and pure culture isolation are difficult. In fact, most microorganisms in nature have not been grown in culture, and it has been estimated that less than 1% of known microbes are culturable with current techniques. Standard culturing often introduces a selective bias toward microorganisms that are able to grow quickly or to utilize substrates provided in the medium more efficiently. Nevertheless, culturing, although highly selective and difficult for unknown or poorly characterized microorganisms, is still a reliable way to determine metabolic capabilities of certain microbial groups. This method is impractical for diverse communities, as it is prohibitive to attempt to grow all possible microbial metabolic groups that may be present in an ecosystem.

If pure or mixed cultures are obtained, then microorganisms can be characterized or identified based on a variety of growth-dependent assays. Biochemical assays can also identify activities in natural samples without cultures. Assays include, but are not limited to, measuring the presence or absence of enzymes involved in degradation of substrates in the growth medium, production of gases (e.g., H_2, CO_2, H_2S), or the utilization of specific carbon substrates, and determining the ability to grow under different physicochemical conditions, such as growth over a range of pH values or temperatures, or even under varying oxygen concentrations. For many organisms, a few key assays are all that is required for identification or characterization. Some more sophisticated assays include measuring enzymatic activity or membrane lipids such as phospholipid-derived fatty acids.

Molecular Methodologies

Molecular techniques that detect nucleic acids can circumvent culture-based problems and provide a culture-independent assessment of which microorganisms are present in a community based on the evolutionary relationships of gene sequences. To characterize microbial taxonomy, most researchers have evaluated the 16S rRNA gene sequence diversity using Sanger-sequencing methods. This approach achieves varying levels of success depending on how many sequences are generated from a sample and the community diversity; sample coverage is generally high for low diversity materials, but samples having high microbial diversity can have very low coverage. Resulting gene sequences are evaluated in a phylogenetic framework to categorize groups. An alternative metagenomics approach has been developed recently that allows for evaluating exceptionally high volumes of gene sequences by using next-generation pyrosequencing chemistries, including an approach referred to as *454 sequencing* that involves high-throughput sequencing platforms (e.g., Roche/454 Life Sciences) (Marsh, 2007).

There are numerous well-known disadvantages to these methods, including the original DNA extraction methods that may bias microbial nucleic acid representation in a sample. Other disadvantages are because the methods rely on the polymerase chain reaction (PCR) to amplify sequences from DNA, so PCR inhibitors within a sample, genomic complexity in a sample, primer choice, chimera generation, and the interpretation of multiple gene copies, can all affect the experimental results. Probably the most significant disadvantage is that the phylogenetic affiliations that can be assigned are based on what is already known. If a unique habitat or microbial community is examined, then phylogenetic assignments of unknown microbes may be weak or incorrect. Another significant disadvantage, especially in terms of understanding the metabolisms of a microbial population, is that phylogeny rarely equals function. So, metabolic activity must be inferred, if not tested separately, from additional ecological, culturing, or molecular methods.

In recent years, and based on whole genome comparative analyses from an ever-increasing number of microbes, the characterization of functional genes is becoming more important. If genes are sequenced from DNA, then this allows for an evaluation of the potential for a specific pathway to be done by certain microbes. However, if genes are sequenced from mRNA, then this allows for an evaluation of the transcripted information, or more specifically, an evaluation of gene expression. This transcriptomics approach can obtain key functional gene information for a community without inferring metabolism from other DNA-based methods.

Microarrays, which are microchips with attached nucleotide probes, can also be used to study metabolic activity and microbial diversity in complex systems (Guschin et al., 1997). Probes can be designed for functional genes, as well as for species-, genera-, group-, or domain-specific targets. The number of probes used in an array is only limited by the size of the microchip. Microarray methods circumvent PCR-based problems, and can be done with DNA to examine gene presence or with mRNA to examine transcripted activities. Although the utility of microarrays is seemingly limitless, at present there are no cave or karst studies incorporating this technique.

Stable Isotope Ratio Analyses and Labeled Substrate Experiments

Microbial activity and ecosystem energetics in natural or cultured samples can be evaluated from ratios of stable isotopes in the sample compared to the surroundings, or from using labeled substrates, either radioisotopes or stable isotopes, such that both carbon fixation and energy-conserving metabolism can be quantified. These experiments can provide activity rate estimates without knowing how many organisms are in the sample, and allow a sensitive, quantitative assessment of biogeochemical transformations in nature (Boschker and Middelburg, 2002).

Stable isotope ratio analysis methods depend on the creation and alteration of compounds by biotic and abiotic processes. For example, the two carbon isotopes of importance in studying carbon cycling are ^{12}C and ^{13}C. Incorporation of carbon into living tissues invokes significant kinetic isotope fractionation, such that biological processes discriminate for the lighter isotope (^{12}C), leaving behind the heavier isotope (^{13}C). Discrimination for ^{12}C is due to kinetic effects caused by the irreversible enzymatic CO_2-fixing reaction that assimilates CO_2 into the carboxyl group of an organic acid. Differences in the isotopic composition are expressed in terms of the delta (δ)-notation of a ratio of $^{13}C/^{12}C$ in a sample relative to a standard, measured in per mil (‰). Biogenic carbon is isotopically lighter than the inorganic reservoir, such as atmospheric CO_2 or dissolved bicarbonate (HCO_3^-), and autotrophic fixation pathways have some of the largest fractionation effects during biological assimilation of carbon. The overall isotopic fractionation is determined by subtracting the carbon isotopic composition of an organism from the isotopic composition of the carbon reservoir, with most autotrophic fractionation values ranging between -20 and -40‰. Excretion, respiration, and heterotrophic carbon cycling are (for the most part) considered negligible processes for carbon fractionation, and the isotopic composition of heterotrophic organic matter will be the same or slightly higher than the source organic carbon. Due to the diversity in carbon-fixation pathways, as well as the level of recycling within a microbial detrital loop, there can be a wide range in carbon isotope values within an ecosystem.

The rate of certain metabolic processes can be determined by measuring the uptake of one or more compounds over time and the subsequent buildup of other compounds, such as metabolic by-products. Electron acceptors are the most commonly radiolabeled substrates. For instance, sulfate reduction can be monitored using radiolabeled $^{35}SO_4^{2-}$ compounds, whereby loss of $^{35}SO_4^{2-}$ would indicate consumption, and a gain of $^{35}S^{2-}$ as the by-product of sulfate reduction would provide sufficient evidence that sulfate reduction is occurring. Carbon fixation rates can also be determined by labeling carbon molecules, such as bicarbonate or acetate, with ^{14}C, and the evolved $^{14}CO_2$ and ^{14}C-biomass are measured. Even when multiple radiolabeled substrates are combined in an experiment, however, there are several problems with carbon radioisotope studies, with the most significant being how to decipher the results. Carbon fixation or heterotrophic consumption of organic matter can be overestimated or underestimated due to sampling perturbation or poor understanding of the community structure. Because natural samples can be extremely sensitive to disturbance, the actual experiment may skew results and demonstrate that one metabolic process is dominant, when in fact, that process is just less sensitive to disturbance than other processes. Additionally, some microbes in a sample could be capable of facultative or mixotrophic growth, acting as chemolithoautotrophs in nature, but when provided with a rich organic carbon molecule, they will use it as chemoorganotrophs or heterotrophs. Detailed culturing or phylogenetic analyses of the microbial communities should be coupled to radioisotope studies to aid interpretation. Recovering and sequencing the DNA or RNA from the experimental biomass that incorporated the labeled substrates can be done to investigate the active microorganisms of interest as well.

New Uses for Microscopes

Cell morphologies and structural relationships between cells and inorganic substrates can be visualized directly by light microscopy or transmission or scanning electron microscopy (TEM or SEM, respectively). But, although microscopy can provide a wealth of information, such as microbial mat structure and the nature of biomineralization, microscopy is not a reliable technique to identify microbes, even with morphologically conspicuous cells. General microscopy can quantify biomass, but does not provide any information about metabolism, although there are basic stains or dyes that can identify actively metabolizing cells. The best use of microscopy is to combine it with geochemical, ecological, and microbiological methodology, such as fluorescence *in situ* hybridization (FISH). FISH probes are similar to those used in microarrays, with a majority of them being based on 16S rRNA targets. This method combines fluorescence microscopy with oligonucleotide fluorescent probes. For low abundance or low rRNA content, catalyzed reporter deposition (CARD) combined with FISH can enhance signal fluorescence for group quantification. FISH combined with microautoradiography (FISH-MAR) identifies not only the microbe, but also what the microbe may be doing by allowing for the incorporation of a labeled substrate (Ouverney and Fuhrman, 1999). New methods have been developed that utilize a high-resolution ion microprobe using secondary ion mass spectroscopy (SIMS) technique combined with FISH. At the single-cell level, this approach simultaneously detects the probe hybridization to inform of phylogeny, as well as the isotopic signal from biomass incubated with labeled substrate (Li *et al.*, 2008).

MICROBIAL DIVERSITY OF CAVES AND KARST

Current research from karst attempts to understand the abundance, diversity, and activities of microbial communities, their ecosystem function, the connection between microbes and geological or geochemical processes. This section highlights some of the implications of microbial diversity and distribution in karst systems, based primarily on the retrieval of 16S rRNA gene sequences retrieved from caves and karst. From about 20 years of data (*e.g.*, Engel and Northup, 2008; Engel, 2010), roughly a third of the recognized bacterial phyla have been identified from caves, and slightly less than half of the recognized archaeal phyla (Fig. 2). Most of what is known of bacterial and archaeal diversity is from just a few systems worldwide, and from focusing on specific types of caves or microbial groups (Engel and Northup, 2008). The diversity of microbial eukaryotes is limited to a few examples for each major group (Fig. 2). We know much more about these organisms from systematics rather than genetics, despite the fact that some, such as fungi and algae, can form structures seen by the naked eye (Fig. 3).

From these different molecular and culture-based studies, the abundance and distribution of microbes in caves and karst settings have to do with, at a fundamental level, how a habitat originated and the source(s) of formational fluids, including epigenic (*i.e.*, sourced from the surface) or hypogenic (*i.e.*, sourced from depth). In general, epigenic systems have been studied more intensely, likely because these are the most common cave type (*e.g.*, Palmer, 2007). Moreover, each major habitat zone (*i.e.*, entrance, twilight, and dark zone) has general physicochemical and nutrient conditions that control the types of organisms present. For instance, dry, air-filled passages can have nearly constant temperatures and stable geochemical conditions, but cave streams or underwater passages are directly influenced by hydrology that can cause seasonal or even episodic habitat variability. Subtle physicochemical variability, speciation and availability of redox-sensitive elements, and the types and loading of organic carbon and other nutrients to a zone, further influence the diversity and metabolism of (micro) organisms that will colonize cave and karst habitats. The juxtaposition of stable physicochemical conditions with extreme geochemistry and physicochemical conditions has provided interesting microbiological research avenues (Barton and Northup, 2007).

Probably the most recognizable macroscopic microbiological feature on moist cave walls in epigenic caves are shiny gold, white, pink, or brown patches of large colonies of filamentous Actinobacteria, commonly and historically referred to as *actinomycetes* (Fig. 4A,B,C). Water condenses onto the cells, which are more hydrophobic than the cave walls, thereby causing a sparkly appearance. Humid lava tubes also contain Actinobacteria, as well as members of the phyla Chloroflexi, Verrucomicrobia, and Proteobacteria. The phylum Actinobacteria includes the genera *Streptomyces*, *Mycobacterium*, and *Nocardia*, and some species can be pathogenic. Some Actinobacteria produce secondary metabolites that are useful antibiotic and antimicrobial agents. One secondary metabolite of *Streptomyces* is geosmin, the substance that causes the earthy smell common to many caves.

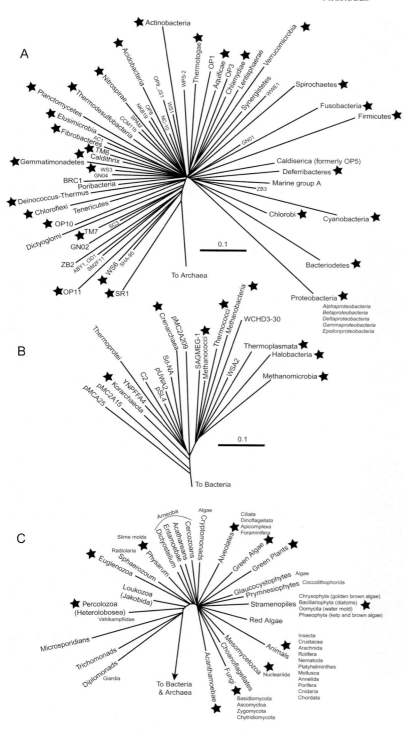

FIGURE 2 Based on Engel (2010), and the representation of microbial phyla, indicated by stars (★), from cave and karst systems within the domains: (A) bacteria, (B) archaea, and (C) eukaryota. The bacterial and archaeal topologies are based on trees created using the NAST alignment of 16S rRNA gene sequences from Greengenes (DeSantis et al., 2006). Bars indicate changes per nucleotide. The schematic phylogeny for eukaryotes is based on the topology from 18S rRNA gene sequences, presented by Dawson and Pace (2002).

Another large group of microbes described from cave and karst settings belongs to the phylum Proteobacteria, which is subdivided into classes, including the *Alphaproteobacteria*, *Betaproteobacteria*, *Gammaproteobacteria*, *Deltaproteobacteria*, and *Epsilonproteobacteria* (Fig. 2). Iron bacteria from the *Betaproteobacteria*, including the chemolithoautotrophic *Gallionella* spp. and heterotrophic *Leptothrix* spp., are common in the allochthonous sediments of noncarbonate rocks in many epigenic caves; the sediments can contain higher ferromanganese content compared to the carbonate host rock. Iron- and manganese-oxidizing bacteria have also been identified from corrosion residues on cave-wall surfaces from Lechuguilla Cave and other caves in the Guadalupe Mountains of West Texas and New Mexico. Among the diverse bacteria, the ferromanganese deposits also consist of Crenarchaeota and Euryarchaeota, which have also

FIGURE 3 (A) Entomopathogenic fungi on *Triphosa dubitata*; moth ~2.5 cm long, Planinska Jama, Slovenia. (B) Fungi growing on organic debris; acorn cap for scale, approximately 2 cm across, Sandy Cave, Kentucky. (C) Algae, cyanobacteria, and other phototrophic organisms growing in the entrance zone.

been retrieved from paleofill sediment deposits in Wind Cave in South Dakota. Poorly crystalline manganese oxide and hydroxide minerals in caves have been attributed to microbial processes. Encrusted sheaths and stalks resembling modern iron-oxidizing bacterial cells have been seen from inside stalactites, known as *rusticles*, as well as within lithified sediments, thick mineral crusts, and boxwork from numerous caves, including those in the Black Hills, South Dakota.

Proteobacteria are also prevalent in active hypogenic caves and karst systems, especially sulfidic caves (Engel, 2007). The classes most commonly identified in sulfidic systems include the *Epsilonproteobacteria*, *Gammaproteobacteria*, and *Betaproteobacteria*, which consist of many different types of sulfur-oxidizing bacteria that are essential to the development of chemolithoautotrophically based ecosystems. These bacteria form the bulk of paper-thin, floating microbial mats and thick, filamentous microbial mats in streams and pools in active caves (Fig. 5A,B,C).

They may have played a role in the formation of the biogenic speleothem pool fingers found from caves such as Lechuguilla Cave and Carlsbad Caverns among others. In active sulfidic caves, these microbes contribute to sulfuric acid speleogenesis. Cave-wall surfaces are another microbial habitat in active sulfidic caves (Fig. 5D). Cave walls can be acidic, due in part to gypsum mineralization and microbial activity, and have condensation droplets of pH <2. Interestingly, the phylogenetic diversity of cave-wall biofilms is similar from active caves around the world, consisting of various members of the Proteobacteria, Actinobacteria, and bacterial candidate divisions, especially TM6, as well as filamentous fungi and protists.

Only recently has the microbial diversity of vermiculations, also known as *biovermiculations*, been investigated (Fig. 4D). These features are broad areas of discontinuous, geometric deposits of organic matter intermixed with clay on cave-wall surfaces, and are found in many different types of caves, including lava tubes, as well as mines. Since the 1960s, they have been considered (micro)biological. The dominant bacteria in biovermiculations from the sulfidic Frasassi Caves, Italy, include *Betaproteobacteria*, *Gammaproteobacteria*, as well as members of the phyla Acidobacteria, Nitrospirae, and Planctomycetes (Jones et al., 2008). Mites and nematodes have also been described. The causes of the geometric structure of biovermiculations are still mysterious, but the patterning is being used by astrobiologists as a way to identify life processes.

Knowing the phylogenetic composition of microbial communities is useful to understand what microbes may be doing in, and to, caves, including being able to determine the health of a cave ecosystem from the microbial community composition. But the problem comes in interpreting the presence and absence of microbial groups in communities. Does absence mean that the microbes are not present in a system, or that the sampling and analytical methods did not detect them? Moreover, there is a big interpretation leap from the information gained from DNA, and specifically 16S

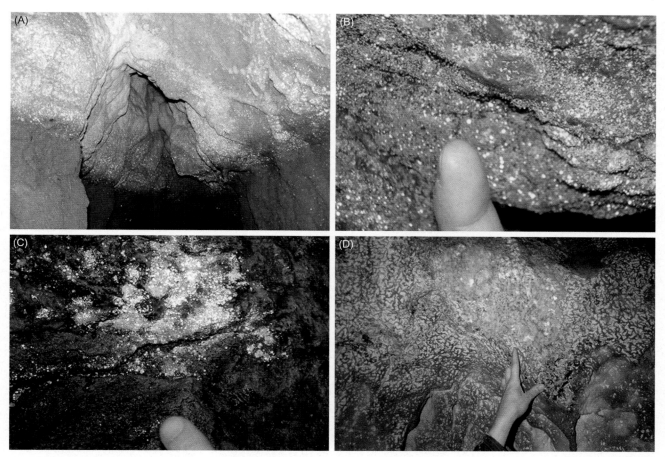

FIGURE 4 (A) Gold and white actinomycete colonies forming in the upper portion of a cave passage, approximately 1 m across, that is not submerged by rising water, Planinska Jama, Slovenia. (B) Close-up of the colonies from (A); water droplets condense on the hydrophobic surfaces of the colonies. (C) White actinomycete colonies with condensation, Cascade Cave, Kentucky. (D) Clay vermiculations on limestone cave wall, being covered in gypsum, Frasassi Caves, Italy.

rRNA genes, to knowing the metabolic roles of most microbes. Attempts to evaluate this dilemma have yielded mixed results. Recent investigations of hydrologically connected caves in different karst basins have indicated that microbial communities in cave streams and the karst aquifers under low-flow conditions include members of the *Deltaproteobacteria*, *Acidobacteria*, and *Nitrospirae*, but high-flow conditions following storm events have more representation from common soil microbial groups, including members of the phyla Firmicutes, Bacteriodetes, and Chlorobi (*e.g.*, Pronk *et al.*, 2009). However, genetically similar 16S rRNA gene sequences to these potential disturbance groups, along with members of the *Betaproteobacteria*, have been considered indigenous communities in pristine and low-impact (from tourism) cave sediments, pristine cave pool waters, and carbonate moonmilk deposits. To complicate matters, the same betaproteobacterial groups have been found in high abundance from high-impact (from tourism) sediments (Ikner *et al.*, 2007). In short, these mixed results highlight that it may be possible to determine the composition of an endemic subsurface microbial community, as long as it is possible to distinguish which microbes may indicate disturbance. This means that broader spatial and temporal distribution studies need to be done, including evaluation of overlying soil, drip waters, air, and so on, as well as studies to understand the metabolism of active microbial groups because the presence of specific DNA sequences does not necessarily mean that the microbes are metabolically active and affecting the system geologically, geochemically, or ecologically.

THE FUTURE

There have been important advances in our understanding of the diversity and roles of microbes in caves and karst, but many challenges wait. It is important that we determine not just the overall habitat biodiversity from many more caves and karst settings, but that

FIGURE 5 (A) Floating microbial mats in Movile Cave, Romania; PVC grids are 10 cm × 10 cm. (B) White filamentous microbial mats in the Frasassi Caves, Italy. (C) White mats in Lower Kane Cave, Wyoming. (D) Cave-wall biofilms on gypsum with condensation droplets with pH <2, Lower Kane Cave, Wyoming. Scale bar is 5 cm.

future research addresses "What are they doing?" and "Where do they come from?" More data are needed to test distribution and dispersal hypotheses, and hypotheses related to how microbes may adapt (or have adapted) to the physicochemical conditions of karst. Is it possible that microbes are endemic to a cave or, more generally, the subsurface? Once these research directions take shape, and we know more about the microbial communities at the base of karst ecosystems, then it should be possible to better protect and conserve the natural habitats through time.

See Also the Following Articles

Chemoautotrophy
Cave Ecosystems

Bibliography

Barton, H. A., & Northup, D. E. (2007). Geomicrobiology in cave environments: Past, current and future perspectives. *Journal of Cave and Karst Studies*, 69(1), 163–178.

Boschker, H. T. S., & Middelburg, J. J. (2002). Stable isotopes and biomarkers in microbial ecology. *FEMS Microbiology Ecology*, 1334, 1–12.

Caumartin, V. (1963). Review of the microbiology of underground environments. *Bulletin of the National Speleological Society*, 25(1), 1–14.

Chen, Y., Wu, L., Boden, R., Hillebrand, A., Kumaresan, D., Moussard, H., Baciu, M., Lu, Y., & Murrell, J. C. (2009). Life without light: Microbial diversity and evidence of sulfur- and ammonium-based chemolithotrophy in Movile Cave. *The ISME Journal*, 3(9), 1093–1104.

Dattagupta, S., Schaperdoth, I., Montanari, A., Mariani, S., Kita, N., Valley, J. W., & Macalady, J. L. (2009). A novel symbiosis between chemoautotrophic bacteria and a freshwater cave amphipod. *The ISME Journal*, 3(8), 935–943.

Dawson, S. C., & Pace, N. R. (2002). Novel kingdom-level eukaryotic diversity in anoxic environments. *Proceedings of the National Academy of Sciences*, 99(12), 8324–8329.

DeSantis, T. Z., Hugenholtz, P., Larsen, N., Rojas, M., Brodie, E. L., Keller, K., Huber, T., Dalevi, D., Hu, P., & Andersen, G. L. (2006). Greengenes, a chimera-checked 16S rRNA gene database and workbench compatible with ARB. *Applied and Environmental Microbiology*, 72(7), 5069–5072.

Engel, A. S. (2007). Observations on the biodiversity of sulfidic karst habitats. *Journal of Cave and Karst Studies*, 69(1), 187–206.

Engel, A. S. (2010). Microbial diversity of cave ecosystems. In L. Barton, M. Mandl & A. Loy (Eds.), *Geomicrobiology: Molecular & environmental perspectives* (pp. 219–238). Springer. doi:10.1007/978-90-481-9204-5_10.

Engel, A. S., & Northup, D. E. (2008). Caves and karst as model systems for advancing the microbial sciences. In J. Martin & W. B. White (Eds.), *Frontiers in karst research* (pp. 37–48). Leesburg, VA: Karst Waters Institute (Special Publication 13).

Faust, B. (1949). The formation of saltpeter in caves. *Bulletin of the National Speleological Society*, 11, 17–23.

Guschin, D. Y., Mobarry, B. K., Proudnikov, D., Stahl, D. A., Rittmann, B. E., & Mirzabekov, A. D. (1997). Oligonucleotide microchips as genosensors for determinative and environmental studies in microbiology. *Applied and Environmental Microbiology*, 63(6), 2397–2402.

Hutchens, E., Radajewski, S., Dumont, M. G., McDonald, I. R., & Murrell, J. C. (2004). Analysis of methanotrophic bacteria in Movile Cave by stable isotope probing. *Environmental Microbiology*, 6(2), 111–120.

Ikner, L. A., Toomey, R. S., Nolan, G., Neilson, J. W., Pryor, B. M., & Maier, R. (2007). Culturable microbial diversity and the impact of tourism in Kartchner Caverns, Arizona. *Microbial Ecology*, 53(1), 30–42.

Jones, D., Lyon, E., & Macalady, J. (2008). Geomicrobiology of biovermiculations from the Frasassi Cave System, Italy. *Journal of Cave and Karst Studies*, 70(2), 76–93.

Li, T., Wu, T. -D., Mazéas, L., Toffin, L., Guerquin-Kern, J. -L., Leblon, G., et al. (2008). Simultaneous analysis of microbial identity and function using NanoSIMS. *Environmental Microbiology*, 10(3), 580–588.

Marsh, S. (2007). Pyrosequencing applications. *Methods in Molecular Biology*, 373, 15–24.

Ouverney, C. C., & Fuhrman, J. A. (1999). Combined microautoradiography-16S rRNA probe technique for determination of radioisotope uptake by specific microbial cell types in situ. *Applied and Environmental Microbiology*, 65(4), 1746–1752.

Palmer, A. N. (2007). *Cave geology*. Dayton, OH: Cave Books.

Principi, P. (1931). Fenomeni di idrologia sotterranea nei dintorni di Triponzo (Umbria). *Grotte d'Italia*, 5, 1–4 (in Italian).

Pronk, M., Goldscheider, N., & Zopfi, J. (2009). Microbial communities in karst groundwater and their potential use for biomonitoring. *Hydrogeology Journal*, 17(1), 37–48.

Sarbu, S. M., Kane, T. C., & Kinkle, B. K. (1996). A chemoautotrophically-based cave ecosystem. *Science*, 272(5259), 1953–1955.

Symk, B., & Drzal, M. (1964). Research on the influence of microorganisms on the development of karst phenomena. *Geographia Polonica*, 2, 57–60.

Whitman, W. B., Coleman, D. C., & Wiebe, W. J. (1998). Prokaryotes: The unseen majority. *Proceedings of the National Academy of Sciences*, 95(12), 6578–6583.

MINERALS

Bogdan P. Onac

University of South Florida, Tampa, U.S.A., and "Emil Racovita" Institute of Speleology, Cluj, Romania

This article presents a short review of the physical, chemical, and crystallographic properties of cave minerals. In addition, their general modes of occurrence and genesis under various cave settings are discussed. To suit the needs of any particular reader, nearly 30 of the most common cave mineral species are described, providing the necessary backdrop for everyone eager to know more about cave mineralogy.

INTRODUCTION

Caves are natural subterranean cavities, fissures, and fragments of conduit systems that are accessible to human exploration. The cave environment typically maintains constant temperature, relative humidity, and carbon dioxide partial pressure over long periods. Solutions entering the caves, according to their primarily chemical composition or following the reaction with different cave deposits (*i.e.*, bedrock, cave sediments, organic deposits) will precipitate a variety of interesting, sometimes unique, cave minerals. Only secondary mineral species (*i.e.*, formed within the cave from a physical process or a chemical process) are considered true cave minerals.

Although several types of reactions may take place in the cave environment (White, 1997), those that are ultimately responsible for the deposition of common cave minerals fall into one of the following categories: (1) dissolution/precipitation (*e.g.*, calcite, gypsum); (2) hetero- or homogeneous acid/base reactions (the $MgO-CO_2-H_2O$ system, carbonic acid); (3) phase transitions (aragonite/calcite inversion); (4) hydration/dehydration (*e.g.*, mirabilite/thenardite or brushite/monetite); (5) microbial processes (*e.g.*, calcite precipitation); and (6) redox reactions involving mainly sulfur and manganese.

Moore (1970) published the first checklist of cave minerals. He included 68 cave minerals formed within and outside of the United States. Six years later, another comprehensive compilation of nearly 80 minerals, primarily a review of U.S. cave mineralogy, was undertaken (Hill, 1976). The first book to provide worldwide coverage on this topic was the first edition of *Cave Minerals of the World* (Hill and Forti, 1986). This edition summed up 173 minerals, 86 of which were either ore-related or being miscellaneous, uncommon cave minerals. In the second edition of the same book published in 1997 were reported 255 cave minerals, 125 of which were precipitated under special cave settings. Since then, 29 new cave minerals have been added and the list is far from being completed (Onac and Forti, 2011). Apart from these two books specifically dedicated to cave minerals, more recent reviews on this topic can be found in White (1988), Moore and Sullivan (1997), Onac and Forti (2011), and Ford and Williams (2007).

This exponential increase in the number of cave minerals over the past 30 years is due to (1) advancements in analytical facilities (X-ray powder diffraction, X-ray fluorescence spectrometry, scanning electron microscopy, electron microprobe, and other techniques); (2) discovery of many new cavities carved by sulfuric acid or hydrothermal ore-derived solutions, skarn-hosted caves, and so on (all displaying a diverse and fascinating mineralogy); and (3) the increasing interest for nondescript weathering crusts and various earthy masses comprising the cave soils.

Minerals most likely to be encountered in a "normal" cave environment (*i.e.*, limestone, gypsum, or salt caves) belong to eight chemical classes. Our approach will emphasize only the chief minerals of each class that are represented in the cave environment following *Dana's New Mineralogy* scheme (Gaines et al., 1997). Less attention will be given to species (ore-related and other miscellaneous minerals) that are rare, or that do not ordinarily appear in caves except in a unique set of conditions. Within each of *Dana's* classes, physical and chemical properties of the more common cave minerals are discussed, along with their general modes of occurrence, genesis, and stability under various cave settings. The crystallographic data for the common cave mineral species are listed in Table 1. Organic minerals are given a brief overview. The presentation is arranged according to the abundance in the cave environment of each chemical subclass.

CARBONATES

Approximately 29 carbonate species are reported from caves worldwide. Of these, calcite and aragonite are the two most common and abundant (Table 1). Calcite is the thermodynamically stable form of calcium carbonate under temperature, pressure, and CO_2 partial pressure found in caves. Crystal habits include rhombohedrons, scalenohedrons (dogtooth), prismatic (nailhead), or combinations of these (Fig. 1). Aragonite is the metastable, orthorhombic polymorph of calcite and occurs as short to long prismatic (along *c*), acicular or tabular (chisel-shaped) crystals, often twinned. With time (*i.e.*, at geologic time scales) aragonite will internally change its crystal structure to the stable one of calcite; however, the external habit of aragonite remains preserved. Both calcite and aragonite may appear as monocrystals. More commonly these crystals are associated in polycrystalline aggregates shaped into a countless number of speleothem types (*i.e.*, soda straws, stalactites, and stalagmites). Mineralogical and isotopic investigations in caves hosting perennial ice deposits enabled the identification of *cryogenic carbonates*. These minerals form during freezing of carbonate-rich water, when expulsion of CO_2 from solution promotes subsequent precipitation of the rare carbonate minerals such as monohydrocalcite and ikaite (*e.g.*, Scărişoara Ice Cave, Romania).

Although dolomite (a mineral containing magnesium) is very common as a rock, it is rare as a secondary cave mineral. Except for huntite, $CaMg_3(CO_3)_4$; dolomite, $CaMg(CO_3)_2$; and hydromagnesite, the Mg-rich cave minerals are quite uncommon. These minerals, along with magnesite, $MgCO_3$; monohydrocalcite, $CaCO_3 \cdot H_2O$; and nesquehonite, $Mg(HCO_3)(OH) \cdot 2H_2O$ are the chief constituents of moonmilk deposits appearing as microcrystalline aggregates of various habits and shapes. They are also identified in crusts and nodules. The Mg/Ca ratio of the solutions governs the deposition of Ca-Mg carbonate species in caves. As the ratio increases, a polymineral sequence precipitates along the pathway best illustrated by coralloid multi-aggregates (Fig. 2).

Some secondary carbonates containing copper and/or zinc, such as malachite; azurite, $Cu_3(CO_3)_2(OH)_2$; and rosasite, $(CuZn)_2(CO_3)(OH)_2$, are extremely colorful. The whole range of speleothems they form or stain have great aesthetic value. These minerals, as well as the relatively rare cerussite, $PbCO_3$; rhodochrosite, $MnCO_3$; and smithsonite, $ZnCO_3$, are the oxidation products of sulfides (chalcopyrite, bornite, sphalerite, *etc.*) dispersed in the bedrock.

Carbonate mineral deposition in caves is dictated by the availability of supersaturated percolation water entering the cave galleries. Figure 3 summarizes the chemical pathway of percolating waters passing through the vertical sequence of air-soil-limestone-cave. The meteoric water is already acidified by atmospheric carbon dioxide, $CO_{2(atm)}$. As these waters percolate through decaying organic soil matter, dissolved $CO_{2(bio)}$ increases further. On entering the cave, degassing of CO_2 occurs, causing the water to become supersaturated with respect to calcium, resulting in the precipitation of the various speleothems of calcite and/or other carbonate minerals. In addition, a growing body of evidence points toward the role of biogenic processes in precipitating calcium carbonates in caves. Through their metabolic activities, a number of bacteria favor $CaCO_3$ deposition by simply modifying the supersaturation of the solution.

The simple linked series of reactions shown in Figure 3 is far more complex in reality; the chemical kinetics and various biogenic processes are discussed in detail by White (1988), Ford and Williams (2007), Palmer (2007), and Jones (2010).

SULFATES

Ca^{2+}, Mg^{2+}, and Na^+ ions, with lesser amounts of K^+, dominate the circulating groundwater. As a result, gypsum, epsomite, and mirabilite are the usual sulfate minerals found in caves (Table 1). By far the most representative of this group and perhaps the second most common cave mineral is gypsum. It forms a huge variety of speleothem types ranging from the less

TABLE 1 Crystallographic Data for Common Cave Minerals

Mineral Chemical formula	Crystal system	Unit cell parameters a_0 (Å), b_0 (Å), c_0 (Å)	$\alpha(°)$, $\beta(°)$, $\gamma(°)$	Observations
CARBONATES				
Calcite $CaCO_3$	Trigonal	4.99 — 17.061		Vigorous effervescence in cold dilute HCl, perfect rhombohedral cleavage; exhibits double refraction; rhombohedral, prismatic (nailhead), scalenohedrons (dogtooth) or twinned crystals.
Aragonite $CaCO_3$	Orthorhombic	4.959 7.964 5.737		Acicular; sinks in bromoform; columnar aragonite has cleavage parallel to elongation; twins are common.
Hydromagnesite $Mg_5(CO_3)_4(OH)_2 \cdot 4H_2O$	Monoclinic	10.11 8.94 8.38	114.5	Common in moonmilk, nodules, earthy masses; feels like cream cheese when rubbed between the fingers.
Malachite $Cu_2CO_3(OH)_2$	Monoclinic	9.48 12.03 3.21	98	Green color and streak; botryoidal habit; typically banded.
SULFATES				
Gypsum $CaSO_4 \cdot 2H_2O$	Monoclinic	5.678 15.20 6.52	118.43	Low hardness; perfect cleavage after (010); swallow-tail twins; flexible; soluble in hot dilute HCl.
Epsomite $MgSO_4 \cdot 7H_2O$	Orthorhombic	11.86 11.99 6.85		Acicular crystals elongated parallel to c, powdery crusts; bitter taste; very soluble in water; vitreous, fibrous, silky luster; frequently associated with gypsum.
Mirabilite $Na_2SO_4 \cdot 10H_2O$	Monoclinic	11.51 10.37 12.85	107.80	Bitter and salty taste, very soluble.
Celestine $SrSO_4$	Orthorhombic	8.37 5.35 6.87		Tabular and fibrous crystals; characteristic pale blue color; brilliant red flame coloration.
Barite $BaSO_4$	Orthorhombic	8.88 5.45 7.15		Tabular and prismatic crystals; high specific gravity; green flame coloration.
Alunite $KAl_3(SO_4)_2(OH)_6$	Trigonal	6.981 — 17.331		Crystals typically pseudocubic or tabular; fibrous to columnar, porcelaneous, commonly granular.
PHOSPHATES				
Hydroxylapatite $Ca_5(PO_4)_3(OH)$	Hexagonal	9.41 — 6.87		Compact to coarse granular in crusts; various shades of brown and red; indistinct cleavage.

(Continued)

TABLE 1 (Continued)

Mineral	Crystal system	Unit cell parameters		Observations
Brushite $CaHPO_4 \cdot 2H_2O$	Monoclinic	6.36		Colorless to ivory yellow; needle-like or prismatic crystals; earthy or powdery.
		15.19	118.45	
		5.18		
Ardealite $Ca_2(HPO_3OH)(SO_4) \cdot 4H_2O$	Monoclinic	5.72		Fine-grained powdery masses; white to pale yellow; Associated with gypsum and brushite; described from Cioclovina Cave, Romania.
		30.95	117.29	
		6.26		
Taranakite $K_3Al_5(PO_3OH)_6(PO_4)_2 \cdot 18H_2O$	Trigonal	8.71		Nodules, flour-like masses; white, gray, yellowish; acidic environment indicator; first reported as *minervite*.
		—		
		95.1		
Variscite $AlPO_4 \cdot 2H_2O$	Orthorhombic	9.822		Earthy masses, nodules, and crusts; grayish-white, pale-yellow to brown.
		9.630		
		8.561		
OXIDES & HYDROXIDES				
Goethite $FeO(OH)$	Orthorhombic	4.62		Blackish brown to yellow or reddish; typically massive, botryoidal, with internal radiating fibrous or concentric structure; pseudomorphs after pyrite.
		9.95		
		3.01		
Ice H_2O	Hexagonal	4.52		Melts at 0°C; form sublimation crystals.
		—		
		7.36		
Birnessite $(Na,Ca,K)_{0.6}(Mn^{4+},Mn^{3+})_2O_4 \cdot 1.5H_2O$	Monoclinic	5.174		Red-brown streak; soluble in concentrate HCl; commonly earthy and admixed with other Fe and Mn oxides and hydrates.
		2.85	103.18	
		7.33		
SILICATES				
Opal $SiO_2 \cdot nH_2O$	Amorphous			Botryoidal crusts, banded; conchoidal fracture; brittle.
Quartz SiO_2	Trigonal	4.91		Prismatic crystals terminated by hexagonal dipyramid or rhombohedra faces; practically any color; high hardness; soluble only in HF; no cleavage.
		—		
		5.40		
Allophane $Al_2O_3(SiO_2)_{1.3-2.0} \cdot 2.5-3H_2O$	Amorphous			Earthy masses, moonmilk-like deposits; white, pale-cream, blue, yellow.
HALIDES				
Halite, $NaCl$	Cubic	5.64		Salty taste, perfect cubic cleavage, low hardness.
Fluorite, CaF_2	Cubic	5.46		Isometric; fluorescent in UV, no effervescence in HCl.
NITRATES				
Niter KNO_3	Orthorhombic	5.41		Cooling taste, soluble in water, violet flame.
		9.16		
		6.43		

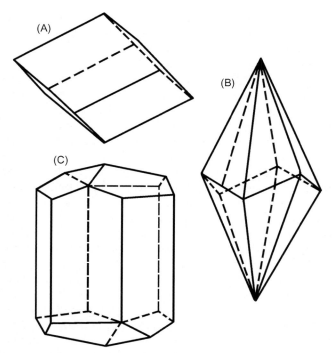

FIGURE 1 Common crystal habits of calcite: (A) rhombohedron, (B) scalenohedron (dogtooth), (C) prismatic (nailhead). *Adapted from Onac, 2000.*

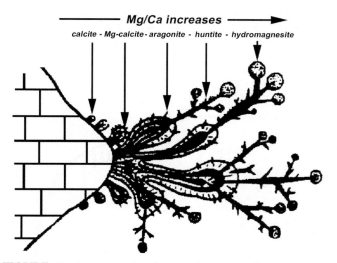

FIGURE 2 Sequence of polymineral precipitation in corraloid multi-aggregates. *Adapted from Maltsev, unpublished manuscript.*

spectacular white, soft moonmilk deposits composed of microcrystals, to curving clusters, to prismatic crystals up to 12 m long (Cave of the Giant Crystals, Naica, Mexico), and to enormous and spectacular stalactites (the so-called chandeliers) exposed in Lechuguilla (New Mexico) and Cupp-Coutunn (Turkmenistan) caves.

Two other common cave sulfates that usually occur as colorless to white fibrous efflorescences or crusts on cave walls and sediments are mirabilite and epsomite.

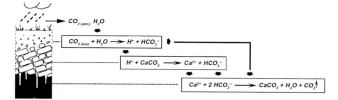

FIGURE 3 Reactions in the air-soil-rock-cave system and carbonates deposition. *Modified from Onac, 2000.*

Both minerals are extremely sensitive to variations in relative humidity and temperature. As the relative humidity drops below ~70% mirabilite will dehydrate to thenardite (Na_2SO_4) whereas epsomite transforms into hexahydrite ($MgSO_4 \cdot 6H_2O$) and kieserite ($MgSO_4 \cdot H_2O$) respectively. Samples of mirabilite and epsomite removed from the cave environment will decompose within minutes into a white milky powder. Cave passages with low relative humidity, high temperatures (above 20°C), or that are well ventilated appear to be ideal locations for the deposition of bassanite ($2CaSO_4 \cdot H_2O$) and anhydrite ($CaSO_4$).

Barite and celestine (Table 1), are relatively insoluble in comparison to other sulfates. Only under special conditions does direct precipitation of these minerals in caves from meteoric waters take place. Both sulfates commonly occur as millimeter- to centimeter-size crystals, crusts, and speleothems in caves invaded by thermal waters or hydrothermal solutions. Famous occurrences of celestine are known from Crystal Cave (Ohio), Lechuguilla Cave (New Mexico), Cango Cave (South Africa), and Valea Rea Cave (Romania). Barite is known from caves in England, Italy, Hungary, France, Russia, and the United States.

Although less common compared to the above cave sulfates, alunite (Table 1) is a key mineral used to trace the origin of caves and their ages. It is a by-product material resulting from the reaction between cave clays and sulfuric acid in the process of cave formation. Because alunite can be dated using radiometric methods (*i.e.*, K-Ar or Ar-Ar), it can provide age estimates for sulfuric acid caves (Polyak *et al.*, 1998).

The other 60 sulfates described from caves (*e.g.*, alunogen, halotrichite, jarosite, chalcantite, melanterite, *etc.*) are known only from warmer caves in which Al, Fe, and Cu were supplied by ore bodies or hydrothermal solutions.

Four mechanisms are ultimately responsible for the deposition of sulfates in caves. Most commonly cited is simple precipitation by evaporation and sulfuric acid reaction with the carbonate bedrock or cave sediments. With respect to the origin of sulfates in caves, when these minerals do not occur in the overlying limestone, the following sources have been proposed: oxidation of sulfides (*e.g.*, pyrite or marcasite), presence of bat guano, or post-volcanic activities (*i.e.*, fumaroles).

PHOSPHATES

Phosphates are the second largest group of cave minerals after sulfates. The $(PO_4)^{3-}$ radical combines with some 30 elements to form over 300 phosphate minerals, out of which 57 have been found in different cave settings. Hydroxylapatite, brushite, ardealite, taranakite, and variscite are the most abundant (Table 1); all the other phosphates are rare. They occur whenever a cave contains fresh or fossil bat guano or significant accumulations of bone deposits. Concentrations of bat guano are found primarily in caves situated in low-latitude humid areas. The most diverse phosphate assemblages are to be found there. Many of the rare phosphate minerals were identified in caves from South Africa, Australia, Namibia, and Romania. Unlike other cave minerals, the phosphates do not form spectacular speleothems, but occur as crusts, nodules, lenses, and earthy or powdery masses.

Depending on whether the percolating water passing through guano reacts with carbonate rocks or clay minerals, Ca-rich or Mg- and Al-rich phosphates will be deposited. Less commonly, the reaction of guano with ore-derived metals or spontaneous combustion of guano can produce some rare phosphate minerals (Fig. 4).

Many cave phosphate minerals include the ammonium ion (*e.g.*, biphosphammite, $H_2(NH_4)PO_4$; struvite, $(NH_4)Mg(PO_4) \cdot 6H_2O$; taranakite, *etc.*), which is derived from decomposition of bat urea. After leaching of alkali ions, the more stable minerals (*e.g.*, hydroxylapatite or brushite) persist. Members of the apatite group, $Ca_5(PO_4)_3(OH,F,Cl)$, are common cave minerals. Fluorine, chlorine, and the hydroxyl ion can mutually replace each other to form almost pure end members: fluorapatite, chlorapatite, and hydroxylapatite. Of these, the last one is the most thermodynamically stable phosphate mineral under ordinary cave conditions, occurring on almost all coatings and crusts. Brushite is stable under acidic (pH <6) and damp conditions, occurring as nodular masses or prismatic crystals. It loses water readily, converting to monetite ($CaHPO_4$). Brushite is isostructural with gypsum, so it is not surprising that ardealite should exist, particularly in view of their frequent association. A decrease of organic content and an increase of the Ca/P ratio characterize the mineral sequence brushite-whitlockite-hydroxylapatite. Like brushite, taranakite is stable in acidic environments and precipitates near contacts between guano and clays, under poorly drained conditions. It usually occurs as white-yellowish soft nodules, veins, or

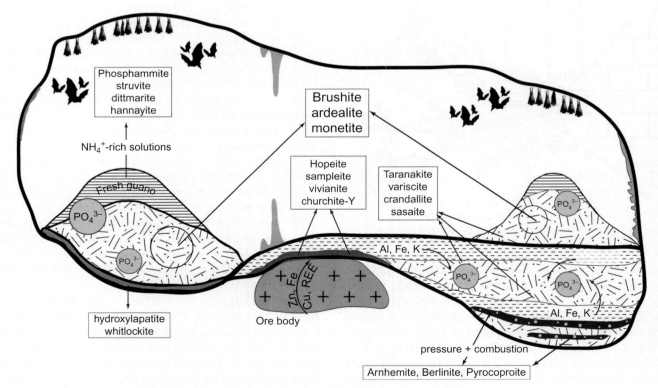

FIGURE 4 Development of phosphate minerals in an idealized and simplified cave environment.

flour-like masses. Variscite, another fairly common phosphate, occurs under acidic conditions as veins and nodules in, or at the contact with, highly phosphatized Al-rich cave sediments. Other products with the interaction of bat guano with various rocks and loose sediments in the cave environment are illustrated in Figure 4.

OXIDES AND HYDROXIDES

Among the most common oxide and hydroxide minerals frequently precipitated in caves are those containing iron (*e.g.*, goethite, hematite, Fe_2O_3; lepidocrocite, $Fe^{3+}O(OH)$) and manganese (*e.g.*, birnessite, pyrolusite, MnO_2; and romanèchite, $(Ba,H_2O)_2(Mn^{4+}, Mn^{3+})_5O_{10}$; Table 1). Of these, goethite occurs in caves worldwide, forming a variety of speleothems colored reddish-brown. Dark-brown to black fine-grained mixtures (chiefly birnessite, romanèchite, and todorokite) have been reported covering cave walls, stream clasts, and other speleothems in caves such as Vântului (Romania), Jewel and Jasper (U.S.A.), and Zbrasov (Czech Republic). The term *wad* is frequently used as a field term for composite mixtures consisting mainly of manganese oxides, in much the same sense as *limonite* is used for hydrous iron oxides. All minerals mentioned above are formed under oxidizing conditions.

Pyrite (FeS_2) and other sulfides are the primary source for iron oxide and hydroxide minerals in caves. Manganese may be released from impure limestone or from within the soil zone where it accumulates from decaying of plants. Some bacterial species are known to catalyze the oxidation and precipitation of Fe and Mn in the cave environment. Furthermore, pH and/or redox conditions within the aqueous solution control this biologically mediated process.

The crystalline forms of solid water (ice) also belong to this group of cave minerals. Ephemeral ice speleothems form in the entrance passages of caves during cold seasons, whereas perennial ice deposits accumulate when particular airflow conditions maintain low temperatures even during the warm season. Freezing of dripping or flowing water creates similar speleothems to those of calcite. Spectacular hexagonal platy ice crystals and hoarfrost may form by sublimation on the cave walls and ceiling from moist, warm air entering the cave.

SILICATES

The silicates comprise about a third of all mineral species. Although many of these minerals are uncommon in caves, others make up some particular cave deposits (*e.g.*, opal, quartz, and allophane) (Table 1). The solidified colloidal silica, opal, is the most common mineral among the cave silicates. It occurs mostly in lava tubes and cavities within volcanic or metamorphic rock, but may also be found in limestone caves. Opal typically forms botryoidal crusts, coralloids, and can be intermixed with calcite in different speleothems. In time it can gradually transform to chalcedony or even quartz. The most abundant opal speleothems are to be found in lava tubes (*e.g.*, Algar do Carvao, Azores). The presence of quartz in caves, usually as euhedral crystals lining the walls, indicates at least one hydrothermal episode in the cave's history. Spectacular quartz crystals are displayed in Wind and Jewel caves (South Dakota), Chiricahua Crystal Cave (Arizona), and Kristallkluft Cave (Swiss Alps). Fluid inclusion studies on quartz allow for estimating the temperature of the hydrothermal solutions from which the crystals were precipitated.

The third most frequent silicate mineral in caves is allophane, an amorphous hydrous aluminum silicate. It has been reported mostly as white-pale cream to yellow-orange moonmilk-like deposits, but also in the form of different speleothems. Among the best displays of allophane are those from Mbobo Mkulu (Transvaal, South Africa) and Iza (Romania) caves, where it forms spectacular blue flowstones. Most allophane occurrences are related to the action of acidic waters on cave sediments.

Most clay minerals (hydrated silicates) in caves are thought to be derived from the bedrock or were transported in along with other clastic sediments by the cave streams or percolating waters. Others, however, are definitely of authigenic origin. Among these are dickite, halloysite, montmorillonite, and sepiolite.

In addition, a long list of other cave silicates can be compiled (*e.g.*, benitoite, clinochlore, epidote, fraipontite, ilvaite, *etc.*); however, their secondary origin is questionable. All these minerals were deposited either as crystals, granular aggregates, or earthy masses mostly under particular settings (*i.e.*, hydrothermal and ore-related environs, lava tubes, *etc.*).

HALIDES

The list of halides, compounds in which a halogen element is the sole or principal anion, comprises some 100 species. Only 11 of these minerals were documented from different cave environments. The more common members are simple in composition and fall into either the anhydrous (halite; sylvite, KCl; fluorite, *etc.*) or hydrous (carnallite, $KMgCl_3 \cdot 6H_2O$; atacamite, $Cu_2Cl(OH)_3$, *etc.*) halide group. They form mostly in caves located within extensive salt beds (Mount

Sedom, Israel; Meledic Plateau, Romania) or in lava tubes from Iceland, Italy, the United States, and Australia.

Halite is by far the most common water-soluble halide mineral (Table 1), displaying an overwhelming variety of speleothem types, especially when precipitated in salt caves. Typically, halite may appear as massive, coarsely granular or compact aggregates, or as cubic crystals, often with cavernous and stepped faces (hopper crystals). It is colorless, white, gray, yellowish (colored by discrete impurities), blue, or purplish (color due to lattice imperfections).

Fluorite is another secondary halide mineral forming various-sized euhedral cubic crystals (sometimes twinned) and coarse-to-fine granular aggregates of blue, purple, green, or violet color (Table 1). Fluorite has been described from a number of ore-related cavities such as Cupp-Coutunn (Turkmenistan), Treak Cliffs (U.K.), Kootnay (British Columbia, Canada), Glori and Blanchard (New Mexico, U.S.A.). In all these locations fluorite was interpreted as being precipitated from hydrothermal solutions circulating along fissures or channels. Rare occurrences are known for fluorite precipitated from low-temperature percolating waters (caves in Arizona and Wyoming) (Hill and Forti, 1997).

Except for the halides occurring in caves within or near ore bodies (fluorite and bromargyrite) or those precipitated from fumarole vapors (*e.g.*, galeite and kainite), all are rather soluble salts that are leached by percolating waters from the overburden rocks and redeposited in caves during evaporation of the solvent. Therefore, these minerals form mainly in caves in dry climate areas that experience little precipitation.

NITRATES

Minerals of this group are structurally similar to the carbonates with plane triangular $(NO_3)^-$ groups in place of $(CO_3)^{2-}$. Since N is more electronegative than C, the anionic group, $(NO_3)^-$, is less stable, restricting the occurrence of nitrate cave minerals much more than the carbonates. Nitrates are highly soluble, forming only in warm, dry (low humidity), and well-ventilated caves. They commonly appear along with gypsum, epsomite, halite, and other salts as efflorescence, tiny crystals disseminated within cave soil and dry guano deposits.

There are 11 nitrate minerals found in caves, with 5 of these being known *only* from particular cave environs (sveite, mbobomkulite, hydrombobomkulite, nickelalumite, and gwihabaite). Although niter; nitratine, $NaNO_3$; nitrocalcite, $Ca(NO_3)_2 \cdot 4H_2O$; and nitromagnesite, $Mg(NO_3)_2 \cdot 6H_2O$, were documented in several caves worldwide, none of these can be considered common for the cave environment.

Most of the nitrate minerals are concentrated in cave soils after being carried inside by water percolating through rich organic soils. Alternatively, bat and rat guano or basic volcanic rocks can also supply nitrates in caves. Ultimately, microorganisms (particularly bacteria) play major roles in the deposition of nitrate minerals in caves. Their precipitation is affected by environmental factors that influence microbial activity, such as temperature, moisture, and organic matter availability.

NATIVE ELEMENTS

Although several chemical elements occur as minerals in the Earth's crust, so far only sulfur has been found in a number of caves from the United States, Italy, Mexico, Iceland, Romania, and Russia. Sulfur occurs as canary-yellow powder, granular to massive aggregates, or it can take a variety of speleothem types (crusts, stalactites, euhedral crystals, cave rafts, *etc.*). Most commonly, native sulfur occurs in sulfuric acid caves, in which its origin was related to either hydrogen sulfide oxidation (caves from the Guadalupe Mountains, United States; Cueva de Villa Luz, Mexico; Sant Cesarea and Grotta dello Zolfo, Italy; Diana, Romania) or reduction of sulfate ion in the presence of sulfur bacteria (Frasassi, Grotta di Cala Fetente, Italy) (Hill and Forti, 1997).

ARSENATES AND VANADATES

The arsenates and vanadates are oxysalts characterized structurally by independent $(AsO_4)^{3-}$ or $(VO_4)^{3-}$ anionic groups linked through intermediate cations. The majority are mostly low-temperature, hydrous minerals. Only eight arsenate and six vanadate minerals were identified in caves known to have sulfuric acid or hydrothermal genesis. Typically, they form at the expense of primary arsenic or vanadium-rich minerals by weathering and oxidation processes; some may have been precipitated in reducing hydrogen sulfide-rich environments. None of these minerals is common for normal cave settings; however, when they occur as small crystals or scaly crusts, they are easily noticed due to their bright-yellow or green color. Common members of each of these two subclasses include: conichalcite and talmessite (arsenate group) and tyuyamunite, carnotite, and calciovolborthite (vanadate group). Most of the above minerals were identified in caves from Tyuya-Muyun region (Kyrgyzstan); Valea Rea Cave (Romania); Sonora Caverns (Texas); Lechuguilla, Carlsbad, Spider, and other caves in the Guadalupe Mountains (New Mexico).

SULFIDES

This group of cave minerals comprises compounds in which the large atoms S, As, and Bi are combined to one or more of the metals (Fe, Zn, Pb, Cu, Hg, and Sb). Thirteen minerals (*e.g.*, pyrite, marcasite, cinnabar, metacinnabar, stibnite, *etc.*) were described from skarn-hosted and hydrothermal-related caves. Most of the later ones are associated with Mississippi Valley-type lead-zinc ore deposits (Hill and Forti, 1997; Ford and Williams, 2007). The usual occurrences of sulfides consist of crusts, inclusions, druses, or crystals. With few exceptions, all sulfides were precipitated in preexisting metasomatic- or meteoric-dissolved karst channels from low-temperature hydrothermal solutions.

ORGANIC MINERALS

The presence of large bat or bird colonies in some caves from arid and semi-arid regions of Western Australia, Namibia, Israel, and the United States is ultimately responsible for the deposition of nine highly soluble organic minerals. These include oxalates, mellates, and purines, which normally form pale-yellow, red, or light brown crusts and efflorescences. Almost all organic minerals were derived from the reaction of urine or animal excreta with carbonate bedrock, clays, or cave detritus.

WHY STUDY CAVE MINERALS?

Cave minerals can yield a wealth of information. Foremost, they simply retained their scientific value as mineral bodies occurring within the cave. Much information, however, is gained by studying speleothems, which are made up of one or more cave minerals. Examinination of such information reveals that cave minerals are valuable data repositories, which allows for the investigation of dozens of interesting problems that cover a broad range of topics; some of these are summarized below:

1. Caves are natural underground laboratories where crystal growth processes can be observed and monitored.
2. Studies of crystallography and mineralogy of cave minerals provide invaluable insight into the chemical and physical (sometimes biological too) conditions existing within given cave environments at various times during mineral formation.
3. Some caves provide a unique set of conditions that allow the deposition of a suite of exotic minerals; many of these are restricted to only such particular cave settings and are never found in the outside world.
4. The mineral assemblage present in a given cave or caves within a region can sometimes be used as diagnostic criterion of the cave speleogenetic pathway.
5. Calcite speleothems in caves are well suited for uranium-series dating based on which an absolute chronology can be derived. Furthermore, carbon and oxygen isotopic variations in carbonate speleothems combined with changes in calcite crystal fabric and luminescence of growth lamina are potentially powerful tracers of changes in Quaternary climate and possibly vegetation. Aragonite and calcite speleothems in littoral caves can resolve sea level changes and tectonic histories.
6. Oxygen isotope analyses ($\delta^{18}O$) and fluid inclusions of hydrothermal-related cave minerals (*e.g.*, quartz, barite, calcite, malachite, *etc.*) provide solution temperature at the time of deposition.
7. The ability to reliably differentiate hypogene from supergene cave sulfates (*e.g.*, gypsum, barite, celestine, *etc.*) by studying their isotopic composition ($\delta^{34}S$) is critical to understanding sulfur source, hydrological, and speleogenetic processes.
8. Some cave minerals (*e.g.*, calcite, alunite, *etc.*) play an important role in reconstructing landscape evolution (*e.g.*, valley incision rate, mountain uplift) or determining the age of caves.
9. Supporting evidence shows some mineral assemblages were precipitated in caves with the aid of various microorganisms. Deciphering the geomicrobiological processes involved on their deposition may help our understanding of the caves as life environments on Earth or beyond.

Apart from the above issues, the cave minerals can provide an accessible window into geochemistry, paleontology, archaeology, soil science, ore deposits, and other Earth science disciplines. It is expected that their use will become even more widespread, stimulating further interest and excellence in future multidisciplinary research carried into these fascinating subterranean worlds.

See Also the Following Article

Speleothems: General Overview

Bibliography

Ford, D. C., & Williams, P. W. (2007). *Karst hydrogeology and geomorphology* (2nd ed.). Chichester, England: John Wiley & Sons.

Gaines, R. V., Skinner, C., Foord, E. E., Mason, B., & Rosenzweig, A. (1997). *Dana's new mineralogy*. New York: Wiley & Sons.

Hill, C. A. (1976). *Cave minerals*. Huntsville, AL: National Speleological Society.
Hill, C. A., & Forti, P. (1986). *Cave minerals of the world*. Huntsville, AL: National Speleological Society.
Hill, C. A., & Forti, P. (1997). *Cave minerals of the world* (2nd ed.). Huntsville, AL: National Speleological Society.
Jones, B. (2010). Microbes in caves. In H. M. Pedley & M. Rogerson (Eds.), *Tufas and speleothems. Unravelling the microbial and physical controls* (336, pp. 7–30). Geological Society of London Special Publication.
Moore, G. W. (1970). Checklist of cave minerals. *National Speleological Society News*, 28(1), 9–10.
Moore, G. W., & Sullivan, N. (1997). *Speleology. Caves and cave environment* (3rd ed.). St. Louis, MO: Cave Books.
Onac, B. P. (2000). *Geology of karst terrains*. Bucuresti: Didactica si Pedagogica.
Onac, B. P., & Forti, P. (2011). State of the art and challenges in cave minerals studies. *Studia UBB Geologia*, 56(1), 33–42.
Palmer, A. N. (2007). *Cave geology*. Dayton, OH: Cave Books.
Polyak, V. J., McIntosh, W. C., Güven, N., & Provencio, P. (1998). Age and origin of Carlsbad Cavern and related caves from $^{40}Ar/^{39}Ar$ of alunite. *Science*, 279(5358), 1919–1922.
White, W. B. (1988). *Geomorphology and hydrology of karst terrains*. New York: Oxford University Press.
White, W. B. (1997). Thermodynamic equilibrium, kinetics, activation barriers, and reaction mechanisms for chemical reactions in karst terrains. *Environmental Geology*, 30(1–2), 46–58.

MODELING OF KARST AQUIFERS

Georg Kaufmann,* Douchko Romanov,* and Wolfgang Dreybrodt†

*Free University of Berlin, Germany, †University of Bremen, Germany

INTRODUCTION

Groundwater flow through a karst aquifer is prone to contamination because of the very nature of the karstified host rock: Fissures and bedding partings in the rock are enlarged by chemical dissolution over time and provide preferential flow paths, through which water is transferred rapidly and almost unfiltered from input points such as sinks and dolines to output points such as large karst springs. The enlarged fractures and bedding partings are responsible for a very heterogeneous distribution of permeability within the karst aquifer. Enlarged passages can be very conductive ($1-10$ ms^{-1}) but have low storage capacity. The surrounding rock is orders of magnitude less conductive (10^{-8} ms^{-1}), but can provide significant storage. This large-scale heterogeneity in conductivity makes it difficult to assess the karst aquifer properties from field studies such as borehole pumping, packer, and slug tests. Monitoring spring discharge, on the other hand, provides only an integral picture of the karst aquifer. A different approach to understanding a karst aquifer and its spatial and temporal evolution are numerical models. This field has evolved dramatically over the past decades, and is described in this article.

CHEMISTRY

In karst landscapes, soluble rocks are dissolved by water enriched with carbon dioxide. In the case of gypsum, the dissolution is a physical dissociation process, in which only water and gypsum are involved. In the case of limestone or dolomite, the water enriched with carbon dioxide, which originates from the atmosphere and possibly the organic soil cover, reacts to form carbonic acid, which then dissociates and chemically dissolves the rock. On the surface, the bedrock is removed under conditions open to the atmosphere, and the dissolved material is then transported away in the aqueous solution. This process, termed *karst denudation*, lowers the surface with time. Within the karst aquifer, bedrock is removed from the walls of narrow fractures and bedding partings, enlarging the voids in which the water circulates. Depending on the situation within the aquifer, dissolution proceeds either under open-system conditions (soil, epikarst, vadose zone) or under closed-system conditions with solution completely filling the voids (phreatic zone). The enlargement of fractures and bedding partings in the aquifer enhances flow by a positive feedback loop, as the enlarged voids provide a much larger conductivity than the initial bedrock.

Calcium Equilibrium

The maximum amount of dissolved limestone or gypsum, described by the calcium concentration c [molm^{-3}], is controlled by the calcium equilibrium concentration, c_{eq} [molm^{-3}]. When the solution reaches this concentration, it is saturated with respect to calcite or gypsum and dissolution stops. For gypsum, the calcium equilibrium concentration is almost independent of the water temperature and amounts to $c_{eq} = 15.4$ molm^{-3}. For limestone in its pure form, calcite, the calcium equilibrium concentration can be derived as an analytical expression (*e.g.*, Dreybrodt, 1988); and it depends strongly on carbon-dioxide pressure and also on water temperature.

As seen in Figure 1, a solution undersaturated with calcium is located in the undersaturated field to the left of the equilibrium lines (white dot on the left axis). Depending on the boundary conditions, dissolution proceeds either along the horizontal line, until the equilibrium is reached (upper white dot on the equilibrium curve), thus keeping the carbon-dioxide

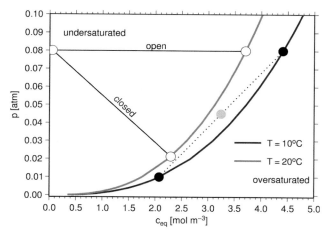

FIGURE 1 Calcium equilibrium as a function of carbon-dioxide pressure. The equilibrium line is shown for two temperatures, $T = 10°C$ and $T = 20°C$. Also shown are the different pathways for solution toward equilibrium (white dots). In the open system, dissolution proceeds along the horizontal line, while in the closed system dissolution proceeds along the sloping line. The effect of mixing two saturated solutions equally (black dots) results in a newly undersaturated solution (gray dot).

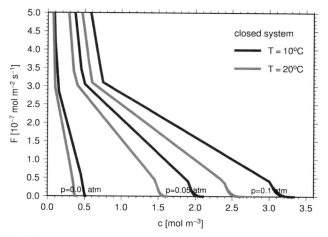

FIGURE 2 Low- and high-order flux rates as a function of calcium concentration. Flux rates are given for different temperatures T and carbon-dioxide pressures P, and a film thickness of $d = 0.001$ cm.

pressure in the solution constant (open system), or if the closed-system condition holds, the carbon-dioxide pressure decreases, and the calcium equilibrium line is reached at a lower value (lower white dot on the equilibrium curve). The dissolution of calcite is also temperature-dependent, and a decrease in temperature results in an increase in dissolved calcium. Thus, a cooling of water through daily variations in surface temperature or cooling of rising thermal water can dissolve additional calcite. While the former process is limited to the first few meters in juvenile karst aquifers, thermal corrosion can be locally important in regions of hydrothermal water supply. Here, large temperature gradients are responsible for a significant amount of additional calcite dissolved. A temperature drop of $T = 10°C$ will result in around 0.5 mol m^{-3} additional calcium dissolved.

The mixing of two solutions saturated with respect to calcite, which have different carbon-dioxide concentrations (black dots), results in a solution which, again, can dissolve calcite (gray dot). This alternative form of corrosion, termed *mixing corrosion* (Bögli, 1980), is a result of the nonlinear relation between the calcium equilibrium concentration and the carbon-dioxide pressure. As a consequence, water with different chemical properties flowing in the karst aquifer through the network of small fissures and cracks, all hydrologically connected, will mix, and results in additional corrosive power. As water with different partial carbon-dioxide pressure is the normal case within a drainage area due to differences in surficial origin of the solution, in surface cover and soil on a scale of meters as well as daily variations of organic activity, mixing corrosion can be a significant process on a local scale, but the additional amount of dissolved calcium rarely exceeds 0.2 mol m^{-3}.

Calcium Flux Rate

The flux rate F (mol/m^{-2} s^{-1}) describes the removal of material from the surface per unit area and per time. It is controlled by several potentially rate-limiting processes on the surface: (1) the surface reaction of water with the bedrock; (2) mass transport of the dissolved species by diffusion; and (3) conversion of carbon dioxide into carbonic acid, in the case of limestone. Flux rates have been measured experimentally for limestone and calcite and for gypsum, and have been predicted numerically (see references in Dreybrodt *et al.*, 2005; Kaufmann and Dreybrodt, 2007). For both gypsum and limestone, they can be approximated by a nonlinear rate law. In the case of limestone, for large undersaturation ($c < 0.3c_{eq}$), an initial fast linear kinetics is present, with very high removal capability. For more saturated solution ($0.3c_{eq} < c < 0.9c_{eq}$), a slower linear kinetics takes over, which removes material about one order of magnitude slower than the initial rate. For concentrations close to the equilibrium ($c > 0.9c_{eq}$), the high-order kinetics is established, and the rate becomes nonlinear and drops fast. The reason is the accumulation of impurities on the reactive surface, which significantly inhibits the dissolution process.

For the flux rate of limestone, several key aspects can be seen (Fig. 2): Firstly, flux rates depend strongly on the carbon-dioxide pressure in the solution. The larger the carbon-dioxide pressure the more calcite can be dissolved. Secondly, flux rates decrease with temperature, with warmer solutions being able to dissolve less calcite. Thirdly, calcium flux rates decrease rapidly with increasing concentration for strong undersaturation ($c < 0.3c_{eq}$), from then on the decrease is slower, but still linear, for calcium concentrations below 90% of the equilibrium concentration. For $c > 0.9c_{eq}$, dissolution rates drop sharply and the higher-order power law controls the flux rates. This nonlinear effect is important for the evolution of karst aquifers during their early phases, as the reduced flux rates close to equilibrium increase the distance undersaturated solution can penetrate into the aquifer significantly. Gypsum shows linear kinetics up to $0.94c_{eq}$ and then becomes nonlinear.

THE SINGLE FRACTURE

Fractures and bedding partings in the karst aquifer are enlarged by chemical dissolution, following the chemical reactions described above (Fig. 3). The fracture aperture d (m) is a function in time of the initial aperture d_i (m) and the calcium flux rate F in a given fracture. A first generation of models introduced conceptual ideas and started developing karst evolution models as a one-dimensional single fracture (see references in Dreybrodt and Gabrovšek, 2003; Dreybrodt et al., 2005).

An example of the evolution of a circular fracture in limestone is shown in Figure 4. The fracture has a length of 1 km and the initial fracture aperture is $d_i = 0.5$ mm; the head difference between the left and the right side is 10 m. The upper panel shows the evolution of the aperture with time. The undersaturated solution ($c_{in} = 0$ mol m^{-3}) enters the fracture from the left side. As dissolution proceeds in the early phase, only the entrance section is enlarged, whereas the downstream section grows only slowly (red lines). This is a result of the slow high-order kinetics, as the calcium concentration quickly reaches levels close to saturation farther downstream, which is shown in the middle panel. Thus, efficient dissolution only occurs within the entrance section. When enlargement proceeds, more highly undersaturated solution is carried farther downstream, and the fracture finally grows at a relatively constant pace (blue lines).

The importance of high-order kinetics close to the saturation of the solution, c_{eq}, becomes obvious when we discuss the calcium-concentration along the fracture. At the onset of evolution, linear kinetics in the entrance section results in solution close to saturation within the first few centimeters. As a consequence, modeling only low-order linear kinetics would result in enlargement limited to a few centimeters to meters from the entrance, depending on fracture geometry and film thickness. By taking high-order kinetics into account, when the solution is close to saturation, the calcium flux rate is reduced by several orders of magnitude, and slightly undersaturated solution can be carried much father downstream. Dissolution is occurring up to several kilometers within a fracture. This mechanism is largely responsible for enlarging fractures

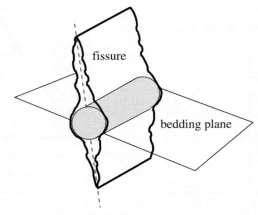

FIGURE 3 Sketch of a single fracture along the junction of a fissure and a bedding plane.

in limestone and gypsum during the early phase of karstification. The widening of the fracture at its exit increases with increasing aperture width. This causes a positive feedback loop. The amount of water flow through the fracture is shown in the lower panel. Firstly, the flow rate increases slowly with time, but due to the increasing widening of the fracture aperture width at the exit, the feedback becomes so strong that suddenly a dramatic increase in the flow rate is observed. This event is called *breakthrough*. After breakthrough the amount of water, which is limited, determines the further evolution of the fracture.

MODELING THE EVOLUTION OF KARST AQUIFERS

Based on the single fracture, more sophisticated two-dimensional karst aquifer network models were assembled (see references in Dreybrodt et al., 2005), which were used to describe the development of karst aquifers both in horizontal and vertical scenarios, the latter with the inclusion of a varying water table.

These two-dimensional models have then been used to describe regional case examples, for example, the gypsum maze caves from the Western Ukraine (Rehrl et al., 2008), and the bathyphreatic loop caves in the Jurassic limestone of the Swabian Alb in southwestern Germany (Kaufmann and Romanov, 2008). Another focus has been the safety and integrity of water reservoirs. Here, karst modeling is used to assess the temporal evolution of leakage and the development of subsurface voids and sinkholes in the vicinity of large hydraulic structures (e.g., Romanov et al., 2003).

The most recent development is the extension of karst aquifer evolution models into the third dimension (Annable, 2003; Kaufmann, 2009; Kaufmann et al., 2010). With these models, it is possible to simulate a karst

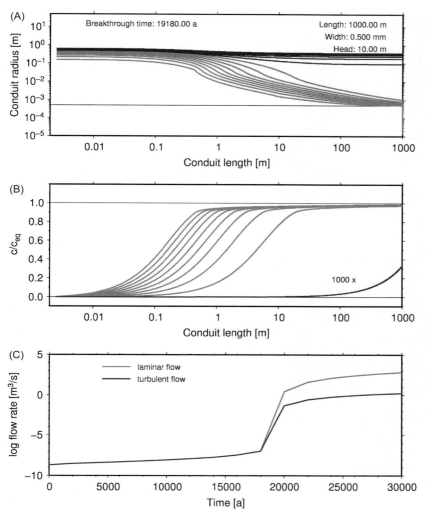

FIGURE 4 Evolution of a single fracture. Flow is driven from left to right. (A) Fracture aperture evolution; the black line indicates the initial fracture width, red lines are enlarged fracture widths before, blue lines after low-order kinetics is established. Lines are plotted at time steps of 2000 years. (B) Calcium concentration along fracture before (red) and after (blue) breakthrough. (C) Flow rate at the exit of the fracture for laminar (red) flow only and for both laminar and turbulent flow (blue).

aquifer in all dimensions, including a time-dependent water table.

As an example, we demonstrate a state-of-the-art karst evolution model for limestone in three dimensions, which depicts an idealized karst aquifer with horizontal dimensions of 742.5-m length and 375-m width, and a vertical extent of 150 m. The aquifer is discretized to provide a fracture spacing of 7.5 m. The model consists of 107,100 nodes, 99,000 matrix elements, and 287,831 fracture elements. The fine fractures have initial aperture values following a log-normal distribution with mean $d_{ave} = 0.15$ mm and standard deviation of $d_\sigma = 0.20$ mm. The rock matrix has conductivities of $K_m = 10^{-10}$ m s^{-1}. Flow in the karst aquifer is driven by a fixed-head boundary condition, with heads fixed on the left and the right sides. The head difference is 100 m. The water infiltrating the karst aquifer is aggressive, $c_{in} = 0$ mol m^{-3}. As climatic conditions we have chosen a temperature of $T = 10°C$ and a carbon-dioxide pressure of $P = 0.05$ atm, resulting in a calcium equilibrium concentration of $c_{eq} = 2.12$ mol m^{-3}.

In Figure 5, the evolution shortly before and some time after breakthrough is shown. Initially, several competing sets of fractures begin evolving from the left side. These evolving fractures push the hydraulic heads toward the right side, as seen in the figures before breakthrough. The right part of the modeling domain is still a bottleneck for flow. One of the enlarged fracture systems will be the first to reach the right side, and the model experiences breakthrough after around 900 years, which can be seen by the pressure lines bouncing back. The enlarged flow path connecting the high-pressure side with the low-pressure side is sinusoidal, because fractures, which initially have a larger aperture, are less frequent. After breakthrough, a zone of enlarged fractures evolves around the continuous flow path, and other branches experience secondary breakthrough events, when they either connect to the existing flow path or manage to continue growing to the low-pressure side, depending on the more favorable flow path.

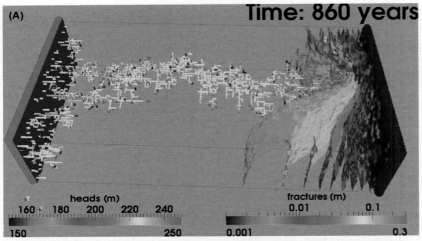
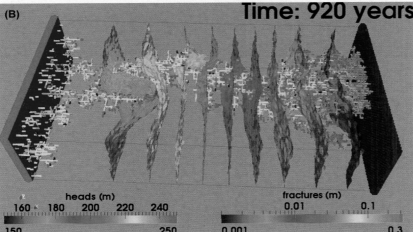

FIGURE 5 Evolution of simple model with statistical fracture distribution. Fixed head boundary conditions are shown as red (high pressure) and blue (low pressure) blocks. Heads are contoured every 10 m (color code as left legend). Enlarged fractures are shown as tubes, with the color indicating the aperture (color code as right legend). (A) Situation before breakthrough; (B) situation after breakthrough.

See Also the Following Article

Hydrogeology of Karst Aquifers

Bibliography

Annable, W. K. (2003). *Numerical analysis of conduit evolution in karstic aquifers* (Ph.D. dissertation). Ontario, Canada: University of Waterloo.

Bögli, A. (1980). *Karst hydrology and physical speleology*. Berlin: Springer-Verlag.

Dreybrodt, W. (1988). *Processes in karst systems*. Berlin & New York: Springer.

Dreybrodt, W., & Gabrovšek, F. (2003). Basic processes and mechanisms governing the evolution of karst. *Speleogenesis and Evolution of Karst Aquifers, 1*(1), 1–26; www.speleogenesis.info.

Dreybrodt, W., Gabrovšek, F., & Romanov, D. (2005). *Processes of speleogenesis: A modeling approach*. Postojna, Slovenia: Karst Research Institute, ZRC Publishing, ZASU.

Kaufmann, G. (2009). Modelling karst geomorphology on different time scales. *Geomorphology, 106*(1–2), 62–77.

Kaufmann, G., & Dreybrodt, W. (2007). Calcite dissolution kinetics in the system $CaCO_3$–H_2O–CO_2 at high undersaturation. *Geochimica et Cosmochimica Acta, 71*(6), 1398–1410.

Kaufmann, G., & Romanov, D. (2008). Cave development in the Swabian Alb, South Germany: A numerical perspective. *Journal of Hydrology, 349*(3–4), 302–317.

Kaufmann, G., Romanov, D., & Hiller, T. (2010). Modeling three-dimensional karst aquifer evolution using different matrix-flow contributions. *Journal of Hydrology, 388*(3–4), 241–250.

Rehrl, C., Birk, S., & Klimchouk, A. B. (2008). Conduit evolution in deep-seated settings: Conceptual and numerical models based on field observations. *Water Resources Research, 44*. doi:10.1029/2008WR006905.

Romanov, D., Gabrovšek, F., & Dreybrodt, W. (2003). Dam sites in soluble rocks: A model of increasing leakage by dissolutional widening of fractures beneath a dam. *Engineering Geology, 70*(1–2), 17–35.

MOLLUSKS

David C. Culver

American University

INTRODUCTION

Clams, snails, and their relatives comprise the phylum Molluska. They occur in subterranean habitats and are sometimes common in aquatic habitats, including caves and springs, and more occasionally in terrestrial habitats. Nearly all obligate cave-dwelling mollusks are gastropods. All gastropods have a muscular foot, visceral mass, and distinct head region. A fleshy mantle covers the viscera and secretes a calcium carbonate shell. Bivalves have hinged calcium carbonate shells. They are represented by a tiny handful of species in caves, and there is but one undoubted cave-limited clam, *Congeria kusceri*. Morphological characteristics of obligate subterranean species include a thin, often translucent shell that is usually white, with a depigmented body, and depigmented and reduced eyes. The aquatic species of gastropods are usually very tiny. This is true not only for species in small cavity habitats such as the underflow of streams, but also for large cavity habitats such as caves. Features associated with the miniaturization of cave snails include complex coiling of the intestine, loss or reduction of gills, simplification of gonadal morphology, and loss of sperm sacs. Although these characteristics represent an interesting case of convergent evolution, they also pose a difficult challenge for any attempt to reconstruct the phylogenetic history of cave snails. Some terrestrial cave species can be quite large. The largest reported terrestrial snail is that of *Paraegopis oberwimmeri*, a pulmonate snail whose shell reaches a diameter of nearly 3 cm.

Almost nothing is known about the ecology of cave mollusks. Because their shells remain long after the death, many collections of cave mollusks consist of dead shells only. This has resulted in a great deal of biogeographic information but a lack of ecological information.

Most gastropods in surface waters are either detritus feeders or algal feeders. Presumably, aquatic subterranean gastropods are detritus feeders. Most bivalves are deposit or filter feeders, and the few cave bivalves are likely also deposit or filter feeders. An amphibious habitat seems especially common among cave snails. Nowhere is this more apparent than in the genus *Zospeum* found in Dinaric caves. Most species appear to be primarily terrestrial but *Z. exiguum*, which is common in Križna Jama in Slovenia, appears to be primarily aquatic.

TAXONOMIC PATTERNS OF CAVE MOLLUSKS

Nearly all subterranean mollusks are gastropods. The only undoubted obligate cave bivalve is the

TABLE 1 Subterranean Families and Genera

Higher Classification	Families	Genera
Bivalvia: Lamellibranches	Dreissensiidae	*Congeria*
	Sphaeridae	*Pisidium*
Gastropoda: Prosobranchia: Archaeogastropoda	Hydrocenidae	*Georissa*
Gastropoda: Prosobranchia: Mesogastropoda	Hydrobiidae	Over 50 genera
	Pomatiopsidae	*Akiyoshia, Moria, Saganoa*
	Assimindeidae	*Cavernacmella*
	Cyclophoridae	*Pholeoteras, Opisthostoma*
Gastropoda: Pulmonata: Basommatophora	Acroloxidae	*Acroloxus*
	Ellobiidae	*Zospeum, Carychium*
Gastropoda: Pulmonata: Stylommatophora	Pupullidae	*Argna, Speleodiscus, Klemmia, Virpazria*
	Orculidae	*Speleodentorcula*
	Enidae	*Speleoconcha*
	Clausiliidae	*Sciocochlea*
	Ferussaciidae	*Ceciloides, Cryptareca*
	Subulinidae	*Opea*
	Helicodiscidae	*Helicodiscus*
	Zonitidae	15 genera
	Trigonochlamydidae	*Troglolestes*
	Polygyridae	*Mesodon, Polygyra*

From Bernasconi and Riedel (1994).

"living fossil" *Congeria kusceri* from the Dinaric karst. Minute clams in the genus *Pisidium* may also be cave-limited species. In this genus it is very difficult to distinguish between stygobionts and surface-dwelling species that occasionally enter caves. Four species of *Pisidium* have only been found in caves: three from caves in the Caucasus Mountains in Abkhazia and one from Turkey (Bole and Velkovrh, 1986). The rarity of bivalves in caves may be the result of the greatly reduced motility of bivalves and their general inability to withstand conditions of reduced oxygen. It seems unlikely that the absence of bivalves in caves is due to the absence of potential surface ancestors. For example, clams in the family Unionidae are extremely diverse in the southeastern United States, which is also an area of extensive cave development. Yet no clams are found in these caves.

Nearly all aquatic subterranean gastropods are prosobranchs. The major morphological difference between prosobranchs and pulmonates is that prosobranchs possess a gill and a horny or calcareous operculum. Pulmonates use a modified portion of the mantle cavity as a lung and lack an operculum. *Acroloxus* from the Dinaric Mountains in Bosnia and Herzegovina, Croatia, and Slovenia and *Hydrophrea* from New Zealand are pulmonates. Conversely, all troglobionts

TABLE 2 Obligate Cave-Dwelling Snails from Caves of the United States

	Order	Family	Species	State
Aquatic	Mesogastropoda	Hydrobiidae	*Amnicola cola*	AR
			Amnicola stygia	MO
			Antrobia culveri	MO
			Antrorbis breweri	AL
			Antroselates spiralis	KY, IN
			Balconorbis uvaldensis	TX
			Dasyscias franzi	FL
			Fontigens antroecetes	IL, MO
			Fontigens proserpina	MO
			Fontigens tartarea	WV
			Fontigens turritella	WV
			Holsingeria unthankensis	VA
			Phreatoceras conica	TX
			Phreatoceras taylori	TX
			Phreatodrobia coronae	TX
			Phreatodrobia micra	TX
			Phreatodrobia nugax	TX
			Phreatodrobia plana	TX
			Phreatodrobia punctata	TX
			Phreatodrobia torunda	TX
			Stygopyrus bartonensis	TX
		Physidae	*Physa spelunca*	WY
Terrestrial	Stylommatophora	Carychiidae	*Carychium stygium*	KY, TN
		Endodontidae	*Helicodiscus barri*	AL, GA, TN
		Zonitidae	*Glyphyalinia pecki*	AL
			Glyphyalinia specus	AL, GA, KY, TN, WV
			Pristiloma cavato	CA

State distributions are shown in the final column.
From Hershler and Holsinger (1990).

are pulmonates, except for the prosobranch *Pholeoteras euthrix*. By far the most common aquatic subterranean species are in the family Hydrobiidae, comprising 97% of all species. There are over 350 described species of aquatic obligate subterranean mollusks. The number of terrestrial cave-limited species is much less, probably less than 50. A summary of molluskan troglobionts and stygobionts is given in Table 1. What is noteworthy is the extraordinarily high diversity of hydrobiid gastropods at both the generic and species levels.

BIOGEOGRAPHIC PATTERNS OF CAVE MOLLUSKS

In the United States, there are 23 described species of aquatic obligate cave snails and 5 species of terrestrial obligate cave snails (Table 2). Within the United States, a hotspot of biodiversity is the Edwards Aquifer in Texas. This deep karst aquifer, which is intersected by numerous wells and caves, includes two endemic genera, *Phreatodrobia* and *Stygopyrus*. Of the 23 described aquatic cave snails, 11 are from the Edwards Aquifer (Hershler and Holsinger, 1990). The reasons for this high diversity are unclear, especially which ecological conditions allow the coexistence of this many snail species. It is known that unlike other U.S. cave areas (but similar to some regions of northern Mexico), snails in the Edwards Aquifer arose not only from old freshwater groups as did other U.S. cave snails, but they also arose from marine ancestors of groups during the late Cretaceous ocean embayments. These species appear to be stranded relicts that resulted from the regression of the Cretaceous seas. The Edwards Aquifer is also a hotspot of subterranean biodiversity for other groups, especially amphipods.

On a worldwide basis, two areas are likely to be hotspots of subterranean snail biodiversity. One is the Dinaric Mountains, ranging along the Adriatic Coast from extreme northeast Italy to Montenegro. In the Slovenian part of the Dinaric Mountains, for example, there are 37 aquatic obligate cave snails, 1 aquatic cave clam, and 11 species of terrestrial obligate cave snails. The entire Dinaric Mountain region has several times that many species. Surface freshwater in this region is also extraordinarily rich in mollusk diversity, especially among aquatic species. A possible explanation for the richness of surface species is the long freshwater history combined with the extremely fragmented nature of surface waters. This fragmentation occurs because of the high frequency of surface streams sinking into subterranean watercourses, resulting in increased chances for speciation. In addition, diversity of mollusks may be higher in regions of carbonate rocks,

FIGURE 1 The hydrobiid snail *Marstoniopsis croatica*. Photo by Boris Sket.

because this greatly increases the availability of calcium carbonate for the construction of shells. The high diversity of subsurface species in the region may be explained by the same factors, as well as the high density of caves and the development of extensive subterranean streams and rivers. A sampling of the diversity of Dinaric cave snails is shown in Figures 1 and 2.

The second hotspot of subterranean snail diversity is East Asia, although this region has been studied less than the Dinaric Mountains. In East Asia it is the terrestrial cave snail fauna that is especially rich in species. The surface terrestrial fauna in karst areas is especially rich as well. A single karst hill in East Asia may contain between 60 and 100 species. A single cave on an island in Halong Bay in Vietnam yielded 17 terrestrial cave snails, 15 of which belong to a single family (Vermeulen and Whitten, 1999).

COMMENTS ON SELECTED SPECIES

The hydrobiid cave snail *Fontigens tartarea* occurs in a dozen or so caves in West Virginia. As far as we know, its habits are typical, at least for the genus. The maximum linear extent of its range is 200 km, and there is considerable variation among populations, enough so that the northern species were at one time considered to be a separate species. Variation includes

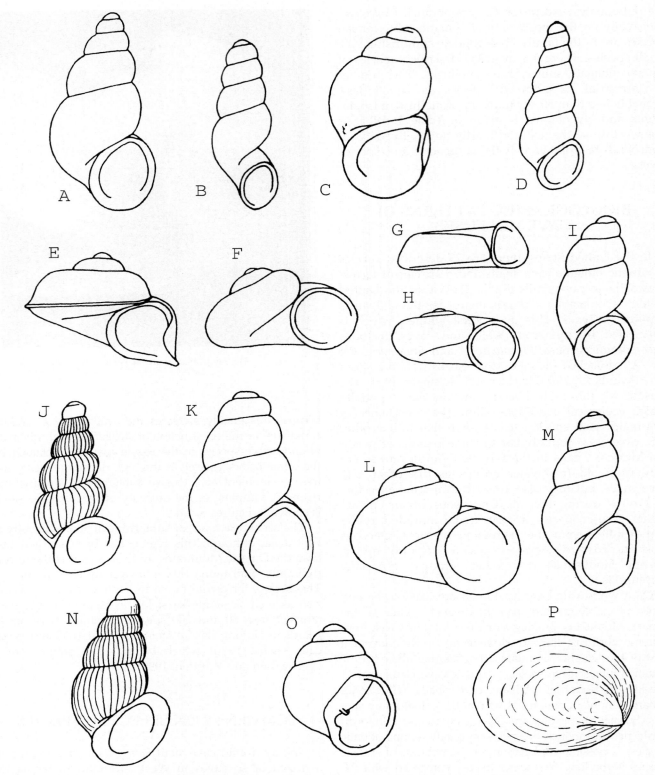

FIGURE 2   A sampling of stygobiotic mollusks from caves in the Dinaric Mountains. (A) *Anagastina hadouphylax*, 4 mm; (B) *Bagalivia karamani*, 1.7 mm; (C) *Belgrandiella kusceri*, 2.4 mm; (D) *Cilgia dalmatica*, 1.8 mm; (E) *Dalmatella sketi*, 2.2 mm; (F) *Erythropomatiana erythropomatia*, 1.5 mm; (G) *Hadziella ephippiostoma*, 1.6 mm; (H) *Hauffenia tellinii*, 1.8 mm; (I) *Istriana mirnae*, 2 mm; (J) *Lanzaia vjetrenicae*, 1.8 mm; (K) *Marstoniopsis croatica*, 3 mm; (L) *Neohoratia subpiscinalis*, 2.4 mm; (M) *Paladilhiopsis robiciana*, 4 mm; (N) *Plagigeyeria mostarensis*, 3.3 mm; (O) *Zoospeum exiguum*, 1.6 mm; (P) *Arcoloxus tetensi*, 4.5 mm. *Adapted from J. Bole and F. Velkorvh (1986).*

both size and shape of the shell, and indeed this may be a complex of species. Shell length ranges from 1 to 2.3 mm, so even large *F. tartarea* are tiny. In place in a cave stream, they look like large sand grains. It is very sporadic in its occurrence, occupying less than 10% of the cave streams within its range. Within cave streams it is also very sporadic in occurrence; it is very common in a few spots and absent in most of the stream. Some of this patchiness may be the result of inadequate collecting, but *Fontigens* has a reputation among cave biologists as being patchily distributed in the extreme. A possible explanation for this patchiness is that they may be associated with patchily distributed biofilms. Concentrations of *F. tartarea* are often on vertical faces of rocks in streams, yet another puzzle in its distribution.

The most unusual and unique cave mollusk is certainly the Dinaric cave clam *Congeria kusceri* (Morton et al., 1990). It is related to the well-known zebra mollusk. The genus *Congeria* was common in the countries of the western Balkans, Hungary, and Romania in the late Miocene, and its near demise occurred when the ancestral Mediterranean Sea dried up (the Messinian salinity crisis). *C. kusceri* escaped extinction by colonizing the subterranean waters that exited into the Mediterranean Sea. Living specimens are only known from Herzegovina and Croatia (shells are known from Slovenia). It attaches to cave walls typically in terminal lakes deep underground. Shell length reaches over 12 mm, and the mollusk often occurs in clusters. Shells are often covered by precipitated calcium carbonate and the tubes of the unique cave polychaete *Marifugia cavatica*. It appears to be able to emerge from the water, probably kept moist by water dripping from above. This seasonal drying is the probable cause of the growth rings observed on the shells of *C. kusceri*. Based on this inference, *C. kusceri* lives to an age of 25 years, compared to 1 or 2 years for surface-dwelling species in the same family. It also shows other demographic adaptations to cave life, possibly including internal fertilization and brooding of eggs.

Bibliography

Bernasconi, R., & Riedel, A. (1994). Mollusca. In C. Juberthie & V. Decu (Eds.), *Encyclopedia biospeologica I* (pp. 54–61). Moulis, France: Société de Biospéologie.

Bole, J., & Velkovrh, F. (1986). Mollusca from continental subterranean aquatic habitats. In L. Botosaneanu (Ed.), *Stygofauna mundi* (pp. 177–206). Leiden, The Netherlands: E. J. Brill.

Hershler, R., & Holsinger, J. R. (1990). Zoogeography of North American hydrobiid snails. *Stygologia, 5*, 5–16.

Morton, B., Velkovrh, F., & Sket, B. (1998). Biology and anatomy of the "living fossil" *Congeria kusceri* (Bivalvia: Dreissenidae) from subterranean rivers and caves in the Dinaric karst of former Yugoslavia. *Journal of Zoology, London, 245*, 147–174.

Vermeulen, J., & Whitten, T. (1999). *Biodiversity and cultural property in the management of limestone resources. Lessons from East Asia.* Washington, D.C.: The World Bank.

MORPHOLOGICAL ADAPTATIONS

Kenneth Christiansen

Grinnell College

INTRODUCTION

Morphological adaptation is any evolutionary modification of the morphology of lineages of organisms associated, in this case, with their existence in caves. Stewart Peck (1998) usefully divided these changes into two groups, *regressive* and *progressive*. Regressive adaptations involve reduction or loss of systems that occur in surface-dwelling organisms. Progressive adaptations involve enlargement, modification, or development of systems not seen in most surface-dwelling organisms.

TROGLOMORPHY

The term *troglomorphy* (Christiansen, 1962) designates both regressive and progressive evolutionary features associated with cave life. Though the term was originally used only for morphological features, subsequent study has shown that it applies equally well to behavioral and physiological features. In aquatic forms, where underground waters are frequently not associated with caves, the equivalent term *stygomorph* is sometimes used. These two terms serve to identify cave-adapted organisms without the unprovable, and sometimes erroneous, designation of troglobite. Troglomorphic features are essentially the same as those discussed under the rubric *Le Facies Cavernicole* (Vandel, 1964). Though characteristic of cave animals, they are by no means limited to cave animals. Many of the regressive features associated with the absence of light are seen in other environments where light is absent or greatly reduced, such as soil and the abyssal benthos. Some environments, such as soil and microcaverns, serve as recruiting grounds for cave organisms. Because of the lack of light, organisms in these environments rely on nonvisual methods of food location that, combined with the common habit of feeding on dead vegetable matter or fungi, preadapt them to survival in caves.

Not all cave organisms display characteristics of troglomorphism. Organisms with troglomorphic characteristics have been called *cave-dependent*, and those organisms without troglomorphic features are called *cave-independent*. The existence of these two types of characteristics facilitates the study of evolution within cave organisms. Cave-independent features enable the determination of the lineages to which different species belong, and the cave-dependent features, which always display polarity, enable the determination of their relative evolutionary position within a lineage.

EVOLUTION OF TROGLOMORPHY

Where troglomorphy does occur, it involves parallel or convergent evolution of different lineages. This convergence can be seen clearly in the Collembolan genera *Sinella* and *Pseudosinella* of the family Entomobryidae, where similar foot modifications occur in many European, North American, and Japanese lineages (Fig. 1). Convergence can also be seen in the head flattening in different fish and amphibian families (Fig. 2) and in physogastry and pseudophysogastry, which occur in beetles (Fig. 3) and at least one Hemipteran.

Troglomorphic changes vary in their generality. Many troglomorphic changes are limited to one order or even to one or a few families, including the foot modifications seen in Collembola (Fig. 1) and in Hawaiian cave planthoppers. Another taxonomically limited troglomorphy is the development of a ridge on the ventral surface of the thorax in the tribe Leptodirini of the beetle family Leiodidae. The clear paedomorphosis seen in the Urodele amphibia and the pseudophysogastry of beetles of the subfamilies Catopocerinae and Cholevinae of the family Leiodidae, and the subfamily Trechinae of the family Carabidae (Fig. 3) are other examples.

Somewhat more widespread is scale loss in some species of four families of teleost fish. The reduction of the swim bladder is more widespread, reduced in some members of three families and totally lost in two families.

Some troglomorphic features are very widespread. Table 1 summarizes some of the major morphological

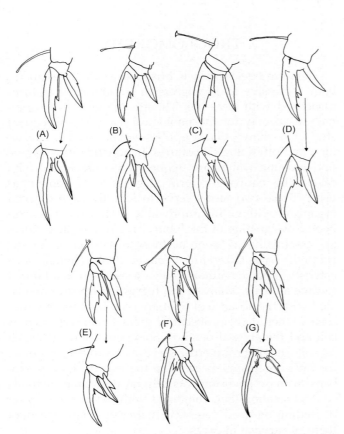

FIGURE 1 Parallel evolution in caves of the foot complex of cave Collembola. (A) *Pseudosinella* in the United States; (B), (C), (D) different lineages of *Pseudosinella* in Europe; (E), (F) different lineages of *Sinella* in the United States; (G) *Sinella* in Japan. Least troglomorphic species above and most troglomorphic species below.

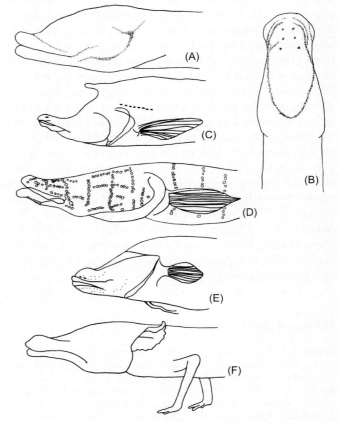

FIGURE 2 Extreme convergent troglomorphic head flattening in cave vertebrates. (A), (B) lateral and dorsal view of the head of *Ophisternon inferniale* (Synbranchidae); (C) *Synocyclocheilus hyalinus* (Cyprinidae); (D) *Spleoplatyrhinus poulsoni* (Amblyopsidae); (E) *Lucifuga subterraneus* (Bythitidae); (F) *Typhlomolge rathbuni* (Plethodontidae). *Adapted from Christiansen, 2004; and Weber, 2000.*

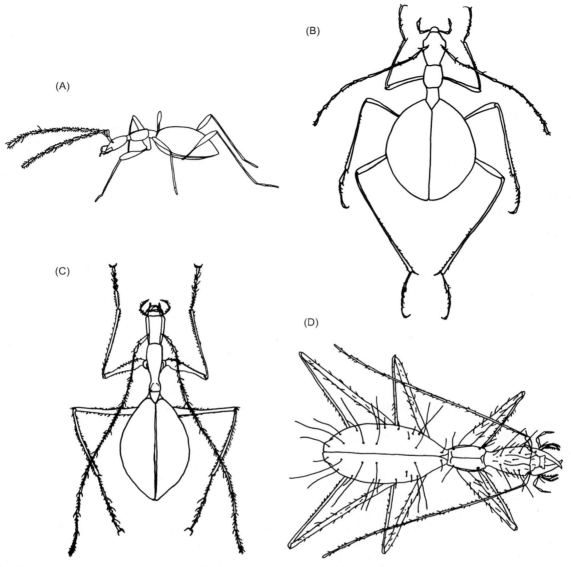

FIGURE 3 Convergent pseudophysogastry in different lineages of cave beetles: (A) Trechinae, *Aphaenops pluto* (after Vandel, 1964); (B) Cholevinae, *Glacicavicola bathysciodes* (courtesy Stewart Peck); (C) Bathysciinae, *Antroherpon scutulatum* (courtesy Pier Mauro Giachino); and (D) Bathsciinae, *Leptodirus hohenwarti*. After Vandel, 1964.

troglomorphic features. These features include head flattening in vertebrates (Fig. 2); thinning of the cuticle in terrestrial arthropods; and the elongation of appendages and body form seen in some fish, amphibia, and arthropods. Size increase is a troglomorphic characteristic in several orders of small arthropods, as is the reduction or loss of wings in most troglobitic pterygote insects. Some troglomorphic characteristics appear to be almost taxonomically unlimited, occurring in virtually all species that show significant troglomorphy. This category includes pigment loss and eye reduction, which are the most common regressive adaptations. An increase in putative nonvisual sensory systems is also virtually universal.

REGRESSIVE EVOLUTION

Much of the controversy surrounding the study of morphological change in caves has been about regressive modifications. Only a few results of evolution have created a wealth of theories and hypotheses comparable to those devoted to explaining the effects of regressive evolution. The exponents of these theories often use cave organisms as primary model systems to support their arguments. At least 14 hypotheses have been advanced to explain the mechanisms of regressive changes. Most of these hypotheses are discussed in more detail in Barr (1968; pp. 69–80). Ten such hypotheses that are of historical or current interest will be dealt with in this article.

TABLE 1 Common Troglomorphic Characteristics

Specialization of sensory organs (*e.g.,* touch, chemoreception)

Elongation of appendages

Pseudophysogastry

Reduction of eyes, pigment, wings

Compressed or depressed body form (Hexapoda)

Increased egg volume

Increased size (Collembola, Arachnida)

Unguis elongation (Collembola)

Foot modification (Collembola, planthoppers)

Scale reduction or loss (teleost fishes)

Loss of pigment cells and deposits

Cuticle thinning

Elongate body form (teleost fishes, Arachnida)

Depressed, shovel-like heads (teleost fishes, salamanders)

Reduction or loss of swim bladder (teleost fishes)

Decreasing hand femur length, crop-empty live weight ratio (crickets)

Five of these hypotheses are primarily of historical interest. Three of them rely on theories that directly oppose the now well-established core tenets of the Neo-Darwinian thesis. These theories are (1) Lamarckism and Neo-Lamarckism, which rely on the inheritance of acquired characteristics; (2) orthogenesis; and (3) Vandel's organicism. Organicism states that lineages go through three stages: birth, specialization, and senescence. Under this model, degenerative changes are characteristic of lineage senescence. Thus, Vandel hypothesizes that all cave species are in the senescent evolutionary stage. They show regressive evolution, not because of environment or selection, but because of their age status in the lineage. No recent evidence has been presented to support Vandel's theory.

Two other hypotheses of historical interest are Lankester's escape hypothesis and Ludwig's trap hypothesis. The escape hypothesis states that weak-eyed organisms falling into caves were unable to escape. The trap hypothesis holds that weakly pigmented surface forms would be sensitive to light and thus would concentrate in caves. These two hypotheses were never widely supported; instead they were dealt a deathblow by the work of Fong and Culver (1985) which showed large-eyed individuals of a highly variable amphipod population were actually more rather than less photophobic than small-eyed individuals. None of these five hypotheses is extensively studied today, but a few are being pursued.

Two explanations still being considered have received no recent support. The first of these is material compensation. This hypothesis comes in many forms but all of them presuppose that, in the food-poor cave environment, selection will favor the most economical use of the energy furnished by food and thus will favor reduction of useless structures or processes. This hypothesis has run into difficulties. Recent discoveries show that regressive evolution can occur in cave environments where there is an abundant supply of food. These environments include some tropical caves and caves where chemoautotrophic bacteria furnish ample food supply. Even in caves with limited food, the evidence to support a theory of material compensation is weak. As Poulson (1963) noted in his detailed study of this problem, the calculations do not clearly support or disprove the idea of selective reduction of "useless" traits as having a positive effect on energy economy. In the cavefish *Astyanax* the reduction of the eyes does not appear to result in energy savings as measured by egg yolk consumption in embryos. Moreover, the caves where blind fish are found are reported to be food-rich. More data on the genetics of cave organisms are needed to settle this debate.

Another explanation is Heuts' (1951) negative allometry hypothesis, in which he posits that organs appearing early in development and growing rapidly at first, such as eyes, are more likely to be negatively affected by lowering growth rates. This thesis is supported by some data but is generally more descriptive than explanatory.

Most recent studies have involved either the increase in the number of neutral genes having a disruptive and/or reductive effect on functionless organs, associative selection with adaptive structures, or some combination of the two. Poulson (1986) and others have strongly supported the disruptive effect of the accumulation of neutral mutations as the most plausible explanation for regressive evolution. The fact that, in the course of increasing troglomorphy, progressive features tend to show less variability than regressive ones would tend to support the accumulation of genes in regressive structures. Sket (1985) challenges the generality of this increase in variability. Culver and Wilkens (2000), in their recent work on the topic, suggested that the accumulation of mutated genes and associative selection reinforce each other in regressive evolution. Hypotheses of this sort afford the best available explanation for regressive evolution.

There are still problems. First, it is questionable whether any single hypothesis can explain all of the aspects of regressive evolution. Moreover, a number of ideas exist concerning regressive evolution that has not been tested in cave animals. One such idea is Regal's

(1977) noise suppression hypothesis. This hypothesis posits, on a basis of cybernetic theory, that natural selection is constantly acting to squelch nonessential messages within developing organisms. This action reduces metabolic noise and hence improves evolutionary fitness. In addition, a few earlier hypotheses, such as orthogenesis, have recently been revisited and revised. Several students of evolution suggest that this concept is misunderstood and actually denotes that there is a limited array of variations, which can occur and be functional in any group of organisms. This results in the fact that very similar characteristics occur under different ecological conditions and independently of selection. Regressive evolution remains a fruitful field for future research. The study of progressive modifications in cave life has produced less controversy.

PROGRESSIVE TROGLOMORPHY AND ADAPTATION

Though some progressive evolutionary modifications may be nonadaptive, they most putatively increase their owner's chances of survival or competitiveness in caves. There have been a number of works dealing with such adaptations in specific groups (Christiansen, 1965; Poulson, 1963; Hobbs, 2000; Niemiller and Poulson, 2010). The general topic of cave adaptation has also received a number of reviews (Vandel, 1964; Barr, 1968; Culver, 1982; Culver et al., 1995; and Culver and Pipan, 2009). External morphological adaptation has been studied far more than internal.

Regressive troglomorphy is commonly found in troglophilic forms, but progressive troglomorphy is usually limited to troglobitic or nearly troglobitic forms (those very rarely found outside caves). Progressive troglomorphy is, however, not universal among such cave organisms. In order for it to occur, three factors have to be present: (1) selection pressure for the development of a particular characteristic; (2) the genetic and physiological or behavioral ability of the organism to respond to this pressure; and (3) sufficient time evolving in caves to develop the adaptations. Many cave organisms lack one or more of these factors. Indeed, the majority of troglophiles and some troglobites show little or no progressive troglomorphy, probably because one or more of the factors necessary to morphological adaptation is absent. Thus, there are many species of fish and crayfish found in caves that show no troglomorphism. Many extremely edaphic Collembola, such as the members of the family Onychiuridae, already unpigmented and eyeless in all surface habitats, rarely show further troglomorphy, although they have many troglobitic species. In many other groups, troglomorphy is questionable or inconsistent (Culver, 1982). Troglomorphy is also usually absent in some cave environments, such as localized guano piles or large masses of organic debris that are extremely energy-rich. Peck (1973) pointed out that troglomorphy should not be normal in these cases, because strong troglomorphy only occurs when organisms are exploiting large-volume spaces such as cave wall or floor surfaces or large bodies of water. The cases where troglomorphy occurs in food-rich regions (some tropical caves and chemoautotrophic caves) most commonly involve only the regressive features of eye and pigment reduction. This fact supports the hypothesis that the mechanisms of regression differ from those of progressive adaptation. It is interesting that Boutin and Coineau (2000) have found something similar to troglomorphy in phreatobites, which live in narrow spaces, where small size and elongate body form are developed convergently.

EFFECTS OF ENVIRONMENTAL FACTORS ON ADAPTATION

There is a strong correlation between the environmental factors found in caves and the development of progressive troglomorphy. While earlier works suggested non Neo-Darwinian explanations for adaptation in cave animals (Vandel, 1964), almost all recent work has been done with the clear assumption of the applicability of the Darwinian core tenets. A possible recurrence of a view similar to that held by Vandel can be seen in the emphasis on phenotypic plasticity espoused by Romero (2009); however, the fact that similar positive troglomorphic features evolve in many different cave regions and lineages implies an adaptive control over their genetic development.

The nature of this adaptive value is sometimes completely unknown, as with the mesosternal carina development in beetles and the flattening of heads in fish and salamanders. In some cases, various untested hypotheses for these adaptive values have been advanced. For example, the large air space under the elytra in pseudophysogastry has been theorized to serve a respiratory function by increasing the air exposure of the thin abdominal membrane and by serving as a bubble lung during flooding. It has also been explained as a flotation device. In other cases the adaptive value appears to be self-evident, as with the increase in motion sensory organs in fish and elongation or hypertrophy of tactile organs generally. In some cases these apparently adaptive functions are associated with specific habits and/or habitats. Thus, the extremely delicate or gracile body form and elongate antennae of the cave crayfish (Fig. 4) *Troglocambarus maclanei* is associated with its normal habitat on walls or ceilings of flooded cave passages.

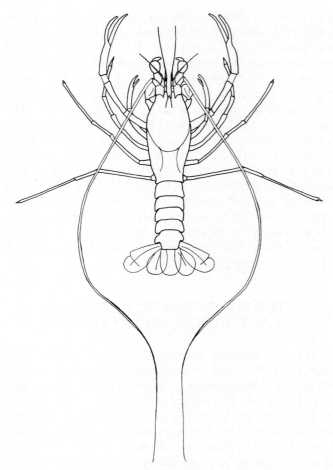

FIGURE 4 Extreme troglomorphy in the crayfish *Troglocambarus maclanei*. Adapted from Hobbs, 2000.

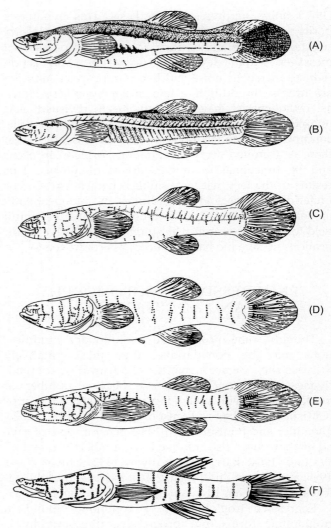

FIGURE 5 The species of Amblyopsidae showing increasing neuromast number: (A) *Chologaster cornutus*; (B) *Forbesichthys agassizi*; (C) *Typhlichthys subterraneus*; (D) *Amblyopsis spelaea*; (E) *Amblyopsis rosae*; and (F) *Speoplatythinus poulsoni*. Adapted from Weber, 2000.

There is general agreement that cave habitats are very specialized and, in some respects, very demanding. The normal scarcity and sporadic availability of food, the absence of light, and the difficulty of long-range dispersal pose problems. On the other hand, the relatively high and constant humidity, absence of wind currents, and limited predation and parasitism represent considerable advantages. All of these lead to selection pressures, which do not occur in surface habitats, and release from some selection pressures, which do occur there. These changes are consistent among caves all over the world.

EVIDENCE OF TROGLOMORPHIC ADAPTATION

Many recent studies have furnished direct evidence to support the adaptive value of troglomorphic changes (Christiansen, 1965; Culver *et al.*, 1995; and Poulson, 1963). These works indicate that those troglomorphic features that have been studied are subject to selection and are, in fact, adaptive.

In spite of all this, few studies have produced data to support the precise adaptive functions of specific troglomorphic developments. One of the best-studied examples of the adaptive function of troglomorphic change is that of the cavefish of the family Amblyopsidae. Here the great increase in neuromast motion sensory organs (Fig. 5) is associated with an increased ability to locate prey and with greater success in capturing it. This adaptation, combined with a variety of behavioral, physiological, and developmental changes, increase the chances of survival in cave waters. A second example occurs in the beetles of the subfamily Trechinae. Here the cave species that are crevice feeders tend to be small and flattened, whereas the cursorial feeders tend to be larger pseudophysogastric types (Fig. 3). Still another example is in the family Entomobryidae of the Collembola, where laboratory studies have shown that the foot adaptations seen in increasing troglomorphy are associated with behavioral changes that first allow for better movement over wet stone surface, then make for increased efficiency in moving over wet clay, finally resulting in an ability to walk on water surfaces (Fig. 6). In the course of this evolution, each change in behavior (the need to walk on wet stone, wet clay, or water) precedes a change in the direction of the development of the foot structure. Thus every troglomorphic characteristic in these cases is associated with the ability to deal with the particular environmental conditions found in caves.

The function of troglomorphic modification is also shown by analogous modifications in other habitats. For example, the extremely troglomorphic form of Collembola is most closely approached in two very different surface habitats. The foot structure is closest to that seen in aquatic Collembola, but the body shape and size are most similar to those seen in forms (largely tropical) that live in trees above the litter or soil. The foot structure is clearly related to the widespread occurrence of wet clay in caves, which makes movement difficult, and to the rimstone or other water pools, which act as traps for many Collembola. The body shape and size changes are probably associated with a method of predator escape. The freedom of movement furnished by open cave surfaces is more similar to conditions found in the arboreal habitat than to the normal surface habitat in soil or leaf litter where cave forms are recruited. Another example of analogous modification is the hypertrophy of organs for nonvisual location of food in both deep-sea and cave-dwelling fish. Wing reduction or loss, which is characteristic of cave insects, is also seen in environments such as soil and high alpine areas where wings are nonadaptive. The reduction of eyes and pigment in cave forms occurs also in dark edaphic and microcavernicole habitats. The loss of the epicuticular wax layers seen in caves is also seen in edaphic habitats where humidity is consistently high, as it is in habitable caves.

OCCURRENCE OF MORPHOLOGICAL TROGLOMORPHY

One of the mysteries of the evolution of cave organisms is the fact that only certain groups of any major taxon evolve morphologically within caves. In some cases the opportunity to evolve is simply lacking due to dietary restrictions or lack of entrance to caves. Yet there are many instances where neither restriction applies. In Collembola, most genera do not reproduce in cave depths, but many occasionally do in entryways where large amounts of organic debris have accumulated. Members of the families Isotomidae and Sminthuridae are often found in deeper reaches of caves where piles of organic debris have accumulated, yet no genus of the Isotomidae and only one genus of Sminthuridae (*Arrhopalites*) have evolved a lineage of cave species. Similarly, the order Psocoptera, with more than 3800 species, has only 8 cavernicole species, none of which is clearly troglomorphic. In beetles, the highly variable families Scarabidae, Ptiliidae, and Tenebrionidae have no troglomorphic species; there is no *a priori* reason to expect this circumstance. In the widely varied family Carabidae, only 4 or 5 of the 40 tribes have developed troglomorphic forms.

The majority of the troglomorphic species of the beetle family Carabidae belong to a single tribe, the Trechini. Similarly, in the 17 tribes of the beetle family Leiodidae, only 3 tribes harbor troglobitic species and most of these belong to the tribe Leptodirini (Bathyscini). In freshwater teleost fish, about half the families found in caves show no species with troglomorphic features or any evidence of derivative

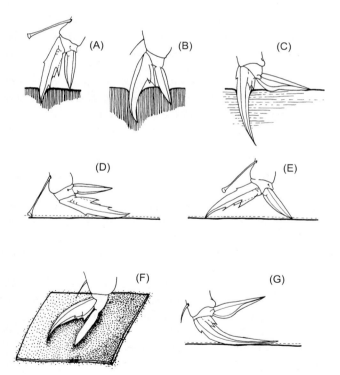

FIGURE 6 Morphological and behavioral changes seen in the process of adaptation of the foot complex of cave Collembola in the genus *Pseudosinella*: (A) typical weak penetration of clay seen in nontroglomorphic species; (B) intermediate troglomorphy showing increased penetration of clay; (C) extreme troglomorphy, capable of walking on water surfaces; (D), (E) nontroglomorphic position on smooth, wet, hard surfaces; (F) sideways position in forms having intermediate troglomorphy; and (G) typical foot position on hard, wet surfaces in highly troglomorphic species.

forms with them. Similarly, in the crustacean order Amphipoda, the large Holarctic family Gammaridae, with over 50 genera and over 300 species, has only 14 genera and 33 species that are troglobitic or stygobitic. In contrast, the family Crangonyctidae, with only 6 extant genera and about 145 species, has 5 troglomorphic genera and 116 such species. Even more striking is the Palearctic family Niphargidae with 8 genera and 207 species, all troglobitic or stygiobitic.

One possible explanation of these cases of failure in the evolutionary exploitation of caves is the absence of necessary exaptations or preadaptations. It has been generally agreed that preadaptation or exaptation is required for successful passage from trogloxene to troglophile. This idea is largely based on inference; little strong evidence to support it has been produced. It is true that many terrestrial troglobite lineages are recruited from habitats such as soil and microcavernicole cavities where suggested preadaptations or exaptations are developed. For example, soil organisms have to deal with survival in the absence of light and, and for most groups, poor food sources. Both of these conditions could serve as preadaptations for cave life. In addition, soil habitats usually have high humidity, as do most biologically active cave habitats.

Four facts argue against this idea as a major explanation. First, each of these environments is significantly different from that of caves. For example, food supply in soil is generally widely and continuously available but also very dispersed. In caves, the supply is usually concentrated and locally available but sporadic or periodic. Second, movement in soil is severely limited, where it is virtually unlimited for small organisms in caves. Third, there are many cases where the most primitive troglobite members of a cave lineage do not show morphological evidence of these preadaptations or exaptations. Fourth and finally, even if these exaptations or preadaptations are adequate to explain the successful invasion of caves, there are many cases of groups with similar potential preadaptations that are not successful. Examples exist in the Carabid beetles and in the Collembola. In the Carabidae, members of the tribe Anillini are eyeless and depigmented. Although most are edaphic, only one species in Alabama appears to be a true troglobite. This contrasts with the tribe Trechini, which has hundreds (if not thousands) of troglobitic species. In the Collembola, the family Entomobryidae has hundreds of troglobites, some showing the highest degree of troglomorphy. In contrast, the Isotomidae, with hundreds of depigmented, reduced-eye, or eyeless edaphobites, has only two troglobitic species, neither of which is strongly troglomorphic.

A possible explanation for these two cases is that physiological and/or behavioral preadaptations rather than morphological changes are crucial for survival in caves. The best evidence to support this explanation comes from the fish families with a large percentage of stygiobiont forms. In the family Amblyopsidae, where 5 of the 6 species are found in caves, the single non-cave species lives in stable habitats, shows low activity level, is nocturnal, and has the ability to feed and orient itself in the dark. All these features could be adaptive in a cave environment. The surface-living forms of the family Synbranchidae include about 17 species, 4 of which are found in underground waters and show similar probable preadaptations. They are mostly light-avoiding species feeding at night and able to move through small spaces. Another piece of evidence to support the idea of nonmorphological preadaptation is the widespread existence of opportunistic troglophile species that lack morphological troglomorphism and are able to survive in caves. These species thrive where they do not face competition from troglomorphic species and where food resources are abundant.

More definitive evidence concerning the role of preadaptation or exaptation in the development of troglomorphy could be obtained by comparing groups successful in exploiting caves with those that are not successful. For instance, compare the behavior and physiology of surface species of the highly troglomorphic fish family Amblyopsidae and Synbranchidae with those of the family Centrarchidae that has 18 species, 11 of which are found in caves, none showing any troglomorphy. In the family Cichlidae, with over 900 species with only 2 reported in caves, there are no troglomorphic species. For a terrestrial example, compare the following: the collembolan genus *Folsomia*, with 190 primarily soil species, many with reduced eyes and pigment, has no troglomorphic cave species, and the genus *Pseudosinella* contains 280 species, the majority of which are troglomorphic troglobites.

Where troglomorphy does occur, it allows for a separate analysis of those features clearly affected by the cave environment (cave-dependent) and those that are unaffected by it (cave-independent). The cave-dependent features show a great deal of convergence and parallelism, giving clear polarity for phylogenetic analysis. This permits a measure, or least an indication, of the degree of cave adaptation of different groups of organisms. An unfortunate corollary is that the large amount of resultant homoplasy makes cladistic analyses of cave animals difficult. Nevertheless, the existence of these two types of features makes caves an excellent natural laboratory for studying evolutionary processes.

The *superficial underground environment* (MSS) is a specialized microcavernicole habitat that is frequently in direct contact with caves. Here we find many, but by no means all, of the fauna of caves. In a sense, this can be looked on as an extension of the cave habitat but with an absence of bodies of water or large open spaces. Because this environment is not well studied, it furnishes a good field for future research.

EXAMPLES OF DETAILED ANALYSIS OF TROGLOMORPHIC EVOLUTION

Wherever progressive troglomorphy occurs, it is clear that understanding the factors that produce it requires detailed study of the organisms involved. In the next section, two such studies will be considered and what they tell us about the development of these adaptations will be discussed.

The first study deals with interspecific troglomorphic changes among the members of the North American teleost fish family Amblyopsidae. This small family has six known species (Fig. 5), all limited to the southeast quadrant of the United States: one surface species (*Chologaster cornuta*), one troglophile (*Forbesichthys agassizi*), and four troglobite species (*Typhlichthys subterraneus, Amblyopsis spelaea, Amblyopsis rosae*, and *Speoplatyrhinus poulsoni*). The single surface species is not found near any region occupied by the troglophile or troglobite species of the family, but it is found in slow streams and swamps on the coastal plain from southern Virginia to southern Georgia. The other species have either overlapping or nearly adjunct ranges (Fig. 7). Although many have studied these fish, Thomas Poulson did the definitive work (1963). There is a clear gradient of increasing troglomorphy as we move from *C. cornuta* to *S. poulsoni*. The troglomorphic changes involve eye, pigment, and optic lobe reduction; increase in neuromast number, size, and exposure; hypertrophy of tactile, olfactory, and equilibrium receptors and their associated brain centers; and enlargement and flattening of the head. The single genetic study of the members of this family points to a multiple origin for different populations of the least troglomorphic of the troglobite species, *T. subterraneus*. This conclusion is reinforced by the wide and disjunctive distribution of this species. Thus, the evidence is strong that troglomorphic features seen in *T. subterraneus* were developed by parallel evolutionary changes. Whether the increasingly troglomorphic features of the remaining troglobite species were developed from some *Typhlichthys*- like ancestor or from now extinct troglophilic forms remains in doubt. The existence of clines in *A. spelaea* from south to north with the most troglomorphic condition occurring in the southern populations would seem to support the former hypothesis. In addition, the extreme troglomorphy of *S. poulsoni* argues for ancestry from some already troglomorphic form. In any case, the widely disjunctive distributions of the highly troglomorphic *A. rosae* and *A. spelaea* strongly support the parallel evolution of their troglomorphic features.

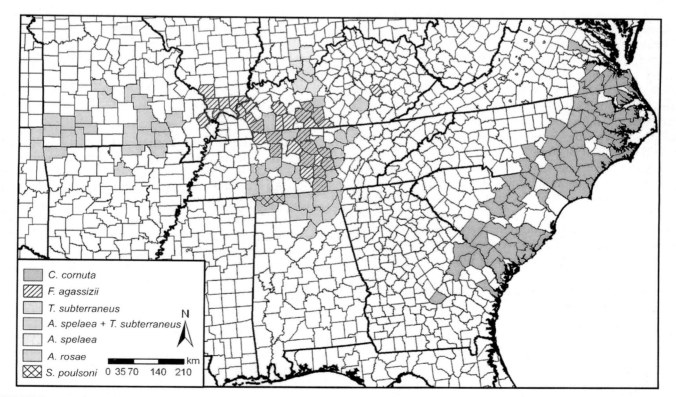

FIGURE 7 Distribution by county of Amblyopsidae in the eastern United States. Only the swampfish *Chologaster cornuta* is found outside the Ozarks and Interior Low Plateaus. *From Niemiller and Poulson, 2010.*

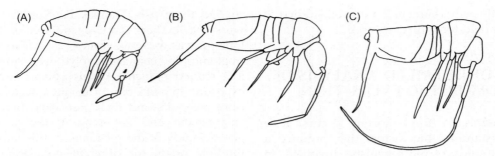

FIGURE 8 Members of the *Pseudosinella hirsuta* lineage: (A) *P. dubia*; (B) *P. hirsuta*; and (C) *P. christianseni*.

A second notable feature is the lack of congruent development of troglomorphic features. The overall gradient for troglomorphy is clear but only one feature, olfactory rosette size, follows this gradient strictly. Eye size shows a sharp reduction from *C. cornuta* to *S. poulsoni*, but the eyes of *T. subterraneus* are smaller than those of the generally more troglomorphic *A. spelaea*. Similarly, although there is a striking increase in neuromast number from *C. cornuta* to *S. poulsoni*, the number in *Amblyopsis* is smaller than that of *Typhlichthys*. A third notable feature is the increasingly limited distribution with increasing troglomorphy in the troglobitic species (Fig. 7). *Typhlicthys subterraneus*, the least troglomorphic of these, is scattered over five states. In contrast, *S. poulsoni*, the most troglomorphic, is found in only one cave system.

The second study involves the troglobitic collembolan *Pseudosinella hirsuta*, studied in great detail by Christiansen and Culver (1968). This study involved the geographic variation of 25 populations of *P. hirsuta* using 12 cave-dependent and 3 cave-independent morphological features as well as one behavioral feature. This species belongs to a lineage of three species (Fig. 8) including *P. dubia* and *P. christianseni*. *P. dubia* shows little progressive troglomorphy and has a widely disjunctive distribution of three caves in Washington County, Arkansas, and one cave in Dent County, Missouri. *Pseudosinella hirsuta* is common and widespread in caves of central Kentucky and Tennessee. It has some populations in caves of northeast Alabama and two in adjacent northwest Georgia. In addition, there are populations in eastern Tennessee and the westernmost part of Virginia (Fig. 9). Assuming that the troglophile ancestor of *P. hirsuta* resembled *P. dubia*, *P. hirsuta* already shows significant increase in troglomorphy. In some other features, such as the foot structure, *P. hirsuta* shows a wide range, varying from one similar to *P. dubia* to one similar to *P. christianseni*. The most troglomorphic of the species, *P. christianseni*, occurs in small, scattered locations in the eastern part of the range of *P. hirsuta*.

All cave-dependent features of each cave population show similar states of advanced or primitive cave adaptation, with the exception of the extremely

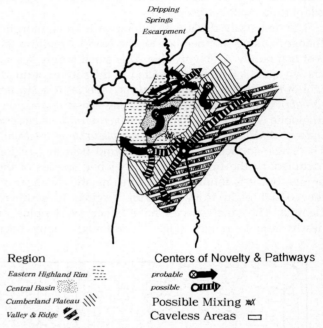

FIGURE 9 Centers of novelty and pathways of evolutionary distribution of *Pseudosinella hirsuta*.

phenotypically plastic mesothorax. On this basis, each population can be placed in one of several sequences of advanced to primitive conditions. Each sequence is largely or entirely limited to a single geological region with a single evolutionarily advanced region and one to five primitive regions, generally marginal (Fig. 9). The populations' cave-independent features do not follow the same pattern, but do give some indication of decreasing variability from advanced to primitive populations. Each sequence of caves contains most or all of the variations seen within the species.

The main conclusions from this study regarding troglomorphism are as follows.

1. The characteristics of *Pseudosinella hirsuta* have evolved independently in at least four different places.
2. *Pseudosinella christianseni* represents a continuation of the evolutionary trends seen in *P. hirsuta*. The two species have never been found in the same

cave, but where they exist in caves near each other, the populations of *P. hirsuta* are usually among the most highly troglomorphic seen in the species.
3. Dispersal of *P. hirsuta* occurs primarily via underground routes. There is no indication of morphological discontinuity between populations in a given cave sequence.
4. There is morphological evidence for genetic discontinuity between different cave sequences.
5. Most importantly, it appears that the extremely troglomorphic *P. christianseni* evolved at four different times from *P. hirsuta*-like ancestors.
6. The ranges of the troglobitic species tend to reduce with increased troglomorphy.
7. The putative ancestral species is no longer found near the ranges of the more troglomorphic species.

These patterns are all seen in many other groups of cave organisms but seldom with the detail and clarity seen in these two examples.

CONCLUSIONS

1. Morphological cave adaptation or troglomorphy is seen in many, but not all, troglobitic animals.
2. The development of troglomorphy requires a number of conditions and is still not clearly understood.
3. Troglomorphy can be either regressive, involving the reduction or loss of features, or progressive, involving alteration or expansion of preexisting features or the development of new features.
4. Progressive and regressive troglomorphy probably involve different combinations of mechanisms.
5. All troglomorphic development involves extensive parallelism and convergent evolution.
6. The occurrence of these troglomorphic (cave-dependent) features alongside nontroglomorphic (cave-independent) features, which show no parallelism or convergence, makes caves an ideal environment in which to study evolution.
7. These trends are associated with the unusual, and often demanding, environmental conditions of caves, which are very different from those in surface habitats but similar in the great majority of caves throughout the world.
8. The common result of these environmental conditions is that the initial development of troglobitic forms results in equivalent levels of troglomorphy evolving in many different places.
9. The higher the level of troglomorphy within troglobitic groups, the more restricted the distribution of the species and the more limited their geographic range.
10. Forms showing low levels of troglomorphy are often troglophilic, whereas those showing high levels are almost always troglobitic.
11. Although it appears likely that development of troglomorphy is facilitated by preadaptations or exaptations for cave life, the reasons why only a few groups of organisms develop this remain unclear.

See Also the Following Articles

Adaptation to Low Food
Astyanax mexicanus: *A Model Organism for Evolution and Adaptation*
Gammarus minus: *A Model System for the Study of Adaptation to the Cave Environment*
Natural Selection

Bibliography

Barr, T. C. (1968). Cave ecology and the evolution of troglobites. *Evolutionary Biology, 2*, 35–102.

Boutin, C., & Coineau, N. (2000). Evolutionary rates and phylogenetic age in some stygobiontic species. In H. Wilkens, D. C. Culver & W. F. Humphreys (Eds.), *Subterranean ecosystems* (pp. 433–451). Amsterdam: Elsevier.

Christiansen, K. A. (1962). Proposition pour la classification des animaux cavernicoles. *Spelunca, 2*, 76–78 (in French).

Christiansen, K. A. (1965). Behavior and form in the evolution of cave Collembola. *Evolution, 19*, 529–537.

Christiansen, K. A. (2004). Adaptation: Morphological (external). In J. Gunn (Ed.), *Encyclopedia of cave and karst science* (pp. 7–9). London: Fitzroy Dearborn.

Christiansen, K. A., & Culver, D. C. (1968). Geographical variation and evolution in *Pseudosinella*. *Evolution, 2*, 237–255.

Culver, D. C. (1982). *Cave life*. Cambridge, MA: Harvard University Press.

Culver, D. C., Kane, T. C., & Fong, D. W. (1995). *Adaptation and natural selection in caves*. Cambridge, MA: Harvard University Press.

Culver, D. C., & Pipan, T. (2009). *The biology of caves and other subterranean habitats*. Oxford: Oxford University Press.

Culver, D. C., & Wilkens, H. (2000). Critical review of relevant theories of the evolution subterranean animals. In H. Wilkens, D. C. Culver & W. F. Humphreys (Eds.), *Subterranean ecosystems* (pp. 389–407). Amsterdam: Elsevier.

Fong, D. W., & Culver, D. C. (1985). A reconsideration of Ludwig's differential migration theory of regression evolution. *Bulletin of the National Speleological Society, 47*, 123–127.

Heuts, M. J. (1951). Ecology, variation, and adaptation of the blind African cave fish *Caecobarbus geertsi*. *Annales de Societe Royal Zoologie Belgique, 82*, 155–230.

Hobbs, H. H. (2000). Subterranean organisms—Crustacea. In H. Wilkins, D. C. Culver & W. F. Humphreys (Eds.), *Subterranean ecosystems* (pp. 95–108). Amsterdam: Elsevier.

Niemiller, M. L., & Poulson, T. L. (2010). Subterranean fishes of North America: Amblyopsidae. In E. Trajano, M. E. Bichuette & B. G. Kapoor (Eds.), *Biology of subterranean fishes* (pp. 169–280). Enfield, NH: Science Publishers.

Peck, S. B. (1973). A systematic revision and the evolutionary biology of the *Ptomaphagus* (*Adelops*) beetles of North America (Coleoptera, Leiodidae, Catopinae), with emphasis on cave-inhabiting species. *Bulletin of the Museum of Comparative Zoology, 145*(2), 29–162.

Peck, S. B. (1998). Cladistic biogeography of cavernicolous *Ptomaphagus* beetles (Leiodidae, Cholevinae: Ptomaphagini) in the United States. In: *Proceedings of the 20th International Congress of Entomology* (pp. 235–260). Torino, Italy: Museo Regionale di Scienze Naturali.

Poulson, T. L. (1963). Cave adaptation in amblyopsid fishes. *American Midland Naturalist, 70*(2), 257–290.

Poulson, T. L. (1986). Evolutionary reduction by neutral mutations: Plausibility arguments and data from amblyopsid fishes and linyphiid spiders. *Bulletin of the National Speleological Society, 47*, 109–117.

Regal, P. J. (1977). Evolutionary loss of useless features: Is it molecular noise suppression? *American Naturalist, 111*, 123–133.

Romero, A. (2009). *Cave biology. Life in darkness*. Cambridge, U.K.: Cambridge University Press.

Sket, B. (1985). Why all cave animals do not look alike: A discussion of the adaptive value of reduction processes. *Bulletin of the National Speleological Society, 47*(2), 78–85.

Vandel, A. (1964). *Biospeleologie—La Biologie des Animaux Cavernicoles*. Paris: Gauthier Villars (in French).

Weber, A. (2000). Subterranean organisms—Fish and amphibia. In H. Wilkens, D. C. Culver & W. F. Humphreys (Eds.), *Subterranean ecosystems* (pp. 109–113). Amsterdam: Elsevier.

MULTILEVEL CAVES AND LANDSCAPE EVOLUTION

Darlene M. Anthony

Roane State Community College, Harriman, TN

MULTILEVEL CAVES

Cave explorers were among the first to note that some caves were found at different levels within a hillside or valley wall. The opportunity to enter a cave via one entrance, journey through extensive horizontal passageways connected to each other by narrow canyons, and exit via a second, lower entrance was considered by many to be the ultimate trip in sport caving. The notoriety of multilevel (or tiered) caves increased with the overall depth of the entire system, or with the number of levels or entrances that could be traversed during a single trip underground. When multilevel cave passages were surveyed and drafted by cartographers, the map often resembled a jumble of ribbon in plan view. However, when viewed from the side, or in vertical profile, the passages fell neatly into narrow zones of horizontal development connected in a few key places by vertical passages. This particular pattern of cave development is found in many of the world's karst landscapes, where scientists are interpreting the role of multilevel caves in landscape evolution.

CAVES AND THE WATER TABLE

Why do some caves develop levels, whereas others do not? What characteristics do multilevel caves share? How many years does it take to form a level as large as a subway tunnel? A basic understanding of the relationship between horizontal conduits and the regional hydrologic system is vital to answering these questions. In a karst terrain, caves typically form where slightly acidic water finds flow paths (such as joints, bedding planes, or faults) in carbonate rocks, directing surface water downward to the regional water table. Caves often provide the fastest (and sometimes shortest) path for surface water to reach the regional base level, which is represented by major regional rivers. Because caves are formed by flowing water, cave development is strongly controlled by base-level elevation. In the *vadose zone* (above the base level), underground streams carve narrow canyons that lead downward in the fastest manner possible until the *phreatic zone* (at or below the base level) is reached. At the phreatic zone, cave streams form nearly horizontal tubes as water flows to a discharge outlet, which in many cases is a spring at or near the regional river. When surface rivers remain at the same elevation for long periods of time, called *stillstands*, cave streams develop large, horizontal passages at grade with the rivers (Fig. 1). When any event (such as a drop in sea level or tectonic uplift) causes the regional rivers to suddenly lower, or incise, the sudden difference in elevation creates disequilibria between the cave stream and the newly lowered base level. The increased hydraulic gradient causes the cave stream (now in the vadose zone) to preferentially downcut narrow canyons in the floor of the horizontal conduit, abandoning the upper level to keep grade with the

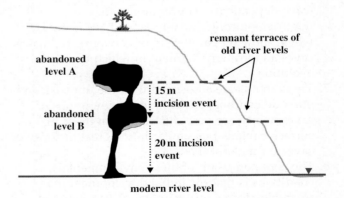

FIGURE 1 Generalized representation of a multilevel cave and its relationship with the regional base level shows the extensive horizontal conduits that form during long periods of base-level stability. These large passages are quickly abandoned in favor of lower levels during periods of accelerated river incision.

new base level. Differences between phreatic tubes and vadose canyons are often striking when exploring multilevel caves, and the vadose–phreatic transition clearly indicates a previous position of the regional base level punctuated by a sudden incision episode.

RIVER INCISION AND THE RECORD IN CAVES

Karst geomorphologists, who examine the relationship between cave levels and the landscape, recognized that as surface rivers incised, the regional base level lowered, leaving formerly active cave passages dry and forming new cave passages beneath the older. Abandoned cave levels were often found to coincide with the scars of ancient terraces and remnants of gravel deposits found on the surface of the river valley. But when did the river incise, abandoning the cave levels? Rivers are by nature a destructive force in the landscape, removing many of the depositional features that mark their passage. Reconstructing the position of a surface river over geologic time is made even more difficult by the weathering and erosion of remnant fluvial deposits and by the lack of appropriate methods (such as carbon-14) for determining the absolute age of unconsolidated gravel and sand.

In sharp contrast to surface streams, multilevel caves contain vast piles of sediments deposited by underground streams (Fig. 2). Subterranean capture of surface streams at sink points directs gravel, sand, and silt from surface highlands into caves, where sediment is transported along developing horizontal conduits to output springs. Banks of sediments are deposited in conduits in the same manner as sediments in surface streams, complete with flood deposits, channel cut-and-fill deposits, and other fluvial features. When an underground stream abandons a conduit in favor of a lower level, sediments left behind may remain undisturbed for millions of years, unaffected by either erosion or weathering. Because sediments in abandoned conduits were deposited by moving water, they represent the last time the conduit was an active part of the local hydrologic system. Conduit elevation and the timing of sediment burial serve as proxies for the paleoelevation of the water table over geologic time.

But how long ago was the sediment washed into the cave? This important question may now be answered with a new method of absolute age dating that compares the radioactive decay of atoms called *cosmogenic nuclides* in quartz-rich sediment. The following examples illustrate how karst geomorphologists are using burial dates from cave sediments to interpret the Plio-Pleistocene history of river incision, tectonic uplift, and erosion rates in various parts of the world.

FIGURE 2 In-place fluvial sediments (Bone Cave, Tennessee) preserve the record of a stream once occupying the abandoned conduit. Cosmogenic nuclides are measured to determine how long ago the cave stream was active and to constrain the timing of the incision event that abandoned the conduit. *Photograph taken by the author.*

LANDSCAPE INTERPRETATION USING MULTILEVEL CAVES

Water Table Positions at Mammoth Cave, Kentucky

The development of cave levels in the Mammoth Cave System of Kentucky (Fig. 3) is now firmly linked to changes in the elevation of the Green River (a modern tributary of the Ohio River), with passage morphology and sediment age determined by cosmogenic nuclides permitting a reconstruction of regional base level over the past 3 million years (myr) and a correlation of passage development and abandonment with regional and global events. Several periods of river stability punctuated by brief periods of accelerated river incision allowed the formation and abandonment of extensive horizontal passages at levels A and B. The most recent river stillstand ended 1.39 million years ago (Ma), with the abandonment of level C. Remnant terraces in the Green River gorge corresponding with cave levels may now be assigned ages, which correlate

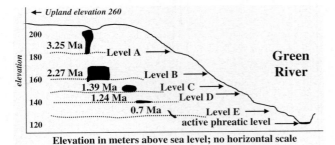

FIGURE 3 Generalized elevations of cave levels in Mammoth Cave, Kentucky, and the age of sediments deposited in each level. Cave levels correlate with remnant terraces and fluvial deposits along the Green River valley walls. Passage morphology and the age of sediments in Mammoth Cave constrain river stillstands and incision events during the past 3 myr.

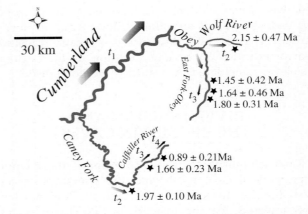

FIGURE 4 Abandonment of multilevel caves along tributaries of the Cumberland River, Tennessee, track the migration of a knickpoint originating on the Cumberland River (arrow) at time t_1 prior to 2 Ma and moving up tributaries, reaching caves at times t_2, t_3, and t_4. Differences in stream power account for the simultaneous abandonment of caves on separate tributaries.

with major ice advances in the Ohio River basin during the Plio-Pleistocene. For example, the abandonment of Mammoth Cave's level C at ~1.4 Ma was initiated by the blocking of major north-flowing rivers by an ice sheet that forced reorganization of North American drainage east of the Mississippi River, creating the modern Ohio River at ~1.5 Ma.

Cosmogenic nuclides in cave sediments may also be used to determine rates of surface processes while the sediments were deposited underground. The dates of abandonment at Mammoth Cave generate an incision rate of ~30 m myr^{-1} for the Green River. Erosion rates indicate that the sandstone-capped uplands surrounding the Green River gorge have maintained a steady rate of ~2–7 m myr^{-1} despite the accelerated river incision rate. Although surface valleys have deepened considerably, landscape features of the Appalachian Plateaus would have been recognizable more than 3 Ma.

For speleologists, the ages of levels B and C abandonment demonstrate that level C passages at Mammoth Cave needed at least 500,000 years to form, defining a minimum number of years for extensive conduits to develop in that particular limestone unit.

Incision Pulses on the Cumberland River, Tennessee

Whereas Mammoth Cave preserved a record of changes in the water table at one location, abandonment of multilevel caves along the valley walls of tributaries to the Cumberland River in Tennessee tracks the migration of incision pulses, or *knickpoints*, along this modern tributary of the Ohio River. Knickpoints are sudden changes in the gradient of a river's long profile, such as rapids or waterfalls, and may form as a response to a sudden drop in the water table. A knickpoint will migrate in the upstream direction at a rate directly related to the size of the river's total drainage area, with pulses moving faster on a larger river than its tributaries. As a knickpoint migrates up a larger river, knickpoints are initiated on the smaller tributaries (much like dominoes). The ages of sediments found in large, abandoned passages across the Cumberland River basin (Fig. 4) at similar elevations above the modern river level show that caves were abandoned sequentially beginning around 1.8 Ma as a knickpoint generated by incision of the Cumberland River moved up the Caney Fork and beyond the cave outlets. This incision pulse is older than the reorganization of the Ohio River basin, indicating that sea level lowstands at either 2.5 or 2.1 Ma are likely candidates for triggering this earlier incision event.

The simultaneous abandonment of passages on two tributaries of different lengths and drainage areas (the Caney Fork and the East Fork Obey rivers) illustrates how the same incision event may initiate knickpoints that migrate upstream of varying stream power, affecting two different caves at the same time. Incision rates for tributarie of the Cumberland River are also ~30 m myr^{-1}, and the erosion rates for the sandstone-capped Cumberland Plateau are a steady 2–7 m myr^{-1}, corroborating those found at the Green River.

Additionally, when knickpoint propagation up fluviokarst tributaries of the Cumberland River is modeled according to the stream power law $E = kQ^m S^n$ and tested using the dated incision events, the ratio m/n is consistent with published parameters but the values for m and n are much higher than those reported in previous field studies.

Incision Rates in the Sierra Nevada, California

Measurement of cosmogenic nuclides in cave sediments and bedrock surfaces, along with studies of landscape morphology, highlight the history of landscape development in the southern Sierra Nevada, California, over the past 5 myr. Cave sediments record moderate to slow river incision rates between ~5 and 3 Ma ($\leq$70 m myr^{-1}), acceleration between 3 and 1.5 Ma (on the order of 300 m myr^{-1}), and a decrease since (on the order of 20 m myr^{-1}). The acceleration of river incision is interpreted to be a response to a discrete tectonic event as opposed to global climate change. Topographic evidence and dated cave positions include the development of narrow inner gorges beginning at ~3 Ma that incised into existing river canyons. Tributary streams have not adjusted to the incising of the inner gorges, indicating that the response to Pliocene uplift has not yet migrated to the upland surfaces. Exposure dating of bedrock shows these surfaces to have eroded slowly over the past 3 myr at rates on the order of 10 m myr^{-1}, whereas adjacent rivers incised rapidly. Thus, climate change is not enough to explain uplift, and a tectonic event is favored.

Erosion Rate Increase in the Southern Rocky Mountains, Colorado

The measurements of cosmogenic nuclides in cave sediment, bedrock, and modern fluvial bedload record a significant increase in Pliocene erosion rates in the Rocky Mountains of southern Colorado. The burial age of cave sediments in Marble Mountain coupled with exposure dating of bedrock on the summit above the cave indicate a sudden erosion rate increase in the Rocky Mountains between 4.9 and 1.2 Ma. Based on passage development and morphology of cave levels in Marble Mountain, the increase likely occurred shortly before 1.2 Ma, coinciding with the transition from a warmer, more humid Pliocene climate to a cooler periglacial Pleistocene climate.

Young Uplift of the Eastern Alps, Austria

The development of multilevel caves in the Central Styrian Karst within the Mur River basin, Austria, records a complex history of changes in the water table. Were these changes due to incision or uplift? Studies indicate that the Mur River acted as an antecedent river in an area of recent uplift of the Eastern Alps, based on the topographic history of the basin, stream piracy by the Mur River, and Plio-Pleistocene incision rates in bedrock on the order of 100 m myr^{-1}. Further correlation of cave levels, stream terraces, and other erosional surfaces along the Mur River constrain the timing of landscape evolution along the Eastern Alps since 5 Ma.

The relationship between multilevel caves and the surrounding landscape has opened the door to many applications in geomorphology due to the ongoing development of burial and exposure dating of sediments and bedrock by measuring cosmogenic nuclides in quartz. Geomorphologists around the globe are continuing to examine multilevel caves in every type of karst terrain, measure the rates of geomorphic and tectonic processes, and test long-established models using these empirical data.

See Also the Following Article

Cosmogenic Isotope Dating of Cave Sediments

Bibliography

Anthony, D. M., & Granger, D. E. (2006). A new chronology for the age of Appalachian erosional surfaces determined by cosmogenic nuclides in cave sediments. *Earth Surface Processes and Landforms, 36*, 874–887.

Anthony, D. M., & Granger, D. M. (2007). An empirical stream power formulation for knickpoint retreat in Appalachian Plateau fluviokarst. *Journal of Hydrology, 343*, 117–126.

Ford, D. C., & Williams, P. W. (2007). *Karst hydrogeology and geomorphology*. New York: Wiley.

Granger, D. E., Fabel, D., & Palmer, A. N. (2001). Pliocene-Pleistocene incision of the Green River, Kentucky, determined from radioactive decay of cosmogenic ^{26}Al and ^{10}Be in Mammoth Cave sediments. *Geological Society of America Bulletin, 113*, 825–836.

Palmer, A. N. (1987). Cave levels and their interpretation. *National Speleological Society Bulletin, 49*, 50–66.

Stock, G. M., Anderson, R. S., & Finkel, R. C. (2005). Rates of erosion and topographic evolution of the Sierra Nevada, California, inferred from cosmogenic ^{26}Al and ^{10}Be concentrations. *Earth Surface Processes and Landforms, 30*, 985–1006.

Wagner, T., Fabel, D., Miebig, M., Hauselmann, P., Sahy, D., Xu, Sheng, & Stuwe, K. (2010). Young uplift in the non-glaciated parts of the Eastern Alps. *Earth and Planetary Science Letters, 295*, 159–169.

White, W. B. (1988). *Geomorphology and hydrology of karst terrains*. New York: Oxford University Press.

MULU CAVES, MALAYSIA

Tony Waltham and Joel Despain

Mulu Caves Project and British Cave Research Association; Gunung Buda Project and Sequoia, Kings Canyon National Park Service

One of the world's great tropical karst regions lies below Gunung Mulu (Mount Mulu) in the Malaysian state of Sarawak on the island of Borneo, off Southeast Asia. Here, the Miocene Melinau Limestone is exposed in a range of precipitous massifs, low mountains, and hills, in a northeast-oriented outcrop, 10 km wide and 35 km long, broken by the floodplains of the Medalam, Melinau, and Melinau Paku rivers. Within the limestone lie cave passages of which more than 440 km have been surveyed to date (Fig. 1). These include the largest cave room in the world (Sarawak Chamber), the largest cave passage in the world (Deer Cave), and perhaps the largest cave in the world (when measured by total volume of Clearwater Cave, with nearly 190 km of passages that are mostly very large).

The terrain around Gunung Mulu is spectacular and dramatic, due to the precipitous rise of the mountains above the coastal plain, the white limestone cliffs more than 100 m high around their lower margins, the prolific dense vegetation, and an almost persistent cloud cover. The three main limestone mountains, Gunungs Api, Benarat, and Buda, lie 75 km from the South China Sea at the edge of the interior highlands of Sarawak. They lie 4° north of the equator, and receive an annual average of more than 5000 mm of year-round rainfall.

The Melinau Limestone is a lenticular body, and the rock is very massive and white or gray with generally less than 1% insoluble material. It is lagoonal in origin, and was deposited within a reef complex. Fossils are locally common and include corals, bryozoans, bivalves, gastropods, and algal balls. The massive beds of the Melinau Limestone dip to the northwest, perpendicular to the trend of the range, and are folded with minor faulting. The limestone is overlain conformably by the Setap Shale, which is 4000–5000 m thick, and is underlain by the Mulu Formation, with 4000–5000 m of shale, sandstone, and orthoquartzite.

Local forest nomads were the first explorers of some of the caves. The Berawan, Tabuan, and Penan peoples still make use of the caves, some of which are burial sites. The caves also house several species of swiftlets, whose nests are used to make bird's nest soup, a highly prized Asian delicacy. Harvesting of nests at Mulu, and elsewhere in Southeast Asia, has not been on a sustainable scale, leading to a steep decline in the numbers of these birds, besides significant damage to some of the caves.

The first significant cave explorations in the rugged and remote forests of Mulu were made by G. E. Wilford of the Malaysian Geological Survey in 1961. He mapped Deer Cave (but both he and the readers of his report underestimated its size), the first part of Cave of the Winds (later connected to Clearwater Cave), and three inlet passages of the Terikan River Caves (but without venturing into the main underground river). His was a pioneering work, but it was only in 1978 that the Mulu caves were again visited, by British cavers on the Royal Geographic Society expedition sponsored by the Malaysian government. This project sought to survey the flora and fauna of the region and to study the geology and hydrology of the area in the newly designated Gunung Mulu National Park.

In 1978, the caves were almost perceived as an expedition sideline, but they proved to be truly spectacular. Consequently, the British cavers followed up with major expeditions in 1980 and 1984, and then with visits that have become annual since the National Park became more accessible; the Mulu Caves Project continues to make new discoveries. Parallel explorations by American cavers revealed many more caves in and around Gunung Buda during four expeditions between 1994 and 2003.

CAVES IN THE SOUTHERN HILLS

Separated from Gunung Api by the Melinau Paku valley, the southern hills contain Deer Cave (Gua Payau in Malay). This could be dismissed as a fragment of an ancient trunk cave that has been left in a limestone hill, now isolated from the Api massif. But it is the largest cave passage in the world, reaching 168 m wide and 125 m high along its southern half (Fig. 2). The very large passage size makes its geomorphology difficult to appreciate; phreatic origins have been lost in progressive roof collapse, and it appears that much of the passage was cut as a giant vadose canyon by the Melinau Paku River, which took this underground route for a very long time. The giant passage is almost entirely within reach of daylight for its kilometer course between the stream entrance in the Garden of Eden doline and the dry exit at the southern end. North of the Garden of Eden, Green Cave is its continuation only on a slightly smaller scale. A network of smaller cave passages lies within the hills parallel to Deer Cave.

The Garden of Eden is an almost enclosed chunk of rainforest within a massive doline on the edge of the limestone outcrop. It may be described as a very large but well-degraded tiankeng, with vertical cliffs still forming its margins on three sides. The concept of development by collapse of a very large cave chamber is reasonable when it is compared to the nearby Sarawak Chamber, which is not much smaller, though any large chamber may always have been open through an inlet from the shale outcrop on its southeast side.

CAVES IN GUNUNG API

Mt. Api is riddled with caves, and nearly 190 km of the passages are now linked to form the Clearwater Cave System (or Gua Air Jernih in its Malay translation), named after the clear karst water which pours from its

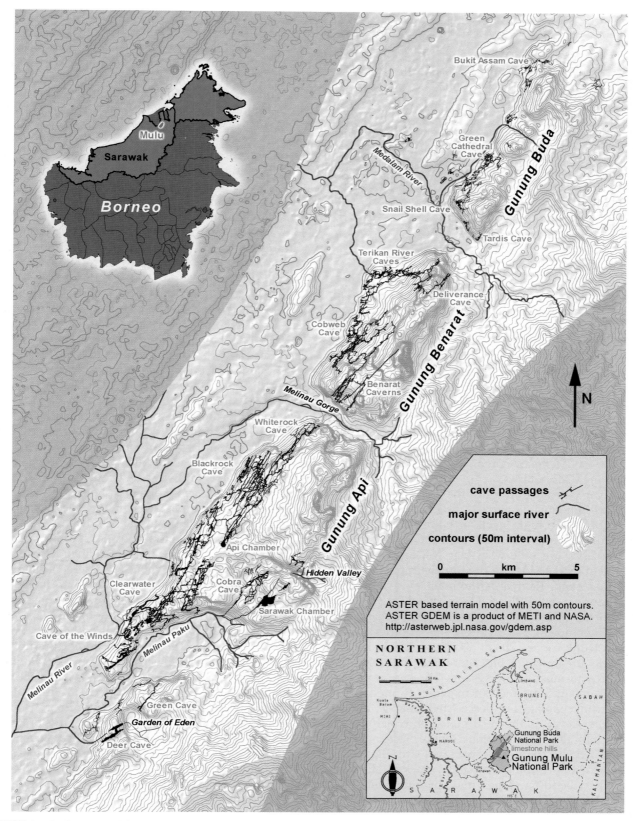

FIGURE 1 Outline map of the known caves within the limestone mountains adjacent to Gunung Mulu.

FIGURE 2 The very large passage in Deer Cave looking out toward the main entrance at its southern end, which is nearly 600 m from the camera. *Photograph taken by Tony Waltham.*

resurgence into the muddy Melinau River. The main Clearwater River Cave, upstream from its resurgence, is a magnificent canyon passage, 30 m high and wide, whose clean walls, cut and undercut in pale limestone, are the epitome of underground sculpture. This one cave river drains most of Api, and has been followed beyond six short sumps. The Firecracker Streamway in Blackrock Cave and the Whiterock River are two further long sections of large and gently graded river passages, all connected by high levels into the sprawling Clearwater system; Whiterock reaches to within a few meters of the Melinau Gorge, where it collects water from sinking distributaries.

The river passages are only the latest phase of the cave development, as a maze of huge abandoned cave tunnels forms the high levels of Clearwater. Underfit streams drain down through some of these, but the main passages are otherwise inactive, and many are truncated at exits hidden in the impenetrable forest. At the southern end, Revival is the main old trunk route above the Clearwater River; it is mostly 30 m high and wide and lies about 100 m above the modern streamway. Farther north, the Big Issue and Borderline are the main trunk routes, 200 m above the Blackrock and Whiterock streamways, but there are multiple high levels through most of the system. The enormous Api Chamber (300 m long and 200 m wide) formed where parallel trunk passages coalesced by collapse, and an even higher level at the resurgence end includes an opening into the forested floor of the Secret Garden tiankeng. All the main cave passages are aligned on the strike of the dipping limestone, stacked obliquely up just a few bedding planes, and covering over 400 m of vertical range. The high levels also continue beyond the resurgence to link to Cave of the Winds, a streamway carrying water through from the Melinau Paku. Lagan's Cave has another segment of large old trunk passage at an intermediate level, but there is no connection to Clearwater as yet.

Parallel to and southeast of Clearwater, two separate cave systems drain from the huge blind gorge of Hidden Valley through to the Melinau Paku lowland. The Cobra-Bridge-Cloud Cave provides a convoluted through route. Gua Nasib Bagus can only be entered at its resurgence, and its high-level route from Prediction Cave is choked with sediment. Midway through the mountain, Sarawak Chamber is the world's largest underground void. It is 700 m long by 300–400 m wide, and its roof arches 100 m over the ramp of fallen blocks that forms its sloping floor. The chamber was formed where the river from Hidden Valley slipped sideways, down the steep bedding, and created the giant undercut; this has subsequently been modified as the roof has collapsed upward to form a stable arch. The river has since found a new entrance passage farther downdip, and now just drains through the boulders at the foot of the chamber, leaving the main part of the chamber and Prediction Cave dry.

Sarawak Chamber almost defies comprehension. The three cavers who explored the cave in 1980 did not know that they were in a chamber; they could only see one wall, which they followed to make their survey. Only when they had gone most of the way around a huge loop did they realize that they might be in a vast cavern, instead of in a very large, curving passage; they calculated a compass bearing and then climbed over the boulders across the middle of the chamber to regain their entry point. Midway, they could see no walls, no roof, nothing—the sense of agoraphobia was acute.

The main trunk passages in the Clearwater Cave System have formed where large flows of water have poured off the sandstone slopes of Gunung Mulu and sunk into the limestone in the Melinau Gorge; these have then flowed underground, to emerge farther south where the limestone mountains drain onto the adjacent plains at lower altitudes. The Gorge has long been at the upper apex of a huge gravel fan that has spread out onto the lowlands, and this fan has continually evolved and reformed at lower levels as both Gunung Mulu and the plains were lowered by surface denudation. The successive falls of the fan, the plains, and the water table within Gunung Api can be recognized in the falling sequence of trunk cave passages, almost along the strike of the dipping beds. Clearer indicators of the past water levels are the deep wall notches which were undercut by the cave rivers whenever they flowed over stable gravel fills (Fig. 3). Abandoned notches are among the most distinctive features of Mulu caves; generally 2–3 m high and cut 4–5 m into the vertical walls, they provide cavers with terraced walkways around deep lakes or large boulder piles in both the old high levels and the modern river passage.

The graded trunk passages, wall notches, and gravel profiles within Clearwater have been correlated with the gravel terraces of the Melinau Plain. The clastic sediments of both the cave and the plain are essentially the lower ends of enormous gravel fans supplied from the slopes of Gunung Mulu and poured through the breaches in the limestone escarpment. Each terrace was formed by aggradation during wetter interglacial periods, followed by incision during drier glacial periods. More than 25 km of passages with traceable wall notches have been mapped within Clearwater Cave, and these fall into a sequence of 20 levels, spread over more than 300 m of altitude. Uranium series dating of calcite on sediments within the lower, more recent, passages indicates a mean rate of base-level decline of 0.19 m per 1000 years. The sequence correlates well with all the interglacial peaks on the oxygen isotope record, and implies that the gravel aggradation was controlled by climatic variations and not by episodic uplift. Preliminary data from Al/Be dating of clastic sediments in some upper levels indicate ages of well over 2 myr, which fit with extrapolation of the erosion rates inferred from the younger stalagmite ages. The constant background rate of base-level lowering is interpreted as the rate of isostatic uplift in response to regional denudation. The limestone blocks have risen almost uneroded on this uplift, while the adjacent plains, and the noncarbonate slopes, have been denuded at rates close to that of the uplift. The limestone mountains of Mulu have taken around 10 myr to evolve to their present topographic relief.

FIGURE 3 Clearwater River Passage, with a conspicuous wall notch at 2–4 m above the present water level. *Photo courtesy of Jerry Wooldridge.*

CAVES IN GUNUNG BENARAT

Massive cave systems within Gunung Benarat largely follow the style of those in Gunung Api. From the high cliffs of the Melinau Gorge (Fig. 4), large abandoned trunk passages head northeast along the strike of the steeply dipping beds; four main levels lie at altitudes

FIGURE 4 The Melinau Gorge, looking upstream toward the cloud-covered slopes of Gunung Mulu; the cliffs along the left side are broken by the dark cave entrances 400 m above the river that are the truncated ends of old trunk passages in Benarat Caverns. *Photograph taken by Tony Waltham.*

roughly 100 m apart. One outlet of these old trunk routes is through Deliverance Cave where 4 km of passage open into the high slopes of the Medalam Valley, but there is a long section choked by sediments and unexplored through the center of Benarat as yet.

From the high-level entrances in the Melinau Gorge, the Benarat Caverns System has just over 50 km of mapped caves that also link, across the beds, to the complex maze of passages in Cobweb Cave. Many of the Cobweb passages are less mature as they still had switchback profiles, up and down the dipping beds, when they were abandoned, though Swift Highway is one high-level trunk with a well-graded profile. Within Cobweb only short sections of low-level streamway have been reached, and these head north toward the Terikan resurgence.

Across the north end of Gunung Benarat, the Terikan River Caves contain another 32 km of passages, with an active cave river largely fed from various sinks along the floodplain margin, and draining out to a resurgence at the head of the Terikan River, a major tributary of the Medalam. Above the river passage, multiple high levels almost parallel the modern river, and also include a complex of caves that reach very close to the northern ends of the Cobweb passages in Benarat Caverns.

It appears that the large phreatic trunk passages through Benarat have, at various times in the past, carried water in both directions, toward the Melinau and toward the Medalam, but the geomorphology of the Benarat caves is still not understood as well as that of Api. Many of these caves did not mature to graded profiles, and consequently wall notches are not as widespread as they are in Api. As yet, there have been no programs of detailed leveling and sediment dating to confirm exactly how the Benarat caves correlate with those in Api. The Terikan caves are altogether different, as they were formed at lower levels by water from a separate alluvial fan built by the River Medalam.

CAVES IN GUNUNG BUDA

Gunung Buda is as rich in caves as the limestone mountains to its south, except that it lacks their very large ancient trunk passages. The most extensive system is Green Cathedral Cave, with 27 km of mapped passages forming a very complex series of old phreatic drains on multiple levels. Some passages switch back up and down to follow steeply dipping bedding planes and available joints, while others have achieved more level profiles that were graded to the contemporary Medalam River. Currently, streams flow from sandstone sinks to the east into the limestone and then drain broadly along the strike toward outlets near the Medalam. To the north, Gua Bukit Assam has 8 km of passages with several small streams and a complex of high levels. Cave hydrology is complex in this part of Buda, with streams forming on adjacent sandstone and within small limestone basins, before draining into the caves. In contrast, passages along the southern edge of Buda (Fig. 5) were probably formed by distributaries of the Medalam; a stream passage in Snail Shell Cave carries water from the Medalam away to the north, possibly to Lower Turtle Cave. These caves have even more local relief where they are aligned directly across the limestone strike; the large phreatic ramp in Snail Shell Cave has been followed over a vertical range of 470 m. Buda's high-level caves are currently being truncated by surface erosion, creating the many entrances in the forest.

The lower levels of the Buda caves contain significant lengths of active streamways. Most of these are flood-prone due to sandstone sinkhole catchments and

FIGURE 5 Giant pinnacles etched into the passage floor by eons of percolation water in Tardis Cave, in the southern tip of Gunung Buda. *Photo courtesy of Dave Bunnell.*

also pulses of rainwater that funnel straight down the fissures from the forest above. In contrast, Lower Turtle Cave contains long lakes that vary little in level or flow within an old phreatic tube.

Buda contains a total of nearly 70 km of known caves, and various outlying limestone blocks contain another 11 km of mapped passage, mostly active stream passages at low level; these include the 5 km of passages in Spirit River Cave away to the northwest of the main limestone ridge. Buda continues the trend set by Api and Benarat that every chunk of limestone in Sarawak's rainforest is riddled with, large, beautiful and spectacular caves that are exceptionally extensive.

GUNUNG MULU NATIONAL PARK

The caves are not the only outstanding feature of Mulu. Gunung Api is one of the world's most spectacular chunks of limestone karst. Except for the vertical cliffs of its margins, its entire surface is fretted into dramatic pinnacle karst. Widely spaced vertical joints have been etched out by dissolution, and are now almost choked by vegetation and organic soils. The remnant blocks of limestone have been carved by direct rainfall into razor-sharp ridges. Most of these pinnacles are just a few meters high, and the terrain is seriously difficult to cross because it consists entirely of steep pinnacles and tapering fissures choked with unstable plant debris. High on Api, pinnacles that are more than 30 m tall rise above the forest canopy to create an amazing landscape, and most of the ground has never been trodden by humans.

The British Royal Geographic Society and other researchers before and since have documented an amazing diversity of plants and animals in the area. The large range in elevations, various soil and rock substrates, and tropical climate have produced what may be the world's most diverse assemblage of tree species, an amazing variety of herbaceous plants, 109 species of palms, and a total of 3500 vascular plants in 17 vegetation zones. More than 270 bird species including all Bornean hornbills, broadbills, and barbets are known from Mulu, and primates found in the area include monkeys, prosimians, and the Bornean gibbon. Other important wildlife includes dozens of snakes, civets, many bats, and more than 200 species of cave-adapted invertebrates.

Cave biology work was undertaken by both British and American expeditions. Four dominant terrestrial habitats (and bio-communities) are recognized within the Mulu caves: bat guano, bird guano, the entrance transition zone, and the deep cave community. Each has unique species of crickets, spiders, beetles, cockroaches, and, in some communities, isopods and millipedes. Some species are found to occur throughout the caves, whereas others are restricted to specific habitats. Swiftlets fly far into many of the larger caves. More than two million bats roost on the roof of Deer Cave, far above any predators. They feed outside the cave, but droppings from their roost accumulate to form great banks of guano on the cave floor. These are both home and food supply for millions of beetles, earwigs, cockroaches, millipedes, and crickets, together with large, predatory spiders and centipedes and an active population of racer snakes. Members of the Mulu cave community include snakes found deep within the caves that predate primarily on birds but also on bats, several species of crabs including some that lack pigment, and at least two species of large centipedes.

The Gunung Mulu National Park was established primarily to conserve forever a substantial section of the Borneo rainforest—which was ever diminishing and under continued threat from Sarawak's powerful logging industry. To save the forest, the park needed economic viability through tourism, and the caves were seen as potential visitor attractions. The fabulous cave discoveries made the initial plan work out very well, and the Mulu Park now offers the dual attraction of the forest environment and the great caves.

The Gunung Mulu National Park was formally established by the Sarawak state government in 1974, with 52,864 ha of forest protected, along with its spectacular caves and its wonderful array of plants and animals. It was inscribed as a World Heritage Site by the United Nations in 2000. The Park includes the sandstone mountain of Gunung Mulu rising to 2377 m, the rugged limestone mountains along its western flank, and the adjacent terraced floodplains that reach to as low as 28 m above sea level. It was opened in 1985, and now boasts a visitor center, forest boardwalks, tourist cave facilities, a large resort hotel, and a small adjacent airport, while the high mountains are preserved as almost inaccessible wilderness. Gunung Buda was established as a separate but adjacent National Park in 2001. It protects another 6235 ha of limestone uplands, swampy floodplains, forested hills, and rivers, but is kept as a wilderness with neither facilities nor easy access.

Bibliography

Brook, D. B., & Waltham, A. C. (Eds.), (1978). *Caves of Mulu*. London: The Royal Geographic Society.

Eavis, A. J. (Ed.), (1981). *Caves of Mulu '80*. London: The Royal Geographic Society.

Eavis, A. J. (Ed.), (1985). *Caves of Mulu '84*. London: The Royal Geographic Society.

Farrant, A. R., Smart, P. L., Whittaker, F. F., & Tarling, D. H. (1995). Long-term Quaternary uplift rates inferred from limestone caves in Sarawak, Malaysia. *Geology*, 23, 357−360.

Hacker, B. (Ed.), (1997). *Caves of Gunung Buda*. Huntsville, AL: National Speleological Society.
Hacker, B. (Ed.), (2000). *Caves of Gunung Buda 1997*. Huntsville, AL: National Speleological Society.
Kirby, M. (Ed.), (2009). *Mulu caves 2009*. Lancaster: Mulu Caves Project.
Meredith, M., Wooldridge, J., & Lyon, B. (1992). *Giant caves of Borneo*. Kuala Lumpur: Tropical Press Sdn. Bhd.
Proffitt, M., & Mosenfelder, J. L. (Eds.), (2003). *Caves of Gunung Buda 2000*. Huntsville, AL: National Speleological Society.

MYRIAPODS

David C. Culver and *William A. Shear*

American University, †Hampden-Sydney College

INTRODUCTION

Myriapods are wingless terrestrial arthropods with at least nine pairs of walking legs and a body not distinctly divided into a thorax and abdomen (Minelli and Golovatch, 2001). Based largely on molecular evidence, myriapods are now regarded as a monophyletic group related to crustaceans and insects. The most familiar myriapods are centipedes and millipedes. Of the four classes of myriapods, three have definitively troglobiotic species. Only the Pauropoda seem to have no troglobionts (Harvey et al., 2000). A variety of cave myriapods is shown in Figure 1. Millipedes are often the most abundant component of terrestrial cave communities at least in terms of biomass. Generally speaking, the myriapods have been less thoroughly studied than other terrestrial arthropod groups (insects and arachnids), with the exception of the mites.

PAUROPODA

Pauropods are the smallest myriapods, rarely more than 1 mm long, eyeless, generally lacking in pigment. Nearly all species have 9 pairs of legs, but a few have 10 or 11. Their bodies are relatively short and stocky, and they have unique 3-branched antennae. Only two species have been observed feeding; both of them bit into fungal hyphae and sucked out the contents. They generally occur in very humid, cool spots, in leaf litter, and soil fissures. Given their general habits, habitats, and morphology, it would seem likely that they should occur in caves. The absence of troglobiontic species may be attributed to the general troglomorphic appearance of all members of the group, especially eyelessness and lack of pigment, which would make it difficult to set troglobionts aside from the others. Large numbers of pauropods have been reported from a few caves, such as Hölloch in Switzerland and several caves in southern France. Pauropods have also been found in mines. It is somewhat surprising that they have not been found to be more widespread and common in caves. This may be due to the fact that they have often been overlooked due to their small size, or that they have food or habitat requirements that are rarely met in caves.

SYMPHYLA

Symphylans are all less than 1 cm long, and have the appearance of small, white centipedes. All species are eyeless, generally lack pigment, and have one pair of elongate multisegmented antennae, 12 pairs of legs, and a pair of posterior unsegmented appendages called *spinnerets*. The number of tergites, or dorsal plates, covering the body is variable and often not correlated with the number of legs. Most symphylans feed on dead and decaying vegetation, but a few species are pests of crops. They are widespread throughout the world in litter and soil. Like pauropods, symphylans are often collected in caves and some species are known only from caves. However, because relatively little surface collecting has been done and the species so far described from caves show few modifications for cave life, it is not known if any of them are really troglobionts. *Scolpendrellopsis pretneri* Juberthie–Jupeau and *Scutigerella hauseri* Scheller are candidate troglobionts from the Postojna-Planina cave system in Slovenia, the most diverse cave community in general.

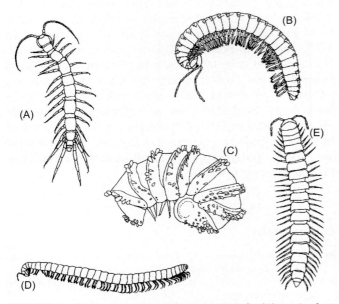

FIGURE 1 Illustrations of troglobiotic myriapods: (A) centipede—Lithobiomorpha, *Lithobius anophthalmus*; (B) millipede—Chordeumatida, undescribed species from China; (C) millipede—Glomerida, *Trachysphaera orientalis*; (D) millipede—Julida, *Antrocoreeana arcuata*; (E) millipede—Polydesmida, *Epanerchodus fontium*.

CHILOPODA

Centipedes exhibit a great range of body lengths from a few millimeters to 30 cm. All have 15 or more pairs of legs. All centipedes are predatory, mostly on insects. The first trunk segment bears a pair of specialized claws that contain a poison gland; the claws are used in the capture of prey. Some centipedes also have the last pair of legs modified as pincers to aid in prey handling. One curiosity of the centipedes is that the number of leg-bearing segments is always odd, ranging from 15 to 191. Centipedes are typically forest litter-dwelling predators although they have also colonized desert habitats. There are five orders of centipedes with about 3300 described species and several times that number of undescribed species. Worldwide, there are slightly more than 50 described troglobiotic species (Negrea and Minelli, 1994).

The smallest order is the Craterstigmatomorpha with 15 pairs of legs and a posterior end shaped like a clam shell. There are only two species, known from the soil in Tasmania and New Zealand. Neither has been found in caves.

Scutigeromorpha is a primarily tropical order characterized by 15 pairs of extremely long legs and large compound eyes, the only centipede order with compound eyes. They are found in caves in Asia, Africa, Australia, and South America. *Scutiera coleoptrata* is a European and North African species that has been spread around the world by man, and is often found in caves in desert regions, though it more commonly occurs in houses and other buildings. Some of the Asian troglophiles are extraordinarily large with leg spans up to 15 cm. None of the species is trogobiontic.

Scolopendromorpha is a widespread order characterized by 21 or 23 pairs of legs. The largest known centipedes, up to 30 cm in length, belong to this group. Members of this order are either eyeless or with a few simple ocelli. Individuals of some of the very large surface-dwelling species are often found in tropical caves. One species has been filmed snatching bats out of the air as they fly by. Two troglobiotic species (*Thalkethops grallatrix* Crabill from New Mexico and *Newportia leptotarsis* Negrea, Matic, and Fundora from Cuba) show morphological modifications for cave life—depigmentation and elongation of appendages. A handful of other species from the Mediterranean and Cuba are known only from caves but show no morphological modifications for cave life. Species such as these may also occur in non-cave habitats but have not yet been found there, or they may recently have been isolated in caves.

Geophilomorpha is the largest centipede order in number of species, common on all continents except Antarctica and in nearly all climates. They are long, thin centipedes with between 35 and 191 pairs of legs and without eyes. The antennae are relatively short and always consist of 14 articles. Some species are capable of producing cyanide as a chemical defense, and still others give off a sticky, luminous secretion. Nearly all species of geophilomorphs are found in moist habitats in leaf litter and in soil, sometimes a meter or more beneath the surface. In spite of the similarities of this habitat to caves, geophilomorphs are quite uncommon in caves. This is probably because geophilomorphs are adapted to small rather than large cavities, are strongly thigmotactic, and have rarely made the transition to large cavities (caves). This explanation is, of course, incomplete because many other groups have made this transition. Although several troglophilic species have been reported in Cuba, there are no known troglobionts.

It is with the Lithobiomorpha that one finds the greatest number of chilopod troglobionts. They have 15 pairs of stout legs, relatively short antennae, and at most a few ocelli. Of the 850 or so described species, nearly 50 are troglobionts. Some troglobiotic species, such as *Lithobius matulicii*, from caves in the Popovo Polje of Bosnia and Herzegovina, have more than 100 segments in their antennae (Fig. 2)! Caves in the Pyrenees Mountains of France and Spain are a hotspot of lithobiomorph diversity with more than 20 species. Europe as a whole has approximately 40 species. By contrast, in the United States, only two species of *Typhlobius*, one from Virginia and one from California, and *Nampabius turbator* Crabill have been described from caves. However, several additional undescribed species from Texas, Colorado, Idaho, and California are known. Curiously, the undescribed Rocky Mountain species belong to a family that is much more dominant in the temperate parts of the Southern Hemisphere, but rare in the Northern Hemisphere. It is likely that the

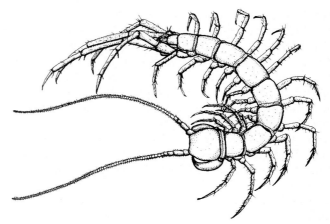

FIGURE 2 Illustration of the cave centipede *Lithobius matulicii* from Vjetrenica Jama in Bosnia and Herzegovina. *Drawing by Slavko Polak.*

disparity in described species is due to the historical lack of centipede specialists in North America. A scattering of troglobionts have been described from elsewhere, including the Near East, Japan, and Korea.

DIPLOPODA

These millipedes range in size from a few millimeters to 35 cm, and are by far the most common of the myriapod orders. The first trunk segment is legless, followed by three or four segments with one pair of appendages. The remaining segments each have two pairs of legs. Millipedes are the "leggiest" animals, and one species from California may have more than 700. They are generally elongate, and the body may or may not be flattened. The exoskeleton is hardened in all but one small group, due to the presence of calcium salts. This requirement for calcium predisposes them to abundance in karst regions. They typically have groups of simple eyes, but all millipedes in the abundant order Polydesmida lack eyes. The mouth parts of many millipedes are adapted for cutting and chewing tough material, such as wood or dead leaves. This may explain in part the remarkable numerical dominance of millipedes in many terrestrial cave communities both in the temperate zone and in the tropics. They are often the dominant detritivores in a cave. It is not unusual to see dozens of millipedes, both troglophilic and troglobiotic, in caves. A single bait with cheese or rotten meat will often attract 100 or more millipedes. Most millipedes have potent chemical defenses that protect them against predators, but these defenses are missing in the order Chordeumatida, the dominant cave millipedes of the Northern Hemisphere (Shear, 1999).

The taxonomic situation with millipedes has been likened to that occurring in entomology at the middle of the nineteenth century (Hoffman, 1999). About 11,000 species have been described and probably 70,000 species await description. At least several hundred troglobionts have been described. Centers of cave millipede richness are in the Dinaric region of Bosnia and Herzegovina, Croatia, and Slovenia, as well as in the Pyrenees of France and Spain. In North America, particular centers of cave millipede richness are in northeast Alabama and adjacent Tennessee, southern Indiana, western Kentucky, and the region along the border between Virginia and West Virginia, extending into Tennessee (Fig. 3). Many species have been described from caves in Mexico and Central America, but in other tropical regions the cave millipede fauna is much less well known. Currently, 16 orders of millipedes are recognized. The Glomerida, or pill millipedes, can roll up into an almost perfect sphere. They are common troglobionts in Europe, Mexico, and Central America (Maruies, 1994).

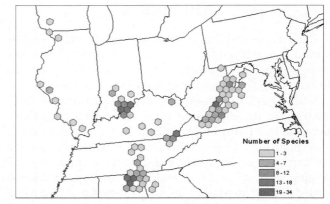

FIGURE 3 Map of troglobiotic millipede species richness in caves in the eastern United States. Each hexagon has an area of 1000 km².

TABLE 1 Number of Species of Troglobiotic Millipedes in U.S. Caves. Modified from Shear (1969, 2008, 2010)

Order	Number of Families	Number of Genera	Number of Species
Callipodida	2	2	3
Chordeumatida	5	15	63
Julida	1	2	2
Polydesmida	1	5	16
Spirostreptida	1	1	1

Polydesmida is the largest millipede order in numbers of species. They are usually easy to recognize by the wing-like lateral extensions on the dorsal side of the animal, though these are absent or inconspicuous in some species. All species in the order are eyeless. Many troglobiont species have been described from Europe, Japan, and Mexico, and they are the dominant invertebrate detritivores in many Mexican caves. The polydesmid *Oxidus gracilis* is a Southeast Asian species that has been spread by man to warm regions all over the world, and may occur in huge numbers in caves. The impact of this species on native troglobionts has not been studied. Relatively fewer troglobiont polydesmids have been described from U.S. caves, mostly in the southwest (Table 1); these species tend to be small and inconspicuous.

The order Julida is restricted to the Northern Hemisphere. Julids are long-bodied, cylindrical, and usually black or gray in color. They are the dominant millipede order in Europe and there are at least two troglobiont species in North America as well. Some troglobiont European julids in the genus *Trogloiulus* have mouth parts modified to comb bacteria from surfaces and to filter food particles from water. As a result, they are amphibious, and often collected in cave pools and streams.

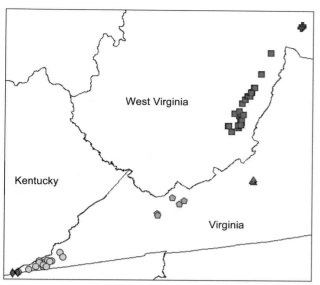

FIGURE 4 Map of troglobiotic *Pseudotremia* species in Virginia and West Virginia caves. Three additional species (see Shear, 2008), not shown, are each known from a single cave.

The most important millipede order in north temperate caves is the Chordeumatida. Most surface-dwelling species are adapted to high humidity and cool temperatures and are found in leaf litter and in the soil. Given similarities of these habitats to the cave environment, it is not at all surprising that there are so many species found in caves. In the tropics, few chordeumatids are found in caves, but those that do occur are often members of families and genera that occur on the surface farther north. Warming postglacial climates may have extinguished surface populations, leaving relicts behind in the cooler conditions of the caves. In the United States alone, there may be more than 100 species, about a third of them undescribed (Table 1). Two genera that include the majority of species in eastern North America are *Pseudotremia* and *Scoterpes*; recent studies have greatly increased the numbers of recognized species in both genera. Most of the ranges of these troglobionts are very small (Fig. 4), and some supposedly widespread described species may actually be a complex of genetically distinct but morphologically very similar species. Examination of the DNA of populations that are indistinguishable on the basis of anatomy have shown that even caves quite close to each other may contain genetically isolated populations that could deserve species rank.

Bibliography

Harvey, M. S., Shear, W. A., & Hoch, H. (2000). Onychophora, Arachnida, Myriapods and Insecta. In H. Wilkens, D. C. Culver & W. F. Humphreys (Eds.), *Subterranean ecosystems* (pp. 79–94). Amsterdam: Elsevier.

Hoffman, R. L. (1999). Checklist of the millipeds of North and middle America. *Virginia Museum of Natural History Special Publication, 8,* 1–584.

Mauries, J.-P. (1994). Diplopoda. In C. Juberthie & V. Decu (Eds.), *Encyclopedia biospeologica I* (pp. 255–262). Moulis, France: Société de Biospéologie.

Minelli, A., & Golovatch, S. I. (2001). Myriapods. In S. A. Levin (Ed.), *Encyclopedia of biodiversity*, Vol. 4 (pp. 291–303). San Diego: Academic Press.

Negrea, S., & Minelli, A. (1994). Chilopoda. In C. Juberthie & V. Decu (Eds.), *Encyclopedia biospeologica I* (pp. 249–254). Moulis, France: Société de Biospéologie.

Shear, W. A. (1969). A synopsis of the cave millipedes of the United States, with an illustrated key to genera. *Psyche, 76,* 126–143.

Shear, W. A. (1999). Millipede *American Scientist, 87,* 232–240.

Shear, W. A. (2008). Cave millipedes of the United States. VII. New species and records of the genus *Pseudotremia* Cope. I. Species from West Virginia, USA (Diplopoda, Chordeumatida, Cleidogonidae). *Zootaxa, 1764,* 53–65.

Shear, W. A. (2010). The milliped family Trichopetalidae, Part 2: The genera *Trichopetalum, Zygonopus* and *Scoterpes* (Diplopoda, Choredeumatida, Cleidogonoidea). *Zootaxa, 2385,* 1–62.

N

NATURAL SELECTION

Peter Trontelj

University of Ljubljana, Slovenia

Natural selection is a major evolutionary force, the only one that consistently leads to adaptation. The process of natural selection is central to Darwinian adaptive evolution that in turn is widely believed to have shaped much of the diversity of life on Earth. A succinct definition is that of differential survival and reproduction. Those individuals that manage to survive and reproduce most prolifically will contribute the highest number of offspring to the next generation, and with them their own variants of heritable traits. In the course of several generations, this process will lead to an increase in frequency for some variants and to the demise of others, which, in a nutshell, means that the population is evolving by natural selection. The key issue is the difference in the net reproductive success between individuals that possess certain heritable variants of traits and those who do not. Survival, mating success, fecundity, longevity, and other factors contribute to it, but ultimately it can be reduced to the number of offspring, technically termed *fitness*. When a population is exposed to a new environment—like at the invasion of a cave—certain variants might prove advantageous over those favored in the old environment, and their bearers will produce more offspring. After some generations, their share in the population will have increased. Consequently, the average fitness of the members of that population will have increased, and the population as a whole will become better adapted to the new environment. The traits contributing to higher survival and reproduction are called *adaptations*.

Cave animals often differ strikingly from their close surface relatives in phenotype (*i.e.*, the totality of biological form and function), but much less so in the heritable information in their genes, or genotype. At the same time they tend to resemble other cave species, even if not closely related. This phenomenon is known as *convergent evolution*. Features like eyelessness, depigmentation, and elongated appendages are among the most common cave-related traits, or troglomorphies. Together with convergent changes in physiology, life history, and behavior (see articles in this volume on behavioral and morphological adaptations), they form a rich source of data demonstrating the power of natural selection. Repeated evolutionary convergence in the same kind of environment but at different places, even continents, can be better explained by natural selection than by random genetic drift simply because the latter is expected to produce random changes (Endler, 1986).

HISTORICAL DEVELOPMENT

The first one to mention the above argument in a slightly modified form was none other than Darwin (1859): "The principle which determines the general character of the fauna and flora of oceanic islands, namely, that the inhabitants, when not identically the same, yet are plainly related to the inhabitants of that region whence colonists could most readily have been derived—the colonists having been subsequently modified and better fitted to their new homes—is of the widest application throughout nature. . . . We see this same principle in the blind animals inhabiting the caves of America and of Europe." He attributed the increase in the length of the antennae or palpi in cave arthropods to natural selection, as a compensation for blindness. The loss of eyes and pigment he explained by disuse, a concept now considered obsolete. In the decades to follow Darwin's discourse on evolution in caves, the notion of natural selection shaping the bizarre appearance of cave animals was increasingly losing ground. The debate centered on regressed and lost organs and structures, chiefly blindness and skin pigment. Speleobiologists working in North America (Arthur M. Banta, Carl H. Eigenmann,

Alpheus S. Packard) and Europe (René G. Jeannel, Armand Viré), although still being evolutionists, put forward nonselectionist and Neo-Lamarckian theories. They were criticized by the Romanian naturalist Emile Racovitza in his famous *Essai sur les problèmes biospéléologiques* (1907). He suggested that both the cave environment and competition between and within species exercised selection on cave-dwelling animals. He disagreed also with Darwin, who was hesitant in ascribing a strong role to these two factors in caves. With regard to the mechanisms of natural selection, Racovitza seemed to adhere to non-Darwinian explanations about the inheritance of traits acquired and lost by the impact of environment. Non-Darwinist views have retained a strong impact on the discipline until the 1960s. The last major proponent of the nonselectionist view of subterranean biology was the French biospeleologist Albert Vandel.

At about the same time, American speleobiologists (Thomas C. Barr, Kenneth Christiansen, and Thomas. L. Poulson) started a new chapter in the understanding of the adaptive and functional significance of the morphology, physiology, life history, and behavior of cave animals. They used laboratory experiments, quantitative and statistical methods, and comparative approaches to demonstrate the adaptive features of aquatic and terrestrial cave life. However, while their work brought an end to metaphysical explanations of evolution in caves, it was still focused on adaptive explanations, not the processes that gave rise to them. The difficult job of showing that natural selection actually takes place in caves, and of explaining its mechanisms, was undertaken by what we may call the second generation of the American selectionist school of biospeleology, represented by David C. Culver, Daniel W. Fong, Ross T. Jones, and Thomas C. Kane.

In the 1940s, the German zoologist and geneticist Curt Kosswig provided an explanation for the regression of structures that had apparently lost their biological function in the subterranean environment, known as the *neutral mutation theory*. This was probably the first evolutionary explanation for cave animals that was completely free of metaphysical speculations. The theory actually accepts the importance of natural selection in that it predicts what happens when it no longer acts, like in the case of sight in total darkness. Mutations that obstruct sight are normally removed by purifying selection, but have a chance to spread in a cave population by random genetic drift. The process of random genetic drift has later became accepted as a major evolutionary force at the level of genes and proteins, formalized in Motoo Kimura's neutral theory of molecular evolution in 1968. The question today is not whether Kosswig's neutral mutation theory is correct, but how important it is compared to natural selection in shaping the subterranean phenotypes (Protas *et al.*, 2007).

Finally, it should be noted that although evolution in caves seemingly provides ideal examples of natural selection, cave fauna never became a really important model for studying natural selection, as did, for example, fruit flies and the Galapagos finches. Moreover, the actual processes of natural selection in caves were explicitly addressed by very few studies, and, in general, remain poorly understood.

THE ACTION OF NATURAL SELECTION IN CAVES

Most of our understanding of subterranean biology rests on the assumption that troglobionts are in some way functionally adapted to the subterranean environment. However, by hypothesizing about the possible functional role and adaptive value of troglomorphic traits we are not studying the evolutionary process. To directly measure and demonstrate selection in subterranean environments is not easy. It requires an accurate knowledge about the variation of the examined trait(s), a functional explanation of the biological performance of different variants, evidence of their heritability, and measurement of the actual difference in fitness (survival and reproduction). The only work so far to explicitly address all of the above points is a large scale monographic account on the amphipod crustacean *Gammarus minus* (family Gammaridae) by Culver *et al.* (1995).

This aquatic amphipod inhabits surface springs and caves in a wide area in the eastern and central United States, usually not showing any morphological differences between habitats. Morphologically distinct, troglomorphic populations have evolved only in the long caves in the karst areas of Greenbrier Valley, West Virginia, and Wards Cove, Virginia. Several cave populations originated from separate cave invasions and acquired the troglomorphic phenotype independently from one another. They are genetically isolated from nearby surface populations. Their cave-related traits—prolonged antennae, reduced eyes and skin pigment, modified brain morphology with increased olfactory and decreased optic ganglia—are less expressed than in many other cave crustaceans, but still distinct enough to be measured. That makes them ideal for a comparison of natural selection between cave and surface environments. Perhaps the biggest difficulty of such studies is to accurately measure fitness of different phenotypes. It is neither possible to follow individual amphipods throughout their life in their natural habitat nor to count their offspring. Therefore, Culver and coworkers measured two out of several components of the overall fitness related to reproductive success: frequency of amplexus (a precopula phase of tight physical contact between a male and female amphipod) and number of eggs carried by the female in her brood pouch. Amplexing animals and those with more eggs were considered to have higher fitness. Using these data, they estimated selection gradients for three traits that are often associated with the transition to subterranean environment: body size (represented by head length), antennal length, and eye size. They found convincing evidence that natural selection in caves favors larger animals, longer antennae, and reduced eyes. This is in line with expectations, because larger animals are, in the absence of predators, stronger competitors, energetically more efficient, and produce more offspring. Likewise, longer antennae enhance the sensory abilities in a lightless environment. The possible advantage of superficially reduced eyes is less clear, and the effect might be caused by genetic correlation with another, unmeasured trait. In surface springs, larger size of all three traits was selected for. It is harder to explain why longer antennae were selected for in this environment.

As important as identifying the traits under selection is the question about the mode, direction, and strength of selection. The view of caves as extreme environments implies monotonous directional selection toward a fully adapted troglobiont (Fig. 1A), one that minimizes variation. Contrary to these predictions, cave populations often show remarkable amounts of morphological variability and genetic polymorphism that would better fit a model of balancing selection. The only ones to explore this problem in caves were Culver et al. (1994) using their model organism *Gammarus minus*. They found that directed selection for larger body, longer antennae, and smaller eyes indeed strongly prevailed in caves. A third mode of selection, pulling trait values apart and thereby disrupting the variability (Fig. 1B) might be involved in colonization and speciation, and is discussed below.

The strength of selection is closely linked to the time required for a new cave invader to evolve its adaptive troglomorphic traits. Culver et al. (1995) estimated that a wild population would need some 100,000 years to become eyeless and otherwise troglomorphic. This estimate agrees well with the facts that most molecular clock datings gave cave species that much time or more, and that very few troglomorphic species are found in areas where the subterranean habitat has only been available since the last glacial retreat about 10,000 years ago. On the other hand, this seems rather slow in the light of several new reports about both vertebrates and invertebrates adapting to new environments and speciating within tens or hundreds of generations. Also, strong directional selection and a consequent fast evolution are to be expected under the sharp ecological gradient from the surface to the subterranean environment.

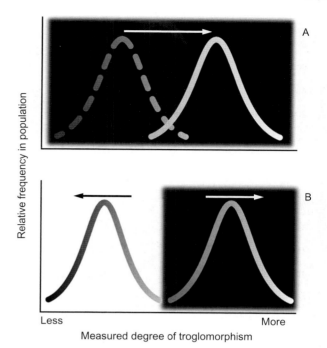

FIGURE 1 Modes of natural selection in caves. (A) *Directional selection*: When isolated from the surface ancestor, for example, by its extinction, a new cave species undergoes directional selection leading to adaptation to the new environment. (B) *Divergent selection*: When the surface ancestor remains in contact with the cave population, divergent selection drives trait values of the cave and the surface form apart. This kind of selection can lead to ecological speciation.

A few cases might nevertheless be pointing to a very fast evolution in caves. One is the troglomorphic form of the aquatic isopod *Asellus aquaticus* (family Asellidae). Several local, independently evolved cave populations and species are known from the Dinaric Karst in Slovenia and Italy and from Dobruja Plateau in Romania. Multivariate morphometric comparisons between historical and recent samples suggest that one of them might have invaded Postojna Cave and evolved a troglomorphic appearance in no more than 80 years (Prevorčnik et al., 2010).

The same species in the same cave demonstrates how natural selection interacts with gene flow to form a cline (Fig. 2; Sket, 1985). A population of nontroglomorphic, normally eyed surface phenotypes of *Asellus aquaticus* lives in the surface freshwaters outside Postojna Cave including the sinking river passing through the cave. Deeper in the cave, a troglomorphic population of depigmented and eyeless individuals inhabits the same sinking river. Individuals from the surface migrate into the cave following the flow direction. Within the first kilometer from the entrance, the nontroglomorphic phenotype strongly predominates. From approximately kilometer two, troglomorphic

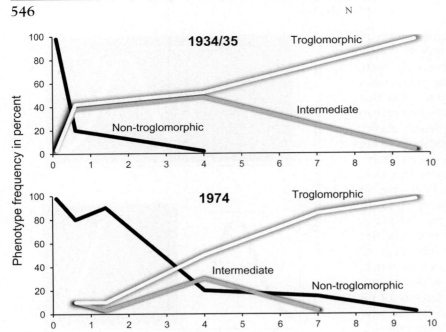

FIGURE 2 Phenotypic clines along the sinking Pivka River in the Postojna-Planina Cave System, Slovenia. The steep cline in 1934–1935 is indicative of strong selection against the surface form. Forty years later, the cline was much flatter, probably because of increased nutrient input by organic pollution of the Pivka River, changing the selective regime in favor of the surface phenotype. *Adapted from Sket (1985), simplified.* Actual counts are represented only at the starting, breaking, and end points of the curves.

Asellus begin to increase in number together with morphologically intermediate animals. Gradually, their frequency increases until at about kilometer four the surface phenotype becomes very rare. The width of the cline is determined by the rate of immigration relative to selection acting against immigrants. Interestingly, in the first half of the twentieth century C. Kosswig observed a much narrower cline with the surface phenotype becoming rare already within the first kilometer. As it is unlikely that the mobility of the animals would have changed during that time, we can assume a change in selection pressure. This was induced by organic pollution introducing nutrients and thus making the subterranean river more hospitable to the surface animals.

THE SELECTIVE REGIME IN CAVES

The selective regime, or selective environment, comprises those environmental and organismic factors that determine how natural selection will act on character variation. It is important to have an idea about the key selective factors of the cave environment if we wish to know whether a troglomorphic trait evolved under the impact of natural selection. The usually mentioned critical factors in caves are complete and constant darkness and food scarcity. Other features, such as absence of predators and competitors, constancy of the environment, and high humidity, are more likely to loosen the selective pressure than to intensify it. Studies on North American amblyopsid cavefish postulated strong selective constrains by food scarceness in a permanently lightless aquatic environment. The flattening of the cline between cave and surface *Asellus* phenotypes following increased availability of food (Fig. 2) suggests a similar selective regime for aquatic crustaceans. Conversely, the selective regime for cave populations of *Gammarus minus* mentioned above was postulated to be determined by darkness alone, as the cave streams have considerable organic input. Culver *et al.* (2010) proposed that different kinds of aquatic subterranean environments (deep groundwater, cave streams, epikarst, superficial subterranean habitats) differ in the availability of food and space, and that these differences should influence the selective regimes in those habitats. Using this prediction they tried to explain the morphological variability within the North American subterranean amphipod genus *Stygobromus* (family Crangonyctidae). Contrary to expectations, they could find no support for the hypothesis that differences in antennal elongation reflect differences in food and nutrient supply. Given that all *Stygobromus* species are eyeless and depigmented, they concluded that absence of light was the single key selective factor responsible for general troglomorphic appearance.

Few examples demonstrate that other features of the cave environment can contribute to the selective regime. One is the permanently wet substratum on which surface collembolans would have trouble moving, and in response to which troglobiotic species have evolved a special foot structure and locomotory behavior. Another one is a special cave microhabitat consisting of films of water moving down cave walls, known as *cave hygropetric* (Sket, 2004). The water contains organic particles, but special adaptations on legs and mouth parts are required to exploit them. Several leptodirine cave beetles have evolved such adaptations convergently. This shows that the cave environment not only poses selective constraints to its colonizers but also offers ecological opportunities, for example, in the form of new trophic niches.

This idea was examined on the Western Palearctic amphipod genus *Niphargus* (family Niphargiadae). The high morphological variation among coexisting *Niphargus* species can be explained by their adaptations to microhabitats within caves (Trontelj *et al.*, unpublished). Up to nine species inhabit a single cave. Large-bodied species with long antennae and legs are consistently associated with phreatic cave waters, stagnant or slowly flowing and with an abundance of available space. Large but slender species with short antennae and legs show a preference for cave streams with high flow velocity. Small-bodied species are associated with either of the two small pore microhabitats: the epikarst and the cave interstitial. The limiting factor in these two habitats appeared to be body diameter rather than body length. This demonstrates that selective regimes in caves can be more complex than simple directional selection for compensation of eyesight and energetic efficiency. What at a first glance seems to be different stages of adaptation might in reality be adaptations to different niches within a cave. Finally, the coexistence of several closely related species means that biotic interactions such as competition and predation add to the selective regime.

THE ROLE OF NATURAL SELECTION IN COLONIZATION AND SPECIATION

Speciation—the origin of new species—often occurs by means of natural selection. One particularly important way for new cave species to evolve is by divergent natural selection operating between the new cave invaders and the part of the population that remains in the original surface environment. Speleobiologists have been aware of this mode of colonization for about thirty years thanks to the work of Francis G. Howarth on Hawaiian cave arthropods. Historically, in subterranean biology, the distinction between active and passive colonization and colonization in the presence or absence of surface ancestors was considered more important than the direct role of selection in this process. Less attention has therefore been paid to the question of how selection leads to speciation than to the question of whether a particular cave fauna had originated by adaptive shift or not. On the other hand, speciation by divergent natural selection, known also as *ecological speciation*, has been in the wider focus of speciation research for many years. Generally, ecological speciation refers to the evolution of reproductive isolation between populations or subsets of a single population by adaptation to different environments or ecological niches.

Natural selection can be directly involved when new cave species are budding off from a surface ancestor. There is an important distinction between the mode of selection involved here and the more familiar directional selection, although both will eventually result in the same troglomorphic adaptations such as longer legs and antennae, smaller eyes, lower metabolism, or a longer life span. This distinction occurs when the cave invaders remain in contact with the ancestral population, for example, through repeated immigration from the surface. Selection now favors one extreme of a trait in the surface environment and the other one in the cave (Fig. 1B). Theory predicts that the population subdivided between two different environments will adapt locally according to the differences in selective regimes. Gene flow between both environments may still persist, but selection will act against it. Traits favored in one environment will be disadvantageous in the other. Intermediate, heterozygous individuals will have lower fitness in both environments. Speciation with ongoing gene flow can only take place under strong divergent selection that results in some form of reproductive isolation.

The reproductive barriers leading to speciation in caves have not been thoroughly studied, but a handful of studies suggest that they may be quite diverse. The most obvious is isolation by habitat where the non-adapted surface species simply cannot penetrate deep enough in sufficient numbers to mate and maintain the contact to the cave population. A case of strong habitat isolation was reported for locally adapted populations of the Atlantic molly (*Poecilia mexicana*, family Poeciliidae) in the Cueva del Azufre system in southern Mexico (Plath *et al.*, 2010). In addition to the lightless cave environment, toxic hydrogen sulfide is another important selective factor there. Adaptation to divergent environmental conditions and selection against immigrants from ecologically different populations drives the genetic differentiation in this system. An exceptionally strong flood with the potential to intermix animals from all parts of the system occurred in 2007. Plath and coworkers used this event as a natural perturbation experiment. They compared the genetic structure of populations from different environments before the flood to the structure encountered two months after the flood. They found that the catastrophic flood displaced individuals among compartments of the same type of environment, but the structure between environments remained unchanged. What this means is that natural selection has eliminated—probably by mortality—all involuntary immigrants from their non-native habitats.

This example might be extreme, but it shows that isolation by habitat is enough to keep the cave and the surface population from admixing even if there are no physical barriers to migration. Similarly, the habitat barrier probably plays a role in keeping some cave populations of *Asellus aquaticus* in the Postojna-Planina Cave System separate from the surrounding surface populations. In captivity, they interbreed successfully, but in the wild there is very little or no genetic exchange between the surface and cave populations (Verovnik *et al.*, 2003; Protas *et al.*, 2011).

Another type of reproductive barrier was suggested by Niemiller *et al.* (2008) to be responsible for the ecological speciation of cave salamanders of the genus *Gyrinophilus* (family Plethodontidae). The cave forms are paedomorphic like many troglobiotic salamanders (see the article "Salamanders" in this volume). Paedomorphosis, and the permanent aquatic lifestyle it is associated with, is believed to be an adaptation to the cave environment, where food and space are more abundant in the aquatic than in the terrestrial part. As a side effect of this adaptive divergence, paedomorphosis may contribute to reproductive isolation. Paedomorphs must court and mate in water while metamorphosed *Gyrinophilus* have only been observed mating on land.

CONCLUSIONS AND PROSPECTS

The evolving cave fauna offers powerful models for the study of natural selection. Although researchers have been aware of this potential ever since Darwin, few have made use of it. Biospeleology has been historically burdened by non-Darwinian explanations of evolution that have left little space for natural selection. Much effort has therefore been put into demonstrating the adaptive value of troglomorphic characteristics, leaving the vast field of the actual evolutionary processes virtually untouched. This has changed since 1990 when natural selection was extensively studied on the amphipod *Gammarus minus*. We know since then that reduced organs—for example, smaller eyes—are selected for in caves, but selected against outside. We know that selection for other troglomorphic traits, such as longer appendages and larger body size, is directional. Individuals with mixed traits probably result from ongoing colonization processes or incomplete reproductive isolation of the cave population, and are selected against, rather than being the result of stabilizing selection. Food scarcity and darkness seem to be the main common factors shaping the selective regime in caves, but interactions between species might be important in caves with communities of ecologically similar or closely related species. The coexistence of several species of niphargid amphipods is possible because they are adapted to microhabitats within a cave, which is likely the result of divergent natural selection. Natural selection is important also for the colonization and speciation under the adaptive shift model, more generally known as *ecological speciation*. Divergent selection between the surface and the cave environment can produce local adaptations that bring about reproductive isolation. Speciation can happen while individuals from the surface continue to immigrate if selection acts against the surface phenotype inside the cave and *vice versa*. Reproductive isolation can also evolve as a by-product of adaptation to the cave environment, for example, by paedomorphosis in cave salamanders.

Cave animals are traditional examples of evolution, but are young models in natural selection research. Many questions remain to be answered or even addressed. Most of our knowledge comes from aquatic cave animals such as cavefishes, amphipods, isopods, and salamanders; the validity of models and concepts needs to be verified on the terrestrials. The strength of selection and the time needed for a troglomorphic species to evolve—anything between 100 and 100,000 years—are still under debate. So is the role of selection in the origin of cave species; here the debate has joined the wider field of speciation research, so examples from caves are likely to attract more general interest. Almost nothing is known about how natural selection shapes the diversity of cave species within caves, that is, the process of adaptive radiation following the initial colonization. Here, the interesting questions include evolutionary tradeoffs between adaptations to various facets of the cave environment, strength and forms of competition and other interactions, and speciation within the subterranean realm without the aid of sharp cline to the surface.

No other scientific debate has marked biospeleology like the debate about the role of neutral mutations and selection in the evolution of reductive traits. The dispute is likely never to be solved in favor of one or the other side, not only because both processes are probably involved but also because the crux of the problem is often not presented clearly enough. If the question is, are neutral mutations involved in eye and pigment degeneration, the answer has to be a clear "yes." And if the question is, are degenerated or reduced eyes adaptive in the cave environment, the answer might be a cautious "probably not in itself." However, if one asks what processes are involved in the developmental background that phenotypically manifests itself as reduced eyes and body pigment, the answer will most likely include natural selection. Through this debate, the most advanced scientific studies in biospeleology have been motivated, ranging from extensive laboratory breeding and classical genetic experiments to developmental genetic and genomic research. It also points the way to the future of evolutionary speleobiology, where genomics methods will be used to identify what genes are under natural selection, and how the changes at these genes repeatedly yield the morphological, physiological, behavioral, and other adaptations to cave life.

See Also the Following Articles

Neutral Mutations
Morphological Adaptations
Behavioral Adaptations

Bibliography

Culver, D. C., Holsinger, J. R., Christman, M. C., & Pipan, T. (2010). Morphological differences among eyeless amphipods in the genus *Stygobromus* dwelling in different subterranean habitats. *Journal of Crustacean Biology, 30*, 68–74.

Culver, D. C., Jernigan, R. W., O'Connell, J., & Kane, T. C. (1994). The geometry of natural selection in cave and spring populations of the amphipod *Gammarus minus* Say (Crustacea: Amphipoda). *Biological Journal of the Linnean Society, 52*, 49–67.

Culver, D. C., Kane, T. C., & Fong, D. W. (1995). *Adaptation and natural selection in caves: The evolution of Gammarus minus*. Cambridge, MA: Harvard University Press.

Darwin, C. (1859). *On the origin of species by means of natural selection, or the preservation of favoured races in the struggle of life*. London: John Murray.

Endler, J. A. (1986). *Natural selection in the wild*. Princeton, NJ: Princeton University Press.

Kimura, M. (1968). Evolutionary rate at the molecular level. *Nature, 217*, 624–626.

Plath, M., Hermann, B., Schröder, C., Riesch, R., Tobler, M., García De León, F. J., Schlupp, I., & Tiedemann, R. (2010). Locally adapted fish populations maintain small-scale genetic differentiation despite perturbation by a catastrophic flood event. *BMC Evolutionary Biology, 10*, 256.

Prevorčnik, S., Trontelj, P., & Sket, B. (2010). Rapid re-invasion and evolution following the mysterious disappearance of Racovitza's *Asellus aquaticus cavernicolus* (Crustacea: Isopoda: Asellidae). In A. Moškrič & P. Trontelj (Eds.), *Organizing Committee of the 20th International Conference on Subterranean Biology*, Abstract book (p. 172). Postojna.

Protas, M., Conrad, M., Gross, J. B., Tabin, C., & Borowsky, R. (2007). Regressive evolution in the Mexican cave tetra. *Astyanax mexicanus. Current Biology, 17*, 452–454.

Protas, M. E., Trontelj, P., & Patel, N. H. (2011). Genetic basis of eye and pigment loss in the cave crustacean, *Asellus aquaticus. Proceedings of the National Academy of Science (U.S.A.), 108*(14), 5702–5707.

Racovitza, E. G. (1907). Essai sur les problémes biospéologiques. *Archives due Zoologie Experimentale et Generale, 6*, 371–488.

Sket, B. (1985). Why all cave animals do not look alike: A discussion of the adaptive value of reduction processes. *National Speleological Society Bulletin, 47*, 78–85.

Sket, B. (2004). The cave hygropetric—A little known habitat and its inhabitants. *Archiv für Hydrobiologie, 160*, 413–425.

Verovnik, R., Sket, B., Prevorčnik, S., & Trontelj, P. (2003). Random amplified polymorphic DNA diversity among surface and subterranean populations of *Asellus aquaticus* (Crustacea: Isopoda). *Genetica, 119*, 155–165.

NEUTRAL MUTATIONS

Horst Wilkens

Zoologisches Institut und Zoologisches Museum, University of Hamburg

The paradigm of selection dominates all modern interpretations of evolutionary processes and provides their most efficient explanation. However, this is in contrast with the conditions characteristic of the basal process of molecular evolution. Here mutation pressure and drift play the major role (Kimura, 1983), whereas as a rule selection only gets involved secondarily, at the phenotypic level. The question is whether these basic principles of molecular evolution, random mutations, and drift may also manifest in phenotypic evolution and can, in some cases, be traced here.

The reduction and rudimentation of structures is very common throughout nature. It can be observed in wingless birds and insects such as ostriches and carabid beetles. Also baleen whales, which have not only reduced their hind legs but also the teeth, provide spectacular examples. The most conspicuous and curious phenotypes are found in cave-dwelling animals. Eye loss and paleness of the so-called troglobites have stimulated the thinking of scientists from the beginning of research. Even Darwin was aware of the bizarre phenotypic appearance of cave animals, which seemed to him "wrecks of ancient life." Cave animals are a central object for the study of the causes of regressive evolutionary processes, because most reduced traits may be no longer influenced by persisting biological functions.

As a unique model system in research of regressive evolution the characid *Astyanax fasciatus* (Cuvier, 1819) (= *A. mexicanus*; Filippi, 1853) from Mexico has served since the middle of the twentieth century (Wilkens, 1988, 2010; Jeffery, 2001). This widely distributed surface fish is the sister form of more than 20 troglobitic populations distributed in different caves within a rather restricted karst area in northeast Mexico (Mitchell et al., 1977). Unlike the majority of surface species entering the hypogean realm, the large-eyed surface *Astyanax* is not a true troglophile, which, like catfish, are usually characterized by extremely negative scotophilia and adaptations to life under light-poor conditions (Langecker, 2000). Specimens of *A. fasciatus* did not colonize caves voluntarily, but got trapped in the underground without the possibility of escape (Wilkens, 1988). They were able to survive in darkness because of preadaptations such as chemically triggered spawning behavior and because they could orientate in darkness on behalf of a well-developed lateral line sense (Wilkens, 2010). Whereas traits such as olfaction, taste, lateral line, yolk content, and fat storage were constructively improved by directional selection in darkness, those having become biologically functionless (*e.g.*, dorsal light reaction, circadian rhythm, shoaling, optically triggered aggressive behavior, pigmentation, eyes, and so on) were subjected to reduction (Langecker, 2000; Wilkens, 1988, 2010).

Surface and cave forms of *Astyanax* are closely related and thus morphological, physiological, and behavioral comparisons can be performed to analyze their adaptations to the cave habitat. As surface and cave forms are interfertile, it was possible to study the genetic basis of regressive as well as constructive traits by crossing analyses. The material examined was derived from nine cave populations (Sótano de la Molino, Cueva de El Pachón, Sótano de la Tinaja, Cueva de los Sabinos, Sótano de Yerbaniz, Sótano de las Piedras, Cueva de la Curva, Cueva Chica, and Cueva del Rio Subterráneo = Micos cave).

One of the most conspicuous traits characterizing cave animals is the loss of coloration. In cave *Astyanax* this relies on several different factors. Albinism, the complete loss of melanin, is known from three populations (Molino, Pachón, Yerbaniz). It is based on one recessive allelic gene, for which all three populations are homozygous. Molecular studies in two cave populations, Pachón and Molino, showed that albinism in them can be attributed to different deletions in the *Oca2* pigment gene, responsible for loss of the corresponding protein OCA2 (Protas *et al.*, 2007). Thus the inability to convert L-tyrosinase to L-dopa and melanin (Jeffery, 2001) seems to rely on the beforementioned mutation and is convergently achieved. With the exception of the Molino and the Micos cavefish, all populations mentioned above are homozygous for the recessive "brown gene," due to which the melanophore color cells as well as the eye pigment epithelium only show partially reduced melanin content. This so-called brown phenotype has arisen independently in geographically separate caves and is mediated by different mutations in the same "brown" gene (Gross *et al.*, 2009).

Also the amount of guanine in the scales, which is camouflaging the fish in the water by its silvery shine under normal light conditions, is slightly reduced in all cave populations. However, in two of them, different recessive genes are found, which reduce the guanine content of the scales in part (Molino) or completely (Piedras). The Piedras population was found to be heterozygous for this gene. Another regressive mutation (yellow gene) causes bright yellow body color. This probably results from the disturbed ability to decompose carotenoids ingested with the food. For this, a recessive gene is responsible, which was only found in the Piedras population (Culver and Wilkens, 2000).

Besides by mere monogenic color mutations, reduction of the dark melanin pigmentation of *Astyanax* cavefish is additionally caused by a decrease of the number of melanophores, of which only about 20% are left. Inheritance of this feature is independent from that of the albino and the brown genes, a fact that is often not considered in studies. Crossings have revealed a polygenic basis for this (Wilkens, 1988; Culver and Wilkens, 2000). Like the color mutations, this feature is already manifested at early embryogenesis, during which fewer precursor cells, the melanoblasts, are developed in the neural crest in the cave forms.

Besides loss of pigmentation, the reduced eyes of cave animals have generated the greatest interest of scientists. In the *Astyanax* cavefish the eye rudiments show varying degrees of differentiation and sizes. In correlation to this, they are sunk below the body surface at differing extents. The degree of differentiation is assumed to be dependent on the phylogenetic age of the cave populations. However, all of them are smaller in size and show less structural differentiation. In adults of the most reduced cavefish, neither lenses nor visual cells are developed. Only those retinal layers not involved in the formation of the visual cells can be found in the eyes best differentiated. In the most reduced ones, the sclera just encloses undifferentiated remnants of nervous, pigment, and chorioid tissues (Wilkens, 1988).

In crossing experiments between surface and cavefish, no genes could be analyzed to be responsible for specific eye structures such as lens or retina. It was only found that size and degree of structural differentiation are in principle correlated and that the larger an eye is, the better its single structures are developed. The genetic factors influencing eye size were called *eye genes*. Crossing experiments and *quantitative trait loci* (QTL) studies revealed that at least six of them were responsible for eye reduction in the cavefish (Wilkens, 1988; Protas *et al.*, 2007). These eye genes seem to have mutated negatively. The defect eye genes are assumed to be randomly distributed in the different cave populations providing a partially diverging genetic basis. They seem to be at least in part heterozygous. This can be concluded from the fact that the hybrid offspring between different cave populations, may develop larger and correlated to this, better differentiated eyes than the parental generation (Wilkens, 1988, 2010).

The genetic character of the eye genes, and which they are, is still being studied. Mapping of developmental control genes such as *sonic hedgehog* (*shh*), *twiggy winkle hedgehog* (*twhh*), and *Pax6* revealed that no eye QTLs are located near these loci. This result makes it unlikely that mutations in any of these control genes are directly responsible for eye regression (Protas *et al.*, 2007). Gene expression and sequencing data showed that destructive loss-of-function mutations have probably not or only in few cases occurred in the cavefish eye gene cascade, including those structural genes that function at the bottom of the regulatory cascades (Jeffery, 2001). This finding is also supported by the observation that all the different-sized eyes developed in hybrids between cave and surface forms show all the structures characteristic of an eye such as lens, lens muscle, pupil, and retinal layers (Wilkens, 1988, 2010). Only the size of these structures generally varies in correlation with overall eye size. This observation furthermore shows that the genes involved in eye development have not mutated and can potentially be expressed. For example, a red-like opsin gene is transitorily transcribed in Piedras cavefish during early ontogeny (Langecker, 2000) and the *a-A-crystalline* gene, which is responsible for the development of a proper lens in surface fish, shows no destructive mutations, but is never activated in strongly reduced cavefish (Wilkens, 2010).

Thus, it can be assumed that eye regression in *Astyanax* seems to be mainly because of down regulation by the expanded *hh* gene expression, which was suggested to have a causal role in eye development of the cavefish (Yamamoto *et al.*, 2004). The question of what causes down regulation of *hh* genes as yet remains unsolved. However, deeper insight into eye genetics can be derived from crossings between different cave populations: whereas hybrids between surface and cave populations contain eyes of all possible intermediate sizes between the parental forms, the crossings between different cavefish populations very often may develop slightly larger and better differentiated eyes than the parental cavefish. Usually, rudimentary lenses are developed in adult cavefish hybrids (Wilkens, 1988, 2010). An exceptional situation is found in the F2-cross between the Pachón and the Molino cavefish populations, in which "back-to-surface eyes" develop. Among a large number of specimens with small eyes that, as in the parental forms and in their F1-cross, are sunken into the orbital cavity, several of the adult specimens have externally visible eyes with wide pupils. Histological sections of the eyes from these latter specimens show the presence of large transparent crystalline lenses as well as retinas containing all optical layers with morphologically intact visual cells including outer segments, pigment epithelium, optic nerve, and lens muscle (Wilkens, 2010). Keeping in mind that expanded expression of the *hh* genes may have a causal role in eye reduction of the parental cavefish, the development of back-to-surface eyes could be explained by the secondary restriction of expression and secondary downregulation of *hh* genes by as yet unidentified genes.

In contrast to the unmutated *hh* genes, it is likely that these as yet unidentified genes show loss-of-function mutations. In back-to-surface eyes they could be expressed again because of the complementary restitution of their function. The complementary restitution then would be brought about by the recombination of different loss-of-function genes in cross hybrids, which derive from different geographically remote cavefish populations that have evolved independently. As shown above, the phenomenon of varying loss-of-function mutations occurring in the same gene in geographically remote cave populations, but nonetheless causing the same phenotypic effect in them, has also been found for the albino gene as well as for the brown gene (Protas *et al.*, 2007; Gross *et al.*, 2009).

The as yet unidentified genes regulating *hh* expression are suggested to be equivalent to those eye genes that were found by crossing experiments to be responsible for eye size (Wilkens, 1988). As mentioned above they were quantitative in character and were originally called "eye genes" because it was not possible to identify specific genes, such as ones responsible for the lens or visual cells, with the crossing experiments.

Histological studies have shown that in the eye two developmentally independent subunits, lens- and retina-dependent structures, exist. Furthermore, it was found that the cavefish growth curve at early ontogeny is characterized by two important subsequent developmental stages contributing to eye regression (Fig. 1). Therefore it was suggested that the eye genes divide into two groups: the first group express upstream of the *hh* genes during the first step of eye regression. These genes included in the first group would determine the size of the primordial eye cup and of the lens through regulation of *hh* expression.

However, whereas the size and the degree of differentiation of the lens is determined during the first step of eye development, that of the retina is induced both during this first step and additionally during the subsequent second step. Thus a second group of eye genes is suggested to act during this second step of eye development, regulating structural genes such as those responsible for the visual cell outer segment formation. *Hh* genes could again be involved as a regulator in the development of the retina, because it is expressed several times during eye development (Wilkens, 2010).

By stating that "It is an important fact that rudimentary organs, such as teeth in the upper jaws of whales and ruminants, can often be detected in the embryo, but afterwards wholly disappear" and that "It is also a universal rule, that a rudimentary part or organ is of greater size relative to the adjoining parts in the embryo, than in the adult," Darwin recognized the process of ontogenetic regression as a characteristic of regressive evolution (Darwin, 1859, p. 430).

Whereas biologically functionless traits such as body color in cave animals are reduced from the very beginning of ontogeny, the eye is subjected to the aforementioned process of ontogenetic reduction. Cavefish are primarily developing smaller but

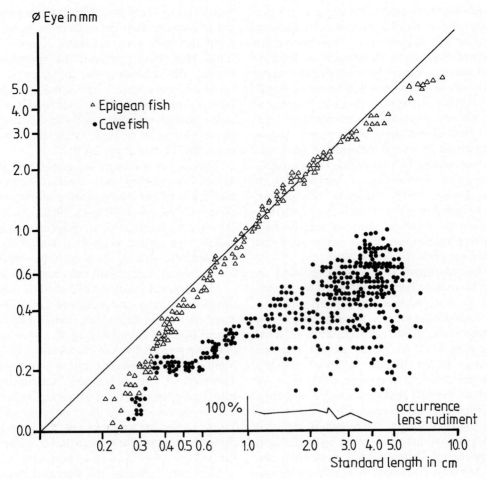

FIGURE 1   Growth of eyes in *Astyanax* surface and cave forms.

structurally almost complete eyes which, for example, contain lenses and visual cells. These are getting reduced during later stages of development and only get completely lost in adult fish. This process can be explained as caused by developmental functions: for example, the eyeball provides by induction the correct formation of the head skeleton. It is this internal selective force that will eliminate all regressive eye structural mutations and which might heavily disturb eye- and, as a consequence, head-building. Disruptions in the formation and evagination of the optic primordia can result in defects in both ocular and craniofacial development. Only those mutations that do not totally disturb eye formation, as, for example, many artificially induced mutations in the zebrafish (*Danio rerio*) or the medaka (*Oryzias latipes*) do, are not subjected to elimination by selection in cavefish in which regression can be studied under "natural conditions." The existence of internal selection is demonstrated by the variability of cavefish eye size during early ontogeny, which is not higher than in the *Astyanax* surface fish until a certain size is reached. Only during the subsequent growth stages the characteristic variability of cavefish eye size develops and enhances, because internal selection has ceased (Fig. 1). Principally similar observations were made in cave-living catfish (*Rhamdia* spp., Pimelodidae) and the cyprinid cavefish *Garra barreimiae* during ontogeny. Summarizing, it can be concluded that only part of the genes involved in eye formation has a neutral character, whereas others are still subjected to internal developmental selection—a characteristic of ontogenetic regression, which, in the case of the cavefish eye, can be coined the *eye paradox*.

The enigmatic disappearance of traits in cave animals has stimulated a large series of different explanations. It is still under dispute, and time and again old theories of eye regression are revived once more and new ones are put forward.

One of these is the *deleterious risk hypothesis*, which suggests that an exposed eye was a deleterious risk in darkness. This hypothesis seems unlikely, because cave animals usually derive from troglophile ancestors, which are already adapted to a partial lifestyle in darkness. Such species are mostly night-active and still have small eyes, because selection is acting against the complete loss of eyes. However, they have already improved senses such as taste, olfaction, or orientation, which enable them to live in complete darkness, too. Furthermore the deleterious risk hypothesis is not apt to explain the reduction of all other biologically functionless traits getting lost in darkness.

Also the *energy economy hypothesis* has long been considered a driving force of eye reduction. Recently, it was speculated that the high energetic costs of visual cell metabolism could be a strong selective force in eye regression. This assumption was based on the finding that neural energy costs of fly visual cells do not lower in darkness when kept in a light−dark 12:12 cycle. These results partially concur with findings of retinal protein synthesis in toads, in which it was found that protein synthesis did not change during the normal light−dark cycle. However, major differences in protein synthesis were evident when animals were stressed with continuous darkness for several days. Under these conditions, a 50% lowering of retinal protein synthesis, which can be considered an indicator of retinal metabolism, took place (Wilkens, 2010).

It was furthermore found in the fish retina that light responsiveness of cone horizontal cells is even suppressed in prolonged darkness. These results show that the effects of darkness on the vertebrate retina are quite complex and it is questionable whether conclusions can be drawn from studies in fly visual cells. To date, no measurements of energy costs of true cave animal retinas have been performed. As the entire ontogenetic development of cave species takes place in complete darkness and thus proceeds without any light stimulus, one can argue that this cannot be compared at all with animals kept in a normal light−dark 12:12 cycle even when measured in darkness. This assumption is substantiated by histological analysis of retinas of cave-living fish sampled under the exclusion of light in their original habitat. The fish bred in darkness showed considerable destructive malformations of the retina, whereas their offspring, hatched and kept at 12:12 light conditions, developed a completely intact and functional retina (Wilkens, 2010). Therefore, physiological deficiencies caused by retinal destruction can be expected in the original cavefish. To date, it is questionable whether energy economy is responsible for cavefish eye reduction.

The hypothesis of energy economy as a selective factor in eye regression is furthermore improbable because behavioral traits, the display of which depends on vision, are reduced although the ancestral colonizer did no longer perform them in darkness and hence they were not subjected to selection. For example, in *Astyanax* cavefish the visually triggered aggressive behavior as well as the dorsal light reaction get reduced in the cavefish, nonetheless (Langecker, 2000; Wilkens, 2010).

Because of the failure to detect specific selective forces responsible for the regression of functionless traits, selection was suggested to be mediated by pleiotropy. The *pleiotropy hypothesis* was recently revived again (Yamamoto et al., 2009). They tried to demonstrate it by conditional overexpression of an injected *shh* transgene at specific times in development of embryonic *Astyanax*, and found that taste-bud amplification and eye degeneration are sensitive to *shh* overexpression during early developmental periods. From this observation, it was concluded that pleiotropy mediated developmental tradeoffs between the regressive eye and oral/pharyngeal constructive cave traits such as jaw size and taste-bud number as well as feeding behavior. However, no reliable experimental proof of this inverse relationship was provided. First, it was attempted to verify it by generating an F3 generation, which is an uncertain tool of genetic crossing analysis, because whereas each individual in an F2 is an unbiased sample of alleles in the original F1, the F3 is a biased sample which may actually create a spurious correlation. Furthermore, the differences in taste-bud numbers of surface and cavefish become significantly notable at much later stages than those studied. Also the assumption of Yamamoto et al. (2009) that the wider cavefish mouth alone accounts for the substantially elevated numbers of taste buds in cavefish was not confirmed. Last but not least, these findings are also in contrast to QTL (Protas et al., 2007, 2008) and morphological studies, which did not reveal correlations between eye size and taste-bud number (Wilkens, 2010).

Independent inheritance of regressive and constructive traits was also shown for the lateral line sense, which compensates the loss of vision in darkness by enlargement of the size and the number of sensory structures sensitive to the slightest water movements, the neuromasts. Counting of the neuromasts on the lateral head side of F2 crossings between surface and cavefish did not reveal any statistical correlation between the number of taste buds and eye size. The distribution of neuromasts is even not expanded into those areas of the cavefish head surface, where the eye was originally located.

The independent assortment of constructive and regressive traits seems to have general validity as was revealed by studies of other constructive and regressive traits in F2-crosses between *Astyanax* surface and cavefish and is corroborated by QTL analyses (Protas et al., 2007). Quite generally the improvement of cave adaptive traits cannot be explained as a tradeoff, pleiotropically mediated by the regression of biologically functionless traits such as the eye or melanin pigmentation (Wilkens, 2010). It is suggested that this independence is even favored by selection because it guarantees evolutionary flexibility, which is an advantage when adapting to a new environment, as cavefish do (Wilkens, 2010).

Darwin (1859, p. 432) observed that "An organ when rendered useless, may well be variable, for its variations cannot be checked by natural selection." The loss of the biological function of those structures for which light is a transporter of information in cave animals was recognized as the key event of their future development by the German biologist Curt Kosswig (Kosswig and Kosswig, 1940). He stated that because of this, stabilizing selection would no longer act on such features. As deleterious mutations became neutral under these instances they were no longer eliminated but accumulated randomly. In the initial phase of reduction, variability of biologically functionless traits arises. For the first time this was experimentally described by him in detail for cave-living isopods of the genus *Asellus* from the Dinarian karst. He showed the high variability of the ocelli these species are convergently developing in three different cave populations—one of them later named *A. kosswigi* in honor of his studies. He found that the crystalline cones may be smaller, completely reduced, or with respect to their number lower or higher compared to the surface sister species *A. aquaticus*. These phenomena also occur on the individual level, generating left–right asymmetry. Similar variability was found for the eyes of other arthropods such as the cave mysid *Heteromysoides cotti*, which additionally even shows variability between the ommatidia of the same eye.

Also in cavefish this variability was observed at the individual and population levels as well as between populations (Wilkens, 1988, 2001). For example, the eye size of the phylogenetically young Micos cave population of *Astyanax* shows half the range of an F2-cross between surface and strongly reduced cavefish (Wilkens, 1988). In the regressive vertebrate fish eye, too, considerable differences in histology and size can be observed between the left and the right side eye rudiment of one single specimen (Wilkens, 2001). Furthermore, in cases in which several cave populations have convergently evolved in geographically remote caves, as in *Astyanax* cavefish or in the *Rhamdia* cave catfish, mean eye size varies among different populations.

The process of reduction, as it is most conspicuously becoming apparent by the variability developed, is perpetuated and finally completed by the continuous accumulation of deleterious mutations in the course of time simply because these mutations are no longer eliminated by selection. The *neutral mutation theory* (NMT) provides the most parsimonious explanation of regression (Wilkens, 2010). In accordance with the generally unquestioned role of selection in constructive evolution, the neutral mutation theory does not contradict the *modern synthetic theory of evolution*, but is embedded in all its rules except for stabilizing selection. It is by no means Lamarckism in disguise, as repeatedly was speculated, the idea that an organism can pass on characteristics that it acquired during its lifetime to its offspring. Furthermore, the neutral mutation theory is until present under dispute because loss of selection is implied to be responsible for troglobitic evolution in general and it is not considered that a difference is made in this theory between constructive troglomorphic traits driven by selection and regressive ones exclusively explained by the above-mentioned theory.

As shown above, strong arguments corroborating the NMT are provided by the reduction of a series of behavioral traits that are no longer performed after the lightless cave environment has been colonized. For example, the optically orientated aggressive behavior of *Astyanax* is getting reduced although not performed by the surface fish in darkness. The same was found for the dorsal light reaction, which helps the surface fish to camouflage under light conditions. Unexpectedly even negative scotophilia, a behavior enabling troglophiles to avoid light, is lost. This behavior is no longer needed, because the hypogean realm is usually so vast that there is no danger of the majority of a cave population to be in contact with the lighted surface environment.

The conditions of absolute darkness in caves can be compared with phenomena arising in domesticated animals, when loss of selection is provided by human care. Breeders allow traits such as body color and shape to no longer be subjected to selection and so variability will arise, by which they are enabled to select for specific features. Similarly, several traits in humans such as the eyes are subjected to a loss of selection, too, because spectacles are used to neutralize the developing variability caused by mutations responsible for short- and wide-sighted phenotypes and many other defects.

Summarizing, the neutral mutation theory and the characteristic variability developed in functionless regressive traits in cave animals would be one of the rare cases in which random mutations are not eliminated by natural selection acting to preserve the functional capability of a trait. Loss of selection allows random mutations as a basic evolutionary principle to manifest in the phenotype.

See Also the Following Articles

Adaptive Shifts
Astyanax mexicanus: A Model System for Evolution and Adaptation
Natural Selection

Bibliography

Culver, D. C., & Wilkens, H. (2000). Critical review of the relevant theories of the evolution of subterranean animals. In H. Wilkens, D. C. Culver & W. F. Humphreys (Eds.), *Ecosystems of the world: Subterranean ecosystems* (pp. 381–389). Amsterdam: Elsevier.
Darwin, C. (1859). *On the origin of species* (Reprint). New York: Gramercy Books.
Gross, J. B., Borowsky, R., & Tabin, C. J. (2009). A novel role for Mc1r in the parallel evolution of depigmentation in independent populations of the cave fish, *Astyanax mexicanus*. *PLoS Genetics*, *5*(1), e1000326.
Jeffery, R. J. (2001). Cavefish as a model system in evolutionary developmental biology. In *Developmental biology* (pp. 1–12). New York: Academic Press.
Kimura, M. (1983). *The neutral theory of evolution*. Cambridge, U.K.: Cambridge University Press.
Kosswig, C., & Kosswig, L. (1940). Die Variabilität bei Asellus aquaticus, unter besonderer Berücksichtigung der Variabilität in isolierten unter- und oberirdischen Populationen. *Revue de la Faculté des Sciences de l'Université d'Instanbul*, *5*, 1–55 (in German).
Langecker, T. G. (2000). The effects of continuous darkness on cave ecology and cavernicolous evolution. In H. Wilkens, D. C. Culver & W. F. Humphreys (Eds.), *Ecosystems of the world: Subterranean ecosystems* (pp. 135–158). Amsterdam: Elsevier.
Mitchell, R. W., Russell, W. H., & Elliott, W. R. (1977). Mexican eyeless characin fishes, genus *Astyanax*: Environment, distribution and evolution. *Special Publications of the Museum of Texas Technological University*, *12*. Lubbock, TX: Texas Tech Press.
Protas, M. E., Conrad, M., Gross, J. B., Tabin, C., & Borowsky, R. (2007). Regressive evolution in the Mexican cave tetra, *Astyanax mexicanus*. *Current Biology*, *17*(5), 452–454.
Protas, M., Tabansky, I., Conrad, M., Gross, J. B., Vidal, O., & Tabin, C. J. (2008). Multi-trait evolution in a cave fish, *Astyanax mexicanus*. *Evolutionary Developments*, *10*(2), 196–209.
Wilkens, H. (1988). Evolution and genetics of epigean and cave *Astyanax fasciatus* (Characidae, Pisces). Support for the neutral mutation theory. In M. K. Hecht & B. Wallace (Eds.), *Evolutionary biology* (pp. 271–367). New York: Plenum Publishing Corporation.
Wilkens, H. (2001). Convergent adaptations to cave life in the *Rhamdia laticauda* catfish group (Pimelodidae, Teleostei). In E. K. Balon (Ed.), *Environmental biology of fishes* (pp. 251–261). Dordrecht: Kluwer Academic Publishers.
Wilkens, H. (2010). Genes, modules and the evolution of cave fish. *Heredity*, *105*(5), 413–422.
Yamamoto, Y., Stock, D.W., & Jeffery, W.R. (2004). Hedgehog signaling controls eye degeneration in blind cavefish. *Nature*, *431*, 844–847.
Yamamoto, Y., Byerly, M. S., Jackman, W. R., & Jeffery, W. R. (2009). Pleiotropic functions of embryonic sonic hedgehog link jaw and taste bud amplification with eye loss during cavefish evolution. *Developmental Biology*, *330*(1), 200–211.

NIPHARGUS: A MODEL SYSTEM FOR EVOLUTION AND ECOLOGY

Cene Fišer

University of Ljubljana, Slovenia

INTRODUCTION

Evolutionary ecology depends on model systems. To understand evolutionary processes in subterranean environments over a broad geographic scale, amphipod genus *Niphargus* (family Niphargidae) seems to be the most appropriate model for focused analyses. With over 300 species it is the largest genus of freshwater amphipod in the world. It is distributed across most of Europe, with the exception of the Pyrenean Peninsula and Northern Europe, and few species are known from the Middle East (Karaman and Ruffo, 1986). The estimated age of three fossilized specimens dates between 30 and 50 Ma BP, and imply synchronous evolution of the genus and geological formation of the continent (*e.g.*, Jazdzewski and Kupryjanowicz, 2010). All species within the genus, including fossils, are eyeless and without pigmentation.

Niphargus is a nightmare for taxonomists. This fact is best illustrated by the five international meetings held between 1969 and 1980 devoted to problematic taxonomy of the genus. Indeed, intraspecific variability, often coupled with sexual dimorphism and allometric growth, may be large compared to interspecific differences. Diagnostic details can easily be overlooked. Furthermore, samples counting a few morphologically unique individuals often remain taxonomically unevaluated (*e.g.*, Fišer et al., 2009).

On the other hand, the high variability of *Niphargus* makes the genus an attractive object for evolutionary ecology if analyzed in spatial and temporal contexts. Until recently, no relevant phylogenetic hypothesis was available; therefore evolutionary hypothesis testing was not possible. Comparative studies are in their early phase and this article mainly reviews the potential of *Niphargus* of being a model system. It consists of three parts: (1) a review of the properties of *Niphargus* considered to be of key importance for comparative studies; (2) studies describing interactions within and between species; (3) a reconsideration of some of these elements within an ecological or evolutionary context.

PART ONE: KEY PROPERTIES

Morphological Variation

The overall morphological variation of *Niphargus* exceeds the variation of most amphipod genera or even families (Fišer *et al.*, 2009). Summarized here are only functionally important morphological characters, or traits otherwise relevant for ecological research. Morphological terminology is clarified in Figure 1.

Body size is the most obvious difference between species (Fig. 2). Body length spans between 2 mm and 35 mm.

Body shape varies from stout to slender, almost vermiform. It determines hydrodynamic properties of an amphipod body. Paired, ventrally extended coxal and epimeral plates form a longitudinal ventral channel. Pleopod action generates water currents, which flow through the channel. The water brings oxygen, food particles, and chemical cues. Currents can be used for jet propulsion in stretched animals while in flexed animals their sum equals zero. Body shape and strength of self-producing currents therefore strongly depend on shape of coxal plates and bases of pereopods V–VII (Figs. 1 and 2).

Maxillae I are the most variable appendages in the mouthpart apparatus. The number of spines on its outer lobes and density of denticles along these spines perhaps indicate different modes of feeding (Fig. 3).

Gnathopods are the two anterior-most appendages on a trunk. Two distal articles are modified for feeding and grooming. Their variable size and shape suggest differences in feeding biology (Figs. 2 and 3).

Antennae and pereopods are of different lengths. Length of the first pair of antennae, for example, measures only one third of the body length in some species, but exceeds the body length by factor 1.3 in others. Similarly variable in length are pereopods V–VII (Fig. 2).

Sexual dimorphism is common in several species. Differences between sexes include body size (males larger), body shape (males slenderer), and elongation of distal articles of male's uropods I and III (Fig. 4).

Allometric growth, that is, nonlinear change of proportions in two body measures during growth (Fig. 4), is another well-known phenomenon in *Niphargus*. Details for most of the species are not known; however, a list of possible allometries has been made. It includes appendage lengths, shapes of coxal plates, and shapes of gnathopods; allometric growth is a proximate mechanism of sexual dimorphism (Fišer *et al.*, 2008a).

Developmental patterns were studied as temporal sequences of post-embryonic developmental events, that is, outgrowth of particular seta or sudden shifts in growth (Fišer *et al.*, 2008a; Fig. 4). To date, only one analysis of developmental events was made on a limited data set. It tentatively implies high independence among individual developmental events, meaning that selection can act on every individual trait without affecting other morphological traits (low pleiotropy).

Physiological Tolerance to Abiotic Factors in Relation to Distributional Data

Niphargus is found in virtually all types of subterranean waters or waters connected with subterranean environments such as epikarst, hypothelminorheic, interstitial, sinking rivers, phreatic habitats, and springs. The overall eurytopic nature of *Niphargus* can be inferred from laboratory and field data. In most cases it is not clear whether individual species tolerate or select for some abiotic conditions. However, on a large scale, when compared to surface crustaceans, niphargids appear to be more tolerant to extreme values for most physical and chemical parameters. This tolerance apparently varies between *Niphargus* species. Here are reviewed known interactions between abiotic parameters and *Niphargus* species, while consequences for local biodiversity are discussed in the last section.

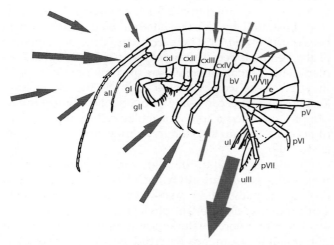

FIGURE 1 Functional model of an amphipod. Arrows indicate water currents created by pleopod action. Abbreviations are aI, aII—antenna I and II; gI, gII—gnathopod I and II; pV, pVI, pVII—pereopods V, VI, and VII; bV, bVI, bVII—basis V, VI, and VII; uI, uIII—uropods I and III; cxI, cxII, cxIII, cxIV—coxal plates I, II, III, and IV; e—epimeral plate (only the first one indicated).

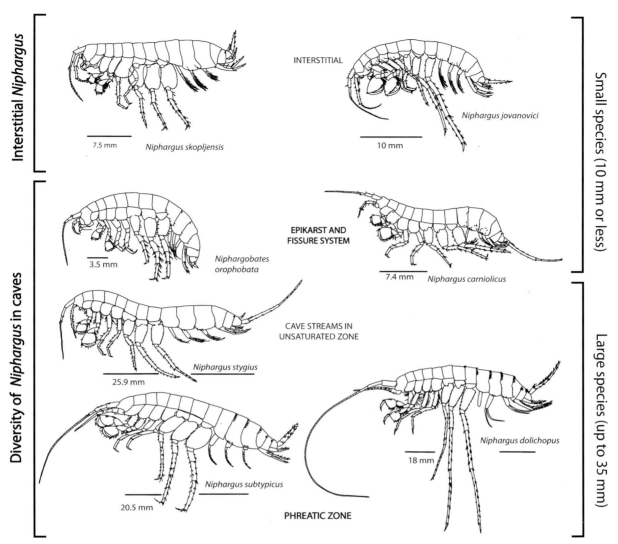

FIGURE 2 Morphological diversity of *Niphargus*: size, shape, and proportions. Species are not drawn to scale: size proportions are indicated by scale bars. Note that size minimally contributes to diversity of interstitial species, which is in stark contrast to the cave species.

Light elicits a stress response in some species, as inferred from measurements of the respiratory electron system and oxygen consumption (Simčič and Brancelj, 2007). Contrary to laboratory evidence, many species are found in springs. Some of them presumably migrate back to darker regions at dawn, having spent the night feeding at springs (Mathieu and Turquin, 1992), but some species constitute permanent populations several hundred meters away from resurgences on the surface (Fišer et al., 2006). It is unclear to what extent the latter tolerate and to what extent they avoid light by usage of dark microrefugees.

Resistance to starvation depends on high fat contents which are 3% and 6% of wet weight in *N. rhenorhodanesis* (hyporheic, more food) and *N. virei* (cave, less food), respectively. Both niphargid species can live without food for more than 200 days. For comparison, the surface amphipod *Gammarus fossarum* and the isopod *Asellus aquaticus* do not survive a starvation period of one month (Hüppop, 2000). Similarly, relying on measurements of the respiratory electron system and oxygen consumption, two other species, *N. stygius* and *N. krameri*, seem to be extremely resistant to starvation compared to surface amphipods (Simčič and Brancelj, 2007). Proximate causes for resistance to starvation include different enzymatic activity and energy-saving behavior such as decreased locomotory and respiratory activity (Hüppop, 2000). At least *N. rhenorhodanesis* can aestivate when water dries up (Mathieu and Turquin, 1992).

Response to hypoxia was studied in *N. virei* and *N. rhenorhodanesis* comparatively to surface amphipods (*Gammarus fossarum*) or isopods (*Asellus aquaticus*). Both hypogean species resist hypoxia significantly longer than surface species. Response to hypoxia

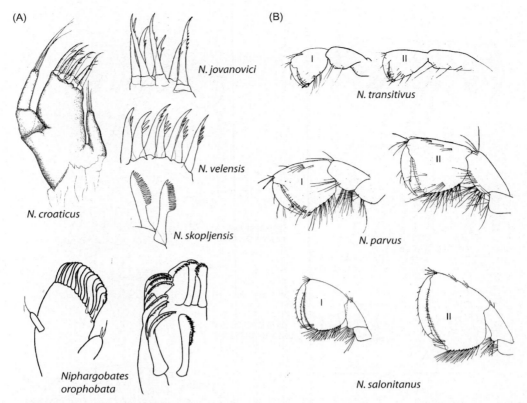

FIGURE 3 Diversity of feeding structures. (A) Maxilla I varies mainly in number and denticulation of spines on the outer lobes (upper part); these spines are in some species greatly modified, such as in *Niphargobates orophobata* (lower part). (B) Gnathopods are used for feeding and grooming. The shape and size of gnathopods I and II greatly vary.

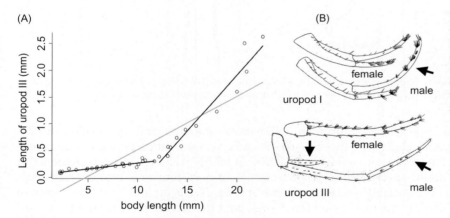

FIGURE 4 Allometry and sexually dimorphic traits. (A) Growth of sexually dimorphic uropod III; nonlinear growth (black lines) evidently better describes growth of the article than linear growth (gray line) *(after Fišer et al., 2008a)*. (B) Sexually dimorphic traits include lengths of inner ramus of uropod I, distal article of uropod III, and inner ramus of uropod III (arrows).

seems to be mediated by different enzyme activities and immediate decrease in locomotory and respiratory activities. Although *N. virei* seems to have lower oxygen consumption and overall lower locomotory and respiratory activities than *N. rhenorhodanesis*, the latter in post-hypoxic recovery more efficiently assimilates metabolites accumulated during hypoxia (Hüppop, 2000). Field evidence implies that at least *N. sphagnicolus* survives in hypoxic waters.

Temperatures in subterranean environments are presumably stabile and subterranean species are presumably stenothermal. Many data support this view. *Niphargus virei* is intolerant to cold temperatures in the laboratory, which agrees with its distribution that is limited to presumable borderlines of Alpine Pleistocene glaciers (Folquier *et al.*, 2008). Furthermore, *N. thermalis* has been reported only from thermal waters. In a cold-adapted mountain complex *N. tatrensis*, post-Pleistocene warming presumably caused extinction of lowland populations and fragmentation of ancestral populations. By contrast, *N. rhenorhodanesis* shows remarkable tolerance to low temperatures in the laboratory. Unexpectedly and difficult to explain, the tolerance for cold in this species varies on population level (Colson-Proch et al., 2009).

Tolerance to water chemistry is less well known. Measurements of the respiratory electron system and oxygen consumption imply high tolerance for acid environment in *N. sphagnicolus* (Simčič and Brancelj, 2007), which agrees with its distributional data (Fišer et al., 2010). Quite a few species live in anchihaline water, where coexisting species apparently distribute along a salinity gradient. Some species survive in mildly to highly sulfide waters and some do not (Flot et al., 2010). Moreover, few species were found exclusively in toxic hydrogen sulfide-rich aquifers, although their preference for sulfide remains unstudied (Dattagupta et al., 2009). Water conductivity may be related to water hardiness or to pollution. *Niphargus sphagnicolus*, for example, was found in a pond where conductivity increased because of pollution. By contrast, allopatric distribution of four species in the western Dinarics corresponds to water conductivity and water hardiness though causal links between physiology and distribution need to be addressed in the laboratory (Fišer et al., 2006). Last but not least, *N. montellianus* showed increased resistance to high concentrations of heavy metals compared to surface *Gammarus* species (Coppellotti Krupa and Guidolin, 2003).

Endemism and Poor Migratory Abilities

Most of niphargid species are known only from few localities, many of these only from their type localities. The overall impression is that the degree of endemism is high, but few species have been reported from large areas. Individual distributional patterns may be biased because of either incomplete sampling or vague taxonomy coupled with misidentifications. Trontelj et al., (2009) addressed the second issue using molecular phylogeny. Indeed several species names in fact cover a number of species, which are not necessarily closely related. Moreover, the authors proposed that the longest diameter of a range only exceptionally exceeds 200 km. The range-size hypothesis was tested by Zagmajster et al. (unpublished) using a dataset of 43 species that could have been unambiguously identified and originated from locally well-explored areas. This analysis indeed showed that most of the species have small ranges, although species with large ranges do exist. Three species were collected in a single spot, about 60% of species had the maximum extent of the range less than 100 km, and about 78% of the species less than 200 km. Extent of range in nine species (21%) exceeded 200 km, reaching up to 600 km. Therefore, it seems plausible to conclude that the majority of species have poor migratory capabilities, a finding that might importantly explain some evolutionary patterns.

PART TWO: INTERACTIONS WITHIN AND BETWEEN SPECIES

Intraspecific Interactions: Cannibalism

A key intraspecific interaction seems to be cannibalism, which affects microdistribution and breeding biology. Observations of *N. timavi* in the cave laboratory unambiguously confirmed that large individuals prey on small conspecifics, even on their own offspring (Luštrik et al., 2011). These observations, although conducted in Petri dishes where juveniles cannot hide, may indicate that niphargids are not selective predators. This also means that juveniles would benefit if they evolved avoidance behavior. Following this idea, Luštrik et al. (2011) observed the distribution of juveniles in aquaria filled with layers of differently sized glass pebbles. Juveniles preferred layers of midsized pebbles, but roamed also through large sized pebbles. As adults were released to aquaria, juveniles withdrew to the midsized layer, evidently inaccessible to adults.

Therefore, cannibalism probably best explains results from field studies using artificial substrates. Juveniles of *N. rhenorhodanesis* are abundant only in the absence of larger (>7 mm) individuals, and rapidly disappear when larger individuals colonize the space (Mathieu and Turquin, 1992 and references therein). Similarly, in the field juvenile *N. timavi* invaded fine-sand boxes while adults preferred gravel and leaves (Luštrik et al., 2011).

Cannibalism in combination with water regime may regulate breeding biology of species. Juveniles in both cave *N. virei* and interstitial *N. rhenorhodanesis* hatch when chances to be eaten by conspecifics are small. In cave *N. virei*, juveniles hatch at the onset of floods, when drifting accelerates dispersion of juveniles. In interstitial *N. rhenorhodanesis*, hatching corresponds to the onset of a dry period, when habitat partially dries up and adult individuals aestivate (Mathieu and Turquin, 1992).

Interspecific Interactions: Predation and Competition

Interactions between different species of *Niphargus* are poorly understood. Perhaps the most informative are results of transplant experiments, in which individuals of *N. virei* were introduced into a cave populated by *N. rhenorhodanesis*. The introduced species exterminated the native one in less than 20 years, implying that competition and predation are important interactions that determine local and regional niphargid diversity (after Folquier et al., 2008). On the other hand, niphargid species frequently coexist, and constitute communities of up to nine species (Table 1). A prerequisite for temporal stable communities is minimization of harmful interactions (competition and predation). Niche separation can be expected. A few cases are reviewed in the final section.

Although various subterranean amphipods may coexist with *Niphargus* species, neither interactions nor niche separation was studied at this taxonomic level. More attention was paid to interactions between subterranean *Niphargus* and surface amphipods. Observations in the Postojna-Planina Cave System in Slovenia implied that competitive strength of surface species increases when food is not a limiting factor. This was extended to the hypothesis that competition influences distribution of troglobionts and prevents their spreading into surface waters (Sket, 1981). Recent work partially supports this hypothesis. Annual observations of selected sites in spring areas revealed high tempor-spatial variation in the microdistribution of *G. fossarum* and *N. timavi*. Boxes with artificial substrates showed that the two species differ in habitat preferences. While *G. fossarum* preferred decaying leaves, *N. timavi* invaded leaves and gravel in similar shares. Laboratory observations suggest that the spread of *N. timavi* into surface waters is restricted by a combination of predation and cannibalism, but not by interspecific competition (Luštrik et al., 2011, and references therein).

TABLE 1 Structure of *Niphargus* Community in Vjetrenica and the Postojna-Planina Cave System

Vjetrenica	Habitat	PPCS
N. boskovici		Niphargobates orophobata
N. factor	epikarst	
N. cvijici		N. wolfi
—	cave streams	N. stygius
		N. spoeckeri
N. zavalanus	cave interstital	N. dobati
N. balcanicus		
N. vjetrenicensis	phratic, deep lakes	
N. hercegovinensis		N. orcinus
N. trullipes		
N. kolombatovici		

Symbiosis with Chemoautotrophic Bacteria

Dattagupta et al. (2009) report on the discovery of abundant filamentous bacteria on the exoskeletons of *Niphargus* species within the sulfide-rich Frasassi cave complex in central Italy. At least two out of four *Niphargus* species living in that complex are infected with filamentous bacteria. Using 16S rDNA sequencing and fluorescence *in situ* hybridization (FISH), they showed that amphipods throughout the cave complex are colonized by a single type of bacteria in the sulfur-oxidizing clade *Thiothrix*. The epibiont phylotype is distinct from *Thiothrix* phylotypes that form conspicuous biofilms in the cave streams and pools inhabited by niphargids. Using a combination of 13C labeling, FISH, and secondary ion mass spectrometry, they show that the epibiotic *Thiothrix* are autotrophic, establishing the first known example of a nonmarine chemoautotroph-animal symbiosis.

PART THREE: COMPARATIVE REVIEW OF KEY ELEMENTS

Phylogeny Reveals Morphological Diversity and Stasis

Introduction of molecular data largely modified our view of the evolution of morphology of *Niphargus*. Allozyme studies and sequencing of mitochondrial COI and 12S, and nuclear 28S and ITS DNA (Trontelj et al., 2009; Fišer et al., 2008b; Flot et al., 2010) yielded several important findings.

First, all members of the family Niphargidae diversified from the common ancestor *Niphargopsis*, and at least two other niphargid genera (*Niphargobates*, *Pontoniphargus*) are nested within *Niphargus*, losing justification for their separate generic status (Fišer et al., 2008b; Flot pers. comm; unpublished data).

Second, several cryptic lineages, which may be tentatively equated with cryptic species, have been discovered. Convergent selection and poor migratory abilities are key factors that explain morphological similarity among some niphargid species. Morphological similarity may be the result of fragmentation of ancestral species range followed by speciation in the absence of morphological change. Alternatively, strong selection operating on entire suits of morphological characters in unrelated species results in morphologically indistinguishable species (Trontelj et al., 2009).

Third, closely related species may be morphologically unexpectedly different (Fig. 5). A tentative conclusion is that not only convergent but also divergent selection operates in subterranean environments (Fišer et al., 2008b).

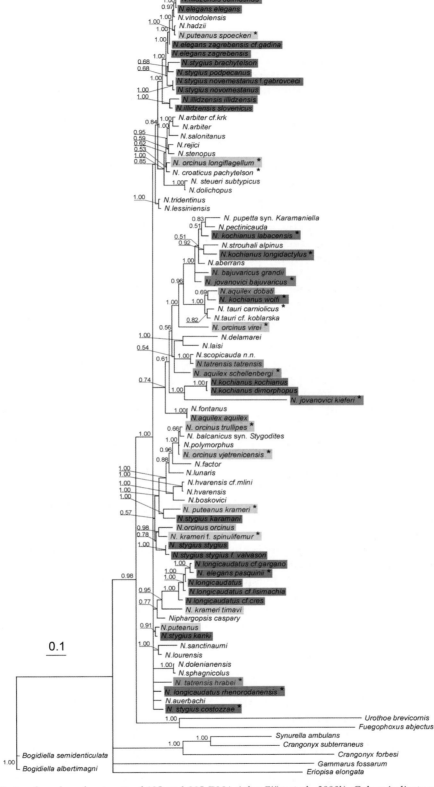

FIGURE 5 Phylogenetic tree based on fragments of 12S and 28S DNA (*after Fišer et al., 2008b*). Colors indicate one of the traditional divisions of the genus based on morphology. There is only little congruence between morphology and phylogenetic relationships.

Communities: Niche Separation and Functional Morphology

Up to nine *Niphargus* species may coexist in one locality. Communities such as these are priceless natural laboratories for evolutionary and ecological research. As expected, coexisting species differ in their ecological niches along chemical and physical gradients. In addition, it appears that microhabitats defined by availability of space and water currents harbor morphologically distinct species.

Hypothelminorheic (seep-dwelling) *N. sphagnicolus* and *N. slovenicus* live in sympatry. More careful examination of distributional maps reveals that the two species may be collected in close proximity, but never together (Fig. 6). It appeared that localities invaded by either of the species differ in oxygen saturation, conductivity, and pH. pH is a key factor that determines which species prosper in a particular locality (Fišer *et al*., 2010).

Both chemical and physical gradients may be involved in niche separation, as is the case in the sulfide-rich Frasassi cave complex in central Italy. *N. ictus* and *N. frasassianus* live in sulfidic water. In localities where both species live together they occupy different microhabitats. *Niphargus frasassianus* is restricted to fast-flowing portions of the stream, whereas *N. ictus* inhabits stagnant parts that are deeper than 20 cm. Two other species seem to be entrapped in nonsulfidic lakes in different parts of the cave complex. *Niphargus montanarius* was collected from the nonsulfidic lake Il Bugianardo. A fourth, yet undescribed, species was found in a remote locality of Lago Primo, a stratified lake with oxygenated water on top and reducing sulfidic waters at depths greater than 3.5 meters (Flot *et al*., 2010).

Most caves have less distinct, if any, chemical gradients. Nevertheless, presence of water currents and availability of space define at least four physically distinct microhabitats: epikarst, interstitial, streams, and lakes. Niche separation, at least in part, includes occupancy of different microhabitats. For instance, in the Postojna-Planina Cave System *Niphargobates orophobata* and *N. wolfi* were found in drips, *N. stygius* and *N. spoeckeri* in sinking rivers and ponds along shores of the river, *N. dobati* in cave interstitial, and *N. orcinus* in phreatic. Moreover, a comprehensive analysis of seven cave communities identified four distinct morphological types (unpublished). Phreatic species are of two size classes with elongated appendages. Species from streams are large with short appendages. Interstitial species cannot be distinguished from epikarstic species. Both are small and of various lengths (Fig. 2). In any case, the communities provide some evidence for divergent selection.

A Case of Adaptation: Female Body Size

Whereas small pore size in interstitial and epikarst habitats may explain the evolution of small species, the question remains: What is the adaptive value of large body size in food-poor phreatic waters? Phreatic niphargids are not sexually dimorphic, and ovigerous females (females carrying eggs) often exceed the size of 20 mm. By contrast, species from streams and springs are sexually dimorphic, where females rarely reach 18 mm in body length (males can reach 30 mm). In a recent study, egg volume, brood volume, and reproductive investment (*i.e.*, what proportion of body volume represents brood volume) were estimated for ovigerous females of seven phreatic species and six species living in springs. Large phreatic females have larger eggs but lower reproductive investment into a single brood compared to species from springs. Moreover, the correlations between body size and egg volume and body size and reproductive investment are significantly positive and negative, respectively. Large body size therefore presumably increases both female and offspring fitness. Egg size in general positively correlates with juvenile size, and larger juveniles presumably more easily cope with environmental challenges, such as avoidance from predators, increased resistance to starvation, and exploitation of a broader range of food. Since phreatic species have a prolonged life span (Mathieu and Turquin, 1992), it is likely that they have increased numbers of broods. Fitness of the female—the product of number of broods and average number of survived offspring per brood—therefore depends on her own survival, which modifies her allocation strategy. Reduction of reproductive effort into one brood—a phenomenon common to species with multiple broods—increases her chances to survive until the next brood. Increase in body size also affects other traits. Larger eggs, for instance, require larger marsupium for two reasons. Not only that more space is needed, but that larger eggs may require increased ventilation due to decreased surface-to-volume ratio. This explains the relatively deeper coxal plates in phreatic species compared to species from springs.

CONCLUSIONS

This short overview illustrates that the genus *Niphargus* is extremely variable considering morphological, developmental, physiological, and life-history traits. This variability, when analyzed in temporal and spatial contexts, may be a precious source of information about

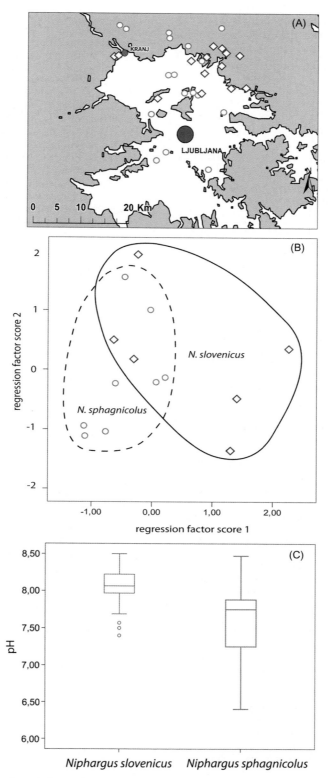

FIGURE 6 Niche separation of two hypothelminorheic species (red circles: *N. sphagnicolus*; blue diamonds: *N. slovenicus*). (A) Distribution of the two species reveals no geographic segregation; white color indicates alluvial plain at altitudes between 290 and 390 m, and most samples were collected below 500 m altitude (*after Fišer et al., 2010*). (B) Principal component analysis of 11 functional characters shows that the two studied species cannot be unambiguously discriminated (the first two principal components explain 98.7% of variance); this may indicate that hydrological conditions in hypothelminorheic species do not allow morphological differentiation of the species. (C) Evidence that distribution of the two species depends on acidity of water. Note that *N. sphagnicolus* tolerates a large range of pH, possibly indicating that this species frequently lives in suboptimal acidic conditions.

processes such as speciation, specialization, adaptation, competition, and migration on different geographic scales (from single locality to Western Palearctics) and different taxonomic levels (from population up to a family).

See Also the Following Articles

Asellus aquaticus: *A Model System for Historical Biogeography*
Gammarus minus: *A Model System for the Study of Adaptation to the Cave Environment*
Invasion, Active versus Passive
Vicariance and Dispersalist Biogeography

Bibliography

Colson-Proch, C., Renault, D., Gravot, A., Douday, C. J., & Hervant, F. (2009). Do current environmental conditions explain physiological and metabolic responses of subterranean crustaceans to cold? *The Journal of Experimental Biology*, 212(12), 1859–1868.

Coppellotti Krupa, O., & Guidolin, L. (2003). Responses of *Niphargus montellianus* and *Gammarus balcanicus* (Crustacea, Amphipoda) from karst waters to heavy metal exposure. *Journal de Physique IV*, 107, 323–326.

Dattagupta, S., Schaperdoth, I., Montanari, A., Mariani, S., Kita, N., Walley, J. W., et al. (2009). A novel symbiosis between chemoautotrophic bacteria and a freshwater cave amphipod. *The ISME Journal*, 3(8), 935–943.

Fišer, C., Bininda-Emonds, O. P. R., Blejec, A., & Sket, B. (2008a). Can heterochrony help explain the high morphological diversity within the genus *Niphargus* (Crustacea: Amphipoda)? *Organisms, Diversity and Evolution*, 8(2), 146–162.

Fišer, C., Konec, M., Kobe, Z., Osanič, M., Gruden, P., & Potočnik, H. (2010). Conservation problems with hypothelminorheic *Niphargus* species (Amphipoda: Niphargidae). *Aquatic Conservation: Marine and Freshwater Ecosystems*, 20(5), 602–604.

Fišer, C., Sket, B., & Stoch, F. (2006). Distribution of four narrowly endemic *Niphargus* species (Crustacea: Amphipoda) in the western Dinaric region with description of a new species. *Zoologischer Anzeiger*, 245(2), 77–94.

Fišer, C., Sket, B., & Trontelj, P. (2008b). A phylogenetic perspective on 160 years of troubled taxonomy of *Niphargus* (Crustacea: Amphipoda). *Zoologica Scripta*, 37(6), 665–680.

Fišer, C., Trontelj, P., Luštrik, R., & Sket, B. (2009). Toward a unified taxonomy of *Niphargus* (Crustacea: Amphipoda): A review of morphological variability. *Zootaxa*, 2061, 1–22.

Flot, J.-F., Woerheide, G., & Dattagupta, S. (2010). Unsuspected diversity of *Niphargus* amphipods in the chemoautotrophic cave ecosystem of Frasassi, central Italy. *BMC Evolutionary Biology*, 10, 171.

Folquier, A., Malard, F., Lefébure, T., Douady, C. J., & Gibert, J. (2008). The imprint of Quaternary glaciers on the present-day distribution of the obligate groundwater amphipod *Niphargus virei* (Niphargidae). *Journal of Biogeography*, 35(3), 552–564.

Hüppop, K. (2000). How do cave animals cope with the food scarcity in caves? In H. Wilkens, D. C. Culver & W. F. Humphreys (Eds.), *Subterranean ecosystems* (pp. 159–188). Amsterdam: Elsevier.

Jazdzewski, K., & Kupryjanowicz, J. (2010). One more fossil Niphargid (Malacostraca: Amphipoda) from Baltic amber. *Journal of Crustaceana Biology*, 30(3), 413–416.

Karaman, G. S., & Ruffo, S. (1986). Amphipoda: *Niphargus*—Group (Niphargidae *sensu* Bousfield, 1982). In L. Botosaneau (Ed.), *Stygofauna mundi* (pp. 514–534). Leiden: E. J. Brill/Dr. Backhuys.

Luštrik, R., Turjak, M., Kralj-Fišer, S., & Fišer, C. (2011). Coexistence of surface and cave amphipods in an ecotone environment (spring area). *Contributions to Zoology*, 80(2), 133–141.

Mathieu, J., & Turquin, M. J. (1992). Biological processes at the population level. II. Aquatic populations *Niphargus* (stygobiont amphipod) case. In A. I. Camacho (Ed.), *The natural history of biospeleology* (pp. 264–293). Madrid: Monografias Museo Nacional de Ciencas Naturales.

Simčič, T., & Brancelj, A. (2007). The effect of light on oxygen consumption in two amphipod crustaceans—The hypogean *Niphargus stygius* and the epigean *Gammarus fossarum*. *Marine and Freshwater Behaviour and Physiology*, 40, 141–150.

Sket, B. (1981). Distribution, ecological character, and phylogenetic importance of *Niphargus valachicus*. *Biološki Vestnik*, 29(1), 87–103.

Trontelj, P., Douady, C., Fišer, C., Gibert, J., Gorički, Š., Lefébure, T., et al. (2009). A molecular test for hidden biodiversity in groundwater: How large are the ranges of macro-stygobionts? *Freshwater Biology*, 54(4), 727–744.

NITRATE CONTAMINATION IN KARST GROUNDWATER

Brian G. Katz

U.S. Geological Survey, Tallahassee, Florida

BACKGROUND INFORMATION

In karst terrain, as well as in other hydrogeologic systems, numerous interrelated factors account for nitrate contamination of groundwater. These factors can be grouped into two main categories: land use (including nitrate sources, land application rates and timing of fertilizers and manure, and waste disposal practices) and hydrogeologic characteristics (such as the degree of connectivity between the aquifer and surface solution features, soil drainage characteristics, the timing and amount of recharge, degree of aquifer confinement, shallow and deep flow patterns, and type of water-supply well). Other conditions that further promote nitrate contamination in karst aquifer systems are the typically oxygenated conditions that contribute to the stability of nitrate (preventing its transformation to reduced nitrogen species), thin soil cover overlying many karst aquifers, and often careless waste-disposal practices. Also, in many areas, an epikarst zone directly above the aquifer can provide temporary storage for nitrate and other contaminants, which can be released from this zone into the aquifer by recharge from storm events. Losses or decreases in nitrate concentrations within a karst aquifer system can result from naturally occurring processes, such as denitrification (microbially mediated transformation of nitrate to reduced nitrogen species) and dilution (*e.g.*, resulting from interactions between surface water and groundwater).

ENVIRONMENTAL CONCERNS AND NITROGEN CYCLING IN GROUNDWATER

Several human and ecological health concerns are associated with elevated nitrate concentrations in groundwater. Infants under 6 months of age who ingest nitrate in drinking water are susceptible to methemoglobinemia, which can lead to reduced blood oxygen levels and can result in death. For these health concerns, the U.S. Environmental Protection Agency established a maximum contaminant level (MCL) for nitrate of 10 mg L^{-1}, as nitrogen (N), for drinking water. Another study found an increased risk of non-Hodgkins lymphoma associated with nitrate concentrations of 4 mg L^{-1} or more in rural drinking-water supplies. Groundwater contributes to base flow in many streams and rivers and elevated nitrate concentrations from groundwater can stimulate the growth of nuisance aquatic vegetation in surface waters. Currently, there are no guidelines for nitrate concentrations in streams to limit the growth of nuisance aquatic vegetation.

Nitrate is one of several species of nitrogen (N) that can exist in the subsurface. The presence of other N species in a karst aquifer depends on oxidation-reduction conditions and microbial reactions in the subsurface. N-cycling processes are complex. For example, oxidation of reduced N species (ammonium, organic N, nitrite) to nitrate or transformation of nitrate to reduced N species generally takes place in the soil or unsaturated zone where organic material and microbial communities are present. These processes are controlled by the presence or absence of oxygen and other terminal electron acceptors (*e.g.*, sulfate, ferric iron, organic carbon). These processes need to be understood to properly assess the fate of nitrate in karst aquifers. It is beyond the scope of this chapter to discuss these processes in detail. Numerous reviews, texts, and other publications have described biologically mediated transformations, storage pools, and nitrate removal.

LAND-USE FACTORS

Activities associated with various land uses can substantially affect nitrate concentrations in groundwater in karst systems. Most cases of elevated nitrate concentrations in karst aquifers are related to agricultural activities. Numerous studies have demonstrated that nitrate concentrations in groundwater are higher beneath agricultural areas compared to forested areas and are related to the percentage of agricultural land used near wells or springs.

Nonpoint or diffuse sources account for nitrate contamination of groundwater on a regional scale in many agricultural areas throughout the world. Elevated nitrate concentrations in many of these areas often are associated with contamination from other agricultural chemicals, such as herbicides and other pesticides. Although several retrospective studies have reported lower nitrate concentrations in groundwater from the southeastern United States compared to other regions, elevated nitrate concentrations exceed the nitrate MCL for drinking water in many parts of the karstic Upper Floridan aquifer in areas of extensive agricultural land use. The application of manure also has contributed to nitrate contamination of groundwater as indicated by statistically significant relations between nitrate concentration in groundwater and the amount of manure applied to soils overlying carbonate aquifers.

Nitrate contamination of groundwater also can originate from localized agricultural point sources, such as concentrated animal feeding areas, poorly designed waste storage areas, and disposal of dead animals in sinkholes. Also, other documented cases of nitrate contamination of groundwater are related to the spreading of organic wastes over agricultural lands, irrigation of pasture land with cheese factory effluent, and spray irrigation of wastewater from a vegetable- and meat-canning facility.

There have been several documented cases of nitrate contamination of groundwater in karst areas that are associated with land uses other than agriculture. In rural areas, contamination from septic tanks and cesspools is fairly common. Effluent from these systems can reach the water table after undergoing little filtration or attenuation due to the typically thin soil cover in many karst areas. Also, nitrate contamination of groundwater has resulted from diversion of untreated sewage into nearby sinkholes or swallow holes. Leaky sewers have contributed to nitrate contamination in some urban areas. In many rural landscapes, the dumping of household refuse, construction materials, and dead livestock has resulted in groundwater contamination problems. Landfills and open dumps in limestone outcrops in the recharge area of a well field have contributed to nitrate contamination of groundwater in China. In an Indiana karst system, nitrate contamination of groundwater was attributed to by-products from ordnance manufacturing and demolition processes at a military institution.

Atmospheric deposition (wetfall and dryfall) is potentially another source for elevated nitrate concentrations in karst groundwater, although very few studies have addressed this issue. Seasonal variability of nitrogen (ammonium and nitrate) in precipitation most likely is related to the application of fertilizers in the spring, but other potential atmospheric sources of nitrate, including industrial waste gases and fossil-fuel combustion, could be responsible. The influence of atmospheric sources has been inferred from isotopically enriched values of $\delta^{18}O$ of nitrate found in winter and summer spring-water samples, when nitrate in groundwater was relatively low and its concentration in precipitation was high.

Most studies have focused on the collection of water from wells or springs in karst aquifers; however, a few studies have collected samples of cave drip water as a novel way of characterizing the chemistry of recharge water moving through the unsaturated zone into caves and groundwater. Caves near and below agricultural land had higher nitrate concentrations in drip water than did groundwater in forested areas. Concentrations of nitrate were about an order of magnitude higher in autogenic recharge waters at a cave located beneath an agriculturally intensive landscape than in recharge at caves beneath less-intensive land uses. This study also found that nitrate varied seasonally with a flush of nutrients during late summer and autumn that corresponded to trends in surface rivers and streams.

HYDROGEOLOGIC FACTORS

Based on an analysis of retrospective data (1972–92) from sites around the United States, the U.S. Geological Survey (USGS) reported that nitrate concentrations were highest in karst bedrock areas compared to other hydrogeologic settings, such as sand and gravel, alluvium, and other bedrock types. Differences reflected the intensive agricultural activity on land overlying the carbonate settings, but also were related to the hydraulic connections to the surface and solution features typical of fractured carbonate rocks.

Similar findings were reported for other karst areas in the midwestern United States. Based on data from more than 6000 water samples collected during 1977–80 from 22 counties in Iowa, nitrate-N concentrations from all well depths were higher in a karst region compared to samples from shallow and deep bedrock. Even though nitrate concentrations decreased with depth in all regions, elevated nitrate concentrations in the karst region extended to 30 m greater depth than in the confined deep bedrock region.

The degree of confinement of the aquifer can also affect nitrate concentrations in groundwater. Several studies have shown that nitrate concentrations in limestone aquifers are substantially higher in water from unconfined aquifers than from confined aquifers. However, in some areas, a rapid response of groundwater to elevated nitrate concentrations after fertilizer applications were attributed to the rapid movement of fertilizer-derived nitrate into macropores.

The density of karst features, such as sinkholes, is another factor that affects nitrate contamination of groundwater. Based on water samples from seven carbonate aquifers in the eastern United States, nitrate concentrations were significantly higher in agricultural areas with a moderate to high density of sinkholes (1–25 sinkholes per 100 km^2) compared to areas with low sinkhole density (<1 sinkhole per 100 km^2).

Nitrate concentrations in groundwater also are related to the type and depth of a sampled well, although very few studies have specifically addressed this issue in karst areas. Based on more than 5600 samples collected throughout the U.S. in both karst and nonkarst settings, the USGS reported that median nitrate-N concentrations ranged from 0.2 mg L^{-1} in public supply wells to 2.4 mg L^{-1} for irrigation and stock wells. Water samples from about 16% of the generally shallow irrigation and stock wells exceeded the nitrate MCL, compared to only about 1% of samples from public supply wells, which are typically open to much deeper parts of an aquifer.

NITRATE CONTAMINATION OF SPRINGS

Springs represent discharge points of groundwater basins and are typically the dominant source of water for surface streams in karst areas. Because springs integrate water temporally, spatially, and vertically from an aquifer, they offer an effective way of assessing sources and time scales of nitrate contamination of groundwater in karst areas. In many areas, nitrate concentrations in spring waters have increased substantially over the past 40–50 years and in most cases, the increasing trends were related to increased amounts of fertilizer application. Nitrate concentrations in spring waters (and nitrate flux) are related to the extent of cultivated zones at the surface.

In several studies, nitrate concentrations were found to be higher in springs than in water from nearby wells. For example, in the Ozarks, higher nitrate concentrations in springs were attributed to groundwater that generally follows shallow flow paths and issues from conduits and, therefore, is more susceptible to contamination from the surface. Elevated nitrate concentrations are found in shallow systems, as well as in springs draining deep groundwater systems (up to 100 m depth) in areas of intensive water exploitation and fertilizer use.

Stable nitrogen isotopes ($\delta^{15}N$) in hydrologic studies have been particularly effective in assessing the sources and fate of nitrate. In studies of karst springs, these and other naturally occurring tracers have been used effectively to test hypotheses about flow-system characteristics and dominant geochemical processes. For example, based on measurements of $\delta^{18}O$ and $\delta^{15}N$ of nitrate in spring waters in the southwest Illinois sinkhole plain, significant denitrification occurred within the soil zone, epikarst, and shallow karst hydrogeologic system and accounted for the loss of nitrate in spring waters from spring to winter. A multitracer approach that included isotopic and other chemical tracers was used to show that nitrate concentrations in Florida spring waters were related to average groundwater residence times determined from age-dating techniques. Springs with lower flow rates had short residence times (around 10 years) and relatively high nitrate-N concentrations (20–35 mg L^{-1}), whereas those with higher flow rates had longer residence times (20–30 years) and lower nitrate concentrations (0.2 to 5 mg L^{-1}). These differences were attributed to the higher contribution of young water (recharged less than 7 years) from shallow parts of the Upper Floridan aquifer in spring waters with low flows (0.028–0.28 $m^3 s^{-1}$), compared to a higher contribution of deep older water (recharged 2–3 decades ago) that accounted for lower nitrate concentrations in spring waters with high flow rates (>2.8 $m^3 s^{-1}$).

TEMPORAL VARIABILITY OF NITRATE

Temporal variations in nitrate concentrations in groundwater and springs reflect the interrelation of several important hydrogeologic, geochemical, land use, and climatic factors. The degree of temporal variation is related to the degree of karstification, mantle thickness, and the proportion of conduit to diffuse flow. For example, nitrate concentrations in groundwater generally increase following seasonal applications of fertilizers and subsequent recharge of the groundwater system. Large fluctuations in nitrate concentrations in springs have been observed during high discharge events related to annual variations in rainfall. In some areas, seasonally high water tables may result in groundwater that is anoxic and if conditions persist, loss of nitrate through denitrification may occur (see below).

PROCESSES THAT RESULT IN LOSSES OF NITRATE IN KARST GROUNDWATER

Although many karst aquifers tend to be oxic, nitrification-denitrification processes can occur within the overlying soil zone, epikarst, and aquifer (rock) matrix. In the relatively immobile pore waters in the rock matrix of the Chalk aquifer in Britain, denitrification accounted for the decrease in nitrate concentrations downgradient in the direction of flow. Other studies have documented nitrate losses in karst groundwater as a result of denitrification.

Dilution of nitrate can result from mixing of surface water and groundwater during following periods of high rainfall in karstic systems. Dilution by runoff transported through fast-flow conduits also has accounted for large decreases in nitrate concentrations in spring waters during high-rainfall periods. In a study in Iowa, wetter than average conditions following a drought period resulted in a substantial increase in nitrate concentrations as a result of the increased water volume moving through the soils and the leaching of nitrate left over from the drought.

PROTECTING GROUNDWATER RESOURCES

Understanding the factors and processes that affect the occurrence, movement, and fate of nitrate is essential for protecting groundwater quality and preventing further nitrate contamination in karst areas. Although the delineation of groundwater contributing areas and (or) protection zones for springs and wells in karst areas remains difficult and challenging, recharge areas are protected in many locales—especially springs or wells used for water supply. Considerable work is being done to evaluate management strategies for reducing nitrate contamination from point and nonpoint sources in karst systems. The time lag between a reduction in land application of fertilizer or manure spreading can vary greatly, depending on site-specific soil characteristics, climatic conditions, and recharge rates. Given the large year-to-year variations in nitrate concentrations at Big Spring, in Iowa, it has been difficult to identify any improvements in water quality resulting from reductions in nitrogen that have occurred incrementally over the past decade. In about one-third of the studied basin, decreases in nitrogen inputs have been overshadowed by large variations in annual recharge. As the worldwide demand for water increases, much work remains to be done to characterize both the residence time of groundwater in varied and complex karst systems, and the travel time from a recharge area to a point of spring discharge or withdrawal of groundwater for public consumption.

Bibliography

DiGnazio, F. J., Krothe, N. C., Baedke, S. J., & Spalding, R. F. (1998). $\delta^{15}N$ of nitrate derived from explosive sources in a karst aquifer beneath the Ammunition Burning Ground, Crane Naval Surface Warfare Center, Indiana, U.S.A. *Journal of Hydrology, 206*, 164–175.

Drew, D., & Hötzl, H. (Eds.), (1999). *Karst hydrogeology and human activities: Impacts, consequences, and implications. International contributions to hydrogeology.* Rotterdam, Netherlands: A. A. Balkema.

Dubrovsky, N. M., Burow, K. R., Clark, G. M., Gronberg, J. M., Hamilton, P. A., Hitt, K. J., et al. (2010). The quality of our nation's waters—Nutrients in the nation's streams and groundwater, 1992–2004. *U.S. Geological Survey Circular, 1350*, 1–174.

Hallberg, G. R., & Keeney, D. R. (1993). Nitrate. In W. M. Alley (Ed.), *Regional ground-water quality.* New York: Van Nostrand Reinhold.

Johnson, C. J., Bonrud, P. A., Dosch, T. L., Kilness, A. W., Senger, K. A., Busch, D. C., et al. (1987). Fatal outcome of methemoglobinemia in an infant. *Journal of the American Medical Association, 257*(20), 2796–2797.

Katz, B. G., Böhlke, J. K., & Hornsby, H. D. (2001). Timescales for nitrate contamination of spring waters, northern Florida, U.S.A. *Chemical Geology, 179*(1–4), 167–186.

Kendall, C., & Aravena, R. (2000). Nitrate isotopes in groundwater systems. In P. G. Cook & A. L. Herczeg (Eds.), *Environmental tracers in subsurface hydrology.* Boston: Kluwer Academic Press.

Lindsey, B. D., Berndt, M. P., Katz, B. G., Ardis, A. F., & Skach, K. (2009). Factors affecting water quality in selected carbonate aquifers of the United States. *U.S. Geological Survey Scientific Investigations Report 2008-5240.*

Lindsey, B. D., Katz, B. G., Berndt, M. P., Ardis, A. F., & Skach, K. (2010). Relations between sinkhole density and anthropogenic contaminants in selected carbonate aquifers in the eastern United States. *Environmental Earth Sciences, 60*, 1073–1090.

Mueller, D. K., Hamilton, P. A., Helsel, D. R., Hitt, K. J., & Ruddy, B. C. (1995). Nutrients in ground water of the United States—An analysis of data through 1992. *U.S. Geological Survey Water-Resources Investigations Report 95-4031.*

Panno, S. V., Hackley, K. C., Hwang, H. H., & Kelly, W. R. (2001). Determination of the sources of nitrate contamination in karst springs using isotopic and chemical indicators. *Chemical Geology, 179*(1–4), 113–128.

Peterson, E. W., Davis, R. K., Brahana, J. V., & Orndorff, H. A. (2002). Movement of nitrate through regolith covered karst terrane, northwest Arkansas. *Journal of Hydrology, 256*, 35–47.

Plagnes, V., & Bakalowicz, M. (2001). The protection of karst water resources: The example of the Larzac karst plateau (south of France). *Environmental Geology, 40*, 349–358.

Scanlon, B. R. (1990). Relationships between groundwater contamination and major-ion chemistry in a karst aquifer. *Journal of Hydrology, 119*, 271–291.

Spalding, R. F., & Exner, M. E. (1993). Occurrence of nitrate in groundwater—A review. *Journal of Environmental Quality, 22*(3), 392–402.

Wells, E. R., & Krothe, N. C. (1989). Seasonal fluctuation in $\delta^{15}N$ of groundwater nitrate in a mantled karst aquifer due to macropore transport of fertilizer nitrate. *Journal of Hydrology, 112*, 191–201.

Zhu, X. Y., Xu, S. H., Zhu, J. J., Zhou, N. Q., & Wu, C. Y. (1997). Study on the contamination of fracture-karst water in Boshan district, China. *Ground Water, 35*(3), 538–545.

NULLARBOR CAVES, AUSTRALIA

Julia M. James, Annalisa K. Contos, and Craig M. Barnes

University of Sydney, Australia

INTRODUCTION

In Australia, the karst land, referred to as the *Nullarbor*, covers an area in excess of 200,000 km^2, although in proportion to its area the karst is poor in known caves. Despite being referred to by their collective name, the Nullarbor Caves, many of the individual caves are world class in their own right. They have played an important role in elucidating Australian prehistory and they have been the focus of many intrepid expeditions—first to find the caves and then to explore them both in the dry waterless passages and in the extensive flooded passages.

The very existence of caves beneath a desert is remarkable, and because of this they have individually and as a group contributed significantly to the understanding of speleogenesis in arid environments. These caves are renowned for their rare secondary minerals, and the great variety of novel cave decorations has provided valuable insight into past surface and climatic conditions. The desiccating environment in these caves has also preserved biological materials in such pristine condition that they are considered outstanding repositories of paleoenvironmental information. Despite being nutrient poor, the dry passages of the Nullarbor Caves host a range of terrestrial invertebrates including several unique troglobitic species. Colonies of microbes have produced abundant biofilms in the saline cave waters and these contain many new species of bacteria.

PHYSICAL SETTING

The Nullarbor Caves are found on one of the largest continuous exposed karst lands in the world in the southwest part of the Australian continent (Fig. 1). These caves are developed in four limestone strata: the Nullarbor (early-middle Miocene), the Mullamullang member of the Nullarbor limestone, Abrakurrie, and Wilson Bluff (middle-late Eocene). Bounded to the north by the Colville Sandstone, the limestone strata are unconformable, almost horizontally bedded and dip to the south-southeast (Fig. 2). Uplift of the limestones probably occurred at the end of the Lower Miocene. Relatively weak jointing and the absence of folding or major tilting suggest that the region has remained relatively stable since uplift. Particularly important to cave development is that all limestone strata are both porous and permeable.

The Nullarbor is a plain that is almost completely flat. The plain in the south is sparsely vegetated and in the north it is treeless. At present, the tree line that delineates the edge of the treeless plain is approximately 80 km north of Eucla (Fig. 1). Pollen studies of sediments and gut contents of extinct fauna have shown that the "tree line" extended to Eucla during the last glacial, a time when the coastline was 160 km farther to the south. These studies have also shown that the vegetation has remained similar throughout and since the last cold period.

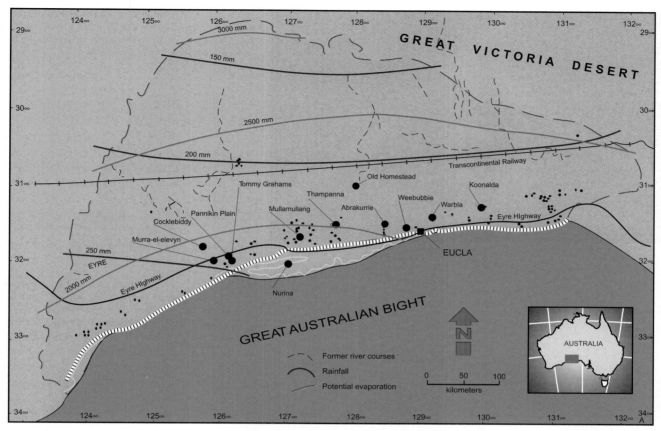

FIGURE 1 Location map for the Nullarbor Caves. The golden yellow color depicts the limestone area. *After Lowry and Jennings (1974).*

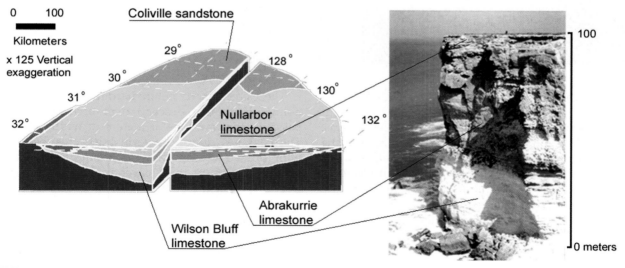

FIGURE 2 The geology of the Nullarbor (after Lowry and Jennings, 1974)—the Mullamullang strata are not shown for clarity—together with a plate showing an outcrop of the limestone strata exposed in the cliffs bordering the Great Australian Bight. *Photograph taken by Annalisa Contos.*

FIGURE 3 The mouth of Pannikin Plain Cave set in the flat plain. The cave entrance is 10 m in diameter and after 8 m leads down through collapse to the main passage. *Photo courtesy of Dirk Stoffles.*

FIGURE 4 The main chamber in Abrakurrie. The chamber is in the white Wilson Bluff limestone. The flat floor consists of fluvial sediments mixed with the products of crystal weathering. The volume of the chamber is ~135,000 m^3. *Photograph taken by Julia James.*

The major surface karst features are ridges and corridors, with relief of 3–5 m amplitude, and large but extremely shallow circular depressions called *dongas*. Set into the plain and distributed throughout the Nullarbor are thousands of blowholes, holes in the ground through which air blows in and out strongly. More frequent in the southern part of the plain are hundreds of abrupt, sharp-edged collapse dolines (Fig. 3). Typical of semiarid to arid karst, the Nullarbor has poorly developed karren features, although there are extensive calcretes and the surface of the limestone is case hardened. Immediately below the indurated surface, the rock mass is extensively weathered with many small cavities.

The caves can be entered from either blowholes or dolines that often descend over breakdown to large passages that are tens of meters in width and height. The passages enlarge to form huge chambers (Fig. 4) with domed ceilings and arch entrances. The most striking feature of these caves is the spectacular lakes (Fig. 5) of considerable depth and clarity. These lakes have helped establish the international reputation of the Nullarbor Caves. In a few of the known caves, kilometers of flooded passage leading from the lakes have been discovered.

Although open to debate, some evidence suggests that the Nullarbor has been semiarid to arid for some 6 myr. At present, the Nullarbor has a warm semiarid

FIGURE 5 Weebubbie Lake has an area of ~7500 m². From the end of the lake a series of large flooded passages continue, reaching a depth of 45 m. *Photo courtesy of Alan Warild.*

climate in the south, but it is hot and arid in the north. Surface temperatures range widely both seasonally and diurnally as a result of continentality, although winters are short and cool to mild. The average temperature of the caves is 18°C. An important meteorological feature is that potential evaporation exceeds precipitation for the entire region (Fig. 1). The rainfall pattern is unreliable; along the coast most precipitation falls in winter as light rain and mist, whereas inland, irregular summer thunderstorms distribute the rainfall. This annual pattern is interrupted every few summers by intense rain-bearing cyclonic depressions. These events can drop hundreds of millimeters of rain in an hour, causing considerable local flooding.

There are no permanent streams on the Nullarbor, but there are numerous intermittently active watercourses. During intense rainfall, water flows into depressions or collapse dolines and rapidly disappears underground. In the caves, the streams flow only short distances before entering lakes or sinking in sediment. Lakes are encountered at similar depths in the Nullarbor Caves and they form part of a regional water table known as the *Nullarbor aquifer*. The surface of the aquifer gently slopes toward the Great Australian Bight and is close to sea level in height. It has both conduit and diffuse flow characteristics, with the flow of water in the caves being from north to south. The caves act as conduits within the unconfined aquifer whose surface marks the top of the phreatic zone. The ultimate destination of the cave waters remains an enigma as there are no large springs issuing from the coastal cliffs and dunes, and no known submarine springs in the Great Australian Bight.

PREHISTORY, HISTORY, AND SPELEOLOGICAL HISTORY

On the Nullarbor, the nomadic Aborigines inhabited the fringes of the plain from the earliest times and entered the caves in a search for flint. Indeed, several of the caves, such as Koonalda (Fig. 1), have been mined for flint. Archaeological excavations 80 m below the surface, on the edge of the twilight zone in Koonalda, the only cave containing a constant supply of potable water, have provided evidence of a period of uninterrupted human "occupation" spanning approximately 30,000 years. The stone artifacts found there show discernible changes in stone implement technology but there are no grinding stones, indicating a lack of grasses and cereals and a purely hunting lifestyle. Several caves are decorated with hand stencils or primitive engravings similar to artwork found in Paleolithic caves in Europe and dating to Late Glacial times.

The presence of caves on the Nullarbor was first recorded in 1866 just prior to European settlement at Eucla in 1867. The local Aboriginal tribe was the Mirning, which at its peak numbered 200 persons but by 1867 numbered only 30. Anthropologists have established that at this time the Mirning no longer used the Nullarbor Caves. Even after European settlement, the population density of the area remained very low on a world scale; thus, until the early twentieth century, many of the Nullarbor Caves formed a pristine underground wilderness, preserved by their remoteness in an inhospitable landscape. However, the change of land use from wilderness to sheep and cattle grazing, coupled with the introduction of other feral animals, especially the rabbit, has subsequently caused considerable damage to many of the caves.

The first expeditions specifically to locate and explore caves took place during the 1930s. On some of these early expeditions, a light aircraft was used to locate the caves. With the formation of Australian speleological societies in the 1950s, major caving expeditions commenced. These benefited from the newly available air photographs, which proved invaluable in locating caves with large dolines and entrances. The highlight of the 1950s exploration was the discovery and survey of Mullamullang, which at 10.8 km was the longest cave in Australia at that time. The flooded passages in many of the caves attracted cave divers, with the first major attempt at underwater exploration occurring in 1972. The 1980s saw cave exploration intensify once more and the use of ultralight aircraft for prospecting. This systematic searching of the surface and depressions led to many new caves being found and explored. In the known caves, passages were extended by dedicated cavers who were prepared to continue even when the way became constricted or required excavation. Improvements in cave diving equipment saw flooded caves explored to world record lengths. This attack on the Nullarbor underground has continued into the new millennium; even in 2002 new caves and flooded maze passages coated in crystal were discovered. Since the 1950s, speleological investigations in the Nullarbor Caves have taken place in parallel with exploration, leading to their being among the most studied arid caves in the world.

ORIGIN AND ENLARGEMENT OF THE CAVES

At first it was anticipated that the number and size of the Nullarbor Caves would be limited by a lack of surface water and biogenic carbon dioxide. This hypothesis was supported by an apparently immature karst surface, lacking in richness and diversity of landforms. In contrast, the enormous caves found indicated that below ground the karst was mature, while the large number of blowholes showed that cavities are numerous beneath the surface, the latter supported by extensive drilling records. In fact, the Nullarbor is not lacking in caves but in entrances to them. Old Homestead (Fig. 1) illustrates this; it has a single entrance, yet on the surface above the cave at least 30 blowholes too narrow for human passage have been identified as having an air connection with the voids below. Within the cave there is evidence of additional entrances that are now blocked with eolian surface materials.

As early as the 1970s, scientists hypothesized that the caves had shallow phreatic origins, developing in zones of fluctuating water tables. This hypothesis was readily accepted as an explanation for the smaller, shallow Nullarbor Caves. For the huge caves, other investigators invoked more pluvial times in the past to supply copious aggressive waters for dynamic phreatic development. The existence of the paleodrainage channels found in the north of the plain (Fig. 1) was regarded as support for this argument. At present, runoff generated by intense rainfall is the only aggressive water to enter the caves. Freshwater lenses form on top of the brackish waters in the cave lakes, allowing corrosion notches (Fig. 6) to develop on the walls around the lake.

The brackish waters in the lakes, flooded passages, and the Nullarbor aquifer are saturated with respect to calcite, evidence that they are dissolving limestone and have a role in the development of the caves. They have large amounts of total dissolved solids, with sodium ions and chloride dominating. The highest levels of totally dissolved solids in caves are found in Nurina, where the water is hypersaline, and the lowest in Mullamullang (Fig. 1). The generalization that salinity in the cave waters decreased with distance from the ocean led to the hypothesis that seawater was back-flooding into the caves. Stable isotope analysis has shown that the source of the salt is actually sea spray and that it is the salt content of the meteoric waters that decreases with distance from the ocean.

The aggressive meteoric waters are implicated in many aspects of karst and cave development on the Nullarbor. The light rains dissolve some limestone on the surface and in the soil profile but when it evaporates calcium carbonate (calcite), sodium chloride (halite), and calcium sulfate-2water (gypsum) crystallize out on the surface. Subsequent cycles of light rain and evaporation increase the salt content on surface and in soil. This concentration of calcium carbonate, where potential evaporation exceeds rainfall, is a recognized mechanism for the formation of calcretes and case-hardening of limestone. On the arid Nullarbor, halite and gypsum are also concentrated in this manner. During periods of intense rain, the soluble salts concentrated through the above mechanism dissolve. The resulting vadose waters are aggressive because

FIGURE 6 Tafoni in Cocklebiddy entrance passage. The tafoni are in lines separated by harder layers in the limestone. Gypsum and halite efflorescence can be seen on the curved surfaces and a fine-powdered deposit of gypsum and calcite sand can be seen on the ledges below the tafoni. Above the surface of the lake is a deeply cut corrosion notch. *Photograph taken by Julia James.*

they have increased ionic strength and are thus able to create cavities immediately below the indurated surface. These cavities, called the *zone of intense phreatic preparation*, were previously interpreted as resulting from shallow phreatic solution at times of past high water tables. However, using the concept of renewed aggressivity from increased soluble salt content, it is possible to propose that the zone is recent, not ancient, and vadose not phreatic, in origin.

The vadose waters continue through the limestone, now saturated with respect to calcite and with a high ionic strength, until they reach the water table. The brackish waters of the cave conduits and lakes and of the Nullarbor aquifer are also saturated with respect to calcite. When these two waters mix, the mixed water is aggressive and mixing corrosion takes place at the mixing surface. The location of the mixing surface in the karst has varied over time as sea levels were lowered during periods of glacial advance. It is likely that this process has been active throughout the history of cave development beneath the Nullarbor, creating shallow phreatic caves.

Whatever the process of creating the first caves, when they drain due to the lowering of the water tables, the major process of enlargement is considered by all to be crystal weathering. It is the scale and intensity of crystal weathering in the Nullarbor Caves that makes them the foremost example in the world of this process of cavern enlargement. The agents of crystal weathering are the meteoric salts in the vadose waters. The Nullarbor Caves are barometric breathers; thus, airflow in the caves reverses in response to pressure change. This breathing removes moisture from the caves and replaces it with desiccating desert air. If the humidity in the caves is sufficiently reduced, the vadose waters will start to evaporate until they become so concentrated that salts they are carrying crystallize in the surface layers of the limestone. Calcite, the first mineral to crystallize, can be observed in thin sections of the wall rock. Following calcite, gypsum and then halite crystallize causing granular disintegration of the wall rock. Gypsum is the more effective weathering agent because it has greater expansion on crystallization. This process produces upward-sloping domes and highly fretted walls. The caves enlarge, sloping upward due to crystal weathering and subsequent collapse, ultimately penetrating to the surface. These collapsed entrances allow aggressive runoff to reach the caves and contribute to their enlargement.

The collapse and crystal weathering products would fill a cave if they were not removed. Beneath the Nullarbor, there are a number of ways for their removal. Mixing corrosion at the vadose and phreatic interface will dissolve them. In Murra-el-elevyn, aggressive runoff waters are contributing to their removal. Solution within the freshwater lens has etched fallen boulders, leaving fossil corals standing up to 3 cm proud of their surface. Eolian processes are the only way the products of crystal weathering can be removed from the dry cave passages. In Mullamullang, crystal weathering has reduced the bedrock to sand below a dome. The sand, carried by the strong cave winds, shifts through the cave.

Crystal weathering masks, modifies, and mimics other geomorphic forms in the caves. Pseudocorrosion notches form around lakes that never receive runoff. These are a result of the brackish waters rising through the surrounding porous rock by capillary action and evaporation. This produces a band of halite and gypsum flowers and causes granular disintegration of the wall rock. Figure 6 shows the multiple rows of tafoni in the entrance chamber of Cocklebiddy. To the advocates of dynamic phreatic solution, tafoni were solution scallops but are actually a product of crystal weathering. They have formed in bands separated by harder indurated limestones. These harder layers were generated when the parent coral reefs that produced the limestones emerged briefly above water.

Crystal weathering is also responsible for the novel "coffee and cream" sediments found in many of the caves. The weathering products are made up of two forms, which appear to flow over one another. Both the light- and dark-colored materials are made up of high magnesium calcite with different impurities. The "cream" contains goethite-dominated iron minerals and some kaolin. The strikingly contrasting dark brown coffee has more gypsum, hematite, and manganese dioxide.

It has been proved that crystal weathering is occurring at observable rates in the caves under the present climatic conditions. Thus, it is easy to accept that these enormous caves could have enlarged continuously over millions of years at a slow but variable pace dominated by this geomorphic process.

Throughout this *Encyclopedia of Caves* article, the words *large*, *enormous*, and *huge* are used. Yet none of the Nullarbor Caves would rate a mention in lists of the deepest and longest caves of the world. Old Homestead is the longest of the Nullarbor Caves at 25 km. Mullamullang is 135 m deep and is the second longest cave (>13 km). Weebubbie is equally deep but its large passage (Fig. 5) is submerged for 45 m of its overall depth. It is the flooded passages of the Nullarbor Caves that have attracted international expeditions in search of world records. Cocklebiddy has the longest underwater section, 6.5 km through large passages with only two modest air spaces to the limit of exploration and has held world records. Pannikin Plain has the largest flooded passages; they are the size of the Abrakurrie main chamber, the largest air chamber in Australia (Fig. 4). Another feature of the Nullarbor Caves that is internationally recognized is the strength of the cave winds. These have been measured at 70 km/h^{-1} issuing from the blowhole entrance of Thampanna.

MINERALS

The Nullarbor Caves are renowned for their secondary minerals. The most unusual are the very rare organic minerals produced within bat guano. These are exceptionably soluble and are thus found only in the driest of caves. Only nine organic minerals have been reported from caves worldwide; of these, five were first collected from the Nullarbor Caves.

Halite is rare in limestone caves around the world, yet beneath the Nullarbor it is abundant and occurs in

FIGURE 7 A halite curtain ~100 mm high. *Photo courtesy of Norm Poulter.*

all the known forms of halite speleothems. Of exceptional beauty are the wide varieties of delicate and erratically shaped halite flowers (Fig. 7). The source of the halite is the meteoric salt in the vadose water. It persists only in places in the caves where there is continuously low relative humidity. One of the most exciting halite speleothems found is a 2.7-m-long fallen column. For its protection this giant, christened Big Salty, was removed to the Western Australian Museum. A number of uranium series dates were obtained from it for growth rate studies and it was concluded that there had been two previous periods on the Nullarbor, dry enough for massive halite deposition. The same cave has five other halite columns, disguised from collectors by their unattractive appearance.

Gypsum is even more widespread and is found in many speleothem shapes. Halite is substantially more soluble than gypsum. This differential solubility is demonstrated when slow-dripping gypsum straws develop halite tips that are lost after subsequent rain increases flow through the straws.

In the Nullarbor Caves calcite speleothems, once regarded as rare, are widespread, although under the current arid conditions subaerial calcite deposition is rare. Conventional calcite deposition by CO_2 degassing is to be found in Nurina where pools are covered with calcite rafts that have coalesced to form fragile floors and have stacked into piles against the cave walls.

Ancient "black" calcite is found in all of the common speleothem shapes. This black calcite, which has a humic compound as the chromophore, is seen both on the surface and throughout the caves to the water table. Uranium series dating has shown that its deposition ceased more than 350,000 years BP. Flowstones of black calcite over a meter thick suggest that its deposition extended over a long period of time. The black calcite must have been deposited when the climate was wetter or at a time when there was a much lower potential evaporation in order to allow the calcite-saturated waters to reach the caves. A more extensive vegetation cover would also have been necessary to provide both the carbon dioxide and the humic chromophore. Some of the massive speleothems of black calcite have been reduced to shards by the relentless attack of the agents of crystal weathering.

Mixing-zone calcites lie submerged in the flooded passages of the Nullarbor Caves. In Tommy Grahams (Fig. 1), these speleothems are being deposited below a halocline at a depth of 22 m. Above the halocline, the limestone wall rock is being eroded, while below it calcite is precipitating as a result of "mixing crystallization." Similar calcite deposits are found in many of the water-filled Nullarbor Caves. Particularly significant is a calcite flowstone over a kilometer in length and covering many meters of the lower passage walls and floor in the flooded passage of Cocklebiddy. It is the nature of their deposition that makes these a valuable source of past environmental data.

Calcite biominerals encapsulated within biofilms have formed speleothems that have been named microbial mantles. Six different mantle morphologies have been observed: icicles (Fig. 8), U-loops, bulbs, feathers, and thin roof and thick floor sheets. Biofilms are implicated in both the nucleation and habit modification of the spindle-shaped microcrystals found within them. The floors below the mantles are either snow fields of loosely packed white crystalline detritus or pavements of cemented calcite.

BIOLOGY PAST AND PRESENT

The Nullarbor Caves contain repositories of biological materials of outstanding paleoenvironmental value preserved in the desiccated sediments. The bone and subfossil material found covers a range of faunal elements, including extinct megafauna—many modern species now extinct on the Australian mainland and species whose range has markedly contracted. The most exciting findings have been the complete skeleton of a marsupial lion Thylacaleo and mummified Thylacines. The latter species is known as the Tasmanian tiger and is now extinct, although at the time of European settlement it was found in Tasmania and the southeastern mainland. Desiccation resulted in mummies being found with soft tissues intact; one carcass has been dated at 4600 BP.

The entrance is the only source of nutrients and energy for the cave ecosystems, excepting those caves inhabited by the bat species *Chalinolobus morio* where modest guano piles add additional nutrients. This limited supply of nutrients restricts the numbers and types of invertebrates. Despite this, the Nullarbor Caves host a range of terrestrial invertebrates, which includes several remarkable and unique troglobitic species that are survivors of local evolutionary, environmental processes. These troglobites have preserved a tiny sample of late Tertiary/Pleistocene surface fauna, now absent from the region, and give the systems their zoogeographic importance. The caves also contain a large number of troglophilic insects (mainly cockroaches, beetles, and crickets) and accidental species.

The almost total absence of aquatic fauna in the cave waters of the aquifer is surprising. Salinity is not the cause of this absence as there is one known aquatic troglobitic species, an amphipod crustacean, in the hypersaline waters of Nurina. Neither is a lack of nutrients responsible, for biofilms are abundant in the cave waters and, like the invertebrate fauna, are remarkable. So far, microbiological studies have been restricted to the calcite precipitating microbial mantles (Fig. 8). They are communities of microorganisms consisting primarily of *Pseudomonas* spp. and *Pseudoalteromonas* spp. as well as nitrite oxidizers. Of the 36 phylotypes, 12 could not be identified to a subdivision level and only 2 showed any relationship to previously described environmental clones.

The presence of chemotrophic bacteria in the Nullarbor introduces another variable into the understanding of the cave origins and their enlargement. They can add and remove carbon dioxide or hydrogen carbonate from the cave waters, thus dissolving limestone or precipitating calcite. They are widespread in the flooded cave passages and thus they may have a considerable impact.

FIGURE 8 Microbial mantles, icicle morphology. The gelatinous biofilm can be seen with its encapsulated calcite crystals reflecting light from the flash gun. In the top right corner, above the mantles, the limestone bedrock is etched. *Photo courtesy of Peter Rogers.*

CONCLUSION

In the world of caves, where in most cases biogenic carbon dioxide and surface waters provide the essential ingredients for the solution of limestone and the precipitation of decorations, the more of both, the bigger and more spectacular the caves. The Nullarbor Caves provide an example of how caves can originate, enlarge, and become decorated with speleothems at times when there is a shortage of both.

Bibliography

Contos, A. K., James, J. M., Heywood, B. R., & Rogers, P. A. W. (2001). Morphoanalysis of bacterially precipitated subaqueous calcium carbonate from Weebubbie Cave, Australia. *Geomicrobiology Journal, 18*, 331−343.

Davey, A. G., Gray, M. R., Grimes, K. G., Hamilton-Smith, E., James, J. M., & Spate, A. P. (1992). *World Heritage significance of karst and other landforms in the Nullarbor region*. University of Canberra, ACT, Australia: Applied Ecology Research Group.

Ford, D. C., & Williams, P. W. (1989). *Karst geomorphology and hydrology*. London: Chapman & Hall.

Gillieson, D. (1996). *Caves processes, development, management*. Malden, MA: Blackwell.

Hill, C. A., & Forti, P. (1997). *Cave minerals of the world* (2nd ed.). Huntsville, AL: National Speleological Society.

Holmes, A. J., Tujula, N. A., Holley, N., Contos, A. K., James, J. M., Rogers, P. A. W., et al. (2001). Phylogenetic structure of unusual aquatic microbial formations in Nullarbor Caves, Australia. *Environmental Microbiology, 3*, 256−264.

James, J. M. (1992). Corrosion par melange des eaux dans les grottes de la Plaine de Nullarbor, Australie. In J. N. Salomon & R. Marie (Eds.), *Karst et evolutions climatiques*. Bordeaux, France: Presses Universitaires de Bordeaux.

Long, J., Archer, M., Flannery, T., & Hand, S. (2002). *Prehistoric mammals of Australia and New Guinea: One hundred million years of evolution*. Sydney, Australia: University of New South Wales Press.

Lowry, D. C., & Jennings, J. N. (1974). The Nullarbor Karst, Australia. *Zeitschrift für Geomorphologie, 18*, 35−81.

P

PALEOCLIMATE RECORDS FROM SPELEOTHEMS

Victor J. Polyak* and Rhawn F. Denniston†
*University of New Mexico, †Cornell College

INTRODUCTION: WHY SPELEOTHEMS FOR PALEOCLIMATE RESEARCH

Stalagmites contain numerous physical and chemical characteristics that can be used to reconstruct past climate records, although stable isotopes of oxygen and carbon are the two most frequently used. Visual characteristics are shown in Figure 1. Oxygen isotope values, once thought to be reliable temperature indicators, are now used primarily as indicators of moisture source or rainfall amount while carbon isotopes express changes in vegetation, soil, and aridity (Denniston et al., 1999; Lachniet, 2009). The roles played by concentrations and isotope values of trace elements such as strontium, magnesium, and uranium have future potential, as does grayscale. This article focuses on published continuous paleoclimate records, most of which were derived from carbon and oxygen stable isotopes.

Most stalagmites grow in environments of total darkness, where temperature and humidity changes are dampened, and this growth may occur continuously for thousands and even hundreds of thousands of years. Water falling as rain or snow finds its way through the soil and into fractures in the limestone rock, and eventually into the caves. This water has a direct link to climate, and oxygen, hydrogen, and carbon that represent that surface environment are transferred by the water to the calcite crystals that form stalagmites and other speleothem types.

Thus far the best records of past climate change are being generated from fine-scale sampling of powders for carbon and oxygen stable isotope analyses tied to high-resolution uranium-series dating of stalagmites. The records included in this chapter illustrate the influence these cave-related studies are having on climate reconstructions from around the globe (Fig. 2).

IMPORTANCE OF ABSOLUTE CHRONOLOGY: DATING TECHNIQUES

In order to be useful, all paleoclimate records must provide for some means of age control, and speleothems are ideally suited for dating by uranium-series techniques. Uranium-238 (^{238}U), which is typically present in 0.1–1 ppm abundances in speleothem calcite, decays through a series of relatively short-lived daughters to lead-206 (^{206}Pb), and U-Pb dating is a cutting-edge component of speleothem-based paleoclimate work. However, disequilibrium of the ^{238}U-^{234}U-^{230}Th portion of this decay chain is the most commonly applied method for dating speleothems. In normal, shallow crustal waters (i.e., well-oxygenated, slightly alkaline to slightly acidic pH), U is highly soluble while thorium (Th) is insoluble, and thus cave drip waters often contain very high U/Th ratios. Crystallization of speleothem calcium carbonate from such solutions satisfies one criterion for dating: that nearly all ^{230}Th contained within the sample was formed by *in situ* decay from ^{234}U and thus ^{234}U/^{230}Th ratios may be used as an absolute chronometer.

The measuring of U and Th isotopic ratios was originally performed by alpha spectrometry, which required large samples and long counting times while offering only limited temporal resolution ($\pm10\%$ of the age), but advances in *thermal ionization mass spectrometry* (TIMS) greatly improved the U-Th dating techniques over the earlier alpha radiation methods, allowing uncertainties at $\pm1\%$ of the age. The latest advancement is the application of the multicollector inductively coupled plasma mass spectrometer (MC-ICPMS) which has led to significant improvements on the measurement of the decay

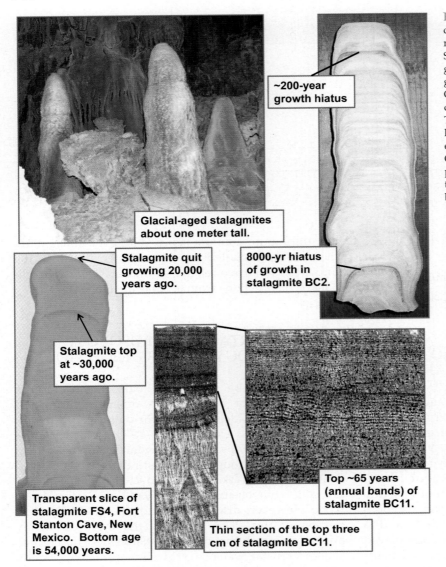

FIGURE 1 Stalagmites are the speleothem of choice for paleoclimate studies. (A) Typical stalagmites in Fort Stanton Cave, southwestern United States. (B) Stalagmite BC2 from Carlsbad Cavern grew during the Late Holocene and exhibits two growth hiatuses. (C) Stalagmite FS4 grew during Glacial 1. Growth of this stalagmite terminated during or immediately after the glacial maximum. Transparency of this slice shows subtle growth layers. (D) A thin section showing the gray differences in the top 300 years of stalagmite BC11, Carlsbad Cavern. The top dark area includes the past 100 years, the period of cave commercialization. (E) Magnification of (D) shows the annual bands in the upper 65 years of growth.

constants (half-lives) of ^{230}Th, ^{234}U, and ^{235}U and which has increased the precision of isotopic measurements. These changes, in turn, have led to improved accuracies of U-series-generated chronologies with absolute ages having errors in sub-1% ranges for samples hundreds of thousand years old (Cheng et al., 2009).

CONTRIBUTIONS TO PALEOCLIMATOLOGY

The Early Studies: Paleotemperature

The early paleoclimate studies using stalagmites were focused on paleotemperature reconstructions over time. The functionality of speleothems as paleotemperature records is founded on two relationships: the temperature dependence of the oxygen isotopic composition of meteoric precipitation ($+0.7‰/°C$) and the temperature-dependent oxygen isotopic fractionation that accompanies mineral crystallization ($-0.23‰/°C$). Because the climate of deep, poorly ventilated caves remains steady throughout the year at approximately mean annual temperature, and because drip waters often reach thermal equilibrium with their surroundings during infiltration into deep caves, the operative relationship linking speleothem calcite $\delta^{18}O$ values to mean annual temperature is approximated by $0.7‰/°C - 0.23‰/°C = 0.4‰/°C$.

However, the magnitude and direction of the slope linking meteoric precipitation with temperature introduces a large degree of uncertainty into this relationship both spatially and temporally. Changes in the seasonality, source, or original isotopic composition of precipitation may not remain constant through time, particularly during glacial/interglacial transitions. And in the low latitudes, temperature-rainwater $\delta^{18}O$

relationships are replaced by "amount effects" in which the intensity of rainfall exerts the strongest control on the oxygen isotopic composition of rainwater. Other complicating influences include enrichment of ^{18}O in water by evaporation prior to infiltration or by kinetic effects on fractionation between HCO_3^- and $CaCO_3$. Because of the complexities involved with extracting absolute temperature records, later studies using stable isotopes evolved to the use of stable isotopes $\delta^{18}O$ and $\delta^{13}C$ as tracers rather than absolute indicators. Lachniet (2009) provides a comprehensive review of stable isotope geochemistry and speleothems.

Devils Hole: The First Long, Continous High-Resolution Speleothem-Based Paleoclimate Record

For decades, the benchmark paleoclimate records have remained the marine foraminifera oxygen isotope curve and the ice records of Greenland and Antarctica. While the information these datasets provide is enormously valuable, each is limited by varying degrees in the precision and accuracy of its chronology. For example, marine sediments cannot be dated by radiocarbon techniques past 50,000 years and paleomagnetism techniques do not offer precise age constraints. And while annual banding in Greenland ice allows precise age determinations on less compacted, Holocene-age sections, the older, more compacted portions of the cores have age errors of hundreds to thousands of years.

The earliest high-resolution paleoclimate study using speleothems with a chronology based on mass spectrometry is the Devils Hole mammillary calcite record (Winograd et al., 1992). Cave mammillaries are subaqueous speleothems that form near the water table. The mammillaries in Devils Hole formed near a water table that apparently fluctuated vertically by only a few tens of meters over the past half a million years. Continuous deposition of the Devils Hole mammillary calcite resultantly preserved the carbon and oxygen stable isotope values of the groundwater paleoclimatic information.

Variations in both oxygen and carbon isotope values roughly mimic those of SPECMAP (the marine record) as shown in Figure 3. But the timing of these variations offered the first significant challenge to the traditional orbital (Milankovitch) theory of climate change. While similar to SPECMAP, glacial-to-interglacial transitions occurred approximately 15,000 years earlier than predicted by orbital theory or as dated in marine sediments. What the timing of glacial to interglacial transitions in the Devils Hole record represents remains the subject of much debate.

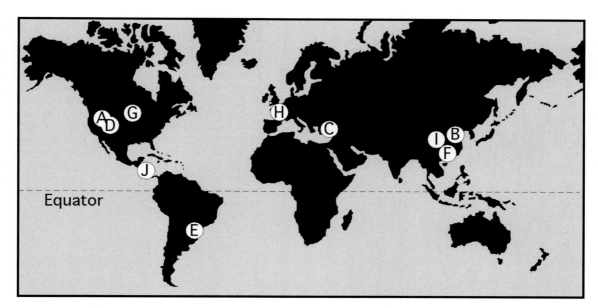

A - Devils Hole, Nevada, USA
B - Hulu Cave, China
C - Sofular Cave, Turkey
D - Fort Stanton Cave and Pink Panther Cave, New Mexico, USA
E - Botuvera Cave, Brazil
F - Dongge Cave, China
G - Cold Water Cave, Iowa, USA
H - Villars Cave, France
I - Shanboa Cave, China
J - Actun Tunichil Muknal Cave, Belize

FIGURE 2 Speleothem records globally are producing impressive accurately dated high-resolution paleoclimate records. Figure 2 shows the location of numerous study sites used in this chapter.

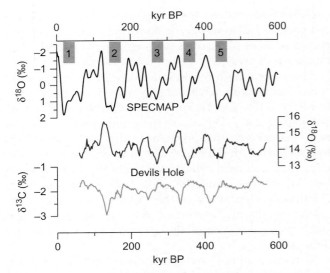

FIGURE 3 Devils Hole, Nevada, produced the first lengthy continuous speleothem-based climate record, going back half a million years and variations in oxygen and carbon isotope values in the speleothem calcite correlated with the marine (SPECMAP) and Antarctic ice (Vostok) core records. The Devils Hole record is shown above in comparison to SPECMAP. These records show that glacial periods (labeled 1–5) have been occurring on a roughly 100,000-yr periodicity over the past million years, which is similar to the cycle of eccentricity of Earth's orbit around the Sun.

The Devils Hole record, unlike almost all other speleothem-derived records, was retrieved from a subaqueous speleothem. Other high-resolution cave-based records are being retrieved almost exclusively from stalagmites.

The Last Glacial: Contribution from Stalagmites

The Greenland ice cores record large shifts in air temperature during glacial periods, referred to as *stadials* and *interstadials*. These include the Bolling-Allerod oscillations and Dansgaard–Oeschger events, the Younger Dryas, and Heinrich events. A major question that speleothem studies are helping to answer is "are stadials and interstadials in the ice core record coincident with similarly profound climate changes in continental regions around the globe?" Many of the stalagmite records documented so far have not only demonstrated the existence of significant climate change coincident with stadials and interstadials, but they are providing increasingly precise timing of these events as well.

Three records from stalagmites showing remarkable agreement with climate changes throughout the last glacial period are from Hulu Cave, eastern China, Sofular Cave, northern Turkey, and Fort Stanton Cave, southwestern United States. Each of these records is shown in Figure 4

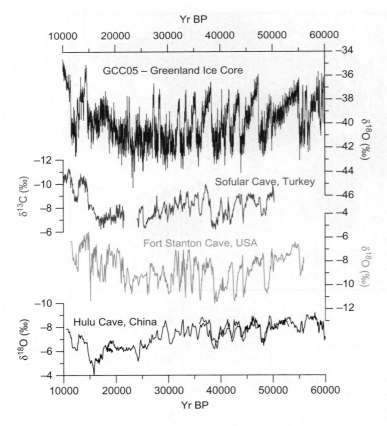

FIGURE 4 Stalagmites are showing that land masses throughout the Northern Hemisphere experienced changes in climate synchronous with the changes recorded in the Greenland ice cores. Records in this figure were retrieved from stalagmites in North America (stalagmite FS2, Fort Stanton Cave, southwestern United States), Europe (stalagmites So-1 and So-2, Sofular Cave, northern Turkey), and Asia (stalagmites H82, MSD, MSL, PD, and YT, Hulu Cave, eastern China), and are compared to the Greenland ice core record GCC05. These records show that changes recorded in the Greenland ice cores drove changes in moisture regime/intensity or vegetation as well as the variations in climate that accompanied these shifts. Increasingly detailed studies will ultimately provide absolute accuracy of the timing of these major shifts in clime.

in comparison to the Greenland ice core record (GCC05, Svensson et al., 2005).

The record from Hulu Cave, China (Wang et al., 2001), reveals precise changes of monsoonal activity related to the last ice-age termination, and to all the stadials and interstadials of the last glacial cycle. This record, based on multiple stalagmites (stalagmites H82, MSD, MSL, PD, and YT) became the stalagmite record to which later speleothem-based studies of the last glacial period compared their records.

The Sofular Cave data, produced from stalagmites So-1 and So-2, yielded a largely continuous 50,000-year record (Fleitmann et al., 2009) through the last glacial. Both the carbon and oxygen values were characterized by isotope variations that matched the Greenland stadials and interstadials. The $\delta^{13}C$ values in stalagmite So-1 (Fig. 4) produced a record more similar to the Greenland ice core $\delta^{18}O$ records. Timing of Dansgaard–Oeschger events in stalagmite So-1 matched those of Hulu Cave and suggested some discrepancies with the timing of events as dated in the Greenland ice core.

The Fort Stanton Cave stalagmite yielded a continuous, 45,000-year-long record of the last glacial from 56,000 to 11,500 yr BP (Asmerom et al., 2010). This record includes large shifts in oxygen isotope values similar to and synchronous with those of the Greenland ice core records and which are related to moisture sources affecting the American southwest. Heinrich events drive more Pacific-dominated winter-like moisture into the southwestern United States resulting in significantly lower stalagmite $\delta^{18}O$ values, while Dansgaard–Oeschger events drive more Gulf of Mexico-dominated summer-like moisture into the region, resulting in significantly higher $\delta^{18}O$ values. The outcome is a record that preserves large shifts in oxygen isotope values, similar to those of the Greenland ice cores. As with the Sofular Cave stalagmites, the Fort Stanton time series also contains some discrepancies with the timing of events in the Greenland ice core.

Tracking Insolation and Paleomonsoons

Kutzbach (1981) described how orbital changes in solar radiation drive monsoon variability over tens of thousands of years. As summer insolation increases, largely due to the precessional (wobble) Milankovitch cycle, so do the land-sea temperature gradients that, in turn, drive monsoon circulation in the subtropics. Thus, according to this so-called orbital monsoon hypothesis, wetter monsoons accompany higher summer insolation. Monsoon strength can be tracked by speleothems because the oxygen in speleothem carbonate is closely tied to the $\delta^{18}O$ values of meteoric precipitation, which in monsoon areas is dominated by the amount effect.

A remarkable relationship between insolation and monsoon variability from a South American stalagmite was reported by Cruz et al. (2005). Soon after, another record from the same cave demonstrated the repeatability and reliability of these records from stalagmites. Figure 5 shows these two stalagmite time series from Botuvera Cave, Brazil, which span the past 116,000 years, and that are characterized by oxygen isotopic ratios that track the sinusoid of austral summer insolation. Periods of elevated summer insolation correspond to the lowest oxygen isotopic ratios (and thus the heaviest rainfall) (Cruz et al., 2005; Wang et al., 2007).

But the longest and perhaps most impressive insolation-monsoon relationship comes from China. This record of monsoonal activity involved the study of numerous stalagmites (Cheng et al., 2009) and tracks the insolation record (Fig. 6) back to 360,000 years.

Ties between Solar Variability and Climate Change

Stalagmites have also been used to test the role played by changes in the Sun's output on the strength of monsoons. Because solar irradiance is tied to radiocarbon production in Earth's atmosphere, radiocarbon measurements made on objects of known age (such as trees and stalagmites) allow the back-calculation of changes in solar irradiance, described as $\Delta^{14}C$ (Stuiver et al., 1998). Overlying $\Delta^{14}C$ values against stalagmite oxygen isotopic time series from monsoonal regions allows an investigation of the role of the Sun on monsoon rainfall. Wang et al. (2005) constructed a stalagmite record from Dongge Cave, China, that spanned the past 9000 years. Their record, which was sampled so as to yield 5-year temporal resolution, found a strong *positive* covariation between the strength of the Asian Summer Monsoon and solar irradiance. On the opposite side of the Pacific, Asmerom et al. (2007) found that a 12,000-year-long stalagmite record from Pink Panther Cave, New Mexico, in the American southwest revealed a strong *negative* covariation between solar output and rainfall. Asmerom et al. (2007) reconciled these seemingly contradictory results by invoking changes in the Walker Circulation of the Pacific Basin, which essentially led to La Niña-like conditions during periods of increased solar output, with increased rainfall in the western Pacific and decreased rainfall in the eastern Pacific creating a see-saw relationship. Figure 7 shows these two records and their comparisons.

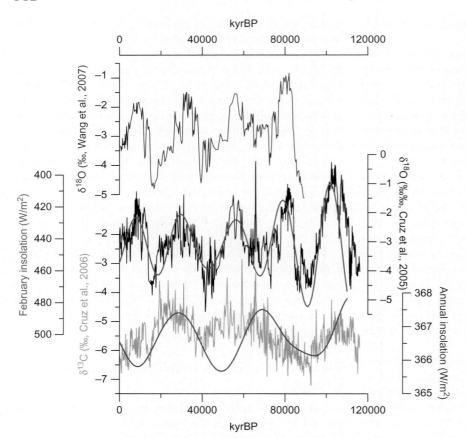

FIGURE 5 Insolation is the measure of solar radiation impacting the Earth's surface (watts per square meter—W/m^{-2}) over time and is generated using Milankovitch's theory. Some stalagmite studies are showing remarkable comparison with insolation. This, figure shows an impressive match between isotope records retrieved from stalagmites BT2 and BTV3a, Botuverá Cave, Brazil, and solar insolation over the past 120,000 years.

Annual Laminations in Stalagmites and Late Holocene Climate Change

Similar to tree rings, annual bands can form in stalagmites, recording the growth rate history of these stalagmites. Numerous studies have been published showing banding history related to general moisture availability or temperature change. Two such studies are offered in Figure 8, one from northeastern China, near Beijing (stalagmite TS90501; Tan et al., 2003), and the other one from the southwestern United States (stalagmites BC2 and BC11; Polyak and Asmerom, 2001; Rasmussen et al., 2006). The relationship of annual band thickness to climate is complex due to the numerous variables involved. However, like the temperature-δ^{18}O problem, such banding records have the potential to accurately reflect climate change variability and to precisely record the timing of climatic events. An example of this is illustrated in Figure 8 where two stalagmites preserve the same, significant change in stalagmite growth between 1400 and 1600 A.D. These records may help resolve climate change interpretations when combined with other annually resolved records coming from tree rings, annually laminated lake sediments, and other high-resolution climate proxies.

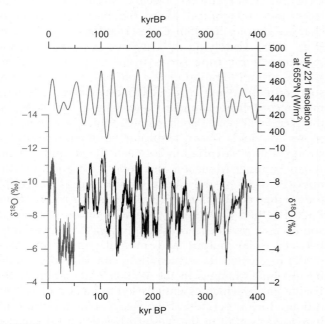

FIGURE 6 A remarkable match between a stalagmite-based record from China and insolation over the past 360,000 years. It took tens of samples to construct this record. The gray isotope values in the left y-axis correspond to the gray curves in the first part of the record. Stalagmites are providing crucial information.

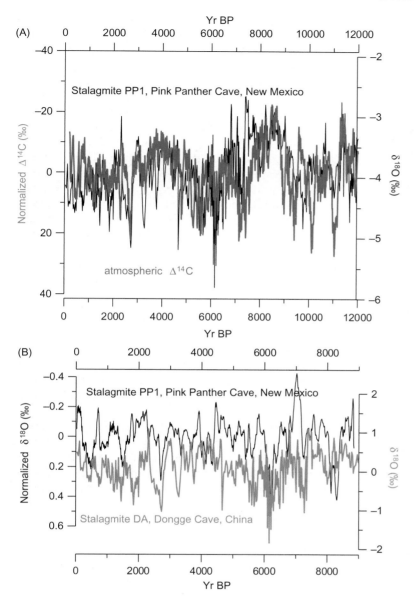

FIGURE 7 Variations in solar radiation are recorded in trees for the entire Holocene expressed through the carbon isotopes, specifically the radioactive carbon isotope ^{14}C. Fleitmann *et al.* (2003) demonstrated that changes in oxygen isotope values in a stalagmite from Oman correlated well with changes in atmospheric ^{14}C ($\Delta^{14}C$) during the Middle Holocene. This figure (A) shows a positive correlation between the atmospheric $\Delta^{14}C$ and the oxygen isotope record from stalagmite PP1, New Mexico, for the entire Holocene, supporting the interpretation that changes in precipitation regimes in the southwestern United States were driven by changes in solar radiation. In (B), stalagmite PP1 is compared to stalagmite DA, China, and the two records express a negative correlation which suggests that monsoonal intensities in these regions are opposite of each other—a "seesaw" effect.

Vegetation and Soil Dynamics

Carbon isotopes, unlike oxygen isotopes, cannot be used to reconstruct past temperature or moisture regimes but still offers important insight into past environmental conditions. Carbon in speleothems is derived from multiple sources. Atmospheric CO_2 and/or soil CO_2 react with water to form carbonic acid which, in turn, dissolves carbonate bedrock. Outgassing of CO_2 or evaporation of water drive precipitation of calcium carbonate via:

$$Ca^{2+} + 2HCO_3^- \rightarrow CaCO_3 + H_2O + CO_2$$

While the $\delta^{13}C$ value of bedrock carbonate is mostly fixed in time, carbon isotopic ratios of soil CO_2 change according to the type of vegetation present. Isotopically lighter C_3 vegetation includes trees, shrubs, and forbs, while isotopically heavier C_4 plants are made up of many prairie grasses, with each type favoring different climatic conditions. Under ideal circumstances, about half of the carbon in speleothem carbonate is derived from soil CO_2 and half from bedrock carbonate (atmospheric CO_2 is typically a trivial source of carbon in water infiltrating through soils that are thick and/or contain dense plant cover). Thus, changes in plant type yield distinct isotopic signals that are transferred into drip water and then into underlying stalagmites.

This relationship was illustrated by Dorale *et al.* (1992) who tied a Middle Holocene increase in $\delta^{13}C$ values in stalagmites from Cold Water Cave, north-central

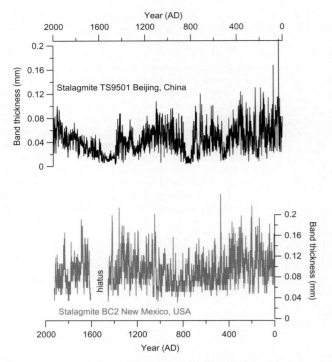

FIGURE 8 Annual banding records in speleothems are similar to those retrieved from trees. This figure offers two annual band records, one from southwestern North America (stalagmites BC2), and one from northeastern China (stalagmite TS90501). There are some similarities in these three records, particularly the period from 1400 to 1600 A.D. Stalagmite BC2 records offered information about general precipitation over the Late Holocene, while stalagmite TS9501 yielded a summer temperature record.

United States, to encroaching dry Pacific air masses that led to replacement of deciduous forest (C_3) by prairie (C_4). However, some complexities of speleothem carbon isotopic dynamics were explored by the Brazilian record of Cruz et al. (2006) who documented large variations in $\delta^{13}C$ values that oscillated in concert with the 40,000-year obliquity (tilt) Milankovitch cycle while $\delta^{18}O$ values oscillated with the precession cycle (Fig. 5). These authors argued that tilt-driven changes in meridional temperature gradients affected local conditions that changed plant activity and thus the amount of soil CO_2 in drip water. Similarly, Genty et al. (2001) documented large swings in $\delta^{13}C$ values in a stalagmite from Villars Cave, France, that occurred synchronously with D–O events and that were tied to vegetation dynamics, with lower stalagmite $\delta^{13}C$ values corresponding to higher contributions from soil CO_2 during wetter intervals when plant density and activity were elevated.

Timing of Climatic Changes

The ice core and marine records document the abrupt termination of glacial periods, but these records cannot examine the full geographic and temporal differences in glacial terminations around the globe. Many stalagmite records are accurately recording the end of glacial periods, however. In the Northern Hemisphere but also in some areas of the Southern Hemisphere, stalagmites are showing that the termination of the last glacial (Termination I) occurred around 14,500 years ago (Fig. 4). The stalagmite records from China (Cheng et al., 2009) and the Dead Sea region (Frumkin et al., 1999) are revealing that the termination of the previous glacial (Termination II) occurred at around 130,000 years ago, and more precisely at 129,000 ± 100 years ago from the Sanboa Cave record (Cheng et al., 2009). The climate record from Chinese caves also reveals the timing of Terminations III and IV at 242,000 and 336,000 years ago, respectively. These high-precision dates allow us to test the orbital (Milankovitch) theory of the ice ages, and to serve as a point of comparison for less precise dates of similar events.

Paleotempestology

Shorter-term climatic variables such as tropical cyclones (hurricanes) may also be assessed using speleothems. In an attempt to extend the record of tropical cyclone activity in the past that are available through direct observation or reconstruction using historical documents, a variety of paleotempestology archives have been developed. The most common method takes advantage of the fact that rain from tropical cyclones is highly depleted in ^{18}O (Lawrence et al., 2002). Values of summer rainfall 6‰ below normal summer rain values serve as markers of severe storms, and thus negative spikes in stalagmite carbonate have been interpreted as marking tropical cyclone activity.

Several studies have compared historical tropical cyclones with ^{18}O-depletion in speleothem (e.g., stalagmite and soda straw) carbonate. Of these, the most high-resolution analysis was that of Frappier et al. (2007) who tied $\delta^{18}O$ values of a fast-growing ($\sim$1 mm yr^{-1}) calcite stalagmite from Belize to historical (1977–2000 A.D.) tropical cyclones that tracked through the region. This study demonstrated unambiguously that through high-precision microsampling (20-µm intervals, averaging $\sim$75 samples per year of growth, subweekly resolution) tropical cyclones can be identified individually by comparison with historical records. Stalagmites will provide valuable information regarding history of hurricane/cyclone activity for the past several thousand years.

Other Studies

While this article focuses primarily on the stable isotope geochemistry of stalagmites, other methods exist.

These include elemental analyses, other isotope systems such as $^{87}Sr/^{86}Sr$, $^{234}U/^{238}U$, magnesium and lithium isotopes, gray and color histograms, growth records defined by growth hiatuses, laser excitation records, and organic materials/substances.

See Also the Following Articles

Cosmogenic Isotope Dating of Sediments
Uranium Series Dating of Speleothems
Paleomagnetic Records in Cave Sediments

Bibliography

Asmerom, Y., Polyak, V. J., & Burns, S. J. (2010). Variable winter moisture in the southwestern United States linked to rapid glacial climate shifts. *Nature Geoscience, 3*(2), 114–117.

Asmerom, Y., Polyak, V. J., Burns, S. J., & Rasmussen, J. (2007). Solar forcing of holocene climate: New insights from a speleothem record, southwestern United States. *Geology, 35*(1), 1–4.

Cheng, H., Edwards, R. L., Broecker, W. S., Denton, G. H., Kong, X., Wang, Y., et al. (2009). Ice age terminations. *Science, 326*(5950), 248.

Cruz, F. W., Burns, S. J., Karmann, I., Sharp, W. D., Vuille, M., Cardoso, A. O., et al. (2005). Insolation-driven changes in atmospheric circulation over the past 116,000 years in subtropical Brazil. *Nature, 434*(7029), 63–66.

Cruz, F. W., Jr., Burns, S. J., Karmann, I., Sharp, W. D., Vuille, M., & Ferrari, J. A. (2006). A stalagmite record of changes in atmospheric circulation and soil processes in the Brazilian subtropics during the Late Pleistocene. *Quaternary Science Reviews, 25*, 2749–2761.

Denniston, R. F., Gonzalez, L. A., Asmerom, Y., Baker, R. G., Reagan, M. K., & Bettis, E. (1999). Evidence for increased cool season moisture during the Middle Holocene. *Geology, 27*(9), 815–818.

Dorale, J. A., Gonzalez, L. A., Reagan, M. K., Pickett, D. A., Murrell, M. T., & Baker, R. G. (1992). A high-resolution record of Holocene climate change in speleothem calcite from Cold Water Cave, northeast Iowa. *Science, 258*(5088), 1626–1630.

Fleitmann, D., Burns, S. J., Mudelsee, M., Neff, U., Kramers, J., Mangini, A., et al. (2003). Holocene forcing of the Indian monsoon recorded in a stalagmite from southern Oman. *Science, 300*(5626), 1737.

Fleitmann, D., Cheng, H., Badertscher, S., Edwards, R. L., Mudelsee, E. M., Göktürk, O. M., et al. (2009). Timing and climatic impact of Greenland interstadials recorded in stalagmites from northern Turkey. *Geophysical Research Letters, 36*, L19707. doi:10.1029/2009GL040050.

Frappier, A. B., Sahagian, D., Carpenter, S. J., Gonzalez, L. A., & Frappier, B. R. (2007). Stalagmite stable isotope record of recent tropical cyclone events. *Geology, 35*, 111–114.

Frumkin, A., Ford, D. C., & Schwarcz, H. P. (1999). Continental paleoclimatic record of the last 170,000 years in Jerusalem. *Quaternary Research, 51*, 317–327.

Genty, D., Blamart, D., Quahdi, R., Gilmour, M., Baker, A., Jouzel, J., et al. (2001). Precise dating of Dansgaard–Oeschger climate oscillations in Western Europe from stalagmite data. *Nature, 421*, 833–837.

Kutzbach, J. E. (1981). Monsoon climate of the Early Holocene: Climate experiment with the Earth's orbital parameters for 9000 years ago. *Science, 214*(4516), 59–61.

Lachniet, M. S. (2009). Climatic and environmental controls on speleothem oxygen-isotope values. *Quaternary Science Reviews, 28*, 412–432.

Lawrence, J. R., Gedzelman, S. D., Gamache, J., & Black, M. (2002). Stable isotope ratios: Hurricane Olivia. *Journal of Atmospheric Chemistry, 41*(1), 67–82.

Polyak, V. J., & Asmerom, Y. (2001). Late Holocene climate and cultural changes in the southwestern United States. *Science, 294*(5540), 148–151.

Rasmussen, J., Polyak, V., & Asmerom, Y. (2006). Evidence for Pacific-modulated precipitation variability during the Late Holocene from the southwestern USA. *Geophysical Research Letters, 33*, L08701. doi:10.1029/2006GL025714.

Stuiver, M., Reimer, P. J., Bard, E., Beck, J. W., Burr, G. S., Hughen, K. A., et al. (1998). INTCAL98 radiocarbon age calibration, 24000–0 cal BP. *Radiocarbon, 40*(3), 1041–1083.

Svensson, A., Andersen, K. K., Bigler, M., Clausen, H. B., Dahl-Jensen, D., Davies, S. M., et al. (2005). The Greenland Ice Core chronology 2005, 15–42 ka. Part 2: Comparison to 15 other records. *Quaternary Science Reviews, 25*(23–24), 3258–3267.

Tan, M., Liu, T., Hou, J., Qin, X., Zhang, H., & Li, T. (2003). Cyclic rapid warming on centennial-scale revealed by a 2650-year stalagmite record of warm season temperature. *Geophysical Research Letters, 30*(12), 1617. doi:10.1029/2003GL017352.

Wang, X., Auler, A. S., Edwards, R. L., Cheng, H., Ito, E., Wang, Y., et al. (2007). Millennial-scale precipitation changes in southern Brazil over the past 90,000 years. *Geophysical Research Letters, 34*, L23701. doi:10.1029/2007GL031149.

Wang, Y., Cheng, H., Edwards, R. L., He, Y., Kong, X., An, Z., et al. (2005). The holocene Asian monsoon: Links to solar changes and north Atlantic climate. *Science, 308*, 854–857.

Winograd, I. J., Coplen, T. B., Landwehr, J. M., Riggs, A. C., Ludwig, K. R., Szabo, B. J., et al. (1992). Continuous 500,000-year climate record from vein calcite in Devils Hole, Nevada. *Science, 258*(5080), 255–260.

PALEOMAGNETIC RECORDS IN CAVE SEDIMENTS

Ira D. Sasowsky

University of Akron

INTRODUCTION

The paleomagnetic record in cave sediments allows us to answer one of the fundamental questions about caves: "How old are they?" This is a hard question because caves are essentially empty space created in rock and, therefore, we are trying to determine the age of something that is no longer there. Hence, we must constrain the age of the cave by other strategies. We know of two conditions that must always apply.

First, the cave must be younger (more recent) than the rock in which it has formed. The only exception is the case of lava tubes, where the cave is the same age as the rock in which it was formed. However, the majority of caves are formed in limestone, a long time after the limestone rock was originally deposited. For example, Mammoth Cave in Kentucky is about 4 million years old, but the rocks in which it is found are more than 300 million years of age. Because of this large difference, dating a cave by the age of the rocks that host it is not very satisfactory.

A second, and more useful, method to determine the age of a cave is to date the deposition of materials that we find within it. This can include things that have been carried in by natural processes such as sediments and speleothems. This provides a minimum age, because the cave must have existed before the material could have been placed in it. The paleomagnetism of cave sediments is one such way to determine age. Other methods (not addressed here) are uranium–thorium radiometric dating of speleothems and cosmogenic isotope dating of quartz particles washed into a cave.

EARTH MAGNETISM AND PALEOMAGNETISM

To understand the paleomagnetism of cave sediments, an understanding of Earth's magnetism and paleomagnetism is required. It has long been known that the Earth has a magnetic field. The field has several sources, but it is mainly generated in the core of the Earth by rotation of the planet. Early natural scientists and explorers used magnetic minerals (lodestones) to measure the direction of the field or to guide them in their travels. This led to the development of compasses, which are precision devices with a narrow magnetic needle that points toward magnetic North (N_m). The Earth also has geographic, or true, North (N) and South Poles that are defined by the axis of rotation of the Earth. The magnetic North and South Poles are in the general region of the geographic poles and are the areas to which a compass needle will point. In addition to a north–south orientation (which is called *declination*), the magnetic field has an up–down component (called *inclination*). At the North magnetic pole, the inclination is $+90°$ (straight down). At the magnetic equator, the inclination is $0°$ (flat), and at the South Pole it is $-90°$ (straight up).

Measurements of the position of the magnetic poles have been made since the early 1600s, and it has been found that the magnetic pole position varies (drifts) with time—a process called *geomagnetic secular variation*, or simply *polar drift*. This drift is on the order of $0.05°$ per year. Then, in the 1900s, it was discovered that the magnetic field had reversed many times throughout geologic history. In these instances, rather than gently drifting though time, the poles had actually (and quite rapidly) flipped: The magnetic North Pole moved to the geographic South Pole and vice versa. These changes are called *field reversals*. When the magnetic North Pole is in the Southern (geographic) Hemisphere, the condition is considered to be one of reverse polarity. When the North Pole is in the Northern Hemisphere, as it is today, it is considered a time of normal polarity.

Many rocks and sediments hold a weak magnetic signal that is "locked in." The signal that they hold points in the direction of magnetic North from the time that they were deposited. Rocks deposited during a time of reversed polarity have a magnetic signal that points to the Southern Hemisphere. Through the study of rocks worldwide, a chronology of the magnetic field has now been constructed. This magnetostratigraphic time scale (Fig. 1) shows the orientation of the magnetic field during different portions of Earth's history. Those times when the fields were reversed are indicated in white.

As a result of the creation of this worldwide time scale, it is now possible to determine the age of other rocks and sediments by measuring their magnetic properties and placing them in their proper position in the magnetostratigraphic time scale. Butler (1992) and Tuaxe (2010) provide comprehensive reviews of the topic.

PALEOMAGNETISM OF CLASTIC CAVE SEDIMENTS

All streams carry small, broken up bits of rock (*clastic sediment*) that come from the weathering of the surrounding landscape. This includes fine materials such as clay and coarser materials such as sand and gravel. When water flow decreases, or stops altogether, these materials are dropped (deposited) from the stream. As the particles slowly settle from the water, tiny magnetic mineral grains (primarily magnetite and hematite) orient themselves with the magnetic field. The process is similar to a compass needle becoming oriented. When the particle reaches the stream bottom, it is buried by additional particles, and the magnetic orientation of the grains is locked in. This is called *detrital remanent magnetism* (DRM). If a sample of this material is analyzed, the direction of the magnetic field during the time of deposition can be determined.

Caves typically have a period of their existence during which a stream is flowing through them (Fig. 2). If this stream comes from the surface of the Earth, it will carry clastic sediments that can then be left in the cave. Such deposits (Fig. 3) can range in thickness from millimeters to tens of meters and can consist of all grain sizes. Virtually all caves have some clastic sediment. To measure the paleomagnetism, researchers

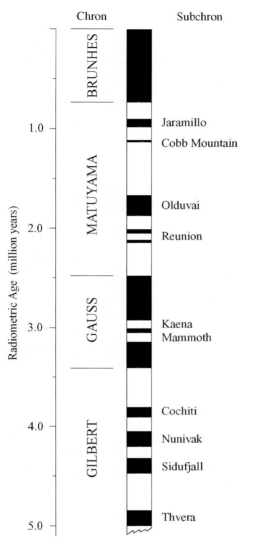

FIGURE 1 Magnetostratigraphic time scale. The polarity of the Earth's magnetic field has switched through time, with the North magnetic pole alternating position near the North geographic and South geographic poles. *Courtesy of U.S. Geological Survey.*

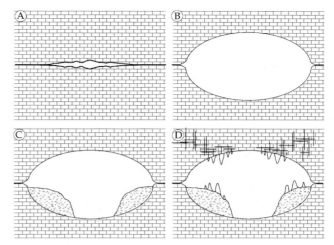

FIGURE 2 Typical stages in the evolution of a cave: (A) Initial rock mass with open bedding plane, (B) enlargement of bedding plane by groundwater circulation, (C) deposition of clastic sediments, (D) deposition of speleothems (chemical sediments).

FIGURE 3 Clastic sedimentary deposit in Windy Mouth Cave, West Virginia, U.S.A. A stream that once flowed through the cave deposited these sediments.

collect a small, oriented sample of the material. Duplicate samples and samples from many locations and levels are usually obtained. Clays are the most desirable material to sample, because they represent a still-water environment, which allows for good orientation of the magnetic particles. But good results have also been obtained from coarse, cross-bedded sands. The samples are then taken to a laboratory and measured in a magnetometer. This instrument determines both the direction and strength (magnetic intensity) of the sample. Oftentimes several steps of magnetic cleaning, or *demagnetization*, are needed to reveal the DRM. On an automated system, such measurement takes about 30 minutes. The resulting plots, which show magnetic direction and intensity (Fig. 4), tell whether the sample has a normal or reverse polarity.

Now comes the tricky part: interpretation. The magnetostratigraphic time scale is a binary one—it only shows normal or reverse. If we have an isolated sample of cave sediment that is normal, how can we know which normal interval (*chron*) it is from? The short answer is that we cannot—at least not with absolute certainty. But we can make educated guesses with an isolated sample. For example, if we find a reversed polarity sample, we know that it must be at least

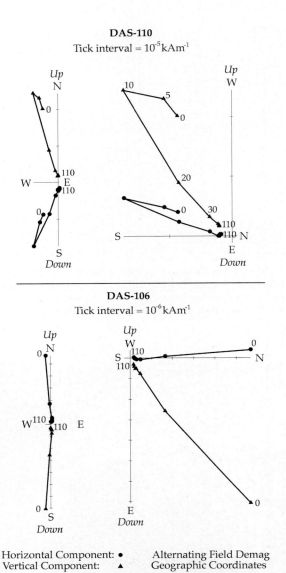

FIGURE 4 Orthogonal vector plot (Zijderveld diagram) of two clastic sediment samples from caves. Each sample has two plots. The lower sample has normal polarity, showing a northward declination and downward inclination. The upper sample is reversed, and shows a characteristic hooking, with southward declination and upward inclination.

780,000 years old (the end of the last reversed interval; also called the Matuyama chron). Likewise, if we find an isolated deposit of normal polarity sediment, we infer that it must be either less than 780,000 years old or more than 990,000 years old. Using these constraints, in context with other information such as U-Th dates of speleothems or position in the landscape, we can narrow down the age range of a cave. The work of Victor Schmidt (1982) at Mammoth Cave, Kentucky, U.S.A., was some of the first to date a cave using this technique.

In many caves, deposits of clastic sediments are discontinuous or complex to decipher. In these cases, interpretation may be limited to that mentioned above. However, in other cases, one or more "stacks" of sediment may be found, and a fairly complete stratigraphy may be constructed. When this occurs, the resulting local magnetostratigraphy may be directly matched with the global column (Fig. 5), and the age range of the method is greater.

PALEOMAGNETISM OF SPELEOTHEMS

Speleothems are deposits of secondary minerals (such as calcite) that form on the ceilings, walls, and floors of caves. Stalactites and stalagmites are the most frequently studied of these features. As with the clastic sediments discussed earlier, speleothems can retain a record of the magnetic field from the time of their formation. In this case, however, the magnetism is a *chemical remanent magnetism* (CRM) rather than a DRM, and the sampling and processing are slightly different.

Because speleothems are hard, drilling is required to collect the oriented sample. This is done using a gasoline-powered drill with a hollow, diamond-tipped bit. Water is pumped on the bit to cool it and remove the rock cuttings. It is a rather onerous procedure to carry out in the difficult setting of a cave, and this may be one reason why there have been relatively few such studies. However, small cores have been taken, and a project conducted by B. Ellwood collected very long cores from large speleothems in Carlsbad Caverns, New Mexico, U.S.A.

IMPORTANCE OF THE PALEOMAGNETIC RECORD IN CAVES

In the preceding sections, the importance of the magnetic record in caves with regard to determining cave age has been discussed. Paleomagnetism provides a relatively quick and inexpensive method for estimating such ages. However, the method relies on establishing a complete record of polarity changes. Therefore, difficulties arise due to discontinuous deposition or erosion of the sediments.

The main advantage of paleomagnetism is that it has a greater range than U-Th disequilibrium dating. The former has been applied back to 4.5 million years, whereas the latter has a range of about 450,000 years. Paleomagnetism also requires less laboratory work. Therefore, both methods have utility for the study of caves.

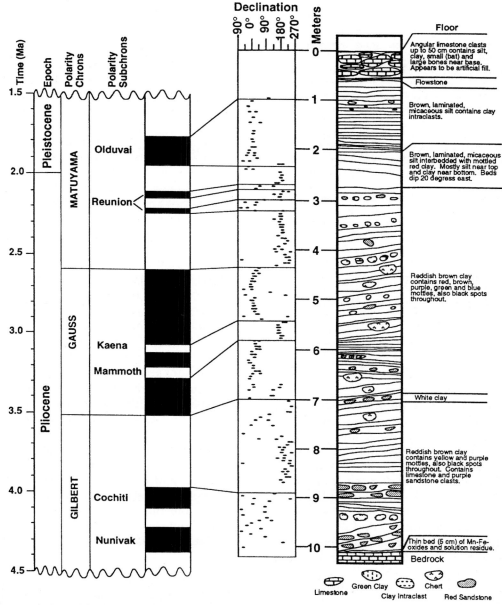

FIGURE 5 Correlation of the magnetostratigraphic column from Cave of the Winds, Colorado, with the global record (Luiszer, 1994). An apparently continuous record of deposition from 4.5 to 1.5 million years ago is shown. *Courtesy of Fred Luiszer and the Karst Waters Institute.*

Through age determination of caves, it is also possible to relate the caves and other features to landscape evolution, such as the incision of rivers. Therefore, the study of caves, which may seem rather esoteric, serves a purpose that is much broader.

The relatively new study of "environmental magnetism" may provide an additional utility for cave sediments. Rather than simply looking at reversals, environmental magnetism examines the magnetic properties of sediments to understand the environmental conditions under which the materials were produced. For example, Ellwood et al. (1998) identified glacial effects on cave sediments. These methods, along with continued development of our understanding of cave processes, suggest that the magnetism of cave sediments will continue to be an important research tool.

See Also the Following Articles

Clastic Sediments in Caves
Paleoclimate Records from Speleothems

Bibliography

Butler, R. F. (1992). *Paleomagnetism: Magnetic domains to geologic terranes*. Boston: Blackwell Scientific Publications. www.geo.arizona.edu/Paleomag/book/.

Ellwood, B. B., Zilhão, J., Harrold, F. B., Balsam, W., Burkhart, B., Long, G. J., et al. (1998). Identification of the last glacial maximum in the Upper Paleolithic of Portugal using magnetic susceptibility measurements in Caldeirão Cave sediments. *Geoarchaeology, 13*, 55–71.

Luiszer, F. G. (1994). Speleogenesis of Cave of the Winds, Manitou Springs, Colorado. In I. D. Sasowsky & M. V. Palmer (Eds.), *Breakthroughs in karst geomicrobiology and redox geochemistry* (pp. 91–109). Charles Town, WV: Special Publication 1, Karst Waters Institute.

Schmidt, V. A. (1982). Magnetostratigraphy of sediments in Mammoth Cave, Kentucky. *Science, 217*, 827–829.

Tauxe, L. (2010). *Essentials of paleomagnetism*. San Diego: University of California Press.

PALEONTOLOGY OF CAVES

Blaine W. Schubert and Jim I. Mead

East Tennessee State University

INTRODUCTION

Paleontology is the study of prehistoric life through the analysis of fossils. Fossils are physical entities that represent evidence of past organisms, and can be preserved plant or animal remains, dung (Fig. 1), molds, casts, or traces such as tracks and claw marks. Remains do not have to be altered or mineralized to be considered a fossil. Generally, caves are subterranean openings or voids in rocks near the surface of the earth. Once caves exist, organisms find their way into these systems in many different ways, and some of them avoid destruction by being cached in caves and become fossils. Caves provide shelter and protection from the primary elements of weathering, and this combined with relatively stable temperatures and humidity make subterranean environments an ideal location for long-term preservation.

Geologically speaking, cave passages are short-lived because the host rock is subjected to weathering and erosion. For this reason, most fossils from extant caves are relatively young, less than ~2 myr old. Thus, the fossil record provided by active caves is essentially equivalent to the Quaternary, the past ~1.8 myr (Schubert et al., 2003). Older cave deposits do exist, but the physical cave is typically gone or filled and the sediments and fossils are often highly mineralized and cemented, forming what are known as *bone breccias*. These more ancient cave or karst features are usually found during construction or quarrying operations of bedrock. Karst deposits with fossils date back to the late Paleozoic, with the earliest example being limestone fissure fills in Iowa where the oldest Mississippian-age vertebrate fauna of North America is preserved (Lundelius, 2006).

Here we focus primarily on the Quaternary, with some mention of the earlier Pliocene as well. It is during the Quaternary that most of our existing traversable caves formed and became accessible to vertebrates, so the fossil record is rich and diverse. The earliest part of the Quaternary is the Pleistocene, which ranges from ~1.8 million years ago until ~12,000 years ago. This time is known as the *Ice Age*, and it is characterized by extreme climatic oscillations, reflected in a series of glacial advances and retreats. Colder (glacial) and warmer (interglacial) periods alternated, with short-term warming during glacials (stadials) and cooling during interglacials (interstadials). The second part of the Quaternary is the Holocene (or Recent)—the past ~12,000 years. Many geologists consider the Holocene to be just another interglacial within the Ice Age.

Researchers have studied fossils from Quaternary cave sites all over the world, particularly in regions where limestone bedrock is common. Most research has focused on mammals, although studies have also occurred on arthropods, mollusks, fish, amphibians, reptiles, and birds. In general, there has been a bias toward studying the mammals because many of them are large and extinct, and the teeth of mammals preserve well and are easier to identify to specific levels than most fossil remains from other taxonomic groups. In addition, mammal teeth can be highly informative about diet and evolution. For these reasons, there has been more interest in the mammals overall. It is important to note that the lack of studies on other animal groups from cave deposits is not a reflection of relative abundance. In fact, most cave collections in major

FIGURE 1 Rampart Cave, Grand Canyon, AZ, U.S.A. Floor of cave covered in sloth dung. *Photograph courtesy of the GRCA National Park Museum Collections.*

natural history museums house a plethora of nonmammalian fossils just waiting to be studied.

Fossil deposits in caves vary widely, ranging from carnivore dens to owl roosts, from natural traps to dung deposits, and packrat middens to early hominid sites. Hominids are the bipedal apes, which includes us, *Homo sapiens*, others members of our genus, and species of *Ardipithecus*, *Australopithecus*, and *Paranthropus*. Here we do not cover human habitation in caves, and leave this to the archaeologists and paleoanthropologists. The overall diversity in site types indicates the potential complexity of interpreting each unique locality, and discerning which taphonomic processes contributed to the fossil deposit. Taphonomy is the study of how fossils come into being. In other words, what happened to the organism from the time of death until it was discovered? This aspect of paleontology is a crucial component in understanding how fossils get into caves.

TYPES OF CAVE DEPOSITS AND EXAMPLES

Natural Traps

Caves that have a vertical entrance can serve as natural traps for organisms that fall in and die. Size of the animals represented depends on the size of the entrance. Natural traps are more effective at accumulating remains if they are deep and lack associated horizontal passages and entrances. These deposits are characterized by relatively complete skeletons with little or no postmortem taphonomic damage, though small rodents often find their way into these areas and gnaw on bones. The trap accumulations are often scattered along a talus cone or rock debris slope, positioned under the vertical entrance. If sedimentation occurs regularly, skeletons become buried more quickly and will be better preserved, more articulated, and a chronology of deposition can be more easily determined. If sedimentation rate is low, more scattering and time averaging will occur.

One of the best examples of an extensive natural trap deposit is Natural Trap Cave in north-central Wyoming, North America. The pit entrance of 4 by 5 m is not visible until the edge, the drop is ~20 m, and the cave is bell-shaped with no other entrances. This pit has accumulated animals for at least the past 110,000 years, including remarkable Pleistocene megafauna such as the American cheetah (*Miracinonyx trumani*), American lion (*Panthera atrox*), giant short-faced bear (*Arctodus simus*), dire wolf (*Canis dirus*), bison (*Bison antiquus*), woodland muskox (*Bootherium bombifrons*), American camel (*Camelops*), extinct horses (*Equus*), and mammoth (*Mammuthus*) (Martin and Gilbert, 1978). Other extraordinary examples of natural traps are found in the Thylacoleo caves on the Nullarbor Plain, south-central Australia. These caves are named after the articulated and well-preserved marsupial lion (*Thylacoleo carnifex*) remains recovered, but many other vertebrates occur in the deposits, including new species of kangaroos.

Alluvial and Colluvial Transport

Many caves have active river systems (alluvial), and these streams can transport carcasses or parts of skeletons into caves, or move them from one place to another in the cave system. This alluvial transport often introduces animals that otherwise would not venture into caves. For example, Powder Mill Creek Cave, Missouri Ozarks, USA, contains numerous remains of late Pleistocene semiarticulated deer. Based on the associated sediments and adjacent stream, these specimens were clearly transported to their current location by water. The level of articulation indicates the carcasses washed to these locations close to the time of death.

Colluvial transport is down-slope movement of sediments without the assistance of water. These types of sediments can introduce fossils into caves through fissure fills, sinkholes, or sloped entrances. Further, existing deposits in caves can change over time as colluvial action occurs, particularly when sediments are on slopes. These sedimentary processes point out some of the potential complexities of cave deposits.

Raptors

Here we refer to raptors, or birds of prey, in the informal sense, meaning those birds that base their diet on scavenging or hunting other vertebrates. Certain raptors, such as owls, hawks, and vultures, often utilize cave entrances as roosts and accumulate bones. Owls are nocturnal hunters, while other raptors hunt and scavenge during the day (diurnal); thus, their prey base can differ dramatically. Raptors do not digest the entire carcass; instead they regurgitate bones and hair (or scales) as pellets. Over time, significant accumulations can develop below roosts. Owl roosts contain the most prolific and undamaged bones because they swallow most prey relatively whole and cast their pellets from a more alkaline digestive system. Diurnal birds of prey have varying degrees of higher acidity in their precasting digestion, and therefore dissolution of bone occurs in most species. Also, some of these raptors tear their prey or scavenged remains into pieces, further disturbing the skeletal material. These differential characteristics, as well as prey size, allow paleontologists to discern the type of raptor responsible for certain fossil deposits (Andrews, 1990).

Skeletal remains collected by raptors and deposited in caves play a significant role in our understanding of the distribution and evolution of small mammals during the late Cenozoic. An example of an extensive raptor roost is Screaming Neotoma Cave in Arizona, along the southern border of the Colorado Plateau, western North America. This cave is a natural trap pit where owls have roosted on and off since the late Pleistocene; thus, at least two different processes are depositing fossils. The deposit has only received minimal excavations, but species such as the sagebrush vole (*Lemmiscus curtatus*) indicate a geographical change in distribution; this species now occurs much farther north. Climatic change and habitat loss are possibilities for this biogeographic change. Other examples of extraordinary raptor roosts in southwestern North America are found in the Red Wall Limestone of the Grand Canyon, Arizona. Here condors (*Gymnogyps*) once roosted and accumulated the remains of scavenged animals during the late Pleistocene. Feathers, skulls, keratin, other tissues, and bones are exquisitely preserved because of persistent arid conditions inside and outside the caves.

Bats

Bats are a widespread group of vertebrates that have been around since at least the early Cenozoic. They are the only mammals with true flight, and this is accomplished by "hands" that are covered in skin membranes that form the wings. With over 1000 living species bats, are one of the most diverse mammalian orders. The majority (>70%) are relatively small insectivores that hunt at night and roost during the day. All bats can see, but night hunters rely on echolocation for navigation and finding their prey.

When caves are available bats use them, particularly for daytime roosting, rearing young, and winter torpor in colder regions. Thus, many species are closely intertwined with cave environments and have evolved adaptations and behaviors that reflect this relationship. Because bat bones are small and fragile, their fossil record is sparse for most of their existence. However, Pleistocene and Holocene deposits from caves are prolific and can give insight into old roosting sites, past distributions of extant species, and climatic and environmental change. Since many living bat species have restricted temperature and humidity requirements for roosts, these ecological parameters are sometimes applied to Quaternary fossils to better understand cave paleoecology.

Not surprisingly, the primary fossil localities for bats are found under abandoned roost sites in caves. Skeletal remains are intermixed with piles or layers of bat guano if the roost was used during a time of feeding. If a roost represents a maternity colony, a high proportion of the bat remains will be juveniles. On the other hand, if a deposit is rich in skeletal remains but lacks an abundance of guano, it may represent a winter die-off during torpor.

Bat deposits in caves are numerous but understudied. The extensive Mammoth Cave system in Kentucky, USA, provides a rare example of a thorough survey for bat fossils. The oldest bat known from the cave belongs to a single vampire bat (*Desmodus*) specimen, and was found in sediments in the upper reaches of the cave, dated over 1 myr old. *Desmodus* today prefers warmer temperatures in the more tropical regions of the Americas, and its occurrence from Pleistocene sites in North America indicates warmer periods than today. One of the most extensive deposits in Mammoth Cave is of free-tailed bats (*Tadarida*), a genus that now has a distribution to the southwest. This deposit is thought to represent an interglacial when it was warmer than today in Kentucky. *Tadarida* are known to form huge colonies; for example, Bracken Cave in Texas houses over 20 million individuals during part of the year. The Historic Entrance section of Mammoth Cave provides a more recent example of environmental change and its impact on bats. Small brown bats of the genus *Myotis* inhabited this region of the cave in vast numbers for roosting during historic times. The primary area of the colony is documented in early journals, and as guano, bat bones, and staining on the ceiling. A survey of this entire region of the cave indicates that modifications to the main entrance altered temperature and humidity patterns, making the roosting site no longer adequate.

There are other ways that bat remains get incorporated into fossil sites. Some species roost alone or in small groups and die while roosting. Others become incorporated into deposits by carnivores such as raptors or mammals. For example, raccoons are known to venture back into caves and eat bats while they are hibernating. Raccoon scat, primarily composed of bat remains in Mammoth Cave, provides evidence for this behavior in the fossil record. In addition, some bats are carnivorous and contribute remains of their prey into cave deposits. An example is the carnivorous ghost bat, *Macroderma gigas*, of Australia.

Large Carnivorans

Many species of large carnivorans (Order Carnivora) utilized caves in the past for denning or other purposes. Carnivorans are not all carnivores; *carnivore* is an ecological term meaning the diet is composed mostly of meat. Evidence of cave use ranges from claw marks to bedding areas, and from foot-prints and coprolites, to chewed-up prey and the remains of carnivorans themselves. Extinct Pleistocene carnivorans that used caves include giant bears, hyenas, wolves, and cats of different types. Indeed, it would not have been the best time to be a caver. Examples of living large carnivorans that use caves include the American black bear (*Ursus americanus*), Asian black bear (*Ursus thibetanus*), giant panda (*Ailuropoda melanoleuca*), lion (*Panthera atrox*), leopard (*Pathera pardus*), cougar (*Puma concolor*), and spotted hyena (*Crucuta crocuta*).

Bears

The most abundant fossils of a cave-dwelling carnivoran are from the cave bear (*Ursus speleus*), making it one of the best-known Ice Age mammals. This species appeared around 300,000 years ago and went extinct near the end of the Pleistocene. Its distribution was limited to Europe, and ranged from sea level to alpine areas. The cave bear was similar to the living Holarctic brown bear (*Ursus arctos*), but its limbs were relatively shorter, its body was more barrel-like, and its teeth were more broad and blunt. It was the largest carnivoran from the Pleistocene of Europe and some caves contain thousands of their bones. These were not mass die-offs, but represent repeated use of caves for hibernation and denning over extended periods of time. The remains that are found in caves are usually of juvenile, old, or diseased animals. Males were much larger than females and sexual differentiation in cave use may have occurred. For example, in Drachenhöhle (Dragons Cave) near Mixnitz, Austria, ~30,000 individuals were estimated to be in the deposit, and the male specimens outnumber the females by 3 to 1. In other sites, the females and associated juveniles are more abundant. Other evidence of the cave bear's use of caves is found in polished cave walls, claw marks, and tracks (Kurtén, 1968; Kowalski, 2005).

In North America, bear fossils are often found in caves but are never represented by the same abundance as the cave bear in Europe. Bears, particularly black bears (*Ursus americanus*), represent the most commonly found articulated remains of large carnivorans in North American cave deposits. Although the geologic age of most of these *U. americanus* has not been determined, a great number of them likely represent the Holocene epoch. In addition to their bones and teeth, bear beds, claw marks, and footprints are relatively common. The brown bear, *Ursus arctos*, is not typically associated with cave deposits, but numerous records exist for Alaskan Caves.

The tremarctine bears (Subfamily Tremarctinae) are known exclusively from the Americas and a number of species (*Tremarctos floridanus*, *Arctodus pristinus*, and *A. simus*) are found in cave deposits in North America. Only one living tremarctine exists, the spectacled bear (*Tremarctos ornatus*) of South America. The extinct Florida cave bear (*T. floridanus*) was a large robust Plio-Pleistocene species that is known from the southern United States, Mexico, and Central America. This bear derives its name from an abundance of cave sites in Florida, where it is often found co-occurring with *U. americanus*. It has been proposed that *T. floridanus* was primarily a vegetarian like *U. speleus* (Kurtén and Anderson, 1980). Reports of *T. floridanus* existing until ~8000 years ago (Kurtén and Anderson, 1980), based on conventional radiocarbon dates from Devil's Den in Florida, should be questioned. While this species certainly went extinct around the end of the Pleistocene, no reliable radiocarbon dates on the species have been reported.

The lesser short-faced bear (*Arctodus pristinus*) occurs in some Pleistocene cave and karst deposits in eastern North America, but it is not abundant. By the late Pleistocene, *A. pristinus* was completely replaced by *A. simus*, the giant short-faced bear, which had a wide distribution from Alaska to Mexico, and coast to coast. The giant short-faced bear was larger, had a relatively shorter face, more crowded teeth with some proportional differences, and longer limbs for its size (Kurtén and Anderson, 1980). These bears were highly sexually dimorphic, with males being much more massive than females. This is a trend that becomes more pronounced among larger bears. Some estimates have placed the largest *A. simus* males around 1000 kg. Well over 100 fossil localities are known for *A. simus*, and many of these are cave sites. Interestingly, none of the large-sized specimens, interpreted to be males, are found in cave deposits but 70% of the smaller, presumably female, adult specimens are from cave sites (Fig. 2). This has been interpreted to indicate denning among the females (Schubert and Kaufmann, 2003). Dietary studies on this bear indicate that it was highly carnivorous in some portions of its range, but additional studies are needed to better understand variation across space and time. *Arctodus simus* is one of the most thoroughly dated late Pleistocene species in North America, and these radiocarbon dates indicate that it overlapped temporally and spatially with humans and may have been one of the last megafaunal species on the continent to go extinct (Schubert, 2010).

In South America, five different species of extinct tremarctines are recognized, all within the genus *Arctotherium*. Overall their fossil record is relatively sparse, but some cave occurrences exist. The most noteworthy is three specimens of the South American

FIGURE 2 Giant short-faced bear forelimbs in Big Bear Cave, MO, U.S.A. *Photo courtesy of James E. Kaufmann. Used with permission.*

giant short-faced bear *Arctotherium angustidens*, recovered during quarry operations in Buenos Aires Province, Argentina. These specimens are interpreted to represent a sow and cubs that perished while denning (Soibelzon et al., 2009). This species represents the largest type of bear ever known, with males as large as 1600 kg. In contrast to North America, where one of the youngest (or latest) tremarctines was the largest (*Arctodus simus*), in South America it was the earliest known species (*Arctotherium angustidens*) that was monstrous, and apparently more carnivorous. This gigantic bear went extinct in the middle Pleistocene and was replaced by smaller and more omnivorous *Arctotherium* species, some of which lasted until the end of the Pleistocene (Soibelzon and Schubert, 2011).

The fossil record of giant pandas (*Ailuropoda*) comes from cave deposits in China and Southeast Asia, where three extinct species are recognized (*Ailuropoda microta*, *Ailuropoda wulingshanensis*, and *Ailuropoda baconi*). The oldest remains are from *A. microta*, and date to the late Pliocene. The first discovered skull of this species, from Jinyin Cave, south China, demonstrates that cranial and dental adaptations associated with bamboo feeding were apparently in place by the late Pliocene (Jin et al., 2007).

Hyena

Fossil hyenas from caves sites in the Old World are relatively common. Perhaps the most famous discovery of cave hyenas is from the Kirkdale Cave deposit in North Yorkshire, England, described by Rev. William Buckland in the early 1820s. The reverend initially set out to show how the deposit was formed by a Biblical deluge, but his thorough analysis instead demonstrated that the bone accumulation, containing an abundance of hyena remains of all ages, and the broken and gnawed bones of extinct hippo (*Hippopotamus*), woolly rhinoceros (*Coelodonta*), and other megafauna, was the result of denning hyenas. He also demonstrated that these hyenas were considerably larger than living species and that coprolites found in the cave were from hyenas (Buckland, 1823; Kurtén, 1968). The cave hyena went extinct near the end of the Pleistocene. While some morphological differences are known between this hyena and living species, there is disagreement over whether or not it is a distinct subspecies or species or represents a larger variant of the extant spotted hyena (*Crocuta crocuta*).

Another historically significant record of hyenas as bone accumulators can be traced to South African early hominid sites. Dr. Raymond Dart, the anatomist who described the first Pliocene hominid, *Australopithecus africanus*, from a South African bone breccia cave deposit, was ostracized for his claims that this was a human ancestor. Dart recognized and described the specimen, known as the Taung Child, as being bipedal with a brain the size of a chimpanzee's brain. This characterization of human ancestors was not to the liking of some British scientists, who conceived that the large brain would have predated bipedalism, and that it should be found in Europe. In the 1940s, Dart began to work on other South African *Australopithecus* sites, particularly Makapansgat Limeworks Cave where the broken up remains of mammals were found in abundance. He imagined that this hominid was a vicious killer and cannibal, and that the remains in the caves were brought there by members of this species. He termed this the *osteodontokeratic* (bone, tooth, and claw) culture and outlined his ideas in a series of papers. Other researchers noted the similarity in the fragmentary faunal remains to hyena dens and Dart vehemently denied this possibility. The fauna has now been carefully analyzed and is considered to be a carnivore den, primarily formed by hyenas. Thus, in this case the hominids appear to be the hunted, not the hunters (Brain, 1981).

Wolves

Of the wolves, the extinct dire wolf (*Canis dirus*) is the only species with an exceptional fossil record from cave sites. This species was widespread, ranging across North America and down into South America. Though it was not necessarily taller than a large gray wolf on average, it was more robust with a large broad head and massive cheek teeth. Based on these morphological attributes, some researchers have suggested that it was hyena-like in its behavior (Kurtén and Anderson, 1980). Eastern and western populations in North America differed significantly in proportions and weight, with the eastern form being larger and heavier.

Most dire wolf sites are from caves and are thought to represent dens. Examples of sites that have been interpreted as dire wolf dens from caves or karst deposits include Guy Wilson Cave in Tennessee and Cutler Hammock in Florida. In addition to relatively intact *C. dirus* remains, these cave sites are littered with the chewed-up skeletons of other vertebrates. In Guy Wilson Cave, the most abundant prey items are flat-headed peccaries (*Platygonus compressus*) and deer (*Odocoileus*).

Cats

During the Pleistocene more types of large cats existed on the landscape, and some extant species had much larger geographic distributions. In Europe there were lions larger than the African lion (*Panthera leo*), and these are sometimes referred to a distinct species, the cave lion (*Panthera spelaea*). Specimens of *P. spelaea* from caves are usually not abundant, but there are exceptions, such as Wierzchowska Cave, Poland, where the remains of ~20 individuals were recovered (Kurtén, 1968). Similarly, in North America there was a large cat known as the American lion (*Panthera atrox*). Its taxonomic position is also debated, and it has been found in few cave localities. However, large cat tracks from a Missouri cave (Fig. 3) have been referred to this species (Graham *et al.*, 1996). Leopards (*Panthera pardus*) and jaguars (*Panthera onca*) ranged much farther north during the Pleistocene; leopards into Eurasia and jaguars over much of the United States. While neither is particularly abundant from cave sites, leopards are known for collecting remains and taking them into caves. Thus, some faunal cave deposits, for example, the South African hominid sites, may have been formed in part by leopards. Jaguar remains are found at Pleistocene sites across North America, and these range from individual skeletons or trackways that represent cave use, to specimens that are part of faunal accumulations (Kurtén and Anderson, 1980). To date there are no fossil bone accumulations in caves attributed to jaguars.

The sabertooth cat, *Smilodon*, is perhaps the most iconic of all American Pleistocene mammals; however, its occurrence in caves is relatively low and there is no evidence of denning and bone accumulation. However, a record from Hurricane River Cave, Arkansas, provides evidence of cave exploration by a sabertooth, which unfortunately for the cat resulted in death at the bottom of a pit. A related species, the scimitar cat (*Homotherium serum*) seems to have had a closer association with caves. In contrast to *Smilodon*, *Homotherium serum* had shorter sabers that are serrated on the front and back, other teeth that were serrated, and exceptionally long forelegs. Friesenhahn Cave in Texas preserves multiple *H. serum* skeletons of varying

FIGURE 3 One of many large cat tracks from a cave in the southern Ozarks of Missouri, USA. The greatest width of this track is ~175 mm, and it is thought to be from the extinct American lion (*Panthera atrox*) (Graham et al., 1996). *Photo courtesy of James E. Vandike. Used with permission.*

individual ages, indicating that it was used as a den by the cats. Associated with these remains are a multitude of mammoth (*Mammuthus*) and mastodon (*Mammut*) milk teeth and other bone fragments. Thus, it has been hypothesized that these scimitar cats focused their hunts on young proboscideans (elephants and relatives) and brought their prey back to the cave (Kurtén and Anderson, 1980). As with other Pleistocene megafauna, it is thought that these cats went extinct near the end of the epoch.

Rodents

Some rodents are collectors and build dens and debris piles (middens) within caves. These middens are accumulations of the surrounding biota, including plants, coprolites, bones, and teeth of whatever they can carry. The collections are continually covered in their own feces and urine, and in arid regions this creates crystallized and cemented masses that can preserve for tens of thousands of years. In North America the appropriately named packrats (or woodrat, genus *Neotoma*) are well known for this peculiar collecting habit, and studies on their middens are common. The paleoecological record in these middens has provided unique and significant information about climatic, ecological, and faunal changes that have occurred over the past 40,000 years. In fact, without these packrat midden deposits, there would not be a faunal record for Quaternary reptiles and amphibians in the American Southwest. Other rodents in South America, Africa, and Australia make similar nests in caves that can preserve fossils but none contains such diverse contents as those made by the packrat (Lundelius, 2006).

Another type of rodent bone collector is the porcupine, both New World (*Erethizon*) and Old World (*Hystrix*) types. In Africa it seems fairly common for porcupines to collect bones, store them in caves, and gnaw on them periodically. Like other rodents, gnawing on hard substances such as bones keeps the incisors from getting too long and provides nutrients. There is the potential that significant fossil accumulations in caves exist from porcupine hoarding behavior (Brain, 1981). Porcupine dens are easy to recognize based on the size and shape of gnawing marks on the bones.

Other Vertebrate Sites

Many other types of vertebrates contribute to cave deposits. For example, small carnivores such as the raccoon (*Procyon lotor*), coati (*Nasua nasua*), ringtail (*Bassariscus astutus*), skunk (*Mephitis mephitis*), marten (*Martes martes*), and red fox (*Vulpes vulpes*), all use caves to some extent today. Some use them as dens, while others specifically seek out cave-based food, such as raccoons and ringtails pursuing roosting bats. Small carnivores often consume prey that is the same size that raptors eat, but their accumulations are marked by higher levels of bone fragmentation, some evidence of digestive dissolution, and often intact coprolites that contain an abundance of fragmented bones (Andrews, 1990). There is no question that small carnivores such as these have contributed substantially to the fossil record in caves. In western Australia, the Tasmanian devil (*Sacrophilus harrisi*), a carnivorous marsupial, used caves as lairs and accumulated massive amounts of faunal material over time during the Pleistocene (Lundelius, 2006).

In dry caves extraordinary preservation of soft tissues, hair, feathers, keratin, and dung occur. North American caves of the Desert Southwest are particularly notable for these deposits. Many types of large herbivores used cave entrances or large rock shelters for latrines, and in some cases massive amounts of dung accumulated over time. Dung attributed to extinct mammals, such as the Shasta ground sloth (*Nothrotheriops shastensis*), Harrington's mountain goat (*Oreamnos harringtoni*), shrub ox (*Euceratherium*), mammoth (*Mammuthus*), and horse (*Equus*), all give insight into the diet of these animals and their paleoecology (Mead and Agenbroad, 1992). The most extensive deposit of fossil *N. shastensis* dung was found in Rampart Cave, Grand Canyon, Arizona (Fig. 1). Extensive deposits of mammoth dung are known from Bechan Cave and Cowboy Cave in Utah, and indicate a diet composed primarily of grass and sedges (Mead et al., 1986a). Dung from the extinct mountain goat, *O. harringtoni*, is found throughout the entrances of caves along the cliffs of the Grand Canyon, attesting to the precipitous lifestyle of this species.

One of the earliest examples of soft tissue preservation for an extinct Pleistocene mammal came from Mylodon Cave in Chile, where skin, hair, and dung of a mylodontid sloth was recovered. In arid western North America, skin, hair, keratin claw sheaths, cartilage, and other soft tissues are known from *N. shastensis* remains from caves. The best known of these is from a lava tube, Aden Crater, in New Mexico, where a relatively complete mummified specimen was recovered. Horn sheaths from bovids, particularly *Oreamnos harringtoni* (Fig. 4), are also well known from cave sites on the Colorado Plateau (Mead et al., 1986b). In eastern North America, caves are typically wet, which is not conducive to soft tissue and coprolite preservation. One example of an exception to this is Big Bone Cave, Tennessee, where keratin and cartilage are preserved on Jefferson's ground sloth (*Megalonyx jeffersonii*) remains. This includes keratinous claws (Fig. 5), intervertebral discs, and other soft tissue components.

Two extinct ground sloths in North America, *N. shastensis* and *Megalonyx jeffersonii*, are often found in caves as relatively intact skeletons without any carnivore damage. Thus, it has been proposed that these sloths used caves for torpor, shelter, and/or maternity denning (McDonald, 2003). One cave in northern Alabama, ACb-3 Cave, contains multiple *M. jeffersonii* skeletons of varying ages, including one female that was apparently pregnant. This site has more known *M. jeffersonii* than any other site and supports the hypothesis that this species utilized caves for denning.

One of the most abundant Pleistocene mammals found in caves is the flat-headed peccary, *Platygonus compressus*. In some cases this peccary clearly found its

FIGURE 4 Extinct Pleistocene Harrington's mountain goat with preserved horn sheaths from a Grand Canyon Cave, AZ, U.S.A. *Photo courtesy of Larry L. Coats. Used with permission.*

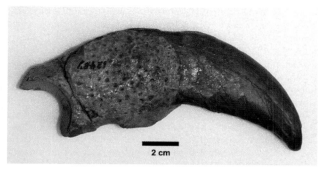

FIGURE 5 Jefferson's ground sloth (*Megalonyx jeffersonii*) claw with preserved keratinous nail from Big Bone Cave, TN, U.S.A. *Photograph taken by Ted Daeschler, courtesy of the Academy of Natural Sciences, Philadelphia.*

way into faunal deposits as the prey of carnivores, such as *Canis dirus*. However, there are many sites where skeletons are intact and abundant, and specimens of all ages are represented. The most extraordinary example is Megenity Peccary Cave, Indiana, where hundreds of individuals have been collected. Extensive peccary trackways are also known from cave sites, for example, the clay floor in River Bluff Cave, Missouri, is covered in hundreds, if not thousands, of their tracks. The abundance of fossil remains and tracks of *P. compressus* in caves indicates that they used caves, probably for shelter, and they apparently lived in large groups. Caves in southwestern North America are actually used today by the living collared peccary, *Pecari tajacu*, as retreats from harsh weather.

IMPORTANCE OF FOSSIL VERTEBRATES FROM CAVES

Caves provide one of the primary sources of faunal remains from the Quaternary fossil record and these fossils are often well preserved (Schubert et al., 2003; Lundelius, 2006; Jass and George, 2010). Thus, the record from caves provides a crucial component for understanding the chronology, distribution, ecology, evolution, and extinction of vertebrates during this relatively recent, yet exciting, time in Earth's history. In sum, this was a period of profound climatic and environmental changes, and the Pleistocene still holds many mysteries, like how did biological communities react to climatic changes, and when and why did the megafauna go extinct?

Near the end of the Pleistocene epoch, most large mammals in North America and South America became extinct. Megafaunal extinctions also occurred in Eurasia and Africa but were minimal in comparison, and Australia's late Pleistocene extinction appears to have occurred earlier than other continents. Many of these species are found in cave deposits, and in some cases caves represent the primary source for fossils. This is especially true for carnivorans that used caves as dens, like the scimitar cat, dire wolf (outside of Rancho la Brea), European cave bear, North American giant short-faced bear, Florida cave bear, and cave hyena. In addition, particular ground sloths, the flat-headed peccary, and Harrington's mountain goat are well known because of their fossil records in caves. Besides providing relatively complete specimens that allow the study of different age individuals, a growth series, the record in caves gives a unique paleoecological view into cave use for torpor, shelter, and denning. Further, the high level of preservation of remains from caves has contributed to better radiocarbon chronologies of species, as well as enhanced knowledge of dietary preferences and evolutionary relationships.

Most fossil sites that are excavated today include careful screenwashing procedures to recover microvertebrates, which typically entered the site as prey. Small mammal remains from screenwashing, particularly rodent teeth, have received the most attention and research for three primary reasons: (1) the teeth are typically well preserved and are often distinct at the species level, (2) they have high reproductive and evolutionary turnover, and (3) they are sensitive to climatic and environmental changes. Because small rodents evolve quickly, they are used as biostratigraphic markers, and assist in age correlation from site to site. However, caution must be applied in making such age assignments because evolutionary turnover can be time transgressive over space, and remnant populations of earlier species sometimes survive in refugia. An example of this is found in Pleistocene deposits from Cathedral Cave, Great Basin, USA. Two extinct voles from the site, *Microtus meadensis* and *M. paroperarius*, represent chronologic range extensions of ~100 ka years. Thus, there is a possibility that this area was a refugium for those species (Jass, 2007).

While many large mammals went extinct at the end of the Pleistocene, small species survived but tended to change their distributions based on ecological preferences. This knowledge is based almost exclusively on cave faunas. For example, several voles, shrews, and moles that now occur much farther to the north are found in Pleistocene cave sites across the south-central and southeastern United States. In these "southern" cave sites "northern" species are found associated with taxa that now occur farther west, and species that still live in the area of the cave; thus, the faunal accumulation is composed of mammals that no longer co-occur and prefer very different environmental settings. These accumulations are known as nonanalog faunas and are thought to reflect climatic and ecological conditions that are quite different from today. Recent advances in radiocarbon dating provided a means to test this hypothesis, and dates on nonanalog taxa show that these associations represent true past communities.

See Also the Following Article

Vertebrate Visitors in Caves—Birds and Mammals

Bibliography

Andrews, P. (1990). *Owls, caves and fossils*. London: Natural History Museum Publications.

Brain, C. K. (1981). *The hunters or the hunted? An introduction to African cave taphonomy*. Chicago, IL: University of Chicago Press.

Buckland, W. (1823). *Reliquiæ diluvianæ; or, observations on the organic remains contained in caves, fissures, and diluvial gravel, and on other geological phenonmena, attesting the action of an universal deluge*. London: Thomas Davison, White Friars.

Graham, R. W., Farlow, J. O., & Vandike, J. E. (1996). Tracking ice age felids: Identification of tracks of *Panthera atrox* from a cave in southern Missouri, U.S.A. In K. M. Stewart & K. L. Seymour (Eds.), *Palaeoecology and palaeoenvironments of late cenozoic mammals: Tributes to the career of C.S. (Rufus) Churcher* (pp. 331–345). Toronto, Canada: University of Toronto Press.

Jass, C. N. (2007). New perspectives on Pleistocene biochronology and biotic change in the east-central Great Basin. Unpublished PhD dissertation, The University of Texas at Austin, Austin, TX.

Jass, C. N., & George, C. O. (2010). An assessment of the contribution of fossil cave deposits to the Quaternary paleontological record. *Quaternary International*, 217, 105–116.

Jin, C., Ciochon, R. L., Dong, W., Hunt, R. M., Jr., Liu, J., Jaeger, M., & Zhu, Q. (2007). The first skull of the earliest giant panda. *Proceedings of the National Academy of Sciences*, 104, 10932–10937.

Kowalski, K. (2005). Paleontology of caves: Pleistocene mammals. In D. C. Culver & W. B. White (Eds.), *Encyclopedia of caves* (1st ed., pp. 431–435). Burlington, MA: Elsevier, Academic Press.

Kurtén, B. (1968). *Pleistocene mammals of Europe*. London: Weidenfield and Nicolson.

Kurtén, B., & Anderson, E. (1980). *Pleistocene mammals of North America*. New York: Columbia University Press.

Lundelius, E. L. (2006). Cave site contributions to vertebrate history. *Alcheringa Special Issue*, 1, 195–210.

Martin, L. D., & Gilbert, B. M. (1978). Excavation at natural trap cave. *Transactions of the Nebraska Academy of Science*, 6, 107–116.

McDonald, H. G. (2003). Sloth remains from North American caves and associated karst features. In B. W. Schubert, J. I. Mead, & R. W. Graham (Eds.), *Ice age cave faunas of North America* (pp. 1–16). Bloomington, IN: Indiana University Press.

Mead, J. I., & Agenbroad, L. D. (1992). Isotope dating of Pleistocene dung deposits from the Colorado Plateau, Arizona, and Utah. *Radiocarbon*, 34, 1–19.

Mead, J. I., Agenbroad, L. D., Davis, O. K., & Martin, P. S. (1986a). Dung of *Mammuthus* in the arid Southwest, North America. *Quaternary Research*, 25, 121–127.

Mead, J. I., Martin, P. S., Euler, R. C., Long, A., Lull, A. J. T., Toolin, L. J., Donahue, D. J., & Linick, T. W. (1986b). Extinction of Harrington's mountain goat. *Proceedings of the National Academy of Sciences USA*, 83, 836–839.

Schubert, B. W. (2010). Late Quaternary chronology and extinction of North American giant short-faced bears (*Arctodus simus*). *Quaternary International*, 217, 188–194.

Schubert, B. W., & Kaufmann, J. E. (2003). A partial short-faced bear skeleton from an Ozark Cave with comments on the paleobiology of the species. *Journal of Cave and Karst Studies*, 65, 101–110.

Schubert, B. W., Mead, J. I., & Graham, R. W. (Eds.), (2003). *Ice age cave faunas of North America*. Bloomington, IN: Indiana University Press.

Soibelzon, L. H., Pomi, L. M., Tonni, E. P., Rodriguez, S., & Dondas, A. (2009). First report of a short-faced bears' den (*Arctotherium angustidens*): Palaeobiological and palaeoecological implications. *Alcheringa*, 33, 211–223.

Soibelzon, L. H., & Schubert, B. W. (2011). The largest known bear, *Arctotherium angustidens*, from the early Pleistocene Pampean region of Argentina: With a discussion of size and diet trends in bears. *Journal of Paleontology*, 85, 69–75.

PASSAGE GROWTH AND DEVELOPMENT

Arthur N. Palmer

State University of New York

Of all cave types, solution caves have the most complex developmental histories. They are formed by the dissolving action of underground water as it flows through fractures, partings, and pores in bedrock. Such caves must grow rapidly enough to reach traversable size before the rock material that contains them is destroyed by surface erosion. Because of their sensitivity to local landscapes and patterns of water flow, solution caves contain clues to the entire geomorphic, hydrologic, and climatic history of the region in which they are located. At the land surface most of this evidence is rapidly lost to weathering and erosion, but in caves these clues can remain intact for millions of years.

STAGES OF CAVE DEVELOPMENT

Solution caves develop in several stages that grade smoothly from one to the next: (1) For the rock to transmit enough water to form caves, it must first contain a network of presolutional openings, such as fractures, partings, and primary pores. (2) Because these openings are narrow, the initial groundwater moves very slowly and becomes nearly saturated after only a short distance of travel. Solutional widening of the openings is likewise very slow. Nevertheless, among the many alternate routes that the water follows, there is a great variation in the amount of flow because of differences in width and length of openings. Those paths with the greatest flow enlarge the fastest. (3) These favored routes eventually become wide enough that groundwater is able to pass all the way to the spring while still retaining much of its solutional capacity. (4) Flow along these preferred routes increases rapidly, enlarging them into cave passages rather uniformly over their entire length. (5) The cave acquires a distinct passage pattern that depends on the nature of groundwater recharge, geologic setting, and the erosional history of the region. (6) Most caves evolve by diversion of water to new and lower routes as surface rivers deepen their valleys and lower the elevations of springs. (7) Caves are eventually destroyed by roof collapse and by intersection of passages by surface erosion. At any given time, different parts of the same cave may be undergoing different stages in this sequence.

THE EARLIEST STAGES

At great depth beneath the surface there is very little groundwater flow because openings in the rock are narrow and few, and hydraulic gradients are feeble. Wide tectonic fractures may penetrate to great depth and can allow substantial water flow, but they are sparse. As rocks are gradually exposed at the surface by uplift and erosion, increasing amounts of groundwater pass through them. At first, the rate at which any opening enlarges by dissolution depends almost entirely on the discharge through it. In turn, the most important controls over the amount of flow are how much water is available to supply the upstream end of the growing caves, and the widths of the initial openings. Also influential, but less so, are the overall hydraulic gradient, length of flow, and water chemistry.

When the openings along a particular flow path enlarge enough that water can pass through all the way to the spring while still retaining much of its solutional capacity, this flow path begins to enlarge much more rapidly than its neighbors' do. From then on, water chemistry becomes the main factor controlling growth rate, and the entire path enlarges rapidly at a maximum wall retreat of about $0.001-0.01$ cm yr^{-1}. Openings that have not reached this stage continue to enlarge only slowly, often at diminishing rates with time. The typical result is a cave with only a few negotiable passages surrounded by openings that have enlarged only slightly.

The time needed to reach this stage of rapid cave enlargement can be considered the "gestation time" through which an incipient cave must pass in order to grow into a true cave. It is difficult to specify exactly when this time begins. Some researchers argue that it should include the entire age of the soluble rock, including early depositional conditions, compaction, and burial by other rocks, as well as later uplift and perhaps folding and faulting. But before cave growth can truly begin, there must be a substantial hydraulic gradient through the rock. This usually requires that the soluble rock be exposed at or near the surface, where zones of groundwater recharge and discharge are well defined. Most people date the age of a cave to the onset of these conditions.

Field evidence and computer modeling show that short fissures in limestone (up to a few hundred meters long) with initial widths of $0.01-0.1$ cm would require no more than a few thousand or tens of thousands of years to reach their maximum enlargement rates. Relatively long and/or narrow paths require much longer times, typically on the order of hundreds of thousands of years.

DEVELOPMENT OF CAVE PATTERNS

Competition between Initial Flow Routes

Most groundwater infiltrates in upland recharge areas and emerges at lower elevations such as river valleys. Patterns of solution caves typically reflect these trends. Such caves can form only along paths that gain discharge with time, which requires one of the following conditions:

- Where water leaks from a stream and drains to a lower outlet, the underground flow increases dramatically as the initial openings enlarge by dissolution. When the incipient cave enlarges enough to carry all of the available flow, the surface stream becomes a sinking stream. From then on the mean-annual flow rate remains roughly the same, but the water level in the cave drops as the passages enlarge. The cave reaches its maximum enlargement rate early in this enlargement history, so the leveling out of the flow rate does not interfere with the growth of the cave.
- Water that seeps through soil into underlying soluble rock can gain discharge only by increasing its catchment area. Because the upstream ends of the major flow paths enlarge most rapidly, soil subsides into these growing voids to form sinkholes. As sinkholes grow they increase their catchment areas, delivering progressively more water to the passages they feed. Water input to the caves increases in an irregular manner, and much less rapidly than in routes fed by leaking streambeds.

The routes fed by leaky surface streams increase their flow much more rapidly, so they are usually the first to form traversable cave passages. Passages fed by sinkholes have a limited catchment area and require more time to form. They usually join the earlier passages as tributaries of a branching cave system. The first passages to form are usually short and direct. With time, as their water levels drop, these early passages serve as targets for later passages that drain more remote areas. Although the growth of any single passage propagates in the downstream direction, the overall system grows in the upstream direction, away from the springs, by the addition of new tributaries. A typical sequence is shown in Figure 1.

Vertical Organization of Cave Passages

Most vadose passages, because their flow is purely gravitational, tend to have a strong component down the dip of the rock strata. This is especially true in well-bedded rocks. Phreatic passages show no consistent relation to the dip, unless that is the only direction to potential outlets, or where prominent fractures also extend in that direction. In well-bedded rocks, the intersections between dipping beds and low-gradient water tables encourage many phreatic passages to develop roughly along the strike of the beds. These relationships tend to be obscure where the geologic structure is complex.

The vertical arrangement of underground water flow is strongly controlled by the deepening of surface valleys by river erosion, and cave passages reflect this control. In general the largest passages form at the elevation of erosional base level when river entrenchment is slow. At such times, rivers develop floodplains, and springs remain at fairly constant elevations for lengthy periods of time. Meanwhile the passages that drain to those springs are able to reach large size. In contrast, most passages that form during rapid river entrenchment are small because the cave-forming water undergoes more frequent diversion to lower routes. The major passages may be arranged in several different levels that decrease in age downward. Partial filling of river valleys with sediment tends to block nearby cave passages, causing them to fill with sediment as well, or at least to become ponded with water.

This conceptual model has been well validated in many karst areas, but several complications can disrupt this simple interpretation. Vadose passages may become perched on insoluble strata and grow to large size well above the local levels of surface rivers. In addition, most phreatic passages contain vertical loops that descend below river levels. Even the ideal cave levels controlled by pauses in river entrenchment are not perfectly level. For this reason, many people prefer to call them *stories* or *tiers*, either of which is preferred in general applications. However, the term *cave level* is still appropriate where there is a clear relation to erosional base level. The critical measurement is not the average elevation of a phreatic passage, but instead the elevation where passages change from vadose to phreatic types (for example, from a canyon to a tube). This transition may not be clear where the geologic structure is complex or in caves that experience severe fluctuations in water level due to flooding.

Figure 2 is an idealized profile through a multistoried cave. Three main tiers of cave development are shown, with decreasing loop amplitudes from the highest tier to the lowest. This is a conceptual ideal that is not characteristic of all multistoried caves. Fissures may be sparse at first, so that passages are constrained to only a few deeply descending loops. As erosional unloading and cave development persist, fissures become more numerous until eventually the passages are able to form more or less along the water table, with minimal phreatic looping. However, in some caves the greater amplitude of loops in upper passages is instead caused by floodwaters, which produce ungraded, looping bypasses around low-flow routes that have more uniform gradients. Furthermore, the depth of phreatic loops also tends to be greatest in

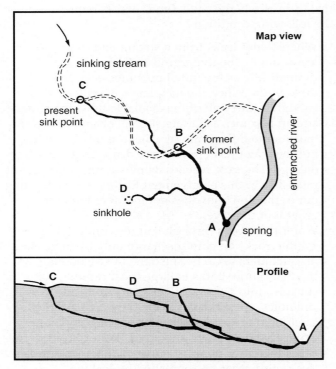

FIGURE 1 Typical stages in the early development of a solution cave. Segment B-A forms first as a short, high-gradient path from a sink point in the upland stream to the entrenched river. Segment C-B forms next, aided by the decrease in head in the previous segment. The passage fed by the sinkhole at D is slower to form because of its smaller catchment and less rapid increase in flow with time.

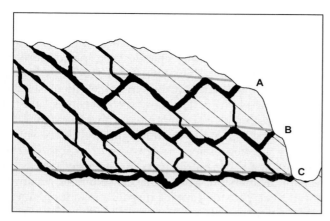

FIGURE 2 Idealized profile through a complex, multitier cave system showing decreased loop amplitude with time as fissure frequency increases from stage A to B to C. Loop amplitude is also proportional to the length of the cave system, to structural deformation, and massiveness of the soluble rock.

highly deformed or massive rocks, and in broad regional flow systems.

Origin of Branching Systems

Branching (branchwork) caves are roughly the underground equivalents of surface drainage. Many small passages join together in the downstream direction to form larger but fewer passages. Of all solution caves, branchwork caves are by far the most common. There are several reasons for the branching pattern:

- As phreatic passages enlarge, the local hydraulic head within them decreases. Groundwater flows from surrounding smaller openings, where the water table is higher, toward the low heads in the passages.
- Vadose passages have no inherent tendency to converge, because their streams are independent of one another. However, the structures that they follow often intersect, forcing their streams to converge as tributaries. Examples of convergent structures include intersecting fractures, and trough-like features in bedding-plane partings.
- Water from broad recharge areas converges toward outlets of limited extent, generally discharging in stream valleys. Thus there is a natural tendency for conduits to join each other simply by competition for space. There is little opportunity for water to diverge farther downstream. The exception is in the vicinity of spring outlets, where local distributary systems may develop because of collapse, flooding, and widening of the main stream routes by erosional stress release. In those areas a tangle of diversion passages may develop like the channels on a river delta.

Development of Maze Caves

Most nonbranchwork caves are mazes in which all passages form more or less simultaneously. A maze cave can form only if the growth rate is similar along many alternate flow paths. This requires that the ratio of discharge to flow distance is large along many alternate flow routes. This condition can be met in any of the following ways:

1. *Floodwater recharge.* A rapid influx of water into caves during floods will fill the caves with high-pressure water and force solutionally aggressive water into all fissures in the surrounding rock. Floodwaters combine steep gradient, high discharge, and short flow paths. As a result, these openings enlarge at approximately the maximum possible rate. This process is most active in the vicinity of constrictions in major stream passages caused by collapse, sediment chokes, or poorly soluble strata.
2. *Short flow paths.* Mazes can also form where many alternate flow paths are short (*i.e.*, from where the water first enters the soluble rock). In this case, all openings except for the narrowest enlarge simultaneously at similar rates. The epikarst is an example. This is a network of enlarged openings, all of which have enlarged simultaneously at comparable rates. The tendency for uniform growth diminishes with flow distance, so only a few major drains lead downward into the aquifer below.
3. *Uniform recharge through a permeable insoluble rock.* Where water enters a soluble rock by first seeping through an insoluble but permeable rock, all fissures in the soluble rock receive roughly comparable amounts of flow, regardless of how wide they are. For example, mazes can be formed by recharge through a permeable but insoluble material such as quartz sandstone. Their uniform enlargement is also aided by the short flow paths. Rising water entering soluble rock through permeable material can produce a similar effect, and the resulting caves tend to be relatively extensive.
4. *Local boosts in aggressiveness.* Where the groundwater gains solutional capacity by mixing of waters of contrasting CO_2 content or salinity, or by oxidation of sulfide-rich water, many alternate routes are enlarged at roughly comparable rates. Mazes are formed along short paths of flow in and around the mixing zone.

The differences in maze patterns depend partly on geologic structure. *Network mazes* consist of intersecting fissures arranged in a pattern like that of city streets. They require many intersecting fractures (joints or faults), which are typical of massive or thick-bedded rock. They can be formed by any of the four processes above. *Anastomotic mazes* have a braided pattern of intersecting tubes, usually arranged two-dimensionally along a single parting or fault. They are nearly all formed by process 1 above. *Spongework mazes* form where primary (matrix) porosity is dominant. In pattern they resemble the intersecting holes in a sponge. Most of them form by process 4, and less commonly by process 1. A two-dimensional variety can form along bedding-plane partings. *Ramiform mazes* consist of rooms with offshoots extending outward from them at various elevations. They usually include areas of network or spongework maze development and are formed mainly by process 4. Many network and anastomotic mazes, and a few spongework mazes, are merely superimposed on a basic branchwork pattern and constitute only part of the entire cave development.

Figure 3 summarizes typical cave patterns and their relation to sources of aggressive water and to dominant structural characteristics.

Adjustment of Caves to Changing Conditions

When an incipient cave finally reaches its maximum growth rate, several other changes take place more or less simultaneously. The flow regime of cave water changes from laminar to turbulent, which increases the solution rate slightly. The flow also becomes able to transport sediment. For example, it can carry away the soil that subsides into caves through karst depressions, allowing the depressions to grow more rapidly. The sediment load can also help to enlarge caves by mechanical abrasion. But in places, sediment accumulates in thick beds that retard dissolution and erosion.

When a cave is able to transmit the entire flow from its recharge area, the average flow can increase no further. Instead the head within the passage decreases as the cross-section continues to enlarge. Much of the upstream part of the cave becomes vadose, and streams may entrench canyons in the passage floors.

Where water first enters caves, it is fairly rich in soil-derived CO_2 and may acquire even more by oxidation of organic materials within the caves. As caves acquire entrances that allow air exchange with the surface, many cave streams lose part of their solutional capacity. Loss of CO_2 through entrances and other openings can drive the stream water to supersaturation with

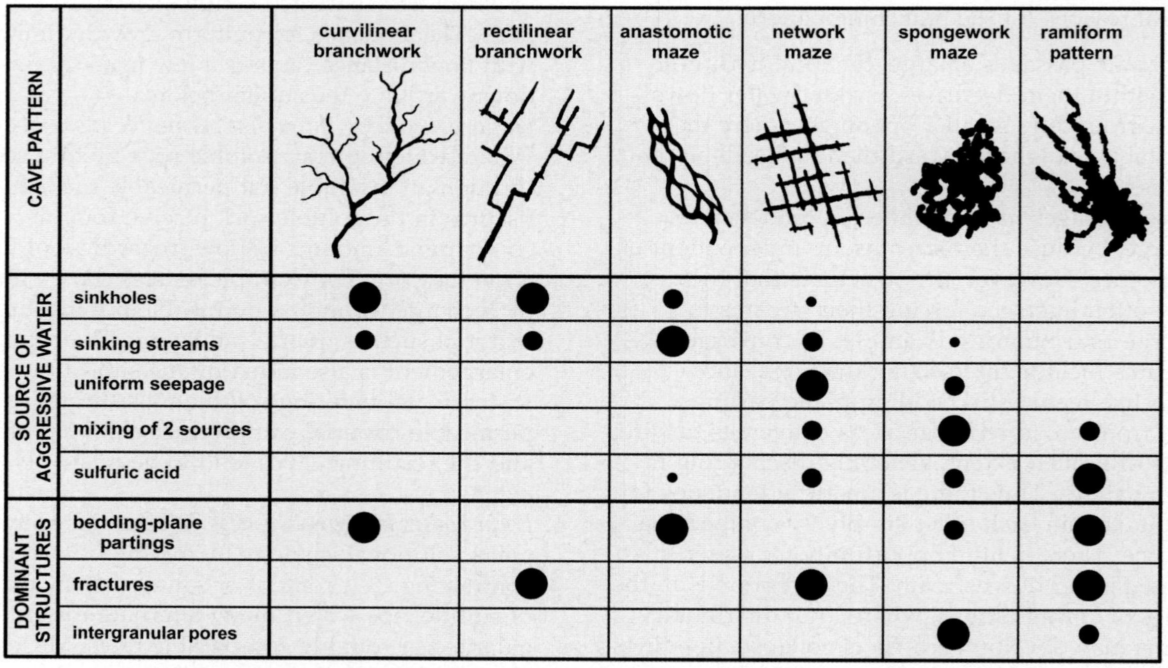

FIGURE 3 Relation of cave patterns to mode of groundwater recharge and geologic structure. Dot sizes show the relative abundance of each pattern within the various recharge and structural categories. Single-passage caves are rudimentary or fragmentary versions of those shown here.

dissolved calcite or dolomite. As a result, many cave streams are aggressive only during high flow.

As the land surface becomes dissected by erosion, patterns of groundwater recharge change. Initial water sources are usually few and large, but with time they may become divided into many smaller ones. Vadose water must travel increasingly greater distances to reach the water table, and extensive complexes of vadose canyons and shafts can form. The resulting pattern of active cave streams tends to become denser with time, and eventually much denser than that of the original surface drainage. Growing dolines eventually merge to form a continuous karst surface. In a mature karst region the only surface streams that retain their flow are the main entrenched rivers and the ephemeral upstream ends of sinking streams.

CAVE DEGENERATION

As the land erodes, the surface intersects underlying cave passages, segmenting them and eventually destroying them entirely. Evidence for the cave may persist for a while as a canyon-like feature or a rubbly zone of collapsed blocks. This final phase in the life of a cave passage usually occupies tens of thousands or even hundreds of thousands of years. However, newer passages continue to develop if the soluble rock extends to lower elevations. In dipping strata, new areas of rock are uncovered by erosion at about the same rate as they are eroded away in the up-dip areas. Thus some of the caves we see today are the descendants of earlier caves that once occupied parts of the soluble rock that are now gone. This process ends when the entire soluble rock in the cave region is eroded away.

CONCLUSION

Patterns of solution caves are highly sensitive to their physical setting. With the aid of dating of cave deposits, cave patterns can shed considerable light on the history of landscape development, hydrology, and climate changes in the surrounding region.

See Also the Following Article

Passages

Bibliography

Dreybrodt, W. (1990). The role of dissolution kinetics in the development of karst aquifers in limestone: A model simulation of karst evolution. *Journal of Geology, 98*(5), 639–655.

Ford, D. C., & Ewers, R. O. (1978). The development of limestone cave systems in the dimensions of length and depth. *Canadian Journal of Earth Sciences, 15*, 1783–1798.

Granger, D. E., Fabel, D., & Palmer, A. N. (2001). Pliocene-Pleistocene incision of the Green River, Kentucky, determined from radioactive decay of ^{26}Al and ^{10}Be in Mammoth Cave sediments. *Geological Society of America Bulletin, 113*(7), 825–836.

Klimchouk, A. B. (2011). *Hypogene speleogenesis: Hydrogeological and Morphogenetic Perspective* (2nd ed.). Carlsbad, NM: National Cave and Karst Research Institute, Special Paper Series 1.

Palmer, A. N. (1991). Origin and morphology of limestone caves. *Geological Society of America Bulletin, 103*, 1–21.

Palmer, A. N. (2007). *Cave geology*. Dayton, OH: Cave Books.

White, W. B. (1969). Conceptual models of carbonate aquifers. *Ground Water, 7*(3), 15–21.

White, W. B. (1977). Role of solution kinetics in the development of karst aquifers. In J. S. Tolson & F. L. Doyle (Eds.), *Karst hydrogeology* (pp. 503–517). International Association of Hydrogeologists, 12th Memoirs. Huntsville, AL: UAH Press.

Worthington, S. R. H. (2004). Hydraulic and geological factors influencing conduit flow depth. *Cave and Karst Science, 31*(3), 123–134.

Worthington, S. R. H. (2005). Evolution of caves in response to base-level lowering. *Cave and Karst Science, 32*(1), 3–12.

PASSAGES

George Veni

National Cave and Karst Research Institute

Voids that transmit or have the capacity to transmit turbulently flowing water in karst are called *conduits*, a term that is used generically to describe all segments of karst drainage networks or to refer specifically to segments too small for human entry. *Turbulent flow*, a key factor that distinguishes conduits from smaller openings in the rock, occurs when an opening achieves a width of about 1 cm. Conduits large enough for human entry are called *caves*. Passages, rooms, pits, and domes are types of conduits within caves that have certain physical characteristics. *Passages* are horizontal to moderately sloping and narrow relative to their lengths. A *room* is a type of passage that is wide relative to its length; rooms often form where passages intersect and are circular, oval, or irregular in shape. Pits and domes are steeply sloping to vertical passages. When encountered from the top, they are called *pits*, and from below, they are called *domes*.

DEFINITIONS AND CONCEPTS

Several factors affect the location, size, and shape of cave passages. Because limestone, dolomite, and other karstic rocks are poorly permeable, passages develop along fractures and bedding planes where permeability is higher. Initially, groundwater will flow through a variety of these structurally guided conduits until a preferential path is developed that enlarges to form a passage. The pattern that the passages form, the morphology or shape that they take, and the sediments and features that they contain, reflect the lithologic, structural, and hydrologic conditions within the karst aquifer. Many of the factors and conditions discussed in this article are shown and summarized in Figure 1. This discussion describes typical conditions; local hydrogeologic factors may produce variations and exceptions. For in-depth discussions of cave and passage development, see Bögli (1980), White (1988), Ford and Williams (1989), Gillieson (1996), Klimchouk et al. (2000), and Klimchouk (2007).

Passages can be classified into four general types according to their position in the karst aquifer. *Vadose passages* occur at the upgradient ends of the aquifer, and they transmit water through the vadose zone from the surface to the water table. They tend to have high gradients and, depending on the gradient and volume of flow, contain small or large accumulations of sediment. *Phreatic passages* occur at and below the water table (the phreatic zone) and transmit groundwater from the vadose to the discharge ends of the aquifer. They have relatively low gradients and usually contain significant accumulations of sediment where intersected by vadose passages. *Tributary passages* feed groundwater into the main vadose and phreatic passages. They decrease in numbers down the hydraulic gradient while generally increasing in size and volume of flow. *Discharge passages* occur near springs where groundwater flows to the surface. They often form a "distributary" pattern that divides the water into multiple outlets that form in response to changing conditions on the surface, such as downcutting of valleys below the springs. Some of these passage types may be hydrologically inactive or active only during flood events, having formed when groundwater levels were higher.

PATTERNS OF CAVE PASSAGES

The distribution and geometry of passages within a cave are dictated by a combination of local lithologic, structural, and hydrologic factors. Variations in the solubility of the strata may determine the particular bed in which a passage forms. Interbedded rocks of very low solubility, such as shales, clays, or chert, may perch groundwater and produce passages at the base of a group of soluble rocks. Fractures that breach those poorly soluble units allow pits to form that may extend down to the next poorly soluble unit, creating a stair-step pattern (Fig. 2).

Passages formed along fractures tend to be narrow and linear, with sharp, often right-angle turns and intersections with other passages (Fig. 1, angular passages). Passages guided by bedding planes are usually

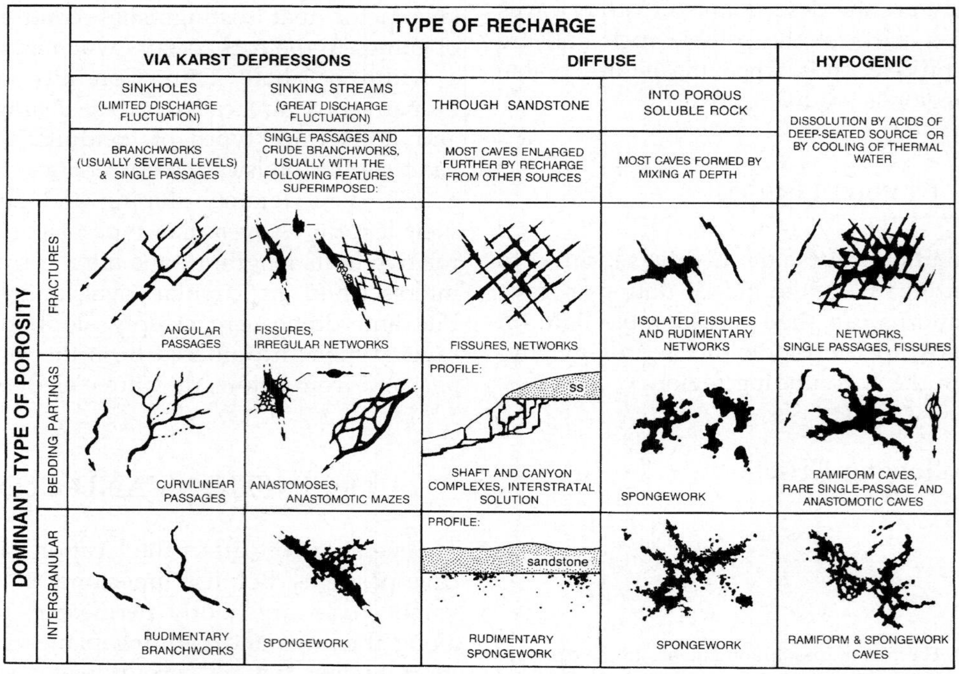

FIGURE 1 Summary of cave patterns and their relationship to types of recharge and porosity. *From Palmer, 1991. Used with permission.*

wide and meandering, with low-angle passage intersections (Fig. 1, curvilinear passages). When the structural grain of the rock, such as the primary fracture direction or dip of the beds, significantly differs from the hydrologic gradient, cave passages may form closely spaced meanders or zig-zags as competing factors alternately control the development of different short segments of passages.

Vadose tributary passages in caves formed in horizontal or nearly level beds join the main passages from both sides in a roughly equal number; however, where beds are dipping, the tributaries enter the main passages from predominantly the up-dip direction, and distributary passages branch off in the down-dip direction. This is most noticeable in steeply dipping strata but has also been observed in beds with <1° of dip. Phreatically formed passages, usually the main passages within a cave, tend to follow the strike of dipping strata (Fig. 3).

The pattern of cave passages will vary depending on the number of sources that contribute most of the groundwater within the cave. Caves recharged by multiple sites of roughly equal contribution will tend to form the tributary systems described previously. Numerous small passages will join and enlarge as they flow downgradient, and their combined flow will form the main drainage passage that leads to a spring (Fig. 1, sinkhole type of cave). In contrast, caves formed by capturing a single large source of water, such as a surface stream, quickly grow into single large passages that maintain a relatively constant size between their source and spring. Tributaries are few and provide relatively little water, and due to their lesser flows those passages are smaller and more recently formed. Occasionally, they will form floodwater mazes next to the main passage (Fig. 1, sinking stream type of cave).

Passages that form down steep hydraulic gradients are more linear than those formed on low gradients. Nearly level gradients will produce anastomotic patterns, where passages divide their flow as they branch, rejoin, and vary in size. Ponded water may dissolve low wide rooms along bedding planes or a criss-crossing network of passages along fractures. Ponding is often the result of a geologically short-lived phenomenon, such as collapse along a cave stream or inundation of the spring. Normal conditions are restored when, for example, groundwater cuts a path through or around the collapse. Flooding may also inject water episodically into fractures and bedding planes, temporarily creating ponded conditions until the floodwaters subside. Flooding from chemically aggressive surface streams is especially effective in creating floodwater mazes. Some maze passages form by vadose water moving down through poorly soluble but permeable strata, such as sandstone, and enlarging fractures in underlying soluble rocks so that they interconnect linearly (Fig. 1, fissure networks).

Passage patterns also reflect their development in shallow to deep portions of the aquifer. Roughly

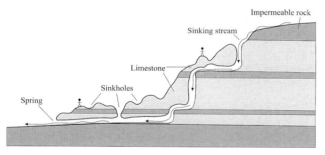

FIGURE 2 Schematic model of stairstep cave development. *Adapted from Crawford, 1996.*

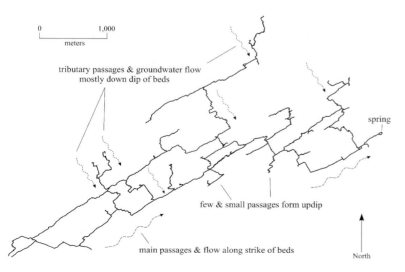

FIGURE 3 Map of Honey Creek Cave, Texas, illustrating passage development and groundwater flow relative to dip of the beds. *From Elliott, W. R., and Veni, G. (Eds.), 1994. Convention Guidebook, National Speleological Society, Huntsville, AL. Used with permission.*

horizontal passages form in the shallow portion of the phreatic zone. They may result from closely underlying poorly permeable strata but also develop without a poorly permeable base in highly fractured rock. In aquifers where the rock is less fractured, fewer permeable zones are available to develop passages between the vadose passages and the springs. Consequently, groundwater pressure will move water through deeper fractures, forming deeper passages. These passages will ascend and descend en route along the more permeable fractures. The greater the distance to the springs, the greater is the potential depth of the conduits (Fig. 4). Springs that discharge from such systems in Mexico have measured depths of more than 300 m.

While most passages form from the top down by water entering the ground and discharging back to the surface at a lower elevation, hypogenic passages form from the bottom up. While initially described as the result of rising gases, typically hydrogen sulfide, that flow into groundwater to form acidic solutions that more aggressively dissolve the rock, recent studies show these hydrologic conditions create a distinct set of morphologies independent of lithology and groundwater chemistry. These passages are characterized by the absence of features that reflect vadose flow from the surface; by decreased passage development with increased elevation; by abrupt terminations of rooms and passage (often with maze passages or honeycombed walls); by the occurrence of irregularly shaped rooms and passages with little apparent relationship to convergent, tributary flow patterns; and, where exposed, by deep, narrow fissures in the floors where groundwater rose (Fig. 1, hypogenic caves). Mineralogical indicators may also be present, such as gypsum deposits, where hydrogen sulfide produced sulfuric acid which dissolved the limestone. Carlsbad Caverns (New Mexico) is the best known example of a cave formed by this process.

PASSAGE MORPHOLOGY AND EVOLUTION

Phreatically formed passages have circular to elliptical cross sections. The ellipses will be vertical, horizontal, or diagonal, depending on whether the passage is enlarged along a fracture or bedding plane and the orientation of those features (Fig. 5). Phreatic walls are relatively smooth. The symmetry of phreatic passages results from their development below the water table where roughly equal water pressure is present on all walls.

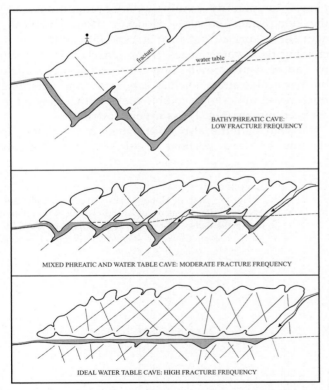

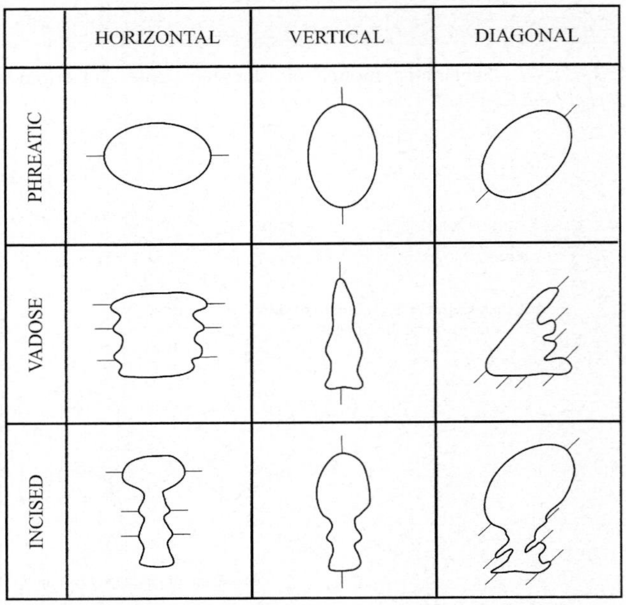

FIGURE 4 Model of cave development below the water table relative to fracture frequency. *Adapted from Ford and Ewers, 1978; and Ford and Williams, 1989.*

FIGURE 5 Cross-sectional patterns of phreatic, vadose, and incised passages relative to bedding and fractures.

Vadose passages are relatively tall and narrow in cross section, especially where the hydraulic gradient is steep and/or the bedding is thick. They form by vadose water downcutting the floor to reach the water table in as short a distance as possible. Water pressure and dissolution are concentrated at their floors and lowermost walls. Pits are the ultimate form of vadose passage. The walls of vadose passages are often rough and irregular due to greater turbulence and differences in water pressure and dissolution, as well as from greater mechanical erosion by rocks and other sediments carried by the streams. Passages formed along vertical fractures tend to be narrow while those formed along horizontal or dipping fractures and beds are wider because they can more easily undercut, collapse, dissolve, and wash away the strata (Fig. 5).

Water levels in an aquifer naturally decline over time as the land surface erodes, allowing groundwater to discharge from progressively lower elevations. Consequently, phreatic passages will often be drained and become modified by the new vadose hydrologic conditions. The most common change is the incision of passage floors to drain water to the descending water table. Keyhole passage shapes are typical, where a broad, smooth-walled phreatic tube sits over a narrow, rough-walled vadose canyon that is cut into its floor (Fig. 5). In caves where phreatic water moved up and down pits, newly vadose passages at the tops of pits will form waterfalls that lengthen the pit by cutting headward, toward the water source, into the wall below the passage.

During phreatic conditions, the buoyancy of the water can support up to 42% of the weight of the passage ceiling. In broad rooms and passages, especially in thin- to medium-bedded limestone of low structural strength, the change to vadose conditions can result in ceiling instability and collapse. The influx of chemically aggressive vadose groundwater may accelerate the collapse. The result after the collapse is a more structurally stable dome shape in the passage's ceiling and a mound of breakdown on its floor. Because the collapsed rock will fill more space than when it was intact, it could extend to the ceiling and prevent further collapse. In some caves, where jagged walls demonstrate extensive collapse but there is no underlying mound of breakdown, groundwater on or under the rubble is dissolving and removing the pile faster than it is accumulating.

Some of the passage patterns, shapes, and modifications described in this article can be hidden by sediments and speleothems. Pits seemingly without passages are the most common example. The passages exist but are buried by sediments deposited on the pit floors where the water that carries them changes from high-energy vertical flow to low-energy horizontal flow conditions. Another somewhat common example is where large passages end in low crawlways that can barely be entered. These often reflect where the once sediment-free phreatic conduit dipped down and up in a V-shaped pattern (Fig. 4); during vadose conditions, the bottom of the "V" was filled by sediments. In some cases, dense deposits of chert, clay, or other poorly soluble materials may armor a passage floor to prevent downward erosion. As more of this material is deposited, groundwater will dissolve the walls and ceiling to enlarge the passage upward by a process called *paragenesis*.

Karst aquifers evolve at different rates according to the geologic factors described in this article, but especially in response to erosion of the overall landscape. The downcutting of surface streams will drain old aquifer levels, create new ones, and change groundwater gradients toward new springs. These changes will

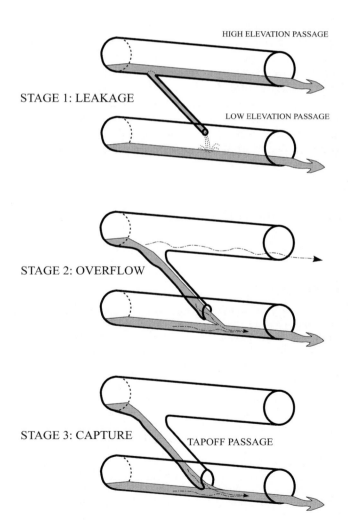

FIGURE 6 Three-stage evolutionary model of tapoff passage development. *Adapted from Veni, 1997.*

be reflected in cave passages. In many cases, water from one passage will be captured or pirated into another passage with a steeper hydraulic gradient. These passages are known as *tapoff passages*. Originally meant to describe a newly formed passage that transmits water from a cave to a newly formed spring, its definition has been expanded to describe a relatively short, high-gradient cave stream passage that pirates flow from a high elevation stream to a lower elevation stream (Fig. 6). Tapoffs can form between any combination of surface and subsurface streams, are usually small relative to the surrounding passages, and have few or no tributaries. With sufficient time, they may enlarge to the point where they may be difficult or impossible to recognize.

CONCLUSIONS

Cave passages form the framework of karst aquifers. Their proper interpretation provides insights into the evolution of the aquifer and land surface, groundwater flow behavior, potential fate and transport of contaminants that might enter the aquifer, and the definition and distribution of habitat for terrestrial and aquatic species living in the cave and aquifer, as well as providing key information on the portion of the conduit system that remains humanly inaccessible.

See Also the Following Article

Passage Growth and Development

Bibliography

Bögli, A. (1980). *Karst hydrology and physical speleology*. Berlin: Springer-Verlag.

Crawford, N. C. (1996). *The karst hydrogeology of the Cumberland Plateau Escarpment of Tennessee*. Nashville: Tennessee Department of Environment and Conservation.

Ford, D. C., & Ewers, R. O. (1978). The development of limestone cave systems in the dimensions of length and depth. *Canadian Journal of Earth Sciences, 15*(11), 1783–1798.

Ford, D. C., & Williams, P. W. (1989). *Karst geomorphology and hydrology*. London: Unwin Hyman.

Gillieson, D. (1996). *Caves: Processes, development, management*. Cambridge: Blackwell Publishers.

Klimchouk, A. B. (2007). *Hypogene speleogenesis: Hydrogeological and morphogenic perspective*. Carlsbad: National Cave and Karst Research Institute.

Klimchouk, A. B., Ford, D. C., Palmer, A. N., & Dreybrodt, W. (Eds.), (2000). *Speleogenesis: Evolution of karst aquifers* Huntsville, AL: National Speleological Society.

Palmer, A. N. (1991). Origin and morphology of limestone caves. *Geological Society of America Bulletin, 103*(1), 1–21.

Veni, G. (1997). Geomorphology, hydrogeology, geochemistry, and evolution of the karstic Lower Glen Rose Aquifer, South-Central Texas, *Texas Speleological Survey Monographs* (Vol. 1).

White, W. B. (1988). *Geomorphology and hydrology of karst terrains*. New York: Oxford University Press.

POPULATION STRUCTURE

Valerio Sbordoni, Giuliana Allegrucci, and Donatella Cesaroni

Tor Vergata University, Roma, Italy

INTRODUCTION

Population structure is an important guideline to understanding the evolution of cave-dwelling animals, because it represents the outcome of their history and adaptation as well as the groundwork for speciation in the cave environment. Population structure can be viewed from two different perspectives. Ecologists usually view the composition of a population according to age and sex of individuals, and population geneticists keep in mind the organization of genetic variation within and between populations, with special emphasis on their spatial arrangement. In this article we address the latter aspect. The advent of molecular techniques, from allozymes to microsatellites and to DNA sequencing analysis, led to significant improvements in understanding the genetics of natural populations. Particularly, the feasibility to estimate parameters, such as the amount of genetic variation, gene flow, time since isolation, provides important evidence to test alternative hypotheses on patterns of colonization and evolution of cave populations. Moreover, mitochondrial and nuclear DNA genealogies have revealed genetic partitions that can be interpreted in terms of habitat distribution and geography. These scenarios appear to be influenced by many factors intrinsic to the life histories of the organism (*i.e.*, the mating system, levels of gene flow) and by historical biogeography (*i.e.*, past distributional ranges) and demographic factors. The most obvious cause of structuring in cave populations is habitat fragmentation. However, as we will discuss, habitat fragmentation may not include the unique factor of population structuring. Generally speaking, caves reflect the historical process of habitat fragmentation quite well. Cave regions in temperate countries have been subjected to profound and repeated changes in climate and vegetation, which are particularly well documented in the Pleistocene. Cold and dry phases corresponding to glacial periods have repeatedly led to changes from mesophyt forest to mountain steppe. Consequently, previously widespread ancestral populations of several cave-dwelling species became confined to small refugial habitats, leaving relictual populations at different steps of isolation, depending on the biological properties and/or the history (i.e., time) of the organisms involved. Extrinsic causes for fragmentation can also be detected in other circumstances such as the structure and history of drainage systems, tectonic changes, or lava flow events in volcanic cave areas. In many cases, these phenomena can be dated to a reasonable extent.

Other evolutionary processes can also account for population structuring, especially where population structure is perceived at much smaller scales, even within a single cave. Double or multiple invasion of a cave habitat at different times by different populations of a given species represents a feasible mechanism. Alternatively, extrinsic factors such as habitat and resource patchiness may interact with the intrinsic biological properties of organisms, like fecundity, dispersal ability, and behavioral characters, leading to a population structure maintained by different forms of natural selection.

In this article, a series of case studies will be discussed to underscore the role of different factors in shaping population structure in a limited sample of cave organisms that have been the object of appropriate population genetic studies.

ESTIMATING GENETIC STRUCTURE

The genetic parameters relevant to investigate population structure are effective population size (N_e), observed and expected heterozygosity (H_o and H_e, respectively), genetic distance (D), amount of structuring between subpopulations (F_{ST}), and gene flow ($N_e m$, where m is the migration rate).

Effective Population Size (N_e)

The number of individuals that effectively participates in producing the next generation is named *effective population size*. Generally, the effective size of a population is considerably less than the census size. Evolutionary processes are greatly influenced by the size of populations.

Heterozygosity

Mean heterozygosity, calculated across a number of loci, is a valuable parameter used to estimate the degree of genetic variation within a population. Population structuring occurs when genotype frequencies deviate from Hardy–Weinberg expected proportions, or panmixia is unfulfilled. If either inbreeding or selection occurs, then populations can be considered "structured" in some way.

Genetic Distance

When two populations are genetically isolated, both mutation and genetic drift lead to differentiation in the allele frequencies at selectively neutral loci. As the amount of time that two populations are separated increases, the difference in allele frequencies between them should also increase, until each population is completely fixed for separate alleles. Therefore, calculation of genetic distance (D) between two populations provides a relative estimate of the time elapsed since these populations have existed as a single panmictic unit. Small estimations of distance among completely isolated populations indicate that they have only been separated for a short period of time. Alternatively, in the absence of isolation, small values of genetic distance may indicate population structure (i.e., subpopulations in which there is random mating, but between which there is a reduced amount of gene flow).

F-Statistics

F-statistics, developed by Wright (1965), represent the basic method to measure the amount of subdivision in populations. *F*-statistics can be viewed as a measure of the correlation of alleles within individuals, and they are related to inbreeding coefficients. An inbreeding coefficient is really a measure of the nonrandom association of alleles within an individual. As such, *F*-statistics describe the amount of inbreeding-like effects within subpopulations, among subpopulations, and within the entire population. In particular, the F_{ST} index (or R_{ST}, as estimated for microsatellite data) is an estimator of the amount of structuring of a population into subpopulations.

Migration

If there is no migration (gene flow) occurring between two populations or demes, eventually alternate alleles will become fixed and will reach 1. Alternatively, it has long been known that if migration, measured in terms of $N_e m$, is >1 (where N_e is the effective population size and m is the proportion of migrants per generation, or migration rate), the allele frequencies in the subpopulations remain homogenized (Wright, 1931). If, however, migration is present but $N_e m < 1$, an equilibrium based on the rate of mutation, migration, and genetic drift will be established.

Phylogeography

Recently, a relatively new discipline named *phylogeography* has been applied to investigate the principles and processes governing the geographic distributions of genealogical lineages within and among closely related extant species. Phylogeographic studies focus on understanding the contribution of historical versus contemporary ecological processes in shaping present-day species distributions. Phylogeographic inferences are based on DNA sequences sampled from the same locus in many individuals collected throughout the geographic range of a species. Statistical analyses are based on coalescence theory that employs a sample of individuals from a population to trace all alleles of a gene shared by all members of the population to a single-ancestral copy. This uses sophisticated model-driven approaches that answer specific questions for inferring population history. Such studies can provide substantially new insights into the processes responsible for shaping the spatial patterns of genetic variation within and among populations as well as their distributions.

POPULATION STRUCTURE AT REGIONAL SCALE

It is well known that troglobitic species (*i.e.*, obligate cave-dwellers) occupy very reduced distribution ranges, sometimes limited to a given karst region or even to a single cave system. Troglophilic species (*i.e.*, species able to live and reproduce in the cave habitat as well as in surface habitats) comparatively show much wider distribution ranges.

Related troglobitic species within a genus are often geographically vicariant in different karst areas or cave systems. A huge number of taxa in terrestrial and aquatic organisms, such as carabid and leiodid beetles, spiders, pseudoscorpions, millipedes, isopods, amphipods, exemplify this situation, particularly in limestone areas of temperate regions. These vicariant species represent the outcome of a geographical speciation process that was initiated with the population genetic structuring of an ancestral troglophilic species at the regional scale. As already outlined, the amount of population structure is strictly dependent on intrinsic factors, such as dispersal ability of the organism implied, and extrinsic factors limiting gene flow between populations, such as geographic distance and the extent of ecological and geographical barriers to migration. Therefore, we cannot automatically expect that different organisms experiencing similar evolutionary pathways show the same geographical pattern of population structure. Careful genetic analyses are required to reveal the occurrence of actual or potential gene exchange between populations. In turn, these analyses represent a tool to disclose historical relationships between populations and to test hypotheses on processes generating spatially structured population systems.

Potential or actual gene exchange between cave populations, and their resultant structuring at a regional scale, can be understood by illustrating study cases involving rhaphidophorid crickets and nesticid spiders living in Italy and the eastern United States. By means of allozyme polymorphisms, levels of gene flow between populations were evaluated for *Nesticus* cave spiders (Cesaroni *et al.*, 1981) and cricket populations belonging to *Dolichopoda laetitiae* from the Italian peninsula (Sbordoni *et al.*, 1987, 1991, 2000) and to *Euhadenoecus puteanus* and *E. fragilis* from the eastern United States (Caccone and Sbordoni, 1987). The enhanced degree of population fragmentation in these organisms is chiefly the result of a gradual reduction of gene flow between populations caused by bioclimatic changes that occurred in the Pleistocene. Hence, levels of gene flow ($N_e m$), as measured on the basis of the present patterns of genetic differentiation (Wright's F_{ST}), are supposed to reflect historical gene flow occurring between cave and surface populations living in a continuum of wet or mesophyllous woody environments. The amount of genetic structuring was estimated for multiple combinations of populations at different geographic scales. F_{ST} and $N_e m$ values were calculated for different groups of populations by considering increasingly wider geographic windows. The size of the geographic window for the population sampling was progressively enlarged from a minimum average population distance of 12–14 km for Italian caves and 30–40 km for

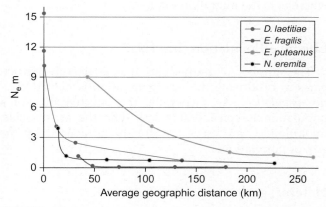

FIGURE 1 The amount of gene flow ($N_e m$), as measured on the basis of the present patterns of genetic differentiation (Wright's F_{ST}), was estimated for multiple combinations of populations at different geographic scales, by considering geographic windows of wider and wider size. The four curves outline the trends of *Dolichopoda laetitiae*, *Euhadenoecus fragilis*, *Euhadenoecus puteanus*, and *Nesticus eremita*.

American caves to a maximum average distance of larger than 200 km (Fig. 1). Moreover, in *D. laetitiae*, estimates of F_{ST} and N_em have also been performed at smaller geographic windows by means of samples caught in sites 50–500 m apart. At the same geographic scale the extent of gene flow among *E. puteanus* populations was higher than among *Dolichopoda* populations, whereas the degree of population structuring in *E. fragilis* is undoubtedly more than in both *D. laetitiae* and *E. puteanus*. Population fragmentation and geographical isolation in *E. fragilis* have already proceeded to a very large extent, because almost irrelevant gene flow is detectable at any scale. This outcome is in agreement with the well-known features of these three species. Both American species are found in caves, but *E. puteanus* also commonly occurs outside caves in forest habitats, whereas *E. fragilis* is a strictly cavernicolous species inhabiting a patchy limestone area. *Dolichopoda laetitiae* shows halfway troglophilic habits with respect to the two American species. Moreover, troglomorphic features are more advanced in *E. fragilis* than in *E. puteanus*, with *D. laetitiae* intermediate both in the degree of leg elongation and in the rate of pigment reduction. The general emerging picture corroborates the idea that gene flow with epigean populations prevents or retards adaptation to the cave habitat, and that isolation is an important prerequisite for the evolution of troglomorphic adaptations.

In the European spider species, *N. eremita*, the relation between N_em estimates and geographic distance was similar to that observed for *D. laetitiae*. *Nesticus eremita* is a widespread troglophilic species, colonizing not only natural limestone caves, but also artificial caves, cellars, and some epigean habitats, just like *D. laetitiae*. Moreover, this spider can move outside caves and disperse to some extent by means of ballooning at juvenile stages. Therefore, both present and past gene flow can explain the pattern of genetic differentiation between populations, which typically reflects a pattern of isolation by distance.

The comparison to the examined cricket species indicates that the dispersal ability for *N. eremita* is lower than that for *E. puteanus*, perhaps lesser than for *D. laetitiae*, but undoubtedly higher than for *E. fragilis*. The study cases examined here proved that different cave species, even if closely related, could show different levels of isolation, fragmentation, and dispersal, depending on both extrinsic obstacles to gene flow and intrinsic properties of organisms. In some cases, different molecular markers, used to investigate population genetic structures, or technical procedure of data analysis may lead to slightly different conclusions or may make the comparison of patterns of structuring difficult.

An example is provided by a species group of Appalachian cave spiders, *N. tennesseensis* "complex," where population genetic structure was investigated using mtDNA sequence data (ND1 gene). Hedin, in a 1997 paper, described high sequence homogeneity within populations with large divergences between populations for all species analyzed.

The revealed pattern was consistent with an ancestral interbreeding population, which, at some time in the past, was fragmented into several habitat-limited subpopulations. Subsequent to fragmentation, the exchange of migrants among subpopulations became extremely restricted to nonexistent. The conclusion was that Appalachian *Nesticus* species are characterized by similar and essentially complete population subdivision regardless of difference in species habitat preference (surface, cave dwelling, troglobitic). However, this outcome is in disagreement with many other studies on cave animals, essentially based on allozyme data, where troglobitic species showed lower rates of gene flow than troglophilic or epigean species. Hedin underlined that mtDNA, which is maternally inherited, could reveal a reduced amount of gene flow with respect to estimates based on nuclear loci. This would explain the discrepancy between levels of gene flow estimated in the American species of *N. tennesseensis* complex and the European *N. eremita*. In addition, Hedin suggested that studies on population structure based on allozymes have to take into account that populations currently exchanging no genes can still share ancestral allozyme polymorphisms, that is, the calculated extent of gene flow could be actually interpreted as the trace of historical gene flow between populations, as already outlined.

Histories of repeated isolation events leading to populations structuring and relictual populations confined in subterranean habitats are frequently invoked to explain the formation of species complexes in caves according to an allopatric speciation model. Conversely, a recent study by Niemiller *et al.* (2008) on Tennessee cave salamander complex belonging to the genus *Gyrinophilus* gave support to the hypothesis of a parapatric speciation mechanism to explain the origin of the cave populations from spring salamanders. The previous classical hypothesis, based on the species distribution, postulated that the subterranean species evolved from an epigean, metamorphosing ancestor similar to present-day *G. porphyriticus* during the Pleistocene as climatic conditions forced surface populations at the periphery of the species' range underground, thus isolating and facilitating speciation and evolution of troglomorphic characters as predicted by the allopatric, climate-relict model. Phylogenetic reconstructions based on mitochondrial and nuclear DNA sequences indicated that all three subterranean forms (*Gyrinophilus palleucus palleucus*, *G. p. necturoides*, and *G. gulolineatus*) are recently derived from the surface-dwelling species

G. porphyriticus and genealogically nested within, but an allopatric speciation model, implying the isolation of populations, should be rejected. In fact, the results of coalescent-based analysis of the distribution of haplotypes among groups supported a specification model implying a continuous or recurrent genetic exchange during divergence (parapatric speciation).

SURFACE VERSUS UNDERGROUND DISPERSAL ROUTES

How can extrinsic and intrinsic factors interact to influence patterns of structuring populations? An additional example is useful. The genetic structures of populations were compared in two pairs of prey–predator species in two regions of Kentucky: (i) the cave cricket, *Hadenoecus subterraneus*, and the carabid beetle *Neaphaenops tellkampfi*, which is a predator of *Hadenoecus* eggs; and (ii) the cave cricket, *H. cumberlandicus*, and its egg predator *Darlingtonea kentuckensis*. The two regions where the species occur show different extents of fragmentation in the karst cover, with the latter much more dissected into relatively smaller limestone fragments. This difference is expected to affect migration via underground routes more than surface routes, because substantial forest cover was almost continuous in both regions, at least in historical times. This is why population fragmentation and divergence could be strongly affected by extrinsic barriers.

Results from allozymic analyses showed a similar overall degree of genetic structuring in the two cave cricket species (average F_{ST}s were 0.58 and 0.46 in *H. subterraneus* and *H. cumberlandicus*, respectively). On the other hand, although *N. tellkampfi* populations showed a degree of genetic structuring ($F_{ST} = 0.56$) similar to cricket species, *D. kentuckensis* populations displayed much higher values of genetic structuring ($F_{ST} = 0.96$, $Nm = 0.01$), suggesting that some of them are already evolving into reproductively isolated species (Kane *et al.*, 1992).

POPULATION STRUCTURE IN AQUATIC TROGLOBITES: HYDROLOGICAL ROUTES VERSUS ADAPTIVE DIVERSITY

In aquatic organisms, hydrologic relationships appear to reflect population connectivity and gene flow much better than geographic distance. In some cases populations from different caves exposed to the same groundwater aquifer may show no trace of genetic differentiation, suggesting a single panmictic gene pool, as reported in several crustaceans and fish. In other cases, strong population subdivisions occur even over very short geographic distances.

When groundwater habitats show high levels of endemism, the most likely cause is strong hydrographical isolation and low dispersal ability of inhabiting species. Therefore, groundwater species are often constituted by highly subdivided populations and it is also not rare to find cryptic species, especially when habitat is fragmented. These general hypotheses can be better understood by describing case studies involving different taxa from Crustacea living in the United States (*Gammarus minus*), in the Balkan Peninsula (*Troglocaris anophthalmus*), and in France (*Niphargus virei*).

The Crustacean, *Gammarus minus*, has been the object of several detailed investigations by Culver *et al.* (1995). This freshwater and cave-dwelling amphipod species is found in caves throughout its range, but morphologically highly modified cave populations are found only in two relatively small regions of Virginia and West Virginia in the United States. Populations of *G. minus* reside in resurgences and related subsurface basins, both in cave stream habitats and occasional openings to the surface (karst windows). Genetic analysis of allozyme polymorphisms pointed out that the hydrological relationships among populations have largely determined the levels of genetic differentiation. Actually, cave and karst-window populations within a given basin are similar to each other. These results have been confirmed by molecular sequence data from two loci, the nuclear gene ITS-I (internal transcribed spacer) and the mitochondrial one, CO-I (cytochrome oxidase I). Molecular data also indicated (a) much lower genetic variation within cave populations than within the surface ones and (b) a codon usage bias significantly lower in caves, suggesting that the cave colonization is rather recent. Overall data suggested limited gene flow among populations and cave populations more prone to bottleneck than spring populations, likely resulting from differences such as temperature variability and drought conditions between the two types of habitats (Carlini *et al.*, 2009).

On the other hand, the analysis of morphological variation in *G. minus* generated a different pattern. In this case, populations are similar by habitat rather than by basin. Populations from cave and resurgence are markedly different, and karst-window populations exhibit a wide range of eye sizes, from very small (a troglomorphic character) to the sizes seen in resurgence populations. Both troglomorphic feature variations and genetic structure of populations suggest independent invasion of subsurface basins, and the overall similarity of eyes in cave populations relative to resurgence populations identifies a role for natural selection.

A significant correlation between allozyme and morphological distances was also found, suggesting that morphological variation among populations is largely influenced by evolutionary history. However, there is no significant path linking habitat and genetic distances, indicating that selection has little influence on genetic structure. The significant path linking habitat distances and morphological distances does indicate that selection has a strong effect on morphological variation among populations. A series of several interrelated analyses led to the conclusion that *G. minus* is actually a species complex; populations in resurgence habitats constitute one species, and populations in different basins may be as many troglomorphic species as independent isolations occurred in different basins.

The Dinaric karst area in the Balkan Peninsula hosts the richest subterranean fauna in the world with several endemisms and cryptic species as shown by different studies based on molecular data. One of these species is represented by *Troglocaris anophthalmus*, a cave shrimp distributed in a large range of more than 500 km. Morphologically, all populations are very uniform, but on a genetic point of view, different studies carried out on allozymes (Cobolli Sbordoni et al., 1990) and on molecular analyses (Zakšek et al., 2009) demonstrated that the taxon is not homogeneous but composed by different genetic units. In particular, considering the study of Zakšek et al., based on molecular sequences from mitochondrial (COI and 16S rRNA) and nuclear (ITS2) genes analyzed in several samples collected throughout the range, several (four—five) monophyletic, geographically defined phylogroups can be observed. Moreover, two of these groups showed biological reproductive isolation in sympatry. These results confirmed the nondispersal hypothesis of subterranean fauna although the southern Adriatic phylogroup showed a pattern of recent dispersal across 300 km of hydrographically fragmented karst area. The authors explain this result by combining the specific geomorphology of the Dinaric Mountains with karst hydrology. Caves inhabited by *Troglocaris* are separated by flat depressions (poljes), on insoluble ground, that are periodically flooded. During periodical floods, aquatic cave animals are frequently washed out of their subterranean habitat and they could reach caves at other parts of a polje and eventually reach a subterranean connection to an adjacent polje. In this way, flooded poljes could act as occasional stepping stones where no permanent hydrological connection exists.

A phylogeographical analysis has been carried out on the karstic groundwater dwelling amphipod *Niphargus virei*, distributed in the Netherlands, Belgium, and eastern France (Lefébure et al., 2006). Two independent molecular loci, one nuclear (28S rRNA) and one mitochondrial (COI), were used in specimens collected throughout the species range. Results from both genes support the same tripartite structure: a northern group from Belgium and the Netherlands, a central eastern group in France, and a more western group including populations from North and South France. These lineages appear to be rather ancient, being separated at least 13 Ma. From morphological point of view all populations belonging to *Niphargus virei* are extremely homogeneous although allometric variation has been observed between the two genetic units in France. These results argue for the occurrence of deep cryptic diversity in *N. virei* and demonstrate either a long morphological stasis in the evolution of the three lineages or that extensive convergent evolution occurred in this taxon. To solve this issue, a phylogenetic study including the best possible sampling of closely related taxa consisting also of surface taxa, a reconstruction of the ancestral morphology of *N. virei*, and subsequent hypothesis about evolutionary trajectories would be necessary.

Although hydrologic relationships are quite determinant for population connectivity and levels of gene flow, in some cases, a similar or even identical habitat fragmentation could produce a different pattern of population structuring and promote differently genetic divergence and speciation in relation to some features of the organisms. Guzik et al. (2009) found a quite interesting evidence of these differences by employing a comparative phylogeographical and population genetic approach to investigate the origins of a sympatric sister species triplet of diving beetles. The three species are found within a single calcrete aquifer in Western Australia and differ in size class: *Paroster macrosturtensis* is $\approx$4 mm, *P. mesosturtensis* $\approx$2 mm, and *P. microsturtensis* $\approx$1.8 mm. Previous morphological and molecular studies showed that these three taxa form a reciprocally monophyletic clade among other stygobiontic and epigean *Paroster* species; thus they are sister taxa. Two plausible hypotheses about the population structures of these beetles' species can be made depending on the role of calcrete environment in limiting gene flow between populations. First, if the groundwater matrix is the same throughout the calcrete and the beetles are highly dispersive, then their populations should be panmictic. Alternatively, if the calcrete environment is heterogeneous with the beetles dispersing locally, then patterns of strong population genetic structure caused by abiotic barriers, isolation by distance, or genetic drift should be observed. The outcome from population analyses, based on mitochondrial DNA sequence data (COI), showed that each species experimented multiple expansion events and that spatial heterogeneity in the distribution of genetic variation occurred both within and among the three taxa. A significant fine-scale differentiation with isolation by distance was found for *P. macrosturtensis* and *P. mesosturtensis*, but not for the smallest species *P. microsturtensis*. The lack of population structure and isolation by distance in *P. microsturtensis* most probably reflects its smaller size and better dispersal capability than the other two species in the calcrete environment. Such calcrete matrix, in fact, varies in physical structure and connectivity (pore size of crevices or holes) and could play a role in influencing levels of gene flow and population genetic structuring.

POPULATION STRUCTURE IN AQUATIC TROGLOBITES: HISTORICAL DETERMINANTS AND SECONDARY CONTACT

In several instances, an apparent population structuring can be determined to some extent by a genetic admixture between already differentiated gene pools. In tropical karst areas rapid evolution of drainage basins combined with tectonic events may alter connections of surface and cave streams, thus producing either isolation or secondary contact between biota adapted to different conditions. The Cueva de Los Camarones in Chiapas, Mexico, offers an excellent example of such situations (Cesaroni et al., 1992; Allegrucci et al., 1992). Two undescribed species of *Procambarus* crayfish belonging to the *P. mirandai* species group inhabit the subterranean stream. They were roughly distinguishable only by comparing extreme phenotypes, ranging from dark, thick, eyed, surface-dwelling-like individuals to light, elongate, micro-ophthalmic, cave-dwelling-like individuals. Analyses of allozyme polymorphisms and morphometry were performed to enlighten evolutionary relationships among individual crayfish and to explain patterns of microgeographic variation previously revealed along the cave stream. Results from multivariate morphometric analyses showed a real discontinuity between the two species mainly determined by the shape of the rostrum, chelae, and telson. Moreover, these same characteristics exhibited clinical variation within the less cavernicolous species. The genetic structure of the two species was investigated at 23 enzyme loci, revealing unusually high levels of heterozygosity in both species. Results of analyses on individual allozymic profiles corroborated morphometric results, yielding a genetic distance of $D = 0.26$ between the two gene pools. Due to the occurrence of alternative alleles, we could quantify patterns of introgression that reveal absence of F1 individuals and asymmetric gene flow between the two species. In the light of these data, the observed microgeographic variation in morphology within one of the two species, as well as the occurrence of aberrant phenotypes, could be interpreted as the outcome of introgression.

Studies on cave organisms based on DNA microsatellites are scantier than those based on sequencing, in part because of the technical difficulties named above. These markers are, however, very informative when it comes to understanding processes at the population level, as attested by the study published by Strecker et al. (2003) on the neotropical fish *Astyanax fasciatus* (Characidae), where both surface and cave populations are known. The authors analyzed simultaneously four cave and four surface populations; cave populations all showed extremely low genetic variability, which most likely resulted from bottleneck events. No appreciable extant gene flow was detected between surface and cave populations, the origin of which is likely the result of independent invasions from surface populations.

POPULATION STRUCTURE AND HABITAT HETEROGENEITY

The hypothesis that habitat heterogeneity affects population structure was investigated in the *D. laetitiae* population inhabiting the Cerveteri's Etruscan necropolis near Rome, Italy. This is a well-known necropolis extending throughout a roughly elliptical area of 70 ha. The necropolis includes approximately 300 tombs of different sizes and locations. Small colonies of *Dolichopoda* inhabit most of these tombs, raising the following questions: How many *Dolichopoda* populations inhabit the necropolis? Is there a case for a metapopulation (*i.e.*, a population of populations) or, alternatively, for a unique population? To what extent does this peculiar habitat structure affect the population genetic structure?

A. Sansotta and the authors of this article faced these questions by means of a long-term series of ecological and genetic investigations to measure the dispersal ability of crickets across tombs, to estimate the effective population size, to determine the amount of heterozygosity of the population, and to assess whether *Dolichopoda* genotypes are randomly distributed throughout the necropolis.

Extensive fieldwork was carried out by means of mark−recapture techniques based on individual tagging. Both allozyme polymorphisms and RAPD-DNA markers were employed to carry out genetic analysis. As a first result from this study, crickets revealed remarkable dispersal ability as they proved to move significantly across tombs to such an extent that the necropolis can be considered as the home range for a panmictic population as a whole. Second, a population sample collected across the necropolis exhibited relatively high values of heterozygosity ($H_o = 0.218$; $H_e = 0.244$)

compared to other *Dolichopoda*. Third, individuals carrying definite allele combinations (genotypes) were located preferentially in some tombs within the necropolis to such an extent that genotypes in different partitions of the necropolis were not randomly distributed, but were significantly associated with some cave ecological descriptors, namely, temperature. In summary, these crickets appeared to respond to the necropolis spatial and environmental heterogeneity by means of habitat choice expressed by different genotypes. Therefore, multiple niche selection associated to habitat choice appears to be the most probable selective process to explain the observed high level of genetic polymorphism.

POPULATION STRUCTURE AND FOOD RESOURCE HETEROGENEITY

The Frasassi Cave System is located in Central Italy, on the Adriatic side of the Apennine Mountains. This karst system is developed in at least four main horizontal levels. Each level shows a complex pattern, strongly influenced by faults and by a hydrogeological setting. The system is characterized by the occurrence of sulfide streams in some sections and by deposits of guano in others. The oniscidean isopod, *Androniscus dentiger*, is the most abundant species in different habitats in the Frasassi caves. Gentile and Sarbu (2004) studied the possible occurrence of genetic structuring in this species in population samples collected at different sites by means of F_{ST} analyses based on 18 allozyme loci. Among other samples collected throughout different caves in this system, three neighbor samples were collected within Grotta del Fiume (*i.e.*, Guano room, Green Lake room, and along a lateral sulfide section). Results indicated departures from the Hardy—Weinberg equilibrium at several loci, generally due to heterozygote deficiencies. The F_{ST} value was 0.180, indicating a level of genetic structuring comparable to values found among geographically distant populations of the same species. However, even by removing the Grotta del Mezzogiorno sample from the analysis, the amount of population subdivision remained unexpectedly high ($F_{ST} = 0.180$), especially considering the neighborhood of sampling locations within the same cave. Such an extent of population structure implies very low levels of gene flow, as shown between the Green Lake room and the lateral sulfide section, although each of these demes shows a limited gene exchange with the subpopulation from the Guano room. Interestingly enough, compared to other samples, individuals collected in the sulfide section were proved to feed on different types of resources such as chemoautotrophically synthesized food, as established by stable isotope ratio analysis. Again we are facing a situation where natural selection could be involved in promoting and maintaining population structuring in *A. dentiger* in the Grotta del Fiume.

Caves are often expected to be very homogeneous environments as a result of their physical parameters and habitat stability. This may not necessarily be true, at least in instances where localized trophic resources may convert a cave into a patchy habitat, constraining colonies to set around resources, in spite of the potential dispersal capability of individuals. If trophic resource heterogeneity remains stable over time, we could expect that genetic polymorphism is maintained by natural selection. This interpretation explains why several cave-dwelling organisms show relatively high heterozygosity levels notwithstanding their reduced population sizes.

GENETIC VARIABILITY, POPULATION SIZE, AND NATURAL SELECTION

The amount of genetic variability expressed by troglobitic populations is a controversial issue. In fish, for example, a recent study based on RAPD markers suggests that hypogean populations of balitorid fish have lower genetic variability than related surface populations. Similar results were obtained from allozymic studies in the Mexican cavefish *Astyanax fasciatus*, amblyopsid fishes from North America, and a trichomycterid species from Venezuela. The decreased genetic variation observed was reputedly consistent with the expectation that the troglobitic fish have smaller population sizes than the epigean species. Limited to single caves and cave systems, they supposedly have small population sizes. However, in a study of hypogean cyprinid fish from Somalia (Fig. 2), Cobolli *et al.*

FIGURE 2 Somali cave fish: *Barbopsis devecchii* and *Phreatichthys andruzzii*. Photos from Ercolini et al. (1982).

(1996) showed that populations of *Phreatichthys andruzzii*, a troglomorphic stygobiont species, are more heterozygous than most epigean cyprinid species studied thus far. In addition, when comparing this species with its closest relative, *Barbopsis devecchii*, a micro-ophthalmic, less specialized fish also occurring in the groundwaters of Somalia, statistically significantly higher heterozygosity values were revealed (H_e ranging from

TABLE 1 Estimates of Heterozygosity in Troglomorphic and Nontroglomorphic Cave Crickets

Orthoptera	No. Species	No. Pops.	Average H_e	Range of H_e
Noctivox	1	1	0.18	—
Longuripes	1	1	0.253	—
Hadenoecus	5	18	0.078	0.020–0.130
Euhadenoecus	4	24	0.064	0.030–0.110
Ceuthophilus	1	7	0.026	—
Dolichopoda	10	52	0.144	0.056–0.209
Troglophilus	9	19	0.06	0.000–0.178

TABLE 2 Estimates of Heterozygosity in Different Troglobitic Taxa

Crustacea	No. Species	No. Pops.	Average H_e	Range of H_e
Amphipoda				
Crangonyx	1	6	0.118	—
Gammarus	1+	8	0.108	0.075–0.130
Niphargus	4	9	0.274	0.104–0.347
Isopoda				
Androniscus	5+	34	0.102	0.027–0.178
Trichoniscus	2+	12	0.089	0.034–0.218
Stenasellus	4	13	0.079	0.000–0.196
Proasellus	1+	3	0.084	0.037–0.148
Typhlocyrolana	2	3	0.04	0.029–0.061
Monolistra	3	3	0.28	0.261–0.316
Coleoptera				
Carabidae				
Duvalius	2	3	0.114	0.100–0.141
Neaphaenops	1	8	0.192	0.173–0.222
Darlingtonea	1	10	0.009	0.000–0.040
Pseudanophthalmus	3	13	0.094	0.053–0.170
Cholevidae				
Bathysciola	1	1	0.121	—
Speonomus	5	44	0.113	0.060–0.192
Orostygia	2	2	0.165	0.133–0.198
Leptodirus	1	1	0.168	—
Ptomaphagus	1	6	0.056	0.012–0.099

0.046 to 0.062 in *Phreatichthys* and from 0.014 to 0.020 in *Barbopsis*). Due to the lack of appropriate population estimates, we cannot test the hypothesis that these differences are related to population sizes. In *Phreatichthys* it has been hypothesized that selective advantage for individuals with high heterozygosity could be involved. Negative correlations have been found in various organisms between individual heterozygosity levels and fitness components such as rate of oxygen consumption, energy requirements for maintenance. These and other metabolic features may represent adaptations to low-energy input conditions in troglobites and susceptibility to affect selective advantage to a considerable extent. Comparing population structures of *Phreatichthys* to *Barbopsis*, it was revealed that the former is strongly structured into genetically isolated populations while consistent gene flow remains genetically homogeneous in even geographically distant populations of *Barbopsis*, indicating an earlier cave isolation of *Phreatichthys* populations. On the contrary, *Barbopsis* could have experienced a longer direct contact with their epigean relatives. Because isolation in the cave habitat by means of breakdown of gene flow with epigean populations is a prerequisite to enhance adaptation of cave populations, it could be speculated that old established cave populations have a higher probability of expressing fitness-related features in their gene pools such as increased heterozygosity. The tenet that troglobites or troglomorphic cave dwellers have low genetic variability may not be generally applicable. A comparison between troglomorphic and nontroglomorphic cave crickets like *Dolichopoda* versus *Troglophilus*, *Hadenoecus* versus *Ceuthophilus*, and *Longuripes* versus *Noctivox* leads to the invariable result that the former have higher heterozygosity than the latter (Table 1). Moreover it can be seen that, among other taxa, several stygobiontic crustaceans such as amphipods and isopods as well as troglobitic beetles show evidence for very high levels of heterozygosity (Table 2).

These findings can hardly be explained by large population sizes, but they could reflect the existence of genetic polymorphisms maintained by balancing selection. Interestingly, high heterozygosity levels have been revealed mainly in small arthropods, which appear to perceive the environmental patchiness as coarse grained and therefore favorable in which to adapt to habitat and/or resource heterogeneity by means of a multiple niche polymorphism strategy.

See Also the Following Article

Life History Evolution

Bibliography

Allegrucci, G., Baldari, F., Cesaroni, D., Thorpe, R. S., & Sbordoni, V. (1992). Morphometric analysis of interspecific and microgeographic variation of crayfish from a Mexican cave. *Biological Journal of the Linnean Society, 47*, 455–468.

Caccone, A., & Sbordoni, V. (1987). Molecular evolutionary divergence among North American cave crickets. I. Allozyme variation. *Evolution, 41*, 1198–1214.

Carlini, D. B., Manning, J., Sullivan, P. G., & Fong, D. W. (2009). Molecular genetic variation and population structure in morphologically differentiated cave and surface populations of the freshwater amphipod *Gammarus minus*. *Molecular Ecology, 18*, 1932–1945.

Cesaroni, D., Allegrucci, G., Caccone, A., Cobolli Sbordoni, M., De Matthaeis, E., Di Rao, M., et al. (1981). Genetic variability and divergence between populations and species of *Nesticus* cave spiders. *Genetica, 56*, 81–92.

Cesaroni, D., Allegrucci, G., & Sbordoni, V. (1992). A narrow hybrid zone between two crayfish species from a Mexican cave. *Journal of Evolutionary Biology, 5*, 643–659.

Cobolli, M., De Matthaeis, E., Mattoccia, M., & Sbordoni, V. (1996). Genetic variability and differentiation of hypogean Cyprinid fishes from Somalia. *Journal of Zoological, Systematic and Evolutionary Research, 34*, 75–84.

Cobolli Sbordoni, M., Mattoccia, M., La Rosa, G., De Matthaeis, E., & Sbordoni, V. (1990). Secondary sympatric occurrence of sibling species of subterranean shrimp in the karst. *International Journal of Speleology, 19*, 9–28.

Culver, D. C., Kane, T. C., & Fong, D. W. (1995). *Adaptation and natural selection in caves*. Cambridge, MA: Harvard University Press.

Ercolini, A., Berti, R., Chelazzi, L., & Messana, G. (1982). Researches on the phreatobitic fishes of Somalia: Achievements and prospects. *Monitore Zoologico Italiano, 17*(9 (N.S. Supplement)), 219–241.

Gentile, G., & Sarbu, S. M. (2004). Trophic sources partition and population genetic structure of the isopod *Androniscus dentiger* from a chemoautotrophy based underground ecosystem. *Subterranean Biology, 2*, 7–14.

Guzik, M. T., Cooper, S. J. B., Humphreys, W. F., & Austin, A. D. (2009). Fine-scale comparative phylogeography of a sympatric sister species triplet of subterranean diving beetles from a single calcrete aquifer in Western Australia. *Molecular Ecology, 18*, 3683–3698.

Hedin, M. C. (1997). Molecular phylogenetics at the population/species interface in cave spiders of the Southern Appalachians (Araneae: Nesticidae: *Nesticus*). *Molecular Biology and Evolution, 14*, 309–324.

Kane, T. C., Barr, T. C., Jr, & Badaracca, W. J. (1992). Cave beetle genetics: Geology and gene flow. *Heredity, 68*, 277–286.

Lefébure, T., Douady, C. J., Gouy, M., Trontelj, P., Briolay, J., & Gibert, J. (2006). Phylogeography of a subterranean amphipod reveals cryptic diversity and dynamic evolution in extreme environments. *Molecular Ecology, 15*, 1797–1806.

Niemiller, M. L., Fitzpatrick, B. M., & Miller, B. T. (2008). Recent divergence with gene flow in Tennessee cave salamanders (Plethodontidae: *Gyrinophilus*) inferred from gene genealogies. *Molecular Ecology, 17*, 2258–2275.

Sbordoni, V., Allegrucci, G., Caccone, A., Carchini, G., & Cesaroni, D. (1987). Microevolutionary studies in Dolichopodinae cave crickets. In B. Baccetti (Ed.), *Evolutionary biology of orthopteroid insects* (pp. 514–540). Chichester, UK: Horwood.

Sbordoni, V., Allegrucci, G., & Cesaroni, D. (1991). A multidimensional approach to the evolution and systematics of *Dolichopoda* cave crickets. In G. M. Hewitt et al. (Eds.), *Molecular techniques in* taxonomy (NATO ASI Series, Vol. H 57, pp. 171–199). Berlin: Springer-Verlag.

Sbordoni, V., Allegrucci, G., & Cesaroni, D. (2000). Population genetic structure, speciation and evolutionary rates in cave dwelling organisms. In H. Wilkens, D. C. Culver & W. F. Humphreys (Eds.), *Subterranean ecosystems* (pp. 453–477). Amsterdam: Elsevier.

Strecker, U., Bernatchez, L., & Wilkens, H. (2003). Genetic divergence between cave and surface populations of *Astyanax* in Mexico (Characidae, Teleostei). *Molecular Ecology, 12*, 699–710.

Zakšek, V., Sket, B., Gottstein, S., Franjević, D., & Trontelj, P. (2009). The limits of cryptic diversity in groundwater: Phylogeography of the cave shrimp *Troglocaris anophthalmus* (Crustacea: Decapoda: Atyidae). *Molecular Ecology, 18*, 931–946.

POSTOJNA–PLANINA CAVE SYSTEM, SLOVENIA

Stanka Šebela

ZRC SAZU Karst Research Institute

PHYSICAL SETTING

Hydrology and Topographic Setting

The Postojna Cave System is the longest (20,570 m long; 115 m deep) known cave in the classic karst of Slovenia. The term *karst* is derived from this region. The Postojna Cave is connected to the Planina Cave (6.656 km long; 65 m deep) by unknown passages. There remains about 2000 m of unexplored separation, although an underground water connection has been known since the explorations of Gruber in 1781.

The Pivka River flows on impermeable Eocene flysch and sinks into the underground system of the Postojna Cave System at an elevation of 511 m on the contact with Cretaceous limestones (Fig. 1). Several kilometers downstream, the underground Pivka joins the Rak River passage in the Planina Cave System and emerges from the cave entrance to form the Unica River at an elevation of 453 m. The average flow of the Pivka at the swallow point is 5.26 $m^3 s^{-1}$. At Postojna the average annual precipitation is 1565 mm (years 1982–2000). Dye-tracing experiments in 1988 provided evidence for the bifurcation of the Pivka, for the surficial river also drains toward the sources of the Vipava River and thus forms part of both the Adriatic and the Black Sea drainage basins. The portion of the Pivka that enters the Postojna–Planina Cave System belongs to the Black Sea drainage basin.

The land surface above the Postojna–Planina Cave System can be divided into two erosional levels. The lower level is above the Postojna Cave System and reaches 632 m. The land surface is higher than 630 m above Planina Cave and can reach 753 m. It is a typical karst surface with numerous dolines and collapse dolines. There are also some horizontal caves, but vertical caves prevail and some end very close to underground Pivka River passages. The landscape was used as pasture about 100 years ago. Today it is covered by forest.

FIGURE 1 Ponor entrance of Pivka River into Postojna Cave System at 511 m above sea level. *Photograph by the author.*

Geologic Setting

Slovenia is situated at the border between the Adria microplate and the Eurasia plate and characterized by complex and neotectonically active geological conditions.

Since the late Miocene to Pliocene paleomagnetic data have indicated about 30° counterclockwise rotation of Adria microplate. During the Miocene to recent, the thrust belts along the Adria margin include the Dinaric thrust system, the South-Alpine thrust system, and the Dinaric faults. The Dinaric faults cut and displace both Dinaric and South-Alpine fold-and-thrust structures. Most of them are characterized by moderate historic and recent seismicity.

The area is part of the Javorniki–Snežnik thrust unit which has been overthrust over the Eocene flysch. The Hrušica thrust unit, which is upper Triassic dolomite, overthrusts the Javorniki–Snežnik thrust unit (Fig. 2). Overthrusting took place after the deposition of the Eocene flysch. During the Miocene and Pliocene, the overthrusting was accompanied by folding. The principal folding deformation in Postojna Cave is the Postojna Anticline (Fig. 3). Between the Postojna and Planina Cave Systems is the Studeno Syncline (Čar and Gospodarič, 1984).

It is important to distinguish older overthrusting and folding deformations from younger faulting deformations. The Postojna–Planina Cave System is situated between two regionally important faults with the NW-SE Dinaric orientation. These are the Idrija Fault on the north and the Predjama Fault on the south (Fig. 2).

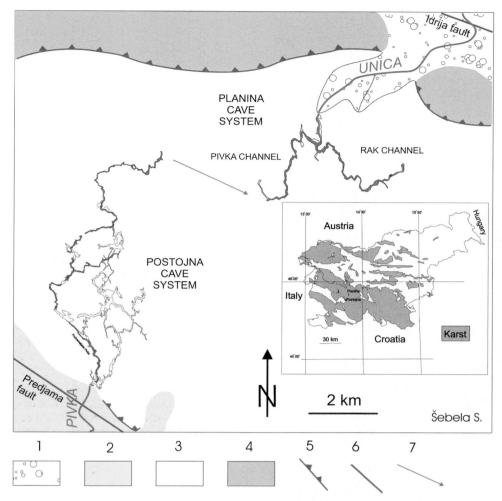

FIGURE 2 Geological position of Postojna–Planina Cave System, Slovenia. (1) Alluvium (Holocene) covering the karst polje of Panina, (2) Eocene flysch, (3) Cretaceous limestone (Javorniki–Snežnik thrust unit), (4) Upper Triassic dolomite (Hrušica thrust unit), (5) thrust, (6) fault, (7) direction of underground water flow.

The tectonic structure of the area between those two faults has all the characteristics of the intermediate zone between two dextral strike-slip faults. The cave passages of the Postojna Cave System follow the strike and dip of the bedding planes, especially those with interbedded slips. They are developed in both flanks of the Postojna Anticline. They also follow Dinaric and cross-Dinaric (NE-SW) oriented fault zones and mostly north–south oriented fissured zones (Šebela, 1998).

During the development stages of Postojna Cave System, some of the geological structures were reactivated and appear to be connected with the formation of collapse chambers. The geological structural elements were used as pathways for cave development. Along the same fault zone in the cave up to four different reactivations can be detected. The fault zone that runs from Pisani Rov and through Velika Gora shows sinistral horizontal movement in Pisani Rov, reverse fault movement at Velika Gora, and dextral horizontal movement and normal fault movement in Lepe Jame (Fig. 3). The same Dinaric oriented fault zone is monitored at two places with TM 71 extensometers. The same fault zone is cut by the cross-Dinaric oriented fault zone in Lepe Jame. This neotectonic fault zone was active post-cave development because it cuts older cave sediments which are at least 0.78–0.99 Ma in age.

On the southwestern flank of Postojna Anticline Dinaric oriented fault zones prevail, while on the northeastern flank the cross-Dinaric oriented fault zones prevail. The underground River Pivka passage follows the strike direction of bedding planes in the southern part and bedding strike dip direction in the northwestern part.

The Postojna–Planina Cave System is developed in Cretaceous carbonate rocks. The passages of Postojna Cave are developed in upper Cretaceous (Cenomanian, Turonian, and Senonian) mostly bedded and thick-bedded limestones (Fig. 4). The Cenomanian and Turonian

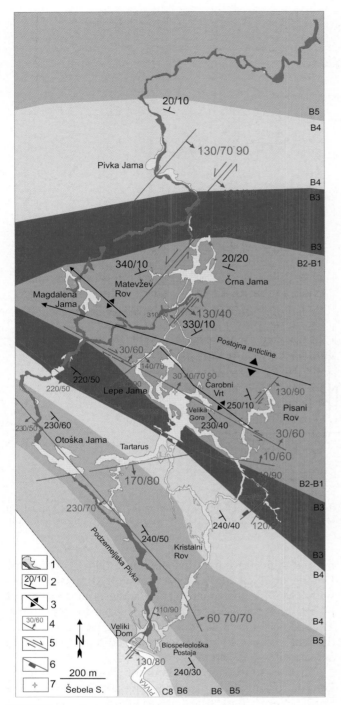

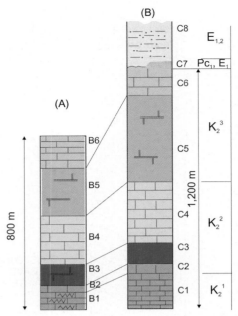

FIGURE 3 Geology of Postojna Cave System. (1) Cave passage with underground River Pivka, (2) strike and dip direction of bedding plane, (3) anticline, (4) strike and dip direction of fault, (5) dextral horizontal movement, (6) vertical movement, (7) TM 71 exstensometer, subsidence of SW block.

FIGURE 4 Lithological columns. (A) Postojna Cave System lithological column. (B1) Thin-bedded limestone with cherts (100 m), (B2) thick-bedded limestone (35 m), (B3) very thick-bedded limestone (100 m), (B4) thick-bedded limestone with rudists (225 m), (B5) very thick-bedded limestone (225 m), (B6) thick-bedded limestone (140 m). (B) Surface lithological column over the Postojna Cave System. (C1) Thin-bedded limestone (165 m), (C2) thick-bedded limestone (50 m), (C3) very thick-bedded limestone with *Chondrodonta* horizon (up to 110 m), (C4) thick-bedded limestone with rudists (290 m), (C5) very thick-bedded limestone with rudists (420 m), (C6) thick-bedded limestone (130 m), (C7) conglomerate (1–2 m), (C8) flysch.

limestones are more thin bedded and can include chert lenses. The Senonian limestones are thick bedded to massive. The cave passages are developed in about an 800-m-thick lithological column. The Planina Cave System is developed in lower and upper Cretaceous limestone and calcarenite with inliers of limestone, which in the northeast part traverses into lower Cretaceous dolomite.

Recent microtectonic deformations have been monitored continuously in 3D in Postojna Cave with TM 71 extensometers since 2004. Two instruments, 260 m apart, were installed on the Dinaric oriented (NW-SE) fault zone that is situated about 1000 m north of the inner zone of the regionally important Predjama Fault.

The locations for installation of the exstensometers in Postojna Cave were selected to evaluate if the monitored fault is tectonically active and to determine whether the tectonic activity has any influence on the speleogensis. The two devices were installed in the same fault zone but on different fault planes. The device at Velika Gora was installed at the contact point between a fault plane (dip angle 70–90° to the NE) and a collapse block, partly covered with flowstone. The second device (Lepe Jame) was installed in a narrow natural passage, which was enlarged artificially 30 years ago, between two fault planes that are about half a meter apart. The northern fault plane dips to the NE at 60° and has a horizontal striation—dextral movement. The southern one constitutes the normal fault, dipping toward the NE by 80–90° and showing vertical striation. The device at Velika Gora is 68 m below the surface and the device in Lepe Jame is 60 m below the surface.

Monitoring on both instruments has shown small tectonic movements, that is, a general dextral horizontal movement of 0.05 mm in 4 years (Velika Gora) and extension of 0.03 mm in 4 years (Lepe Jame). Between the longer or shorter calm periods, eleven extremes have been recorded regarding characteristic changes in displacement. The largest short-term movement was a compression of 0.04 mm in 7 days, detected in March 2005, which coincided with the 25-km distant Ilirska Bistrica earthquake ($M_L = 3.9$). About two months before the earthquake, an extension of 0.05 mm occurred and one month before the earthquake the strain changed into a compression of 0.05 mm. The largest permanent peak was detected at the end of 2004. Along the y-axis (Velika Gora), there was a dextral horizontal movement of 0.075 mm in one month (November 10 to December 15, 2004). After the sinistral horizontal movement of 0.02 mm (December 15–27, 2004), the y-axis retained its permanent position on 0.05 mm where it remained until today. According to speleogenesis, the monitored fault zone represents a stable cave environment (Šebela et al., 2010).

Other Interesting Research in the Postojna Cave System

In Kristalni Rov more than 20-year-long studies of percolation water from the surface through the 100-m-thick limestone cave roof are going on. Polluted waters rich in nitrates and chlorides continued for 11 years after the source removal, while sulfates and phosphates continued even longer (Kogovšek and Šebela, 2004).

In the catchment area of the Unica River, more combined tracer tests with fluorescent dyes have been performed to characterize the properties of groundwater flow and transport of contaminants through the karst system under different hydrologic conditions. During the tracer test when injection was followed by rain, the flow velocities were 88–640 m h^{-1} (Gabrovšek et al., 2010).

Since 1932 in Postojna Cave in Tartarus passage, an underground station with horizontal pendulums was in operation to measure Earth tides. The study that stopped before 1945 was important for both karst hydrology and seismology.

In the past few years, the idea to establish an underground geophysical station in Postojna Cave has been prevalent. The cave passages provide good underground conditions for such studies. First steps started in 2004 when two extensometers TM 71 were installed to measure microtectonic displacements in three dimension. Since December 2009, data on Velika Gora have been automatically taken from the TM 71 extensometer. Parallel with microtectonic displacements, the radon air measurements are organized. There are connections between extension periods and higher radon concentrations (Šebela et al., 2010).

Since July 2010, the tilt data on a vertical 20-m-long pendulum in Magdalena Jama have been collected by Czech explorers. The two-dimensional optical measurement of tilt of the rock mass and continous digital evaluation of the results are important to understand the behavior of the stress changes within the rock mass.

In the same place as the 1932 station with horizontal pendulums, the accelerometer instrument was installed within the common project with Italian colleagues in the beginning of 2010. Our aim is to establish the first underground seismological station in Slovenia with online registration.

The Postojna Cave preliminary underground geophysical station includes two TM 71 extensometers (with air temperature, air pressure, and radon monitoring), one vertical pendulum in Magdalena Jama, and an accelerometer instrument in Tartarus passage.

Paleomagnetic properties and magnetostratigraphy of karst sediments in Slovenia (Zupan Hajna et al., 2008) indicated that the age of the oldest samples ranged from about 1.8 to more than 5.4 Ma. The study indicates that part of Postojna Cave System has sediments being much older than 0.78 Ma. Sediments from Planina Cave System are younger than 0.78 Ma.

All karst caves in Slovenia are natural worths and are the property of the state. The Postojna Cave System is three-fourths managed by a private company and one-fourth by the Comune of Postojna. Both need to accomplish expert control and recommendations for management of cave system and climatic and biologic monitoring of the cave system. Expert control and recommendations for sustainable management of natural worths, formulation of suitable directives for sustainable use of natural worths, and climatic and biologic monitoring of cave system need to be fulfilled. Besides this the cave guardian collects data on old and actual scientific and popular researches of the cave system.

The first studies of the Postojna Cave underground climate were published by Crestani and Anelli in 1939. Air temperature in different parts of the cave, underground River Pivka temperature, and also temperature of the rock were obtained. The ventilation and humidity in the cave were studied. The connections between blowholes on the surface and cave passages were described.

A systematical study of underground meteorological conditions in Postojna Cave has been started in 2009 to determine human impact on natural worth. Besides two meteorological stations (air temperature, water temperature, humidity, wind direction, wind speed, CO_2), 15–20 other places have been selected in the Postojna Cave as monitoring sites where air temperature, air pressure, humidity, and wind are continuously measured. The cave is well ventilated deep inside. The average air temperature (2009–2010) is 11.10°C on the top of Velika Gora and 10.66°C for Lepe Jame (Fig. 5). For the same period the average air temperature on the surface outside Otoška Jama was 9.20°C.

HISTORY OF EXPLORATION

The oldest inscription on a cave wall is from 1213. In 1818, the native Luka Čeč discovered new parts of the cave and this is the official year of discovery of the tourist section of the Postojna Cave System. The Postojna Cave System has a long history of tourism (Shaw, 2000; 2010).

Valvasor in 1689 described the cave as one of the most remarkable caves in the world. Hacquet provided a description of the cave in his work *Oryctographia Carniolica* written from 1778 to 1779. Nagel produced the first cave map in 1748. After the first guidebook written by Agapito in 1823, Hohenwart in his 1830/32 guidebook also explained the geology of the cave. Schaffenrath provided cave illustrations. In 1854 Schmidl described the importance of the discovery of new cave passages by Luka Čeč. Schmidl is also the author of the first monograph on the Postojna Cave System in 1854. He was one of the first explorers of some parts of the underground Pivka River in the Postojna Cave System. In the Planina Cave System, he accomplished the exploration of the Pivka Passage to the sump and also part of the Rak Passage in 1850. In 1889 the first Slovene caving club, named "Anthron," was established in Postojna.

The first underground post office in Postojna Cave was situated close to Kongresna Dvorana in the years 1899–1927. In 1927 it was removed to Koncertna Dvorana.

Since 1872 the railway was in operation for tourists in Postojna Cave. A small man-powered railway was introduced for all who could afford it. In 1924 man-power was replaced by locomotive power (Fig. 6).

Permanent electric lighting has been installed since 1884, which means that Postojna Cave was the third in the world to be lighted electrically.

The French speleologist E. A. Martel in his book *Les Abîmes* (1894) included many interesting explorations and studies of the Postojna Cave System. In the beginning of the twentieth century, Perko was one of the leading names connected with the Postojna Cave System. He was a manager of the cave from 1909 to 1941.

The Speleological Institute was established in Postojna in 1929. Its first task was to find new natural caves and passages between Postojna and Planina Cave systems and to connect the caves. Gallino and Petrini surveyed the Postojna Cave System by theodolite in 1924–1934. In 1947 the Karst Research Institute of the Slovene Academy of Sciences and Arts was established. Systematic speleological, geological, morphological, and hydrological studies of the Postojna Cave System and other caves were carried out.

FIGURE 5 Lepe Jame is rich in speleotheme decorations. *Photograph taken by the author.*

FIGURE 6 The tourist railway runs under fallen stalagmite, Postojna Cave System. *Photograph taken by the author.*

DESCRIPTION OF THE CAVE

The Postojna Cave System has several known entrances:

- Ponor entrance of the Pivka River (511 m)
- Entrances from the artificial platform (529.5 m)
- Entrance to the Otoška Jama (531.7 m)
- Magdalena shaft entrance (562 m)
- Črna Jama entrance (531 m)
- Pivka Jama entrance (550 m)

Besides known entrances, there are also other, less known, accesses from the surface. Some are just widened fissures not big enough for a man to pass, representing breathing holes; others are small entrances used 100 or more years ago that are no longer accessible (e.g., Matevžev Rov). There are also some old entrances that are still completely filled with sediments as was the case with entrances east of Biospeleološka Postaja. And there are examples where the karstological processes on the surface draw the dolines or collapse dolines near the underlying cave passages as is the case with Pisani Rov.

On the area of Postojna karst there are some horizontal caves and shafts that do not have the connection with the Postojna Cave System but at least some of them probably belong to the older stages of the speleogenesis, meaning that in the past they were connected with active passages of Postojna Cave.

The area between the Postojna and Planina Cave systems was always of great interest due to the unknown underground connection between both caves. The diving explorations in 1998 and 1999 provided sufficient information to understand the underground Pivka flow toward the Planina Cave System, where about 2 km of unknown passages still exist.

Postojna Cave is developed in two principal levels. The higher level begins at an altitude of 529.5 m and is represented by dry passages (Figs. 5 and 6) which are filled with cave sediments, flowstone, and collapse blocks. The lower level (Pivka sink at 511 m) contains the underground channels of the Pivka River which are shifted more to the southwest and northwest with respect to the higher levels. Both levels are connected by side passages. The river channels of Postojna Cave are 10–20 m lower than the older dry passages. In some parts, the phreatic passages are still preserved. There are also paragenetic passages and collapse chambers. The largest collapse chambers are aligned to principal fault zones and developed in thick-bedded limestone. Velika Gora with the volume of 240,000 m^3 is the biggest collapse chamber in Postojna Cave.

Studies of the cave sediments show that there were ten principal development stages of the cave system between the first sedimentary deposits and the present. The stages have been determined by absolute dating of flowstone and the relative ages of the cave sediments (Gospodarič, 1976). The Postojna Cave System was active at least 0.99 Ma ago. Samples from a small natural passage which is accessible from an artificial passage showed reverse polarity and are at least 0.78–0.99 myr old belonging to the Matuyama Reversed Epoch.

Most of the passages in the Planina Cave System are active water channels with some higher elevation dry passages. Paleomagnetic analyses of sediment in smaller passages showed that the cave sediments were deposited during the Brunhes Normal Epoch (younger than 0.78 Ma) (Šebela and Sasowsky, 1999).

SPECIAL ATTRIBUTES

Biology

In 1831, one of the first guides in the Postojna Cave System, Luka Čeč, found the first beetle in the cave. Hochenwart gave it to Schmidt who described it as a new species *Leptodirus hochenwarti*. Because the first specimens of cave animals were described from Postojna Cave, the cave is called the birthplace of biospeleology. The first specimen of the cave amphibian *Proteus anguinus* was found in Črna Jama Cave, which is part of the Postojna Cave System, by Josip Jeršinovič Löwengreif in 1797. The only known cave vertebrate in Europe *Proteus anguinus* was otherwise first described in 1768 by Laurenti from another cave in southeast Slovenia. The Postojna Cave System is the type locality for 37 species.

Mineralogy

The Postojna–Planina Cave System is richly decorated with flowstone (Fig. 5), which is the most common form of calcite in Slovene caves. Allochthonous mechanical minerals, composed of cave sands and loams, were transported to the cave from elsewhere. These minerals reflect the composition of the original noncarbonate rock, mostly Eocene flysch.

Paleontology/Archaeology

The first traces of human settlement (remains of the cave bear *Ursus spelaeus*, cave lion, wolf, deer, etc.) in the valley of the Pivka River derive from the Postojna Cave System. Cultural layers from Postojna Cave belong to the last half of the Middle Paleolithic, which is the end of the Mousterian. In the beginning of the Würm glacial period, the presence of Paleolithic man in the cave can be proved reliably.

See Also the Following Article

(Brodar, 1969).

Bibliography

Brodar, S. (1969). Latest Palaeolithic Discoveries in the Postojna Caves. *Acta Archaeologica, 20*, 141–145.

Čar, J., & Gospodarič, R. (1984). About geology of karst between Postojna, Planina and Cerknica. *Acta Carsologica, 12*, 93–106.

Gabrovšek, F., Kogovšek, J., Kovačič, G., Petrič, M., Ravbar, N., & Turk, J. (2010). Recent results of tracer tests in the catchment of the Unica River (SW Slovenia). *Acta Carsologica, 39*(1), 27–37.

Gospodarič, R. (1976). The Quaternary caves development between the Pivka basin and polje of Planina. *Acta Carsologica, 7*, 7–13.

Kogovšek, J., & Šebela, S. (2004). Water tracing through the vadose zone above Postojnska Jama, Slovenia. *Environmental Geology, 45*, 992–1001.

Shaw, T. R. (2000). *Foreign travellers in the Slovene karst 1537–1900*. Ljubljana: ZRC Publishing.

Shaw, T. R. (2010). *Aspects of the history of Slovene karst 1545–2008*. Ljubljana: ZRC Publishing.

Šebela, S. (1998). *Tectonic structure of Postojnska jama cave system* (Vol. 18, 112 pp.). Ljubljana: ZRC Publishing.

Šebela, S., & Sasowsky, I. D. (1999). Age and magnetism of cave sediments from Postojnska jama cave system and Planinska jama cave, Slovenia. *Acta Carsologica, 28*, 293–305.

Šebela, S., Vaupotič, J., Košťák, B., & Stemberk, J. (2010). Recent measurement of present-day tectonic movement and associated radon flux in Postojna cave, Slovenia. *Journal of Cave and Karst Studies, 72*, 21–34.

Zupan Hajna, N., Mihevc, A., Pruner, P., & Bosák, P. (2008). Palaeomagnetism and magnetostratigraphy of karst sediments in Slovenia. In *Carsologica* (Vol. 8, 266 pp.). Ljubljana: ZRC Publishing.

PROTECTING CAVES AND CAVE LIFE

William R. Elliott

Missouri Department of Conservation

PROTECTING CAVE LIFE

Threats to Cave Life

Caves are often thought of as unchanging environments, but with modern instruments in the farthest reaches of large caves there are detectable, annual changes in air and water. Some caves are naturally perturbed by flooding or temperature shifts, and these events influence the types of communities found there.

Caves that flood violently usually lack truly cave-adapted species, called *troglobites* or *troglobionts*, but may have *troglophiles* (cave-loving species) and *trogloxenes* (animals, such as bats, that roost in caves but exit to feed or migrate). Indiana bats (*Myotis sodalis*) and gray bats, (*Myotis grisescens*) hibernate in the near-freezing zone of certain cold-air-trap caves, which have deep or large entrances. Such entrances allow cold air to flow inward during strong cold fronts. In contrast, in summer gray bats and Mexican free-tailed bats (*Tadarida brasiliensis mexicana*) prefer warm caves with high ceiling domes, which, combined with bat body heat, help to create an incubator effect for young bats in the summer, >30°C in grays and >40°C for free-tails.

Different animals are adapted to the natural extremes of caves and other habitats. Human-caused threats, however, can severely tax the ability of wildlife to adjust. Some of the most destructive changes to caves were brought about by quarrying and water projects that completely destroyed, or flooded caves. Human disturbance of bat roosts caused severe declines in numerous bat species (Elliott, 2000), but since 2006 white-nose syndrome (WNS) has killed more than 1 million bats in the eastern United States. Bats are discussed in more detail below.

Amphibian chytrid fungus, *Batrachochytrium dendrobatidis* (Bd), was found in southern Missouri caves by Rimer and Briggler (2010). This disease has killed many amphibians worldwide, but this was the first report from caves. Eight of 12 caves sampled were positive in all five counties visited. Five of the 7 species of salamanders and frogs sampled were positive, with an individual infection rate of 31%. Chytrid was confirmed in the grotto salamander (*Eurycea spelaea*), the first report in a troglobite. It is not known if humans spread the chytrid spores into caves, but once in an area it can spread via groundwater, humans, or pickerel frogs, which use caves seasonally. Chytrid and WNS in caves point to the need to practice "clean caving" and decontamination as a precaution against the spread of wildlife diseases.

Other pressures on cave life act over long time spans and are more difficult to measure. They include hydrological changes caused by land development, which can alter the normal hydrological cycle and increase sedimentation. For example, residential developments may cause an increase in nitrogenous wastes and sediments washed into caves. Sediments can be harmful to aquatic creatures with gills and/or soft body tissues, such as cavefishes, salamanders, and "cave snails" (eyeless, subterranean snails).

The enrichment of caves with wastes, such as ammonia and other nitrogen-based compounds, in infiltrating waters, can bring an increase in bacterial growth. Bacteria in groundwater are not killed by sunlight and they can be transmitted for many miles. Residential developments may also bring exotic species, such as the aggressive red imported fire ant, which has caused many problems in caves in Texas (Elliott, 2000).

Regional overpumping of wells can lower karst groundwater to the point where important springs and wells run dry, endangering species that live there and threatening water sources. In Texas, endangered species of salamander, amphipod, and wild rice have been affected by such trends.

Dramatic chemical spills can harm caves if the contaminants seep into streams or other routes into the cave. Once chemicals are in the groundwater, they are difficult to remove. Nutrient loss seems to happen less frequently than nutrient enrichment, but can cause severe problems (see example of Shelta Cave, below).

Extinct and Endangered Species

Although about six North American troglobitic species are thought to be extinct, it is likely that others became extinct before they could be discovered or described. Local populations of invertebrates, fishes, salamanders, and bats have been extirpated. Since some troglobitic species are endemic to a single cave or a small cluster of caves, and many caves have been disturbed, filled, quarried, mined, or polluted, it is possible that some species have disappeared recently without our knowledge.

Bats

Bats are important contributors to the world's ecological health. Caves harbor numerous bat species, which consume night-flying insects, some of which are agricultural, forest, or public health pests (Cleveland et al., 2006). In Missouri, about 800,000 gray bats, which roost in caves year-round, consume 490 metric tons of insects per year (Missouri Department of Conservation, 2010). In the tropics, bats that eat fruit and pollinate plants often roost in caves. So, even though some bats do not use caves, the bat-cave connection is important.

Declines in North American cave bats became noticeable in the 1950s. Six of the 47 continental U.S. bats are currently on the U.S. endangered species list and additional species have been petitioned for listing. The 6 are dependent on caves for part of their life cycle, and human disturbance has been the major factor in their decline. For example, Indiana bats (*M. sodalis*) lost significant numbers through disturbance of their hibernacula and improper gating decades ago. If such bats cannot hibernate deeply, they use up their body fat too fast, which results in starvation or death before spring.

Large water projects can drown caves under reservoirs, or use them as recharge wells. A recharge project caused violent flooding of Valdina Farms Sinkhole, a large cave near San Antonio, Texas. In 1987, a large flood pulse cleaned out the cave. The cave lost a colony of four million Mexican free-tailed bats (*Tadarida brasiliensis mexicana*) and a rare colony of ghost-faced bats (*Moormoops megalophylla*). A salamander (*Eurycea troglodytes*) which lived only in that cave, probably is extinct as a result.

The mining of caves for saltpeter, bat guano, or other minerals can have a drastic effect on bat colonies and other fauna. Mexican free-tailed bats have been disturbed by some guano mining in Texas, while some miners may have aided the colonies by mining out rooms that would have filled with guano. The better operations mine only in the winter when the bats are gone. The opening of large, second entrances can severely alter the meteorology of a cave, causing bats to vacate. Marshall Bat Cave, Texas, lost its free-tail colony after 1945, when a large, 40-m-deep shaft was dug into the rear of the cave to hoist out guano, causing too much ventilation and cooling of the cave.

Mammoth Cave once harbored Indiana bats before the entrance was modified to block incursions of cold winter air. The National Park Service is currently trying to reinstate the natural temperature profile of the cave.

Nutrient loss from the loss of gray bats apparently caused a domino effect in Shelta Cave, in Huntsville, Alabama. Shelta had the most diverse cave community known in the southeastern United States, but land development encroached on the cave in the 1960s and the townspeople were concerned about youths entering the cave. The cave harbored a large colony of gray bats. The National Speleological Society purchased the cave in 1967, and they moved their headquarters to a building nearby. They gated the cave in 1968 with a strong, cross-barred gate that had been taken from an old jail. This gate, in hindsight, was inappropriate for bats, which abandoned the cave within two years. Urbanization of the area probably affected the colony too. In 1981 a horizontal-bar, bat-friendly door was put on the gate, but no bats returned to the cave. The loss of bat guano to the lake in the cave probably contributed to the decline of cave crayfishes there (Elliott, 2000). The gate was replaced with a high fence around the area in the early 2000s, and some bats returned.

WNS is the most severe threat to cave- and mine-hibernating bats to date, killing more than 1 million bats of nine species from 2006 to 2010 in the eastern United States and Canada. As the fungal infection, probably originating from Europe, spread from the northeastern United States into Canada and south and west as far as Oklahoma, many agencies and landowners prepared for the worst and wrote response plans. Thousands of caves were closed or restricted to human access to avoid possible spread of the infectious, microscopic spores of *Geomyces destructans*, the apparent causative agent. Bats spread the spores, but it is not clear how much was spread on unwashed caving gear and clothing. Strict decontamination rules are required for anyone to enter many caves now, and an effective treatment is not yet in sight. It is clear that more and better cave gates and alarm systems will be needed for important bat caves, but also more research, bat monitoring, and disease surveillance. It is impossible and undesirable to gate all caves containing bats, which might number 90,000 or perhaps 100,000 caves in the United States. Even noncolonial bats are affected, such as *Perimyotis subflavans*, the tri-colored or eastern pipistrelle bat, and sites remain infectious for long periods after bats have died off. Captive holding of bats for study, treatment, or even propagation, may be the last resort for some species that are rapidly declining, such as *Myotis lucifugus*, the little brown bat. It is likely that WNS will spread to the western United States and even into the highland tropics, wherever bats hibernate in chilly caves and mines.

Many cave bats produce small amounts of guano during migration and hibernation. If WNS causes a decline in gray bat maternity caves in summer, there could be a decline in nutrient input to some cave systems in the southeastern United States. Some conservationists have proposed replacing lost gray bat guano in caves, but there are problems in that strategy, as outlined below:

1. Many caves with high biodiversity lack gray bats (Elliott, 2007), so loss of gray bat guano is not a general threat to all cave biodiversity.
2. Large-scale guano transfer is an untested idea. A search of 25,000 references in the American cave biology literature found 32 reports on guano, but none was an experimental transfer of bat guano from one cave to another.
3. Large-scale guano transfer from one cave to another might introduce infectious *Geomyces destructans* spores, whereas many are trying to practice contamination control within caves, as well as decontamination before and after entering caves.
4. Large-scale guano transfer from a gray bat cave would disrupt the source cave. Other types, such as free-tail guano, contain different fauna and flora, and it might be biochemically harmful to the recipient community of guanophiles.
5. The total energy contribution of gray bat guano to a cave ecosystem has not been measured well. Many nutrients come from rotting detritus, organics in infiltrating groundwater, bacteria, fungi, dead pickerel frogs and salamanders, cave crickets, and so on. Most of the carbon flux in two Missouri cave streams that were studied is from allochthonous sources (Lerch *et al.*, 2000, 2002).
6. Wildlife agencies and researchers do not have the manpower for large-scale guano transfer in a scientifically controlled fashion.

Further discussion of bats may be found below, in the subsections "Cave Gating Criteria" and "Cave Restoration."

CAVE PRESERVE DESIGN

Cave conservation encompasses many aspects, which can be applied to developed show caves as well as to "wild" (unmodified) caves. Show caves, however, have special problems, such as growth of cyanobacteria and plants near trail lights, accumulation of "cave lint" and trash, and general disruption of the cave's ecology. Some show caves avoid the use of trail lights but still have good trails; the visitors are provided with electric hand lanterns, and viewing native wildlife along the trail is a goal of the tour. Such show caves usually provide a more educational experience for the public.

Good cave management includes rules of access to the cave. Many publicly owned caves can be left open to the public as long as they do not vandalize or disturb cave life.

Besides WNS considerations, some caves require a permit system for entry, based on flooding hazards or other considerations of safety or sensitivity. Usually a permit or restricted access cave would be gated to control access, but appropriate signs are needed to inform people of the availability of permits (Fig. 1). Certain caves may be considered closed for recreation, but not for monitoring and research. Examples include a few caves that are especially pristine and rich in multiple resources, or which have overlapping seasons for endangered bats. Heretofore, a few eastern American caves harboring gray bats in the summer and Indiana bats in the winter were accessible between bat seasons in May and September, but now few are managed that way.

Cave preserves have been set aside for the protection of endangered, nonbat species. Too often, however, such preserves surround only the entrance area and do not include the entire extent of the cave, much less the recharge area to the cave (often called the *watershed* in American usage). It is essential to have good scientific information about the cave: an accurate map, a description, inventories of the cave's resources, and a hydrogeologic assessment. The assessment may require a dye-tracing study in which tracer dyes are put into streams, sinkholes, and other input points. The dyes are recaptured with charcoal traps placed in cave streams, springs, or wells. Maps can then be

FIGURE 1 A 2010 version of a cave sign with information on white-nose syndrome.

drawn that delineate the cave's water sources, which makes it possible to more scientifically manage the landscape.

Lack of detailed information should not stall conservation planning, however. For example, foresters in Missouri try to maintain water quality to ensure a pesticide-free food supply for gray bats. They maintain a continuous forest canopy 60 m wide along streams, in the 8 ha around and above gray bat cave entrances and as travel corridors 60 m wide from gray bat caves to riparian foraging areas. This canopy provides protection from predators and a substrate for insect production.

Buffer zones around small caves in Texas lacking streams have been as small as one hectare just for the protection of terrestrial invertebrates. It is important to maintain native vegetation and drainage patterns on the epigeum (surface). Even intermittent cave streams may have sources beyond a few hectares, so the preserve would necessarily be larger in such cases. Krejca et al. (2002) recommended a minimum preserve size of 28–40 ha (69 to 99 acres) around a small cave or cave cluster, as well as maintenance and adaptive management against other threats, such as red, imported fire ants (Solenopsis invicta). If the cave preserve is adjacent to undeveloped lands, then occasional visits to the cave by raccoons and other nonresident species may continue to provide necessary nutrient inputs in the form of droppings. If the preserve is isolated by developed lands, then it probably should be larger to maintain native flora and fauna. Camel crickets and harvestmen may exit the cave at dusk and forage for carrion and feces in the surrounding area, but these arthropods may not travel more than 100 m from the cave entrance. Pesticide use is banned or limited in cave preserves to avoid poisoning the cave fauna directly or indirectly. Even a small cave may have to be protected with a strong cave gate to prevent heavy visitation and vandalism, which can alter the cave habitat. Such cave gates should be designed to freely allow bats and small animals to pass back and forth. Preserve designs are discussed in various species recovery plans and in guidelines issued by the U.S. Fish and Wildlife Service and other agencies.

Buffer zones around large caves may be more difficult to achieve. For example, Tumbling Creek Cave in Missouri has high biodiversity and a recharge area of 2331 ha. The cave was under two private properties for many years with extensive forest cover, losing streams, light farming, and cattle production. Even though the cave was managed by the Ozark Underground Laboratory, gray bats gradually declined for several reasons. In the 1990s sedimentation from poor land usage by a neighbor contributed to the decline of a cave snail unique to the cave. A drought of several years probably contributed to the decline too. Today, more of the cave's recharge area is under careful land use, but it remains to be seen if the cave snail will increase again (USFWS, 2001; Elliott and Aley, 2006; Elliott et al., 2008).

In southeast Alaska's Tongass National Forest, studies found that some old-growth forest areas had to be protected on intensely karsted terrain. The limestone was so pure that little mineral soil had developed, and trees grew out of a thin moss blanket. When clear-cut, the thin soils washed off into the numerous sinkholes, which fed cave streams, which fed salmon streams. In a sense, the karst served as three-dimensional stream banks feeding the local streams, and could be considered as stream buffers removed at a distance, and protected under existing laws and forestry standards. Some cave entrances and sinkholes received slash and runoff from logging and roads, in violation of the Federal Cave Resources Protection Act. Many of these areas are now protected from road building and timber harvest (Elliott, 1994).

Some states publish "Karst BMP" sheets, providing the "best management practices" for construction and other ground-disturbing projects on karst. These BMPs should be applied to prescribed burns and other conservation projects. Different types of karsts may require different BMPs. The essentials of a karst BMP are summarized below:

1. A karst BMP should generally describe karst geology, hydrology, and biology, and emphasize that karst resources are easily damaged by sediments, spills, dumping, and construction activities.
2. Identification of sinkholes, recharge areas, sinking streams, gaining streams, caves, and springs is needed to avoid problems. Consult geological, speleological, and state natural heritage surveys. Cave maps are useful for determining the extent of a cave. Dye-tracing studies can delineate the recharge area for a cave, sinkhole, or spring. Some karst features are hidden under certain soils.
3. Staging and storage areas with fuel and oil should be located away from karst features, and must have erosion controls. Concrete and wash water is disposed off-site. Temporary roads should not be steep, and should have runoff controls.
4. Buffer zones of at least 30 m (100 ft) should be maintained around caves, sinkholes, and springs. Sediment controls are installed upslope of buffer zones. Pesticides and prescribed burning are generally banned within buffer zones.
5. Disturbed areas should be revegetated or allowed to regrow after a project. Annual nonnative grasses such as rye or wheat may be planted in conjunction with native species to provide short-term erosion control. Areas subject to erosion may be planted with nonnative mixtures for rapid establishment and erosion control. Follow-up inspections are required and continued controls may be necessary.
6. Avoid altering drainage patterns, and properly dispose of debris and trash away from karst features. Temporary erosion controls should be removed after serving their purpose, but some permanent controls may be needed.

Cave Gating Criteria

Cave gates are steel structures built to protect cave resources by keeping out human intruders while allowing air, water, and wildlife to pass freely in and out. Cave gates have locking doors or removable bars so that authorized persons can gain access during appropriate seasons for necessary work. Poor cave gates can harm wildlife and cave resources, but good ones can improve conditions for bats and other wildlife. Cave gating is not an automatic solution to cave conservation problems and there are many reasons for not gating a cave. Technical knowledge and experience are needed to gate a cave; for example, it cannot be done properly by a general welding contractor without providing specifications, a design, and onsite supervision by an experienced cave gater. Knowledge of cave ecology, especially bats, is necessary before a gate is considered. Similar techniques are used for gating abandoned mines, which often harbor bats of great value. Some governmental agencies and cave conservation groups assist cave owners in cave gating, but first a decision guide must be followed (below).

A few rules of thumb can be followed for cave gates. Natural entrances should not be sealed, but opening a long-sealed cave can also cause problems for the cave unless some means of protection is devised. Gates should not be made of reinforcing bar, or *rebar*; it is much too weak. Chain link fences are easily breached, but can be used around sinkholes if necessary; stronger "vertical bar fences" are now preferred. Do not construct any raised footings, stone work, or concrete walls on the floor or around a gate because they can hinder air exchange and change the temperature at the bats' favorite roosts. Gates should be tailored for the wildlife and other resources in the cave. A cave gate is not a substitute for good land management, but it is a last resort.

Limited space does not allow a full discussion here of the many construction techniques that have been developed for cave gating. Cave gate designs and specifications by the American Cave Conservation Association (ACCA), Bat Conservation International (BCI), Missouri Department of Conservation, and National Speleological Society (NSS) are available on the World Wide Web and in publications.

Cave Gate Styles

Depending on the needs of the cave, the type of entrance, bats and other wildlife, the design could specify a full gate (Fig. 2), half gate or fly-over gate (Fig. 3), window or chute gate (particularly for maternal gray bats), cupola or cage gate (Fig. 4), bay window gate, enclosure, fence, or no gate at all. For example, some bat caves that may need a gate for protection are not feasible to gate for certain physical reasons. Many caves that are feasible to gate ought not to be gated because other modes of protection may work better.

It is important to rely on an experienced cave-gating expert. Leading organizations are ACCA, BCI, U.S. Forest Service, U.S. Fish and Wildlife Service, the Missouri Department of Conservation, NSS, and others. ACCA's designs were adopted by BCI and many government agencies, and have become the industry standard.

FIGURE 2 The National Cave Gating Workshop, 2009, built a bat-friendly, full cave gate at Cliff Cave, Missouri, to protect multiple resources, including endangered Indiana bats.

FIGURE 3 A large half gate or fly-over at Great Spirit Cave, Missouri, completed in 2002. The gate weighs 18 metric tons and measures 31 m wide by 5 m high. It protects gray bats (*Myotis grisescens*) in summer, Indiana bats (*Myotis sodalis*) in winter, and multiple resources in this Natural Area cave.

FIGURE 4 A cupola or cage gate over a sinkhole entrance, which allows bats to gain altitude and exit laterally, easier than flying up through a horizontal grate.

The above organizations teach annual, regional cave-gating workshops to demonstrate the proper decision-making process, design, and construction techniques for ecologically sound cave gates (Fig. 2). These gates have resulted in significant protection and increases of colonies of endangered bats, such as grays, Indianas, and others. Protection of other irreplaceable cave resources is another benefit of properly built gates. Such workshops may also lead to the formation of regional cave-gating groups from different organizations. The workshop comprises lectures supplemented by building a real cave gate under the direct supervision of cave-gating experts. Each gate is somewhat different, and various problems in funding, logistics, design, safety, teamwork, and construction must be solved.

Gates are now made stronger than in the past, but it is important to check and repair them because any cave gate can be breached by determined vandals. Designs have evolved to foil attempts to tunnel under a gate or to destroy the door or lock. Specifications vary for different gate styles. Gates are usually made of mild or modified steel "angle iron," although stainless and manganal steels may be used in corrosive environments. The latter are more expensive and are unnecessary in most applications. The service life of most steel gates may be 30 years.

Cave Gating Decision Guide

Are There Poor Reasons Not to Gate the Cave?

For example,

- Purely aesthetic objections to a gate while the cave's resources are being degraded anyway.
- It may "start a trend" toward too much gating.
- Because a few people consider themselves above the rules and may threaten the gate.

Score no points for any poor reasons not to gate.

Are There Poor Reasons for Gating the Cave?

For example,

- For fear of liability, which probably is nonexistent. Cave owners are protected by law in some states.
- For administrative convenience (instead of having a comprehensive conservation program).
- To keep wild animals or competing explorers out.

Score no points for any poor reasons to gate.

Are There Good Reasons Not to Gate the Cave?

For example,

- The gate, as designed, will not comply with current ACCA and BCI standards.
- A vigilant owner or manager lives nearby.
- Other controls can be used—road gates, signs, surveillance.
- Visitors probably will comply with a good permit system.
- Cave management experts are opposed to the gate.
- The cave gaters are inexperienced and overconfident.
- No one will commit to checking and maintaining the gate.
- Technical reasons: The entrance is too small for a proper gate (*e.g.*, half gate for gray bat maternity colony), or the environment or budget will not allow a good design.

Score one point each *against* gating if any good reasons against gating hold true.

Are There Good Reasons to Gate the Cave?

- The cave is hazardous to casual visitors and no other controls (permits and signs) are adequate.
- Endangered species inhabit the cave and can be bolstered by protection.

- The cave is a target for vandals, looters, and trespassers. A "better clientele" is needed.
- The cave has high value, is threatened, and it can best be studied and appreciated with a good permit system combined with a gate.

Score one point each *for* gating if any good reasons hold true.

Final Results

Add up the points for and against gating, and determine which seems more important. Other criteria may have to be considered.

Security Systems

Electronic technology now allows cave conservationists to deploy alarms and surveillance equipment in lieu of cave gates, or to supplement gates that are frequently attacked. Light, vibration and magnetic sensors are becoming available that can distinguish human intruders from bats and other small animals.

An option for a cave manager would be an audible alarm versus a silent alarm. The former may frighten off intruders or anger them, possibly leading to vandalism of the equipment. A silent alarm would alert an authority, who could apprehend or warn the intruders. The key question to consider in designing a silent cave alarm system is, "Who will respond, and how quickly?" If law officers or managers cannot respond within one hour to apprehend the intruders, then the alarm system probably will not be a deterrent.

There are many options for these security systems, a few of which are outlined below.

Sensors for detecting humans could include attractive objects, pressure mats, light detectors, infrared light beams, motion detectors, vibration detectors, magnetic detectors, and infrared trail cameras with invisible flash.

The cheapest intrusion detector is an attractive, inexpensive object, such as a coin placed in a known place in the cave. If it disappears since the previous visit, then the manager knows that an unauthorized person or a packrat has been there.

Pressure mats were designed for indoor use. Such mats are switches using no electrical power until they are triggered by someone stepping on them. They are not reliable sensors unless they are placed on a flat, hard surface, and then covered with soil. They can be linked together to create a security zone. It may help to place something just beyond the mats that will attract an intruder and lead him to step on the mat unknowingly. A simpler method might be a sign with a momentary switch to the system labeled "Do not touch this."

Light detectors are available in custom alarms or as data loggers, which record the time and date that a light illuminated the detector in the dark zone of the cave. Such data loggers may be used to measure the amount of traffic in well-traveled caves, or to detect if anyone has trespassed into a closed cave, but the data are downloaded later and cannot determine anyone's identity.

Custom light sensor alarms can alert a manager or law officer via radio, telephone, or satellite communication to the Internet. However, it is usually necessary to install such sensors in the dark zone, leaving the entrance and twilight zone unprotected. They require continuous electrical power, as do motion detectors, which can be accomplished with solar panels and batteries or other power sources and cables, which must be concealed.

Invisible infrared light beams can be positioned at a certain height so that bats and wildlife are unlikely to trip them, and they can be deployed in the entrance, but false alarms may occur. Arrays of infrared beams or "beam breakers" are now used to monitor elevated bat activity during WNS-related activity.

Vibration detectors can be concealed and tuned to respond only to the seismic shaking caused by a human walking, or by earthquakes. They are more effective when attached to a cave gate, but like all cave electronics, they must be hardened against the harsh, damp, mineral-rich cave atmosphere, and flooding.

All of the above sensors must be linked to a system that conveys information out of the cave, either via a cable that must be concealed, or by wireless relays. The system can then trigger an automated telephone call, radio call, or satellite signal to an authority.

Concealed video and still cameras are routinely used by law officers to obtain evidence leading to the arrest and conviction of law breakers. The image must be clear enough to identify the suspect. They are somewhat labor-intensive and vulnerable to attack if the culprit notices the equipment, but they are easily deployed to new sites as needed.

MANAGEMENT AND EDUCATION

Cave Management

A general goal of land management over a cave is to not alter the landscape much or build infrastructure, such as sewer lines, pipelines, roads, and the like, especially if the cave contains streams and species of concern. Paving over the top of a cave cuts off much of its water supply. Installing septic systems may relegate the cave to a sewer.

In prescribed burns one should insure that smoke does not enter caves, especially those occupied by endangered bats. For example, if one burns a hillside under certain meteorological conditions, a cold front may carry the smoke down the slope and into a cave, especially if it is a cold-air trap. This could be harmful for hibernating bats. Conversely, smoke could rise up a hill and into a cave that serves as a summer roost, because it may be a warm-air trap.

Cave Restoration

Many cavers and cave managers favor cave cleanup and restoration projects for multiple reasons. Graffiti removal may seem to be only an aesthetic pursuit, and it usually does not directly help restore wildlife in the cave unless toxic materials are removed. However, one has to consider how some caves become targets for vandalism. If a government agency or cave owner allows people to vandalize and litter a cave, then many visitors assume that it is all right to do so, and they will continue. Inaction is considered condoning the bad behavior. Vandalism sometimes extends to harassment or killing of bats and other wildlife. Trashy caves become training grounds for ignorant behavior, places to act up, and this behavior spreads to other caves. The same behavior is seen in illegal dumpgrounds, where a few leave some household trash on vacant land. If no one objects, others then opportunistically dump there and the problem accelerates. Signage does not seem to help at that point.

We can involve volunteers in photographing, documenting, then carefully removing graffiti. Such a conservation project helps to restore public respect for a cave, especially if we publicize the work and explain the effort. An important educational need is filled in this way. We can also take the opportunity to educate volunteers about wildlife during the project, which they appreciate.

Most graffiti that this author has seen is not historically significant, but important signatures and markings are usually left in place or else documented in a photographic file or database before removal. Certainly aboriginal and other valuable markings and art must be protected. Various rules and laws define the types and ages of markings that must be preserved in place, so it is best to check with the appropriate state historical program office about such requirements. More stringent requirements may be practiced by conservationists on their own initiative.

An example of restoration is Little Scott Cave in Missouri, managed by the Missouri Department of Conservation (MDC). This cave is near a highway, and it became a target for repeated vandalism, including extensive and gross graffiti, breakage, spray painting two bats to death, hundreds of beer cans from underground parties, and candles from occult ceremonies.

Cavers repeatedly tried to clean up the mess and notified MDC. Finally, MDC gated the cave in April 2000 to stop the cycle. MDC simultaneously instituted an easy procedure for getting access to the cave by telephone interview. MDC awarded grants to two grottoes (caving clubs) to clean up the cave using low impact techniques in which no chemicals were used. This work was noted in local newspapers and in Elliott and Beard (2000).

Before restoring a cave to a more natural state, one should consider how altered the cave is and define realistic goals for the restoration project. Is it a show cave with many years of accumulated change and little hope of complete restoration, or is it a wild cave that is not so ecologically disturbed? In a show cave nuisance species may be present: cyanobacteria near electric lights, exotic species in cave lint along trails, or epigean (surface) wildlife that is attracted to artificial food sources in the cave. The following questions should be asked before planning projects:

1. Do we know what native species should be in the cave?
2. How can we restore the cave to a more normal aesthetic and ecological state without harming the native species?
3. To what historic or prehistoric period should we restore the cave?

The historic period is an important question in ecological restoration for bats, many of which species have declined drastically within the past few decades. Some species of bat can be protected by bat-friendly gates. Some roosts may return to the "maximum past population" as measured by ceiling stains. Most colonies do not fully recover. Very large bat colonies (*e.g.*, free-tails) may not tolerate a full gate at all because of the bottle neck, traffic jam, and acoustical confusion it may cause.

It is important not to remove decayed wood and organic matter from a cave without first checking it for cave life, which may have colonized it over many years (Elliott, 2006).

No cave is ever completely restored to its former aesthetic or ecological state. Proper biological inventory and project planning will improve the success of restoration efforts.

Prioritizing Caves

Cave protection usually involves crisis management: caves and cave life that are under the greatest threat receive the most attention. Rarely do we look at caves over a large region and consider which ones deserve protection before they become degraded.

Caves can be very different, and comparing and ranking them seems difficult. In some states adequate records exist to allow cavers, managers, and scientists to evaluate and prioritize caves for protection. Ideally, a cave would receive a numerical score for each of its resources. A composite score can then be derived for the whole cave and used for ranking and prioritizing caves or karst regions. A cave could be scored for its length, depth, hydrology, biodiversity, value as a bat cave, geology, paleontology, archaeology, history, speleothems, aesthetics, recreational value, and the threats against it. There are various ways to score each aspect; for instance, for biodiversity one can consider the total number of species present, how rare or endemic the cave's troglobitic species are, the cave's importance to endangered species, and its overall biodiversity. Small numerical differences between caves are not important, and the rankings are used as a general guide.

Most states have a Natural Heritage Database, usually within a state agency, and many states have a state cave database, usually managed by a nonprofit cave survey organization. Increasingly these databases are being carefully shared among organizations for the purpose of protecting significant caves from road and land development and other threats. The locations of caves, especially those on private lands, are protected in each system so as not to draw the attention of potential intruders. Coordinates can be truncated to coarse values for mapping over large areas without revealing cave names or precise locations.

Public Education

Cave conservation includes publications, videos, and educational programs for the public. MDC and other organizations offer workshops on cave ecology for biologists and teachers.

It is not necessary for the public to become experienced cavers to achieve conservation goals, but it is helpful to inform the public about all the resources connected to caves. Two concepts are important to convey to the public: (1) caves operate on a much longer time scale than surface landscapes, and are essentially non-renewable resources. (2) Our caves have already lost so much—when we visit a cave we should give something back to it. For example, we can pick up trash, teach others about cave conservation, or advocate for the cave and the bats to those who have authority over it. Even for those who do not enjoy caves, it is important to know that bats provide ecosystem services to us in the form of insect control, and that valuable karst groundwater resources and caves are highly related.

See Also the Following Articles

Exploration of Caves—General
Recreational Caving

Bibliography

Cleveland, C. J., Betke, M., Federico, P., Frank, J. D., Hallam, T. G., & Horn, J., et al. (2006). Economic value of the pest control service provided by Brazilian free-tailed bats in south-central Texas. *Frontiers in Ecology and the Environment, 4*, 238–243.

Elliott, W. R. (2000). Conservation of the North American cave and karst biota. In H. Wilkens, D. C. Culver & W. F. Humphreys (Eds.), *Subterranean ecosystems* (pp. 665–689). Amsterdam: Elsevier.

Elliott, W. R. (1994). Alaska's forested karstlands. *American Caves 7*, 8–12.

Elliott, W. R. (2006). Biological dos and don'ts of cave restoration and conservation. In J. Hildreth-Werker & J. Werker (Eds.), *Cave conservation and restoration* (pp. 33–44). Huntsville, AL: National Speleological Society.

Elliott, W. R. (2007). Zoogeography and biodiversity of Missouri caves and karst. *Journal of Cave and Karst Studies, 69*, 135–162.

Elliott, W. R., & Beard, J. B. (2000). Cave restoration. *Missouri Conservationist, 61*, 14–16 (also available at *http://mdc.mo.gov/conmag/2000/10/cave-restoration*).

Elliott, W. R., & Aley, T. (2006). Karst conservation in the Ozarks: forty years at Tumbling Creek Cave. In G.T. Rea (Ed.), *Proceedings of the 2005 National Cave and Karst Management Symposium, Albany, NY, Oct. 30–Nov. 4, 2005.* (pp. 204–214). Albany, NY: Northeastern Cave Conservancy.

Elliott, W. R., Echols, K., Ashley, D. C., Aley, T., Leary, A., & McKenzie, P. (2008). Waterborne contaminants in Tumbling Creek Cave, Missouri. In W.R. Elliott (Ed.), *Proceedings of the 18th National Cave and Karst Management Symposium, October 8–12, 2007, St. Louis, Missouri* (pp. 107–123). Jefferson City, MO: Missouri Department of Conservation.

Fant, J., Kennedy, J., Powers, R., Jr., & Elliott, W. R. (2009). *Agency guide to cave and mine gates.* Austin, TX: American Cave Conservation Association, Bat Conservation International.

Krejca, J., O'Donnell, L., Kennedy, K., Knapp, S. M., & Shull, A. (2002). How much surface habitat is enough? Preserve design and application for cave-limited species. In G. T. Rea (Ed.), *Proceedings of the 2001 National Cave and Karst Management Symposium, October 16–19, 2001* (p. 141). Tucson, AZ: USDA-Forest Service, Coronado National Forest.

Lerch, R. N., Erickson, J. M., & Wicks, C. M. (2002). Intensive water quality monitoring in two karst watersheds of Boone County, Missouri. In G. T. Rea (Ed.), *Proceedings of the 2001 National Cave and Karst Management Symposium. Tucson, Arizona, October 16–19, 2001* (pp. 158–168). Tucson, AZ: USDA-Forest Service, Coronado National Forest.

Lerch, R. N., Erickson, J. M., Wicks, C. M., Elliott, W. R., & Schulte, S. W. (2000). Water quality in two karst basins of Boone County, Missouri (Abstract). *Journal of Cave and Karst Studies, 62*, 187.

Missouri Department of Conservation (2010). *White-nose syndrome action plan.* Jefferson City, MO: Missouri Department of Conservation.

Rimer, R. L., & Briggler, J. T. (2010). Occurrence of the amphibian chytrid fungus (*Batrachochytrium dendrobatidis*) in Ozark caves, Missouri, U.S.A. *Herpetological Review, 41*, 175–177.

U.S. Fish and Wildlife Service. (2001). Endangered and threatened wildlife and plants, list the Tumbling Creek cave snail as endangered [Emergency rule] *Federal Register, 66*, 66803–66811.

Q

QUARTZITE CAVES OF SOUTH AMERICA

Augusto S. Auler

Instituto do Carste

INTRODUCTION

A significant portion of South America comprises ancient geological units, especially in the highlands east of the Andes, where Proterozoic carbonates and quartzite host several significant caves. Quartzite rocks, being usually very resistant to weathering, tend to form the highest and most inaccessible ridges and plateaus. Poor sand-rich soil does not favor agriculture and these areas are usually little populated. Due to the difficulty of access, quartzite ridges have received limited attention from speleologists. Although the potential for major caves has been recognized for several decades, only in recent years has systematic exploration started to reveal outstanding caves.

The total number of quartzite caves is still very small when compared to the number of carbonate caves. It is likely that fewer than 1000 quartzite caves have so far been documented in South America. This is due to the lack of exploration, access difficulties as well as the smaller density of caves in quartzite rocks when compared to carbonate. Only one cave, 16.1-km-long Cueva Roraima Sur (Ojos de Cristal) in Venezuela, is known to surpass 10 km in length, but at least twenty caves are now over 1 km long. Large cave volumes have been recognized from a few sites, both in Venezuela and Brazil. Very few caves can be considered as possessing significant depth, the area containing the largest collection of deep caves being the Inficionado Peak in southeastern Brazil, where two caves are known to surpass −400 m in depth.

Many quartzite caves occur in scenic mountain ridges and plateaus, where the rocky landscape rends it as of limited commercial value and the lack of settlements made the establishment of protected areas easy. As a result, some of the more remarkable quartzite caves are within the boundaries of state and national parks.

LITHOLOGY AND OCCURRENCE

Quartzite caves occur in a number of different geological formations throughout South America. The large majority represent ancient (Late to Mid Proterozoic) fluvial and eolic sandstones metamorphosed at various degrees of intensity. Terms such as *quartzite*, *metasandstones*, and even *sandstones* have been applied to these rocks. In southern Venezuela, close to the border with Brazil and Guyana, the Gran Sabana area hosts several quartzite plateaus known by the Indian denomination of tepuis. These flat-topped plateaus have vertical walls that may exceed 1000 m in relation to the surrounding plain. Quartzite belongs to the >1.5-Gyr-old Mataui Formation that lies subhorizontally and displays limited tectonic disturbance. Annual rainfall exceeds 3500 mm. Only a few of the tepuis have been subjected to systematic speleological exploration, the best known being Roraima, Chimanta, Auyan, Kukenan, Akopan, Autana, and Sarisariñama. Roraima tepui hosts the best-known cave, Roraima Sur (also known as Ojos de Cristal), currently the longest quartzite cave in the world at 16.1 km in length (Fig. 1). Recent exploration at Chimanta tepui has revealed very extensive caves such as Cueva Charles Brewer (7.3 km long) and Muchimuk-Colibri system (8 km long), both containing 50-m-wide passages and spacious chambers representing some of the largest cave volumes in quartzite recorded so far (Fig. 2). Auyan tepui contains a vast depression that hosts the vertically significant Aonda Cave at −383 m in depth and over 2 km in length. Autana tepui was among the first to be explored in the 1970s, being home to the Autana Cave, a large tunnel that cuts through the upper third of the tepui vertical wall. The Sarisariñama plateau contains some vast circular depressions, some of which (Sima Mayor) are over 300 m deep. Several tepuis await exploration, there being considerable potential for further major discoveries. South of the tepui area, the little explored quartzite plateau of Aracás Ridge in Brazil has recently yielded at least one major vertical cave.

Proterozoic quartzite comprises much of the Espinhaço ridge, running north—south along eastern Brazil. The well-lithified rocks of the Espinhaço Supergroup comprise numerous cave-bearing quartzite units at various degrees of tectonism and metamorphism. The 1.1-Gyr Chapada Diamantina Group in south-central Brazil is characterized by both conglomerate and quartzite, giving rise to tabletop mountain scenery similar to that in the tepui area, although much smaller in scale. Torras Cave, 3.8 km long, is one among a series of major caves within and around Chapada Diamantina National Park. The very large Lapão Cave is entirely within a conglomerate containing quartzite pebbles and a silica-rich matrix. The Espinhaço ridge is over 1000 km long and many other relevant caves occur throughout its length.

In the tectonically complex Iron Quadrangle area in south-central Brazil, metamorphosed quartzite occurs together with several other rock types. Quartzites of both the Moeda Formation (2.4 Gyr) and Itacolomi Group (~1.8 Gyr) contain major caves. The most significant area is the 0.9-km^2 top of the Inficionado Peak, where several fissures open into major vertical caves (Fig. 3). Three caves are over 350 m in depth, Gruta do Centenário being 3.7 km long and −484 m deep.

The Andrelândia Group (~1 Gyr) quartzites at the Ibitipoca ridge contain a large density of caves, including the 2.7-km-long Gruta das Bromélias. In the surroundings dozens of significant horizontal or gently sloping caves occur, sometimes associated with cuesta relief.

FIGURE 1 River passage at Cueva Roraima Sur (Ojos de Cristal). *Photo courtesy of Marek Audy. Used with permission.*

FIGURE 2 Spacious river passage at Muchimuk Cave. *Photo courtesy of Marek Audy. Used with permission.*

FIGURE 3 Vertical fissures at the top of the Inficionado Peak. Some of these fissures give access to very deep and long caves such as Gruta do Centenário. *Photograph taken by the author.*

Important quartzite and sandstone caves occur in all major biomes, from equatorial Amazonia to temperate southern Brazil and semiarid northeastern South America, comprising a large variety of silicate rocks of widely different age and tectonic history. Paleozoic and Mesozoic sandstone sequences, although devoid of the tectonic processes that affected older quartzites, display similar lithological features and geomorphological processes. Sandstone caves can, thus, be discussed in the same context as quartzite caves. Devonian sandstone of the Maecuru Formation in Eastern Amazonia hosts at least four major caves over 1 km in length. Devonian to Silurian sandstones of the Alto Garças and Vila Maria Formations in western Brazil (Chapada dos Guimarães National Park and surroundings) contain two important and large caves. The Furnas Formation in southern Brazil is home to several shafts and important caves, although few tend to surpass 1 km in length.

CAVE MORPHOLOGY

Quartzite caves display a variety of patterns, some of which find equivalents in traditional carbonate karst settings (Fig. 4). Branchwork caves, with many passages containing active streams, are common. Gruta de Torras (Fig. 4A) is a good example, in which a small stream collects water from a few tributaries and runs for a few kilometers along the dip and the strike of a quartzite bed. Passages following the dip tend to form laterally wide passages with a uniform slope. Passages along the strike display a series of waterfalls followed by deep pools. Interestingly, although it is very common for quartzite cave passages to contain streams, cave entrances are rarely associated with swallets, streams originating inside the cave as localized outlets of the water table. There is no clear relationship between underground stream discharge and surface catchment areas, especially considering that many of the upstream entrances lie close to the top of the cuesta slope.

Anastomotic caves are well represented by Cueva Roraima Sur (Ojos de Cristal) (Fig. 4B). At the top of the Venezuelan tepuis, a much more friable and unconsolidated unit of the Mataui Formation is sandwiched between more hardened quartzites. This easily eroded unit allows for the development of a series of interconnected passages sometimes divided by remarkable quartzite pillars.

Another common type of quartzite cave is represented by enlarged deep fissures. Major fissure caves, such as Gruta do Centenário (Fig. 4C) and Gruta da Bocaina (3.2 km long and −404 m deep), occur at the top of Inficionado Peak (Fig. 3). They display a number of curious features not found elsewhere, such as absence of a proper ceiling (twilight being provided by the curving vertical profile of the fissures) and cave levels generated by breakdown edged in the fissures, voids being filled with smaller pebble and sand sediments, some even containing cave streams. The common presence of cave drainage in such a small area is due to the very wet climate characteristic of these high peaks.

Many quartzite caves are shallow features, closely paralleling the slope of the surface. Cueva Roraima Sur is seldom more than 20 m deep in relation to the surface. Gruta de Torras is still shallower. These caves exhibit frequent upper entrances due to

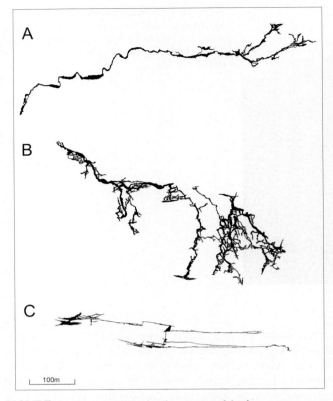

FIGURE 4 Cave patterns (at the same scale) of some representative quartzite caves. (A) Gruta de Torras, south-central Brazil, a branchwork cave with few tributaries. (B) Cueva Roraima Sur (Ojos de Cristal) in southern Venezuela, an anastomotic cave. *Map by Sociedad Venezolana de Espeleologia.* (C) Gruta do Centenário, a vertical network cave in southeastern Brazil. *Map courtesy of Grupo Bambuí de Pesquisas Espeleológicas.*

collapse and interception by surface fissures. Although some caves contain active river passages that can be followed downstream from the upper entrance to resurgence, they appear to represent perched streams. At the Venezuelan tepuis as well as at Inficionado Peak, river passages may terminate at the cliff face, hundreds of meters above the base of the mountain. At the gently sloping caves of eastern Brazil, some caves, such as Gruta de Torras, end at narrow impassable fissures, but a characteristic type of passage termination, especially common at Ibitipoca Ridge, is the "sand siphon," in which a passage not being able to transport sand produced by the disaggregation of walls and ceilings ends up being choked with sand.

Small-scale morphological features in quartzite caves may include pendants, upper levels, canyon-shaped passages, and remnants of phreatic tubes. The friable nature of the rock may prevent preservation of original speleogenetic landforms. Biokarst has also been reported for Charles Brewer Cave in the Chimanta tepui.

CONTROLS ON QUARTZITE CAVE GENESIS AND DEVELOPMENT

Quartzite tends to be a relatively homogeneous rock, comprised of quartz grains with (or without) a matrix composed of amorphous silica. A series of other minerals such as micas (biotite) and kaolinite are common constituents. Dissolution of these rocks proceeds slowly but tends to increase at higher temperatures. Amorphous silica, in particular, is quite soluble under tropical climates (>100 mg L^{-1}). The geochemistry of quartzite cave waters in many areas in Brazil and Venezuela has demonstrated that surface streams and percolation waters are very acidic but also has shown very low content of dissolved silica, suggesting that dissolution must proceed very slowly. Timescales for speleogenesis may be orders of magnitude higher than in carbonate areas. The very old nature of the rock and stable tectonic condition of most of the quartzite cave areas allow for such expanded timescales for speleogenesis.

Quartzite cave genesis appears to involve a sequential process comprising both chemical and erosional mechanisms. The most applied hypothesis involves dissolution of the matrix, prompting release of quartz grains. Since many quartzite rocks do not possess matrix, dissolution at the boundaries of quartz grains may also occur. A number of alternative mechanisms have been proposed. Removal of kaolinite has been shown to favor quartz removal at Furnas Formation in southern Brazil. Chemical alteration of micas (mostly biotite) may also provide enough space for the loosening quartz grains. Field observations in many caves in Brazil and Venezuela have shown that quartzite rock inside caves tends to be very porous, water soaked, and friable. The initial process of quartz loosening appears to create favorable conditions for erosional removal of the rock, a process known as *arenization*. Some authors have even suggested that dissolution is not required to explain the formation of quartzite caves, although most sources tend to agree that dissolution, although responsible for a small portion of rock removal, plays an essential initial role in karstification and conduit generation.

Some caves appear to have developed upstream from a scarp through sapping processes. These caves usually contain a perennial or temporary stream and tend to progressively narrow as one goes upstream.

Many quartzite caves exhibit a clear lithological control. Tepui caves develop mostly along the more unconsolidated unit of the Mataui Formation, the harder overlying quartzite providing ceiling support. At Chapada Diamantina area, some caves, such as Torras Cave, show passage control by a laterite-rich alteration zone. Kaolinite- and mica-rich layers also appear to provide a more favorable zone for conduit initiation. In general, the absence of deep phreatic passages or springs and the close relationship between surface topography and cave profiles may suggest that most quartzite caves may originate close to the water table. At present many caves, such as in the tepui area or Inficionado Peak, although containing active streams, lie well above the present water table.

Fissure caves appear to display other additional speleogenetic mechanisms. These caves tend to be concentrated close to very steep scarps, being less common at the inner zone of tepuis. They appear to be related to unloading joints due to stress release associated to the proximity to the scarp line. These caves are, thus, at least in part tectonic features enlarged by dissolutional and erosional processes. Observations at the very deep entrance shafts of Inficionado caves show that cave walls (and thus fissure width) enlarge through flaking processes, water-soaked slabs of quartzite undergoing disaggregation, and gravitational spalling. Fissures tend to become progressively narrower at depth.

Breakdown processes are very common in quartzite caves. The quartzite conglomerate Lapão Cave in the Chapada Diamantina area is a massive cave passage, now entirely filled with breakdown. In other caves, cave streams will rapidly disaggregate the breakdown and transport sand downstream.

CAVE DEPOSITS AND AGE

Clastic sedimentation in quartzite caves tends to be monotonous, comprising basically quartz sand and breakdown. The higher relief of these areas prevents the occurrence of overlying formations and soil. In areas where kaolinite forms the rock matrix, red fine-grained sedimentation may be ubiquitous. The profusion and abundance of speleothems commonly found in carbonate karst is largely absent in quartzite caves. Speleothems are mostly represented by small opal coralloids and crusts on cave walls. Flowstone of gelatinous composition is sometimes observed. An exception is provided by the remarkable speleothems found in some of the caves in the tepui area, especially at the Chimanta tepui. Large globular or filamentous speleothems are unlike anything described elsewhere. Genesis is interpreted as being associated with condensed cave moisture mediated by biological processes. Some speleothems have been described as stromatolites formed by silicified filamentous microbes (either bacteria or cyanobacteria). These opalline-laminated speleothems have been successfully dated by the U-series method. Their slow growth and chemical composition may yield useful paleoclimate information.

Low rates of dissolution, low content of dissolved silica in the water, and the old age of the landscape have led to the widespread assumption that quartzite caves may be extremely old. This may well be the case, although direct dating evidence is lacking. On the other hand, erosional processes in unconsolidated quartzite may proceed extremely fast once an underground route is wide enough to evacuate the frequent sand residue. Initiation of quartzite caves may proceed in timescales an order of magnitude slower than in carbonate areas, although cave evolution may be quite fast.

Bibliography

Aubrecht, R., Brewer-Carias, C., Smída, B., Audy, M., & Kovacik, L. (2008). Anatomy of biologically mediated opal speleothems in the world's largest sandstone cave: Cueva Charles Brewer, Chimantá Plateau, Venezuela. *Sedimentary Geology, 203*, 181–195.

Aubrecht, R., Lánczos, T., Smída, B., Brewer-Carias, C., Mayoral, F., Schlog, J., et al. (2008). Venezuelan sandstone caves: A new view on their genesis, hydrogeology and speleothems. *Geologia Croatica, 61*, 345–362.

Galán, C., Herrera, F., & Astort, J. (2004). Génesis del Sistema Roraima Sur, Venezuela, com notas sobre El desarrollo del karst em cuarcitas. *Boletín de la Sociedad Venezolana de Espeleologia, 38*, 17–27.

Lundberg, J., Brewer-Carias, C., & McFarlane, D. A. (2010). Preliminary results from U-Th dating of glacial–interglacial deposition cycles in a silica speleothem from Venezuela. *Quaternary Research, 74*, 113–120.

Martini, J. E. J. (2000). Dissolution of quartz and silicate minerals. In A. Klimchouk, D. C. Ford, A. N. Palmer & W. Dreybrodt (Eds.), *Speleogenesis: Evolution of karst aquifers* (p. 527). Huntsville, AL: National Speleological Society.

Piccini, L., & Mecchia, M. (2009). Solution weathering rate and origin of karst landforms and caves in the quartzite of Auyan-tepui (Gran Sabana, Venezuela). *Geomorphology, 106*, 15–25.

Wray, R. A. L. (1997). A global review of solutional weathering forms on quartz sandstones. *Earth Science Reviews, 42*, 137–160.

R

RECREATIONAL CAVING

John M. Wilson
Marks Products, Inc.

RECREATIONAL CAVING DEFINITION

Recreational caving may be defined as the activity of entering a void, such as a cave, for the pure joy of the activity. When entering a cave, cavers would expect it to be totally dark, even during normal daylight, for the experience to be truly considered as caving. Most cavers generally agree that the definition of *cave* is "an underground void large enough for a person to get into." Although some measure of fun can be had checking out small cave-like openings, true recreational caving involves exploring longer cave passages, which also might include manmade voids such as mines and tunnels. A few deep pits and caves have long, relatively straight entrances that allow daylight to penetrate the entire cave, but cavers still consider these to be valid caving caves. Deer Cave in Malaysia and Neversink Pit in Alabama are examples of exceptions to the total darkness rule. The information in this article is based on caving in true recreational caves or voids.

Recreational caving as defined here includes sport caving as well as relatively easy cave recreation that might involve only walking passages and visiting caves in large groups. Recreational caving may overlap with more avocational or professional caving purposes when recreational cavers add speleological activities to their cave visits or *vice versa*. These avocational and professional activities may include photography, mapping, and various cave-related scientific specialties. Cave activities or caving have become a subordinate occupation for avocational cavers, who give many volunteer hours for the joy of it, as material compensation for caving is rare.

TYPES OF CAVES USED

The following five types of natural caves used in recreational caving are ranked by popularity, based on information obtained from cave publications and other sources. There are significant differences in cave popularity in some geographic areas, depending on the types of caves available for caving.

Solution Caves

Limestone and other solution caves are by far the most used by cavers for recreation. They are formed by several types of solution processes described elsewhere in this book. Availability is the most important factor in determining the popularity of types of recreational caves in a given geographic area. This is why limestone caves top the list of popular caves. This group also includes caves formed in marble, dolomite, and gypsum. These three types make up a small part of the recreational caving in solution caves.

Lava Caves

Lava caves, which are often referred to as *lava tubes*, are the second most common type of cave used for recreational purposes. Cavers cave in lava tubes as well as lava voids that are not in tube form. These caves result from flowing lava and other processes described elsewhere in this book. Many lava caves have relatively flat floors and multiple entrances, making them suitable for those with limited caving skills. Lava caves do have their own special risks. They often have dark, very rough-textured surfaces with sharp edges that absorb light, reducing the effectiveness of caving light systems. Many novices have been amazed to discover the extent of torn clothing after traversing small passages in lava caves, in contrast to an equivalent-sized passage in a limestone cave.

Sea Caves

Sea caves are formed primarily by wave action at the shores of oceans and large lakes. They are the most common type of erosion cave. These caves are less common than limestone or lava caves and are always located at past or present sea levels. The relative predictability and attractiveness of their locations make sea caves the third most popular type of cave for recreational caving. Some sea caves have considerable incidental visitation by people primarily doing other forms of recreation. Many sea caves have no totally dark areas; however, this apparent deficiency is more than made up for by other factors. The interaction of tides and waves and the often abundant variety of life forms in and near sea caves are appealing to cavers who would not normally visit these comparatively small caves if they were not next to the ocean. Some partially flooded caves at sea level were formed primarily by solution processes or flowing lava. Technically, these are not sea caves but are solution or lava caves. Recreational cavers may not be able to make this distinction, and visitation information in some reports indicates visits to sea caves when it was in fact that caving occurred in solution or lava caves.

Talus and Tectonic Caves

Talus caves are formed when very large rocks fall from mountains or cliffs. The spaces between the rocks are sometimes large enough to allow a person to enter. Tectonic caves are formed by tectonic processes along faults near the Earth's surface. Few people use talus and tectonic caves for recreation, as they are usually small and have few caving features of interest to most recreational cavers, making these the fourth most popular type of cave. Little documented caving occurs in these types of caves; therefore, ranking them separately is not statistically meaningful.

Glacier Caves

Caves are formed in glaciers and are the least popular type of the five natural voids used for recreational purposes. These caves are cold, wet, and unstable. Specialized skill and equipment are necessary to explore ice caves successfully. These caves are located only in glaciers and icefields; thus, they are rare and remote from population centers. Sometimes lava caves with permanent ice deposits are referred to as ice caves, but these ice deposit caves are still considered lava caves, despite the common mistake of referring to them as ice caves.

Mines and Tunnels

These and other voids in the Earth created by people are rarely used for recreational caving purposes because of their instability. In terms of absolute visitation, mines are used by mine workers far more intensively than caves are used by recreational cavers. Miners use some of the same equipment as cavers. Once abandoned, mines tend to become unstable, and owners are often obligated to construct gates or other barriers and post entrance restrictions. Among the few interesting exceptions to these generalizations about mines is the Wieliczka Salt Mine near Cracow, Poland. Part of the mine is developed as a show cave, but cavers should not expect to go caving in the mine, as visitors are limited to the show cave area.

OTHER VOIDS AND URBAN CAVING

Some manmade voids are occasionally used by people for caving. They might even use some of the same type of equipment, but in most places it is not recommended by caving organizations due to safety and legal considerations. As with all caving, the permission of the property owner should be obtained before entering. Most owners of abandoned mines and buildings will not grant permission for recreational purposes. However, some urban areas have mines, aqueducts, tunnels, and quarries that have become popular with people for recreation. The location of these voids near major population centers probably contributes to their use. One of the best examples is "Les Carrieres de Paris" or Catacombs of Paris. These tunnels under a portion of the city of Paris were originally Roman-era quarries that were modified in the eighteenth century as mass tombs to resolve severe overcrowding in cemeteries. Beside human bones, visitors will find considerable upscale graffiti in an extensive underground maze.

ACTIVITIES NOT CONSIDERED CAVING

Some activities that appear to be caving are usually not considered recreational caving by cavers. Visiting a show cave and show caving are activities done by many more people than recreational caving. Some surface activities, such as looking for caves (which cavers call *ridge walking*), are not considered caving in this section. This section presents a more precise description of traditional recreational caving by not including data from show caving and cave-related surface activities.

Shelter Caves

Shelter caves are small, naturally occurring voids that may be formed by any of the previously described cave formation processes. In addition, aeolian caves formed by wind are usually considered shelter caves. Many people visit shelter caves as an incidental part of other surface recreational activities, so there is little documentation of their visits. Because shelter caves are penetrated by daylight and require no caving equipment, visiting them is usually not considered recreational caving.

Show Caves

Show caves have manmade improvements that allow for easier passage; thus, almost anyone can go into one of these caves without any special skills or equipment. Show caves are managed as a commercial business that charges an admission fee or are maintained by a governmental agency, often as part of a park system, and may also require a user fee. Show caving is not considered caving by most cavers because of the cave improvements. The major disqualifying improvement is the artificial light provided, usually as a permanently installed electric light system or as a communal light carried by a tour guide. Some show caves have supplemental offerings and are referred to as *wild caving*. These cave trips may or may not be led by a guide. In either case, wild caving in show caves is considered recreational caving.

Manmade Show Caves

Manmade show caves are usually built to provide a hands-on educational experience for the public. The historic cave paintings in Lascaux, France, are shown to the public in a manmade cave that was created because the originals were being threatened because of changes in environmental conditions associated with visitation. Some such manmade structures do include the kinds of caving challenges that appeal to some recreational cavers and provide an opportunity to practice caving skills. Visiting these manmade show caves might be considered recreational caving, depending on the specific caving conditions.

Abandoned Buildings

Little information is available on the recreational exploration of these types of structures, perhaps because it is often illegal. Manmade surface enclosures usually do not appeal to the traditional caver, although some of the same caving skills and equipment are used by the few people who do enter these structures. Most cave organizations do not recommend caving in abandoned buildings because of structural instability and legal considerations.

Factors Contributing to Recreational Cave Visitation

Some caves become very popular, with visitation averaging several hundred people a week; however, most caves rarely see humans, and the few who do enter have other purposes, such as mapping or research. Cave popularity can be affected by word of mouth, the media, availability of other recreational options, and cultural changes. Recent increases in the popularity of other forms of recreation and extreme sports may have provided alternatives for potential cavers; thus, the number of recreational cavers is most likely lower than it might be otherwise. Cave visitation can be influenced by the following six criteria, listed in a plausible order of importance:

- *Known existence.* The cave must be known to exist, and such knowledge is often spread by those taken to caves by previous visitors. The location of a cave can also be spread by word of mouth and by various print and electronic media. Keeping the location of a cave secret, at least in the short run, has significantly delayed the advent of recreational caving to these caves.
- *Physical access.* For visitation, a cave must have physical access. Factors affecting physical access to a cave might include gates, barriers, or proactive access control, as well as caves being difficult to find or requiring long hikes from the nearest road to reach the entrance. Other physical factors include vertical entrances, which reduce visitation of nonvertically equipped cavers to near zero but attract those with vertical caving capabilities. Water barriers or in-cave sumps usually keep all but properly equipped cave divers from entering.
- *Distance to cave.* With all other factors being equal, the farther a cave is from a given starting point, the less likely it is to be visited. The percentage of the population who has caved is higher in communities located in karst areas than in areas more distant from suitable caves; however, even moderate distances may change the nature of how people organize in order to cave. It appears that cavers living in urban areas tend to organize more effectively, perhaps to accommodate their transportation needs; thus, these organized cavers may continue to remain actively involved longer than those people who live close to caves but are not organized into a cave group. It is true that many very experienced cavers travel all over the world to go caving; however, most

recreational cavers are not experienced and will go on only a few cave trips in their lives. Recreational cave trips tend to be to caves that are relatively nearby. It is reasonable to assume that a higher percentage of existing caves have been discovered in areas that are easy to access.

- *Caver appeal.* Popular caves have one or more interesting features that appeal to the caver, such as significant size and complexity, aesthetics, unique or unusual geologic features, interesting biota, fun caving features, caving challenges, or cultural mystique. The attractiveness of a cave to cavers often depends on the degree to which cavers can do the things they like to do in a cave, such as climbing through complex passageways, crawling into different parts of the cave, and exploring and seeing interesting cave features.
- *Suitability for group caving.* Almost all recreational caving is done in groups of 3 to 30 people. Caves without places where people can stop and socialize or at least communicate are noticeably visited less often by recreational cavers.
- *Legal access.* Anarchy exists in a few places in the world, and social deviance occurs to some degree in all cultures. These factors partially account for the fact that the single act of legal posting or closing of a cave is not always effective. Cave owners usually take additional actions if they intend to achieve total compliance with their cave restrictions. Some posted caves with little or no enforcement may continue to have cave visitation; the extent of the trespassing depends on the respect for property rights among the area's cave visitors. All respected cave organizations recommend that cavers should never trespass in a cave that has been gated or posted. Serious consequences of ignoring such postings include lawsuits, serious injuries, and fatalities, partially due to spelunkers disregarding cave closings.

WHY PEOPLE BEGIN CAVING

There is no single reason why people enter caves. Attempts to sum up caving using a short, catchy phrase have resulted in oversimplified explanations. People who cave do so for many different reasons. Non-cavers may find it difficult to imagine why anyone would crawl in the mud, climb in the dark, and go into tight places. Most people start caving to have an enjoyable experience. The most common reasons why people cave are listed below. They apply to novice as well as avocational cavers.

- *To enjoy the company of others in a fun group activity.* The social aspect is a significant factor for novice recreational cavers. Schools, youth groups, and community groups often sponsor cave trips.
- *To explore or have an adventure.* (Zuckerman, 1994) Maze exploration is considered great fun by many. The excitement of cave exploration can be compared to exploring a complex maze. The curiosity to see what is around the bend in the passage or beyond the extent of one's light leads the caver on. The increased adrenaline and other hormone production in the stimulating cave environment provides a pleasurable sensation. Some people enjoy this experience and will return for more. Psychologists (Archer and Birke, 1983) have provided strong evidence from studies of animals and people that exploratory behavior is an innate characteristic. If so, this behavior may explain why people cave; however, it does not explain why they cave as opposed to choosing other forms of exploration. It is unlikely that an innate exploratory need completely accounts for caving behavior, as most people explore other things and never experience the inside of a cave.
- *To accomplish something unique, to see things that most people have not seen or done, or to learn something about caves and nature.* For example, some cavers have the goal of discovering a "virgin" cave, a cave passage thought to have had no prior entry by humans.
- *To engage in a physical or sporting activity for the personal challenge or just for the exercise.*

THE OCCASIONAL DOWNSIDE

Sometimes a person goes on a cave trip with mistaken assumptions, believing that it will be an exciting adventure full of glamorous achievements and revealing a hidden world of wondrous vistas, not to mention the possibility of finding something of value such as a hidden treasure. It is a considerable disappointment when the new caver discovers reality. It is often hard work getting to and through caves. Much of the time is spent traversing passages that are anything but glamorous, and the wondrous sites are few and far between. No crowds are cheering the caver on. Actually, only a few other people care about these exploits. On top of all that, there is no buried pirate or other kind of treasure in caves, and in most places laws prohibit the removal of mineral formations and artifacts. In fact, most cavers help enforce these laws, and the National Speleological Society (NSS) even offers a reward of up to $1000 for the arrest and conviction of anyone guilty of cave vandalism (check with the NSS about the current status of this award). It is not unusual for a new caver to emerge from a cave wet, muddy, bruised, and tired with sore muscles he did not know he had and nothing much to show for his efforts, with memories being mostly of mud and rock seen from a distance of 20 cm while crawling. He has probably decided this will be his first and last cave trip. In fact, he has a lot of company, as between one fourth and one half of the people who have gone caving once will never go on another cave trip. This figure does vary, depending on the type of cave trip in which the person participated and the types of future cave trips available. Cave trips that are competently led in caves appropriate for the first-time caver will lead to higher repeat caving rates.

NOVICE CAVERS

Since the founding of the National Speleological Society in 1941, there have been significant improvements in caving techniques, equipment, and safety procedures. Many recreational cavers are among the best-equipped and most experienced cavers. As a result, recreational cavers have penetrated farther into caves and coped with more challenging situations.

Inexperienced cavers sometimes try to emulate competent recreational cavers with bad results when they exceed their skill level and have inadequate equipment for the situation. Novice cavers make up a significant portion of all recreational cavers and sometimes can be identified by their inappropriate equipment, misdirected motivation, or lack of cave knowledge. Some novices who have not had a good orientation and/or have had no contact with the mainstream of the organized caving community may do strange and sometimes dangerous things in caves. For example, while there may be some instances when the following items might be appropriate in a cave, the possession of some of these items in a cave and the kinds of reasons given for having them usually identify the person as being either unprepared for cave trips or someone who has plans to vandalize the cave. In general, one should not take any of the following into a cave on a recreational trip:

- Large knives for killing snakes, bats, or other wildlife
- Guns for shooting bears or outlaws
- Handheld flashlight or candle as the primary or only light source
- Inadequate or unreliable light sources or no light source
- Clothesline or other light-duty rope for rappelling down a drop or pit
- Hemp rope for climbing hand over hand out of a pit
- Balls of string to unwind and mark one's route in order not to get lost
- Hand-carried coolers for beverages, especially alcoholic beverages
- Alcoholic beverages or mind-altering drugs
- Paint, especially a can of spray paint
- Equipment for collecting cave mineral formations

Except when it is clear that such equipment has been brought along for certain professional or ceremonial activities, observing someone with any of this equipment should cause enough concern to take appropriate action. Some cavers might try to advise such a person of the inappropriateness of the items and recommend corrective action before going caving. Cave conservation and appropriate equipment and techniques are discussed in other chapters of this book.

Unfortunately, a few recreational cavers have engaged in destructive activities such as painting graffiti in a cave, breaking speleothems, stealing mineral formations and artifacts, and harming cave life. Many states and some countries have cave protection acts that provide for punishment under the law for harming caves or their natural inhabitants. Cavers will sometimes refer to cave vandals as spelunkers.

The term *spelunker* is often applied to recreational cavers. The word is derived from the Latin word *spelunca* ("cave") or the even earlier Greek word, *spelaion*. However, in the organized caving community today, comprised of those who have learned proper techniques for cave exploration and developed an awareness of the fragile nature of this underground resource, the term *spelunker* has become a derogatory term. The majority of English-speaking cavers would not refer to another caver as a spelunker, although there are a few exceptions to this rule. Two cave clubs in Missouri are composed of competent cavers with an independent spirit who refer to themselves as spelunkers.

WHY DOESN'T EVERYONE CAVE?

Some avocational cavers have wondered why everyone does not go caving, while other cavers are glad everyone doesn't. Millions of people living in North America have caved at least once. Professional and avocational cavers number in the tens of thousands. For these thousands, caving is the epitome of experiences life has to offer. The stereotypical cave trip involves crawling or wiggling through tight places, sometimes no more than 20 cm high, where mobility is severely restricted. Sometimes cavers must crawl through a stream or slide through a passage on their back, with only a few inches of air space between the water and ceiling. Some cavers thrive on these challenges; others do not. Anyone with even mild claustrophobia will be unlikely to return to this type of activity. The same is true of people with acrophobia, as many caves have floors that are little more than a series of rocks, boulders, crevices, and pits. Bats and a variety of invertebrates do not appeal to everyone. Erroneous beliefs and imaginary threats stop others from even considering going into a cave. The vast majority of the population does not enjoy these types of experiences. They either avoid exposure to such situations or accept them as a one-time experience in life. Some people enjoy a more sedentary lifestyle that is incompatible with sport caving. Most caving requires effort, sometimes strenuous effort, to participate.

Other limiting factors include vertical caves with entrance drops or pits. These entrance drops may be anywhere from a few meters to more than 300 meters deep. Special skills and equipment are needed for this type of caving. Many difficult cave rescues have been a result of poorly informed and inadequately equipped recreational cavers attempting to descend into vertical caves.

A CAVING LEVEL FOR ALMOST EVERYONE

People interested in caving are pleased to discover that they can choose caving activities appropriate to their training and interests. Caving varies widely in its intensity, required skills, and equipment. Exploring some caves requires little more effort than taking a short hike; such caving requires no special equipment beyond a good light system. Diversity is the very nature of caves, and cavers can pursue their interests in many different cave environments. Caves present a continuum of difficulty requiring more organization, planning, and stamina. Cave divers require special training and underwater breathing equipment.

Someone seeking an athletic challenge or someone who is an extreme sports aficionado can find caves suited to their goals. Cavers who search for new caves, draft cave maps, or take cave photographs have a seemingly endless opportunity to find and record nature's handiwork. A cave is usually a low-energy environment that is often isolated from outside environments and hence spawns many different life forms. Caves are a natural attraction for geologists and biologists. This variety of activity attracts people with widely varying interests, nowhere more pronounced than at a National Speleological Society annual convention, where the program includes numerous special-interest sessions, contests, and workshops stressing various cave-related skills, as well as hundreds of different presentations.

WHY DO PEOPLE CONTINUE TO CAVE?

The transition from recreational caving status to being an avocational caver or speleologist often occurs as these cavers find that caving is more interesting if it has a purpose. Caving merely to do informal exploration of a cave that many others have seen and studied before may no longer satisfy the needs of the more adventurous. A few cavers have adopted caving as a life-long avocation. The careers of most speleologists were preceded by some recreational caving. Others who were recreational cavers at one time may have stopped caving for various reasons but maintained their support of cave organizations and goals, and are involved in other cave-related activities.

The hierarchy of human needs model proposed by Maslow (1954) describes the sequence in which people focus their efforts on meeting their various needs—physiological, safety, belonging and love, esteem, self-actualization, cognitive, and aesthetic. People meet their more basic needs first and then proceed to address their higher needs. Considering the characteristics of large, complex, and scientifically interesting caves, it can be easily understood how caving meets the self-actualization, cognitive, and aesthetic needs of an avocational caver. Participation in cave-related organizations may also meet some belonging and esteem needs. A person mainly focused on basic needs and trying to survive is unlikely to participate in caving.

DEMOGRAPHIC FACTORS

Contemporary studies conducted by members of the National Speleological Society have discovered several significant differences between cavers and the general population. None of these studies has been published outside local caving newsletters, but the results are intriguing:

- *Age.* The mean age of recreational cavers, as determined by information gathered from cave registers between 1975 and 1985, was 21 years (mode, 18 years). The frequency of caving decreases significantly with increasing age. Anecdotal information obtained recently indicates that the mean age of cavers may be greater by several years.
- *Sex.* Males make up 75% of cavers and an even higher percentage of cavers on cave trips, which indicates that males have taken a greater average number of cave trips. There does not appear to be any component of recreational caving that would favor one sex over the other. Most people say that cultural factors account for this difference. While

this may be true, it is not an adequate explanation. A more explicit hypothesis suggests that caving activity is closer to traditional male activities such as outdoor physical activity, exploratory behavior, and risk taking. These activities, plus being covered with mud and dirt when caving, are not those normally associated with traditional feminine activities. A test of this hypothesis would occur in a society where the roles of women have veered away from the traditional feminine roles; if caving is primarily a culturally linked phenomenon, then one would expect the percentages of female cavers in such a society to increase.

- *Race*. In North America, Caucasians are the most numerous ethnic group, representing at least 95% and probably more of all cavers. There are far fewer African-American cavers than would be expected, given their percent of the population. There is anecdotal information that caving activity among Hispanic populations is lower than for non-Hispanic Caucasians. Recreational caving exists in most parts of the world, but data indicate that the percentage of the population engaging in recreational caving is highest in Europe and North America, although no definitive studies on this issue have been conducted. One plausible explanation that may account for some of these differences could be that recreational caving is similar to other forms of recreation that require significant leisure time and equipment. While the cost of basic caving equipment is not considered expensive by Western standards, it might be beyond the reach of many. Maslow's hierarchy of needs may also help explain this difference, as people will focus most of their attention on meeting basic needs before acting on higher needs. The hypothesis is most plausible if one assumes that in Europe and North America a greater percentage of the population is focused on meeting their higher needs, and recreational caving fulfills these needs for people who are active cavers. While Maslow's hierarchy of needs may account for some caving frequency, other factors are at work here that have not been studied in relation to recreational caving.
- *Length of cave trips*. The length of cave trips can last from a few minutes to several days. Most recreational cave trips fall within 2 to 8 hours, and the average is about 4 hours. The length of cave trips is mostly a factor of the size of the cave, difficulty traversing a particular cave, and the interest and ability of each caver.
- *Cycles in cave visitation*. Cycles in the frequency of caving do occur (*e.g.*, seasonal variations), with the warmer months being more popular than the colder ones. Because most caves have a relatively constant temperature year-round, it appears that factors other than temperature are more important in determining when to go caving. Also, weekends are generally significantly the most popular time to visit caves.

CONTEMPORARY TRENDS

The Outdoor Foundation (2010) published its most extensive survey ever on outdoor recreation. The results show a continual decline in participation in many forms of outdoor recreation from 2006 through 2009. A few activities such as adventure racing, snowshoeing, and kayaking did have significant increases in participation. Caving was not included as a category and the report does not mention caving or caves at all. However, an activity such as sport climbing could serve as a proxy for recreational caving in lieu of any other systematic study. The survey showed a modest drop in participation in the four years of the study. This study gave some of the reasons young people gave for not participating in outdoor recreational activities. They include lack of interest, no time, no access, prefer TV and movies, computer or video games, and hanging out with friends.

From 1950 to 1980, young people from the age of 15 to 30 dominated participation in caving. That demographic is relatively much less common now. This change in society away from outdoor activities is dramatically on display at many gatherings of cavers in which the older ones far outnumber the young. Historically, the vast majority of older cavers and members of groups such as the National Speleological Society started caving when they were young.

Social norms evolve over time and can be driven by technological change. Today, the ever-present Internet and other electronic media offer an alternative method for work, recreation, and finding information. The reality is that people now have more choices in how to use their time. If one wanted to learn about caves or have a visual cave experience, a case could be made that the most efficient method would be to use the Internet. For example, there are now many times more cave photos on the Internet than there were available in all printed material published more than 30 years ago.

SUMMARY

While recreational caving has parallels with other forms of recreation, its diverse nature allows people to enjoy an activity in the same places and sometimes in the same organizations as professionals with similar interests. People cave for many different reasons, at many different skill levels, and in several different types of caves requiring a variety of equipment and techniques. While young males are the most common cavers, all types of people can be found caving. People have clear preferences in the features and types of caves they choose and in the distances they choose to travel to cave. Recreational caving can be a risk to both the cave environment and the caver. The effort required to navigate through restricted passageways filled with water and mud, drops, rough terrain, and assorted other obstacles is enough to keep most people from trying it or discourage them from returning. These same features are part of the appeal of a cave and contribute to the cognitive challenges, diverse types of participation, and spectacular aesthetics associated with caves. These features may also be less unique with the advent of newer forms of recreation and electronic media which may offer alternatives that are contributing to a declined in participation in recreational caving.

See Also the Following Article

Exploration of Caves—General

Bibliography

Archer, J., & Birke, L. (1983). *Exploration in animals and humans*. New York: Van Nostrand-Reinhold.

Maslow, A. H. (1954). *Motivation and personality*. New York: Harper.

The Outdoor Foundation (2010). Outdoor recreation participation report. Boulder, Colorado. www.outdoorfoundation.org.

Zuckerman, M. (1994). *Behavioral expressions and biosocial bases of sensation seeking*. New York: Cambridge University Press.

RESCUES

John C. Hempel

EEI Geophysical

From time to time those who explore caves do not return to the surface under their own power. Trained teams of cavers take it on themselves to rescue cavers who are in trouble. Rescues are required for many reasons: there may have been an accident and a caver is injured, cavers may be trapped by rising water, and cavers may simply have lost their lights and are unable to move or become lost. Cave rescue is a highly organized activity in the United States and Europe (Hempel and Fregeau-Conover, 2001).

INTRODUCTION

Since humans began using caves for habitat, recreation, or religious reasons, accidents have been a real threat to those exploring the netherworld. Prior to electronic communication, cave rescues went almost unnoticed outside of the community in which they occurred. Since the 1920s, media coverage of cave rescues has turned unfortunate ordeals into public entertainment. Cave rescues first attracted national attention when Floyd Collins, a young caver and explorer, became trapped in Sand Cave, Kentucky, in 1925. His attempted rescue and death have been the subject of records, movies, books, and documentaries. During the attempt to free Floyd, hourly radio news reports held the nation captive for days. Newspaper articles documenting the efforts of his rescuers won a young journalist, Skeets Miller, a Pulitzer Prize.

In 1925 few individuals practiced caving as a sport and recreation in caves was a novel idea. As a fairly new sport, experienced cavers were rare and few could be found to help at Sand Cave. No one at the rescue had been trained for cave rescue emergencies and they lacked the specialized equipment to save Floyd's life. The rescue team had no idea how to help Floyd remove the rock that pinned him. Miners, farmers, relatives, and hundreds of other people came to the scene and tried to help. Eventually there were so many bystanders that the rescue took on a circus-like appearance, complete with hot-dog stands and vendors selling souvenirs. After days of entrapment and bitter cold, Floyd died, imprisoned by a 26-pound rock.

HISTORY OF RESCUE

Since 1925, cave exploration has increased in popularity and frequency. Caving has become one of the "new outdoor sports." With this increase, people of many backgrounds have become attracted to caves. They come for adventure, for recreation, and for science. Unfortunately, not all visitors come with the proper equipment or training. It is often these poorly prepared "spelunkers" that rescuers are called to help. Currently almost 90% of cave rescue emergencies result from poorly prepared or inexperienced people entering caves. As the number of people entering caves has increased, so has the need for rescue.

Like any sport, people can be injured in various ways while exploring caves, and many rescues require rescuers to possess special skills to reach and treat the caver. However, most missions are initiated for simple problems, such as a light failing or the group lost in an unfamiliar area of the cave.

Cave rescues take many forms. The simplest rescue mission might involve locating a lost person, and a more complex rescue could involve negotiating multiple vertical drops or water problems while carrying an injured patient. These missions may involve as many as 200 rescuers.

The National Speleological Society (NSS) first began recording cave rescue information around 1950 and each year since has published the rescue information gathered. Currently, these accident reports are collected in a single yearly publication entitled *American Caving Accidents*.

Additional record keeping is done by the National Cave Rescue Commission (NCRC) and by individual regions of the NSS. One example is the eastern region of the NCRC, which has maintained its regional mission reports on file, on the web, and available for research. These are available on the eastern region website or by accessing its files. The NCRC records are compiled each year as part of the publication *American Caving Accidents*. These data banks have made it possible to take a comprehensive look at the growth of cave rescue in America. By studying mission reports compiled over the past 30 years, rescuers can better prepare for future rescue situations and determine training needs for an area.

DEVELOPMENT OF RESCUE TEAMS

In order to better understand the growth of cave rescue in the United States, the historical records were categorized and divided into several time periods. The mission types, frequency, and occurrence showed that there was an almost proportional increase in the number of rescue callouts and the growth of the caving population as determined by increases in NSS membership. From this, one can infer that the growth of the NSS reflects an overall increase in total hours people spend sport caving each year. As more people take up caving, more accidents occur.

The increase in total accidents graphs as a parallel linear trend when compared to NSS growth but when accidents are graphed by type, the growth trend shows several sudden peaks followed by a gradual return to a more linear trend. An example of this was the sharp increase in the vertical and water rescue accident rates when new equipment and skills were introduced in the 1970s. It is theorized that this surge in vertical and water missions reflects a lack of training or experience in using the new equipment, resulting in higher accident rates. As training and experience were gained, the rate declined and eventually returned to a more normal growth line.

As accidents increased over the 50 years of study, more rescuers were needed to help on rescue missions. This resulted in the formation of local cave rescue teams. During the 1960s and 1970s, teams such as the National Capital Cave Rescue Team and the Hondo Underground Rescue Team (HURT) responded to rescues all over the country. These teams were soon featured in men's magazines such as *Saga*, which ran an article entitled "Get the Heroes from HURT." The backlash from all of this publicity set cave rescue team training back a decade.

IN THE BEGINNING

Cave rescue teams were a relatively new concept in the United States during the 1950s. Organized training in cave rescue was done informally by local clubs but not on a national scale. Before 1976, no structured rescue classes were available to cavers. Because of this, cavers learned rescue skills by doing rescues.

From training done by early clubs, the first true American cave rescue teams emerged. The earliest teams were formed in response to accidents occurring on club trips or when club members were nearby. During the early years an injured caver often waited for hours as one of the established cave rescue teams was summoned to the site. This often took many hours, resulting in a number of body recoveries instead of rescues.

Cavers recognized the need for more formal training. Grottos (local NSS chapters) began training and new teams were formed around experienced leaders. As these teams gained in mission experience, they began to amass information on which techniques worked and which did not.

In the 1960s. the NSS formed the first Cave Rescue Commission and began planning for a national policy on cave rescue. The first commissioners met at NSS conventions and began the first formal exchange of information between teams. This effort continued with several false starts and slowdowns until 1976 when the first formal certification in cave rescue was offered by the West Virginia Fire Extension Service. This class, taught by NSS instructors, was the beginning of the society's first organized effort to train cave rescuers.

Since 1977, the NCRC of the NSS has taken the lead in cave rescue training throughout the United States and the Caribbean. Standards for cave rescue developed by the NCRC are derived from mission experiences, mission data analysis, and current rescue training standards.

LEARNING FROM THE PAST

Mission records reveal some interesting facts about the types of missions each decade spawned, including how training and experience influence the outcome of many missions. As the years progressed, better trained teams succeeded in saving lives more often than untrained teams and experienced fewer injuries to team members.

Accident Frequency Rate during the 1950s

National Speleological Society cavers in the United States spent a maximum of 1,440,000 hours caving per year. There were 13 rescue missions or incidents reported during those 1,440,000 hours of caving in the 1950s. For NSS caving this translates to one accident per roughly 111,000 hours underground during the decade.

Accident Frequency Rate during the 1960s

Caving in the 1960s started off with a bang. New rappelling systems, such as brake bars and carabiners, replaced the body rappel for descending. Prusik knots were replaced by new and faster jumars. Ascending devices and improved cable ladders opened previously unexplored passages to a new generation of cavers. All over the country cavers were off to push greater depths and longer drops. With this quest to push the limits, rescues became harder and more frequent.

By the beginning of the 1960s, the NSS was recording several rescues per year. As a result the NSS became more interested in defining its rescue capabilities. In 1960 the NSS polled its members to determine who had rescue experience. The survey found that only 81 members nationwide had cave rescue experience, about 8% of the active members of the NSS at that time.

Accident Frequency Rate during the 1970s

From 1970 to 1980 there were 198 incidents reported. Caving was getting safer but people were pushing harder and accidents were increasing. The NSS had about 3200 members and cavers had one accident per 23,272 hours underground. For the first time chronic medical problems became a factor and rockfall turned up as a cause of 8% of the rescue missions.

The 1970s saw additional improvements in equipment and technique. Descending devices now included racks and figure eights in addition to brake bars and carabiners. The introduction of better equipment was not complete until braided nylon ropes appeared and Charlie Gibbs introduced his Gibbs ascending system at the 1969 NSS Convention. During this period southeastern cavers were experimenting with racks for descending long drops and with new abrasion-resistant rope designs. Vertical caving took a giant leap forward and vertical incidents took a leap upward. The percentage of vertical missions dropped back from 42% in the 1960s, but the number of vertical problems increased.

Accident Frequency Rate during the 1980s

The 1980s found cavers well equipped to push the hardest caves and "world class caving" was introduced. Cavers took the new equipment, mastered it, redesigned it, and even manufactured it in order to explore the world's largest caves. With equipment mastery and vertical skills training a part of every caver gathering, the vertical accident rate started to trend back toward the linear caver-population growth curve.

From 1980 to 1990, accidents continued to increase as did membership in the NSS. However, vertical accident rates began to drop. This decade saw 300 incidents with an NSS membership of 5200. The calculated rate of accidents per hour underground remained near that of the previous decade with one accident occurring per 24,960 hours of caving. Analysis of incidents indicated that for the first time vertical accidents were not the leading type. Water-related accidents surpassed vertical accidents this decade because sport diving became more popular and the great cave springs claimed many lives. Chronic medical problems doubled in response to the aging of the caving populations and the availability of treatments that allowed persons with chronic medical problems to become more active.

THE PRESENT

The years from the 1990s and into the twenty-first century were ones of consolidation. The NCRC sponsored cave rescue training courses at beginner and advanced levels in most parts of the country at regular intervals. Rescue certification was well established. Many involved with rescue had advanced from basic first aid to emergency medical technician (EMT) and paramedic status. Most caving groups have a local cave rescue committee and relations with other search-and-rescue organizations such as police and fire departments have greatly improved. For the unfortunate victim of a caving accident this means a shorter wait for rescue personnel to arrive, better medical training when they do arrive, and an effective organizational structure that deals with everything from proper packaging and transport of the victim from the cave to news conferences with the media.

Most countries with active caving groups have found it necessary to organize formal rescue teams. These groups have interacted and exchanged information through the International Congresses of Speleology which are held every four years.

See Also the Following Article

Exploration of Caves—General

Bibliography

Hempel, J. C., & Fregeau-Conover, A. (2001). *On call.* Huntsville, AL: National Speleological Society.

RESPONSES TO LOW OXYGEN

Frédéric Hervant and Florian Malard

UMR CNRS 5023, Laboratoire d'Ecologie des Hydrosystèmes Naturels et Anthropisés, Université Lyon 1

INTRODUCTION

Before the late 1970s, ecological studies were carried out mainly in the unsaturated zone of karst aquifers, particularly in caves. Because cave water bodies are exposed to the atmosphere, they are usually saturated with oxygen, *dissolved oxygen* (DO) was not measured routinely, and oxygen availability in subterranean biotopes was rarely considered as a key ecological factor that governed the occurrence and spatiotemporal distribution of hypogean animals. Biological activity, animal density, and organic matter content in the groundwater were assumed to be too low to induce oxygen deficiency; therefore, the possibility that subterranean organisms may have to face hypoxic stress was not considered. Ecophysiological studies essentially concerned the adaptive responses of hypogean animals to low food supply including their higher food-finding ability, strong resistance to starvation, and reduced metabolism (Hervant and Renault, 2002; Issartel et al., 2010; Salin et al., 2010).

Since the 1970s, this view has been reevaluated by many researchers who have frequently reported low dissolved oxygen concentrations from shallow groundwater in unconsolidated sediments, particularly in the hyporheic zone of streams; the groundwater environment has been described as being hypoxic or weakly oxygenated. Also, several authors have conducted laboratory studies to test the resistance and adaptive strategies of animals in response to low oxygen (Malard and Hervant, 1999; Issartel et al., 2009).

The ensuing material begins by examining the oxygen status of different groundwater systems including deep and shallow-water-table aquifers and the hyporheic zone of streams. Then, based on laboratory studies conducted by one of the authors (Hervant et al., 1996, 1997, 1998; Malard and Hervant, 1999; Issartel et al., 2009), we examine the behavioral, respiratory, and metabolic responses of groundwater organisms, especially crustaceans and amphibians, to low oxygen concentrations. Finally, we suggest that the selection of organic-matter-rich habitats in groundwater increases the risk of facing hypoxic stress.

SUPPLY AND CONSUMPTION OF DISSOLVED OXYGEN IN GROUNDWATER

Because of permanent darkness, there is no photosynthesis in groundwater, thus no production of oxygen; therefore, the oxygen status of groundwater is determined by the rate of oxygen transport from the surface environment and by the rate of oxygen consumption in the subsurface. Replenishment of dissolved oxygen occurs by air diffusion from the unsaturated zone or by recharge with rainwater or river water. Fluctuations of the groundwater table enhance air entrapment, thereby increasing dissolved oxygen transfer from entrapped air. Oxygen transport within groundwater may occur as a result of oxygen diffusion, convection currents caused by heat transfer, and advection of water in response to hydraulic gradients. The diffusion of oxygen in water is negligible, and convective currents are limited in groundwater by sediment and weak thermal gradients (the geothermal gradient is usually about $0.01°C\ m^{-1}$) (Malard and Hervant, 1999); therefore, oxygen transport in groundwater is primarily due to advective movement of water in response to hydraulic gradients. Because groundwater velocity is usually low (i.e., $10^{-6}-10^{-4}\ m\ s^{-1}$), the available flux of DO in groundwater is much lower than in surface waters. On the other hand, oxygen consumption by microorganisms is limited in many aquifers by the availability of biodegradable organic carbon. Consequently, DO may persist at considerable distances from the recharge zone in deep-water-table aquifers where soil-generated dissolved organic carbon is completely degraded during the transit of infiltrating water in the unsaturated zone. In confined aquifers of the Ash Meadows basin (south central Nevada), Winograd and Robertson (1982) sampled groundwater with $2\ mg\ L^{-1}\ O_2$ at a distance of 80 km from the recharge zone. In contrast, dissolved oxygen may be totally consumed over very short distances (i.e., a few meters or even centimeters) in shallow-water-table aquifers or in the hyporheic zone of rivers because of the input of soil- or river-labile dissolved organic carbon. Malard and Hervant (1999) reported strong variability among groundwater systems in the length of underground pathways for dissolved oxygen. Based on cross-system comparison from literature data, these authors suggested that differences among hyporheic zones reflect variation in the contact time of water with sediment, whereas differences among confined aquifers are primarily a result of differences in the rate of DO consumption (see also Foulquier et al., 2010).

SMALL-SCALE HETEROGENEITY IN DISSOLVED OXYGEN IN GROUNDWATER

Small-scale investigations of oxygen distributions in subsurface waters revealed strong variations over distances of a few centimeters or meters. This heterogeneity, an essential feature of the groundwater environment, was observed in a number of subsurface water habitats, including the saturated zone of karst aquifers, the water-table region of deep- and shallow-water-table porous aquifers, the halocline of anchihaline caves, the hyporheic zone of rivers, and the interstitial environment of marine and freshwater beaches (Malard and Hervant, 1999) (Fig. 1). Small-scale spatial heterogeneity in DO reflects changes in sediment composition and structure, subsurface water flow velocity, strength of hydrological exchanges with the surface environment, dissolved and particulate organic matter content, and activity of microorganisms. Strong temporal changes in DO may also occur in the hyporheic zone of streams as well as in the recharge zone of aquifers, but these fluctuations are strongly damped with increasing distance from the stream and the recharge zone. Whereas diminished oxygen concentration is

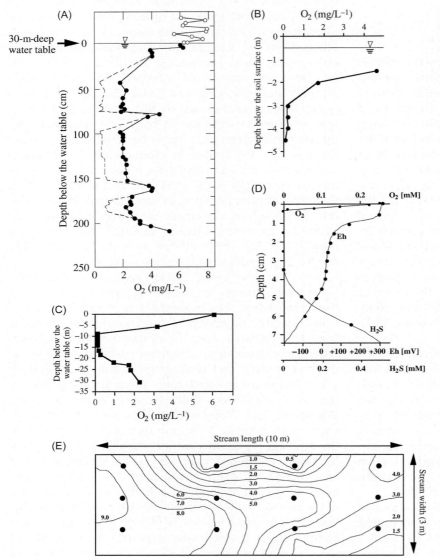

FIGURE 1 Meso- (m) and micro- (cm) scale heterogeneity in dissolved oxygen concentrations in: (A) a deep-water-table aquifer; (B) a shallow-water-table aquifer; (C) an anchihaline cave; (D) the interstitial environment of a sandy marine beach; and (E) the hyporheic zone of a backwater (Malard and Hervant, 1999, and references therein.).

typically not a rule for the groundwater environment, the high spatial heterogeneity of DO at meso- (meter) and micro- (centimeter) scales is considered a peculiarity of groundwater habitats (Malard and Hervant, 1999). This implies that animals living in groundwater have to experience highly variable oxygen concentrations as they move through a mosaic of patches with contrasted DO concentrations.

FIELD EVIDENCES OF SURVIVAL IN LOW OXYGENATED GROUNDWATER

Results of field studies reveal that most animals can be found living in a wide range of DO, even anoxia in some cases (Malard and Hervant, 1999). The blind salamander *Proteus anguinus*, the only European vertebrate that lives exclusively in caves, has to frequently cope with low DO (Issartel et al., 2010, and references therein). Based on 700 faunal samples collected in the hyporheic zone of several desert streams in Arizona, Boulton et al. (1992) examined the field tolerance to DO of 23 common invertebrate taxa. These authors showed that most taxa could be encountered in subsurface waters with less than 1 mg L^{-1} O_2. Strayer et al. (1997) obtained similar results based on 167 samples of hyporheic invertebrates collected at 14 sites in the eastern United States; however, in both studies, several animal groups, particularly crustaceans and insects, occurred more frequently in well-oxygenated sediments than where oxygen was scarce. Several crustaceans (including the amphipod *Niphargus hebereri* Schellenberg, the thermosbaenacean *Monodella argentarii* Stella, and some species of remipedes) are known to develop dense populations in the sulfide zone of anchihaline caves (Malard and Hervant, 1999). It is not yet clear, however, whether these crustaceans permanently live in deoxygenated waters or if they temporarily seek shelter in a more aerated environment. A diversified aquatic fauna was also found to live in the uppermost hypoxic layer (DO <0.3 mg L^{-1} O_2) of sulfidic groundwater in Movile Cave, Romania (Malard and Hervant, 1999). On the other hand, the paucity of fauna in extensive areas of hypoxic groundwater suggests that most hypogean taxa are probably not able to survive a lack of oxygen for very long (more than 3–4 days). Reports of groundwater animals among poorly oxygenated groundwaters are equivocal because their presence may be due to either tolerance to low oxygen concentrations by hypogean invertebrates or the existence of microzones of high DO.

Recent detailed laboratory studies on the ecophysiological (*i.e.*, behavioral, respiratory, and metabolic) responses of several subterranean crustaceans and amphibians to anoxia have enabled us to define more precisely the degree of tolerance of groundwater animals to low oxygen concentrations and to elucidate some of the mechanisms responsible for extended survival under oxygen stress (Hervant et al., 1996, 1997, 1998; Issartel et al., 2009).

SURVIVAL TIMES UNDER ANOXIA

Most data available on adaptations to low oxygen in groundwater organisms arise from comparisons of the adaptive responses of three hypogean crustaceans (the amphipods *Niphargus virei* Chevreux and *Niphargus rhenorhodanensis* Schellenberg and the isopod *Stenasellus virei* Dolfus) and two epigean crustaceans (the amphipod *Gammarus fossarum* Koch and the isopod *Asellus aquaticus* L.) to anoxia (Hervant et al., 1996, 1997). The responses of these five crustaceans were examined in darkness (at 11°C) under three different experimental conditions: anoxia, normoxia following an anoxic stress (postanoxic recovery), and declining DO concentration (from 10.3 to 0.7 mg L^{-1} O_2). Information gained from these experiments concerned (1) the lethal time for 50% of the population (LT 50%) in anoxia; (2) the locomotor activity (number of periods of locomotion per minute) and the ventilation activity (number of pleopod beats per minute) of animals; (3) the oxygen consumption rates in normoxia and under declining PO_2 (in order to determine the critical partial pressure of oxygen, Pc); and (4) changes in the concentrations of key metabolites such as high energy compounds (*e.g.*, adenosine triphosphate [ATP], phosphagen), anaerobic substrates (glucose, glycogen, amino acids), and anaerobic end products (lactate, alanine, succinate, malate, propionate).

Groundwater crustaceans show survival times (LT 50%) of about 2 to 3 days under anoxia (Table 1) and several months in moderate hypoxia (Malard and Hervant, 1999). They are much more resistant to oxygen deprivation than morphologically closed epigean species whose LT 50% values range from a few hours to one day. Hypogean fishes, crayfishes, and urodel amphibians also display much higher survival times than their related epigean species. For example, as far as we know, the cave-dwelling urodel *P. anguinus* is the most anoxia-tolerant amphibian (exhibiting LT 50% in anoxia of about 12 h at 12°C) and one of the most anoxia-tolerant vertebrates. At high temperatures, only the goldfish *Carassius auratus* presents a higher half-lethal time (22 h at 20°C; Issartel et al., 2009). *Proteus anguinus* could thus represent an excellent model organism for the study of key mechanisms in anoxia tolerance.

TABLE 1 Comparison of Locomotory and Ventilatory Activities, Oxygen Consumption (SMR), and Intermediary and Energetic Metabolism during Normoxia, Anoxia and Postanoxic Recovery in Two Surface-Dwelling and Three Groundwater Crustaceans

	Epigean		Hypogean		
	G. fossarum	A. aquaticus	N. rhenorho.	N. virei	S. virei
Survival during anoxic stress					
• Lethal time for 50% of the population	6.3 h	19.7 h	46.7 h	52.1 h	61.7 h
Activity					
• Number of periods of locomotion per minute (in normoxia)	19.3	6.7	12.5	1.4	4.5
• Changes during anoxia	↗	↗	↘	↘	↘
Ventilation					
• Number of pleopod beats per minute (in normoxia)	60	43	44	48	38
• Changes during anoxia	↗	↘	↘	↘	↘
Respiratory metabolism					
• Normoxic O_2 consumption (SMR; $\mu L\ O_2/g\ dw/h$)	940	1325	605	305	425
• O_2 critical pressure (mg L^{-1})	3.6	3.8	2.0	1.8	2.0
Body energy stores (in normoxia)					
• Stored glycogen ($\mu mol/g\ dw$)	110	165	356	243	307
• Stored arginine phosphate ($\mu mol/g\ dw$)	7.0	10.5	26.5	26.5	31.5
Post-anoxic recovery					
• Glycogen re-synthesis rate ($\mu mol/g\ dw/min$)	0.003	0.01	0.07	0.03	0.04
• % of glycogen re-synthesis	8	19	47	49	53

After Hervant et al., 1996, 1997, 1998; Malard and Hervant, 1999.

BEHAVIORAL RESPONSES

Every spontaneous or stress-induced activity increases energetic expenditures and, therefore, oxygen consumption. Some hypogean species show a very low activity rate when compared to morphologically closed epigean animals (Table 1). This reduced activity has been interpreted to be a result of (1) food scarcity, stable environmental conditions, and general lack of predators and territorial behavior; (2) cramped interstitial habitats; and/or (3) low oxygen availability (Hervant et al., 1996; Hervant and Renault, 2002; Salin et al., 2010). A low activity rate in groundwater animals which is often associated with an increase in food-finding ability is an efficient adaptation to low food supply because it results in a reduced energy demand.

Epigean *Gammarus fossarum* and *Asellus aquaticus* respond to experimental anoxia by a marked hyperactivity (Table 1) that, in natural conditions, probably corresponds to an attempt to move to more oxygenated habitats (i.e., to a behavioral compensation). In contrast, all three hypogean crustaceans show a drastic reduction in locomotory activity under lack of oxygen (Table 1). This adaptive behavior reduces energy expenditure during oxygen deprivation (by decreasing the oxygen consumption) and therefore increases survival time in deoxygenated groundwater. This "sit-and-wait" strategy, very advantageous under hypoxic (and/or food limited) environments, was also observed in starved stygobite crustaceans (Hervant and Renault, 2002) and cave salamanders (Hervant et al., 2001). The hypogean isopod *Proasellus slavus* actively explored its interstitial environment but slowed down exploratory movement under severe hypoxic conditions (i.e., 0.1 mg L^{-1} O_2). The lack of an escape behavior is probably not universal for hypogean aquatic organisms, however, because active animal migration in response to fluctuating DO concentrations has already been documented in natural biotopes and in a laboratory flume.

RESPIRATORY RESPONSES

Subterranean animals generally have reduced *standard metabolic rates* (SMRs) and reduced *routine metabolic rates* (RMRs) when compared to their surface-dwelling counterparts (Hervant et al., 1998; Issartel et al., 2010). For example, the cave salamander *Proteus anguinus* Laurenti (Hervant et al., 2001), the hypogean fish *Astyanax fasciatus mexicanus* (Salin et al., 2010), and the groundwater crustaceans *Niphargus virei*, *Niphargus rhenorhodanensis*, and *Stenasellus virei* show SMRs in normoxia from 1.6 to 4.5 times lower than their epigean relatives (Table 1) (Hervant et al., 1998). Low SMRs and RMRs among hypogean animals have long been interpreted only as an adaptation to low food availability; however, reduced SMRs in some populations of subterranean animals may also reflect an adaptation to low oxygen concentrations in the groundwater (Hervant and Renault, 2002). A reduced metabolic rate (*i.e.*, low energetic requirements) results in a lower oxygen removal rate and an increased survival time in anoxia.

The hypogean amphipods *N. virei* and *N. rhenorhodanensis* and the hypogean isopod *S. virei* reduce their ventilatory activity during anoxic stress (Table 1), thereby limiting their energy expenditure. Several animals contract an oxygen debt during oxygen deprivation that is repaid on return to normoxia (Hervant et al., 1998). The repayment of this debt during recovery from anaerobic stress involves a significant increase in SMR. The oxygen debt of hypogean *N. virei* and *N. rhenorhodanensis* is 2.2 to 5.3 times lower than that of epigean *G. fossarum* and *A. aquaticus* (Hervant et al., 1998). This lower oxygen debt indicates a reduced energetic expenditure (*i.e.*, an energy sparing) which is probably linked to lower locomotory and ventilatory activities during anoxia.

The resistance to anoxia (and/or long-term fasting) displayed by numerous hypogean species may be explained by their ability to remain in a prolonged state of torpor. This state enables hypogean organisms to tolerate a prolonged reduction in oxygen (and/or food) availability by maximizing the time during which metabolism can be fuelled by a given energy reserve (or a given food ration). This supports the classic suggestion that difficulties in obtaining food in stressful environments may select for conservative energy use (Hervant and Renault, 2002). These adaptive responses may be considered for numerous subterranean organisms as an efficient energy-saving strategy in a harsh and unpredictable environment where hypoxic (and/or starvation) periods of variable duration alternate with normoxic periods (and/or sporadic feeding events). Hypoxia-tolerant (and/or food-limited) groundwater species appear to be good examples of animals representing a low-energy system (Hervant et al., 2001; Hervant and Renault, 2002), although starvation tolerance is not universal for cave organisms (Salin et al., 2010).

Theoretically, aerobic organisms can be described as metabolic conformers or regulators if their oxygen consumption varies directly with or is independent of PO_2, respectively; however, these two types are merely the two ends of a large spectrum of respiratory responses of species. Animals never fully belong to one or the other type, and below a critical PO_2 (Pc) a regulator becomes a conformer (Hervant et al., 1998); therefore, Pc is a good indicator of the tolerance or adaptation of an organism to low PO_2. *Gammarus fossarum*, *A. aquaticus*, *N. virei*, *S. virei*, and *N. rhenorhodanensis* are able to maintain a relatively constant rate of oxygen consumption that is independent of PO_2 and between normoxia and Pc (Hervant et al., 1998). The blind salamander *P. anguinus* presented the same respiratory pattern (Hervant, unpublished data). However, the PO_2 at which respiratory independence is lost is significantly lower for hypogean species than for surface-dwelling ones (Table 1). This implies that groundwater species are able to maintain an aerobic metabolism for a longer time in declining PO_2 (progressive hypoxia) instead of partly switching to a low-energy anaerobic metabolism. The maintenance of an aerobic metabolism (*i.e.*, survival at a lower energetic cost) under hypoxic conditions is partly due to the lower SMR of hypogean animals. The occurrence of respiratory regulation over such a wide range of DO concentrations had, until then, not been observed among epigean crustaceans and had rarely been observed in other invertebrates (Malard and Hervant, 1999, and references therein).

METABOLIC RESPONSES

During Anoxia

A number of biochemical adaptations that permit extended survival under prolonged hypoxia or anoxia have been identified in various well-adapted epigean groups, especially marine annelids and intertidal mollusks. These include the maintenance of high reserves of fermentable fuels (such as glycogen and amino acids) in some tissues under normoxic conditions, the use of anaerobic pathways to enhance ATP yield and to maintain redox balance during low-oxygen conditions, and mechanisms for minimizing metabolic acidosis often associated with anaerobic metabolism. Nevertheless, epigean crustaceans have been recognized as being of a "modest anaerobic capacity" without special and efficient mechanisms of anaerobic metabolism (Hervant et al., 1996, and references therein).

Examination of biochemical (*i.e.*, intermediary and energy metabolism) responses of hypogean crustaceans *N. virei*, *N. rhenorhodanensis*, and *S. virei* during anoxic stress show that, similarly to epigean crustaceans, anaerobic metabolism does not lead to a high ATP production rate (Hervant et al., 1996, 1997). The five crustaceans studied respond to severe experimental anoxia with a slight improvement of a classical anaerobic metabolism that is characterized by a decrease in ATP and phosphagen (arginine phosphate, representing an immediate source of ATP), a coupled utilization of glycogen and some amino acids (mainly glutamate), and the accumulation of lactate and alanine as end products. The only difference is that both *Niphargus* species also accumulate a low proportion of succinate, which slightly enhances ATP yield during anaerobic respiration. Lactate is largely excreted by all five crustaceans. This excretion, which is unusual for crustaceans, can be considered a simple way to fight against metabolic acidosis linked to anaerobic end-product (including H^+) accumulations (Hervant et al., 1996).

There is a striking difference in the respective amounts of glycogen and phosphagen stored by epigean and hypogean crustaceans. Glycogen body reserves (Table 1) are 1.5 to 3.2 times higher in the three hypogean crustaceans studied than in the epigean *G. fossarum* and *A. aquaticus* (Hervant et al., 1996, 1997) but are also higher than those reported for all surface-dwelling crustaceans, even those most tolerant of anoxia or hypoxia (Malard and Hervant, 1999). High amounts of fermentable fuels result in a more sustained supply for anaerobic metabolism, thereby increasing survival time during oxygen deprivation. Moreover, glycogen utilization rates and lactate production rates are significantly lower in hypogean crustaceans. This finding is probably linked to lower SMRs and to the reduction of locomotory and ventilatory activity during anoxia (Table 1) (Hervant et al., 1996, 1998; Issartel et al., 2009).

During a Postanoxic Recovery Phase

It is ecologically very important for organisms to recover quickly and completely from hypoxic or anoxic stress when oxygen is available once more. This recovery implies a restoration of high energy compounds (mainly ATP, phosphagen, and glycogen), as well as disposal of anaerobic end products (mainly lactate, alanine, and succinate). End products can be disposed of in three different ways during a postanoxic recovery phase: complete oxidation, conversion back into storage products such as glycogen (via the glyconeogenesis pathway, such as glycogen de novo synthesis from lactate, amino acids, and/or glycerol), and excretion into the medium (Hervant et al., 1996). Excretion is an important mechanism for the disposal of lactate during aerobic recovery in the epigean *G. fossarum* and *A. aquaticus*. This is a costly strategy because it implies a loss of energy-rich carbon chains. In contrast, hypogean crustaceans preferentially use glyconeogenesis to convert lactate into glycogen stores. The existence of glyconeogenesis has already been demonstrated in several crustaceans, although the organ sites for this metabolic pathway have not been identified clearly. Experiments using injections of labeled glucose and lactate (see Malard and Hervant, 1999) revealed that the gluconeogenesis rate in *N. virei* during postanoxic recovery was higher than any rate measured previously for epigean crustaceans. Glycogen reserve restoration was indeed 2.5 to 6.6 times greater in hypogean species than in *G. fossarum* and *A. aquaticus* (Table 1). This ability to quickly resynthesize during recovery periods the body stores depleted during lack of oxygen allows groundwater organisms to successfully fuel an ensuing hypoxic or anoxic period; therefore, groundwater species are well adapted to live in habitats showing frequent and unpredictable alternations of normoxic and hypoxic/anoxic phases.

In addition to the immediate harmful effect of anaerobiosis, anoxia-tolerant animals must also cope with postanoxic damage that occurs during reoxygenation: a burst of *reactive oxygen species* (ROS) occurs during the recovery of activity of the mitochondrial respiratory chain. Thus, because of the damaging effects of ROS overproduction, animals that experience anoxia have developed both enzymatic and non-enzymatic antioxidant mechanisms (Issartel et al., 2009, and references therein). Although we demonstrated exceptional anoxia tolerance in the cave-dwelling urodel *P. anguinus*, this species paradoxically does not present the "ordinary" antioxidant responses (showing no increase in liver antioxidant enzyme activities and a general low control of ROS damage; Issartel et al., 2009). These surprising results open new research perspectives in the comprehension of anoxia tolerance.

IMBRICATION OF ADAPTIVE STRATEGIES TO LOW FOOD RESOURCES AND LOW OXYGEN SUPPLY

The occurrence of adaptive strategies in response to low oxygen among animals living in an oligotrophic environment may seem paradoxical. Notwithstanding the fact that the supply of organic matter is typically lower in groundwater than in surface water, field ecological studies carried out for the past 30 years have shown that several groundwater habitats (particularly unconsolidated sediments) have reduced DO concentrations. Although a few subterranean habitats are known to be organic-matter rich (Salin et al., 2010), low DO concentration in many groundwater habitats is most likely attributable to the lack of oxygen production and low transport rate of DO than to elevated concentrations of organic matter. If food availability drives habitat selection in groundwater, hypogean animal populations would preferentially occur in groundwater biotopes receiving higher fluxes of organic matter from the surface environment. These habitats are also more likely to exhibit reduced DO concentrations because of increased respiration rates associated with the input of organic matter. Thus, the selection of habitats having increased food supply increases the probability of facing hypoxic stress among hypogean animals. Meantime, the development of behavioral, respiratory, and metabolic strategies to low food supplies also selects for higher resistance to a low oxygen supply. Clearly, the role of food availability and the significance of low oxygen supply in determining the development of adaptive strategies and distribution patterns of animals in groundwater are overlapping aspects that can hardly be treated independently.

Some metabolic pathways, therefore, that are specifically linked to the response to long-term starvation and/or adaptation to oxygen stress are associated with energy-limited subterranean organisms. Hervant and Renault (2002) and then Issartel et al. (2010) demonstrated that the groundwater crustacean *S. virei* and a cave population of the urodel *Calotriton asper* preferentially utilizes lipids during food shortage, in order to (1) save carbohydrates and phosphagen, the two main fuels metabolized during oxygen deficiency, and (2) save proteins (and therefore muscular mass) for as long as possible. Thus, this species can (1) successfully withstand a hypoxic period subsequent to (or associated with) an initial nutritional stress, and (2) rapidly resume searching for food during short-term, sporadic, nutrition events.

A general adaptation model for groundwater animals involves their ability to withstand prolonged hypoxia and/or long-term starvation (Hervant et al., 2001; Hervant and Renault, 2002) and to utilize in a very efficient way the high-energy body stores. Because the three hypogean crustaceans studied lack a high-ATP-yielding anaerobic pathway (such as observed in permanent anaerobic organisms), their higher survival time in anoxia is mainly due to the combination of four mechanisms: (1) high storage of fermentable fuels (glycogen and phosphagen); (2) low SMR in normoxia; (3) further reduction in metabolic rate by lowering energetic expenditures linked to locomotion and ventilation during hypoxia; and (4) high ability to resynthesize the depleted body stores during subsequent recovery periods. The ability to maintain and rapidly restore (without feeding) high amounts of fermentable fuels for use during lack of oxygen can be considered an adaptation to life in a patchy environment. Through their efficient exploratory behavior in a moving mosaic of patches of low and high DO concentration (Malard and Hervant, 1999), numerous groundwater animals probably experience highly variable DO and/or food concentrations. The behavioral, physiological, and metabolic responses of numerous hypogean animals partly explain why they occur in groundwater systems with a wide range of DO (Malard and Hervant, 1999).

A high resistance to lack of oxygen (and/or to food deprivation) is not universally found in subterranean organisms but is probably more related to oxygen availability and/or to the energetic state of each subterranean ecosystem, as demonstrated by Salin et al. (2010) with hypogean fish living in energy-rich tropical caves. Indeed, groundwater ecosystems are far more complex and diverse than earlier presumed. The aquatic amphipod *Gammarus minus* (Hervant et al., 1999) and the tropical fish *Astyanax fasciatus* (Hervant, unpublished data) showed no significant difference in ecophysiological responses to experimental anoxia and subsequent recovery among a surface and a cave population. The ability to withstand and recover from long periods of nutritional and/or oxygen stress is clearly a critical adaptation for the colonization of energy-limited subterranean environments, but troglomorphism is not necessarily linked to starvation capacities and/or hypoxia tolerance in some energy-rich caves. Despite a strong tendency toward morphological and ecophysiological convergence, subterranean organisms do not form a homogeneous group.

Bibliography

Boulton, A. J., Valett, H. M., & Fisher, S. G. (1992). Spatial distribution and taxonomic composition of the hyporheos of several Sonoran Desert streams. *Archiv für Hydrobiologie, 125,* 37–61.

Foulquier, A., Malard, F., Mermillod-Blondin, F., Datry, T., Simon, L., Montuelle, B., et al. (2010). Vertical change in dissolved organic carbon and oxygen at the water table region of an aquifer recharged with stormwater: Biological uptake or mixing? *Biogeochemistry, 99,* 31–47.

Hervant, F., Mathieu, J., Garin, D., & Fréminet, A. (1996). Behavioral, ventilatory, and metabolic responses of the hypogean amphipod *Niphargus virei* and the epigean isopod *Asellus aquaticus* to severe hypoxia and subsequent recovery. *Physiological Zoology, 69,* 1277–1300.

Hervant, F., Mathieu, J., & Messana, G. (1997). Locomotory, ventilatory and metabolic responses of the subterranean *Stenasellus virei* (Crustacea: Isopoda) to severe hypoxia and subsequent recovery. *Compte Rendus de l'Académie des Sciences de Paris, Life Sciences, 320,* 139–148.

Hervant, F., Mathieu, J., & Messana, G. (1998). Oxygen consumption and ventilation in declining oxygen tension and posthypoxic recovery in epigean and hypogean aquatic crustaceans. *Journal of Crustacean Biology, 18*(4), 717–727.

Hervant, F., Mathieu, J., & Culver, D. C. (1999). Comparative responses to severe hypoxia and subsequent recovery in closely related amphipod populations (*Gammarus minus*) from cave and surface habitats. *Hydrobiologia, 391,* 97–204.

Hervant, F., Mathieu, J., & Durand, J. P. (2001). Behavioural, physiological and metabolic responses to long-term starvation and refeeding in a blind cave-dwelling salamander (*Proteus anguinus*) and a facultative cave-dwelling newt (*Euproctus asper*). *Journal of Experimental Biology, 204*, 269–281.

Hervant, F., & Renault, D. (2002). Long-term fasting and realimentation in hypogean and epigean isopods: A proposed adaptive strategy for groundwater organisms. *Journal of Experimental Biology, 205*, 2079–2087.

Issartel, J., Hervant, F., De Fraipont, M., Clobert, J., & Voituron, Y. (2009). High anoxia tolerance in the subterranean salamander *Proteus anguinus* without oxidative stress nor activation of antioxidant defenses during reoxygenation. *Journal of Comparative Physiology B, 179*, 543–551.

Issartel, J., Voituron, Y., Guillaume, O., Clobert, J., & Hervant, F. (2010). Selection of physiological and metabolic adaptations to food deprivation in the Pyrenean newt *Calotriton asper* during cave colonisation. *Comparative and Biochemical Physiology A, 155*(1), 77–83.

Malard, F., & Hervant, F. (1999). Oxygen supply and the adaptations of animals in groundwater. *Freshwater Biology, 41*(1), 1–30.

Salin, K., Voituron, Y., Mourin, J., & Hervant, F. (2010). Cave colonization without fasting capacities: An example with the fish *Astyanax fasciatus mexicanus*. *Comparative and Biochemical Physiology A, 156*(4), 451–457.

Strayer, D. L., May, S. E., Nielsen, P., Wollheim, V., & Hausam, S. (1997). Oxygen, organic matter, and sediment granulometry as controls on hyporheic animal communities. *Archiv für Hydrobiologie, 140*, 131–144.

Winograd, I., & Robertson, F. (1982). Deep oxygenated groundwater. Anomaly or common occurrence? *Science, 216*(4551), 1227–1229.

ROOT COMMUNITIES IN LAVA TUBES

Fred D. Stone, * *Francis G. Howarth,*[†] *Hannelore Hoch,*[‡] *and Manfred Asche*[‡]

University of Hawai'i at Hilo, [†]Bernice P. Bishop Museum, [‡]Museum für Naturkunde, Germany

Plant roots were not considered an important food resource in cave ecosystems until the discovery by Francis Howarth of a planthopper and other cave-adapted animals on tree roots in a lava tube in Hawai'i Volcanoes National Park, Howarth (1973). Furthermore, very few troglobites (obligate cave-dwellers) had been reported from lava tubes or from tropical caves. In the subsequent decades, cave-adapted species have been discovered in many areas of the tropics and in lava tubes as well as in other suitable subterranean habitats. Advances in knowledge of tropical cave communities and of the Adaptive Shift hypothesis for cave species evolution are described by Hoch and Howarth (1989, 1993) elsewhere in this encyclopedia. The potential for discovery of new cave species in the tropics is great, as only a tiny part of the potential underground habitat has been studied to date (Fig. 1).

HABITAT AVAILABLE FOR TROGLOBITE EVOLUTION

Lava tubes are a small subset of the total habitat available to troglobites; other habitats include:

- Mesocaverns, which are small voids, about 0.5–25 cm in width, beneath the surface in many kinds of substrates. These mesocaverns include fractures, cracks, and vesicles in many kinds of rock; cracks, animal burrows, and tree root holes in soil; spaces between rocks in talus slopes or other types of rock piles; spaces in rocks and gravel from alluvial deposits or along stream beds; and solution cavities. Substrates, in which the mesocaverns are interconnected to form a vast anastomosing system, can provide ideal habitat for troglobitic species. The mesocavernous environment is most likely the predominant habitat for cave species because it has such a large potential area and is near enough to the surface to contain abundant energy resources from root penetration, migrating surface organisms, and water-transported organic matter.
- Caves formed in limestone or other soluble rock are generally the best studied cave communities because of their attraction to cavers and their accessibility, but most, if not all, obligate cave species in caves also live in the surrounding mesocaverns. Limestone caves may also contain important root communities, particularly in areas near the surface or with ready access to the water table.
- Tree root habitats are far more widespread than those found in lava tubes. Roots can penetrate through soil layers and enter the mesocavernous habitat in many kinds of surface materials. Lava tube root areas are particularly important because they allow access to the underground habitat by scientists, who otherwise have a difficult time gaining access to the cave species. Lava tubes have the additional advantage that they form in lava flows that flow along the surface of the ground, so they generally tend to be fairly shallow throughout their length.

THE CAVE ENVIRONMENT IN RELATION TO CAVE SPECIES

Energy Sources

Tree root communities gain their energy from photosynthesis on the ground surface and therefore have the same trophic structure as surface communities with primary producers, herbivores, carnivores, detritivores, decomposers, and fungivores. Sugar produced

FIGURE 1 Lava tube passage with tree roots.

in the tree leaves is transported downward through the trunk and into the roots, where it becomes the energy source for root growth and for the species that feed on the roots (Fig. 2). Species that live on and around tree roots include many surface species, such as ants, that venture underground but are not necessarily adapted to survive entirely in the deep cave zone. Other species spend a portion of their life cycle underground on or near tree roots, such as immature stages of planthoppers (Hemiptera: Fulgoromorpha; e.g., Cixiidae) and cicadas (Hemiptera: Cicadidae). Cicada nymphs spend up to 17 years underground feeding on tree roots until they crawl up the tree trunk and emerge as surface-dwelling adults. In spite of their long sojourn underground, no cicada species has yet been discovered that spends its entire life cycle in the subterranean environment. This adaptive shift has occurred many times by planthopper species, however.

Moisture/Humidity

Troglobites live in air saturated with water vapor and have a number of adaptations to cope with high humidity, such as loss of pigments and thinning of their skin or cuticle. Species that feed directly on the watery sap in tree roots can survive in somewhat drier areas due to the constant supply of moisture through the tree roots. Examples include nymphs of cave planthoppers in the family Cixiidae (Table 1).

Air Flow

Restricted airflow is essential to survival of cave-adapted species, as airflow is usually desiccating. Certain types of air motion may actually introduce moisture into caves in dry areas. For example, the interaction of surface air with cave air often causes moisture to condense in the zone of contact. This results in a drip zone just inside the entrance of caves. Lava tubes in dry areas of Hawai'i sometimes act to remove moisture from air flowing through them, if they are sufficiently deep and have restrictions to prevent the loss of water vapor. These cave passages also capture percolating rain water. Hawaiians took advantage of these water traps as sources of water for drinking and agriculture. Tree roots follow the moisture, so often on dry lava flows on the leeward side of Hawai'i, native pioneering trees, *Metrosideros polymorpha* ('ōhi'a), survive by sending their roots into these water-trap caves.

O_2 and CO_2

In the root zone deep in the soil, or in mesocaverns, oxygen is depleted and carbon dioxide increased by respiration of the roots and of soil organisms. Air exchange with the surface is restricted in these zones by the small, labyrinthine nature of the spaces, allowing the oxygen/carbon dioxide imbalance to be maintained. Species living in this zone need to adapt to the low oxygen and high carbon dioxide levels. In deep caves with restricted airflow, a similar oxygen/carbon dioxide imbalance can occur for the same reasons. Areas with animal guano or

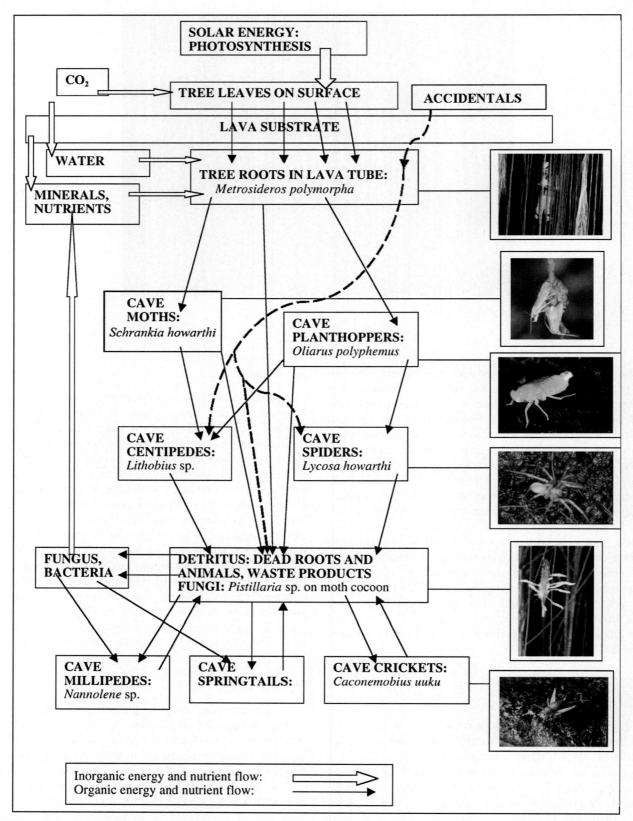

FIGURE 2 Cave root food web.

TABLE 1 Cave Tree Root Communities in Hawai'i and Australia

	Hawai'i Lava Tubes Howarth (1981, 1991)	Australia: Bayliss Lava Tubes Howarth & Stone (1990)
Producers		
	Metrosideros polymorpha ('Ōhi'a tree)	*Brachychiton* spp., *Eucalyptus* spp., *Ficus* spp.
Primary consumers		
Insecta: Hemiptera: Cixiidae (cave planthoppers)	*Oliarus polyphemus* and other *Oliarus* species	*Solanaima baylissa*
Insecta: Lepidoptera: Noctuidae (cave moths)	*Schrankia howarthi*	
Insecta: Coleoptera (beetles)		Curculionidae: Rhytirhininae
Secondary consumers		
Aranea (spiders)	Lycosidae: *Lycosa howarthi*; Linyphiide; Theridiidae	Pholcidae: *Spermophora* sp.; Nesticidae: *Nesticella* sp.; Zodariidae: *Storena* sp.
Acari (mites)	Rhagidiidae: *Foveacheles* sp.	
Chilopoda (centipedes)	Lithodiidae: *Lithobius* sp.	Scutigeridae
Hemiptera	Reduviidae: *Nesidolestes ana*	Reduviidae: *Micropolytoxus*
Scavengers, detritivores, fungivores		
Insecta: Dermaptera: Anisolabidae	*Anisolabis howarthi*	
Insecta: Orthoptera (crickets)	Gryllidae: *Caconemobius varius*, *C. uuku*, *Thaumatogryllus cavicola* (cave tree cricket)	
Insecta: Blattodea: Nocticolidae, Blatellidae (cave cockroaches)		Nocticolidae: *Nocticola* sp.; Blatellidae: *Neotemnopteryx baylissensis*
Insecta: Hemiptera: Mesoveliidae	*Cavaticovelia aaa*	
Diplopoda (millipedes)	Cambalidae: *Nannolene* sp.	Polyxenidae; Polydesmidae; Cambalidae (all undetermined species)
Crustacea: Isopoda	*Littorophiloscia* sp.	Oniscomorpha: undetermined species
Insecta: Collembola	*Sinella yoshia*; *Protanura hawaiiensis*	Entomobryidae: *Pseudosinella* sp.
Decomposers		
Fungi	*Pistillaria* sp.	

washed-in organic matter can have enhanced carbon dioxide and lowered oxygen due to respiration by bacteria and other decomposers. Carbon dioxide is denser than air; thus, if no mixing occurs, CO_2-laden air will tend to flow downward and accumulate. With restricted air flow, the air in caves may become stratified with higher carbon dioxide concentrations in lower areas, Deharveng & Bedos (1986).

Influence of Cave Geomorphology on Ecological Conditions

The shape of caves and lava tubes is critical for the occurrence of habitat suitable for cave-adapted species. The most important elements are the following: (1) restriction of airflow to reduce the impact of surface air, which can cool and dry the cave; (2) passages that trap water vapor, such as low ceiling segments followed by higher ceiling areas, upsloping, or upper-level, dead-end passages; and (3) passages that serve as carbon dioxide sumps (*e.g.*, low floor areas or downsloping dead-end passages) Howarth (1981, 1991).

CAVE ZONES IN RELATION TO ROOT GROWTH

Tree Roots

Tree roots penetrate the substrate to obtain water and nutrients for growth. In lava flows and limestone areas with fractured or cavernous substrate, surface water rapidly percolates to the water table at some depth below the surface. Some tree species are adapted to send their roots deep beneath the surface to reach the water table. Areas with fractured or cavernous rock often have little or no surface soil, due to its erosion into the underground habitat, or, in the case of lava tubes, due to the recent lava flows. Trees may also have adaptations that allow them to survive on nutrients that leach from the cave rocks. Where the water table is too far below the surface for roots to penetrate, the voids in the substrate may accumulate water vapor that the roots can utilize. In Hawai'i, trogloxenes such as bats and crickets do not occur and therefore do not contribute nutrients in guano to the cave community. In mainland areas, such as the Undara lava tubes in Queensland, Australia, bats bring significant amounts of guano into the lava tubes that provide nutrients for the tree roots as well as the scavengers (guanobites).

Trees that can grow on bare rock have adaptations for sending roots deep to a water source and may penetrate into caves more than 30 m belowground, Jackson et al (1999). This is especially true in the tropics, where the high evapotranspiration rate imposes a severe stress on trees. In the wet/dry tropics, these trees often stay green while those growing in soil often lose their leaves in the dry season. In Hawai'i, varieties of 'ōhi'a are adapted to colonize bare rock in new lava flows. Roots appear to have special adaptations for growing on bare rock. The 'ōhi'a tree is one of the first colonists on new lava flows, and it also grows in wet soil areas in the rainforest, Cordell et al (1998). Research in lava tubes in Hawai'i Volcanoes National Park showed that trees in wetter areas had roots with thinner outer layers and larger vessels in the xylem, McDowell (2002). This adaptation would make the roots easier to penetrate by the piercing mouthparts of the cave planthoppers. In Australia, *Brachychiton* species and some *Eucalyptus* species send roots into caves, and *Ficus* species do this wherever they occur throughout the tropics.

Entrance

Entrance zones of lava tubes are often more densely vegetated than are the surrounding areas. This is partly because the entrance pits accumulate dust and debris from the surface, providing nutrients, and partly due to the ready access to moisture. A drip zone often occurs just inside cave entrances, due to the mixing of surface and cave air which results in condensation of water vapor. This can often be seen as a fog zone when conditions are right. Cave-adapted species that generally occur in the deep cave zone can sometimes be found under rocks, in cracks, or on roots due to the constant moisture in this zone. Tree roots that hang from the ceiling in the entrance zone may be more similar to aerial roots that occur on the surface, due to their periodic exposure to desiccating surface air. They tend to be bushier and have a thicker cortex than roots in the deep cave environment.

Twilight

This is the zone just inside the entrance where some light penetrates. It is strongly impacted by surface air, so it is generally not suitable for cave-adapted species. In cases where there is a restricted entrance or an upsloping dead-end passage, the twilight zone may be small or absent, and cave-adapted species can then be present very near the entrance.

Transition/Mixing Zone

Beyond the zone of light penetration is an area that still receives some impact from surface air. This zone is drier than the deeper cave and has a more variable temperature. In many temperate caves, this zone expands in the winter when the surface temperature remains below cave temperature. This winter effect is due to cool surface air moving into the cave, where it is warmed by the latent heat of the cave. The warming process evaporates moisture from the cave surface, desiccating the passage. In the tropics, surface temperatures generally rise above and fall below cave temperatures nearly every day creating a diurnal winter effect. This zone often has troglophilic and trogloxenic species that can cope with the drier conditions and more variable temperature. A zone of mixing occurs between the transition zone and the deep cave air, which can sometimes be seen as a foggy area due to condensation of water vapor into droplets.

Deep Cave (Saturated Humidity) Zone

This zone is beyond the effects of surface air. Humidity remains near 100%, and the temperature remains relatively constant. Air motion is reduced. Due to the constant moisture and lack of desiccation, tree roots proliferate in this zone and provide the basis for some of the most diverse communities of cave-adapted species.

Bad Air (Saturated Humidity Plus High Carbon Dioxide/Low Oxygen) Zone

In areas with low dead-end passages, virtually no air motion, and an energy source for respiration, carbon dioxide can increase and oxygen decrease. Carbon dioxide readily mixes with the air, so there is no buildup where there is air exchange with other zones; however, where air is stagnant and biogenic carbon dioxide is being produced, cave air can stratify, resulting in higher carbon dioxide levels in low areas. If these areas also have high humidity and food energy sources, they are home to the most highly cave-adapted species. Bayliss Cave is a lava tube in Australia with a large bad air zone, high humidity, and abundant tree roots penetrating the cave ceiling and growing through the soil on the floor. It is home to 24 highly cave-adapted species, among the highest diversity of any cave community, Atkinson et al (1976), Howarth & Stone (1990).

MORPHOLOGY AND ADAPTATIONS OF CAVE SPECIES

Obligate cave species respond to cave conditions of total lack of light, high humidity, constantly wet substrate, and potentially high carbon dioxide and low oxygen by adaptive alteration of their morphology, physiology, and behavior. Morphological changes include loss of eyes and pigment; flightlessness; enhanced senses of smell, hearing, and touch and ability to detect air motion; elongation of appendages such as legs and antennae; loss of pulvilli and lengthening of claws; and rotation of legs to keep the body upright. Physiologically, cave species lose their circadian rhythms and diurnal activity patterns; they often have fewer but larger offspring and a slower metabolic rate than surface species. Though energy sources may be abundant in caves, they are often more dispersed than on the surface and more difficult to find in the dark, three-dimensional maze, so cave species have adapted to survive long periods without feeding.

Behaviorally, cave species must develop special adaptations for finding food, finding a mate, and avoiding predators. In the labyrinthine spaces of caves and mesocaverns, cave species may resort to a constant random search for food. Tree roots can assist the search by providing a pathway that cave species can follow. Because the roots are also an energy source, predators and scavengers may use the root pathways to assist in locating prey. Tree roots also provide hiding places for predators such as the small-eyed/big-eyed hunting spiders.

Cave planthoppers use tree roots for their food source as well as to assist them in finding mates. Female planthoppers sit on the roots and produce a substrate-borne sound that travels readily along the roots. Male planthoppers move randomly over the cave walls until they find a root. When a male detects a signal from a female, he returns the call and the pair "duet" until the male locates the female. The female lays its eggs on the roots, and the nymphs insert their mouthparts into the xylem to feed. Planthopper nymphs produce wax filaments from a gland on their abdomen and make a loose wax "cocoon" that protects them from predators while they feed, Hoch & Howarth (1989, 1993)

Cave moths use a different strategy for finding their mates. Female moths sit on the roots and often have reduced wings and are flightless. Male moths fly erratically through the cave with their legs held out in front. They have extremely sensitive touch and can feel a small root and even alight on a strand of spider web without alerting the spider. The female moth releases a pheromone, and when the male detects the chemical it flies upwind until it reaches the female. The female lays its eggs on the root, and the inchworm caterpillar feeds on the small tree roots. It also uses the roots to build a hanging cocoon to protect it from predators while it pupates.

Male cave crickets and cockroaches have tergal glands that release pheromones to attract the females, as do their surface relatives. The female crawls onto the male to the tergal gland, and the male attaches to her with a special genital hook. In Thai species of the cockroach family Nocticolidae, the tergal glands are highly modified into large horn-like structures, whose function has not been studied.

CONSERVATION OF CAVE ROOT COMMUNITIES

As with other cave communities, cave root communities include many rare and highly specialized species that may be threatened by impacts to their habitat. However, their survival requires more than just protection of caves or cave passages. It is dependent on protection of the plant roots that are the main energy source for the community. This requires protection of the overlying surface ecosystems. Identification of the plant species that supply the roots in the caves is important for a management plan that will protect the critical plant species and allow their re-introduction in adjacent areas where they have been lost. Invasive deep-rooted species that compete with the native plant species and alter the cave root composition also need to be identified. Roots deep in caves are difficult to identify visually, as their growth form differs significantly from that of roots near entrances or in soil. Recently, use of DNA analysis has become an important conservation and management tool by making accurate root identifications possible, Howarth et al (2007).

See Also the Following Articles

Adaptive Shifts
Food Sources
Volcanic Rock Caves

Bibliography

Atkinson, A., Griffin, J. J., & Stephenson, P. J. (1976). A major lava tube system from Undara Volcano, North Queensland. *Bulletin of Vulcanology, 39*, 1–28.

Cordell, S., Goldstein, G., Meuller-Dombois, D., Webb, D., & Vitousek, P. M. (1998). Physiological and morphological variation in *Metrosideros polymorpha*, a dominant Hawaiian tree species, along an altitudinal gradient: The role of phenotypic plasticity. *Oecologia, 113*, 118–196.

Deharveng, L., & Bedos, A. (1986). *Gaz carbonique* (pp. 144–152). *Thai-Maros 85: Rapport Speleologique et Scientifique to Thailand and Sulawesi*. Toulouse, France: Association Pyreneenne de Speleologie.

Hoch, H., & Howarth, F. G. (1989). Six new cavernicolous cixiid planthoppers in the genus *Solonaima* from Australia (Homoptera: Fulgoroidea). *Systematic Entomology, 14*, 377–402.

Hoch, H., & Howarth, F. G. (1993). Evolutionary dynamics of behavioral divergence among populations of the Hawaiian cave-dwelling planthopper *Oliarus polyphemus* (Homoptera: Fulgoroidea: Cixiidae). *Pacific Science, 47*, 303–318.

Howarth, F. G. (1973). The cavernicolous fauna of Hawaiian lava tubes. 1. Introduction. *Pacific Insects, 15*, 139–151.

Howarth, F. G. (1981). Community structure and niche differentiation in Hawaiian lava tubes. In D. Mueller-Dombois, K. W. Bridges & H. L. Carson (Eds.), *Island ecosystems: Biological organization in selected Hawaiian communities* (Vol. 15, US/IBP Synthesis Series (pp. 318–336). Stroudsburg, PA: Hutchinson Ross Publishing.

Howarth, F. G. (1991). Hawaiian cave faunas: Macroevolution on young islands. In E. C. Dudley (Ed.), *The unity of evolutionary biology* (Vol. 1). Portland, OR: Dioscorides Press.

Howarth, F. G., James, S. A., McDowell, W., Preston, D. J., & Imada, C. T. (2007). Identification of roots in lava tube caves using molecular techniques: Implications for conservation of cave arthropod faunas. *Journal of Insect Conservation, 11*, 251–261.

Howarth, F. G., & Stone, F. D. (1990). Elevated carbon dioxide levels in Bayliss Cave, Australia: Implications for the evolution of obligate cave species. *Pacific Science, 44*, 207–218.

Jackson, R. B., Moore, L. A., Hoffman, W. A., Pockman, W. T., & Linder, C. R. (1999). Ecosystem rooting depth determined with caves and DNA. *Proceedings of the National Academy of Science USA, 96*, 11387–11392.

McDowell, W. M. (2002). An ecological study of *Metrosideros polymorpha* Gaud. (Myrtaceae) roots in lava tubes. Master's thesis, University of Hawai'i, 91 pp.

S

SALAMANDERS

Špela Goricki,* Matthew L. Niemiller,† and Danté B. Fenolio‡

*University of Maryland, †University of Tennessee, ‡Atlanta Botanical Garden

Salamanders are a diverse group of vertebrates, exploiting moist cool habitats in a variety of ways. Several lineages have colonized subterranean habitats, particularly in regions of climatic extremes. Salamanders associated with karst exhibit differences in the amount of time they spend in subterranean habitats, their dependence on these resources, and morphological and behavioral adaptations to life underground (troglomorphisms). Differences are manifested within and across taxonomic groups as well as between geographical regions. North America and Europe host the greatest number of known cave-dwelling species. Cave-dwelling salamanders may inhabit subterranean environments for significant portions of their life cycle but not all of it, others migrate to the surface periodically, and still others are exclusively found underground. Sources of general information on salamanders are Amphibia Web (2011), Grossenbacher and Thiesmeier (1999), and Lannoo (2005).

Salamanders are the only tetrapods that exhibit an exclusive subterranean existence. Such troglobitic salamanders belong to the families Plethodontidae (in North America) and Proteidae (in Europe). Troglomorphisms seen in troglobitic salamanders vary from species to species and even from locality to locality, but they frequently include reduced eyes and pigmentation, reduced number of trunk vertebrae, elongated appendages, and modified head shape with increased dentition. Salamanders are opportunistic predators, feeding on a variety of small invertebrates and will often also consume their own kind. There are records of notable dietary oddities in troglobitic salamanders, including feeding on bat guano and silt. Troglobitic salamanders are usually found in caves where fishes do not occur, and function as top predators in those subterranean waters. Because of the constant water temperatures in their habitat, they are likely active year-round. For orientation, feeding, and mating, they rely on mechano- and chemosensory cues, which corresponds to progressive development of extraoptic sensory systems. Salamanders from subterranean environments usually have lower metabolism, efficient energy (fat) storage, and a longer life span than their surface counterparts. They have fewer, larger eggs and offspring compared to related surface-dwelling species. The rate of development in amphibians strongly depends on environmental factors, most notably ambient temperature. When compared with related epigean species that reproduce at the same water temperature, troglobitic salamanders usually develop more slowly. All begin their life inside a gelatinous egg capsule, deposited in water. Eggs hatch into free-swimming aquatic larvae, with visible bushy gills and a tailfin. Instead of metamorphosing into mature adults, however, most troglobitic salamanders attain sexual maturity while retaining these and other larval characteristics—a condition known as *neoteny* or *paedomorphism*.

We can only speculate about the origins and time required for the evolution of obligate subterranean existence. While the ancestors of some troglobitic species may have colonized caves as early as in the Miocene, it is more likely that subterranean colonization occurred much later, even as late as during the Pleistocene. Salamander fossils are rare and the oldest found that may belong to a cave-dwelling species date to the Pleistocene.

The following sections portray all ten currently described troglobitic salamanders. Accounts also are provided for their close relatives that show affinities to subterranean habitats.

OLMS AND WATERDOGS (PROTEIDAE)

The family includes six species of surface-dwelling mudpuppies or waterdogs (*Necturus* sp.) from eastern and central North America and the troglobitic olm (*Proteus anguinus*) found in southeastern Europe. These

Olm (*Proteus anguinus*)

The olm (*Proteus anguinus*) is the only European obligate cave-dwelling vertebrate. It has been found in over 250 springs and caves in the Dinaric karst on the western Balkan Peninsula, between the rivers Isonzo/Soča (northeastern Italy) and Trebišnjica (Bosnia and Herzegovina). Several populations, however, have been destroyed or severely diminished by dam construction, earth fill, or pollution. The salamander inhabits karst waters from close to the surface to deep in fissures. During flooding or at night it may be found in cave entrances or short distances away from springs.

This elongated salamander reaches 25 cm in total length, although 40-cm-long individuals have also been found. The salamander's predominant mode of locomotion is swimming with eel-like undulations of the body ending in a short, finned, laterally compressed tail (Fig. 1A). The very short limbs, with digits reduced in number to three on the forelegs and two on the hindlegs, are used for crawling on the water bottom and occasionally outside the water. Its muscular head ends with a flattened, blunt snout.

The species is currently divided into two subspecies based on the extent of troglomorphism. Analyses of molecular data suggest, however, that *P. anguinus* may be a complex of several species that independently evolved troglomorphic traits. Mitochondrial DNA data suggest that *P. anguinus* is divided into six divergent lineages. These are about 5 to about 15 million years old, that is, they originated in the Pliocene or Miocene. Conversely, it has been hypothesized that troglomorphism might have evolved in less than 500,000 years (Trontelj *et al.*, 2007).

Morphologically, the extremely rare subspecies *Proteus anguinus parkelj*, found only in the springs of two streams near Jelševnik in Bela Krajina (southeastern Slovenia), is believed to resemble the supposed epigean ancestor of both forms. Despite having an entirely subterranean life, this salamander is darkly pigmented (Fig. 1B). Its small but functional eyes are covered with a transparent skin and have well-differentiated lens and retina. Conversely, the pineal gland, which in amphibians controls circadian rhythms, gamete development, and pigmentation changes, is greatly reduced.

In contrast, the widespread, troglomorphic *Proteus anguinus anguinus* is characterized by a yellowish to pinkish-white skin, with few scattered melanophores (pigment cells) invisible to the naked eye (Fig. 1A). Melanin synthesis can be light-induced, and after a prolonged exposure to light, the animals turn dark. The skin is thinner than in the pigmented form, and contains fewer multicellular glands. The head and snout of *P. a. anguinus* are elongated and flattened, and

FIGURE 1 The olm (*Proteus anguinus*). (A) Described in 1768, this blind salamander was the first scientifically documented cave-dwelling animal, but its existence had already been known long before. Its earliest representation may be a Venetian stone carving from the tenth or eleventh century. (B) A rare, pigmented individual with eyes, from southeastern Slovenia. This unique population was discovered toward the end of the twentieth century. (C) Female protecting her clutch of eggs from predation by conspecifics. *Photos courtesy of G. Aljančič, Tular Cave Laboratory. Used with permission.*

salamanders are aquatic throughout their life, retaining external gills, but are also able to breathe air through their skin and with their sac-like lungs. All representatives of this family are obligate paedomorphs; paedomorphism in *Proteus* is thus seemingly not a result of conditions particular to the subterranean environment. Thyroid glands, which control metamorphosis, are functional in all species of this family. Metamorphosis cannot be induced artificially, although changes in the skin structure toward a metamorphosed form have been observed in *Proteus*. Furthermore, it has been shown that *Necturus* possesses functional thyroid hormone receptors and its tissues are not generally irresponsive to the hormone (Safi *et al.*, 2006). Rather, the absence of metamorphosis may be due to loss of function of thyroid hormone-dependent genes required for tissue transformation.

the cranial bones are longer and the teeth more numerous than in the pigmented form. Cervical vertebrae are often elongated as well. Compared to the pigmented form, *P. a. anguinus* has a shorter trunk, but a proportionately longer tail and extremities. Morphometric variability among genetically and hydrologically isolated populations can be quite high.

The eyes of this salamander are greatly reduced in size and structure, and lie embedded deep in the hypodermal tissue. While vision is lost, the eye is still capable of detecting light. The pineal gland is also greatly reduced, but may still be capable of light detection and hormonal activity. As is common in salamanders, the skin and midbrain are also photosensitive. Extraoptic sensory systems, including the lateral line, inner ear, ampullary electroreceptors, olfactory epithelium, and taste buds, are well developed and believed to play a crucial role in prey detection and communication among individuals. Furthermore, behavioral experiments suggest that the animals may use the Earth's magnetic field to orient themselves.

Because of its very low metabolic rate and efficient fat storage, predominantly in the liver and tail, *P. anguinus* can survive prolonged periods of food deprivation. When food in the form of small subterranean crustaceans, mollusks, or surface insect larvae is abundant, it becomes a voracious and efficient predator.

P. anguinus is a social animal, recognizing and communicating with conspecifics through scent. Both males and females use water-borne chemical signals to locate each other, but in order to recognize sex and reproductive state, they need to come into direct contact. Outside of the breeding period, aggressive behavior is reduced and the animals aggregate at communal resting places under stones and in cracks. Long periods of inactivity are punctuated by sessions of foraging or exploratory behavior. The animals show no apparent daily rhythm of activity and resting.

Little is known about timing of reproduction. A female is thought to reproduce only every 12 years on average. Up to 70 eggs are attached under rocks, and are guarded by the female until hatching (Fig. 1C). Larvae hatch after 4 to 6 months. The newly hatched larvae are 2.5 cm long and have incompletely formed limbs, but well-differentiated eyes and scattered, visible melanophores in the skin. During early larval development, the eyes only slightly increase in size without further differentiation, and then gradually sink into the surrounding tissue. After four months, major degenerative changes appear which cannot be prevented by illumination. Sexual maturity is reached at a size of about 20 cm, which is attained after about 15 years. *P. anguinus* is thought to have the longest life span of all amphibians, living for 70 years or more.

LUNGLESS SALAMANDERS (PLETHODONTIDAE)

Salamanders of the family Plethodontidae are widely distributed in North and Central America, with populations also in South America, southern Europe, and on the Korean Peninsula. Lacking lungs, these salamanders breathe primarily through their skin (cutaneous respiration). Pheromones play an important role in communication, and the characteristic nasolabial grooves present in metamorphosed adults allow fine-tuning in chemoreception. A great diversity of lifestyles has evolved in this group, from terrestrial, arboreal, to semi-aquatic and aquatic; life histories range from biphasic and larval reproduction to direct development and possibly even vivipary. It is no surprise to find in this family species associated with caves. The family includes nine described troglobitic species, all found in central and eastern North America. They are grouped in the genera *Eurycea*, *Haideotriton*, and *Gyrinophilus*. Formerly used genera *Typhlotriton* and *Typhlomolge* have been synonymized with *Eurycea*.

Brook Salamanders (*Eurycea*)

This is a diverse genus of 26 recognized species, five of which are troglobites. Except for *Eurycea spelaea*, the troglobitic species inhabit subterranean waters of the Edwards Plateau and Balcones Escarpment in central Texas, in the United States. Many species of *Eurycea* in this region are spring-dwelling troglophiles, and additional troglophilic *Eurycea* are encountered in caves outside the region. Paedomorphism is common in both cave- and spring-dwelling salamanders. *Eurycea spelaea* is the only metamorphosing troglobitic member of the genus.

Texas Cave and Spring-Dwelling *Eurycea*

All species of *Eurycea* from the Edwards Plateau in Texas share a common ancestor. It has been hypothesized that about 15 million years ago a split into two lineages occurred, corresponding to the geographic divide imposed by the Colorado River. The resulting lineage south of the divide later gave rise to a group of extremely troglomorphic cave-dwelling species collectively called Typhlomolge and its sister lineage comprised of less troglomorphic cave- and spring-dwelling species called Blepsimolge. For the spring-dwelling species north of the Colorado River the name Septentriomolge has been coined. Each troglobitic species has a surface counterpart inhabiting the spring outflows of the same watershed or geological formation. The complex interspecies relationships along with

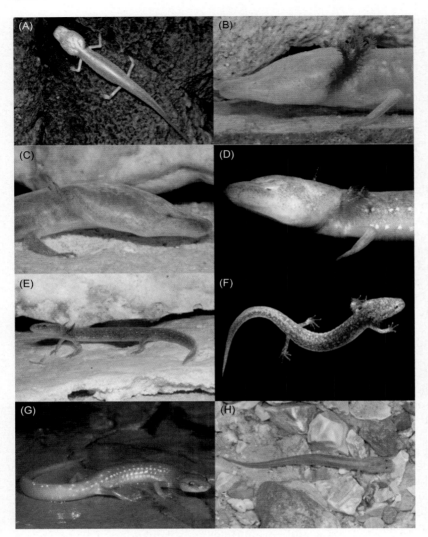

FIGURE 2 Brook salamanders (*Eurycea*). Troglobitic *E. rathbuni* (A) and (B), *E. waterlooensis* (C), and *E. tridentifera* (D) from the Edwards Plateau have degenerate eyes and reduced pigmentation compared to spring-dwelling populations of *E. sosorum* (E) and *E. latitans* (F). Adult (G) and larva (H) of *Eurycea spelaea* from the Ozark Plateau. This species is the only troglobitic brook salamander to readily undergo metamorphosis in nature. *Photos by D. Fenolio.*

their geographic and ecological segregation indicate that an ancient speciation event giving rise to the subterranean and largely epigean species was followed by a rapid adaptive radiation and more recent colonization of subterranean habitats by still more species. Evidence of intermittent gene flow between species that diverged millions of years ago has been detected. Conversely, in some of the species with relatively broad ranges, there is generally very little migration between geographically distant populations, with "distant" sometimes referring to as little as a few hundred meters (Chippindale *et al.*, 2009; Lucas *et al.*, 2009).

Typhlomolge (Blind Salamanders)

All species are neotenic troglobites with very long and thin limbs, short trunk, broad and flattened snout, and a virtually pigmentless and translucent skin (Fig. 2A). They have a shimmering white appearance due to the reflective connective tissue that lies beneath the skin. The rudimentary eyes may be visible as tiny dark spots under the skin, and possibly still detect light. These salamanders are moderately small with the total length of adults ranging from 7–14 cm. Their diet consists of a variety of small aquatic subterranean invertebrates, including crustaceans and snails. When feeding, the salamanders usually probe the bottom using lateral movements of their spatulate head.

The Texas blind salamander (*Eurycea rathbuni*) has been found in caves, wells, and pipes that intersect the San Marcos Pool of the Edwards Aquifer, although it may have a wider geographic range than previously thought. Molecular analyses revealed high levels of genetic variation and the possibility that two distinct species occur within *E. rathbuni*. In springs immediately above the subterranean waters occupied by *E. rathbuni*, *E. nana* (see below) is locally abundant. A captive breeding program has been established for both species.

Artificial induction of transformation in *E. rathbuni* resulted in only partial metamorphosis, with some tissues (*e.g.*, skin) continuing to show only larval features. This salamander has an extremely depressed and broad anterior part of the head (Fig. 2B), with elongated anterior cranial bones, and the longest limbs of all troglomorphic *Eurycea*. The eyespots are barely visible. Juveniles have darker pigmentation and proportionally larger eyes.

In contrast, the olfactory system is well developed and plays a very important role in the social behavior of this salamander (Epp *et al.*, 2010). Nothing is known about social interactions or breeding of this species in nature; however, they have been observed in captivity. These salamanders rarely show aggressive or social behavior. The male of *E. rathbuni* lacks mental (chin) and caudal hedonic glands used during courtship in other *Eurycea* salamanders. Nevertheless, the species has an elaborate tail-straddling walk similar to that which has been observed in other plethodontid salamanders, albeit somewhat simplified. In captivity, these salamanders can live for over 10 years.

The Blanco blind salamander (*Eurycea robusta*) was described based on a single specimen collected from a well drilled in the bed of the Blanco River east of San Marcos. More specimens had been collected at the time, but birds consumed two individuals while they were in a bucket and one preserved specimen has vanished. Molecular data are not available for this species, but the geological formation in which it occurs is hydrologically isolated from that in which the geographically proximal *E. rathbuni* is found. The epigean salamander found in the springs of the Blanco River drainage is *E. pterophila* (see below). Compared to *E. rathbuni*, the body of *E. robusta* is stockier, with longer trunk, and shorter, relatively robust limbs. The head also is wider posteriorly and more massive. The eyes are extremely reduced and invisible through the skin.

The Austin blind salamander (*Eurycea waterlooensis*) is known only from the outflows of Barton Springs in Austin, where individuals presumably are washed out from the Barton Springs segment of the Edwards Aquifer. This species is much more rarely encountered than the sympatric surface-dwelling *E. sosorum* (see below), both of which are included in a captive breeding program. This species is similar to *E. rathbuni*, but has a wider head (Fig. 2C). The eyespots are visible and superficially resemble those of *E. rathbuni*. The limbs are shorter and the skin is slightly more pigmented than in *E. rathbuni* and *E. robusta*. The tail fin is weakly developed.

Blepsimolge (Sighted Salamanders)

A large majority of these salamanders are neotenic and completely aquatic. Besides those listed below, the group includes several populations of yet undescribed species. The salamanders of this group are primarily surface-dwelling, found in the immediate vicinity of spring outflows, but they also inhabit caves, fissures, and sinkholes. Various species have independently evolved cave-associated morphological traits, which has become more apparent as new populations are discovered and genetically analyzed (Bendik *et al.*, 2009).

These small salamanders (6–10 cm long) usually have a dark, finely patterned dorsal coloration and translucent ventral parts (Figs. 2E,F). The intensity and pattern of dorsal coloration may vary due to an irregular pattern of melanophores and reflective white iridophores. Single or double dorsolateral rows of light spots extend along the body. The species differ in eye size, head shape, tail fin shape, and color pattern on the tail and flanks. Certain cave-dwelling populations exhibit marked reduction of eyes and pigmentation. The skin of the troglomorphic animals has a light yellowish color, with diffuse gray or brown spots dorsally and lighter spots laterally (Fig. 2D). Juveniles are darker than adults. The snout is flattened but truncated. The eyes are regressed, sometimes lacking a lens, but are usually still visible through the skin as dark spots; they are larger than the rudimentary eyes of *E. rathbuni* and other members of the Typhlomolge group. The trunk is short, containing fewer vertebrae, and the limbs are somewhat elongated.

The life history of most species is poorly known and based predominantly on observations of *E. nana* and *E. sosorum* in captive breeding programs (e.g., Najvar *et al.*, 2007). The spring-dwelling *Eurycea* are fairly sedentary, although they may seasonally move between surface and cave habitats. On the surface, these secretive salamanders usually hide under rocks, in gravel substrate, or among aquatic plants. They can survive temporary drying of ephemeral springs, presumably by retreating to subsurface refugia. They are active throughout the year and appear to reproduce during all seasons. Up to 70 eggs are laid singly and do not receive parental care. In captivity, the salamanders usually live for about 4 years, but an individual of *E. sosorum* survived to at least 12 years.

Historically, many spring-dwelling populations from throughout the Edwards Plateau were assigned to the Texas salamander (*Eurycea neotenes*) based on morphological similarity, but they were found to differ genetically. This species is now restricted to the springs in the vicinity of Helotes, near San Antonio. A disjunct and genetically divergent population inhabits the area of Comal Springs in New Braunfels.

One salamander that has been elevated to species from *E. neotenes* is the Barton Springs salamander (*Eurycea sosorum*), known from the Barton Springs pool and adjacent springs in Austin. This species is primarily surface-dwelling, but appears to reproduce in subterranean conduits. Skin pigmentation in *E. sosorum* varies considerably, but all individuals have very small eyes and somewhat elongated limbs. The flattened, slightly elongated head ends with a truncated snout (Fig. 2E). The subterranean portion of the Barton Springs Aquifer is inhabited by the troglobitic *E. waterlooensis* (see above).

The San Marcos salamander (*Eurycea nana*) is abundant in the Spring Lake pool at the source of San Marcos River, the only site where it occurs. Metamorphosis has been induced artificially in this species.

The Fern Bank salamander (*Eurycea pterophila*) inhabits springs and caves of the Blanco and Guadalupe River drainages. Most populations are morphologically similar to *E. neotenes*, to which they were formerly assigned, but cave-dwelling individuals with reduced eyes and pigmentation have also been identified. This species is genetically different from *E. neotenes*, but similar to, perhaps even conspecific with, *E. latitans* and/or *E. tridentifera* (see below).

The Cascade Caverns salamander (*Eurycea latitans*) is possibly a complex of several species found in caves and springs of the Cibolo Creek basin and south of the Guadalupe River in the southeastern part of the Edwards Plateau. For some time before initial molecular analyses were completed, most populations were regarded as *E. neotenes*, and the population at the type locality (Cascade Caverns) was considered to be hybrid between *E. neotenes* and *E. tridentifera* (see below). This population includes individuals with a spectrum of morphological features, ranging from highly troglomorphic, most similar to those of *E. tridentifera*, to surface-like (Fig. 2F), most similar to what was historically considered *E. neotenes*. Recent molecular analyses suggest that at least part of *E. latitans*, including the Cascade Caverns population, is conspecific with *E. tridentifera*. Furthermore, a few cave populations were determined genetically to be hybrids between *E. latitans* and *E. neotenes*, and in two cases individuals of both species were located within the same cave.

As currently recognized, the Comal blind salamander (*Eurycea tridentifera*) is a troglobitic species. It exhibits morphological modifications similar to the blind salamanders of the Typhlomolge group, although generally not to the same extreme extent (Fig. 2D). The salamander is found in caves of the Cibolo Creek basin and south of the Guadalupe River, in the southeastern part of the Edwards Plateau. These populations may actually be cave populations of *E. latitans* that have become troglomorphic. At the entrance to the type locality (Honey Creek Cave), individuals intermediate in morphology between *E. tridentifera* and the surface species have been found.

The Valdina Farms salamander (*Eurycea troglodytes*) is likely a complex of several species found in the southwestern part of the Edwards Plateau. Most populations in this species complex are neotenic. However, natural metamorphosis has been observed in populations from several springs and caves. Two groups of populations within this complex are troglomorphic, with enlarged head, and reduced eyes and pigmentation. The morphologically variable population from the type locality (Valdina Farms Sinkhole) was for some time considered a hybrid between *E. neotenes* and *E. tridentifera*. This population is now extinct, but analyses of others have revealed that *E. troglodytes* is genetically distinct from both *E. neotenes* and *E. tridentifera*.

Septentriomolge (Northern Salamanders)

Most populations of these spring-dwelling species from north of the Colorado River were discovered after 1995. The few known prior to 2000, when these species were formally described, had been considered peripheral isolates of *E. neotenes*. Morphologically they resemble spring-dwelling species south of the Colorado River (Blepsimolge), described above. They can be seen year-round, most easily during the spring and summer months, but small juveniles are rarely observed.

The San Gabriel Springs salamander (*Eurycea naufragia*) is known from springs and caves in the drainage of the San Gabriel River, in the vicinity of Georgetown. The adults have prominent eyes with melanophores concentrated around them. The range of the Jollyville Plateau salamander (*Eurycea tonkawae*) is limited to a few drainages on the Jollyville Plateau segment of the Edwards Aquifer. This species makes extensive use of subterranean aquatic habitat, especially when surface spring flow decreases. Adults have well-developed eyes, broad jaws, and blunt snouts. The rare Chisholm Trail salamander (*Eurycea chisholmensis*) is known only from springs at Salado. Compared to *E. tonkawae* and *E. naufragia*, the eyes of this salamander are reduced, and the head is flattened and elongated.

Other Troglobitic and Troglophilic *Eurycea*

Several species of *Eurycea* with overlapping ranges occur syntopically in caves and spring-fed headwaters within the Ozark and Appalachian Highlands and the Interior Lowlands of North America. These include *E. tynerensis*, *E. lucifuga*, *E. longicauda*, and the troglobitic *E. spelaea*.

The grotto salamander (*Eurycea spelaea*) is restricted to the Springfield and Salem Plateaus in the Ozark region of southern Missouri, extreme southeastern Kansas, and adjacent areas in Arkansas and Oklahoma (U.S.A.). DNA sequence divergence within *E. spelaea* is relatively large, and some of the populations may represent distinct species. This is the only known troglobitic member of the genus that undergoes complete metamorphosis. Adults are known only from caves, but a large fraction of the larval population develops in surface springs and streams. Here they displace the larvae of sympatric *Eurycea* species to more ephemeral parts of the stream and where predatory fishes and crayfish are present. Both adults and larvae are sensitive to floods, which can extirpate a significant proportion of a population.

Adults grow to 7–14 cm in length and are white, pinkish white, or light brown in color (Fig. 2G). The eyes are reduced in size and structure, visible as raised dark spots through the fused eyelids. The larvae are tan dorsally, often spotted or mottled; those developing inside caves can be pale to pink (Fig. 2H). They have relatively small but functional eyes, which degenerate in old larvae or during metamorphosis. The extent of retinal degeneration in adults is related to postmetamorphic age but there is variability in each age group. In its terminal stage, reduction encompasses the entire area of the retina and vision is lost. Individuals raised in light maintain some pigmentation or become pigmented. The larvae may retain vision longer than those kept in darkness and the eyelids in adults do not fuse.

Adults congregate in the main caverns where colonial bats roost, most likely for the purposes of feeding and breeding. They are most active during spring and summer, when moisture levels are high and food is abundant. They may be found climbing moist vertical rock walls outside water, but commonly return to the water to hunt aquatic invertebrates. Besides aquatic and terrestrial invertebrates they also feed on bat guano (Fenolio et al., 2006). In caves where colonial bats have declined, emaciated individuals have been observed. Mating in late spring is followed by oviposition through summer and fall. Larvae metamorphose 2–3 years after hatching, at the size of 5–12 cm, and mature soon afterward.

The sister species of *E. spelaea*, the troglophilic Oklahoma salamander (*Eurycea tynerensis*), is also found on the Ozark Plateau. The species comprises both strictly aquatic neotenic populations and more terrestrial metamorphic populations. Neotenic animals that inhabit caves are often pale, whereas those on the surface retain normal larval pigment patterns. The metamorphs rarely venture far away from water. Usually they inhabit cool, moist habitats both on the surface and in the twilight zone of caves. Larvae may use the karst cavities to move within or between streams.

The troglophilic cave salamander (*Eurycea lucifuga*) occurs in the broad region between southern Indiana, northwestern Virginia, northern Alabama, and northeastern Oklahoma (U.S.A.); this includes the karst areas of the Ozarks, Interior Low Plateau, and the Appalachians. The twilight zone of caves is the preferred habitat of this species, although it can also be found deeper into caves and, despite its name, outside of caves in forested limestone ravines, springs, and spring-fed streams. The species is broadly sympatric with the long-tailed salamander (*Eurycea longicauda*), which is distributed throughout a similar area. This species is also frequently associated with caves and mines. These slender salamanders have very long tails and grow to 18–19 cm. They are very conspicuous, colored yellow or orange, and covered with black spots (Figs. 3A,B). *E. lucifuga* has long limbs and a prehensile tail used in climbing rock walls and ledges. The eyes of this salamander are large and capable of twilight vision. Adults can be found in caves year-round, but their numbers fluctuate seasonally. They aggregate there between spring and fall, when surface temperatures are high. During this time the females deposit their eggs in subterranean waters. Larvae hatch after rains, when water flow increases and more food becomes available (Ringia and Lips, 2007).

Georgia Blind Salamander (*Haideotriton wallacei*)

The geographically isolated troglobitic Georgia blind salamander (*Haideotriton wallacei*) has been found in a few caves pertaining to the Floridan Aquifer below Georgia, Florida, and Alabama (U.S.A.). At least two historic localities have been destroyed by human activities, but cave divers have since spotted this species in the corresponding portion of the aquifer. This species is sometimes included in the genus *Eurycea*.

Adults of this neotenic salamander measure 5–8 cm in length and have little pigmentation, being mostly pink or whitish with scattered melanophores on the back and sides. The head is broad and the snout is long, but not flattened. The limbs are slender and the eyes are almost invisible, embedded in a mass of adipose tissue below the skin. Juveniles are slightly more pigmented and the eyespots are visible.

Haideotriton wallacei can be seen in cave pools and streams especially in caves where bats defecate over or near the water. Salamanders move about slowly, resting on bottom sediments or climbing on submerged limestone sidewalls and ledges. Deeper in subterranean tunnels the salamanders are much less common, presumably because their food (benthic invertebrates) is

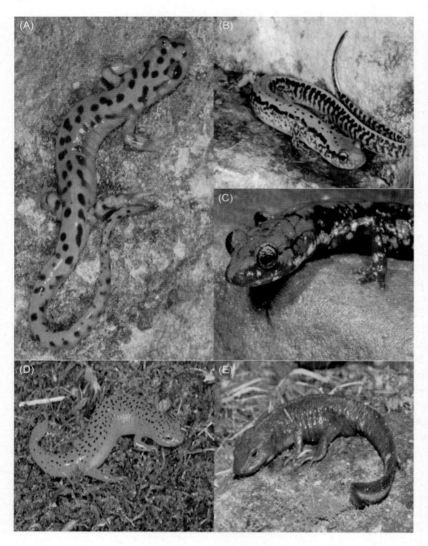

FIGURE 3 A few salamanders frequently found in caves. In eastern North America, *Eurycea lucifuga* (A) and *E. longicauda* (B) frequently inhabit the same cave. (C) *Plethodon petraeus* is associated with karst and caves on Pigeon Mountain in Georgia. (D) *Pseudotriton ruber* also frequently uses caves for reproduction (Miller et al., 2008). (E) This undescribed species of *Paramesotriton* inhabits pools in the twilight zone of caves in China. *Photos (A)–(C) by D. Fenolio; (D)–(E) by M. Niemiller.*

scarce. Cave stream sediments have been found in the digestive tracts of *H. wallacei*. One hypothesis attempting to explain silt feeding holds that *Haideotriton* intentionally ingests the material to digest biofilm and microorganisms in the sediments (the same has been observed in young *Proteus anguinus*). Another hypothesis posits that the silt represents failed feeding attempts.

Spring Salamanders (*Gyrinophilus*)

Four North American species are currently recognized. Three species are troglobites, found in caves of southern and central Appalachian Highlands (U.S.A.). In *Gyrinophilus*, paedomorphosis is not universal and has appeared after colonization of caves, although the metamorphosing nontroglobitic species already has an extremely long larval period.

The troglophilic spring salamander (*Gyrinophilus porphyriticus*) is common in and around small streams and springs in eastern North America. Next to *Eurycea lucifuga* this is the most frequently encountered salamander in caves. Both larvae and adults can be found in caves throughout the Appalachian Highlands, even considerable distances from the entrance. These large salamanders, which can grow over 20 cm in total length, are notorious for supplementing their invertebrate diet with conspecifics and other salamanders. Metamorphosed animals are usually orangish-red to salmon and turn darker with age (Fig. 4B), but some individuals of cave populations are pale. The eyes of cave-dwelling individuals do not differ from those from surface populations. A much higher occurrence of ingested debris found in cave-dwelling individuals compared to individuals from surface streams may indicate a lower feeding efficiency in the dark. Larvae, which are also often pale in color, can grow up to 16 cm. They metamorphose after 3–5 years in surface populations but perhaps after 10 years or more in cave populations.

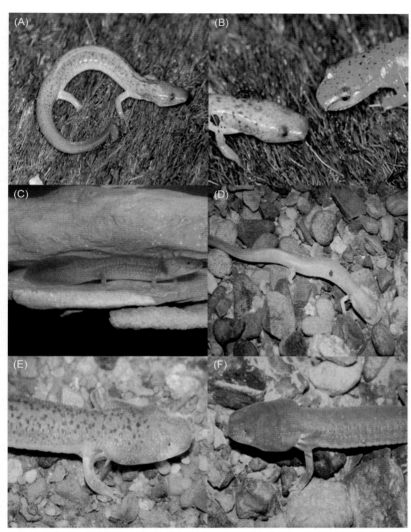

FIGURE 4 Spring salamanders (*Gyrinophilus*). Metamorphosed adults of *G. palleucus* (A) and left in (B) are pale and have smaller, degenerate eyes compared to the troglophile *G. porphyriticus* (right in (B)). The larvae of *G. subterraneus* (C), *G. palleucus palleucus* (D), *G. palleucus necturoides* (E), and *G. gulolineatus* (F) have small but functional eyes. In addition, the numerous neuromast mechanoreceptors on the head and flanks enable them to detect vibrations in the water. *Photos (A)–(B) and (D)–(F) by M. Niemiller; photo (C) by D. Fenolio.*

The three troglobitic species are believed to have arisen independently from a single epigean ancestor similar to *G. porphyriticus* as early as 2.5 million (Pliocene) to as recently as 60,000 years ago (Pleistocene), but precise phylogenetic relationships are obscured due to recent speciation (Niemiller et al., 2008, 2009). Nonetheless, other evidence supports their independent origins and DNA data indicate that divergence has occurred in the presence of continuous or periodic gene flow between subterranean populations and their surface-dwelling progenitor.

The Tennessee cave salamander (*Gyrinophilus palleucus*) is found in caves in the Central Basin, Highland Rim, and Cumberland Plateau of central Tennessee, northern Alabama, and extreme northwestern Georgia. Its range is on the periphery of that of *G. porphyriticus*. Two subspecies of *G. palleucus* are recognized: the pale salamander (*Gyrinophilus palleucus palleucus*; Fig. 4D), which inhabits caves in the Lower Tennessee River watershed in Alabama and Tennessee, and the big mouth cave salamander (*Gyrinophilus palleucus necturoides*), found in the Collins, Duck, Elk, and Stones River drainages in Tennessee. This form is tan with dark spots dorsally and laterally (Fig. 4E). The subspecies also differ slightly in head width, leg length, eye size, and the number of trunk vertebrae. Except for the lost ability for eye accommodation (focusing), vision is not greatly impaired in *G. palleucus* (Besharse and Brandon, 1973). Exposure to light has no effect on eye development and does not induce any change in skin pigmentation.

G. palleucus inhabits sinkhole-type caves that are rich in nutrients, which support its invertebrate prey base. Besides cave streams, where it is easily observed, the salamander is believed to inhabit subterranean waters inaccessible to humans. It has occasionally been found in springs outside caves. Little is known about the life history of this species. Breeding is most likely seasonal, occurring in late autumn or early winter. This salamander is typically paedomorphic, although metamorphosed individuals have been encountered in at least four caves. Metamorphosed individuals of

G. palleucus differ from adult *G. porphyriticus* in their pale skin, gaunt appearance, longer and narrower snout and smaller eyes (Fig. 4A,B) and in retaining the larval characteristic of an undivided premaxillary bone.

The Berry Cave salamander (*Gyrinophilus gulolineatus*) is known from caves in the Ridge and Valley of eastern Tennessee. Its range is entirely contained within the range of *G. porphyriticus* and the two species sometimes occur in the same cave (Miller and Niemiller, 2008). Like *G. palleucus*, *G. gulolineatus* is usually paedomorphic and generally resembles the former (Fig. 4F), but can be distinguished from it by having a darker pigmentation, wider head, more spatulate snout, and by attaining a greater adult size. The eyes are small and degenerate, comparable in size to the eyes of *G. palleucus*. Unlike *G. palleucus*, however, transformed individuals have fully divided premaxillae, in which they resemble *G. porphyriticus*.

The West Virginia spring salamander (*Gyrinophilus subterraneus*) is known only from General Davis Cave in West Virginia where it co-occurs with *G. porphyriticus*. Unlike *G. palleucus* and *G. gulolineatus*, *G. subterraneus* regularly metamorphoses, although at an exceedingly large size (up to 18 cm). Large larvae have fully developed, mature gonads. Both larvae and adults have a light brownish-pink skin color, overlaid dorsally and laterally by a darker reticulate pattern (Fig. 4C). The head of larval *G. subterraneus* is broader than the heads of either *G. porphyriticus* or *G. palleucus*, but does not have the spatulate snout typical of *G. gulolineatus*. Metamorphosed adults retain fused premaxillae (as in *G. palleucus*). Both larvae and adults have smaller eyes than *G. porphyriticus* and, unlike *G. palleucus*, only weak visual perception.

Although *G. porphyriticus* also occurs in General Davis Cave, only adults of this species have been found (Niemiller *et al.*, 2010). *G. subterraneus* has been observed as far as 2 km into the cave; however, in the first 300 m of the stream passage, predominantly larvae have been encountered, suggesting that its main habitat may be located deeper into the cave. The larvae prefer shallow, calm pools and are found in water depths of 1–20 cm. Adults can be found in shallow water as well, whereas *G. porphyriticus* almost never enters the water. The banks of the cave stream contain thick deposits of decaying leaf litter, washed into the cave by floods. This leaf litter is the source of nutrients for cave invertebrates that adults of both species putatively prey on.

DIVERSITY PATTERNS OF SALAMANDERS FOUND IN CAVES

Including the salamanders described above, over 90 species (approximately 15% of all known salamander species) belonging to five families have been reported from natural or manmade subterranean environments. Their dependence on the cave environment varies not only between species, but also among populations of the same species.

Over half of all species found in caves are reported from eastern North America. Several woodland salamanders (*Plethodon* sp.), especially *Plethodon albagula*, *P. glutinosus*, *P. dorsalis*, and *P. petraeus* (Fig. 3C) are frequently associated with caves. Also reliant on caves are the red salamander (*Pseudotriton ruber*; Fig. 3D), a few dusky salamander species (*Desmognathus* sp.), and the green salamander (*Aneides aeneus*). To the south, in the Mexican Sierra del Madre Oriental and Sierra Madre del Sur, two splayfoot salamanders (*Chiropterotriton* sp.) and two false brook salamanders (*Pseudoeurycea* sp.) occur in caves. Of the salamanders that occur in western North America, the web-toed salamanders (*Hydromantes* sp.) in California, especially the shasta salamander (*Hydromantes shastae*), are often found in caves. Apart from the plethodontids, the mole salamanders (*Ambystoma* sp.) of the family Ambystomatidae are regularly encountered in caves, particularly in the region from Arizona to Alabama and Tennessee. Neotenic cave-dwelling populations of the barred tiger salamander (*Ambystoma mavortium*) have been reported from New Mexico and in gypsum caves of northeastern Oklahoma.

Europe's salamander fauna is dominated by the family Salamandridae, but seven species of the plethodontids also occur here. The European cave salamanders (*Speleomantes* sp. and *Atylodes* sp.) of Sardinia, northwestern mainland Italy, and southern France are closely related to the Californian *Hydromantes* and share many life history features with them. They are strongly associated with caves, and will aggregate in the twilight zone. Although they have prominent eyes, they can also detect prey in darkness using olfactory information. Several surface species of the family Salamandridae frequent natural or manmade subterranean refugia in Europe, especially in the arid circum-Mediterranean region. The widespread fire salamander (*Salamandra salamandra*) not only seeks refuge in caves but also frequently favors them as reproduction sites even when surface water is available nearby. The Pyrenean mountain newt (*Calotriton asper*) inhabits streams, lakes, and ponds along the Pyrenees Mountains of France and Spain, but is also found in subterranean waters, especially on the periphery of its range. Except for one pale-colored cave population, no morphological modifications have been described in this species. Instead, modifications related to cave life involve the reproductive cycle and its periodicity, including prolonged or continuous gamete production, slower egg deposition, and delayed metamorphosis. Neotenic individuals have also been found.

Despite extensive karst areas in Central America and Asia, and great salamander diversity (especially in Central America), the number of salamanders that inhabit caves and other subterranean habitats is low. The caves on the Yucatán Peninsula in Mexico and adjacent Guatemala are the temporal habitat of the plethodontid Yucatán salamander (*Bolitoglossa yucatana*), (perhaps) the nimble long-limbed salamander (*Nyctanolis penix*), and potentially two more species. In western Asia, only one species, of the family Hynobiidae, has been found in a cave. The almost fully aquatic Gorgan mountain salamander (*Paradactylodon gorganensis*) is known from two localities in the Alborz Mountains of northern Iran, one of which is a cave. In eastern Asia, another hynobiid species, the oki salamander (*Hynobius okiensis*) from the Dogo Island of Japan, is believed to reproduce in hypogean waters. The salamandrid wart newts (*Paramesotriton* sp.) have been observed in the twilight zone of caves in Guizhou Province, China.

CONCLUSION

Very few salamanders are obligate cave-dwellers, which is in contrast to the great number of species that temporarily inhabit caves. Ranges of troglobitic salamanders are usually small, limited by hydrological or geological barriers. Due to the limitations such as inaccessibility or rarity, only the more common species have been studied in detail. A fair amount of attention has been given to their phylogenetic relationships, observations on regressive morphological modifications associated with cave life, development, and behavior, but limited knowledge exists of their life histories and ecology. Continuing research in these areas is essential for gaining insight into the origins and processes involved in evolution of obligate subterranean existence.

Because of suspected small population sizes, limited distribution, and high specialization, troglobitic salamanders are threatened by habitat and water quality degradation, caused mainly by urbanization, agriculture, and deforestation. Their position in the food web and the tendency to accumulate energy reserves in their bodies make them particularly vulnerable to the effects of toxic chemicals in contaminated water. Most troglobitic salamanders are legally protected. However, in the face of increasing demands for groundwater, pollution, and alterations to the landscape above cave systems, their numbers continue to decline. Not only troglobitic species but also all salamanders that use caves during part of their life history, are important components of subterranean ecosystems and greatly depend on the preservation of these special habitats.

Bibliography

AmphibiaWeb: Information on amphibian biology and conservation (2011). Berkeley, CA: AmphibiaWeb. Available: <http://amphibiaweb.org>.

Bendik, N. F., Gluesenkamp, A. G., & Chippindale, P. T. (2009). The biogeography and rapid radiation of central Texas neotenic salamanders. *Proceedings of the 15th International Congress of Speleology*, 1, 219–225.

Besharse, J. C., & Brandon, R. A. (1973). Optomotor response and eye structure of the troglobitic salamander *Gyrinophilus palleucus*. *American Midland Naturalist*, 89, 463–467.

Chippindale, P. T., Gluesenkamp, A. G., & Bendik, N. F. (2009). Texas cave and spring salamanders (*Eurycea*): New discoveries and new surprises. *Proceedings of the 15th International Congress of Speleology*, 1, 227.

Epp, K. J., Gonzales, R., Jr., & Gabor, C. R. (2010). The role of waterborne chemical cues in mediating social interactions of the Texas blind salamander, *Eurycea rathbuni*. *Amphibia-Reptilia*, 31(2), 294–298.

Fenolio, D. B., Graening, G. O., & Stout, J. F. (2006). Coprophagy in a cave-adapted salamander; the importance of bat guano examined through stable isotope and nutritional analyses. *Proceedings of the Royal Society B*, 273(1585), 439–443.

Grossenbacher, K., & Thiesmeier, B. (Eds.), (1999). *Handbuch der Reptilien und Amphibien Europas. Band 4/I: Schwanzlurche (Urodela) I*. Wiesbaden: AULA (in German).

Lannoo, M. (Ed.), (2005). *Amphibian declines: The conservation status of United States species*. Berkeley: University of California Press.

Lucas, L. K., Gompert, Z., Ott, J. R., & Nice, C. C. (2009). Geographic and genetic isolation in spring-associated *Eurycea* salamanders endemic to the Edwards Plateau region of Texas. *Conservation Genetics*, 10, 1309–1319.

Miller, B. T., & Niemiller, M. L. (2008). Distribution and relative abundance of Tennessee cave salamanders (*Gyrinophilus palleucus* and *G. gulolineatus*) with an emphasis on Tennessee populations. *Herpetology Conservation and Biology*, 3(1), 1–20.

Miller, B. T., Niemiller, M. L., & Reynolds, R. G. (2008). Observations on egg-laying behavior and interactions among attending female red salamanders (*Pseudotriton ruber*) with comments on the use of caves by this species. *Herpetological Conservation and Biology*, 3(2), 203–210.

Najvar, P. A., Fries, J. N., & Baccus, J. T. (2007). Fecundity of San Marcos salamanders in captivity. *Southwestern Naturalist*, 52(1), 145–147.

Niemiller, M. L., Fitzpatrick, B. M., & Miller, B. T. (2008). Recent divergence-with-gene-flow in Tennessee cave salamanders (Plethodontidae: *Gyrinophilus*) inferred from gene genealogies. *Molecular Ecology*, 17(9), 2258–2275.

Niemiller, M. L., Miller, B. T., & Fitzpatrick, B. M. (2009). Systematics and evolutionary history of subterranean *Gyrinophilus* salamanders. *Proceedings of the 15th International Congress of Speleology*, 1, 242–248.

Niemiller, M. L., Osbourn, M. S., Fenolio, D. B., Pauley, T. K., Miller, B. T., & Holsinger, J. R. (2010). Conservation status and habitat use of the West Virginia spring salamander (*Gyrinophilus subterraneus*) and spring salamander (*G. porphyriticus*) in General Davis Cave, Greenbrier Co., West Virginia. *Herpetological Conservation and Biology*, 5(1), 32–43.

Ringia, A. M., & Lips, K. R. (2007). Oviposition, early development and growth of the cave salamander, *Eurycea lucifuga*: Surface and subterranean influences on a troglophilic species. *Herpetologica*, 63(3), 258–268.

Safi, R., Vlaeminck-Guillem, V., Duffraisse, M., Seugnet, I., Michelina Plateroti, M., Margotat, A., et al. (2006).

Pedomorphosis revisited: Thyroid hormone receptors are functional in *Necturus maculosus*. *Evolution and Development*, 8(3), 284–292.

Trontelj, P., Gorički, Š., Polak, S., Verovnik, R., Zakšek, V., & Sket, B. (2007). Age estimates for some subterranean taxa and lineages in the Dinaric Karst. *Acta Carsologica*, 36, 183–189.

SALTPETRE MINING

David A. Hubbard Jr.
Virginia Speleological Survey and Virginia Department of Mines, Minerals and Energy

DEFINITIONS AND IMPORTANCE

Historically, saltpetre is one of the most strategic of commodities. It occurs naturally in caves and rockshelters, but it is rare. A suite of related nitrates occurs in many caves. The mining and processing of cave nitrate-enriched sediments is a relatively simple endeavor, although labor intensive. The tendency of these sediments to contain a suite of nitrates rather than just potassium nitrate is one reason the archaic spelling *saltpetre* is used in reference to the mining of cave nitrates and the caves in which they occur. This convention is followed throughout this article.

The invention of gunpowder revolutionized weaponry and warfare. Gunpowder, also referred to as black powder, consisted of a mixture of saltpetre, sulfur, and charcoal. Although saltpetre was used in the preservation of meats, the greatest historic demand for saltpetre was during times of insurrection and war. Nowhere has the quest for saltpetre contributed to historic events more than in the United States, where this commodity contributed to both the formation of a country and almost its destruction.

The mineral *niter* (synonym, *saltpeter*) is potassium nitrate (KNO_3). Like many other nitrate compounds, niter is deliquescent; that is, it has a natural tendency to draw water to itself and dissolve into a solution. Although deliquescent minerals can absorb moisture from humid air, they occur naturally in sheltered locations under conditions of low humidity or during periods of reduced humidity. The deliquescent nature of saltpeter is the reason for the old warning of soldiers and frontiersmen, who depended on their firearms for survival, to "keep your powder dry!"

Caves and rockshelters, also termed *rockcastles*, are locations where nitrates may accumulate. Analyses of cave sediments, which were mined historically for saltpetre, commonly reveal no nitrate minerals. The reason is that most of the classical saltpetre caves are in regions where the humidity typically is too high for niter and the even more deliquescent minerals *nitromagnesite*, $Mg(NO_3)_2 \cdot 6H_2O$, and *nitrocalcite*, $Ca(NO_3)_2 \cdot 4H_2O$, to crystallize into their solid mineral forms. Instead, the saltpetre-rich sediments, historically termed *petre dirt*, contain concentrated viscose nitrate solutions in the form of sediment moisture.

Because nitrate minerals rarely crystallize in most of the known saltpetre caves, other clues to the accumulation of nitrates in cave sediments are important. In the absence of niter, the foremost evidence of significant nitrate concentrations in cave sediment is the presence of *efflorescent crusts*. These white or light-colored powdery crusts commonly are composed of a mixture of soluble salts and minerals, such as gypsum and calcite, that accumulate on cave sediment and rock surfaces as a result of evaporation. Efflorescent incrustations signify locations where periodic atmospheric conditions allow evaporation and the concentration of the minute amounts of dissolved solids in interstitial soil and rock moisture. Precipitation and concentration drive the wicking action of the dissolved solids through soil and rock pores from their respective remote sources. The sources of most saltpetre cave nitrates are the surface ecosystems overlying saltpetre caves.

Recent microbiological work in caves and karst has shown that bacteria are important in cave development (*speleogenesis*) and in the development of the secondary cave mineral forms (*speleothems*) that were thought to be the result of physiochemical reactions (Taylor, 1999). The importance of nitrifying and other bacteria in the accumulation of efflorescent crusts and nitrate accumulations in sediments (petre dirt) is unknown but probably is not trivial.

SALTPETRE MINING

The mineral niter (KNO_3) is rarely found in caves, but when observed it occurs as clear to white lint-like fibers, acicular (needle-like) crystals, powder, crusts, coralloid, or flowstone forms. The most extensive form observed in saltpetre caves is the lint-like fibers that occur in dense carpets on bedrock walls (Fig. 1) and sediment-covered walls and floors. These niter fiber occurrences can be harvested with the use of a thin wooden spatula or paddle-like scraper, leaving little or no evidence of extraction. Such a wooden scraper was observed high on a Virginia saltpetre cave ledge before this author had observed an efflorescent niter occurrence.

The majority of the documented saltpetre caves does not normally contain crystalline niter. At the humidities typically found in these caves, the deliquescent nitrate accumulations occur as viscose nitrate solutions in efflorescent crusts on rock and sediment of the effects of niter in the preservation of meat may predate gunpowder, no evidence is known that niter was extracted from caves for preservation of meat remote from caves. The importance of gunpowder in revolutionizing armed conflict resulted in intense periods of demand for niter. In addition to mining in rockshelters and caves, saltpetre was obtained from artificial niter beds and from soils collected under buildings. In Western societies, the earliest documented search for niter is the 1490 quest of Hans Breu for saltpetre in Sophienhöhle (Sophie's Cave) in Germany. The French Revolution and Napoleonic wars of Europe (1792–1815) and the Revolutionary (1775–1783), 1812 (1812–1815), and Civil (1861–1865) wars of North America were periods of intense demand for saltpetre. The written record of actual saltpetre mining processes is sparse, as is typical for many mundane tasks. The most extensive known distribution of saltpetre caves and the best-preserved evidence of mining can be found in the southeastern United States. Although the mining evidence in each saltpetre cave is as different as caves are from one another, similar patterns of marks, disturbances, and artifacts convey some information about basic mining and processing methodologies.

Tangible evidence of saltpetre mining includes the principal physical evidence of wall and floor sediment removal, the secondary evidence of lighting and names and artifacts used in mining or as bracing that date to the mining era, evidence of simple sediment processing and separation of rocks and clay balls, and modified pathways and conveyances of miners and mined sediment. Nitrate processing evidence is important in distinguishing saltpetre mining from sediment extraction for other purposes and is discussed in the section on saltpetre processing.

Principal Physical Evidence of Mining

Physical evidence of sediment removal includes tool marks, wall discolor marks of former sediment levels, and tunnels and pits within sediments. Tool marks on worked sediment faces are of wide- and narrow-bladed tools. Sediment mining tool marks predating the 1860s are typically hoe-like and wide bladed (10 to 13 cm), while many of those of the American Civil War era are narrow bladed (5 to 7 cm) and attributed to mattocks. The blades of these tools were metal, and rarely were such tools left in caves, while pointed digging and prying sticks and paddle-like spatulas are more commonly found. Pick and mattock impact marks on rock walls may relate to efflorescent wall crust or sediment removal. The discoloration marks of old sediment levels on bedrock walls provide an

(A)

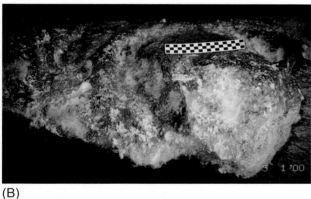

(B)

FIGURE 1 (A) Fibrous form of the mineral saltpeter carpeting cave walls in a Virginia saltpetre cave (*photograph by John C. Taylor*; image width approximately 12 cm). (B) Fibrous form of the mineral saltpeter carpeting cave surfaces in a Virginia saltpetre cave; scale is in centimeters.

surfaces and within sediments and bedrock proximal to the cave.

The earliest mining of saltpetre may have been in China, dating from the development of gunpowder approximately 1000 years ago. Although the discovery indication of the thickness of mined nitrate-rich sediment. An important distinction is that not all sediment in a saltpetre cave is equally enriched or desirable. Slightly worked deposits of difficult to process clay sediments with well-preserved mattock marks and well-developed white efflorescent crusts are evidence that the difficulty in extracting nitrates from clay outweighs the effort. Similarly, pits and tunnels through thick sediments may indicate low yield deposits, which were partially exploited, but such excavations may have had more utility as passage and haulage routes.

Secondary Evidence of Mining

Indirect evidence of mining includes soot stains above torch perches and the stubs of pine faggots (torches) at worked sediment faces and along well-traveled routes between entrance and mining areas. Rock piles in wall niches or backfilled in small passages are evidence of hand-sorting of rock from the sediment. Less common are piles of small rocks and clay balls from sieving near the site of sediment mining or near locations where sediment was transported in cloth or burlap sacks. The combination of old sediment level stains and names dating to the mining era stranded high on walls, as well as remote clusters of tally and torch marks, may differentiate periods of mining. Mining artifacts include pointed wooden digging and prying sticks, wooden scrapers, and wood hoops for holding sacks open during filling. Stone and wood bracing stabilize undermined large rocks in some saltpetre caves. In breakdown mazes, the use of rock for slab and ceiling stabilization may not be obvious. The breakup of rock slabs to access the underlying sediment has resulted in labor-intensive passage enlargement and rock walls and fills that may appear as having other significance. Other modified pathways and conveyances include cut steps in steep sediment banks, stone steps, wooden stairs, boardwalks, and plank ramps across chasms and canyons; various types of ladders; and windlasses.

Perhaps the most rewarding evidence of saltpetre mining is the correlation of names etched into cave walls with written mine-era records of payrolls, equipment receipts, or saltpetre production or sales receipts. Such corroboration sometimes allows the matching of different historic cave names with present-day names as well as verification of the nature of the sediment mining.

SALTPETRE PROCESSING

In some saltpetre caves, evidence of the processing is intermingled with the mining evidence. Large hoppers or smaller vats were commonly used in saltpetre processing, but barrels and sections of hollow logs also were employed. Two of the best-preserved examples of sediment processing equipment are displayed on cave tours: rectangular hoppers in Mammoth Cave in Kentucky and V- or wedge-shaped vats in Organ Cave in West Virginia. Intact hoppers and vats are rare in most present-day saltpetre caves. All that remains in many saltpetre caves are the internal casts of nitrate-depleted mine spoils. Surrounding such casts occur the discarded piles of leached cave sediments, discarded from the vats prior to their last loads of sediment. The wooden remains of many hoppers and vats have rotted or been destroyed by vandals.

Processing requires a source of water, which is not usually located close to the nitrate-rich cave sediments that require at least a periodic reduction in the humidity to accumulate. Processing of the majority of saltpetre cave sediment has entailed the puddling of the nitrate-rich sediments in a hopper or vat. After the nitrates are leached from the cave sediments, the liquid, termed *beer*, is decanted. To this beer another leachate, from wood ashes (potash), is added until a white precipitate ceases to form from the mixing of the two solutions. This process exchanges the potassium cation of the potash solution for the calcium and magnesium cations of the cave nitrate solution. The resulting potassium-rich nitrate solution is transported to large iron kettles for further processing and eventual fractional crystallization of the niter crystals. The kettles are usually hemispherical and without legs or other attachment points that would serve as heat sinks, resulting in differential heating within the kettle. Most kettles have a wide lip or rim by which they are supported over a heat source. The fractional crystallization process enables workers to selectively crystallize specific soluble salts, while leaving more soluble salts in solution.

The processing of sediment, especially for smaller caves, occurred outside of many saltpetre caves for lack of a suitable water source. Some small caves were worked as satellite locations to a nearby major saltpetre works. No evidence of kettle processing is known from within any saltpetre cave, although kettles are known to have been hidden in and recovered from caves.

Perhaps the most impressive examples of evidence of in-cave saltpetre processing are preserved at Mammoth Cave in Kentucky. Some of these War of 1812 era workings may be observed on tours and include large hoppers and a complex system of hollow log plumbing, whereby freshwater was pumped into and saltpetre leachate was pumped out of the cave.

The processing of the nitrate deposits of the sandstone rockcastles of eastern Kentucky is abbreviated. The nitrates from these shelter caves are typically potassium nitrate and do not require the ionic exchange step required for the calcium- and magnesium-rich nitrates of the carbonate solutional caves.

Evidence of saltpetre processing in caves is important in establishing that the mining evidence is of saltpetre mining. Cave sediments have been excavated from caves worldwide for uses other than saltpetre extraction.

CONTRASTING SALTPETRE MINING WITH OTHER CAVE SEDIMENT EXTRACTION

The cave sediments of many European caves are enriched not only in nitrates but also phosphate as a result of the bones of mega fauna, which used these caves during the Ice Ages of the Pleistocene. Locally, nitrate- and phosphate-enriched sediments have been extracted from European caves for fertilizing gardens. In China, cave sediments have historically been worked for fertilizer and for bones to be used in the apothecary trade. Marketed as "dragon bones," these fossil remains have been processed for use in folk remedies and as aphrodisiacs. Bat guano has been utilized as fertilizer in numerous areas within the United States, but the extraction of other nitrate-rich cave sediments for garden use is also known from at least one U.S. locality. It is likely that European immigrants to the United States continued the time-honored traditional exploitation of enriched cave sediments as garden fertilizer at other southeastern U.S. sites. Evidence of the extraction of cave sediments without associated saltpetre processing evidence, written historic records of saltpetre mining, or local saltpetre mining lore may represent other uses of cave sediments, such as for garden fertilizer, chinking for a log home, ceramics, fossil or artifact pilferage, or other usage.

In summary, saltpetre mining is a historic extractive industry tied to the development and usage of black-powder-charged firearms. The archaic spelling of the mineral, saltpetre, is retained as a descriptor because, in most cases, this mineral has only rarely been encountered in saltpetre caves; rather, a suite of the deliquescent nitrates typically occurs as viscous solutions within cave sediments, which were mined and the nitrates extracted by leaching and chemically converted to the crystalline commodity, potassium nitrate (saltpeter), the major constituent of black powder.

Bibliography

DePaepe, D. (1985). *Gunpowder from Mammoth Cave: The saga of saltpetre mining before and during the War of 1812.* Hayes, KS: Cave Pearl Press.

Duncan, M. S. (1997). Examining early nineteenth century saltpeter caves: An archaeological perspective. *Journal of Cave and Karst Studies, 59*(2), 91–94.

Faust, B. (1964). Saltpetre caves and Virginia history. In H. H. Douglas (Ed.), *Caves of Virginia* (pp. 31–56). Falls Church, VA: Virginia Cave Survey.

Hill, C. (Ed.), (1981). Saltpeter. *National Speleological Society Bulletin, 43*(4), 83–133.

Hill, C., & Forti, P. (1997). *Cave minerals of the world* (2nd ed.). Huntsville, AL: National Speleological Society.

Shaw, T. R. (1992). *History of cave science.* Sydney, Australia: Sydney Speleological Society.

Smith, M. O. (1990). *Saltpeter mining in East Tennessee.* Maryville, TN: Byron's Graphic Arts.

Taylor, M. R. (1999). *Dark life.* New York: Scribner.

SCALLOPS

Phillip J. Murphy

University of Leeds, UK

INTRODUCTION

Studies of the solutional sculpturing of cave walls can provide information on both the direction and discharge of water flow in a cave passage. Scallops are the most well-studied type of solutional sculpturing. Their asymmetry indicates the direction of groundwater flow and their wavelength are inversely proportional to the flow velocity. Laboratory and field investigations have enabled the calculation of mean flow velocity from scallop wavelength data and thus the calculation of discharge at the time of scallop formation.

SCALLOPS AND FLUTES

Scallops are asymmetrical, cuspate, oyster-shell-shaped dissolution depressions in cave walls. The term *scallop* was first proposed by Coleman (1949) to replace the term *flute* used prior to this. The term *flute* has been used for the elongate, nearly parallel crested forms seen in some cave passages. They are scallops of infinite width and are much rarer than true scallops. Vertical grooves seen in vadose shafts formed by water streaming down in thin sheets have also been called *flutes*.

GROUNDWATER FLOW DIRECTION

Scallops are asymmetrical in the direction of flow, with a smooth slope on the downstream side of the scallop and a steep cusp on the upstream side. Coleman (1949) realized they could be used as a quick and simple indicator of the direction of flow in a conduit. They occur in packed patterns on cave floors, walls, and ceilings. They are measured from cusp to cusp and can vary from 1 cm to several meters in length (where they become difficult to distinguish from bends in the passage walls).

The simplest way to ascertain the direction of asymmetry in a scallop population is to shine a light along the scalloped surface and then to position yourself so that you are looking directly toward the scallops. If the light is shining upstream, the steep slopes of the scallops are brightly illuminated and the shallow slopes are in darkness, resulting in the majority of each scallop not being illuminated (Fig. 1). If the light is shining downstream, the steep slopes are in darkness and the shallow slopes are brightly illuminated, resulting in the majority of each scallop being illuminated. If the scallop asymmetry is not sufficiently distinct, their profiles may be transferred to paper using a profile template and analyzed graphically. A large sample of scallops can then be analyzed statistically for asymmetry.

GROUNDWATER FLOW VELOCITIES

The size of the scallops is inversely proportional to the flow velocity of the water that formed them. Scallop populations usually form unimodal well-sorted, log-normal distributions. In the right

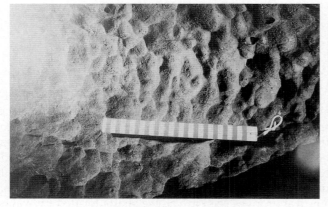

FIGURE 1 Scallops illuminated with the light source pointing upstream, Joint Hole, North Yorkshire, UK. *Photograph taken by the author.*

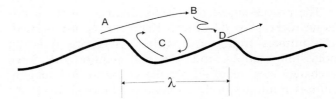

FIGURE 2 Fluid motion in the vicinity of a scallop: (A) detachment of the laminar boundary, (B) transition to turbulence, (C) the lee eddy, and (D) flow reattachment. *After Curl (1974).*

conditions, patterns of scallops of characteristic length cover all available surfaces. They are the stable form of these surfaces in the prevailing conditions and can be considered as solutional analogs to current ripples in unconsolidated sediments.

Curl (1974) studied the process of scallop and flute formation and was able to show that they are a hydraulic phenomenon. Curl proposed that scallops are formed where there is detachment of the saturated boundary layer at a specified Reynolds number, occurring in the subcritical turbulent flow regime. Where the main flow separates, it passes above an area of slower, recirculating flow. Within a short distance the fast flow becomes irregular and turbulent. Because the turbulence produced causes mixing between the fast stream and the lee eddy, fluid is entrained out of the lee eddy, causing the jet to turn toward the surface and reattach (Fig. 2). Detachment allows aggressive waters to come into contact with the bedrock, increasing the rate of direct erosion. The rate of solution is the highest in the area of reattachment where the turbulent flow impinged the most on the surface. This means scallops migrate downstream. The more the frequency of detachment, the more the flow velocity, hence reducing the erosion length available to each individual scallop.

Being a hydraulic phenomenon, scallop formation and the conduit flow conditions that gave rise to them are described by fluid dynamic equations. Previous work has shown that scallops form at a stable scallop Reynolds number (Re^*) of ~2200 where Re^* is related to the mean boundary shear velocity, $\bar{u}^*$, the mean scallop wavelength $\bar{\lambda}$, fluid density ρ_f, and fluid dynamic viscosity μ by

$$Re^* = \frac{\bar{u}^* \bar{\lambda} \rho_f}{\mu} \quad (1)$$

Measuring a statistically viable sample allows $\bar{\lambda}$ to be calculated, and thus an estimate of $\bar{u}*$ to be made using Eq. (1). A water temperature appropriate for the study area needs to be chosen. Then a modified Prandtl's universal velocity distribution equation for a parallel-sided conduit [Eq. (2)] or a circular conduit

[Eq. (3)] can be used to calculate the mean flow velocity at the time of creation of the scallops:

$$\bar{u} = \bar{u}^* \left[2.5 \left\{ \ln\left(\frac{d}{2\bar{\lambda}}\right) - 1 \right\} + B_L \right] \quad (2)$$

$$\bar{u} = \bar{u}^* \left[2.5 \left\{ \ln\left(\frac{d}{2\bar{\lambda}}\right) - 1.5 \right\} + B_L \right] \quad (3)$$

where d is the hydraulic diameter of the conduit and B_L is Prandtl's bed roughness constant. Blumberg and Curl (1974) showed through experimental flume work on scallops that $B_L = 9.4$.

An approximation to the groundwater flow velocity (in meters per second) indicated by a scallop population can be calculated by dividing 3.5 by the mean length of the scallops in centimeters.

CALCULATION OF DISCHARGE AND OTHER PARAMETERS

Following the approach of Gale (1984), further hydraulic parameters can be estimated from the values obtained above and the following equations:

$$f = \frac{8}{(\bar{u}/\bar{u}^*)^2}$$

$$Q = \bar{u} a$$

$$Re = \frac{\bar{u} d \rho_f}{\mu}$$

$$F = \frac{\bar{u}}{\sqrt{gd}}$$

$$\bar{\tau} = (\bar{u}^*)^2 \rho_f$$

$$P = \bar{u} \bar{\tau}$$

$$\omega_{max} = 0.96 \bar{u}^*$$

where f is the Darcy–Weisbach friction factor, Q the discharge, a the cross-sectional area of the conduit, Re the conduit Reynolds number, F the Froude number, g the gravitational acceleration, $\bar{\tau}$ the mean boundary shear stress, P the power of flow per unit area of boundary, and ω_{max} the maximum settling velocity of material in suspension.

USES AND PITFALLS OF SCALLOP DISCHARGE DATA

Curl's theory of scallop formation has been tested in laboratory studies and in the field both in vadose and phreatic conduits. Studies in network cave systems have shown that paleodischarges calculated from

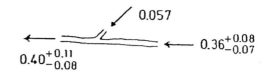

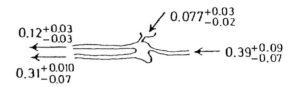

FIGURE 3 Flow through the junctions (m^3 s^{-1}) with estimates of errors of flow. *From Lauritzen (1982).*

scallop data fit the requirements of the continuity equation, whereby the same amount of incompressible liquid has to pass through each cross-section of a tube or tube system (Fig. 3).

Working in relict conduits where enough scallop lengths can be measured, the paleovelocity and paleodischarge of the conduit can be calculated by assuming a temperature of the water and measuring the conduit dimensions. A number of other factors, however, have to be considered. One problem is "Under what flow conditions in the conduit does scallop formation occur?" Field studies in phreatic conduits have shown that the velocities indicated by the mean scallop length (the scallop dominant discharge) correspond to the upper 5% of the flow regime (Fig. 4). For vadose conduits, however, a scallop dominant discharge is difficult to define because of the unconstrained flow depth. A complex relationship is seen to exist between the velocity and depth of flow as discharge changes (Charlton, 2003). Thresholds occur over discrete depth ranges where there is little or no change in velocity; these are observed during both rising and falling stages. These thresholds may be related to changes in hydraulic radius, and hence flow resistance at different depths of flow.

Scalloped surfaces exhibit a number of small depressions, which appear to be related to the intersection of the rims of the scallops and therefore not related to the flow velocity. The use of the Sauter mean (λ_{32}) rather than the arithmetic mean is used by

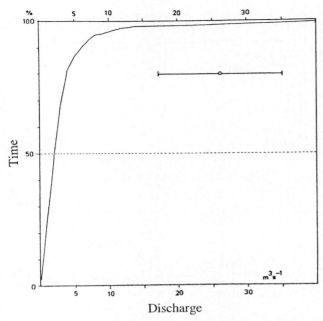

FIGURE 4 Flow duration curve for the period from September 1983 to April 1985 with the flow rate deduced from scallops for the underground outlet for Lake Glomdal, Norway. The flow rate deduced from scallops represents only 2% or less of the time of flow. *From Lauritzen et al. (1985).*

some workers to suppress the importance of these smaller features.

$$\lambda_{32} = \frac{\sum_{i=1}^{n} \lambda_i^3}{\sum_{i=1}^{n} \lambda_i^2}$$

Being hydraulic phenomena, the process of scallop formation is independent of substrate; however, field observation has shown that bedrock variation does exert some control on scallop size. Homogeneous rock with a uniform grain size and lack of small cavities provides the best substrate for development. Scallops are rarely developed in dolomite caves.

Studies of scallop distributions in an active phreatic conduit have shown that distinct and adjacent scallop distributions indicative of very different flow regimes can coexist (Murphy *et al.*, 2000). Contrast in size between ceiling and lower wall scallop populations in relict conduits have been explained by a partial draining of the conduit with a corresponding change in discharge. Such a situation could occur in the epiphreatic zone. Workers in relict conduits have often assumed that the presence of small scallops on the floor contrasting with large scallops on the walls and ceiling is due to invasion of the relict conduit by a vadose stream. This may not be the case and the contrasting scallop sizes could be a product of the phreatic phase of the conduit's history.

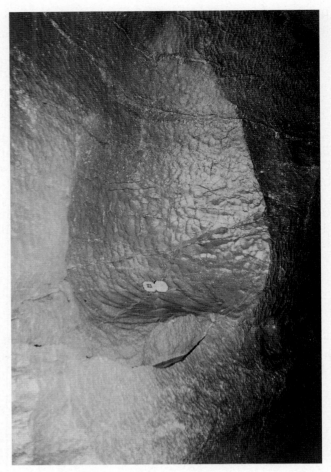

FIGURE 5 A lateral pothole on Buckeye Creek Cave, West Virginia. *Photograph courtesy of G. Springer.*

Sims (2004) compared mean velocities required to entrain sediment particles and to produce scallops. Velocities which would entrain sand grains of 1-mm diameter would produce scallops with a wavelength of approximately 10 cm. Scallop formation still occurs at velocities two orders of magnitude below the minimum velocity required to entrain any particles.

The occurrence of scalloping superimposed on speleothems means the scallops must postdate initial draining of the conduit and initiation of speleothem growth. Such occurrences allow for the possibility of constraining the timing of paleohydraulic events by dating the speleothems. This approach has been used by Murphy *et al.* (2001) to date the occurrence of a second phreatic event in a conduit's history.

OTHER PHENOMENA

Pockets resemble hollow, bisected hemispheres or hemiellipses (Springer and Wohl, 2002). They are readily distinguishable from scallops by their lack of oversteepened upstream faces. The flow field in pockets can be forward, reversed, upward, or downward. All four flow fields were recorded in Buckeye Creek Cave in West Virginia, U.S.A. In contrast to scallops, pockets do not migrate but are developed on defects such as joints and stylolites, the morphology of the pocket being dependent on the nature and orientation of the defect. The presence of pockets within fields of scallops suggests that some pockets may have originated as preferentially enlarged scallops.

Lateral potholes may resemble hollow bisected cylinders, hemispheres, hemiellipses, or upright teardrops in vertical channel walls (Fig. 5). They are eroded by sediment-laden vortical flow near the air–water interface and are potential indicators of paleoflow depths and velocities. They are described from a vadose canyon passage in Buckeye Creek Cave, West Virginia, by Springer and Wohl (2002) where scallops within lateral potholes record recirculating flows, with flows descending at an angle of 10–30° from the horizontal.

The preferential development of small-scale solutional etching on one set of passage walls in gull rifts in southwest England has been used by Self (1995) to infer paleoground water flow.

See Also the Following Article

Karren, Cave

Bibliography

Blumberg, P. N., & Curl, R. L. (1974). Experimental and theoretical studies of dissolution roughness. *Journal of Fluid Mechanics, 75*, 735–742.

Charlton, R. A. (2003). Towards defining a scallop dominant discharge for vadose conduits: Some preliminary results. *Cave and Karst Science, 30*(1), 3–7.

Coleman, J. C. (1949). An indicator of water flow in caves. *Proceedings of the University of Bristol Speleological Society, 6*(1), 57–67.

Curl, R. L. (1974). Deducing flow velocity in cave conduits from scallops. *National Speleological Society Bulletin, 36*, 1–5.

Gale, S. J. (1984). The hydraulics of conduit flow in carbonate aquifers. *Journal of Hydrology, 70*, 309–327.

Lauritzen, S.-E. (1982). The paleocurrents and morphology of Pikhaggrottene, Svartisen, North Norway. *Norsk Geographischen Tidsskrift, 36*, 183–209.

Lauritzen, S.-E., Abbot, J., Arnessen, R., Crossley, G., Grepperud, D., Ive, A., et al. (1985). Morphology and hydraulics of an active phreatic conduit. *Cave Science, 12*, 139–146.

Murphy, P. J., Hall, A. M., & Cordingley, J. N. (2000). Anomalous scallop distributions in Joint Hole, Chapel-le-Dale, North Yorkshire, UK. *Cave and Karst Science, 27*(1), 29–32.

Murphy, P. J., Smallshire, R., & Midgley, C. (2001). The sediments of Illusion Pot, Kingsdale, North Yorkshire, UK: Evidence for subglacial utilisation of a karst conduit in the Yorkshire Dales. *Cave and Karst Science, 28*(1), 29–34.

Self, C. A. (1995). The relationship between the gull cave Sallys Rift and the development of the River Avon east of Bath. *Proceedings of the University of Bristol Speleological Society, 20*(2), 91–108.

Sims, M. (2004). Tortoises and hares: Dissolution, erosion and isostasy in landscape evolution. *Earth Surface Processes and Landforms, 29*, 477–494.

Springer, G. S., & Wohl, E. E. (2002). Empirical and theoretical investigations of sculpted forms in Buckeye Creek Cave, West Virginia. *Journal of Geology, 110*, 469–481.

SHALLOW SUBTERRANEAN HABITATS

Tanja Pipan * *and David C. Culver* †

*Karst Research Institute ZRC-SAZU, †American University

INTRODUCTION

What are subterranean habitats? Speleobiologists (*e.g.*, Botosaneanu, 1986; Juberthie, 2000) generally recognize two primary categories of subterranean habitats: large cavities (caves) and small interstitial cavities (gravel and sand aquifers, and the soil). These habitats share two important characteristics—the absence of light and the presence of species both limited to and modified for subterranean life. While species in both habitat types are typically without eyes and pigment, large cavity species have elongated appendages while animals limited to interstitial habitats are often miniaturized with shortened appendages (Coineau, 2000). This difference between large and small cavity species makes it relatively easy to separate blind and depigmented species into the appropriate habitat category. An example of these differences is shown for ingolfiellid amphipods in Figure 1.

However, there is one other category—shallow subterranean habitats (SSHs). SSHs are aphotic habitats such as seeps, talus slopes, solution pockets, and tubes in epikarst. They are shallow (less than 10 m from the surface) which means that they are much more intimately connected with the surface environment than deeper cavities. They are typically more variable and with more organic carbon than deeper subterranean habitats. The habitable spaces within these habitats are considerably larger than their inhabitants (Culver and Pipan, 2008). They share with large cavities a habitable space large enough that organisms are not in contact with solid surfaces in all three dimensions. This contrasts with small cavity habitats where cavity size is a major constraint in the evolution of morphology. On the other hand, animals limited to caves and SSHs have similar morphological and ecological characteristics, such as appendage elongation, specialization of extraoptic sensory organs, increased

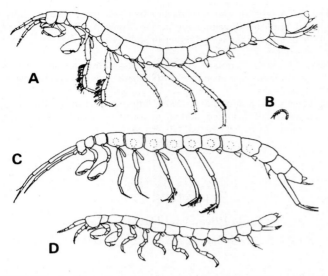

FIGURE 1 Diagrams of stygobiotic ingolfiellid amphipods. (A) *Trogloleleupia leleupi* (12–20 mm) from a cave; (B) *Ingolfiella* sp. (<1 mm) from interstitial habitats, shown at same scale as (A); (C) *Trogloleleupia opisthodorus* (24–28 mm) from a cave; and (D) *Ingolfiella petkovskii* (1 mm) from an interstitial habitat. Note the relatively short appendages of *I. petkovskii* as well as its small size. *From Coineau (2000). Used with permission of Elsevier Ltd.*

life span, resistance to starvation and reduced aggressive behavior. This set of traits, especially the morphological part, was termed *troglomorphy* by Christiansen, and this term is widely used to describe the morphological convergence of obligate cave dwellers.

SSHs have a number of unique features relative to other subterranean habitats, including (1) the areal extent of an individual habitat is often small, usually <0.1 km², but many replicates exist; (2) they are rich in organic matter relative to other subterranean habitats, in part because of their close proximity to the surface; (3) they have intimate connections to the surface, resulting in greater environment variation than other subterranean habitats; and (4) they have fauna which includes troglomorphic species, some of which are limited to SSHs.

Among terrestrial SSHs are the spaces between rocks in areas of moderate to steep slope typically stabilized by moss, spaces in bedrock caused by weathering, similar spaces in volcanic terrains, clinker in lava, air-filled epikarst spaces, and sometimes even leaf litter. Juberthie (2000) used the term *milieu souterrain superficiel* or *mesovoid shallow substratum* (MSS) for this type of habitat, especially the spaces between rocks, and we use *MSS* as a collective term for these habitats since it is a well-established term in the literature and is self-descriptive.

Among aquatic SSHs are epikarst, the uppermost layer of karst with poorly integrated solution cavities;

seepage springs, also called the *hypotelminorheos*; and the underflow of streams and rivers, the hyporheos and associated groundwater. For each habitat, we discuss a representative example, give data on environmental variability, summarize available information on organic carbon, and enumerate the fauna.

SEEPAGE SPRINGS

The hypotelminorheic habitat, or seepage spring (Fig. 2), is (1) a persistent wet spot, a kind of perched aquifer fed by subsurface water in a slight depression in an area of low to moderate slope; (2) rich in organic matter; (3) underlain by a clay layer typically 5–50 cm beneath the surface; (4) with a drainage area typically <10,000 m²; and (5) with a characteristic dark color derived from decaying leaves which are usually not skeletonized (Culver et al., 2006). The habitat can occur in a wide variety of geologic settings anywhere outside of arid regions where there is a layer of impermeable sediment, but it is probably less common in karst landscapes because of the extensive occurrence of an impermeable clay layer which prevents the downward movement of water and the development of karst landscapes. Most of the available habitat for the animals comprises spaces between decomposing leaves and sediment, and the animals literally live in their food.

Chemical and physical conditions vary considerably between sites, but conductivity tends to be high, indicating that the water had been underground for some time. Although oxygen concentrations varied considerably, the fauna does not seem to be especially sensitive to this parameter. There was about 3.0 mg dissolved organic carbon (DOC) L^{-1} in seepage springs in

FIGURE 2 Photograph of the authors at a seepage spring at Scotts Run Park, near Washington, D.C., U.S.A. *Photo courtesy of W. K. Jones. Used with permission.*

Nanos, Slovenia, several times higher than most cave water. Based on a 10-month monitoring period (March 2007 to January 2008) of a hypotelminorheic habitat in Prince William Forest Park in Virginia, U.S.A., the habitat was temporally variable (Fig. 3). From May to September, hypotelminorheic temperatures were depressed compared to the nearby surface stream, and approximated surface water temperatures for the rest of the year. In spite of the variability, the amplitude of variation in hypotelminorheic temperatures is less than that of surface waters. The maximum recorded temperature in the hypotelminorheic was 22°C compared to 28°C in a nearby (<10 m) stream (Table 1). This is a remarkable difference, given the superficial nature of seepage springs. The differences may become more important, given predictions of climatic variability and change.

Based on a study of 50 seepage springs in the lower Potomac River basin that drain hypotelminorheic habitats within a radius of 45 km, a total of 15 macroinvertebrates have been recorded, including 12 amphipod, 2 isopod, and 1 gastropod species (Fig. 4, Table 2). Four of the amphipods were probably accidentals—they were uncommon and showed no evidence of reproduction. Of the remaining 11 macroinvertebrate species, 7 were species living exclusively in subterranean habitats. Of these seven, five are exclusively found in seeps: they are hypotelminorheic specialists. Five of the seven stygobionts were troglomorphic. Hypotelminorheic sites in Croatia and Slovenia have a similar mixture of specialized and nonspecialized amphipods.

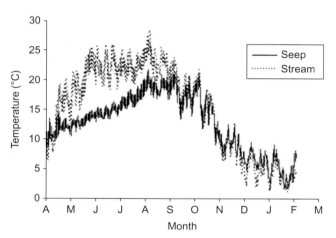

FIGURE 3 Hourly temperature from April 7, 2007, to February 4, 2008, in a seepage spring and adjoining stream in Prince William Forest Park, Virginia, U.S.A. Because of the scale, line thickness indicates the extent of daily fluctuations.

FIGURE 4 Photograph of the stygobiotic hypotelminorheic specialist *Stygobromus tenuis*, from Scotts Run Park, Fairfax County, Virginia. *Photo courtesy of William K. Jones. Used with permission.*

TABLE 1 Statistical Properties of Temperature (in °C) Time Series for a Seepage Spring in Prince William Forest Park, Virginia, U.S.A. Data from Culver and Pipan (2008)

	Seepage Spring	Surface Stream
Mean	12.80	15.10
Standard error	0.06	0.09
Standard deviation	4.88	7.51
Coefficient of variation	38.15	49.75
Range	20.03	27.47
Minimum	1.79	1.02
Maximum	21.82	28.49
Count	7274	7274

TABLE 2 List of Species Found in Seepage Springs (Hypotelminorheic) in the Lower Potomac River Basin According to Ecological Category, Whether They Are Limited to Seepage Springs, and Whether They Show Troglomorphy

Group	Ecological Category	Habitat Specialist	Troglomorphic	Number of Species
Amphipoda	Stygobiont	Yes	Yes	3
Amphipoda	Stygobiont	No	Yes	2
Amphipoda	Stygophile	No	No	3
Amphipoda	Accidental	No	No	4
Isopoda	Stygobiont	Yes	Weakly	1
Isopoda	Stygophile	No	No	1
Gastropoda	Stygobiont	Yes	Weakly	1

Source: Data from Culver and Pipan (2008).

EPIKARST

Epikarst is a perched aquifer, the uppermost layer of rock in karst regions, and the major point of contact and transmission between surface and subterranean water. Water is transmitted vertically either through conduits or through small fissures to the water table. Lateral transmission occurs through poorly integrated cavities. Epikarst stores considerable volumes of water which explains why, during a drought, cave streams take much longer to dry up than surface streams in the same area. Its vertical extent is usually 10 m or less. Its principal characteristic is its heterogeneity, with many solution pockets whose water chemistry is also quite variable. Water dripping from epikarst (Fig. 5) typically contains about 1.0 mg DOC L^{-1}, and while this is relatively low, it is an important carbon source in caves both in the establishment of the biofilm in cave streams and in cave passages without active streams. Epikarst is a nearly universal feature of karst areas, except in arid zones and glaciated areas. Karst areas cover approximately 15% of the earth's surface, and epikarst is likely to occur over most of this area. Although epikarst water occurs throughout karst areas, it can usually only be sampled by filtering the dripping and seeping water of caves (Pipan, 2005).

Epikarst water shows considerable temperature buffering relative to surface water, and more temperature buffering than hypotelminorheic water. An example is the cave Županova jama in Slovenia (Fig. 6) where temperature varied by approximately 4°C over a 2-year period. Drip rates of water, however, were highly variable, ranging from <1 mL min^{-1} to >100 mL min^{-1}; with the seasonal pattern dependent on rainfall, but displaying a lagged response. Even though epikarst water is only a few meters below the surface, its residence time may be weeks or even months.

In all of the places where epikarst fauna has been sampled, stygobionts have been found, but the Slovenian assemblages are by far the most diverse (Pipan, 2005). In Županova jama in Slovenia, a total of 16 copepod species were found in five drips and their associated pools, and remarkably 14 of these are stygobionts (Table 3). Seven of these species are epikarst specialists, including five undescribed species, two of which are unique to Županova jama. Other caves in Slovenia have similar overall epikarst copepod species richness, although with more stygophiles (species that can complete their life cycles in either subterranean or epigean habitats) than Županova jama.

HYPORHEIC

The hyporheic zone comprising water-filled spaces between the grains of unconsolidated sediments beneath

FIGURE 5 Photograph of water dripping from epikarst in Organ Cave, West Virginia, U.S.A. *Photo courtesy of Horton H. Hobbs III. Used with permission.*

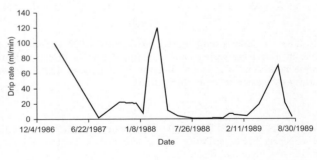
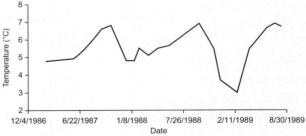

FIGURE 6 Discharge (upper panel) and temperature (lower panel) variation of an epikarst drip in Županova jama, Slovenia, based on monthly measurements taken over a 2-year period. *Data from Kogovšek (1990).*

and lateral to streams, the best studied of all interstitial habitats is the surface–subsurface hydrological exchange zone beneath and alongside the channels of rivers and streams. The hyporheic zone of rivers is an ecotone between surface and groundwater. The connection between the hyporheic and permanent groundwater (phreatic water) can be very direct or without any direct connection at all. In the case of direct connections between the hyporheic and permanent groundwater, stygobiotic species are often found. Even though the hyporheic appears to be highly uniform, it actually has a series of upwellings and downwellings. Downwellings typically have higher oxygen levels and more organic matter. The exact position of these upwelling and downwelling zones along the stream course depends on the relative pressure of the subsurface and surface waters, and other hydrological details. When there are unconsolidated sediments along the stream bank, the hyporheic can extend laterally tens of meters from the stream bank.

Formed by a meander arm, hyporheic habitats in the Lobau wetlands are part of the floodplain of the Danube River near Vienna, Austria, and comprise the Danube Flood Plain National Park. This UNESCO Biosphere Reserve, with an area of 0.8 km^2, has been extensively sampled for decades. Bou-Rouch pumps, specialized pumps designed for sampling in these habitats, and minivideo cameras in shallow wells revealed a complex habitat with areas of differing porosity and permeability, variable oxygen levels, and a rich fauna. A small 900 m^2 component of this floodplain, called "Lobau C" is a recent terrace of the backwater system "Eberschüttwasser–Mittelwasser" and is a self-contained ecosystem with clear inputs and outputs because of its position between two channels and a dam. Loosely packed gravel, alternating with a thin layer of finer sediments, extends several meters beneath a thin soil cover. Temperature at a depth of 2 m varied between 3°C and 21°C over a two-and-a-half-year period, while surface water temperatures varied between 1°C and 26°C over the same period (Fig. 7). DOC averaged 3.71 mg CL^{-1} (Danielopol et al., 2000). Animals were found throughout the depth of gravel, but were most common 0.5 m beneath the surface, and rare below 2 m. In Lobau C, at least 27 species have been found, 11 of them stygobionts (Table 4). While these species are stygobiotic, they do not in general show the troglomorphic syndrome of size increases and appendage elongation, but rather are miniaturized with shortened appendages (see Fig. 1). Thus, they show more morphological kinship with

TABLE 3 List of Species of Copepods Found in Epikarst Drips and Pools in Županova Jama, Slovenia, According to Ecological Category, Whether They Are Limited to Seepage Springs, and Whether They Show Troglomorphy

Copepod Group	Ecological Category	Habitat Specialist	Troglomorphic	Number of Species
Cyclopoida	Stygobiont	Yes	Yes	1
Cyclopoida	Stygobiont	No	Yes	1
Harpacticoida	Stygobiont	Yes	Yes	6
Harpacticoida	Stygobiont	No	Yes	6
Harpacticoida	Stygophile	No	No	2

Source: Data from Pipan (2005).

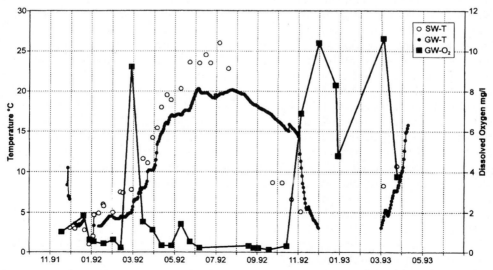

FIGURE 7 Variation in temperature and oxygen from 1991 through 1993 at site D10 in the Lobau wetlands: surface water temperature (SW-T); groundwater temperature at 3.5 m (GW-T), and oxygen concentration at 3.5 m (GW-O$_2$). *Data from Danielopol et al. (2000).*

TABLE 4 Number of Species and Stygobionts from the "Lobau C" Area of Danube Flood Plain National Park, Austria

Group	Number of Species	Number of Stygobionts
Rotatoria	>1	1
Mollusca	>2	2
Copepoda: Cyclopoida	14	3
Copepoda: Harpacticoida	7	2
Amphipoda	1	1
Isopoda	2	2

Source: Data from Danielopol and Pospisil (2001).

deep interstitial species rather than species from other SSH habitats.

The hyporheic is an atypical habitat compared to other SSHs because the animals are not troglomorphic, but rather have the morphology of small body size and small appendages characteristic of interstitial habitats, and because its linear reach can be extensive. Whether it is useful to continue to include these as SSHs remains to be seen.

MILIEU SOUTERRAIN SUPERFICIEL (MSS)

Shallow subterranean terrestrial habitats include the spaces between rocks in areas of moderate to steep slope, spaces in bedrock caused by weathering, similar spaces in volcanic terrains, air-filled epikarst spaces, and even leaf litter. Initially, the importance of MSS habitats was thought to be as dispersal corridors between cave regions and did in fact account for some disjunct distributions, but soon it was recognized that the primary habitat for some troglobionts was the MSS, rather than caves. MSS habitats can be thought of as relatively large cavities embedded in a matrix of small cavities, that is, the soil (Gers, 1998). An example is Barranco de los Cochinos, an erosional habitat in lava in a laurel (*Laurus*) forest at 940 m a.s.l. in the Teno area, northwest Tenerife, Canary Islands. This site comprises fragmentized basaltic rocks several million years in age, and covered by 40–60 cm of soil. The surface is characterized by a dense tree covering, low exposure to the sun, and rather high humidity. Temperatures at a depth of 50 cm ranged from 9.2°C to 16.4°C, compared to a range of 7.9–19.4°C on the surface (Table 5). In fact, the annual pattern of temperature fluctuation was quite similar between surface and MSS (Fig. 8), although there was no detectable daily temperature cycle at the MSS site.

TABLE 5 Statistical Properties of Temperature (in °C) Time Series for an MSS Site in Barranco de los Cochinos, Tenerife

Description	Surface	MSS
Mean	12.90	12.50
Standard error	0.025	0.021
Standard deviation	2.51	2.12
Coefficient of variation	19.45	17.00
Range	10.50	7.19
Minimum	7.93	9.24
Maximum	19.43	16.43
Count	10,055	10,055

Source: Data from Pipan et al. (2011).

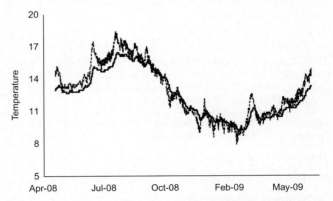

FIGURE 8 Temperature profiles at hourly intervals for an MSS site (solid line) and nearby surface site (dotted line) in a laurel forest in Teno in northwest Tenerife, Canary Islands. *Data from Pipan et al. (2011).*

The Barranco de los Cochinos site has been exceptionally well studied and a total of 73 invertebrates species have been found. Of these, 41 are generalist species, found both in surface and subterranean habitats. Of the 32 species known from subterranean habitats, 22 are soil specialists with shortened appendages and small size, and 10 are troglomorphic species, with elongated appendages and larger size. The 10 troglomorphic species are not usually found in soil, and the 22 soil specialists may not necessarily be permanent inhabitants of the MSS. The troglobionts include spiders, cockroaches, beetles, millipeds, and pseudoscorpions.

GENERALITIES

These SSHs share several features. First, all of these habitats, with the exception of the hyporheic, support highly modified troglomorphic species, which have a morphology that in many of its characters is convergent with that of the morphology of related deep cave-dwelling species. Second, all of the superficial subterranean environments have species that are not only obligate subterranean species but also SSH specialists. Third, SSHs are not generally resource-poor habitats. This is especially true for seepage springs and interstitial habitats. Resources are not as abundant in epikarst or MSS, but at least in the case of epikarst, there is more DOC than in epikarst-fed streams (Simon *et al.*, 2007). Fourth, with the exception of light, environmental conditions in SSHs are intermediate between surface conditions and deep subterranean conditions. Fifth, what all these habitats share in common is that they are aphotic. In the case of interstitial and epikarst habitats, this is self-evident. In the case of seepage springs, light may be present at the exit of the seep, but the habitat itself—the hypotelminorheic—is without light. For the MSS long-term measurements indicate that no light penetrates even 50 cm.

ORIGIN OF THE SSH FAUNA

It might seem curious that any species are ever found in SSHs except for an occasional transient. The dominant paradigm of colonization of deeper subterranean habitats such as caves is that animals were forced into these habitats by the vicissitudes of climate change, such as glaciation, sea level change, and aridity. This may be relevant to SSHs as well, given that temperature extremes are never as high in SSHs relative to surface habitats. If it is temperature extremes that drive species underground, SSHs may be a relatively favorable environment in this respect.

In the case of aquatic SSHs (epikarst and hypotelminorheic) both habitats offer an aquatic environment that is less likely to dry out than some surface habitats. Epikarst retains water long after surface water and some cave streams have dried up and the clay that underlies the hypotelminorheic also retains water during periods of drought. Both habitats may thus be a refuge against climate change. Aside from darkness, it seems likely that conditions in SSHs do not present a formidable barrier to colonization, and may even present some advantages. For example, predators, at least large ones, are absent in most SSHs because the size of the habitat does not provide access. Organic carbon is not scarce, or at least not as scarce as in caves, and all of these habitats have a virtual rain of organic matter—falling and decaying leaves in the hypotelminorheic, DOC in epikarst, organic debris in MSS, and detrital accumulation in the hyporheic.

EVOLUTIONARY AND BIOGEOGRAPHIC CONNECTIONS WITH OTHER SUBTERRANEAN HABITATS

Depending on the proximity of SSHs to deeper subterranean habitats, especially caves, colonization of and subsequent adaptation to SSHs may be an evolutionary pathway to colonization of deeper, more extreme subterranean environments. If this is the case, then a phylogenetic tree of a group with both SSH and deep cave species should have a topology with SSH species more basal than deep cave species. Unfortunately, no detailed phylogenies are available for groups with both SSH and subterranean species. Part of the problem may be classification of the habitat. At first glance, the hypotelminorheic might appear to be a surface habitat, while epikarst species might simply appear to be cave species.

In many studies of subterranean fauna, the degree of troglomorphy has been equated with the degree of adaptation to subterranean environments and to the length of time a lineage has been isolated in subterranean environments. We suggest an alternative view—the degree of troglomorphy reflects site-specific differences in the subterranean habitat the species occupy. Extreme appendage elongation, reduction in metabolic rate, and increased longevity, to cite only a few troglomorphic characters, may occur in organisms in extreme, isolated subterranean environments—deep caves and deep groundwater. More typical troglobionts and stygobionts may be the inhabitants of SSHs. They are not necessarily phylogenetically younger, they are just in a different habitat, one that is dark but neither constant nor extremely food-poor.

See Also the Following Articles

Epikarst
Epikarst Communities

Bibliography

Botosaneanu, L. (Ed.), (1986). *Stygofauna mundi*. Leiden, The Netherlands: E. J. Brill.
Coineau, N. (2000). Adaptations to interstitial groundwater life. In H. Wilkens, D. C. Culver & W. F. Humphreys (Eds.), *Subterranean ecosystems* (pp. 189–210). Amsterdam: Elsevier.
Culver, D. C., & Pipan, T. (2008). Superficial subterranean habitats— gateway to the subterranean realm? *Cave and Karst Science, 35*, 5–12.
Culver, D. C., Pipan, T., & Gottstein, S. (2006). Hypotelminorheic—a unqiue freshwater habitat. *Subterranean Biology, 4*, 1–8.
Danielopol, D. L., & Pospisil, P. (2001). Hidden biodiversity in the groundwater of the Danube Flood Plain National Park, Austria. *Biodiversity and Conservation, 10*, 1711–1721.
Danielopol, D. L., Pospisil, P., Dreher, J., Mösslacher, F., Torreiter, P., Geiger-Kaiser, M., et al. (2000). A groundwater ecosystem in the Danube wetlands at Wien (Austria). In H. Wilkens, D. C. Culver & W. F. Humphreys (Eds.), *Subterranean ecosystems* (pp. 481–511). Amsterdam: Elsevier.

Gers, C. (1998). Diversity of energy fluxes and interactions between arthropod communities from soil to cave. *Acta Oecologica*, 19, 205–213.
Juberthie, C. (2000). The diversity of the karstic and pseudokarstic hypogean habitats in the world. In H. Wilkens, D. C. Culver & W. F. Humphreys (Eds.), *Subterranean ecosystems* (pp. 17–39). Amsterdam: Elsevier.
Kogovšek, J. (1990). The properties of the precipitations seeping through the Taborska Jama roof. *Acta Carsologica*, 19, 143–156.
Pipan, T. (2005). *Epikarst—A promising habitat.* Postojna–Ljubljana: ZRC Publishing at Karst Research Institute ZRC-SAZU.
Pipan, T., López, Oromí., P., Polak, S., & Culver, D. C. (2011). Temperature variation and the presence of troglobionts in terrestrial shallow subterranean habitats. *Journal of Natural History* 45, 253–273.
Simon, K. S., Pipan., T., & Culver, D. C. (2007). A conceptual model of the flow and distribution of organic carbon in caves. *Journal of Cave and Karst Studies*, 69, 279–284.

SHOW CAVES

Arrigo A. Cigna

Union Internationale de Spéléologie

A SHORT HISTORY OF SHOW CAVES

Caves have always attracted the attention of humans since prehistory, but at that time the interest was mainly quite practical, that is, to have a shelter, a sanctuary, or a burial place. Later, until the Middle Ages, caves were associated with the devil or hell in general and people feared going into them. For this reason, bandits could use caves as hiding places without problems from undesired visitors.

Some historical show caves were known already in ancient times. About 2000 year ago Plinius, a Roman writer, described the "Dog's Cave" near Naples, Italy, being visited by people because of the peculiar release of carbon dioxide close to the floor which killed small animals (hence its name) while standing people were not affected. Other caves were visited not purely for tourism but mainly for religious purposes: such shrines may be found everywhere.

In Postojna Cave (Slovenia), on the walls of the so-called Passage of the Ancient Names were found old signatures left by occasional visitors, the most ancient ones dating back to 1213, 1323, and 1393 according to some authors of the nineteenth century. Around 1920, such signatures were scarcely visible on account of the seepage; presently the oldest signature which can be read easily dates 1412 and from the sixteenth century onward they became rather abundant. This means that from the sixteenth century the cave was visited more frequently by many persons attracted by the underground world. Therefore, this period may be considered the start of cave tourism.

In a more recent time the Cave of Antiparos in Cyclades, Greece, became a great attraction in the seventeenth century as documented by the many prints reproducing the cave. The Kungur Cave, 100 km SE of Perm, near Kungur (Urals), Russia, is an ice show cave already visited in the eighteenth century. Probably it is the largest show cave in gypsum. On August 13, 1772, the scientist Joseph Banks landed on Staffa Island and in November he wrote in the *Scots Magazine*: "... there is a cave in this island which the natives call the Cave of Fingal." Since that time this cave became one of the best known caves of the world, inspiring poets and musicians. Its fame was so great that it became the natural cave most represented in paintings and engravings all over the world.

The Cango Cave (Oudtshoorn, South Africa) was discovered around 1780 and the first recorded visit was made in 1806. A few years later, a farmer bought the land around the cave with the exclusion of the entrance. The Governor included into the deeds the condition that the farmer was obliged to leave, perfectly free and undisturbed, the entrance of the cave, to be considered as public property, with a road in his land to reach the cave. This document has a historical importance because it is probably the first attempt in the world to legislate for cave protection.

In the United States, the Mammoth Cave, Kentucky, was defined "the stellar attraction of the Mammoth Cave National Park" by Gurnee and Gurnee (1990). Already known in prehistory, in the late eighteenth century the cave was mined for saltpeter to make gunpowder. But only at the beginning of the following century, Mammoth Cave became a tourist attraction.

If a show cave is defined as a cave where a fee is paid in order to have access and to visit it, then the oldest one is the Vilenica Cave in Slovenia. The cave is close to the village of Sezana, just a few kilometers from the Italian border. At the beginning of the seventeenth century, the Count of Petac began to invite the people of Trieste and some noble friends to visit the cave. On certain holidays, a hundred meters from the entrance, an area for the orchestra and a dance floor were set up and the entire dripstone passage was illuminated with torches and candles. Probably already in 1633, the Count Benvenut Petac charged admission to visit the cave. Part of the money was donated to the local church of Lokev where masses were dedicated to "greater safety" of the people in the cave.

THE ENVIRONMENTAL PROTECTION OF SHOW CAVES

A cave is an environment with little contact with the outside. For this reason its equilibrium may be easily changed when additional energy is introduced

(Cigna, 1993). Obviously, such changes may occur more frequently when the whole energy budget of the cave is small, but in case of show caves the energy budget is often not very small, because of their size which is generally large. A river or a subterranean lake plays an important role in keeping the natural equilibrium because they may absorb any further input of energy more easily than rock.

In a show cave both the visitors and the electric lighting system release energy into the environment. A person who is walking releases nearly as much energy as a 200-W bulb at a temperature of about 37°C. Therefore, the total energy released by hundreds or thousands of visitors in a day is not negligible as an absolute amount. The heat released by the electric lighting system has the same order of magnitude.

There are different ways to keep the additional energy input into the cave as low as possible. A limit of the number of visitors is given by the so-called visitors' capacity, which is defined as the maximum number of visitors acceptable in a time unit under defined conditions which does not imply a permanent modification of a relevant parameter. Otherwise, instead of reducing the number of persons, the time they spend in the cave may be reduced. This result may be easily achieved when people enter the cave through one entrance and exit along another passage, instead of returning along the same pathway where they got in.

Using high-efficiency lamps can reduce the contribution by the electric lighting system. A further reduction can be obtained if the lamps are switched on only when visitors are in the vicinity.

Another perturbation of the cave environment is the lint (hair, dry-flaking skin, dust from shoes, and lint from clothing) left by visitors. In caves visited by a large number of people, the accumulation of lint becomes a real problem to be solved by an accurate removal. In fact, such lint would cause deterioration of formations and reduce their pristine white beauty to a blackened mess.

Lint released into a cave might be reduced by means of air curtains at the entrance. Such a solution would wash people entering the cave and, at the same time, isolate the cave environment from outside since an air curtain acts as an invisible door and prevents airflow through it.

The protection of the environment of a show cave is fundamental both from the point of view of avoiding any damage to a nonrenewable patrimony and the conservation of the source of income for the cave management. Therefore, such a common interest may have an important role in the implementation of any action aiming to safeguard the cave environment.

Visitors also release carbon dioxide as a result of their breathing. Until few years ago, such carbon dioxide was considered a threat to the cave formations since it could have increased the water acidity and, consequently, the corrosion instead of the deposition of new formations. Further accurate studies (Bourges et al., 1998) have shown that in many instances the carbon dioxide produced by natural processes (oxidation of organic matter in the soil above a cave) may introduce through the water percolating into the cave amounts very much larger than the carbon dioxide released by visitors.

When the water, with a relatively high concentration of carbon dioxide, reaches the cave environment it releases immediately part of such carbon dioxide, which is not in equilibrium with the carbon dioxide in air. Therefore, the chemical reaction moves toward the deposition of calcium carbonate and the formations continue to grow. In general, rather small caves with a high visitor flux and without any input of natural carbon dioxide might have formations corroded because the chemical reactions would be reversed when the carbon dioxide of air dissolves into water, particularly when water vapor condenses on the cave walls.

Another form of environmental pollution may occur through a joint contribution by visitors and light. Persons release into the cave spores or seeds of plants and they may grow in the vicinity of lamps if the light flux is high enough. The result is the so-called lampenflora, that is, green plants (generally algae, fern, moss) developing on cave walls or formations close to a light source. Such plants cover the surfaces with a greenish layer, which can become included into the calcite deposition and no longer removable. In fact, the lampenflora may be washed away by bleach or hydrogen peroxide if it is not covered by any calcite. Special care must be taken to avoid any nuisance to the cave fauna.

The growth of lampenflora can be avoided by employing light sources with a very low emission of light useful for the chlorophyllian process and low light flux at the rock surface.

THE DEVELOPMENT OF A SHOW CAVE

A correct development of a show cave must take into account both the protection of the environment and the safety of the visitors. As it has been already pointed out in the previous paragraph, the physical and chemical equilibria of the environment should be modified outside the range of the natural variations.

At the same time, any undue source of harm to the visitors must be avoided. This means that the pathways must be strong enough to withstand the very high humidity and, sometimes, also floods. In the past, wooden structures were often used, but they had to be replaced frequently; currently, some "green" people would still use wood because this material is natural. Nevertheless, the rather short life of a wooden structure in the cave environment implies an additional cost which is not justified by any advantage. On the contrary, the rotten wood supplies large amounts of food modifying the equilibrium of the cave life.

In particular at present, the criterion to use only structures which can be easily decommissioned is substantially wrong because once it is no longer convenient to manage a show cave, no one will spend money to take out structures inside the cave. Only when show cave managers will be obliged to deposit a given amount of money to assure the future decommissioning of any structure, it is possible to use structures to be easily disassembled.

In the meanwhile, it is preferable to use materials which are compatible with the cave environment and will not release pollutants in the long run. A material with these characteristics and not expensive is concrete. It may be conveniently used for pathways in general.

The handrails in stainless steel are also a convenient solution, particularly when they are also used as pipes to provide water in different parts of the cave to wash out the pathways. In fact, a higher cost of stainless steel is justified by a lack of any maintenance after many years of operation. Sometimes plastic may be used under the condition that it does not contain any contaminant (e.g., heavy metals or organic compounds which may be released).

When an artificial entrance is needed in order to give an easy access to the cave or to establish a circuit by avoiding the return of visitors on the same pathway, it is absolutely necessary to install a system of doors to stop any additional airflow in the cave. Up until now, doors operated mechanically or manually are normally used, but it would be most preferable to install air curtains. This solution (suggested many years ago by Russell Gurnee and Jeanne Gurnee) is less expensive, quite safe, and has the great advantage of avoiding any sense of claustrophobia to visitors. In addition, it also decreases the release of lint by people as reported previously.

The surveillance of the main parameters (temperature, humidity, carbon dioxide, radon, etc.) can be achieved by a monitoring network, which should always be installed in any show cave. Presently, it is possible to install networks at a very reasonable cost, which are reliable and require little care, as, for instance, data loggers, which can be discharged every month, and the data transferred into a computer for any further evaluation. Automatic networks directly connected to a computer are operated more easily but, of course, their cost is higher. In any case, it must be stressed that any kind of monitoring network always requires some attention to avoid malfunctioning and calibration, possibly once a year.

Such monitoring networks have also another important advantage because they contribute much interesting data, which have greatly enlarged the knowledge of the behavior of cave environment. A rather widespread feeling among speleologists, and people in general, that a cave is lost to science when it is developed as a tourist attraction is not at all supported by the important scientific results obtained just within many show caves. Sometimes the borderline between use and abuse may be difficult to define; nevertheless, a careful development continuously monitored may be the most efficient way to protect a cave.

It is evident that the economy of a region around a show-cave-to-be can be radically modified by the cave development. Therefore, strenuous opposition to any tourist visitation appears to be rather unfair also toward the local people, particularly when a suitable compromise between strict conservation and a sound development can be found. But in any case, as it was previously reported, cave development cannot be accepted if it is not supported by appropriate preliminary research.

An evaluation of the number of show cave visitors all around the world (Cigna and Burri, 2000), based on data obtained for about 20% of all show caves indicated a number of more than 150 million visitors globally per year. By assuming a budget per person as reported in Table 1, the total amount of money spent to visit the show caves is around US$ 2.3 billion. The number of the local people directly involved in the show cave business (management and local services) can be estimated to be several hundred per cave, that is, some hundreds of thousands of individuals in the world.

By taking into account that there are several hundred other people working indirectly to each person directly connected with a show cave (Forti and Cigna, 1989), a

TABLE 1 Rough Estimation of the Annual Direct and Local Budget of a Show Cave per Visitor (2008 US$)

Direct income	6.5
Other local income:	
Souvenirs & snacks	2.0
Meals	6.5
Transportation	2.5
Travel agency	2.5
Total	20.0

gross global figure of about 100 million people receive salaries from the show cave business. Therefore, it can be roughly assumed that behind each tourist in a show cave there is about one employee directly or indirectly connected.

In addition to show caves, the existence of karst parks, which include a cave within their boundaries, must also be considered. As reported by Halliday (1981), the number of visitors of three top karst national parks in the United States (Mammoth Cave, Carlsbad Caverns, and Wind Cave) amounted to about 2500,000 tourists each year. Therefore, karst parks give a further increase to the number of people involved in the whole "karst" business.

There are many other human activities which involve a larger number of people; nevertheless, the figure reported above is not negligible and gives an indication of the role that show caves play in the global economy.

GUIDELINES FOR THE DEVELOPMENT OF SHOW CAVES

Some guidelines aiming to supply a recommendation to be endorsed for the development of show caves were drafted over the past few years and received strong recommendations from the UIS Department of Protection and Management at both the 14th International Congress of Speleology held in Kalamos, Greece, in August 2005 and the 15th International Congress of Speleology held in Kerrville, TX, in July 2009. These guidelines are here reported.

1. Development of a Wild Cave into a Show Cave

The development of a show cave can be seen as a positive financial benefit to not only itself but also the area surrounding the cave. The pursuit of these antici-pated benefits can sometimes cause pressure to be applied to hasten the development of the cave. Before a proposal to develop a wild cave into a show cave becomes a physical project, it is necessary to carry out a careful and detailed study to evaluate the benefits and risks, by taking into account all perti-nent factors such as the access, the synergy, and possible conflict with other tourism-related activities in the surrounding area, the availability of funds, and many other related factors. The conversion should only take place if the results of the studies are positive. A wild cave that is developed into a show cave, and is subsequently abandoned, will inevitably become unprotected and be subject to vandalism in a very short time. A well-managed show cave assures the protection of the cave itself, is a source of income for the local economy, and may also contribute to a number of scientific researches.

1-1 A careful study of the suitability of the cave for development, taking into account all factors influencing it, must be carried out, and must be carefully evaluated, before physical development work commences.

2. Access and Pathways within the Cave

In many caves it has been found to be desirable to provide an easier access into the cave for visitors through a tunnel, or a new entrance, excavated into the cave. But such an artificial entrance could change the air circulation in the cave, causing a disruption of the ecosystem. To avoid this, an air lock should be installed in any new entrance into a cave. On the other hand, it must be mentioned that in some very exceptional cases a change in the air circulation could revitalize the growth of formations. A decision not to install an air lock must be taken only after a special study.

2-1 Any new access into a cave must be fitted with an efficient air lock system, such as a double set of doors, to avoid creating changes in the air circulation within the cave.

Caves are natural databases, wherein an incredible amount of information about the characteristics of the environment, and the climate of the cave, are stored. Therefore, any intervention in the cave must be carried out with great care to avoid the destruction of these natural databases.

2-2 Any development work carried out inside the cave should avoid disturbing the structure, the deposits, and the formations of the cave as much as possible.

When a wild cave is developed into a show cave, pathways and other features must be installed. This invariably requires materials to be brought into the cave. These materials should have the least possible impact on both the aesthetics of the cave and its underground environment. Concrete is generally the closest substance to the rock that the cave is formed in, but once concrete is cast it is extremely expensive and difficult to modify or decommission. Stainless steel has the distinct advantage that it lasts for a long time and requires little to no maintenance but it is expensive and requires special techniques to assemble and install. Some recently developed plastic materials have the advantage of a very long life, are easy to install, and are relatively easy to modify.

2-3 Only materials that are compatible with the cave, and have the least impact on the cave, should be used in a cave. Cement, concrete, stainless steel, and plastics are examples of such materials.

The environment of a cave is usually isolated from the outside and therefore the introduction of energy from the outside will change the equilibrium balance of the cave. Such changes can be caused by the release of heat from the lighting system and the visitors and also by the decay of organic material brought into the cave, which introduces other substances into the food chain of the cave ecosystem. In ice caves, the environmental characteristics are compatible with wood, which is frequently used for the construction of pathways, as it is not slippery.

2-4 Organic material, such as wood, should never be used in a cave unless it is an ice cave where, if necessary, it can be used for pathways.

3. Lighting

The energy balance of a cave should not be modified beyond its natural variations. Electric lighting releases both light and heat inside the cave. Therefore, high-efficiency lamps are preferred. Discharge lamps are efficient, as most of the energy is transformed into light, but only cold cathode lamps can be frequently switched on and off without inconvenience. Lightemitting diode (LED) lighting is also very promising. As far as possible, the electric network of a cave should be divided into zones to enable only the parts that visitors are in to be lit. Where possible a noninterruptible power supply should be provided to avoid problems for the visitors in the event of a failure of an external power supply.

3-1 Electric lighting should be provided in safe, well-balanced networks. The power supply should preferably be noninterruptible.

It is essential to ensure the safety of the visitors in the cave, particularly in the event of a failure of the main power supply. Emergency lighting should always be available whether it is a complete noninterruptible power supply or an emergency lighting system with an independent power supply. Local code requirements may be applicable and these may permit battery lamps or a network of LEDs or similar devices.

3-2 Adequate emergency lighting should be available in the event of a power outage.

Lampenflora is a fairly common consequence of the introduction of an artificial light supply into a cave. Many kinds of algae, and other superior plants, may develop as a result of the introduction of artificial light. An important method to avoid the growth of green plant life is to use lamps that do not release a light spectrum that can be absorbed by chlorophylll.

3-3 Lighting should have an emission spectrum with the lowest contribution to the absorption spectrum of chlorophyll (around 440 nm and around 650 nm).

Another way to prevent the growth of lampenflora is the reduction of the energy reaching any surface where the plants may live. The safe distance between the lamp and the cave surface depends on the intensity of the lamp. As a rough indication, a distance of one meter should be safe. Special care should also be taken to avoid heating the formations and any rock paintings that may exist.

3-4 Lighting sources should be installed at a distance from any component of the cave to prevent the growth of lampenflora and damage to the formations and any rock paintings.

The lighting system should be installed in such a way that lights only in the portions of the cave occupied by visitors are switched on, leaving the lighting in the portions of the cave that are not occupied switched off. This is important from the aspects of reducing the heating of the cave environment and preventing the growth of lampenflora, as well as decreasing the amount of energy required and its financial cost.

3-5 Lighting should be installed in a manner to enable only the portions of the cave that are occupied by visitors to be illuminated.

4. Frequency of Visits and Number of Visitors

The energy balance of a cave environment can be modified by the release of heat by visitors. A human being, moving in a cave, releases about 150 W— approximately the same as a good incandescent lamp. Consequently, there is also a limit on the number of visitors that can be brought into a cave without causing an irreversible effect on the climate of the cave.

4-1 A cave visitor capacity, per a defined time period, should be determined and this capacity should not be exceeded. Visitor capacity is defined as the number of visitors to a given cave over a given time period, which does not permanently change the environmental parameters beyond their natural fluctuation range. A continuous tour, utilizing an entrance and another exit, can reduce the time that visitors spend in a cave, compared to the use of a single entrance/exit.

In addition to the normal tours for visitors, many show caves have special activities, sometimes called adventure tours, where visitors are provided with speleological equipment for use in wild sections of the cave. If such a practice is not properly planned, it may cause serious damage to the cave.

4-2 When visits to wild parts of a cave are arranged, they must be carefully planned. In addition to providing the participants with the necessary speleological safety equipment, there must be experienced guides to show the wild caves. The pathway that visitors travel along, must be clearly defined, for

example, with red and white tape, and the visitors should not be allowed to walk beyond this pathway. Special care must be taken to avoid any damage to the cave environment, and the parts beyond the pathway must be maintained in a clean condition.

5. Preservation of the Surface Ecosystem When Developing Buildings, Parking, Removal of Surface Vegetation, and Waste Recovery

It is important that the siting of the aboveground facilities, such as the buildings, parking, and waste recovery, be well planned. There is a natural tendency to try to place these development features as close as possible to the cave entrance. Sometimes these features are built over the cave itself, or relevant parts of it. The hydrogeology above the cave must not be modified by any intervention such as the watertight surface of a parking area. Any change in the rainwater seepage into a cave can have a negative influence on the cave and the growth of its formations. Care should be exercised also when making any change to the land above the cave, including the removal of the vegetation and disturbance of the soils above the bedrock.

5-1 Any siting of buildings, parking areas, and any other intervention directly above the cave must be avoided in order to keep the natural seepage of rainwater from the surface in its original condition.

6. Monitoring

After the environmental impact evaluation of the development, including any other study of the cave environment, it is necessary to monitor the relevant parameters to ensure that there is no deviation outside acceptable limits. Show caves should maintain a monitoring network of the cave environment to ensure that it remains within acceptable limits.

6-1 Monitoring of the cave climate should be undertaken. The air temperature, carbon dioxide, humidity, radon (if its concentration is close to or above the level prescribed by the law), and water temperature (if applicable) should be monitored. Airflow in and out of the cave could also be monitored.

When selecting scientists to undertake studies in a cave, it is very important that only scientists who have good experience with cave environments be engaged for cave-related matters. Many otherwise competent scientists may not be fully aware of cave environments. If incorrect advice is given to the cave management, then this could result in endangerment of the cave environment. Cave science is a highly specialized field.

6-2 Specialized cave scientists should be consulted when there is a situation that warrants research in a cave.

7. Cave Managers

The managers of a show cave must never forget that the cave itself is "the golden goose" and that it must be preserved with great care. It is necessary that persons involved in the management of a show cave receive a suitable education, not only in the economic management of a show cave, but also about the environmental issues concerning the protection of the environment at large.

7-1 Cave managers should be competent in both the management of the economics of the show cave and its environmental protection.

8. Training of the Guides

The guides in a show cave have a very important role, as they are the "connection" between the cave and the visitor. Unfortunately, in many instances the guides have not been trained properly and, notwithstanding that they are doing their best, the overall result will not be very good. It is very important that the guides receive proper instructions about the environmental aspects of the cave as well as dealing with the public. It is important that guides are skilled in tactfully avoiding entering into discussions that can have a detrimental effect on the overall tour. The guides are the guardians of the cave and they must be ready to stop any misbehavior by the visitors that could endanger the cave environment.

8-1 Cave guides should be trained to correctly inform the visitors about the cave and its environment.

INFORMATION ON SHOW CAVES IN THE WORLD

There are many books published in different countries providing guides to the local caves. On one hand they report a rather large amount of information but, on the other hand, they are fully reliable only for a short time after their publication. Information on visitor details and even the existence of the show caves can change frequently.

Recently, a rather useful way to obtain up-to-date information became available. "Showcaves of the World" is a website (Duckeck, 2000) that can be found at www.showcaves.com/. This site changes and grows continually, so the latest version may always be seen on the web Figures 1–7.

FIGURE 1 Kartchner Caverns (AZ, U.S.A.) are probably the show caves in the world, managed according to the most up-to-date criteria. An electric train from the office building to the cave is available for visitors, avoiding both acoustic and environmental pollution. *Photograph taken by the author.*

FIGURE 3 Kartchner Caverns (Arizona). A monitoring network should always be envisaged in a show cave in order to avoid relevant changes of the cave environment. In the case of Kartchner Caverns, such monitoring is particularly important since the relative humidity within the cave is close to 100%, while outside the climate is very dry. During the construction of a pathway, many pipes are placed to be covered by concrete. Such pipes are used for circuits connecting any sensor within the cave with a computer in the main building for present and future monitoring networks. *Photograph taken by the author.*

FIGURE 2 Kartchner Caverns (AZ, U.S.A.). Sometime in a cave there are joints which may imply possible rock fall. In order to assure the safest condition in a show cave, it is necessary to detect well in advance any minor movement. A detector, monitoring possible displacement of a limestone layer, can be seen in this photo. *Photograph taken by the author.*

FIGURE 4 Cango Caves (Oudtshoorn, South Africa) was discovered around 1780 and the first recorded visit was made in 1806; now it is the most important show cave of Africa. Here a flowstone deposited over a clay filling, which has been washed away successively, can be seen. Iron handrails along the pathway in the forefront: Nowadays such handrails are generally substituted by stainless steel or plastic handrails, to avoid iron contamination of the cave environment. *Photograph taken by the author.*

FIGURE 5 Cango Caves (Oudtshoorn, South Africa). In one of the innermost parts of the cave, Cango 3, which is not open to tourism, the floor is covered by calcite crystals; a plastic ribbon fixes the boundaries of the trail, to avoid any destruction of the crystals due to scientists moving around. *Photograph taken by the author.*

FIGURE 7 Frasassi Caves (Ancona, Italy) developed mainly in the shallow phreatic zone, where rising hydrogen sulfide–rich groundwater mixes with oxygenated seepage water. Presently, these are the most important show caves in Italy. Details may be found in www.frasassi.com. Shown here is Niagara Falls. *Photo courtesy of Ente Consorzio Frasassi.*

See Also the Following Articles

Lampenflora
Recreational Caving

Bibliography

Bourges, F., D'Hults, D., & Mangin, A. (1998). *Étude de l'Aven d'Orgnac*. Rapport final, Laboratoire Souterrain de Moulis C.N.R.S. Géologie Environnement Conseil, pp. 1–84.

Cigna, A. A. (1993). Environmental management of tourist caves. *Environmental Geology, 21,* 173–180.

Cigna, A. A., & Burri, E. (2000). Development, management and economy of show caves. *International Journal of Speleology, 29B* (1–4), 1–27.

Duckeck, J. (2000). Showcaves of the world. <www.showcaves.com/>.

Forti, P. & Cigna, A.A. (1989) Cave tourism in Italy: an overview. Proceedings of the International Symposium on the 170th Anniversary of Postojnska Jama. Postojna, Nov. 10-12, 1988, Centre for Scientific Research SAZU & Postojnska Jama Tourist and Hotel Organization, p. 46–53.

Gurnee, R., & Gurnee, J. (1990). *Gurnee guide to American Caves*. Closter, NJ: R. H. Gurnee Inc.

Halliday, W.R. (1981) Karstic national parks: International economic and cultural significance. Proceedings of the International Symposium on Utilization of Karst Areas, Trieste, March 29–30, 1980, Commissione Grotte Boegan, CAI Trieste, pp. 135–144.

FIGURE 6 Hwanseon Cave (Daei, South Korea) is a very important show cave of Korea. Tourists reach the vicinity of the cave by car or bus and then they have to climb a slope of a couple of kilometers to reach the cave entrance, which is here depicted. *Photograph taken by the author.*

SIEBENHENGSTE CAVE SYSTEM, SWITZERLAND

Pierre-Yves Jeannin and Philipp Häuselmann

Höhlenforschergemeinschaft Region Hohgant (HRH) and Swiss Institute for Speleology and Karst Studies (SISKA), Switzerland

GEOGRAPHICAL AND GEOLOGICAL SETTING

With a length of more than 156 km and a depth range of 1340 m, the Réseau Siebenhengste-Hohgant Cave System is one of the most important on Earth. Close to this network, several other large caves have been found, but they remain unconnected to the main system so far. The total length of explored conduits within the area is about 320 km, and this ensemble is referred to as the *Siebenhengste Cave System* in this article.

The system is located in the frontal Alpine range (Helvetic Nappe), directly facing the Swiss Plateau (Fig. 1). From the edge of Lake Thun, where the main spring of the system (Bätterich) is located, it extends more than 20 km northeastward to the Schrattenfluh, crossing the deep Emme Valley (Bitterli, 1988).

The mountain range enclosing the caves is the southeast-dipping limb of an anticline. Its northwestern limit is formed by high cliffs. Elevations of the summits located along the cliffs range between 1950 and 2190 m a.s.l. Patches of denuded karren fields occur above 1700 m a.s.l., where limestone is exposed. Below this limit, sandstone mainly crops out; therefore, forest, meadow, and swamp cover most of the area. The climate is humid and temperate, dominated by a western wind. The average annual temperature is about 2°C at 1800 m a.s.l., and annual precipitation ranges between 1500 and 2000 mm. Southeastward, the limestones/sandstone series disappears below the thick flysch deposits of the Pennine Nappes (Fig. 2).

Karst features are developed predominantly within the Schrattenkalk formation (Barremian to Aptian, Cretaceous, Urgonian facies), which is usually 150–200 m thick. Detailed geological studies identified 6 different formations within the Schrattenkalk, which is underlain by the Drusberg Marls (lower Barremian). These marls are 30–50 m thick. Most of the underground rivers follow the dip on the top of this impervious layer, often along the main faults. The Hohgant sandstones (Eocene), which are locally up to 200 m thick, overlie the Schrattenkalk (Fig. 2). In most of the region, faults enable the surface waters to flow through the sandstones down into the Urgonian limestones.

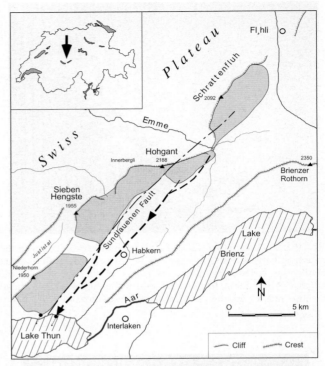

FIGURE 1 Location of the Siebenhengste region with the water catchments of the St. Beatus Spring (bottom left) and of the Bätterich/Gelber Brunnen springs (Siebenhengste-Hohgant-Schrattenfluh). *Adapted from Jeannin et al., 2000.*

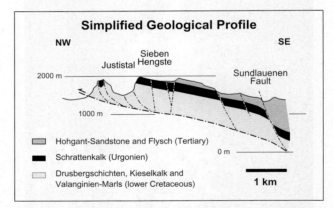

FIGURE 2 Geological cross-section of the Siebenhengste range; see Figure 3 for profile location.

An important longitudinal normal fault, stretching from Lake Thun to Schrattenfluh (the Hohgant-Sundlauenen Fault), disrupts the continuity of the southeastward-dipping monocline of the Siebenhengste range (Figs. 2 and 3). The offset on the fault is 200–1000 m, depending on the location. Several parallel normal faults are also present. The normal faults in the Siebenhengste region mainly developed during the Lower Cretaceous to the Eocene. Another set of faults is extensively developed in the area: the dextral strike-slip

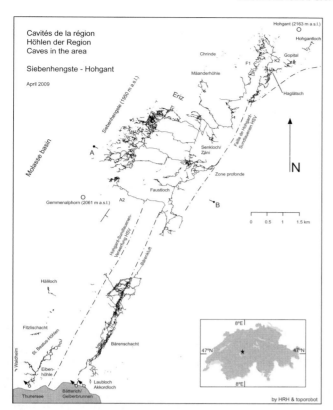

FIGURE 3 Plan view of the cave systems.

faults and, to a lesser degree, the related sinistral strike-slip faults. These faults are related to alpine tectonics and are of Miocene age. They appear to have been partly reactivated very recently.

West of the Siebenhengste, the frontal folds of the Helvetic Nappe are observed in the Sigriswilergrat Range, west of Justistal (Figs. 1 and 2), allowing placement of the original crest of the anticline between 1 and 2 km in front of (northwest of) the Siebenhengste cliffs.

HYDROLOGY

Most of the region is drained by underground karst systems emerging at two main locations: the St. Beatus spring and the Bätterich/Gelberbrunnen springs.

The St. Beatus spring with a discharge ranging between 10 L s^{-1} and 2–3 m^3 s^{-1} (average 72 L s^{-1}) has a catchment area of about 10.5 km^2 that lies in the southeast part of the region. The Bätterich/Gelberbrunnen spring catchment extends at least 21 km to the northeast, reaching the Schrattenfluh massif, as proven by a tracing experiment (Fig. 1). As the main spring (Bätterich) lies below Lake Thun, the discharge of the system is very difficult to measure. It probably exceeds 20 m^3 s^{-1} during floods. The catchment area is around 32 km^2 and is largely covered by sandstone where surface flow may occur locally. Recharge rates are therefore difficult to assess.

OVERVIEW OF THE CAVE SYSTEMS AND HISTORY OF THEIR EXPLORATION

The different caves known at present sum to a total length of more than 320 km of passage, lying between 480 and 2000 m a.s.l. The system has a distinct labyrinth character, whose complex geometry can only be explained by the development of several superimposed cave systems that correspond to different and more or less independent times and conditions.

Seven large caves enclose most of the known passages in the area (Figs. 3 and 4). Many passages of the caves are surveyed and mapped in detail, as shown in Figure 5.

St. Beatus Cave

St. Beatus Cave is a spring cave that can be followed from the spring upstream for approximately 2 km as a crow flies (Häuselmann et al., 2004). The explored length of the cave is 12 km with a height of +353 m with respect to the entrance. Most of the cave is of phreatic origin (elliptic passage), although canyons are found in many places (keyhole cross-section). Most of the passages lie on the top of the Drusberg marls. A detailed study of the morphology of this cave and its sediments was conducted (Häuselmann 2002), allowing the identification of several speleogenetic phases (see later discussion). Some smaller caves are known in the vicinity of St. Beatus Cave.

The entrance of the cave has been known since prehistoric times and seems to have been inhabited by the hermit called St. Beatus in the Early Middle Ages. The first caving explorations were made in the nineteenth century, and already in 1904 the entrance part of the cave had become a show cave. Maybe one-third of today's known passages were explored before World War II. The evolution of the caving techniques (especially cave diving) allowed the exploration of the other two-thirds of the cave over the past 50 years. A complete and systematic resurvey of the cave was conducted in the 1990s, allowing the exploration of about 1 km of new passages. Cave exploration is almost complete by today's standards.

Bärenschacht ("Bear's Shaft")

Bärenschacht is the second largest cave in the region with a length of 71 km and a depth of −979 m (as of

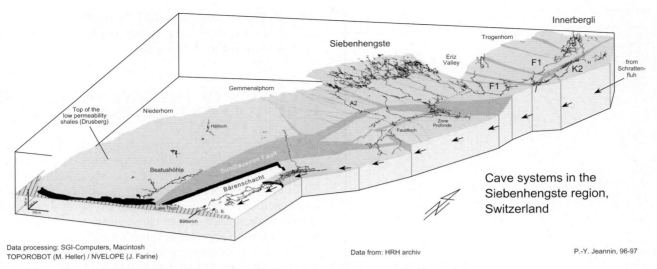

FIGURE 4 Three-dimensional view of the cave system. *From Jeannin et al., 2000. Used with permission.*

October 2011). The name of the cave comes from three bear skeletons found at the bottom of the first shaft, some hundreds of meters from the entrance. The entrance part presents a series of shafts leading to the bottom of the limestone (Schrattenkalk) at a depth of 150 m (Bitterli et al., 1991). From this point down to −550 m, the passage is steep, with small shafts and a cross-section on the order of 5 m². Most of this section has developed along a thin calcareous bed within the Eocene sandstone series (Hohgant series). Due to large faults, deep shafts make it possible to reach an incredible labyrinth of large elliptic conduits at a depth of −900 m. The bottom part of the cave floods, as it is located at the level of Lake Thun (Bätterich spring). Higher levels remain dry and are richly decorated by dripstones and flowstones, gypsum crystals, and some aragonite formations.

With the exception of 2 km in the entrance part, Bärenschacht is almost exclusively of phreatic origin. Direct infiltration into the cave is prevented by a thick cover of marls and flyschs. In some places, the depth of the conduits below the land surface reaches as much as 1000 m. Therefore, the large cross-sections found in the labyrinth (between 12 and 25 m²) clearly indicate that Bärenschacht is the downstream collector of the Bätterich/Gelberbrunnen catchment area, that is, the downstream part of most of the caves of the Siebenhengste region.

Bärenschacht was first discovered in 1965 and explored in 1973 and 1974 down to a sump at −565 m. Explorations were conducted by large teams recalling the conquest of the Himalayan summits in the 1950s. The sump was dived and considered too tight. In 1986, an audacious caver dug underwater and was able to squeeze through a very narrow flooded conduit for about 40 m. A steep and large passage ended in the darkness of very large shafts. This was the key to the labyrinth at −900 m and the beginning of incredible explorations by a very restricted group of cavers, able to cope with the dive of both a nasty siphon and difficult vertical caving. Nearly 40 km were explored at that time (Funcken, 1994). In 1995, a tunnel was dug on top of the sump, making it possible also for nondivers to reach the deep part of the cave (Funcken et al., 2001). Today's exploration is driven by the Coordination Group of Bärenschacht, which is part of the body of HRH.

The Réseau Siebenhengste-Hohgant

The Réseau Siebenhengste-Hohgant is located below the Siebenhengste karren field. In the international scene of speleology, it is the best-known system of the region, because its exploration was quite spectacular. This is also the reason why its name has been used for the entire cave region. This cave network was connected to F1 and Faustloch in the 1980s, giving today a system of 156 km of connected passages with a depth of 1340 m. The Siebenhengste labyrinth itself is nearly 108 km long.

Schematically, most of the entrances to the Réseau start with a series of shafts and short meanders down to a depth of 150−250 m (Hof, 1984). The Réseau ("network" in French) itself is reached at that depth, where a labyrinth of elliptic or keyhole passages can be followed for kilometers, upstream, downstream, or horizontally. This labyrinth is clearly visible on the map of the caves (Fig. 3). Similar to antennas, underground streams have been explored downstream along large and straight canyons that follow some of the main strike-slip faults

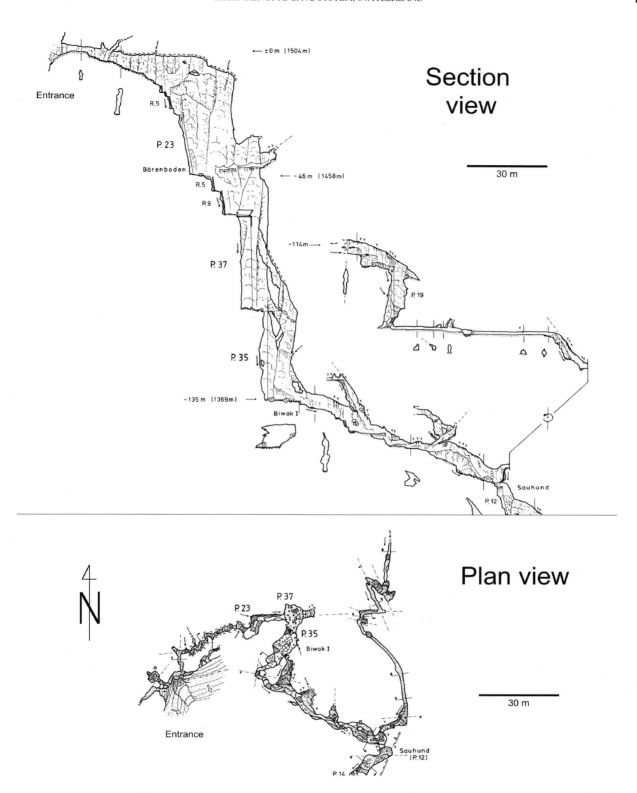

FIGURE 5   Example of a detailed map of the cave system. *Adapted from Bitterli et al., 1991.*

of the region. These all reach a perpendicular major conduit coming from F1. Due to geological complications (crossing a normal fault), the major conduit splits into three-dimensional anastomoses in the so-called *Zone Profonde*.

Exploration of the Siebenhengste massif started in 1966, but it was in 1972 and 1973 that the existence of such a large cave system was revealed. Exploration was then quick, with a length of 20 km being reached in 1976, 32 km (to a depth of −800 m) in 1978, and 45 km in 1982. It was at that time that the F1 (Innerbergli area) was connected, giving a length of 60 km to the Réseau. This quick exploration was not systematic at all, leaving many unexplored passages. A more systematic exploration of these side passages was begun in 1982 and is still ongoing today. As a consequence, the overall aspect of the Réseau Siebenhengste-Hohgant did not change much during this period, but the density of the conduits has more than doubled. The exploration of this incredible cave system is still far from finished, and new kilometers of passage are found every year. Exploration and survey are achieved in a very systematic way in the Réseau as well as at the surface. Hundreds of small caves have been explored in the Siebenhengste karren field, some of them reaching a length of 1 or 2 km; however, only a restricted subset of them could be connected to the Réseau.

The Faustloch

The Faustloch is a 14-km-long cave (reaching −930 m), which was connected to the Réseau Siebenhengste-Hohgant in 1987. The entrance part, down to −200 m, developed in the sandstone overlying the Urgonian limestone and is followed by two huge shafts (80 m and 60 m) crossing the limestone down to its bottom. From here onward, an underground stream can be followed along a canyon. At −450 m, the stream crosses the Drusberg marls and goes deeper into a series of small and wet shafts and meanders, developed in a siliceous limestone (Hauterivian). At the top of the descent across the Drusberg marls, a major elliptic conduit is found (cross-section of $10-20\ m^2$). This conduit is the continuation of the F1-Siebenhengste major collector (fossil phreatic conduit). It can be followed downstream over several kilometers toward the Bärenschacht and divides into several branches. Two of them end on deep siphons (one of them has been dived down to −38 m for a length of 300 m), which lie slightly higher than those of the Bärenschacht. The other branches are clogged by thick sediments. Despite strong digging efforts, the continuation has not yet been opened.

The Faustloch was discovered in 1970, the entrance being as large as a fist ("faust" in German). It took several years to descend the shaft series down to −350 m—first, because of large quantities of water cascading down the shafts, and, second, because it was very difficult to set anchors in the sandstone. The first third of the cave was explored between 1971 and 1978. In 1987, a large flood event occurred in the Siebenhengste region, and the fossil passage coming from the Réseau Siebenhengste-Hohgant was reactivated by a stream of several cubic meters per second. This opened the passage upstream of Faustloch, allowing the Réseau to be connected. It also opened the downstream end of the cave (Funcken and Moens, 2000). After a long digging effort, almost 10 km of passage could be found toward the Bärenschacht. In 1997, a new flood clogged this passage again, and the most remote part of the cave was not accessible until 2009, when the access was dug open again.

Due to strong cold water cascades, the entrance shaft series of Faustloch are a dangerous section of the cave, where three cavers have lost their lives (one in 1976 and two in 1998).

F1 (Innerbergli Area)

F1 cave has a typical dendritic pattern. Its length is nearly 32 km and its depth is 650 m. It was connected in 1982 with the Réseau Siebenhengste-Hohgant. The entrance shaft series is comprised of small pits and meanders, with some tight passages. At a depth of 150 m, the Drusberg marls are found, and the passage gets larger, thanks to the presence of some tributaries converging at that point and to the collapse of the conduit wall where it contacts the Drusberg marl. A large active canyon can be followed for about 3 km before disappearing into a large collapse room. From this room onward, a phreatic fossil conduit continues for another 3 km. It is joined by four main tributaries but ends (clogged) very close to the Réseau. The connection could be managed only through tight and wet passages. The cave entrance was found in the summer of 1981 and the main conduit grew over 15 km in the 18 months prior to the connection to the Réseau at Christmas 1982. A systematic exploration of the tributaries has been undertaken over the past 30 years, but the complete exploration of F1 will require many more years.

K2

This cave is parallel and quite similar to F1. Its length is 14 km and its depth is −750 m. The entrance shaft series, found in 1980, is followed by a severe squeeze and a long and narrow meander down to a room at −300 m. The passage out of this room is very tight, and many cavers have had to turn back at this point. After a short maze of small phreatic conduits, a large canyon passage is reached and can be followed for several kilometers. This section is very dark and slippery. At a depth of −600 m, the main passage splits into a complex three-dimensional labyrinth of elliptic conduits (fossil phreatic passages) reaching a siphon at its lowermost point. In 1991, a dive of the siphon led to the deeper point of the cave (−750 m), where any continuation does not seem possible. At that time, this point was probably one of the remotest parts of Switzerland. In 1992, a lower entrance was found, giving access to the lower part of the cave and reactivating exploration in this region. Many tributaries were explored, but no major continuation could be found. The intermediate section of K2 is still being explored with the hope of being able to connect it to F1 or to the Haglätsch cave (7.5 km), which develops about 80 m straight above K2.

A2 (Hohlaub Area)

This 11-km-long cave (−690 m) can be considered as the southwestern continuation of the Siebenhengste labyrinth, but no connection has been achieved. The entrance was found in 1973 and explored down to a depth of 30 m. There, a very tight meander was found. At least four groups have tried to squeeze through the meander, where a strong draft indicated a continuation. Only in 1986 was the 35-m-long passage traversed. A mining action was organized in order to be able to explore the galleries found beyond. A series of shafts and meanders was explored down to a depth of 220 m. A labyrinth of horizontal elliptic passages (fossil phreatic conduits), corresponding to the Siebenhengste labyrinth, crosscuts the shaft zone. Downstream, an active canyon stretches toward the east to a collapse room, where the water disappears. A fossil phreatic conduit goes on toward Faustloch, but turns abruptly to the right some hundreds of meters before reaching it. This region, called *Keller* (Cellar), develops along a fault and gives access to the lowermost point of the cave (−690 m). It is still nearly 100 m above the nearby Faustloch. Most of the cave was explored between 1986 and 1994; however, there are still a lot of passages to discover (Gerber, 1994).

GENESIS AND AGE OF THE CAVE SYSTEMS

Theories regarding the genesis of the Siebenhengste cave system began to be developed immediately on exploration of the cave, as a result of the exceptional density of conduits. Various contradicting hypotheses were discussed. In the middle of the 1980s, Hof (1984) laid the groundwork for the current ideas. Jeannin *et al.* (2000) and Häuselmann *et al.* (2002) refined this model. The ideas are based on the observation of the conduit morphology that allows recognition of paleowater tables.

Two major observations help to identify the position of paleowater tables:

1. The change of a vadose conduit morphology (canyon or meander) into a phreatic morphology (elliptic tube) gives the exact height of the paleowater table. However, it has been shown that this paleowater table is not horizontal (Häuselmann *et al.*, 2003).
2. In looping passages, the position of the paleowater table can be inferred to be higher than the top of the loops, if they do not present any indication of free-surface flow morphology (entrenchment).

By applying this concept, several phases (or levels) of conduits have been recognized. The inferred paleowater tables are considered to be linked to the position of paleosprings (*i.e.*, of paleovalley bottoms; regional base level). Phases are labeled with elevations inferred for the paleosprings. Due to the Alpine uplift, real elevations may have been different when the systems were active.

The oldest speleogenetic phases are found in the uppermost caves. Five phases could be identified at 1950, 1800, 1720, 1585, and 1505 m a.s.l., respectively. The conduit morphology (scallops) indicates that water flowed to the northwest and that the springs of those hydrological systems were located in the Eriz Valley (Fig. 6). Based on additional data, we hypothesize that the Aare Valley (today's main valley) did not exist when this part of the cave system was active and that the inferred paleosprings were located along the (mostly disappeared) Eriz paleovalley.

During the next series of speleogenetic phases, water flowed to the southeast to springs located in the Aare Valley (Fig. 7), meaning that a significant geomorphic event happened between phases 1505 and 1440, capable of turning the flow direction 180° within the karst system. Evidence for springs at 1440, 1150, 1050, 890, 805, 760, 700, 660, and 558 m a.s.l. has been found so far.

Cave sediments were investigated, and the following ages of the cave systems could be obtained by applying the U-Th and the cosmogenic burial age methods (Häuselmann *et al.*, 2007, 2008):

- Phase 1950: must be older than 4350 ky
- Phase 1800: oldest found deposition of quartz sand at 4350 (±600) ky

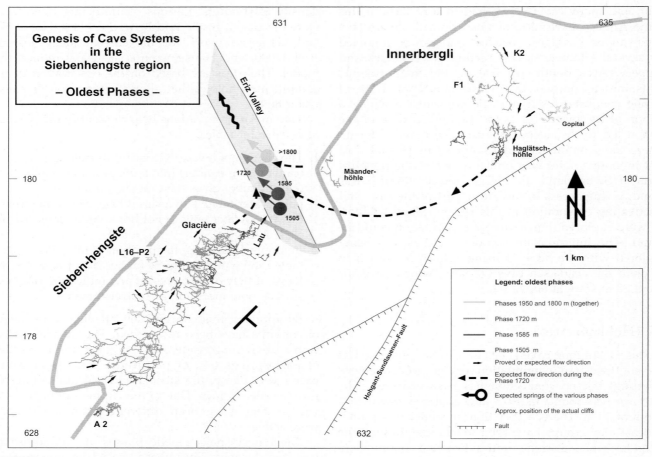

FIGURE 6 Older phases of the cave system genesis.

- Phase 1800: first deposition of glacial pebbles at 1870 (±210) ky
- Phase 1585: active system (not necessarily phreatic) between 1570 and 760 (±420) ky
- Phase 1505: active phreatic system between 740 and 600 (±170) ky
- Phase 1050: last deposition of quartz sand at 440 (±160) ky
- Phase 760: active phreatic system between more than 350 ky and 207 (±27) ky
- Phase 700: active phreatic system between 207 (±27) and 146 (±11) ky
- Phase 660: active phreatic system between 146 (±11) and 28 (±11) ky
- Phase 558: active phreatic system since 28 (±11) ky until today

These datings are novel for the Alps and made it possible not only to date valley deepening phases with relatively high precision, but they also contribute to document the drain switch from Eriz to the Aare Valley (due to its glacial incision), as well as incision velocity of the valleys (0.12 mm a^{-1} in the Eriz, 1.2 mm a^{-1} in the Aare Valley) and erosion rates at the Siebenhengste plateau surface (roughly 0.12 mm a^{-1} throughout, therefore proof of enhancing relief).

Further studies in the area proved that the oldest glacial pebbles found in the Siebenhengste (1.87 Ma BP) were transported from southern Valais to the north, before the present-day valley pattern of the Alps was established. All these data together make the Siebenhengste a prime site for the geomorphological evolution of this area of the Alps.

CAVE MINERALS

Many sections of the caves in the Siebenhengste region are nicely decorated. White-to-yellowish flowstones are present in many passages, producing a strong contrast with the dark limestone walls. Some of the flowstones can be followed over hundreds of meters, up to almost 1 km.

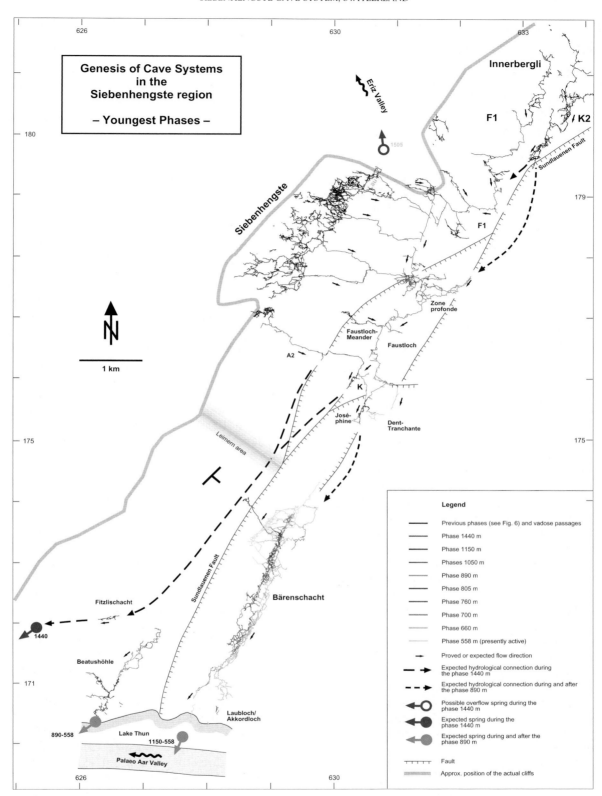

FIGURE 7 Younger phases of the cave system genesis.

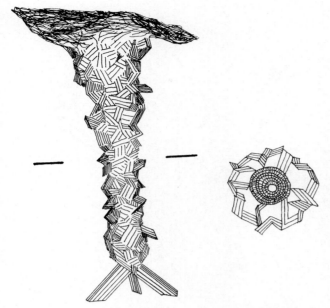

FIGURE 8 Sketch of a gypsum mantle covering and stalactite. *Illustration courtesy of A. Hof, 1984.*

Stalactites and stalagmites are common features, but they are never very large and are usually quite dispersed. Some particular speleothems are calcite stalactites mantled in a gypsum crust that remain at a few centimeters from the calcite core (Fig. 8). Larger gypsum crystals are sometimes present at the tip of the stalactite.

Soda straws are found in many places, some of them reaching more than 3 m in length. Helictites, draperies, cave pearls, rimstone dams, cave shields, moonmilk, and some spars are found in all parts of the cave system, but mainly in the lower levels. Although frequent, speleothems are generally not set in a very dense pattern in the Siebenhengste caves, but really decorate the conduit walls, roof, and floor.

Beside these very classical calcite formations, aragonite is observed as frostwork or is interbedded in stalagmites or stalactites. Rarely, soft fibrous aragonite aggregates looking similar to asbestos have also been found. Gypsum is very frequently found in the caves of the Siebenhengste region. Gypsum stars and needles (up to 15 cm long) are the most common formations. Gypsum crusts, snow, and mantles (Fig. 8) are also quite common. Gypsum hair up to 30 cm is quite rare, but can be seen in some cave parts. In Faustloch, a floor is covered by a 20-cm-thick carpet of mirabilite crystals, probably coated by some thenardite white crust.

In fact, very few data have been collected on minerals and speleothems of the Siebenhengste region. More systematic investigations would probably reveal many other types of minerals and formations.

PRESENT AND PAST FAUNA DISCOVERED IN THE CAVES

Many investigations of cave biology have been undertaken in the Siebenhengste region, but have never been brought to the attention of a wider public.

Investigations in St. Beatus Cave (Häuselmann *et al.*, 2004) revealed the existence of 26 mosses, 4 ferns, and 3 algae as well as 4 mushroom species; also, 11 bat species have been observed, sometimes quite far from the cave entrances. Many invertebrates have been collected and determined. The current numbers follow: *Collembola* (11), Niphargus (3), Dytiscides (1), Asellus (2), Diptera (4), Turbellaria (2), Ostracoda (1), Gasteropoda (8), Copepoda (1), Diplopoda (1), Opiliones (1), Arachnia (2), Acaria (2), Trichoptera (4), Coleoptera (4), Thysanura (1), Lepidoptera (3), Oligochaeta (1), and rainworms (2), giving a total of 54 species (24 of which are described from St. Beatus Cave). *Bythiospeum alpinum* (Mollusca: Gasteropoda: Prosobranchia: Hydrobiidea), found in Bärenschacht, is a new animal species for science. The same is valid for *Onychiurus dunarius*, a collembol first described from St. Beatus Cave.

Attention has also been paid to remains of past fauna. Bone fragments of a cave bear (*Ursus spelaeus*) have been found in one cave, but these are the only remains of Pleistocene fauna.

A large spectrum of Holocene animals (46 species total) have been found, usually at the bottom of entrance shafts, which acted as natural traps. The Brown Bear (*Ursus Arctos*), who was extinct in Switzerland by around 1850, was present in many caves. Bear nests have been found in two caves. Moreover, a moose (*Alces alces*) found in the Innerbergli area dated back to 424 to 633 A.D. Many species of small mammals and small bovines common to the region were found in the caves as well and also provide interesting information. For instance, the presence of domestic animals and of species typical for a forest environment is surprising in caves that open in a denuded limestone pavement. In fact, these animals are relicts of warmer periods, when forest still covered much of the Siebenhengste region almost to the top of the mountains.

The same type of observation has been made from bat bones found in the caves. The presence of species living today at least 500 m below their finding location indicates relicts of a warmer climate (Morel, 1989). Today, bats are still present but scarce. The very large number of skeletons found in the caves seems to indicate that the bat population was much higher in the past. Most of these bones probably date from a period extending between 4000 B.C. and 1500 A.D. as confirmed by 14C datings of bat bones from a nearby karst area. Although still poorly investigated, bones found in caves provide a rich record of the past fauna and climate of Central Europe.

Acknowledgments

The English was reviewed by Monique Hobbs. We also thank all the cavers who explored this huge system and provided very valuable work of exploration and cave mapping. Marco Filipponi and Duccio Malinverni kindly revised Figures 6 and 7.

Bibliography

Bitterli, T. (1988). Das Karstsystem Sieben Hengste-Hohgant-Schrattenfluh: Versuch einer Synthese. *Stalactite, Journal de la Société Suisse de Spéléologie, 38*(1−2), 10−22.

Bitterli, T., Funcken, L., & Jeannin, P.-Y. (1991). Le Bärenschacht: Un vieux rêve de 950 mètres de profondeur. *Stalactite, Journal de la Société Suisse de Spéléologie, 41*(2), 71−92 (in French).

Funcken, L. (1994). Bärenschacht: Plus de 36 km post-siphon. Une exploration hors du commun. *Stalactite, Journal de la Société Suisse de Spéléologie, 44*(2), 55−82 (in French).

Funcken, L., & Moens, M. (2000). Synthèse des explorations au Faustloch depuis 1987. *Stalactite, Journal de la Société Suisse de Spéléologie, 50*(1), 3−22 (in French).

Funcken, L., Moens, M., & Gillet, R. (2001). Bärenschacht: L'interminable exploration. *Stalactite, Journal de la Société Suisse de Spéléologie, 51*(1), 9−22 (in French).

Gerber, M. (1994). Le A2-Loubenegg (région des Sieben Hengste). *Stalactite, Journal de la Société Suisse de Spéléologie, 44*(1), 3−8 (in French).

Häuselmann, P. (2002). *Cave genesis and its relationship to surface processes: Investigations in the Siebenhengste region (BE, Switzerland)*. Switzerland: University of Fribourg.

Häuselmann, Ph., Jeannin, P.-Y., Monbaron, M. & Lauritzen, S. E. (2002). Reconstruction of alpine Cenozoic paleorelief through the analysis of caves at Siebenhengste (BE, Switzerland). *Geodinamica Acta, 15*, 261−276.

Häuselmann, P., Bitterli, T., & Höchli, B. (2004). *Die St. Beatus-Höhlen: Entstehung, Geschichte, Erforschung*. Allschwil, Switzerland: Speleo Projects.

Häuselmann, P., Granger, D. E., Lauritzen, S.-E., & Jeannin, P.-Y. (2007). Abrupt glacial valley incision at 0.8 Ma dated from cave deposits in Switzerland. *Geology, 35*(2), 143−146.

Häuselmann, P., Jeannin, P.-Y., & Monbaron, M. (2003). Role of epiphreatic flow and soutirages in speleogenesis: The Bärenschacht example (BE, Switzerland). *Zeitschrift für Geomorphologie N.F. 47*(2), 171−190.

Häuselmann, P., Lauritzen, S.-E., Jeannin, P.-Y., & Monbaron, M. (2008). Glacier advances during the last 400 ka as evidenced in St. Beatus Cave (BE, Switzerland). *Quaternary International, 189*, 173−189.

Hof, A. (1984). Sieben Hengste-Hohgant Höhle: Le Réseau. *Le Trou, Bulletin du Groupe Spéléo Lausanne, 34*, 18−39 (in French).

Jeannin, P.-Y., Bitterli, T., & Häuselmann, P. (2000). Genesis of a large cave system: The case study of the north of Lake Thun system (Canton Bern, Switzerland). In A. B. Klimchouk, D. C. Ford, A. N. Palmer & W. Dreybrodt (Eds.), *Speleogenesis, evolution of karst aquifers* (pp. 338−347). Huntsville, AL: National Speleological Society.

Morel, P. (1989). Ossements de chauves-souris et climatologie: Collecte systématique de squelettes de chiroptères dans des systèmes karstiques des Préalpes et Alpes Suisses—Premiers résultats. *Stalactite, Journal de la Société Suisse de Spéléologie, 39*(2), 59−72 (in French).

SINKING STREAMS AND LOSING STREAMS

Joseph A. Ray

Crawford Hydrology Laboratory, Western Kentucky University

The purpose of this chapter is to discuss the characteristics and significance of influent streams, creeks, and rivers that drain underground and to evaluate their integral function in karst hydrology. Losing streams are sometimes observed recharging sand and gravel aquifers and volcanic material such as basalt. This chapter, however, will restrict discussion to the more common karst phenomena developed within soluble rocks such as limestone. Many of the terms associated with influent streams are often interchanged depending on the practice of the user.

INFLUENT RIVERS AND STREAMS

The unique nature of losing and sinking influent streams is their development and evolution of conduit-flow routes and caves through soluble rocks. Large volumes of concentrated recharge from losing and sinking rivers are central to the evolution of most of the world's largest and most significant caves and springs in karst (Fig. 1). Caves formed by influent streams may contain a major stream or active floodwater route or these may be long abandoned high above active base level. Abandoned flood routes often retain fluvial sediment and features left by the former stream.

Where major rivers cross soluble rocks they generally carve a low-gradient, continuous surface valley through the region and usually function as the master drainage or base level for local tributaries and springs; however, certain rivers and most smaller streams become influent and lose water where they traverse soluble rocks. For example, influent water from the upper Danube River resurfaces at the Aach Spring, Germany's largest resurgence. Because of early conflicts over water rights, this subsurface river was the site of the first quantitative water trace to prove the spring's source. The pioneering 2-day study, conducted by Knop in 1877, used hourly salt content measurements of Aach Spring to calculate a concentration curve totaling nearly 10,000 kg of salt, about half of the quantity deliberately injected into the Danube River sinks (Käss, 1998).

The Takaka River in New Zealand begins losing base flow into sediment-covered marble, 20 km from its

FIGURE 1 Steep-gradient stream sinking into a large-capacity cave opening in southeastern Kentucky. The flood capacity of this cave may exceed 20 m^3 s^{-1}. *Photo courtesy of James C. Currens. Used with permission.*

resurgence at Waikoropupu Spring. This large spring maintains a low-flow discharge of >5 m^3 s^{-1} (Ford and Williams, 1989). The Reka River in Slovenia sinks into Skocjanska Cave and resurges 30 km south near Trieste, Italy, forming a group of 16 springs with a mean discharge of >26 m^3 s^{-1} (Jennings, 1985). The Trebinjcica River in the Dinaric karst is one of the largest rivers in the world to sink into limestones and contributes to the Ombla River Spring near Dubrovnik. One of the largest springs in the world, the Ombla has a discharge ranging from 4 to 140 m^3 s^{-1}.

River rises, supplied primarily by nearby losing rivers, are a common type of large spring in Florida. The typical losing river reach is less than 5 km in length. The Santa Fe River, a major tributary of the Suwannee River in north-central Florida, may be the largest losing stream in the United States. The losing reach has developed where the Santa Fe River incises a regional escarpment or linear bluff. The influent conduit capacity ranges from about 5 to 16 m^3 s^{-1} at O'Leno State Park, 4.8 km northeast of its resurgence at Santa Fe Rise. Subsurface flow is locally visible in a karst window about 3 km from the sink area, and when river stage exceeds 14 m, the dry river channel is reoccupied by flood waters (Martin and Dean, 1999).

Big Spring, Missouri's largest spring with a discharge ranging from 6 to 36 m^3 s^{-1}, receives water from several losing tributaries of the Eleven Point River. This regional-scale recharge was demonstrated by groundwater tracer tests conducted by Aley over distances of 60 km. Elk River is possibly the largest losing stream in West Virginia, with 8 km of dry river channel, under-drained by caves and conduits feeding Elk River springs (Jones, 1997).

The Edwards Aquifer in south-central Texas is one of the largest, most important karst aquifer systems in the United States. It contains a 5- to 25-km-wide, 350-km-long recharge zone that is traversed by numerous losing rivers and creeks. The Blanco, Medina, Frio/Dry Frio, and Nueces/West Nueces rivers are four of the larger losing-stream systems that contribute to the karst aquifer with a combined average recharge of

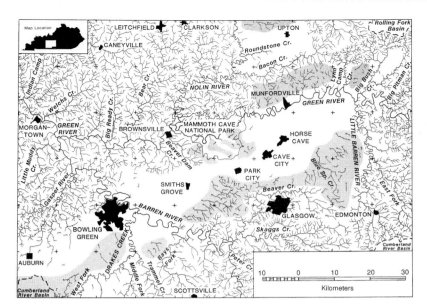

FIGURE 2 Stream patterns and urban centers in an 8250-km^2 region surrounding Mammoth Cave National Park in Kentucky. A consistent stream density occurs over most of the region except for the streamless karst plain, oriented southwest to northeast. Numerous sinking streams drain beneath this sinkhole plain and contribute to large springs on the Green and Barren rivers. These sinking stream watersheds are identified with a gray tone.

20 m^3 s^{-1}. Large springs, including Comal, San Marcos, and Leona springs, and high-volume artesian wells, yielding groundwater flow under hydrostatic pressure, are supplemented by this extensive network of losing streams.

With more than half of the state covered by soluble rocks, Kentucky contains many losing and sinking streams. Figure 2 shows stream patterns in the region surrounding Mammoth Cave National Park. The light-shaded diagonal band showing an absence of streams corresponds with the sinkhole plain. This gently rolling landscape was an early settlement-transportation corridor because of the streamless, prairie terrain. The main portion of the sinkhole plain, between the Green and Barren rivers, is flanked to the southeast by dozens of sinking streams. These influent streams comprise the headwaters of the three largest karst drainage basins in Kentucky, including groundwater in the National Park. Blue Spring Creek, with a watershed of 37 km^2, is the largest of these sinking streams. Its basin comprises nearly 10% of the 390-km^2 watershed of Gorin Mill Spring, Kentucky's largest spring. Figure 3 shows the sequence of multiple stream swallets that are activated over a 5-km reach of Blue Spring Creek as discharge increases from low flow to moderate and high flow.

Losing streams and subterranean meander cutoff caves and springs are common in southwestern Kentucky. These features frequently occur along the West Fork of Red River and Sinking Fork of Little River (Mylroie and Mylroie, 1991). The most significant losing stream is probably the Sinking Creek/Boiling Spring system in north-central Kentucky. Although 19 km of Sinking Creek is shown on topographic maps upstream of Boiling Springs, this 20- to 30-m-wide stream is actually a dry channel for most of the year.

FUNCTION OF INFLUENT STREAMS

In order to comprehend the hydrology of influent streams, one must also understand the structure of normal streams. An idealized stream basin may be viewed as having three basic components: the headwaters, or allogeneic source; the main stem, or zone of transfer; and the mouth, or discharge point (Schumm, 1977). Allogeneic waters are defined as stream flow derived from nonlocal sources. In karst drainage, influent allogeneic streams may be derived from adjacent insoluble rock terranes or soluble rocks that include perching units such as shale, chert, or sandstone. This allogeneic flow comprises the headwaters of most losing and sinking streams. Karst watersheds that contain a significant amount of stream or fluvial recharge are often termed *fluviokarst* (White, 1988).

Likewise, the concept of base level is important in understanding influent streams. *Base level* refers to sea level or the lowest level of master stream erosion. Influent streams develop when they cross soluble rocks along their transfer route to base-level rivers or seas. As potential groundwater discharge locations (down-valley or along master rivers) deepen over time, tributary headwaters tend to keep pace by developing advantageous subsurface flow routes. Enlarging conduits typically follow preferential pathways along bedrock fractures and bedding planes and function as efficient flow routes to local base level. Because karst conduit development can occur some depth below base level, losing reaches may occur along local base-level streams.

As incipient conduit-flow routes are established, convergent recharge from losing and sinking streams tends to rapidly accelerate conduit enlargement. This preferential and aggressive conduit erosion controls the location of main flow routes and caves within the developing

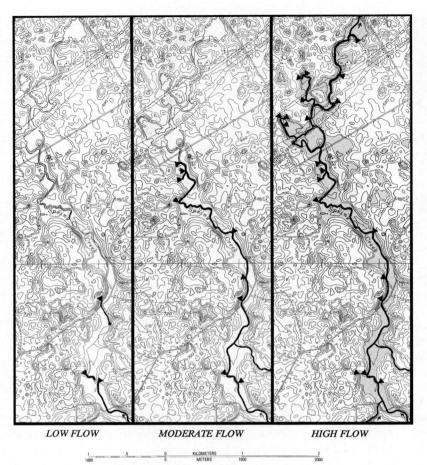

FIGURE 3 The sinking-stream complex of Blue Spring Creek in central Kentucky. Flow variations are compared during low-, moderate-, and high-flow conditions on three images of the same topographic map segment. Active swallets or stream sinks are identified with black triangles tangent to the flowing stream. The influent capacity of successive swallets is exceeded as stream discharge increases. Stream flooding and intermittent lakes are identified by a gray tone in the high-flow map. More than 40 additional sink points are located up to 2 km beyond the map border.

karst drainage basin. Similar in function to local headwater stream networks, diffuse percolation from the shallow weathered zone (epikarst) continually feeds the under-draining trunk-flow routes created by losing and sinking streams. This integrating process often creates an efficient subsurface network, generally dendritic or trellised in pattern, that functions as a unified karst drainage basin or watershed.

Jakucs (1977) pointed out that much classic theory on the development of karst ignored the erosional and hydrologic influence of through-flowing and losing streams derived from adjacent nonkarst terrane. He described several varieties of erosion within karst terrane by allogeneic or "alien" streams and illustrated the evolution of a chain of swallow-holes formed as these streams incise soluble rocks.

In many cases, the largest karst caverns are created in the zones where allogeneic rivers pass through the subsurface. Because gravel, cobbles, boulders, and other debris can be swept underground by sinking streams, features resembling normal surface landforms such as floodplains, alluvial bars, and stream meanders can be formed within flood-route caves. This transported sedimentary load contributes to mechanical abrasion (corrasion) along with the solutional enlargement (corrosion) of passages by aggressive flood waters. Exceptional flood-route caves and mazes are often formed where influent tributaries converge, where floodwater routes diverge around obstacles, and within effluent distributaries (Palmer, 1975). Accordingly, sinkhole collapse and karst-window development tend to focus along the trunk route and near the boundary zones of subsurface rivers.

Losing and sinking streams can concentrate large amounts of floodwater runoff and contaminants into a karst aquifer at specific points. Obviously, this type of rapid recharge makes karst groundwater extremely vulnerable to pollution. Education of the agricultural industry, urban developers, and the public about the dynamics of karst drainage is vital if we are to protect and restore these important groundwater resources.

EVOLUTIONARY SEQUENCE OF LOSING AND SINKING STREAMS

The subsurface flow route of influent streams may range from minor cutoffs to regional trunk conduits.

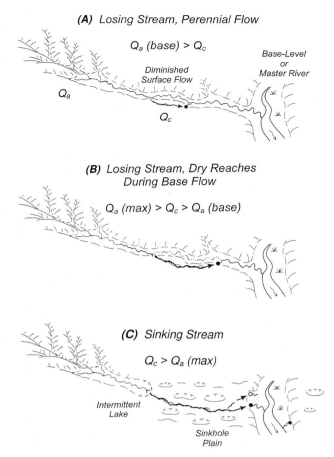

FIGURE 4 Three basic types of influent streams draining to a base-level or master river. These generalized illustrations utilize the terms Q_a for allogeneic discharge or headwater source and Q_c for conduit-flow capacity in the subsurface transfer zone. Arrows show flow direction from under-draining conduits to springs in the discharge zone. Case (A) shows a losing stream with perennial flow, where base-flow allogeneic discharge is greater than conduit-flow capacity. Case (B) shows a losing stream where maximum allogeneic discharge is greater than conduit-flow capacity, but base-flow allogeneic discharge is less than conduit-flow capacity; therefore, a dry reach occurs during base flow. Case (C) shows a terminal sinking stream where conduit-flow capacity is greater than maximum allogeneic discharge. *Adapted from White (1999).*

As an evolving surface drainage network initially incises soluble rocks, subterranean meander cutoffs and short-distance detours of flow into the subsurface may begin to develop. Where the exposure of soluble rocks is limited, the depth and extent of conduit flow are accordingly minimized. Such watercourses usually regain lost conduit flow at downstream springs.

This intravalley losing and gaining karst flow is probably the most common type of stream behavior in soluble rocks. In addition to flow in highly soluble rocks, losing and gaining hydrology can occur in thin or interbedded limestone/shale units that lack the potential for deep conduit flow. Likewise, less soluble rocks such as dolomite locally develop this type of interrupted stream flow. Often, normal stream morphology remains largely unaffected by the stream loss, and many topographic maps do not reveal the subsurface diversions.

The idealized diagrams in Figure 4 illustrate a reasonable evolutionary sequence in the development of losing and sinking streams. Three types of influent streams tributary to a base-level river are shown and are compared by means of a generalized water balance. Evapotranspiration and artificial withdrawals are ignored in these simplified relationships.

Case (A) illustrates karst flow in a losing stream where influent conduit capacity is exceeded during all conditions. The base flow from the allogeneic watershed, Q_a (base), is greater than the conduit-flow capacity (Q_c); consequently, surface flow is only diminished within the losing-stream reach. Due to subtle flow variations, some losing reaches can be difficult to identify. This less common (and less recognized) situation may occur during incipient karst development, because of persistent conduit clogging, or along broad rivers with large base flows. Streams with abundant loads of coarse sediment often lose a portion of surface flow entirely through its alluvium. This type of intergranular flow may be locally indistinguishable from karst diversions.

Case (B) shows a losing stream that develops a dry reach during base flow. Maximum allogeneic discharge during flood, Q_a (max), is greater than the conduit-flow capacity (Q_c), causing a continuous overflow channel to be fluvially eroded and maintained. However, the conduit-flow capacity (Q_c) is greater than the allogeneic base flow, Q_a (base), which causes a dry reach below the losing point. This is probably the most common type of influent stream. Conduit flow may return to the surface at a down-valley spring within the same watershed (intrabasin flow) or it may divert beneath a surface divide into a neighboring watershed (interbasin flow). Because of their frequency and significance in karst hydrology, groundwater systems or spring basins dominated by losing streams may be classified as type I or overflow allogeneic karst basins.

Case (C) illustrates a sinking stream that is pirated or diverted entirely underground. Conduit-flow capacity Q_c is greater than the maximum allogeneic discharge, Q_a (max). Because of the exceptional conduit-flow capacity, a surface escape route or overflow channel is not maintained. An abandoned karst valley or sinkhole-dominated plain often separates the stream sink and its resurgence. Groundwater systems dominated by sinking streams in blind valleys may be classified as type II or underflow allogeneic karst basins.

These three cases are simplified. Actual influent streams are usually more complex, as they may exhibit overlap or transitional situations based on weather trends or unique conditions. For example, case (A) may resemble case (B) during severe drought, whereas case (B) may resemble case (A) in wet years. Also, during extreme flood events some sinking streams, case (C), may yield intermittent lakes and overland flow through degraded gaps or hanging valleys that are progressively being abandoned.

Some karst areas contain relatively thick surficial deposits of soil or sediment. Streams that lose water through appreciable bedrock cover tend to maintain surface overflow routes. Influent soil pipes or channels formed in sediment frequently collapse, partially clogging flow routes. Cobbles and gravel concentrated from the cover material tend to collect within these unstable conduits and may help to maintain a limited capacity of leakage into the under-drain system.

Depending on season or flow condition, losing and sinking stream flow may advance to, or retreat from, a series of sink points along a considerable stream reach. The 32-km-long dry channel of Lost River in southern Indiana is an exceptional example. Such streams have a tendency of adjustment between the influent capacity of under-draining conduits and loss of base flow within a stream reach. Because winter base flow exceeds summer low flow, the influent front tends to migrate between the two conditions. Figure 3 shows a channel reach of 3 km over which flow migrates between low and moderate conditions, which generally corresponds to the two base-flow conditions. Also, excessive runoff from developed or agricultural areas increases the magnitude of peak runoff and may alter the expected base flow discharge. Knowledge of these potential flow dynamics is important during hydrogeologic mapping and karst inventories for dye tracing studies.

In summary, many rivers drain through the karst regions of the world and are altered by interaction with underlying conduit flow developed within soluble rocks. Countless allogeneic tributary streams lose water or sink entirely and contribute to the primary groundwater circulation routes through karst watersheds and drainage basins. These main conduit-flow routes function as transfer paths from headwater influent streams to discharge points at springs. Many of the largest and most important caves in the world were developed by influent recharge. The active trunk-flow routes serve as master groundwater collectors of diffuse recharge from infiltration and runoff into sinkholes. Due to limited filtration of influent stream recharge and rapid groundwater velocities, most karst aquifers are extremely vulnerable to pollution from surface activities. Knowledge of the recharge sources and variability inherent to karst drainage is vital for protecting and restoring threatened cave resources, water supplies, and many perennial spring-fed streams.

See Also the Following Articles

Hydrogeology of Karst Aquifers
Water Tracing in Karst Aquifers

Bibliography

Ford, D. C., & Williams, P. W. (1989). *Karst geomorphology and hydrology*. London: Unwin Hyman.
Jakucs, L. (1977). *Morphogenetics of karst regions*. New York: John Wiley & Sons.
Jennings, J. N. (1985). *Karst geomorphology*. Oxford: Basil Blackwell.
Jones, W. K. (1997). *Karst hydrology atlas of West Virginia*. Charles Town, WV: Karst Waters Institute.
Käss, W. (1998). *Tracing technique in geohydrology*. Rotterdam: A.A. Balkema.
Martin, J. B., & Dean, R. W. (1999). Temperature as a natural tracer of short residence times for groundwater in karst aquifers. In A. N. Palmer, M. V. Palmer & I. D. Sasowsky (Eds.), *Karst modeling* (pp. 236–242). Charles Town, WV: Karst Waters Institute (Special Publication 5).
Mylroie, J. E., & Mylroie, J. R. (1991). Meander cutoff caves and self piracy: The consequences of meander incision into soluble rocks. *National Speleological Society Bulletin, 52*, 33–44.
Palmer, A. N. (1975). The origin of maze caves. *National Speleological Society Bulletin, 37*, 56–76.
Schumm, S. A. (1977). *The fluvial system*. New York: John Wiley & Sons.
White, W. B. (1988). *Geomorphology and hydrology of karst terrains*. New York: Oxford University Press.
White, W. B. (1999). Conceptual models for karstic aquifers. In A. N. Palmer, M. V. Palmer & I. D. Sasowsky (Eds.), *Karst modeling* (pp. 11–16). Charles Town, WV: Karst Waters Institute (Special Publication 5).

SISTEMA HUAUTLA, MEXICO

C. William Steele[*] and James H. Smith, Jr.[†]

[*]Boy Scouts of America, [†]Environmental Protection Agency

Sistema Huautla, the second deepest cave in the Western Hemisphere, is located in the southern Mexican state of Oaxaca. It is among the 12 deepest caves in the world, and one of the longest of the world's deep caves. The exploration of the Huautla caves is still ongoing, nearly 47 years since cavers first discovered them. The stories connected to the exploration are well documented in magazines, books, caving publications, photography, maps, and film (Steele, 2009; Stone et al., 2002). Sistema Huautla was the first cave outside of Europe to be explored deeper than 1000 m.

CAVE DESCRIPTION

The integrated Sistema Huautla is comprised of 20 entrances and 62,099 m of passage to a depth of 1475 m. This most complex vertical system has more independent deep routes than any other cave in the world: three routes over 1000 m deep, one over 900 m, two over 800 m, one over 700 m, and two over 600 m. The connection of Nita Nanta with the rest of the system in 1987 marked the first and, still one of the very few times two caves each over 1000 m deep have been joined together. The world's deepest down and up traverse, a spectacular crossover trip, can be made: 1225 m down via Nita Nanta's highest entrance, through spectacular stream galleries of San Agustin, and 1107 m back up Li Nita, all without retracing a single footstep. This down and up traverse would be 11 km in traverse length, although the two entrances are less than a kilometer apart on the surface. Such a trip is likely never to be done due to the logistics of rigging well over 100 pitches and diving two widely separated sumps at the bottom.

The most remote recess of Sistema Huautla is beyond a series of water-filled tunnels at −842 m in San Agustin. Sump dives of 450 and 170 m in length reach air-filled tunnels that measure 20 m wide and 20 m tall that ultimately lead to Perseverance Hall, the second largest chamber in Sistema Huautla. From Perseverance Hall, measuring at 120 m wide and 120 m long and 30 m tall, the passage dips steeply to galleries that lead to the deepest level and final sump of Sistema Huautla, 1475 m below the highest entrance. The final sump and farthest exploration is located 2.5 km from the 842-m sump or 6.6 km from the San Agustin entrance.

Hidden in the depths of the Sierra Mazateca, 640 m below the entrance of San Agustin, is the jewel of Sistema Huautla, Anthodite Hall. Anthodite Hall, at 300 m long and 200 m wide and 70 m high, is one of the largest chambers in the region. The ceiling and walls are covered with milky white to clear speleothems, the majority of which are large anthodites. These radiating pin-cushions are half a meter in length.

The caves of Huautla were first discovered in 1965 by cavers from Austin, TX. The caves were initially explored through the rest of the 1960s by Canadian cavers and cavers from assorted American states such as Texas, Georgia, Indiana, and Tennessee. In 1967 the distinction of having the deepest cave in the Western Hemisphere came to the Huautla area caves with the

FIGURE 1 Sistema Huautla is located in the northeast corner of the state of Oaxaca, Mexico. *Map by James H. Smith.*

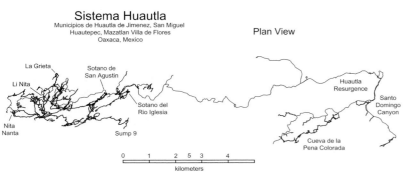

FIGURE 2 Sistema Huautla is a complex network of cave streams with 20 known entrances. The pattern of stream flow is determined by the structural geology of the karst groundwater basin. *Map by Stone (2003).*

exploration of Sotano del Rio Iglesia to a depth of 531 m. This was surpassed the following year when Sotano de San Agustin was explored to a depth of 612 m.

Following a hiatus from 1970 to 1976, expeditions to Huautla began again in 1976 and carried on steadily into the twenty-first century. These were organized

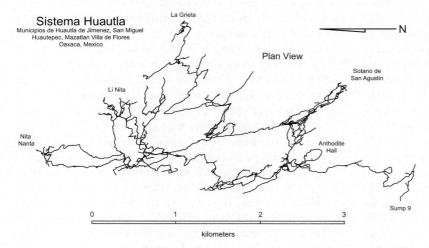

FIGURE 3 Sistema Huautla is a cave composed of many separate streams that descend vertically to a base level and then converge to form one large stream before discharging out a spring entrance. *Map by Stone (2003).*

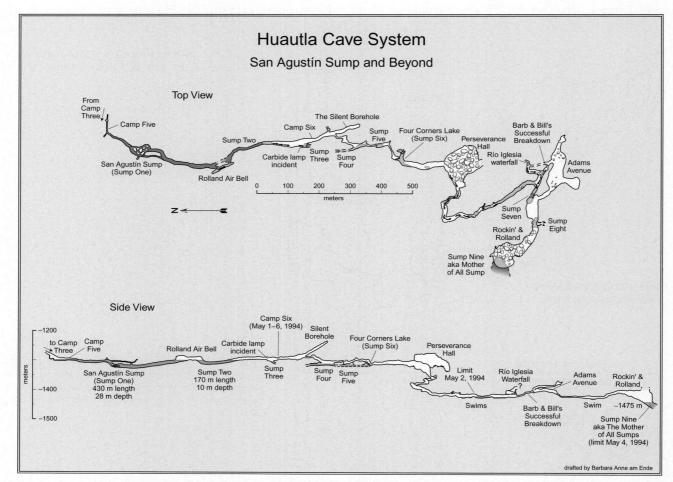

FIGURE 4 Exploration beyond the 840 Sump to Sump 9 in Sotano de San Agustin is the greatest single feat of U.S. deep cave exploration ever accomplished. *Map by Barbara Anne Am Ende.*

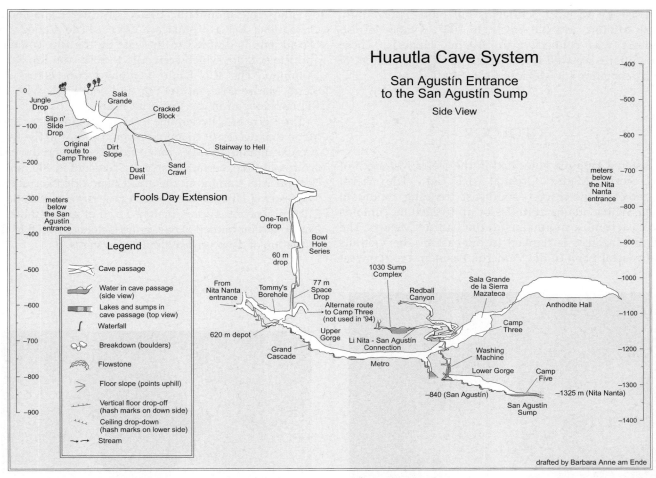

FIGURE 5 Anthodite Hall, the jewel box of Sistema Huautla, contains the most spectacular anthodite speleothems found in the cave system. Anthodite Hall is the largest chamber in Sistema Huautla. *Map by Barbara Anne Am Ende.*

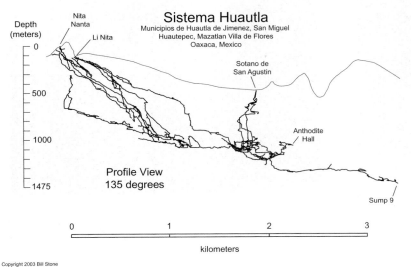

FIGURE 6 Sistema Huautla is a vertical drainage system composed of separate streams that descend through the mountain down steeply descending passages and a multitude of vertical shafts. More than 62 km of cave passages are represented with 1475 m of vertical relief. *Map by Bill Stone.*

and led by Richard Schrieber, Bill Stone, Jim Smith, Mark Minton, and Bill Steele. In 2007, Sotano del Rio Inglesia was connected to Sistema Huautla. These expeditions are well documented in the newsletters of the Association for Mexican Cave Studies.

GEOLOGY

Sistema Huautla is located in the Sierra Mazateca of the Sierra Madre Oriental del Sur. The Sierra Mazateca, a northwest—southeast trending mountain range which attains altitudes of up to 2200 m, contains the easternmost mountains to the Gulf of Mexico. The Sierra Mazateca is flanked to the east by the foothills and coastal plain of the Veracruz Basin and to the west by the intermountain basin of the Cañada de Oaxaqueña (a.k.a. Tehuacan Valley). The Cañada de Oaxaqueña is drained to the east by the Rio Tomellin and Rio Salado, which converge to form the Rio Santo Domingo. The Rio Santo Domingo cuts across the Sierra Madre Oriental del Sur and converges with the Rio Papaloapan in the Veracruz Basin.

The Sierra Mazateca range receives orographic precipitation that decreases in amount from the east to the west. Annual rainfall averages range from 6 m along the eastern front of the Sierra Mazateca at Tenango to 400 mm at Teotitlan del Camino in the arid Cañada de Oaxaqueña. The Sistema Huautla area, which ranges to an altitude of 2100 m, receives approximately 2.5 m of annual rainfall. The abundant rainfall is one of the factors essential to the formation of deep vertical drainage systems.

FIGURE 7 Anthodites are speleothems that are characterized by radiating clusters of needles or quills and generally are composed of aragonite, one of the crystalline forms of calcium carbonate ($CaCO_3$). Anthodites may grow from the ceiling, walls, and floor. They may form single clusters, huge clusters, or chandeliers. This anthodite was found in Anthodite Hall, Sistema Huautla. *Photograph taken by James H. Smith.*

FIGURE 8 Laura Smith is descending a waterfall on a rope during the exploration of Cueva Agua Carlota, Huautla. In order to follow the stream, cave explorers must use rope techniques perfected specifically for cave exploration. *Photograph taken by James H. Smith.*

HYDROGEOLOGY

The Sistema Huautla karst groundwater basin is an elongate basin, 4 km wide and 14.5 km long, oriented north—south, and is defined by its structural geology. The basin is bordered on the west by allochthonous sandstones and shale of Jurassic Age and to the north and east by cretaceous limestones. The drainage of the basin is characterized as subterranean with more than 100 km of active and paleo conduits that have been explored and surveyed. The drainage outlet or spring for the Sistema Huautla Cave is to the south in the Rio Santo Domingo Canyon at an elevation of 300 m above sea level. Subterranean groundwater flow is determined by regional strike and plunge of dipping limestone strata. The vertical extent of the subterranean drainage was verified by dye tracing. From the highest mapped entrance of Sistema Huautla, Nita Nanta, to the Sistema Huautla Resurgence, the vertical extent of 1760 m is second only to Sistema Cheve, which dye traced to 2525 m (Smith, 2020).

BIOLOGY

The cave life in the Huautla System is highly diverse and rich in cave-adapted animals. Of the 48 species of invertebrates known from the caves of Huautla, 10 are highly adapted troglobitic forms. Many groups of tropical arachnid have been found inhabiting the caves. The most striking animal is the scorpion *Alacran tartarus* (Francke, 1982). Found deep in the system, it is usually found near or in the water. One of the rarest troglobitic tarantula spiders, *Schizopelma grieta* is also found deep in the caves. Another, as yet unnamed, spider of the genus *Pholocophora* has also been found in the caves. An amblypigid with very reduced eyes, *Paraphrynus grubbsi*, is known to be in four of the caves. An undescribed schizomid of the genus *Schizomus* has also been found. Two cave-adapted millipedes, *Cleidogona baroqua* and *Mexicambala fishi*, have been found. Among the insect groups the collembola *Pseudosinella bonita* and the beetle *Platynus urqui* have been described, while a large nicoletiid thysanuran, common on the mudbanks of the streams, remains undescribed. Closer examination of the caves will no doubt reveal more species of cave-adapted life. A new species of the spider genera, Ctenus, Maymena, Coryssocnemis, Metagonia, and Modismus await description, as do opilionids of the genera Hoplobunus and Karos, millipedes of the genus *Sphaeriodesmus*, and campodeid diplurans.

FIGURE 9 At −740 m deep in lower La Grieta, Sistema Huautla, explorers search for a potential connection to Sotano de Agua Carrizo, an 848-m-deep cave, the bottom of which is located only a few meters from La Grieta's stream passage. This photograph was taken near the Triple Connection between Sotano San Agustin, La Grieta, and Nita Nanta (the highest entrance). *Photograph taken by James H. Smith.*

ARCHAEOLOGY

Few archaeological remains have been discovered in the passages integrated into the Sistema Huautla; however, a site in an area cave, named Blade Cave, suggests use by ancient Mazatecs. Evidence from the cave includes 60 ceramic vessels that date from the Early Urban stage (1—300 A.D.) through the Early City—State stage (750—1250 A.D.). Many of the vessels appear to have held perishable offerings. Frequently, wealth items such as beads and pendants of jade, stone, shell, and coral were included in these offerings or added to the vessels later. Evidence of blood sacrifice is present in the form of remains of animals, in particular dog mandibles. Four bifacially chipped blades, one 18 cm in length, were likely used as knives for such sacrifices. Auto-sacrificial bloodletting was possibly occurring as smaller prismatic obsidian blades were found. Human skeletal material is abundant in the Blade Cave assemblage. It is not certain what these individuals represent. The investigations have amassed an abundance of detailed contextual information on ancient Mazatec cave ritual. Ethnographic research leads to the belief that the ceremonies conducted in Blade Cave were primarily petitions for rain (Smith, 1987).

Bibliography

Francke, O. F. (1982). Studies of the scorpion subfamilies Superstitioninae and Typhlochactinae, with descriptions of a new genus (Scorpiones, Chactoidea). *Association for Mexican Cave Studies Bulletin, 8*, 51−61.

Smith, J. H. (2002). *Hydrogeology of the sistema huautla karst groundwater basin, bulletin 9*. Austin, TX: Association for Mexican Cave Studies.

Steele, C. W. (2009). *Huautla: Thirty years in one of the world's deepest caves*. Dayton, OH: Cave Books.

Steele, J. F. (1987). *Blade Cave: An archaeological preservation study in the Mazatec Region, Oaxaca, Mexico*. San Antonio: University of Texas.

Stone, W. C., Am Ende, B., & Paulsen, M. (2002). *Beyond the deep: The deadly descent into the world's most treacherous cave*. New York: Warner Books.

SOIL PIPING AND SINKHOLE FAILURES

Barry Beck

P.E. LaMoreaux & Associates, Inc.

If you visit Mammoth Cave National Park in Kentucky, you may hike down into huge forested sinkholes like Cedar Sink. These low areas are completely enclosed by steep rock walls, and they are many hectares in size and more than 100 meters deep, possibly with a stream flowing out of a cave on one side and into a cave on the other. If you live in Central Florida you may have seen a house or a road that collapsed into a sinkhole—one that is smaller than those mentioned in Kentucky, but significantly more devastating. If you are a farmer in some areas of Slovenia, the only land with sufficient soil to grow crops is at the bottom of the numerous bowl-shaped basins that dot the landscape, thousands of feet across and tens of feet deep; these are locally termed *dolinas*, the original term for sinkholes. If you visit West Texas, you may go to see the Devil's Sinkhole, a gaping collapse in the ground surface more than 30 meters in diameter that drops vertically into an even larger cave room extending down to 100 meters below the ground surface; whatever cave lies below is completely plugged with the debris from the upward-stoping roof collapse. All of these different features have been termed *sinkholes*, and they are part of the same overall process, even though each one is a little bit different.

Sinkholes are generally thought of as "natural enclosed depressions found in karst landscapes." In European terminology they are generally called *dolines*, an Anglicized form of the Serbo-Croatian word *dolina*. Although they vary over a wide range of sizes and shapes, they are generally subcircular in plan, and they vary in cross-section from saucer shaped through bowl shaped to cylinder shaped. Depending on the type of sinkhole under discussion, they may vary from 0.3 m in diameter and 1 m deep for small cover collapses, to a kilometer or more in diameter and hundreds of meters deep for the polygenetic sinkholes of mature karst landscapes. In the following description of the origin and types of sinkholes, it is assumed that the reader is generally familiar with karst processes and terminology. Moreover, this discussion generally refers to traditional karst developed on limestones or dolostones (dolomites). Some useful sources are given in the bibliography.

Although five different types of sinkholes are generally identified by geologists (Beck, 1988), these are the result of only two different processes: the transport of surficial material downward along solutionally enlarged channels, or collapse of the rock roof over large bedrock cavities (Fig. 1). The natural solution process dissolves the limestone most rapidly at its surface. If the limestone is bare, or almost bare (that is, exposed at the ground surface), water flows over the top of the limestone toward the points of easiest infiltration, usually at joint intersections, and then downward. As water converges toward these points of infiltration, vertical solution pathways develop, and the converging flow dissolves and lowers the limestone surface, forming a bowl-shaped depression. This is the classic *solution sinkhole* (Fig. 1A).

Because the solution process is imperceptibly slow, solution sinkholes are not generally hazards to the engineering of human-made structures. Their role is to act as drainage inlets into a conduit drainage network. They are environmental hazards in that they are a direct input to the deeper karstic groundwater. Contaminants introduced into an open sinkhole (of any type) will flow rapidly into the karst drainage network with little or no degradation or filtering of the contaminants.

If the limestone is mantled (covered) by an accumulation of residual soil (insoluble residue from the dissolution of the limestone), the terrain is termed a *subsoil karst*. If the limestone is mantled by sediments of an outside origin (for example, marine sands or glacial sediments) the terrain is termed a *mantled karst*. In terms of the processes that will form sinkholes, the two settings are generally the same. The overlying unconsolidated sediments may simply be termed *cover*. Water infiltrates through the cover to the limestone surface,

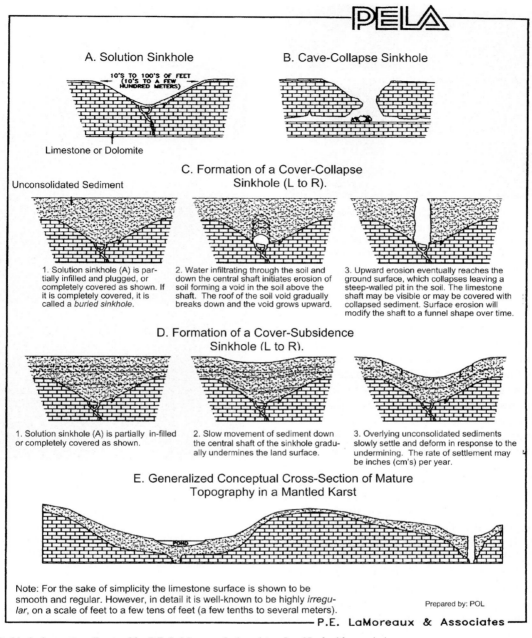

FIGURE 1 Sinkhole formation. *Prepared by P.E. LaMoreaux & Associates, Inc. Used with permission.*

then downslope to the vertical drains and downward. In the process, the cover sediment is eroded down the vertical conduits. This process, internal erosion of the cover sediment, forms what have been termed *subsidence sinkholes*. Jennings (1971) describes them as follows: "Where superficial deposits or thick residual soils overlie karst rocks, dolines (sinkholes) can develop through spasmodic subsidence and more continuous piping of these materials into widened joints and solution pipes in the bedrock beneath. They vary very much in size and shape. A quick movement of subsidence may temporarily produce a cylindrical hole which rapidly weathers into a gentler, conical or bowl-shaped depression" (p. 126). The process is diagrammatically shown in Figures 1C and 1E. These have also been called *alluvial dolines*. However, that term is somewhat misleading and has not been generally accepted.

If the covering sediment is somewhat cohesive, an arched void may form in the sediment immediately above the limestone drain, as the sediment is eroded

FIGURE 2 A cover-collapse sinkhole that developed rapidly (overnight) near Orlando, Florida. It is approximately 10 meters in diameter and 5 meters deep. Note that only surficial sand (cover) is exposed in the hole. The limestone is approximately 30 meters below the surface in this area. *Photo by the author.*

down the drainage conduit. As upward erosion of the cover sediment continues, the void may grow larger over time while the eroded sediment is removed down the conduit or shaft. Or, if collapse of the soil is relatively rapid, the void may simply migrate vertically upward, as the crumbling sediment from the arched roof accumulates on the floor (Fig. 1C, parts 1–3). More cohesive strata within the cover, such as a clay layer, may impede the upward erosion and cause the cavity to widen, rather than grow upward for a period of time. Eventually the upward growth of the void may leave only a thin roof of soil that is not strong enough to support its own weight, resulting in ground collapse. At that moment a hole suddenly appears in the ground surface; this is termed a *cover-collapse sinkhole* (Fig. 2). If the limestone is not too far below ground, and if the previously collapsed sediment was removed down the drain as the collapse grew upward, then the limestone and drain may be visible in the hole. More often, all that is visible in the hole is collapsed sediment. A cover-collapse sinkhole is a subtype of a subsidence sinkhole.

The process of localized internal erosion of the soil is often called *soil piping* by geologists, in this case caused by subsurface karstic erosion. Soil piping was described by White (1988): "Concentrated runoff ... can form channelized paths through the soil to a solution opening in the underlying bedrock. The increased hydraulic gradient and increased flow velocities speed the removal of soil from the subsurface. Because the pavement or the building itself, or even the root-bound soil zone, acts as a supporting structure, the loss of soil may not be noticed until a substantial cavity has formed. At some point structural support is lost and the building collapses into the hole" (p. 357). This downward

FIGURE 3 The remnants of a home engulfed as a large cover-collapse sinkhole developed over several hours, Keystone Heights, Florida (east of Gainesville). The occupants were able to evacuate safely, although their Christmas tree and presents were lost. The surficial sediment is clayey sand, which spans the upward eroding subsurface void allowing it to grow to a significant size before it collapsed. The limestone at this location is approximately 30 meters below the land surface. *Photo by the author.*

erosion of unconsolidated sediment has been termed *ravelling* by engineers; it is the same process.

When the earth suddenly collapses beneath a building, a roadway, or a railroad, this is an obvious disruption to human-made infrastructure (Fig. 3). Such unexpected damages due to cover-collapse sinkholes are common in karst areas, although the actual rate of incidence is not high. Where detailed records have been compiled and averaged, a corrected incidence of 1.7 sinkholes per square mile per year is the highest reported. Rates of less than one sinkhole per square mile per year are commonly reported in high incidence areas. Inasmuch as most sinkholes are less than 60 meters in diameter, the actual risk of a sinkhole damaging an individual 180-square-meter home in

Pinellas County, Florida, has been calculated as on the order of 1 chance in 5000.

The incidence of cover-collapse sinkholes is often increased by the ponding or damming of surface-water flow due to human construction and drainage. Storm water retention basins or waste treatment lagoons have frequently emptied rapidly into sinkholes formed in the bottom. This is obviously a source of groundwater contamination. Large reservoirs also cause sinkholes to collapse in their bed, often causing unacceptable leakage and costing hundreds of thousands of dollars to repair. In some cases, episodes of intense groundwater withdrawal due to well pumping also trigger the collapse of cover-collapse sinkholes. This is particularly true in Florida where high-intensity irrigation for freeze protection of crops often causes tens of sinkholes to form overnight, some extremely damaging.

The surface expression of a sinkhole may also develop imperceptibly slowly. In karst terrain a generally circular area of the ground may slowly subside at the rate of inches to feet per year due to subsurface karstic erosion, not local compaction. These features have been termed *cover-subsidence sinkholes*. The downward erosion may take place slowly, grain by grain, if the sediment is loose and granular, or it may take place by plastic flow into cavernous voids. It is also possible that the surface sediment is gradually "sagging" downward as the area below it is undermined by karstic erosion removing the structural support. The timeframe involved is years. Cover-subsidence sinkholes have been documented due to their damage to human-made rigid structures (Figs. 4 and 5). Whereas an inch of settlement spread out over a year is unnoticeable, after 10 years a 25-cm-deep depression has formed.

FIGURE 4 Damage to a home in East Tennessee caused by a cover-subsidence sinkhole. This damage developed slowly over several years. However, the area is underlain by limestone and has other characteristics of karst topography. Drilling showed a depression in the limestone surface in the area of the damage, and a slight depression in the ground surface could also be seen. *Photo by Arthur J. Pettit, P.E. LaMoreaux & Associates, Inc. Used with permission.*

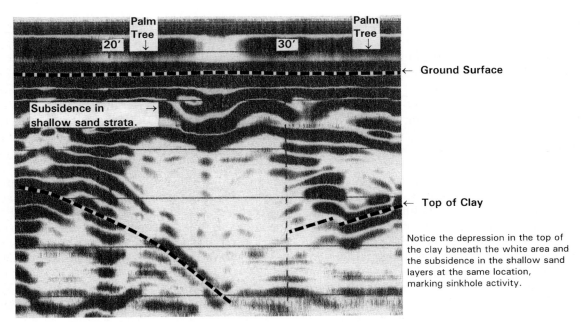

FIGURE 5 Ground-penetrating radar graph showing a cross-section of shallow sediment layers that have settled due to a cover-subsidence sinkhole in Tampa, Florida. The surficial sediment is loose sand and a clay layer occurs approximately 2 to 3 meters below ground surface. Note that the clay stratum has subsided as it has been undermined by deeper karstic erosion. Note the small, but obvious, area of settlement in the surface sands directly over the clay subsidence. The limestone is approximately 6 meters below ground surface and cannot be seen on this graph. This cover-subsidence sinkhole developed slowly over several years, partially beneath a home, damaging it to such an extent that it was uninhabitable. *Radar graph by P.E. LaMoreaux & Associates. Used with permission.*

If this is beneath a foundation, significant damage may occur. We do not yet have a complete understanding of the mechanisms by which cover subsidence occurs. Both cover-collapse and cover-subsidence sinkholes are subtypes of subsidence sinkholes.

Sinkholes are generally thought of as bowl-shaped, funnel-shaped, or cylindrical depressions in the ground surface. However, during and after the development of these basins, depending on their origin, natural processes will begin to fill them. The sides will erode outward and the eroded sediment will accumulate in the basin, creating a broader, shallower depression. This low area will also collect runoff and sediment from the surrounding area. It may be wet and swampy, possibly even forming a small lake, where organic growth will fill the basin over time. These processes, and others, may fill the sinkhole completely, eventually leaving no sign of its previous existence. Such a feature is now termed a *buried sinkhole* (Fig. 1C, part 1).

It is also possible for external sedimentation processes, which operate more rapidly than the rate at which karst develops, to completely cover a karst terrain and fill all the surface depressions. In the northern United States and Europe, many karst areas have been blanketed by a thick mantle of glacial debris. In some areas of Florida, the limestone is covered by a blanket of marine or coastal sands, obscuring the underlying karst until the moment when the ground surface collapses due to ongoing karstic erosion. Figure 1C, part 1, shows a buried solution sinkhole. It is also possible to have cover-collapse or cover-subsidence sinkholes buried. As Waltham (1989) points out, "... these phenomena develop either by pre-burial subaerial erosion or subsequent subsoil erosion ..." (p. 35). That is, the karstic erosion process will continue to operate while the terrain is covered, creating modifications to the unconsolidated strata overlying the limestone.

All four of the aforementioned types of sinkholes—solution, cover-collapse, cover-subsidence, and buried—are the result of the downward movement of material along solutionally enlarged channels through the limestone. However, the dissolution of the limestone by laterally flowing groundwater at depth may also produce conduit drainage systems—caves. These caves continue to grow as the groundwater flowing through them continually dissolves the limestone, over time making the passages larger and larger. Simultaneously, the upper surface of the limestone is also being lowered by solutional attack. These processes may eventually result in a situation where a large cave is close to the limestone surface with only a thin rock roof over the void. If this rock roof suddenly collapses into the cave, a *bedrock-collapse* or *cave-collapse sinkhole* is formed (Fig. 1B).

Bedrock-collapse sinkholes are generally very steep walled, or even overhung. They are often deeper than they are wide and relatively cylindrical. Because the walls are in solid rock, they will maintain this profile for a long time, in terms of a human life span. Thus, examples of bedrock-collapse sinkholes are not difficult to find. Karst geologists generally agree that bedrock-collapse sinkholes are rarely observed in the act of collapsing because the rate of surface down-wasting is extremely slow on a human time scale. That is, bedrock-collapse sinkholes are rare, except on a geologic time scale. The Florida Sinkhole Research Institute collected data on more than 1700 sinkholes that developed in Florida, and not one of these was definitely a bedrock-collapse sinkhole.

The Mitchell Plain of southern Indiana is an example of the long-term product of karstic erosion. It is pockmarked with sinkholes. The largest and deepest of them are generally formed over major, active cave passages due to roof collapse. One of the rare examples of the formation of a cave-collapse or bedrock-collapse sinkhole within historic time occurred at the Colglazier entrance to Blue Spring Cave, on the Mitchell Plain southwest of Bedford, Indiana. What had formerly been a shallow marshy depression grew into a cave-collapse sinkhole nearly 2800 cubic meters in volume during a heavy rainstorm in 1941. Although other cave-collapse sinkholes occur across the Mitchell Plain over large cave passages, the incidence of collapse in the human timeframe is extremely rare, as mentioned above.

Europeans favor a different terminology. They favor using the Anglicized Serbo-Croatian term *doline*, and they include all collapse (a rapid process) dolines in one subgroup, within which they differentiate cave or rock collapse from overburden collapse. Sinkholes due to the slow karstic erosion of cover (herein cover subsidence) are called *suffosion dolines*. The suffosion process is defined as the gradual winnowing and downwashing of fines by a combination of physical and chemical processes. According to European usage, suffosion dolines are small depressions usually only a few meters in diameter and depth. This specific example does not match the features herein called *cover-subsidence sinkholes*, although it could describe one possible example. The term *suffosion* is little used in the United States, and the process of winnowing and downwashing of fines leaving coarser materials in place does not accurately describe the processes occurring in the development of cover-subsidence sinkholes. Therefore, we will eschew the European terminology in favor of the American nomenclature, which appears to more correctly characterize the erosion process.

Subtypes of sinkholes (dolines) are routinely defined as if they occurred in isolation, but, in fact, in the evolution of a complex karst landscape these features are stages in time rather than subtypes. Most sinkholes in a mature, mantled karst landscape are polygenetic, and most karst landscapes will eventually be mantled, at least by residual soil, except for those few cases where the area is underlain by hundreds of meters of very pure limestone.

Solution and internal drainage are the basic paradigm underlying karst landscape development. As the landscape evolves and cover sediments are deposited (either as residuum from the insoluble residue in the limestone or transported into the area from outside sources), other erosional processes will also play a role, integrated with the underlying paradigm. As the cover sediments are eroded downward through the internal drainage network, the solution process continues accompanied by episodes of cover collapse, infilling of karstic depressions, and possible cover subsidence under the appropriate conditions. Where enlarging subsurface karst features approach the top of solid rock, cavern roof collapse can also impact the landscape. The Mitchell Plain of southern Indiana is a good example of such a complex karst landscape.

According to this concept, a sinkhole is part of a long-term process and it may have different forms and surface expressions at different times. When the sinkhole is completely filled, there is no surface indication of its presence, but the epikarstic drainage feature still exists and continues to function below ground. This has been termed a *buried sinkhole*, which is a valid and appropriate concept despite the fact that no depression exists during this stage of sinkhole development. In recognition of the complex evolutionary process that characterizes the development of karstic drainage and subsurface erosion, and in view of the varying surface expressions that this process may have as it evolves and the landscape matures, and considering that under some conditions there may be no ground surface expression of the underlying karst drainage, the traditional definition of a sinkhole as "a depression" appears inappropriate. The following is a functional, process-oriented definition of a sinkhole, or doline:

Sinkholes (dolines) are the surface (and near-surface) expressions of the internal drainage and erosion process in karst terrain, usually characterized by depressions in the land surface. Formative processes include bedrock solution, downward transport of overburden sediment, and/or bedrock collapse. Most large, mature sinkholes have a complex origin involving all three processes. At some stages in sinkhole development, a surface depression may not be present.

See Also the Following Articles

Closed Depressions in Karst Areas
Epikarst

Bibliography

Beck, B. F. (1988). Environmental and engineering effects of sinkholes—The processes behind the problems. *Environmental Geology and Water Science, 12*(2), 71–78.
Beck, B. F. (1991). On calculating the risk of sinkhole collapse. In E. H. Kastning & K. M. Kastning (Eds.), *Appalachian karst* (pp. 231–236). Huntsville, AL: National Speleological Society.
Ford, D., & Williams, P. W. (1989). *Karst geomorphology and hydrology*. London: Chapman & Hall.
Jennings, J. N. (1971). *Karst*. Cambridge, MA: The MIT Press.
Parise, M., & Gunn, J. (2007). *Natural and anthropogenic hazards in karst areas*. Geological Society of London.
Sowers, G. F. (1996). *Building on sinkholes*. New York: ASCE Press.
Sweeting, M. M. (1972). *Karst landforms*. London: Macmillan.
Waltham, A. C. (1989). *Ground subsidence*. Glasgow: Blackie & Sons.
Waltham, T., Bell, F., & Culshaw, M. (2005). *Sinkholes and subsidence*. Chichester, U.K.: Springer.
White, W. B. (1988). *Geomorphology and hydrology of karst terrains*. New York: Oxford University Press.
Williams, P. W. (2003). Dolines. In J. Gunn (Ed.), *Encyclopedia of caves and karst science* (pp. 304–310). New York: Fitzroy Dearborn.

SOLUTION CAVES IN REGIONS OF HIGH RELIEF

Philipp Häuselmann
Swiss Institute for Speleology and Karst Studies
La Chaux-de-Fonds, Switzerland

INTRODUCTION

The topic of solution caves in high-relief areas, although they are caves constrained by two conditions, is remarkably broad and variable. It is therefore difficult to choose good examples of the many existing caves. We restrict ourselves voluntarily to epigenic caves, formed by rainwater percolating from above, because, although recent research in hypogenic caves revealed that they are much more numerous than previously thought (Klimchouk, 2007), virtually all large cave systems in mountainous ranges are epigenic and may show only some minor hypogenic chambers or passages.

It is apparent that there is no definite geomorphological trait that characterizes all these caves; on the contrary, they seem even more diverse than the countryside that surrounds them. Therefore, it is not possible to define what caves in mountains have to look like. However, caves in high-relief areas share some common traits that are not directly linked to, but do influence, morphology. They are summarized below.

It is broadly known that caves are natural archives that register events in the Earth's history. It is less known that this conservation of events is not only reflected in sediments (deposition), but also in the cave's morphology (erosion). Both are equally important if one wants to know what happened in time. The difficulty is often to link sedimentological and morphological observations. An article on the importance of studying caves therefore gives a broad recipe. Examples of widely known caves then illustrate the diversity of settings and morphology. In many cases it is apparent that the cave's morphology is not sufficient to understand the morphogenesis of the landscape without taking into account other geological parameters. However, other examples show that insight into cave-forming processes is the only tool to reconstruct the evolution of the surface.

Literature research reveals that the geomorphology of many alpinotype caves is only known in its broad outlines; in many cases, the morphological description amounts to one or two phrases in the text. This fact makes it very difficult to review their geomorphological traits and significance, but on the other hand, it shows how much research is still needed to understand these caves. Consequently, the last paragraph, although entitled "Conclusions," also gives ideas for future research.

GENERAL CHARACTERISTICS OF CAVES IN HIGH-RELIEF AREAS

Caves in high-relief areas differ from caves in the lowlands in several aspects. Infiltrating water tries always to flow down vertically until it reaches either impervious strata or the base level. In high-relief areas, large and long shafts and steep meandering canyons are created this way. In a general perception, a cave with many shafts and meandering canyons is commonly regarded as *alpinotype*, although it is by no means restricted to the Alps. On the contrary, in lowlands, caves are most usually more or less flat, with small shafts between, and meandering canyons tend to be subhorizontal.

Due to orogenetic processes, the limestone strata are seldom subhorizontal; most of the time, the strata are more or less inclined. Because many caves preferentially develop on bedding planes (Lowe, 1992; Filipponi, 2009), eventually also at the intersection of bedding planes with faults and fractures, cave passages almost invariably are steeply inclined. This is evident for vadose passages following stratal dip, but it is commonly also the case for phreatic passages following strike. The permeability of the bedding plane being inhomogeneous, the water seeks its way along the most permeable places (Ford and Williams, 2007) and thus wiggles along the bedding plane, creating the so-called *looping passages* (Fig. 1). "Looping" in this case does not necessarily mean "making a closed loop," although lateral passages that do loop often exist.

The temperature of the infiltrating water is lower in mountain chains. Colder water can contain more CO_2 (and thus may be more aggressive), whereas the lower temperature should slow down the reaction kinetics. Therefore, it would be expected that the water in high-relief areas retains more aggressivity over a longer distance, thus enhancing speleogenetic processes in depth. However, the availability of CO_2 is not only due to water temperature, but also strongly influenced by vegetation at the surface (Bögli, 1978). This vegetation is reduced or absent in high altitudes. Moreover, investigations (Jeannin, 1990) have shown that the water temperature relatively quickly adjusts to the cave temperature. The subject has still to be investigated.

Glaciers at the surface in high mountains produce a high load of sediment suspended in water, which then is often called *glacier milk*. This enriched water has two effects on caves: where it flows torrentially through passages, it increases mechanical erosion and may dig out large canyons with polished walls. Where the water velocity gets very low, the sediment may settle out and eventually obstruct the passages, thus creating the silty fillings that are found in many caves in the Alps and elsewhere (Bini et al., 1998). The absolute effect of glacier milk on passage enlargement is still unknown.

Climatic variation in mountain chains is usually much larger than in lowlands. In wintertime, precipitation falls as snow, and there is almost no water available. In springtime, snowmelt releases huge amounts of water in a comparably short time, the caves are often flooded, and discharge is near its maximum. In summertime, thunderstorms form, releasing huge precipitation. There is an ongoing debate as to whether climate and its variability has a notable influence on cave-forming processes (and whether such influence can be seen in passage morphology).

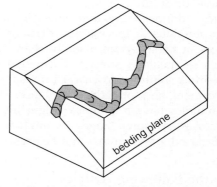

FIGURE 1 A *looping passage* formed along an inclined bedding plane. Such phreatic and epiphreatic passages enlarge at those places where permeability of the bedding plane is largest.

WHY IS IT IMPORTANT TO STUDY CAVES IN HIGH-RELIEF AREAS?

In the lowlands, surface erosion is commonly slow and, in some cases, even negligible, Therefore, geomorphological markers of the past are often recognizable and present. Commonly, these superficial forms can be linked to cave levels and cave sediments. Dating of one sedimentary deposit then gives the age range of the other one. The interrelation between surface and caves is often relatively clear in the lowlands.

On the other hand, mountain chains generally have large erosion rates. In midlatitudes, glaciation is another way of erosion that often leaves no trace of prior events. The overall result is that mountainous regions do very rarely present traces of older evolutionary stages. In order to reconstruct the geomorphic evolution of the surface in high-relief areas, caves are therefore the only means to reach this goal. It also becomes evident that caves are a valuable tool for geomorphology and Quaternary geology (Audra et al., 2006).

THE RELATIVE CHRONOLOGY

A considerable problem exists with caves being a hollow form within the parent rock, so they cannot be dated in themselves. On the other hand, the exact morphology of the void contains much information about processes that took place at the surface. It is thus very important to date the void. This can be done by applying the concept of relative chronology.

Relative chronologies are based on the interrelation between cave morphology (which expresses the removal of rock) and the sediments contained within the cave (which expresses depositional events). Within a sedimentary pile, the laws of stratigraphy can be applied, which in the principle of superposition indicate that a younger sediment overlies an older one. Morphological indications, on the other hand, also give chronological information. A keyhole passage in an epigenic cave informs us that a phreatic phase was followed by a vadose one. Successions of speleogenetic phases are found in many cave systems. This in itself is also chronological information.

The task now is to connect the sediments of several, basically independent, sedimentary profiles and to link them with the morphological succession of the cave passages. Thus, the sedimentary profiles are not independent from each other, and a relative chronology of erosional and depositional events over the whole cave can be made.

Example of a Relative Chronology

Figure 2 shows a real situation encountered in St. Beatus Cave, in Switzerland.

To the right side is a typical keyhole passage which proves that a phreatic initiation of the ellipse on top was followed by a canyon incision. In the middle part of the figure, the meander gradually disappears and is replaced by a more or less elliptic passage that continues toward the left side of the figure. We see therefore a transition of a vadose feature into a phreatic one, and thus an old water level. In the profile to the right, we observe flowstone deposition that was truncated by the river incising the meander. Therefore, the flowstone predates the canyon, but postdates the initial genesis of the elliptic passage to the right. The meander changes into an elliptic passage; thus the two forms are contemporaneous. Consequently, the older flowstone disappears in the area of this transition. Within all the passages, silts were deposited. They are younger than the meander incision, and younger than the passage to the left, and proof of an inundation of the whole cave. Stalagmites grow on the silts and are partially still active. This example can be written as a table. For practical reasons, that table is not rewritten with each sedimentary succession found. Instead, the single sedimentary sequence is coupled with morphology, and is written as a column in the table. The next

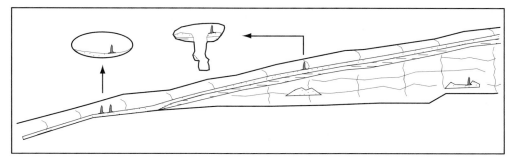

FIGURE 2 A schematic (but real) example to establish a relative chronology between morphology (erosional events) and sediments (depositional events) within a cave. (For explanation see text.)

TABLE 1 Chronological Table with Columnar Writing of Figure 2

Sequence at Left	Sequence at Right
	Phreatic genesis of top ellipse
	Water level lowering
Phreatic genesis	Deposition of flowstone
Phreatic genesis	Erosion of flowstone
Phreatic genesis	Erosion of meander
Water level lowering	
Silt deposition	Silt deposition
Stalagmite growth	Stalagmite growth

sedimentary sequence, again coupled with morphology, is written as another column. Thus, the above example would then look like Table 1. If we continue upstream and downstream of that profile, we find several other morphological indications and sedimentary successions, each of them having a link with our initial profile until we encounter the next paleowater level and thus the next morphological change. There, the links have to be established again. The table thus slowly grows and gets more complete.

When the basic table is established, absolute dating is used to assign ages to the respective formations. This has a twofold benefit: on the one hand the dates ascertain that the relative chronology is consistent, because the absolute ages should be concordant with the relative ones. On the other hand, the relative chronology can be used to assess whether a date is to be trusted or whether it might have experienced changes, for example, U leakage from speleothems.

EXAMPLES OF CAVES

Caves in high-relief areas can be divided into three main groups: (1) large, multiphase labyrinths, often of considerable age; (2) vadose shafts and meandering canyons leading steeply down; and (3) long inclined passages. Of course, there is almost always some mix of these types. Nevertheless, these three groups cover virtually all cave types in mountain ranges.

EXAMPLES OF CAVES: LARGE LABYRINTHS

Considering that relief and thus hydrogeological gradients are steep, one might expect mostly vertical to steeply inclined caves in mountainous regions. However, some of the densest labyrinths of the world's known caves develop in the mountains. Most of these labyrinths contain passages of different ages, and typically represent floodwater mazes that are often intersected by younger vadose passages.

Hölloch Cave, Switzerland

Hölloch Cave was, in the 1960s, considered to be the longest cave in the world. At about the same time, Alfred Bögli executed considerable scientific work within this cave. A summary of his work around Hölloch Cave is published in a well-known book on karst hydrography and physical speleology (Bögli, 1978). Hölloch Cave is situated in central Switzerland. The main entrance is at an altitude of 733 m asl. The cave develops in Schrattenkalk (Urgonien, Lower Cretaceous) of the Helvetic realm.

Hölloch Cave represents the lower end of a vast catchment area extending between 700 and 2300 m in altitude. Whereas the entrance of Hölloch is dry in normal seasons, it may flood during exceptionally wet times. This fact indicates that Hölloch Cave has a huge flooding zone that may reach more than 320 m in the far parts of the cave. In fact, 2/3 of the currently explored 195 km can be flooded during exceptional events (Bättig and Wildberger, 2007; Fig. 3).

As it follows from the floods, and as shown on Figure 3, Hölloch can be considered an anastomotic floodwater maze following Palmer (1991). This is very nicely seen also by its passage morphology: elliptical tubes of phreatic or floodwater origin abound everywhere (Fig. 4); entrenched meandering canyons occur only very rarely.

Bögli (1970), on the base of distribution of the large passages, concluded that Hölloch was made by three speleogenetic phases that were conditioned by valley deepening in the spring area. However, in Bögli's time, no absolute dating was possible. A recent paper (Wildberger et al., 2010) tried to couple different obtained age information (absolute and relative) with a general geomorphological evolution of the surface, in order to get ideas about the possible onset of karstification in the region of Hölloch. The results are yet inconclusive; however, karstification of the area should have begun at around 3 Ma BP.

Réseau Siebenhengste-Hohgant, Switzerland

Réseau means network—and this is a good description of the cave system of Siebenhengste. Together with many caves that are still not connected to the main system, the area comprises 320 km of cave passages, to which the Réseau itself contributes 156 km with a depth of 1340 m. This makes the area of Siebenhengste one of the large cave areas of the world.

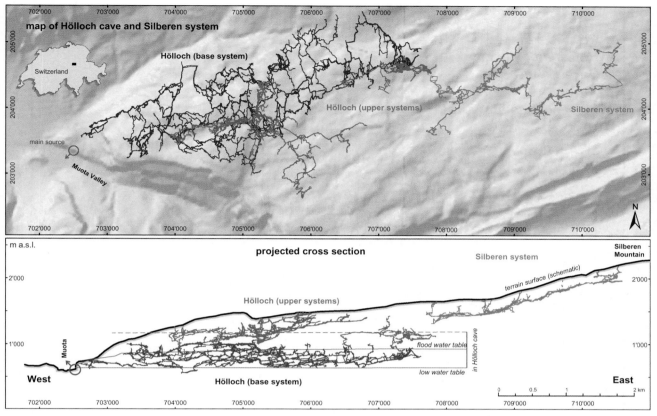

FIGURE 3 Map and projected section of Hölloch Cave (Switzerland) and neighboring Silberen System. Plan pattern as well as heights of flood (projection) clearly show that Hölloch Cave can be qualified as an epiphreatic floodwater maze. *Adapted from Wildberger et al., 2010.*

FIGURE 4 A passage in Hölloch cave reveals its *phreatic* form that is still enlarged in the epiphreatic zone. This passage gets regularly flooded, especially in snowmelt and in summer thunderstorms. *Photo courtesy of J. Godat.*

The Siebenhengste area is located in western-central Switzerland, north of Lake Thun. The caves occur between 460 m asl (100 m below the level of the lake), and the summit of Siebenhengste at 1950 m. The system develops in Schrattenkalk (Urgonian, Lower Cretaceous). A general view of the cave area is given in Figure 5.

The Siebenhengste caves are quite diverse in their character due to different genesis and age. In the following paragraphs, we describe the three main morphological parts.

The caves F1 and K2 (Fig. 5) show a typical dendritic pattern (Palmer, 1991). They are caves with very minor phreatic development: most of the passages are of vadose origin.

The labyrinth of the Siebenhengste itself represents a dense maze of old, phreatic tubes interlaced with many younger shafts and meandering canyons (Hof et al., 1984). Morphological research revealed that there were five speleogenetic phases responsible for the genesis of the different phreatic passages.

From the labyrinth, mostly vadose rivers flow downward toward the *zone profonde* (deep part) of the Siebenhengste, which again forms a 3D labyrinth with vadose shafts and meandering canyons coming from the surface. From there onward, an old phreatic tube that has its origin in the downstream part of F1 continues toward Faustloch and farther to the southwest.

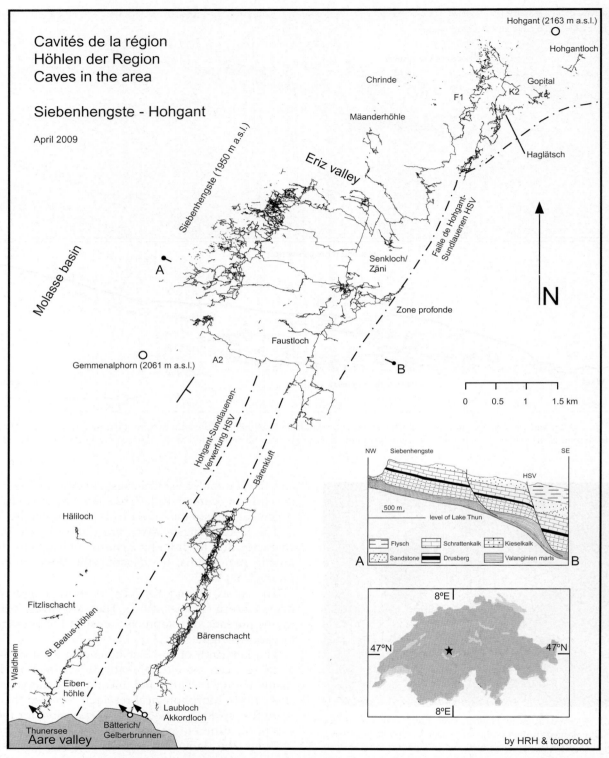

FIGURE 5 Caves in the Siebenhengste area, Switzerland. The map shows a total of 320 km of passages, of which 156 km are currently interconnected. The difference in plan pattern between dendritic (F1, K2) and phreatic multilevel mazes (Bärenschacht) can clearly be seen.

Bärenschacht is, like Hölloch, a floodwater maze cave. While several speleogenetic phases can be recognized in Bärenschacht (Häuselmann et al., 2002), there are almost no vadose meandering canyons and affluents from the surface that is covered by impermeable flysch. Therefore, all the maze is of (epi)phreatic origin.

The Réseau Siebenhengste-Hohgant is one of the few well-explored cave systems where a lot of scientific work has been done. Geomorphological investigations of the passage morphology revealed the existence of 14 speleogenetic phases (Häuselmann et al., 2002). All these phases can be linked with ancient valley floors, and therefore give valuable information on the evolution of the surface outside. The uppermost (and therefore oldest) of the 5 levels had their spring in the northeast of the Siebenhengste labyrinth in the Eriz valley (Fig. 5). The following 9 speleogenetic phases, on the contrary, had their spring in the Aare valley. Dating of the several cave phases by cosmogenic isotopes and U-Th series indicated that the uppermost 5 phases were active between more than 4.4 and 0.8 Myr BP, and that the valley incision rate during that period was around 0.12 mm yr^{-1} (Häuselmann et al., 2007). The subsequent incision of the Aare valley, between 0.8 Myr BP and today, happened 10 times faster with an incision rate of roughly 1.2 mm yr^{-1}.

EXAMPLES OF CAVES: VERTICAL TO SUBVERTICAL SHAFTS AND MEANDERING CANYONS

Although the labyrinths presented above all occur within the mountains, the shaft-and-canyon type is considered a *typical alpine cave*. By number, these caves exceed all other types by far. For instance, all but three of the 176 caves in the Innerbergli karren field (Siebenhengste, Switzerland) contain mostly shafts and meandering canyons.

Many of these shaft systems may be smallish. To develop vertical or subvertical systems of more than 500 m in altitudinal difference, the limestone thickness (or its folds) has to be sufficiently large to allow that. This is not everywhere the case.

Krubera, Georgia

The currently most famous of these shaft systems is Krubera (Voronja) cave, Georgia. By its depth of currently −2191 m, it is the world's deepest cave (Klimchouk et al., 2009). A succession of shafts and steep canyons leads down to several sumps which were dived to show that the passages continue behind. This is therefore a typical *alpine shaft system*, albeit of exceptional dimensions (Fig. 6).

Krubera Cave is situated in the Arabika massif (Western Caucasus, Georgia). The entrance opens at 2256 m asl within a glacial trough. The rocks are composed of Upper Jurassic and Lower Cretaceous limestones that are folded and complexly faulted and that dip continuously down to the Black Sea shore.

The cave map makes very quickly clear that the cave is essentially vertical: a total depth of −2191 m and a length of about 14 km lies within an area of only *circa* 700 × 400 m. Geomorphological observations point out that most of the cave was formed in the vadose zone; only minor passages seem to have their origin in an ancient phreatic zone (Fig. 7). This in itself is not very surprising for the given geologic setting. However, two facts make the cave interesting from the geomorphological point of view: first, the very low gradient between the water table occurring in the cave (110 m asl) and the spring area distant around 12−13 km at the Black Sea coast, and second, U-Th dating of stalagmites occurring at a depth of 640 and 436 m asl that yielded an age of between 200 and 300 ka. These age dates reveal that these parts of the cave were already in the vadose zone at that time. The current knowledge indicates that the initiation of the upper phreatic tubes could have happened during the Messinian salinity crisis, and the vadose cave formed during the Plio-Pleistocene, more or less concordant with uplift, which was very huge in the central parts of the massif (estimated to be 2−2.5 km during Plio-Pleistocene; Fig. 7).

Krubera Cave shows that not only phreatic tubes connected to paleosprings can be used for reconstructing the paleogeomorphology of a landscape, but in ideal cases, also vadose caves are very valuable. Therefore, scientific observations in all kinds of alpinotype caves are to be promoted.

EXAMPLES OF CAVES: LONG INCLINED PASSAGES

Caves that present long inclined passages are almost invariably caves that develop along a geological boundary with impermeable rocks; they therefore always contain an active river. Generally, these caves are characterized by a vertical shaft zone where the waters transect the limestone in the most direct way before reaching the impervious layers. Often, the limestone thickness is rather small (150−250 m)—as a matter of fact, many of the deepest caves of the world actually belong to this cave style. The most famous of them is Gouffre Berger (Vercors, France), which was, in 1956, the first cave known to reach −1000 m (Caillault et al., 1997).

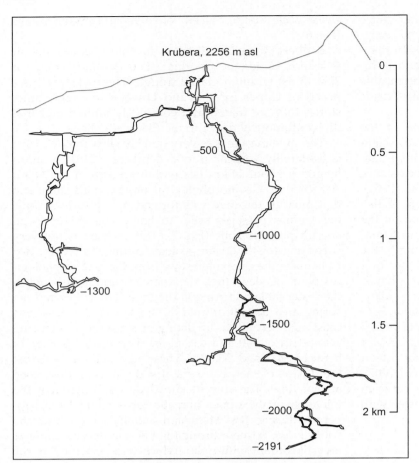

FIGURE 6 Projected section of the deepest cave of the world, Krubera (Voronja), Caucasus, Georgia. The cave is mainly a succession of steep meandering canyons and shafts. Almost all the cave is of vadose origin, with the exception of the subhorizontal branch at around −300 m, which is interpreted as being a possible Messinian cave.

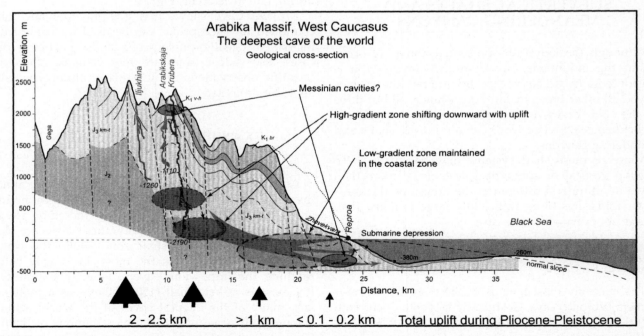

FIGURE 7 Genetic interpretation of the genesis of Krubera Cave in connection with recent uplift of the Caucasus range. *From Klimchouk et al., 2009.*

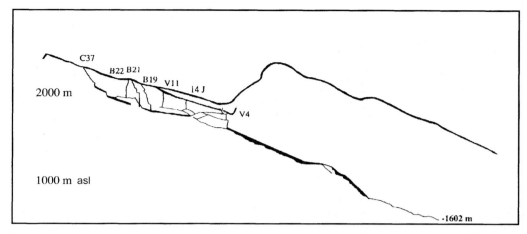

FIGURE 8 Projected cross-section of Jean-Bernard Cave (France). Several entrances (with numbers) connect to partly vadose, partly phreatic channels. The phreatic cave had its spring in the area of the entrance V4. The rest of the cave is essentially vadose and formed by a canyon which is sometimes very high (80–100 m). *Adapted from Lips et al., 1993.*

The genesis of this cave type is actually favored by orogenetic processes that overthrust (and fold) limestone nappes onto other rocks and so create the possibility of large altitudinal differences within the same (thin) limestone sequence.

The Jean-Bernard Cave (France)

The *Gouffre Jean-Bernard* was the world's deepest cave in the beginning of the 1990s. It is situated in the Pre-Alps in the east of France. Its several entrances are between 1840 and 2320 m asl. The cave develops in Urgonian limestones of 150- to 200-m thickness, which are overthrust, folded, and faulted. The cave follows a minor syncline within the limestone series.

For many reasons, the cave is again considered an intermediate system between the simple shaft-canyon type (as is Gouffre Berger) and a complex labyrinth like the Siebenhengste: its upper part consists not only of vadose shaft systems rapidly reaching the impervious base level (Fig. 8), it also contains ancient phreatic tubes that actually had their paleospring in the lower entrances (Lips *et al.*, 1993). The lower part of the cave contains one large single meandering canyon, sometimes up to 100 m high, where the river flows. This river sinks between breakdown; the following passage leading farther downward is of much smaller size.

Jean-Bernard Cave has the peculiarity that the lowermost end of the cave (716 m asl) is lower than the altitude of the resurgence (780 m asl). This caused a vigorous debate as to whether this would be possible or whether this is simply caused by topographic errors. The current idea is that the mapping is correct and that indeed the cave is deeper than its spring, because the river diverts laterally from the cave and resurfaces on the flank of the syncline, whereas the small annex passage goes farther downward along the synclinal axis and was drained by a now-abandoned spring downstream from the present one.

There is at present no comprehensive dating of the cave that permits determination of its age. A hypothesis is that the ancient phreatic passages occurring on top could have formed before the Messinian salinity crisis. The only proven fact is that when these passages form, the valleys and the overall geomorphology of the Alps in that region has to have been completely different from their aspect of today.

Cassowary System (Papua New Guinea)

The Cassowary System is situated in the island of New Britain (Papua New Guinea), at an altitude of 1445 m asl (Audra *et al.*, 2001). The system is, with −1178 m in altitudinal difference, the deepest cave of the Southern Hemisphere; it can be followed from the ponor entrances all the way to the spring at 267 m asl (Fig. 9). The cave is formed within pure, only slightly compacted, Miocene limestone, overlain by alteritic residues of former volcanoclastic rocks. The area was subject to vigorous uplift beginning in the Upper Pliocene.

The cave character differs from the general notion of shafts going down to an impervious layer and a meandering passage following it to the spring: first, there is no impervious layer present and then the cave presents only very few shafts: it is mainly composed of a long, inclined tube going from ponor to resurgence that was later incised to form a canyon. No ancient phreatic system is present; the cave was therefore formed in a single phase, which suggests a recent

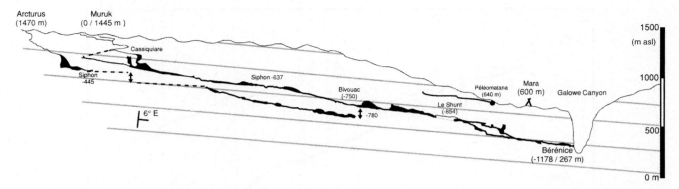

FIGURE 9 Projected cross-section of Cassowary System (Papua New Guinea). The cave monotonously dips down toward its spring in the Galowe Canyon. Dating as well as morphological consideration led to the hypothesis that the cave must be very young. *Adapted from Audra et al., 2001.*

development. The plan view supports this hypothesis: the cave goes practically straightforward from ponor to spring with basically no lateral wiggles.

The Nakanai mountains, where the cave is situated, present special conditions for speleogenesis and karstification: the presence of pure, but only slightly consolidated, limestone offers high potential for both active corrosion as well as mechanical erosion. The rainfall in the area (10 to 12 m yr^{-1}!) provides huge amounts of waters, which furthermore are concentrated at the surface due to the argillaceous alteritic cover of the former volcanoclastic rocks. The presence of a rainforest with a highly productive growth increases the CO_2 content of the water to a considerable extent, and the vigorous recent uplift of the area is responsible for a high hydraulic gradient.

Dating of some speleothems gave ages of less than 50 ka BP, and paleomagnetic analyses of clay sediments showed normal polarity. The age of the cave system is therefore less than 780 ka BP. Audra *et al.* (2001) suppose an age of 100 to 200 ka BP. The cave system is therefore an example of a recent, single-phase ponor-to-spring system. In that respect, it is a good example of an extreme end-member of cave formation.

CONCLUSIONS

From the above examples, it is clear that there is no simple geomorphological trait that links all the caves in high-relief areas together. While the basic cave-forming processes are always the same, every cave is dependent on local conditions.

When reading speleological literature about alpino-type caves, it appears that very many caves actually are vertical shafts without fossil passages. The question is now whether this is speleological reality (there are actually quite few fossil passages) or whether this is exploration bias (many cavers are not geomorphologists and simply want to go as deep as possible, so all possible horizontal passages are neglected). My personal experience rather tends toward the second explanation: very many shaft caves I know do in reality intersect (or follow) fossil phreatic passages. As a first conclusion, one can thus say that caves have to be explored more thoroughly, and complete and well-drawn maps (and longitudinal sections) that contain all relevant information must be made.

When literature is searched specifically for scientifically interesting information on caves in high-relief areas, the reader is left with some few articles compared to the mass of such caves that are explored. In a general sense, there is still much fieldwork to be done almost everywhere. Rewards are most certain, because caves now are known to be archives of the past, recording not only valley deepening phases, but also in registering climatic variations, both in speleothems and in other sediments (and even in their passage morphology). Therefore, the second conclusion follows: almost everywhere there is still a huge amount of work to be done.

As stated above, caves everywhere, but especially in high-relief areas, are very important for reconstructing the evolution of the landscape outside the cave. They are actually, in very many places, even the single best possible means to get insight into valley deepening processes, glaciations, and climatic variations. Such research is not purely academic: caves may offer insight into incision rates of valleys, which may directly condition the potential suitability of, for instance, a deposit of radioactive waste planned by the government. Now, much of this work cannot be done by "traditional" geomorphologists, simply because they cannot reach the cave. Either cavers become geomorphologists, or the geomorphologists teach the cavers to bring as much information as possible back to the surface. Experience has shown that it may work! If "normal" scientists regard caves no longer as "dirt-loving and strange," but as collaborators, general knowledge will improve tremendously. The third conclusion: *do collaborate!*

While caves in high-relief areas are often not very easily accessible, they may offer some advantages to caves in the lowlands. For instance, due to the generally more diverse climate, looping caves present larger loops, which make it easier to decipher speleogenetic processes and to gain insight into basic principles of speleogenesis. In following years, it is expected that there will be important discoveries.

See Also the Following Article

Solution Caves in Regions of Moderate Relief

Bibliography

Audra, P., Bini, A., Gabrovsek, F., Hauselmann, Ph., Hoblea, F., Jeannin, P.-Y., Kunaver, J., Monbaron, M., Sustersic, F., Tognini, P., Trimmel, H., & Wildberger, A. (2006). Cave genesis in the Alps between the Miocene and today: A review. *Zeitschrift für Geomorphologie N.F., 50*(2), 153–176.

Audra, P., de Coninck, P., & Sounier, J.-P. (2001). *Nakanai—20 ans d'exploration*. Antibes: Association Hemisphère Sud.

Bättig, G., & Wildberger, A. (2007). Ein Vergleich des Hölloch-Hochwassers vom August 2005 mit seinen Vorgängern. *Stalactite, 57*(1), 26–34 [in German].

Bini, A., Tognini, P., & Zuccoli, L. (1998). Rapport entre karst et glaciers durant les glaciations dans les vallées préalpines du Sud des Alpes. *Karstologia, 32*(2), 7–28 [in French].

Bögli, A. (1970). *Das Hölloch und sein Karst*. Neuchâtel: La Baconnière.

Bögli, A. (1978). *Karsthydrographie und physische speläologie*. Berlin: Springer Verlag.

Caillault, S., Haffner, D., & Krattinger, T. (1997). *Spéléo dans le Vercors*. Aix-en-Provence: Edisud.

Filipponi, M. (2009). *Spatial analysis of karst conduit networks and determination of parameters controlling the speleogenesis along preferential lithostratigraphic horizons.* Switzerland: Ecole Polytechnique Fédérale de Lausanne.

Ford, D. C., & Williams, P. (2007). *Karst hydrogeology and geomorphology*. Chichester, U.K. Wiley.

Häuselmann, P., Granger, D. E., Jeannin, P.-Y., & Lauritzen, S.-E. (2007). Abrupt glacial valley incision at 0.8 Ma dated from cave deposits in Switzerland. *Geology, 35*(2), 143–146.

Häuselmann, P., Jeannin, P.-Y., Monbaron, M., & Lauritzen, S.-E. (2002). Reconstruction of Alpine Cenozoic paleorelief through the analysis of caves at Siebenhengste (BE, Switzerland). *Geodinamica Acta, 15*, 261–276.

Hof, A., Rouiller, P., & Jeannin, P.-Y. (1984). Le Réseau. *Höhlenforschung im Gebiet Sieben Hengste-Hohgant, 0*, 1–106.

Jeannin, P.-Y. (1990). Températures dans la zone vadose du karst. *Bulletin du Centre d'Hydrogéologie de l'Université de Neuchâtel, 9*, 89–102 [in French].

Klimchouk, A. B. (2007). *Hypogene speleogenesis: Hydrogeological and morphogenetic perspective*. Socorro, NM: National Cave and Karst Research Institute.

Klimchouk, A. B., Samokhin, G. V., & Kasian, Y. M. (2009). The deepest cave in the world in the Arabika massif (Western Caucasus) and its hydrogeological and paleogeographic significance. In *Proceedings of the 15th International Congress of Speleology* (pp. 898–905). Kerrville, TX.

Lips, B., Gresse, A., Delamette, M., & Maire, R. (1993). Le Gouffre Jean-Bernard (−1602 m, Haute-Savoie, Fr.). *Karstologia, 21*(1), 1–14.

Lowe, D. J. (1992). *The origin of limestone caverns: An inception horizon hypothesis.* : Manchester Polytechnic.

Palmer, A. N. (1991). Origin and morphology of limestone caves. *Geological Society of America Bulletin, 103*, 1–21.

Wildberger, A., Geyh, M., Groner, U., Hauselmann, Ph., Heller, F., & Ploetze, M. (2010). Dating speleothems from the Silberen Cave system and surrounding areas: Speleogenesis in the Muota Valley (Central Switzerland). *Zeitschrift für Geomorphologie, 54*(2), 307–328.

SOLUTION CAVES IN REGIONS OF MODERATE RELIEF

Arthur N. Palmer

State University of New York

Caves in regions of moderate relief are widely regarded as the standard to which all others are compared. Tectonic stability is their most significant characteristic. The presence of moderate relief implies that uplift of the land is slow and erosional processes are able to keep pace. As a result, rivers easily erode to their local base levels, and base-level control is reflected in cave passages. The relation between caves and regional geomorphic history is stronger than in any other setting.

CHARACTERISTICS OF KARST REGIONS OF MODERATE RELIEF

In this article, the standard cave-forming process is considered to be dissolution of carbonate rocks by groundwater rich in carbonic acid derived from the soil. Some solution caves in moderate-relief karst are the product of quite different mechanisms (*e.g.*, by sulfuric acid or by gypsum dissolution) and are discussed in other articles.

Relief by itself has little influence on cave origin. For example, alpine caves form by the same mechanisms as those in moderate relief, but they show different relationships to the landscape and to the geologic structure. Most regions of moderate relief have a well-developed soil cover, except where it has been lost through glacial scour, deforestation, or overgrazing. Soil is the source for most of the carbonic acid that allows groundwater to dissolve caves, and it enhances the deposition of carbonate speleothems where groundwater seeps into aerated caves. It also affects the nature of groundwater recharge, and therefore cave patterns.

GEOGRAPHIC DISTRIBUTION

Regions of moderate relief include plateaus of relatively undeformed rock, as well as the roots of deformed mountains that uplifted hundreds of millions of years ago and are tectonically nearly stable today. These landscapes are well illustrated by the low karst plateaus of the east-central United States and the Appalachian Mountains (Fig. 1). These were the main field areas for the classic American karst studies of the late nineteenth century through the mid-twentieth century. Base-level and water-table control of cave development were topics of great weight among American karst scientists, but were far less conspicuous in European studies. Many of these differences can be traced to the strong focus on alpine karst by most European karst scientists. Nevertheless, many important karst areas throughout the world are in plateaus and ancient mountains of moderate relief.

The most cavernous states in the United States, ranked by number of known caves, are Tennessee, Kentucky, West Virginia, Alabama, Virginia, and Missouri. Each has more than 3000 significant caves, and all are in the moderate-relief interior plateaus or Appalachian Mountains. The extensive Florida karst shares many of the same cave-forming processes, although it has much lower relief and is influenced by the unusually high porosity of its carbonate rock and in local areas by mixing of freshwater and seawater. In the mountainous and semi-arid western United States, most caves are of alpine, thermal, sulfuric acid, and volcanic origin, or have formed in gypsum.

ORIGIN OF SOLUTION CAVES

Most solution caves in carbonate rocks are formed by the circulation of groundwater from upland surfaces through a network of fractures, partings, and intergranular pores, to springs at lower elevations. Groundwater recharge takes place in several ways, including diffuse seepage through soil, small trickles delivered by sinkholes, and concentrated inflow as sinking streams.

As water passes through the atmosphere and soil, it absorbs carbon dioxide, which combines with the water to form carbonic acid. This acid, though weak, is abundant and is the most common agent of carbonate-rock dissolution. Where the water first enters limestone or dolomite, it dissolves all available openings in the rock at nearly uniform rates, producing a maze of widened fissures and pores (the epikarst). In areas of moderate relief, most of these openings are soil-filled. Water drains from the base of the epikarst through a relatively small number of major conduits, which deliver solutionally aggressive water deep into the soluble rock. It is along these paths that caves form. As the openings enlarge, soil subsides into them, forming sinkholes at the surface that funnel increasing amounts of water into the ground as their catchment areas expand.

Water descends by gravity along the steepest available openings until it reaches the zone where all openings are filled with water. The top of this zone is the *water table*. Above the water table is the *vadose zone*, and below it is the *phreatic zone*. (Most groundwater specialists prefer to call them the *unsaturated* and *saturated* zones, but these terms blur the distinction between hydrologic and chemical saturation.) Cave passages that form in the vadose zone have continuously descending profiles that typically consist of canyon-like passages interrupted by vertical shafts, so that the resulting pattern is step-like. Canyons develop mainly along the partings between rock strata and grow by entrenchment of their floors by flowing streams (Fig. 2). Shafts form where the descending water follows steeply inclined or vertical openings, mainly joints or faults.

At and below the water table, gravity is offset to varying degrees by the downward increase in hydrostatic pressure. Phreatic water follows the paths of least resistance to the nearest available outlets, usually river valleys. Passages of phreatic origin are tubular or fissure-like conduits with low overall gradients (Fig. 3), and most have looping profiles that fall and rise irregularly along their length. Because

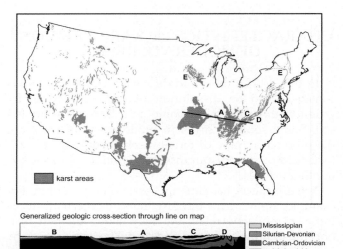

FIGURE 1 Map of karst regions of the 48 conterminous United States, showing areas discussed in the text. A = margins of the Illinois Basin (Indiana and Kentucky); B = Ozark Plateaus of Missouri and Arkansas; C = Appalachian Plateaus; D = Ridge and Valley Province (folded Appalachians); E, E = karst areas modified by continental glaciation. The cross-section shows the distribution, simplified structure, and ages of the three major cavernous rock sequences.

FIGURE 2 A typical canyon passage in McFail's Cave, New York, formed by vadose flow.

FIGURE 3 A tubular passage in Sullivan Cave, Indiana, formed by phreatic flow (*i.e.*, at or below the water table). Although the main stream has since diverted to a lower route, periodic overflow still utilizes the passage shown here.

fractures and partings tend to become narrower with depth, most phreatic passages follow rather shallow paths at, or not far below, the water table. This characteristic is especially common in regions of moderate relief.

As surface rivers entrench their valleys, groundwater seeks progressively lower outlets, and new cave passages form along those paths. Much of this entrenchment is caused by uplift of the Earth's crust, while stream erosion tries to keep pace. Therefore, younger passages are not necessarily lower in absolute elevation above sea level. Changes in the rate of river entrenchment leave their mark on both valleys and caves. Steep-walled valleys are produced by rapid entrenchment. Wide flat-floored valley bottoms (or their remnant terraces) indicate slow entrenchment, static conditions, or even rising river levels, and major cave passages tend to form at those elevations. Sediment accumulates in valleys and caves when river levels rise.

The erosional-depositional history of stream valleys is controlled by significant events such as climate changes, sea-level fluctuations, rearrangement of river patterns, changes in sediment supply, changes in uplift rate, or depression of the continent by the weight of glaciers. At the surface, much of the evidence for this history is destroyed by later erosion. A great deal more evidence is preserved in caves.

Cave origin is not as thoroughly understood as the above paragraphs might imply. There is still debate on many topics, such as: When does cave development actually begin? Of all the original openings in soluble rock, which ones evolve into caves? How deep can caves form below the surface? How does the local setting affect cave patterns? What do caves tell us about past geologic history and climates? Each question is best examined in karst regions of moderate relief, because they not only show the greatest geologic and hydrologic variety, but also are tectonically stable enough that cave features can be preserved for lengthy periods of time.

FIELD EXAMPLES

Several well-documented karst areas are briefly described here, with emphasis on the details that help to answer the questions posed above. These areas are located in the low plateaus of the east-central United States and in the Appalachian Mountains and have three major karst-bearing sequences in common: limestones and dolomites of Cambrian-Ordovician, Silurian-Devonian, and Mississippian age. The same strata, arranged in different settings, produce a great variety of cave types.

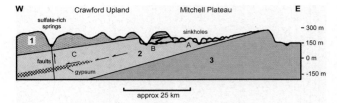

FIGURE 4 Cross-section through the karst area of southern Indiana. 1 = mainly insoluble sandstones and shales, 2 = Mississippian limestones, 3 = insoluble siltstones. A = Caves in exposed limestones in the Mitchell Plateau. B = Caves beneath insoluble caprock along the eastern edge of the Crawford Upland. C = Deep flow routes guided by interbeds of gypsum, with outlets through the caprock along faults.

Dissected Plateaus of Southern Indiana and Western Kentucky

Perhaps no other karst area in the world shows the relation between river erosion and cave development as clearly as these low-relief plateaus, where carbonate rocks are exposed at the surface around the perimeter of the Illinois structural basin (Fig. 1). In general, the rocks dip gently at less than half a degree toward the center of the basin, which is located in southern Illinois (Fig. 4). The main cavernous units are carbonate rocks of Mississippian age with a thickness of 100–200 m. Most of them exhibit prominent bedding. At considerable depth below the surface, gypsum is interbedded with limestone and dolomite in the middle part of the sequence. At depths of less than about 100 m, dissolution by groundwater has removed most of the gypsum.

Where the carbonate rocks are exposed at the surface they form a vast sinkhole plain covering thousands of square kilometers (the Pennyroyal Plateau of Kentucky and Mitchell Plateau of Indiana). Many river caves are located in this region, including some of the longest caves in the United States. Binkley's Cave, Blue Spring Cave, and the Lost River Cave System (all in Indiana) and Hidden River Cave (Kentucky) all have mapped lengths of 30 to 40 km. The erosional relief of the sinkhole plain is generally less than 50 m, and so the caves rarely contain more than one or two distinct levels of passage development. In river valleys, postglacial sediment has accumulated to thicknesses as much as 25 m, flooding and choking some cave passages that lie below present base level.

Farther in the down-dip direction, the carbonate rocks are capped by insoluble sandstone and shale, which form a dissected upland of irregular limestone ridges capped by the more resistant insoluble rocks (the Chester Upland of Kentucky and Crawford Upland of Indiana). Multilevel caves are numerous, because the protective caprock inhibits erosional destruction of the underlying caves. The most extensive caves are the Mammoth Cave System (628 km of surveyed length), the Fisher Ridge Cave System (182 km), and the Whigpistle Cave System (56 km), all of which are close neighbors at the southeastern edge of the Illinois Basin in Kentucky. (A *cave system* is a group of several interconnected caves explored separately and later joined by the discovery of interconnecting passages.) Mammoth is the world's longest known cave, more than twice as long as any other, and it is likely that future exploration will connect it with one or both of the neighboring caves.

Still farther down-dip, the insoluble caprock becomes continuous. Confined groundwater moving through the confined underlying carbonate rocks is able to escape upward through the caprock only along sparse high-angle faults. Springs fed by this water have high sulfate contents, showing that the interbedded gypsum is hosting deep groundwater flow through the carbonates. Hydrogen sulfide in the spring water indicates that sulfate reduction is also present at depth.

By scanning these three regions from their down-dip to up-dip ends, it is possible to envision how the karst landscape must have evolved with time. Presumably the deep gypsum dissolution in the confined part of the karst aquifer has prepared the way for later cave development at shallower depths. In places, the exposed carbonate rocks are fragmented and contorted and contain veins of calcite and quartz, all of which are indicators of former gypsum bodies. However, extensive geologic mapping of caves in the capped ridges and sinkhole plain shows that the cave patterns are closely adjusted to present-day shallow groundwater flow and must have had very little (if any) inheritance from earlier deep-seated dissolution.

The clues that indicate passage development by shallow groundwater flow are fairly clear. In the prominently bedded rocks of this region, most vadose water preferentially follows the dip of the strata, except where it jogs downward along joint-controlled shafts. Where it reaches the water table, the water loses its dip tendency, and most of the water follows paths that have no dip trend. Instead, many phreatic passages closely parallel the strike of the beds, which is approximately the line of intersection between the dipping beds and the less steeply sloping water table. Most passages consist of a vadose upstream section and a phreatic downstream section. At the present or former vadose-phreatic transition point, the passages change from down-dip canyons to low-gradient tubes with irregular profiles. In Indiana the lower parts of the carbonate sequence are massive and prominently jointed, so that narrow, fissure-like passages are formed in which the vadose-phreatic transition is not so clear.

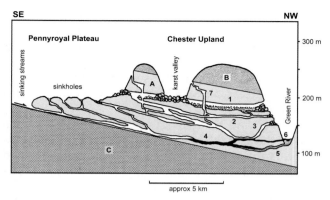

FIGURE 5 Idealized profile through the Mammoth Cave System, Kentucky, and its relation to the surrounding landscape. A = Cavernous limestones (from bottom to top: St. Louis Limestone, Ste. Genevieve Limestone, and Girkin Formation). B = Resistant caprock, mainly of sandstones and shales. C = Impure, poorly cavernous limestone. 1 = large Tertiary-age canyons, with thick sediment fill, 180–210 m above sea level; 2, 3 = major cave levels with vadose-phreatic transitions, 168 and 152 m above sea level; 4 = Late Pleistocene passages; 5 = sediment fill from postglacial rise in base level; 6 = springs fed by active cave streams; 7 = vadose passages (canyons and shafts).

Mapping of these transition points in any part of the region shows a tight clustering at a few consistent altitudes. At Mammoth Cave these are 200–210 m, 180–190 m, 168 m, and 152 m (Fig. 5). These same levels of development can be traced through much of the region, although their altitudes and spacing vary. In each of the lowest two levels in Mammoth Cave, the vadose-phreatic transitions in different passages fall within a meter of each other. If deep-seated dissolution had prepared the way for later cave development, this tight correlation would be absent. No good example of control of cave development by ancient, deep-seated dissolution has yet been demonstrated in either Kentucky or Indiana, although it is likely that a few may yet be revealed.

The drainage history of the region has been greatly clarified by cave evidence. The uppermost passages are large canyons partly or completely filled with sand and gravel deposited by cave streams. These passages represent slow dissection by surface rivers, alternating with periods when river valleys and cave passages were partly filled with sediment. This alternating sequence may reflect climate changes from humid (entrenchment) to relatively dry (sediment accumulation), but it may also have been the result of sea-level changes. Throughout this period the Pennyroyal Plateau and Mitchell Plateau formed a low-relief surface near base level with little or no karst development. The largest cave passages in the Chester and Crawford Uplands are located at roughly the same elevations, and many of them appear to have formed by water draining through the upland ridges and fed by runoff from the adjacent plains.

A widespread episode of sediment accumulation followed, covering the plains and filling the caves to depths of up to 30 m. Soon afterward, rivers began to entrench rapidly, allowing the Pennyroyal Plateau and Mitchell Plateau to develop many sinkholes and caves. This rapid entrenchment was apparently triggered by rerouting of rivers by glacial advances, which led to the development of the present Ohio River. The large canyon passages and sediment fill in the upper levels of the caves date from the Late Tertiary Period, which predates the first significant continental glaciation. Rapid entrenchment and development of the sinkhole plain took place during the Quaternary Period, when glacial ice masses periodically covered much of high-latitude North America. Glaciers fell short of the present karst areas, but their effect on the caves was strongly felt in the form of changes in river entrenchment and valley filling.

Dating of the passages in Mammoth Cave has recently been accomplished with cosmogenic aluminum and beryllium isotopes in quartz-rich sediment (Granger et al., 2001). The oldest cave sediments in the large upper-level canyons are up to 4 million years old, and the canyons themselves are presumably much older. The period of widespread sedimentation took place about 2.6 million years ago, just before the onset of extensive North American glaciation. The 168-m and 152-m passage levels date from roughly 1.5 and 1.0 million years ago and correlate with adjustments of the Ohio River drainage pattern. Some of these changes were caused by diversion of surface rivers to more southerly routes. The contribution of caves to the interpretation and timing of the regional geomorphic and climatic history is clearly demonstrated.

The Ozark Plateaus

Karst of the Ozark Plateaus of Missouri and Arkansas (Fig. 1) resembles that of the sinkhole plains of Indiana and Kentucky, except that most of its caves are in dolomite. This rock type is widely considered to be inferior to limestone in cave-forming potential, but the caves seem not to have noticed. Ordovician and Mississippian carbonates are the main cavernous units. In general they dip gently away from the St. François Mountains of southeastern Missouri at less than half a degree, but with local broad, gentle warps in the regional structure. Chert is so abundant in some of the carbonate rocks that residual chert fragments have accumulated to depths as great as 40 m, subduing the surface karst topography. The chert fragments also retard flood runoff into caves, and some surface rivers

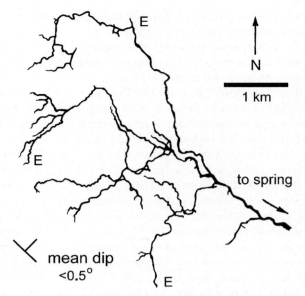

FIGURE 6 Map of Crevice Cave, Missouri, showing an ideal branching passage pattern guided by prominent bedding. The relation between dip-oriented canyons and the strike-oriented main tubular passage is shown. E = entrance. *From map by Paul Hauck.*

fed by karst springs have relatively low flood peaks as a result.

Caves are numerous and widespread in the Ozark Plateaus. The longest are sprawling branchworks composed of sinuous, low-gradient passages guided by bedding. Vertical shafts are sparse, and the low topographic relief prevents them from being very deep. Crevice Cave is the largest cave in the region, with 46 km of mapped passages (Fig. 6). The Ozarks are also well known for large karst springs. Some of these are fed by deep phreatic tubes that extend more than 60 m below base level.

In the 1930s through 1950s the well-known geologist J Harlan Bretz (1956) used qualitative field observations in this region to support his view that caves originate deep in the phreatic zone and are later invaded by vadose streams only late in their history. This hypothesis has gone out of favor in the past half century. In recent years, however, certain aspects of this model are being seen in more favorable light. The geomorphic history of Ozark caves has yet to be studied in detail using quantitative methods.

Appalachian Plateaus

This region stretches all the way from central New York to northern Alabama (Fig. 1). Its karst is developed mainly in Mississippian carbonates, with maximum thicknesses of about 500 m. At the northeastern end the cavernous rocks are much thinner and of Silurian-Devonian age. The Appalachian Plateaus contain the largest caves of the east, including the Friar's Hole Cave System, Organ Cave System, and Scott Hollow Cave (West Virginia); Blue Spring Cave and Cumberland Caverns (Tennessee); and Sloan's Valley Cave (Kentucky), all of which have surveyed lengths between 40 and 70 km.

In many ways the geology and caves of this region resemble those of the Mammoth Cave area of Kentucky, but the Appalachian Plateaus have greater relief and more obvious control by faults and insoluble beds. The strata have been warped into broad folds, typically with dips of only a few degrees. In many places the carbonate rocks are overlain by a thick impermeable cap and are exposed only in narrow bands along escarpments and entrenched river valleys. Perhaps the most dramatic example is the Obey River Gorge of northeastern Tennessee, where a broad, nearly flat plateau of sandstone and shale has been breached by abrupt V-shaped canyons up to 300 m deep, in which carbonate rocks are exposed only in the lower 25–30%. Many complex caves have developed along the valley bottoms parallel to the rivers. These cave patterns reflect the widening of joints and partings by stress release from erosional unloading along valleys (Sasowsky and White, 1994). Many caves in northern West Virginia have a similar setting. A conspicuous exception in the latter area is Simmons-Mingo Cave, which crosses beneath a prominent surface divide along a linear fault zone for a distance of 3.2 km.

In parts of southeastern West Virginia, the insoluble caprock is of irregular thickness and forms only the tops of hilly plateaus. It has been breached by small surface streams high above base level, and caves can extend from these points of focused recharge to remote springs along the plateau edges. Caves do not follow the valley bottoms as faithfully as they do in the examples cited above. The Friar's Hole Cave System is the best example.

Along the plateau edges, many caves are perched on underlying insoluble beds. In southeastern West Virginia, hundreds of kilometers of vadose passages have formed at the contact between limestones of the Greenbrier Group and underlying shale of the Maccrady Formation (Fig. 7). Erosion by vadose flow has entrenched canyons into the shale as much as 15 m deep. Flow along the contact is mainly in the dip direction, but is locally deflected from that direction by intersecting fractures. As a result, vadose passages are not oriented so relentlessly down the dip as in the Mammoth Cave area. The vadose passages feed conduits of phreatic origin that follow the strike of the beds, often along synclinal troughs. Dye tracing shows that some of these conduits feed springs at distances up to 20 km.

FIGURE 7 Cave development at the contact between limestones of the Greenbrier Group (ceiling) and shales of the Maccrady Formation (walls and floor) in Ludington Cave, southeastern West Virginia.

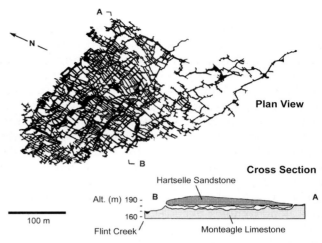

FIGURE 8 Map of Anvil Cave, in northern Alabama, a joint-controlled network cave beneath a thin permeable sandstone cap. *From map by William Varnedoe.*

Faults exert a strong effect on some caves in the Appalachian Plateaus. The deepest vertical shaft in any cave of the eastern United States, in Ellison's Cave, northwestern Georgia, is located along a large strike-slip fault. The fault zone features slickensides, pulverized rock, displacement of solutional rock surfaces, and occasional mild earthquakes. Upper-level stream passages in the cave are perched on shaly beds hundreds of meters above the local valley floors, but where they intersect the fault the water plunges down shafts up to 180 m deep.

Some caves in the region consist of extensive networks of intersecting fissures, in which all major fractures within a large area of soluble rock have enlarged more or less simultaneously by dissolution. Network caves of this type can form in various ways, but, in the areas described here, the largest are located where seepage takes place through a thin overlying cap of quartz sandstone. Water infiltrates thorough the sandstone as if through a sponge, and enlarges each fracture at or near the top of the underlying carbonate rock at fairly uniform rates. This process can take place in either phreatic or vadose conditions, and often both in sequence. The best example of a sandstone-capped network in the Appalachian Plateaus is Anvil Cave, Alabama (Fig. 8). Twenty kilometers of tightly labyrinthine fissure passages are constrained to the top of the Monteagle Limestone beneath a thin caprock of Hartselle Sandstone no more than 10 m thick. This and other similar caves have undergone some enlargement by backflooding from nearby rivers.

Ridge and Valley Province, Appalachian Mountains

The intensely folded and faulted eastern region of the Appalachian Mountains (Ridge and Valley Province) consists of numerous subparallel ridges and valleys that follow the trends of the folds (Fig. 1). Resistant quartz sandstones and conglomerates are the main ridge-formers. Carbonate rocks up to 2 km thick are exposed in linear bands over much of the region, mostly in valleys and along the flanks of ridges. Caves in this region contain the same general passage types as those in the plateaus farther west, but the overall cave patterns show some differences. Because the average dips are much steeper than in the plateaus, vadose water tends to follow short, steep paths interrupted by many shafts and fissures along discordant fractures. Major phreatic passages usually have elongate patterns extending along the strike of the beds.

A good example is the group of caves at Trout Rock, in northeastern West Virginia (Figs. 9 and 10). These include several levels of strike-oriented passages in steeply dipping limestone, with a network maze at the crest of a small anticline. Nearby Rexrode Cave and Smoke Hole Caverns have linear patterns tightly constrained along the strike of beds that are almost perfectly vertical.

The Burnsville Cove System of Virginia illustrates a variant of the tendency for strike orientation. The main passage drains approximately along the axis of a syncline, with many joint-controlled tributaries, overflow routes, and local networks superimposed on the main trend. Gravel and large boulders carried through the cave during peak flow suggest that part of the cave's complexity is due to enlargement by periodic floodwaters.

In the western part of the Ridge and Valley Province most of the caves are in Silurian-Devonian limestones only about 100–200 m thick. The thinness of these rocks helps to constrain many of the caves to strike-oriented trends. Along the eastern margin of the region, prolonged weathering of the thicker Cambrian-Ordovician carbonate rocks has formed a broad, rolling lowland that extends through western Virginia and southeastern Pennsylvania. Caves in this area are not as large as might be expected from the thickness of soluble rock. The largest accessible caves tend to occupy small residual knobs (*e.g.*, Luray Caverns, Virginia). Elongation of caves along the strike of the beds is less prominent than in the thinner rocks to the west, because the massive nature of the carbonates and the intense structural deformation allow groundwater to flow in a variety of somewhat unpredictable directions along complex fractures.

Caves in the folded Appalachians have been used to validate some of the most widely accepted views on cave origin. The idea that levels of cave development correlate with periods of rather static base level in nearby river valleys had been proposed by various geologists in the first half of the twentieth century. Many of these studies were not entirely convincing because some "levels" were caused simply by perching of water on insoluble beds, and correlations between caves in mountain ranges were clouded in places by irregular rates of tectonic uplift. The elevations of major cave passages in the Appalachians correlate well with river terraces, which are remnants of former floodplains produced when rivers pause in their downward entrenchment and widen their valleys. The caves are not perched on insoluble beds, because most of them are oriented along the strike of steeply dipping beds or cut across geologic structures. Most caves in the region could be attributed to dissolution at the top of the phreatic zone, because those in dipping beds concentrate in narrow horizontal bands that terminate abruptly in both the up-dip and down-dip directions. Although these observations do not hold true for every cave in the region, this general view is still valid for the majority.

Caves in the Potomac River basin cluster vertically in a few groups whose elevations decrease in the downstream direction at roughly the same rate as the present river gradients (White, 2009). This suggests that the caves, while forming, were adjusted to river patterns not much different from those of today. The various cave levels become more distinct in the downstream direction, apparently because base-level control is better defined in the mature lower parts of the river valley.

Several excellent examples of sandstone-capped mazes are located in the Ridge and Valley Province where a thin cap of Oriskany Sandstone overlies jointed limestones of the Devonian Helderberg Group, and where the dip is locally rather low. Some of these mazes have also been enlarged partly by periodic flooding from nearby rivers, so the zone of maximum

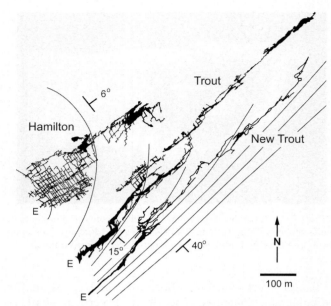

FIGURE 9 Composite map of caves at Trout Rock, West Virginia, showing their relation to the geologic structure. Hamilton Cave is a network maze along the crest of a local anticlinal flexure. Trout Cave and New Trout Cave are located along the anticlinal flank. Structural contours are drawn at intervals of 10 m. The local dip of the strata is shown in degrees. E = entrance. *Cave maps by David West. Geology by A. and M. Palmer.*

passage development is not necessarily at the very top of the limestone. In fact, flooding has produced local mazes superimposed on the branching pattern of many Appalachian caves, even where a sandstone cap is absent. Mazes are especially common along the crests of anticlines where fractures have been enlarged by tension (Fig. 9).

Glaciated Plateaus of Northeastern and North-Central United States

Extensive caps of glacial ice covered the north-central and northeastern states several times during the past 2 million years. Where glacial deposits are thick (as in northern Illinois, Indiana, and Ohio) preexisting karst surfaces were completely buried, and there is no substantial field evidence for caves in these areas. However, where the glacial deposits are thin or absent, preglacial caves have survived with only minor disturbance.

Howe Caverns and related caves in New York, located in gently dipping Silurian-Devonian carbonates, are examples of preglacial caves that have survived at least one phase of continental glaciation. Ice sheets were more than a kilometer thick over the region, and yet the caves remain more or less intact. Some of the overlying surface karst was partly or completely masked by glacial sediments, and patterns of groundwater flow were altered slightly by blockage of inputs and springs by these deposits.

During the waning phases of the latest glaciation, about 14,000 years ago, the local north-flowing river (Schoharie Creek) was blocked by ice to form a proglacial lake up to 150 m deep. Thinly laminated clays were deposited on the surface during this event, and they are also present in every major cave in the Schoharie Valley (Fig. 11) but absent in caves in surrounding drainage basins. In Howe Caverns and nearby caves the clays occupy the lowest passage levels, and postglacial entrenchment has extended at most only 10–20 cm below their base. Radiometric dating of speleothems in these caves includes ages of more than 350,000 years, showing that the caves had reached their present state of maturity well before the onset of the latest (Wisconsinan) glaciation about 70,000 years ago. Only a few minor cave passages can be demonstrated to have formed in the 14,000 years since the last glacial retreat. Some are diversions from upper levels, and others are floodwater passages formed under the most favorable conditions of aggressive water, short flow path, and steep hydraulic gradient. A few caves in the Schoharie Valley contain narrow passages up to 0.5 m^2 in cross-sectional area that were formed by water diverted around passages that are blocked by glacial deposits.

Sinking streams are most common in karst of moderate relief, where runoff from large areas of insoluble material feeds swallow holes in adjacent cavernous rock. Like most surface streams, they tend to exhibit severe flooding. Floods have their greatest effect on caves that are at least partly air-filled. Cave enlargement is rapid during floods because the incoming water tends to have a low concentration of dissolved solids. Solutional growth is aided by abrasion by stream-borne solid particles. Caves that have been

FIGURE 10 Strike-oriented passage in steeply dipping Devonian limestone, New Trout Cave, West Virginia.

FIGURE 11 Laminated glacial lake clays in Howe Caverns, New York, were deposited in proglacial Lake Schoharie roughly 14,000 years ago. The bedrock floor beneath the clay lies about 1 meter below the present level of the main cave stream (in foreground). Width of photo is approximately 50 cm.

influenced greatly by floodwater exhibit complex patterns, with diversion mazes, dead-end fissures, and solution pockets superimposed on the basic pattern of stream passages. Such caves occur in all the regions described above, but they are fairly sparse, accounting for an average of about 10% of all caves. The percentage is much higher in glaciated regions because of diversion of surface water into caves by low-permeability glacial deposits, which also block many springs.

An example is Mystery Cave, in southeastern Minnesota, located in a low-relief plateau of highly fractured Ordovician dolomites and limestones. It developed as a subterranean cutoff of several meanders in a surface river, across a reach of particularly steep gradient. The entire cave consists of a fissure network with 21 km of mapped passages. The local dip is 0.3–0.8° to the west, and the groundwater flow is to the east, against the dip, along fractures discordant to the bedding. The maze pattern is a result of periodically aggressive water following steep hydraulic gradients through a network of fairly wide initial fissures. The stresses of glacial loading and unloading may have aided the mechanical widening of joints, although much of the cave predates the latest glaciation. Uranium-thorium radiometric dates for calcite speleothems cluster at 100,000–150,000 and 10,000–15,000 years, which are respectively pre-Wisconsinan and post-Wisconsinan. Flowstone in a high-level passage dates to more than 350,000 years. The close relation of the cave to the present river pattern is intriguing, because the cave probably began to form at least half a million years ago.

Isotopic analysis of calcite speleothems in this midcontinental region has provided paleoclimatic evidence that is both more precise and geographically more specific than that of other methods such as coring of marine sediments. The fact that these speleothems can be dated fairly accurately to ages of at least 350,000 years makes them doubly useful for paleoclimatic studies. Variations in oxygen isotopes are temperature-related, and although there are difficulties in translating them directly into former temperatures, they are useful in detecting paleoclimatic trends. Variations in carbon isotopes help to determine the type of vegetation in the area at any time.

SUMMARY

The examples described here represent the typical range of caves found in karst regions of moderate relief. Even with such a brief overview it is possible to answer the questions posed earlier:

- *When does cave development begin?* This depends on the definition of terms as much as on geologic setting. Slow flow far below river levels can dissolve primitive routes in highly soluble materials like gypsum (as in Indiana), but there is evidence for well-developed deep-seated conduit flow through carbonates as well (as in Missouri). However, in most karst areas of moderate relief there is little evidence that deep flow has determined the patterns of presently traversable caves. The close association between caves and the present cycle of landscape development argues against a widespread influence by older deep-seated processes. This conclusion is supported by the correlation of vadose-phreatic transition points between cave passages and with surface features such as river terraces.

- *Which of the original fractures and pores in a soluble rock evolve into caves?* Those that are widest, have the steepest hydraulic gradients, and shortest paths are favored. Mystery Cave, Minnesota, illustrates rapid development in such a setting, even though the cave cuts diagonally across the strata in a direction opposite that of the stratal dip. The limited depth of phreatic loops in most caves, and the tendency for phreatic tubes to follow the strike of the strata, show that shallow paths are the most favorable, owing to a decrease in fissure width with depth. Concentrations of caves in zones of stress release in deep canyons also show the importance of initial fracture width, as well as the close association with river levels.

- *How far below the surface can caves form?* The flow systems that extend deepest below river levels appear to follow major faults (even in relatively undisturbed strata such as those of Indiana and Kentucky). Although the depth of cave development is limited in the moderate-relief regions described here, phreatic caves at least 100 m below the present base level are known from geochemical studies, water tracing, diving in water-filled caves, and measurement of high-amplitude phreatic loops in presently air-filled caves.

- *What is the effect of local geologic setting on cave development?* The nature of groundwater recharge controls the overall cave pattern. Most caves fed by typical karst recharge are branchworks (*e.g.*, Fig. 6). Thick overlying insoluble rocks restrict recharge to a few points, which makes the cave pattern deviate from the ideal branchwork pattern. Perching of vadose water along partings in well-bedded rocks, or on insoluble beds, produces dominantly down-dip passages (Fig. 2). Most phreatic passages are oriented roughly parallel to the strike, except where the host rocks are massive and prominently fractured. All areas discussed above exhibit these trends, especially the Ridge and Valley Province. Network mazes are formed either by uniform seepage through overlying or underlying permeable but insoluble rock or by severe discharge fluctuations during floods. Worldwide, many large network caves are formed by the rise of deep-seated water, with dissolution performed by sulfuric acid or by mixing of deep high-CO_2 water with shallow low-CO_2 water. Although some caves in the Appalachians have patterns and geologic settings that suggest such an origin, there is so far little supporting documentation.

- *What information do caves provide about past climates and geologic history?* Their close association with fluvial erosion relates them to the developmental history of the entire surrounding region. Dating of sediments and speleothems provides a reliable time scale, and isotopic analysis of speleothems provides information on former climates.

Although this section is limited to karst areas of moderate relief, and describes only caves formed by the carbonic acid reaction, the conclusions drawn from it are valid for the majority of karst regions throughout the world.

See Also the Following Article

Solution Caves in Regions of High Relief

Bibliography

Bretz, JH. (1956). *Caves of Missouri* (Vol. 39, Second Series). Rolla, MO: State of Missouri, Division of Geological Survey and Water Resources.

Davies, W. E. (1960). Origin of caves in folded limestone. *National Speleological Society Bulletin*, 22(1), 5–18.

Granger, D. E., Fabel, D., & Palmer, A. N. (2001). Pliocene-Pleistocene incision of the Green River, Kentucky, determined from radioactive decay of 26Al and 10Be in Mammoth Cave sediments. *Geological Society of America Bulletin*, 113(7), 825–836.

Jones, W. K. (1997). *Karst hydrology atlas of West Virginia.* Charles Town, WV: Karst Waters Institute.

Kastning, E. H., & Kastning, K. M. (Eds.) (1991). Appalachian karst. In *Proceedings of Appalachian Karst Symposium*, Radford, VA. Huntsville, AL: National Speleological Society.

Palmer, A. N. (2007). *Cave geology.* Dayton, OH: Cave Books.

Palmer, A. N., & Palmer, M. V. (Eds.), (2009). *Caves and karst of the USA.* Huntsville, AL: National Speleological Society.

Sasowsky, I. D., & White, W. B. (1994). The role of stress release fracturing in the development of cavernous porosity in carbonate aquifers. *Water Resources Research*, 30(12), 3523–3530.

Stone, R. W. (1953). Caves of Pennsylvania. *National Speleological Society Bulletin*, 15, 51–137.

White, W. B., & White, E. L. (Eds.), (1989). *Karst hydrology—Concepts from the Mammoth Cave Area.* New York: Van Nostrand Reinhold.

White, W. B. (2009). The evolution of Appalachian fluviokarst: Competition between stream erosion, cave development, surface denudation, and tectonic uplift. *Journal of Cave and Karst Studies*, 71(3), 159–167.

SPECIES INTERACTIONS

David C. Culver

American University

INTRODUCTION

Species can affect each other in a variety of ways. They can be predator and prey, they can be competitors, or they can be mutualists in which case each species has a beneficial impact on the other. When two species interact in an ecologically significant way, the interaction affects population size and population growth. If species A decreases the number of individuals of species B and species B decreases the number of individuals of species A, then A and B are competitors. If species A increases population size of species B but species B decreases population size of species A, then B is a predator of A. In this demographic sense, parasites also fall into the category of predators. All interactions can be classified according to the effect that an encounter of an individual of species A with species B has (Fig. 1). Two-way interactions are quite familiar—competition, mutualism, and predation. One-way interactions are less so—amensalism and commensalism. Ecologically, amensalism is a very one-sided case of competition, and commensalism is a very one-sided case of mutualism. The interactions can occur between two free-living species, or between individuals, one of which is permanently attached to or embedded in another—symbioses. Symbiotic (literally "living together") organisms can have mutualistic, commensal, or parasitic population interactions.

It is likely that many animals initially enter the subterranean realm in order to escape predators (including parasites) and competitors. Visually oriented predators cannot effectively find their prey in the darkness of a cave. If a species has a competitor that is more visually oriented rather than, for example, olfactory oriented, the visually oriented competitor will be at a disadvantage in subterranean habitats. Parasites may also find it difficult to make the transition to subterranean environments perhaps because it is more difficult to locate a host in the absence of light.

Caves have proved to be useful ecological laboratories for the study of interspecific interactions (Culver, 1982). The number of species in any one cave is quite small compared to most surface communities. This makes analysis much simpler because the number of pairwise interactions is much smaller in smaller

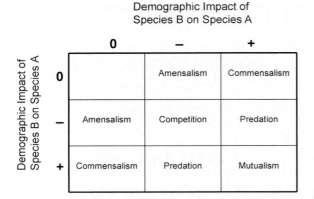

FIGURE 1 Classification of pairwise interspecific interactions. When both species have a negative effect on each other this is competition; when both species have a positive effect on each other this is mutualism; when one has a positive effect and the other a negative effect this is predation. Less familiar are the highly asymmetric interactions of commensalism and mutualism.

communities. In a community of 4 species there are 6 pairs of possible interactions, whereas in a community of 10 species there are 45 such pairs. (In general, for a community of S species there are $\{S(S-1)/2\}$ interaction pairs.) Many cave streams have only two or three macroscopic invertebrates compared to dozens in surface streams. This makes the job of determining major competitors, predators, and prey much simpler. Second, cave communities typically have a simple trophic structure with only two or three trophic levels. Food webs are simpler and easier to analyze. Third, because of the patchy nature of the habitat and the resulting patchy distribution of most species, nearby caves often have different species combinations present. This allows these "natural experiments" to be used to study the effects of species additions and removals. Because of the relatively large number of caves in cave-bearing areas (there are more than 45,000 known caves in the United States), the number of such natural experiments can be quite large.

In subterranean environments, predation and competition are relatively well known. The other two-way interaction—mutualism—has been little studied in subterranean environments. Hobbs (1975) studied ostracods living on the exoskeletons of cave crayfish. Ostracods gain an advantage from the interaction because they feed on microorganisms and detritus that accumulate on the host exoskeleton. Crayfish may gain some advantage from the ostracods directly from exoskeleton cleaning. Hobbs found that the interaction seemed to decrease in intensity with increasing cave adaptation. In Pless Cave, Indiana, U.S.A., stygobiotic *Orconectes inermis inermis* had an average of 13.5 ostracods per individual. *Cambarus tenebrosus*, a stygophile also common in surface streams, had an average of 31 ostracods in Pless Cave, more than twice as many. Strictly surface-dwelling species, such as *Cambarus bartoni*, have even more ostracod ectocommensals—sometimes more than 100.

Relative to free-living predators, there are few studies of parasitic interactions in subterranean environments. There are a few parasites that have specialized on subterranean species, in some ways the *ne plus ultra* of extreme specialization. One of the most spectacular examples are Temnocephala, parasitic flatworms which parasitize cave shrimp living in Balkan caves, probably feeding on the haemolymph of the shrimp that they access through thin parts of the exoskeleton (Matjašič, 1994). Seven species and several genera of Temnocephala are found only on the cave shrimp *Troglocaris schmidti*, with each species specializing in a particular region of the body. For example, *Subtelsonia perianalis* is only found around the anus of *T. schmidti*.

Although many interactions have been remarked on in the literature, and complete food webs have been constructed for a few communities, two sets of species have been studied extensively. One is the predator–prey system of carabid beetles and their cricket egg prey, which occurs extensively in caves throughout the Interior Low Plateaus of Kentucky, Tennessee, and Alabama and in the Balcones Escarpment of Texas. The other is the intricate systems of interactions between amphipods and isopods in cave streams in the Appalachian Mountains of Virginia and West Virginia.

BEETLE PREDATORS AND THEIR CRICKET EGG PREY

Because of the scarcity of food in caves, especially for predators, the great Romanian biospeleologist Emil Racovitza pointed out that many cave animals are "carnivores by predilection but saprovores by necessity." This has the effect of both shortening food webs and making predatory behavior difficult to observe because predatory events themselves are usually quite rare. An exception to both the rarity of observation and tendency of predators toward omnivory is the predator–prey pair of cave crickets (in particular, their eggs) and beetles found in many caves in North America.

The best studied pair is that of the beetle *Neaphaenops tellkampfi* and the cave cricket *Hadenoecus subterraneus* in Mammoth Cave, Kentucky, and other nearby caves (Kane and Poulson, 1976). Very similar predator–prey interactions occur with other beetles (*e.g.*, *Rhadine subterranea*) and cave crickets in Texas (*e.g.*, *Ceuthophilus cunicularis*) and with other beetles and crickets in other parts of Kentucky.

FIGURE 2 Concentration of cave crickets, *Ceuthophilus stygius*, on the ceiling of Dogwood Cave, Hart County, Kentucky, U.S.A. *Photo by H. H. Hobbs. Used with permission.*

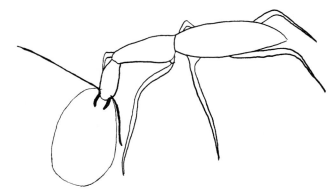

FIGURE 3 Sketch of *Rhadine subterranea* eating a cricket egg. *Adapted from a photograph by Robert W. Mitchell.*

The key to this predator–prey interaction begins with the cave crickets (Lavoie et al., 2007). Species such as *Hadenoecus subterraneus* are often quite common in caves, typically reaching numbers into the hundreds and even thousands (Fig. 2). Morphologically, they show signs of both subterranean and surface life—they have eyes (sometimes reduced) but appendages are elongated relative to surface species. Many individuals, especially those near the entrance, leave the cave at night to forage for food during the warmer times of year. Some species may leave every night during the summer months but *H. subterraneus* probably leaves the cave to feed at 10- to 20-day intervals. During the day in the summer, and during day and night in winter months, they stay in caves, where they typically lay their eggs. Presumably the advantage they gain from cave life is avoidance of most predators and some relief from high daytime temperatures in the summer. Crickets and other relatively large-bodied orthopterans have many enemies—many birds during the day and small rodents at night.

Whatever the frequency of outside feeding, it represents an important food source entering the cave. Many species feed on the splattering of cricket guano present in the caves and a few beetles have specialized on the eggs crickets bury in the cave. In caves with sand or uncompacted silt, crickets bury their eggs in this soft substrate. Cricket eggs, rich in protein and lipids, and nearly one-third the size of an adult *N. tellkampfi* (Fig. 3), are a bonanza to the beetles if they can find them. Successful cricket egg predation behavior has evolved several times among cave beetles. Most of these beetle species are 6.5 to 8 mm long and actively forage for eggs, locating the holes where the eggs were laid by a combination of chemoreception and mechanoreception. Beetles dig a hole to find an egg, and when they are successful, they remove the egg, pierce it with their mandibles, and the egg contents are pumped into the gut. In caves without soft substrates, beetles are unable to locate and dig up the eggs. The size of a cave cricket egg relative to that of *N. tellkampfi* (other cricket egg predators are of similar size) indicates the potential energetic importance of cave cricket eggs (Fig. 3). The dry weight of an *H. subterraneus* egg (2.26 g ± 0.03) is nearly ¾ that of the dry weight of an adult *N. tellkampfi* (3.02 g ± 0.07). A single egg is sufficient food for a beetle for weeks and it takes approximately 50 days for the beetle to return to its prefeeding weight.

The intensity of egg predation is high. In sandy areas in Mammoth Cave beetles consume over 90% of the cricket eggs laid throughout the year, and more than half are found and eaten by beetles within 15 days of being deposited. Beetles dig in the substrate to find eggs, and when they are successful, they remove the egg, pierce it with their mandible and the egg contents are pumped into the gut. The impact of this resource on beetle population dynamics is striking, and reproduction in beetles closely follows the time of maximum rate of egg depositions by crickets (Fig. 4). The quantitative impact of egg predation on *H. subterraneus* populations is unknown but it must be considerable.

An evolved response on the part of *H. subterraneus* has been to increase the ovipositor length in populations experiencing egg predation by cave beetles. The increase in ovipositor length from 12 to 13 mm is sufficient to reduce egg predation. Beetles often do not dig deep enough. Most if not all populations of cricket-egg-eating beetles must switch to other prey because cricket eggs are not available year-round. In periods of cricket egg scarcity, *N. tellkampfi* and other species become "ordinary" predators, feeding on a variety of prey. They also must switch habitats since the sandy

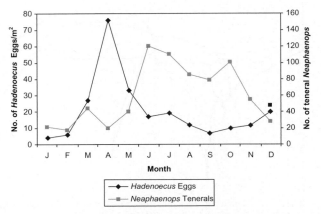

FIGURE 4 Seasonal changes in number of eggs per square meter of the cave cricket *Hadenoecus subterraneus* and a visual census of newly emerging adults (tenerals) of the beetle *Neaphaenops tellkampfi* in Edwards Avenue, Great Onyx Cave, Kentucky. *Data from Kane and Poulson, 1976.*

substrates and uncompacted silt where cricket eggs are laid have few other species. Interestingly, they are apparently not very efficient general predators, much less so than related species in the genus *Pseudanophthalmus* as well as conspecific populations that do not prey on cricket eggs.

COMPETITION AND OTHER INTERACTIONS IN APPALACHIAN CAVE STREAMS

A thoroughly studied case of interaction within a trophic level is that of the amphipod and isopod species that occupy many gravel-bottomed streams in Appalachian caves (Culver and Pipan, 2009). In these streams, there is an alternation between deeps (pools) and shallows (riffles). The amphipods and isopods are highly concentrated in riffles as a result of the concentration of food (especially leaf detritus), increased oxygen, and the absence of salamander predators, which live in pools. In this habitat, the three obvious kinds of interactions are as follows: (1) Species may compete for food, (2) species may compete for space (the underside of gravels), and (3) species may serve as food for other species. All can and do occur in particular situations, but the most universal (and easiest to analyze) is competition for space on the underside of riffles (Culver, 1994).

There is a general risk associated with life on the underside of rocks—the risk of washing out into the current. Individuals that wash out into pools run the risk of being eaten by salamanders, the primary predators living in pools. Individuals also run the risk of damage from buffeting by the current. As with many subterranean species, the amphipods and isopods

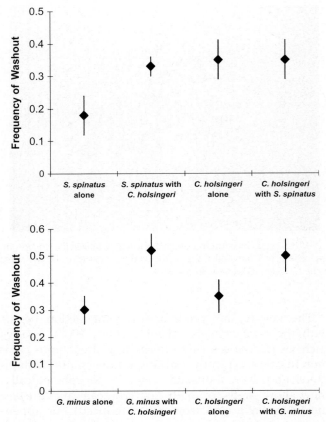

FIGURE 5 Mean frequency, together with standard errors, of washout rates from riffles of *Gammarus minus*, *Caecidotea holsinger*, and *Stygobromus spinatus* in various combinations in laboratory stream experiments.

have long, thin appendages that are easily broken. Amphipods and isopods with broken appendages are not easy prey for salamanders but are also attacked and eaten by other amphipods and isopods. Amphipods may be at more risk than isopods in current. The cave stream amphipod species are not good swimmers and their laterally compressed body shape, compared to the dorso-ventrally flattened body shape of isopods, is not hydrodynamically efficient in moving water. It is very easy to observe the behavioral response to most encounters even in a small dish in the lab—one or both individuals rapidly move away. More realistic laboratory experiments can be done in a small artificial riffle, where the washout rate of individuals put in the riffle in various combinations can be measured.

An example is shown in Figure 5, involving two pairs of species from Organ Cave. When the amphipod *Stygobromus spinatus* is in an artificial riffle with the isopod *Caecidotea holsingeri*, *S. spinatus* has nearly double the washout rate that it has when it is alone, indicating that *C. holsingeri* is its competitor (Fig. 5A). However, the converse is not true. *S. spinatus* seems to have no effect on *C. holsingeri*. Thus the overall interaction is one of amensalism (Fig. 1), a very one-sided competition in this case. This asymmetry was not the result of size differences because these two species are roughly the same size, about 5 mm. The isopod is the superior competitor perhaps of its more hydrodynamic body shape.

Another pair of species, the amphipod *Gammarus minus* and the isopod *C. holsingeri*, displayed competitive behavior, and both species had a higher washout rate in the presence of the other (Fig. 5B). This would seem to be a simple interaction, but it proved to be more complex than first thought. The complication was that some *C. holsingeri* "disappeared" during the day-long washout experiments and where they disappeared to was the gut of *G. minus*! The interaction between these two species had elements of predation (hence the "disappearing" isopods) and of competition (hence the mutual avoidance by both species when in a riffle).

The next step was to construct a model of competition relating relative washout rate to the interaction coefficients of the standard competition equations:

$$dN_1/dt = r_1 N_1 (K_1 - N_1 - \alpha_{12} N_2)$$

$$dN_2/dt = r_2 N_2 (K_2 - N_2 - \alpha_{21} N_1)$$

where r is the intrinsic rate of increase of species 1 and 2, N is population size, K is carrying capacity, and α_{ij} represents the effect of species j on species i. For competition for space in a riffle α_{ij} is approximately equal to the ratio of the washout rate of species i when species j is present to the washout rate of species i when species j is not present. The resulting estimates of competition can then be compared with field data on the amount of overlap among species in caves and within riffles within a cave.

Three different cave stream communities show the range of interactions. The first involves a trio of species found in Thompson Cedar Cave and other nearby caves in Lee County, Virginia: the amphipod *Crangonyx antennatus* and the isopods *Caecidotea recurvata* and *Lirceus usdagalun*. The intensities of competition measured in the laboratory were such that, if they accurately reflect the situation in the field, only one of the three pairs (*Crangonyx antennatus* and *Caecidotea recurvata*) should be able to persist in the same stream, as well as the triad of species itself. In a survey of seven cave streams, no "unstable" species pairs were found, suggesting that laboratory measurements reflected the situation in the field. This was further supported by the distribution within Thompson Cedar Cave (Fig. 6). In various parts of this small cave stream, all three species occurred together, *Lirceus usdagalun* occurred by itself, and *Crangonyx antennatus* and *Caecidotea recurvata* occurred together, but in

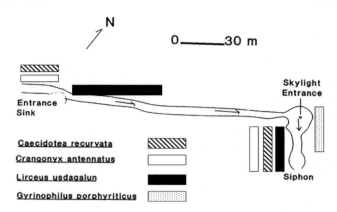

FIGURE 6 Map of distribution of *Crangonyx antennatus*, *Caecidotea recurvata*, *Gyrinophilus porphyriticus*, and *Lirceus usdagalun* in Thompson Cedar Cave, Lee County, Virginia.

no place did the pair *Lirceus usdagalun* and *Caecidotea recurvata* or the pair *Lirceus usdagalun* and *Crangonyx antennatus* occur. The final wrinkle is the occurrence of predaceous larvae of the spring salamander *Gyrinophilus porphyriticus*. Its preferred prey, *Caecidotea recurvata*, is reduced in abundance, and very few amphipods and isopods occur in pools. When *Gyrinophilus porphyriticus* finds isolated rimstone pools, it decimates any amphipod or isopod population that might be there.

The third example of species interactions that has been intensively studied is the isopod community in Alpena Cave, West Virginia, U.S.A., and one that demonstrates that competition is not universal in cave stream communities. Two species occur in the same stream—the isopods *Caecidotea cannulus* and *Caecidotea holsingeri*. Superficially they would seem to be competitors. Neither laboratory stream studies nor field perturbation experiments detected any evidence of competition between these two species. They both occur in the same riffles throughout the cave stream. It is possible that competition between the two species did exist in the past. Typically, *Caecidotea cannulus* is larger than *Caecidotea holsingeri* and this difference is enhanced in Alpena Cave when the two species occur together, thus reducing competition. However, the size of the isopods is strongly correlated with the size of the rocks in the streams and it turns out that Alpena Cave has a bimodal distribution of gravel sizes. So, it can only be said that at present competition is not occurring.

COMPETITION AS A RESULT OF EUTROPHICATION

Species interactions also play a key role in the replacement of a specialized cave fauna in a cave stream subjected to eutrophication (Sket, 1977). In the late 1950s, the Pivka River in Slovenia carried a heavy load of organic pollutants into Postojnska Jama. Where the Pivka River entered the cave, oxygen concentrations (a measure of eutrophication) in the river were 10% of saturation. One kilometer into the cave, oxygen levels recovered to nearly 50% of saturation. Six kilometers into the cave oxygen concentration was nearly at saturation and other measures of eutrophication indicated it had largely disappeared. The very rich obligate cave-stream fauna, including the very interesting and intensively studied isopod *Asellus aquaticus cavernicolus*, was largely extirpated in the first several kilometers of stream passage. The fauna was replaced by surface-dwelling aquatic insects, surface-dwelling amphipods in the genus *Gammarus*, and the surface-dwelling isopod species *Asellus aquaticus aquaticus*. These species were able to invade because of the high energy and nutrient levels in the eutrophic Pivka River. As a result of their invasion, the stygobionts (such as *Asellus aquaticus cavernicolus*) were pushed farther into the cave, not because they couldn't survive under the higher food conditions, but because they are out-competed. In extreme cases, such as probably occurred near the entrance to Postojnska Jama, low oxygen levels can prevent the survival of stygobionts, but most areas did not have extremely low oxygen values. Especially interesting are those areas of the underground Pivka River that came to be dominated by aquatic insects and *Gammarus*, both of which are not species of highly polluted waters, but rather stygophiles, species that can live in both surface and subterranean habitats. Stygophiles were probably better competitors of surface-dwelling species than were stygobionts in polluted cave streams. Of course, stygophiles can survive and reproduce in nonpolluted surface streams.

It also seems likely that competition with surface-dwelling species and predation by surface-dwelling species is a major factor in preventing the movement of subterranean species onto the surface. Sket showed that in the absence of competitors and predators, stygobionts may forage in surface environments. Of course, other factors prevent movement of subterranean species to the surface. Many subterranean species are sensitive to light and also unable to cope with environmental fluctuations.

CONCLUSION

Because of the highly replicated nature of caves, they can be important ecological laboratories for the study of species interactions. The above studies indicate some of the potential for such work. Many more cave communities await such an analysis.

See Also the Following Article

Natural Selection

Bibliography

Culver, D. C. (1982). *Cave life: Evolution and ecology*. Cambridge, MA: Harvard University Press.
Culver, D. C. (1994). Species interactions. In G. Gibert, D. L. Danielopol & J. A. Stanford (Eds.), *Groundwater ecology* (pp. 271–281). San Diego: Academic Press.
Culver, D. C., & Pipan, T. (2009). *The biology of caves and other subterranean habitats*. Oxford: Oxford University Press.
Hobbs, H. H., III (1975). Distribution of Indiana cavernicolous crayfishes and eco-commensal ostracods. *International Journal of Speleology, 7*, 273–302.
Kane, T. C., & Poulson, T. L. (1976). Foraging by cave beetles: Spatial and temporal heterogeneity of prey. *Ecology, 57*, 793–800.
Lavoie, K. H., Helf, K. L., & Poulson, T. L. (2007). The biology and ecology of North American cave crickets. *Journal of Cave and Karst Studies, 69*, 114–134.
Matjašič, J. (1994). Turbellaria, Temnocephala. In C. Juberthie & V. Decu (Eds.), *Encyclopaedia biospeologica (Tome I)* (pp. 45–48). Moulis, France: Société Internationale de Biospéologie.
Sket, B. (1977). Gegenseitige Beeinflussung der Wasserpollution und des Höhlenmilieus. *Proceedings of the 6th International Congress of Speleology, Olomouc, ČSSR, 4*, 253–262.

SPELEOGENESIS, HYPOGENIC

Alexander Klimchouk

Ukrainian Institute of Speleology and Karstology

INTRODUCTION

Historically, scientific notions of karst evolved from regions where it is most noticeable, where soluble rocks are exposed to the surface, and specific geomorphic and hydrologic features such as dolines, sinking streams, and caves are well represented. Karst and cave development in such regions is driven by recharge from the surface, either diffused authogenic or concentrated allogeneic. Dissolution of carbonate rocks occurs mainly due to carbonic acid generated by reactions with soil CO_2.

Karst developed in these settings is termed *epigenic*. The traditional paradigm of karst science originally evolved from studying epigenic karsts. The adoption of the very scientific term from the Germanized name of the geographic area in Slovenia further strengthened the epigenic bias in karst researches, dominating till now. Traditional concepts and models for the origin of caves have been mainly developed based on the geomorphic and hydrogeologic context of epigenic karst.

Epigenic speleogenesis, directly related to the infiltration of meteoric water, is predominantly associated with local flow systems and/or recharge regimes of intermediate to regional meteoric flow systems. Epigenic caves form along paths (connected joints and/or pores) where the greatest flow occurs between recharge points at the surface and discharge points in topographic lows. The recharge regime is characterized by highly variable input parameters, relatively high hydraulic heads a decreasing with depth, downward and divergent flow, chemically aggressive groundwaters with low dissolved solids content, oxidizing conditions, and negative anomalies of geothermal heat and gradient (Fig. 1). The types of recharge, the relationships of recharge and discharge points, and the initial percolation patterns are controlled by the geomorphic settings and distribution of primary porosity and fractures. The amount of recharge is controlled by climate. The positive feedback loop between discharge and the rate of conduit growth is fully operational in epigenic speleogenesis, which leads to the competitive development of conduits, progressive concentration of flow, and domination of branching cave patterns. Epigenic conduit systems favor lateral flow and respective porosity diminishes with depth.

Abundant instances of karst cavities encountered by boreholes and mines at greater depths have been commonly interpreted as paleo(epigenetic) karst, simply because of the overwhelming dominance of the epigenic concepts of karst and speleogenesis. However, there are many examples of deep-seated cavernous porosity in strata, in which epigenic karst development can definitely be ruled out, for example, in rocks that were buried continuously after deposition and have never been exposed. The advancements in karst and cave science during past 20–30 years led to the growing recognition of the possibility and wide occurrence of cave development in deep-seated conditions, without direct recharge from the surface, by recharge to the cave-forming zone coming from depth. This type of speleogenesis is termed *hypogenic* (or *hypogene*).

The idea that some caves could form at depth by rising waters was introduced as early as the nineteenth century with regard to hydrothermal systems. More significant attention has been given to hydrothermal speleogenesis since the mid-twentieth century, particularly in Eastern Europe. Later on, another chemical mechanism of cave development by rising waters was recognized, the oxidation of H_2S and dissolution by sulfuric acid. By the end of the twentieth century the term and concept of *hypogene speleogenesis* had been firmly established, although remained largely limited to caves formed by the above dissolution mechanisms (Ford and Williams, 1989; Palmer, 1991; Dublyansky, 2000).

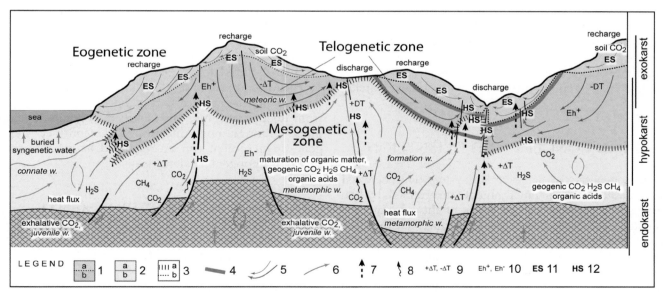

FIGURE 1 Karst and speleogenesis in the context of diagenetic zones and groundwater flow regimes. The diagram is out of scale and the vertical dimension is greatly exaggerated. 1 = meteoric, topography-driven regime: a = local systems (unconfined), b = regional and subregional systems (confined); 2 = expulsion (exfiltration, basinal) regime, commonly overpressured, driven by compaction and tectonic compression: a = in newly deposited sediments, b = in older rocks; 3 = interfaces between groundwater regimes and systems: a = meteoric/expulsion regimes, b = local/regional-subregional meteoric systems; 4 = poorly permeable beds (only a few are shown on the diagram); 5 = meteoric flow paths; 6 = basinal flow paths; 7 = enhanced cross-formational communication; 8 = intense gas inputs; 9 = temperature and gradient anomaly: positive, negative; 10 = redox conditions: oxidizing, reducing; 11 = epigenic speleogenesis; 12 = hypogenic speleogenesis.

The recent past two decades have witnessed a rapid growth in the number of in-depth studies of different kinds of hypogene speleogenesis in various places around the world. It was recognized that cave development by various processes in deep-seated environments is much more common than previously thought (Palmer, 1995; Klimchouk, 2000, 2007; Klimchouk and Ford, 2009; Stafford et al., 2009).

At the same time, sedimentologists and industry geologists concerned with the origin of porosity began to realize limitations of the heavily used model of subaerial meteoric diagenesis in freshwater, which implied that deep-seated porosity in carbonates is related to unconformities and is mainly due to dissolution in paleovadose and paleophreatic freshwater zones (i.e., is paleokarst). They came to realize that deep-burial diagenesis in the mesogenetic environment can contribute significantly to secondary dissolution porosity and permeability evolution in many carbonate hydrocarbon reservoirs and ore deposits (e.g., Mazzullo and Harris, 1992; Machel, 1999).

The ongoing burst of interest in hypogene speleogenesis has been largely related to the establishment of a hydrogeological rather than geochemical approach to its definition. This approach highlighted the common hydrogeological genetic background and explained the multifaceted similarity of caves formed by upwelling flow, previously seen as unrelated because of their attribution to different chemical processes involved. The recent reviews, special symposia, and paper collections (Klimchouk, 2007; Klimchouk and Ford, 2009; Stafford et al., 2009) have demonstrated that hypogene speleogenesis is not just an aberrant curious phenomenon within the otherwise predominantly epigenic karst paradigm but is one of the fundamental categories of karst, at least of equal importance with epigene karst.

BASIC CONCEPT AND DEFINITION

Speleogenesis is a result of interaction between groundwater and its environment, determined by the various components and attributes of the two systems seeking equilibration. Geochemical mechanisms that cause the dissolution effect on the host rocks are many and varied. To cause speleogenetic development, the dissolution effect of disequilibria has to accumulate over sufficiently long periods of time and/or to concentrate within relatively small rock volumes or areas. The systematic transport and distribution mechanism that creates and supports the required disequilibrium conditions and hence controls the distribution of the dissolution effect is the groundwater flow, which suggests that the latter has to be a primary consideration.

Sedimentology considers (among other alteration processes) any dissolution in the rock during diagenesis, most of which occurs in a dispersed mode in the pore media. Speleogenesis deals with the creation and evolution of *organized permeability structures on the macro-scale* ($>ca.$ 5–15 mm) due to dissolutional enlargement of an earlier porosity.

The concept of hypogene speleogenesis does not necessarily mean cave development at great depth but, rather, refers to the origin of the cave-forming agency from depth. Hypogene speleogenesis is defined as *the formation of solution-enlarged permeability structures by water that recharges the cavernous zone from below, independent of recharge from the overlying or immediately adjacent surface* (Ford, 2006; Klimchouk, 2007). Hypogene caves form by a variety of geochemical mechanisms including those that do not rely on acids (such as dissolution of evaporites). At the same time, hypogene caves formed in different lithologies and by different dissolution processes demonstrate remarkable similarity in their patterns, morphologies, hydrostratigraphic occurrence, and current or inferred hydrogeologic functioning, which suggests their common hydrogeologic backgrounds.

Hypogene speleogenesis is associated with upwelling flow characteristic of discharge regimes of deeper (subregional and regional) meteoric flow systems, either terminal or intervening, or those of basinal flow systems (Fig. 1). It also tends to be related with interfaces between different flow systems. An upwelling flow inherently implies a certain degree of hydrogeological confinement, which is almost always characteristic of deep flow systems because of the almost ubiquitous layered heterogeneity in sedimentary sequences. When hypogenic solution porosity structures are shifted to the shallower, unconfined situation due to uplift and denudation but their further development continues to depend on upwelling flow from deeper systems, this is still hypogenic development although now unconfined. Unconfined hypogene development can be regarded as an extinction phase of hypogene speleogenesis.

DIAGENETIC ENVIRONMENTS, HYDROGEOLOGICAL CYCLES, AND HYDRODYNAMIC ZONES OF THE EARTH'S CRUST

Studies of hypogene speleogenesis have emphasized the need to approach the development of karst porosity and permeability from a perspective that goes far beyond contemporary geomorphological epochs. To set a framework for characterization and understanding of hypogene speleogenesis, one has to view karst from the perspective of the whole geologic evolution of soluble deposits, particularly of the evolution of diagenetic and hydrogeologic environments.

Diagenetic Environments

The realm of diagenesis comprises all alterations of sedimentary deposits that occur since ultimate deposition till the onset of metamorphism. Being the result of dissolution, speleogenesis can be viewed as a specific kind of diagenetic changes. However, in discussing aspects of diagenesis in this account we are concerned mainly with localized effects of dissolution processes rather than with dissolution that may occur diffusely throughout the micropore media of sediments.

Various classifications have been proposed to distinguish diagenetic environments and respective stages and zones. Interested readers should consult the abundant existing literature. The scheme of Choquette and Pray (1970) is one of the most widely used, which divides diagenetic history into three stages:

1. *Eogenesis*, when the shallow marine sediment is first laid down and may be altered by the effects of periodic drying or exposure to rainwater. The eogenetic zone grades downward into the mesogenetic zone where the newly deposited sediment is buried to depth where near-surface processes become ineffective.
2. *Mesogenesis* denotes alteration of sediments in the conditions of increasing burial depth (and hence increasing temperature and pressures) by the processes unrelated to the surface. Probably the most of hypogene speleogenesis occurs in the mesogenetic domain.
3. *Telogenesis* replaces mesogenesis when buried deposits, being shifted to a shallower position due to uplift and denudation, are affected again by processes related to the exposed surface, particularly to invading meteoric groundwater. This notion implies that the rock has been subjected to burial diagenesis to some degree. The lower limit of the telogenetic realm is not clearly defined. Some researchers draw this limit at or near water table (thus, telogenesis would roughly coincide with the realm of epigenic karst in this connotation) whereas others put it to the lower boundary of penetration of meteoric groundwaters. With the latter meaning, hypogene speleogenesis in confined deep but still meteoric flow systems falls in the domain of telogenesis.

Machel (1999) classifies the diagenetic realm into near-surface, shallow-, intermediate-, and deep-burial diagenetic settings, hydrocarbon-contaminated plumes, and fractures. Near-surface settings and the upper part of the shallow burial settings in this scheme roughly correspond to the eogenetic zone of Choquette and Pray (1970) whereas intermediate and deep burial settings, with a boundary between them at depths varying from 2000 to 3000 m (the top of liquid oil window), represent the mesogenetic realm. Deep-burial diagenetic settings merge down into the metamorphic realm at temperatures around 200°C (at depth of about 6–9 km). Hydrocarbon-contaminated plumes and fractures, which are particularly important for hypogene speleogenesis, may crosscut other settings.

Most sedimentary rocks have been buried to various depths (typically several hundreds or thousands of meters) for geologically long periods. Some of them experienced subsequent uplift, exposure, and new burial (with possible repetitions of the cycle) while others have never been exposed. The occurrences of large cavities at great depths in the rocks that have never experienced exposure are common and constitute a strong evidence for hypogene speleogenesis in deep-seated settings.

Aqueous fluids in the mesogenetic environment can be of different origins and they mix in varying proportions to constitute formation waters. *Connate waters* dominate in pores in shallow and intermediate settings of subsiding basins. With increasing burial depth they are expelled from clayey sediments into coarsely grained reservoirs. During the continental exposure, *meteoric waters* can penetrate to depths of several km in geologically and topographically favorable conditions. In deep burial settings, the increasing temperature causes dehydration of minerals containing combined water and the consequent release of substantial volumes of *dehydration waters* that partially replace connate waters in the sediments. Water is also generated during maturation of organic matter and subsequent hydrocarbon degradation, including thermochemical sulfate reduction. Prograde metamorphism and devolatilization reactions at still greater depth can liberate water as well (*metamorphic waters*), and *volcanogenic (juvenile) waters* can rise into the sedimentary cover from the deeper parts of the Earth's crust.

Groundwater Systems and Regimes

In subsiding basins the dominant flow drive in progressively buried strata is compaction due to the increasing load, which requires an expulsion of the pore waters from the sediments. This is the *expulsion regime*, also termed *elision* or *exfiltration* in the East European literature. Flow is directed upward, and on the regional scale from areas of greatest subsidence to the margins of basins. Varying susceptibility of sediments to compaction and pressure solution, as well as to chemical water-rock interactions, lead to increasing inhomogeneity in porosity and permeability. In more homogeneous strata and strata comprising evaporites, large-scale *density-driven* (convection) flow systems may develop. With still deeper burial and further rise of temperature and lithostatic load a *thermobaric regime* develops in which the fluid pressures are caused by the thermal expansion of water or by the release of water by mineral dehydration in a low-permeability environment. Yet another flow regime can be generated by tectonic strain in the vicinity of collision and uplift areas (a *compression regime*). Because of the ubiquity of some pressure-generating mechanisms in the progressive burial environment and the commonness of others, the elision regime is commonly characterized by the abnormal (higher than hydrostatic) pressures.

Following uplift and continental exposure, the *meteoric circulation regime* driven by topography evolves, also termed *infiltration regime* in the East European literature. Meteoric waters increasingly flush out the formation waters from a basin, and the meteoric regime substitutes the expulsion in the upper part of the crust, although expulsion circulation may still predominate in deep environments (Fig. 1). The pattern of the meteoric circulation is controlled by the basin geometry and relief, and by geological inhomogeneities that determine permeability distribution.

In the meteoric regime, hierarchical flow systems of increasing scales and depths are distinguished: local, intermediate (subregional), and regional, the last two being almost universally confined due to the layered heterogeneity of strata. Despite the confinement, basin-scale hydraulic continuity is a ubiquitous feature of the meteoric regime (Tóth, 1995). Cross-formational hydraulic communication between the systems of different scales occurs through both the pore media in low-permeable strata and (in more localized and intense mode) stratigraphic windows and transverse disruptions such as major fractures and conductive faults. This communication in the meteoric regime is directed downward beneath highlands and upward below topographic depressions. The upward communication is generally much more intense as topographic depressions often coincide with fault zones and other conductive geologic weaknesses.

The meteoric water circulation (regional systems) can penetrate to depths of several km affecting even deeply buried strata in geologically and topographically favorable conditions, especially in tilted and faulted basins adjacent to mountainous areas and in intermountain basins where the deep plunge of meteoric waters is favored by descending limbs of large convection circulation systems. In several regions fresh, meteoric groundwater flow was found by ocean drilling far offshore. However, the water flux through deeper systems is often significantly less than that through shallower systems.

The meteoric regime is perched on ubiquitously ascending waters of the expulsion regime, commonly overpressured (Fig. 1). Zones of interaction between the two regimes, either crosscutting or lateral, are particularly favorable for hypogene speleogenesis. Hypogene speleogenesis is commonly a part of mixed flow systems, where topography-driven flow interacts with the deeper compaction- or density-driven regimes. The nature and the geometry of the transition between the two regimes (and hence of the transition from mesogenesis to telogenesis, in the extended connotation of the latter) are controlled by respective fluid potentials and geological heterogeneities, especially sedimentary windows and conductive faults. The boundary may be blurred but it is more distinct when it coincides with some low-permeability strata of a regional extent. In the course of the geological evolution, the deeper strata migrate upward through this boundary with the onsets of uplift and denudation, and the nature and the geometry of this transition adjusts to the structure of rising strata and changing potentials of the two regimes. The meteoric systems at shallower depths are often pierced by upwelling cross-formational plumes of basinal waters.

In many generalizing schemes the hydrogeodynamic circulation on the continents (especially in cratons) is described as displaying two distinct stories. The upper story has active meteoric water circulation signified by the presence of considerable amounts of dissolved oxygen and nitrogen of meteoric origin. It corresponds to local and subregional flow systems of Tóth (1995). The notion of telogenesis should probably be restricted to this story. The lower story, which corresponds to regional flow systems, is dominated by formation waters lacking such gases.

Hydrogeologic Cycle

The hydrogeologic cycle starts with subsidence and marine sedimentation, continues through uplift, denudation, and exposure, and closes with the new marine transgression. During the second cycle, the new marine waters may enter any older lithified rocks in the mature basin, partly replacing meteoric or formation waters in them inherited from the previous continental regime (Fig. 2). In deeper parts of a basin the meteoric regime can be "trapped" by the new sediment pile, and the expulsion regime from the first burial stage may still be continuing. There can be significant time lags in the adjustment of the flow systems to the changing conditions and mingling

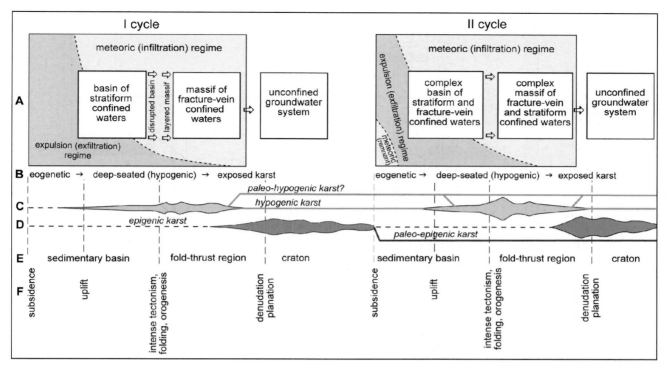

FIGURE 2 Evolution of geohydrodynamic systems and hydrogeological structures in the upper part of the hydrosphere (A) and karst development in the context of this evolution (B–D). B = direction of changes of evolutionary types of karst in newly deposited rocks in the course of diagenetic and geomorphic evolution; C–D = relative intensity of karst development in the course of evolution (the superposition of bars also demonstrates a potential for inheritance of preformed karst features by evolving karst systems): C = hypogenic karst; D = epigenic karst; E = tectonic settings; F = dominant geodynamic/geomorphic regimes.

groundwaters of differing ages and origins. The complexity is further added by increasing deformation, disruption, and subdivision of originally unified immature groundwater basins through their history. This leads to the directed evolution of large-scale hydrogeologic structures (Fig. 2) from artesian basins with prevailing porous stratiform groundwater bodies (layered aquifers) through disrupted basins and layered massifs to block-fault massifs with a prevalence of fracture and vein (conduit) types of porosities, which are often crosscutting.

Evolution of Karst Settings

A useful framework to characterize the changes in karst and speleogenetic environments in the upper part of the Earth's crust is provided by the evolutionary classification of karst types (Klimchouk and Ford, 2000; Fig. 3). It distinguishes major stages in the geologic and geomorphic history of a soluble rock as the types of karst, marked by distinct combinations of the diagenetic and structural prerequisites for groundwater flow and speleogenesis and different potential for inheritance from earlier conditions. These types are (in the order they potentially evolve) *syngenetic/eogenetic karst* in freshly deposited rocks; *deep-seated karst*, which develops during mesogenesis, particularly during its ascending limb (when the rocks are being shifted toward the surface); *subjacent karst*, where the cover is locally breached by erosion; *entrenched karst*, in which valleys incise below the bottom of the karst aquifer and drain it, but where the soluble rocks are still covered by insoluble formations for the most part; and *denuded karst*, where the insoluble cover materials have been completely removed. If karst bypasses burial, or if the soluble rock is exposed after burial without having experienced any significant karstification during burial, it represents the *open karst* type. Later on, karst may become *mantled* by a cover that develops contemporaneously with the karst, or *reburied* under younger rocks to form paleokarst, and be reexposed (*exhumed karst*).

Although this classification does not directly specify the origin of caves, it characterizes dominant speleogenetic environments and their evolutionary changes. The types of karst correlate well with the three major types of speleogenetic settings distinguished now (Klimchouk *et al.*, 2000). Coastal and oceanic speleogenesis in diagenetically immature rocks falls into the syngenetic/eogenetic karst domain.

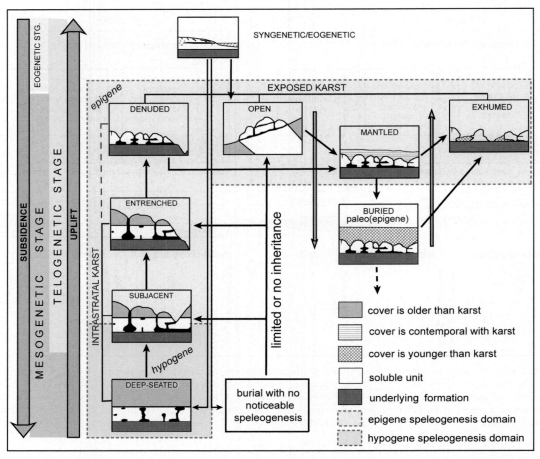

FIGURE 3 Evolutionary types of karst and speleogenetic environments (*modified from Klimchouk and Ford, 2000*). Background colors indicate the domains of hypogenic and epigenic speleogenesis.

Deep-seated karst is represented exclusively by hypogenic speleogenesis. In subjacent karst both hypogenic and epigenic speleogenesis may operate, depending on the dominant groundwater regime and interaction of flow systems, but hypogene speleogenesis still dominates. Exposed karst types (entrenched and denuded) are overwhelmingly epigenic, with an inheritance of hypogenic features that can be reworked by epigenic processes or get fossilized. In both karst types, however, ceasing hypogene systems may still operate. Open karst is marked by exclusively epigenic speleogenesis.

Very Deep Perspective: The Endokarst Realm

The realm of mesogenesis grades downward into the metamorphic realm. Water and CO_2 are generated there through prograde metamorphism and devolatilization reactions, and deeper juvenile fluids rise from still greater depths. Little is known about fluid flow and dissolution pattern at great depth.

There is mounting evidence for extensive fluid circulation in very deep boreholes in many regions, for example, large flows of hot brine detected at depths to greater than 9 km in the deep borehole of the Kola Peninsula, which proves the existence of fluid-filled fracture zones at these depths in the crust. In many instances, mineralization of water at great depths is comparatively low (5–10 g L^{-1} and less) and the total amount of dissolved gases, mainly CO_2 and H_2S, is high (up to 60–70 vol %).

An important conceptual framework about the hydrodynamics of the Earth's crust has been developed by Russian scientists in the 1980s–1990s (S.N. Ivanov, Y.A. Ezhov, G.A. Lysenin, and others; see a summary publication by Ivanov and Ivanov, 1993), based on observations from very deep boreholes, theoretical calculations and experimental results of rock deformation at high temperatures and pressures, and seismic data (Fig. 4). These ideas were discussed from the perspective of karst in Dublyansky (2000), Klimchouk (2000), and Andreychouk *et al.* (2009).

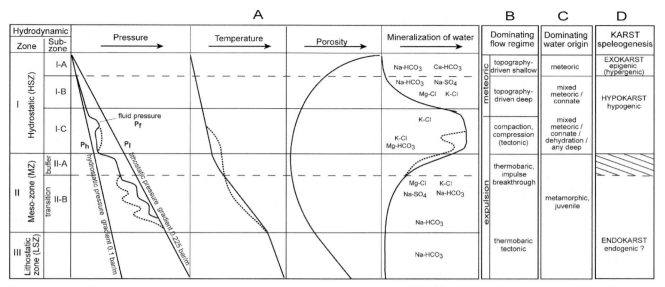

FIGURE 4 Vertical hydrodynamic zoning (A) and karst stories (D) of the Earth's crust. (B) and (C) show, respectively, dominating flow regimes and the origins of groundwater in different zones and stories. (A) also shows changes of some important parameters with depth: solid line—parameters before the impulse breakthrough of fluids through buffer subzone II-A; dotted line—parameters after the breakthrough. See text for definitions of zones and subzones. *(A) is from Ezhov and Lysenin, 1990, as reproduced in Andreychouk et al., 2009.*

According to this model, the almost complete closure of fissures, cavities, and open pores in rocks at a certain depth (commonly of about 7–15 km but sometimes less), creates a dense "buffer" interval of maximum compaction almost impermeable to fluids. This interval (subzone II-A; the upper part of the mesozone) serves as a planetary-scale regulator of defluidization of the deepest parts. Above it lies a hydrostatic zone (I), in which fluid pressures are close to the hydrostatic pressures, not exceeding them by more than 10–25%.

Below the buffer interval is a transition subzone (II-B) marked by variable suprahydrostatic pressures, decreasing mineralization, reduced rock density, and high porosity and fracturing. This grades downward into the deepest zone where liquid-vapor fluids released from thermal breakdown of hydrous minerals and arriving from the lower crust and mantle are under lithostatic pressure (lithostatic zone III in Fig. 4).

The buffer subzone is the base of a variety of tectonic structures, including the overwhelming majority of deep faults and hydrothermal conduit systems. Under seas and oceans, due to additional pressure from the water column, it is closer to the sea floor than it is to the land surface on the continents. When pressure of fluids beneath the buffer subzone reaches values sufficient for hydrofracturing, the fluids break through it into the above hydrostatic zone. Such periodic breakthroughs of pulses of deep-seated fluids are believed to be responsible for local aquifer pressure anomalies in the hydrostatic zone and associated thermal, hydrochemical, and gas anomalies. In favorable structural conditions, such plumes may transect the hydrostatic zone to various heights and up to the surface, causing respective anomalies in all shallower groundwater systems (Fig. 5). Apparently, these aggressive fluxes play the important role in generating hypogene speleogenesis in the hydrostatic zone along and around these paths, especially hydrothermal speleogenesis.

Based on the above model, a new notion of *endokarst* has been suggested (Ezhov and Lysenin, 1990; Andreychouk et al., 2009) that radically differs from the convenient usage of this term in the karst literature (*i.e.*, simply meaning underground karst) but correlates well with a broader geologic usage. The endokarst realm encompasses the deep parts of the crust below the buffer subzone. Endokarst processes involve metamorphic and igneous fluids and act at elevated temperatures (>80–100°C) and pressures much exceeding the hydrostatic ones. In such conditions the fluids are highly aggressive with respect to many sedimentary, metamorphic, and igneous rocks, not only to conventional soluble rocks such as carbonates and sulfates. In these conditions, solubility of quartz, for instance, becomes comparable to that of gypsum or anhydrite in near-surface conditions (Fig. 6H). Cavities may exist in the endokarst story because fluids there are under lithostatic pressures so that effective pressure is zero. However, they can be preserved while passing through the above buffer interval only if filled with some secondary mineral such as calcite. This could be removed in the course of further dissolution in the mesogenetic zone.

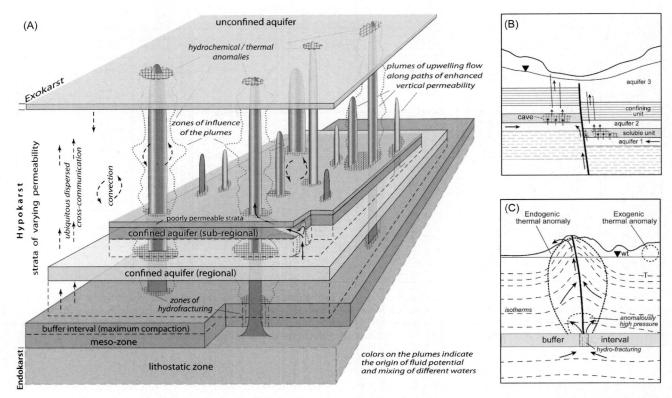

FIGURE 5 Highly idealized conceptual representation of the cross-formational nature of hypogene speleogenesis in the upper part of the Earth's crust (A), and of typical hypogene speleogenesis settings: (B) artesian settings; (C) deep-rooted settings. Hypogene speleogenesis is related to upwelling plums of water originated at varying depths, extending upward from the source of the potential along zones of enhanced vertical permeability. The upwelling plums interact with waters in any upper aquifers and develop zones of influence, which lateral extension varies depending on permeabilities and the groundwater regime in the aquifers crossed. *(C) is from Andreychouk et al., 2009.*

At least some of the known caves in silicate rocks could owe their origin to endokarst processes (Andreychouk et al., 2009).

The above cited works also suggested the term *exokarst* be used for all karst processes operating above the buffer interval, that is, in the entire mesogenetic and telogenetic zones. The present author, however, recommends limiting the use of this term by the meaning equivalent to epigenic karst, that is, karst that develops in the shallowest groundwater systems with direct meteoric recharge. The zone between the buffer interval and the exokarst zone, the arena of hypogenic speleogenesis by upwelling fluids in the hydrostatic zone, is termed here the *hypokarst* zone (Figs. 4D and 5), the main focus of this article.

HYPOGENE SPELEOGENESIS

Chemical Mechanisms

There was a long belief in the karst literature that confined conditions generally offer limited dissolution potential for karstification as slow artesian and basinal flow would be chemically equilibrated with the soluble rocks. A new perspective emerged from the recognition of the variety of deep-seated sources of the aggressiveness and of widespread leakages across strata that were supposed to be "impermeable" within the old artesian paradigm. The fundamental feature of hypogene speleogenesis is that it is mainly driven by transverse flow across boundaries between different strata, diagenetic zones, and flow systems (Fig. 5), whereas the boundaries commonly coincide with major contrasts in water chemistry, gas composition, and temperature. This causes disequilibrium conditions and supports diverse mineral reactions, including those of dissolution with the spatially concentrated effects. The variety of processes that can potentially cause the aggressiveness and dissolution in deep-seated settings has been characterized in Mazullo and Harris (1992), Surdam et al. (1984), Palmer (1991, 2007), Klimchouk (2000, 2007), and Dublyansky (2000). The most important processes are briefly overviewed below (Fig. 6).

Carbonic acid dissolution dominating epigenic carbonate speleogenesis also operates as a hypogenic agent,

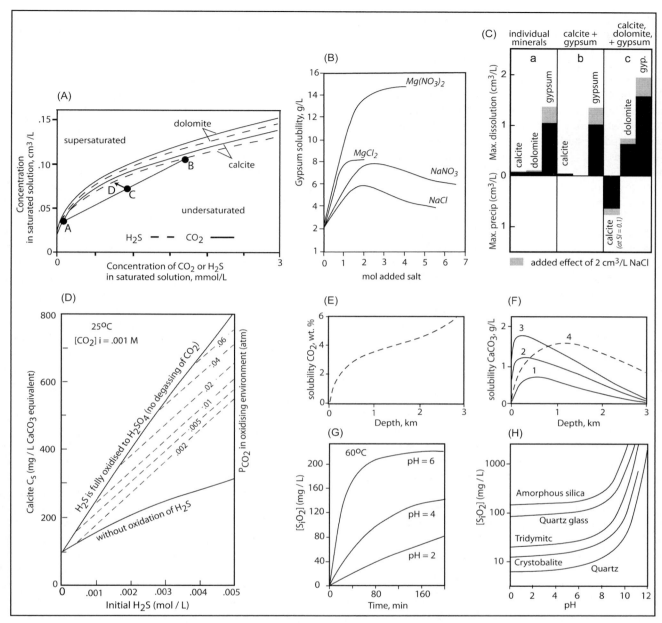

FIGURE 6 Diagrams illustrating the potential for aggressiveness of groundwater in different conditions with respect to various rocks. (A) The effect of mixing of water differing in CO_2 and H_2S content on dissolution of carbonates (*from Palmer, 1991*). (B) Gypsum solubility as a function of concentration of other salts. Magnesium chloride and nitrate have a particularly strong effect on gypsum solubility. (C) Dissolution in mixed carbonate/sulfate strata (*from Palmer, 2007*). The diagram shows solubilities of calcite, dolomite, and gypsum: (a) as individual minerals at $15°C$ and $P_{CO2} = 0.01$ atm; (b) in a mixture of calcite and gypsum (both become less soluble, although the effect on gypsum solubility is minor); (c) where water first encounters calcite and dolomite, and then gypsum (both dolomite and gypsum become more soluble as calcite is forced to precipitate). The shaded areas show the effect of $2\ cm^3\ L^{-1}$ of dissolved salt (NaCl). In (c), calcite is held at $SI = 0.1$ to account for the residual supersaturation required to precipitate it. (D) Change in calcite solubility when H_2S in solution oxidizes to sulfuric acid (*from Palmer, 1991*). Dissolution is greatest if the CO_2 generated by the reaction is retained in solution. (E) and (F) Solubility of CO_2 (E) and $CaCO_3$ (F) in ascending thermal water as function of temperature. Temperature changes from depth toward the surface: on (E) from 200 to $25°C$; on (F) lines indicate the following ranges: 1—from $225-100°C$, 2—from $225-50°C$, 3—from $225-25°C$, 4—from $200-100°C$ for solution with ionic strength $I = 1$. (G) and (H) Dissolution of SiO_2 in water in different conditions: (G) the dissolution rate of amorphous SiO_2 at $T = 60°C$ and varying pH; (H) solubility of different varieties of silica with change of pH in standard conditions (*from Andreychouk et al., 2009*).

though the origin of the acidity is different. CO_2 is produced at depth mainly by prograde metamorphism and devolatilization reactions at great depths, thermal degradation and oxidation of deep-seated organic compounds by mineral oxidants, and by several other reactions. Efflux of CO_2 of the deep origins into upper aquifers can be massive, point-wise or disperse, depending on geologic/structural conditions. CO_2 from oxidation of deep-seated organic compounds is especially common in the vicinity of hydrocarbon fields, where waters characteristically contain high CO_2 concentrations, and in the areas of young volcanism.

Dissolution by cooling thermal water can occur along ascending flow paths, even at constant CO_2 levels. In a closed system, the solubility of calcite increases with decreasing temperature, so that more calcite can be dissolved (Fig. 6F). Near the surface, CO_2 concentration decreases and the solubility of calcite drops (Fig. 6E). Carbonates dissolve at depth along the rising thermal flow paths but calcite precipitates near the surface. The dissolution effect increases with increasing CO_2 partial pressure. This mechanism is known as *hydrothermal speleogenesis*, occurring in high-gradient zones where ascending flow is localized along some highly permeable paths (Dublyansky, 2000). However, dissolution by cooling of thermal water alone can produce substantial caves only where high thermal gradients are sustained through long periods of time. Most caves thought to be thermal actually owe their origin to the mixing of high-CO_2 thermal water with low-CO_2 water of shallower flow system (Palmer 1991, 2007).

Hydrogen sulfide is another common hypogene source of acidity, abundant in deep meteoric groundwater and basinal brines, generated at depth by microbial or thermal reduction of sulfates in the presence of organic carbon. When dissolved in water, hydrogen sulfide forms a mild acid, but it can cause carbonate dissolution if this water flows elsewhere and mixes with water that has little or no H_2S content (Palmer, 1991, 2007), or escapes from reducing zones as a gas and is reabsorbed in freshwater. H_2S can also react with dissolved metals, such as iron, to produce sulfide ores and dissolve carbonate rocks.

As both CO_2 and H_2S are commonly present together in deep groundwater systems, and their proportion changes in the course of basinal evolution, the particular contribution of the above dissolution mechanisms in hypogene speleogenesis is difficult to discriminate.

Mixing of waters of contrasting chemistry, particularly those differing in CO_2 or H_2S content or salinity, can renew or enhance solutional aggressiveness to carbonates due to the concavity of solubility curves (Fig. 6A), the effect widely referred to in the karst literature as *mixing corrosion*. Gases have a greater effect than salinity in controlling mixing dissolution, especially when differences in equilibrium gas contents are large (Palmer, 1991). Mixing corrosion is highly relevant for hypogene systems, where water that rises from depth encounters shallower meteoric water along paths of the cross-formational flow.

Dissolution by sulfuric acid, a very strong speleogenetic agent (Fig. 6D), occurs in shallower conditions where H_2S-bearing waters rise to interact with oxygenated shallower groundwaters. It is recognized as the main speleogenetic process for certain large caves (*e.g.*, caves in the Guadalupe reef complex in the United States and in the Apennines in Central Italy) and many smaller ones. The dissolution by sulfuric acid is most readily observable in caves that are now in unconfined settings but continue receiving rising H_2S-bearing water from depth. Intense production of sulfuric acid near the water table and in the subaerial (above the water table) conditions causes pronounced speleogenic and mineralogical effects, which led some researchers to believe that most of the cave origin is due to subaerial sulfuric acid dissolution. Substantial sulfuric acid dissolution can also be caused by oxidation of iron sulfides such as pyrite and marcasite, localized in ore bodies or along certain horizons or bedding planes in carbonates (Lowe and Gunn, 1997).

Organic acids generated in great quantities during the maturation of organic matter in the mesogenetic environment can significantly enhance the solubility of carbonates (Surdam *et al.*, 1984; Palmer, 2007) and therefore be a powerful agency for hypogenic speleogenesis. Their dissolution effect appears to be the most pronounced under high (above 80°C) temperatures, as at lower temperatures they are readily metabolized by bacteria. At still higher temperatures, organic acids break down to CH_4 and CO_2, and the latter forms carbonic acid to dissolve more carbonates. Although some of the reactions involved may lead also to precipitation, the onsets of dissolution and precipitation can be spatially separated.

Dissolution of evaporites is a simple dissociation of the ions and diffusional mass transport from the surface into the solution. The solubility of gypsum increases significantly (up to 3–6 times) with the presence of other salts in the solution (Fig. 6B). Gypsum dissolution can be rejuvenated by reduction of sulfates, which removes sulfate ions from the solution and allows more sulfates to dissolve, and by dedolomitization, which generates further dissolutional capacity with respect to gypsum because Ca^{2+} is removed from solution and the sulfate ions react with Mg (Fig. 6C). Because mixed evaporite (gypsum/anhydrite and salt) and evaporate/carbonate formations are common in sedimentary successions, the above effects are extremely relevant to hypogene speleogenesis.

Dissolution in mixed carbonate/sulfate strata is a kind of synergetic process, more complex than in each lithology alone (Palmer, 2007), which does not require an acid source. In a mixed system, up to 1.5 times more gypsum can dissolve, and up to 7 times more dolomite, compared to the solubilities of gypsum and dolomite alone (Fig. 6C). The effect increases with salinity of water. In a simple carbonate-sulfate system, the volume of precipitated calcite is less than 30% of the combined total volume of dissolved gypsum and dolomite, which results in considerable solution porosity. The process is mediated by the relatively slow dissolution of dolomite, so that it is a viable mechanism for the development of dissolution paths in gypsum in deep-seated settings.

There may be other chemical conditions and reactions operative at depth, poorly known so far or still overlooked in karst science. Low-mineralization groundwater anomalies, frequently documented at great depth, especially in the vicinity of hydrocarbon fields, bring about chemical disequilibrium conditions that can be localized and extended upward along cross-formational flow paths. Chemical interaction of matrix pore and conduit waters is still poorly studied. Radiolysis of groundwater can locally modify redox conditions and therefore the speciation and the solubility of the compounds. It possibly accounts for the recorded instances of the presence of oxygen in considerable amounts (up to 50 vol % of dissolved gases) in water at depths of 2 to 3 km, which can generate reactions involving oxidation of hydrocarbons, sulfides, and other compounds.

There is mounting evidence suggesting that more than one process could be involved in most cases of hypogene speleogenesis, operating either in combination or sequentially in time.

Distribution and Settings

Recent studies made it apparent that hypogene speleogenesis is much more widespread in nature than previously thought. It operates in various geological and tectonic settings, at different depths (ranging from a few tens of meters to several kilometers), due to different dissolutional mechanisms operating in various lithologies. According to the scale and depth of flow systems that drive upwelling flow, hypogene speleogenesis settings can be classified into three types:

1. *Artesian hypogene speleogenesis*, related to upwelling limbs of regional or subregional (intermediate) meteoric systems, roughly corresponding to the subzone I-B (Figs. 4 and 5B). This type of speleogenesis is most common for the post-orogenic tectonic regime marked by massive and deep penetration of topography-driven meteoric waters.
2. *Deep-rooted hypogene speleogenesis*, related to upwelling plumes generated by deep basinal systems in the subzone I-C or fluxes from the mesozone II-B through the buffer interval (Figs. 4 and 5C). It is common for both the extensional tectonic regimes of continental rifting and the compressional regimes of active margins and collision fold-thrust zones, favored by intense and deep dislocations and related thermal and geochemical cross-formational anomalies.
3. *Combined artesian/deep-rooted speleogenesis*, where the settings 1 and 2 interact intensely. It typically occurs in the wedge-top and forebulge settings of foreland basins, as well as in disturbed cratonic zones.

A common misconception about hypogenic karst is that it is thought to be peculiar to arid climates and less common in humid regions. In reality, hypogenic speleogenesis is largely independent of climate, but the epigenic overprint of hypogenic caves, inherited by entrenched and denuded karst types, strongly depends on climate. The overprint is particularly strong in regions of moderate and high runoff. In arid and semi-arid regions epigenic speleogenesis and surface karst morphologies are normally subdued due to limited infiltration and short supply of CO_2 and organic acids from the soil, but hypogenic features are often abundant in the subsurface, resulting in a strong contrast in the degree of karstification between the surface and subsurface.

Hydrogeologic and Structural Controls

Cross-formational flow in a heterogeneous (layered, deformed, and fractured) sequence may have very complex patterns. They are determined by complex interplay of driving forces of different flow systems and highly variable permeabilities, which vary with classes of heterogeneities (*e.g.*, layered and discontinuous) and types of elementary porosity structures (*e.g.*, pore, vugs, and fractures) and their connectivity. Soluble rocks, with their own variety of geological heterogeneities and primary porosity structures, may have different positions and play different functions within the overall hydrostratigraphic framework and flow patterns.

Distinct integrated stratiform permeability structures, relatively independent of each other (hydrostratigraphic units, or hydrofacies), can be related to single beds or be comprised of several rock units. They can be separated by beds of distinctly lower permeability (aquitards, or hydrostratigraphic barriers) or by less obvious boundaries along which the vertical

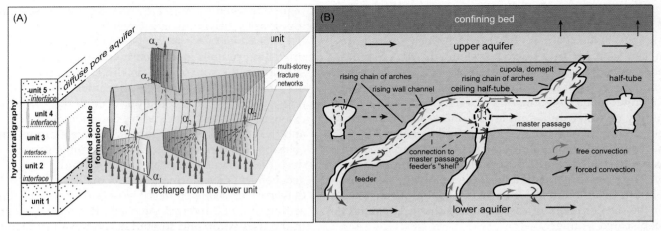

FIGURE 7 Transverse flow across several hydrostratigraphic units and interfaces in a soluble bed (A) and the morphologic suite of rising flow (MSRF), diagnostic of hypogenic transverse origin of caves (B). In (A) α_n denotes exchange terms between different permeability structures. *Modified from Klimchouk, 2007.*

hydraulic connection between contrasting and discordant permeability structures in the adjacent beds is limited (hydrostratigraphic interfaces). Leakage across hydrostratigraphic barriers occur along major crosscut fractures and sedimentary windows. Leakage through hydrostratigraphic interfaces occurs through points where permeability features of adjacent units (stratiform pores and vugs, or minor fractures) vertically intersect (Fig. 7A).

Zones of enhanced vertical permeability that cross thick sequences and interfaces between major flow systems and diagenetic subdomains (the fracture diagenetic setting of Machel, 1999) are rarely comprised of single fractures. More common are combinations of several major fractures in echelon, associated feathering fracture systems and lithological variations within different intervals. The vertical permeability of such zones varies at different levels. Flow along such structures is invariantly vertical within hydrostratigraphic barriers, but it may diverge laterally where high-permeability beds are intersected, especially if the guiding vertical structure terminates upward at one location but continues into the upper strata in the other nearby location.

Though both insoluble and soluble units can initially (prior to speleogenesis) serve as hydrostratigraphic barriers in an aquifer system, their further behavior is different. The vertical permeability of the former remains conservative, while in the latter the conduit permeability develops in the course of speleogenesis, so that they lose their separating function (Fig. 7). The same happens with hydrostratigraphic interfaces within the heterogeneous soluble sequence (Fig. 7A). In this way, hypogene speleogenesis serves to enhance cross-formational hydraulic communication across initially less pervious soluble beds or hydrostratigraphic interfaces. The net result is that multiple aquifer systems achieve greater integration.

In epigenic karst (exokarst), directly recharged from the surface, the mode of recharge evolves in coordination with the development of solution conduits. The evolution is characterized by a switch from the hydraulic control of flow (by size of a conduit itself) to the catchment control (available recharge) when the conductivity of a growing conduit becomes larger than available recharge (Palmer, 1991).

In contrast, in hypogene settings, recharge to a soluble sequence comes from the adjacent strata below and discharge occurs through the overlying strata. Recharge is independent from the surface conditions. Both recharge and discharge can be diffuse or point-wise, laterally coinciding or shifted, which determines to a large extent the pattern of speleogenesis in the soluble sequence.

The mode of recharge is conservative, although its amount and chemical characteristics may change on the long run of the geologic time scales as the deep driving force or the regional flow system changes.

Discharge conditions are more dynamic as they change with erosion and breaching of the cover and the major confinement. The breaching can be by external geomorphological processes, such as erosional entrenchment, or by the internal development of the hypogene cave system, such as vertical stoping of breakdown structures above cavities across the confining strata. Where there is no major geologic confinement and semiconfined conditions are caused by the layered heterogeneity within the soluble sequence itself, changes in the discharge conditions are most dynamic and occur mainly through the internal development of a cave system.

As recharge to and discharge from the soluble rock occurs through the adjacent strata with relatively conservative permeabilities, flow rates are controlled by the permeability of the least permeable member. Thus there is an external conservative hydraulic control on the amount of flow through the evolving conduits. When conduits have grown to the breakthrough conditions and the hydraulic gradient across the unit is diminished, their further growth is not accelerated significantly as flow across the soluble bed is now controlled not by the hydraulic resistance of the conduit system but by the permeability of the least permeable member, and by the boundary conditions of the system. This suppresses the positive flow-dissolution feedback and speleogenetic competition, the main mechanism acting in unconfined (epigenic) speleogenesis, and promotes more pervasive and uniform enlargement of initial permeability structures in the cave-forming zone where recharge diffuses.

The presence of hydrostratigraphic interfaces within the soluble formation, as well as the restricted outflow, also favors pervasive conduit development. If recharge and discharge are both through a single major crosscutting fracture, the conduit development in the soluble unit is localized, although it can get dispersed below the upper insoluble barrier that restricts outflow despite the growth of the conduit in the soluble unit. In hydrothermal speleogenesis in carbonates there is another specific mechanism, caused by the thermal coupling between the fluid and rock, that also suppresses speleogenetic competition.

Large, deep, and complex transverse structures supporting cross-formational upwelling flow are commonly geologically long-lived, although periods of intense and abated activity may alternate. Geochemical environments may migrate along them and change many times, promoting alternating dissolution and precipitation conditions. Hypogene speleogenesis at different depths along such structures can operate through various processes, either simultaneously or sequentially. The resultant solution porosity structures, aligned to the same transverse zone, can have multiphase evolution and complex spatial organization, although they can be represented by certain elementary patterns within particular intervals.

Porosity Patterns

Solution porosity patterns produced by hypogenic speleogenesis are the result of the complex interaction of structural, hydraulic, and geochemical conditions, all varying in the course of geological evolution. The overall position of cave-forming zones is controlled by the 3D distribution of soluble rocks, their position and hydraulic function within the flow systems, the nature and geometry of major crosscutting fractures, and the pattern of geochemical environments in a given setting. Overall patterns of cave systems are strongly guided by the initial (prespeleogenetic) permeability features in a sedimentary sequence, that is, by the spatial distribution of the permeability structures and hydrostratigraphic interfaces within the soluble and adjacent units; by the mode of water input to, and output from, cave-forming zones; and by the overall recharge-discharge configuration in the multiple aquifer system. The presence of crosscutting permeability features such as major fractures can exert strong effects on cave patterns through their inflow, throughflow, and outflow controls. In contrast to epigene settings where initial effective permeability structures are exploited by speleogenesis in a very selective manner, hypogene speleogenesis tends to exploit most of them within cave-forming zones, provided the aggressiveness is maintained. Geochemical interactions of flow components guided by transverse and lateral permeability pathways determine zones of pronounced speleogenetic development and influence the resultant patterns. General evolutionary factors, such as regional tectonic and geomorphic developments that control rates and architecture of flow and timing of speleogenesis, also affect cave patterns forming in hypogene settings.

Hypogene caves demonstrate a variety of patterns, as classified and briefly described below. For more extended discussion refer to Palmer (1991, 2007), Ford and Williams (2007), Klimchouk (2000, 2007, 2009), and Audra *et al.* (2009). Same patterns are known to form in different lithologies and by different dissolution mechanisms. Certain patterns common for hypogene speleogenesis can also be formed locally in epigenic environments. Conversely, branchwork patterns, the most common one for epigenic speleogenesis, never form in hypogenic settings. This reflects the fundamental difference between the mechanisms of epigenic speleogenesis, largely competitive, and of hypogenic speleogenesis, in which the competition between alternative flow paths is subdued.

The following elementary cave patterns are typical (although not necessarily exclusive) for hypogene speleogenesis:

- Single passages or rudimentary networks of passages
- Cavernous edging along transverse hypogene conduits
- Network maze
- Spongework maze
- Irregular isolated chambers
- Rising, steeply inclined passages or shafts
- Collapse shafts over large hypogenic voids and breccia pipes

Single isolated passages or rudimentary networks of passages are probably the most common cave structures of hypogene origin. They form by transverse flow across a single soluble bed where guiding cross-bed fractures are isolated or have only limited lateral connection. The passages are typically slot-like, sometimes with rounded sections in the middle or upper parts of the cross-section, single or combining in small clusters of few intersecting passages according to the pattern of guiding fractures. Their notable characteristic is that they are dead-end in every direction. They can be observed directly when exposed in cliffs or encountered by underground mines. The cross dimensions vary from several meters to a few centimeters (dissolutionally enlarged fractures), the latter being the most common.

Cavernous edging is represented by zones of cavernous porosity comprised of closely spaced irregular touching vugs and small conduits (up to 2–5 cm in diameter but sometimes much larger) which frequently border walls of crosscutting fracture conduits up to depths of a meter or more. The size of voids diminishes in the direction normal to walls, and farther into the rock they pinch out. The cavernous edging can be stratiform or distributed by clusters along the fracture/conduit walls. The formation of this type of cavernousity is apparently related to the mixing corrosion effect, where pore water in more diffusely permeable beds or zones converges toward a conductive fracture.

Network maze patterns are common for hypogenic caves. Passages are strongly fracture-controlled and form more or less uniform networks, which may display either systematic or polygonal patterns, depending on the nature of the fracture networks. Systematic, often rectilinear, patterns are most typical, displaying networks of regular cross-bed fractures. Networks of different density and orientation may be present at different stories of a single cave. Network maze caves of hypogene origin are known in limestones, dolomites, and gypsum, in mixed limestone-dolomite-gypsum strata, and also in conglomerates. A common feature of network mazes is a very high passage density.

Spongework maze patterns are much less common than networks. Highly irregular passages develop through enlargement and coalescing of pore- and vuggy-type initial porosity in those horizons of the cave-forming zone that have no fractures but interconnected pores and vugs. Clusters of spongework-type mazes are commonly combined with other patterns in the adjacent intervals to form complex cave structures.

Isolated chambers commonly occur at the bottoms of soluble sequences but they are also frequent within particular beds in stratified carbonate sequences. They form in two situations: (1) by buoyant dissolution at the bottom of a soluble formation, commonly evaporites, where there is a major underlying aquifer; (2) where aggressive recharge from below is localized along major crosscutting fractures and aggressiveness is enhanced by mixing of rising deep flow with a shallower flow system. Similar chambers form where there is a point-wise input of geogenic gases into the basal aquifer. Irregular chambers of the hypogene origin can attain very large dimensions, such as directly documented cavities in evaporates of southern Harz, Germany (cavities of the *schlotten* type), the Big Room in Carlsbad Cavern, or a huge cavity encountered by boreholes in the Archean and Proterozoic marbles in southern Bulgaria with a maximum vertical dimension of 1340 m and an estimated volume of 237.6 million m^3, probably the largest known, although not accessible, cave chamber on Earth. Smaller isolated chambers are abundant everywhere. Where large irregular chambers are complicated by blind branches extending outward, the pattern is termed *ramiform*.

Rising, steeply inclined passages or shafts are outlets of deep hypogene systems in which the "root" structure remains unknown in most cases. A type example is the 392-m-deep Pozzo del Merro near Rome, Italy, presumably formed by rising thermal water charged with CO_2 and H_2S.

Collapse shafts over large hypogenic voids can form mega-sinkholes at the surface, locally termed *cenote* (Mexico) and *obruks* (Turkey). The type examples are Sistema El Zacatón in Mexico (several features with depth up to 329 m from the surface and ~317 m below the water table), obruks in the Central Anatolia, Turkey (71 features with depths up to 125 m from the surface), and cenotes near Mt. Gambier in southeastern Australia. These collapse features indicate giant chambers formed at the base around localized inputs of fluids charged with volcanogenic CO_2 in the areas of young volcanism.

Fossil vertically extended collapse features known as *breccia pipes, collapse columns*, or *geologic organs*, and also termed *vertical through structures*, are a common by-product of hypogene speleogenesis in a variety of geological settings. They originate from collapse of large cavities, such as described above, and propagate upward by stoping across sedimentary successions that include soluble rocks. Such features may reach many hundreds of meters in the vertical extent. They are not merely breakdown structures but complex hydrogeologic structures whose development depends on (and favors) focused cross-formational groundwater circulation and continuing dissolution of intercepted soluble beds and infallen clasts.

Composite 3D systems are comprised of various elementary patterns at different levels, such as irregular chambers, clusters of network or spongework mazes, and rising subvertical conduits and other morphs connecting them. Composite 3D systems include many of the world's largest documented caves. Their organization reflects vertical heterogeneity in distribution of initial permeability structures that guided, redistributed, or impeded rising flow within a soluble formation, and the interplay between structural, hydrogeological, and chemical factors.

Composite 3D cave structures may develop within a rather thin bed and be laterally extended (*e.g.*, two- to four-story giant mazes in the western Ukraine confined within the 16–30-m-thick gypsum strata), or be vertically extended through a range of several hundred meters (*e.g.*, Monte Cucco system, central Italy: 930 m; Lechuguilla Cave in the Guadalupe Mountains, New Mexico, U.S.A.: 490 m).

Multistory mazes are "layered" variants of composite 3D systems. In multistory mazes, stories (up to 5–6) are commonly comprised of stratiform networks differing in patterns, connected via rising conduits or *chimneys*, or cross-cutting rift passages. The stories can superpose each other within the same area or display a staircase arrangement within a system, with cave areas at different stories shifted relative to each other because areas of discharge through the cover are shifted with respect to areas of preferential recharge from below.

Some vertically extensive 3D caves have prominent feeders as large isolated, steeply ascending passages, or clusters of rift-like passages connecting to some master levels of passages and chambers, and prominent outlet segments rising from the latter. Many such caves are composed of network and spongework mazes at various levels connected through rising vertical conduits, coalescing with large chambers and passages. Other notable variants include complex bush-like upward-branching structures, composed of rising sequences of chambers and large spherical cupolas, and network maze clusters at the base of a soluble sequence with ascending staircase limbs of vertical ascending pits and subhorizontal passages at different levels (outlet components). Composite 3D structures dominated by large irregular chambers, from which blind branches extend outward, are distinguished as ramiform patterns.

Because of the transverse nature of hypogene speleogenesis, caves and their parts tend to have a clustered distribution in the plan view, although clusters may merge to extend over considerable areas. Laterally extensive multistory maze caves such as in the western Ukraine, or in the Black Hills, South Dakota, or vertically extensive 3D structures such as in the Guadalupe Mountains and the Apennines, are in fact combinations of many clusters representing relatively independent transverse flow subsystems.

Mesomorphology Features

In spite of the great variability in geological and chemical conditions of hypogene speleogenesis and in size and patterns of hypogene caves, their mesomorphological features exhibit the remarkable similarity at all scales. This arises due to several reasons: (1) generally sluggish conditions of rising flow in which hypogene speleogenesis occurs; (2) the great role of buoyant convection due to density differences and the related abundance of upward-oriented morphs; and (3) the clustered nature of cavities resulting from transverse flow along multiple closely spaced paths, and occasional lateral merging of individual clusters.

Individual occurrences of some features characteristic for hypogene caves, such as cupola-form solution pockets or ceiling half-tubes, also occur in epigenic caves where they form in the unconfined phreatic zone or subaerial conditions. Specific to hypogene speleogenesis is, however, that different morphs commonly occur in spatially and functionally related groups where fluid flow paths, including distinct buoyant convection components, can be traced from rising inlet conduits, through transitional wall and ceiling features (rising wall channels and ceiling half-tubes), to outlet features (cupolas and domepits). This regular combination has been distinguished as the morphologic suite of rising flow (MSRF; Klimchouk, 2007, 2009; Fig. 7B). It provides diagnostic evidence for hypogene speleogenesis. MSRF has been recognized in hundreds of hypogene caves across the globe. The diagram on Fig. 7B is generic and elastic; it can be stretched vertically, and complexity can be added to allow for multiple stories. The pattern of the forms will repeat itself on each story. Hypogene caves may consist of a few elementary segments such as the one depicted in Fig. 7, or combine hundreds and thousands of laterally merged segments within a single system.

Among elementary morphs in this suite, inlet features (termed *feeders*, *vents*, or *risers* in recent publications) are the most indicative of recharge from below and hence the hypogene origin for a cave. Original feeders are basal input points to hypogenic cave systems, the lowermost components, vertical, or subvertical conduits, through which fluids rise from the source aquifer. The vertical extension of feeders can be a few meters, such as in the gypsum mazes in the western Ukraine fed by a basal artesian aquifer, but they can rise through several hundred meters, as documented in mines that intercepted feeder conduits below a currently shallow-lying relict multistory maze cave in dolomites of the Neoproterozoic Vazante Group, Minas Gerais, Brazil.

Where the water table is established at a given level within a former confined hypogenic cave system, significant lateral widening and merging of passages into large chambers may occur, with characteristic speleogens such as horizontal notches and corrosion tables. This is typical for caves in evaporites and limestones subjected to back-flooding from a nearby river, and for limestone caves where deep recharge by sulfidic deep waters continues during the unconfined development. In the latter case, distinct upward-oriented morphological effects of subaerial sulfuric-acid dissolution (*i.e.*, condensation corrosion above the water table) may also develop, such as air-convection cupolas, which are difficult to distinguish from speleogens formed under submerged conditions. Their distribution, however, is restricted to the zone immediately above the water table, whereas in composite 3D systems the elements of the MSRF are systematically distributed through various stories in large vertical ranges.

For a more extended recent discussion of the morphology of hypogene caves the reader is referred to Klimchouk (2007, 2009), Palmer (2007), and Audra et al. (2009).

Hypogene Speleogenesis and Paleokarst

Within the conventional, predominantly epigenic, karst paradigm, instances of deep-seated karst have commonly been interpreted as paleo-(epigenetic) karst because the possibility of karstification in deep environments without recharge from the immediately overlying or adjacent surface has been neglected for a long time.

Paleokarst is not a particular type of karst but is, rather, a fossilized condition. Features become paleokarst as they get hydrologically decoupled from contemporary systems, in contrast to relict features that exist within contemporary systems but are removed from the environment in which they developed (Ford and Williams, 2007). True paleokarst is buried karst (see Fig. 2), which is a complete infilling and burial of exposed karst by later materials such as transgressive marine sediments. Paleokarst horizons are reliably recognized where they underlie unambiguous stratigraphic unconformities related to subaerial exposure.

With growing recognition of hypogenic speleogenesis, it becomes increasingly obvious that in many cases features previously interpreted as paleo-(epigenetic) karst, including coastal/oceanic karst, can be better explained as active or relict hypogenic features. Review of the international literature, especially concerned with carbonate-hosted hydrocarbon reservoirs, reveals that in many cases, because of their occurrence beneath distinct formational contacts, paleokarst features have been dubiously interpreted as evidence of subaerial exposure by a process of reciprocal reasoning. Other common cases of problematic paleokarst are stratiform breccia horizons which are commonly the ultimate result of hypogenic speleogenesis, namely, the collapse of laterally extensive, stratigraphically conformable maze caves.

Hypogene speleogenesis tends to operate over long time spans, intermittently or being repeatedly suspended and reactivated. It may even continue uninterruptedly from one hydrogeologic cycle through another (Fig. 2). Hypogenic features may become relict but still remain within the contemporary systems, for example, in a system where original confinement was breached and the flow pattern reversed from upwelling to descending. Hypogenic features are not paleokarst unless their evolution is completely halted by the removal of the cave-forming units from the geological section and their substitution by stratiform breccia horizons, complete sealing by cementation (mineralization), or lithification of the fill material. Establishing the paleostatus of hypogenic features and their distinction from epigenic paleokarst requires additional discussion.

Many important hydrocarbon and mineral deposits are karst-related, and commonly thought to be related to paleo-(epigenic) karst. Better understanding of hypogene speleogenesis is of paramount importance for their proper genetic interpretation, which in turn is crucial for the development of more adequate approaches to the prediction, prospecting, and development of these resources.

See Also the Following Articles

Speleogenesis, Telogenetic
Sulfuric Acid Caves

Bibliography

Andreychouk, V., Dublyansky, Y., Ezhov, Y., & Lysenin, G. (2009). *Karst in the Earth's crust: Its distribution and principal types.* Sosnowiec–Simferopol: University of Silesia–Ukrainian Institute of Speleology and Karstology.

Audra, P., Mocochain, L., Bigot, J.-Y., & Nobecourt, J.-C. (2009). Morphological indicators of speleogenesis: Hypogenic speleogens. In A. B. Klimchouk & D. C. Ford (Eds.), *Hypogene speleogenesis and karst hydrogeology of artesian basins* (pp. 17–22). Simferopol: Ukrainian Institute of Speleology and Karstology (Special Paper 1).

Choquette, P. W., & Pray, L. C. (1970). Geologic nomenclature and classification of porosity in sedimentary carbonates. *American Association of Petroleum Geologists Bulletin, 54,* 207–250.

Dublyansky, Y. V. (2000). Hydrothermal speleogenesis—Its settings and peculiar features. In A. B. Klimchouk, D. C. Ford, A. N. Palmer & W. Dreybrodt (Eds.), *Speleogenesis: Evolution of karst aquifers* (pp. 293–297). Huntsville, AL: National Speleological Society.

Ezhov, Y. A., & Lysenin, G. A. (1990). Vertical zonation of karst development. *Izvestija AN SSSR Serija Geologii, 4,* 108–116 (in Russian)

Ford, D. C. (2006). Karst geomorphology, caves and cave deposits: A review of North American contributions during the past half century. In R. S. Harmon & C. W. Wicks (Eds.), *Perspectives on karst geomorphology, hydrology and geochemistry* (pp. 1–14). Geological Society of America (Special Paper 404).

Ford, D.C., & Williams, P.W. (1989). *Karst geomorphology and hydrology.* London, Unwin Hyman.

Ford, D. C., & Williams, P. W. (2007). *Karst hydrogeology and geomorphology.* Chichester, U.K.: Wiley and Sons.

Ivanov, S. N., & Ivanov, K. S. (1993). Hydrodynamic zoning of the Earth's crust and its significance. *Journal of Geodynamics, 17*(4), 155–180.

Klimchouk, A. B. (2000). Speleogenesis under deep-seated and confined settings. In A. B. Klimchouk, D. C. Ford, A. N. Palmer & W. Dreybrodt (Eds.), *Speleogenesis: Evolution of karst aquifers* (pp. 244–260). Huntsville, AL: National Speleological Society.

Klimchouk, A. B. (2007). *Hypogene speleogenesis: Hydrogeological and morphogenetic perspective.* Carlsbad, NM: National Cave and Karst Research Institute.

Klimchouk, A. B. (2009). Morphogenesis of hypogenic caves. *Geomorphology, 106*, 100–117. doi: 10.1016/j.geomorph.2008.09.013.

Klimchouk, A. B., & Ford, D. C. (2000). Types of karst and evolution of hydrogeologic settings. In A. B. Klimchouk, D. C. Ford, A. N. Palmer & W. Dreybrodt (Eds.), *Speleogenesis: Evolution of karst aquifers* (pp. 45–53). Huntsville, AL: National Speleological Society.

Klimchouk, A. B., & Ford, D. C. (Eds.), (2009). *Hypogene speleogenesis and karst hydrogeology of artesian basins.* Simferopol: Ukrainian Institute of Speleology and Karstology.

Klimchouk, A. B., Ford, D. C., Palmer, A., & Dreybrodt, W. (Eds.), (2000). *Speleogenesis: Evolution of karst aquifers.* Huntsville, AL: National Speleological Society.

Lowe, D. J., & Gunn, J. (1997). Carbonate speleogenesis: An inception horizon hypothesis. *Acta Carsologica, 26*(2), 457–488.

Machel, H. G. (1999). Effects of groundwater flow on mineral diagenesis, with emphasis on carbonate aquifers. *Hydrogeology Journal, 7*, 94–107.

Mazzullo, S. J., & Harris, P. M. (1992). Mesogenetic dissolution: Its role in porosity development in carbonate reservoirs. *American Association of Petroleum Geologists Bulletin, 76*, 607–620.

Palmer, A. N. (1991). Origin and morphology of limestone caves. *Geological Society of America Bulletin, 103*(1), 1–21.

Palmer, A.N. (1995). Geochemical models for the origin of macroscopic solution porosity in carbonate rocks. In D. A. Budd, P. M. Harris & A. Saller (Eds.), Unconformities in carbonate strata: their recognition and the significance of associated porosity (pp. 77–101). *American Association of Petroleum Geologists Memoir 63.*

Palmer, A. N. (2007). *Cave geology.* Dayton, OH: Cave Books.

Stafford, K., Land, L., & Veni, G. (Eds.), (2009). *Advances in hypogene karst studies.* Carlsbad, NM: National Cave and Karst Research Institute (NCKRI Symposium 1).

Surdam, R. S., Boese, S., & Crossey, C. (1984). The chemistry of secondary porosity. *American Association of Petroleum Geologists Memoir, 37*, 127–149.

Tóth, J. (1995). Hydraulic continuity in large sedimentary basins. *Hydrogeology Journal, 3*(4), 4–15.

SPELEOGENESIS, TELOGENETIC

Franci Gabrovšek

Karst Research Institute ZRC SAZU, Slovenia

INTRODUCTION

We describe basic principles of *meteoric* (as opposed to hypogenic) *speleogenesis* in *telogenetic* (as opposed to eogenetic) rocks. In other words, we discuss cave genesis in mature rocks with dominant fracture porosity caused by normal circulation of meteoric water. A dividing line between each of the above-mentioned systems is often not clear, for example, speleogenesis in the telogenetic stage can use horizons developed in the early eogenetic stage. Also a clear distinction between meteoric and hypogenic cave development is often not possible. We primarily discuss speleogenesis in limestone, although many of the principles are valid for other soluble rock.

Speleogenesis is a result of dissolution of soluble rocks by groundwater. It is a complex interplay of many factors and processes leading to the evolution of conduit networks in karst aquifers. The dynamics of speleogenesis and the resulting geometry of the conduit network strongly depend on local hydrology, lithology, structure, geochemical settings, and topography. Each of these imposes different parameters which can, as well, change in time.

Cave genesis is a major topic in speleology, with implications in other fields, particularly in karst hydrology. Many important papers discuss speleogenesis in various settings. They compose an invaluable base of knowledge of which only a part is summarized in this entry. An interested reader is referred to a few textbooks that contain most of what we know today on speleogenesis and contain many references to original works. The book edited by Klimchouk *et al.* (2000) provides information about history, basic principles, and a broad review of speleogenesis in different settings. For a comprehensive review of numerical modeling of speleogenesis in the past two decades, the volume by Dreybrodt *et al.* (2005) is recommended. Ford and Williams (2007) is considered to be the major reference in karst by many; it dedicates one of its 12 chapters to speleogenesis. Probably the best source to learn about speleogenesis is the book by Palmer (2007), which successfully links the basic principles with many field examples.

BASIC DYNAMICS OF SPELEOGENESIS

The simplest scenario includes a single flow path via one or a series of fractures connecting the points of recharge and discharge at different hydraulic heads (Fig. 1). Hydraulic head, h, is defined as $h = z + p/(\rho g)$, where the first term, z, presents elevation head and the second, $p/(\rho g)$, the pressure head. ρ is the density of water and g is the gravitational acceleration. Water tends to flow from higher to lower head. Water that is aggressive to rock-forming minerals (*e.g.*, calcite) dissolves fracture walls. The concentration of ionic species, and correspondingly the saturation ratio of water with respect to the mineral, increases along the flow path.

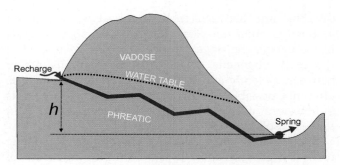

FIGURE 1 A single pathway evolving from the point of recharge elevated by h above the spring.

If dissolution rate (the rate of wall retreat is directly proportional to dissolution rate) decreases linearly with saturation ratio, it decreases exponentially with the flow distance along the fracture. If the rate law is nonlinear, the dissolution rate drops hyperbolically along the fracture. One can introduce penetration length, λ, which defines how rapidly from the entrance the dissolution rates drop below some value. For the nonlinear rate law, the penetration length λ is proportional to

$$Q\,P^{-1}\left(1-\frac{c_{in}}{c_{eq}}\right)^{(1-n)}$$

where P is the fracture perimeter, Q is the flow rate, c_{in} is the concentration at the input, c_{eq} is the equilibrium concentration, and n is the kinetic order. Because the fracture is being dissolutionally widened, the flow rate and penetration length increase with time. Both quantities are locked in a feedback loop, that is, an increase of one causes an increase of the other. They accelerate in time until an abrupt increase of both occurs. The terms *breakthrough* and *breakthrough time* (denoted as T_B) have been used for the event and the time when it happens. After the breakthrough, λ becomes large compared to the fracture length L, and the whole fracture widens uniformly at dissolution rates defined mainly by the saturation state at the entrance.

Such growth cannot last long, as there is no catchment large enough to sustain a constant head difference between the input and the output. The flow rates no longer depend on fracture resistance, but on the amount of water that a catchment can supply to the input. If the position of the flow path is above the base level the full pipe flow conditions will change to open channel flow and a transition to canyon will start.

Figure 2 schematically presents some of the features discussed above. The evolution of a fracture is divided into three stages: initiation, breakthrough, and growth. The curves present the evolution of a 1-km-long fracture with initial an aperture of 0.02 cm and width 1 m. Hydraulic head at the input is 50 m higher than at the

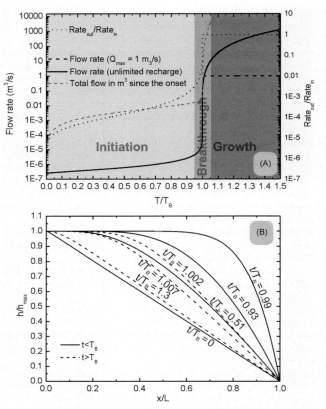

FIGURE 2 Evolution of a single fracture. (A) Flow rates for the case of limited (black dashed line) and unlimited (full black line) recharge. Total flow passed through the fracture since $t = 0$ (dashed gray line). The ratio between output and input dissolution rates (gray dashed line with scale on the right axis). (B) Hydraulic head along the fracture at different times in units of T_B (breakthrough time).

spring. Black lines show the evolution of flow rates when the recharge to the system is unlimited (full lines) or limited to 1 m³/s (dashed). The gray dashed line shows the total flow that has passed through the fracture from the beginning to a certain time. Note that only about 20 L of water passes the fracture until the breakthrough occurs. The dotted line (scale on the right side) presents the ratio between dissolution rates at the output and dissolution rates at the input of the fracture. This ratio is initially small and asymptotically approaches 1 at the breakthrough.

The hydraulic head distribution along the evolving fracture shows interesting dynamics. Initially the head drop is linear. As the fracture evolves, the high head from the input is shifted toward the tip of the fracture, and a high gradient is built up between the tip and the output (solid lines on Fig. 2B). After the breakthrough (dashed lines on Fig. 2B), the fracture is widened uniformly, which makes the head drop increasingly uniform. When recharge dominates the evolution, the head at the input decreases, and when vadose canyon starts to evolve, the head becomes equal to elevation.

The dynamics of head distribution as presented is important for understanding the evolution of more complex systems.

The natural conditions are usually far more complex than those described above. However, by having in mind a feedback mechanism and the concept of penetration length, the processes in more complex settings can be more easily understood.

EVOLUTION OF CONDUIT NETWORKS

Conduit systems in nature are far more complex than a single flow path. Caves rarely look linearly, and even if they do, it is likely that most of their complexity is hidden to explorers. Local geological and hydrological conditions determine the dynamics of speleogenesis and the final geometry of the conduit system. Cavers have mapped thousands of kilometers of caves around the world, exhibiting a variety of geometries in various settings. However, a careful look into cave maps allows us to organize cave patterns into several categories, which can be related to the type of recharge and geological structure (Palmer, 2007). In telogenetic settings *branchwork* and *maze* patterns are most common.

To demonstrate simple scenarios that lead to the formation of one or the other of the two above-mentioned patterns, we apply a numerical model of a 2D fracture network as shown in Figure 3. The domain consists of a network of 100×100 fractures with initial apertures distributed statistically between 0.007 cm and 0.015 cm. Into this network, a coarse network of 10×10 fractures, 100 m long with initial apertures of 0.03 cm, is embedded. Only fractures with aperture widths equal to or larger than 0.03 cm are shown. Four inputs of calcite aggressive water on the left side are denoted as I1–I4. On the right side are the outputs denoted by O1–O4. The head difference between input and output is 50 m. The upper and lower boundaries are impermeable. Aperture widths are presented by the line thickness and flow rates by colors, as shown on the legend. Flow rates are shown in relative units Q/Q_{max}, where Q_{max} is the flow rate in the fracture with the maximal flow in the net. Line thicknesses are shown in logarithmic scale. Lines of equal heads with values in cm are also shown.

During the initiation stage several competing flow paths have grown from the inputs to the outputs (Fig. 3A). The flow path from I4 to O3 has evolved most efficiently. Shortly before the breakthrough, about 80% of its length is widened considerably. Its wide part offers low resistance to flow, resulting in only a small head drop along its length. Almost complete head drop occurs between the tip of the evolved part and the outputs. The head distribution in the leading pathway influences the head distribution in its vicinity. It causes low gradients in the wider upstream region of the network and thus prevents the evolution of the competing pathways emerging from inputs I1–I3.

After the breakthrough (Fig. 3B), the flow path I4-O3 is widened by uniform rates defined by the solution at the input. Consequently the head drop along it becomes increasingly uniform. The gradients in the upstream part reestablish so that other flow paths start growing again. Meanwhile, the new distribution of hydraulic heads diverts some of the flow from inputs I1 to I3 toward the flow path I4–O3.

From this point, further evolution of the network can take diverse routes depending on the recharge. Figure 3C shows the situation at $2\ T_B$ when the recharge can sustain constant head at the inputs. I1 first connects to the output O1 and later to the conduit system evolved from the input I4, to which I2 also connects. As long as the gradients in the network are maintained, a system of bypasses evolves, shortcutting and connecting the prominent flow paths. This results in a maze pattern (Fig. 3C). Input I3 is isolated from the prominent network and cannot develop its own flow path.

In most relevant situations, the catchment cannot supply enough water to the input. Figure 3D presents the outcome when the recharge to the inputs is limited to 100 L/s. After the breakthrough, the head at I4 and along the flow path I4–O3 drops. After $1.2\ T_B$, the head at I4 falls below 10 m and drops along the flow path toward I3. This results in high gradient between the flow paths originating from I1 to I3, which are still at high heads, and the flow path I4–O3. Therefore, inputs I1 and I2 integrate to I4–O3. Later on (not shown), I3 also integrates, because of the extreme gradients in its vicinity. Although some bypasses have evolved, the scenario clearly leads to a branchwork conduit system.

Although the presented model is simple and just one case, it clearly shows some basic rules of pattern development, which can be observed in the field and in other modeling results:

> Maze patterns evolve when the recharge is high enough to sustain high head in the network. In this case the maze evolves by the successive growth of bypasses emerging from the leading pathways. The whole conduit network thus evolves simultaneously with high dissolution rates.
> When recharge is limited, the head at the input and along the whole path decreases over time. This attracts the flow and growth of the neighboring flow paths emerging from other recharge points. The branchwork pattern evolves. The branchwork pattern continues to develop when one or more of the passages become vadose.

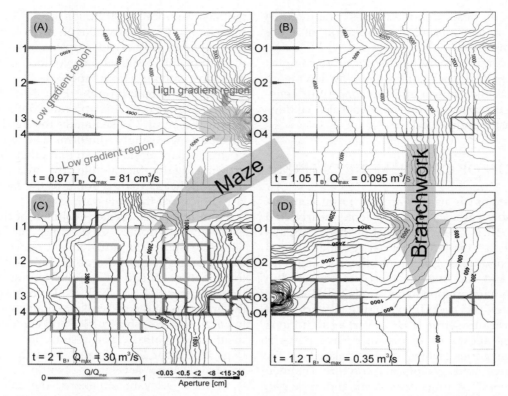

FIGURE 3 Evolution of a 2D fracture network with inputs I1–I4 on the left side of the domain and outputs O1–O4 on the right side. (A) Initiation phase, (B) situation shortly after the breakthrough, (C) growth phase (post-breakthrough) with recharge sustaining initial heads at the inputs, (D) growth phase with limited recharge to the inputs. Aperture width is represented by line thickness. Color presents the flow rates in units Q/Q_{max}, where Q_{max} is the maximal flow rate in the whole network.

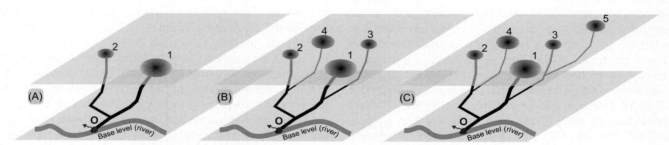

FIGURE 4 Upward evolution of branchwork pattern. See text for discussion.

The recharge conditions in natural systems are not constant. A branchwork pattern can develop during low flow and a maze pattern during high flow, when high head conditions allow expansion of the network by bypassing. These conditions give rise to the so-called floodwater mazes (Palmer, 2007).

The mechanisms revealed by this simple model are not the only ones leading to one or the other pattern. Maze caves can develop in settings where the input solutions are close to saturation. In this case the long penetration lengths, proportional to $(1 - c_{in}/c_{eq})^{1-n}$, result in slow but simultaneous widening of many alternative flow paths. Maze caves also develop when recharge occurs through a thin insoluble layer to many openings in limestone or directly into soluble rock, as in epikarst (Palmer, 2007).

However, in most telogenetic cases the maze development is not sustained. If and when a flow path evolves to the stage when the head along it drops, the neighboring flow paths (still at high heads) will grow toward it and a branchwork will develop.

Figure 4 shows a conceptual model where a branchwork pattern develops upward from the spring. In the example shown, input is distributed through the sinkholes on a plain (denoted 1 to 5 in Fig. 4A,B,C). The flow path connecting Sinkhole 1 evolves first. Sinkhole 2 initially evolves toward the valley, but is later redirected to the flow path 1—O, when its head becomes low enough (Fig. 4A). The evolution proceeds upward from the spring when the head in the evolved flow paths drops or they become vadose. Sinkholes 3 to 5 connect to one of the two primary flow paths (1—O, 2—O).

The parameter space that determines speleogenesis has many dimensions. It seems impossible to embrace even the most common speleogenetic scenarios in a short text. Conditions during speleogenesis continuously change; valleys are down cut and/or filled up; glaciations and deglaciations, transgressions and regressions, surface denudation and other processes continuously alter the evolution of cave systems. The role of sediments, which is important, has also not been discussed here. For a comprehensive picture, the reader is once more referred to the cited textbooks. However, there are many open questions in speleogenesis, which, when resolved, will surely add many new aspects to the present state-of-the-art.

See Also the Following Articles

Speleogenesis, Hypogenetic
Solution Caves in Regions of Moderate Relief
Solution Caves in Regions of High Relief

Bibliography

Dreybrodt, W., Gabrovsek, F., & Romanov, D. (2005). *Processes of speleogenesis: A modeling approach.* Ljubljana, Slovenia: ZRC Publishing.
Ford, D. C., & Williams, P. (2007). *Karst geomorphology and hydrology.* London: John Wiley & Sons.
Klimchouk, A., Ford, D. C., Palmer, A., & Dreybrodt, W. (Eds.), (2000). *Speleogenesis: Evolution of karst aquifers* Huntsville, AL: National Speleological Society.
Palmer, A. N. (2007). *Cave geology.* Dayton, OH: Cave Books.

SPELEOTHEM DEPOSITION

Wolfgang Dreybrodt

University of Bremen, Germany

INTRODUCTION

Precipitation of calcite from supersaturated H_2O–CO_2–$CaCO_3$ solutions sculptures fascinating cave decorations of stalagmites, stalactites, and other speleothems and also creates flowstone and sinter terraces. Rainwater seeping through vegetated soils on its way down to the limestone bedrock can absorb large amounts of carbon dioxide, which is present in the soil in much higher concentrations than in the atmosphere. When this water moves down, it dissolves limestone in the soil and in the fissures of the bedrock and comes very close to saturation with respect to calcite. If such a solution enters a cave, outgassing of CO_2 creates a supersaturated solution from which calcite is deposited. This causes the growth of calcite speleothems in their various shapes (White, 1988). The chemical composition of this solution depends on the CO_2 content of the soil atmosphere, but also on the conditions under which CO_2 is absorbed.

Two extremes can be envisaged. If the soil is free of calcite, only CO_2 is absorbed there, and dissolution of limestone later proceeds exclusively in the fractures of the rock under conditions closed to carbon dioxide. For each $CaCO_3$ unit dissolved, one CO_2 molecule is consumed. When this solution enters a cave, its CO_2 concentration may have become so low that carbon dioxide is taken up from the cave atmosphere and the solution becomes aggressive. This can cause the formation of Rinnenkarren extending below bedding planes in many caves. If, however, the partial pressure of carbon dioxide in the soil atmosphere is sufficiently high, outgassing of CO_2 creates a supersaturated solution precipitating calcite. The other extreme occurs when the soil contains limestone particles, and rainwater seeping down dissolves limestone under conditions open to the atmosphere such that for each $CaCO_3$ unit dissolved, one molecule of CO_2 is absorbed from the soil air. If such a solution reaches equilibrium with respect to calcite and later on encounters a cave, it will deposit calcite, provided the P_{CO_2} of the cave atmosphere is lower than that in the soil above. Intermediate cases, for example, when dissolution first proceeds in the soil under open conditions and then continues in the fissures of the rock under closed conditions, are also possible (Dreybrodt, 1999). These considerations show that supersaturation and consequently growth rates of speleothems are not simply related to soil P_{CO_2}, but also in a complex way to the hydrochemical pathway of the water on its way down to the cave.

THE CHEMISTRY OF PRECIPITATION AND DISSOLUTION OF CALCITE

To address the issues of how fast speleothems grow and which shape they take, one has to know the chemistry of a carbonate solution. Figure 1 illustrates the chemical reactions in the system H_2O–CO_2–$CaCO_3$. Gaseous CO_2 from the air enters into the water and is dissolved as aqueous CO_2, which reacts with water to form

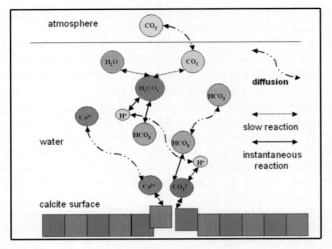

FIGURE 1 Visualization of the chemical reactions and transport by diffusion during dissolution/precipitation of calcite.

carbonic acid H_2CO_3. This is a slow reaction. The following reactions are fast. The complete reaction chain is:

$$CO_2^{atm} + H_2O \overset{K_H}{\leftrightarrow} H_2O + CO_2^{aq} \overset{K_0}{\leftrightarrow} H_2CO_3 \overset{K_1}{\leftrightarrow} H^+ + HCO_3^- \overset{K_2}{\leftrightarrow} H^+ + H^+ + CO_3^{2-}$$

CO_2^{aq} is in equilibrium with the partial pressure, $P_{CO_2}^{air}$, of CO_2 in the soil air by Henry's law $(CO_2) = K_H \cdot P_{CO_2}^{air}$. The mass action laws control the concentrations of the other species:

$$K_0 \cdot (H_2CO_3) = (CO_2); \quad K_1 \cdot (H_2CO_3) = (H^+) \cdot (HCO_3^-);$$
$$K_2 \cdot (HCO_3^-) = (H^+) \cdot (CO_3^{2-})$$

The Ks are the corresponding mass action constants. Parentheses denote concentrations. Furthermore dissociation of water, $H_2O \leftrightarrow H^+ + OH^-$ with $(H^+) \cdot (OH^-) = K_W$, must also be considered.

Ca^{2+} ions and CO_3^{2-} ions are released from the calcite surface by three simultaneous reactions:

$$CaCO_3 + H_2O \Leftrightarrow Ca^{2+} + CO_3^{2-} + H_2O; \quad CaCO_3 + H^+ \Leftrightarrow Ca^{2+} + HCO_3^-$$

and

$$H_2CO_3 + CaCO_3 + H_2O \Leftrightarrow Ca^{2+} + 2HCO_3^- + H_2O$$

These can be summarized by the global reaction $CO_2 + CaCO_3 + H_2O \Leftrightarrow Ca^{2+} + 2HCO_3^-$. Stoichiometry requires that for each Ca^{2+} released from the calcite surface, one molecule of CO_2 is converted to HCO_3^- and H^+ and is thus removed from the solution. On the other hand, if Ca^{2+} is deposited, one molecule of CO_2 is delivered into the solution. Since calcite dissolution and precipitation are ubiquitous in nature, this plays an important role in the global carbon cycle. During the chemical reactions at the surface all participating reactants must be transported to the surface or away from it by molecular diffusion. In this way concentration gradients arise and the concentrations at the calcite surface differ from those in the bulk of the solution.

The dissolution/precipitation rates F (mol cm^{-2} s^{-1}) at the surface have been found experimentally by Plummer et al., 1978. They are given by

$$F = k_1 \cdot (H^+)_s + k_2 \cdot (H_2CO_3)_s + k_3 - k_4 \cdot (Ca^{2+})_s \cdot (HCO_3^-)_s \quad (1)$$

The suffix s denotes the concentrations at the reacting surface, which are unknown. The rate constants of dissolution k_1, k_2, k_3 depend on temperature only, whereas the rate constant k_4, which comprises the three backward reactions of precipitation, depends also on the CO_2 concentration in the solution. The state of saturation of the solution is given by the ratio $\Omega = (Ca^{2+})(CO_3^{2-})/K_C$. K_C is the solubility constant. For $\Omega < 1$ the solution is undersaturated and calcite is dissolved; for $\Omega > 1$ the solution is supersaturated and calcite precipitates. Chemical equilibrium is attained if $\Omega = 1$. The equilibrium concentration of Ca^{2+} is $C_{eq} = (K_1 \cdot K_C \cdot K_H/4 \cdot K_2 \cdot \gamma_{Ca} \cdot \gamma_{HCO_3}^2)^{\frac{1}{3}} \cdot (P_{CO_2}^{air})^{\frac{1}{3}}$ (Dreybrodt, 1988). The γ are ionic activity coefficients and $P_{CO_2}^{air}$ is the partial pressure of CO_2 in equilibrium with the solution by Henry's law.

PRECIPITATION RATES FROM THIN WATER LAYERS

To calculate the rates, one needs the values of the concentrations at the surface. To obtain them from the bulk concentrations, which can be measured, one has to solve a system of coupled reaction–diffusion equations. These must take into account the reaction rates F as a boundary condition at the surface, the transport of reactants by diffusion, the slow reaction $H_2O + CO_2 \leftrightarrow H_2CO_3$, and the condition that in a system open to CO_2 for each calcium released one molecule of CO_2 must de-gas from the solution. Buhmann and Dreybrodt (1985) have solved the transport reaction equations numerically. In their numerical model they assumed a planar calcite surface covered by a stagnant water layer of thickness δ. The free surface of this layer is in contact with an atmosphere with partial pressure, $P_{CO_2}^{air}$, of carbon dioxide, and equilibrium between the CO_2 concentration in the water and its surrounding atmosphere is established. Figure 2 shows the precipitation rates for various thicknesses of the water film as a function of the Ca concentration in the solution at a temperature of T = 10°C and in equilibrium with a

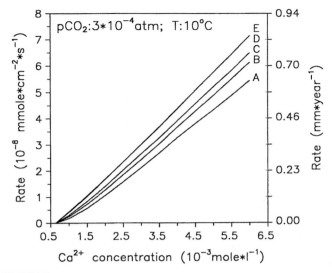

FIGURE 2 Precipitation rates of calcite from thin solution films of various depth: A: $\delta = 0.005$ cm; B: $\delta = 0.0075$ cm; C: $\delta = 0.01$ cm; D and E: 0.02 cm $< \delta < 0.04$ cm. The right-hand scale gives growth rates in mm yr^{-1}.

cave atmosphere with $P_{CO_2} = 3 \cdot \times 10^{-4}$ atm. All the curves in Figure 2 can be approximated within an error of less than 10% by the equation:

$$F = \alpha \cdot (C - C_{eq}) \quad (\text{mol cm}^{-2}\,\text{s}^{-1}) \qquad (2)$$

α (cm s^{-1}) is a reaction constant; C is the calcium concentration (mol cm^{-3}) in the solution, and C_{eq} its equilibrium concentration with respect to the partial pressure of CO_2 in the cave air. Values of α have been obtained numerically (Buhmann and Dreybrodt, 1985; Baker et al., 1998) and verified experimentally by Dreybrodt et al., 1997. The values of α depend on the film depth δ. For $\delta < 0.005$ cm they decrease with decreasing δ, because the slow reaction $H_2CO_3 \leftrightarrow H_2O + CO_2$ becomes rate controlling. The number of CO_2 molecules forming per time due to precipitation of calcite is $V \cdot R_{CO_2}$, where the rate function R_{CO_2} depends only on the carbonate and H^+ concentrations in the solution with volume V. Stoichiometry requires this number to be equal to the number of Ca^{2+} ions deposited per time to the surface area A, which is $A \cdot F$. This gives $F = R_{CO_2} \cdot (V/A)$. For planar water films $V/A = \delta$ and for sufficiently low values of δ the precipitation rates decrease linearly with δ. By use of the enzyme carbon anhydrase, R_{CO_2} can be enhanced. Therefore adding this enzyme to a calcite-precipitating solution enhances precipitation rates by an order of magnitude (Dreybrodt et al., 1997). As seen from Figure 2 the rates are only weakly dependent on δ for depths from 0.0075 cm to 0.04 cm. This is the thickness of films covering stalagmites. It turns out that α is almost independent of the P_{CO_2} in the cave air in the range from 3×10^{-4} atm up to 5×10^{-3} atm, which covers most of the natural conditions. Values of α in this range between 0°C and 30°C are reported by Baker et al. (1998). These can be fitted by the relation

$$\alpha = (0.52 + 0.04 \cdot T_C + 0.004 \cdot T_C^2) \cdot 10^{-7} \text{ m/s} \qquad (3)$$

where T_c is the temperature in centigrade (Romanov et al., 2008). α increases by a factor of ten with increasing temperature from 0°C ($\alpha = 5.2 \times 10^{-6}$ cm s^{-1}) to 30°C ($\alpha = 5.3 \times 10^{-5}$ cm s^{-1}). For $C = 2 \times 10^{-6}$ mol cm^{-3} and $C_{eq} = 0.5 \times 10^{-6}$ mol cm^{-3} one obtains growth rates between 0.2 mm yr^{-1} up to 2 mm yr^{-1}. Such values have been observed on many actively growing stalagmites (Baker et al., 1998; Genty et al., 2001), measuring growth rates by annual lamina counting, using stalagmites where drip-water samples were analyzed for Ca. When comparing theoretically predicted values, they observed good agreement between theoretical and measured stalagmite growth rates.

PRECIPITATION FROM WATER FILMS IN TURBULENT FLOW

Deposition of cave sinter in many cases occurs from turbulently flowing water. Relatively thin films of a few millimeters flow down from stalactites or stalagmites when water supply from drip points is sufficiently high. Supersaturated water flowing down cave walls precipitates flowstone. Finally cave rivers can shape sinter terraces and rimstone dams.

When water flow is turbulent, mass transport is enhanced by eddies in its bulk. This turbulent bulk, however, is separated from the mineral surface by a laminar diffusion boundary layer (DBL) with thickness ε, which depends on the hydrodynamic conditions, such as flow velocity, surface roughness, and depth of the flowing fluid. In this layer, mass transport proceeds by molecular diffusion solely. Because diffusional mass transport increases with decreasing ε, deposition rates are strongly dependent on the hydrodynamics of flow. In all the cases of interest here, ε is about a tenth of a millimeter (Dreybrodt and Buhmann, 1991).

Rates also depend on the depth δ of the flowing water, because $\delta = V/A$ determines the influence of HCO_3^- on the CO_2 reaction. For small $\delta < 0.2$ cm this reaction becomes rate limiting. As in laminar flow, the rates exhibit a linear relation of the Ca concentration as given by Eq. (2). Values of α have been listed by Liu and Dreybrodt (1997) for various values of δ, ε, P_{CO_2}, and temperature.

Figure 3 depicts precipitation rates calculated by use of these values and Eq. (2) for various thicknesses ε and δ as listed on the curves for a partial pressure

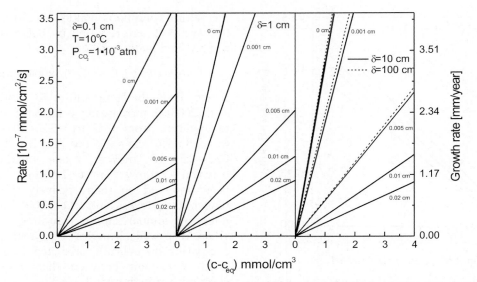

FIGURE 3 Precipitation rates of calcite from thin films of solution in turbulent flow with film depth δ of 0.1 cm, 1 cm, 10 cm, and 100 cm. The numbers on the lines depict the values of the thickness ε of the DBL.

$P_{CO_2} = 1 \times 10^{-3}$ atm of CO_2 in the solution at 10°C. These conditions apply to many cave rivers, from which flowstone is deposited.

In all cases one visualizes a strong decrease of the rates with increasing ε. The uppermost curves show the rates for $\varepsilon = 0$. In this case, due to complete mixing concentrations at the mineral surface equal those in the bulk. Due to the rate-limiting slow reaction of HCO_3^- conversion, the rates increase with increasing δ. There is a thickness δ (10 cm in Fig. 3) above which the rates become independent of δ. At that thickness the V/A ratio is sufficiently large and HCO_3^- conversion is no longer controlling the rates. This thickness increases with decreasing P_{CO_2} in the solution. Its value is about 1 cm at 5×10^{-3} atm, 10 cm at 1×10^{-3} atm, and 100 cm at 3×10^{-4} atm. For $\varepsilon = 0$ and $\delta > 10$ cm, neither mass transport nor HCO_3^- conversion is rate limiting and the rates can be obtained from the PWP equation, Eq. (1). These are the maximal possible rates for a given chemical composition of the solution.

The values of α depend on the P_{CO_2} in the solution. Figure 4 depicts as an example values of α in their dependence on P_{CO_2} for $\delta = 1$ and 10 cm, and $\varepsilon = 0.01$ and 0.02 cm. α depends almost linearly on temperature increasing by about a factor of 3, from 5°C to 20°C. Details can be found in Liu and Dreybrodt (1997).

In summary, rates from turbulently flowing water can be significantly higher than those from thin laminar films. This is the case when flow is highly turbulent and DBL thickness is less than 0.005 cm. When a DBL with $\varepsilon \approx 0.01$ cm is present rates drop, and for water films less than 1 mm deep they are close to rates in laminar or stagnant films of about 0.01 cm, because in both cases diffusion proceeds through a layer (DBL) or a laminar film of comparable thickness, and also HCO_3^-—CO_2 conversion occurs mainly in the DBL. This has been verified by field observations (Liu et al., 1995; Bono et al., 2001).

MORPHOLOGY OF REGULAR STALAGMITES

Franke (1965) was the first who recognized the principle of the growth of ideal regular stalagmites. He proposed the following mechanism assuming growth conditions constant in time. A drop of solution, supersaturated with respect to calcite falling to the ground spreads out radially when it impinges at the surface. Precipitation rates are maximal at the center and decrease with distance outward. This way a thin layer of precipitated calcite is created. It has a lenticular shape, thick at the center and thinning out at the edges. The next layer is built in the same way on the top of the previous one and a vertical candle-shaped stalagmite grows upward. After it has grown to a height, which is equal to its diameter its shape on the top becomes stable and also the diameter of the stalagmite does not change any more. In other words, the growth rates in the vertical direction are constant everywhere and the shape of the stalagmite is shifted vertically in its growth direction. Figure 5 shows this on a natural stalagmite, where the vertical distances between two growth layers are constant everywhere. Dreybrodt (1988, 1999) and Romanov et al. (2008) derived this growth

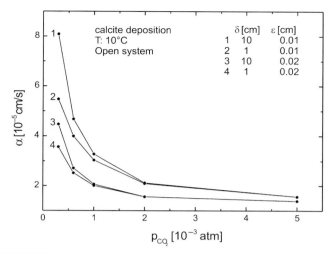

FIGURE 4 Values of the rate constants α in dependence on P_{CO_2} at 10°C for various values of the thickness ε of the DBL and the depth δ of a solution in turbulent flow. Filled circles are calculated points. *Redrawn after Bono et al., 2001.*

principle from three simple generally valid assumptions by use of numerical modeling, but also by strict mathematical proof. These are (1) the point of drop impact, the drip interval, and temperature are constant in time; (2) growth by deposition of calcite is directed normal to the actual surface of the stalagmite; (3) growth rates W(l) (cm yr^{-1}) decrease monotonically with distance l from drop impact along the actual surface and become zero at some large but finite distance l_0.

The basic principle of stalagmite growth can be formulated as follows: under constant growth conditions in time the top of a stalagmite attains an equilibrium shape, which during further growth is shifted up vertically. The part of the stalagmite below is a vertical cylinder with constant equilibrium radius R_{eq}. This is illustrated schematically in Figure 6, which shows two growth surfaces of a stalagmite in equilibrium, which are separated by some distance $W_0 \cdot \Delta t \cdot W_0$ (cm yr^{-1}) is the growth rate at the apex and Δt (year) the time distance when the two layers originated. The total volume V_c between these two layers is composed of circular rings (yellow) with radius x and width Δx, all with equal height $W_0 \cdot \Delta t$. They form a circular disc with radius R_{eq} and height $W_0 \cdot \Delta t$ with volume:

$$V_c = W_0 \pi R_{eq}^2 \cdot \Delta t \quad (4)$$

W_0 is the growth rate at the apex in cm yr^{-1}. The precipitation rates F are given by Eq. (2).

They are converted to a growth rate W in cm yr^{-1} (Dreybrodt, 1988) by

$$W = 1.17 \cdot 10^9 \alpha (c - c_{eq}) \text{ cm yr}^{-1} \quad (5)$$

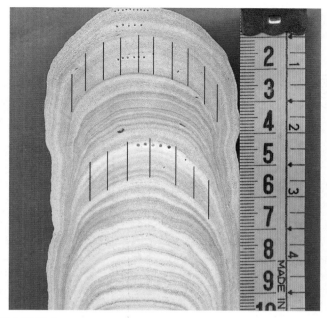

FIGURE 5 Longitudinal section through an ideal stalagmite. The black lines between growth lines, representing earlier surfaces, are of equal length. The stalagmite has grown by shifting its surface into the vertical direction. *Photo courtesy of D. Genty. Used with permission.*

Now the volume V_d deposited to the stalagmite in 1 year ($\Delta t = 1$) must be equal to the volume V_L of the amount of calcite lost during this time from the water, which enters with flow rate Q (cm^3 s^{-1}) and concentration C_{in} at the apex and leaves the stalagmite with equilibrium concentration C_{eq}:

$$V_L = \frac{V_{drop}}{\tau} \cdot (C_{in} - C_{eq}) \cdot 1.168 \cdot 10^9$$
$$V_d = W_0 \pi R_{eq}^2 = 1.168 \cdot 10^9 \alpha \cdot (C_{in} - C_{eq}) \cdot \pi R_{eq}^2 \quad (6)$$

V_{drop} is the volume of the drop and τ the drip interval. Therefore by equating $V_L = V_d$ one finds

$$R_{eq} = \sqrt{V_{drop}/\pi \alpha \tau} \quad (7)$$

This result relates in a simple way the equilibrium radius to the kinetic constant α and the drip interval τ. Note that the radius does not depend on the initial supersaturation $(C_{in} - C_{eq})$.

As an example, with $\alpha = 1.3 \times 10^{-5}$ cm s^{-1}, corresponding to a temperature of 10°C, a drop volume $V_{drop} = 0.1$ cm^3, and $\tau = 30$ s the equilibrium radius is $R_{eq} = 10$ cm, a very realistic number.

In this approach one assumes that the water supplied to the apex spreads out to a water film, which expands radially, flowing down the stalagmite in laminar flow. A water film covers the stalagmite with radial symmetry, as shown in Figure 7. One has to

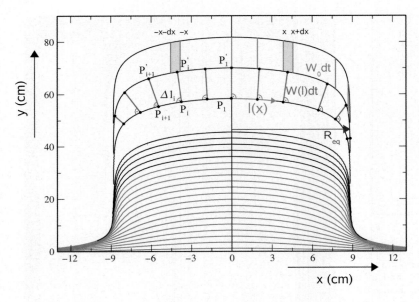

FIGURE 6 Computer simulation of a regular stalagmite. The red lines show the surface every 200 years during the initial stage. The black lines depict the equilibrium shape. The top four lines are separated by 1000 years. The green lines depict the growth $W(l) \cdot dt$ normal to the actual surface decreasing with distance l from the apex. The new surface after time step dt is generated by approximating the actual surface by a polygon P_1—P_i and calculating the new positions of the points P_i' from the points P_i by adding the vector $\to W(l_i)dt$. The red lines between the top two surfaces show vertical growth at all points equal to growth $W(0) \cdot dt = W_0 \cdot dt$ at the apex.

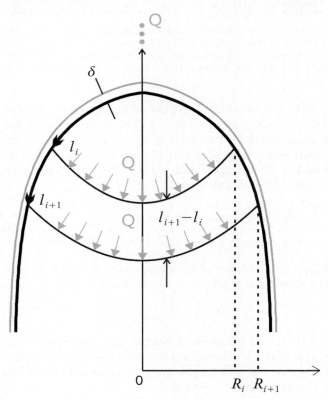

FIGURE 7 Water impinging on the apex with flow rate Q spreads out radially forming a water film with depth δ. An annulus of this film between distance ℓ_i and ℓ_{i+1} is shown. The water supply Q flows radially from the apex. At distance ℓ_i along the growth surface the water enters into the annulus with concentration c_i and leaves it at distance ℓ_{i+1} with concentration c_{i+1}.

consider an annulus between the points i and i + 1 at the surface with radial distance R_i and R_{i+1}, respectively. The distance between these two points along the surface is $l_{i+1} - l_i = \Delta l_i$. Mass balance requires that the amount of calcite deposited to the area of the annulus between points i and i + 1 must be equal to the loss of $CaCO_3$ in the solution when passing over the annulus. Because the total flow $Q = V_{drop}/\tau$ enters the annulus at R_i with concentration C_i and leaves it at R_{i+1} with concentration C_{i+1}, the mass of calcite lost per second from the solution covering the surface of the annulus is

$$M_s = Q(C_i - C_{i+1}) \qquad (8)$$

M_s must be equal to the mass M_p of calcite precipitated per second to the surface $S_a = 2\pi \cdot \Delta l_i \cdot R_i$ of the annulus. Note that M_S and M_P in this context are in mol s^{-1}.

$$M_p = 2\pi \cdot \Delta l_i \cdot R_i \cdot \alpha \cdot (C_i - C_{eq}) \qquad (9)$$

Equating $M_p = M_s$, one obtains

$$C_{i+1} = C_i - 2\pi \cdot \alpha \cdot (C_i - C_{eq}) \cdot R_i \cdot \Delta l_i/Q \qquad (10)$$

From this the deposition rates W_{i+1} can be calculated by subtracting C_{eq} from both sides of the equation and then multiplying both sides by α

$$W_{i+1} = W_i(1 - 2\pi \cdot \alpha \cdot R_i \cdot \Delta l_i/Q) \qquad (11)$$

Remembering that $R_{eq} = \sqrt{V/\pi\alpha\tau} = \sqrt{Q/\pi\alpha}$ Eq. (11) can be rewritten as

$$W_{i+1} = W_i \cdot (1 - 2 \cdot R_i \cdot \Delta l_i/R_{eq}^2) \qquad (12)$$

This recursive equation shows that growth rates decline with distance l_i. Equation 11, which contains only defined parameters and the somewhat idealistic assumption of continuous flow spreading out radially, can be used for numerical modeling. This has been performed as FLOW-model by Romanov et al. (2008). They employed the following procedure: the actual surface of the stalagmite at time t is approximated by the polygon P_1 to P_n, where P_1 is the point of drip impact. The surface at later time $t + \Delta t$ is obtained by drawing lines with lengths $W(l_i) \cdot \Delta t$ from points P_i normally to the actual surface (green lines in Fig. 6). This way the three basic assumptions on stalagmite growth are fulfilled and the new surface is approximated by the polygon P'_1 to P'_n, as shown in Figure 6.

The results are also shown in Figure 6. It depicts growth layers of a stalagmite growing on a plane surface. Note that the x and y axes have different scales. Therefore the picture presents a shape compressed in height. Clearly the shape of the stalagmite changes in its initial stage of growth (red lines) until an equilibrium shape is attained (black lines). From then on this shape is maintained during further growth and the vertical distance between any two equilibrium growth layers remains constant on all points of the surface. The vertical red lines, all of equal lengths, between the two upper surfaces indicate this. The equilibrium shape is attained after a growth height close to its diameter. Furthermore it is independent of the shape of the surface from where it started to grow (Dreybrodt, 1988).

A SIMILARITY RULE OF REGULAR STALAGMITES

An important rule can be obtained from Eq. (12). By introducing dimensionless coordinates $\tilde{x} = x/R_{eq}$ and $\tilde{y} = y/R_{eq}$ one finds

$$W_{i+1} = W_i \cdot (1 - 2 \cdot \tilde{R}_i \cdot \Delta \tilde{l}_i) \quad (16)$$

This, however, means that any regular stalagmite scaled by the factor $1/R_{eq}$ into this new coordinate system must exhibit a unique shape independent of its former dimensions. This proves the similarity rule: scale the stalagmites in both coordinates by the factor $1/R_{eq}$ and choose their growth axis as y-coordinate and their apex as origin. Then all regular stalagmites have the same shape.

To prove this by observation one has to select regular stalagmites, which show vertical walls about 1 diameter below their apex. An example is shown in Figure 5. Seven longitudinal sections of stalagmites exhibited in the Deutsche Archiv für Sinterchronologie (DASC) in the Schillat-cave, Germany, are used.

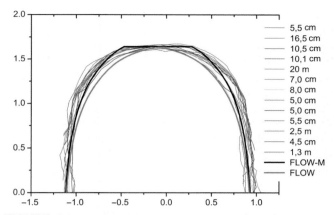

FIGURE 8 Rescaled shapes of all stalagmites compared to the shape obtained from the FLOW-model (fat red line) and a modified FLOW-model (fat black line). The numbers give the diameters of the stalagmites.

Furthermore four stalagmites have been taken from the literature. For details see Dreybrodt and Romanov (2008). They have also performed a search of regular large stalagmites on the Internet. The largest stalagmite in the world is reported from Cueva San Martin Infierno in Cuba. This giant stalagmite exhibits a height of about 70 m and 20-m diameter (www.goodearthgraphics.com/virtcave/largest.htm). Other ones have been found in Carlsbad Caverns, New Mexico, U.S.A., with a diameter of 4 m (www.fingerlakesbmw.org/visual/az/cccolumn1.jpg), and finally from Lost World Caverns, West Virginia, U.S.A., the Bridal Veil with 2-m diameter (www.wonderfulwv.com/archives/june01/fea1.cfm). Suitable growth layers, or the surface of those stalagmites, have been digitized. To compare the shapes all stalagmites were rescaled to the same radius $\tilde{R}_{eq} = 1$. Figure 8 shows the plots of the complete selection of samples and compares them to the theoretically obtained rescaled shape of the FLOW-model depicted in Figure 7. Although not perfect, the theoretical shape (fat red line) is quite close to the natural shapes. The deviation could be explained in the following way: whenever a drop impinges to the apex of a stalagmite, splashing occurs in a water layer covering the central part of the stalagmite's surface. This acts like a reservoir from which outward flow is supplied. In this layer due to mixing, when a new drop impinges, a constant calcium concentration will be maintained and the growth rate W(l) is constant within the radius. To model the shape of this stalagmite requires $W_i = W_0$ for all points i with $R_i \leq R_c$. For $R_i > R_c = 0.33 \cdot R_{eq}$ the recursive equation, Eq. (11), is valid. The shape obtained this way is closer to nature and is shown by the fat black line.

Also ideally shaped cone-like stalactites obey an analogous similarity rule (Short et al., 2005). These authors have modeled growth of stalactites, considering a thin film of water flowing evenly down the stalactite. Because of the decreasing diameter of the stalactite, its depth increases downward. As stated already, for very thin water films precipitation rates increase with film thickness. Consequently growth rates increase downwards the stalactite. Using this their model predicts similar shapes for all ideal stalactites. It has been verified by comparing the rescaled shapes of stalactites of various sizes in Kartchner Caverns (Benson, AZ, U.S.A.) to the predicted shape.

GROWTH OF STALAGMITES UNDER CONDITIONS VARIABLE IN TIME

Only a few stalagmites grow in their ideal shapes under conditions constant in time. Changes in temperature affect the rate constant α and may also influence meteoric precipitation, thus altering the drip intervals. In many caves one finds cone-shaped stalagmites, which form when the equilibrium radius $R_{eq}(t) = \sqrt{V/\pi\tau(t)\alpha(t)}$ decreases in time. Club-like stalagmites grow when $R_{eq}(t)$ increases in time. From this, one understands why, in spite of the simple growth mechanism, stalagmites exhibit such a variety of shapes.

As an example, one assumes a periodic change of drip intervals τ as shown in Figure 9B. The temperature remains constant. The shape of a stalagmite growing under such conditions can be modeled by adjusting in each time step $\alpha(t)$ and $\tau(t)$. The result is shown in Figure 9A. The stalagmite grows with periodic change of its diameter. To illustrate which effect climatic changes could imprint into the shape of a stalagmite, one can use the temperature variations from the Vostok ice core rescaled to moderate climates. Additionally one assumes a linear relation between temperature and meteoric precipitation, which cause low drip intervals at warm climatic conditions and higher ones for cooler climate. This is illustrated in Figure 9D. Furthermore P_{CO_2} is low at cold climates and because of stronger vegetation it becomes high at warmer temperatures. The resulting stalagmite is shown in Figure 9C. The growth lines are drawn every 1000 years. At high temperatures τ is small and α is moderately large. For high P_{CO_2} the equilibrium concentration C_{eq} is high, causing large growth rates. Large diameters and wide distances between the growth lines reflect this. The regions where growth is slow and growth lines become dense (black parts) exhibit small diameters because drip intervals are long. This rather crude simulation shows as an example how climatic changes might affect the shapes of stalagmites. More details can be found in Kaufmann (2003).

Recently Tan et al. (2006) have measured the width of annual lamina, as they occur in many stalagmites, on 2000 lamina in a stalagmite of China. They suggest that the time series of growth rates obtained this way may be used as a proxy of paleoclimate. The theoretical

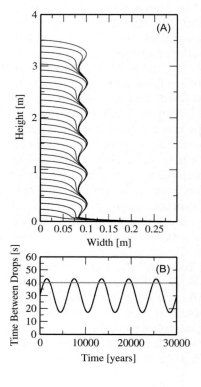
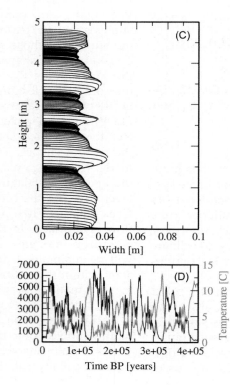

FIGURE 9 Model of stalagmite growth under conditions varying in time. (A) and (B) Growth at constant temperature; drip rate varies periodically. $\alpha = 1.3 \cdot 10^{-5}$ cm s^{-1}. (C) and (D) Growth under climatic conditions (temperatures) as obtained from the Vostok ice core.

and experimental knowledge on calcite precipitation is an important tool to decipher such proxies.

Bibliography

Baker, A., Genty, D., Dreybrodt, W., Barnes., W. L., Mockler, N. J., & Grapes, J. (1998). Testing theoretically predicted stalagmite growth rate with recent annually laminated samples: Implications for past stalagmite deposition. *Geochimica et Cosmochimica Acta*, 62, 393–404.

Bono, P., Dreybrodt, W., Ercole, S., Percopo, C., & Vosbeck, K. (2001). Inorganic calcite precipitation in Tartare karstic spring (Lazio, central Italy): Field measurements and theoretical prediction on depositional rates. *Environmental Geology*, 41, 305–313.

Buhmann, D., & Dreybrodt, W. (1985). The kinetics of calcite dissolution and precipitation in geologically relevant situations of karst areas: 1. Open system. *Chemical Geology*, 48, 189–211.

Dreybrodt, W. (1988). *Processes in karst systems. Physics, chemistry and geology.* Berlin: Springer.

Dreybrodt, W. (1999). Chemical kinetics, speleothem growth and climate. *Boreas*, 28, 347–356.

Dreybrodt, W., & Buhmann, D. (1991). A mass transfer model for dissolution and precipitation of calcite from solutions in turbulent motion. *Chemical Geology*, 90, 107–122.

Dreybrodt, W., Eisenlohr, L., Madry, B., & Ringer, S. (1997). Precipitation kinetics of calcite in the system $CaCO_3-H_2O-CO_2$: The conversion to CO_2 by the slow process $H^+ + HCO_3^- \rightarrow CO_2 + H_2O$ as a rate limiting step. *Geochimica et Cosmochimica Acta*, 61(18), 3897–3904.

Dreybrodt, W., & Romanov, D. (2008). Regular stalagmites: The theory behind their shape. *Acta Carsologica*, 37, 175–184.

Franke, H. W. (1965). The theory behind stalagmite shapes. *Studies in Speleology*, 1, 89–95.

Genty, D., Baker, A., & Vokal, B. (2001). Intra- and inter-annual growth rate of modern stalagmites. *Chemical Geology*, 176, 191–212.

Kaufmann, G. (2003). Stalagmite growth and paleoclimate: The numerical perspective. *Earth and Planetary Science Letters*, 214, 251–266.

Liu, Z., & Dreybrodt, W. (1997). Dissolution kinetics of calcium carbonate minerals in H_2O-CO_2 solutions in turbulent flow: The role of the diffusion boundary layer and the slow reaction $H_2O + CO_2 \rightarrow H^+ + HCO_3^-$. *Geochimica et Cosmochimica Acta*, 61, 2879–2889.

Liu, Z., Svensson, U., Dreybrodt, W., Yuan, D., & Buhmann, D. (1995). Hydrodynamic control of inorganic calcite precipitation in Huanglong Ravine, China: Field measurements and theoretical prediction of deposition rates. *Geochimica et Cosmochimica Acta*, 59(15), 3087–3097.

Plummer, L. N., Wigley, T. M. L., & Parkhurst, D. L. (1978). The kinetics of calcite dissolution in CO_2-water systems at 5°C to 60°C and 0.0 to 1.0 atm CO_2. *American Journal of Science*, 278, 537–573.

Romanov, D., Kaufmann, G., & Dreybrodt, W. (2008). Modeling stalagmite growth by first principles of chemistry and physics of calcite precipitation. *Geochimica et Cosmochimica Acta*, 72, 423–437.

Short, M. B., Baygents, J. C., Beck, J. W., Stone, D. A., Toomey, R. S., & Goldstein, R. E. (2005). Stalactite growth as a free-boundary problem: A geometric law and its platonic ideal. *Physical Review Letters*, 94, 018501.

Tan, M., Baker, A., Genty, D., Smith, C., Esper, J., & Cai, B. (2006). Applications of stalagmite laminae to paleoclimate reconstructions: Comparison with dendrochronology/climatology. *Quaternary Science Reviews*, 25(17–18), 2103–2117.

White, W. B. (1988). *Geomorphology and hydrology of karst terrains.* New York: Oxford University Press.

SPELEOTHEMS: GENERAL OVERVIEW

William B. White

The Pennsylvania State University

Speleothems are secondary deposits in caves. *Speleothem* is a made-up word. The term was invented in 1952 to replace the obsolete term *cave formations*, which could be easily confused with *bedrock formations*. Speleothems are the primary attraction in show caves and provide much of the esthetic attraction for cave explorers. This article is intended as a broad overview of speleothems. More detail is provided in other articles on specific speleothem groups. For a comprehensive discussion of cave minerals and speleothems see Hill and Forti (1997).

SPELEOTHEM MINERALS

Although speleothems can be composed of many minerals, in fact the vast majority of speleothems is composed of one (or a mixture) of only three minerals: calcite, $CaCO_3$; aragonite, also $CaCO_3$; and gypsum, $CaSO_4 \cdot 2H_2O$. Of these, calcite is, by far, the most common. Each of these minerals has a different crystal structure and also different habits of crystal growth.

Calcite is trigonal, space group $R\bar{3}c$. The structure consists of alternating layers of Ca^{2+} ions and CO_3^{2-} ions stacked perpendicular to the c-crystallographic axis. Calcite occurs in nature in a great variety of forms but the most common in caves is the scalenohedron. The c-axis is the fast growth direction in calcite and this is responsible for much of the observed microstructure in speleothems.

Aragonite is orthorhombic, space group Pmcn. Although it has the same chemical composition as calcite, the calcium and carbonate ions are more densely packed in aragonite, giving the mineral a somewhat higher density. Crystal growth in the [001] direction is much faster than other directions, giving many aragonite speleothems an acicular habit.

Gypsum is monoclinic, space group A2/a (or I2/a). Layers composed of Ca^{2+} ions and SO_4^{2-} ions are separated by a layer of H_2O molecules so that these layers are held together only by hydrogen bonds (Fig. 1). The result is that gypsum has a perfect cleavage on {010}. Large crystals can be easily cleaved into transparent slabs, a form of gypsum known as *selenite*. The strongest bonding is in the [001] direction which is also the fast crystal growth direction. Gypsum grows readily as [001] fibers, a habit that produces the form of gypsum

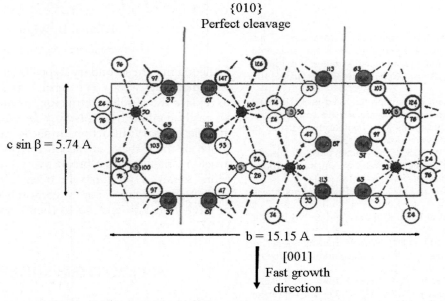

FIGURE 1 The crystal structure of gypsum, $CaSO_4 \cdot 2H_2O$, showing the direction of perfect cleavage and also the fast growth (fiber) direction.

known as *satin spar* and is responsible for the widely occurring gypsum flowers in caves.

MECHANISMS FOR SPELEOTHEM DEPOSITION

The essential condition for the deposition of any speleothem is that water bearing mineral matter in solution enters the cave out of chemical equilibrium with the cave environment. Some reaction will then take place in which the water chemistry strives toward equilibrium with the concurrent deposition of minerals.

Carbonate minerals, both calcite and aragonite, are formed when water, supersaturated with respect to these minerals is exposed to the cave atmosphere.

$$Ca^{2+} + 2HCO_3^- \leftrightarrow CaCO_3 + CO_2 + H_2O$$

If the CO_2 pressure of the aqueous solution is less than the CO_2 pressure of the cave atmosphere, CO_2 will de-gas and calcite (or aragonite) will precipitate. This is the overall mechanism. The details are much more complicated with multiple subreactions depending on temperature and other ions present in the water. However, the most important variable is the CO_2 pressure in the infiltrating water compared with the background CO_2 pressure of the cave atmosphere. Calcite and aragonite can also be precipitated by evaporation of the solutions. Evaporation becomes important in well-ventilated caves and near cave entrances. In general, evaporation produces very small carbonate crystals; the largest and best-developed crystals are found in water-saturated caves where CO_2 loss is the only mechanism.

In contrast, the sulfate minerals, gypsum, and also mirabilite, epsomite, and others, are deposited mainly by evaporating water. Dissolved sulfate ions recombine with Ca^{2+}, Mg^{2+}, Na^+, and other cations as evaporation drives the solution to supersaturation.

SPELEOTHEM SHAPES

The shapes of speleothems are the result of a competition between the hydraulics of the fluids carrying the nutrient to the growing speleothem and the tendency of the crystals to take on their natural habits. Because of the great variety of combinations of flow regimes and growth habits there are a very large number of speleothem shapes. These shapes have been given a great variety of names and there have been numerous attempts at classifications, none of them entirely satisfactory. Some classifications are very elaborate, attempting to provide a slot for every possible speleothem. Others consist only of an alphabetical

TABLE 1   A Classification of Speleothems

- A. Forms created by flowing water
  1. Flowstone
  2. Rimstone dams
  3. Botryoidal forms (globulites, cave popcorn, cave coral)
- B. Forms created by dripping water
  1. Stalactites
  2. Stalagmites
  3. Draperies
  4. Shields
- C. Forms created by seeping water
  1. Crusts
  2. Helictites
  3. Oulopholites (gypsum flowers)
  4. Anthodites/frostwork
- D. Forms created in standing water
  1. Pool spar
  2. Concretions (including cave pearls)

FIGURE 2   Flowstone formed on the bed of a small in-feeder stream, Marvel Cave, Missouri. *Photo by the author.*

FIGURE 3   Flowstone in Furong Dong, Chongqing Province, China. The height of the flowstone is about 10 meters. The colors are due to the show cave lighting. *Photo by the author.*

listing of the names. The classification given in Table 1 is a compromise based on the flow characteristics of the fluids that deposit the speleothems.

Speleothems Created by Flowing Water

Flowing water in caves may occur as streams flowing along the cave floor and it may take the form of thin sheets of water flowing down cave walls. Of these, thin sheets of flowing water, draining through the vadose zone from the epikarst, contain higher concentrations of dissolved carbonates, and are at higher CO_2 pressures. Thin sheets of flowing water are responsible for a great deal of speleothem deposition. In contrast, caves streams are usually undersaturated with respect to carbonates and speleothem deposition on stream beds is relatively uncommon although there are exceptions (Fig. 2).

CO_2 degassing from moisture films on cave walls produces sheets of calcite known as *flowstone* (Fig. 3). Flowstone consists of superimposed layers of calcite crystals which typically grow with their c-axes perpendicular to the sheet.

Cave streams near chemical saturation have a remarkable ability to pond themselves in the form of *rimstone dams* (Fig. 4). The mechanism for the formation of dams is not known with certainty and it is possible that microorganisms play a role. They are found in small and large flowing streams where the chemistry is correct and are also found in surface rivers of which the Krka River and the Plitvice Lakes in Croatia are outstanding examples.

More enigmatic speleothems are the nodular or bead-like forms variously known as *cave beads, cave coral, cave grape, cave popcorn,* or *globulites.* The speleothems are small, typically a few millimeters to a few centimeters, and grow from cave walls and from other speleothems. Some consist entirely of calcite and form on rough cave walls that have been etched by descending streams of water (Fig. 5) or on the sides of stalactites (Fig. 6). These are layered structures with the growth originating from small projections on the cave wall (Fig. 7). Some nodular speleothems contain aragonite and other minerals and are not as obviously associated with flowing water.

Speleothems Created by Dripping Water

Water infiltrating downward from the epikarst follows available pathways along fractures until it emerges from the ceiling of an underlying cave. If the seepage rates are low, drops form and hang for a period of time during which CO_2 is degassed from the drop, thus supersaturating it with respect to calcite.

FIGURE 4 Rimstone dams formed in a shallow pool, a form sometimes called "lily pads." *Photo by the author.*

FIGURE 6 Nodular speleothems on the sides of stalactites in Swago Pit, West Virginia. *Photo by the author.*

FIGURE 5 Cave beads formed on a bare bedrock wall in Harlansburg Cave, Pennsylvania. The top scale is in inches; the bottom scale in centimeters.

FIGURE 7 Cross-section of nodular speleothem showing banded structure. The diameter is about 1 cm. *Photo by the author.*

FIGURE 8 Straw stalactites and other speleothems in Whisper Rocks (Lincoln Caverns), Huntingdon County, Pennsylvania. Note the fracture that provides the pathway for the seeping vadose water. *Photo courtesy of Lincoln Caverns. Used with permission.*

The calcite deposits as a thin ring on the surface from which the drop is suspended. Eventually, the drop falls, leaving the ring of calcite behind and a new drop forms. This process is repeated and the deposit takes the form of a straw-shaped *stalactite* (Fig. 8). Straw stalactites are hollow tubes, about 5 mm in diameter, and are often very coarsely crystalline with the straw axis corresponding to the c-axis direction of the calcite crystals. Stalactites continue to grow laterally and vertically from water flowing down the outside as well as through the central canal (Fig. 9). Stalactites can grow very large and indeed reach a weight that exceeds the tensile strength of their attachment and they spontaneously break loose and fall. Cross-sections of stalactites typically consist of concentric rings of calcite around the central canal.

Water drops from cave ceilings or from stalactite tips usually fall before depositing their entire load of calcite. Additional deposition takes place when these falling drops hit the floor, thus forming upward-growing *stalagmites*. Stalagmites grow as a sequence of superimposed layers. There is no central canal. Stalagmites growing beneath a single drip site with a uniform drip rate and uniform chemistry take the form of vertical cylinders with varying diameters depending on the drip rate and drip water chemistry (Fig. 10). Because stalagmites rest on cave floors, they can grow to sizes limited only by the size of the cave passage.

Stalactites and their counterpart stalagmites eventually grow together to form columns. Multiple drip sites and shifting drip sites combine to generate complex speleothems that can become very massive (Fig. 11). Water dripping along fractures can produce a stalactite form known as the *drapery*, an intermediate form between flowstone and an isolated stalactite.

Related to dripstone and flowstone forms are the relatively rare speleothems known as *shields* (Fig. 12). Shields appear to be formed by waters moving along fractures under pressure so that shields can form at any angle to the vertical. Shields contain a medial crack rather than a medial canal. Shields often have a fringe of stalactites formed by water seeping from the medial crack and dripping off the edges. Because shields require unusual hydraulic conditions, they are found on only a few caves such as Grand Caverns, Virginia, and Lehman Caves, Nevada.

FIGURE 10 Stalagmites in Postojna Cave, Slovenia. *Photo by the author.*

FIGURE 9 Carrot-shaped stalactites grown by enlargement of straw stalactites in Ohio Caverns, near West Liberty, Ohio. *Photo by the author.*

FIGURE 11 Massive stalactite-stalagmite combination. Buckhill Cave, Virginia. *Photo by the author.*

FIGURE 12 Shields located in Grand Caverns, Virginia. Note that the medial planes are at different oblique angles to the vertical. *Photo by the author.*

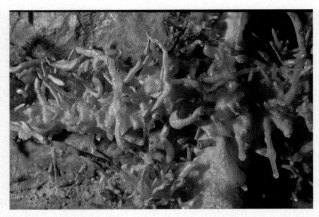

FIGURE 13 Helictites in Sites Cave, Pendleton County, West Virginia. *Photo by the author.*

FIGURE 14 Gypsum flower from Martin Ridge Cave System, Kentucky. *Photo by Joe Kearns. Used with permission.*

Speleothems Created by Seeping Water

The carbonate and gypsum rocks in which caves form tend to have a very low permeability so that most of the groundwater movement is through joints and bedding plane partings. However, low permeability is not zero permeability and there can be seepage through the bulk rock in addition to seepage along fractures. Fluids seeping through pores in the rock carry mineral matter which is deposited as crusts or more fully developed speleothems when the solutions evaporate into the cave passage. The high supersaturations achieved by these evaporating solutions often result in the deposition of aragonite rather than calcite and thin crusts are revealed under the microscope to be composed of mats of very small acicular crystals of aragonite. The growth of larger aragonite crystals in the form of *frostwork* or *anthodites* is not well understood but they appear to be formed from seeping fluids. These are masses of aragonite fibers or radiating sprays of aragonite crystals that may reach lengths of many centimeters.

Helictites are an erratic stalactite-like form that occurs when seepage rates are too slow for complete drops to form. Instead, the speleothem follows the whims of calcite growth (Fig. 13). Helictites have a central canal and grow primarily from the tip.

Sulfate-bearing solutions seeping from pores in the bedrock deposit crusts of gypsum and other sulfate minerals. Continued seepage produces additional crystal growth. If there is a zone of higher permeability, gypsum growth takes the form of fibrous crystals which push outward. Because the fastest growth rate is at the center, the fibers curl to the outside producing the gypsum flower (Fig. 14). "Flowers" can also form from other sulfate minerals such as mirabilite, $Na_2SO_4 \cdot 10H_2O$, and epsomite, $MgSO_4 \cdot 7H_2O$.

Speleothems Created in Standing Water

There is a marked contrast between speleothems that grow at the interface between air and a water surface and speleothems that grow entirely immersed in water. Subaerial speleothems have smooth rounded surfaces dictated by the characteristics of the flowing water. Subaqueous speleothems exhibit the growth habits of their component minerals.

Cave pearls are near-spherical, unattached speleothems that form in pools agitated by water dripping from considerable heights (Fig. 15). They form in concentric layers around a nucleus which can be a sand grain or even a bat bone. Drip pools in former show caves sometimes contain cave-pearl–like objects formed as calcite coatings built up over bits of gravel that had been used in the

FIGURE 15 A pool with cave pearls. Barberry Cave, Bath County, Virginia. *Photo by Nevin W. Davis. Used with permission.*

FIGURE 16 Pool spar from Helictite Cave, Highland County, Virginia. *Photo by Philip C. Lucas. Used with permission.*

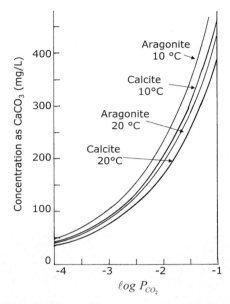

FIGURE 17 Solubility curves for calcite and aragonite as a function of CO_2 partial pressure. Calculated from thermodynamic data.

tourist trail. Sufficient agitation from dripping water is needed to keep the pearls from becoming attached and to roll them into roughly spherical forms. There are a few rare examples of cave pearls in the form of cubes tightly nested in the drip pool.

Standing water in cave pools can become supersaturated by the gradual degassing of CO_2 from the pool surface. Calcite crystals may precipitate on the surface of the pool and aggregate to form *calcite rafts*. These may be suspended by the surface tension of the water until they finally accumulate sufficient mass to sink. They collect on the bottoms of pools as loosely aggregated plates.

Crystals forming in pools are unconstrained and can adopt the natural habit of the precipitating mineral. Calcite is the most common and takes the form of pool spar. Cave pools are frequently crystal-lined, often with a coarse mass of small crystals but sometimes with near perfect scalenohedra (Fig. 16).

SPELEOTHEM PROPERTIES

The Calcite/Aragonite Problem

Calcite is the thermodynamically stable form of $CaCO_3$ at atmospheric pressure over the entire temperature range from absolute zero to the decomposition temperature. The thermodynamic stabilization of aragonite at ambient temperature requires pressures in excess of 3000 atmospheres. But oblivious to thermodynamics, aragonite is a common mineral in caves where it is precipitated from seepage and drip waters at cave temperature and atmospheric pressure. How to explain this observation has occupied many researchers for many years and is known as the "calcite/aragonite problem."

Aragonite is more soluble than calcite at any given temperature and CO_2 pressure but only 10–15% more soluble (Fig. 17). What is needed are mechanisms that will either inhibit the precipitation of calcite so that the concentration of ions in solution can increase past the aragonite solubility curve or that will enhance the nucleation of aragonite which can then grow metastability in competition with calcite. Two important mechanisms that have been identified involve the common minor elements in karst waters: Mg^{2+} and Sr^{2+}. Mg^{2+} ions have been shown to poison the growth centers on calcite crystal, thus inhibiting calcite growth. $SrCO_3$ has the same crystal structure as aragonite. Early precipitation of $SrCO_3$ provides a nucleus on which aragonite can grow. There may be other mechanisms and other contributing ions.

In many caves, the carbonate speleothems consist entirely of calcite. But when aragonite speleothems are found, there are usually many of them. This suggests that the presence (or absence) of aragonite is controlled by local chemical composition of the bedrock or chemical composition of the vadose seepage water. Aragonite is often identified by its crystal habit since it has a tendency to grow as tufts of radiating acicular crystal. Flowstone and dripstone forms of aragonite do occur but are uncommon (Fig. 18).

Color and Luminescence of Speleothems

Chemically pure calcium carbonate and calcium sulfate are colorless. Most calcite speleothems are colored various shades of tan, orange, and brown. Although a few speleothems are pigmented by incorporated iron oxides and hydroxides, most speleothem colors arise from humic substances extracted from the overlying soil and incorporated in the speleothems. The large and complex humic and fulvic acid molecules seem to fit into the calcite structure but not the aragonite structure. Aragonite and gypsum speleothems tend to be colorless (Fig. 18).

Calcite speleothems are usually luminescent under excitation by ultraviolet light. The emission occurs as a broad band in the range of 400–450 nm, giving the luminesence a blue-white color (Fig. 19A). Occasionally, a more yellow emission is observed on organic-rich speleothems (Fig. 19B). The luminescence, like the color, is due to humic substances. The fulvic acid fraction produces luminescence; the humic acid fraction is responsible for the color. For this reason, nearly white speleothems tend to have the strongest luminescence.

SPELEOTHEMS IN VOLCANIC CAVES

In solution caves, those in carbonate rocks and gypsum, the deposition of speleothems is the reverse of the process that formed the caves. Speleothem deposition in volcanic caves is a more complicated sequence of events (Forti, 2005). Lava tube caves result from the draining of tubular flows of lava beneath the surface. Such caves form very rapidly and many of the explored lava tube caves are only years to centuries old. The speleothems that form in lava tube caves develop in a sequence as the tube drains, cools, and becomes subject to rainwater infiltration.

In the final stages of the draining of the tube, still-molten lava can drip from the tube ceiling and then

FIGURE 18 Massive column of aragonite in Carroll Cave, Missouri. Note color contrast between the white aragonite and the yellow and brown calcite speleothems. *Photo by the author.*

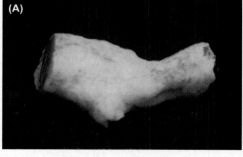 

FIGURE 19 (A) Typical blue-white luminescence from a segment of stalactite. (B) Oily yellow luminescence from a stalactite fragment from Cueva del Guacharo, Venezuela. *Photos by the author.*

freeze to produce lava stalactites. Lava stalagmites and other forms can also develop. Some have questioned whether these lava forms should be called speleothems. Purists have argued that speleothems are secondary deposits and that lava stalactites are primary features.

As cooling progresses, the still hot lava can be impacted by water either from rainfall or—in coastal areas—by saltwater from wave action or storm surges. The tubes and associated fractures can be infused with superheated steam which can extract and transport heavy metals from the lava. The resulting mineral deposits take the form of coatings and small nodules and stalactites sprayed on the tube walls as brightly colored silicates of copper, vanadium, and other metals. The chloride content of seawater is particularly effective in mobilizing metals. Elemental sulfur is also found.

The most dramatic speleothems form in tubes only a few years old that have cooled enough to permit some circulation of liquid water. The few such caves that have been examined (e.g., Porter, 2000) contain suites of complex sulfate minerals, many of which are highly hydrated and water soluble. The sequence of mineralization is highly transitory. Further cooling of the tube and infiltration of rainwater quickly removes the water-soluble minerals.

The final stage is a lava tube at ambient temperature with the same dripping water found in other caves. The resulting speleothems are also very similar to those found in solution caves. Gypsum is common as crusts and nodules. Calcite occurs as nodules and rarely as dripstone.

SPELEOTHEMS AS PALEOCLIMATE ARCHIVES

Although speleothems and the processes that form them are interesting in their own right, the most important speleothem study is their role as archives of paleoclimate (Fairchild *et al.*, 2006). Stalagmites, in particular narrow cylindrical stalagmites, are built up layer by layer and can be regarded as a drill core available without the necessity of drilling. A longitudinal section through such a stalagmite provides a microstratigraphy that can span thousands to tens of thousands of years of time. U-Th dating provides an absolute chronology than can, in principle, extend back to the mid-Pleistocene.

Profiles can be measured along the stalagmite axis of oxygen isotope ratios, H/D ratios, carbon isotope ratios, strontium isotope ratios, and the concentration profiles of chosen trace elements such as Mg, Sr, and any other trace elements that might be climate

FIGURE 20 Sequences of speleothem deposition in Ruddle Cave, Pendleton County, West Virginia. Note first-deposited dark brown speleothems that were coated by a later flooding event, followed by more recent deposition of the orange stalagmite. *Photo by the author.*

signatures. Likewise, the color banding, luminescence banding, and crystal growth textures can be measured. Because the water that deposits speleothems begins its journey as rainfall which then passes through the soil, the epikarst, and the vadose zone to reach the site of deposition in the cave, it should carry a signature of the surface climate. The climate signatures at each time interval are stored in the growth layers of the stalagmite. There are certainly signals present and much current research is focused on the interpretation of these signals.

Because paleoclimate studies require removing stalagmites from caves and making measurements in the laboratory, attention has focused on climate changes during the closing phases of the last ice age and on climate changes during the Holocene. What awaits are investigations of entire caves and of larger and older speleothems. Consider Figure 20. There was clearly an earlier phase of speleothem deposition, following by a period of flooding that coated the speleothems with silt and clay, and finally the most recent stage when the clear orange stalagmite was deposited. The size of the final stalagmite suggests that the previous events are quite old. There are many cave sites similar to Figure 20, just waiting for the investigators.

See Also the Following Articles

Gypsum Flowers and Related Speleothems
Helictites and Related Speleothems
Paleoclimate Records from Speleothems
Speleothem Deposition
Stalactites and Stalagmites
Uranium Series Dating of Speleothems

Bibliography

Fairchild, I. J., Smith, C. L., Baker, A., Fuller, L., Spötl, C., Mattey, D., et al. (2006). Modification and preservation of environmental signals in speleothems. *Earth Science Reviews, 75*, 105–153.

Forti, P. (2005). Genetic processes of cave minerals in volcanic environments: An overview. *Journal of Cave and Karst Studies, 67*, 3–13.

Hill, C., & Forti, P. (1997). *Cave minerals of the world.* Huntsville, AL: National Speleological Society.

Porter, A. (2000). The initial exploration of Lower Lae'apuki Cave System, Hawai'i Volcanoes National Park. *National Speleological Society News, 58*, 10–17.

SPIDERS AND RELATED GROUPS

James R. Reddell
The University of Texas at Austin

INTRODUCTION

The class Arachnida is second only to the Insecta class in number of species and diversity (Savory, 1977). It is also among the more important groups of invertebrates occurring in caves, with 9 of the 11 extant orders containing troglobites. The windscorpions (order Solifugae) do not occur in caves except as accidentals. The whipscorpions (order Uropygi) rarely occur as trogloxenes in tropical caves.

DISTRIBUTION

The class Arachnida is distributed worldwide and, with the exception of many Acarina, is entirely terrestrial. Of the orders containing troglobites, three (Schizomida, Amblypygi, and Ricinulei) are predominantly tropical and subtropical in distribution. Troglobitic Ricinulei occur only in the New World tropics and subtropics and troglobitic Schizomida are known only from North America, South America, and Australia. With the exception of one species each in Israel and Sarawak, unquestionably troglobitic Scorpiones occur only in Mexico and Ecuador. The remaining orders include troglobites in both Old and New Worlds and in tropical and temperate regions.

With some exceptions, genera containing troglobites belong to families also known from the surface. The spider family Holarchaeidae is known only from two troglobitic species in New Zealand. Several families of aquatic mites are apparently restricted to subterranean waters. Although some genera contain only troglobites, they typically are closely related to other genera occurring on the surface in the same area. Relict species are known in several groups. The troglobitic telemid spider *Telema tenella* from Europe is the only species in the family occurring in temperate regions. Other isolated species now inhabit caves in areas outside the general range of the genus or family. These include species that have survived in the moist habitat of caves in desert regions. Examples of this type of distribution include several species of Schizomida, Ricinulei, and Opiliones in northern Mexico and the southern United States.

Early studies of the distribution of troglobites in general indicated that troglobites were extremely rare in tropical lowlands. In recent years extensive studies in the tropics have demonstrated a rich troglobitic fauna. Troglobitic arachnids are especially abundant in areas previously thought to contain a depauperate fauna. It is typical of most tropical areas for the insect fauna to include few troglobites. For example, in the Yucatan Peninsula of Mexico, only one true insect (a phalangopsid orthopteran) is considered to be troglobitic, whereas about 18 species of arachnids are troglobites. Arachnids to a large extent appear to occupy the predatory positions held by beetles and other insects in temperate caves.

Most species of troglobitic Arachnida have very limited distributions, with many occurring in a single cave. A few widely distributed species may represent newly evolved troglobites that have not been isolated long enough for morphological differentiation to have occurred. In some cases, more detailed analyses of populations have revealed the existence of cryptic species.

BIOLOGY

The biology of few species of troglobitic arachnids has been studied. In numerous cases only a single sex or even specimen is known and the rarity of most troglobites precludes anything more than cursory observations. Except for many mites that may feed on living or decaying organic matter or are parasites, most arachnids are predators. Some Opiliones are known to also feed on organic material.

Most troglobitic arachnids are preadapted for life in caves. All species of Ricinulei and Palpigradi are eyeless and in the Schizomida the ocular structures consist of one small eyespot on each side of the head. Arachnid families containing troglobites typically live in leaf litter, soil, or other sheltered habitats. Scorpions of the genus *Typhlochactas* are all blind, but the troglobitic species have longer appendages than the litter-dwelling surface species.

Adaptations for the cave environment include loss or reduction of eyes and pigment, elongation of appendages, and an increase in sensory structures. For the few species where data are available, fewer and larger eggs are produced and metabolic rates are lower. They may also have longer life spans. Troglobitic species of the dictynid spider genus *Cicurina* have been maintained in the laboratory for as much as 2 years before reaching sexual maturity.

Epigean Amblypygi are nocturnal, spending the day under rocks, ground debris, and crevices. Many cavernicolous species have extremely elongate pedipalps and exhibit various degrees of eye reduction. A few species are totally eyeless.

A study of the Hawaiian troglobitic spider *Lycosa howarthi* Gertsch revealed greater water loss and a lower metabolic rate than a related epigean species. Oxygen consumption in the troglobitic species remained constant over a 12-hour period, whereas it increased in the epigean species during darkness.

The biology and environment of the Australian troglobite *Draculoides vinei* (Harvey) have been studied in detail. The troglophilic species *Stenochrus portoricensis* Chamberlin of southern Mexico and the West Indies is of special interest in that it is facultatively parthenogenetic. With only a few exceptions, all insular populations are parthenogenetic. In the Yucatan Peninsula of Mexico there is an equal ratio of males and females from the surface, but essentially all specimens from deep within caves are females. This presumably explains the extreme abundance of this species in the caves of the peninsula.

Epigean Ricinulei inhabit leaf litter and the underside of rocks. They are primarily predators, but cavernicole species have been found feeding on dead arthropods and the feces of bats and millipedes.

Endogean Palpigradi are found under rocks and in leaf litter. All are eyeless, depigmented, and with slender appendages. Troglobitic species are distinguished by longer appendages and larger numbers of trichobothria and other sensory structures.

CONSERVATION

Troglobitic Arachnida, as in all groups of cave organisms, are particularly susceptible to loss of populations and even extinction. Small population sizes, long life span, slow reproductive rates, and a high degree of endemicity all place troglobitic arachnids at risk. Deforestation in many parts of the world has resulted in the loss of habitat for many species of invertebrates and bats that provide essential nutrients into the cave ecosystem. The loss of vegetation may result in environmental degradation, especially of small, shallow caves. Modifications of natural vegetation in the vicinity of the cave may result in invasion of introduced species. Urbanization in many areas has resulted in the destruction of many caves and impacted many others.

The actual number of species threatened by extinction is unknown but it must be a very high number. The U.S. Fish and Wildlife Service has placed 11 species of troglobitic arachnid on the official list of endangered species. Ten of these are from two limited areas of Central Texas where active urbanization threatens essentially every cave with destruction. Furthermore, the red imported fire ant *Solenopsis* (*Solenopsis*) *invicta* Buren has invaded hundreds of caves with devastating effect on the cavernicole fauna. The listed species are the neobisiid pseudoscorpion *Tartarocreagris texana* (Muchmore); the phalangodid harvestmen *Texella cokendolpheri* Ubick and Briggs, *T. reddelli* Goodnight and Goodnight, and *T. reyesi* Ubick and Briggs; and the dictynid spiders *Cicurina* (*Cicurella*) *baronia* Gertsch, *C.* (*C.*) *madla* Gertsch, *C.* (*C.*) *venii* Gertsch, and *C.* (*C.*) *vespera* Gertsch; and the leptonetid spiders *Neoleptoneta microps* (Gertsch) and *N. myopica* (Gertsch). In addition to these species, the Hawaiian lycosid spider *Adelocosa anops* Gertsch is also considered endangered.

The listing of these species has slowed destruction of caves in these very limited areas, but only an aggressive program of conservation throughout the world can preserve the remarkable diversity of troglobitic arachnids.

SYSTEMATICS

The classification of the Arachnida is far from settled, especially with respect to the phylogenetic relationships of the orders. Many specialists now consider the Acari to constitute a separate subclass of arachnids, with some placing the order Ricinulei within that subclass. Others consider the Acari to be polyphyletic and provide a different arrangement for arachnid orders and suborders. For convenience the Acari is considered here as an order and the classic suborders maintained.

Order Scorpiones (Scorpions)

This order of about 1300 species in nine families is worldwide in distribution (Fig. 1) (Volschenk and Prendini, 2008). Scorpions where they occur in caves are the top predator in the system. There are now 22 unquestionable troglobites known worldwide. An additional six species known only from caves show slight reduction in pigmentation or eyes but are probably not troglobitic. There have been no studies, however, on the biology or ecology of any of the species. All of the troglobitic species are extremely rare, most known only from a few specimens.

Family Akravidae. The family includes the troglobitic *Akrov israchanani* Levi from Israel.

Family Buthidae. The only unquestioned troglobite in the family is *Troglorhopalurus translucidus* Lourenço et al. from Brazil. *Troglotityobuthus gracilis* (Fage) from Madagascar has reduced pigment and may be troglobitic.

FIGURE 1 *Alacran tartarus* from Sótano de San Agustin, Oaxaca, Mexico. *Photo courtesy of Robert and Linda Mitchell. Used with permission.*

Family Chactidae. *Broteochactas trezzii* (Vignoli and Kovarik) from Venezuela has reduced eyes and pigmentation and may be a troglobite.

Family Chaerilidae. Two species, *Chaerilus chapmani* Vachon and Lourenço from Sarawak and *C. sabinae* Lourenço from Indonesia, are troglobites. A third species, *Chaerilus cavernicola* Pocock from Indonesia, is probably a troglophile.

Family Diplocentridae. The genus *Diplocentrus* includes six species from caves in southern Mexico, of which three are unquestioned troglobites (*D. actun* Armas and Palacios-Vargas, *D. anophthalmus* Francke, and *D. cueva* Francke). *Diplocentrus mitchelli* Francke contains reduced pigmentation but is probably not a troglobite. *Heteronebo clareae* Armas from the U.S.-owned Navassa Island in the Caribbean has reduced eyes and pigmentation but is probably not troglobitic.

Family Euscorpiidae. Two species of the genus *Troglocormus* (*T. willis* Francke and *T. ciego* Francke) have been described from caves in Mexico.

Family Liochelidae. *Liocheles polisorum* Volschenk et al. from Christmas Island has reduced eyes and pigment and is certainly a troglobite.

Family Troglotayosicidae. The only troglobite in the family is *Troglotayosicus vachoni* Lourenco from Ecuador. A closely related species is known from litter in adjoining Colombia. The European *Belisarius xambeui* Simon was once considered troglobitic because of its reduced eyes. It is now known to be a soil-dwelling form occasionally found in caves.

Family Typhlochactidae. This family includes seven troglobites. The Mexican genus *Typhlochactas* includes four troglobites (*T. cavicola* Francke, *T. poncho* Sissom, *T. reddelli* Mitchell, and *T. rhodesi* Mitchell) from the Mexican states of Tamaulipas and Veracruz. This genus also contains an eyeless species from leaf litter in southern Mexico. *Sotanochactas elliotti* (Mitchell) is known only from San Luis Potosi. The genus *Alacran* includes two troglobitic species *A. tartarus* Francke and *A. chamuco* Francke from Oaxaca. *Alacran tartarus* is especially interesting in that it has been found more than 900 m below the surface and has been collected from under-water on at least one occasion.

Family Urodacidae. *Aops oncodactylus* Volschenk and Prendini from Barrow Island, off the coast of Western Australia, is the only troglobite in the family.

Family Vaejovidae. The only unquestioned troglobite in the family Vaejovidae is *Vaejovis gracilis* Gertsch and Soleglad from Veracruz, Mexico. Many other species of *Vaejovis* have been found in caves but most are at best considered troglophiles. *Uroctonus grahami* Gertsch and Soleglad with reduced eyes is known only from caves in California, U.S.A., and is probably a troglobite. The genus *Pseudouroctonus* includes the troglobitic *P. sprousei* Francke from Mexico. Other species of troglophile are known from Mexico. *Pseudouroctonus reddelli* (Gertsch and Soleglad), is an extremely abundant troglophile in the caves of Texas, U.S.A.

Order Schizomida (Short-Tailed Whipscorpions)

This small order contains about 200 species in two families (Reddell and Cokendolpher, 1995). About 80 species, some awaiting description, have been recorded from caves. They range in size from 2 to 12.4 mm in total length. Epigean species are found under rocks, in leaf litter, and some species are associated with termites and ants.

Family Hubbardiidae. This family occurs worldwide in the tropics and subtropics. A large number of species of Hubbardiidae have been recorded from caves, but unquestioned troglobites are known only from caves in Cuba, Jamaica, Belize, Mexico, California (U.S.A.), Ecuador, and Australia. Two monotypic genera, *Cokendolpherius* and *Reddellzomus*, have recently been described from caves in Cuba. Two species of *Draculoides* have been found in Western Australia. *Hubbardia* from California and Arizona is a genus most closely related to the Asian fauna. The only troglobite is *H. shoshonensis* (Briggs and Hom) from a single cave surrounded by desert. The genus *Rowlandius* includes numerous species, including several troglophiles. Troglobites are known from caves in Cuba and Jamaica. The genus *Sotanostenochrus* contains only two troglobitic species from Mexico. The large genus *Stenochrus* contains eight described and several undescribed troglobites from caves in Mexico. The genus *Stewartpeckius* contains only the troglobite *S. troglobius* (Rowland and Reddell) from Jamaica. The genus *Tayos* contains only the troglobite *T. ashmolei* (Reddell and Cokendolpher) from Ecuador.

Family Protoschizomidae. Eleven species have been described from Mexico. *Protoschizomus* includes four species of troglobite from the states of San Luis Potosí and Tamaulipas; an undescribed species is from Texas (U.S.A.). *Agastoschizomus* with five species from caves in Guerrero, Hidalgo, San Luis Potosí, and Tamaulipas includes *A. lucifer* Rowland, at 12.4 mm in length the largest species in the order.

Order Amblypygi (Tailless Whipscorpions)

This order of about 100 species in three families is distributed worldwide in tropical and subtropical habitats (Weygoldt, 1994). Many species are large (up to 4.5 cm in body length) with extremely elongate pedipalps and antenniform first legs (Fig. 2). Cavernicolous species occur in all families and virtually every species found in karst regions also inhabits caves. About 40 species have been recorded from caves.

Family Charontidae. This family is largely Asian in distribution but the genus *Charinides* includes three troglobitic species in Cuba. Five or six species of the genus *Stygophrynus* inhabit caves in Malaysia, Burma, Borneo, and Java. An undescribed eyeless *Charinus* is known from caves in Belize. Other genera containing cavernicolous species include *Phrynichosaurus* and *Sarax* from Malaysia, *Tricharinus* from Jamaica and Surinam, and *Charinus* from Africa, Venezuela, and Cuba.

Family Phrynichidae. This tropical family contains cavernicolous species in the genera *Phrynichus* (Africa), *Damon* (Africa), and *Trichodamon* (Brazil). All are probably troglophiles.

Family Phrynidae. The family Phrynidae is the dominant family in the New World tropics and subtropics and is well represented in caves. All genera contain cavernicolous species, but the genus *Paraphrynus* is notable for containing six described troglobites and several troglophiles in Mexico. A large undescribed troglophile (previously recorded as *P. raptator*) is sympatric with the troglobitic *P. chacmool* (Rowland) in Yucatan, Mexico. Both species are found on cave walls, but *P. chacmool* has been found only in total darkness far from the cave entrance. The genus *Acanthophrynus* includes *A. coronatus* Kraepelin from southern Mexico. One species of *Heterophrynus* has been found in Colombian caves, but other species of the genus may occur in caves. *Phrynus* contains several troglophiles in Mexico and Texas (U.S.A.)

Order Araneae (Spiders)

This is the largest order of Arachnida with more than 50,000 described species (Ribera and Juberthie, 1994). Thousands of species, including large numbers of troglobites and troglophiles, have probably been recorded from caves. The diversity of troglobitic spiders is great with numerous families being recorded in essentially every karst region worldwide.

Suborder Mesothelae

Family Liphistiidae. This primitive family is tropical in distribution. Troglobitic species of the genus *Liphistius* are known from Malaysia and Thailand.

Suborder Mygalomorphae

Family Barychelidae. The only troglobite in this family is *Troglothele coeca* fage from Cuba.

Family Cyrtaucheniidae. The only troglobitic species in this family is *Acontius stercoricola* (Denis) from Guinea.

Family Dipluridae. Troglobitic and troglophilic species in the genus *Euagrus* have been found in Mexico. The genus *Masteria* includes troglobites in Jamaica and the Philippines. *Troglodiplura lowryi* Main is a troglobite in Australia.

Family Hexatelidae. The only troglobite in this family is *Hexathele cavernicola* Forster from New Zealand.

Family Microstigmatidae. The only troglobite in this family is *Spelocteniza ashmolei* Gertsch from Ecuador.

Family Theraphosidae. This family of tarantulas includes six troglobites in the genus *Hemirrhagus* in Mexico. One species is eyeless while the others have reduced eyes and pigment.

Suborder Araneomorphae. At least 25 families of Araneomorphae are considered to contain troglobites, although some are better represented in caves than others.

Family Agelenidae. The genus *Tegenaria* includes troglobitic and troglophilic species in Europe and the United States.

FIGURE 2 *Paraphrynus chacmool* from Actun Ziz, Yucatan, Mexico. *Photo courtesy of Robert and Linda Mitchell. Used with permission.*

Family Anapidae. This largely tropical family includes troglobites in the genera *Cruzetulus* in South Africa and *Conculus* from Korea.

Family Austrochilidae. This primitive tropical family contains only one troglobite, *Hickmania troglodytes* (Higgens and Peters) from Tasmania.

Family Cybaeidae. This family includes troglobitic species of the genus *Cybaeus* from Japan and Korea.

Family Cycloctenidae. This family includes troglobitic species in the genera *Cycloctenus* and *Toxopsiella* from Australia and Tasmania.

Family Dictynidae. The genera *Blabomma* and *Chorizomma* contain troglobitic species in Europe. The genus *Cicurina* includes troglobites in the United States, Mexico, and Japan. The radiation of this genus in the caves of Texas (U.S.A.) is especially remarkable with more than 50 eyeless species known.

Family Dysderidae. This family includes a large number of troglobitic species from caves in the Mediterranean basin. The most abundant genus is *Dysdera* with numerous species recorded. Other significant genera include *Folkia*, *Harpactea*, *Harpactocrates*, *Minotauria*, *Rhode*, *Rhodera*, *Speleoharpactea*, and *Stalita*.

Family Gnaphosidae. Troglobitic species in this family include two species of *Lygromma* from the Galapagos Islands and one of *Herpsyllus* from Cuba.

Family Hahniidae. This family of small spiders includes troglobites in the genera *Hahnia* and *Iberina* from Europe.

Family Holarchaeidae. This family includes only two troglobitic species, *Holarchaea globosa* (Hickman) from Tasmania and *H. novaeseelandiae* (Forster) from New Zealand.

Family Leptonetidae. This family is primarily holarctic in distribution with a few species in southern Mexico. It is well represented in caves, with numerous troglobitic and troglophilic species in many genera having been described. About 60 species of troglobites in several genera, including *Falcileptoneta*, *Leptoneta*, *Marisana*, and *Sataurana*, have been described from caves in China, Korea, and Japan. The fauna of Europe and northern Africa include many troglobites in the genera *Barusia*, *Leptoneta*, *Leptonetela*, *Paraloptoneta*, *Protoleptoneta*, *Sulcia*, and *Teloleptoneta*. Cave-associated species in the United States include representatives of the genera *Appaleptoneta*, *Archoleptoneta*, *Callileptoneta*, and *Neoleptoneta*. The genus *Darkoneta* includes a troglobitic species in Mexico. Two species of *Archeoleptoneta* and eight of *Neoleptoneta* are known from caves in Mexico, but only five of the latter genus are considered troglobites. Other species in the United States await description.

Family Linyphiidae. This large family contains numerous troglobitic and troglophilic species in caves throughout the Northern Hemisphere. The European fauna includes many troglobites in the genera *Caviphantes*, *Centromerus*, *Icariella*, *Lepthyphantes*, *Porrhomma*, *Thyphlonyphia*, and *Troglohyphantes*. Species of *Lepthyphantes* have also been found in caves in Korea and South Africa. Other genera represented by troglobites include *Anthrobia* in the United States and *Phanetta* in the United States and Mexico, *Erigone* and *Meioneta* in the Hawaiian Islands, *Allomengea* and *Jacksonella* in Korea, *Dunedinia* in Australia, and *Metopobactrus* and *Walckenaeria* in the Canary Islands.

Family Liocranidae. Genera represented by troglobites include *Brachyanillus* from Algeria and Spain, *Liocranum* from Spain, and *Agraecina* from the Canary Islands and Romania.

Family Lycosidae. This family of wolf spiders is represented in caves only by accidentals, with the exception of two remarkable species, *Lycosa howarthi* Gertsch and *Adelocosa anops* Gertsch from the Hawaiian Islands.

Family Mimetidae. The only troglobitic species in this family is *Mimetus strinatii* Birgnoli from Ceylon.

Family Mysmenidae. This family is represented from caves by troglophiles in the genus *Maymena* in Mexico and troglobites in the genus *Trogloneta* in Tasmania.

Family Nesticidae. The family Nesticidae is closely associated with caves, with numerous species of troglobite and troglophile having been described. The genus *Nesticus* includes troglobites and troglophiles in the United States, Mexico, China, Korea, Ceylon, Japan, and Europe. *Eidmannella* includes several troglobites from Texas (U.S.A.). Other genera containing troglobites include *Canarionesticus* and *Typhlonesticus* from Europe. The genus *Nesticiella* is abundant in caves in Asia, central Africa, Hawaii, and Fiji.

Family Ochyroceratidae. This is a tropical family of small spiders that contains cavernicolous species in North and South America, the Antilles, Asia, and Africa. The genus *Ochyrocera* includes troglobites and troglophiles in Mexico, Guatemala, Peru, and Cuba. Troglobitic and troglophilic species of *Theotima* are known from Mexico, Belize, Hawaii, Jamaica, and Cuba. Two other genera, *Fageicera* and *Speocera*, include troglophiles in Cuba. Troglophilic species in the genera *Athepus*, *Psiloderces*, *Simonicera*, *Speocera*, and *Theotima* have been recorded from caves in Asia. The genus *Speloderces* includes a troglophile in South Africa.

Family Oonopidae. This tropical family of small spiders includes several cave-associated species, including several troglobites. Troglobitic species include species of *Wanops* in Mexico, *Oonopsides* in Cuba, *Gamasomorpha* in Ecuador and Ceylon, and *Dysderoides* in India.

Family Pholcidae. This cosmopolitan family is well represented in caves, especially in the New World tropics. More than 90 troglophiles and troglobites in the genera *Anopsicus*, *Coryssocnemis*, *Ixchela*, *Metagonia*, *Modisimus*, *Physocyclus*, and *Psilochorus* have been collected from caves in Mexico. Troglobitic species of *Metagonia* have been described from caves in Belize, Cuba, Jamaica, and the Galapagos Islands. Other genera containing cave-associated species include *Artema*, *Pholcus*, and *Spermophora* in New Guinea; *Aymaria* in the Galapagos Islands; *Blancoa*, *Chebchea*, and *Mesabolivar* in Venezuela; *Priscula* in Peru; *Spermophora* in Tasmania; and *Spermophoroides* in the Canary Islands.

Family Prodidomidae. This small family includes only one troglobite, *Lygromma gertschi* Planick and Shadab. *Lygromma anops* Peck and Shear from the Galapagos Islands was considered troglobitic but it has since been found on the surface

Family Stiphidiidae. This small African and Australasian family includes troglobites and troglophiles in the genera *Baiami*, *Stiphidion*, and *Tartarus* from caves in Australia and Tasmania.

Family Synotaxidae. Three species of the genus *Tupua*, one troglophile and two troglobites, have been described from caves in Tasmania.

Family Telemidae. Cave-associated species, including some troglobites, of this family include species of *Telema* in Japan, and Guatemala; of *Telemofila* in New Caledonia; *Usofila* in the United States; *Cangoderces* in South Africa; and *Apneumonella* in Tanganyika. Of special interest is the troglobitic *Telema tenella* Simon from Europe. This is considered a "living fossil" and is the only representative of the family from temperate regions.

Family Tetrablemmidae. This small tropical family includes cave-associated species in the genus *Tetrablemma* from Mexico, *Caraimatta* from Mexico, and *Ablemma* from Okinawa and Sumatra.

Family Tetragnathidae. Genera in this family containing cave-associated species include *Meta* in Canada, the United States, Europe, Russia, and Tasmania; *Metellina* in Europe; and *Orsinome* in Tasmania.

Family Theridiidae. This large family includes numerous troglophiles in caves in many parts of the world. Troglobitic species include representatives of the genera *Achaearanea*, *Coscinidia*, and *Stemops* in New Guinea; *Pholeomma* and *Steatoda* in Austrialia; *Icona* in Australia and Tasmania; *Robertus* in Europe; *Theridion* in the Azores and Galapagos Islands; and *Thymoites* in the United States.

Family Theridiosomatidae. Cave-associated species in this family include representatives of the genera *Plato* in Venezuela, *Wendilgarda* in Guadeloupe and Ceylon, and *Andasta* in Ceylon.

Order Palpigradi (Micro Whipscorpions)

This is a small order containing 78 species arranged in two families (Conde, 1998) - note there should be an accent over the e in Conde. It is remarkable that 27 described species are known only from caves. Several others have also been reported from caves but also occur on the surface. Palpigrades range in body length from about 1.9 to 2.8 mm.

Family Eukoeneniidae. The genus *Koenenides* is represented in caves only by *K. leclerci* Condé from Thailand. Troglobitic species of *Eukoenenia* have been found in Europe (22 species), Cuba (1 species), Thailand (3 species), and Indonesia (1 species). Additional troglobitic species in the genera *Allokoenenia*, *Eukoenenia*, and *Koenenides* have been found in India, Sulawesi, and Thailand.

Family Prokoeneniidae. The only described troglobitic species of this family is *Prokoenenia celebica* Condé from Sulawesi, Indonesia. Two additional species of the genus from Southeast Asia and one from California await description.

Order Pseudoscorpiones (Pseudoscorpions)

This order of about 3000 species in 20 families is worldwide in distribution and is extremely well represented in caves (Fig. 3) (Heurtault, 1994). More than 400 species in 15 families have been described from caves.

Family Atemnidae. Cave-associated species have been described from the genera *Atemnus* in the Philippines, *Oratemnus* in Australia, and *Catatemnus* and *Titanatemnus* from Africa.

Family Bochicidae. This family is primarily restricted to the neotropics. The northernmost populations are apparent relicts in the caves of Texas (U.S.A.) and northern Mexico. Most species are troglobitic. Five genera that contain only troglobites are *Antillobisium* with two species in Cuba, *Troglobochica* with two species in Jamaica, *Troglohya* with two species in southern

FIGURE 3 *Tartarocreagris infernalis* from Electro-Mag Cave, Williamson County, TX, U.S.A. *Photo courtesy of Robert and Linda Mitchell. Used with permission.*

Mexico, *Paravachonium* with five species in Mexico, and *Vachonium* with seven species in southern Mexico and one in Belize. The genus *Leucohya* includes two troglobites in northern Mexico and one in Texas. The genus *Mexobisium* includes one troglobite from Guatemala, two from Belize, one from the Dominican Republic, and two from southern Mexico.

Family Cheiridiidae. This family contains few species associated with caves but species of *Cheiridium* from Namibia and Cuba; *Cryptocheiridium* from Australia, Malaysia, Philippines, and Cuba; and *Neocheiridium* from Curaçao are probably troglophiles.

Family Cheliferidae Genera containing cave-associated species in this family include *Lissochelifer* in Kenya, *Mexichelifer* in Mexico, *Protochelifer* in Australia, and *Stygochelifer* in Java.

Family Chernetidae. Many species of this family have been recorded from caves where they are frequently present in bat guano in very large numbers. The ecological status of these species is unknown but most are probably troglophiles. Genera recorded from caves include *Bitulochernes* and *Epactiochernes* from Cuba; *Neoallochernes* from Antigua in the Lesser Antilles; *Hesperochernes*, *Neoallochernes*, *Tejachernes*, and *Dinocheirus* from the United States; *Chernes*, *Lasiochernes*, and *Pselaphochernes* from Europe; *Chelanops*, *Dinocheirus*, *Lustrochernes*, *Parachernes*, and *Tejachernes* from South America; *Dinocheirus* and *Megachernes* in Asia; *Caffrowithius* and *Nudochernes* from Africa; and *Sundochernes* and *Troglochernes* from Australia.

Family Chthoniidae. This is one of the more important families of pseudoscorpions inhabiting caves worldwide. An estimated 200 species of the family have been described from caves, of which a large number in many genera are troglobites. The European fauna include about 80 species of *Chthonius*, as well as species of the genera *Microchthonius*, *Paraliochthonius*, *Spelyngochthonius*, *Troglochthonius*, and *Tyrannochthonius*. The Asian fauna include representatives of the genera *Allochthonius*, *Lagynochthonius*, *Mundochthonius*, and *Pseudotyrannochthonius*. *Austrochthonius*, *Lagynochthonius*, *Pseudotyrannochthonius*, *Sathrochthonius*, and *Tyrannochthonius* have been recorded from Australia. *Chthonius* has been recorded from northern Africa, *Selachthonius* from South Africa, and *Tyrannochthonius* from Kenya. The South American fauna include species of *Lechytia* from Trinidad, *Pseudochthonius* from Brazil, and *Tyrannochthonius* from Peru. *Tyrannochthonius* and *Kleptochthonius* are especially abundant in the caves of the United States, with other genera including species of *Aphrastochthonius*, *Apochthonius*, *Mexichthonius*, *Mundochthonius*, and *Neochthonius*. The fauna of Mexico and Central America include species of *Aphrastochthonius*, *Lechytia*, *Mexichthonius*, *Pseudochthonius*, and *Tyrannochthonius*. *Lagynochthonius* and *Tyrannochthonius* have been recorded from caves in Jamaica. Species of *Tyrannochthonius* and *Vulcanochthonius* are troglobites in Hawaiian lava tubes.

Family Garypidae. The family Garypidae includes species of *Archeolarca* and *Larca* in the United States and of *Larca* in Europe.

Family Ideoroncidae. The genus *Typhloroncus* includes several species of troglobite in Mexico. *Albiorix* includes three troglobites in Mexico and one in Arizona (U.S.A.). Species of *Dhanus* have been described from caves in Malaysia and one of *Negroroncus* from the Congo.

Family Neobisiidae. This is a dominant family of pseudoscorpions in the caves of the Holarctic region, with particular radiation in the United States and Europe. Species in the genera *Microcreagris*, *Pararoncus*, and *Parobisium* are troglobites in Japan. The genera *Neobisium* with more than 70 species and *Roncus* with more than 40 species are the dominant genera in Europe and northern Africa. Other genera with troglobites in Europe include *Acanthocreagris*, *Balkanoroncus*, *Insulocreagris*, and *Roncobisium*. The fauna of the United States include species in the genera *Alabamocreagris*, *Austrolinocreagris*, *Lissocreagris*, *Minicreagris*, *Novobisium*, *Parobisium*, *Tartarocreagris*, and *Trisetobisium*.

Family Pseudochiridiidae. This small tropical family includes cave-associated species in the genera *Pseudochiridium* from Malaysia and *Paracheiridium* from Madagascar.

Family Pseudogarypidae. This small family includes three troglobitic species of *Pseudogarypus* in California and Arizona (U.S.A.).

Family Sternophoridae. This small family includes a species of *Afrosternophorus* described from a cave in Papua New Guinea.

Family Syarinidae. This family contains cave-associated species in the caves of Europe, North America, South America, and the West Indies. The genus *Chitrella* contains several species, including possible troglobites, from caves in the United States. *Chitrellina chiricahuae* Muchmore, the only member of the genus, is an eyeless species from a cave in Arizona (U.S.A.). *Ideoblothrus* includes cave-associated species from caves in Mexico and Australia. The genus *Ideobisium* contains species in Ecuador, Venezuela, and Puerto Rico. The genera *Hadoblothrus*, *Pseudoblothrus*, and *Troglobisium* include troglobites from Europe.

Family Tridenchthoniidae. The only cave-associated species of this family is *Tridenchthonius juxtlahuaca* Chamberlin and Chamberlin from a cave in Guerrero, Mexico.

Family Withidae. Cave-associated species of this family have been recorded for the genera *Parawithius* in Venezuela and *Pycnowithius* in Kenya.

Order Ricinulei (Hooded Tickspiders)

This is the smallest order of arachnids with about 50 extant species in one family and three genera (Fig. 4) (Juberthie, 1994). One genus, *Ricinoides*, is exclusively African and contains no cave-associated species. Three species from northern Mexico inhabit caves at higher elevations of isolated mountain ranges surrounded by desert and are apparent relicts. Once thought to be extremely rare, populations of both troglophiles and troglobites may be present in vast numbers.

Family Ricinoididae. *Cryptocellus* is known from South and Central America, whereas the genus *Pseudocellus* occurs in Texas (U.S.A.), Mexico, Central America, and Cuba. The only described troglobitic species of *Cryptocellus* is *C. bordoni* (Dumetresco and Juvara-Bals) from Venezuela. Cave-associated species of *Pseudocellus* occur in Cuba, Guatemala, and Mexico. The troglobitic *P. silvai* (Armas) and the troglophilic *P. paradoxus* (Cooke) have been described from Cuba *Pseudocellus krejcae* Cokendolpher is a troglobite in Belize. The Mexican cavernicole fauna include the following troglobitic species: *P. osorioi* (Bolívar y Pieltain) from Tamaulipas and San Luis Potosí, *P. reddelli* (Gertsch) from Durango, and *P. sbordonii* (Brignoli) from Chiapas. Two additional undescribed troglobites are also known from Mexico.

Order Acari (Mites and Ticks)

This order of about 40,000 described species is probably represented in subterranean habitats by thousands of species, but comparatively few are terrestrial troglobites (Dusbabek, 1998; Palacios-Vargas, 1998). Numerous species of mites found in the interstitial habitat below streams and in cave waters exhibit adaptations to subterranean conditions.

Suborder Notostigmata

Family Opilioacaridae. This rare primitive group includes two troglophilic species of *Opilioacarus* from caves in Cuba. The genus *siamacarus* includes two species, one of which is troglomorphic, from caves in Thailand.

Suborder Mesostigmata. Large numbers of species, both free living and parasitic, in this suborder have been recorded from caves. Numerous species in the families Macronyssidae and Spinturnicidae are parasites of bats and have been recorded from caves. The free-living members of the suborder inhabit a wide variety of biotopes, including litter and soil, and they easily colonize caves. Many species have been recorded from caves in all parts of the world, but only a few exhibit troglomorphic adaptations.

Family Macrochelidae. This family contains a large number of cavernicolous species; some are known only from caves but all are probably troglophiles. Among the more significant genera that inhabit caves are *Macrocheles* in Japan, Europe, Mexico, Venezuela, and the West Indies; *Holostaspis* in Europe and Algeria; and *Geholaspis* in Europe. Most of these have been taken from bat guano.

Family Parasitidae. This family includes a large number of cave-associated species, many abundant in bat guano. Many species of *Eugamasus* have been recorded from caves in Afghanistan, Japan, and Europe. One European species is considered a probable troglobite. Other important cave-associated genera include *Pergamasus* and *Parasitus* in Europe and Mexico; *Paracarpais* in Europe and the United States; and *Vulgarogamasus* in Europe and the United States.

Family Uropodidae. This family contains a large number of cave-associated species. They are frequently present in bat guano with populations in the millions. Troglobitic species are known in the following genera: *Chiropturopoda* (3 species from Romania, South Africa, and Zaire), *Nenteria* (5 species from Europe and Trinidad), *Oplitis* (2 species from Cuba, Venezuela, and Trinidad), *Trichouropoda* (5 species from Cuba, Java, Afghanistan, and Europe), *Uroobovella* (12 species from Europe, Java, Cuba, and Trinidad), *Uropoda* (2 species

FIGURE 4 *Pseudocellus osorioi* from Sótano del Tigre, San Luis Potosí, Mexico. *Photo courtesy of Robert and Linda Mitchell. Used with permission.*

from Mexico, the Antilles, and Austria), and *Uroseius* (3 species from Japan, Italy, and Bosnia–Herzegovina).

Family Zerconidae. This family of predatory mites includes cave-associated species of the genera *Dithnozercon* from the United States, *Prozercon* from Japan, and *Zercon* from Europe. *Paleozercon cavernicolus* Blaszak et al. is a late Pleistocene fossil recovered from calcite in Hidden Cave, New Mexico (U.S.A.).

Suborder Metastigmata. This suborder includes the ticks, many species of which have been found as parasites of bats.

Family Argasidae. Many species of this family are parasites of bats. Species of the genus *Carios* occur in vast numbers in bat guano in the caves of Cuba and Mexico and apparently feed on the guano. A few species of the genus *Argas* have unusually long legs and sensory organs that indicate a degree of adaptation to cave life. Numerous species of *Carios* have been taken from caves. Some are parasites of bats but others have been taken from a variety of ground-dwelling mammals. *Ornithodoros turicata* (Dugès) in the caves of the southwestern United States is a known vector of relapsing fever and many cases of this disease have been documented in cave explorers following visits to caves inhabited by it.

Family Ixodidae. Few species of this family have been recorded from caves except as accidentals. Several species of the genus *Ixodes* in Europe are parasites of bats and frequently found in caves. One species, *Ixodes conepati* Cooley and Kohls, a parasite of small mammals, is frequently taken from caves in Texas and New Mexico (U.S.A.).

Suborder Prostigmata (Terrestrial). This suborder includes the majority of mites considered troglobitic. The families Myobiidae and Psorergatidae include numerous bat parasites that have been recorded from caves. The majority of aquatic mites recorded from subterranean habitats also belong in the suborder.

Family Cunaxidae. The family includes one troglobitic species of *Bonzia* in England. Troglophiles in the genera *Cunaxa* and *Cunaxoides* are known from Spain and Mexico, respectively.

Family Ereynetidae. This family includes, in addition to bat parasites in the genera *Neospeleognathopis* and *Speleochir*, a troglobitic species of *Ereynetes* from lava tubes in Japan. Other cave-associated species include representatives of the genera *Ereynetes* from Mexico and *Riccardoella* from Europe.

Family Leeuwenhoeckiidae. Four species of troglobite belonging to the genera *Heterotectum, Ischnothrombium, Pentagonotectum,* and *Tectumpilosum* have been described from bat guano in Cuban caves.

Family Proterorhagiidae. This family was described for the troglomorphic species *Proterorhagia oztotlica* Lindquist and Palacios-Vargas from caves in Colima, Mexico.

Family Rhagidiidae. This is the most important family of mites inhabiting caves. A large number of species have been recorded as troglophiles and troglobites. The ecological status of some species is uncertain, but many are doubtless troglobitic. Troglobites and troglophiles have been described in the genera *Coccorhagidia* from Cuba; *Flabellorhagidia* from the United States; *Foveacheles* from Europe, Mexico, United States, and Hawaii; *Poecilophysis* from Europe, Mexico, and United States; *Rhagidia* from Europe, Mexico, United States, South Korea, and Japan; *Robustocheles* from Canada, United States, and Mexico; *Traegardhia* from Italy; and *Troglocheles* from Europe.

Family Trombiculidae. This family includes numerous species of bat parasites that have been recorded from caves. Species in the genera *Cubanothrombium* and *Heterothrombium* from caves in Cuba appear to be troglobitic. Two species of the genus *Trombicula* from Afghanistan, Java, and Morocco are considered troglobites. Species of *Microtrombicula* and *Nycterinastes* have been collected from caves in Mexico.

Family Trombidiidae. This family includes parasites of bats, but others are predators of insects. Many species have been recorded from caves. Cave-associated species, including some probable troglobites, are known in the genera *Anomalothrombium* from Madagascar; *Hannemania* from Mexico; *Leptothrombium* and *Neotrombicula* from Romania; *Spelaeothrombium* from Europe and Africa; *Speothrombium, Dolichotrombicula,* and *Discotrombidium* from Cuba; and *Typhlothrombium* from Belgium.

Suborder Prostigmata (Aquatic). An estimated 4000 species in 100 families and more than 300 genera have been recorded from aquatic habitats (Schwarz et al., 1998). A large percentage of these are known only from subterranean habitats that include caves and phreatic and interstitial habitats. Subterranean species show reduction or loss of eyes and pigment and many species possess elongated bodies and short legs. Only a few of the more important families are discussed here. Other families containing subterranean species include the Anisitsiellidae, Arenohydracaridae, Arrenuridae, Athienemanniidae, Bogatiidae, Chappuisididae, Feltridae, Halacaridae, Hungarohydracaridae, Hydrovolziidae, Kantacaridae, Lebertiidae, Mideopsidae, Momoniidae, Neocaridae, Nipponacaridae, Omartacaridae, Piersigiidae, Pionidae, Sperchontidae, Torrenticolidae, Trombidiidae, and Unionicolide.

Family Aturidae. This family contains a large number of species adapted for subterranean existence. Among the more important genera are *Aturus* from Europe, Mexico, and Japan; *Axonopsella* from Cuba, South America, Australia, and Tasmania; *Axonopsis* from Europe, North America, Asia, North Africa, and Cuba; and *Frontipodopsis* from Europe, Asia, North America,

Central America, and South America. Numerous genera are known only from subterranean waters.

Family Hydryphantidae. This is an abundant family in underground waters, with a large number of described species. Genera containing subterranean species include *Clathrosperchon* from the southern United States, Mexico, and South America; *Cowichiana* from Canada; *Cyclothyas* from the United States; *Euwandesia* from Chile and New Zealand; *Thyasella* from Europe; and *Wandesia* from Europe, North America, South America, Cuba, India, Australia, Tasmania, and Siberia.

Family Hygrobatidae. This family contains numerous species recorded from subterranean waters. Among the more important genera are *Atractides* from Europe, South America, and Asia; *Australiobates*, *Callumobates*, *Camposea*, and *Decussobates* from Chile; *Corticacarus* from South America; *Gondwanobates* from Australia and New Zealand; and *Hygrobates* from India.

Family Limnesiidae. This family includes a large number of subterranean species, including many depigmented, eyeless species. Among the more important genera are *Kuwamuraarus* from North America, Mexico, India, Indonesia, and Japan and *Neomamersa* with many species in North America, Central America, South America, and the West Indies.

Family Limnohalacaridae. Two genera, *Lobohalacarus* and *Soldanellonyx*, contain species recorded from caves in Japan. Other genera with species recorded from caves include *Troglohalacarus* and *Parasoldanellonyx* from Europe and *Homohalacarus* from the United States. Numerous other genera have been recorded from other subterranean habitats.

Family Stygotrombidiidae. This family is extremely abundant in subterranean waters of Europe, with numerous species of *Stygotrombidium* having been described. Other genera associated with underground waters include *Cerberotrombidium*, *Charonotrombium*, *Hydrotrombium*, and *Victatrombium*.

Suborder Astigmata. Most of the species in this suborder are parasites of bats and other mammals. Representatives of the families Acaridae, Chirodiscidae, Chirorhynchobiidae, Rosensteiniidae, Sarcoptidae, Teinocoptidae, and Gastronyssidae have been taken from bats in caves.

Family Acaridae. Species of this family feed on dead insects and other organic material. The ecological status of cave-associated species is generally unknown. Genera recorded from caves include *Aellenella* from Kenya, *Acotyledon* from Puerto Rico, *Caloglyphus* from Mexico and Trinidad, *Sancassania* from Mexico, and *Schwiebea* from Europe and Kenya.

Suborder Cryptostigmata. This large group contains more than 7000 species arranged into 140 families and 700 genera. Epigean species are taken from soil, litter, and other endogean habitats. They are extremely abundant in caves but remain poorly studied. A few of the more important families recorded from caves are discussed below.

Family Damaeidae. The genus *Belba* includes one troglobite from Europe and one apparent troglophile from Mexico. Other genera with cave-associated species include *Damaeus* from Algeria, Europe, and Korea; *Dameosoma* from Europe; *Epidamaeus* from the United States and Korea; and *Hypodamaeus* from Scandanavia.

Family Galumnidae. Species of the genus *Galumna* occur in the caves of Yucatán, Mexico, and other neotropical areas. Populations in the millions may be present in some caves.

Family Haplozetidae. The frugivorous species *Rostrozetes foveolatus* Sellnick has been taken from caves in Cuba and Trinidad, where it is extremely abundant on bat guano.

Family Hypochthoniidae. Genera with cave-associated species include *Eohypochthonius* from Cuba and Trinidad, *Hypochthonius* from Europe, and *Malacoangelia* from Cuba.

Family Microzetidae. The cave-associated species of this family have been found in the guano of birds and bats. Genera taken from caves include *Acaroceras* from Cuba and Mexico, *Gymnozetes* from Cuba, and *Microzetes* from Trinidad.

Family Oppiidae. This is one of the more important families of cryptostigmatid mites found in caves. Most species are probably troglophiles. Many genera have been recorded from caves, including *Amerioppia* and *Multioppia* from Fiji; *Brachioppia* from Puerto Rico; *Hypogeoppia*, *Kunoppia*, *Leuroppia*, *Medioppia*, *Micropipa*, *Oppia*, and *Serratoppia* from Europe; *Lasiobella* from the Canary Islands; *Amerrioppia* and *Oppia* from Mexico; and *Oppia* from Cuba.

Family Scheloribatidae. This important family includes cave-associated species in the genera *Monoscheloribates* from Mexico; *Poroscheloribates* from the Canary Islands; and *Scheloribates* from Mexico, Cuba, Puerto Rico, and Fiji.

Family Sphaerochthoniidae. The genus *Sphaerochthonius* is frequently found in caves, with species having been recorded from Fiji, Cuba, Puerto Rico, and Mexico.

Order Opiliones (Harvestmen)

This order contains an estimated 5500 species divided into four suborders (Fig. 5) (Rambla and Juberthie, 1994).

Suborder Cyphophthalmi. The Cyphophthalmi is a small group of about 100 species. These mite-like harvestmen are considered the most primitive members of the order. Many epigean species are blind, but the

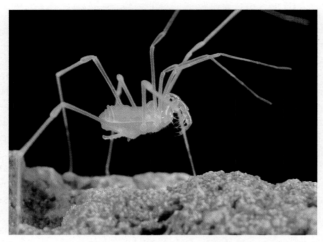

FIGURE 5 *Texella reyesi* from Electro-Mag Cave, Williamson County, TX, U.S.A. *Photo courtesy of Robert and Linda Mitchell. Used with permission.*

cave species possess longer legs and other adaptations for a subterranean existence.

Family Neogoveidae. The only troglobitic species in this small family is *Neogovea mexasca* Shear from Mexico.

Family Pettalidae. Although all species of this family are blind, the only troglobite is *Speleosiro argasiformis* from South Africa.

Family Sironidae. Eight troglobitic and one troglophilic species of *Siro* have been described from caves in Europe. The most highly troglomorphic species is *Tranteeva paradoxa* from Bulgaria. The only species recorded from tropical caves is *Marwe coarctata* from Kenya.

Family Stylocellidae. Two species in the genus *Stylocellus* have been described from caves in Sarawak.

Family Troglosironidae. This family was created for the troglobitic species *Troglosiro aelleni* Juberthie in New Caledonia.

Suborder Eupnoi. This suborder includes the delicate long-legged harvestmen commonly found in cave entrances. Most cavernicolous species belong to the family Sclerosomatidae with numerous species of *Leiobunum* having been recorded as trogloxenes in the caves of Europe, Canada, Mexico, and the United States.

Suborder Dyspnoi. This small suborder includes troglobites in three families.

Family Ischyropsalidae. This monotypic family from Europe includes a large number of troglophiles and troglobites in the genus *Ischryopsalis*. Ten species are troglobitic. Some species are notable for occurring in caves with a temperature as low as 1.5°C.

Family Nemastomatidae. Troglobitic species in the following genera have been described: *Buresiola* (four species from Europe and Tunisia), *Nemaspela* (two species from Europe), *Mitostoma* (two species from Europe) and *Nemastoma* (three species from Europe and the United States). *Ortholasma sbordonii* Brignoli is a possible troglobite in Mexico.

Family Sabaconidae. The only troglobite is *Sabacon picosantrum* Martens from caves at high altitude in Spain. A second species is a troglophile in the caves of France.

Suborder Laniatores. This suborder occurs worldwide but is particularly abundant in the tropics and subtropics. The taxonomy of the group is still poorly known with the limitations of some families not yet settled. Many species from the tropics are large and heavily spinose.

Family Agoristenidae. Two cavernicole species of the family Agoristenidae have been described from caves in Venezuela. *Phalangozea bordoni* Munoz-Cuevas is a troglobite.

Family Assamiidae. This family is restricted to tropical Africa and Asia. Blind, depigmented species have been described from forest litter and termite nests, but some from caves appear to be troglobitic and troglophilic. The following genera include species of troglobites: *Typhlobunellus* from Tanzania, *Sijucavernicus* and *Metassamia* from Assam, and *Calloristus* from India.

Family Biantidae. The only troglobite in this family is *Decuella cubaorientalis* Avram from Cuba. Other species from Cuba and Ceylon are probably troglophiles.

Family Cladonychiidae. This Holarctic family contains a few troglophiles in Europe. *Speleomaster lexi* Briggs and *S. pecki* Briggs are troglobites from lava tubes in the northwestern United States. *Erebomaster flavescens* Cope includes two subspecies of troglophile and one of troglobite from caves in the eastern United States.

Family Gonyleptidae. This neotropical family includes numerous cavernicolous species, but the only apparent troglobite is *Vima chapmani* Rambla from Venezuela. Genera containing troglophiles or trogloxenes include *Pachyloides* from Uruguay and Venezuela; *Rhopalocranaus, Mendellina, Vima,* and *Santinezia* from Venezuela; *Ancistrotellus* from Brazil; and *Aucayacuella* from Peru.

Family Phalangodidae. This family is especially abundant in the caves of North America and Europe. The genus *Banksula* occurs in the caves of California (U.S.A.). The genus *Texella* includes troglobitic and troglophilic species in California and Texas (U.S.A.). Other troglobites from caves in the United States belong to the genera *Bishopella, Calicina, Crosbyella, Goodnightiella, Phalangodes,* and *Phalangomma*. Two troglobites in the genus *Guerrobunus* are known from Mexico. The genus *Jimeneziella* includes two troglobitic species in Cuba. *Panopiliops inops* Goodnight and Goodnight is a troglobite in Costa Rica. The genus *Scotolemon* includes several troglobites from caves in Europe. Other troglobitic species in Europe include representatives of the genera

Lola and *Paralola*. The only troglobitic member of the family from Africa is *Conomma troglodytes*.

Family Stygnommatidae. The genus *Stygnomma* includes troglobites and troglophiles from Mexico, Belize, and the Galapagos Islands.

Family Stygnopsidae. This family includes numerous cavernicolous species, many undescribed, in the caves of Mexico. Troglobites include representatives of the genera *Hoplobunus*, *Mexotroglinus*, and *Troglostygnopsis*. Troglophiles are found in the genera *Hoplobunus*, *Karos*, *Sbordonia*, and *Stygnopsis*. Two species of *Chinquipellobunus* are troglobites in Texas (U.S.A.). South American troglobites include *Pachyspeleus strinatii* from Brazil and *Galanomma microphthalma* Juberthie from the Galapagos Islands.

Family Travuniidae. This family includes troglobites from caves in Europe, the United States, and Japan. The genus *Peltonychia* includes numerous species with troglomorphic adaptations ranging from eyed to totally eyeless from caves in Europe. Other troglobitic European species occur in the genera *Dinaria* and *Travunia*. The remaining troglobitic travuniid species are *Speleonychia sengeri* Briggs from lava tubes in Washington (U.S.A.) and *Yuria pulcra* Suzuki from Japan.

Family Triaenonychidae. This family is largely tropical but a few species are known from the United States. The genus *Hendea* includes several species of troglobite and troglophile in New Zealand. The Australian troglobite fauna include species of the genera *Calliuncus* and *Holonuncia*. The cavernicole fauna of Tasmania include species of troglobitic *Hickmanoxyomma*, *Lomanella*, and *Picunchenops*. Other troglobitic genera containing troglobites include *Spelaeomontia* in South Africa, *Picunchenops* in Argentina, and *Cryptobunus* in the northwestern United States.

Family Zalmoxidae. The genus *Ethobunus* includes troglobites from Belize and Jamaica.

Bibliography

Condé, B. (1998). Palpigradida. In C. Juberthie & V. Decu (Eds.), *Encyclopaedia biospeologica* (Vol. II, pp. 913–920). Moulis, France: Société de Biospéologie.

Dusbabek, F. (1998). Acari parasiti. In C. Juberthie & V. Decu (Eds.), *Encyclopaedia biospeologica* (Vol. II, pp. 921–928). Moulis, France: Société de Biospéologie.

Heurtault, J. (1994). Pseudoscorpions. In C. Juberthie & V. Decu (Eds.), *Encyclopaedia biospeologica* (Vol. I, pp. 185–196). Moulis, France: Société de Biospéologie.

Juberthie, C. (1994). Ricinulei. In C. Juberthie & V. Decu (Eds.), *Encyclopaedia biospeologica* (Vol. I, pp. 231–235). Moulis, France: Société de Biospéologie.

Palacios-Vargas, J. G., Decu, V., Iavorski, V., Hutzu, M., & Juberthie, C. (1998). Acari terrestria. In C. Juberthie & V. Decu (Eds.), *Encyclopaedia biospeologica* (Vol. II, pp. 929–952). Moulis, France: Société de Biospéologie.

Rambla, M., & Juberthie, C. (1994). Opiliones. In C. Juberthie & V. Decu (Eds.), *Encyclopaedia biospeologica* (pp. 215–230). Moulis, France: Société de Biospéologie.

Reddell, J. R., & Cokendolpher, J. C. (1995). *Catalogue, bibliography, and generic revision of the order Schizomida (Arachnida)*. Austin: Texas Memorial Museum.

Ribera, C., & Juberthie, C. (1994). Araneae. In C. Juberthie & V. Decu (Eds.), *Encyclopaedia biospeologica* (Vol. II, pp. 197–214). Moulis, France: Société de Biospéologie.

Savory, T. (1977). *Arachnida* (2nd ed.). New York: Academic Press.

Schwarz, A. E., Schwoerbel, J., & Gruia, M. (1998). Hydracarina. In C. Juberthie & V. Decu (Eds.), *Encyclopaedia biospeologica* (Vol. II, pp. 953–976). Moulis, France: Société de Biospéologie.

Volschenk, E. S., & Prendini, V. (2008). *Aops oncodactylus*, gen. et sp. nov., the first troglobitic urodacid (Urodacidae: Scorpiones), with a re-assessment of cavernicolous, troglobitic and troglomorphic scorpions. *Invertebrate Systematics*, 22, 235–257.

Weygoldt, P. (1994). Amblypygi. In C. Juberthie & V. Decu (Eds.), *Encyclopaedia biospeologica* (Vol. I, pp. 241–247). Moulis, France: Société de Biospéologie.

SPRINGS

William B. White

The Pennsylvania State University

INTRODUCTION

A spring is a localized orifice where groundwater returns to the surface. All aquifers must ultimately drain somewhere. If the drainage occurs at a single location, as distinguished from a line of seeps, a wetland, or discharge into the bed of a surface stream, the place where the groundwater appears is called a *spring*. The sizes of springs vary greatly from barely perceptible trickles to full-sized rivers gushing forth from the earth.

Springs have been of immense importance to humankind throughout history. Most have been used as water supplies and most farmsteads were sited next to a spring which provided water for both household and livestock. Indeed, in arid regions, springs may have been the only water supply. The city of Jerusalem was founded at the site of the sacred Gihon Spring about 3800 years ago (Amiel et al., 2010). Many other towns were located at springs so that the water supply would be safely inside the city walls. Hot springs and mineral springs have been valued for their curative properties with many health resorts constructed around springs. Many of the largest springs come from cave passages or from fractures in carbonate rock and are known as karst springs. Large springs are also found in volcanic rocks and springs of various sizes in nearly all rocks. This article is concerned only with karst springs. For a comprehensive discussion of springs including descriptions of many individual springs, see Kresic and Stevanovic (2010).

TYPES OF SPRINGS

Springs can be categorized and described in terms of the nature of the spring orifice and by the characteristics of the discharge (Fig. 1).

Conduit and Diffuse Flow Springs

Springs that drain from cave passages (conduits) have distinctive characteristics. They tend to be flashy; that is, flow increases rapidly in response to storms in the watershed. The springs often become turbid, sometimes muddy, in response to storms. The chemistry of the spring water also varies with seasons and in response to storms. These springs are known as *conduit springs* even if the conduit that feeds the spring is not visible at the spring orifice. Other springs, called *diffuse flow springs*, drain from faults, networks of fractures, or other localized zones of high permeability. These springs tend to flow at a constant rate with at most only sluggish response to storms, although the flow may rise and fall with wet and dry seasons. The water usually remains clear. The temperature and chemical composition of the water remain constant throughout the year. Some very large springs, those in Florida, for example, do drain from conduits but show little storm response.

Gravity Springs

In a sense, all springs are gravity springs because it is the action of gravity that causes water to drain from storage in the aquifer. However, the term *gravity spring* is reserved for springs that represent the emergence of an underground stream from within the cave system that feeds the spring. Some gravity springs are open cave mouths (Fig. 2). One can enter the cave by simply walking (or crawling) into the spring mouth. In a few of these springs, one can follow the cave stream upstream for considerable distances; in others the underground stream is found rising from a sump at varying distances upstream. In other cases, the spring mouth is obscured by surface debris that has slumped across the passage. In these cases, the spring water

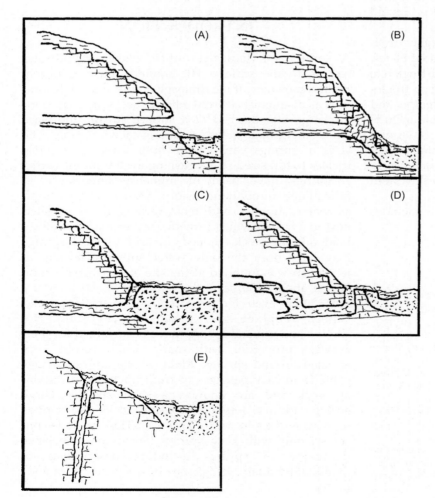

FIGURE 1 Sketches of various types of springs: (A) Gravity spring draining from an open cave mouth. (B) Gravity spring draining through rubble. (C) Alluviated spring. (D) Shallow artesian spring. Water rises from flooded cave passages at depths of a few meters to a few tens of meters. (E) Deep artesian spring. Water rises along a deep channel in the bedrock. Depths may be hundreds of meters.

FIGURE 2 Overholt Blowing Cave Spring, Pocahontas County, West Virginia, a gravity spring draining from an open cave conduit. *Photograph taken by the author.*

emerges from hillside rubble with no obvious cave passage.

Alluviated Springs

Spring mouths are frequently located on the banks of rivers or large creeks. Over time, the beds of the rivers and creeks accumulate thick deposits of sand, silt, and clay, materials known as *alluvium*. The sediments may block the spring mouths, thus forcing the water to rise through a channel between the hillside and the alluvial river sediments (Fig. 3). Such blockages have occurred in the Mammoth Cave (Kentucky) area caused by changes in the Ohio River and its tributaries during glacial and postglacial periods. Similar springs are found in New York where spring mouths have been choked with glacial drift. The spring orifices of alluviated springs are rise pools, sometimes called *blue holes*, where water wells upward and then flows out to the surface stream. Divers have explored some alluviated springs and found the flooded cave passage feeding the spring at depths of 5–10 m.

Offshore Springs

During periods when the glaciers of the Pleistocene ice ages had their maximum advance, so much water was stored in the glaciers that sea levels were lowered by as much as 100 m. Cave systems discharged their water through springs located near the shoreline. When the glaciers retreated and sea levels rose again, these springs were drowned but continued to discharge freshwater. Freshwater is less dense than seawater. As a result, the location of drowned offshore springs is marked by a plume of freshwater rising to the surface, sometimes with a pronounced boil. Offshore springs are found in many coastal areas where carbonate rocks extend below sea level.

FIGURE 3 An alluviated spring in Mammoth Cave National Park, Kentucky. *Photograph taken by the author.*

Artesian Springs

Artesian springs are those in which the water is forced upward under pressure from channels in the bedrock. Some artesian springs are shallow, fed by cave passages below the water table. Shallow artesian springs sometimes well up with a pronounced boil, but with water coming from depths of only a few tens of meters. Many cave streams end in sumps from which the connection to the spring is through flooded passages. Many shallow artesian springs have been penetrated by divers who re-emerged into air-filled caves after relatively short distances underwater.

Deep artesian springs are fed by channels that extend to depths of hundreds of meters. Usually these channels are guided by geological structures such as major fracture systems or faults. These springs are sometimes known as *Vauclusian springs*, named after the Fontaine de Vaucluse in southern France. Exploration by divers and remote sensing equipment has established the depth of the Fontaine de Vaucluse at 308 m. Deep artesian springs are found along the Sierra de El Abra in Tamaulipas, Mexico. The Nacimiento del Rio Mante was explored to 252 m without reaching bottom.

A recent discovery is Cenote de Zacatón, also in Tamaulipas, Mexico, with an explored depth of 329 m (Gary, 2010). Exploration of these deep underwater systems is at (or beyond) the limits of diving techniques and as a result little is known about the feeder system of these springs.

Overflow, Underflow, and Distributary Springs

Springs may have more than one orifice. During dry season, only the orifice at the lowest elevation, the underflow spring, may be flowing. Underflow springs are often low on the banks of surface streams or in the channel itself and may not be obvious to observers. When flow through the aquifer increases during wet seasons or in response to storms, other orifices at higher elevation may become active. Overflow springs may appear as dry cave entrances during dry seasons and may discharge a large volume of water during wet seasons. The Planina Cave in Slovenia is among the largest springs in Europe, discharging the combined flow of the Pivka and Rak rivers. It is, however, the overflow spring and the discharge varies by several orders of magnitude. It is the more constant underflow spring, the Malenščica, that is used for municipal water supplies.

Cave systems sometimes reach the surface as a single conduit leading to a single spring. However, it is common to find downstream distributaries where the passages split into multiple branches each of which may reach the surface to produce a spring. Distributary springs have been observed scattered along several kilometers of surface stream, all discharging water from the same karst groundwater basin. Distributary springs are usually not at exactly the same elevation. As a result, only one or a few springs may be flowing during the dry season but many more become active during storm flow or during the wet season.

Thermal Springs

The water temperature in springs draining shallow groundwater basins is typically the same as the mean annual temperature of the region, although water temperature from some open and fast-flowing conduit systems fluctuates with the seasons. The water is warmer than average in the summer and colder than average in the winter. The transit of water from inlet to spring in these very open systems takes place faster than the water can come into thermal equilibrium with the surrounding rock. The seasonal average, however, is close to the mean annual temperature.

There exist limestone springs that discharge water at temperatures higher than the local average temperature (Goldscheider et al., 2010). These thermal springs are called warm springs or hot springs depending on the temperature. Not all thermal springs are limestone springs. Hot springs are common in volcanic areas where groundwater is heated by volcanic activity. In karstic regions, the source of heat is usually geothermal. The temperature of the earth increases with depth at rates ranging from 15 to $40°C\,km^{-1}$. If the groundwater circulation path takes the water to great depths before reaching the spring, the water temperature will be increased. Thermal springs draining from karst aquifers have observed temperatures from just above ambient to as high as $70°C$, although most springs are somewhat cooler.

Mineral Springs

Springs can also be categorized by the chemical composition of their waters. Water from a typical karst spring will contain mainly Ca^{2+}, Mg^{2+}, and HCO_3^- ions with minor concentrations of Na^+, K^+, SO_4^{2-}, Cl^-, NO_3^-, and possibly a few other ions. Total ion concentrations are typically in the range of a few hundred milligrams per liter. Mineral springs are those that contain much higher concentrations of dissolved substances, sufficient to give the water a strong taste and possibly an odor. Mineral springs have a great range of both chemical composition and concentrations of dissolved substances and may be broadly subdivided according to the dominant chemistry. Saline springs contain high concentrations of NaCl and other dissolved salts of magnesium, calcium, and sodium, with magnesium sulfate often being an important constituent. Sulfur springs contain dissolved hydrogen sulfide, H_2S, which gives the water a rotten egg odor. Iron springs, also called *chalybeate springs*, have high iron concentrations. Mineral springs may or may not also be thermal springs. Mineral springs were prized for their medicinal properties, particularly in the 19th century. Resort hotels and baths were built on the sites of many of these springs, although only a few remain in operation.

Seeps and Wetlands

Marshy areas often result from the discharge of groundwater from any type of aquifer. Karst aquifers in which the movement of water is through networks of fractures may return water to the surface through an extended wetland region rather than through a single large spring. Other karst regions may be mantled with glacial drift, volcanic ash, or other recent permeable cover, thus spreading and diffusing the groundwater discharge. Seeps and wetlands are found where the karstic carbonate rocks are underlain by impenetrable layers. Water may emerge from the bottom of the aquifer along the line of contact with the impermeable layer as a continuous zone of small seeps and springs rather than as one large spring.

SPRING DISCHARGES

Magnitudes of Spring Discharges

The volume of water discharged from karst springs varies by many orders of magnitude. Table 1 lists a selection of some of the largest karst springs of the world. Where data were available, maximum, minimum, and mean discharges are listed. Some spring discharges vary only slightly about the mean; others vary by several orders of magnitude between base flow and peak flow. None of the largest karst springs are in the United States. The largest reported springs are in Papua New Guinea where several large rivers flow directly from caves. For the United States, large springs occur in Florida, Missouri, and Texas (Table 2). Although there are hundreds (probably thousands) of karst springs in the eastern United States, many of the Appalachian groundwater basins are relatively small as are their springs. They provide spring discharges typically an order of magnitude smaller than the large springs listed in Table 2.

A plot of discharge as a function of time is known as a *hydrograph*. Hydrographs are routinely measured for surface streams and provide a graphic picture of the variations in stream discharge with seasons and with storm flow. Hydrographs can also be measured for springs and provide some insight into the characteristics of the groundwater basin and the conduit system that feeds the spring. Figure 4 shows two examples. The hydrograph of Alley Spring contains many sharp peaks, which represent the spring response to individual storms. The hydrograph of Rainbow Spring is smooth. The contributions of individual storms have been averaged out and only the rise and fall of discharge with seasonal wet and dry periods is reflected in the hydrograph. Aquifers with well-developed conduit systems drain rapidly. Little water is retained in storage. Spring discharge has time to fall to base flow before the next storm arrives. The ratio of peak flow to base flow gives a measure of the "flashiness" of storm response. For Alley Spring, this parameter is 4.7. For Rainbow Spring, it is 1.2. Ratios as high as 100 have been observed for Davis Spring, West Virginia.

Ebb-and-Flow Springs

Ebb-and-flow springs are those in which the discharge varies in a cyclical fashion. Discharge drops to some minimum value, then rises rapidly to a peak, and

TABLE 1 Some Representative Large Karst Springs Excluding the United States

Spring	Location	Maximum Flow	Mean Flow	Minimum Flow
Tobio	New Guinea	—	85–115	—
Matali	New Guinea	>240	90	20
Dumanli	Turkey	—	50	25
Buna	Bosnia	440	40	2
Ljubljana	Slovenia	132	39	4.2
Ras el Ain	Syria	—	39	—
Chingshui	China	390	33	4
Vaucluse	France	200	29	4.5
Frio	Mexico	515	28	6
Coy	Mexico	200	24	13
Timava	Italy	138	17.4	9
Niangziguan	China	15.8	12.1	8.1
Aachquelle	Germany	24.1	8.2	1.3
Blautopf	Germany	26.1	2.2	0.35

Data from various sources, especially Ford and Williams (2007) and Kresic and Stevanovic (2010). All flow measurements are in $m^3\ s^{-1}$.

TABLE 2 A Selection of Large Karst springs in the United States

Spring	Location	Maximum Flow	Mean Flow	Minimum Flow
Silver Spring	Florida	36.3	23.0	15.3
Rainbow Spring	Florida	28.9	19.8	13.8
Wakulla Spring	Florida	54.1	11.1	0.71
Comal Springs	Texas	15.1	9.2	—
San Marcos Spring	Texas	8.5	4.2	2.3
Big Spring	Missouri	36.8	12.1	6.7
Greer Spring	Missouri	25.5	9.4	2.9
Maramec Spring	Missouri	18.4	4.2	1.6
Alley Spring	Missouri	—	3.6	1.5
Davis Spring	West Virginia	28.3	3.1	0.6

Data from various sources. All flow measurements are in $m^3 \, s^{-1}$.

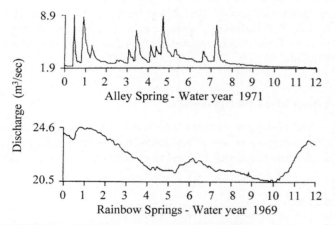

FIGURE 4 Hydrographs, each drawn for a single water year from U.S. Geological Survey water records for Alley Spring, Missouri, and Rainbow Spring, Florida.

then decreases again. Usually this flow-and-flush cycle occurs on a timescale of minutes to hours. Ebb-and-flow behavior has been ascribed to a siphon action in the conduit system that feeds the spring. The deeper part of the conduit slowly fills with water until the level reaches a spillover route or the pressure head becomes sufficient to force water out of the system. Draining through the spillover route triggers a siphon action so that the lower, flooded parts of the conduit are also drained. Once the conduit system has been drained, siphon action ceases and the system slowly fills to begin the next cycle. One of the best studied examples is Big Spring, draining the Lilburn Cave System in Kings Canyon National Park, California, U.S.A. Detailed analyses of hydrographs measured at the spring and in the lowest levels of the cave substantiate a forced siphon mechanism.

Analysis of Spring Hydrographs

Some information concerning the groundwater basin that provides the spring discharge can be obtained from a more careful analysis of the spring hydrograph. Figure 5 shows a schematic hydrograph of a single storm to illustrate the various features. If there has been no rain in the watershed for a long time, the flow from the spring will decrease to a minimum value called the base flow, Q_B (of course, if the spring completely dries up during droughts, the base flow is zero). The storm precipitation is represented by the bars on the left side of the diagram (drawn upside down). After a period of time, the storm lag, the flow of the spring will begin to increase and quickly rise to a peak value, Q_{max}. The rising limb of the hydrograph is generally very steep. After the storm passes, the flow from the spring will begin to decrease, but the rate of decrease is slower than the rate of rise. Often the recession limb of the hydrograph can be represented by an exponential function with two adjustable parameters, Q_0 and τ. The parameter τ has units of time and is a measure of the response time of the aquifer. Exponential fits to karst spring recession curves often produce two or more segments. The fast response segment, with τ on the order of 3–10 days, may represent the draining of the open conduits, while the slow response segment, with τ on the order of 20–50 days, may represent the draining of fractures and other small tributaries. Sometimes spring hydrographs have multiple segments or do not fit exponential equations at all. These responses do not lend themselves to straightforward hydrologic interpretation. More detailed analysis of spring hydrographs can provide considerable information about the aquifer feeding the spring (Geyer et al., 2008).

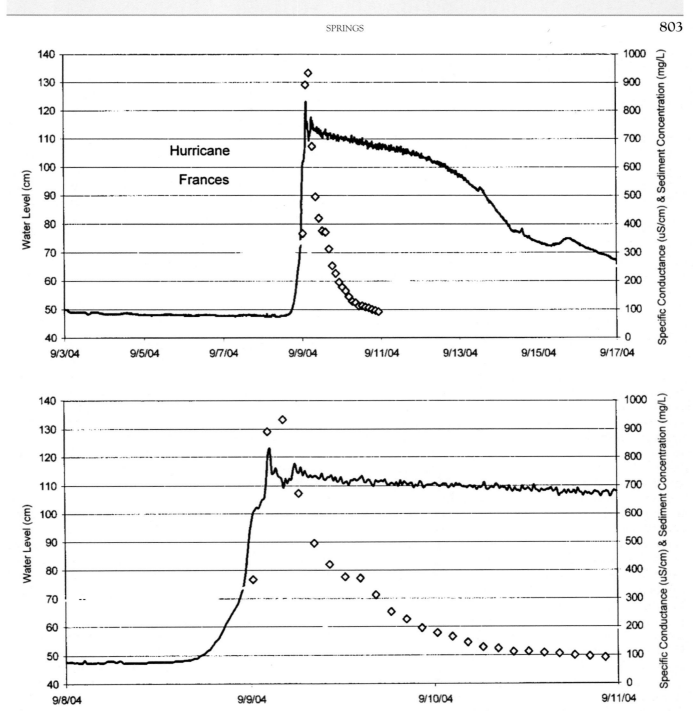

FIGURE 5 Hydrograph for Hurricane Francis at Arch Spring, Blair County, Pennsylvania. Solid line is the water level at the spring; individual points are measured sediment load. Upper figure shows ten-day record; lower figure is expanded three-day record. *From Herman et al. (2008).*

Turbidity and Chemical Variability

Flow velocities in the conduit system of karst aquifers, especially during storm flow, are often sufficient to take clays, silts, and other small particles into suspension and carry the suspended particles to the spring. As a result, some springs become turbid or muddy following storms with the water gradually becoming clear as the discharge decreases back toward base flow conditions (Fig. 5). The presence of muddy water during storm flow is often an indication that the groundwater basin is recharged by sinking streams or open sinkholes that can carry sediment into the system.

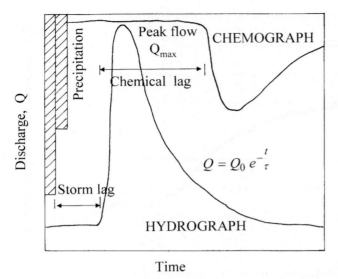

FIGURE 6 Sketch of a single storm peak on a spring hydrograph showing the rapid rising limb and the more drawn out recession limb.

The chemical composition of springs draining from fracture systems (diffuse flow springs) tends to remain constant throughout the year regardless of storms and wet and dry periods. The chemical composition of springs draining from conduit systems tends to be much more variable. The variability in such components as Ca^{2+}, Mg^{2+}, and HCO_3^- is due to dilution of groundwater by storm water, and by changes in recharge chemistry due to winter/summer changes in vegetation. The bulk chemistry of the spring water can be easily monitored by measuring the electrical conductivity. This produces a curve called a *chemograph*, which can be superimposed on the hydrograph as indicated in Figure 6. The dip in the chemograph indicates dilution due to the input of storm water. In Figure 6, the dip in the chemograph is shown offset from the peak in the hydrograph by a time period called the *chemical lag*. If the chemical lag is zero, that is, if the dip in the chemograph is directly superimposed on the peak of the hydrograph, it indicates that the storm water has moved directly to the spring. In this case, the storm lag is a rough measure of the travel time between the storm water input points and the spring. If the dip in the chemograph is offset, it indicates that the conduit system is at least partially flooded. Rising water levels at the upstream end of the system force water out of the flooded conduits to produce the peak in the hydrograph and it is only later that the actual storm water reaches the spring.

Chemographs can be constructed for individual chemical constituents if enough samples are collected and analyzed during the storm flow. These may also be interpreted in terms of the movement of various constituents through the system.

Saturation State of Spring Waters

From a chemical analysis of spring water, it is possible to calculate how the measured concentrations of dissolved carbonate minerals deviate from chemical equilibrium. Springs fed by open conduit systems receiving a considerable fracture of surface water can then be undersaturated with respect to carbonate minerals. The water moved through the aquifer in less time than the time required for reaction with the bedrock. Springs discharging from fracture aquifer tend to be close to chemical equilibrium. Water from aquifers that are recharged at high CO_2 partial pressures can discharge the excess CO_2 at or just below the spring mouth, resulting in the precipitation of the form of calcium carbonate known as *tufa*.

SPRINGS AS WATER SUPPLIES

Water gushing from the earth has seemed an obvious source of freshwater for human consumption for millennia. Villages and towns have grown up around springs because of the convenient supply of high-quality water. As a result, there is a popular myth that spring water is intrinsically pure. Indeed, spring water is often chosen by water bottling companies (LaMoreaux and Tanner, 2001). In the case of karst springs, the myth of pure spring water is exceedingly dangerous.

Karst springs frequently receive their water from cave passages, which in turn receive their water from surface streams that sink underground or from storm flow into sinkholes. Sinkholes are often repositories for household garbage and farm waste such as empty pesticide cans, other trash, and dead animals. Sinking surface streams and sinkholes may admit spilled petroleum hydrocarbons (gasoline, diesel fuel, and home heating oil) as well as industrial solvents such as perchloroethylene (PCE) and trichloroethylene (TCE) and many others. Springs may be impacted by discharge from mining activities in the catchment area. Any spring that becomes turbid after storms or which shows a high degree of chemical variability should be immediately suspected. Even fracture flow springs with clear water and constant chemistry may require extensive testing as well as filtration before being approved as public water supplies. Regulations now often require an elaborate test for surface water influence. Minor debris such as bits of plant material or insect parts is sufficient to require a filter plant.

SPRINGS AS HABITAT

Springs provide a stable habitat for a variety of aquatic organisms. Spring waters, with their nearly constant temperature, remain warm and ice-free during winters. Springs are a transitional environment between true cave conditions upstream and normal surface stream conditions downstream. The temperature and water chemistry of springs are comparable to cave stream conditions, but springs are also under the influence of sunlight. Aquatic plants can grow in springs and provide sources of nutrient that are not available in the cave system. However, springs are generally more sheltered and subject to smaller swings in microclimate that are completely open surface streams.

See Also the Following Article

Hydrogeology of Karst Aquifers

Bibliography

Amiel, R. B., Grodek, T., & Frumkin, A. (2010). Characterization of the hydrogeology of the sacred Gihon Spring, Jerusalem: A deteriorating urban karst spring. *Hydrogeology Journal, 18*, 1465—1479.

Ford, D., & Williams, P. (2007). *Karst hydrogeology and geomorphology*. Chicheser, UK: John Wiley & Sons.

Gary, M. O. (2010). Karst hydrogeology and speleogenesis of Sistema Zacatón. *Association for Mexican Cave Studies Bulletin, 21*, 1—114.

Geyer, T., Birk, S., Liedl, R., & Sauter, M. (2008). Quantification of temporal distribution of recharge in karst systems from spring hydrographs. *Journal of Hydrology, 348*, 452—463.

Goldscheider, N., Mádl-Szönyi, J., Eröss, A., & Schill, E. (2010). Review: Thermal water resources in carbonate rock aquifers. *Hydrogeology Journal, 18*, 1303—1318.

Herman, E. K., Toran, L., & White, W. B. (2008). Threshold events in spring discharge: Evidence from sediment and continuous water level measurement. *Journal of Hydrology, 351*, 98—106.

Kresic, N., & Stevanovic, Z. (Eds.), (2010). *Groundwater hydrology of springs*. Amsterdam: Elsevier.

LaMoreaux, P. E., & Tanner, J. T. (Eds.), (2001). *Springs and bottled waters of the world*. Berlin: Springer-Verlag.

STALACTITES AND STALAGMITES

Silvia Frisia[*] and Jon D. Woodhead[†]

[*]The University of Newcastle, Australia,
[†]The University of Melbourne, Australia

INTRODUCTION

Stalactites and stalagmites are the most common speleothems, the morphology of which is basically controlled by dripping; therefore, both speleothems can be considered as gravitational forms. *Stalactites* are centimeter to meter in scale, hanging from the ceiling and growing toward the cave floor. *Stalagmites* grow from the cave floor upward and are commonly fed by water dripping from an overhead stalactite (Fig. 1). *Stalagmitic flowstones* are a particular type of stalagmite formed by a thin flowing film of water itself fed by groups of dripping stalactites, and coat the cave floor and walls. When a stalagmite and the overhanging stalactite merge, they form a *column* (Fig. 1). Most stalactites and stalagmites are composed of calcite, a few of aragonite, the rhombohedral and orthorhombic phases of calcium carbonate ($CaCO_3$), respectively. Rare stalactites and stalagmites consisting of huntite (a Mg-carbonate), halite (NaCl), gypsum ($CaSO_4 \cdot 2H_2O$), and even opal (amorphous hydrated SiO_2) have been found.

Stalactites and stalagmites likely started to develop in caves when the first carbonate rocks had been subaerially exposed and eroded well over 1 billion years ago. Most speleothems that have been extensively studied date from the Quaternary, and the genesis of these is commonly driven by the process of degassing, which occurs when drip waters having a high carbon dioxide concentration (pCO_2) interact with the cave atmosphere that has a relatively low pCO_2. It is, therefore, believed that occurrence of stalagmites and stalactites greatly increased since the rise of vascular plants in the Devonian, which led to an acceleration of chemical weathering, greater availability of soil CO_2, and a decline in global atmospheric CO_2 concentration (Alonso-Zarza and Tanner, 2010). Chemical weathering of Ca-bearing silicate minerals by acidic waters generated in peat soil, for example, is very effective in yielding calcite stalactites and stalagmites in caves cut into granite and gneiss; in such cases karst dissolution does not play a role in the genesis of these speleothems, but the presence of vascular plants does. The focus of the following sections is, therefore, on the genesis, structure, and chemical properties of calcium carbonate stalagmites and stalactites since the Devonian.

MINERALOGY AND PROCESSES OF FORMATION

Carbonate stalactites and stalagmites are commonly composed of calcite, the calcium carbonate ($CaCO_3$) phase that is thermodynamically stable at surface temperature (T) and pressure (P). They may be also composed of aragonite (Fig. 2A), the high pressure polymorph of $CaCO_3$, thermodynamically unstable at surface P and T. The formation of aragonite in caves is most probably related to a combination of factors, some of which are well understood, others more speculative. Aragonite precipitation in stalactites and stalagmites appears to be favored by high Mg/Ca ratio of the parent water, high pH, presence of ions and

FIGURE 1   Stalactites, both soda straws and cone stalactites, candle-shaped stalagmites, columns (stalactites and stalagmites merged), and stalagmitic flowstone coating the cave floor. The plastic containers host glasses onto which *in situ* calcite precipitation experiments have been carried out to determine the processes that influence the development of different crystals.

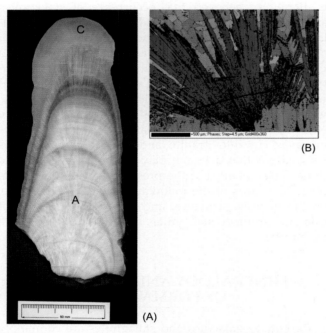

FIGURE 2   (A) Stalagmite consisting of aragonite layers (A) capped by calcite layers (C). (B) Backscattered electron diffraction image of calcite (colored in red) replacing aragonite crystals (colored in blue). *Photo courtesy of I. J. Fairchild.*

not be the same as that of the feeding stalactite. It is not uncommon, in fact, to observe calcite stalactites feeding aragonite stalagmites. Aragonite, however, being thermodynamically unstable, could dissolve and be replaced by calcite through dissolution and reprecipitation, yielding diagenetically modified stalagmites and stalactites (Fig. 2B).

The precipitation of calcium carbonate in caves occurs from solutions that commonly have low ionic strength and reach supersaturation with respect to calcite (or aragonite). The capability of a solution to precipitate calcite is commonly measured by its Saturation Index (SI):

$$SI_{calcite} = \log\ \{(Ca^{2+})(CO_3^{2-})/K_{calcite}\}$$

where $(Ca^{2+})(CO_3^-)$ is the ion activity product and $K_{calcite}$ is the temperature-dependent solubility constant for calcite. Calcite will precipitate when $SI_{calcite} > 0$, and dissolve when $SI_{calcite} < 0$.

Monitoring studies have shown that active calcite growth may occur in stalagmites even if the initial drip waters are barely at saturation, which is the case for most high-altitude, high-latitude caves. The mechanism producing supersaturation of an already saturated solution and resultant precipitation of calcite in caves is CO_2-degassing, a set of reactions that involve the loss of carbon dioxide from the drop to the cave atmosphere. The precipitation of calcite (or aragonite) is described by the following equilibrium reaction:

$$Ca^{2+} + 2HCO_3^- \Leftrightarrow CaCO_3 + CO_2 + H_2O$$

If the Ca^{2+} ion concentration in the parent solution does not increase, the loss of CO_2 from the solution through degassing shifts the equilibrium reaction to the left, and calcite (or aragonite) precipitates. Surface area and timing for CO_2 degassing varies for stalactites and stalagmites, and is dependent on drip rate: the slower the drip rate, the longer the degassing process may proceed. Degassing may also be forced by kinetic (irreversible) processes, in particular, in ventilated caves where the pCO_2 of the cave air may attain values similar to those of surface air (slightly less than 400 ppm), while drips may have a CO_2 concentration at least ten times higher. Cave ventilation modulated by the temperature difference between the cave (warm in winter and cool in summer) and the surface (cold in winter and warm in summer) may, thus, force stalagmite and stalactite growth in alpine and high-latitude cold caves, where calcite precipitation is not favored given that the solubility of calcite (in this case, that of the host rock) decreases with increasing temperature, and soil CO_2 production is restricted to a short, warm season.

complexes in solution that may inhibit calcite growth sites, low drip rate, and forced degassing (which, by promoting degassing, increases the pH of the solution). The mineralogical composition of a stalagmite need

STALACTITES

The most common variety of stalactite is the tubular soda straw, consisting of thin translucent, tubular layers of calcite crystals surrounding a central canal. Soda straws are speleothems generated by axial feeding, and characterized by a central canal with constant diameter and by a wall structure that is controlled by geometric selection during growth on the meniscus of a drip. The term *soda straw*, therefore, is not to be applied to any stalactite with a central, hollow tube.

Commonly, the outer diameter of a soda straw varies from 5 to 10 mm, and the diameter of the inner canal ranges from 2 up to 6 mm. Scanning electron microscopy (SEM) observations indicate that soda straws grow by linear, downward accretion of coalescent rhombohedra crystallites with their c-axis (the threefold symmetry axis of the rhombohedral unit cell) dimension nearly vertically oriented. When a droplet of water at saturation with respect to calcium carbonate flows out of pores or fissures at the roof of the cave it degasses and deposits a small rim of calcite (or, less commonly, aragonite), which has the diameter of the droplet. Successive layers of crystals then deposit on the previous layer, maintaining the original diameter. Soda straws, thus, commonly show growth bands typically 0.05 to 0.5 mm thick, which is the linear annual extension rate expected for these speleothems. Growth bands identify a periodic thickening of the inner channel wall. Field observations and sampling reveal that the dimension of the central tube is a function of drip rate. In some cases, air is present within the channel and degassing occurs within the straw, causing calcite precipitation and, eventually, its complete obstruction. The straw then would not be capable of feeding an underlying stalagmite, unless rupture followed by development of a new straw at the same drip point occurs. Aragonite soda straws are less common and consist of acicular crystals.

Conical stalactites characteristically taper downward, and may be several tens of centimeters wide. Their length can reach the meter-scale and, eventually, the stalactite coalesces with the underlying stalagmite. Conical stalactites consist of concentric layers of crystals elongated perpendicular to a central growth axis (the axis extending from the ceiling toward the floor of the cave), which may be hollow to allow the passage of water (Fig. 3). With time, the center of the stalactite may be clogged by calcite and once the central conduit is closed, water flows along the outer surface of the speleothem. A conical stalactite, thus, grows through deposition of crystals in radial layers developing on the sides of a remnant soda straw, and spreading outward. The crystals of the radial layers have their c-axes perpendicular to the original straw. Growth on a cone stalactite, therefore, occurs both in vertical and lateral directions.

FIGURE 3 Thin section perpendicular to the vertical growth axis of a cone stalactite showing the hollow central canal and concentric layers of columnar calcite (crossed polarizers). Base of the photo: 8 mm.

STALAGMITES

Stalagmites are convex deposits that grow upward, commonly fed by an overhanging stalactite, and show flat, rounded, or slightly hollow tops, but not a central canal. Calcite and aragonite stalagmites form through degassing of a saturated thin film of fluid (about 0.1 mm thick), which slowly moves down from the stalagmite tip to its flanks. The morphology of stalagmites is mostly determined by the drip rate, and in particular by the interannual variability of the drip rate and by the distance from the tip of the feeding stalactite. Given similar $SI_{calcite}$, drips that show a high interannual variability in the time elapsed between the fall of two drops, that is, a high coefficient of variation, produce cone-shaped stalagmites. By contrast, stalagmites fed by drips that show interannual constant time elapsed between two drops yield candle-shaped stalagmites. Development of candle-shaped stalagmites is favored also by drips with high $SI_{calcite}$, whereas, given the same drip rate variability, a decrease in $SI_{calcite}$ may result in the passage from candle to cone-shaped stalagmites (Fig. 4). Relatively high drip rates and waters with low $SI_{calcite}$ form stalagmitic flowstones, where calcite precipitation is controlled by degassing over a large surface area, possibly aided by a flow regime that is more turbulent than the laminar flow on top of stalagmites. The relationships between $SI_{calcite}$ and drip rates reported in Figure 4, do not account for

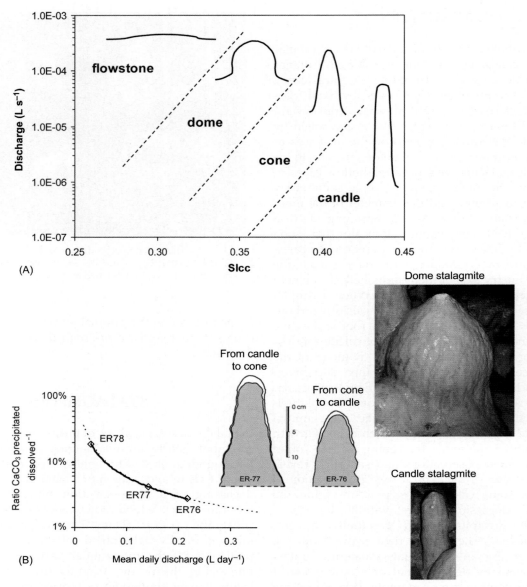

FIGURE 4   Relationship between stalagmite morphologies, discharge, and $SI_{calcite}$. Stalagmite ER 77 has become cone-shaped in the last 150 years (top layer), whereas ER76 has become candle-shaped. (Redrawn from Miorandi et al., 2010).

the effect of the height of the drop. When this exceeds 2 m, there is fragmentation of the impinging drop at the impact point with the stalagmite surface, which tends to be flat. A great fall height may also give rise to stalagmites whose morphology resembles stacked dishes or cylindrical stalagmites characterized by a sunken splash cup.

The rate of stalagmite growth was theoretically calculated to be in the range of 0.9 to 2.1 nmol min^{-1} for a water film thickness of 50 μm; laboratory experiments and observations support these theoretical calculations. The mechanisms of crystal growth, and the presence of ions or compounds that may poison crystal growth sites, however, may have the overall effect of promoting or retarding stalagmite growth rates. A stalagmite most commonly grows under interannual variable physical and chemical parameters, which may result in periodic drying, fluctuations in the supersaturation state of the solution, or peaks in the numbers of organic molecules transported by the drip water onto the speleothem. Stalagmites, therefore, can be considered polycrystalline aggregates (Sunagawa, 2005) and rarely consist of single crystals. Stalagmite growth can be explained as occurring through the "addition" of growth units arranged in successive, well-stacked layers, which resemble miniature strata when the stalagmite is cut along the vertical growth axis. The way each subsequent growth unit attaches to the units below and adjacent to it is commonly syntaxial, that is, the new unit keeps the same crystallographic orientation. The composite crystals, thus, are formed of crystalline calcite in perfect optical continuity, one with respect to the other, and show uniform extinction. There are stalagmites where the stacking of successive growth units is not perfectly syntaxial, due to fluctuations in drip rate and degree of supersaturation (or in the amount of impurities), which results in the development of diverse stalagmite fabrics (*e.g.*, Frisia and Borsato, 2010).

Columnar is the most common fabric type observed in stalagmites consisting of calcite. Macroscopically, it is composed of crystals ≥ 1 mm wide and ≥ 2 mm long, elongated perpendicular to the growing speleothem surface, with a length to width ratio of about 6:1, uniform extinction under crossed polarizers, and serrated or open crystal boundaries. In the latter case, intercrystalline pores may host fluid inclusions or impurities. Each macroscopically visible crystal consists of the regular stacking of rhombohedra crystallites typically 50 to 100 μm wide and ≥ 100 μm long, which are deposited during the hydrological cycle that characterizes each cave (*e.g.*, Miorandi *et al.*, 2010). Stalagmites composed of columnar fabrics are commonly translucent unless the crystals do not fully coalesce and there are intercrystalline voids elongated perpendicular to the substrate. In this case, the fabric becomes open columnar and the stalagmite calcite has an opaque macroscopic appearance. When the presence of impurities in the drip water or physical phenomena perturb the regular stacking of the crystallites, the composite crystals attain a milky, porous aspect, and show irregular crystal boundaries and patchy extinction when viewed under crossed polarizers of a petrographic microscope. This subtype of the columnar fabric has been termed *microcrystalline* (Frisia and Borsato, 2010) and distinguished from the columnar fabric because it has an environmental significance: it forms under highly variable drips rates ranging from 30 mL to less than 0.1 mL per minute, and in the presence of impurities (P ions, fulvic acids, trace metals). Stacking of crystallites into composite crystals formed by crystallographic dendrites (Jones and Renaut, 2010) gives rise to a dendritic fabric, which is typical of stalagmites formed near cave entrances, subject to air currents and evaporation, or in ventilated caves. The presence of dendritic fabric in stalagmites indicates that they formed under the influence of kinetic processes, and, most probably, under highly variable drip rates (Frisia and Borsato, 2010). The role of cave bacteria in the growth of stalagmites, and on their fabrics, is still largely unknown, but it must not be excluded. Lipids related to cave microorganisms, in fact, have been found within stalagmite calcite.

Stalagmites consisting of aragonite are commonly characterized by acicular crystals, which have a length to width ratio $\gg 6:1$, are elongated along the c-axis, and show needle-like terminations. Acicular aragonite in stalagmites has been interpreted as being caused by precipitation from alkaline solutions with high pH and under low drip rates. Alternating layers of columnar calcite and acicular aragonite in stalagmites within the same cave thus clearly indicate that the hydrology of the cave has varied periodically. In monsoonal settings, for example, calcite layers form in the wet season, and aragonite develops in the dry season.

ARCHIVES OF THE DEEP PAST

Changes in the morphology and mineralogy of stalagmites clearly respond to climate-related phenomena, such as drip rate variability and temperature-modulated cave ventilation. The physical and chemical properties of stalagmites, therefore, provide unparalleled information, often at very high temporal resolution (annual to seasonal), on past climate and environment changes in continental regions. These records are especially powerful because stalagmites can be dated precisely with radiometric techniques and by counting annual growth layers.

Although most of the speleothem-based paleoclimate research to date has been undertaken on samples from the Late Quaternary, recent developments have now extended their use for paleoclimate studies back to the Paleozoic. For example, a speleothem with the morphology of a conical stalactite was dated 289 ± 0.68 Ma BP by U-Pb radiometric determination. The stalactite shows well-preserved concentric layers of columnar and mosaic calcite. Due to its old age, diagenetic alteration such as dissolution and reprecipitation cannot be excluded, despite the pristine appearance under the microscope of the calcite crystals. The distribution of trace elements in the speleothem, however, is coherent with its concentric structure and, thus, consistent with a primary geochemical signal in unaltered calcite. The chemical properties of this ancient stalactite were used to gain insight into climate on the supercontinent Pangaea at an important period for early terrestrial vertebrate evolution (Woodhead *et al.*, 2010).

The preservation of ancient stalagmites is seriously endangered by erosion, tectonics, and, in the Boreal hemisphere, the effects of aggressive meltwaters during periodic deglaciations in the Quaternary. There are, therefore, few karst regions where ancient stalagmites are likely to be extremely well preserved; one of

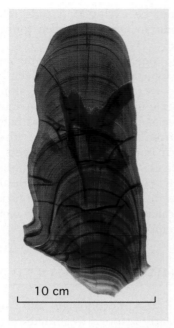

FIGURE 5 Polished, thin slab of a mid-Pliocene stalagmite from the Nullarbor showing well-preserved layers of columnar calcite indicative of pristine preservation of its chemical and physical properties.

these is the Nullarbor Plain of southwestern Australia. This large, relatively flat karst area is, at present, characterized by arid conditions, but its caves preserve dark, layered, candle-shaped stalagmites consisting of columnar calcite (Fig. 5), most of which formed between 3 and 4 million years ago, that is, during the mid-Pliocene climatic warm period when global surface temperatures are believed to have been several degrees higher than today, with relatively high atmospheric CO_2 concentration. The Nullarbor Caves also preserve stalagmites that are as old as the Late Miocene, when northward movement of Australia and New Guinea are believed to have caused the gradual closure of the Indonesian seaway, precipitating a major reorganization of oceanic circulation. It is entirely possible, then, that the age distribution, geochemistry, and morphology of the Nullarbor stalagmites preserves a wealth of information about the influence of plate tectonics of regional climates, and the timing and effects of warm periods in Earth's history, which may provide analogues of the projected climate scenarios for the next 100 years.

In conclusion, the combination of radiometric dating, petrography, and geochemical analyses of speleothems has the potential to become the most powerful tool to reconstruct climate and environmental changes in continental settings from at least the Paleozoic through to the Quaternary.

Bibliography

Alonso-Zarza, A. M., & Tanner, L. H. (2010). Preface. In A. M. Alonso-Zarza & L. H. Tanner (Eds.), *Developments in sedimentology, Vol. 61. Carbonates in continental settings—Facies, environments and processes* (pp. xi–xiii). Amsterdam: Elsevier.

Frisia, S., & Borsato, A. (2010). Karst. In A. M. Alonso-Zarza & L. H. Tanner (Eds.), *Developments in sedimentology, Vol. 61. Carbonates in continental settings—Facies, environments and processes* (pp. 269–318). Amsterdam: Elsevier.

Jones, B., & Renaut, R. B. (2010). Calcareous spring deposits in continental settings. In A. M. Alonso-Zarza & L. H. Tanner (Eds.), *Developments in sedimentology, Vol. 61. Carbonates in continental settings—Facies, environments and processes* (pp. 177–224). Amsterdam: Elsevier.

Miorandi, R., Borsato, A., Frisia, S., Fairchild, I. J., & Richter, D. V. (2010). Epikarst hydrology and implications for stalagmite capture of climate changes at Grotta di Ernesto (NE Italy): results from long-term monitoring. *Hydrological Processes, 24*, 3101–3114.

Sunagawa, I. (2005). *Crystals: Growth, morphology and perfection*. Cambridge, U.K.: Cambridge University Press.

Woodhead, J. D., Reisz, R., Fox, D., Drysdale, R., Hellstrom, J., Maas, R., et al. (2010). Speleothem climate records from deep time? Exploring the potential with an example from the Permian. *Geology, 38*, 455–458.

SULFURIC ACID CAVES

Arthur N. Palmer* and Carol A. Hill†

State University of New York, †University of New Mexico

Most caves owe their origin to carbonic acid generated in the soil. In contrast, sulfuric acid caves are produced by the oxidation of sulfides beneath the surface. Although sulfuric acid caves are relatively few, they include some large and well-known examples, such as Carlsbad Cavern, New Mexico. They also provide evidence for a variety of deep-seated processes that are important to petroleum geology, ore geology, tectonic history, and the nascent field of karst geomicrobiology.

GEOLOGIC SETTING

Cave origin by sulfuric acid requires the reaction of oxygen with either solid or aqueous-phase sulfides. The most common sulfide source is hydrogen sulfide (H_2S) which originates in reducing environments such as petroleum-rich sedimentary basins. Sulfuric acid is generated when rising H_2S-rich water encounters oxygen-rich water at or near the water table (Fig. 1). The resulting caves are examples of *hypogenic caves*, the origin of which depends on deep-seated processes and rising groundwater. Oxidation of metallic sulfides (*e.g.*, pyrite) accounts for only a few minor caves and generally produces only scattered solutional pores. This

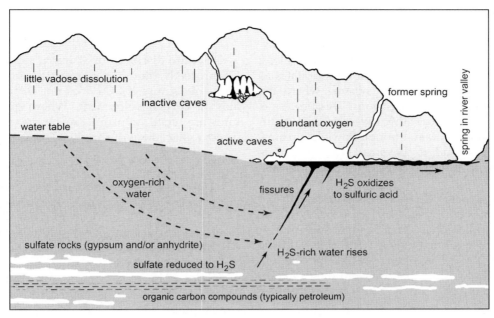

FIGURE 1 Typical setting for sulfuric acid cave origin. Alternatively, in many areas the H_2S is generated within an adjacent basin and driven in aqueous solution laterally or upward into carbonate rocks by favorable hydraulic gradients or by tectonic or sedimentary overpressuring. The H_2S source may also be volcanic.

chapter focuses only on caves produced by H_2S oxidation.

The source of cave-forming H_2S is usually a sedimentary basin bordering a carbonate upland, or deep sulfate deposits interbedded within the carbonate rock itself. Water from basinal sediments can be expelled by compaction, overpressuring by rapid sediment accumulation, as well as by flow down a regional hydraulic gradient. Because of their higher permeability, carbonate rocks often have lower hydraulic heads than surrounding materials and serve as the zones of groundwater discharge to the surface. Cave origin by these rising fluids requires oxidation of H_2S beneath the surface in carbonate rocks. This requires either convergence with oxygenated groundwater fed by meteoric infiltration, or exchange of air through openings to the surface. Either way, oxidation is greatest at and near the water table. Most H_2S-rich groundwater simply discharges at springs without encountering enough oxygen below the surface to form caves.

DIAGNOSTIC FEATURES OF SULFURIC ACID CAVES

The typical sulfuric acid cave consists of a central area of irregular rooms, with local mazes of intersecting fissures or sponge-like solution pockets and sparse outlet passages that lead to active or relict springs. Some caves consist of only a single passage. From these main zones of cave development, narrow fissures commonly descend and pinch downward. In active examples these fissures are the main inlets for H_2S-rich water.

Much of the solutional enlargement of a sulfuric acid cave takes place above the water table. H_2S escapes from the rising water into the cave air and is absorbed by moisture droplets and films, along with oxygen from the cave atmosphere. The moisture can be supplied by infiltration and/or condensation. Production of sulfuric acid within this water dissolves the adjacent carbonate rock and typically produces a rind of replacive gypsum. Fractionation during the redox reactions makes the sulfur isotopes in the secondary gypsum relatively "light" compared to those of primary gypsum. Sulfuric acid can also alter clays, yielding several by-products including alunite $(KAl_3(SO_4)_2(OH)_6)$ and hydrated halloysite (or endellite, $Al_2Si_2O_5(OH)_4 \cdot 2H_2O$), while liberating silica, which precipitates as intergranular cements and irregular chert beds. Native sulfur can accumulate by incomplete oxidation of H_2S at low pH, especially where acids are locally shielded from the carbonate rock by gypsum or siliciclastic rock.

Cave enlargement can continue even after sulfuric acid activity has ceased. Through-flowing meteoric water easily removes the secondary gypsum. Condensation of moisture in cool upper levels and on cave ceilings is also prominent, owing to circulation of warm humid air from lower levels. Because it absorbs CO_2 from the cave air, condensation moisture readily dissolves carbonate rocks. This moisture consists mainly of capillary films, which produce smooth, rounded surfaces and cupolas. The dissolved material is carried downward by seepage along the walls and within pores, and it often reprecipitates as cave popcorn and needle-like speleothems where the water descends into evaporative regions. This process is more noticeable than in most other caves, because sulfuric acid caves rarely contain through-flowing streams, and their entrances tend to be small and few, thus enhancing thermally driven convection within the cave atmosphere.

Many of the diagnostic features of sulfuric acid cave origin can be removed by near-surface processes after the caves cease to be active. This is especially true in humid climates, where the evidence of former sulfuric acid cave development is often effaced by shallow meteoric groundwater. Cave patterns and their relation to the geologic setting may be the only remaining genetic clues.

CHEMICAL REACTIONS

Most H_2S is generated by reduction of sulfates by organic carbon compounds. Gypsum and anhydrite are the typical sulfate sources involved in sulfuric acid speleogenesis. The simplest form of the reduction reaction is

$$SO_4^{2-} + 2CH_2O \rightarrow H_2S + 2HCO_3^- \quad (1)$$

where CH_2O represents a generic organic compound.

H_2S is highly soluble in water, and except at depths of only a few meters below the water table it rarely reaches concentrations high enough to produce gaseous bubbles. Where H_2S is carried by groundwater into an oxidizing environment, sulfuric acid is produced:

$$H_2S + 2O_2 \rightarrow 2H^+ + SO_4^{2-} \quad (2)$$

usually with other sulfur species as intermediate products. Microbial processes greatly speed the reactions and help to determine the intermediate reactions. At low pH, the following equilibrium becomes important:

$$2H^+ + SO_4^{2-} \leftrightarrow H^+ + HSO_4^- \quad (3)$$

Within carbonate rocks, sulfuric acid is a potent cave-former. For limestone, the dissolution reaction can be stated:

$$CaCO_3 + 2H^+ + SO_4^{2-} \rightarrow Ca^{2+} + SO_4^{2-} + H^+ + HCO_3^- \quad (4)$$

$$H^+ + HCO_3^- \leftrightarrow H_2CO_3 \leftrightarrow CO_2 + H_2O \quad (5)$$

Dolomite dissolves in a similar manner. As reaction (5) proceeds, the partial pressure of carbon dioxide (P_{CO_2}) rises, and if CO_2 does not escape, carbonic acid aids in cave development. Where groundwater flow rates are small, gypsum can be deposited, often as a direct replacement of carbonate rock:

$$Ca^{2+} + SO_4^{2-} + 2H_2O \leftrightarrow CaSO_4 \cdot 2H_2O \quad (6)$$

In groundwater, buffering of the acids by dissolution of carbonate rock generally keeps the pH moderate ($\sim$6–7). Lower pH can develop if the acids are shielded from the carbonate rock by a nonreactive material. Gypsum crusts tend to provide this shield so that above the water table the pH of water films and droplets on gypsum can decrease considerably. Build-up of acid in contact with gypsum is limited by reaction (3), which usually holds pH to values of about 1.5 (depending on temperature and other dissolved species), while gypsum dissolution consumes H^+ to produce HSO_4^-. However, droplets and films of moisture on relatively nonreactive materials such as chert, quartz, and microbial filaments can reach extremely low (sometimes negative) pH values, as in Cueva de Villa Luz (described below).

MICROBIAL INTERACTIONS

Reduction of sulfate in the presence of organic carbon can take place spontaneously at temperatures above about 85°C. At lower temperatures the process depends on bacterial sulfate reduction. Thus in many cases (perhaps most), H_2S production requires microbial mediation. Sulfate reducers include the bacterium *Desulfovibrio*, which ingests sulfate and excretes H_2S.

In the oxidizing zone, where the caves are formed, sulfur-oxidizing bacteria (*e.g.*, *Thiothrix*) can speed reactions by several orders of magnitude. Also, certain species are able to facilitate the conversion of sulfides directly to sulfuric acid without intermediate by-products. The oxidation reactions are strongly exothermic, providing an energy source for microbial growth. Bundles of bacterial filaments are common in active sulfuric acid caves, and their mineral-coated fossilized forms have also been observed in relict caves. Microbes are also active mediators of redox reactions in the deeply weathered walls of certain sulfuric acid caves.

FIELD EXAMPLES

Some of the best-documented examples of sulfuric acid caves are described here.

Kane Caves, Wyoming

Lower Kane Cave, in north-central Wyoming, is where sulfuric acid cave development was first described in detail (Egemeier, 1981). The cave consists of a single linear stream passage in the Madison Limestone of Mississippian age, where the limestone is breached by the Bighorn River along the Sheep Mountain Anticline. The cave spring is located at the present river level, and the passage extends at a low gradient for approximately 350 m. The cave stream is fed by H_2S-rich water that discharges from floor fissures at several places along the passage, including the upstream terminus (Fig. 2). Bundles of filaments of sulfur-oxidizing bacteria form conspicuous tendrils in and around the fissures. A gypsum rind coats most of the walls and ceilings around the inlets as the result of oxidation of H_2S within moisture films. The gypsum flakes off the limestone bedrock and drops to the floor, where it either falls into the stream or builds mounds adjacent to it. The stream is undersaturated with gypsum, and any gypsum that comes in contact with it is carried away in solution. This process contributes greatly to the enlargement of the cave.

About 30 m directly above is Upper Kane Cave, a relict precursor of the active lower cave. The upper cave is a dry passage similar in pattern and length to the lower one, but with a larger cross-section and a breakdown-strewn floor. Breakdown and dissolution by vadose seepage have obscured or removed much of the gypsum crust that must have once lined most of its surfaces. Discharging along both banks of the Bighorn River are other similar caves that contain H_2S-rich water, but they are smaller than the Kane Caves and present more difficult access.

The water that feeds the active caves in this area comes from two large regional groundwater systems with infiltration as much as 150 km away. The apparent source of the H_2S is reduction of sulfates interbedded within the Madison in the presence of basinal hydrocarbons. Sulfates have been removed by shallow meteoric groundwater in most exposed areas of the Madison, but it is still abundant within inter-mountain basins.

FIGURE 2 Fissure inlet midway in Lower Kane Cave, Wyoming. Note the mat of white bacterial filaments in foreground. *Photo by A. N. Palmer.*

Cueva de Villa Luz, Mexico

Of all well-studied sulfuric acid caves, this one has the most intense H_2S-H_2SO_4 activity. It is located in the semitropical jungle of southern Tabasco, in a low upland of folded and faulted Cretaceous limestone (Fig. 3; Hose *et al.*, 2000). It lies several tens of kilometers south of a basinal oilfield and 50 km east of the active El Chichón volcano. The cave atmosphere is highly toxic, with fluctuating H_2S concentrations that sometimes exceed 200 ppm. (The OSHA safety standard

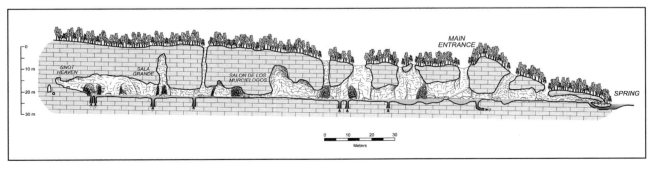

FIGURE 3 Profile of Cueva de Villa Luz, Mexico. *From map by Bob Richards and Louise D. Hose.*

for H$_2$S was once 10 ppm but has recently been decreased to zero.) Levels of atmospheric carbon monoxide and carbon dioxide are also dangerously high at times. Gas masks are mandatory for safe entry, although visitors often unwisely explore the near-entrance regions without protection.

The cave consists of braided stream passages that discharge to a single spring, which is the head of the surface stream Rio Azufre ("Sulfur River"). Total discharge is a rather uniform 200–300 L s^{-1} and shows little response to rainfall variation. The cave water is a sulfate-chloride-bicarbonate brine compatible with a basinal origin, presumably from the oilfield to the north. Light sulfur isotopes in the H$_2$S support this hypothesis. Water flows into the cave through approximately 26 inlets, all too small for human entry. Most are located at the upstream end of the cave, with a few about midway in the cave. The cave streams have moderate and rather uniform gradients, with a few scattered deep pools up to several meters deep. The ceiling rises and falls abruptly, with many skylights to the surface.

Each stream inlet delivers one of two types of water (Table 1). Some are anoxic and high in H$_2$S. Others are oxygenated, with no detectable H$_2$S, and have apparently encountered aerated conditions farther upstream. Otherwise the two water types are chemically rather similar. The anoxic water releases H$_2$S to the cave atmosphere, and white filaments of sulfur-fixing bacteria coat the floors over which it flows. The water is roughly at equilibrium with dissolved limestone and dolomite. The oxygen-rich water precipitates iron hydroxide, which coats filaments of iron-fixing bacteria that grow on the floors of these streams. This water is slightly supersaturated with respect to limestone and dolomite, and in a few places it deposits calcite travertine. The two waters mix to produce streams that are white with colloidal sulfur and which have a nearly neutral pH.

Infiltration moisture on the cave walls and ceilings absorbs H$_2$S and oxygen from the cave atmosphere, and the resulting sulfuric acid has produced a thick coating of crystalline gypsum over most limestone surfaces, especially in the vicinity of sulfide-rich inlets. Sulfur accumulates on the gypsum-coated walls above these inlets. The gypsum and sulfur are highly depleted in ^{34}S, with δ^{34}S ranging from -26 to $-22‰$ CDT, showing the influence of microbial mediation. The cave walls are coated in many places by organic slime intermixed within the gypsum. Bacterial filaments hang from many of the gypsum surfaces, and drops that accumulate on them have an average pH of 1.4 (Fig. 4). However, the drops that linger longest before falling have measured pH values as low as zero. They can burn skin and eat holes in clothing.

The filaments form white, gelatinous, elastic bundles. Their mucus-like texture has inspired a vivid informal name, but the term *biothem* is gaining favor. The bacteria form the base of a complex food chain, including midges, spiders, gastropods, and fish (*Poecilia mexicana*). In an annual spring ceremony, the local Soque Indians harvest the fish just inside the cave entrance to provide food for a community feast.

Although it is compelling to think of the entire cave community as based solely on the energy from sulfide oxidation, the system is complicated by an influx of organic material and light from the many openings to the surface. A more likely candidate for pure chemoautotrophy is the less diverse underground community of Movile Cave in southeastern Romania, which occupies an air-filled H$_2$S-rich room that is almost

TABLE 1 Water Chemistry in Lower Kane Cave, July 1970 (Egemeier, 1981) and Cueva de Villa Luz, January 1998 (Hose et al., 2000)

	Lower Kane Inlets	Villa Luz H$_2$S-Rich Inlet	Villa Luz O$_2$-Rich Inlet	Villa Luz Spring
Temp (°C)	23.5	27.5	28.3	28.0
pH	7.0	6.61	7.23	7.14
Ca (mg L^{-1})	97	396	383	393
Mg	31	81	97	88
Na	11	484	477	–
HCO$_3$	218	498	451	477
SO$_4$	189	940	980	910
Cl	12.5	814	792	803
H$_2$S (aq)	6 ± 1.5	500 ± 50	N.D. (<0.1)	17.5 ± 3
O$_2$ (aq)	0	N.D. (<0.1)	4.3 ± 0.7	1.3 ± 0.2
P$_{CO_2}$ (atm)	0.0005–0.0014	0.11	0.023	0.030

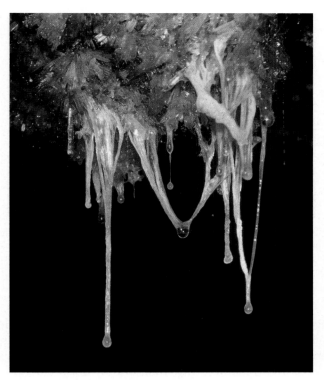

FIGURE 4 Bundles of bacterial filaments in Cueva de Villa Luz, Mexico. The droplets are dilute sulfuric acid with pH as low as zero. Height of photo is ~15 cm. *Photo by A.N. Palmer.*

FIGURE 5 Massive gypsum in the main upper level of the Frasassi Cave System, Italy. Note vertical grooves formed by dripping meteoric water. The dark stain at the bottom is from late-stage clay fill. *Photo by A.N. Palmer.*

totally isolated from external nutrients (Sarbu *et al.*, 1996). However, Movile Cave has an H_2S activity far lower than that of Cueva de Villa Luz.

Much of the sediment on the floor of the main stream of Cueva de Villa Luz is an organic-rich muck in which sulfate reduction and iron sulfide precipitation are actively taking place. A large percentage of the sediment consists of tiny gastropod shells a few millimeters in diameter, as well as fragments of allogenic metamorphic rock.

The cave enlarges in several ways. The most vigorous enlargement is by the same gypsum-replacement mechanism described above for the Kane Caves. It also accounts for the irregular ceiling profile. The skylights apparently originated as narrow fissures communicating with the surface, which have enlarged as escaping H_2S is absorbed by moisture on their walls. Although the main streams are slightly supersaturated with both calcite and dolomite during dry periods, they are able to dissolve the bedrock during periods of high infiltration, when the dripping of H_2SO_4-rich water is most abundant. Beneath skylights, drips from the surface form cylindrical holes in the gypsum piles. Where acidic water drains through the porous gypsum of the cave walls, it forms solutional rills in the limestone near stream level, where the gypsum is absent.

Frasassi Cave System, Italy

Perhaps the largest known active sulfuric acid cave, the Frasassi Cave System is one of Europe's best-known show caves. It is located in folded and faulted Jurassic limestone along the eastern flank of the Apennine mountain chain. Its lowest level is a stream passage with high concentrations of H_2S and active sulfuric acid cave enlargement. The source of the sulfides is thought to be reduction of underlying gypsum beds. Passages and rooms at higher levels are much larger, and although presently inactive they show evidence for a sulfuric acid origin, such as massive bodies of gypsum derived from replacement of the native limestone (Fig. 5). Passages have highly irregular patterns, as is typical for caves in highly deformed host rock, but the overall passage profiles are fairly horizontal and reflect present and past fluvial base levels.

The processes and passage character at the stream level are similar to those described above in Cueva de Villa Luz, although present H_2S concentrations in Frasassi are only about 10% as great, and there are fewer known inlet points for sulfide-rich water. Infiltrating water dilutes the sulfide-rich water by 30–60%. Measurements by Galdenzi *et al.* (2008) show that limestone dissolution by H_2S oxidation is most intense in the wet season, when the infiltration of oxygen-rich water is greatest.

Caves of the Guadalupe Mountains, New Mexico

The Guadalupe Mountains of southeastern New Mexico contain some of the world's most spectacular caves, of which Carlsbad Cavern and Lechuguilla Cave are the largest and best known (Figs. 6 and 7). Although presently inactive relics, these caves have been the focus of extensive studies of sulfuric acid cave origin. Many of the processes recently observed in Cueva de Villa Luz were anticipated by studies of now-inactive caves in the Guadalupes.

The Guadalupes are composed of a Permian reef complex, which today stands at altitudes up to 2600 m as the result of Cretaceous and Late Cenozoic uplift. The Capitan Reef, which forms the southeastern escarpment, grades northwestward into bedded back-reef limestones and dolomites. An apron of fore-reef talus extends to the southeast and forms the boundary with the Delaware Basin, which consists of Permian carbonates and quartz-rich sandstones capped by sulfates and halides. Hydrocarbons are abundant at depth within the Delaware Basin and in smaller concentrations beneath the Guadalupe Mountains.

The typical Guadalupe cave has a ramifying pattern consisting of irregular rooms and mazes with passages branching outward from them (Fig. 6). Overlapping

FIGURE 7 The Chandelier Maze, Lechuguilla Cave, shows the sponge-like pattern of dissolution in the massive Capitan reef rock, vivid wall colors produced by mineral oxidation, and accumulations of speleogenetic gypsum. *Photo by A.N. Palmer.*

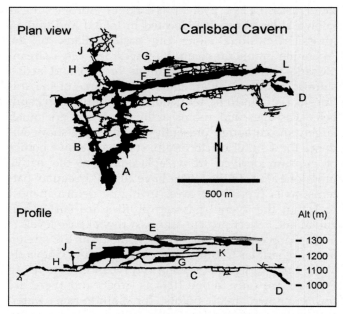

FIGURE 6 Map and profile of Carlsbad Cavern, New Mexico. E = entrance. Other letters show correspondence between locations on map and cross-section. *From map by Cave Research Foundation.*

tiers of rooms and passages are common. Branches do not converge as tributaries, but instead serve as distributary outlets at successively lower elevations. Some caves involve simple widening of only a few fractures. Many caves, or parts of caves, have complex network or sponge-like patterns. Because of the large primary porosity in the reef, spongework is most abundant in caves in that rock unit.

As in the Frasassi Caves described above, caves of the Guadalupe Mountains have considerable vertical relief. Some of this may be the result of mixing between H_2S-rich water and oxygenated water deep beneath the water table, but successive stages of water-table lowering appear to be the main cause. Most cave enlargement takes place above the active passages, as in Cueva de Villa Luz, so that there is great integration between levels. H_2S-rich water rising from depth along prominent fractures and partings (either from the Delaware Basin or the Guadalupe block itself), converged with fresh groundwater that infiltrated at higher elevations to the west and southwest. Although sulfuric acid production may have begun as much as 100–200 m below the contemporary water table when the caves were actively forming, the greatest burst of oxidation took place at and just above the water table, producing some of the largest cave rooms in North America. The Big Room of Carlsbad Cavern is up to 80 m high and covers an area of 33,200 m^2.

Large cave rooms have irregular outlines rimmed by dead-end galleries and blind alcoves. Some rooms and adjacent passages have nearly horizontal floors that disregard stratigraphic boundaries, indicating dissolution at former water tables. The oxygen demand in producing a large room requires plentiful air exchange with the surface. Most rooms connect to the surface via older ascending phreatic passages. In the absence of an open cave passage, air can also be exchanged in limited quantities through narrow fissures in the overlying bedrock. The ceilings of major cave rooms are typically smooth and arched, and they show evidence of gypsum replacement. Many are still lined by gypsum rinds (Fig. 7). Some rooms are floored by thick gypsum up to 10 m thick, which accumulated in pools by a combination of subaqueous precipitation and accumulation of fallen material from the ceiling. The $\delta^{34}S$ of the cave gypsum is highly negative, with a mean of $-16.8‰$ CDT, compared to $+10.3‰$ for the primary marine gypsum and anhydrite of the region, which shows that the H_2S that formed it is derived from reduction of primary sulfates (Hill, 1987).

Many passages have steep gradients that connect different levels or serve as entrance galleries. Although entrances of this type give the impression that they were originally surface-water inlets, closer inspection shows that they may have been initiated as groundwater outlets while the caves were forming. The entrance passages contain no stream entrenchment, and almost no vertical shafts or coarse sediment. Instead, the passage ceilings rise in a series of smooth convex-upward arcs quite distinct from the abrupt stair-step pattern typical of vadose passages in humid climates. Local confinement of rising water by resistant beds forced some passages to follow an up-dip course. However, much of the passage enlargement probably took place by absorption of H_2S and oxygen in the cave atmosphere. In places, dripping sulfuric acid produced dissolution rills and potholes in the bedrock. Freshwater seepage that has entered the cave since the sulfuric acid phase has formed drip holes that extend through gypsum blocks.

The caves also contain a wide variety of carbonate speleothems. Besides the common dripstone, flowstone, and pool deposits are several rare forms. Underwater helictites, produced by the common-ion effect between dissolved calcite and gypsum, were first discovered in Lechuguilla Cave. They are among the very few known in the world. Finger-like *pool fingers* and *u-loops* are calcite-coated bacterial filaments. The filaments are now fossilized, but the resulting forms are remarkably like the bundles of bacterial filaments in Cueva de Villa Luz (Fig. 4).

Other cave deposits include clay alteration products such as alunite and hydrated halloysite (endellite). Argon dating can be applied to the potassium-bearing alunite, which is generated by sulfuric acid attack of montmorillonite and therefore provides the dates of active cave enlargement (Polyak *et al.*, 1998). The highest-elevation caves in the Guadalupes are about 12 million years old, and the lowest caves, which apparently represent the latest phase of sulfuric acid development, are about 4 million years old. This was the first application of this procedure to cave dating.

Because of the close association of major cave development with former water tables, the vertical arrangement of Guadalupe caves must relate to the erosional history of nearby rivers. One might expect major passages and rooms to correlate with pauses in base-level lowering, and therefore to cluster at similar elevations. However, correlations between levels of development are somewhat obscure from one cave to another, and even within different sections of the same cave. As the water table dropped, bursts of cave enlargement occurred when and where H_2S happened to rise to the water table in significant quantity. Major rooms and passages were produced by episodic influx of H_2S, probably during periods of uplift and faulting. When these releases coincided with periods of rather static water table, distinct horizontal levels resulted. Alunite dating suggests that the main episodes lasted at least 10^5 years.

Recent studies suggest that the Guadalupe caves were probably fed by water from a large recharge area to the west, which has since dropped in elevation because of extensional tectonics and no longer serves as a catchment area for the present Guadalupe Mountains (DuChene and Cunningham, 2006). If so, the caves provide a record not simply of local water-table lowering, but of the Late Tertiary tectonic history of the entire region.

Caves of the Grand Canyon, Arizona

Grand Canyon National Park, while not primarily a "*cave park*", does contain a widespread karst system. Most of the caves in this region are located in the uppermost unit (Mooney Falls Member) of the Late Mississippian Redwall Limestone. Minor cave development also occurs in the Permian Kaibab Limestone and Cambrian Muav Limestone. There are two types of caves in the canyon: (1) older paleocaves in the Redwall that drained across the Colorado Plateau before the downcutting of the Grand Canyon, and (2) younger vadose caves that drained (or are still draining) to the Colorado River during or after the downcutting of the Grand Canyon. The older paleocaves contain a number of deposits that can be dated with respect to the paleo-watertable. This information can then be used to help determine the age of downcutting of the canyon.

Various analyses relate the cave deposits to the evolution of the Grand Canyon: $^{40}Ar/^{39}Ar$ dating of the minerals alunite and jarosite; U-series and U-Pb dating of cave spar, cave mammillaries, and other speleothems; and stable isotope analysis, X-ray diffraction, and chemical analysis of various cave deposits. From these analyses, the following sequence of events has been established for the Redwall paleocaves: (1) Mississippian Redwall Limestone/Stage 1 karst episode, which is represented by paleokarst breccias; (2) Stage 2 (Laramide?) phreatic karst and iron oxide episode represented by hematite-goethite cave deposits formed at redox boundaries; (3) a calcite spar episode characterized by crystal linings overlying the hematite-goethite; (4) a shallow phreatic, or water-table, episode of cave origin, including some sulfuric acid processes, characterized by solution domes, alunite, mammillaries, and replacement gypsum; and (5) subaerial speleothem deposition.

The sulfuric acid episode (3) in Grand Canyon caves probably represents renewed dissolution at or near the water table in the Pliocene-Pleistocene, with the source of H_2S possibly derived from hydrocarbons in the underlying Precambrian Chuar Group. In the Grandview mine, which is at the same level as the Cave of the Domes on Horseshoe Mesa, an $^{40}Ar/^{39}Ar$ date of ~700,000 YBP for alunite suggests that water-table cave development in that part of the canyon occurred around that time (Fig. 8A). Mammillaries collected from Tse'an Bida Cave, Mother Cave, and Grand Canyon Caverns gave U-series dates of >600,000 YBP. One stalactite from Bat Cave, western Grand Canyon, gave a U-series date of ~460,000 YBP, indicating that the level of the water table (river level) must have been somewhere below the level of Bat Cave ~0.5 Ma BP and that the rate of downcutting for that part of the western Grand Canyon must have been <0.67 mm yr^{-1} (Fig. 8B). More recent data suggest that canyon entrenchment may have extended as far back as 17 Ma BP (Polyak et al., 2008).

RELATION OF SULFURIC ACID CAVES TO MISSISSIPPI VALLEY–TYPE ORE DEPOSITS, URANIUM DEPOSITS, HYDROCARBONS, AND RESERVOIR POROSITY

Sulfuric acid caves are important not only because they represent a mode of speleogenesis recognized only recently, but also because they can be genetically related to native sulfur deposits, Mississippi Valley– type (MVT) ore deposits, uranium deposits, and hydrocarbons and reservoir porosity. All of these deposits and features are interconnected via the generation and migration of hydrogen sulfide within and around intracratonic basins. The general model is that H_2S is generated at depth by the reaction of hydrocarbons with evaporite (sulfate) rock according to reaction (1). The H_2S can then either stay within the source beds or it can move into neighboring or overlying carbonates.

Within a basin, aqueous H_2S can react with oxygen to form native sulfur deposits. Examples include the massive economic sulfur deposits at the Culberson sulfur mine, Delaware Basin, southeastern New Mexico, and west Texas (south of Carlsbad Cavern). Also,

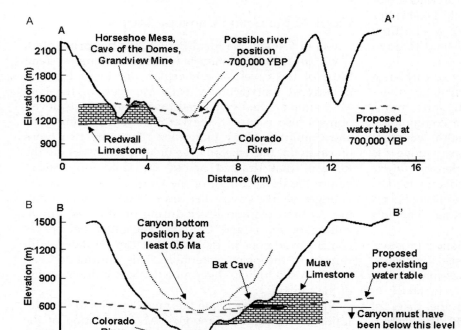

FIGURE 8 Relationship of dated cave and mine deposits in the Grand Canyon to the history of downcutting of the canyon. *From Hill et al., 2004.*

porosity can be created where dissolution is produced by the mixing of waters of different H_2S content or by the oxidation of H_2S. *Sulfuric acid oil-field karst* is a specific kind of H_2S-related porosity where carbonate reservoirs of cavernous size have been dissolved by a sulfuric acid mechanism. In an H_2S system, porosity can be produced entirely in the deep subsurface and does not have to represent a paleokarst surface formed by shallow-phreatic or vadose dissolution.

Mississippi Valley–type (MVT) ore deposits commonly occur in carbonate rock around the edges or margins of intracratonic basins. This is due to the migration of H_2S from hydrocarbon basins into structural and stratigraphic traps along the basin margins. As groundwater moves through these traps, metal ions such as lead, zinc, and copper that precipitate readily in the H_2S-rich environment. Thus the H_2S migration mechanism applies not only to the small MVT deposits around the Delaware Basin, but also to huge lead-zinc deposits such as those of the Viburnum Trend, Missouri.

Uranium is another metal that precipitates readily in the presence of H_2S. Such precipitation is typical of *roll-front* deposits where groundwater encounters and percolates through a reducing environment, allowing uranium to precipitate preferentially at the redox interface. In the case of the Grand Canyon, it is possible not only that the isotopically light cave gypsum was ultimately derived from hydrocarbons in the subsurface, but also that H_2S from hydrocarbons was the source of reduced sulfur for the copper and uranium mineralization in the mines of the Grand Canyon. Some of the highest-grade uranium deposits in North America are located in breccia pipes of the Grand Canyon, and these breccia pipes and paleokarst breccias may have acted as stratigraphic and structural traps.

Sulfuric acid caves are an integral part of this *intracratonic basin* model of H_2S generation and migration. Where H_2S migrates from a basin into carbonate rock surrounding the basin and encounters oxygenated groundwater, then sulfuric acid forms via reactions (2) and (3), and cave dissolution occurs via reactions (4) and (5). As a by-product of these reactions, gypsum and native sulfur form within these caves, hydrated halloysite (endellite) and alunite are produced under low pH sulfuric acid solutions, and the uranium-vanadium minerals tyuyamunite and metatyuyamunite precipitate in an H_2S-rich environment.

CONCLUSIONS

The role of sulfuric acid in cave origin, which was virtually unrecognized only a few decades ago, is now recognized as one of the most significant cave-forming processes. It also provides a view of deep-seated processes that integrates cave studies with other fields, such as economic geology and microbiology.

See Also the Following Articles

Speleogenesis, Hypogenetic
Lechuguilla Cave, New Mexico

Bibliography

DuChene, H. R., & Hill, C. A. (Eds.), (2000). The caves of the Guadalupe Mountains. *Journal of Cave and Karst Studies* (62). pp. 51–158.

DuChene, H. R., & Cunningham, K. I. (2006). *Tectonic influences on speleogenesis in the Guadalupe Mountains, New Mexico and Texas* (pp. 211–218) (Guidebook to 57th Annual Field Conference). New Mexico Geological Society.

Egemeier, S. J. (1981). Cavern development by thermal waters. *National Speleological Society Bulletin*, 43, 31–51.

Galdenzi, S., Cocchioni, M., Morichetti, L., Amici, V., & Scuri, S. (2008). Sulfidic ground-water chemistry in the Frasassi Caves, Italy. *Journal of Cave and Karst Studies*, 70(2), 94–107.

Hill, C. A. (1987). *Geology of Carlsbad Cavern and other caves in the Guadalupe Mountains, New Mexico and Texas* (Bulletin 117). New.

Hill, C. A. (1995). Sulfur redox reactions: Hydrocarbons, native sulfur, Mississippi Valley-type deposits and sulfuric acid karst in the Delaware Basin, New Mexico and Texas. *Environmental Geology*, 25, 16–23.

Hill, C. A., Polyak, V. J., McIntosh, W. C., & Provencio, P. P. (2001). Preliminary evidence from Grand Canyon caves and mines for the evolution of the Grand Canyon and Colorado River System. In R. A. Young & E. E. Spamer (Eds.), *The Colorado River: Origin and evolution* (pp. 141–145). Grand Canyon, AZ: Grand Canyon Association (Monograph 12).

Hose, L. D., Palmer, A. N, Palmer, M. V., Northup, D. E., Boston, P. J., & DuChene, H. R. (2000). Microbiology and geochemistry in a hydrogen-sulphide-rich karst environment. *Chemical Geology*, 169, 399–423.

Klimchouk, A. B. (2007). *Hypogene speleogenesis: Hydrogeological and morphogenetic perspective*. Carlsbad, NM: National Cave and Karst Research Institute.

Northup, D. E., Dahm, C. N., Melim, L. A., Spilde, M. N., Crossey, L. J., Lavoie, K. H., et al. (2000). Evidence for geomicrobiological interactions in Guadalupe caves. *Journal of Cave and Karst Studies*, 62(2), 30–40.

Palmer, A. N., & Palmer, M. V. (2000). Hydrochemical interpretation of cave patterns in the Guadalupe Mountains, New Mexico. *Journal of Cave and Karst Studies*, 62(2), 91–108.

Polyak, V. J., McIntosh, W. C., Güven, N., & Provencio, P. (1998). Age and origin of Carlsbad Cavern and related caves from $^{40}Ar/^{39}Ar$ of alunite. *Science*, 279(5358), 1919–1922.

Polyak, V. J., Hill, C. A., & Asmerom, Y. (2008). Age and evolution of the Grand Canyon revealed by U-Pb dating of water table-type speleothems. *Science*, 319(5896), 1377–1380.

Queen, J. M., & Melim, L. A. (2006). *Biothems: Biologically influenced speleothems in caves of the Guadalupe Mountains, New Mexico, USA* (Guidebook to 57th Annual Field Conference) (pp. 167–173). New Mexico Geological Society.

Sarbu, S. M., Kane, T. C., & Kinkle, B. K. (1996). A chemoautotrophically based groundwater ecosystem. *Science*, 272(5270), 1953–1955.

T

TIANKENG

Xuewen Zhu, Weihai Chen, and Yuanhai Zhang

Institute of Karst Geology, Chinese Academy of Geological Sciences, Guilin, China

DEFINITION

The word *tiankeng* is a transliteration from two Chinese characters, "天坑" (tiān kēng in *pinyin*), that roughly mean sky hole or heaven pit or some similar variation on that double theme. Tiankeng are very large dolines formed in carbonate rocks that are more than 100 m deep and wide, and have volumes of more than 1 million cubic meters, steep profiles with vertical cliffs around all or most of the perimeters, and are connected by underground cave rivers.

TYPES OF TIANKENGS

Tiankengs may be classified as *collapsed tiankeng*, *erosional tiankeng*, and *water tiankeng* by their origin and development (Fig. 1) or classified as *immature tiankeng*, *mature tiankeng*, and *degraded tiankeng* by their evolution (Fig. 2).

Collapsed tiankengs are dominant in number, distribution, and dimension as well as difference of varied ages. The development of a collapsed tiankeng is distributed relative to the underground river trace instead of surface topography of the karst, and is the result of intensive erosion and dissolution of underground water. However, the formation of a tiankeng is the reconstruction of large collapses. The appearance of tiankengs with different ages may be the symbol of mature karst development, such as Dashiwei Tiankengs Group in the Bailang Underground River System in Guangxi, China (Fig. 3).

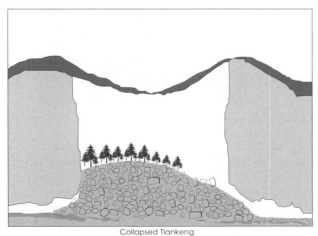 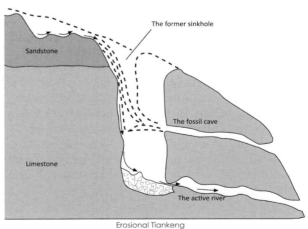

FIGURE 1 Collapsed tiankeng and erosional tiankeng.

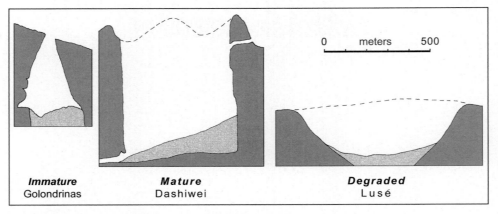

FIGURE 2 Immature tiankeng, mature tiankeng, and degraded tiankeng.

FIGURE 3 Dashiwei tiankeng.

An erosional tiankeng develops at the site of a sinking allogenic stream, as distinct from collapse tiankengs that are independent of surface drainage. The erosional type therefore has a cave stream draining from it, but not into it. Qingkou Tiankeng in Chongqing is therefore an excellent example (Fig. 4), and there may be erosional tiankengs in the Muller Plateau and in Nakanai, New Britain, Papua New Guinea.

A water tiankeng has a pool inside, and is generally of the collapsed type. Some are the skylights of active underground rivers, such as the Sanmenhai Karst Windows in Leye-Fengshan Global Geopark. Some are connected with underground water in a phreatic zone, such as Grveno Jezero in Croatia and Lago Azul in Brazil. Some are collapsed tiankengs submerged by water. The tiankeng of Grveno Jezero is 528 m deep with a lake of 281 m deep and 16 Mm³ water in volume. Its altitude is −6 m, and there is an underground stream flowing at the bottom.

FEATURES OF TIANKENGS

Tiankengs are unusually large as karst landforms and they have clear collapse triangular profiles. They

FIGURE 4 Qingkou tiankeng.

usually destroy any other landforms of the surface karst, including dry valleys, depressions, dolines, or cone karst, either individually or collectively, which indicate that they are young karst landforms. That is the significant difference between tiankengs and dolines.

There are four main stages of development in a tiankeng: (1) a cave with large river passage; (2) a large cave chamber; (3) a tiankeng that is opened to the surface; (4) and finally a degraded tiankeng (Fig. 5). A collapsed tiankeng has the following critical factors:

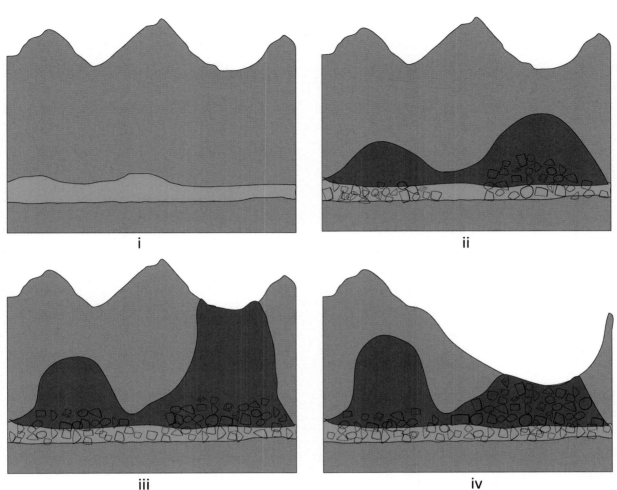

FIGURE 5 The four main stages of development in a tiankeng.

1. A great and continuous thickness of carbonate rocks and a deep vadose zone in the karst
2. Tropical-subtropical humid climate and abundant precipitation
3. Slow tectonic uplifting and deep trunk river incision
4. Developed underground water system, particularly a subterranean river flowing through the tiankeng bottom
5. Gently dipping beds and networks of joints or faults in carbonate rocks
6. In contrast to the development of collapse tiankengs where the surface hole opened abruptly from beneath, erosional tiankengs are formed gradually from the surface to the floor; excavation downward by stream erosion, and fluvial removal of debris and solutes, complement each other (Fig. 6)
7. An erosional tiankeng develops at the boundary between soluble and nonsoluble rocks; it has formed within underlying limestone that has propagated to the ground surface through a cover of insoluble rock

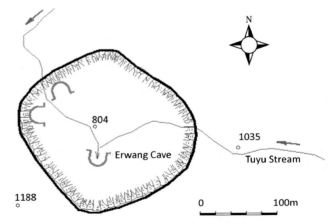

FIGURE 6 Erosional tiankeng and underground river.

8. An erosional tiankeng develops in abundant thickness of a vadose zone or continuous descent of an underground water table with tectonic uplifting and river incision

TABLE 1 Dimensions of the Main Tiankengs in the World and Some Comparable Features

Tiankeng	Length × Width (m)	Area (m²)	Depth (max. m)	Volume (Mm³)	Type	Location
Xiaozhai	625 × 535	274,000	662	119.3	C, M	China
Dashiwei	600 × 420	167,000	613	75	C, M	China
Xiaoyanwan	625 × 475	200,000	248	36	C, D	China
Datuo	530 × 380	149,000	290	32.7	C, D	China
Longgang	350 × 170	53,000	350	9.2	C, M	China
Huangjing	320 × 170	51,700	161	6.3	C, E, M	China
Qingkou	250 × 220	40,700	295	9.2	E, M	China
Yanzi	100 × 60	14,340	250	1.69	C, I	China
Minyé	350 × 350	75,000	510	26	C, E, M	Papua New Guinea
Lusé	800 × 600	350,000	250	61	C, D	Papua New Guinea
Kavakuna	380 × 300	80,000	480	24	C, D	Papua New Guinea
Naré	150 × 120	13,000	310	4.7	M	Papua New Guinea
Garden of Eden	1200 × 800	750,000	300	150		Malaysia
Sendirian	115 × 90	12,000	240	2	M	Malaysia
Crveno Jezero	200 × 280	56,000	528	30	W	Croatia
Modro Jezero	700 × 400	19,000	290	22	D	Croatia
Verika Dolina	300 × 170	30,000	165	3.5	M	Slovenia
Lisicina	400 × 200	60,000	115	2	D	Slovenia
El Sotano	440 × 210	70,000	445	16	M	Mexico
Golondrinas	300 × 130	26,000	400	5	I	Mexico
Peruacu North	450 × 200	85,000	170	10		Brazil
Peruacu South	400 × 180	55,000	150	5	D	Brazil
Lago Azul	200 × 140	18,000	280	4.3	W	Brazil
Mangilg	700 × 500	280,000	140	25		Madagascar
Styx2	400 × 300	88,000	140	8		Madagascar
Tres Pueblos Sink	190 × 180	25,000	120	2.5		Puerto Rico

C = collapsed; E = erosion; W = water; I = immature; M = mature; D = degraded.

9. An erosional tiankeng develops in regions with a humid climate and abundant precipitation

Different water tiankengs at different sites may have different ages, but most of them are young, formed after the Quaternary.

DISTRIBUTION OF TIANKENGS

Tiankengs are the result of intensive karst evolution and are mainly distributed in the highly active hydrodynamic karst region, which have the essential conditions including continuous great thick and gently dipping carbonate rocks, a humid climate, and abundant precipitation. There are two main tiankeng regions: South China Karst-North Vietnam Karst and Muller-Nakanai Karst in Papua New Guinea. Both these regions are distinguished by unusually large numbers of tiankengs, especially in the fengcong karst area of south China (Table 1). In addition, tiankengs have been discovered in Malaysia, Mexico, Brazil, Croatia, Slovenia, Madagascar, and Puerto Rico.

Bibliography

Garasic, M. (2001). New speleo-hydrogeological research of Crveno Jezero (Red Lake) near Imotski in Dinaric karst area (Croatia Europe). *Proceedings of the 13th International Congress of Speleology, Brazil*, 2, 168–171.

James, J. (2005). Tiankengs of the Muller Plateau, Papua New Guinea. *Cave and karst Science*, 32(2–3), 85–91.

Waltham, A. C. (2005). Tiankengs of the world, outside China. *Cave and Karst Science*, 32(2–3), 67–74.

Waltham, A. C., Bell, F. G., & Culshaw, M. G. (2005). *Sinkholes and subsidence*. Chichester, U.K. Praxis Publishing.

Zhu, X. W. (2001). Karst tiankeng and its value for science and tourism in China. *Science & Technology Review*, 10(160), 60–63 (in Chinese).

Zhu, X. W., & Chen, W. H. (2005). Tiankengs in the karst of China. *Cave and Karst Science*, 32(2–3), 55–66.

Zhu, X. W., & Waltham, A. C. (2005). Tiankeng: Definition and description. *Cave and Karst Science*, 32(2–3), 75–77.

Zhu, X. W., Zhu, D. H., Baojian, H., & Chen, W. H. (2003a). *Dashiwei Tiankeng Group in Leye, Guangxi: Discoveries, exploration, definition and research*. Guangxi Scientific & Technical Publishers.

Zhu, X. W., Zhu, D. H., & Chen, W. H. (2003b). A brief study on karst tiankeng. *Carsologica Sinica*, 22(1), 51–65 (in Chinese).

U

UKRAINE GIANT GYPSUM CAVES

Alexander Klimchouk
Ukrainian Institute of Speleology and Karstology, Simferopol, Ukraine

REGIONAL GEOLOGY AND HYDROGEOLOGY

The host gypsum bed, ranging from a few meters to more than 40 m in thickness, is the main component of the Miocene evaporite formation that girdles the Carpathians to the northeast, from the Nida River basin in Poland across the Western Ukraine and Moldova to the Tazleu River basin in Romania. The gypsum occurs on the southwestern edge of the Eastern European platform, where it extends along the Carpathian foredeep for over 300 km in a belt ranging from several kilometers to 40–80 km wide (Fig. 1A,B). It occupies over 20,000 km^2, together with some separated areas that occur to the northeast of the unbroken belt.

Most Miocene rocks along the platform margin rest on the eroded terrigenous and carbonate Cretaceous sediments. The Miocene succession comprises deposits of Badenian (Tortonian) and Sarmatian age. The Lower Badenian unit, beneath the gypsum, includes mainly carbonaceous, argillaceous, and sandy beds (30–90 m thick) adjacent to the foredeep, and these grade into rocks of calcareous biothermal and sandy facies (10–30 m thick) toward the platform interior. The Miocene gypsum bed is variable in structure and texture. Most commonly it grades from microcrystalline massive gypsum in the lower part through to variably grained bedded gypsum in the middle, to coarsely crystalline rock in the upper horizon. A layer of evaporitic and epigenetic limestone, locally called *Ratynsky*, commonly overlies the gypsum. This layer ranges from half a meter to more than 25 m in thickness. The gypsum and the Ratynsky limestone comprise the Tyrassky Formation, which is overlain by the Upper Badenian unit, comprised either by argillaceous and marly limestones and sandstone or, adjacent to the foredeep, by marls and clays of the Kosovsky Formation. In the platform interior parts of the gypsum belt the Kosovsky Formation is replaced by the Lower Sarmatian clays, up to 40–50 m thick. The total thickness of the capping marls and clays ranges from 40–60 m in the platform interior to 80–100 m or more in the areas adjacent to the regional faults that separate the platform edge from the foredeep.

The present distribution of Miocene formations and the levels of their denudation stripping vary in a regular manner from the platform interior toward the foredeep (Fig. 1C). The Tyrassky Formation dips 1–3° toward the foredeep and is disrupted by block faults in the transition zone. To the south and southwest of the major Dniester Valley, large tectonic blocks drop down as a series of steps, the thickness of the clay overburden increases, and the depth of erosional entrenchment decreases. Along the tectonic boundary with the foredeep the Tyrassky Formation drops to a depth of 1000 m or more. This variation, the result of differential neotectonic movements, and incision of the Dniester Valley and its north subparallel tributaries during the Plio-Pleistocene, played an important role in the hydrogeological evolution of the Miocene aquifer system, and resulted in the differentiation of the platform edge into four zones (Fig. 1A,C). The gypsum was entirely removed by denudation within the first zone, but the other three zones represent distinct types of intrastratal karsts: entrenched, subjacent, and deep-seated (for more details on the evolutionary classification of karst types see this book's "Gypsum Caves" and "Speleogenesis, Hypogenic" entries). The gypsum bed is largely drained in the entrenched karst zone, is partly inundated in the subjacent karst zone, and remains under artesian confinement in the deep-seated karst zone.

In hydrogeological terms the region represents the southwestern portion of the Volyno-Podolsky artesian

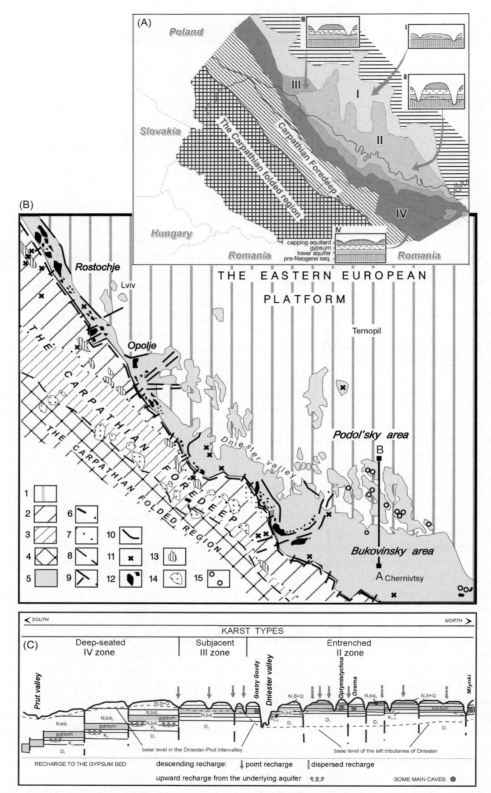

FIGURE 1 Distribution, geological conditions, and types of gypsum karsts in the Western Ukraine. (A) Zones of different evolutionary types of karsts: I = the gypsum entirely denuded; II = entrenched karst; III = subjacent karst; IV = deep-seated karst. (B) Distribution of the gypsum stratum and large caves: 1 = the fringe of the Eastern European platform; 2–3 = the Carpathian foredeep: 2 = outer zone; 3 = inner zone; 4 = the Carpathian folded region; 5 = sulfate rocks on the platform. Tectonic boundaries include: 6 = platform/foredeep; 7 = outer/inner zone of the foredeep; 8 = foredeep/folded region; 9 = other major faults; 10 = flexures; 11 = sulfur mineralization; 12 = sulfur deposits; 13 = gas deposits; 14 = oil deposits; 15 = large maze caves in the gypsum. (C) Geological profile and zones of different types of karsts along the line A−B on the panel B.

TABLE 1 Morphometric Parameters of Large Gypsum Caves in the Western Ukraine

No.	Cave	Length, km	Average Cross-section Area, m^2	Density of Passages, km/km^2	Areal Coverage, %	Cave Porosity, %
1	Optymistychna	232.0	2.8	147	17.6	2.0
2	Ozerna	131.4	6.0	150	44.6	5.0
3	Zoloushka	92.0	8.0	142	48.4	3.8
4	Mlynki	36.0	3.3	141	37.6	3.4
5	Kristal'na	22.6	5.0	169	29.2	6.0
6	Slavka	9.1	3.7	139	27.6	3.4
7	Verteba	8.5	6.0	118	34.7	12.0
8	Mushkarova Jama	5.05	n.a.	n.a.	n.a.	n.a.
9	Gostry Govdy	3.58	1.7	270	17.5	4.0
10	Atlantida	2.52	4.5	168	30.0	4.0
11	Bukovinka	2.4	2.5	120	21.5	4.4
12	Ugryn	2.12	3.8	177	33.3	5.7
13	Jubilejna	1.5	2.3	278	37.0	4.0
14	Komsomol'ska	1.24	2.1	177	24.3	3.0
15	Dzhurinska	1.13	2.4	126	17.8	2.0

basin. The Sarmatian and Kosovsky clays and marls serve as an upper confining sequence. The lower part of the Kosovsky Formation and the limestone bed of the Tyrassky Formation form the original upper aquifer (above the gypsum), and the Lower Badenian sandy carbonate beds, in places together with Cretaceous sediments, form the lower aquifer (below the gypsum), the major regional one. The hydrogeologic role of the gypsum unit has changed with time, from initially being an aquiclude, intervening between two aquifers, to a karstified aquifer with well-developed conduit permeability. The regional flow is from the platform interior, where confining clays and the gypsum are largely denuded, toward the large and deep Dniester Valley and the Carpathian foredeep. In the northwest section of the gypsum belt the confined conditions prevail across its entire width. In its wide southeast section the deeply incised valleys of the Dniester and its left-hand tributaries divide the Miocene sequence into a number of isolated, deeply drained interfluves capped with the clays (Podolsky area; Fig. 1B). This is the entrenched karst zone where most of the explored, presently relict, maze caves are located. To the south-south-east of the Dniester (Bukovinsky area) the gypsum remains largely intact and is partly inundated (the subjacent karst zone). Further in this direction, as the depth of the gypsum below the clays increases and entrenchment decreases, the Miocene aquifer system becomes confined (the deep-seated karst zone). In this zone the groundwater flow pattern includes a lateral component in the lower aquifer (and in the upper aquifer, but to a lesser extent) and an upward component through the gypsum in areas of potentiometric lows, where extensive cave systems develop, as evidenced by numerous data from exploratory drilling.

PATTERNS AND MESOMORPHOLOGY OF CAVES

Fifteen large caves over 1 km in length are known in the region (Table 1). Most of them are located north of the Dniester River. Two other large caves, Zoloushka and Bukovinka, occur in the Bukovinsky region, near the Prut River and the border with Moldova and Romania, generally in the area of artesian flow within the Miocene aquifer system but within local, particularly uplifted blocks, where entrenchment into the upper part of the gypsum caused unconfined (water-table) conditions to be established in the Holocene.

Most of the caves have only one entrance, either through collapse dolines and swallow holes at the interfluves or from gypsum outcrops in the slopes of the major valleys. Some caves and their entrance series were known to local people since long ago (e.g., Ozerna, Kristal'na, Mlynki, Verteba), but others were

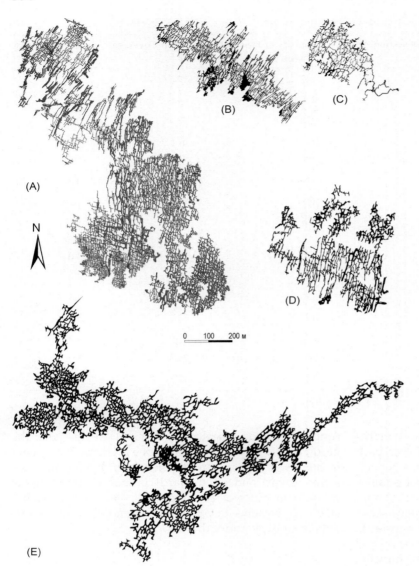

FIGURE 2 Maps of some of the main maze caves in the gypsum of the Western Ukraine: (A) Ozerna Cave, (B) Kristal'na Cave, (C) Slavka Cave, (D) Mlynki Cave, (E) Zoloushka Cave. *Courtesy of the speleological clubs of Ternopil, Chernivtsy, and Kiev.*

discovered by cavers via digs (*e.g.*, Optymistychna, Slavka, Mushkarova Jama, Atlantida). Two caves (Zoloushka and Bukovinka) became accessible when opened by gypsum quarries. Systematic cave exploration and mapping in the region began in the 1960s.

All of the large gypsum caves in the region are mazes developed along vertical and steeply inclined fissures arranged into multistory laterally extensive networks. Aggregating passages form lateral two- to four-story systems that extend over areas of up to 2 km^2 (Fig. 2). A notable feature of the mazes is the exceptionally high passage network density, which is characterized conveniently by using the ratio of a cave length to an area occupied by a cave system. This parameter varies from 118 (Verteba Cave) to 270 km/km^2 (Gostry Govdy Cave), with the average value for the region being 164 km/km^2. Values of areal coverage and cave porosity (fractions of the total area and volume of the rock within a cave field, occupied by passages) vary for individual caves from 17.5 to 48.4% (average 29.5%) and from 2 to 12% (average 4.5%), respectively, being roughly an order of magnitude greater than these characteristics for typical epigenic caves. Optymistychna Cave (Optimisticheskaya in Russian spelling) is the longest gypsum cave, and the second-longest cave of any type known in the world, with more than 232 km of passages surveyed (Fig. 3). By area and volume the largest caves are Ozerna (330,000 m^2 and 665,000 m^3) and Zoloushka (305,000 m^2 and 712,000 m^3), followed by Optymistychna Cave (260,000 m^2 and 520,000 m^3).

Maze caves in the region are hypogenic in origin, developed under confined conditions, due to upward transverse groundwater circulation between aquifers below and above the gypsum beds (Klimchouk, 2000).

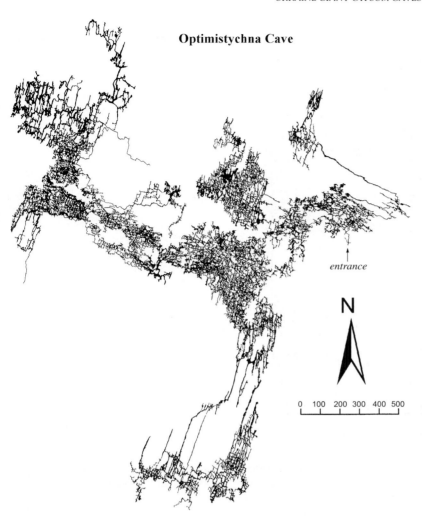

FIGURE 3 Map of the Optymistychna Cave, the largest gypsum cave in the world. *Courtesy of the speleological club of Lviv.*

They richly demonstrate morphological features characteristic for hypogene speleogenesis (Klimchouk, 2009). These features are related in suites indicating rising flow through caves, with considerable buoyancy dissolution effects (see Fig. 7 in the article "Speleogenesis, Hypogenic"). According to the morphology, arrangement, and hydrologic function of cave mesoforms during the main (artesian) speleogenetic stage, three major components can be distinguished in the cave systems:

1. *Feeding channels.* These are the lowermost components in a system: vertical or subvertical conduits through which water rose from the subgypsum aquifer to the master passage networks. Such conduits are commonly separate but sometimes they form small networks at the lowermost part of the gypsum. The feeding channels join master passages located at the next upper level and scatter uniformly through their networks (see Fig. 3 in the article "Gypsum Caves").

2. *Master passages.* These are horizontal passages that form laterally extensive networks within certain horizons in the middle part of the gypsum bed. They receive dispersed recharge from numerous feeding channels and conduct flow laterally to the nearest outlet feature.

3. *Outlet features.* These include domes, cupolas, and vertical channels (dome pits) that rise from the ceiling of the master passages to the bottom of the overlying bed. They discharge water from cave systems to the overlying aquifer.

Also common are raising wall channels above feeders, continued through ceiling half-tubes to nearest outlets. Where many smaller ceiling channels are braiding closely, they leave solutionally sculptured pendants in-between as they curve into the ceiling. All of these features are the result of buoyancy dissolution in the natural convection system.

In areas where water table conditions remained stable within the gypsum bed for prolonged time

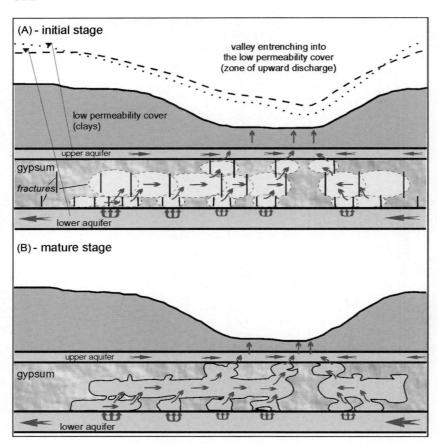

FIGURE 4 Conceptual model of hypogene transverse speleogenesis in gypsum in the Western Ukraine.

during the subjacent karst stage, enhanced dissolution along the water table caused extensive horizontal notching in passage walls and considerable widening of passages.

The modern dissolution in the entrenched karst zone is limited to points of leakage from the aquifer perched on the cover clays, which creates vertical solution pipes, and also to courses of streams swallowed via ponors.

CAVE SEDIMENTS AND FORMATIONS

The predominant sediments in the maze caves of the region are successions of fine clays, with minor beds of silty clays. These fill passages to a variable extent and can reach 5–7 m in thickness. Breakdown deposits are also common. They include chip, slab, and block breakdown material from the gypsum, as well as more massive breakdown from the overlying formations. Calcite speleothems (stalactites, stalagmites, flowstones, and helictites) occur locally in zones of vertical water percolation from the overlying formations. Gypsum crystals of different habits and sizes are the most common cave decorations. They are largely of subaerial origin. Hydroxides of Fe and Mn occur as powdery layers within the clay fill of many caves, indicating repeated transitional cycles from a reducing to an oxidizing geochemical environment. Massive deposition of Fe/Mn compounds in the form of powdery masses, coatings, stalactites, and stalagmites has occurred in Zoloushka Cave, where a rapid dewatering caused by groundwater abstraction during the past 50 years gave rise to a number of transitional geochemical processes, some of which appear to show considerable microbial involvement.

SPELEOGENESIS

The Western Ukrainian maze caves are the model examples of the hypogene transverse speleogenesis (Fig. 4) (Klimchouk, 2000, 2007). This type of speleogenesis in the Podolsky region took place mainly during the late Pliocene through to the middle Pleistocene. It was induced by initial incision of the Dniester valley and its left-hand tributaries into the confining clays, and respective activation of the upward cross-formational groundwater flow in the underlying artesian system. Breaching of artesian confinement and further incision of the valleys during the middle Pleistocene caused substantial acceleration of groundwater circulation within the Miocene artesian system. The majority of passage growth probably occurred during this transitional period. Where the water table was established in the gypsum for a prolonged time, further widening of passages occurred due to horizontal notching at the water table. With the water table dropping below the lower gypsum contact, cave systems in the entrenched karst zone became largely relict. Hypogene cave development under confined and semiconfined conditions continues today within the zones of deep-seated and subjacent karst.

See Also the Following Article

Gypsum Caves

Bibliography

Klimchouk, A. B. (2000). Speleogenesis of great gypsum mazes in the Western Ukraine, in Klimchouk, A. B., Ford, D. C., Palmer, A. and Dreybrodt, W., Eds., *Speleogenesis: Evolution of karst aquifers*, Huntsville, Alabama: National Speleological Society, p. 261–273.

Klimchouk, A. B. (2007). Hypogene Speleogenesis: Hydrogeological and Morphogenetic Perspective. National Cave and Karst Research Institute. National Cave & Karst Research Institute Special Paper 1, Carlsbad, New Mexico.

Klimchouk, A. B. (2009). Morphogenesis of Hypogenic Caves. *Geomorphology* 106, 100–117.

UNDERWATER CAVES OF THE YUCATÁN PENINSULA

James G. Coke, IV

The Woodlands, Texas, U.S.A.

The geologic history of the Yucatán Platform is difficult to establish due to the paucity of deep well data, exposed anatomic strata, and relative inaccessibility of the inland terrain. In a quest for potential oil reserves, PEMEX (Petróleos Mexicanos) drilled 10 exploratory wells into this platform during the 1970s. Most of these wells were drilled in the state of Yucatán, often in excess of −2000 m. Data from the test wells summarize the stratigraphy of the isthmian portion of the Yucatán Peninsula as that of a great limestone platform (Ward and Weidie, 1978). Paleozoic metasediments form the basement strata at varying depths exceeding −2400 m.

The Cretaceous period launched the onset of numerous marine transgressions, submerging much of the isthmus under warm shallow seas until the Pleistocene. Over 2000 m of limestone strata were formed in the period, including the uppermost (Miocene–Pliocene) Carrillo Puerto Formation (>300 m thick). Uplift in the south-central area of the isthmus began in the Oligocene, encouraging the Carrillo Puerto Formation to be deposited in stages over the ever-expanding margins of the peninsula. This slow uplift continued until the Pleistocene. Pleistocene deposits are the result of marine transgressions during the Illinois and Wisconsin ice ages. Most Quaternary deposits on the isthmus are restricted to the present margins of the peninsula. These deposits are typically thin, extending one to three kilometers inland from the Caribbean. Unconsolidated Holocene deposits are restricted to the present shoreline. As a final consequence, a near-uniform deposition of carbonate strata has produced a low relief plain over much of the peninsula. Both north and northeastern areas of the isthmus display a steady increase in elevation from the coastal periphery (3 m above sea level) toward the interior flatlands (over 30 m above sea level). This shared region of the isthmus is generally acknowledged to contain the highest concentration of known dry and underwater caves.

As a relatively stable karst platform, the peninsula contains very few surface drainage systems or lakes. It is also highly vulnerable to the dynamics of speleogenesis. Capped by a scrub tropical forest, the peninsula receives over 150 cm of precipitation per year (Tulaczyk et al., 1993). Approximately 10% of this rainwater filters rapidly through the topmost calichified zone and into lower fractured strata; the remainder is lost to evapotranspiration. Reaching the shallow freshwater aquifer, it begins a journey toward the Caribbean and Gulf of Mexico that is guided in part through bedding fractures and joints. The process of freshwater aquifer recharge in Yucatán has facilitated dissolution within the parent limestone since its initial uplift from the ocean. However, local dissolution is complicated through interaction with the base saltwater intrusion that buoys the lens of freshwater. The underwater cave environment can embody as many as 3 chemically dissimilar water layers. This includes a thin upper fresh-cap layer that is independent of the lower freshwater column during normal meteorological conditions. A middle freshwater layer contains varying amounts of dissolved chlorides. Depending on tunnel depths and distance from the coast, this brackish stratum inundates horizontal underwater caves, deep underwater pits, and coastal or offshore discharge vents to different depths within the caves' parent limestone beds. The base of the water column is occupied by a near-static anoxic saltwater intrusion.

Today the Mexican Yucatán Peninsula is divided into three administrative states: Campeche, Yucatán, and Quintana Roo. Each of these states lays claim to a variety of dry and underwater caves that display a multiplicity of environmental and physical conditions. Underwater cave explorations are not well documented in Campeche. Exploration teams must be self-sufficient when providing for compressed gas supplies and rechargeable electrical devices. Access to the interior of

Campeche is also a challenge due to a limited network of roads. With few expeditions exploring the interior of Campeche, it is understandable why limited underwater cave discoveries are reported to this date.

The state of Yucatán in the north is well respected for both its dry caves and cenotes (from the Maya word *dzonot*). Unlike many of the explored cenotes in Quintana Roo, the majority of cenotes in Yucatán takes the shape of flooded classic sinks, or pits. They are recognized for their vertical relief rather than horizontal scope, due to the mechanism of their formation. Dissolution of deep strata and past sea-level fluctuations stimulated an eventual collapse of surface strata, creating near-circular pools of water at the jungle surface. In profile, most Yucatán cenote entrances crown an ever-widening bell-shaped pit which is floored by a central organic debris and breakdown mound. The size of the collapse is symbolized by the shape, size, and ultimate depth of the underwater pit. The deepest sinks in Yucatán offer a direct descent to great depths from the cenote surface. As an explorer descends through the freshwater layer, the walls of the cave pit recede and disappear into a gloomy distance. At great depths, the explorer's only visual reference within the void is the dull glow of sunlight from the cenote entrance above. At -70 m, explorers arrive at the boundary zone between freshwater and salt water layers. The halocline is a distinctive density interface. When disturbed, the mixing of freshwater and saltwater layers create temporary, yet dazzling visual distortions within the immediate mixing area. A white or red hydrogen sulfide cloud can also be encountered at the interface. This corrosive layer is a by-product of organic digestion by anaerobic bacteria that live in the lower anoxic saltwater intrusion. A "smell" of rotten eggs is often detected by divers as they descend through this thin chemical zone. Absorbed through the skin, high concentrations of hydrogen sulfide will produce headaches and nausea for the diver. Should a hydrogen sulfide boundary be passed for greater depths, the chemical cloud will effectively block any remaining natural light from the cenote entrance. The deepest areas in Yucatán's classic sinks are attained at the junction of the bottom of the debris cone and the surrounding cave wall.

Classic sinks in Yucatán contain small horizontal freshwater tubes within the first 10 m of the freshwater column. Substantial horizontal cave development, such as that found in coastal caves, is not recorded at any depth within the classic sink environment. Modern explorers are exploring many of these deep-pit cenotes for the first time. The deepest explored classic sink cenotes in Yucatán include Cenote Sabak Ha (-147 m), Cenote Xkolac (-121 m), and Cenote Ucil (-118 m). The final depth of Cenote Sabak Ha sink ("smoke-water" in Maya) remains a mystery. Only two underwater caves in Yucatán display significant horizontal passage development. Further explorations in Cenote Chacsinikché (670 m) and Cenote Kankirixché (438 m) are serious undertakings due to the depth and dimensions of their outlying tunnels.

The state of Quintana Roo has received the greater share of attention from cave researchers during the past 27 years. The state contains three of the five longest caves in Mexico, along with the longest cave in Mexico. Sea-level fluctuations during the Sangamon Interglacial through the Wisconsin postglacial periods expedited numerous developments within subterranean drainage networks, as demonstrated by the underwater caves on the eastern coast. Receding sea levels during active glaciation resulted in a considerable drop in the local water table and, thus, water in the subsurface drainage channels. Unsupported cave ceilings immediate to the surface collapsed, creating karst windows at irregular intervals along the course of the conduits. Streambed erosion and dissolution gradually expanded the existing caves while eroding new passages into deeper limestone beds (currently explored to -120 m within horizontally developed caves). About 18 ka BP sea level began a slow rise from 125 m below its present level as the Wisconsin glaciation period declined. As the freshwater table was elevated by a rising salt water intrusion, deeper conduits were rapidly immersed by this intrusion.

Holthuis (1973) coined the term *anchialine* (also known as *anchihaline*) to define specific attributes of inland, seemingly isolated aqueous pools. Many cave systems in eastern Quintana Roo contain anchihaline pools; those that contain salt or brackish waters that fluctuate with oceanic tides yet lack a surface connection to the ocean. Anchihaline caves in Quintana Roo contain an upper freshwater layer that can flow over a near-static saltwater intrusion. Unlike the near-static freshwater layer in classic sinks, a freshwater flow courses through horizontal cave tunnels as it travels toward the ocean. A dynamic mixing zone occurs (enlivened by water current and passage configuration) at the halocline density interface. The depth of the halocline and thickness of the freshwater layer increases with distance from the ocean, while periods of sustained precipitation, ocean tides, and the pervasiveness of exposed mixing zones in cave passages instigate random fluctuations of the halocline. Investigations conclude that preferential dissolution of limestone at the mixing zone (Back et al., 1986), allied with native fracture zone speleogenesis, has produced an arena for vast growth and passage complexity within these cave systems. Volume of recharge, the availability of coastal discharge centers, and a constant supply of freshwater for the mixing zone also

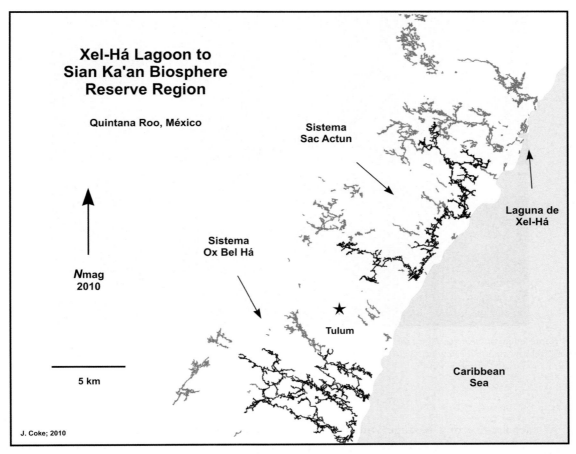

FIGURE 1 Map of underwater caves Xel-Ha to Sian Ka'an Biosphere Reserve region, Quintana Roo. *Drawing by J. Coke, from 2010 data supplied by the Quintana Roo Speleological Survey.*

appears to be decisive for the substantial growth of Quintana Roo caves (Smart et al., 2006). Coupled with Pleistocene sea-level fluctuations, a snapshot of historical speleogenesis would incorporate both solutional dissolution mechanisms with preferential dissolution by halocline migration through the depths of the cave passage. With the addition of karst windows along the systems' corridors, injections of organic debris enter the caves within a wide range of locations to enrich both chemical and nutrient microenvironments in the caves.

A persistent assemblage of speleogenic processes is responsible for creating a multifaceted coastal region that is riddled by underwater caves and microenvironments. A section of this cave region surrounds the city of Tulum on all flanks, while paralleling the coastline for a total distance of 36 km (Fig. 1). The present extent of inland cave development is presently limited to a 12-km distance from the coast. Cave passage configuration within the coastal region of Quintana Roo is oriented from the northwest to southeast. This orientation is also representative of the established inland-to-coastal flow of freshwater drainage. Why the 12-km boundary exists has perplexed underwater explorers for many years. Coastal horizontal cave formation appears to surrender to interior classic sink patterns at this frontier. The first explorers to confront this boundary reported large, fracture controlled passages just as they reached the terminal end of the tunnel. Swimming against strong freshwater currents, they would describe 100-m-wide tunnels decorated by huge flowstone columns, calcite rafting floors, and slender stalactites hanging from 6-m-high ceilings (Fig. 2). Without warning, this trunk passage would end abruptly at a vertical wall of limestone. As exploration efforts continued in other areas, a collection of unassociated cave systems were surveyed to an equal distance from the coast. Each of their cave passages terminated at an all-too-familiar limestone barricade. The eastern edge of the Holbox Fracture Zone has proved to be a formidable opponent for coastal cave explorers. This fracture zone extends south near the island of Holbox to Laguna Chunyaxché near the town of Muyil. We have discovered that this fracture appears to play a significant role within the local

FIGURE 2 Representative cave passage and speleothems in Sistema K'oox baal, Quintana Roo, Mexico. Speleothems deposited during Pleistocene sea-level low stands. *Photographs courtesy of Radoslav Husák. Used with permission.*

hydrology. At each intersecting passage with the fracture zone, today's explorers are quick to note an abrupt change in the freshwater currents. In each instance, a strong flow of freshwater is recorded, emanating from the south. This flow anomaly can be tracked over a 25-km distance through five widely separated cave systems. It appears that the Tulum Region's aquifer is partially supplied from a yet-to-be-determined freshwater source south of Laguna Chunyaxché. The implications concerning the nuances of the local freshwater aquifer are fascinating. It is a worthy complement to the pristine microenvironments that are supplied by these unknown freshwater supplies.

As the aquifer begins its 12-km traverse to the Caribbean through limestone, cave passage configurations and their environmental niches change over the course of the drainage pattern. Hundreds of natural openings are encountered along this watercourse. In earlier times these natural openings were vital centers of activity for pre-Maya populations. Radiocarbon (^{14}C) analyses date one set of human remains in Sistema Naranjal (an underwater cave in Quintana Roo) to over 10 ka BP in age. Three additional pre-Maya skeletons are documented from the Naranjal Cave, while another four skeletal sets are under study in nearby underwater caves. During a period of lowered sea levels, proto-explorers gained entrance to many caves due to their position above the water table. Whether they used caves for shelter, a source for drinking water, or as a spiritual focal point is a matter of speculation. It was the rise of the Maya civilization that brought a collective essence to Yucatán's caves. The Maya were relentless dry cave explorers, at times traversing hundreds of meters of intricate dry crawlways and galleries with little more than a smoldering torch. Leaving a legacy of huge stone structures, cave paintings, rock art, and pottery to mark their explorations, modern researchers are convinced the Maya used caves and cave entrances as centers for spiritual vitality and political advantage into the twentieth century. Sacrificial pots, small structures, and petroglyphs from past Maya activities continue to be documented at cenote and dry cave sump entrances to underwater caves.

Today cenotes play another role within the environment of the underwater caves. As a window to the jungle, terrestrial organic materials are able to invade limited sections of the watercourse to enrich the caves' tenuous ecological niche. Each of these openings to the heart of underwater caves is analogous to bat populations' stimulus to the biological web in dry caves. Cenotes are a point source for organic fuels that sustain many of the 41 species of cave-adapted stygobionts on the Yucatán isthmus. They also nourish jungle mammal, reptile, amphibian, and avian populations. As underwater passages progress toward the ocean, each cenote opening supports a succession of terrestrial, stygophile, and stygobiont species. The aquifer's water chemistry is altered during this journey. Flowing at an approximate velocity of 250 m per day, the aquifer is transformed from a substantial brackish body of water to a thin and near-saline layer at coastal discharge centers. Not only does this transformation affect local biological populations, it can degrade the parent limestone strata of the conduits. Distinct underwater cave passage begins to disintegrate into a Quaternary mix of breakdown and unconsolidated strata. Erosion by the halocline can also give the impression of "bedding plane" dissolution. Daily tidal fluctuations encourage the halocline to focus its authority on a 45-cm section of strata, creating a sharply defined horizontal void.

Two extensive caves located in the Tulum and Xel-Ha regions best illustrate the complex nature of underwater caves in Quintana Roo. Sistema Ox Bel Ha (in Maya, "Three Paths of Water") is the longest cave in Mexico, and the longest underwater cave in the world. It contains over 231 km of surveyed passage that integrates four known ocean-discharge vents, and over 137 inland cenote entrances within 52 km^2. Extending nearly 9 km inland from a series of small offshore vents, the cave traverses beneath a dense coastal mangrove zone that rises to a scrub jungle interior that is peppered by karst windows and *aguadas* (temporary pools of surface water). However, the highest density of cave development is found in a centralized maze along major drainage conduits within a 38-km^2 area.

As the Ox Bel Ha Cave spreads inland from the Caribbean Sea, it transforms its overall structure to reflect two of three general classes of geological and environmental conditions that typify most underwater caves of the region. These classes may be subjectively ordered (progressing inland) as caves found in recent coastal strata, caves that are developed in older strata 1 to 6 km from the Caribbean, or those underwater caves that impinge and terminate on the Holbox Fracture Zone (10–12 km inland). The sections of Ox Bel Ha within a kilometer of the coast tend to settle into four independent maze areas. Their passages are formed in unconsolidated Recent deposits, giving way to more substantial Quaternary strata farther inland. Each maze is arranged around one or two dedicated conduits that are engaged in directing freshwater drainage toward the ocean. An extensive mixing zone between the freshwater and saltwater layers creates an elevated concentration of chlorides with these sections. Passages are often barren of speleothems, given the advanced degree of dissolution and limited geological history in this area. They are likely formed along small fractures and joints; however, conduit erosion makes this difficult to discern. Fine clay-like silts cover the floors and ceilings of the cave, while the parent limestone is often frangible and unstable. Unlike other underwater caves proximal to the Caribbean, few if any conduits attempt to parallel extant coastal topography through linear fractures.

Fracture-based speleogenesis is evident within the primary discharge conduits as they advance farther inland. At 1 km from the ocean, Ox Bel Ha assumes an entirely different character than that of its coastal temperament. Cave formation in this section is based on a northwest–southeast pattern of conduit evolution that is present in many of the area's underwater caves. The parent cave strata become more massive and stable, the mixing zone between water layers evolves into a more distinct halocline, and speleothems become more common within the well-defined passages. This area transitions into older cave passage and limestone strata. This part of the cave is also thought to have longer dry periods during Pleistocene glaciations. Many of the cave tunnels are situated within 20 m of the land surface; deeper conduit development is not unusual although it is not widespread.

Lateral branching of fault-controlled tunnels occurs in this section. Secondary tunnels may rejoin their original passage or become an independent subsidiary section. In rare instances, they can connect two trunk passage sections of Ox Bel Ha that are separated by over a kilometer. It is not clear if these distant connections serve as conduits to pirate water from other areas of the cave, or if they are a product of simple fracture evolution. Numerous karst windows become highly engaged within this lateral passage development. Inline flowing sinks such as Cenote Esmeralda or Cenote So'sook are focal points in the fracture patterns.

Sistema Sac Actun (in Maya, "the White Cave") is the second longest cave in Mexico, and the second longest underwater cave in the world. It contains over 215 km of surveyed underwater passage, and an additional 2 km of dry passage. The Sac Actun cave integrates two known discharge vents to the ocean, and over 162 inland cenote entrances. Unlike the more centralized development of Sistema Ox Bel Ha, Sistema Sac Actun parallels the east coast of Quintana Roo through a complex set of interconnected high-density cave centers. Reaching over 8 km inland through fracture- and dissolution-controlled passages, Sac Actun extends laterally nearly 12 km between its most northern and southern survey stations. It incorporates a highly complex inland drainage system that plunges toward the coast at established freshwater discharge points. Given its geographical range, the cave traverses the first two of the three general classes of geological and environmental conditions that typify most underwater caves of the region.

Two coastal sections of Sac Actun tend to settle into localized maze configurations once the immediate discharge tunnels are passed. Conduits in the Cenote Abejas area attempt to parallel coastal topography through linear fractures in Recent depositional strata. Similar to the coastal sections of Sistema Ox Bel Ha, the high concentration of chlorides in the mixing zone produces low horizontal tunnels ("bedding plane") and ill-defined conduits that are broken by periodic fracture-controlled rooms. Also as in Sistema Ox Bel Ha, these passages are barren of speleothems, covered in fine clay-like silts, and developed in unstable limestone. Hydrologically active passages may be complicated by deformations in a nearby Pleistocene ridge, or through young Holocene deposits that are positioned between the coastline and the Pleistocene ridge.

As the cave migrates inland from its discharge areas, it passes beneath the Pleistocene ridge (~200 m to 1 km from the coast). The ridge produces a rise in topographic elevation (~10 m) that accompanies a change to older more consolidated limestone strata. Clear of local strata disturbances created by the ridge, much of Sac Actun begins to adopt an entirely different character from that of the coastal region. Fracture and fault-based speleogenesis becomes evident within primary and some secondary discharge conduits as they progress inland. Cave formation in this section turns to a northwest–southeast pattern of conduit evolution. The parent cave strata become more massive and stable, the cave passage becomes clearly defined, and lateral branching of fault-controlled tunnels occurs frequently. Most conduits are situated within 15 m of the land surface in this area of the cave.

At 3 to 6 km from the coast, both northern (Nohoch Nah Chich) and southern (Sac Actun-Naval) arms of Sac Actun enter or impinge on Tertiary (Miocene-Pliocene) strata. Although some lateral branching of fault-controlled tunnels occurs in these areas, cave development is often limited to one or two primary discharge tunnels. Most secondary tunnels are generally limited in length and small in size; they often end in impassable cracks or tubes that discharge freshwater.

Explorations in the northern section of Sistema Sac Actun are now within 5 m of connecting to Sistema Dos Ojos (68 km). Due to the wandering nature of the cave, the possibilities for expanding Sac Actun are far greater than those in Sistema Ox Bel Ha.

See Also the Following Article

Exploration of Caves—Underwater Exploration Techniques

Bibliography

Back, W., Hanshaw, B. B., Herman, J. S., & Van Driel, J. N. (1986). Differential dissolution of a Pleistocene reef in the ground-water mixing zone of coastal Yucatán, Mexico. *Geology, 14*, 137–140.

Coke, J. G., IV, Perry, E. C., & Long, A. (1991). Sea level curve. *Nature, 353*(6339), 25.

Holthuis, L. B. (1973). Caridean shrimps found in land-locked saltwater pools at four Indo-West Pacific localities (Sinai Peninsula, Funafuti Atoll, Maui and Hawaii Islands), with the description of one new genus and four new species. *Zoologische Verhandelingen, 128*, 1–48.

Lutz, W., Prieto, L., & Sanderson, W. (Eds.), (2000). *Population, development, and environment on the Yucatán Peninsula: From ancient Maya to 2030*. Laxenburg, Austria: International Institute for Applied Systems Analysis.

Rissolo, D. (2005). Tancah Cave revisited. *Association for Mexican Cave Studies Activities Newsletter, 28*, 78–82.

Smart, P. L., Beddows, P. A., Coke, J., Doerr, S., Smith, S., & Whitaker, F. F. (2006). Cave development on the Caribbean Coast of the Yucatán Peninsula, Quintana Roo, Mexico. In R. Harmon & C. Wicks (Eds.), *Perspectives in karst geomorphology, hydrology, and geochemistry* (pp. 105–128) (Special Paper 404). Boulder, CO: Geological Society of America.

Tulaczyk, S. M., Perry, E. C., Duller, C. E., & Villasuso, M. (1993). Influence of the Holbox fracture zone on the karst geomorphology and hydrogeology of northern Quintana Roo, Yucatán Peninsula, Mexico. In B. F. Beck (Ed.), *Applied karst geology* (pp. 181–188). Rotterdam: Balkema.

Ward, W. C. (1985). Quaternary geology of northeastern Yucatán Peninsula, Part 2. In W. C. Ward, A. E. Weidie & W. Back (Eds.), *Geology and hydrogeology of the Yucatán and quaternary geology of northeastern Yucatán Peninsula* (pp. 23–53). New Orleans, LA: New Orleans Geological Society.

Ward, W. C., & Weidie, A. E. (Eds.), (1978). *Geology and hydrogeology of northeastern Yucatán* New Orleans, LA: New Orleans Geological Society.

Weidie, A. E. (1985). Geology of the Yucatán Platform, Part 1. In W. C. Ward, A. E. Weidie & W. Back (Eds.), *Geology and hydrogeology of the Yucatán and quaternary geology of northeastern Yucatán Peninsula* (pp. 1–19). New Orleans, LA: New Orleans Geological Society.

URANIUM SERIES DATING OF SPELEOTHEMS

Christoph Spötl and Ronny Boch

University of Innsbruck

INTRODUCTION

Entering a cave decorated with various sorts of speleothems commonly prompts questions about the age of these formations. It is intuitive to suggest that tall stalagmites or thick flowstone took a long time to form, but until the middle of the twentieth century no method existed that allowed the quantitative determination of how old speleothems are. First attempts were made by physicists in the 1950s who applied the radiocarbon (^{14}C) method (discovered in the late 1940s) to

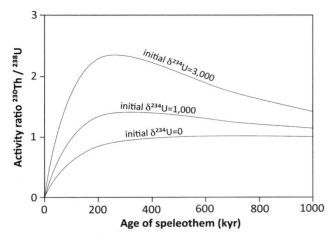

FIGURE 1 ^{230}Th/^{238}U activity ratio versus age. A high initial ^{234}U/^{238}U ratio (expressed in delta units) accounts for an extended dating range as secular equilibrium is reached after a longer time period (after Richards and Dorale, 2003). Most speleothems show δ^{234}U values between ca. 100 and 1200.

dripstones. It was soon realized, however, that these results must be viewed with great caution, because the origin of the carbon atom (utilized for age dating) of dripstone carbonate is different from that in organic remains, such as wood. In addition, radiocarbon dates require a calibration to obtain true calendar ages (which was not available in the early days of radiocarbon dating). In short, ^{14}C dates of speleothems provide only a rough estimate of their formation age and samples older than ca. 50,000 years cannot be dated using this method.

The dating method of choice for speleothems, known as the *uranium-thorium* (U-Th), or *thorium-uranium* (Th-U) *method*, belongs to the family of uranium-series methods and was first applied to speleothems in the early 1970s, although its physical principles were known since the beginning of the twentieth century. The past four decades have seen dramatic improvements of this method and today U-series dating of speleothems is among the most highly respected chronological techniques, providing invaluable time information for studies of the environment and climate of the past.

This article provides a concise summary of the method for cavers, speleologists, and other people interested in finding out how the age of speleothems and their growth rate can be measured, and which practical aspects are relevant when selecting samples for analysis.

PRINCIPLES OF U-SERIES DATING

From the three naturally occurring U isotopes (99.28% ^{238}U, 0.72% ^{235}U, 0.005% ^{234}U). ^{238}U and ^{235}U are the starting points for two different radioactive decay chains comprising numerous unstable isotopes toward stable Pb isotopes. The rate of radioactive decay is statistically constant (*decay constant*) for a particular isotope, but different isotopes decay at different rates (they have different *half-lives*). As a consequence, different isotope ratios can be measured and are equivalent to a specific time elapsed, that is, the isotopic abundances depend on the decay laws and this accounts for an absolute chronology. An accurate knowledge of the decay constants is essential to calculating the age, and these constants are subject to ongoing reevaluation and refinement.

U-Th (disequilibrium) dating determines the relative abundances of unstable U and Th isotopes and is one of several methods based on the radioactive decay starting with the parent isotope ^{238}U. This particular method, however, is by far the most important one for determining the age of speleothems, and it lays the foundation for their prominent role in paleoclimate research. Depending on the geochemical characteristics of the sample (e.g., the initial ^{234}U/^{238}U ratio, Fig. 1) and the laboratory instrumentation, the U-Th method allows dating of samples as old as 600,000–700,000 years, after which radioactive decays of the isotopes ^{230}Th and ^{234}U are in a state of secular equilibrium. An initial state of disequilibrium between the U and Th isotopes is a precondition for this method, that is, U and Th are separated from each other prior to speleothem formation due to the vastly different chemical properties of these elements: Th is insoluble in water and tends to be adsorbed on particles (e.g., clay minerals); in contrast, U is highly soluble, transported in solution (drip water) and coprecipitated during speleothem deposition. As soon as the U atoms are incorporated into the lattice of the carbonate crystals radiogenic ^{230}Th starts to build up from U decay, that is, the U-Th "clock" starts to run. Under ideal conditions, the ^{230}Th content originates solely from the decay of ^{234}U and its concentration increases systematically with time. No gain or loss of these isotopes, other than the intrinsic decay, should have occurred since the formation of the speleothem, that is, the system must have remained closed. For this reason, speleothems showing recrystallization or dissolution should be avoided. Some speleothems contain an additional, nonradiogenic Th fraction resulting from the incorporation of mineral or organic particles. In these cases, the isotope ratios have to be corrected for the initial ^{230}Th contribution using the measured concentration of ^{232}Th and an initial ^{230}Th/^{232}Th activity ratio which is either assumed (e.g., the mean Th/U ratio of the upper continental crust) or determined using isochron methods. The extent of detrital contamination may vary significantly within a cave and also within an individual speleothem.

Regarding U-Th technology, the first laboratories started in the 1970s and the subsequent evolution was characterized by two major steps prompted by breakthroughs in analytical science. The first revolution occurred in 1986 and marked the introduction of mass spectrometric techniques (instead of week-long counting of alpha particles from radioactive decay). Thermal ionization mass spectrometry (TIMS) allowed measurement of U and Th isotope abundances with much better precision and smaller sample sizes and extended the dating range from ca. 350,000 years to ca. 500,000 years. The second improvement involved the introduction of a new type of mass spectrometer which was successfully coupled to an ionizing plasma source. These machines are referred to as inductively coupled plasma-mass spectrometers (ICP-MS) with the prefix "single-collector" or "multi-collector" depending on the number of particle collectors. Introduced in 1997, these instruments now dominate many fields of geochemistry and geochronology. Compared to TIMS, the ICP-MS technology holds the advantages of a reduced sample size, shorter analysis time (increased sample throughput), and an increased sensitivity due to higher ionization efficiency. Using ideal samples and optimum instrumental conditions, an upper dating limit of ca. 700,000 years can today be achieved. Modern developments include the application of *in situ* laser ablation ICP-MS which increases the spatial resolution, decreases the sample size, and facilitates sample preparation, but currently suffers from poorer precision and accuracy.

Before measuring the relevant isotopic ratios on a mass spectrometer of the carbonate samples—preferentially taken from pristine crystal fabrics along the central speleothem growth axis and in powder form—rather time-consuming, wet-chemical preparation steps have to be conducted in a dedicated clean laboratory. U and Th are enriched and isolated from each other and from the carbonate matrix using an ion-exchange resin. Ultra-pure acids and other chemicals including well-characterized radioactive materials (spikes) are used and this work therefore requires extreme care.

REPORTING AND ASSESSING U-TH AGES

Individual speleothem U-Th ages are quoted either in years (*annum*; a) or, in the case of older samples, in thousands of years (ka), for example, 34,200 a is equivalent to 34.2 ka. Traditionally, U-Th ages have been referred to the year 1950 AD (the reference point for radiocarbon ages), but more recently other reference years have also been used, for example, the year 2000 AD (abbreviated as b2k for "before 2 ka"). This information has to be stated clearly when reporting speleothem ages to avoid confusion, in particular when dealing with young (Holocene) samples.

In general, each measurement has two kinds of uncertainties ("errors"): accuracy and precision. The former determines the deviation from the "true" value, and requires careful evaluation in the laboratory involving standard reference materials. Using different decay constants, for example, also results in systematically different results (which may still be extremely precise). Precision is easier to assess and provides statistical information on how far replicate measurements deviate from the mean value. Recent years have seen an impressive improvement in the precision of U-Th data, although it is important to bear in mind that this does not include the uncertainty associated with accuracy. Dating precision is traditionally reported at the 2-sigma uncertainty level, for example, an age of 123.3±1.2 ka means that 95% of the sample reruns fall between 122.1 and 124.5 ka. Some other geochronological methods report their results at the 1-sigma level, such as ^{14}C and burial ages. For example, a bone fragment yielded a (calibrated) ^{14}C age of 37,140±300 a BP and is embedded in flowstone whose U-Th age is 36,400±300 a BP. Note that (1) the latter date is twice as precise as the radiocarbon date, and (2) the speleothem is statistically younger than the bone at the 1-sigma level; on the 2-sigma level, however, the two dates are indistinguishable from each other. Speleothem ages determined by alpha-spectrometry were also reported at the 1-sigma level; U-Th data produced prior to 1987/88 are quoted in this way. In addition, these dates were calculated using decay constants that have since been updated, resulting in an age-dependent deviation.

What precision can be expected for speleothem samples? This question depends on a number of factors which range from the type and geochemical quality of the sample to the instrumentation and calibration of the particular laboratory. For typical speleothem samples, most laboratories are capable of measuring ages with relative errors of 1–2% for samples younger than ca. 200 ka. For example, a flowstone that formed 120 ka ago (during the last interglacial) can be routinely dated to ±1.2 to ±2.4 ka. Some of the leading laboratories report precisions as low as 0.2–0.3%. So, the analytical uncertainty of the same flowstone can be reduced to ±240 to ±360 a and a very small number of these labs have recently begun reporting U-Th ages at "epsilon precision," that is, better than per-mil level. Even taking further technological advances into account, it

FIGURE 2 Drilling a small-diameter core near the base of *in situ* stalagmites allows the determination of the approximate starting date of deposition at this drip site, thereby minimizing the impact on the cave environment.

is highly questionable if such ultra-high precision will be routinely possible within the next five years. After all, the most commonly encountered problem is impurities in the sample. Although various approaches exist to correct for the nonradiogenic Th fraction, they are all based on assumptions and therefore result in larger age uncertainties. Regardless of the instrumentation used, the overall uncertainty increases with increasing age—for example ±30–50 a at 10,000 a, ±250–400 a at 50,000 a, ±700–1500 a at 100,000 a, ±1000–3000 a at 200,000 a, ±3000–15,000 a at 400,000 a—and reaches infinite values at the upper dating limit.

Finally, it should be noted that the analytical precision achievable using state-of-the-art instrumentation approaches the resolution imposed by sampling a finite thickness of speleothem material. Given a stalagmite with a slow vertical growth rate of 25 μm/a, a 2-mm-thick sample represents some 80 a of growth. If sampling is not done carefully, for example, if growth layers are not visible, these 2 mm might encompass significantly more than 80 a.

PRACTICAL ISSUES

Who Is Allowed to Sample Speleothems?

Each cave and each dripstone is a unique part of nature which took literally thousands of years to form. In most countries speleothems are therefore protected by law. As a consequence, sampling of speleothems requires special permits and should always be done with great caution and in close collaboration with local authorities and/or speleologists.

How to Sample Speleothems in Caves

Most paleoenvironmental studies require the entire stalagmite to be removed from the cave. As significant parts of the dripstone are expended during laboratory analyses it is rarely possible to restore the original stalagmite after the scientific study has been completed. Although there are ways to replace the stalagmite with a replica, most scientific studies have not made such an attempt. Consequently, the removal of a stalagmite from a cave is an invasive act which should be evaluated very carefully.

One way to minimize the impact on the cave and to screen for relevant stalagmites (*e.g.*, covering a time period of interest) is to take small *in situ* samples by drilling horizontally into the base of a stalagmite (Fig. 2). U-Th dating of such small drill cores (*e.g.*, 5 mm in diameter) from the inner part of a stalagmite provides useful information on the starting time of growth, and the drill hole can be easily sealed, thereby causing hardly any impact on the appearance of the stalagmite. Drilling is also an option for sampling flowstone deposits, but in this case larger diameter cores (*e.g.*, 2–3 cm) are preferable so that a whole series of analyses (in addition to dating) can be performed. Again, such holes can be sealed so that there is no impact on the appearance of the dripstone formation to visitors of the cave.

How Much Sample Material Is Needed for an Age Determination?

It is possible to obtain a meaningful age from as little as 10–20 mg of speleothem material. Most laboratories, however, prefer larger samples (*e.g.*, 50–200 mg) to improve precision and possibly run duplicates. As a rule of thumb, the younger the sample and the lower its U concentration, the more sample is required to obtain a good result. Aragonitic speleothems have significantly higher U contents than the more common calcitic ones, so smaller sample amounts are needed. The subsamples for U-Th dating are typically drilled from a polished slab cut from the axial part of a stalagmite (Fig. 3). Some researchers prefer to use a thin diamond-coated blade- or wiresaw to cut discrete aliquots from the stalagmite instead of using powdered drill samples. This, however, requires cutting the axial slab in half as it is advisable to stay close to the axial zone where the growth layers are rather thick, as opposed to the flanks of the stalagmite where they are thinner, and impurities may be present at higher concentrations.

FIGURE 3 Polished slab of a small (13.8 cm high) stalagmite with several elongated holes drilled to sample calcite for U-Th age determinations. Note that the sampling holes follow individual growth layers in order to increase the age resolution. The vertical band along the growth axis is a milling trench—only fractions of a millimeter deep—along which numerous subsamples for stable isotopic analyses were taken. Small holes along growth layers are samples to check for stable isotopic equilibrium.

How Much Does an Age Determination Cost?

In contrast to, for example, radiocarbon analyses of organic material, only a few laboratories worldwide offer U-Th analyses on a commercial basis. Prices typically range from *ca.* US$200–400. The majority of the U-Th facilities, however, belong to universities or other academic institutions and their pricing may deviate significantly.

Why Focus on Stalagmites?

Speleothems, despite their uniform mineralogy (calcite, less commonly aragonite), occur in a great variety of forms, dimensions, and colors, and any of these can be dated by U-series methods provided their age is within the dating limit of the particular method. Still, with some exceptions, the majority of scientific studies have so far relied on stalagmites rather than stalactites, flowstones, or other formations. The main reason for this is that regularly shaped (candlestick) stalagmites are the only speleothem growth variety that shows a well-defined internal structure. Flowstone also commonly shows a regular internal growth structure, but is often characterized by growth interruptions due to switching of the flow paths on the surface. Stalactites have also been investigated, but given their complex (chevron-like) internal structure and the presence of an internal drip-water feeder channel most researchers tend not to work with them.

What Is a "Good/Bad" Sample for Dating?

Along the lines of the preceding paragraphs, it is advisable to stay away from samples that macroscopically suggest impure calcite (or aragonite), that is, those showing a brown or gray stain. A clean, white, outer surface of actively forming (*i.e.*, drip-water fed) speleothems is a strong indication of low impurities, but such a diagnosis may fail on older, inactive formations, whose outer surface often disguises the interior composition. Drilling a basal small-diameter core is a way of screening potentially interesting formations (see above). Flowstone embedded in fine-grained clastic sediments should also be viewed with caution as it likely contains clay particles which compromise the quality of the dating result.

Many dripstone caves contain abundant broken speleothems, locally reworked by a cave stream. Sampling such fragments is clearly easier to justify than removing *in situ* stalagmites and still provides valuable insights into the history of a cave or the regional paleoclimate. Speleothem fragments reworked by cave streams, however, are sometimes problematic as the water can alter the chemical composition of the samples due to undersaturation with respect to calcite.

GROWTH DYNAMICS OF SPELEOTHEMS

U-Th dating offers the opportunity of using speleothems as archives of environmental change in the past. To this end, the internal age structure of a given stalagmite (or flowstone) needs to be known in as much detail as possible. This typically requires a set of some 10–30 individual U-Th subsamples, depending on the size and age range of the speleothem. The relationship between distance (measured from the top or base of a stalagmite) and the age provides fundamental insights into the growth history (Fig. 4), including possible growth interruptions (hiati). Various approaches exist to mathematically fill the gaps between individual U-Th data points and to assess the degree of uncertainty

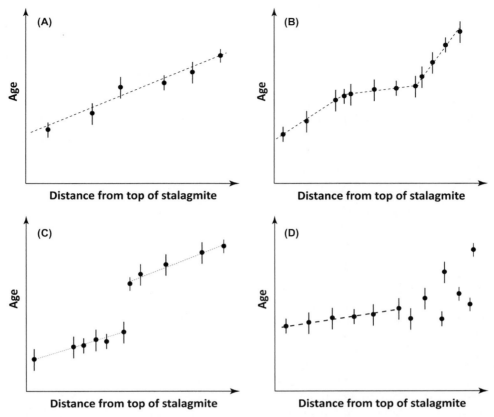

FIGURE 4 Schematic depth-age relationships commonly found in stalagmites. Filled circles and vertical lines represent individual U-Th samples and their associated analytical uncertainties, respectively. (A) the gradual progression of individual U-Th ages suggests an overall linear growth model (stippled line). Note, however, that given the relatively large uncertainties associated with some of the ages, more complex growth histories cannot be ruled out. (B) Stepwise growth model which demonstrates that the rate of vertical extension increased strongly in the middle part of the speleothem. Such increases in growth rate commonly reflect a change to higher drip rate and hence a shift to more humid (and possibly warmer) outside climate. (C) This speleothem shows a distinct growth interruption (hiatus) indicated by the jump in the growth curve, that is, during this time interval the drip above the stalagmite presumably dried up. (D) The older part of this speleothem shows a large scatter in the U-Th dates (including age inversions), strongly suggesting that this part has not behaved as a closed system. In contrast, the ages in the younger segment are in correct stratigraphic order and hence provide a robust chronology.

resulting from modeling the growth history. The most conservative way of linking up individual data points along the extension axis of a stalagmite is a piecewise linear interpolation. Alternatively, polynomial regressions or smooth-spline functions may be used.

Thanks to a large number of stalagmites analyzed in the past two decades, we have a robust database at our disposal which shows that stalagmites often show a near-linear growth pattern, that is, the growth rate (commonly expressed in $\mu m/a$) remained rather constant over millennia. These growth rates, however, may vary over 2–3 orders of magnitude in stalagmites and show a strong relationship with climate: rapidly growing stalagmites (fractions of a mm/a) are often found in temperate and warm regions with sufficient net infiltration of meteoric precipitation, for example, regions in Asia affected by summer monsoon rains, or in humid circum-Mediterranean regions. At the lower end of the spectrum are small stalagmites from cold caves located at high latitudes or high elevation. In these locations, growth rates do not exceed a few tens of $\mu m/a$. Very slow growth also occurs in caves located in semi-arid regions, where caves are dry most of the year. By far the lowest rate was measured in some subaqueous deposits which precipitated under extremely stableconditions of low calcite supersaturation. A thick subaqueous calcite deposit present in a warm (34°C) aquifer underlying the desert of southern Nevada formed at an astonishingly slow average rate of $0.8\,\mu m/a$ over the past half million years.

U-Th dating has also provided insights into the typical "lifetime" of speleothems. Again, there is a large variability ranging from stalagmites in semi-arid regions, where growth phases are confined to humid climate intervals separated by hiati, to large or very slowly forming formations where deposition occurred uninterruptedly over tens or even hundreds of thousands of years. It is worth emphasizing that these findings are also relevant for hydrological studies and show that individual drip sites in caves remain stable over these very long periods.

U-PB DATING OF ANCIENT SPELEOTHEMS

Even with today's most advanced instrumentation, it is impossible to date samples beyond ca. 600−700 ka using the U-Th "clock." Under very special geochemical circumstances it may be possible to use a variant of this method, the ^{238}U-^{234}U *decay scheme*, to extend the datable time range to ca. 1.5 million years. There is also another method within the U-series family which is capable of dating speleothems, in principal, as old as the age of the Earth, based on the two radioactive decay series from ^{238}U and ^{235}U to stable ^{206}Pb and ^{207}Pb, respectively. This is not a routine technique, however, because it is very time-consuming and expensive and requires samples of exceptional chemical composition. In essence, speleothem samples must be rich in U (ideally at least ca. 1 ppm) and contain very little initial (so-called common) Pb. If initial Pb was present at the time of speleothem deposition, significant corrections have to be applied, increasing the age uncertainty, in particular, if the speleothem is less than a few millions of years old. Moreover, elements such as the noble gas radon, which is part of both decay series, are volatile and therefore tend to disturb closed-system conditions. Despite these difficulties, U-Pb remains a very promising technique to explore, for example, Tertiary-age cave deposits, provided that clean, U-rich samples are available.

See Also the Following Article

Paleoclimate Records from Speleothems

Bibliography

Dorale, J. A., Edwards, R. L., Alexander, E. C., Shen, C. C., Richards, D. A., & Cheng, H. (2007). Uranium-series dating of speleothems: Current techniques, limits, and applications. In I. D. Sasowsky & J. Mylroie (Eds.), *Studies of cave sediments. Physical and chemical records of paleoclimate, revised edition* (pp. 177−197). Dordrecht: Springer.

Richards, D. A., & Dorale, J. A. (2003). Uranium-series chronology and environmental applications of speleothems. In B. Bourdon, G. M. Henderson, C. C. Lundstrom & S. P. Turner (Eds.), *Uranium-series geochemistry, Reviews in Mineralogy and Geochemistry* (Vol. 52, pp. 407−460). Washington, D.C.: Mineralogical Society of America.

Scholz, D., & Hoffmann, D. (2008). ^{230}Th/U-dating of fossil corals and speleothems. *Eiszeitalter und Gegenwart, 57,* 52−76.

Spötl. C., & Mattey, D. (2012). Scientific drilling of speleothems − a technical note. *International Journal of Speleology, 41,* 29−34.

van Calsteren, P., & Thomas, L. (2006). Uranium-series dating applications in natural environmental science. *Earth Science Reviews, 75,* 155−175.

Woodhead, J., Hellstrom, J., Maas, R., Drysdale, R., Zanchetta, G., Devine, P., et al. (2006). U-Pb geochronology of speleothems by MC-ICPMS. *Quaternary Geochronology, 1,* 208−221.

V

VERTEBRATE VISITORS—BIRDS AND MAMMALS

Nikola Tvrtkovic
Croatian Natural History Museum, Croatia

CAVE-DWELLING SPECIES

There are no troglobites, permanent inhabitants of caves, in the group of warm-blooded animals. Most birds and mammals documented in connection with caves are random guests, although some species are not accidental visitors. They prefer to stay in the cave for a shorter or a longer period of time and have adapted in some ways to cave habitats, such as in the case of mammals that have developed special pads for climbing, long vibrissae, partial heterothermy, and certain types of echolocation. Like troglobites, these types of cave visitors are referred to as *cave-dwelling animals* or *cavernicoles* (Vandel, 1965), with different levels of modification for cave life. Most of them originate from the climbing type of animal, especially from rock-dwellers (petricola). According to the ecological classification of cavernicola modified from the scheme of Schiner–Racovitza, cavernicolous birds and mammals are *habitual trogloxenes* after the Hazelton and Glennie system (Ford and Cullingford, 1976), or *troglophiles* (eutroglophiles and only partly subtroglophiles) after the Pavani system (Vandel, 1965)—cavernicola by choice, not by chance.

In the first group of cavernicolous birds and mammals are *facultative visitors*: they only visit caves, but they are also common in different surface habitats. Facultative cave visitors by choice are the pack rat (*Neotoma* spp.) in America, the edible dormouse (*Glis glis*) in Europe, the Egyptian fruit bat (*Rousettus aegyptiacus*) in North Africa and western Asia, and the bird origma (*Origma* spp.) in Australia. Their nests or roost sites have been found many hundreds of meters from the entrances, but they are common in other habitats too, such as in tree holes or shallow depressions in cliffs.

The second group are *regional cavernicoles*—they are permanent cave-dwellers only in some parts of their range. The cave swallow (*Hirundo = Petrochelidon fulva*) is the first example: it breeds in caves only in southern parts of its range, in the border area between the United States and Mexico, and in Cuba. In this group there are several owls, such as the little owl (*Athene noctua*) and large eagle owls like *Bubo bubo* in the karst area of Europe, or *Bubo africanus* and *Bubo ascalaphus* in Saudi Arabia. The monk seal (*Monachus monachus*) rears its young on beaches along the North African coast, but most of their breeding sites along the northern Mediterranean coast were (before extinction) in the sea caves. Two types of crows (*Pyrrhocorax graculus* and *P. pyrrhocorax*) are high mountain birds of Eurasia, nesting in crevices in cliffs. But in the Mediterranean karst, they nest only in deep cold potholes (pits). The Geoffroy bat (*Myotis emarginatus*), primarily a Mediterranean species, has large nursery colonies in sea caves or warmer caves together with the greater horseshoe bat (*Rhinolophus ferrumequinum*). In Central Europe both species are found in the nonkarst areas. Their small roost colonies can be found in attics, and they migrate in the fall to karst areas with their traditional hibernacula.

The last group is *regular cavernicoles*. According to Brosset (1998) and other sources, the birds among these regular cave visitors include the guacharo or oilbird (*Steatornis caripensis*) in Central and South America, the Guinean cock of the rock (*Rupicola rupicola*) in South America, the waterfall swift (*Hydrochous gigas*) in Southeast Asia, the swiftlets (*Aerodramus* spp. and *Collocalia* spp.) in Southeast Asia and the neighboring islands from the Seychelles to West Oceania, pallid swifts (*Apus pallidus*) on Mediterranean coasts, and two rockfowls from central Africa (*Picathartes oreas* and *Picathartes gymnocephalus*). Regular cave visitors among mammals include the Dinaric or Martino's vole

FIGURE 1 The Martino's vole *(Dinaromys bogdanovi)*, subadult female. *Photograph taken by the author.*

(Dinaromys bogdanovi; Fig. 1), restricted only to the western Balkans in Europe, and the dawn bat *(Eonycteris spelea)* from the group of fruit bats of Southeast Asia. The largest group of regular cavernicola are microchiropteran bats restricted to caves, like the gray bat *(Myotis grisescens)* in North America, the long-fingered bat *(Myotis capaccinii)* from Mediterranean karst habitats, and Schreiber's bat *(Miniopterus schreibersi)* from Europe, North Africa, Asia, and Australia.

GAPS IN KNOWLEDGE

Lists of cave-dwelling birds and mammals are incomplete. Most of the data have been documented accidentally and we have not yet had the ability to compare the numbers of cave-dwelling species in different karst regions of the world. Some species have been documented with only one specimen, like the shrew *Soriculus salenskii* in Sichuan (China) inside a cave with water, or the fruit bat *Aproteles bulmerae* from only one cave in Western Papua (Nowak, 1994). The ecology and behavior of known cavernicolous birds and mammals are also poorly known. Of more than 20 species of the Asian bird genus *Aerodrama*, until now echolocation is documented in only three species (del Hoyo et al., 1999). How successful is the first flight of nestlings of the Alpine chough from a 70-m-deep nest site in the darkness of a pothole? Many similar questions about cavernicolous birds and mammals are without answers even today.

CAVES AS SHELTER FOR RESTING AND MATING

Caves offer a wide spectrum of microhabitats (Ginet and Decou, 1977), because they are always a suitable shelter from rain, snow, sun, and wind and from most predators. Isolation from surface habitats and from most of the predators, low level of disturbance, darkness and a stable temperature, humidity, and airflow regime are important characteristics for resting and mating. Caves provide daytime shelter for nighttime active species such as owls, hyenas, hyraxes, wild boars, peccaries, porcupines, bats, and various species of rock-dwelling rodents, or nighttime shelter suitable for digesting food in the case of owls and bats. Croatian speleologists observed the karst mouse *(Apodemus epimelas)* in many Mediterranean caves and pits, some specimens were seen at a depth of up to 70 m. In one of the deepest pits in Dinaric karst, Slovakia pit, bats were observed in flight up to 900 m deep in summer. Most bats mate in wintering or transitional places in caves before and after hibernation. Males of the lesser mouse-eared bat *(Myotis blythii)* in Europe and Asia position themselves on the ceiling and walls of caves from spring to fall in isolated solitary places marked with urine, waiting for females. Schreiber's bat forms small harems in different parts of caves early in spring, as well as in August and September.

CAVES AS DEN SITES

Most cave birds breed colonially, like the oilbird *(Steatornis caripensis)*, the swifts *(Apus pallidus)*, the swiftlets *(Aerodramus* spp. and *Collocalia* spp.), the crows *(Pyrrhocorax graculus* and *P. pyrrhocorax)*, and the rock doves *(Columba livia)*. Their nests have been found many hundreds of meters from the cave entrances and in pits up to 100 m below the entrance. In limestone caves many traces were documented of the edible dormouse *(Glis glis)* such as nests, feces, footprints, and urine marks. It has been reported that the nests of the dormouse have been found deep (several hundred meters) in the caves of Italy and Slovenia.

Desert caves in West Asia are regular den sites for striped hyenas *(Hyaena hyaena)*. Hyenas enter caves to avoid temperature extremes, to eat large scavenged prey undisturbed, to store large bones for later consumption, to give birth and to raise their offspring, and to die. Jackals *(Canis aureus)* in the Mediterranean part of eastern Europe, western Asia, and northern Africa use caves as den sites too. The famous Ghost Cat of the Himalayas—snow leopard *(Pardus = Unciauncia)*—can rear cubs for extended periods of time in a cave den in the mountains in winter. Its cave dens are lined with the mother's fur.

Warm chambers in caves with temperatures greater than 12–25°C and high humidity are the summer home for nursery colonies of a great number of bats. Relative high temperatures are important for juveniles, because they have no possibility to regulate their own temperature. Females of some bat species like the long-fingered bat (*Myotis capaccinii*) are extremely philopatric to caves in which they had been born: they usually return to their natal colony after hibernation (Papandatou et al., 2009). Different bat species usually form mixed nursery colonies in caves. The most common were nursery roosts with up to several thousands of specimens. The world's largest extant nursing bat colony is that of the guano bat (*Tadarida brasiliensis*) in Bracken Cave in Texas with about 20 million individuals (Hutson et al., 2001).

CAVES AS WINTERING SITES

In the temperate region or in the upper extremes of the subtropical regions, caves are suitable wintering places for many bats. They hibernate alone or in colonies, in crevices inside caves or on the ceiling of chambers. Most bat species gather to hibernate even in larger concentrations in a small number of sites. From all of the population of gray bat (*Myotis grisescens*) in the southeastern United States, 90% of bats hibernate in only three caves. The parti-colored bat (*Vespertilio murinus*) migrates more than 1700 km each year from mountains and steppe regions in northern Europe to wintering sites in the Alps (Hutterer et al., 2005). Schreiber's bat forms thick clusters in several layers, with an average of 2000 specimens per square meter of roof surface. The lesser mouse-eared bat prefers sites at only 2–3°C, and the gray bat prefers sites with temperatures between 10°C and 15°C. The long-fingered bat forms wintering clusters only above cave waters. In the Pleistocene, cave bears (*Ursus spaeleus*) wintered in caves all over the Eurasia. Recent species of bears usually use small caves for wintering, too.

CAVES AS FORAGING HABITATS

Some snakes and owls such as the barn owl (*Tyto alba*) catch bats at cave entrances. Mammals such as the ringtail (*Bassariscus astustus*), the red fox (*Vulpes vulpes*), the fennex (*V. zerda*), the coyote (*Canis latrans*), the badger (*Meles meles*), the stone marten (*Martes foina*), and the striped skunk (*Mephitis mephitis*) visit caves too, but mostly only for eating young bats or birds that fall to the floor of the roosting site, or to eat bats during their hibernation (Juberthie et al., 1998). Throughout the year predators in tropical caves include carnivores such as *Herpestes*, *Viverra*, and *Felis*, and shrews such as *Crocidura malayana* and *C. cinerea*. *Desmodus rotundus* and other vampire bats are a special case of bat predation, preying on other bats in some South American caves. In the Late Quaternary the Eurasian spotted hyena (*Crocuta crocuta* spelaea) and the steppe lion (*Panthera leo spelaea*) hunted cave bears especially during hibernation times in caves in low to medium high mountainous elevations, and in high alpine regions, leopard (*Panthera pardus*) seems to have used the ecological niche of the absent hyenas (Diedrich, 2010).

The most dangerous cave predator of cave birds and bats is man (*Homo sapiens*). Native hunters from Papua New Guinea collect fruit bats in caves for food, and modern hunters in Asia collect nests made of swiftlets' saliva to make a special soup. Local people on the Mediterranean and along other coasts harvest hatchlings of sea birds from small sea caves: one man from Socotra Island confessed that they could take 30 nestlings of rare Jounin's petrel (*Bulweria fallax*) from a single cave in a night. Local people of islands of the Vis archipelago in the Adriatic Sea traditionally harvested eggs and nestlings of Cory's shearwater (*Calonectris diomedea*), a social sea bird which breeds in sea caves and large coastal crevices.

IMPACT OF BIRDS AND MAMMALS ON CAVE ECOLOGY

Birds and mammals have an important impact on cave ecology. Juberthie et al. (1998) noted that the feces of porcupines in two caves in Africa are an important food for endemic stygobitic isopods. The input of the droppings of birds and bats is sometimes extremely high. Invertebrate guano fauna is formed in guano piles, an important food resource for cave-dwelling invertebrates. Rodents such as the Martino's vole take green plants and mice and dormice take seeds inside caves for storage. Rodents such as edible dormice and birds build subterranean nests of plant particles. Hyenas accumulate remains of its prey and bones for later consumption in caves. The remains of nests and food storage, and carcasses of cave visitors are an energy source for troglobites, although in tropical caves an abundance of food input attracts more surface competitors (Mohr and Poulson, 1969). In cases of colonial breeding in relatively small cave chambers, bats have changed habitat factors such as temperature and humidity. In extreme situations, millions of bat specimens decomposing guano produce a high concentration of ammonia and/or carbon dioxide, which is dangerous for man (Hill and Smith, 1984).

CAVES AS IMPORTANT HABITAT FOR SOME RARE SPECIES

The rare sea bird Jounin's petrel (*Bulweria fallax*) known from the NW Indian Ocean has breeding sites only in the sea-facing caves of and fissures on Socotra Island. The Saint Helena petrel (*B. bifax*) was endemic to Saint Helena Island, but went extinct in historical times. In western Balkans the endemic Martino's vole (*Dinaromys bogdanovi*) is known as a regular cave visitor in the rough karst landscape. It is a habitat specialist that generally depends on underground shelters in limestone karst (Kryštufek and Buan, 2008). It is documented by several speleologists to be seen deep in caves and pits. It is also an altitude generalist, known from sea level to mountain peaks at 2200 m above, in ecologically very different plant belts. Paleontological evidence suggests that the range of the genus *Dinaromys* was historically small and restricted mainly to the Dinaric karst in the western Balkans. This small-range species is presumably the only surviving member of the Late Tertiary genus *Pliomys*. Caves as shelter were crucial in survival for this paleoendemic taxon. Mitochondrial DNA sequencing has linked *Dinaromys* to *Prometheomys*, a monospecific subterranean genus endemic to the Caucasus in Asia Minor with divergence estimated to have occurred 6.2 Ma. The available evidence suggests that recent populations are small and frequently isolated due to high habitat specificity. Last results of molecular investigation with nuclear markers suggest that Martino's vole consists of two from phylogeographical lineages, dating from about 1 Ma, which may represent distinct cryptic species, and have different threat statuses.

MAN AND CAVERNICOLOUS BIRDS AND MAMMALS

In the past, ancient man and his relatives in temperate and northern areas, especially in glacial stages from Middle to Late Pleistocene, had to compete hard for caves as dens with large carnivores like the cave bear and the cave or Eurasian spotted hyena (*Crocuta crocuta spelaea*). Scientists suppose that ancient hyenas were probably the most dangerous animal for men in the Late Pleistocene period. It has been estimated that they weighed up to 102 kg and have hunted in packs as large as 40 to 50 animals. Some have hypothesized that the domestication of dogs might finally have aided humans to successfully defend against marauding hyena packs. Cave hyena populations began to shrink after roughly 20,000 years ago, completely disappearing from Western Europe between 14,000 and 11,000 years ago with decline of grasslands (Stiner, 2004). Competition and predation of hyenas in Siberia were probably significant factors in delaying human colonization of Alaska with crossing the Bering Strait earlier. The oldest Alaskan human remains coincide with the same time when cave hyenas became extinct.

In time, *Homo sapiens* created artificial caves: houses, mines, and other human-made shelters, which are suitable habitats for most of cave birds and mammals. Tin roofs provide a microclimate that is very similar to that of warm caves, which proves to be a good place for nursery colonies of bats or the nesting places of rock doves (*Columba livia*). Rock doves nest in limestone pits in Mediterranean karst and in lava tubes in Saudi Arabia, for example, in the pit of Dharb al Najem. Recently, the most numerous populations of this bird are in cities. The most of European bats change their former roost sites in caves to lofts of houses. There has, therefore, been a co-evolution of man and formerly cave-dwelling animals.

Some cave-dwelling birds and mammals are of direct economic importance (Nowak, 1994). Local people in Southeast Asia collect bats for food. Swiftlet nests, mostly of the black-nest swiftlets (*Aerodrama maximus*) and the edible-nest swiftlets (*A. fucifugus*), are high-priced export goods to some Southeast Asia countries. Cave guano harvesting was important in the past for the gunpowder industry (Hutson et al., 2001), while today it is an important source of nitrates for the fertilizer industry. Cave-dwelling birds and bats are consumers of pest insects or insect vectors of different diseases. Fruit bats are pests in orchards, but the role of cave bats as pollinators of some fruits and their role in tree seed dispersion in subtropical areas are very important.

On the other hand, in some cases, cave bats are problematic for public health as potential vectors of rabies. In northern areas we know of some lethal cases in humans: the death of one bat worker bitten by a wintering Daubeton's bat (*Myotis daubentonii*) in northern England in 2001. UNEP/EUROBATS recommends gloves for bat handling. In South America, vampire bats (*Desmodus rotundus*) are vectors of some diseases. Jackals (*Canis aureus*) are accused of being the natural reservoir hosts of protozoan *Leishmania donovani*, the cause of Kala azar disease in areas from the eastern Mediterranean to west India. Sand flies (*Phlebotomus* spp.) live in jackals' dens in caves and are vectors of *Leishmania* to humans. Dry caves inhabited by birds or bats are dangerous places for humans because they can act as reservoirs of the fungus *Histoplasma capsulatum*. The fungus grows in soil rich with bat or bird droppings (guano) and men were inoculated by inhalation of fungal spores from disturbed guano. The acute phase of this "cave disease" is characterized by nonspecific respiratory symptoms like cough or flu, which cause damage to the retinas of the eyes or affect diverse organ systems, occasionally with fatal results.

The tendency of some species of birds and bats to aggregate in large numbers only in caves makes them very vulnerable. The best example is the white-nose syndrome. This cutaneous fungal infection caused by the fungus *Geomyces destructans* was responsible for massive mortality of hibernating bats in the northeastern United States (Biehert *et al.*, 2009). Since the winter of 2006/2007, over a million bats died: bat declines exceeding 75% have been observed at surveyed hibernacula. In 2009 and 2010 *G. destructans* was for the first time identified in European caves too, but here bats remain healthy.

There is a long list of threats for cave-dwelling birds and mammals today. In combination with threats outside the caves, formerly high regional populations of cave bats are in drastic decline. In Eagle Creek Cave in Arizona, the population has declined from more than 25 million guano bats in 1963 to 30,000 in 1969. Some formerly large colonies of this species in Mexico are now extinct. On the IUCN Red Data List of Threatened Animals, there are 11 species of cave birds (Stattersfield and Capper, 2000), snow leopard, Martino's vole, and many more cave bats, which make up the vast majority of the 370 threatened species of bats (Hutson *et al.*, 2001).

See Also the Following Articles

Bats
Paleontology of Caves

Bibliography

Biehert, D. S., Hicks, A. C., Behr, M., Meleyer, C. U., Berlowski-Zier, B. M., Buckles, E. V., et al. (2009). Bat white-nose syndrome: An emerging fungal pathogen? *Science*, 323(5911), 227.

Brosset, A. (1998). Caves. In C. Juberthie & V. Decu (Eds.), *Encyclopaedia biospeleologica* (Vol. II, pp. 1249–1256). Moulis, France: Société de Biospéologie.

del Hoyo, J. E., Elliot, A., & Sargatal, J. (1999). *Handbook of the birds of the world, Vol. 5, Barn-Owls to Hummingbirds*. Barcelona: Lynx Edicions.

Diedrich, C. G. (2010). Spotted hyena and steppe lion predation behaviours on cave bears of Europe—Late quarternary cave bear extinction as result of predator stress? *Geophysical Research Abstracts*, 12, EGU2010–2111.

Ford, T. D., & Cullingford, C. H. D. (1976). *The science of speleology*. London: Academic Press.

Ginet, R., & Decou, V. (1977). *Initiation a la biologie et a l'écologie souterraines*. Paris: Jean-Pierre Delarge.

Hill, J. E., & Smith, J. D. (1984). *Bats, a natural history*. London: British Museum of Natural History.

Hutson, A. M., Mickleburgh, S. P., & Racey, P. A. (Eds.), (2001). Microchiropteran bats: Global status survey and conservation action plan. Gland, Switzerland: IUCN/SSC Chiroptera Specialist Group.

Hutterer, R., Ivanova, T., Meyer-Cords, C., & Rodriques, L. (2005). *Bat migrations in Europe*. Bonn: Naturschutz und Biologische Vielfalt, Federal Agency for Nature Conservation.

Juberthie, C., Decu, V., & Radulescu, C. (1998). Mammalia (Marsupialia, Insectivora, Artiodactyla, Rodentia et Fissipedia). In C. Juberthie & V. Decu (Eds.), *Encyclopaedia biospeleologica* (Vol. II, pp. 1257–1261). Moulis, France: Société de Biospéologie.

Kryštufek, B., & Bužan, E. V. (2008). Rarity and decline in palaeoendemic Martino's vole *Dinaromys bogdanovi*. *Mammal Revue*, 38, 267–284.

Mohr, C. E., & Poulson, T. L. (1969). *The life of the cave*. New York: McGraw-Hill.

Nowak, R. (1994). *Walker's bats of the world*. Baltimore, MD: The Johns Hopkins University Press.

Papandatou, E., Butlin, R., Pradel, R., & Altringham, J. D. (2009). Sex-specific roost and population dynamics of the vulnerable long-fingered bat, *Myotis capaccinii*. *Biological Conservation*, 142, 280–289.

Stattersfield, A. J., & Capper, D. R. (2000). *Threatened birds of the world*. Barcelona and Cambridge: Lynx Edicions and Bird Life International.

Stiner, M. C. (2004). Comparative ecology and taphonomy of spotted hyenas, humans, and wolves in Pleistocene Italy. *Revue de Paléobiologie*, 23, 771–785.

Vandel, A. (1965). *Biospeleology, the biology of cavernicolous animals*. Oxford: Pergamon Press.

VICARIANCE AND DISPERSALIST BIOGEOGRAPHY

John R. Holsinger

Old Dominion University, Norfolk, Virginia, U.S.A.

OVERVIEW OF BIOGEOGRAPHY

Biogeography is a subdiscipline of biological systematics briefly defined as the science that documents and attempts to explain the geographic distribution of organisms. Two conceptually different approaches have developed over the years. They include historical or large-scale biogeography and ecological or small-scale biogeography. *Historical biogeography* is the oldest and most commonly practiced approach and explains distributional patterns of organisms on the basis of their dispersal and/or vicariance over evolutionary time. It is essentially large scale, inasmuch as it examines the distribution of biotas from a regional, continental, or global perspective. In comparison, *ecological biogeography*, sometimes called *deterministic biogeography*, explains distribution patterns on the basis of relatively short-term, ecological effects. It is generally regarded as small-scale biogeography because it focuses on the effects of environmental differences that have developed over ecological time. An example of the latter would be the Case and Cody (1987) application of island biogeography theory to explain distributional patterns on islands in the Sea of Cortez in Mexico.

Historical biogeography had its beginnings in Europe in the eighteenth century, with the early, largely botanical, works of Linnaeus, Buffon, Forster, Humboldt, and de Candolle (Lomolino et al., 2005). Modern historical biogeography began in the following century and owes many of its basic principles to Alfred Russel Wallace, a colleague of Charles Darwin, who is often called the "father of zoogeography."

As a science, historical biogeography is very different now from what it was in the 1800s or, for that matter, in the first six or seven decades of the twentieth century. Significant changes came in the 1960s and 1970s with wide acceptance of the theory of plate tectonics and continental drift, coupled with the rapidly developing popularity and utilization of phylogenetic analysis or cladistics. With subsequent application of these exciting new theories and concepts to the interpretation of distribution patterns, historical biogeography has become more focused and analytical. Some workers now refer to it as *cladistic biogeography* (Humphries and Parenti, 1999). Prior to the 1960s, biogeographic discussions consisted largely of lengthy narratives and scenarios, sometimes highly subjective, to explain distribution patterns in terms of centers of origin and outward dispersal. Dispersalist models identified centers of origin and then speculated on the subsequent peripheral movement of organisms across potential dispersal barriers, such as over imaginary land bridges. In retrospect, many of these barriers appear to have been insurmountable.

The application of cladistics to biogeography has placed significant emphasis on construction of biological-area cladograms, combined with subsequent examination of relationships between terminal taxa and areas of endemism (Fig. 1). In these analyses, congruence between area cladograms of different monophyletic taxa suggests that these organisms share a common evolutionary history. This procedure has led ultimately to the acceptance of vicariance as an alternative hypothesis to explain heretofore many otherwise inexplicable distribution patterns. However, this is not to say that dispersalist models have diminished in importance. On the contrary, dispersal is still considered very important in explaining distributions. But in recent studies, centers of origin are rarely evoked, and dispersal often provides an alternative hypothesis rather than the sole explanation for a given pattern of distribution (see Wiley, 1988). Another analytical method called *panbiogeography* (Craw et al., 1999), which places greater emphasis on congruent distribution tracks (*i.e.*, generalized tracts) than area cladograms, has also gained popularity as an analytical methodology. When used in combination with phylogenetics, panbiogeography has proven effective for explaining distribution patterns of subterranean animals (Christiansen and Culver, 1987).

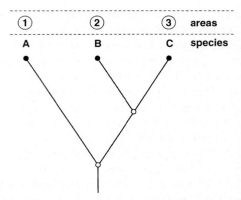

FIGURE 1 Example of biological or taxon area cladograms for a hypothetical genus composed of species A, B, and C that occur in discontinuous areas 1, 2, and 3, respectively. These cladograms are produced by combining the name of the species with the area(s) it inhabits. In this simple example, each species is endemic (restricted) to a different geographic area, suggesting a strong relationship between species distributions and geographic isolation. Open circles are internal nodes, which indicate splits in lineages leading to terminal species. Internal nodes can also correspond to vicariant events.

BIOGEOGRAPHY OF CAVE ANIMALS

To date, between 10,000 and 15,000 species of animals have been described from caves and similar subterranean habitats throughout the world. However, an exact count is not available and would be meaningless nevertheless because new species from subterranean environments are being discovered and described on a regular basis and the number continues to increase. A majority of these species typically lack eyes and pigmentation and are said to be troglobites (troglomorphic) if living in terrestrial environments or stygobites (stygomorphic) if living in aquatic ones. In either event these organisms are apparently well adapted for life underground. From a biogeographic perspective, subterranean environments typically include cave passages in limestones and other soluble carbonate rocks, lava tubes in basalt, and interstitial spaces in a variety of unconsolidated sediments. However, virtually any habitable space beneath the surface of the Earth can be a subterranean environment.

Troglomorphic and stygomorphic animals make excellent candidates for biogeographic studies and are increasingly studied in this research for five reasons. First, they are taxonomically diverse and widespread geographically. Second, they are becoming well known taxonomically, largely through the diligent collecting efforts and taxonomic research of the past 50 years, coupled in the past decade with phylogenetic analyses. Third, dispersal abilities, although undoubtedly variable, are generally limited because these organisms are adapted to the restricted confines of subterranean habitats. Fourth, ranges are commonly, but not invariably, narrowly circumscribed, resulting in numerous locally endemic species. Fifth, many taxa apparently represent old phylogenetic lineages that have persisted in subterranean refugia for long periods of time, and both aquatic and terrestrial taxa are represented by distributional and phylogenetic relicts in temperate and tropical karst regions. What roles have vicariance and dispersal played in shaping the distribution patterns of subterranean fauna? Is one of these processes more important than the other in explaining distributions, or are they approximately equal in explanatory value? *Dispersal* is defined as the movement of an organism from one area to another, which changes the natural distribution of that organism, whereas vicariance is defined as the occurrence of closely related taxa in disjunct areas, which have been separated by the development of a natural barrier (Humphries and Parenti, 1999). The formation of a natural barrier that results in splitting a formerly continuous range is a *vicariant event*. As pointed out below, both dispersal and vicariance have played significant roles in the development of cave faunal distribution patterns, and constructing both dispersalist and vicariance models is necessary for deeper insight into the complexity of these patterns. Therefore, because most present-day ranges probably reflect aspects of both dispersal and vicariance, teasing out these two components and delineating their contribution to the formation of distribution patterns is one of the principal tasks of large-scale cave biogeography.

Dispersal and vicariance biogeography of terrestrial troglobites and aquatic stygobites are considered separately below, primarily because of important differences between the two with respect to origins, habitats, and dispersal potentials. Although most of the examples used here to illustrate various distribution patterns are species from North America, many comparable examples could be cited from Europe and other parts of the world where subterranean fauna have also been studied in detail.

TERRESTRIAL TROGLOBITES

Two levels of dispersal have played roles in shaping the distribution patterns of troglobites. The first level involves the actual invasion of caves from the surface by putative preadapted founder populations. This may occur under constraint when climatic changes precipitate invasion, which apparently occurred in temperate karst during climatic vicissitudes of the Pleistocene. It may also occur when founders invade newly developing subterranean food niches, which is apparently common and ongoing in tropical karst. These two models for invasion and colonization are termed *climatic relict* and *adaptive shift*, respectively. They were reviewed recently in detail (Holsinger, 2000) and both are largely dispersalist in character because they involve the active movement of an organism from one environment to another, effectively changing its natural distribution. However, because climatic change envisioned under the climatic relict model would likely preclude reinvasion of surface habitats and isolate founders in caves, it might arguably be regarded a vicariant event.

The second level of dispersal involves movement of troglobitic organisms within the subterranean realm, typically, but not invariably, between caves in karst areas or lava flows. Both the extent of a given karst area or lava flow and the dispersal vagility (mobility) and niche breadth of a given species determine potential limits of dispersal. For many troglobites this potential is greatly limited and species ranges are often determined by the extent of interconnected cave passages and solution channels within a contiguous karst area. In a biogeographic sense, contiguous karst areas are analogous to islands, inasmuch as they are physically isolated from each other by noncavernous rocks, and often contain locally endemic cave species.

Dispersal of terrestrial troglobites includes movement through larger cave passages (>20 cm) and numerous smaller mesocaverns (<20 cm) within karstic limestone or other soluble carbonate rocks or basaltic lava. It may also include movement through shallow underground compartments (=*milieu souterrain superficiel* or *MSS*), which occur in some areas in loose rocks at the base of scree slopes or in cracks and fissures in mantle rock just beneath the lower layer of soil, both within and outside karst areas (Juberthie and Decu, 1994). Moreover, it is likely that some of the more widely distributed troglobites, such as species of linyphiid spiders and certain collembolan and dipluran insects in eastern North America, commonly disperse through cracks and crevices in epikarstic habitats above caves and even occasionally through damp leaf litter or deep soil outside karst areas. At the opposite end of this spectrum are troglobitic pseudoscorpions of the genus *Kleptochthonius* and pselaphid beetles, of which a majority of species are highly localized endemics restricted to single caves (Barr and Holsinger, 1985). Presumably the low vagility and limited niche breadth of these species prevent them from dispersing far from very small areas or even a single cave habitat.

Trechine beetle species of the genus *Pseudanophthalmus* (family Carabidae) in eastern North America are a classic example of well-studied troglobites, whose ranges are often largely determined by the dispersal limits of contiguous karst (Barr and Holsinger, 1985). Many species appear to be limited in their distribution to areas that typically encompass only interconnected caves in a contiguous karst area. The effect on dispersal by the extent of contiguous karst is nicely illustrated by the distribution of species of

Pseudanophthalmus in the Appalachian Valley and Ridge physiographic province, where ranges vary in extent from a single cave to many caves that occur over a linear distance of approximately 95 km. In the Shenandoah Valley of northwestern Virginia, where belts of cavernous limestones are strongly folded, discontinuous, and separated from each other by insoluble sandstones and shales, four closely similar species of the monophyletic *hubbardi* group occur in caves within relatively short distances of each other (Fig. 2). Extensive searching in caves of this region over the years suggests that the range of each species is restricted to an isolated belt of cavernous limestone, which for all intents and purposes constitutes a small contiguous karst area. Further evidence of the extreme isolation of these species is that three of the four species are recorded from single caves. Elsewhere in the Appalachian Valley and Ridge, belts of cavernous limestone are more extensive and continuous over large areas, and several species of *Pseudanophthalmus* have much larger ranges that cover linear distances of 50–95 km and include many caves (Fig. 3). In the Mississippian plateaus of Indiana, Kentucky, and Tennessee, where limestone outcrops are typically broader and less disturbed by folding than in the Appalachian Valley and Ridge to the east, ranges of trechine beetles are often more extensive (Barr and Holsinger, 1985) and provide further evidence that ranges of these troglobites are generally defined by the limits of contiguous karst.

The effect of vicariance on the distribution patterns of troglobites is more difficult to document than that of dispersal, primarily because barriers to subterranean dispersal develop gradually over extended periods of time, making it difficult to assess their impact. However, vicariant events are probably relatively common in the subterranean realm and thus are significant factors in isolating cave populations and promoting speciation. Physical barriers to dispersal may develop externally from the surface

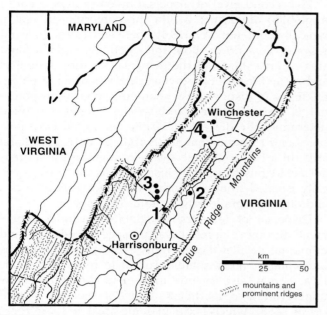

FIGURE 2 Distribution of four troglobitic beetle species (numbers 1–4) of the *hubbardi* group of *Pseudanophthalmus* in caves of the northern Shenandoah Valley of Virginia. Cave localities are indicated by solid circles. Each species inhabits a cave or small cluster of caves in a physically isolated limestone belt (contiguous karst).

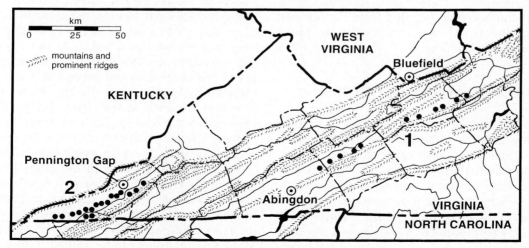

FIGURE 3 Distributions of the troglobitic beetles *Pseudanophthalamus hoffmani* (1) and *P. delicatus* (2) in southwestern Virginia karst areas. Cave locations are indicated by solid circles. Both species occur in continuous belts of cavernous middle Ordovician limestone. The former is recorded from caves developed in a long, relatively narrow belt of moderately dipping limestone along the northwestern flank of Big Walker Mountain. The latter is recorded from caves that are developed in a relatively broad exposure of low-dipping limestone, which extends throughout much of the Powell Valley in central Lee County, Virginia.

though sinkhole collapse into underlying cave passages or by deep erosional gorges that eventually completely bisect subterranean channels. Physical barriers may also develop internally and destroy or close off subterranean dispersal corridors by filling them with silt, flowstone deposits, or rubble from a ceiling collapse. Moreover, physical barriers may develop vertically as well as laterally, inasmuch as new generations of caves are actively forming beneath old ones. Thus, it is assumed that animals will move downward to colonize newly developing living space and available niches, and that the progressive isolation of populations in newly developing, lower-level passages may result from this activity.

AQUATIC STYGOBITES (STYGOFAUNA)

The subterranean aquatic environment in general is more diverse and interconnected than its terrestrial counterpart, and the extent to which groundwater is continuous over wide areas plays a significant role in the dispersal of stygobites. It is generally accepted that the pervasive nature of subterranean groundwaters provides stygobites, on average, greater dispersal potential and therefore wider ranges than their terrestrial troglobite counterparts. Available groundwater habitats occur both within and outside karst regions, and many areas far beyond the boundaries of karst terranes are inhabited by an interesting diversity of stygobites (Ward *et al.*, 2000). However, it is not uncommon to find the same species, or closely related species, simultaneously living in caves of a karst area and adjacent subterranean aquifers in a noncavernous area. As pointed out below, dispersal between karst areas is often facilitated by movement through the interstitial spaces of shallow groundwater aquifers that are not directly associated with the caves in karst.

Subterranean groundwater habitats in karst caves typically include streams, drip and seep-fed pools, and phreatic lakes. They are often more accessible and easier to observe than those outside karst areas. In addition to typical karst caves with freshwater habitats, anchihaline caves are a special cave subset that develop in limestones or basalt in coastal areas near the sea and are connected to nearby marine waters by subterranean channels (see Iliffe, 2000). As one might expect, the water in anchihaline caves varies from nearly fresh to fully marine and, depending on depth, may be horizontally partitioned by a halocline. With some notable exceptions, anchihaline caves are inhabited by stygofaunas having morphological affinities with marine faunas living in the nearby sea.

Outside karst areas, the more common subterranean groundwater habitats include the saturated interstitial spaces of gravels and coarse sands situated beneath stream beds (hyporheic) or in coral rubble, coarse sand, and so on, in littoral and sublittoral zones (macroporous sediments), unconsolidated sediments beneath the water table (*nappes phréatiques*), and the outflow of small springs and seeps (through fine sediments or leaf litter) from perched water tables (hypotelminorheic). The movement of stygobites through the saturated media of unconsolidated sediments is often termed *interstitial dispersal* and is apparently the principal reason why many species that live in the shallow groundwater aquifers of unconsolidated sediments are typically widely distributed. For many species, interstitial dispersal appears to be a very successful mechanism for range expansion, and probably accounts for the relatively extensive ranges of some of the common and widespread stygobitic amphipods in the genus *Stygobromus* that live in shallow groundwater aquifers along the outer margin of the Piedmont and on the Coastal Plain of eastern North America. These species, along with those in other crustacean groups (*e.g.*, bathynellids, copepods, isopods), are excellent examples of stygobites that apparently rely exclusively on interstitial dispersal to maintain their distributions.

The distribution patterns of stygofaunas suggest that dispersal is often not limited to a single type of groundwater habitat but more frequently involves two or more such habitats. For example, an important aspect of cave drip pools is their close relationship with epikarstic habitats that occur in open spaces and bedrock fractures between cave passages and the Earth's surface. Aquatic epikarstic habitats are in the form of perched aquifers or perched water tables situated well above the true groundwater table. A number of studies have demonstrated passive dispersal of small (<10 mm in length) stygobitic species (*e.g.*, planarians, amphipods, and isopods) from epikarstic aquifers via ceiling drips into shallow cave pools on the floor below. Passive dispersal may also occur when the water in ceiling drips or wall seeps passes into a cave stream. There is little doubt that epikarstic aquifers facilitate dispersal of small stygobitic crustaceans both within and between caves.

A second example of dispersal involving more than one cave habitat is illustrated by species that move actively or passively between pools and small streams. Studies on Appalachian cave amphipods in the genus *Stygobromus* indicate that, in general, the larger, more widely distributed species are recorded from both pools and streams, whereas in contrast, smaller species with greatly delimited ranges are generally found only in tiny drip pools. Finally, it should be noted that some species living in cave streams may be passively washed downstream and into springs at the surface, where a few can apparently survive for long periods of time with adequate cover. It is conceivable that the outwash from springs might occasionally facilitate limited passive dispersal through surface water between caves.

Larger stygobites, such as crayfish, shrimps, salamanders, and fishes, and amphipods and isopods exceeding 20 mm in length, are rarely if ever encountered in cave drip pools but are commonly found in cave streams or lakes. In the relatively shallow, wet caves of the Florida lime sink region, decapods (primarily crayfish) also commonly inhabit submerged cave passages. Because of the larger size of cave crayfish, which excludes them from interstitial dispersal, their distribution patterns are usually defined by the limits of contiguous karst.

Dispersal over relatively great distances through deep phreatic water in karst areas is another dispersalist mechanism available to some stygobites, although it is probably utilized primarily by taxa that inhabit deep groundwater aquifers. Recent investigations suggest that deep phreatic water habitats are more common and interconnected than previously believed, but they remain relatively poorly known because of limited accessibility to cave biologists. Most observations have been made where lower level cave passages allow access to bodies of phreatic water or through water wells in limestone bedrock that access deep groundwater aquifers. Recent observations on amblyopsid cavefishes in the Ozarks and cirolanid isopod crustaceans in the Appalachians provide strong evidence that deep phreatic aquifers can serve as important avenues for dispersal. According to a study in southern Missouri by Noltie and Wicks (2001), the principal habitats of the cavefishes *Amblyopsis rosae* in the Springfield Plateau and *Typhlichthys subterraneus* in the Salem Plateau are numerous small channels, solution tubes, and vugs beneath the groundwater table in these karst regions. Interestingly, because a majority of these organisms live in areas that are inaccessible to cave biologists, their population size has been underestimated. Lateral movements in the phreatic zone beneath the water table facilitate dispersal of the fishes over a relatively wide range. However, further observations on *A. rosae* indicate the existence of four semi-isolated genetic groupings within its overall range, suggesting a limited degree of physical impediment to dispersal and gene flow. Dispersal through deep phreatic water is probably also responsible for the extensive distribution of *Typhlichthys subterraneus* east of the Mississippi River in caves of the Interior Low Plateaus in southern Kentucky, central Tennessee, and northern Alabama. However, there is some evidence in this region for differentiation into several local "biological species" separated by extrinsic dispersal barriers (Barr and Holsinger, 1985).

Biogeographic studies by Holsinger *et al.* (1994) and Hutchins *et al.* (2010) on the stygobitic cirolanid isopod crustacean *Antrolana lira*, currently recorded from phreatic water habitats in 17 localities (12 caves and 5 water wells) in the Shenandoah Valley karst region of northwestern Virginia and eastern West Virginia (Fig. 4), suggest that this species, which may reach 21 mm in

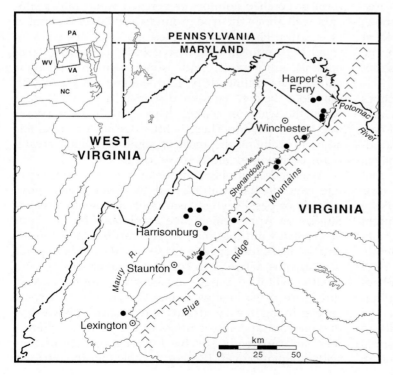

FIGURE 4 Distribution of the stygobitic cirolanid isopod *Antrolana lira* in the Shenandoah Valley of the Virginias. Solid circles represent collecting sites (mostly caves except for five water wells and one deep pit in the northern part of the range). Note that the range of this species extends beneath a major surface drainage divide between Staunton and Lexington, Virginia. The question mark indicates a sight record for specimens pumped from a water well.

length, has utilized deep phreatic water for dispersal. But unlike the near flat-lying strata in the Missouri plateaus, the limestone layers in this region are steeply folded and separated from each other by nonsoluble layers of sandstone and shale. However, fracturing of the sandstones and shales makes it likely that the limestone beds are not fully isolated from each other at variable depths below the water table and that minimal lateral movement of water and isopods has occurred across the insoluble beds. Therefore, tenuous gene flow has conceivably occurred throughout the range of this species in the past and probably accounts for the morphologically identical populations. However, a recent genetic study by Hutchins *et al.* (2010) yielded 14 haplotypes, and a phylogenetic analysis of the haplotypes revealed the presence of three distinct clades. The clades, in turn, correspond to three geographic groups throughout the presently known range of the species.

Dispersal through deep phreatic water also appears to be utilized by some of the species that inhabit the Edwards Aquifer in south-central Texas. This deep, extensive aquifer is associated with the Balcones Fault Zone and forms a biological mixing zone where karst groundwater of the Edwards Plateau comes into contact with interstitial groundwater of the Coastal Plain. The intense faulting and fracturing of the bedrock in this region has also resulted in compartmentalization and development of phreatic subaquifers within the main aquifer. Thus, although dispersal obviously occurs freely in some parts of the aquifer, barriers prevent it in other parts, and it is probable that a combination of these two effects has contributed significantly to the evolution of the highly diverse stygofauna of the region. More than 31 stygobitic species have been sampled from the artesian well in San Marcos, where specimens are coming from a depth of approximately 60 m. Sampling has also been carried out from very deep artesian wells located approximately 90 km southwest of San Marcos near San Antonio. Deep springs and caves have also been carefully investigated in this region. That part of the aquifer extending from San Marcos southwest to San Antonio appears to be a viable dispersal corridor for a number of stygobites. Four of the 10 stygobitic amphipod species recorded from the artesian well in San Marcos have also been flushed out of two of the artesian wells near San Antonio, and a fifth species has been found in phreatic waters of two nearby caves and possibly additional caves west of San Marcos. The cirolanid isopod *Cirolanides texensis* is also recorded from the San Marcos artesian well. Like the amphipods, it has been sampled from deep artesian wells near San Antonio as well as from caves to the north and west on the Edwards Plateau.

As mentioned above, dispersal between karst areas or even between caves is often facilitated by movement through shallow groundwater aquifers that may or may not be developed in limestone terranes. Thus, dispersal through these aquifers, which can occur in a variety of terranes, appears to be a very common means of range expansion for relatively widely distributed stygobites. Studies on stenasellid isopod crustaceans in groundwaters on the Arabian Peninsula by Magniez and Stock (1999) illustrate how interstitial groundwater in the gravel sediments of a hyporheic zone and the "free" water in nearby limestone cavities can be linked through an ecotone, which is cohabitated by representatives of species from the different habitats. Many species of isopods and amphipods occur in caves and to a lesser extent in springs and seeps at the surface and they apparently utilize interstitial dispersal to broaden their ranges. The amphipod crustaceans *Crangonyx antennatus* and *C. packardi* and the isopod *Caecidotea pricei* are three good examples of North America stygobitic crustaceans that frequently inhabit drip pools and small streams in caves but utilize dispersal through epikarstic aquifers and hypotelminorheic habitats (seeps) outside karst terranes to maintain extensive ranges. The geographic distribution of *C. antennatus*, which extends for approximately 700 km through a number of different karst areas in parts of several states in the Appalachians, is shown on the map in Figure 5. The range of *Crangonyx packardi* extends from southern Indiana and central Kentucky west to eastern Kansas over a linear distance of approximately 1000 km, whereas that of *Caecidoeta pricei* extends from Rockbridge County, Virginia, northeast to Montgomery County, Pennsylvania, covering a linear distance of approximately 450 km.

Most of the same vicariant events that form physical barriers to the dispersal of troglobites mentioned above apply equally to stygobites. For example, sinkhole collapse into underlying cave passages and bisection of subterranean channels by deep erosional surface gorges would both serve to effectively block potential stygofaunal dispersal corridors. Moreover, siltation, flowstone deposition, and ceiling collapse can also destroy or close off aquatic dispersal corridors.

However, three vicariant phenomena are unique to the geographic distribution of stygofaunas: stranding, stream capture, and spring failure. In a cladistic biogeographic analysis, any one of these events might coincide with the nodes on the area cladogram of the taxon being investigated (Fig. 1). The nodes, in turn, reflect splits in the lineages that gave rise to terminal taxa. The possibilities for applying this kind of analysis to the interpretation of biogeographic patterns of various groups of stygobites is exciting and will increase as more phylogenies of these organisms become available.

Stranding involves the isolation of marine or brackish water organisms in gradually freshening

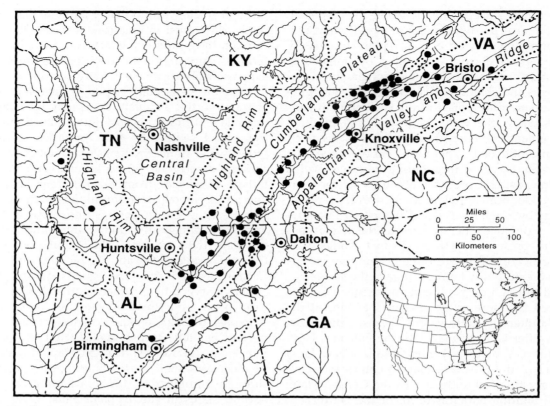

FIGURE 5 Distribution of the stygobitic amphipod crustacean *Crangonyx antennatus* in the southern Appalachians. The solid circles represent one to ten closely proximate localities. Although most of these localities are caves, the species has also been observed to enter cave pools through ceiling drips from epikarstic zones and occasionally has been found in seeps and small springs at the surface.

subterranean groundwaters following marine regressions (see Parenti and Ebach, 2009). Founding populations adapt to freshwater conditions as the salinity gradually decreases in concert with regression of marine waters. This process is gradual, but it is vicariant because ultimately the stranded populations become isolated from their marine ancestors when ecological conditions are changed and genetic continuity is eliminated. The evidence for stranding comes from the large number of freshwater stygobitic crustaceans that inhabit subterranean groundwaters in parts of the world that once bordered on or were covered by the putative warm, shallow waters of the Tethys Sea. Most of these taxa have obvious taxonomic affinities with modern marine species and are apparently relicts of former marine embayments and island emergence that occurred in different parts of the Tethyan realm at different times between the Cretaceous and Middle to Late Tertiary. Marine relict stygobites are especially common in Texas, Mexico, and the West Indies in southern North America, and the Mediterranean region of southern Europe and northern Africa. They are also recorded from Eastern Africa, the Arabian Peninsula, and parts of the western Pacific region including Australia.

Stream capture involves the piracy or capture of one stream by another, and in karst areas it includes diversion of surface streams through sinkholes into subterranean channels. Initially, preadapted freshwater organisms enter subterranean waters passively when their surface habitat stream is diverted into a subterranean channel. Founder populations become permanently established in caves only after they are isolated in underground streams and contact is cut off with surface ancestors. The capture of surface streams by sinkholes is a relatively common phenomenon in karst areas and is vicariant when remnant surface streams recede up-gorge or dry up completely. In accordance with this model, ancestors of cave-adapted populations of the widely distributed characid fish *Astyanax fasciatus* invaded caves in east-central Mexico when surface streams were captured by sinkholes in karst areas and were unable to reestablish contact with epigean forms. Similarly, stygobitic crayfish in north-central Florida are apparently derived from surface ancestors that gained entry into subterranean waters via stream capture in karst areas, presumably becoming isolated underground by the same processes.

Spring failure involves the retreat of preadapted surface populations into cave streams from surface springs during episodic droughts. Subsequent isolation of founder populations in underground streams is vicariant when the springs are eliminated by erosional processes or when stream flow to the surface ceases due to lowering of the groundwater table. Spring failure is a fairly common phenomenon in karst areas where the climate is becoming progressively more arid. Apparently, it has been a significant factor in the evolution of troglomorphic salamanders in central Texas, which are derived from spring-dwelling ancestors that migrated into cave streams to survive severe surface droughts since the Pleistocene. Spring failure may also be implicated in the evolution of cave-adapted hydrobiid snails in parts of North America.

A final note regarding vicariance includes a brief discussion of groundwater calcrete aquifers, which constitute a unique subterranean environment recently identified from the Western Australian arid zone by Humphreys (2001). Calcretes are isolated subterranean carbonate deposits with cavelike, freshwater habitats. They have developed under arid conditions by evaporation and became isolated in the upper tributaries of paleodrainage systems by the progressive upstream movement of salinity, perhaps beginning in the Eocene. Calcretes are numerous and inhabited by a stygofauna consisting predominantly of locally endemic crustaceans, many of which are relict species. The intervals of hypersalinity/calcrete discontinuities have acted as barriers to the dispersal of the stygofauna, and long-term isolation of crustacean populations in upstream pockets of freshwater has resulted in the evolution of a remarkably diverse regional stygofauna. In a biogeographic sense, calcretes are discontinuous groundwater habitats, which, like isolated belts of contiguous karst, are analogous to islands.

CONCLUSION

The biogeography of organisms living in subterranean environments is primarily concerned with explanation and interpretation of the distribution patterns of troglomorphic species. Both vicariance and dispersal have played important roles in shaping the distribution patterns of species restricted to a subterranean environment, which include terrestrial troglobites and aquatic stygobites. Whereas vicariance delimits ranges through the development of dispersal barriers, dispersal increases ranges by movement of organisms into new areas. The effect of these processes has been different for troglobites and stygobites, largely because the two groups inhabit ecologically different subterranean habitats and have different dispersal potentials. In general, subterranean aquatic environments are more pervasive and diversified than their terrestrial counterparts and, as a result, stygobites typically inhabit a wider variety of habitats and have more extensive ranges. Stygobites not only exploit karst groundwaters but also inhabit the extensive interstitial groundwater media that extends far beyond the boundaries of karst terranes. Some karst species apparently also use deep phreatic aquifers for dispersal. Troglobites may also exploit terrestrial habitats outside karst, such as "shallow underground compartments," but these are generally limited in scope, and the ranges of a majority of terrestrial species are restricted to caves in karst or basaltic lava.

Although several forms of vicariance, such as the destructive geomorphic processes that close subterranean passages, are similar for both troglobites and stygobites and often establish the outer limits of distribution in karst areas, several important vicariant events are unique to stygobites. They include stranding, stream capture, and spring failure, and each has played a significant role in the distribution of stygobites. The newly discovered groundwater calcretes in Western Australia are also vicariant in character, inasmuch as they restrict the distribution of stygobitic crustaceans in this arid region.

Acknowledgments

I am grateful to David C. Culver, Daniel W. Fong, David A. Hubbard, Jr., Glenn Longley, and Wil Orndorff for providing useful information during the preparation of this article; Donald K. Emminger and Justin Shafer for preparation of the figures; and Lynnette J.M. Ansell for editorial assistance.

See Also the Following Articles

Adaptive Shifts
Invasion, Active versus Passive

Bibliography

Barr, T. C., Jr., & Holsinger, J. R. (1985). Speciation in cave faunas. *Annual Reviews of Ecology Systematics, 16*, 313–337.

Case, T. J., & Cody, M. L. (1987). Testing theories of island biogeography. *American Scientist, 75*(4), 402–411.

Christiansen, K. A., & Culver, D. C. (1987). Biogeography and the distribution of cave collembola. *Journal of Biogeography, 14*, 459–477.

Craw, R. C., Grehan, J. R., & Heads, M. J. (1999). *Panbiogeography: Tracking the history of life*. Oxford, U.K.: Oxford University Press.

Holsinger, J. R. (2000). Ecological derivation, colonization, and speciation. In H. Wilkins, D. C. Culver & W. F. Humphreys (Eds.), *Subterranean Ecosystems* (pp. 399–415). Amsterdam: Elsevier.

Holsinger, J. R., Bowman, T. W., & Hubbard, D. A., Jr. (1994). Biogeographic and ecological implications of newly discovered populations of the stygobiont isopod crustacean *Antrolana lira* Bowman (Cirolanidae). *Journal of Natural History, 28*, 1047–1058.

Humphreys, W. F. (2001). Groundwater calcrete aquifers in the Australian arid zone: The context to an unfolding plethora of stygal biodiversity. *Records of the Western Australia Museum, Supplement No. 64*, 63–83.

Humphries, C. J., & Parenti, L. R. (1999). *Cladistic biogeography.* Oxford, U.K.: Oxford University Press.

Iliffe, T. M. (2000). Anchialine cave ecology. In H. Wilkins, D. C. Culver & W. F. Humphreys (Eds.), *Subterranean ecosystems* (pp. 59–76). Amsterdam: Elsevier.

Juberthie, C., & Decu, V. (1994). Structure et diversite du domaine souterrain; particularites des habitats et adaptations des especes. In C. Juberthie & V. Decu (Eds.), *Encyclopaedia biospeologica* (Vol. I, pp. 5–22). Moulis, France: Société de Biospéologie (in French).

Lomolino, M. V., Riddle, B. R., & Brown, J. H. (2005). *Biogeography.* Sunderland, MA: Sinauer Associates.

Magniez, G., & Stock, J. H. (1999). Consequence of the discovery of Stenasellus (Crustacea, Isopoda, Asellota) in the underground waters of Oman (Arabian Peninsula). *Contributions to Zoology, 68* (3), 173–179.

Noltie, D. B., & Wicks, C. M. (2001). How hydrology has shaped the ecology of Missouri's Ozark cavefish, *Amblyopsis rosae*, and southern cavefish, *Typhlichthys subterraneus*: Insights on the sightless from understanding the underground. In A. Romero (Ed.), *Environmental biology of fishes* (Vol. 62, pp. 171–194). Dordrecht: Kluwer Academic Publishers.

Parenti, L. R., & Ebach, M. C. (2009). *Comparative biogeography: Discovering and classifying biogeographical patterns of a dynamic Earth.* Berkeley, CA: University of California Press.

Ward, J. V., Malard, F., Stanford, J. A., & Gonser, T. (2000). Interstitial aquatic fauna of shallow unconsolidated sediments, particularly hyporheic biotopes. In H. Wilkins, D. C. Culver & W. F. Humphreys (Eds.), *Subterranean ecosystems* (pp. 41–58). Amsterdam: Elsevier.

Wiley, E. O. (1988). Vicariance biogeography. *Annual Reviews of Ecology Systematics, 19*, 513–542.

VJETRENICA CAVE, BOSNIA AND HERZEGOVINA

Ivo Lučić

Speleological Association Vjetrenica–Popovo Polje, Bosnia and Herzegovina

For centuries, Vjetrenica was perceived as a magic hole in the ground, with a freezing breeze coming out even during the hottest days of summer. The wind always came with sounds that aptly illustrated its supernatural force. This marked the colorful history of Vjetrenica and its rich scientific history—it also gave it the name (*vjetar*—wind in South Slavic; *vjetrenica*—a windy place). It is the largest cave in Bosnia and Herzegovina, of total length of 7014 m, with a rich and versatile karst inventory. It is one of the two leading world hotspots of underground fauna, a very specific paleontological site, a prominent shrine, and a protected site of natural heritage. It is located in the far south of the country, specifically at the southwestern edge of Popovo Polje, in the vicinity of the village of Zavala (at 42°50′45″ S, 17°59′02″ I, 268 m above sea level). It is 10.1 km from the nearest town on the Adriatic Coast, Slano, Republic of Croatia.

THE KARST AREA OF VJETRENICA

It comprises the end section of the basin of the River Trebišnjica, with total estimated surface of 2800 km^2, and an altitude range from the top of Mt. Lebršnik (1985 m) to sea level. The average annual precipitation in the basin is 1780 mm, of which only 40% can be traced through surface flows. Trebišnjica runs through several types of karst *poljes* (Gatačko p., Cerničko p., Fatničko p., Popovo p., Mokro p., Hutovo blato) and receives water from rivers from other *poljes*: Brova (Ljubomorsko p.) and Bukov potok (Ljubinsko p.); it resurfaces several times at powerful springs: Obod (max. 35 m^3 s^{-1}), springs of Trebišnjica (max. >800 m^3 s^{-1}), Ombla (max. 112 m^3 s^{-1}), and several others; it runs through large capacity *ponors*: Srđevići (60 m^3 s^{-1}), Pasmica (25 m^3 s^{-1}), Tučevac (>20 m^3 s^{-1}), Doljašnica (55 m^3 s^{-1}), and several more; it changes the name from the first spring of Dobra voda, to Vrba, then Mušnica, Ključka Rijeka, and Obod, to Trebišnjica, Ombla, and Krupa; and ends partly in a series of submarine springs (some 35 have been reported in a narrow coastal strip off the coast near Vjetrenica, at the site of Doli). The most prominent part of the network of flows is the final surface stretch named Trebišnjica, of total natural length of 94.5 km, and with a subterranean section with a total areal distance of almost 34 km from the point of sinking to the point of resurfacing (Srđevića ponor and Dejanova pećina Cave). Popovo Polje is the final stretch of the blind valley of Trebišnjica, which is, at 68 km^2, some 40 times smaller than the actual basin. In a matter of days, it would turn into a lake with depths reaching 40 m and with water staying there for an average of 248 days, eventually exiting in several directions, via a web of subterranean flows of maximum total capacity of 250 m^3 s^{-1}, including well-developed examples of bifurcations and intersecting subterranean flows. For that reason, for most of the twentieth century, Vjetrenica was explored as the key to understanding the peculiar functioning of the karst, not only in terms of how its subterranean areas relate to the surface topography, but also as an important drainage canal of the Trebišnjica basin and Popovo Polje.

SPELEOMORPHOLOGY OF VJETRENICA

The cave is mildly forked, with the Main Channel (Glavni kanal; see Fig. 1, item 1) as the central level, angling horizontally toward the southeast and south, of total length of 2465.9 m. The Main Channel is spacious, with width varying from 5 m to an extreme width of 25 m (the average width is 9.6 m), with an inclination of 5.6°, and with several pronounced chambers. Exceptions include the first section of 160 m, with a lower ceiling, usually of average human height, and several narrow

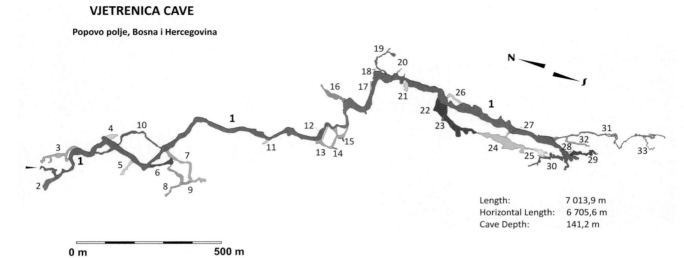

FIGURE 1 An overview of Vjetrenica with key sites. *Map by Darko Bakšić.*

stretches inside the cave. Those are the Siphon (Sifon, 1007.4 m from the entrance; see Fig. 1, item 11), a lowered area covered in water for most of the year, a sediment deposit at the entrance to the Cvijić Chamber (Cvijićeva dvorana, 1755.4 m from the entrance), as well as a collapse in the Cvijić Chamber (1919.3 m from the entrance). The last section is crossed diagonally by the Main Channel, with a recently discovered narrow passage through the blocks named Ass (Prkno—a traditional name for narrow passages in estavelles used by visitors seeking potable water; see Fig. 1, item 26), which is why the rest of the Main Channel is called the Hidden Main Channel (Skriveni glavni kanal, see Fig. 1, item 27). In addition to the Cvijić Chamber (see Fig. 1, item 22), there are also the Golden Chamber (Zlatna dvorana, 363.2 m from the entrance; see Fig. 1, item 4, Fig. 4), the Deep Chamber (Duboka dvorana, 565.4 m from the entrance; see Fig. 1, item 6), the Tall Chamber (Visoka dvorana, 1506.5 m from the entrance; item 16 in Fig. 1), and the Welsh Channel (Velški kanal, 2133.3 m from the entrance; items 24–25 in Fig. 1).

The floor of the Main Channel is covered with blocks reaching as far as 650 m from the entrance, occasionally masked by spacious sequences of calcite flowstones and rimstone pools, and the rest of the channel is covered with sediments of clay, loam, and occasionally sand, with stone blocks fitted over them in several areas. The actual width and height of the channels in Vjetrenica reach almost 90 m. The Hidden Main Channel is almost entirely closed off by a collection of breakdown blocks and is the actual end of Vjetrenica, as it is known today.

Different channels rise from the Main Channel. Most of them are impassable and insufficiently explored vertical channels, chimneys, connecting the Main Channel with the topographic surface above. Their actual connection with the surface has not been demonstrated, although there are several pits and caves on the surface, and the places where some of them connect with the Main Channel are marked by large calcite flowstones, some of them exceeding several tens of meters. The largest and best explored channels above the level of the Main Channel are the Welsh Channel and the Leopard's Channel (Leopardov kanal; item 19 in Fig. 1), the Ravno Channel (Ravanjski kanal; item 31 on Fig. 1) and the Clay Figurines Channel (Kanal glinenih figurica; 2410 m from the entrance; item 30 on Fig. 1), the last three complex expansions and vistas, attractive beds of permanent streams, two of which flow toward the Main Channel, and the Clay Figurines Stream runs southward, not connecting with the Main Channel. The Leopard's Channel and the Ravno Channel run as occasionally very narrow meanders, with sections layering out, creating a labyrinth of passages. These four channels are located deep in the interior of Vjetrenica, two with orientation toward the entrance, and one, the Leopard's Channel, runs perpendicular to the Main Channel. All have stone bases. Several networks of channels are concentrated around Lower Vjetrenica (Donja Vjetrenica; item 3 on Fig. 1) and the Absolon Channel (Absolonov kanal; items 8 and 10 on Fig. 1), farther from the Main Channel. They are younger, narrower, with clean stone edges, with active water flow, though located close to the entrance of the cave and oriented toward it. This means that almost all the peripheral channels are inclined due north, that is, toward the entrance.

In addition to layers of clay and loam there are numerous examples of calcite sediments, with particularly prominent rimstone pools called *kamenica* (traditionally also called *pjati*, Italian for plate, Fig. 2) and

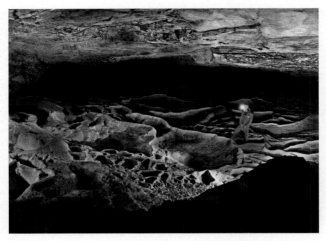

FIGURE 2 Rimstone pools—the so-called plates which open the northern catchment of Vjetrenica, at some 1500 m from the entrance. *Photograph taken by Darko Bakšić.*

FIGURE 3 The Great Lake, view of the exit: to the left is the Hadezija Cone, the site where *Hadesia vasiecki* was first found. *Photograph taken by Ivo Lučić. After Lučić (2003).*

FIGURE 4 The Golden Chamber, view of the exit: yellow flowstone to the right, mother rock to the left, with numerous erosion and corrosion formations. *Photograph taken by Ana Opalić. After Lučić (2003).*

flowstones. A sequence of *pjati* covers the front part of the Main Channel, some 150 m. Of the flowstones, the most prominent are the Yellow Flowstone (Žuti saljev, in the Golden Chamber, Fig. 4, right side), the White Flowstone (Bijeli saljev, 2100 m from the entrance), and several others. There are smaller formations, often aragonites. Since karstification of the area is well advanced and has led the cave into its mature stage, there are numerous examples of broken ceilings and distorted sediments. This is particularly visible on calcite formations, which have been destroyed in some places and with some sections of the Main Channel with no calcite sediments at all. There are also different forms of subterranean corrosion (anastomoses) and erosion (facets), as well as examples of destruction caused by changing hydrology.

Vjetrenica has a network of water bodies, which includes flows (permanent streams) such as the Absolon Stream (Absolonov potok), the Leopard's Stream (Leopadov potok), the Ravno Channel Stream (Potok Ravanjskog kanala), the Hanging Lakes Stream (Potok visećih jezera), and the Clay Figurines Stream (Potok glinenih figurica), flowing in different directions and at different heights, along with lakes; in addition to the Great Lake (Veliko jezero, Fig. 3), 180 m long, there are many smaller ones. The interior of the cave reacts quickly to precipitation and water bodies also increase into a series of occasional lakes. The Main Lake expands and floods most of the Main Channel for as far as 400 m, thus closing of the Siphon. Sources indicate that interior parts of the cave, inaccessible in winter, also merge into a single large lake. There are numerous occasional flows as well.

The hydrology of Vjetrenica is organized in two interior catchment areas, delimited by the Dividing Flowstone (Vododjelnica, at 1505.5 m from the entrance) one evacuates the water toward the north, via the Great Lake, the Great Lake Stream and the Absolon Stream to the Lukavac spring, located at the edge of the slope and the floor of the Popovo Polje near the entrance of the Vjetrenica cave; the second runs toward the interior of the cave, due south. There are instances of successive reduction of subterranean flows: the Great Lake Stream has four visible ponors in the Main Channel, from 650 to 800 m from the entrance. The last is close to the most likely local erosion basis, the Absolon Stream.

MICROCLIMATE

In addition to a powerful stream of air at the entrance area, milder currents are present in some narrow sections, particularly at the Siphon and the entrance to the Leopard's Channel. At higher outdoor temperatures, the air current is steady as it runs toward the entrance and maintains a steady climate. At lower outdoor temperatures, the direction of the air current from inside the cave changes. Since the cave faces the north, the strong northerly winds enter the front of the cave and reduce the temperature. The speed of the air current can increase to 10 m s^{-1} or greater. The air temperature averages $11°C$, with seasonal variations. For example, on February 7, 2008, at 07:20 AM, the recorded temperature was $-3.85°C$, relative humidity 38.75% and wind speed 9.83 m s^{-1}; at the same time, an outdoor station at some 300 m from Vjetrenica showed the outdoor temperature of $-4.33°C$, relative humidity of 33.25% and wind speed of 3.15 m s^{-1}. The water temperature corresponds to the air temperature with very little deviation, though there is a notable decrease in the winter, below major streams. The soil temperature below the surface may drop to $10.2°C$. During average air circulation, CO_2 levels vary from 400 to 600 ppm. All the values were registered by monitoring equipment placed close to the entrance.

HABITATS

Vjetrenica has many habitats. Terrestrial habitats include rocks, calcite sediments, rock, clay and loam surfaces, crevices, and areas under rocks. Aquatic habitats include fast streams in pure rock beds, slow streams running over clay surfaces, sand deposits in lakes, as well as sedentary water: larger lakes and small pools over clay surfaces, occasional lakes on clay surfaces, small lakes with sediments of sand, lakes with erosion depressions on rocks surfaces, and small occasional surfaces on rimstone dams and pools. There are several hygropetric habitats—flowstones with a thin water flow enveloping them (Fig. 3, left). There is considerable variation in the level of potassium, sometimes as low as 0.1 mg L^{-1} in some lakes, though somewhat higher in most of the water, and sometimes as high as several tens of milligrams. Organic particles, often brought by streams dripping from wide cracks in the ceiling, can be found all along the cave, including the deep interior (Sket in Lučić, 2003).

FAUNA

Due to its past geological development and abundant karst elements that offer versatile subterranean habitats, Vjetrenica (together with the Bjelušica Cave and the Lukavac spring) is rich with fauna. It comprises a total of 219 taxa, of which 37 protists and 182 animals with 101 cave-dwelling taxa: 49 troglobiotic (obligate terrestrial subterranean species) and 52 stygobiotic (obligate aquatic subterranean species). Sorted by taxonomic groups, they are: Turbellaria (5), Hydrozoa (1), Gastropoda (12), Bivalvia (1), Nemertina (1), Polychaeta (1), Oligochaeta (6), Hirudinea (1), Palpigradi (1), Araneae (6), Opiliones (2), Pseudoscorpiones (3), Copepoda (6), Ostracoda (4), Decapoda (4), Isopoda (7), Amphipoda (12), Chilopoda (2), Diplopoda (5), Collembola (4), Diplura (1), Thysanura (1), Coleoptera (14), and Vertebrata (1). The fauna is highly localized with 57 taxa endemic to Popovo Polje region, 25 of them only in Vjetrenica and related Bjelušica and Lukavac Spring. The distribution is still unknown for 37 protists and 24 animals, determined only up to family or genus levels, so levels of endemism may be even higher.

Vjetrenica is the type locality for 38 taxa: *Scutariella stammeri* Matjašič, 1958,* *Stygodyticola hadzii* Matjašič, 1958 (Temnocephalida, Scutariellidae), *Lanzaia vjetrenicae* Kuščer, 1933,* *Zavalia vjetrenicae* Radoman,* 1973 (Gastropoda, Hydrobiidae), *Vitrea kiliasi* Pinter, 1972 (Gastropoda, Pulmonata, Zonitidae), *Eukoenenia remy* Conde, 1974* (Palpigradi, Eukoeniidae), *Stalagtia (Stalagtia) hercegovinensis* (Nosek, 1905), *Stalitella noseki* Absolon & Kratochvil, 1933 (Araneae, Dysderidae), *Tegenaria conveniens* Kulczynski, 1914 (Araneae, Agelenidae), *Lephtyphantes spelaeorum* Kulczynski, 1914* (Araneae, Linyphiidae), *Chthonius (Chthonius) occultus* Beier, 1939 (Pseudoscorpiones, Chthoniidae), *Neobisium vjetrenicae* Hadži,* 1933 (Pseudoscorpiones, Neobisiidae), *Dinaria vjetrenicae* (Hadži, 1932)* (Laniatores, Travuniidae), *Diacyclops karamani* (Kiefer, 1932), *Eucyclops inarmatus* Kiefer, 1932, *Acanthocyclops troglophilus* (Kiefer, 1932) (Crustacea, Cyclopidae), *Pseudocypridopsis hartmanni* Petkovski et al., 2009 (Ostracoda, Cyprididae), *Troglocaris hercegovinensis* (Babić, 1922) (Crustacea, Decapoda, Atyidae), *Troglomysis vjetrenicensis* Stammer, 1936* (Crustacea, Mysidacea, Mysidae), *Monolistra (Pseudomonolistra) hercegoviniensis* Absolon, 1916 (Crustacea, Isopoda, Sphaeromatidae), *Proasellus hercegovinensis* (S. Karaman, 1933) (Crustacea, Isopoda, Asellidae), *Armadillidium absoloni* Strouhal, 1939* (Crustacea, Isopoda, Armadillidiidae), *Niphargus balcanicus* (Absolon, 1927), *Niphargus boskovici* S. Karaman, 1952, *Niphargus vjeternicensis* S. Karaman, 1932, *Niphargus trullipes* Sket, 1958, *Niphargus factor* G. Karaman & Sket, 1991,* *Niphargus cvijici* S. Karaman, 1950, *Niphargus zavalanus* S. Karaman, 1950* (Crustacea, Amphipoda, Niphargidae), *Hadzia fragilis* S. Karaman, 1932 (Crustacea, Amphipoda, Hadziidae), *Lithobius (Troglolithobius) sketi* Matic & Darabantu, 1968 (Chilopoda, Lithobiidae), *Typhloiulus (Attemsotyphlus) edentulus* Attems, 1951* (Diplopoda, Julidae), *Plusiocampa (Stygiocampa) remyi* Conde, 1947 (Diplura, Campodeidae), *Hadesia vasiceki* (J. Müller, 1911), *Nauticiella stygivaga* Moravec & Mlejnek, 2002,* *Speonesiotes (Speonesiotes) schweitzeri* Jeannel, 1941 (Coleoptera, Cholevidae, Leptodirini), *Aphaenopsis (Adriaphaenops) pretneri* Scheibel, 1935*, *Aphaenopsis (Scotoplanetes) arenstorffianus* Absolon, 1913 (Coleoptera, Carabidae, Trechini).

An asterisk indicates taxa found in the Vjetrenica system only.

Some animals belong to monotypic genera: *Spelaeoconcha paganettii* (Gastropoda), *Marifugia cavatica* (Polychaeta), *Velkovrhia enigmatica* (Hydrozoa), *Zavalia vjetrenicae* (Gastropoda), *Stalitella noseki* (Araneae), *Dinaria vjetrenicae* (Opiliones), *Typhlogammarus mrazeki*, *Spelaeocaris pretneri*, *Troglomysis vjetrenicensis* (Crustacea), *Nauticiella stygivaga* (Coleoptera), and *Proteus anguinus* (Vertebrata). The cave-dwelling fauna of Vjetrenica includes some fauna of special interest. The hydrozoans *Velkovrhia enigmatica* is a unique species of a single genus, in a unique freshwater species of the family Bougainvilliidae. The clam *Congeria kusceri* is a relict, a living fossil, the only survivor of a lineage of 100 *Congeria* species present during the Upper Miocene. *Marifugia cavatica* is the only freshwater serpulid worm and the only stygobiotic polychaete. *Pholeoteras euthrix* (Gatropoda, Prosobranchia) is the only known cyclophorid in Europe, a relic of old, tropical fauna. *Proteus anguinus* is the only European stygobiotic salamander. With ten species of genus *Niphargus* (Amphipoda) Vjetrenica holds a world record as a single cave; seven of which also have their type locality there.

PALEONTOLOGY

Vjetrenica is a paleontological site which acted as a natural trap in the Pleistocene for the leopard (*Panthera pardus*). Skeletons of four leopards have been found in different parts of the cave, one of them very well preserved. All the findings of the species, now extinct in Europe, were almost 2 km of the modern-era entrance, either found on the surface or under shallow sediment. The level of preservation of the skeletons, the number of samples, and the fact that they were found deep in the interior make Vjetrenica a unique Pleistocene paleontological site.

CULTURAL HISTORY

Legends, myths, and material cultural remains testify that throughout recorded history, Vjetrenica has been considered a prominent spiritual site. This fact gave a primary determination to its place and function in the cultural communities it served. Myths say that Vjetrenica was the place where fairies gathered, danced, and sang, and that the "song" coming from inside the cave was, in fact, a lullaby for the fairies' children, and so on. Sounds from the Vjetrenica cave were often compared with bagpipes. A written source from between 1580 and 1584 claims that visible tracks of "cracked" footprints were visible on the floor (according to legend, fairies had cloven hooves instead of feet). Until as late as early twentieth century, peasants from Zavala swore they had seen the fairies with their own eyes. Vjetrenica also had a system of sounds used for telling the future. They appeared in several places in the ceiling of one of the chambers, called the Millstone, the Drum, and the Mill. Particular sounds, varying from one to three, could be heard at the entrance to the cave. In short, its own mythology always rested on the interpretation of phenomena related to air currents.

Only traces of material artifacts remain, as they were gradually removed by new cultures: the prehistoric human habitat was at the very entrance, which was also the site of discovery of Roman coins as well as coins minted in Dubrovnik, as well as a medieval tomb, destroyed in modern times. The only reminder of the tomb is an interesting ornament, typical of medieval tombstones in Bosnia, called *stećak*. It is unusual that instead of the commonly used freestanding monolith, the *stećak* used a rock at the entrance, with two shallow carvings. The visual expression of the carvings is also original and authentic, with integrity that fits perfectly with the cave itself. Just like elsewhere, it depicts old motifs of hunting and tournaments, symbolizing mystic rituals striving toward immortality, present since ancient Trace. Here, however, two opposing horsemen are not separated by the commonly found image of a woman or a symbol representing female deity (a circle, a rosette, a crescent moon, etc.) (Wenzel, 1961). The reason is simple: whoever stood in front of the relief could confirm the presence of a female deity (a fairy), since her voice spoke through the wind from Vjetrenica.

According to some sources, the sacred status of the cave evident in pagan times continued also in the Christian era, though this remains only marginally explored. Namely, there are traces of an edifice at the entrance to Vjetrenica. Some interpret it as the presence of a shrine of the medieval Bosnian Church, and some as the castle of an unknown nobleman. Historical sources indicate that the interior of the edifice used the cave wind to keep cool, which is an interesting and rare example of early air-conditioning. Vjetrenica seems to have been abandoned as human or animal shelter for a long time, though remains of domestic animals indicate that this period may not have been all that long. Vjetrenica has also been the generator of sanctity of the surrounding area, as certified by material remains of different cultures. There are the foundations of an early Croatian Church of St. Peter (with cultural elements specific to the ninth to twelfth centuries), subsequently surrounded by a necropolis of late medieval tombstones, as well as a medieval Eastern Orthodox Monastery of Presentation of the Blessed Virgin Mary, with the church built on an inferior cave, partly connected to live rock. Its interior is richly decorated with frescos dating back to 1619. The construction of this church disturbed the cultural sediments where a Mesolithic tool was found.

EXPLORATION OF VJETRENICA

The known history of Vjetrenica is a web of mythology and science conditioned by the ever-present fascination with the "wind." The oldest written sources are those from *Historia Naturalis* by Pliny the Elder, from the year 77 BC. When speaking about the sources of wind (Plinio Secondo, 1884:245) the source speaks of an unnamed cave on the Dalmatian coast, which the Renaissance tradition of Dubrovnik names as Vjetrenica. The Dubrovnik writer Benko Kotruljević wrote in 1464 that the wind at its entrance was colder in the summer than any winter wind in Italy. The duke of the nearby town of Ston (in the medieval Dubrovnik Republic), Jakov Sorkočević, wrote (1580– 1584) to the natural scientist Ulise Aldrovandi and gave its first description. In his *Sopra le Metheore d' Aristotile* (1584), Nikola Gučetić Gozze, a leading intellectual, scientist, and politician of the then Dubrovnik Republic, analyzed the wind and how it differed from the Šipun cave in Cavtat. He explained it via Aristotle's natural philosophy, on the basis of the four elements—earth, water, wind, and fire—and four features—warm, cold, dry, and wet. Inside the cave, the sun triggered the evaporation of the element of earth, which was warm, dry, and thick. Once it reached the cold interior surface of the cave, it stopped and as it found an exit at the entrance, which is where it created the great wind. In Šipun, however, there was no evaporation, since the interior was damp, hence no wind, according to Gučetić. In 1830, a monk from the Eastern Orthodox Monastery of Zavala named Joanikije Pamučina wrote that, in his opinion, the phenomenon of air currents and the creation of sediments should be interpreted as a phenomenon of physics. In 1850, a Catholic missionary named Antonio Ayala compared the wind with the sound of a thousand steam engines on the move.

The first research exploration was organized in the late nineteenth century, following the establishment of the Austrian rule. In July 1887, Joseph Riedel conducted the first modern temperature measurements. The *hegumen* (head) of the Zavala Monastery, Hristifor Mihajlović, described Vjetrenica in 1889 and 1890 and this description was accepted all across the monarchic realm. He also produced the first sketch of the cave, and after that (in 1893) Josip Vavrović, a railway engineer, produced an even more detailed description and a map of the first 600 m. During his explorations in the period 1912–1914, Karel Absolon, a Czech explorer who conducted a total of 27 visits to Vjetrenica, discovered most of the segments known today, but he never produced a map. He described his expedition as the most difficult in the history of speleology, and he insisted that Vjetrenica surpassed anything he had ever seen inside any cave. He also suggested that it was there that he established the measurements to assess limestone formations in the Balkan region. Under instruction by his mentor Jovan Cvijić, Mihajlo Radovanović conducted detailed explorations throughout the 1920s, including detailed mapping. He presented his explorations in a dissertation, which he then published as a monograph in 1929. In 1958, Bosnian geographers Ratimir Gašparović and Orhan Zubčević produced a detailed geodetic map of the Main Channel (up to 2084 m). In 1968, members of the South Wales Cave Club found two previously unknown channels of a total length of some 500 m.

New speleological explorations were launched in 2000, eventually producing a monograph entitled *Vjetrenica—A View into the Soul of the Earth* (2003) and a totally new and modern topographic map, published in 2002 (Lučić *et al.*, 2002). Regular summer speleology camps were organized from 2002 to 2009 by the newly established Speleological Association Vjetrenica– Popovo Polje. The camps led by Darko Bakšić (Velebit Speleology Section from Zagreb) brought together a host of clubs and individuals from the region. The camps uncovered new channels of total length of 1668 m. In 2004 the camp also established a biospeleology program, led by Roman Ozimec from the Croatian Biospeleological Society from Zagreb. In an unexplored channel, the SWCC speleologists found a complete leopard skeleton; in the late 1990s Mirko Malez found remains of a leopard and a brown bear; the Velebit Speleology Unit explorers found remains of a leopard. During the 2007 paleontological excavations led by Kazimir Miculinić from the Croatian Biospeleological Society from Zagreb, skeletal remains of three leopards and one bear were excavated and handed over to the National Museum in Sarajevo.

The most intensive exploration of Vjetrenica has been by biospeleologists, though their findings were available to a limited scientific community only. Vjetrenica became attractive for biospeleologists after L. Vašiček and Lucijan Matulić had found cave-dwelling insects, described by Giuseppe (Josef) Müller as *Antroherpon apfelbeckia* and *Hadezia vasiceki*. Exploration was boosted by K. Absolon (research from 1908 to 1922) who unfortunately never published his list of the taxa. It was then explored by members of the Slovenian Society for Cave Research (1931), by Stanko Karaman in the 1950s, and by Slovene biologists Janez Matjašič, Jože Bole, and Boris Sket throughout the second half of the twentieth century, as part of the visits organized by the Slovene Academy of Arts and Science or the School of Biotechnology of the Ljubljana University. More than 30 researchers have explored Vjetrenica and described its fauna or submitted information for description. Most of the new species were described by S. Karaman (9 taxa), followed by K. Absolon (*et al.*) and K. Verhoeff (8 taxa),

J. Matjašič (7), F. Kiefer and L. Kulczynski (5), and others. They thus created a database that allowed Boris Sket to prepare an integral description of the fauna of Vjetrenica (2003). It was also of help in synthesizing conclusions on the international prominence of Dinaric subterranean fauna. Vjetrenica was known for 75 cave-dwelling taxa at the onset of the biospeleological camp program, and by 2009 101 taxa were described. The recent exploration, which includes the first year-round climate monitoring and new environment-sensitive lighting for tourist access, has been followed by a host of events aimed at promoting the cave, such as the publication of the monograph, a web page, documentary films, and numerous texts. This is also a new presentation of karst, based on an environmental outlook, as well as a new view of contemporary organization and management of Vjetrenica.

VISITOR ACCESS

The first facilities to allow access to visitors date back to the late nineteenth century, when the entrance to Vjetrenica was expanded. The old entrance was low and narrow, barely allowing a crouching adult to enter, though often blocked due to strong wind from the inside. A railway track was laid immediately in front of Vjetrenica in 1901 and the entrance was fitted with a metal frame and a gate with vertical bars. The first visitor trail of 1250 m was built between the two world wars. Construction included flattening of the footsteps over small rock, placement of a fine stone layer over clay beds, placement of gravel over rimstone pools to create paths. In the early 1960s, the path was expanded to 110 cm and extended to 1700 m; electric lighting was introduced on 1045 m of the path in 1964, along with a full-time guide service and a motel in the vicinity. Official reports of 7000–10,000 visitors per year seem conservative. All the equipment and facilities for visitors were destroyed during the 1991–1995 war, and real interest in the revival of tourism was triggered by the new speleological research at the onset of the twenty-first century.

ECOLOGICAL CHANGES

Following the massive hydropower construction on the River Trebišnjica (1965–1979), which reduced this karst river to just 67 km and gradually converted almost all of its flow into a concrete canal, thus reducing its environment of some 4 billion cubic meters of water, Vjetrenica was at the very edge of the endangered area, though not physically included. The villages above it have no industrial facilities, but there is also no garbage disposal system. Also, the 1991–1995 war left behind damaged military technology based on PCB compounds, which no one has yet investigated. Conversion of Vjetrenica to accommodate visitors included several expansions of the entrance—which terminated the magical sounds—excavation to set paths, introduction of construction materials and metal, tiling and laying of concrete, defragmentation of rock and sediments, gravelling and placement of gravel over *plates* and flowstones, disintegration of several lakes, introduction of powerful lighting, which led to the generation of lampenflora, as well as collection of garbage. And sources identify a disposed camera battery as the principal cause of the disappearance of several crustaceans from some of the ponds of the Main Channel. The extinction of the entire fauna in the Third Calcite Lake was noted in the early autumn of 1996, but no causes were identified. Vjetrenica is threatened by the post-socialist transition of a war-torn country, where a cave is used for commercial exploitation with absolutely no systematic state-backed protection of the natural system. The struggle to protect Vjetrenica culminated in early 2009, with a 14-day hunger strike of an activist from a local NGO. Although the struggle to protect it enjoys the support of many prominent international professional speleologists, there has been no response from Bosnian authorities.

PROTECTION

Vjetrenica was declared a natural monument (1950) and was subsequently declared a special geological reservation (1965 and 1985). In 2004, the BiH Academy of Arts and Science nominated it for the UNESCO List of Natural Heritage Sites and in 2007 the National Commission for Monuments expanded the nomination to include the architectural ensemble of the village Zavala. However, at the time of this writing, in late autumn of 2010, it remains only on the tentative list.

Translated into English by Amira Sadiković.

See Also the Following Article

Diversity Patterns in the Dinaric Karst

Bibliography

Lučić, I. (2003). Vjetrenica—pogled u dušu zemlje. Ravno, Zagreb, Croatia.

Lučić, I., Bakšić, D., Mulaomerović, J., & Ozimec, R. (2005). Recent research into Vjetrenica cave (Bosnia–Herzegovina) and the current view of the cave regarding its candidature for the World Heritage List. *Proceedings of the 14th International Congress of Speleology*, Athens, August 21–28, 2005 (Vol. 2, pp. 430–435) Athens: Hellenic Speleological Society.

Ozimec, R., & Lučić, I. (2010). The Vjetrenica cave (Bosnia & Herzegovina)—One of the world's most prominent biodiversity hotspots for cave-dwelling fauna. *Subterranean Biology, 7*, 17−23.

Wenzel, M. (1961). A medieval mystery cult in Bosnia and Herzegovina. *Journal of the Warburg and Courtauld Institutes, 24*, 89−107.

VOLCANIC ROCK CAVES

Stephan Kempe

Institute of Applied Geosciences, University of Technology Darmstadt, Germany

INTRODUCTION

Exploration of volcanic rock caves, termed *vulcanospeleology* by William R. Halliday, the Nestor of the field, did not receive much attention until the 1970s. Before, it remained a topic of local caving interest, for example, in Australia, in the western United States on Tenerife, Sicily, and Iceland, or a topic of ecologists looking at biological diversity. One of the reasons for this general lack of attention may have been the question of accessibility but even more so the term *lava tube* may have suggested "tubular caves" of little morphological variety and interest ("you have seen one, you have seen them all"). Even in volcanological textbooks, the important role of lava tubes for the lateral transport of lava and the shape and functioning of shield volcanoes has rarely been acknowledged. These textbooks would publish at most one picture of a lava tube (very often of Thurston Lava Tube, Hawai'i), state that they form by "crusting over of channels" and then move on.

The beginning of lava cave research on Hawai'i in the 1970s and the consecutive acknowledgment of the enormous Hawai'ian cave potential spurred activity there and worldwide, leading to the foundation of the Hawaii Speleological Survey in 1989 during the International Congress of Speleology in Budapest, Hungary, and the International Commission of Vulcanospeleology in 1993. Furthermore, the Vulcanospeleological Symposia, initiated by W.R. Halliday in 1972 (Halliday, 1976) and held every other year (the 14th was in Australia, August 2010), is attracting many speleologists and other scientists. The proceedings of these symposia form the largest body of professional papers available in this field. They are now accessible at www.vulcanospeleology.org/symposia.html. At the International Congresses of Speleology, lava cave symposia are also held; the papers of the latest one in Kerrville, Texas, August 2009, are available at http://www.karstportal.org/FileStorage/IntlCongressofSpeleology/15thproceedings-v002.pdf. Other sources are the Newsletters of the Hawaii Speleological Survey and the Hawaiian Grotto of the National Speleological Society. U.S.-centered terminology of vulcanospeleology was summarized by Larson and Larson (1993).

In general lava caves, or better, *caves in lava*, can be formed by many different processes. Most of them are primary caves, but secondary caves occur as well. Kant (or one of the editors of his work in 1803) was probably the first to differentiate between primary and secondary caves, albeit on the basis of quite abstruse surmises.

SECONDARY VOLCANIC ROCK CAVES

Secondary caves in volcanic rocks, that is, created long after the lava was deposited, form tectonically, by collapse or by erosion.

Tectonic caves are still poorly documented. Along the Great Crack, the southwest rift zone of the Kilauea, several essentially tectonic fissure caves have been explored, including Pit H (183 m deep) and the Wood Valley Pit Crater (90 m deep).

Pit craters form by cold collapse of the roof of small sections of a magma chamber (collapse of the entire roof of a magma chamber leads to the formation of large-scale depressions, called *calderas*). One recent example is Devil's Hole on the Island of Hawai'i (Hawai'i National Park, Chain of Craters Road), that started as a small hole in the ground leading into a conical chamber about 50 m deep. Further collapse of ceiling and walls led to an open pit over the course of a few decades. Many other older pits occur in the vicinity. Pit craters have recently obtained more attention because they may be the first cave-like objects detected on Mars. Collapse of hypogene karst caves and the consecutive stoping upward of the cavity can also result in lava caves. An example is the Basalthöhle near Ortenberg, Germany, a chamber in columnar basalts that was intercepted by quarrying for basalt.

Erosional caves in lava may either be formed by waves along coasts or by running water. Wave-cut caves in volcanic rocks are quite common and may be of substantial size. The most famous is Fingal's Cave of the Island of Staffa, Scotland. Some of the most spectacular examples occur along the cliffs of the Hawai'ian Islands, specifically on the Na-Pali Coast of Kauai. Sometimes basaltic dykes are eroded by waves leading to long and narrow fissure-like sea caves such as the >40-m Sorte Gryde on Bornholm, Denmark. Quite common are also small caves that are produced by the lateral erosion along river valleys that cut down into stacked series of lava flows. The flowing water can then preferentially remove loose a'a rubble from underneath the solid cores of the a'a flows that serve as cave roofs. For example, such caves occur along the sides of Wadi Rajil in Jordan near the famous Bronze Age city of Jawa. Water entering primary lava caves can cause internal water erosion such as documented for the Pa'auhau Civil Defense Cave on Mauna Kea, Hawai'i (Kempe et al., 2003). In this cave, several spectacular waterfalls eroded the bottom of the cave, polishing the rocks and gravel to a surprising extent. It is difficult to understand how water can erode a cave in lava without following an existing lava cave, such as the kilometer-long and 100-m-deep Kuka'iau Cave (Kempe and Werner, 2003) which is formed in a series of pāhoehoe flows and diamict layers in weathered Mauna Kea lavas. It not only pirates the tributaries of a parallel valley but it also forms a series of sumps and chutes where the water moves upward, similar to karstic caves under phreatic conditions. At flood, this cave transports large amounts of water, gravel, rounded blocks, and trees and may pond up to 60 m above its sump.

PRIMARY VOLCANIC ROCK CAVES OF LARGE EXTENT (PYRODUCTS)

Tunnels formed by the lateral transport of lava through interior conduits of pāhoehoe lava flows comprise the most important group of primary caves worldwide. Within the tunnel, heat is lost only conductively while surface lava loses heat convectively at the upper and conductively at the lower interface. Tunnel transport is thus the reason why basaltic shield volcanoes have low flank-slopes, often <2° (compare Table 1) and why multiple eruptions can form extensive intracontinental lava plateaus. What appears later on the geological map as an individual *pāhoehoe flow* is actually a low ridge that grew only at the downhill end being supplied by the internal duct that in most cases is not visible to the outside. In fact, one can walk across an active tunnel without even noticing it. A'a flows, on the other hand, move like glaciers, gliding downhill with their entire mass, moving like a caterpillar on a bed of rubble that falls down at the front and is overridden. On Hawai'i a'a flows do not develop tunnels.

The first to describe a lava tunnel having inspected it himself was probably Eggert Olafsen (1774–1775, § 358, p. 130) who visited Iceland from 1752 to 1757. He wrote about Surtshellir: "... the running lava flowed through this channel like a river...." Uno von Troil (1779, p. 225), who visited Iceland with Joseph Banks and Daniel Solander in 1772, wrote (transl. from German by author):

> The upper crust sometimes cools and solidifies, even though the molten matter keeps running underneath; in this way large caves form, the walls, floors and ceiling of which are composed of lava and where a lot of dripstones of lava occur.

TABLE 1 Comparison of Some Morphological Indices of Some of the Hawai'ian Lava Tunnels

Cave	Total Length, km	Main Trunk Length, km	End-to-End Bee-Line, km	Sinuosity	Vertical Distance, m	Slope	Volcano
Kazumura Cave	65.50	41.86	32.1	1.30	1101.8	1.51°	K, A
Ke'ala Cave	8.60	7.07	5.59	1.25	186	1.51°	K, A
J. Martin/Pukalani System	6.26						K, A
Epperson's Cave	1.93	1.13	0.80	1.41	—	—	K, A
Thurston Lava Tube	0.490	0.490	432	1.13	20.1	2.4°	K, A
Ainahou Ranch System	7.11	4.82*	4.27	1.13	323	3.83°	K, A
Ke'auhou Trail System	3.00	2.27	1.99	1.13	213.3	5.36°	K, A
Charcoal System	1.5		1.4		60	2.6°	K
Earthquake System	0.34				33	4.7°	K
Huehue Tube	10.8	6.17	5.13	1.2	494.6	4.58°	H, HH
Clague's Cave**	2.73	1.39	1.18	1.15	157.1	6.49°	H, HH
Pa'auhau Civil Defense C.	1.00	0.58	0.50	1.14	49	4.87°	MK
Whitney's Cave	0.651	0.509	0.438	1.15	17.3	1.97°	ML
Manjang Gul, Jeju, Korea		4.304	3.197	1.32	32.4	0.4°	Jeju

*horizontal
**upper part of Huehue
For sources of data see Kempe, 2002.

The first to report having seen an active tunnel was Coan (1844) who, in 1843, scaled Mauna Loa:

> But we soon had ocular demonstration of what was the state beneath us; for in passing along we came to an opening in the superincumbent stratum, of twenty yards long and ten wide, through which we looked, and at the depth of fifty feet, we saw a vast tunnel or subterranean canal, lined with smooth vitrified matter, and forming the channel of a river of fire, which swept down the steep side of the mountain with amazing velocity. The sight of this covered aqueduct—or, if I may be allowed to coin a word, this *pyroduct*—tiled with mineral fusion, and flowing under our feet at the rate of twenty miles an hour, was truly startling.

Even though Coan used the more general term *tunnel* to start with, he coined a new one: *pyroduct*. This highly specific term should take precedence over younger terms describing the same phenomenon. Early geologists such as James Dana continued to use the word "tunnel"; J.W. Powell introduced the term *volcanic pipes* and Tom Jaggar used "tunnel" as well as "tube"; only after 1940 did the term *lava tube* become standard (Loowood & Hazlett, 2010). For reasons of scientific priority, the term *pyroduct* should be used. Avoiding the word "tube" is also advisable, because it has invoked the picture of pipes in which lava can flow up and down under pressure like in plumbing and it implies a circular cross-section that is only rarely found.

The discovery of extensive lava flows on Venus, Mars, and the Moon and even active volcanoes on Io has increased the interest in pyroducts in the past few decades. On Earth, the longest surveyed lava tunnel is Kazumura Cave (main trunk length 41 km) (Hawai'i, Kilauea Volcano) (Allred *et al.*, 1997), and the longest Quaternary pyroduct-fed flow on Earth is that of Undara/Australia (Atkinson and Atkinson, 1995). The author's group explored and surveyed many other caves on Hawai'i (*e.g.*, Kempe, 2002) and in Jordan that give the opportunity to study formation and evolution of pyroducts from the inside. Many other areas are under investigation as well: The islands of Galapagos, Rapanui, Jeju, Mauritius, Comores, Sicily, Iceland, Canaries, Azores, and in the intracontinental lava fields of Syria, Saudi Arabia, East Africa, the western United States, Mexico, and the Andes. Petrographically, almost all pyroducts documented (yet) occur in tholeiitic and alkali basalts, only the caves in Mount Susua, Kenya, are formed in phonolites (pers. com. C. Wood).

Formation of Pyroducts

Even though in many text books lava tunnels are described as having formed by "crusting over of channels" (*e.g.*, Francis, 1993), there is a totally different mechanism by which they can form, and that is by *inflation* (Hon *et al.*, 1994). This is an incremental process starting at the distal tips of pāhoehoe flows where hot lava rapidly covers the ground in a thin sheet. This sheet cools quickly, causing the dissolved gases to form vesicles that diminish the overall density of the lava. This sheet will float on top of the next pulse of advancing melt (the lava flow is "inflated" from below because of buoyancy) before forming the next distal surface sheet. Multiple advances can occur, forming a primary roof composed of several sheets, separated by sheer interfaces (only the first or top sheet will display the subaerially formed typical ropy pāhoehoe structure) (Fig. 1). The "oldest" lava sheet is therefore on top of the stack, contradicting normal stratigraphic rules. Below the primary roof the lava can stay hot and can keep flowing. This is the initial conduit. Inflation caves are characterized by roofs built of one or several sheets, sometimes more than ten continuous sheets of lava. This roof structure can be studied at roof collapses, called *pukas*, in Hawai'i. Many of the long Hawai'ian caves are inflationary in origin (Kazumura, Keala, Huehue, Ainahou, Keauhou Trail; just to name a few). However, the "crusting-over" of channels can also lead to long caves. Inspection of cross sections of the roof of the second longest cave on Hawai'i, the Kipuka Kanohina System (37 km of interconnected passages), has shown that it consists of welded, irregular fragments of pāhoehoe plates, stabilized by inserted lava fingers (*squeeze balls*) and a 10–50-cm-thick lining welding the roof from below. Such structure suggests that the roof is formed by agglomeration of floating lithoclasts that are wedged into each other similar to a logjam on a river. Sometimes roofing-over may also occur by accretion of vertical to subvertical thin lava layers growing from the sides inward and having a central vertical parting where the growing lateral shelves meet. The levees of open-surfaced lava channels grow by overbank events of thin lava sheets or by stranded lava floats. Sometimes large sections of the bank break loose and float downstream. These can then jam and form short roofs that are stabilized by spatter. Examples have been documented for the channels of the Puhia Pele eruption of 1801 on Hualalai.

Internal Development

If the area to be covered by the first advance is rather flat, many small, parallel conduits can develop (Fig. 2). Each of them can start to erode down soon after. One of the threads will, however, erode fastest and attract most of the lava. This passage will then drain the other parallel ducts, one by one, often leaving them as small-scale labyrinths high above the final floor. Since these mazes are drained when they are still

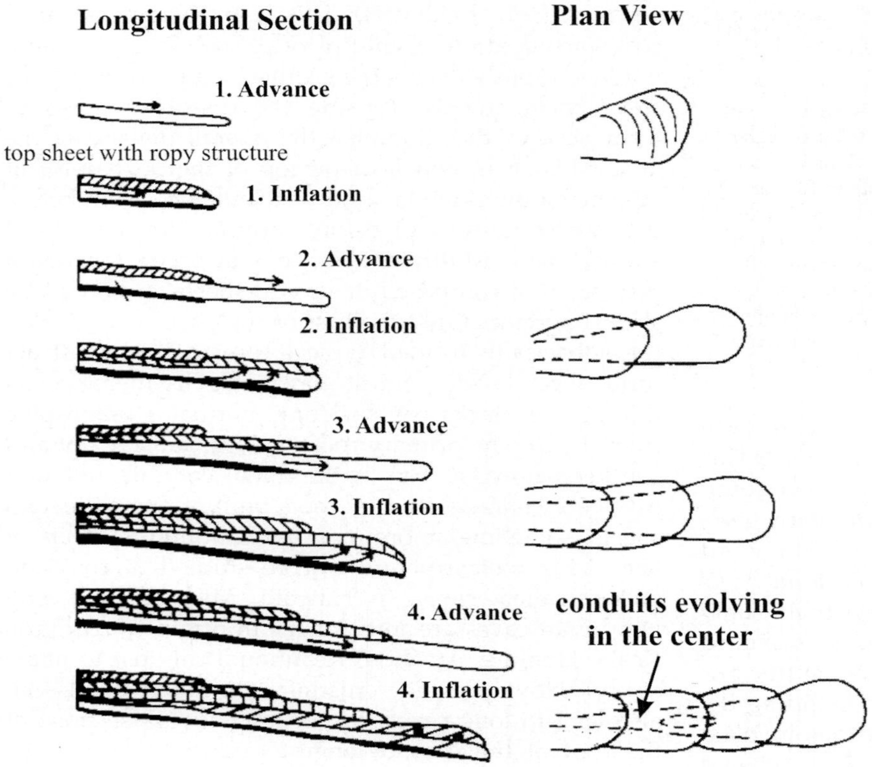

Longitudinal Section   **Plan View**

FIGURE 1 Sketch illustrating pyroduct (lava tunnel) formation. At the tip of a pāhoehoe flow lava advances quickly in the form of a delta of thin, ropy lava. The next pulse of lava lifts the first sheet up (inflation). This process is repeated until a stack of lava sheets (the primary roof) is formed, below which the hottest flow thread later becomes the main conduit.

very hot, their floors are mostly smoother than that of the later main tunnel where the terminal flow can convert to a'a rubble.

As the erosion continues, the lava flows with an open surface in a self-generated underground canyon cut into older rocks, not associated with the current eruption. This fact can be studied at places, where the thin lining of the side walls has fallen away (*e.g.*, Greeley *et al.*, 1998; Kempe, 2002). Often we find a'a blocks behind the lining or even ash horizons, both clearly not integral parts of pāhoehoe flows. Downcutting is facilitated by a variety of processes; one of the more spectacular is backcutting of lavafalls (*e.g.*, Allred and Allred, 1997) (Fig. 3). These are quite common in long Hawai'ian caves, but none has been found yet in Jordan. The falling lava acts like a sledge hammer, forcing loose rocks from the floor. These are less dense than the lava itself and float up and are transported on the surface of the river. Thus, in contrast with a water river, the bed is not protected by bedload and is therefore prone to continued erosion. The mobilized blocks are cold and receive a coating of lava, forming lavaballs. Some of the lavafalls seem to be stationary forming large plunge-pools and chambers (Allred and Allred, 1997). Due to these erosive mechanisms the cave grows in depth and width in an uphill direction. The passage above the lavafalls is quite small in contrast. As one proceeds uphill, the canyon becomes larger and larger until one enters the next plunge pool chamber. It is not quite understood how much mechanical erosion and how much melting of the river bed occurs (*e.g.*, Greeley *et al.*, 1998; and citations in Kempe, 2002). Other enlarging processes may occur, such as small phreatic explosions, blowing out sections of the wall or floor as intersected groundwater is vaporized.

As the downcutting continues the lava river meanders, undermining walls and destabilizing the roof. Breakdown falling into the flowing lava is also carried away. If the primary ceiling collapses entirely, a skylight or *puka* opens up. If the flow is still active, the rubble can be carried away and we speak of a *hot puka*. If the collapse occurs after the flow terminated, breakdown will remain, sometimes giving easy access to the cave below, sometimes sealing it completely; this is termed a *cold puka*. Through a hot puka gases can escape from the tunnel, triggering convective cooling. Heat loss is specifically efficient if two pukas open up: cold air will be drawn into the lower one and hot gases will escape from the upper one, freezing out a *secondary ceiling* above the flowing lava in the canyon in between the pukas, splitting the passage into two levels on top of each other (Fig. 3). This process can be repeated, forming even more internal ceilings. Later spills from below through breakdown holes or from upstream can reinforce these ceilings. In Ke'ala Cave, is a block about 12 m wide, 8 m long, and 5 m thick welded into the ceiling of the cave. The cold, oxygen-containing external air that is drawn into the cave can oxidize lava surfaces that are still hot. The iron, contained in the volcanic glass, is oxidized to fine-grained hematite, tinting the surfaces of secondary ceilings in various hues of red. If steam is present, goethite or limonite can also be formed, introducing ocher or yellow colors.

Hot pukas can also serve as temporary rootless vents when the tunnel below is obstructed or even closed entirely. Lava erupting from these pukas can form rapidly cooling, thin, ropy pāhoehoe, reinforcing the primary roofs from above. The Puka 17 Flow out of the lower part of the Huehue tunnel (Fig. 4) is an example. Pukas, cold or hot, can also serve as entrances for lava of later flows, such as at the upper and lower ends of Ke'ala Cave which are plugged by later invasive lava.

All these processes act to form caves of complex patterns, morphologically not representing "tubes" at all. Also the total length of the caves is usually much larger than the simple distance along the main tunnel. Table 1 gives some basic morphometric data, such as sinuosity and slope for some pyroducts.

General Types of Pyroducts

Overall, we can differentiate several general types of lava tunnels. These are (1) single-trunked systems; (2) double (or multiple)-trunked systems; and (3) superimposed-trunked systems

Most of the lava tunnels so far documented in enough detail appear to belong to the *single-trunked* category. They are fed by one eruption vent and the mesomorphological internal structure can be explained by the processes discussed above. The tunnel size depends on the lava discharge rate and on the length of activity (days, weeks, months, and possibly even years), that is, on the time available for erosion. If the eruption stops or the cave collapses or is blocked, the pyroduct will cool. The next or even—in case of a blocked tunnel—the same eruption will then create a new pyroduct. Normally, it will be situated to either side of the previous flow because that now forms a topographic ridge (flow lobe). Typical examples of single-trunked systems are Kazumura, Ainahou Ranch, Ke'ala, and others of the long Hawai'ian caves. If lava from the new tunnel should spill through a puka, or break (because of its overburden) into any older, underlying tunnel, then the older tunnel will be filled by lava cooling in the same pattern as at the surface.

Double-trunked systems are comprised of two lava tunnels, active side by side at the same time and fed by two separate eruption points. Such tunnels can interact and cause more complex morphologies than described above.

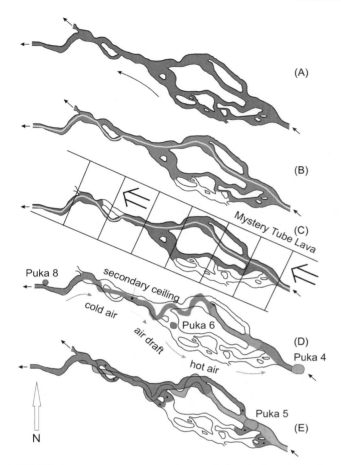

FIGURE 2 Detail of the ground plan of the Huehue Cave illustrating how a primary maze of parallel lava threads (A) was drained until the master trunk that cut down fastest remained (E). (A) Initial pattern of lava tunnels; flow was from right to left in up to five parallel conduits. (B) The southern-most conduits were drained first. (C) Further downcutting reduced the number of active conduits to three. At the same time, lava from a parallel flow (Mystery Flow) covered the area. (D) Only one tunnel remained (following the thick line in B and C), its bed 2 m below the original surface. The added overburden caused collapse of pukas (labeled 8, 6, and 4), the breakdown of which was removed by the lava river. Now external air began flowing uphill from puka 8 to 4. As a consequence a secondary ceiling froze out uphill and downhill of puka 8. (E) Various spill-events reinforced the secondary roof, even spilling into the already drained upper conduits, closing the northern most one. These spilled lavas were oxidized by the passing surface air and attained various shades of red caused by the crystallization of very fine-grained hematite. The last event was the collapse of puka 5 while the tube cooled. Its material is still in place.

Hawai'i, one section of secondary ceiling is over 1 km long. Very often the upstream end of the secondary ceiling is sealed. This is caused by lavaballs floating on the lava river that are too buoyant to be dragged below the secondary ceiling. Instead they strand on the upper edge of the secondary ceiling. The accumulated blocks are then welded together by splashed-up lava. Floating blocks can be very large: in Waipouli Makai Cave there

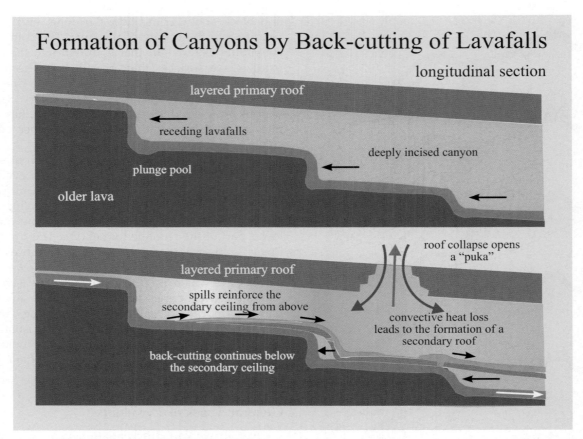

FIGURE 3 Longitudinal section of an evolving lava tunnel. Top: Erosional enlargement of the underground canyon by backcutting lavafalls. Bottom: On static failure and partial collapse of the primary roof, a skylight (puka) opens up, allowing cold air to enter the tunnel, freezing an internal secondary roof below which erosional enlargement can continue. Spills from uphill or through holes reinforce the secondary roof.

One example is the interaction between the Huehue Flow and the secondary Mystery Flow (Fig. 4; Kempe, 2002). The Huehue Flow established its tunnel first, then a second vent (the very inconspicuous, low *Mystery Shield*) erupted lava, forming a small tunnel in parallel. Part of the Mystery lava quickly cooled, forming a'a flows. These superseded the upper part of the Huehue tunnel. Once thick enough, the primary, sheeted roof of Huehue collapsed and left a roof composed of Mystery a'a lava. The resulting breakdown was removed with the active Huehue lava river. Due to the large, hall-like cavity that formed, a secondary roof froze out over the active flow of Huehue. Later rockfall covering the newly formed *false floor* gives the upper passage the appearance as if the tunnel was formed in a'a, an impossibility near to a vent issuing very hot basaltic lava.

The least understood and documented category is the *superimposed-trunked* system. It is defined as a set of lava tunnels superimposing and crossing each other, all being active at the same time. The upper tunnels stop their activity first, so that the lower ones carry on for some time before they also stop operating and become evacuated. There may even be connecting openings between the levels exchanging lava between crossovers. In such systems at least the lower ducts must have been filled to the ceiling until very late in their development. Such systems could arise when a volcanic vent increases its output volume during an ongoing eruption. Then the already established pyroducts cannot accommodate the increased flow volume and a new level of independently operating tunnels is built on top of the already active ones. The Kipuka Kanohina System on Hawai'i (Coons, 2009) is the largest example of such a superimposed-trunked system. Sistima Tlacotenco (16 km long), a segmented system of superimposed passages, most probably also belongs to this category of conduits.

PRIMARY VOLCANIC ROCK CAVES OF LIMITED EXTENT

Primary volcanic rock caves also occur in many different types: hollow imprints of trees and animals, partings along the central plane of lava sheets, hollow tumuli,

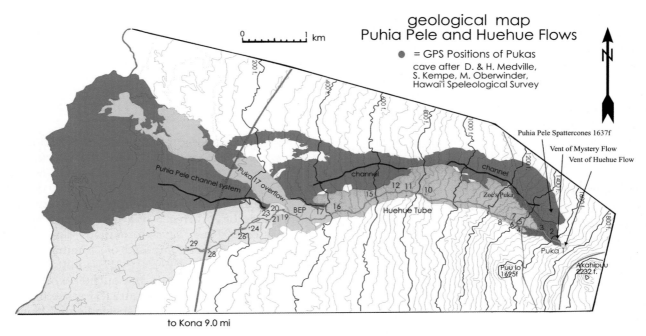

FIGURE 4 Geological map of the Huehue Flow (Hualalai Volcano, Hawai'i) of 1801 according to surveys of the author's group. Flow was from right to left, vertical distance 500 m. Oldest are the lavas (blue) of the Puhia Pele vent (a series of spectacular spatter cones) that was gas-rich and formed an open channel system (with a few roofed-over sections). After its termination the Huehue Flow erupted (numbers label pukas on the Huehue Cave, thick line; lava light yellow-green at left and at puka 1) accompanied by the Mystery Shield eruption that was active for only a short time (upper right) and covered much of the upper part of the Huehue Flow (brown). Zoe's Puka is a tunnel belonging to the Mystery Flow. From puka 17 a surface flow issued (light green) and the terminal lava from the mystery shield formed several short a'a flows (dark red-brown at the right).

drained lava tongues, pressure ridge cavities, and volcanic vents, and there may be more to be discovered.

Imprints

One of the most astounding caves in any respect is the hollow imprint of a diceratherium in Miocene pillow lavas (Rhino Cave, U.S. Quadrangle Park Lake, Grant Country, Washington; C. Holler, pers. comm.). Not only is it extremely rare that an animal gets encased in lava, even more unlikely is the fact that the cave is just now opened by erosion, so that it can be entered through "the rear." Imprints of trees are, compared to those of animals, more common. A Mikado-like jumble of trees was encased in lava of Mount St. Helens and is now publicly accessible to the joy of children who can easily crawl through the hollow imprints from one tree to the other. A large, accessible tree trunk is also encountered in Pa'auhau Civil Defense Cave on Mauna Kea, Hawai'i (Kempe et al., 2003). Often tree trunks are still standing, encased with lava that cooled around them. When the lava flow is subsiding around them they may be left standing, such as in Lava Tree State Park, Hawai'i, which includes an accessible pit-like imprint of the former tree.

Partings

In the process of cooling, gas exsolves from the lava, forming small vesicles. The more time is available, the larger they become. Cooling is fastest from the surface downward and slower from the bottom of the sheet upward: therefore the largest vesicles are mostly found in the lower third of the sheet. Sometimes they become dense enough to cause a parting along which the upper section of the sheet can be separated and bent upward by lateral pressure. These caves are low, but can be quite wide; they are closed on all sides and only accessible if opened in a road cut or by erosion.

Hollow Tumuli, Peripheral Lava Rise Caves, and Drained Lobes

Tumulus is a morphological term describing a variety of hummocks or small hills rising above the general lava surface (Walker, 1991). Some appear to be pressed up by lateral forces; others may result from lava being injected from below under pressure, resulting in the extrusion of lava from the tumulus. A few of these tumuli are hollow, forming dome-like cavities. Some of the largest occur in Kilauea Caldera in the Postal Rift

Flow of 1919 (*e.g.*, Tumulus E1; Walker, 1991). Other caves follow the perimeter of larger lava rises (*e.g.*, lava rise E5) that deflated in their centers once the lava drained from them. These can be rich in rock speleothems, specifically cylindrical stalactites. Other caves in the same flow appear to be drained lava flow lobes and lava tongues. All in all about 250 mostly shallow caves have been recorded of various genetic origins in the 1919 Caldera Flow (W.R. Halliday, pers. comm.).

Pressure Ridge Caves

This class of caves is much wider and longer than tumuli caves. In Jordan we know of ten caves of this type, all occurring in the lava field of the Qais/Makais eruption. The longest of these caves is Al-Ameed, with a total horizontal extent of 150 m. It actually consists of two low, 30-m-wide (now centrally collapsed) and 15-m-wide chambers connected by a low passage. These caves are not related to tumuli nor do they show flow features suggestive of drainage. Rather they seem to be associated with low ridges that are thought to be created by the lateral compression of the upper lava layers, already solidified, caused by the general movement of the lower, still plastic layers, thereby pressed upward and forming low, arched domes. On Hawai'i, Eclipse Cave is of similar type, forming a 70-m-long, up to 2.5-m-high hall, perpendicular to the direction of the flow.

Volcanic Vents

Volcanic vents form pit-like or slanted caves, which are potentially very deep. However, solidified lava and wall collapse normally limit the accessible depth of such caves. Kaukako Crater on Molokai, probably over 350 m deep (100 m above and >250 m below water), is one of the deepest open vents on record. Its diameter narrows down to about 15 to 20 m, *circa* 30 m below the water level. The lake is anaerobic below 4 m of depth and was dived by M. and S. Garman to a depth of 140 m, possibly one of the deepest cave dives in anoxic waters. Of similar depth is the Na-One pit on Hualalai, a vent explored to a depth of over 268 m. The current eruption on Kilauea, at the wall of the Halemaumau Crater, has opened up a 40-m-wide and 160-m-deep pit, at the bottom of which the top of the magma chamber is seen boiling and through which gases, clasts, and ash are ejected. Many more vents exist on Hualalai and Kilauea that have not yet been explored due to the high risk of rockfall. On Iceland, the Þríhnúkagígur is 120 m deep (200 m total), funneling out to a width of 49–70 m below its orifice (Stefánsson, pers. comm.). On Terceira, Azores, the Algar do Carvão, a 90-m-deep vent, has been made accessible to the public. The vent leads into a large chamber hollowed out by convecting basalt magma in a body of trachyte. Into this class of caves we also must count caves in hollow dikes that have been reported from several places.

ROCK SPELEOTHEMS

The term *speleothem* is composed of the Latin word *spelaeum* (poetical for "cave" after the Greek root τό σπήλαιον; to *spaelaion* = "the cave") and the Greek word ὁ θημών (*ho thaemon* = "the pile, deposit"). According to Hill and Forti (1997, p. 13) the term "refers to the mode of occurrence of a mineral and not to its composition." Thus, lava formations are excluded because they are not composed of "a mineral" but of "rock." Hill and Forti (1997, p. 217) admit that "there are deposits in caves or in the outside world which, while not speleothems in the strictest sense, nevertheless mimic the forms taken by speleothems." However, since the term *speleothem* is just indicating a "deposit in a cave" it is logical to use the term *rock speleothem* for those deposits in lava caves that are strikingly homomorphous in appearance to their cousins, the *mineral speleothems* (Kempe, 2011).

Among the most common forms of rock speleothems in lava caves are stalactites and stalagmites, of which several types can be differentiated. Allred and Allred (1998) have investigated cylindrical stalactites. They appear to be extrusions of the ceiling of residual melt. It is not clear if they are growing like soda straws at their tips or if they are extruded at the ceiling interface. Below these stalactites often stalagmitic driplet spires occur, composed of discrete drips of lava melt. In other cases, where lava cascades into pukas, large curtains, stalagmites, and columns can form. Rarely, spattering inside a cave is observed; in Manu Nui (A. and P. Bosted, pers. comm.) apparently water-bearing strata were eroded into and steam explosions threw up spatter that was oxidized surficially, showing all colors from black to tan.

CONCLUSIONS

In spite of the tremendous progress made in lava cave exploration, we still are far from understanding all the features and processes that interact to form caves in volcanic rocks, specifically the large and extensive pyroduct systems. It is clear that the concept of a "tube," simply piping lava downhill, is far too simple to explain the observed morphologies. Furthermore, many published lava cave maps are of limited value because they are not linked to a geological map; many of them lack morphological details and cross-sections often do not show structural information of the lava flow itself. Thus, much more process-oriented analysis is needed in order to advance lava cave research.

See Also the Following Articles

Kazumura Cave, Hawaii
Root Communities in Lava Tubes

Bibliography

Allred, K., & Allred, C. (1997). Development and morphology of Kazumura Cave, Hawai'i. *Journal of Cave Karst Studies, 59*(2), 67–80.

Allred, K., & Allred, C. (1998). Tubular lava stalactites and other related segregations. *Journal of Cave and Karst Studies, 59*(2), 131–140.

Allred, K., Allred, C., & Richards, R. (1997). *Kazumura Cave Atlas, Island of Hawai'i*. Hilo, HI: Hawaii Speleological Survey.

Atkinson, V., & Atkinson, A. (1995). *Undara Volcano and its lava tubes: A geological wonder of Australia in Undara Volcanic National Park, North Queensland*. Brisbane, Queensland, Australia (privately published).

Coan, T. (1844). Letter of March 15, 1843, describing the Mauna Loa eruption of 1843. *Missionary Herald*, 1844.

Coons, D. (2009). The Kipuka Kanohina Cave System, southwestern Hawai'i. In A. Palmer & M. Palmer (Eds.), *Caves and karst of the USA* (pp. 320–321). Huntsville, AL: National Speleological Society.

Francis, P. (1993). *Volcanoes, a planetary perspective*. New York: Oxford University Press.

Greeley, R., Fagents, S. A., Harris, R. S., Kadel, S. D., & Williams, D. A. (1998). Erosion by flowing lava, field evidence. *Journal of Geophysical Research, 103*(B11), 27325–27345.

Halliday, W. R. (Ed.), (1976). *Proceedings 1st International Symposium of Vulcanospeleology and its Extraterrestrial Applications, White Salmon, WA., 16 Aug. 1972* Vancouver, WA: Western Speleological Survey, ABC Publishing.

Hill, C., & Forti, P. (1997). *Cave minerals of the world* (2nd ed.). Huntsville, AL: National Speleological Society.

Hon, K., Kauahikaua, J., Denlinger, R., & Mackay, K. (1994). Emplacement and inflation of pāhoehoe sheet flows: Observations and measurements of active lava flows on Kilauea Volcano, Hawai'i. *Geological Society of America Bulletin, 106*, 351–370.

Kempe, S. (2002). Lavaröhren (Pyroducts) auf Hawai'i und ihre Genese. In W. Rosendahl & A. Hoppe (Eds.), Angewandte Geowissenschaften in Darmstadt, *Schriftenreihe der deutschen Geologischen Gesellschaft, 15*, 109–127 (in German).

Kempe, S. (2011). Morphology of speleothems in primary (lava) and secondary caves. In A. Frumkin (Ed.), *Karst geomorphology*. Amsterdam: Elsevier.

Kempe, S., Bauer, I., & Henschel, H. V. (2003). The Pa'auhau Civil Defense Cave on Mauna Kea, Hawai'i, a lava tube modified by water erosion. *Journal of Cave and Karst Studies, 65*(1), 76–85.

Kempe, S., & Werner, M. S. (2003). The Kuka'iau Cave, Mauna Kea, Hawai'i, created by water erosion, a new Hawaiian cave type. *Journal of Cave and Karst Studies, 65*(1), 53–67.

Lockwood, J. P., & Hazlett, R. W. (2010). *Volcanoes, a global perspective*. New York: John Wiley.

Olafsen, E. (1774–1775). *Des Vice-Lavmands Eggert Olafsens und des Landphysici Bianre Povelsens Reise durch Island, veranstaltet von der Königlichen Societät der Wissenschaften in Kopenhagen* (from the Danish 1st ed., 1772). Kopenhagen und Leipzig: Heinecke und Faber (in German).

von Troil, U. (1779). *Briefe welche eine von Herrn Dr. Uno von Troil im Jahr 1772 nach Island angestellte Reise betreffen*. Upsala und Leipzig: Magnus Swederus (in German).

Walker, G. P. L. (1991). Structure, and origin by injection of lava under surface crust, of tumuli, "lava rises," "lava rise pits," and "lava-inflation clefts" in Hawaii. *Bulletin of Volcanology, 53* (546–558).

W

WAKULLA SPRING UNDERWATER CAVE SYSTEM, FLORIDA

Barbara Anne am Ende

Deep Caves Consulting, South Riding, VA, USA

The source of the Waculla forms a large circular basin of great depth, in which the water appears to be boiling up from a fathomless abyss, as colorless as the air itself.

Charles Latrobe, 1833

INTRODUCTION

Wakulla Spring, located 30 km south of Tallahassee, Florida, is a compelling source of freshwater. An average of nearly 1 million cubic meters of water discharges every day—enough to fill a thousand Olympic-sized swimming pools! A multikilometer-long cave system, with passages up to 30 m in diameter at an average depth of nearly 90 m, provides water to the spring vent. However, during the Pleistocene, sea level was lower and mastodons and other terrestrial animals found their way deep into the passage that currently feeds the spring.

The earliest evidence of human visitation to Wakulla Spring comes from Paleo-Indian artifacts dated from 8500 to 12,000 years BP (Revel, 2002). The first European explorer in the area was Panfilo de Navarez, who in 1528 was searching for the province of Apalachen with its purported gold and riches. He and his Spanish crew ran out of provisions, diverting to the village of Aute for resupply. The exact location of Aute is not certain, though it was probably at or near Wakulla Spring. A recent archaeological investigation at the spring found red, unglazed earthenware olive jar fragments, suggesting Hernando de Soto (retracing Navarez's route) visited Wakulla Spring in 1539. The beauty of this spring has caused much speculation that it might have been the Fountain of Youth sought by Ponce de Leon, though there is no evidence that he visited Wakulla.

The spring's name is a European language bastardization of the Muskhogean *Wacara* (pronounced "wakala") where *cara* meant "spring of water," which is similar to the Creek word *wahkola* meaning "loon." Over the years, the name has been published variously as *Wackhulla*, *Wakully*, *Waculla*, *Wahkula*, and *Wakulila*. In 1843, the name *Wakulla* (pronounced "Wah-cuh-lah") was officially given to the county. This large spring, and the nearby smaller one, Sally Ward, are grouped together and termed the *Wakulla Springs*.

Beginning in the early nineteenth century, American settlers started moving into the Tallahassee area. The Wakulla River, which suddenly appears in the midst of a pine, live oak, and cypress forest, was a draw for local tourists. Its remote location with poor roads helped protect the natural environment. The trip from Tallahassee through the pines was long and monotonous, prompting the rumor that "young people never went to Wakulla Springs without coming back engaged." Enterprising inhabitants provided boat tours for visitors. Several published reports commented on the vertigo induced in visitors who boated over the spring basin because of the clarity of the water. The county held an annual picnic on the site on the first Saturday in May for many years. Several plans and attempts were devised to commercialize the spring given the success of other Florida springs such as Weeki Wachee and Silver Springs, but none of these plans materialized.

In 1934, Edward Ball, while working for his brother-in-law, Alfred I. du Pont, began purchasing land around Wakulla Springs. A year later, he began constructing a hotel next to the spring basin. Though rumored it was to become a private lodge, the hotel was always a commercial operation. However, local hunters were excluded from the land. A fence built across the river denied access to the spring and caused much contention. The county could no longer hold its annual picnic. Nevertheless, Ball resisted the notion of expanding the commercial operation to a gaudy, highly developed tourist trap. Over the years, a number of movies were filmed at Wakulla including two Tarzan movies, *Creature from the Black Lagoon*, and *Airport '77*, in which a 70-foot model of a 747 was perched on the ledge over the cave entrance in the spring basin. Edward Ball died in 1981 and the land went to the Nemours Foundation. It was realized that Wakulla Springs would never make a profit, and the operation was sold to Florida, in 1986 becoming the Edward Ball Wakulla Springs State Park.

Today, the park continues its boat tours of the river where wildlife such as alligators, turtles, deer, and squirrels can be seen at close proximity. Birds abound, including anhingas, which are frequently seen perched, holding their wings out to dry after diving for fish. Glass-bottom boats circle the spring basin when the visibility of the water is good. Swimming from a white sand beach is encouraged, though scuba diving is generally not allowed except by special permit.

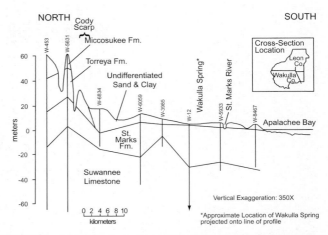

FIGURE 1 Geologic cross-section from the Tallahassee Hills through the Woodville Karst Plain. The approximate location of Wakulla Spring has been projected onto the cross-section line. *After Rupert, 1993.*

GEOLOGY

Geologic Framework

Wakulla Spring lies within the Woodville Karst Plain, a low-lying area underlain by Oligocene- and Miocene-aged limestone overlain by a thin blanket of unconsolidated sands (Fig. 1). The Eocene Ocala Group is a widespread unit that is an important part of Florida's aquifer system. However, near Wakulla Spring it lies well below all known cave passages; the top of the Ocala was encountered between 120 and 180 m below the land surface in oil test wells near Wakulla (Rupert, 1988). The Lower Oligocene Suwannee Limestone unconformably overlies the Ocala Group. The Suwannee Limestone is a calcarenite containing miliolid foraminifera, mollusks, bryozoans, echinoids, and corals. Chert is common and was used by Paleo-Indians for tools and weapons. The passages in Wakulla Cave are developed in the Suwannee Limestone. The Lower Miocene St. Marks Formation unconformably overlies the Suwannee and is the uppermost bedrock unit at Wakulla Spring. The ledge above the Wakulla Spring vent is St. Marks Formation. The unit is primarily a calcilutite-containing quartz sand, clay stringers, and mollusks. The lower contact of the St. Marks Formation lies at an approximately 27-m water depth.

Samples collected during exploration in 1987, as well as visual descriptions and video footage from within Wakulla Cave, indicate a distinct lithologic and color change within the Suwannee Limestone at a depth of about 65 m (Rupert, 1988). Above lies a soft biocalcarenite, whereas below the contact is a less soluble, recrystallized dolomitic calcarenite. The dolomitic rock may have retarded additional downward dissolution in the conduits. The 3D map produced during the 1998–1999 diving project shows that the floor level for much of the cave is uniform at a depth of about 90 m (Fig. 2).

During the Pleistocene, the shoreline transgressed across this area, reworking sands from older formations and depositing the sediment over the limestone. Five marine terraces are recognized in Wakulla County, and Wakulla Spring lies within the Pamlico Terrace, ranging from 3 to 8 m above sea level. The Cody Scarp (Fig. 1) is an escarpment marking the boundary between the northern Tallahassee Hills and the Coastal Lowlands to the south. The boundary represents the ancient shoreline location.

Paleontology

The first correct identification of the enormous bones seen through the clear water on the bottom of the Wakulla Spring basin was made by Sarah A. Smith for the *Tallahassee Floridian and Journal* in 1850. This publication prompted a local professor to collect a significant portion of a mastodon skeleton, but the bones were lost in a shipwreck on their way to a museum on the Atlantic coast.

Additional bones were collected during the next decades. In 1930, more mastodon bones were discovered in shallow water during construction of a swimming area at Wakulla Spring (Rupert and Spencer, 1988). The U.S. Geological Survey was enlisted in its collection and the skeleton remains on display today

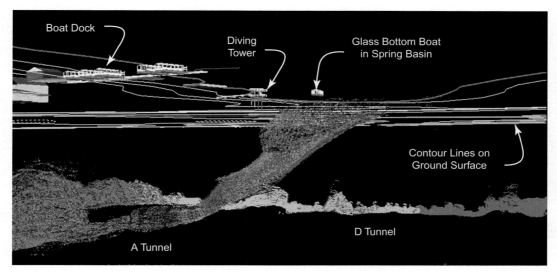

FIGURE 2 Perspective view of the entrance area to Wakulla Spring Cave. Each colored dot represents a wall point surveyed with the digital wall mapper. Different colors of dots are from different data files. Topographic contours of the ground surface are shown in yellow and the boundary of the Wakulla River is shown in blue.

FIGURE 3 Mastodon skeleton in the Museum of Natural History, Tallahassee, assembled with bones found at Wakulla Spring, Florida. *Photo by the author.*

in the Museum of Natural History in Tallahassee (Fig. 3).

Dives by Wally Jenkins, Garry Salsman, and their buddies in 1955 and 1956 (see exploration section below) resulted in the discovery of mastodons, mammoths, deer, camels, giant ground sloths, and bears, along with many spear points from Paleo-Indians. The divers used pillowcases lined with plastic bags inflated by air from their tanks to lift the heavy bones to the surface from depths as great as 60 m. During later exploration Pleistocene mammal bones were discovered as far back as 366 m from the entrance.

Hydrology

Wakulla Spring is one of 33 first-magnitude springs in Florida. Average discharge from 1907 to 1974 was 11 $m^3 s^{-1}$ (Rosenau et al., 1977). Wakulla Spring displays the greatest range of discharge of any Florida spring. A minimum flow of 0.7 $m^3 s^{-1}$ was recorded on June 18, 1931, whereas a maximum flow 54 $m^3 s^{-1}$ was reported on April 11, 1973.

The presence of fossil mammal bones located deep within the cave (Fig. 4), coupled with lower Pleistocene sea level (and by extension, base level on land), constrains hydrologic models. It has been suggested that mammals wandered into the dry cave entrance looking for water. Further, it was hypothesized that the Wakulla Spring may have been a sink where water entered the aquifer at the time. Indeed, a report from another Florida spring noted the presence of an extinct land tortoise with a wooden stake stuck in its shell at 26 m below the current water level. A less likely explanation published for the presence of mastodon bones in Wakulla spring was that "In winter, mastodons crossing frozen pools broke through the ice and drowned."

Directly north of the Wakulla area, an unconfined surficial aquifer system is found in the unconsolidated sands and gravels of the Tallahassee Hills. Recharge occurs by direct precipitation. An intermediate aquifer

FIGURE 4. Mastodon bone lying nearly 200 m inside Wakulla Spring Cave at a water depth of about 80 m. *Photo courtesy of Wes Skiles.*

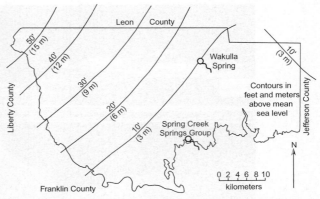

FIGURE 5 Potentiometric surface map of the Upper Floridan aquifer in Wakulla County. *After Meadows, 1991.*

system lies below the surficial sediments ranging from about 15 to 46 m thick. The bedrock contains interlayered clayey sediments, limestone, and dolomite resulting in discontinuous water-bearing zones. Water recharges into the intermediate aquifer by leakage from the surficial aquifer and from sinkhole-drained lakes. A dramatic example of flow into the aquifer from a lake in the Tallahassee Hills occurred on September 16, 1999. Much of the 1620-hectare Lake Jackson in the Tallahassee Hills suddenly drained down Porter Hole, a 5-m-deep sinkhole in the lake bed. The pit leading down from the sink swallowed the lake's southern half in a single day. Similar drainage events have occurred in the past as well.

Below the surficial and intermediate aquifers lies the Floridan aquifer system, which is a major carrier of water and extends through much of the northern part of the state. In the Woodville Karst Plain, the Floridan aquifer is comprised of the St. Marks Formation, Suwannee Limestone, and Ocala Group. Transmissivities are high and range from 465 to 11,613 m^2 per day. Recharge comes from downward leakage from the intermediate aquifer in the Tallahassee Hills and through sinkholes. The Woodville Karst Plain also forms a recharge area via direct rainfall and through sinkholes. Four streams sink underground, also contributing to recharge, though one reemerges.

Regional groundwater flow of the Upper Floridan aquifer is to the southeast across Wakulla County (Fig. 5). In the Woodville Karst Plain the Floridan aquifer is unconfined and no low-permeability units lie between the surface and carbonate aquifer units.

The primary source of Wakulla Springs was determined, through the use of uranium isotopes, to be southward-flowing Floridan aquifer water. Further, strontium isotopic ratios were used to conclude that the spring water has not come solely from local recharge, but rather comes from water that has been in contact with Floridan aquifer bedrock for considerable time. Conversely, tritium age dating for shallow and deep groundwater gave recharge times of less than 30 years.

The conduit flow of groundwater in the Wakulla Spring cave and the Woodville Karst Plain is remarkably complex considering the relatively uniform piezometric surface of the area (Fig. 5). A groundwater divide has been reported 1 to 2 km inside the main tunnel of Wakulla Spring. The divide marks a divergence between water flowing north to the Wakulla Spring vent and water flowing approximately along the regional groundwater gradient south, presumably to the Spring Creek Springs Group of 13 submarine springs. This flow divergence is perplexing because one would not expect diffuse aquifer percolation to supply water to a tunnel approximately 30 m in diameter.

The Spring Creek Springs Group has pulsating changes in flow where the surface of the water alternates between flat quiescence and boiling surface turbulence (Lane, 2001). The alternating surges generally last for several minutes and are thought to be a result of flushing through complex, tortuous passages. Since the Spring Creek Springs Group is likely to be connected, at least indirectly, with the nearby Wakulla Spring, much remains to be learned about the flow within the Woodville Karst Plain.

The importance of understanding the sources of water for Wakulla Spring is apparent when considering the clarity of water discharging from the spring. The vertigo-inducing clarity of the water in the nineteenth and early twentieth centuries has diminished during the past few decades. One of the first reports of low visibility came in 1894: "The water has been stirred up by the heavy rains, and we could only see down 80 feet [24 m]," less than the maximum of 38 m. Reports in 1945 and 1946 by the commercial operation run by Ed Ball noted that visibility was affecting the ability to run glass bottom boat tours and turn a profit. The visibility is diminished primarily from tannic and humic acids in surficial swamp or river waters that enter the karst. Particulate matter may also contribute to lower visibility, but the dark color of the tannic water is a bigger problem. Boat guides at Wakulla report that dark water has become a more frequent, longer duration problem. Records of dark water indicated water visibility was poor for 58% of the time during the past 12 years and that the poor visibility is correlated to rainfall events. The best chance for crystal clear water is during the dry months of May and June.

Another significant problem with water quality at Wakulla Springs is the increase in nitrates, presumably the result of runoff from fertilizers used on lawns and agriculture. Simultaneous has been the introduction of exotic algae, particularly hydrilla. A virtual explosion of algae growth has choked much of the Wakulla Spring basin and river in recent years, probably enhanced by the high nitrate levels. The state park has attempted to manage the hydrilla using mowers and divers to remove as much as they can, but it has been a losing battle. Recently, the park applied an herbicide to the spring basin to improve visibility of the spring by glass bottom boat tours and to regain a healthy ecologic balance in the river system. It is hoped that native plants will reestablish themselves and that the hydrilla will not be reintroduced.

Speleogenesis

Little work has been conducted on the speleogenesis of Wakulla Spring cave system considering its magnitude. The reason is probably that the cave system is entirely flooded and averages a nearly 90-m water depth.

One study suggested that karstification began approximately 9000 years BP based on the regional structural geology, sea-level fluctuations, climatic change, and groundwater flow of the Woodville Karst Plain and its offshore extension. A model for the origin of Wakulla Spring described the tunnels as a branching, flow-dominated, saturated cave. A four-stage sequence of development started through self-initiation. Small, random variations in permeability created positive feedback loops. The upgradient process was based on geochemical feedback. Downgradient, hydrodynamic processes governed the feedback loop and was enhanced by corrosion from mixing of waters. The largest conduits were developed in the cave when sea level was lower during the Pleistocene. Discharge occurred in springs that are presently submarine, while recharge entered through a sinkhole that now is the current Wakulla Spring.

EXPLORATION AND MAPPING OF WAKULLA SPRING CAVE

A filming crew arrived at Wakulla Spring in 1955 for one of the many movies that needed a clear water setting. The movie company hired Garry Salsman, a Florida State University (FSU) student, to assist because he owned an air compressor to fill diving tanks (Burgess, 1999). He and a buddy, Wally Jenkins, made their initial dive into the cave after completing the first day's filming. On their second dive into the cave, the two explorers discovered a mastodon long bone some 60 m inside and carried it out of the cave. The manager of the spring was so excited about the find that he permitted Salsman and his FSU buddies to continue diving in Wakulla.

Equipment at that time was quite primitive and techniques for diving at such depths (55 m and heading down) were not well established. Early dives in the 69°F water were made in nothing more than swimming trunks. Hardware-store flashlights wrapped in plastic bags were used for lighting. The scuba tanks did not have pressure gauges, so the students would venture in on each dive just a little bit farther than the previous one, to ensure an adequate air supply. The team of divers diligently created safety plans and avoided decompression sickness. However, by 1957 their diving ended, having reached nearly 300 m into the cave at a water depth of 73 m. In part because of limitations in technology, in part because of the students graduating and moving on, and because of liability concerns by the springs manager, diving in Wakulla Springs ended in 1961 for two decades. In 1981 three Florida cave divers used diver propulsion vehicles and stage bottles to penetrate 335 m into the cave at 79-m depth. Nitrogen narcosis was a significant impairment factor, as it had been for the FSU students.

When Wakulla Springs was sold to the state of Florida, cave divers hoped to obtain access to the underground tunnels once again. In 1987 the U.S. Deep Caving Team, Inc. (USDCT), received a permit to dive in the cave for a 2-month period in the fall (Stone, 1989). State officials were particularly interested in obtaining a map of the cave, as well as geologic, hydrologic, and biologic sampling and studies. Bill Stone, leader of the project, was in the process of designing a mixed-gas rebreather, the MK-1, and wanted the opportunity to test the device. During this project, the rebreather was successfully tested to a depth of 43 m and Stone was able to remain underwater continuously for 24 hours. However, the rebreather was not used for missions into the cave. For this, the divers used clusters of scuba tanks filled with mixed gas attached to the diver propulsion vehicles, as well as tanks on the divers' backs. Dives were made using dry suits, and a specially designed habitat was built to allow divers to decompress in a "bubble" of air. The total surveyed length of the cave was 3310 m at a maximum depth of 110 m. The culmination of diving was due to time and technology constraints. The intention was to return to explore the tunnels that beckoned onward into the darkness.

A group of cave divers known as the Woodville Karst Plain Project (WKPP) who had been diving in various sinks and springs in the area began diving in Wakulla Spring in 1991. During the next few years, the team explored and surveyed extensively, culminating in an impressive dive with a reported maximum penetration in Wakulla Spring of 5488 m using semiclosed rebreathers. The most recent maps published by the WKPP extend in maximum distance from the entrance to about 3000 m.

In December 1998, the USDCT returned to Wakulla Springs for the Wakulla 2 project (am Ende, 2000). Stone's rebreather was no longer a prototype and the MK-5 was commercially available. Diver propulsion vehicles were powered by electric automobile batteries. Decompression was conducted in a rented decompression chamber deployed in the spring basin. The purpose of the expedition was not exploration (although 1.1 km of new passage was discovered), but rather mapping. A special device, the digital wall mapper (DWM) was designed and built for the project. Also powered by electric auto batteries, the DWM kept track of its location and measured distances to the walls of the cave using sonar (Fig. 6). A total of about 10 million points on the walls of the cave were surveyed during the 3-month expedition. A total of 6.4 km of tunnels were imaged this way (Fig. 7). The point cloud generated during the project shows exquisite detail of passage morphology that had not heretofore been measured in caves. Current efforts are aimed at meshing the point cloud into solid polygons (Fig. 8) to render a more realistic cave.

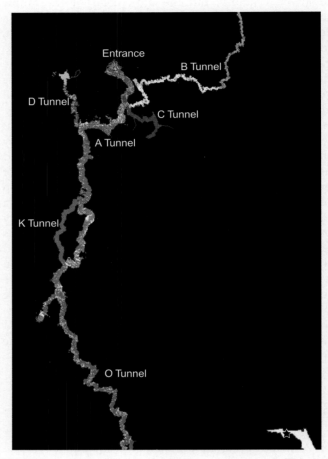

FIGURE 7 Map of Wakulla Spring Cave showing 3 million wall points. In B, C, and D Tunnels, segments of passage mapped by hand are shown by lines.

FIGURE 6 Divers entering the tunnel with DWM equipment. *Photo courtesy of Wes Skiles.*

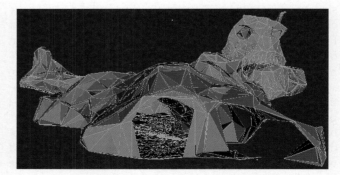

FIGURE 8 Image of a meshed portion of the Wakulla point cloud. The points are connected by cream-colored lines to form solid blue triangles. This section of cave is at the junction of A and C Tunnels.

In a separate effort, a 3D map of the Wakulla Spring cave was built based on a traditional survey by the WKPP (Kinkaid, 1999). Solid walls were created from survey station data of four wall distance estimates taken about 10 to 100 m apart (269 stations and 1076 estimated wall points). Although this map lacks the detail of the USDCT map, the sparseness of data makes it more easily viewable on the average personal computer. In July, 2007, Jarod Jablonski and Casey McKinlay of the WKPP made the impressive connection between the Leon Sinks underwater cave system and passage they explored in Wakulla Spring. The total cave system stands at 51 km in length and 107 m in depth.

CONCLUSION

The end of Wakulla Spring Cave is nowhere in sight and exploration will probably continue. However, even with the best technical diving equipment available, the current frontier is extremely remote. It may require yet greater advances in diving technology to make significant additions to the exploration of Wakulla Spring.

Long periods of time with poor visibility of the water also hampers exploration. The source of low-quality water in terms of clarity and nitrates has not been directly identified. It is probably from nonpoint source pollution such as excessive runoff from cleared and/or paved areas and residential and agricultural contaminants. The state of Florida is currently taking many positive steps to preserve and restore the quality of water in its many springs. Time will tell if the state can overcome the ecological pressure created by the modern lifestyle of a large population in the area around Wakulla Spring.

See Also the Following Article

Exploration of Caves—Underwater Exploration Techniques

Bibliography

am Ende, B. (2000). Wakulla 2—Building the first fully 3D cave map. *NSS News*, 58, 244–260 (270).

Burgess, R. F. (1999). *The cave divers*. New York: Aqua Quest Publications.

Kincaid, T. R. (1999). *Morphologic and fractal characterization of saturated karstic caves*. Laramie: University of Wyoming.

Lane, E. (2001). *The Spring Creek Submarine Springs Group, Wakulla County, Florida* (Special Publication No. 4+). Tallahassee, FL: Florida Geologic Survey.

Meadows, P. E. (1991). *Potentiometric surface of the Upper Floridan aquifer system in the Northwest Florida Water Management District, Florida, May, 1990* (U.S. Geological Survey Open File Report 90–586).

Revel, T. J. (2002). *Watery eden: A history of Wakulla Springs*. Tallahassee, FL: Sentry Press.

Rosenau, J.C., Faulkern, G.L., Hendry, W.W., Jr. & Hull, R.W. (1977). *Springs of Florida*. Tallahassee: Florida Geological Survey, Bulletin 31 (revised), 461 p.

Rupert, F. R. (1988). *The geology of Wakulla Springs* (FGS Open File Report No. 22). Tallahassee, FL: Florida Geological Survey.

Rupert, F. R. (1993). Karst features of northern Florida. In S. A. Kish (Ed.), *Geologic field studies of the coastal plain in Alabama, Georgia, and Florida: Southeastern section* (Meeting Guidebook, Tallahassee, FL, April 1998, pp. 49–61).

Rupert, F., & Spencer, S. (1988). *Geology of Wakulla County, Florida* (Bulletin No. 60). Tallahassee, FL: Florida Geological Survey.

Stone, W. C. (1989). *The Wakulla Springs project*. Derwood, MD: The U.S. Deep Caving Team.

WATER CHEMISTRY IN CAVES

Janet S. Herman
University of Virginia

INTRODUCTION

Caves would not exist in limestone bedrock if it were not for water. Interestingly, water is not only the agent of formation of the natural openings in the Earth that we call caves, but it is also the agent of cave decoration. The duality of water's role in bedrock dissolution to form the cave and in mineral precipitation to decorate the cave can be understood through an examination of the chemistry of cave waters. Just as any natural water acquires its chemical composition, cave waters acquire solutes from many different sources, including gases from the atmosphere, weathering and dissolution of the minerals of soils and bedrock, and human activities.

LIMESTONE DISSOLUTION AND THE CHEMICAL EVOLUTION OF CAVE WATERS

Calcite, $CaCO_3$, is the principal component of limestone. The solubility of calcite in natural waters is established by chemical equilibria among the solid calcite, the dissolved calcium and carbonate-bearing species in aqueous solution, and the carbon dioxide in the gas phase in contact with the solution. Indeed, the pH of most natural waters is controlled by reactions involving the carbonate system, calcium is the predominant cation in most freshwaters, and most groundwater in contact with limestone is near equilibrium with respect to calcite solubility. These facts allow us to understand the chemical evolution of cave waters using the tools of chemical thermodynamics.

TABLE 1 Chemical Composition for Water in Equilibrium with Respect to Both Calcite and CO_2 Gas at Two Different P_{CO_2} Values and at Two Different Temperatures

P_{CO_2}	Temperature (°C)	pH	Ca^{2+} (mg L^{-1})	HCO_3^- (mg L^{-1})
$10^{-3.5}$	12	8.3	24	72
$10^{-3.5}$	25	8.3	19	58
$10^{-1.5}$	12	7.0	120	372
$10^{-1.5}$	25	7.0	96	298

These values for Ca^{2+} and HCO_3^- concentration were obtained in a hypothetical calculation of simultaneous equilibria.

Role of CO_2 Gas in Calcite Solubility

We can describe the dissolution of calcite in natural waters by

$$CaCO_3(s) + H_2O + CO_2(g) \leftrightarrow Ca^{2+} + 2HCO_3^- \quad (1)$$

where the role of carbon dioxide is to generate acid. Dissolved carbon dioxide is the most common acid in natural waters, and the reactions that give rise to free acid are the formation of carbonic acid from the hydration of CO_2 gas:

$$CO_2(g) + H_2O \leftrightarrow H_2CO_3^0 \quad (2)$$

and the subsequent dissociation of carbonic acid according to

$$H_2CO_3^0 \leftrightarrow H^+ + HCO_3^- \quad (3)$$

$$HCO_3^- \leftrightarrow H^+ + CO_3^{2-} \quad (4)$$

A natural water in contact with a gas phase containing CO_2, therefore, has the dissolved carbonate-bearing species (i.e., $H_2CO_3^0$, HCO_3^-, and CO_3^{2-}) in solution regardless of contact with a solid carbonate phase such as calcite.

The first fact of the chemistry of cave waters is therefore evident. The extent to which CO_2 gas is dissolved in infiltrating waters determines the extent to which acid is available to carry out limestone dissolution to form caves out of previously solid bedrock. Caves in limestone bedrock in continental settings are typically formed by the circulation of shallow groundwater. These waters derive from the infiltration of dilute, fresh rainfall into soil. Microbial respiration in the soil biodegrades natural organic matter (presented simply here as CH_2O from dead vegetation) and generates CO_2 gas according to

$$CH_2O(s) + O_2(g) \leftrightarrow CO_2(g) + H_2O \quad (5)$$

Infiltrating water dissolves the CO_2 gas available in the soil environment, thereby generating the acidity for dissolution of limestone bedrock as the water continues along its flow path.

The measure of the amount of CO_2 gas in the atmosphere is its partial pressure (P_{CO_2}), where the total atmospheric pressure is 1 bar. The value of the partial pressure of CO_2 ranges from the low value of $10^{-3.5}$ bar in the open atmosphere at the Earth's surface, to the higher values of $10^{-2.2}$ to $10^{-2.0}$ bar in the typical soil atmosphere of a humid, temperate climate, to a high value of $10^{-1.5}$ bar in a notably CO_2-rich cave. This range in partial pressure of CO_2 gas is an important factor in the starting point for calcite dissolution. A pure water in equilibrium with $10^{-3.5}$ bar CO_2 has a pH of 5.7, whereas a water in equilibrium with $10^{-1.5}$ bar CO_2 has a pH of 4.7 at 25°C. The greater amount of CO_2 generates a distinctly more acidic solution.

The greater the amount of CO_2 dissolved in the groundwater, the greater the extent of dissolution of calcite at equilibrium [see Eq. (1)]. The solubility of a mineral is the mass of mineral that can dissolve in a volume of aqueous solution. At equilibrium between solution and mineral, no further dissolution occurs nor does mineral precipitate from solution. An idealized calculation of pure water in equilibrium with respect to calcite at each of two different partial pressures of CO_2 and two different temperatures illustrates the impact these two variables have on solution composition (Table 1). The availability of CO_2 gas is clearly an important factor in determining the extent of calcite dissolution. Temperature is the other. Like all gases, CO_2 solubility is greater at lower temperatures than at higher temperatures. The greater CO_2 gas solubility that is possible at 12°C in groundwater compared to a warm 25°C surface water further promotes calcite dissolution.

Chemical Composition of Groundwater in Limestone Terrain

The resulting chemical character of the groundwater in caves is one rich in dissolved calcium (Ca^{2+}) and bicarbonate (HCO_3^-). The pH of cave waters tends to be relatively high because the protons released from carbonic acid are consumed in the dissolution of calcite. Equation (1) demonstrates the molar relationships of Ca^{2+} and HCO_3^- in this system: 2 mol HCO_3^- are generated for every mole of Ca^{2+}. The typical

TABLE 2 Chemical Composition of Groundwater Collected in Limestone Terrain

Sample Location	pH	Ca^{2+}	Mg^{2+}	Na^+	K^+	HCO_3^-	SO_4^{2-}	Cl^-	NO_3^-
State College, PA, U.S.A.	7.36	54.9	23	6	1.4	238	16.5	13.5	22
Mallorca, Spain	7.25	119	25	31	2.9	334	99	58	4.7
Warm River Cave, VA, U.S.A.	6.72	180	32.7	4.3	18.2	341	3.6	ND	
Huntsville, AL, U.S.A.	7.5	48	3.6	2.1	ND	152	3.2	8.0	ND
Manatee County, FL, U.S.A.	7.59	93.3	48.9	47.0	8.1	159	352	47.0	ND
Yucatán Peninsula, Mexico	7.23	94	28	16	0.6	382	5.9	34	14

Some samples were collected in caves; some samples were collected in wells drilled in limestone bedrock. All chemical concentrations are reported in units of milligrams per liter. ND = not determined.

groundwater collected from limestone terrain, then, has very large concentrations of dissolved HCO_3^- compared to groundwater in other geological settings (Table 2).

Unlike the hypothetical case of pure water in contact with pure calcite and a CO_2-containing atmosphere (e.g., Table 1), Ca^{2+} and HCO_3^- are not the only dissolved chemicals in natural water. Limestone bedrock, although predominantly made up of calcite, ordinarily has some other minerals present. Rainwater infiltrating through soil and groundwater migrating to greater depths come in contact with these various solid phases.

A mineral commonly associated with calcite is dolomite, $CaMg(CO_3)_2$, whose dissolution is described by

$$CaMg(CO_3)_2(s) + 2H_2O + 2CO_2(g) \leftrightarrow Ca^{2+} + Mg^{2+} + 4HCO_3^- \quad (6)$$

In addition to the common occurrence of dolomite in limestone bedrock, some caves are developed in dolostone, bedrock made up primarily of the mineral dolomite. Dissolved Mg^{2+} is, therefore, a major constituent in many cave waters (e.g., Table 2). Gypsum, $CaSO_4 \cdot 2H_2O$, may be associated with calcite and dolomite in some sequences of sedimentary rock. If present, this highly soluble sulfate mineral can have a big impact on the chemical composition of cave waters

$$CaSO_4 \cdot 2H_2O(s) \leftrightarrow Ca^{2+} + SO_4^{2-} + 2H_2O \quad (7)$$

For instance, in the sample from Manatee County, Florida, U.S.A. (Table 2), the groundwater circulation put it in contact with a gypsum layer at depth, and the resultant groundwater had higher dissolved SO_4^{2-} concentrations than HCO_3^- concentrations.

The other common rock-forming elements of Na^+ and K^+ may enter solution from dissolution of minerals or from ion-exchange reactions whereby the base cations occupying the surface sites of clay minerals in the limestone bedrock or in the sediments of the cave stream can exchange by

$$Ca^{2+} + 2Na^+ - clay \leftrightarrow 2Na^+ + Ca^{2+} - clay \quad (8)$$

This mass action expression illustrates the shift in the composition of the ion-exchange complex on the clay mineral surface with shifting concentrations of dissolved cations. When more dissolved Ca^{2+} is present from calcite dissolution, it tends to shift toward sorption onto the clay [i.e., Eq. (8) shifts to the right]; however, when more dissolved Na^+ is present from salt water intrusion, Na^+ will move toward occupying more surface sites [i.e., Eq. (8) shifts to the left].

The anions commonly present in cave waters may derive from any of several sources. Neither Cl^- nor NO_3^- is a major rock-forming element. Both chemicals have atmospheric sources, especially in regions where the atmospheric chemistry is impacted by industrial emissions, and, thus, they may enter with rainfall. Cl^- may be introduced to the groundwater of near-coastal limestone terrain by the dissolution of sea spray or marine aerosols into rainfall or by salt water intrusion in limestone aquifers. Several samples in Table 2 have been influenced by the proximity of marine waters (Mallorca, Spain; Manatee County, Florida, U.S.A.; Yucatán Peninsula, Mexico). Furthermore, both chemicals may also come in as contaminants in waters influenced by animal fecal wastes, road salts, and agricultural chemicals.

Saturation State of the Aqueous Solution

The chemical status of a water capable of dissolving calcite is termed *undersaturated*. That is, the solution is *undersaturated* with respect to calcite and will dissolve calcite until equilibrium solubility is obtained [i.e., Eq. (1) will shift toward the right]. An undersaturated solution has less dissolved Ca^{2+} and HCO_3^- in it than it would have at equilibrium. Such a solution is also described as being aggressive toward calcite in that it is capable of dissolving more calcite in the cave-forming process. A water initially at equilibrium with respect to

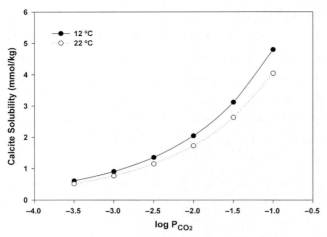

FIGURE 1 Solubility of calcite as a function of P_{CO_2} at two different common groundwater temperatures, 12° and 22°C, for 1 bar total pressure. For a given P_{CO_2} value (in bars), water samples with total dissolved calcium concentrations (in mmol kg^{-1}) plotting below the line are undersaturated with respect to calcite and capable of dissolving calcite; solution compositions on the curves are in equilibrium with respect to calcite and no further reaction can occur without a change in environmental conditions; solution compositions plotting above the line are supersaturated with respect to calcite and capable of precipitating calcite. The retrograde solubility of calcite can also be observed by noting that at a given P_{CO_2} value, the equilibrium solution at 12°C is capable of holding more total dissolved calcium (greater calcite solubility) than is the 22°C solution.

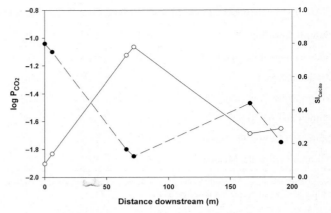

FIGURE 2 Plot of the log partial pressure of CO_2 (dashed line) and saturation index with respect to calcite (solid line) as a function of distance along the flow path for a cave stream, Warm River Cave, Virginia. From the point of entry into the cave room, the stream water is outgassing CO_2 from an initial high concentration in the groundwater toward a lower value in equilibrium with the cave atmosphere. The decreased P_{CO_2} thereby drives the solution from an initial condition near equilibrium with respect to calcite toward greater and greater supersaturation. Precipitation of calcite begins at some greater distance (delayed time) from the point of water entry into the cave. As calcite precipitation proceeds, the solution approaches equilibrium with respect to calcite at a new, lower P_{CO_2} level.

calcite can be shifted toward undersaturation by changes in the two important factors discussed above: P_{CO_2} and temperature. An increase in P_{CO_2} or a decrease in temperature will cause a water to again become undersaturated with respect to calcite and to further the dissolution of limestone bedrock. Conversely, an equilibrium solution can be shifted toward being *supersaturated* and capable of precipitating calcite [*i.e.*, Eq. (1) will shift toward the left] with a loss of CO_2 gas or an increase in temperature. A supersaturated solution has more dissolved Ca^{2+} and HCO_3^- in it than it would have at equilibrium. An adjustment toward equilibrium is achieved by precipitating calcite from solution.

A summary of the influences of P_{CO_2} and temperature is illustrated in a solubility diagram (Fig. 1). Along each solid curve, the dissolved Ca^{2+} concentration and the pH are the equilibrium values for the P_{CO_2} and temperature indicated. A groundwater with lower dissolved Ca^{2+} concentration or with lower pH than the values on the curve is undersaturated and capable of calcite dissolution. A groundwater with higher dissolved Ca^{2+} concentration or with higher pH than the values on the curve is supersaturated and capable of calcite precipitation.

The saturation state of a solution is usually quantified through the calculation of the saturation index. The saturation index is the logarithm of the ratio of the actual dissolved Ca^{2+}, HCO_3^-, and H^+ concentrations to those expected at equilibrium. The log of that ratio is greater than 0 for a supersaturated solution, less than 0 for an undersaturated solution, and equal to 0 at equilibrium.

The process of CO_2 gas exchange in cave waters often controls the saturation state of the solution. Groundwater rich in dissolved CO_2 picked up during infiltration through an organic-rich soil and circulation in the confined subsurface will have a high P_{CO_2}. When such a groundwater enters a cave, the water begins to equilibrate with the cave atmosphere that is usually less rich in CO_2 by virtue of circulation and exchange with the open atmosphere on the Earth's surface. The outgassing of CO_2 from a water entering the cave can drive the pH up and the saturation state of the water from being near equilibrium with respect to calcite to being significantly supersaturated with respect to calcite. In essence, the loss of CO_2 to the atmosphere acts to drive Equation (1) back to the left. If calcite does not immediately precipitate, the result is a supersaturated solution capable of precipitating calcite. A longitudinal sampling of a cave stream in Virginia showed just such a trend in saturation state of the solution along the flow path (Fig. 2). Until the solution can reach equilibrium by precipitating calcite, it remains supersaturated with saturation indices significantly greater than 0.

Environmental Factors Influencing Calcite Dissolution

The critical role of CO_2 gas in shifting the chemical equilibria toward more calcite dissolution in the presence of more CO_2 gas or toward less calcite dissolution in the presence of less CO_2 gas is the basis for understanding some of the climatic effects on cave development. In a moist, warm environment, the growth and decay of terrestrial vegetation occurs to greater extents and at greater rates than in cool, dry environments. The overall result is greater availability of CO_2 gas to generate carbonic acid in a warm, organic-rich environment. Cave development in tropical settings, therefore, is recognized to be far greater than in arctic or desert settings. Some of the most dramatic caves we know from locations such as Puerto Rico, Mexico, and Belize developed under tropical climatic conditions.

There are other hydrogeological and geochemical settings that generate waters undersaturated with respect to calcite and capable of dissolving limestone to form caves. The mixing of freshwater and saline water in coastal regions gives rise to an aggressive solution that can dissolve more calcite than either individual type of water can. The chemical characteristics of groundwater in coastal regions are those of a mixture of two solutions: (1) a dilute, fresh groundwater such as might be generated by the infiltration of rain into a soil and into an underlying limestone where calcite dissolution occurs, and (2) seawater that is intruding into the limestone aquifer. The dense saline water that encroaches into coastal aquifers forms the downgradient boundary of the fresh groundwater circulation system where discharging freshwater mixes with intruding seawater to form a zone of brackish groundwater.

The chemical reactivity of the brackish-water mixture results from the nonlinearity of mineral solubility as a function of the relative concentrations of dissolved constituents in the mixture. The great contrasts in solute concentrations between the mixing end members make it possible to generate a mixture that is capable of further dissolution of calcite from two solutions that were individually in equilibrium, or even supersaturated, with respect to calcite (Fig. 3). This renewed potential for calcite dissolution in the coastal mixing zone has been recognized as an important contribution to the formation of caves in the near-coastal setting. Caves developed along the coasts of Florida, Mexico, Spain, Greece, and the Bahamas have been influenced by calcite dissolution in the freshwater–saline water mixing zone. The chemical composition of water in the resulting coastal caves once again has Ca^{2+} and HCO_3^- as major constituents, but the overall character of the water could be described as diluted seawater with sodium (Na^+), chloride (Cl^-), and sulfate (SO_4^{2-}) concentrations also being high; however, the source of these solutes is not in the local bedrock.

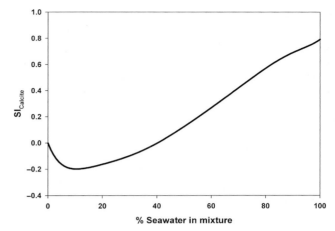

FIGURE 3 Saturation state of a solution derived from mixing fresh and saline waters. The mixing curve was generated for a fresh groundwater in equilibrium with respect to calcite at $P_{CO_2} = 10^{-2.2}$ bar and 22°C. The freshwater was mixed with a standard composition of surface seawater for which $P_{CO_2} = 10^{-3.3}$ bar. The mixture is undersaturated where the mixing curve plots below value $SI_{Calcite} = 0$; the mixture is supersaturated where it plots above the value $SI_{Calcite} = 0$. Starting with a saturated freshwater and a supersaturated saline water, mixing produces an undersaturated brackish water which is capable of calcite dissolution that contributes to the formation of caves in coastal limestone.

Role of Strong Acids in Limestone Dissolution

Although the abundant natural acid in hydrogeological systems is carbonic acid, there are settings in which other natural acids are important. The migration of hydrogen sulfide into shallow, oxygenated groundwater environments from greater depths where reducing conditions prevailed sets up the potential for the generation of strong acid in the subsurface environment. For instance, caves of the Guadalupe Mountains of New Mexico have been formed by sulfuric acid derived from the oxidation of hydrogen sulfide migrating from underlying oil and natural gas reservoirs. As dissolved H_2S gas comes in contact with dissolved O_2, oxidation generates acid in the groundwater:

$$H_2S(g) + 2O_2(g) \leftrightarrow 2H^+ + SO_4^{2-} \quad (9)$$

Once acid is present, the dissolution of calcite looks just as it did for the carbonic acid source:

$$CaCO_3(s) + H^+ \leftrightarrow Ca^{2+} + HCO_3^- \quad (10)$$

but the chemical composition of the cave water has a high concentration of SO_4^{2-} that is not seen in the more typical solution caves.

The oxidation of other sulfide phases may also generate sulfuric acid for limestone dissolution. Iron sulfide minerals, mainly pyrite (FeS_2), and other base-metal sulfide minerals are common in the rocks of the Earth's crust. Although stable under reducing conditions, with exposure to O_2-containing air and water, the sulfide minerals undergo oxidation as in this reaction with pyrite:

$$FeS_2(s) + O_2(g) + H_2O \leftrightarrow Fe^{2+} + 2H^+ + 2SO_4^{2-} \quad (11)$$

Just as in the oxidation of hydrogen sulfide gas, acidity is released to solution that can further react with limestone bedrock according to Equation (10). Although natural weathering of bedrock containing sulfide minerals does occur, the acceleration of this reaction due to the human activity of mining and processing minerals far outstrips the natural occurrence of acidity from such reactions. Today, the contact of acid mine drainage with limestone bedrock in active or abandoned mining districts may be contributing to the formation of new conduits and caves in the regional bedrock.

One environment fosters both reduction and oxidation reactions involving sulfur. In the saline waters of coastal caves and blue holes, density stratification of the groundwater in the cave or aquifer may limit the mixing of shallow oxygenated waters to greater depth. At depth, modest concentrations of dissolved and suspended organic matter can result in reduction of SO_4^{2-} as the organic matter undergoes microbially mediated oxidation in the absence of dissolved O_2. The use of SO_4^{2-} as an alternate electron acceptor results in the formation of dissolved sulfide:

$$SO_4^{2-} + 2CH_2O \leftrightarrow H_2S(g) + 2HCO_3^- \quad (12)$$

where the organic matter, CH_2O, may be dissolved or particulate. Further migration of H_2S along a groundwater flow path may bring the water into contact with O_2 where the H_2S oxidizes to sulfate, generating acidity [Eq. (9)]. The increased capacity to dissolve limestone has contributed to the formation of caves and blue holes on the islands of the Caribbean.

PRECIPITATION OF MINERALS AND THE FORMATION OF SPELEOTHEMS FROM CAVE WATERS

Not all cave waters derive their final composition from the process of calcite dissolution. Just as the increased abundance of CO_2 gas was a factor in shifting the extent of calcite dissolution in undersaturated subsurface waters, the same factor in reverse can shift cave water to being supersaturated and capable of mineral precipitation. Carbonate minerals precipitate as a result of Eq. (1) shifting toward the left due to CO_2 outgassing when soil water or groundwater with elevated P_{CO_2} enters the open room of a cave. Although CO_2 levels were increased in the restricted environment of the soils, when soil water or groundwater with high P_{CO_2} comes in contact with a cave atmosphere with lower P_{CO_2}, sudden outgassing can occur. Precipitation of calcite from drip waters forms dripstone and flowstone on cave ceilings, walls, or floors, and precipitation from stream waters forms travertine. In each case, the resulting chemical composition of the cave water is somewhat reduced in dissolved Ca^{2+} and HCO_3^- concentrations and the pH of the solution is slightly lower; that is, the solution composition shifts down to the left on Figure 1 as the supersaturated solution (saturation index greater than 0) establishes equilibrium (saturation index equal to 0).

Although CO_2 outgassing can drive the cave water to high levels of supersaturation from which calcite precipitation ensues, the rate of calcite precipitation may be slow enough to maintain a supersaturated solution for some time. In the example from a cave stream in Virginia, for instance, the solution reaches a high degree of supersaturation before beginning to adjust toward equilibrium through calcite precipitation as the stream moves through the cave (Fig. 2). The principles of thermodynamics are useful in determining the ultimate equilibrium composition of the calcite–water–CO_2 system, but the rates at which the reactions occur are less known and fall under the purview of geochemical kinetics.

All cave water is not confined to a stream channel in which travertine or rimstone dams may be the product of calcite precipitation. Diffuse infiltration waters commonly emerge from the ceiling or walls of caves where rapid outgassing of CO_2 from small water droplets with large surface area drives the solution to become supersaturated. Generally, the precipitated mineral is calcite, although in some instances the polymorph aragonite is formed. Being formed from a solution with a variety of dissolved chemicals present, some trace incorporation of cations other than Ca^{2+} contributes to some chemical impurity of the carbonate speleothems. The chemical and morphological variety of the precipitated carbonates makes the study of cave speleothems an endlessly fascinating endeavor.

In addition to the transport and reaction of Ca^{2+} and HCO_3^- in the cave environment, chemical reactions in the soil water and groundwater environment can mobilize iron and manganese. Ordinarily found in their oxidized form in insoluble oxide minerals, iron and manganese can be reduced in much the same way that O_2 is reduced coupled to the oxidation of organic matter. In the water-saturated subsurface, when biodegradation of organic matter consumes all of the available O_2, bacteria will turn to the use of alternate electron acceptors. Oxidized iron (Fe(III)) and manganese (Mn(IV)) may be present in the form of goethite (FeOOH) and birnessite (MnO_2) as grain coatings in the soil. Here shown for birnessite, an oxidized manganese mineral is reduced when coupled to microbially mediated oxidation of organic matter to CO_2:

$$2MnO_2(s) + CH_2O + 4H^+ \leftrightarrow 2Mn^{2+} + 3H_2O + CO_2(g) \tag{13}$$

When these elements are used as electron acceptors, they are reduced to Fe(II) and Mn(II), respectively, and are then soluble in groundwater. Dissolved Fe^{2+} and Mn^{2+} enter the cave in flowing groundwater where they may remain dissolved in cave waters or they may be reoxidized to precipitate Fe(III) and Mn(IV) oxide and carbonate minerals. The presence of oxidized iron and manganese minerals can be recognized as coatings on cave walls or on speleothems as black or reddish brown coatings. For instance, birnessite can be a black, dendritic deposit on the surface of calcite speleothems.

HUMAN IMPACT ON CAVE WATERS

Just as the activities of crop and animal farming, industrial production, and fecal, municipal, and hazardous waste disposal have had deleterious effects on the quality of surface water and groundwater in many regions of the world, some cave waters have been negatively impacted. The variety of chemical contaminants is too numerous to elucidate in detail, but everything from fertilizers, pesticides, metals, bacteria, petroleum hydrocarbons, and chlorinated solvents to exotic synthetic organic compounds and pharmaceuticals have been reported to contaminate cave waters. In these situations, cave waters are not unique but rather are like other groundwater. Human activities, sometimes immediate and sometimes at remote times and places, can impact the chemical composition of groundwater along its flow path.

There are, however, some dramatic examples in which the physical characteristics of limestone aquifers have distinguished the resulting groundwater contamination from that typical in other types of bedrock. The large conduits and caves in the subsurface of karst terrains are openings many orders of magnitude larger than are present in the subsurface in other geological settings. These relatively huge openings lend themselves to rapid transmission of high concentrations of contaminants and allow the migration of large particles of contaminants without the benefit of filtration by the porous media of the subsurface environment. Some examples of the dramatic contamination of cave waters include the presence of free-phase gasoline floating on cave streams and the fecal colliform bacterial contamination of rural drinking water supplies. Once introduced to groundwater, the rapid flow through limestone aquifers makes it particularly difficult to limit the spread of contamination. The hard lessons learned about the contamination of water supplies in karst regions drive home the need to avoid contamination of groundwater in the first place rather than to try to remediate contamination after it occurs.

Bibliography

Back, W., Hanshaw, B. B., Pyle, T. E., Plummer, L. N., & Weidie, A. E. (1979). Geochemical significance of groundwater discharge and carbonate solution to the formation of Caleta Xel Ha, Quintana Roo, Mexico. *Water Resources Research, 15,* 1521–1535.

Butler, J. N. (1982). *Carbon dioxide equilibria and their applications.* Reading, MA: Addison-Wesley.

Drever, J. I. (1997). *The geochemistry of natural waters* (3rd ed.). Upper Saddle River, NJ: Prentice Hall.

Garrels, R. M., & Christ, C. L. (1965). *Solution, minerals, and equilibria.* San Francisco: Freeman, Cooper & Co.

Hem, J. D. (1989). *Study and interpretation of the chemical characteristics of natural water.* U.S. Geological Survey.

Krauskopf, K. B., & Bird, D. K. (1995). *Introduction to geochemistry* (3rd ed.). New York: McGraw-Hill.

Langmuir, D. (1971). The geochemistry of some carbonate groundwaters in central Pennsylvania. *Geochimica et Cosmochimica Acta, 35,* 1023–1045.

White, W. B. (1988). *Geomorphology and hydrology of karst terrains.* New York: Oxford University Press.

White, W. B., Culver, D. C., Herman, J. S., Kane, T. C., & Mylroie, J. E. (1995). Karst lands. *American Scientist, 83,* 450–459.

Wigley, T. M. L., & Plummer, L. N. (1976). Mixing of carbonate waters. *Geochimica et Cosmochimica Acta, 40,* 989–995.

WATER TRACING IN KARST AQUIFERS

William K. Jones

Karst Waters Institute, Leesburg, VA, USA

DEFINITIONS AND OBJECTIVES OF TRACER TESTS IN KARST

Water tracer tests are usually conducted to establish the hydrologic connections between two or more points. The tracer is an identifiable label or marker added to flowing water that establishes the links between the injection point of the tracer and the monitoring points where the tracer reappears. Fluorescent dyes are the most commonly used tracers in karst aquifers, but a wide range of substances has been used successfully. The experimental design of a tracer test may be *qualitative* to simply establish if a hydrologic connection exists between two points, or *quantitative* to measure the time-concentration series (breakthrough curve) generated by the recovery of the tracer. Water tracer tests usually work well in karst areas because of the fast groundwater flow rates and the prevalence of flow paths restricted to discrete conduits.

The level of effort and the cost of conducting a tracer test are functions of the question being asked. Many tracer tests are an extension of cave exploration and are conducted to "see where the water goes." The tracer test may have the relatively simple objective of establishing a connection between two cave passages separated by an impassable reach. Tracer tests may be used to determine the destination of water flowing into a sinkhole or blind valley. The karst drainage basin contributing water to a spring or resurgence may be estimated based on the results of tracer experiments. The interbasin transfer of water is frequently demonstrated by tracer tests. The travel rate of a tracer may be used to establish some of the hydrologic properties of an aquifer. Examination of the breakthrough curve of a tracer test can aid in the interpretations of some of the internal flow characteristics of the aquifer. The movement of a tracer may mimic the subsurface behavior of a pollutant introduced into the aquifer at the same point. Tracers may be used to identify the source of pollution at a spring. Tracer tests are sometimes used to demonstrate the "vulnerability" of karst aquifers to chemical spills and pollution associated with inappropriate land use practices and to establish "wellhead protection areas" for public water supplies. A study of the Unica River basin in Slovenia (Gabrovsek *et al.*, 2009) provided a demonstration of the range of hydrologic information that can be obtained from tracer tests and the applicability to the protection and management of karst water resources.

Naturally occurring chemicals or isotopes may be used to determine the age or residence time of water at sampling points within the aquifer. The sources of the water in terms of percentages contributed by conduit flow versus drainage from the overlying epikarst or deeper fracture zones may be studied. These chemical markers are considered tracers, but they identify the various components or storage areas of the aquifer rather than establish direct hydrologic connections or paths between specific points in the aquifer.

HISTORY OF WATER TRACING IN KARST AQUIFERS

A considerable amount of folklore concerning underground water connections has accumulated throughout the world's cave areas. Tests have been reported using tracers such as wheat chaff, duck feathers, marked logs, tagged eels, and muddy water from storm events. A 2000-year-old tracer test is attributed to tetrarch Philippus who supposedly established the source of the springhead of the Jordan River in 10 A.D. Chaff was thrown into Phialo Pond (Berekhat Ram Crater Lake) and reemerged from the cave spring Panium, the head of the Jordan River. Although the results of this test have been questioned, it does establish that the basic idea of tracing underground water is quite old.

A scientifically planned tracer test was conducted in 1872 by A. Hagler to determine the origin of a typhoid fever outbreak in the village of Lausen in Switzerland. About 800 kg of salt (NaCl) was injected in a sinking stream on a farm south of the village and the water supply spring for the village showed a strong reaction for chloride the following day (Käss, 1998).

The start of modern water tracing studies followed the discovery of fluorescein by A.V. Baeyer in 1871. The sodium salt of fluorescein, called *uranine* in Europe, rapidly became the most used and probably the most successful groundwater tracer for cave and karst studies to this day (Fig. 1). The first reported use of sodium fluorescein as a tracer was an experiment conducted in southwestern Germany on October 9, 1877, by Professor Albert Knop. Ten kg of sodium fluorescein supplied by C. Ten Brink was injected at infiltration points in the bed of the Danube River and reappeared two days later and 12 km away at the Aach Spring, a tributary of the Rhine River. A test conducted a couple of weeks earlier from the same location used almost 10,000 kg of salt and may have been the first quantitative tracer test with hourly sampling at the Aach Spring.

FIGURE 1 Fluorescein sodium powder is red but turns bright fluorescent green when dissolved in water. Here the dye is being added to an underground stream just upstream of an impassible sump. *Photo by Philip C. Lucas. Used with permission.*

Work on the development of methods for tracing underground waters continued in Europe. Many of the early tracer tests used large quantities of dye, because detection depended on visual coloring of the water at the resurgence. An interesting example of an accidental tracer experiment occurred in the French Franche-Comte in 1901. As the result of a fire at the Pernod distillery at Pontarlier, a quantity of absinthe poured into the Doubs River. A significant amount of the alcohol seeped through the bed of the river and reemerged the following day 10 km away at the source spring of the Loue River. The test of the connection between the Doubs and the Loue was repeated by E.A. Martel in 1910 using 100 kg of sodium fluorescein. The Loue was brightly colored two days later for a distance of 100 km downstream. The source of the Garonne River in southern France was determined by a tracer test in 1931 by Norbert Casteret. He injected 60 kg of sodium fluorescein at the Trou du Toro on the Spanish side of the Pyrenees and it emerged 3.7 km away and 10 hours later at the Goueil de Joueou on the French side. The Garonne River was colored a bright green for over 50 km downstream. Quantitative tests at the same site in the early 1990s used less than 2 kg of sodium fluorescein and yielded travel times for the tracers of about 11 hours during high-flow conditions in the summer and 180 hours for a low-flow test in the winter (Freixes et al., 1997).

Various tracer techniques were tried in Europe through the 1920s. A "fluoroscope" was developed which allowed a rough estimation of dye concentrations in water samples by visual comparison with prepared laboratory standards. A number of different salts, dyes, and even radioactive compounds were tried. Bacteria were tried as a particulate drift tracer as early as 1896. Spores from the club moss *Lycopodium calvatum* were suggested as a possible tracer in 1910, and the first reported tracer test using spores is from Europe in 1940 (Käss, 1998). Spores became a popular tracer in Europe in the 1950s because they could be dyed different colors and several sinkpoints could be traced simultaneously. Plankton nets were suspended at the springs to passively collect the spores, so constant surveillance of a number of springs was no longer required. Spores dyed with fluorescent dyes proved to be easier to identify under a microscope. Drifting tracers such as spores and bacteria are still used to a certain extent in Europe but have never been popular in North America.

One novel tracing experiment in Slovenia in 1929 used marked eels. The dorsal fins of 494 eels were notched and the eels were released in the Reka River that sinks into Skocjanske Cave and resurges 34 km to the west at the Timavo Springs in Trieste Bay (Italy). Twenty-nine of the eels were caught in eel pots at the Timavo Springs during the one-year observation period.

The Yorkshire Geological Society carried out a series of tracer tests in the Ingleborough area of Great Britain in 1904. One principle from this work has been to test "key sinks" and to guess the remainder in lieu of testing everything. This approach is still practiced in many karst areas, but more intensive studies at later dates sometimes produce surprising results. Much of the subsequent work on tracing underground drainage in the British Isles has been conducted by caving groups associated with various universities. A study of the caves of northwest County Clare, Ireland, in the early 1960s by E.K. Tratman and the University of Bristol Speleological Society, involved a number of tracer tests using sodium fluorescein and passive carbon detectors to monitor the resurgences.

Just as professional cave and karst studies in North America lagged behind Europe in general, so did the interest in tracing underground water. The first description of tracer tests from North America was a U.S. Geological Survey water supply paper by R.B. Dole in 1906. This paper discussed the use of sodium fluorescein as a tracer, but few tracer tests are reported from North America before the late 1950s. Whereas most of the water tracing studies in Europe were conducted by professional hydrologists or speleologists, often at considerable expense, the real beginning of karst water tracing studies in North America was the contribution of unpaid cavers as an extension of cave exploration. A brief description of a method to recover sodium fluorescein sorbed onto activated coconut charcoal granules was published by J.R. Dunn in a cave club newsletter in 1957. This discovery offered an inexpensive technique to monitor different resurgences without the necessity of constant surveillance. The researchers could place carbon packets (called *Dunn Bugs*) in the springs, inject the dye in a sinking stream or cave, and return in a few weeks to collect and test the carbon packets. The concentration of the dye on the carbon increases with the exposure time to the dye, so recovery concentrations of sodium fluorescein below the normal visual threshold could be detected. The testing relied on visual identification of the dye as a fluorescent yellow-green sheen floating on top of the carbon in an elutant. Apparently a very similar technique was described by Mayrhofer in 1904 (Kass, 1998), but the activated carbon system was used in Europe more as a technique for enriching the dye concentration in a sample than as an unattended monitoring system. The first field test of the passive detector system in North America was described by T.D. Turner in 1958. Turner reported a tracer test using carbon detectors and 4 g of sodium fluorescein over a distance of 4.8 km near Pine Grove Mills, Pennsylvania.

Water tracing in North American cave studies really began in the early 1960s using sodium fluorescein and passive detectors of activated carbon. Hermine Zotter began a dye tracer study in Pocahontas County in east-central West Virginia in 1960. Thomas Aley conducted a tracer test for a court case in California in 1963. This was probably the first use of underground water tracing to resolve litigation in the United States. Aley worked on a study for the U.S. Forest Service in the Ozarks in Missouri in 1966 and conducted a 64-km-long trace from the Eleven Point River Basin to Big Spring on the Current River Basin. William Jones conducted a study for the U.S. Geological Survey in Greenbrier County, West Virginia, in 1966. The results of the tracing tests were coupled with the positions of surveyed cave streams and topographic divides to delineate karst drainage basins. Charles Brown was conducting tracer tests in the Maligne basin in Alberta, Canada, by 1969. James Quinlan became the park geologist for Mammoth Cave National Park in Kentucky in 1973 and started an intensive series of water tracing tests. The results of the tracing tests were combined with data from cave surveys and potentiometric contours to produce a detailed map of drainage basins and subbasins around Mammoth Cave (Quinlan and Ewers, 1989).

The development in the early 1960s of fluorometers that control or measure light emissions in both the excitation and emission wavelengths of different fluorescent compounds enabled dye analytical methods to become much more quantitative. Quantitative water tracing using fluorometers dates from the early 1960s. Much of the early quantitative tracer work was for surface water time-of-travel studies. Fluorometers also allowed several different fluorescent compounds to be used simultaneously because the instruments could be set to narrow specific wavelengths.

Much of the development of quantitative tracer techniques was presented in a series of conferences in Europe beginning with the Specialists Conference on Tracing of Subterranean Waters in Graz, Austria, in 1966. This conference series has continued at about four-year intervals as the Symposium on Underground Water Tracing (SUWT) conferences. The published proceedings from the series contain a wealth of information on various tracer techniques for both karst and porous media water tracing. Much of the material from these conferences is distilled in the book by Käss (1998). A U.S. Geological Survey paper by Wilson (1968) and a paper on the evaluation of fluorescent tracers by Smart and Laidlaw (1977) were probably the most influential papers on quantitative water tracing published in North America.

TRACERS

Groundwater tracers are generally classified as: (1) naturally occurring (often accidental in their addition to the water); (2) artificial tracers deliberately introduced into the aquifer; and (3) pulses. Naturally occurring tracers include the chemical constituents and isotopes naturally found in water or present due to human activities. Tritium, fluorocarbons, pollutants from waste sites, and thermal waters may all serve as tracers in some circumstances. Isolated thunderstorms may generate a distinctive flow response at springs some distance from the storm. Organisms unique to a particular subsurface drainage basin may be considered tracers if their distribution can be used to define the boundaries of the basin. Naturally occurring tracers are often used to "date" the water, artificial tracers are used to label a specific sink or cave, and pulses send an identifiable signal through the conduit or sometimes a large part of the aquifer.

Most tracer tests in karst areas are conducted using artificially injected tracers. These tracers may be broadly classified as (1) water-soluble tracers (dyes, salts); (2) particulate or drift material tracers (club moss spores); and (3) physical pulses (flood waves). The ideal tracer should be:

1. Nontoxic
2. Not normally present in the study area (low background)
3. Detectable at very low concentrations
4. Conservative (minimal sorptive and decay losses)
5. Have the same density of water (or be neutrally buoyant)
6. Relatively inexpensive
7. Easy to sample using passive collectors
8. Easy to analyze and quantify in the laboratory

The vast majority of tracer tests in cave and karst areas are conducted using fluorescent dyes. All fluorescent dyes have some tendency to adsorb to clay minerals and some have high photochemical decay rates. The relative fluorescent intensity of many dyes is pH- and temperature-dependent. These considerations are usually not a problem for underground tracing in cavernous aquifers with high flow rates. Tracer tests in diffuse-flow aquifers often involve the use of more conservative isotopes where the longer residence times make fluorescent dyes less suitable. A list of the principal fluorescent tracer dyes commonly used in North American karst studies is presented in Table 1.

Tracer tests from Europe have largely been conducted quantitatively and required direct sampling of the resurgences. Studies in the British Isles and North America through 1990 have mostly been qualitative and used passive collectors to monitor the resurgences. Qualitative tracing is still used for reconnaissance-type studies and work in remote areas, but quantitative tracing techniques are now the norm for most scientific studies throughout the world.

TABLE 1 Principal Fluorescent Tracers Used in North American Karst Studies

Tracer Colour Index Name Color	Excitation Emission Maximum	Passive Detector	Remarks
Phorwite AR Solution FB # 28 UV blue	349 nm 430 nm	Unbleached cotton	Invisible in solution. High photochemical decay rate. Often high background fluorescence. May be present due to household detergents.
Solophenyl Direct Yellow 96 UV yellow	≈397 nm ≈490 nm (pH sens.)	Unbleached cotton	Sensitive to pH. No coloring of water. Usually low background.
Sodium Fluorescein Acid Yellow 73 Yellow-green	491 nm 512 nm	Activated carbon 6–14 mesh	High photochemical decay and sensitive to pH. Probably the most frequently used tracer for karst studies. Low sorption.
Eosin Acid Red 87 Red	516 nm 538 nm	Activated carbon 6–14 mesh	High photochemical decay. Can overlap with sodium fluorescein if both are present in the same sample.
Rhodamine WT CI Acid Red 388 Red	554 nm 580 nm	Activated carbon 6–14 mesh	Low background. Good stability in sunlight. Moderate sorption. Generally requires a fluorometer for detection. Some environmental concerns.

The fluorescent dyes commonly used as tracers are believed to be quite harmless to people and the environment. Reviews of the toxicity data for tracer dyes were presented by Smart (1984) and Behrens et al. (2001). Several dyes in the rhodamine group, especially rhodamine B, are not recommended due to environmental concerns, but the high sorptive tendency for some of these dyes makes them a poor choice for groundwater tracing under any circumstances.

A lot of confusion exists over the names of many of the tracers. Sodium fluorescein is known as uranine in Europe and commonly called fluorescein in the North American literature. Only the sodium salt of fluorescein is soluble enough to make it an efficient tracer. When describing a dye, the Color Index name should be cited at least once. Sodium fluorescein has a Color Index constitution number CI 45350 and a generic name of acid yellow 73. Also, all dye shipments should be checked when they are received to make certain the correct product is being used.

Note that a few words of caution are in order for anyone planning on conducting tracer tests. Specialists now do most dye tracing, and water tracing by untrained people may be very damaging. In the United States, some states require a permit for conducting a tracer test, and the local water inspectors should always be notified in advance of any test so a hazardous materials team is not called out at great expense. A few states do not allow any water tracing tests. Care should be taken to avoid coloration at water intakes. All work in a given area should have one coordinator so various tests cannot cross tracers. Some epikarst aquifers may take several years to clear a given dye from an injection point, so careful planning is needed to avoid confounding future tests using the same tracer. The following description of the basic procedures for conducting tracer tests in karst areas is not intended to serve as a training manual. Tracers, such as antibiotics, will lose their usefulness if overused or improperly used through time.

QUALITATIVE TRACING USING PASSIVE DETECTORS

Passive collectors such as packets of activated carbon or unbleached cotton are placed in the monitoring points and collected at regular intervals. The dye, if present, is then eluted from the carbon or exposed by examining the cotton under ultraviolet light. The use of a fluorometer to analyze the elutant increases the minimum detectable concentration of the tracer by several orders of magnitude. It is not possible to reliably obtain the dye concentrations at the resurgences from this technique, because exposure time of the detector to the dye and other variables affect the elutant dye concentration. Time of travel for the tracer can only be approximated based on the changing interval for the detectors. Detailed instructions for conducting qualitative tracer tests are presented in Aley and Fletcher (1976), Jones (1984), and Alexander and Quinlan (1996).

The principal groups of passive detectors and tracers are:

1. *Activated carbon* (activated charcoal)
   a. Sodium fluorescein
   b. Eosin
   c. Rhodamine WT, Sulforhodamine B

2. *Cotton*
   a. Tinopal CBS-X (FB 351)
   b. Phorwite
   c. Calcophor white (FB number 28)
   d. Direct yellow 96
3. *Plankton netting* (about 25-micron mesh openings)
   a. Colored spores of *Lycopodium calvatum*

The fluorescent dyes have been used extensively in North America and Europe while tracing using spores has been primarily a European technique. The following discussion outlines the basic procedure for conducting a qualitative tracer test using sodium fluorescein and carbon detectors.

The execution of a tracer test should involve an initial survey of the study area to identify the hydrologic boundaries and all possible springs and resurgences. Any previous tracer tests from the area and all cave survey data should be studied. All landowners potentially affected by the test should be contacted and access secured to injection and monitoring points. Local and state agencies should be notified about the study and any required permits obtained. Measurements of background fluorescence from the springs should be used to help determine the most suitable tracers. Initial tests in an area should be conducted during average flow conditions; tests under minimal discharge should generally be avoided.

The amount of dye to use for any given test is an educated guess at best. The quantity of tracer to be injected is a function the hydrologic conditions of the study area and the characteristics of the tracer and the analytical procedure. The possible travel distances and discharges of the resurgences may be used to help estimate the travel time and tracer quantity required for the test. Several formulae to estimate dye quantities are presented in Käss (1998), but these provide only rough guidelines. Tests in conduit aquifers involving distances less than 1 km and discharges less than 30 L s^{-1} should require less than 200 g of sodium fluorescein. Most of the other commonly used tracer dyes probably require two to four times the mass used for a sodium fluorescein test. Less dye is usually needed for qualitative tests than for quantitative tests. If the possible coloration of the resurgences is not objectionable, the estimated amount of dye should be increased. Care must be taken to keep maximum tracer recovery concentrations below the visible threshold (about 30 μg L^{-1} for sodium fluorescein) if the tracer could affect a water supply. The U.S. Geological Survey suggested a maximum dye concentration of 10 μg L^{-1} at water intakes, but this is a conservative number and is not based on toxicological considerations (Alexander and Quinlan, 1996).

The monitoring locations should include all possible resurgences and one or two "impossible" ones to serve as controls for the test. Carbon detectors are usually made of envelopes of plastic screening material containing a few grams of activated carbon granules (about 6–14 mesh). Cotton detectors are usually balls of unbleached cotton. The passive detectors should be changed at regular intervals, usually between 2 and 14 days. The detectors should be in place for at least one changing interval prior to dye injection to establish background information. They should be placed to remain submerged with changing water levels, be readily recoverable, and remain out of sight to casual observers. An interesting "hanger" for detectors in deeper water was developed by James Quinlan and is shown in Figure 2. The detectors should be placed so medium velocity current flows through them and they are not exposed to direct sunlight. Detectors lose their ability to adsorb the tracers through time, but they will retain the dye for periods of at least one year following exposure when kept in total darkness. Caution must be used to prevent contamination of the detectors and they should be placed before the dye is injected.

Injection points for the tracer are ideally flowing streams at surface sinkpoints or upstream of impassable reaches in a cave. Powdered dyes are often mixed with water prior to injection to avoid unwanted spread of the dye. The injection points should be chosen to minimize the surface exposure of the tracers to sunlight.

FIGURE 2 A "Quinlan Gumdrop" anchor for passive dye collectors. The base is concrete molded in a coffee can and the detectors are suspended above the bottom of deeper springs or streams. The screen envelops contain unbleached cotton (right arm) and activated carbon (left arm and bottom). The packets can also be secured with wire to rocks or other solid anchors such as old railroad spikes.

Tests involving injection points without rapid continuous flow such as a dry sinkhole or water well are much more difficult and more prone to failure. Tracer tests from sinkholes or wells are usually only conducted by professional hydrologists. The basic procedure is to use a tank truck to deliver water to the site. Several thousand liters of water are run into the sinkhole before the dye is added. The dye is then flushed into the aquifer system with about an additional 10,000 liters of water. Injection sites that use water wells often provide unsuccessful tests, which means that the dye injected as the tracer will not be usable in the area for a number of years. Caution and careful planning are needed for this type of test.

Another dye injection method involves leaving the dye in a safe storage area in a presently dry sink or streambed and letting the next storm wash the tracer into the flow system. This "dry set" system works best in remote areas where people or animals are unlikely to encounter the tracer before it is removed by precipitation. The analysis of the carbon for the presence of sodium fluorescein involves the following steps:

1. Rinse the carbon in clean water to remove sediment.
2. Fill a test tube about half full with the carbon.
3. Pour in the test solution, called the *eluent*, to cover the carbon about 3 cm deep (5–10 mL).
4. Let the samples sit undisturbed in the dark for between 30 min and 24 hr.
5. Shine a concentrated white light source such as a focusable flashlight through the test tube and look for the characteristic fluorescent green sheen on top of the carbon granules (do not shake or disturb the solution for a visual examination).

Some workers prefer to use a larger diameter vial to increase the length of the light path through the eluent. The most commonly used eluent for sodium fluorescein is 5% potassium hydroxide (KOH) in 70% isopropyl alcohol. Many other eluents have been used including the "Smart Solution" (50% 1-propanol, 20% NH_4OH, and 30% distilled water by volume). It takes some practice to distinguish weakly positive sodium fluorescein tests from some naturally occurring background. Numerous trials at varying dye concentrations should be run under controlled conditions before moving on to actual field tests. One of the advantages of the passive collector system is that the samples are time-integrated for the exposure period of the detectors at each monitoring point.

QUANTITATIVE WATER TRACING IN KARST AREAS

Many of the professionally conducted water tracing tests are quantitative. Resurgences are sampled at short discrete time intervals and the water is analyzed using a calibrated fluorometer to determine the concentration of the dye in each sample. Several different dyes may be injected simultaneously so multiple sink points may be tested at the same time and under the same flow conditions. The results are plotted on a graph showing the concentration of the dye through time (breakthrough curve). The expense and complexity of quantitative tests are much higher, but considerably more information is obtained about the nature of the subsurface flow systems. The combination of quantitative dye tracing results with discharge measurements allows a more certain delineation of karst drainage basins and the calculation of the percentage of discharge from different sources at a given resurgence. Good results are dependent on using an appropriate tracer, adequate sampling frequency, and accurate analysis of the samples. A paper by Field (2003) provides formulas for determining dye mass and sampling frequency, but a good estimation of hydraulic and transport characteristics of the system is needed. This usually requires advance knowledge of the flow conditions and the resurgence(s). An estimate of the expected travel time from sink to resurgence is also helpful in planning tests. Flow velocities for conduit karst aquifers are typically in the range of 0.2 to 7 km per day, but some tracer tests have shown results far outside of these values.

The planning and dye injection procedure is essentially the same as described for qualitative traces. Qualitative tests are often conducted first to identify which resurgences to monitor and to estimate the travel time for the tracer. Most tracer tests in karst areas use an instantaneous injection of the dye, but continuous (over several hours) releases are occasionally used to facilitate the calculation of discharge at the recovery point. Fluctuations in background fluorescence are often a problem in the blue and green wavelengths, so the red dye Rhodamine WT is the most commonly used fluorescent dye for quantitative studies. Quantitative tests generally require about twice as much dye as qualitative tests. The much greater sampling effort required for quantitative testing is the major reason for the increase in cost.

The instrument used to determine fluorescent dye concentrations is a fluorometer (Fig. 3). Fluorometers pass light through the sample at a controlled excitation wavelength and measure the relative fluorescent intensity of the output signal at the emission wavelength. Laboratory standards at different dye concentrations must be prepared for each batch of dye to calibrate the fluorometer to determine dye concentrations. Filter fluorometers may be laboratory or field instruments and use various filters to set the excitation and emission wavelengths, while scanning spectrofluorophotometers are laboratory instruments that are better

FIGURE 3 A scanning spectrofluorometer used for laboratory quantification and identification of fluorescent dyes and a battery-powered hand held fluorometer (lower left) used for field monitoring of a dye pulse.

suited to analyzing samples containing multiple dyes or having high fluorescent backgrounds. A continuous scan through the wavelengths of interest with about a 20-nm separation between the excitation and emission settings allows a single scan to be analyzed for several different dyes. Some workers use computer enhancement of the instrument output signal to help identify and separate overlapping fluorescent peaks. Spectrofluorophotometers are more selective in their ability to isolate specific fluorescent compounds, but both types of fluorometers are highly sensitive to low concentrations of dye. Recording filter fluorometers provide a high resolution record of the passage of the dye pulse, but spectrofluorophotometers are more reliable in the identification and quantification of individual tracers.

Filter fluorometers may be fitted with a flow-through door which circulates water past the instrument's light source. This provides a continuous record of fluorescent intensity in one set of wavelengths, but this is usually only practical for tests with travel times of less than 24 hours. Fluorometric probes compatible with data loggers make continuous fluorometric measurements more cost-effective for many field studies and record a nearly continuous picture of the dye pulse through time. Handheld fluorometers with several "channels" set for specific tracer dyes are robust enough to take into caves, but the samples must be taken manually. Most quantitative tracer studies at present involve the use of automatic water samplers at several resurgences over a period of days to weeks. The sampling interval is dependent on the expected travel time of the tracer and may range from 15 min to one per day. Some resolution in the shape of the dye recovery curve may be lost if the sampling interval is too long. Sampling should continue for a time sufficient to ensure that most of the dye pulse has been recovered. Samples should be stored in the dark in glass containers to minimize photochemical decay and sorption of the dye to the sample bottles.

INTERPRETATION OF WATER TRACING TESTS

Qualitative tracer tests are positive if the dye is detected at one or more monitoring points. A successful tracer test may prove a hydrologic connection between the injection point and recovery point(s) of the tracer, but the exact flow path(s) will remain unknown. Depending on the changing interval for the detectors, some information of the travel time for the tracer may be available. If the tracer is not detected at a monitoring station, it is listed as "none detected" rather than "negative." Negative tests do not prove that a connection does not exist and the percentage of the injected tracer actually recovered can only be determined by quantitative measurements of the dye concentration and discharge at the resurgences during the recovery period of the dye pulse.

Tracer tests are representative of the karst flow system for the hydrologic conditions prevailing at the time of the test. Overflow routes may exist that function only during high-flow conditions, so a thorough study of a karst

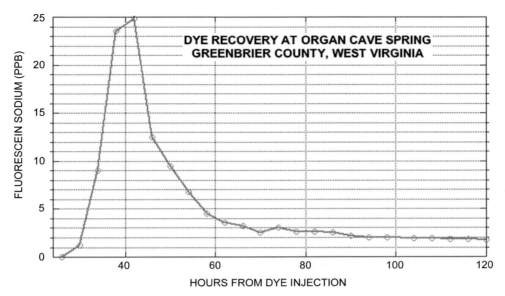

FIGURE 4 Graph showing a simple breakthrough curve from a tracer test from a cave to a spring. The principal components of the curve are (A) first arrival of the dye at 30 hours; (B) peak (maximum) recovery concentration at 42 hours; (C) dye centroid (center of mass and median recovery point) at 50 hours; and (D) total time of dye cloud passage from 30 through 120+ hours. Note the four-hour sampling interval and that sampling was stopped before the complete passage of the dye cloud.

region will require multiple tests at different water levels. Some resurgences may only function during high-flow conditions. The travel times for the tracers will be less at high water levels, and the dye will move much faster if the conduits are completely flooded. Quantitative tests between the same injection and recovery sites under a range of flow conditions may be used to develop the relationship between travel time and groundwater discharge and to predict solute transport characteristics. Discharge-dependent thresholds may be identified.

The graph of dye recovery concentration plotted through time is called a *dye recovery* or *breakthrough curve* (Fig. 4). This curve presents a picture of the passage of the dye "cloud" or pulse through time at the stationary sampling point. Breakthrough curves are typically skewed to the right with rapid rise to peak (maximum) dye concentration and a long trailing edge of declining dye concentrations. The shape or form of the breakthrough curve is a function of the characteristics of the tracer, the flow conditions during the test, and the characteristics of the aquifer or conduit system. The mean flow velocity is usually computed using the straight-line distance between sink and resurgence and the time between dye injection and recovery of half of the tracer. This is the *centroid* or center of recovered mass on the breakthrough curve. Some workers use the time to peak concentration if the resurgences are not completely sampled through the long trailing period of declining tracer concentrations. If discharge is known throughout the dye recovery period, the amount of dye recovered can be calculated by integrating the area under the breakthrough curve and multiplying by the discharge:

$$M = Q \int_0^\infty C \, dt$$

where M = mass of dye recovered, Q = discharge, C = dye concentration at time t.

Discharge is usually based on current-meter measurements during the test. If the tracer is conservative (not subject to decay or sorptive losses) and all of the water from the injection point is accounted for, the mass of dye recovered should be equal to the mass injected. This is the dye budget for the test. None of the fluorescent tracers is completely conservative, but tracer tests with travel times of less than a few days should balance reasonably well if all of the resurgences have been monitored. Assuming complete mixing and a conservative tracer, the area under the time-concentration breakthrough curve should be constant even as the shape of the curve changes in the downstream direction. A significant loss of the tracer from the system suggests that additional resurgences are present or some of the tracer has moved into storage areas along the flow route. Less conservative tracers usually produce a longer travel time estimate due to retardation caused by sorption and desorption along the flow route.

Breakthrough curves for many tracer tests in karst areas exhibit multiple peaks, and the time between the peaks may be dependent on discharge. The interpretation of these tests is somewhat subjective, but much insight into the internal flow characteristics of the karst aquifer and the conduit system can be gained from these tests. The best breakthrough curve data are obtained using continuous flow fluorometry during the passage of the dye pulse. The shape or form of the dye cloud is dependent on the interaction of dispersion, dilution, divergence and convergence, and storage (Smart, 1988). The quantitative interpretation of breakthrough curves is generally based on solute transport equations. An introduction to this type of analysis was presented in Benischke et al. (2007) and a computer program to analyze tracer recovery data was developed by Field (2002).

Traces involving longer distances and longer travel times show lower peak concentrations and longer persistence of the dye pulse as the tracer becomes increasingly more dispersed with time and distance. The shape of the tracer cloud becomes more asymmetrical as dispersion is influenced by the tracer moving in and out of pockets of *dead zone storage* along the flow route. Dispersion is also affected by the retardation factor for tracers that are less than completely conservative.

Divergence occurs where the underground flow routes divide or break away from the main conduit. Convergence is where the routes rejoin the main conduits. The divergent routes generally are longer flow paths than the main conduit, so the portion of the flow and tracer diverted may reappear as second or multiple pulses at the sampling station.

Dilution is caused by unlabeled tributary water mixing with the water containing the original tracer injection. This will cause a lower concentration of dye in the downstream water samples.

Storage of the tracer in the conduit system may be very complex. Storage may just represent very low velocity flow through one section of a branching conduit, called *inline storage*. A test done under conditions of diminishing flow may leave some of the dye abandoned in upper level pockets to be remobilized when the flow increases at a later date. Some dye may become stored offline in lateral areas along the route such as small voids or in pores in the sediments. Inline storage tends to create dye pulses that correlate with flushing due to storm events. Dye from offline storage tends to drain slowly during groundwater recessions. Some conceptual models of breakthrough curves for different flow levels and conduit geometries are presented in Figure 5 (adapted from Smart and Ford, 1982).

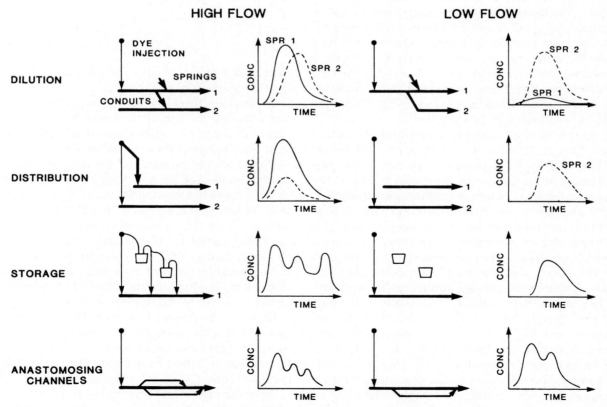

FIGURE 5 Sketches showing conceptual models of possible dye breakthrough curves. The pattern of the recovery curve is a function of prevailing water levels at the time of the test and the geometry of the conduit system. *From W.K. Jones, 1984; after C.C. Smart and D.C. Ford, 1982. With permission.*

The best picture of the conduit aquifer system for any karst area is from the actual surveying of the cave passages. Tracer tests, interpreted in the context of the regional geology and hydrologic setting, can be used to fill in the blank place on the map where actual exploration of the caves is not possible. The rather unusual mixture of quantitative techniques and subjective interpretation of the unseen part of the aquifer makes water tracing a fascinating endeavor.

See Also the Following Articles

Hydrogeology of Karst Aquifers
Springs

Bibliography

Alexander, E. C., Jr., & Quinlan, J. F. (1996). Practical tracing of groundwater with emphasis on karst terranes. In *Guidelines for wellhead and springhead protection area delineation in carbonate rocks* (EPA 904-B-97-003, Appendix B, pp. 1–38). Nashville, TN: Eckenfelder, Inc.

Aley, T., & Fletcher, M. W. (1976). The water tracers cookbook. *Missouri Speleology*, 16(3), 1–32.

Behrens, H., Beims, U., Dieter, H., Dietze, G., Eikmann, T., Grummt, T., et al. (2001). Toxicological and ecotoxicological assessment of water tracers. *Hydrogeology Journal*, 9, 321–325.

Benischke, R., Goldscheider, N., & Smart, C. (2007). Tracer techniques. In N. Goldscheider & D. Drew (Eds.), *Methods in karst hydrogeology* (pp. 147–170). London: Taylor and Francis.

Field, M. S. (2002). *The QTRACER2 program for tracer-breakthrough curve analysis for tracer tests in karstic aquifers and other hydrologic systems.* U.S. Environmental Protection Agency.

Field, M. S. (2003). A review of some tracer-test design equations for tracer-mass estimation and sample collection frequency. *Environmental Geology*, 43, 867–881.

Freixes, A., Monterde, M., & Ramoneda, J. (1997). Tracer tests in the Joeu karstic system (Aran Valley, Central Pyrenees, NE Spain). In A. Kranjc (Ed.), *Tracer hydrology 97* (pp. 219–225). Rotterdam, The Netherlands: A. A. Balkema.

Gabrovsek, F., Kogovsek, J., Kovacic, G., Petric, M., Ravbar, N., & Turk, J. (2009). Recent results of tracer tests in the catchment of the Unica River (SW Slovenia). *Acta Carsologica*, 39(1), 27–37.

Jones, W. K. (1984). Analysis and interpretation of data from tracer tests in karst areas. *National Speleological Society Bulletin*, 46(2), 41–47.

Käss, W. (1998). *Tracing technique in geohydrology.* Rotterdam, The Netherlands: A. A. Balkema.

Quinlan, J. F., & Ewers, R. O. (1989). Subsurface drainage in the Mammoth Cave area. In W. B. White & E. L. White (Eds.), *Karst hydrology: Concepts from the Mammoth Cave Area* (pp. 65–103). New York: Van Nostrand Reinhold.

Smart, C. C. (1988). Artificial tracer techniques for the determination of the structure of conduit aquifers. *Ground Water*, 26(4), 445–453.

Smart, C. C., & Ford, D. C. (1982). Quantitative dye tracing in a glacierized alpine karst. *Beitrage Geol. Schweiz-Hyrdrologie*, 28(1), 191–200.

Smart, P. L. (1984). A review of the toxicity of twelve fluorescent dyes used for water tracing. *National Speleological Society Bulletin*, 46(2), 21–33.

Smart, P. L., & Laidlaw, I. M. S. (1977). An evaluation of some fluorescent dyes for water tracers. *Water Resources Research*, 13, 15–33.

Wilson, J. F. (1968). Fluorometric procedures for dye tracing. *Techniques of water-resources investigations of the U.S. Geological Survey* (Book 3, Chapter A12, pp. 1–34) (Revised).

WETLANDS IN CAVE AND KARST REGIONS

Tanja Pipan[*] *and David C. Culver*[†]

[*]*Karst Research Institute ZRC SAZU,* [†]*American University*

INTRODUCTION

A general definition of wetlands is an area of land where the soil is saturated with moisture either permanently or seasonally, which may also be covered partially or completely by shallow bodies of water. They include swamps, marshes, bogs, fens, and peatlands. A more restricted and legalistic definition is that of the U.S. Environmental Protection Agency: "those areas that are inundated or saturated by surface or groundwater at a frequency and duration sufficient to support, and that under normal circumstances do support, a prevalence of vegetation typically adapted for life in saturated soil conditions." The emphasis of this definition on vegetation makes identification of wetlands easier. Many wetlands, both freshwater and saltwater, play an important role in water purification, are highly productive in terms of carbon fixation, and play a key role in the life cycle of many vertebrates, such as anadromous fish and migratory waterfowl.

At the same time, wetlands are highly susceptible to destruction by human activities, particularly by draining. Wetlands are of great interest and concern to both international and national agencies. The UNESCO Ramsar Convention on Wetlands is designed to highlight important wetlands internationally. In the United States, wetlands as defined by the U.S. Environmental Protection Agency are probably the most regulated and protected (and threatened) general landscape features.

In karst areas, wetlands have received less recognition and attention than elsewhere, probably because wetlands are not especially common, due in part to the nature of karst and cave areas. Since precipitation rapidly sinks into the ground, wetlands are not usually formed by the accumulation of surface runoff in areas of recharge. Euliss *et al.* (2004) suggest a continuum of wetlands along a groundwater axis (ranging from recharge to discharge) and atmospheric water axis (precipitation). In karst areas, wetlands usually only occur at the discharge portion of the groundwater continuum, near springs and resurgences. Secondly, since most streams are underground, there is less opportunity for development of surface wetlands.

The broad patterns of distribution of major wetland areas (Mitsch and Gosselink, 2007) are nearly the reverse of the distribution of karst areas (Fig. 1). Wetlands are especially common at far northern latitudes (50°–70° N) and near the equator. These are regions with poor soil drainage where hydric soils form. In contrast, karst regions are concentrated at intermediate northern latitudes (30°–50° N). Of the 65 major wetlands enumerated by Mitsch and Gosselink in their textbook on wetlands, only two are unequivocally in karst areas. One is the Hudson Bay lowlands in Canada, one of the most extensive wetlands in the world. However, it is probably the relatively low relief and Pleistocene glaciations that are the defining features of this wetland. The other large karst wetland listed by Mitsch and Gosselink is Cuatro Ciénegas in Mexico, a 1500-km^2 wetland in a valley of the Sierra Madre Oriental formed by more than 200 springs and home to a number of endemic obligate aquatic subterranean (stygobiotic) invertebrates.

At smaller geographic scales, karst wetlands fit uneasily into wetland classification schemes. Leibowitz (2003) suggests that isolated wetlands (ones without direct surface connection to rivers and streams) be defined as wetlands "completely surrounded by uplands." By this definition every sinkhole and dolina is potentially an isolated wetland and it almost becomes a synonym for sinkhole and dolina. The number of dolinas and sinkholes in karst areas can easily exceed 100 per km^2. Karst wetlands are also different because large connections (e.g., cave entrances) exist between surface water and groundwater, and these connections allow for the rapid movement of water onto or away from the surface. Hence, rapidly disappearing lakes and ponds can be found in many karst regions. Finally, there is a complex terminology of landscape features in karst wetlands (e.g., poljes, turloughs, estavelles), one that is almost completely different from the terms used by wetland scientists. Mitsch and Gosselink list 35 common terms for wetlands (e.g., billabongs, bogs, bottomland), only one of which (turlough) is a karst feature. The general wetland type that is most useful in framing discussion about karst wetlands is the isolated wetland, especially temporary freshwater wetlands. Temporary wetlands are categorized on the basis of

1. Periodicity of flooding, whether it occurs every year, occasional years, or several times throughout the year
2. Seasonality of flooding
3. Duration of flooding, the average length of time the wetland persists
4. Predictability of flooding

Numerous authors have pointed out that not only is the vegetation of wetlands adapted to saturated soil, but the animals of wetlands must cope with the periodic drying of the habitat (Wiggins et al., 1980). Among the adaptations to life in temporary water are burrowing, resistant resting stages, and migration. In spite of the differences in distribution, definition, and focus, wetlands and karst systems do share some key features which make their joint consideration important. Both are highly dependent on the hydrologic cycle, both involve groundwater/surface water interactions, and both are very fragile systems that are especially susceptible to human impacts.

Depending on the definition used (see above), wetlands in karst areas could include not only relatively large wetlands of the types we review below, but also smaller features, often less than one hectare in size, that are sometimes covered with water but may lack a characteristic flora adapted to life in saturated soil. These smaller features include some sinkholes, dolinas, uvalas, and springs. We limit our description to the larger surface wetlands, as the logical starting point for the incorporation of wetlands concepts into the physical geography of karst.

The Dinaric karst, ranging from extreme northeast Italy south to Albania, is both the "type locality" and the site of an extensive variety of landscape features that are the benchmark against which other karst areas, especially in temperate regions, are compared. A series of karst wetlands in the Dinaric karst of Slovenia, although they have not been called wetlands, are well described, and serve as a useful starting point of analysis of karst wetlands. Karst wetlands, especially poljes, are the dominant wetland type throughout the Dinaric karst. They comprise about 4% of the 72,000 km^2 of the Dinaric karst and are critical habitat

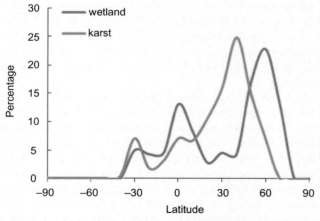

FIGURE 1 Qualitative view of the distribution of wetlands and karst areas with respect to latitude. *Data for wetlands was based on a graph in Mitsch and Gosselink (2007) and data for karst areas was taken from Balász (1977), and values estimated at 10° intervals.*

for many waterbirds (Stumberger, 2010). Undoubtedly, other kinds of large karst wetlands occur in tropical and polar regions, but they have been little studied.

CERKNIŠKO JEZERO/POLJE

Called *Cerknica Lake* in English, the site is given the name *jezero* (lake) when flooded and *polje* (field) when it is dry (Gaberščik, 2003). This is perhaps the best-known seasonal lake, and scientists have sought explanations for its sudden appearance and disappearance since Valvasor's diagram from the seventeenth century (Fig. 2). The lake featured prominently in Valvasor's epic description of the Slovenian landscape, *Die Ehre dess Herzogthums Crain*. The entire closed depression (Cerkniško polje) in which Cerkniško jezero occurs is 38 km^2. The lake itself can occupy up to 27 km^2 and reach a depth of more than 2 m, or be completely dry. When the lake is present, it is used for recreation. When it is dry, the land is used for grazing and cropland. Typically, the lake is completely dry for 80 days in the summer (August to October). Occasionally (about once a decade), the lake does not completely dry out anytime during the year. It reaches its maximum extent in April and May, and again in December. The transition from wet to dry (and *vice versa*) is rapid, with the lake changing level of up to 8 cm per day.

One of the water exits, a spectacular ponor (swallow hole), appears during drying (Fig. 3). In the past there have been attempts both to permanently drain the lake for agricultural purposes and to make the lake permanent to enhance recreational opportunities. All these projects failed.

Not surprisingly, the flora of Cerkniško jezero is very diverse. Aquatic plant communities are very restricted because of the intermittent nature of the lake. Most species die back during the dry phase. Among such species are *Charetea fragilis* and several species of *Potametum*. Amphibious species, ones where the species occur in both wet and dry phases and where the duration of the aquatic phase is at least 3–4 months, include *Rorippa amphibian* and *Sium latifolium*. Typical wetland species include *Phragmites australis*, *Equisetum fluviatile*, many species of *Carex*, and *Nymphaea alba*. There are also species living on the edge of the inundation zone, including *Deschampsia caespitosa* and *Plantago altissima*. Many of these wetland species are rare or endangered in Slovenia.

The Crustacea reported from Cerkniško jezero include a fairy shrimp, *Chirocephalus croaticus*, a rare species limited to temporary karst waters (Fig. 4). The population was destroyed because pools were filled with debris, and its recent occurrence has not been demonstrated. A total of 21 cladoceran species and 15 copepod species have been recorded from the lake.

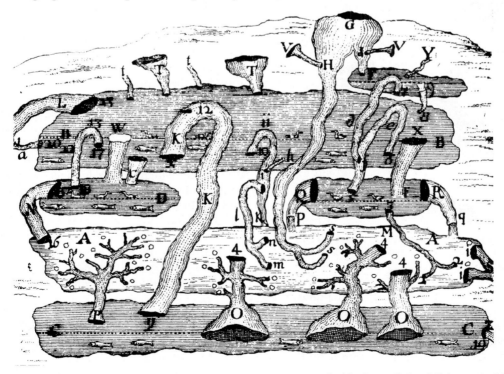

FIGURE 2 Diagram of the subterranean drainage of Cerkniško jezero as conceived of by Janez Vajkard Valvasor in 1689. It was the first attempt to provide a rational explanation for the disappearance and appearance of Cerkniško jezero. *From J.V. Valvasor (1689) Die Ehre dess Herzogthums Crain. I. Th., Laibach-Nurnberg.*

FIGURE 3   Ponor in Cerkniško polje, a major drain for the lake. *Photo courtesy of S. Polak. Used with permission.*

FIGURE 4   The fairy shrimp (Anostraca), *Chirocephalus croaticus*, a rare species limited to temporary karst waters. *Photo courtesy of S. Polak. Used with permission.*

While copepods are ubiquitous, cladocerans tend to be most common in temporary waters. One of the copepod species, *Diacyclops languidoides*, is an obligate subterranean dweller (stygobiont). The lake is also a hotspot for both Odonata and Trichoptera, with several rare and endangered species in Slovenia.

In the seventeenth century, three fish species were known to inhabit Cerkniško jezero—pike (*Esox lucius*), tench (*Tinca tinca*), and burbot (*Lota lota*). Unfortunately, as in many other lakes worldwide, several species have been introduced. In a stream in the polje, 11 species of fish are known, the majority of which were introduced. The cave salamander *Proteus anguinus* has been reported from Cerkniško jezero, where it washes in from subterranean channels as the lake fills.

PLANINSKO POLJE

Planinsko polje is the second most studied polje. In comparison to Cerkniško jezero/polje, it is much drier, and the period of flooding is much shorter, up to a maximum of 2 months. It is typically referred to as a polje, not a jezero. The geological setting for Planinsko polje and poljes in general is quite complex, typically aligned along tectonic axes, and covered with alluvium. Two main springs drain into Planinsko polje (Fig. 5) from the Javornik and Snežnik massifs, a string of poljes including Cerkniško polje, and from the Pivka River (Petrič, 2010). One of the springs, Unica, drains into the large Postojna–Planina Cave System. Unica River flows across Planinsko polje and plunges underground in a series of ponors at the northern end of the polje.

Since the polje is dry for most of the year, it is an important terrestrial habitat for mammals such as the wildcat, *Felix sylvestris*, many birds, butterflies, and beetles. The Postojna–Planina Cave System, whose waters exit at Unica spring, is the richest cave in the world in terms of the number of obligate cave-dwelling species. Especially during high water, the cave salamander *Proteus anguinus* washes out of the cave and can be found in the surface part of the Unica River. A number of fish live in the Unica River, including the brown trout *Salmo trutta* and seven other fish species native to the river, and is an important recreational fishery.

Since there is a permanent stream across the polje, there is a persistent zonation of plants, from terrestrial to aquatic. Much of the land is used for grazing and some crops. Its position at the junction of the Dinaric and Alpine regions accounts in part for the high floristic diversity. A species of hyacinth, *Scilla literdierei*, is limited to Dinaric karst wetlands and is common in Planinsko polje. Other species characteristic of Planinsko polje are *Gladiolus illyricus*, *Iris sibirica*, *Iris pseudacorus*, and *Ophioglossum vulgatum*.

PIVKA INTERMITTENT LAKES

In the drainage of Pivka River, which flows for only 26 km across a classic karst landscape before it empties into Postojnska jama, a total of 17 intermittent lakes of widely varying sizes appear with different periodicities

FIGURE 5 Planinsko polje, with the two major springs (Malenščica and Unica) indicated. *Photo courtesy of M. Petrič. Used with permission.*

(Fig. 6). Unlike Cerkniško jezero, none of these lakes is associated with a polje, but rather are in closed depressions (dolinas). The largest, Palško jezero, is 1.03 km^2, about 3% of the size of Cerkniško jezero. The actual number of lakes is difficult to determine because there are a number of very small dolinas (<0.01 km^2) that occasionally fill with water. For the entire area there is a close connection between underground and surface water. The karst aquifer is a rather shallow one, probably the result of underlying flysch rocks. During most of the year water flows underground but after periods of intensive or long-lasting precipitation, the water table rises and emerges at the surface in different locations (Petrič and Kogovšek, 2005). Some of them appear annually and last for up to 6 months, as is the case for Palško jezero and Petelinjsko jezero, the two largest lakes. In Palško jezero, the lake is present for an average of 3 months, but may fill more than once a year if heavy rainfall occurs. Most of the other lakes appear rarely, and only fill up in times of big floods. The water table fluctuates more than 40 m at the estevelle that both fills and drains Palško jezero (Fig. 7). In the smaller lakes, they recharge and discharge mainly through the alluvial sediments which cover their limestone lake bed. Some of the smaller lakes probably fill only once a decade on average, although there are few long-term data on the presence or absence of these lakes. During and after the Second World War the area was an important military base. In the past, the area was used primarily for seasonal sheep grazing. At present, agriculture is less intensive than in the past, and forests are gradually covering the landscape, including some of the intermittent lakes (Mulec *et al.*, 2005). In 1985, in the Pivka valley as a whole, 44% of the surface was pasture and meadows, 26% forest, 9% cultivated fields, and 2% wetlands.

Vegetation of the intermittent lakes has an interesting pattern. Soil is thin and rocky on the lake margins and becomes thicker toward the center of the lake, making the center more suitable for cultivation. In general, open meadow-like flooded areas and dry meadows at the margin of the lakes are important habitats. Species characteristic of dry margins of the lakes include *Galium verum* and *Lotus corniculatus*. In wetter areas with thicker, nutrient-poor soil, species such as *Allium angulosum* and *Carex panacea* are found. Some of the species characteristic of wetter areas, such as *Carex hirta* and *Ranunculus repens*, are becoming less common because of drying out of some areas in recent decades. In the areas that are flooded for longer periods, willows (*Salix*) overgrow meadows and pastures in the absence of cultivation or grazing. The impact of reforestation on the hydrological regime and the appearance of the temporary lakes are not precisely known but apparently lead to reduced frequency of lake formation and reduced residence time of the lake.

The crustacean fauna of the Pivka lakes includes two species of fairy shrimp (*Branchypus schafferi* and *Chirocephalus croaticus*; Fig. 4), three Cladocera species, nine Copepoda species, two Ostracoda species, one Amphipoda, and one Isopoda. While the species list is very incomplete and nonexistent for insects, it includes stygobionts, generalist species, and species adapted to temporary waters (Pipan, 2005). One of the copepod species (*Diacyclops charon*) and one isopod subspecies (*Asellus aquaticus cavernicolus*) are stygobionts. The European cave salamander *Proteus anguinus* can also be found in the lakes as a result of washing out from springs and estavelles. Typical of temporary wetlands, fairy shrimp have thick-shelled winter eggs that are resistant to desiccation. They also have a very high reproductive rate, and in the case of *C. croaticus*, they can complete their life cycle in 14 days, an obvious advantage in temporary waters such as the Pivka lakes.

Pivka intermittent lakes may be representative of karst wetlands in general than either poljes or disappearing lakes. While Cerkniško jezero and Planinsko polje are large wetlands of major importance in Slovenia and elsewhere in the Dinaric karst, the geological conditions needed for polje formation (large valley along tectonic axes) are not that common (Bonacci, 2004). Pivka lakes, on the other hand, occur in a landscape of dolinas (sinkholes) in a river valley, an extremely common feature of karst landscapes. Intermittent lakes in the Sinkhole Plain of central

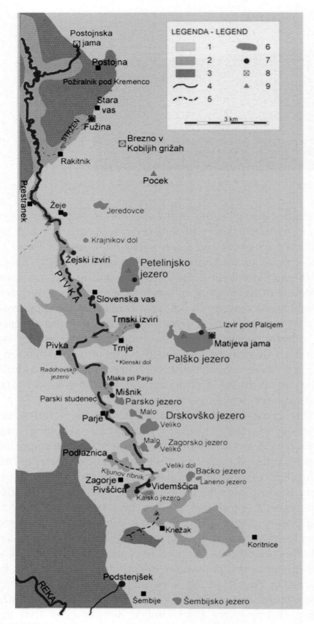

FIGURE 6 Hydrogeological map of the area of Pivka intermittent lakes. Legend: (1) karst aquifer; (2) porous aquifer; (3) rock of very low permeability; (4) permanent surface flow; (5) intermittent surface flow; (6) intermittent karst lake; (7) karst spring; (8) cave; (9) ponor. Map courtesy of M. Petrič. Used with permission.

Kentucky have many of the features of the intermittent Pivka lakes. Vessels and Jack (2001) recognize these intermittent lakes as wetlands, and note that the fauna of the lakes includes components similar to what Pipan (2005) reported for Pivka Lakes.

Sheehy Skeffington and Gormally (2007) and Gunn (2006) discuss a karst wetland feature in Ireland, called *turloughs*, found in grasslands in Ireland. They are also intermittent lakes and differ in some details from the Pivka intermittent lakes (Sheehy Skeffington et al., 2006). They suggest that the term *turlough* be used for the Pivka intermittent lakes, but Gunn (2006) on the other hand emphasizes the importance of glaciation in the development and functioning of turloughs. Turloughs have priority status under the European Habitats Directive and therefore the term is a useful tool for conservation. Perhaps the term *intermittent lakes* should also be retained for these features to recognize their connection to other kinds of wetlands.

PROTECTION OF KARST WETLANDS

There are a variety of country- or region-specific laws that are important in wetlands protection, such as the European Union's Habitat Directive and the United States' Clean Water Act. Internationally, the most important structure is the Ramsar Convention on Wetlands, adopted in 1971. Utilizing a broad definition of *wetlands* and a process by which individual countries nominate a site, nearly 1900 freshwater and marine wetlands have been designated as of March 2010. One of the wetland types is "karst and other subterranean hydrological systems—freshwater." A total of 69 sites include this designation as one of the wetland types, and for 23 it was listed as the dominant type (Table 1). The 46 sites for which the karst designation is not the primary one serve as a reminder that a variety of wetland types can occur on karst, but are not unique to karst. Of the 23 designated as primarily karst wetlands, 6 correspond to the poljes and intermittent lakes described above. Ramsar sites of this type include Cerkniško polje/jezero and Bayerische Wildalm and Wildalmfilz in Austria. Another six focus primarily on surface features, but permanent bodies of water rather than intermittent lakes. These include a groundwater-fed lake in Turkey—Kizören Obrouk—and alpine karst streams in Austria—Nationalpark Kalkalpen. To a lesser extent than intermittent lakes, these kinds of wetlands are not necessarily unique to karst. Several large ecoregions were also designated as karst wetland types, including a large region in Yucatan, Mexico.

The final category is that where caves are the primary wetlands feature in the designation. These include the binational Baradla Cave/Domica Cave System in Hungary and the Slovak Republic, and Škocjanske jame in Slovenia. Exactly what can be considered as a subterranean wetland is not clear, but the Ramsar Convention has an extraordinarily broad definition of *wetland*, basically including any temporary or permanent body of water less than 6 m deep. Under the Ramsar definition, almost all subterranean waters, except for deep phreatic lakes, could be called wetlands. All of the caves on the Ramsar list are

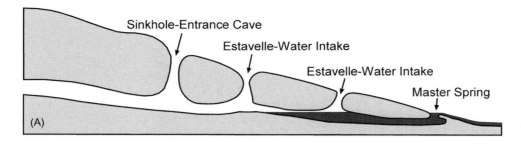
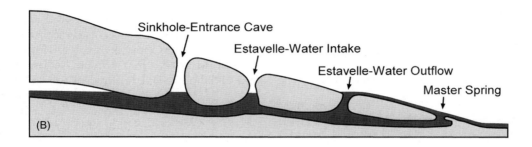
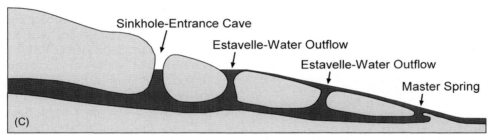

FIGURE 7 Hypothetical stream profile showing sequential reversal of flow in estavelles: (A) low flow, only lower section of karst drainage system flooded and the estavelles act as ponors (swallow holes); (B) medium flow, headward flooding of the system and the lowest estavelle acts as a spring; (C) high flow, storage capacity of karst system exceeded and all estavelles become springs. *From J.D. Vineyard and G.L. Feder (1982). Springs of Missouri. Missouri Department of Natural Resources, Division of Geology and Land Survey.*

TABLE 1 Primary Wetland Type for Ramsar Sites on Karst

Wetland Type	Number
Marine/saline	14
Permanent lakes	11
Tree or shrub-dominated wetland	3
Marsh	3
Peatlands	4
Rivers	3
Intermittent lake	2
Karst—primarily cave	8
Karst—primarily surface	6
Karst—intermittent lakes	6
Karst—large-scale landscapes	3
Other/unknown	6

Source: From www.ramsar.org. (Accessed 10 March 2010).

substantial caves with unique biological or hydrological features, often more than 1 km long, and typically including either insurgence or resurgences of water.

An alternative way to use the term *wetland* for subterranean water bodies is to look for analogies with surface wetlands. For example, there can be underground floodplains and even oxbows (abandoned stream channels). Of course, subterranean floodplains and oxbows will lack any vegetation and any migratory waterfowl. Their biological significance is that these are areas of significant organic carbon that was deposited by lateral transport. Many obligate cave-dwelling terrestrial species depend on this resource as their primary source of organic carbon.

See Also the Following Articles

Closed Depressions in Karst Areas
springs

Bibliography

Balászs, D. (1977). The geographical distribution of karst areas. *Proceedings of the Seventh International Congress of Speleology, Sheffield*, 1, 13–15.

Bonacci, O. (2004). Poljes. In J. Gunn (Ed.), *Encyclopedia of caves and karst science* (pp. 599–600). New York: Fitzroy Dearborn.

Euliss, N. H., LaBaugh, J. L., Fredrickson, L. H., Mushet, D. M., Laubhan, M. K., Swanson, G. A., et al. (2004). The wetland continuum: A conceptual framework for interpreting biological studies. *Wetlands*, 24, 448–458.

Gaberščik, A. (Ed.), (2003). In A. Gaberščik (Ed.), *Jezero, ki izginja Ljubljana: Monografija o Cerkniškem jezeru*. Društvo Ekologov Slovenije.

Gunn, J. (2006). Turlough and tiankengs: Distinctive doline forms. *Speleogenesis and Evolution of Karst Aquifers*, 4, 1–4.

Leibowitz, S. G. (2003). Isolated wetlands and their function: An ecological perspective. *Wetlands*, 23, 517–531.

Mitsch, W. J., & Gosselink, J. G. (2007). *Wetlands* (4th ed.). New York: Wiley.

Mulec, J., Mihevc, A., & Pipan, T. (2005). Intermittent lakes in the Pivka basin. *Acta Carsologica*, 34, 543–565.

Petrič, M. (2010). Characterization, exploitation, and protection of the Malenščica karst spring, Slovenia. In N. Kresic & Z. Stevanovic (Eds.), *Groundwater hydrology of springs* (pp. 428–441). Amsterdam: Elsevier.

Petrič, M., & Kogovšek, J. (2005). Hydrogeological characteristics of the area of intermittent karst lakes of Pivka. *Acta Carsologica*, 34, 599–618.

Pipan, T. (2005). Fauna of the Pivka Intermittent Lakes. *Acta Carsologica*, 34, 650–659.

Sheehy Skeffinton, M., Moran, J., O'Connor, A., Regan, E., Coxon, C. E., Scot, N. E., et al. (2006). Turloughs—Ireland's unique wetland habitat. *Biological Conservation*, 133, 265–290.

Sheehy Skeffinton, M., & Gormally, M. (2007). Turoughs: a mosaic of biodiversity and management systems unique to Ireland. *Acta Carsologica*, 36, 217–222.

Stumberger, B. (2010). A classification of karst poljes in the Dinarides and their significance for waterbird conservation. In D. Denac, M. Schneider-Jacoby & B. Stumberger (Eds.), *Adriatic flyway – clasoing the gap in bird conservation* (pp. 151–154). Radofzell, Germany: Euronatur.

Vessels, N., & Jack, J. D. (2001). Effects of fish on zooplankton community structure in Chaney Lake, a temporary karst wetland in Warren County, Kentucky. *Journal of the Kentucky Academy of Science*, 62, 52–59.

Wiggins, G. B., MacKay, R. J., & Smith, I. M. (1980). Evolutionary and ecological strategies of animals in annual temporary pools. *Archieves de Hydrobiologie Supplement*, 2, 97–206.

WHITE-NOSE SYNDROME: A FUNGAL DISEASE OF NORTH AMERICAN HIBERNATING BATS

Marianne S. Moore and Thomas H. Kunz

Boston University

INTRODUCTION

White-nose syndrome (WNS) is an epizootic currently affecting several species of hibernating bats in North America. The most devastating wildlife disease in recorded history, WNS is causing unprecedented mortality and threatening regional extinction in at least one previously common bat species (Frick et al., 2010). The syndrome is named for a white, filamentous growth of fungal hyphae and conidia (spores) on the nose, ears, wings, and tail membranes of affected bats (Blehert et al., 2009; Fig. 1). The first evidence of this fungal growth on hibernating bats was photographed on 16 February 2006 at Howes Cave, located approximately 50 km west of Albany, New York (Blehert et al., 2009). In the four years since it was first observed, the putative fungal pathogen, *Geomyces destructans* (*Gd*), associated with WNS, has spread rapidly from its epicenter, southward to North Carolina, westward to Missouri and Oklahoma, and northward to Canada. As of April 2011, *Gd* has been reported from bats in 17 states (Connecticut, Delaware, Indiana, Maryland, Massachusetts, Missouri, New Hampshire, New Jersey, New York, North Carolina, Ohio, Oklahoma, Pennsylvania, Tennessee, Vermont, Virginia, and West Virginia) and three Canadian provinces (New Brunswick, Ontario, and Quebec), although confirmed infections based on histopathology have not been determined for bats from Missouri and Oklahoma (United State Fish & Wildlife Service, 2011).

In contrast to stable or increasing pre-WNS bat populations, declines of hibernating bats in the northeastern United States, ranging from 30–99% annually and averaging 73%, have been documented (Frick et al., 2010). Assuming that current rates of mortality continue, the once common little brown myotis (*Myotis lucifugus*) is expected to be extinct in the northeastern United States by the year 2026 (Frick et al., 2010), and at least five other hibernating bat species (*M. septentrionalis*, *M. sodalis*, *M. leibii*, *Eptesicus fuscus*, and *Perimyotis subflavus*) in this region are at risk (Blehert

FIGURE 1 Hibernating little brown myotis (*Myotis lucifugus*) infected with a putative fungal pathogen, *Geomyces destructans*, associated with white-nose syndrome. *Photo by A.C. Hicks, New York Department of Environmental Conservation. Used with permission.*

et al., 2009; Turner and Reeder, 2009; Courtin et al., 2010; K.E. Langwig, pers. comm). Based on morphological and molecular (polymerase chain reaction, PCR) criteria, Gd was recently reported from three other hibernating bat species in North America (*M. grisescens* [listed as federally endangered], *M. austroriparius*, and *M. velifer*) although infections have not been confirmed in *M. grisescens* from Missouri or *M. velifer* from Oklahoma based on histopathology. Two federally endangered subspecies of Townsendi's long-eared bat, *Corynorhinus townsendii* (*C. t. virginianus* and *C. t. ingens*), are within the geographic range of hibernating bats affected by Gd, but to date there is no evidence that this fungus has spread to these two species.

No introduced pathogen has caused such a precipitous decline in any mammal species in recorded history. Low reproductive rates and long life span of bats (Kunz and Fenton, 2003) will make their recovery extremely slow unless the syndrome can somehow be controlled. The recent decline of bat populations in the northeastern United States caused by WNS is likely to have adverse economic consequences on the ecology of both natural and human-derived agroecosystems.

BAT MORTALITY AND THE PUTATIVE PATHOGEN

Beginning in late winter 2007/2008, wildlife biologists began to document unusual physical conditions, behaviors, and mortality in bats hibernating in caves and abandoned mines in New York State. Subsequently, postmortem histopathological examinations revealed a distinct cutaneous fungal infection where hyphae invaded the epidermis, hair follicles, sebaceous and sweat glands, and breached the basement membrane of hibernating bats (Blehert et al., 2009). Examination of these bats with cutaneous fungal infections revealed a previously undescribed fungus with an asymmetrically curved conidia (Blehert et al., 2009). Laboratory studies later documented that this fungus is psychrophilic (*i.e.*, cold loving) and grows optimally between 5 and 10°C, well within the seasonal temperature range of most hibernacula (2–14°C) (Blehert et al., 2009). Phylogenetic analysis has placed these isolates within the inoperculate ascomycetes (order Helotiales) and near the anamorphic genus *Geomyces* (teleomorph *Pseudogymnoascus*). The closest relative of the fungal isolates from bats was identified as the ubiquitous *G. pannorum*, a keratinophylic species of which a subspecies, *G. pannorum* var. *pannorum*, has been identified in superficial skin and fingernail infections in humans. The combination of a unique morphology and genetic composition (specifically at the identical internal transcribed spacer region and small subunit ribosomal RNA gene) distinguishes the bat isolate from other members of *Geomyces*. This discovery led to the recognition of a new species within the genus, and subsequently the fungus associated with WNS was aptly named *Geomyces destructans*. To date, it is not entirely clear whether Gd is the causative agent of WNS, but it appears that this fungus alters the physiology of hibernating bats (Boyles and Willis, 2009) and thus may have a direct pathogenic role (Meteyer et al., 2009; Cryan et al., 2010). Aside from the histopathological evidence that Gd causes severe cutaneous lesions (Meteyer et al., 2009), secretory proteases involved in tissue digestion have been identified from Gd supporting the etiologic role of the fungus in WNS (Chaturvedi et al., 2010). Additionally, genotyping of five genetic markers using Gd isolates collected from bats within a 200-km survey distance from its epicenter provide preliminary evidence that a single strain was introduced into New York and a clonal population is infecting all bats within the affected region (Chaturvedi et al., 2010).

DIAGNOSTICS

Histopathology is currently the gold standard of diagnostic tools used to confirm WNS, which with specialized training and appropriate staining and visualization of tissues, allows the identification of Gd and associated cutaneous lesions (Meteyer et al., 2009). Culture, isolation and morphological examination can also be used to identify presence of the fungus from tissue or swab samples (Blehert et al., 2009; Gargas et al., 2009; Chaturvedi et al., 2010); however, a number of researchers have noted low rates of isolation (*e.g.*, 54%) (Lorch et al., 2010) despite an abundance of available Gd-affected tissue, a complication that may be due to competition from other microorganisms colonizing bat skin (Chaturvedi et al., 2010; Lorch et al., 2010). PCR amplification and sequencing of genetic material is a promising, economical, and rapid alternative that can also be used to identify the presence of Gd on skin, hair, and other tissue samples (Chaturvedi et al., 2010; Lorch et al., 2010), and trials using this method on the skin of bats inoculated with various concentrations of Gd showed 96% diagnostic sensitivity and 100% diagnostic specificity (Lorch et al., 2010). However, although this technique can be used to less invasively screen individual bats for the presence of Gd, small tissue samples usually collected for DNA extraction and PCR amplification may not provide accurate results because the fungus is not equally distributed across all tissue surfaces (Lorch et al., 2010). Additionally, the use of PCR amplification and culture/isolation/examination methods only allow for confirmed presence of the fungus, which may occur in the absence of cutaneous lesions, and do not provide the ability to determine the extent of *Gd* invasion into cutaneous tissues as compared with histopathological examination.

BAT SPECIES AFFECTED BY WNS

To date, six hibernating cave-roosting bat species have experienced varied rates of mortality associated with WNS, including the little brown myotis (*Myotis lucifugus*), northern long-eared myotis (*M. septentrionalis*), the federally listed endangered Indiana myotis (*M. sodalis*), the small-footed myotis (*M. leibii*), the tricolored bat (*Perimyotis subflavus*), and the big brown bat (*Eptesicus fuscus*) (Blehert *et al.*, 2009; Courtin *et al.*, 2010; A.C. Hicks and K.E. Langwig, pers. comm.). More recently, *Gd*, the fungus associated with WNS, has also been identified morphologically and genetically (using PCR) from the southeastern myotis (*Myotis austroriparius*), the federally listed endangered gray myotis (*M. grisescens*), and cave bat (*M. velifer*), although to date no other symptoms associated with WNS, including mass mortality, have been observed in these three species.

CHARACTERISTICS OF WNS

Fungal Infections and Tissue Damage

The primary characteristic of WNS is the presence of cutaneous lesions from *Gd*, where the invasion of fungal hyphae causes cup-like epidermal erosions, ulcers, and degradation of regional connective tissue (Meteyer *et al.*, 2009; Fig. 2). Invasion by *Gd* leads to the replacement of cutaneous glands, blood, and lymphatic vessels and muscle as the fungus digests tissues and also causes a reduction in tone, tensile strength, and elasticity of wing membranes (Cryan *et al.*, 2010). Additionally, tissue damage and necrosis, apparently due to oxygen depletion (*i.e.*, infarcted tissue), has been observed in regions of wing membrane distant from the area of fungal invasion (Cryan *et al.*, 2010). Given the importance of bat wings in many physiological processes, these observations suggest that a massive homeostatic imbalance caused by the destruction of wing tissue may be a possible proximate cause of death in hibernating bats (Cryan *et al.*, 2010).

Depleted Fat Reserves

WNS is also manifested in hibernating bats by an apparent premature mobilization of fat reserves (Blehert *et al.*, 2009; Meteyer *et al.*, 2009; Courtin *et al.*, 2010; J.D. Reichard, pers. comm.) suggesting that starvation due to depleted fat reserves may also be a proximate cause of death in bats affected by WNS. Depletion in energy reserves may be due to the potential homeostatic imbalance noted above or an initial dearth of fat reserves in bats as they enter hibernation possibly due to a lack of fatty acids in the diet because of changes in insect abundance. It is also possible that depleted fat reserves are the result of accelerated energy consumption if bats are arousing from torpor more frequently to groom the fungus from their skin, or to mount immune responses. Arousing from torpor and maintaining euthermic body temperatures at typical hibernaculum temperatures are energetically expensive activities (Boyles and Willis, 2009). If normal torpor-arousal patterns are disrupted during the development of WNS, increases in metabolic rates could easily require more fat reserves than bats typically store in preparation for hibernation and result in the observed depletion of these reserves and death by starvation.

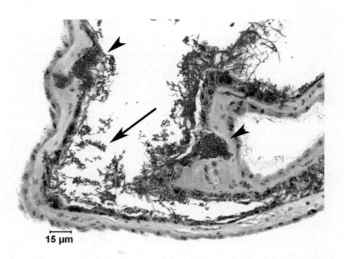

FIGURE 2 Histological section of a wing membrane collected from a hibernating little brown myotis (*Myotis lucifugus*) affected by white-nose syndrome immediately following euthanasia. Exuberant growth of *Geomyces destructans* and conidia (spores) are present on the skin surface (arrow) as well as penetrating wing membrane (arrowheads) without associated inflammation. PAS stain. Bar = 15 μm. *Photo by C.U. Meteyer, U.S. Geological Survey; Meteyer et al., 2009. Used with permission.*

Atypical Winter Behavior

In addition to wing damage, depleted fat reserves, and fungal infections, bats affected by WNS may also exhibit atypical behavior including flying outside hibernacula in midwinter (A.C. Hicks, pers. comm.) and the inability to arouse from torpor when disturbed by human activity (A.C. Hicks, J.G. Boyles, and J.D. Reichard, pers. comm.). In fact, infrared thermal imaging of clusters of hibernating little brown myotis shows that many torpid bats affected by WNS do not elevate their body temperatures upon anthropogenic torpor are due to irritation caused by the cutaneous fungal infection, or to elevate immune responses in an attempt to resist fungal invasion. It is also possible that the depletion of fat reserves prompts bats to arouse from torpor to feed, which is supported by the observations of bats emerging from hibernacula in midwinter potentially in an attempt to forage (Blehert et al., 2009; A.C. Hicks, pers. comm.; J.D. Reichard, pers. comm.).

Changes in Immune Response during Hibernation

As true hibernators, bat species affected by WNS may experience levels of immunocompetence constrained by prolonged periods of deep torpor, given that research on other hibernating mammals suggests that deep torpor depresses numerous mechanisms associated with immunity (Bouma et al., 2010). Because of attempted resistance against fungal invasion, the immune function in hibernating bats affected with WNS may be altered, with some aspects apparently elevated and others reduced in comparison to unaffected bats (M.S. Moore, unpubl. data; R. Jacob and D.M. Reeder, unpubl. data). Based on histopathological examination, the little brown myotis does not appear to mount a morphologically detectable inflammatory response to *Gd* (Meteyer et al., 2009); however, the extent to which WNS-affected bats attempt to resist invasion by *Gd* through other immunological responses remains unknown.

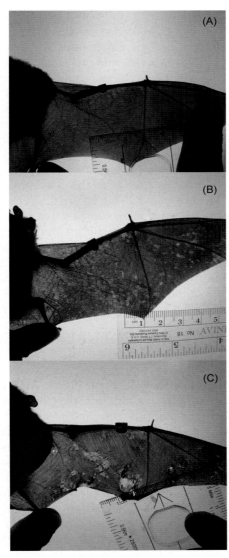

FIGURE 3 Wings of little brown myotis (*Myotis lucifugus*) unaffected by and affected by white-nose syndrome captured in early spring from a maternity colony. (A) Undamaged wing (wing-damage index 0); (B) wing showing moderate scarring and spotting (wing-damage index 1+); (C) wing showing severe scarring and necrosis (wing-damage index 3). *Wing-damage index after Reichard and Kunz, 2009. Photo by N.W. Fuller, Boston University. Used with permission.*

Wing Damage

Bats that survive hibernation while being affected by WNS may return to maternity colonies with ulcerated, necrotic, and scarred wing membranes (Reichard and Kunz, 2009; N.W. Fuller, pers. comm.). In fact, severe wing damage was observed in the majority of little brown myotis roosting in two New Hampshire maternity colonies during the summer of 2008 (Fig. 3). Damage to wings caused by wounds or infections can impair the ability of wings to perform their functions and may result in reduced foraging success and increased vulnerability to predation. However, the results from a recent study indicate that complete healing from WNS-associated damage can occur in some individuals during summer months (N.W. Fuller, pers. comm.). If bats with damaged wings experience reduced foraging success during the active season, further reductions may occur in the relative body condition of bats arriving at maternity colonies in an already compromised state. This reduction in body condition may also result in an inability to properly care for young, avoid predation, and enter the following

disturbance (J.D. Reichard, unpubl. data; J.G. Boyles, unpubl. data).

Atypical behavior of hibernating bats infected with *Gd* also appears to manifest increased frequencies of arousal from torpor (D.M. Reeder, pers. comm.). Because elevating body temperature (T_b) from near ambient to ~37°C and sustaining this elevation of T_b during arousals is extremely energetically costly (Boyles and Willis, 2009), this behavioral change is likely to incur extreme costs in the form of fat depletion. It is possible that increased frequencies of arousal from hibernation period with sufficient fat reserves. In fact, reduced body condition and lower recapture rates have been observed in relation to WNS-induced wing damage (Reichard and Kunz, 2009). In addition to the massive mortality that has been observed during hibernation, this physiological challenge incurred during the active season also has presumably contributed to the approximately 73% decline in summer activity of little brown myotis within regions affected by WNS (Dzal et al., 2011; Brooks, 2011).

FUNGAL TRANSMISSION

The pattern of expansion away from its epicenter, the known behavioral patterns of bats throughout the year, and the fact that some gated caves and mines have had no human visitors for several years prior to becoming affected, support the hypothesis that bat-to-bat contact is the primary mode of transmission. However, the possibility of anthropogenic transmission was acknowledged when researchers first became aware that the majority of newly affected sites in 2008 had been visited by humans, either as biologists or recreational users (Turner and Reeder, 2009). However, in 2009, bats infected with *Gd* appeared in areas distant from other affected sites, suggesting that human transmission of the fungus was a possibility. These infected sites have high recreational use and small hibernating populations or may have been visited by people with gear used at other affected sites. Thus, it appears that while bat-to-bat transmission facilitates the movement of *Gd* into new sites adjacent or near to affected hibernacula, anthropogenic transmission may accelerate the spread and introduce the fungus into relatively distant sites and areas.

A EUROPEAN CONNECTION?

As early as the 1980s, repeated observations of white fungal growth on the muzzles of hibernating bats were reported in Europe; however, no unusual mortality events have been documented in association with these reports (Martínková et al., 2010; Wibbelt et al., 2010). Since the advent of WNS in North America and the identification of the associated fungus, *Gd*, in nine species of bats in Europe, sampled in Germany, Switzerland, Hungary, the Czech Republic, Slovakia, and France (Martínková et al., 2010; Puechmaille et al., 2010; Wibbelt et al., 2010), no mass mortality has been associated with these isolations. Some researchers have postulated that the fungus may be native to Europe or at least that it predates its presence in North America (Martínková et al., 2010; Wibbelt et al., 2010). Alternatively, it is also possible that the fungus had its origin in North America and was recently introduced to Europe. However, because considerable genetic variability was observed in *Gd* isolates from the Czech Republic and Slovakia (Martínková et al., 2010), in contrast to the identical clones found in New York (Chaturvedi et al., 2010), these results suggest that the fungus had its origin in Europe. One hypothesis is that *Gd* caused mass mortality in European bats long before human observations were recorded and that they evolved resistance while being exposed to this fungus. Different hibernation strategies may also affect the susceptibility of various species to the syndrome. As suggested by Wibbelt et al. (2010), competition between *Gd* and other microbial flora colonizing bat skin or roost surfaces and soils in hibernacula in Europe may have contributed to the evolution of a nonpathogenic form in Europe, while a lack of similar competition in North America may have led the evolution of a pathogenic variant of *Gd* that affects hibernating bats.

CURRENT AND FUTURE RESEARCH

A multicontinent research effort has been mobilized to study the effects of WNS on North American and European bats, and includes researchers from numerous academic, governmental and nonprofit institutions. One of the primary areas of focus is in determining if the fungus is the cause of mortality or is a secondary symptom. Controlling the spread and mitigating the effects of WNS are also high priority areas of research. As an example, based on modeling efforts, Boyles and Willis (2009) suggested that localized "thermal refugia," or warm areas inside hibernacula, could reduce heat loss and energy expenditure during periodic arousals, possibly providing a stopgap measure to employ while other control measures are being developed. However, this unconventional practice would require that bats somehow could detect and travel to refugia and that these practices would not alter conditions within the rest of hibernacula where refugia are installed. Culling has also been suggested as a possible means to reduce transmission (Arnold Air Force Base, 2009), although Hallam and McCracken (2011) modeled the effects of culling on transmission cycles and concluded that this mitigation practice will have little effect on the spread of the syndrome. Some groups are also conducting trials to test the ability of fungicides to eradicate *Gd* from contaminated gear and the efficacy of compounds to treat affected bats (H. Barton, pers. comm.; D.M. Reeder, pers. comm.; A.H. Robbins, pers. comm.).

Additional studies initiated shortly after the emergence of the syndrome include investigations into patterns of arousal and torpor, thermoregulatory changes, variation in metabolic rates, behavioral changes, determination of variation in body condition and types of fat throughout the hibernation period, immunological correlates, and examination of the bacterial flora of digestive systems in WNS-affected bats (Turner and Reeder, 2009). The possible role of environmental contaminants has been investigated to a degree, showing high levels of persistent organic pollutants (Kannan et al., 2010), lead, and arsenic (Courtin et al., 2010) in WNS-affected bats. However, high levels of organic pollutants were also found in bats hibernating in unaffected hibernacula (Kannan et al., 2010), making the interpretation of these results ambiguous. Primers developed for the detection of Gd on bat tissues have been tested on environmental samples to better understand the geographic distribution of this newly described fungus, although high levels of cross-reaction with other species of fungi have prevented this method from being used to successfully identify Gd in samples other than those directly collected from bat tissue (Lorch et al., 2010). Modeling efforts have compared demographics in the affected little brown myotis to predict the risk of extinction as a result of mortality caused by WNS (Frick et al., 2010), and demographic models are being developed to help explain the differential rates of mortality among affected species (K.E. Langwig, pers. comm.; A.P. Wilder, pers. comm.).

More recently, researchers have begun to investigate additional components important to the development of WNS. Quantitative PCR methods are being developed to determine fungal load on affected bats. Investigations are being conducted into the pathogenesis, microbial ecology and phylogeography of the fungus, including sequencing the genome of Gd (J.T. Foster, pers. comm.). Population genetic structure and gene flow in the little brown myotis is being evaluated to help predict the further spread of the syndrome (C.M. Miller-Buttersworth, pers. comm.; A.P. Wilder, pers. comm.). Studies are also under way to compare the North American and European strains of the fungus and differences in susceptibility between North American and European species.

CAVE CLOSURES AND DECONTAMINATION PROTOCOLS

Because of the likelihood that human visitations facilitate the transmission of Gd to unaffected sites, a number of states have closed caves on state-owned property and the U.S. Fish and Wildlife Service (USFWS) recommends that all cavers observe cave closures, advisories, and decontamination protocols. Decontamination procedures and a list of current cave closures can be found at www.fws.gov/WhiteNoseSyndrome/cavers.html. For the most current list of WNS-affected and adjacent states, visit www.fws.gov/WhiteNoseSyndrome/. Most importantly, bats should not be handled by people unless authorized to do so. If live or dead bats are encountered in caves and mines with the appearance of WNS, contact the nearest state wildlife agency, USFWS Ecological Services Field Office, or email WhiteNoseBats@fws.gov. State office listings can be found at www.fws.gov/offices/statelinks.html. USFWS office listings can be found at: www.fws.gov/offices/.

Acknowledgments

We wish to thank H.A. Barton, J.G. Boyles, N.W. Fuller, A.C. Hicks, R. Jacob, K.E. Langwig, D.M. Reeder, J.D. Reichard, A.H. Robbins, and A.P. Wilder for sharing unpublished information referenced in this chapter. We are grateful to the American Society of Mammalogists, Bat Conservation International, the Eppley Foundation for Research, National Science Foundation, Morris Animal Foundation, National Speleological Society, U.S. Fish and Wildlife Service, and the Woodtiger Fund for supporting our research on WNS.

See Also the Following Article

Bats

Bibliography

Arnold Air Force Base (2009). *White nose syndrome cooperative monitoring and response plan for Tennessee 2009*. Tullahoma, TN: Arnold Air Force Base.

Blehert, D. S., Hicks, A. C., Behr, M., Meteyer, C. U., Berlowski-Zier, B. M., Buckles, E. L., et al. (2009). Bat white-nose syndrome: An emerging fungal pathogen? *Science, 323*(5911), 227.

Bouma, H. R., Carey, H. V., & Kroese, F. G. M. (2010). Hibernation: The immune system at rest? *Journal of Leukocyte Biology, 88*, 619–624.

Boyles, J. G., & Willis, C. K. R. (2009). Could localized warm areas inside cold caves reduce mortality of hibernating bats affected by white-nose syndrome? *Frontiers of Ecology and the Environment, 8*, 92–98.

Brooks, R. T. (2011). Declines in summer bat activity in central New England 4 years following the initial detection of white-nose syndrome. *Biodiversity and Conservation, 20*(1), 2537–2541. doi:10.1007/s10531-011-9996-0.

Chaturvedi, V., Springer, D. J., Behr, M. J., Ramani, R., Li, X., Peck, M. K., et al. (2010). Morphological and molecular characterizations of psychrophilic fungus *Geomyces destructans* from New York bats with white nose syndrome (WNS). *PLoS ONE, 5*(5), e10783. doi:10.1371/journal.pone.0010783.

Courtin, F., Stone, W. B., Risatti, G., Gilbert, K., & Van Kruiningen, H. J. (2010). Pathogenic findings and liver elements in hibernating bats with white-nose syndrome. *Veterinary Pathology, 47*, 214–219.

Cryan, P. M., Meteyer, C. U., Boyles, J. G., & Blehert, D. S. (2010). Wing pathology of white-nose syndrome in bats suggests life-threatening disruption of physiology. *BMC Biology, 8*, 135.

Dzal, Y., McGuire, L. P., Veselka, N., & Fenton, M. B. (2011). Going, going, gone: The impact of white-nose syndrome on the summer activity of the little brown bat (*Myotis lucifugus*). *Biology Letters, 7* (3), 392–394. doi:10.1098/rsbl.2010.0859.

Frick, W. F., Pollock, J. F., Hicks, A. C., Langwig, K. E., Reynolds, D. S., Turner, G. G., et al. (2010). An emerging disease causes regional population collapse of a common North American bat species. *Science, 329*(5992), 679–682.

Gargas, A., Trest, M. T., Christensen, M., Volk, T. J., & Blehert, D. S. (2009). *Geomyces destructans* sp. nov. associated with bat white-nose syndrome. *Mycotaxon, 108*, 147–154.

Hallam, T. G., & McCracken, G. F. (2011). Management of the panzootic white-nose syndrome through culling of bats. *Conservation Biology, 25*, 189–194.

Kannan, K., Yun, S. H., Rudd, R. J., & Behr, M. J. (2010). High concentrations of persistent organic pollutants including PCBs, DDT, PBDEs and PFOS in little brown bats with white-nose syndrome in New York, U.S.A. *Chemosphere, 80*, 613–618.

Kunz, T. H., & Fenton, M. B. (2003). *Bat ecology*. Chicago, IL: University of Chicago Press.

Lorch, J. M., Gargas, A., Meteyer, C. U., Berlowski-Zier, B. M., Green, D. E., Shearn-Bochsler, V., et al. (2010). Rapid polymerase chain reaction diagnosis of white-nose syndrome in bats. *Journal of Veterinary Diagnostic Investigations, 22*, 224–230.

Martínková, N., Bačkor, P., Bartonička, T., Blažková, P., Červený, J., Falteisek, L., et al. (2010). Increasing incidence of *Geomyces destructans* fungus in bats from the Czech Republic and Slovakia. *PLoS ONE, 5*(11), e13853. doi:10.1098/rsbl.2010.0859.

Meteyer, C. U., Buckles, E. L., Blehert, D. S., Hicks, A. C., Green, D. E., Shearn-Bochsler, V., et al. (2009). Histopathologic criteria to confirm white-nose syndrome in bats. *Journal of Veterinary Diagnostic Investigations, 21*, 411–414.

Puechmaille, S. J., Verdeyroux, P., Fuller, H., Gouilh, M. A., Bekaert, M., & Teeling, E. C. (2010). White-nose syndrome fungus (*Geomyces destructans*) in bats, France. *Emerging Infectious Diseases, 16*, 290–293.

Reichard, J. D., & Kunz, T. H. (2009). White-nose syndrome inflicts lasting injuries to the wings of little brown myotis (*Myotis lucifugus*). *Acta Chiroptera, 11*, 457–464.

Turner, G. G., & Reeder, D. M. (2009). Update of white-nose syndrome in bats, September 2009. *Bat Research News, 50*, 47–53.

United States Fish & Wildlife Service (USFWS). (2011). White-nose syndrome: Something is killing our bats. www.fws.gov/whitenosesyndrome/. Accessed 02.11.11.

Wibbelt, G., Kurth, A., Hellmann, D., Weishaar, M., Barlow, A., Veith, M., et al. (2010). White-nose syndrome fungus (*Geomyces destructans*) in bats, Europe. *Emerging Infectious Diseases, 16*, 1237–1242.

WORMS

Elzbieta Dumnicka

Institute of Nature Conservation, Polish Academy of Sciences

GENERAL CHARACTERISTICS

Several types (phyla) of small invertebrates are usually included under the common term *worms*: flatworms (Turbellaria), roundworms (Nematoda), ribbon worms (Nemertea), and annelids (Annelida). Sometimes the wormlike Onychophora are also included. The shape of worms is rather simple: flat, filiform, or tube-like. The body is usually small: from half a millimeter to 2–3 cm, exceptionally more. Nevertheless true "giants" can be found among worms—some reaching lengths in the meters. The majority of worm species do not show morphological changes connected with life in the cave environment, because they are usually unpigmented and blind. Only in flatworms, leeches, and a few ribbon worms, which are normally occulated and pigmented, do these changes occur. The body coloration is determined principally by the contents of the digestive tube and by blood pigments (as in annelids). Features characteristic of typical cave fauna other than morphological ones have been observed, including aperiodicity of breeding and prolongation of all stages of the life cycle (Dumnicka, 1984).

The majority of worms living in caves are the same species that live in the soil or in marine or freshwater environments. Nevertheless, in all phyla there are also species found exclusively in various kinds of underground habitats (including caves), and they are classified as troglobionts (terrestrial forms) or stygobionts (aquatic forms). The distribution of particular stygobiotic species is often wider than that of species found only in caves since stygobionts can live in various types of underground waters (*e.g.*, interstitial waters, hyporheic waters of rivers, *etc.*; (Botosaneanu, 1986). They can even migrate from the drainage area of one river to the other—due to so-called *stream capture* (or *stream piracy*)—via an underground connection between streams from different river catchments. Stygobionts also live in regions without caves; in such a situation they inhabit fissures and small spaces between the grains of sediments filled with water. Their presence in underground waters of such areas is noticed during studies made in wells or springs, where they have been caught after being washed out from underground waters following heavy rains.

HISTORY OF STUDIES

Though studies on cave worms started many years ago (*e.g.*, Jeannel, 1926; Hyman, 1937), there is still a lot that is unknown. World data on the distribution, biology, and ecology of various taxonomic groups were summarized in *Encyclopaedia Biospéologica* (Juberthie and Decu, 1994, 1998) and *Stygofauna Mundi* (Botosaneanu, 1986) (for water fauna). From the United States, Culver *et al.* (2000) listed 927 species living obligatory in caves, but Nematoda were omitted from this list, whereas other worms were represented by only a small number of species: Tricladida, 24 species; Oligochaeta, 5 species; and Branchiobdellida, 7 species. From that time a few new stygobiotic worms were described in this region but more studies of this group are needed. Since the publication of those books, many new species or even higher taxa have been described from other regions, especially Europe. For example, 66 stygobiotic oligochaete species were cited in *Stygofauna Mundi* (1986) but that number is now more than 130. Recently, a special European research project (PASCALIS—Protocols for the Assessment and Conservation of Aquatic Life in the Subsurface) was carried out in five Mediterranean countries and Belgium that allowed the finding of many new species as well as to elucidate the origin of some stygobiotic taxa. Moreover, in the studied karstic regions, mapping of biodiversity was performed to aid in the choice of the best conservation strategies. Thus the majority of published information on the number and distribution of troglobiotic and stygobiotic species is incomplete. According to our present state of knowledge, troglobiotic and stygobiotic worms are most numerous in Europe and North America, but these are also the continents with the most intensive biospeleological studies. Studies of worms in karstic regions of Malaysia and South America started only recently, and their results should increase our knowledge about this group. Recently in Australia, the most interesting discoveries of the underground fauna were made in various regions, mainly in the western part of the country (Austin et al., 2008). Troglobiotic and stygobiotic fauna inhabiting small fissures up to 100 m deep were sampled from boreholes but worm fauna in this country has been little studied.

BRIEF CHARACTERIZATION OF INVERTEBRATE TYPES CONSIDERED WORMS

Phylum: Turbellaria (Flatworms)

The body of flatworms is elongated with a delicately ciliated surface, with only one alimentary canal opening (no anal opening). Representatives of two orders (discussed next) have been found in cave waters, whereas so-called *microturbellaria* (small representatives of various turbellarian groups) are known predominantly from interstitial waters.

Class: *Tricladida*

The Tricladida are almost exclusively aquatic forms feeding on living or dead animals. The triclad engulfs food particles extruding a long and muscular pharynx through its mouth (Fig. 1) and sucks up the food into the gastrovascular cavity. These invertebrates are very resistant to lack of food—the starvation period can reach some months. Morphological adaptations to cave life are clearly visible in this group of worms, expressed in depigmentation of the body and disappearance or reduction of the eyes. They have been found mainly in Europe and North America, but recent discoveries have been made in Tahiti, Mexico, and Australia. Of

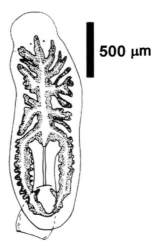

FIGURE 1 *Dinizia sanctaehelenae* (Tricladida). *From Botosaneanu, 1986.*

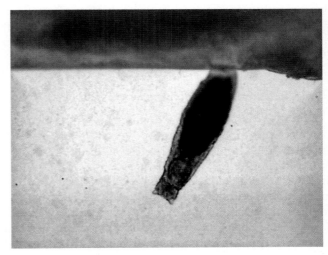

FIGURE 2 The Temnocephalid, an external parasite of crustaceans.

the approximately 400 triclad species currently known, about 150 are stygobionts. In recent systematics of this group, a suborder named Tricladida Cavernicola was separated (Sluys, 1990).

Class: *Temnocephalida*

The Temnocephalida are external parasites of crustaceans (Decapoda and Amphipoda), attaching to the host's gills, antennae, or legs (Fig. 2). The body is simple, with one or more (up to six) acetabula allowing for attachment to the host animal (Fig. 3). They produce a viscous substance to better attach and glue their eggs, from which small specimens, similar to adult forms, hatch. Those worms feed mainly on secretions or body fluids of the host; only the predatory genus *Bubalocerus* uses crustaceans as a means of dispersal.

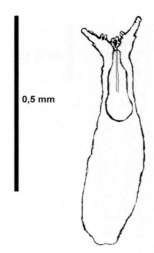

FIGURE 3 *Troglocaridicola capreolaria* (Temnocephalida). *Redrawn from Juberthie and Decu, 1994.*

So far only 15 species of these highly specialized worms are known from caves in southern Europe, but one undetermined species was found on a cave crab in Papua New Guinea.

Phylum: Nematoda (Roundworms)

Roundworms show high homogeneity in their morphology and anatomy. The body is filiform, usually small in free-living forms (length 0.2–3 mm), pointed at one or both ends, covered by smooth chitin cuticle, and without appendages (Fig. 4). This very large group (about 100,000 known species) lives in all habitats: terrestrial and aquatic (sea bottom and all kinds of freshwater), and as parasites on plants, animals, and humans. The parasitic species living in cultivated plants or breeding animals are well known whereas the free-living nematodes are poorly known. Many free-living species are eurytopic; such species can be found in different environments (terrestrial or aquatic) as well as in various climatic zones. Their high resistance to unfavorable life conditions (the eggs are resistant to desiccation and extreme temperatures) and ability to form cysts means that they can be easily transported for long distances by wind or water. That is why common species of nematodes are abundant in cave environments where they feed on bacteria, vegetal debris, and bat guano; some of them are predators. Among nematodes the troglobiotic species are extremely rare; only about 20 are known.

Phylum: Nemertea (Ribbon and Proboscis Worms)

In Nemertea, the body is unsegmented, strongly elongated, slender, soft, and usually without appendages

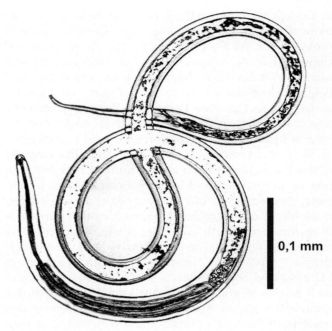

FIGURE 4 *Dorylaimus* sp. (Nematoda).

(Fig. 5). A retractable, muscular proboscis apparatus at the anterior body end is characteristic of these worms only. It is used for catching the prey—almost all ribbon worms are predators or scavengers. The length of the majority of species does not exceed a few centimeters but the longest marine species reach 30 m. Almost all species of ribbon worms (about 1200) live in marine environments (Sundberg and Gibson, 2008), but there are some exceptions: about 20 freshwater species are known and among them 3 or 4 are blind, depigmented inhabitants of various kinds of subterranean waters. Their dimensions are smaller than these of epigean species. Stygobiotic species have been found mainly in Europe (France, Austria, Switzerland, Germany) but one species has been described from New Zealand also. Moreover some marine species inhabiting deep layers of sandy beaches could be treated as stygobionts, but the morphological adaptations for subterranean life have not been observed in them.

Phylum: Annelida

The Annelida are tube-like invertebrates with a regularly segmented body. Usually they are bilaterally symmetrical, elongated, mainly hermaphroditic (bisexual) with gonads situated in some segments only.

Class: Polychaeta

Segments have muscular, paired, laterally situated "pseudolegs" (parapodia) with setae on their ends.

WORMS

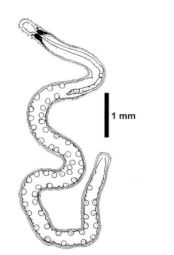

FIGURE 5 *Ototyphonemertes* sp. (Nemertina). *From Botosaneanu, 1986.*

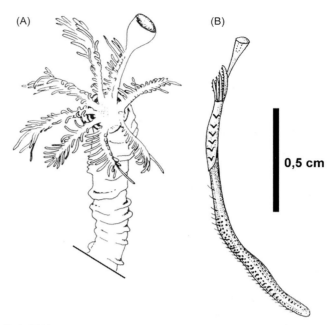

FIGURE 6 *Marifugia cavatica* (Polychaeta): (A) Anterior part of the body *(from Botosaneanu, 1986)*. (B) Habitus of a female individual *(from Juberthie and Decu, 1998)*.

Polychaeta are an almost exclusively marine group represented by about 9000 known species, but the degree of knowledge of this group is insufficient. Many polychaetes are sessile forms, living in tubes; others are mobile and move actively. Numerous marine species live in conditions similar to cave conditions: inside the bottom material, under stones, and in the rock fissures. In submarine (anchihaline) caves, stygobiotic forms very tolerant to changes of environmental conditions were found (Brito et al., 2009). In Central and South America rich polychaete fauna was found in a special type of sinkhole, *cenotes*, which connect with the ocean by flooded cave corridors. Nevertheless, due to their lack of morphological adaptations to cave environment, only on the basis of their habitat can they be classified as stygobionts.

About 2% of all polychaete species live in freshwater and some curious stygobionts are known from this habitat. The most famous one is *Marifugia cavatica* (Fig. 6), known from caves situated in Dinaric karst (northeastern Italy, Slovenia, Croatia, and Bosnia and Herzegovina) but occurring in small, isolated areas (Kupriyanova et al., 2009). Even though some localities are situated a few kilometers from the seashore, this species has never been found in brackish or saltwater. Individuals live in calcareous tubes which are attached to the walls, roofs, and bottoms of water-filled corridors forming colonies of various dimensions. In recent years live animals were collected in a few caves only, whereas aggregations of their broken tubes were found in many localities (Kupriyanova et al., 2009). Other freshwater stygobiotic polychaetes belong to the genus *Namanereis*. *Namanereis cavernicola* is known from Mexican and Peruvian caves, while *N. tiriteae* from New Zealand and Fiji (Gray et al., 2009) lives in interstitial waters. The disjunct distribution of these species according to the Gondwana continent indicates their ancient origin (Glasby and Timm, 2008).

Class: Oligochaeta

All segments (except for the first and last ones) of Oligochaeta have four bundles of setae (Fig. 7). They live in both aquatic and terrestrial (earthworms) environments. About 5000 species have been described, with only 130 from caves and subterranean waters. The smallest ones (Fig. 8) live among the grains of sand and gravel saturated with water (interstitial waters) at the seashore as well as along streams and in stagnant water bodies in caves. The bigger ones "walk" on the bottom surface or swim using setae as locomotory organs. The representatives of naidids and lumbriculids generally are "anchored" inside the sediments with the anterior part of the body, waving the posterior end for better oxygenation of the surrounding water. They move slowly when the food is used up or environmental conditions deteriorate, which is when oxygenation decreases or products of metabolism accumulate. Like nematodes, some stygobiotic oligochaete species can form cysts, which help them to survive in small, periodically desiccated cave pools. These invertebrates are bisexual. Normally cross-fertilization occurs but self-fertilization is also known. The majority of oligochaetes feeds on organic matter from sediments or soil. In the cave environment, organic matter usually forms only a small part of the sediment, so oligochaetes ingest big

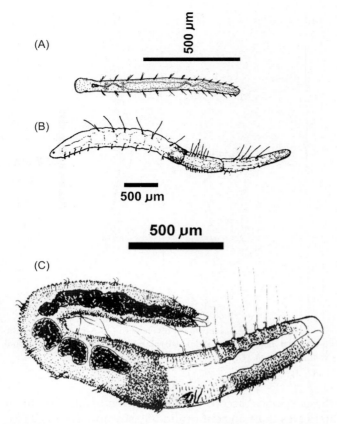

FIGURE 7 Representatives of cave oligochaetes: (A) *Aeolosoma* sp. (B) *Nais* sp.; (C) *Haber turquini* (from Botosaneanu, 1986).

Species of this group (classified as oligochaetes or leeches, or even a separate class among annelids) are usually ectoparasites of crustaceans, some of them (similarly to flatworms Temnocephalida) were found on 1–3 host species only. Troglobiotic and stygobiotic oligochaetes are known from the caves of all continents (except for Antarctica), but the most diversified fauna have been found in the Mediterranean region (Creuzé des Chatelliers et al., 2009).

Class: Hirudinea (Leeches)

The Hirudinea have segments without parapodia and setae. All leeches have two suckers: usually a small one around the mouth and a big one at the end of the body (Fig. 9). About 500 species of leeches are known; they live principally in freshwater, but some are terrestrial and these live in humid habitats. Usually the species from surface waters are found in caves, but they can form separated cave populations, depigmented to various degrees, and with eyes partially or totally reduced (e.g., *Dina* sp.) (Fig. 9A). An unusual species of leech, *Croatobranchus mestrovi*, was found in deep caves in Croatia (Sket et al., 2001). This species has head morphology not found in other Hirudinea. The rim surrounding the oral sucker forms four pairs of conical lobes, with five finger-shaped papillae on each lobe. Their role is not known yet. Moreover, the posterior half of the body has 10 or more pairs of lateral processes (Fig. 9B) not serving as locomotory organs, but most probably as respiratory organs. This species lives in deep vertical

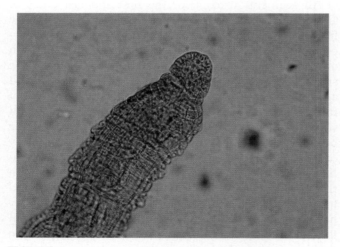

FIGURE 8 *Gianius aquedulcis* (Oligochaeta), anterior segments.

amounts of substrate but utilize only a small part of it. Some species feeding on algae in surface waters can change their food preference in subterranean waters and eat small particles of organic matter. Only a few stygobiotic species are predators. Stygobiotic crayfish-worms (Branchiobdellida) are also known from caves.

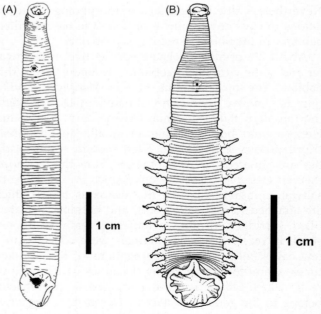

FIGURE 9 Representatives of cave leeches: (A) *Dina* sp. (from Botosaneanu, 1986). (B) *Croatobranchus mestrovi*, ventral side (from Sket et al., 2001).

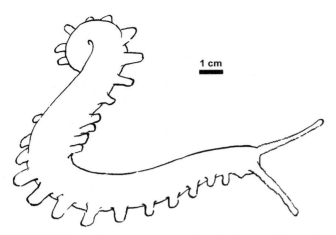

FIGURE 10 *Speleoperipatus speleus* (Onychophora). *Redrawn from Juberthie and Decu, 1994.*

caves, usually near their bottom, where temperatures are low and constant (4–6°C). Most leeches were seen on the rocky walls in a thin layer of flowing water; rarely they can be found on the bottom of cave pools. All cave leeches are predators, feeding on other small invertebrates; bloodsuckers have not been found in subterranean waters. Leeches are known from caves of the northern hemisphere.

Phylum: Onychophora

Onychophora are worm-like, predatory invertebrates with one pair of antennae and 13–43 pairs of locomotory appendages (Fig. 10). This is a small group (about 100 species) of "living fossil" terrestrial animals, which are very interesting from the systematics and phylogenetic point of view. They are found in tropical and subtropical regions but in very restricted areas. Cave species are known from the Southern Hemisphere except for *Speleoperipatus speleus*, which was found in caves in Jamaica.

ECOLOGICAL AND PHYLOGENETIC REMARKS

As mentioned above, among worms living in caves there are both common, widely distributed species and highly specialized forms known exclusively from very restricted areas or from only a few localities. This first group is not very interesting to biologists working on the systematics and evolution of cave fauna, but their role in the functioning of cave biocenoses is usually important. Very often detritivorous, small worm species occur at high densities and sometimes their biomass is also high so they constitute a good source of food for predators.

Troglobiotic and stygobiotic worm species originated from soil, freshwater, and marine forms. Their ancestors colonized the underground environment in various geological epochs and in various places. Some of them are very old; for example, Onychophora formed in the Paleozoic. Among the oligochaetes there are cave forms at various systematic levels, from the exclusively troglobiotic or stygobiotic families (Ocnerodrilidae in South America, Dorydrilidae in a restricted area of Europe, and Parvidrilidae from North American and European caves) and genera (*Delaya*, *Krenedrilus*), and genera with mainly subterranean species (*Rhyacodrilus*, *Trichodrilus*), to genera with only a few or single stygobiotic species. Troglobiotic or stygobiotic species belong to almost all oligochaete families of various origins (marine, freshwater, and terrestrial) and of various ages—both old and recent. This tells us that multiple migrations to the subterranean environment took place at various times and places whereas in other groups such as Temnocephalida and Hirudinea the settlement in caves was an exceptional event. Among stygobiotic worms many species seem to be endemic to one cave or one karstic region, but with advances in research sometimes it turns out that the distribution of some species previously considered as endemic is actually quite—for example, an oligochaete *Gianius aquedulcis* that was originally known only from interstitial waters of Weser River in Germany was recently found in some countries of southern Europe, Poland, and even in the United States.

Bibliography

Austin, A. D., Cooper, S. J. B., & Humphreys, W. F. (Eds.), (2008). Subterranean connections: Biology and evolution in troglobiont and groundwater ecosystems. *Invertebrate Systematics*, 22, 1–310.

Botosaneanu, L. (Ed.), (1986). *Stygofauna mundi. A faunistic, distributional, and ecological synthesis of the world fauna inhabiting subterranean waters (including the marine interstitial)*. Leiden, The Netherlands: E. J. Brill.

Brito, M. C., Martinez, A., & Núñez, J. (2009). Changes in the stygobiont polychaete community of the Jameos del Agua, Lanzarote, as a result of bioturbation by the echiurid *Bonellia viridis*. *Marine Biodiversity*, 39(3), 183–187.

Creuzé des Chatelliers, M., Juget, J., Lafont, M., & Martin, P. (2009). Subterranean aquatic Oligochaeta. *Freshwater Biology*, 54(4), 678–690.

Culver, D. C., Master, L. L., Christman, M. C., & Hobbs, H. H. (2000). Obligate cave fauna of the 48 contiguous United States. *Conservation Biology*, 14(2), 386–401.

Dumnicka, E. (1984). Laboratory studies on biology of cave-dwelling Enchytraeidae: *Enchytraeus dominicae* Dumnicka and *Fridericia bulbosa* (Rosa). *Mémoires de Biospéologie*, 11, 199–205.

Glasby, C. J., & Timm, T. (2008). Global diversity of polychaetes (Polychaeta; Annelida) in freshwater. *Hydrobiologia*, 595(1), 107–115.

Gray, D. P., Harding, J. S., & Winterbourn. M. J. (2009). *Namanereis tiriteae*, New Zealand's freshwater polychaete: New distribution records and review of biology. www.biol.canterbury.ac.nz/ferg/people/Duncan/Natural%20history/Polychaete%20paper.pdf.

Hyman, L. H. (1937). Studies on triclad Turbellaria, VIII. Some cave planarians of the United States. *Transactions of the American Microscopal Society, 44*, 51–89.

Jeannel, R. (1926). *Faune Cavernicole de la France, avec une Etude des Conditions d'Existance dans le Domain Souterrain*. Paris: Lechevallier.

Juberthie, C., & Decu, V. (Eds.), (1994). *Encyclopaedia biospeologica* (Vol. I). Moulis, France: Société de Biospéologie (in French).

Juberthie, C., & Decu, V. (Eds.), (1998). *Encyclopaedia biospeologica* (Vol. II). Moulis, France: Société de Biospéologie (in French).

Kupriyanova, E. K., ten Hove, H. A., Sket, B., Zaksek, V., Trontelj, P., & Rouse, G. W. (2009). Evolution of the unique freshwater cave-dwelling tube worm *Marifugia cavatica* (Annelida: Serpulidae). *Systematics and Biodiversity, 7*(4), 389–401.

Sket, B., Dovc, P., Jalzic, B., Kerovec, M., Kucinic, M., & Trontelj, P. (2001). A cave leech (Hirudinea, Erpobdellidae) from Croatia with unique morphological features. *Zoologica Scripta, 30*(3), 223–229.

Sluys, R. (1990). A monograph of the Dimarcusidae (Platyhelminthes, Seriata, Tricladida). *Zoologica Scripta, 19*(1), 13–29.

Sundberg, P., & Gibson, R. (2008). Global diversity of nemerteans (Nemertea) in freshwater. *Hydrobiologia, 595*, 61–66.

Index

A

A2, 703
Abandoned buildings, 643
Abandonment suite, 137
Acari, 793–795
Acaridae, 795
Acarina, 214
Accelerated corrosion, 144
Accidentals, 276–277
Acid mine drainage (AMD), 161–162
Actinomycete colonies, 494, 497f
Activated carbon passive detector, 890
Active colonization model, 275, 405
Active dispersal models
  of invasion, 405
    active colonization model, 405
    adaptive-shift model, 405
    climatic-relict model, 405
    competitive pressure, 405
    deep sea colonization, 405
    epigean limnic system model, 405
    escape from epigean predators, 405
    marine shallow-water ecotones model, 405
Adaptation. *See also* Behavioral adaptations
  to cave environment, 341
  Cavefish of China, 113–121
    adipose storage, 117, 118f, 119f
    body shape changes, 108f, 113–114, 113f
    eye degeneration, 114f, 119f, 117–120, 121f
    humpback and horn, 114–116, 114f, 115f
    pigmentation loss, 108f, 120
    scale disappearance, 120–121
    sensory apparatus, 113f, 116–117, 116f
    specialized appendages, 108f, 117
  Gammarus minus
    cave population origin, 345, 346f
    cave streams and, 344
    compound eyes and, 345–346, 347f
    foods of, 343–344
    general ecology of, 343–344
    idealized subterranean drainage basin and, 343, 343f
    karst window and, 344–345
    as model system, 347
    population density of, 343–344
    population structure of, 345
    reproduction, 343–344
    springs and, 344
    variation in, 344–345, 344f
  to low food
    energy demand and, 4–6
    factors concerning, 2f
    food resources and, 1–3
    food scarcity types and, 1
    food-finding and, 3–4
    life history characters and, 6–8
    overview, 1, 2f
    periodic starvation and, 8–9
  mechanism hypotheses, 341–342
    directly adaptive hypothesis, 342
    indirectly adaptive hypothesis, 342
    nonadaptive hypothesis, 341–342
  model organisms and, 37
  preadaptation, 10
  regressive evolution and, 341
  study criteria, 342
Adaptive shift
  case studies, 12–16
    cixiid planthoppers, 13–14, 13f
    continents, 16
    crickets, 15
    Hawaii, 12–16, 12t
    moths, 15–16
    other islands, 16
    terrestrial Isopoda, 14–15
    wolf spiders, 15
  conclusions about, 16–17
  factors underlying, 10–12
    ancestral habitats, 12
    cavernicolous habitats, 11
    environmental stress, 12
    extrinsic, 11–12
    food resources, 11–12
    founder events, 10–11
    genetic repertoire, 10
    intrinsic, 10–11
    mating behavior and hybridization, 11
    preadaptation, 10
    stress response, 11
  future regarding, 17
  theory of, 9–10
Adaptive-shift model, 405
Aeolian caves, 105f, 106
Aerobes, 490f, 491
Agagonite, 777
Agelenidae, 789
Aggradation, 136f, 137
Agoristenidae, 796
Ailuropoda, 594
Akravidae, 787
Allochthonous sediments, 135
Allogenic recharge, 387
Allophane, 501t, 505
Alluviated springs, 799, 799f
Alpine
  karrenfields, 430
  karst, 436–437
Alpinotype caves, 724
Alunite, 501t, 503
Amarnath Cave, 322
Amblyopsidae, 522–523, 522f, 525, 525f
Amblyopsis rosae
  life history characters of, 7–8, 8t
  life history evolution of, 465, 466, 466t, 467–468
Amblyopsis spelaea, 465, 466, 466t, 467–468
Amblypygi, 214, 789, 789f
AMD. *See* Acid mine drainage
Amensalism, 743, 744f
Ammonia, chemoautotrophy and, 129
Amphibia, 260, 261t
Amphibian chytrid fungus, 624
Amphipoda, 178t, 186, 187t, 257
  anchihaline habitats and, 22
  Appalachian stream competition and, 746–747, 746f, 747f
  in Australia, 207
  in Dinaric karst, 233
AMT. *See* Audio-magnetotelluric soundings
Anaerobes, 490f, 491
Anapidae, 790
Anaspidacea, 178t, 184, 184f
Anastamose, 422, 423f
Anastomose, 422
Anastomotic caves, 395–396, 637, 638f
Anastomotic maze, 602f
Anchialine pools, 18
Anchihaline and marine caves, 240
Anchihaline habitats
  in Australia, 209, 210, 214f
  biogeography, 22–23
  biology and inhabitant distribution, 23–24
  cave morphology and hydrology, 18–20
    horizontal caves, 18
    U-shaped tube effect, 19
    vertical caves, 18–19
  Crustacea, 20–22, 209, 210, 214f
    Amphipoda, 22
    Copepoda, 21
    Decapoda, 21
    Isopoda, 22
    Leptostracans, 21
    Mictaceans, 22
    Mysids, 21–22
    Niphargus, 22, 22f

Anchihaline habitats (*Continued*)
  Ostracods, 21
  Remipedia, 21
  Thermosbaenaceans, 21
 fauna and humans, 25
 geography, 18
 noncrustacean groups, 20
  filter feeders, 20
  fishes, 20
  gastropods, 20
  sessile chaetognaths, 20
  sponges, 20
 overview about, 17
 trophic relations, 24
Ancient cavers, eastern North America
 archaeology and, 26
 conclusions about, 30
 Mammoth Cave
  archaeological evidence, 27–28, 27f, 28f
  archaeology, 26–27
  chronology and, 27
  evidence interpretation, 28–30, 29f
  prehistoric archaeology, 27–30
 overview about, 26
Androniscus dentiger, 615
Angel hair, 375
Angular passages, 604–605, 604f
Anhydrite, 374
Annelida, 911–914
 Hirudinea, 913–914, 913f
 Oligochaeta, 912–913, 913f
 Polychaeta, 911–912, 912f
Anomura, 178t, 192
Anophthalm beetles, 56–57, 56f
Anoxia survival times, 653
Anthodite Hall, 715f
Anthodites, 379–380, 380f, 782
Anthuridea, 178t, 188, 189t
Antroherpon scutulatum, 519f
Antrolana lira, 854–855, 854f
Anvil Cave, 739, 739f
Aphaenops Pluto, 519f
Aphenopsian beetles, 56–57, 56f
Appalachian Mountains, 734f, 739–740, 740f, 741f
Appalachian Plateaus, 734f, 738–739, 739f
Aqua Cave, 74, 77
Aquatic cave animals, energy demand and, 4, 4t
Aquatic interstitial subterranean habitats, 240
Aquatic stygobites distribution, 853–857, 854f, 856f
 Antrolana lira, 854–855, 854f
 Crangonyx antennatus, 855, 856f
 fishes, 854
 Stygobromus, 853–854
Aquifers. *See also* Calcrete aquifers; Edwards Aquifer; Karst aquifer, hydrology of; Karst aquifers, modeling of; Karst aquifers, water tracing in
 NAPLs storage in aquifers, 171
  fractures, 171
  matrix and vugs, 171

  pools in conduits, 171
  sediment, 171
 NAPLs transport into aquifers, 168–171
  climate and, 169, 170
  epikarst and, 169–170
  fractures and, 170–171
  nonwetting fluid and, 169
  open drains and, 170–171
  sinkhole formation and, 170
  sinking streams and, 170–171
  soil and regolith pores and, 169
  volume and rate of release and, 168–169
Arabika Massif
 deep cave evolution, 449–450
 deep caves in, 443–447
 geology, 447–448, 447f
 hydrogeology, 448–449
 location and physiography, 443, 444f
 Ortobalagan Valley in, 443, 445f
Arachnida, 256–257
 mites of, 256
 Opiliones of, 256
 Pseudoscorpions of, 256–257
 Schizomida of, 257
 scorpions of, 257
 spiders of, 256
 tropical diversity patterns, 245t
Aragonite, 500, 501t
 calcite problem, 783–784, 783f
 massive column of, 783, 784f
 stalagmites, 809
Araneae, 256, 789–791
 in Australia, 215
 in Dinaric karst, 232
Araneomorphae, 789
Arctodus pristinus, 593
Arctodus simus, 593, 594f
Arctotherium, 593–594
Argasidae, 794
Arsenates, 506
Artesian hypogene speleogenesis, 759
Artesian springs, 799–800
Ascending vertical exploration techniques, 307, 316f, 317–318
Asellota, 178t, 188–190, 189t
Asellus aquaticus, 653, 654t
 Asellus kosswigi and, 34–35, 34f
 broader geographic distribution, 33–34, 34f
 continental scale, 35–36, 35f, 35t
 future prospects, 36
 habitats, 30–31, 31f
 natural selection and, 545
 overview, 30–31
 phenotypic clines and, 545–546, 546f
 PPCS and, 32–33
 respiratory area of, 31, 31f
 subspecies, 31
 troglomorphies development, 31–32, 32f
Asellus kosswigi, 34–35, 34f
Assamiidae, 796
Astacidea, 178t, 192, 192t
Astigmata, 795
Astyanax fasciatus

 aggressive behavior of, 65–66, 66f, 67f
 egg size, 6–7, 7f
 energy demand, 4, 4t
 excitement and aggression and, 5
 feeding behavior of, 63
 neutral mutations
  albinism and, 550
  dark melanin and, 550
  deleterious risk hypothesis and, 552–553
  energy economy hypothesis and, 553
  eyes and, 550, 552f, 554
  guanine and, 550
  neuromasts and, 553
  overview, 549–550
  pleiotropy hypothesis and, 553
  surface and cave forms regarding, 550
 reproductive behavior of, 64–65
Astyanax mexicanus
 developmental stages, 39f
 melanophore differentiation and, 41–42, 41f
 as model organism, 37
 natural history of, 37–38
 populations and distribution, 37–38, 38f
 troglomorphic traits, 38–40
  constructive, 38t, 39, 40f
  developmental basis of, 40–42
  evolution of, 42–43
  genetic basis of, 42
  regressive, 38–39, 38t
Atemnidae, 791
Atlantida Cave, 829t
Aturidae, 794–795
Audio-magnetotelluric soundings (AMT), 349–350
Austin blind salamander, 669
Australia. *See also* Diversity patterns, in Australia; Nullarbor Caves, Australia
 root communities in, 661t
Austrochilidae, 790
Autochthonous sediments, 135
Autotroph, 490–491
Avens, 423
Aves, 261

B

Back to network, 6
Bacteria
 conservation and, 624
 cyanobacteria, 453
 iron and manganese, 133
 Proteobacteria, 495–496
 saltpetre mining and, 676
 symbiosis, 560
Bad air zone, 662–663
Balitoridae, 107
Ball, Edward, 874–875
Bamboo, 402, 403f
Banta, A. M., 404–405
Barberry Cave, 80, 80f
Bärenschacht, 699–700
Barite, 501t, 503
Barometric airflow, Jewel Cave, 411

Barr, T. C., 276
Barton Springs salamander, 669–670
Barychelidae, 789
Basic Cave Diving: A Blueprint for Survival (Exley), 310
Basins, 427–428
Basswood Cave, 81–82
Bathynellacea, 178t, 184, 184f, 257
Batrachochytrium dendrobatidis, 624
Bats
  in Australia, 218
  bell holes and, 423–424, 424f
  cave climate influenced by, 357
  conservation, 625
    decline of, 625
    mining and, 625
    overview, 625
    Shelta Cave, 625
    water projects and, 625
    WNS and, 625–626
  in Dinaric karst, 232–233
  distribution of, 46
  echolocation and, 45
  ecosystem services provided by, 51–54
  flight and, 45
  food sources and, 331t, 333
  guano, 358
    carnivorous, 358–359, 359f
    community structure and succession in, 360–361
    conservation and, 242–243, 249, 626
    frugivorous, 358–359, 358f
    hematophagous, 358–359, 359f
  habitat destruction and alteration and, 52
  hibernation of, 48–49, 48f
  human disturbance and, 52
  Jewel Cave, 415–416
  life-history traits, 45
  living in caves
    benefits, 49–50
    costs, 49–50
  management, 50–51
  overview about, 45–46
  paleontology and, 592
  roosting habits, 45, 46–48
    courtship and mating, 47
    rearing young, 47–48
    roost functions, 47
  Rousette fruit, 46
  Townsend's big-eared, 46f
  tropics and, 242–243, 249
  in United States, 261–263, 262t
  vampire, 53–54
  wind turbines and, 53
  WNS and, 52–53, 625–626
    atypical winter behavior, 905–906
    cave closures and decontamination, 908
    characteristics, 905–907
    conservation and, 625–626
    current and future research, 907–908
    depleted fat reserves, 905
    diagnostics, 904–905
    European connection, 907
    fungal infections and tissue damage, 905, 905f
    fungal transmission, 907
    immune response changes, 906
    mortality and, 904
    overview, 903–904
    pathogen, 903f, 904
    species affected by, 905
    wing damage, 906–907, 906f
Battered Bar Cave, 80–81
Battery operated light sources, 305–306
Beaded helictites, 379, 380f
Bears, 593–594
  Ailuropoda, 594
  Arctodus pristinus, 593
  Arctodus simus, 593, 594f
  Arctotherium, 593–594
  tremarctine, 593
  Ursus americanus, 593
  Ursus speleus, 593
Bear's Shaft, 699–700
Bedrock-collapse sinkhole, 720f, 722
Beetles
  anophthalm and aphenopsian, 56f, 56–57
  biogeography and, 851–852, 852f
  cricket egg prey of, 744–746
  development cycles of, 56, 56f
  ecology and, 59–61
  fecal food sources and, 331t, 333
  importance and protection of, 61
  Leptodirini group, 57, 58f
    Leptodirus hochenwarthii, 55, 55f, 231f, 519f
    phylogenetic Mediterranean tree, 59, 60f
  other types of, 57, 58f
  overview about, 54–55
  Pholeuon
    flooding periods influencing, 60, 61f
  proportion of cave, 55, 55f
  systematics of, 56–58, 56f, 58f
  troglomorphic features of, 56–57, 57f
  White Cave food web and, 60–61, 61f
Behavioral adaptations
  aggressive, 65–68, 66f, 67f
  Astyanax fasciatus
    aggression, 65–66, 66f, 67f
    feeding, 63
    reproduction, 64–65
  discussion, 68
  food and feeding, 63, 64f
  Munidopsis polymorpha
    aggression, 66–67, 67f
    reproduction, 63, 64f
  overview, 62–63
  Poecilia mexicana
    aggression, 67–68
    reproduction, 64, 65f, 67–68
  Proteus anguinus
    aggression, 67–68, 67f
    feeding, 63, 64f
    reproduction, 64–65
  reproductive, 63–65, 64f, 65f
Bell holes, 423–424, 424f
Benarat Caverns System, 536

Berchilskaja Cave, 443, 444f
Berry cave salamander, 673f, 674
Best management practices (BMPs), 627–628
Better Forgotten Cave, 77–78
Biantidae, 796
Bible, 322
Big river caves, 308
Biogeography
  anchihaline habitats, 22–23
  aquatic stygobites distribution and, 853–857, 854f, 856f
    Antrolana lira, 854–855, 854f
    Crangonyx antennatus, 855, 856f
    fishes, 854
    Stygobromus, 853–854
  of cave animals, 850–851
  cavefish of China, 121–124
  cladistic, 850, 850f
  epikarst communities and, 293–294
  overview of, 849–850
  panbiogeography, 850
  Sinocyclocheilus, 121–124
  summary about, 857
  terrestrial troglobites distribution and, 851–853, 852f
    beetles, 851–852, 852f
  of thalassoid limnostygobionts, 483, 484f, 485f
  types of, 849–850
  vicariant phenomena, 855
    calcrete aquifers and, 857
    spring failure, 856–857
    stranding, 855–856
    stream capture, 856
Biospeologica, 219–220
Birds, as cave visitors
  in Australia, 218
  cave ecology impact of, 847
  cave-dwelling species, 845–846
    facultative visitors, 845
    regional cavernicoles, 845
    regular cavernicoles, 845–846
  den sites and, 846–847
  foraging habitats and, 847
  guano and, 358
  knowledge gaps, 846
  man and, 848–849
  rare species, 848
  resting and mating and, 846
  wintering sites and, 847
Bird's nest soup, 532
Black Hills caves, 417. *See also* Jewel Cave, South Dakota
Blanco blind salamander, 669
Blarney Stone Cave, 79
Blattodea, 216–217
Blepsimolge, 668f, 669–670
Blind Faith Cave, 80, 280
Blind salamanders, 668–669, 668f
Block breakdown, 69
Block-fault structure, 445–447
Blowholes, 570
Blue holes, 159–160
  defined, 160
  exploration, 160

Climate (*Continued*)
  Vjetrenica and, 860–861
Climatic-relict model, 405
Clint, 428–429, 429f
Closed circuit rebreathers (CCRs), 313
Closed depressions, 434–435
  compound hollow, 152–153
  dolines
    collapse, 150–151, 150f, 151f
    cover, 150f, 152
    intersection, 150f, 152
    other types of, 150–152, 150f
    subsidence, 150f, 151–152
  dolines, solution
    conclusions about, 150
    defined, 140
    drainage and, 140
    drawdown, 143f, 144
    evolution examples, 144f, 145–147
    Eyre Peninsula, 147
    filling material evacuation and, 145
    frost shattering and, 145
    inception, 143f, 144
    as karst surface diagnostic forms, 140–150
    locations, 141
    midlatitude, 149
    Monte Baldo, 147
    Montello, 145–147, 146f, 148f
    Monti Lessini, 144f, 145
    morphology and size, 140–141
    other evolution processes, 145–147
    overview about, 140
    point-recharge, 142–144, 143f, 145
    Santa Ninfa Plateau, 145
    slope processes and, 145
    soil and, 145
    specific environmental conditions and, 147–149
    structure, 141–142, 142f
    subsoil, 141
    tropical landscape and, 147–149, 149f
    Velebit Mountains, 147
    Waitomo district, 145
  poljes, 140, 153–155
    classification, 154
    defined, 153
    Dinaric karst and, 153, 199–200
    lakes and, 153, 153f
    lithological contact, 154, 154f
  polygenetic sink, 152–153
  types of, 140
  uvala, 152–153
Coastal caves
  blue holes, 159–160, 160f
  categories of, 155
  conclusions about, 161
  flank margin caves, 156–159, 157f, 158f, 159f, 160f
  overview, 155
  sea caves, 155–156, 156f
  sea level and, 155
Coastal karren, 425f, 430
Coastal karst, 437

Cobitidae, 107
Cobra-Bridge-Cloud Cave, 534
Cockroaches, 241, 242f, 362, 663
Cold air trapping, 400–401
Coleoptera, 230, 259–260. *See also* Beetles
  Carabidae, 230–232
  Cholevidae, 230
  Dystiscidae, 217–218
  Pselaphidae, 232
Collapse dolines, 150–151, 150f, 151f
Collapse shafts, 761, 762
Collapsed tiankeng, 821, 821f, 822f
Collembola
  Australian diversity patterns, 216
  European diversity patterns, 221–222, 224f
  morphological adaptations in, 523, 523f
  tropical diversity patterns, 245t
Collins, Floyd, 85–86, 648
Colonization
  active model of, 275, 405
  deep sea, 405
  Dinaric karst human, 229
  ecological classification and, 275
  natural selection and, 547–548
Comal blind salamander, 670
Combined artesian/deep-rooted speleogenesis, 759
Commensalism, 743, 744f
Competition, 743, 744f
  Appalachian stream, 746–747, 746f, 747f
  eutrophication and, 747–748
  initial flow routes, 599–600, 600f
  Niphargus and, 559–560, 560t
Competitive pressure, 405
Components, 269–270
Composite 3D systems, 762–763
Compound depressions, 140, 152–153
Compound hollow, 152–153
Condensation
  corrosion hollows, 425, 425f
  rills/flutes, 425
Conduit, 104
  diffuse flow springs and, 798
  hypogene, 367f
  NAPLs and, 171
  networks evolution, 767–769, 768f
  permeability, 384, 385–387, 385f, 385t
  single-conduit caves, 394
  systems carrying capacity, 389
  tubular, 734–735, 735f
Cones, 435–436, 436f
Congeria kusceri, 512–513, 517
Conical stalactite, 807, 807f
Conn, Herb, 411
Conn, Jan, 411
Conservation, 309
  bats, 625
    decline of, 625
    guano and, 242–243, 249, 626
    mining and, 625
    overview, 625
    Shelta Cave, 625
    water projects and, 625
    WNS and, 625–626

  of cave life, 624–626
    bacteria and, 624
    chemical spills and, 625
    extinct and endangered species and, 625
    fungus and, 624
    human impact and, 624
    threats, 624–625
    water pumping and, 625
  cave preserve design, 626–631
    gate styles, 628–630, 629f, 630f
    gating criteria, 628
    karst BMPs and, 627–628
    large cave buffer zones, 627
    security systems, 631
    signs and, 626, 627f
    small cave buffer zones, 627
  gating decision guide, 630–631
    final results, 631
    good reasons not to gate, 630
    good reasons to gate, 630–631
    poor reasons to gate, 630
    poor reasons to not gate, 630
  management, 631–633
    education and, 631–633
    prioritizing caves, 632–633
  Vjetrenica, 864
Contamination. *See* Heavy metals; Nonaqueous phase liquids
Copepoda, 178t, 181–182, 291–293
  anchihaline habitats and, 21
  in Australia, 211
  Calanoida, 178t, 181, 181t
  Cyclopoida, 178t, 181–182
  in Dinaric karst, 233
  Gelyelloida, 178t, 182
  Harpacticoida, 178t, 182
  Misophrioida, 178t, 181, 182t
  Platycopioida, 178t, 181
Coral Sea, 464
Coralloid multiaggregates, 500, 503f
Corkscrews, 423
Corridors, 428–429
Corrosion hollows, 425, 425f
Corrosion notch, 425, 429–430
Cosmogenic isotope dating
  burial dating and, 173–176, 174f
    caves and human evolution and, 176
    conclusions, 176
    Mammoth Cave and, 175–176
    requirements for, 174–175
  dating techniques
    relative *vs.* absolute, 173
  overview, 172–173
Cotton passive detector, 891
Cover dolines, 150f, 152
Cover-collapse sinkhole, 718–719, 719f, 720f
Cover-subsidence sinkhole, 719f, 721–722, 721f
Crangonyx antennatus, 855, 856f
Crash cycles, 10–11
Craterstigmatomorpha, 539
Crawford Upland, 736
Crayfish, 521, 522f
Creep closure, 354–355, 355f

Crete, 321
Crevice Cave, 738, 738f
Crickets, 15, 333, 331t
　eggs, 330, 744–746
　guano and, 358
　heterozygosity in, 615–617, 616t
　root communities and, 663
CRM. See Chemical remanent magnetism
Cross-bedding, 135–136, 136f
Crustacea. See also Gammarus minus
　anatomical variety, 177
　in anchihaline habitats, 20–22, 209, 210, 214f
　　Amphipoda, 22
　　Copepoda, 21
　　Decapoda, 21
　　Isopoda, 22
　　Leptostracans, 21
　　Mictaceans, 22
　　Mysids, 21–22
　　Niphargus, 22, 22f
　　Ostracods, 21
　　Remipedia, 21
　　Thermosbaenaceans, 21
　in Australia, 209–214
　　anchihaline habitats, 209, 210, 214f
　　Copepoda, 211
　　Decapoda, 213–214
　　Isopoda, 212–213
　　Ostracoda, 211–212
　　overview, 209
　　Spelaeogriphacea, 213
　　Syncarida, 209–211
　low oxygen and, 653
　overview about, 177
　stygobitic taxa synopsis, 177–193, 178t
　　Branchiopods, 177–179, 178t
　　Hexapoda, 258–259
　　Insecta, 259–260
　　Malacostraca, 178t, 183–193, 257–258
　　Maxillopoda, 178t, 180–182, 257
　　Ostracoda, 178t, 182–183, 182f
　　Remipedia, 178t, 179–180, 180f, 180t
　taxonomy, 177, 178t, 290, 290t
　troglomorphy, 177
　tropical diversity patterns, 245t
Crusts, gypsum, 374, 375
Cryogenic cave micropearls, 401–402
Cryptostigmata, 795
Crystal Cave Expedition (C-3 Expedition), 85–86
Crystal wedging, 72–73
Cueva de Villa Luz, Mexico, 813f, 813–815, 814t, 815f
Culturing, 492
Cumacea, 190, 191t
Cumberland River incision pulses, 530, 530f
Cunaxidae, 794
Cupola, 395, 424, 425f
Curl, Rane, 278
Curvilinear branchwork, 602f
Curvilinear passages, 604–605, 604f
Cusps, 425, 425f
Cut-and-closure caves, 354–355, 355f

Cutter, 429
Cvijič, Jovan, 433
Cyanobacteria, 453
Cybaeidae, 790
Cycloctenidae, 790
Cyclopoida, 178t, 181–182
Cyphophthalmi, 795–796
Cyprinidae, 234
Cyrtaucheniidae, 789

D

Damaeidae, 795
Danielopolina kornickeri, 214f
Dan's Cave, 311f
Danube River, 707
Darwin, Charles, 543–544
Das Karstphanomen, 433
Dating. See also Uranium series dating, of speleothems
　burial, 173–176, 174f
　　caves and human evolution and, 176
　　conclusions, 176
　　Mammoth Cave and, 175–176, 737
　　requirements for, 174–175
　Castleguard Cave, 93
　ice in caves, 402
　paleoclimate records from speleothems, 577–578
　relative vs. absolute, 173
　from stygofauna, 484–486
Davis, Nevin W., 76–77
Deal, Dwight, 411
Decantation runnel, 428
Decapoda, 178t, 191, 191f, 258
　anchihaline habitats and, 21
　Anomura, 178t, 192
　Astacidea, 178t, 192, 192t
　in Australia, 213–214
　Brachyura, 178t, 192, 193t
　Caridea, 178t, 191, 192t
Deep sea colonization, 405
Deep Secrets, 463
Deepest caves, 443
Deep-rooted hypogene speleogenesis, 759
Deep-seated karst, 365, 367
Deer Cave, 532, 534f
Degeneration, cave, 603
Delaya bureschi, 231f
Deleterious risk hypothesis, 552–553
Density-driven flow, 751–752
Denuded karst, 365
Dermaptera, 259
Descending, 317
Deterministic biogeography, 849–850
Detrital remanent magnetism (DMR), 586
Devils Hole, 579–580, 580f
Diagenetic environments, 750–756
　eogenesis and, 751
　mesogenesis and, 751
　telogenesis and, 751
Dictynidae, 790
Diffuse flow springs, 798
Dinaric karst. See also Diversity patterns, in Dinaric karst
　boundaries of, 195, 196f

　caves and, 201
　as classical karst, 195
　climate regarding, 195–197, 229
　contact karst and, 201
　depressions, 199–200
　described, 228–230
　dolines, 199–200
　flora and fauna and, 229–230
　formation of, 229
　human colonization of, 229
　hydrological characteristics of, 197, 197f
　karst features of, 199–202
　land use, 202
　leveled surfaces, 200–201, 201f
　lithology, 198–199, 199f
　location, 195, 228–229, 229f
　poljes and, 153, 199–200
　relief forms, 199
　roofless caves, 201–202, 202f
　structural geology, 198
　surface of, 201–202, 202f
　tectonic map of, 196f
　today, 229
　traits, 195–197
　uvalas, 199–200
　vegetation and, 197, 197f
Diplocentridae, 788
Diplopoda, 258
　in Australia, 216
　in Dinaric karst, 232
　Millipedes, 540–541
　　Chordeumatida, 540t, 541
　　Julida, 540, 540t
　　map of, 540f
　　overview, 540
　　Polydesmida, 540, 540t
　　Pseudotremia, 541, 541f
　　Scoterpes, 541
Diplura, 216, 245t
Dipluridae, 789
Diptera, 260
Directional selection, 545, 545f
Directly adaptive hypothesis, 342
Discharge passages, 604
Dispersed infiltration, 387
Dissolution basins, 420, 421f
Dissolutional pavements, 427, 427f, 429f, 430
Dissolutional potholes, 420, 422f
Dissolved organic carbon (DOC), 289
Distributary springs, 800
Divergent selection, 545, 545f
Diversity patterns, in Australia
　amphipods, 207
　Chelicerata, 214–215
　　Acarina, 214
　　Amblypygi, 214
　　Araneae, 215
　　Opiliones, 215
　　Palpigradi, 214
　　Pseudoscorpiones, 215
　　Schizomida, 215
　　Scorpiones, 215
　conclusions about, 218–219
　deficiency of, 203
　geographic factors

Diversity patterns, in Australia (*Continued*)
- cave atmosphere, 205–208
- energy supply, 208–209
- groundwater calcretes, 207–208, 213f
- overview, 204
- shield regions and Cretaceous marine transgressions, 204–205, 205f
- stagnant-air zone, 205
- tropical winter effect, 205
- insects, 216–218
  - Blattodea, 216–217
  - Coleoptera Dystiscidae, 217–218
  - Collembola, 216
  - Diplura, 216
  - Orthoptera, 217
  - Planthoppers, 216
  - Zygentoma, 216
- lack of study of, 204
- Myriapoda, 216
  - Chilopoda, 216
  - Diplopoda, 216
- overview, 203–204, 203f
- stygofauna and Crustacea, 209–214
  - anchihaline habitats, 209, 210, 214f
  - Copepoda, 211
  - Decapoda, 213–214
  - Isopoda, 212–213
  - Ostracoda, 211–212
  - overview, 209
  - Spelaeogriphacea, 213
  - Syncarida, 209–211
- subterranean animal examples, 206f
- subterranean invertebrates distribution, 212t
- vertebrates, 218
  - bats, 218
  - birds, 218
  - fishes, 218
  - snakes, 218

Diversity patterns, in Dinaric karst
- Amphipoda, 233
- Araneae, 232
- bats, 232–233
- biogeographical, 234–236
  - endemism, 235–236
    - holodinaric distribution, 235
    - merodinaric distribution, 235
    - smaller distribution areas, 235
    - transdinaric distribution, 235
  - widely spread taxa, 234
- Bougainvilliidae, 234
- Coleoptera, 230
  - Carabidae, 230–232
  - Cholevidae, 230
  - Pselaphidae, 232
- Copepoda, 233
- Cyprinidae, 234
- Diplopoda, 232
- Dreissenidae, 234
- fauna composition and diversity, 230–234
  - aquatic, 233–234
  - examples, 231f
  - history, 230
  - terrestrial, 230–233

- Gastropoda, 232, 233
- hotspots, 236
  - Hercegovina, 236
  - Postojna-Planina Cave system, 236
- Isopoda, 233–234
- overview, 228–230, 229f
- pollution and protection and, 237–238
- *Proteus anguinus*, 234
- Pseudoscorpiones, 232
- Scutariellidae, 234
- Serpulidae, 234
- special assemblages, 236–237
  - cave hygropetric, 236
  - crevice systems, 237
  - ice caves, 236
  - sinking streams, 237
  - thermal waters, 237
- Spongillidae, 234
- Thermosbaenacea, 234

Diversity patterns, in Europe
- Carabidae, 221, 222f
- Collembola, 221–222, 224f
- conservation and, 226–228
- different habitats and, 226, 227t
- factors influencing, 221
- geographic patterns and, 224–226, 225f, 226f, 227f, 227t
- hotspots, 225, 226f
- Leiodidae, 221, 223f
- midlatitude biodiversity ridge and, 224–225, 225f
- molluscs, 221, 224f
- Niphargidae, 221
- other continents compared with, 220–221, 220f
- overview, 219–220
- troglobionts to stygobionts ratio and, 225, 227f
- troglomorphy and relictness and, 221–223, 222f, 223f, 224f

Diversity patterns, in tropics
- background regarding, 238, 239f
- conservation issues, 249–250
  - bats and guano, 242–243, 249
  - endemism and vulnerability, 249–250
- current knowledge state about, 238–240, 239f
- historical context, 238
- relictual *vs.* nonrelictual taxa, 247–249
- subterranean habitats, 240–243
  - anchihaline and marine caves, 240
  - aquatic interstitial, 240
  - cave freshwater, 240–241
  - guano and guanobionts, 242
  - oligotrophic terrestrial, 241–242
  - troglophilic species, 242–243
- temperate subterranean biodiversity *vs.*, 239f, 247
- troglobiont broad-scale, 243–247
  - along environmental gradients, 247
  - geographical patterns, 243–247, 245t

Diversity patterns, in United States
- cave ecology and, 252

- cave/karst regions and, 252–254, 253f, 253t, 254f
- discussion about, 263
- frequency histograms by karst region, 256f
- frequency histograms by state, 255f
- habitat types and, 252
- invertebrates, 254–260
  - Arachnida, 256–257
  - Crustacea, 257–258
  - Hexapoda, 258–259
  - Insecta, 259–260
  - Malacostraca, 257–258
  - Maxillopoda, 257
  - Mollusca, 255
  - Oligochaeta, 254–255
  - Turbellaria, 254
- maps regarding, 253f, 254f
- overview, 251–252, 251t
- species richness comparison and, 253t
- vertebrates, 260–263
  - Amphibia, 260, 261t
  - Aves, 261
  - Mammalia, 261–263, 262t
  - Osteichthyes, 260
  - overview, 260
  - Reptilia, 260–261

DMR. *See* Detrital remanent magnetism
DNAPLs. *See* Heavy nonaqueous phase liquids
DOC. *See* Dissolved organic carbon
Documentation
- databases, 269–271
  - advantages, 271
  - components, 269–270
  - data transfer and, 271
  - disadvantages, 271
  - entities and, 269
  - fields and, 269
  - foreign key and, 270
  - forms and, 270
  - future needs, 271–272
  - issues, 270–271
  - linkages and, 270
  - overview, 269
  - queries and reports and, 270
  - records and, 270
  - tables and, 270
- defined, 264
- GIS and, 264–265
- history, 264–269
- information safety, 268–269
- information storage, 266–267
  - in digital form, 266–267
- issues, 265
  - access policy, 265
  - cave identification, 265
  - clerical, 265
  - credibility, 265
  - information retrieval, 265
  - information transfer, 265
  - record safety, 265
  - secrecy, 265
  - standards, 265

Krubera Cave, 443

overview, 264
publication, 269
recording and, 265–266
    coordination, 266
    external issues, 266
    helpful references, 266
    information consolidation, 266
    results publication, 266
    storage and distribution, 266
    updating, 266
retrieval, 267
standards, 267
    existing, 267
    planned, 267
surveying and mapping, 267–268
Voronja Cave, 443
Dolichopoda, 614
Dolichopoda laetitiae, 610–611, 610f
Dolines, 140. *See also* Tiankeng
collapse, 150–151, 150f, 151f
cover, 150f, 152
defined, 434–435, 723
Dinaric karst, 199–200
intersection, 150f, 152
Nullarbor Caves, 570, 570f
other types of, 150–152, 150f
in Slovenia, 435f
solution
    conclusions about, 150
    defined, 140
    drainage and, 140
    drawdown, 143f, 144
    evolution examples, 144f, 145–147
    Eyre Peninsula, 147
    filling material evacuation and, 145
    frost shattering and, 145
    inception, 143f, 144
    as karst surface diagnostic forms, 140–150
    locations, 141
    midlatitude, 149
    Monte Baldo, 147
    Montello, 145–147, 146f, 148f
    Monti Lessini, 144f, 145
    morphology and size, 140–141
    other evolution processes, 145–147
    overview about, 140
    point-recharge, 142–144, 143f, 145
    Santa Ninfa Plateau, 145
    slope processes and, 145
    soil and, 145
    specific environmental conditions and, 147–149
    structure, 141–142, 142f
    subsoil, 141
    tropical landscape and, 147–149, 149f
    Velebit Mountains, 147
    Waitomo district, 145
subsidence, 150f, 151–152
suffosion, 722
Dolonization, deep sea, 405
Double-trunked lava tunnels, 869–870
Downstream Complex, Castleguard Cave, 89–90

Dowsing, 352
Draculoides vinei, 206f
Drainage grooves, 422
Drained lobes, 871–872
Drapery, 781
Drawdown doline, 143f, 144
Dreissenidae, 234
Drip line, 283
Drip pits, 420, 421f
Dye recovery, 894, 894f, 895f
Dysderidae, 790
Dyspnoi, 796
Dystiscidae, 217–218
Dzhurinska Cave, 829t

E

Earthworms, 362
Eastern Alps, young uplift, 531
Ebb-and-flow springs, 801–802
Echolocation, 45
Ecological biogeography, 849–850
Ecological classification
accidentals, 276–277
Barr and, 276
colonization and genetic isolation and, 275
eutroglophile, 276–277
Gilbert and, 276
Joseph and, 275
occasional visitors, 275
overview about, 275
Pavan and, 276–277
Racovitza and, 275–276, 277
Schiner and, 275, 276, 277
Schiodte and, 275
selection and, 275
Sket and, 276
source and sink population and, 277
subterranean organisms, 276
subtroglophile, 276–277
Trajano and, 276
troglobites, 275–276, 277
troglophiles, 275–276, 277
trogloxenes, 275–276, 277
true *vs.* false cavernicoles, 276
Ecosystems
bats and, 51–54
chemoautotrophy and, 133–134
connectivity and disturbance in, 102
definition and boundaries, 99
energy flux and limitation, 100–101, 100f
nutrients, 101–102
physical environment and habitat zones, 99–100, 100f
Ecotones, 283
Ectothermy, 5–6
Edwards Aquifer, influent streams and, 708–709
Effective population size, 609
Efflorescent crusts, 676
Electrical resistivity imaging (ERI), 348–349, 349f
Elision, 751–752
The Elk Horn, Bobcat Cave, 79f
Ellison's Cave, 86

Embryonic cryptodolines, 141
Endokarst realm, 754–756, 755f, 756f, 757f
Energy demand
activity and, 5
aquatic cave animals and, 4, 4t
back to network and, 6
body size conflicts and, 5
character reduction and, 6
ectothermy and neoteny and, 5–6
excitement and aggression and, 5
food scarcity, 4–6
hypoxic conditions and, 6
overview, 4
periodic starvation and, 9
terrestrial cave animals and, 4
Energy economy, 1
hypothesis, 553
Englacial caves, 354–356
Entities, 269
Entranceless caves
background regarding, 277–278
Bobcat Cave and, 280
Buckwheat Cave and, 280
Chestnut Ridge Cave System, 280
deciduous trees and, 279
entrance lifetime and, 278–279
environmental settings of, 278–279
examples of, 280
finding, 279–280
    clastic ridges and, 279
    digging bar and, 279
    melted snow and, 279
    sinks and, 279
obstacles to, 278
statistical theory regarding, 278
trash and, 279
variables that produce, 278
virgin passages and, 277
Entrances
as habitat, 283
human created, 281, 282, 282f
intrinsic, 281
lifetime, 278–279
locations of, 280
overview, 280
as paleontological/archaeological sites, 283–284
pit and shaft, 281
sinkhole, 281, 282f
sinking stream, 281
size scales of, 281, 282f
at springs, 281, 281f
statistics of, 282–283, 283f
talus, 284
truncation and collapse and, 281
types of, 281–282
valley deepening and, 281
Entrenched karst, 365, 372
Eogenesis, 751
Eogenetic karst, 437
Epigean limnic system model, 405
Epikarst. *See also* Shallow subterranean habitats
as carbon dioxide reservoir, 286
concept, 284–285

Epikarst (*Continued*)
    disagreement about, 285–286
    in ecology, 284–285
    in hydrogeology, 285
    in karstology, 285
  defined, 284, 434
  erosion, 287
  evapotranspiration and, 285
  GPR and, 287
  how it works, 286, 286f
  hypotelminorheic zone and, 284
  illustration of, 285f
  karst evolution and morphology and, 286–287
  karst landforms and, 434
  as karst skin, 287
  Mangin and, 284
  NAPLs and, 169–170
  overview about, 288
  SSHs and, 685t, 686, 686f
  subcutaneous karst and, 284
  superficial karst and, 285
  underground shallow medium and, 284–285
  water
    calcium and, 289
    discharge and, 288
    DOC and POC and, 289
    temperature, 289
Epikarst communities. *See also* Shallow subterranean habitats
  biogeography and, 293–294
  Copepodas, 291–293
  ecology and, 294–295, 294f
  environmental conditions, 288–289
  morphological features, 293
  sampling techniques, 288, 289f, 290f
  taxonomy, 290–291
    aquatic genera, 290–291, 291t, 292f
    crustaceans, 290, 290t
    terrestrial species, 291
Epikarstic aquifer, 284
Epsomite, 374, 501t, 503
  stalactites, 375, 376f
Ereynetidae, 794
ERI. *See* Electrical resistivity imaging
Erosional caves, 865–866
Erosional tiankeng, 821f, 822, 822f
Erosion/surface water flow model, 406–407
Escape from epigean predators, 405
Escape hypothesis, 520
Eucarida, 178t, 191–193
  Decapoda, 178t, 191, 191f
    Anomura, 178t, 192
    Astacidea, 178t, 192, 192t
    Brachyura, 178t, 192, 193t
    Caridea, 178t, 191, 192t
Euhadenoecus fragilis, 610–611, 610f
Euhadenoecus puteanus, 610–611, 610f
Eukoeneniidae, 791
Eumalacostraca, 178t, 183–193
  Eucarida, 178t, 191–193
  Peracarida, 178t, 184–191
  Syncarida, 178t, 183–184
Eupnoi, 796

European diversity patterns. *See* Diversity patterns, in Europe
Eurycea, 667
  blepsimolge, 668f, 669–670
  E. chisholmensis, 670
  E. latitans, 670
  E. longicauda, 671, 672f
  E. lucifuga, 671, 672f
  E. nana, 670
  E. naufragia, 670
  E. neotenes, 669
  E. pterophila, 670
  E. rathbuni, 668, 668f
  E. robusta, 669
  E. sosorum, 669–670
  E. spelaea, 670–671
  E. tonkawae, 670
  E. tridentifera, 670
  E. troglodytes, 670
  E. tynerensis, 671
  E. waterlooensis, 669
  other, 670–671
  septentriomolge, 670
  Texas cave and spring-dwelling, 667–668
  typhlomolge, 668–669, 668f
Euscorpiidae, 788
Eutroglophile, 276–277
Evaporites dissolution, 758
Evapotranspiration, 285
Evolution of lineages
  background about, 295
  genetic differentiation models and, 297–298
  molecular data overview, 301–303
  subterranean habitat colonization, 295–297
    hypotheses, 296
    preadaptations and, 295–296
    routes, 296
  troglobites
    single *vs.* multiple origin of, 298–299, 298f
  troglomorphic traits, 299–301
    constructive *vs.* regressive, 299–301
    isolation variation and time, 299
Evolutionary ecology, 408–409
Evolutionary ecology, *Niphargus* and
  conclusions about, 562–564
  distribution, 555
  interactions within and between species
    cannibalism, 559
    chemoautotrophic bacteria symbiosis, 560
    predation and competition, 559–560, 560t
  key elements comparative review, 560–562
    communities: niche separation, 562, 563f
    female body size, 562
    morphological diversity and stasis, 560–561, 561f
  key properties, 556–559
    abiotic factors tolerance, 556–559
    allometric growth, 556, 558f

    antennae and pereopods, 556, 557f
    body shape, 556, 556f, 557f
    body size, 556, 557f
    developmental patterns, 556, 558f
    endemism/poor migratory abilities, 559
    gnathopods, 556, 557f, 558f
    light and, 557
    maxillae I, 556, 558f
    morphological variation, 556, 557f
    response to hypoxia, 557–558
    sexual dimorphism, 556, 558f
    temperatures, 558–559
    water chemistry tolerance, 559
  overview about, 555–556
  taxonomy challenges of, 555
Exaptations, 524
Exfiltration, 751–752
Exley, Sheck, 310, 312
Expedition caving, 308
Exploration, 103. *See also* Entranceless caves
  basic equipment, 305–307
    battery operated light sources, 305–306
    carbide lamps, 305
    clothing, 306
    hard hats, 306, 306f
    LEDs, 306
    light sources, 305–306
    packs, 306–307
  blue holes, 160
  Burnsville Cove, 75–77
  Butler Cave Camp, 86
  conclusions about, 309
  conservation and, 309
  Crystal Cave Expedition, 85–86
  demands of, 85
  Ellison's Cave, 86
  European model for, 85
  first camps used in, 85
  Friars Hole System, 336
  of glacier caves, 357
  Jewel Cave, 411
  Kazumura cave, 439
    entrances, 439, 439f
    local visitation, 439
    survey and systematic exploration, 439
  Krubera Cave, 444–445
  large systems
    expedition and project caving, 308
  Lechuguilla Cave, 462–464
    cartographic effort, 464
    Chandelier Maze, 464
    Coral Sea, 464
    Deep Secrets, 463
    Eastern Branch, 463–464
    guano mining and, 462–463
    before May 1986, 462
    1970s, 463
    1986, 463
    Southwest Branch, 463
    Rift, 463
  Mammoth Cave System, 469–470
  Mulu caves, 532

need for, 277–278
Nullarbor Caves, 572
overview about, 304
PPCS, 622
Project SIMMER and, 86
reasons for, 644
safety, 309
skill levels, 304–305
    large cave systems, 305
    small, near-horizontal caves, 304
    underwater caves, 305
    vertical caves, 304
Sullivan Cave, 86
vertical caving, 307
    ascending, 307, 316f, 317–318
    from bottom up, 319–320
    changing over, 318
    descending, 307, 317
    disclaimer, 314
    overview regarding, 314–315
    rappel, 307
    rigging and, 315–316, 315f
    rope and, 315
    single-rope, 307
    sit harness and, 316–317, 316f
    traverses and tyrolians, 318–319, 319f
Vjetrenica, 863–864
Wakulla Spring underwater cave system, 878–880, 879f
water, 308–309
    big river caves, 308
    under water caves, 308–309
Wind Cave, and, 86
Expulsion regime, 751–752
Extraterrestrial caves, 404
Eyre Peninsula dolines, 147

F

F1, 702
Facultative organisms, 491
Facultative visitors, 845
Fage, Louis, 219–220
Fantail, 438, 439f
Fat accumulation, 8
Faustloch, 702
Fern Bank salamander, 670
Ficco, Mike, 76
Fields, 269
Filter feeders, 20
Fishes, 20, 218, 854. *See also specific fish*
Fissure caves, 155
Fissure networks, 604f, 605
Fixed-beam model, 70, 71f
Flabellifera, 178t, 188, 189t
Flank margin caves, 156–159
  dissolutional morphology of, 158f
  formation of, 157, 158–159
  freshwater lens and, 157, 157f
  Hamilton's Cave, 160f
  Isla de Mona and, 159, 159f
  sea level and, 159
  Sistema del Faro, 159f
  typical, 158–159, 159f
  water density and, 157
Flatworms. *See* Turbellaria
Flies, 362
Floor ice, 401–402
Florida. *See also* Wakulla Spring underwater cave system, Florida
  rivers, 708
Flowstone, 779f
Fluorescent dyes, 889, 890t
Fluorite, 501t, 506
Flutes, 420, 421f, 428
  condensation, 425
  defined, 679
  transverse, 422
Fluviokarst, 436
Fluvio-karstic hollow, 153
Folklore. *See* Myth
Fontigens tartarea, 515–517
Food
  adaptation to low
    energy demand and, 4–6
    factors concerning, 2f
    food resources and, 1–3
    food scarcity types and, 1
    food-finding and, 3–4
    life history characters and, 6–8
    overview, 1, 2f
    periodic starvation and, 8–9
  adaptive shift and, 11–12
  behavioral adaptations and, 63, 64f
  finding, 3–4
    appendages and sensory equipment, 3
    back to network and, 4
    behavior, 3
    other factors of, 4
  Gammarus minus diet, 343–344
  general scarcity of, 1
  guano and, 363, 363f
  Organ Cave and, 101
  resources, 1–3
    cave type influence, 3
    chemoautotrophy and, 3
    input, 1–3
  scarcity types, 1
  White Cave and, 60–61, 61f
Food sources
  fecal types of, 331–334, 331t
    bats and, 331t, 333
    beetles and, 331t, 333
    crickets and, 331t, 333
    pack rats and, 331t, 332
    quality, quantity, predictability, 331t, 332
    raccoons and, 331t, 332
  limited, 323–328
    adaptations predicted from, 324–326
    aquatic and terrestrial differences from, 326
    costs of, 324, 325t
    exceptions proving the rule about, 326–328
    glacial cycles and, 326
    limestone *vs.* lava tube caves and, 327
    observations and principles, 323–324
    pollution and, 327
    short time scale and, 327
    specializations predicted from, 324–326, 325t
  scientific method use
    caveats and tradeoffs, 323
  terrestrial habitats, 328–331
    aquatic *vs.*, 328–329, 328t
    corpses and, 330
    cricket eggs and, 330
    extreme food types and, 329–331, 329t
    leaf and twig litter and, 330–331
Forbesichthys agassizi, 465, 466t
Foreign key, 270
Forms, 270
Forti, P., 499
Founder events, 10–11
Fracture permeability, 384, 385f, 385t
Frasassi Cave System, Italy, 696f, 815–816, 815f
Freshwater cave, 240–241
Freshwater lens, 157, 157f
Friars Hole System
  cave hydrology, 337–338, 337f
  exploration history, 336
  overview about, 334
  paleohydrology, 338–339
    cave, 338–339
    cave age, 339
    surface, 338
  setting, 334–336
    geology, 336
    hydrology and hydrogeology, 334–336
    location, 334, 335f
Frost shattering, 145
Frostwork, 380, 381f, 782
F-statistics, 609
Fuhl's Paradise Pit, 82
Fungus, 624

G

Galumnidae, 795
Gammarus fossarum, 653, 654t
Gammarus minus, 612
  cave population origin, 345, 346f
  cave streams and, 344
  compound eyes and, 345–346, 347f
  foods of, 343–344
  general ecology of, 343–344
  idealized subterranean drainage basin and, 343, 343f
  karst window and, 344–345
  as model system, 347
  natural selection and, 544–545
  population density of, 343–344
  population structure of, 345
  reproduction, 343–344
  springs and, 344
  variation in, 344–345, 344f
Garden of Eden, 532
Garypidae, 792
Gastropoda, 20, 232, 233
Gate styles, 628–630, 629f, 630f

Gating criteria, 628
Gating decision guide, 630–631
  final results, 631
  good reasons
    to gate, 630–631
    not to gate, 630
  poor reasons
    to gate, 630
    to not gate, 630
Gelyelloida, 178t, 182
General food scarcity, 1
Genetic differentiation models, 297–298
Genetic distance, 609
Genetic isolation, 275
Genetic repertoire, 10
Genrikhova Bezdna Cave, 443, 444f
Geographic information systems (GIS), 264–265
Geoid, 348
Geophilomorpha, 539
Geophysics of location
  AMT and, 349–350
  cave radiolocation, 351–352, 351f
  conclusions about, 352
  dowsing and, 352
  ERI and, 348–349, 349f
  GPR and, 350, 350f
  interferometry and, 350, 351f
  microgravity and, 348, 349f
  overview about, 348
  seismic modeling techniques and, 350
  targets of, 348f
  techniques, 348
  TEM and, 352
  thermal variation and, 350–351
  tradeoffs, 348
Georgia Blind Salamander, 671–672
Ghost stories, 323
Gigalere Cave, 85
Gilbert, Janine, 101, 276
GIS. *See* Geographic information systems
Glaciated plateaus, 740–742, 741f
Glacicavicola bathysciodes, 519f
Glacier caves, 105f, 106
  basics about, 353
  creep closure in, 354–355, 355f
  cut-and-closure, 354–355, 355f
  englacial, 354–356
  exploration of, 357
  formation processes, 354
  keyhole and, 355–356, 356f
  overview about, 353
  recreational caving and, 642
  subglacial, 356, 356f
  temperature and, 353
  thrust faults and, 354, 354f
Glacier milk, 724
Glaciokarst, 437
Glacio-karstic hollow, 153
The Glop Slot, Butler Cave, 78f
Gnaphosidae, 790
Gnathopods, 556, 557f, 558f
Golden Chamber, 859–860, 860f
Gonyleptidae, 796
Gostry Govdy Cave, 829t

Gouffre Berger Cave, 85
GPR. *See* Ground penetrating radar
Grand Canyon caves, 817–818, 818f
Gravity
  sediments, 139
  springs, 798–799, 799f
Green Cathedral Cave, 536
Grike, 428–429, 429f
Grotto, Nittany, 74, 75
Ground penetrating radar (GPR), 287, 350, 350f
Guadalupe Mountains caves, 816–817, 816f
Guano
  bat, 358
    carnivorous, 358–359, 359f
    community structure and succession in, 360–361
    conservation and, 242–243, 249
    frugivorous, 358–359, 358f
    hematophagous, 358–359, 359f
  cave communities and science and, 357–358
  conclusions about, 363–364
  conservation and, 363
  eutrophic caves and, 357
  fauna associated with, 359–360
  fauna examples, 360–361
    basic taxa, 361
    bugs, 362
    cockroaches, 362
    earthworms, 362
    flies, 362
    mites, 361–362, 361f
    overview, 361
    pseudoscorpions, 362
  food web examples, 363, 363f
  replacement, 626
  sources, 358
  studies of, 358
  tropics diversity patterns and, 242
  types of, 358–359
    bats, 358
    birds, 358
    crickets, 358
Guanophages, 359
Gunung Api, 532–535, 537
  Clearwater River Cave, 532–534
  Clearwater River Passage, 534, 535f
  Cobra-Bridge-Cloud Cave, 534
  Sarawak Chamber, 534
Gunung Benarat, 535–536
  Benarat Caverns System, 536
  Melinau Gorge, 535–536, 535f
  Terikan River Caves, 536
Gunung Buda, 536–537
  Green Cathedral Cave, 536
  Tardis Cave, 536f
Gunung Mulu National Park, 537
  cave biology, 537
  Gunung Api, 537
  plant/animal species, 537
Gypsum, 500–503, 501t. *See also* Ukraine giant gypsum caves
  angel hair, 375
  bulk crystals, 374–375

cave
  conclusions about, 372–373
  lengths and locations, 369t
  modification of hypogene maze, 371f, 372
  Ordinskaya, 372, 372f
  patterns and morphology, 372, 373f
  sketches, 366f
cave cotton, 375
chandeliers, 374–375
crusts, 374, 375
as evaporite mineral, 374
hypogene conduits and, 367f
karst types, 364–365
  deep-seated karst, 365, 367
  entrenched karst, 365, 372
  intrastratal karst, 365, 367
  open karst, 365, 366–367
  subjacent karst, 365, 371
  syngenetic karst, 365
needles, 374–375, 375f
Nullarbor Caves and, 574, 574f
occurrence of, 364–365
speleogenesis
  chemistry and kinetics, 365–366
  karst types and, 366–372
  transverse hypogene, 367–368, 368f
speleothems and, 374–375
structure of, 777–778, 778f
Gypsum flowers
  formation of, 374, 375, 375f, 782, 782f
  overview about, 374
  sources and depositional mechanisms, 375–377
Gyrinophilus, 672–674, 673f
  G. gulolineatus, 673f, 674
  G. palleucus, 673, 673f
  G. porphyriticus, 672, 673f
  G. subterraneus, 673f, 674

H

Hadenoecus subterraneus, 744–746, 745f, 746f
Hahniidae, 790
Haideotriton wallacei, 671–672
Half-tubes, 422, 423, 424f
Halides, 501t, 505–506
  fluorite, 501t, 506
  halite, 501t, 506
  overview, 501t, 505–506
Halite, 501t, 506
  Nullarbor Caves and, 574, 574f
Halocline, 157
Halosbaena tulki, 214f
Hamilton's Cave, 160f
Haplozetidae, 795
Hard hats, 306, 306f
Harpacticoida, 178t, 182
Harvestmen, 795–797, 796f
Hawaii. *See also* Kazumura cave, Hawaii
  adaptive shift and, 12–16
    cixiid planthoppers, 13–14, 13f
    crickets, 15
    moths, 15–16
    Oliarus polyphemus, 14, 14f
    species-pairs occurring, 12t

terrestrial Isopoda, 14–15
wolf spiders, 15
root communities in, 661t
volcanic rock caves, 866t
Head flattening, 518, 518f
Headwater Complex, Castleguard Cave, 89–90
Heavy metals
  chemistry, 162–164
    general, 162–163
    interactions, 162, 163f
    iron and manganese, 163–164
    pH, 162–163
    redox, 162–163
    trace and contaminant metals, 164
  conclusions about, 165–166
  defined, 161, 162t
  forms of, 162
  potentially toxic, 162t
  sources, 161–162
    anthropogenic, 161–162, 163t
    in karst settings, 161, 162f
    natural, 161, 163t
  spring water concentrations, 165f
  storage and transport, 164–165
    schematic scenarios, 164
    speleothems and cave deposits, 164
    storm-enhanced, 165
    suspended and bed sediments, 164–165
  storage compartments, 163f
Heavy nonaqueous phase liquids (DNAPLs), 391
Helictite Cave, 81, 81f
Helictites
  beaded, 379, 380f
  bushes, 382–383, 382f
  formation of, 782, 782f
  subaerial, 379
  subaqueous, 381–383, 381f, 382f
Hemiptera, 259
Hercegovina, 236
Heteroptera, 259
Heterotrophs, 126–127, 490–491
Heterozygosity, 609
Hewitt, Bevin, 74, 75
Hexapoda, 258–259
Hexatelidae, 789
Hibernation, of bats, 48–49, 48f
Hill, C. A., 499
Hirudinea, 245t, 913–914, 913f
Historical biogeography, 849–850
Hjulstrom's Diagram, 138, 138f
Hohlaub area, 703
Holarchaeidae, 790
Hölloch Cave, Switzerland, 85, 726, 727f
Homoptera, 259
Hon Chong hills, 249–250, 250f
Hondo Underground Rescue Team (HURT), 649
Hooded tickspider, 793, 793f
Hosterman Pit, 278
Howe Caverns, 740–741, 741f
Hubbardiidae, 788–789
Huehue Cave, 869f

Huehue Flow, 869–870, 871f
HURT. See Hondo Underground Rescue Team
Hwanseon Cave, 696f
Hybridization, 11
Hydrogen
  chemoautotrophy and, 127–128
  sulfide, 758
Hydrogeology of karst aquifers. See Karst aquifer, hydrology of
Hydrographs, 388–389, 389f
Hydromagnesite, 500
  balloons, 416–417, 416f
Hydrothermal caves
  chemistry, 393–394
    mixing water dissolution, 393
    rising thermal water dissolution, 393
    subaqueous and subaerial dissolution, 393–394, 394f
    sulfides oxidation, 393
  defined, 391–392
  deposits, 396
    character of, 396
    mineralogy, 396
  identification, 397
  life, 396–397
  morphology, 394–396
    anastomotic caves, 395–396
    bush-like caves, 395, 395f
    caves with cupolas, 395
    individual chambers, 394
    isometric rooms, 394
    network caves, 395
    phreatic maze caves, 395–396
    single-conduit caves, 394
    small-scale, 396
    solution porosity, 394
    spongework caves, 396
  regional extent, 397
Hydrothermal karst, 437
  defined, 391–392
  in noncarbonate rocks, 393
  oxidation of sulfide ores and, 393
  settings, 392–393
    deep seated, 392
    endokarst, 392
    shallow, 392–393
Hydroxides, 501t, 505
Hydroxylapatite, 501t, 504–505
Hydryphantidae, 795
Hyena, 594
Hygrobatidae, 795
Hygropetric cave, 236
Hypochthoniidae, 795
Hypogenetic karst, 437
Hypogenic speleogenesis, 756–764
  artesian, 759
  basic concept, 750
  chemical mechanics, 756–759, 756f, 757f
    carbonic acid dissolution, 756–758
    evaporites dissolution, 758
    hydrogen sulfide, 758
    mixed carbonate/sulfate strata dissolution, 759
    organic acids, 758

sulfuric acid, 758
thermal water cooling, 758
water mixing, 758
combined artesian/deep-rooted, 759
defined, 750
diagenetic environments, 750–756, 751
  eogenesis and, 751
  mesogenesis and, 751
  telogenesis and, 751
distribution and settings, 759
endokarst realm, 754–756, 755f, 756f, 757f
groundwater systems and regimes, 751–752
hydrogeologic and structural controls, 759–761, 760f
hydrogeologic cycle, 752–753, 753f
karst settings evolution, 753–754, 754f
mesomorphology features, 763–764
overview, 748–750, 749f
paleokarst and, 764
porosity patterns, 761–763
  cavernous edging, 761, 762
  collapse shafts, 761, 762
  isolated chambers, 761, 762
  network maze, 761, 762
  single or rudimentary networks, 761–762
  spongework maze, 761, 762
  steeply-inclined passages or shafts, 761, 762
Hyporheic zone, 686–688, 687f, 688t
Hypotelminorheic zone, 284
Hypoxic conditions, 6
Hypsimetopidae, 212

I

Ice blocks, 401–402
Ice caves, 236
Ice Crawls, Castleguard Cave, 92f
Ice in caves
  bamboo, 402, 403f
  cold air trapping and, 400–401
  cryogenic cave micropearls, 401–402, 403
  crystallography and mineralogy and, 403
  extraterrestrial caves and, 404
  floor ice, 401–402
  high latitude/altitude and, 399–400
  ice blocks, 401–402
  ice stalagmites, 402
  isotope analysis and, 402–403
  lake ice, 401–402
  mechanisms causing, 399, 400f
  overview about, 399
  pollution and, 403
  radiocarbon dating and, 402
  Scàrisoara Ice Cave, 403
  scientific significance of, 402–404
  snow and, 400
  speleothems, 401f
  tourism and, 403–404
  ventilation cooling and, 401
Ice wedging, 72
Icicles, 575, 575f
Idealized subterranean drainage basin, 343, 343f

Ideoroncidae, 792
Illinois structural basin, 734f, 736
Imprints, 871
Inception doline, 143f, 144
Incised passages, 606f
Indiana, southern, 736–737
Indirectly adaptive hypothesis, 342
Inficionado Peak, 637, 637f
Infiltration regime, 752
Inflation, 867, 868f
Influent streams, 707–709
    cave evolution and, 707, 708f
    Danube River and, 707
    Edwards Aquifer and, 708–709
    evolutionary sequence of, 710–712
        case A, 711, 711f
        case B, 711, 711f
        case C, 711, 711f
    Florida, 708
    function of, 709–710
    Kentucky, 709, 709f, 710f
    overview, 707
Innerbergli area, 702
Insecta, 216–218, 259–260
    Blattodea, 216–217
    Coleoptera, 259–260
        Dystiscidae, 217–218
    Collembola, 216
    Dermaptera, 259
    Diplura, 216
    Diptera, 260
    Hemiptera, 259
    Heteroptera, 259
    Homoptera, 259
    Orthoptera, 217, 259
    Planthoppers, 216
    tropical diversity patterns, 245t
    Zygentoma, 216
Insolation, 581, 582f
Interferometry, 350, 351f
Internal runoff, 387–388
Interpolators, 477–479, 477f
Intersection dolines, 150f, 152
Intrastratal karst, 365, 367
Invasion
    active dispersal models, 405
        active colonization model, 405
        adaptive-shift model, 405
        climatic-relict model, 405
        competitive pressure, 405
        deep sea colonization, 405
        epigean limnic system model, 405
        escape from epigean predators, 405
        marine shallow-water ecotones model, 405
    conclusions about, 409
    ecological evidence, 407–408
    evolutionary scenarios, 405–407
    overview about, 404–405
    passive dispersal models, 406–407
        erosion/surface water flow model, 406–407
        rafting model, 406
        regression model, 406, 407f
        surface water infiltration model, 406

    unique vs. multiple, 407
Iron
    bacteria, 133
    chemistry and, 163–164
    chemoautotrophy and, 129, 133
Ischyropsalidae, 796
Isla de Mona, 159, 159f
Isolated chambers, 761, 762
Isometric rooms, 394
Isopoda, 178t, 187–188, 189t, 258
    adaptive shift and, 14–15
    anchihaline habitats and, 22
    Anthuridea, 178t, 188, 189t
    Appalachian stream competition and, 746–747, 746f, 747f
    Asellota, 178t, 188–190, 189t
    in Australia, 212–213
    Calabozoidea, 178t, 189t, 190
    Cirolanidae, 212–213
    in Dinaric karst, 233–234
    Flabellifera, 178t, 188, 189t
    Hypsimetopidae, 212
    Microcerberidea, 178t, 188, 189t
    Oniscidea, 178t, 189t, 190, 213
    Phreatoicidea, 178t, 188, 189t
Ixodidae, 794

J

Japanese folklore, 322–323
Jean-Bernard Cave, 731, 731f
Jeannel, René, 219–220, 404–405
Jefferson's ground sloth, 596, 597f
Jewel Cave, South Dakota
    barometric airflow, 411
    biology, 415–416
        bats, 415–416
        cave life, 415
        microbiology, 416
    calcite crystals of, 411, 412f
    conclusion about, 417
    geology, 413–415
        cave fill, 414–415
        Minnelusa Cap/caves relationship, 414, 414f
        recent mapping, 414
    history, 411
        early, 411
        early exploration, 411
        later exploration, 411
    hydrology, 415
    levels, 413f
        basement level, 412
        chert level, 412
        loft level, 412
        lower level, 412
        main level, 412
        subchert level, 412
    origin, 415
        working hypothesis, 415
    other Black Hills caves, 417
    physical characteristics, 411–413
        breakdown modification, 412–413
        passage sizes, 413
        three-dimensional system, 411–412, 413f

    speleothems, 416–417
        hydromagnesite balloons, 416–417, 416f
        popcorn stalagmites, 416
Jollyville Plateau salamander, 670
Joseph, G., 275
Jubilejna Cave, 829t
Julida, 540, 540t

K

K2, 702–703
Kakuk, Brian, 311f
Kamenica, 859–860, 860f
Kamenitza, 140
Kane caves, Wyoming, 813, 813f, 814t
Karren, cave
    biological activity influencing, 419–420, 420f
    clastic material and, 420
    development, 419
    feature descriptions, 420–425
        anastamose, 422, 423f
        anastomoses, 422
        avens, 423
        bell holes, 423–424, 424f
        bypass tubes, 423
        chimneys, 423
        condensation corrosion hollows, 425, 425f
        condensation rills/flutes, 425
        corkscrews, 423
        corrosion notch, 425
        cupola, 424, 425f
        cusps, 425, 425f
        dissolution basins, 420, 421f
        dissolutional potholes, 420, 422f
        drainage grooves, 422
        drip pits, 420, 421f
        half-tubes, 422, 423, 424f
        laugdecke, 425
        paragenesis, 422
        pendants, 422, 423f
        rills/flutes, 420, 421f, 425
        scallops, 420–422, 422f
        scour marks, 422
        solution pockets, 423, 424f
        spongework, 422
        transverse flutes, 422
        waterline notch, 425
    flow and, 419
    form controls, 419–420
    overview about, 419
    photokarren and, 419–420, 420f
    rock properties influencing, 419, 420f
    splitkarren, 428–429
Karren, surface
    assemblages, 430
        alpine karrenfields, 430
        coastal karren, 425f, 430
        pinnacle karst, 430
        ruiniform karrenfield, 430
    basic controls, 426–427
        biological activity, 427
        climate, 427
        hydrodynamic, 426, 426f

lithological factors, 427, 427f
dissolutional pavements, 427, 427f
features
    basins, 427–428
    cavernous karren, 429, 429f
    channels, 428
    corrosion notch, 429–430
    cutter, 429
    decantation runnel, 428
    flute, 428
    grike, 428–429, 429f
    pans, 427–428, 428f
    pits, 427–428
    rillenkarren, 428, 428f
    solution runnels, 428, 429f
    spitzkarren, 429
overview about, 425–426
Pinnacles of Mulu, 426, 426f
positive forms, 426
San Salvador Island, 426f
Karst
    coastal, 437
    denudation, 508
    ecosystems
        connectivity and disturbance in, 102
        definition and boundaries, 99
        energy flux and limitation, 100–101, 100f
        nutrients, 101–102
        physical environment and habitat zones, 99–100, 100f
    endokarst realm, 754–756, 755f, 756f, 757f
    eogenetic, 437
    history surrounding, 432–433
    hypogenetic, 437
        hydrothermal, 437
        sulfuric acid, 437
    landforms, 433–436
        caves and, 436
        closed depressions, 434–435
        cones and towers, 435–436, 436f
        overview, 433
        pavements and epikarst, 434
        poljes, 435
    mixed origin landscapes, 436–437
        alpine karst, 436–437
        fluviokarst, 436
        glaciokarst, 437
    mixing-zone, 437
    overview about, 430–431
    settings evolution, 753–754, 754f
    telogenetic, 437
    window, 344–345
    word origin, 431–433
        karst triangle and, 432, 432f
        Kras plateau and, 431–432, 432f
        location and, 431–432
    world distribution, 431, 431f, 432t
Karst aquifer, hydrology of, 384–388
    characteristics, 388–390
        base flow/area relationships, 389–390
        clastic sediments, 390
        conduit systems carrying capacity, 389
        hydrographs, 388–389, 389f
    contaminant transport, 390–391
        DNAPLs, 391
        LNAPLs, 391
        metals, 391
        pathogens, 391
        water-soluble compounds, 391
    groundwater basins/flow systems, 387–388
    overview, 383
    piracies and spillover routes, 388
    porous media aquifers and, 383–384
    standard model, 387–388, 387f
        allogenic recharge, 387
        dispersed infiltration, 387
        internal runoff, 387–388
    triple porosity model, 384–385
        conduit permeability, 384, 385–387, 385f, 385t
        fracture permeability, 384, 385f, 385t
        matrix permeability, 384, 385f, 385t
    types, 385, 386f
    water quality, 390
    water supply, 390–391
        wells and, 390
    water table and, 388
Karst aquifers, modeling of
    chemistry, 508–510
        calcium equilibrium, 508–509, 509f
        calcium flux rate, 509–510, 509f
    evolution, 510–512, 512f
        three-dimensional, 510–511, 512f
        two-dimensional, 510
    overview about, 508
    single fracture, 510, 510f
Karst aquifers, water tracing in
    definitions and objectives of, 886–887
    dry set system, 892
    dye amounts to use, 891
    history, 887–889, 887f
    passive detector types
        activated carbon, 890
        cotton, 891
        plankton netting, 891
        Quinlan Gumdrop, 891, 891f
    qualitative
        using passive detectors, 890–892
    quantitative, 892–893
    test interpretation, 893–896, 894f
        dye recovery or breakthrough curve and, 894, 894f, 895f
    tracers and, 889–890
        categories of, 889
        cautions, 890
        fluorescent dyes, 889, 890t
        ideal characteristics, 889
Kartchner Caverns, 694f
Kazumura cave, Hawaii
    description, 439–442
        braided passages, 440–441, 441f
        canyon-like passages, 439–440, 440f
        extraneous tubes, 441
        intermediate roofs, 441–442, 441f
        multilevel/lavafall development, 441–442
        plunge pools, 441–442, 442f
        speleogenesis, 439–440
        thermal erosion, 440–441, 440f
    exploration history, 439
        entrances, 439, 439f
        local visitation, 439
        survey and systematic exploration, 439
    fantail and, 438, 439f
    lava cave speleothems, 442
        primary, 442
        secondary, 442
    overview about, 438
    physical setting, 438
        human impacts, 438
        location, 438, 438f
        surface relationship, 438
Kebara Cave, 95f, 98, 98f
Keller Well, 471f
Kentucky. *See also* Mammoth Cave System, Kentucky
    influent streams, 709, 709f, 710f
    solution caves, 736–737
Keyhole, 355–356, 356f
Komsomol'ska Cave, 829t
Kras plateau, 431–432, 432f
Kriging, 477–479
Kristal'na Cave, 829t
Krosswig, Curt, 544
Krubera Cave
    Arabika Massif
        deep cave evolution, 449–450
        geology, 447–448, 447f
        hydrogeology, 448–449
        location and physiography, 443, 444f
        Ortobalagan Valley in, 443, 445f
    block-fault structure and, 445–447
    branches, 445, 445f
    combined cave profile, 446f
    depth of, 443
    evolution of, 449–450
    exploration since 1999, 444–445
    first documentation of, 443
    genesis of, 729, 730f
    hydrology, 447
    overview about, 443
    as shaft and canyon cave, 729, 730f
    temperatures, 447
Kujbyshevskaja Cave, 443, 444f

L

Labeled substrate experiments, 493–494
Labyrinth caves, large, 726–729
    Hölloch Cave, 726, 727f
    Réseau Siebenhengste-Hohgant system, 726–729, 728f
Lake ice, 401–402
Lamarckism, 520
Lampenflora
    appropriate lighting, 455, 455f
    composition and ecology, 452–453
    control and restriction of, 454–455, 454f
    cyanobacteria, 453
    definitions, 451–452
    dispersion modes, 452–453

Lampenflora (Continued)
  diversity, 453
  factors influencing, 452
  future management, 455
  habitat alteration and, 453–454
  killing, 455
  mosses, 453
  overview, 451
  in Postojnska Jama, 452f
  problems associated with, 454
  surfaces and, 453
  temperature and, 452
  tourism and, 451–452
Landscape evolution
  multilevel caves, 529–531
    Cumberland River incision pulses, 530, 530f
    eastern Alps, Austria, young uplift, 531
    Mammoth Cave, 529–530, 530f
    Sierra Nevada incision rates, 531
    southern Rocky Mountains erosion rates, 531
Laniatores, 796
Lankester, E. R., 404–405
Lanzaia vjetrenicae, 231f
Large labyrinth caves, 726–729
  Hölloch Cave, 726, 727f
  Réseau Siebenhengste-Hohgant system, 726–729, 728f
Lateral potholes, 682f, 683
Lattice data, 475
Laugdecke, 425
Lava cave, 641. *See also* Root communities, in lava tubes
  double-trunked tunnels, 869–870
  rise, 871–872
  single-trunked tunnels, 869
  speleothems, 442
    primary, 442
    secondary, 442
  superimposed-trunked tunnels, 870
  tubes, 327, 658
Lavafall development, 441–442
Lechuguilla Cave, New Mexico, 816–817, 816f
  biology, 459
  Capitan Formation, 457–458, 457f
  conservation and, 456
  description, 460–462
    boneyard maze, 461–462, 462f
    fissures, 461, 462f
    horizontal passages, 461, 462f
    vertical tubes, 461
  discovery of, 456
  exploration history, 462–464
    cartographic effort, 464
    Chandelier Maze, 464
    Coral Sea, 464
    Deep Secrets, 463
    Eastern Branch, 463–464
    guano mining and, 462–463
    before May 1986, 462
    1970s, 463
    1986, 463
    Southwest Branch, 463
    Rift, 463
  folds within vicinity of, 458
  four units spanned by, 457–458, 457f
  geologic setting, 457
  as hypogene cave system, 458
  map, 461f
  microbial influence in, 456
  mineralogy, 459–460
  paleontology, 460
  physical setting, 456–457, 457f
  Queen Formation, 457–458, 457f
  regional/local hydrogeology, 458–459
  rooms and passages, 458, 459f
  Seven Rivers Formation, 457–458, 457f
  stratigraphy and structure, 457–458
  sulfuric acid and, 456, 460–461
  as world-class cave, 456
  Yates Formation, 457–458, 457f, 458
LEDs. *See* Light-emitting diodes
Leeches, 913–914, 913f
Leeuwenhoeckiidae, 794
Legend. *See* Myth
Leiodidae, 221, 223f
Lepe Jame, 622, 622f
Leptodirini group, of beetles, 57, 58f
  Leptodirus hochenwartii, 55, 55f, 231f, 519f
  phylogenetic Mediterranean tree, 59, 60f
Leptodirus hochenwartii, 55, 55f, 231f, 519f
Leptonetidae, 790
Leptostracans, 21
Life history characters
  adaptation and, 6–8
  back to network and, 8
  case study, 7–8, 8t
  egg size and, 6–7, 7f
  growth rate and, 7
  longevity and, 7
Life history evolution
  of Amblyopsidae, 465–466
    Amblyopsis rosae, 465, 466, 466t, 467–468
    Amblyopsis spelaea, 465, 466, 466t, 467–468
    Chologaster cornuta, 465, 466t
    Forbesichthys agassizi, 465, 466t
    life history characteristics, 466–467, 466t
    Typhlichthys subterraneus, 465, 466t
  high reproductive effort paradox and, 467–468
Light nonaqueous phase liquids (LNAPLs), 391
Light-emitting diodes (LEDs), 88, 306
Lighting, show cave, 694–695
Limbodessus eberhardi, 206f
Limestone caves, 658. *See also specific cave*
Limnesiidae, 795
Limnohalacaridae, 795
Limonite, 163–164
Linkages, 270
Linyphiidae, 790
Liochelidae, 788
Liocranidae, 790
Lithobiomorpha, 539–540, 539f
Lithological contact poljes, 154, 154f
Littoral cave, 155
LNAPLs. *See* Light nonaqueous phase liquids
Long inclined passages, 729–732
  Cassowary System, 731–732, 732f
  Jean-Bernard Cave, 731, 731f
Looping passages, 724, 724f
Losing streams, 707–709
  cave evolution and, 707, 708f
  Danube River and, 707
  Edwards Aquifer and, 708–709
  evolutionary sequence of, 710–712
    case A, 711, 711f
    case B, 711, 711f
    case C, 711, 711f
  Florida, 708
  function of, 709–710
  Kentucky, 709, 709f, 710f
  overview, 707
Loubens, Marcel, 85
Low food adaptation
  energy demand and, 4–6
  factors concerning, 2f
  food resources and, 1–3
  food scarcity types and, 1
  food-finding and, 3–4
  life history characters and, 6–8
  overview, 1, 2f
  periodic starvation and, 8–9
Lungless salamanders, 667–674
Lycosidae, 790

M

Machaerites spelaeus, 231f
Macrochelidae, 793
Magnetostratigraphic time scale, 586, 587f
Makua Cave, 322
Malachite, 500, 501t
Malacostraca, 178t, 183–193, 257–258
  Eumalacostraca, 178t, 183–193
    Eucarida, 178t, 191–193
    Peracarida, 178t, 184–191
    Syncarida, 178t, 183–184
  Phyllocarida, 183f, 183, 178t
Mammalia, 261–263, 262t
Mammals, as cave visitors
  cave ecology impact of, 847
  cave-dwelling species, 845–846
    facultative visitors, 845
    regional cavernicoles, 845
    regular cavernicoles, 845–846
  den sites and, 846–847
  foraging habitats and, 847
  knowledge gaps, 846
  man and, 848–849
  rare species, 848
  resting and mating and, 846
  wintering sites and, 847
Mammoth Cave System, Kentucky
  ancient cavers
    archaeological evidence, 27–28, 27f, 28f
    archaeology, 26–27
    chronology and, 27

evidence interpretation, 28–30, 29f
  prehistoric archaeology, 27–30
 biology, 473–474
 Cleaveland Avenue, 472f
 dating, 175–176, 737
 entrances, 470, 470f
 exploration, 469–470
 geographic and geologic setting, 471
 history, 470
 idealized profile through, 737, 737f
 as international research magnet, 474
 Keller Well, 471f
 landscape evolution and, 529–530, 530f
 minerals, 472–473, 473f
 origin, 473
 overview about, 469, 470f
 passage patterns, 471–472, 471f, 472f
 prehistoric explorers and miners, 469–470
 recent/ongoing exploration, 469
 as solution caves, 737, 737f
Manganese
 bacteria, 133
 chemistry and, 163–164
 chemoautotrophy and, 129, 133
Mangin, A., 284
Mangkurtu mityula, 206f
Manmade show caves, 643
Mapping subterranean biodiversity
 data types, 475
  lattice data, 475
  point patterns, 475
 interpolators and, 477–479, 477f
 kriging and, 477–479
 modeling issues, 479–481
  aggregation and scale, 479–481, 480f
  excess zeroes, 481
 overview about, 474–475
 spatial pattern investigation, 476–477
  spatial autocorrelation, 476–477
 spatial process decomposition, 475–476
  example, 476, 476f
  large-scale trend, 475
  mesoscale or small-scale variation, 475
  random error or unexplained variability, 475–476
 zero-inflated spatial models and, 478–479
Marginal cave, 18
Marine caves, 240
Marine regressions
 background, 482
 change rate evaluation, 486
 dating from stygofauna, 484–486
 diverse habitat adaptability, 482
 from sea to freshwater and land, 482
 thalassoid limnostygobionts
  biogeography of, 483, 484f, 485f
  Morocco and, 483–484
  as sole evidence, 483–484
Marine shallow-water ecotones model, 405
Martel, Edouard A., 103, 433
Martino's vole, 845–846, 846f
Masodon bones, 875–876, 876f, 877f
Material compensation, 520
Mating behavior, adaptive shift and, 11

Matrix permeability, 384, 385f, 385t
Mats, microbial, 496, 498f
Maxillopoda, 178t, 180–182, 257
 Copepoda, 178t, 181–182
  Calanoida, 178t, 181, 181t
  Cyclopoida, 178t, 181–182
  Gelyelloida, 178t, 182
  Harpacticoida, 178t, 182
  Misophrioida, 178t, 181, 182t
  Platycopioida, 178t, 181
 Mystacocarida, 178t, 180–181
 Tantulocarida, 178t, 180
Maya, 322, 836
Maya caves
 ancient utilization
  antiquity of, 487
  cave art, 488, 489f
  cave burial, 487–488
  cave modifications, 488, 488f
  general characteristics of, 487
 artificial, 489f, 490
 community and, 489–490
 ethnohistory and ethnography, 488–489
 map, 487f
 overview about, 486–487
Maze caves, 395–396
Maze development
 aggressiveness boosts, 601
 floodwater recharge and, 601
 short flow paths and, 601
 uniform recharge and, 601
Melinau Gorge, 535–536, 535f
Melinau Limestone, 532
Mesocaverns, 658
Mesogenesis, 751
Mesostigmata, 793
Mesothelae, 789
Metabolic assays, 492
Metabolic span, 7
Metastigmata, 794
Meteoric circulation regime, 752
Methane cycle, 133
Mexico. See also Sistema Huautla, Mexico
 camps and, 86–87, 87f
 Cueva de Villa Luz, 813–815, 813f, 814t, 815f
Micro whipscorpions, 791
Microaerophilic organisms, 491
Microbes. See also Chemoautotrophy
 actinomycete colonies, 494, 497f
 cave and karst diversity of, 494–497, 495f
 diversity and metabolism, 490–491
  aerobes and anaerobes and, 490f, 491
  energy-yielding metabolic reactions, 490–491, 490f
 future regarding, 497–498
 history of studies on, 491–492
  early studies, 491
  1960s through 1980s, 491
  1980s and 1990s, 491–492
  technology and, 492
 methodological approaches and, 492–494
  culturing and metabolic assays, 492
  labeled substrate experiments, 493–494

 microscopes, 494
 molecular, 492–493
 stable isotope ratio analysis, 493–494
 microbial mats and, 496, 498f
 naked eye and, 494, 496f
 overview about, 490
 Proteobacteria, 495–496
 vermiculations, 496, 497f
Microbial mats, 496, 498f
Microcerberidea, 178t, 188, 189t
Microcracks propagation model, 71
Microgravity, 348, 349f
Microstigmatidae, 789
Microzetidae, 795
Mictacea, 22, 178t, 185–186, 186f
Middle East
 anthropogenic deposits, 96
 cave occupations, 97–99
  Chalcolithic and Bronze age, 98
  Middle Paleolithic, 98
  Neolithic, 98
  prehistoric, 97–98
  Upper and Epi-Paleolithic, 98
 deposits and processes, 96–97
 Kebara Cave, 95f, 98, 98f
 location, 95f, 96
 overview about, 94–96
 prehistoric sites, 95f
 present environment, 96
 Qafzeh Cave, 95f, 98f
 Tabun Cave, 95f, 97–98, 97f
Midnight Cave, 86
Migration, 609
Milieu souterrain superficiel (MSS), 686f, 688, 688f
Millipedes, 540–541
 Cambalopsidae and, 242, 244f
 Chordeumatida, 540t, 541
 Julida, 540, 540t
 map of, 540f
 overview, 540
 Polydesmida, 540, 540t
 Pseudotremia, 541, 541f
 Scoterpes, 541
Milyeringa veritas, 214f
Mimetidae, 790
Mineral springs, 800
Minerals. See also specific mineral
 arsenates and vanadates, 506
 books on, 499
 carbonates, 500, 501t
  aragonite, 500, 501t
  calcite, 500, 501t, 503f
  coralloid multiaggregates and, 500, 503f
  deposition of, 500
  hydromagnesite, 500
  malachite, 500, 501t
 halides, 501t, 505–506
  fluorite, 501t, 506
  halite, 501t, 506
  overview, 501t, 505–506
 hydrothermal caves and, 396
 ice in caves and, 403
 Lechuguilla Cave, 459–460

Minerals (*Continued*)
　Mammoth Cave System, 472–473, 473f
　native elements, 506
　Nullarbor Caves, 574–575
　　calcite, 574
　　gypsum, 574, 574f
　　halite, 574, 574f
　organic, 507
　overview, 499–500, 501t
　oxides and hydroxides, 501t, 505
　phosphates, 504–505
　　brushite, 501t, 504–505
　　development of, 504, 504f
　　hydroxylapatite, 501t, 504–505
　　overview, 501t, 504
　　taranakite, 501t, 504–505
　PPCS, 623
　reaction types creating, 499
　reasons to study, 507
　Siebenhengste cave system, 704–706, 706f
　silicates, 501t, 505
　　allophane, 501t, 505
　　opal, 501t, 505
　　origins, 505
　　other, 505
　　quartz, 501t, 505
　speleothems and, 777–778, 885–886
　　agagonite, 777
　　calcite, 777
　　gypsum, 777–778, 778f
　stalactites and, 805–806, 806f
　stalagmites and, 805–806, 806f
　sulfate mineral crystals
　　overview about, 374
　　sources and depositional
　　　mechanisms, 375–377
　　types of, 374
　sulfates, 500–504, 501t
　　alunite, 501t, 503
　　barite and celestine, 501t, 503
　　deposition mechanisms, 503–504
　　epsomite, 501t, 503
　　gypsum, 374, 500–503, 501t, 574, 574f, 777–778, 778f
　　mirabilite, 501t, 503
　sulfides, 507
Mines, 642
Minnelusa Cap, 414, 414f
Mirabilite, 374, 501t, 503
　stalactites, 375, 376f
Misophrioida, 178t, 181, 182t
Mississippi Valley-type (MVT) ore deposits, 818–819
Mitchell, Simon, 313
Mitchell Plain, 722
Mitchell Plateau, 736
Mites, 256, 361–362, 361f, 793–795
Mixed carbonate/sulfate strata dissolution, 759
Mixing-zone, 157
　karst, 437
Mixotrophs, 126–127, 490–491
Mlynki Cave, 829t
MMS. *See* Superficial underground environment

Model organisms
　Astyanax mexicanus, 37
　evolution and adaptation and, 37
Modern synthetic theory of evolution, 554
Mollusca
　biogeographic patterns, 514t, 515
　　Dinaric Mountains, 515, 515f, 516f
　　United States, 514t, 515
　Europe, 221, 224f
　overview about, 512–513
　Paraegopis oberwimmeri, 512–513
　Pisidium, 513–514
　prosobranchs, 514–515
　pulmonates, 514–515
　selected species, 515–517
　　Congeria kusceri, 512–513, 517
　　Fontigens tartarea, 515–517
　taxonomic patterns, 513–515, 513t
　tropical diversity patterns, 245t
　United States diversity patterns, 255
　Zospeum, 513
Monolistra monstruosa, 231f
Monte Baldo dolines, 147
Montello dolines, 145–147, 146f, 148f
Monti Lessini dolines, 144f, 145
Morocco, 483–484
Morphine Waterfalls, Barberry Cave, 80f
Morphological adaptations
　in Amblyopsidae, 522–523, 522f, 525, 525f
　analysis examples, 525–527
　　Amblyopsidae, 522f, 525, 525f
　　Pseudosinella hirsuta, 526, 526f
　in Collembola, 523, 523f
　conclusions about, 527
　in crayfish, 521, 522f
　environmental factors effects, 521–522
　evidence, 522–523
　exaptations or preadaptations and, 524
　MMS and, 524–525
　occurrence of, 523–525
　overview about, 517
　progressive, 521
　　factors needed for, 521
　regressive evolution hypotheses, 519–521
　troglomorphy, 517–518
　　common characteristics of, 518–519, 520t
　　convergence and, 518, 518f, 519f
　　evolution of, 518–519
Mosses, 453
Moths, 15–16, 663
MSS. *See* Milieu souterrain superficiel
Muchimuk-Colibri system, 635–636, 636f
Multilevel caves
　landscape interpretation using, 529–531
　　Cumberland River incision pulses, 530, 530f
　　eastern Alps, Austria, young uplift, 531
　　Mammoth Cave, 529–530, 530f
　　Sierra Nevada incision rates, 531
　　southern Rocky Mountains erosion rates, 531
　overview about, 528
　representation of, 528f

　river incision and cave record, 529
　sediments and, 529, 529f
　water table and, 528–529
Mulu caves, Malaysia
　bird's nest soup and, 532
　exploration, 532
　Gunung Api, 532–535
　　Clearwater River Cave, 532–534
　　Clearwater River Passage, 534, 535f
　　Cobra-Bridge-Cloud Cave, 534
　　Sarawak Chamber, 534
　Gunung Benarat, 535–536
　　Benarat Caverns System, 536
　　Melinau Gorge, 535–536, 535f
　　Terikan River Caves, 536
　Gunung Buda, 536–537
　　Green Cathedral Cave, 536
　　Tardis Cave, 536f
　Gunung Mulu National Park, 537
　　cave biology, 537
　　Gunung Api, 537
　　plant/animal species, 537
　Melinau Limestone and, 532
　overview, 531–532, 533f
　Pinnacles of Mulu, 426, 426f
　southern hills, 532
　　Deer Cave, 532, 534f
　　Garden of Eden, 532
Munidopsis polymorpha
　aggressive behavior of, 66–67, 67f
　reproductive behavior of, 63, 64f
Mushkarova Jama Cave, 829t
Mutualism, 743, 744f
MVT. *See* Mississippi Valley-type ore deposits
Mygalomorphae, 789
Mysmenidae, 790
Myriapoda
　in Australia, 216
　　Chilopoda, 216
　　Diplopoda, 216
　Centipedes, 539–540
　　Craterstigmatomorpha, 539
　　Geophilomorpha, 539
　　Lithobiomorpha, 539–540, 539f
　　overview, 539
　　Scolopendromorpha, 539
　　Scutigeromorpha, 539
　Millipedes, 540–541
　　Chordeumatida, 540t, 541
　　Julida, 540, 540t
　　map of, 540f
　　overview, 540
　　Polydesmida, 540, 540t
　　Pseudotremia, 541, 541f
　　Scoterpes, 541
　overview about, 538, 538f
　Pauropods, 538
　Symphylans, 538
　tropical diversity patterns, 245t
Mysida, 21–22, 178t, 185, 186t
Mysmenidae, 790
Mystacocarida, 178t, 180–181
Mystery Cave, 742
Myth
　Amarnath Cave and, 322

Bible and, 322
Chinese, 322
creatures of, 321
Crete and, 321
ghost stories and, 323
Japanese, 322–323
Makua Cave and, 322
Maya, 322
Sibyl's Cave and, 321–322
Underworld in, 321
Vjetrenica and, 862

N

NAPLs. See Nonaqueous phase liquids
Natural selection. See also Species interactions
  Asellus aquaticus and, 545
  cave regime, 546–547
  in caves, 544–546
  colonization and speciation role, 547–548
  conclusions and prospects, 548–549
  Darwin and, 543–544
  Gammarus minus and, 544–545
  historical development, 543–544
  modes of, 545, 545f
  NMT and, 544
  overview about, 543
  phenotypic clines and, 545–546, 546f
  population structure and, 615–617
  Racovitza and, 543–544
  time and, 545
Natural traps, 591
Neaphaenops tellkampfi, 744–746
Needles, gypsum, 374–375, 375f
Neff Canyon Cave, 85
Negative allometry hypothesis, 520
Nemastomatidae, 796
Nematoda, 911, 911f
Nemertea, 911, 912f
Neobisiidae, 792
Neogoveidae, 796
Neoteny, 5–6
Nesticidae, 790
Nesticus eremita, 610–611, 610f
Network caves, 395
Network maze, 602f, 761, 762
Neuromasts, 553
Neutral mutation theory (NMT), 544, 554
Neutral mutations
  Asellus, 554
  Astyanax fasciatus
    albinism and, 550
    dark melanin and, 550
    deleterious risk hypothesis and, 552–553
    energy economy hypothesis and, 553
    eyes and, 550, 552f, 554
    guanine and, 550
    neuromasts and, 553
    overview, 549–550
    pleiotropy hypothesis and, 553
    surface and cave forms regarding, 550
  modern synthetic theory of evolution, 554
  NMT and, 554

reduction and, 549
selection *vs.*, 549
summary, 554–555
Ngamarlanguia luisae, 206f
Nicholson, David, 76
Nicholson, Ike, 74, 75
Niphargidae, 221
Niphargus
  anchihaline habitats and, 22, 22f
  conclusions about, 562–564
  distribution, 555
  interactions within and between species
    cannibalism, 559
    chemoautotrophic bacteria symbiosis, 560
    predation and competition, 559–560, 560t
  key elements comparative review, 560–562
    communities: niche separation, 562, 563f
    female body size, 562
    morphological diversity and stasis, 560–561, 561f
  key properties, 556–559
    abiotic factors tolerance, 556–559
    allometric growth, 556, 558f
    antennae and pereopods, 556, 557f
    body shape, 556, 556f, 557f
    body size, 556, 557f
    developmental patterns, 556, 558f
    endemism/poor migratory abilities, 559
    gnathopods, 556, 557f, 558f
    light and, 557
    maxillae I, 556, 558f
    morphological variation, 556, 557f
    response to hypoxia, 557–558
    sexual dimorphism, 556, 558f
    temperatures, 558–559
    water chemistry tolerance, 559
  N. orcinus, 231f
  N. rhenorhodanensis, 653, 654t
  N. virei, 613, 653, 654t
  overview about, 555–556
  taxonomy challenges of, 555
Nitrate contamination, in Karst groundwater
  background information, 564–565
  environmental concerns and nitrogen cycling, 565
  groundwater resource protection and, 567
  hydrogeologic factors, 566
  nitrate loss processes, 567
  springs and, 566–567
  temporal variations and, 567
Nitrates, 506
Nitrites, 129
Nitrogen cycle
  chemoautotrophy and, 132–133
  environmental concerns and, 565
NMT. See Neutral mutation theory
Nocticolidae cockroach, 241, 242f
Nodules, 779, 780f
Nonadaptive hypothesis, 341–342

Nonaqueous phase liquids (NAPLs). *See also* Heavy nonaqueous phase liquids; Light nonaqueous phase liquids
  characteristics and sources, 166–168, 167f, 167t
  conclusions about, 172
  density and, 166, 167f, 167t
  detection, 171–172
    in soils, 172
    in springs and caves, 171
    in wells, 171–172
  distribution of, 168f
  overview about, 166
  solubility and, 166, 167f, 167t
  storage in aquifers, 171
    fractures, 171
    matrix and vugs, 171
    pools in conduits, 171
    sediment, 171
  transport into aquifers, 168–171
    climate and, 169, 170
    epikarst and, 169–170
    fractures and, 170–171
    nonwetting fluid and, 169
    open drains and, 170–171
    sinkhole formation and, 170
    sinking streams and, 170–171
    soil and regolith pores and, 169
    volume and rate of release and, 168–169
  types of, 166, 168
  vapor pressure and, 166–167, 167f, 167t
  viscosity and, 167–168
Northern salamanders, 670
Notch, 425
Notostigmata, 793
Novice cavers, 645
Nullarbor Caves, Australia
  arid environments and, 568
  biology, 575
  blowholes of, 570
  chambers of, 570, 570f
  climate and, 570–571
  conclusions about, 575–576
  dolines of, 570, 570f
  geology, 568–571, 569f
  history, 571–572
    early, 571–572
    exploration, 572
  icicles in, 575, 575f
  lakes of, 570, 571f
  minerals, 574–575
    calcite, 574
    gypsum, 574, 574f
    halite, 574, 574f
  Old Homestead Cave, 574
  origin and enlargement of, 572–574, 573f
  overview, 568
  physical setting, 568–571, 569f
  water flow and, 571

O

Occasional visitors, 275
Ochyroceratidae, 790
Offshore springs, 799–800

Oklahoma salamander, 671
Old Homestead Cave, 574
Oliarus polyphemus, 14, 14f
Oligochaeta, 254−255, 912−913, 913f
Oligotrophic terrestrial, 241−242
Oniscidea, 178t, 189t, 190, 213
Onychophora, 914, 914f
Onychostoma macrolepis, 107, 109t
Oonopidae, 790
Opal, 501t, 505
Open karst, 365, 366−367
Ophisternon candidum, 206f
Opiliones, 215, 256, 795−797, 796f
Oppiidae, 795
Optymistychna Cave, 829t, 830, 831f
Ordinskaya Cave, 372, 372f
Organ Cave, 101
Organic acids, 758
Organic minerals, 507
Organicism, 520
Orthogenesis, 520
Orthogonal vector plot, 586−587, 588f
Orthoptera, 217, 259
Ortobalagan Valley, 443, 445f
    combined cave profile, 446f
    major caves in, 445f
Osteichthyes, 260
Ostracoda, 178t, 182−183, 182f
    anchihaline habitats and, 21
    in Australia, 211−212
Overflow springs, 800
Ox Bel Ha, 837
Oxides, 501t, 505
Oxygen, responses to low
    anoxia survival times, 653, 654t
    behavioral responses, 654
    Crustacea, 653
    groundwater and
        DO small-scale heterogeneity in, 652−653, 652f
        DO supply and consumption in, 651−652
        survival evidence in, 653
    metabolic responses, 655−656
        during anoxia, 655−656
        during postanoxic recovery, 656
    overview, 651
    respiratory responses, 655
Ozark Plateaus, 734f, 737−738, 738f
Ozerna Cave, 829t

P

Pack rats, 331t, 332
Packs, cave, 306−307
Pairwise species interactions, 743, 744f
Paleoclimate records, from speleothems
    dating techniques, 577−578
    global keeping of, 577, 579f
    overview, 577
    paleoclimatology contributions, 578−585
        climatic changes timing, 584
        Devils Hole, 579−580, 580f
        early studies: paleotemperature, 578−579
        last glacial, 580−581, 580f
        late Holocene, 582, 584f
        other studies, 584−585
        paleotempestology, 584
        solar variability and climate change, 581, 583f
        tracking insolation, 581, 582f
        tracking paleomonsoons, 581, 582f
        vegetation and soil dynamics, 583−584, 584f
    stalagmites use for, 577, 578f
Paleofill, 414−415
Paleokarst, 764
Paleomagnetic records
    clastic cave sediments and, 586−588, 587f
    CRM and, 588
    DMR and, 586
    earth magnetism and, 586
    interpretation, 587−588
    local/global column results and, 588, 589f
    magnetostratigraphic time scale and, 586, 587f
    orthogonal vector plot and, 586−587, 588f
    overview, 585−586
    speleothems and, 588
    value of, 588−589
Paleomonsoons, 581, 582f
Paleontology of caves
    deposit types and examples
        alluvial and colluvial transport, 591
        bats, 592
        bears, 593−594
        cats, 595, 595f
        hyena, 594
        Jefferson's ground sloth, 596, 597f
        large carnivorans, 592−595
        large herbivores, 596
        natural traps, 591
        other vertebrate sites, 596−597
        peccary, 596−597
        raptors, 591−592
        rodents, 595−596
        small carnivores, 596
        soft tissue preservation, 596, 596f
        wolves, 594−595
    dung and, 590, 590f
    fossil vertebrates importance, 597−598
    Lechuguilla Cave and, 460
    overview about, 590−591
    PPCS and, 623
    Quaternary and, 590
    research focus, 590−591
    screenwashing and, 597
    Vjetrenica and, 862
    Wakulla Spring underwater cave system and, 875−876, 876f
Paleotempestology, 584
Palpigradi, 214, 791
Panbiogeography, 850
Pans, 427−428, 428f
Paraegopis oberwimmeri, 512−513
Paragenesis, 422, 607
Paragenetic ceiling tube, 423, 424f
Paragenetic pendants, 422, 423f
Parasitidae, 793
Paroster species, 613−614
Particulate organic matter (POC), 289
Partings, 871
PASCALIS (Protocol for the Assessment and Conservation of Aquatic Life in the Subsurface), 220
Passage growth and development
    cave degeneration, 603
    cave patterns and, 599−603
        anastomotic maze, 602f
        branching systems origin, 601
        changing conditions, 602−603
        curvilinear branchwork, 602f
        initial flow routes competition, 599−600, 600f
        maze development, 601−602, 602f
        network maze, 602f
        ramiform, 602f
        rectilinear branchwork, 602f
        spongework maze, 602f
        summary, 602f
        vertical organization, 600−601, 601f
    conclusion, 603
    overview, 598
    stages, 598−599
        earliest, 599
Passages
    angular passages, 604−605, 604f
    below water table and, 605−606, 606f
    braided, 440−441, 441f
    canyon-like, 439−440, 440f, 734, 735f
    conclusions about, 608
    curvilinear, 604−605, 604f
    definitions and concepts, 603−604
    discharge, 604
    fissure networks, 604f, 605
    hypogenic, 604f, 606
    incised, 606f
    Jewel Cave, 413
    Lechuguilla Cave, 458, 459f, 461, 462f
    long inclined, 729−732
        Cassowary System, 731−732, 732f
        Jean-Bernard Cave, 731, 731f
    looping, 724, 724f
    Mammoth Cave System, 471−472, 471f, 472f
    morphology and evolution of, 606−608, 606f, 607f
    overview, 603
    patterns of, 604−606, 604f
    phreatic, 72, 604, 605, 605f, 606f
    stairstep development, 604, 605f
    steeply-inclined, 761, 762
    tapoff, 607−608, 607f
    tributary, 604
    types of, 604
    Ukraine giant gypsum caves, 831
    vadose, 604, 604f, 605, 606f
    virgin, 277
Passive detectors
    water tracing and
        activated carbon, 890
        cotton, 891
        plankton netting, 891
        Quinlan Gumdrop, 891, 891f
Passive dispersal models
    of invasion, 406−407
        erosion/surface water flow model, 406−407

rafting model, 406
regression model, 406, 407f
surface water infiltration model, 406
Patchy food scarcity, 1
Pauropods, 538
Pavan, T., 276–277
Pavements, 434
Peccary, 596–597
Pendants, 422, 423f
Pennyroyal Plateau, 736
Peracarida
  Amphipoda, 178t, 186, 187t
  Bochusacea, 178t, 186
  Cumacea, 190, 191t
  Isopoda, 178t, 187–188, 189t
    Anthuridea, 178t, 188, 189t
    Asellota, 178t, 188–190, 189t
    Calabozoidea, 178t, 189t, 190
    Flabellifera, 178t, 188, 189t
    Microcerberidea, 178t, 188, 189t
    Oniscidea, 178t, 189t, 190, 213
    Phreatoicidea, 178t, 188, 189t
  Mictacea, 178t, 185–186, 186f
  Mysida, 178t, 185, 186t
  Spelaeogriphacea, 178t, 184–185, 185f
  Tanaidacea, 190, 190t
  Thermosbaenacea, 178t, 185, 185t
Peracarida, 178t, 184–191
Periodic food supply, 1
Periodic starvation, 8–9
  back to network and, 9
  energy demand and, 9
  fat accumulation and, 8
  overview, 8
Pettalidae, 796
Phalangodidae, 796–797
Pholcidae, 790–791
Pholeuon, 60, 61f
Phosphates, 504–505
  brushite, 501t, 504–505
  development of, 504, 504f
  hydroxylapatite, 501t, 504–505
  overview, 501t, 504
  taranakite, 501t, 504–505
Photokarren, 419–420, 420f
Phreatic looping, Castleguard Cave, 90, 92f
Phreatic maze caves, 395–396
Phreatic passages, 72, 604, 605, 605f, 606f
Phreatic water, 157
Phreatoicidea, 178t, 188, 189t
Phreatoicoides gracilis, 206f
Phrynichidae, 789
Phrynidae, 789
Phyllocarida, 178t, 183, 183f
Phylogeography, 610
Pinnacle, 428–429
  karst, 430
Pinnacles of Mulu, 426, 426f
Pisces, 245t
Pisidium, 513–514
Pit craters, 865
Pits, 427–428
Pivka intermittent lakes, 899–901, 901f, 902f
Pivka River, 618, 618f
Planinsko polje, 899, 900f
Plankton netting passive detector, 891

Planthoppers, 216, 663
  cixiid, 13–14, 13f
Platycopioida, 178t, 181
Pleiotropy hypothesis, 553
Plethodontidae, 667–674
  Eurycea, 667–668
  Gyrinophilus, 672–674, 673f
  Haideotriton wallacei, 671–672
Plunge pools, 441–442, 442f
POC. See Particulate organic matter
Poecilia mexicana
  aggressive behavior, 67–68
  reproductive behavior, 64, 65f, 67–68
Point patterns, 475
Point-recharge doline, 142–144, 143f, 145
Poljes, 140, 153–155
  Cerknica, 433, 434f
  Cerkniško, 898–899, 898f, 899f
  classification, 154
  defined, 153, 435
  Dinaric karst and, 153, 199–200
  lakes and, 153, 153f
  lithological contact, 154, 154f
  Planinsko, 899, 900f
Pollution, 237–238, 327, 403
Polychaeta, 911–912, 912f
Polydesmida, 540, 540t
Polygenetic sink, 152–153
Pool spar, 783, 783f
  Helictite Cave, 81f
Population flush, 10–11
Population structure
  in aquatic troglobites
    Gammarus minus, 612
    historical determinants and
      secondary contact, 614
    hydrological routes vs. adaptive
      diversity, 612–614
    Niphargus virei, 613
    Paroster species, 613–614
    Troglocaris anophthalmus, 613
  cave crickets heterozygosity and,
    615–617, 616t
  dispersal routes
    surface vs. underground, 612
  food resource heterogeneity and, 615
  genetic structure estimation, 609–610
    effective population size, 609
    F-statistics, 609
    genetic distance, 609
    heterozygosity, 609
    migration, 609
    phylogeography, 610
  genetic structure estimation and, 609–610
  genetic variability and natural selection,
    615–617
  habitat heterogeneity and, 614–615
  overview, 608–609
  at regional scale, 610–612
    gene flow study cases, 610–611, 610f
    vicariant species, 610
  troglobitic heterozygosity comparisons,
    615–617, 616t
Porous media aquifers, 383–384
Postojna-Planina Cave System (PPCS)
  Asellus aquaticus and, 32–33

as biotically diverse, 236
description, 623
entrances, 623
exploration history, 622
exstensometers used in, 620–621
Lepe Jame in, 622, 622f
physical setting, 618–622
  faults, 618–619, 619f
  geologic, 618–621, 619f, 620f
  hydrology and topography, 618
  lithological columns, 619–620, 620f
  other research, 621
  Pivka River, 618, 618f
special attributes, 623–624
  biology, 623
  mineralogy, 623
  paleontology/archaeology, 623
tourist railway, 622, 622f
Postojnska Jama, Slovenia, 452f
Potassium nitrate, 676. See also Saltpetre
  mining
Potholes
  dissolutional, 420, 422f
  lateral, 682f, 683
Poulson, Thomas, 465
PPCS. See Postojna-Planina Cave System
Preadaptation, 10, 295–296, 524
Predation, 743, 744f
Pressure ridge caves, 872
Proboscis worms, 911, 912f
Prodidomidae, 791
Progressive adaptations, 517
Project caving, 308
Project SIMMER, 86
Prokoeneniidae, 791
Prosobranchs, 514–515
Prostigmata, 794
Protection. See Conservation
Proteidae, 665–667
  Proteus anguinus, 666–667, 666f
Proteobacteria, 495–496
Proterorhagiidae, 794
Proteus anguinus, 219, 230, 653, 666–667, 666f
  aggressive behavior, 67–68, 67f
  in Dinaric karst, 234
  feeding behavior, 63, 64f
  reproductive behavior, 64–65
Protocobitis, 107
Protocol for the Assessment and
  Conservation of Aquatic Life in the
  Subsurface. See PASCALIS
Protoschizomidae, 789
Pselaphidae, 232
Pseudochiridiidae, 792
Pseudogarypidae, 792
Pseudokarst caves, 155
Pseudoscorpiones, 791–792, 791f
  in Australia, 215
  in Dinaric karst, 232
  guano and, 362
  in United States, 256–257
Pseudosinella, 518, 518f
Pseudosinella hirsuta, 526, 526f
Pseudotremia, 541, 541f
Puka, 868–869
Pulmonates, 514–515

Pygolabis humphreysi, 206f
Pyroducts, 866–870, 866t
  canyon formation, 868–869, 870f
  double-trunked, 869–870
  formation of, 867, 868f
  Huehue Cave and, 869f
  Huehue Flow and, 869–870, 871f
  inflation and, 867, 868f
  internal development, 867–869, 870f
  puka and, 868–869
  single-trunked, 869
  superimposed-trunked, 870
  types of, 869–870

Q

Qafzeh Cave, 95f, 98f
Quartz, 501t, 505
Quartzite caves of South America
  deposits and ages, 639
  genesis and development controls, 638–639
  Inficionado Peak, 637, 637f
  lithology and occurrence, 635–637
  morphology and, 637–638
    anastomotic caves, 637, 638f
    branchwork caves, 637, 638f
    vertical network caves, 637, 638f
  Muchimuk-Colibri system, 635–636, 636f
  overview, 635
  Roraima Sur, 635–636, 636f
Quaternary, 590
Queen Formation, 457–458, 457f
Quinlan Gumdrop, 891, 891f
Quintana Roo State, 834

R

Raccoons, 331t, 332
Racovitza, E., 219–220, 275–276, 277, 404–405, 543–544
Radio detection and ranging (RADAR), 350
Radiolocation, 351–352, 351f
Rafting model, 406
Rain pits, 140
Ramiform pattern, 602f
Rappel, 307
Raptors, 591–592
Ravelling, 720
Rebelays, 307
Rebreathers, 312, 313f
  CCRs, 313
  SCRs, 313
Records and, 270
Recreational caving
  activities not considered
    abandoned buildings, 643
    overview, 642–644
    shelter caves, 643
    show caves, 643
  cave types used for, 641–642
    glacier caves, 642
    lava caves, 641
    mines and tunnels, 642
    sea caves, 642
    solution caves, 641
    talus and tectonic caves, 642
  contemporary trends, 647
  definition, 641
  demographic factors, 646–647
    age, 646
    frequency cycles, 647
    length of trips, 647
    race, 647
    sex, 646–647
  discouraged items for, 645
  downside, 644–645
  factors contributing to, 644
    appeal, 644
    distance to cave, 643–644
    group caving suitability, 644
    known existence, 643
    legal access, 644
    physical access, 643
  levels for everyone, 646
  novice cavers and, 645
  other voids and, 642
  people avoiding, 645–646
  people continuing, 646
  reasons people begin, 644
    challenge/exercise, 644
    exploration/adventure, 644
    social aspect, 644
    unique accomplishment, 644
  spelunker regarding, 645
  summary, 647–648
  urban, 642
Rectilinear branchwork, 602f
Redox, 162–163
Regional cavernicoles, 845
Regression model, 406, 407f
Regressive adaptations, 517
Regressive evolution hypotheses, 519–521
Regular cavernicoles, 845–846
Remipedia, 21, 178t, 179–180, 180f, 180t
Reports, 270
Reptilia, 260–261
Rescues
  background about, 648
  Collins and, 648
  history of, 648–649
    early, 649
  HURT and, 649
  learning from past, 650
    accident frequency, 1950s, 650
    accident frequency, 1960s, 650
    accident frequency, 1970s, 650
    accident frequency, 1980s, 650
  overview, 648
  present and, 650–651
  team development, 649
Réseau Siebenhengste-Hohgant system, 700–702, 726–729, 728f. *See also* Siebenhengste cave system
Restoration, 632
Rhagidiidae, 794
Ribbon worms, 911, 912f
Ricinoididae, 793
Ricinulei, 793, 793f
The Rift, 463
Rigging, vertical exploration, 315–316, 315f
Rillenkarren, 428, 428f
Rills, 420, 421f, 425
Rimstone dams, 779, 780f
Robins Rift, 78
Rock shelters, 105f, 106
Rocky Mountains erosion rates, 531
Rodents, 595–596
Roofless caves, 201–202, 202f
Roosts functions, bat, 46–48
Root communities, in lava tubes
  cave zones, 661–663
    bad air zone, 662–663
    deep zone, 662
    entrance zones, 662
    transition/mixing zone, 662
    tree roots and, 661–662
    twilight zone, 662
  conservation of, 663
  environmental factors, 658–661
    air flow, 659
    energy sources, 658–659
    geomorphology influence, 661
    moisture/humidity, 659
    oxygen and carbon dioxide, 659–661
  food web, 660f
  in Hawaii and Australia, 661t
  overview, 658
  species morphology and adaptation, 663
    crickets and cockroaches, 663
    moths, 663
    planthoppers, 663
Roraima Sur, 635–636, 636f
Rouch, Raymond, 99
Roundworms, 911, 911f
Rousette fruit bats, 46
Ruiniform karrenfield, 430
Runnels, 428, 429f

S

Sabaconidae, 796
Sac Actun, 837–838
Safety, 309
Salamanders, 260, 261t. *See also specific salamander*
  blepsimolge, 668f, 669–670
  conclusions about, 675
  diversity patterns of, 674–675
  overview, 665
  Plethodontidae, 667–674
    Eurycea, 667–668
    Gyrinophilus, 672–674, 673f
    Haideotriton wallacei, 671–672
  Proteidae, 665–667
    Proteus anguinus, 666–667, 666f
  septentriomolge, 670
  typhlomolge, 668–669, 668f
Saltpetre mining, 676–678
  bacteria and, 676
  definitions and importance, 676
  efflorescent crusts and, 676
  evidence of, 677
    principle, 677–678
    secondary, 678
  gunpowder and, 676
  history, 677

other sediment extraction contrasted with, 679
potassium nitrate and, 676
rockshelters and, 676
saltpetre processing, 678–679
evidence, 678
outside, 678
steps, 678
summary, 679
white lint-like fibers and, 676, 677f
San Gabriel Springs salamander, 670
San Marcos salamander, 670
San Salvador Island karren, 426f
Santa Ninfa Plateau dolines, 145
Sarawak Chamber, 534
Scallops, 420–422, 422f
defined, 679
discharge calculation, 681
discharge data uses and pitfalls, 681–682, 681f, 682f
groundwater flow
direction and, 680, 680f
velocities and, 680–681, 680f
other phenomena, 682–683, 682f
pockets, 682–683
potholes, 682f, 683
overview about, 679
scour marks and, 422
transverse flutes and, 422
Scàrisoara Ice Cave, 403
Scheloribatidae, 795
Schiner, J. R., 275, 276, 277
Schiodte, J. C., 275
Schizomida, 215, 257, 788–789
Schmidl, Adolf, 103
Schoharie Valley, 741, 741f
Schwartz, Ben, 76
Scolopendromorpha, 539
Scorpiones, 257. *See also specific Family*
in Australia, 215
Families, 787–788, 788f
Scoterpes, 541
Scour marks, 422
SCR. *See* Semi-closed circuit rebreather
Screenwashing, 597
Scutariellidae, 234
Scutigeromorpha, 539
Sea caves, 105f, 106
as coastal caves, 155–156
human interaction with, 156
last interglacial and, 156
modern, 156
overview about, 155–156
recreational caving and, 642
ubiquity of, 156
variations of, 156, 156f
Sea level, 155
flank margin caves and, 159
Security systems, 631
Sediments, Castleguard Cave, 93
Seepage spring, 684–685
characteristics, 684, 684f
chemical and physical conditions, 684–685, 685f
species found in, 685, 685t

temperature and, 684–685, 685f, 685t
Seeps, 800–801
Seismic modeling techniques, 350
Semi-closed circuit rebreather (SCR), 313
Septentriomolge, 670
Serpulidae, 234
Sessile chaetognaths, 20
Seven Rivers Formation, 457–458, 457f
Shaft and canyon caves, 729
Krubera Cave, 729, 730f
Shallow subterranean habitats (SSHs)
epikarst, 685t, 686, 686f
fauna origin, 689
hyporheic zone, 686–688, 687f, 688t
MSS, 686f, 688, 688f
other subterranean habitats and, 689
overview, 683–684
seepage spring, 684–685
Shelta Cave, 625
Shelter caves, 643
Shields, 781, 782f
Shield's Diagram, 138f, 139
Shifflett, Tom, 76–77
Short-tailed whipscorpions, 788–789
Show caves, 643
budget estimate of, 692t
Cango caves, 695f
development guidelines, 693–697
aboveground facilities, 696–697
access and pathways, 693–694
guide training, 697
lighting, 694–695
managers, 697
monitoring, 697
visitor frequency and numbers, 695–696
of wild cave into show cave, 693
development of, 691–693
environmental protection of, 690–691
Frasassi caves, 696f
history, 690
Hwanseon Cave, 696f
Kartchner Caverns, 694f
obtaining information about, 697
Sibyl's Cave, 321–322
Side-mount diving, 311, 312f, 313f
Siebenhengste cave system
detailed map of, 701f
fauna discovered in, 706–707
genesis and age, 703–704, 704f, 705f
geographical and geological setting, 698–699, 698f, 699f
hydrology, 698f, 699
minerals, 704–706, 706f
overview and exploration, 699–703, 699f, 700f
A2, 703
Bärenschacht, 699–700
F1, 702
Faustloch, 702
K2, 702–703
Réseau Siebenhengste-Hohgant, 700–702
St. Beatus Cave, 699
Sierra Nevada incision rates, 531

Sighted salamanders, 668f, 669–670
Silicates, 501t, 505
allophane, 501t, 505
opal, 501t, 505
origins, 505
other, 505
quartz, 501t, 505
Similarity rule, 775–776, 775f
SIMMER, Project, 86
Simmons, Ron, 76
Simmons-Mingo Cave, 86
Sinella, 518, 518f
Single fracture, 510, 510f
Single or rudimentary networks, 761–762
Single-conduit caves, 394
Single-rope techniques, 307
Single-trunked lava tunnels, 869
Sinkholes, 140
building damage by, 720–721, 720f, 721f
causes of, 718, 719f
collapse, 150–151, 150f, 151f
cover, 150f, 152
defined, 434–435, 723
Dinaric karst, 199–200
entrances, 281, 282f
evolution of, 723
examples of, 718
formation, 718, 719f
NAPLs and, 170
intersection, 150f, 152
Mitchell Plain and, 722
other types of, 150–152, 150f
overview about, 718
ponding or damming and, 721
reservoirs and, 721
soil piping and, 720
solution, 718, 719f
conclusions about, 150
defined, 140
drainage and, 140
drawdown, 143f, 144
evolution examples, 144f, 145–147
Eyre Peninsula, 147
filling material evacuation and, 145
frost shattering and, 145
inception, 143f, 144
as karst surface diagnostic forms, 140–150
locations, 141
midlatitude, 149
Monte Baldo, 147
Montello, 145–147, 146f, 148f
Monti Lessini, 144f, 145
morphology and size, 140–141
other evolution processes, 145–147
overview about, 140
point-recharge, 142–144, 143f, 145
Santa Ninfa Plateau, 145
slope processes and, 145
soil and, 145
specific environmental conditions and, 147–149
structure, 141–142, 142f
subsoil, 141

Sinkholes (*Continued*)
  tropical landscape and, 147–149, 149f
  Velebit Mountains, 147
  Waitomo district, 145
 subsidence, 150f, 151–152
 suffosion dolines and, 722
 types of, 718, 719f
  buried sinkhole, 719f, 722
  cover-collapse sinkhole, 719f
  cover-subsidence sinkhole, 719f, 721–722, 721f
  subsidence sinkhole, 718–719, 719f
Sinking streams, 707–709
 cave evolution and, 707, 708f
 Danube River and, 707
 Dinaric karst and, 237
 Edwards Aquifer and, 708–709
 entrances and, 281
 evolutionary sequence of, 710–712
  case A, 711, 711f
  case B, 711, 711f
  case C, 711, 711f
 Florida, 708
 function of, 709–710
 Kentucky, 709, 709f, 710f
 NAPLs and, 170–171
 overview, 707
Sinocyclocheilus, 108
 adipose storage, 117, 118f, 119f
 biogeography, 121–124
 eye degeneration, 117–120, 119f, 121f
 humpback and horn of, 114–116, 114f, 115f
 phylogenesis, 121–124, 123f
 pigmentation loss, 108f, 120
 reproduction, 108–112
 research and conservation of, 124–125
 S. hyalinus, 107, 108f, 109t
 scale disappearance, 120–121
 sensory apparatus, 113f, 116–117, 116f
 specialized appendages, 108f, 117
 speciation mechanisms, 121–124
Sironidae, 796
Sistema del Faro flank margin cave, 159f
Sistema Huautla, Mexico
 Anthodite Hall, 715f
 archaeology, 717–718
 biology, 717
 cave description, 712–716
 exploration, 714f
 geology, 716
 hydrogeology, 713f, 714f, 717
 location, 713f, 716
 overview, 712
 speleothems, 715f, 716f
 stream network, 713f
Sistema K'oox baal, 836f
Sistema Ox Bel Ha, 837
Sistema Sac Actun, 837–838
Sit harness, 316–317, 316f
Sket, B., 276
Skiles, Wes, 312
Škocjanske, 433, 434f
Slab breakdown, 69

Slavka Cave, 829t
Slickenslides Room, Helictite Cave, 82f
Sloth, Jefferson's ground, 596, 597f
Slovenia
 dolines in, 435f
 Postojnska Jama, 452f
Snakes, 218
Soda straw, 807
Soil piping, 720
Solution caves, 105f, 106–107
 alpinotype, 724
 Chester and Crawford Uplands, 736
 in high relief regions
  Cassowary System, 731–732, 732f
  cave examples, 726
  characteristics of, 724
  climate and, 724
  conclusions about, 732–733
  glacier milk and, 724
  Hölloch Cave, 726, 727f
  Jean-Bernard Cave, 731, 731f
  Krubera Cave, 729, 730f
  large labyrinths, 726–729
  long inclined passages, 729–732
  looping passages, 724, 724f
  overview about, 723–724
  relative chronology and, 725–726
  relative chronology example, 725–726, 725f, 726t
  Réseau Siebenhengste-Hohgant system, 726–729, 728f
  shaft and canyon caves, 729
  St. Beatus Cave, 725–726, 725f
  study of, 725
  water temperature and, 724
 in moderate relief regions
  Appalachian Mountains, 734f, 739–740, 740f, 741f
  Appalachian Plateaus, 734f, 738–739, 739f
  canyon-like passages, 734, 735f
  characteristics of, 733
  dissected plateaus and, 736–737
  drainage history and, 737
  field examples, 735–742
  geographic distribution of, 734, 734f
  glaciated plateaus, 740–742, 741f
  Illinois structural basin and, 734f, 736
  Mammoth Cave and, 737, 737f
  northeastern/northcentral U.S., 740–742, 741f
  origins of, 734–735
  overview about, 733
  Ozark Plateaus, 734f, 737–738, 738f
  Pennyroyal and Mitchell Plateau, 736
  sediment accumulation and, 737
  southern Indiana and, 736–737
  summary, 742–743
  tubular conduits, 734–735, 735f
  western Kentucky and, 736–737
 recreational caving and, 641
Solution dolines
 conclusions about, 150
 defined, 140
 drainage and, 140

 drawdown, 143f, 144
 evolution examples, 144f, 145–147
 Eyre Peninsula, 147
 filling material evacuation and, 145
 frost shattering and, 145
 inception, 143f, 144
 as karst surface diagnostic forms, 140–150
 locations, 141
 midlatitude, 149
 Monte Baldo, 147
 Montello, 145–147, 146f, 148f
 Monti Lessini, 144f, 145
 morphology and size, 140–141
 other evolution processes, 145–147
 overview about, 140
 point-recharge, 142–144, 143f, 145
 Santa Ninfa Plateau, 145
 slope processes and, 145
 soil and, 145
 specific environmental conditions and, 147–149
 structure, 141–142, 142f
 subsoil, 141
 tropical landscape and, 147–149, 149f
 Velebit Mountains, 147
 Waitomo district, 145
Solution pockets, 423, 424f
Solution runnels, 428, 429f
Solution sinkhole, 718, 719f
Solutions pans, 140
South America, quartzite caves of
 deposits and ages, 639
 genesis and development controls, 638–639
 Inficionado Peak, 637, 637f
 lithology and occurrence, 635–637
 morphology and, 637–638
  anastomotic caves, 637, 638f
  branchwork caves, 637, 638f
  vertical network caves, 637, 638f
 Muchimuk-Colibri system, 635–636, 636f
 overview, 635
 Roraima Sur, 635–636, 636f
Sparassidae spiders, 242, 243f
Spatial pattern investigation, 476–477
Spatial process decomposition, 475–476
Species interactions
 Appalachian stream competition
  amphipod and isopod, 746–747, 746f, 747f
 beetle predators
  cricket egg prey of, 744–746
 conclusion, 748
 eutrophication and competition and, 747–748
 overview about, 743–744
 pairwise, 743, 744f
Spelaeogriphacea, 178t, 184–185, 185f, 213
Speleogenesis, hypogenic, 756–764
 artesian, 759
 basic concept, 750
 chemical mechanics, 756–759, 756f, 757f
  carbonic acid dissolution, 756–758
  evaporites dissolution, 758
  hydrogen sulfide, 758

mixed carbonate/sulfate strata dissolution, 759
organic acids, 758
sulfuric acid, 758
thermal water cooling, 758
water mixing, 758
combined artesian/deep-rooted, 759
deep-rooted, 759
defined, 750
diagenetic environments, 750–756
eogenesis and, 751
mesogenesis and, 751
telogenesis and, 751
distribution and settings, 759
endokarst realm, 754–756, 755f, 756f, 757f
groundwater systems and regimes, 751–752
hydrogeologic and structural controls, 759–761, 760f
hydrogeologic cycle, 752–753, 753f
karst settings evolution, 753–754, 754f
mesomorphology features, 763–764
overview, 748–750, 749f
paleokarst and, 764
porosity patterns, 761–763
cavernous edging, 761, 762
collapse shafts, 761, 762
isolated chambers, 761, 762
network maze, 761, 762
single or rudimentary networks, 761–762
spongework maze, 761, 762
steeply-inclined passages or shafts, 761, 762
Speleogenesis, telogenetic
basic dynamics, 765–767, 766f
branchwork pattern upward evolution, 768–769, 768f
breakthrough and, 766
conduit networks evolution and, 767–769, 768f
fracture stages and, 766
overview, 765
Speleothems, 103–104. *See also* Gypsum flowers; Paleoclimate records, from speleothems; Stalactites; Stalagmites
anthodites, 379–380, 380f, 782
blue holes and, 160
Castleguard Cave, 93
cave shields, 381
classification of, 778–783, 779t
deposition mechanisms, 778
dripping water created, 779–781
carrot-shaped stalactite, 779–781, 781f
drapery, 781
shields, 781, 782f
stalactite-stalagmite combination, 781, 781f
stalagmites, 781, 781f
straw-shaped stalactite, 779–781, 780f
flowing water created, 779, 779t
cave beads, 779, 780f
flowstone, 779f

nodules, 779, 780f
rimstone dams, 779, 780f
frostwork, 380, 381f
growth dynamics
uranium series dating and, 842–844, 843f
gypsum, 374–375
heavy metals and, 164
helictites
beaded, 379, 380f
bushes, 382–383, 382f
subaerial, 379
subaqueous, 381–383, 381f, 382f
ice, 401f
Jewel Cave, 416–417
hydromagnesite balloons, 416–417, 416f
popcorn stalagmites, 416
Kazumura cave, 442
primary, 442
secondary, 442
mineral precipitation and, 885–886
minerals and, 777–778
agagonite, 777
calcite, 777
gypsum, 777–778, 778f
overview, 777
as paleoclimate archives, 785
paleomagnetic records and, 588
properties, 783–784
calcite/aragonite problem, 783–784, 783f
color and luminescence, 784, 784f
seeping water created, 782
frostwork, 782
gypsum flower, 782, 782f
helictites, 782, 782f
shapes of, 778–783
Sistema Huautla, 715f, 716f
standing water created, 782–783
calcite rafts, 783
cave pearls, 782–783, 783f
pool spar, 783, 783f
in volcanic caves, 784–785, 872
Speleothem deposition
chemistry
calcite precipitation/dissolution, 769–770, 770f
overview, 769
regular stalagmite
morphology, 772–775, 773f, 774f, 774f
similarity rule, 775–776, 775f
stalagmite growth
under variable time conditions, 776–777, 776f
thin water layers and, 770–771, 771f
turbulent water flow and, 771–772, 772f, 773f
Speleothems, paleoclimate records from
dating techniques, 577–578
global keeping of, 577, 579f
overview, 577
paleoclimatology contributions, 578–585
climatic changes timing, 584

Devils Hole, 579–580, 580f
early studies: paleotemperature, 578–579
last glacial, 580–581, 580f
late Holocene, 582, 584f
other studies, 584–585
paleotempestology, 584
solar variability and climate change, 581, 583f
tracking insolation, 581, 582f
tracking paleomonsoons, 581, 582f
vegetation and soil dynamics, 583–584, 584f
stalagmites use for, 577, 578f
Speleothems, uranium series dating of
ancient speleothems and, 844
disequilibrium dating, 839–840, 839f
overview, 838–839
practical issues, 841–842
cost, 842
in-cave sampling, 841, 841f
permits, 841
sample material quantity needed, 841, 842f
sample quality, 842
stalagmite focus, 842
precision and, 840–841
principles of, 839–840
reporting and assessing, 840–841
speleothem growth dynamics and, 842–844, 843f
technology, 840
uncertainties and, 840
Spelunker, 645
Sphaerochthoniidae, 795
Spiders. *See also* specific Family, Order or Sub
Acari, 793–795
Amblypygi, 789, 789f
of Arachnida, 256
Araneae, 789–791
biology, 786–787
conservation, 787
distribution, 786
Opiliones, 795–797, 796f
overview, 786
Palpigradi, 791
Pseudoscorpiones, 791–792, 791f
Ricinulei, 793, 793f
Schizomida, 788–789
Scorpiones, 787–788, 788f
Sparassidae, 242, 243f
systematics, 787–797
wolf, 15
Spitzkarren, 429
Splitkarren, 428–429
Sponges, 20
Spongework, 422
caves, 396
maze, 602f
patterns, 761, 762
Spongillidae, 234
Spring salamanders, 672–674, 673f
Springs. *See also* Seepage spring; Wakulla Spring underwater cave system, Florida
Barton Springs salamander, 669–670

Springs (*Continued*)
    discharges, 801–804
        ebb-and-flow springs, 801–802
        hydrographs analysis, 802
        magnitudes of, 801, 801t, 802f, 802t
        saturation state, 804
        turbidity and chemical variability, 803–804, 803f, 804f
    entrances at, 281, 281f
    Eurycea and, 667–668
    failure, 856–857
    Gammarus minus and, 344
    as habitat, 804–805
    heavy metals and, 165f
    NAPLs and, 171
    nitrate contamination and, 566–567
    overview about, 797
    salamanders, 669–670, 672–674, 673f
    San Gabriel Springs salamander, 670
    types of, 798–801, 798f
        alluviated springs, 799, 799f
        artesian springs, 799–800
        conduit and diffuse flow springs, 798
        gravity springs, 798–799, 799f
        mineral springs, 800
        offshore springs, 799–800
        overflow, underflow, and distributary springs, 800
        Seeps and wetlands, 800–801
        thermal springs, 800
    as water supplies, 804
    West Virginia spring salamander, 673f, 674
SSHs. *See* Shallow subterranean habitats
St. Beatus Cave, 699, 725–726, 725f
Stable isotope ratio analysis, 493–494
Stagnant-air zone, 205
Stalactites
    carrot-shaped, 779–781, 781f
    column and, 805, 806f
    conical, 807, 807f
    defined, 805
    epsomite, 375, 376f
    mineralogy and formation processes, 805–806, 806f
    mirabilite, 375, 376f
    overview about, 805
    soda straw, 807
    stalagmite combination, 781, 781f
    straw-shaped, 779–781, 780f
Stalagmites. *See also* Paleoclimate records, from speleothems
    aragonite, 809
    as archives of past, 809–810
    column and, 805, 806f, 809
    creation of, 781, 781f
    defined, 805
    formation of, 807–809
    growth
        rate, 808–809
        under variable time conditions, 776–777, 776f
    ice, 402
    Jewel Cave, 416
    microcrystalline fabric, 809
    mid-Pliocene, 809–810, 810f
    mineralogy and formation processes, 805–806, 806f
    morphologies, 807–808, 808f
    overview about, 805
    paleoclimate records from, 577, 578f
    popcorn, 416
    regular
        morphology, 772–775, 773f, 774f
        similarity rule, 775–776, 775f
    stalactite combination, 781, 781f
    uranium series dating and, 842
Stalita taenaria, 231f
"A Statistical Theory of Cave Entrance Evolution" (Curl), 278
Steeply-inclined passages or shafts, 761, 762
Stenasellus virei, 653, 654t
Sternophoridae, 792
Stillstands, 528–529
Stiphidiidae, 791
Stone, Bill, 312
Straw-shaped stalactite, 779–781, 780f
Stygiocaris stylifera, 214f
Stygnommatidae, 797
Stygnopsidae, 797
Stygobiotic ingolfiellid amphipods, 684f
Stygobites, 245t
Stygobromus, 853–854
Stygotrombidiidae, 795
Stylocellidae, 796
Subaerial helictites, 379
Subaqueous helictites, 381–383, 381f, 382f
Subcutaneous karst, 284
Subglacial caves, 356, 356f
Subjacent karst, 365, 371
Subsidence dolines, 150f, 151–152
Subsidence sinkhole, 718–719, 719f
Subterranean organisms, 276
Subtroglophile, 276–277
Subway, Castleguard Cave, 92f
Suffosion dolines, 722
Suffosional caves, 105f, 106
Sulaplax ensifer, 241, 241f
Sulfate mineral crystals. *See also* Gypsum flowers
    overview about, 374
    sources and depositional mechanisms, 375–377
    types of, 374
Sulfates, 500–504, 501t
    alunite, 501t, 503
    barite and celestine, 501t, 503
    deposition mechanisms, 503–504
    epsomite, 501t, 503
    gypsum, 500–503, 501t
    mirabilite, 501t, 503
Sulfides, 507
Sulfur compounds, reduced inorganic, 128–129
Sulfur cycle, 130–132, 132f
Sulfuric acid
    hypogenic speleogenesis and, 758
    karst, 437
    Lechuguilla Cave and, 456
Sulfuric acid caves
    chemical reactions, 812
    conclusions about, 819
    diagnostic features, 811–812
    field examples, 812–818
        Carlsbad Cavern, 816–817, 816f
        Cueva de Villa Luz, 813–815, 813f, 814t, 815f
        Frasassi Cave System, 815–816, 815f
        Grand Canyon, 817–818, 818f
        Guadalupe Mountains, 816–817, 816f
        Kane caves, 813, 813f, 814t
        Lechuguilla Cave, 816–817, 816f
    geologic setting, 810–811, 811f
    microbial interactions, 812
    MVT ore deposits and, 818–819
    overview, 810
    uranium and, 818–819
Sullivan Cave, 86
Superficial karst, 285
Superficial underground environment (MMS), 524–525
Superimposed-trunked lava tunnels, 870
Surface water infiltration model, 406
Swiftlets, 242, 243, 243f
Syarinidae, 792
Symphylans, 538
Syncarida, 178t, 183–184
    Anaspidacea, 178t, 184, 184f
    in Australia, 209–211
    Bathynellacea, 178t, 184, 184f
Syngenetic karst, 365
Synotaxidae, 791
Synurella ambulans, 231f

T

Tables, 270
Tabun Cave, 95f, 97–98, 97f
Tafoni caves, 155
Tailless whipscorpions, 789, 789f
Talus, entrance, 284
Talus caves, 105f, 106, 155
    recreational caving and, 642
Tanaidacea, 190, 190t
Tantulocarida, 178t, 180
Tapoff passages, 607–608, 607f
Taranakite, 501t, 504–505
Tardis Cave, 536f
Tecto-karstic hollow, 153
Tectonic caves, 105–106, 105f, 865
    recreational caving and, 642
Telemidae, 791
Telogenesis, 751
Telogenetic karst, 437
Telogenetic speleogenesis
    basic dynamics, 765–767, 766f
    branchwork pattern upward evolution, 768–769, 768f
    breakthrough and, 766
    conduit networks evolution and, 767–769, 768f
    fracture stages and, 766
    overview, 765
TEM. *See* Transient electromagnetics
Temnocephalida, 910–911, 910f, 911f

Tennessee cave salamander, 673, 673f
Terikan River Caves, 536
Terminal breakdown, 69
Terrestrial cave animals, 4
Terrestrial Isopoda, 14–15
Terrestrial troglobites distribution, 851–853, 852f
    beetles, 851–852, 852f
Tetrablemmidae, 791
Tetragnathidae, 791
Texas
    cave and spring-dwelling *Eurycea*, 667–668
    salamander, 669
        blind, 668
Thalassoid limnostygobionts
    marine regressions and
        biogeography of, 483, 484f, 485f
        Morocco and, 483–484
        as sole evidence, 483–484
Tharp model, 71
Theraphosidae, 789
Theridiidae, 791
Theridiosomatidae, 791
Thermal erosion, 440–441, 440f
Thermal springs, 800
Thermal variation, 350–351
Thermal waters
    chemistry, 393
    cooling, 758
    Dinaric karst and, 237
Thermobaric regime, 751–752
Thermoindicator, 402
Thermosbaenacea, 178t, 185t, 185, 257
    anchihaline habitats and, 21
    in Dinaric karst, 234
Thrust faults, 354, 354f
Tiankeng, 151
    critical factors, 822–824
    defined, 821
    development stages, 822–824, 823f
    distribution of, 824
    features, 822–824, 824t
    major world, 824t
    types, 821–822, 821f, 822f
        collapsed, 821, 821f, 822f
        erosional, 821f, 822, 822f
        water, 822
    underground river and, 823, 823f
Ticks, 793–795
Tourism, 403–404, 451–452, 622, 622f
Towers, 435–436, 436f
Townsend's big-eared bat, 46f
Trajano, E., 276
Transient electromagnetics (TEM), 352
Transverse flutes, 422
Trap hypothesis, 520
Traverses and tyrolians, 318–319, 319f
Travuniidae, 797
Tree roots, 661–662
Tremarctine bears, 593
Triaenonychidae, 797
Tributary passages, 604
Tricladida, 910, 910f
Tridenchthoniidae, 792

Triple porosity model, 384–385
Troglobites, 107. *See also specific troglobite*
    ecological classification and, 275–276, 277
    habitats
        lava tubes, 658
        limestone caves, 658
        mesocaverns, 658
        tree roots, 658
    single *vs.* multiple origin of, 298–299, 298f
    tropical diversity patterns, 245t
Troglocaris anophthalmus, 613
Troglomorphy, 517–518
Troglophiles, 107, 275–276, 277
Troglosironidae, 796
Troglotayosicidae, 788
Trogloxenes, 107. *See also* Bats
    ecological classification and, 275–277
Trombiculidae, 794
Trombidiidae, 794
Tropical landscape, sinkholes and, 147–149, 149f
Tropical winter effect, 205
Tropics diversity patterns. *See* Diversity patterns, in tropics
Trout Rock caves, 739, 740f, 741f
Tubular conduits, 734–735, 735f
Tumuli, 871–872
Tunnels, recreational caving and, 642
Turbellaria, 254, 910–911
    Temnocephalida, 910–911, 910f, 911f
    Tricladida, 910, 910f
    tropical diversity patterns, 245t
Turbulent water flow, 771–772, 772f, 773f
Twilight zone, 283
Typhlichthys subterraneus, 465, 466t
Typhlochactidae, 788
Typhlogammarus mrazeki, 231f
Typhlomolge, 668–669, 668f
Tyrolians, 318–319, 319f

U

Ugryn Cave, 829t
Ukraine giant gypsum caves
    feeding channels and, 831
    hypogene transverse speleogenesis, 832–833, 832f
    master passages and, 831
    as mazes, 830, 830f
    Optymistychna Cave, 830, 831f
    patterns and morphology, 829–832, 829t
    regional geology, 827–829, 828f
    regional hydrogeology, 827–829, 828f
    sediments and formations, 832
Underflow springs, 800
Underground shallow medium, 284–285
Underwater exploration techniques
    background about, 310
    carbon dioxide and, 313
    equipment development, 310–311
    factors contributing to, 310
    fatalities and, 313–314
    rebreathers, 312, 313f
        CCRs, 313
        SCRs, 313
    side-mount diving, 311, 312f, 313f

    survey tools, 312f, 314
Underworld, in myth, 321
United States diversity patterns. *See* Diversity patterns, in United States
Uranium, 818–819
Uranium series dating, of speleothems
    ancient speleothems and, 844
    disequilibrium dating, 839–840, 839f
    overview, 838–839
    practical issues, 841–842
        cost, 842
        in-cave sampling, 841, 841f
        permits, 841
        sample material quantity needed, 841, 842f
        sample quality, 842
        stalagmite focus, 842
    precision and, 840–841
    principles of, 839–840
    reporting and assessing, 840–841
    speleothem growth dynamics and, 842–844, 843f
    technology, 840
    uncertainties and, 840
Urban caving, 642
Urodacidae, 788
Uropodidae, 793–794
Ursus americanus, 593
Ursus speleus, 593
U-shaped tube effect, 19
Uvalas, 152–153, 199–200

V

Vadose passages, 604, 604f, 605, 606f
Vadose water, 157
Vaejovidae, 788
Valdina Farms salamander, 670
Vampire bats, 53–54
Vanadates, 506
Vandel, A., 404–405, 520
Variscite, 501t, 504–505
Velebit Mountains dolines, 147
Ventilation cooling, 401
Vermiculations, 496, 497f
Verteba Cave, 829t
Vertebrate visitors, birds and mammals
    cave ecology impact of, 847
    cave-dwelling species, 845–846
        facultative visitors, 845
        regional cavernicoles, 845
        regular cavernicoles, 845–846
    den sites and, 846–847
    foraging habitats and, 847
    knowledge gaps, 846
    man and, 848–849
    rare species, 848
    resting and mating and, 846
    wintering sites and, 847
Vertical exploration techniques, 307
    ascending, 307, 316f, 317–318
    from bottom up, 319–320
    changing over, 318
    descending, 307, 317
    disclaimer, 314
    overview regarding, 314–315

Vertical exploration techniques (*Continued*)
  rappel, 307
  rigging and, 315–316, 315f
  rope and, 315
  single-rope, 307
  sit harness and, 316–317, 316f
  traverses and tyrolians, 318–319, 319f
Vertical network caves, 637, 638f
Vicariant phenomena, 855
  calcrete aquifers and, 857
  spring failure, 856–857
  stranding, 855–856
  stream capture, 856
Vietnam, Hon Chong hills in, 249–250, 250f
Virgin passages, 277
Vjetrenica
  cultural history, 862
    artifacts, 862
    myths, 862
    sacred status, 862
  ecological changes to, 864
  exploration, 863–864
  fauna, 861–862
  Golden Chamber of, 859–860, 860f
  habitats, 861
  hydrology of, 860
  kamenica of, 859–860, 860f
  karst area of, 858
  microclimate, 860–861
  overview about, 858
  paleontology, 862
  protection, 864
  speleomorphology of, 858–860, 859f
  visitor access, 864
  water bodies of, 860, 860f
Volcanic rock caves, 105f, 106
  conclusions about, 872–873
  limited extent
    drained lobes, 871–872
    imprints, 871
    lava rise, 871–872
    partings, 871
    pressure ridge, 872
    tumuli, 871–872
    volcanic vents, 872
  morphological indices of Hawaiian, 866t
  overview about, 865
  pyroducts, 866–870, 866t
    canyon formation, 868–869, 870f
    double-trunked, 869–870
    formation of, 867, 868f
    Huehue Cave and, 869f
    Huehue Flow and, 869–870, 871f
    inflation and, 867, 868f
    internal development, 867–869, 870f
    puka and, 868–869
    single-trunked, 869
    superimposed-trunked, 870
    types of, 869–870
  secondary, 865–866
    erosional caves, 865–866
    pit craters, 865
    tectonic caves, 865
  speleothems in, 784–785, 872
Volcanic vents, 872

Voronja Cave
  Arabika Massif
    deep cave evolution, 449–450
    geology, 447–448, 447f
    hydrogeology, 448–449
    location and physiography, 443, 444f
    Ortobalagan Valley in, 443, 445f
  block-fault structure and, 445–447
  branches, 445, 445f
  combined cave profile, 446f
  depth of, 443
  evolution of, 449–450
  exploration since 1999, 444–445
  first documentation of, 443
  hydrology, 447
  overview about, 443
  temperatures, 447

W

Waitomo district dolines, 145
Wakulla Spring underwater cave system, Florida
  Ball and, 874–875
  early exploration, 874
  exploration and mapping, 878–880, 879f
  future of, 880
  geology, 875–878
    geologic framework, 875, 876f, 877f
    hydrology, 876–878, 877f
    paleontology, 875–876, 876f
    speleogenesis, 878
  history, 874
  name derivation, 874
  overview about, 874–875
Water caves, 308–309
Water chemistry in caves
  human impact on, 886
  limestone dissolution, 880–885
    aqueous solution saturation state, 882–883, 883f
    carbon dioxide and, 881, 881t
    environmental factors influencing, 884, 884f
    groundwater composition and, 881–882, 882t
    strong acids role in, 884–885
  mineral precipitation and speleothems formation, 885–886
  overview about, 880
Water Sinks Cave, 82
Water table, 528–529
Water tracing, in karst aquifers
  definitions and objectives of, 886–887
  dry set system, 892
  dye amounts to use, 891
  history, 887–889, 887f
  passive detector types
    activated carbon, 890
    cotton, 891
    plankton netting, 891
    Quinlan Gumdrop, 891, 891f
  qualitative
    using passive detectors, 890–892
  quantitative, 892–893
  test interpretation, 893–896, 894f

  dye recovery or breakthrough curve and, 894, 894f, 895f
  tracers and, 889–890
    categories of, 889
    cautions, 890
    fluorescent dyes, 889, 890t
    ideal characteristics, 889
Waterdogs, 665–667
Waterline notch, 425
West Virginia spring salamander, 673f, 674
Wetlands, 800–801
  Cerkniško jezero/polje, 898–899, 898f, 899f
  defined, 896
  distribution of major, 897, 897f
  overview about, 896–898
  Pivka intermittent lakes, 899–901, 901f, 902f
  Planinsko polje, 899, 900f
  protection, 901–902, 902t
  temporary, 897
White Cave food web, 60–61, 61f
White-nose syndrome (WNS), 52–53, 625–626
  cave closures and decontamination, 908
  characteristics, 905–907
    atypical winter behavior, 905–906
    depleted fat reserves, 905
    fungal infections and tissue damage, 905, 905f
    immune response changes, 906
    wing damage, 906–907, 906f
  current and future research, 907–908
  diagnostics, 904–905
  European connection, 907
  fungal transmission, 907
  mortality and, 904
  overview, 903–904
  pathogen, 903f, 904
  species affected by, 905
Wild caving, 643
Wilford, G. E., 532
Wind Cave, South Dakota, 86
Wind turbines, 53
Wishing Well Cave, 82
Withidae, 792
WNS. *See* White-nose syndrome
Wolf spiders, 15
Wolves, 594–595
Worms
  Annelida, 911–914
    Hirudinea, 913–914, 913f
    Oligochaeta, 912–913, 913f
    Polychaeta, 911–912, 912f
  characterization of invertebrate types, 910–914
  earthworms, 362
  ecological and phylogenetic remarks, 914
  general characteristics, 909
  Nematoda, 911, 911f
  Nemertea, 911, 912f
  Onychophora, 914, 914f
  studies history, 909–910
  Turbellaria, 910–911
    Temnocephalida, 910–911, 910f, 911f
    Tricladida, 910, 910f

X

Xel-Há to Sian Ka'an Biosphere Reserve region, 835–836, 835f

Y

Yangpterochiropera. *See* Bats
Yates Formation, 457–458, 457f
Yinchiroptera. *See* Bats
Yucatán Peninsula, underwater caves of
   Campeche State and, 833–834
   cenotes and, 836–837
   deep well data paucity regarding, 833
   geologic history, 833
   hydrology, 833
   pre-Mayans/Mayans and, 836
   Quintana Roo State and, 834
   Sistema K'oox baal, 836f
   Sistema Ox Bel Ha, 837
   Sistema Sac Actun, 837–838
   Xel-Há to Sian Ka'an Biosphere Reserve region, 835–836, 835f
   Yucatán State and, 834

Z

Zalmoxidae, 797
Zerconidae, 794
Zero-inflated spatial models, 478–479
Zoloushka Cave, 829t
Zospeum, 513
Zospeum kusceri, 231f
Zygentoma, 216, 245t